Wörterbuch Antriebstechnik
Dictionary of Drives

Wörterbuch Antriebstechnik
Dictionary of Drives

Thomas Antoni

Publicis MCD Verlag

Die Deutsche Bibliothek – CIP-Einheitsaufnahme
Ein Titeldatensatz für diese Publikation ist bei Der Deutschen Bibliothek erhältlich

Die Deutsche Bibliothek – CIP-Cataloguing-in-Publication-Data
A catalogue record for this publication in available from Die Deutsche Bibliothek

Autoren und Verlag haben alle Texte in diesem Buch mit großer Sorgfalt erarbeitet. Dennoch können Fehler nicht ausgeschlossen werden. Eine Haftung des Verlags oder der Autoren, gleich aus welchem Rechtsgrund, ist ausgeschlossen. Die in diesem Buch wiedergegebenen Bezeichnungen können Warenzeichen sein, deren Benutzung durch Dritte für deren Zwecke die Rechte der Inhaber verletzen kann.

This book was carefully produced. Nevertheless, author and publisher do not warrant the information contained therein to be free of errors. Neither the author nor the publisher can assume any liability or legal responsibility for omissions or errors. Terms reproduced in this book may be registered trademarks, the use of which by third parties for their own purposes may violate the rights of the owners of those trademarks.

ISBN 3-89578-156-8

Herausgeber: Siemens Aktiengesellschaft, Berlin und München
Verlag: Publicis MCD Verlag

© 2000 by Publicis MCD Werbeagentur GmbH, München
Das Werk einschließlich aller seiner Teile ist urheberrechtlich geschützt. Jede Verwendung außerhalb der engen Grenzen des Urheberrechtsgesetzes ist ohne Zustimmung des Verlags unzulässig und strafbar. Das gilt insbesondere für Vervielfältigungen, Übersetzungen, Mikroverfilmungen, Bearbeitungen sonstiger Art sowie für die Einspeicherung und Verarbeitung in elektronischen Systemen. Dies gilt auch für die Entnahme von einzelnen Abbildungen und bei auszugsweiser Verwendung von Texten.

Printed in Germany

Vorwort

Der große Erfolg der ersten Auflage hat mich ermutigt, das Wörterbuch ständig weiter auszubauen und zu verbessern. Der Umfang ist um fast 40% angewachsen. Das Wörterbuch enthält jetzt in beiden Sprachrichtungen zusammen über 60.000 Stichpunkte und ca. 120.000 Übersetzungen – wobei die neue deutsche Rechtschreibung bereits berücksichtigt ist.

Die Schwerpunkte der Erweiterungen sind:
- Automatisierungstechnik
- Feldbustechnologien
- Elektrische Maschinen

Gerade in der Automatisierungs- und Feldbustechnik hat sich seit der ersten Auflage viel getan.

Viele Übersetzungen sind durch Nennung des Fachgebietes nun besser zuzuordnen. Eine Auflistung der entsprechenden Abkürzungen finden Sie nach dem Vorwort.

Das Wörterbuch richtet sich an Ingenieure und Techniker, die sich in Entwicklung, Fertigung, Dokumentation, Ausbildung und Vertrieb mit elektrischen Antrieben und Automatisierungstechnik befassen. Im Laufe meiner langjährigen Tätigkeit auf dem Gebiet der Antriebs- und Automatisierungstechnik hat sich durch die vielen internationalen Kontakte ein großer deutsch-englischer Wortschatz angesammelt, der in dieses Wörterbuch eingeflossen ist. Außerdem habe ich eine Vielzahl von Fachzeitschriften und -büchern sowie Betriebsanleitungen, Kataloge und Druckschriften zahlreicher Weltfirmen ausgewertet.

Schwerpunkte dieses Wörterbuchs sind die Themengebiete
- Elektrische Maschinen
- Stromrichtergeräte
- Regelungstechnik
- Steuerungs- und Automatisierungstechnik
- Anlagen- und Prozesstechnik
- Feldbussysteme

Darüber hinaus enthält das Wörterbuch grundlegende Begriffe aus den Gebieten
- Allgemeine Elektrotechnik
- Datenverarbeitung
- Maschinenbau
- Geschäfts- und Wirtschaftsenglisch
- Marketing
- Verhandlungswortschatz
- Technische Aus- und Weiterbildung
- Werbung

Außerdem sind in dem Wörterbuch eine Reihe von Vokabeln enthalten, die häufig in technischen Beschreibungen und im beruflichen Alltag benötigt werden.

Die einzelnen Übersetzungen sind mit einer Vielzahl technischer Erläuterungen und Hinweisen auf relevante Normen kommentiert, so dass das Wörterbuch gleichzeitig ein kleines technisches Nachschlagewerk darstellt.

Im deutsch-englischen Teil des Wörterbuchs sind die Übersetzungen eines Stichpunktes in der Reihenfolge ihrer Geläufigkeit angeordnet; im Zweifelsfalle ist die zuerst stehende Übersetzung zu verwenden. Dies begünstigt es, dass in einzelnen Abteilungen oder Firmen immer die gleiche Übersetzung verwendet wird.

Der englisch-deutsche Teil ist durch einen automatischen Umkehrvorgang aus dem deutsch-englischen Wörterbuch entstanden. Dadurch ergibt es sich, dass die Übersetzungen eines Stichpunkts streng alphabetisch geordnet sind und nicht nach der Geläufigkeit und dass die Kommentare teilweise mehrfach aufgelistet sind.

Wenn Sie Fehler oder Lücken im Wörterbuch gefunden haben, scheuen Sie sich bitte nicht, mir eine entsprechende E-Mail an thomas@antonis.de zu senden. Ich freue mich über jeden Verbesserungsvorschlag. Aktuelle Informationen finden Sie auch unter http://www.antonis.de.

Erlangen, August 2000   Thomas Antoni

Preface

The great success of the first edition has encouraged me to extend and improve the dictionary permanently. The scope has been increased by nearly 40%. Now the dictionary contains a total of over 60,000 keywords and approx. 120,000 translations – and the new German orthography has already been taken into account.

The main fields of extension are:
- Automation
- Field bus technology
- Electrical machines

Especially in the automation and field bus technologies, many innovations and new terms have appeared since the first edition.

By quoting the special fields, many translations can now much better be assigned to the appropriate field of knowledge. The German-language designations are noted as abbreviations at the beginning of the comment. You will find a list of these abbreviations after the preface.

The dictionary is addressed to engineers and technicians who deal with electrical drives and automation in the fields R&D, manufacturing, documentation, technical training and sales/marketing. During many years' work in the field of electrical drives and automation and the stimulated by many international contacts I have collected a great many of German-English technical terms, which have been incorporated into this dictionary. Furthermore a great variety of technical journals and books as well as operating instructions, catalogs, and other publications from many companies of the world market have been taken as sources for this book.

Main fields of this dictionary are:
- Electrical machines
- AC and DC Converters
- Closed-loop controls
- Control engineering and automation
- Industrial systems and process engineering
- Field busses

Furthermore the book contains basic terms of the following sectors:
- General electrical engineering
- Data processing
- Mechanical engineering
- Business and economics
- Sales and marketing
- Terms often used in meetings, conferences and negotiations
- Technical training
- Advertising

Additionally, a range of terms is integrated into the dictionary which are often used in technical documentation and discussions as well as during the everyday telephone, letter and FAX contacts.

The translations are supplemented by a wealth of technical explanations and references to applicable international standards. Thus the dictionary also serves as a small reference book.

The translations of a keyword in the German-English dictionary are arranged into the order of the most common use. In case of doubt you should therefore always use the translation which is noted first. This principle of arrangement supports the uniform translation of a certain term over all documents of a department.

The English-German part of the dictionary was generated by an automatic reversing process out of the German-English dictionary. This leads to the consequences that the translations of an individual keyword are arranged in strictly alphabetic order and not in order of the most common use and that multiple notations of one and the same comment cannot be avoided.

Have you found any error or missing information in the dictionary? Then please do not hesitate to send me an e-mail to thomas@antonis.de. I am very happy about any suggestion for improvement. New informations concerning the book will be presented at http://www.antonis.de.

Erlangen, August 2000   Thomas Antoni

Abkürzungen zur Kennzeichnung der Fachgebiete

Abkürzung	Fachgebiet
{(Ab)Wasser}	Wasser- und Abwasserversorgung
{Agrikultur}	Land- und Forstwirtschaft
{Aufzüge}	Aufzugstechnik
{Auto}	Automobiltechnik
{Bahnen}	Eisenbahnwesen, Stadtbahnen
{Bautechnik}	Bautechnik
{Bergbau}	Bergbau
{Buchbinderei}	Buchbinderei
{Chemie}	Chemie
{Computer}	Computer-Hard- und Software, Informatik
{Draht}	Draht, Drahtziehmaschinen
{Druckerei}	Druckerei, Druckmaschinen, graf. Gewerbe, DTP
{el.Masch.}	elektrische Maschinen
{Elektr.}	Elektrotechnik, elektrisch, Elektronik
{Fördertechnik}	Fördertechnik
{Glas}	Glasherstellung
{Gummi}	Gummi, Reifen, Kautschuk
{Hebezeuge}	Hebezeuge, Kräne
{Holz}	Holzverarbeitung
{Kabel}	Kabel, Leitungen, Kabelherstellmaschinen
{Klimatechn.}	Heizung-, Klima-, Lüftungstechnik
{Kraftwerk}	Kraftwerkstechnik
{Kunststoff}	Kunststofferzeugung und -verarbeitung, Folien
{Masch.bau}	Maschinenbau
{Mathem.}	Mathematik
{mechan.}	mechanisch
{Medizin}	Medizintechnik
{Meßtechn.}	Meßtechnik
{Metallurgie}	Metallurgie und Eisenhüttenwesen
{Nahrungsmittel}	Nahrungsmittel- und Getränkeindustrie
{NC-Steuerung}	numerische Werkzeugmaschinensteuerung
{Network}	Datennetze, Datenübertragung, Feldbusse
{Optik}	Optik
{Papier}	Papier- und Zellstoffindustrie
{Physik}	Physik
{Rechtswesen}	Rechtswesen
{Regel.techn.}	Regelungstechnik, Regelungstheorie
{Schifffahrt}	Schifffahrt
{Schweißtechnik}	Schweißtechnik
{SPS}	Speicherprogrammierbare Steuerung
{Stromrichter}	Stromrichtertechnik
{Textil}	Textilmaschinen, Textilien
{Verkehr}	Verkehrstechnik
{Verpack.techn.}	Verpackungs- und Flaschenabfülltechnik
{Walzwerk}	Walzwerke, Bandbehandlungsanlagen
{Werkz.masch.}	Werkzeugmaschinen, Metallbearbeitung
{Wirtschaft}	Betriebs- und Volkswirtschaft, Handelswesen, Logistik

Sonstige Abkürzungen

Abkürzung	Bedeutung
Adj.	Adjektiv
Adv.	Adverb
amerikan.	amerikanisch (-es Englisch)
brit.	britisch (-es Englisch)
e.	einer, eines, ein, einen, einem
etw.	etwas
figürl.	figürlich, im übertragenen Sinne
Ggs.	Gegensatz
i.	in, im
jdm.	jemandem
jdn.	jemanden
konkret	konkret, dinglich
m.	mit
o.	oder
Plural	Plural, Mehrzahl
s.	sich
salopp	salopp, umgangssprachlich
Subst.	Substantiv, Hauptwort
u.	und
v.	von
zw.	zwischen

Abbreviations designating the special fields and branches

Abbrevation	Special Field
{(Ab)Wasser}	Water supply and waste water treatment
{Agrikultur}	Agriculture and forestry
{Aufzüge}	Elevators and Lifts
{Auto}	Automotive technology
{Bahnen}	Railways
{Bautechnik}	Civil and construction engineering
{Bergbau}	Mining
{Buchbinderei}	Book-binding
{Chemie}	Chemistry
{Computer}	Computers and information technology
{Draht}	Wires and wire-drawing machines
{Druckerei}	Printing technology and trade, desktop publishing
{el.Masch.}	Electrical machines (motors and generators)
{Elektr.}	Electrical engineering, electricity, electronics
{Fördertechnik}	Conveyor technology
{Glas}	Glassmaking
{Gummi}	Rubber and tires
{Hebezeuge}	Cranes and hoists
{Holz}	Woodworking
{Kabel}	Cables and cable machines
{Klimatechn.}	HVAC (heating, ventilation and air conditioning)
{Kraftwerk}	Power plants
{Kunststoff}	Plastics and plastic films industry
{Masch.bau}	Mechanical engineering
{Mathem.}	Mathematics
{mechan.}	Mechanical
{Medizin}	Medical engineering
{Meßtechn.}	Measuring and instrumentation
{Metallurgie}	Metallurgy and ironworks
{Nahrungsmittel}	Foods and beverages
{NC-Steuerung}	Numerical controls for machine tools
{Network}	Data networks and field busses
{Optik}	Optics
{Papier}	Pulp and paper industry
{Physik}	Physics
{Rechtswesen}	Legal terms
{Regel.techn.}	Closed-loop controls and automatic control theory
{Schifffahrt}	Shipping
{Schweißtechnik}	Welding technology
{SPS}	PLCs (programmable logic controllers)
{Stromrichter}	Drive and power converter technology
{Textil}	Textile machines and textiles
{Verkehr}	Transport systems
{Verpack.techn.}	Packaging and bottling
{Walzwerk}	Rolling mills and strip processing lines
{Werkz.masch.}	Machine tools and metalworking
{Wirtschaft}	Business administration and economics

Other abbreviations

Abbrevation	Meaning
Adj.	Adjective
Adv.	Adverb
amerikan.	American, American English
brit.	British, British English
e.	a, an
etw.	something
figürl.	in a figurative/ methaphorical sense
Ggs.	opposite, antithesis
i.	in
jdm.	someone, somebody
jdn.	someone, somebody
konkret	concrete
m.	with
o.	or
Plural	Plural
s.	oneself
salopp	Colloquialism, nonchalant term
Subst.	Noun, substantive
u.	and
v.	from
zw.	between

time onwards: *ab diesem Zeitpunkt* || above * *oberhalb (auch figürlich)* || and above * *ab einschließlich (Reihenfolge; nachgestellt)* || starting * *starting in November: ab November; starting immediately: ab sofort* || from and above * *ab einschließlich* || away * *nicht (mehr) vorhanden (from: von), weg* || off * *weg, entfernt, ab; washed off: abgewaschen; rubbed off: abgerieben* || as of * ~ *einem bestimmten Datum; as of 1996:* ~ *1996*
ab Herstellerwerk   ex works
ab Lager   ex-stock || from stock || off-the-shelf || available off the shelf * *ab Lager lieferbar*
ab Lager lieferbar   available from the stock || off-the-shelf || available off the shelf
ab sofort   effective immediately * *mit sofortiger Wirksamkeit/Gültigkeit* || immediately
ab Version x   from version x onwards || with version x and more recent versions * *ab inklusive Version x*
ab Werk   ex works
abändern   modify
abarbeiten   execute * *Programm* || process || run * *ein Programm* || work off
Abarbeitung   execution * *Ausführung, z.B. eines Programms* || processing
Abarbeitungsgeschwindigkeit   processing speed || processing rate
Abarbeitungsreihenfolge   processing sequence
Abbau   disassembly || dismantling * *z.B. einer Maschine* || stripping || mining * *Bodenschätze* || cut * *Löhne, Preise* || cut down * *Personal* || reduction * *Reduzierung* || exhaustion * *Erschöpfung* || breaking-down * *Niederreißen, Abbrechen, Zerlegen, Auflösen* || decay * *Rückgang, Schwund* || build-down * *z.B des Durchmessers beim Abwickler* || exploitation * *Abbau/Gewinnung (von Bodenschätzen), Verwertung/Ausnutzung/-beutung (auch figürl.)* || downsizing * *Belegschaft, Produktion usw.*
abbauen   dismantle * *demontieren* || disassemble * *auseinanderlegen, demontieren, zerlegen* || dismount * *abmontieren, ausbauen, auseinandernehmen* || remove * *abnehmen, abmontieren, ausbauen* || suppress * *abschaffen* || reduce * →*reduzieren*, →*verringern* || extinguish * *abschaffen, aufheben, (aus)löschen, ersticken* || disconnect * *eine Verbindung* ~ || clear * *eine Verbindung* ~ || win * *Bodenschätze* ~ || mine * *Bodenschätze* ~ || cut down * *Personal, Kosten usw.* ~ || make redundant * *Personal* ~ || exploit * *Bodenschätze* ~
Abberufung   recall
abbestellen   cancel || countermand || discontinue * *Abonnement usw.*
abbiegen   turn off * *in; auch Straße usw.; into: in.* || turn aside || turn right * *nach rechts* ~ || turn left * *nach rechts* ~ || bend * *krümmen* || avoid * *(figürl.) Konflikt, Nachteil usw. verhindern* || deflect * *(sich) biegen, krümmen, neigen* || bend off || branch off * *Straße usw.*
Abbild   image || copy || replica * *genaue Nachbildung* || process image * *Prozessabbild* || thermal image * *thermisches Abbild*
abbilden   copy || map * *map an input: einen Eingang* ~ || show * *zeigen (in e. Abbildung)* || depict
Abbildung   diagram || figure * *besonders als Bildunterschrift mit Zahl, z.B. Fig. 23: Abb. 23* || illustration || graph * *grafische Darstellung, Schaubild, Kurvenbild/blatt* || picture * *Bild* || image * *Abbild* || display * *an Bildschirm, Anzeigeeinrichtung* || chart * *Diagramm, Tabelle, Schaubild, Kurvenblatt, grafische Darstellung* || drawing * *Zeichnung*
Abbildungsmaßstab   object-to-image ratio
Abbrand   burn-off || contact erosion * ~ *von Kontakten*

abbrechen   stop * *aufhören* || interrupt * *unterbrechen* || abort * *Programm/Vorgang/Ablauf z.B. auf Grund eines Fehlers* ~ || truncate * *Rechenvorgang/Programmablauf* ~; *abrunden* || pull down * *Gebäude* ~ || cut short * *plötzlich* ~ || terminate * *abschließen, beenden* || break off * *Beziehung usw.* ~ || call off
abbremsen   brake || slow down * *verzögern*
Abbremsung   braking || slowing down * *Verzögerung* || deceleration * *Verzögerung*
abbrennen   burn away || burn down
Abbruch   abort * *eines Programms* || pulling down * *Gebäude* || truncation * *eines Rechenvorganges* || break off || dismantlement
abbuchen   debit
Abbundanlage   timber crafter * *{Holz}* || trimming machine * *{Holz}* || cross-cutting and trenching machine * *{Holz}*
Abdeckblech   cover sheet || sheet-metal cover || cover
abdecken   cover * *auch figürlich, z.B. einen Markt, einen Leistungs/Anforderungsbereich, ein Risiko*
Abdeckplatte   cover plate || cover
Abdeckung   cover || covering || guard || insulating cover * *Isolierung*
abdichten   seal * *versiegeln, dichtmachen* || make tight * *stuff* || pack * *waterproof* * *gegen Wasser* || plug up * *Loch* ~
Abdichtung   sealing
abdrehen   turn down * *an der Drehmaschine* || face * *Stirnfläche mit einer Drehmaschine* || skim down * *an Drehmaschine leicht* ~ *zum Beseitigen der Unebenheiten*
abdruckbares Zeichen   printable character
abdrücken   force off * *z.B. Lager von Welle* || lift off * *z.B. Lager von Welle* || force away || perform the pressure test * *Druckprüfung durchführen (z.B. für Pressluftkessel, Rohrsystem usw.)* || pull the trigger * *Gewehr usw.* || fire off * *Gewehr usw.* || overpressure test * (Substantiv) *Überdruckprüfung*
Abdrückschraube   jack screw * *zum Demontieren/Abdrücken eines Maschinenteils (z.B. Nabe von einer Welle)* || extracting screw || lifting screw || puller screw || withdrawer screw || forcing-off screw * *zum Abziehen eines im Presssitz montierten Maschinenteils* || setscrew
Abendessen   supper || evening meal || dinner * *Hauptmahlzeit, Mittagessen, Abendessen*
Abendkurs   evening class
Abfall   drop * *z.B. Spannungsabfall an einem Widerstand, Druckabfall, Abfall e. Relaiskontakts* || depression || droop * *Herabhängen; Statik eines Reglers* || dip * *kurzzeitig, z.B. Spannung* || waste * *zum Wegwerfen* || litter * *herumliegender Abfall* || decay * *Verfall, Schwächerwerden* || drop-out * *eines Relais-/Schützkontakts* || dropout * *eines Relais-/Schützkontakts* || refuse * *Abfall, Ausschuss, Abraum {Bergwerk}, Müll-* || spoilage || scrap
abfallen   drop * *Relais, Schütz, Druck, Spannung an einem Widerstand usw. (by: um)* || drop out * *Relais, Schütz* || release * *Relais, Schütz usw.* || be de-energized * *Relais* || decrease * *sich vermindern (wert-/zahlenmäßig)* || decline * *abnehmen, nachlassen, zurückgehen, sich verschlechtern, schwächer werden, verfallen* || fall off * *abnehmen, hinunterfallen* || slope * *Gelände usw.* || compare badly with * ~ *gegenüber, im Vergleich zu ...* ~ || decrease * *Geschwindigkeit* || slacken * *Geschwindigkeit* || be a spin-off * *als (zufälliges) Nebenprodukt*
abfallend   falling || descending * *in descending order: in* ~*er Reihenfolge*
abfallende Flanke   falling edge * ~ *eines Impulses* || trailing edge * *Rückflanke eines Impulses*

abfallende Hochlaufgeber-Rampe ramp from run mode
Abfallentsorgung waste disposal
abfallverzögert with drop-out delay ‖ drop-out delayed ‖ with OFF delay
abfallverzögertes Schütz dropout delayed contactor
Abfallverzögerung dropout delay ‖ switch-off delay
Abfallverzögerungszeit dropout time ‖ drop-out time
Abfallzeit drop-out time * *Schütz, Relais* ‖ dropout time
abfangen clamp * *z.B Kabel mit Hilfe einer Schelle festklemmen*
abfeilen file off
abflachen flatten * *auch 'sich ~'* ‖ level * *auch 'sich ~'* ‖ slope down * *sich ~*
abfließen flow off ‖ run off ‖ drain off * *besonders Flüssigkeiten; auch 'abfließen lassen'* ‖ leak off * *durch ein Leck*
Abfolge sequence ‖ succession
Abfrage scan * *auf einen Signalzustand* ‖ interrogation * *Frage, Befragung, Vernehmung* ‖ sample * *Abtastung eines Sinalzustands* ‖ test * *Prüfung* ‖ check * *Überprüfung, Prüfung* ‖ polling * *Wählen, zyklische Abfr. (im Computer)*, einen Partner nach d. anderen ‖ inquiry * *Erkundigung, Nachforschung, Untersuchung, Auskunft-Einholen* ‖ scanning * *eines Signalzustandes/Impulses usw.; Abtastung v. Messwerten in e. Digitalsystem*
Abfrageergebnis scan result ‖ sample * *Abtastwert einer Messung*
abfragen scan * *abtasten, z.B. einen Signalzustand* ~ ‖ sample * *Messgröße/wert abtasten* ‖ check * *überprüfen; against a limit: auf eine Grenze* ~ ‖ interrogate * *be/ausfragen (auch einen Binärausgang), vernehmen* ‖ query * *aus/befragen* ‖ sense * *Messwert/Signalzustand usw.* ~ ‖ poll * *(zyklisches) Abfragen "auf Änderung" durch einen Rechner*
Abfuhr leading off ‖ dissipation * *of heat: Wärmeabgabe* ‖ removal ‖ hauling off * *von Sachen* ‖ discharge * *Leerung eines Behälters* ‖ draining off * *Wasser usw.* ‖ carrying off * *einer Sache, von Wärme usw* ‖ extraction
abführen dissipate * *z.B. Verlustwärme; auch '(sich) auflösen/zerstreuen/verflüchtigen'* ‖ lead off ‖ lead away ‖ carry off * *eine Sache/Hitze* ~ ‖ pay over * *to: Geldbetrag an ...* ~ ‖ discharge * *(Behälter) leeren* ‖ drain off * *Wasser usw.* ~ ‖ haul off * *eine Sache* ~ ‖ pass to a persons credit * *jemandem gutschreiben* ‖ branch off * *abzweigen* ‖ exhaust * *Gas/Dampf* ~ ‖ withdraw * *wegnehmen, abziehen* ‖ conduct outwards * *nach außen* ~ *(z.B. Kühlluft)*
Abführung dissipation * *von Verlustwärme usw.* ‖ leading off
Abfüllanlage filling line * *(Verpack. techn)* *z.B. für Flaschen* ‖ bottling plant * *(Verpack. techn.)* Flaschenabfüllanlage ‖ filling plant * *(Verpack. techn.) z.B. für Flaschen*
Abfüllmaschine bottling machine * *zur Flaschenfüllung, z.B. in d. Getränkeindustrie* ‖ filling machine * *(allg.)*
Abfüllung filling * *von Behältern, Flaschen usw.* ‖ bottling * *von Flaschen*
Abfüllwaage filling weigher * *(Verpack· technik)* ‖ weigh filler * *(Verpack. technik)*
Abgabe duty * *z.B.Steuer,Zoll* ‖ delivery
Abgabeleistung output power ‖ output ‖ power output
Abgang outgoing section * *Stromkreis* ‖ outgoing feeder * *Ausleitung/Abzweig z.B. zur Leistungsversorgung* ‖ outlet * *Herausführung; Anschluss für Stromverbraucher*

Abgas waste gas ‖ exhaust gas ‖ stale air * *verbrauchte/abgasgeschwängerte Luft*
abgeben deliver * *auch Drehmoment, Leistung, Strom usw.* ~ ‖ supply * *power: Leistung* ‖ deposit * *Gepäck usw. deponieren* ‖ emit * *Wärme usw.* ~ ‖ output * *Leistung usw.* ~ ‖ produce * *erzeugen* ‖ supply * *z.B. Spannung/Strom/Leistung* ~ ‖ dissipate * *z.B. heat: (Verlust)Wärme* ~ ‖ give off * *z.B. heat: Verlustwärme* ~ ‖ issue * *a declaration: eine Erklärung* ~ ‖ pass on * *weitergeben (to: an)* ‖ return * *wieder ~, zurückgeben*
abgebrochen aborted * *vorzeitig abgebrochen (z.B. Ablauf e. Programms, z.B. wegen eines Fehlers)* ‖ cancelled * *gestrichen, annulliert, widerrufen, ungültig gemacht, storniert* ‖ truncated * *z.B. Programmablauf/Rechenalgorithmus usw.* ‖ canceled * *(kann mit 'l' oder mit 'll' geschrieben werden)*
abgebunden cured * *verfestigt (z.B. Klebeverbindung), ausgehärtet*
abgedeckt covered * *(auch figürlich: durch Versicherung, Garantie usw.)*
abgedichtet sealed ‖ gasketed * *mit →Dichtung* ~ *(z.B. Schranktür mit Gummiwulst)* ‖ impermeable * *undurchlässig, undurchdringlich; impermeable to water: wasserdicht*
abgefallen dropped out * *Relais, Schütz* ‖ deenergized * *Relais, Schütz* ‖ deenergized * *Relais, Schütz*
abgeflacht flattened
abgegeben delivered ‖ output
abgegebene Leistung output * *z.B. eines Stromrichters/Motors*
abgeglichen balanced ‖ trimmed
abgehen go off ‖ go away ‖ leave * *Zug, Post usw.* ‖ start * *Zug usw.* ‖ depart * *Zug usw.* ‖ sail * *Schiff (for: nach)* ‖ come off * *sich loslösen* ‖ branch off * *Weg usw.* ‖ digress * *from: von einem Thema* ~/*abschweifen* ‖ abondon * *von einem Vorhaben/einer Lösung* ‖ be missing * *fehlen* ‖ drop * *von einem Thema usw.*
abgehend outgoing ‖ leaving * *Zug usw.*
abgeknickte Kennlinie bending characteristic ‖ fold-back characteristic * *zurückgebogene Kennlinie (bei Netzteil: Spannungsreduzierung b.Überstrom)*
abgekündigt discontinued * *nicht mehr gefertigtes Produktes* ‖ to be discontinued
abgelaufen elapsed * *Zeit* ‖ expired * *Frist, Überwachungszeit*
abgelöstes Ventil preceding valve * *ausgehendes Ventil beim Kommutierungsvorgang eines Stromrichters*
abgerundet curved ‖ round * *Zahl* ‖ well-rounded * *Leistung, Stil, Bildung* ‖ in round figures * *(Adv.) ganze Zahl*
abgeschaltet switched off ‖ off
abgeschirmt shielded ‖ screened * *screened enclosure: ~es Gehäuse*
abgeschirmte Leitung shielded cable ‖ screened cable ‖ shielding cable
abgeschlossen completed * *beendet* ‖ secluded * *abgesondert* ‖ isolated * *abgesondert, isoliert* ‖ locked * *abgesperrt* ‖ self-contained * *(in sich) geschlossen, unabhängig, selbstständig, autark* ‖ autonomous * *autark* ‖ terminated * *beendet, aufgehoben (z.B. Vertrag), abgeschloss. (Leitung m.Wellenwiderstand)* ‖ settled * *endgültig abgeschlossen* ‖ brought to a close * *zum Abschluss gebracht* ‖ ended * *beendet* ‖ over * *vorbei (zeitlich)*
abgeschrägt beveled * *(Kante)* ‖ chamfered * *(Kante)* ‖ bevelled * *(Kante)* ‖ bevel
abgesehen davon dass not to mention that ‖ let alone that ‖ quite apart from the fact that * *ganz abgesehen davon dass*
abgesehen von with the exception of * *mit Ausnah-*

me von || apart from || aside from || except for || exclusive of || leaving out || beside the
abgesichert   fused * *z.B. mit Schmelzsicherung* || guarded against * *geschützt gegen* || protected against * *geschützt gegen* || checked * *überprüft*
abgestimmt   harmonized * *harmonisiert*
abgestuft   graded || graduated || scaled || classified * *in Klassen eingeteilt*
abgetastet   sampled * *(Signal)*
Abgleich   adjustment || trim || balancing * *Austarieren* || compensation || matching * ~ *von mehreren (meistens 2) Dingen aneinander* || adjust * *Kurzform für Beschriftung e.Bedienelements* || alignment * *Anpassung, Ausrichtung* || intervention * *Einstellung des Eingriffs, das Abgleichen* || scaling * ~ *auf Nennwert (z.B. Tachometersignal), Normierung, Gewichtung* || balance * *(Null)Abgleich, Ausgleich, Kompensation* || calibration * *Eichung, Kalibrierung, Abgleich einer Messeinrichtung*
abgleichen   balance * *austarieren, (auch Konten ~)* || adjust * *anpassen, einstellen, justieren* || trim || compensate * *kompensieren* || calibrate * *eichen, z.B. Messinstrument* || match * *mehre Teile aneinander anpassen* || tune * *einstellen, abstimmen, anpassen, optimieren, in Übereinstimmung/Resonanz bringen* || align * *abgleichen/-fluchten, ausgleichen/-fluchten, symmetrieren, zum Fluchten bringen* || equalize * *ab-/ausgleichen, egalisieren, entzerren, gleichmachen/-stellen, nivellieren* || square * *Konten ~* || synchronize * *in Übereinstimmung bringen (auch Datenbestände), abstimmen, synchronisieren*
Abgleichkondensator   trimming capacitor || tuning capacitor || trimmer capacitor || trimmer
Abgleichpotentiometer   trimming potentiometer || trimming pot || trim-pot || trimmer
Abgleichwiderstand   trimming resistor || trimmer || adjusting resistor
abgraten   deburr || clip * *Gussnaht*
abgreifen   pick off * *z.B. Spannung von einem Potentiometer, Messsignal usw.*
Abgrenzung   delimitation || demarcation
Abgriff   pick-off * *z.B. an Potentiometer, Schiebetrafo* || tap * *Anzapfung, z.B. an Spule, Transformator* || slider * *Schleifer an Potentiometer, Schiebetrafo usw.* || tap-off point * *Abgreifpunkt*
abhaken   check off
abhandeln   deal with
abhängen   depend * *(up)on: von* || be dependent * *finanziell usw.* || be subject * *to: von einer Zustimmung, Vorschrift usw.abhängen* || hinge * *hinge (up)on: letztlich abhängen von* || unhang || unhook * *vom Haken*
abhängig   dependent * *on/upon: von* || depending * *(up)on: von/davon; on whether...or not: davon ob...oder nicht* || contingent * *(up)on: von Umständen* || subject * *to: von Zustimmung usw.* || dependant * *siehe 'dependent'* || conditional * *bedingt (durch), abhängig (von); make something conditional on: etwas abh. machen von* || be subject * *to: von einer Zustimmung, Vorschrift usw. abhängig sein*
abhängig davon ob   depending on whether * *or not: oder nicht* || depending as to whether
abhängig von   depending on || in relation to * *im Verhältnis zu* || dependent on || contingent on * *von Umständen* || subject to * *von Zustimmung* || dependent upon
Abhängigkeit   dependence || sensitivity * *Empfänglichkeit, Empfindlichkeit* || relationship * *z.B. mathematische Funktion; between: zwischen; to: zu/mit* || characteristic * *Kennlinie, math.Funktion* || interdependence * *gegenseitige ~*
Abhaspel   uncoiler * *Abwickler für Blech, Draht,*

Kabel || pay-off reel * *in Walzstraße/Blechbehandlungsanlage/Draht-/Kabelmaschine* || payoff reel * *in Walz-/Blech-/Draht-/Kabelstraße*
abheben   lift off
Abhilfe   remedy * *take remedial measures: Abhilfe schaffen, Abhilfemaßnahmen ergreifen* || relief * *afford relief: Abhilfe schaffen* || counter-measure * *Abhilfe-/Gegenmaßnahme (possible: mögliche)*
Abhilfemaßnahme   corrective measure || countermeasure * *Gegenmaßnahme* || counter-measure * *possible: mögliche* || remedy * *(Plural: remedies)* →*Abhilfe*
Abhilfemöglichkeit   possible counter-measure
abholen   pick up || fetch || come for || collect || collection * *(Substantiv)* || retrieve * *heraus/zurückholen/fischen, wiedererlangen*
abisolieren   strip || bare || skin
Abisolierwerkzeug   stripper || wire stripper || stripping tool
Abisolierzange   wire stripper
Abkantbank   folding press * *zum Biegen/Abkanten von Blech*
Abkantpresse   press brake * *zum Biegen von Blech* || power break * *motorisch angetriebene Abkantbank zum Biegen von Blechen* || folding press || bending press * *zum Biegen von Blech (-profilen) usw.* || trimming press || press break || brake * *zum Biegen von Blech ("brake" und "break" ist beides richtig)*
Abkantvorrichtung   bending device
abklemmen   disconnect
abklingen   die away * *schwächer werden, nachlassen, sich verlieren* || fade away * *dahinschwinden, verschwinden, vergehen, zerrinnen* || subside * *(figürlich) einsinken, absacken, sich setzen* || ebb * *(figürlich) verebben, versiegen, abnehmen, (dahin-) schwinden* || dissipate * *sich zerstreuen/auflösen/verflüchtigen* || decay * *schwinden, abnehmen, schwach werden, (herab)sinken, zerfallen, verfallen, zugrundegehen* || die out
abklingend   decaying * *z.B. Schwingungen*
Abklingzeit   decay time
abknickende Kennlinie   inflected characteristic
Abkommen   agreement * *Vereinbarung (make/enter into: treffen)* || settlement || accord || pact * *Pakt, strategisches Abkommen*
abkoppeln   uncouple || decouple
abkratzen   scrape off || scratch off
abkühlen   cool-off || chill * *schlagartig ~*
Abkühlphase   cooling-off period
Abkühlung   cooling || damper * *(figürlich)* || cooling down * *z.B. eines Motors nach dem Abschalten*
Abkühlungskurve   cooling-down curve
Abkündigung   announcement of discontinuation * *dass ein Produkt nicht mehr gefertigt/geliefert wird* || notification of delivery-discontinuation * →*Produkt~, Liefer*~
Abkündigung eines Produktes   announcement that a product will be discontinued
abkürzen   shorten || short || bypass * *einen Umweg ~* || abbreviate * *(Zeit, Wörter, Rede usw.) abkürzen* || kürzen
Abkürzung   abbreviation || acronym * *(kryptisches) Kürzel* || code * *Kode(nummer); siehe auch 'Kurzbezeichnung'* || shortcut * *(mnemotechnisches) Kürzel*
Ablage   depot * *Lagerstelle, Ablegestelle* || deposition * *das Ablegen; Aufbewahrungsort, Magazin* || place of deposition * *Aufbewahrungs-/Ablageort* || filing * *das Ablegen von Akten, Dokumenten usw.* || place of deposit * *Aufbewahrungsort, ~ort*
Ablagerung   deposition || accumulation || deposit * *das Abgelagerte; dust: Staub; moisture: Feuch-*

ablängen

tigkeit; dirt: Schmutz || sediment * *{Geologie}* || residue * *Rückstand, Überbleibsel*
ablängen cut to length
Ablängsäge crosscut saw * *{Holz}* || cross-cut saw * *{Holz}*
Ablass drain * *z.B. für Flüssigkeiten* || drain hole * *Ablauf-/Ablassöffnung* || discharge device * *Entleerungsvorrichtung*
ablassen drain * *Wasser, Öl* ~ || let off || blow off * *Dampf* ~
Ablassschraube drain plug
Ablauf sequence * *['si: kwens] Abfolge* || program flow * *Programmablauf* || procedure || outlet * *Vorrichtung* || expiration * *einer Frist* || termination * *eines Vertrages* || process || drain * *Ablass (vorrichtung), z.B. für Flüssigkeiten* || drain hole * *Ablassöffnung* || discharge device * *Ablassvorrichtung, z.B. an einem Behälter* || expiration * *einer Frist, Zeit usw.*
Ablauf der Überwachungszeit timeout
Ablaufdiagramm sequence chart || flow chart * *Flussdiagramm*
Ablaufebene task execution level * *in einem Echtzeit-Computersystem*
ablaufen expire * *Frist, Zeit, Garantie usw.* || run * *z.B. Programm, Prozedur* || be in progress * *im Gange sein, fortschreiten, in Entwicklung befindlich sein* || flow off * *abfließen* || drain off * *auch Flüssigkeit* || lapse * *z.B. Frist, Vertrag* || terminate * *z.B. Frist, Vertrag* || end * *Ausgang nehmen* || turn out * *Ausgang nehmen* || come to an end * *sein Ende nehmen* || run down * *Uhr* || wear out * *verschleißen* || run off * *Flüssigkeit* || elapse * *vergehen, verstreichen von Zeit* || snub * *snub a person off: jemanden ablaufen lassen* || flow down * *abfließen* || run through * *Gegend* || run out * *eines Zeitgliedes* || be executed * *Programm*
ablaufen lassen run * *z.B. ein Programm/Projekt*
ablauffähig ready to run * *z.B. Softwareprogramm* || runnable * *z.B. Programmcode auf einem Rechner* || executable * *ausführbar, z.B. Programmcode auf einem Rechner*
Ablaufglied sequential logic element || step logic element || sequential control element
Ablaufkette sequence cascade || sequencer || sequencer cascade || drum sequencer || drum controller || sequential control * *Ablaufsteuerung* || sequence control || control sequence
Ablaufkettensteuerung sequential control || sequence cascade control
Ablaufkontrolle debug function * *{SPS}* || program check * *{SPS}* || debugging * *{SPS}*
Ablaufplan sequence control chart || control system flowchart * *{SPS}*
Ablaufschritt sequence step || sequencer step
Ablaufsteuerung sequence control * *['si: kwens]* || sequencing control || operating system * *Betriebssystem* || state machine * *elektronisches oder softwaremäßig realisiertes Zustandssteuerwerk* || sequencing controller * *als Einrichtung/Gerät* || sequential control || sequence control system * *bei SPS* || sequencer || sequence controller
Ablauftisch discharge table * *zur Materialabfuhr*
Ablaufumgebung runtime environment * *Hard-/Softwareumgebung, auf der eine Software ablaufen soll* (→*Zielsystem*)
Ablaufzeit ramp-down time * *Rücklaufzeit eines Hochlaufgebers* || runtime * *eines Programms* || run-time * *eines Programms*
Ablegeeinrichtung offloading equipment
ablegen arrange * *anordnen* || locate || deposit || lay down || file * *Akten, Briefe* || set down * *absetzen*
Ableger stacker * *Einrichtung zum Ablegen von Produkten (z.B. von bedruckten Papierbögen) auf*

einen Stapel || branch * *eines Unternehmens/einer Firma*
ablehnen refuse || decline || reject * *auch* →*abweisen*
Ablehnung rejection * *Ablehnung, Zurückweisung, Verwerfung, Einwand* || refusal || objection * *against/to: von*
Ableit- bypass- || leakage || deflection || discharge
ableiten derive * *herleiten (auch Gleichung, Formel), zurückführen (from: von), mathematisch differenzieren* || discharge * *Strom, Ladung* || dissipate * *Wärme* || shunt * *Strom* || drain off * *z.B. Waser* || deduce * *folgern; schließen (from: aus); herleiten z.B. einer Formel* || leak off * *Strom* || abduct * *Hitze* || dissipate * *Hitze* || divert * *z.B. Spannungsspitzen; auch ablenken,abwenden; from: von, to: nach* || arrest * *aufhalten, hemmen* || deduce * *ab-/herleiten (from: von); folgern, schließen (from: aus); vermuten*
Ableiter conductor * *Leiter* || surge suppressor * *z.B.Überspannungsableiter* || arrester || diverter
Ableitkondensator bypass capacitor * *Querkondensator*
Ableitung leakage * *Streuung, Auslaufen, Lecken, Schwund, Entweichen* || derivation * *mathematisch* || derivative * *{Mathem.} das Abgeleitete/das Differenzierte (z.B Beschleunigung: dv/dt)* || deduction * *(Schluss-) Folgerung,* →*Herleitung, (logischer) Schluss* || outgoing cable * *elektr. Kabel/Leitung* || outgoing leads * *z.B. in einer Schaltanlage*
Ableitwiderstand bleed resistor || discharge resistor || bleeder
ablenken divert * *from: von, to: nach* || deflect * *Licht, Feld, Strahlen usw.* || take off * *Aufmerksamkeit, Gedanken usw.* || turn away || turn off || arrest * *aufhalten, hemmen, ableiten, (Aufmerksamkeit usw.) fesseln/festhalten* || turn aside
Ablenkung deflection * *Verbiegung, Abweichung, Ausschlag, Ablenkung (auch eines Magnetfeldes)* || deflexion || diffraction || refraction || averting * *das Abwenden/-wehren* || turning away || turning off || diversion || distraction
Ablesbarkeit readability * *z.B. einer Anzeige*
ablesen read * *Skala* || read out * *Messgerät, Anzeigewert usw.* ~ || read off * *Messgerät, Anzeigewert usw.* ~ || take a reading from * *Messwert von einem Messgerät*
Ablesung read-off * *eines Mess/Anzeigewerts* || eadout * *von e. Messgerät*
Ablösedrehzahl cut-in speed * *oberhalb derer die Feldschwächung beginnt* || basic speed * *oberhalb derer die Feldschwächung beginnt* || base speed * *oberhalb derer die Feldschwächung beginnt* || cut-out speed * *oberhalb derer die Feldschwächung beginnt*
Ablösefrequenz cut-out frequency * *Frequenz bei der die Spannungsanhebung b.e.Frequenzumrichter f/U-Kennlinie endet* || end point of the voltage boost * *Frequenz,bei d.die Spannungsanhebung bei e. Frequenzumrichter endet*
ablösen replace * *ersetzen* || supersede * *einem Vorgänger nachfolgen* || take the place of * *einem Amtsvorgänger nachfolgen* || take over from * *übernehmen von z.B. von Vorgänger* || discharge * *Schuld* ~ || (sich) blättrig ~ *(z.B. Farbe)*; peel off * *(sich) blättrig* ~ *(z.B. Farbe); (Klebeetikett usw.) abziehen; abschälen* || alternate * *sich* ~ *(in: bei)* || relieve one another * *sich (gegenseitig)* ~ *(at: bei)* || relieve * *Schicht, Wache, Einheit usw.* ~ || work in shifts * *bei der Arbeit* ~ || detach * *lösen, abtrennen/montieren* || take turns * *sich (gegenseitig)* ~ *(at: bei)* || transfer * *übergeben, übertragen, übertreten (to: zu/an/nach)*
ablösende Regelung transfer control || changeover

control ‖ overriding control ‖ overrule control ‖ override control ‖ overrun control
ablösender Hochlaufgeber self-bypassing ramp function generator * *überbrückt sich nach erfolgtem Hochlauf*
ablösender Regler override controller ‖ controller with transfer characteristic * *z.B. n-Regler mit Steuerung d. Ausgangsbegrenzg (Moment).* ‖ transition-type controller ‖ controller with override characteristic ‖ controller with dynamically acting limit signals * *Regler m.ablös. Begrenzg. (Grenze dynam. geführt)* ‖ controller with constricting output limiting signals * *Regler mit ablösender Begrenzung*
Ablösepunkt switch over point * *z.B. für Feldschwächung* ‖ cut-out point ‖ crossover point ‖ transfer point ‖ change-over point ‖ field weakening point * *bei diesem (EMK-)Wert beginnt die Feldschwächung des Motors*
Ablöseregelung override control * *siehe auch →ablösende Regelung*
Ablöseschaltung transfer circuit ‖ takeover circuit ‖ override circuit ‖ overrule circuit ‖ transition-type circuit
Ablösespannung switch over voltage * *für Feldschwächung* ‖ CEMF crossover voltage * *oberhalb dieser Spanng.beginnt d.Feldschwächbereich b. DC-Motor* ‖ voltage at which field weakening starts * *Einsatzpunkt f. Feldschwächung*
Ablösung substitution * *Substitution* ‖ replacement * *Ersetzen* ‖ detachement * *(Ab)Lösung, Abtrennung, Abbau* ‖ peeling off * *(Sich-) Abschälen, Abblättern, Abbröckeln*
ablöten unsolder
Abluft exhaust air * *ausgestoßene/ausströmende Luft, Abluft, Abgas* ‖ extracted air * *abgesaugte Luft* ‖ hot air * *ausströmende erwärmte (Kühl-) Luft* ‖ stale air * *Abgase, verbrauchte Luft* ‖ discharged air
Abluftreinigung exhaust air purification
Abmachung arrangement ‖ term arranged * *Einzelpunkt einer Abmachung* ‖ settlement ‖ agreement ‖ conventional agreement * *vertragliche*
abmagnetisieren demagnetize ‖ neutralize
Abmagnetisierung demagnetization
abmanteln strip * *the insulation: Kabel ~*
Abmaße dimensions * *WxHxD: BxHxT (Breite x Höhe x Tiefe)* ‖ measurement dimensions
abmelden log off * *im Computernetzwerk, Steuerfunktion*
abmessen measure ‖ measure off ‖ gauge * *genau abmessen (hauptsächlich Längen/Dickenmaß)*
Abmessung dimension ‖ size * *Größe*
Abmessungen dimensions * *of smaller dimensions: mit kleineren ~*
Abminderung reduction
Abnahme acceptance * *z.B. Abnahmeprüfung eines Produktes durch Kunden oder Prüfinstitut* ‖ acceptance test * *Abnahmeprüfung durch Kunden bzw. Prüfinstitut/Kontrollbehörde (z.B. TÜV, PTB)* ‖ final inspection * *Abnahmeprüfung* ‖ removal * *Demontage* ‖ decrease * *Verringerung* ‖ web transfer * *einer Papierbahn vom Sieb in d. Papiermaschine* ‖ depression * *Fallen (z.B. von Preisen), ~, Schwäche, Schwächung, Flaute, Depression, Wirtschaftskrise*
Abnahmebeauftragter inspector ‖ testing officer * *offizieller*
Abnahmebericht acceptance report
Abnahmefilz pick-up felt * *in Papiermaschine (übernimmt das vom Sieb kommende Papier)*
Abnahmeprüfprotokoll inspection report ‖ acceptance test report
Abnahmeprüfung conformance test * *Normprüfung eines Produkts; siehe auch →Abnahme* ‖ acceptance test
Abnahmeprüfzeugnis inspection certificate ‖ acceptance-test certificate
Abnahmesystem take-off system * *Abtransport/Abfuhr von Teilen/Material aus einer Maschine* ‖ discharge system
Abnahmeverpflichtung commitment to take delivery * *of: von Lieferungen*
Abnahmevorschrift quality specifications
abnehmbar removable ‖ detachable ‖ demountable ‖ easy-to-remove * *leicht ~/demontierbar*
abnehmen accept * *(offiziell) anerkennen, annehmen, akzeptieren* ‖ approve * *genehmigen, an erkennen, bestätigen* ‖ inspect ‖ test ‖ decrease * *reduzieren, kleiner werden* ‖ diminish * *vermindern, verringern, verkleinern, herabsetzen* ‖ shrink * *schrumpfen* ‖ slow down * *an Geschwindigkeit ~* ‖ go down ‖ validate * *zertifizieren* ‖ taper * *sich verjüngen; z.B. b. hoher Drehzahl abgesenkte Strombegrenzung, abnehmend Wickelhärte* ‖ decay * *schwinden, sinken, verfallen* ‖ pick off * *z.B. Strom von einem Stromabnehmer/Schleifkontakt ~/abgreifen* ‖ detach * *lösen* ‖ remove * *lösen* ‖ take off * *lösen* ‖ decay * *schwinden, abnehmen, (ab)sinken, schwach werden, ver-/zerfallen*
abnehmend decreasing ‖ degressive ‖ decelerating * *Geschwindigkeit, Drehzahl*
abnehmende Wickelhärte tension taper * *beim Aufwickler* ‖ taper tension reduction * *beim Aufwickler* ‖ taper tension * *Zugreduzierung m.steig. Durchm. bei Aufwickler z.Schonen d. Innenlagen*
Abnehmer buyer * *Käufer* ‖ purchaser * *Käufer* ‖ customer * *Kunde* ‖ client ‖ consumer * *Verbraucher* ‖ current collector * *Stromabnehmer* ‖ pickup * *Abtastgerät, Messwertaufnehmer, Aufnahmeapparatur,* ‖ receiver * *Empfänger* ‖ doffer * *{Textil} Abnahmewalze in Krempel-/Kardenmaschine*
abnutzen wear ‖ wear out ‖ use up ‖ wear up
Abnutzung wear ‖ abrasion * *Abrieb* ‖ erosion * *Verschleiß, Abnützung, Schwund, Zerfressen* ‖ rate of wear * *bezogen auf die Betriebs/Nutzungszeit*
Abonnent subscriber * *auch eines Telekommunikations-Dienstes*
abordnen delegate ‖ assign ‖ send
abpumpen evacuate
Abräumer discharging machine * *z.B. in der Verpackungstechnik*
Abrechnung statement * *Kosten-* ‖ account * *Rechnung* ‖ settlement * *of accounts: von Konten* ‖ accountancy * *Buchhaltung* ‖ clearing * *Abrechnungs/Bankverkehr* ‖ invoicing * *{Wirtschaft}*
Abrechnungseinheit invoicing department * *Dienststelle*
Abreise departure * *for: nach; on my departure: bei meiner ~*
abreisen leave ‖ depart ‖ start * *for: nach* ‖ lift off * *mit Flugzeug*
abreißen break ‖ break off ‖ snap * *zerspringen, zerreißen, entzweigehen* ‖ come to a dead stop * *ein Prozess, eine Beziehung usw.* ‖ tear off * *außeinanderreißen (aktiv)* ‖ tear down ‖ niederreißen ‖ pull off * *auseinander reißen (aktiv)* ‖ pull down * *Gebäude* ‖ strip * *Werksanlage, Maschine* ‖ dismantle * *Werksanlage, Maschine*
Abreißspannung extinction voltage * *Löschspannung*
Abreißverschluss tear-off closure * *{Verpack. techn.}*
Abricht-Hobelmaschine surface planer * *{Holz}*
abrichten trim * *Schleifscheibe usw.* ‖ dress * *Schleifscheibe egalisieren, Säge usw. scharfmachen* ‖ plane * *{Holz} mit Hobel(maschine)* ‖ true

* *z.B. Schleifscheibe ~, einpassen, einschleifen, abziehen, nachschleifen*
Abrichthobel surface planer * *{Holz}*
Abrichthobelmaschine surface planing machine || surface planer * *{Holz}*
Abrichtwerkzeug dressing tool * *{Werkz.masch.} zum Scharfmachen von Sägen, Schleifscheiben usw.*
Abrieb abrasion || wear * *Verschleiß* || abraded particles * *abgeriebenes Material* || dust * *(abgeriebener) Staub, carbon dust: Kohlebürstenabrieb*
abriebbeständig wear-resistant
abriebfest abrasion-proof
Abriebfestigkeit wear resistance || abrasion resistance
Abriss breakage * *Draht, Faden, Materialbahn usw.*
abrollen pay out * *Kabel* || unwind * *abwickeln* || uncoil * *Blech, Draht, Kabel, Seil* || unreel || roll off * *mit der Hand*
Abroller unwinder || unwind stand
Abrollung rollstand * *Rollenständer/Gerüst; shaftless: achslos* || unwind || unwind stand * *Abroller*
Abruf call * *on call: auf ~~* || poll || polling
abrufen call || recall || poll || call up * *z.B. Diagnosedaten* || call-up * *z.B. einzelne Lieferungen bei Rahmenvertrag* || fetch * *herbeiholen, ab-/hervorholen*
abrunden round || round off * *eine Zahl* || round down * *eine Zahl nach unten* || round up * *eine Zahl nach oben* || bring up to a round figure * *eine Zahl nach oben aufrunden* || bring down to a round figure * *eine Zahl nach unten abrunden* || make perfect * *vollkommen machen* || complete * *vollenden* || perfect * *perfektionieren*
Abrundung rounding || rounding off
ABS ABS * *'American Bureau of Shipping': amer. Schiffsklassifikationsgesellsch., erlässt Prüfvorschriften*
absacken sag * *z.B.Last am Kran* || sink || drop suddenly
Absage rejection || cancellation || refusal * *Ablehnung*
Absatz paragraph * *in einem Text* || clause * *kurzer Abschnitt in einem Text* || ledge * *vorstehender Rand* || sales * *von Produkten/Waren auf dem Merkt* || distribution * *das Absetzen auf dem Markt* || selling * *das Verkaufen*
Absauganlage exhauster || exhaust system || extractor
absaugen suck off || exhaust * *Gas* || vacuum * *mit Staubsauger usw.* || vacuum off * *mit Unterdruck*
Absauggerät extractor || suction box || exhaust fan * *Absauggebläse* || exhaust
Absaugung suction * *removal by suction* || exhauster * *Gasabsauger* || extraction
abschaben scrape off || abrade * *abreiben, abschürfen, abscheuern, aufscheuern* || wear off * *abnutzen* || shabby * *abgeschabt, z.B. Textilstoff* || threadbare * *abgeschabt, z.B. Textilstoff*
abschaltbar switched || interruptible || interruptible || disconnectible || capable of being switched off || which can be turned off * *z.B. Thyristor ($\rightarrow$GTO)* || gate-turn-off * *Thyristor ($\rightarrow$GTO)* || which can be switched off
Abschaltdrehmoment tripping torque
Abschaltdrehzahl tripping speed * *Auslösedrehzahl einer Drehzahlüberwachung* || cut-off speed
abschalten switch off || turn off || break * *z.B. durch Relaiskontakt ~* || disconnect * *Stromkreis/Verbindung unterbrechen* || stop || shut down * *z.B. Antrieb stillsetzen* || trip * *Störabschaltung* || put out of action || disable * *Freigabe für Ein/Ausgang (z.B. e.SPS)/Überwachung usw.* ~ || de-energize * *mit Low-Signal ansteuern* || stall * *stehenbleiben* ||

deactivate * *deaktivieren, z.B. eine Überwachung* || switch out || cut off || power off * *die Versorgungsspannung ~* || shutdown * *(Substantiv) Abschaltung*
Abschaltschwelle shutdown threshold || tripping threshold * *für Fehlerabschaltung/Auslösung*
Abschaltspannung voltage induced on circuit interruption * *beim Ausschalten e.induktiven Last* || gate turn-off voltage * *(Mindest-) Abschaltspannung e. Thyristors*
Abschaltstrom turn-off current * *z.B. bei Thyristor*
Abschaltthyristor gate turn-off thyristor * *Thyristor, der sich über einen Gate-Anschluss abschalten lässt* || GTO * *Abkürzung für 'Gate Turn-Off Thyristor'* || turn-off thyristor
Abschaltung switching off || shutdown || cut off || stopping * *Maschine* || opening * *Kontakt, Stromkreis* || tripping * *durch Überwachungskontakt* || interruption * *Unterbrechung* || disconnection * *eines Stromkreises* || trip * *z.B. auf Grund eines Fehlers oder des Ansprechens einer Überwachung* || shut-down || disengagement * *einer Kupplung* || switch off
Abschaltvermögen breaking capacity * *eines Leistungsschalters, einer Sicherung usw.*
Abschaltvorgang switch-off operation || tripping sequence * *Ablauf einer Störabschaltung (Abschaltung auf Grund e. Fehlers)*
Abschaltzeit break time || clearing time || turn-off time * *Thyristor* || interruption time * *~ einer Sicherung/eines Leistungsschalters usw.*
abschätzen estimate * *schätzen* || value * *bewerten*
Abschätzung estimation || evaluation * *auch Berechnung* || valuation * *Bewertung, (Ab-) Schätzung, Wertbestimmung, Taxierung, Veranschlagung* || estimate * *superficial estimate: überschlagsmäßige/grobe ~* || analysis * *Analyse* || appraisal * *(Ab-) Schätzung, Taxierung, Schätzung, Schätzwert, Bewertung* || assessment * *(steuerliche usw.) Einschätzung, Veranlagung* || assessing * *Einschätzen, Bewertung, Beurteilung*
abscheiden separate out * *abtrennen*
Abscheider separator * *z.B. oil separator: Ölabscheider; centrifugal separator: Zentrifugal~*
abscheren shear off
Abscherfestigkeit shearing strength
abschicken send off || issue * *z.B. ein Telegramm auf Bus* || dispatch * *ab/versenden*
Abschieber deflector * *{Verpack.techn.}*
Abschirmblech shield plate || screening plate
abschirmen shield * *from: gegen* || screen
Abschirmhaube shield cover || screening cover
Abschirmplatte shield plate
Abschirmung shield || screen
abschleifen grind off
abschließbar lockable
abschließen lock * *ver-/zusperren* || terminate * *z.B. eine Leitung mit Wellenwiderstand ~* || close * *zum Ende bringen* || settle * *endgültig ~* || transact * *einen Handel ~* || complete * *beenden, komplettieren, vervollständigen* || conclude * *abschließend beenden* || finalize * *beenden, vollenden, endgültig erledigen, endgültige Form geben*
abschließend final * *endgültig* || concluding || definitive || in conclusion * *(Adv.)* || finally * *(Adv.)*
Abschluss closing * *Beendigung* || termination * *auch Leitungs~ z.B. durch Widerstand oder Bus~stecker* || deal * *~ eines Geschäfts* || end || completion * *Fertigstellung* || termination * *~ eines Geschäfts* || qualification * *~ einer (Berufs)Ausbildung* || terminator * *Leitungs~, Bus~widerstand, ~Stecker*
Abschlussstecker terminating connector * *zum Leitungsabschluss* || terminator * *zum Leitungsabschluss*

Abschlusswiderstand terminating resistor
Abschlusszeugnis diploma || leaving-certificate
abschmieren lubricate || grease
Abschneidemaschine cut-off machine
* →*Querteilanlage z.b. für Blech*
abschneiden cut off || cut || clip || shear off
Abschneidmaschine cutting machine || shear
* →*Schere*
Abschnitt section * *Kapitel, Abschnitt/Teil einer Maschine/Anlage, Bereich* || cut * *abgeschnittenes Stück* || paragraph * *z.B eines Buches* || period * *Zeit~* || offcut * *das Abgeschnittene* || stage * *Stufe (eines Prozesses usw.)* || segment * *Abschnitt, Teil, Segment*
abschnittweise sectionally * *(Adv.)* || section by section * *(Adv.)* || by sectors || in stages
* →*stufenweise* || by steps * →*schrittweise*
abschrägen slope || bevel || chamfer
Abschrägung slope || bevelling || taper
abschranken fence || barrier
Abschrankung fence || barrier || guard
abschrauben unscrew || screw off || screw out
* *herausschrauben*
Abschreibung depreciation || write-off
abschwächen weaken * *z.R. Motorfeld ~* || diminish || reduce || attenuate
Abschwächer attenuator
Abschwächung attenuation || weakening * *z.B. Motorfeld* || decrease || reduction || fading * *Schwund* || reduction * *Reduzierung*
Abschwung downturn * *economic: wirtschaftlicher*
Absender sender * *z.B.eines Briefs* || transmitter
* *einer Nachricht, von Daten*
absenken ower || reduce * *reduzieren* || diminish
* *reduzieren* || drop down * *(figürl.) wert-/zahlenmäßig ~* || slow down * *Drehzahl, Leistungsfähigkeit usw. ~*
Absenkung reduction * *Reduzierung* || lowering
* *Niedriger-Machen* || decrease * *Herabminderung, (in: von)* || diminution * *Verminderung, Verringerung, Verkleinerung*
Absetzbecken settling tank * *bei der Wasseraufbereitung*
absetzen set down * *auf eine Unterlage* || sell * *verkaufen* || deposit * *ablegen, sich ~* || strip-off * *Kabelisolierung* || recess * *aussparen, einsenken, vertiefen, ausbuchten, zurücksetzen*
Absetzlänge stripping length * *Abisolierlänge, entlang derer das Kabel von der Isolierung befreit wird*
absichern fuse * *mit Sicherung/Kurzschlussschutz* || guard against || provide security for * *z.B. Kredit ~* || protect by fuse * *mit Sicherung ~* || give security * *sicherstellen* || qualify * *eine Aussage ~*
Absicherung guarding * *Schutz* || fencing * *Einzäunung, Abschrankung* || protection * *gegen Gefahr, Störung und Zerstörung* || fusing * *z.B. mit Schmelzsicherung* || fuses * *mit meheren Schmelzsicherungen* || fused link || fuse protection * *mit Schmelzsicherung usw.*
Absichtserklärung letter of intent * *schriftliche* || agreement of understanding
absinken go down * *z.B. Last am Kran* || sink * *sinken* || subside * *z.B. Boden* || give way * *z.B. Boden* || sag * *zusammensacken, herunterhängen* || fall
* *fallen, z.B. Preise* || decrease * *sich vermindern* || drop * *sinken, (herunter/ab)fallen, z.B. Spannung; schwächer werden*
absolut absolute
absolute Abweichung absolute error
absoluter Fehler absolute error
absoluter Gleichlauf absolute synchronism * *z.B.*
→*Winkelgleichlauf*
absoluter Sprung unconditional jump
Absolutwert absolute value

Absolutwertbildner absolute-value generator || absolute-value function || absolute-value module
Absolutwertbildung absolute-value generator || absolute-value function
Absolutwertgeber absolute encoder * *erzeugt e.absolute Winkel-/Lageinformation z.B. i. Gray-Code (einschrittiger Code)* || absolute position encoder * *Lagegeber*
absondern segregate
absorbieren absorb
Absorption absorption
abspecken downsize
abspeichern store || save * *retten* || memorize * *in Speicher festhalten, aufbewahren*
Absperrhahn stopcock || cock
Absperrventil shut-off valve || check valve
* →*Rückschlagventil* || stop valve || gate valve
absprechen coordinate * *koordinieren* || arrange
* *verabreden, arrangieren* || agree * *verabreden* || disallow * *Ansprüche (z.b. auf Schadenersatz) negativ bescheiden*
absprengen crack off
Abstand distance || space * *(auch Verb: in einen definierten ~ bringen)* || interval * *zeitlich; in short intervals: in kurzen Abständen* || clearance * *lichter Raum, Zwischenraum, Spiel(raum)* || gap * *Lücke* || difference * *Unterschied* || apart * *125 mm apart: im ~ von 125 mm* || margin * *vom Rand e.Druckseite; Vorsprung (z.B. zu einem Mitbewerber auf dem Markt); (Handels)Spanne* || spacing * *(einzuhaltender, vorgesehener, vorzusehender) Abstand* || pitch * *in einem Raster; z.B. auf dem Bildschirm, Zähne auf Zahnstange usw.*
Abstand nehmen von desist from || refrain from || stand off from * *Abstand halten/sich entfernt halten von*
Abstandhalter spacer || stand-off
Abstandshalter stand-off || spacer washer * *Abstandsscheibe* || spacer * *auch in einer Batterie* || cage * *in einem Wälzlager (Kugelkäfig)* || standoff
Abstandsmessgeber distance measurement transducer * *z.B. nach d. Ultraschallprinzip*
Abstandsstück spacer * *siehe auch* →*Abstandshalter*
Abstapelung destacking
abstechen cut off * *mit Drehbank ~* || tap off
* *Schlacke/Stahl ~* || run off the iron * *Hochofen ~* || contrast with * *~ gegen(-über)* || kill * *totmachen (auch figürl.)*
Abstechmaschine cutting-off machine * *Werkzeugmaschine*
abstecken unplug * *Stecker*
absteigender Ast descending branch
abstellen stop * *Maschine* || switch off || shut down
|| stall * *Motor* || deposit * *ablegen* || set down * *auf eine Unterlage absetzen* || turn off * *(Wasser, Gas) abdrehen, (Licht, Radio usw.) ausschalten, (Schlag usw.) abwenden/-lenken*
abstimmen harmonize * *in Einklang bringen, harmonisieren* || tune * *z.B. Radio, Schwingkreis; to a frequency: auf e. Frequenz* || balance * *austarieren* || attune to * *z.B. auf eine Norm* || coordinate * *koordinieren, aufeinander abstimmen, richtig anordnen* || vote * *bei einer Wahl* || match * *anpassen (to: an/auf)* || jointly agree upon * *etwas gemeinsam vereinbaren*
Abstimmung harmonization || coordination * *Koordierung* || tuning * *eines Schwing-/Regelkreises usw.*
abstrahlen emit || radiate
Abstrahlung radiation * *Strahlung* || emission
* *Ausstrahlung, Ausströmung*
abstrakt abstract
Abstreifer wiper || blade * *doctor blade:* →*Rakel* || stripper * *z.B. in der Gießerei*

abstufen grade || graduate || group * *in Gruppen einteilen* || classify * *in Klassen einteilen*
Abstufung graduation || grading || grouping * *Aufteilung in Gruppen* || classification * *Aufteilung in Klassen*
Absturz crash * *Programm, Computer, Flugzeug usw.; system crash: System~ (eines Comuters usw.)* || fall || sudden fall * *plötzlicher ~* || drop * *Fallen, Sinken, Sturz, Rückgang; drop in prices: Preisrückgang* || precipice * *Abgrund (auch figürlich)* || shoot down * *zum ~ bringen* || force down * *zum ~ bringen* || plunge * *Sturz, Stürzen, Ein/Untertauchen* || deadlock * *"Auf-Der-Stelle-Treten" z.B. eines Softwareprogramms* || ockup * *~/Sich-Verklemmen eines Softwareprogramms*
abstürzen fall down || drop * *z.B. Last vom Kran* || crash * *Programm, Flugzeug usw.* || go down || be precipitated * *(figürlich) hinein/hinabgestürzt sein (into/in)* || descend steeply * *abschüssig sein, steil abfallen*
abstützen support
Abstützung support
Abt Dpt. * *Abk. für 'department'; Abteilung* || dept * *Abk. für 'department'; Abteilung*
Abtast-Halteglied sample and hold element || SH * *Abk.f. 'Sample and Hold element'*
Abtast-und-Halte Verstärker sample-and-hold amplifier || sample-and-hold
Abtast-und-Halte-Glied sample-and-hold circuit || sample-and-hold
Abtast-und-Halte-Verstärker sample-hold amplifier || sample-and-hold amplifier || sample-and-hold * *Abtast-und-Halteglied*
Abtast-und-Halteglied sample-and-hold
abtasten scan || sample * *Regler* || sense * *Messwert von Messaufnehmer, Sensor*
Abtaster sensor || pickup || scanner * *optischer/berührungsloser, z.B. für Barcode*
Abtastfehler sampling error
Abtastfrequenz sampling frequency || scanning frequency || sampling rate * *Abtastrate* || scan rate * *Abtastrate*
Abtastgeschwindigkeit sampling rate * *Abtastrate* || sampling frequency * *Abtastfrequenz* || scan rate * *Abtastrate*
Abtastimpuls sampling pulse
Abtastintervall sampling interval || sample interval
Abtastkopf scanning head
Abtastperiode sampling period
Abtastpunkt sampling point
Abtastrate sampling rate || scan rate
Abtastregelung sampling control || digital control
Abtastregler sampling controller
Abtastsignal sampled signal
Abtastsystem sampled-data system || sampling system
Abtasttheorem sampling theorem || Nyquist theorem
Abtastung scanning * *z.B. optisch* || sample * *(Abfragen eines) Augenblickwert(s)/Messwert(s)* || sampling * *z.B. eines Reglers* || sensing * *das Abtasten, Messen, Erfassen eines Messwerts*
Abtastzeit sampling time || scan time || sampling interval
Abtastzeitintervall sampling time interval
Abtastzeitpunkt moment of sampling || sampling instant || point of sampling
Abtastzeitüberwachung sampling time monitoring * *erkennt Überlauf eines Softwarezyklus*
Abtastzyklus sampling cycle || sampling interval * *Abtastintervall, Tastperiode*
Abtauung defrost
Abteilung department || Dept. || group * *Gruppe von Abteilungen, Unternehmensbereich* || center * *Zentrum* || division * *Unternehmensbereich* || section || shop * *Werkstatt, Fertigungsabteilung*

Abteilungsbezeichnung department name
Abteilungsleiter head of department || departmental chief || division head * *Leiter eines Unternehmensbereichs*
abtragen erode * *wegfressen, erodieren* || remove || carry off
Abtransport removal * *Fortschaffen, Wegräumen* || evacuation * *Aus/Entleerung, Verlegung, Räumung, Umsiedlung* || transport * *Ab/Antransport, Beförderung, Versand, Verschiffung* || shipment * *Versand, Verladung*
abtrennen disconnect || isolate || clear * *a connection: eine Nachrichtenverbindung* || cutt-off || separate
Abtrennung disconnection
Abtrieb output * *eines Getriebes* || driven end of shaft || outdrive || output shaft * *Abtriebswelle eines Getriebes, einer Kupplung usw.* || downward pressure * *nach unten wirkender Druck* || power take-off * *Kraftausleitung* || drive-out || transmission * *Kraft-/Drehmoment-übertragung*
Abtriebs- output * *z.B. eines Getriebes, einer Kupplung usw.*
Abtriebsdrehmoment output torque * *eines Getriebes*
Abtriebsdrehzahl speed of the driven side || r.p.m. of the driven side || output speed * *eines Getriebes*
Abtriebsflansch output flange * *z.B. eines Getriebes, Durchtriebs usw.*
Abtriebsmoment output torque * *eines Getriebes usw.*
Abtriebsseite output side
abtriebsseitig on output side * *z.B. bei Getriebe, Durchtrieb usw.*
Abtriebsstufe secondary gear * *bei Getriebe*
Abtriebswelle outgoing shaft * *bei Getriebe* || slow speed shaft * *bei Untersetzungsgetriebe* || output end * *z.B. einer Kupplung/eines Getriebes* || output shaft * *eines Getriebes/einer Kupplung*
Abwärme waste heat
abwärts downwards || downward || down
abwärts fahren descend * *{Aufzüge}*
abwärtskompatibel downwards compatible
Abwärtsregler step-down regulator * *Schaltregler für Schaltnetzteil mit Ausgangsspannng. < Eing.spanng.*
abwärtszählen count down
Abwasser waste water || sewage
Abwasserbehandlung waste water treatment
Abwasserreinigung waste water purification
abwechseln alternate
abwechselnd alternating || in turns * *(Adv.)* || by turns * *(Adv.)* || alternately * *(Adv.)*
abwehren block || beat back * *Angriff usw.* || repulse || parry * *Stoß usw. (auch figürl.)* || ward off * *Stoß usw.(auch figürl.)* || avert * *Unheil* || head off * *Unheil* || be defensive * *abwehrend/in der Defensive sein*
abweichen deviate * *from: von; by: um* || differ * *from (one another): von(einander)* || vary || yaw * *gieren, ~ (from: von; z.B. ungewolltes ~/-kommen e.Schiffs v.Kurs auf Grund d.Seegangs)*
abweichend deviating || varying || irregular
Abweichung deviation * *transient deviation: vorübergehende ~, frequency deviation: Frequenz~* || error * *Regel~, mean error: mittlere ~* || variation || difference * *anomaly* || tolerance * *zulässige ~* || departure * *from: von einer Regel* || discrepancy * *z.B. in einem Vertragsablauf* || violation * *Verletzung einer Vorschrift* || change * *Änderung* || deflection * *Verbiegung, Ablenkung, Ausschlag* || excursion * *Abweichung; Abschweifung; Ausflug*
abweisen refuse * *ablehnen* || decline * *zurückweisen* || reject * *ablehnen, abweisen* || repulse * *Angriff*

abwenden divert * *from: von* || avert * *Gefahr/Unheil ~*
abwerfen throw off * *the load: Last ~* || drop * *z.B. Behälter ~* || render * *Gewinn/Profit usw. ~* || shed * *abstoßen (Wasser, Hörner usw.), abwerfen (auch Last), verbreiten (Geruch, Laune), ablegen*
abwesend absent || away || not in || missing
Abwesenheit absence
Abwickelbremse tension brake
Abwickelhaspel uncoiler || decoiler
abwickeln unwind * *Material v. einem Abwickler ~* || reel off * *Material ~* || uncoil * *Blech, Kabel, Draht von einer Haspel ~* || transact * *Geschäfte ~* || handle * *durchführen* || administer * *verwalten, handhaben, durchführen* || manage * *verwalten, durchführen* || wind off * *Material von einer Rolle/Spule/Haspel/Trommel ~* || wind up * *Geschäfte usw. ~* || process * *(figürl.) einen Vorgang, einen Auftrag usw. ~; behandeln, einem Verfahren unterwerfen* || de-coil * *Draht/Blech ~*
Abwickelrolle unwind reel || payoff reel
Abwickeltrommel pay-out reel
Abwickelwerkzeug unwrapping tool * *für Wire-Wrap Verbindung*
Abwickler unwinder || unwind || uncoiler * *Abhaspel für Blech, Draht, Kabel oder Seil* || pay-off reel * *Abhaspel in Walzwerk/Blechbehandlungsanlage* || pay-off * *z.B. für Metallband, Kabel, Draht* || pay-off station * *Abwickelstation, z.B. für Kabel* || pay-off stand * *Abwickelmaschine, z.B. für Kabel* || unwind stand * *→Rollenständer*
Abwicklung control * *Management* || management || winding-up * *das Abwickeln von Geschäften usw.* || wind-up * *(amerikan.) das Abwickeln von Geschäften usw.* || unwinder * *Material-Abroller/Abwickler* || unwind * *Material-Abrollung/Abwicklung/Abwickler* || carrying out * *~ eines Auftrags* || transaction * *~ einer Bestellung usw.* || processing * *Erledigung* || administration * *verwaltungsmäßige ~* || uncoiler * *Abhaspel für Blech, Draht, Kabel* || winding off * *das Abwickeln von Material von e. Rolle/Spule/Haspel* || winding * *das Abwickeln von Material von e.Rolle/Spule/Trommel/Haspel* || settlement * *Erledigung* || pay-off * *Abwickeleinrichtung, z.B. für Blechband, Kabel* || pay-off station * *Abwickelstation, z.B. für Kabel* || pay-off stand * *Abwickler, z.B. für Kabel; Rollenständer*
abwinkeln bend * *down: nach unten; upwards: nach oben*
Abwurf dropping
abwürgen stall * *Motor* || choke * *Motor* || strangle
abziehbar withdrawable || removable
abziehbare Klemme removable terminal * *z.B. Schraubsteckklemme*
abziehen draw || pull || pull off * *z.B. Kupplung ~* || substract * *subtrahieren* || extract || remove || withdraw * *Personal, Wälzlager, Stecker, Steckbaugruppe usw. ~* || move off || detach * *entfernen, z.B. ein Steckkabel* || remove * *z.B.Lager/Kupplung/Zahnrad ~* || unplug * *Stecker/Stöpsel ~* || peel off * *abschälen, z.B. Schutzfolie* || draw off
Abziehen des Lagers bearing extraction || bearing remove * *Entfernen des Lagers von der Welle mit Abziehwerkzeug*
Abzieher bearing extractor * *für Kugellager* || extraction tool * *Werkzeug* || extractor
Abziehkraft withdrawing force
Abziehvorrichtung bearing extractor * *z.B. für Kugellager* || puller tool * *Abziehwerkzeug; z.B. für Lager, Zahnrad* || puller || extracting tool
Abziehwerkzeug extraction tool || extractor || puller tool * *z.B. für Lager, Zahnrad* || puller
Abzisse abcissa * *X-Achse* || axis of x * *X-Achse* || X-coordinate

abzuführende Verlustleistung heat to be dissipated
Abzug exhaust * *Lüftung* || withdrawal || departure || rebate * *Rabatt* || deduction * *Geld, Preisnachlass; charges deducted: nach ~ der Kosten* || exit feed * *dort wird Material herausgefördert* || discount * *Skonto, Preisnachlass, Rabatt, Abschlag (on: auf)* || substraction * *Substraktion; when substracting: nach ~ von* || exit section * *fördert (bandförmiges) Material aus einer Verarbeitungsstraße heraus* || allowance * *Preis-~/-Nachlass, Rabatt; Vergüt., (zahlen-/wertmäßige) Berücksichtigung (for: von)* || outlet * *Auslass* || pull-off * *z.B. Antrieb i.Kabelmaschine, in d. d.Kabel geklemmt u.mit konstanter Geschw. geförd.wird* || drawing off * *das Abziehen* || trigger * *bei Gewehr usw.*
abzüglich less || minus || deducting || charges deducted * *abzüglich der Spesen*
Abzugsgeschwindigkeit take-up speed * *z.B. in Textilmaschine*
Abzugskraft withdrawal force || retention force * *eines Steckers*
abzulösendes outgoing * *zu löschendes (Ventil in Stromrichterschaltung)*
Abzweig branch || branch circuit || distribution || tap * *Stichleitung* || feeder * *einer Energieverteilungsanlage*
Abzweigdose branch box || distribution box || tapping box || branching box || junction box
abzweigen tap off * *→anzapfen*
abzweigend tapping * *Leitung usw.*
Abzweigung branch || branching || branching off || junction || setting apart
abzwicken nip off || pinch off * *auch figürlich*
Achsabstand distance between achses || wheelbase * *Radabstand* || center distance
Achsantrieb axle drive || center drive * *(amerikan.) bei Wickler* || centre drive * *(brit.)*
Achse axle || shaft || axis * *geometrisch* || spindle
Achsen axes * *(Plural von 'axis') auch bei Werkzeugmaschine*
Achshöhe shaft height * *(el.Masch.) Abstand zw. d. Längsachse der Welle u. der Fußauflagefläche (DIN 42939)* || motor frame * *Motor-Baugröße* || frame size
Achslager journal bearing * *Zapfenlager* || axle bearing || shaft bearing
achslos shaftless * *z.B. Papierroller*
Achsmodul axis module * *bei Modular- oder Servoumrichter*
achsparallel axially parallel || parallel-axis
Achsschenkel axle journal * *{Auto}* || axle neck * *{Auto}* || steering knuckle * *{Auto}* || stub axle * *{Auto}*
Achstacho shaft tachometer
Achswickler center winder * *(amerikan.)* || centre winder * *(brit.)* || core winder || axial winder || shaft-driven winder
achteckig octagonal
achten auf pay attention to * *siehe auch →beachten* || mind || take care of || take care that * *darauf achten, dass* || bear in mind * *daran denken, nicht vergessen* || regard * *beachten, berücksichtigen, respektieren; disregard: nicht achten auf, nicht beachten* || give consideration to
Achterbahn roller-coaster
achtfach octal
Achtkantschraube octagon bolt
Achtung attention || caution * *als Hinweis z.B. auf Schild/in Betriebsanleitung* || important * *vor Hinweis*
AD-PE explosion-proof enclosure * *Schutzart druckf.Kapselg.i.Italien: Antideflagranti Prove di Esplosione*
Adapter adapter || converter || adaptor

Adapterbaugruppe adapter module || adapter board * *Leiterkarte*
Adapterkabel adapter cable || adaptor cable
Adapterplatte adapter plate
Adapterstecker adapter connector || adaptor plug || adaptor connector
adaptieren adapt
adaptiert adapted
Adaption adaption || adaption function
Adaptionskapsel adapter casing || adapter module
adaptive Regelung adaptive control
adaptiver Regler adaptive regulator
Adaptivregler adaptive controller
addieren add || sum up * *auf-, zusammenaddieren*
Addierer adder || summing element || sum block * *Funktionsbaustein* || summator
Addierpunkt summing point
Addierstelle summing point * *{Regel.techn.}*
Addierverstärker summing amplifier
Additionsglied summing element || adder
additiv additive * *(Adjektiv), Zusatz (Substantiv)* || adding
Ader core * ~ *einer mehradrigen Leitung/Kabel* || wire || conductor || strand * *Litze(nleiter)*
Aderendhülse ferrule || wire-end ferrule || pin-end connector * *Stiftkabelschuh* || conductor end sleeve || end sleeve || connector sleeve
Aderkennzeichnung core identification
Aderpaar core pair
Aderquerschnitt core cross-section
Aderzahl number of conductors
Adhäsion adhesion
Adhäsionsbelag adhesion lining * *z.B. bei einem Treibriemen, z.B. aus Chromleder*
Adhäsionskraft adhesive power || adhesiveness
adiabatische Erwärmung linear temperature rise * *Temperaturanstieg über d.Zeit b.reiner Wärmespeicherg. (keine Wärmeleitg.)*
administrativ administrative
Administrator administrator * *auch Verwalter eines Computernetzes*
Admittanz admittance
Adressat addressee
Adressbus address bus
Adressdekodierung address decoding
Adresse address
Adressfehler address error
adressierbar addressable
adressieren address
Adressierfehler addressing error
Adressierung addressing
Adressiervolumen address range || address space || addressing range
Adressleitung address line
Adressliste address map * *{Computer}* || memory map * *{Computer}* || mailing list * *für Postversand* || distribution list * *für die Verteilung eines Informationsdienstes*
Adressraum address space || memory space || memory address space
Adressversatz address displacement || offset address
ADU ADC * *Abkürz. f. 'Analog to Digital Converter': Analog/Digital-Umsetzer* || AD converter
AEE AEE * *'Asociación Electrotécnica Espanola': Spanischer Elektrotechn.Verband (ähnlich VDE)*
aerodynamisch aerodynamic
AFNOR AFNOR * *Association Francaise de Normalisation': französischer Normenverband ähnlich DIN/VDE*
AG PLC * *programmable logic controller: Speicherprogrammierbare Steuerung* || programmable controller * *speicherprogrammierbare Steuerung* || Inc. * *Incorporated: Aktien/Kapitalgesellschaft* ||
incorporated company * *eingetragene Gesellschaft, Aktiengesellschaft*
Aggregat aggregate || unit || generating set * *zur Stromerzeugung*
aggressiv aggressive || elbow-pushing * *gegenüber Mitbewerbern mit 'Ellenbogen' vorhehend* || corrosive * *chem. zersetzend, zerfressend, ätzend, angreifend; z.B. Gase, Atmosphäre*
aggressive Atmosphäre corrosive atmosphere
aggressive Gase corrosive gases * *Korrosion verursachend* || aggressive gases
aggressive Medien aggressive substances
agieren act * *as: als*
Ah Ah * *Amperestunde* || shaft height * *Achshöhe AH* || frame size * *Baugröße/Achshöhe AH bei Motor*
ähneln resemble * *ähnlich sein/sehen, gleichen, ähneln* || be similar to * *ähnlich sein* || be like * *ähnlich sein* || be comparable to * *vergleichbar sein mit* || correspond to * *entsprechen* || be analogous to * *entsprechen* || be a parallel to * *entsprechen* || look like
ähnlich similar * *to: zu/wie* || like || alike || close together * *dicht beieinander* || resembling * *ähnlich sehend/seiend, ähnelnd* || analogous to * *entsprechend, analog/parallel zu* || corresponding to * *entsprechend*
Ähnlichkeit similarity * *Ähnlichkeit, Gleichartigkeit* || resemblance * *to/between: Ähnlichkeit mit/zwischen* || likeliness || analogy * *Analogie, Ähnlichkeit, Übereinstimmung*
akademischer Grad university degree
akkreditieren accredit * *offiziel/als berechtigt anerkennen/bestätigen/beglaubigen*
Akkumulator accumulator * *Ergebnisregister in Computer-Rechenwerk* || storage battery * *ladbare Batterie*
akkumulieren accumulate
akquirieren canvass * *Kunden bereisen, Aufträge hereinholen*
Akquisiteur canvasser * *Vertriebsmann im Außendienst, (Handels-) Vetreter*
Akquisition sales/marketing || promotion * *Verkaufsförderung* || canvassing * *Kundenwerbung* || acceptance * *Hereinnahme von Aufträgen* || acquisition * *(käuflicher) Erwerb, (An-) Kauf, Erwerbung, Erfassen, Aneignung* || sales promotion || sales/marketing activities
Akquisitionshilfe sales aid || selling tool
Akquisitionsunterlagen promotion material
akquisitorisch sales/marketing * *Marketing-, vertrieblich* || promotional * *werblich, Werbe-*
Akte file * *Aktenordner, abgelegte Akten; put on file: zu den Akten legen* || document
Aktenkoffer file case || attaché case || briefcase * *Aktentasche* || portfolio * *Aktenmappe* || executive case
Aktennotiz memo || memorandum
Aktenordner file || ring binder * *Ringbuch, dünner Aktenordner* || folder * *Ringbuch, dünner Aktenordner* || arch file || binder
Aktentasche brief-case || portfolio * *Mappe* || document case * *Aktenmappe* || brief-bag
Aktenvermerk memo
Aktenzeichen reference number
Aktie share * *shares are at a premium: die ~n stehen gut; prospects are fine: die ~n stehen gut (figürl.)* || stock * *(amerikan.)*
Aktion action * *Handeln, Handlung, Tat, Tätigkeit, Gang, Funktionsweise; enter into action: in Aktion treten* || measure * *Maßnahme* || campaign * *z.B. Werbeaktion, Kampagne, Feldzug* || drive * *Vorstoß, (Sammel-) Aktion, Kampagne,* || activity * *Tätigkeit, Betätigung, Unternehmung, Rührigkeit, Wirksamkeit, Betrieb* || scheme

* *Plan, Projekt, Programm, Komplott, Intrige* || project * *Projekt, Plan, (Bau)Vorhaben, Entwurf*
Aktionär   shareholder
Aktionsradius   operating range * *Fahrzeug* || radius of action * *Reichweite*
aktiv   active || effective
Aktiva   assets
aktiver Zustand   active state
aktives Filter   active filter
aktivierbar   can be activated * *kann aktiviert werden* || able to be activated * *möglich, aktiviert zu werden*
aktivieren   activate || enable * *freigeben* || mobilize * *mobilisieren, z.B.Kräfte* || speed up * *(Arbeit usw.) beschleunigen* || bring into action * *ins Spiel bringen, einsetzen* || invoke * *aufrufen (Programm usw.)*
aktiviert   activated
Aktivierung   activation
Aktivität   activity
Aktivkohlefilter   active carbon filter * *z.B. in der Wasser-/Abwasseraufbereitung*
Aktivteile   core-and-winding-asssembly * *Teile einer el.Masch., in denen sich die Energieumwandlung vollzieht* || active parts * *[el.Masch.]*
Aktor   actuator
aktualisieren   update || refresh || renew || prolong * *Vertrag* || up-date || actualize || bring up-to-date * *auf den neuesten Stand bringen*
Aktualisierung   updating || update
Aktualoperand   actual parameter * *bei SPS*
Aktuator   actuator * *Stellglied* || final control element * →*Stellglied*
aktuell   current || up-to-date * *modern, gängig, dem letzten Stand entsprechend* || acute * *Problem* || actual * *wirklich, tatsächlich* || present * *gegenwärtig* || fashionable * *modisch* || momentary * *momentan, augenblicklich* || topical * *von aktueller Bedeutung/ aktuellem Interesse, z.B. Frage, Problem, Film* || up-to-the-minute * *in letzter Minute verfügber (Informationen, Nachrichten usw.)*
aktueller Wert   actual value
Akustikkoppler   acoustic coupler * *zur Datenübertragung über Telefon*
akustisch   acoustic
akustische Meldung   audible signal || audible alarm * *Warnsignal*
akzeptabel   acceptable * →*tragbar; (to: für)* || fair
Akzeptanz   acceptance * *enjoy a high level of acceptance: eine hohe Akzeptanz haben*
Akzeptor   acceptor
Alarm   alarm * *auch Programmunterbrechung in einem Computer* || interrupt * *[SPS | Computer]*
Alarm freigeben   enable interrupt
Alarm sperren   disable interrupt
Alarmbearbeitung   interrupt service routine * *durch Computer* || interrupt processing * *durch Computer, SPS usw.* || interrupt procedure * *durch Computer* || alarm processing * *Verarbeitung von Fehler- und Warnungsmeldungen, z.B. durch ein Leitsystem*
Alarmeingang   interrupt input
Alarmereignis   interrupt event || alarm event
alarmfähig   interrupt-capable * *z.B. Digital-/Binäreingang*
alarmgesteuert   interrupt-controlled || interrupt-driven
Alarmhupe   alarm horn
Alarmliste   interrupt list * *z.B. bei SPS*
Alarmmeldung   alarm message || alarm signal
Alarmprogramm   interrupt service routine || interrupt handler
Alarmverarbeitung   alarm processing || interrupt processing

alfabetisch   alphabetic
Algorithmus   algorithm || law * *Bildungsgesetz, Rechenvorschrift*
Alkydharz   alkyd resin
alle   all || every * *every 4 sec: alle 4 Sekunden* || everything * *alles; everything new: alles neue; everything else but: alles andere als*
alle mögliche   all sorts of || all possible means * *try all possible means: alles Mögliche versuchen (auch 'try everything')*
allein   alone || unassisted * *ohne Hilfe* || singlehanded * *ohne Hilfe* || by oneself * *ohne Hilfe, automatisch* || separately * *jedes Teil für sich* || individually * *jedes Teil für sich* || only * *nur* || merely * *(Adv.) nur, bloß* || exclusively * *(Adv.) ausschließlich* || but * *aber* || however * *jedoch*
Alleinschutz   motor protection exclusively by PTC thermistors * *[el.Masch.]*
Alleinstellungsmerkmal   unparalleled feature * *einzigartiges Merkmal* || prestigious feature * *mit Prestige behaftetes Merkmal* || unique feature
allen Ansprüchen gerecht werden   meet all requirements || come up to expectations * *die Erwartungen erfüllen*
Allen Bradley   A-B * *Kurzform*
aller Art   all kinds of * *(vorangestellt) all kinds of machines: Maschinen aller Art* || of all kinds || all sorts of * *(vorangestellt)* || of different makes * *verschieder Marken/Hersteller*
alles aus einer Hand   one stop shopping || from a full-liner * *vom Komplettanbieter*
alles in einem   all in one
alles mögliche   all sorts of things
allgemein   general || generic * *oberbegrifflich; generic term: Oberbegriff; generic standard: übergreifende Norm* || popular * *allgemein verbreitet* || common * *üblich* || overall * *umfassend* || universal * *universell, (all-) umfassend* || generally * *(Adv.) im Allgemeinen* || in general * *(Adv.)* || generically * *(Adv.) oberbegrifflich*
allgemein gültig   universally valid
allgemein verwendbar   general-purpose
allgemeine Anforderungen   general requirements || normal applications * *for normal applications: für allgemeine Anforderungen (z.B. Kondensatoren)* || basic requirements
allgemeine Anwendung   general use
allgemeine Daten   general data || general specifications * *technische Daten*
allgemeine Hinweise   general information
allgemeine Lieferbedingungen   general conditions of sale and delivery || General Supply Conditions for Products and Services of the Electrical Industry * *allg."grüne" Liefb.*
allgemeine Übersicht   general view * *of: über*
allgemeine Übersicht über   general view of
allgemeiner Hinweis   general note
allgemeiner Maschinenbau   general mechanical engineering
allgemeiner Trend   general trend || mainstream
Allgemeines   general || general information * *Kapitelüberschrift in Dokument*
Allgemeingültigkeit   universal validity || universality || generality * *Allgemeingültigkeit, allgemeine Regeln*
allmählich   gradual || by degrees * *(Adv.).* || little by little * *(Adv.).* || gradually * *(Adv.) nach und nach, allmählich.*
allmähliche Änderung   gradual change
allokieren   allocate * *(Speicherplatz) zuweisen/-teilen (to: an), anweisen, d.Platz bestimmen für*
allozieren   allocate * *(Speicherplatz) zuteilen*
allpoliges Abschalten   all-pole disconnection
allpoliges Schalten   all-pole switching

Allradantrieb all-wheel drive || four-wheel drive
* *bei Vierrad-Fahrzeug*
allseitig geschlossen closed on all sides
Allstrom universal current || AC-DC || d.c.-a.c.
Allstrom Relais universal relay || AC-DC relay || a.c.-d.c. relay
Allstrommotor universal motor * *mit Kommutator* || a.c.-d.c. motor || AC-DC motor * *mit Kommutator, z.Anschluss an Gleich- und Wechselstrom (Hauptschlussschaltung)* || universal AC/DC motor
Allstromrelais universal relay || AC-DC relais || a.c.-d.c. relay
Allzweck- general-purpose || all-purpose * *universal* || universal || all-in-one
alpha g rectifier stability limit * *Gleichrichtertrittgrenze* || minimum firing angle * *Gleichrichtertrittgrenze* || zero delay angle
alpha w inverter stability limit * *Wechselrichtertrittgrenze* || maximum firing angle * *Wechselrichtertrittgrenze* || maximum delay angle || inverter limit
alpha-G rectifier stability limit || zero delay angle || minimum firing angle
alpha-W inverter stability limit || maximum delay angle || maximum firing angle || inverter limit
alphabetisch alphabetic || alphabetically * *(Adv.)*
alphabetische Reihenfolge alphabetic sequence || alphabetic order
alphanumerisch alphanumeric
als bisher than previously || than up to now
als Funktion von in dependance of * *in Abhängigkeit von, z.B. Wert auf der X-Achse einer Kennlinie*
als Option optional
also thus * *(Adv), auch auf diese Weise* || so * *(Adv.).* || therefore || consequently * *(Adv.)* || hence * *folglich, daher, deshalb* || logically * *(Adv.) logischerweise* || i.e. * *d.h.*
alt old * *(auch Substantiv: "das Alte")* || aged * *bejahrt, gealtert (auch technisch)* || used * *gebraucht* || second-hand * *gebraucht, aus zweiter Hand* || old-time * *zurückliegend* || ancient * *geschichtlich alt* || same old * *unverändert, das gleiche alte* || unchanged * *unverändert* || long-standing * *seit langem bestehend (Freudschaft, Beziehungen usw.)* || long-established * *firm: alt eingeführte Firma* || former * *ehemalig* || worn * *verschlissen* || worn out * *verschlissen* || ex- * *ehemalig* || as it was * *wie üblich/letztes Mal/immer; everything remains as it was: es bleibt alles beim Alten* || past * *(Substantiv) das Alte, die Vergangenheit* || old things * *(Substantiv) das Alte* || the customary * *(Substantiv) das Alte/Gewohnte* || old ways * *(Substantiv) die alte (Art und) Weise* || original * *ursprünglich*
alteingesessen long-established
altern age * *Maschinenteil* || wear out * *verschleißen* || fatigue * *ermüden* || season * *z.B. durch Wettereinflüsse/künstlich* ~ || grow old * *Lebewesen*
alternativ alternative
Alternative alternative * *to: zu/für; have no alternative: keine Alternative haben* || choice * *Wahlmöglichkeit; have no choice: keine Alternative haben* || option * *Wahlmöglichkeit; have no option: keine Alternative haben*
Alterung ageing || seasoning || weathering || wear * *Verschleiß* || fatigue * *[fe'tieg] Ermüdung* || aging
alterungsbeständig nonageing || resistant to ageing || nonaging || resistant to aging || non aging || nonageing
Alterungsbeständigkeit resistance to ageing || resistance to aging || aging resistance
Alterungserscheinungen ageing effects || ageing symptoms || ageing phenomenon || aging effects || aging symptoms || aging phenomena

Alterungsriss season crack || fatigue crack
* *[fe'tieg ...] Ermüdungsriss*
Altmaterial salvage * *verwertbares* ~ || junk || scrap
Altöl used oil
Altpapier waste paper || recycled paper
ALU ALU * *Abkürzg. für 'Arithmethic and Logic Unit': Rechenwerk eines Computers*
Alu-Profil extruded aluminium
Aluminium aluminium || aluminum * *(amerikan.)*
Aluminium-Druckguss die-cast aluminium || cast aluminium
Aluminium-Folienkondensator aluminium foil capacitor
Aluminium-Spritzguss die-cast aluminium alloy
Aluminium-Strangguss extruded aluminium || extruded aluminium alloy * *aus Aluminium-Legierung*
Aluminiumfolie aluminium foil
Aluminiumgehäuse aluminium housing
Aluminiumlegierung aluminium alloy || aluminum alloy * *(amerikan.)*
Aluminiumprofil extruded aluminium
am at * *at the input: am Eingang* || the * *(in Zusammenhang mit Superlativ) z.B. the fastest: am schnellsten*
am Anfang initially * *(Adv.)*
am Beispiel von exemplified by
am besten the best * *(Adv.)*
am gebräuchlichsten most common
am häufigsten verwendet most commonly used || most popular
am Markt einführen introduce to the marketplace
am Netz on the supply system
am Telefon on the phone || on the telephone || on the end of the line * *am (anderen) Ende der Leitung* || speaking * *(salopp) 'am Apparat'*
amagnetisch non-magnetic
Amboss anvil
amorph amorphous
Amortisation amortization || liquidation || payback || return on investment
amortisieren amortize || pay off || redeem * *(Anleihe usw.)* || write off * *abschreiben*
Amortisierung payback || amortization || return on investment
AMP-Steckverbinder AMP blade-type terminal * *Flachsteckverbinder, z.B. von der Fa. AMP*
Ampere amps || amp * *1 Ampere*
Amperemeter ammeter || amps meter
Amperestunde ampere-hour
Amperewindungen ampere-turns
Amplitude amplitude || magnitude
Amplitudenfaktor crest factor
Amplitudengang amplitude response
Amplitudenspektrum amplitude spectrum
amtlich anerkannt officially approved
Amtsblatt official journal
an at || on || upon || by || against || in respect to * *hinsichtlich* || in the way of * *hinsichtlich* || to || onward * *(Adv.)* || up * *(Adv.)* || from now on * *(Adv.) von nun* ~ || along * *entlang (an)*
an Bedeutung gewinnen become increasingly significant
an der Momentengrenze along the torque limit
an erster Stelle stehen be at a premium * *in der Priorität* || have top priority * *höchste Priorität haben* || be the leader * *führend sein, z.B. am Markt* || be the main player * *Marktführer sein*
an Spannung legen energize
analog analog || analogue || analogous * *to/with: zu* || analogously * *(Adv.)* || by analogy * *(Adv.)*
analog zu analogous to || similar to * *ähnlich* || corresponding to * *entsprechend*

Analog-Ausgabebaugruppe analog output module
Analog-Digital-Umsetzer A/D converter || ADC || analog-to-digital converter || analog-digital converter || A-to-D converter || analog-digital convertor * *(veraltet)*
Analog-Digital-Umsetzung A/D conversion || analog-to-digital conversion
Analog-Digital-Wandler A/D converter || ADC || analog-digital converter || analog-to-digital converter || A-to-D converter
Analog-Digitalwandler analog to digital converter
Analog-Eingabebaugruppe analog input module
Analogausgabe analog output || analog output channel * *Ausgabekanal*
Analogausgang analog output
analoge Regelung analog closed-loop control
Analogeingabe analog input
Analogeingabebaugruppe analog input module
Analogeingang analog input
analoger Wahlausgang analog select output || programmable analog output || selectable analog output
analoger Wahleingang analog select input || programmable analog input || selectable analog input
analoges Messinstrument analog meter
Analoggröße analog value || analog variable
Analoginstrument analog meter
Analogmessgerät analog meter
Analogmultiplexer analog multiplexer
Analogregelung analog control
Analogschalter analog switch
Analogsignal analog signal
Analogsollwert analog setpoint
Analogtacho analog tachometer
Analogtechnik analog technology * *in: in der*
Analogwert analog value
Analysator analyzer
Analyse analysis || analytical process
Analysegerät analyzer
analysieren analyze
Anbau fitment || mounting || annex * *an Gebäude* || attachment * *Anbringung, Befestigung*
Anbau- built-on
anbauen attach * *Maschinenteil* || add * *Gebäude* || annex * *Gebäude(to: an)* || fit || retrofit * *nachträglich anbauen, nachrüsten*
Anbauflansch fixing flange || mounting flange || attaching pad
Anbaugeräte mounted equipment || mountes accessories * *z.B. Tacho, Bremse, Kupplung usw. an einem Motor*
Anbaugetriebe gearhead * *mit dem Motorflansch verschraubtes* ~ || built-on gearbox
Anbaumaße fixing dimensions * *z.B. für AC-Motoren: siehe IEC 72, DIN 42673 u. DIN 42677, jew. Blatt 1*
Anbausatz mounting kit || extension kit
Anbauschalter built-on circuit breaker * *z.B. am Motor angebauter EIN/AUS, Polum- oder Stern/Dreieckschalter* || built-on switch || built-on contactor * *{el.Masch.} direkt an den Motor angebautes Motorschütz*
Anbauten attachments || built-on accessories
anbieten offer || tender || quote * *Preis ansetzen*
Anbieter vendor * *Verkäufer, Lieferfirma* || supplier * *Lieferant* || provider * *Lieferant, Anbieter (auch von Online-Diensten und Netzzugängen)*
anbinden link to
Anbindung interfacing * *durch Signale, serielle Schnittstelle usw.* || interlinking * →*Verknüpfung* || connecting-up * →*Verbindung* || connection * *Verbindung; siehe auch* →*Verknüpfung; (to: an)* || linkage
anbrennen burn || catch fire
anbringen attach || fit || mount * *montieren, befestigen (at: an/auf)* || apply * *to: an, auf* || fix * *befestigen* || affix * *befestigen, anbringen, anheften, ankleben, beilegen, hinzufügen (to: an)*
andauernd lasting * *dauernd, anhaltend, beständig, nachhaltig; dauerhaft, haltbar* || continuous * *ununterbrochen, (fort)laufend, zusammenhängend, unaufhörlich, andauernd, fortwährend* || persistent * *ständig, nachhaltig, anhaltend, beharrlich, ausdauernd, hartnäckig* || incessant * *unaufhörlich, unablässig, ständig* || ceaseless * *unaufhörlich* || sustained * *anhaltend, Dauer-, ungedämpft, aufrechterhalten*
andenken contemplate * *ins Auge fassen, erwägen, nachdenken (über), überdenken, erwarten, rechnen mit* || schedule tentatively * *provisorisch einplanen, ungefähr zeitlich planen (for: für)*
ander other || different || second * *zweiter* || next * *folgender* || opposite * *gegenüberliegend*
änderbar alterable || changeable || variable || writeable * *beschreibbar, z.B. Speicherzelle, Datei, Parameter*
Änderbarkeit changeability || adaptability * *Anpassbarkeit*
andere als other than
andererseits on the other hand || in turn
ändern change || modify * *modifizieren* || alter || vary * *variieren, verändern, sich verändern (with: mit)* || revise * *revidieren, überarbeiten (Buch usw.), (wieder/nochmals) durchsehen, überprüfen*
anders otherwise * *(Adv.)* || differently * *(Adv.)* verschieden || else * *z.B. somebody else: jemand* ~ || other * *(Adj.)* || different * *(Adj.)* verschieden || second * *(Adj.)* zweit || opposite * *(Adj.)* gegenüberliegend, z.B. Seite || another * *be another story:* ~ *(gelagert) sein*
anderthalb one and a half
Änderung change * *in: in einem Ab/Verlauf; of: von einer Größe* || variation * *Veränderung, Wechsel, Schwankung* || degradation * *Verminderung, Schwächung, Verschlechterung* || drift * *z.B. temperatur/langzeitabhängig* || modification || update * *neue Version* || enhancement * *Verbesserung* || alteration * *(Ab/Um)Änderung, Veränderung, Umbau* || amendment * *Berichtigung, Verbesserung, Neufassung, Nachtrag, Zusatzartikel; (für Norm, Gesetz usw.)*
Änderungen vorbehalten subject to change without notice * *ohne Benachrichtigung d. Betroffenen* || alterations reserved || object to change without prior notice * *ohne vorhergehende Benachrichtigung* || subject to change without prior notice * *kann geändert werden ohne vorherige Benachrichtigung* || we reserve the right to make modifications * *wir behalten uns das Recht auf Änderungen vor*
Änderungs- of change * *(nachgestellt) z.B.: rate of change:* ~*geschwindigkeit*
Änderungsdatei file of the changes || change-log file
Änderungsgeschwindigkeit change rate || rate of change * *time derivative* * *Differenzialquotient nach der Zeit* || slew rate * *Anstiegsgeschwindigkeit eines Signals* || response * *bezüglich der Reaktion eines Systems auf einen Stimulus* || rate of rise * *Anstiegsgeschwindigkeit* || gradient * *Neigung, Steigung*
Änderungsmitteilung change note
andeuten hint at * *einen Hinweis geben auf* || intimate * *zu verstehen geben, andeuten/-kündigen, mitteilen* || indicate * *kurz erwähnen* || outline * *umreißen, in groben Zügen darstellen, skizzieren* || signify * *an-/bedeuten, kundtun, ankündigen*
andocken hook up
Andruckmaschine proofing press * *{Druckerei}*
Andruckrolle pinch roll

Andruckwalze pressing-on roller ‖ snubber roll * z.B. in Walzstraße ‖ pinch roll ‖ nip roller * (An)-Press-/Haltewalze
aneinander together * (Adv.) ‖ at each other
aneinander fügen join ‖ butt ‖ butt-join * auf Stoß ‖ link together * miteinander verbinden, verknüpfen, verketten
aneinander hängen chain ‖ cascade * kaskadieren, hintereinanderschalten ‖ adhere * aneinandergehängt sein ‖ stick together * aneinandergehängt-/geklebt sein ‖ daisy-chain * mit Hilfe v.Signalen, die jew.von einem Element zum nächsten durchgeschleift sind
aneinander reihen butt * (stumpf) aneinander stoßen (lassen) ‖ butt-mount * aneinander stoßend befestigen/montieren ‖ mount side by side * aneinander stoßend befestigen/montieren ‖ gang * z.B.Klemmen ‖ arrange side by side ‖ line up ‖ string together ‖ join together ‖ butt together
aneinander stoßen abut
aneinander stoßend butt-jointed
aneinanderreihbar buttable
anerkennen acknowledge ‖ recognize * as: als; auch durch eine Normenbehörde ‖ accept * als gültig ~ ‖ appreciate * lobend ~ ‖ approve * billigend/offiziell ~
Anerkennung acknowledgement ‖ certification * Zertifikation ‖ approval * z.B. durch eine Normenbehörde
anfachen excite * Schwingung ‖ induce
Anfahrautomatik automatic starting control ‖ automatic start up
Anfahrbedingung starting condition
anfahrbereit ready to start ‖ ready to go ‖ ready to run
Anfahrdrehmoment starting torque
anfahren start * losfahren, starten (Maschine, Fahrzeug) ‖ start-up * Anlage, Maschine, Kraftwerk, Reaktor ‖ approach * annähern, sich nähern, nahekommen, (fast) erreichen ‖ run up * hochfahren ‖ carry up * Material antransportieren ‖ initiate * beginnen, einleiten, einführen, aufnehmen ‖ starting * (Substantiv) ‖ reach * erreichen ‖ arrive * ankommen (at: bei) ‖ run into * rammen ‖ run-up * (auch Substantiv)
Anfahrfrequenz start-up frequency ‖ starting frequency
Anfahrmoment starting torque ‖ start-up torque
Anfahrsteuerung start control ‖ starting sequence control
Anfahrstrom starting current ‖ starting-up current
Anfahrstromstoß in-rush current surge * beim Einschalten eines (induktiven) Verbrauchers (Motor, Drossel usw.)
Anfahrt travel ‖ arrival * Ankunft
Anfahrumrichter start converter ‖ starting converter
Anfahrwarnung starting alarm
Anfahrwiderstand starting resistor
anfallen result * sich ergeben ‖ occur * auftreten, sich ergeben, passieren ‖ accrue * Gewinn, Zinsen usw. ‖ be incurred * Kosten, finanzielle Verluste; losses incurred: angefallene Verluste
anfällig sensitive * to: gegenüber ‖ prone * anfällig sein (to: für) ‖ susceptible * auch 'empfindlich, empfänglich' (to: für/gegen) ‖ unreliable * unzuverlässig ‖ undependable * unzuverlässig ‖ vulnerable * to: gegenüber ‖ delicate * →empfindlich, zart, zerbrechlich, fein, filigran, heikel, schwierig
Anfälligkeit susceptibility * auch 'Empfindlichkeit, Empfänglichkeit' (to: für/gegen) ‖ unreliability * Unzuverlässigkeit ‖ undependability * Unzuverlässigkeit ‖ predisposition * to: für/gegen ‖ prone-

ness * to: für, gegen ‖ sensitivity * Empfindlichkeit, Sensibilität, Sensitivität (to: für/gegen)
Anfang beginning ‖ start ‖ lead * z.B. einer Wicklung ‖ commencement * →Beginn ‖ early stage * früher Zeitpunkt (eines Projekts usw.)
Anfänger beginner ‖ greenhorn * (salopp) Grünschnabel ‖ non-professional * Laie ‖ novice * →Neuling ‖ novices * (Plural)
anfänglich initial * (Adj.) ‖ in the beginning * (Adv.) ‖ at first * (Adv.) ‖ initially * (Adv.)
anfangs in the beginning ‖ at first ‖ originally * ursprünglich
Anfangs- initial ‖ starting ‖ first ‖ start-up
Anfangsadresse start address ‖ restart address ‖ initial address
Anfangsbedingung initial condition
Anfangsdurchmesser start diameter * bei Wickler
Anfangskurzschlussleistung initial symmetrical short-circuit power
Anfangslage initial position
Anfangsverrundung initial rounding-off * für Hochlaufgeber-Rampe ‖ ramp start rounding * bei Hochlaufgeber ‖ ramp begin rounding ‖ initial rounding
Anfangswert initial value ‖ start value
Anfangswindung leading turn
anfasen chamfer
anfassen touch ‖ grip
Anfasung chamfer ‖ bevel
anfertigen make ‖ manufacture ‖ fabricate ‖ prepare ‖ draw up * schriftlich ‖ assemble * zusammenbauen/-setzen, montieren
anfeuchten moisten ‖ wet ‖ damp
anflanschbar able to be flange-mounted
anflanschen flange on ‖ flange-mount
anfordern demand ‖ call for ‖ request * bitten, ersuchen; auch Zugriffsrechte, Speicherplatz usw. ~
Anforderung requirement * regarding: bezüglich; depending on individual requirements: je nach ~; auch e.Norm ‖ demand * place higher/strigent demands on: höhere/strenge Ansprüche stellen an ‖ request * Wunsch, Ersuchen, Nachfrage, Bitte; (for: von/um) ‖ call for * Anforderung von, z.B. von Tagungsbeiträgen (call for papers) ‖ needs * Erfordernisse, Bedürfnisse
Anforderung erfüllen meet the requirements ‖ match the requirements
Anforderungen requirements * meet the requ.: die ~ erfüllen;be subject to stringent requ.: hohe ~ stellen (Aufgabe) ‖ demands * make high demands on: hohe Ansprüche stellen an; increased: erhöhte ‖ conditions * (Betriebs-) Bedingungen
Anforderungen stellen place demands * place extreme demands on: hohe ~ an ‖ place requirements * on: an
Anforderungsalarm user interrupt * bei SPS
Anfrage inquiry * z.B. nach e. Lieferangebot, auch 'Nachfrage, Erkundigung'; make: stellen ‖ question ‖ request * Anforderung/Frage; (up)on request: auf ~ ‖ survey * Umfrage
anfragen ask for ‖ inquire * for: nach ‖ consult * um Rat fragen
anfressen corrode * zer/anfressen, angreifen, korrodieren, rosten, s.einfressen ‖ pit * an/zerfressen, Löcher/Grübchen/Narben/Vertiefgn. bilden z.B. i.Lager, Zylinderlaufbahn ‖ erode * an/zer/wegfressen, verschleißen, abnützen; z.B.Kontakte
anführen quote * zitieren ‖ mention * erwähnen ‖ lead * vorne stehen
Angabe information * Information (on: über) ‖ specification * (technische) Spezifikation ‖ description * Bezeichnung, Beschreibung
Angaben information ‖ details ‖ statement ‖ data ‖ specifications * z.B. in einem Datenblatt
Angaben zu information regarding

Angaben zur Person personal data
angebaut built-on * *onto: an*
angeben specify * *im Einzelnen ~, spezifizieren* || declare * *erklären* || show * *ausweisen* || indicate * *z.B.Richtung; e.Einzelheit in/durch e. Diagramm/Grafik/Spezifikation; Größe i.e. Dimension* || ;inform * *informieren* || pretend * *etw. falsches vorgeben; protzen* || quote * *z.B. eine Größe in einer Maßeinheit, Schreibweise ~; in Newton: in Newton* || give * *z.B. Grenzwerte in einer Norm, Kenndaten in einem Datenblatt ~* || define * *by: durch (Definition)* || rate * *with: einen Wert für eine Kenngröße e.Produkts ~/zusichern* || note * *notieren, aufschreiben* || state * *angeben, darlegen, dar-/feststellen; z.B. Daten, (Sonder)Wünsche (i.Bestellformular)* || stipulate * *festsetzen, vereinbaren* || tabulate * *in Tabelle ~*
Angebot offer * *on offer: im Angebot* || quotation * *Preis/Verkaufs/Lieferangebot; generate a quotation: ~ ausarbeiten/erstellen* || business offer || bid * *~ bei öffentlicher Ausschreibung; Gebot (bei Versteigerung), Bewerbung (for: um), Bemühung* || quote
Angebotsausarbeitung preparation of offer
Angebotsbearbeitung handling of quotations || processing of quotations || administration of inquiries
Angebotsstadium quatation phase || quotation stage
angebracht advisable * *ratsam* || appropriate * *passend, gut ~; inappropriate/out of place/ill-timed: un~* || indicated * *angebracht, angezeigt* || mounted * *montiert, befestigt (siehe auch →anbringen)*
angedacht sein be tentatively scheduled * *unverbindlich eingeplant sein*
angeflanscht flanged on || flange-mounted
angefordert requested || demanded * *z.B. Momentenrichtung*
angeforderte Momentenrichtung demanded torque direction
angegeben stated * *z.B. in einer Beschreibung* || specified * *z.B. in Datenblatt ~, spezifiziert*
angegossen cast integrally * *z.B. Aufstellfüße eines Motors*
angegossene Füße integrally cast feet * *z.B. am Motor*
angehängt trailing * *hinten/am Ende angehängt*
angehen go on * *Lampe, Licht, Anzeige*
Angelegenheit matter || affair || business || concern * *Sache (that's his concern: d.ist seine ~), Geschäft, Bedenken, Interesse, Beziehung*
angelegte Spannung applied voltage
angelenkt hinged * *mit Scharnier/Gelenk/Angel versehen, (zusammen)klappbar* || pivoted
angelernt trained on the job * *Arbeiter* || job-trained * *Arbeiter*
angelernter Arbeiter semiskilled worker || job-trained worker
Angelpunkt pivot * *Drehzapfen* || cardinal point * *(figürlich)* || pivot point
angemessen appropriate || proper || pertinent * *passend, schicklich, sachdienlich, einschlägig* || reasonable * *vernünftig* || fair * *reell, z.B. Preis* || due * *gebührend, angemessen, fällig, passend, vorschriftsmäßig, recht(zeitig), erwartet* || commensurate * *in Einklang stehend (with/to: mit), angemessen, entsprechend*
angenähert approximate * *annähernd*
angenehm pleasant || agreeable * *gefällig, liebenswürdig, von angenehmen Wesen* || comfortable * *komfortabel* || cosy * *behaglich* || welcome * *willkommen*
angenommen supposed * *(Adj.)* || vermutet || let us assume that * *(Adv.)*
angepasst matched || adapted
angeregte Schwingung sympathetic vibration * *mechanische* || sympathetic oscillation

angeschraubt bolted || bolted on
angesichts in view of * *in view of the fact that: angesichts der Tatsache, dass* || considering || in face of * *konfrontiert mit*
angesprochen responded * *z.B. Regelung, Messung, Überwachung*
Angestellter employee || white-collar worker || clerk * *Büro~*
angestelltes Lager preloaded bearing * *Lager m.axialer Vorspannng.,die.e.Vorzugslage bewirkt u.so Vibrationen vermeidet* || spring-loaded bearing * *axial federnd vorgespanntes Kugellager*
angesteuert fired * *Thyristor* || energized * *Klemme/Steuereingang mit HIGH-Signal*
angetrieben driven || powered
angewandte Wissenschaften applied sciences
angezapft tapped
angezeigt indicated || displayed
angezogen pulled in * *Relais* || energized * *Relais* || tightened * *Schraube*
angreifen act * *z.B. eine Kraft an einem Punkt* || touch * *anfassen, (Kapital) angreifen* || handle * *anfassen* || tackle * *(figürlich) e. Aufgabe usw.*
angreifen || set about * *(figürlich) Aufgabe usw.* || approach * *(figürlich) Aufgabe usw.* || weaken * *schwächen* || exhaust * *erschöpfen* || attack ⊥ *(auch figürlich) feindlich* || assail * *(auch figürl.) feindlich* || corrode * *korrodieren, chemisch zersetzen* || bite * *Säge usw.*
angrenzend adjacent * *to: an* || contiguous || adjoining * *to: an*
Angriffspunkt point of application * *z.B. einer Kraft* || point of engagement * *Eingriffspunkt* || point of action || point of attack || place of application * *einer Kraft usw.*
Angriffsrichtung line of action * *auch einer Kraft* || direction of force * *einer Kraft*
anhalten stop || bring to a stop || come to a stop || halt || stall * *abwürgen, blockieren, auslaufen lassen* || retain * *beibehalten, halten (z.B. I-Anteil, Durchmesserrechner), bewahren* || hold * *(an)halten (z.B. Durchmesserrechner, SPS-Programm usw.)* || coast * *austrudeln, im Leerlauf fahren*
Anhaltezeit deceleration time * *Rücklaufzeit*
Anhaltswert approximate value || rough value
anhand by * *durch* || with the help of * *mit Hilfe von* || by means of * *mittels* || by the example of * *am Beispiel von*
Anhang appendix * *eines Buches,einer Beschreibung, eines Berichts* || annex * *auch einer Beschreibung* || supplement * *Ergänzung, Zusatz, Nachtrag, Anhang*
Anhänge appendices
anhängen affix || attach
Anhänger trailer * *Fahrzeug*
Anhängevorrichtung lifting eyebolt * *beim Krahn* || lifting pin * *beim Krahn*
Anhängsel suffix * *an ein Wort,eine Bestellbezeichnung,einen Variablennamen* || appendix * *auch Anhang, Ansatzteil*
Anhäufung accumulation || increase * *Anstieg* || aggregation || piling-up || aggregate * *Anhäufung, (Gesamt)Menge, Summe*
anheben lift * *heben* || raise * *emporheben, erhöhen* || accentuate * *anheben (z.B. Frequenzkurve), hervorheben, akzentuieren, betonen* || jack up * *mit Vorrichtung* || boost * *Strom/Spannung/Drehmoment durch eine bestimmte Maßnahme* || increase * *erhöhen*
Anhebung increase || boost * *z.B. Spannung/Leistung/Drehmoment in einem bestimmten Arbeitsbereich* || elevation * *in die Höhe*
anhören listen to || sound * *sich ~ (well: gut, badly: schlecht)* || lend an ear to * *jdn. anhören; jdm. zuhören*

Anker armature * *['a: matje] an elektr.Maschine, Relais, Bremse; Magnetanker* || anchor * *Befestigungselement* || brace * *Befestigungselement* || rotor * *Rotor*
Ankerbandage armature band
Ankerblech armature lamination
Ankerdrossel armature reactor || armature choke || smoothing reactor * *Glättungsdrossel*
Ankerdurchflutung armature ampere-turns || armature flux || rotor flux * *Rotordurchflutung*
Ankereisen armature iron
Ankereisenverlust armature core loss
Ankerfeld armature field
Ankerflussdichte armature flux density || armature induction
Ankergesteuerter Wickler armature winder * *DC-Auf/Abwickler mit Durchmesseranpassg. i.Ankerkreis*
Ankerinduktion armature induction
Ankerinduktivität armature inductance
Ankerkreis armature circuit
Ankerkreisinduktivität armature circuit inductance || armature inductance
Ankerkreisumschaltung armature reversal * *Momentenumkehr b.1Q-Stromrichter über Schütze i.Ankerkreis*
Ankerkreiswiderstand armature resistance || armature circuit resistance
Ankerkreiszeitkonstante armature-circuit time constant * *typ. 20 msec bei Gleichstrommaschinen*
Ankerkurzschlussbremsung braking by armature short-circuiting
Ankerlängsfeld armature direct-axis field
Ankernennspannung rated armature voltage
Ankernennstrom rated armature current
Ankernutschrägung skewed armature * *schräg genuteter Anker*
Ankerplatte armature * *bei Bremse*
Ankerquerfeld armature quadrature-axis field
Ankerreaktanz synchronous reactance
Ankerrückwirkung armature reaction
Ankerscheibe armature plate * *bei einer (elektromagnet.) Bremse/Kupplung* || armature ring * *einer elektromagnet. Bremse/Kupplung* || armature disk * *bei einer Bremse*
Ankerspannung armature voltage || arm voltage * *(Kurzform)*
Ankerspannungserfassung armature voltage sensing
Ankerspannungsistwert armature voltage actual value || armature voltage feedback
Ankerspannungsregler armature voltage controller
Ankerstellbereich armature control range
Ankerstern armature spider
Ankersteuerbereich armature control range
Ankersteuersatz armature gating unit
Ankerstreufluss armature leakage flux
Ankerstreuinduktivität armature leakage inductance
Ankerstreureaktanz armature leakage reactance
Ankerstrom armature current * *['a: matje...]*
Ankerstrombelag armature ampere conductors
Ankerstromgrenzen armature current limits
Ankerstromistwert armature current actual value
Ankerstromregelung closed-loop armature current control
Ankerstromregler armature current controller || armature current regulator
Ankerstromrichter main converter || armature-circuit converter
Ankerstromsollwert armature current setpoint
Ankerwickelkopf armature winding overhang
Ankerwicklung armature winding || rotor winding * *Läuferwicklung einer elektr. Maschine*
Ankerwiderstand armature resistance
Ankerwinde anchor winch || capstan || anchor winding gear || anchor winder
Ankerzeitkonstante armature time constant
ankleben glue on || paste on || stick * *angeklebt sein*
anklemmen connect * *Draht (to a terminal: an eine Klemme)*
anklicken click-on * *mit der Computermaus*
ankommen arrive
ankommen auf count * *zählen, wichtig sein*
ankommend incoming * *Telegramm, Strom, Signal*
ankommende Leitung incoming line
ankoppeln couple * *to: an; with: koppeln mit/ankoppeln an* || interface * *verbinden, anschalten (to: an)* || link * *to: an; z.B. über serielle Schnittstelle* || tie in || connect || hook up * *anschließen, "andocken", anhaken*
Ankopplung connection * *to: an* || coupling || interface * *to: an* || interfacing * *Signalankopplung, Datenkopplung (to: an)*
ankörnen punch
ankündigen announce
Ankündigung announcement
Ankunft arrival * *upon arrival.: bei Ankunft*
ankurbeln crank up * *(konkret u. figürl.)* || stimulate * *(figürl.)* || boost * *(figürl.) ankurbeln, in die Höhe treiben, Auftrieb geben, anpreisen* || ginger up * *anfeuern, ankurbeln, scharfmachen; aufmöbeln* || step up * *Produktion usw.* ~ || pep * *(amerikan.; salopp) Pepp geben*
Anlage plant * *z.B. Produktionsanlage* || field * *praktischer Einsatz, Einsatzfeld, Praxis, Wirklichkeit, Branche* || installation * *Einrichtung, Anlage; Installierung, Errichtung, Einbau* || station * *(Funk/Forschungs)Station, Kraftwerk* || equipment || system * *(Energieversorgungs-) System, Anlage, Aggregat* || facility * *Betriebs-* || works * *Produktionsanlage* || investment * *Kapitalanlage* || enclosure * *zu einem Brief, einer Sendung* || attachment * *zu einem Handbuch, Bericht, Brief* || project * *in Planung befindliches Projekt* || structure * *z.B. eines Romans* || site * *Baustelle, Industriegelände* || attached * *in der Anlage (z.B. eines Briefes)* || line * *(Fertigungs-, Walz-, Verarbeitungs-, Behandlungs-) Straße/Linie* || arrangement * *Art der ~, →Konzept*
anlagenabhängig process-dependent
Anlagenabteilung systems department || industial systems department
Anlagenbau plant engineering || plant construction
Anlagenbauer plant manufacturer
anlagenbedingt for plant-specific reasons
Anlagenbedingungen plant conditions || site conditions
Anlagenbild system display * *auf einem Bildschirm* || plant display * *auf Bildschirm* || process display * *auf Bildschirm* || plant diagram || mimic diagram * *grafische Nachbildung der Anlage auf dem Bildschirm*
Anlagendaten plant data
Anlagenerfordernisse plant requirements
Anlagenfehler plant fault
Anlagenfrequenz plant frequency
Anlagenkennzeichen higher level designation || higher level assignment || location designation * *Ortskennzeichen* || plant code || plant identification * *siehe DIN 40719 Teil 2* || plant designation || plant identifier
Anlagenkennzeichnungssystem plant designation system
Anlagenkonfiguration site configuration || installation configuration

Anlagenmoment system torque
Anlagenparameter plant parameter
Anlagenpersonal plant personnel
Anlagenplanung system design
Anlagenprojekteur plant engineer || project engineer || system designer
Anlagenprojektierung system engineering || plant-projecting || project engineering || plant engineering || plant design
Anlagenschema plant diagram || layout plan || plant schematic
Anlagenseite plant side || installation side
anlagenseitig on the plant side || at site * *auf der Baustelle* || at the installation || on the system side || plant-side
Anlagensignale external signals || plant signals
anlagenspezifisch system-specific || process-specific || process-related || site-specific || plant-specific
Anlagenstillstand plant outage || plant downtime || plant standstill || plant shutdown || plant stoppage || production standstill
Anlagentechnik plant technology || plant engineering || industrial engineering || industrial systems engineering || systems engineering || industrial systems || system technology || plant construction || industrial technology
Anlagenverbund process network || inline operation || combined operation
Anlagenverfügbarkeit plant availability
Anlagenverhältnisse plant conditions
Anlagenwirkungsgrad system efficiency
Anlagenzustand plant status
Anlagevermögen fixed assets * *unbewegl. (Immobilien), bewegliches(Maschinen..) u. immateriell. Vermög. (Patente..)* || invested capital
Anlassdrossel starting reactor * *{el.Masch.} 3phasige ~ mit Eisenkern, wird beim Anlassen vor d.Wicklg. geschaltet.* || starting choke
anlassen start || temper * *Stahl* || anneal * *glühen*
Anlasser starter * *während d.Hochlaufs zur Energie/Momenten-Entlastg.i.d.Motorkreis geschaltete Widerstände* || motor starter
Anlasshäufigkeit starting frequency * *{el.Masch} zuläss. aufeinander folgende Anz. d. Anlassvorgänge mittels Anlasser*
Anlasshilfe starting equipment * *z.B. für einen Motor (Softstarter, Anlaßwiderstand usw.)*
Anlassschützkombination combination starter
Anlassschwere starting-duty rating * *{el.Masch.} Leistungsaufnahme beim Anlassen/Leist.aufn. im Bemessungspunkt* || starting load * *{el.Masch.}* || starting load factor * *{el.Masch.} mittlerer Anlassstrom/Läufernennstrom*
Anlassspitzenstrom peak starting current * *{el. Masch.}*
Anlassstrom locked-rotor current * *wird vom Motor beim Anlaufen aufgenommen*
Anlasstransformator starting transformer * *für Sanftanlauf eines Motors* || autotransformer starter * *{el.Masch.} als Spartrafo ausgelegte Starthilfe für Motor*
Anlassvorgang starting process || starting operation || starting cycle * *Zyklus* || starting sequence
Anlasswiderstand starting resistor
Anlasszahl permissible number of starts in succession * *{el.Masch} zulässige Anz. aufeinanderfolg. Startvorgänge*
Anlasszeit starting time * *{el.Masch.}*
Anlauf start || start-up || starting || initialization * *Anlaufprozedur für Computer* || starting cycle * *Anlaufvorgang* || startup || initial start * *bei SPS*
Anlauf- starting
Anlaufart starting duty * *{el.Masch.}*
Anlaufbaustein start-up block * *bei SPS*
Anlaufbedingungen starting conditions

Anlaufdauer starting duration * *Zeit zum Hochlaufen z.B. eines Motors*
Anlaufdrehmoment starting torque
Anlaufeigenschaften starting properties * *eines Motors* || starting performance * *eines Motors*
anlaufen start || start-up || start running || tarnish * *Metall* || run up * *z.B. eine Maschine* || come online * *einer Produktionslinie, z.B. nach der Montage*
Anlaufgerät starting unit
Anlaufgüte starting performance * *Anlaufeigenschaften e. Drehstrommotors (gewünscht: mögl. niedriger Anlaufstrom)* || starting quality * *Anlaufeigenschaften e. Drehstrommotors*
Anlaufhilfe starting aid
Anlaufkäfig starting cage * *in einem Läufer eines (Synchron-) Drehstrommotors (z.B. SIEMENS SI-MOSYN-Maschine (R))*
Anlaufkondensator starting capacitor
Anlaufkupplung starting clutch * *reduziert das Anlaufdrehmoment beim Starten eines Motors* || starting coupling * *für AC-Motor zum schnellen Hochlauf u. thermisch. Entlastg. (reduz. Anlaufmoment)* || speed-controlled coupling * *{el.Masch.} Kupplung, die dem Motor einen Leerlauf ermöglicht*
Anlaufmoment starting torque * *z.B.Drehmoment e.Asynchronmotors bei Drehzahl Null*
Anlaufphase starting period
Anlaufprogramm start program || initialization || start-up || start-up program || initialization procedure || power-on routine
Anlaufreibmoment starting friction torque
Anlaufspannung starting voltage * *wird für Motor häufig erhöht z. Überwindg. d.Losbrechmoments*
Anlaufsperre starting lockout || starting inhibit circuit
Anlaufsteuerung starting control
Anlaufstrom starting current || warm-up lamp current * *für Lampe* || locked-rotor current * *wird z.B. von AC-Motor beim Anlaufen benötigt*
Anlaufverhalten starting characteristic || starting peformance * *für Käfigläufermotoren: siehe DIN 57530, VDE 0530/IEC 34 jeweils Teil 12* || starting behavior || starting behaviour
Anlaufverzögerung stop time at start * *Wartezeit bevor e.Antrieb nach d.Startbefehl hochläuft*
Anlaufvorgang starting-up sequence
Anlaufwärme starting heat * *Verlustwärme, die beim Anlaufen eines Motors entsteht*
Anlaufwärme im Läufer rotor starting heat
Anlaufwärme im Motor motor starting-heat * *{el. Masch.}*
Anlaufwärme im Ständer stator starting heat * *{el.Masch.}*
Anlaufwicklung starting winding || auxiliary winding * *Hilfswicklung*
Anlaufzeit starting time * *benötigt ein Motor zum Anfahren d. Nenndrehz. aus d. Stillstand heraus* || acceleration time * *(mechan.) Hochlaufzeit* || running-up time * *{el.Masch.} Dauer eines Anlaufs v. Einschalten bis zum stationären Betrieb*
anlegen apply * *a voltage across/to: eine Spannung zwischen/an; apply a standard: einen Maßstab ~* || design * *planen* || inject * *(Signal) einprägen, anlegen* || create * *erzeugen, schaffen* || generate * *erzeugen* || invest * *Kapital usw.* || lay out * *auslegen* || join to || put to || dock * *Schiff* || install * *einrichten* || feed * *{Druckerei} Papierbögen usw.* || lay alongside * *Schiff {of: an}* || dock * *Schiff* || lay in * *Vorrat* || provide with * *versorgen mit*
Anleger feeder * *{Druckerei}*
Anlegetacho surface tachometer * *zur Messung der Bahngeschwindigkeit*

anleimen glue on || gum * *gummieren* || paste * →*kleben* || glue * →*leimen*
Anleitung instructions || operating instructions * *Betriebsanleitung* || guidance * *Belehrung, Richtschnur* || introduction * *Einführung* || instruction book * *Handbuch* || manual * *Handbuch* || instruction manual * *Betriebsanleitung* || user manual * *Anwenderhandbuch*
anlenken hinge * *mit Scharnier/Gelenk/Angel versehen* || pivot * *dreh/schwenk/klappbar lagern*
anliefern deliver || supply
Anlieferung delivery
Anlieferungszustand as-delivered condition * *siehe auch* →*Auslieferungszustand* || as shipped * *im ~*
Anliegen wish * *Wunsch* || presence * *Vorhandensein* || be present * *(Verb) vorhanden sein, z.B. Signal, Warnung, Meldung* || be available * *(Verb) z.B. Versorgungsspannung* || be applied * *(Verb) angelegt sein, z.B. eine Spannung* || prevail * *(vor)herrschen, maß-/ausschlaggebend sein. maßgebend/gültig sein*
Anmeldebestätigung confirmation of enrolment * *für Vorlesung, Kurs, Tagung usw.*
Anmeldeformular registration form * *z.B. für Schulungskurs*
anmelden announce || log on * *als Steuerfunktion ; im Computernetz ~* || book * *buchen* || apply for * *Patent, Konkurs* || register * *Lehrgang, Tagung*
Anmeldeprozedur logon procedure * *(Computer) für Betriebssystem, Netzwerk, Mailbox usw.*
Anmeldung announcement || registration * *auch zu Seminar, Kongress* || booking * *application* * *Patent, Trainingskurs usw.* || reception * *Empfangsbüro* || enrolment * *Einschreibung*
Anmerkung remark || comment * *Kommentar* || note * *Hinweis (schriftlicher)*
annähern approach || approximate
annähernd approximate * *siehe auch* →*näherungsweise* || rough || fairly exact * about * *(Adv.)* || not far from || almost * *fast*
annähernd stromloses Schalten switching of negligible currents
Annäherung approach || approximation * *z.B. mathematisch*
Annahme acceptance * *Empfang; auch figürlich; Akzeptanz* || reception * *Entgegennahme* || assumption * *Vermutung, Voraussetzung* || adoption * *Übernahme* || passing * *eines Gesetzes*
Annahme verweigern refuse || refuse to accept || reject
Annahmeprüfung acceptance test
Annahmeverfahren acceptance procedure
annehmbar acceptable * *to: für* || fair * *ehrlich, fair* || reasonable * *vernünftig* || tolerable * *leidlich*
annehmen accept || take * *z.B. Form/Gestalt/Farbe; take various forms: verschiedene Formen annehmen* || receive || adopt * *Antrag, Haltung* || undertake * *Auftrag* || pass * *Gesetz* || assume * *Gestalt, vermuten, auch Wert annehmen* || think * *vermuten* || guess * *vermuten* || take on * *z.B. eine Farbe* || attain * *erreichen; auch 'Wert erreichen'* || pick-up * *aufnehmen, (Telefonanruf usw.) annehmen*
Annehmlichkeit convenience * *Bequemlichkeit, Komfort, Vorteil, Nutzen, Nützlichkeit* || goody * *(salopp) Zusatznutzen, 'Bonbon'*
annullieren annul || nullify || declare null and void * *für null und nichtig erklären* || cancel * *Auftrag* || set aside * *Urteil*
Annullierung annulment || cancellation
Anode anode || anode terminal * *Anodenanschluss*
Anoden-Kathoden-Spannung anode-cathode voltage

anordnen arrange || design || group * *gruppieren* || order * *befehlen* || instruct * *befehlen* || direct * *lenken, steuern, leiten, jdn. anweisen* || provide * *sorgen für, vorsehen* || locate * *örtlich* || place * *örtlich; platzieren* || position * *in die (richtige) Lage bringen, aufstellen, positionieren*
Anordnung arrangement || design || layout * *räumliche, konstruktive ~* || grouping * *Gruppierung* || structure * *Aufbau* || pattern * *System* || scheme * *Schema* || order * *Befehl* || instruction * *Befehl* || rule * *Vorschrift* || direction * *Anweisung* || location * *räumliche* || configuration * *Konfiguration, Gestalt(ung), Struktur* || coordination * *richtige An/Bei/Zuordnung* || positioning * *das In-Die-Richtige-Lage-Bringen, Aufstellung*
Anordnungsplan layout diagram * *(auch bei SIMADYN D (R) von SIEMENS)* || arrangement drawing || location diagram || disposition diagram
anorganisch inorganic * *{Chemie}*
anormal abnormal || non-standard * *nicht dem normalen Standard entsprechend* || anomalous || special * *spezial, Sonder-*
Anpass- matching || interface
anpassbar adaptable || customizable * *an Kundenwünsche*
Anpassbarkeit adaptability
Anpassbaugruppe adaptor board * *Leiterkarte* || adaptor module * *Baustein, Modul* || adaption module * *siehe auch* →*Anpassmodul*
Anpasselektronik adaptor || interface circuit || interface electronics
anpassen adapt * *auch 'sich ~' (to: an); adapt oneself to: sich ~ an* || fit * *on: an* || match * *mehrere (meist 2) Dinge aneinander; to: an; closely to: weitgehend an; auch 'sich ~'* || adjust * *einem Zweck ~; adjust oneself to: sich ~ an* || tune * *einem Zweck ~* || proportion * *im Verhältnis ~* || suit * *to: an* || tailor * *to: an (Bedürfnisse, Kundenwunsch ~)* || rengineer * *umprojektieren, umkonstruieren*
anpassen an Kundenwünsche customize
anpassfähig adaptable || customizable * *an Kundenanforderungen*
Anpassfaktor adjustment factor || scaling factor * *Normier/Skalierfaktor* || adjustable factor
Anpassmodul matching module || interface modul
Anpassschaltung matching circuit || adapting circuit
Anpassstufe matching unit || adaptor
Anpasstrafo matching transformer
Anpasstransformator matching transformer
Anpassung adaption * *(to: an)* || adjustment || matching * *mehrere Dinge aneinander* || conditioning * *Signal~* || tuning || body conformity * *an die Oberfläche eines Körpers* || trim * *Abgleich* || accommodation * *mehrere Dinge aneinander* || customization * *(amerikan.) ~ an Kundenbedürfnisse/-spezifikationen* || customisation * *(brit.) ~ an Kundenanforderungen*
anpassungsfähig adaptable || customizable * *an Kundenbedürfnisse* || versatile * *['wörsetail] vielseitig* || flexible
Anpassungsfähigkeit adaptability
Anpassungsschaltung matching circuit
Anpassungstrafo matching transformer
Anpassverstärker matching amplifier || signal conditioner
Anpress- pressure || contact || clamping * *{Druckerei}* || tightening * *zum Straffen* || holding down * *zum Niederhalten* || nip || squeezing * *Abquetsch-, Press-* || squeegee * *Gummiquetsch-* || nip * *Press-, Abkneif-*
Anpressdruck contact pressure || surface pressure ||

bearing pressure || working pressure * *in Druckluftsystem, für Pneumatikkolben usw.*
anpressen press against || press down * *niederdrücken* || force down * *niederhalten* || tighten * *fest-/anziehen (Schraube, Zügel usw.),* spannen *(Feder, Gurt, Kette), straffen (Seil)*
Anpresskraft applied force * *z.B. bei Bremse, Kupplung usw.* || contact pressure * *Anpressdruck* || surface pressure || bearing pressure
Anpresswalze pressure roll || nip roll * *Haltewalze*
anquetschen crimp * *z.B. Kontakt an Kabel usw.* || crimp on * *z.B. Kontakt an Kabel usw.*
Anregelzeit rise time * *benötigt d.Regelgröße nach e.Sollwertsprung z. erstmaligen Erreichen d. Sollwerts* || step rise time * *bei Sprungantwort*
anregen suggest * *vorschlagen* || encourage * *ermuntern* || stimulate * *stimulieren* || induce * *Schwingung* || excite * *Schwingung, einspeisen eines Stimulations-Signals* || activate || initiate || energize
Anregung stimulation || encouragement * *Ermutigung* || impulse || suggestion * *Vorschlag* || excitation * *einer Schwingung* || input * *Signal*
Anreicherung enhancement || enrichment || concentration
anreihbar mountable side by side || butt-mountable
anreihen arrange in a series * *in (einer) Reihe anordnen* || align || butt * *auf Stoß anordnen, aneinander stoßen/fügen* || gang || arrange side by side || line up || butt-mount * *angereiht/auf Stoß befestigen/montieren*
Anreihklemme modular terminal
Anreihklemmenblock modular terminal block
Anreihmontage butt mounting || mounting side by side
anreisen travel || arrive * *ankommen*
anreißen mark out * *markieren* || touch * *Thema* || scribe * *mit Reißnadel*
Anruf call || telephone call
Anrufbeantworter telephone answering machine || answering machine
anrufen call * *call about someth.: wegen etwas anrufen* || call up || phone * *phone somebody about: jdn. anrufen wegen; be on the phone: am Telefon sein*
Anrufer caller * *am Telefon*
Ansammlung agglomeration * *Anhäufung, Zusammenballung* || collection || accumulation || concentration || accrual || heap
Ansatz extension * *Ansatzstück* || shoulder * *Ansatzstufe* || approach * *logisches Herangehen (to: an)* || disposition * *Anlage (figürl.)* || base * *logischer*
Ansaugdruck intake pressure
Ansaugdurchmesser intake diameter * *z.B. bei Lüfter*
ansaugen suck in * *z.B. Kühlluft, Frischluft, Flüssigkeit, Wasser, Gas* || take in * *auch Luft* || draw in * *Luft* || prime * *durch Pumpe*
Ansaugfilter intake filter || suction filter * *für Pumpe, Hydrauliksystem usw.*
Ansaughöhe suction lift
Ansaugleistung suction capacity
Ansaugluft intake air || induction air * *Verbrennungsmotor* || suction air
Ansaugseite air inlet end * *{el.Masch.}* || inlet side * *Pumpe* || suction side * *Pumpe, Lüfter* || intake side * *Lüfter* || air-inlet end * *{el. Masch.} (für Kühlluft)*
Ansaugstutzen intake flange * *z.B. für (Kühl)Luft bei elektr. Maschine* || intake stub * *Pumpe*
Ansaugung suction * *z.B. durch Pumpe, Lüfter*
Ansaugventil intake valve || suction valve
anschaffen acquire || buy || purchase || procure * *besorgen* || provide * *besorgen*

Anschaffungskosten purchase cost || prime cost || first cost || initial outlay
Anschaltbaugruppe interface module * *z.b. für ein externes Gerät, einen Feldbus usw.* || interface board * *Leiterkarte zum Anschluss eines externen Gerätes, eines Busses usw.*
anschalten connect * *mit Draht* || interface * *Schnittstelle* || wire up * *mit Draht* || turn on * *einschalten* || switch on * *einschalten*
Anschaltung interface module * *als Baugruppe/Baustein* || interface || adapter * *{Computer} für Monitor, Netzwerk usw.* || interface card * *als Leiterkarte; z.b. network interface card: Netzwerk~*
Anschaltungsbaugruppe interface module || interface board * *Leiterkarte*
anschauen look at * *(auch figürl.)* || view * *auch auf dem Bildschirm; (auch figürl.)* || browse * *durchschauen/-stöbern v.Datenbeständen, Dateien, Verzeichnissen, Datennetz-Inhalten usw.*
anschaulich clear * *→deutlich, übersichtlich, →klar, eindeutig* || vivid * *deutlich, lebendig, intensiv, klar, lebhaft* || graphic * *plastisch, lebendig, grafisch, zeichnerisch* || concrete * *konkret, greifbar, wirklich, dinglich* || understandable * *→verständlich*
anschellen clamp-on * *Befestigung mittels einer Schelle*
Anschlag stroke * *Schlag; auch bei Schreibmaschine* || impact * *Anprall* || depression * *einer Taste* || key-stroke * *Tasten~* || stop * *mechan. End~* || dead stop * *~ einer Messeinrichtung* || limit stop * *End~ z.B. ~ einer Messeinrichtung* || account * *Betracht; take into (leave out of) account: (nicht) in ~ bringen* || endstop * *mechanischer End~, z.B. an Potentiometer* || end stop * *End~*
anschlagen strike || fasten * *befestigen* || fix * *befestigen* || estimate * *(ver)anschlagen, schätzen* || take effect * *wirken, z.B. Arznei, Abhilfemaßnahme* || attach * *Hebezeug*
Anschlagplatte stop plate * *z.B. für Säge*
anschleifen rub * *(ab)schleifen; with emery: abschmirgeln* || rub down
anschließbar which can be connected * *was angeschlossen werden kann*
anschließbarer Leiterquerschnitt nominal conductor cross-section
anschließen connect-up || wire-up * *hinverdrahten* || plug in * *mit Stecker* || add * *anfügen* || join * *anfügen* || follow * *einem Beispiel folgen, nachfolgen* || connect * *verbinden, anschließen, auflegen* || wire * *verdrahten* || connecting-up * *(Substantiv)* || hook up * *z.B. Signal/Leitung an Empfänger/Klemme/Stecker ~/andocken/anstecken/auflegen*
anschließend adjacent * *räumlich* || next * *(räuml. u. zeitl.) nächstfolgend, nächststehend, (gleich) neben, nächst-, gleich nach* || neighbouring * *räumlich benachbart* || subsequent * *zeitlich; to: an* || following * *zeitlich* || thereupon * *(Adv.) danach, (zeitlich) anschließend*
Anschluss connection * *Verbindung, Leitungs-~, auch Netz~* || terminal * *Klemme, Schraub-~* || terminal connection * *~ an Klemme* || joining * *(allg., mechanisch, auch figürlich)* || plug-in connection * *Steck-~* || supply * *Gas-/Wasser~* || wiring * *Verdrahtung* || connector * *~vorrichtung, Stecker* || connecting lead * *~leitung* || connecting-up * *das Anschließen, der Vorgang des Anschließens* || line * *Telefon~* || subscriber's station * *Telefonapparat* || lead * *Zuleitung* || termination * *besonders für einen flächigen ~punkt (auch in Motorklemmenkasten)* || pin * *~stift*
Anschluss verlieren miss the bus * *(figürl.)*
Anschluss- connecting
Anschluss- und Abzweigdose junction and tapping box

Anschlussanordnung pinout * *IC, Stecker* || pinning * *IC, Stecker* || pinning diagram * *IC- oder Stecker-Anschlussbild* || terminal assignment * *Klemmenbelegung*
Anschlussauftrag follow-up order
Anschlussbeispiel wiring example || connecting example || example for wiring up
Anschlussbelegung terminal assignment || pin assignment * *Stecker* || pinning * *IC, Stecker* || pinning diagram * *IC- oder Stecker-Anschlussbild* || pinout * *IC, Stecker*
Anschlussbezeichnung terminal marking || terminal designation * *Klemmenbezeichnung; für E-Motoren: siehe DIN 57530/VDE 0530/IEC 34 jew.Teil 8* || connection designation || connecting-up designation
Anschlussbild connection diagram || wiring diagram
Anschlussbolzen terminal stud
Anschlussdose junction box || outlet box
Anschlussdraht connecting lead || solid wire connecting lead * *steifer Draht, kein Litzenleiter* || lead wire
Anschlusseinheit connecting unit
Anschlussfahne terminal lug || connecting lug
Anschlussfeld connecting panel
anschlussfertig ready-to-connect || prewired || plug-in * *steckbar* || factory-assembled || pre-wired || ready-wired || ready for connection
Anschlussfläche terminal face || connecting surface * *(mechanisch)*
Anschlusshinweise connecting-up instructions
Anschlussimpedanz terminal impedance
Anschlusskabel connecting cable || power lead * *Stromversorgungskabel* || supply cable * *Stromversorgungskabel* || power cord * *flexible Stromversorgungsleitung, z.B. für Kleingerät* || cable set * *konfektionierter Kabelsatz* || service cable * *für Hausanschluss* || communication cable * *Kommunikationsleitung, z.B. für serielle Schnittstelle*
Anschlusskasten terminal box || terminal housing || cable box
Anschlussklemme terminal || supply terminal * *für Stromversorgung* || connecting terminal
Anschlussklemmen terminals
anschlusskompatibel terminal-compatible * *Klemmen-kompatibel* || pin compatible * *IC, Stecker* || plug-in compatible * *Stecker-kompatibel*
Anschlusslasche connecting lug || terminal lug
Anschlussleiste terminal block || terminal strip || connector strip
Anschlussleistung installed load || power supply capacity || connected load || input power * *z.B. eines Motors*
Anschlussleitung connecting cable || connecting lead || power lead * *für Stromversorgung* || service cable * *Hausanschluss* || supply cable * *für Stromversorgung, Motor, Stromrichter usw.* || power cord * *flexible Netzanschlussleitung* || lead-wire * *einer Wicklung* || communication cable * *für (serielle) Kommunikationsverbindung* || supply pipe * *Anschlussrohr (z.B. in Wasserversorgungs-, Pneumatiksystem usw.)* || cord * *flexible Netz~ z.B. für Kleingeräte* || connecting line * *Kommunikations-/Telefon~* || subscriber's line * *Telefon-Endanschlussleitung*
Anschlussmaße fitting dimensions || connecting dimensions
Anschlussmöglichkeit für capability of connecting
Anschlussmöglichkeiten wiring facilities || interface facilities * *z.B. für Peripheriegeräte*
Anschlussöse connecting tag * *z.B. an Lötfahne*
Anschlussplan terminal connection diagram * *für Klemmenanschlüsse* || connection diagram || connecting-up diagram

Anschlussplatte terminal plate || terminal board || conduit mounting plate
Anschlusspunkt termination point || connecting point || terminal || connection point
Anschlussquerschnitt connection cross-section || wire cross section || wire range || wire gauge || conductor size || terminal capacity * *~ der anschließbaren Leitungen bei Klemme*
Anschlussraum wiring space || wiring area
Anschlussschaltbild connection diagram
Anschlussschaltung connection circuit
Anschlussschelle terminal clamp
Anschlussschema connection diagram
Anschlussschiene connecting bar || terminal bar
Anschlussschnur cord || flexible cord
Anschlussschraube terminal screw
Anschlussseite terminal end
Anschlusssicherung mains fuse * *Netzsicherung* || service fuse
Anschlussspannung supply voltage * *z.B. eines Stromrichters/Motors* || input voltage || system voltage * *Netzspannung* || incoming voltage || line voltage * *Netzspannung*
Anschlussstecker plug || cable connector || connecting plug
Anschlussstelle connection point || terminal || termination point
Anschlussstück connecting adapter * *für Schlauch, Rohr usw.* || tube nozzle * *für Rohr* || hose nipple * *für Schlauch* || union || connecting piece || joint coupling
Anschlussstutzen entry fitting || connecting piece * *Rohr* || duct adapter * *für Luft*
Anschlusstechnik wiring method || wiring technique || termination technology || connection system
Anschlussverteiler multi-point connector
Anschlusswiderstand terminal resistance
Anschnitt phase angle control * *Phasenanschnitt (steuerung)*
Anschnittsteuerung phase angle control * *Phasenanschnittsteuerung* || phase control
Anschraub- bolt-on
Anschraubbefestigung bolt-on fixing
Anschraubbohrung mounting bore
anschrauben screw on * *siehe auch →festschrauben* || bolt-on || bolt || bolt to
Anschreiben cover note
Anschrift address
Anschwingzeit response time || built-up time
ansehen view * *siehe auch →anschauen* || look at
ansenken counterbore * *z.B. einer Bohrung mit Spitzsenker*
ansetzen quote * *(Preis) angeben, berechnen* || put on || set on || add * *anfügen* || piece on * *anstücken (to: an)* || fasten * *befestigen (to: an)* || apply * *anlegen, ansetzen (Hebel, Meißel usw.) (to: an)* || fix * *Frist, Termin, Preis* || appoint * *Termin* || schedule * *Termin* || set * *a date: einen Termin* || rate * *abschätzen* || value * *abschätzen, bewerten* || overstate * *zu hoch ansetzen* || understate * *zu niedrig ansetzen* || put up * *(mathem.) Gleichung* || begin to do * *zu einer Handlung* || start doing something * *mit etwas den Anfang machen* || lead off * *den Anfang machen* || join * *anfügen*
Ansetzstelle jacking lug * *zum Heben* || lifting lug * *zum Heben*
ANSI ANSI * 'American National Standards Institute': nationales USA-Normeninstitut (Ähnlich DIN)
Ansicht view * *optische; Blick; side view: Ansicht von der Seite* || sight || opinion * *Meinung; in my opinion: meiner ~ nach; be of the opinion that: der Meinung sein, dass*
Ansicht im Schnitt sectional view

Ansicht vertreten be of the opinion that || take the view that || hold that * *meinen, der Ansicht sein*
Ansicht von der Seite side view
Ansicht von hinten rear view
Ansicht von links view from the left
Ansicht von oben top plan view
Ansicht von rechts view from the right
Ansicht von unten inverted-plan view || worm's eye view
Ansicht von vorn front view
anspitzen point || sharpen || goad * *(salopp) goad someone (into action): jdn. ~/scharfmachen/anstacheln/-treiben*
Ansporn motivation
Ansprechbereich operating range
Ansprechempfindlichkeit sensitivity || response threshold * *Ansprechschwelle* || responsiveness || resolution * *Auflösung* || response sensitivity
Ansprechen response * *z.B. Messinstrument, Überwachung, Begrenzung, Regeleinrichtung* || give response * *(Verb) z.B.Messinstrument* || trip * *(Verb und Substantiv) Überwachungseinrichtung* || fault trip * *Fehlerauslösung, Fehlerabschaltung* || be released * *(Verb) Überwachungseinrichtung* || respond * *(Verb) z.B. Regelung auf Sollwertänderung; Messgerät* || pick up * *(Verb) Relais* || treat * *(Verb) Thema usw. behandeln* || operate * *(Verb) Relais, Sicherung, Überspannungsableiter* || spark over * *(Verb) Überspannungsableiter* || blow * *(Verb) Sicherung* || rupture * *(Verb) Sicherung* || contact * *(Verb) jemanden* || get in touch * *(Verb) jemanden* || access * *zugreifen auf* || come into effect * *(Verb) wirksam werden*
ansprechend appealing * *gefallend* || attractive * *attraktiv*
Ansprechgeschwindigkeit response rate
Ansprechgrenze operating limit || response threshold * *Ansprechschwelle*
Ansprechleistung pickup power * *Relais*
Ansprechpartner contact person || contact partner
Ansprechpegel sparkover level * *eines (Überspannungs-) Ableiters* || response threshold * *Ansprechschwelle*
Ansprechschwelle response threshold * *z.B. e. Mess/Regel/Überwachungs/Schutzeinrichtung* || sensitivity level * *z.B. einer Schutzeinrichtung*
Ansprechspannung threshold voltage || response voltage
Ansprechstrom threshold current || pick-up current * *Relais*
Ansprechtemperatur response temperature
ansprechverzögertes Relais ON-delay relay
Ansprechverzögerung switch-in delay || pickup delay * *z.B. Relais* || switch-on delay * *Einschaltverzögerung* || pick-up delay * *Relais* || ON-delay || response delay * *z.B. einer Kupplung/Bremse* || engagement delay * *einer Kupplung/Bremse*
Ansprechverzug response delay * *→Ansprechverzögerung z.B. einer Bremse, Kupplung*
Ansprechwert switching value * *Relais, Schalter* || pickup value * *z.B. Relais* || pull-in value * *z.B. Relais* || sensitivity * *kleinster erfassbarer Messwert* || response threshold * *Ansprechschwelle* || response level * *response value*
Ansprechzeit response time || opening time * *eines Öffnerkontakts* || closing time * *eines Schließerkontakts* || pick-up time * *Relais* || tripping time * *z.B. e. Auslösegeräts, Überwachungseinrichtung, Leistungsschalters*
Anspruch claim * *auch 'für s. in ~ nehmen'; auch Patent~; (to: auf); for damages: auf Schadeners.* || demand * *Forderung (for: nach); make high demands: hohe Ansprüche stellen* || requirement

* *(An)Forderung, Bedingung; meet all requirements: allen Ansprüchen gerecht werden*
Ansprüche demands * *rising: steigende; place demands on: ~ stellen an*
anspruchslos unpretending || unpretentious || unassuming * *schlicht* || modest * *auch 'bescheiden'* || simple * *einfach* || plain * *einfach, schlicht* || nearly maintenance-free * *wartungsarm* || easy to maintain * *wartungsfreundlich* || low performance * *mit niedriger Leistungsfähigkeit* || low end * *am unteren Ende des Typenspektrums* || economy * *preisgünstig*
anspruchsvoll ambitious * *von Sachen* || hard to please * *schwer zufriedenzustellen* || fussy * *['fassi] übertrieben anspruchsvoll, wählerisch, heikel* || demanding * *technische Anwendung/Anforderung; geistig ~; highly demanding: höchst-/ sehr ~* || high-brow * *hochgestochen, betont intellektuell* || sophisticated * *hochentwickelt, raffiniert* || exacting * *streng, genau; z.B. Kunde; be exacting: hohe Anforderungen stellen*
Ansprung branch
anstatt instead of || as opposed to
Ansteck- clip-on
anstehen be present * *Spannung, Signal* || be applied * *angelegt sein, z.B.Spannung* || be up * *for decision: zur Entscheidung* || pend * *z.B. Interrupt, Warnung* || be available * *Fehler/Warnsignal* || exist * *vorhanden sein; fault condition exists: Fehler steht an*
anstehend pending * *z.B. Meldung, Warnung, Interrupt* || under consideration * *Entscheidung usw.*
anstehende Meldung pending message || pending signal || existing message
ansteigen rise * *Gelände, Spannung, Strom usw.; auch 'zunehmen'* || slope * *Gelände* || increase * *zunehmen (by: um)* || ascend * *auf/empor/hinaufsteigen, (schräg) in die Höhe gehen* || surge * *vorübergehend/stoßartig ~, z.B. Spannung, (Luft-) Druck* || mount * *zunehmen* || sky-rocket * *(salopp) jäh/raketenartig ~* || zoom * *(salopp) jäh ~* || be increased || be increasing * *im Ansteigen begriffen sein* || be on the rise * *im Anstieg begriffen sein* || be rising * *~d sein*
ansteigend increasing || progressive * *fortlaufend, zunehmend, progressiv* || rising * *Gelände, Fläche; zunehmend* || ascending * *Rang, Töne, Zahlen, Kennlinie usw.* || surging * *plötzlich, z.B. Druck, Spannung; hochbrandend* || accelerating * *Geschwindigkeit, Drehzahl*
ansteigende Flanke rising edge * *eines Impulses* || leading edge * *Vorderflanke eines Impulses*
ansteigende Hochlaufgeber-Rampe ascending speed ramp * *für Drehzahlsollwert* || ramp to run mode * *für Drehzahlsollwert*
ansteigende Rampe ascending speed ramp * *e.Drehzahlhochlaufgebers* || ramp to run mode * *e. Drehzahlhochlaufgebers*
ansteigender Ast ascending branch
anstelle instead * *of: von*
anstellen hire * *Personal* || preload * *Lager z.B. mit Federring zur Vermeidung von Axialspiel*
Anstellmotor screw-down motor * *{Walzwerk} zur Walzenanstellung*
Anstellung employment || job || screwing down * *{Walzwerk} →Walzenanstellung in Walzgerüst*
Ansteuerbaugruppe firing circuit module * *Steuersatz für Thyristor-Stromrichter* || trigger module * *Steuersatz für Thyristor-Stromrichter* || driver module * *Verstärkerbaugruppe* || control module * *Steuerbaugruppe* || gating unit * *zur Ansteuerung von Leistungshalbleitern* || gating module * *zur Ansteuerung von Leistungstransistoren/Thyristoren* || gating board * *Leiterkarte*

Ansteuerbaustein energizing unit * z.B. zur Schützansteuerung
Ansteuereinrichtung gating unit * für Thyristorschaltung
Ansteuerelektronik gating electronics * z.B. für Transistor/Thyristor-Leistungsteil
Ansteuerimpuls input pulse || control pulse || trigger pulse * z.B.für Thyristor || firing pulse * für Thyristor || gate pulse * z.B. für Thyristor/Transistor || drive pulse * Treiberimpuls
Ansteuerleistung control power || driving power || triggering power * zur Ansteuerung von Thyristoren, Pneumatikventilen usw.
Ansteuerleitung firing cable * für Thyristor(en) || gating cable * für Leistungstransistoren/Thyristoren
Ansteuerlogik control logic || trigger logic * z.B. für Thyristor || gating logic * z.B. für Transistor
ansteuern control || drive || trigger * Stromrichter || set * e. Ein/Ausgang mit High-Signal (z.B. bei SPS) || energize * e. Ein/Ausgang mit High-Signal; eine Relaispule; de-energize: mit LOW-Signal || activate || gate * z.B. Transistor/Thyristor-Leistungsteil mit Ansteuer/Zündimpulsen || feed * z.B.Motor || phase-control * bei Phasenanschnittsteuerung || power * mit Leistung ansteuern, Leistung zuführen || actuate * betätigen, z.B. Magnetventil || fire * Thyristor, Schneidmesser bei Rollenwechsel usw.
Ansteuerschaltung gating circuit * für Transistor/Thyristor-Leistungsteil || gate driving circuit * für Transistor/Thyristor-Leistungsteil || trigger circuit * für Thyristoren
Ansteuersignal triggering signal * für Transistor, Leistungstransistor usw.
Ansteuerung control || driving || driver || driving circuit || triggering * Thyristor-Stromrichter || activation || selection * Auswahl || gate control * Transistor/Thyristor || gating * Thyristor, Leistungstransistor, Schütz || firing * eines Leistungshalbleiters bei Phasenanschnittsteuerung || delay angle setting * Einstell. d. Steuerwinkels für Phasenanschnittsteuerg. || energizing * z.B. eines Steuereingangs mit HIGH-Signal || phase control * z.B. Phasenanschnittsteuerung
Anstieg rise || increase || progress * Fortschritt, (Weiter-)Entwicklung, Fortschreiten, Umsichgreifen || ascent * Aufstieg, Steigung, Gefälle, Auffahrt, Rampe || gradient * Gradient, Steigung (auch physikal.), Neigung, Gefälle || grade * Steigung, Gefälle, Neigung || surge * vorübergehender, plötzlicher, hochbrandender ~ (z.B. Spannungs~, Druck~) || growth * Wachstum
Anstiegsantwort ramp response * eines Regelkreises bei Rampe als Eingangsgröße
Anstiegsbegrenzer velocity limiter || change rate limiter || ramp generator * Hochlaufgeber
Anstiegsbegrenzung slew rate limiting * Begrenzung der Anstiegsgeschwindigkeit (einer Spannung) || rate limitation
Anstiegsfunktion ramp function generator * Rampenbaustein, Hochlaufgeber-Funktionsbaustein
Anstiegsgeschwindigkeit rate of rise || derivative * (math.) erste Ableitung || slew rate * Halbleitertechn.: eines Impulses || change rate * Änderungsgeschwindigkeit
Anstiegssteilheit rate of rise || rate-of-rise || slew rate
Anstiegsverzögerung ramp response time * bei Rampe als Eingangsgröße || rise delay
Anstiegsverzögerungszeit ramp response time * bei Rampe als Eingangsgröße || ramp delay
Anstiegsvorgang ramp * entlang einer Rampe || unit ramp * Anstiegsfunktion mit Anstiegszeit 1 sec ('Einheitsrampe')
Anstiegszeit rise time * eines Impulses || rate of rise || build-up time
Anstoß impulse * (figürl.) || start * (auch Verb) || initiate * (Verb) den ersten ~ geben || point of contact * mechanisch || triggering * Auslösung || iniative * Iniative, Anstoß, Anregung, Unternehmungsgeist || incentive * Ansporn, Antrieb, (Leistungs-) Anreiz
anstoßen push || strike || knock || butt * aneinander stoßen, (stumpf) aneinander fügen, (an)grenzen, auf Stoß aneinandergefügt sein || trigger * auslösen || initiate * einleiten, initiieren || activate * aktivieren || bump * (heftig) anstoßen/-prallen
anstreben aim at || aspire to * er-/anstreben, streben/trachten nach || strive for * streben nach, (erbittert) ringen um, sich mühen um || aim for * beabsichtigen, im Sinne haben, hinzielen auf
anstreichen paint || coat || mark * Fehler, Textstelle ~ || underline * Fehler, Textstelle ~ || check off * abhaken
Anstrengung effort * Bemühung || strain * Strapaze || stress * Strapaze || exhaustion * Erschöpfung || attempt * Versuch, Bemühung || endeavour * Bemühung, Anstrengung, Bestreben || endeavor
Anstrich painting || coating || film * hauchdünn || appearance * Aussehen || paint finish || coat || paint system
Anteil fraction * Bruchteil || percentage * Prozent~ || content * Inhalt, Gehalt, z.B. an Oberwellen || ratio * Verhältnis || part * Teil, Pflicht; do one's part: seinen ~ leisten || term * Teil einer Formel || component * Komponente, z.B. integral component: Integral~, direct component: Gleich~ || proportion * (An)Teil, Verhältnis, Ausmaß, Umfang || share * (An)Teil, Geschäfts/Kapital/Aktien~ e. Unternehmens, Teilhaberschaft || contribution * Beitrag, Mitwirkung; make a contribution: einen Beitrag liefern || constituent * (Bestand-) Teil, Komponente || quota * Quote || interest * {Wirtschaft} Beteiligung
Antenne aerial || antenna * (amerikan.) ferrite-bar antenna: Ferrit~
Antichronstutzen cable gland * für PG-Verschraubung
Anticompoundwicklung differential compound winding
Antidröhnmittel anti-vibration compound
antiferromagnetisch antiferromagnetic
Antikorrosiv anticorrosive
antimagnetisch antimagnetic || non-magnetic
antiparallel anti-parallel * anti-parallel connected: ~ geschaltet || inverse-parallel
Antiparallelschaltung antiparallel connection || back-to-back connection || anti-parallel connection
antippen touch || press * Taste ~
Antistatikbekleidung anti-static clothes
Antistatikmittel antistatic agent
antistatisch antistatic || anti-static
Antivalenz exclusive OR || non-equivalence * z.B. logische Exklusiv-Oder Funktion || EXOR * Exklusiv-Oder || non-coincidence
Antrag submission * Unterbreitung, Vorlage || offer * Antrag; Vorbringen (e. Vorschlags, e. Meinung usw.) || proposal * Vorschlag || proposition || petition * Gesuch, Eingabe, Petition || application * Gesuch
antreffen encounter
antreiben drive * Maschine || power * mit Antriebskraft versehen || actuate * betätigen, auslösen, anstoßen || hurry * zur Eile || propel * Fahrzeug || stimulate * stimulieren || urge * (figürlich) vorwärtstreiben, ansporen, drängen, nötigen || impel * (figürlich) treiben, zwingen, nötigen
Antrieb drive * Maschine, Motor-Stromrichterkombination; im amerikan. häufig nur d.Strom-

richter gemeint || motive power * *Antriebskraft* || propulsion * *Antriebs-/Fortbewegungskraft, besonders für Fahrzeug; auch figürlich* || actuator * *Stell~* || magnet system * *Relais~, Schütz~* || impulse * *Impuls* || motive * *Beweggrund* || incentive * *Anreiz, Ansporn, Antrieb, Leistungsanreiz* || impetus * *innerer ~* || inducement * *Beweggrund* || electric drive * *elektrischer ~* || initiative * *Initiative* || solenoid * *Magnet~* || operating mechanism * *~/Betätigungsteil eines Leistungsschalters usw.*

Antrieb am Bus   drive connected to the bus
Antrieb Betriebsbereit   drive healthy * *Steuersignal* || drive ready || drive ready-to-run
Antrieb großer Leistung   high-power drive
Antrieb hoher Leistung   high-power drive
Antrieb mit Drehzahleinstellung   adjustable speed drive
Antriebelement   power transmission element * *Kraftübertragungselement*
Antriebkonzept   drive concept
Antriebs- und Regelungstechnik   drives and controls
Antriebsaufgabe   drive task
Antriebsausrüstung   drive equipment || drive system
Antriebsdaten   drive data
Antriebsdrehmoment   driving torque || output torque * *am Ausgang e. Getriebes/Motors/Kupplung* || drive torque
Antriebsdrehzahl   drive speed * *z.B. Drehzahl n2 einer Getriebeausgangswelle* || input speed * *eines Getriebes*
Antriebseinheit   drive unit
Antriebselement   driving component || drive element || transmission element * *Kraftübertragungs-Element, z.B. Kette, Ritzel, Riemen, Getriebe*
Antriebsenergie   motive power || operating energy * *für Betätigungsantrieb*
Antriebsgeräusche   drive noise
Antriebsgruppe   drive section || drive group
Antriebskonzept   drive configuration || drive concept
Antriebskraft   motive power || motive force || driving force || propulsion force * *bei Schiff, Fahrzeug, Linearmotor*
Antriebskurbel   crank
Antriebsleistung   driving power || drive power || motive power || input * *Generator* || mechanical input power * *eines Generators* || propulsion power * *bei Schiff, Fahrzeug, Linearmotor*
Antriebslösung   drive solution
Antriebsmaschine   driving maschine || drive motor || driving motor || prime mover
Antriebsmechanik   drive mechanics || driving mechanism * *Antriebsmechanismus* || motion mechanism * *Antriebsmechanismus*
Antriebsmoment   drive torque || driving torque
Antriebsmotor   drive motor || propulsion motor * *bei Fahrzeug* || traction motor * *Fahrmotor bei Schienenfahrzeug*
antriebsnah   drive-related * *in engem Zusammenhang mit dem Antrieb stehend (z.B. Steuerung, Regelung)* || drive-specific
antriebsnahe Steuerung   drive-related open-loop control || drive-specific open-loop control || drive-specific control
Antriebspaket   drive package * *z.B. Stromrichter und Motor*
Antriebsprojektierung   drive engineering || layout of drive equipment * *Auslegung der Antriebsausrüstung* || engineering of the drive system * *für eine Maschine/Anlage*
Antriebsrad   drive wheel * *z.B. eines Fahrwerks*
Antriebsregelung   drive control || drive regulation

Antriebsregelungssystem   drive control system || closed-loop drive control system
Antriebsritzel   driving pinion || drive pinion
Antriebsscheibe   driving pulley || driving sheave
Antriebsseite   drive end || driving end || D-end * *Abkürzung für drive end* || coupling end * *Kupplungsseite* || pulley end * *Seite, an der die Riemenscheibe angebracht ist* || back * *Rückseite, hintere Seite* || drive side * *z.B. eines Motors* || driving side || driven side * *angetriebene Seite*
antriebsseitig   drive-end || at the drive end || D-end
Antriebsspindel   jack shaft * *im Walzwerk* || actuator stem * *eines Ventils* || drive spindle
Antriebsstaffel   sectional drive * *auch 'Antrieb innerhalb einer Antriebsstaffel'* || drive group || drive sequence
Antriebsstelle   drive station || drive location || drive position
Antriebssteuerung   drive-related control * *antriebsnahe Steuerung* || drive control
Antriebssteuerungsebene   drive control level
Antriebsstrang   drive train || power train || drive line
Antriebsstromrichter   drive converter || motor controller || drive controller || drive * *(amerikan.)*
Antriebssystem   drive system || actuating system * *~ für Schalter, Relais*
Antriebstechnik   drive engineering || drive technology || drives || drive systems * *Spektrum antriebstechnischer Produkte* || power transmission engineering * *Techn. d.mechan. Kraftübertragg. (Getriebe, Kupplungen,Bremsen usw.)* || power transmission technology * *Technik d.mechan. Antriebselemente, z.B. Bremsen, Kupplgn., Getriebe*
Antriebstechnik mit System   System-Based Drive Technology * *SIEMENS-Slogan*
Antriebstrommel   driving drum || drive drum
Antriebsverbund   drive system
Antriebswelle   drive shaft || driving shaft || motor shaft * *Motorwelle* || input shaft * *Getriebeeingangswelle* || actuating shaft * *Schalter, Relais* || propshaft * *eines Fahrzeugs/Schiffs (z.B. Kardanwelle)* || high-speed shaft * *bei einem Untersetzungsgetriebe* || input end * *z.B. eines Getriebes/ einer Kupplung* || output shaft * *Ausgangswelle (eines Getriebes)*
Antwort   answer * *to: auf* || reply * *to: auf* || response * *~ einer Mess- o. Regeleinrichtung (z.B. Sprung-/Anstiegs~)* || reaction * *Reaktion* || return letter * *~schreiben, ~brief*
antworten   answer * *to: auf* || reply * *to: auf* || respond * *von einer Mess- oder Regeleinrichtung; to: auf; with: mit* || react * *reagieren* || give an answer
Antwortkanal   reply channel
Antwortkennung   reply specifier * *('PROFIBUS-Profil; für Parameter-Schreib-/Leseaufträge)*
Antworttelegramm   response telegram || response message * *Inhalt*
Antwortzeit   response time
Anwahl   selection || selecting
anwählbar   selectable
anwählen   select || preselect || call for
Anwahltaste   selection key || selector button
Anwalt   lawyer
anwärmen   preheat * *vorwärmen* || warm up
Anwärmzeit   preheat period || warming up period || warm-up period || preconditioning time * *Messgerät*
anweisen   allocate * *z.B. Speicherplatz*
Anweisung   instruction * *auch Computerbefehl, besonders für Assembler/Maschinensprache* || statement * *Computerbefehl, besonders für Hochsprache* || order * *Befehl, Auftrag, Instruktion* || direction * *Weisung, Anordnung, Vorschrift, Richtlinie, Belehrung* || regulation * *Vorschrift*

Anweisungsliste 34

specification * *Festlegung* || assignment * *Zahlung* || transfer * *Zahlungsüberweisung* || command * *Befehl, Kommando*
Anweisungsliste statement list * *{SPS} Programmiersprache für Steuerungen mit mnemotechn. Befehlen nach DIN 19239* || instruction list
anwendbar applicable * *to: auf; auch 'passend, zutreffend'* || practicable * *ausführbar* || relevant * *einschlägig* || suitable * *geeignet*
anwendbar sein can be applied || be applicable * *to: auf* || be relevant * *einschlägig/relevant sein* || be practicable * *ausführbar sein* || be suitable * *geeignet sein* || apply * *to: Anwendung finden bei; passen/zutreffen auf*
anwendbar sein auf apply to
Anwendbarkeit applicability || feasibility * *Durchführbarkeit* || application * *Anwendungsmöglichkeit*
anwenden use * *for: für, zu; benutzen, verwenden* || apply * *to: auf; auch Gesetz, Prinzip* || employ * *for; einsetzen* || utilize * *sich zunutze machen, ausnutzen* || make use of * *Gebrauch machen von*
Anwender user || customer * *Kunde*
Anwender-definiert user-defined
Anwender-Schnittstelle user interface
Anwenderbedürfnisse user's needs || user needs
Anwenderberatung user advice
Anwenderebene user layer * *z.B. bei einem Bussystem*
anwenderfreundlich user friendly || easy-to-apply * *leicht anwendbar* || easy-to-operate * *einfach zu bedienen* || user convenient * *bequem* || easy-to-use * *anwendungsfreundlich* || user-friendly
Anwenderfreundlichkeit user convenience || user-friendliness || usability || ease of use || ease of operation
Anwendernutzen user advantages || user benefits
Anwenderoberfläche user interface || human interface
anwenderorientiert user-oriented || application-oriented
Anwenderprogramm user program || application program || application
Anwenderschicht application layer * *Schicht 7 im ISO 7-Schichten-Kommunikationsmodell*
Anwenderspeicher user memory * *für Anwenderprogramme verwendbarer Speicher, z.B. in einer SPS*
anwenderspezifisch application-specific || customized * *auf einen Kunden zugeschnitten*
Anwenderunterstützung user support
Anwendung application * *Einsatzfall, Einsatzmöglichkeit* || use * *Gebrauch* || utilization * *Nutzbarmachung* || employment * *Verwendung, Beschäftigung, Einsatz* || app * *(salopp) Kurzform für 'application' (Computer-Anwendungsprogramm)*
Anwendung finden apply * *to: bei*
Anwendung finden bei apply to
Anwendungsbeispiel application example || example of application
Anwendungsbereich applications || field of application || scope * *Geltungsbereich z.B. für Vorschrift* || range of application || area of application || application scope || field of use
anwendungsbezogen application-related
Anwendungsfall application || applicability * *Anwendungsmöglichkeit* || use * *Verwendung(szweck), Verwendbarkeit, Gebrauch, Benutzung*
Anwendungsgebiet field of application || range of application * *Anwendungsbereich* || application area || area of application
Anwendungshinweise application notes || application guidelines || application hints * *Tipps/Ratschläge für die Anwendung* || application instructi-

ons || application information || user's instructions
Anwendungsklasse DIN climatic category * *Klimaklasse nach DIN*
Anwendungsmakros application macros
Anwendungsmöglichkeiten application possibilities || application spectrum * *Anwendungsspektrum* || applications || application potential
anwendungsorientiert application-oriented || application oriented
Anwendungspektrum application spectrum
Anwendungsprogramm application program
Anwendungsrichtlinien user's guide || application guide || operating principles
Anwendungsschwerpunkt main application
anwendungsspezifisch application-specific || task-specific || user-specific
Anwendungstechnik applications engineering
Anwendungszweck application || intended use * *Bestimmung* || function * *Funktion*
anwerfen start || start up
Anwerfschalter centrifugal starting switch
Anwesenheit presence * *auch 'Vorhandensein'* || attendance * *at: bei*
Anwurfmotor starting motor * *{el.Masch.} Anwurfs-Hilfsmotor* || pony motor
Anzahl number * *Zahl* || quantity * *auch Stück Ware* || q'ty * *Kurzform für 'quantity'* || count
Anzahlung deposit
Anzapfdrossel tapped variable inductor
anzapfen tap * *Trafo/Drossel/Bierfass* ~
Anzapfung tap * *tapped: mit ~ versehen* || tapping || tapping point || wiretap * *~ einer Daten-/Telefonleitung*
anzeichnen mark || scribe * *anreißen (z.B. mit Reißnadel)* || stroke * *mit Strich(en)*
anzeigbar displayable
Anzeige indication * *auch v. Messinstrument* || display * *z.B. auf LED-~ oder Bildschirm* || reading * *Ablesung e.Instruments* || readout * *Ablesewert, z.B. von einem Messinstrument* || announcement * *Ankündigung* || annunciation * *Anzeige (z.B. von Fehler, Zustand), Verkündigung* || annunciator * *~gerät, z.B. Störleuchte* || advertisement * *Inserat, Werbeanzeige* || advert * *(salopp) Werbeanzeige* || condition code * *bei SPS* || indicator * *~gerät, ~element (z.B. Lampe, LED, LCD-Anzeige)* || readout * *Ablesewert* || Ad * *(salopp) Abkürzung für 'advertisement': Werbe~*
Anzeige- indicating || indicator || display
Anzeigebereich display range || display limits * *(Anzeigegrenzen)* || instrument range * *eines Messinstruments*
Anzeigedaten display data
Anzeigeeinheit display panel * *Anzeigefeld* || display unit
Anzeigeelement display element || display control || annunciator
Anzeigefeld display panel
Anzeigegenauigkeit accuracy of indication
Anzeigegerät display unit
Anzeigeinstrument indicator || meter * *Messinstrument* || indicating instrument
Anzeigelampe pilot lamp || indicator lamp || indicating lamp
Anzeigemodus display mode
anzeigen display || indicate || record * *registrieren, aufzeichnen* || read * *Ablesewert an einem Messgerät ~* || annunciate * *z.B. Fehlermeldung ~* || show * *Ablesewert auf Messgerät ~; show no reading: nichts ~*
anzeigend indicating
Anzeigeparameter display parameter || monitoring parameter * *Beobachtungsparameter* || read-only parameter * *Nur-Lese-Parameter*

anzeigepflichtig notifiable || reportable
Anzeiger indicator || annunciator || gauge * *z.B. für Druck, Batterie-Restkapazität*
Anzeigeschwelle threshold of indication
Anzeigetableau annunciator || annunciator panel || annunciator board
Anzeigetafel annunciator panel || annunciator board
Anzeigewert reading || displayed value || indicated value || readout value
Anziehdrehmoment tightening torque * *für Schrauben; siehe* →*Anzugsmoment*
anziehen tighten * *Schraube; in diagonal-opposite sequence: über Kreuz* || draw * *(allg.)* || pull * *(allg.) auch Bremse* || pull off * *abfahren, sich in bewegung setzen* || stretch * *spannen* || break away * *Motor* || start up * *Motor* || pick up * *Relais* || put on * *Kleider* ~ || attract * *Magnet/Kapital* ~ || rise * *von Preisen usw.* || advance * *von Preisen usw.* || pull in * *Relais, elektromagnet. Bremse usw.* || tighten up * *Schraube* ~ || apply * *Bremse* ~ || operate * *Relais, Bremse usw.*
Anziehschema bolt tightening scheme * *für Schrauben*
Anziehung attraction * *(auch figürl.) magnetic: magnetisch; electrostatic: elektrostatisch*
Anziehungskraft attractive power * *(physikal.)* || attraction * *(figürl.)* || attractive force * *z.B. magnetische*
Anzug starting power * *Anzugskraft, Anziehvermögen bei Fahrzeug/Motor* || starting tractive effort * *Anzugskraft b.Fahrzeug* || starting drawbar pull * *Anzugskraft bei Fahrzeug* || getaway power * *Anzugskraft bei Fahrzeug* || zip * *(salopp) Anzugskraft bei Fahrzeug usw.*
Anzugsdrehmoment stud torque * *für Schrauben* || breakaway torque * *Motor* || starting tractive torque * *Bahn* || tightening torque * *für Schrauben*
Anzugsmoment stud torque * *beim Anziehen einer Schraube* || breakaway torque * *Motor* || starting tractive torque * *Bahn* || locked-rotor torque * *Anlaufmoment e. Motors aus d. Stillstand heraus* || tightening torque * *Schraube* || locked-motor torque * *Motor* || starting torque * *Anlaufmoment eines Motors (kleinstes gemessen. Drehmomente e.festgebremsten Motors)* || accelerating torque * *Beschleunigungsmoment* || tightening torque * *zum Anziehen von Schrauben*
Anzugsspannung pickup voltage * *Relais* || operate voltage * *Relais* || breakaway voltage * *Motor* || pick-up voltage * *Relais*
Anzugsstrom locked-rotor current * ~ *eines Motors im Startmoment* || pickup current * *Relais* || breakaway starting current * *Motor* || starting current * *eines Motors (gemessener Effektivwert I1 d.Stroms bei festgebremstem Läufer)* || inrush current * *erhöhter Einschaltstrom b.Schalten induktiver Lasten (Relais, Schütz, Magnetventil)*
anzugsverzögert with ON delay || pickup-delayed || with pull-in delay
Anzugsverzögerung pickup delay * *Relais* || ON-delay * *Relais*
Anzugswicklung pickup winding * *Relais*
Anzugszeit pickup time * *Relais* || operate time * *Relais*
anzuverzögert with ON delay || pickup-delayed
anzutreffen encountered
aperiodisch aperiodic || aperiodical || deadbeat * *auch 'aperiodisch gedämpft'*
Apotheke chemist's shop || pharmacy || apothecary * *(amerikan.)* || drug store
Apparat apparatus * *(auch figürlich)* || instrument * *feinmechanischer* ~, *Messeinrichtung* || device * *Gerät, Vorrichtung* || appliance * *z.B. Haushaltsgerät* || machine * *Maschine* || mechanism * *Mechanismus* || organization * ~ *einer Verwaltung*

Apparate-Stückliste device parts list || equipment parts list
Apparatur equipment || apparatuses * *Apparate* || device || outfit * *Gerät(e), Ausstattung, Werkzeug(e), Ausrüstung, Ausstattung* || mechanical outfit || installation || system
Applikation application
Applikationshandbuch application manual || application guide
Applikationshinweis application note
Applikationshinweise application notes || application guidelines
Applikationsingenieur application engineer
Applikationsschrift application note || application document
Approbation listing * *z. B.UL* || certificate || certification * *Zertifizierung/Approbierung* || approval * *z.b. Zulassung/Anerkennung bei einer Normenbehörde*
approbiert listed * *z.B. UL listed: beim UL-Normengremium approbiert* || approved * *z.B. bei Normenbehörde UL, CSA usw.* || recognized * *anerkannt z.B. bei Normenbehörde UL, CSA usw.*
Approximation approximation
äquidistant equidistant
Äquipotenzial- equipotential
äquivalent equivalent
Arbeit energy * *auch als physikalische Große in [Ws] oder [J]* || work * *auch als physikalische Größe in [Ws] oder [J]* || paper * *schriftliche Ausarbeitung* || effort * *Mühe* || employment * *Berufstätigkeit* || job * *Berufstätigkeit, Arbeitsplatz, Aufgabe* || task * *Aufgabe* || occupation * *Berufstätigkeit, Tätigkeit*
arbeiten work || operate * *z.B.Antrieb/Verstärker/ Gerät/Einrichtung* || act * *handeln, tätig sein, wirken, eingreifen* || act as * *fungieren/agieren/ amtieren/dienen als* || run * *laufen (Maschine)* || be operated * *betrieben werden*
Arbeiter worker || workman * *Hand~* || unskilled worker * *ungelernter* ~ || operator * ~ *an der Maschine* || attendant * ~ *an der Maschine* || semi-skilled worker * *angelernter* ~ || labourer
Arbeitgeber employer
Arbeitnehmer employee
Arbeitsbelastung work load * *bezügl. der Menge der zu leistenden Arbeit*
Arbeitsbereich operating range || working area * *Bereich in dem jemand arbeitet (Sicherheitszone)* || operating area
Arbeitsbreite working width * *einer Maschine zur Verarbeitung von durchlaufenden Warenbahnen (z.B. Papiermaschine)*
Arbeitsdiskette working disk
Arbeitsgang pass || step || cycle of operation * *Maschine* || operating cycle || production step * *working process* || phase of operation * *in a single operation: in einem* ~
Arbeitsgemeinschaft working group || working pool || study group || team || union * *Vereinigung* || alliance * *Bund, Bündnis, (Interessen-) Gemeinschaft, Allianz*
Arbeitsgeschwindigkeit production speed * *einer Produktionsmaschine/-linie/-anlage* || working speed || operating speed
Arbeitsgruppe working group * *siehe auch* →*Arbeitskreis, Gremium*
arbeitsintensiv labor-intensive
Arbeitskennlinie operation characteristic
Arbeitskleidung work clothes
Arbeitsklima working atmosphere
Arbeitskontakt normally open contact || NO contact * *Abkürzg. für 'Normally Open contact'* || make contakt * *Schließerkontakt* || A contact || form A contact

Arbeitskraft manpower || human resources * *menschliche*
Arbeitskreis working committee || study group || working group
Arbeitsleistung manpower || work rate * *einer Maschine*
Arbeitslosigkeit joblessness
Arbeitsluftspalt operating airgap * *z.B. bei Kupplung, Bremse*
Arbeitsmaschine driven machine * *angetriebene Maschine* || prime mover || production machine
Arbeitsmittel tools || working materials || auxiliary materials || aids * →*Hilfsmittel*
Arbeitspferd workhorse * *(auch figürl.) robustes, erprobtes Produkt*
Arbeitsplan work schedule || working chart * *als Diagramm auf Papier*
Arbeitsplatz working place * *z.B. Schreibtisch, Werkbank* || place of work * *Stelle* || employment * *Anstellung, Arbeits/Beschäftigungsverhältnis* || job * *Stelle; change ones job: ~ wechseln; maintenance: Erhaltung; security: Sicherheit* || operator's position * *z.B. eines Maschinenführers* || seat * *~ im Büro, z.B. an einem Computer* || vacancy * *freier ~/Stelle* || working area * *Arbeitsbereich* || working space * *Arbeitsbereich* || workplace * *Arbeitsplatz, Arbeitsbereich* || work station * *~ für Bedienpersonal einer Maschine, Anlage usw.*
Arbeitsplatzrechner workstation computer || workstation
Arbeitspunkt operating point
Arbeitsrollgang working roller table * *{Walzwerk}*
Arbeitssicherheit industrial safety
Arbeitssicherheitseinrichtung industrial safety equipment
Arbeitsspeicher main memory * *Hauptspeicher, z.B. eines PC* || working memory || RAM memory * *RAM-Speicher* || operating memory || RAM * *RAM-Speicher*
Arbeitsstelle post || job || employment
Arbeitsstellung operating position || run position
Arbeitsstrom operating current * *z.B. bei Bremse, Magnetventil* || working current * *siehe z.B VDE 0160*
Arbeitsstrombremse magnetically operated brake * *Bremsen nur mit Strom (d.h. nicht bei Spannungsausfall) möglich*
Arbeitsstromprinzip make-current principle of operation * *b. el.mechan. Gerät (z.B. magn. Bremse schließt bei Stromfluss)*
Arbeitsstunde man-hour || work hour
Arbeitstage working days
Arbeitsteilung division of labour || task distribution
Arbeitstitel working title * *eines Buches, Projekts usw.*
Arbeitsumgang extent of work
Arbeitsunfall industrial accident || industrial injuries
Arbeitsunterlage working document
Arbeitsverhältnis employment * *Anstellung, Stellung*
Arbeitsvorbereitung production scheduling * *Fertigungsvorbereitung* || production planning * *Fertigungsplanung* || work scheduling
Arbeitsvorgang procedure
Arbeitswalze worker roll * *{Textil}* in →*Kardenmaschine*
Arbeitsweise principle of operation || mode of operation || practice * *einer Person/Abteilung usw.* || operation || working method
Arbeitswelle power synchro-tie * *ein Funktionsprinzip d.elektr. Welle mit Asynchron-Schleifringläufermotoren*
Arbeitswinkel working angle * *z.B. einer Bremse*
Arbeitszeit working hours

Arbeitszyklus operating cycle
Arbitrierung arbitration * *Verwaltung u. Zuteilung der Bus-Zugriffsrechte bei Multiprozessorsystem*
Arbitrierungseinheit arbiter * *organisiert die Zusteilung d. Buszugriffsrechte bei Multiprozessorsystem* || arbitration unit * *organisiert Zusteilg. d. Buszugriffsrechte bei Multiprozessorsystem*
Architektur architecture * *(auch figürl.) auch eines Rechners*
Archiv archive * *['a:kaiv] zur Ablage/Sicherung von Software, Akten*
Archivdiskette archiving floppy disk
archivieren archive * *['a:kaiv] Software, Akten, usw.* || file * *Dokumente, Akten usw.* || put into the archives
Archivierung archiving || filing * *von (Papier-) Dokumenten, Akten*
Ärgernis scandal * *skandalöses Ereignis, Schande, (öffentliches) Ärgernis* || offence * *(brit.) Ärgernis, Anstoß, Vergehen, Verstoß, Delikt; give offence: ~ erregen* || offense * *(amerikan.)* || annoyance * *Störung, Belästigung, Plage(geist), Ärger(nis)* || vexation * *Verdruss, Ärger, Plage, Qual, Belästigung, Schikane* || bother * *Last, Plage, Mühe, Ärger, Schererei, Aufregung, Getue* || nuisance * *Ärgernis, Plage, etwas Lästiges/Unangenehmes, Missstand, Unfug, Quälgeist, Landplage*
Argument argument || point * *(springender) Punkt* || contention * *Argument, Behauptung, Streitpunkt*
Argumenteheft argumentation brochure
Argumentenheft promotional-arguments booklet * *mit Nennung der Produktvorzüge zur Vertriebsunterstützung* || arguments booklet || acquisitional argumentation booklet || argumentation booklet || argumentation brochure
Arithmetikbaustein arithmetic block
Arithmetikfunktionen arithmetic functions || math functions || mathematics || computation functions
arithmetisch arithmetic
Arithmetische Funktion arithmetic function || arithmetic operation
arithmetischer Befehl arithmetic instruction * *Rechner, SPS*
arithmetischer Mittelwert arithmetic mean value || arithmetic average value || arithmetic average || arithmetic mean || average || av * *Abkürzg. für 'Average'*
arithmetisches Mittel arithmetic mean
Armatur fitting || fittings || mountings
Armaturen valves and fittings
Ärmel hochkrempeln roll up the shirt sleeves * *(auch figürl.)*
Armiereisen reinforcing steel || reinforcing mesh * *in Mattenform*
Armierung sheath * *z.B. eines Kabels*
Armierungsgarn armoring yarn * *{Kabel}*
arretieren lock || block || arrest || clamp || lock in place
Arretierung lock || arresting element || clamping device || blocking || locking
Art kind * *Sorte, Wesen, Gattung, Natur* || method * *Methode* || type * *Typ, Modell, Ausführung, Baumuster* || nature * *Beschaffenheit* || category * *Gattung, Klasse* || mode * *Art u.Weise, Modus, Betriebsart* || fashion * *Art u.Weise, Stil, Brauch, Sitte,* || style * *Stil, Typ, Art u.Weise, Manier* || manner * *Art u.Weise, von 'Verhalten, Betragen, Auftreten'* || class * *Klasse, Gruppe* || character * *Wesen, Natur, Eigenart* || breed * *(Menschen-) Schlag, Art* || ART * *{el.Masch.} Abkürzg. für 'Allowable Running-up Time: zulässige Anlaufzeit*
Art und Weise manner || way * *(Lösungs/Realisierungs) Weg; in this way: auf diese ~* || fashion * *Stil, Brauch, Sitte* || mode * *Wirkungs/Arbeits-*

weise, Methode || style * *Stil, Typ, Manier* || method * *Verfahren, Methode* || procedure * *Verfahren* || kind of doing * *des Handels, Vorgehensweise*
Artikel goods * *(Plural) (Transport)Güter, Waren, Gegenstände, Frachtgut* || item * *Stück Ware* || part || article * *Aufsatz/Bericht z.B. in Zeitschrift*
Artikel-Nr item No. || order No. * *Bestellnr.* || part No.
Arzt physician || doctor || medical man * *(salopp)*
AS drive side * *Antriebs-Seite eines Motors* || drive end * *Antriebs-Seite eines Motors* || D-end * *Abkürzg. für 'Drive end': Antriebsseite* || AS * *Australian Standard (Australische Norm ähnlich DIN und VDE)*
AS-Wellenende drive shaft end * *{el.Masch.} antriebsseitiges Wellenende (z.B. eines El.motors)*
Asbest asbestos
asbestfrei asbestos-free || non-asbestos
Asche ash * *auch als Bestandteil von Papier*
Aschemessung ash measurement * *zur Messung des Füllstoffgehalts bei Papier*
Aschenbecher ash tray
ASCII ASCII * *Abk.f.'American Standard Code for Information Interchange': Code zur Textdarstellg, ISO 646*
ASCII-Zeichen ASCII character
ASIC ASIC * *Abk. f. 'Application-Specific Integrated Circuit': kundenspezifischer Integrierter Schaltkreis*
Aspekt aspect || issue * *(Kern)Punkt, Sachverhalt, (Streit/Kern)Frage* || point of view * *Gesichtspunkt, Blickwinkel; from the...point of view: aus dem Blickwinkel der...* || viewpoint * *siehe 'Gesichtspunkt'*
Assembler assembler * *Übersetzungsprogramm für Maschinenspracheprogramme; →Assemblersprache*
Assemblercode assembly code
Assemblersprache assmbly language * *Maschinensprache in mnemotechn. Notation (1 Statement entspr. 1 Maschinenbefehl)*
Assemblierer assembler * *Übersetzungsprogramm für Maschinenspracheprogramme*
Assistent assistant || personal assistant * *persönlicher ~* || assistant * *Assistent(in), Helfer, Mitarbeiter, Gehilfe, Angestellte(r), Verkäufer, Hilfs-* || wizard * *kleiner Helfer (Hilfe-Funktion in Softwareprogramm), Zauberer, Hexenmeister*
Assoziation association * *Zusammenhang, Gedankenverbindung/verknüpfung, Vereinigung, Zusammenschluss* || connotation * *Nebenbedeutung, 2. Bezeichnung, Begriffsinhalt* || partnership * *Partnerschaft*
assoziativ associative * *z.B. (Cache-) Speicher*
assoziativer Speicher associative memory || content-addressable memory || CAM * *Abk.f. 'Content-Addressable Memory': Speicher, dess.Inhalt als Adresse dient*
Ast branch * *einer Kurve; ascending: aufsteigender; descending: abfallender*
Astloch knothole * *{Holz}*
asymmetrisch asymmetric || asymmetrical
asymptotisch asymptotical || asymtotically * *(Adv.)*
asynchron asynchronous || asynchronously * *(Adv.)*
Asynchron-Frequenzumformer asynchronous frequency converter * *rotierender*
Asynchron-Käfigläufermotor squirrel-cage induction motor
Asynchron-Schleifringläufermotor slipring induction motor
Asynchrongenerator asynchronous generator || induction generator

Asynchronmaschine induction machine || asynchronous induction machine
Asynchronmotor induction motor * *siehe auch →Käfigläufermotor* || AC-induction motor || asynchronous motor
Äthylen ethylene
atmen breathe || vary * *variieren, sich ändern; auch Preise, Kurse* || oscillate * *(hin und her) schwingen*
Atmosphäre atmosphere
Atmung respiration * *Ausdehnung und Schrumpfung bei Temperaturänderungen*
attraktiv attractive || eye-catching * *das Auge anziehend*
Attrappe dummy || mock-up || display model * *für Ausstellungszwecke*
Attribut attribute * *{Computer} (Zusatz-) Kennzeichen einer Variablen/einer Datei usw.*
ätz- und säurefeste Ausführung corrosion-resistant design
ätzen etch
auch also * *gleichfalls* || too * *(nachgestellt)* || as well * *(nachgestellt) gleichfalls* || likewise * *eben/ gleichfalls desgleichen, ebenso* || even * *selbst, sogar* || at that * *(nachgestellt) übrigens* || even if * *wenn auch* || although * *obwohl, wenn auch* || either * *[am. i: the; engl. aithe] (immer nachgestellt) auch i Zusamm.hang m. Verneinung(not either)*
auch nicht neither * *neither can I: ich kann es auch nicht* || not either * *I cannot do it either: ich kann es auch nicht*
Audit audit
auf on || at || in * *in the market: auf dem Markt* || to * *(mit Akkusativ)* || towards * *(mit Akkusativ)* || open * *offen* || up * *(Adv.) hinauf; up and down: auf und ab* || upwards * *(Adv.) hinauf* || on top of * *auf ... oben 'drauf'*
auf Abruf on call
auf alle Fälle at all events || in any case || by all means || to be on the safe side * *zur Sicherheit, sicherheitshalber, lieber*
auf Anforderung on request || on demand
auf Anfrage upon request || on request || please inquire * *'bitte fragen Sie an'*
auf dem Laufenden up-to-date
auf dem Markt on the market || in the market || on sale * *käuflich*
auf dem richtigen Weg sein be on the right track
auf den Markt bringen launch * *z.B. ein Produkt* || put on market || put on sale
auf den neuesten Stand bringen upgrade || update * *Software*
auf der Basis von based on
auf der Gleichstromseite on the DC side
auf Display ausgeben display
auf Drehzahl Null herunterfahren ramp down to zero * *entlang einer Rampe*
auf eine Grenze abfragen check against a limit
auf einen Blick at a glance
auf einfache Art und Weise simply
auf einmal all at once * *gleichzeitig* || suddenly * *plötzlich* || at once * *zugleich* || in one step * *in einem Schritt/Arbeitsgang* || in one go
auf Grund due to * *verursacht durch; due to the fact that: ~ der Tatsache, dass* || as a result of * *als Ergebnis von* || because of * *because of the fact that: ~ der Tatsache, dass* || for the sake of * *um (einer Sache) willen* || as a result of * *als Ergebnis von, ~ von* || owing to * *infolge, wegen, zurückzuführen auf* || on account of * *um ... willen, wegen* || by virtue of * *kraft (Gesetz, Vollmacht usw.), auf Grund von, vermöge*
auf Grund von as a result of || due to
auf halbem Weg halfway
auf HIGH schalten switch to HIGH

auf keinen Fall by no means || on no account || in no case
auf Knopfdruck just hitting a key * *siehe auch →Knopfdruck* || at a mouse click * *mit einem Mausklick*
auf Kommando upon command
auf Lager in stock * *have stocked: ~ haben* || in store || on hand * *bei der Hand, greifbar* || on the shelf
auf Lager halten keep in stock || store * *auf Lager nehmen*
auf Lager legen stock || stock up
auf lange Sicht in the long run * *auf die Dauer* || on a long-term basis || long-term || long-range
auf LOW schalten switch to LOW
auf Lücke taggered
auf Masse bezogen single ended * *z.B. Analogsignal (im Gegensatz zum Differenzialsignal)* || to ground
auf Null zurückfallen fall to zero
auf Null zurückgehen go back to zero
Auf Wiedersehen good bye || bye bye || bye || see you later * *bis später* || have a nice day * *einen schönen Tag noch* || have a nice evening * *einen schönen Abend noch*
auf Wunsch on request * *auf Anfrage* || when requested
Auf-Abwickler winding gear
Auf-Putz surface-mount
Aufbau build-up * *z.B. Spannung, Wickel beim Aufwickler, Wärme in e. Wärmenest, Magnetfeld* || design * *Konstruktion* || arrangement * *Anordnung* || assembly * *Zusammensetzung, Zusammenbau* || construction * *Bauweise, Konstruktion* || erection || building || building-up * *z.B. Spannung, Durchmesser beim Aufwickler* || disposition * *Anlage, Anordnung* || mounting * *Montage* || structure * *Struktur, Gefüge* || system * *Gefüge, System* || built-up * *das Aufgebaute, z.B. Spannung, Durchmesser beim Aufwickler* || buildup * *siehe 'build-up'* || installation * *Errichtung, Einbau, Intallierung* || configuration * *Anordnung, Konfiguration* || design * *Konstruktion, konstruktiver/schaltungstechnischer* ~ || formation * *(Aus)Formung*
aufbauen build up * *auch Strom, Spannung, Feld, Druck, Wickeldurchmesser, Kundenstamm, Theorie usw.* || build-up * *auch Feld, Druck, Wickeldurchmesser, Kundenstamm, Theorie usw.* || erect * *errichten, bauen, aufstellen, montieren, einrichten* || construct * *gestalten, formen, entwerfen, konstruieren* || assemble * *zusammensetzen, montieren* || mount * *errichten, aufstellen, montieren, anbringen, einbauen* || set up * *aufstellen, montieren, gründen, etablieren, kräftigen, wiederherstellen* || establish * *z.B. Strom, Spannung, Feld, eine Organisation, eine Kommunikationsverbindung* || base * *gründen, stützen, sich verlassen (on:* ~*)* || generate * *aufbringen, erzeugen* || install * *intallieren, montieren, aufstellen, einbauen* || configure * *(auf)bauen, gestalten, konfigurieren*
Aufbaulüfter top-mounted fan
Aufbaurichtlinie installation guideline * *zum störsicheren Einbau u.Verdrahtung v. Steuer- u. Regelsystemen-/Geräten*
Aufbaurichtlinien mounting guidelines
Aufbausystem mechanical rack system * *mit Baugruppenträgern und Steckbaugruppen* || rack system || packaging system
Aufbautechnik packaging system || mounting technology
Aufbauzeit build-up time * *z.B. für Motor-Feld*
aufbereiten condition * *konditionieren, i.e.bestimmten Zustand bringen; z.B. Signale* || prepare * *vorbereiten, zurecht/fertigmachen, ausarbeiten, anfertigen, (Schriftstück) abfassen* || work up * *ausarbeiten, entwickeln, erweitern (into: in), (Thema) bearbeiten, sich einarbeiten* || refine * *z.B. in e. Produktionsschritt, auch 'verfeinern, ausklügeln'* || process * *in Podunktions/Verarbeitungsschritt* || dress * *Erz* || upgrade * *Kohle* || wash * *Erz*
Aufbereitung preparation * *Aufbereitung (von Sollwerten, Erz, Kraftstoff usw.), Vorbehandlung, Vorbereitung* || processing * *Verarbeitung* || conditioning * *Signale* || preparing * *Bereitmachen, Präparieren, Zu-/Herrichten, Ausarbeitung (auch von Dokumenten)* || reprocessing * *Wiederaufbereitung* || treatment * *z.B. von Wasser* || dressing * *Nachbearbeitung, Zurichtung, Aufbereitung (auch von Erz)* || washing * *Waschen, Wässern, Nasswäsche (auch von Erz, Schlamm)* || upgrading * *von Kohle, Erz usw.*
Aufbewahrung storage
aufblasen blow up || inflate
aufblenden display * *anzeigen* || pop up * *sich entfalten, aufspringen (z.B. Menü auf dem Bildschirm)* || open up * *aufblättern, öffnen (auch Anzeigefeld auf dem Bildschirm)*
Aufbohrwerkzeug boring tool
aufbrauchen exhaust * *erschöpfen*
aufbrechen break up * *(auch figürl.)*
aufbringen apply * *anwenden, anbringen, bestreichen usw.* || coat on * *Farbe, Auftrag, Beschichtung* || get open * *öffnen, offen bekommen* || raise * *Geld* || meet * *Kosten* || introduce * *Mode* || provoke * *erzürnen, provozieren* || generate * *erzeugen, z.B. Drehmoment*
Aufbruchstimmung spirit of moving forward * *'zu neuen Ufern'* || atmosphere of departure * *'Abschieds/Abreisestimmung'*
Aufdrückvorrichtung pusher tool || pusher || fitting tool
aufeinander one on top of the other || one after another * *nacheinander* || one by one * *nacheinander* || one upon the other * *aufeinander, gegenseitig; matching: ~ abgestimmt* || on top of one another || on top of each other || against one another * *gegeneinander*
aufeinander folgen succeed * *one another: einer auf den anderen*
aufeinander folgend consecutive || successive || sequenced * *in Reihe angeordnet* || butted * *aneinandergereiht*
Aufeinanderfolge sequence * *['siequens]* || succession * *['saxeschen]*
auffahren drive against * *auf etwas ~, rammen* || run into * *auf etwas ~* || get close to * *sicht an etwas annähern; get too close to: zu dicht auf etwas ~ (im Verkehr)* || run on * *Schiff* || run aground * *(Schiff) auf Grund fahren* || dish up * *Speisen, Getränke usw. ~ lassen*
auffangen capture * *einen Messwert, Zählerstand usw. in einen Zwischenspeicher ~* || collect || catch up || snatch || intercept * *Funkspruch, Brief usw.* || cushion * *abfedern/puffern, z.B. Fall, Stoß* || pick up * *Neuigkeiten usw.*
Aufforderung prompt * *Eingabe-Aufforderung durch einen Computer auf d.Bildschirm* || call || request || invitation * *Einladung*
Aufforderungstext prompting text * *auf einem Computerbildschirm* || prompting message * *auf einem Computerbildschirm*
auffrischen refresh * *auch Daten, Prozessabbild, Ein/Ausgänge* || update * *z.B. einen Puffer, Speicher*
Auffrischkurs refresher course
Auffrischung refresh
Auffrischzeit refresh time
aufführen give * *z.B. Angabe in einer Tabelle* || enumerate * *aufzählen* || enter * *eintragen* || state

* *in einer Liste* || show * *in einer Liste, Grafik* || list
* *in einer Liste* || specify * *im Einzelnen* ~ || itemize
* *(amerikan.) im Einzelnen* ~ || set out * *in einer Liste* || thread * *der Papierbahn in einer Papiermaschine* || tabulate * *in einer Tabelle*
auffüllen replenish * *Vorräte usw. (siehe auch →füllen)* || fill up || refill * *nachfüllen* || restock * *Lager* || top up * *(Öl usw.) nachfüllen*
Aufgabe task || job || duty * *auch Pflicht* || responsibility * *Verantwortlichkeit* || function || mission * *Reisezweck/auftrag, Sendung, taktischer Auftrag* || problem || challenge * *schwierige/lockende Aufgabe, Herausforderung* || exercise * *Übungs-/Schulaufgabe*
Aufgabe haben be dedicated to || be responsible for * *verantwortlich sein für*
Aufgabenklärung technological analysis * *technologische* ~ || clarification * *(Ab)Klärung*
Aufgabenlösung task solution
Aufgabenstellung task || objective * *Zielsetzung* || target * *Ziel(setzung)* || task definition
Aufgabenumfang scope of tasks
Aufgabenverwalter task scheduler * *in einem (Multitask)-Betriebssystem* || task manager
aufgebaut built-on * *(oben) angebaut* || mounted * *montiert* || top-mounted * *oben montiert, auf der Oberseite 'draufgebaut'*
aufgeben place * *order: Bestellung* || give up * *übergeben, preisgeben, z.B. Geschäft*
aufgedampft vapor-deposited
aufgeführt given * *z.B. Angabe in einer Tabelle* || enumerated * *aufgezählt* || stated * *in einer Liste* || shown * *in Liste, Grafik usw.* || specified * *im Einzelnen* || listed * *in einer Liste* || itemized * *(amerikan.) im Einzelnen* ~
aufgehängt hung * *z.B. Programm*
aufgeladen charged * *z.B. Kondensator, Batterie, Akku*
aufgelegt connected * *z.B. ein Signal auf eine Klemme* || hooked up * *z.B. ein Signal an eine Klemme*
aufgelöst decomposed * *zerlegt, auseinandergespaltet, auseinandergelegt, vereinzelt*
aufgenommen absorbed * *z.B. Wärmeaufnahme/Leistungsaufnahme eines Motors* || consumed * *verbraucht, z.B. Leistung* || inserted * *Klausel* || started * *Betrieb* || entered into * *Verhandlungen*
aufgenommene Leistung input * *z.B. eines Motors*
aufgenommene Wirkleistung input * *eines Motors*
aufgeräumt neat * *ordentlich, sauber, übersichtlich* || clean * *ordentlich, sauber* || uncluttered * *z.B. Bildschirmdarstellung, Bedienungselement usw.*
aufgeschrumpft shrunk-on
aufgliedern break up || split up || subdivide || break down * *"herunterbrechen", aufgliedern, aufschlüsseln, analysieren* || analyze
Aufgliederung subdivision || breakdown * *Aufschlüsselung/gliederung, Analyse* || analysis || structure * *Aufbau* || departmental classification * *eines Unternehmens*
aufhängen hang * *auch Programm* || put off * *Mäntel, Garderobe ablegen* || hang up || suspend * *from: an* || lock up * *→sich aufhängen (Softwareprogramm)*
Aufhängepunkt suspension point
Aufhängung suspension
Aufhaspel coiler * *für Blech, Draht, Kabel* || recoiler * *Wiederaufhaspel, z.B. bei Metallband-Behandlungsanlage*
aufheben cancel * *each other: einander; auch Erlass/Verbot* ~ || neutralize * *each other: sich gegenseitig* ~; *ausgleichen* || compensate * *aufwiegen, ausgleichen, kompensieren* || take up || pick up

* *vom Boden* ~ || lift up * *emporheben* || raise * *hochheben; Maßnahme usw.* ~ || keep * *aufbewahren* || preserve * *aufbewahren* || store * *lagern* || warehouse * *['wärhaus'] einlagern, auf Lager halten* || dissolve * *auflösen (Organisation usw.)* || reduce * *in einer Bruchrechnung* || abolish * *abschaffen* || revoke * *widerrufen* || supersede * *ersetzen* || declare null and void * *für ungültig erklären* || suspend * *zeitweilig~, vorläufig* ~ || terminate * *Vertrag* ~ || release * *Verriegelung* ~, *Sperre* ~
Aufhebung cancellation
aufheizen heat up * *auch sich aufheizen*
Aufheller brightener
aufholen take-up * *durchhängendes Material aufholen, z.B. in Papierbahn* || slack take-up * *Durchhang, Lose aufholen, z.B. in Papierbahn* || catch up * *Vorsprung, Verspätung usw.* ~ || make up for lost time * *Verspätung* ~ || make up for * *Zeit, Verspätung usw.* ~
Aufholen-Nachlassen slack control * *{Papier}*
Aufholtaste slack-take-up button * *z.B. bei Papiermaschine*
aufhören cease * *zu Ende gehen; cease to/from: aufhören zu (mit Infinitiv)/mit* || stop || leave off || quit * *(amerikan.)* || have done * *(mit Gerundium)* || discontinue * *abbrechen* || subside * *allmählich* || ebb * *allmählich verebben* || cut out
aufkaufen buy out * *Firma usw.* || buy up * *Firma usw.*
Aufkleber sticker || label || adhesive label
aufladbar rechargeable
aufladen charge * *z.B. Kondensator, Batterie, Akku*
Auflage edition * *eines Druckerzeugnisses* || support * *Stütze* || rest * *Auflage, Stütze, Lehne* || seat * *Sitz, Auflager, Fundament* || lining * *Futter* || coat * *Anstrich, Beschichtung* || layer * *Schicht* || base * *Unterlage*
Auflagefläche seating face || bearing face || seat || contact area || rest * *Auflage, Stütze* || bearing area || bearing surface || bedding
auflegen connect-up * *"hinverdrahten/verbinden", z.B. Leitungen an Klemme* || clear the line * *Telefon* ~ || hang up * *Telefon* ~ || apply * *Beschichtung/Pflaster usw.* ~ || publish * *Buch usw.* ~; *reprint/republish: wieder~*
aufleuchten light up || go on * *angehen, eingeschaltet werden, z.B. Lampe/Leuchte* || be lit || flash up || shine || flash
aufliegen seat * *seinen Sitz haben, liegen, sitzen* || lie * *ruhen, lasten, liegen, gelegen sein, sich befinden (upon: auf)* || rest * *ruhen, sich stützen, lehnen (upon: auf); sich stützen/beruhen auf* || weigh * *lasten ((up)on: auf)* || make contact * *z.B. Kohlebürsten auf d. Kommutator einer Gleichstrommaschine*
auflisten list * *z.B. in Aufzählung/Tabelle/Liste* ~ || tabulate * *Listen-/Tabellenförmig zusammenstellen* || enlist * *in Liste eintragen*
auflösen solve * *Rätsel, Gleichung, Klammer usw.* ~; *for: nach* || reduce * *(mathemat.) Bruch, Gleichung* ~ || dissolve * *Verein, Ehe, Versammlung, Stoffe in Flüssigkeit usw.* || cancel * *Vertrag usw.* ~ || annul * *Vertrag usw.* ~ || liquidate * *Firma* ~ || disentangle * *entwirren* || unravel * *entwirren* || decompose * *zerlegen, zerspalten, vereinzeln*
Auflösewalze opening roll * *{Textil} bei Open-End-Spinnmaschine*
Auflösung resolution || dissolution * *Firma, Stoff in einer Flüssigkeit usw.*
aufmagnetisieren magnetize || remagnetize
aufmerksam machen auf draw attention to
Aufmerksamkeit attention * *pay attention to: jmd./e.Sache Aufmerksamkeit schenken* || attentiveness || watchfulness * *Wachsamkeit* || alertness * *Wachsamkeit* || vigilance * *Wachsamkeit* || cour-

aufmodulieren

tesy * *Höflichkeit* || small gift * *kleines Geschenk* || small token * *kleines Mitbringsel*
aufmodulieren   modulate upon
Aufnahme   consumption * *Verbrauch* || absorption * *Aufsaugung, auch Wärme* || uptake * *(auch figürl.: Waren vom Markt usw.)* || intake * *Zustrom, aufgenommene Menge, Ein/Ansaugen* || accommodation * *Beherbergung, Unterkunft, Platz für...* || starting * *Beginn* || initiation * *Beginn* || assumption * *~ einer Tätigkeit* || integration * *Eingliederung (within: in)* || incorporation * *Eingliederung (into: in)* || inclusion * *Einbeziehung (into: in/bei)* || reception * *Empfang* || admission * *Zulassung* || enrolment * *Einschreibung* || entry * *~ in eine Liste* || listing * *~ in eine Liste* || assessment * *~ eines Schadens* || recording * *Aufzeichnung, z.B. von Signalen, eines Protokolls (auch 'plot': Kennlinie aufnehmen)* || establishing * *~ von Beziehungen* || input * *eingespeiste/zugeführte Menge/Spannung/Leistung*
Aufnahmeleistung   input * *z.B. eines Motors*
Aufnahmeort   pick-up point * *z.B. für Kran*
Aufnahmeprüfung   entrance examination
aufnehmen   record * *aufzeichnen, z.B. Messwerte, Kennlinie* || pick up * *z.B. vom Boden, Signal durch e. Funkempfänger/e. Messaufnehmer usw. ~* || absorb * *absorbieren, in sich ~, "schlucken", aufsaugen, z.b. Wärme, Stöße* || take up * *vom Markt, auch "take"* || comprehend * *geistig erfassen, verstehen, begreifen, umfassen, einschließen* || receive * *empfangen* || accept * *annehmen; auch eine Baugruppe/ein IC in e. Einbauplatz* || accommodate * *räumlich ~ (z.B. Zusatzbaugruppe), beherbergen* || contain * *enthalten* || include * *eingliedern, into: in* || integrate * *eingliedern, integrieren, within: in* || incorporate * *eingliedern, in: in* || embody * *eingliedern, in: in* || insert * *Klausel/Passus ~* || list * *eintragen* || enter * *eintragen* || admit to * *in einen Verein ~* || enrol * *in einen Verein ~* || start * *starten, the operation: den Betrieb ~* || establish * *Beziehungen ~*
Aufnehmer   transducer * *Mess~/Geber* || sensor * *Sensor, Mess~*
aufprägen   impress * *(on: auf)* || imprint * *stamp*
Aufpreis   extra price || additional price || surcharge || extra charge * *zusätzlich Verrechnung/Gebühr/Belastung* || extra cost * *zusätzliche Kosten, zusätzlicher Preis* || price adder
aufquellen   swell * *anschwellen* || soak * *einweichen* || steep * *durchtränken (in/with: in), einweichen, eintauchen, imprägnieren*
aufrauen   roughen || nap * *{Textil} Textilstof, Tuch usw. ~* || card * *{Textil} Wolle ~*
aufräumen   clear off * *ausräumen, reinigen; beseitigen, loswerden, erledigen* || tidy up * *z.B. Zimmer* || straighten up * *(amerikan.) z.B. Zimmer* || clear away * *wegräumen, wegschaffen* || put in order * *ordnen* || do away with * *(figürl.) mit etwas aufräumen* || make a clean sweep of * *mit etwas aufräumen*
aufrechterhalten   maintain * *(einen Zustand aufrecht-) erhalten, beibehalten, (be-) wahren* || adhere to * *festhalten an, bleiben bei, (Urteil usw.) bestätigen, etw./jdm. treu bleiben* || uphold * *Brauch, Lehre, Urteil ~* || sustain * *Brauch, Lehre, Urteil ~* || retain * *beibehalten*
Aufrechterhaltung   preservation * *Bewahrung, Schutz, Konservierung, Vorbeugung* || maintenance || support
aufreißen   tear up || rip up || tear open || split open || burst * *bersten* || crack * *reißen* || disintegrate * *Schmierfilm* || break down * *auch Schmierfilm*
Aufreißstreifen   tear-open strip * *{Verpack. techn.}*
Aufreißverschluss   tear-off closure * *{Verpack. techn.}*

Aufrichtemaschine   erecting machine * *{Verpack. techn.} z.B. für Kartons*
aufrichten   set-up * *von der Horizontalen in die Vertikale*
Aufrichter   erector * *{Verpack.techn.}*
Aufriss   upright projection * *Zeichnung* || vertical plan * *Zeichnung* || vertical section * *senkrechter Schnitt* || front view * *Vorderansicht* || profile section * *Seitenriss* || elevation * *Zeichnung*
aufrollen   reel up
Aufroller   reel-up || winder
Aufrollung   reeling up || reeler * *Aufwickler* || wind stand * *Aufroller*
aufrücken   close up || move up * *→aufschließen (to/with: zu)* || be promoted * *im Rang ~* || advance to a higher position * *im Rang ~*
Aufruf   call * *z.B. eines Unterprogramms*
aufrufen   call * *z.b. Unterprogramm* || call-up * *auch 'abrufen'; z.b. Hilfetext, Diagnosedaten* || initiate * *initiieren* || invoke * *Programm usw.*
aufrunden   bring up to a round figure * *eine Zahl nach oben* || round off * *eine Zahl nach oben oder unten* || round up * *eine Zahl nach oben (to: auf)*
aufrüsten   upgrade * *hochrüsten (Hard-/Software)*
Aufrüstsatz   upgrade kit
Aufrüstung   upgrade
Aufsatz   article * *in Zeitschrift* || paper * *Abhandlung, Referat, Aufsatz* || treatise * *Abhandlung* || headpiece * *Oberteil* || top * *Oberteil*
aufschalten   switch to || switch into || switch-in * *(auch Substantiv) z.B. (Zusatz)Sollwert* || inject * *hineinleiten, z.B. Signale (siehe auch →vorsteuern), Rechtecksprünge, Feldstrom*
aufschalten auf einen laufenden Motor   start into a running motor
Aufschaltung   switching-in * *z.B. eines Zusatzsollwertes* || injection * *Einspeisung e. (Zusatz- Stimulations-/Kompensations-) Signals*
aufschaukeln   amplify * *→verstärken* || build up * *anschwellen*
aufscheuern   rub * *abreiben, reiben an, (ab)scheuern, (ab)schaben, abschleifen* || chafe * *(sich) durch-/wundreiben, scheuern (against: an), verschleißen (z.B. Kabelisolierung)* || scour * *scheuern, schrubben, polieren* || sore * *Haut, Wunde usw.*
aufschieben   defer * *zeitlich* || postpone * *zeitlich* || adjourn * *zeitlich* || reschedule * *(Termin) neu festsetzen* || put off || push open * *z.B. einen Schieber öffnen* || slip on * *ein Teil auf ein anders "draufschieben"* || delay * *verzögern*
Aufschlag   impact * *Aufprall (on: auf)* || thud * *dumpfer Aufprall* || additional charge * *zusätzliche Berechnung, Verrechnung, Gebühr* || extra charge * *zusätzliche Berechnung* || extra price * *Zusatzpreis, Preisaufschlag* || rise * *Erhöhung, Hochsetzung (Preis, Kurs usw.)* || advance * *Erhöhung, Hochsetzung (Preis, Kurs usw.)* || extra costs * *supply at small extra costs: zu einem kleinen (Preis)Aufschlag liefern* || striking * *Auftreffen*
aufschließen   unlock * *Tür usw. ~* || open * *Tür usw. ~* || desintegrate * *{Chemie}* || open up * *Markt/Bodenschätze usw. auf-/erschließen* || close up * *zum Vordermann ~, einholen* || join up * *zum Verband* || move up * *zur Spitzengruppe ~ (to/with: zu)* || catch up * *zur Spitzengruppe ~*
aufschlüsseln   subdivide || break down * *statistisch aufgliedern/schlüsseln* || discriminate * *(scharf) unterscheiden, einen Unterschied machen*
aufschlussreich   informative || instructive || insightful

Aufschmelzlöten reflow soldering * *von* →*SMD-Bauelementen*
Aufschnapp- clip-on || snap-on
aufschnappen snap on * *snap onto:* ~ *auf* || snap onto * *z.b. auf eine Installations/Hutschiene* || snap-mount * *montieren per Aufschnappmontage* || snap into place * *auf den Einbauplatz* ~
Aufschnappgehäuse rail-mount housing * *zum Aufschnappen auf* →*Installationsschiene*
Aufschnappmontage snap-on mounting || clip-on fixing
aufschrauben screw off * *abschrauben* || unscrew * *abschrauben* || screw on * *anschrauben* || bolt on * *anschrauben*
aufschrumpfen shrink on * *z.b. Aufziehen eines Lagers mit Wärme*
Aufschwung upturn * *economic: wirtschaftlicher*
Aufsichtsbehörde controlling authority * *z.B. TÜV*
Aufsichtsrat supervisory board
Aufsichtsratmitglied member of the supervisory board
Aufsichtsratsmitglied member of the supervisory board || member of the board
Aufsichtsratsvorsitzender chairman of the supervisory board || chairman of the board
Aufsichtsratvorsitzender chairman || chairman of the supervisory board
aufspalten split up || break up || cleave || disintegrate * *{Chemie}* || divide * *(ein)teilen (in(to): in)* || divide up * *zerteilen, zerlegen, zerstückeln, spalten*
Aufspulmaschine winding machine || take-up machine
aufstapeln pile || stack
Aufsteck- clip-on * *z.B. clip-on heatsink: ~-Kühlkörper*
Aufsteck-Getriebemotor slip-on geared motor
aufstecken insert * *z.B. Stecker/Steckbaugruppe* || snap on * *aufschnappen* || clip on * *rastend* || plug on || slip on * *ein Teil auf ein anderes*
Aufsteckgetriebe shaft mounted gearbox || floating-type gear || shaft-mounted speed reducer * *als Untersetzungsgetriebe* || gearhead * *mit Motorflansch verschraubtes Getriebe*
Aufsteckkarte piggyback board * *Subleiterkarte*
aufsteigen ascend * *Rang, Zahlen, Kennlinie, Töne, Ballon usw.* || rise * *Fläche, Gelände usw.; zunehmen* || increase * *zunehmen* || take off * *Flugzeug* || climb * *klimmen, auch Flugzeug* || loom * *drohendes Ereignis* || well up * *Gefühl* || come up * *Gewitter, Unheil* || be promoted * *beruflich be/gefördert werden*
aufsteigend ascending * *Rang, Zahlen, Kennlinie, Töne usw.* || rising * *Gelände, Fläche; zunehmen* || increasing * *zunehmend* || in ascending order * *(Adv.)*
aufsteilen steepen * *Impulsflanken usw.*
aufstellen set up * *auch Maschine* ~ || put up || erect * *Bauten* ~ || install * *Maschine* ~; *auch 'mount' (Motor, Maschine ~)* || assemble * *Maschine* ~ || make * *an assertion: eine Behauptung* ~ || set * *Beispiel* ~ || make up * *Bilanz* ~, *Rechnung* ~ || lay down * *Grundsatz* ~ || nominate * *als Kandidaten* ~ || specify * *Kosten* ~ || propound * *Lehre* ~, *Theorie* ~ || make out * *Liste, Rechnung usw.* ~ || prepare * *Liste usw.* ~ || organize * *Einheit, Streikräfte usw.* ~; *(arrange: anordnen, arrangieren)* || establish * *System* ~, *Rekord* ~ || state * *Problem, Regel* ~ || compile * *zusammentragen, z.B. tabellarisch* || nominate * *benennen, z.B. Teammitglieder* || annunciate * *verkünden, z.B. Lehrsatz, Axiom, Richtlinie*
Aufstellfläche floor space * *auf dem (Fuß-) Boden* || footprint * *für die Montage benötigte Grundfläche*
Aufstellhöhe site altitude * *above sea level: über dem Meeresspiegel* || altitude of site || operating altitude || installation altitude
Aufstellung listing || list || installation * *Installation, Montage* || table * *Tabelle* || survey * *Übersicht* || installation * *Installation, Montage; indoor: Innenraum-~; outdoor:* ~ *im Freien* || mounting * *Montage* || tabulation * *tabellarische* ~
Aufstellungshöhe installation altitude * *above sea level: über dem Meeresspiegel* || erection altitude * *above sea level: über dem Meeresspiegel* || site altitude || altitude of site || site elevation
Aufstellungsort site || mounting location
Aufstieg ascent || ascension * *(amerikan.)* || take off * *Flugzeug* || rise || promotion * *Beförderung* || advancement * *sozialer* ~ || social climbing * *sozialer* ~
auftauchen arise * *(figürl.) Frage, Problem usw.* || emerge || rise up || pop up * *(salopp) plötzlich auftauchen*
aufteilen divide || split-up * *aufspalten; be splitted up in(to): sich aufteilen in* || divide up || partition || distribute * *verteilen* || subdivide * *unterteilen* || split * *(auf)spalten* || rip * *{Holz} auf-/zertrennen, (der Länge nach) auseinandersägen* || size * *{Holz} zuschneiden; nach Größe sortieren*
Aufteilung distribution * *Verteilung* || alotting * *in Teile, Austeilung, Verteilung (auch durch Los)* || division * *Teilung (into: in), Gliederung, Verteilung* || partitioning * *in Teile* || allotment * *Ver/Zuteilung, Anteil, das Zugeteilte* || sharing * *untereinander* || splitting up * *into subassemblies: in Unterbaugruppen* || grouping * *in Gruppen* || classification * *in Klassen* || share * *z.B. ~ e. Last,* ~ *v. Last/Drehmoment auf mehrere Motoren; (An)Teil, Beitrag, Beteilig.*
Auftrag task * *Aufgabe* || order * *Bestellung, Befehl, Instruktion; by order of: im ~ von; be in the order backlog: in ~ sein* || request * *Anforderung* || instruction * *Weisung* || commission * *Beauftragung* || charge * *Befehl, Pflicht* || mission * *Reisezweck, Sendung, taktischer Auftrag* || contract * *Liefervereinbarung, öffentlicher Auftrag, Vertrag* || message * *Botschaft* || application * *von Farbe, Leim, Klebstoff usw.*
Auftrag bearbeiten process an order * *siehe auch* →*Auftragsabwicklung*
Auftrag erteilen place an order * *with: an* || give an order
Auftrag erteilt bekommen win the order * *gegen Konkurrenz* || be awarded the order || receive the order
auftragen spread * *verstreichen (z.B. Klebstoff)* || apply * *aufbringen* || coat * *beschichten. eine Schicht auftragen*
Auftraggeber orderer * *Besteller* || customer * *Kunde* || employer * *Arbeitgeber* || client * *Kunde* || party who issues the order * *{Rechtswesen}*
Auftragnehmer contractor || supplier * *Lieferant*
Auftragsabwicklung processing of order || order transaction || contract administration || order handling || order administration || order processing
Auftragsbearbeitung processing of orders || handling of orders || administration of orders || order processing
Auftragsbestand orders in hand || unfilled orders
Auftragsbestätigung confirmation of order
auftragsbezogene Fertigung order-driven manufacturing
Auftragseingang orders received || incoming orders || volume of orders received * *geldmäßiger Betrag, Volumen* || order entry * *Bestelleingang, Eintreffen von Bestellungen* || new orders || order intake || receipt of orders * *[ri'ßiet] Empfangen, Erhalt von Aufträgen*

Auftragserhalt contract award * *gegen Konkurrenz* || receipt of orders * *[ri'ßiet ...]*
Auftragserteilung placing of order || award * *bei einer Ausschreibung* || contract award * *Auftrags-Gewährung/Vergabe (meist gegen Konkurrenzangebote)* || award of contract * *bei Ausschreibung/gegen Konkurrenz* || conferring of contract * *das "Zuerteilt-Bekommen" eines Auftrags*
Auftragsfertigung contract manufacturing
auftragsgebunden on contract-specific basis
auftragsgemäß as ordered || as per order || according to instructions * *den Anweisungen entsprechend* || according to directions * *anweisungsgemäß*
Auftragskanal request channel
Auftragskennung request specifier * *('PROFI-BUS-Profil; für Parameter-Schreib/Leseaufträge)*
Auftragskennzeichen contract reference number || contract reference || contract ref. || order reference || contract ID || suppliers ref. || contract number || contract ref. No.
Auftragslage order situation || order inflow
Auftragsnummer contract No
Auftragsplanung job planning
Auftragspolster wealth of orders || abundance of orders in hand || orders in hand * *Auftragsbestand* || backlog of unfilled orders * *Auftragsüberhang*
Auftragsrückgang falling off of orders
Auftragsrückstand backlog of unfilled orders
auftragsspezifisch order-related
Auftragsvorbereitung production planning * *Fertigungsplanung* || job planning
Auftragswalze inking roller * *[Druckerei]* || spreading roller || applicator roll * *bei Streich-/Beschichtungsmaschine* || coating roller * *in (Papier-) Streichmaschine, Folien-beschichtungsmaschine usw.* || application roller
Auftragswerk coater * *bei Streichmaschine*
Auftragswert contract value || contract volume
auftrennen separate * *(mechan.)* || interrupt * *(elektr.) Stromkreis* || disconnect * *(elektr.) Stromkreis, Leitung* || open
auftreten occur * *vorkommen, eintreten, geschehen; z.B. Fehler* || arise * *erscheinen, entstehen; z.B. Nebeneffekt, Verluste in e.Motor* || occurrence * *(Substantiv) Vorkommen, Auftreten; Ereignis, Vorfall, Vorkommnis* || appearance * *(Substantiv) Erscheinen* || behavior * *(Substantiv) Verhalten, Benehmen* || result * *sich ergeben, resultieren, anfallen* || be incurred * *Kosten, finanzielle Verluste usw.*
Auftrieb lift || buoyancy * *eines Körpers in einer Flüssigkeit, e.Flugzeuges, Tragkraft*
Aufwand expenditure * *~ an Geld, Zeit, Kraft; of: an; considerable: großer* || costs * *Kosten* || expense * *an Geld; Auslagen, Kosten; invoice according to the effective expenses: nach ~ verrechn.* || effort * *Mühe, Anstrengung, Kraftanspannung* || extent of work * *Arbeits~* || complexity * *Kompliziertheit*
aufwändig costly * *kostspielig, teuer* || expensive * *teuer* || largescale * *ausgedehnt, in großem Maßstab* || troublesome * *mühsam* || difficult * *schwierig* || hard * *schwierig* || tough * *schwierig, unangenehm, ein saures Stück Arbeit darstellend* || complicated * *kompliziert* || elaborate * *ausführlich, kompliziert, sorgfältig ausgearbeitet/-geführt; (sorgfältig) ausarbeiten*
aufwandsarm labor-saving * *Arbeit einsparend* || time-saving * *zeitsparend* || easy to handle * *leicht hantierbar* || at little expense * *(Adv.) mit geringen Kosten verbunden* || economical * *→kostengünstig*
aufwärts upward
aufwärtskompatibel upwards compatible
Aufwärtsregler step-up regulator * *(Schalt-) Regler (z.B. für Schaltnetzteil) mit Ausgangsspanng. > Eingangsspanng.*
Aufwärtssteuerung upward control * *beim Aufzug*
aufwärtszählen count up
aufweisen feature * *als Besonderheit/Merkmal haben, sich auszeichnen durch* || present || show || have || boast * *etwas aufzuweisen haben* || exhibit * *aufweisen, zeigen, an den Tag legen, entfalten, ausstellen (Waren usw.)*
aufwenden expend || spend || use * *anwenden* || take * *pains: Mühe*
Aufwickelhaspel coiler
aufwickeln wind up || coil up * *Blech, Draht, Kabel* || wind || coil
Aufwickler winder || rewinder * *Wiederaufwickler, auch Umroller* || coiler * *Aufhaspel für Blech, Draht, Kabel* || recoiler * *Wiederaufhaspel für Blech, Draht, Kabel* || take-up * *z.B. für Kabel* || take-up station * *Aufwickelstation, z.B. für Kabel* || take-up stand * *Aufwickler, z.B. für Kabel* || take-up winder
Aufwicklung winding up * *das Aufwickeln* || winder * *Aufwickler* || coiler * *Aufhaspel* || wind * *Aufwickelstelle* || take-up * *z.B. für Kabel* || take-up station * *Aufwickelstation, z.B. für Kabel* || take-up stand * *Aufwickelmaschine, z.B. für Kabel*
Aufwind up-wind || tail wind * *(auch figürl.) with tail wind: im ~ (z.B. Geschäfte)*
aufwirbeln stir up
aufzeichnen record * *Messwerte ~, Mitschrift/Niederschrift erstellen* || log || note * *notieren* || plot * *Messkurve, Messwerte usw. ~ mit Mess-/Registriereinrichtung*
Aufzeichnung recording || plot * *Messkurve* || tracing * *in einem Messwert/Diagnosespeicher* || record || entry * *Eintrag* || record * *das Aufgezeichnete* || logging * *von Betriebsdaten* || log * *aufgezeichnete Betriebsdaten usw.*
aufziehen fit * *z.B. Lager, Kupplung, Zahnrad, Riemenscheibe ~* || shrink * *mit Hitze ~, aufschrumpfen* || shrink on * *mit Hitze ~, aufschrumpfen* || mount || wind * *Feder ~* || charge * *Feder ~*
Aufziehvorrichtung fitting tool * *z.B. für Lager* || pusher tool * *Aufdrückvorrichtung, z.B. für Lager*
Aufzug lift * *(siehe EN 81)* || elevator * *(amerikan.; siehe EN 81)* || hoist * *Lastenaufzug*
Aufzuggruppe elevator bank
Aufzugmotor elevator motor || lift machine
Aufzugsantrieb lift drive || elevator drive * *(amerikan.)*
Aufzugschacht elevator shaft || lift shaft || hoistway
Aufzugseil elevator hoisting cable || hoisting rope || lift hoisting rope
Aufzugskabine lift car || elevator cage
Aufzugsmotor lift motor || elevator motor || hoisting motor
Aufzugsregelung elevator control * *(amerikan.)* || elevator closed-loop control * *(amerikan.)* || lift control || elevator/lift control
Aufzugsseil lift rope || elevator hoisting cable * *(amerikan.)* || hoisting rope
Aufzugssteuerung elevator control system || elevator control || lift control
Aufzugsteuerleitung elevator control cable
Aufzugsteuerung elevator control system || elevator control || lift control
Aufzugwinde lift winch || hoisting winch
Augenblick moment * *at the moment: im ~* || instant || point of time
augenblicklich instantaneous || immediate || momentary * *vorübergehend* || present * *gegenwärtig* || at the moment * *(Adv.)* || just now * *(Adv.)*
augenblicklicher Zustand instantaneous status * *z.B. der Zustandsgrößen eines Steuer/Regelsystems i. e. Moment*

Augenblickswert instantaneous value || instantaneous power * ~ *der Leistung*
aus off || stop * *Halt-Kommando z.B. für Antrieb* || coast stop * *Kommando z.spannungsfrei machen u. Austrudeln eines Antriebs (AUS2 im →PROFIBUS-Profil)* || ramp stop * *Kommando zum Stillsetzen eines Antriebs über Hochlaufgeber (AUS1 im →PROFIBUS-Profil)* || quick stop * *Kommando z.schnellstmögl. Stillsetzen e. Antriebs a.d. Momentengrenze (AUS3 i.'PROFIBUS)* || emergency stop * *Nothalt (entspricht meist 'coast stop', siehe dort)* || E-stop * *Abk.f. 'Emergency stop': Nothalt (entspr. meist 'coast stop'; siehe dort)* || from * *von e.Ort, aus Material; z.B.: from cast iron: ~ Grauguss; from plastic material: ~ Kunststoff* || manufactured from * *hergestellt aus (... Material)* || out of || of || by || through || on || upon
aus dem Rahmen fallen be out of place || stick out || be unsuited to the occasion * *Benehmen*
aus dem Verkehr ziehen withdraw from service
aus dem Vollen from a block * *{Werkz.masch.} z.b. ~ drehen/fäsen* || with abundant resources
aus dem Vollen schöpfen draw on ample resources || have plenty
aus der Sicht from the perspective * *from the price perspective: aus der Preis-Sicht* || from the point of view
aus Gründen der for reasons of the || for reasons concerning the * *den ... betreffend*
aus Versehen erroneously * *siehe auch →versehentlich*
aus- und wiedereinschalten switch-off and on again
Aus-Zustand OFF condition
AUS1 OFF1 * *Steuerbit im →PROFIBUS-Profil; Motor über Rampe stillsetzen, bei Drehz.Null Spannungsabschaltg* || ramp stop * *Steuerbit im →PROFIBUS-Profil; Motor über Rampe stillsetzen, dann elektr. AUS*
AUS2 OFF2 * *Steuerbit im →PROFIBUS-Profil: Motor mögl.schnell spannungslos machen, Antrieb trudelt aus* || coast stop * *Steuerbit im →PROFIBUS-Profil: Motor mögl.schnell spannungslos mach., Antrieb trudelt aus* || emergency stop * *Not Aus* || E-stop * *Abk. für 'emergency stop' (Not Aus)* || electrical OFF * *'Elektrisch AUS', Steuerbit im →PROFIBUS-Profil 'Drehzahlveränderbare Antriebe'*
AUS3 OFF3 * *Steuerbit im →PROFIBUS-Profil: Motor m.max. mögl. Bremsmoment stillsetz.,dann Spanng. abschalt* || quick stop * *Schnellhalt (max. Bremsmom.), Steuerbit im →PROFIBUS-Profil 'Drehzahlveränderb. Antriebe'*
ausarbeiten elaborate * *sorgfältig ~; auch herausarbeiten (siehe auch →erarbeiten)* || prepare * *entwerfen* || draw up * *entwerfen* || work out * *auch herausarbeiten* || compose * *schriftlich* || formulate * *schriftlich* || write * *schriftlich* || finish * *vollständig ~* || perfect * *vollständig ~* || refine * *verfeinern, 'ausklügeln', aufbereiten* || work up * *erweitern (into: zu), entwickeln; (Thema usw.) ~, bearbeiten, s. einarbeiten, aufbauen*
Ausarbeitung preparation * *Abfassung (z.B. e. Dokuments), Aufbereitung, Vorbereitung, Ausfüllen (z.B. Formulare)* || working out * *Ausarbeitung, Entwicklung* || elaboration * *Ausarbeitung, genaue Darlegung, (Weiter-) Entwicklung, Vervollkommnung* || composition * *Abfassung, Entwurf, Aufsatz, (Schrift-) Werk* || compilation * *Zusammenstellung, Zusammentragung von (schriftl.) Material* || paper * *schriftliche ~*
Ausbau development * *Erschließung, Ausbau* || extension * *Erweiterung* || completion * *Fertigstellung* || dismounting * *Abmontieren, Herausbauen*

ausbaubar expendable || upgradable || upgradeable
Ausbaubarkeit upgradability * *bezüglich der Leistungsfähigkeit; Höchrüstbarkeit* || expendability * *Ausbaubarkeit* || capability of development * *Entwicklungsfähigkeit, z.b. eines Marktes* || expansion capability
ausbauen upgrade * *auf/hochrüsten* || complete * *fertigstellen* || extend * *vergrößern, erweitern* || expand * *erweitern* || enlarge * *vergrößern* || develop * *entwickeln* || enhance * *verbessern* || improve * *entwickeln, verbessern* || cultivate * *pflegen* || consolidate * *(ver)festigen* || remove * *abbauen (z.b. Maschinenteil)* || disassemble * *auseinanderbauen* || detach * *abbauen (z.B. Maschinenteil)* || intensify * *intensivieren, z.b. Geschäft(sbeziehungen)* || dismantle * *de/abmontieren, auseinandernehmen, zerlegen, niederreißen*
ausbaufähig extendable || expendable || upgradable * *hoch/aufrüstbar* || extensible
Ausbaufähigkeit expandibility || upgradability * *bezüglich der Leistungsfähigkeit*
Ausbaugrad degree of expansion || degree of extension
Ausbaustufe expansion stage || project stage || step || extension step || development step * *Entwicklungsschritt* || extension stage || stage
ausbeulen flatten
Ausbeute yield * *Ertrag, Ernte, Ausbeute, Gewinn, Ergiebigkeit, Metallgehalt (von Erz)* || output * *(Ertrag)* || profit || gain || yield rate * *Gut/Schlecht-Verhältnis, z.B. bei einem Produktionsprozess*
ausbilden train * *schulen* || educate * *Schule, Hochschule* || instruct * *belehren* || drill * *streng schulen* || form * *formen, gestalten* || develop * *entfalten, bilden, wachsen*
Ausbilder instructor * *auch Lehrer, Dozent* || trainer * *Schulungsleiter* || coach * *Trainer, einarbeitender Kollege, Einpauker* || teacher * *Lehrer*
Ausbildung education * *Bildung (Schule, Hochschule, Universität)* || training * *Schulung, innerbetriebliche/fachliche Aus-/Weiterbildung* || schooling * *Unterricht, Ausbildung* || instruction * *Belehrung, Ausbildung* || practice * *Übung* || drill * *strenge Schulung* || formation * *Formung, Gestaltung, Bildung* || development * *Entfaltung, Entstehen, Bildung, Wachstum* || indoctrination * *Unterweisung, Belehrung, Schulung*
Ausbildung am Arbeitsplatz on-the-job training
Ausbildungsaufwand training requirements * *z.B. zum Lernen der Maschinenbedienung*
Ausbildungskosten training costs * *für berufliche Aus/Weiterbildung*
Ausbildungsleiter education manager
Ausbildungsstätte training facility * *für berufliche/betriebliche Aus- und Weiterbildung*
Ausbildungssystem training system * *für berufliche Aus/Weiterbildung*
Ausbildungsweg line of training * *z.B. in beruflicher Aus- und Weiterbildung*
ausblasen blow out * *mit Druckluft usw.*
Ausblendband frequency inhibit band * *Ausblendung von Frequenzsollwerten zur Vermeidung mechan. Resonanzen* || skip frequency zone * *zur Ausblendung von (mechan.) Resonanzfrequenzen im Sollwert* || frequency avoidance zone * *zur Ausblendung von (mechan.) Resonanzfrequenzen im Sollwert* || lock-out band || critical speed rejection * *Ausblendung v. Drehzahlsollwertbereichen zur Vermeidung mechan. Resonanzen* || frequency jumping * *Ausblendung bestimmter Frequenzsollwerte zur Vermeidung mechan. Resonanzen* || suppression band || frequency suppression band
ausblenden mask * *maskieren, z.B. Impulse, Daten* || mask out * *ausmaskieren; z.B. Impulse, Daten* ||

Ausblendfrequenz

suppress * *unterdrücken* || extract * *herausholen, (her)ausziehen (from: aus/von)*
Ausblendfrequenz jump frequency * *ausgeblendeter Frequ.sollwert f.Frequenzumrichter zur Vermeidg. mechan. Resonanzen* || frequency jumping * *ausgeblendeter Frequ.sollwert f. Frequenzumrichter z. Vermeidg. mechan. Resonanzen* || skip frequency * *ausgeblendeter Frequ. sollwert f. Frequenzumrichter zur Vermeidg.mechan. Resonanz* || critical frequency rejection * *ausgeblendet. Frequ. sollw. f. Frequ. umrichter z. Vermeidg. mechan. Resonanz*
ausbohren bore || bore out || drill out
Ausbohrmaschine boring machine
ausbreiten spread || spread out || extend || expand || circulate || propagate * *sich ~, z.B. Ladungsträger i.Halbleiter, Signale, Verlustwärme usw.*
Ausbreitung propagation * *das Sich-Ausbreiten/Fortpflanzen, z.B. von Ladungsträgern, Signalen, Verlustwärme* || spreading || circulation
Ausbruch cutout * *→Ausschnitt, z.B. in Blechkonstruktion* || aperture * *→Öffnung*
ausdehnbar extensible
ausdehnen expand * *erweitern, verlängern* || extend || stretch || elongate || enlarge
Ausdehnung elongation * *Streckung* || expansion || extent * *Ausmaß* || stretching * *Strecken* || volume expansion * *des Volumens z.B. bei Erwärmung*
ausdenken think out || think up * *(amerikan.)* || invent * *contrive* || decise || devise || cook up * *(salopp) auskochen* || imagine * *vorstellen* || conceive * *erdenken, ersinnen, planen, in Worten ausdrücken*
ausdrehen bore * *z.B. mit Drehbank, Ausbohrmaschine*
Ausdruck printout * *von Drucker* || listing * *Programm* || term * *Fachausdruck,Teil einer Formel* || expression || word || print-out * *durch einen Drucker* || hardcopy * *durch einen Drucker (z.B. Bildschirminhalt, Parametersatz)* || engineering term * *technischer (Fach)Ausdruck*
ausdrucken print out || print-out
ausdrücklich explicit || express || strict * *Befehl usw.* || expressly * *(Adv.)*
auseinanderbauen disassemble
auseinandernehmen disassemble || strip || dismount
außen out || outside * *außerhalb; from the outside: von ~; outside diameter: ~durchmesser; auch "nach (dr)außen"* || on the outside * *(Adv.) außerhalb* || without * *außerhalb; from without: von außen* || out of doors * *im Freien* || outdoors * *im Freien* || outwards * *nach außen hin* || externally * *außerhalb, nach außen hin, extern*
Außen- external || outer
Außenabmessungen outline dimensions
Außenansicht external view || external appearance * *äußeres Erscheinungsbild*
Außenbüro branch office
Außendienst field service * *z.B. von Servicepersonal* || field sales service * *von Vertriebspersonal*
Außendurchmesser outside diameter || outer diameter * *z.B. eines Kabels*
Außenfläche outer surface
Außenfühler sensor in the ambient * *Außen-Temperaturfühler für Heizungszwecke*
Außengewinde external thread
Außenkante outside edge
Außenkreis outer circuit * *z.B. Kühlkreislauf*
Außenläufer external rotor
Außenläufer-Motor external rotor motor
Außenläufermotor external-rotor motor || friction-drum motor * *für Förderband usw.*
Außenleiter outer conductor || phase conductor || phase

Außenleiterspannung phase-to-phase voltage * *verkettete Spannung*
Außenleiterstrom phase current
Außenluft surrounding air * *z.B. bei Kühlsystem* || ambient air * *Umgebungsluft*
Außenlüfter external fan
Außenmantel outer sheath * *eines Kabels*
Außenraumaufstellung outdoor installation * *Aufstellung/Installation im Freien*
Außenring outer ring
Außenseite outside
Außenstände accounts receivable
Außenverzahnung external gearing || external teeth
Außenwand outside wall || end wall * *Stirnwand*
außer except * *ausgenommen; except that: außer dass* || but * *außer, als, ohne dass* || other than * *anders als, verschieden von* || with the exception of * *mit Ausnahme von* || beyond * *außer, abgesehen von, über ... hinaus, jenseits* || apart from * *abgesehen von* || not counting * *nicht mitgezählt* || aside from * *(amerikan.) abgesehen von* || out of * *außer (Betrieb, Dienst usw.)*
außer Acht lassen disregard || pay no heed to || leave out of account || omit * *aus/unter/weglassen (from: aus/von), versäumen (doing/to do: etw. zu tun)*
außer Betrieb out of operation || out of service || inoperative || out of function * *defekt* || not operational
außer Betrieb nehmen put out of operation || shut down
außer Betrieb setzen put out of operation || shut down
außer Schritt fallen fall out of step || fall out of synchronism
außer Tritt fallen fall out of step || fall out of synchronism * *Sychronität verlieren* || pull out of step * *Kippen (z.B. Synchronmotor)*
außerdem besides * *außerdem, ferner, überdies* || moreover * *außerdem, überdies, ferner* || what is more || furthermore * *ferner, überdies, außerdem, weiterhin* || further * *ferner, überdies, außerdem, weiter, entfernter* || amongst others * *→unter anderem*
äußere Abmessungen outer dimensions
äußere Einflüsse external influences
äußere Oberfläche outer surface
äußere Umstände external conditions || external forces
äußerer external || exterior || outer || outward || outside * *außen; outside diameter: Außendurchmesser*
äußerer Regelkreis outer control loop
äußeres external * *außen befindlich* || maximum * *maximales* || outer * *am äußeren Rand befindlich*
äußeres Erscheinungsbild external appearance
außergewöhnlich extraordinary * *→außerordentlich* || unusual * *→ungewöhnlich, außergewöhnlich, ungewohnt, selten, äußerst* || unusually * *(Adv.)*
außerhalb outside * *the value lies outside: der Wert liegt außerhalb; siehe auch →außen* || off- * *z.B. off-chip: außerhalb des integrierten Schaltkreises* || external * *äußer(lich), Außen-, external to: ~ von* || externally * *(Adv.)*
außerhalb der external to
außerhalb der erlaubten Grenzen outside the permitted limits
außerhalb der Toleranz out of tolerance
außerhalb des external to || outside the
außerhalb des erlaubten Bereichs outside the permitted range

außerhalb des zulässigen Bereiches outside the permissible range
außermittig off-center * *(amerikan.)* || off-centre * *(brit.)*
außerordentlich extraordinary * →*ungewöhnlich*, →*außergewöhnlich* || astonishing * *erstaunlich* || outstanding * →*hervorragend* || enormous * *enorm, ungeheuer* || immense * *immens, ungeheuer* || extreme * →*extrem* || uncommon * *außergewöhnlich* || unusual * *ungewöhnlich, außergewöhnlich, ungewohnt, selten, äußerst* || exceptional * *außergewöhnlich, Ausnahme-, Sonder-, ungewöhnlich (gut)* || singular * *einzigartig, außergewöhnlich, einmalig* || remarkable * *bemerkenswert* || extraordinarily * *(Adv.)* || exceptionally * *(Adv.)*
außerplanmäßig non-scheduled
äußerster Termin final date || deadline || time-limit * *Frist*
Ausfall loss * *Wegbleiben e. Signals, Spannung oder Strom* || failure * *Ausbleiben, Fehlen, Versagen, Zusammenbruch, Fehler*,
Ausfalldauer down-time
ausfallen fail * *auch Spannung, Gerät, Maschine, Netz* || break down || become inoperative || get out of commission || get out of order || turn out * *Ergebnis usw. (well: gut; badly: schlecht)* || prove * *Ergebnis usw.*
Ausfallerscheinungen failure phenomena
Ausfallrate failure rate
ausfallsicher fail-safe
Ausfallzeit downtime * *z.B. einer Maschine, Anlage* || outage
Ausflug excursion || trip || outing
ausfügen delete * *auf PG-Tastatur für SPS-Programmierung*
ausführbar executable * *z.B. Softwareprogramm*
Ausführbarkeit practicability || feasibility
ausführen carry out * *z.B. a work: eine Arbeit, a repair: eine Reparatur* || perform * *durchführen, z.B. eine Aufgabe* || execute * *Auftrag, Aufgabe, Computerbefehl/Programm/Prozedur* || realize * *Plan/Konzept ~/verwirklichen* || export * *exportieren* || implement * *durchführen, vollenden, (Vertrag) erfüllen* || design * *konstruktiv ~/ auslegen*
Ausfuhrgenehmigung export license * *(amerikan.)* || license to export || export authorization || export licence * *(brit.)*
ausführlich in detail || detailed || comprehensive * *umfassend* || extensive || at length || verbose * *ausführlich, vollständig, wortreich, geschwätzig, langatmig* || verbosely * *(Adv.) in voller Ausführlichkeit*
Ausführung execution * *z.B. Programm, Befehl* || performance * *einer Verrichtung* || realization * *Verwirklichung eines Plans, Konzepts* || implementation * *~ eines Gesetzes/Befehls, Softwarerealisierung* || design * *Konstruktion, Auslegung, Bauart* || type * *Typ* || model * *Modell* || version * *Version* || workmanship * *Verarbeitungsgüte, (handwerkliche) Qualität* || quality * *Qualität* || explanation * *Darlegung, Erklärung* || method * *Methode, Verfahren* || style * *Art, Typ, Art und Weise, Manier, (Bau-) Stil* || construction * *Aufbau, Konstruktion* || type of construction * *Bauart*
Ausführungszeit execution time * *für Rechnerbefehl/-programm*
ausfüllen fill out * *auch Formular*
ausfüttern line * *(aus) füttern, auskleiden, ausschlagen, (auf der Innenseite) überziehen* || pad * *polstern* || upholster * *polstern*
Ausgabe edition * *Auflage einer Druckschrift* || issue * *Herausgabe/Folge/Nummer, z.B. einer Zeitschrift* || copy * *Exemplar, z.B. eines Buches* ||
expense * *Geldausgabe* || expenditure * *Geldausgabe* || cost * *Unkosten* || handing out * *Aushändigung, z.B. von Besprechungsunterlagen*
Ausgabebaugruppe output module
Ausgabebefehl output command
Ausgabedatum revision date * *von Softwareprogrammen, Druckerzeugnissen usw.*
Ausgabegerät output device
Ausgaben outgoings * *Geldausgaben* || expenses * *von Geld* || expenditures * *von Geld*
Ausgabesperre output inhibit * *z.B. für Digital-/ Analogausgaben*
Ausgabestand version || release
Ausgang output * *z.B. Digital-,Analog-, eines Stromrichters/Motors/Getriebes* || exit * *eines Raums/Gebäudes/Programms* || outlet * *Auslass* || ending * *Ende*
result * *Ergebnis* upshot * *Ergebnis* || coil * *in Kontaktplandarstellung eines SPS-Programms*
Ausgang rücksetzen reset output || unlatch output * *in Kontaktplandarstellung eines SPS-Programms*
Ausgang setzen set output || latch output * *in Kontaktplandarstellung einer SPS*
Ausgangsbaugruppe output module || output board * *Leiterplatte*
Ausgangsdrossel output reactor * *z.B. für Pulsumrichter* || output choke
Ausgangsfilter output filter * *z.B. für Stromrichter (siehe auch 'Sinusfilter' u. 'dV-dt-Filter')*
Ausgangsfrequenz output frequency
Ausgangsgröße output signal * *Ausgangssignal* || output quantity * *Betrag/Höhe des Ausgangssignals* || output variable || output value * *Ausgangswert*
Ausgangsimpedanz output impedance || internal impedance * *Innenwiderstand z.B. einer Spannungs- oder Signalquelle*
Ausgangsklemme output terminal
Ausgangskondensator output capacitor
Ausgangsleistung output power || output capacity || horse power output * *in Pferdestärken*
Ausgangsmaterial basis material
Ausgangsmerker output flag
Ausgangsnennleistung rated output capacity * *auch eines Stromrichters* || output rating
Ausgangspegel output level
Ausgangspunkt starting point || starting position
Ausgangsschaltung output circuit
Ausgangssiebdrossel output filter reactor * *z.B. für Pulsumrichter* || output smoothing reactor
Ausgangssignal output signal
Ausgangsspannung output voltage
Ausgangsspannungsbereich output voltage range
Ausgangsstellung starting position || initial position || initial condition || normal position * →*Ruhestellung*
Ausgangsstrom output current
Ausgangsstufe output stage * *z.B. eines Leistungsverstärkers*
Ausgangswelle outgoing shaft * *eines Getriebes*
Ausgangswiderstand output resistance || internal resistance * *z.B. einer Spannungs- oder Signalquelle* || internal source resistance * *Quellwiderstand e. Spannungs-/Signalquelle*
Ausgangswort output word * *bei SPS*
ausgeben display * *anzeigen* || direct * *leiten, führen; z.B.auf eine Analogausgabe* || output * *z.B. ein Signal auf eine Digital/Analogausgabe* || issue * *ein (Warn/Zustands) Signal*
ausgefallen faulty * *auf Grund eines Fehlers/einer Störung* || out of order * *defekt* || broken down * *auf Grund eines Fehlers/einer Störung* || unusual * *unüblich* || eccentric * *exzentrisch* || odd * *sonderbar, seltsam, merkwürdig, kurios*

ausgegeben output * *z.B. Signal,Leistung* || displayed * *angezeigt* || read out * *ausgelesen*
ausgeglichen balanced
ausgehen go out * *z.B. Licht, Lampe, LED, Thyristor* || be switched off * *ausgeschaltet werden, z.B. Lampe*
ausgehend outgoing * *zu löschendes Ventil i.Stromrichterschaltung*
ausgeklügelt sophisticated || clever || ingenious || well-contrived
ausgelegt laid out * *z.B. konstruktiv ausgelegt, dimensioniert* || dimensioned * *dimensioniert* || sized * *(in Wortzusammensetzg.) von...Größe, -dimensioniert, z.B.full/over/under/small-sized*
ausgelöst triggered || fired * *z.b. Messer bei fliegendem Rollenwechsel* || released
ausgenommen with the exception of
ausgenutzt utilized
ausgeprägt distinct * *entschieden, klar, deutlich, merklich, unverkennbar, eigen, selbstständig* || marked * *markant, deutlich, merklich,* || pronounced * *deutlich, stark ~, sichtlich, ausgesprochen, bestimmt, entschieden* || salient * *herausragend, (her)vorspringend, hervorstechend, ins Auge springend*
ausgerastet disengaged
ausgeregelter Zustand steady state
ausgereift perfected * *technische Lösung, Konstruktion usw.* || fully developed * *technische Lösung, Konstruktion usw.* || mature * *reif, voll entwickelt, reichlich erwogen; fällig, zahlbar, zur Reife bringen* || maturing || matured * *(aus)gereift; abgelagert* || service-proven * *betriebs-/praxiserprobt*
Ausgereiftheit degree of perfection
ausgereizt developed as far as it can go * *eine techn. Konstruktion* || largely perfected
ausgerichtet aligned * *zum Fluchten gebracht* || targeted * *auf ein Ziel ~*
ausgerüstet furnished * *auch technisches Gerät* || fitted * *mit Zubehör, Lager, Beschlag usw.* || equipped * *with: ausgestattet mit*
ausgeschaltet disconnected from the supply * *vom Netz getrennt; when disconnected from the supply: im ~en Zustand* || switched off * *siehe →ausschalten*
ausgeschlossen excluded
ausgesetzt sein be subject of * *z.B. einer Belastung/Gefahr*
ausgestattet furnished * *auch technisches Gerät* || equipped * *with: mit* || provided || supplied || staffed * *mit Personal* || fitted out
ausgestellt sein be on show * *auf Messe, Ausstellung usw.*
ausgestellte Produkte exhibits * *auf Messe*
ausgesucht selected
ausgewogen well-balanced * *gut/wohl ~*
ausgezeichnet excellent * *hervorragend, vorzüglich* || outstanding || first-class || splendid * *großartig* || fine * *perfect* || award-winning * *mit einem Preis/einer Auszeichnung ~* || awarded * *mit einer Auszeichnung/einem Preis ~*
Ausgießer nozzle * *Schnauze, Tülle, Rüssel, Mundstück (an Gefäßen usw.)*
Ausgleich compensation || balancing || equalization * *z.B. energy equalization: Energieausgleich* || self-regulation * *selbsttätig eintretender ~*
ausgleichen compensate * *kompensieren* || get across * *vermitteln* || balance * *ins Gleichgewicht bringen, ausbalancieren, auswuchten, Konto ausgleichen* || equalize * *ausgleichen, kompensieren, stabilisieren, egalisieren* || compensate for * *etwas ~* || level up * *aufs gleiche Niveau heben*
ausgleichend compensative || compensatory

Ausgleichs- compensating || equalizing || balancing || counter- || balance
Ausgleichscheibe end-float washer * *zur Einstellung des axialen Lagerspiels* || spacer washer * *Abstandsscheibe* || washer * *Unterlegscheibe* || shim * *z. Reduzierg./Einstellg. d. Axialspiels, Unterlegscheibe z. Motorausrichtg. usw.* || spacer ring * *Distanzring, Abstandsscheibe* || float limiting shim * *zur Reduzierung des axialen Lagerspiels*
Ausgleichsgewicht counterbalance weight || damper weight * *"Wuchtgewicht" zur Vermeidung unrunden Laufs* || counter-weight || comensating weight
Ausgleichskupplung alignment coupling
Ausgleichsofen soaking furnace * *Walzwerk*
Ausgleichsscheibe end-float washer * *zur Einstellung des axialen Lagerspiels* || spacer washer * *Abstandsscheibe* || washer * *Unterlegscheibe* || shim * *z. Reduzierg. d. Axialspiels, Unterlegscheibe z. Ausrichtg. e. Motors usw.* || spacer ring * *Distanzring, Abstandsscheibe* || equalizing ring * *z.B. zum Ausgleichen des axialen Lagespiels*
Ausgleichsstrom equalizing current || compensating current
Ausgleichstrom equalizing current
Ausgleichsvorgang equalization process || equalization effect || equalization operation || transient phenomenon || transient effect || transient reaction || transient * *make/break transient: ~ bei Stromkreis-Schließung/-Unterbrechg.* || transient recovery || transient
Ausgleichswelle differential selsyn * *Ausführungsform d. elektr.Welle mit Asynchron-Schleifringläufermotoren*
ausgliedern separate out || eliminate
aushändigen hand over * *to a person: jemandem* || deliver up || surrender
aushärten harden * *non-hardening: nicht ~d* || bond * *abbinden* || cure
Aushärtung hardening * *z.B. von Isolier-Tränkharz, Kunstharz* || curing || bonding * *(chem.) von Klebstoff usw.*
Aushaumaschine nibbling machine * *zur Blechbearbeitung*
auskennen be familiar * *with: sich auskennen mit*
Ausklammerfrequenz suppression frequency * *unterdrückter Frequenzsollwert e. Umrichters zur Vermeidung mechan.Resonanzen* || skip frequency * *siehe auch →Ausblendfrequenz, →Ausblendband*
ausklappbar able to be swung-out || swung-out type || hinged * *angelenkt, mit Scharnier versehen*
Auskleidung lining || lining material * *Auskleidungsmaterial*
auskoppeln extract * *z.B. Signal* || tap * *z.B. Energie* || uncouple * *auskuppeln, entkoppeln* || retrieve * *(wieder) herausholen, →herausfischen"; (from: aus)*
Auskunft information
Auskunft erteilt for further information
auskuppeln declutch * *eine Kupplung* || release the clutch * *eine Kupplung* || disengage the clutch * *eine Kupplung* || uncouple * *entkuppeln/koppeln* || disconnect * *trennen, lösen, auskuppeln/-rücken, ab/ausschalten* || de-clutch * *eine Kupplung* || disengage || throw out * *auskuppeln, ausrücken* || decouple
Ausladung offload * *Entladung von Frachtgütern* || unloading * *das Entladen* || projection * *das Nach-Außen-Ragen eines Teils, →Vorsprung* || overhang * *z.B. einer Werkzeug-maschine* || length of jib * *eines Drehkrans* || swing * *eines Schwenkkrans* || depth of throat * *~ z.B. einer Blechschere, Überhang* || protrusion * *Hervorstehen/-treten/-*

springen (Maschinenteil), vorstehender Teil, Vorwölbung/-sprung
Auslagerungsdatei swapfile * *zur vorübergehenden Auslagerung von RAM-Daten auf d. Festplatte*
Ausland foreign country || abroad * *ins Ausland, im Ausland; from abroad: aus dem Ausland; trip abroad: Auslandsreise* || outside Germany * *außerhalb Deutschlands* || other countries
ausländisch foreign
Auslandsniederlassung foreign subsidiary || foreign branch || overseas branch
Auslandsvertretung representation abroad
Auslass outlet * *outlet valve: Auslassventil*
auslassen omit * *weglassen* || skip * *überspringen, auslassen, übergehen*
Auslasskanal outlet duct * *z.B. für Kühlluft*
Auslastung work load || utilization * *Ausnutzung* || load * *Belastung* || loading
Auslauf exit section * *Auslaufteil z.B. bei Blechstraße* || exit || outflow * *von Flüssigkeit* || discharge * *von/für Flüssigkeit* || outlet * *Auslass* || drain * *Ableitung, Abfluss, Entwässerung* || sailing * *{Schifffahrt} das ~en eines Schiffes* || run down * *{el.Masch.}*
auslaufen coast down * *austrudeln (Maschine, Motor)* || coast to standstill * *bis zum Stillstand austrudeln (Maschine, Motor)* || run down * *z.B. Motor* || coast * *→trudeln* || running down * *(Substantiv)* || sail * *{Schifffahrt}* || clear the port * *{Schifffahrt} aus dem Hafen ~* || run out * *Produktion, Motor usw.* || wear out * *{Masch.bau} Lager usw.* || be discontinued * *Produktion eines Modells/einer Type; discontinue: ~ lassen* || be phased out * *allmählich ~; phase out: ~ lassen* || taper off * *allmählich ~ lassen*
auslaufendes Produkt discontinued product
Auslaufrollgang exit roller table * *Walzwerk*
auslaufsicher leak-proof
Auslaufteil exit section * *z.B. einer Walzstraße*
Auslaufversuch run-down test * *{el.Masch.} Aufnahme des zeitl. Drehzahlverlaufs nach Abschalten*
Auslaufwalze take-off roll * *z.B. bei Textilmaschine*
Auslaufzeit stopping time * *z. Stillsetzen e. Antriebs*
ausleeren empty || clear || drain * *Flüssigkeit leeren/ableiten/abführen* || deplete * *(auch figürlich) entleeren, entblößen, erschöpfen*
auslegen design * *entwerfen, auch bemessen* || plan * *planen, projektieren, entwerfen* || lay * *Kabel* || cover * *auskleiden* || size * *bemessen, for: für* || configure * *anordnen, konfigurieren, gestalten, strukturieren, bauen* || lay out * *konstruktiv ~, dimensionieren* || dimension * *dimensionieren* || rate * *z.B. Stromrichter/Motor ~* || interprete * *interpretieren, z.B. Norm, Vorschrift*
Auslegung design || layout || interpretation * *Interpretation* || dimensioning * *Dimensionierung* || evaluation * *Abschätzung* || rating
Auslegungskriterien design criteria
Auslegungskriterium dimensioning criterium || rating criterium * *für die Leistung (z.B. des Motors/Stromrichters; Plural: criteria)*
Ausleitsystem rejection system * *zum Aussondern fehlerhafter Produkte aus einer Transportlinie*
auslenken deflect
Auslenkung deflection || displacement
auslesen read out * *z.B. Daten ~* || read-out || upload * *herauf laden von Daten/Parametern aus einem Gerät heraus* || upread * *herauflesen von Daten/Parametern aus einem Gerät heraus*
Auslieferdatum shipping date
ausliefern ship || supply || dispatch * *(schnell) ver/absenden*

Auslieferungtermin delivery date || shipping date
Auslieferung shipment || shipping
Auslieferungsdatum shipping date || delivery date
Auslieferungslager distributing warehouse || picking warehouse * *Kommissionierlager*
Auslieferungstermin delivery date || shipping date
Auslieferungszustand delivered condition || as delivered * *im Auslieferungszustand* || as-delivered condition || condition when supplied || 'as supplied' condition || status as supplied || as shipped * *im Auslieferungszustand*
ausloggen log-off * *Arbeitssitzung in einem Computer-Netzwerk beenden*
auslöschen extinguish
Auslöse- trip * *Ansprechwert bei Überwachungseinrichtung* || trigger
Auslösegerät tripping unit * *Schutz/Überwachungseinrichtung/-Relais z.B. für Thermofühler im Motor* || releasing device || release device || release * *Auslöser, Auslösevorrichtung* || tripping device || trip unit || tripping control
Auslösekennlinie tripping characteristic * *z.B. eines Motorschutzschalters*
Auslösekraft release force
auslösen trigger * *(auch figürl.) ein Ereignis* || trip * *z.B. Überwachungseinrichtung, Störabschaltung* || initiate * *veranlassen, bewirken, initiieren* || fire * *z.B. knife: Messer beim fliegenden Rollenwechsel* || cause * *verursachen* || release * *freigeben, freiklinken, z.B. Überwachungseinrichtung/Schnappverschluss, Kameraverschluss* || operate || activate * *aktivieren (auch Stör/Warnmeldung)*
Auslösepunkt trip point * *z.B. zum Ansprechen einer Überwachung, eines Grenzwertmelders usw.*
Auslöser release * *Auslöse-/Schutzeinrichtung* || trigger || trip element * *Schutzeinrichtung* || tripping device * *Schutzeinrichtung*
Auslösestrom trip current * *einer Überwachungs-/Schutzeinrichtung* || tripping curren * *einer Überwachungs-/Schutzeinrichtung*
Auslösetemperatur tripping temperature * *einer Temperaturüberwachung* || releasing temperature * *eines thermischen Auslösegerätes*
Auslösezeit tripping time * *eines Auslösegeräts*
Auslösung trip * *einer Überwachungs/Schutzeinrichtung* || tripping * *einer Überwachungseinrichtung* || severance pay * *Trennungsgeld* || redemption * *Abzahlung/-lösung, Tilgung, Freikauf, Wiedergutmachung, Ausgleich, Wiederherstellung*
auslöten unsolder
ausmachen put out * *Feuer/Licht usw. ~* || switch out * *Licht usw. ~* || turn out * *Licht usw. ~* || amount to * *(zahlen-/wertmäßig) betragen* || total to * *in der Summe ~, sich belaufen auf* || locate * *orten* || agree * *vereinbaren* || arrange * *vereinbaren* || make up * *einen Teil bilden* || form * *einen Teil bilden/formen; ausbilden* || matter * *that does not matter: das macht nichts aus; it matters a great deal: es macht viel aus* || be of consequence * *it is of no consequence: das macht nichts aus* || mind * *never mind: das macht nichts; would you mind if: macht es Ihnen etwas aus, wenn?* || make out * *sichten, feststellen* || sight * *beobachten, sichten, zu Gesicht bekommen, anvisieren* || stipulate * *ausbedingen, (vertraglich) vereinbaren, festsetzen* || agree upon * *vereinbaren, verabreden, sich einigen auf/über* || make it a condition * *ausbedingen, zur Bedingung machen*
Ausmaß extent * *to a large extent: in hohem Ausmaß* || dimensions * *Abmessungen* || amount * *Betrag, Summe, Höhe, (Geld)Wert, Inhalt, Ergebnis* || level * *(figürl.) Niveau, Stufe, Stand, Ebene*
Ausnahme exception
Ausnahmefall exceptional case

Ausnahmesituation exceptional situation * *exception handling: Behandlung von ~en {Computer}*
Ausnahmezustand exceptional condition * *exception handler: Programm(teil) zum Reagieren auf ~/Sonderfälle* || exceptional status
ausnutzen take advantage of * *vorteilhaft* || utilize * *nutzbar machen; under-utilize: schlecht ausnutzen* || econimize * *sparsam umgehen mit, (so gut wie möglich) ausnützen* || exploit * auswerten, (kommerziell) verwerten, ausbeuten (auch figürl.), ausnutzen, ausschlachten
Ausnutzung utilization
Auspackmaschine depacker || unpacking machine
Ausprägung characteristic * *Eigentümlichkeit, Kennzeichen, Eigenschaft, charakteristisches Merkmal* || characteristics || kind * *Art, Sorte, Wesen, Gattung, Natur* || form * *Form, Gestalt, Fasson, Manier, Art u.Weise, Schema, System* || shape * *(auch figürl.) Form, Gestalt* || way of realization * *Art u.Weise der Realisierung/Verwirklichung/Ausführung* || way of implementation * *Art u.Weise der Ausführung/Durchführung* || embodiment * *Verkörperung, Darstellung, Inbegriff* || incarnation * *Verkörperung, Inbegriff* || version
Auspuff exhaust * *exhaust box: Auspufftopf* || muffler * →*Schalldämpfer*, →*Auspufftopf*
Auspuffkrümmer exhaust manifold
Auspufftopf silencer * *Schalldämpfer* || exhaust box || muffler * *Schalldämpfer*
Ausputzwalze cleaning roller * *{Textil} z.B. bei* →*Kratzenraumaschine*
ausräumen remove * *z.B. Ladungsträger b. Abschalten e. Thyristors* || discharge * *entleeren* || empty || evacuate || clear || remove * *Möbel usw. (auch figürl.: Bedenken usw. ~)* || clear off * *Waren usw. ~* || broach * *{Werkz.masch.} z.B. mit Räumnadel/Räumeisen* || dispel * *(figürl.) Bedenken usw. ~* || clear up * *(figürl.) Missverständnis usw. ~* || settle * *(figürl.) Meinungsverschiedenheit usw. ~*
Ausräumung removal * *z.B. von Ladungsträgern beim Abschalten e. Thyristors*
ausrechnen figure out * *(amerikan.)* || calculate || compute || reckon out * *(auch figürl.)*
ausregeln respond * *ansprechen, reagieren auf; respond to: e. Störgröße ausregeln* || compensate * *for: etwas* || correct || adjust
Ausregelzeit settling time * *verstreicht nach.e.Sollwertsprung bis d.Regelgröße endgültig i.Toleranzbereich bleibt* || correction time
ausreichen be sufficient || be enough || be adequate * *angemessen sein* || suffice * *genügen, hinreichen, reichen, (aus)reichen, jdm. genügen* || do * *that will hardly do: das wird kaum* ~ || last
ausreichend sufficient || enough * *genug* || adequate * *passend, hinreichend, entsprechend, genügend (to: dem)* || sufficiently * *(Adv.)* || reasonable * *angemessen, zumutbar, annehmbar*
Ausreißer irregularity || outlier || maverick
ausrichten align * *ausfluchten, z.B. Motor, Objekte an einer Linie* || straighten * *gerade* ~ || true || adjust * *(auch figürl.: Benehmen usw.)* || effect * *bewirken* || deliver * *Botschaft usw. ~* || take a message * *Botschaft* ~ || orient * *(Lage, geistig) ausrichten, (sich) orientieren, sich* ~ || justify * *(Text usw.) am Rand ~; richten,justieren, richtigstellen; rechtfertigen, jdm.Recht geben*
Ausrichtung alignment * *z.B. eines Motors* || orientation * *(auch figürl.) Orientierung (towards: an), Ausrichtung, auch eines Materialgefüges* || corporate orientation * ~ *eines Unternehmens*
ausrücken disengage * *Kupplung* || throw out of gear * *außer Eingriff bringen* || unmesh * *Zahnräder usw. außer Eingriff bringen* || declutch * *auskuppeln, Kupplung ~* || march out * *Feuerwehr usw.* || disengagement * *(Substantiv) Kupplung* ||

throw out * *Kupplung* || declutch * *Kupplung* || release * *Kupplung, Bremse*
Ausrückfeder release spring * *z.B. in Kupplung/Bremse*
Ausrückkraft disengagement force * *zum Öffnen einer Kupplung*
Ausrückmoment moment of disengagement * *einer (Sicherheits-) Kupplung* || disengagement torque * *einer (Sicherheits-) Kupplung*
Ausrückweg disengagement travel * *einer Kupplung*
ausrüsten equip || furnish || provide || supply || fit out || finish * *Papier, Tuch*
Ausrüstung equipment * ~/-stattung, ~sgegenstände, techn. Material/Gerätschaft/Einrichtung, Rüstzeug* || gear * *Werkzeug, Gerät, Zubehör; besonders in zusammengesetzten Wörtern* || fitting out || finishing * *Weiterverarbeitung/Endbehandlung z.B. von Papier, Textilstoffen*
Ausrüstungsmaschine finishing machine
ausrutschen slip
Aussage statement || assertion * *Behauptung, Erklärung, Geltendmachung* || declaration * *Erklärung, Aussage* || message * *Miteilung, Bescheid, Botschaft* || deposition * *schriftliche (rechtlich verwertbare) Erklärung* || evidence * *give evidence: eine ~ machen* || information * *give information about: Aufschluss geben über* || affirmation * *Behauptung, Bestätigung, Beteuerung* || assurance * *Ver/Zusicherung, Garantie* || prediction * *Vorhersage, Vor*~
aussagekräftig informative || instructive
Ausschalt-I2t-Wert total operating I2t value * *Für Sicherung: Summe aus Schmelz- und Lösch-I2t-Wert*
ausschalten switch off || turn off || shut down || cut off || disable * *sperren, deaktivieren* || break || cut out || interrupt || disengage * *Kupplung trennen* || let go out * *Lampe, Licht, LED-Anzeige ~* || open * *(Kontakt) öffnen* || switch out || disconnect * *Stromkreis ~* || power off * *die Stromversorgung ~* || cut out * *(sich) ausschalten, aussetzen, ausschneiden* || power down * *Stromversorgung usw.* ~
Ausschaltfolge shutdown sequence
Ausschaltleistung rupturing capacity * *e. Leistungsschalters, e. Sicherung*
Ausschaltlogik switch-off logic || power-down logic
Ausschaltprüfung disconnecting test * *z.B. für Sicherungen*
Ausschaltreihenfolge shut-down sequence || topping sequence * *Motor* || power-down sequence
Ausschaltsteuerung shut-down control || stopping sequencing control * *für Antrieb* || stop control * *Antrieb* || shut-down sequencing || power-down sequence control || shutdown control
Ausschaltsteuerwerk shut-down control || stopping sequence control * *für Antrieb ~* || stop control * *für Antrieb usw.* || shut-down sequencing || power-down sequence control
Ausschaltstrom breaking current * *eines Leistungsschalters*
Ausschaltvermögen breaking capacity * *z.B. Sicherung, Leistungsschalter* || interrupting capacity || interrupting rating || clearing capacity
Ausschaltverzögerung switch off delay || OFF delay || off-delay time * ~*szeit*
Ausschaltzeit disengagement time * *einer Kupplung* || operating time * *einer Schmelzsicherung usw.* || turn-off time * *eines Transistors, Thyristors usw.* || opening time * *eines Relais usw.* || breaktime * *eines Leistungsschalters* || tripping time * *einer Überwachungseinrichtung*

ausschäumen foam-up
ausschlachten takt to pieces for reutilization ‖ cannibalize * z.B. Fahrzeug ‖ make the most of * ausnutzen
Ausschlag deflection * z.b. eines Zeigers ‖ decisive factor * ausschlaggebender Punkt ‖ reading * Anzeigewert eines Messgeräts
ausschlaggebend decisive * entscheidend, maßgebend, endgültig; be decisive in: entscheidend beitragen zu ‖ determining * determining factor: entscheidender Umstand/Faktor ‖ material * von Belang, sachlich wichtig (to: für) ‖ determinative * entscheidend ‖ final * endgültig ‖ crucial * entscheidend, kritisch; crucial point: springender Punkt
ausschlaggebend sein be decisive ‖ govern * z.B. Auslegungskriterium
ausschleusen remove ‖ feed aside
Ausschleusung discharging ‖ diversion * Umleitung, Ableitung ‖ deflection turn exit * durch Umlenken ‖ lift transfer * durch Hubbewegung ‖ cross transfer * seitliche ~ ‖ bypassing * am Hauptstrom vorbei
ausschließen eliminate * eliminieren ‖ exclude ‖ lock out ‖ regard as highly unlikely * für höchst unwahrscheinlich halten ‖ make highly unlikely * weitgehend unwahrscheinlich machen
ausschließlich exclusive * of: von ‖ exclusively * (Adv.) ‖ excluded * (nachgestellt) z.B. packing excluded: ausschließlich Verpackung
Ausschluss exclusion ‖ disclaimer * Haftungs-/Gewährleistungs~(erklärung)
ausschneiden cut out ‖ clip ‖ cut * ~ und in die Zwischenablage einfügen (z.B. unter WINDOWS)
Ausschnitt cut-out * z.B. in einem Blech ‖ cutout ‖ recess * Aussparung, Vertiefung, Einschnitt ‖ notch * Kerbe, Einschnitt, Aussparung, Nute ‖ cut ‖ aperture * Öffnung ‖ sector * Kreisausschnitt ‖ segment * Kreisausschnitt ‖ part * Teil ‖ section * Teil
Ausschreibung call for tenders * für Lieferangebote ‖ invitation to bid * für Angebote ‖ competitive procurement procedure * (meist staatliches) Ausschreibungsverfahren für Beschaffung ‖ tender invitation * für Angebotsabgaben
Ausschuss waste * Schrott, verdorbenes Material ‖ scrap * auch Schrott ‖ low-quality goods * schlechte Ware(n) ‖ rejects * nicht verwendbare/ zurückgewiesene Teile ‖ broke * z.B. bei der Papierherstellung
ausschwenken swing out ‖ swing away ‖ swing over
ausschwitzen sweat out ‖ exude ‖ exudation * (Subst.) Ausschwitzen, Auschwitzung (von Schmiermittel usw.) ‖ seep ‖ glue penetration * {Holz} (Subst.) ~ von Leim ‖ floating * (Subst.) ~ von Farbe
Aussendung emission
Aussetzbetrieb intermittent operation ‖ intermittent periodic duty type * Betriebsart S3 nach VDE0530/DIN41756 für Motor/Stromrichter ‖ intermittent duty ‖ periodic duty
Aussetzbetrieb mit Einfluss des Anlaufvorgangs intermittent periodic duty type with starting * Motor-Betriebsart S4 nach VDE0530/IEC34-1
Aussetzbetrieb mit Einfluss von Anlaufvorgang und elektrischer Bremsung intermittent periodic duty type with starting and electric braking * Motorbetriebsart S5 nach VDE0530
Aussetzbetrieb ohne Einfluss des Anlaufvorgangs intermittent periodic duty type without starting * Motorbetriebsart S3 nach VDE0530
aussetzen expose to * Hitze, Wind, Wetter, Unbill, Gefahr usw. ‖ subject * unterwerfen, unterziehen, subject to: (etw. Hitze usw.) ~ ‖ intermit * unterbrechen ‖ interrupt * unterbrechen ‖ discontinue * mit (Tätigkeit usw.) aussetzen ‖ stop * (Tätigkeit, Maschinenlauf usw.) unterbrechen, anhalten; stoppage: das Aussetzen/Anhalten ‖ fail * versagen; failure: das Versagen ‖ arrest * an/aufhalten, hemmen, hindern ‖ suspend * Zahlung usw. ~ ‖ defer * aufschieben ‖ postpone * verschieben ‖ adjourn * vertagen ‖ skip * überspringen, überschlagen ‖ find fault with * etwas auszusetzen haben ‖ object to * etwas einzuwenden haben ‖ criticize * kritisieren, etwas auszusetzen haben ‖ pause * unterbrechen, pausieren ‖ break off * unterbrechen ‖ be irregular * öfter aussetzen ‖ stall * Motor
Aussparung notch * Kerbe, Einschnitt, Aussparung, Nute ‖ cutout * Ausschnitt ‖ recess * Aussparung, Vertiefung, Einschnitt ‖ cut-out * Ausschnitt ‖ opening * →Öffnung
Ausspritzapparat rinsing machine * z.b. zur Reinigung von Flaschen
ausspülen flush out ‖ rinse out ‖ wash out
ausstanzen punch out ‖ blank out
ausstatten provide ‖ equip ‖ furnish * versehen, aufrüsten, einrichten, möblieren ‖ supply * with: mit ‖ fit out ‖ staff * mit Personal
Ausstattung equipment
Ausstattungsmaschine finishing machine * z.B. in der Verpackungstechnik
aussteigen exit * ein SPS Programm verlassen ‖ get out * auch aus einem Fahrzeug/Unternehmen/ einer Firma (of: aus) ‖ drop out ‖ give up * aufgeben
ausstellen show * auf Messe ~ ‖ expose * zur Schau stellen, z.B. auf Messe ‖ showcase * in einem Schaukasten, auf einem Messestand usw. ‖ present * präsentieren, (erstmalig) zeigen
Aussteller exhibitor * auf Ausstellung/Messe
Ausstellerverzeichnis list of exhibitors * bei einer Messe/Ausstellung
Ausstellung exhibition ‖ show ‖ fair * Messe ‖ exposition * (amerikan.) ‖ display * von Waren
Ausstellungsfläche exhibition area * auf einer Messe
Ausstellungsgelände fair grounds * Messegelände ‖ exhibition park
Ausstellungsraum exhibition room ‖ showroom
Ausstellungsstand exhibition booth ‖ exhibition stand ‖ booth ‖ exhibition space * Fläche
Aussteuerbarkeit control range * Aussteuerbereich
Aussteuerbegrenzung firing limits * Steuerwinkelbereich für netzgeführten Stromrichter
Aussteuerbereich control range
Aussteuergrad pulse control factor ‖ firing angle * Steuerwinkel besonders bei Phasenanschittsteuerung ‖ control setting ‖ control factor ‖ gating ratio
aussteuern control * z.B.Thyristor bei Phasenschnittsteuerung; to give the voltage x: auf die Spannung x
Aussteuerung gating * Ansteuerung, z.B. eines Thyristorsatzes ‖ firing angle setting * Aussteuerwinkel bei Thyristorsatz ‖ control setting * Aussteuergrad ‖ gating ratio * Aussteuergrad ‖ saturation degree * Drossel ‖ modulation * in der Mess-/Nachrichtentechnik usw. ‖ firing angle * Aussteuerwinkel ‖ output voltage ratio * Verhältnis max. zu aktueller Ausgangsspannung bei Stromrichter
Aussteuerungsgrad control factor ‖ pulse control factor
Aussteuerwinkel firing angle * für netzgeführten Stromrichter (Phasenanschnittsteuerung)
ausstoßen push out ‖ thrust out ‖ eject ‖ expel * auch (Kühlluft) 'ausblasen'
ausströmen emit ‖ stream out ‖ flow forth * hervor-

strömen || issue || exhaust * *Gas, Dampf* || escape * *Gas, Dampf* || issue * *herauskommen, herausströmen, hervorbrechen, entspringen* || discharge * *ausströmen (lassen), z.B. Wasser; ausstoßen* || drain off * *ausströmen lassen, z.b. Wasser*
Austausch replacement * *Ersetzen* || exchange * *das ~en (for: gegen)* || swap * *gegenseitiger ~; z.B. Nibbles in einem Byte* || interchange * *z.B. ~ von Signalen, Telegrammen usw.* || substitution * *Substitution, ersatzweise Verwendung, Ersatz-*
Austausch- replacement
austauschbar interchangeable * *untereinander* || exchangeable * *wechselbar* || replaceable * *easily: leicht*
Austauschbarkeit interchangeability
austauschen interchange || exchange * *vertauschen* || replace * *ersetzen* || swap * *(miteinander ver-) tauschen, z.b. Bytes in einem Datenwort, Daten in einem Speicher*
Austauschgerät replacement unit
Austrag discharge * *Herausförderung aus einem Behälter, einem Ofen usw.*
austragen discharge * *entladen; Material/Teile aus Behälter/Ofen/Fertigungslinie austragen*
Austragmaschine discharging machine * *Entlademaschine z.B. für Ofen* || exit machine * *z.B. für Ofen*
Austragspumpe discharge pump
austreten escape * *entweichen, z.B. von Kühlluft* || wash one's hand * *Bedürfnis verrichten* || emerge * *auftauchen, herauskommen, erscheinen, hervortreten, zum Vorschein kommen* || discharge * *ausströmen (lassen), z.b. Kühlluft*
austretende Kühlluft dicharged cooling air
Austrittsöffnung outlet * *z.B. für Kühlluft*
austrudeln coast down || coast to stop * *bis zum Stillstand* || rotate freely back to zero * *bis Drehzahl Null* || windmilling * *(Substantiv; salopp)* || coast to rest
ausüben produce * *erzeugen* || exert * *gebrauchen, anwenden, z.B. Druck, Einfluss* || impose * *(Kraft, Druck usw.) ~ (upon/on: auf); beanspruchen; (Pflicht, Steuer usw.) auferlegen*
Auswahl selection || sampling * *z.B. der wichtigsten Produkte in e. Übersichtskatalog* || choice * *(Aus)Wahl*
Auswahldaten selection data
auswählen select * *from: aus* || choose
Auswahlfenster select window * *am Bildschirm*
Auswahlkriterien selection features
Auswahlkriterium selection criterion * *(Plural: criteria)*
Auswahlliste selection list || selection sheets * *Auswahlblätter*
Auswahlmenü select menu
Auswahlmöglichkeiten select possibilities || selection capabilities || choices
Auswahlparameter select parameter
Auswahlschalter selection switch
Auswahltabelle selection table
auswechselbar replaceable || interchangeable * *one with another: miteinander* || exchangeable
auswechseln exchange || interchange || replace * *ersetzen* || change * *Verschleißteil (Reifen, Batterie, Öl)*
Ausweg way out * *auch 'Ausflucht'* || loophole * *Ausflucht, Ausweichmöglichkeit* || alternative * *(figürl.) Ausweichmöglichkeit, Alternative* || expedient * *Notbehelf* || shift * *Notbehelf* || last resort * *letzter Ausweg* || workaround * *Notbehelf*
ausweiten expand * *erweitern* || extend * *erweitern* || stretch * *ausweiten*
auswendig by heart * *know by heart: ~ wissen/können; learn by heart: ~ lernen* || from memory * *aus dem gedächtnis; memorize: auswendig lernen, sich einprägen* || by memory * *aus dem Gedächtnis*
auswerfen eject * *auch 'hinauswerfen, ausstoßen'*
Auswerfer ejector || throw-out
Auswertebaustein evaluation block * *Funktionsbaustein*
Auswerteelektronik evaluation electronics
Auswertegerät evaluation unit * *z.B. für (Temperatur-) Messgeber, Näherungsschalter usw.* || tripping unit * *für Überwachungszwecke (Auslösegerät), z.b. Temperaturüberwachung*
auswerten evaluate || analyze || interprete
Auswerteprogramm analysis program
Auswerter evaluator || interpreter
Auswertung evaluation * *Abschätzung, Bewertung, Auswertung, Beurteilung* || analysis * *Analyse* || interpretation
auswirken take effect * *Auswirkung haben* || effect * *bewirken* || affect * *sich auswirken auf, beeinflussen* || work out || result * *Ergebnis*
Auswirkung effect * *Wirkung* || result * *Ergebnis* || consequence * *Rückwirkung, Folge* || impact * *Beeinflussung* || implication * *Konsequenz, Begleit-/Folgeerscheinung; by implication: als (natürliche) Folge*
auswuchten balance || counterbalance || balance out
Auswuchtmaschine balancing machine
Auswuchtring balancing ring
Auswuchtung balancing
Auswurf ejecting || ejection
Auswurfhebel eject lever * *z.B. für Flachbandkabel-Steckverbinder* || eject * *Auswerfer*
ausziehbar sliding || telescopic || extensible * *verlängerbar* || removable * *herausnehmbar*
Ausziehvorrichtung withdrawal device
Auszubildender apprentice * *Lehrling*
Auszug extract || excerpt * *aus einem Dokument*
autark stand-alone * *allein arbeitend* || autonomous * *für sich allein lebens-/betriebsfähig* || self-contained * *in s.geschlossen, selbstständig (arbeitend), unabhängig* || independent * *unabhängig* || self-supplying * *selbstversorgend*
autogen autogenous * *z.B. Schweißverfahren*
Autogenschweißen autogenous welding || acetylene welding || gas welding
Autoindustrie automotive industry
Autokran mobile crane
Automat automatic machine || automat * *automatic lathe: Drehautomat; siehe auch →Sicherungsautomat* || robot * *Handhabungsautomat, Roboter*
Automatenstahl machining steel
Automatik automatic control || automatism || automatic operation || automatic transmission * *automatisches Getriebe*
Automatik- automatic || auto-
Automatikbetrieb automatic mode * *~sart* || automatic operation
Automatiksollwert automation setpoint
automatisch automatic * *(Adj.)* || automatically * *(Adv.)* || auto- * *(als Vorsilbe)* || automated * *automatisiert* || computerized * *rechnergesteuert* || self-acting * *selbsttätig* || unattended * *unbeaufsichtigt, ohne Beteiligung von Bedienpersonal*
automatische Anpassung an die Netzspannung automatic supply voltage matching || automatic source matching
automatische Belichtung computer-to-plate technology * *{Druckerei} bei Offsetdruck* || computer-to-cylinder technology * *{Druckerei} bei Tiefdruck*
automatische Entflechtung auto-routing * *einer Leiterkarte*
automatische Nachführung auto tracking

automatische Netzfrequenzanpassung automatic adaption to the system frequency * z.B. bei Steuersatz e. netzgeführten Stromrichters || automatic line frequency tracking * aut. Netzfrequ.nachführg. b. Steuersatz e.netzgef. Stromrichters
automatischer Wiederanlauf auto restart * z.B. nach einem Netzspannungseinbruch || automatic warm restart
automatisieren automate
automatisiert automated
Automatisierung automation || automation system * Automatisierungssystem
Automatisierungsaufgabe automation task || control problem
Automatisierungsebene level of automation || automation level * lower/medium/higher: untere/mittlere/obere
Automatisierungsgerät automation unit || programmable controller * AG, SPS, speicherprogrammierbare Steuerung || PLC * Abk. f. 'Programmable Logic Controller': speicherprogrammierbare Steuerung
Automatisierungsgrad degree of automation || level of automation
Automatisierungskonzept automation concept
Automatisierungslandschaft automation environment || automation scenario
Automatisierungslösung automation solution
Automatisierungsschnittstelle automation interface
Automatisierungssollwert automation setpoint
Automatisierungssystem automation system || automation control system
Automatisierungstechnik automation || automation engineering * als Fachrichtung || automation technology || automatization
Automobilbau car manufacturing || automotive industry
Automobilelektronik automotive electronics
Automobilindustrie automotive industry
autorisiert authorized * siehe auch 'befugt' || designated * ernannt, bestimmt || licensed || empowered * bevollmächtigt, ermächtigt
Autoverleih car rental
Avivagerolle preparation roll * zum Befeuchten der Fasern bei Kunstfaser-Herstellung
AWL statement list * Anweisungsliste als Programmiersprache für SPS || STL * Abkürzung für 'STatement List' Anweisungsliste als Programmiersprache für SPS
axial axial || axially * (Adv.)
axial bedrahtet axial-lead
Axialbelastung axial load
Axialbewegung axial movement
Axialdichtring axial sealing ring * z.B. Simmerring
Axialdruck axial pressure
Axialfluss axial flux
Axialgebläse axial blower
Axialkraft axial force || axial load * axiale Belastung || thrust
Axiallast axial load || axial thrust || thrust
Axiallüfter axial-flow fan || axial fan
Axialschub axial force || thrust
Axialspiel end play * z.B. an einem Wellenende || axial clearance * freier Abstand || axial play
Axialventilator axial fan || vane-axial fan || axial-flow fan
AZNIE AZNIE * Associazione Nazionale Industrie Elettrotecniche: ital. Verband d. Elektroindustrie (wie ZVEI)
azyklisch acyclical || acyclic || aperiodic * nicht periodisch || acyclically * (Adv.) || event-driven * ereignisgesteuert

B

B x H x T width x heigth x depth || W x H x D
B-Bauform B type of construction * Motor
B-Seite non-drive end * Nicht-Antriebsseite eines Motors || NDE * Abkürzg. f. 'Non-Drive End': non-drive end || ND end * Abk. für 'Non-Drive end'
B2-Schaltung single-phase bridge connection * [Stromrichter] || H-bridge * [Stromrichter]
B2C-Schaltung B2C connection * vollgesteuerte Einphasen-Gegenparallel-Brückenschaltg.für 4Q-DC-Stromrichter || fully controlled single-phase antiparallel bridge connection
B2H-Schaltung B2H connection * halbgesteuerte Einphasen-Brückenschaltung für 1Q-DC-Stromrichter || half-controlled single-phase bridge connection * halbgesteuerte 1-Phasen-Brückenschaltung
B6-A-B6-C-Schaltung (B6)A(B6)C connection * vollgesteuerte Drehstrombrücke i.Antiparallelschaltung für 4Q-DC-Stromrichter || fully-controlled inverse-parallel three-phase bridge connection * vollgest. gegenpar.Drehstrombrücke
B6-Schaltung three-phase bridge connection * [Stromrichter]
B6C-Schaltung B6C connection * vollgesteuerte Drehstrombrückenschaltung für 1Q-DC-Stromrichter || fully-controlled three-phase bridge connection
Backenbremse shoe brake || drum brake * Trommelbremse
Bäckereimaschine bakery machine
Badewannenkurve bathtub curve * Kurve d. Ausfallwahrscheinlichkt. über d.Gebrauchsdauer (hohe Früh- u.Spätausfälle)
Badlöten bath soldering
Bagger excavator * z.B. Trocken-/Greifbagger || dredger || power shovel * Schaufelbagger || dredge
Baggerantrieb excavator drive
Baggereimer scoop * Schöpfeimer/-Kelle, (Wasser-) Schöpfer, Schippe || bucket || dredger bucket * Baggerschaufel (auch eines Schwimmbaggers)
Bahn railway * Eisen~ || train * Eisen~, Zug || track * Pfad, Spur, auch Transport~ in Transferstraße || web * Material~ (aus Papier, Kunststoff, Textil) || railroad * (amerikan.) Eisen~ || tram * Stadt~, Straßen~ || trajectory * [NC-Steuerung] Verfahr~/-kurve im 2-/3-dimensionalen Raum bei Positioniersteuerung usw. || pathway
Bahn- traction * [Verkehr] || railway * [Verkehr] || railroad * [Verkehr] || orbital * umlaufend || trajectory * [NC-Steuer.] || path * [NC-Steuerg.]
Bahnantrieb traction drive * siehe VDE 0535 und IEC Publ.411
Bahnbeobachtungsanlage web viewer * für durchlaufende Warenbahn (z.B. in Druckmaschine)
bahnbrechend pioneering * z.B. Erfindung || poineer || epoch-making * epochemachend || revolutionary * revolutionär
Bahnfolgeregelung tracking control
Bahnführung web guide * Einrichtung zur seitl. Führung einer durchlaufenden Warenbahn
Bahngeschwindigkeit line speed || speed || web speed * einer Warenbahn (Papier, Textil usw) || material speed * Material || surface speed * Oberflächengeschwindigkeit || web velocity * einer durchlaufenden Papier/Textil/Kunststoffbahn
Bahninterpolation path interpolation * bei →NC-Steuerung
Bahnkurve curve of trajectory || trajectory

Bahnmotor

Bahnmotor rail traction motor || traction motor * *Traktionsmotor*
Bahnriss web break || web breakage
Bahnschweißen seam welding
Bahnspanngerät web tensioning unit
Bahnspannung web tension * *Zug in einer Materialbahn*
Bahnspannungsregelsystem web tension control * *meist Abwickler mit Magnetbremse/Kupplung* || web tensioner * *meist Abwickler mit Magnetbremse/Kupplung*
Bahnsteuerung path control * *(NC-Steuerung) Position entlang einer Kontur Steuern* || contouring control * *(NC-Steuerung) Bewegung entlang einer Kontur Steuern* || continuous-path control * *(NC-Steuerung)* || track control || contour control * *(NC-Steuerung)*
Bahnstromversorgung traction power supply || traction supply
Bahntacho web-speed tachometer || surface tachometer * *Anlegetacho*
Bahnumrichter traction converter
Bahnzug web tension * *Zug in einer Materialbahn (z.B. Papier, Textil)*
Bahrzahlung cash || cash payment
Bajonett bayonet
Bajonettfassung bayonet socket
Bajonettverschluss bayonet joint || bayonet catch
Balken bar * *auf Anzeige, Messgerät, Bildschirm usw.* || bargraph * *zur Messwertanzeige auf Messinstrument/Bildschirm*
Balkenanzeige bargraph display || bargraph || bar display || bar graph
Balkendiagramm bargraph || bar diagram
Balkenhobel beam planer * *(Holz)*
Balkon patch * *behelfsmäßige Modifikation eines Programms durch Ansprung einer ausgelagerten Befehlssequenz* || binary patch
Ballastschaltung ballast circuit
Ballen bale || pack || bundle
Ballenpresse baling press * *(Verpack.techn.)* || bale press * *(Verpack.techn.)*
ballig curved || convex || cambered || barelled * *fassförmig* || bareled * *fassförmig* || crowned
Ballon balloon || carboy * *für Chemikalien* || demijohn * *Korbflasche*
Bananenstecker banana plug * *z.B. 4mm-Stecker* || split plug * *Büschelstecker*
Band band * *Frequenzband, Metallband* || tape * *schmales Band, Streifen, Textil/Klebeband* || strip * *Walzband, Metallband, Bandmaterial* || belt * *Gurt, (Förder-) Band* || assembly line * *→Fließband* || sliver * *(Textil) Flor~, Kammzug in der Spinnerei*
Bandabzug belt take-off
Bandage bandage || banding * *Wicklung* || taping
Bandagierung banding * *Wicklung*
Bandantrieb belt drive * *z.B. Antrieb für Förderband; Gurtantrieb* || capstan drive * *für Magnetbandgerät* || line drive * *z.B. in Blechstraße, Bandbehandlungsanlage, Walzwerk*
Bandbearbeitungsmaschine strip finishing line * *Nachbearbeitungsstraße für Walz-/Blechband* || strip working machine * *für Blechband*
Bandbehandlungsanlage strip processing line * *für Metallband* || strip finishing plant * *für Metallband*
Bandbreite bandwidth * *z.B. eines Frequenzbandes/Bandpasses/Datenübertragungskanals* || strip width * *(Walzwerk) ~ eines Walzbandes*
Bandbremse band brake || strap brake
Banddicke strip thickness * *z.B. von Walzband*
Bandeinlauf entry of strip * *im Walzwerk: Stelle, an der das Walzband einläuft* || strip entering * *Walzwerk* || mill entry section * *im Walzwerk*

Bandende strip tail * *(Walzwerk) Ende des Walzbandes*
Banderoliermaschine banding machine * *(Verpack.technik)* || banderoling machine * *(Verpack.technik)* || taping machine * *(Verpack.techn.) zum Verpacken mit (Klebe-) Streifen*
Bänderregelung conveyor control * *für Transportband*
Bandfilterpresse belt press * *(Nahrungsmittel) z.B. zur Saftherstellung*
Bandförderer belt conveyor || band conveyor
bandförmig strip-type || band-shaped
Bandgeschwindigkeit strip speed * *z.B. von Walzband*
Bandglühanlage strip annealing line || band annealing plant || band annealing line
Bandkabel ribbon cable
Bandkabelverbindung ribbon cable connection
Bandkern wound core
Bandkerntrafo wound-core transformer
Bandkopf strip head * *(Walzwerk) Beginn des Walzbandes*
Bandlaufwerk tape drive * *für Magnetbandgerät*
Bandmaß measuring tape || tape measure
Bandpass band-pass
Bandpassfilter band-pass filter
Bandrichtmaschine strip-levelling machine * *→Richtmaschine zum Planbiegen von Blechband*
Bandriss strip breakage * *(Walzwerk)*
Bandrücklauf belt return * *eines Transportbandes*
Bandsäge band saw || belt saw || band iron || ribbon saw
Bandsägeblatt endless saw blade || band saw blade
Bandsägemaschine band-sawing machine
Bandschleifmaschine abrasive-band grinding machine || abrasive-belt grinding machine || belt sander * *(Holz)* || belt sanding machine * *(Holz)*
Bandspeicher strip accumulator * *für Metallband, wird zur Überbrückung der Coilwechselzeit gefüllt, dann geleert* || looper * *für Metallband*
Bandsperre band-stop filter || rejection band || rejector circuit * *Sperrkreis*
Bandvorschub belt feed * *bei Fördersystem*
Bandwaage conveyor-type weigher || belt weigher || belt conveyor scale || belt scale || belt conveyor scale * *in Förderband integrierte Waage* || conveyor belt scales
Bandwalzwerk strip mill
Bandzug strip tension
Bandzugregelung strip-tensioning control * *in Walzstraße*
Bank bank * *Geldinstitut; Speicherbank, Kondensatorbank, Reihenanordnung*
Barren ingot * *Metallbarren, auch 'Stange, Block'*
BASIC BASIC * *Beginner's All-purpose Symbolic Instruction Code, einfache höhere Programmiersprache*
BASIC-Interpreter BASIC interpreter
basieren base * *(up)on: auf* || be based * *beruhen, sich stützen ((up)on: auf)* || be founded * *sich gründen, beruhen (on: auf)* || rest * *beruhen, sich stützen ((up)on: auf)*
basieren auf be based on
Basis basis * *auch Grundlage, Fundament, Basis eines Transistors; on the basis of: auf d. ~ von* || base * *auch Grund, Grundlage, Fundament, ~ e.Transistors, Sockel, Unterbau, Ausgangspunkt* || radix * *Grundzahl einer potenzierten Größe* || gate * *Steueranschluss bei Thyristor, Transistor usw.*
Basis-Emitter-Strecke base-emitter circuit
basisch basic * *(chem.)*
Basismaterial basis material || base material
Basisstecker backplane connector * *auf e. Leiter-*

karte zur Rückwandplatine hin angeordnet || base connector

BASP output inhibit * *Befehlsausgabe-Sperre für Digital-/Analogausgaben z.B. bei SPS*

Batchdatei batch file * *Textdatei, die eine Folge von Betriebssystem-Kommandos enthält*

Batterie battery || bank * *Reihe von Geräten/Bauelementen, z.B. von Kondensatoren*

Batterie-gepuffert battery-backed

Batterie-Lebensdauer battery life

Batterieausfall battery low * *Meldung '(Puffer-) Batterie entladen' (z.B. bei SPS)*

Batteriebetrieb battery operation

batteriebetrieben cordless * *ohne Netzanschluss (z.B. Elektrowerkzeug)*

batteriegepuffert battery backed || with battery backup

batteriegestützt battery-backed * *z.B. nichtflüchtiger Datenspeicher*

Batteriekästchen battery box * *z.B. zur Vorgabe v. Sollwerten/Sollwertsprüngen für d. Antriebsregler-Optimierung*

Batterieladegerät battery charger

Batterielebensdauer battery life

batterielose Pufferung data retention without battery * *nichtflüchtige Datenspeicherung ohne Batterie (z.B. in NOVRAM)* || non-volatile storage without backup battery

Batterien batteries

Batteriepufferung battery back-up || battery buffer

Bau manufacture * *Herstellung, Erzeugung* || production || erection * *~ eines Gebäudes, einer Anlage usw.* || building * *~ eines Gebäudes* || construction * *~ von Maschinen usw.* || manufacturing * *das ~en, z.B. einer Maschine*

Bau- construction || structural

Bauart type || design || type of construction * *Bauform*

bauartbedingt as a result of the design || inherent in the type of construction

Baud Baud * *['bod] bits pro sec.* || bps * *bits pro sec*

Baudrate baud rate * *Maßeinheit bits pro sec, Übertragungsgeschwindigkeit bei einer seriellen Kommunikation*

Baueinheit package || sub-assembly * *Baugruppe*

Bauelement component || device || construction component * *(mechanisches/maschinenbautechnisches) Konstruktionselement*

Bauelement-Bestückung insertion of components || mounting of components

Bauelemente-Bestückung mounting of components || insertion of components

bauen build || construct || erect * *errichten* || raise * *errichten* || manufacture * *herstellen* || make * *herstellen* || design * *entwerfen* || trust * *vertrauen (in: auf)* || count * *sich verlassen, zählen; (on: auf)* || base * *Hoffnung, Urteil usw.; (upon: auf)*

Bauform type of construction * *für elektr. Maschinen: siehe DIN IEC 34, Teil 7 u. DIN 42950* || design || type

Bauform-Kennziffer figure denoting type of construction * *Bestellnummer-Ergänzung bei SIEMENS Motor* || type-of-construction code

baugleich of the same type of construction || of the same rating * *mit gleichen Nenn-/Kennwerten/ Typenschildangaben* || of the same construction

Baugröße size || construction size || frame * *bei Elektromotor (Achshöhe)* || motor frame * *für Elektromotors (ident.m. Achshöhe; z.B. BG225: Achshöhe 225mm; s.IEC Publ. 72 (1971))* || motor size * *{el. Masch.} eines Motors (Achshöhe plus Buchstabe für Gehäuselänge; IEC72-1/2)*

Baugruppe module || plug-in module * *Steck~* || board * *Leiterkarte* || PCB * *Leiterkarte* || printed circuit board * *Leiterkarte* || card * *Leiterkarte* || component * *Komponente, Bauteil* || sub-assembly || hardware module * *Hardware-~* || assembly * *aus Einzelteilen zusammengebaute ~*

Baugruppen-Identifikation board identification * *(automatische) Baugruppen-Erkennung*

Baugruppentausch replacement of the printed-circuit board * *Tausch der Leiterkarte*

Baugruppenträger rack || subrack || cage || card frame * *für Leiterkarten* || card rack* *für Flachbaugruppen/Leiterkarten* || card cage

Bauholz structural timber || lumber * *(amerikan.)* || bilding lumber || structural lumber || timber

Bauindustrie construction industry

Baujahr year of manufacture || date of manufacture || year of construction

Baukastenprinzip modular principle || modular design * *modular entworfen/konstruiert* || unitized construction * *Baukastensystem, System aus abgeschlossenen Baueinheiten*

Baukastensystem modular system || unitized construction || modular design * *modular konstruiert/ entworfen* || modular-assembly system * *einer Maschine, einer Konstruktion usw.*

Baukran building crane || construction-site crane

Baukreissäge construction circular saw * *{Holz}* || builder's saw bench * *{Holz}*

Baulänge construction length * *z.B. eines Motors*

Bauleistung rated power * *z.B. eines Transformators* || design rating * *z.B. eines Transformators*

Bauleiter site manager

Baum tree || beam * *{Textil} Wickel, aufgewickelter Textilstoff*

Baumaschinen construction machinery

baumförmig tree-type

Baumstruktur tree-type structure

Baumuster prototype

Baumusterprüfbescheinigung prototype test certificate || prototype test certification

Baumwollspinnerei cotton-spinning plant

Baureihe series || line || range || class || product range || type series || series of products || type of construction * *→Bauart*

Bausatz assembly set || construction kit || component set

Bauschaltplan wiring layout

bauschig swelled || puffy || baggy || bulky * *massig, sperrig; (sehr) umfangreich* || flossy * *flaumig, seidig*

Baustein module * *Hardware/Software-Baugruppe* || block * *Software-Funktionsbaustein* || building block || modular assembly * *Baugruppe in Modulbauweise* || function block * *Funktionsbaustein* || chip * *integrierter Baustein* || IC * *Integrierter Baustein*

Baustein- modular

Baustein-Bibliothek function block library * *mit Software-Funktionsbausteinen*

Bausteinaufruf block call || function block call || module call

Bausteinende block end * *Ende eines Funktionsbausteins bei SPS* || end module * *Befehl z. Beenden eines Funktionsbaustein bei einer SPS*

Bausteinmaske function block mask * *Bildschirmmaske für Funktionsbaustein; z.B. b.SIEMENS Regelsystem SIMADYN D (R)*

Bausteinpaket function block package || block package * *Paket aus Software-/Funktionsbausteinen*

Bausteinparameter block parameter

Bausteinsystem modular system || modular assembly system * *modularer Aufbau bei Geräten, Maschinen usw.*

Bausteinverbindung function block interconnection * *zwischen Funktionsbausteinen*
Baustelle site || building site * *Haus~, Anlagen~* || construction site || erection site
Baustoff construction material || building material || material for construction || structural material
Baustoffindustrie building material industry
Bautechnik construction engineering * *Hochbau* || civil engineering * *Hoch- und Tiefbau*
Bauteil component || device
Bauteilprüfung component inspection || components inspection
Bauteilseite component side * *einer Leiterkarte*
Bauvolumen construction volume || unit volume * *eines Geräts* || volume
Bauweise design || type || type of construction * *Bauform*
Bauwesen building and civil engineering * *Hoch- und Tiefbau* || civil engineering * *Tief- und Hochbau* || building construction * *Hochbau*
BBS BBS * *Abk.f. 'Bulletin Board System': Mailbox-System, z.B. Support-Mailbox mit Zugriff über Modem*
BCD BCD * *Abkürzg.f. 'Binary-Coded Decimal': binär codierte Dezimal- (Zahl).*
beabsichtigen intend
beabsichtigt intended
beachte note
beachten pay attention to || note * *kindly note: bitte beachten* || notice || take notice of || observe * *befolgen; z.B.Gesetze* || take care * *warnend* || mind * *warnend; that: dass* || consider * *berücksichtigen* || bear in mind * *berücksichtigen, nicht vergessen* || take into account * *berücksichtigen* || take note of || follow * *befolgen*
beachtlich noticeable || considerable || remarkable * *bemerkenswert* || respectable * *ansehnlich, (recht) beachtlich, respektabel*
Beachtung observance * *Befolgung; z.B. einer Vorschrift* || attention * *pay attention to: etwas ~ schenken* || notice * *be noticed: ~ finden* || consideration * *Berücksichtigung; give due consideration to sth.: etwas gebührend →berücksichtigen* || regard * *Berücksichtigung; ~ schenken (Verb); disregard: keine ~ schenken*
Beamter officer || official || civil servant * *Berufsgruppe* || public servant * *Berufsgruppe*
Beanspruchbarkeit stressability || load capability * *Belastbarkeit*
beanspruchen stress * *belasten, (Material, Maschinenteil, Bauelement) beanspruchen* || impose * *auferlegen, aufbürden ((up)on): e. Ding, e. Person), beanspruchen, z.B. mit Oberwellen* || claim * *ein Recht ~* || require * *z.B. Mühe, Sorgfalt, Zeit, Platz ~* || take * *z.B. Mühe, Sorgfalt, Zeit, Platz ~* || call for * *verlangen, abrufen, (er)fordern* || make use * *Gebrauch machen (of: von)* || strain * *ermüden* || take-up * *(Platz, Zeit usw.) ~, in Anspruch nehmen, beanspruchen*
Beanspruchung stressing * *von Material* || stress || strain * *(starke) Inanspruchnahme, Belastung, Last* || demand * *Anspruch (on: an), Anforderung* || wear * *Verschleiß* || conditions * *(Betriebs-) Beanspruchung; for most stringent conditions: für härteste Beanspruchung* || loading * *Belastung* || duty * *das zu Leistende, Nutzleistung*
beanstanden object * *to: etwas ~* || complain * *of: etwas ~; reklamieren; sich beklagen/beschweren (of/about: über; that: dass)* || frow upon * *etwas ~* || demur to * *Forderung usw. ~* || reject * *Waren, Lieferung usw. ~* || refuse * *Waren, Lieferung usw. ~; refuse acceptance: Annahme verweigern* || be critical of * *etwas ~/kritisieren* || take exception to
Beanstandung complaint

beantworten answer * *with: mit* || respond to || reply to
bearbeiten work at || process * *verarbeiten, bearbeiten* || treat * *behandeln* || deal with * *behandeln* || handle * *erledigen, hantieren, handhaben, abwickeln* || act upon * *mit etw. befasst sein* || consider * *bedenken, überlegen* || treat with * *behandeln* || work out * *ausarbeiten* || prepare * *ausarbeiten* || re-edit * *Buch ~* || revise * *überarbeiten* || work * *Metall ~, spanlos ~* || tool * *Metall ~, spanabhebend ~, z.B.auf Dehbank* || machine * *Werkstück(e) maschinell ~ (meist spanabhebend)* || edit * *Text, Artikel, Buch, Gafik usw. ~/redigieren/druckfertig machen*
Bearbeiter official in charge * *Sachbearbeiter* || engineer * *Ingenieur, Techniker* || contact person * *Kontaktperson, z.B.in Vertriebsniederlassung* || person responsible * *for: für* || responsible person * *zuständiger Bearbeiter*
Bearbeitung treatment * *(allg.) auch Akten~, Werkstück~, chemische ~ usw.* || mechanical treatment * *mechanische ~* || working * *spanlose ~* || machining * *zerspanende ~ (durch Werkzeugmaschine)* || tooling * *spanabhebende ~* || processing * *auch Akten, Werkstück usw. ~* || preparation * *Akten usw. ~* || revision * *~ eines Druckerzeugnisses; Überarbeitung* || revised edition * *überarbeitete Auflage eines Druckerzeugnisses* || dressing * *Zurichtung* || handling * *~ eines Falles, Vorgangs usw.* || management * *Verwaltung, Handhabung, Behandlung* || administration * *Handhabung, Durchführung, Verabreichung* || forming * *Formung, Gestaltung*
Bearbeitungsgeschwindigkeit processing speed || processing rate
Bearbeitungskontrolle debug * *bei der Abarbeitung eines SPS-Programms* || program check * *bei SPS*
Bearbeitungsmaschine processing machine
Bearbeitungsreihenfolge processing sequence
Bearbeitungsstation machining station * *{Werkz. masch.} mit spanabhebender Bearbeitung*
Bearbeitungszeit execution time * *Ausführungszeit eines (SPS-/Computer-) Programms/-Befehls* || processing time
Bearbeitungszentrum machining center * *{Werkz. masch.}(amerikan.)* || machining centre * *(brit)*
beaufschlagen act upon || admit * *zuführen, einlassen* || load * *(elektr.) belasten* || apply * *auflegen/-tragen, anbringen (to: an/auf), zur Anwendung bringen*
Beaufsichtigung supervision || superintendence || surveillance
beauftragen contract * *vertraglich* || charge * *entrust* || direct || order || put in charge || appoint * *berufen* || authorize * *ermächtigen*
Beauftragter officer * *mit einer offiziellen Aufgabe betraut* || delegate * *Abgeordneter, Bevollmächtigter, Delegierter*
Becher cup || beaker * *großer Becher, Becherglas (chem.)* || tumbler || mug || bucket * *bei Bagger*
Becherkondensator metal-cased capacitor
Becherwerk bucket conveyor * *Förderer*
bedacht sein auf be concerned about || be tender of || be careful of || be anxious to
Bedämpfung dampening || damping || attenuation * *Dämpfung, Abschwächung*
Bedarf requirement * *auch an Stromversorgung; as required: nach Bedarf* || demand * *auch an Stromversorgung; meet the demand for: dem ~/der Nachfrage nach ... nachkommen*
bedarfsgerecht according to requirement
bedauern regret || be sorry || sympathize with * *a person: jemanden*
bedecken cover

bedeuten mean ‖ signify ‖ imply * *in sich schließen* ‖ represent * *kennzeichnen*
bedeutend considerable * *(Adj.)* ‖ considerably * *(Adv.)* ‖ important ‖ major ‖ eminent * *hervorragend* ‖ remarkable * *bemerkenswert*
Bedeutung meaning ‖ significance * *becoming increasingly significant: an ~ gewinnen* ‖ importance * *Wichtigkeit; gain in importance: an ~ gewinnen* ‖ consequence * *Folge, Resultat, Wirkung, Wichtigkeit, Bedeutung; of no consequence: ohne ~* ‖ acceptation * *gebräuchlicher Sinn (eines Wortes)* ‖ concern * *Wichtigkeit, ~; Unruhe, Sorge, Bedenken (at/about/for: um); be a concern: von ~ sein*
Bedien- und Beobachtungsfunktionen operator control and visualization functions
Bedienablauf operational sequence
Bedieneingriff operator intervention
Bedieneinheit operator station ‖ operator control unit ‖ operator's station ‖ operator panel * *Bediengerät*
Bedienelement operator control element ‖ control element ‖ operator * *(amerikan.)* ‖ operating device ‖ actuating element * *an einem Stellglied/Aktutor* ‖ actuator * *Stell-/Betätigungsglied, Aktuator* ‖ operator device
bedienen operate * *auch 'handhaben, betätigen'* ‖ control * *steuern, führen, lenken, beaufsichtigen, überwachen* ‖ work * *arbeiten an, beschäftigen (Fabrik, Maschine), arbeiten lassen* ‖ manipulate * *handhaben, geschickt behandeln* ‖ handle * *handhaben, hantieren mit, umgehen mit, führen, leiten* ‖ serve * *z.B. einen Markt/Kunden/Gäste, versorgen, dienen, nützen* ‖ attend * *sich kümmern/sorgen um, Acht geben/achten auf, Dienst tun* ‖ help oneself * *sich (selbst) bedienen* ‖ run * *fahren, laufen lassen, führen* ‖ service * *softwaremäßig*
Bedienen und Beobachten operator control and monitoring ‖ operating and monitoring ‖ operator communication and observation ‖ man-machine dialog
Bediener operator
Bedienereingriff operator intervention
Bedienerführung operator prompting * *z.B. am Bildschirm* ‖ user prompting
Bedieneroberfläche man-machine interface * *siehe auch →Bedienungsoberfläche*
Bedienfeld operator control panel ‖ operator panel ‖ control panel ‖ keyboard unit * *Tastenfeld* ‖ panel operator * *(amerikan.)* ‖ operator * *(amerikan.)* ‖ operational panel ‖ operating control panel
Bedienfunktionen operator control functions
Bediengerät operator unit ‖ operator panel * *Bedienfeld, z.B. für SPS* ‖ operator control unit ‖ operator station * *z.B. Bedienkasten/Pult* ‖ operation panel * *Bedienfeld* ‖ operator * *(amerikan.)* ‖ keyboard unit * *mit Tasten* ‖ operating unit
Bedienhoheit parameter change rights * *(→PROFIBUS-Profil)*
Bedienkomfort user convenience * *Anwenderfreundlichkeit* ‖ user-friendliness * *Anwenderfreundlichkeit* ‖ ease of operation * *einfache Bedienung, leichte Bedienungsweise* ‖ operator friendliness
Bedienoberfläche operator control interface ‖ user interface ‖ operator interface ‖ man-machine interface
Bedienort control location
Bedienpersonal operators staff
Bedienplatz operator control station
Bedienpult control desk ‖ operator console ‖ operator desk ‖ control console
Bedienschnittstelle operator interface
Bedienseite operator side * *einer Maschine*

Bedienstation operator's panel ‖ operator's station ‖ operating station ‖ operating panel ‖ operator control station
Bedientableau operator control panel * *siehe auch →Bedientafel*
Bedientafel operator panel ‖ operator's panel ‖ operator control panel
Bedientasten control buttons * *Druckknöpfe* ‖ control keys * *Steuertaster*
Bedienteil operating element ‖ operator control element
Bedienterminal control terminal
Bedienung operator control ‖ operator input * *Eingabe* ‖ operative control ‖ operation ‖ operations ‖ handling * *Hantierung, Handhabung* ‖ manipulation ‖ working ‖ control ‖ control and operation ‖ operator communication * *bei SPS usw.*
Bedienungs- operating ‖ operator control ‖ control ‖ operator ‖ service
Bedienungs- und Beobachtungsstation operator's station
Bedienungsanleitung operating instructions ‖ operators guide ‖ operator's guide
Bedienungsbeispiel operational example
Bedienungseintrag initialization * *{SPS} (amerikan.)* ‖ initialisation * *{SPS} (brit.)*
Bedienungselement operator device ‖ control device ‖ control element ‖ operator's control element ‖ operator control element ‖ operator * *(amerikan.)* ‖ actuating element * *an einem Stellglied/Aktuator* ‖ actuator * *Aktuator, Stellglied, Betätigungselement* ‖ pilot device * *z.B. Taster, Schalter, Meldeleuchte*
Bedienungselemente operator devices ‖ controls ‖ control elements ‖ operator's control elements
Bedienungsfehler operator mistake ‖ incorrect operation ‖ operating error ‖ faulty operator control ‖ attendance mistake
bedienungsfreundlich easy to use ‖ easy-to-use
Bedienungsfreundlichkeit ease of operation ‖ simple operations ‖ user-friendliness * *Anwenderfreundlichkeit*
Bedienungshebel control lever
Bedienungsklappe service cover ‖ servicing cover * *siehe auch →Serviceöffnung*
Bedienungskomfort operator convenience ‖ ease of operation
Bedienungsmann operator ‖ attendant
Bedienungsoberfläche human communication interface ‖ human interface ‖ operator interface ‖ user interface ‖ man-machine interface ‖ front-end * *{Computer} Software, die e. (komfortablere) ~ für e. anderes Programm erzeugt*
Bedienungspersonal operating personnel ‖ operators staff ‖ operating staff
Bedienungspult control desk * *siehe auch →Bedienpult*
Bedienungsschnittstelle operator interface
Bedienwert operating value * *bei SPS*
bedingen involve * *mit sich bringen* ‖ cause * *verursachen* ‖ call for * *erforden, verlangen* ‖ presuppose * *→voraussetzen*
bedingt caused * *verursacht; by: durch* ‖ conditional * *abh. von e. Bedingung, z.B. Programmsprung (on: durch), Funktionsbaustein usw.* ‖ dependent * *abhängig* ‖ limited * *beschränkt*
bedingter Sprung conditional jump * *im Programm* ‖ conditional branch * *im Programm*
Bedingung condition * *on condition that: unter der Bedingung, dass; fulfill/comply with: erfüllen* ‖ requirement * *Anforderung* ‖ conditions * *Verhältnisse* ‖ restriction * *Einschränkung* ‖ provision * *in Vertrag; auch 'Vorbehalt'*
bedrahtet through-hole type * *Bauelement zur Be-*

bedrahtete Ausführung

stückung auf Leiterkarte || leaded * *mit Anschlussdraht versehen*
bedrahtete Ausführung through-hole version * *Bauelement zur Bestückung auf Leiterkarte*
Bedruckbarkeit printability
beeinflussen influence * *on: auf; with: bei* || affect * *einwirken auf; oft negativ* || have influence on || interfere * *s.einmischen, (be)hindern, eingreifen, stören, beeinträchtigen* || control * *beherrschen* || exercise an influence on * *Einfluss ausüben auf* || sway * *beherrschen* || have a bearing on * *einwirken auf* || have an effect * *einen Einfluss ausüben; have an adverse effect on: nachteilig/schädlich* ~
beeinflusst influenced
Beeinflussung influence * *Einfluss* || control * *Steuerung* || interference * *[inter]fierens] störende, schädliche ~ (Störung, Beeinträchtigung, Übersprechen)* || interaction * *gegenseitige ~; between: zwischen* || interdependence * *gegenseitige Abhängigkeit*
beeinträchtigen impair || injure || affect || affect adversely || have an adverse affect on || prejudice * *jemandes Rechte ~* || infringe * *jemandes Rechte ~* || handicap * *behindern* || diminish * *verringern, herabsetzen, (ab)schwächen* || obstruct * *behindern, verstellen, verstopfen*
beenden end || abort * *abbrechen* || finish || complete * *abschließen, beenden, vollenden, fertigstellen, erledigen* || bring to an end || terminate * *abschließen* || discontinue * *aufhören mit, aufgeben* || truncate * *Programmablauf, Digitalisierg., Rechenalgorithmus abbrechen/-schneiden (Restfehler mögl.)* || close * *(ab)schließen*
beendet completed * *abgeschlossen*
Beendigung completion * *Fertigstellung* || termination
beengt confined || restrained || narrow * *eng, räumlich beschränkt, knapp* || choked * *zurückgedrängt, abgewürgt, zusammengedrängt* || hampered || cramped * *eng, beengt*
beengter Platz restricted space
befassen be engaged * *(figürl.) be engaged in: sich befassen mit* || touch * *(auch figürl.)* || handle || deal with * *(figürl.) sich befassen mit* || occupy oneself with * *sich befassen mit* || concern oneself with * *sich befassen mit* || study * *sich prüfend befassen mit* || examine * *sich prüfend befassen mit* || consider * *sich prüfend befassen mit* || deal with * *sich befassen mit, behandeln (z.B. in Aufsatz, Abhandlung)*
Befehl command * *Steuerbefehl/Kommando* || instruction * *[Computer] (Assembler-/Maschinen)befehl* || opcode * *[Computer] Computerbefehl in Binärdarstellg.* || order * *Auftrag* || statement * *[Computer] Anweisung in Hochsprache* || operation * *[Computer | SPS] Operation*
Befehl freigeben command enable * *bei SPS*
Befehl sperren command disable * *bei SPS*
Befehlsausführung instruction execution
Befehlsausführungszeit instruction execution time * *[Computer]* || command execution time * *[SPS]* || operation execution time * *[SPS]*
Befehlsausgabe command output * *bei SPS*
Befehlsausgabe-Sperre output ihibit * *"BASP" bei SPS (Sperren der Digital-/Analogausgaben)*
Befehlscode instruction code * *Rechner* || command code * *Bediengerät* || op code * *in Binärdarstellung* || opcode * *in Binärdarstellung* || operation code * *in Binärdarstellung*
Befehlsregister instruction register * *(Rechner)*
Befehlssatz instruction set * *Rechner* || command set * *Bediengerät*
Befehlszähler program counter
Befehlszyklus instruction cycle
befestigen mount || attach * *to: an* || fasten * *fest machen* || fix || mount on || secure * *sichern* || clamp * *mit Klammer(n)* || cleat * *mit Klammer(n)* || couple * *aneinanderkoppeln* || connect * *miteinander verbinden* || affix * *befestigen, anbringen, anheften, ankleben*
befestigt mounted || attached
Befestigung fixing || fastening || mounting * *Montage* || clamping * *mit Klammer(n)* || fastener * *Verschluss, Befestigungsteil* || attachement * *Befestigung, Anbringung, Verbindung, Anschluss* || suspension * *Aufhängung, Federung* || fitting * *Zubehör(teil), Beschlag, Armatur (für Rohre, Kabel usw.)* || fixture * *(Befestigungs-/Aufspann-/Halte-)Vorrichtung, Befestigung, Spannzeug*
Befestigungs- fixing || mounting
Befestigungsart type of mounting || mounting style
Befestigungsbohrung mounting hole * *tapped mounting hole: ~ mit Innengewinde zur Schraubbefestigung* || fixing hole || fastening hole || retaining hole
Befestigungselement mounting element || fixing element * *Halteelement* || retaining element * *Halteelement* || fastener * *Verschluss, Halter, Befestigungsvorrichtung, Sicherung* || fastening element || connecting element * *→Verbindungselement*
Befestigungselemente mounting elements || mounting accessories * *als Zubehörsatz ("Satz loser Teile")* || fixing accessories || fixing material
Befestigungsflansch fixing flange * *für Motor: siehe DIN 42984* || mounting flange
Befestigungslasche mounting lug || fixing lug || fitting plate || mounting strap
Befestigungsloch mounting hole
Befestigungsmaße mounting dimensions * *für Trafos/Drosseln: siehe EN 60852-4*
Befestigungsmaterial mounting material || mounting accessories * *Montagezubehör*
Befestigungsmutter retaining nut || lock nut
Befestigungsschiene fixing rail || mounting rail * *Montageschiene* || DIN rail * *DIN-Montageschiene (Hutschiene)*
Befestigungsschlitz retaining slot * *z.B. in Montageblech oder Spulen-/Trafo-Befestigungswinkel z. Schraubbefestig.*
Befestigungsschraube fixing bolt || mounting screw || retaining screw * *zum Festhalten/Niederhalten; verhindert das Herunterfallen* || fastening screw || fastening bolt || retaining bolt
Befestigungsteile fastening parts || mounting hardware || fixing parts || fittings * *für Rohre usw.* || fastenings
Befestigungswinkel L-bracket * *z.B. Blechwinkel*
befeuchten wet || moisten
Befeuchtung moisturization || moistening || humidification || wetting * *nass machen* || humidifying
Befeuchtungsgerät moistening unit
befindlich located * *örtlich*
befolgen adhere to * *Grundsatz~, Regel ~* || follow * *Ratschläge ~* || take * *Ratschläge ~* || comply with * *Vorschrift ~* || observe * *Vorschrift ~* || obey * *Vorschrift ~, Gesetz ~; gehorchen, Folge leisten* || abide by * *festhalten an, z.B. Grundsatz*
Befolgung compliance * *with: von* || observance * *of: von* || adherence * *to: von* || following * *Befolgung, Beachtung, 'Sich-Richten-Nach'*
Beförderung carriage * *von Frachtgut* || conveyance * *Materialtransport, Transport, Beförderung, Spedition* || advancement * *im Rang* || promotion * *beruflich, im Rang*
befreien relieve * *of: von; Sorge, Last usw.* || free * *from: von*
Befugnis authority || power || right || privilege * *durch Vorrecht/Privileg* || competence * *Zuständigkeit* || jurisdiction * *Zuständigkeit* || warrant
befugt authorized * *to: zu* || entitled || empowered

* *bevollmächtigt, ermächtigt* || have the right * *befugt sein; have no right to do so: dazu nicht befugt sein* || competent * *zuständig; for a thing: für etwas; to do a thing: etwas zu tun* || have jurisdiction * *over: zuständig sein für*
befüllen   charge * *Behälter, Zentrifuge usw.* ~
Befüllung   filling
Befüllungsanlage   charging unit
Befund   findings
begegnen   encounter * *jemanden ~,einer Sache ~* || meet with
Begehung   walkaround tour
Beginn   beginning || commencement * *Anfang, Beginn, (Tag der) Feier* || outset || start || opening * *~ einer Verhandlung, Schul~ usw.* || inception * *Beginn, Anfang, Einsatz* || early stage * *früher Zeitpunkt (eines Projekts usw.)*
beginnen   begin || start || commence
Begleiterscheinung   secondary phenomenon * *Sekundäreffekt, Nebeneffekt* || side-effect * *Nebeneffekt* || attendant symptom || concomitant phenomenon || accompaniment || concomitant
Begleitkarte   job card * *z.B. für ein Produkt in der Fertigung*
Begleitpapiere   accompanying paperwork * *z.B. Versandpapiere* || way-bills || pass-bills || permits * *Zollpapiere* || accompanying documents || way bills || customs permit * *zollamtliche ~* || bond note * *zollamtliche ~*
begrenzen   limit * *to: auf (einen Wert) ~* || restrict || determine * *festlegen*
Begrenzer   limiter || limiter module
begrenzt   limited || restricted * *eingeschränkt* || narrow * *eingeschränkt, beschränkt (Mittel, Verhältnisse, Platz, Raum usw.)*
begrenzt setzen   set conditionally * *bei SPS*
Begrenzung   limiting * *before/after the limiting: vor/nach der Begrenzung* || limit * *Grenze* || restriction * *Einschränkung* || limitation
Begrenzungsdiode   limiting diode
Begrenzungsregelung   limiting control
Begrenzungsregler   limit controller
Begriff   term * *Ausdruck* || word * *Wort* || conception * *Konzeption* || idea * *Vorstellung* || notion * *Vorstellung*
Begriffe   terminology || nomenclature || terms * *Fachausdrücke* || technical terms * *techn. Fachbegriffe*
Begriffsdefinitionen   definitions of terms
behalten   keep || retain * *auch im Gedächtnis, Speicher usw.* || remember * *im Gedächtnis*
Behälter   container || case * *Kasten* || box * *Kasten* || tank * *für Flüssigkeiten* || basin * *für Flüssigkeiten* || receptacle || reservoir * *Vorrats-/Zwischen-/Ausgleichs~, Sammel-/Staubecken, Bassin; reservoir of: Vorrat an* || bin * *für Schüttgut*
Behältersteuerung   tank filling and discharge control * *für Flüssigkeiten*
behandeln   treat || handle * *auch in einem Buch ~* || deal with * *auch Thema* || manage * *lenken, meistern* || manipulate * *betätigen* || finish * *nach-/end~*
Behandlung   treatment * *~ von Material* || therapy * *~ zur Besserung* || processing || handling * *Handhabung* || management * *Verwaltung, Handhabung, Behandlung* || usage * *Gebrauch, Verwendung, Behandlungsweise*
Beharrungstemperatur   steady-state temperature * *im eingeschwungenen Zustand/therm. Gleichgewicht* || permanent temperature * *im eingeschwungenen Zustand*
Beharrungszustand   steady state * *stationärer/eingeschwungener Zustand* || equilibrium * *Gleichgewichtszustand*
behaupten   claim * *geltend machen* || state * *erklä-*
ren || declare * *erklären* || assure * *versichern* || assert * *behaupten, erklären, (Anspruch, Recht) geltend machen, sich behaupten/durchsetzen*
beheben   remove * *Fehler ~* || fix * *instand setzen, (Software-) Fehler ~, in Ordnung bringen* || overcome * *Fehler ~* || repair * *Schaden ~* || debug * *Softwarefehler ~* || rectify * *berichtigen, korrigieren, richtigstellen, verbessern*
Behebung   removal * *auch ~ von Fehlern* || fixing * *~ von Fehlern* || settlement * *Problem ~, Fehler ~* || fix * *Fehler ~, Problem ~* || elimination || relief
beheizbar   heatable
behelfsmäßig   makeshift || provisional
beherrschen   master * *meistern* || cope with * *fertigwerden mit* || tackle with * *fertig werden mit, lösen*
beherrschend   dominating * *(be)herrschend, dominierend, emporragend über* || pervasive * *durchdringend, überall vorhanden/anzutreffen, beherrschend* || governing * *bestimmend (Prinzip usw.), leitend* || determinant * *bestimmend, entscheidend; entscheidender Faktor* || dominant * *(be)herrschend, vorherrschend, entscheidend; emporragend, weithin sichtbar* || predominant * *vorherrschend, überwiegend, vorwiegend; überlegen, das Übergewicht innehabend*
Beherrschung   mastery * *z.B. Prozess, Technik, Sprache* || rule || domination || control * *auch des Marktes; self-control: Selbst~*
behindern   hinder * *aufhalten, hindern, verhindern (from: zu tun), abhalten (from: von), im Wege sein* || hamper * *(be-) hindern, hemmen, verstopfen* || handicap * *benachteiligen, behindern, belasten* || impede * *(be-) hindern, verhindern, erschweren* || restrain * *zurückhalten, i.Schranken halten, Einhalt gebieten; restrain s.o. from: jmd. abhalten von* || obstruct * *versperren, verstopfen, blockieren, behindern, nicht durchlassen, auch Sicht, Verkehr*
Behinderte   disabled persons
Behinderung   obstruction
Behörde   authority || government body * *Regierungs-/staatliche Behörde* || body || regulatory body * *Vorschriften erlassende ~*
behutsam   cautious || careful || gentle * *sachte*
bei   with * *z.B. Abteilung, Firma, Person* || in case of * *im Falle von, bei Eintreten von* || at * *z.B. Spannung, Drehzahl, Betriebsbedingung; at high frequencies: bei hohen Frequenzen* || under conditions of * *z.B. bei bestimmten Betriebsbedingungen* || in the case of * *im Falle des* || near * *(räumlich)* || close to * *(räumlich)* || next to * *(räumlich)* || at * *in Verbindung mit Institutionen und Personen* || during * *(zeitlich) während* || on * *(zeitlich)* || upon * *(zeitlich)* || on the occasion of * *gelegentlich des ...* || for * *für, bezogen auf*
bei Bedarf   when required || if required
bei einer Stückzahl von   for a quantity of
bei hohen Geschwindigkeiten   at high speeds
bei niedrigen Geschwindigkeiten   at low speeds
bei Null anfangen   start from ground zero
bei Teillast   under partial load condition
beibehalten   retain || maintain || keep up || adhere to * *Grundsatz beibehalten, sich halten an* || keep
Beiblatt   supplement sheet * *z.B. zu einer Norm/einer Betriebsanleitung* || supplementary sheet || supplement * *to: für*
beidseitig   at both ends * *an beiden Enden, z.B. Kabel, Welle* || on both sides || mutual * *→untereinander, gegen-/wechsel-/beiderseitig, gemeinsam*
beifügen   add || attach || annex || append || join * *to: an* || enclose * *einem Brief (with: in/bei)*
beige   beige || tan * *gelbbraun, hellbraun*
Beilaufpapier   separation paper * *Papierzwischenlage*

beilegen adjoin * *z.B. in eine Verpackung* || bundle with * *zusammenpacken mit*
beim Kunden at the customer's
beinhalten include * *einschließen* || comprise || incorporate * *eingebaut haben* || integrate * *in sich tragen*
Beipack enclosed * *in der Anlage beiliegend, beigefügt, im ~* || included loose * *lose im ~ beiliegend*
beipacken enclose * *loose: lose* || include || add * *hinzufügen* || pack-up with * *mit/dazu einpacken*
Beirat advisory council || advisory board
Beispiel example * *for example/e.g.: zum Beispiel; exemplified by/by the example of: am Beispiel von* || paradigm * *Paradigma, (Muster)Beispiel*
beispielhaft exemplary
beistellen supply * *sorgen für* || provide * *beliefern/versorgen (with: mit), z. Verfügung stellen, bereitstellen* || provide from own sources * *durch den Kunden usw.*
Beitrag contribution * *(allg.) auch zu einem Projekt, Geldbeitrag usw.* || subscription * *Mitgliederbeitrag* || share * *Beitragsanteil* || premium * *(Versicherungs-) Prämie* || paper * *zu Kongress, Fachzeitschrift usw.*
beitragen contribute * *to: zu*
Beiwert coefficient * *Koeffizient*
Beizanlage pickling plant * *z.B. für Blech* || staining plant * *{Holz}*
Beize mordant * *{Chemie}* || pickle || caustic * etching acid || etching solution || stain * *{Holz}* *Holz~, Färbemittel*
beizen pickle * *Metall* || stain * *{Holz| Textil}* *Holz beizen, Glas bemalen, Stoff färben; Flecken verursachen, beflecken* || mordant * *{Textil}* || bate * *Häute ~* || dip * *Metall ~* || sauce * *Tabak*
bekannt known * *to: jemandem* || well-known * *gut bekannt* || noted * *berühmt; for: wegen* || acquainted * *be acquainted with: mit etwas/jmd.bekannt sein* || familiar * *be familiar with: mit etwas bekannt/vertraut sein* || wide-spread * *weit verbreitet* || renowned * *nahmhaft, berühmt*
bekannt geben announce
Bekanntgabe announcement * *Verkündigung* || publication || notification || proclamation * *Verkündigung* || disclosure * *Verlautbarung* || communique * *(offizielle) Verlautbarung* || advertisement * *in den Medien*
bekanntmachen make known || introduce * *einführen, vorstellen (auch eine Person)* || establish * *z.B. ein Produkt/eine Marke am Markt* || report * *berichten* || notify * *anzeigen, melden, amtlich mitteilen* || give notice of || make public * *öffentlich* || publish * *öffentlich* || announce * *verkünden, ankündigen* || advertise * *durch Werbung, in der Zeitung*
Bekanntmachung announcement * *Verkündigung* || proclamation * *Verkündigung* || publication || notification || disclosure * *Verlautbarung* || communique * *(offizielle) Verlautbarung* || advertisement * *in den Medien*
Bekleidungsindustrie clothing industry
bekommen get || receive || win * *Auftrag, Preis* || be awarded * *Preis, Auszeichnung, Auftrag* || obtain * *Anstellung, Ware usw.* || acquire * *erwerben*
Beladeeinrichtung intake equipment * *z.B. für Chargenprozess* || charging system || filling system || loading equipment || loading system || loading facility
Belademaschine charging machine
Belag coat * *Überzug* || coating * *Überzug, Auftrag* || film * *Häutchen, Überzug* || lining * *z.B. Reib~, Brems~, Kupplungs~* || layer * *Schicht*
belastbar loadable || stressable * *beanspruchbar* || can be loaded

Belastbarkeit rating * *Nennleistung/-Strom/-Drehmoment* || load capacity * *Belastungsvermögen (zahlen/wertmäßig)* || load capability * *z.B. ~ eines Analog/Digital-Ausgangs* || stressability * *Beanspruchbarkeit* || loadability || load rating * *spezifizierte~ , zugelassene ~* || allowable load * *zulässige/maximale ~* || maximum load * *maximale ~* || loading capability || capacity * *z.B. thermal capacity: Wärme~*
belasten load || impose * *z.B. mit einer Pflicht, Steuer, Oberwellen ~* || stress * *beanspruchen* || set back * *finanziell mit einem Betrag von ... ~* || burden * *mit (Trag)Last/Bürde ~; burden someone with: jdn. mit etwas ~* || invoice * *Konto ~* || debit * *debit an account: ein Konto ~* || bring onto load * *auf die Last aufschalten (z.B. Motor)*
belastet loaded || stressed * *beansprucht* || on load * *unter Last*
belästigen molest || annoy || trouble * *stören* || pester * *plagen*
Belästigung molestation || annoyance || nuisance
Belastung load * *Last; (maximum) permissible: zulässige; per unit load: pro Flächeneinheit* || loading || duty * *das zu leistende, Nutzleistung* || burden * *Traglast, Tragfähigkeit, (auch finanzielle) Bürde, Ladung* || stress * *Beanspruchung* || drag * *(figürlich) Hemmschuh* || handicap * *Erschwernis* || strain * *Anstrengung* || charge * *finanzielle ~ (besonders Steuern, Zinsen, Abgaben) auch e. Kontos, Grundstücks* || burden * *financial: finanzielle ~* || drain * *finanzielle ~* || debit * *~ eines Kontos* || weight * *Gewicht*
Belastungs- load
Belastungsänderung variation in load
Belastungsänderungen load changes || load fluctuations * *Lastschwankungen* || variations in load
Belastungsbereich load range
Belastungsfaktor load factor * *{el.Masch.} kennzeichnet die erlaubte Schalthäufigkeit e. Motors*
Belastungsgrenze loading limit
Belastungskennlinie load characteristic || service characteristic * *{el.Masch.} →Betriebskennlinie, Belast.abhängigkeit v.Drehzahl, Strom usw.*
Belastungsklasse duty class * *Betriebsart für elektr.Maschinen (VDE 0530) u.Stromrichter (EN 60141, IEC 146-1-1)* || rating class * *Betriebsart eines Stromrichters (definierte Lastspiele), siehe DIN 41756*
Belastungsmaschine load machine || dynamometer * *mit Drehmomenten-Messeinrichtung (für Prüfstände usw.)* || loading machine
Belastungstest load test * *z.B. ~ für el. Maschine, Stromrichter* || stress test * *Beanspruchungsprüfung (mit Temperatur-/Schwingungsbeanspruchung)*
belastungsunabhängig load-independent
Belastungsverlauf load cycle characteristic * *zeitlicher Verlauf eines Lastzyklus, eines Lastdrehmoments usw.*
Belastungswiderstand load resistor
Belastungswinkel load angle
Belastungszustand load condition
Beleg voucher * *['wautscher] (Rechnungs-) Beleg, Quittung, Gutschein, Eintrittskarte, Verrechnungsscheck* || proof * *Beweis* || authentic record || document * *Unterlage* || receipt * *[ri'ßiet] Quittung*
belegen prove * *beweisen* || assign * *zuweisen, z.B. Speicherplatz ~, Klemme mit einer Funktion ~* || occupy * *besetzen, (für sich) beanspruchen, z.B. Speicherplatz* || line * *Kupplung, Bremse usw. mit Belag ~/auskleiden* || coat * *mit (Schutz-) Überzug usw. ~* || reserve * *einen Platz usw. ~* || enter one's name for * *Vorlesung usw. ~* || inscribe for * *Vorlesung usw. ~* || verify * *beweisen* || illustrate * *durch*

Beispiele ~ || exemplify * *durch Beispiele ~* || register * *sich eintragen lassen*
beleglose Bestellabwicklung   electronic ordering procedure
beleglose Bestellung   automatic ordering || electronic ordering || voucherless ordering || computerized ordering
Belegschaft   work force || personnel * *Personal* || staff
Belegung   assignment * *Zuordnung z.B. für Stecker-Kontakte, IC-Pins, Ein/Ausgänge usw.* || pinout * *~ von IC-Anschlüssen, Steckerstiften usw.* || designation * *Bezeichnung, Kennzeichnung, Bestimmung, z.B. von Klemmen* || allocation * *→Zuordnung* || occupancy * *Inanspruchnahme (von Raum, Förderkapazität usw.), Innehaben, Besitz(ergreifung)*
Belegungsplan   I/Q/F reference list * *Abk. f. 'Input/Output/Flag': Beleg.plan für Ein-/Ausgänge und Merker bei SPS* || input/output/flag reference list * *Belegungsplan für Ein-/Ausgänge und Merker bei SPS* || input/output/memory map * *Speicherbelegungsliste eines Rechners/einer SPS* || terminal connection diagram * *(Klemmen-) Anschlussplan* || terminal diagram * *(Klemmen-) Anschlussplan*
Belegungstabelle   reference list * *z.B. für Ein-/Ausgänge bei SPS* || memory map * *Speicherbelegungsliste bei Rechner/SPS*
beleimen   glue || gum * *gummieren*
Beleimgerät   glue applicator * *{Holz}*
Beleimmaschine   adhesive applicator * *{Holz}* || glue-applicating machine * *{Holz}*
Beleimung   gluing || adhesive application * *{Holz} Aufbringung v. Klebstoff, z.B. z. Furnier- oder Kantenanleimung*
Beleuchtung   lighting
Beleuchtungstechnik   lighting technology
Belgien   Belgium
Belichter   image setter * *{grafisches Gewerbe}*
beliebig   any * *z.B. ein Beliebiger* || arbitrary * *willkürlich* || random * *wahlfrei, wahllos, zufällig, Zufalls ..., aufs Geratewohl* || any given * *jedes beliebig herausgegriffene Ding* || at pleasure * *(Adv.)* || as desired * *(Adv.) wie es beliebt, nach Belieben*
beliebig oft   as often as required
beliebig viele   as many as you like
beliebt   popular * *with: bei* || favorite * *Lieblings-* || wide-spread * *weit verbreitet*
Beliebtheit   popularity
beliefern   provide * *with: mit*
Belohnung   reward || award * *Verleihung, Zuerkennung*
belüften   ventilate || aerate || vent
Belüfter   aerator * *z.B. bei Wasseraufbereitungsanlage*
Belüftung   ventilation || aeration || ventilation system
Belüftungsanlage   aerating plant || ventilation system
Belüftungskanal   ventilation duct
Belüftungsöffnung   ventilation opening
Belüftungsventilator   ventilation fan
bemaßen   dimension
Bemerkung   comment * *Kommentar; on: über* || remark * *on: über* || note
bemessen   dimension * *dimensionieren (for: nach)* || rate * *z.B. Stromrichter/Motor (bezüglich des Nennwertes) auslegen* || size * *Größe festlegen, nach Größe ordnen* || specify * *bestimmen, (i. einzelnen) festsetz., genaue Angaben machen, spezifizieren, (einzeln) angeben* || rate * *Leistung bemessen; abschätzen* || design * *(for: nach)* || calculate * *berechen, auslegen (for: nach)* || estimate * *abschätzen* || assess * *abschätzen, einschätzen, veranlagen, bewerten* || measure * *bewerten (by: nach)* || judge * *bewerten (by: nach)* || proportion * *bemessen (to: nach), anpassen, i.d. richtige Verhältn. bringen, verhältnismäßig verteilen* || time * *zeitlich bemessen* || dimensioned * *(Adj.)* || measured * *(Adj.)* || adjusted * *(Adj.)* || rated * *(Adj.)*
Bemessung   dimensioning * *Dimensionierung* || rating * *festgelegter Nennwert* || design
Bemessung für Maschinen mit mehreren Drehzahlen   rating of machines with several speeds * *{el. Masch.} es gibt mehrere Bemessungsdrehzahlen*
Bemessung für Maschinen mit veränderlichen Größen   rating of machines for variable values * *{el. Masch.} Bem.größen dürf. mehrere Werte annehm.*
Bemessungs-   rated * *Nenn-* || nominal
Bemessungs-Ankerspannung   rated armature voltage || design armature voltage * *eines Stromrichters*
Bemessungs-Ankerstrom   rated armature current || design armature current * *eines Stromrichters*
Bemessungsbetrieb   rated duty * *{el. Masch.} Betrieb mit allen vom Hersteller festgelegten Bemessungsgrößen*
Bemessungsbetriebsbedingungen   nominal operating conditions
Bemessungsdaten   rating * *continuous rating: Bemessungsdaten für Dauerbetrieb*
Bemessungsdrehzahl   rated speed
Bemessungsfeldstrom   rated field current || design field current * *eines Stromrichters*
Bemessungsgleichstrom   rated DC current
Bemessungsgröße   rated quantity || rated value
Bemessungsklasse   class of rating * *{el. Masch.}*
Bemessungslast   rated load * *at rated load conditions: bei*
Bemessungsleistung   rated power || rated output * *z.B. eines Motors, Stromrichters usw.*
Bemessungsluftspalt   rated air gap * *bei einer Bremse usw.* || nominal air gap * *bei einer Bremse usw.*
Bemessungslüftweg   nominal release travel * *bei einer Bremse*
Bemessungsmoment   rated torque
Bemessungsscheinleistung   rated apparent power
Bemessungsspannung   rated voltage * *für Motoren: siehe VDE 0530 und EN 60034-1* || design voltage
Bemessungsspannungsbereich   range of rated voltages * *{el.Masch.}*
Bemessungsstrom   rated current
Bemessungswert   rated value
Bemühung   effort
bemustern   supply samples * *of: mit* || sample || send samples to
Bemusterung   supplying samples || sending samples
benachbart   adjacent * *to: zu/an* || adjoining * *angrenzend, aneinander stoßend* || butted * *(stumpf) aneinander gefügt* || neighbouring
Benadelung   pinning
Benchmark   benchmark * *Vergleichstest*
benennen   name * *einen Namen geben* || call * *nennen, bezeichnen als, heißen (after: nach), halten für* || designate * *bezeichnen, (be)nennen; kennzeichnen* || denominate * *(be)nennen, bezeichnen* || term * *mit einem Fachausdruck ~* || fix * *einen Termin usw. ~* || lable * *bezeichnen, mit Namen/Symbol/Schild usw. versehen* || nominate * *ernennen*
Benennung   designation * *Bezeichnung* || denomination || name * *Name* || term * *Fachausdruck*
Benetzbarkeit   wettability
Benetzung   wetting
benötigen   require || need || want * *wünschen* || stand in need of || be necessary * *notwendig sein* || be in need of

benutzen use * share: gemeinsam ~ || employ || utilize * nutzbar machen
Benutzer user
Benutzeranleitung user's guide
benutzerdefiniert user-defined
Benutzereingriff user interaction || operator interaction * Bedienereingriff
Benutzerfreundlichkeit user convenience || user-friendliness || usability || ease of use
Benutzerhandbuch user manual || user's manual || instruction book || reference book * mit Beschreibung aller/Bezugnahme auf alle Funktionen
Benutzeroberfläche user interface
Benutzung use
Benzin gas * (amerikan.) || petrol * (brit.)
Benzinmotor gasoline engine || petrol engine
Beobachtbarkeit observability * {Regel.techn.}
beobachten observe || watch * genau ~ || keep an eye on || scan * dauernd ~, systematisch ~ || survey * genau betrachten, sorgfältig prüfen, mustern, überblicken, abschätzen, begutachten || monitoring * (Substantiv) || viewing * (Substantiv) || visualization * (Substantiv) Visualisierung || monitor * zur Überwachung ~
Beobachter observer * auch in der {Regel.techn.}
Beobachtermatrix observer matrix * {Regel.tech.}
Beobachtung monitoring || visualization * Visualisierung
Beobachtungsfunktionen monitoring function || visualization functions * Visualisierung, z.B. über Bildschirm
Beobachtungsparameter monitoring parameter || display parameter * Anzeigeparameter || read-only parameter * Nur-Lese-Parameter
Bepalettiermaschine pallet loading machine || palletizing machine * siehe auch →Palletiermaschine
bequeme Bedienung ease of operation
beraten advise || counsel || discuss * etwas beratschlagen
Berater advisor || consultant
Beratung advising * Beratung, Raterteilung, (An-)Empfehlung || discussion * Diskussion || conference * Besprechung || consultation * Besprechung mit e. Fachmann (with a person: mit jemandem) || advice * Rat, Raterteilung || counsel * Rat (to a person: jemanden beraten), Raterteilung || guidance * (Berufs-, Ehe- usw.) -Beratung || deliberation * Beratschlagung (on: über) || consideration * Beratschlagung (of: über) || technical advice * technische ~ || advisory service * als Dienstleistung, Dienststelle || consulting * Industrieberatung(sfirma) || consultancy service * (Industrie-) Beratung als Dienstleistung || technical support * technische Unterstützung || consultancy || support * Unterstützung
Beratungsfirma consultancy || consultant
berechenbar calculable
berechnen calculate || invoice * Kosten in Rechnung stellen, fakturieren || charge * in Rechnung stellen, Preis berechnen, (Konto) belasten || pass to a person's discount * jemandem in Rechnung stellen || compute || bill * jdm. einen Geldbetrag ~ || figure out * (amerikan.) →ausrechnen, →rauskriegen
berechnet calculated || invoiced * in Rechnung gestellt || charged * in Rechnung gestellt, belastet
Berechnung calculation || computation || estimate * (Ab)Schätzung || valuation * Bewertung || determination * Bestimmung || charge * in Rechnung stellen (Preis, Kosten) || quotation * Preisstellung || invoicing * in Rechnung stellen || debit * Belastung (eines Kontos) || design calculation * zur Auslegung/Dimensionierung e. Konstruktion/Schaltung usw. || computing * das Berechnen
Berechnungsbeispiel calculation example

Berechnungsfaktor calculation factor
Berechnungshinweise notes for calculation
Berechnungsunterlagen calculation documents
berechtigen authorize
Berechtigung authorization * Ermächtigung, Vollmacht, Genehmigung (to: zu) || right * (An)Recht (to: zu), (Rechts)Anspruch, Berechtigung || title * Rechtsanspruch, Rechtstitel, Recht, Berechtigung || warrant * Vollmacht, Befugnis || license * (amerikan.) von Behörde zugesprochene/offizielle ~ || qualification * Befähigung (for: zu) || eligibility * Befähigung (for: zu) || justification * Rechtfertigung || competence * Zuständigkeit || licence * (brit.) offizielle ~ || empowerment * →Befugnis, Ermächtigung || entitlement * Anrecht, Anspruch || validity * Rechtskraft, Rechtsgültigkeit, Gültigkeit, Gültigkeitsdauer || warranty * Berechtigung, Vollmacht, Rechtsgarantie
Beregnungsanlage sprinkling system
Bereich range * in the range from x to y/in the x to y range: im ~ von x bis y; low/high: klein/großer || area * (örtlich und figürl.) || field * z.B. Industrie~, Wirtschafts~, Branche, ~ im Markt, Anwendungs~ || branch * Branche || zone * {soun] Gebiet, Zone, Gegend, Raum, Bereich || location * Stelle, Platz || domain * Gebiet, Sphäre, Reich, Domäne || sector * Sektor, Teil || division * eines Unternehmens || section * →Abschnitt einer Maschine/Anlage usw.
Bereich-Anschaltung interface data area * {SPS} Datenbereich für Anschaltungen
Bereich-System system data area * {SPS} Datenbereich für Systemdaten
Bereichsleiter division director
Bereichsumschaltung range selection
Bereinigung settlement * siehe auch 'Fehlerbereinigung' || restoration of healthy conditions * (figürl.) || clearance * Räumung, Beseitigung, Leerung
bereit ready * be ready for: bereit sein zu/für
bereits already || previously * zuvor
Bereitschaft readiness * for: zu || stand-by mode * Betriebsart || standby
Bereitschaftsdienst on-call service
bereitstellen provide || prepare || keep ready * in Bereitschaft halten
Bergbau mining
Bergbaumotor mining motor
Bergfahrt uphill operation
Bergwerk mine
Bergwerks- mining || mine-type || underground mining * Untertage- || open-pit mining * Tagebau- || surface mining * Tagebau-
Bergwerksausrüstung mining equipment
Bericht report || minutes of meeting * Besprechungsbericht
Berichtigung correction
BERO BERO * Markenzeichen für einen 'BErührungslosen Näherungsschalter mit Rückgekoppeltem Oszillator' || proximity switch * Näherungsschalter || proximity limit switch * Endschalter || non-contacting limit switch * berührungsloser Endschalter || non-contact proximity detector * Berührungsloser Abstandssensor, 'Analog-BERO'
bersten burst || break * Eis, Glas usw || crack * einen Riss bekommen || explode * explodieren || detonate * explodieren, detonieren
Berstfestigkeit bursting strength
Berstschutz bursting protection
berücksichtigen take into account || allow for * auch 'in Betracht ziehen', z.B. e. Umstand || make allowance for * berücksichtigen, bedenken, im Budget reservieren || have regard to || have respect to || consider * beachten, i. Betracht ziehen, bedenken, erwägen, genehmigen || take into consideration * in Erwägung ziehen || bear in mind * beach-

ten, nicht vergessen || heed * beachten, Acht geben auf || take special account of * (einem Umstand) Rechnung tragen || regard * beachten, berücksichtigen, betrachten, respektieren; disregard: nicht ~
Berücksichtigung   consideration * be considered: ~ finden; in consideration of: unter ~ von || regard * with (due) regard to: unter (angemessener) ~ von; disregard: nicht berücksichtigen || view * Hinblick; with a view of: in Hinblick auf; in view of: unter ~ von
Berufsbezeichnung   job title
Berufsgenossenschaft   Safety Authority || Employer's Liability Assurance Association || professional association * Fachvereinigung || trade association * Wirtschaftsverband
berufsständische Organisation   professional organisation || trade association
beruhen   be based * basieren (on: auf), zur Grundlage haben
beruhen auf   be based on || depend on * abhängen von || be due to * zurückzuführen sein auf || be owing to * zurückführbar sein auf || be founded on * sich gründen auf
Beruhigung   stabilization || smoothing || damping * Dämpfung
berühren   touch || contact * z.B. mit der Hand, von Maschinenteilen untereinander usw. || affect * betreffen, (ein-) wirken auf, beeinflussen, beeinträchtigen, bewegen
berührgeschützt   protected against accidental contact
Berührung   contact || touching
berührungsfrei   non-contacting || contactless
berührungsgefährdete Teile   parts at hazardous voltage levels || components at hazardous voltage level
berührungsgeschützt   protected against accidental contact || safe-to-touch
berührungslos   no-contact * kontaktlos || contact-free || non-contacting || contactless
berührungsloser Näherungsschalter   proximity switch
Berührungsschutz   protection against accidental contact * siehe z.B. DIN VDE 0106 und 0113) || shock-hazard protection * gegen Personengefährdung/elektr. Schlag || shock protection * gegen Stromschlag || protection against contact with live or moving parts
Berührungsschutzmaßnahmen   protective measures against contact with moving or live parts
berührungssicher   safe-to-touch * siehe auch →berührungsgeschützt
Berührungssicherheit   shock-hazard protection * gegen elektr. Schläge
Beryllium   Beryllium
Besäumautomat   automatic edger * {Holz} || optimizing edger * {Holz}
besäumen   edge * {Holz} Kanten von Brettern/Platten gerade- bzw. glattschneiden || trim * {Holz} || square * (Holz)Platten usw.
Besäumer   edger * {Holz} zum Kantenabschneiden bei Holzplatten
Besäumkreissäge   edging circular saw * {Holz} zum Beschneiden der Brettkanten
Besäumsäge   edger * {Holz} || edging saw * {Holz}
Besäumschere   trimming shear * zum seitlichen Abschneiden der Blechränder in Blech-/Walzstraße
besäumt   edged * {Holz}
beschädigen   damage
beschädigt   damaged
beschädigt werden   incur damage
Beschädigung   damage * incur damage: beschädigt werden
beschaffen   provide * sich || make available * sich beschaffen, verfügbar machen || procure * auch

'sich beschaffen, jemandem etwas beschaffen' || constituted * gebildet, gestaltet, geartet || obtain * erlangen || secure * erlangen || furnish * liefern || supply * liefern || raise * z.B. Kapital || find * z.B. Arbeit, Kapital
Beschaffenheit   nature * Art, Sorte, Beschaffenheit, Naturt; Veranlagung, Charakter || character * Wesen, Natur, (Eigen-) Art, Merkmal, Kennzeichen, Eigenart || property * Eigenschaft || characteristic * charakteristisches Merkmal, Eigenschaft, Eigenschaft, Eigentümlichkeit, Kennzeichen || composition * Wesen, Natur, Anlage; (chemische) Zusammensetzung/Verbindung || structure * Struktur
Beschaffung   procurement || acquisition || purchase * Anschaffung, Einkauf
beschäftigt sein   be on the payroll * angestellt sein, auf der Gehaltsliste stehen
Beschäftigtenzahl   numbers employed
beschaltet   damped * Relais/Schützspule mit RC-Glied oder Diode zum Überspannungsschutz
Beschaltung   wiring * z.b. mit Bauelementen || circuitry || external circuitry || circuit
bescheinigen   certify * z.b. Erfüllung einer Norm ~ || attest || verify * beglaubigen || vouch for * beglaubigen || authenticate * amtlich ~ || receipt * z.B. mit Hilfe einer Quittung ~
Bescheinigung   certificate * Dokument, (Prüf)Zeugnis, Prüfbescheinung einer Abnahmestelle (z.B. TÜV, PTB) || certification * Vorgang des Bescheinigens || acceptance certification * über eine Abnahmeprüfung z.b. durch Prüfbehörde (z.B. TÜV, PTB)
beschichten   coat || laminate * (eine von mehreren) Teilschichten auftragen
beschichtet   coated || laminated
Beschichtung   coating || lamination * Schichtung, Auftrag (einer) von (mehreren) Teilschichten || facing * Außenschicht, Belag, Überzug || lining * Belag (Bremse, Kupplung usw.), Auskleidung, Überguss, Ausfütterung || deposition * Schicht, Belag, Ablagerung, Niederschlag, (Boden)satz, Aufdampfung || covering
Beschichtungsanlage   coating plant
Beschichtungsmaschine   coating machine
Beschickung   charging || feeder system * ~sanlage || feeding
Beschickungsanlage   charging equipment
Beschickungseinrichtung   charging unit * z.B. für Ofen; siehe auch →Einschieber, →Lademaschine || loading equipment
Beschickungssystem   charging system || infeed system
Beschlag   metal fitting || hardware * {Holz} (eingelassener) ~ an Möbeln usw. || band * an Kisten || clasp * Klammer, Haken, Schnalle, Spange, Schließe, Schloss; Umklammerung || fixture * Befestigs., fester Gegenstand, festes Zubehör, (Aufspann-/Halte-) Vorrichtg., Befestigung || absorb * mit ~ belegen, ganz in Anspruch nehmen
Beschlageinlassmaschine   hardware recessing machine * {Holz} in der Möbelindustrie
Beschlagsetzmaschine   fitting-insertion machine * {Holz}
beschleunigen   accelerate || ramp up * entlang einer Rampe || speed up * schneller machen/bearbeiten || accel * Kurzform für 'accelerate'
beschleunigt   speedy * zügig || accelerated || expeditious * schnell, zügig
Beschleunigung   acceleration * [m/s²] || acceleration rate * wertmäßig
Beschleunigungsaufnehmer   acceleration transducer || acceleration sensor || accelerometer
Beschleunigungsaufschaltung   acceleration feed forward * Beschleunigungsvorsteuerung || acceleration compensator * →Beschleunigungsausgleich

Beschleunigungsausgleich (als Einrichtg., Schaltg., Funktionsbaustein) || acceleration compensation
Beschleunigungsausgleich inertia compensation * Trägheitsmoment-Kompensation: Vorsteuerg. d. Beschleun./Verzög.moments || inertia precontrol * Trägheitsmoment-Vorsteuerung || acceleration compensation || acceleration compensator * als Schaltung, Funktionsbaustein, Einrichtung
Beschleunigungsband acceleration belt * bei Bandförderer || intermediate belt * bei Bandförderer
Beschleunigungskompensation inertia compensation * Vorsteuerg. d. Schwungmassen-abhäng. Beschleun./Verzögerungsmoments || acceleration compensation || inertia precontrol
Beschleunigungskraft force of acceleration || acceleration force
Beschleunigungsleistung acceleration power
Beschleunigungsmesser accelerometer
Beschleunigungsmoment acceleration torque || accelerating torque
Beschleunigungsstrom accelerating current
Beschleunigungsvorgang acceleration process
Beschleunigungsvorsteuerung inertia compensation * Trägheitsmoment-Vorsteuerung || inertia precontrol || acceleration compensation * →Beschleunigungsausgleich
Beschleunigungsweg acceleration distance
Beschleunigungszahl acceleration figure * {el. Masch.| Walzwerk} bei Rollgangsmotor
Beschleunigungszeit acceleration time || duration of acceleration
beschließen end * beenden, abschließen || close * abschließen || conclude * beschließen, beendigen, abschließen || finish * beenden || terminate * beenden || decide * sich entschließen, entscheiden || determine * bestimmen, entscheiden, sich entschließen, festsetzen || make up one's mind * sich ~; to do: etwas zu tun || pass * eine Vorlage ~, einen Vorschlag ~
Beschluss decision || resolution * Entscheidung || close * Ende, Beendigung, Abschluss || conclusion * Beendigung, Abschluss, Folgerung
beschneiden trim * z.B. Papierbogen, Papierbahn (seitlich)
beschnittene Breite cut width * {Papier}
beschränken restrict * to: auf || confine * confine oneself to: sich beschränken auf || limit || restrain * einengen || narrow * einengen || clip * beschneiden || cut * beschneiden
beschränkt confined || limited || restricted * to: auf || restrained * eingeengt || narrow * eingeengt
Beschränkung limitation || constraint || restriction
beschreiben describe * schildern, (Kreis, Bahn usw.) ~; indescribable/beyond all description: nicht zu ~ || give a description of * schildern || characterize || picture * anschaulich || relate * erzählend ~ || portray * anschaulich ~ || depict * darstellen, schildern, (anschaulich) ~, darstellen, (ab)malen || go into detail about * genau ~ || particularize * genau ~ || specify * spezifizieren, vorschreiben || trace * eine Bahn usw. ~
beschreibend descriptive
Beschreibung manual * Handbuch || instructions * Betriebsanleitung || operating instructions * Betriebsanleitung || instruction book * Handbuch || guide * Führer || description * Erläuterung, Beschreibung || descriptive information * reference manual * ausführliches/umfassendes Handbuch
beschriftet labelled
Beschriftung legend * z.B. an einem Messinstrument || labelling * z.B. auf einem Etikett || labeling * (amerikan.) auf einem Etikett || label * mittels Aufkleber || mark * Markierung || marking * z.B.

von Steckern, Kabeln, Klemmen usw. || lettering * z.B. Werbeaufschrift
Beschriftungsschiene labelling bar * z.B. bei einem Baugruppenträger
Beschriftungsschild legend plate * z.B.für Bedienelement
Beschriftungsstreifen lettering strip || marking label || labelling strip * z.B. bei einer Flachbaugruppe/eines Baugruppenträgers || labeling strip * z.B. eines Baugruppenträgers
beseitigen remove || eliminate * auch Fehler, Problem ~ || abolish * abschaffen || dispose of * Abfälle usw. ~ || redress * Übel ~ || remedy * Nachteil, Unrecht, Fehler, Schaden ~ || cure * Fehler ~ || clear away * Hindernisse, Fehler usw. ~ || overcome * Hindernisse, Problem usw. ~ || settle * Auseinandersetzng/Streit aus dem Weg räumen ~ remove * Gegner ~ || get rid of * Problem, Gegner ~ || liquidate, ausschalten || purge * säubern, reinigen, entleeren || rectify * (Fehler, Störung usw.) berichtigen, korrigieren, verbessern, richtigstellen, beseitigen
Beseitigung removal || disposal || elimination || liquidation * (figürl.) || redress || purge || remedy * Abhilfe, Behebung, (Problem-) Beseitigung; siehe auch →Fehlerbeseitigung
Besen broom
besetzen occupy * Markt, (Sitz-)Platz, Land usw || fill up * offene Stelle || man * durch Bedienungspersonal
besetzt manned * bemannt, mit Personal ~ || engaged * Telefon || busy * (amerikan.) Telefon
Besichtigung visit || sightseeing * Besichtigungsrundgang/-tour || examination * prüfende ~ || inspection * prüfende ~ || review * prüfende ~ || tour * Rundgang, Rundreise (of: durch)
Besitz property
besitzen possess * auch eine Eigenschaft ~ || have * eine Eigenschaft ~ || be in possession of || have * auch Eigenschaft, Talent usw. ~ || own * innehaben, beitzen; auch Eigenschaft, Talent usw. ~ || be endowed with * Talent, Gaben usw. ~ || be provided with * ausgestattet sein mit, verfügen über || be equipped with * ausgestattet sein mit, verfügen über
Besitzer possessor || holder * Inhaber, auch von Aktien, Kreditkarten, Rechten usw. || owner * Eigentümer; change hands: den Besitzer wechseln || proprietor * Eigentümer
besondere special
besondere Anforderungen special demands || special requirements
besondere Merkmale special features || characteristics of special note
besonderen Wert legen auf place considerable significance on
Besonderheit particularity || special feature * besonderes Merkmal || highlight * herausragendes Merkmal || feature * Merkmal
besonders special * speziell || particular || specific || exceptional * außergewöhnlich, Sonder- || exceptionally * (Adv.) ausnahmsweise, außergewöhnlich || particularly * (Adv.) in besonderem Maße || extreme * sehr, extrem || extensive * umfangreich, ausgedehnt; siehe auch 'umfassend'
besonders geeignet especially suited
Besprechung meeting || discussion * Gespräch || conference * Konferenz || talk * Gespräch || interview * Befragung || deliberation * Beratung || negotation * Verhandlung || review * Buchbesprechung, Zwischenbesprechung, Entwicklungsbesprechung
Besprechung abhalten hold a meeting
Besprechung einberufen convene a meeting || convene a conference

Besprechungsbericht minutes of meetings || meeting minutes || memorandum
Besprechungsprotokoll minutes of meeting * *Besprechungsbericht*
Besprechungsthema subject of conversation || topic of conversation
besser better * *all the better: umso besser* || improved * *verbessert* || superior * *über dem Durchschnitt stehend, überragend* || performing better * *in der Leistungsfähigkeit überlegen sein*
best best * *(Adj. und Adv.) best: am ~en; at best: im ~en Falle; in the best (possible) manner: ~ens* || optimum * *optimal, am ~en* || at the most * *(Adv.) im ~en Falle* || prime * *in prime condition: im ~en Zustand* || cream * *(Subst., salopp) the cream: das Beste*
beständig resistant * *widerstandsfähig (against: gegen)* || protected * *geschützt* || immune * *geschützt, unempfänglich, gefeit (from/against: gegen)* || proof || constant * *(an)dauernd, gleich bleibend* || steady * *ständig, stetig (auch Börse, Nachfrage usw.)* || unchanging * *unveränderlich* || lasting * *dauerhaft, lang andauernd* || permanent * *dauerhaft* || stable * *dauerhaft, stabil (auch Börse, Nachfrage)* || continual * *ausdauernd* || persistent * *ausdauernd, beharrlich* || persevering * *beharrlich, ausdauernd* || steadfast * *treu* || of resistant material * *be of resistant material: aus ~em Material bestehend*
Beständigkeit immunity || stability || resistance
Bestandsaufnahme inventory * *take: machen*
Bestandteil component || constituent * *Teil, Bestandteil, (physikal.)* Komponente, *Auftrag-/Vollmachtgeber* || part * *Einzelteil; essential part: wesentlicher ~* || constituent part * *(wesentlicher) Bestandteil*
bestätigen confirm || acknowledge * *den Empfang bestätigen, quittieren* || certify * *bescheinigen* || legalize * *amtlich* || validate * *für (rechts)gültig erklären,bestätigen (z.B.durch Prüfungsbehörde), rechtswirksam machen*
Bestätigung confirmation || acknowledgement
bestehen consist * *of: aus* || stand * *sich behaupten* || stand-up * *widerstehen, z.B.einer aggressiven Substanz* || comprise * *of: aus* || be composed * *zusammengesetzt sein (of: aus)*
bestehen aus comprise * *enthalten, umfassen* || consist of || be made of * *hergestellt sein aus (Material)* || be made up of * *sich →zusammensetzen aus, zusammengesetzt sein aus*
Bestellabwicklung order administration || processing of orders || order processing || order procedure || ordering procedure || contract administration
Bestellangaben ordering information || ordering data
Bestellannahme acceptance of orders
Bestellbeispiel ordering example || oder example || example for order
Bestellbezeichnung order designation
Bestelldaten ordering data
Bestelleingang order entry * *Zugang/Eintreffen von Bestellungen* || orders placed
bestellen order || give an order for || place an order for || book * *buchen (Platz, Zimmer usw.)* || reserve * *(amerikan.) reservieren (auch Zimmer)* || ask to come * *kommen lassen* || send for * *kommen lassen* || make an appointment with * *eine Verabredung treffen mit* || give * *Grüße usw.*
Besteller orderer || customer || buyer || purchaser * *Abnehmer, Käufer*
Bestellformular order form || order blank * *(amerikan.)*
Bestellhinweise ordering instructions || ordering information || order data
Bestellnummer order No || part No * *Artikelnummer, Ersatzteilnummer* || item No * *Artikelnummer* || catalog No * *Artikelnummer im Katalog* || order code
Bestellnummerergänzung order No suffix || order suffix || order number supplementary code
Bestellnummern order numbers || Order Nos. * *Abkürzg. f. 'Order Numbers'*
Bestellort ordering location || ordered from * *zu bestellen bei*
Bestellort ist ordered from
Bestellschein order form || order blank * *(amerikan.)* || purchase order * *Bestellung, ausgefülltes Bestellformular*
Bestellschlüssel ordering code
bestellt ordered || in hand * *vom Lieferer hereingenommener Auftrag*
Bestellung order * *Auftrag, Bestellung; give/place orders with: Bestellung bei ... aufgeben* || ordering * *Auftragserteilung, das Bestellen* || purchase order
Bestellung an send your order to
Bestellzettel order form * *Bestellformular* || order blank * *(amerikan.) Bestellformular* || purchase order * *ausgefüllter Bestellzettel, Bestellung*
bestenfalls at best || at the very best
bestens geeignet ideally suited || admirably suited
bestimmen determine * *festlegen, ermitteln* || govern * *maßgeblich sein für* || define * *pinpoint* * *genau festlegen* || quantify * *quantitativ ~*
bestimmend determining
bestimmt determined * *festgelegt* || certain * *gewiss; up to a certain limit: bis zu einer ~en Grenze* || bound * *for: zu/für; be bound to: dazu ~ sein, etwas tun zu müssen* || intended * *intended for: vorgesehen für* || specified * *festgelegt* || destined * *vorgesehen; be destined for: ~ sein für* || dedicated * *vorgesehen, gewidmet (für einen ~en Zweck)* || specific * *spezifisch, speziell, bestimmt, eigen (tümlich). besonders, wesentlich, definitiv, präzise*
bestimmte certain
bestimmter certain * *gewisser* || specific * *spezifisch, speziell, eigen(tümlich), typisch, kennzeichnend, besonderer*
Bestimmung regulation * *Vorschrift, Norm; legal/statutory regulation: gesetzliche ~* || specification * *Spezifikation, Vorgabe* || ascertainment * *Ermittlung, Feststellung*
bestimmungsgemäße Verwendung specified application
bestmöglich best possible || in the best possible manner * *(Adv.)*
Bestrahlung irridation || exposure to radiation || radiation treatment * *Strahlenbehandlung*
bestromen energize || power on
bestromt current carrying || current conveying || live * *Draht oder Leiter* || energized
bestücken mount
bestückt mounted || populated * *(salopp) ein vorbereiteter Einbauplatz, z.B. für Speicherbänke im Rechner* || equipped * *ausgerüstet, ausgestattet (with: mit)*
bestückte Leiterplatte assembled board
Bestückung insertion * *von Bauelementen auf einer Leiterkarte* || mounting || complement * *eines Baugruppenträgers, z.B. bei SPS* || tooling * *einer Werkzeugmaschine mit Werkzeug(en)* || assembling * *Bestücken, Einbau, Zusammenbau* || component set * *Satz von zum Einbau vorgesehenen Teilen* || placement * *→SMD-Bauteile auf Leiterkarte* || placing * *→SMD-Bauteile auf Leiterkarte* || equipping * *das Bestücken (z.B. eines Baugruppenträgers), Ausrüsten*
Bestückungsautomat PCB assembly machine * *für Leiterkarten* || pick-and-place machine || automatic pick-and-place machine || automatic component insertion equipment

Bestückungsplan component mounting diagram * z.B. für Leiterkarte
Bestückungsplatz component mounting position * für Bauelemente auf Leiterkarte || mounting position
Bestückungsseite component side * einer Leiterkarte
betätigen press * Taste || hit * Taste || depress * Taste || actuate * auch Taste, Bremse, Kupplung || manipulate * bedienen || operate * z.B. Schalter/Bremse; auch bedienen, laufen lassen,handhaben, einwirken (on/upon: auf) || control * steuern || set in motion * in Gang setzen || set going * in Gang setzen || participate * sich betätigen (in: an/bei) || act * sich betätigen (as: als) || work * sich betätigen (as: als) || energize * Relais, Klemme usw.
betätigt operated || actuated * hydraulically: hydraulisch; pneumatically: pneumatisch; electro-magnetically: elektromagnet.
Betätigung actuation || operation * Bedienung
Betätigungsart actuation method * z.B. bei Kupplungen
Betätigungseinheit actuating unit || actuation unit
Betätigungselement operator control element * zur Bedienung
Betätigungshebel operating lever
Betätigungsmagnet actuating magnet || actuating solenoid
Betätigungsmechanismus actuating mechanism
Betauung moisture condensation * Flüssigkeitskondensation || dew * Tau
Betauung nicht zulässig moisture condensation non-permissible * siehe auch →keine Betauung
Betauungsschutz protection against dew
beteiligen take part * sich beteiligen (in: an/bei) || participate * sich beteiligen, teilhaben (in: an/bei) || contribute * Beitrag leisten, sich beteiligen (to: zu/an) || cooperate * sich helfend beteiligen (in: an) || be concerned * beteiligt/betroffen sein (in: an) || have an interest * Geschäftsanteile haben (in: an) || have a share in * geschäftlich beteiligt sein (in: an) || share in profits * am Gewinn beteiligt sein || be involved in * verwickelt/involviert sein in
beteiligt involved
Beteiligung participation * in: auf/bei || partnership * geschäftliche ~, z.B. an einem Unternehmen || share * Anteil || interest * Anteil, Kapitalbeteiligung || involvement * das "Verwickelt-/Beteiligt-/Betroffen-Sein" || investment * durch Kapitalanlage || holdings * (Plural) durch Aktienbesitz || cooperation * Mitwirkung || support * Unterstützung (of: an/von) || contribution * (to: an/von) || attendance * Teilnehmerzahl
Beteiligungsgesellschaft associated company
Beton concrete
Betonmischer concrete mixer
betrachten consider * as: als || look at || have a look at || view * (auch figürl.) || inspect * genau ~ || examine * genau, prüfend ~ || observe * beobachten || watch * (aufmerksam) beobachten || reflect on * sinnend ~ || contemplate * sinnend ~ || regard * as: als || look upon * as: als || strictly speaking * genau betrachtet
beträchtlich considerable * (Adj.) auch 'erheblich, ansehnlich, bedeutend' || considerably * (Adv.) || important * wesentlich, bedeutend, wichtig || substantial * wesentlich (Fortschritt, Unterschied usw.), wesentlich, substanziell, namhaft || heavy * Kosten, Verluste || essential * wesentlich || severe * ernsthaft, schwer, schlimm (Verluste, Kritik usw.)
Betrag absolute value * Absolutwert || amount * Menge, Summe, Höhe, Geldbetrag, quantitative Größe || quantity * Menge || magnitude * Größe, Höhe, Betrag (auch eines Vektors) || amplitude * Amplitude
betragen amount to * wert/zahlenmäßig ~ || come to * zahlen/wertmäßig ~ || run up * zahlen/wertmäßig ~ || run up to || total * mehrere Posten zusammen; sich belaufen (to: auf) || aggregate * mehrere Posten || behave * (sich) benehmen (towards: gegenüber) || be of * zahlen/wertmäßig ~ || number * zahlenmäßig ~ || be * zahlen/wertmäßig ~ || be equal to
Betragsbildner absolute-value generator
Betragsbildung absolute-value generation
Betragskennlinie value characteristic
Betragsoptimum optimum of magnitude * Regleroptimierg. so das Betrag d.Amplitud.gangs gleich 1 über mögl. weit. Frequ.
beträgt is
betrauen entrust * with: mit; auch 'anvertrauen'
Betreff Ref * Abkürzung für 'reference': Abkürzung in Briefkopf
betreffen affect || be concerned * betroffen sein || pertain to || concern * angehen, interessieren, v. Belang sein für, beunruhigen; c. oneself with: sich befass. mit
betreffend concerning || involving * einbeziehend || in terms of || pertaining to || associated * zugeordnet || regarding || relevant || involved * betroffen
betreiben operate * Maschine, Anlage ~ (at: bei (e. Betriebspunkt, Frequenz usw.)) || run * Maschine, Anlage, Unternehmen, Verkehrslinie usw. ~ || work * Maschine, Anlage ~ || pursue * Beruf, Geschäft, Hobby, Studien, Angelegenheit usw. ~ || be engaged in * Beruf, Geschäft usw. ~ || manage * Unternehmen, Verkehrslinie usw. ~ || work for * hinarbeiten auf || aim at * hinarbeiten auf || practise * ausüben, (Beruf, Geschäft usw.) betreiben, tätig sein als/in, praktizieren || carry on * (Geschäft usw.) betreiben; fortführen, fortsetzen
Betreiber operator || user
betreuen care for || attend || support * unterstützen || take care of
Betreuung care * of/for: von/für || support * Unterstützung || looking after || supervision * Beaufsichtigung
Betrieb operation * Betreiben; out of op.: außer ~; start op.: - aufnehmen; go into op.: ~ in - gehen || mode * ~sart (auch 'duty': Lastspiel bei einer elektr. Maschine usw.) || management * Betrieb, Leitung, Verwaltung || working * Arbeit(sweise) || running * enterprise * Unternehmen || business * Unternehmen, Firma || firm * Firma || concern * Konzern || production unit * produzierender ~ || public utility * Versorgungs~ || factory * Fabrik (anlage) || manufacturing plant * Fabrikanlage || works * Werk(e) || workshop * Werkstatt, Fertigungs/Werkhalle, Werk, Betrieb || manufacture * Herstellungsgang || plant * Maschinen(anlage) || facility * Einrichtung || RUN * ~sart einer SPS, einer Maschine usw. (z.B. als Lampenbeschriftung) || service * Dienst (auch e.Maschine/Fahrzeugs), Betrieb (in: in; out of: außer)
Betrieb am 60 Hz-Netz operation from a 60 Hz supply
Betrieb am Drehstromnetz operating from a three-phase system
Betrieb am Netz direct on-line operation * Betrieb eines Motor ohne Stromrichter direkt am Netz || direct online supply
Betrieb am Umrichter converter operation * Motor || converter duty * Motor || converter-fed operation * eines Motors
Betrieb am ungeerdeten Netz operation from ungrounded supply * z.B. eines Stromrichters

Betrieb eines Motors am Umrichter operating a motor from a converter
Betrieb mit wechselnder Belastung varying duty
Betrieb mit wechselnder Last changing-load operation
Betrieb ohne Last no-load operation
Betrieb unter Last on-load operation
Betriebsart S3 duty type S3 * *{el.Masch.} periodischer Aussetzbetrieb* || intermittent periodic duty * *{el. Masch.} periodischer Aussetzbetrieb*
betrieben operated
betrieblich operational || internal * *intern* || company's * *innerbetrieblich*
Betriebs- operating * z.B. *Strom, Spannung, Bedingungen* || operational || working
Betriebs-Bezugsbedingungen reference serviceconditions
Betriebsablauf operational sequence
Betriebsanleitung operating instructions || instructions || manual * *Handbuch* || instruction manual * *Handbuch*
Betriebsanzeige operating display || operational display
Betriebsart operating mode || mode || mode of operation || duty class * *genormtes Lastspiel elektr. Maschinen lt. VDE 0530 (z.B.Kurzzeit/Dauerbetrieb)* || duty type * *genormtes Lastspiel el. Maschinen n.VDE0530/IEC34-1 u. v. Stromrichtern nach DIN41756* || rating class * *Belastungsklasse/ Betriebsart für Stromrichter, siehe DIN 41756* || duty * *Kurzform für duty class* || type of load * *Belastungsklasse* || duty cycle * *Lastpiel, Art des Lastspiels* || run mode
Betriebsart B12 duty type S12 * *{el.Masch.} wie →Betriebsart S3,jedoch Motor während Stillstandszeit unter Spannung*
Betriebsart S1 continuos running duty * *{el. Masch.} Dauerbetrieb* || duty type S1 * *{el. Masch.} Dauerbetrieb*
Betriebsart S10 duty type S10 * *{el. Masch.} Betrieb mit einzelnen konstanten Belastungen* || duty with discrete constant loads * *{el. Masch.}*
Betriebsart S2 duty type S2 * *{el. Masch.} Kurzzeitbetrieb* || short-time duty * *{el. Masch.} Kurzzeitbetrieb*
Betriebsart S4 duty type S4 * *{el. Masch.} periodischer Aussetzbetrieb mit Einfluss des Anlaufvorgangs* || intermittent periodic duty with starting * *{el.Masch.} period. Aussetzbetrieb mit Einfluss d. Anlaufs*
Betriebsart S5 duty type S5 * *{el. Masch.} Periodischer Aussetzbetrieb mit elektr.Bremsung* || intermittent periodic duty with electric braking * *{el. Masch.}*
Betriebsart S6 duty type S6 * *{el. Masch.} Ununterbrochener periodischer Betrieb* || continuousoperation periodic duty * *{el. Masch} Ununterbrochener periodischer Betrieb*
Betriebsart S7 duty type S7 * *{el. Masch.} Ununterbrochener periodischer Betrieb mit elektrischer Bremsung* || continuous-operation periodic duty with electric braking * *{el. Masch.}*
Betriebsart S8 duty type S8 * *{el. Masch.} ununterbrochener periodischer Betrieb mit Last-/Drehzahländerungen* || continuous-operation periodic duty with related load/speed changes * *{el. Masch.}*
Betriebsart S9 duty type S9 * *{el. Masch.} Betrieb mit nichtperiodischen Last- und Drehzahländerungen* || duty with non-periodic load and speed variations * *{el.Masch.}*
Betriebsartanwahl mode selection
Betriebsartenänderung change of operating mode || mode change

Betriebsartenschalter mode selector || mode switch || function selection switch
Betriebsartenwechsel switch-over of operating mode || change of operating mode * *Betriebsartenänderung*
betriebsbedingt operational * *betriebsmäßig vorkommend*
Betriebsbedingung operating condition || operation condition || service condition * *{el. Masch.} Betriebsverhältnisse, die der Festlegung d.Betriebsart zugrundeliegen* || environmental conditions * *(Plural) Umgebungsbedingungen* || site condition * *~ am Aufstellort, auf der Anlage*
Betriebsbedingungen working conditions
betriebsbereit ready || healthy || in standby mode * *in Bereitschaft stehend* || ready to go || ready-to-run || ready-to-start * *Motor* || ready for operation || operable * *siehe auch 'betriebsfähig'* || serviceable * →*betriebsfähig*
Betriebsbremse operational brake * *im Gegensatz zur →Haltebremse* || service brake
Betriebsdaten operating data || operation data || process data * *Prozessdaten*
Betriebsdatenerfassung industrial data acquisition || manufacturing data recording || operational data acquisition || production data acquisition || operating data acquisition
Betriebsdauer operational duration || duration of operation || running time
Betriebsdrehzahl operating speed || rated speed * *Nenndrehzahl z.B. eines Motors* || operation speed
Betriebsdrehzahlbereich operating speed range
Betriebsdruck operating pressure * z.B. *in Hydrauliksystem* || working pressure
Betriebseigenschaften operating characteristics
Betriebserfahrung field experience || operating experience
Betriebserfahrungen experiences gained in operation || practical experience || operating experience
Betriebserhaltung plant maintenance
betriebsfähig operational || in working condition || seviceable * *siehe auch* →*betriebsbereit* || fully functional
Betriebsfaktor service factor * *Betriebsbeiwert/ Servicefaktor: Verhältn. zuläss. Motordauerleistg./ Motornennleistg.* || overload factor
Betriebsferien business vacation || annual shut period
Betriebsfreigabe operational enable * *Steuerkommando* || operating enable * *Steuerkommando* || enable signal * *als Signal*
Betriebsfrequenz operating frequency
Betriebsgeräusch operating noise || running noise
Betriebsgüte performance value * *Produkt (Wirkungsgrad mal cos phi) eines Motors* || efficiency/ power-factor value * *Produkt (Wirkungsgrad mal cos phi) eines Motors*
Betriebshöhe altitude * *above sea level: über d. Meeresspiegel* || installation altitude * *above sea level: über dem Meeresspiegel* || site altitude
Betriebskapital working capital || stock in trade
Betriebskennlinie service characteristic * *{el. Masch.} Belastungabhängigkeit v. Drehzahl, Strom, Schlupf, Leistung usw.*
Betriebsklasse application class * *für Sicherung (siehe DIN VDE 0636)* || duty class * *siehe →Betriebsart (nach VDE0530/IEC 34-1)*
Betriebsklima working atmosphere
Betriebskondensator running capacitor * *für Einphasenbetrieb eines Drehstrommotors (Steinmetzschaltung)* || motor-run capacitor * *permanent-split capacitor * *Phasenschieberkondensator z.Betrieb e. Drehstrommotors an 1-phasigem Netz* || motor operating capacitor || running capacitor

Betriebskontrolle operation check
Betriebskosten operating expenses || operating costs || running costs
Betriebslast running load * *eines Motors* || working load || consumable load
Betriebsleitung plant management || works management * *Werksleitung*
betriebsmäßig actual * *tatsächlich auftretend* || operational * *den Betrieb betreffend, während des Betriebs auftretend* || in operation
Betriebsmittel equipment || device || manufacturing equipment || factory equipment
Betriebsmittelkennzeichnung equipment designation || equipment identification
Betriebsparameter operating parameter
Betriebspause shutdown period
Betriebsprotokoll operation log * *für eine Produktionsanlage/Maschine*
Betriebspunkt operating point || working point
Betriebsquadrant operating quadrant
Betriebsrat worker's council
Betriebsschaltung running connection * *für Motor/Trafo, z.B. Stern- od. Dreieckschaltung* || operating connection * *z.B. Stern-, Dreieckschaltung*
Betriebssicherheit safety in operation || reliability * →*Zuverlässigkeit* || reliability in operation * *Zuverlässigkeit* || operational reliability * *Zuverlässigkeit* || operational safety || operational dependability
Betriebsspannung operating voltage || service voltage * *z.B. eines Netzes*
Betriebsstellung operating position
Betriebsstoff working material
Betriebsstörung stoppage || breakdown || interruption of service * *Stockung/Unterbrechung des Betriebs* || failure || down time * *Stillstandszeit*
Betriebsstrom on-load current
Betriebsstunden operating hours || running hours || service hours || hours run
Betriebsstundenzähler hour counter || hours-meter || hours-run meter || elapsed-hour meter || elapsed-time meter || elapsed-time counter
Betriebssystem operating system * *Software. die den Ablauf u.die Resourcen-Zuteilung für Anwenderprogramme verwaltet* || OS * *Abkürzung für 'Operating System'* || operating environment * *Umgebung*
Betriebssystem-Speicherbereich system data memory * *bei SPS*
Betriebssystemschnittstelle API * *{Computer}* Application Program Interface: Schnittst. z. Anwenderprogr.* || application program interface * *{Computer} zum Anwenderprogramm* || application-specific interface * *{Computer}*
betriebstechnisch operational
Betriebstemperatur operating temperature || working temperature
betriebstüchtig operational * *siehe auch* →*betriebsfähig* || reliable * *zuverlässig*
Betriebstunden operating hours
Betriebsüberwachung operating control || plant supervision
Betriebsumrichter main converter * *Übernimmt nach d.Hochfahren; siehe auch* →*Synchronisierzusatz,* →*Hochfahrumrichter*
betriebsunfähig out of order
Betriebsunterbrechung interruption of operation || shutdown || downtime * *Stillstandszeit einer Maschine/Anlage (auch auf Grund eines Fehlers)* || operation stoppage || interruption of service
Betriebsunterhaltung plant maintenance
Betriebsverhalten operating characteristics || performance || operational behaviour
Betriebsverhältnisse plant conditions || service conditions || operating environments

Betriebsvermögen operating assets * *Betriebskapital, betriebsnotwendiges Kapital* || movable property * *bewegliches Vermögen (Maschinen, Geschäftsausstattung usw.)* || movables * *bewegliches Vermögen (Maschinen, Geschäftsausstattung usw.)*
betriebswarm in the warm state || at operating temperature || under hot running condition * *Verbrennungsmotor usw.*
betriebswarmer Zustand warm condition || at rated-load operating temperature
Betriebswert operating value || operating specification * *z.B. Nennwert bei Bemessungsbedingungen*
Betriebswerte operating characteristics || characteristics
Betriebswirt business economist
Betriebswirtschaft applied economics || managerial economics || science of business * *Betriebswirtschaftslehre* || business administration
Betriebszeit operating period
Betriebszustand operating status * *(Plural: operating statuses)* || condition of operation * *Betriebsbedingungen (auch an einem bestimmten Arbeitspunkt)* || operating condition * *in: in* || operating point * *Betriebs/Arbeitspunkt* || condition || operating state || operational status * *einer Maschine/Anlage*
Betriebszustände operating statuses
Betriebszyklus cycle of operation
Betrifft Subject * *'Thema' im Kopf eines Berichts, Briefes*
betroffen concerned * *berührt, beteiligt, verwickelt* || affected * *auch 'beeinflusst, beeinträchtigt'* || respective * *jeweilig* || relevant * *relevant, einschlägig* || afflicted * *belastet, behaftet, geplagt, leidend* || visited * *by: von; heimgesucht, befallen* || stricken * *with: von; heimgesucht, schwer betroffen* || shocked * *bestürzt* || startled * *bestürzt, erschrocken* || taken aback * *verblüfft, überrascht* || involved * *beteiligt, verwickelt, berührt, involviert (in/with: in/an)*
Bettfräsmaschine bed-type milling machine
Beugungsachse deflection axis
Beugungswinkel diffraction angle * *{Optik}* || angular misalignment * *{Masch.bau} Fluchtfehler bei der Ausrichtung von Wellen usw.*
beurteilen judge * *by: nach; of: über; misjudge: falsch beurteilen* || criticize * *fachmännisch* || comment on * *kommentieren, Meinung abgeben* || discuss * *besprechen, diskutieren* || rate * *Leistung (e. Mitarbeiters, von Produkten im Vergleich usw.), bewerten, (Arbeits)Qualität* || estimate * *abschätzen* || assess * *(ein)schätzen (z.B. Schaden), bewerten, veranschlagen, (z.B. Steuer) festsetzen* || view * *betrachten*
Beurteilung judgement * *auch 'Urteil, Ansicht'; in my judgement: meines Erachtens* || opinion * *Meinung; of: über* || review * *kritische Besprechung, Rezension (z.B. eines Buches), Kritik* || assessment * *Einschätzung, Bewertung* || rating * *Leistung (auch eines Mitarbeiters), Wert* || confidential report * *in Personalakte* || efficiency report * *Leistungsbeurteilung in Personalakte* || view * *Ansicht, Auffassung, Urteil, in my view: meines Erachtens* || judgment * *siehe 'judgement'* || evaluation * *Abschätzung, Bewertung, Beurteilung, Auswertung*
Beutel bag || sack * *(amerikan.) Sack, Beutel, Tüte*
Beutelmaschine bags packaging machine * *{Verpack.techn.}* || bag filling machine * *{Verpack. techn.} Beutelfüllmaschine*
bevorrechtigt privileged
bevorzugt favoured || favored * *(amerikan.)* || preferred * *bevorzugt, Vorzugs-* || preferable * *vorzuziehen(d), to: gegenüber* || preferably * *(Adv.) vor-*

zugsweise, lieber, am besten || by preference * (Adv.) mit Vorliebe, begünstigt || chiefly * (Adv.) hauptsächlich || mostly * (Adv.) hauptsächlich || preferential * bevorzugt, bevorrechtigt, Vorzugs- || preferentially * (Adv.) vorzugsweise
bewähren prove * they have proven themselves: sie haben sich bewährt
bewährt proven || well-proven * gut bewährt || highly-proven * sehr bewährt || field-proven * im breiten (Feld)Einsatz bewährt, →felderprobt || tried-and-tested || well-established * gut eingeführt, eine Referenz darstellend, unzweifelhaft
bewegen move * auch sich; freely; frei; to and fro: hin und her || stir * unruhig; auch 'sich bewegen' || set in motion * in Bewegung setzen || set going * in Bewegung setzen || carry * befördern,fördern || convey * fördern || fluctuate * schwanken, z.B. Preise || vary * sich ändern || range * in einer Spanne (z.B.Preise) || induce * jemanden/etwas zu etwas bewegen || travel * sich (hin- und her-) bewegen, laufen; auch Maschinenteil, z.B. Kolben || ascend * nach oben || descend * nach unten
beweglich movable || moving || mobile * z.B. auf Rädern || flexible * elastisch, beweglich (auch figürl.) || portable * tragbar || agile * (figürl.) behände, wendig, beweglich, agil || nimble * (figürl.) flink, hurtig, gewand, behend, nimble mind: rasche Auffassungsgabe, ~er Geist || versatile * (figürl.) wendig, vielseitig
bewegliche Teile moving parts
bewegliches Vermögen movable property * Maschinen, Geschäftsausstattung usw. || movables * Maschinen, Geschäftsausstattung usw.
Beweglichkeit mobility * auch 'Mobilität'
bewegtes Ziel moving target
Bewegung motion || movement || move * mit einer bestimmten Absicht || stir * unruhige ~, Aufregung, Aufruhr; Betriebsamkeit, reges Treiben || jerk * ruckweise || trend * Tendenz || travel * eines Maschinenteils, auch 'Lauf, Hub'
Bewegungsablauf sequence of movement || sequence of motion || motional process || motional action || motion process
Bewegungsenergie kinetic energy
Bewegungsgleichung equation of motion
Bewegungssteuerung motion control || motion controller * als Einrichtung/Gerät
bewehrtes Kabel armoured cable
Bewehrung armour * z.B. für Kabel || reinforcement
Beweis proof * of: für || evidence
beweisen prove || show || evidence * zeigen, beweisen || verify * verifizieren, nachweisen || demonstrate * demonstrieren
Beweismittel evidence
bewerben apply * for a job: sich um eine Stelle; to a person: bei jdm. || have an interview * ein Bewerbungsgespräch führen
Bewerber applicant * for: für/um || candidate || aspirant * to: für/um || bidder * bei einer Ausschreibung || competitor * Mitbewerber bei einer Ausschreibung
Bewerbung application * for: um || candidature * for: um || interview * Bewerbungs-/Vorstellungsgespräch
bewerkstelligen manage || accomplish || realize || bring about * by: mittels/durch || effect || contrive || engineer || bring off || make
bewerten rate * klassifizieren || grade * einstufen, klassifizieren || assess * abschätzen || estimate * abschätzen || appraise * abschätzen || scale * skalieren || normalize * normieren || value * (allg. at: auf; by: nach) || price * preislich || evaluate * abschätzen, berechnen, beurteilen, bewerten, auswerten ||

benchmark * bezüglich der Leistungsfähigkeit/Messgrößen im Vergleich zu anderen ~
bewertet weighted * gewichtet || valued * at: auf, by: nach || estimated * abgeschätzt
Bewertung valuation * (Ab)Schätzung, Wertbestimmung || estimation * (Ein)Schätzung, Meinung, Urteil || assessment * (Wert)Einschätzung, Bewertung (auch figürl.), (Schadens)Feststellung || rating * (Ab)Schätzung, Beurteilung, (Zeugnis)Note, Leistungsbeurteilung e.Person || judgement * Urteil, Beurteilung, Ansicht, Einschätzung || scaling * eines Signals mittels Anpassungsfaktor || intervention * Eingriff eines Signals (z.B. Zusatzsollwert) || evaluation * Abschätzung, Bewertung, Beurteilung, Auswertung || weighting * Gewichtung || appraisal * (Ab)Schätzung, Taxierung, Schätzungswert, Bewertung || judgment * siehe 'judgement' || benchmarking * der Leistungsfähigkeit (im Vergleich)
Bewertungsfaktor evaluation factor || scaling factor * Skalierfaktor || intervention factor * Eingriff-Faktor
Bewetterung ventilation * Belüftung im Bergbau; mine ventilation: Gruben~
bewilligen grant
bewirken effect || ensure * sicherstellen || cause * veranlassen,verursachen || result in * als Resultat haben, resultieren in
bewusst conscious * be conscious of: sich einer Sache ~ sein || beware * be beware of: sich einer Sache ~ sein || known * bekannt || deliberate * überlegt, wohlerwogen, bewusst, absichtlich, vorsätzlich; besonnen, vorsichtig || intentional * absichtlich || sensible * bei Bewusstsein; be sensible of: sich einer Sache ~ sein; fühl-/spürbar, vernünftig
bezahlt machen pay for itself * sich ~
bezeichnen term * mit einem Wort/Ausdruck || class * klassifizieren || designate * mit Namen, Markierung (auch Klemme, Kabel usw.) || mark * mit Markierung, Marke
Bezeichner designator
Bezeichnung marking * Kennzeichnung, z.B. Anschlussbezeichnung an einer Klemme || term * (Fach-)Ausdruck || notation * Aufzeichnung, Schreibweise, Notierung || name * Name || designation * Name, Kennzeichnung || naming * Namensgebung || identifier * Kennzeichen, Kennung
beziehen have reference to * sich ~ auf || pertain * to: betreffen, gehören zu || relate to * sich ~ auf, z.B. auf eine Nenngröße || move into * Gebäude, Wohnung, Stellung usw. ~ || obtain * Ware ~ || procure * Ware ~ || buy * Ware ~ (from: von) || suscribe to * z.B. Zeitschrift ~ || draw * Gelder, Gehalt, Einkommen, Energie aus dem Netz usw. ~ || connect with * ~ auf || apply to * ~ auf || refer to * (sich) ~ auf || use a person's name as a reference * sich auf jemanden ~
beziehen auf apply to || refer to * sich ~ || relate to
Beziehung reference || relation * between: zwischen; to: zu/mit || connection * with: zu Personen, Firmen || relationship * between: zwischen; to: zu/mit || respect * in this/some/every respect: in dieser/mancher/jeder ~ || concern * Geschäft, Unternehmen; Beziehung; have no concern with: nichts zu tun haben mit
beziehungsweise respectively * nachgestellt || resp. || or || rather * eher
Bezirksvertreter local representative * örtlicher Vertreter
bezogen related * to: auf || referred * to: auf || per-unit * auf eine Nenngröße/Bemessungsgröße || specific * spezifisch || percentage * →prozentual || percental * →prozentual
bezogen auf corresponding to || as compared with * verglichen mit || related to || referring to * Bezug

haben auf || referred to || with respect to * *bezüglich* || relative to || with reference to
bezogene thermische Lebensdauererwartung relative thermal life expectancy * *{el.Masch.}Lebensdauerveränderg. b.veränd. Betriebsbeding.*
bezogenes Drehmoment per-unit torque
Bezug reference * *['referens] (figürlich) Bezugnahme* || purchase * *Erwerb von Ware* || procurement * *von Ware* || supply * *von Ware* || cover * *Überzug* || covering * *Überzug* || regard * with regard to: in *Bezug auf*
Bezug nehmen auf refer to * *verweise/hinweisen auf* || make reference to || reference to * in *Dokument/Buch Verweise anbringen auf*
bezüglich relative to * *relativ zu* || regarding * *bezogen auf, wenn man ... betrachtet* || with respect to || with regard to || in terms of * *im Sinne von* || as far as * *so weit ... betrifft, betreffend* || concerning * *betreffend* || relating to
Bezugs- reference * *reference line: Bezugslinie; reference frequency: Bezugsfrequenz*
Bezugsfrequenz reference frequency
Bezugsgröße base value || reference quantity || reference variable * p.u. quantity * *diejenige Größe, die das "Einheitsnormal" darstellt* || reference value
Bezugsklemme reference terminal * *0V* || ground terminal * *Massepotenzial*
Bezugsmaßsystem absolute measuring system * *Fahren auf absolute (immer auf d.gleichen Wegnullpunkt bezogene) Position*
Bezugspotenzial reference potential * *z.B. für ein(e) Signal(klemme)* || ground * *Masse* || GND * *Abkürzg. für 'GrouND'*
Bezugsprofil basic profile * *einer Verzahnung* || reference profile * *einer Verzahnung*
Bezugspunkt reference point * *z.B. für Potenzial/Signal* || specified point
Bezugsquelle source of supply
Bezugsquellenverzeichnis list of suppliers
Bezugsschalldruck reference sound pressure * *{el.Masch.} Eff.wert p0 d. Schallwechseldrucks an d.Hörschwelle bei 1 kHz*
Bezugsschalleistung reference sound power * *{el. Masch.} Bezugsgröße P0 für Schalleistungen (10^-12 W)*
Bezugssignal signal common * *gemeinsames Bezugssignal, "Wurzel"* || reference signal
Bezugsspannung reference voltage * *Bezugspotenzial* || common ground * *gemeinsame Masse* || reference voltage level * *Bezugspotenzial* || reference potential
Bezugswert reference value || per unit value * *auf den sich andere Angaben beziehen (Bezugswert entspr. 1 oder 100%)*
Bezugszeitpunkt reference instant
BG frame size * *Motor-Baugröße, Achshöhe* || motor frame * *Motor-Baugröße, Achshöhe* || size * *Sicherung, Trafo usw.*
Bibliographie bibliography
Bibliothek library * *auch von Softwaremodulen (Plural: libraries)*
bichromatisiert bichromated
bidirektional bi-directional || bidirectional
Biegefestigkeit bending strength || bending resistance
Biegekraft bending force
Biegemaschine bending machine || bender
Biegemoment bending moment || bending torque
biegen bend
Biegeradius bending radius || bend radius
Biegeschwingungen bending vibrations
Biegesteifigkeit bending resistance || flexural stiffness || flexural strength
Biegewiderstand bending resistance

biegsam flexible || pliable
biegsame Welle flexible shaft
Biegung bend * siehe auch →*Krümmung* || bending || curve * *Kurve, Wölbung* || turn || turning || curvature * →*Krümmung* || flexure * *Krümmung* || deflection * *Biegung, Krümmung, Faltung* || arch * *Wölbung* || sagging * *Durchhang, Durchbiegung*
bieten offer * *anbieten, bereitstellen* || afford * *gewähren, liefern* || give * *schenken, geben, (dar)bieten, gewähren, liefern* || provide * *sorgen für, bereitstellen, z. Verfügung stellen, (be)liefern (with: mit), befriedigen* || render * *möglich machen, ermöglichen, wiedergeben*
bifilar gewickelt double-wound
bifilare Wicklung double-wound winding || bifilar winding
Bilanz balance || balance sheet * *Aufstellung* || result * *(figürl.)*
bilanzieren balance * *{Wirtschaft} Konto ausgleichen, aufrechnen, saldieren, abschließen*
Bild picture * *allg.* || figure * *Abbildung (in Dokument)* || illustration * *Illustration, Abbildung* || image * *Abbild, optisches Bild* || sketch * *Skizze* || diagram * *Diagramm, grafische Darstellung* || graph * *grafische Darstellung, Schaubild, Kurve(ndarstellung), Kurve* || chart * *Diagramm, Schaubild, Tabelle, Kurve(nblatt), Karte, Skizze* || photo * *Foto* || photograph * *Fotografie* || draft * *Skizze* || drawing * *Zeichnung, Skizze* || rough draft * *Grob-/Entwurfsskizze* || rough drawing * *Roh-/Grobskizze* || hardcopy * *z.B. ~ von Oszillografenkamera, Computerausdruck d. Bildschirminhalts* || plot * *aufgezeichnete Messkurve* || screen * *Bildschirm~*
Bildaufbau display buildup || display building * *manuelle Erstellung eines Bildschirmbildes* || picture formatting || display construction * *manuelle Erstellung eines Bildschirm-Bildes* || screen structure * *Anordnung der Grafikelemente auf d. Bildschirm*
Bildbereich complex variable domain
bilden form || make-up || establish || generate * *erzeugen* || constitute * *ausmachen, bilden, ernennen, einsetzen, in Kraft setzen, einsetzen*
Bilderkennung automatic image recognition * *automat. Bilderkennung* || automatic imaging || imaging || machine vision analysis || automatic pattern recognition * *automatische Mustererkennung* || automatic picture recognition || picture recognition
Bilderneuerungsfrequenz frame rate * *Häufigkeit d. Ausgabe bewegter Videobilder (flüssige Bewegg. ab 25 frames/s)*
bildgebend imaging
bildlich graphic || pictorial * *bildmäßig, Bild-, Bilder-*
Bildpunkt pixel * *{Computer} z.B. auf einem Monitor*
Bildqualität image quality * *eines Computer-Monitors*
Bildröhre cathode-ray tube * *Kathodenstrahlröhre* || CRT * *Abkürzg. f. 'Cathode-Ray tube': Kathodenstrahlröhre* || display tube
Bildschirm screen || display * *Anzeige* || monitor screen || monitor || display monitor
Bildschirmanweisung on-screen instruction
Bildschirmflimmern screen flicker
Bildschirmfoto screen shot * *Hardcopy eines Monitorbildes*
Bildschirmgerät video display unit || VDU * *Abk. für 'Video Display Unit'* || CRT unit * *Abk. für 'Cathode-Ray Tube': Kathodenstrahl-Bildröhre*
Bildschirmmaske screen form
Bildschirmprogrammiergerät CRT-based programmer * *für SPS*
Bildschirmseite screen page || screen

Bildschirmspeicher video memory
Bildung education * *durch Schule/Hochschule* || development * *Entstehung, Entfaltung* || generation * *Erzeugung* || forming * *Formung*
Bildunterschrift caption || legend * *Legende, Bildinschrift*
Bildwiederholfrequenz refresh rate * *Bilder/sec bei Monitor (ab ca. 70 Hz kein Flimmern mehr sichtbar)* || image refresh rate || frame rate * *softw.- mäßige Ausgabehäufigkt. v.Videobildern; flüssige Bewegg. ab 25 frames/s* || vertical frequency
Bildwiederholspeicher refresh memory || picture refresh memory || refresh buffer
billig cheap || low-cost * *siehe auch →preisgünstig* || low-priced || inexpensive
billigen agree to || approve || approve of || consent to || sanction * *sanktionieren, billigen (auch nachträglich), gutheißen, dulden, bindend machen* || condone * *stillschweigend billigen, verzeihen, (Fehltritt usw.) entschuldigen* || accept * *akzeptieren, anerkennen, gelten lassen, bejahen, hin/annehmen, sich mit ... abfinden* || assent to * *zustimmen, billigen, beipflichten, genehmigen* || authorize * *gutheißen, billigen, genehmigen*
Bimetall bimetal
Bimetall- bimetallic
Bimetallauslöser bimetallic release || bimetallic trip
Bimetallrelais bimetal relay
Bimetallschalter bimetal switch * *z.B. zum thermischen Motorschutz* || bimetallic-element switch * *z.B. zum thermischen Motorschutz* || bimetallic trip * *Bimetallauslöser* || bimetal release * *{el. Masch.} therm. Schutzschalter*
Bimsstein pumice block * *z.B. zum Reinigen des Kommutators bei Gleichstrommaschine* || pumice
binär binary || boolean * *boolsch*
Binäranweisung binary statement * *z.B. für SPS*
Binärausgabe binary output
Binärausgang binary output || digital output || status output * *für (Betriebs-) Zustandssignal*
Binärcode binary code
binäre Funktion binary function || logic function || binary operation * *SPS-Befehl*
Binäreingabe binary input
Binäreingang binary input || digital input || logic input || command input * *für Steuerbefehl*
binärer Beobachter binary observer * *{Regel.-techn.}*
Binärfunktion binary function || logic function || binary operation * *SPS-Befehl*
Binärgröße binary value || binary variable || logic variable || binary quantity
Binärsignal binary signal
Binäruntersetzer binary scaler
Binärvektor binary vector
Bindeglied intermediate element || link
Bindemaschine binding machine * *Walzwerk*
Bindemittel binding agent * *{Chemie}* || bonding agent || cementing agent
binden link * *z.B. Softwaremodule zusammen~ ("Befriedigung" der externen Adressbeziehungen)* || bond * *ab~ (Lack, Harz, Klebestelle usw.)* || tie * *(figürl.) hemmen, fesseln, lahm legen, (Produktion usw.) stillegen, (Resourcen) blockieren* || truss * *bündeln, schnüren (up: fest), zusammen~* || bind * *(an-/um-/fest)binden, Bücher (ein)~, binden {Chemie}, verpflichten (figürl.)* || cord * *mit Stricken ~* || pack * *Ballen usw. ~* || bundle * *bündeln* || tie up * *Geldmittel ~*
bindend binding || obligatory * *auch verpflichtend, (rechts)verbindlich, obligatorisch*
Binder linker * *zum Zusammenlinken von Softwaremoduln ("Befriedigung" der gegenseitigen Adressreferenzen)*

Bindestrich hyphen * *auch 'mit Bindestrich versehen'; hyphenated: mit Bindestrich versehen* || dash * *Gedankenstrich*
Binnenmarkt domestic market
biologisch biological
biologisch abbaubar bio-degradable
BIOS BIOS * *'Basic Input Output System': D. Betriebssystem unterlagerte →Firmware z. Ansprechen d. Hardware*
bipolar bipolar
bis up to * *maximal, bis hinauf zu* || not exceeding * *nicht größer als* || to || until * *(zeitlich)* || up until * *(zeitlich) bis (zu)* || up to and including * *bis einschließlich; z.B. Versionsnummer, Baugröße* || down to * *bis herunter zu* || thru * *(salopp) bis einschließlich*
bis auf except * *außer* || with the exception of
bis auf weiteres until further notice || for the time being * *momentan* || until you are informed otherwise * *(an jemanden gerichtet) bis Sie weitere Nachrichten erhalten*
bis einschließlich through || thru * *(amerikan.)* || up to and including
bis zu up to
bisher up to now || till now || hitherto || so far * *bis jetzt* || thus far || as in the past * *wie bisher* || previously * *vorher, früher; than previously: als bisher* || until now || up until now * *bis jetzt*
bisherig former || present || prevailing || hitherto existing || past * *ehemalig, vergangen* || previous * *vorherig*
bisherige Ausführung previous version
bisheriger Stand status up until now
bislang hitherto
bistabil bistable
bistabiles Relais bistable relay || latching relay * *Haftrelais*
Bit * *Abkürzg. für 'Binary Digit' (Informationseinheit)*
Bitbelegung bit assignment
Bitebene bit level
Bitfolge bit string
Bitmuster bit pattern || bit sequence * *Bitfolge*
Bitprozessor bit processor * *Spezialprozessor zur schnellen Bitverarbeitung in SPS*
bitte please * *anfragend* || pardon * *wie bitte?* || yes, thank you * *Bitte sehr!* || you are welcome * *als Antwort auf "Danke!"* || welcome * *als Antwort auf "Danke!"* || don't mention it * *als Antwort auf "Danke!"* || never mind * *es macht nichts* || here you are * *nach e. erbetenen Handreichung, z.B. auf haben Sie Feuer?* || it's all right * *(salopp) 'schon gut/in Ordnung' als Antwort auf Entschuldigung*
bitte zu beachten kindly note
bitten ask * *for: um*
Bitumen bitumen
bitweise bitwise || bit by bit || bit for bit
blank bare * *nicht isoliert (z.B. Draht), nicht abgedeckt, bloß* || bright * *(metallically) bright: metallisch blank* || shining || polished * *poliert, geputzt* || naked * *bloß* || clean * *sauber*
Blankstahl cold-drawn steel * *{Draht}*
Blase bubble || blister * *an der Oberfläche* || blow * *im Inneren*
blasen blow
Blasenbildung bubble formation || blistering || pimpling
Blasmaschine blowing machine * *z.B. für Glas*
Blaswalze blowing roll
Blatt sheet * *Papier, Holzfurnier* || leaf * *in Buch; Folie, Blatt einer Blattfeder* || blade * *Klinge* || page * *Seite in einem Dokument/Druckerzeugnis*
blättern scroll * *z.B. über Bildschirmseiten* || turn over the leaves * *in Buch* || flake off * *abblättern* ||

Blattfeder 70

scale off * *abblättern* || flip * *flip through: (ein Buch) durchblättern* || browse * *{Computer} durch Datenbestand/Bildschirm-/Internetseiten* ~
Blattfeder leaf spring || plate spring
blau blue
Blaupause blue-print
Blech sheet metal || rolled plate * *(dickes) Walz~* || sheet steel * *dünnes Stahl~* || steel plate * *dickes Stahl~* || sheet || steel sheet * *[el.Masch.]* Elektro~, →*Magnetblech*
Blech-Formmaschine sheet-section machine * *erzeugt Formprofile*
Blech-Richtmaschine sheet-straightening machine
Blech-Rundbiegemaschine sheet metal bending machine
Blechbearbeitung sheet metal working * *Fein~* || metal plate working * *Grob~* || sheet working
Blechbearbeitungsmaschine sheet metal working machine || sheet-working machine
Blechdosen-Herstellungsmaschine can-making machine
Blechform shape of the lamination * *Form/Querschnitt der Magnetbleche in einem Blechpaket*
Blechkante plate edge
Blechpaket core assembly || core || laminated core || lamination
Blechprofil sheet metal section
Blechschere shear || shears || plate-cutting machine || sheet-iron shears || plate-shearing machine || tinner's snips * *Handblechschere* || snips * *Handblechschere* || plate shears * *für Grobblech*
Blechschnitt shape of the lamination * *Form der Magnetbleche in einem Blechpaket (z.B. e. Motorläufers)* || cross section of the core stack * *Querschnitt eines (Motorläufer-/-stator-) Blechpakets*
Blechschraube self-tapping screw || tapping screw
Blechstärke sheet thickness
Blechung lamination
Blechwender plate turner * *im Walzwerk*
Blei lead
Bleiakkumulator lead-acid battery
Bleibatterie lead-acid battery
bleiben remain || stick * *to something: bei etwas (z.B. Meinung)* || stay || hold * *hold the line, please: bleiben Sie bitte am Telefon* || persist * *verharren (in: bei), beharren (in: auf; z.B. Meinung), weiterarbeiten (with: an)*
bleibend lasting * *ever-lasting: ewig* ~ || enduring * *(an-, fort-) dauernd, bleibend* || permanent * *(fort-) dauernd, bleibend, permanent, ständig, dauerhaft, Dauer-* || persistent * *nichtflüchtig (Speicherinhalt), ständig, anhaltend, beharrlich, ausdauernd, hartnäckig*
bleibende Regelabweichung permanent control error || steady-state control deviation || permanent offset error
bleibender Fehler permanent error || residual error * *Restfehler* || steady-state error * *Fehler im stationären Zustand* || steady-state deviation * *Abweichung im stationären Zustand*
bleibender Schaden permanent damage
Bleiche bleaching * *das Bleichen, z.B. von Papier*
bleichen bleach
Bleicher bleaching apparatus * *{Papier} in der Zellstoff-/Papierindustrie* || bleacher * *{Papier} in der Zellstoff-/Papierindustrie*
Bleicherei bleach plant * *z.B. für Papier* || bleaching plant * *z.B. in der Textilindustrie*
Bleistift pencil
Bleistiftskizze pencil sketch || pencil drawing * *Bleistiftzeichnung*
Blendschutz glare protection
Blickfang eye catcher

Blickwinkel perspective * *(auch figürlich)* || aspect * *Gesichtspunkt*
Blind- reactive * *(elektr.)* || dummy * *z.B. dummy plug: Blindstopfen*
Blindleistung reactive power || VAr * *'reactive' Volt-Ampere*
Blindleistungsanteil reactive-power component || VAr component
Blindleistungsaufnahme reactive power consumption
Blindleistungsbedarf reactive-power demand
Blindleistungskompensation reactive power compensation || power factor compensation || VAr compensation || power-factor correction || harmonic compensation
Blindleistungskompensationsanlage VAr compensator
Blindstecker dummy plug
Blindstopfen filler plug * *z.B. für Kabeleinführungsloch* || blanking plug || hole plug * *z.B. für Kabeleinführungsloch* || dummy plug
Blindstrom reactive current || idle current || wattless current
Blindstromkompensation power factor compensation || reactive-current compensation * *durch Kondensatoren in der Nähe induktiver Verbraucher*
Blindstromkompensationsanlage reactive current compensation system || reactive current compensation plant
Blindstromkompensationseinrichtung power factor compensation equipment || VAr compensator
Blindwiderstand reactance
blinken flash * *Leuchtdiode, Meldeleuchte, Anzeige, Cursor, Verkehrsampel* || blink * *Leuchtdiode, Meldeleuchte, Anzeige, Cursor* || gleam * *glänzen, leuchten, schimmern* || twinkle * *blitzen, glitzern, funkeln, blinken, zwinkern*
blinkend flashing * *z.B. Anzeigelampe, LED-Anzeige* || blinking || twinkling * *glimmend*
Blinklicht flashing light || flashlight
Blinksignal flashing signal || blinking signal
Blistermaschine blister machine * *{Verpack.-techn.}*
Blisterverpackung blister package
Blitz lightning
Blitzableiter lightning conductor
Blitzschlag lightning stroke
blitzschnell lightning * *(Adj.)* || split-second * *(Adj.) in Bruchteilen einer Sekunde* || with lightning speed * *(Adv.)* || like a shot * *(Adv.) wie aus der Pistole geschossen* || in a flash * *(Adv.) wie der Blitz* || abruptly * *(Adv.) plötzlich* || all of a sudden * *(Adv.) plötzlich* || with blazing speed * *(Adv.; salopp) verteufelt schnell* || lightning-fast
Blitzstoßspannung lightning withstand voltage
Block block
Blockbandsäge log band saw * *{Holz} Holzsäge mit umlaufendem Sägeblatt*
Blockbauform block design * *z.B. bei SPS*
blockförmig square-wave * *rechteckförmig, z.B. Impulse, Spannung, Strom, Modulation*
Blockhobelmaschine log planer * *{Holz}*
blockieren block || stall * *aussetzen (Motor), sich festfahren, stecken bleiben* || disable * *sperren* || blocking * *(Substantiv)* || interlock * *verriegeln*
Blockierschutz stall protection || blocking protection || locked-rotor protection * *für Motor* || stalled-motor protection * *für Motor*
blockiert stalled * *abgewürgt, festgefahren, steckengeblieben* || obstructed * *blockiert, gehemmt, lahmgelegt, versperrt* || blocked || locked * *versperrt*
blockierter Läufer locked rotor
Blockierung blockage
Blockierzeit stall time

Blockkraftwerk district power station || distributed power station
Blockprüfungszeichen block check character * *Datensicherungszeichen, z.B. bei serieller Datenübertragung* || BCC * *Abk. f. 'Block Check Character': Datensicherungszeichen b.seriell. Übertragg.*
Blockprüfzeichen block-check character * *sorgt für die Datensicherung bei einer Datenübertragung (z.B. Quersumme)* || BCC * *Abkürzung für "Block-Check Character"*
Blockschaltbild block diagram || block circuit diagram * *Schaltbildübersicht* || schematic diagram
Blockstrom square-wave current * *rechteckförmiger Strom ;z.B. z. Ansteuerg. bürstenloser Servomotoren* || block current * *rechteckmodulierter Strom z. Ansteuerung →bürstenloser Servomotoren (SIEMENS 1FT5)*
bloß merely * *lediglich* || simply * *nur, einfach* || only * *nur* || just * *nur* || but * *aber* || bare * *unbedeckt* || naked * *nackt* || uncovered * *unbedeckt* || sheer * *bloß, →rein, nichts als, nackt; sheer nonsense: nichts als/bloß Unsinn*
BNC BNC * *Abkürzg. für 'BajoNet Connector': Bajonett-Steckverbindung für 50 Ohm Koax-Leitung usw.*
Bock stand || pedestal * *Untergestell, Sockel, Bock, →Lagerbock* || support frame || trestle * *Gestell, Gerüst, Bock, Schragen* || jack || support
Bocksprünge machen cut capers * *Kapriolen/Luftsprünge machen* || gambol * *herumtanzen, Luftsprünge machen* || caper * *Kapriolen/Luftspünge machen* || frolic * *herumtoben, (ausgelassen) herumtollen* || frisk * *herumtollen*
BOD BOD * *Break-Over Diode: Diode zum Thyristor-Schutzzünden bei Überschreiten d.max. zuläss. Sperrspanng.* || break-over diode * *zum Schutzzünden v.Thyristoren bei Überschreiten der max. zulässigen Sperrspannung*
Bode-Diagramm Bode diagram
Bodediagramm Bode diagram
Boden soil * *Erde, Grund* || floor * *eines Raums* || bottom * *eines Gefäßes, Schrankes, Gehäuses usw.* || ground * *(Unter)Grund; gain ground: an ~/ Bedeutung gewinnwn*
Bodenbearbeitung soil cultivation * *{Agrikultur}*
Bodenbelastbarkeit bearing capacity of floor
Bodenbelastung floor loading
Bodenblech bottom plate
Bogen arc * *auch Licht~* || sheet * *Papier~*
Bogenableger layboy * *für Papierbögen* || stacker * *Stapler; z.B. für Papierbögen* || sheet piling device * *für Papierbögen*
Bogenanleger sheet feeder * *{Druckerei} für Bogendruckmaschine* || automatic feeder * *{Druckerei} automatischer ~ für Bogendruckmaschine*
Bogenausleger sheet delivery unit * *für Bogendruckmaschine*
Bogendruckmaschine sheet-fed press * *{Druckerei} druckt vorgeschnittene Papierbögen (i. Ggs. z. Rollendruckmaschine)* || sheet-fed printing machine || sheet-fed printing press
Bogenentladung arc discharge
Bogenmaß arc measure || circular measure * *Maßeinheit rad (1 rad entspr. 360 Grad/2 Pi entspr. 57 Grad 17 min)* || radian * *Maßeinheit für Winkel (1 rad entspr. 360 Grad/2 x Pi entspr. 57 Grad 17 min)*
Bogenoffsetmaschine sheet-fed offset press * *{Druckerei}*
Bogenstapler sheet piling device * *für fertig geschnittene oder bedruckte Papierbögen* || layboy * *für Papierbögen*
Bogenzahnkupplung curved-tooth coupling
Bohle plank

Bohr- und -Fräswerk milling and boring machine * *{Werkz.masch.}*
Bohr- und Fräszentrum boring and milling center * *{Werkz.masch.}* || boring and moulding center * *{Holz}*
Bohrantrieb drilling drive
Bohrautomat automatic drilling machine * *{Werkz.masch.}*
bohren drill || bore * *ausbohren, mit Bohrmeißel ~, mit Holzbohrer ~*
Bohrer drill || drill bit * *Bohrereinsatz (s.B. Spiral-, Steinbohrer)* || boring bit * *(Aus)Bohreinsatz, z.B. für Holz*
Bohrfutter drilling jig || chuck || boring jig || drill chuck || boring socket
Bohrfutterschlüssel chuck key
Bohrgestänge drill string * *z.B. für Erdöl-/Erdgasbohrung*
Bohrkopf boring head * *{Holz}*
Bohrloch drill-hole * *Bohrung ins Volle, Vollbohrung; auch zur Erdölgewinnung* || borehole * *durch Aufbohrung; auch in Holz und in der Erde (zur Öl-/Wasser-/Gasgewinnung)*
Bohrlochpumpe borehole pump || down hole pump
Bohrmaschine drilling machine || drill machine
Bohrmeißel drill bit * *z.B. bei Erdöl-Bohreinrichtung*
Bohrmilch drill milk * *{Werkz.masch.}*
Bohrplattform offshore drilling platform || mobile offshore drilling unit || drilling platform || oil rig * *zum Bohren nach Öl auf dem Lande oder im Meer*
Bohrschablone drilling template
Bohrständer drill stand
Bohrstange drill pipe * *für Erdölbohrung*
Bohrung hole || drill-hole * *Bohrloch* || bore || drill hole || drill * *nach Bodenschätzen (Erdöl, Gas); lower a drill: eine ~ niederbringen* || drilling
Bohrungsdurchmesser bore diameter || hole diameter
Bohrwerk boring machine || boring mill
Bohrzentrum multi-way drilling machine
Bolzen bolt || pin || stud * *Steh~, Stift(-Schraube), Zapfen*
Bolzenbrett stud board * *zum Leitungsanschluss im Motorklemmenkasten*
Bolzenklemme threaded-stud terminal * *zum Anschluss von Leitungen* || stud terminal * *zum Anschluss von Leitungen*
Bolzenkupplung pin coupling
Bolzenschraube screwed rod || bolt screw
Bondierung bonding * *Befestigung der Anschlussdrähte eines Halbleiterchips auf dem Trägermaterial*
Bonus bonus || premium || discount * *Rabatt (on: auf)* || abatement * *Rabatt* || allowance * *Rabatt* || price reduction * *Preisreduzierung*
boolsch boolean
boolsche Größe boolean variable
Bördelmaschine flanging machine
bördeln flange * *flanging press: Bördelpresse; double-flanged butt weld: Bördelschweißung* || bead * *mit Wulst/mit Perlen versehen* || edge * *durch Schmieden* || curl * *kräuseln; seal-curling: verschließen durch Bördeln (z.B. Kondensator, Dose ...)* || border * *einfassen, (um-)säumen* || seal curling * *(Substantiv) Verschließen durch Bördelung, z.B. Kondensator, Dose usw.* || rim
Bordnetzversorgung on-board power supplies
Bordstromversorgung on-board power supply
Boxpalette box pallet
Brackwasser brackish water || contaminated water * *verschmutztes Wasser*
Bramme slab
Branche branch || application * *Anwendungsbe-*

reich || field * *Bereich, (Arbeits)gebiet, Fach* || industrial sector * *in der Indutrie* || industry * *Industrie* || sector * *Bereich* || line * *Fach, Gebiet, Tätigkeitsfeld, Sparte, Geschäftszweig* || trade * *Gewerbe, Geschäftszweig* || market sector * *Marktsektor* || branch of industry
branchenbedingt due to the conditions in the particular trade || branch-specific
Branchenkenntnis knowledge of the trade
branchenkundig experienced in the trade
branchenorientiert branch-oriented || sector-oriented || application-oriented * *anwendungsorientiert*
Brandgasentlüfter flue gas exhauster * *zum Absaugen der Brandgase im Falle eines Gebäudebrands*
Brandgefahr fire hazard || fire risk
Brandschutz fire protection
Brandstelle burn mark * *z.B. an Kommutator* || sign of burning * *z.B. am Kommutator e. Gleichstrommaschine nach Bürstenfeuer*
brauchbar usable * *siehe auch* →*verwendbar*
brauchen need || require * →*benötigen* || be in need of * *nötig haben,* →*benötigen*
Brauerei brewery
braun brown
Braunkohle lignite coal
Braunkohlentagebau lignite open-cast mining * *Abbauverfahren* || lignite open-pit mine * *Bergwerk*
Brechanlage crushing plant * *z.B. für Erz* || breaker plant * *z.B. für Steine, Koks* || pulverizing plant
brechen break * *auch Steine/Vertrag/Rekord/Versprechung/Widerstand/Knochen/Kanten; break down: zusammen*~ || crack * *bersten, platzen* || snap * *(zer)springen, (zer)reißen, entzweigehen* || rupture * *reißen, ab*~ || smash * *zertrümmern; to pieces: in Stücke* || crush * *mahlen* || refract * *{Optik} Lichtstrahl* || violate * *Vertrag usw.* ~ || fracture * *(zer)brechen* || fold * *Papier* || crease * *Papier* || burst * *bersten, (zer)platzen, (auf-/zer-) springen* || fractionize || collapse * *zusammen*~, *kollabieren*
brechersicherer Lüfter icing fan * *{Schifffart} Lüfter, d.e. plötzl.Blockade durch Brecherschlag verkraftet*
Brechung refraction * *{Optik} Lichtstrahl; plane/ angle of action: ~sebene/-Winkel; refractive index: ~szahl* || breaking
Brechungszahl refractive index
Brei paste || pulp * *Zellstoff-/Papier*~ || slurry * *Aufschlämmung*
breit broad || wide || large * *ausgedehnt* || vast * *ausgedehnt* || spacious * *ausgedehnt* || diffuse * *weitschweifig, langatmig, diffus* || long-minded * *weitschweifig*
Breitband wide-strip * *im Walzwerk*
breitbandig high-bandwidth || broadband || wideband || broad-band
Breitbandstraße wide-strip line * *Walzwerk*
Breite width * *['with]*
Breite x Höhe x Tiefe width x height x depth || W x H x D
Breitenmessung width gauge * *Breitenmessanlage/-Einrichtung*
breiter Einsatz use in a wide range of applications
Breithobelmaschine wide planer * *{Holz}*
Breitstreckmaschine width drawing machine * *z.B. zur Herstellung von Kunststofffolien*
Breitstreckwalze spreader roll * *mit gebogener Achse zur Vermeid.v. Falten bei Papiermaschine* || curved roll * *mit gebogener Achse zur Vermeidg.v. Faltenbildg. b.Papier(verarbeitungs)maschine* || width drawing roll || banana roll * *(salopp) Bananen-förmige Walze* || stretch roll || stretching roll || web-spreading roller
Breitziehpresse wide-frame drawing press || double-crank drawing press

Bremsarbeit braking energy
Bremsbacke brake shoe
Bremsbelag brake lining
Bremsbetrieb braking operation
Bremschopper brake chopper || braking copper
Bremsdruck braking pressure
Bremse brake
Bremseinheit braking module || braking unit
Bremseinrichtung braking facility || braking equipment || braking gear || braking unit || braking module
Bremseinsatzpunkt deceleration point || deceleration-initiation point || brake-application point
bremsen brake || decelerate * *verzögern* || slowdown * *langsamer machen/werden*
bremsend braking
Bremsenergie braking energy
Bremsenöffnung brake release
Bremsenschließzeit brake closing time
Bremsensteuerung braking control || brake control
Bremsgenerator brake generator * *z.b. Abwickler*
Bremsgerät braking unit || braking device
Bremskennlinie braking characteristic
Bremskonus brake cone
Bremslamelle brake plate || brake disc || brake disk
Bremsleistung braking power
Bremslüfter brake fan
Bremsmodul braking module
Bremsmoment braking torque || deceleration torque * *Verzögerungsmoment*
Bremsmotor brake motor * *Motor mit Anbau-/Einbaubremse* || brakemotor * *Motor mit Anbau-/Einbaubremse* || motor with brake
Bremsnabe brake hub
Bremsrelais braking relay
Bremssattel brake caliper * *an Scheibenbremse*
Bremsscheibe brake plate || brake disk || brake disc
Bremsschütz brake contactor
Bremsspannung braking voltage
Bremsspule brake coil
Bremsstrom braking current
Bremsstufe brake control circuit
Bremsung braking || braking operation * *Bremsvorgang* || braking effect * *Bremswirkung*
Bremsvorgang braking operation
Bremswächter brake monitor || zero-speed relais
Bremswärme im Läufer rotor braking heat
Bremswärme im Motor motor braking heat
Bremswärme im Ständer stator braking heat
Bremsweg stopping distance || length of brake path || braking distance
Bremswiderstand braking resistor || brake resistor || brake resistance || braking rheostat || load rheostat || pulsed braking resistor * →*Pulswiderstand* || bleeder || bleed resistor
Bremswinkel braking angle * *bis zum Stillstand noch zurückgelegter Winkel ab Einschalten einer Bremse*
Bremswirkung braking effect
Bremszeit braking time || braking interval || stopping time * *zum Stillsetzen benötigte Zeit* || deceleration time * *Zeit, während der verzögert wird* || braking time * *{el.Masch.} Dauer einer Bremsung vom Ausschalten d.Motors bis zum Stillstand*
brennbar combustible || flammable * *entflammbar*
Brennbarkeit combustibility || flammability * *(amerikan.) Entflammbarkeit* || inflammability * *Entflammbarkeit*
brennen burn
Brenner burner || torch * *Schweißbrenner* || blowpipe * *Schweißbrenner* || power burner * *für Heizungszwecke*
Brennpunkt focus
brennschneiden flame-cut

Brennschneidmaschine flame-cutting machine || gas cutting machine
Brennstoff fuel
Brennstoffzelle fuel cell
Brett board || bulletin board * *Anschlagbrett, Diskussionsforum (in Mailbox usw.)*
Bretterverleimmaschine board gluing machine
Brettverleimanlage board gluing line * *[Holz]*
Briefumschlag envelope
Brikettiermaschine briquetting machine
Brikettpresse briquetting press
brillant brilliant || splendid || superlative
Briseur licker-in * *[Textil] bei →Krempelmaschine*
Broadcast broadcast * *[Network] (Telegramm] an alle*
Bronze bronze || gun metal
Bronze-Sinterlager sintered bronze bearing
Bronzegewebebürste woven-bronze brush * *z.Stromübertragg. auf bewegte Teile (el.magn.Bremse/Kupplg.) b. Naslauf (in Öl)*
Bronzegleitlager bronze sleeve bearing
Broschüre brochure || booklet || pamphlet || folder * *Werbeblatt, Faltblatt, Broschüre, Heft*
Browser browser * *Programm zur Anzeige verschiedenster Daten/Dateien in einer Schnell-/Inhaltsübersicht*
Bruch breakage * *Draht, Material; auch 'Bruchstelle, Bruchschaden'* || loss * *Wegbleiben eines Signals, z.B. Tachobruch* || fracture || crack * *Riss, Sprung* || fraction * *(mathem.)* || burst * *eines Rohrs, Tanks usw.*
Bruchdehnung stretch at break
bruchfest breaking resistant || resistant to fracture || breakproof || breakage-proof
Bruchkraft tensile strength * *z.B. von Papier*
Bruchlochwicklung fractional-slot winding * *[el. Masch.] Nutenzahl je Pol/Nutenzahl je Wicklungsstrang ist gebroch. Zahl*
Bruchstrich fraction bar * *(mathem.)*
Bruchteil fraction
Bruchwiderstand tensile strength * *z.B. von Papier*
Brücke jumper * *Steck-, Klemmenbrücke* || bridge * *auch Dioden/Transistor/Thyristorbrückenschaltung* || link * *Verbindungs(stück)* || gantry * *Portal (bei Kränen; Füll~ bei der Flaschenabfüllung usw.)*
brücken jumper * *über~, z.B. mit Löt-/Klemmen-/Drahtbrücke*
Brückengleichrichter rectifier bridge || bridge rectifier
Brückenhälfte converter section * *eines Gleichrichters* || half bridge * *z.B. einer Stromrichter-Brückenschltung*
Brückenkran gantry crane
Brückenschaltung bridge connection * *Stromrichter* || bridge circuit * *Messbrücke* || bridge configuration * *Anordnung*
Brückenstecker jumper plug || plug-in jumper
Brumm hum * *Geräusch, elektr. Brummen*
Brummen hum * *Geräusch, elektr.Brummen*
brummfrei hum-free
Brummspannung hum voltage || ripple voltage * *Spannungswelligkeit*
brünieren brown * *Metall* || burnish * *polieren*
Brunnen well || fountain || spring * *Quelle*
Brushless DC-Motor brushless DC motor * *permanenterregter Synchron-Servomotor mit Block- oder Sinusmodulation* || AC servo motor * *(salopp)*
brutto gross
Bruttogewicht gross weight
Bruttosozialprodukt gross national product || GNP * *Abkürzung für 'Gross National Product'*
BS non-drive end * *B-Seite, Nicht-Antriebsseite eines Motors* || NDE * *Abkürzg. für 'Non Drive End': Nicht-Antriebsseite eines Motors (B-Seite)* || N-end * *Abkürzg. für 'Non-Drive end': Nicht-Antriebsseite* || BS * *Abk. für 'British Standard': britische Norm*
BS-Wellenende non-drive shaft end * *auf der Nicht-Antriebsseite eines Motors* || ND shaft end * *Abkürzg.f. 'Non-Drive': nicht antreibend*
Buchbinderei book-binding || bookbinding || bookbindery * *Buchbinderwerkstatt/-Firma* || binding department * *als Abteilung*
Buchdruck offset printing * *Offsetdruck, Druck auf Gummi, Platten-Druckverfahren* || bookprinting || printing || letterpress printing || typography * *Buchdruckerkunst, (Buch-) Druck*
Buchdruckerei printing office || printing plant * *(amerikan.)* || typography * *Buchdruckerkunst, (Buch-) Druck*
buchen book || register || list || enter into the books
Bücherei library
Buchführung bookkeeping || accounting * *Rechnungswesen* || accountancy * *Buchhaltung*
Buchhalter bookkeeper
Buchse socket * *(elektrisch)* || jacket * *(elektrisch)* || jack * *Steckdose, Buchse (z.B. für 4 oder 2mm Prüfstecker), Klinkenbuchse* || test socket * *Prüfbuchse* || measuring socket * *Messbuchse* || bushing * *Muffe, Spannhülse, Durchführungshülse* || bush * *Maschinenteil: Büchse, Buchse* || sleeve * *Muffe, Hülse, Buchse, Manschette* || liner * *Zylinderbuchse* || cup * *Fettbuchse* || receptacle * *Netz-)Steckdose, (Kaltgeräte-/Bananenstecker-) Buchse*
Buchsenklemme sleeve terminal || pillar terminal || tunnel terminal
Buchsenleiste female connector * *"weiblicher" Steckverbinder* || socket connector
Buchsenstecker female connector * *"weiblicher" Stecker* || socket connector || receptacle
Buchstabe letter || character * *Zeichen*
Budget budget
Bügel bracket || bail * *Henkel, Bügel, (Hand-) Griff; Schranke; Bürge, Kaution; (Wasser) schöpfen* || strap || clamp * *Klammer, Schelle* || bow * *federnder Bügel (z.B. für Brille); bow collector: Stromabnehmer~*
bügeln iron
Bügelsägemaschine hack sawing machine || hacksaw || framed crosscut saw
Bühne platform * *Arbeits~, Steuer~* || stage * *im Theater* || gallery * *Galerie, Laufgang, Empore; Stollen im Bergbau* || trestle * *Gestell, Gerüst, Bock*
Bund coil * *aufgewickeltes Material (Draht, Blech, Seil usw.), Blechwickel, Spule* || collar * *z.B. einer Welle* || shoulder * *Vorsprung, Schulter* || seat * *Auflager* || rod-stop * *Anschlag* || cording * *[Buchbinderei]* || bundle * *Bündel, Bund, Paket, Ballen*
Bundabtransport coil conveyor * *Fördereinrichtung im Walzwerk* || coil removal * *Funktion im Walzwerk*
Bündel bundle
Bündelmaschine bundling machine
bündeln bundle * *(zusammen-) bündeln, zusammenknöpfen, zusammenfassen (zu Gruppe, Haufen), kombinieren* || focus * *konzentrieren, vereinigen (sich) sammeln, fokussieren* || bunch * *zusammen~, zusammenfassen (zu Gruppe, Haufen usw., z.B. beim Verseilen von Kabeln)* || truss * *bündeln, schnüren (up: fest), zusammen'binden*
Bündelpresse bundling press || bundle press || baling ress * *Ballenpresse*
Bundgehäuse shoulder housing * *Gehäusebauform z.B. einer Kupplung/Bremse*
bündig flush * *fluchtend/fluchtrecht (with: mit); be flush with: bündig abschließen mit* || to the point

bündigziehen 74

* *kurz und bündig* || conclusive * *überzeugend* || concise * *Stil, Rede* || precise * *genau* || curt * *schroff* || point-blank * *geradeheraus, unverblümt, klipp und klar* || flush-mounted * *bündig abschließend montiert* || justified * *ausgerichtet; right-/left-justified: rechts-/linksbündig*
bündigziehen pull flush with floor level * *{Aufzüge}*
Bundwagen coil car
Bunker bunker * *Behälter z.B. für Kohle* || bin * *(großer) Behälter, (Vorrats-) Bunker, Fülltrichter, Lagerfach, Kasten, Tasche* || silo * →*Silo für Getreide usw.* || hopper * *Schüttgutbehälter, Vorratsbehälter, Fahrzeug mit Schnellentlade-Bodentrichter*
Bürde burden || load * *Last, Belastung* || burden resistor * *Bürdenwiderstand* || shunt resistor * *Bürdenwiderstand zur Strommessung*
Bürdeneinheit load unit * *Belastungseinheit aus Widerständen zur Strommessung*
Bürdenwiderstand shunt resistor * *zur Strommessung* || load resistor * *Lastwiderstand* || burden resistor
Burn-In burn-in
Büro office
Büroautomatisierung office automation
Büroklammer paperclip
bürokratisch bureaucratic || bureaucratically * *(Adv.)* || red-tape
Büroleiter head || senior clerk * *bei Behörde, Bank*
Büromaschine office apparatus
Büroumgebung office environment
Bürste brush * *auch Kohlebürste in Motor/Generator*
Bürstenabhebevorrichtung brush lifting device * *{el.Masch.}* || brush lifting and short-circuiting device * *{el.Masch.} Kurzschließ- und ~ für Schleifringläufermotor*
Bürstenabrieb brush wear * *Bürstenverschleiß* || carbon dust * *Abriebpulver, Grafitstaub* || carbon dust deposits * *abesetzter Kohlebürstenstaub (am Kommutator einer Gleichstrommaschine usw.)*
Bürstenapparat brushgear * *z.B. einer Gleichstrommaschine* || brush unit || brush mechanism || brush assembly * *einer DC-Maschine*
Bürstenbrücke brush rocker * *für Kohlebürsten z.B. in Gleichstrommaschine*
Bürsteneinrichtung brush gear * *an Gleichstrommaschine usw.*
Bürstenfeder brush spring * *für Kohlebürste*
Bürstenfeuer brush sparking * *Funkenbildung am Kommutator* || brush flashover || sparking under the brushes
Bürstenführung brush guide * *z.B. bei Gleichstrommaschine*
Bürstenhalter brush holder * *für Kohlebürsten in Motor/Generator*
Bürstenhalterbolzen brush-holder stud * *z.B. bei Gleichstrommaschine* || brush-holderarm * *z.B. bei Gleichstrommaschine* || brush spindle
Bürstenhebel brush arm
Bürstenhebevorrichtung brush raising device || brush lifting gear
Bürstenlänge brush length
Bürstenlebensdauer brush life || brush service life
Bürstenlineal brush-holder arm * *z.B. im Gleichstrommaschine* || brush-holder stud || brush spindle
bürstenlos brushless
bürstenloser Gleichstrommotor brushless d.c. motor * *AC Synchronmotor mit "elektron. Kommutator" (Blockstromansteuerung)* || brushless DC motor * *AC Synchronmotor mit "elektron. Kommutator" (Blockstromansteuerung)* || AC servo motor * *(salopp) AC Synchronmot. m."elektron. Kommutator" (Blockstromsteuerg.)*

Bürstenrundfeuer heavy brush flashover || brush flashover
Bürstensatz brush set * *z.B. in Gleichstrommaschine*
Bürstenstandzeit brush service life || brush life
Bürstenüberwachung brush monitoring * *z.B. der Restlänge der Kohlebürsten in Gleichstrommotor* || brushwear monitoring * *Überwachung d. Bürsten e. Gleichstrommaschine auf Verschleiß* || brush monitoring device * *als Gerät/Einrichtung*
Bürstenverschleiß brush wear
Bürstenverstellapparat brush-rocker gear * *bei Drehstromkommutatormotor*
Bürstenverstelleinrichtung brush-rocker gear * *bei Drehstromkommutatormotor*
Bürstenwalze brush roller
Bürstenwechsel brush-replacement
Bürstenzugänglichkeit brush access * *external: von außen, easy: leichte*
Bürstmaschine brush machine * *z.B. zur Reinigung in Blechstraße*
Bürstwalze brush roller * *z.B. in Bandbehandlungsanlage zum Reinigen des Walzbandes* || cleaning roller * *{Textil} bei* →*Kratzenraumaschine*
Bus bus || data highway || multi-drop connection || trunk * *(amerikan.) Kommunikations(haupt)bus, Hauptnervenstrang*
Bus-Hauptstrang backbone * *von dem die Vor-Ort-Kommunikationsstränge über Stichleitungen abzweigen*
Busabschlußmodul bus terminating module
Busabschlußmöglichkeit bus terminating possibility
Busabschluss bus termination * *z.B. mit Wellenwiderstand der Busleitung (z.B. 120 Ohm)*
Busabschlussstecker bus terminating connector || bus termination * *Busabschluss* || bus terminator * *Busabschluss-Element*
Busabschlusswiderstand bus terminating resistor * *zum Abschluss der Busleitung mit dem Wellenwiderstand (z.B. 120 Ohm)* || bus termination resistor
Busaktivität bus activity
Busanschaltung bus interface || bus interface module * *Baugruppe* || bus interface board * *Leiterkarte*
Busanschaltungsmodul bus-coupling module || bus-interfacing module
Busanschluss bus connection || bus connector * *Busanschlussstecker*
Busanschlussstecker bus connector
Busarbitrierung bus arbitration * *Verwaltung und Zuteilung der Zugriffsrechte auf einen (Multiprozessor-) Bus*
Busbetrieb bus operation
Busbrücke bus jumper
busfähig bus-capable * *Gerät* || multi-drop * *Kommunikationsverbindung* || with bus capability || busable || networkable * *netzwerkfähig, vernetzungsfähig*
busfähige Schnittstelle bus-capable interface || multi-drop interface || interface with bus capability || busable serial link * *serielle* || bus interface
Busfehler bus error || bus fault
busförmig multidrop || bus-type || in a bus-type configuration || multi-drop
busförmig verbinden multidrop || connect in a bus-type configuration
busförmige Kommunikation multi-drop communication
Busfreigabe bus enable
busgekoppelt bus-coupled || multi-drop
Buskabel bus cable
Buskoppelbaugruppe bus coupling module || bus-interfacing module
Buskoppler bus coupler || bus transceiver * *Sende-/Empfangsschaltung*

Buskopplung bus interface || bus-type connection * *busförmige Verbindung* || multidrop communication
Busleitung bus cable || bus line * *logisch gesehen*
Busplatine backplane || bus PCB * *Bus-Leiterplatte* || wiring backplane
Busprotokoll bus protocol
Busse buses * *auch Kommunikationsbusse*
Bussystem bus system
Busteilnehmer bus node * *Knoten* || node || bus station || master * *aktiver* || slave * *passiver* || master station * *aktive Station, Leitstation* || slave station * *passive Station, Unterstation* || station
Busübertragung bus transmission
Busumlaufliste bus circulating list || bus polling list || poll list for slave stations || job table
Busumlaufzeit bus cycle time || bus circulating time * *wird insgesamt benötigt, alle Busteilnehmer einmal anzusprechen*
Busumsetzer bus converter || gateway * *Protokollumsetzer* || data bridge * *Protokollumsetzer*
Busverbindung bus interconnection || bus linkage
Busverstärker repeater
Busverwalter bus arbiter
Busverwaltung bus management || bus arbitration * *kollisionsfreie Buszuteilung z.B bei einem Multiprozessorbus*
Buszugriff bus access
Buszugriffsperre bus access inhibit
Buszuteiler bus arbiter
Buszuteilungslogik bus arbitration
Bütte tank || butt || tub || vat || pulp vat * *Zellstoffbütte in der Zellstoff-/Papierherstellung* || chest * *z.B. in der Zellstofferzeugung*
Büttenregelung chest level control * *Füllstandsregelung für Bütte in der Zellstoffaufbereitung*
BxHxT WxHxD * *width x height x depth: Breite x Höhe x Tiefe*
Byte byte || bytes * *(Plural)* || octet * *{Network}* z.B. *beim →PROFIBUS*
BZ order form * *Bestellformular* || purchase order * *Bestellung, ausgefülltes Bestellformular* || order blank * *(amerikan.) Bestellformular*
BZ-Empfänger order form receiver * *Bestellzettel-Empfänger* || order address * *Bestelladresse*
BZ-Nummer order No * *Bestellnummer* || departmental order number
bzw resp * *Abkürzung für 'respectively'.* || respectively * *nachgestellt* || or || rather * *eher*

C

C-Achse C-axis * *{Werkz.masch.} Hauptspindel bei Werkzeugmaschine* || main spindle axis * *{Werkz.masch.} Hauptspindel*
C-Schiene C-type symmetrical mounting rail according to DIN EN 50 022 * *35mm breite Aufschnapp-Montageschiene* || C-type mounting rail * *Aufschnappmontageschiene nach DIN EN 500222 (35 mm breit)*
ca appr. || approx. || approximately
Cachespeicher cache memory * *schnell.Pufferspeicher z. Zwischenspeichern v. Befehlen u. Daten f. CPU-Schnellzugriff*
CAD CAD * *Abkürzg.für 'Computer Aided Design': Rechnergestützte Konstruktion/Entwicklung*
CAD-Arbeitsplatz CAD workstation
cadmiert cadmium plated

CAE CAE * *Abkürzg. für 'Computer-Aided Engineering': rechnergetütztes Konstruieren*
CAM CAM * *Abkürzg.für 'Computer Aided Manufacturing': rechnergestützte Fertigung* || CAM * *Abkürzg.für 'Content-Addressable Memory': assoziativer Speicher (siehe dort)*
CAN CAN * *Abk.f. 'Controller Area Network': seriell. Bussystem n. ISO DIS 11898 u.11519-1 (urspr.f.Auto)*
Cannon-Stecker Sub-D connector || D-type conector * *mit "D"-förmigem Querschnitt*
CANOpen CANOpen * *{Network} erweitertes →CAN Busprotokoll mit Anwenderprofilen (z.B. DSP-402 für el.Antriebe)*
Carborundumstein carborundum block * *zum Schleifen*
CASE CASE * *{Computer} Abkürzg. für 'Computer-Aided Software Engineering'*
cc c.c. * *Kopie an* || copy to * *Kopie an*
CE CE * *französ. Abk.für 'Communautéen Europeéen': Europäische Gemeinschaft*
CE-Kennzeichen CE mark
CE-Kennzeichnung CE marking
CE-Kennzeichnung explosionsgeschützter Elektromotoren CE-marking of explosion protected eletrical motors
CEB CEB * *Abk.für Comité Électrotechnique Belge: Belgisches Elektrotechn Komitee (ähnl. d.deutschen DKE)*
CEF CEF * *Abk.für Comité Électrotechnique Francais: französ. Eletektrotechn. Komitee (ähnl. dem DKE)*
CEI CEI * *Abkürzg. für 'Comitato Electrotechnico Italiano': italienisches Normengremium* || CEI * *französ. Abk.für Comité Électrotechnique International: Internat.Elektrotechn. Kommission (IEC)*
Cellulose cellulose
CEMA CEMA * *Canadian Electrical Manufacturers Association: Verband d. canad, Elektroindustrie (ähnl. ZVEI)*
CEMEP CEMEP * *Abk. f. 'Comité des Constructeurs d'Machines Electriques et d'Electronique de Puissance'*
CEN CEN * *Abkürzg.f. 'Comité Européen de Coordination des Normes': Europäisch. Komitee für Normung*
CENELEC CENELEC * *European Commitee for Electrotechnical Standardization (europäisches Normengremium)*
CEP CEP * *Abk.für 'Comissao Electrotécnica Portoguêse' portugiesische Elektrotechn. Kommission (wie DKE)*
CES CES * *Abk.für 'Comité Électrotechnique Suisse': Schweizer. Elektrotechn. Komitee (ähnlich DKE)*
CFC CFC * *'Continuous Function Chart': Grafische Beschreib.sprache für kontinuierl. Steuer-/Regelaufgaben*
Changierantrieb traversing drive * *in Textilfaserverarbeitung*
Changiereinrichtung traversing device * *{Textil}* || traversing unit * *{Textil}*
changieren traverse * *Quer- und Herverlegen beim Aufwickeln von Textilfäden, Kabel, Draht usw.*
Changiersteuerung traverse control || traversing control
Changierung traversing * *seitliche Hin- und Herbewegung z.B. von Textilfäden/Kabel bei der Aufwicklung* || traverse unit * *Einrichtg. z. seitlichen Hin- u. Herbewegen z.B. v. Textilfäden/Kabel b. d. Aufwicklg.*
chaotisch chaotic
charakteristisch characteristic * *feature/property:*

charakteristische Gleichung

Eigenschaft || typical * *of: für* || distinguished * *heraus-/hervorragend, bemerkenswert, charakteristisch*
charakteristische Gleichung characteristic equation
Charge batch * *auch Fertigungslos*
Chargenprozess batch process * →*diskontinuierlicher Prozess, z.B. in der Zement-, Zucker-, chemischen Industrie*
Chassis-Gerät chassis unit * *Baueinheit auf offenem Tragblech ohne besondere Schutzart*
Checkliste checklist
Chef boss * *salopp* || chief * *Oberhaupt, Vorgesetzter* || head * *Führer, Leiter, Vorstand, Direktor* || principal * *Direktor* || employer * *Arbeitgeber* || manager * *Verwalter* || head manager * *"Obermanager"* || chief engineer * *Ingenieur* || chief designer * *Konstrukteur, Entwickler* || senior * *besonders Rang/Dienstälterer* || senior engineer * *rang-/dienstälterer Ingenieur*
Chefredakteur editor in chief || chief editor
Chemie chemistry
Chemiefaser man-made fiber * *(amerikan.)* || man-made fibre * *(brit.)* || synthetic fiber * *(amerikan.)* || synthetic fibre * *(engl.)*
Chemiefaserindustrie synthetic fiber industry * *(amerikan.)* || synthetic fibre industry * *(brit.)* || man-made fiber industry
Chemiefaserzellstoff dissolving pulp * *{Papier}* || chemical processed pulp * *{Papier}*
Chemikalie chemical
Chemikalien chemicals
Chemikalienaufbereitung chemicals preparation
Chemikalienrückgewinnung chemical recovery
chemisch chemical || chemically * *(Adv.)*
chemisch aggressiv chemically aggressive
chemische Industrie chemical industry
chemische Reaktion chemical reaction
Chip-Karte chip-card
Chip-Kondensator chip-type capacitor * *z.B. für SMD-Bestückung*
Chip-Widerstand chip-type resistor * *z.B. für SMD-Bestückung*
Chlor chlorine * *chlorine-free: ~frei (z.B. ungebleichtes oder mit anderen Stoffen gebleichtes Papier)*
Chopper chopper
Chopper-Umrichter chopper converter * *mit nachgeschaltetem Gleichspannungssteller im Zwischenkreis*
Chrom chrome * *chromic: Chrom-* || chromium
chromatisiert chromated
chronologisch chronological
chronologische Reihenfolge chronological order
CIGRE CIGRE * *Conference Internationale des Grandes Réseaux Electriques: Intern. Hochspanngs.normenkonferenz*
CIM CIM * *Abkürz. für: 'Computer Integrated Manufacturing': rechnerbasierte Fertigung*
CISC CISC * *'Complex Instruction Set Computer': Computer m.komplexem Befehlssatz (mächtige Befehle;* →*RISC)*
Client client * *(Personal)Computer, d. als Anwenderstation i.e. 'Client-Server-Netzwerk' dient (s. 'Server')*
Clip clip
Cluster cluster * *kleinste logische Einheit auf der Festplatte (1 Festplatte hat unter MS-DOS 65536 Cluster)*
cm cm * *Zentimeter; 1 cm entspr. 0,393701 in (inch)* || in * *inch; 1 cm entspr. 0,393701 in*
CNC CNC * *{Werkz.masch} Abk. für "Computer Numerical Controller": Computergestützte numerische Steuerung*
CNC-gesteuert CNC-controlled

CNC-Interpreter CNC interpreter * *interpretiert Programme für Bewegungssteuerungen, z.B. nach DIN 66025*
CNC-Steuerung CNC control * *Abk. f. 'Computerized Numeric Control': numerische Steuerung (Bewegungssteuerung)*
Code code * *1-out-of-8 code: 1-aus-8 Code*
CODEC CODEC * *COmpressor/DECompressor: Einrichtg. z.Echtzeit-Komprimieren/Dekomprimieren v. Videosignalen*
Codegenerator code generator * *erzeugt z.B. Hochsprache- oder Assemblerquellcode z.B. aus einer graf. Darstellung*
Codewort password
Codier- coding
codierbar codable
Codierschalter coding switch || decode switch || DIP switch * *Dual-In-Line-Schalter (auf Leiterkarte)* || thumb-wheel switch * *Daumenradschalter, Zifferneinsteller*
Codierstecker coding plug
Codierung coding || keying * *von Steckern zur Vertauschungssicherheit*
Codierzapfen keying pin * *an einem unverwechselbaren Steckverbinder* || coding pin * *an einem unverwechselbaren Steckverbinder*
COMEL COMEL * *'Comité des Coord. d. Contr. d. Machines Tournantes Elect. d. Marché Commun.': El.masch.verband*
Compiler compiler * *Übersetzungsprogramm für Hochspracheprogramme*
Compoundwicklung compound winding * *{el. Masch.}*
Container container
Containerkran container crane
cos phi cos phi * *Leistungsfaktor (Verhältnis Wirk-/Scheinleistung)* || power factor * *Leistungsfaktor (Verhältnis Wirk-/Scheinleistung)*
cosinus cosine
CPU CPU * *Abkürzung für 'Central Processing Unit': Zentraleinheit eines Rechners* || central processing unit
CPU-Baugruppe CPU module * *Zentraleinheit eines Rechners/einer SPS* || CPU card * *Flachbaugruppe* || CPU board * *Flachbaugruppe*
CRC CRC * *Abk.f. 'Cyclic Redundancy Check': Datensicherungsverfahren mit zykl.berechneten Prüfmustern*
Crimpanschluss crimp termination || crimp snap-on connection || crimp snap-on connector * *Crimpanschlussverbinder*
Crimpkontakt crimp contact || crimp snap-in contact * *zum Einrasten z.B. in Steckergehäuse*
Crimpverbinder crimp connector
Crimpzange crimping tool * *zum Anquetschen von Kabelschuhen und Kabel-Quetschverbindern*
CSA CSA * *Abkürzg. für 'Canadian Standard Association': kanadisches Normengremium*

D

D-Anteil derivative component || derivative term || differential term || D-component
D-Sub Stecker Sub-D connector || D-type connector * *mit "D"-förmigem Querschnitt*
D-Verhalten derivative characteristic || derivative action
D-Wellenende D-shaft end * *Drive side, Antriebsseite, A-Seite eines Motors*
dabei near * *nahe* || near by * *nahebei* || close by

* *nahebei* || about to * *im Begriffe; do a thing: etwas zu tun* || going to * *im Begriffe* || at the same time * *gleichzeitig* || in doing so * *gleichzeitig* || in so doing * *wenn/indem man das tut, hierdurch, gleichzeitig* || besides * *überdies* || nevertheless * *dennoch* || yet * *dennoch* || for all that * *dennoch* || present * *anwesend* || there * *anwesend* || part * *take part in: teilnehmen an, dabei sein* || on the occasion * *anlässlich* || then * *dann, anlässlich* || as a result * *dadurch* || thereby * *dadurch, auf diese Weise, daran, davon, nahe dabei*
Dach roof * *roof section: Dachteil eines Schaltschranks* || hood || top
Dachaufsatz top assembly * *eines Schaltschranks* || roof section * *z.B. eines Schaltschranks; siehe auch →Dachhaube*
Dachblech top cover || top cover sheet || top sheet cover || top plate * *aus dickerem Blech* || roof section * *z.B. eines Schaltschranks*
Dachgesellschaft holding company || holding
Dachhaube roof cover assembly * *Über einem Schaltschrank angebrachte ~ zum Schutz der Kühlluft-Austrittsöffnung* || roof chamber * *z.B. eines Schaltschranks* || roof section * *Dachteil (z.B. eines Schaltschranks)*
Dachventilator roof fan || roof-mounted fan
dadurch by it * *auf solche Weise* || through it * *auf solche Weise* || thereby * *auf diese/solche Weise* || in this manner * *auf diese Art und Weise (am Satzanfang)* || in this way * *auf diese Art und Weise (am Satzanfang)* || by that means * *auf diesem Weg, hierdurch* || thus * *somit, folglich, demgemäß, so, folgendermaßen* || through there * *(örtlich) da hindurch* || that way * *(örtlich) auf diesem Weg* || owing to * *dadurch dass, wegen* || owing to the fact that * *dadurch dass, wegen; auf Grund d.Tatsache ,dass* || thanks to * *dank* || by || indem || as * *weil* || because * *weil*
dagegen against it * *(Adv.)* || on the other hand * *andererseits* || however * *andererseits, jedoch* || on the contrary * *indessen* || but then * *indessen* || whereas * *während* || whilst * *während* || while * *während*
daher in consequence of this || therefore || hence * *folglich,deshalb,also* || consequently * *folglich*
Dahlander-Schaltung Dahlander-connection * *für polumschaltbaren Motor, ermöglicht 2 Drehzahlen*
damit thereby * *dadurch* || hence * *daher, deshalb, somit*
Damm dam
dämmen insulate * *z.B. Schall, Wärme ~* || absorb * *absorbieren*
Dämmstoff insulating material
Dampf vapor * *(amerikan.)* || steam || vapour * *(brit.)*
dampfbeheizt steam-heated
Dampfdruck steam pressure
Dampfdüse steam nozzle
Dämpfe vapours
dämpfen damp * *(elektrisch, mechanisch, akustisch); Schwingung, auch Stimmung ~* || dampen * *(elektrisch, mechanisch, akustisch) auch Stimmung; dampen down: abdämpfen* || deaden * *(akustisch)* || absorb sound * *Geräusche ~* || cushion * *(mechanisch) puffern, abfedern, weich betten, (Stoß, Fall) dämpfen* || absorb * *Stöße, Schall, Schwingungen absorbieren/dämpfen* || attenuate * *(elektr.) abschwächen* || muffle * *(akustisch) Ton ~* || soften * *Licht ~* || stabilize * *Schwingungen ~* || suppress * *unterdrücken* || weaken * *(ab)schwächen*
Dämpferkäfig damping cage * *im Läufer eines AC-Motors*
Dämpferring short-circuit ring
Dämpferwicklung damper winding

Dampferzeuger steam generator
Dampffeuchte steam moisture
Dampfkessel steam boiler
Dampfturbine steam turbine
Dämpfung damping * *(elektr., mechan., akust.) abschwächen, drosseln, auch Schwingungen* || attenuation * *Abschwächg.(z.B. Signal auf Leitg./an Messgerät.), Dämpfg. elektr./opt./mech. Schwingg.* || loss * *von Energien* || stabilization * *Stabilisierung* || suppression * *Unterdrückung* || slowing down * *Nachlassen* || cushioning * *Abfederung, weiche Aufhängung* || muffling * *Geräusche, Ton (siehe auch 'dämpfen')* || damping effect * *Dämpfungseffekt*
Dämpfungselement damper * *mechanisch, elektrisch und akustisch*
Dämpfungsglied attenuator * *(elektr.) Abschwächer* || damper * *(mechan., elektr. und akustisch)*
Dämpfungsgrad damping coefficient
Dämpfungskreis damping circuit
danach after that || afterwards * *später* || later on * *später* || subsequently * *anschließend* || thereupon * *anschließend* || according to it * *gemäß* || accordingly * *entsprechend* || about * *ask about: danach fragen*
daneben adjacent * *angrenzend/-liegend/-stoßend; benachbart* || beside it * *~liegend* || near it * *in der Nähe liegend* || next to it * *nächstliegend, angrenzend* || close by it * *dicht ~* || hard by it * *dicht ~* || besides * *außerdem* || moreover * *außerdem* || at the same time * *gleichzeitig* || parallel to it * *gleichzeitig* || beside * *vorbei; beside the mark: am Ziel vorbei* || missing * *verpassend; miss: vorbeigehen; miss the mark: danebentreffen/ schlagen* || failing * *verpassend; fail to hit: am Ziel vorbeigehen/nicht treffen*
Dänemark Denmark
dank thanks to || owing to || thanks * *(Substantiv)*
dankbar thankful * *in Wort und Tat* || grateful * *innerlich* || obliged * *verpflichtet* || worthwhile * *lohnend* || satisfactory * *befriedigend*
danke thanks || many thanks * *dankeschön!* || thank you || thank you very much * *dankeschön* || no, thank you * *bei Ablehnung* || you are welcome * *nichts zu danken!*
danken thank * *for: für* || return thanks || decline with thanks * *ablehnend*
Dankschreiben letter of thanks || thank you letter
dann then
daraufhin thereupon * *zeitlich*
daraus folgt hence it follows * *that: dass*
dargestellt shown * *z.B. in einer Zeichung, einer Tabelle* || illustrated
darlegen lay down * *in einer Schrift* || outline * *in Umrissen, in groben Zügen*
Darlingtonstufe Darlington stage * *Transistorschaltung (aus zwei "hintereinandergeschalteten" Transistoren)*
darstellen represent * *allg.* || show * *veranschaulichen, z.B.in Bild/Tabelle/Anhang* || describe * *beschreiben* || illustrate * *illustrieren* || figure * *grafisch, rechnerisch ~* || plot * *grafisch, zeichnerisch ~* || chart * *grafisch ~, mit Diagrammen* || outline * *in Umrissen, in groben Zügen ~* || be * *sein* || present || render * *wiedergeben, interpretieren* || depict * *(bildlich) darstellen, veranschaulichen, zeichnen, (ab)malen, schildern*
Darstellung representation * *z.B. Zahlendarstellung* || notation * *Notation*
Darstellungsart way of representation || method of representation || representation mode || representation type
Darstellungsform representation method
darüber over that || over it || over them * *(Plural)* || above it * *(auch figürl.: over and above: darüber*

hinaus) || on top of it || on that point * *in dieser Hinsicht* || about that * *in dieser Hinsicht* || on that matter * *bezüglich dieser Angelegenheit/Sache*
darüber hinaus moreover || above that * *bei mehr/größerem* || what is more || beyond it || past it || in addition to it * *(figürl.)* || over and above it * *(figürl.)* || over it * *darüber hinweg*
darüberhinaus furthermore || further || in addition
darunter underneath || under that || under it || under them * *(Plural)* || beneath it || below || among them * *unter einer Anzahl von Dingen* || including * *einschließlich* || less * *weniger*
das ist sinnvoll that makes sense
das vorliegende this
dasselbe the same
Datei file
Dateibelegungstabelle FAT * *'File Allocation Table' (siehe →FAT) zur Dateibuchhaltung in Betriebssystem*
Daten data * *(Plural, nie mit End-s verwenden: "datas" ist falsch!)* || facts * *Tatsachen* || features * *Eigenschaften* || datum * *(Singular von 'data') ein Einzeldatum*
Daten-Rettroutine data-save routine
Datenaustausch data communication || data transmission * *Übertragung* || data exchange || data interchange || data communications || data transfer * *Datenübertragung*
Datenautobahn data highway
Datenbank database
Datenbasis data base
Datenbaustein data block * *z.B. bei SPS* || data module * *z.B. bei SPS*
Datenbereich data area || data block || data array * *Datenfeld* || data record * *Datensatz* || data segment * *Datensegment (im Gegensatz zum Programmcode-Segment)* || data storage area * *im Speicher einer SPS*
Datenbit data bit
Datenblatt data sheet || data form * *Formular zum Eintragen individueller Werte (Rechenblatt)*
Datenblattangabe data sheet information || published data * *vom Hersteller veröffentlichte Daten*
Datenbreite data width
Datenbus data bus
Datendarstellung data representation || method of data representation * *Art d. Datendarstellung*
Dateneingabe data entry
Datenerfassung data acquisition * *von Messdaten* || data recording
Datenfernübertragung remote data transmission
Datenfluss data flow
Datenflussplan data flowchart
Datenhaltung data storage * *Speicherung* || data management * *Verwaltung*
Dateninhalte data contents
Datenkonsistenz data integrity * *keine Durchmischung von Teildaten aus verschiedenen Zyklen in einem Datensatz* || data consistency
Datenleitung data line
Datenmenge extent of data || volume of data
Datenmengen data quantities || amounts of data
Datenpaket data packet
Datenpuffer data buffer
Datenrettroutine data save routine
Datensatz data set || data record || data block * *Datenblock* || data area * *Datenbereich* || record * *[Computer | NC-Steuerg.] in einer Datenbank (-Tabelle) oder in einem NC-Programm*
Datensicherheit data security || data integrity * *Unverfälschtheit der Daten*
Datensicherung data protection || data security || data save || data backup
Datenspeicher data memory || data register * *Datenregister bei SPS*

Datenstrom data stream || data flow * *Datenfluss*
Datenträger storage medium || data storage medium || data carrier * *Diskette, Magnetband usw.; mobile data carrier: mobiler* ~ *(z.B. in Materialflussystem)* || data media * *(Plural)* || data medium
Datentransfer data transfer
Datentransferprogramm data transfer program
Datentyp data type || variable type * *Variablentyp*
Datenübermittlung data transmission || data communication
Datenübertragung data communication || data transmission || data transfer
Datenübertragungsfehler data transmission error || communication error * *Kommunikationsfehler*
Datenübertragungsgeschwindigkeit data-transfer rate
Datenübertragungsrate data-transfer rate
Datenübertragungssteuerung data transfer control || data transmission control
Datenverarbeitung data processing || computer science * *als Wissenschaftszweig* || computer engineering * *als Fachrichtung* || DP * *Abk. f. 'Data Processing'* || computing * *Rechnertechnik (als Fachgebiet)* || information technology || IT * *Abk. für 'Information Technology'*
Datenverarbeitungssystem data processing system
Datenverkehr data interchange || data exchange || data communication || data transfer || data transmission || data traffic
Datenverlust data loss || memory loss * *in einem Datenspeicher* || loss of data
Datenwort data word * *meist 16-bit-Information*
Datiereinrichtung date-stamping system * *z.B. in der Nahrungsmittel- u. Verpackungsindustrie*
Datum date * *Tag/Monat/Jahr* || datum * *(Singular von 'data') ein Einzeldatum* || data item * *(Singular von 'Daten')*
DAU DAC * *Abk. für 'Digital to Analog Converter'* || DA converter
Dauer duration || period * *Zeitraum* || length of time * *Zeitlänge*
Dauer- continuous || permanent || long-term || standing
Dauer-Ausgangsstrom permanent output current || continuous output current
Dauerbelastbarkeit continuous loading capability
Dauerbelastung continuous loading || continuous duty * *das dauernd zu leistende (Betriebsart)* || continuous load * *Dauerlast* || continuous stressing * *Dauerbeanspruchung*
Dauerbetrieb continuous operation * *für Motor: siehe EN 60034-1* || continuous duty * *Belastung, Betriebsart* || continuous running || continuous running duty type * *{el.Masch.}Betriebsart S1* || continuous duty type * *Betriebsart S1 nach VDE 0530/DIN 41756 für Motoren/Stromrichter* || permanent service * *continuous duty cycle* || continuous-running duty * *Betriebsart S1 für Motoren/Stromrichter nach VDE 0530* || permanent operation
Dauerbremsleistung continuous braking power || continuous brake power
dauerelastisch permanently flexible
Dauerfestigkeit long-term endurance
dauergeschmiert permanently lubricated || prelubricated
Dauergleichstrom continuous direct current || continuous DC current || continuous DC forward current * *einer Diode* || continuous DC on-state current * *eines Thyristors*
Dauergrenzspannung category voltage * *Kategoriespannung, z.B. nach IEC*
Dauergrenzstrom maximum mean forward current * *Diode* || maximum mean on-state current * *Thyristor*

dauerhaft durable * *dauerhaft, haltbar, beständig* || lasting * *langwährend, nachhaltig, bleibend* || long-term * *(zeitlich)* || stable * *fest* || enduring * *nachhaltig, bleibend* || persistent * *hartnäckig* || permanent * *bleibend, (fort-) dauernd, ständig, Dauer-, dauerhaft, massiv* || permanently * *(Adv.)*
Dauerimpuls continuous pulse
Dauerladen float charge * *Batterie*
Dauerlast continuous load || continuous loading || permanent service * *Dauerbelastung*
Dauerlastgrenze continuous load limit
Dauerlaststrom continuous load current
Dauerleistung continuous power * *maximum permissible: maximal erlaubte* || continuous output * *z.B. ~ eines Motors/Stromrichters* || continuous rating * *Nenn~, z.B. eines Motors*
Dauerlicht steady light * *z.B. einer Diagnose-Leuchtdiode* || continuous light * *z.B. einer Diagnose-Leuchtdiode*
Dauermagnet permanent-magnet || permanent magnet
Dauermagnetbremse permanent-magnet brake
dauermagneterregt permanently excited || permanent-magnet excited || permanent-field
Dauermagnetmotor permanent magnet motor
dauern last
dauernd permanent || continuous * *auch ständig* || continuously * *(Adv.)* || continually * *(Adv.)* || fortwährend, dauernd, immer wieder, ständig || continual * *immer wiederkehrend, (sehr) häufig, oft, wiederholt, fortwährend, dauernd*
dauernd belastbar can be continuously loaded * *with: mit*
Dauernennstrom continuous rated current
Dauerprüfung endurance test
Dauerschleife idle loop * *Untätigkeitsschleife* || endless loop * *Endlosschleife* || continuous loop
Dauerschmierung permanent lubrication || long-term lubrication * *Langzeitschmierung* || continuous lubrication
Dauerschwingprüfung vibration fatigue test
Dauersignal continuous signal || maintained signal
Dauerstrom continuous current
Dauertest long-time test
Dauerversuch long-time test || continuous test
Daumenregel rule of thumb * *siehe auch* →*Faustformel*
Daumenwert rough value || recommended value || guide value * *Richtwert, empfohlener Wert* || empirical value * *Erfahrungswert* || practically proven value * *in der Praxis bewährter Wert* || pragmatical value
davon of that * *(Adv.)* || thereof * *(Adv.) z.B. i. Zusammenhang mit Zahlenangabe* || off * *hinweg/-fort* || away * *hinweg/-fort* || thereby * *thereby affected: davon betroffen*
davorliegend preceding || upstream * *im Signal-/Leistungs-/Energiefluss* || situated in front of
dazu bestimmt sein be intended for * *bestimmt/vorgesehen sein für* || be destined for * *bestimmt/vorgesehen sein für* || be meant for * *bestimmt/vorgesehen sein für* || be bound to * *etwas tun zu müssen*
dazugehörig belonging to it || forming part of it || pertinent * *(zur Sache) gehörig, einschlägig, sachdienlich, passend, angemessen* || related * *in Beziehung/Zusammenhang stehend* || associated * *zugeordnet*
dazwischenkommen intervene || supervene * *auch unvermutet eintreten*
dazwischenliegend interposed || intermediate * *(zeitlich u. räumlich)*
DCS DCS * *abkürzg.f. 'Digital Control System': Digital-Regelsystem, vorzugsw. f.verfahrenstechn. Prozesse*
deaktivieren deactivate || disable * *sperren* || obstruct * *blockieren, hemmen, lahm legen, hindern, versperren* || block * *sperren* || suppress * *unterdrücken* || shut down * *stillsetzen* || put out of action * *außer Betrieb/Gefecht setzen*
Deck deck * *Schiffsdeck*
Deckanstrich final coat
Deckblatt cover sheet * *z.b.eines Berichts/einer Telefax-Nachricht/eines Dokuments* || title sheet * *z.B. eines Buches*
Decke ceiling * *eines Raums* || blanket * *Stoffdecke* || awning * *Plane*
Deckel cover * *Abdeckung, auch Buch~; side/top cover: Seiten-/obenliegender ~* || lid * *hinged lid: Klapp~ (bei Behälter usw.), lid catch: ~verschluss* || top * *screw top: Schraub~* || cap * *Kappe* || top cover * *obenliegender ~*
Deckelschraube cover screw
decken cover * *auch figürlich*
Deckscheibe cover washer * *z.B. bei Wälzlager* || cover plate * *z.B. zur Abdichtung bei Wälzlager*
Deckung alignment * *Fluchtung; bring into alignment: zur ~ bringen* || covering || cover * *~ der Kosten usw.; Sicherheit*
deckungsgleich congruent * *(geometr.)* || non-overlapping * *nicht überlappend*
deckwassergeschützt deckwater-proof * *z.B. Motor-Schutzart für Einsatz auf Schiffen (Schutzart IP56)*
Defaultwert default value *' Vorbesetzungswert* || initial value
defekt out of order * *nicht in Ordnung, defekt* || broken || damaged || faulty || imperfect * *mangel/fehlerhaft* || fault * *(Substantiv) in: an* || defect * *(Substantiv) in: an* || flaw * *(Substantiv) fehlerhafte Stelle, Fehler, Defekt, Bruch, Riss, Sprung* || defective
definieren define
definiert defined || in a defined fashion * *(Adv.) auf ~e Art und Weise*
Definition definition
deformieren deform
Defuzzifizierung defuzzification * *bei Fuzzy-Regelung*
degressiv tapered || deminishing
dehnbar extendible || elastic || flexible
dehnen stretch || expand * *ausdehnen, z.B. Impulse* || draw * *ziehen, verstrecken* || tension || elongate * *in die Länge ziehen*
Dehnung stretching * *elastische ~* || extension * *permanent: bleibende* || expansion * *Ausdehnung* || stretch * *elastische ~* || elongation * *verformende/bleibende ~* || dilatation * *z.B. Wärme~* || longitudinal stress * *Längsspannung/-beanspruchung*
Dehnungsmessstreifen strain gauge || SG * *Abkürzung für 'strain gauge'* || strain gage * *(amerikan.)*
Dehnungsschlaufe expansion loop * *in (Halbleiterschutz-) Sicherung*
Dejustage misalignment
DEK DEK * *Abk.f. Dansk Elektrotekinsk Komite': Dänisches Elektrotechn. Komitee (ähnlich DKE)*
Dekade decade
dekadisch decade
Dekanter decanter * *Zentrifuge zur Trennung von Substanzen z.B. Fest- u.Flüssigstoffe i. d. Getränkeindustrie*
Deklaration declaration
dekodieren decode
Dekodierer decoder
dekomprimieren decompress * *auch Daten* || unzip * *(salopp) Daten*
dekontaminierbar can be decontaminated || decontaminable
Dekorpapier decorating paper
Dekotextilien decorative textiles
Delta-I-Regelung delta-I control

Demagnetisierung demagnetization
Demontage disassembly || dismantling * *ganzer Werksanlagen* || release * *"Entschnappen" eines Geräts von einer Montage-Hutschiene*
demontierbar dismountable || easy-to-remove * *leicht ~; siehe auch →abnehmbar*
demontieren disassemble || dismantle * *auseinanderbauen* || detach * *abbauen/-montieren* || gut * *ausweiden (auch techn. Anlage usw.)*
der vorliegende this
derartig such || of that kind || of such kind || of this nature * *a voltage notch of this nature: ein ~er Spannungseinbruch* || such as this * *(nachgestellt)* || of this kind || in such a manner || in such a way
Derivat derivate
derjenige he who || the one who || that which * *(auf eine Sache bezogen)*
derselbe the same || one and the same * *ein und ~*
derzeit presently * *gegenwärtig* || at the present time
derzeitig present
deshalb therefore * *als Folge davon, infolgedessen* || for that reason * *aus diesem Grund, als Folge davon* || for this account * *deswegen* || that is why || for that purpose * *für diesen Zweck* || to that end * *für diesen Zweck* || this is why * *aus diesem Grunde* || on that account * *als Folge davon, infolgedessen* || that's why * *deswegen* || so || hence * *so, daher, folglich*
Design styling * *äußeres Design, stilistische Formgebung* || design * *Konstruktion, Auslegung, Entwicklung*
dessen whose * *für Personen, umgangssprachlich auch für Sachen* || of which * *(nachgestellt) für Sachen*
deswegen therefore * *→daher* || for that reason * *→daher*
Detail detail
Detaildarstellung detail representation || zoom || detail display * *am Bildschirm* || detailed drawing * *Detailzeichnung*
Detailinformationen detailed information
detailliert detailed || with full details || in detail * *im Einzelnen* || precise * *genau*
Determinante determinant * *{Mathem.}*
deterministisch deterministic * *vorausbestimmbar/-berechenbar*
deutlich clear * *klar* || distinct * *klar, ausgeprägt, entschieden, merklich, spürbar* || plain * *klar, leicht verständlich, einfach, schlicht* || intelligible * *verständlich* || obvious * *einleuchtend* || evident * *offensichtlich* || considerable * *beträchtlich* || definite * *bestimmt, fest, klar, eindeutig, genau* || marked * *merklich, auffällig, markant, ausgeprägt*
significant * *bedeutsam, spürbar* perceptible * *wahrnehmbar* || noticeable * *wahrnehmbar* || visible * *sichtbar* || audible * *hörbar* || significantly * *(Adv.)*
deutsch German
Deutschland Germany
deutschsprachig German-language
dezentral distributed * *verteilt angeordnet* || decentralized || remote * *entfernt angeordnet, ferngesteuert* || local
dezentrale Peripherie decentralized peripherals * *ferngesteuerte Klemmenleiste, z.B. SIMATIC (R) ET100, ET200 von SIEMENS* || distributed periphery * *ferngesteuerte Klemmenleiste, z.B. SIMATIC (R) ET100, ET200 von SIEMENS* || distributed peripherals * *ferngesteuerte/intelligente Klemmenleiste, z.B. ET 100/200 von SIEMENS* || remote I/O * *entfernt angeordnete Ein-/Ausgaben, elektronische Klemmenleiste*
dezentrale Steuerung distributed control
dezentraler Aufbau distributed configuration

dezentrales Peripheriegerät decentral peripheral device || remote I/O unit
Dezibel decibel
dezimal decimal
Dezimalpunkt decimal point
Dezimalsystem decimal system
Dezimalzahl decimal || decimal number
dh i.e. * *das heißt*
Diagnaose-LED diagnostic LED
Diagnose diagnostics || diagnosis * *[daieg'nousis]* || troubleshooting * *→Fehlersuche, →Fehlerbeseitigung*
Diagnoseanzeige diagnostic indicator || diagnostic display
Diagnosebaustein diagnostic block * *Funktionsbaustein* || diagnosis function block * *Funktionsbaustein* || debug function block * *Funktionsbaustein*
Diagnoseeinrichtung diagnostic aid system || diagnostic equipment
Diagnosefunktion diagnostic function
Diagnosegerät diagnosis unit || diagnostic unit
Diagnosehilfe diagnostic aid || diagnostic help || diagnostic tool * *Diagnosehilfsmittel/Werkzeug* || diagnostic support * *Unterstützung*
Diagnosemeldung diagnostic signal || diagnostic message
Diagnosemittel diagnostic aids || diagnostic tool
Diagnosespeicher diagnostic memory || trace memory * *speichert Signalverlauf*
Diagnosestecker diagnosis connector || diagnostic plug
Diagnosesystem diagnostic system
Diagnosetechnik diagnostics technology
Diagnosezwecke diagnostics purposes
diagnostizieren diagnose
diagonal diagonal || transverse * *schräg, quer- (verlaufend), sich überschneidend; (to: zu)* || transversal * *schräg, quer- verlaufend), sich überschneidend; (to: zu)*
Diagonale diagonal
Diagramm diagram || chart * *grafische Darstellung, Kurvenblatt, Schaubild* || plot * *Kurvendiagramm, Messschrieb* || graph * *['græ: f] Schaubild, grafische Darstellung, Kurvenbild/-blatt*
Dialog dialogue || dialog * *(amerikan.)*
dialogfähig interactive
dialoggesteuert menue driven || interactive || with operator prompting
Dialogprogramm interactive program
Diamant diamond
diamantbestückt diamond-tipped * *{Werkz. masch.\ Holz} Werkzeug usw.*
Diamantwerkzeug diamond tool
dicht dense || tight
Dichte density * *high-/low-density mit geringer/hoher ~* || concentration * *[Chemie]* || consistency * *Konsistenz, Beschaffenheit, →Dichtheit, Dicke, Übereinstimmung, Folgerichtigkeit* || volumetric mass
Dichtelement sealing element || sealing
dichten seal * *abdichten*
Dichtfläche sealing face || sealing surface || seal face
Dichtgummi sealing rubber
Dichtheit tightness * *z.B. gegen Luft, Wasser; test for leaks: auf ~ prüfen* || compactness * *Kompaktheit, Festigkeit, Gedrängtheit* || density * *→Dichte, Dichtheit, Gedrängtheit, Enge* || closeness * *Dichte, Festigkeit, Knappheit, Verschlossenheit, Enge*
Dichtigkeitsprüfung leak test * *leak-tested: auf Dichtigkeit geprüft* || sealing test || high-pressure test * *~ mit Überdruck*

Differenzverstärker

Dichtlippe seal lip * *z.B. bei einem Dichtring* || sealing lip
Dichtmanschette U-seal || cup packing || U packing ring
Dichtmasse sealing compound || sealing agent || sealant
Dichtmittel sealant * →*Dichtmasse*
Dichtpropil sealing strip * *z.B. für Schranktür*
Dichtring sealing ring || seal ring || sealing washer || rubber sealing washer * *Gummidichtring*
Dichtscheibe sealing washer * *z.B. Gummidichtscheibe in PG-Verschraubung* || side plate * *eines Wälzlagers* || sealing disc || packing washer || sealing disk
Dichtung seal || gasket * *Dichtungsring/-manschette/-zwischenlage*
Dichtungselement sealing element
Dichtungsmasse sealing compound || sealing agent
Dichtungsmittel sealant
Dichtungssatz sealing kit
dick thick || heavy-gauge * *Draht, Kabel, Blech* || big * *massig* || large * *groß, massig* || bulky * *massig* || voluminous * *voluminös, umfangreich* || stout * *umfangreich* || viscid * *zähflüssig, dickflüssig, klebrig* || sirupy * *zähflüssig* || fat * *fett*
Dicke thickness || gauge * *Stärke von Draht, Blech, einer Rohr/Gehäusewandung usw.* || caliper * *von Papier*
Dicken-Messanlage thickness gauge
Dickenhobel thicknessing planer * *{Hobel}*
Dickenhobelmaschine thicknessing planer * *{Holz}* || thicknessing machine * *{Holz}*
Dickenmessanlage thickness gauge || gauge
Dickenmesser thickness gauge
Dickenmessgeber thickness gauge
Dickenmessgerät thickness gauge
Dickenmessung thickness gauge * *Dickenmessanlage/-Einrichtung* || caliper measurement * *z.B. für Papier*
dickflüssig viscous || semifluid || viscid
Dickschichtschaltung thick-film circuit
Dickschichttechnik thick-film technology * *z.B. zur Herstellung v. Hybridbauteilen auf Keramikträger*
Dickstoff slush * *Schlamm, Matsch* || slurry * *Schlamm, Matsch* || solid * →*Feststoff* || thick mud * *Dickschlamm* || thick mash * *Dickmaische*
Dickungsmittel thickener
die Oberhand gewinnen über overrule
die vorliegende this
Dielektrikum dielectric * *allg.Ausdruck f. e. festen, flüssigen, gasförmigen Isolierstoff o. Vakuum als Isolierst.*
dielektrisch dielectric
dielektrische Beanspruchung dielectric stress * *Beanspruchung eines Isolierstoffs (Dielektrikum) durch ein elektr. Feld*
Dielektrizitätskonstante dielectric constant || permittivity
dienen serve * *as: als; for: (da)zu; it serves the purpose of: es dient dazu, ...* || be used * *in: verwendet werden für* || act * *fungieren, arbeiten, tätig sein, agieren, wirken, eingreifen (as: als)* || be conducive to * *dienlich/förderlich sein*
dienen als serve as || act as * *fungieren/agieren/arbeiten/tätig sein/eingreifen als*
Dienst service * *Dienstleistung, Dienstleistungseinrichtung, ~ eines Feldbussystems usw.*
Dienstleistung service
Dienstleistungsbetrieb service company || service provider
Dienstplan personnel schedule
Dienstprogramme system software || utilities
Dienstreise official trip || official journey || sales trip * *zu Verkaufsgesprächen bei Kunden* || business trip * *Geschäftsreise* || call on customers * *Kundenbesuch (at: an einem Ort)*
Dienststelle department || Dept. || authority * *bei einer Behörde*
Dienstzeit working hours * *Arbeitszeit*
diesbezüglich referring to this || relating thereto || relevant * *to this: in dieser Sache* || pertinent * *to this: in dieser Sache* || appropriate * *passend*
Dieselaggregat Diesel generator || Diesel-generator set
Dieselbetrieb diesel-engine operation
dieselelektrisch diesel-electric
dieselelektrischer Antrieb diesel-electric drive
Dieselgenerator Diesel generator || Diesel-generator set
dieselhydraulisch diesel-hydraulic
Dieselkraftwerk Diesel power station
Dieselsstromversorgung diesel-generator supply
Differenz difference * *between: zwischen*
Differenzdruck differential pressure || pressure drop * *Druckabfall* || DP * *Abkürzg. für 'Differential Pressure'*
Differenzdruckgeber differential pressure transmitter
Differenzeingang differential input * *z.B. eines Verstärkers (nicht massebezogen)*
Differenzempfänger differential receiver || differential line receiver * *Leitungsempfänger*
Differenzial differential * *differential gear· ~getriebe; differential equation: ~gleichung*
Differenzial-Leitungsempfänger differential line receiver
Differenzial-Leitungstreiber differential line driver
Differenzialempfänger differential receiver || differential line receiver * *Leitungsempfänger*
Differenzialgetriebe differential gear
Differenzialgleichung differential equation * *solve: lösen*
Differenzialgleichungssystem differential equation system
Differenzialglied derivative element || derivative-action element
Differenzialleitungsempfänger differential line receiver
Differenzialleitungstreiber differential line driver
Differenzialquotient derivative * *of the third order: dritter Ordnung* || differential coefficient
Differenzialregler derivative-action regulator || derivative regulator
Differenzialsignal differential signal * *z.B. für Analogeingangsklemme, RS485 (Gegensatz: single ended signal)*
Differenzialtreiber differential driver || differential line driver * *Leitungstreiber*
Differenzier-Zeitkonstante differentiation time constant
differenzieren differentiate * *unterscheiden, einen Unterschied machen* || derive * *erste Ableitung bilden (math.)* || make a distinction * *unterscheiden, einen Unterschied machen*
Differenzierer derivative element * *Differenzierglied liefert am Ausg.die 1.Ableit. (nach d.Zeit) der Eingangsgröße* || differentiator * *~ als Funktionsbaustein*
Differenzierglied derivative-action element || derivative element || differentiator
Differenzierzeit differentiation time || derivative-action time
Differenzsignal differential signal * *z.B. für Analogeingangsklemme, Verstärker, RS482-Signal*
Differenzspannung differential voltage * *z.B. an Differenzverstärker-Eingang* || potential difference * *Potenzialdifferenz*
Differenzverstärker differential amplifier

diffundieren diffuse
Diffusion diffusion
Diffusionsstrom diffusion current
digital digital || digitally * *(Adv.)*
Digital-Analog-Wandler digital-analog converter || DAC
Digital-Analogwandler digital to analog converter || digital-analog convertor * *(veraltet)*
Digital-Ausgabebaugruppe digital output module
Digital-Eingabebaugruppe digital input module
Digitalanzeigegerät digital readout system || digital display unit
Digitalausgang digital output || binary output * *Binärausgang* || logic output * *Binärausgang* || status output * *für (Betriebs-) Zustandssignal*
digitale Drehzahlistwerterfassung digital speed actual value evaluation || digital speed evaluation || digital speed measurement || digital speed sensing
digitale Funktion digital function || digital operation
digitale Regelung digital control || direct digital control || DDC * *Abkürzg. für 'Direct Digital Control'*
digitale Signalverarbeitung digital signal processing
digitale Wegerfassung digital position sensing || digital position decoding module * *als Baustein*
Digitaleingang digital input || binary input * *Bináreingang* || logic input || command input * *für Steuerbefehl*
digitales Regelsystem digital control system || digital control * *Einrichtung*
Digitalisiereinrichtung digitizer
digitalisieren digitize * *digitizer: Digitalisiereinrichtung; digitizing tablet: Digitalisiertablett* || digitalize
digitalisiert digitized || digitalized
Digitalisierung digitizing || digitization || analog-to-digital conversion * *Analog-Digital-Wandlung* || digitalization
Digitalregler digital regulator || digital controller
Digitalspeicher-Oszilloskop digital storage oscilloscope || DSO * *Abkürzg. für 'Digital-Storage Oscilloscope'*
Digitaltacho pulse tachometer || pulse tacho || pulse tach || digital tachometer
Digitaltechnik digital technology || digital technique
DIL-Schalter DIP-type switch * *Abkürzg. für 'Dual In Line package': im DIL-Gehäuse*
Dimension engineering dimension * *physikalische* ~ || dimension * *auch physikalische* ~
dimensionieren dimension * *bemessen, auch einer Schaltung* || rate * *z.B. Stromrichter/Motor auslegen* || design * *eine Konstruktion; for: nach einem Zielkriterium* || lay out * *auslegen, entwerfen* || size * *Größe festlegen*
dimensioniert dimensioned || laid out * *ausgelegt* || rated * *von der Baugröße her* || sized * *(i. Wortzusamm.setzg.) von...Größe, -dimensioniert; z.B. full-/under-/over-/small-*
Dimensionierung dimensioning * *Bemessung, auch einer Schaltung* || engineering * *Entwurf, konstruktive/technische Auslegung* || sizing || rating
Dimensionsangabe dimension information
DIN DIN * *Abkürzg. für 'Deutsche Industrienorm bzw.Deut.Institut für Normg.': German Industry Standard(s)* || German Industry Standard
DIN EN DIN EN * *Deutsche Fassung einer Europäischen Norm*
DIN IEC DIN IEC * *Deutsche Fassung einer IEC-Norm*
DIN ISO DIN ISO * *Deutsche Fassung einer internationalen ISO-Norm*
DIN VDE DIN VDE * *VDE-Bestimmung, die gleichzeitig eine Deutsche Elektrotechnische Norm ist*
DIN-Schiene DIN mounting rail * *Aufschnapp-Installationsschiene, z.B. →C-Schiene oder →G-Schiene*
Diode diode
Diodenbrücke diode bridge
Diodengleichrichter diode rectifier || diode converter
Diodensatz diode set || diode stack
Diplomarbeit degree thesis work || diploma thesis || defence of thesis
Diplomingenieur Bachelor of Engineering || diploma'd engineer || B.Sc. * *(brit.) Abk. f. 'Bachelor of Science'* || graduated engineer * *→graduierter Ingenieur*
Dirac-Funktion Dirac pulse || unit pulse
Dirac-Impuls Dirac pulse * *"unendlich hoher" Nadelimpuls, Sprungantwort eines Differenzierers* || unit pulse * *Sprungantwort eines Differenzierers auf "Einheitssprung"*
direkt direct || directly * *(Adv.)*
direkt am Netz direct-on-line * *z.B. Betrieb eines Motors ohne Stromrichter/Sanftanlaufhilfe* || DOL * *Abkürzg. für 'Direct On Line'*
direkt eingelötet directly soldered-in * *z.B. ohne Sockel*
direkt gekoppelt directly coupled * *z.B. ohne Getriebe/Riementrieb*
direkt gespeist directly fed
direkt neben right next to
direkt proportional directly proportional
direkt proportional zu in direct proportion to
direkt-leitergekühlte Wicklung direct-cooled-conductor winding * *{el.Masch.} Kühlmittel umfließt direkt die Wicklungsleiter*
Direktantrieb direct drive || ungeared drive * *ohne Getriebe* || gearless drive * *ohne Getriebe*
direkte Adresse immediate address * *im Befehlscode steht direkt e.(konstanter)Wert, nicht d.Adresse einer Konstanten*
direkte Kühlung direct cooling * *ohne Sekundärkühlkreislauf*
direkte Leiterkühlung direct cooling of the conductors * *z.B. Direktkühlung der Wicklungsleiter bei Großmotor*
direkte Zugregelung closed loop tension control * *mit Zugerfassung durch Tänzer oder Zugmessdose* || closed loop tension regulation
Direkteinbau inline installation * *z.B. Magnetventil in e. Rohrleitung*
Direkteinfahrt direct approach * *{Aufzüge}*
Direkteinschaltung direct-on-line starting * *Direktstart e. AC-Motors am Netz ohne Anlasswiderst. oder Anlasshilfe*
direkter Draht direct line
direkter Netzanschluss direct connection to the line * *z.B. ohne Trafo oder Stromrichter*
direkter Speicherzugriff Direct Memory Access * *Peripherelement trennt CPU v.Speicher u.greift auf diesen direkt zu* || DMA * *Abk.f. 'Direct Memory Access'*
direktes Einschalten direct-on-line starting * *{el. Masch.}e.Motors direkt am Netz ohne Anlasshilfe/Umrichter/Softstarter* || DOL * *{el.Masch.}Direct-On-Line Starting: Start e.AC-Motors ohne Umrichter/Anlasser*
Direktor president || director || managing director * *Geschäftsführer* || principal
Direktstart direct-on-line starting * *eines Motors direkt am Netz ohne Umrichter/Sanftanlaufgerät*
Direktumrichter cyclo-converter * *setzt ohne Zw.kreis e. AC-Eingangs- i.AC-Ausg.spanng. mit max 0,5-facher Frequ.um*

Direktzugriff direct access
Disjunktion disjunction * *Schaltalgebra*
Diskette floppy disk || floppy || disk || diskette
Diskettenlaufwerk floppy disk drive || FDD * *Abkürzung für 'Floppy Disk Drive'* || diskette drive
diskontinuierlich discontinuous || discontinuously operating || batch * *Chargen-; batch process: Chargenprozess/~er (Produktions-) Prozess*
diskontinuierlicher Prozess batch process
diskontinuierlicher Regler discontinuous-action controller * *z.B. Digitalregler mit relativ langsamer Abtastung*
diskret discrete * *einzeln, einzeln/getrennt aufgebaut; z.B. Vorgang, Regelung, Bauelement*
Diskussion discussion
diskutieren discuss
disponieren plan ahead * *vorausplanen* || make arrangements * *Vorkehrungen treffen* || make dispositions || arrange matters || dispose * *over: über etwas/jdn.* || place orders * *Aufträge erteilen* || schedule orders * *zeitliche Verteilung von Aufträgen planen*
Disposition layout || disposition || coordinating * *Koordinierung* || purchasing schedule * *~ des Wareneinkaufs* || forward planning * *Vorausplanung* || arrangement * *Anordnung* || placing of orders * *Auftragserteilung* || plan * *Entwurf, Anlage, Plan* || disposal * *Verfügung* || scheduling * *(zeitliche) Planung*
Dispositionsplan disposition diagram || location diagram * *→Anordnungsplan*
Dispositionszeichnung layout drawing * *mit Darstellung einer Baugruppenträger/Schrankbelegung*
Dissertation dissertation || thesis
Distanzbuchse spacer sleeve
Distanzhalter spacer || stand-off
Distanzhülse spacer bush
Distanzring spacer ring
Distanzscheibe spacer washer
Distanzstück spacer * *Abstandshalter* || distance link || separator piece * *zum Trennen/Auseinanderhalten von Teilen* || distance piece
Distributor distributor
Disziplin discipline * *Wissenschaftszweig, Selbstdisziplin, Zucht*
Dithering dithering * *{Computer} Erzeugung von Zwischenfarben durch über-/nebeneinanderlegen von Farbpunkten*
dito ditto || same || the same || do * *Abkürzung für 'ditto': dito*
Divergenz divergence
diverse miscellaneous || sundry * *(Adj.) verschiedene, diverse, diverse, allerhand* || various
Diverses miscellaneous || sundries * *Diverses, Verschiedenes, diverse Unkosten, allerlei Dinge*
Diversifikation diversification
Dividend dividend || numerator * *Zähler, das über dem Bruchstrich stehende*
dividieren divide * *by: durch*
Dividierer divider
Division division
Divisor divisor || denominator * *Nenner, das unter dem Bruchstrich stehende*
DKE DKE * *Abk.f. 'Deutsche Elektrotechn. Kommission' im DIN und VDE* || DKE * *Abk. für 'Deutsche Elektrotechnische Kommission im DIN und VDE': Normengremium aus VDE u. DIN*
DLL DLL * *'Dynamic Link Library': Unterprogrammbibliothek, die erst bei Programmaufruf "gelinkt" wird*
DM DMK * *Deutsche Mark* || Deutschmark * *Einzahl* || Deutschmarks * *Mehrzahl*
DMS SG * *Abkürzung für 'Strain Gauge': Dehnungsmessstreifen* || strain gauge * *Dehnungsmessstreifen*

DNV DNV * *Abk. für 'Det Norske Veritas': Schiffsklassifizierungsgesellschaft*
Doffer doffer * *{Textil} Abnehmerwalze in bei Krempel-/Kardenmaschine*
Doktorarbeit doctoral thesis || dissertation || thesis
Dokumentation documentation * *bezüglich techn. Dokumentation siehe DIN-Entwurf 8418, 66055 u. VDI-Richtl. 4500*
dokumentieren document
Dokumentiersatz documenting package
Dolmetscher interpreter
Domäne domain * *Sphäre, →Gebiet, →Bereich, Reich; übergeordnete Adresskennung im WWW*
dominant dominant * *factor: Umstand, Faktor* || dominating * *dominierend* || preponderant * *→überwiegend, →entscheidend*
Dominanz predominance * *over: über* || predominancy
dominieren dominate || predominate * *vorherrschen* || prevail * *das Übergewicht haben, die Oberhand gewinnen (over, against), maß-/ausschlaggebend sein* || be predominant || have the upperhand
doppel double || dual || twin
Doppel- double- || twin- * *zweifach, Zwillings-* || duplex * *zweifach* || tandem * *aus 2 gleichen Teilen bestehend (z.B. Kompressor)* || dual
Doppel-Europaformat double-height Eurocard format * *Leiterkartenformat, T x H: 160 x 233,4 mm oder 220 x 233,4 mm*
Doppelantrieb twin drive * *z.B. Walzmotor-Paar*
Doppelbremse double brake
Doppeldreieck double-delta * *z.B. Schaltung für polumschaltbaren Motor*
Doppeleuropaformat double-Europe format * *T x H: 160 x 233.4 mm oder 220 x 233.4 für Leiterkarte* || double-Euroformat * *T x H: 160 x 233.4 mm oder 220 x 233.4 für Leiterkarte*
Doppeleuropakarte double-Eurocard * *Leiterkarte T x H: 160 x 233.4 mm oder 220 x 233.4 mm* || double-size Eurocard * *Leiterkarte T x H: 160 x 233.4 mm oder 220 x 233.4 mm*
Doppelfehler double fault
Doppelgehrungssäge double mitre cutting saw * *{Holz}* || double mitre saw * *{Holz}*
Doppelimpuls double pulse
doppelklicken double-click * *mit Computermaus (on: auf)*
Doppelkontakt double break contact * *z.B. bei Schütz*
Doppellager duplex bearing || double bearing
doppellogarithmisch log-log
Doppellüfter double fan
Doppelmotor twin motor * *{el.Masch.} 2 konzentrische ineinandergebaute Käfigläufermotoren (→doppelte Drehz.)*
Doppelpuffer double-buffer * *{Computer} zur Datenverwaltung/-übergabe*
Doppelpunkt colon
Doppelschaltstück double-moving contact
Doppelscheiben- double-disk || double-disc
Doppelschenkelbürste caliper-type brush * *zur Stromübertragung auf bewegte Maschinenteile* || double-leg brush * *zur Stromübertragung auf bewegte Maschinenteile*
Doppelschlussmotor double shunt-wound motor * *DC-Motor*
Doppelschutzart double explosion-safety * *Kombinat.d. Zündschutzarten d- Druckfeste Kapselg. u. e- Erhöhte Sicherheit*
doppelseitig double sided || at both ends * *beidseitig*
doppelseitiger Antrieb double-ended drive || bilateral drive

Doppelspuler dual take-up * *Doppelaufwickler (meist mit fliegendem Spulenwechsel) z.B. in Kabelmaschine*
Doppelstern double-star * *z.B. Schaltung für polumschaltbaren Motor*
Doppelsternschaltung double-star connection
Doppelstockklemme double-level terminal
Doppelstocktrafo double-tier transformer * *z.B. 3-Wickl.~ z.Speisg. e. 12puls-Stromricht., 2 um 30 versetzte Sek.wickl.* || phase-shifted transformer * *{Stromrichter} für 12-pulsige Stromrichterschaltung usw.*
Doppelstromrichter double converter || twin converter
doppelt double * *(Adj.u.Adv.)* || double the * *das Doppelte von, doppelt so viel wie* || twice the * *wert/zahlenmäßig; twice the amount: der doppelte Betrag/das Doppelte des Betrages* || twofold * *(Adj.)* || duplicate * *(Adj.)* || twin * *Zwillings-* || doubly * *(Adv.vor Adj.)* || twice * *(Adv.)* || twice as * *twice as much: doppelt so viel*
doppelt hoch double-height
doppelt wirkende Presse double-action press
Doppeltacho double tachometer * *z.B. zur Gewährleistung einer gewissen Sicherheitsstufe*
doppelter Genauigkeit double-precision
Doppeltragwalzenroller two drum surface winder * *Umfangswickler mit 2 Tragwalzen (Tragw. 1 drehz-, Tragw.2 momentengeregelt)* || two drum rewinder || twin drum winder
Doppelwickler double winder * *z.B für fliegenden Rollenwechsel*
Doppelwort double word * *z.B. Datenwort mit einer Länge von 32 Bit* || DWORD * *Abkürzung für 'double word'*
Dorn mandrel * *Wickeldorn, →Spanndorn, Dreh~, Hauptspindel* || arbor * *Aufsteck~, Aufnahme~, Aufsteckhalter; z.B. zum Aufstecken eines Schleif-/Fräswerkzeugs* || pin * *Nadel, Stift, Bolzen* || reamer * *Reib~, Reibahle, Räum~, Räumahle* || spike * *Spitze* || thorn * *allg. (auch figürl.)* || bolt * *Bolzen* || stem * *Stängel, Stiel, (Ventil-) Schaft, (Aufzieh-) Welle* || mandril * *Wickel~, Spann~, Dreh~* || prickle * *Stachel* || triblet * *Ausweite~*
Dornantrieb mandrel drive * *für Wickeldorn z.B. in Walzwerk, Blechverarbeitungsstraße*
DOS-Ebene DOS level
Dose can || tin * *Blechdose*
Dosenfüllmaschine can filling machine
Dosier-Füllmaschine volumetric metering filling machine * *in der Verpackungstechnik*
Dosieranlage dosing plant * *z.B. in der Getränkeindustrie* || proportioning plant || proportioning system
Dosiereinrichtung dosing equipment || proportioning device || metering unit || metering system
Dosieren dosing || dose * *(Verb)* || measure out * *(Verb)* || meter * *(Verb)*
Dosiergenauigkeit metering accuracy
Dosiermaschine dosing equipment || metering machine || auger doser * *mit Dosierschnecke* || proportioning plant * *Dosieranlage*
Dosierpumpe dosing pump || metering pump || proportioning pump
Dosierschnecke metering screw * *in der Verpackungstechnik* || proportioning screw || dosing auger
Dosierung metering || dosing || dosage
Dosierwaage dosing scales
dotieren dope * *Halbleiter*
Dotierung doping * *Halbleiter*
Doubliermaschine doubling machine * *zum Doublieren von Papier*
Download download * *Transfer von Dateien zum eigenen Computer (z.B. über d. →FTP-Protokoll im Internet)*

Dozent lecturer || instructor * *Kursleiter*
DP DP * *Abkürzg.für 'drip-proof': tropfwassergeschützt (Schutzart IP22)* || drip-proof * *tropfwassergeschützt (IP22)* || DP * *Abk. für 'Data Processing': Datenverarbeitung*
Draht wire * *solid: massiver, steifer* || lead * *Leitung* || conductor * *Leiter* || wire strand * *(Draht)-Litze, Litzendraht, Drahtseil* || strand * *Litze(nleiter)* || wire rope * *Drahtseil*
Draht-Fertigblock wire rod finishing block * *in Drahtwalzwerk*
Drahtbruch wire breakage || conductor breakage || cable breakage * *Kabel/Leitungsbruch*
Drahtbrucherkennung wire breakage sensing || conductor breakage detection
drahtbruchfest protected against wire breakage
drahtbruchsicher protected against wire breakage || fail-safe
Drahtbruchüberwachung wire-break monitoring || wire breakage monitoring || open-circuit monitoring
Drahtbrücke jumper
Drahtbürste wire brush
Drahtdicke wire gauge
Drahtende wire end
Drahtgeflecht wire mesh
Drahtgitter wire grate || wire grating
Drahtgitterpalette wire-lattice pallet
Drahtglühen wire annealing
Drahthaspel wire wheel || wire coiler
Drahtindustrie wire industry
Drahtisolierung strand insulation * *für Motorwicklung und elektr. Wickelgüter*
Drahtlitze wire strand
drahtlos wireless
Drahtpotentiometer wire-wound potentiometer
Drahtrichtmaschine wire straightener || straightening machine
Drahtseil wire rope
Drahtwalzwerk wire mill
Drahtwickelautomat automatic wire winder
Drahtwickeltechnik wire-wrap technique * *Verdrahtung über Wire-Wrap-Stifte mit Wickelpistole*
Drahtwiderstand wire-wound resistor
Drahtziehen wire drawing * *Drahtherstellung mittels →Drahtziehmaschine*
Drahtzieherei wire drawing mill
Drahtziehmaschine wire-drawing machine
Drainage drainage
Drall twist || spin || moment of momentum || spiral
DRAM DRAM * *Dynamic Random Access Memory, dynamisch aufzufrischender Schreib/Lesespeicher*
Dränage drainage
dranbleiben hold * *am Telefon* || hold the line * *am Telefon* || pursue * *eine Idee, Politik usw. verfolgen* || follow up * *seine Idee, Stoßrichtung usw. verfolgen; energetically: mit Nachdruck*
drängen press * *(auch figürl.)* || push || shove || urge * *(figürl.) auch zur Eile ~* || hurry * *zur Eile ~* || crowd * *sich ~* || be pressing * *dringend sein* || be urgent * *dringend sein*
drastisch drastic || dramatic || drastically * *(Adv.)* || dramatically * *(Adv.)* || drastical
Draufsicht top view * *Sicht von oben*
Dreh- rotational || swivel * *Dreh-, Schwenk-, dreh/schwenkbar (z.B. über Gelenk, Drehzapfen)* || rotating * *sich drehend* || oscillating * *(sich) hin- und herdrehend*
Drehachse axis of rotation || rotational axis || center line of rotation
Drehautomat automatic lathe
Drehband rotating belt
Drehbank lathe || turning lathe
drehbar turnable || rotatable || hinged * *angelenkt, mit Scharnier versehen* || can be turned * *drehbar*

sein || swivelling * *dreh/schwenkbar (z.B. über Gelenk, Drehzapfen)* || pivoted * *drehbar gelagert* || which can be rotated * *(nachgestellt)*
Drehbewegung   rotary motion || rotational motion || rotating movement
Drehbolzen   pintle * *z.B. bei Kabel-Auf-/Abwickler (greifen beidseitig in die Achse d. Kabeltrommel ein)*
Dreheiseninstrument   soft-iron instrument * *Weicheiseninstrument (misst Effektivwert)* || moving-iron type instrument
drehelastisch   torsionally flexible * *z.B. Kupplung*
Drehelastizität   angular flexibility
drehen   turn * *auch mit Drehmaschine* ~ || rotate * *rotate freely/unobstructed: (sich) frei/ungehindert* ~ || twist || strand * *verseilen (Kabel, Seil)* || revolve * *sich* ~ || spin * *(sich schnell) drehen*
drehend   rotating
drehende Teile   rotating parts
Drehfeld   phase sequence * *Phasenfolge: clockwise: Rechts-, counterclockwise: Links-* || direction of phase rotation * *Richtung des Drehfeldes* || rotating field * *umlaufendes Feld*
Drehfeld-unabhängig   phase sequence tolerant || phase rotation insensitive * *Netzphasen können beliebig angeschlossen werden*
Drehfelddrehzahl   synchronous speed
Drehfelderkennung   phase sequence identification
Drehfeldleistung   rotor power * *bei Motor*
Drehfeldmagnet   torque motor * *Drehmomentmotor mit hohem Stillstandsmoment*
Drehfeldmaschine   polyphase machine * *Oberbegriff für alle Drehstrommaschinen* || rotating-field machine || AC machine
Drehfeldmesser   phase-sequence indicator
Drehfeldmotor   polyphase motor * *Oberbegriff für alle Drehstrommotoren*
Drehfeldrichtung   rotating-field direction
Drehfeldüberwachung   phase sequence monitoring || rotating field monitoring
Drehfeldumkehr   field reversal || phase sequence reversal || field rotation reversal || rotating-field reversal
Drehfeldumschaltung   phase reversal
drehfeldunabhängig   phase rotation insensitive * *Netzphasen können beliebig angeschlossen werden* || phase sequence tolerant
Drehfilter   rotary filter
Drehfutter   lathe chuck
Drehgeber   rotary transducer || selsyn || rotary encoder * *digitaler*
Drehgeschwindigkeit   rotational velocity || angular velocity * *Winkelgeschwindigkeit*
Drehgestell   bogie * *{Bahn}* || wheel frame * *{Bahn}* Radgestell
Drehimpuls   angular momentum
Drehimpulsgeber   rotary pulse encoder || rotary incremental encoder || incremental encoder || incremental pulse encoder || pulse encoder
Drehknopf   control knob || knob || rotary knob
Drehkondensator   variable capacitor
Drehkreuz   turret * *Drehkreuz bei Wickler mit fliegendem Rollenwechsel* || turret butt * *Drehkreuz-Auf/Abrollung bei Wickler mit fliegend. Rollenwechsel (butt: aneinander fügen)* || turret winder * *Drehkreuzwickler* || carrousel || *Karussell* || spider * *z.B. bei Non-Stop-Wickler* || turnstile * *zur Zugangskontrolle an Durchgängen/Toren*
Drehkreuz-Rollenwechsler   turret butt splice * *für fliegenden Rollenwechsel bei Auf/Abwickler*
Drehkreuzwickler   turret winder * *Nonstop-Wickler mit fliegendem Rollenwechsel; automatic: automatischer* || carousel winder
Drehling   lathe tool * *Werkzeug für Drehmaschine* || tool bit * *Werkzeugeinsatz*

Drehmagnet   solenoid || torque motor * *{el.Masch.}* →*Drehmomentmotor, Drehfeldmagnet*
Drehmaschine   lathe * *{Werkz.masch.}* Drehbank || turning lathe * *{Werkz.masch.} Drehbank* || turning machine * *{Werkz.masch.\ Holz}*
Drehmatrix   rotary matrix
Drehmelder   synchro-generator * *Drehgeber für elektrische Welle* || synchro * *drehender Umformer, wandelt Wellenwinkelpositionen i.sinusförm. Signal um u.umgekehrt* || resolver
Drehmoment   torque * *['to:k] transmittable: übertragbares*
Drehmoment-Drehzahl-Ebene   torque/speed chart * *Koordinatensystem*
Drehmoment-Drehzahlkennlinie   torque-speed characteristic
Drehmoment-Drehzahlverlauf   torque/speed characteristic
Drehmoment-Messeinrichtung   torque meter || dynamometer * *Pendelmaschine, Momentenwaage*
Drehmoment-Überlastbarkeit   excess-torque capacity * *{el.Masch.} kurzzeitige Überlastbarkeit, ohne dass d.Motor stehenbleibt*
Drehmomentanhebung   torque boost
Drehmomentaufbau   torque build-up
Drehmomentbegrenzer   torque limiter * *siehe auch* →*Rutschkupplung,* →*Sicherheits-Rutschkupplung*
Drehmomentbereich   torque range
Drehmomentenaufbau   torque build-up
Drehmomentenbegrenzer   torque limiter * *z.B.* →*Rutschkupplung*
Drehmomentenbegrenzung   torque limiting
Drehmomentenrechner   torque calculator * *z.B. bei Motor-Prüfstand*
Drehmomentenregelung   torque control
Drehmomentenrichtung   torque direction || direction of torque
Drehmomentenschlüssel   torque spanner || torque wrench
Drehmomentenspitze   torque peak
Drehmomentenstoß   sudden change in torque
Drehmomentenstütze   torque arm || torque bracket
Drehmomentenumkehr   torque reversal
Drehmomentenverlauf   torque characteristic
Drehmomentenwaage   dynamometer || torque meter
Drehmomentenwandler   torque converter
Drehmomentenwechsel   torque reversal
Drehmomentenwelligkeit   torque ripple
drehmomentfreie Pause   zero-torque interval || interval of zero torque
Drehmomentkennlinie   speed-torque characteristic * *Abhängigkeit des Drehmoments von d. Drehzahl* || speed-torque curve
Drehmomentkonstante   torque constant * *eines Elektromotors [Nm/A]*
drehmomentlose Pause   zero-torque interval || interval of zero torque
Drehmomentmessgerät   dynamometer
Drehmomentmotor   torque motor * *Motor mit hohem Stillstandsmoment ('Drehfeldmagnet')*
Drehmomentrechner   torque calculator * *z.B. bei Motor-Prüfstand*
Drehmomentregelung   torque control
Drehmomentrichtung   torque direction || direction of torque
Drehmomentschlüssel   torque spanner
Drehmomentverlauf   torque characteristic * *z.B. über der Drehzahl*
Drehmomentwandler   torque converter
Drehmomentwelligkeit   torque ripple
Drehofen   rotary kiln * *z.B. zur Zementherstellung*
Drehpunkt   center of rotation || pivot * *(auch figür-*

Drehrichtung

lich) einer Zapfenverbindung, eines Scharniers, eines Hebels || pivot point || fulcrum point
Drehrichtung direction of rotation * *clockwise: rechts; counterclockwise: links; forward: vorwärts; reverse: rückwärts* || speed direction * *siehe auch →Drehsinn* || motor direction * *eines Motors* || running direction || rotational direction || sense of rotation * *Drehsinn*
Drehrichtungs-Umkehr speed direction reversal
drehrichtungsabhängig depending on the direction of rotation
drehrichtungsabhängiger Lüfter unidirectional fan
Drehrichtungsauswahl selection of the speed direction || selection of the rotational direction
Drehrichtungspfeil direction arrow || rotation arrow * *zur Kennzeichnung d.Drehsinns v. Maschinen, die nur in eine Drehrichtg. laufen*
Drehrichtungssperre lock of rotational direction || rotation lock
Drehrichtungssteuerung forward/reverse control
Drehrichtungsumkehr speed reversal || reversing || rotation reversal
Drehrichtungsumschalter direction-of-rotation switch || rotational-direction switch
drehrichtungsunabhängig regardless of the speed direction || regardless of the direction of rotation || bidirectional * *in beiden Drehrichtungen arbeitend/wirkend* || independent of the direction of rotation * *z.B. Lüfter, der i. beiden Drehrichtungen arbeitet*
drehrichtungsunabhängiger Lüfter bidirectional fan
Drehrichtungsvorgabe rotation setting || setting of rotational direction
Drehrichtungswahl selection of the speed direction || selection of the rotational direction
Drehrollen turn-table rollers * *"Um-Die-Ecke"-Rollgang (z.B. im Walzwerk)*
Drehschalter rotary switch
Drehschieber-Vakuumpumpe rotary-vane vacuum pump
Drehschwingung torsional vibration
Drehschwingungen torsional vibrations
Drehsinn direction of rotation * *für elektr. Maschinen: siehe DIN VDE 0530 Teil 8 (IEC 34-8)* || sense of rotation
Drehsinnumkehr speed reversal * *siehe auch →Drehrichtungsumkehr*
Drehspannung three-phase voltage * *im dreiphasigen System*
Drehspulinstrument moving-coil instrument
Drehstahl turning tool * *für →Drehbank* || cutting tool * *für →Drehbank* || lathe tool * *für →Drehbank*
drehstarr torsionally stiff * *z.B. Kupplung* || torsionally rigid * *coupling: Kupplung*
drehsteif torsionally stiff * *z.B. Kupplung* || non-twist || anti-rotation
Drehsteifigkeit torsional stiffness || torsional rigidity
Drehstomnetz three-phase system
Drehstrom three-phase current || 3-phase current || AC * *auch 'Wechselstrom'*
Drehstrom- three-phase || 3-phase || AC * *auch 'Wechselstrom-'*
Drehstrom-Asynchronmotor three-phase induction motor
Drehstrom-Kommutatormaschine three-phase shunt wound machine || polyphase commutator machine || AC commutator machine
Drehstrom-Kommutatormotor three-phase shunt-wound motor || polyphase commutator motor || AC commutator motor
Drehstrom-Nebenschlussmotor three-phase shunt-wound motor || polyphase commutator motor || AC commutator motor || three-phase commutator motor || three-phase commutator motor with shunt characteristic
Drehstrom-Schleifringläufermotor three-phase slipring motor
Drehstromantrieb AC drive || three-phase drive || three-phase AC drive
Drehstromasynchronmaschine three-phase induction machine
Drehstrombrücke three-phase bridge * *z.B. Gleichrichterschaltung*
Drehstrombrückenschaltung three-phase bridge connection || 3-phase bridge connection || three-phase bridge circuit
Drehstromkommutatormotor polyphase commutator motor || AC commutatormotor || three-phase shunt-wound motor
Drehstrommotor AC motor || a.c. motor || three-phase motor * *dreiphasig* || 3-phase motor * *dreiphasig* || polyphase motor * *drei- und mehrphasig* || three-phase induction motor
Drehstromnebenschlussmaschine three-phase commutator shunt-type machine
Drehstromnebenschlussmotor three-phase commutator shunt-type motor
Drehstromnetz 3-phase line || three-phase supply system || three-phase system
Drehstromsammelschiene three-phase busbar
Drehstromseite three-phase side * *on the: auf der*
drehstromseitig on the three-phase side
Drehstromsteller three-phase AC controller || 3-phase AC controller || three-phase power converter || 3-phase AC voltage controller || three-phase AC power controller * *{Stromrichter} AC-AC-Stromrichter mit Phasenanschnittsteuerung*
Drehstromsynchronmaschine three-phase synchronous machine
Drehstromsystem three-phase system
Drehstromtransformator three-phase transformer
Drehstromverbraucher three phase load
Drehstromverteilung three-phase power distribution || three-phase distribution
Drehstromwickler AC winder || three-phase winder * *z.B. mit Spezial-Wickelmotor der Fa. Lenze*
Drehstromwicklung three-phase winding
Drehstromzuleitung three phase leads || three-phase supply cable
Drehteil turned part
Drehteller rotating disk || rotary table * *Drehtisch*
Drehtisch turntable || rotary table * *z.B. bei Werkzeug-/Verpackungsmaschine* || indexing table * *Positioniertisch mit einstellbarer Winkellage* || rotating table || carrousel table || rotary transfer station * *{Fördertechnik}* || rotary indexing table * *Positionier~*
Drehtrafo rotary transformer
Drehtransformator rotary transformer || induction regulator * *ähnl. Asynchronmotor m. fixiertem Läufer f.stetige Spanngs.-u. Phasenänderg.* || variable transformer * *Stelltrafo*
Drehung turn || revolution * *[rewol'uschen] Umdrehung* || rotation || twist * *Ver~*
Drehvektor rotational vector
Drehverschluss rotary latch * *~ mit Verriegelung*
Drehverteiler carousel distributor * *in Förder-/Dosiersystem*
Drehvorrichtung turning unit || turnover unit * *Wendevorrichtung (zum Umdrehen von Teilen, Werkstücken, Holzplatten usw.)*
Drehwerk slewing gear * *{Hebezeuge} →Schwenkwerk eines Krans*
Drehwiderstand potentiometer || variable resistor || rotary rheostat
Drehwinkel angle of rotation || rotation angle

Drehzahl speed || r.p.m. * in [upm] || rotational speed * in upm (i. Gegensatz zu speed i. Sinne v.(Bahn)Geschwindigkeit) || RPM * Abkürzung für 'Revolutions Per Minute': Drehzahl in [upm]
Drehzahl erreicht at speed || drive at speed || drive at set speed
Drehzahl Null zero speed * at: bei || standstill * Stillstand
Drehzahl-Drehmoment-Kennlinie torque/speed characteristic
Drehzahl-Drehmomentenkennlinie speed-torque characteristic
Drehzahl-Festsollwert pre-set speed || preset speed
Drehzahl-Leistungsdiagramm speed-output diagram * z.B. bei DC-Motor || speed-power diagram * z.B. bei DC-Motor
Drehzahl-Messgeber speed sensor
Drehzahl-Messgerät speed meter || speedometer
Drehzahl-Null-Erfassung zero speed sensing
Drehzahl-Null-Meldung zero speed signal
Drehzahlabfall speed drop
drehzahlabhängig speed-dependent
drehzahlabhängige Feldschwächung speed-dependent field weakening * erlaubt bei Kranhubwerk schnelle Fahrt bei leichter Last
drchzahlabhängige Strombegrenzung speed-dependent current limitıng * z.Einhaltg. d. Kommutierungs-Grenzkurve || tapered current limit * Stromabsenkung bei hoher Drehzahl z. Kommutatorschutz || taper current limit * Stromabsenkung b. hoher Drehz. schützt Kommutator b.DC-Motor
Drehzahlabweichung speed error * Regelabweichung || speed deviation || speed control error * Regelabweichung im Drehzahlregelkreis
Drehzahländerung speed changing || speed variation || speed fluctuation * Schwankung || speed regulation * bei Laständerung
Drehzahlanpassung speed match || speed adaption || speed adjustment || tachometer scaling * siehe auch →Drehzahlistwertanpassung
Drehzahlanregelzeit speed rise time
drehzahlbestimmend speed-defining
drehzahlbestimmende Größen factors which determine the speed
Drehzahldifferenz speed difference
Drehzahleinbruch speed drop || drop in speed || speed undershoot
Drehzahlerfassung speed sensing || speed measurement
Drehzahlgeber tachometer || tacho-generator || tacho || pulse tachometer * Impulsgeber || pulse tach * Impulsgeber || speed encoder * digitaler || tach * (salopp) Kurzform für 'tachometer' || speed transducer || speed sensor || speed transmitter || tachogenerator
drehzahlgeregelt variable-speed || adjustable-speed || speed-controlled
drehzahlgeregelter Antrieb variable-speed drive
drehzahlgesteuert variable-speed
Drehzahlgleichlauf speed synchronism || speed match || speed slaving
Drehzahlgrenze speed limit
Drehzahlgrenzwertmelder speed comparator || speed limit monitor
Drehzahlhub speed range || speed control range * Drehzahlstellbereich
Drehzahlistwert speed actual value || speed feedback || speed signal || actual speed || actual-speed value || actual speed value || instantaneous value of speed * Augenblickswert der Drehzahl || actual value of speed
Drehzahlistwertabgleich speed feedback adjustment || speed signal adaption || speed trim || tachometer scaling
Drehzahlistwertanpassung speed signal adaption || speed feedback adjustment || speed trim || actual-speed signal adapter * als Einrichtung/Gerät || tachometer scaling
Drehzahlistwerterfassung speed sensing || actual speed sensing || speed actual value evaluation || speed feedback evaluation || speed actual value evaluator * Baustein, Einrichtung || actual speed value acquisition
Drehzahlistwertgeber tachometer || speed actual value encoder * digitaler || speed sensor || pulse tachometer * Impulsgeber als Tachometer || pulse tach * Impulsgeber als Tachometer
Drehzahlkonstante speed constant * eines Elektromotors [upm/V] (Kehrwert der →Spannungskonstanten)
Drehzahlkonstanz speed stability || speed constancy
Drehzahlkorrektur speed trim * durch Korrektur-Zusatzsollwert (z.B. von Tänzerlage- oder Zugregler) || speed correction
Drehzahlmesser revolution counter || revolution indicator || tachometer
Drehzahlmessung speed measurement
drehzahlproportional proportional to speed
Drehzahlregelung speed control || speed loop control || closed-loop speed control || speed regulation || speed variation * Veränderung der Drehzahl || speed adjustment * Drehzahl(ein)stellung
Drehzahlregler speed controller || speed regulator
Drehzahlreglerausgang speed controller output || speed loop output
Drehzahlreglereingang speed controller input
Drehzahlregulierung speed control || rpm control
Drehzahlrelais speed monitor || tachometric relay || tacho-switch
Drehzahlrelation speed ratio
Drehzahlschwankungen speed fluctuations || speed hunting * Pendeln, Instabilität || hunting * periodische Schwankungen der Drehzahl (Kenngröße: Ungleichförmigkeitsgrad)
Drehzahlschwelle speed threshold
Drehzahlschwellwert speed threshold
Drehzahlsollwert speed setpoint || speed reference || preset speed * Festsollwert || desired speed * gewünschte Drehzahl || setpoint value of speed || speed reference value || set speed
Drehzahlsollwertpotentiometer speed setting potentiometer || speed control potentiometer
Drehzahlsollwertvorgabe speed setting || speed setpoint setting || speed reference setting || speed demand * Drehzahlanforderung, geforderte Drehzahl
Drehzahlsprung step change in speed
Drehzahlstabilität speed stability
Drehzahlstatik speed droop * Abfall d. Drehzahl bei Belastungssteigerung, z.B. durch weich gemachten Drehz.regler
drehzahlstellbar variable-speed || adjustable-speed
Drehzahlstellbereich speed setting range || speed control range * Regelbereich (wide: großer) || speed range
Drehzahlsteuerung open loop speed control || speed control
Drehzahlsuche speed search * zum Aufschalten eines Stromrichters auf einen laufenden Motor
drehzahlsynchron speed-synchronized || speed-sychronous
drehzahlveränderbar variable-speed || variable speed || adjustable-speed
drehzahlveränderbarer Antrieb variable-speed drives * siehe VDI/VDE 2185
drehzahlveränderbarer Antrieb variable-speed drive || adjustable speed drive || variable-frequency drive * Drehstomsntrieb || VFD * Abk.f. 'Variable-

Frequ. Drive': AC-Antrieb mit veränderbarer Frequenz || VSD * Abk.f. 'Variable-Speed Drive'
Drehzahlverhalten speed regulation characteristic * Drehzahl i.Abhängigkeit v. Lastmoment, gespiegelte Drehmom.kennl..
Drehzahlverhältnis speed ratio || draw * →relative Geschwindigkeitsdifferenz, z.B. zw.mehreren Papiermaschinen-Antrieben
Drehzahlverhältnisregelung speed ratio control
Drehzahlverstellbereich speed setting range || speed range || speed control range * Drehzahlregelbereich
Drehzahlverstellung speed adjustment || speed control || speed variation || speed setting
Drehzahlvorwahl speed pre-selection || rpm pre-selection
Drehzahlwächter speed monitor || tachometric relay || zero-speed monitor * Stillstandsüberwachung, spricht bei Drehzahl Null an || zero-speed switch * Stillstandsüberwachung, spricht bei Drehzahl Null an || zero-speed relay * Stillstandsüberwachungsrelais, löst bei Drehzahl Null aus || overspeed monitor * Überdrehzahlüberwachung || overspeed tripping unit * Auslösegerät zum Überdrehzahlschutz || tacho-switch
Drehzapfen pivot
drei- triple-
Drei-Exzess-Code excess-three code
Drei-Exzess-Gray-Code Gray excess three code || Gray-coded excess three code
dreidimenional three-dimensional
Dreieck triangle || delta * Schaltgruppe z.B. für Trafo/Stromrichter/Motor; connected in delta: in ~ geschaltet
Dreieck- triangular * z.B. triangular wave: periodisches ~ssignal || delta * Schaltgruppe, Schaltung || triangle || V-
dreieckförmig triangular || triangular shaped * z.B. Impulse, Kennlinie
Dreieckfunktion delta function
Dreieckmodulation delta modulation || triangular modulation
Dreieckschaltung delta connection * z.B. Trafo, Motor; delta-connected: in ~
Dreiecksignal triangle signal || triangular signal
Dreieckspannung delta voltage || phase-to-phase voltage
dreifach triple
dreifach- triple-
Dreikantschlüssel triangular wrench
dreiklemmiger Klemmenkasten 3-terminal terminal box * bei explosionsgeschütztem Motor
dreiphasig 3-phase || three-phase
Dreipuls- triple-pulse || three-pulse
dreipulsig 3-pulse || triple-pulse
Dreipunktaufhängung three-point fixing * {el. Masch.} eines Motors mit →Pratzenanbau
Dreipunktbefestigung three-point fixing
Dreipunktregelung three-step action control || three-step control
Dreipunktregler three-step controller || three-step action controller || three-position controller || three-term controller || three-point controller || double-circuit controller * z.B. Temperaturregler || two-stage controller * z.B. Temperaturregler
Dreipunktwechselrichter 3-step inverter
Dreischichtarbeit three shift working
Dreischichtbetrieb three-shift operation
Dreistellungsschalter 3-position switch || three-position switch
dreistufig 3-stage || three-stage
dreiweg three-way
Dreiwegeventil three-way valve
Dreiwicklungstrafo three-winding transformer * siehe auch →Doppelstocktrafo

Dreiwicklungstransformator three-winding transformer
Dremoment-Drehzahlkennlinie torque-speed characteristic * siehe auch →Lastkennlinie
Dressiergerüst temper mill * zur Nachbehandlung in Walzstraße/Bandbehandlungsanlage
Dressiergrüst temper mill * in Blechstraße
Drift drift
driftfrei drift-free || droopless
Driftfreiheit no drift
dringend urgent * be in urgent need of: ~ brauchen || urgently * (Adv.) || imperative * ~/zwingend notwendig || pressing * auch dringlich || priority * vordringlich || badly * (Adv.) want badly: ~ brauchen || hard * Bitte usw. || strongly * (Adv.) advise/warn strongly: ~ abraten || emergency * Notfall, Notlage, Notruf
dringender Fall urgent case
Dringlichkeit urgency || priority * Vor~ (top: höchste, lowest: niedrigste), Priorität || exigency * Dringlichkeit, dringlicher Fall, (dringendes) Erfordernis
Dringlichkeitsstufe priority * top: höchste || priority class
dritt third || 3rd * Abkürzung für 'third'
Dritt- third party * Drittersteller, dritte Kraft, dritte Partei/Person || third * -klassig, -bester usw.
Drittanbieter third-party vendor
dritte Potenz cube * increase (proportionally) with the cube of: mit der dritten Potenz von ... anwachsen
dritte Wurzel cubic root
Drittel third * one third: ein ~; two-thirds: zwei ~; a third of the amount: ein ~ des Betrages || third part * der dritte Teil
drittens thirdly || in the third place
Drittfirma third-party company || third party
Dritthersteller third party
drittletzt third from last || third last
drittletzter third from last || third last
Dröhnen rumbling noises
Drossel choke * kleine || reactor * größere || inductor || choke coil * Drosselspule || throttle * (mechan.) ~ventil, ~klappe || throttle valve * (mechan.) Drosselventil/klappe || restrictor * (mechan., strömungstechn.) z.B. in Hydrauliksystem || damper * strömungstechnische ~, z.B. in Rohrleitung
drosseln throttle
Drosselorgan throttling element * (mechan.) z.B. in Rohrleitung
Drosselspule reactor * siehe VDE 0532; größere || choke * kleinere || choke coil
Drosselverluste throttling losses * bei Durchflussregelung über Schieber/Drosselventile
Druck pressure * low/light/high: niedriger/leichter/hoher || printing * {Druckerei} Druckvorgang || compression * Zusammendrücken/-pressen, Verdichtung, Druck, Kompression || presswork * {Druckerei} Druckvorgang
Druckabfall decompression || drop in pressure || pressure drop || fall in pressure
Druckakkumulator pressure accumulator * in Hydrauliksystem
Druckaufnehmer pressure transducer * Drucksensor || pressure sensor
Druckausgleichsbehälter pressure equalizing reservoir || pressure compensating air reservoir * in einem Druckluftsystem
Druckbeanspruchung compressive stress || pressure load || compression stress
Druckbegrenzungsventil pressure regulating valve || relief valve jet * Überdruckventil
Druckbehälter pressure tank
Druckbild printing style || printing quality
Druckbogen printed sheet

Druckdifferenz pressure difference || fan pressure * *bei Lüfter*
Druckeinheit printing unit * *{Druckerei}*
druckempfindlich pressure-sensitive || sensitive to pressure
drucken print || publish * *veröffentlichen*
drücken press * *auch eine Taste* || depress * *niederdrücken, auch Taste, Druckknopf* || hit * *eine Taste* || actuate * *betätigen (z.B. eine Taste)*
Drucker printer * *auch Beruf* || line printer * *Zeilendrucker* || page printer * *Seitendrucker* || event recorder * *Meldedrucker*
Druckerei print shop || printing shop || printing office || printing company || printing factory || printing firm
Druckerhöhung pressure difference * *Differenzdruck, Druckdifferenz* || fan pressure * *bei Ventilator* || pressure increase * *z.b. bei Lüfter, Ventilator*
Druckerpresse printing press || letter press
Druckerschnittstelle printer interface || printer port
Druckerschwärze carbon black || printing ink
Druckerzeugnis printed product * *{Druckerei}*
Druckfarbe ink * *{Druckerei}* || printing ink * *{Druckerei}* || printing color * *{Textildruck}* || carbon black * *{Druckerei} Druckerschwärze* || paste * *{Druckerei} bei Siebdruck*
Druckfeder pressure spring || release spring * *einer Kupplung, Bremse usw.* || compression spring * *z.b. zur Lagervorspannung (siehe →vorgespannt (-es Lager))* || compressing spring
Druckfehler typographic error || erratum * *(Singular)* || errata * *(Plural)* || misprint || typographical error
druckfest pressure-proof || explosion proof * *(amerikan.) Explosionsschutzart* || pressure-resistant || flameproof * *(brit.) →explosionsgeschützt*
druckfeste Kapselung flame-proof enclosure * *Zündschutzart 'd' nach EN50018 für Ex-geschützte el. Betriebsmittel* || explosion-proof enclosure * *(amerikan.) Zünd-/Explosionsschutzart* || pressurized enclosure * *Zünd-/Explosionsschutzart*
Druckfestigkeit compressive strength || pressure strength
Druckform printing form * *{Druckerei}* || stencil * *{Druckerei}*
Druckgeber pressure transmitter || pressure pick-up || pressure sensor
Druckgewerbe printing trade
Druckgießmaschine die-casting machine
Druckgießwerkzeug die-casting die
Druckguss die-cast || die-cast metal * *Druckgussmetall/-material*
Druckgussläufer die-cast rotor * *{el.Masch.}*
Druckhebel pressure lever || printing lever * *{Druckerei} bei Bogendruckmaschine* || pressure finger * *kurzer Hebel*
Druckindustrie printing industry
Druckknopf pushbutton || button || key * *Drucktaste*
Druckkolben pressure piston * *z.B. in einem pneumat./hydraul. Aktuator/Zylinder*
Druckkontakt pressure contact * *z.B. bei Scheibenthyristoren*
Druckkopf printhead * *eines Computerdruckers*
Druckleitung compressed-air pipe * *für Druckluft* || compressed-air line * *für Druckluft* || delivery pipe * *Pumpenausgangsleitung, Zuführleitung* || pressure line || high-pressure line * *Hochdruckleitung* || discharge line * *am Ausgang eines Kompressors*
Druckluft compressed air
Druckluft- compressed-air || air-pressure || pneumatic || pneumatically operated * *[nju'mätik ...]* || air-operated
Druckluftanlage compressed-air system

Druckluftanschluss compressed-air connection
Druckluftbehälter compressed-air tank || compressed-air reservoir * *Zwischenbehälter/Ausgleichsbehälter für Pressluft* || compressed-air container
Druckluftbremse air brake || pneumatic brake || pneumatically operated brake || compressed-air brake || pneumatically actuated brake
Druckluftfilter compressed-air filter
Druckluftkupplung pneumatic clutch || compressed-air clutch
Druckluftleitung compressed-air line || compressed-air pipe
Druckluftschlauch compressed-air hose
Druckluftwerkzeug pneumatic tool
Druckluftzylinder pneumatic cylinder * *[nju'mätik ...]* || compressed-air cylinder
Druckmarke register mark * *{Druckerei}*
Druckmaschine printing machine
Drückmaschine spinning machine * *"Drück-Drehbank" zur Blechformung (z.B. zur Erzeugung von Hohlgefäßen)*
Druckmesser pressure gauge || pressure meter
Druckmessgerät pressure gauge || pressure meter || liquid manometer
Druckminderer pressure reducer || pressure-reducing valve
Druckminderungsventil pressure regulating valve || pressure reducer || pressure reducing valve
Druckminderventil pressure-reducing valve || pressure reducer
Drucköl pressure oil * *in Hydrauliksystem*
Druckpapier printing paper || newsprint * *für Zeitungen*
Druckplatte platen * *{Druckerei} auch Drucktiegel, Druckzylinder in Druckmaschine* || pressure plate * *z.B. bei einer Kupplung* || printing plate * *{Druckerei}* || stereotype * *{Druckerei}* || cliché * *{Druckerei} Druckstock, Bildstock, Klischee*
Druckpunkt tactile touch * *an Taste/Tastatur*
Druckregelung pressure control || pressure closed-loop control
Druckregler pressure regulator || pressure controller
Druckring clamping ring * *übt axialen Druck auf ein e.Lager aus (siehe →angestelltes Lager)* || pressure ring * *z.B. bei einer Reibkupplung*
Druckrohr pressure tube
Druckschalter pressure switch
Druckscheibe thrust washer * *zur →Vorspannung eines Lagers*
Druckschrift publication || pamphlet || brochure * *Broschüre* || printed publication || documentation
Drucksensor pressure transducer
Druckspannung compressive strength
Druckspeicher pressure reservoir * *z.B. in Hydrauliksystem* || print buffer * *für Computerdrucker*
Druckstoss pressure surge
Drucktaste push-button || key
Druckaster pushbutton * *['puschbatn]* || key * *Taste(r)*
drucktechnisch typographic || typographical
Drucktiegel platen * *{Druckerei}* →*Druckplatte (z.B. einer Bogendruckmaschine)*
Drucktuch printing blanket * *{Druckerei}* || back cloth * *{Druckerei}*
Druckverminderung decompression
Druckvorgang process of printing
Druckvorspannung pre-compressive stress || compressive pre-tensioning
Druckvorstufe prepress level * *{Druckerei}*
Druckwächter pressure switch
Druckwalze rider roll * *z.B.obere 3.Rolle bei Doppeltragwalzenroller,dient auch zur Durchmessererfassung* || roller * *im Walzwerk* || pressing on rol-

Druckwerk

ler * *Andruckwalze* || snubber roll * *z.B. in Walzstraße* || printing roller * *{Druckerei}*
Druckwerk   printing unit || printing mechanism || printing element || printing engine * *bei Laserdrucker*
Druckzylinder   pressure-cylinder * *z.B. Pneumatikzylinder* || impression cylinder * *{Druckerei}* || rubber cylinder * *{Druckerei} Gummi-Druckzylinder einer Druckmaschine* || platen * *{Druckerei} einer Rotationsdruckmaschine*
DSP   DSP * *Digitaler SignalProzessor*
DSS1   DSS1 * *Abk.f. 'Digital Subscriber System No. 1': Europäisches Übertragungsprotokoll für →ISDN*
DT1-Glied   DT1-function || DT1-block * *Funktionsbaustein*
DTP   DTP * *{Computer} Abk. f. 'DeskTop Publishing': Layout von Druckerzeugnissen per Computer*
du-dt-Filter   dv/dt-filter * *zur Spanngs.anstiegsbegrenzg. z.B. hinter Umrichter, schützt Motor vor Spannungsspitzen*
Dual-Code   binary code
Dual-Port-RAM-Anschaltung   dual-port RAM interface || dual port RAM interface board * *Leiterkarte, Flachbaugruppe*
Dualcode   binary code
Dualzahl   binary number
Dübel   dowel || peg || wall plug * *Wanddübel* || treenail || plug
Dübeleinschießpistole   dowelling gun
Dübeleintreibmaschine   doweldriving machine * *{Holz}* || dowel-inserting machine * *{Holz}*
Dübelloch   dowel hole * *{Holz}*
Dübellochbohrmaschine   dowel hole boring machine * *{Holz}*
dunkel   dark
dünn   thin || light-gauge * *Draht, Kabel, Blech* || fine * *fein, zart* || delicate * *zart* || flimsy * *Gewebe* || slight * *schlank* || slim * *schlank* || slimline * *in schlanker Bauart* || low-profile * *in flacher Bauart* || slender * *schlank* || lean * *mager* || spindly * *mager* || weak * *Flüssigkeit* || diluted * *verdünnt*
Dünnbramme   thin-slab
Dünnfilmtechnik   thin-film technology
dünnflüssig   thinly-liquid || low-viscosity || watery || light * *Öl*
Dünnschlamm   thin sludge
dünnwandig   thin-walled
Dunst   vapour
Dunstabsauganlage   vapour extractor
Duo-   two-high * *Walzstraße, Walzwerk* || duo- || two- || double
Duplexpapier   two layer paper
Duplikat   duplicate || copy * *Kopie* || identical copy * *Kopie* || replica * *Nachbau* || dupe * *(salopp)*
duplizieren   duplicate
durch   by means of * *mittels* || due to * *auf Grund* || through * *durch hindurch, z.B. Stromfluss; auch mittels* || by the way of * *mittels* || by * *z.B. by pressing the key: durch Drücken der Taste* || as a result of * *als Ergebnis von* || per * *per, durch, laut, gemäß, pro, je, für*
durcharbeiten   work through * *z.B. Lehrbuch*
Durchbiegung   bending || deflection * *Abbiegung, Ablenkung, Durchbiegung (einer Antriebswelle usw.); Ausschlag* || sagging * *z.B. eines stark belasteten Regalbodens* || sag * *Durchhang, Durch-/Absacken, Senkung* || bend * *Biegung, Krümmung* || flexure * *Biegung, Krümmung* || deflexion * *Ablenkung, Abbiegung*
durchbrechen   break through * *(figürl. und gegenständl.)* || break in two * *entzweibrechen* || force one's way * *mit Gewalt seinen Weg machen* || break out * *zum Vorschein kommen*

durchbrennen   fuse * *Sicherung* || blow * *Sicherung* || melt * *schmelzen (Sicherung usw.)* || burn out * *z.B. Wicklung* || burn through * *z.B. Wicklung*
Durchbruch   breakthrough * *(auch figürl.)* || penetration || opening * *Öffnung* || aperture * *Öffnung*
Durchbruchspannung   breakdown voltage || flashover voltage
Durchbruchstrom   breakdown current || flashover current || avalanche current * *z.B. bei Halbleiter (Thyristor usw.)*
durchdacht   well devised * *gut durchdacht (z.B. Plan, Konstruktion, techn. Lösung)* || well-reasoned || well weighed
durchdringen   penetrate * *auch Markt/geistig; auch eindringen, durchbohren, durchstoßen, geistig ergründen*
durcheinander   in confusion * *(Adj.)* || promiscuously * *(Adv.) wahllos* || misorder * *(Subst.)* || confusion * *(Subst.)* || muddle * *(Subst.)* ~, *Unordnung, Wirrwar, Unklarheit; make a muddle of sth.:* ~*bringen*
durchfahren   pass
durchfließen   flow through || pass through
durchflossen werden von   carry * *führen, z.B. Strom*
Durchfluss   flow || flow rate * *Durchflussrate, z.B. in Kubikmeter pro h* || flow through
Durchflussgeber   flow sensor || flow transducer * *Messumformer* || flow meter * *Durchflussmesser* || flow transmitter
Durchflussgeschwindigkeit   flow velocity || flow rate * *Förderrate z.B.in [Volumen/h]* || delivery rate per hour * *Förderrate einer Pumpe, z.B. in Kubikmeter pro h* || delivery rate * *Förderrate einer Pumpe*
Durchflussmenge   rate of flow * *pro Zeiteinheit* || flow
Durchflussmess-Füllmaschine   flow-metering filling machine * *in der Verpackungstechnik*
Durchflussmesser   flow transducer || flow sensor || flow transmitter || flow meter || volume flow meter * *Volumendurchfluss-Messgerät* || flowmeter
Durchflussregelung   flow rate control || flow control
Durchflussregler   flow controller || mass flow controller * *Mengenstromregler* || flow rate regulator || flow rate controller
Durchflusssteuerung   flow control
Durchflusswächter   flow monitor
Durchflusswandler   forward converter * *bei Schaltnetzteil*
durchflutet   magnetized * *magnetisch* ~
Durchflutung   magnetic flux || ampere-turns || magnetomotive force * *Summe aller Ströme (Ampere-Windungen), die ein bestimmtes Magnetfeld erzeugen*
durchführbar   practicable * *durch-/ausführbar; möglich, tunlich, brauchbar, anwendbar* || feasible * *ausführbar, durchführbar, möglich* || workable * *durch-/ausführbar (Plan, Vorhaben usw.); bearbeitungsfähig* || viable * *machbar, lebensfähig*
Durchführbarkeit   feasibility || practicability || workability || implementability
Durchführbarkeitsstudie   analysis of feasibility || feasibility analysys || feasibility study
durchführen   perform || carry out * *ausführen, z.B. Reparatur, Anlagenprojektierung* || pass through * *(durch etwas) hindurchführen* || implement || run * *z.B. einen Versuch* || do * *tun* || make * *bilden, formen, bilden, machen* || conduct * *führen/leiten, z.B. Kurs, Vorlesung, Tagung* || execute * *aus-/durchführen* || manage * *organisieren* || feed through * *hindurchführen* || undertake * *(Sache) in die Hand nehmen, (Aufgabe, Verantwortung)*

übernehmen, (Reparatur) ~ || subject * unterwerfen, unterziehen; someth.is subjected to: an etwas wird ... durchgeführt || hold * Kurs/Seminar usw. ~
Durchführung carrying-out * Aus/Durchführung || execution * Ausführung, z.B. eines Projektes || performance || completion * Fertigstellung, Vollendung, Abschluss, Vervollständigung, Erfüllung || realization * Realisierung, Verwirklichung, Ausführung || implementation * Erfüllung, Ausführung, (Software) Realisierung || passing through * einer Leitung || wall entrance * Wanddurch/einführung von Leitungen || inlet * Enlass(stelle), z.B. von Leitungen || entry * Einlass-/Einführungs(stelle) z.B. von Leitungen || enforcement * Durchsetzung, Vollzug (Gesetz, Vorschrift) || bushing * Leitungs-Durchführungshülse, Muffe, Spannhülse || insulating bushing * isolierte Leitungsdurchführungshülse || penetration * (Kabel)Eintritt, Durchbohrung, Durchtritt || feed through * z.B. von Kabeln || feed-through * z.B. Kabel~
Durchführungshülse bushing * z.B. für Kabel || strain-relief bushing * mit Zugentlastungsfunktion, z.B. für Kabel
Durchführungsisolator insulating bushing
Durchführungskondensator bushing-type capacitor || feed-through capacitor
Durchgang pass * Arbeitsgang, Durchlauf, Passage || continuity * Stromdurchgang, geschlossene Verbindung || passage * für Personen, Flüssigkeiten, Gase || gateway * (Eingangs)Tor, Zugang (auch figürl.), Einfahrt, Eingang || gate * eines Ventils || connecting passage * für Flüssigkeiten, Gase || run * Durchlauf, Folge, Arbeitsgang || heat * Runde, Lauf, Endrunde, Entscheidungskampf
durchgängig unified * vereinheitlicht, z.B. e. techn. Lösung (bei e. Produktfamilie) || general * allgemein, umfassend, gängig, allgemein üblich || uniform * einheitlich, gleichmäßig, übereinstimmend || universal * (all)umfassend, universal, allgemein(gültig), allgemein üblich || consistent * konsequent, folgerichtig, logisch, übereinstimmend, stetig, gleichmäßig || conductive * (den Strom) leitend || sustained * aufrecht erhalten, durchgehalten, anhaltend; ungedämpft
Durchgängigkeit uniformity * z.B. einer techn. Lösung über eine Produktfamilie || continuity * ununterbrochener Zusammenhang, Kontinuität, zusammenhängendes Ganzes || universality * Allgemeingültigkeit, Universalität, Vielseitigkeit || conducting state * stromdurchlässiger/-führender Zustand || unified system concept
Durchgangsbohrung through-hole
Durchgangsloch through-hole
Durchgangsprüfer continuity tester
Durchgangsverdrahtung through-wiring
Durchgangsverluste conductive losses * z.B. bei Halbleiter
durchgehärtet completely hardened || through-hardened
durchgehen run away * eines Motors (Überdrehzahl) || overspeed * eines Motors (Überdrehzahl) || runaway * (Substantiv) z.B. eines Motors
durchgehend continuous || unified * durchgängig
durchgehende Welle through-shaft
durchgeschaltet switched through || connected || completed * completed circuit: durchgeschalteter/geschlossener Stromkreis || enabled * (Signal) freigegeben || signal flow * Zustand eines Strompfades in der Kontaktplandarstellung eines SPS-Programms
durchgezogene Linie solid line
Durchhang sag || slack
durchhängen sag
Durchhangregelung sag control * z.B. in für Metallband in Schlingengrube oder in Kabelumman-
telungsanlage || slack control || catenary control * z.B. b. Kabelummantelungsanlage; catenary: →e.mathem. Seil-/Kettenlinie folgend".
Durchkontaktierung plated through-hole * in Leiterkarte
Durchlass passage || outlet * Auslass || opening * Öffnung || gate || port
Durchlassbelastungsgrenze on-state loading limit * z.B. bei Diode, Thyristor
durchlassen let pass || allow to pass || let through || be pervious to * (physikal.) durchlässig sein für || be permeable to * (physikal.) durchlässig sein für || transmit * Licht ~ || leak * Flüssigkeit ~ || filter * filtern || strain * filtern
durchlässig permeable || pervious * durchlässig, durchdringbar, gangbar (to: für), zugänglich, offen, undicht || porous * porös || leaky * leck, ~ für Flüssigkeiten || translucent * ~ für Licht || conductive * ~ für Strom || diaphanous * ~ für Licht || conducting * (Strom-) leitend
Durchlässigkeit permeance || permeability || perviousness || porosity || leakiness || translucence || transmission factor * otische || electric constant * elektrische
Durchlasskennlinie forward characteristic * (Diode) || on-state characteristic * (Thyristor)
Durchlassrichtung forward direction * z .B. Thyristor/Diode || on-direction
Durchlassspannung forward voltage * (Diode) || on-state voltage * (Thyristor) || conducting-state voltage drop * z.B. einer Diode || forward voltage drop * z.B. bei Diode/Thyristor/Transistor
Durchlassstrom cut-off current * ~ einer Sicherung/eines Leistungsschalters im ausgeschalteten Zustand || on-state current * bei Diode, Thyristor, Transistor usw. || forward current * bei Diode, Thyristor usw.
Durchlassverluste forward losses * z.B. bei Diode/Transistor/Thyristor || losses due to forward current * bei Diode, Thyristor usw.
Durchlasswiderstand ON resistance * z.B. Feldeffekttransistor, FET-Schalter || on-resistance || on-state resistance * eines elektronischen Schalters (Diode, Transistor, Thyristor usw.)
Durchlasszustand on-state * z.B. Thyristor || conducting state
Durchlauf pass || repetition * Wiederholzyklus || cycle * Zyklus || run * siehe auch 'Durchgang'
Durchlauf- continuous || through-feed * z.B. ~- Presse für Spannplatten usw.
Durchlauf-Feuchtemessanlage continuous moisture metering equipment
Durchlaufbetrieb continuous operation * bei {el. Masch.}: →Betriebsarten S6 bis S8 || continuous running || uninterrupted operation
Durchlaufbetrieb mit Aussetzbelastung continuous-operation periodic duty type * Motorbetriebsart S6 nach VDE0530/IEC34-1 || continuous-operation periodic duty
Durchlaufbetrieb mit Kurzzeitbelastung continuous operation with short-time loading
durchlaufen pass through || run continuously * ohne Unterbrechung/kontinuierlich ~ || run nonstop * ohne Unterbrechung ~ || run 24 hours a day * rund um die Uhr ~
durchlaufend continuous || continuously running || flowing || continually running * ohne Unterbrechung/fortwährend ~; immer wiederkehrend
durchlaufende Warenbahn continuous web
durchlaufender Betrieb continuous operation
Durchlaufofen continuous furnace || continuous oven
Durchlaufpresse continuous press * z.B. für Spanplatten (im Gegensatz zur →Taktpresse) || through-feed press * {Holz}

Durchlauftrockner continuous-flow dryer || conveyor dryer
Durchlaufwiderstandsglühofen continuous resistance annealer * *{Draht}*
Durchlaufzeit throughput time * *z.B. in einer Fertigungslinie* || production time * *in der Fertigung* || processing time * *für Bestellungen usw.*
Durchlegieren breakdown * *bei Halbleiterbauelement* || shooting-through * *z.B. bei Thyristor* || break down * *bei Halbleiterbauelement; siehe auch →Durchschlag*
durchmessen check * *überprüfen* || take the measurements of * *Messwerte (z.B. eines Bauelements) aufnehmen*
Durchmesser diameter
Durchmesser-abhängig diameter-dependent
Durchmesser-Sensor diameter sensor * *z.B. mit Ultraschall für Auf/Abwickler*
Durchmesser-Setzwert diameter preset value
Durchmesseraufbau diameter buildup * *Durchmessererhöhung bei Aufwickler während des Wickelvorgangs*
Durchmesserbereich diameter range * *bei Auf-/Abwickler* || diameter ratio * *→Durchmesserverhältnis (bei Auf-/Abwickler)*
Durchmessererfassung diameter sensing
Durchmesserrechner diameter calculator * *bei Auf/Abwickler z.B. nach der Formel [Durchm. entspr. const x V/n)]*
Durchmessersetzwert diameter preset value
Durchmesserverhältnis diameter ratio * *beim Wickler*
durchnummerieren number consecutively * *fortlaufend*
durchreichen pass || pass through || let pass * *durchlassen*
durchreißen tear || get torn || break || tear asunder * *entzweireißen* || tear in two * *entzweireißen* || rend * *entzweireißen*
Durchreißwiderstand tearing resistance * *z.B. von Papier*
durchrutschen slip * *z.B. Kupplung, Bremse*
durchsacken sag
Durchsatz throughput || throughput rate * *Rate* || flow rate * *Durchflussrate* || production rate * *Fertigung*
durchschalten switch through || connect through || connect || complete the circuit * *Stromkreis schließen* || enable * *(Signal) durchschalten/freigeben* || put through * *Signal usw. ~*
durchschieben shift through
Durchschlag breakdown * *Isolationsversagen* || puncture * *Isolationsversagen, (Ein-)Stich, Loch, Reifenpanne; puncture-proof: durchschlagsicher* || copy * *Kopie; c.c.: Durchschlag an ("carbon copy")* || break down * *Isolationsversagen (z.B. bei Überspannung)* || brown-out * *Isolationsversagen*
Durchschlag an c.c. * *Abk. f. 'Carbon 'Copy': Schreibmaschinendurchschlag/Fotokopie an*
Durchschlagfestigkeit dielectric strength || breakdown strength * *Widerstandsfähigkeit gegen Isolierversagen*
Durchschlagsicherung protective break down * *{el.Masch.} Überspannungsableiter z. Schutz v. Meskreisen i.Hochspannungsmasch.*
Durchschlagspannung breakdown voltage || flashover voltage * *Überschlagspannung* || arc-over voltage * *Überschlagspannung*
durchschleifen loop through
Durchschnitt average * *on (an) average: im Durchschnitt* || mean
durchschnittlich average || mean || medium * *Größe, Preis, Qualität usw.* || on the average * *(Adv.) im Durchschnitt*
durchsetzen carry through * *durchhelfen/bringen,*

durchführen || put through * *durch/ausführen* || succeed with * *erfolgreich* || enforce * *erzwingen* || force * *zwingen* || carry one's point * *seine Meinung ~; with: bei* || win through * *erfolgreich sein* || succeed * *erfolgreich sein* || prevail * *erfolgreich sein; auch Erzeugnis am Markt ~* || mix * *durchmischen* || saturate * *sättigen* || realize * *realisieren* || establish * *establish oneself: sich etablieren/durchsetzen, z.B. am Markt/eine neue Technik*
durchsichtig clear * *Farbe*
Durchsichtigkeit transparency
Durchsprache review || discussion
durchsprechen discuss || review
Durchsteck- through-hole
Durchsteckmontage through-hole mounting || thru-the-door mounting * *(amerikan.) in einen Türausbruch montiert/montierbar*
Durchsteckstromwandler winding-type current transformer * *ohne Primärwicklung* || bar-type current transformer * *mit integriertem Durchsteck-Primärleiter*
Durchsteckwandler winding-type transformer * *ohne Primärwicklung* || bar-type transformer * *mit integriertem Durchsteck-Primärleiter* || straight-through transformer || ring-type transducer * *z.B. Strom-Messwandler* || ring-type current transformer * *Stromwandler*
durchstellen connect * *jemanden am Telefon weitervermitteln* || put through * *(amerikan.) jemanden am Telefon weitervermitteln*
durchströmen flow through || run through
durchströmt current-carrying * *elektr. Strom führend*
Durchtrieb drive-through mechanism * *sealed: abgedichteter; z.B. z.Antreiben e.in Vakuum befindl. Antriebsstelle*
Durchtritt exit * *z.B. einer Welle*
Durchtrittsöffnung exit opening
durchverbinden connect through
Durchwahl extension * *Haus-Telefonapparat*
durchziehen pull through
Durchziehwicklung pull-through winding * *Wicklungsart bei Motor: Drähte werden einzeln durch die Nut hindurchgezogen*
durchzugsbelüftet open-circuit air-cooled * *→eigen- oder →fremdbelüfteter Motor mit Luftstrom durch die Maschine* || open-circuit cooled || enclosed ventilated * *durchzugsbelüftet, innengekühlt (el. Maschine mit Gehäuse-interner Lüftung)* || with open-circuit cooling * *Motorkühlart mit Luftstrom durch das Gehäuse (niedrige Schutzart)* || open-circuit ventilated * *Motorkühlart mit Luftstrom durch das Gehäuse (Schutzart z.B. IP23)*
Durchzugsbelüftung open-circuit cooling * *z.B. →eigen- oder →fremdgekühlter Motor mit Luftstrom durch d.Maschine* || open-circuit air-cooling || end-to-end cooling || drawing-through ventilation || open-circuit ventilation || enclosed ventilation * *{el.Masch.}*
Durchzündung breakthrough * *(Thyristor)*
dürfen be allowed to || be permitted to || have the right to
dürftig poor * *→ungenügend, →mangelhaft, →schwach, →unbefriedigend* || inadequate || scanty * *(amerikan.) spärlich, mager* || meagre * *spärlich, mager* || meager * *(amerikan.) spärlich, mager* || skimpy * *shabby* || schäbig || flimsy * *fadenscheinig, →schwach*
Düse nozzle
Düsenrohr spray pipe || jet tube
DV data processing * *Datenverarbeitung* || DP * *Abkürzg. für 'data processing': Datenverarbeitung*
dv-dt dv/dt
dV-dt-Filter dV/dt filter * *am Umrichterausgang z. Schaltflanken-Verschliff (zur EMV u. Motorschonung)*

DV-System data processing system ‖ DP system
Dynamik dynamic performance * *Leistungsfähigkeit bezügl. der (Regel-) Dynamik* ‖ dynamic response ‖ fast response * *hohe Dynamik* ‖ response ‖ response characteristic * *dynam. Eigenschaften* ‖ dynamic behaviour * *dynam. Verhalten* ‖ dynamics ‖ responsiveness ‖ speed of response ‖ impetus * *(Auf-) Schwung (z.B. der Wirtschaft, einer Entwicklung(stendenz) usw.)*
dynamisch dynamic ‖ transient * *vorübergehend, kurzzeitig; transient error: dynamische Regelabweichung* ‖ high response * *mit hoher Dynamik* ‖ quick-response * *hochdynamisch, schnell reagierend* ‖ responsive * *schnell ansprechend, mit hoher Dynamik (z.B. Regelung)*
dynamisch auswuchten balance dynamically
dynamische Anforderungen response requirements
dynamische Beanspruchung dynamical loading
dynamische Bremsung dynamic braking
dynamische Eigenschaften dynamic response characteristic ‖ dynamic performance characteristic
dynamische Momentenkennlinie dynamic speed-torque characteristic * *{el.Masch.} M(n)-Kennl. e. AC-Motors b. schneller Drehz.änderg.*
dynamische Regelabweichung dynamic control deviation
dynamische Überlastbarkeit dynamic overload capability
dynamische Verluste dynamic losses
dynamischer Datenaustausch dynamic data exchange * *feste Verknüpfung e."Wirtsdatei" mit Elementen anderer Dateien* ‖ DDE * *Abkürzg.f. 'Dynamic Data Exchange' (siehe dort)*
dynamisches Verhalten dynamic behavior ‖ dynamic behaviour ‖ dynamic characteristics
Dynamoblech electrical sheet steel ‖ magnetic sheet steel
Dynamometer dynamometer * *Drehmomentvorgabe- und Messeinrichtung, z.B. Pendelmaschine, z.B. für Prüfstand*

E

e-Funktion e-function ‖ exponential characteristic
E-Haus electrical equipment house
E-Modul Young's modulus * *Elastizitätsmodul* ‖ elastic modulus
E-Raum electrical equipment room
eben plane * *plan, flach* ‖ level * *waagerecht* ‖ even * *flach, waagerecht* ‖ flat * *flach* ‖ horizontal * *horizontal* ‖ planar * *Oberfläche*
Ebene layer * *Schicht* ‖ level * *Grad, Stufe, Klasse* ‖ chart * *im Koordinatensystem, z.B. Strom/Spannungsebene* ‖ plane
ebene Fläche flat surface
ebenfalls likewise ‖ also ‖ too ‖ as well
echt genuine
Echtzeit real-time * *Rechnerfunktionen mit schnellen u.vorherbestimmbaren (deterministischen) Reaktionszeiten* ‖ online * *z.B. bei laufender Maschine* ‖ on the fly * *während des Betriebes, ohne anzuhalten* ‖ real-time clock * *Echtzeituhr*
Echtzeit-Betriebssystem real-time operating system
Echtzeitbetriebssystem real-time operating system
echtzeitfähig real time * *z.B. Rechner, Betriebssystem* ‖ realtime-capable ‖ fast response * *mit kurzen Reaktionszeiten*
Echtzeitgröße real-time quantity

Echtzeitregelung real-time control
Echtzeituhr real-time clock
Echtzeitverarbeitung real time processing
Ecke corner * *(auch figürl.)* ‖ angle * *spitzer Punkt* ‖ edge * *Kante* ‖ nook * *Nische* ‖ recess * *Aussparung, Vertiefung, Einschnitt, Nische, das Innere*
Eckfrequenz base frequency * *z.B.e.U/f-Kennlinie beim Pulsumrichter, bei dieser Frequ.wird die max.Spanng.erreicht* ‖ corner frequency * *z.B. ~ einer U/f-Kennlinie (bei dieser Frequenz wird die max. Spannung erreicht)* ‖ cutoff frequency
eckig angular ‖ cornered
eckige Klammern square brackets
Eckpunkt transition point * *Übergangspunkt* ‖ vertex * *Scheitelpunkt* ‖ corner point ‖ joint * *Verbindungspunkt*
Eckverbindung edge joint * *{Holz}* ‖ squaring-up * *{Holz}*
ED switch-on duration * *→Einschaltdauer* ‖ CDF * *Abk. f. 'Cyclic Duration Factor': relative Einschaltdauer (e. el. Maschine: s. VDE 0530 Teil 1)* ‖ cyclic duration factor * *relative →Einschaltdauer* ‖ load factor ‖ duty cycle * *Abk. für →Einschaltdauer; relative Einschaltdauer*
Edelmetall noble metal ‖ precious metal
Edelstahl stainless steel
editieren edit
EDV computer ‖ DP * *Abk. für 'Data Processing': (elektron.) Datenverarbeitung* ‖ computing ‖ EDP * *Abk. für 'Electronic Data Processing'* ‖ IT * *Abk. für 'Information Technology'*
EEMAC EEMAC * *Abkürzg. f. 'Electrical Electronic Manufacturers Association of Canada': Canad. Norm.gremium*
EEPROM EEPROM * *Electrically Erasable Programmable Read Only Memory, elektrisch löschbarer Nur-Lese-Speicher*
EEx d flameproof enclosure type d * *Schutzklasse EEx d (druckfeste Kapselung)*
EExe Motor motor in hazardous locations
Effekt effect * *Effekt, (Ein-) Wirkung, Einfluß, Erfolg, Folge* ‖ efficiency * *Wirkungsgrad, Leistungsfähigkeit, Tauglichkeit, Tüchtigkeit* ‖ phenomenon * *(Physikal.) Phänomen, Erscheinung*
effektiv effective * *wirksam, effektvoll; effectively required: ~ benötigt* ‖ rms * *Effektiv- (Wert, Spannung, Leistung usw.)*
Effektivität effectiveness
Effektivspannung RMS voltage
Effektivstrom RMS current
Effektivwert root-mean-square value ‖ rms value ‖ r.m.s. ‖ RMS * *Abkürzg. für 'Root Mean Square value': quadrat. Mittelwert, Effektivwert*
Effektivwert der Spannung RMS voltage
Effektivwert des Stroms RMS current
Effektivwertregelung RMS control * *'Abkürzg. für 'Root Mean Square': Quadratischer Mittelwert, Effektivwert*
effizient efficient ‖ effective * *wirsam, erfolgreich, wirkungsvoll, kräftig, effektvoll; wirklich, tatsächlich*
Effizienz efficiency
EFTA E.F.T.A. * *European Free Trade Association*
EG EC * *Abkürzg.f. 'European Community': Europäische Gemeinschaft* ‖ EEC * *Abkürzg.f. 'European Economic Community': Europ. Wirtschaftsgemeinsch.: EWG*
EG-Konformitätserklärung EC Declaration of Conformity * *Herstellererklärung der Übereinstimmung mit EG-Normen/Richtlinien*
EG-Niederspannungsrichtlinie EC Low-Voltage Directive
EG-Richtlinie ECC Guideline * *siehe →EG* ‖ EU Directive * *Abk. für 'Union European'* ‖ European Council Directive ‖ EC Directive

Egalisierapparat side dresser || shaping unit * *{Holz}* zum Schärfen von Sägen
egalisieren equalize * *siehe auch 'ausgleichen'*
Egalisiergerüst skin-pass stand * *{Walzwerk}* im Kaltwalzwerk || skin-pass edging stand * *{Walzwerk}* im Warmwalzwerk
Egalisiermaschine dressing machine * *{Holz}* zum Pflegen/Schärfen von Sägen, Schleifscheiben usw.
Egalisierwalze evener roll * *bei der Papierherstellung*
Egalisierwalzwerk equalizing rolling mill * *{Walzwerk}*
EGB ESD * *Abk f. 'Electrostatically Sensitive Devices': elektrostatisch gefährdete Bauteile*
Egoutteur dandy roll * *bei Papierherstellung* || egoutteur * *{Papier} bei Papiermaschine hinter dem Sieb*
EIA EIA * *Abkürzg. f. 'Electronic Industries Association': US-Normengremium*
eichen calibrate
Eichfaktor calibration factor
Eichgerät calibrator || calibration device
Eichung calibration
eifrig eager || keen || passionate || enthusiastic || studious || anxious * *be very anxious to: eifrig bestrebt sein, zu* || ambitious * *strebsam, ehrgeizig*
eigen own || of one's own || dedicated * *speziell zugeordnet, zugeeignet, gewidmet* || individual || dedicated * *speziell zugeordnet/zugeschnitten; zugeeignet, gewidmet* || private * *persönlich, privat, allein, vertraulich, geheim* || intrinsic * *innewohnend, inner, wahr, eigentlich, wirklich* || inherent * *innewohnend, zugehörend, eigen*
Eigen- intrinsic * *innewohnend* || self- * Selbst- || natural * *-Frequenz, -Kapazität, -Schwingung usw.* || inherent * *innewohnend, Selbst-*
Eigenart peculiarity || characteristic || particularity * *Besonderheit, Umstand*
eigenartig peculiar || strange
eigenbelüftet self-ventilated * *mit durch die Motorwelle angetriebenem Lüfter* || self-cooled * *selbstgekühlt (unüblicher Ausdruck für Motorkühlart)* || with natural air cooling * *luftselbstgekühlt (z.B. Leistungshalbleiter, für Motor unüblich)*
Eigenbelüftung self-ventilation * *{el.Masch.}* (veraltet) →*Eigenkühlung*
eigener own || of one's own || one's own || dedicated * *speziell zugeordnet, zugeeignet, gewidmet* || individual
Eigenfertigung own manufacture || own production || in-house production
Eigenfrequenz natural frequency
eigengekühlt self-ventilated * *z.B. Motor mit vom Läufer angetriebenem Lüfterrad* || self-cooled * *(für Motorkühlart unüblicher Ausdruck)* || self ventilated * *with fan: mit Lüfter* || with shaft-mounted fan * *Motorkühlart mit durch die Motorwelle angetriebenem Lüfterrad* || TEFC * *'Totally Enclosed Forced Ventilated': vollgekapselte Motorschutzart m. Fremdlüftg.*
eigengetaktet self-clocked
Eigenheit particularity || peculiarity || characteristic
Eigeninduktivität self-inductance
Eigeninitiative individual initiative
Eigenkapazität intrinsic capacitance
Eigenkapital equity capital * *~ einer Gesellschaft; Wert nach Abzug aller Belastungen, "reiner Wert"* || capital stock || capital stock and reserve || capital resources || privately owned capital || equity * *Wert nach Abzug aller Belastungen, reiner Wert* || equity capital || owner's equity || equity
Eigenkapitalrendite return on equity * *Verhältnis (Gewinn/eingesetztes Kapital) x 100%* || ROE * *Abkürzg. für 'Return On Equity': Verhältnis Gewinn/eingesetztes Kapital*

Eigenkühlung self ventilation * *Motorkühlart* →*"eigenbelüftet" (Lüfter durch Motorwelle angetrieben)* || self-cooling * *z.B. eines Motors mit am Läufer angebrachten oder von ihm angetriebenen Lüfter* || self circulation
Eigenlüfter main-shaft mounted fan * *in elektr. Maschine* || shaft-mounted fan * *durch (Motor)welle selbst angetriebener Lüfter* || shaft-mounted fan * *z.B. durch die Welle angetriebener ~ bei eigenbelüftetem Motor*
Eigenschaft feature * *auch 'Eigenschaft haben/sich auszeichnen durch'* || characteristic || quality || status || points * *good/bad: gute/schlechte* || property * *Eigenart, Merkmal, Werkstoffeigenschaft, Vermögen (z.B.Isolationsvermögen)* || particularity * *Eigenheit, Umstand* || nature * *Beschaffenheit* || peculiarity * *Eigenart*
Eigenschaften features * *Merkmale* || characteristics || properties
eigensicher intrinsically safe * *Ex-Schutzart 'i' nach EN50020*
Eigensicherheit intrinsic safety * *Schutzart 'i' nach EN50020 für Ex-geschützte el. Betriebsmittel*
eigenständig stand alone || independent * *unabhängig* || self-contained * *z.B. Gerät* || autonomous
eigentlich real * *wirklich* || true * *wirklich, echt* || actual * *wirklich, tatsächlich* || essential * *wesentlich* || basic * *grundlegend, die Grundlage bildend* || precise || genau || proper * *genau, richtig* || basically * *(Adv.) im Grunde grundsätzlich* || really * *(Adv.) tatsächlich* || actually * *(Adv.) tatsächlich* || as a matter of fact * *(Adv.) tatsächlich* || originally * *(Adv.) ursprünglich* || exactly * *(Adv.) genau* || strictly speaking * *(Adv.) genaugenommen* || by rights * *von Rechts wegen* || to tell the truth * *(Adv.) offen gesagt* || anyhow * *(Adv.) what do you want anyhow: was willst Du eigentlich* || exact * *genau* || intrinsic * *innewohnend*
Eigentum property
eigenverantwortlich self-controlling
Eigenverbrauch internal consumption || own consumption
Eigenwert eigen value || characteristic number
Eignung suitability * *Eignung, Angemessenheit (to: für, zu)* || applicability * *Eignung, Anwendbarkeit (to: für, auf, zu)* || qualification * *~ einer Person (for: zu)* || fitness * *(auch einer Person) Eignung, Fähigkeit, Tauglichkeit, Zweckmäßigkeit, Angemessenheit* || aptitude * *(einer Person) Befähigung, Fähigkeit, Tauglichkeit, Begabung (for: für, zu)*
Eignungstest aptitude test * *für Bewerber, Personal* || qualification test * *für Bewerber, Personal* || suitability test * *für anzuschaffendes Investitionsgut*
Eilablauf high sequence * *bei SPS*
Eilgang rapid taverse * *{NC-Steuerung}* || rapid feed * *{NC-Steuerung}*
EIN on
Ein- single- || mono-
Ein- oder Ausschaltkontakt single-throw contact * *hat nur 2 Anschlüsse (kann entweder nur schließen oder nur öffnen)*
Ein-Aus-Schalter on-off switch || power switch * *für Stromversorgung*
Ein-Ausgabe IO
Ein-Ausgabe-Baugruppe IO-module || I/O module || I/O board * *Leiterkarte, Flachbaugruppe*
Ein-Ausgabe-System I/O-system
Ein-Ausgabebaugruppe I/O module || I/O board * *Leiterkarte/Flachbaugruppe*
Ein-Ausgabedaten in/output data
Ein-Ausgabemodul I/O module || input-/output module
Ein-Ausschaltautomatik automatic switch-on and switch-off control || automatic switch-on and switch-off circuit || automatic start-stop control

Ein-Ausschalter on/off switch || power switch * *für die Stromversorgung*
Ein-Ausschaltlogik switch-on and switch-off control || switch-on and switch-off logic || start-stop sequencing || start-stop logic || sequence control || sequence logic || start-stop control
Ein-Ausschaltsteuerung start-stop sequencing * *Ablaufsteuerung zum Starten u. Anhalten e. Motors* || start-stop control || switch-on and switch-off control
Ein-Chip-Computer single-chip computer || embedded controller || microcontroller
Ein-Chip-Mikroprozessor embedded controller || microcontroller || single-chip processor
Ein-Rückspeiseeinheit regenerating rectifying unit * *Umkehrstromrichter zur Zw. kreisspeisg. f. (mehrere) Wechselrichter* || regenerative rectifier unit * *zur Zw.kreispeisg. v. Antriebswechselrichtern m. Bremsenergie-rückspeisg.* || regenerating front end || active front end * *selbstgeführt. Umkehrstromrichter z.Zwischenkreisspeis. (z.B. m. Transistoren)* || regenerative front end || rectifier/regenerative feedback unit || rectifier and regenerative feedback unit
Einachs-Antrieb single axis drive
Einachs-Positionierung single-axis positioning
Einachspositionierung single-axis positioning
einarbeiten incorporate *⁰ einfließen lassen, einbauen, hin-* || become familiar with * *sich in etwas ~* || get familiar with * *sich in etwas ~* || train * *someone for a job: jemanden in eine Aufgabe ~* || initiate * *someone in a job: jemanden in eine Aufgabe ~* || familiarize * *someone with his new work: jmd. in seine neue Aufgabe; oneself: sich* || work into * *sich in eine Sache ~; neue Erkenntnisse in eine Arbeit ~* || make up for * *by extra work: einen Rückstand durch Zusatzarbeit* || work in * *by extra work: einen Rückstand durch Zusatzarbeit* || coach * *jdm. einarbeiten, ausbilden, trainieren, instruieren (meist durch älteren Kollegen)*
Einarbeitung training || coaching * *von einem erfahrenen Kollegen*
Einarbeitungsphase familiarization period
Einarbeitungszeit learning time || training period || breaking-in period
Einbau installation || mounting * *Anbau, Befestigung, Bestückung* || accommodation * *Unterbringung* || building in || fitting || insertion * *das Einfügen* || incorporation * *Eingliederung, Einverleibung*
Einbau-Messgerät panel meter * *Schalttafel-Instrument*
Einbauanleitung installation instructions
Einbaubreite mounting width
Einbaueinheit chassis module * *auf offenem Tragblech montiert, ohne besondere Schutzart, für Schrankeinbau*
einbauen build in || build into * *in ... hinein* || incorporate * *in: in* || install * *installieren* || mount * *montieren* || fit * *into: in* || insert * *einfügen, into: in* || assemble * *zusammenbrechen/bauen, montieren* || house * *befestigen, verzapfen/schrauben* || retrofit * *nachrüsten, nachträglich einbauen* || integrate * *eingliedern, integrieren*
Einbaufläche mounting surface
Einbaugerät chassis unit * *auf offenem Tragblech montiert, m.niedriger Schutzart (meist IP20), für Schrankeinbau* || open-chassis unit * *zum Schrankeinbau ohne besondere Schutzart, z.B. auf offenem Tragblech (IP20)* || built-in unit
Einbauhinweise installation instructions || installation recommendations * *Empfehlungen den Einbau betreffend* || mounting instructions
Einbauhöhe installation height above sea level * *über d. Meeresspiegel*

Einbauhülse fitting sleeve
Einbaulage mounting position || installation position
Einbaumaß mounting dimension
Einbaumotor built-in motor * *"nackter" Motor, nur aus Ständer- und Läuferpaket bestehend* || shell-type motor * *{el.Masch.}*
Einbauort installation location || mounting location
Einbauplatte mounting plate * *Montageplatte*
Einbauplatz mounting location || slot * *zum Einschieben einer Steckbaugruppe/Leiterkarte* || build-in position || module location * *für Baugruppe* || component mounting position * *z.b. für Einbauplatz auf Leiterkarte* || mounting position || SPS * *Abk. f. 'Standard Packaging Slot': →Standard~ (SEP, EP)*
Einbaurahmen mounting rack * *z.b. für Baugruppen*
Einbausatz mounting set || mounting kit
Einbausystem packaging system || modular packaging system
Einbauten built-in components || fittings || built-in accessories * *eingebaute Zubehörteile*
Einbautiefe mounting depth
Einbautoleranz mounting tolerance
Einbauvariante mounting version
Einbauvorschrift mounting specification
einberufen call * *eine Sitzung, Versammlung, Besprechung usw. ~; for elghi o'clock: für 8 Uhr* || convene * *Versammlung usw. ~* || call together * *Teilnehmer (zu einer Versammlung) zusammenrufen* || convoke * *z.B. Sitzun ~g* || summon * *z.B. Versammlung ~* || hold a meeting * *eine Besprechung abhalten*
einbetonieren set in concrete || embed in concrete || let into concrete
einbetten embed
einbeziehen involve || include * *einschließen* || cover * *abdecken* || incorporate * *einverleiben, vereinigen, verbinden, aufnehmen, eine Körperschaft bilden*
Einbeziehung incorporation * *into: in* || inclusion
einbezogen involved
einbinden link into || integrate * *integrieren; into: in* || tie in * *(auch figürl.)* || tie to * *jdn. verpflichten (zu), jdn. binden an*
Einblockziehmaschine single block drawing machine * *[Draht] Drahtziehmaschiene*
einbrechen break in || collapse * *kollabieren, zusammenbrechen* || break into * *gewaltsam ~* || drop * *z.B. Spannung*
Einbrennofen burn-in oven
einbringen bring in * *auch Vorlage, Vorschlag* || work into * *hineinarbeiten* || bring forward * *auch Vorlage, Vorschlag* || introduce * *z.B. Vorschlag, Gesetz; auch 'Thema/Frage anschneiden/aufwerfen/vorstellen'* || present * *präsentieren* || propose * *vorschlagen*
Einbruch notch * *z.B. Kommutierungseinbruch* || dip * *kurzzeitiger, z.B. Spannungseinbruch* || sag * *Absacken z.B. der Netzspannung* || fall * *wirtschaftlicher ~* || decline * *Nieder-/Rückgang, Abnahme, Verschlechterung* || burglary * *krimineller ~, ~sdiebstahl (besonders b. Dunkelheit)* || housebreaking * *~sdiebstahl* || undershoot * *Unterschwinger*
Eindampfanlage evaporator
Eindampfung evaporation || vaporization || concentration
eindeutig definite || clear * *klar, deutlich; clearly identified: eindeutig gekennzeichnet* || unambiguous * *unzweideutig* || determinate * *~ bestimmt (Fehlerzustand usw.), bestimmt, fest(gesetzt), entschieden*
eindicken thicken

Eindicker thickener
Eindickung thickening || concentration || livering * *von Farbe usw.,* || bodying * *von Öl usw.*
eindrähtig single-core * *einadrig (Kabel)* || solid * *steifer Leiter (keine Litze)* || single-conductor
eindringen penetrate * *durchdringen, eindringen (in) (auch geistig), durchbohren, durchstoßen* || migrate * *hinein/abwandern* || infiltrate * *hinein, auch 'eindringen lassen, einsickern, unterwandern'* || enter * *eintreten, eindringen in* || enter by force * *mit Gewalt* || break in * *(hin)einbrechen* || invade * *in einen Markt/ein Land* || soak in * *einer Flüssigkeit* || invasion * *(Substantiv) Eindringen, Einfall, Angriff* || infiltration * *(Substantiv) auch 'Einsickern'* || ingress * *(Substantiv) Eintritt, Eintreten, Zutritt, Zugang*
eindrucksvoll impressive || spectacular * *spektakulär* || striking, appealing
Eindruckwerk imprinting unit * *{Druckerei}*
eine Fülle von an abundance of || a wealth of * *ideas: Einfälle/Ideen; knowledge: Wissen*
eine Reihe von a series of
eineinhalb one and a half
einerseits on the one hand * *on the other hand: andererseits* || on the one side
einfach single * *einmal vorhanden* || simple * *leicht* || easy * *leicht* || in an easy way * *(Adv.) auf einfache Art* || plain * *klar, leicht verständlich, schlicht* || clear * *klar* || uncomplicated * *unkompliziert* || straight-forward * *→unkompliziert, direkt, geradlinig, gerade(her)aus* || foolproof * *kinderleicht, narrensicher, idiotensicher, ungefährlich, betriebssicher*
einfach inbetriebzunehmen easy to install
Einfach- single- || one-
Einfachbedienfeld simple operator control panel
einfache Bedienbarkeit ease of operation * *bequeme Bedienbarkeit; siehe auch 'Bedienungsfreundlichkeit'* || usability
einfache Bedienung simple operations || easy operation * *siehe auch 'Bedienungsfreundlichkeit'*
Einfachfehler single fault * *bei SPS* || single-chance fault * *bei SPS*
Einfachheit simplicity * *to simplify matters/for reasons of simplicity/to save trouble: d. Einfachheit halber* || ease * *Leichtigkeit, Lockerheit, Bequemlichkeit* || plainness * *Einfachheit, Schlichtheit, Deutlichkeit, Klarheit, Offenheit, Ehrlichkeit* || frugality * *Genügsamkeit, Einfachheit*
Einfachrohrkühler single-tube heat exchanger
einfachst simplest
einfachster Fall simplest case * *in the simplest case: im einfachsten Fall*
Einfachstromrichter one-way converter * *Einquadrant-Stromrichter*
einfädeln thread || set afoot * *(figürl.)* || contrive * *(figürl.) zustande bringen, ermöglichen, es einrichten (to: zu)*
einfahren drive in || enter || come in || arrive || pull in * *{Bahnen}* || approach * *sich nähern, heranfahren (z.B. Aufzug, Zug), ein-/anfliegen, nahekommen (auch figürl.)* || descend * *{Bergwerk}* || bring in * *Ernte* || drive into place || transport into place || run in * *neues Auto, Getriebe usw. ~* || break in * *(amerikan.) neues Auto usw.*
Einfahrt entrance || arrival || approach * *Annäherung, (Heran-)Nahen, Anmarsch, Nahekommen (auch figürl.); auch bei Zug, Aufzug usw.*
einfallen come to mind * *Gedanke* || be applied * *Bremse* || be put on * *Bremse* || engage * *Bremse* || application * *(Substantiv) einer Bremse*
Einfallzeit application time * *einer Bremse* || brake delay time * *einer Bremse*
einfetten grease * *mit Fett* || oil * *mit Öl*

Einflächen- single-face * *z.B. Bremse, Kupplung*
Einfluss influence || effect * *Wirkung, Ein/Auswirkung* || power * *Macht* || control * *Beherrschung* || flowing in * *das Einfließen* || influx * *das Einfließen* || interference * *Eingreifen (with: in), Beeinflussung, Beeinträchtigung, Störung,Überlagerung* || impact * *Einwirkung* || action * *(Ein-)Wirkung, Wirksamkeit, Einfluss, Vorgang, Prozess, Handeln, Tätigkeit* || leverage * *Mittel/Hebel zur Einflussnahme, Einfluss, Hebelwirkung*
Einflussgröße influencing variable || factor * *(allg.)* * *auch figürl.: Meßwert usw.*
einfrieren freeze * *auch figürl.: Meßwert usw.*
einfügen insert || put in || fit in * *Sache und Person; (well: gut)* || insertion * *(Substantiv) Einfügung, Einfügen* || paste * *Inhalt von Zwischenablage, Textpuffer usw. "einkleben" (Cut and Paste-Funktion)*
Einfügung insertion
einführen thread * *z.B. eine Materialbahn ~* || introduce * *(auch figürl.) in eine Öffnung ~; well introduced: gut eingeführt (Person, Produkt)* || launch * *Mode, neues Produkt usw. ~* || initiate * *Maßnahmen usw. ~; initiate someone into: jdn. einweihen in* || adopt * *Maßnahmen usw. übernehmen, annehmen, einführen, ergreifen* || establish * *Einrichtungen usw.; well established: gut eingeführt (Firma, Produkt)* || set up * *Einrichtungen, Methode usw. ~* || import * *Waren aus dem Ausland ~* || present at * *eine Person in einen neuen Kreis ~* || insert into * *in eine Öffnung ~* || lead in * *eine elektr. Leitung ~* || feed in * *zuführen*
Einfuhrgenehmigung license to import || import license
Einführung introduction * *~ in Wort oder Schrift; ~einer Person; ~ von Maßnahmen* || presentation * *Vorstellung* || establishment * *~ von Einrichtungen* || importation * *Import aus d.Ausland* || entry * *z.B. Kabel~* || insertion * *Einsetzen eines Bauteils, Markt~* || lead-in * *~ von Leitung, Kabel* || threading * *~ einer Wahrenbahn in eine Maschine (z.B. Papier, Folie, Blechband, Textil usw.)* || coupling * *~ eines Rohrs*
Einführungskurs introductory training course
Einführungsplatte entry plate * *z.B. zur Kabeleinführung in einen (Motor-) Klemmenkasten* || cable entry plate * *zur Kabeleinführung in Klemmenkasten*
einfüllen fill in || pour in * *eingießen* || fill into || bottle * *in Flasche* || barrel * *in Fass* || charge * *(Behälter usw.) befüllen; Metall (-guss), Granulat usw. einfüllen*
Einfüllöffnung filling hole || feed opening
Einfüllschacht filling funnel || feed tube
Einfüllschraube filler screw || filler plug || filling plug
Einfüllstutzen filler stub || filler cap || filler socket
Einfülltrichter infeed hopper || feed hopper || hopper || funnel tube || funnel
Eingabe entry * *per Bedienung, z.B. in einen Computer, einem Bedienfeld* || setting * *per Bedienung* || entering * *z.B. über Tastatur*
Eingabeaufforderung promt * *eines Betriebssystems am Bildschirm (z.B. "C:\>" bei MS-DOS)* || user prompt
Eingabefehler input error
Eingabefeld entry field * *auf einer LCD-/Bildschirmanzeige* || dialog box * *Dialog-Feld auf einem Bildschirm*
Eingabegerät input device
Eingabeinformation input information
Eingabekontrolle input check
Eingabemaske input mask * *am Bildschirm*
Eingabesequenz input sequence
Eingabetaste ENTER key || ENTER
Eingang input || inlet * *Einlassöffnung*

Eingang begrenzt setzen   set input conditionally * *bei SPS*
Eingangs-   input || incoming || receiver || driving-point
Eingangs-Nennspannung   nominal input voltage
Eingangsbaugruppe   input module || input board * *Leiterkarte*
Eingangsbyte   input byte
Eingangsfunktion   input function
Eingangsgleichrichter   incoming rectifier || line-side rectifier * *netzseitiger ~*
Eingangsgröße   input variable || input value * *Eingangswert* || input quantity * *zahlen-/wert-/pegelmäßig* || input signal
Eingangsimpedanz   input impedance
Eingangsklemme   input terminal
Eingangsleistung   input power
Eingangsspannung   input voltage * *to: für/an*
Eingangspegel   input level
Eingangsprüfung   incoming inspection || acceptance test * *Abnahmeprüfung bestellter Waren*
Eingangsschaltung   input circuit
eingangsseitig   input-side || input || supply-side * *Netz-/Versorgungs(spannungs)-seitig*
Eingangssicherung   line-side fuse * *netzseitig angeordnete ~*, →*Netzsicherung* || incoming fuse
Eingangssignal   input signal || source signal * *Quellsignal*
Eingangsspannung   input voltage
Eingangsspannungsbereich   input voltage range
Eingangsstrom   input current
Eingangsstufe   input stage
Eingangstaktfrequenz   input clock frequency
Eingangsvermerk   receive note * *z.B. auf einem Brief* || Received * *z.B. auf einem Brief*
Eingangswelle   ingoing shaft * *bei Getriebe*
Eingangswiderstand   input resistance || input impedance * *Eingangsimpedanz*
Eingangswort   input word
eingebaut   built-in || incorporated || embedded * *eingebettet* || integral * *integriert* || mounted || integrated * *integriert*
eingebaut haben   incorporate
eingebaut sein   be incorporated
eingeben   enter * *z.B. über Tastatur/Bediengerät/Klemme* ~ || type in * *über Tastatur* ~ || edit * *Text/Daten in einen Rechner* ~ || input || type * *über Tastatur* ~
eingebettet   embedded
eingebracht   embedded * *z.B. Wicklung in Motornut*
eingefallen   switched-in * *z.B. Schützkontakt* || applied * *z.B. Bremse* || put on * *z.B. Bremse*
eingefroren   frozen * *auch Messwerte, Speicherinhalt usw.*
eingegeben   entered || input
eingehend   incoming * *hereinkommend* || exhaustive * *gründlich, erschöpfend* || thorough * *gründlich, sorgfältig, eingehend, genau; vollkommen, völlig, echt, durch und durch* || detailed * *detailliert, in Einzelne gehend* || throrougly * *(Adv.) gründlich* || in detail * *(Adv.) detailliert, im Einzelnen* || thoroughgoing * *kompromisslos, durch und durch*
eingelötet   soldered in
eingeprägt   load-independent * *z.B. Strom/Spannung* || impressed * *auch Strom/Spannung*
eingeprägte EMK   impressed EMF
eingeprägte Spannung   impressed voltage || load-independent voltage
eingeprägter Strom   impressed current || load-independent current
eingepresst   press-fit
eingerastet   engaged
eingeschaltet   switched on || lit * *Leuchtdiode, Meldeleuchte* || energized * *mit High-Signal angesteuert*

eingeschalteter Zustand   ON-state * *z.B. Thyristor* || energized state * *z.B. Klemme, Relais, Motor* || on-condition * *auch eines Thyristors*
eingeschlossen   included * *inbegriffen; siehe auch* →*einschließen*
eingeschränkt   restricted * *(to: auf)* || limited || to a limited extent * *(Adv.) in eingeschränktem Maße*
eingeschwungener Zustand   steady-state || stabilized condition * *thermal: thermisch* ~ || steady-state condition
eingestellt   adjusted || set || discontinued * *z.B. Fertigung* || preset * *voreingestellt, vorgegeben*
eingestellter Wert   set value
eingetragenes Warenzeichen   registered trademark
eingießen   cast in * *z.B. in Kunstharz* || pot * *z.B. in Becher* || encapsulate * *(ein-) kapseln, umgießen*
eingreifen   engage * *in/with: in* || gear * *Getriebe, Maschinenteile; in(to): in* || take action * *(figürlich)* || step in * *(figürlich)* || come into action * *taktisch* || mesh * *Maschinenteile ; in(to): in; with: ineinander* || intervene * *vermittelnd* ~ *(in: in)* || interfere * *störend* ~ || act * *on: in/auf* || intervention * *(Substantiv) Eingriff; manual intervention: Eingreifen von Hand* || engagement * *(Substantiv)* || action * *(Substantiv) figürlich* || access * *zugreifen auf* || influence * *beeinflussen*
Eingreifen von Hand   manual intervention
Eingriff   intervention * *auch: eines Sıgnals/Sollwerts* || weighting * *Gewichtung, auch eines Signals* || engagement * *von Zahnrädern, Kupplung usw.* || meshing * *von Zahnrädern usw.*
einhaken   hook in || hook into || hook * *an-/ein-/fest-/zuhaken* || hook up * *anhaken, (mit Haken) aufhängen, anschließen*
einhalten   observe || adhere to * *sich halten an, einhalten (z.B. an eine Regel)* || follow || maintain * *(aufrecht)erhalten, beibehalten, unterstützen, (Grenzen, Spezifikation) beachten* || satisfy * *z.B. Bedingung/Gleichung* || keep within * *limits: Grenzen; time limit: Frist; appointed time: Termin* || keep to * *term: Frist* || comply with * *Vorschriften, Norm* || meet * *Vorschriften, Norm* || keep with * *Forderung usw.*
Einhaltung   compliance * *Erfüllung; with: von* || observance * *of: von* || adherence * *to: von*
Einhand   one-hand * *z.B. ~bedienung, ~hobel usw.*
einhängen   hook in * *into: in* || suspend * *into: in (Konstruktionselement, Bauteil usw.)* || hang up * *aufhängen* || put on its hinges * *in Scharnier* ~ *(z.B. Tür)* || hang in
Einheit   unit * *Gerät, Maß~* || unit of measure * *Maß~* || unity * *zusammengehörende* ~ || engineering unit * *physikalische* ~ || package * *Bau-* || module * *Bau-, Baustein, Modul*
einheitlich   uniform || standardized * *genormt* || undivided * *ungeteilt* || unitarily * *(Adv.)* || standard || harmonized * *ver~t* || unified * *ver~t*
Einheitsprung   unit step
Einheitssprungfunktion   unit step
Einheitsstoß   unit pulse || Dirac pulse
einhergehend mit   accompanying
einholen   catch up * *with: jemanden*
Einhüllende   envelope || envelope curve * *Hüllkurve*
einigen   settle * *on: sich einigen auf*
einigermaßen   to some extent || to a certain extent || somewhat || rather * *ziemlich* || fairly * *ziemlich*
einkanalige Steuerung   single-channel control * *{SPS}*
einkapseln   encapsulate || enclose * *einschließen*
Einkauf   purchasing department * *Einkaufsabteilung* || purchasing * *das Einkaufen* || purchasing management

Einkäufer purchasing manager * *in einem Unternehmen*
Einkaufsabteilung purchasing department
einkerben notch
einklammern put in brackets || parenthesize * *[pe'renthisais]* || put in parentheses
einklinken latch * *on the latch: eingeklinkt* || engage || catch || clinch
Einkopfsäge single-head saw * *{Holz}*
einkoppeln inject || couple into
Einkopplung injection * *z.B. von (Stimulations-) Signalen* || coupling || interference * *von Störungen*
Einkünfte income
einladen invite * *jemanden zu Besuch, Besprechung usw.* ~ || load in * *Transportgüter* || entruck * *Lastwagen beladen*
Einladung meeting request * *zu einer Besprechung* || invitation * *auch zu einer Besprechung*
Einlage intermediary * *{Kabel | Draht}*
einlagern store || store up || warehouse || put into stock || stock * *vorrätig haben, führen* || stock up * *auf Lager legen, speichern*
Einlagerung warehousing || storage
einlagig single-layer
Einlass inlet * *inlet valve: ~ventil* || intake
einlassen let in || open the door to || let into * *in Holz/Metall (z.B. Möbelbeschlag)* || fit into * *in Holz/Metall* || sink into || imbed in || recess * *aussparen, einsenken, vertiefen (z.B. Fläche für Möbelbeschlag)* || agree to * *sich auf Vorschlag usw.* ~
Einlasskanal inlet duct * *z.B. für Kühlluft*
Einlauf entry * *z.B. einer Walzstraße, einer Warenbahn in e. Verarbeitungsmaschine* || inlet * *Einlass* || entry section * *Einlaufteil z.B. bei Blechstraße* || infeed * *Zufühг-/Zufördereinrichtung*
einlaufen come in || arrive * *{Bahnen}* || enter * *{Schifffahrt} eines Schiffes, einer Wahrenbahn* || shrink * *{Textil} schrumpfen; unshrinkable/ Sanforized: nicht ~d*
Einlaufteil entry section * *z.B. einer Walzstraße*
Einlaufwerk draw-in unit * *in Faserverstreckanlage usw.*
einlegen insert * *Werkstück, Batterie usw.* || engage * *Kupplung, Bremse, Gang usw.*
einleiten initiate * *anstoßen, initieren*
einlesen read in || download * *hinunterladen von Daten/Programmen/Parametern in ein Gerät hinein* || upload * *herauflanden von Daten/Parametern aus einem Gerät heraus* || upread * *herauflesen von Daten/Parametern aus einem Gerät heraus*
einleuchtend clear || evident || obvious || plausible || insightful * *aufschlußreich*
einloggen log-in * *Starten einer Arbeitssitzung in einem Computer-Netzwerk* || log-on * *in Computer-Netzwerk*
einlöten solder in * *z.B. Lötbrücke, Bauteil auf Lötstützpunkte* ~
einmal once
einmalig unique || once * *einmal* || unsurpassed * *unübertroffen*
einmalige Kosten initial cost * *nur zu Beginn einmal aufzubringende (Initialisierungs-) Kosten*
Einmannbedienung one-man operation
einmessen calibrate
Einmotorenantrieb single-motor drive
Einnahme occupation * *Besetzung* || receipts * (Plural) *[ri'ßiets]* (Plural) *Geschäftseinnahmen* || takings * (Plural) *Geschäftseinnahmen* || return * *Geschäftseinnahme* || proceeds * (Plural) *Erlös* || earnings * (Plural) *Gewinn, Ergebnis, Verdienst* || revenue * ~ *des Staates*
einnehmen take-up * (Platz, Zeit usw.) *beanspruchen, in Anspruch nehmen, beanspruchen*
einordnen arrange in proper order * *in der richti-*

gen Ordnung/Reihenfolge ~ || arrange * *anordnen* || put in its proper place * *an der richtigen Stelle* ~ || file * *Akten usw.* ~ || classify * *in Klasse(n)* ~ (*as: unter)* || categorize * *in Kategorie* ~ (*as: unter)* || class * *in Klasse(n)* ~ (*with/as: unter)* || rank * *im Rang* ~ (*among: unter)* || take one's place * *sich* ~
Einpackmaschine packing machine
einpassen fit in || fit into place || fit into || trim in || true
Einphasenbetrieb single-phase operation * *z.B. eines Motors*
Einphasenbrückenschaltung single-phase bridge connection
Einphasenmotor single-phase motor * *{el. Masch.} Asynchronmotor mit Phasenschiebung der Spannung über Kondensator(en)* || single-phase capacitor motor * *mit Kondensator zur Erzeugung einer Hilfsphase (Steinmetzschaltung)*
einphasig single phase || single-phase
Einplatinencomputer single-board computer
einpolig single-pole || single-line * *Darstellung im Übersichtsschaltbild mit nur einer Phase des Drehstromsystems*
einpolige Darstellung single-pole representation
einprägen impress * *(auch figürl.: jemd. etwas einprägen, siehe auch 'merken') auch Strom, Spannung usw.* || imprint || stamp || inject * *z.B. Strom* || memorize * *→auswendig lernen, sich* ~
einprägsam easy to remember || easily remembered || catchy * *z.B. Motto, Slogan* || impressive * *eindrucksvoll* || eye-catching * *optisch*
Einpress- press-in || press-fit
Einpressmutter knock-in nut
Einquadrant- single-quadrant || one-quadrant || unidirectional * *mit einer Drehrichtung*
Einquadrantbetrieb single-quadrant operation || one-quadrant operation
Einquadrantenantrieb one-quadrant drive || single-quadrant drive
Einquadrantenbetrieb single-quadrant operation
einrasten click into place || lock home || catch || latch * *(sich) einklinken/einschnappen; on the latch: eingeklinkt* || engage || clinch || engage * *in: an/in* || snap into place
einrastend snap- || catching type || snapping
Einrichtdrehzahl thread speed * *Einziehdrehzahl/-geschwindigkeit*
einrichten install || equip * *ausrüsten* || set * *Werkzeugmaschine* ~ || arrange * *anordnen, aufstellen, einteilen* || organize * *gründen, ins Leben rufen, gestalten, organisieren* || regulate * *regulieren, anpassen, einstellen (Gerät)* || establish * *z.B. Firma, Geschäft, Niederlassung* ~ || set-up mode * *(Substantiv) Betriebsart bei NC-Steuerung und SPS* || wetting-up mode * *(Substantiv) Betriebsart bei NC-Steuerung und SPS* || presetting * *Maschine usw.* ~ || set-up * *Maschine usw.* ~
Einrichter setter * *Maschinen-* || fitter * *Installateur, Maschinenschlosser*
Einrichtung facility * *Ausrüstung, Institution, (Instrie-) Anlage, Werkstatt, Labor, (öffentliche) Einrichtung* || device * *Gerät, Vorrichtung* || equipment * *Ausrüstung* || institution * *öffentliche Einrichtung* || arrangement || organization || set-up || disposition * *Anordnung* || design * *Bauart* || establishment * *Gründung; öffentliche/private Einrichtung/Institution* || setting-up * *Gründung* || furniture * *z.B. eines Hauses* || furnishings * *z.B. eines Hauses* || fittings * *z.B. eines Ladens* || installation * *Einbau, Anlage* || setting * *Einstellung, z.B. einer Maschine* || plant * *Anlage* || apparatus * *Vorrichtung* || appliance * *Vorrichtung* || mechanism * *Vorichtung*
einrücken throw in * *Gang, Kupplung usw* || engage * *Kupplung usw.* || throw into gear * *Kupp-*

lung usw. || indent * *Absatz in einem Dokument, Text auf dem Bildschirm usw.*
Einrückkraft engagement force * *zum Schließen einer Kupplung*
eins one || unity * *unity power factor: Leistungsfaktor von eins* || one thing * *das eine*
eins-zu-eins one-for-one
Eins-Zu-Eins-Verbindung 1-to-1 connection
Einsattelung dip * *in einer Kurve/Kennlinie*
Einsatz use * *Verwendung; in use/at work: im ~; be put to use: zum ~ kommen* || application * *Anwendung* || utilization * *Nutzbarmachung* || insert * *Teil zum Einsetzen in etwas, Einsatzstück* || cartridge * *Patrone, (Filter-) Einsatz* || dedication * *Hingabe, Arbeits~* || struggle * *Kampf, Mühen* || working moral * *Arbeitsmoral* || action * *be in action: im ~ sein; be brought into action: zum ~ kommen* || start * *das Beginnen* || mission * *Auftrag (für Dienstreise), taktischer Auftrag* || assignment * *Anweisung, Arbeitsauftrag, taktischer Auftrag* || adapter * *(An-) Passstück* || task * *Aufgabe* || efforts * *Anstrengung* || hard work * *harte Arbeit* || deployment * *von Arbeitskräften* || service * *Dienst (auch von Maschinen)* || bit * *zum Einsetzen in Werkzeughalter (z.B. Bohr-/Schraub~)* || beginning * *Anfang, Beginn; (initiation: Einleitung, Beginn, Aufnahme)*
Einsatzbedingungen operating conditions * *most severe: härteste* || working conditions || application conditions
Einsatzbeispiel example of application
Einsatzbereich range of application || application area
Einsatzbereiche range of applications
einsatzbereit operational * *betriebsfähig, arbeitend* || ready for operation || ready for action || ready for service || usable * *einsatzfähig* || workable * *einsatzfähig* || ready-to-run * *startbereit*
Einsatzbereitschaft readiness for action || readiness for service
Einsatzbericht service-call report * *von Servicepersonal·*
Einsatzerfahrungen practical experience
Einsatzfrequenz der Feldschwächung frequency at which field weakening starts
Einsatzgebiet application area
Einsatzgebiete fields of application * *z.B. eines Geräts* || application areas || range of applications
einsatzgehärtet case hardened
Einsatzhinweise application information * *siehe auch 'Anwendungshinweise'* || application instructions || application hints * *Tipps/Ratschläge für die Anwendung*
Einsatzmöglichkeiten possible applications
Einsatzschwerpunkt main application
Einsatzspektrum scope of applications * *wide: breites* || range of applications
Einsatzstahl hardening steel
Einschalt-Ausschalt-Logik on-off logic sequencing || on-off logic module || start-stop sequencing || switch-on and switch-off control || start-stop control || start-stop logic
Einschalt-Reset power-on reset * *wirkt beim Einschalten der Versorgungsspannung*
Einschalt-Rush inrush * *z.B. erhöhte Leistungsaufnahme b. Einschalten einer induktiven Last*
Einschaltbefehl switch-on command
Einschaltbereit ready-to-switch-on || ready to start * *Motor* || switch-ON readiness * *(Zustandsbit im →PROFIBUS-Profil)* || ready to start * *Motor, Maschine* || ready for energizing
Einschaltbereitschaft switch-on readiness
Einschaltdauer switch-on duration || conduction interval * *Stromflusszeit* || running time * *z.B. bei Motor* || cyclic duration factor * *{el.Masch.} relati-*

ve ~, z.B. eines Motors bei Aussetzbetrieb || duty ratio * *relative ~* || load time ratio * *relative ~* || operating time || duty cycle * *relative ~ (→ED) in % des gesamten Lastzyklus* || duty factor * *relative ~*
einschalten switch on || energize * *an Spannung legen, HIGH-Signal an eine Klemme anlegen* || power up * *an Spannung legen, hochfahren* || activate * *aktivieren* || switch-on || pull-in * *anziehen, z.B. Schütz, Relais* || close * *Kontakt schließen* || light * *(auf)leuchten lassen, z.B. LED, Meldeleuchte; lit: eingeschaltet* || switch in * *hineinschalten, einlegen, aktivieren* || turn on || power on * *die Stromversorgung* || engage * *Kupplung, Bremse* || engagement * *(Substantiv) das Einschalten einer Kupplung/Bremse*
einschaltfeste Erdung make-proof earthing * *Personen-Schutzeinrichtung/-Maßnahme für Servicearbeiten an Hochspannung*
Einschaltfolge starting sequence || start-up sequence || power-up sequence
Einschaltleistung inrush * *erhöhte Leistung im Einschaltaugenblick* || making capacity * *Schalter, Relais usw.*
Einschaltlogik starting logic || starting sequencing control || start-up control || power-up sequencing control || switch-on sequencing * *→Einschaltsteuerwerk*
Einschaltmoment starting torque * *Drehmoment* || instant of turn-on * *Augenblick des →Einschaltens, z.B. eines Thyristors*
Einschaltreihenfolge start-up sequence * *z.B. Motor* || starting sequence * *z.B. Motor* || power-up sequence
Einschaltscheinleistung locked-rotor apparent power * *Kurzschluss-Scheinleistg. e. Asynchronmotors bei festgebremst. Läufer*
Einschaltschwelle switch-in threshold
Einschaltsperre switch-on inhibit || lock-out function
Einschaltsteuerung start-up control * *z.B. Motor* || power-up sequencing control || starting sequencing control || starting logic * *Einschaltlogik*
Einschaltsteuerwerk start-up control * *z.B. für Antrieb* || starting sequence control * *z.B. für Motor* || switch-on control circuit
Einschaltstoß inrush * *Stoßstrom/Spannung, z.B. beim Einschalten induktiver Last*
Einschaltstoßstrom inrush current * *z.B. beim Einschalten induktiver Last*
Einschaltstrom inrush current * *hoher ~; z.B. beim Schalten induktiver Lasten* || starting current * *~ eines Motors beim Starten* || making current * *~ eines Leistungsschalters*
Einschaltstrombegrenzung inrush-current limiting * *begrenzt den Überstrom beim Einschalten induktiver Verbraucher*
Einschaltstromstoß inrush current * *beim Einschalten induktiver Lasten* || power-up current surge
Einschaltverhältnis switch-on ratio * *Verhältnis Einschaltdauer zur Gesamt-Schaltperiode z.B. bei Bremswiderstand*
Einschaltverlustleistung turn-on loss
Einschaltverriegelung starting interlock * *~ für Motor, Maschine* || start preconditioning circuit * *~ für Motor, Maschine*
Einschaltverzögerung ON delay || pickup delay * *Relais, Schütz* || closing delay * *Schalter, Kontakt* || switch-on delay || pulse delay block * *Funktionsbaustein* || on-delay
Einschaltvorgang switch-on-sequence
Einschaltwiderstand on resistance * *z.B. eines FET-Transistors im On-Zustand* || turn-on resistance * *z.B. eines FET-Transistors*
Einschaltzeit turn-on time

Einschätzung evaluation
Einscheiben- single-disc * *Kupplung, Bremse usw.*
Einscheibenbremse single-disk brake
Einscheibenkupplung single-disc clutch || single-plate clutch
Einschichtwicklung single-layer winding * *Motor-Wickl.art: jede Nut enth.nur Drähte einer Spule; Einziehtechnik-geeignet*
einschieben slide in || insert
Einschieber feeder * *Beschickungseinrichtung, z.B. für Ofen* || charging unit * *Beschickungseinrichtung, z.B. für Ofen* || charger * *Beschickungseinrichtung*
einschlägig pertinent || relevant || respective * *entsprechend*
Einschlagmaschine wrapping machine * *{Verpack.techn.}*
einschleifen loop-in * *z.b. Signal, Leitung* || bed in * *Kohlebürsten* || loop into * *loop into a circuit: in einen Stromkreis einschleifen*
Einschleusung charging || infeed || chanelling in || deflection turn entry * *durch Umlenkung* || passing in
einschließen include || encompass * *auch figürlich* || comprise * *umfassen, enthalten* || enclose * *einschließen, einkapseln* || encapsulate * *einkapseln* || allow for * *berücksichtigen, in Betracht ziehen, einberechnen* || lock in * *mit Schlüssel, Riegel usw.* ~ || lock up * *mit Schlüssel, Riegel usw.* ~ || encase * *in Gehäuse* ~ || house * *in Gehäuse* ~ || be inclusive of * *im Preis usw. eingeschlossen sein*
einschließlich including || inclusive of || comprising * *beinhaltend*
einschnappen snap in || latch * *auch 'einrasten'*
Einschnitt cut || notch * *Kerbe, Scharte, Loch* || cut-out * *Aussparung, Ausschnitt*
einschnüren constrict * *zusammenziehen, zusammenschnüren, einengen, zusammenpressen* || confine * *beengen, begrenzen, einschränken, beschränken* || narrow down * *einengen* || cramp * *einzwängen* || strangle * *strangulieren, (er-)würgen*
Einschnürung contraction * *magn.* || pitch * *magnet.* || constriction * *mechan.* || pinching * *Einzwängung, Einzwicken; auch von Ladungsträgern in Halbleiter*
Einschränkung restriction || limitation
Einschränkungen limitations
einschrauben screw in
Einschraubtiefe reach of screw
Einschreibung enrolment * *z.B. zu Vorlesung, Kurs, Tagung* || enrollment
Einschub slide-in || slide-in module * *Einschubmodul* || plug-in module || slide-in unit || draw-out section * *bei Leistungsschalter usw.*
Einschubstator press-in stator core * *{el. Masch.} Ständerblechpaket, das außerh.d. Motors gewickelt wird*
Einschwallen flow-soldering
Einschwingdauer settling time
Einschwingvorgang settling process || transient || transient condition
Einschwingzeit settling time || response time || built-up time * *z.B. zum Aufbau der Versorgungsspannung* || transient time
einseitig single-ended * *z.B. Antrieb* || one-sided * *(auch figürlich)* || unilateral * *Vertrag usw.* || partial * *parteiisch* || biased * *parteiisch* || exclusive * *ausschließlich* || unbalanced * *unausgeglichen* || at one end * *auf einer Seite* || single-sided || at one side * *auf einer Seite* || at one end only * *nur auf einer Seite* || single-end
einseitig saugend single-inlet * *Ventilator*
einseitiger Anschluss single-ended termination * *z.B. bei einem Kondensator*

einseitiger Antrieb single-ended drive || unilateral drive
Einseitigkeit one-sidedness * *(auch figürlich)*
einsetzbar applicable * *anwendbar* || usable * *verwendbar; universally: universell*
einsetzen employ * *verwenden, beschäftigen* || set-up * *z.B. eine Kommission, ein Gremium* || use * *verwenden, for: für* || apply * *verwenden, anwenden (to: bei/auf), in eine Formel einsetzen* || insert * *einfügen* || fit into * *einbauen, einmontieren* || enlist * *jmd. heranziehen/gewinnen/engagieren/anheuern (in: für)* || begin * *anfangen*
einsparen save up || economize * *on: etwas* || not incur the expenses of * *die Kosten/den Aufwand für etwas* ~
Einsparung saving || savings * *(Plural) cost saving(s): Kosteneinsparung* || economizing
Einsparungen cost savings || savings
Einspeiseeinheit rectifying unit * *netzseitiger Gleichrichter z.B. zur Speisung mehrerer Wechselrichter* || rectifier unit * *netzseitiger Stromrichter für mehrere Wechselrichter an gemeins.Zwischenkreis* || line-side rectifier unit || supply unit * *z.B. z. Erzeugen der Zwischenkreisspannung für Antriebswechselrichter* || feeder module || DC rectifier unit || rectifying front end || front end * *zur Speisung einer Gleichstrom-Zwischenkreisschiene vom Netz* || feeder unit || incoming unit * *(allg.)* || infeed module
Einspeiseklemmen supply terminals
Einspeisemodul feeder module
einspeisen supply || feed || inject * *z.B. einen Sollwert, ein Stimulus-Signal* || feed in || apply to
Einspeiseschrank incoming supply cubicle
Einspeisestrom infeed current
Einspeisetrafo infeed transformer
Einspeisetransformator infeed transformer
Einspeisung incoming powerline * *Strom/Energieversorgung* || supply * *Versorgung* || incoming feeder * *Schalteinrichtung zur Netzeinspeisung* || feeder * *Leistungs/Netzeinspeisung* || incomer section || incoming supply || incoming section || feeding section
Einspindel- single-spindle * *z.B. Werkzeugmaschine*
Einspritzung injection
Einspruch opposition * *z.B. gegen Patentanmeldung* || objection
Einspruch einlegen object || enter a protest * *against: gegen* || veto * *gegen etwas* || file an objection * *against: gegen (schriftlich)*
Einspruch erheben raise objection
einsteckbar plug-in || slide-in * *einschiebbar*
einstecken plug in || insert || can take a knock or two * *etwas – können* || pocket * *in die Tasche stecken, einstecken, einheimsen, mitgehen lassen*
Einsteckmontage through-hole mounting
Einstecktiefe insertion depth
einsteigen get in * *in ein Fahrzeug* || get into * *in ein Fahrzeug* || board * *an Bord eines Schiffes* || enter
Einsteiger first-time user || beginner * *Anfänger* || non-professional * *Laie* || newcomer
Einsteiger- entry-level
Einstell- single * *z.B. -schraube,-knopf, ~ring* || set * *z.B. ~schraube* || adjustment * *Einstell-, Regulier-* || adjusting * *z.B. ~schraube, ~knopf, ~hebel, ~markierung*
Einstellanweisung adjusting instructions
Einstellarbeiten setting up || adjustments
einstellbar adjustable || can be adjusted * *einstellbar sein* || can be set * *einstellbar sein* || programmable * *digital*
Einstellbereich setting range * *z.B. für Parameter* || adjustment range

einstellen stop * beenden || adjust * justieren; (auch figürl.: adjust oneself to: sich einstellen auf) || set up * einstellen, regulieren; z.B. Parameter, Potentiometer || set * z.b. Parameter, Sollwert ~ || tune * auch abstimmen (to: auf) || select * auswählen || discontinue * beenden/abbrechen, z.b. Fertigung eines Produktes || be obtained * sich einstellen || position * in die richtige Lage bringen || recruit * (Personal) rekrutieren || engage * Arbeitskräfte usw. engagieren || program * Parameter usw. ~ || gear * (figürl.) einrichten, anpassen, einstellen (to/ for: auf/an); gear with: passen zu
Einstellfehler setting error || incorrect setting
Einstellgenauigkeit setting accuracy
Einstellgerät setting instrument || adjusting device
Einstellhinweise guideline for setup || adjustment instructions
einstellig single digit
Einstellknopf adjustment knob
Einstelllehre adjusting gauge || setting gauge
Einstellmöglichkeiten setting possibilities
Einstellregel setting rule * siehe auch →Einstellvorschrift
Einstellring setting ring
Einstellschraube set screw
Einstellung setting * ~ von Parametern, Potentiometern || adjustment * ~ von Parametern || set up * v.Parametern; auch f.Messgerät (Oziiloskop, Logikanalysator) u.PC-Konfiguration (BIOS) || discontinuation * Aufgabe; z.b. product discontinuation: Einstell. d.Fertigung/Lieferung e.Produkts || set-up * z.B. ~ von Einstellparametern || stoppage * Anhalten, Stillstand, Betriebsstörung, Stockung, Verstopfung || suspension * vorübergehende Beendigung, Aufschieben || attitude * mental: geistige ~, personal: persönliche ~ || view * Ansicht (of: über) || approach * "Herangehen", Haltung (to: gegenüber/zu), Ansatz
Einstellungsbereich setting range
Einstellvorschrift adjustment instructions || setting procedure * Vorgehensweise zum Einstellen || setting rule
Einstellwert setting value || setting
Einstich groove * mit Drehbank erzeugte Rille
Einstiegslösung entry-level solution
Einstrahlung radiation
Einstreuungen noise || EMI noise || interference || parasitics
einstufig single-stage || single-reduction * bei Untersetzungsgetriebe
Eintakt- single clocking * z.B. Netzteil
eintauchen immerse * in Flüssigkeit || submerge * untertauchen, eintauchen, überschwemmen, unterdrücken, übertönen || dip in || plunge in || dive in || dip into * ~ in || dive * (intransitiv, aktiv) into: in || plunge * (intransitiv, aktiv) into: in || immersion * (Substantiv)
Eintauchtiefe depth of immersion * in Flüssigkeit; z.B.Getrieberad in Ölbad || submergence * Untertauchtiefe, Eintauchung
einteilen divide * in/into: in || subdivide * unterteilen, in/into: in || arrange * anordnen; in: in Gruppen || distribute * verteilen || parcel out * verteilen in Paketen, paketieren || graduate * in Grade || grade * in Klassen einteilen, abstufen || classify * in Klassen ~ || group * in Gruppen ~ || plan * planen || map out * (im Einzelnen voraus)planen, Zeit ~, mittels einer Karte/Grafik ~ || budget * (Geld)Ausgaben ~ || time * zeitlich ~ || dispose of * Zeit ~ || assign to * Arbeit, Aufgaben ~ || detail * Arbeit ~ || schedule * zeitlich ~, durch Zeit/Terminplan ~
einteilig single-part || one-part || one-piece
Einteilung classification * z.B. in Klassen || division || arrangement * Anordnung || distribution

* Verteilung || planning * Planung || schedule * zeitliche ~ || scheduling * zeitliche ~ || budget * ~ der Finanzen || grouping * ~ in Gruppen, Klassen || graduation * ~ in Grade, Abstufung || scale * ~ in Grade || subdivision * Unterteilung
eintourig single-speed * z.B. Motor mit Festdrehzahl
Eintrag entry * in Tabelle, Liste usw.; Buchung || item || charging * z.B. in Ofen, Presse
Einträge entries
eintragen enter * buchen usw.; into: in || book || list || record || register * amtlich ~ (z.b. Warenzeichen), sich amtlich ~ lassen; with: bei || insert * einfügen || fill in * ausfüllen; z.B. Angaben in einen Fragebogen/ein Formular ~ || enrol * (sich) eintragen/ einschreiben (für Kurs, an Hochschule usw.), sich immatrikulieren || load * beschicken (z.B. Ofen) || charge * beschicken (z.B. Ofen)
Eintragmaschine entry machine * z.B. für Ofen || charging machine * Beschickungs-/Lademaschine, z.B. für Ofen, Presse
eintreffen arrive * ankommen || come in || happen * geschehen || come true * sich erfüllen || be fulfilled * sich erfüllen
eintreiben drive in || run into || dowel * Dübel ~ || insert || collect * Schulden, Steuern ~ || recover * Zahlung ~
eintreten enter * hin~ (auch Dienst, Beruf), into: in Verhandlungen, als Teilhaber usw. || step in * hin~ || come in * hinein-/hereinkommen || join * in Verein/Verband usw. ~ || happen * sich ereignen, geschehen || occur * auftreten, sich ereignen, geschehen || take place * stattfinden || come about * geschehen, passieren, entstehen, umspringen (Wind, Markt) || arise * Fall, Notwendigkeit, Umstände || accrue * Versicherungs/Gewährleistungsfall || stand up for * für jdn. einstehen || intercede for * für jdn. Fürsprache einlegen, intervenieren || intervene * ein Ereignis; auch 'plötzlich ~, unerwartet dazwischenkommen, eingreifen'
Eintriebsdrehzahl input speed * eines Getriebes
Eintritt inlet * Einlassöffnung || ingress * z.B. von Fremdkörpern || entry || intake || entrance * Eintritt (zu einer Veranstaltung), Einlass, Zutritt, Beginn || admission * zu einer Veranstaltung usw.
Eintrittsöffnung inlet * z.B. für Kühlluft
Eintrittstemperatur inlet temperature * z.B. von Kühlluft/Wasser
einverstanden sein agree * to: mit || be agreeable * to: mit || consent * to: mit || approve * of: mit || be satisfied * zufrieden sein (with: mit)
Einwand objection || opposition
einwandfrei unobjectionable || contestable * unanfechtbar || unassailable * unanfechtbar || completely accurate * genau || blameless * tadellos || perfect * fehlerfrei, vollendet, vollkommen || O.K. * (auch Adv.) vermutl. Abkürzg. für 'all correct': alles in Ordnung || faultless || correct * satisfactory * zufrieden stellend
Einweg- one-way || single-way * z.B. bei Stromrichterschaltung || disposable * throw-away
Einweggleichrichter one-way rectifier || half-wave rectifier || single-wave rectifier
Einweggleichrichtung one-way rectification || half-wave rectification || single-wave rectification
Einwegschaltung single-way connection * (Stromrichter)
Einweisung guidance * Anleitung, Belehrung, Beratung, Führung || briefing * (genaue) Anweisung, Instruktion, Einsatzbesprechung, Befehlsausgabe || training * Schulung, Ausbildung, Üben || assignment * Arbeits/Aufgabenzuteilung, Anweisung || coaching * jdm. einarbeiten, trainieren, ausbilden, instruieren (meist durch älteren Kollegen) || on-the-job training * Vor-Ort-Schulung

Einwickelmaschine wrapping machine * *{Verpack.techn.}*
einwickeln wrap up
einwirken influence * *beeinflussen,einwirken auf* || act * *(up)on: auf; through: durch; wirksam werden, be-/hinwirken(for: auf), verursach., betätigen* || operate * *upon: auf* || have effect on * *Ein/Auswirkung haben auf* || have an effect upon * *Ein/ Auswirkung haben auf* || affect * *angreifen, einwirken auf, beeinflussen, beeinträchtigen* || work * *on: auf eine Person ~*
Einwirkung effect || influence * *Einfluss (of: von)* || action || operation
einzeilig single-line * *auf Anzeige oder im (Aus)- Druck* || single-tier * *Baugruppenträger usw.* || single-row * *Baugruppenträger usw.*
Einzel- single || individual * *einzeln, individuell, eigen(tümlich), bestimmmt, charakteristisch* || one- || mono- || stand-alone * →*autark*
Einzelanschluss point-to-point wire connection * *Anschlusstechnik ohne mehrpoligen Stecker bei SPS*
Einzelantrieb single drive || individual drive * *bei mehreren Antriebsstellen* || single motor drive * *mit nur einem Motor* || single-motor drive * *z.B. mit nur einem Motor am Stromrichter*
Einzelantriebe single drives
Einzelarbeitsplatz individual workplace
Einzelbediengerät individual operating unit || remote operating unit * *Vor-Ort-~*
Einzelfertigung single unit production * *Ein- Stück-Fertigung* || single-batch production || one- off production || custom-specific production * *kundenspezifische Fertigung*
Einzelfunktion individual function
Einzelhalbleiter discrete semiconductor
Einzelhandelsgeschäft retail store
Einzelhändler retailer
Einzelheit detail * *(on: über), with full details: mit allen ~en; go into details: sich mit ~en befassen* || item * *Einzelgegenstand, Stück, Posten, Position, Artikel* || particular * *Einzelheit, besonderer/näherer Umstand; with full particular: mit allen ~* || particularity * *Besonderheit, besonderer Umstand, Eigenheit, eigentümliche Beschaffenheit* || particular point * *besonderer Punkt* || isolated fact * *abgesonderter/abgeschlossener Tatbestand*
Einzelheiten details * *go into details: sich mit ~ befassen* || specifics || particulars
Einzelimpuls solid pulse * *z.B. zur Thyristor-Zündung i. Ggs. zum Kettenimpuls*
Einzelkalander offline calender * *einzeln stehendes Glättwerk, nicht in Papiermaschine integriert*
Einzelkartenhalterung printed-board holder || plug-in card holder * *für Steckkarte* || PCB holder || single-PCB receptacle
Einzellizenz single licence || single license * *(amerikan.)* || single-user licence || single-user license * *(amerikan.)* || individual license
einzeln single || individual * *eigen, für sich allein, eigenständig* || discrete * *getrennt, getrennt betrachtetes Teil, eins von mehreren Teilen* || solitary * *isoliert, einsam* || particular * *besonder, speziell* || special * *besonder* || isolated * *für sich allein* || separate * *getrennt* || detached * *abgetrennt* || each * *each one: jeder ~e (aus einer Gruppe)* || singly * *(Adv.)* || individually * *(Adv.)für sich/~ genommen* || separately * *(Adv.) getrennt* || severally * *(Adv.) getrennt, einzeln* || one by one * *(Adv.) einer nach dem andern, jeder für sich* || detail * *(Substantiv) Einzelheit; in detail: im Einzelnen*
einzeln aufgebaut discrete * *z.B. Schaltung aus Einzelbauteilen*
einzelner single || individual
einzelnwirkend individually-acting

Einzelpreis unit price || retail price * *Einzelhandelspreis, Ladenpreis, Kleinmengenpreis*
Einzelschritt single step * *{Computer\NC-Steuerung} Ablauf e. Programms mit Stopp nach jeder Anweisung zum Test*
Einzelspeisung individual supply
Einzelsteuerebene individual control level
Einzelsteuerung individual control * *bei SPS*
Einzelsteuerungsebene individual control level
Einzelsteuerungsglied individual control module * *bei SPS*
Einzelteil component part || component * *Komponente, Bauteil* || part * *Teil* || member * *Glied* || machine element * *Maschinenteil, Konstruktionselement* || item * *(Einzel)Gegenstand, Stück* || individual part
Einzelton single sound * *hörbare Schallschwingung mit einer einzigen Frequenz*
Einzelverlustverfahren segregated-loss method * *zur Bestimmung der Nenndaten einer elektr. Maschine (siehe VDE 0530,Teil 2)* || loss-summation method * *zur Ermittlung der Verluste und des Wirkungsgrads*
Einzelverpackung single packing
Einzelzyklus single scan * *bei SPS*
Einziehbetrieb threading operation * *Einziehen der Materialbahn mit langsamer (Kriech-) Geschwindigkeit*
einziehen feed in || pull in * *z.B. (Motor)wicklung in eine Nut* || thread in * *z.B. Warenbahn in eine Maschine* || thread || retract * *(Fahrwerk, Klauen usw.) ~, (Behauptung usw.) zurückziehen/widerrufen, zurücktreten*
Einziehgeschwindigkeit thread speed || threading speed || crawl speed * *Kriechgeschwindigkeit*
Einziehscheibe threading capstan * *erste* →*Ziehscheibe einer Drahtziehmaschine (gleich hinter der Abwicklung)*
Einziehtechnik pull-in technique * *Einziehen vorgefertigter Spulen in Motor (Herstellverfahren)*
Einziehvorgang threading procedure * *für eine Warenbahn*
Einziehwicklung pull-in winding * *{el.Masch.} vorgewickelte Spulen aus Runddrähten, die maschinell eingezogen werden*
einzig only * *the only one: der ~e* || single * *einzeln; not a single person: kein ~er* || sole * *alleinig* || solely * *(Adv.) ~ und allein* || once * *(Adv.) never once: nicht ein ~es Mal* || one * *my one thought: mein ~er Gedanke* || unique * *~artig; be unique: ~ darstellen*
einzigartig unique
Einzug entry feed * *fördert Material hinein* || entry section * *Materialeinzugsystem bei Verarbeitungsstraße* || feeding unit || entry || infeed
Eisbrecher ice-breaker
Eisen iron
Eisenbahnsignaltechnik railway signalling systems
Eisenbrand core burning * *{el.Masch.} Zerstörung der Blechisolation und Verschmelzen der Bleche*
Eisendrossel iron-core reactor
Eisenfüllfaktor stacking factor * *{el.Masch.}* →*Stapelfaktor*
Eisenkern iron core
Eisenschluss short circuit to core * *{el.Masch.} unerwünschter galvan. Kontakt zw. Blechen in e. Blechpaket*
Eisenverlust iron-loss || core loss
Eisenverluste iron losses
Ejektor ejector
eklatant striking * *bemerkenswert, auffallend, eindrucksvoll, schlagend, überraschend, verblüffend, treffend* || brilliant * *ausgezeichnet, glänzend* || sensational || blazing * *schreiend, auffallend, offen-*

kundig, eklatant, (salopp) verteufelt || blatant
* *brüllend, marktschreierisch, lärmend, eklatant (Beispiel, Unsinn, Lüge usw.)*
elastisch   elastic || flexible
elastische Kupplung   flexible coupling
Elastizität   elasticity * *(auch figürlich)* || angular flexibility * *Dreh~, z.B. einer Kupplung* || elastic constant * *~skonstante* || flexibility || resilience || springiness
Elastizitätsgrenze   elastic limit || yield strength
Elastizitätskoeffizient   modulus of elasticity || coefficient of elasticity
Elastizitätsmodul   elastic modulus || modulus of elasticity || Young's modulus
Elastomer   elastomer
elegant   smart
elegante Lösung   smart solution
Elekomagnet-Kupplung   electromagnetic clutch
Elektriker   electrician
Elektrikerschraubendreher   electrician's screwdriver
Elektrikerschraubenzieher   electrician's screwdriver
elektrisch   electrical * *meist nicht direkt mit Elektrizität behaftet, z.B. electrical engineer: Elektroingenieur* || electric * *meist direkt mit Elektrizität behaftet, z.B. electric field: elektr. Feld* || electrically * *(Adv.)*
elektrisch AUS   electrical OFF * *entspr. 'AUS2' beim Profibusprofil 'Drehzahlveränderbare Antriebe'*
elektrisch leitende Beschichtung   conductive coating
elektrisch leitender Anstrich   conductive coating
elektrische Betriebsmittel   electrical equipment
elektrische Bremse   electric brake
elektrische Bremsung   electric braking
elektrische Energietechnik   power engineering
elektrische Feldlinie   electric line of force
elektrische Feldstärke   electric field strength || electric field intensity
elektrische Festigkeit   electrical loadability * *z.B. eines Isoliersystems im Motor*
elektrische Flussdichte   electric flux density
elektrische Grade   electrical degrees
elektrische Größen   electrical quantities
elektrische Lebensdauer   electrical life
elektrische Maschine   electrical machine
elektrische Maschinen   electrical machines
elektrische Störung   electrical disturbance
elektrische Welle   synchro system * *Gleichlaufeinrichtung mit Asynchron-Schleifringläufermotoren* || electrical simulated shaft || self-synchronous system || selsyn
elektrischer Antrieb   electrical drive
elektrischer Schlag   electric shock
elektrisches Feld   electric field
Elektrizität   electricity
Elektrizitäts-Versorgungsunternehmen   electricity supply company || power utility * *EVU* || power supply company || power company
Elektrizitätsgesellschaft   power utility * *EVU* || electricity supply company || power supply company || power company
Elektrizitätslehre   electrical science
Elektrizitätsverbrauch   electricity consumption
Elektroakustik   electroacoustics
Elektroantrieb   electrical drive || electric drive
Elektroblech   magnetic sheet steel * *{el.Masch.} Ausgangsmaterial der Bleche für Läufer und Ständer*
elektrochemisch   electro-chemical
Elektrode   electrode
Elektroerosion   electro-erosion * *z.B. Verschleißeffekt bei elektr. Kontakten*

Elektrofachmesse   electrical trade fair
Elektrofahrzeug   electric vehicle
Elektrofilter   electric filter * *z.B. zur Abgasreinigung*
Elektroflotation   eloctroflotation
Elektrohandwerk   electrical trades || electrical craftsmen
Elektroindustrie   electrical industry
Elektroinstallation   electrical installation
Elektrokonstruktion   electrical engineering * *auch als Abteilungsbezeichnung* || electrical design
Elektrolyse   electrolysis
Elektrolyseanlage   electrolysis plant
Elektrolyt   electrolyte
Elektrolyt-Kondensator   electrolytic capacitor
elektrolytisch   electrolytic
Elektrolytkondensator   electrolytic capacitor || lytic capacitor * *Kurzform*
Elektromagnet   electromagnet
Elektromagnet-Bremse   electromagnetic brake
Elektromagnet-Einscheibenkupplung   electromagnetic single-disc clutch
elektromagnetisch   electromagnetic
elektromagnetisch betätigt   electromagnetically actuated
elektromagnetisch gelüftete Federkraftbremse   spring loaded brake with electromagnetic release
elektromagnetische Bremse   electromagnetic brake
elektromagnetische Kupplung   electromagnetic clutch
Elektromagnetische Störung   electromagnetic interference || EMI * *Abkürz. für 'electromagnetic interference'* || noise * *Störung*
elektromagnetische Verträglichkeit   electromagnetic compatibility || EMC * *Abkürzg.f. 'Electro-Magnetic Compatibility':* →*EMV*
elektromagnetisches Feld   electro-magnetic field
Elektromagnetismus   electromagnetism
Elektromagnetventil   solenoid valve
elektromechanisch   electro-mechanical || electromechanical
Elektromotor   electric motor || electromotor
elektromotorisch   electromotive || electric-motor driven || electric-motor actuated * *Stellantrieb*
elektromotorisch betrieben   motor-operated || electric motor driven
Elektronenstrahl   electron beam || cathode ray * *Kathodenstrahl* || E beam
Elektronik   electronics
Elektronik-Masse   electronics ground
Elektronik-Stromversorgung   electronics power supply
Elektronikbaugruppe   electronics board * *Leiterkarte* || electronic modul * *Baustein, Modul* || electronics module || electronic board * *Leiterkarte* || control card
Elektronikgehäuse   electronics housing
Elektronikmotor   electronic motor * *regelbarer Kleinmotor 0,6 bis 60 W und 6 bis 60 V*
Elektronikstromversorgung   electronics power supply
Elektronikteil   control unit * *Elektronikeinheit* || electronic module * *Baugruppe* || electronics section
elektronisch   electronic || electronically * *(Adv.)*
elektronische Klemmenleiste   electronic terminal block || electronic terminal || remote I/O terminal block || electronic terminal strip
elektronische Schaltung   electronic circuit
elektronischer Kommutator   electronic commutator
elektronisches Getriebe   electronic gearbox || electronic gearing

Elektropersonal electrical personnel
elektrostatisch gefährdete Bauteile electrostatic sensitive devives || ESD * *Abkürzg. für 'Electrostatic Sensitive Devices':* EGB
elektrostatische Aufladung electrostatic charging
Elektrotechnik electrical engineering
elektrotechnisch electrotechnical
Elektrowärme electric heat || electroheat || induction heating * *induktiv erzeugte* || electroheating
Elektrowärmeanlage electric heating equipment
Elektrowerkzeug power tool || portable electric tool
Element element
elementar elementar || fundamental
eliminieren eliminate
Elko electrolytic capacitor || lytic capacitor * *Kurzform*
Ellipse ellipse
elliptisch elliptical || elliptic
elliptisches Drehfeld elliptical field * *{el.Masch.} läuft entlang d. Umfangs d.Ständerbohrg.mit lageabhäng. Amplitude*
eloxieren anodize || eloxadize
eloxiert anodized
emaillieren enamel
Emission emission
Emitter emitter
Emitterfolger emitter follower * *Transistorschaltung*
Emitterschaltung emitter connection || common emitter circuit
EMK EMF * *Abküzg. für 'Electro-Motive Force' oder 'ElectroMagnetic Force': elektromagnetische Kraft* || counter-EMF || motor back EMF || CEMF * *Abkürzung für 'counter electromagnetic force': Gegen-EMK*
EMK-Erfassung EMF sensing
EMK-geregelter Antrieb voltage feedback drive * *tacholoser Betrieb eines DC Motor*
EMK-Istwert EMF actual value
EMK-Messung EMF measurement
EMK-Regelung EMF closed-loop control * *Abkürzg.f. 'ElectroMagnetic Force': elektromagnetische Kraft* || EMF control
EMK-Regler EMF controller
EMK-Sollwert EMF setpoint || EMF reference value
EMK-Vorsteuerung EMF pre-control || EMF feedforward
Empfang receipt * *[re'ßiet] Erhalt (von Sendungen, Briefen usw.)* || reception * *z.B. Radioempfang, Empfang einer Nachricht, eines Telegramms*
empfangen receive || welcome * *freundlich* ~ || see * *Zutritt gewähren* || accept * *annehmen* || draw * *Geldmittel usw.* ~
Empfänger receiver || recipient || addressee * *Adressat, z.B. eines Briefes*
Empfängermaschine receiver machine * *bei elektr. Welle*
Empfangsantenne receiving aerial
Empfangsbaustein receive block * *Funktionsbaustein* || receiver block * *Funktionsbaustein* || receiver
Empfangsbestätigung confirmation of receipt * *für Ware, Brief, Zahlung* || acknowledgement * *Qittierung, auch für Datentelegramm*
Empfangsdaten receive data
Empfangskanal receiving channel || receive channel
Empfangsrichtung receive direction
Empfangszeichen received characters || receive character
empfehlen recommend
Empfehlung recommendation
Empfehlungsschreiben letter of recommendation

empfindlich sensitive * *allg.; auch leicht zerstörbar* || susceptible * *auch anfällig, empfänglich, zugänglich (to: gegen/für)* || delicate * *zart, fein, zerbrechlich, leicht, dünn, fein, genau, heikel, schwierig* || unreliable * *unzuverlässig* || fragile * →*zerbrechlich*
Empfindlichkeit sensitivity * *auch 'Sensibilität, Sensitivität, Veranlagung' (to: für/gegen)* || susceptibility * *to: für/gegen; Anfälligkeit, Empfindlichkeit, Empfänglichkeit* || sensibility * *to: für/gegen; Empfindlichkeit, Empfänglichkeit, Sensibilität, Empfindsamkeit*
empfohlen recommended
empirisch empirical || experimental * *experimentell* || by experiments * *durch Versuche* || by tests * *durch Versuche* || by way of experiments * *durch Versuche*
Emulations- und Testadapter in-circuit emulator || ICE * *Abk. f. 'In-Circuit Emulator'*
Emulator emulator || in-circuit emulator * *online-Emulator als Mikroprozessor-Entwicklungshilfsmittel*
emulieren emulate * *nachbilden, simulieren, nachahmen (z.B. ein Computer-Zielsystem)*
Emulsion emulsion
EMV EMC * *ElectroMagnetic Compatibility: elektromagnet. Verträglichkeit (z.B. EN 61000)* || noise * *Störeinkopplung* || EMI * *Abk. f. 'ElectroMagnetic Interference': elektromagnetische Störung*
EMV-Maßnahmen EMC measures
EMV-Messung EMC measurement
EMV-Richtlinie EMC directive * *z.B. der EG (EG-Richtlinien 89/336/EWG, 92/31/EWG, 93/68/EWG)* || EMCD * *Abkürzg. für 'ElektroMagnetic Compatibility Directive'*
EMV-Verhalten EMC behaviour
EN * *Abkürzg. für 'Europanorm': European Standard (zuständiges elektrotechn. Normengremium:* →*CENELEC)* || European Standard
End- final * *endgültig, Schluss-, letzt* || end || limit * *Grenz-* || ultimate * *äußerst, (aller)letzt, endlich, End- (Ergebnis, Verbraucher usw.)*
Endabnehmer end user * *z.B. einer Ware/eines Geräts* || ultimate buyer * *einer Ware* || final customer * *Endkunde* || final user
Endanwender end user
Endarretierung end lock
Endausbau final extension stage * *letzte Ausbaustufe*
Endausschlag full-scale * *(Messgerät)* || full-range * *(Messgerät)*
Endbehandlung final treatment * *z.B. bei Faserherstellung* || finishing * *z.B. bei Papier, Metall*
Enddruckregler discharge pressure controller * *für eine Turbine*
Enddurchmesser end diameter * *z.B. bei Wickler*
Ende end * *auch Nahlenende, Ende einer Liste, einer Wicklung usw.* || by the end of * *z.B. by the end of September: Ende September*
enden end
Endergebnis final result || upshot || end result
Endezeichen end character * *kennzeichnet z.B. das Ende eines Datentelegramms (häufig ETX entspr. 03hex)* || final character || delimiting character || delimiter
Endfertigung end finishing || final assembly * *Endmontage*
Endfrequenz end frequency * *z.B.der f/U-Kennlinie eines Frequenzumrichters*
Endgerät terminal * *in Nachrichtenübertragungssystem*
endgültig final * *schließlich* || definite * *eindeutig, genau, definitiv, bestimmt, fest* || permanent * *dauernd*
Endkontrolle final inspection

Endkunde final customer || end user * *letztendlicher (eigentlicher) Anwender* || ultimate buyer * *Endabnehmer einer Ware* || end customer
Endlage end position || limit position
endlich finite * *(Adj.) begrenzt, nicht unendlich* || final * *(Adj.)* || finally * *(Adv.)* || at last * *(Adv.)* || after all * *(Adv.) endlich doch*
endlos endless
Endlosschleife endless loop || continuous loop * *Dauerschleife* || idle loop * *Untätigkeitsschleife*
Endmaß caliper * *Tastvorrichtung z.B. zur Messung von Innendurchmessern* || caliper gauge * *Tast-/Rachenlehre* || precision-gauge block * *zur Messung von Durchmessern usw.*
Endmontage final assembly
Endprodukt end product || finished procuct * *Fertigprodukt* || final product
Endprüfung final test || final inspection
Endpunkt end point * *z.B. einer Kennlinie*
Endschalter limit switch || proximity switch * *Näherungsschalter*
Endstellung end position
Endstufe power stage * *Leistungs-Endstufe* || output stage * *z.B. für Stromrichter* || power module * *Leistungsstufe* || final stage * *letzte/abschließende Stufe*
Endtemperatur final temperature || steady-state temperature * *im thermisch eingeschwungenen Gleichgewichtszustand* || stable temperature
Endverrundung final rounding-off * *für Hochlaufgeber-Rampe* || ramp end rounding || final rounding
Endverschluss sealing end * *an einem Leistungskabel*
Endwert full scale * *eines Meßbereichs* || final value || full value || maximum value * *Maximalwert* || end value || steady-state value * *im eingeschwungenen Zustand*
Energie energy
Energieanwendung electric power utilization * *elektrische ~*
Energieausgleich energy equalization * *z.B. durch gemeinsamen Zwischenkreis bei Umrichtern* || energy balancing * *z.B. zw. mehreren Umrichtern am gemeinsamen Zwischenkreis* || power balancing
Energieaustausch exchange of energy || energy transfer || power balancing * *Energieausgleich*
Energiebedarf energy demand || requirement for energy
Energiebilanz energy balance
Energieeinsparung saving of energy || energy saving
Energieerzeugung power generation || generation of electricity * *Stromerzeugung* || energy production
Energiefluss energy flow
Energieführungskette cable drag chain * *(für Kabel)* || cable and hose drag chain * *(für Kabel und Schläuche)*
Energiekennziffer energy index
Energiekette cable drag chain * *für (Schlepp-)Kabel bei Werkzeugmaschine usw.* || cable and hose drag chain * *für Kabel und Schläuche*
Energiekosten power costs
Energiekreis power circuit
Energiepufferung energy buffering || power back-up
Energiequelle energy source
Energierückspeisung return of energy * *z.B.bei Nutzbremsung (to: in)* || power recovery || energy recovery || regenerating || regeneration || regenerative energy feedback * *z.B. vom Motor ins Netz beim Bremsen* || regeneration of energy * *back to the mains: →Netzrückspeisung* || regenerative feedback

Energierückspeisung ins Netz return of energy to the supply system
energiesparend energy-saving || energy efficient || low-power
Energiespeicher energy storage
Energietechnik power engineering
Energieumwandlung energy conversion * *z.B. von elektrischer in mechanische Energie*
Energieverbrauch energy consumption
Energieverlust energy loss
Energieversorgung power supply
Energieversorgungs-Unternehmen power utility * *siehe auch 'EVU'*
Energieversorgungsanlage power distribution installation
Energieversorgungsunternehmen power utility * *EVU* || power company
Energieverteilung power distribution || distribution of electricity * *von elektr. Energie* || energy distribution
Energiewandler energy transducer
eng narrow || close * *dicht, nah; close-tolerance: ~ toleriert* || intimate * *innig* || clinging * *~ anliegend* || tight * *Termin(plan), Etat, Geldmittel usw.*
eng verzahnt closely meshed * *(figürl.)*
eng zusammenhängend closely related
engagiert sein be committed * *to: bei, an* || engage * *engage oneself: sich engagieren (in: bei)* || commit * *commit oneself to: ~ bei* || be engaged * *beschäftigt sein/arbeiten (in: mit/an), nicht abkömml.s., in Anspruch genomm.s.*
enge Toleranzen close tolerances
enger Raum limited room * *beengte Raumverhältnisse* || narrow space || cramped space
englisch English
englischsprachig English-language
Engpass bottleneck * *(auch figürlich) delivery bottleneck: Liefer~* || shortage * *Knappheit, Mangel (of: an)* || squeeze * *wirtschaftlicher, finanzieller ~, Klemme*
Engstelle bottleneck * *Flaschenhals*
enorm enormous || tremendous * *gewaltig, ungeheuer, "toll", kolossal* || huge * *sehr groß, rieseg, ungeheuer, gewaltig*
Ent- de * *Vorsilbe für Substantiv* || un * *Vorsilbe*
entfallen be inapplicable * *nicht in Frage kommen* || be allotted to * *auf jemanden/etwas (anteilmäßig) ~* || fall to a person's share * *auf jemanden (anteilmäßig) entfallen* || slip a person's memory * *dem Gedächtnis ~* || can be dispensed with * *~ können, weggelassen werden können, verzichtbar/entbehrlich sein*
entfällt not applicable * *in Formularen* || N.A. * *Abkürz. f. 'Not Applicable': 'nicht anwendbar', in Formularen, Tabellen, Listen* || not required * *nicht benötigt (for: bei/für)* || is eliminated * *wird überflüssig (gemacht)*
entfernen remove * *beseitigen*
entfernt remote || distant || far || far-away || apart * *voneinander ~*
Entfernung distance
entfetten degrease
Entfeuchter dehumidifier
entflammbar inflammable || flammable * *(amerikan.)*
Entflammbarkeit flammability * *(amerikan.)* || inflammability || ignitability * *Entzündbarkeit (Explosion verursachend)*
Entflechtung routing || artworking
entgegen in opposition to * *im Gegensatz zu* || contrary to * *Gegensatz; contrary to all expectations: ~ allen Erwartungen* || in the face of * *gegenüberstehend, angesichts* || against * *Gegensatz, Richtung* || towards * *Richtung* || opposite * *~gesetzt (gerichtet), gegenüberliegend*

entgegen dem Uhrzeigersinn   counterclockwise || ccw || anti-clockwise
entgegengerichtet sein   oppose * *zu etwas*
entgegengesetzt   opposite || contrary * *(figürlich)* || opposed * *to: zu* || in opposition * *in entgegengesetzter Anordnung* || opposing * *unvereinbar (to: mit), gegenüberliegend* || reverse * *Bewegung*
entgegengesetzt gepolt   of opposite polarity
entgegengesetzt gerichtet   opposite in direction || reverse * *Bewegung*
entgegengesetztes Vorzeichen   opposite sign || inverse sign
entgegenkommen   accommodate * *z.B. jemanden bei einer Verhandlung* || co-operate * *mitwirken, beitragen, helfen; to: an/bei/zu* || come to meet * *treffen* || meet * *treffen*
entgegenkommend   oncoming * *z.B. Fahrzeug, Verkehr* || coming the other way * *z.B. Fahrzeug, Verkehr* || co-operative * *(figürl.)* || obliging * *(figürl.) verbindlich, gefällig, zuvorkommend, entgegenkommend*
entgegensetzen   oppose * *auch entgegentreten,sich widersetzen*
entgegenwirken   counteract * *to: gegen* || act in opposition to * *z.B. einer Kraft*
Entgiftung   decontamination
entgraten   deburr || clip * *Gussnaht*
enthalten   contain || imply * *beinhalten* || comprise * *einschließen, umfassen, beinhalten* || included * *beinhaltet* || include * *einschließen, beinhalten* || hold * *fassen (Gefäß), enthalten, Platz haben für, in sich schließen*
entionisiert   de-ionized * *z.B. Wasser (für Kühlkreislauf, zur Erzielung einer geringen Leitfähigkeit)*
entkoppeln   neutralize * *neutralisieren* || uncouple * *aus/los-koppeln/kuppeln* || tune out * *z.B. bei Rundfunkempfänger* || decouple || buffer * *puffern durch Treiber- oder Speicherstufe* || treat separately * *getrennt behandeln*
Entkopplung   decoupling
Entkopplungsdiode   decoupling diode * *z.B. bei Stromversorgung: schaltet d.höhere Spannung z.Verbraucher durch*
Entladeeinrichtung   discharging device * *z.B. für Chargenprozess* || discharger || unloading equipment || unloading system || unloading facility
entladen   discharge * *Kondensator (auch "sich ~"), Behälter; oneself: sich (Person b.Hantierung m.el.Bauteilen)* || unload * *Fahrzeug, Waggon, Ladung, Gewehr usw. ~* || dump * *Schüttgut ~*
Entladeort   dircharge point * *für Kran usw.*
Entladeschlussspannung   final discharge voltage * *bei Batterie/Akku*
Entladestrom   discharge current
Entladung statischer Elektrizität   static discharge
entlang   in line with * *z.B. entlang einer Rampe* || along * *Weg, Straße (vorangestellt)*
Entlassung   dismissal
entlasten   unload || unburden || relieve * *befreien (of/from: von)* || discharge * *Vorstand ~* || disburden * *(meist figürl.) von einer Bürde befreien, entlasten (of/from: von)* || offload
Entlastung   unloading || relief || discharge
entleeren   empty || evacuate || deplete || drain * *Flüssigkeit* || deflate * *Luft usw. ablassen, aus-/entleeren*
Entleerung   emptying || drainage * *Entleerung (v. Flüssigkeiten), Drainage, Abfluss, Ableitung, Entwässerung, Trockenlegung* || deflation * *Ablassen von Luft; {Wirtschaft} Deflation* || depletion * *(Ent)Leerung, Erschöpfung, Entblößung* || evacuation * *Aus-/Entleerung, Abtransport, Räumung*
Entleerungszeit   discharge time
entlöten   unsolder
entlüften   evacuate the air from || de-aerate * *(chem.)* || bleed * *Bremse, Hydraulikanlage* || deaerate || ventilate || vent * *z.B. ölgekühlten Trafo ~*
Entlüftung   de-aerating || bleeding * *einer hydraulischen Bremse* || evacuation of air || venting || air vent || stale air extraction * *Abfuhr der Abgase, z.B. bei Tunnel* || de-aeration || pressure release * *~ zum Druckausgleich* || ventilation || vent || deaeration
Entlüftungsbohrung   breather hole || vent hole
Entlüftungsschraube   vent plug || venting screw
entmagnetisieren   demagnetize
Entmagnetisierung   demagnetization
Entnahme   extraction || removal || withdrawal * *von Geld usw.* || taking out
entnehmen   draw * *z.B. Strom aus dem Netz ~ (from: von/aus)* || derive * *ab/herleiten* || take * *a sample: eine Probe ~* || remove * *herausnehmen, entfernen (from: von/aus)* || quote from * *zitieren* || infer from * *schließen/folgern/ableiten aus* || take from * *(auch figürlich) wegnehmen; auch Angabe aus einer Tabelle/e.Buch usw. ~* || draw from * *Geld ~; aus einem Buch usw. ~* || borrow from * *aus einem Buch usw. ~* || gather * *schließen (from: aus), folgern, (auf)lesen, sich zusammenreimen, sich denken* || retrieve * *herausholen*
entpacken   decompress * *Daten ~* || unpack * *eingepackte Produkte, Daten ~* || unzip * *(salopp) Daten ~*
Entpalettiermaschine   pallet unloading machine || depalletizer
Entprellen   debouncing
Entprellung   debouncing
entriegeln   unlock || release
Entriegelung   reset * *logische ~, Rückstellen* || release * *Freigabe, mechanische ~* || unlocking * *mechanische oder steuerungstechnische ~*
Entriegelungs-   release
Entrindung   debarking
Entrindungsmaschine   debarker * *{Holz}* || barker * *{Holz}* || wood peeling machine * *Holzschälmaschine*
Entrindungstrommel   debarking drum * *{Holz}*
entrosten   derust
Entsalzung   desalination
Entsalzungsanlage   desalination plant
Entsättigung   desaturation
entscheiden   decide * *on: über; in favo(u)r of: für/zugunsten; against (doing): gegen/nicht zu tun* || determine || settle * *endgültig*
entscheidend   decisive * *z.B. Kriterium, Eigenschaft* || conclusive * *z.B. schlüssig* || crucial * *kritisch; crucial point: ~der Punkt* || final * *endgültig* || critical * *Augenblick* || essential * *wesentlich, (lebens)wichtig, unentbehrlich* || basic * *grundlegend, Grund-, die Grundlage bildend* || vital * *grundlegend wichtig* || at a premium * *be at a premium: ausschlaggebend/-sein* || determinant * *bestimmend, entscheidend; ~der Faktor*
entscheidende Rolle spielen   play a decisive role
entscheidender Faktor   major factor
Entscheider   decision maker
Entscheidung   decision * *on: über; of: des; take a decision: eine ~ treffen* || determination || ruling || decree || judgement * *gerichtliche ~* || selection * *Auswahl*
Entscheidungsbefugnis   competence || jurisdiction * *rechtlich*
Entscheidungsprozess   decision-making || decision-making process
Entscheidungsspielraum   freedom in decision-making || room to move * →*Spielraum* || freedom of choice * *Entscheidungsfreiheit* || decision-making power * *Entscheidungsgewalt*
Entscheidungsträger   decision maker

entschieden definitely * *(Adv.)* || decidedly * *(Adv.)* *entschieden, fraglos, bestimmt, deutlich*
entschlossen sein be determined * *to do: etwas zu tun*
Entschraubmaschine unscrew machine
entschuldigen excuse || apologize * *oneself: sich* || justify * *rechtfertigen*
Entschuldigung excuse * *it is unexcusable/there is no excuse for it: dafür gibt es keine* ~ || apology * *Abbitte* || pretext * *Ausrede, Vorwand* || sorry * *(gesproch.) um Verzeihung bittend; (I'm) sorry (about that):* ~ *(f.d.Ungemach)* || excuse me * *(Ausspruch) in Verbindung mit einer Bitte*
Entschwefelung desulpherization
entsenden send off || delegate * *abordnen; auch als Vertreter* ~ || assign * *abordnen* || send
entsorgen dispose * *of: etwas* || discharge * *entladen, austragen*
Entsorgung disposal * *z.B. von Schrott* || waste disposal * *Abfallentsorgung* || dispersing * *(Weiter-) Verteilung von Daten z.b. bei SPS*
entspannen relieve * *z.B. eine Feder* ~ || tension-relieve * *z.b. Zugfeder, Zugspannung* ~ || slacken * *Seil* ~ || relax * *Muskeln, Nerven, Geist* ~*; sich* ~ || ease * *sich* ~ *(Lage usw)*
entsprechen correspond to || meet * *einer Bitte, einer Vorschrift* ~ || match * *einer Anforderung, einem zweiten Teil* ~ || conform to * *einer Norm/ Vorschrift* ~ || equal * *identisch sein mit* || comply with * *einer Vorschrift, Norm, Regel, Bedingung, Bitte* ~ || coincide * *übereinstimmen, sich decken (mit), etwas genau* ~ || be equivalent to * *gleichkommen* || reflect * *widerspiegeln, sich widerspiegeln, seinen Niederschlag finden in, s.auswirken auf* || be as per * *wie bei ... sein* || amount to * →*betragen*
entsprechend corresponding to * *in Bezug stehend mit* || in accordance with * *z.B. gemäß einer Norm/ Vorschrift* || conforming to * *z.B. einer Norm/Vorschrift (auch "in line with":* →*in Übereinstimmung mit)* || equivalent to * *gleichwertig* || analogous to * *sinngemäß* || proportionate with * *im Verhältnis* || commensurate with * *im Verhältnis stehend zu, Einklang stehend mit, angemessen* || suitable for * *passend* || respective * *jeweilig, betreffend* || according to * *(Adv.) gemäß (auch 'accordingly')* || corresponding || in compliance with * *(Adv.) gemäß, entsprechend (e. Vorschrift/Wunsches usw.)* || following * *(Adv.) entsprechend* || appropriate * *passend (auch "adequate":* →*angemessen); (Adv.: appropriately)* || suitable * *passend* || particular * *besonder, einzeln, speziell, jeweilig* || pursuant to * *gemäß, zufolge, entsprechend, laut (einer Vorschrift, einem Gesetz usw.)* || per * *laut, gemäß* || as per * *laut, gemäß*
Entsprechung equivalent * *Gegenstück, Entsprechung, Äquivalent* || counterpart * *Pendant* || analogy * *Übereinstimmung, Analogie, Ähnlichkeit* || opposite number * *Gegenstück*
entspricht corresponds to
entstandene Kosten costs incurred
Entstauber dust remover || dust extractor || dust exhauster * *Staub-Absauger*
Entstaubung dust removal || dust removing * *siehe auch* →*Staubabsauggerät*
Entstaubungsanlage dust removal system || dust separation system
Entstaubungseinrichtung dust extraction system
Entstaubungsgerät dust extractor * *z.B. bei Holzbearbeitungsmaschine*
entstehen come into being || spring up || grow * *erwachsen; out of: aus* || develop * *(sich) entwickeln, z. Vorschein kommen, auftreten, s. zeigen* || arise * *erscheinen, auftreten, hervorgehen, herrühren, stammen (from/out of: von)* || take its rise || originate * *stammen, herrühren (from/in: aus)* || be due to * *zurückzuführen sein auf* || result * *resultieren (from: aus)* || be incurred * *Kosten, Nachteile usw. (by: durch)* || accrue * *Kosten usw. (from: durch)* || development * *(Substantiv) Bildung, Entstehung, Entfaltung* || occur * *eintreten, sich ereignen, vorkommen, auftreten* || emerge * *auftauchen, zum Vorschein kommen, herauskommen (from: aus)* || be caused * *verursacht sein (by: durch)* || break out * *Feuer usw.* || be in the making * *im Entstehen begriffen sein* || be in process of development * *in Entwicklung befindlich* || nascent * *anfänglich/erstmalig entstehen*
Entstehung origin * *Ursprung, Quelle, Herkunft* || beginning * *Anfang, Beginn, Ursprung* || coming into being * *das Entstehen, das Erschaffen-Werden* || rise * *Ursprung, Entstehung, Anlass* || emergence * *Heraus-/Hervorkommen, Auftreten, Erscheinen, Auftauchen, Entstehen* || birth * *Geburt, Ursprung, Entstehung* || formation * *Bildung, Formung, Entstehung, Gestaltung* || genesis * *Ursprung, Beginn, Entstehung, Werden*
Entstördrossel interference suppression choke || EMI suppression choke * *gegen elektromagn. Beeinflussung von außen u. nach innen* || interference suppression coil || noise suppression choke || RFI suppression reactor * *Funkentstördrossel* || RFI reactor
entstören suppress interference || take EMI-suppression measures * *Entstörmaßnahmen ergreifen* || take anti-noise measures * *Entstörmaßnahmen ergreifen*
Entstörfilter interference filter || EMI filter * *ElectroMagnetic Interference: elektromagnetische Störung* || noise filter || interferenco suppression filter || EMI suppression filter * *ElectroMagnetic Interference: elektromagnetische Störung*
Entstörkondensator interference-suppression capacitor || EMI suppression capacitor * *zur Sicherstellung der* →*EMV* || radio interference suppression capacitor * *zur Funkentstörung* || RFI suppression capacitor * *zur Funk'entstörung*
Entstörmaßnahmen noise suppression measures || EMI-suppression measures || anti-noise measures
Entstörmittel interference suppressor || EMI-suppressing components || noise suppression equipment
Entstörung interference suppression * *Störunterdrückung* || EMI suppression * *Unterdrückung v. 'ElectroMagnetic Interference': Störunterdrückung* || interference suppression measures * *Entstörmaßnahmen* || EMI suppression measures * *Entstörmaßnahmen* || radio interference suppression * *Funkentstörung* || RFI suppression * *Unterdrückung von 'Radio Frequency Interference': Funkentstörung* || noise suppression equipment * *Entstörmittel, Entstörausrüstung*
Entwärmung heat discharge || heat dissipation || cooling
entwässern drain * *ablaufen lassen* || dewater * *dehydrate* * *Wasser entziehen*
Entwässerung drainage * *z.B. von Papier* || dewatering
entweder oder either or
entwerfen design || lay out * *auslegen* || project * *planen, entwerfen* || sketch * *flüchtig, grob* ~ || outline * *in groben Zügen* ~ || draw up * *Vertrag, Schriftsatz, Regelstruktur usw.* ~ || draft * *Vertrag, Vorschrift, Norm usw.* ~
entwickeln design * *konstruieren, auch Software; newly designed: neu entwickelt* || devise * *ausdenken, ersinnen, erfinden, konstruieren, auch Software* || develop * *auch 'sich* ~*'* || dissipate * *(Verlust-) Wärme* ~ || evolve * *entfalten, herausbilden, hervorrufen, erzeugen, ausscheiden, (sich)* ~*/-falten, entstehen* || process * *Film* ~

Entwickler design engineer * *Konstrukteur, Entwerfer* || designer * *Entwerfer, Konstrukteur* || developer || software designer * *Softwareentwickler* || hardware designer * *Hardwareentwickler* || application engineer * *Projekteur, Anwendungsentwickler* || development engineer
Entwicklung development * *eines Produktes; be in/under development: in ~ sein* || design || layout || evolution * *Entfaltung, Entwicklung*
Entwicklungsabteilung development department || Development
Entwicklungsaufwand development costs * *Entwicklungskosten* || engineering effort
Entwicklungsgruppe development group
Entwicklungshilfsmittel design tool || development tool || design tools * *(Plural)*
Entwicklungsingenieur design engineer * *siehe auch 'Entwickler'*
Entwicklungskosten development costs
Entwicklungsland less developed country || LDC * *Less Developed Country* || developing country
Entwicklungsmuster engineering sample || prototype * *Prototyp*
Entwicklungsplattform development platform
Entwicklungssprung development jump
Entwicklungsstand state of development
Entwicklungsstufe stage of development
Entwicklungssystem development system * *z.B. für Mikroprozessor-Software*
Entwicklungstendenz development trend
Entwicklungtrend developing trend || development trend
Entwicklungsumgebung development environment * *die Gesamtheit aller Entwicklungswerkzeuge für eine Softwareentwicklung*
Entwicklungsvorhaben development project || development programme
Entwicklungswerkzeug design tool || development tool
Entwicklungszeit time to market * *bis ein Produkt marktfähig ist* || period of development
Entwicklungsziel development target || objective of development || design target || aim of development || development aim
entwirren disentangle * *auch 'sich ~'; befreien, auf-/loslösen* || unravel * *auch 'sich enträtseln/~'* || unsnarl * *(auch figürl.)* || unclutter * *Ordnung in etwas bringen, Wirrwar "geradeziehen"* || declutter * *Ungeordnetes übersichtlich gestalten (z.B. Bildschirmaufbau/-darstellung)*
Entwurf draft * *fist/rough: Grobentwurf* || design * *Konstruktion* || scheme * *Schema, System, Anlage, Übersicht, schemat.Darstellg., Aufstellung, Plan, Projekt, Programm*
Entwurfsprüfung design review * *Qualitätsprüfung an einem (Konstruktions-) Entwurf (DIN 55350 T.17)*
entzündbar inflammable * *highly: leicht* || ignitable * *eine Explosion verursachend* || combustable
entzundern descale * *von Stahl* || pickle || scour || scale off || scale removal * *(Substantiv)* || descaling * *(Substantiv)*
Entzundungsmaschine descaling machine * *[Walzwerk]*
EP SPS * *Abk. für 'Standard Packaging Slot': Standard-Einbauplatz*
epitaxial epitaxial
Epoxid epoxy
epoxidbeschichtet epoxy coated
Epoxidharz casting epoxy * *Gießharz* || epoxy resin || epoxy casting resin * *Gießharz* || epoxy
Epoxyd epoxy
Epoxydharz glass epoxy || epoxy-resin
Epoxydharzlack epoxy-resin paint
EPROM-Löschgerät EPROM eraser

EPROM-Programmiergerät EPROM burner || EPROM programmer || prommer * *(salopp)*
erarbeiten acquire * *z.B. Wissensstoff, Problemlösung ~* || make one's own * *sich einen Wissensstoff/ e. Thema ~/aneignen* || extract * *herausarbeiten, ab/herleiten* || collect * *zusammentragen* || compile * *zusammentragen* || work out * *herausarbeiten, 'in die Mache nehmen'* || elaborate * *sorgfältig aus-/durcharbeiten* || work out * *→ausarbeiten* || gain by working * *mühsam ~*
Erdanschluss grounding terminal * *Klemme* || earth terminal
Erdbebenfestigkeit seismic withstandibility || resistance to earthquakes || seismic withstand capability
Erdbebenprüfung seismic test
erdbebensicher earthquake-proof || aseismic
Erde earth * *auch elektr.; protective earth: Schutz~* || ground
erden ground || earth || connect to earth
erdfrei not-grounded || floating || earth-free || ungrounded
erdfreies Netz floating supply system || earth-free network
Erdgas natural gas
Erdklemme earthing terminal
Erdöl mineral oil || petroleum || crude oil * *Rohöl*
Erdschleife ground loop
Erdschluss earth fault || GND fault || ground fault
erdschlussfest protected against ground faults || earth-fault-proof || ground fault protected || earth-fault resistant || ground-fault resistant
Erdschlussfestigkeit earth-fault resistance || resistance against ground-faults
Erdschlussschutz ground fault protection || earth fault protection
erdschlusssicher ground fault proof || protected against ground faults
Erdschlusssicherheit ground fault protection
Erdschlussüberwachung ground fault monitoring || ground fault monitoring device || *als Gerät* || earth-fault monitoring
Erdschlusswächter earth-leakage monitor || earth leakage relay || ground fault monitor
Erdstrom earth current || earth leakage current * *Fehlerstrom*
Erdung earthing || grounding || earth connection * *Erdverbindung*
Erdungsanschluss earthing terminal || earth terminal
Erdungsanschlusspunkt earth terminal || ground terminal
Erdungsbürste grounding brush * *z.B. zum Erden eines Motorläufers* || earthing brush * *z.B. zum Erden eines Motorläufers*
Erdungsdraufschalter grounding make-proof switch || make-proof earthing switch
Erdungsgarnitur earthing accessories
Erdungsklemme earthing terminal || grounding terminal
Erdungsleitung earthing conductor || earth
Erdungsschraube earth connection screw
Erdverbindung earth connection || earthing connection
Ereignis event || occurrence * *Vorfall* || incident * *Vorfall* || affair * *Angelegenheit* || phenomenon * *Erscheinung*
ereignisgesteuert event-driven || event-controlled
Ereigniszähler event counter
erfahren experienced || skilled * *in: in; geübt, Fertigkeiten besitzend, geschickt (meist für Facharbeiter)* || expert * *in: in* || seasoned * *langgedient* || well versed * *bewandert (in: in)* || at home * *bewandert, vertraut (in: in)* || proficient * *tüchtig* || come to know * *(Verb)* || learn * *(Verb)* || be told * *(Verb)* || be informed * *(Verb)* || experience

* *(Verb) erleben* || go through * *(Verb) erleben, durchmachen* || suffer * *erleiden* || be subject to * *(einem Effekt) ausgesetzt sein* || knowlegeable * *(salopp) klug, kenntnisreich*
Erfahrung experience || expertise * *[expe'ties] Fachwissen, Expertenwissen* || know-how * *"gewußt wie"* || wealth of experience * *~sschatz*
Erfahrungsaustausch exchange of experience * *interdepartmental: zwischen Abteilungen eines Unternehmens* || sharing of experience
Erfahrungskurve know-how development characteristic
Erfahrungsträger experienced person || experienced personnel * *(Plural)*
Erfahrungswert empirical value || practically-proven value || pragmatical value || 'experience' value
erfassen sense * *Messwert (auch von Messgeber)* || record * *aufzeichnen, statistisch erfassen* || capture * *auffangen, z.B. in Tracespeicher* || acquire * *z.B. Istwerte, Messwerte, (Analog-) Signale* || sample * *(Messwert) abtasten* || detect || measure * *messen* || meter || seize * *greifen (auch figürl.)* || grasp * *greifen (auch figürl.)* || catch * *greifen (auch figürl.)* || comprehend * *geistig* || catch hold of * *ergreifen* || clutch * *(fest) packen; (mechanisch)* || grip * *packen (auch figürl.)* || realize * *erkennen* || register * *statistisch ~, registrieren, anmelden, verzeichnen, (handelgerichtlich usw.) eintragen* || consider * *berücksichtigen* || collect * *ein-/aufsammeln* || scan * *abtasten (z.B. Messwert)*
Erfassung sensing * *von Messwerten* || acquisition * *z.B. von Messwerten* || sampling * *Abtastung* || measurement * *Messung* || recording * *Aufzeichnung* || registration * *statistische ~* || listing || consideration || capture * *~ eines Messwertes oder Zählerstandes im laufenden Betrieb in einen Zwischenspeicher*
Erfassungsbereich sensing range * *für Messgeber, Näherungsschalter usw.*
Erfassungssystem sensing system * *für Messwerte* || acquisition system * *für Messwerte*
Erfindung invention
Erfolg success * *great: großer/voller; successfully/ with success. mit ~; be unsuccessful: keinen ~ haben* || achievement || result * *(End)Ergebnis* || consequence * *Folge* || effect * *Wirkung*
erfolgen happen * *sich ereignen* || take place * *stattfinden* || occur * *geschehen, auftreten* || come * *sich einstellen* || arrive * *sich einstellen* || be made * *payment must be made: Zahlung muss ~* || be provided * *zur Verfügung gestellt werden (by: von)* || be realized * *realisiert/durchgeführt werden* || be executed * *aus-/durchgeführt werden*
erfolgreich successful * *in (doing): in/bei* || effective || crowned with success * *von großem Erfolg gekrönt*
Erfolgsfaktor success factor
Erfolgsrezept recipe for success || formula for success || rule for success
erforderlich essential * *wichtig* || required * *benötigt, gefordert (to: für)* || needed * *benötigt* || necessary * *notwendig*
erfordern necessitate || call for || require * *benötigen* || demand * *verlangen, fordern, bedürfen*
erfrischen refresh
erfüllen fulfill * *Bedingung, Aufgabe ~, requirements: Anforderung, Spezifikation* || accomplish * *Aufgabe* || meet * *Bedingung, Erwartung, Norm, Vorschrift, Anforderung, Bedürfnis ~* || comply with * *erfüllen, entsprechen (z.B. Norm, Vorschrift, Bedingung, Bitte, Regel)* || carry out * *Pflicht ~* || keep * *Versprechen ~* || satisfy * *z.B. eine Bedingung/Formel/Gleichung/Vorschrift ~* || conform to * *einer Norm/Vorschrift usw. entspre-*

chen || serve * *serve the needs/demands: Bedürfnisse/Anforderungen ~*
erfüllt met * *z.B. Vorschrift, Voraussetzung*
Erfüllung compliance
ergänzen complete || supplement * *auch durch einen Anhang, z.B. an eine Bestellnummer* || expand * *erweitern* || suffix * *durch eine angehängte Zusatzangabe, z.B. an Bestellnummer* || augment * *(sich) vermehren, vergrößern, zunehmen, ergänzen* || add * *hinzufügen* || amend * *(Gesetz, Vorschrift, Norm usw.) abändern, ergänzen; verbessern, berichtigen*
ergänzend complementary * *ergänzend, sich ergänzend, Ergänzungs-; Komplementär-* || supplementary
ergänzende Hinweise supplementary information
Ergänzung supplement * *Nachtrag, Anhang, Zusatz, das Ergänzte* || completion * *Vervollständigung, Vollendung, Fertigstellung, Abschluss* || restoration * *Wiederherstellung, Instandsetzung* || supplementation * *Ergänzung, Nachtragen, Nachtrag, Zusatz* || replenishment * *Auffüllung, Ersatz, Ergänzung* || complement * *das Ergänzte, Vervollständigung, Ergänzung; full complement: volle Anzahl/Menge* || amendment * *zu Gesetz, Norm, Vorschrift* || appendix * *Anhang* || suffix * *an Bestell-/Codenummer*
ergeben result * *in· sich ~; from: sich ~ aus; to: Ergebnis einer Berechnung* || can be calculated * *can be calculated from: ergibt sich rechnerisch/ physikal.aus* || can be obtained * *kann man erhalten/erlangen/bekommen* || arise * *entstehen, erscheinen, auftreten* || result in || be obtained * *sich ergeben* || yield * *(Resultat,Ertrag, Ausbeute) ~; ein-/er-/hervorbringen; (Gewinn, Zinsen usw.) hervorbringen*
Ergebnis result || earnings * *Geschäftsergebnis, Gewinn, Profit* || profit * *Profit, Gewinn, Geschäftsergebnis* || score * *erzielte Punkte, Treffer (zahl), Ergebnis, (Be-)Wertung, Spielstand*
Ergebnisanzeige result bit * *bei SPS*
Ergonomie ergonomics
ergonomisch ergonomical
ergreifen take * *auch Maßnahmen*
Erhalt receipt * *[ri'βiet] Empfang* * *Sendung, Brief*
erhalten receive * *empfangen, bekommen* || get * *bekommen* || obtain * *erlangen, bekommen, erwerben, sich verschaffen* || fetch * *eine Sache ~* || preserve * *bewahren, aufrecht~, beibehalten* || keep * *bewahren, beibehalten, aufrecht~* || conserve * *erhalten, bewahren, beibehalten* || maintain * *aufrecht~* || retain * *aufrecht~* || support * *unterstützen* || keep in good condition * *in gutem Zustand halten* || keep in good repair * *in gutem Zustand halten* || be given * *gegeben/verliehen bekommen* || be awarded * *Auftrag/Auszeichnung erteilt/verliehen bekommen; z.B. 'he was awarded the prize'*
erhältlich available
Erhaltung preservation * *Bewahrung, Vorbeugung, Konservierung* || upkeep * *~ von Bauten, Anlagen usw.* || conservation * *Bewahrung, Instandhaltung, Schutz; Haltbarmachung, Konservierung* || maintenance * *~ von Maschinen, Frieden usw.* || retention * *Bewahrung, Festhalten, Beibehaltung, Zurückbehalten, Einbehaltung*
Erhaltungsladung trickle charge * *einer Batterie* || equalizing charge || trickle charging * *einer Batterie/eines Akkus*
erheblich considerable * *beträchtlich, bedeutend* || serious * *Schaden, Aufwand* || grave * *Schaden, Verlust* || heavy * *Kosten, Verluste* || important * *wichtig* || substantial * *substanziell, z.B.Sachschaden* || considerably * *(Adv.) beträchtlich, be-*

deutend || significant * *bedeutend, bedeutsam* || severe * *schwer, schlimm (z.B. Verluste, Kritik)* || noticeable * *beachtlich*
Erhitzer heater
erhöhen increase * *auch 'sich ~'; (by: um; to: auf)* || raise * *steigern (auch Preise)* || boost * *Spannung/Strom/Drehmoment in einem bestimmten Arbeitspunkt/durch eine besondere Maßnahme* || enhance * *Wirkung, Leistungsfähigkeit, Zuverlässigkeit usw.* ~ || lift * *in die Luft heben; auch Preise usw.* ~ || elevate * *emporheben (in die Luft)* || augment * *steigern (to: auf, by: um)* || intensify * *verstärken, intensivieren* || deepen * *Eindruck* ~ || advance * *Preise usw.* ~ || mark up * *Preise usw.* ~ || lift up * *(amerikan.) Preise usw.* ~ || speed up * *die Geschwindigkeit* ~
erhöht increased * *gesteigert* || higher * *höher* || raised * *angehoben, gesteigert* || lifted * *angehoben (in die Luft)* || elevated * *emporgehoben* || intensified * *verstärkt* || advanced * *Preise usw.* || enhanced * *Wirkung usw.* || boosted * *(salopp) aufgeblasen, z.B. Leistungsvermögen, Wirkung usw.*
erhöhte Anforderungen stricter requirements || high-reliability requirements || high rel requirements * *Kurzform für 'high reliability ..'* || high requirements
erhöhte Genauigkeit increased accuracy
erhöhte Leistung increased output * *Motor/Stromrichter*
erhöhte Sicherheit increased safety * *Schutzart 'e' nach EN50019 f. Ex-geschützte el. Betriebsmitt. (keine Funkenbildg.)*
Erhöhung increase || boost * *Leistungs~, Spannungs~, Druck~* || build-up * *Aufbau, z.B. Temperatur in e. Wärmenest, Durchmesser bei e. Aufwickler* || rise * *Anstieg* || over- * *(Vorsilbe) z.B.: overvoltage: Spannungs~ (über die Nenn-/erlaubte Spann. hinaus)* || gain * *Zunahme, Steigerung, (Wert-/Leistungs-) Zuwachs*
Erholung recovery
Erholzeit recovery time
erkaufen buy * *(auch figürl.)* || purchase * *(auch figürl.) at the expense of: mit* || have to pay a heavy price for * *(auch figürl.) etwas teuer ~ müssen*
erkennbar noticeable || perceptible * →*wahrnehmbar*
Erkennbarkeit perceptibility
erkennen recognize * *auch 'wahrnehmen, sehen'* || identify * *identifizieren (by: an/durch)* || know * *make oneself known: sich zu ~ geben* || discern * *wahrnehmen* || distinguish * *wahrnehmen* || detect * *entdecken, detektieren; z.B. Fehler* || discover * *entdecken* || realize * *sehen, begreifen, erfassen, wahrnehmen* || diagnose * *diagnostizieren, z.B. Fehler* || find out * *herausfinden* || see * *(er/ein/nach/ab)sehen, erkennen, entnehmen, herausfinden, erleben, verstehen*
Erkenntnis knowledge * *Wissen* || finding * *etwas Herausgefundenes*
Erkennung detection || sensing * *eines Signals/Zustandes* || identification * *Identifizierung*
erklären explane * *erläutern* || interprete * *deuten* || define * *definieren* || illustrate * *veranschaulichen* || demonstrate * *veranschaulichen, zeigen* || declare * *aussprechen, kundtun* || state * *(aus-)sagen, vorbringen/-tragen, (Einzelheit usw.)angeben, behaupten, erwähnen, bemerken*
erklärend explanatory
Erklärung explanation * *Erläuterung (of: für/zu)* || interpretation * *Deutung* || definition * *Begriffsbestimmung* || reasons * *Gründe* || comment * *Kommentar* || illustration * *Veranschaulichung* || declaration * *Aussage, Feststellung (auch rechtsverbindliche)* || statement * *Aussage, Feststellung (auch offizielle)*

erlangen acquire || attain * *erreichen, erzielen*
erlauben permit * *gestatten* || allow
Erlaubnis permit || allowance * *make allowance for: berücksichtigen*
erlaubt legal * *per Gesetz* ~*, den Vorschriften entsprechend* || permissible * *zulässig* || allowed || permitted
erläutern explain || illustrate * *illustrieren anhand eines Beispiels/eines Bildes* || outline * *in Übersichtsform*
Erläuterung explanation || note * *Anmerkung*
Erläuterungen explanatory notes
erledigen carry through * *durchführen* || execute * *aus/durchführen* || dispose of * *aus der Welt schaffen* || settle * *Geschäft; Problem lösen* || finish * *beenden* || bring to a close * *beenden* || be settled * *sich erledigen* || take on * *(Arbeit) annehmen, übernehmen* || carry out * *aus-/durchführen, erfüllen*
Erledigung settlement
Erledigungsliste to-do list
erleichtern make easy * *einfach machen* || facilitate * *auch 'fördern'* || ease * *vereinfachen* || relieve * *erleichtern, entlasten, befreien (of: von)* || disburden * *von einer Last befreien* || make easier * *gegenüber vorher* ~ || lighten * *eine Bürde* ~ || simplify * *vereinfachen*
erleiden sustain * *Schaden usw.* ~ || incur * *erleiden (damage: Schaden; losses: Verluste), auf sich laden, geraten in*
erlöschen go out * *z.B. Lampe, Lichtbogen* || go off * *z.B. Lampe, Lichtbogen*
ermitteln determine * *bestimmen,festsetzen* || ascertain * *feststellen* || establish * *feststellen, festsetzen, nachweisen, etablieren* || find out * *herausfinden* || discover * *entdecken, ausfindig machen* || locate * *Aufenthaltsort* || identify * *indentifizieren* || determining * *(Substantiv)* || evaluate * *berechnen, abschätzen, auswerten* || figure out * *(amerikan.) ausrechnen, herausbekommen* || quantify * *quantitifizieren, quantitativ bestimmen*
Ermittlung determination * *Bestimmung* || discovery * *Fund, Entdeckung, Enthüllung* || ascertainment * *Feststellung* || investigation * *Untersuchung* || findings * *Feststellungen* || evaluation * *durch Berechnung, Messung* || estimation * *durch Abschätzung* || determining
ermöglichen make possible || make feasable || enable || give || permit * *gestatten, erlauben* || allow * *gewähren, möglich machen, ermöglichen (of: etwas), zulassen* || facilitate * *erleichtern, fördern*
ermüden tire * *ermatten* || fatigue * *von Material* || exhaust * *erschöpfen* || wear down * *abnutzen*
Ermüdung fatigue * *Materialermüdung*
ernennen nominate || appoint * *zum Amtsinhaber; he was appointed chairman: er wurde zum Vorsitzenden ernannt* || constitute
erneuern refresh * *auffrischen, z.B. Daten, Farben* || renew * *auch Vertrag* ~*; auch 'wiederholen'* || renovate * *wiederherstellen, erneuern* || recondition * *(wieder) instand setzen, überholen, erneuern* || change * *Verschleißteil, Öl usw. (aus)wechseln* || replace * *auswechseln* || repeat * *wiederholen* || revive * *(sich) neu beleben* || reinstate * *Patent* ~ || prolong * *verlängern, z.B. Vertrag* || recharge * *wieder auffüllen; with grease: Fett* ~
Erneuerung renewal * *renewal rate: ~rate* || renovation || reconditioning || restoration || replacement || revival || reinstatement || reiteration * *Wiederholung*
erneut re- * *als Vorsilbe für Verb* || renewed * *erneut, erneuert* || repeated || fresh || anew * *(Adv.)* || again * *(Adv.)* || once more * *(Adv.) noch einmal, von neuem* || once again * *(Adv.) noch einmal*
erneut starten restart

erneutes re- * *Vorsilbe für Substantiv/Verb*
erniedrigen lower * *kleiner machen, reduzieren* ‖ reduce * *reduzieren* ‖ decrease * *kleiner machen*
ernst zu nehmend serious
eröffnen open up * *z.b. Möglichkeiten* ‖ open-up
erproben test ‖ try ‖ try out ‖ prove
erprobt proven * *['pru: ven]* ‖ well proven * *wohl erprobt, gut bewährt* ‖ highly proven * *wohl erprobt, gut bewährt*
Erprobung test ‖ trial ‖ testing ‖ try-out
Erprobungsphase proof-of-concept phase ‖ testing stage ‖ trial phase
errechnen calculate ‖ compute ‖ reckon out
erregen excite * *z.B. elektr. Maschine* ‖ energize * *z.b. die Spule eines Magneten, einer Bremse*
Erreger- field ‖ excitation
Erregerdurchflutung field ampere turns ‖ excitation strength
Erregergerät excitation control unit * *z.B. für Bremse*
Erregergleichrichter field rectifier ‖ excitation rectifier ‖ field converter
Erregerkreis field circuit
Erregerkreisumschaltung field reversal * *Feldumkehr z.B. zur Momentenumkehr bei DC 1Q Antrieb*
Erregerkreiszeitkonstante field circuit time constant * *ca 1 bis 4,5 sec bei Gleichstrommaschinen*
Erregerleistung field rating * *z.B. Nennerregerleistung eines DC-Motors* ‖ excitation power
Erregermaschine exciter ‖ rotating exciter * *z.B. bei Leonard-Umformer* ‖ excitor
Erregerpol field pole
Erregerspannung excitation voltage ‖ field voltage
Erregerstrom field current * *Feldstrom* ‖ excitation current
Erregerstromregelung excitation current control ‖ field current control ‖ excitation control
Erregerstromrichter field converter ‖ exciter supply converter ‖ excitation converter * *z.B. für Synchronmaschine*
Erregerstromversorgung field power supply ‖ field current supply ‖ field supply
Erregerverluste excitation losses ‖ field losses
Erregerversorgung excitation supply ‖ field supply ‖ excitation power supply
Erregerwicklung field winding ‖ excitation winding ‖ exciter winding
Erregerwicklungsinduktivität field winding inductance
Erregerwicklungswiderstand field winding resistance
erregt excited * *z.B. (Motor)Feld, Relaisspule, Bremse*
Erregung excitation * *Motor, Bremse usw.* ‖ field * *Feld* ‖ energization * *einer Federdruck-Magnetbremse*
erreichbar attainable * *erzielbar* ‖ achievable ‖ obtainable * *ereichbar, erlangbar, erhältlich, zu erhalten* ‖ within reach * *(örtlich) within easy reach: leicht* ~ ‖ available * *verfügbar* ‖ reachable * *z.B. am Telefon*
erreichen attain * *erlangen, erzielen, z.B. Maximalwert erreichen* ‖ get * *bekommen* ‖ obtain * *erlangen, erhalten, sich verschaffen* ‖ achieve * *hingelangen, erzielen* ‖ realize * *realisieren, wahr machen* ‖ reach * *z.B. einen Wert, ein Ziel* ‖ acquire * *erlangen* ‖ gain * *gewinnen* ‖ range-up to * *im (Leistungs- usw.) Bereich* ‖ arrival * *(Subst.)* →*Ankunft, Eintreffen, Gelangen (auch figürl.; at: zu); Erreichen e. Wertes usw.* ‖ reaching * *(Subst.) Erreichung* ‖ achievement * *(Subst.) Erreichung eines Ziels usw.*
erreicht reached
errichten erect

Ersatz replacement * *for: für* ‖ exchange * *Auswechslung* ‖ interchange * *Auswechslung* ‖ changing * *Auswechslung (besonders eines Verschleißteils, z.B. Batterie, Reifen, Bremsbelag)* ‖ substitution * *Substitution. ersatzweise Verwendung, Ersatz-, Ersetzung (durch neue Technik usw.)* ‖ substitute * *Ersatz-, ~mann, ~produkt*
Ersatz- spare ‖ replacement * *(Aus-) Tausch* ‖ standby * *in Reserve/Bereitschaft stehend* ‖ substitute * *ersetzend, Austausch-, als Ersatz/Stellvertreter dienend* ‖ reserve ‖ equivalent
Ersatzbaugruppe replacement module ‖ replacement board * *Leiterkarte* ‖ spare module
Ersatzbürste replacement brush
Ersatzgröße replacement value * *für eine nicht direkt messbare Größe* ‖ substitution variable
Ersatzkohlebürsten replacement carbon brushes ‖ spare brushes
ersatzlos without replacement
ersatzlos streichen cancel and not replace
Ersatzschaltbild equivalent circuit ‖ equivalent circuit diagram
Ersatzschaltung equivalent circuit
Ersatzsperrschichttemperatur virtual junction temperature * *bei Halbleiterbauelement*
Ersatzteil spare part ‖ replacement part * *Austauschteil*
Ersatzteilbestellung spare parts order ‖ ordering of spare parts
Ersatzteildienst spare parts service
Ersatzteile spare parts ‖ spares
Ersatzteilhaltung spare parts inventory ‖ spares inventory ‖ spare parts stocking ‖ spares back-up ‖ spare parts stockage ‖ stockkeeping of spares
Ersatzteilkatalog spare parts catalog
Ersatzteillager stock of spare parts ‖ parts department ‖ spare parts warehouse * *beim Hersteller/Lieferanten* ‖ spare parts store
Ersatzteilliste list of spare parts ‖ spare parts list
Ersatzteilversorgung spare parts supply ‖ spares supply
Ersatztyp replacement type
ersatzweise by substitution ‖ by the way of substitution ‖ in exchange for * *im (Aus-) Tausch für*
erscheinen appear * *auch auf einer Anzeige, einem Bildschirm, vor Gericht* ‖ be published * *Druckerzeugnis* ‖ advent * *(Substantiv) das Erscheinen, das Kommen, die Ankunft* ‖ appearance * *(Substantiv)* ‖ publication * *(Substantiv) eines Buches (forthcoming: im ~ begriffen; when published: bei ~)*
Erscheinung appearance * *äußere* ‖ phenomenon * *Natur-*
erschließen open up * *z.B. neue Anwendungen, Märkte usw.* ~ ‖ open-up * *auch Absatzgebiete, Märkte usw.* ~ ‖ make accessible ‖ open ‖ develop * *entwickeln, nutzbar machen, (sich) entfalten; auch "Baugelände usw. ~"* ‖ exploit * *nutzbar machen; auch "Bodenschätze usw. ~"* ‖ disclose * *aufdecken, enthüllen, offenbaren* ‖ reveal * *offenbaren, enthüllen, erkennen lassen, aufdecken* ‖ unfold * *(sich) entfalten, ausbreiten, öffnen, enthüllen, offenbaren* ‖ acquire * *erwerben, erlangen, erreichen, gewinnen*
Erschließung development
erschlußfest protected against earth faults
erschöpfen deplete * *entleeren*
erschöpfend exhaustive * *z.B. Information, Auskunft*
Erschütterung vibration ‖ shock ‖ jolt ‖ blow * *Schlag* ‖ shaking * *das Schütteln* ‖ concussion * *Erschütterung, Aufschlag, Stoß*
erschütterungsfrei vibration-free ‖ free from vibration ‖ shock-free
erschweren obstruct * *hemmen, behindern, ver-*

erschwerte Umgebungsbedingungen

sperren, verstopfen, blockieren, verhindern || complicate * (ver-) komplizieren, verwickelt machen, erschweren || render difficult * schwer/schwierig machen; render more difficult: noch schwerer machen || impede * (be-) hindern, erschweren, verhindern || aggravate * erschweren, verschlimmern, verschärfen, verstärken
erschwerte Umgebungsbedingungen severe environments
erschwinglich affordable
ersetzen replace * by: durch || substitute * ersetzen/austauschen/an die Stelle setzen (for: durch/gegen/von) || supersede * etwas/jdm. ersetzen/ablösen (by: durch), etw. abschaffen, beseitigen || upgrade * durch eine neue/leistungsfähigere Version ~
Ersparnis saving * in/of: an
ersponnen spun * {Textil} z.B. Chemiefasern
erst first * zu~, zuvor || only * bloß || just * bloß || but * bloß || not before * nicht früher als || not until * nicht bevor || at first * anfangs || 1st * Abkürzung für "first" || one * (nachgestellt) step one: ~er Schritt
Erstabfrage first input bit scan * bei SPS || pulse edge evaluation * Flankenerfassung
Erstanlauf cold restart
Erstanwender first-time user || newcomer * →Neuling
Erstattung refund || repayment || reimbursement || crediting * z.B. Gewährleistungsansprüche
Erste Hilfe first aid
Erste Hilfe leisten render first aid || give first aid
erstellen provide || make available || supply || erect * Gebäude ~ || construct * Gebäude ~ || build * Gebäude ~ || generate * erzeugen, herstellen, z.B. Dokumente, Unterlagen, Zeichnungen || create * erschaffen, erzeugen, z.B. Dokumentation || write up * etwas Schriftliches ~ || draw up * Dokumentation, Zeichnungen usw. ~
Erstellung generation * Erzeugung, Erstellung (auch von Dokumenten) || erection * ~ einer Fabrik, einer Anlage, eines Gebäudes usw.
erstens firstly || first || in the first place || to begin with * zunächst einmal || for one thing || first of all * an erster Stelle, zunächst || primarily * an erster Stelle || at first * anfangs, zuerst
erster Ordnung first order * (vorangestellt) z.B. Oberwelle || 1st order * (vorangestellt) || of the 1st order
erstes first || 1st
Erstfehler first fault
Erstfehlererkennung first-fault identification
Erstfehlerspeicher first fault memory
Erstinbetriebnahme initial start-up || first commissioning
erstklassig first-class || top-quality || first-grade
Erstlauf initial run || first run || first cyclical run * eines zyklisch abgearbeiteten Programms
erstmalig for the first time || first-ever
erstmaliges Auftreten first occurrence
Erstmeldungserkennung first message identification
erstrecken extend * auch 'sich ~' (to: bis zu; over: über); extend up to: sich ~ bis (zu) || stretch * auch 'sich ~' (from ... up to: von ... bis) || reach || range || refer to * (figürlich) sich ~ auf || concern * (figürlich) sich ~ auf, betreffen || be concerned with * (figürlich) sich ~ auf || cover * sich ~ über, z.B. Garantie, Versicherungsschutz || span * sich ~ über
Erstwerterkennung first-up indicator * Funktionsbaustein
Erstwertmeldung first-up signalling
erteilen award * z.B. (einen Lieferauftrag) vergeben, gewähren, zuteilen, zuerkennen, zusprechen
Ertrag profit * Gewinn, Profit || revenue * Geldeinnahmen || earnings * Gewinn, (Geschäfts-) Ergebnis, Profit || return * Ertrag, Einnahme, Verzinsung, Gewinn; yield/bring returns: Nutzen abwerfen, ertragreich s. || return of investment * Nutzbringung des eingesetzten Kapitals
ertüchtigen upgrade * auf/hochrüsten || improve * verbessern || strengthen * stärker machen || reinforce * verstärken. kräftigen
erwähnen mention
erwähnenswert noteworthy * especially noteworthy: besonders ~ || worth mentioning || worthy of note || worthy of special mention
erwähnt mentioned * above: oben; below: unten; earlier: bereits/früher || named * named above: oben genannt
erwärmen heat up * auch 'sich erwärmen' || warm * auch 'sich erwärmen' || heat || grow warm * sich erwärmen
erwärmte Luft warmed air
Erwärmung heating * (Auf)Heizung, Erhitzung || heating up * Aufheizen, Sich-Aufheizen || warming * Wärmung || temperature rise * Temperaturerhöhung/Anstieg (z.B. gegenüber der Umgebungs/Kühlmitteltemp.) || heat build-up * Entwicklung höherer Temperaturen z.B. bei Motor/Getriebe i.Dauerbetrieb
Erwärmungsmessung measuring of the temperature rise
Erwärmungsprobe temperature rise test
Erwärmungsprüfung heat run * Erwärmungslauf || heat test || run in * Erwärmungslauf zur Erkennung von Frühausfällen || burn in * Hitzebehandlung zur Voralterung und Erfassung von Frühausfällen || temperature rise test * z.B. für Elektromotor (siehe VDE 0530 Teil 1)
Erwärmungsverluste heat losses
Erwärmungszeit temperature rise-time * {el. Masch.} benötigt e. blockierter ex.gesch. Motor z.Erreichen d.Grenztemp.
erwarten expect || anticipate * erwarten, erhoffen, voraussehen, vorausahnen, vorausempfinden
erwartet expected
Erwartung expectation * (Gegenstand der) Erwartung; come up to/exceed expectations: ~n erfüllen/übertreffen || expectancy * Erwartung (-shaltung), Hoffnung, Aussicht
Erwartungen voll erfüllen come fully up to expectation
Erwartungswert expectation value
erweisen prove * sich erweisen (to be: als) || turn out * turn out to be difficult: sich als schwierig ~ (siehe auch →herausstellen)
erweiterbar expandable
Erweiterbarkeit expandability * modular: modulare
erweitern expand || extend || enhance * verbessern || complete * vollständig machen || widen * ausweiten (auch figürlich) || upgrade * in der Funktion hochrüsten || broaden * z.B. Wissen || add * erweitern um, hinzufügen
erweiterte Peripherie extended I/O area * bei SPS
erweiterter Bereich extended range || extended area * z.B. Datenbereich bei SPS
Erweiterung expansion || extension || enlargement || upgrade * Aufrüstung (Hard- und Software) || option * wählbare, zusätzlich erhältliche
Erweiterungs- extension || expansion || upgrade
Erweiterungs-Baugruppenträger expansion rack
Erweiterungsbaugruppe expansion board || extension board
erweiterungsfähig extendable || upgradable * hochrüstbar || extensible
Erweiterungsgerät expansion unit || extension unit
Erweiterungskarte extension board * Leiterkarte
Erweiterungssteckplatz extension slot * für Leiterkarte || extra slot * für Leiterkarte || expansion slot

Erwerb acquisition || acquiring || buying || purchase * *Kauf*
erwerben acquire * *erwerben, erlangen, gewinnen, bekommen, (er-) lernen, erfassen (z.B. Messwerte)* || purchase * *käuflich* ~ || earn * *durch Arbeit, auch Achtung usw.* ~ || gain * *(sich) Reichtum, Achtung usw.* ~ || make * *ein Vermögen, Einkommen* ~ || win * *Achtung, Vertrauen usw.* ~
Erz ore
Erzaufbereitung ore dressing
erzeugen generate || produce || create * *(er)schaffen, hervorbringen, erzeugen; verursachen, ein-/errichten, ernennen (to: zu)* || exert * *ausüben*
Erzeugnis product
Erzeugnis-Identnummer product identification code
Erzeugnisbeschreibung product description
Erzeugnisbezeichnung product designation
Erzeugung generation * *auch von Programmen, Dokumenten, Zeichnungen usw.* || production || manufacture
erzielbar attainable || achievable
erzielen obtain * *erlangen, erhalten* || attain * *erlangen, erreichen* || achieve * *zustandebringen, ausführen; (Ziel) erreichen, (Erfolg) erzielen* || score * *Erfolg/Treffer* ~ || reach * *Verständigung usw.* ~ || come to * *Übereinkunft usw.* ~ || realize * *Gewinn usw.* ~ || make * *Gewinn usw.* ~
Erzmühle ore mill || ore crusher
Erzrivale archrival * *auch auf dem Markt*
erzwingen force || enforce * *(besonders gesetzlich)*
erzwungen forced
ESD * *Abk.für 'ElectroStatic Discharge': Elektrostatische Entladung (bezüglich EMV: siehe EN61000-4-2*
eskalieren escalate
etablieren establish
Etage floor || storey || story * *(amerikan.)* || tier * *Reihe, z.B. eines Baugruppenträgers* || deck * *auf Schiff*
Etat budget
Etikett label * *adhesive label: Selbstklebe-Etikett* || docket * *Adresszettel, Etikett, Vermerk; Liste; Zollquittung, Lieferschein*
Etiketten-Druckmaschine label printer
etikettieren label
Etikettiermaschine labeling machine || labelling machine
ETSI ETSI * *Akürzg.f. 'European Tele-communications Standards Institute': europ. Norm.grem. f.Telekommunikat*
Etui etui || little case * *siehe auch* →*Kästchen*
etwa about * *ungefähr* || approximately * *ungefähr* || approx. * *ungefähr (Abkürzg. für approximate(ly))* || perhaps * *vielleicht* || possibly * *möglicherweise* || for example * *zum Beispiel* || almost * *beinahe, fast* || roughly * *grob, annähernd, ungefähr* || roughly speaking * *grob gesprochen, etwa, ungefähr*
etwaig any || whatever || any possible || eventual * *eventuell, möglich, möglich* || possible * *möglich*
etwas something || some * *(Adj.)* || any * *irgendwas* || somewhat * *(Adv.) ein wenig, ein bisschen* || rather * *(Adv.)* || a little * *(Adv.)* || a bit * *(Adv.)* || a certain something * *ein gewisses Etwas* || thing * *(Substantiv) das Etwas* || slight * *gering(fügig), leicht, unbedeutend* || slightly * *(Adv.) gering(fügig), leicht, unbedeutend*
EU EU * *Abkürzung für 'European Union': Europäische Union*
EURO-Flansch EURO flange
Euro-Netzspannungen Euro-supply voltages * *siehe* →*Eurospannungen*
Europa Europe
Europaformat Europe format * *100 x 160 mm für Leiterkarte* || Euroformat * *100 x 160 mm für Leiterkarte* || Eurocard standard format * *100 x 160 mm für Leiterkarte*
europäisch European
Europäische Norm European Standard || EN * *Abkürzg. für European Standard: Europäische Norm*
Europakarte Euro-card * *Leiterkarte 100 x 160 mm*
Europanorm European Standard || EN
Eurospannung Euro voltage * *genormte Netzspannungen, z.B. 230 V, 400 V, 690 V usw.*
Eurospannungen Euro-voltages * *Netz-Normspannungen nach DIN IEC 38/5.87 (z.B. 230V, 400V, 690V usw.)* || European standard supply voltages
evakuieren evacuate * *auch Luft*
eventuell possible || eventual || perhaps * *(Adv.) vielleicht* || if necessary * *(Adv.) falls nötig* || eventually * *(Adv.)* || possibly * *(Adv.) möglicherweise* || if existing * *falls vorhanden* || if need be * *(Adv.) falls nötig, notfalls* || in that case * *(Adv.) gegebenenfalls* || should the occasion arise * *(Adv.) gegebenenfalls* || contingent * *eventuell, möglich, zufällig, ungewiss, gelegentlich*
Evolution evolution
Evolvente evolvent || involute * *z.B. Flankenform einer Verzahnung*
EVU power utility * *Energieversorgungsunternehmen* || power supply company || power company || power supply utility
EVU-Netz public network
Ex e increased-safety type * *Zünd/Explosionsschutzart "Druckfeste Kapselung" nach EN 50019*
Ex-Bereich seriously hazardous environment
Ex-geschützt explosion protected * *siehe EN50014* || flameproof * *siehe EN50014*
Ex-Schutz explosion protection
Ex-Schutzzone explosion-protected zone
exakt exact || accurate * *genau* || proper * *ordentlich, ordnungsgemäß* || precise * *präzise, klar, genau, exakt, korrekt, peinlich genau*
Examen examination
Exemplar copy * *Exemplar/Kopie eines Druckerzeugnisses/Buchs/Dokuments* || number * *einer Zeitschrift usw.* || issue * *Ausgabe (einer Zeitschrift usw.)* || specimen * *Einzelstück* || piece * *(Einzel-) Stück* || sample * *Einzelstück*
existent existent || real || being in existence * *vorhanden*
Existenz existence * *Vorhandensein, Leben, Lebensunterhalt*
existenzgefährdend throat-cutting * *Wettbewerb usw.*
existieren exist || be in existence || live * *leben* || be present * *vorhanden sein* || be available * *vorhanden/verfügbar sein*
existierend existing
Expansionsventil expansion valve * *zum Verdampfen, z.B. bei einer Kältemaschine*
expedieren dispatch
experimentell experimental || by way of an experiment
experimentieren experiment * *with: mit*
Experte specialist
Expertensystem expert system
Expertenteam specialist team
Expertenwissen expertise * *[expe'ties]* || knowhow * *"gewusst wie"* || expert knowledge
explizit explicit * *ausdrücklich, deutlich, bestimmt*
explodierte Darstellung exploded view
Explosionsdarstellung exploded view
explosionsfähig explosive
explosionsgefährdet hazardous || potential explosive
explosionsgefährdete Umgebung hazardous location || hazardous area

explosionsgefährdeter Bereich hazardous area || potentially hazardous area * *siehe DIN 57165/ VDE 0165*
explosionsgefährdeter Ort hazardous location
explosionsgefährdeter Raum hazardous area || space liable to contain explosive atmospheres
explosionsgeschützt explosion-proof || explosion protected * *siehe EN50014* || flameproof * *siehe EN50014* || hazardous type || with explosion-proof enclosure || explosions-protected || suitable for harzadous duty || for use in hazardous locations || hazardous-duty
explosionsgeschützte Ausführung hazardous-duty design || hazardous-duty type || design suitable for use in explosive atmospheres || explosion-protected design
explosionsgeschützter Motor hazardous-duty motor || explosion-protected motor
Explosionsgruppe explosion group * *kennzeichnet jew. Gase ähnlichen Zünddurchschlagvermögens, siehe EN50014 bis 50020* || class of inflammable gases and vapours * *kennzeichn.Gase ähnlich. Zünddurchschlagvermögens (EN50014)*
Explosionsklasse explosion class * *Explosionsgruppe nach alter Norm VDE0170/0171*
Explosionsschutz explosion protection || protection against explosion || protection against firedamp * *Schlagwetterschutz*
Explosionsschutz-Richtlinie explosion-protection directive
Explosionsschutz-Vorschriften hazardous-location regulations
Explosionsschutzart protection against explosion * *siehe IEC-Publikation No.79, EN50014 bis 50020* || type of protection
Exponate exhibits * *auf Messe,Ausstellung*
exponentiell exponential || according to an exponential function * *entspr. e. Exponentialfunktion*
Exportanteil export share
exportorientiert geared up for export * *{Wirtschaft}* || export-minded
Exportvorschrift export regulation
extern external || externally * *(Adv.)* || off-chip * *außerhalb des integrierten Schaltkreises*
externe Steuerung remote control * *Fernsteuerung, z.B. über seriellen Bus* || external control system || higher-level control system * *überlagertes Steuersystem*
extra extra || separate * *getrennt* || additional * *zusätzlich*
extrahieren extract * *herausziehen*
extrapolieren extrapolate
extrem extreme || excessive * *übermäßig* || extremely * *(Adv.)* || exceptional * *außergewöhnlich, Ausnahme-, ungewöhnlich*
Extremwert extreme-value || extremum
Extremwertauswahl extreme-value selector * *als Funktionsbaustein/Einrichtung* || extreme value selection * *als Funktion* || maximum value selector * *Maximalwertauswahl-Baustein*
Extruder extruder
extrudieren extrude
extrudiert extruded
Extrusion extrusion || extrusion * *'Strangpressen' von erhitztem Gummi, Kunststoff.., durch Schnecken-/Spalt-/Blasextruder*
Extrusionsbeschichtung extrusion coating * *z.B. zur Erzeugung kunststoffbeschichteten Papiers*
exzellent excellent || superior * *hervorragend* || outstanding * *hervor-/herausragend*
Exzenter eccentric movement || eccentric element || eccentric cam * *auf Welle*
Exzenterantrieb excenter drive
Exzentergetriebe eccentric gear reducer * *Untersetzungs-~*
Exzenterpresse eccentric press * *mit Schwungrad zur Verringerung der Laststöße*
Exzenterscheibe eccentric disk
Exzenterschneckenpumpe progressive cavity pump
Exzenterwelle eccentric shaft
exzentrisch eccentric
Exzentrizität eccentricity * *z.B. von Motorwelle u. Anbauflansch (DIN 42995)* || out of balance * *Rundlauffehler, z.B. bei Motorwelle (siehe DIN 42955)*

F

f-u-Wandler F/V-converter || voltage-to-frequency converter
F-und-E DE * *Abkürzg. für 'Development and Research': Entwicklung und Forschung*
Fa Co. * *Company: Firma*
Fabrik factory || works * *Werk(sanlagen), Betriebe, Fabrikationsanlagen* || production plant * *Produktionsanlage* || manufacturing plant * *Fabrikationsanlage* || installations * *Fabrikationsanlagen* || mill * *in der Grundstoffindustrie (Papier, Stahl, Textil)* || work * *Werk, Fabrik, Betrieb* || fab * *(salopp) Abk. f. 'FABrication plant': Produktionsanlage; nofab company: Firma ohne Fabrik*
Fabrikant manufacturer
Fabrikat brand * *Herstellermarke* || manufacture * *Erzeugnis* || product * *Produkt* || make * *Marke, Typ, Bauart, Machart, Fabrikat; Anfertigung, Herstellung*
Fabrikatebezeichnung product designation
Fabrikatedatenbank product data base
Fabrikationsnummer serial number * *Seriennummer*
Fabrikautomatisierung factory automation || manufacturing automation
Fabrikbus factory bus || mill bus * *in Papierfabrik, Walzwerk usw.*
fabriklos nofab * *z.B. nofab company: Unternehmen ohne eigene Fertigungseinrichtg.*
Fabriknetz factory network
fabrikneu brand-new || fab new
Fabriknummer serial number * *Seriennummer*
Fabrikumgebung factory environments || factory floor * *(salopp) at the factory floor: in der (rauen) ~*
fabrikweit factory wide || millwide * *über die ganze Papierfabrik/d.ganze Walzwerk (z.B. einheitl. Automatisierungssystem)* || fab-wide * *(salopp) über die ganze Fabrikationsanlage*
facettenreich multifaceted || complex * *komplex* || many-sided || many-layered * *vielschichtig* || intricate * *verwickelt, verschlungen, vielschichtig, knifflig, schwierig* || multifacetted
Fach compartment * *Abteil, Ablage~* || partition * *Abteil, auch in Schrank, Aktentasche usw.* || division * *Abteil* || drawer * *Schub~* || shelf * *Regalbrett* || department * *Fach, Gebiet, Geschäftsbereich, Abteilung* || province * *(Wissens)Gebiet, Fach* || branch * *Branche* || field * *(Betätigungs/Sach/Arbeits)Gebiet, Branche* || activity * *Tätigkeit, Betätigung(sfeld)* || business * *Geschäft(stätigkeit), Aufgabe, Angelegenheit* || trade * *Geschäftszweig, Gewerbe, Branche, Handel* || line * *Gebiet, Tätigkeitsfeld, Sparte, Fach, Berufszweig* || speciality * *Spezial-~, Spezialgebiet, Spezialität* || subject * *Lehr/Schul/Studien-~, Unterrichtsthema* || profession * *Beruf, Fach* || bay * *Fach (in Regal,*

Hochregallager), Abteil(ung), Nische, Laufwerks-Einbauschacht in Computer
Facharbeiter skilled worker || trained worker || expert worker || expert || specialist
Fachausbildung special training || professional training
Fachausdruck technical term || trade term * ~ *in der Geschäftswelt/Branche/Gewerbe* || engineering term * *technischer* ~ || term
Fachausdrücke terminology || nomenclature || engineering terms * *technische* || trade terms
Fachbegriffe terminology || nomenclature || technical terms || engineering terms * *technische* || trade terms || terms
Fachberater technical adviser || consultant * *von externem Unternehmen*
Fachberatung expert consultation
Fachbuch technical book || professional book || specialized book || specialist book || handbook * →*Handbuch*
Fachbücher trade literature
Fächer fan * *siehe auch 'Fach'*
Fächerscheibe serrated lock washer * *Schraubensicherungsscheibe*
Fachgebiet special field * *siehe auch* →*Fach* || special subject || specialty * *(amerikan.)* || field || trade * *siehe auch* →*Fach* || speciality * *Spezialfach, Spezialgebiet, Spezialität*
fachgerecht professional
Fachgroßhandel wholesalers
Fachgrundnorm generic standard
Fachhandel specialized dealers || specialized trade || branch trade
Fachhändler specialized dealer || dealer || local dealer * *am Orte* || reseller
Fachhochschule technical college || technical academy || professional school || college || academy
Fachingenieur specialist engineer || specialist || expert
Fachkenntnisse technical knowledge || experience * *Erfahrung* || expertise * *[expe'ties] Expertenwissen* || know-how * *"gewußt wie"*
Fachkönnen proficiency || workmanship * *handwerkliches* ~ || skill || knowledge || efficiency
fachkundig competent || expert || technically knowledgeable
Fachliteratur technical literature || trade and technical literature || trade literature || specialized literature || specialist literature || specialist publications
Fachmann specialist || specialist engineer || expert * *Experte (on: für); in this field: auf diesem Gebiet* || professional * *'Profi'; qualified: qualifizierter*
fachmännisch professional || expert || competent * *Arbeit* || workmanlike * *Arbeit* || professionally * *(Adv.)* || expertly * *(Adv.)*
Fachmesse trade fair
Fachorientierte Leitstelle specialist service department
Fachpresse trade press * *siehe auch* →*Fachliteratur*
Fachpublikationen technical and trade publications || technical and trade literature
Fachschule technical school
Fachsprache terminology || nomenclature * *Fachbegriffe*
Fachtermini technical terms || engineering terms * *technische* ~
Fachterminus technical term || engineering term * *technischer* ~
Fachwissen technical knowledge || experience * *Erfahrung* || knowledge * *Wissen* || expertise * *[expe'ties] Expertenwissen* || know-how * *"gewusst wie"*

Fachwörterbuch technical dictionary || specialist dictionary
Fachzeitschrift technical journal * *technische* ~ || journal || magazine * *populärwissenschaftliche* ~ || trade magazine * ~ *für Industrie, Handel und Handwerk* || trade journal * ~ *für Industrie, Handel und Handwerk* || special periodical || technical magazine * *technische* ~ || technical publication || technical periodical || specialist publication
Faden thread || twine * *gezwirnter* || fibre * *Faser* || fiber * *(amerikan.) Faser* || filament * *Faden, Faser, Glüh-/Heizfaden* || hairline * *Haarlinie*
Fadenbruch yarn breakage || thread break
Fadenriss yarn breakage
Fadenschar filament group * *{Textil} bei der Chemiefaserherstellung*
Fadenschärspannung yarn warping tension * *{Textil} in* →*Schärmaschine*
Fadenstreckung fiber stretching || filament stretching
fähig capable * *to/of: zu etwas; auch 'in der Lage'* || able * *to: zu* || qualified * *qualifiziert, geeignet, befähigt (for: für)* || fit * *passend*. geeignet, fähig, tauglich, in (guter) Form; (for: für/zu) || competent * *tüchtig, wissend* || efficient * *tüchtig* || clever * *gewitzt, gescheit* || ingenious * *einfallsreich, genial, erfinderisch, klug* || enabled * *fähig gemacht, in die Lage versetzt (to: für/zu)* || apt * *passend, geeignet (to: für/zu)*
Fähigkeit capability || ability || skill * *eines Menschen*
Fahrantrieb traction drive * *Bahnantrieb* || travelling drive * *Kran* || traversing drive * *Kran* || propelling drive * *Fahrzeug, Bagger usw.* || travel drive || crawler drive * *eines Baggers*
fahrbar mobile
Fahrbetrieb travel operation
Fahrdraht contact wire
Fahreigenschaften riding characteristics * *eines Fahrzeugs/Aufzugs usw.*
fahren go * *in einem (beliebigen) Fahrzeug* || travel * *reisen (by: mit), sich bewegen (auch Maschinenteil), laufen (auch Maschinenteil, Kolben)* || drive * *selbstlenkend* ~ || ride * *mit einem (beliebigen) Beförderungsmittel; auch laufen/gleiten/schweben/aufliegen auf* || sail * *mit Schiff* || run * *laufen, (be/durch)fahren, laufen lassen (Maschine, Fahrzeug)* || be moving * *in Fahrt sein*
fahrerlos driverless * *z.B. Transportfahrzeug, Kran* || automated guided * *Fahrzeug*
fahrerloses Transportsystem automated guided vehicle system
Fahrgeschäft fairground ride || merry-go-round * *Karussel* || coaster * *Achterbahn usw.*
Fahrgeschwindigkeit driving speed * *Fahrzeug* || travelling speed * *z.B. Kran, Förderzeug, Fahrzeug, Positionierantrieb* || traveling speed
Fahrkomfort riding comfort * *eines Fahrzeugs/Aufzugs usw.* || ride comfort
Fahrkurve track curve || traversing profile || speed-versus-time characteristic * *Geschwindigkeitskurve über der Zeit* || target curve * *{Aufzüge}*
fahrlässig careless || reckless || negligent * *rechtlich relevant*
Fahrlässigkeit carelessness || negligence * *gross/culpable: grobe* || recklessness
Fahrleitung traction supply system * *{Bahn}* || contact wire * *{Bahn}* || overhead conductor * *{Bahn}* || conductor rail * *als Schiene ausgelegte* ~
Fahrmotor traction motor * *{Bahn}*
Fahrständer moving column * *{Werkz.masch.}*
Fahrstrecke travel distance
Fahrstuhl lift || elevator
Fahrt drive * *mit Wagen* || ride * *mit Wagen* || journey * *Reise* || tour * *Reise* || trip * *Reise; official*

Fahrtrichtung

trip: Dienstreise || voyage * *Seereise* || excursion * *Ausflug* || speed * *Tempo; at full speed: in voller ~; gather speed: ~ aufnehmen* || travel * *Reise, Bewegung, Lauf, Hub (e. Kolbens usw.), Hin-und-her-Bewegung* || motion * *Bewegung*
Fahrtrichtung travel direction || running direction || riding direction * *bei Zug, Bus, Aufzug usw.* || travelling direction * *Verfahrrichtung* || driving direction * *{Verkehr} facing the front/the engine: in ~* || direction of motion * *Bewegungsrichtung*
Fahrtwind air draught
Fahrweise operating regime * *Art und Weise der Maschinenführung durch das Personal*
Fahrwerk running gear || travelling gear * *z.B. eines Krans* || traversing gear * *Kran* || travel drive || traction gear * *bei Kran, Bagger usw.*
Fahrwerksantrieb travel drive * *z.B. eines Krans* || travelling drive * *Kran* || traversing drive * *Kran*
Fahrzeit travel time || running time || duration of a trip * *Fahrtdauer*
fahrzeitoptimiert with optimal ride times * *{Aufzüge}*
Fahrzeug vehicle * *allg. für Landfahrzeug* || rail car * *{Bahn}*
Fahrzeugantrieb vehicle drive || traction drive * *Traktionsantrieb, Bahnantrieb* || propelling drive
Fahrzeugbau vehicle engineering * *als (Ingenieur-) Fachrichtung* || vehicle industry * *als Industriezweig*
Fahrzeuge vehicles || rolling stock * *Fuhrpark*
Fahrzeugsteuerung vehicle control
Fahrzeugwaschanlage vehicle washing system
fair fair
Faktor factor || determinant * *bestimmender ~* || variable * *veränderlicher ~*
fakturieren invoice
Fakturierung invoicing
Fall case * *be never the case: nie der ~ sein; in that/either/no case: in diesem/auf jeden/keinen ~* || event * *Ereignis, Vor~, Begebenheit* || matter * *Angelegenheit* || affair * *Angelegenheit* || instance * *Einzel~* || fall * *das ~en; auch von Preiseen; give a fall: zu ~ bringen* || drop * *Sturz; auch von Preisen* || tumble * *Sturz* || decay * *Niedergang* || slump * *(Börsen-/Preis) Sturz, Geschäfts/Produktionsrückgang* || instance * *(einzelner) Fall, Beispiel*
Fallbeschleunigung acceleration due to gravity || gravitational acceleration
fallen fall * *auch hin~; auch Preise/Bemerkung usw; fall within the scope of: in eine Kategorie ~* || blow * *Sicherung* || be blown * *Sicherung* || drop * *abfallen; auch Preise usw.; auch ~ lassen (Bemerkung, Gegenstand, Plan, Ansprüche)* || tumble * *hinfallen, purzeln, stolpern (over: über), taumeln, Saltos machen, plötzlich ~* || trip * *Sicherung, Schutzschalter usw.* || be falling * *Luftdruck usw.* || abate * *(figürl.) nachlassen* || decline * *(figürl.) nachlassen, sich neigen/senken, abnehmen, nachlassen (auch Preise), zurückgehen* || go down * *Preise usw.* || slump * *plötzlich ~* || come under * *unter Gesetz, in eine Kategorie ~* || be covered of * *in eine Kategorie ~*
fallen lassen drop
fallend falling || trailing * *hintere (Impulsflanke usw.)*
fallende Flanke falling edge * *eines Impulses* || trailing edge * *Rückflanke eines Impulses* || negative-going edge * *negative Flanke eines Impulses*
Fallregister FIFO memory * *Abk.f.'First-In-First-Out': ältester Eintrag wird als Erster geles. (Schieberegister)*
falls erforderlich if required || if need be || if necessary
falls vorhanden if any || if available * *falls verfügbar*

Fallschacht tumble box * *{Metallurgie}*
Fällung precipitation * *{Physik} Chemie}*
Fallunterscheidung differentiation of cases || distinction between cases || distinction of cases
falsch incorrect * *unrichtig* || false || wrong || erroneous * *irrig, unrichtig, irrtümlich, versehentlich* || erroneously * *(Adv.) irrtümlicherweise, zu Unrecht, aus Versehen* || incorrectly * *(Adv.) unrichtig* || spurious * *falsch, unecht, Fehl- (z.B. Messung)*
falsch angeschlossen incorrectly connected
falsch herum upside down * *die Oberseite nach unten* || back to front * *die Rückseite nach vorn* || in the wrong order * *in der falschen Reihenfolge*
falsche Behandlung misuse
falsches Drehfeld incorrect phase sequence
Falschlieferung wrong delivery
Falschpolung incorrect poling || incorrect polarity
Falt- collapsible * *faltbar, zusammenlegbar* || folding
faltbar collapsible || foldable
Faltblatt folder * *Faltprospekt, Faltblatt, Broschüre, Heft* || leaflet || flyer * *(amerikan.) Prospekt, Reklamezettel* || flier * *(amerikan.) Prospekt, Reklamezettel* || advertising leaflet * *Werbe-Faltblatt* || fold-out
Falte wrinkle || crease
falten fold || plait * *{Textil} Matte, Stoff, Geflecht usw.) falten, (ver)flechten, gefaltet ablegen*
Faltenbalg bellows || bellow || siphon
Faltenbalgkupplung bellow-type coupling || bellows coupling
Faltenfreiheit crease-freedom * *z.B. beim Aufwickler (kann z.B. durch abnehmende Wickelhärte erreicht werden)*
Faltkiste folding box
Faltschachtel folding-box || folded box
Faltschachtelkarton boxboard * *Kartonagenpappe* || folding boxboard * *Faltschachtelkartonpappe*
Falz fold || notch * *Auskehlung, Kerbe, Falz* || welt * *Rollsaum in der Metallbearbeitung, Klempnerei* || turned-over edge || edge || rebate * *{Holz}*
Falzapparat folder * *am Ende einer Druckmaschine zum Sammeln und Falten der Exemplare* || folding machine *.am Ende einer (Zeitungs-) Druckmaschine*
falzen fold || welt * *in der Klempnerei* || rabbet * *zusammenfugen* || rebate * *zusammenfugen* || groove * *{Holz} nuten, auskehlen*
Falzhobel rebating cutter * *{Holz}*
Falzklappe folder shutter * *{Druckerei} im Falzapparat*
Falzmaschine seaming machine * *zur Blechbearbeitung* || folding machine
Falzwiderstand folding endurance * *von Kartonpapier*
Familienbetrieb family-owned company * *Unternehmen in Familienbesitz* || family-owned business || family business || family-run company
Fangaufhebung grip release * *{Aufzüge}*
Fangen flying-restart * *Aufschaltg. e. Frequenzumrichters auf laufenden Motor z.B. nach Spannungseinbruch* || auto-restart || restart on the fly * *Aufschaltung auf laufenden Motor* || speed search * *'Suchen',erlaubt Zuschalt.e. Frequ.umrichters auf lauf. Motor z.B.nach Spanngs.einbr.* || catch * *(Verb)*
Fangschaltung flying restart * *Einschalten e.Umrichters auf e.s. drehende Maschine(frequ./phasen/amplitud.gerecht)* || flying restart module || flying restart circuit || flying restart function * *als Funktion* || flycatcher || auto-restart || restart on the fly || restart-on-the-fly capability || restart-on-the-fly circuit || trap circuit * *z.B. zum "Dingfest-Machen" eines Fehlers oder eines selten auftretenden Ereignisses*

Fangvorrichtung grip device * *{Aufzüge}* || gripping device * *{Aufzüge}*
FAQ FAQ * Abk. *f.. 'Most frequently Asked Questions'*: die am häufigsten gestellten Fragen
Farbanstrich coat of paint || paint || paint finish
Farbbildschirm color monitor || color VDU * *Abk. f. 'Video Display Unit'*: Prozessvisualisierungsstation
Farbe color || paint * *Lack, Mal~ usw.* || coating * *Lack, Beschichtung* || ink * *Druck~* || hue * *Farbton* || pigment * *Farbkörper*
Färbeeinrichtung ink dosing equipment
färben color || colour || tinge || stain * *Glas, Papier usw.*
Färberei dye plant || dyeing
Färbereimaschine dyeing machine * *für Textilien*
Farbgebung color structuring * *z.B. Verwendung v. Farben am Bildschirm*
Farbgrafik colour graphics || color graphics
Farbkennzeichnung color code || color coding system * *Farbkennzeichnungssystem*
Farbmonitor color monitor || colour monitor
Farbnebel paint mist
Farbspritzanlage paint spraying system
Farbtiefe color depth * *bei Monitor; z.B.* ~ 8 bit →256 *Farben darstellbar (16/24Bit: High/True Color)*
Farbvoreinstellsystem ink pre-setting system * *{Druckerei}*
Farbwalze inking roller * *{Druckerei}*
Farbwerk inking unit * *{Druckerei}*
Farbzone ink zone * *{Druckerei}*
Farbzonenvoreinstellung ink-zone presetting * *{Druckerei}*
Faser fibre * *(brit.) virgin fibre: Primärfaser (zur Papierherstellung)* || fiber * *(amerikan.)* || thread * *Faden* || strand * *Strähne, Gewebe* || wool * *metallische Faser, Schlackenfaser* || grain * *Holzfaser; crosscut: quer zur ~ sägen* || filament * *(allg.) feine Faser* || staple * *{Textil} Woll~, Rohwolle, Faser; Fadenlänge-/Qualität; of short staple: kurz~ig/-staplig*
Faserflocken fiber flocks * *(amerikan.) {Textil}* || fibre flocks * *(brit.) {Textil}*
Faserflug fiber fly * *{Textil}*
Faserholz pulpwood
Faserlänge staple length * *{Textil} z.B. in der Spinnrei; of short staple: kurzstaplig/mit kurzer Faden-/-*
Faserlinie fiber line * *z.B. in Zellstoff- oder Textilfabrik*
Faserlunte fiber roving * *{Textil} (amerikan.) bei der Fadenherstellung* || fibre roving * *{Textil} (brit.) bei der Fadenherstellung*
Fasermaterial fibrous material * *{Textil}*
Fasern fibres * *(amerikan.)* || fibres || filaments
Faserplatte fiberboard * *{Holz}* || MDF board * *{Holz} Abkürzg. für 'Medium Density Fiber board': mittelharte ~*
Fasersichter fiber grader * *zum Trennen unterschiedlicher Faserpartikel*
Faserstoff fibrous material * *siehe auch* →*Zellstoff (Papierfaserstoff)* || pulp * *{Papier} (Faser)Zellstoff*
faserverstärkt fibre-reinforced * *(brit.)* || fiber reinforced * *(amerikan.)*
Faserverstreckanlage fiber stretching plant * *(amerikan.)* || fibre stretching line
Faserverstreckantrieb fiber stretching drive * *(amerikan.)* || fibre stretching drive
Faserverstreckstraße fiber stretching line * *(amerikan.)* || fibre stretching line * *(brit.)*
Fass cask * *(auch als Verb: 'in ein Fass füllen')* || barrel * *(auch als Verb: 'in ein Fass füllen')* || tun * *großes Fass Tonne* || keg * *kleines Fass, z.B. für Getränke* || vat * *Bottich* || tub * *Bottich*

fassen seize * *(er-)greifen, packen, fassen, befallen* || grasp * *packen, fassen, (er-)greifen (at: nach), an sich reißen, zugreifen/-packen, streben, trachten* || take hold of * *zu ~ bekommen* || catch * *fangen* || hold * *aufnehmen können, Fassungsvermögen von ... haben* || have a capacity of * *Fassungsvermögen von ... haben* || accomodate * *unterbringen/aufnehmen können* || bite * *Werkzeug usw.* ~ || grasp * *packen, greifen (Werkzeug usw.)*
Fassondrehmaschine profiling lathe * *{Holz}*
Fassonfräsmaschine profile moulder * *{Holz}*
Fassung edition * *Ausgabe eines Druckerzeugnisses* || socket * *für Lampe, IC usw.*
Faston-Anschluss faston connector || blade contact * *Messerkontakt*
Faston-Stecker FASTON tab connector || tab connector || FASTON tab receptacle * *Flachsteckhülse, "weiblicher" Faston-Stecker*
FAT FAT * *'File Allocation Table': enthält für jeden →Cluster d.Festplatte e. Verweis auf d. Folgecluster*
faul rotten * *ver~t* || foul * *schmutzig, schmutzig, verdorben, stinkend, widerlich* || worthless * *(figürl.) Wechsel, Versprechungen usw.* || shady * *verdächtig* || lazy * *träge* || indolent * *träge* || idle * *träge* || slothful * *träge* || lame * *lahm, mangelhaft, stockend; (auch figürl.: lame excuse: ~e) Ausrede*
Faustformel rough formula || rough and ready formula || approximation formula * *Näherungsformel* || rule of thumb * *Faustregel*
Faustregel rule of thumb || general rule || rough formula * *Faustformel* || rough and ready formula * *Faustformel*
Fax fax
Faxgerät fax machine
Fazit result * *draw one's conclusions from something: das Fazit aus etwas ziehen* || upshot * *(End)Ergebnis, Ende, Ausgang, Fazit; sum something up: das Fazit aus etw. ziehen* || conclusions * *Folgerungen*
FCC FCC * *Flux Current Control, Regelverfahren f.Asynchronmotor für optimalen Wirkungsgrad u.hohe Dynamik*
Feder key * →*Passfeder, Nutkeil* || spline * *Passfeder für Keilnut* || spring * *z.B. Spiralfeder* || feather key * *Passfeder* || tongue * *{Holz} bei einer Holzverbindung* || feather key * *Passfeder Keil*
Federbalg spring bellows
Federband spring band
Federbandbremse wrap-spring brake
federbelastet spring-applied * *z.B. Bremse/Kupplung*
Federbolzen spring bolt
Federdruck spring pressure
Federdruck-Bremsmotor brake motor with fail-to-safe brake * *mit Anbaubremse nach d.Ruhestromprinzip (bremst,wenn kein Strom)*
Federdruckbremse spring-applied brake || spring-operated brake || fail-safe brake * *nach dem Ruhestromprinzip arbeitend (schließt bei Spannungsausfall per Federdruck)* || spring-loaded brake || spring-actuated brake
Federdruckkupplung spring-pressure clutch
Federkeil key * *Nutkeil, Passfeder* || spline * *Passfeder* || featherkey
Federkennlinie spring characteristic || spring curve
Federkonstante spring constant || spring rate || elasticity constant * *Elastizitätskonstante*
Federkraft spring force || spring tension
Federkraftbremse spring-loaded brake * *zum Öffnen muss Strom fließen (Ruhestromprinzip)*
Federkupplung spring-loaded coupling
Federleiste female connector * *'weiblicher' Steckverbinder, Buchsenstecker* || socket connector || fe-

male socket connector * 'weiblicher' Steckverbinder, Buchsenstecker
federnd   resilient || springy || elastic
Federring   spring washer * →*Federscheibe, Unterlegscheibe mit axialer Federwirkung* || resilient preloading disc * *Federscheibe für axiale Vorspannung eines Lagers* || spring lock washer * *Feder-(Unterleg)Scheibe*
Federrückdruck   spring return pressure
Federscheibe   spring washer * *Unterlegscheibe mit axialer Federwirkung* || spring lock washer * *Unterlegscheibe zur Schraubensicherung* || resilient preloading disc * *zur axialen Vorspannung eines Lagers* || spring disc || spring shim || cup spring * *Tellerfeder (z.B. zur axialen Vorspannung eines Lagers)*
Federung   springing || springs * *(Plural)* || resilience || cushioning || spring suspension * *an einem Fahrzeug*
Federwickelmaschine   spring-coiling machine
Federzugklemme   cage clamp terminal * *Käfigzugfederklemme (mit Befestigung ohne Schraube)*
Fehl-   mis- || erroneous || incorrect || faulty * siehe auch →*fehlerhaft* || mal- || wrong
Fchlauslösung   spurious tripping * *z.B. eines Überwachungsgeräts*
Fehlausrichtung   misalignment
Fehlbedienung   mishandling || maloperation || incorrect operator control || incorrect operator command * *bei einer Kommandoeingabe*
fehlen   be missing || be absent * *abwesend/nicht vorhanden sein* || not be provided * *nicht vorgesehen/zur Verfügung gestellt sein*
fehlend   missing
Fehler   error * *auch Mess~, Regelabweichung (Abweichung auch 'deviation'); typographic error: Druck~* || fault * *Fehlfunktion, Vergehen, Schuld, (Charakter-) Schwäche; diagnose a fault: e. ~ erkennen* || defect * *Mangel, Defekt, Ausfall, hidden defect: versteckter ~* || imperfection * *Unvollkommenheit, Material~* || weakness * *Schwäche, Mangel* || drawback * *Beeinträchtigung, Nachteil, Hindernis, Haken* || shortcoming * *Unzulänglichkeit, Pflichtversäumnis, Verknappung* || blemish * *Makel* || breakdown * *Zusammenbruch; zum Maschinenstillstand führender ~* || failure * *Versagen, Fehlschlag, Zusammenbruch* || bug * ~ *in Softwareprogramm* || mistake * ~ *beim Rechnen, Schreiben usw.; Versehen, Missgriff, Irrtum* || slip * *leichter ~, Flüchtigkeits~ beim Rechnen, Schreiben usw.* || incorrectness * *Unrichtigkeit, ~haftigkeit, Ungeschicklichkeit* || faux pas * *Taktlosigkeit* || trouble * *Störung, Defekt, Verdruss, Schwierigkeit(en)* || disadvantage * *Nachteil* || misspelling * *orthographischer ~* || misprint * *Druck~ (auch 'typographic error')* || erratum * *Druck~ (Plural: "errata")*
Fehler erkennen   diagnose a fault
Fehler finden   diagnose a fault || find a fault
Fehler lokalisieren   isolate faults || locate a fault
Fehler steht an   fault condition exists
Fehlerabschaltung   fault trip || disconnection on faults * *eines Stromkreises*
Fehleranalyse   fault analysis || postmortem review * *rückblickende Analyse nach Auftreten eines (fatalen) Fehlers*
fehleranfällig   failure prone * *siehe auch 'störanfällig'*
Fehleranzeige   error display || error flag * *Fehlermerker* || error indication || fault indication
Fehleraufzeichnung   error logging
Fehleraugenblick   fault instant
Fehlerausblendung   fault suppression
Fehlerauswertung   fault evaluation

Fehlerbehandlung   fault handling || error management || fault management
Fehlerbehebung   fault correction || fault fixing || fault fix * *in (bereits ausgelieferter) Software* || error elimination || troubleshooting * *Fehlerbeseitigung* || debugging * *beim Test (noch nicht ausgelieferter) Software* || workaround * *provisorische Maßnahme zur Umgehung/Behebung e. Fehlers* || error fix || fixing || fix || bugfix * ~ *in (bereits ausgelieferter) Software*
fehlerbereinigt   debugged * *Software* || corrected * *korrigiert* || fixed * *in Ordnung gebracht*
Fehlerbereinigung   debugging * *in Software* || fault correction || bug fixing * *in Software* || fixing * *das In-Ordnung-Bringen* || fault fixing || error correction
Fehlerbeschreibung   fault description
Fehlerbeseitigung   troubleshooting || debugging * *Software* || bugfix * *in (bereits ausgelieferter) Software* || error elimination || fault fixing || error fixing || error fix || fix || workaround * *provisorische Fehlerumgehung* || fault correction
Fehlerbild   error profile || error characteristic || error phenomenon || fault profile
Fehlercode   error code
Fehlerdetektor   flaw detector * *erfasst Fehler im Material* || fault detector * *erfasst Funktionsfehler* || error detector * *erfasst Funktionsfehler*
Fehlerdiagnose   fault diagnosis * *[daieg'nousis]* || post-mortem diagnostics * *nachträgliche Diagnose der Fehler-Vorgeschichte*
Fehlerdiagnosespeicher   fault diagnostics memory
Fehlererfassung   fault sensing
Fehlererkennung   error detection
Fehlerfall   fault condition * *under: im* || case of failure * *in: im*
fehlerfrei   faultless || perfect || correct || trouble-free || flawless || error-free || fault-free || free from error || free of defects * *Holz usw.*
fehlerhaft   faulty || incorrect * *nicht richtig* || invalid * *ungültig* || erroneous * *falsch, verkehrt, irrtümlich; z.B. Signalzustand, Daten* || defective * *defekt, kaputt, ausgefallen* || garbled * *verstümmelt (Daten, Telegramme usw.)* || buggy * *(salopp) Software*
Fehlerhäufigkeit   fault rate || fault frequency
Fehlerhistorie   fault history || historical fault chronology
Fehlerkennung   error code || fault code || fault identification || fault indication
Fehlerkorrektur   error correction || error recovery * *bei Datenübertragung/-Speicherung*
Fehlerliste   fault list
Fehlermelderelais   fault alarm relais
Fehlermeldung   fault message || error message || fault alarm || error signal
Fehlernummer   error code || fault code || alarm code * *Warnungsnummer*
Fehlerortung   fault location * *z.B. auf einer Leitung*
Fehlerprotokoll   fault log
Fehlerquittierung   fault acknowledgement || fault clear * *Signal* || fault acknowledge * *als Kommando*
Fehlerrate   fault rate || fault frequency || failure rate * *Ausfallrate*
Fehlerschlüssel   fault code
fehlersicher   fail-safe || fault-tolerant || error-proof * *es können kaum Fehler auftreten* || fail-safe * *Fehler können keine unsicheren Zustände herbeiführen*
Fehlerspeicher   fault memory || fault log || fault storage
Fehlerstatistik   error statistics
Fehlerstrom   leakage current || earth leakage cur-

rent * ~ *gegen Erde* || earth leakage * ~ *gegen Erde* || fault current * *auch bei ~ Kurzschluss, Wechselrichterkippen, Isolationsversagen usw.* || ground fault current
Fehlerstromschalter earth leakage breaker * *siehe auch →FI-Schutzschalter*
Fehlerstromschutzeinrichtung RCD * *Abk. für Residual Current Device; siehe auch →FI-Schutzschalter* || residual current device
Fehlersuche troubleshooting || fault finding || debugging * *in Software* || error fixing * *Fehlerbeseitigung* || fault localization * *Lokalisierung, Findung, Ortung von Fehlern* || fault locating || fault search || fault identification * *"Dingfest-Machen" von Fehlern* || fault diagnosis * *Fehlerdiagnose*
Fehlertext fault text
fehlertolerant fault-tolerant
Fehlertoleranz fault tolerance
Fehlerüberprüfung error check
Fehlerunterdrückung fault suppression
Fehleruntersuchung fault analysis
Fehlerursache fault cause || error cause
Fehlervorgeschichte fault history || historical fault chronology
Fehlerwahrscheinlichkeit probability of error || error probability
Fehlerzähler error counter
Fehlerzustand fault condition || error condition || error status
Fehlfunktion malfunction || misfunction || failure * *Fehler* || erroneous function
Fehlinformation misinformation
Fehlinterpretation misinterpretation * *siehe auch →Missverständnis*
Fehljustierung misalignment
Fehlmessung wrong measurement || false measurement || spurious measurement
Fehlpolung incorrect poling || incorrect polarity
Fehlrollenschneider salvage reel cutter * *{Papier}* || salvage winder * *{Papier}* || faulty-reel cutter * *{Papier}* || clinic roll cutter * *{Papier}*
Fehlschlag miss || flop * *(salopp)*
Fehlschnitt off-cut * *z.B. bei Querschneider*
Fehlsignal spurious signal * *'unechtes' (Mess) Signal*
Feiertag holiday || public holidy * *offizieller*
Feigabetaste release button
Feile file || rasp * *grobe*
feilen file
fein fine || dignified * *gediegen, edel* || fine-grained * *feinkörnig, z.B. Schleifmittel, Bimsstein*
Feinabgleich fine adjustment || fine adjust || fine tuning
Feinanpassung fine adjustment || streamlining
Feinblech sheet || light sheet || metal sheet * *thin metal sheet: Feinstblech*
Feindarstellung detailed representation
feindrähtig flexible || stranded * *mit Litzenleiter* || flexible stranded || finely stranded || flexible-conductor
feindrähtige Leitung flexible conductor
feindrähtiger Leiter flexible conductor || stranded conductor
Feineinsteller vernier
Feineinstellung fine tuning || fine adjustment
feinfühlig sensitive * *sensibel* || tactful * *taktvoll* || delicate * *zartfühlend* || precise * *exakt, →genau*
Feingewinde fine thread
Feinheit fineness * *z.B. von Fäden; finer points: die ~en (figürl.)* || size * *{Textil} von Garn* || grist * *{Textil} von Garn*
Feinheitsgrad degree of fineness
Feinimpuls vernier pulse || fine pulse
Feininterpolation fine interpolation * *{NC-Steuerung}*

Feinjustierung fine setting || fine adjustment
feinkörnig fine-grained * *z.B. Schleifmittel, Bimsstein*
Feinmechanik precision mechanics || fine mechanics || precision engineering * *als Fachrichtung*
Feinmessgerät precision measuring instrument
Feinmesszeug precision measuring instrument
Feinpapier fine paper || woodfree paper * *holzfreies Papier*
Feinpappe fiberboard * *(amerikan.)* || fibreboard * *(brit.)*
feinporig fine-pore
Feinputzhobel fine-finishing planer * *{Holz}*
Feinsicherung microfuse * *super-rapid: superflinke* || minifuse
Feinstaub fine dust
Feinstaubfilter micro-filter || fine-dust filter
Feinstblech thin sheet || thin metal sheet
Feinstoff fines
feinstufig high-resolution * *mit hoher Auflösung* || in small steps
Feinwasser deionized water * *z.B. im Kühlwasserkreislauf e. Stromrichters/Motors*
Feld field * *el./magn. ~, Bereich, Sachgebiet; zero-divergence: quellenfreies; irrotational: wirbelfreies* || exciting field * *Erreger~* || excitation * *Erregung* || array * *{Computer} Daten~, mehrdimensionale Variable* || factory floor * *(figürl.) in der untersten Ebene der Fertigungs/Produktion, (harter) Praxiseinsatz* || bay * *~ in einer (Freiluft-) Schaltanlage* || panel * *~ in einer Schaltanlage* || section * *auch ~ in Schaltanlage, Aufbausystem usw.* || tile * *~ in der Wartentechnik*
Feldausfall field loss
Feldbereich field area * *untere Ebene der 'Automatisierungs-Pyramide'*
Feldbus field bus || device network * *z.Anschlus v. Feldgeräten (Sensoren, Aktuatoren, Remote-I/O, dezentrale Intelligenz)*
Felddrossel field reactor || field choke
Feldeffekt-Transistor field-effect transistor || FET
Feldeffekttransistor field-effect transistor || FET
Felderfahrung experience gained in the field
felderprobt field-proven
Felderprobung field test || field trial
Feldgerät field device * *Endgerät an einem Feldbussystem* || field supply unit * *zur Feldstromversorgung z.b. einer DC-Maschine*
Feldgleichrichter field rectifier || bridge exciter * *Brückengleichrichter* || controlled bridge exciter * *gesteuerter* || field-circuit rectifier
Feldkennlinie field characteristic
Feldkennlinienaufnahme field characteristic plotting || recording of the field characteristic
Feldkreis field circuit
Feldkreisumschaltung field reversal * *Feldumkehr z.B. zur Momentenumkehr bei IQ Gleichstromantrieben* || field-circuit reversal
Feldkreisunterbrechung field loss * *Feldausfall*
Feldkreiswiderstand field resistance
Feldlinie line of force * *magnetic: magnetische, electric: elektrische* || field line
Feldlinienbild field pattern
Feldliniendichte field density
Feldlinienverlauf flux distribution characteristic
feldorientiert field-oriented * *z.B. Vektorregelung*
feldorientierte Regelung field-oriented closed-loop control * *für Asynchronmotor* || vector control * *für Asynchronmotor* || TRANSVEKTOR control * *(R) Variante des Erfinders dieses Verfahrens, der Fa. SIEMENS* || field-oriented control * *für Asynchronmotor*
Feldorientierung field orientation || field-oriented principle

Feldplattenpotentiometer magneto-resistive potentiometer
Feldplattenpoti magneto-resistive potentiometer
Feldregelung field control
Feldregler field controller || excitation controller
Feldschwächbereich field-weakening range || field weakened speed range * *Drehzahlerhöhung durch Feldreduzierung* || field weakening operating range || constant-power range * *Drehzahl steigt u. max. Drehmoment sinkt → annähernd const. Leistung*
Feldschwächbetrieb field weakening operation || above base speed operation * *Betrieb oberhalb der Ablösedrehzahl (bei der d. Feldschwächung einsetzt)*
Feldschwächdrehzahl field weakening speed * *Drehzahl bei max. Feldschwächung eines AC- oder DC-Motors* || field-weakening speed
Feldschwächebereich field-weakening range * *siehe →Feldschwächbereich*
Feldschwächregelung field-weakening control
Feldschwächregler field weakening controller
Feldschwächstellbereich field-weakening range || field weakening control range
Feldschwächung field weakening
Feldspeisegerät field supply unit || field supply
Feldstärke field strength || electric field strength * *elektrische ~*
Feldsteuerbereich field control range * *Feldschwächbereich eines AC-Asynchron- oder DC-Motors*
Feldsteuersatz field gating unit
Feldstrombegrenzung field current limiting
Feldstromgrenze field current limit
Feldstromreduzierung field current reduction * *z.b. bei Stillstand, um eine Aufheizung des Motors zu vermeiden* || field economizing * *DC-Motor beim Stillstand*
Feldstromregelung closed-loop field current control
Feldstromregler field current controller
Feldstromsollwert field current setpoint || excitation setpoint
Feldstromversorgung field-current supply * *controlled: geregelte* || field supply || field voltage supply
Feldumkehr field reversal || field reversing
Feldumschaltung field reversal * *Feldumkehr z.B. zur Momentenumkehr bei 1Q Gleichstromantrieb*
Feldversorgung field supply || field-current supply * *controlled: geregelte*
Feldversuch field test || field trial
Feldverteilung field distribution * *z.B. in einem Motor*
Feldwicklung field winding
Fell sheet * *Zwischenprodukt bei der Gummiherstellung*
Fenster window * *auch auf dem Bildschirm* || opening * *Öffnung* || band * *Band (z.B. Toleranzband)*
Fensterdiskriminator window discriminator * *Grenzwertmelder m. 2 Schwellen (erkennt "im oder außerhalb Fenster")*
FEPROM FEPROM * *Abkürzg. für 'Flash EPROM': elektrisch schnell löschbarer Festwertspeicher*
Fern- remote * *z.B. remote control: Fernsteuerung* || tele- * *drahtlos* || long-distance * *im Verkehrswesen, in der Telekommunikation usw.* || long-haul * *{Bahnen}* || main-line * *{Bahn}* || trunk * *z.B. -Kabel, -Leitung, -Straße, -Verbindung*
Fernanzeige remote display || remote indication
Fernbahn-Lokomotive long-haul locomotive
Fernbedien- remote
Fernbediengerät remote operator control unit

Fernbedienung remote control * *als Funktion* || external operator * *als Gerät* || remote operator control
Ferndiagnose remote diagnostics || teleservice diagnostics * *z.b. über Modem*
Ferndrehwelle synchro-system * *Ausführungsform d. elektr. Welle mit Asynchron-Schleifringläufermotoren*
ferner furthermore || further
ferngesteuert remote-controlled || wireless-trolled * *funkferngesteuert* || radio-controlled * *funkferngesteuert*
fernhalten isolate * *from: von; z.B. Störungen vom Netz ~* || keep away * *from/of: von; auch 'sich ~'*
Fernmeldeanlage telecommunications system
Fernmeldestromversorgung telecommunication power supply
Fernmeldesystem telecommunications system
Fernmeldetechnik telecommunications
Fernmeldewesen communication
Fernmessung remote measurement || telemetry * *drahtlose ~* || telemetering * *drahtlose ~*
Fernprogrammierung remote programming * *bei SPS*
Fernschreiben teletype message || teleprint
Fernschreiber teletype writer || teletype || TTY * *Abk.f. 'TeleTYpe writer': Fernschreiber (mit serieller Stromschleifen-Schnittstelle)*
Fernservice teleservice * *über Telefon/Modem*
Fernsprechtechnik telephony * *als Fachgebiet*
Fernsteuerbarkeit remote controllability
Fernsteuerung remote control || master control * *durch ein überlagertes System* || telecontrol * *z.B. drahtlos*
Fernthermometer distant-reading thermometer
Fernverkehr long-distance traffic || heavy rail traffic * *Bahn* || main line traffic
Fernverkehrsfahrzeug heavy rail vehicle * *Schienenfahrzeug*
Fernwärme district heating
Fernwärmeanlage district heating plant
Fernwirkstation telecontrol station
Fernwirktechnik telecontrol
Ferrit ferrite
Ferritmagnet ferrite magnet
ferromagnetisch ferromagnetic
fertig ready * *bereit* || finished * *abgeschlossen, beendet, ~ bearbeitet* || done * *abgeschlossen* || prefabricated * *vorgefertigt* || ready-built * *vorgefertigt*
fertig montiert factory-assembled || pre-assembled * *vormontiert*
fertig werden mit tackle || cope with
Fertig- finishing || final || finished
fertigbearbeiten finish
fertigbohren finish-bore
fertigen assemble * *zusammenbauen, montieren* || make || produce || manufacture
Fertigerzeugnis finished product
Fertiggerüst finishing stand * *Walzwerk*
Fertighaus prefab house
Fertighausbau prefab house construction
Fertigkeit skill
Fertigprodukt finished product
fertigstellen complete
Fertigstellung completion
Fertigstraße finishing mill * *Walzwerk*
Fertigteil finished part || prefabricated part
Fertigung production || manufacturing || manufacturing shop * *Fertigungsabteilung* || manufacturing facilities * *Fertigungseinrichtung(en)* || manufacture * *Erzeugung, Fertigung, Herstellung, Fabrikation, Industrie (-zweig)*
Fertigung einstellen cease production || discontinue the production

Fertigungs- und Vertriebsergebnis production and sales profit
Fertigungsanlage production plant ‖ manufacturing facility
Fertigungsanlagen manufacturing facilities ‖ production facilities
Fertigungsautomatisierung manufacturing automation
Fertigungsbereich manufacturing area ‖ shop floor ‖ factory floor * on: im
Fertigungsbetrieb production plant ‖ factory ‖ production facility ‖ manufacturing plant ‖ works
Fertigungseinrichtungen manufacturing facilities ‖ manufacturing equipment ‖ production facilities
Fertigungseinstellung end of production ‖ discontinuation of production
Fertigungsfehler manufacturing defect ‖ manufacturing deficiency
Fertigungsfluss production flow
Fertigungsgenauigkeit workpiece accuracy * *[Werkz.masch.]*
Fertigungskosten production costs
Fertigungsleitrechner production control computer ‖ production host computer
Fertigungsleitsystem production control system
Fertigungslos batch ‖ lot
Fertigungsmaschinen production machinery
Fertigungsmethode manufacturing method
Fertigungsnummer serial No. * *Seriennummer*
Fertigungspalette production spectrum * →*Fertigungsprogramm*
Fertigungsplanung production scheduling ‖ production planning ‖ manufacturing engineering ‖ production engineering
Fertigungsprogramm line of merchandise * *lieferbare Produkte* ‖ sales program * *lieferbare Producte* ‖ line of products ‖ sales programme * *Verkaufsprogramm* ‖ manufacturing program ‖ product range * *Produktpalette* ‖ production spectrum
Fertigungsprozess production process
Fertigungsqualität production quality
Fertigungsreife readiness for production ‖ production stage
Fertigungsspektrum production spectrum
Fertigungssteuerung production control ‖ manufacturing control ‖ production management and control
Fertigungsstraße production line ‖ manufacturing line
Fertigungssystem manufacturing system * *flexible: flexibles*
Fertigungstechnik manufacturing technology ‖ production technology ‖ production engineering
Fertigungstechnologie production technology ‖ manufacturing technology
Fertigungsverfahren production technique
Fertigungsvorbereitung production planning ‖ production scheduling
Fertigungszentrum production center * *(amerikan.)* ‖ production centre * *(brit.)*
fertigwalzen finish-roll
fertigwerden get ready ‖ manage * *mit jdm./etwas ~* ‖ cope with * *mit etwas ~* ‖ get over * *mit Problem, Kummer usw. ~*
Fertigwert linearized value * *linearisierter Messwert, z.B. von Temperaturgeber*
fest fixed * *orts~, be~igt, starr, unveränderlich, ~gesetzt* ‖ solid * *von ~er Beschaffenheit, massiv* ‖ firm * *fest, stark, beständig; auch 'unerschütterlich', auch Markt, Kurse usw.* ‖ compact * *von ~er Beschaffenheit* ‖ hard * *hart von ~er Beschaffenheit* ‖ rigid * *starr, stabil* ‖ stationary * *orts~, stationär* ‖ permanent * *dauernd, unveränderlich; z.B. Wohnsitz, Struktur, Beschaffenheit* ‖ constant * *standhaft* ‖ stable * *dauerhaft, stabil, auch*

(Geschäfts)Beziehung ‖ durable * *dauerhaft; auch (Geschäfts)Beziehung* ‖ lasting * *dauerhaft, bleibend* ‖ steady * *von Markt, Kursen usw.* ‖ binding * *verbindlich, z.B. Angebot* ‖ regular * *z.b. Kunden* ‖ tight * *fest(sitzend), stramm, straff, (an)gespannt, dicht (gedrängt)* ‖ secure * *sicher, geschützt (from: vor), ~ (Grundlage, Hoffnung, Schaubverbindung), gesichert (Existenz)*
fest angezogen fully tightened * *z.b. Schraubenverbindung*
fest anziehen tighten fully * *z.b. Schraubverbindung* ‖ tighten
fest belegt permanently assigned
fest definiert permanently defined
fest eingebaut stationary mounted ‖ fixed-mounted ‖ non-withdrawable
fest einstellen auf set permanently to
fest montiert permanently fixed ‖ permanently mounted ‖ fixed-mounted ‖ non-detachable * *nicht demontierbar/lösbar* ‖ stationary-mounted * *fest eingebaut*
fest verdrahtet hardwired
fest verlegt permanently installed * *Kabel, Leitungen usw.*
fest zugeordnet individually assigned ‖ dedicated * *zugeschnitten, speziell zugeordnet*
Fest- fixed
festbremsen lock the rotor * *Motor*
festdrehen tighten * *Schraube*
Festdrehzahl fixed speed ‖ pre-set speed * *fester Drehzahlsollwert*
Festdrehzahl-Antrieb fixed-speed drive
Festdrehzahlantrieb constant speed drive ‖ fixed-speed drive
Festeinbau permanent mounting
fester Zustand solid state
Festfrequenz fixed frequency * *fester Frequenzsollwert* ‖ preset frequency * *Frequenz-Festsollwert*
festfressen seize * *sich festfressen* ‖ freeze * *sich festfressen* ‖ jam * *klemmen, hemmen, stocken, verstopfen, blockieren*
festgebremst with the rotor locked * *Motor* ‖ locked ‖ under locked-rotor conditions * *Motor* ‖ locked-rotor * *Motor*
festgebremster Läufer locked rotor
festgelegt specified * *spezifiziert* ‖ fixed * *festgesetzt, unveränderlich* ‖ determined * *bestimmt*
festhalten hold ‖ hold tight * *packen* ‖ freeze * *einfrieren (Integrator, Messwerte usw.)* ‖ retain * *einbehalten* ‖ adhere to * *(figürl.) an Meinung, Grundsatz, Vorgehen usw. ~* ‖ grip * *greifen, packen (firmly: fest/sicher)* ‖ withhold * *einbehalten, zurückhalten, abhalten (withhold someone from: jdn. ~ von), vorenthalten* ‖ fix in place * *in Lage/ Stellung ~* ‖ keep in position * *in Lage/Stellung ~* ‖ record * *in Wort/Bild/Ton ~* ‖ put down * *schriftlich ~* ‖ record in writing * *schriftlich ~* ‖ commit to paper * *schriftlich ~* ‖ take down in writing * *schriftlich genau ~* ‖ make a note of * *Gedanken usw. schriftlich ~*
Festigkeit stressability * *Belastbarkeit, z.B. thermische, mechanische* ‖ strength * *auch einer Konstruktion; Bruch/Zerreiß~, Stärke, Kraft* ‖ resistance * *Widerstandskraft/-fähigkeit, Beständigkeit, (Biegungs-/Stoß-/Verschleiß-) ~* ‖ ruggedness * *Robustheit* ‖ stability * *auch einer Konstruktion; Beständigkeit, Stand~* ‖ firmness * *Beständigkeit, Entschlossenheit* ‖ solidity * *kompakte/massive Struktur/Beschaffenheit; Dichtigkeit, Gediegenheit, Zuverlässigkeit* ‖ compactness * *Kompaktheit, Gedrängtheit* ‖ immunity * *Immunität, Geschützt-/ Sicher-/Unempfänglich-Sein, Unempfindlichkeit (against: gegen)* ‖ tenacity * *Zähigkeit* ‖ material

Festigkeitsklasse

strength * ~ *eines Werkstoffs* || mechanical strength * *mechanische* ~
Festigkeitsklasse strength class * *eines Werkstoffs, einer Schraube usw.*
Festkondensator fixed capacitor
Festkörper solid || solid particles
Festkörperreibung solid friction
Festlager location bearing || locating bearing || locked bearing
festlegen fix || determine * *bestimmen* || establish * *festsetzen, einführen, etablieren, erlassen, durchsetzen* || mark out * *bestimmen, festsetzen, ausersehen(for: für,zu), bezeichnen, markieren* || set * *Termin* ~ || schedule * *Termin* ~ *(on: auf)* || stipulate * *vertraglich* ~ || lay down * *Grundsatz/Regel* ~ *z.B.* in einer Norm/Vorschrift || define * *definieren, genau bestimmen, kennzeichnen* || specify * *spezifizieren, genau angeben/vorgeben* || pinpoint * *auf den Punkt genau* ~
Festlegung appointment * *besonders für Termin, Verabredung, Zusammenkunft* || fixing * *auch Festsetzung, Termin*~ || establishment * *Festsetzung, Feststellung* || arrangement * *Vereinbarg., Verabredg.(make: treffen), Einteilg., Ab/Übereinkommen, Auf/Zusammenstellg.* || regulation * *Regelung, Verordnung, Verfügung, Vorschrift, Satzung, Statuten* || laying down * *Festsetzen einer Regel/Bedingg.; i.e. Vertrag niederlegen* || stipulation * *(vertragliche) Abmachung, Übereinkunft, Festsetzung, Klausel,Bedingung* || provision * *Bestimmung, Vorschrift, Vorkehrung,* || assessment * *Festsetzung einer Zahlung/Steuer, Bewertung* || agreement * *Abkommen, Vereinbarung, Übereinkunft, Vertrag* || fixation * *Festsetzung* || determination * *Festsetzung, Entschluss, Beschluss, Bestimmung* || specification * *Spezifikation, Festlegung*
Festplatte hard disk || hard disk drive * *Festplattenlaufwerk* || HDD * *Abkürzg. für 'Hard Disk Drive':Festplattenlaufwerk* || HDU * *Abkürzg. für 'Hard Disk Unit': Festplattenspeichereinheit* || hard drive
Festplattenkomprimierprogramm hard disk data compression program
Festplattenlaufwerk hard disk drive || HDD * *Abkürzg. für 'Hard Disk Drive'* || hard drive
Festpreis fixed price
Festpunktwert fixed-point value
Festpunktzahl fixed-point number
Festpunktzahl mit Vorzeichen integer
festschrauben bolt into place * *Maschinenteil an seinen Einbauplatz, Schaltgerät an Montageplatte* || screw tight || screw on || tighten up * *(Schraube) festziehen* || screw home || bolt || bolt down * *z.B. Deckel* || screw in place * *am vorgesehenen Einbauplatz* ~
festsitzen fit tightly || sit fast || be frozen in place * *klemmen (z.B. Passsitz)* || be stuck * *festgefahren sein, steckengeblieben sein* || be stalled * *bestgefahren sein, steckengeblieben sein*
Festsollwert fixed setpoint || pre-set value || pre-set speed * *für Drehzahl* || preset value || preset
festspannen clamp tight * *z.B. Werkstück in Schraubstock/Spannvorrichtung*
feststehend stationary || *ortsfest* || established * *feststehend (Tatsache, Brauch usw), planmäßig, fest begründet, unzweifelhaft.* || fixed * *ortsfest, befestigt, fest (auch figürl.: Preis, Grundsatz, Termin), laufend (Ausgaben)*
feststellen establish * *festsetzen, nachweisen, begründen* || state * *konstatieren, erklären, behaupten, erwähnen, bemerken, anführen, darlegen, aussagen* || declare * *erklären* || find out * *herausfinden* || ascertain * *ermitteln* || detect * *herausfinden, ermitteln, entdecken, wahrnehmen, aufdecken, enthüllen* || determine * *bestimmen, festlegen, ermitteln, herausfinden* || assess * *einschätzen, bewerten, veranschlagen,*

festsetzen, schätzen (besonders Schaden) || locate * *Ort, Lage, Fehler lokalisieren* || identify * *identifizieren* || notice * *bemerken, beobachten, wahrnehmen, merken* || observe * *beobachten* || lock * *verriegeln* || secure * *sichern; in position: in (s)einer Lage* || set * *festsetzen, bemessen, festlegen* || set out * *(ausführlich) darlegen, angeben* || pinpoint * *auf den Punkt genau* ~
Feststellknopf locking button
Feststellschraube set-screw || setscrew || grubscrew * *Madenschraube* || lock-bolt || locking screw || anchoring screw
Feststellung findings * *Tatsachenfeststellung, Befund, Fund, Entdeckung* || establishment * *Feststellung, Festsetzung* || statement * *Aussage* || locking * *mechan. Arretierung* || observation * *Beobachtung, Wahrnehmung* || determination * *Ermittlung, Feststellung, Bestimmung, Festsetzung*
Feststellvorrichtung locking device
Feststoff solid
Feststoffe solids * *granular solids: kornförmige* ~ || solid matter || solid substances || hard substances
festverdrahtet hardwired
Festwert fixed value || preset value * *vorbesetzter Wert*
Festwertregelung fixed-command control
Festwertspeicher read-only memory || ROM * *Abkürzg. für 'Read Only Memory': Nur-Lesespeicher*
Festwiderstand fixed resistor
festziehen tighten up || tighten
Fett grease || bold * *(Adj., typografisch) in bold print: fettgedruckt*
Fettdichtigkeit greaseproofness
Fettdruck bold print
Fetterneuerung grease replacement || regreasing * *Nachschmieren* || grease renewal
Fettfüllmenge quantity of grease
Fettfüllung grease packing * *z.B. eines Lagers* || packing of grease
Fettgebrauchsdauer service life of grease || grease stability time || period between lubrications || lubrication interval || grease life time
fettgedruckt in bold print
fettgeschmiert grease-lubricated
Fettmenge grease quantity
Fettmengenregler grease slinger * *Fettschleuderscheibe (zum Abschleudern überschüssigen Lagerfetts)*
Fettschleuderscheibe grease slinger
Fettschmierung grease lubrication * *grease lubricated bearing: Lager mit* ~
Fettsorte type of grease
Fettstandzeit grease life * *Schmiermittel-Gebrauchsdauer*
Fettverbrauch grease consumption * *von Schmierfett, z.B. in einem Lager*
Fettvorratsraum grease chamber * *in Schmiersystem, Lager usw.*
Fettwechselfrist grease change interval
feucht moist * *with: von* || damp * *feucht, dunstig (z.B. Luft)* || humid * *(physikal.) Luft* || wet * *nass* || dank * *nasskalt, (unangenehm) feucht*
Feuchte humidity * *absolute humidity: absolute* ~ *[g/m^3], relative humidity: relative Luftfeuchte [%]* || moisture || moisture content * *Feuchtegehalt (z.B. von Papier)*
feuchte Wärme damp heat || moist heat
Feuchtebeanspruchung moisture conditions * *Betriebsbeanspruchung; high-: hohe* || humidity rating * *zugelassene, spezifizierte*
Feuchtegehalt moisture content
Feuchteklasse humidity classification * *siehe DIN 40040* || humidity class || humidity rating
Feuchteregelung moisture control * *z.B. bei Papiermaschine*

Feuchtesensor humidity sensor
Feuchtigkeit humidity || moisture
Feuchtigkeitsmessgerät humidity measuring equipment || moisture measuring instrument
Feuchtwalze damping roller * *[Druckerei]*
Feuchtwerk damping unit * *[Druckerei]*
Feuerbeschichtung hot-dip galvanizing * *Galvanisierung, z.B. Verzinkung*
Feuerbeständigkeit flame resistance
feuergefährlich inflammable
feuerhemmend fire-retarding || fire-retardant
Feuerraum combustion chamber * *in Heizkessel usw.*
feuerverzinken hot-galvanize
feuerverzinkt hot galvanized || hot-dip galvanized
Feuerverzinkung hot-dip galvanizing
FI Schutzschalter ELCB * *Earth Leakage Circuit-Breaker; Sicherh.einrichtg., schaltet b.Erdfehlerstrom ab* || earth leakage circuit-breaker * *Sicherheitseinrichtg., schaltet bei Erdfehlerstrom den Stromkreis ab*
Fi-Schutzschalter ELCB * *Earth Leakage Circuit-Breaker; Sicherh.einrichtg., schaltet b. Erdfehlerstrom ab* || earth leakage circuit-breaker * *Sicherheitseinrichtg., schaltet bei Erdfehlerstrom ab* || GFCI * *Abk.für 'Ground Fault Circuit Interrupter'* || ground fault circuit interrupter
Fibel primer * *z.B. Einfachbetriebsanleitung/Nachschlagewerk*
FIFO FIFO * *Abk. f. 'First In First Out': Schieberegister mit einer Breite von mehreren Bits*
FIFO-Speicher FIFO memory * *Abk. für "First In First Out": zuerst eingegebene Daten werden als Erstes ausgegeben*
fiktiv fictitious
Filament filament || yarn
filigran delicate || sophisticated * *(figürl.)*
Filter filter * *auch Glättungsglied, Staubfilter* || low pass filter * *Tiefpassfilter* || strainer * *(mechan.) Sieb, Seiher, Filter*
Filteranbau filter mounting
Filteranlage filtration plant
Filterkondensator filter capacitor || filtering capacitor
Filterkreis filter network || filter circuit
Filtermatte filter mat || filter pad
filtern filter || smooth * *glätten*
Filterpresse filter press * *z.B. bei Wasseraufbereitungsanlage*
Filterung smoothing * *(elektr.) Glättung* || filtering || filtration * *(mechan.)*
Filterzeitkonstante filter time constant
Filtration filtration * *Filterung*
Filz felt * *auch z.ur Führung noch nassen Papiers in der Papiermaschine*
Filzlaufregler felt guide * *Filzregulier-/Führungswalze in Papiermaschine*
Filzleitwalze felt roll * *in der Entwässerungspartie einer Papiermaschine* || felt guide roll
Filzregulierwalze felt-guide roll * *in Papiermaschine*
Filzring felt sealing ring * *zum Abdichten (z.B. eines Wälzlagers)*
Filzschreiber felt-tip || felt-tipped pen || felt tip pen
finden find || meet with * *antreffen* || discover * *entdecken* || come across * *entdecken, stoßen auf* || locate * *auffinden, lokalisieren, ausfindig machen, "dingfest machen"* || think * *denken* || consider * *dafürhalten*
findig ingenious || clever || resourceful
fingersicher safe from finger-touch * *siehe VDE 0106 Teil 100*
Finite-Elemente-Methode finite-element method * *CAD-Berechnungsverfahren: Annäherung e. Kontur durch kleinste Teilflächen*

Finite-Elemente-Verfahren finite-elements method * *CAD-Berechnungsverfahren: Annäherung einer Kontur durch kleinste Flächen*
FIP FIP * *Abkürz.. für 'Field Instrumentation Protocol': französisches Feldbusprotokoll*
Firma company || firm
Firma gründen establish a company || found a company
firmeneigen proprietary || in-house * *auch ~e technische Lösung* || company-specific * *firmenspezifisch* || company owned * *der Firma gehörend*
Firmengelände company premises
Firmengründer company founder
Firmengruppe group of companies
firmenintern in-house || proprietary * *z.B. technische Lösung*
firmenneutral company neutral * *z.B. Bussystem* || open * *offen, z.b. Bussystem* || non-proprietary * *nicht firmeneigen/spezifisch* || manufacturer-neutral * *z.b. offenes Bussystem*
Firmenpolitik company policy
Firmenprofil company profile
Firmenschild name plate * *an Maschine, Gerät usw.* || name-plate * *an Maschine, Gerät usw.* || sign-board || fascia
firmenspezifisch company-specific || in-house * *proprietary* * *siehe auch →firmeneigen*
firmenübergreifend open * *offen(gelegt) z.B. Bussystem/protokoll* || non-proprietary * *nicht firmeneigen/herstellerspezifisch* || manufacturer-neutral * *z.B. offenes Bussystem* || company-neutral * *firmenneutral*
Firmenvertretung representative office || sales agency * *Handelsvertretung, →Industrievertretung*
Firmware firmware * *in einem Festwertspeicher (z.B. EPROM) hinterlegte Software*
Firmwarestand firmware release * *Ausgabestand einer in Festwertspeicher hinterlegten Software* || firmware version
fixieren secure * *mechanisch sichern*
Fixkosten overheads * *Gemeinkosten*
flach flat || low-profile * *flach bauend* || slim * *schlank, flach* || slim-line * *schlank/flach bauend* || book-size * *flach/schmal bauend wie ein Buch*
Flach- low-profile
Flachanschluss flat connector * *gelochtes ~blech zum Anschluss der Leitung, z.B. bei Trafo, Drossel usw.* || flat connector * *Flach(steck)verbinder*
Flachbandkabel ribbon cable
Flachbandleitung ribbon cable
Flachbandstecker ribbon cable connector
Flachbauform flat type || low-profile type of construction || flat design
Flachbaugruppe board || PCB || printed circuit board || card || p.c. board || card module || PC board || printed-circuit module * *Leiterkarte*
Flachbildschirm flat-screen || flatscreen
Fläche area * *(begrenzte) Fläche, ~ninhalt, Ober/Grund~* || space * *Raum, Platz, Stelle, Zwischenraum* || surface * *Ober~; machined surface: bearbeitete ~; mating surfaces: aufeinander arbeit. ~n* || face * *Stirnfläche, Hammer-/Ambossbahn usw.* || surface area * *(geometr.) Ober~*
Flacheisen flat iron * *flat iron section: ~profil* || flat bar || rectangular bars || flats
Flächendruck unit pressure * *Druck pro Flächeneinheit, z.B. in einem Gleitlager* || surface pressure
Flächeneinheit unit area * *energy per unit area: Energie pro ~*
Flächengewicht mass per unit area || basis weight * *Papier* || grammage * *Flächenmasse, z.B. von Papier [g/m^3]*
Flächenintegral surface integral
Flächenisolierstoff insulating sheet * *z.B. als*

Flächenisolierstoffe

Wicklungsisolation für el.Maschine, Trafo, Drossel || insulating sheeting material || wrapper insulation material * z.B. zur Isolation von Wicklungen für Motor, Transformator usw.
Flächenisolierstoffe insulation sheets
Flächenmasse mass per unit area || grammage * z.b. von Papier [g/m^3]
Flächenpressung contact pressure * siehe auch →Spannkraft
Flächenschwerpunkt center of area
Flachglas float glass * Glasschmelze wird über ein Bett aus flüss.Zinn (>600°C) geleitet u. härtet dort aus || flat glass || plate glass
flächig flat
Flachkabel ribbon cable
Flachkopfschraube flat-headed screw || flat-head srew
Flachlederriemen flat leather belt
Flachleiter flat core * {Kabel}
Flachleitung ribbon cable || flat cable
Flachleitungs-Steckverbinder ribbon cable connector || ribbon-cable flat connector
Flachmotor flat-frame motor * {el.Masch.} || pancake motor * {el.Masch.} || face-mounting motor * {el.Masch} || flattened motor * {el.Masch.}
Flachriemen flat belt
Flachschleifmaschine surface grinding machine * {Werkz.masch.} || flat grinding machine * {Werkz. masch.}
Flachstecker tab connector || FASTON tab connector || FASTON terminal * Faston-Flachsteckeranschluss/Verbindung || tab * Messerkontakt-Steckzunge || FASTON tab * Faston-Steckzunge || FASTON tab receptacle * weiblicher Faston-Stecker, Flachsteckhülse || tab receptacle * "weiblicher" Flachstecker, Flachsteckhülse || flat connector * für Flachbandleitung || flat plug
Flachsteckhülse tab receptacle * Gegenstück zur Flachstecker-Messerzunge || FASTON tab receptacle * 'Weiblicher' Faston-Stecker
Flachsteckkontakt tab connector * Flachstecker || blade-type terminal
Flachsteckverbinder fast-on connector || flat connector || ribbon cable connector * für Flachleitung
Flachstelle flat spot
Flachwalzwerk strip mill * zur Erzeugung von Metallband
Flachzange flat pliers || flat nose pliers
flackern flicker * z.B. Licht, Lampe
flammbeständig flame resistant
flammhemmend flame-retardant
Flammpunkt flash point * Temperatur, bei der eine Substanz (z.B. Luft-Lösungsmittelgemisch) zündfähig wird
Flammschutzmittel flame retardant
flammwidrig flame resistant
Flammwidrigkeit flame retardance
Flanke edge * eines Impulses || slope * schräg abfallende/steigende || flank * Flanke (auch von Getriebezähnen), Weiche (bei Tier usw.), Seite, Flügel (beim Militär usw.)
Flankenabstand distance between signal edges * von (Impuls-) Signalen || edge displacement * von (Impuls-) Signalen || edge clearance * Zwischenraum zwischen den (Impuls-) Flanken
Flankenauswertung edge evaluation * z.B. von Impulsgebersignalen || edge detection * Erkennung von Impulsflanken, z.B. von e. Impulsgeber
flankengetriggert edge-triggered
Flankenmerker edge trigger flag * bei SPS
Flankenmodulation pulse-edge modulation * z.B. b.Wechselrichter (Pulsweitenmodul. mit nichtäquidistanten 0>1-Flanken)
Flankenspiel backlash * im Getriebe zw. Zahnflanken

Flankensteilheit rate of voltage rise * eines Spannungsverlaufes
Flankenwechsel edge change
Flansch flange * (bei Elektromotoren: siehe DIN 42948) || muff * Muffe, Stutzen, Flanschstück
Flanschausführung flange-mounted design * z.B. Motor mit Flanschbefestigung (i.Ggs.z. Fußausführung) || flange design || flanged type of construction * z.b. Motor mit Flansch- (statt Fuß-) Befestigung
Flanschbauform flanged type of construction * z.B. Motorbauart mit Flanschbefestigung || flange-mounting design
Flanschfläche flange face * z.B. an Motor-Wellenseite
Flanschgehäuse flange housing * Gehäusebauform z.B. einer Kupplung
Flanschgenauigkeit flange accuracy * für Elektromotoren: siehe DIN 42955
Flanschlagerschild flange bearing end shield * z.B. bei Motor || flanged endshield || flange endshield
Flanschmontage flange mounting * z.B. eines Motors (im Gegensatz zur Fußmontage)
Flanschmotor flange-mounted motor
Flanschnabe flange hub
Flanschplatte shovel plate * in Form einer Schaufel, z.B. zur Verbindung v. Motor und Getriebe
Flanschverbindung flanged joint
Flanschwanne flange pan || flange shovel * schaufelförmige Flanschplatte z.B. zur Verbindung von Motorfuß u.Getriebe || shovel plate * schaufelförmige Flanschplatte
Flanschzentrierrand flange centering rim
Flanschzentrierung flange centering
Flaschenabfüllanlage bottle filling machine || bottling line
Flaschenabfüllmaschine bottle-filling machine || bottle filler
Flaschenhals bottleneck * (auch figürlich: Engstelle, kritischer Weg, den Durchsatz bestimmendes Element)
Flaschenkasten bottle crate || bottle box
Flaschenreinigungsmaschine bottle cleaning machine || bottle rinsing machine
Flaschensortieranlage bottle sorting plant
Flaschenzug pulley block || block and tackle || electric chain hoist * mit elektr. Antrieb || trolley block * mit eingebauter Laufkatze || block || tackle || tackle block
Flash-Speicher flash memory * nichtflüchtiger Flash-'EPROM-Speicher
flattern flutter || wobble * z.B. Räder || jitter * z.B. Impulse || oscillate * schwingen || chatter * klappern, z.B. Relais || rattle * rattern, klappern (z.B.Relais), rasseln, rütteln
Flechtmaschine braiding machine * {Textil} || netting machine * zur Herstellung von Drahtgeflecht || plaiting machine * z.B. zur Herstellung von Moniereisen-Geflechtmatten
fleckig stained || blotted * mit Schmierflecken || spotted * mit Schmutz, Rostflecken, Pusteln, Pickeln befleckt/getüpfelt/gesprenkelt
flexibel flexible
Flexibilität flexibility
flexible Kupplung flexible coupling
Flexodruckmaschine flexographic printing machine
Flexodruckwalze flexographic roller * {Druckerei}
fliegen fly || go by plane * per Flugzeug reisen
fliegend on the fly * während des Betriebes, ohne anzuhalten || hot * unter Spannung; hot plugging: Ziehen/Stecken unt. Spanng.; hot fixing: Reparatur u.Spanng. || flying * fliegend, flüchtig, kurz, be-

weglich; flying (re-) start: fliegender (Neu-) Start || overhung * *fliegend angeordnet (Montage, Lagerung usw.), freitragend, überhängend*
fliegend angeordnet in overhung position
Flaschenreinigungsmaschine bottle cleaning machine || bottle rinsing machine
Flaschensortieranlage bottle sorting plant
Flaschenzug pulley block || block and tackle || electric chain hoist * *mit elektr. Antrieb* || trolley block * *mit eingebauter Laufkatze* || block || tackle || tackle block
Flash-Speicher flash memory * *nichtflüchtiger Flash-'EPROM-Speicher*
flattern flutter || wobble * *z.B. Räder* || jitter * *z.B. Impulse* || oscillate * *schwingen* || chatter * *klappern, z.B. Relais* || rattle * *rattern, klappern (z.B.Relais), rasseln, rütteln*
Flechtmaschine braiding machine * *{Textil}* || netting machine * *z.B. zur Herstellung von Drahtgeflecht* || plaiting machine * *z.B. zur Herstellung von Moniereisen-Geflechtmatten*
fleckig stained || blotted * *mit Schmierflecken* || spotted * *mit Schmutz, Rostflecken, Pusteln, Pickeln befleckt/getüpfelt/gesprenkelt*
flexibel flexible
Flexibilität flexibility
flexible Kupplung flexible coupling
Flexodruckmaschine flexographic printing machine
Flexodruckwalze flexographic roller * *{Druckerei}*
fliegen fly || go by plane * *per Flugzeug reisen*
fliegend on the fly * *während des Betriebes, ohne anzuhalten* || hot * *unter Spannung; hot plugging: Ziehen/Stecken unt.Spanng.; hot fixing: Reparatur u. Spanng.* || flying * *fliegend, flüchtig, kurz, beweglich; flying (re-) start: fliegender (Neu-) Start* || overhung * *fliegend angeordnet (Montage, Lagerung usw.), freitragend, überhängend*
fliegend angeordnet in overhung position
fliegende Säge flying saw * *läuft während des Sägevorgangs mit dem Material mit*
fliegende Schere flying shears * *läuft während des Schneidvorgangs mit d. Material mit (Materialfluss stoppt nicht)*
fliegende Umschaltung flying changeover
fliegender Anbau overhung mounting
fliegender Master flying master * *z.B. bei Multiprozessor- oder Feldbussystem*
fliegender Rollenwechsel flying splice
fliegender Start flying start
fliegender Wechsel hot swapping * *z.B. fliegender Tausch von Baugruppen während des Betriebes/ unter Spannung*
Fliehkraft centrifugal force || flywheel effect
Fliehkraftabhebung centrifugal lift * *z.B. bei Klemmzahn-'Rücklaufsperre/-'Freilauf*
Fliehkraftkupplung centrifugal clutch || dry-fluid coupling
Fliehkraftregler centrifugal switch
Fliehkraftschalter centrifugal switch * *centrifugal starting switch: ~ z. Abschalten e. Anlas-Hilfswicklung bei AC-Motor* || centrifugal contactor || centrifugal overspeed switch * *Überdrehzahlschalter* || overspeed switch * *Überdrehzahlauslöser/-schalter*
Fließanlage fleecing system * *{Textil}*
Fließband belt conveyor * *Förderband* || conveyor belt * *Förderband* || assembly line * *come off the assembly line: vom Band laufen* || production line || manufacturing line
Fließbild flow diagram || flow chart
fließen flow * *Strom, Programm*
Fließfett semi-fluid grease
Fließkomma floating point
Fließkommaarithmetik floating point arithmetic

Flussregler

flimmerfrei flicker-free * *Bildschirm, Monitor*
flimmern flicker * *eines Bildschirms*
flink quick-acting * *Sicherung* || fast-acting * *Sicherung* || fast
Floatglas float glass * *siehe auch →Flachglas*
Flockung flocculation * *{Physik}*
Flor nap * *{Textil} Gewebeflor in der Weberei*
Flotation flotation || floating || floatation
fluchten align * *in e. gerade Linie bringen, ausrichten, ausfluchten* || be aligned * *in Linie gebracht/ausgerichtet sein* || be in line * *in einer Linie/Achse ausgerichtet sein*
fluchtend be aligned * *in Linie gebracht/ausgerichtet sein* || in line * *in einer Linie/Achse ausgerichtet*
Fluchtfehler misalignment * *Fehler in der Ausrichtung*
flüchtig transient * *vorübergehend* || volatile * *z.B. Speicher*
Fluchtstellung alignment * *Ausrichtung (z.B. einer Welle)* || aligned position
Fluchtungsfehler misalignment
Flug Nr plane No. || flight No.
Flügel wing * *auch eines Gebäudes; lend wings to a person: jdm. ~ verleihen* || blade * *z.B. eines Lüfters* || vane * *z.B. eines Lüfters* || airfoil * *eines Flugzeugpropellers* || casement * *eines Fensters* || leaf * *einer Tür* || aisle ⊥ *eines Gebäudes*
Flügelmutter wing nut
Flügelrad impeller * *z.B. eines Lüfters*
Flughafen airport
Flugwesen aviation
Flugzeug airplane
Flurförderzeug in-house vehicle || ground conveyor * *driverless/automated-guided: fahrerlos* || stock conveyor * *Lagerförderzeug*
flurgesteuert ground-controlled * *{Hebezeug}*
Fluss flux * *magnetischer* || flow * *von Signalen/ Daten/Programm, Durchfluss, Fließen*
Fluss-Stromregelung flux current control || FCC * *Flux Current Control*
Flussabsenkung flux reduction
Flussbremsung flux braking
Flussdiagramm flow chart || flow diagram
Flussdichte flux density * *magnetic: magnetische; electric: elektrische*
flüssig liquid || non-solid * *als Gegensatz zu "fest"*
Flüssiggas liquid gas
Flüssigkeit liquid || fluid
Flüssigkeitsabscheider liquid separator
Flüssigkeitsanlasser liquid-resistor starter * *Flüssigk.-Anlasswiderst. f.Großmotor z.B.mit Sodalauge u.Verstellelektroden*
flüssigkeitsgekühlt liquid-cooled
Flüssigkeitskristall-Anzeige liquid crystal display || LCD * *Abkürzg. für 'liquid crystal display'*
Flüssigkeitskristallanzeige liquid-crystal display || LCD * *Abkürzg. für 'Liquid Crystal Display'*
Flüssigkeitskühler liquid cooler
Flüssigkeitskühlung liquid cooling
Flüssigkeitskupplung fluid coupling
Flüssigkeitsläufer wet-rotor * *Nassläufer (bei Pumpe)*
Flüssigkeitspumpe liquid pump
Flüssigkeitsreibung fluid friction
Flüssigkeitsring-Vakuumpumpe liquid-ring vacuum pump
Flüssigkeitsstand liquid level
Flüssigkeitswiderstand liquid resistor * *mit Flüssigkeit als Widerstandsmedium, z.B.Wasser* || liquid resistor * *mit Flüssigkeit als Widerstandsmedium, z.B.Wasser*
Flusskomponente field component
Flussmittel soldering flux || flux
Flussregler flux controller

Flussrichtung forward direction * *bei Halbleitern*
Flussvektor flux vector
flüsterleise whisper quiet
FM FM * *Frequenzmodulation* || frequency modulation
FMS FMS * *Abkürzg.für 'Field bus Message Specification': Busprotokoll, b. PROFIBUS genormt (DIN 19245,T.1)*
Föhn hair drier
Folge sequence * *['siequens] Aufeinander~, Ab~* || succession * *Aufeinander~; continuous: ununterbrochene/fortlaufende* || consequence * *Wirkung, logische ~, ~erscheinung* || result * *Resultat* || set * *~ von zusammengehörigen Teilen* || sequel * *~zeit; in the sequel: in der ~; ~erscheinung, (Aus)Wirkung, (Aufeinander-) Folge* || future * *Zukunft* || effect * *Wirkung* || after-effect * *Nachwirkung, ~erscheinung* || aftermath * *ernste ~* || order * *Reihen~* || series * *Reihe, Serie*
Folge- slave * *(Gegensatz: 'master': Leit-)* || subsequent * *nachfolgend*
Folgeachse slave axis
Folgeantrieb slave drive || speed follower * *erhält seinen Drehzahlsollwert vom Leitantrieb* || following drive || follower * *z.B. bei Leit-Folgebetrieb, Lastausgleichsregelung usw.*
Folgeauftrag future order || follow-up order * *Anschlussauftrag*
Folgeerscheinung consequence || result || after-effect || sequel || consecutive effect || consecutive symptom * *Symptom* || side effect * *Nebeneffekt* || secondary phenomenon * *Nebeneffekt* || side benefit * *vorteilhafter Nebeneffekt* || unintentional effect * *ungewollter Effekt, unbeabsichtigte Folge*
Folgefehler secondary fault || consecutive fault || sequential errors * *(Plural)*
Folgefrequenz repetition frequency || recurrence frequency || repetition rate
Folgekosten secondary costs
folgen follow * *auch logisch*
folgend following || subsequent * *nach~, später, nachträgl., Nach-, i. Anschluss an; subs'ly described: i. ~den beschrieben* || subsequently listed * *nach~ aufgeführt (in e. Aufzählung, Liste)*
folgendermaßen as follows
folgendes gilt the following is valid
folgern conclude || deduce || infer || gather * *from: aus*
Folgerung conclusion * *draw a conclusion: eine ~ ziehen* || implication * *stillschweigende ~* || inference * *['inferens] (Schluß-) ~, (Rück-) Schluss; make inferences: Schlüsse ziehen* || deduction
Folgeschaden secondary damage
Folgestart serial start-up * *→zeitlich versetztes Einschalten*
Folgesteuerung sequential control || sequential gating * *{Stromrichter}*
Folgetelegramm response message * *Antworttelegramm* || subsequent telegram
Folgeventil oncoming valve * *das nächste zu zündende Ventil in e. Stromrichter*
folglich consequently || hence * *daher, deshalb, also*
Folie film * *meist aus Kunststoff* || foil * *meist aus Metall*
Foliendruckmaschine film printing press * *für Kunststofffolie* || foil printing press * *für Metallfolie*
Folieneinschweißmaschine foil sealing machine
Folienherstellung plastic film production * *Kunststofffolie* || film production
Folienindustrie plastic film industry * *Kunststofffolie*
Folienmaschine plastic film machine * *für Kunststofffolie* || foil producing machine || film manufacturing line || foil machine

Folienproduktion plastic film production || film production
Folienreckanlage foil stretching machine
Foliensatz set of overhead transparencies || collection of overhead slides
Folienschrumpfverpackung shrink-film wrapping || shrink wrapping
Folienschweißmaschine foil sealing machine
Folientastatur touch pad keyboard || touch pad panel * *Tastenfeld* || membrane-type keyboard || membrane keyboard * *sealed: hermetisch dicht* || membrane keypad * *kleines Tastenfeld* || sealed membrane keyboard * *mit hoher Schutzart* || membrane panel || foil-covered keyboard
Folienwickler stretch-wrapping machine * *{Verpack.techn.} z.Verpacken v. Gegenständen, Paletten usw. i.Stretchfolie* || film winding machine * *Auf-/Abwickler für Kunststofffolie*
Foliermaschine foiling machine * *{Verpack.techn}*
forcierte Luftkühlung forced air cooling * *z.B. bei Leistungshalbleitern*
Förderanlage mine winder * *Schachtförderanlage für Bergwerk* || winding machine * *Haspel für Bergwerk* || conveyor system || conveyor || haulage system
Förderantrieb conveyor drive
Förderband conveyor belt || conveyor * *Förderzeug* || belt conveyor || band-conveyor || moving belt || band conveyor
Förderberg inclined conveyor * *{Bergbau} Schrägfördereinrichtung/Rollenbahn im Bergbau*
Förderbergband inclined conveyor
Fördereinrichtung conveyor system * *Transport/Fördersystem* || conveyor equipment || conveyor system
Förderer conveyor * *Förderzeug, Materialförderer* || backer * *Gönner* || patron * *Gönner, Förderer, Schirmherr*
Förderhöhe head * *einer Pumpe; head of water: Wassersäule* || delivery head * *einer Pumpe* || suction lift * *Ansaughöhe einer Pumpe* || pumping head * *einer Pumpe*
Förderkapazität feed rate * *siehe auch →Förderleistung* || conveyor capacity * *eines Fördersystems* || delivery rate * *(pumpe, Lüfter usw.)* || volumetric capacity * *z.B. eines Lüfters* || pump delivery capacity * *z.B. eines Lüfters* || pump delivery * *einer Pumpe*
Förderkette chain conveyor
Förderkorb mine cage || drawing cage
Förderleistung delivery rate * *einer Pumpe/eines Ventilators; siehe auch →Förderkapazität* || conveying capacity * *eines Förderzeugs* || elevating capacity * *eines Vertikalförderers/Aufzugs* || feed performance || delivery * *einer Pumpe* || flow rate * *z.B. eines Lüfters* || haulage capacity * *max. Förderleistung z.B. einer Bandförderanlage {Bergwerk}* || conveying rate * *eines Fördersystems* || feed rate || lift capacity * *{Aufzüge}*
förderlich conducive * *[kondju: siw] dienlich, förderlich (to: für)* || promotive * *of: für* || useful * *nützlich* || profitable * *vorteilhaft, nützlich (to: für); Gewinn bringend, einträglich, lohnend, rentabel* || effective * *wirksam* || beneficial * *heilsam* || healthy * *gesund (auch figürl.), heilsam, förderlich, fehlerfrei funktionierend*
Fördermaschine conveyor belt * *Förderband* || mine hoist * *Bergwerks-Förderaufzug* || conveyor * *Förderer* || mine winder * *Haspelmaschine für Bergwerk* || winding machine * *im Bergbau*
Fördermedium pumped medium * *(Plural: "media") durch eine Pumpe gefördertes ~*
Fördermenge flow rate * *z.B. eines Lüfters* || volumetric capacity * *z.B. Pumpe*
fordern demand || require || call for || claim * *rechtlich ~/beanspruchen* || charge * *Preis ~*
fördern convey * *Material transportieren, z.B.*

durch ein Förderzeug || route * *be~, leiten, weiterleiten* || transport * *transportieren* || promote * *Vorhaben/Plan/berufliches Fortkommen usw. unterstützen* || advocate * *befürworten, eintreten für* || make easy * *erleichtern* || facilitate * *erleichtern* || deliver * *z.B. Gase, Flüssigkeit durch Pumpe* ~ || charge * *in einen Behälter hinein~* || discharge * *aus einem Behälter heraus~* || feed * *zuführen* || patronize * *als Gönner* ~ || support * *unterstützen* || speed up * *beschleunigen* || back up * *den Rücken stärken*
Förderschnecke spiral conveyor
Förderstrom medium flow || flow medium
Fördersystem conveying system * *z.B. zum Gütertransport* || conveyor system
Fördertechnik conveyor technology
Forderung demand * *for: nach; on: an* || call * *for: nach* || claim * *Anpruch (for: auf)* || debt claim * *Schuldforderung* || title to a debt * *Schuldforderung* || requirement * *for: nach; auch 'Anforderung'*
Förderung conveyance * *Transport, Beförderung, Überbringung, Zufuhr* || promotion * *Verkaufsförderung, Werbung, Befürwortung, Unterstützung, berufliche (Be)Förderung* || encouragement * *Ermutigung* || support * *Unterstützung* || dispatch * *Absendung, Versand, Abfertigung, Beförderung, schnelle Erledigung* || drawing * *von Bodenschätzen* || extraction * *von Bodenschätzen* || hauling * *Ziehen, Schleppen, Transport, auch Fördern von Bodenschätzen* || haulage * *Ziehen, Schleppen, Transport, auch Fördern von Bodenschätzen* || transport * *Transport, Beförderung* || assistance * *Unterstützung, Mithilfe* || transport * *Transport* || movement * *Bewegung, z.B. von Luft durch e. Ventilator* || delivery * *(Be-)Förderung (durch Pumpe usw.), (Zu-)Leitung, Zuführung, Aus-/Übergabe, Lieferung*
Förderzeug conveyor
Form type * *Art, Typ* || form * *Gestalt, Förmlichkeit; pro forma: der ~ halber; be off form: nicht in ~ sein* || shape * *Gestalt, Form (auch e. Impulses); take shape: Form annehmen; be in bad shape: nicht i.form sein* || appearance * *(äußeres) Erscheinung(sbild); (to) keep up appearances: die Form wahren/der Form halber* || figure * *Figur, (geometr. Körper)Form, Gestalt, Aussehen, Erscheinung, Charakter* || style || design || model || profile * *Profil* || mould * *Guss~, Press~* || mold * *(amerikan.) Guss-/Press~* || die * *Spritzguss~, Strangpress~* || mode * *Art und Weise* || manner * *Art und Weise* || condition * *(z.B. körperliche) Verfassung/Leistungsfähigkeit* || ceremony * *Förmlichkeit* || usage * *Brauch, Gepflogenheit, Usus* || notation * *{Mathem.} Darstellungsart/Schreibweise*
formal formal
Formaldehyd formaldehyde
formaler Parameter formal parameter || dummy parameter
formalisieren formalize
Formalitäten formalities
Formaloperand formal operand || assignable parameter
Formalparameter formal parameter || dummy parameter
Format format * *auch von Daten* || size * *Größe, z.B. eines Papierbogens (untrimmed size: Rohformat)* || sheet size * *eines Papierbogens usw.* || sheet length * *bei Querschneider* || format size || size
Formatbreite sheet width * *bei* →*Längsschneider*
formatieren format
formatiert formatted
Formation formation
Formatkreissäge sliding-table circular saw * *{Holz}* || panel sizing circular saw * *{Holz}*

Formatlänge sheet length * *bei* →*Querschneider*
formbeständig dimensionally stable
Formblatt form || blank form || blank
Formel formula * *using/according to the following formula: nach folgender* ~
Formeln formulae
Formelzeichen symbol * *für {el.Masch}: siehe z.B. DIN 1304 Teil 7*
Formfaktor form factor * *Effektiv-/Arithmet. Gleichricht-Mittelwert (1 bei Gleichspanng.; 1,11 bei sin-Spanng.)* || harmonic factor
formieren reform * *z.B. Elektrolytkondensator*
Formierung forming * *z.b. langsame Aufladung eines (Elektrolyt-) Kondensators nach langer Lagerung* || forming process * *Vorgang der* ~, *z.B. eines Elektrolytkondensators*
Formkreissäge sliding-table circular saw * *{Holz}*
formlos formless
Formmaschine forming machine * *z.B. für Pralinen*
Formpresse mould press * *{Holz} Formteilpresse* || shaping press * *{Holz}*
Formpressen molding * *(amerikan.)* || moulding * *(brit.)*
Formschluss closing shape || locking || non-slip engagement || rigid coupling * *starre Kopplung*
formschlüssig form-locking || form-closed || positive || non-slipping || non-slip || form-fit || keyed * *mit Federkeil (auf Welle) befestigt* || geared * *über Zahnräder verbunden* || no-slip || rigidly engaged || rigid * *starr*
Formschneiden scroll corner cutting
formschön pleasing the eye || eye-catching * *das Auge auf sich ziehend, auffällig*
Formspule preformed coil * *{el.Masch.}* || form-wound coil * *{el.Masch.}*
formstabil rigid
Formteil moulded part * *(Kunststoff, Metall usw.)* || shaped workpiece * *{Holz}* || shaped part * *{Holz}* || shaped component * *{Holz}*
Formular form || blank
Formulardruckmaschine business forms press || stationery printing machine * *(stationery: Schreib-/Papierwaren, Büro-/Schreibmaterial, Briefpapier)*
Formularvordruck printed form
Formveränderung forming * *bei Verarbeitungsmaschine*
Forschung research
Forschungsgebiet area of research
Forschungsinstitut research institute
Forschungslabor research laboratory || research lab
Forschungsstätte research facility
Forstwirtschaft forestry || forestry industry
fortfahren go ahead || resume * *wiederanfangen, wieder (an der alten Stelle) aufsetzen*
fortführen lead away * *weg bewegen* || remove * *entfernen* || go on with * *fortsetzen* || continue * *fortsetzen* || keep on * *fortsetzen* || resume * *wieder fortsetzen*
fortgeschritten advanced
Fortgeschrittenenkurs advanced training course
fortlaufend consecutive * *aufeinander folgend*
fortlaufende Nummern consecutive numbers
fortleiten lead away * *fortführen, hinwegführen* || propagate * *fortpflanzen (Ton, Licht usw.), aus-/verbreiten*
fortpflanzen propagate * *auch 'sich fortpflanzen, verbreiten, ausbreiten'*
Fortschaltbedingung step criterion * *(Plural: criteria) bei einer Ablaufsteuerung* || step enabling condition * *bei Schrittsteuerung*
Fortschaltkriterium step criterion * *(Plural: criteria) in Ablaufsteuerung*

fortschreiten advance || advancement * *(Substantiv)*
Fortschritt progress * *in:* in; *make progress:* Fortschritte machen || advance || advancement || improvement * *Verbesserung* || enhancement * *Verbesserungen* || headway * *make headway:* Fortschritte machen
Fortschritte machen advance || make headway
fortschrittlich advanced || progressive || modern * *modern* || up-to-date * *dem neuesten Stand entsprechend* || state-of the art * *modern, dem neuesten techn. Stand entsprechend* || innovative * *innovativ* || forward-looking
fortsetzen continue || pursue || proceed on || resume * *wieder fortsetzen nach Unterbrechung* || go ahead * *weitermachen*
Fortsetzung continuation || resumption * *Wieder~, Wiederaufnahme*
Fortsetzungsstart warm restart * *Warmstart*
Foto photo || photograph || picture * *take a picture of: fotografieren, ein Foto machen von*
Fotokopierer photo copier || copier
Fototransistor phototransistor
Fotowiderstand photoresistor
Fotozelle photocell
Fourier-Analyse Fourier analysis || harmonic analysis
Fourier-Reihe Fourier series
Fourier-Transformation Fourier transformation || Fourier transform || FFT * *Abk.f. 'Fast Fourier Transform': schnelle ~ z.b. mit Digitalrechner*
Fourier-Transformierte Fourier transform
FPGA FPGA * *'Field-Programmable Gate Array': On-Board/im Anwendersystem programmierbarer Logikbaustein*
Fracht freight || cargo || load || goods * *Frachtgut*
Frachtbrief bill of lading * *(amerikan.)* || waybill || consignment note
Frachtgüter freight goods || goods || cargo || ship load * *bei Schiffstransport*
Frachtverkehr goods traffic || freight transportation * *(amerikan.)* || freight traffic
Frage question * *ask/settle/pose a question: e.~ stellen/klären/aufwerfen; FAQ: →oft gestellte ~n* || inquiry * *Erkundigung* || problem * *Problem* || query * *(besonders 'anzweifelnde, unangenehme') Frage*
Fragebogen questionnaire
Fragesammlung FAQ * *Frequently Asked Questions: Kollektion häufig gestellter Fragen u. ihrer Anworten*
Fragezeichen question mark
Framegrabber framegrabber * *[Computer] Einrichtung z. Echtzeit-Digitalisierung u.Speicherung analoger Videosignale*
Frankreich France
französisch French
Fräseinheit routing unit * *[Holz]* || moulding unit * *[Holz] zum Profilieren*
fräsen mill || face * *~ von Stirnflächen* || shape * *form~* || cut || mortize * *[Holz] (amerikan.) ~ von Fugen/Nuten* || mortise * *[Holz] (brit.) ~ von Fugen/Nuten* || mould * *[Holz] ~ von Formteilen/Kehlen/Leisten usw.*
Fräser milling cutter * *[Werkz.masch.]* || moulding cutter * *[Holz]*
Fräser-Dorn milling arbor
Fräserdorn cutter arbor * *[Werkz.masch.]*
Fräsersatz cutter set * *[Holz]* || set of cutters * *[Holz]*
Fräskopf milling head || cutter block || cutterhead || milling cutter || inserted-blade milling cutter || router head * *[Holz] einer Oberfräse*
Fräsmaschine milling machine * *[Werkz.masch.]* || für Metall || moulding machine * *[Holz] Fräsma-*

schine, Kehl(hobel-/fräs-)maschine || shaper * *[Holz] Form~* || mortizer * *[Holz] Fugen-/Nut-~* || moulder * *[Holz]* || spindle moulder * *[Holz]*
Fräswerk milling machine
Fräszentrum milling center * *[Werkz.masch.]* || routing center * *[Holz]*
frei free * *from/of: von; free (of charge): unentgeltlich* || freely * *(Adv.)* || vacant * *Stelle, Leitung* || independent * *unabhängig* || exempt * *from: von Steuer, Zoll, Gebühren usw.* || frank * *freimütig, unumwunden* || open * *offen, freimütig* || unrestrained * *unbehindert* || unobstructed * *unversperrt, ungehindert, ohne Hindernisse* || clear * *Straße, von Schulden usw.; swing clearly: frei schwingen* || free and easy * *ungezwungen* || open * *Markt, Feld, Himmel, Stelle usw.* || freelance * *freiberuflich* || off * *off-day: freier Tag* || plainly * *(Adv.) unumwunden* || bluntly * *(Adv.) unverblümt* || voluntarily * *(Adv.) freiwillig* || spontaneously * *(Adv.) von selbst, unwillkürlich, selbsttätig*
frei belegbar freely assignable * *z.B. Tasten, Ein/Ausgangsklemmen mit einer Funktion* || user-definable * *z.B. Tasten*
frei beschaltbar with free components * *z.B. mit Bauelementen auf Leiterkarte ("Spielwiese")*
frei herausgeführt accessible * *with accessible cable ends: mit ~n Leitungen*
frei herausgeführtes Kabel motor-fixed cable * *[el.Masch.] aus d.Motor herausgeführte Wicklgs. enden; kein Klemmenkasten*
frei parametrierbar sein can be freely parameterized
frei projektierbar freely configurable * *the control system can be freely configured: d. Regelsystem ist frei projektierb.* || user-configurable * *vom Anwender projektierbar* || freely-programmable
frei verschaltbar freely interconnectable || freely configurable * *frei konfigurierbar/projektierbar*
frei- freely
Freiberufler freelance || self-employed person || consultant * *freier Berater*
freiberuflich freelance * *work freelance: freiberufl. tätig sein* || self-employed * *be self-employed: freiberufl. tätig sein* || in private praxis * *Arzt, Rechtsanwalt usw.*
freibleibend not binding * *z.B. Preise, Kurse* || subject to change * *Änderungen unterworfen*
freidefinierbar freely-definable || freely-assignable * *frei zuzuordnen (z.B. eine Funktion zu einer Klemme)*
Freie outdoor * *das Freie, im Freien*
freie Konvektion natural convection || natural air circulation * *z.B. bei Luftselbstkühlung von Leistungshalbleitern*
freie Kühlung natural cooling * *→Konvektionskühlung*
freie Länge free length
freie Marktwirtschaft free market economy || free enterprise || free economy || free competitive system
freier Mitarbeiter freelance * *Freiberufler* || consultant * *Berater, Spezialist*
freies Wellenende free shaft end * *z.B. ~ an der Nicht-Antriebsseite eines Motors (z.Anbau von Bremse/Handrad ...)*
Freiformfläche sculptured surface * *z.B. zur Bearbeitung mit Werkzeugmaschine*
Freigabe enable * *(Verb) als Kommando* || unblocking || permit * *Erlaubnis, auch Aus-/Einfuhrerlaubnis* || release * *z.B. eines Produktes zur Fertigung oder Lieferung; durch Schutzvorrichtung* || enabling * *das Freigeben* || approval * *Genehmigung, Anerkennung*
Freigabesignal enable signal
Freigabeuntersuchung conformance test * *zur Si-*

cherstellung der Übereinstimmung mit einer Norm/Vorschrift || approval test * Abnahmeprüfung || pre-release tests * vor der Freigabe durchgeführte Tests
freigeben enable * eine Funktion, ein Signal, einen Regler ~ || release * ein Produkt ~; for sale: zum Vertrieb || unlock * Sperre aufheben
freigegeben enabled
freigestellt optional
freihalten keep free * from: von || keep clear * z.B. Straße || keep open * z.B. Angebot
Freihaltezeit hold-off interval
Freiheit freedom * from: von; degree of freedom: →Freiheitsgrad || exemption * from: von Lasten usw. || liberty * politische/persönliche ~; take the liberty of: sich die ~ nehmen zu || scope * Spielraum || latitude * Spielraum || absence * Nicht-Vorhandensein, Abwesenheit
Freiheitsgrad degree of freedom
Freilauf freewheel || coasting * das Freilaufen, Trudeln, ungeführter/nicht angetriebener Lauf || overrunning clutch || freewheel clutch || freewheeling * im Freilauf fahrend || one-way clutch || sprag freewheel * Klemmkeil-/Klemmkörper-Freilauf || roller freewheel * Klemmrollen-Freilauf
Freilaufdiode free-wheeling diode * z.B. bei halbgesteuerter Thyristorbrücke || surge absorber diode * z.B. parallel zu Relaisspule, vermeidet Überspanng.b.Schalten induktiver Last || bypass diode || freewheel diode
freilaufend free-wheeling || coasting || asychronous * asynchron
Freilaufkupplung freewheel clutch
Freilaufstrom free-wheeling current
Freilaufventil free-wheeling valve * meist Diode in einem Freilaufzweig
Freilaufzweig free-wheeling arm || bypass arm || free-wheeling path
Freileitung overhead line || overhead power transmission line || electric power line || open line
Freiluft- outdoor- * outdoors: im Freien || outdoors * for outdoor use: für ~aufstellung/Verwendung im Freien || overhead * Leitung
Freiluftanlage outdoor plant
Freimaß free size * ohne Toleranzangabe || dimension with unspecified tolerance * Maßangabe ohne Toleranz
freiprogrammierbar freely-programmable * z.B. Funktion einer Klemme
freiprojektierbar user-configurable || freely configurable * z.B. Regelsystem SIEMENS SIMADYN D (R)
Freiraum clearance * unobstructed: unbehindert; z.B.erforderlich unter- u.oberhalb e.Stromrichtergeräts || free space * freier Raum || space that is kept free * freigehaltener Raum || private area * (figürlich) || reserved space * reservierter/freigehaltener Platz; Platz, der nicht belegt werden darf || headroom * lichte Höhe, Platz für Erweiterungen
freischalten disconnect * Stromkreis ~ || isolate * Stromkreis ~; siehe VDE 0100 Teil 2 || isolate from the supply * vom Netz trennen || de-energize || unlock * Software ~ (z.B. durch Eingabe eines Freischaltcodes) || enter the license code * Software ~ (z.B. durch Eingabe der Registriernummer)
Freischaltmöglichkeit isolating facility * Auftrennbarkeit des Stromkreises
Freischaltung disconnection || safety disconnection * VDE 0100 Teil 200 || safety isolation * VDE 0100 Teil 200 || disconnect
Freistich relief groove * {Werkz.masch.} "Hinterdrehung" an Drehteil || undercut * {Werkz.masch.}

mit Drehmaschine, an Drehteil || relieving cut * mit Drehmaschine
freitragend self-supporting
freitragende Konstruktion cantilevered construction || self-supporting construction
Freiwerdezeit recovery time * braucht ein Thyristor beim Umkommutieren,um richtig zu sperren || turn-off time * Abschaltzeit eines Thyristors
Freizeit leisure time || free time || spare time || time off
freizügig free * frei, zwanglos, ungezwungen, ungebunden, nach Belieben, unbeengt, unbelastet || freely * (Adv.) || unhampered || free to move || unrestricted || generous * großzügig
fremd strange || external * extern || extraneous * nicht dazugehörig || unknown * unbekannt || unfamiliar * ungewohnt, nicht vertraut || third-party * von einem anderen Hersteller/Fremdhersteller
Fremd- external || externally * (Adv.) || separate || outside || parasitic * parasitär (Strom usw.)
fremdangetrieben externally driven
fremdbelüftet separately-ventilated * Motorkühlart; with duct connection: mit Rohranschluss || forced-ventilated * z.B. Motorkühlart mit getrennt angetriebenem Lüfter || separately cooled * fremdgekühlt || force-ventilated || externally ventilated || duct-ventilated * mit Kühlluftzufuhr/-abfuhr über Rohranschluss || separately fan-ventilated || forcecooled
fremdbelüftet über Rohranschluss duct-ventilated * z.B. Motorkühlarten IC05/06/17/37 nach IEC34 Teil 6 || separately-ventilated with pipe-connection || pipe-ventilated
Fremdbelüftung forced ventilation * z.B. Motor || separate ventilation * with duct connection: mit Luftzu-/abfuhr über Rohranschluss || forced-air cooling || external ventilation
Fremdbelüftung über Rohranschluss duct-ventilation || separate ventilation with duct-connection * z.B. Motorkühlarten IC05/06/17/37 nach IEC 34 Teil 6 || separate ventilation with pipe-connection
Fremdeinspeisung external supply
fremderregt separately excited * z.B. Gleichstrommaschine ohne Permanenterregung || separatefield
fremderregte Gleichstrommaschine separately-excited DC machine
fremderregter Gleichstrommotor separately-excited DC motor
Fremderregung separate-field excitation || separate field || separate excitation
Fremdfabrikat product from another manufacturer || third-party product || outside make
Fremdfertigung outside manufacture || outside production || manufacturing by a subcontractor * von Unterlieferant gefertigt
Fremdfirma other company || other manufacturer || third party
Fremdführung external commutation * für Stromrichterschaltung
fremdgefertigt manufactured by a subcontractor * durch einen Unterlieferanten
fremdgeführt externally commutated * Stromrichter; netz- oder lastgeführt i.Ggs.zu selbstgeführt
fremdgeführter Stromrichter externally commutated converter * netz- oder lastgeführt i.Ggs.zu selbstgeführt
fremdgekühlt separately cooled * Motorkühlg. durch getrennt angetriebenen Lüfter (Kühlwirkg.unabh.von Motordrehz.) || separately ventilated * ~fremdbelüftet || forced-ventilated * Motorkühlung durch getrennt angetriebenen Lüfter || forcecooled || forced-cooled || force-cooling
fremdgekühlt über Rohranschluss ventilated by duct system

Fremdgeräusch   background noise
Fremdhersteller   other manufacturer || third-party manufacturer
Fremdkapital   borrowed capital || capital from outside sources || outside capital || outside funds
Fremdkörper   foreign body || foreign bodies * *(Plural)* || foreign matter || foreign particles * *(Plural)* || foreign objects * *protection from ingress of foreign objects: Schutz gegen Eindringen von ~n* || solid foreign bodies * *(Plural) körnförmige/feste ~*
Fremdkühlung   separate cooling * *z.B. Motor durch fremdangetriebenen Lüfter oder Flüssigkeitskühlg* || separate ventilation * *Motorkühlart m.Luftkühlung (z.B. durch nicht v.d.Motorwelle angetrieb. Lüfter)* || forced cooling || forced-air cooling * *Fremdbelüftung über fremdangetriebenen Lüfter oder Rohranschluss* || duct-ventilation * *Motorkühlart mit Luftkühlung über Rohranschluss*
Fremdlüfter   external fan || external blower || separately-driven fan * *fremdangetriebener Lüfter (sitzt nicht auf Motorwelle)* || separate fan || separate fan unit * *inklusive Lüfterzusatzaggregate (Antriebsmotor, Filter usw.)* || independent blower || externally-driven fan || externally-mounted fan
Fremdlüftermotor   separately-driven fan motor * *Motor mit Fremdlüfter*
Fremdrechner   computer from other manufacturers || third-party computer || foreign computer
Fremdschwingungen   external vibrations * *(mechan.)*
Fremdspannung   interference voltage || parasitic voltage || noise voltage * *Störspannung* || disturbing voltage * *Störspannung*
Fremdspannungen   parasitic voltages
Fremdsprache   foreign language
Fremdsysteme   systems from other manufacturers
Fremdträgheitsmoment   external moment of inertia * *z.B. Trägheitsmoment der Last/der Arbeitsmaschine bei e. Motor*
Frequenz   frequency * *Schwingungen, Impulse* || rate * *Häufigkeit des Auftretens von Ereignissen*
Frequenz erreicht   at frequency || frequency arrival * *Substantiv*
Frequenz-Ausblendband   critical speed rejection band * *siehe 'Frequenzausblendband'*
Frequenz-Spannungs-Kennlinie   voltage/frequency curve || voltage/frequency characteristic
Frequenz-Spannungs-Wandlung   frequency-voltage conversion
Frequenz-Spannungswandler   frequency to voltage converter
frequenzabhängig   frequency-related || frequency-dependent
frequenzabhängige Größe   frequency-dependent variable
Frequenzänderung   frequency change || frequency variation || change in frequency
Frequenzanlauf   synchronous starting * *{el. Masch.} Anlassen eines Asynchronmotors mit stetig wachsender Frequenz*
Frequenzausblendband   critical speed rejection band * *Ausblendung von (mechan.) Resonanzfrequenzen im Sollwert* || skip frequency zone || frequency avoidance zone || frequency jumping band || critical frequency lockout band
Frequenzausblendung   frequency jumping * *Ausblendung kritischer (mechan.) Resonanzfrequenzen im Sollwert* || frequency skipping || frequency avoidance || critical speed rejection
Frequenzausgang   frequency output
Frequenzausklammerung   frequency jumping * *Ausblendung (mechan.) Resonanzfrequenzen im Sollwert* || frequency skipping || frequency avoidance || critical speed rejection
Frequenzausklammerungsband   skip frequency zone || frequency avoidance zone || frequency jumping band || critical speed rejection band || jump frequency band || critical frequency lockout band
Frequenzband   frequency band
Frequenzbandausklammerung   frequency band inhibit * *Ausblendband für Frequenzsollw. z. Vermeidg. mechan. Resonanzen* || frequency jumping || critical speed rejection || skip frequency zone || frequency avoidance zone
Frequenzbereich   frequency range || frequency band || frequency domain * *{Regel.techn.}*
Frequenzeingang   frequency input
Frequenzen   frequencies
Frequenzgang   frequency response
Frequenzgeber   frequency generator || waveform generator * *Funktionsgenerator*
Frequenzgenerator   frequency generator || waveform generator * *Funktionsgenerator*
frequenzgestellter Antrieb   variable-frequency drive
Frequenzistwert   frequency actual value || actual frequency || running frequency
Frequenzkennlinie   frequency characteristic || frequency response * *→Frequenzgang*
Frequenzmesser   frequency meter
Frequenznachführung   frequency tracking
Frequenzregelung   closed-loop frequency control || frequency control
Frequenzschwankungen   frequency fluctuations
Frequenzsignal   frequency signal
Frequenzspektrum   frequency spectrum
Frequenzstellbereich   frequency control range || frequency setting range
Frequenzsteuerung   open-loop frequency control * *im Gegensatz zur Regelung (closed-loop)*
Frequenzteiler   frequency divider || frequency scaler || rate multiplier * *Frequ.multiplizierer (lässt nur e.bestimmten %-Anteil d.Eingangsimpulse passieren)*
Frequenzthyristor   frequency thyristor * *für über der Netzfrequ. liegende Schaltfrequenzen (für fremdgeführt.Schaltgn.)*
Frequenzumformer   frequency inverter || frequency converter * *auch rotierender Umformer (z. Erreichen extrem hoher Antriebdrehzahlen)* || PWM Inverter * *Um-/Wechselrichter mit Pulsweitenmodulation* || frequency changer set * *{el.Masch.} rotierender Drehfeldumformer (Maschinensatz)* || frequency changer * *{el.Masch.}*
Frequenzumrichter   frequency converter || PWM inverter * *Pulse Width Modulation, mit Pulsbreitenmodulation* || variable frequency inverter || frequency inverter
frequenzumrichtergespeist   frequency-converter fed || frequency-converter powered
Frequenzwählverfahren   touch-tone dialing * *Telefonwählverfahren: für jede Ziffer wird e. bestimmte Tonfrequenz übertragen* || tone dialing
Frequenzzähler   frequency counter
fressen   freeze * *z.B. Kolben* || stick * *Lager* || pit * *Grübchen/Narben/Löcher bilden; siehe auch →anfressen und →zerfressen* || seize * *(sich) fest~* || seize up * *(sich) fest~* || corrode * *zer~, an~, korrodieren, wegätzen, angreifen* || erode * *an-/zer-/weg~, erodieren*
Frigen   cyrogen || frigen
frigenfest   resistant against frigen || resistant against cyrogen
Friktion   friction * *Reibung*
Friktionsantrieb   friction drive
Friktionswalze   friction roller
frisch machen   fresh up * *z.B.sich waschen u. umziehen*
Frischdampf   live steam

Frischluft fresh air || intake air * *angesaugte Luft; siehe →Kühlluft*
Frischöl fresh oil
Frist time-limit * *vorgegebener Zeitpunkt* || deadline * *maximal akzeptierte/äußerste ~* || final date * *äußerste ~* || interval * *Zwischenzeit/raum* || prolongation * *Aufschub* || appointed time * *festgesetzte Zeit* || prescribed period * *vorgeschriebener Zeitraum* || date of completion * *Fertigstellungs/Erledigungstermin*
fristgerecht in time || to schedule * *termingerecht* || according to schedule * *termingerecht*
Front- front-
Frontansicht front view
Frontblende bezel * *Frontblende (z.B. f. Messinstrument), Skalenfenster (f. Messinstrument, Radio), Guckloch*
Frontplatte front panel || frontplate || faceplate || frontpanel
Frontplatten-Einbauelement front-panel element * *z.b. Schalter, Potentiometer*
Frontplattenausbruch panel aperture || panel cut-out
Frontplattenauschnitt front panel cut-out
Frontplattenaussparung opening in the front panel || front-panel cut-out
Frontplattenöffnung panel aperture
Frontseite front side || front || front panel * *Frontplatte*
frontseitig front || on front
Frontstecker front connector || front plug
Frostschutz frost protection || antifreezing protection || antifreezing agent * *Frostschutzmittel* || anti-frost
Frühausfall early failure
früher previous * *vorhergehend* || former * *ehemalig* || earlier * *zeitiger* || sooner * *zeitiger*
Frühwarnung early warning
FTP FTP * *Abk.f. 'File-Transfer Protocol': Datenübertrag.protokoll z. Dateitransfer von/zu einem Rechner*
FuE RD * *Research and Development: Forschung und Entwicklung*
Fuge joint * *Verbindung(s~)* || seam * *→Naht* || slit * *→Schlitz* || groove * *→Nut, →Falz, →Rille, Furche, Rinne, →Hohlkehle, Kerbe, Vertiefung, Aushöhlung* || parting line * *→Trennfuge* || rabbet * *Fuge, →Falz, Nut, Falzverbindung; rabbet plane: Falzhobel* || mortise * *Falz, Fuge; Zapfenloch*
Fügemaschine jointing machine * *{Holz}*
fugen joint * *(zusammen)fügen, fugen, verlaschen* || groove * *nuten, auskehlen* || point up * *verfugen, verstreichen, Fugen glatt streichen* || rabbet * *einfügen, (zusammen)fugen {Holz}, falzen* || rebate * *einfügen, (zusammen)fugen {Holz}, falzen*
Fugendichtungsmittel joint sealant
Fugenverleimanlage joint-gluing line * *{Holz}*
fühlbar tactile * *mit Hand/Finger* || sensible || tangible * *körperlich* || palpable * *körperlich fühl/greifbar (auch figürl.: augenfällig, deutlich)* || perceptible * *geistig; wahrnehmbar* || noticeable * *geistig* || distinct * *deutlich* || marked * *deutlich* || considerable * *beträchtlich* || appreciable * *beträchtlich* || significant * *deutlich, spürbar* || tangible * *greif-/fühlbar, (figürl.) handfest*
Fühler sensor * *Messgeber*
Fühlerleitung sense wire * *z.B. b.Netzteil nach d.4-Drahtprinzip z.Kompensation d. Leitungs-Spanngs.abfälle*
führen direct * *ein Signal ~* || connect * *ein Signal ankoppeln, schalten, verbinden* || feed * *ein Signal ~; to: auf/zu; through: hindurch, über* || switch-through * *ein Signal durchschalten* || assign * *zuweisen, zuteilen, to: an* || carry * *z.B. Signal, (gefährliche) Spannung, Strom ~* || route * *leiten, be-*

fördern, verlegen (Leitung, Leiterbahn) || take * *take a signal to the terminal: ein Signal auf die Klemme ~* || guide * *lenken, führen, leiten, geleiten* || have stocked * *Artikel im Einzel-/Großhandel ~* || have on stock * *Artikel im Einzel-/Großhandel ~* || result * *resultieren (in: zu)* || channel * *durch einen Kanal/ein Ventil usw. hindurchleiten/-lenken* || duct * *durch Kanal/Röhre ~* || prompt * *{Computer} den Anwender ~ durch ein dialoggesteuertes Programm*
führen zu result in || lead to * *~ nach/zu* || end in * *enden mit* || entail * *mit sich bringen, zur Folge haben, nach sich ziehen, erfordern*
führend leading || leading-edge * *z.B. Firma, Produkt* || top-of-the-market * *am Markt*
führende Nullen leading zero digits
führende Position leadership position * *z.b. eines Herstellers am Markt*
führerloses Fahrzeug automated-guided vehicle || driverless vehicle || AGV * *Abkürzg. für 'Automated-Guided Vehicle'*
Fuhrpark transport park || vehicle pool || rolling stock || fleet
Führung leadership * *Leitung* || guide * *mechanische; auch '~ von Luft' usw.* || conduction * *Leitung (von Strom), (Zu-) Führung, Übertragung* || direction * *Leitung* || management * *Leitung, Verwaltung* || command * *~ einer Regelung/Steuerung* || control * *~ eines Unternehmens/Prozesses/Antriebs* || guidance * *Leitung, Führung (einem Ziele zu), Anleitung, Belehrung, Beratung, Richtschnur* || showing round * *~ in einer Ausstellung, bei einer Besichtigung* || use * *~ eines Titels* || slide * *mechanische ~, Schlitten* || controlling * *der ~sgröße in einem Regelkreis* || duct * *Kanal, Weg, Gang, Röhre; z.B. für Kabel, Kühlluft* || government * *Herrschaft, Kontrolle, Leitung, Verwaltung, Regierung* || guide slide * *mechan. (Gleit-) Führung* || conduct * *Führung, Leitung, Verwaltung, Handhabung, Haltung, Verhalten, Betragen* || conducting * *Leitungs-, das Leiten (auch von Strom)* || guideway * *{mechan.}*
Führung gefordert control requested * *(→PROFIBUS-Profil)* || call for host control * *(→PROFIBUS-Profil)*
Führung übernehmen take over the control * *(→PROFIBUS-Profil)*
Führungs- guide * *(nachgestellt) z.B. Rolle, Schiene* || leading || managing || master * *Leit-* || pilot || controlling * *die Steuerung ausübend* || guiding || lead * *Führung, Leitung, Spitze; Vorsprung*
Führungsbahn slide-way * *z.B. für Verfahrschlitten bei Werkzeugmaschine* || slideway
Führungsbolzen guide bolt
Führungsebene management || managing body || executives || managerial class || management level || top management * *obere ~*
Führungsgröße command variable * *Sollwert in e. Regelkreis* || reference variable * *setpoint * *Sollwert* || reference value * *Sollwert*
Führungskräfte executives || management
Führungslager location bearing * *→Festlager* || guide bearing
Führungsleiste guide || guide rail * *Führungsschiene* || guide arm
Führungsregler master controller
Führungsrolle guide pulley || guide roller || contact roller || idle wheel || snatch block * *zum Regeln des Drahtdurchhangs* || idler * *z.B. in Fördersystem*
Führungsschiene guide rail || raceway * *{Holz} für Kettensäge* || cutter bar
Führungsübertragungsfunktion reference transfer function * *{Regel.techn.}*
Führungsverhalten response to setpoint changes * *{Regel.techn.}* || control performance * *Regelverhalten* || setpoint response * *{Regel.techn.}* ||

management behaviour * ~ *eines Vorgesetzten/ einer Führungskraft* || management style * ~ *einer Führungskraft*
Funktionstaste function key || function button
Fülle abundance || wealth * *of ideas/knowledge: an Ideen/Wissen*
füllen fill || inflate * *aufpumpen (Reifen usw.)* || stuff * *vollstopfen* || cram * *vollstopfen* || charge * *beladen, aufladen* || load * *beladen* || pack * *mit Schmierfett* || replenish * *wieder(auf)füllen (Vorräte usw.)* || bottle * *auf/in Flaschen* ~ || barrel * *in Fässer* ~ || sack * *in Säcke* ~ || put into bags * *in Beutel* ~ || fill-up * *auffüllen* || refill * *nachfüllen* || restock * *Lager wieder auf~*
Füller charger * *Befülleinrichtung* || filler * *Füllmaterial*
Füllgarn stuffer yarn * *{Kabel}*
Füllmaschine filling machine * *{Verpack.techn.}* *(siehe →Dosier~, →Kolben~, →Durchfluss~, →Schlauchbeutel~)*
Füllschnecke metering screw * *{Verpack.techn.}*
Füllstand level || liquid level * *einer Flüssigkeit* || fluid level * *einer Flüssigkeit* || filling level
Füllstandsanzeiger level indicator || liquid level indicator * *für Flüssigkeiten* || oil level indicator * *für Öl* || oil level gauge * *für Öl*
Füllstandskontrolle level monitoring || level monitoring system * *als Einrichtung*
Füllstandsmesser level meter || level gauge
Füllstandsmessgerät level gauge
Füllstandsregelung liquid level closed-loop control * *für Flüssigkeitsbehälter*
Füllstandssensor level sensor || liquid level sensor * *für Flüssigkeiten*
Füllstoff filler || ash * *{Papier} bei der ~messung erfasste Bestandteile*
Füllstoffe filler
Füllung filling || charging || bottling process * *Vorgang des Flaschen-Abfüllens* || stuffing || packing || charge * *Beschickung, Befüllung, Ladung, Einsatz (metall.)*
Fundament foundation * *eines Gebäudes, einer Maschine* || ground || base
Fundamentbelastung foundation load * *für el. Masch: siehe: DIN ISO 3945, VDI 2060 und VDE 0530 Teil 14*
Fundamentklötzchen foundation block * *z.B. zur Motorbefestigung*
Fundamentschwingungen foundation vibrations || ground vibrations
Fundbüro lost property office
fundiert well-founded * *z.B. Kenntnisse, Auskunft*
fungieren act * *as: als* || function * *as: als*
Funk-Empfang wireless reception || radio reception
Funk-Entstörfilter RFI suppression filter * *z. Unterdrückung von 'Radio Frequency Interference'* || RFI filter * *z. Unterdrückung von 'Radio Frequency Interference'* || radio interference filter || radio interference voltage filter
Funk-Entstörung radio frequency interference suppression || radio interference suppression || RFI suppression * *Unterdrückung von 'Radio Frequency Interference'*
Funkempfang radio reception || wireless reception
Funkempfänger radio receiver
Funken spark || sparks * *(Plural)* || flashes * *(Plural)*
funken Sprühend emitting sparks
Funkenbildung sparking
Funkenentladung spark discharge
Funkenerosion spark erosion
Funkenerosionsmaschine spark-erosion machine * *{Werkz.masch.}*

funkenfrei arc-free * *ohne Lichtbogen* || sparkless * *z.b. Kommutierung einer Gleichstrommaschine*
funkenfreie Kommutierung arc-free commutation * *z.B. bei Gleichstrommotor*
Funkenlöschglied quenching element * *zur Beschaltung von Induktivitäten (z.B. bei Relais, elektromagnet. Bremse usw.)* || supressor
Funkenlöschkondensator spark suppression capacitor || spark-quenching capacitor
Funkenlöschung spark quenching || spark suppression
Funkenstrecke spark gap
Funkentstördrossel radio interference suppression reactor || RFI suppression reactor || RFI reactor || interference suppression choke
Funkentstörfilter RFI suppression filter * *z. Unterdrückg.v. 'Radio Frequency Interference': Funkfrequenzstörungen* || radio interference filter || radio interference voltage filter || filter for radio interference suppression || RFI filter || input filter against EMI * *am Eingang e.Geräts geg. 'Electro-Magnetic Interference': el.magn. Störungen* || radio interference suppression filter || noise suppression filter
Funkentstörgrad radio interference level * *class B radio interference level:* ~ *B* || noise suppression level || RFI suppression class * *(Abk.f. 'Radio Frequency Interference') siehe VDE 0871, 0875*
Funkentstörkondensator noise suppression capacitor || RFI suppression capacitor
Funkentstörmittel radio interference suppression devices
funkentstört radio-interference suppressed || RFI suppressed
Funkentstörung RFI suppression * *Unterdrückung von 'Radio Frequency Interference': Funkfrequenz-Störungen* || radio frequency interference suppression || radio interference suppression * *bei Elektromotor: siehe VDE 0875* || interference suppression || radio and television interference suppression
Funkenüberschlag spark-over || sparking || flashover
Funkschutzgrad radio interference level
Funkstör-Grenzwert radio inteference limit value
Funkstörgrad radio-interference level * *siehe z.B. DIN VDE 0875 Teil 11 und EN 55011* || degree of radio interference || siehe z.B. 'Radio Frequency Interference' level * *siehe VDE 0871, EN 55011* || RFI level
Funkstörgrenzwert radio interference limit
Funkstörung radio interference * *hochfrequente Störg.,die d.Funkempfang stören kann* || radio frequency generated interference || radio interference
Funktion function * *as a funct.of: als math.* ~ *von; variation of y as a function of x : Verlauf v. y als* ~ *v. x* || dependance * *in dependance of: als* ~ *von (z.B. bei Kennlinie: auf X-Achse aufgetragene Variable)* || characteristic * *Kennlinie* || role * *(figürlich) Rolle, Part* || operation * *Operation (eines Rechners)*
funktional functional
Funktionalität functional scope * *Funktionsumfang* || functionality
funktionell functional
funktionieren function * *correctly: richtig* || work * *Art u.Weise d. Funktionierens; this is how it works: so funktioniert es* || be functional * *funktionsfähig sein* || be in working order * *funktionsfähig sein* || be O.K. * *funktionsfähig sein, in Ordnung sein* || do the job * *Aufgabe/Funktion erfüllen*
funktionierend workable * →*funktionsfähig*
Funktions- functional
Funktionsablauf functional sequence || sequence of functional activities
Funktionsabschnitt function section

Funktionsaufruf function call
Funktionsaufteilung distribution of functions * *over: zwischen*
Funktionsbaustein function block || block || function module || functional block
Funktionsbausteinbibliothek function block library
Funktionsbereich functional area || functional scope
Funktionsbeschreibung function description || functional description
funktionsbezogen function-related
Funktionseinheit functional unit || functional module
Funktionselement functional element
Funktionserweiterung functional expansion || function expansion || functional upgrade * *Hochrüstung* || upgrade * *Hochrüstung*
funktionsfähig functional || working || O.K. || in order * *in Ordnung* || repaired * *wieder in Ordnung, wiederhergestellt, repariert* || functioning || in working order * *in betriebsfähigem Zustand* || efficient || able to operate perfectly || operable || ready for use * *gebrauchstüchtig/-fertig, funktionsbereit* || functioning correctly * *ordnungsgemäß/richtig funktionierend*
Funktionsfähigkeit correct functioning || functioning
Funktionsfehler fault || malfunction * *Fehlfunktion, Funktionsstörung*
Funktionsgeber function generator
Funktionsgenerator waveform generator || function generator
funktionsgeprüft function-tested
funktionsgerecht very functional * *siehe auch →zielführend*
Funktionsglied function element
Funktionsgruppe functional group
funktionsidentisch functionally identical
Funktionsklasse function class || performance level
Funktionskontrolle function check || function checking || operational test * *→Funktionsprüfung*
Funktionspaket function package * *(SIEMENS SIMADYN D (R))*
Funktionsplan function chart * *[SPS] 'FUP'* || control system function chart * *[SPS]* || CSF * *[SPS] Control System Function chart: FUP (mit Grafiksymbolen; DIN 40700, IEC 117-5)* || function diagram || function plan * *beim SIEMENS Regelsystem SIMADYN D (R)* || control system flowchart * *beim SIEMENS Regelsystem SIMADYN D (R)*
Funktionsprinzip operating principle || method of functioning || principle of operation || mode of functioning || mode of operation * *Betriebsweise, Betriebsart*
Funktionsprüfung functional test || verification of operation * *Nachprüfung auf Einhaltung d. Spezifikationen* || operating test || test for correct functioning || performance test || general operating test || operational test * *siehe auch →Funktionskontrolle* || operation test
Funktionsschaltbild function diagram || functional circuit diagram
Funktionsschaltplan function-oriented diagram
funktionssicher fail-safe || reliable * *zuverlässig*
Funktionsstörung malfunction
Funktionsstufe function stage
Funktionssymbol function symbol || functional symbol
Funktionstabelle function table
Funktionstaste function key || function button
Funktionstest function test
funktionstüchtig functional || functioning || in working order || efficient || able to operate perfectly || working || O.K.

Funktionsüberwachung watchdog monitor * *für Rechner/SPS usw.*
Funktionsumfang functional scope || scope of functions || function scope || functionality * *Funktionalität* || functional extent || functional characteristics * *funktionelle Eigenschaften* || function range
Funktionsvielfalt functional diversity
Funktionsweise operating principle || mode of functioning
FUP CSF * *'Control System Function chart': Funktionsplan für SPS (siehe DIN 40700, IEC117-17)* || function chart * *Funktionsplan für SPS* || control system function chart
für for || pertaining to * *betreffend, gehörend zu* || in exchange for * *als Ersatz für* || in favour of * *zugunsten von* || instead of * *anstatt* || by * *step by step: Schritt für Schritt, piece by piece: Stück für Stück, day by day: Tag für Tag* || pros * *(Substantiv) the pros and cons: das Für und Wider* || considering * *in Anbetracht von* || in return for * *(Gegenwert ausdrückend)* || to * *gegenüber*
für allgemeine Zwecke general-purpose
für etwas sein be in favour of
für spätere Verwendung for use later || for future use
Furnier veneer
furnieren veneer
Furnierholz veneers || plywood
Furniermesser veneer knife * *[Holz]*
Furniermessermaschine veneer slicer * *[Holz]*
Furnierpresse veneering press * *[Holz]*
Furniersäge veneer saw * *[Holz]*
Furnierschälmaschine veneer peeler * *[Holz]* || veneer peeling lathe * *[Holz]*
Furnierschere veneer clipper
Furnierschneidemaschine veneer jointing guillotine * *[Holz] zum Schneiden von lückenlos anfügbaren Furnierblättern*
Furnierung veneering || inlaying || inlayed work
Fuß base * *Maschinenteil* || foot * *auch Befestigungs/Aufstellfuß eines Motors*
Fuß fassen establish a foothold * *(auch figürl.: Unternehmen usw.)* || gain a foothold || become established || settle down permanently
Fußanbau foot mounting || base mounting
Fußausführung foot mounting type of construction * *Motor mit Fußbefestigung (Ggs.: Flanschbefestigung)* || foot-mounted design * *z.B. (Motor mit) Fußbefestigung*
Fußbauform foot-mounting type * *z.B. Motor mit Aufstellfüßen (→angegossen oder →angeschraubt)* || foot-mounted version * *z.B. Motor/Getriebe mit Aufstellfüßen*
Fußbefestigung foot-mounting * *Motor*
Füße feet * *auch Aufstell~ eines Motors*
Fusion merger * *auch von Firmen* || amalgamation || consolidation * *(amerikan.)* || merging * *auch von Firmen*
fusionieren merge * *auch Firmen* || amalgamate || consolidate * *(amerikan.) (with: mit)*
Fußkreisdurchmesser diameter of root circle * *eines Zahnrades* || root diameter * *eines Zahnrades*
Fußloch hole in foot * *z.B. in Motorbefestigungsfuß*
Fußmontage foot mounting * *eines Motors (im Gegensatz zur Flanschmontage)*
Fußnischen feet recesses * *z.B. zur E-Motor-Befestigung*
Fußnote footnote || note
Fußpedal foot pedal
Fußpedalpotentiometer foot-operated potentiometer
Fußplatte base plate * *z.B. an Getriebemotor zur Befestigung*

Fußschraube   holding-down bolt * *zum Festschrauben eines Maschinen-/Motorfußes*
Fussel   lint || fluff || fuzz
Fußzeile   footer
Futter   chuck * *für Drehbank, Bohrmaschine usw.* || lining * *Ausfütterung, Futter, Auskleidung, Verkleidung*
Fuzzifizierung   fuzzification * *bei Fuzzy-Regelung*
Fuzzy-Regelung   fuzzy control

G

g   g * *Gramm; 1 g entspr. 0,0352740 oz* || oz * *Unzen; 1 g entspr. 0,0352740 oz*
G-Schiene   G-type mounting rail according to DIN EN 50045 * *asymmetr. Aufschnapp-Install.schiene (32 mm breit)* || asymmetrical DIN rail * *asymmetrische Aufschnapp-Install.-Schiene (32 mm breit)*
Gabelschlüssel   open-ended spanner || open-end wrench || fork wrench || fork spanner
Gabelstapler   fork-lift truck
Galette   godet * *[Textil] Rolle für Fäden z.B. in Textilfaserstraße* || feed-wheel
Galliumarsenid   gallium arsenide || GaAs
Galvanik   electroplating || galvanizing || galvanization || galvanizing shop * *Galvanikwerkstatt/-Anlage*
galvanisch getrennt   isolated || metallically separated || galvanically isolatet
galvanische Trennung   isolation || electrical isolation || metallic isolation || galvanic isolation
galvanische Verbindung   direct electrical connection
galvanisieren   galvanize
galvanisiert   galvanized
Gang   gear * *Getriebe (low/high: niedriger/hoher); shift into: wechseln in* || motion * *Bewegung, set in motion: in ~ setzen/bringen* || running * *Laufen (smooth: ruhig, swift: schnell)* || movement * *Bewegung* || operation * *Betrieb* || action * *Wirkungsweise* || way * *Weg* || passage * *Durch~, Verbindungsgang* || corridor * *~ im Haus* || gangway * *~ zwischen Sitzreihen* || aisle * *Durchgang (zwischen Bänken, Tischen usw.); Schneise (figürl.)* || speed * *Getriebe~ (first/second/top: erster/zweiter/höchster)* || duct * *Röhre* || worm * *Gewinde(gang)* || thread * *Gewinde(gang)* || backlash * *toter Gewinde~* || dead travel * *toter ~/Spiel/Lose bei Maschinenteilen* || tunnel * *~ unter Tage* || course * *~iner Mahlzeit; Bahn, Lauf, Verlauf; course of business: Geschäftsverlauf* || routine * *gewohnheitsmäßiger ~*
gängig   usual * *gebräuchlich* || movable * *gangbar, bewegbar verschiebbar, verstellbar* || major * *hauptsächlich* || commonly used * *→weit verbreitet*
Gänsefüßchen   quotation-marks || inverted commas
ganz   all || whole * *vollkommen, vollständig, heil, unversehrt; whole number: ~e Zahl* || entire * *vollständig* || undivided * *ungeteilt* || complete * *vollständig* || full * *voll(er), voll(ständig), ganz, völlig, gänzlich; full of: erfüllt mit/reich an* || total * *gesamt, ganz, völlig, gänzlich* || all over * *über das ~e (hinweg)* || overall * *über alles (z.B. Länge)* || quite * *(Adv.) völlig; quite another thing: etw.~ anderes; not quite th.same thing: nicht ~ dasselbe* || all * *(Adv.) sehr; not at all: ~ und gar nicht* || entirely * *(Adv.) vollständig, völlig* || very * *(Adv.) sehr* || by no means * *(Adv) ~ und gar nicht, auf keinen Fall* || throughout * *(Adv.) ~ durch* || most * certainly * *(Adv.) ~ gewiss* || not bad * *(Adv.; salopp) ~ gut* || in full * *(Adv.) (voll)ständig, "in Gänze"; write in full: etw. ausschreiben* || in whole * *(Adv.) vollständig; in whole or in part: ~ oder teilweise*
ganz oder teilweise   in whole or in part
Ganzbereichs-   full-range
ganze Zahl   integer || whole number
ganzheitlich betrachtet   considered as parts of a cohesive whole
ganzjährig   the year round
Ganzlochwicklung   integral slot winding * *[el. Masch.] AC-Motorwicklg., mit ganzzahlig. Nutzahl je Pol u. Wicklungsstrang*
ganzzahlig   integer || integral
ganzzahlige Oberwelle   integer-frequency harmonic
ganzzahliges Vielfaches   integer multiple
Garantie   guarantee * *Garantie, Gewähr, Bürgschaft* || warranty * *des Verkäufers* || guaranty
Garantiebedingungen   terms of guarantee || terms and conditions of warranty
Garantiefrist   warranty interval
garantieren   guarantee * *auch 'sicherstellen, (sich ver) bürgen (für), Garantie leisten'* || warrant * *auch 'zusichern, haften für, bestätigen'* || grant * *bewähren, bewilligen, zugestehen* || ensure * *sicherstellen, gewährleisten für, für etwas sorgen* || vouch * *['wautsch] bürgen; for: (sich ver-) bürgen für; that: dafür bürgen, dass*
Garantiereparatur   warranty repair
garantiert   guaranteed || warranted
Garantiezeit   warranty period || guarantee period
Garn   yarn || thread * *Faden, Zwirn, Garn* || cotton * *Baumwollgarn* || twine * *gezwirntes ~* || worsted * *wollenes ~, Kammgarn, Woll-, Kammgarnstoff; worsted wool: Kammgarnwolle*
Gärung   fermentation
Gas   gas || throttle * *Drosselklappe; open the throttle: Gas geben (auch figürl.); throttle down: Gas wegnehmen*
Gasabsaugung   gas exhausting || gas extraction
gasdicht   gas-tight
Gase   gases
Gasentladung   gaseous discharge
gasförmig   gaseous
Gasgebläse   gas blower
Gasringverdichter   gas-ring compressor
Gasstrahler   air ejector
Gasturbine   gas-turbine
Gatter   gate * *auch logisches/elektronisches ~*
Gattersäge   frame saw * *[Holz] zersägt Baumstämme zu Brettern* || reciprocating saw || gang saw || log frame saw * *[Holz] für Holzstämme* || gangsaw * *[Holz]*
Gattungsname   generic term * *auch 'Oberbegriff'*
Gauß-Verteilung   Gaussean process || Gaussean distribution
Gautscher   coucher * *in der Papierindustrie*
Gautschpresse   couch press * *zur Papier-Entwässerung* || couch roll * *zur Papier-Entwässerung*
Gbaud   Gbaud * *Gigabits pro sec.*
GDI   GDI * *Abk. f. 'Graphics Device Interface': Bindeglied zw. Betriebssystem u. Grafikkarte/ Drucker usw.*
geändert   altered || modified * *modifiziert*
Gebäude   building || house
Gebäudeautomatisierung   building automation
Gebäudetechnik   building systems
Geber   sensor * *Messgeber* || transducer * *Messumformer* || encoder * *digitaler Drehzahl/Lagegeber* || transmitter * *Messgeber* || pickup * *(Messwert-) Aufnehmer* || detector * *pick-up*
Geber-Interface   encoder interface * *Anschaltung für Drehzahl-/Lagegeber*

geberlos sensorless * *ohne Messgeber (arbeitend)* || encoderless * *ohne Drehzahl-/Lagegeber*
geberlose Regelung sensorless control * *z.b. Drehzahlregelung ohne Drehzahl-Messgeber/Tacho*
Gebermaschine transmitter-machine * *bei elektrischer Welle*
Geberrückführung encoder feedback * *Istwert von Drehzahl/Lagegeber wird zur Regelung verwendet* || sensor feedback
Gebiet territory * *räumliches ~* || zone * *Zone, (territorialer) Bereich* || field * *Fach~*, →*Branche* || domain * *Domäne*, →*Bereich, Sphäre, Reich* || industry * *Industrie*, →*Branche* || realm * *Reich, Sphäre, Bereich, (Fach-)~*
gebildet made up * *zusammengesetzt (of: aus)* || generated * *erzeugt, abgeleitet (from: aus)*
Gebläse blower || fan * *Lüfter, Ventilator*
Gebläseantrieb blower drive || fan drive * *Lüfterantrieb*
geblecht laminated
gebleicht bleached * *bleached chemical pulp: ~er Zellstoff*
gebogen bent * *backwards: nach hinten* || curved * *(ab)gerundet* || contoured * *Fläche*
Gebot rule * *Vorschrift, Regel* || obligation * *Verpflichtung* || bid * *Angebot (z.B. bei Versteigerung)*
Gebrauch use * *Anwendung, Gebrauch; in use: in ~* || employment
gebräuchlich current || commonly used || ordinary * *gewöhnlich* || common * *gewöhnlich, üblich* || customary * *herkömmlich* || usual * *with: bei* || in use * *(Adj.) in Gebrauch/Benutzung* || commonly encountered * *häufig anzutreffen* || popular * *populär, (allgemein) beliebt, weit verbreitet* || generally used
gebräuchlichst most commonly encountered * *am häufigsten anzutreffen*
Gebrauchsanleitung instructions for use || directions for use * *siehe auch 'Betriebsanleitung'* || instructions
Gebrauchsanweisung instructions for use || instructions * *siehe auch* →*Betriebsanleitung*
Gebrauchsbedingungen conditions of use
Gebrauchsdauer service life || life || time of service
Gebrauchsmuster design patent || registered design or pattern || utility-model patent
Gebrauchsmuster-Eintragung registration of a utility or design
Gebrauchsmustereintragung registration of a utility or design
gebraucht used || second-hand
gebremst braked
Gebühr fee || charge || surcharge * *zusätzliche ~, Zuschlag, zusätzliche Belastung (Konto)*
Gebühr erheben charge a fee
Gebühren charges
gedämpft damped * *z.B. elektr./mechan./akust. Schwingung; (siehe auch 'dämpfen')*
gedämpfte Schwingung damped oscillation
Gedanke thought || idea || intention * *Absicht* || concept * *Begriff* || notion * *Gefühl, Ahnung* || reflection * *Betrachtung* || supposition * *Mutmaßung*
Gedankengebäude edifice of thoughts || edifice of ideas
Gedankenstrich dash
gedanklich intellectual || mental || in ideas || of thought * *depth of thought: gedankliche Tiefe* || notional * *begrifflich, rein gedanklich, spekulativ, eingebildet, imaginär*
gedecktes Lager sealed bearing * *Rillenkugellager mit eingebauten Lagerdichtungen* || double-sealed bearing * *Rillenkugellager mit eingebauten Lagerdichtungen*
gedreht twisted

gedruckte Schaltung printed circuit
gedrückter Zustand retained condition * *z.b. bei einer Taste*
geeicht calibrated * *Messgerät*
geeignet proper || suitable * *passend* || suited * *passend; well/especially suited: gut/besonders geeignet (to/for: für)* || applicable * *anwendbar* || convenient * *gut geeignet, günstig*
geerdetes Netz grounded supply network * *aus der Sicht eines Verbrauchers (z.B. Motor, Stromrichter usw.)* || earthed-neutral supply system
Gefachesteckmaschine separator inserting machine * *Verpackungsmaschine, packt Abtrennpappen in Karton*
Gefahr danger || hazard || risk * *Wagnis*
gefährden endanger || imperil || expose to danger || risk * *aufs Spiel setzen, riskieren* || jeopardize * *in Frage stellen*
gefährdet sein be subject to potential injury * *Verletzungsgefahr* || be endangered || be in danger
Gefahrenbereich danger area
Gefahrenhinweise safety notes
Gefahrenübergang risk transfer * *(Wirtschaft) Übergang des Risikos vom Hersteller auf den Käufer*
gefährlich dangerous || perilous || risky * *riskant* || hazardous * *riskant, gewagt, gefährlich* || ticklish * *heikel* || serious * *ernst, schwerwiegend* || grave * *ernst, schwer* || critical * *kritisch, ernst* || unsafe * *unsicher*
gefährliche Spannungen hazardous voltage levels
Gefahrstoff dangerous substance
Gefälle gradient || slope || incline || head * *z.B. bei Wasserkraftanlage*
gefärbt coloured
Gefäß vessel || receptacle * *Behälter, Gefäß; Steckdose* || container || basin * *Bassin, Schüssel* || tank * *für Flüssigkeiten*
geflanscht flanged
Geflecht mesh * *z.B. aus Draht; fine: feines; coarse: grobes* || plait * *aus Textil* || network * *(auch figürl.)* || reticulation * *Netz(werk)* || braid * *Litze, ~ aus feindrähtigen Drähten*
Geflechtdraht netting wire
Geflechtschirm braided screen * *bei abgeschirmtem Kabel*
geflutet saturated * *gesättigt, z.B. Diode*
gefordert required || demanded
Gefriergerät freezer || freezer unit
Gefüge structure || framework || system || texture * *(auch figürl.) Struktur, Beschaffenheit, Gefüge, Gewebetextur, Maserung*
Gefühl feeling || sense
geführt controlled * *gesteuert* || fed * *geleitet; fed out to terminals: auf Klemmen herausgeführt (z.B. Signal)* || guided * *auch durch eine mechan. Führung*
gegebenenfalls if necessary * *falls notwendig/nötig* || if required * *falls erforderlich* || if applicable * *falls zutreffend/anwendbar* || if need be * *falls nötig* || should the occasion arise
gegen towards * *örtlich, zeitlich, in Richtung auf; towards zero speed: ~ Drehzahl Null* || against * *~sätzlich* || opposed to * *~sätzlich* || in the face of * *~sätzlich* || about * *ungefähr* || nearly * *ungefähr* || in the neighbourhood of * *ungefähr* || by * *(amerikan.) zeitlich* || for * *Mittel ~, z.B. ein Leiden* || compared with * *verglichen mit* || in exchange * *als Entgelt (for: für)* || versus * *juristisch z.B. Prozessgegner* || with respect to * *bei Spannungen* || to * *bei Spannungen, z.B. voltage to neutral: Sternspannung (~ den Sternpunkt)* || in return * *als Entgelt (for: für)*
gegen Erde to ground * *Spannung, Kapazität usw.*

gegen M schaltend switching to ground * *z.B. Open Collector-Digitalausgabestufe*
gegen P schaltend switching to P potential * *z.B. Open Emitter-Digitalausgabestufe*
Gegen- counter * *z.B. Gegenmaßnahme, Gegenmoment/spannung* || anti- * *entgegen, z.B.anti-clockwise: entgegen d. Uhrzeigersinn* || opposite * *entgegengesetzt, in entgegengesetzter Richtung* || back * *in Gegen-/Rückwärtsrichtung*
Gegen-EMK counter EMF || motor-back EMF || back EMF
Gegendrehmoment counter torque
Gegendruck counter-pressure || back-pressure
Gegendruckzylinder impression cylinder * *{Druckerei}*
gegeneinander against one another || against each other || towards one another || towards each other || back-to-back * *Rücken an Rücken, aneinander* || reciprocally * *gegenseitig, wechselseitig, umgekehrt, reziprok* || mutual * *gegenseitig, wechselseitig, beiderseitig, gemeinsam* || with respect to each other * *aufeinander bezogen*
gegeneinander versetzt staggered
gegeneinanderschalten connect back to back
Gegenelektrode counterelectrode
Gegenerregung counter-excitation * *z.B. bei elektromagnet. Kupplung/Bremse zum Erreichen schneller Ansprechzeiten*
Gegengewicht counterweight * *z.B. bei Aufzug* || balance weight || counterpoise * *(auch figürl.) Gleich-/Gegengewicht (auch Verb: als ~ wirken zu, ausgleichen)* || counterbalance * *(auch Verb: ein ~ bilden, ausgleichen, aufwiegen, die Waage halten)*
Gegeninduktivität mutual inductance
Gegenkomponente negative phase-sequence component * *{el.Masch.}* || negative-sequence component * *{el.Masch.} Anteil mit entgegengesetzter Phasenfolge e. 3-Phasensystems*
Gegenkopplung negative feedback
gegenläufig in opposite direction || counter-rotating * *in Gegenrichtung rotierend* || counterrotating * *in Gegenrichtung rotierend* || contra-rotating * *in Gegenrichtung rotierend*
Gegenläufigkeit counterrotation
Gegenmaßnahme counter-measure || countermeasure || answer * *Gegenmittel, (Problem-) →Lösung*
Gegenmaßnahmen ergreifen counter-act || take countermeasures
Gegenmoment counter torque * *Moment, das die Arbeitsmaschine d.Motor entgegensetzt (Last/Beschleunigungsmoment)* || load torque * *Lastmoment*
Gegenmomentverlauf load torque characteristic * *~ über der Drehzahl*
Gegenmutter lock nut || check nut
gegenparallel anti-parallel || inverse parallel || double way
Gegenparallelschaltung double way bridge connection * *z.B. Gleichrichterbrücke* || antiparallel bridge connection * *z.B. Gleichrichterbrücke* || inverse-parallel connection * *4-Quadranten DC-Doppelstromrichter in Brückenschaltung* || anti-parallel connection
gegenphasig in opposition
Gegenrichtung opposite direction
Gegensatz contrast * *to: zu; in contrast to/with: im Gegensatz zu* || opposite * *of: zu* || contrary * *of: zu/von* || opposition * *Widerspruch; in opposition to/as opposed to: im Gegensatz zu* || antagonism * *Widerspruch (between: zwischen)* || antithesis * *Gegensatz, Widerspruch*
gegensätzlich contrary || opposite || opposed * *entgegengesetzt, gegensätzlich, gegenüberliegend* || opposing * *gegenüberliegend, entgegengesetzt, opponierend*

gegenseitig mutual * *gegen-/wechselseitig, gemeinsam* || reciprocal * *wechsel-/gegenseitig, umgekehrt, reziprok* || bilateral * *zweiseitig* || each other * *einander; praise each other: sich ~ loben; neutralize each other: sich ~ aufheben* || one another * *einander*
gegenseitige Beeinflussung interaction || interference * *störende* || mutual effect
gegensinnig in the opposite sense
Gegenspannung back-voltage || back voltage || back EMF * *EMK eines Motors (EMF: ElectroMagnetic Force)* || back-EMF || motor-back EMF * *Gegen-EMK eines Motors* || counter EMF * *Gegen-EMK*
Gegenstand object * *Objekt, Gegenstand, Absicht, Ziel, Zweck* || thing * *Ding, (Haupt)Sache, Angelegenheit* || item * *Punkt (der Tagesordnung usw.), (Einzel)Gegenstand, Stück, Posten, Artikel* || subject * *(Gesprächs- usw.) Gegenstand, Thema, Stoff, Objekt, Subjekt, Substanz* || theme * *Thema* || topic * *Thema, Gegenstand* || subject-matter * *Inhalt* || matter * *Angelegenheit* || affair * *Angelegenheit* || issue * *Diskussionspunkt, Streitfrage* || motif * *(Leit)Motiv (auch künstlerisches), Leitgedanke,* || body * *(Fest)Körper, Masse, Substanz* || foreign body * *Fremdkörper* || fixture * *fester ~, Installationsteil, Inventarstück, zum Inventar gehörend (auch Mitarbeiter)*
Gegenstrom opposing current || reverse current || contra-flow * *von Luft oder einer Flüssigkeit*
Gegenstrombremsung plug braking * *e. Asynchronmotors durch Drehfeldumpolg. (Vertausch.2er Phasenanschlüsse)* || reversal braking * *eines Asynchronotors durch Umpolung des Drehfeldes* || braking by reversal || countercurrent braking
Gegenstück counterpart || antitype * *Gegenbild, Vorbild* || equivalent || matching piece * *zugehöriges ~* || companion * *zugehöriges ~* || fellow || mirror image * *identisches Teil, getreues Abbild* || cousin || opposite number || mating piece || mating connector * *~ für Stecker/Buchse*
Gegensystem negative phase-sequence system * *{el.Masch.}* || negative-sequence network * *{el. Masch.}*
Gegentakt push-pull
Gegentakt- push-pull * *-Verstärker, -Eingang, -Ausgang* || normal-mode * *-Spannung, -Signal, -Unterdrückung* || series-mode * *-Spannung, -Signal, -Unterdrückung*
Gegentaktunterdrückung series-mode rejection || normal-mode rejection
Gegenteil contrary * *(to: von)quite on the contrary to: ganz im Gegensatz zu* || reverse * *of: von* || opposite * *of: von* || antithesis * *Gegensatz, Widerspruch*
gegenteilig contrary || opposite
gegenüber opposite * *to: von* || facing || in front of || vis-à-vis || face to face with * *Personen* || compared with * *im Vergleich zu* || compared to * *im Vergleich zu* || as against * *im Vergleich zu* || relative to * *relativ zu, bezogen auf* || over * *(dar)über, mehr (verglichen mit); over against: gegenüber, im Gegensatz zu* || contrary to * *im Gegensatz zu* || in view of * *in Anbetracht von* || in the face of * *angesichts* || considering * *angesichts* || confronted * *be confronted with: sich gegenübersehen, konfrontiert sein mit* || with respect to * *hinsichtlich, bezüglich, in Anbetracht von, Gegensatz zu, as opposed to * *verglichen mit, im Vergleich zu, im Gegensatz zu* || versus * *verglichen mit* || vs * *Kurzform für 'versus': verglichen mit*
gegenüberliegen face || be opposite || lie opposite
gegenüberliegend opposing || opposite || facing || back to back * *Rücken an Rücken*
Gegenüberstellung comparison * *Vergleich*

Gegenwalze backing roll * *z.B. bei Streichmaschine, Beschichtungsmaschine*
gegenwärtig at the present time * *siehe auch 'momentan'*
Gegenwert value || equivalent || proceeds * *Erlös*
geglättet smoothed || smooth || filtered * *durch Glättungsfilter* ~
gegliedert arranged * *angeordnet* || organized * *organisiert* || subdivided * *unterteilt, unter~ (into: in)* || structured * *strukturiert* || grouped * *gruppiert* || classified * *in Gruppen/Klassen eingeteilt* || divided * *eingeteilt (into: in)*
gegossen cast
Gehalt salary * *Einkommen* || content * *Inhalt*
Gehaltsliste payroll
gehärtet hardened
Gehäuse case || housing || enclosure * *auch b. Motor* || frame * *Motorgehäuse* || chassis * *tragende Konstruktion, Tragblech* || body * *äußere Gestalt* || package * *z.B. für Halbleiterbauelement, integrierten Schaltkreis usw.* || casing || box || chassis * *Gestell (-Rahmen), Grundplatte*
Gehäuse-Erde chassis earth
Gehäuseabmessungen housing dimensions
Gehäuseausführung design of the frame * *Motor* || housing construction || housing design * *konstruktive Ausführung des Gehäuses* || housing version
Gehäusebreite housing width
Gehäusedach housing roof
Gehäusedeckel top housing cover * *obere Abdeckung* || front cover * *vorderer* ~
Gehäuseform housing type || type of housing
Gehäusefüße housing feet * *z.b. an Motor; cast: angegossen*
Gehäusegröße housing size || frame size * *eines Motors*
Gehäuseinneres case interior
gehäuselos with no housing || chassis- * *z.b. offenes Gerät, nur auf Tragblech montiert* || without enclosure || frameless * *bei Elektromotor: Blechpaket dient gleichzeitig als Gehäuse*
Gehäuseschwingungen housing vibrations * *[el. Masch.]*
Gehäusetemperatur case temperature * *z.B. eines Getriebes, Halbleiterbauelements usw.*
geheim secret * *in secret: im Geheimen; top/most secret: streng ~; secret agreement: Geheimabkommen* || confidential * *vertraulich* || private * *vertraulich* || hidden * *versteckt* || concealed * *verborgen* || underground * *unerlaubt* || clandestine * *heimlich, verborgen, verstohlen, unerlaubt* || surreptitious * *unerlaubt, erschlichen, betrügerisch, heimlich, verstohlen* || mysterious * *~nisvoll, mysteriös* || occult * *Lehre, Geheimwissenschaft* || secretly * *(Adv.) ins~* || privately * *(Adv.) im Geheimen* || classified * *classified matter: -e Dienstsache* || restricted * *Vermerk auf Dokumenten* || cryptic * *verschlüsselt, verborgen, schwer entzifferbar/entschlüsselbar* || in strict confidence * *streng vertraulich*
Geheimhaltungsvereinbarung non-disclosure agreement
gehen walk * *zu Fuß* ~ || leave * *weggehen (for: nach), aus dem Dienst* ~ || quit * *aus dem Dienst* ~ || resign * *aus dem Amt* ~ || be fired * *(salopp) 'rausgeschmissen werden'* || work * *bei Maschine usw.; doesn't work: geht nicht* || run * *bei Maschine usw.* || operate * *bei Maschine usw.* || function * *bei Maschine/Mechanismus usw.* || sell * *sich verkaufen (Ware)* || sell well * *sich glänzend verkaufen* || do well * *bei Geschäften, Absatz usw.* || pass through * *gehen (hin)durch* || follow a rule * *nach einer Regel* ~
gehen auf go * *z.B. signal goes High/Low: Signal geht auf High/Low*
gehende Meldung outgoing message

gehören belong * *to: zu* || be owned by || form part of * *einen Teil bilden von* || be among * *zählen zu* || fall under * *zählen zu* || be associated with * *~ zu, zusammen~ zu, zugeordnet sein zu*
gehörig pertinent * *zur Sache gehörend, sachdienlich, passend, angemessen*
Gehrung mitre * *auch ~sfläche, ~sfugen gehren, auf ~ verbinden; mitre-welded: auf ~ geschweißt* || miter * *(amerikan.) siehe 'mitre'; miter joint: ~sverbindung*
Gehrungsfräsmaschine mitre moulder * *[Holz]* || mitre cutting machine * *[Holz]*
Gehrungssäge mitre saw * *[Holz]*
Gehrungsverbindung mitre joint * *[Holz]*
geistig intellectual * *intellektuell* || mental || spiritual * *unkörperlich* || immaterial * *unkörperlich*
geistiger Horizont thought horizon
geistiges Eigentum intellectual property
gekapselt encapsulated * *auch 'vergossen'* || enclosed || canned * *'eingedost'* || sealed * *abgedichtet, hermetisch dicht* || potted * *vergossen* || moulded-case || flame-proof * *→explosionsgeschützt*
gekennzeichnet typified * *charakterisiert* || characterized * *charakterisiert; by: durch* || marked * *markiert* || identified || labeled * *beschriftet*
geklebt pasted
gekoppelt coupled || connected via a serial link * *über serielle Schnittstelle* ~ || networked * *über ein Datennetz* ~ || linked-up
gekreppt crimped || crêpe * *Papier usw.*
gekrümmt curved || bended || twisted || crooked
Gelände premises * *Grundstück, Häuser nebst Zubehör, Lokal(ität), Räumlichkeiten*
Gelände- offroad * *[Verkehr]*
Geländer railing || rails || balustrade || banisters || hand-rail * *Handgeländer, Handlauf* || guard rail * *Schutzgeländer*
geläufig easy * *fließend* || smooth * *fließend* || familiary * *bekannt, vertraut* || current * *allgemein bekannt* || generally used * *→gebräuchlich*
gelb yellow
Geldautomat cash dispenser || automated teller machine * *Bankautomat* || ATM * *Abk. für 'Automated Teller Machine': Bankautomat*
Gelder monies * *grant monies: ~ bewilligen*
Geldgeschäft financial transaction
Geldhahn zudrehen pull the plug
Gelegenheit opportunity
Gelegenheit geben allow to
Gelegenheitsgesellschaft joint venture * *vorübergehende Zusammenarbeit mehrerer Firmen zu einem Geschäftszweck*
gelegentlich occasional * *gelegentlich, zufällig* || casual * *unregelmäßig, ~; zufällig, unerwartet; ungenau, ungezwungen, zwanglos, beliebig* || temporary * *zeitweilig* || now and then * *(Adv.)* || at times * *(Adv.)* || on occasion * *(Adv.) bei Gelegenheit* || when there is a chance * *(Adv.) bei (passender) Gelegenheit* || at your leisure * *(Adv.) wann es Ihnen passt* || occasionally * *(Adv.)* || from time to time * *(Adv.) von Zeit zu Zeit*
geleimtes Papier sized paper
Gelenk hinge * *Scharnier* || hinged joint * *articulation* * *Gelenkverbindung* || joint || link || universal joint * *Kardangelenk* || pivot * *Dreh-/Angelpunkt, Drehzapfen*
Gelenk- hinged || hinge || joint || articulated * *z.B. bei Stadtbahn-~wagen* || swivel-
Gelenkwelle cardan shaft || universal shaft || universal-joint shaft || propeller shaft * *bei Fahrzeug, Schiff usw.*
gelernter Arbeiter skilled worker
gelesen read
geliefert delivered * *supplied* || shipped * *ausgeliefert, versandt*

gelocht holed || punched * *mit ausgestanzten Löchern*
gelöscht deleted || cancelled * *ungültig/rückgängig gemacht*
gelten be of value || be valid * *gültig sein; become valid: gültig werden; remained valid unchanged: unverändert ~* || count * *zählen* || be effective * *Gesetz* || be in force * *in Kraft sein, z.B. Vorschrift* || be in operation || matter * *wichtig sein* || cover * *abdecken, einschließen,* || apply * *Anwendung finden (to: bei); passen, zutreffen (to: für/auf/bei); z.B. Vorschrift, Norm* || be true * *Formel, techn. Angabe usw. (for: für)* || be considered as * *betrachtet werden als* || hold * *gelten; (sich) halten, festhalten, standhalten; hold to: festhalten an*
gelten als be considered as || be looked upon as * *betrachtet werden als* || rate as
gelten für apply to * *sich anwenden lassen auf* || be applicable to * *anwendbar sein für* || be true for * *Formel, Technische Angabe usw.* || be right for * *richtig sein für*
gelten lassen let pass || allow || admit of
Geltungsbereich scope
gemäß according to || pursuant to * *gemäß, entsprechend, laut (z.B. e. Norm, Vorschrift, Gesetz)* || in accordance with || in conformity with || in agreement with || in compliance with * *gemäß, befolgend* || in consequence of * *zufolge* || as a result of * *zufolge* || in pursuance of * *gemäß e. Norm, Vorschrift, Gesetz* || following * *(Vorschrift, Betriebsanleitung usw.) beachtend/folgend* || conforming to * *(eine Norm, Vorschrift usw.) erfüllend*
Gemeinkosten overhead costs || overheads || overhead expenses
gemeinsam common * *to: mehren Dingen/Personen; common to all: allen ~; common denominator: ~er. Nenner* || combined * *kombiniert, zusammengenommen* || collective * *kollektiv, zusammengenommen* || mutual * *gegenseitig* || joint * *joint/ concerted action: ~e Aktion; joint property: ~es Eigentum* || jointly * *(Adv.)* || together * *(Adv.)* || in a body * *(Adv.)* geschlossen || in concurrence * *(Adv.) act in concurrence/conjointly/in concert with: ~ handeln mit* || shared * *~ benutzt, untereinander aufgeteilt* || divisional * *~ innerhalb einer Abteilung, eines Unternehmensbereichs*
gemeinsam benutzen share * *teilen* || use commonly || pool * *Resourcen in einer Interessengemeinschaft vereinen*
gemeinsam betreibbar interoperable * *z.B. Geräte verschiedener Hersteller an einem Feldbussystem*
gemeinsame Benutzung sharing || pooling * *von Resourcen in einer Interessengemeinschaft*
gemeinsame Welle common shaft
gemeinsamer Zwischenkreis common DC link bus || common DC bus || DC common bus || power equalizing bus * *zum Energieausgleich*
Gemeinsamkeit commonness || mutuality || common interest || mutuality
Gemeinschaftsstand common booth * *gemeinsamer (Messe-) Stand*
Gemeinschaftsunternehmen joint venture
gemeldet signaled || signalled
gemessen measured
gemessen an measured against
Gemisch mixture * *siehe auch →Mischung; mixture control: ~regelung* || medley * *(figürl.)*
gemischt mixed || combined * *kombiniert* || composite * *(aus Verschiedenen Elementen) zusammengesetzt, gemischt, vielfältig, Misch-; Mischung*
gemischter Stern-Dreieck-Anlauf combined stardelta starting * *[el.Masch.] Kombination v. Stern-Dreieck- und Wicklungsumschaltung*
gemittelt averaged * *over 10 sec; über 10 sec*
genannt called * *mit e.Ausdruck bezeichnet* || termed * *mit e.Fachausdruck bezeichnet; termed as: bezeichnet als* || referred to * *z.B. mit einem Fachausdruck; be referred to as: bezeichnet werden als* || mentioned * *erwähnt*
genau accurate * *with centimeter accuracy: zentimeter~* || exact || precise * *präzise, exakt, genau, klar, korrekt, klar umrissen, richtig (Betrag, Augenblick usw.)* || true * *echt* || strict * *streng* || detailed * *ins Einzelne gehend* || just * *(Adv.) just as good: ~ so gut* || definite * *klar, deutlich, eindeutig, definitiv, bestimmt ,endgültig, fest* || proper * *ordnungsgemäß, angemessen, korrekt, einwandfrei, genau* || precisely * *(Adv.)* || exactly * *(Adv.)* || due * *(Adv.) genau, gerade; due west: ~ westlich*
genau dasselbe just the same thing
genau festlegen pinpoint
genauer more detailed * *(Komparativ) detaillierter, mehr ins Einzelne gehend*
genauere Angaben more detailed data || more delailed information
Genauigkeit accuracy * *steady state accuracy: im eingeschwungenen Zustand* || precision || exactness || fidelity * *Wiedergabetreue* || constancy * *Konstanz* || tolerance * *Toleranz* || preciseness
Genauigkeitsgrad degree of accuracy
genauso just as || exactly as
genauso wie the same as * *(Adv)*
genehmigen grant * *bewilligen* || agree to * *gutheißen* || assent to * *gutheißen* || consent to * *gutheißen (of: etwas/jemanden)* || approve * *billigen, gutheißen, anerkennen, annehmen, akzeptieren* || authorize * *gutheißen, billigen, rechtfertigen, ermächtigen, bevollmächtigen, berechtigen* || license * *amtlich ~, lizensieren*
Genehmigung approval * *Billigung, Erlaubnis (of: von/des)* || grant * *Bewilligung* || assent * *Billigung (to: von/des)* || acceptance * *Billigung (of: von/des)* || permission * *Erlaubnis; give a person the permission: jdm. die Erlaubnis erteilen* || authorization * *Ermächtigung; authorize a person to: jdm. zu etw. ermächtigen* || licence * *(behördliche/offizielle) Zulassung* || license * *(amerikan.) offizielle Zulassung* || permit * *behördliche Zulassung*
Genehmigungsverfahren approval procedure || licencing procedure || licensing procedure * *(amerikan.)*
geneigt inclined * *schräg, abschüssig, (auch figürl.; to: zu)*; downward/upward: *nach unten/oben* || sloping * *abschüssig* || well disposed towards * *(figürl.) jdm./zu etwas ~* || incline * *geneigt sein (auch figürl.)* || be disposed to * *etwas begünstigen, zu etw. ~ sein* || well disposed * *(figürl.) be well disposed towards somebody/something: jemandem/zu etwas ~ sein*
Generalist allrounder || full-liner * *Vollanbieter*
Generallizenz general license
Generalunternehmer general contractor
Generation generation * *z.B. einer Gerätereihe, auch 'Erzeugung'*
Generator generator * *electric: elektrischer* || alternator * *für Drehstrom/Wechselstrom*
Generatorbetrieb generator operation
generatorisch generating * *Energie erzeugend; z.B. Drehmoment, Bremsbart eines Stromrichters* || regenerating * *zurückspeisend* || regenerative * *zurückspeisend* || when generating * *(Adv.) im Energie-erzeugenden Zustand* || braking * *im Bremsbetrieb* || operating as a generator * *~ arbeitend*
generatorische Bremsung over-synchronous braking * *[el.Masch.] Bremseffekt bei Asynchronmotor wenn Drehzahl > Synchr.drehz.* || regenerative braking * *mit Rückspeisung der Bremsenergie ins Netz*
generatorische Energie regenerative energy * *wird z.B. beim Abbremsen eines Antriebs frei*

generatorisches Bremsen regenerative braking || rheostatic braking * *mit Bremswiderstand*
generatorisches Moment torque when generating
generell general * *(Adj.)* || generally * *(Adv.)* || generic * *(Adj.)* || generically * *(Adv.)*
generieren generate
Generierung generation * *z.B. eines (ablauffähigen) Softwarepakets z.B. durch Zusammenbinden mehrer Module*
genial ingenious
genormt standardized
Genossenschaft cooperative || society
genug enough
genügen be sufficient * *aus-/hinreichend sein* || conform to * *einer Vorschrift ~; siehe auch →entsprechen* || satisfy * *einer Bedingung/Vorschrift ~* || suffice * *ausreichen sein, ausreichen, hinreichen* || be adequate * *angemessen/ausreichend sein*
genügend sufficient || enough || satisfactory * *befriedigend, zufrieden stellend* || ample * *reich(lich), (vollauf) genügend, weit, groß, geräumig, weitläufig, ausführlich, umfassend* || sufficiently * *(Adv.) ausreichend, hinreichend*
genutet slotted || keywayed * *mit Paßfedernut versehen (z.B. Motorwelle)*
geöffnet opened * *auch Schaltkontakt*
Geometrie geometry
geometrisch geometric || geometrical
geometrische Summe geometrical sum || vector addition * *Zeigeraddition*
geordnet orderly * *(Adj. und Adv.; auch figürl.)* || systematic || structured || disciplined
gepflegt well kept || kept in good order || kept in good service || kept in good condition || well cared for || kept in good repair
geplottet plotted
gepolt polar * *z.B. (Elektrolyt-) Kondensator* || polarized * *z.B. (Elektrolyt-) Kondensator*
gepolter Kondenstor polarized capacitor
geprägt embossed
geprüft tested || checked * *überprüft*
gepuffert buffered || battery-backed * *batteriegepuffert*
gepulst pulsated
gepunktet dotted * *line: Linie*
gerade even * *Zahl* || linear slope * *(Substantiv) gerade Linie* || straight * *straight line: gerade Linie* || upright * *Haltung usw.* || direct * *direkt* || just * *(Adv.)* || exactly * *(Adv.)* || precisely * *(Adv.)* || currently * *(Adv.) momentan* || straight line * *(Substantiv) [Mathematik]*
gerade Linie straight line
Geradeaus- in-line || straight in-line || straight || straight-line * *[Draht]*
Geradengleichung linear equation
geradlinig straight-lined || rectilinear || straightforward * *(figürlich)*
geradlinige Bewegung linear movement || straight-lined motion || translatory motion || moving in a straight line
geradzahlig even-numbered || even
gerändelt knurled
Gerät unit || device * *electrical: elektrisches* || equipment * *→Ausrüstung (-sgegenstände), techn. Material/~schaft, Apparatur(en)* || appliance * *z.B. Haushalts~* || instrument * *Mess~* || apparatus * *Apparat, Gerät, Vorrichtung, Apparatur, Maschinerie* || gear * *Ausrüstung* || gadget * *(amerikan.) Apparat, Gerät, Vorrichtung, techn.Spielerei,; (auch Verb: mit ~en ausstatten)*
Geräte units || equipment
Geräteanzeige device condition code * *z.B. bei SPS*
Geräteausfall equipment failure
Gerätebedienfeld operator control panel

Gerätefehler equipment error
Gerätehandbuch equipment manual
Geräteklasse equipment class * *(→PROFIBUS-Profil)*
Gerätekoppling device link || unit interfacing
Gerätekopplung unit interfacing || device link
Geräteleistung converter output * *eines Stromrichters*
Gerätelüfter equipment fan || converter fan * *eines Stromrichters*
Gerätenennanschlussspannung rated converter supply voltage * *eines Stromrichters*
Gerätenenndaten converter ratings * *eines Stromrichters*
Gerätenennstrom rated current of converter * *eines Stromrichters*
Gerätereaktion drive response * *Fernsteuermöglichktn. ü.d.serielle Automatis.-Schnittstelle e. Antriebsstromrichters* || converter response * *(→PROFIBUS-Profil)*
Geräteschaltplan equipment circuit diagram
Geräteschutz device protection
Gerätesicherheitsgesetz equipment safety law
Gerätetausch unit replacement || converter replacement * *von Stromrichtergeräten*
Geräteteil component
Gerätetreiber device driver * *Softwaretreiber für Rechner-Peripheriegeräte*
Gerätetür converter door * *eines Stromrichters*
Gerätetyp device type
geräteübergreifend global * *(PROFIBUS)*
geräumig spacious || amply-sized || roomy
Geräumigkeit spaciousness || roominess
Geräusch noise || noise level * *~pegel* || audible noise * *hörbares ~*
geräuscharm low-noise || silent * *leise* || whisper quiet * *flüsterleise* || quiet || smooth * *bei Motor; smooth running: ~er/geschmeidiger Lauf*
Geräuschdämmhaube noise-reducing cover
geräuschdämpfend noise-absorbing
Geräuschdämpfer muffler || silencer || sound absorber
Geräuschdämpfung noise suppression
Geräusche noise
Geräuschemission noise emission * *Grenzwerte für Motoren: DIN 57530/VDE 0530/IEC 34 jeweils Teil 9*
Geräuschentwicklung noise development || noise emission * *Abgabe, Emission* || noise level * *Geräuschspiegel*
Geräuschgrenzwert noise limit * *[el.Masch.] max. erlaubter Schallleistungspegel (VDE 0530 Teil 5)*
geräuschlos noiseless || silent
Geräuschmessung noise measurement * *für elektr. Maschinen: siehe DIN 45635* || noise level test * *[el.Masch.] siehe DIN 45635 Teil 10*
Geräuschminderung noise reduction
Geräuschniveau noise level * *low: niedriges; high: hohes*
Geräuschpegel noise level * *Messmethode: siehe DIN 45633 u. 45635; Motorgeräusch: siehe DIN VDE 0530* || sound level * *Schallpegel*
Geräuschspektrum noise spectrum
Geräuschstärke noise level
Geräuschwert noise-level value * *[dB (A)]; Messung: siehe DIN 45635, Motorgeräusch: DIN VDE 0530*
Gerbfass tanning vat
geregelt controlled || regulated || closed-loop controlled || settled * *Problem, Verfahren usw.* || stabilized * *Stromversorgung, Versorgungsspannung*
geregelte Feldstromversorgung controlled field-current supply
geregelter Antrieb variable-speed drive || adjusta-

geregeltes Feld 140

ble-speed drive || closed-loop-controlled drive || servo-controlled drive
geregeltes Feld controlled field || closed-loop controlled field * *mit Feldstromrückführung*
Gerichtsstand revenue * *Gerichtsort* || legal domicile * *Gerichtsort*
geriffelt rippled || having a ripple || knurled * *gerändelt* || fluted * *gerieft, gerillt* || scored * *(ein)gekerbt, gerillt* || corrugated * *gewellt*
gering little || small || poor * *unzureichend, mangelhaft* || low * *niedrig* || trifling * *unbedeutend* || slight * *leicht, unbedeutend* || unimportant * *unbedeutend, unwichtig* || light * *gering (z.B. Strom, Gewicht), leicht, Schwach-, leicht beladen*
gering halten keep down || keep low || keep to a minimum * *so gering wie möglich halten*
geringfügig little || slight * *leicht, gering, unbedeutend, schwach, ein wenig, oberflächlich* || negligible * *unwesentlich, nebensächlich, vernachlässigbar, unbedeutend* || insignificant * *bedeutungs/belanglos, unwichtig, unbedeutend, nichtssagend* || unimportant * *unwichtig, nicht von Gewicht* || trifling * *unbedeutend, oberflächlich* || trivial * *unbedeutend, gering, trivial, banal, oberflächlich* || petty * *Betrag, Vergehen* || a little bit * *(Adv.)* || slightly * *(Adv.)*
gerippt ribbed * *ribbed frame; geripptes (Motor)-Gehäuse* || finned * *z.B. Kühlkörper, Motorgehäuse*
Germanischer Lloyd German Lloyd * *Schiffs-Klassifizierungs- und -Versicherungsgesellschaft*
Germanium germanium
Germaniumdiode germanium diode
Gerüst frame * *z.B. Schalt~* || stand * *z.B. Walz~* || rack * *Gestell, Gerüst* || scaffold * *{Bautechnik} Bau/Montage~* || rack * *Gestell, Gerüst, Schalt~; open frame: offenes Schalt~* || gantry * *Gerüst, Stütze, Bock, (Signal-, Bedienungs-) Brücke, Portal-* || truss * *Gitterwerk, Gerüst, Träger {Bautechnik}, Fachwerk*
gesamt over-all * *über-alles* || entire || total * *über-alles, vollkommen* || complete * *komplett*
Gesamt- total || overall * *über alles* || whole * *das ganze* || global * *mit weitem Geltungsbereich* || entire || aggregate * *angehäuftes, angesammeltes, vereinigtes* || general * *allgemein, umfassend, General-, →Haupt-*
Gesamtanlage overall plant
Gesamtanzahl total number || total || aggregate figure
Gesamtbelastbarkeit overall load capacity || overall output rating || derated loading * *verminderte*
Gesamtbelastung total load
Gesamtbetrag total amount
gesamter total || overall
Gesamtgeräusch mehrerer Quellen noise addition of several sources * *{el.Masch.} Gesamt-Schalldruckpegel mehrerer Maschinen*
Gesamtkosten overall costs
Gesamtlebensdauer lifetime
Gesamtmenge aggregate || total amount * *Gesamtbetrag/summe* || overall quantity
Gesamtpreis total price
Gesamtprojekt over-all project
Gesamtwirkungsgrad overall efficiency
Gesamtzahl total number * *siehe auch →Gesamtanzahl*
gesättigt saturated
geschachtelt nested * *z.B. Unterprogramme, Interrupts usw.*
Geschäft operation * *~stätigkeit eines Unternehmens* || operations * *~stätigkeiten eines Unternehmens* || business * *(allg.)* || transaction * *Unternehmung, Verhandlung, Abwicklung* || deal * *Handel* || affair * *Angelegenheit* || trade * *Handel* || commerce

* *Handel* || firm * *Firma* || enterprise * *Unternehmmen* || concern * *Konzern* || shop * *Laden* || store * *Laden* || bargain * *(~s-) Abkommen, Handel, Geschäft, vorteilhaftes ~, günstiger Kauf*
Geschäft ist geplatzt deal is off
Geschäfte machen make deals || do business
Geschäfte tätigen transact business || do business || make deals
Geschäftsabschluss business transaction || business contract || orders secured || contract secured
Geschäftsbedingungen terms of trade
Geschäftsbereich business unit || company division || business group * *Abteilung* || sphere of action * *(allg./unternehmerischer) Tätigkeitsbereich* || business area
Geschäftsbericht fiscal report || business report || annual report * *jährlicher ~* || market report * *~ über die Marktlage* || financial results * *Ergebnisbericht*
Geschäftsbeziehung business * *do business with: in ~ stehen mit* || business relations
Geschäftsbeziehungen business connections || business relations
Geschäftsbeziehungen haben mit do business with || have business relations with
Geschäftseinheit business unit
Geschäftsfeld field of activity
geschäftsführend executive
Geschäftsführer executive || executive president || general manager || chief executive || managing director
Geschäftsführung executive management
Geschäftsgebiet division * *Abteilung* || sphere of action * *(allg./unternehmerischer) Tätigkeitsbereich*
Geschäftsgeheimnis trade secret || business secret
Geschäftsinhaber holder || owner of a business || proprietor || principal * *Direktor, Chef, Vorsteher, Geschäftsherr*
Geschäftsinteresse interest in business || business interest
Geschäftsjahr fiscal year
Geschäftslage business situation || business outlook * *Geschäftsaussichten*
Geschäftsleben business life
Geschäftsmann businessman
Geschäftspartner business partner || party to the transaction * *bei Geschäftsverhandlungen*
Geschäftsräume business premises || office * *Büro* || shop * *Laden* || store * *Laden* || business space
Geschäftsreise business trip
Geschäftsrisiko business risk
Geschäftsrückgang decline || falling off of business * *recession*
Geschäftssinn business sense || business acumen
Geschäftssitz place of business || location of business
Geschäftsstelle office || agency || secretariat
geschäftstüchtig efficient in business || capable in business || clever * *gewieft* || smart * *gewieft*
Geschäftsverbindung business connection || business contact || business relation
Geschäftsverbindung haben mit do business with
Geschäftszweig branch of business || line of business || sector || sub-division * *in einer Unternehmensorganisation*
geschaltet switched
geschalteter Reluktanzmotor switched reluctance motor
geschehen occur
geschickt clever * *gewandt, geschickt, raffiniert, gescheit, klug, begabt* || skilful * *at: zu, in: bei; geschickt, gewand, geübt, sachkundig; be skilful at: s.verstehen auf* || cleverly * *(Adv.) geschickt ausge-*

dacht, genial || ingenious * *geschickt ausgedacht* || play one's cards well * *geschickt handeln*
geschirmt screened * *siehe auch →abgeschirmt* || shielded
geschliffen ground || polished * *(figürl.) Vortrag usw.*
geschlitzt slotted
geschlossen closed * *auch Schaltkontakt* || shut * *verschlossen* || locked * *verschlossen, verriegelt, versperrt* || enclosed * *(durch Gehäuse) umschlossen, ab~* || fully-enclosed * *voll ~ (Gehäuse-Schutzart)* || self-contained * *in sich ~* || compact * *(figürlich) z.B.Stil* || consistent * *(figürlich) Arbeit, Leistung usw.* || round * *(figürlich) Arbeit, Leistung usw.* || united * *vereint* || uniform * *einheitlich* || en bloc * *(Adv.) als Gesamtheit, alles in einem Rutsch* || compact whole * *(Substantiv) ~es Ganzes* || private * *private party: ~e Gesellschaft* || in a body * *(Adv.) alle gemeinsam* || unanimously * *(Adv.) einstimmig* || closed-circuit * *Kreislauf-* || encased * *mit Gehäuse versehen*
geschlossene Ausführung enclosed version || close version
geschlossene Bauart totally enclosed type * *IP44* || totally enclosed type of protection * *IP 44* || totally enclosed type of enclosure * *IP44*
geschlossene Kapselung sealed enclosure
geschlossene Maschine closed machine || totally enclosed machine * *{el.Masch.} Maschine mit geschlossenem Gehäuse*
geschlossener Kontakt closed contact
geschlossener Kreislauf closed circuit || closed circulation
geschlossener Kühlmittelkreislauf closed coolant circuit * *siehe auch →Kühlmittel* || closed circuit of the heat exchanger
geschlossener Raum enclosed space * *wettergeschützter Bereich/Ort/Platz*
geschlossener Regelkreis closed-loop control circuit || control loop || closed control loop
geschmeidig smooth
geschmiedet forged
geschmolzen molten
geschützt protected || secured * *gesichert* || shielded * *abgeschirmt* || sheltered * *überdacht; in sheltered areas: in ~en Räumen*
geschwächt weakened
Geschwindigkeit speed * *Drehzahl, Bahn~* || velocity * *Oberflächen~, Bahn~* || pace * *Fahrt, Tempo, Schwung, Marsch~ z.B. eines Fahrzeugs* || rate * *Ereignisse/Menge pro Zeiteinheit*
Geschwindigkeitsdifferenz speed difference * *relative: relative ~; z.B. zw. Antrieb 1 und 2: [(v2 - v1)/v1] x 100%*
Geschwindigkeitsgleichlauf speed match || speed synchronism
Geschwindigkeitskorrektursignal speed correction signal * *z.B. von einem Tänzerlage- oder Zugregler* || velocity correction signal || speed trim signal * *z.B. von einem Tänzerlage- oder Zugregler*
Geschwindigkeitsregelung speed control || velocity control
Geschwindigkeitsregler speed controller
Geschwindigkeitsrelation speed ratio || draw * *→relative Geschwindigkeitsdifferenz [(v2-v1)/v1] x 100% zw. 2 Antrieben*
geschwungen curved
gesehen viewed * *right/left viewed: von rechts/links aus ~*
Geselle journeyman * *Handwerker*
Gesellschaft mit beschränkter Haftung limited company
Gesellschafter associate || companion || partner || shareholder

Gesenk die || swage
Gesenkfräsmaschine die-sinking machine
Gesetz law * *auch Natur~, naturwissenschaftliches ~; enact: erlassen* || rule * *Vorschrift, Regel* || regulation * *gesetzliche Vorschrift* || principle * *Naturgesetz, naturwissenschaftliches ~* || code * *~buch, Vorschriftenbuch*
Gesetzgebung legislation
gesetzlich vorgeschrieben statutory || determined by law || prescribed by law
gesetzliche Bestimmung legal issue || statutory regulation
gesetzliche Vorschrift legislative provision || legal regulation
Gesetzmäßigkeit regularity * *Regelmäßigkeit, Gleichmäßigkeit, Stetigkeit* || principle * *Prinzip, Grundsatz, Regel, (Natur)Gesetz*
Gesichtspunkt aspect || issue || point of view * *from the ... point of view: aus/unter dem ~ des ... betrachtet* || viewpoint || motive * *Beweggrund* || angle * *(amerikan.) Blickwinkel* || factor * *Faktor* || criterion * *Merkmal, Kriterium*
gesintert sintered || baked || clinkered || trickled
gesondert separate * *getrennt; to be ordered separately: ~ zu bestellen* || discrete * *einzeln* || individual * *einzeln* || apart * *einzeln, für sich; apart from: (ab)~ von; siehe auch →trennen*
gespeichert stored
gespeist fed
gesperrt disabled || locked || blocked || inhibited * *z.B. Regler*
Gespräch discussion * *Diskussion* || meeting * *Besprechung* || conference * *Konferenz* || conversation || talk || dialog * *Zwiesprache* || colloquy * *gelehrtes ~, Kolloquium*
Gesprächseinladung meeting request
Gesprächspartner interlocutor || person to talk to || contact person * *Ansprechpartner* || those present * *(Plural) die Anwesenden* || attending persons * *Anwesende* || opposite number * *Partner am Telefon, Gegenspieler*
Gesprächsprotokoll minutes of meeting || meeting report
Gesprächsthema topic * *(of conversation)* || subject || item * *Einzelpunkt (von mehreren)*
gestaffelt staggered * *gegeneinander versetzt angeordnet (auch zeitlich)* || grouped * *in Gruppen aufgeteilt* || graded * *abgestuft; siehe 'staffeln'* || cascaded * *kaskadiert* || in steps * *gestuft*
Gestalt shape * *take shape: ~ annehmen*
gestalten conceive * *er/ausdenken, ersinnen, planen, in Worten ausdrücken* || design * *entwerfen* || implement * *realisieren* || shape * *in Umrissen ~ (z.B. ein Programm, Ideen); formen* || form || plan
Gestaltung configuration * *Konfiguration, Anordnung*
gestanzt punched
gestatten allow || permit || tolerate * *dulden, tolerieren, zulassen*
Gestell frame * *auch 'Gerüst'* || rack || support || rack * *auch 'Schaltgerüst'*
gesteuert controlled || open-loop controlled * *nicht geregelt* || controllable * *steuerbar, z.B. Gleichrichterbrücke*
gesteuerter Gleichrichter controlled rectifier
gestört faulted * *fehlerhaft geworden* || disturbed * *durcheinander/in Unordnung gebracht* || interfered * *behindert, belästigt; auch durch elektromagnet. Störung, Netzrückwirkung usw.* || out of order * *nicht in Ordnung, defekt* || incorrect * *nicht richtig, nicht korrekt*
gestörter Programmablauf faulty program execution || incorrect program run
gestrichelt dash-lined
gestrichelte Linie dashed line

gestrichen machine-coated * *Papier* || coated * *Papier*
gestrichenes Papier coated paper || LWC paper * *Abk. f. 'Light-Weight-Coated paper': leicht ~*
gestuft graduated || graded || in steps * *(Adv.)*
gesunder Menschenverstand common sense
getaktet clocked || switched-mode * *Netzteil, Netzgerät* || pulsating * *pulsierend, gepulst*
getaktetes Netzteil switch-mode power supply
geteilt split * *aufgespalten, (in zwei Hälften) geteilt (z.B. Gehäuse)* || divided
geteuertes Ventil controlled valve
Getränk beverage
Getränkeabfüllmaschine bottling machine
Getränkeindustrie beverage industry
getrennt separate * *voneinander* || discrete * *einzeln* || individual * *einzeln* || segregated * *getrennt, abgesondert, abgespaltet, isoliert*
getrennt angetrieben separately driven
getrennt einstellbar can be separately adjusted * *kann getrennt eingestellt werden*
getrennt erregt separately excited * *Motor*
getrennt verlegen lay in seperate ducts * *z.B. Motor- und Signalleitungen in unterschiedlichen Kabelkanälen/-pritschen* || route separated
getrennt zu bestellen to be ordered separately
getrennte Wicklungen separated windings * *(el. Masch.) galvan. voneinand. getrennte Wicklungen bei polumschaltbarem Motor*
Getriebe gearbox || gear unit || gearing || gear * *primary/secondary gear: ~ein-/ausgangsstufe* || gearhead * *an Getriebemotor* || speed variator * *stufenlos verstellbar, z.B. PIV-Getriebe* || speed reducer * *→Untersetzungsgetriebe* || step-up gearing * *Hochsetzgetriebe* || gear-box || transmission * *Transmission, Übersetzung (-sgetriebe)* || gear transmission
Getriebeausgang gearbox output || gear output
Getriebeausgangsdrehzahl gear output speed || gearbox output speed
Getriebeeinbausatz gear component set
Getriebegehäuse gear case || gear frame || gear housing
Getriebekasten gearcase || gearbox || gear frame
getriebelos gearless || ungeared || direct driven * *direkt angetrieben*
Getriebemotor gearbox motor || geared motor || gearhead motor || reducer motor * *mit Untersetzungsgetriebe* || reducer * *mit Untersetzungsgetriebe* || gearmotor || gearmotor unit
getrieben driven || output * *Maschinenteil*
Getriebeöl gearbox oil
Getriebeprüfstand gear testing machine || gearbox test stand || gearbox test bed || gear wheel testbed
Getriebeschonung low gearbox stressing * *z.B. durch Hochlaufgeber vor oder hinter dem Momentenregler* || gearbox protection || gearbox protection ramp * *durch Hochlaufgeber vor oder hinter dem Momentenregler*
getriebeseitig gearbox-side
Getriebesteifigkeit gear stiffness
Getriebestufe gearbox stage || gear stage || reduction stage || speed * *Gang*
Getriebestufenumschaltung gearbox stage changeover
Getriebetechnik transmission engineering * *als Fachrichtung*
Getriebeübersetzung gear ratio * *"i" entspr. Getriebe-Eingangs- zu Ausgangsdrehzahl* || reduction ratio * *bei Untersetzungsgetriebe*
Getriebeumschaltung gearshift
Getriebeuntersetzung gear reduction || speed reduction || gear reduction ratio * *~sverhältnis*
Getriebeverhältnis gear ratio
gewachsen sein cope with * *fertigwerden mit* || be equal to || be a match for || be up to || measure up to || tackle * *fertig werden mit, lösen (z.B. Aufgabe, Problem)*
gewählt selected || chosen
gewähren grant * *bewilligen* || allow * *bewilligen, einräumen* || accord * *bewilligen* || concede * *einräumen* || give * *geben, darbieten* || yield * *zugestehen, einräumen, hergeben* || furnish * *liefern, beschaffen, bieten, ausstatten, ausrüsten, versehen* || afford * *gewähren, spenden* || let alone * *allein/in Ruhe lassen, sich selbst überlassen* || admit * *zulassen, erlauben* || offer * *anbieten, sich bereit erklären zu (z.B. zu Sonderpreis, Vorteil)* || let have his way * *gewähren lassen* || give full play * *gewähren lassen* || leave alone * *sich selbst überlassen, allein/in Ruhe lassen*
gewährleisten guarantee * *garantieren, sicherstellen, (sich ver-) bürgen (für), Grantie leisten* || warrant * *garantieren, (sich ver)bürgen (für), haften für, bestätigen* || ensure * *sicherstellen, Gewähr bieten für, garantieren, für etw. sorgen* || vouch for * *['wautsch] (sich ver-) bürgen für* || vouch that * *['wautsch] dafür bürgen, dass*
Gewährleistung guarantee * *Garantie, Gewähr, Bürgschaft* || warranty * *~ des Verkäufers; under: unter/in der* || guaranty
Gewährleistungsanspruch warranty claim * *warranty credits: Erstattung/Gutschrift von Gewährleistungsansprüchen*
Gewährleistungsausschluss warranty exclusion || warranty disclaimer * *~erklärung*
Gewährleistungsbedingungen terms and conditions of warranty || terms of guarantee
Gewährleistungsfall warranty case
Gewährleistungsfrist warranty interval || warranty period
Gewährleistungsreparatur warranty repair
Gewährleistungsverfahren warranty procedure
Gewährleistungszeit warranty period
Gewalt force * *zwingende Kraft, Gewalttätigkeit; by force/forcibly: mit Gewalt; force majeure: höhere Gewalt* || power * *of/over: über; judicial power: richterliche Gewalt* || sway * *Herrschaft (over: über)* || control * *Herrschaft (of: über); beyond one's control: höhere Gewalt* || dominion * *Herrschaft (over: über)* || might * *zwingende Kraft* || restraint * *Zwang, Beschränkung, Hemmnis, Einschränkung* || violence * *Gewalttätigkeit* || vehemence * *Wucht* || impact * *Wucht* || authority * *z.B. gesetzgebende*, || sheer force * *nackte Gewalt* || with might and main * *mit aller Gewalt* || at all costs * *um jeden Preis* || by hook or crook * *mit aller Gewalt* || government * *Kontrolle, Leitung, Verwaltung, Regierung*
Gewalt anwenden use force
gewaltig tremendous * *enorm*
gewaltsam by force
gewalzter Stahl rolled steel
Gewebe woven material || woven fabric || textile || web || cloth * *Tuch, Stoff, Leinwand (Buchbinderei), Lappen* || netting * *Netz*
gewellt corrugated || rippled
Gewerbe trade
Gewerbeaufsichtsbeamter factory inspector
Gewerkschaft union || trade-union || labour-union
Gewicht weight * *low: geringes* || emphasis * *Betonung, Schwerpunkt, Nachdruck, Deutlichkeit; lay emphasis on: ~/Wert legen auf* || focus * *Brenn-/Schwerpunkt (des Interesses, der Bemühungen, der Arbeit usw.)*
gewichten weight || scale * *skalieren, normieren eines Signals*
gewichtet weighted || scaled * *skaliert* || valued * *bewertet*
Gewichtsersparnis weight saving

Gewichtung weighting ‖ scaling * *Skalierung, Normierung* ‖ intervention * *Eingriff*
gewickelt wound
Gewinde thread * *right-/lefthand(ed): rechts/linksgängiges* ‖ worm ‖ tapped hole * *Gewindeloch*
Gewinde bohren tap
Gewinde- thread ‖ srewed * *mit (Außen-) Gewinde versehen* ‖ threaded * *mit Gewinde versehen* ‖ threading * *Gewinde-herstellend* ‖ tap * *Gewindebohr-, Gewindebohrer-*
Gewindeanschluss thread connection
gewindebohren tap
Gewindebohrer screw tap ‖ tap
Gewindebohrung tapped hole ‖ threaded hole ‖ threaded bore
Gewindebolzen threaded bolt
Gewindedrehbank threading lathe
Gewindeeinsatz helicoil * *zur Reparatur eines ausgerissenen Gewindes*
Gewindegang thread
Gewindegröße thread size
Gewindeherstellungsmaschinen threading and tapping machines ‖ screwing and threading machines
Gewindekommutator threaded-type commutator * *bei DC-Maschine* ‖ skewed commutator * *geschrägter Kommutator bei DC-Maschine*
Gewindelänge length of thread
Gewindelehre thread pitch gauge
Gewindeloch tapped hole
Gewindesackloch tapped blind hole
Gewindeschneideisen screw die
Gewindeschneiden thread cutting ‖ thread hobbing * *mit Wälzfräser* ‖ thread * *(Verb) Gewinde schneiden in, mit Gewinde versehen*
Gewindeschneidkopf screwing chuck
Gewindeschneidmaschine threading machine
Gewindespindel screw spindle ‖ threaded spindle ‖ screwed spindle ‖ elevating screw * *Hubspindel* ‖ worm-gear spindle
Gewindestange screwed rod ‖ threaded rod
Gewindesteigung lead * *in axialer Richtung (Höhe einer vollen Umdrehung)* ‖ pitch * *achsparalleler Abstand, Gewindesteigung* ‖ lead angle * *Steigungswinkel*
Gewindestift grub screw * *Madenschraube* ‖ headless screw ‖ screwed pin ‖ threaded pin ‖ stud bolt * *Stehbolzen* ‖ threaded rod * *Gewindestange* ‖ grubscrew * *Madenschraube*
Gewindestutzen threaded spud * *zum Anbau eines Rohrs* ‖ threaded pipe ‖ threaded adapter nipple * *kleiner ~* ‖ gland * *z.B. zur Kabeldurchführung (bei →PG-Verschraubung)* ‖ threaded flange * *Gewindeflansch* ‖ threaded pipe adapter * *für Rohranschluss*
Gewindetiefe depth of thread
Gewindewalzen thread rolling
Gewindewalzmaschine thread rolling machine
Gewindezapfen threaded stud ‖ threaded pin ‖ threaded journal * *Dreh-/Lager-/Wellenzapfen* ‖ threaded stem
gewinkelt angular
Gewinn gain * *{Elektr.} Verstärkung* ‖ earnings * *Profit,Geschäftsergebnis* ‖ profit * *Profit* ‖ win * *Sieg* ‖ yield * *Ertrag* ‖ returns * *Ertrag, Verzinsung, Gewinn, Rückfluss* ‖ proceeds * *Erlös* ‖ rate of yield * *prozentualer ~ (Ergebnis bezogen auf den Umsatz)*
Gewinn- und Verlustrechnung profit-and-loss statement * *(amerikan.)*
gewinnen gain * *z.B. Erfahrung* ‖ acquire * *erlangen, erreichen* ‖ win * *(Sieg, Preis, Unterstützg. usw.) ~; zu Auftrag usw. gelangen; s.durchsetzen, siegen* ‖ retrieve * *(Wieder-) gewinnen/-erlangen, (Verlust usw.) wettmachen, der Vergessenheit entrei-*

ßen ‖ obtain * *erhalten, (Bodenschätze, Zwischenprodukte usw.) gewinnen* ‖ recover * *aus Altmaterial usw. ~* ‖ extract * *Bodenschätze, Stoff durch chemische Prozess usw. ~* ‖ derive * *durch chem. Prozess ~* ‖ be winner * *siegen* ‖ gain in * *(an Bedeutung, Ansehen usw.) ~; gain ground: an Boden ~*
Gewinnspanne profit margin
gewissenhaft conscientious ‖ scrupulous * *peinlich genau* ‖ careful * *→sorgfältig*
gewisser certain
gewissermaßen so to speak ‖ in a manner of speaking ‖ as it were ‖ to some extent * *in gewissem Maße* ‖ in a way * *in gewissem Maße*
gewöhnlich usually * *(Adv.)* ‖ normal * *normal* ‖ standard * *standardmäßig* ‖ usual * *gebräuchlich* ‖ ordinary * *gewöhnlich (siehe auch →in gewohnter Weise)* ‖ regular * *üblich*
gewölbt arched ‖ vaulted ‖ domed ‖ convex ‖ cambered
gewollt intended * *beabsichtigt* ‖ intentional ‖ intentionally * *(Adv.)*
gewünscht intended * *beabsichtigt, bezweckt* ‖ required * *erforderlich, benötigt, gefordert* ‖ desired * *erwünscht*
gewünschte Funktion desired function
gewurzelt grouped * *mit gemeinsamem Bezugspotenzial (z.B. Digital-Ein-/Ausgänge)* ‖ connected to common potential * *mit gemeins. Bezugspotenzial (z.B. Digital-Ein-/Ausgänge)*
gezackt serrated ‖ intended * *eingekerbt* ‖ scalloped ‖ jagged ‖ ragged ‖ notched * *gekerbt*
gezahnt toothed ‖ cogged ‖ notched * *gekerbt* ‖ serrated * *gezackt*
gezielt carefully directed ‖ controlled ‖ well-aimed ‖ purposive * *Maßnahmen usw.* ‖ directed to specific objectives * *Maßnahmen usw.* ‖ well-calculated ‖ selective * *selektiv, auswählend, Auswahl-, trennscharf* ‖ directed * *specific* ‖ well-placed * *gut plaziert* ‖ precisely placed * *genau plaziert*
gezielte Werbung specific advertising ‖ selective advertising
gezündet fired * *z.B. Thyristor*
gieren yaw * *(um d.Hochachse) gieren/scheren (Flugzeug, Kran); vom Kurs abkommen (Schiff); abweichen*
Gießanlage casting plant * *Stahl/Walzwerk*
Gießen casting * *z.B. Gusseisen* ‖ pour * *(Verb) Flüssigkeit ~* ‖ cast * *(Verb) Metall, Gussstücke* ‖ mould * *(Verb) spritzgießen (Kunststoff, Glas usw.)* ‖ mold * *(Verb; amerikan.) spritzgießen (Kunststoff, Glas usw.)*
Gießerei casting plant ‖ foundry * *siehe auch →Hüttenwerk*
Gießharz cast resin ‖ moulding resin ‖ cast-resin
gießharzisoliert cast-resin insulated ‖ resin-encapsulated
Gießmaschine casting machine ‖ caster
Gießtiegel ladle * *z.B. im Stahlwerk*
Gießverfahren casting process ‖ curtain coating process * *zum Auftragen von Leim, Klebstoffen usw. auf eine Fläche*
Gießwalze chill roll * *wassergekühlte Walze in der Kunststoff-Folienherstellung (hinter Extruder angeordnet)*
Giftigkeit toxity
gigantisch giant
Gips gypsum
Gipsfaserplatte gypsum fiberboard * *{Bautechnik} zum Innenausbau*
Gipskartonplatte plasterboard
Gipsplatte gypsum board * *{Bautechnik} für den Innenausbau* ‖ plasterboard
Gitter grating ‖ lattice ‖ grille * *an Tür, Fenster usw.* ‖ grate * *Rost* ‖ wire-lattice * *Drahtgitter* ‖

Gitterboxpalette

grid * *auf Diagramm/Landkarte; Raster* || fence * *Zaun*
Gitterboxpalette wire-lattice pallet || grid-box pallet
gitterförmig latticed || grated || trellised || in a grid pattern * *z.B. Halbleiterstruktur*
GL GL * *Abkürz. für 'Germanischer Lloyd': deutsche Schiffsklassifikationsgesellschaft*
Glanz gloss
Glas glass
Glasfaser fibre-optic * *Lichtleiter-* || glass-fibre || glassfibre
Glasfaser-Element fibre-optic device
Glasfaser-Lichtleiter glas-fibre optocable
Glasfaserkabel fibre-optic cable || glassfibre cable
glasfaserverstärkt glass-fibre reinforced * *(brit.)* || glass-fiber reinforced * *(amerikan.)*
glasfaserverstärkter Kunststoff glass-fibre reinforced plastic
Glasindustrie glass industry
Glasseide glass silk
glatt smooth || plain || slippery * *schlüpfrig, glatt, glitschig, aalglatt, gerissen (Person), heikel (Thema)*
glatte Welle keyless shaft * *Welle ohne Federkeil*
Glätte smoothness * *Glätte, Reibungslosigkeit, Geschmeidigkeit, glatter Fluss, Eleganz, Gewandheit* || sleckness || slipperiness * *Schlüpfrigkeit* || polish * *Politur* || gloss * *Glanz, Politur*
glätten smooth || filter * *filtern, sieben* || glaze * *Papier kalandrieren; (Papier/Leder) satinieren*
Glättung smoothing || filtering * *Filterung über Glättungsfilter*
Glättungsdrossel smoothing reactor * *z.B. i. Ankerkreis eines DC Motors* || smoothing choke
Glättungsglied filter element || first order filter element * *PT1-Glied* || smoothing element * *Glättungsglied*
Glättungsinduktivität smoothing reactor * *{Stromrichter}*
Glättungskondensator smoothing capacitor || filter capacitor || filtering capacitor
Glattwalzmaschine roller finishing machine * *{Werkz.masch.}*
Glättwerk calender * *Kalander, glättet Papier, Textil usw. durch Zusammenpressen mittels Walzen/Dampfzylinder*
Glättzylinder calender * *Kalander zur Glättung/ Satinierung von Papier und Textilstoffen* || glazing cylinder * *zur Satinierung von Papier, Textilstoffen, Leder usw.* || M. G. cylinder * *Abk. für 'Machine Glazing cylinder'* || yankee cylinder * *großer Trocken- und Glättzylinder in Papiermaschine* || yankee dryer * *großer Trocken- und Gättzylinder in Papiermaschine* || machine glazing cylinder
glauben think * *vermuten* || guess * *annehmen*
Glaubwürdigkeit credibility || authenticity || trustworthiness
gleich equal to * *identisch, mathematisch gleich mit (something equals ...: etwas ist ~ ... (in Gleichung)* || same * *ähnlich, identisch* || synonymous * *with: gleichbedeutend mit* || the same * *(Adv.)*
gleich bleibend permanent * *(fort)dauernd, bleibend, permanent, ständig, ständig* || constant * *unveränderlich, konstant* || continuous * *ununterbrochen, fortlaufend* || always the same || unchangeable * *unveränderlich* || invariable * *unveränderbar/lich* || even * *gleich/regelmäßig, ausgeglichen, glatt, ruhig, gerade* || steady * *gleich bleibend/mäßig, ausgeglichen, fest, stabil* || consistent * *gleichmäßig, stetig, ausgeglich., ständig, fest, unveränderl., stabil (Preise)*
gleich sein be equal to || equal * *(Verb)*
Gleichanteil direct component || DC component * *bei Strom/Spannung/Leistung* || constant component

gleichartig of the same kind || homogeneous || similar * *ähnlich* || analogous * *analog, ähnlich* || uniform * *einheitlich*
gleichbedeutend equivalent * *to: mit* || tantamount * *to: mit; be tantamount to: etwas gleichkommen* || synonymous * *bedeutungsgleich, sinnverwand, gleichbedeutend (with: mit)*
gleichberechtigt equal-priority * *mit gleicher Priorität* || having equal rights
gleichen equal || be equal to || be similar to * *ähnlich sein* || resemble * *ähnlich sein* || be comparable to * *vergleichbar sein mit* || correspond to * *entsprechen* || be analogous to * *entsprechen* || be a parallel to * *entsprechen* || be like * *ähnlich sein*
gleichermaßen in like manner || likewise || equally
Gleichfluss unidirectional flux * *magnetic: magnetischer ~*
gleichförmig uniform || equal || steady * *→unveränderlich, gleich bleibend, gleichmäßig, stetig, regelmäßig* || invariable * *→unveränderlich* || monotonous * *monoton (auch steigend/fallend), eintönig* || steady * *unveränderlich, fest, stabil, stetig, regelmäßig*
gleichförmige Bewegung uniform movement
gleichgerichtet rectified
Gleichgewicht balance * *unbalance: aus dem ~ bringen; loose one's: verlieren; balance: ins ~ bringen* || equilibrium * *~szustand*
Gleichgewichtszustand state of equilibrium || steady state * *stationärer/eingeschwungener Zustand*
Gleichheit equality || equivalence * *Äquivalenz, Gleichwertigkeit*
gleichkommen be equivalent to
Gleichlauf synchronous operation || synchronism || running in synchronism
gleichlaufen run in synchronism
Gleichlauffehler synchronism deviation
Gleichlaufgenauigkeit synchronous accuracy
Gleichlaufgüte degree of synchronism || synchronous accuracy
Gleichlaufregelung synchronous speed control || synchronized drives control || synchronous control loop * *Gleichlaufregelkreis* || synchronizing control || synchro control || accurate synchronism control * *Winkelgleichlauf-Regelung* || synchronization control || synchronous control
Gleichlaufregler synchronized drives controller || synchronisation controller || synchronizing controller
Gleichlaufschwankungen fluctuations in synchronism || wow and flutter * *bei Magnetbandberät usw.* || synchronism deviation * *Gleichlauffehler*
Gleichlaufsteuerung synchronization control
gleichmäßig proportionate * *ebenmäßig* || symmetric * *ebenmäßig* || symmetrical * *ebenmäßig* || even * *ausgeglichen* || uniform * *gleichmäßig; uniform in texture: ~ strukturiert (Oberfläche)* || regular * *gleichmäßig* || constant * *konstant, gleich bleibend* || steady * *stetig, gleich bleibend, unveränderlich, ausgeglichen, fest, stabil* || smooth * *ruhig laufend* || consistent * *gleichmäßig, stetig, übereinstimmend, →homogen, ebenmäßig im Gefüge*
Gleichmäßigkeit uniformity || evenness || regularity || continuity || equableness
gleichphasig in phase || equiphase
gleichpolig homopolar
gleichrichten rectify
Gleichrichter rectifier
Gleichrichterbetrieb rectifier operation || rectifier range * *bei Umkehrstromrichter*
Gleichrichterblock rectifier block
Gleichrichterbrücke rectifier bridge
Gleichrichterdiode rectifier diode * *DIN 41781/82*
Gleichrichtersatz rectifier set
Gleichrichtertrittgrenze rectifier stability limit

Glimmentladung

* *minimaler Steuerwinkel, oft 30 Grad bei 4Q-Stromrichter* ‖ minimum firing angle ‖ minimum delay angle ‖ zero delay angle ‖ rectification limit ‖ minimum gating angle ‖ converter limit
Gleichrichtmittelwert   rectified average value
Gleichrichtung   rectification * *half-wave/one-way: Einweg~; fullwave/two-way: Zweiweg-/Vollwellen~* ‖ DC conversion ‖ rectifying * *das Gleichrichten*
Gleichrichtwert   rectified mean value ‖ rectified value
gleichsam   almost * *beinahe* ‖ so to speak * *sozusagen*
gleichsinnig   in the same sense
Gleichspannung   DC voltage ‖ direct-current voltage ‖ d.c. voltage
Gleichspannungs-Zwischenkreis   DC bus
Gleichspannungsanteil   DC voltage component
Gleichspannungshilfsschiene   DC voltage auxiliary bus
Gleichspannungsnetz   DC supply network
Gleichspannungsquelle   DC supply * *Versorgungspannung*
Glelchspannungsschiene   DC busbar * *Sammelschiene* ‖ DC voltage bus
Gleichspannungsversorgung   DC supply
Gleichspannungswandler   DC-DC converter ‖ dc-to-dc converter
Gleichspannungszwischenkreis   DC link ‖ DC intermediate circuit ‖ d.c. intermediate circuit
Gleichspannungszwischenkreis-Schiene   DC bus
Gleichspannungwandler   DC-DC converter
Gleichstrom   DC current ‖ direct current ‖ DC ‖ d.c.
Gleichstrom-Fahrleitung   DC traction supply system
Gleichstrom-Nebenschlussmotor   DC shunt-wound motor ‖ shunt wound DC motor
Gleichstrom-Normalerregung   normal DC excitation * *für Magnetbremse*
Gleichstrom-Permanentmagnetmotor   DC permanent magnet motor
Gleichstrom-Schnellerregung   high-speed DC excitation
Gleichstromanteil   DC component
Gleichstromantrieb   DC drive ‖ d.c. drive
Gleichstrombahn   DC traction system
Gleichstrombremsung   DC injection braking * *durch Gleichstrom-Einprägung in vom Netz getrennte Ständerwicklg. e.AC-Motors*
Gleichstromerregung   DC excitation * *z.B. Motor, Magnetbremse*
Gleichstromkreis   DC circuit
Gleichstrommaschine   DC machine ‖ D.C. machine
Gleichstrommotor   DC motor ‖ d.c. motor
Gleichstromnebenschlussmaschine   DC shunt-wound motor
Gleichstromnetz   DC supply system
Gleichstromschiene   DC busbar ‖ DC-bus * *on: an* ‖ common DC-bus * *gemeinsame*
Gleichstromseite   DC side * *on the: auf der*
gleichstromseitig   on the DC side
Gleichstromsicherung   d.c. side fuse ‖ DC side fuse link * *Sicherungseinsatz* ‖ DC fuse
Gleichstromsteller   DC chopper * *erzeugt aus konst. Gleichsp. e.variable Gleichsp.,DIN 57558/VDE 0558 jew.Tl.3* ‖ DC chopper regulator ‖ direct DC converter ‖ DC chopper converter ‖ d.c. chopper
Gleichstromsystem   DC system
Gleichstromtacho   DC tachometer ‖ DC tacho ‖ DC tach
Gleichstromtachogenerator   DC tachogenerator ‖ DC tacho ‖ DC tach

Gleichstromzwischenkreis   DC link ‖ DC link circuit ‖ d.c. link
Gleichtakt   common mode
Gleichtaktaussteuerung   common-mode control
Gleichtaktunterdrückung   common-mode rejection ‖ common mode rejection ‖ CMR * *Abk. für 'Common-Mode Rejection'*
Gleichteiligkeit   uniformity of components * *von Bauteilen*
Gleichung   equation * *mathematical: mathematische; boolean: Boolsche/logische*
gleichwertig   equivalent * *to: mit* ‖ of the same value * *~ bezüglich des Geld-/Nutzwertes* ‖ equal to ‖ on a par with
gleichwertiger Dauerbetrieb   equivalent continuous rating * *[el.Masch.] Dauerbetrieb mit gleicher Erwärmung wie reale Betriebsart*
Gleichwertigkeit   equivalence
gleichzeitig   simultaneous * *simultaneously occurring: gleichzeitig auftretend* ‖ synchronous ‖ coincident * *zusammenfallend* ‖ at the same time * *(Adv.)* ‖ together ‖ at one blow * *(Adv.)* ‖ all at once ‖ concurrent * *gemeinsam ablaufend, z.B.Programme*
Gleichzeitigkeit   coincidence ‖ concurrence ‖ concomitance ‖ simultaneitity
Gleichzeitigkeitsfaktor   coincidence factor ‖ simultaneity factor
Gleitbahn   slideway ‖ slide ‖ slipway ‖ guideway * *zur Führung* ‖ shoot * *Rutsche, Rutschbahn* ‖ chute * *Rutschbahn, Gleitbahn, Rutsche*
Gleitbelag   pad covering ‖ sliding coat
Gleiteigenschaften   antifriction property * *Reibarmut, →Leichtgängigkeit*
gleiten   slide ‖ glide * *(leicht) gleiten, dahingleiten* ‖ slip * *(aus)gleiten* ‖ make use of flexible working hours * *Gleitzeit in Anspruch nehmen*
gleitend   sliding ‖ flexible * *(figürl.)* ‖ slipping ‖ gliding
gleitende Mittelwertbildung   sliding-type mean value generation ‖ shifting-mode averaging ‖ sliding-mode mean value generation
Gleitfläche   sliding surface
Gleitgeschwindigkeit   sliding speed
Gleitkomma   floating point
Gleitlager   slide bearing ‖ sleeve bearing ‖ friction bearing ‖ plain bearing ‖ journal bearing ‖ sliding bearing
Gleitmittel   slip additive * *als Zusatz(mittel)* ‖ lubricant * *Schmiermittel*
Gleitpunkt   floating point * *(mathem.) Gleitkomma, Fließkomma*
Gleitpunkt-Arithmetikeinheit   floating point arithmetic unit
Gleitpunktarithmetik   floating point arithmetic
Gleitpunkteinheit   floating-point unit
Gleitpunktzahl   floating-point number
Gleitreibung   viscous friction ‖ running friction ‖ sliding friction
Gleitring   sliding ring
Gleitstein   sliding block * *z.B. an einer mechanisch betätigten Kupplung*
Gleitverschluss   sliding closure * *in der Verpackungstechnik*
Gleitzeit   flexitime ‖ flexible working hours
Glied   element * *active: aktives* ‖ component ‖ link * *Verbindungsglied* ‖ block * *Funktionsbaustein*
gliedern   arrange * *anordnen* ‖ organize * *organisieren* ‖ subdivide * *unterteilen (into: in)* ‖ structure * *strukturieren* ‖ group * *gruppieren, in Gruppen aufteilen* ‖ classify * *in Gruppen/Klassen aufteilen* ‖ divide * *(ein-) teilen*
Glimmeinsatzspannung   discharge inception voltage ‖ corona inception voltage
Glimmentladung   corona discharge * *Teilentla-*

dung; Mikrolichtbögen i.Dielektrikum b. Überschreit. d.Durchschl.festigk. || glow discharge || partial discharge * →Teilentladung
Glimmer mica * z.B. bei Kondensator, Isolierung (zwischen Kommutatorlamellen)
Glimmerblättchen mica lamina * zur Isolierung Leistungshalbleiter ↔Kühlkörper/Leiterkarte || mica lamella || mica splitting || mica flake
Glimmerplättchen mica lamina * zur Isolierung Leistunghalbleiter ↔Kühlkörper/Leiterkarte || mica splitting || mica flake || mica lamella
Glimmerscheibe mica lamina * zur Isolierung Leistungshalbleiter ↔Kühlkörper/Leiterkarte || mica lamella || mica splitting || mica flake
Glimmerzwischenlage mica segment * zur Isolierung zwischen den Kommutatorlamellen || mica separator * zur Isolierung zwischen den Kommutatorlamellen
Glimmlampe glow lamp || glim lamp
Glimmschutz corona shielding * {el.Masch.} Schutz vor →Glimmentladg. an durch Feldstärkespitzen gefährd. Stellen
Glimmtemperatur smouldering temperature * Temp., bei der explosionsfähige Staub-Luftgemische explodieren können
Globalisierung globalization
Glockenanker bell-shaped rotor * z.B. für Mikromotor (System "Faulhaber")
glockenförmig bell-shaped
Glossar glossary * Liste von Fachausdrücken mit kurzen Erklärungen
glücklicherweise fortunately
Glühanlage annealing plant * z.B. für Blech
Glühbehandlung annealing * zur Vergütung von Metall
Glühbirne electric bulb || incandescent bulb
Glühe annealing furnace * Glühofen für Blechbehandlung im Walzwerk
glühen anneal * Metallveredelung || glow || twinkle * glimmen
glühend heiß red hot
Glühfaden filament
Glühlampe incandescent lamp || electric bulb || filament lamp || tungsten filament lamp
Glühofen annealing furnace * z.B. in Walzwerk/ Bandbehandlungsanlage || annealer
Glut glow || heat
GMA GMA * Gesellschaft für Mes- u.Automatisierungstechnik: Assoc. for Instrumentation Automation Engin.
Goldgrube gold mine * (auch figürl.) || cash cow * (figürl.) Melkkuh, sehr gute Einnahmequelle (z.B. Produkt oder Unternehmensbereich)
Goldkontakt goldplated contact || gold contact
Goldspitzkontakt gold-point contact
Göpel animal-powered mechanism * Vorrichtung zum Antrieb landwirtsch. Maschinen, Pumpen usw. durch Zugtiere || drawing engine
graben dig
Grad degree * Winkelgrad, Grad elektr. || rate * Quote, Maßstab, Verhältnis, Ziffer || electrical degree * Grad elektrisch, z.B. Steuerwinkel bei Phasenanschnittsteuerung || class * z.B. Klasse, Funkstörgrad || level * z.B. Funkstörgrad, Schärfegrad || degrees * (Plural) Winkelgrad, Grad elektr.
Grad elektrisch electrical degrees
Gradient gradient
graduierter Ingenieur graduated engineer * undergraduated engineer: noch etwas ~
Grafik graphics || graph * z.B. Kurvendarstellung || chart * Diagramm, grafische Darstellung, Schaubild, Kurvendarstellung
Grafikbeschleuniger graphics accelerator * Grafikkarte mit eigener "Bildaufbau-Intelligenz" || graphics accelerator card

Grafikeditor graphical editor
Grafiker artist * advertising artist: Werbe~
Grafikkarte graphics card || video card || graphic adapter || video board
Grafikleistung graphics performance * z.B. Geschwindigkeit eines Computers b. Bildaufbau auf e. Monitor
grafikorientiert grahics-oriented
Grafiksymbol graphical symbol
Grafiktreiber graphic driver || video driver
grafisch graphic * (amerikan.) || graphical * (engl.) || graphically * (Adv.)
grafisch darstellen chart * in Schaubild, Diagramm || picture || visualize * visualisieren || illustrate
grafische Anwenderoberfläche graphical user interface * (z.B. MS-WINDOWS (R)) || GUI * Abk.f. 'Graphical User Interface' (z.B. WINDOWS (R))
grafische Darstellung graph || graphic representation || diagram || chart || illustration || drawing * Zeichnung || graphical representation || graphic presentation
grafische Oberfläche graphic operator interface || graphic user interface * für Rechnerbenutzer || GUI * Abk.f. 'Graphic User Interface': graf. Benutzeroberfläche (z.B. MS-Windows)
grafische Projektierung graphical configuring * z.B. m.d.SIEMENS Projektiersprache STRUC G für d.Regelsystem SIMADYN D (R) || graphic configuring
grafisches Gewerbe printing trade
Grafit graphite
Grammatik grammar * auch einer Programmiersprache
Granulat granulate * plastic granulate: Kunststoff~ || granulated material || granular material || pellets * kleine Kügelchen; z.B. zur Kunststoff-Verarbeitung || granules || granular matter
Granulator pelletizer * zur Herstellung von Kunststoffgranulat
granulieren granulate
Granuliermaschine pelletizing machine * zur Herstellung von Kunststoffgranulat || pelletizer
Granulierung granulation
Grat burr
gratis free of charge
grau grey
Grauguss cast iron * kohlenstoffhaliges Gusseisen (Kohlenst. liegt als Lamellengrafit vor; DIN 1691) || cast-iron || grey cast-iron
Graugussgehäuse cast-iron housing * z.B. für Elektromotor
Graupappe chipboard
gravieren engrave || inscribe
gravierend serious || aggravating * (auch jurist.) erschwerend
graviert engraved
Gravur engraving
Gray-Code Gray code
greifbar ready on hand * zur Hand || within reach * in Reichweite || available * verfügbar || palpable * (figürl.) augenfällig, handgreiflich, fühl-/greifbar; impalpable: nicht ~ || tangible * (figürl.) greifbar, fühlbar, real || concrete * (figürl.) konkret (siehe auch dort)
greifen grip * packen, (er)greifen, festhalten, festklemmen || seize || grasp || catch hold of || gear into each other * ineinander~ || mesh * ineinander~ || grab * (hastig) ~ (auch mit Greiferkran), an sich reißen, fassen, packen, schnappen, einheimsen
Greifer pick-up * Aufnehmer || clamshell * ~ eines Baggers || clam * ~ eines Baggers || grab * ~ eines Krans/Baggers || gripper * Greifer, Halter; gripping lever: Spannhebel; gripping Tool: Spannwerkzeug; auch b.Roboter || claw * Klaue, Kralle; siehe auch

Größe

"Klaue" || shuttle * ~ zum Hin- und Herbewegen || hook * Haken || engaging dogs * Greifhaken || gripping device * Greifvorrichtung || bucket * ~ eines Baggers/Krans || grab bucket * ~schaufel eines Krans/Baggers || hooking-on arm * ~ eines Krans || grapple claw * ~ eines Fördermechanismus
Greiferhubwerk grab hoisting gear * {Hebezeuge}
Greiferkran grab crane
Greifvorgang grabbing operation * {Hebezeuge} bei Greiferkran/Bagger
Greifwerkzeug gripping tool * z.b. für Roboter
Gremium group * siehe auch →Arbeitsgruppe || board || authoritative body || working committee * siehe auch →Arbeitskreis
Grenzbeanspruchung maximum limit stress || tolerated stress
Grenzbedingung boundary condition
Grenzbelastung boundary load * z.B. bei einem Walzgerüst || critical load || limiting load
Grenzdauerstrom limiting continuous current
Grenzdrehzahl critical speed * bei Gleichstrommotor z.b.durch Kommutator begrenzt || limit speed || limiting speed * mechanical: mechanische
Grenze limit * auch 'Begrenzung'; upper/lower/utmost: obere/untere/äußerste; (at: an) || boundary || margin * Toleran~, Spielraum || border * Landes~; border on: grenzen an (auch figürl.) || barrier * Barriere || limiting * Begrenzung || extremity * äußerstes Ende || border line * Grenzlinie || threshold * Schwelle || bound * Grenze, Schranke, Bereich; within wide bounds: in weiten ~n; keep within b.: in ~en halten || frontier * (Landes)~, Grenzbereich (figürl.); frontier town: Grenzstadt; new frontiers: neue Ziele
Grenzerwärmung temperature-rise limit
Grenzfrequenz limiting frequency * z.B. von AC-Motor || cut-off frequency || critical frequency * kritische Frequenz || maximum frequency || maximum operating frequency
Grenzkennlinie limiting characteristic
Grenzkurve limit curve || limiting characteristic
Grenzleistung limit rating || power limit
Grenzsignalgeber limit signal transmitter * z.B. Endschalter
Grenzspaltweite maximum width of gap * oberhalb besteht Zündgefahr; Kriterium für Einstufg. i. Explosionsklasse
Grenzstrom limit current
Grenztemperatur temperature limit * {el. Masch.} höchstzulässige Dauertemperatur (VDE 530 Teil 18.32, DIN 46416) || temperature-rise limit * Grenzübertemperatur || safety temperature || limiting temperature || permissible temperature * zulässige Temperatur || maximum permissible temperature
Grenzübertemperatur maximum permissible temperature rise * max. zulässige Temperaturerhöhung (z.B. VDE0530 Teile 1 u. 22) || temperature-rise limit || maximum temperature rise * z.B für Motor || limiting temperature rise
Grenzwert limit value || limit || limiting value || boundary value
Grenzwert überschreiten violate a limit
Grenzwertklasse limit class * siehe VDE 0875, EN 55011; DIN VDE 0871 bezügl. Funkentstörung
Grenzwertmelder limit value monitor || limit monitor || limit detector || comparator * Komparator || limit value indicator || signal comparator
Grenzwertmeldung limit indication || limit signal
Grenzwertüberschreitung off-limit condition || limit violation
Grenzwertüberwachung limit monitoring
Griechenland Greece
Griff handle * Halte~, Stiel; auch "Anfasseigenschaften" von Papier, Textilstoff usw. || grip * Zu~,

Halte~ || grasp * (Zu-) Griff, Macht,Gewalt, Verständnis; within one's grasp: in Reichweite/in jmdes. Gewalt || hold * Halt, Griff, Stütze; Gewalt, Macht (on/over/of: über); catch hold of: zu fassen kriegen || snatch * (auch figürl.) schneller ~ (at: nach) || clutch * klammernder ~ (at: nach) || handhold * Festhalte~ || feel * "Anfasseigenschaften" von Stoff, Papier usw. || stranglehold * würgender ~ || knob * Tür~ || pull * Zieh~ || lever * Hebel
griffbereit ready to hand || handy
grob rough || coarse * auch für Einstell/Justierelement, Schleifmittel
Grobabgleich coarse adjustment || coarse adjust || coarse setting || coarse tuning
Grobabschätzung rough estimation
Grobblech heavy plate || plate || rough plate
Grobblechwalzwerk plate rolling mill
grobe Fahrlässigkeit culpable negligence || gross negligence || severe carelessness
Grobeinstellung coarse setting * z.B. an einem Potentiometer || coarse adjustment
Grobentwurf draft
Grobimpuls coarse pulse * z.B. zur Positionserfassung || rough pulse
Grobjustierung rough adjustment || coarse adjustment
Groblech heavy plate || plate
Grobregelung coarse control
groß great || large * umfangreich, large motor: ~motor || big * umfangreich; grow big: ~ werden || tall * ~ von Wuchs (auch Baum, Haus usw.) || high * zahlen-/wertmäßig hoch, z.B. Spannung, Strom, Qualität || spacious * geräumig || vast * weit || extensive * weit || huge * riesig || enormous * enorm, riesig || immense * immens || large-scale * in großem Maßstab/Stil, ~ angelegt || bad * (Fehler) || intense * (Hitze) || severe * (Kälte, Schaden, Nachteil, Unwetter usw.) || capital * capital letter: ~buchstabe || high-rating * (Adj.) z.B. Motor, Stromrichter, Pumpe mit hoher Leistung || heavy * schwer (Maschine, Fahrzeug, Motor, Last, Verlust, Unwetter), stark, heftig, umfangreich || bulky * (sehr) umfangreich, massig, sperrig: bulky goods: Sperrgut || ample * weit (-läufig/-räumig), geräumig, reichlich (bemessen) ausführlich
groß herauskommen go big
Groß- heavy-duty * Schwerlast || large || large-capacity * z.B. Ventilator, Pumpe, Zentrifuge
Großanbieter big player
Großantrieb large drive || high-power drive * Antrieb mit hoher Leistung || high-rating drive || large hp drive * 'mit vielen horsepowers'
Großbritannien Great Britain || U.K. * United Kingdom: England, Schottland, Wales u.Nordirland || United Kingdom * England, Schottland, Wales u.Nordirland
Großbuchstabe capital letter
Großcomputer mainframe || mainframe computer || large computer
Größe size * Umfang, Format; to DIN: nach DIN || largeness * Umfang, Format || stature * Gestalt || dimensions * Abmessungen || width * Geräumigkeit, Weite || quantity * Menge, Höhe, Pegel, Wert e.Signals, auch Einflussgröße, physikal.Größe, Informationsmenge || volume * Rauminhalt, Höhe des Umsatzes usw. || bulk * Rauminhalt || cubic contents * eines Gefäßes || format * Format || variable * sich ändernde mathem./physikal.Größe, Variable || factor * Einflußgröße,Faktor || magnitude * Betrag, Ausmaß, Bedeutung || vector quantity * vektorielle Größe || scalar quantity * skalare Größe || variable quantity * variable Größe || amount * Betrag, Summe, Höhe, Ergebnis, Wert, Bedeutung || entity * Einheit, Grundelement, (Zeichnungs-) Objekt, Gebilde || parameter * Einflussgröße, Parameter

Größen

Größen quantities || variables * *Variablen*
Größenattribut quantity attribute * *(→PROFIBUS-Profil)*
Größenindex unit of measure * *(→PROFIBUS-Profil)*
Größenordnung order of magnitude || order * *Art, Rang, Klasse* || volume * *Umfang, Inhalt, Masse* || arrangement as to size * *Ordnung/Einteilung nach der Größe* || terms * *in terms of minutes: in einer ~ von Minuten*
großer Leistung high-rating * *(Adj.) z.B. Motor, Stromrichter, Pumpe, Ventilator*
größer greater * *(Komparativ) than: als* || larger || exceeding * *wert/betragsmäßig größer sein als* || higher-rating * *mit höherer Leistung, z.B. Motor, Stromrichter, Pumpe*
größer als larger than || greater than
größer gleich greater than or equal to || equal to or greater than
größer sein als exceed
größerer Teil greater part * *of: von* || larger half * *von zweien* || better half * *besserer Teil von zweien* || majority * *einer größeren Menge*
großflächig large surface of * *(Adj.)* || large area of * *(Adj.)* || with large surface area * *(Adv.) z.B. Auflagefläche*
großflächig mit Erde verbinden connect to earth with the greatest possible surface area
Großhandel wholesale || wholesalers * *Großhändler*
Großhändler wholesaler || distributor * *Vertragshändler* || wholesale dealer
Großkunde key-account customer || key-account
Großmaschine heavy machine
Großmotor large motor
Großraum- large-capacity * *mit großem Fassungsvermögen*
Großrechner mainframe || mainframe computer || large computer
Großserienfertigung large batch production || mass production || large-scale production || large-series production
größt greatest || largest || highest-rating * *mit der größten Leistung, z.B. Motor, Stromrichter, Pumpe* || largely * *(Adv.) zum größten Teil* || maximum * *wert-/zahlenmäßig* || biggest || highest || utmost * *äußerst, höchst, größt, entlegenst, fernst*
Großteil large portion
größtenteils for the most part || mostly || chiefly
größtmöglich maximum || largest possible || greatest possible || best || utmost
Großunternehmen large-scale enterprise || big company
großzügig dimensioniert generously-dimensioned || oversized || generously sized || oversize || amply dimensioned * *mit reichlich Platz ausgestattet (z.B.Klemmenkasten)*
Grübchenbildung pitting * *Grübchen-/Narbenbildung: Verschleiß-/Korrosionseffekt, z.B. in einem Lager* || grooving * *Riefenbildung z.B. an einem Kommutator*
Grube mine * *Bergwerk* || pit * *Bergwerk, Schacht; Montagegrube* || hollow * *Höhlung* || cavity * *Höhlung*
grün green
Grund reason * *Ursache, Vernunfts~; for this reason: aus diesem ~; not unreasonably: nicht ganz ohne ~* || perspective * *Blickwinkel, Perspektive* || cause * *Ursache* || occasion * *Anlass, Anstoß, Ursache, Gelegenheit* || motive * *Beweg~* || argument * *Beweis~, arguments for and against: Gründe für u. wider* || bottom * *~ von Gefäßen* || foundation * *Fundament* || ground * *Boden, Erde, Fläche, Gebiet, Motiv, Beweg~, ~lage, on grounds of: auf ~ von* || soil * *Erdboden* || land * *Land, ~ u. Boden* || estate * *Land, ~ u. Boden* || real estate * *~ u. Boden*
Grund- basic || base
Grund-Reserve-Umschaltung basic-reserve changeover
Grundabtastzeit base sampling time * *oft als "T0" bezeichnet* || basic sampling time
Grundachse basic axis * *eines Roboters*
Grundanstrich primary coat * *z.B. mit Vorstreichfarbe* || undercoat
Grundausbau basic configuration
Grundausführung basic type || basic design
Grundbauform basic type of construction
Grundbaugruppe basic module || basic board * *Leiterkarte* || base board * *Leiterkarte*
Grundbaustein basic block || basic building block || basic module
Grundbegriffe fundamental terms || basic terms || basic concepts
Grundbestückung basic complement
Grunddrehzahl base speed * *Drehz. bei Nennspannung/Nennfrequenz, oberhalb derer die Feldschwächung beginnt*
Grunddrehzahlbereich base speed range * *Drehzahlen unterhalb des Feldschwächbereichs* || base speed control range
Gründe reasons * *for/because of technical reasons: aus technischen ~n*
Grundeinstellung initial setting * *z.B.von Potentiometern/Parametern*
gründen set up * *z.B. eine Firma* || establish * *z.B. eine Firma, Niederlassung; auch sich niederlassen* || found * *z.B. eine Firma*
Grundfeld fundamental field * *{el.Masch.} räuml. sin.förmig. Drehfeld mit gleicher Polzahl wie die Wicklung* || first harmonic field * *{el.Masch.}*
Grundfläche footprint * *benötigte Aufstell-/Einbaufläche (z.B. eines Geräts, Schaltschranks usw.)* || basal surface * *base* || basis || floorspace * *Bodenfläche*
Grundfrequenz base frequency
Grundfunktion basic function || essential function || basic logic funktion * *logische*
Grundgerät basic unit || base unit || basic converter * *Stromrichter (aus Sicht einer Zusatzbaugruppe)*
Grundgerätebedienfeld basic converter operator control panel * *eines Stromrichters* || base unit operator panel
grundieren prime * *mit Vorstreichfarbe streichen*
Grundierung primer * *Grundierfarbe/-lack*
Grundkenntnisse basic knowledge
Grundlage base || foundation || basis * *on the basis of: auf ~ von* || groundwork || elements * *~ einer Wissenschaft usw.* || fundamentals * *~ einer Wissenschaft usw.* || basic criterion * *grundlegendes Kriterium (Plural: 'criteria')*
Grundlagen fundamental principles * *einer Wissenschaft/Technik/Fachrichtung* || basics * *z.B. Wissensgrundlagen eines Fachgebiets* || foundations * *Unterbau, Fundamente, Grundfesten, Stützen* || fundamentals * *Grundlagen, Grundprinzip, Grundbegriffe* || basic principles
Grundlagenwissenschaft basic science
Grundlast base load * *wird bei Definition einer Überlastbarkeit zugrundegelegt* || basic load * *'Sockelbetrag' der dauernd erlaubten Belastung bei einer Überlastbarkeits-Spezifikation*
Grundlastbetrieb base load duty
Grundlastdauer base load time
Grundlaststrom base load current * *wird bei der Definition einer Überlastbarkeit zugrundegelegt*
grundlegend substantial || essential || principal * *hauptsächlich* || basic || fundamental * *als Grundlage dienend, grundlegend, wesentlich* || prime

* *primär, ~, wichtigst, erst(klassig); Primzahl; anlassen (Pumpe), grundieren* || *radically* * *(Adv.)*
grundlegend, von Grund auf, radikal
grundlegendes Konzept basic concept
Grundleiterplatte base board || motherboard
gründlich comprehensive * *umfassend* || thorough * *vollständig* || in-depth * *tiefgehend*
gründliche Analyse rigorous analytical process
Grundmaske basic screen mask * *am Bildschirm*
Grundnorm generic standard
Grundoperation basic operation
Grundplatine main board || master board
Grundplatte baseplate || panel || chassis * *Tragblech, tragende Konstruktion*
Grundpreis basic price
Grundprinzip fundamental principle || basic principle
Grundrahmen base frame * *z.B. für die Aufstellung einer elektr.Maschine*
Grundraster basic grid
Grundregel fundamental rule || basic principle
Grundriss ground-plan * *z.B. eines Gebäudes* || plan view || layout diagram * *Anordnungsplan* || floor plan * *eines Raums, Gebäudes* || horizontal projection
Grundsatz principle || axiom * *unbestreitbarer ~* || maxim * *Maxime, Lebensregel, Leitsatz* || rule * *Regel; make it a rule: es (sich) zum ~ machen*
grundsätzlich fundamental * *(Adj.)* || fundamentally * *(Adv.)* || basically || on principle || as a general principle || always * *(Adv.) immer*
grundsätzlich gilt the following is generally valid
Grundschaltung basic circuit || basic converter connection * *[Stromrichter]*
Grundschwingung fundamental wave || fundamental frequency * *Grundfrequenz* || fundamental component * *~santeil* || first harmonic
Grundschwingungsanteil fundamental-frequency content || fundamental component
Grundschwingungsblindleistung fundamental-frequency reactive power
Grundschwingungsgehalt fundamental factor * *Anteil* || fundamental frequency content * *Verhältnis Effekt.wert der Grundschw./Gesamteff.wert (DIN40110)* || fundamental component * *Grundschwingungsanteil* || fundamental frequency component
Grundschwingungsleistung fundamental power
Grundschwingungsleistungsfaktor fundamental power factor * *%-Verhältnis Grundschw.wirk-/Grundschw.scheinleistung (cos Phi_1)* || power factor of the fundamental wave
Grundschwingungsscheinleistung fundamental-frequency apparent power
Grundschwingungsstrom fundamental-frequency current
Grundschwingungswirkleistung fundamental-frequency active power
Grundstoffindustrie basic industry * *z.B. Stahl-, Zellstoff-, Papierindustrie*
Grundtyp basic design * *z.B. eines Geräts* || standard type * *Normalausführung z.B. eines Geräts*
Gründung foundation * *~ einer Firma, Vereinigung usw.* || formation * *auch ~ einer Firma* || establishment * *Errichtung*
Grundwelle base frequency sine wave || base frequency
Grundwissen basic knowledge || basic understanding
Grundzustand initial state
Grüne Lieferbedingungen General Supply Conditions for Products and Services in the Electrical Industry || Green Supply and Delivery Conditions
Grünspan verdigris
Gruppe group || category * *Kategorie, Gattung,*

Klasse; fall in the category: zur ~ gehören || class * *Klasse*
Gruppenantrieb group drive || sectional drive * *ein Antrieb in einem Mehrmotorenverbund* || parallel drive * *bei Parallelschaltung von Motoren an einem Stromrichter* || multi-motor drive * *Antrieb mit mehreren Motoren z.B.an einem Stromrichter*
Gruppeneinspeisung group supply
Gruppenführer team leader || section leader || squad leader * *Truppführer* || head of section || senior engineer * *Ober-/Chefingenieur*
Gruppenspeisung common supply * *z.B. von parallelgeschalteten Motoren an einem Umrichter*
Gruppensteuerung group control
Gruppenumrichter multi-motor-drive converter * *Umrichter, der mehrere parallelgeschaltete Motoren speist* || group converter * *speist mehrere parallelgeschaltete Motoren*
gruppieren group * *siehe auch →einteilen*
Gruppierung grouping || classification * *Aufteilung in Klassen*
Grußwort welcome address
GTO GTO * *Abkürzg. f. 'Gate Turn-Off thyristor': abschaltbarer Thyristor* || GTO-thyristor * *GTO-Thyristor* || gate turn-off thyristor * *GTO-Thyristor*
GTO-Thyristor gate turn-off thyristor || GTO thyristor
gültig valid * *validate/legalize; ~ machen; remain valid: ~ bleiben* || effective * *wirksam, in Kraft; from: ab* || in force * *in Kraft* || admissible * *zulässig* || legal * *zulässig* || applicable * *anwendbar, zutreffend (to: auf)*
gültig bleiben remain valid || remain effective * *wirksam bleiben*
Gültigkeit validity || legal force * *~ e. Gesetzes* || legality * *Zulässigkeit*
Gültigkeitsdauer validity period
Gummi rubber
Gummi-Spritzanlage rubber injection molding system
Gummiband rubber band || elastic band
gummibereift rubber-tired * *(amerikan.)* || rubbertyred * *(brit.)*
Gummibolzenkupplung rubber-bushed pin coupling * *zur Welle-Welle-Verbindg. (kann kleine Fluchtfehler ausgleichen)*
Gummibuchse rubber-bushing
Gummidichtring rubber sealing ring || rubber sealing washer
Gummidichtung rubber seal
gummiert gummed || rubberized
gummiisoliert rubber insulated * *cable: Leitung*
Gummiklotz rubber pad * *z.B. in elastischer Kupplung*
Gummiklötzchen rubber pad * *z.B. in Kupplung*
Gummikneter rubber kneader
Gummimanschette rubber sleeve
Gummiquetschwalze squeegee || squeegee roll * *['skwiedjie 'roul]*
Gummiring rubber bush * *z.B. Dichtring für Kabeldurchführungs- (PG-) Verschraubung* || rubber band * *Gummiband*
Gummischlauchleitung rubber insulated cable
Gummituch blanket * *[Druckerei] auch Stoffdecke, Filzunterlage*
Gummitülle rubber sleeve
Gummiwalze rubber roller || squeegee * *['skwiedjie]* →*Gummiquetschwalze, Gummischrubber*
Gummiwalzwerk rubber rolling mill
günstig favourable * *to: für; be favourable/favour to: ~ sein für* || favorable * *(amerikan.)* || cheap * *billig* || opportune * *Moment; opportunity: ~e Gelegenheit* || encouraging * *ermutigend* || promising * *viel versprechend* || suitable * *passend, zu-*

günstiger Preis   150

träglich || advantageous * *vorteilhaft* || profitable * *vorteilhaft, einträglich, profitabel* || beneficial * *vorteilhaft* || satisfactory * *befriedigend* || agreeable * *befriedigend* || at best * *im ~sten Falle* || on easy terms * *im ~en Bedingungen* || propitious * *geeignet, vorteilhaft (to: für); z.B. Moment* || favourably * *(Adv.)* || favorably * *(Adv., amerikan.)* || favourably priced * *preis~*
günstiger Preis   favourable price || low price || favorable price * *(amerikan.)*
Gurt   belt || tape * *z.B. für gegurtete Bauteile*
Gurtantrieb   belt drive
Gurtbandförderer   belt conveyor
Gurtförderer   belt conveyor
Gurtspanner   belt stretcher || belt tensioner
Gurtung   tape packaging * *für elektr. Bauteile*
Guss   casting * *z.B. Herstellung von Gusseisen*
Guss-   cast || die cast * →*Druckguss-* || extruded * →*Strangguss-*
Gussbronze   cast bronze
Gusseisen   cast iron
Gussgehäuse   cast housing
Gussglas   cast glass
Gussstreichverfahren   cast coating * *z.B. zur Papierbeschichtung*
Gussteil   casting || cast piece
gut   good * *for: für; (Adv.: 'well')* || suitable * *passend, geeignet* || fine * *fein, prächtig* || excellent * *ausgezeichnet (auch 'perfect')* || favourable * *günstig* || lucrative * *lukrativ; business: Geschäft* || profitable * *profitabel; business: Geschäft* || capable * *fähig, tüchtig* || efficient * *tüchtig, effizient* || high-quality * *von hoher Qualität (auch 'high-grade')* || reliable * *zuverlässig* || right * *richtig* || useful * *nützlich* || correct * *richtig, korrekt* || proper * *angebracht* || adequate * *angemessen* || respected * *gut angesehen* || conducive * [kondju: siw] *dienlich, förderlich, nützlich, ersprießlich, zielführend (to: für)* || advantageous * *vorteilhaft* || material * *(Substantiv) Material (auch 'goods': Güter, Waren, Gegenstände, Frachtgut)*
gut durchdacht   well thought-out || well-reasoned || well-devised * *z.B. Plan* || well-weighed * *gut ausgewogen*
gut geeignet   well suited
gut lesbar   easy to read
gut sortiert   well-stocked * *Händler, Lager usw.*
Gutachten   expert's report || expert opinion || certificate * *Dokument* || survey report
Güte   quality * *auch eines Kondensators/Filters* || quality factor * *eines Kondensators/Filters* || grade * →*Qualitätsstufe, Handelsklasse,* →*Qualität*
Gütefaktor   quality factor * *cos Phi x Eta (Leistungsfaktor x Wirkungsgrad) z.B. eines El.-Motors* || Q factor
Güteklassensortiermaschine   grading system * *z.B. für Holz*
guten Appetit   bon appétit * *("guten Appetit" in England u. USA ungebräuchlich)* || enjoy it || I hope you like it
Guten Tag   How are you || Thank's, fine * *Antwort hierauf* || Nice to see you
Gütesicherung   quality assurance
gutgeführt   well-conducted || well-patronized * *Firma*
Guthaben   credit || balance * *my balance stands at 5 $: mein ~ beträgt 5 $; balance standing for my favour: mein ~* || assets
gutschreiben   credit
Gutschrift   credit * *for credit: zur ~*

H

H-Bahn   overhead-cabin railway
Haarriss   hairline crack || capillary fissure
Haartrockner   hair dryer
haben   have || possess * *besitzen* || dispose of * *verfügen über* || have at ones's disposal * *zur Verfügung haben* || be provided with * *ausgestattet sein mit* || be equipped with * *ausgestatet sein mit* || lack * *nicht ~, Mangel ~ an*
Habilitation   habilitation || appointment as a university lecturer
habilitieren   qualify as a university lecturer || habilitate
Hackerei   chipping plant * *in Zellstoff-Fabrik*
Hackmaschine   hogger * *{Holz} in der Spanplattenherstellung*
Hackmesser   hogging knife
Hackschnitzel   chips * *{Holz}* || hogged wood * *{Holz}* || coarse chips * *{Holz}*
Häcksler   shredder * *z.B. für Holzabfälle*
Hadern   rags * *Lumpen (zur Papierherstellung)*
Hafen   port || seaport * *Seehafen*
Hafenmeister   habour master
Haftbild   sticker || adhesive symbol
haften   cling * *(an-) kleben; to: an* || adhere * *(an-) kleben; to: an* || stick * *(an-) kleben; to: an* || be fixed * *befestigt sein; (auch figürl.: Gedanken usw.)* || be liable * *bürgen; for: für; without limitation: unbeschränkt; personally: persönlich* || be responsible * *verantwortlich sein, bürgen* || be held responsible * *zur Haftung gezogen werden* || assume liability * *Haftung übernehmen* || guarantee * *garantieren, gewährleisten, verbürgen, sicherstellen, Garantie leisten* || warrant * *garantieren; warrant a thing: für etwas ~*
Haftreibung   static friction || friction of rest || stiction
Haftrelais   latching relay || bistable relay * *bistabiles Relais* || remanent relay
Haftung   liability * *z.B. für Schäden; limited: beschränkte; personal: persönliche; exemption from: Ausschluss* || adhesion * *Adhäsion* || responsibility * *Verantwortlichkeit* || guarantee * *Gewährleistung*
Haftungsausschluss   suspension of liability
Haken   hook || hook pin * *Hakenbolzen*
Hakenschlüssel   hook wrench || hooked wrench || hooked spanner || hook-spanner
Hakenstellung   hook position * *{Hebezeuge}*
halb   half * *half an apple: ein ~er Apfel; halfway: auf ~em Wege; half as much: ~ so viel* || middle * *at the middle height: auf ~er Höhe*
Halb-   semi- * *z.B. semi-public company: ~öffentliche Gesellschaft; semi-annual: ~jährlich* || half- * *z.B. half-controlled bridge: halbgesteuerte Brücke*
halbautomatisch   semi-automatic
Halbaxiallüfter   semi-axial-flow fan * *Mischung aus Axial- und Radiallüfter (drehricht.abh. Lüfter zur Motorkühlg.)*
halbduplex   half duplex || half-duplex
halber Federkeil   half key
halbgesteuert   half-controlled * *z.B. Brückenschaltung* || semi-controlled * *z.B. Brückenschaltung, zur Hälfte aus Thyristoren, zur anderen Hälfte aus Dioden*
halbgesteuerte Einphasen-Brückenschaltung   half-controlled single-phase bridge connection * *B2H-Schaltung für 1Q-DC-Stromrichter*
halbgesteuerte Schaltung   half-controlled circuit
halbieren   halve || cut in half || divide into halves * *divide into equal halves: in gleich große Hälften (auf-) teilen* || chop by half * *Zeitaufwand, Kosten usw. ~*

Halbkeil   half-key * *zur Motorauswuchtung (Federkeil wird nur zur Hälfte mit einbezogen)*
Halbkeilwuchtung   half-key balancing * *Federkeil wird beim Auswuchten des Motorläufers zur Hälfte berücksichtigt*
Halbleiter   semiconductor || semiconductors * *(Plural)*
Halbleiter-Relais   semiconductor relay
Halbleiter-Sicherung   fuse for protection of semiconductors * *Halbleiterschutz* || fast 'semiconductor' type fuse * *zum Schutz v. Leistungshalbleitern* || SCR type high speed fuse * *zum Thyristorschutz*
Halbleiterbauelement   semiconductor device || semiconductor || semiconductor component
Halbleiterelektronik   solid state electronics
Halbleiterrelais   solid-state relay
Halbleiterschutz-Sicherung   fuse for protection of semiconductors || semiconductor protective fuse || semiconductor protecting fuse || fuse for semiconductor protection
Halbleitersicherung   fast 'semiconductor' type fuse * *zum Schutz von Leistungshalbleitern* || SCR type high speed fuse * *zum Schutz von Thyristoren* || fuse for protection of semiconductors * *schützt Leistungshalbleiter* || semiconductor fuse || semiconductor protecting fuse
Halbportalkran   semi-portal crane || semiportal crane
Halbprodukt   intermediate product * *Zwischenprodukt* || half-finished product
Halbschwingung   half wave
Halbstoff   pulp * *für die Papierherstellung (z.B. Zellstoff)*
Halbwelle   half wave
Halbzeug   semifinished product || semiproduct || preform * *vorgeformtes Teil (z.B. aus Kunststoff)* || first stuff * *Papier usw.* || prefabricated part
Halde   waste-heap * *z.B. für Erz, Abraum usw.* || dump * *Schutt-/Müllhaufen, (Müll-) Abladeplatz, Stapel-/Lagerplatz (für Schüttgut)* || stock-pile * *Vorratshaufen*
Hälfte   half * *half the amount: die ~ des Betrages; half as much: halb so viel; halfway: auf halbem Wege* || one-half * *one-half the: die ~ des*
Hälften   halves
Hall-Effekt   Hall effect
Hall-Generator   hall generator
Halle   hall || workshop hall * *Werkshalle* || workshop * *Werkshalle* || manufacturing shop * *Werkshalle* || shop * *Werkshalle* || hangar * *(Flugzeug)~, Schuppen*
Hallenkran   overhead crane * *mit obenliegenden Laufschienen*
Hallsensor   Hall sensor
Halogen   halogen
halogenfrei   halogen-free
HALT   STOP || HOLD mode * *Betriebsart bei SPS*
Haltbarkeit   durability || life * *Lebensdauer* || service life * *Lebensdauer (von Verschleißteilen)* || resistance to wear * *Verschleißfestigkeit* || stability || shelf life * *~ von Lebensmitteln, Schmierstoffen usw.*
Halteblech   retaining bracket * *z.B. für PC-Karte*
Haltebremse   holding brake * *fail-safe holding brake: Ruhetrom-~* || standstill brake * *→Stillstandsbremse* || brake for static duty || parking brake
Haltegenauigkeit   stopping accuracy * *{Aufzüge}*
Haltemoment   holding torque
halten   hold * *fest-/auf-/zurück-/an-/stand-/ent~, (Versammlung) ab~* || keep * *(bei)be-/fest-/an-/zurück-/ver~; Versprechg. ~; bleiben; keep to/low: s. ~ an/niedrig ~* || support || *stützen* || contain * *enthalten* || deliver * *Rede ~* || give * *Vorlesung ~* || hold out * *stand~* || stand firm * *stand~* || stand

* *aus~* || withstand * *widerstehen* || conduct * *z.B. einen Kurs, eine Vorlesung ~* || adhere * *adhere to: sich an etwas ~/fest~ an/treubleiben* || retain * *zurück-/fest~, (zurückbe-) halten, sichern, stützen, befestigen, (Wasser usw.)* stauen || deem * *halten, erachten für, betrachten als, denken, meinen*
Halteposition   stop position || stop station
Haltepunkt   breakpoint * *{Computer} zum Programmtest*
Halter   holder * *auch Haltevorrichtung* || carrier * *Träger eines mechan. Bauteils* || support * *Stütze* || clip * *Festhalteklammer, Festklemmer* || clamp * *Festklemmer* || bracket * *(Wand)Konsole, Winkelstütze, Stützbalken, offenes Lagerschild* || rack * *Gerüst, Baugr.träger, Einb./Stützrahmen, Gestell/Ständer für Zeitschriften, Prospekte usw.* || L-bracket * *Haltewinkel* || fastener * *Befestigung, Verschluss, Halter* || toolholder * *Werkzeughalter*
Halterahmen   retaining frame * *z.B. für Leiterkarte*
Halterung   mounting support || holding device || fixture || support || carrier * *Träger* || clamp * *Klemme, Klammer* || mount || holder * *Halter, Halterung, Fassung (für Lampe, Sicherung, Werkzeug usw.)* || retaining unit || receptacle * *Aufnahme (für Stecker, Behälter ..), Fach, Fassung, (Sammel-) Gefäß, Behälter; Steckdose* || mounting bracket * *biacket * Halter, Konsole, Träger, Winkelstütze, Wandarm*
Halteschiene   mounting rail * *z.B. Installations-/Aufschnappschiene*
Haltestelle   stopping place || halting place || stop * *für Straßenbahn, Bus usw.* || stopping point * *für Straßenbahn, Bus, Aufzug usw.* || stopping position * *{Aufzüge}* || station
Haltestrom   holding current * *benötigt ein Thyristor, um den Strom im Ein-Zustand aufrechtzuerhalten*
Haltewerk   grab-holding gear * *{Hebezeuge} bei Greiferkran*
Haltezeit   hold time
Hammer   hammer
Hammermühle   hammer mill * *für Steine, Erz usw.*
Hammernut   T-slot * *zur variablen Schraubbefestigung in einem Profilelement (mit →Nutsteinen)*
Hammerschraube   T-head bolt * *für Schienenbefestigungssystem (siehe auch →T-Nut, →Nutstein)*
Hamming-Distanz   Hamming distance * *Maß HD für Datensicherheit: HD - 1 Bitfehler pro Datenblock können erkannt werd.* || Hamming metric * *Maß für Datensicherheit: HD - 1 Bitfehler pro Datenblock können erkannt werd.* || HD * *Abkürzg. f. 'Hamming Distance': Maß für Datensicherheit eine Codes/Protokolls*
Hammingdistanz   Hamming distance * *Maß →HD für Datensicherheit b.serieller Übertragung (HD-1 Fehlbits sind erkennbar)* || Hamming metric * *→HD* || HD * *Maß für Datensicherheit z.B. b. seriell.Übertragg. (HD-1 Falschbits erkennb.)*
Hand   hand * *handy: →zur Hand* || manually * *von ~* || source * *Quelle; from a single source: aus einer ~/von einem Anbieter*
Hand-   manual || manually actuated || hand-held * *~gehalten, ~geführt (z.B. Werkzeug, Bediengerät, Computer)* || manually operated || hand- || portable * *tragbar*
Hand-Automatik-Umschalter   manual-automatic mode selector switch
Hand-Sollwert   manual setpoint
Handachse   wrist axis * *eines Roboters*
Handarbeit   manual workmanship
Handbediengerät   hand-held operator * *tragbares*
handbetätigt   manually operated || hand-operated
Handbetrieb   manual mode

Handbohrmaschine hand drilling machine || hand-drill || drill-gun
Handbuch manual || instruction book || operating instructions * *Betriebsanleitung* || instructions || guide * *Leitfaden* || user's manual * *Anwender~* || reference manual * *technisches Referenz~* || primer * *Fibel, Leitfaden, Anfangslehrbuch* || handbook
Handebene manual control level || local control level
Handeingabegerät handheld terminal || hand-held controller || portable control operator || hand-held operator panel || hand-held keybord unit * *mit Tasten* || manual control
Handeingriff manual intervention
Handel trade * *Geschäftsverkehr* || commerce * *in großem Maßstab* || transaction * *Geschäft, Abwicklung, Verhandlung* || business * *Geschäft* || bargain * *Geschäft(sabkommen), vorteilhaftes Geschäft*
handeln act || proceed * *verfahren* || take action * *in Aktion treten* || trade * *Handel treiben (with: mit einer Person, in: mit Waren)* || deal * *Handel treiben (in: mit Waren); deal with: ~ von/über* || bargain * *feilschen (for: um)* || treat * *(figürl.) ~ von/über* || he a question of * *sich um etwas ~* || be a matter of * *sich um etwas ~* || be concerned * *sich ~* || the question is if * *es handelt sich darum, ob* || involve * *~ um, angehen, betreffen, berühren, mit sich bringen*
Handelsbedingungen trade conditions
Handelsgesellschaft trading company
Handelskammer chamber of commerce
Handelsmarke brand label
Handelsniederlassung trading post
Handelsschiff commercial vessel
Handelsschranken trade barriers
Handelsspanne sales margin
handelsüblich usual in trade || commercial || off-the-shelf * *lagermäßig, von ~r Qualität; Teile* || available on the market || conventional * *herkömmlich*
Handgerät hand-held unit || portable unit || hand-held device
Handgriff handle * *Griff zum Anfassen/Bedienen* || grip * *Griff zum Anfassen, auch Art des Zugreifens* || manipulation * *(meist Plural) Bedienung* || movement * *Bedienungsgriff, Bewegung*
handhabbar manageable || controllable * *kontrollierbar, beherrschbar*
Handhabbarkeit ease of operation * *leichte Bedienung, Bedienungsfreundlichkeit* || manageableness || usability
handhaben handle || wield * *(Macht, Einfluss)* ausüben, (Waffe, Werkzeug) handhaben, führen, schwingen* || manage || operate * *Maschine, Werkzeug usw.* || apply * *Methode usw.*
Handhabung handling * *ease/simplicity of use: einfache ~* || manipulation * *auch durch Roboter, Transfereinrichtung usw.*
Handhabungsgerät manipulator || handling device
Handhabungssystem handling system
Handhebel hand lever || lever handle
händisch manual || manually * *(Adv.)*
Handkurbel hand crank || crank-handle
Händler trader || traders * *(Mehrzahl)* || dealer || reseller * *Wiederverkäufer* || retailer || *Einzelhändler* || merchant * *(Groß)Händler*
handlich handy
Handlüfteinrichtung manual release device * *zum Lösen einer Bemse von Hand (z.B. über einen Hebel)*
Handlüftung manual release * *einer Bremse (z.B. mittels Handhebel bei Aufzugs-/Hebezeugantrieb)*
Handlung act || handling * *Handhabung* || action * *das Handeln*
Handoptimierung manual optimizing

Handpotentiometer manually actuated potentiometer
Handprogrammiergerät hand-held programming unit || hand-held programmer
Handrad handwheel * *(auch bei Aufzug); elevating: zur Höhenverstellung; traversing: zur Seitenverstellung*
Handskizze freehand sketch || rough sketch * *Grobskizze, flüchtige Skizze* || rough drawing * *Grobzeichnung*
Handsollwert manual setpoint
Handtacho hand tacho || hand speed counter * *digitaler* || portable tachometer
Handumgehung manual bypass * *handbetätigter Überbrückungsschalter für unterbrechungsfreie Stromversorgung* || manual override
Handwerk handicraft || craft || skilled trade * *auch 'handwerklicher Beruf'* || trade * *(auch figürlich: z.B. e.Politikers) follow: betreiben* || craftsmen's trade * *Berufsstand; craftsmen enterprises. ~sbetriebe* || métier * *(figürlich)* || job * *(figürlich)* || small trade
Handwerker craftsman || workman || mechanic * *Mechaniker* || artisan * *(Kunst)Handwerker, Mechaniker*
Handwerksbetrieb craftsman's establishment || craftsman's business || craftsmen's workshop || craftshop || craft business
Handwerkzeuge hand-held tools
Handwicklung hand-made winding * *[el.Masch.] Wicklung, der. Spulen v.Hand i.d. Nuten eingebracht/geträufelt werd.*
Handy mobile phone || cellular phone || cell phone
Hanf hemp
Hängebahn overhead conveyor * *Fördersystem* || suspended railway
Hängeförderer overhead conveyor
Hängekabel suspension cable
hängende Last suspended load * *am Kran usw.* || hanging load * *am Kran usw.* || overhauling load * *am Seil hängend* || over-hauling load
Hannover Messe Hanover Exhibition || Hanover Fair
Hannover-Messe Industrie Hanover Industrial Fair
hantierbar manageable
Hantierung handling
Hantierungsbaustein data handling block * *bei SPS* || driver software module * *Treibersoftware-Baustein*
Hardware hardware
Hardware-Plattform hardware platform
Hardwareadresse hardware address
Hardwareänderung hardware change || hardware modification
Hardwareentwicklung hardware design || hardware development
Hardwarefehler hardware fault
Hardwarekomponenten hardware components
Hardwarekonfiguration hardware configuration * *[SPS]*
hardwaremäßig hardware || by hardware * *(Adv.) durch/mittels Hardware* || from the hardware point of view * *(Adv.) aus dem Blickwinkel der Hardware* || from the hardware side * *(Adv.) hardwareseitig* || hardwired * *hardwaremäßig/fest verdrahtet*
Harmonische harmonic
harmonisieren harmonize || coordinate * *koordinieren, aufeinander abstimmen, richtig anordnen*
Harmonisierung harmonization
hart hard * *(auch figürl.) (auch Adv.) auch 'schwierig'* || firm * *fest* || tough * *zäh, widerstandsfähig, robust, schwierig, unangenehm, 'bös', eklig, grob* ||

severe * *streng* || harsh * *streng (z.B. Test), grob, schroff* || troublesome * mühevoll
hart arbeiten work hard || work madly * *(salopp) "wie verrückt" arbeiten*
Härte hardness * *mechan. Härte, ~ von Material* || temper * ~ *von Stahl* || tenacy * *Zähigkeit, Zugfestigkeit, Zähfestigkeit* || harshness * *(figürl.) Strenge, Schärfe* || toughness * *(figürl.) Zähigkeit, Brutalität, Aggressivität* || severity * *(figürl.) Strenge* || vigor * *(amerikan., figürl.) Nachdruck, Stärke, Tatkraft, Vitalität, Strenge, Härte, Rauheit* || vigour * *(brit., figürl.) Nachdruck, Stärke, Tatkraft, Vitalität*
härten harden || temper * *Stahl* || caseharden * *einsatzhärten* || grow hard * *hart werden* || cure * *aus~ (mit Wärme, Wartezeit usw.; z.B. Gummi, Kunststoff)* || solidify * *fest werden (lassen), erstarren* || bond * *(chem.) (ab-) binden, aushärten (z.B. Klebstoff)*
Härteofen tempering furnace
harter Wettbewerb stiff competition
hartlöten braze || brazing * *(Substantiv)*
Hartmetall carbide metal || carbide * *cementedcarbide. gesintertes ~* || hard alloy || hard metal || tungsten carbide * *Wolframkarbid*
Hartmetallbesatz carbide tips * *{Werkz.masch.} bei (Schneid-) Werkzeug*
hartmetallbestückt carbide-tipped * *{Werkz.masch.}* || TC tipped * *{Werkz.masch.} Abkürzg. für 'Tungsten-Carbide tipped'* || tungsten-carbide tipped * *{Werkz.masch.}*
Hartmetallbohrer carbide-tipped drill || tungsten carbide drill bit * *Bohreinsatz*
Hartmetallplättchen carbide tip * *{Werkz.masch.} an Werkzeug*
Hartmetallschneidplatte carbide tip
Hartmetallwerkzeug carbide-tipped tool * *Werkzeug mit (Schneid-/Bohr-/Fräs-) Einsatz aus Hartmetall*
Hartpapier laminated paper || epoxy laminated paper * *mit Epoxidharz verstärkt/getränkt* || resinbounded paper
Hartpappe fibreboard
Hartschaum rigid foam
Härtungseinrichtung curing equipment * *zum Aushärten*
hartverchromt hard chromium plated || hard chrome plated
Harz resin
Harzkochung resin boiling * *i.b. in der Zellstoff-Herstellung*
Haspel coiler || wire winder * *Draht-Auf-/Abwickler* || reeler
Haspelanlage coiler unit
Haspelantrieb coiler drive
Haube cowl * *z.B. für Lüfter; Motorhaube* || cover * *Abdeckung; top cover: obere Abdeckhaube/Deckel* || hood
hauchdünn filmy || wafer-thin
Haufen heap || pile * *Haufen, Stoß, Stapel, (auch Geld-) Menge/Masse* || accumulation * *Ansammlung, Häufung* || cluster * *Ansammlung, Haufen, Menge, Schwarm, Gruppe* || mass * *Häufung* || crowd * *Menge, Ansammlung, Haufen, Häufung, Fülle, Schwarm*
häufig frequent || frequently * *(Adv.)* || prevalent * *vorherrschend, überwiegend, weit verbreitet*
Häufigkeit frequency || rate * *Ereignisse pro Zeiteinheit*
Häufigkeitsverteilung frequency distribution
Häufung heaping * *Anhäufung; (auch figürl.: Überhäufung/Überschüttung)* || accumulation * *(auch figürl.)* || increase * *Ansteigen* || spreading * *Verbreitung* || frequent occurring * *Wiederholung, häufiges Vorkommen*

Haupt- main || primary * *primär* || basic * *grundlegend, die Grundlage bildend* || major * *bedeutend, größer, hauptsächlich; z.b. Hauptfach, Hauptachse, Hauptanbieter* || key * *Schlüssel-, entscheidend; z.B. Kunde, Rolle* || prime * *erst, wichtigst, wesentlichst, of prime importance: von größter Wichtigkeit* || principal * *hauptsächlich, Erst-, Haupt-, Stamm-, Chef-* || chief * *Ober-, Höchst-, Haupt-, in chief: leitender; chief part: ~rolle*
Haupt-Busstrang backbone * *in Computer-Netzwerk*
Hauptanbieter main provider || major player * *(salopp)* || major supplier
Hauptanteil main portion
Hauptantrieb main drive
Hauptanwendung main application || principal application
Hauptanwendungen main applications
Hauptanwendungsgebiet main field of application
Hauptaspekt main issue
Haupteigenschaft principal characteristic || main property || leading feature || outstanding feature * *herausragendes Merkmal* || key feature
Hauptfeld main field * *{el.Masch.} magn.Feld, das Ständer- u.Läuferwicklg. einer Asynchr.masch. gemeinsam haben*
Hauptgeschäftsführer president
Hauptgeschäftsstelle principal office || place of principal business
Hauptinduktivität mutual inductance * *Gegeninduktivität* || magnetizing inductance || main inductance
Hauptinspektion main inspection
Hauptistwert main actual value
Hauptkomponenten main components
Hauptkontakt main contact
Hauptkreis main circuit
Hauptkritikpunkt main criticism
Hauptkunde major customer || key account * *Schlüsselkunde mit hohem Umsatz* || major account * *~ mit hohem Umsatz*
Hauptleiter main conductor * *z.B. L1, L2, L3 bei Drehstromnetz* || phase conductor * *Außenleiter im Drehstromnetz*
Hauptlinie trunk line * *{Bahnen} Hauptstrecke/-Linie z.b. bei der Eisenbahn*
Hauptmenü main menu
Hauptnervenstrang backbone * *(figürl.) Haupt-Busstrang, von d.Neben-/Vor-Ort-Busse/Stichleitungen abzweigen* || trunk * *(amerikan.) eines Kommikationsnetzes*
Hauptplatine main circuit board * *Leiterkarte* || main board * *Leiterkarte, auch für PC* || motherboard * *trägt Zusatzbaugruppen ,z.B. bei PC*
Hauptpol main pole
Hauptpolkern main pole core
Hauptprogramm main program
Hauptreaktanz magnetizing reactance * *{el. Masch.} Blindwiderst. e. Wicklungsstrangs mit Berücksicht. d. →Hauptfelds* || air-gap reactance || armature-reaction reactance
Hauptsache essential || most important thing || main point || main issue * *Hauptaspekt, Kernpunkt* || focal question * *zentrale Frage* || crucial point * *springender Punkt*
hauptsächlich principal || main || essential || most important || basic * *grundlegend, die Grundlage bildend* || major * *größer, bedeutend, hauptsächlich* || mainly * *(Adv.)* || primary * *primär* || primarily * *(Adv.)*
hauptsächlich verwendet most widely used
Hauptschalter line circuit breaker * *Leistungsschalter in Netzeinspeisung* || mains switch * *in Netzeinspeisung* || master switch || mains isolating switch * *in Netzzuleitung* || main switch || main cir-

Hauptschlussmotor

cuit breaker * *Leistungsschalter* || main breaker * *Leistungsschalter* || line switch || line breaker
Hauptschlussmotor   series-wound motor
Hauptschütz   main contactor
Hauptschützansteuerung   main contactor control
Hauptsitz   headquarters || headquarter || head office * *Stammhaus, Hauptniederlassung* || registered office * *(eingetragener/gerichtlicher) Firmensitz*
Hauptsitz haben   be headquartered
Hauptsollwert   main setpoint || main reference || main reference value
Hauptspeicher   main memory * *z.B. eines PC*
Hauptspindel   main spindle * *{Werkz.masch.}* || main shaft * *{Werkz.masch.}*
Hauptspindelantrieb   main spindle drive * *Werkzeugmaschine* || main shaft drive * *Werkzeugmaschine*
Hauptspindelmotor   main spindle motor
Hauptstromkreis   main circuit
Haupttransformator   main transformer
Hauptverteilung   main distribution
Hauptverwaltung   head office
Hauptzweig   main arm * *{Stromrichter} eines Leistungsteils* || principal arm * *{Stromrichter}*
Hausapparat   extension * *interne Telefonnummer*
hauseigen   inhouse * *→firmeneigen, allein einer Firma zugeordnet/gehörend* || proprietary * *→firmenspezifisch*
Hausgerät   domestic appliance
Haushaltsanwendung   household application * *Anwendung für den Privathaushalt* || domestic application || application in domestic premises
Haushaltsgerät   appliance for domestic purposes * *VDE 0730* || household appliance || household device || domestic appliance || appliance
Haushaltsgeräte   household appliances || household equipment
Haushaltsnetz   domestic supply network || domestic supply
Hauspost   internal mail
HD   Hammig distance * *→Hammingdistanz, Maß für Fehlererkenn- u. korrigierbarkeit eines Codes* || HD * *→Hammingdistanz* || Hamming metric * *→Hammingdistanz*
HDF-Platte   HDF board * *{Holz} Abk. für 'High-Density Fiberboard': hochdichte Faserplatte*
HDLC   HDLC * *Abk.f. 'High-Level Data Link Control': Datenübertragungsprotokoll mit Fehlerkorrektur*
Hebebühne   lifting platform || scissor lift
Hebel   lever || handle * *Handgriff* || crank * *Kurbel* || arm * *Arm (auch Maschinenteil)* || finger * *kurzer Hebel, Klinke, Sperrhaken, Finger usw.*
Hebelarm   lever arm
hebeln   lever * *hebeln, stemmen; lever out (of): aus-/herausstemmen aus*
Hebelverschluss   lever closure * *in der Verpackungstechnik*
Hebelwirkung   leverage * *Hebelkraft/-Wirkung, Hebelanordnung/-Anwendung* || lever * *Hebelübersetzung, Hebelkraft/-wirkung; Einfluss (figürl.)*
heben   raise * *auch durch Kran* || hoist * *durch Winde/Aufzug/Kran* || lift || elevate * *hoch~, empor~, aufrichten, er~, befördern (auch Mitarbeiter)*
Hebeöse   lifting eye * *z.B. an Motor, Schaltschrank* || lifting eyebolt * *z.B. an Motor, Schaltschrank* || lifting lug * *in Form einer Lasche* || hoisting lug * *in Form einer Lasche*
Hebewerk   draw work * *z.B. bei Bohrplattform* || hoisting device
Hebezeug   hoisting gear || hoisting device || crane * *Kran* || hoisting equipment || hoist * *electric hoist: elektrisch angetriebenes ~* || lifting equipment || lifting appliance || lifting unit * *→Hubwerk* || lifting gear
Hebezeuge   lifting gear and cranes || hoisting equipment || cranes and hoisting systems
Hebezeugmotor   crane-type motor * *z.B. AC-Schleifringläufermotor bemessen für period. Aussetzbetrieb*
Hefe   yeast || leaven || barm || dregs * *Bodensatz*
Heft   booklet * *Büchlein, Broschüre* || pamphlet * *Druckschrift, Broschüre* || brochure * *Broschüre* || copy-book || handle * *Griff*
heftig   vehement || violent * *Sturm usw.* || impetuous * *stürmisch, mit Macht* || fierce * *wild, heftig (auch Konkurrenz usw.), hitzig, verbissen* || strong * *stark* || intense * *Schmerz usw.* || ardent * *lebhaft*
Heftklammer   staple
Heftmaschine   stapler * *auch im Büro*
heikel   delicate * *Angelegenheit usw.* || ticklish || critical || sensitive * *Thema, Punkt usw.* || sticky * *schwierig, heikel, kritisch, zäh*
Heim-   domestic
helmisch   homegrown * *Pflanzen usw.*
Heimtextilindustrie   domestic textile industry
Heimwerker-   Do It Yourself || DIY * *Abkürzg. für 'Do It Yourself'*
Heimwerkermarkt   do-it-yourself store
heiß   hot * *(auch figürl.: hot topic: ~es/~ diskutiertes Thema)*
heißester Punkt   hottest spot
heißlaufen   run hot
Heißleiter   NTC resistor * *Abkürzg. für 'Negative Temperature Coefficient'* || NTC thermistor * *als Temperaturfühler* || NTC || thermistor
Heißluftgebläse   hot-air blower
Heißpunkt   hot spot * *heiß(est)er Punkt i. Motorwicklung usw., auch Feld in Grafikelement z. Anklicken mit d.Maus*
Heißwasserkessel   hotwater boiler
heizbar   heatable
Heizkessel   heating boiler
Heizkraftwerk   heating power station || combined generating and heating plant || district heating power station || district heating power plant
Heizlüfter   fan heater
Heizplatte   hot platen || heating plate
Heizpresse   hot press * *{Holz} Gummi}*
Heizstab   heating rod || rod heater || tubular heater
Heizung   heating * *auch 'Heizungsanlage'* || firing || heater * *Heizeinrichtung* || radiator * *Heizkörper*
Heizung-Klima-Lüftung   heating-ventilation-air-conditioning || HVAC * *Abkürzg.f. 'Heating, Ventilation and Air Conditioning'*
Heizungspumpe   heating pump
Heizungsregelung   heating control
Heizungsregler   heater control
Heizwert   caloric value || heating power
Heizzone   heating zone * *['hieting 'soun] z.B. bei Extruder, Kunststoffspritzmaschine*
helfen   help || assist * *assistieren, Beistand leisten; assist us in doing: ~ Sie uns bei*
Helligkeit   brightness
Helligkeitssteuerung   brightness control
herabgesteuert   at reduced output voltage * *z.B. Stromrichter*
herabgesteuerter Betrieb   operation at high delay angles * *niedriger Aussteuergrad bei Thyristorbrücke*
herabsetzen   lower * *erniedrigen, verringern* || reduce * *verringern; auch Geschwindigkeit, Preis* || diminish * *vermindern, verringern, verkleinern* || decrease * *vermindern, verringern, reduzieren; abnehmen, sich verringern* || derate * *absenken, heruntersetzen; z.B. e. Spezifikation (by: um) bei bestimmt. Betriebsbedinggn.* || put down * *heruntersetzen, z.B. Preise, Ausgeben; hinstellen/setzen* || take down

* *herunternehmen* || cut * *kürzen* || cut down * *kürzen* || slash * *stark kürzen, zusammenstreichen* || depriciate * i.*Preis/Wert ~; Abschreibungen machen von; verächtlich machen, herabwürdigen* || disparage * *verächtlich machen, in Verruf bringen* || degrade * *(im Rang) herabsetzen, vermindern, heruntersetzen, absinken, (sich) verschlechtern*
Herabsetzung lowering * *Verringerung, Erniedrigung* || reduction * *auch ~ von Preis/Geschwindigkeit* || cut * *Kürzung* || derating * *Absenkung; z.B. ~ einer Spezifikation bei bestimmten Umgebungsbedingungen* || depreciation * *Verächtlichmachung*
heraufsetzen increase || raise || mark up * *z.B. Preise*
herausarbeiten elaborate * *sorgfältig; auch ausarbeiten* || work out * *auch ausarbeiten*
herausbekommen figure out * *(amerikan.)* ermitteln, ausrechnen || find out * *herausfinden* || discover * *herausfinden, entdecken* || puzzle out * *Rätsel usw.* ~ || worm out * *Geheimnis usw.* ~ || get wise to * *herausfinden* || work out * *Rätsel, Antwort usw.* ~ || make out * *Sinn usw.* ~ || get back * *Geld, Investition usw.* ~ || get change * *Wechselgeld ~* || get out || get back * *wiederbekommen*
herausbilden appear * *sich herausbilden* || develop * *(sich) entwickeln, entstehen, sich zeigen, bekanntwerden, erweitern, ausdehnen* || grow up * *auf-/heranwachsen (into: zu), entstehen, sich einbürgern*
herausbrechen break off
herausfallen fall out
Herausforderung challenge * *auch technische*
herausführen bring out * *z.B. Signale aus einem Gerät (to terminals: auf Klemmen ~)* || feed out * *auch Signale, Leitungen usw.* || take out * *z.B. Signal aus einem Gerät* || connect to terminal * *auf Klemme* || carry out
herausgeben publish * *Druckerzeugnis ~* || issue * *z.B. Material, Buch ~* || hand out * *aushändigen*
Herausgeber editor * *Herausgeber, Redakteur* || publisher * *z.B. eines Buches/einer Zeitschrift*
herausgehen exit
herauskommen come out * *auch 'ruchbar werden'* || appear * *erscheinen* || emerge * *erscheinen* || get out * *entfliehen* || come up * *hervor-/heraufkommen, Mode werden* || result * *resultieren*
herauslocken extract
herausnehmen remove * *from: aus etw. heraus* || take out * *from: aus etw. heraus*
herausragen jut out * *vorspringen, →herausragen* || project * *vorspringen, vorstehen, herausragen, vorragen (over: über)* || stand out * *(figürl.) from: aus* || stand proud * *(figürl.) of: über* || protude * *herausstehen, (her)vorstehen, herausragen, hervortreten*
herausragend outstanding * *(figürl.) hervor-/überragend, hervorstechend, prominent, offen stehend (Rechnung usw.)* || jutting out * *z.B. Maschinenteil* || excellent * *ausgezeichnet, hervorragend* || worth highlighting * *wert, hervorgehoben zu werden* || salient * *(her)vorspringend, hervorstechend, ins Auge springend* || superior * *hervorragend* || protruding * *hervorstehend, →hervorragend (z.B. Maschinenteil)* || projecting * *vorspringend, vorstehend*
herausragendes Merkmal key feature || highlight
herausschrauben screw out
herausstellen underline * *unterstreichen, herausstreichen, betonen; z.B. in einem Bericht* || emphasize * *(figürl.), z.B. Gedanken ~* || point out * *(figürl.), z.B. Gedanken, Vorzüge* || make public * *an die Öffentlichkeit bringen* || feature * *z.B. in der Werbung, Presse ~* || distinguish * *erkennbar machen* || turn out * *sich ~* || prove * *sich ~/erweisen (to be: als)* || appear * *sich ~ (to be: als)* || become

apparent * *sich ~* || highlight * *hervorheben, betonen, herausstreichen*
hereinnehmen accept * *entgegen/annehmen, z.B. Auftrag* || book * *(ver)buchen (z.B. Auftrag)* || take in * *Geld, Ware, Auftrag ~* || take in stock * *auf Lager nehmen*
hergestellt manufactured * *from: aus (...-Material)*
herkömmlich traditional || conventional * *(auch eine techn. Konstruktion); in the conventional sense: im ~n Sinn* || customary * *gebräuchlich, herkömmlich, üblich, Gewohnheits-* || usual * *→üblich* || orthodox * *anerkannt, üblich, →konventionell*
Herkunft origin || source
herleiten derive
Herleitung derivation || development * *→Ableitung einer Formel usw.* || inference * *(Schluss-) Folgerung, (Rück-) Schluß*
hermetisch hermetic
hermetisch abgeschlossen hermetically sealed
hermetisch dicht hermetically sealed
hermitisch dicht hermetically sealed
herrschen prevail * *vor~, herrschen, maß-/ausschlaggebend/verbreitet sein* || rule * *over: über* || dominate * *dominieren, be~* || govern * *regieren, lenken, regeln, steuern, bestimmend/maßgebend sein für, leiten* || predominate * *vor~* || be vogue * *in Mode sein* || exist * *bestehen* || be * *bestehen*
herstellen manufacture * *produzieren, herstellen, erzeugen* || realize * *realisieren* || produce || process * *verarbeiten* || establish * *errichten, durchsetzen, schaffen; z.B. (Betriebs-) Zustand, Leitungs-/Telefonverbindung* || create * *(er)schaffen* || generate * *erzeugen, bilden, entwickeln, bewirken, verursachen, hervorrufen* || make
Hersteller manufacturer * *on the manufacturer's premises: beim ~* || producer || supplier * *Lieferant* || vendor * *Verkäufer, Lieferer* || maker
Herstellerangabe manufacturer's specification
Herstellererklärung manufacturer's declaration
herstellerspezifisch proprietary || in-house || manufacturer-specific
Herstellerwerk factory
Herstellerzeichen manufacturer's designation
Herstellfehler manufacturing defect
Herstellkosten production costs || manufacturing costs
Herstellung production || manufacture || making || fabrication || output * *Ausstoß* || manufacturing
Herstellungskosten production costs
Herstellungsprozess manufacturing process
Herstellungstoleranz manufacturing tolerance
Hertz Hertz || Hz || cycles per second || c.p.s. * *Cycles Per Second* || cps * *Cycles Per Second*
herum around || round || about * *um herum (in Zusammenhang mit Zahl)*
herunter down || downward || downwards || down there
herunterbremsen brake || brake down || slow down || throttle * *(auch figürl.: Computer usw.)*
herunterfahren ramp down * *entlang einer Rampe* || slow-down * *verlangsamen* || decelerate * *verzögern* || shut down * *stillsetzen* || bring to a stop * *stillsetzen*
herunterladen download * *von Dateien*
herunterlaufen ramp down * *entlang einer Rampe* || slow-down * *verlangsamen* || decelerate * *Verzögern*
heruntertakten step down
herunterteilen attenuate * *z.B. Spannung(ssignal) über einen Spannungsteiler ~*
heruntertransformieren step down
hervor forth || forward || out || out of * *hervor aus* || from behind * *hinter ... hervor* || from under * *unter ... hervor*
hervorgehen aus arise from * *stammen von* || come

hervorheben 156

from * *stammen von* || emerge from * *stammen von* || spring from * *stammen von* || result from * *sich als Folge ergeben aus* || follow from * *sich als Folge ergeben aus; from this follows that: daraus geht hervor, dass* || shown by * *z.B. aus einer Abbildung/einem Kapitel* ~ || come off winner * *als Sieger* ~ || come out winner * *als Sieger* ~ || emerge a winner * *als Sieger* ~
hervorheben highlite * *hervorheben, betonen, herausstreichen; auch in einem Dokument, z.B. durch Fettschrift* || point out * *herausstreichen* || show off * *herausstreichen* || accentuate * *hervorheben, Wertlegen auf, betonen, akzentuieren* || emphasize * *(nachdrücklich) betonen, hervorheben, unterstreichen* || stress * *betonen* || render prominent * *become prominent: sich hervorheben (from: aus)* || make stand out
hervorkommen come up
hervorragend superb || excellent * *exzellent, vorzüglich, ausgezeichnet* || outstanding * *herausragend* || superior * *bezüglich der Leistung(-sfähigkeit), Qualität usw.* || jutting out * →*herausragend (z.B. Maschinenteil)* || projecting * →*herausragend (z.B. Maschinenteil)* || protruding * →*herausragend (z.B. Maschinenteil)* || very well * *(Adv.)* || outstandingly * *(Adv.)* || excellently * *(Adv.)* || ingenious * *klug erdacht*
hervorstehen protrude || project || stand out || jut out
Herzstück centre-piece || center-piece * *(amerikan.)* || kingpin * *(figürl.)*
heterogen heterogenous * *aus verschiedenartigsten Teilen zusammengesetzt*
heutig today's || this day's || on this day || present * *gegenwärtig* || modern * *modern* || contemporary * *zeitgenössisch* || of today || ongoing
heutzutage nowadays || these days || in these days || in our times
Hex-Monitor hex monitor * *Hilfsprogramm z. Testen v.Anwenderprogrammen mit hexadezimaler Anzeige d.Speicherinhalte*
hexadezimal hexadecimal
Hexadezimalsystem hexadecimal system
Hexazahl hexadecimal number || hex number
Heylandkreis Heyland diagram * *Ortskurve der Ströme der Drehstrom-Induktionsmaschine*
HF- HF- * *High Freqency, Hochfrequenz-*
HGÜ high voltage DC power transmission * *Hochspannungs-Gleichstromübertrag. elektr. Stroms über Leitg.* || high voltage DC transmission || HVDC transmission * *Abkürzg. für "High Voltage Direct Current"*
Hierarchie hierarchy
Hierarchieebene hierarchical level || hierarchy level || hierarchical layer
hierarchisch hierarchical * *hierarchical layered: ~ aufgebaut* || hierarchic
hierbei at this || in this || with this || on this occasion * *bei dieser Gelegenheit* || in this connexion * *in diesem Zusammenhang* || herewith * *hierdurch, hiermit* || enclosed * *ein/beigefügt* || attached * *angeschlossen, zugeteilt, dazugehörig* || annexed * *beigefügt, beifolgend* || therewith * *damit, darauf, hierauf, danach, daraufhin, demzufolge, darum* || thereupon * *darauf, hierauf, danach, daraufhin, darum* || in doing so * *dabei, gleichzeitig* || in so doing * *dabei, gleichzeitig* || at the same time * *gleichzeitig* || hereby * *hierdurch, hiermit*
hiermit with this || herewith || effective immediately * *mit sofortiger Wirkung* || therewith * *damit, darauf, hierauf, daraufhin, demzufolge* || hereby * *hierdurch, hiermit* || with it
High-aktiv High active
HIGH-Signal anlegen energize * *Klemme*
Hilfe help || assistance * *Beistand, Unterstützung,*

Hilfestellung, Assistenz || aid * *with the aid of: mit Hilfe von*
Hilfetext help text || help message
Hilfetexte help texts
Hilfs- auxiliary || helper * *z.B. Antrieb* || supplementary || standby * *Reserve-* || assistant || ancillary * *untergeordnet (to: zu); Hilfs-, Neben-*
Hilfs-Reihenschlusswicklung stabilizing series winding * *kompensiert Lastabhängigk. d. Drehz. e.DC-Motors bei konst. Ankerspanng.* || compound winding * *kompensiert Lastabhängigk. d. Drehz. e.DC-Motors bei konst. Ankerspanng.*
Hilfsantrieb helper drive
Hilfsarbeiter unskilled worker
hilfsbereit ready to help || eager to help || co-operative
Hilfsbetriebe auxiliaries
Hilfseinrichtung auxiliary device
Hilfsenergie auxiliary power || auxiliary energy
Hilfsfunktion auxiliary function
Hilfsklemme auxiliary terminal * *z.B. für Temperaturgeber und Tacho in Motorklemmenkasten*
Hilfskontakt auxiliary contact * *z.B. an Schütz, Relais, Leistungsschalter*
Hilfsmaschine auxiliary machine
Hilfsmerker auxiliary flag * *bei SPS* || relay equivalent * *bei SPS in Kontaktplandarstellung*
Hilfsmittel auxiliary material * *i. weitesten Sinne* || supplementary device * *Hilfsvorrichtung, Gerät, Behelf* || aid * *Hilfe, Hilfsmittel, Hilfsgerät, Mittel* || means * *(Plural) Mittel, Werkzeug, Wege, (Geld)-Mittel* || tool * *(auch figürlich) Werkzeug im weitesten Sinne* || device * *Vorrichtung, Gerät, Behelf* || remedy * *(Gegen)Mittel, Abhilfe* || resource * *(Hilfs)Quelle/Mittel/Geldmittel, Aktiva* || expedient * *(Hilfs-)Mittel, (Not)Behelf, Ausweg* || stopgap * *Notbehelf, Lückenbüßer, Überbrückung,* →*Notlösung* || shift * *Ausweg, Hilfsmittel, Notbehelf, Kniff, List, Ausflucht* || appliance * *(Hilfs)Mittel, Gerät, Vorrichtung, Apparat* || auxiliary * *Hilfs-/Behelfs-/Ausweich- (Mittel)* || resources * *(Plural) (Hilfs-) Quellen/Mittel, Geldmittel* || auxiliary device * *Gerät, Einrichtung* || supplementary equipment || aids * *(Plural)* || auxiliary equipment
Hilfsmotor auxiliary motor || helper motor || pony motor * *Anwurf-/Hilfsmotor (pony enspr. "klein, Zwerg-, Mittelklasse, Rangier-, Vorstreck-")*
Hilfsphase auxiliary phase * *z.B. bei Einphasen-/* →*Kondensatormotor*
Hilfsprogramm utility || subroutine * *Unterprogramm* || auxiliary program
Hilfsreihenschlusswicklung series stabilizing winding * *in DC-Motor* || compound winding * *in DC-Motor*
Hilfsschütz contactor relay || control relay
Hilfssollwert supplementary reference value || supplementary reference || supplementary setpoint || additional reference value || correcting setpoint * *Korrektursollwert*
Hilfsspannung auxiliary voltage || auxiliary supply * *Versorgungsspannung*
Hilfsstoff additive || auxiliary material * *siehe auch* →*Hilfsmittel* || auxiliary equipment
Hilfsstoffe additives
Hilfsstromkreis auxiliary circuit
Hilfsstromversorgung auxiliary power supply
Hilfsthyristor auxiliary thyristor
Hilfswickler auxiliary winder * *z.B. zum Auf-/Abwickeln des Trägermaterials*
Hilfswicklung auxiliary winding || auxiliary starting winding * *z.B. zur Anlaufhilfe von Einphasenmotoren* || starting winding * *Anlaufwicklung in e. Motor* || start winding * *Anlaufwicklung in e. Motor*

Hilfszweig auxiliary arm * *z.B. eines Stromrichters*
hin und her back and forth || to and fro
Hin- und Herbewegung reciprocating || oscillating * *schnelle* ~
hin- und herschalten toggle
hinab down * *right down to 0 V: bis hinab zu 0 Volt* || downward || downwards || down there
hinaufrollen scroll up * *Anzeige, Bildschirm*
hinausgehen über go beyond * *Zahl, Maß, Wert usw.* || surpass
hinausragen über project beyond || tower above * *(figürl.)* || stand out from * *(figürl.)*
Hindernis obstacle || barrier * *Barriere* || difficulty * *Schwierigkeit* || stumbling-block * *(figürl.) Stolperstein, Stein des Anstoßes, Hindernis (to: für)*
Hinderniserkennungssystem obstacle detection system * *für führerloses (Transport-) Fahrzeug* || ODS * *'Obstacle Detection System'; für führerloses (Transport-) Fahrzeug*
hindurchführen pass through
hindurchleiten feed through || route through || guide through
hineinstecken insert
hinnehmen accept * *annehmen* || take || put up with * *sich gefallen lassen, in Kauf nehmen* || tolerate * *tolerieren*
hinreichend sufficient || reasonable * *angemessen, zumutbar, annehmbar*
hinreichend genau with sufficient accuracy
hinreichende Bedingung sufficient condition
hinsichtlich regarding || on the level of || with respect to || from the perspective * *from the performance perspective: ~ der Leistungsfähigkeit* || as far as concerned * *as far as the performance is concerned: ~ der Leistungsfähigkeit*
hinten behind * *from behind: von* ~ || at the back || in the rear * *am Ende, auf der Rückseite* || rearmost * *am Ende* || backwards * *nach hinten* || at the end * *am Ende* || to the back * *nach hinten* || to the rear * *nach hinten; shifted towards the rear: nach ~ versetzt*
hinter rear * *(Adj.)* || behind || at the back of || in the rear * *am Ende* || back of * *(salopp)* || after * *in der (Reihen/Rang) Folge, z.B. des Signalflusses* || downstream of * *im Signal-/Daten-/Material-/Kühlmittelfluss weiter hinten liegend*
hinter dem Hochlaufgeber post-ramp || after ramp || after the ramp-function generator
Hinter- back || rear * *hinten angeordnet, hintenliegend* || trailing
Hinterachse back axle
hinterdrehen relieve * *z.B. Drehteil auf Drehmaschine*
Hinterdrehmaschine relieving lathe
hintere Flanke falling edge * *fallende Flanke* || trailing edge || negative edge * *negative Flanke*
hintere Tragwalze back drum * *Tragwalze 1 eines Doppeltragwalzenrollers*
hintereinander one after the other || one by one || in succession || in series || successively || running * *3 days running: 3 Tage hintereinander; 5 times running: 5-mal hintereinander* || in tandem arrangement * *hintereinander angeordnet* || connected in series * *(elektr.) hintereinandergeschaltet* || inline * *entlang einer (gedachten) Linie angeordnet*
hintereinander anordnen cascade
hintereinander schalten cascade * *kaskadieren, in Kaskade schalten* || connect in series * *in Reihe schalten (with: zu)* || chain * *verketten* || join in series
Hintereinanderschaltung series connection
Hintergrund background * *auch langsame Programmablaufebene (Programme mit niedriger Priorität)*
hintergrundbeleuchtet backlit * *z.B. LCD-Anzeige*

Hintergrundbeleuchtung backlight * *z.B. bei einer LCD-Anzeige*
Hintergrundprozess background process || daemon * *vom Betriebssystem in Auftrag gegebener ~ bei UNIX*
Hintergrundspeicher background memory
Hintergrundverarbeitung background processing * *Rechnerprogramme, die mit niedriger Priorität ablaufen (as: in)* || background task * *Rechnerproramm, das nicht auf externe Ereignisse reagieren muss*
hinterlegen deposit * *deponieren, hinterlegen, übergeben* || store * *speichern, Daten in einer Datei hinterlegen*
Hinterschleifen relief grinding
hinunter down || downward || downwards || down there
hinunterrollen scroll down * *Anzeige, Bildschirm*
hinverdrahten connect-up
Hinweis reference * *['referens] Bezug, (Ver-) Weis (to: auf); with reference to: mit ~ auf* || note * *in Beschreibung; concerning: betreffend* || advice * *Rat, Richtlinie* || hint * *Rat, Wink, Tipp, Fingerzeig* || instruction * *Anleitung (for: zu)* || tip * *Tipp, Fingerzeig* || indication * *Anhaltspunkt* || announcement * *Ankündigung, of: von* || remark * *Anmerkung; on: zu* || information * *on/about: zu/über; relevant information: sachdienliche ~e* || direction * *(An-) Weisung, Anleitung, Anordnung, Belehrung* || recommendation * *Empfehlung*
Hinweise guidelines * *als Richtschnur dienende ~*
hinweisen notify * *of: auf* || refer to * *jdn. auf etw. ~; verweisen auf* || draw a person's attention to * *jdn. auf etwas ~ ; point at * ~ auf* || indicate || point out || hint at * *anspielen auf* || point out that * *darauf ~, dass*
hinweisen auf refer to || draw attention to * *draw a person's attention to: jdn. ~ auf* || call attention to * *call a person's attention to: jdn. ~ auf* || point out * *point out that/to: darauf ~, dass* || point at || point to || indicate * *hinweisen auf, angeben, bezeichnen* || make mention of * *erwähnen* || hint at * *anspielen/hinweisen auf* || allude to * *anspielen auf* || stress * *betonen/darauf hinweisen, dass* || emphasize * *betonen*
hinzufügen add || attach * *beilegen, beifügen, anheften, befestigen, verbinden, verknüpfen, angliedern* || append * *einen Nachtrag ~, eine Anlage ~* || affix * *beiheften/-kleben, beilegen*
Hitze heat
hitzebeständig heat-resistant
Hitzebeständigkeit heat resistance
HM- TCT * *[Werkz.masch.] Abkürzg. für 'Tungsten Carbide Tipped': hartmetallbestückt (Werkzeug)*
HM-Werkzeug T.C.T. tool * *Abk. für 'Tungsten Carbide Tipped' mit Hartmetallplättchen bestückt*
Hobby- hobby || do-it-yourself || DIY * *Abkürzg. für 'Do It Yourself'*
Hobelbank planing bench * *[Holz]*
Hobelmaschine planer || planing machine || shaping machine
Hobelmesser planing knife * *[Holz]* || planing blade * *[Holz]* || planer knife * *[Holz]*
hobeln plane || shape
Hobelstahl planing tool * *Schneidwerkzeug für Hobelmaschine* || cutting tool
Hobelwerk planing system * *[Holz]* || planing mill * *[Holz]*
Hobelzahnriemen chipper chain * *[Holz] für Kettensäge*
hoch high * *auch wert/zahlenmäßig* || elevated * *Lage* || great * *groß* || power of * *zur Potenz von* || important * *wichtig, bedeutsam* || heavy * *Steuer, Gebühr, Zoll, Strafe, Unkosten* || severe * *Schaden,*

Strafe, Gebühr || high-ranking * hochrangig (Person) || high degree of * ein hohes Maß an, hoher Grad an || with a high level of * mit einem hohen Grad an/von || added * vermehrt, erhöht, zusätzlich
hoch 2 squared * (vorangestellt) sqared speed: Drehzahl hoch 2
hoch empfindlich highly sensitive * feinfühlig, empfindsam, hochauflösend (Messgerät) || highly delicate * zart, zerbrechlich
hoch- highly * z.B. highly-integrated: hochintegriert; highly-flexible: hochflexibel
Hochachtungsvoll Regards || Yours respectfully || Yours sincerely || Yours faithfully || Yours truly || Best Regards
hochauflösend high-resolution || high-definition * Bildschirm || hires * (salopp) Abkürzg.f. 'high-resolution'
hochbelastet high-load * z.B. durch mechan./elektr. Last
hochdicht high-density
Hochdruck high pressure * auch figürl.: work at high pressure: mit ~arbeiten || full speed * (figürl) work at full speed/at full blast: mit ~ arbeiten || letter press * (Druckerei) Druckverfahren
Hochdruck- high-pressure
Hochdrucksammelschiene high-pressure steam mains * zur Dampfversorgung
hochdynamisch highly-dynamic || high response || quick-response || having a high dynamic response || fast-response || high-dynamic
hochdynamische Regeleigenschaften highly dynamic response
hochdynamischer Antrieb quick-response drive || high dynamic response drive
hochelastisch highly flexible
Hochenergie- high energy || high-energy
Hochenergiebatterie high-energy battery
hochentwickelt advanced || highly developed
hochfahren run-up || start * anfahren || accelerate * beschleunigen || ramp-up * entlang einer Rampe || boot * einen Computer
Hochfahrintegrator ramp-up integrator
Hochfahrt run up * z.B. auf End-/Betriebsgeschwindigkeit/-Drehzahl || ramping up * über eine (Hochlaufgeber-) Rampe
Hochfahrumrichter starting converter * übergibt den Motor nach erfolgtem Hochlauf ans Netz oder e. →Betriebsumrichter
hochflexibel highly-flexible || highly flexible
hochfrequent high-frequency
hochfrequente Einstrahlung high-frequency radiation
Hochfrequenz- high-frequency || radio-frequency || high-bandwidth * mit großer Bandbreite (bei Nachrichtenübertragung)
Hochfrequenzstörung high-frequency interference
hochgenau with high accuracy || highly accurate || extremely accurate || high-accuracy || high-precision || ultra-precise
Hochgeschwindigkeit high speed
Hochgeschwindigkeits- high-speed
Hochgeschwindigkeitsläufer high-speed rotor * siehe auch →Massivläufer, →Turboläufer
Hochgeschwindigkeitszug high-speed train
hochglanzverchromt highly-polished chromium-plated
hochintegriert highly integrated || VLSI * Abkürzg.für 'Very Large Scale Integration': (IC) mit sehr hohem Integrationsgrad || LSI * Abkürzg.für 'Large Scale Integration': (IC) mit hohem Integrationsgrad
hochisolierend high-insulating
hochkant on edge
hochkoerzitiv high-coercive

hochkomplex highly complex || highly sophisticated
Hochlauf ramp-up * über eine (~geber-)Rampe || run-up || running up || ramping-up * entlang Rampe || acceleration * Beschleunigung; current limit acceleration: ~ an der Stromgrenze || initialization * Rechner || initialization procedure * Anlaufprozedur eines Rechners || runup * auch ~ eines Rechners
hochlaufen run up * z.B. eines Motors || accelerate * beschleunigen || ramp up * entlang einer Rampe ~
Hochlaufgeber ramp-function generator || ramp generator || speed ramp * für Drehzahl, Geschwindigkeit || reference integrator * Sollwertintegrator || ramp
Hochlaufgeber-Rampe drive-ramp * für Drehzahlsollwert
Hochlaufgeber-Verrundung S-curve rounding * Rampe hat die Form des Buchstabens "S" || rounding-off of the ramp generator
Hochlaufgeberfreigabe ramp enabling || enable ramp-function generator
Hochlaufgebernachführung ramp function generator tracking
Hochlaufintegrator run-up integrator
Hochlaufkennlinie run-up characteristic
Hochlaufmoment running-up torque * Dehmoment, das zum Hochlaufen benötigt wird
Hochlauframpe acceleration ramp
Hochlaufverzögerung run-up delay
Hochlaufzeit ramp-up time || acceleration time * auch mechanische Hochlaufzeit e. Arbeitsmaschine bei konstantem Antriebsmoment || increase rate || ramp time * Oberbegriff für Hoch/Rücklaufzeit || ramping time || accel time * (salopp) Abk. f. 'acceleration time'
hochlegiert high-alloy
Hochleistungs- high performance * mit hoher Leistungsfähigkeit || high end * Spitzen- || front end * Spitzen- || high tech * hochtechnologisch || high-rating * mit hoher Abgabeleistung/hohem Strom/hoher Spannung, z.B. Stromrichter || heavy duty * für Schwerlast- oder rauen Betrieb || high-power * mit hoher elektr./mechan.Leistung || high-capacity * mit hoher Leistung/hohem Leistungs-/Fassungsvermögen || high-energy || high-efficiency * mit hoh. Leistungsfähigkeit/Leistung/Wirkungsgrad/Funktionsgüte/Effizienz || high-end * der Spitzenklasse zuzuordnen
Hochleistungsstromrichter high-rating converter
hochmodern ultra-modern
Hochofen blast furnace
hochohmig high-resistance || high impedance || floating * ohne festes Potenzial || high-resistive || having a low Ohm value
Hochpass high pass
Hochpassfilter high-pass filter
Hochpegellogik HTL * Abk. f. 'High-Threshold Logic': z.B. mit 24V-Signalen arbeitende Logikschaltungen || high-threshold logic
hochperformant high-performance
hochpermeable Magnetmaterialien highly permeable magnet materials
hochprior top-priority
Hochrechnung extrapolation
Hochregallager high bay racking system || high-bay storage || vertical-stack warehouse || high-bay warehouse
hochrein ultra-pure * mit geringem Anteil an Fremdstoffen || ultra-clean * sauber
hochrüsten upgrade * Hard- und Software ~; from; von; to: auf
Hochrüstung upgrade || upgrading
Hochschulabschluss university degree

Hochschule university
Hochsetzsteller step-up controller || step-up converter * ~-*Stromrichter*
Hochspannung high voltage || HV * *Abkürz. für 'High Voltage'*
Hochspannungs- high voltage || HV * *Abkürzung für 'High Voltage'*
Hochspannungsanlage high-voltage system
Hochspannungsgleichrichter high-voltage rectifier
Hochspannungskondensator high-voltage capacitor
Hochspannungsmotor high-voltage motor * *ca. 1 bis 12 kV und 200 bis 16000 kW* || HV motor * *ca. 1 bis 12 kV und 200 bis 16000 kW*
Hochspannungsprüfung high-voltage test
Hochspannungsschalter high-voltage switch || HV switch || h.v. switch || HV circuit breaker || high-voltage circuit breaker
Hochspannungsverteilung high-voltage distribution || HV distribution || h.v. distribution
hochsperrend highly-blocking
Hochsprache high level language * *Programmiersprache* || HLL * *Abkürzung für 'high level language' (Programmiersprache)*
höchst highest || uppermost * *oberst, höchst, ganz oben, obenan, zuoberst* || topmost * *höchst, oberst* || superior * *überlegen* || greatest * *größt* || supreme * *höchst, oberst, größt, äußerst* || extreme * *äußerst, weitest, höchst, außergewöhnlich, radikal* || utter * *äußerst, höchst, völlig* || highest-ranking * *am rang~en (Person)* || highly * *(Adv.)* || greatly * *(Adv.)* || very * *sehr* || much * *sehr* || extraordinary * *außerordentlich* || exceptional * *ungewöhnlich* || remarkable * *bemerkenswert* || enormous * *enorm* || top * *z.B. top priority: ~ Priorität, top speed: Höchstgeschwindigkeit/Drehzahl* || maximum * *zahlen-/wertmäßig am ~en, maximal*
höchste Anforderungen extreme requirements
höchste Dynamik ultra high-speed response
höchste Priorität top priority
höchstens not exceeding || at the most * *b. Zahlenangaben nachgestellt* || at best * *im besten Falle; b. Zahlenangaben nachgestellt* || at most
höchster Gang high gear * *Getriebe* || high * *Getriebe*
Höchstfrequenztechnik micro-wave technology || micro wave engineering
Höchstgeschwindigkeit maximum speed || top speed
Höchstmaß maximum dimension * *größte Abmessung* || highest level * *(figürl.) of: an* || highest degree * *(figürl.) of: an* || extremely high level * *an extremely high level of: ein ~ von*
höchstwertiges Bit most significant bit || MSB * *Most Significant Bit*
höchstwertiges Zeichen most significant digit * *einer Ziffernanzeige* || MSD * *Most Significant Digit; einer Ziffernanzeige*
höchstzulässig maximum permissible
hochtakten step up
Hochtemperaturfett high-temperature grease
hochtourig high-speed
hochtransformieren step up || boost * *anheben, verstärken (z.B. Druck, Spannung)*
hochverfügbar high-MTBF * *mit hoher MTBF ('Mean Time Between Failures': Durchschnittl. fehlerfreie Betriebszeit)* || fault-tolerant * *fehlertolerant* || high-availability || redundant * *redundant aufgebaut*
hochviskos high-viscosity
hochwertig high quality * *qualitativ hochstehend* || high-grade || high-tensile * *hochfest, z.B. Stahl* || of high quality || high value || high performance * *bezüglich der Leistungsfähigkeit*

Hof yard * *Hof(raum), (Arbeits-/Bau-/Stapel-) Platz*
höflich courteous
hohe Anforderungen stringent demands || exacting requirements
hohe Ansprüche heavy demands * *make heavy demands on: ~ stellen an* || stringent requirements
hohe Ansprüche stellen make heavy demands * *on: an* || set a high standard * *of: an* || expect a great deal * *from: an jdm.* || tax severely * *an etwas/jemanden* || be demanding || be exacting * *streng/genau/anspruchsvoll sein (z.B. Anwendung, Kunde)*
hohe Aussteuerung low delay angle setting * *kleiner Steuerwinkel bei Phasenanschnittsteuerung* || high control setting || high output voltage ratio || high modulation * *in der Mess-/Nachrichtentechnik usw.*
hohe Dynamik high dynamic performance || fast response
hohe Genauigkeit high accuracy
hohe Standzeit long life
Höhe height * *['hait]* || altitude * *geographisch, above sea level: über dem Meeresspiegel* || amplitude * *e.elektr.Größe, z.B.Spannung Strom,Leistung* || magnitude * *auch e.Stromes/Spannung/Impulses/Drehzahl/Signal* || amount * *Betrag, Summe, Höhe, Ergebnis, Wert*
Höhe über dem Meeresspiegel altitude above sea level
Höhen-Füllmaschine level filling machine
hoher Leistungsbereich upper performance range * *bezügl. der Leistungsfähigkeit* || high-end * *bezügl. der Leistungsfähigkeit* || high-rating range * *bezüglich der Nennleistung, z.B. Motor/Stromrichter* || high-output range * *bezügl. der Ausgangsleistung, z.B. Motor/Stromrichter*
höher increase * *Kommando für Motorpotentiometer* || raise * *Kommando, Taste* || higher * *(auch Adv.)* || higher up * *(Adv.) weiter oben* || wider * *Bereich usw.*
Höher-Taste increase button || raise key || raise button || RAISE pushbutton
Höher-Taster raise button * *z.B. für Motorpotentiometer* || increase key
Höher-Tiefer-Taster Up-/Down button
höhere Gewalt flooding, lightning etc. * *(versicherungstechnisch) Überschwemmung, Unwetter usw.* || influence beyond one's control || force majeure || acts of God || compulsary circumstances
höhere Programmiersprache high level language || HLL * *Abkürzung für 'High-Level Language'*
höherer Ordnung higher-order * *z.B. Oberwelle*
höherprior of higher priority || major * *höherrangig* || higher-ranking * *im Rang höherstehend (Person)*
höherprior sein override
höherpulsig having a higher pulse number * *Stromrichterschaltung*
höherwertig higher-quality * *von höherer Qualität* || more significant * *Ziffer, Datenbit usw.*
höherwertige Stelle upper digit * *z.B.an Zahlenschalter oder Ziffernanzeige*
hohes Maß an Zuverlässigkeit high degree of reliability
hohl hollow || concave * *vertieft* || arched * *gewölbt* || tubular * *röhrenförmig* || empty * *leer*
Hohlglas hollow glass
Hohlglasmaschine hollow-glass machine || glass container production machine || bottle-making machine * *zur Herstellung von Glasflaschen*
Hohlkehle hollow groove || flute || fillet
Hohlprofil hollow section
Hohlraum cavity || hollow || hollow space || air spacing || excavation * *Aushöhlung* || empty space
Hohlwelle hollow shaft

Hohlwellenausführung hollow-shaft design
Hohlwellentacho hollow-shaft tacho
Hohlzylinder hollow cylinder
hohnen hone
holen fetch || pick up * *ab~ (auch Informationen)* || get || go for || come for * *ab~* || call for * *ab~*
Holländer beater * *{Papier} zum Feinmahlen der Faserstoffe bei der Zellstoff-/Papierherstellung* || pulp engine * *{Papier}* || pulper * *{Papier} Breimühle in der Zellstoffindustrie*
Holm beam * *Balken, Tragbalken, Holm* || crossbeam * *Querträger/-balken* || transom * *Querbalken* || spar * *z.B. bei Flugzeug, Schiff* || longeron * *Längs-/Rumpfholm usw.* || brace
Holz wood * *(made) of wood/wooden: aus ~; wood pile: ~stapel; wood yard: ~platz* || timber * *Nutzholz; (brit.) Schnitt~; round timber: Rund~* || lumber * *(amerikan.) Nutz~* || leaf-wood * *Laub~* || coniferous * *Nadel~* || firewood * *Brenn~* || piece of wood * *~stück* || green wood * *grünes ~* || seasoned wood * *abgelagertes ~* || dead wood * *abgelagertes Holz,* totes ~ || plastic wood * *flüssiges ~* || hardwood * *Hart~, Laub~* || foliage trees * *Laub~* || coniferes * *Nadelhölzer* || log * *Rundholz, unbehauener ~klotz, gefällter Baumstamm* || softwood * *Nadel~, Weich~*
Holz- wooden
Holzbearbeitung woodworking || wood working
Holzbearbeitungsindustrie woodworking industry || woodworking trade
Holzbearbeitungsmaschine wood-working machine
Holzbearbeitungsmaschinen woodworking machinery || wood machinery
holzfrei woodfree * *Papier; fine paper/woodfree paper: ~es Papier*
holzhaltig wood containing * *(Papier)*
Holzhandwerk woodworking trade
Holzindustrie timber and woodworking industry
Holzkiste wooden crate
Holzleimbau gluelam beam construction || gluelam construction || glued timber construction
Holzplatte wood plate || wood panel || board
Holzplatz wood-handling yard * *z.B. in Zellstofffabrik, Sägewerk* || timber yard * *im Sägewerk usw.* || woodyard
Holzrücken skidding * *(Agrikultur)*
Holzschälmaschine wood peeling machine || debarker * *Entrindungsmaschine* || barker * *Entrindungsmaschine*
Holzschleifer wood grinder * *zur Zellstofferzeugung*
Holzschliff mechanical woodpulp || groundwood pulp || mechanical pulp
Holzschraube wood screw
Holzspäne wood chips
Holzstäbchen wooden stick
Holzstoff mechanical pulp * *{Papier} Zellstoffrohprodukt* || woodpulp * *{Papier} Zellstoffrohprodukt* || pulp * *{Papier}*
Holzstoffrefiner pulp refiner * *{Papier} zur Zellstofferzeugung*
Holzwerkstoff wood-based material
Holzwolle wood-wool || excelsior * *(amerikan.)* || woodwool
homogen homogeneous * *[homo'dschi: njes]* || consistent * *→gleichmäßig*
homogenisieren homogenize
Homogenität homogeneity || homogeneousness
homokinetisch homokinetic
honen hone
Honmaschine honing machine * *Werkzeugmaschine*
Honorar honorarium || payment || fee * *eines Arztes usw.* || royalties * *eines Buchautors usw.* || commission * *Provision* || brokerage * *Provision eines Maklers usw.*

Hopfen hop
hörbar audible * *inaudible: nicht hörbar*
Hörbereich audible range
Hörgrenze threshold of hearing || perception threshold * *Wahrnehmungsschwelle*
horizontal horizontal
Horizontalfrequenz horizontal frequency * *{Computer} Anzahl Zeilen/sec b. Monitor (Anz.Zeilen x Bildwiederholrate x 1,1)*
Hörsaal auditorium
Hörschwelle threshold of hearing || perception threshold * *Wahrnehmungsschwelle*
Hotelreservierung hotel reservation
HSS HSS * *Abk. für 'High-Speed Steel': Hochleistungs-Schnell(schnitt)-Stahl*
HTL HTL * *Abkürzg. für 'High-level Transistor Logic': hochpegelige Transistorlogik (mit 24V-Pegel)*
HTML HTML * *'HyperText Markup Language': Seitenbeschreib.-Sprache mit Hyperlinks, Grafik, Sound für →WWW*
HTTP HTTP * *'HyperText Transfer Protocol': Protokoll zur Übertragung von →HTML-Seiten im →WWW des Internet*
Hub stroke * *e. Werkzeugmaschine, ~spindel, Kompressors, Kolbens in Verbrennungsmotor* || travel * *~ e. Maschinenteils, Stellventils, Positioniersteuerung* || range * *Bereich* || lift * *~ e. Ventils* || throw * *~ eines Exzenters, eines Kolbens, einer Presse usw.*
Hubantrieb lift drive || lifting drive
Hubbalken walking beam * *z.B. im Walzwerk*
Hubbühne elevated platform || lifting platform * *siehe auch →Hebebühne*
Hubelement lifting element || lift element
Hubgeschwindigkeit speed of lift || lifting speed || hoisting speed * *{Hebezeuge}*
Hubhöhe lifting height || hoisting height * *{Hebezeuge}* || travel height * *{Aufzüge}*
Hubkolbenverdichter piston compressor || reciprocating compressor
Hubkraft lifting force
Hubkreissäge stroke circular saw * *{Holz}*
Hublänge stroke length * *z.B. bei Hubspindeltrieb*
Hubleistung hoisting capacity * *eines Hebezeugs, einer Vertikalförderanlage usw.* || lifting capacity * *eines Hubwerks*
Hubmagnet lifting solenoid || lifting magnet
Hubraum displacement * *z.B. Verbrennungsmotor, Kompressor*
Hubspindel jackscrew || lifting spindle || elevating screw || thrust spindle * *Schubspindel* || jacking spindle
Hubspindeltrieb linear actuator
Hubtisch lifting table || elevating platform || lift table
Hubvolumen stroke volume * *eines (Hydraulik-)Kolbens*
Hubwagen lift trolley || hand-pallet truck * *handbedienter Paletten-Hub-/Rollwagen* || lift-truck || lifting trolley
Hubwerk hoist * *→Hebezeug, Winde, Kran, Flaschenzug, Lastenaufzug* || hoist drive * *~santrieb z.B. eines Krans/Förderzeugs* || hoisting gear * *Hubeinrichtung z.B. eines Krans* || lifting gear * *~ eines Krans/Förderzeugs* || winch drive * *Seil-/Windenantrieb* || lifting unit || crane hoist * *~ eines Krans*
Hubwerksantrieb lifting drive * *bei Kran* || hoisting drive * *bei Kran*
Huch Oops * *(salopp)*
Huckepack- piggyback * *auch Leiterkarte, IC usw.*
Huckepackbaugruppe piggyback board * *Leiterkarte* || daughtercard * *Leiterkarte*
Hülle wrapper || covering || wallet * *Futteral*

Hüllkurve envelope curve || envelope * *Einhüllende*
Hüllkurvenanalyse eveloping measuring method * *zur Ermittlung des Körperschalls (z.B. zur Erkennung von Lagerschäden)*
Hüllkurvenumrichter cycloconverter * *Direktumrichter* || direct AC converter
Hüllmaße enveloping dimensions * *Abmessung eines gedachten Quaders, der den Umriss e. Gegenstands kennzeichnet*
Hülse test socket * *Prüfsteck~* || test jack * *Prüfsteck~* || test receptacle * *Hülse/Verbindungsstück z.b. für Bananenstecker* || sleeve * *Hülse, Buchse, Büchse, Muffe, Manschette, Spann~* || case * *Einfassung, Mantel, Gehäuse, Behälter, Kapsel, Schutzhülle* || bush * *Buchse, Büchse, Lagerfutter* || bushing * *Muffe, Spann~, Durchführungs~ (für Kabel)* || shell * *Schale, Kapsel, Gehäuse, das (bloße) Äußere* || tube * *Röhre, Tubus, Tube, Kanal* || socket * *Steck~, Muffe, Rohransatz, Fassung* || cap * *Kappe; slip-on cap: (Kugelschreiber-/Füllhalter-) Aufsteck~* || hull * *Hülse (besonders von Feldfrüchten), Schale, Rumpf* || husk * *Schale, (wertlose) Hülle (auch figürl.); Hülse, Schote (besonders von Feldfrüchten)* || capsule * *Kapsel* || core * *Wickelkern; z.B. paper core: ~ zum Aufwickeln von Papier* || jacket * *Ummantelung, Hülle; tubular jacket: tubusförmige Hülle*
Hülsendurchmesser core diameter * *z.b. einer Papphülse bei Wickler*
Hunderterstelle hundreds digit
hundertfach hundred times || hundred fold
Hupe horn
Hutmutter cap nut
Hutprofilschiene C-type mounting rail according to DIN EN 50022 * *C-Schiene zur Aufschnappmontage (35 mm breit)*
Hutschiene mounting rail * *Montageschiene zum Aufschnappen von Geräten* || DIN rail * *z. Aufschnappmontage: C-Schiene DIN EN50022 (35mm), G-Schiene DIN EN 50035 (32 mm breit)* || DIN mounting rail * *Normschiene zur Aufschnappmontage z.B. nach DIN EN 50022 (35 mm breit)* || C-type mounting rail according to DIN EN 50022 * *C-Schiene zur Aufschnappmontage, 35 mm breit* || top-hat rail * *zur Aufschnappmontage*
Hüttentechnik metallurgy
Hüttenwerk iron and steel plant || steelworks * *Stahlwerk* || ironworks * *Stahlwerk* || metallurgical plant || smelting house * *Schmelzhütte* || foundry * *Gießerei, Hüttenwerk*
Hüttenwesen metallurgy
hybrid hybrid
Hybridbaustein hybrid module * *z.B. in Dickoder Dünnschichttechnik auf Keramikträger* || hybrid device * *z.B. in Dick- oder Dünnschichttechnik auf Keramikträger*
Hybridmodul hybrid module
Hybridschaltung hybrid circuit
Hybridtechnik hybrid technology * *z.B. Schaltung in Dickschichttechnik auf Keramikträger*
Hydraulikflüssigkeit hydraulic fluid
Hydraulikmotor hydraulic motor
Hydrauliköl hydraulic fluid
Hydraulikpresse hydraulic press
Hydraulikpumpe hydraulic pump * *zur Druckerzeugung in Hydrauliksystem*
Hydraulikzylinder hydraulic cylinder
hydraulisch hydraulic || hydraulically * *(Adv.)*
hydraulisch betätigt hydraulically actuated || hydraulically operated
hydrodynamisch hydrodynamic
Hydromotor hydraulic motor * *axial piston type: in Axialkolbenbauweise*
hydrostatisch hydrostatic

Hydrotechnik hydraulic engineering
hyperbolisch hyperbolic
Hypoidgetriebe hypoid gear
hypothetisch hypothetical
Hysterese hysteresis || dead band * *Totband/zone*
Hysteresebremse hysteresis brake
Hysteresekupplung hysteresis clutch
Hz Hertz || Hz || cycles per second || c.p.s. * *Cycles Per Second* || cps * *Cycles Per Second*

I

I-Anteil I component || integral term || integral component
I-mal-R-Kompensation IR compensation * *z.B. Komp. d. Ankerwiderstands bei tacholoser n-Regelg.*
I-quadrat-t Überwachung I-square-t monitoring
I-quadrat-t-Wert I-square-t-value
I-Regler I controller || integral-action controller
I-Umrichter current fed inverter * *Stromzwischenkreisumrichter* || current source inverter || current-source DC link converter || current DC-link converter * *→Stromzwischenkreisumrichter*
I-Verhalten integral action
I2t-Wert I2t-value * *"I-Quadrat t Wert", Grenzlastintegral, (thermische) Belastungsgrenze*
IBM-kompatibler PC IBM-compatible PC
IBN commissioning * *Inbetriebnahme* || start-up * *Inbetriebsetzung*
IBS start-up * *Inbetriebsetzung* || commissioning * *Inbetriebnahme*
IC IC * *Abkürzg.f. 'Integr. Circuit': integr. Schaltg.' u. 'Internat. Cooling': Kühlsyst. nach DIN IEC 34* || chip
IC-Sockel IC socket
ich glaube I think || I guess * *(amerikan.)*
Icon icon
ideal ideal * *(auch Substantiv) for: zu/für* || model * *(auch Substantiv: Vorbild)* || perfect
Idealfall ideal case || ideal conditions * *under ideal conditions/ideally: im ~*
idealisiert idealized
Idee idea || notion || conception * *Vorstellung* || trace * *Spur* || brain wave * *Geistesblitz, guter Einfall* || inspiration * *Einfall, Eingebung, Erleuchtung, Anregung*
ideell ideal || hypothetical * *hypothetisch, angenommen, mutmaßlich* || idealized * *idealisiert*
ideelle Gleichspannung ideal no-load direct voltage * *~ am DC-Stromrichter-Ausgang ohne Ohmsch. Spann.abfälle o. Schwachlast* || ideal no-load DC voltage * *an DC-Stromrichterausgang ohne Ohmsche Spann.abfälle u.Schwachlast*
ideelle Leerlaufgleichspannung ideal no-load DC voltage
ideeller Wechselspannungsgehalt ideal AC content
ideeller Wert ideal value
Identifikation identification * *auch einer Regelstrecke/Baugruppe*
Identifikationssystem identification system
identifizieren identify
Identifizierung identification || identifier * *Bezeichner*
identisch identical
identisch sein mit be identical with || equal
Identität identity
Identnummer code number || identification code
IEC IEC * *Abkürzg. für 'International Electrotech-*

nical Commission'; internationales Normengremium
IEEE IEEE * Abkürzg.f. 'Institute of Electrical and Electronic Engineers': amerikan. Ingenieursverband
IGBT IGBT * Insulated Gate Bipolar Transistor, schneller Leistungstransistor (Schaltzeit 50...100ns)
ignorieren ignore
Ilgner-Umformer Ward-Leonard-Ilgner set || Ilgner flywheel equalizing set || Ilgner system
illustrieren illustrate || illuminate || demonstrate * (figürl.) || exemplify * (figürl.) durch Beispiele
IM IM * {el.Masch.} 'International Mounting': Bauform und Aufstellung (VDE 0530 Teil7) || type of construction and mounting arrangement * {el. Masch.} 'International Mounting' (VDE 0530 T.7)
im Abstand von apart * (nachgestellt) 125 mm apart: im Abstand von 125 mm || at a distance of
im Allgemeinen generally || in general || commonly * üblicherweise || normally * normalerweise || usually * gewöhnlich || generally speaking
im Anschluss an following || subsequent to || in connexion with || in connection with * in Zusammenhang mit, mit Bezug auf
im Auftrag von by order of || on behalf of * auch für jemanden
im Augenblick at the moment || at present || presently * (Adv.)
im Bereich in the range
im Bereich von in the range || in the range of
im Besonderen in particular
im Detail in detail || detailed
im einfachsten Fall in the simplest case
im Einzelnen in detail || with full details * mit allen Einzelheiten
im engeren Sinne in a narrow sense
im Falle von in the case of
im Folgenden subsequently * subsequently described: im Folgenden beschrieben || hereafter || in the following text * in Schriftstück
im Freien outdoors || outdoor * designed for outdoor services: geeignet für den Betrieb ~
im Gegensatz zu in contrast to || in opposition to || unlike * a/the: zu || as opposed to || contrary to
im Griff haben have the knack * of a thing: etwas || have the feel * of a thing: etwas || master * meistern || tackle * fertig werden mit (Aufgabe, Problem usw.) || cope with * bewältigen, meistern, fertig werden mit, gewachsen sein
im Großen und Ganzen on the whole
im Laufe der Zeit in course of time
im Leerlauf fahren coast || run idle || idle || run in no-load condition
im Moment currently || momentary * momentan || at present * gegenwärtig || at the moment || for the time being || for the present * gegenwärtig
im Prinzip in principle || on principle * aus Prinzip | essentially * im Wesentlichen, eigentlich, im Prinzip || basically * grundsätzlich, im Grunde
im Rahmen von within the boundaries of * innerhalb der Grenzen von || within the scope of * innerhalb des (Geltungs)Bereichs von || within the framework of * innerhalb des Gefüges/der Struktur || within the || in the course of * während, des Verlaufes/der Abwicklung von || within the limits of * innerhalb der Grenzen von || under * im Rahmen eines Vertrages, einer Vorschrift usw. || for the purpose of * für Zweck/Ziel/Vorhaben
im schlechtesten Fall under worst case conditions * unter den ungünstigsten Umständen
im Sekundenbereich in the vicinity of seconds || in the range of seconds || in the seconds range
im Uhrzeigersinn clockwise || cw * Abkürzg. für 'clockwise'
im Verbund networked

im Vergleich zu as compared to || compared to || in comparison with
im Voraus in advance || beforehand
im weitesten Sinne in a broad sense || in the widest sense
im Wesentlichen mainly * hauptsächlich || essentially
im Zuge des in the course of
im Zweifelsfall in doubt
Image image
Imaginärteil imaginary part || imaginary component || complex component
Immissionsschutz pollution protection
immitierte Schwingungen vibrations from external sources * mechanische ~ || external vibrations * von außen eingeleitete mechanische ~
immittierte Schwingungen vibrations from external sources * mechanische ~
Impedanz impedance
Impedanzwandler impedance transformer || impedance converter || voltage follower * Spannungsfolger (Operationsverstärkerschaltung zur Signalentkopplung)
implementieren implement
imprägnieren impregnate || soak * durchtränken/nässen/-feuchten, ~ (with: in); durchtränkt werden, sich vollsaugen || proof * Textilien usw. (wasser-) dicht machen, imprägnieren || steep * eintauchen, durchtränken, imprägnieren (in/with: in)
Imprägnierharz impregnating resin
Imprägniermittel impregnating agent * z.B. für Motorwicklung || impregnant || impregnating agent || impregnating compound || impregnating material
imprägniert impregnated
Imprägnierung impregnation * auch einer Wicklung mit Isolierlack oder Tränkharz
Impuls pulse || impetus * Stoß/Triebkraft z.B. für Wirtschaftswachstum || stimulus * (Plural: stimuli) Anstoß, (An)Reiz, Stimulus, Antrieb, Ansporn
Impuls-Schalldruckpegel impulse sound-power level
Impulsamplitude pulse amplitude
Impulsantwort pulse-forced response
Impulsart pulse shape || pulse waveform
Impulsaufbereitung pulse conditioning || pulse shaping * Impulsformung
Impulsauswertung pulse evaluation
Impulsbelastbarkeit pulse strength * von Kondensatoren || pulse handling capacity * von Kondensatoren
Impulsbildner monoflop * Zeitstufe || timer * Zeitglied || pulse generator * Impulsgenerator || one-shot * Zeitstufe
Impulsbildung pulse generating * als Funktion || pulse generator * als Einrichtung/Schaltung
Impulsbreite pulse width || pulse duration
Impulsdauer pulse duration || pulse width
Impulsdiagramm timing diagram || pulse diagram
Impulse pro Umdrehung pulses per revolution || PPR * Abkürzg. für 'Pulses Per Revolution' (z.B. bei Impulsgeber)
Impulseingang pulse input
Impulserfassung pulse sensing
Impulsfestigkeit pulse strength * von Kondensatoren
Impulsflanke pulse edge
Impulsfolge pulse train
Impulsform pulse shape || pulse waveform || pulse waveshape
Impulsformer pulse conditioning circuit || pulse shaper
impulsförmig in pulse form || pulse-shaped || pulse || rectangular * rechteckförmig
Impulsformung pulse shaping * z.B. Aufsteilung

Impulsfreigabe pulse enable ∥ firing pulse enable * *für Thyristor(satz)* ∥ pulse enabling
Impulsfrequenz pulse frequency ∥ pulse rate
Impulsgeber pulse encoder * *z.b. an Motor* ∥ pulse tach * *['pals 'täk] Impulstacho* ∥ pulse generator * *allg.* ∥ shaft encoder * *Inkrementalgeber* ∥ incremental encoder * *Inkrementalgeber* ∥ pulse tachometer * *['pals 'täkomiete]*
Impulsgeberauswertung pulse encoder evaluation
Impulsgebernachbildung pulse encoder simulation * *Ausgabe v.Impulsketten durch e. Resolver-/ Sin-Cos-Geber-Auswerteschaltung*
Impulsgeberrückführung encoder feedback
Impulsgenerator pulse generator
Impulsgruppe pulse group ∥ pulse run ∥ burst * *von Nadelimpulsen*
Impulskette pulse train ∥ burst * *Paket von Nadelimpulsen*
Impulskondensator pulse capacitor * *mit hoher Impulsbelastbarkeit*
Impulslänge pulse duration ∥ pulse width * *Impulsbreite* ∥ pulse length
Impulslöschung trigger pulse blocking * *für Thyristor/Transistor-Zündimpulse* ∥ pulse suppression * *Impulsunterdrückung* ∥ pulse blocking * *Impulssperre*
Impulslücke interpulse period
Impulsmuster pulse pattern ∥ switching pattern
Impulspaket burst * *~ von Nadelimpulsen* ∥ pulse train * *Impulskette* ∥ pulse packet
Impulspaketsteuerung multi-cycle control
Impulspause interpulse period
Impulsperiode pulse period ∥ pulse repetition period
Impulsraster pulse pattern
Impulssperre pulse blocking ∥ pulse suppression * *Impulsunterdrückung* ∥ firing pulse blocking * *für Thyristorsatz* ∥ trigger pulse blocking * *für Thyristorsatz* ∥ pulse inhibit ∥ gate pulse blocking * *für Ansteuerimpulse* ∥ pulse disable
Impulsspur pulse track * *z.B. eines Impulsgebers*
Impulstacho pulse tachometer ∥ pulse tacho ∥ pulse tach ∥ pulse encoder * *Pulsgeber* ∥ shaft encoder * *Winkelschrittgeber* ∥ incremental encoder * *Inkrementalgeber, Winkelschrittgeber*
Impulsteiler pulse divider ∥ pulse scaler ∥ rate multiplier * *Impulsmultiplizierer (läßt einen einstellbaren Anteil d. Eingangsimpulse passieren)*
Impulsübertrager pulse transformer ∥ firing pulse transformer * *zur Thyristoransteuerung*
Impulsuntersetzer pulse divider
Impulsverkürzer pulse contracting block * *Funktionsbaustein*
Impulsverlänger pulse stretching block * *Funktionsbaustein*
Impulsverschiebung pulse displacement ∥ pulse shift
Impulsverstärker pulse amplifier ∥ gate pulse amplifier * *zur Ansteuer.v.Leistungsthyristoren/ Transistoren*
Impulsverteiler pulse distribution module
Impulsverteilung pulse distribution
Impulszahl number of increments * *bei (Inkremental)Impulsgeber*
Impulszähler pulse counter
in Abhängigkeit von dependent on ∥ over * *auf der X-Achse aufgetragene Variable einer Kurvendarstellung* ∥ as a function of * *(mathem.)* ∥ versus * *bei Kurve/Kennlinie/Funktion (z.B. Größe auf der X-Achse)*
in Analogtechnik analog based
in andern Worten in other words
in Anschlag bringen take into account * *leave out of account; nicht ~*

in Anspruch nehmen utilize * *(aus)nutzen* ∥ lay claim to * *Anspruch erheben auf*
in Benutzung in use
in Bereitschaft stehen stand by ∥ be ready
in Betracht ziehen take into account ∥ take into consideration
in Betrieb in operation ∥ working ∥ running ∥ online * *während des Betriebes, in Echtzeit* ∥ dynamically * *dynamisch, (Änderung) während des Betriebes* ∥ operational * *~ befindlich, betriebsfähig*
in Betrieb gehen start working ∥ go online * *Kraftwerk ans Netz* ∥ begin operation ∥ become operational ∥ come on-line ∥ go on stream ∥ go into operation
in Betrieb sein be operational
in Betrieb sein in operation ∥ start ∥ actuate * *in Gang bringen* ∥ put into operation ∥ start-up * *inbetriebnehmen* ∥ install * *intallieren* ∥ open * *eröffnen, Betrieb aufnehmen* ∥ put into service
in Bewegung setzen start ∥ set going ∥ set in motion
in den Griff bekommen get the knack * *of a thing: etwas ~* ∥ master * *meistern* ∥ overcome * *Herr werden, lösen, meistern (z.B. e. Problem)*
in den Kinderschuhen in its infancy * *be still in its infancy: noch in den ~ stecken*
in der Lage sein be capable * *fähig/im Stande sein (of: zu)* ∥ be in the position * *to do someting: ~ etwas zu tun*
in der Praxis in praxis ∥ in real life
in der Regel as a rule ∥ ordinarily ∥ generally speaking * *im Allgemeinen* ∥ generally
in die Wege leiten set on foot ∥ initiate ∥ start ∥ prepare * *vorbereiten* ∥ pave the way for * *den Weg bahnen für, in die Wege leiten*
in Dreieck geschaltet connected in delta ∥ in delta connection ∥ delta-connected ∥ D-connected
in einem Schritt in one step
in einer Art und Weise dass such that
in Einklang mit in accordance with
in entgegengesetzter Richtung in opposition ∥ in the opposite direction
in erster Linie primarily ∥ first of all ∥ above all ∥ in the first place
in Funktion treten become functional
in Gang setzen start ∥ set going ∥ set in motion ∥ throw into gear * *Maschine ~* ∥ put into operation * *Maschine ~, inbetriebsetzen* ∥ launch * *(figürlich) lancieren, Starthilfe geben, unternehmen, beginnen, vom Stapel lassen* ∥ set on foot * *(figürlich) in Gang bringen, auf den Weg bringen*
in Gebrauch in use
in Geschäftsbeziehung stehen mit do business with
in geschützten Räumen in sheltered areas
in gewohnter Weise as customary ∥ as usual * *wie gewöhnlich, gewohntermaßen* ∥ by habit * *gewohnheitsmäßig* ∥ habitually * *gewohnheitsmäßig*
in groben Zügen in broad terms ∥ in brief outlines ∥ along general lines
in großem Umfang on a large scale ∥ large-scale ∥ wholesale
in Hinblick auf regarding * *betreffend* ∥ from the perspective * *aus der ... Perspektive*
in Klammern in parantheses
in Kraft setzen put into force
in Kraft treten come into force
in Kürze lieferbar available soon
in nicht zu ferner Zukunft not too far down the road ∥ in near future * *in naher Zukunft*
in Ordnung O.K. ∥ correct
in Phase sein be in phase
in Reihe geschaltet connected in series * *with: zu* ∥ series-connected
in Reihe schalten connect in series * *with: zu*

in Richtung auf toward || towards
in Schritten von in steps of
in Serie schalten connect in series * with: zu
in Stern geschaltet in star connection || connected in star || star-connected || Y-connected
in Übereinstimmung bringen make agree * with: mit || synchronize * with: mit || reconcile * with/to: in Einklang bringen/abstimmen mit || harmonize * harmonisieren
in Übereinstimmung mit in accordance with || pursuant to * Vorschrift || in conformity with || in agreement with || in keeping with || in harmony with || in line with * (amerikan.) in Übereinstimmung/Einklang mit, einhergehend mit
in Unordnung bringen disturb
in Verbindung mit in conjunction with || combined with * in Kombination mit || in connection with
in vollem Betrieb in full action
in vollem Umfang in its entirety
in voraus ahead of time
in Vorbereitung being prepared || under preparation
in Vorbereitung befindlich being prepared
in Zusammenhang mit in conjunction with * siehe auch →in Verbindung mit || connected with * be connected with: ~ stehen mit; have no connexion with: nicht ~ stehen mit
Inbegriff substance * Wesen, Substanz, das Wesentliche, wesentlicher Inhalt/Bestandteil, Kern, Gehalt || essence * Substanz, Wesen, Geist, das Wesentliche, Kern; of the essence: von wesentlicher Bedeut. || quintessence * Kern, Quintessenz, höchste Vollkommenheit || aggregate * Summe, Anhäufung, Zusammenfassung || totality * Summe, Gesamtheit, Vollständigkeit || embodiment * Verkörperung, Darstellung, Inbegriff || incarnation * Verkörperung, Inbegriff (auch figürl.) || paragon * Muster, Vorbild, Ausbund || synonym * Synonym, sinnverwandtes/bedeutungsgleiches Wort
inbegriffen included || implicit * (mit/stillschweigend) inbegriffen, implizit
Inbetriebnahme commissioning || start-up || set-up || start-up procedure || opening * Eröffnung || commencement of operation * Einweihung, Produktionsbeginn || field service * Abteilungsbezeichnung || putting into operation * eine Maschine/Anlage || system start-up * Rechner/SPS softwaremäßig || startup || initial operation * erstmaliger Betrieb, z.B. eines Kleingeräts || initial start-up * erstmalige ~
Inbetriebnahme-Ingenieur commissioning engineer || start-up engineer || service engineer * Service/Wartungsingenieur || specialist * Fachmann
Inbetriebnahme-Personal commissioning personnel
Inbetriebnahmeanleitung start-up guide || installation guide
Inbetriebnahmeaufwand start-up costs || commissioning expenses
Inbetriebnahmehinweise information on commissioning
Inbetriebnahmeingenieur start-up engineer || commissioning engineer || system start up engineer || service engineer
Inbetriebnahmeschritte start-up steps
Inbetriebnahmezeit start-up time || commissioning time
inbetriebnehmen start-up || put into operation * z.B. Elektromotor || install * installieren || set in operation || start || actuate * in Gang bringen || open * eröffnen, Betrieb aufnehmen || commission || put into service * z.B. eines vom Lager entnommenen Motors || set to work * ingangsetzen
Inbetriebnehmer start-up engineer || commissioning engineer || system start up engineer

inbetriebsetzen set in operation || put into operation || start || actuate * in Gang bringen || start-up * inbetriebnehmen || install * installieren || open * eröffnen, Betrieb aufnehmen || commission
Inbetriebsetzer start-up engineer * Inbetriebnahmeingenieur
Inbetriebsetzung commissioning || field service * als Tätigkeitsfeld/Abteilungsbezeichnung || start-up || putting into operation || starting-up || putting into service || commencement of operations * Aufnahme des Betriebes einer (Fabrik)anlage || beginning of work * Aufnahme des Betriebes einer (Fabrik)Anlage || opening * Aufnahme des Betriebes (of: von einer (Fabrik)Anlage usw.) || initial start-up * Erstinbetriebnahme
Inbusschlüssel hexagon-socket spanner || socket spanner || socket wrench || box spanner || socket screw wrench
Inbusschraube socket-head cap screw || hexagon-socket screw
indem by
Index index || subscript * tiefgestellte(s) alphanumerisches Zeichen
Indien India
indirekt indirect * (Adj.) || indirectly * (Adv.)
indirekt gekühlte Wicklung indirect-cooled winding * {el.Masch.} nicht →direkt-leitergekühlte Wicklung
indirekte Kühlung indirect cooling * mit Sekundär-Kühlkreislauf
indirekte Zugregelung open loop tension control * ohne Tänzer oder Zugmessdose || open loop tension regulator
indizieren index
indiziert indexed * z.B. Variable, Parameter
indizierter Parameter indexed parameter
Induktion induction * elektromagn. ~: Erzeugung e. Spanng. in e. Leiter, d. sich i.e.magn. Wechselfeld befindet || flux density * Feldliniendichte, Flussdichte (Dichte der magnet. Feldlinien)
induktionsarm low induction
Induktionsgesetz law of induction || Faraday's law
Induktionshärtung induction hardening
Induktionsmaschine induction machine
Induktionsmotor induction motor
Induktionsofen induction furnace
Induktionsspannung induction voltage
induktiv inductive
induktiv gekoppelt inductively coupled
induktive Belastung inductive load
induktive Last inductive load
induktiver Blindwiderstand inductive reactance
induktiver Messaufnehmer inductive sensor
induktiver Spannungsabfall inductive voltage drop
induktives Bauelement inductor
Induktivität inductance * Fähigkeit e. stromdurchflossenen Spule, magnet. Energie zu speichern || inductance value * ~swert
induktivitätsarm low-inductance || low-impedance
Industrie industry
Industrie- industrial || for industrial purposes
Industrie- und Handelskammer Chamber of Industry and Commerce
Industrie-PC industrial PC computer || IPC * Abkürzg.f. 'Industrial PC' || industrial PC
Industrieanlage industrial plant
Industrieanlagenbau industrial plant construction || plant construction
Industrieanwendung industrial application
Industrieausrüstungen industrial equipment
Industrieautomatisierung industrial automation
Industriebereich branch of industry || sector of industry || area of industry
Industriebetrieb industrial enterprise || manufactu-

ring plant || industrial establishment || engineering operation
Industrieelektronik industrial electronics
Industriegebiet industrial area || industrial district || manufacturing district * *Fabrikgebiet* || industrial estate * *z.B. am Rande eines Ortes*
industriell industrial
Industriemesse industrial fair
Industrienetz industrial network * *Energieversorgungsnetz*
Industrieofen industrial furnace || industrial oven
Industriequalität industrial quality * *i. Ggs. zur kurzlebigen Konsumqualität*
Industrieroboter industrial robot
Industriestandard industrial standard || industry standard
Industriestaubsauger industrial vacuum cleaner
Industrieumgebung industrial environments * *rugged: raue*
Industrievertretung representative office || sales agency
Industriezone industrial zone || industrial region
Industriezweig branch of industry || sector of industry || field of industry
induzieren induce
ineinander greifen engage || mesh
infolge due to * *auf Grund* || as a consequence of
infolgedessen consequently || as a result || owing to this || owing to which || accordingly * *folglich, demgemäß, entsprechend* || so
Informand trainee * *Praktikant, Auszubildender*
Informatik informatics || computer science * *Computertechnik/-wissenschaft* || information schience || information technology * →*Informationstechnik*
Information information * *(Singular und Plural) on/about: über; auch 'Auskunft'* || piece of information * *(Singular) Einzel-/Teil~* || details * *nähere Einzelheiten* || particulars * *nähere Auskünfte* || inquery * *(Ergebnis von) Nachforschungen* || info * *(salopp) Kurzform für 'Information'*
Informationen information * *nie in d.Mehrzahl verwenden ("informations" ist falsch!)* || literature * *schriftliche* || data * *Daten*
Informations- informative * *informativ*
Informationsbeschaffung information procurement
Informationsdienst info || information || bulletin * *Folge eines ~es* || periodical technical information * *regelmäßig erscheinender techn. ~*
Informationsfluss information flow
Informationsgespräch informative discussion
Informationsmaterial information material
Informationsmenge amount of information
Informationsschrift bulletin * *kleines Informationsblatt aus aktuellem Anlass*
Informationsstand information counter * *z.B. bei Messe*
Informationstechnik information technology || IT * *Abkürz. f. 'Information Technology'* || information science * →*Informatik*
Informationsveranstaltung information event
Informationsverarbeitung data processing || information processing
informativ informative || information-rich || informatory
informieren inform * *of/on/about: über; oneself: sich* || feed information * *to somebody: jdm. Informationen zukommen lassen* || notify * *jemanden benachrichtigen, (amtlich) mitteilen, melden* || advise * *of: über; benachrichtigen (of: von, that: dass), beraten* || acquaint * *with: über; mitteilen (with a thing: etwas, that: dass)* || instruct * *anweisen* || brief * *eine Kurzinformation/Einweisung geben* || gather information * *sich informieren, Informationen sammeln* || orient * *sich informieren, sich*

orientieren || provide information * *to somebody: jdn.*
infragekommend respective || relevant
Infrarot infrared
Infrarot-Trocknung infrared drying
Infrastruktur infrastructure
Ingenieur engineer
Ingenieurbüro engineering firm || consultant firm * *Beraterfirma* || software company * *Softwarehaus* || engineering consultant || consultant company || engineer's office || consulting engineer's office || engineering office
Ingenieurwissen engineering expertise
Inhaber holder * *einer Firma, Lizenz, Genehmigung usw.* || owner * *Eigentümer* || possessor * *Besitzer, Inhaber* || proprietor * *Eigentümer*
Inhalt content || table of contents * *z.B. Inhaltsverzeichnis am Anfang eines Buches*
Inhaltsübersicht summary || statement of contents || list of contents * *~ in Listenform* || synopsis * *kurze*
Inhaltsverzeichnis contents || list of contents || table of contents || index * *Stichwort-/~, Register; (An)-Zeiger; (Hand)Zeichen, Index, Kennziffer*
inhomogen inhomogeneous * *[inhomo'dschi: njes]*
initialisieren initialize || default * *[Computer] auf den Vorbesetzungswert setzen (Variable, Parameter usw.)*
Initialisierung initialization || power-up procedure * *e. Rechners beim Einschalt. d. Versorgungsspann.*
Initialisierungsfehler initialization error
Initiative initiative * *take the initiative: d. Initiative ergreifen; on one's own initiative: aus eigen. Antrieb*
initieren initiate
initiieren initiate * *einleiten, anstoßen*
inkonsistent corrupted * *verfälscht, z.B. Text, Dokumentation, Daten* || unhealthy || mixed up * *durcheinandergebracht*
Inkrement increment
Inkrementalgeber incremental encoder * *rotary: Dreh-* || pulse encoder || shaft encoder
inkrementell incremental
inländisch domestic
Inlandsmarkt domestic market
Inline-Pumpe inline pump * *Kreiselpumpe mit integriertem Motor*
innen inside * *auch "nach (dr)innen"* || internal || interior || indoor * *im Innenraum (nicht im Freien)* || on the inside * *(Adv.)*
Innen- internal || inner || inside
Innenansicht internal view
Innenausbau interior construction * *[Bautechnik]* || interior finishing * *[Bautechnik]* || interior decoration * *[Bautechnik]* || interior fittings
innenbelüftet open, ventilated || with open-circuit cooling || enclosed-ventilated
Innendurchmesser inside diameter
innengekühlt open, ventilated * *Motorkühlart* || with open-circuit cooling * →*durchzugsbelüftet (Motorkühlart)* || enclosed ventilated * →*durchzugsbelüftet (el. Maschine mit Kühlluftstrom durchs Gehäuse)* || enclosed self-ventilated * *innengekühlt,* *eigenbelüftet; Motorkühlart* || open, self-ventilated * *innengekühlt mit Motorwelle-getriebenem Lüfter;Kühlart IC01/11/21/31 IEC34.6* || open-circuit cooled || open-circuit ventilated || open-circuit air-cooled || internally cooled || internally ventilated * *elektr. Maschine*
innengekühlt durch Fremdbelüftung open, separately ventilated * *z.B. Motorkühlarten IC05/06/ 17/37 nach DIN/IEC 34 Teil 6*
Innengewinde internal thread || female thread || tapped hole

Innenkühlung open-circuit cooling * *z.B. Motor mit innen durchströmender Kühlluft, Kühlarten IC01,IC11, IC21,IC31* || open-circuit ventilation * *Motorkühlart mit innen durchströmender Kühlluft (Durchzugsbelüftung)* || open-circuit air cooling * *Durchzugsbelüftung (Motorkühlart)*
Innenläufer internal rotor
Innenläufermotor internal-rotor motor
innenliegend internal || interior || inboard * *{Schifffahrt} innenbords*
Innenlüfter internal fan
Innenölung internal lubrication
Innenpolerregung inner pole excitation
Innenraum inside room * *im Gebäude* || interior * *das Innere* || interior space * *der innere Raum* || indoor * *Innenraum-; für die Aufstellung/Verwendung in geschlossenen Innenräumen*
Innenraum- indoor- * *nicht im Freien*
Innenraumaufstellung indoor installation
Innensechskant hexagon-socket || internal hexagon
Innensechskantschraube hexagon-socket screw
Innenteile interiors || inner parts || inside parts || interior parts
Innenverzahnung internal gear || internal gearing || internal teeth
Innenwiderstand internal resistance || internal impedance || output resistance * *Ausgangswiderstand einer Signal- oder Spannungsquelle* || output impedance * *Ausgangsimpedanz* || source resistance * *Ausgangswiderstand*
inner interior || internal
innerer inner || internal || inherent * *innenwohnend, →zugehörig, angeboren, →eigen, eingewurzelt*
innerer Kühlluft-Kreislauf primary cooling air circulation
Inneres interior
innerhalb inside * *auf der Innenseite, im Inneren* || within * *~ eines Bereiches, within a period: ~ einer Periode* || interior * *innen gelegen*
innewohnend inherent * *in: dem*
Innovation innovation
Innovationsanstöße incentives of innovation
Innovationskraft innovative strength
Innovationsstrategie innovation strategy
Innovationszyklus innovation cycle
innovativ innovative
innovieren innovate
innoviert innovated
Innung guild
ins Netz into the supply
insbesondere in particular || especially * *(Adv.)* || particularly * *(Adv.)*
Insel- cell- || local || separated || solitary || sattelite-
Inselbetrieb isolated operation * *z.B. Inselnetz, Dieselgenerator* || solitary operation
Inselkraftwerk isolated power station
Insellösung insular solution || singular solution
Inselnetz private power system || island power system
Inserat advertisement || ad * *(salopp) Abkürzg. für 'advertisement'*
Inserent advertiser || advertizer * *(amerikan.)*
insgesamt total * *gesamt, total* || altogether * *gänzlich, ganz und gar, im Ganzen genommen, völlig* || in a body * *zusammen, geschlossen* || in total || a total of * *in Zusammenhang mit Zahl* || in whole * *(Adv.) als Gesamtes/Gesamtheit*
Inspektion inspection || overhaul work * →*Überholungsarbeiten* || check-up
Inspektionsintervall service interval
Inspektionsklappe servicing cover * *Wartungsklappe; siehe auch* →*Wartungsöffnung*
Inspektionsmaschine inspection machine * *Sortierroller z.Hin- u. Herwickeln v. Papier usw. z.*

Aussortieren v.Schadstell. || paper inspection machine * →*Sortierroller für Papier*
inspizieren inspect || examine * *prüfen* || superintend * *beaufsichtigen* || visit * *besuchen, aufsuchen, inspizieren, in Augenschein nehmen*
instabil unstable
Instabilität instability || hunting * *Reglerschwingen/pendeln* || unstable condition || chattering * *Reglerschwingen, Rattern (bei Antriebsregelung)* || runaway * *Durchgehen (eines Motors usw.)*
Installation installation * *auch eines Computer-Programms*
Installations-Bezugsbedingungen installation reference-conditions * *bei Installation e. Motors zu beachtende ~, VDE 0530 Beiblatt 1*
Installationsmaterial wiring accessories || installation material
Installationsschiene mounting rail || DIN rail * *C-bzw.G- Montageschiene n.DIN EN50022 (35 mm breit) bzw.DIN EN50035 (32mm)* || C-type mounting rail according to DIN EN 50022 * *zur Aufschnappmontage, 35 mm breit* || miniature C-type mounting rail according to DIN EN 50045 * *Mini-Aufschnappmontageschiene, 15mm breit* || G-type mounting rail according to DIN EN 50035 * *asymmetr. G-Aufschnappmontageschiene, 32 mm breit*
Installationstechnik installation engineering || installation technology * *als Fachrichtung oder Geschäftszweig*
installieren install * *siehe auch* →*einbauen*
instand setzen enable * *in die Lage versetzen* || repair * *reparieren* || mend * *ausbessern, flicken, reparieren, richten* || restore * *wiederherstellen* || fix * *(amerikan.)* || recondition * *überholen* || overhaul * *überholen*
Instandhaltung maintenance * *siehe z.B. DIN 31051* || service || servicing || upkeep
instandsetzen restore || repair || recondition * *überholen* || overhaul * *überholen*
Instandsetzung restoration || reconditioning * *(Wieder-) Instandsetzung, Überholung, Erneuerung* || repair * *Reparatur* || repairing * *Reparatur* || corrective maintenance || repair work * *Reparaturarbeiten*
Instandsetzungsarbeiten repair works || repairs * *to: an*
Instanz instance * *(auch in der Datenverarbeitg.: Exemplar eines mehrfach ablaufenden Programms)* || authority * *Dienststelle*
Instrument instrument || tool || measuring instrument * *Mess-*
Instrumentierung instrumentation * *Mess- und Regelausrüstungen für verfahrenstechnische Prozesse*
intakt intact || in working order * *siehe auch 'funktionsfähig'* || unhurt || healthy
Integral integral * *[in'ti: grel] of: von/über*
Integralanteil integral component * *[in'ti: grel]* || integral term
Integralregler integral-action controller || I controller
Integration integration
Integrationsfähigkeit ability to be integrated
Integrationsgrad degree of integration
Integrationszeit integral-action time * *[inti: grel]* || integral time || integral time constant
Integrationszeitkonstante integral time || integral gain || integral-action time || integral time constant
Integrator integrator
Integratorsetzwert integrator setting value
integrieren integrate * *of: von; over: über* || incorporate * *einbauen (in/into: in), (in sich) aufnehmen, einverleiben*
integrierend integrating

integrierende Messung integrating measurement * *z.B. bei A/D-Umsetzer nach d."Dual-Slope-Verfahren" (langsam aber störarm)*
Integrierer integrator
Integrierglied integrator || integrating element || I-element
integriert integrated || on-board * *auf der Leiterkarte* || integral * eingebaut, beinhaltet, beinhaltend; ganz, vollständig || built-in * eingebaut || on-chip * *Zusatzfunktion auf einem integriertem Schaltkreis* || embedded * *eingebettet, eingebaut*
integrierte Schaltung integrated circuit || IC * *Abkürzung für 'integrated circuit'* || chip * *(salopp) Halbleiter-Chip*
integrierter Schaltkreis integrated circuit || IC || chip
Integrierzeit integral-action time * *[in'ti: grel]* || integral time * *[in'ti: grel]* || integral gain || integral time constant
Integrierzeitkonstante integral-action time * *[in'ti: grel]* || integral time || integration time || integral gain || integral time constant
Intel-Format Intel format * *Datenformat: niederwertiges Byte/Wort auf niederer, höherwertig. auf höherer Adresse* || little endian format * *Datenformat: niederwertig.Byte/Wort auf niederer, höherwert.auf höher.Adresse*
intelligent intelligent || smart
intelligente Klemmenleiste remote I/O module * *entfernt vom Rechner angeordnete Ein/Ausgaben* || remote I/O || remote IO module
intelligente Peripheriebaugruppe intelligent I/O-module * *z.B. bei SPS* || intelligent peripheral
Intelligenz intelligence
intensiv intensive || comprehensive * *umfassend*
intensivieren intensify
Intensivkurs intensive course
interaktiv interactive || menu driven * *menügesteuert* || menu-prompted * *menügesteuert*
interdisziplinär interdisciplinary
interessant interesting || of interest * *to: für* || attractive * *attraktiv* || fascinating * *faszinierend*
Interesse interest * *auch (An)Teilnahme, Reiz, Wichtigkeit; be of (little) interest: v.(geringer) Bedeutg. sein* || concern * *~ (for: für; in: an), Unruhe, Sorge, Bedenken (at/about/for: um/wegen), Wichtigkeit*
interessierend interesting || relevant * *sachdienlich, von Bedeutung/Belang* || of interest
Interface interface
Interfacemodul interface module
Interferenz interference
intermittierend intermittent
intern internal * *(Adj.)* || internally * *(Adv.)* || interior * *innen, innengelegen* || integrated * *eingebaut* || built-in * *eingebaut* || integral * *eingebaut, beinhaltet* || secret * *geheim* || confidential * *vertraulich* || private * *vertraulich* || hidden * *verborgen* || on-chip * *auf Halbleiterchip integriert*
interner Gebrauch internal use
Interoperability interoperability * *Betreibbarkeit von Geräten mehrerer Hersteller an einem Bussystem*
Interpolation interpolation * *linear interpolation: Linear~*
Interpolationsalgorithmus interpolation algorithm
interpolieren interpolate * *linearly: linear*
interpoliert interpolated
Interpretation interpretation
interpretationsfähig admitting different interpretations || construable || interpretable * *ambiguous* * *→mehrdeutig*
Interpreter interpreter * *z.B. Hochspracheübersetzer, der während des Programmablaufs arbeitet (z.B. →BASIC)*

interpretieren interpret
Interrupt interrupt
Interrupt-Reaktionszeit interrupt response time
Interrupt-Verzugszeit interrupt response time || interrupt latency
Interruptadresse interrupt address || interrupt vektor * *indirekte Interruptadresse, Zeiger auf das Interruptprogramm*
Interruptanforderung interrupt request || IRQ * *Abk. f.'Interrupt Request'*
Interruptbearbeitung interrupt processing || interrupt service routine * *~sprogramm* || interrupt handler * *~sprogramm* || interrupt program || interrupt procedure
Interruptebene interrupt level
Interruptereignis interrupt event
interruptfähig interrupt-capable * *z.B. Binär-/Digitaleingang* || having interrupt capability || interrupt generating * *alarmbildend, einen Interrupt erzeugend (z.B. Digitaleingang)*
interruptgesteuert interrupt-driven || interrupt-controlled
Interruptkennung interrupt identification
Interruptprogramm interrupt program || interrupt task || interrupt service routine || interrupt handler
Intervall interval
intervenieren intervene || interpose * *vermittelnd eingreifen, "dazwischengehen"*
intuitiv intuitive
invariant invariant
invers inverse || inverted * *invertiert* || complementary * *komplementär* || negated * *logisch invertiert*
invertieren invert
invertierend inverting
Invertierer inverter || sign inverter * *Negierer, Vorzeichenumkehr*
invertiert complementary * *Impulse* || inverted * *Analogsignal (Vorzeichenumkehr) oder Digitalsignal* || negated * *Binärsignal* || toggled * *Binärsignal ('umgeklappt')*
invertiertes Signal inverted signal * *Analog- und Digitalsignal* || negated signal * *Binärsignal* || complement * *Binärsignal, z.B. Impulsgeber-Signal*
Invertierung inversion * *für Binär- und Analogsignal (logische ~ bzw.Vorzeichenumkehr)* || reversal * *Umkehr, z.B. der Drehrichtung* || inverting * *für Binär- und Analogsignal*
investieren invest
Investition investment || first cost * *einmalige Kosten, Anschaffungskosten*
Investitionen capital goods * *Investitionsgüter* || capital investments
Investitionsgüter capital goods || investment goods || industrial and capital investment goods
Investitionskosten investment expenses || investment || purchase costs * *Anschaffungskosten* || capital cost
Investor investor
inwieweit how far || to what extent * *bis zu welchem Ausmaße* || in what way * *in welcher Hinsicht* || in what respect * *in welcher Hinsicht*
Ionenaustauscher ion exchange unit * *z.B. zum →Entionisieren von Kühlwasser in einem geschlossenen Kühlkreislauf*
Ionentauscher ion exchanger * *z.B. zum →Entionisieren des Wassers i.e. Kühlwasserkreislauf (zur Leitfähik.reduz.)*
Ionisierung ionization
IP IP * *'International Protection': kennzeichnet Schutz geg.Berührg. Fremdkörper, Wasser, Staub (IEC 34)* || intelligent IO-module * *(SPS) intelligente Peripheriebaugruppe* || intelligent I/O module
IP21 degree of protection IP21 * *Schutzart geg.*

Fingerberührg., Fremdkörper >12mm u. senkrecht. Tropfwass.
IP22 degree of protection IP22 * *Schutzart geg. Fingerberührg., Fremdkörper > 12mm u. Tropfwasser bis 15°* || ODP enclosure * *Abk. f. 'Open Drip-Proof': offen, tropfwassergeschützt nach IEC und DIN*
IP23 degree of protection IP22 * *Schutzart geg. Fingerberührg., Fremdkörper >12mm u. Sprühwasser bis 60°*
IP44 degree of protection IP44 * *Schutzart geg. Berühr.m.Werkzeug, Fremdkörper >1mm und Spritzwasser* || TEFC * *Abkürzg. für 'Totally Enclosed Fan Cooled': vollk. geschlossen u. lüftergekühlt (Motorschutzart)*
IP54 degree of protection IP54 * *Schutzart mit vollständ Berührschutz, geg. Staubablagerg. u. Spritzwasser* || TENV enclosure * *Abk.f. 'Totally Enclosed Non Ventilated': vollk.geschloss. ohne Lüfter;Motorschutzart* || totally enclosed type of enclosure
IP55 degree of protection IP55 * *Schutzart mit vollständ. Berührschutz, geg. Staubablagerg. u. Strahlwasser*
IP56 degree of protection IP56 * *Schutzart m. vollständ. Berührschutz, geg.Staubablagerg. u. schwere Seen*
IP65 degree of protection IP65 * *Schutzart mit vollständ. Berühr- u.Staubschutz u. gegen Strahlwasser*
IP67 degree of protection IP67 * *Schutzart mit vollständ. Berühr und Staubschutz u. für Unterwasserbetrieb*
IPM IPM * *'Intelligent Power Module': hochintegrierter Leist.halbleiter, z.B. in Dick-/Dünnschichttech.*
IRQ IRQ * *{Computer\ SPS} Abkürzg. für 'Interrupt ReQuest': Interupt-Anforderung*
irregulär irregular
irregulärer Zustand irregularity || abnormal condition
irrelevant irrelevant * *to: für* || insignificant * *bedeutungslos*
irrtümlich erroneous || mistaken || false || inadvertent * *unbeabsichtigt* || erroneously * *(Adv.) ~erweise* || mistakenly * *(Adv.) ~erweise* || by a mistake * *(Adv.) ~erweise, versehentlich* || through oversight * *(Adv.) versehentlich* || inadvertently * *(Adv.) unbeabsichtigt* || by mistake
irrtümlicherweise erroneously * *(Adv.) siehe auch →versehentlich*
ISDN ISDN * *Abkürzg.für 'Integrated Services Digital Network': dienstintegrierendes digit. Nachrichtennetz*
ISO ISO * *Abkürzg. f. 'International Organization for Standardization': Internationale Organis. für Normen*
ISO-Passung ISO tolerance
Isolation insulation || isolation * *Potenzialtrennung, (Sicherheits)Abschaltung e. Stromkreises*
Isolationsbemessung insulation rating * *~sspannung z.B. einer Wicklung* || rated insulation voltage * *~sspannung einer Wicklung*
Isolationsfehler insulation fault
Isolationsfestigkeit insulation strength || dielectric strength
Isolationsgruppe insulation group || insulation class * *Isolationsklasse, Isolierstoffklasse*
Isolationsklasse insulation class || class of insulation system
Isolationskoordinaten insulation coordination * *z.B. Luft- und Kriechstrecken (für Stromrichter; siehe IEC 664)* || coordination of insulation || isolation coordinates
Isolationsmessgerät insulation tester
Isolationsmessung isolation resistance measurement || insulation test * *→Isolationsprüfung* || insulation testing * *→Isolationsprüfung*
Isolationsprüfung insulation test || insulation testing || high-voltage test
Isolationsspannung insulation voltage
Isolationsüberwachung insulation monitoring
Isolationswächter insulation monitor
Isolationswiderstand insulation resistance * *z.b. von Kabel, Leitung, Wicklung (siehe z.B. IEC 79-15 für el. Masch.)*
Isolator insulator
Isolierband insulating tape
isolieren insulate * *durch Isolierstoff* || isolate * *potenzialmäßig trennen, spannungs-/stromlos machen, (sicherheitsmäßig) abschalten*
Isolierglas insulating glass
Isolierklasse insulation class
Isolierlack insulating varnish || insulating enamel
Isolierscheibe insulating washer * *siehe auch →Glimmerplättchen*
Isolierschicht insulating layer || insulation layer
Isolierschlauch insulating sleeving
Isolierstoff insulating material || insulant
Isolierstoffgehäuse moulded case
Isolierstoffhülse insulated sleeve
Isolierstoffklasse insulation class * *siehe VDE 0530 Teil 1/DIN IEC 85; class F insulation: ~ F* || class of insulation system
Isolierstoffklasse B insulation class B || class B insulation * *Isolierung entsprechend Isolierstoffklasse B*
Isoliersystem insulating system * *z.b. eines Motors* || insulation system * *z.B. für eine Motorwicklung*
isoliert insulated * *z.B. Kabel* || isolated * *potenzialgetrennt, (zur Sicherheit) abgeschaltet* || segregated * *getrennt, abgesondert, abgespaltet, isoliert*
isoliertes Lager insulated bearing * *{el.Masch.} zur Vermeidung von Lagerströmen, die zum Lagerschaden führen können*
isoliertes Netz isolated network
Isolierumhüllung insulating sleeve * *buchsen-/schlauchförmige ~*
Isolierung insulation * *(für {el.Masch.}: siehe VDE 530, Teile 18-32)*
Isolierungsfehler insulation failure
Isoliervermögen insulation strength * *(eines →Isoliersystems, kann mit (Stoß)Spannungsprüfung ermittelt werden)*
Isolierverstärker isolation amplifier
ISP ISP * *Abkürzg.f. 'Interoperable Systems Project': Feldbus-Normungsvorhaben*
ist act * *Kurzform für Istwert* || f.b. * *Abkürzg. f. 'feedback'*
Ist- actual * *z.B. Istwert, Istfrequenz, Istdrehzahl*
Istweraufbereitung actual value conditioning
Istwert actual value * *['äktjuel 'wälju]* || feedback || act. value
Istwertabfrage scanning of actual value || actual value sensing
Istwertabgleich actual-value adjustment || actual-value matching
Istwertanpassung actual-value matching || actual value matching circuit * *hardwaremäßig* || feedback-signal matching
Istwertaufbereitung actual value conditioning || feedback signal processing
Istwerteingang actual-value input || feedback-signal input
Istwerterfassung actual value sensing || actual value evaluator * *als Einrichtung* || actual value acquisition
Istwertgeber actual value transmitter || actual value transducer || actual value sensor || actual value encoder * *digitaler Drehzahl~*

Istwertglättung actual value smoothing || actual value filtering
Istwertinformation actual value information
Istwertkanal actual value channel
Istwertleitung feedback cable * *{Regel.techn.}* || actual-value cable * *{Regel.techn.}*.
Istwertpolarität actual value polarity
Istwertschätzung actual-value estimation || state estimation
Istwertsiebung actual value filtering || actual value smoothing * *Istwertglättung*
Istwertumpolung actual value polarity reversal
Istzustand actual condition || current state
IT-Netz IT system * *erdfreies Netz (Sternpunkte bei Einspeisung u.Verbraucher ungeerdet)* || isolated supply * *(VDE 0100 Teil 410)* || ungrounded system
Italien Italy
italienisch Italian
Iterationsverfahren iteration method || iteration technique * *Rechen~ mit Annäherung ans Endergebnis durch wiederholte gleiche Rechenschritte*
iterativer Vorgang process of iteration
IxR Kompensation IxR compensation || IR compensation
IxR-Kompensation IR compensation * *Ankerwiderstandskomp. für tacholosen Betrieb e. DC Motors* || IxR compensation
I²T-Wert I²T Value * *"I-Quadrat t Wert", Grenzlastintegral, (thermische) Belastungsgrenze*

J

Jahres- annual * *jährlich, pro Jahr* || yearly
Jahresbedarf annual demand || annual requirements || annual purchasing volume * *jährl. Einkaufsvolumen eines Kunden*
Jahresbericht annual report || fiscal report * *(Jahres-) Geschäftsbericht*
Jahresdurchschnitt annual average
Jahresumsatz annual sales || annual sales volume || annual turnover
Jahresurlaub annual holidays
Jahresvolumen annual sales volume * *z.B. Bestellvolumen eines Kunden*
Jahreszeit time of year || season
jährlich annual || annually * *(Adv.)*
jahrzehntelang during decades
Jalousie louver * *auch an Lufteinauslassöffnung* || louvered cover * *mit "überdachten" Kühlschlitzen versehene Abdeckung (einer Luftein-/Auslassöffnung)* || shutter
je per * *pro* || ever * *jemals* || at a time * *immer; three at a time: je drei* || by * *three by three: je drei* || each * *jedes* || the * *the lower the price the greater the demand: je niedriger der Preis desto höher die Nachfrage*
je nach depending on || according to
je nachdem depending on * *abhängig davon; depending on how you do it: je nachdem, wie Du es machst* || as the case may be || it depends on || according to || in proportion as || according as
je nachdem ob dependent on whether * *abhängig davon ob*
jederzeit at any time || at all times || as many times as required * *beliebig oft, sooft wie gewünscht/erforderlich*
jedes each * *~ Einzelne; each thing: ~ Ding* || every * *~ insgesamt; every day: mit jedem Tage; everyone: jeder* || any * *~ Beliebige; without any doubt: ohne jeden Zweifel; at any time: z u jeder Zeit* || either * *~ von zweien*
jedes beliebige any
jedes Mal each time || every time
jedoch however || still || yet || nevertheless || but || though * *(am Satzende) aber, dennoch, trotzdem, allerdings*
jemals zuvor ever before
jenseits beyond
jetzig present-time || present * *bestehend* || actual * *aktuell* || existing * *bestehend* || current * *laufend* || of today * *heutig* || prevailing * *vorherrschend* || nowadays * *in der jetzigenZeit*
jeweilig respective * *(Adj.)* || in each case * *(Adv.)* || respectively * *(Adv.)* || of the day * *die jeweilige Person* || applicable * *anwendbar, passend, geeignet* || current * *aktuell* || corresponding * *entsprechend, gemäß* || correspondingly * *(Adv.) entsprechend, demgemäß* || particular * *einzeln, individuell, eigentümlich, speziell, besonder* || individual * *einzeln*
Joch yoke * *Magnetjoch, Poljoch; z.B. eines Trafos (wicklungsloser Teil e. magnet. Kreises)*
jungfräulich virgin * *(figürl.) jungfräulich, unberührt* || blank * *(figürl.) leer, unbeschrieben* || fresh * *frisch* || unused * *unbenutzt, ungebraucht, (noch) nicht verwendet/beansprucht*
jungfräulicher Zustand initial state || original state * *Urzustand, ursprünglicher Zustand* || factory setting * *Werkseinstellung, z.B. von Parametern* || 'as delivered' condition * *Lieferzustand*
Justage adjustment || alignment * *Ausrichtung, Ausfluchtung; Ausrichten (in einer geraden Linie usw.)*
Justiereinrichtung adjusting device || calibrator * *zum Eichen von Messgeräten* || aligner * *zum mechanischen Ausrichten*
justieren adjust || trim

K

Kabel cable
Kabel-Außendurchmesser outside cable diameter
Kabel-Verseilanlage cable twisting plant || cable stranding plant
Kabelabfang-Einrichtung cable clamping device * *zur Zugentlastung und Schirmauflegung*
Kabelabfangschiene cable-clamping bar
Kabelabgangsrichtung cable outlet direction * *z.B. bei Motorklemmenkasten*
Kabelabschirmung cable shield || cable screen
Kabelader cable core || cable strand
Kabelanschluss cable connection || cable termination
Kabelausführung cable exit
Kabelbaum cable harness || cable form || wire bundle
Kabelbefestigung cable clamping * *durch Schellen*
Kabelbelegung assignment of cable cores * *Zuordnung der Kabeladern zu (Signal-) Funktionen* || assignment of conductors
Kabelbinder cable tie || cable ties * *(Plural)* || twist-lock tic * *mit Knebel-Drehverschluss*
Kabelbruch cable breakage || interrupted cable
Kabeldurchführung wiring hole * *Loch* || cable entry * *Kabeleinführung* || wall entrance * *Wandeinlass* || bushing * *~shülse* || insulating bushing * *isolierte ~shülse*
Kabeldurchführungsplatte cable bushing plate

Kabeleinführung 170

Kabeleinführung cable entry * z.B. in Klemmenkasten (für el.Masch.: siehe z.B. DIN 42925)
Kabeleinführungsöffnung cable entry-hole * (z.B. in Klemmenkastenwand)
Kabelendverschluss sealing end * Schmutz-/Feuchteschutz am Kabelende an der Verzweigung d. Einzelleiter
Kabelführung cable routing || cable duct * z.B. Kabelkanal
Kabelgehäuse cable housing
Kabelkanal cable conduit || wiring duct || cable duct || conduit || cable channel
Kabelkapazität cable capacitance * zwischen den Adern oder zwischen Ader und Schirm || capacitance per unit length * →Kapazitätsbelag
Kabelklemme cable clamp * siehe auch →Kabelschelle || cable cleat * Kabel-Klemmschelle aus Isoliermaterial
Kabellänge cable length
Kabellegung cable laying
Kabelpritsche cable rack || cable support || cable shelf || cable bearer || raceway
Kabelquerschnitt cable cross-section || cable size || cable gauge
Kabelreste cable remains * siehe auch →Leitungsreste
Kabelschelle cable clamp || cable clip * meist aus Kunststoff || cable cleat * Kabel-Klemmschelle aus Isoliermaterial
Kabelschirm cable screen || cable shield
Kabelschuh cable lug || cable shoe || crimp connector * Quetschverbinder || spade terminal * offener ~ (schaufelförmig) || lug connector * Flach-/Messer~ || crimp-on lug * Flach-Quetschverbinder
Kabelstecker cable plug connector
Kabelstutzen cable gland
Kabeltrasse cable route || cable run
Kabeltrommel cable drum || cable reel
Kabeltülle cable sleeve * →Knickschutz bei Kabelausführung
Kabelumhüllung cable sheathing
Kabelummantelungsanlage cable sheathing line
Kabelumwicklung cable sheathing * Armierung
Kabelverbindung cable connection
Kabelverlegung routing of cables * Kabelführung, Art und Weise der Kabelverlegung || cable installation || cable laying || cable routing
Kabelverseilanlage cable twisting plant || cable stranding line || twister
Kabelverseilmaschine cable stranding machine || cable twisting machine
Kabelverseilvorrichtung twister || cable twister || cable stranding device
Kabelzuführung cable running || cable feeder
Kabelzusammenführung tow stacker * (Textil) z.B. in Faserverstreckanlage
Kabine cabin || cage * (Aufzüge) || cockpit * Führerraum eines Fahrzeugs || cubicle * abgeteilter Raum || compartment * (Zug-) Abteil || car * (Aufzüge) Fahrkorb
Kachel page frame * Speicherbereich im Rückwandbus einer SPS (bei SIEMENS SIMATIC (R) S5)
Kaffeepause coffee break
Käfig cage * z.B. für Käfigläufermotor oder Kugelkäfig Wälzlager
Käfigform kind of cage * (el.Masch.) bei Käfigläufermotor (Rundstab-, Hochstab-, L-Stab-, Keilstabkäfig usw.)
Käfigläufer squirrel cage rotor || cage rotor
Käfigläufermotor squirrel-cage induction motor || squirrel-cage motor || cage motor
Käfigläufermotor mit Anlauf- und Betriebswicklung double-deck squirrel-cage motor
Käfigstab cage bar * in einem Motorläufer

Käfigwicklung squirrel-cage winding * (el. Masch.) Kurzschlusswicklung bestehend aus den Stäben des Läuferkäfigs
Käfigwiderstand cage resistance
Käfigzugfederklemme cage clamp terminal * schraubenlose Kabelanschlussklemme
Kalander calender * beheizter Glättzylinder zur Glättung/Satinierung v.Papier oder Textilstoff durch Pressen || calenders * (Plural) siehe auch →Maschinenkalander, →Superkalander || calender stack * Gruppe senkrecht übereinander angeordneter Kalanderwalzen || paper mangle * Glättwerk für Papier
Kalanderwalze calender roll
kalandrieren calender * (z.B. Papier/Tuch/Stoff) pressen/glätten/satinieren mit Hilfe von Kalanderwalzen || calendize * mit Hilfe eines →Kalanders glätten (z.B. Papier, Textil)
Kalender calendar
Kalenderwoche calendar week
Kalibrator calibrator
Kalibriereinrichtung calibration device
kalibrieren calibrate
kaliometrisch caliometric
Kalkofen lime kiln
Kalkül calculus
Kalkulation calculation
kalorimetrisch calorimetric
kalorimetrisches Messprinzip calorimetric measuring principle
Kalotte cup || calotte || cap
kalt cold
kalte Lötstelle faulty solder joint || faulty solder connection
Kälte cold
Kälte- und Klimatechnik refrigeration and air conditioning
Kälteanlage refrigerator system
Kälteerzeugung cold production
Kältekompressor refrigerating compressor
Kälteleistung refrigerating capacity * einer Kältemaschine (DIN 8928, ISO 9303) || cooling capacity * eines Kompressors/einer Kältemaschine (DIN 8928, ISO 9309)
Kältemaschine refrigerating system || refrigerator
Kältemittel refrigerant * für Kälteanlage || cooling agent * für Kälteanlage
Kältemittelverdichter refrigerant compressor * bei Kältemaschine
Kältetechnik refrigeration technology
Kaltflusspressteil cold extruded part
kaltgewalzt cold rolled * z.B. Blechband
Kaltleiter PTC resistor * Widerstand mit 'Positive Temperature Coefficient': positivem Temperaturkoeffizient || PTC thermistor * bei Verwendung als Temperaturfühler || PTC * Positive Temperature Coefficient, Widerst. dess. Ohmwert s.m.steig. Temperatur erhöht || PTC sensor * als Temperaturfühler
Kaltleitertemperaturfühler PTC-thermistor
Kaltpresse cold press
Kaltstart cold start || cold restart * kompletter Neuanlauf (eines Computers usw.)
Kaltwalzen cold rolling
Kaltwalzwerk cold rolling mill
Kaltwassersatz waterchiller * Kältemaschine zur Erzeugung kalten Wassers
Kaltwiderstand resistance when cold || cold resistance * z.B. Lampe (ist wesentlich niedriger als der Warmwiderstand) || resistance in the cold state
Kamin chimney * →Schornstein || flue * Abzug (rohr)
Kamm comb || reed * (Textil) in der Weberei
Kammer chamber * auch mechan. Gehäusekammer

Kammrelais cradle relay
Kampagne season * *Zuckerrübenkampagne* || campaign * *(Werbe-) Feldzug*
Kanal channel || duct * *(Zu-/Ab-) Führungskanal, z.B. für Kabel, Luft; ventilation ductwork: Lüftungskanäle*
Kanalverwaltung channel control * *für Datenübertragungskanal*
Kanister canister
Kanne jug
Kannenablage can coiler * *(Textil) zur Ablage von Fäden*
kannibalisieren cannibalize * *aus-/abschlachten (auch Auto usw.)*
Kante edge * *lower: Unter~; upper: Ober~-; sharp: scharfe* ~ || brim * *Rand (auch eines Gefäßes), Krempe* || corner * *Ecke* || ledge * *Leiste, vorstehender* ~, *Sims, Anschlag~* || rim * *Rand, Randwulst, Felge* || selvage * ~ *eines Tuches, Webkante*
Kantenanleimmaschine edge-gluing machine * *(Holz) bei der Möbelherstellung* || edge-banding machine * *(Holz) in der Möbelindustrie*
Kantenbearbeitung edge working * *(Holz)*
Kantenbearbeitungsmaschine edge working machine * *(Holz)*
Kantenbearbeitungswerkzeug edge-trimming tool * *(Holz)*
Kantenführung edge guide * *Einrichtung* || edge guiding * *Funktion*
Kantenglättung anti-aliasing * *Glättung von Treppeneffekten bei Computergrafik, z.B. durch Interpolieren*
Kantenschutz edge protection * *als Funktion* || edge protector * *als Bauteil*
Kantholz square timber || squared timber
Kantine canteen
Kaolin kaolin || china clay
Kapazität capacitance * *elektr. Kapazität* || capacity * *(Fassungs)Vermögen, (Leistungs)Fähigkeit, (Nutz)Leistung, Höchstmaß* || capacitor * *Kondensator* || capability * *Fähigkeit (of something: zu etwas), Vermögen, Tauglichkeit*
Kapazitätsbelag capacitance per unit length * *Kabel, Leitung* || capacitance per unit length * *Kapazität pro Längeneinheit einer Leitung/eines Kabels (z.B. 100 pF/m)*
kapazitiv capacitive
kapazitive Last capacitive load
kapazitiver Blindwiderstand capacitive reactance
Kapillarwirkung capillary effect
Kapital resources || finance || capital
Kapitalbindung capital tied up || inactivation of capital
kapitalintensiv requiring a considerable amount of capital || capital-intensive
Kapitalrückfluss return on investment * *einer Investition* || capital returns || investment returns
Kapitel chapter * *in einem Buch* || paragraph * *Absatz, Abschnitt* || section * *Abschnitt* || topic * *Thema, Gegenstand* || passage * *Abschnitt, (Text-) Passage*
Kappe cap
kappen trim * *an den Enden abschneiden* || cut * *Tau usw.* || cross-cut * *(Holz) (an den Enden) quersägen*
Kappsäge cut-off saw * *(Holz)* || cross-cut saw * *(Holz)* || cross-cutting saw * *(Holz)* || undercut saw * *(Holz)*
Kapsel encapsulated module * *gekapselte Baugruppe* || case || box || capsule || chill * *Gusskapsel* || cap * *(Verschluss-) Kappe, Deckel* || module holder * *Baugruppenhalter* || module casing * *Baugruppengehäuse*
Kapselgebläse positive-displacement blower

kapseln cap * *mit Deckel versehen*
kaputt defective * *defekt* || broken * *entzwei, zerbrochen* || faulty * *schadhaft, fehlerhaft* || out of order * *nicht in Ordnung, defekt*
Kardangelenk cardan joint || universal joint
Kardanwelle cardan shaft || propeller shaft * *bei Fahrzeug, Schiff usw.*
Karde carding plant * *(Textil)*
Kardenmaschine carding machine * *(Textil) zum Aufrauen von Fasern*
kardieren card * *(Textil) krempeln/kämmen/kardieren von Wolle, Fasern, Vliesstoffen usw.*
Karte card * *auch Leiterkarte* || board * *Leiterkarte* || PC board * *Leiterkarte* || PCB * *Leiterkarte; Abkürzg. für 'printed circuit board'* || map * *Landkarte*
Kartei index file || card index || cardbox * *Karteikasten*
Karteikarte index card || file card
kartesisch Cartesian * *z.B. Koordinatensystem*
Karton cardboard box * *Schachtel* || carton * *~pappe, Schachtel* || board * *~papier* || cardboard * *~papier* || paste-board * *starke ~pappe* || box board * *~agenpappe* || folding-box * *Faltschachtel* || paperboard * *~agenpappe* || pasteboard box * *Schachtel*
Kartonage box board * *~npappe*
Kartonfabrik cardboard factory
Kartonmaschine cardboard machine || board machine
Kartusche cartridge * *z.B. für Tinte*
Karussel roundabout || merry-goround || turret * *Karussel- (Drehbank), Wende-/Drehkreuz-(Wickler); vertical turr.boring machine: ~drehbank* || coaster * *Fahrgeschäft (Achterbahn usw.)*
kaschieren laminate * *mit Schicht/Plättchen belegen* || line
Kaschiermaschine laminating machine
kaschiert lined
kaschierte Pappe pasted lined board
Kaschierung lining || lamination
Kaskade cascade * *connected in cascade: in Kaskade geschaltet, kaskadiert* || tandem
Kaskadenregelung cascade control * *Regelungseinrichtung mit mehreren einander unterlagerten Regelkreisen*
Kaskadenschaltung cascade connection || concatenated connection || concatenation connection || cascade arrangement || cascade circuit || tandem connection
kaskadieren connect in cascade || concatenate * *auch 'verketten,verknüpfen'* || cascade || connect in tandem
kaskadiert cascaded
kaskadierte Regelung cascade control * *Regeleinrichtung mit mehreren einander unterlagerten Regelkreisen* || cascaded control
Kaskadierung cascading || concatenation * *Schaltung in Kaskade, Verkettung* || tandem arrangement
Kaskode cascode * *cascode amplifier: ~verstärker*
Kasse cash register
Kassette cassette * *[ke'sset] z.B. Video-/Audio-/Magnetband-/Pneumatikventilkassette* || cartridge * *Patrone, Einsatz* || box || holder * *Halter* || case * *Kasten, Kästchen, Koffer, Kapsel, Schachtel, Behälter, Gehäuse, Hülle*
Kästchen small box || little box || square * *auf Rechenpapier usw.* || casket
Kasten box * *auch auf dem Bildschirm* || crate * *Lattenkiste, (Getränkeflaschen-) Kasten*
Kastenentstapler crate unloading machine * *in der Getränkeindustrie*
Kastenstapler crate loading machine * *in der Getränkeindustrie*

Katalog catalog || catalogue || list * *Liste* || directory * *Verzeichnis*
katalogmäßig catalog * *siehe auch* →*listenmäßig*
Katalysator catalyst
Kategorie category * *(Begriffs-) Klasse, Gattung; fall in the category: in die Kategorie fallen*
Kategoriespannung category voltage * *Dauergrenzspannung z.B. nach IEC*
Kathode cathode
kathodenseitig on the cathode side
Katze trolley * *{Hebezeuge}*
Katzfahrwerk trolley drive * *{Hebezeuge}* || cross-traversing gear * *{Hebezeuge}* || overhead-travelling gear * *{Hebezeuge}* || overhead travelling hoist * *{Hebezeuge} mit Hebewinde* || trolley hoist * *{Hebezeuge} mit Hebewinde* || trolley carriage * *{Hebezeuge}* || crab drive * *{Hebezeuge}* || cross-travel gear * *{Hebezeuge}* || trolley cross travel gear * *{Hebezeuge}* || trolley travel gear * *bei Portalkran*
Kauf purchase * *complete a purchase: einen ~ abschließen* || buy * *good buy: guter ~* || bargain * *guter ~, "Schnäppchen"* || purchasing * *das ~en* || buying * *das ~en* || acquisition * *Erwerb, (An-)Kauf, (Neu)Anschaffung, das Erworbene* || put up with * *etwas (mit) in ~ nehmen* || have to accept * *etwas in ~ nehmen müssen*
kaufen purchase * *erwerben* || buy || acquire * *erwerben*
Kaufentscheidung buying decision
Käufer purchaser || buyer || customer * *Kunde*
Kaufkraft purchasing power
kaufmännisch commercial
kaufmännische Leitung business administration * *auch als Abteilungsbezeichnung*
Kaufmannssund ampersand * *Zeichen " auf Tastatur/Bildschirm/Drucker*
kaum hardly * *schwerlich, fast nicht, mühsam, schwer* || barely * *nur gerade* || with difficulty * *mit Mühe* || hardly any * *kaum irgendwelche*
kaum jemals hardly ever
Kaustifizierer caustisizer * *{Papier} in der Zellstoffindustrie*
Kaustizierung causticizing * *z.B. in der Zellstoff-Herstellung*
Kautschuk rubber || india-rubber || gum elastic || caoutchouc || unvulcanized rubber
Kbaud Kbaud * *Kilobits pro sec* || kbs * *Kilobits pro sec*
Kbd kbs * *Kilobaud, Kilobits pro sec*
Kegel cone || taper * *verjüngtes Teil*
Kegel-Stirnradgetriebe helical bevel gearbox
kegelförmig taper || tapered || conical * *konisch* || coniform
kegelig taper || tapered || conical * *konisch*
Kegelmühle conical-roller mill || perfecting engine || conical refiner * →*Refiner mit konischen Mahlwalzen in der Zellstoffindustrie (mahlt Holzstoff)*
Kegelrad bevel gear || mitre gear || miter gear * *(amerikan.)* || bevel-gear wheel || conical wheel
Kegelradgetriebe bevel gear || bevel gear unit
Kegelrollenlager taper roller bearing || tapered-roller bearing || conical-roller bearing
Kegelstirnradgetriebe helical bevel gearbox
Kegelstumpf truncated cone || frustrum of cone
Kehle flute || channel || chamfer * *{Bautechnik}* || hollow groove * →*Hohlkehle* || throat * *{anatomisch}* || gullet * *Schlund*
Kehlmaschine moulding machine * *{Holz} zum Fräsen von Kehlen* || moulder * *{Holz}*
Kehrgewindewelle traverse cam * *z.B. zur* →*Changierung von Textilfäden bei der Aufwicklung*
Kehrmaschine sweeping machine

Kehrseite der Medaille flip side of the coin
Kehrwert reciprocal value
Keil key * *auch* →*Federkeil* || wedge || shim * *Klemmstück, Ausgleichsscheibe*
Keilnut keyway || keyseating || spline * *~ einer Keilverzahnung* || key seat
Keilnuten-Stoßmaschine key-seating machine
Keilriemen V-belt
Keilriemenantrieb V-belt drive
Keilriemenscheibe V-belt pulley
Keilstabläufer keyed-bar cage rotor * *{el. Masch.}* *Stromverdrängungsläufer bei Käfigläufermotor*
Keilverzinkung finger jointing * *{Holz}*
Keilwelle splined shaft * *mit* →*Federkeil versehene Welle* || spline shaft || integral-key shaft || spline * *Keilwelle, Nutwelle, Welle mit Keilverzahnung*
Keilzinkenmaschine finger jointing machine * *{Holz}*
Keilzinkenverbindung finger jointing * *{Holz}* || finger joint * *{Holz}*
Keilzinkenverleimpresse finger jointing gluing press * *{Holz}*
kein no * *(Adj.)* || not any * *(Adj.)* || none * *(Subst.)* * *none of us: keiner von uns* || no one * *(Subst.) keiner* || nobody * *(Subst.) keiner* || nothing * *(Subst.) kein*
keine Betauung non-condensing * *als Datenblattspezifikation* || no moisture condensation || moisture condensation non-permissible * *Betauung nicht zulässig*
keinesfalls on no account || on no condition || in no case || by no means * *als Antwort* || never * *niemals*
Kellerspeicher stack * *in Rechner (Stapelspeicher, Stack)*
Kennbit ID bit || identifier bit || flag
Kennbuchstabe identification letter || code || letter symbol || code letter
Kenndaten characteristic data
Kenndatensatz characteristic data set * *z.B. wählbarer umschaltbarer Parametersatz im Stromrichter*
Kenngröße essential quantity || characteristic value * *Kennwert* || parameter
Kennlinie characteristic * *falling/rising: fallende/(an)steigende ~* || curve * *Kurve* || function curve * *eine math. Funktion darstellend* || characteristic curve || characteristic line || diagram * *Kurvenbild*
Kennlinienbaustein polygon curve characteristic * *Polygonzug-Funktionsbaustein* || polyline function block * *Polygonzug-Baustein*
Kennlinieneinstellung characteristic setting
Kenniniengeber function generator || signal characterizer
Kennlinienneigung characteristic droop * *siehe* →*Statik*
Kennlinientyp characteristic type
Kenntnis knowledge || know-how * *"Gewusst Wie"* || expertise * *[expe'ties] Expertenwissen*
Kenntnisstand knowledge level
Kennung reference || identification || code || identifier || specifier || ID * *Abkürzg. für 'identifier': Kennung*
Kennwert characteristic value || parameter || characteristic
Kennzahl code number
Kennzeichen identifier || designation || characteristic feature * *Haupt/Schlüsseleigenschaft* || abbreviation * *Abkürzung* || code designation * *Kurzbezeichnung* || code * *Kurzbezeichnung* || parameter characteristics * *(→PROFIBUS-Profil)* || assignment * *Zuordnung, Zuweisung*
kennzeichnen tag * *mit Etikett/Kabelbezeichner versehen, (Ware usw.) auszeichnen* || identify * *mit e. Erkennungsmerkmal versehen* || designate * *be-*

zeichnen, (be)nennen, kennzeichnen; berufen, ernennen, ausersehen || characterize ** charakterisieren* || typify ** charakterisieren* || class ** klassifizieren, z.B.* in Schärfegrade beim Funkstörgrad || mark ** markieren, bezeichnen, kennzeichnen* || denote ** bezeichnen, kennzeichnen, anzeigen, andeuten, bedeuten*
Kennzeichnung   designation ** Bezeichnung, Name, Benennung, Kennzeichnung, Bestimmung* || mark of distinction ** zur Unterscheidung dienende ~* || identifier ** Kennzeichen* || mark ** Kennzeichen, Markierung* || labelling ** ~ mit Schild; Beschriftung* || indication ** Angabe/Anzeige z.b. auf Typenschild* || identification ** ~ zur Unterscheidung/Vertauschungssicherheit* || identification symbol || marking ** Markierung, Kennzeichnung (auch von Anschlussklemmen usw.)* || code of identification ** ~ durch Kennbuchstaben, Ziffern, Balken-/Farbcode usw.*
Kennzeichnungssystem   system of designation || code system ** mit Kurzbezeichnungen* || designation system || identification system
Kennziffer   code || code number || code No. || reference number || index ** (Plurel: indices)* || numeral ** Ziffer* || coefficient ** Koeffizient, Verhältniszahl* || identifier || identification number || code digit
Keramik   ceramic
Keramik-Vielschichtkondensator   multilayer ceramic capacitor
Keramikkondensator   ceramic capacitor
keramisch   ceramic
Kerbe   notch || cut ** Einschnitt*
Kerbverbinder   crimp connector
Kerbwirkung   notch effect || notching effect
Kerbzange   crimping tool ** Quetsch-/Presszange für Quetschkontakte, Kabelschuhe usw.*
Kern   core ** das Innere; auch Spulen/Transformator/Wickel~; (auch figürl.: get to the core: z. Kern kommen)* || essence ** Wesen, das Wesentliche; in essence: im ~* || kernel ** Kern(punkt), das Innerste, das Wesen, Guss~, Getreidekorn* || nucleus ** ~ eines Teams, einer Partei; Atom~ usw.* || main issue ** der wichtigste Punkt* || heart ** of the matter: ~ einer Sache* || centre ** z.B. Stadt~, Zentrum*
Kern-   essential ** wesentlich, Kernstück, Kernthema* || core ** Inneres* || key ** wichtigst, Schlüssel-* || nuclear ** Atom-*
Kerngruppe   core group || core team ** von Personen*
Kernkompetenz   core competence || core strength
Kernkraftwerk   nuclear power plant || NPP ** Abk. f. 'Nuclear Power Plant'*
kernlos   coreless ** z.B. Induktionsofen, Motor usw.*
Kernpunkt   issue || basic issue
Kerntechnik   nuclear technology
Kernteil   essential part
Kerntyp   core type ** bevorzugter Typ im Produktspektrum* || standard type
Kessel   boiler
Kesselspeisepumpe   boiler feed pump ** für Kraftwerk*
Kette   chain || train ** Kolonne (von Fahrzeugen, Personen usw.), Impulskette, Reihe (von Ereignissen, Gedanken usw.)* || warp ** {Textil} Kette(nfäden) in der Weberei; warp and woof: ~ und Schuss*
Kettenantrieb   chain drive
Kettenbaum   warp beam ** {Textil}* || yarn beam ** {Textil}*
Kettenelement   cascade element ** in Ablaufkette* || sequence cascade element ** in Ablaufkette*
Kettenförderer   chain conveyor
Kettenfräse   chain mortizing machine ** {Holz}* || chain mortizer ** {Holz}*
kettengetrieben   chain-driven
Kettenglied   chain link

Kettenimpuls   pulse train ** Impulskette, z.B. zur Thyristorzündung i.Ggs. z. Kettenimpuls*
Kettenimpulse   burst pulses ** z.B. zur Thyristor-Ansteuerung*
Kettenmaß   incremental dimension ** Schrittmaß/Positionsdifferenz bei Numerik-/Positioniersteuerung*
Kettenmaßsystem   incremental measuring system ** Fahren auf relative Positionen bei Positioniersteuerung ("Fahren um")*
Kettenrad   sprocket wheel || sprocket || cog wheel ** Ritzel* || chain-sprocket
Kettensäge   chain saw
Kettensatz   incremental data set ** Satz mit Relativ-Lagesollwerten für Lageregelung/Positioniersteuerung*
Kettenschaltung   daisy chain ** z.B. v. Kommunikationspartnern (i.Gegensatz. z. stern-/busförmigen Kommunikation)*
Kettenschleifer   chain grinder ** Stetigschleifer z.B. bei der Zellstoffherstellung*
Kettenschlepper   chain transfer ** z.B. in Walzwerk*
Kettenschutz   chain guard
Kettenspanner   chain stretcher || chain tightener || chain tensioner
Kettenstahl   chain steel
Kettenwirken   warp knitting ** {Textil}*
Kettenwirkmaschine   warp knitting machine ** {Textil}*
Kettung   chaining
kg   kg ** Kilogramm; 1 kg entspr. 2,2046 lb (pounds)* || lb ** pounds; 1 kg entspr. 2,2046 lb* || lb (av) ** pounds; 1 kg entspr. 2,2046 lb (av)* || lb (tr) ** pounds; 1 kg entspr. 2,6792 lb (tr)*
Kiemenblech   louvered cover ** Abdeckblech mit Schlitzen, z.B. für Be-/Entlüftungsöffnung* || fan louvre ** Lüfterjalousie, Abdeckblech mit Schlitzen für Be-/Entlüftungsöffnung* || hooded louvre plate ** zur Abdeckung einer Belüftungöffnung*
Kies   gravel
Kilowatt   kilowatt ** ['kilewot]*
Kilowattstunde   kilowatt hour ** ['kilewot 'auer]*
Kilowattstunden   kilowatt hours
Kilowattstundenzähler   kilowatt hour meter
Kinderkrankheiten   teething trouble ** auch von techn. Geräten, Verfahren und Einrichtungen* || initial defects ** von technischen Geräten, Einrichtungen*
kinderleicht   very easy || as easy as ABC ** (salopp)* || foolproof ** →narrensicher* || child's play ** it's mere child's play to them: es ist ein Kinderspiel für sie*
Kinderschuhe   infancy ** be still in its infancy: noch in den Kinderschuhen stecken*
Kinderspiel   child's play ** (auch figürl.) it's mere child's play to us: das ist ein ~ für dich*
Kinetik   kinetics
kinetisch   kinetic
kinetische Energie   kinetic energy || inertia energy
kinetische Pufferung   kinetic buffering ** Überbrückg. v. Netzausfällen durch Ausnutzg. d. kinet. Energie e.Antriebs* || kinetic ride through ** Überbrückg.v. Netzausfällen durch Nutzung d. kinet.- Energie e.Antriebs* || flywheel backup || kinetic backup
kinetische Schwungenergie   kinetic moment of inertia ** Schwungmoment*
Kippen   stall ** z.B. ~ eines Asynchronmotors* || pullout ** (auch Verb) z.B. ~ einer Asynchronmaschine* || breakdown ** z.B. ~ eines Asynchronmotor* || fall out of step ** (Verb) ~ einer el. Maschine* || become instable ** (Verb) ~ einer el. Maschine* || pull out of step ** (Verb) ~ einer el. Maschine* || topple ** (Verb) (um)~ durch Kopflastigkeit* || tilt ** (Verb) (Verb) ~ einer el. Maschine* || fall out of synchronism ** (Verb) ~ einer el. Maschine* || pulling out of step

Kippgrenze

* ~ *einer Synchronmaschine* || toggle * *(Verb)* ~ *eines Bits, eines Flipflop usw.*
Kippgrenze   stall limit * *z.B. bei Asynchronmotor*
Kipphebel   rocker || rocker arm
Kippmoment   stalling torque * *Asynchronmotor (größtes stationäres Drehmoment unter Bemessungsbedingungen)* || breakdown torque * *bei Synchron- u. Asynchronmotor* || pull-out torque * *Synchronmotor und Asynchronmotor* || pullout torque * *Synchron- und Asynchronmotor* || stall torque
Kippschalter   toggle switch
Kippschutz   stall protection * *z.B. Asynchronmotor* || pull-out protection * *z.B. Asynchronmotor* || stall prevention * *"Kipp-Verhinderung" z.B. bei Asynchronmotor* || anti-stall protection
kippsicher   stable || stall-protected * *(elektr.) mit Kippschutz versehen, z.B. Motor mit Umrichter* || tilt-free * *mechanisch* || non-tilting * *mechanisch* || stall-resistant
Kippsicherheit   stall protection * *Kippschutz (bei Asychronmotor usw.)* || stall stability * *Sicherheit gegen Kippen (bei Asynchronmotor usw.)* || stability * *siehe auch →Kippsteifigkeit*
Kippsteifigkeit   tilting rigidity * *mechan. Steifigeit, die das (Um)Kippen verhindert*
Kipptisch   tilting table
Kippverstärker   bistable amplifier
Kirchhoffsches Gesetz   Kirchhoff's law
Kiste   box || crate * *aus Holzlatten* || case * *Kasten, Kiste, Behälter, Schachtel*
Klammer   bracket * *Bauteil und Schriftzeichen (meist eckige ~); bracketed: in ~n (gesetzt)* || square bracket * *eckige ~* || brace * *geschweifte ~* || cramp * *Krampe, Klammer; Schraubzwinge; Fessel, Einengung* || clamp * *Klammer, Krampe, Klemmschraube, Zwinge, Erdungsschelle* || staple * *Heft~* || parenthesis * *[pe'renthisis] Schriftzeichen '(', ')'; (Plural: parentheses); paranthesize: i.~n setz.* || paper clip * *Büro~*
Klammerebene   nesting level || bracket level
Klammergerät   stapler * *für Papier, Holz usw.; Tacker*
Klammern   paranthases * *als Schriftzeichen* || brackets * *auch als Schriftzeichen*
Klammerstackpointer   nesting stack pointer
Klammertiefe   bracket depth || nesting depth
Klammerung   clamping
Klang   sound || tone || timbre * *Klangfarbe* || rattle * *eines Papierwickels (beim 'Draufklopfen' hörbar)*
Klapp-   hinged * *angelenkt, mit Scharnier versehen* || collapsible * *zusammenklappbar* || tipping || swing-out * *ausschwenkbar* || pivoted * *mit Scharnier/Zapfen versehen* || folding-type * *faltbar*
klappbar   hinged * *mit Scharnier/Gelenk/Angel versehen, (zusammen)klappbar* || folding || collapsible * *zusammenklappbar* || tipping || swing-out * *ausschwenkbar* || pivoted * *mit Zapfen/Scharnier versehen*
Klappe   flap
klappen   fold * *falten; fold away: weg-/zurückklappen* || clap || flap || work well * *gut funktionieren* || swing * *schwenken*
Klapprahmen   hinged rack * *Baugruppenträger für Flachbaugruppen*
klar   clear || definite * *eindeutig, genau, deutlich, fest, bestimmt* || distinct * *ausgeprägt, deutlich, entschieden, merklich, spürbar* || plain * *leicht verständlich, einfach,schlicht* || evident * *augenscheinlich, offenbar*
klar strukturiert   clearly structured
Kläranlage   purification plant || sewage treatment plant * *zur Abwasserbehandlung*
Klärbecken   setting-basin || filterbed
klären   clarify || clear || settle || clear up

Klärschlamm   sewage sludge
Klarsichtverpackung   transparent package
klarstellen   clarify
Klartext   plain text
Klartextangabe   plain text specification || plain text description * *Klartext-Beschreibung/-Bezeichnung*
Klartextanzeige   plain text display || plaintext display || plain language display
Klartextmeldung   plain-text message || plaintext message
Klärung   clarification || clearing up || settling
Klärwerk   sewage treatment plant || sewage work || purification plant
Klasse   level * *Ebene* || category * *Kategorie* || class * *auch 'Typdeklaration' in der objektorientierten Programmierung (z.B. bei C)* || classification * *Klassifizierung, z.B.in einer Norm*
klassifizieren   class * *z.B. in Schärfegrade beim Funkstörgrad* || classify
Klassifizierung   classification
klassisch   classical * *konventionell, üblich* || conventional * *konventionell, herkömmlich* || traditional * *wie gewohnt* || classic * *klassisch, althergebracht, anerkannt*
Klaue   claw * *Kralle* || carrier * *Nase, Mitnehmer* || catch * *Nase* || driver * *Mitnehmer, Nase* || dog * *Mitnehmer, Nase* || jaw * *(Klemm-)Backe, Backen, Klaue; jaw clutch: Klauenkupplung* || grip * *Greifer, Klemmer* || clip * *Klammer, Klemme, Halter, Spange* || pawl * *Sperrhaken, Sperrklinke, Klaue*
Klauenkupplung   dog clutch || claw coupling || jaw clutch || jaw coupling || claw clutch
Klebeband   adhesive strip || adhesive tape
Klebebild   sticker || sticker diagram
kleben   glue || stick || paste * *z.B. Papier* || adhere to * *kleben bleiben* || stick to * *kleben bleiben* || cement * *kitten* || bond * *Metall kleben, chem./techn. binden*
Kleber   adhesive || glue
Klebestelle   splice * *bei Abrollung/Abwickler*
Klebestellenverfolgung   splice tracking * *z.B. in Papierbearbeitungs-/Druckmaschine*
Klebestreifen   adhesive strip
Klebstoff   adhesive || glue * *Leim; auch 'umgebende Logik (m.niedr.Integrationsgrad) bei e. Mikroprozessor-Schaltung'* || cement * *Kitt* || paste * *Kleister*
Klebung   pasting
klein   little || small * *small letter: ~buchstabe; small motor: ~motor; keep/grow small: ~ halten/~er werden* || low * *niedrig bezügl. Zahlenwert/Höhe, z.B. Spannung, Leistung, Reaktanz* || compact * *mit hocher Dichte; kompakt* || small-scale * *mit/in ~em Umfang/Maßstab, z.B. Unternehmen, Integrationsgrad* || tiny * *winzig* || minute * *winzig* || insignificant * *unbedeutend* || trifling * *unbedeutend* || minor * *geringfügig* || microscopic * *verschwindend ~* || miniature * *~st* || downsized * *ver~ert, in Baugröße/Volumen reduziert* || low-rating * *mit ~er Leistung, z.B. Motor, Stromrichter, Pumpe* || small-footprint * *mit ~er Grundfläche/Stellfläche*
Klein-   small || compact || miniature || mini || micro || pony * *z.B. pony motor: →Hilfsmotor, Anwurfmotor*
kleine Drehzahlen   low speeds
kleiner   less * *(Komparativ) zahlenmäßig; than: als* || smaller * *(Komparativ) bezüglich Umfang* || minor * *(Komparativ) geringer, untergeordnet, geringfügig, von geringerer Umfang* || of smaller size * *(Komparativ) von kleiner Größe* || lower * *(Komparativ) niedriger* || lower-rating * *mit kleiner Leistung, z.B. Motor, Stromrichter, Pumpe*
kleiner gleich   equal to or less than
kleiner Leistung   low power || fractional horse-power * *fractional horse power motor: Motor ~*

Kleinheit smallness || littleness || minuteness ||
compactness * *Kompaktheit*
Kleinigkeit minor matter
Kleinleiterplatte compact PCB || small PC board ||
mini PCB
Kleinmotor miniature motor || fractional horsepower motor * *unter 1 PS (0,74 kW) Leistung* ||
small motor || FHP motor * *Abkürzg.f. 'Fractional-Horsepower motor': Motor bis 1 hp (entspr. 0,746 kW)* || fractional-hp motor * *Motor bis 1 hp Leistung (entspr. 0,746 kW)*
Kleinsignal- low-level
Kleinspannungsausführung low-voltage version
kleinst lowest || smallest || minimum || subminiature || microscopic || lowest-rating * *mit ~er Leistung, z.B. Motor, Stromrichter, Pumpe*
Kleinst- miniature || micro- || subminiature || microscopic
Kleinsteuerung mini PLC * *speicherprogrammierbare ~*
kleinstmöglich minimum * *zahlen/wertmäßig* ||
smallest possible * *Größe, Volumen*
Kleinstmotor sub-miniature motor
Kleinstumrichter micro-converter
Kleinteile small parts
Kleinteilelager small parts store
Klemmblock terminal block
Klemmbügel clip || terminal bracket * *Teil in einer Schraubklemme* || clamp strap || bracket * *(eckige) Klammer* || terminal clip * *Teil einer Schraubklemme* || clamping saddle * *an einer Schraubklemme*
Klemmdiode clamping diode
Klemme terminal * *(elektr.)* || screw terminal
* *Schraubklemme* || clamp * *(mechan.)*
klemmen jam * *festsitzen, festklemmen* || clamp ||
squeeze * *drücken, quetschen, zwängen* || pinch
* *(ein)klemmen, quetschen, einengen, einzwängen, drücken, kneifen, zwicken* || nip * *kneifen, zwicken (off: ab-), klemmen (Maschine)*
Klemmenanordnung terminal arrangement || terminal assignment * *Zuordnung der Klemmen(nummern) zu bestimmten Funktionen* || terminal layout
* *(siehe DIN46289, Leistungsklemmen: z.B. DIN 46206 T.2, DIN46223, DIN46260/64/65)*
Klemmenbelegung terminal assignment
Klemmenbelegungsplan terminal connection diagram
Klemmenbezeichner terminal identifier
Klemmenbezeichnung terminal designation * *für E-Motoren: siehe DIN 57530/VDE 0530/IEC 34 jew. Teil 8* || terminal marking
Klemmenblock terminal block || terminal strip
* *Klemmenleiste*
Klemmenbolzen terminal bolt * *z.B. im Motorklemmenkasten*
Klemmenbrett terminal block || terminal board
* *z.B. im Motorklemmenkasten* || terminal insulator
Klemmenbrücke terminal jumper || jumpered terminals * *gebrückte Klemmen*
Klemmenerweiterung terminal expansion || additional terminal modules
Klemmenfach terminal compartment * *Klemmenanschlussraum in einem Gerät*
Klemmenkasten terminal box || cable box || conduit box || terminal housing
Klemmenkastendeckel cover of terminal box
Klemmenkastengehäuse terminal box housing
Klemmenkastenlage terminal box position * *z.B. oben, links rechts bei einem Elektromotor* || location of terminal box * *z.B. eines Motorklemmenkastens (siehe z.B. DIN 42673 Teil 1)*
Klemmenkastenoberteil terminal box housing
Klemmenkastensockel terminal box base plate

Klemmenleiste terminal strip || terminal block
* *Klemmenblock*
Klemmenmodul terminal module
Klemmennummer terminal number || terminal No
Klemmenplatte terminal board * *z.b. in einem (Motor-) Klemmenkasten*
Klemmenschraube terminal screw
Klemmenspannung terminal voltage
Klemmenstein terminal block
Klemmenstützer terminal post insulator * *z.b. im Motor-Klemmenkasten* || terminal insulator * *in Klemmenkasten usw.*
Klemmenträger supporting insulator * *für eine (Schraub-) Klemme*
Klemmhülse locking sleeve
Klemmkasten terminal box
Klemmkörper sprag * *keilförmiger Klemmkörper, z.b. bei Rücklaufsperre, Freilauf*
Klemmkörperfreilauf sprag freewheel * *mit Klemmkeilen*
Klemmlasche clamping strap
Klemmleiste terminal strip || terminal block
* *Klemmenblock*
Klemmnabe clamping hub
Klemmnase clip
Klemmpunkt nip * *Stelle, an der d. Geschwind. e. durchlaufend. Warenbahn zwangsmäßig konstantgehalten wird*
Klemmring clamping ring || locking ring || clamp collar || collet * *Klemmhülse, Klemmring, Kragen, Konushülse, Spannhülse*
Klemmrollenfreilauf roller freewheel
Klemmscheibe clamping washer
Klemmschelle cable clamp * *zur Kabelbefestigung*
Klemmschraube setscrew || clamping screw || terminal screw * *Klemmenschraube*
Klemmstelle nip * *Stelle, an d.die Geschwind. e.durchlauf. Warenbahn zwangsmäßig konstanthalten wird* || speed master * *Antrieb, der die Geschwindigkeit einer Produktions-/Verarbeitungsstraße vorgibt*
Klemmung clamping
Klickfeld button * *auf dem Bildschirm* || active field
* *auf dem Bildschirm*
Klima climate
Klima- und Kältetechnik air conditioning and refrigeration
Klimaanlage air conditioning || air conditioner
* *Gerät* || air conditioning system
klimabeständig climate-proof || all-climate-proof
klimafest climate-proof
Klimafolge climatic sequence
Klimagerät air conditioner
Klimagruppe climate group || climate classification
* *siehe z.B. IEC-Publikation 721-2-1* || class of climate * *Klassifizierung v.natürl. Umwelteinflüssen (Temp., Luftfeuchte... nach IEC721-2-1)*
Klimakategorie climatic categorie * *z.B. nach IEC*
Klimaklasse climatic category * *z.B. nach IEC*
Klimaschutz weatherproofing || weather protection
Klimatechnik air conditioning
klimatisch climatic
klimatische Bedingungen climatic conditions
klimatisieren air-condition
Klimatisierung air conditioning
Klimaverpackung climate package * *siehe auch →Tropenschutzverpackung*
Klimazone climatic zone
Klimmzüge machen do gymnastics * *(figürl., salopp)* || do mental acrobatics * *(figürl., salopp) geistige ~*
Klinke jack * *(elektr.)* || pawl * *Sperr~* || catch
* *Sperr~* || latch * *Rast~*
Klinkenstecker jack plug || plug switch || jack
Klirrfaktor distortion factor * *Verhältn. Effekt.-*

klopfen 176

wert d. Oberschwingungen/Gesamteff.wert (DIN40110) || harmonic distortion || total harmonic distortion || THD * Abkürz. für 'total harmonic distortion' || DF * Abkürzung für 'Distortion Factor' || harmonic factor || distortion * (salopp)
klopfen knock || rap || beat * schlagen || drive * Nagel in die Wand usw. || tap * sanft klopfen
Kloss'sche Gleichung Kloss' equation * {el. Masch.} Formel für die Momentenkennlinie einer idealen Asynchronmaschine
Kluppe die-stock * {Werkz.masch.} an Drehbank || screw-plate * {Werkz.masch.} (amerikan.) an Drehbank || slide caliper * Gabelmaß || screwstock * zum Festschrauben eines Teils
knacken crack * (auch figürl.: harte Nuss, Software usw.) || click * metallisch || break * (figürl.) Software, Zugangscode usw. || hack * (salopp) Software, Zugangscode usw.
knapp scarce * spärlich, selten, rar; scarce commodities: Mangelwaren; make oneself scarce: sich rar machen || scanty * dürftig, kärglich, spärlich, knapp, unzureichend, beengt (Raum usw.) || tight * eng (Termin), dicht (gedrängt), mulmig, angespannt (Marktlage), geizig || spare * kärglich, dürftig, sparsam; a bare mile: eine ~ Meile; bare majority: ~ Mehrheit || meagre * dürftig, kärglich, mager || meager * (amerikan.) dürftig, kärglich, mager || bare * knapp, kaum hinreichend, arm (of: an); a bare mile: e.~ Meile; bare majority: ~ Mehrheit || stringent * knapp (Geld), gedrückt (Geldmarkt), streng || limited * beschränkt; my time is limited: meine Zeit ist ~ bemessen || short * of money: an Geld; of cash: bei Kasse; be in short supply: ~ sein; run short: ~ werden || hard up * (salopp) ~ an Geld || critical * critical items: ~ Waren || barely * (Adv.) mit ~er Not; barely sufficient: ~ ausreichend || just * (Adv.) gerade eben || just under * (Adv.) vor Zahlen || a little less than * (Adv.) vor Zahlen || give short measure * (Adv.) ~ bemessen || cut it fine * (Adv.) ~ berechnen (Preis) || keep short * ~ halten || poor * mangelhaft, unzureichend, ungenügend, schlecht
Knappheit shortage
Knarren- ratchet * mit Sperrklinke versehen, z.B. Schraubenschlüssel/-dreher
Knebelschalter knob-operated switch
kneten knead
Kneter kneader
Knetmaschine kneading machine || kneader
Knetstraße kneading line * in der Backwarenindustrie
Knetwerk kneading machine
Knick kink * in Draht, Kennlinie usw. || angle * Winkel || sharp bend * in Kurve || knee * Knie (-Stück, Rohr-), Winkel, Ausbeulung; auch ~ in Kennlinie usw. || bend * Biegung, Krümmung; auch ~ in Kennlinie usw. || break * Abbruch, Wechsel, Umschwung, Sturz (z.B. Preise, Wetter, Wirtschaft, Geschäft), || fold * Faltung in Papier Blech usw. || elbow * Biegung, Krümmung, Ecke, Knie, Kniestück, Krümmer, Winkel(stück)
Knickarmroboter articulated-arm-type robot
knicken kink * knicken (z.B. Kabel, kräuseln || break * brechen || crack * brechen, zerspringen, ausreißen, einbrechen, (zer-) spalten || buckle * sich krümmen, verbiegen || fold * Papier ~ || snap * Zweig usw. ~
Knickfrequenz cutoff frequency
Knickpunkt kneepoint * z.B. e. Kurve, Kennlinie, Polygonzuges usw. || breakpoint * einer Kurve, Kennlinie usw. || corner point * Stützpunkt einer Kurve, eines Polygonzuges usw.
Knickschutz kink protection
Knickschutztülle kink-protection sleeve || anti-kink sleeve

Knie knee * auch Rohr~; bring to its knees: in die ~ zwingen || elbow * Rohr~ || angle * Winkel
Kniehebelpresse toggle-lever press || toggle press
Kniff trick * Kunstgriff, List || knack * Dreh, Kniff, Apparätchen, "Dingsda"
Knopf button || pushbutton * Drucktaster || knob * auch Dreh/Druckknopf
Knopfdruck hitting of a key * just hitting a key: auf Tastendruck/lediglich durch Betätigen einer Taste || touch of a button * at the: auf ~ || mouse click * at a mouse click: mit einem Mausklick
Knopfzelle button cell * Batterie
Knoten node * auch eines Computernetzwerk; Teilnehmer an einem Bussystem
Knotenblech junction plate || gusset plate || fishplate || connection plate
Knowhow know-how || expertise || experience
Knüppel billet * Metall/Walzknüppel (Stahl/Walzwerk) || stick * Holz~, Steuer~ usw.
Knüppelwalzwerk billet mill
koaxial concentric || coaxial * Kabel, Buchse usw. || in-line * z.B. Armatur
Koaxialität concentricity * auch Fluchtigkeit; für Motor: siehe DIN 42955, IEC 72-1
Koaxialkabel coaxial cable || coax cable
Koaxialleitung coaxial cable || coax cable
Kobald-Samarium cobald-samarium * permanentmagnet. Material, z.B. für Elektromotor
Kocher digester * z.B. für Zellstoff || boiler || cooker
Köcherbürste plug-type brush || cartridge-type brush * zur Stromübertragung auf bewegte Maschinenteile
Kodierelement coding element * z.B. bei vertauschungssicherem Stecker || keying element * z.B. bei vertauschungssicherem Stecker
kodieren encode || code * Programmcode erstellen
Kodierer encoder || coder * Programmierer, der Programmcode erstellt
Kodiernut keying slot * an Stecker zur Gewährleistung der Vertauschungs-/Verpolungssicherheit
Kodierstift keying plug * z.B. für vertauschungssicheren Stecker
kodiert coded
Kodierung coding || keying * von Steckern zur Vertauschungssicherheit
Koeffizient coefficient * z.B. coefficient of friction: Reibzahl
Koeffizientenschreibweise coefficient notation
koerzitiv coercive
Koerzivitäts- coercive * ~feldstärke, ~kraft usw || coercivity
Kohärenz coherence * Zusammengehörigkeit in Ort und/oder Zeit
Kohäsion cohesion
Kohäsionskraft cohesiveness || cohesive force
Kohle coal
Kohlebürste carbon brush || brush
Kohlebürstenabrieb carbon dust || brush wear * Bürstenverschleiß
Kohlebürstenhalter carbon brush holder || brush holder
Kohlekraftwerk coal power station
Kohlendioxid carbon dioxide
Kohlendioxyd carbon dioxide * CO_2
Kohlensäure carbon dioxide || carbonic acid gas * dry ice * feste ~, Trockeneis
Kohlenstaubgebläse pulverized coal blower
Kohlenstoffgehalt carbon content
Kohlepapier carbon paper
Koks coke
Koksofen coke oven || coke furnace
Kokstrommel coke drum
Kolben piston || plunger * Tauchkolben (auch in der Hohlglasherstellung), Tauchbolzen, Tauchspule

Kolbenbolzen wrist-pin ‖ gudgeon pin
Kolbenfüllmaschine metering-piston filling machine * *in der Verpackungsechnik*
Kolbenkompressor reciprocating compressor ‖ piston compressor
Kolbenmaschine reciprocating engine
Kolbenpumpe reciprocating pump ‖ positive-displacement pump
Kolbenstange piston rod ‖ connecting rod * *Verbindungs-/Pleuelstange*
Kolbenverdichter reciprocating compressor ‖ piston compressor
Kollege colleague ‖ co-worker * *Mitarbeiter* ‖ fellow * *Gefährte, Kamerad* ‖ companion * *Kamerad, Gefährte*
Kollektor commutator * *z.B. bei Gleichstrommaschine* ‖ collector * *Anschluss eines Transistors*
Kollektorschaltung collector connection ‖ common collector circuit
Kollergang pan grinder * *Mahlwerk mit schweren Walzen, die auf einer waagerechten Kreisbahn rollen* ‖ edge mill * *Walzen-Mahlwerk* ‖ pug mill * *Walzen-Mahlwerk* ‖ Chile mill * *Walzen-Mahlwerk* ‖ edge-runner mill * *Kollermühle* ‖ edge mill * *Kollermühle*
kollidieren collide * *kollidieren, zusammenstoßen, im Widerspruch stehen (with: mit/zu)* ‖ conflict * *(figürl.)* ‖ clash * *(figürl.)*
Kollision collision * *auch von Interessen, Interrupts, Datenzugriffen* ‖ clash * *von Interessen* ‖ conflict ‖ coincidence * *zeitliche* ‖ crash * *Zusammenstoß*
kollisionsfrei collision-free * *z.B. Speicherzugriff von 2 Teilnehmern*
Kolloquium colloquium
Kombination combination
Kombinationsmöglichkeiten combination possibilities ‖ possible combinations
kombinierbar can be combined * *freely: frei*
kombinieren combine ‖ hook up * *ein Teil an ein anderes "andocken"* ‖ mate * *zwei Dinge passend miteinander ~*
kombiniert combined * *with: mit*
Kombizange combination pliers
Komfort ease * *Leichtigkeit, Einfachheit, Bequemlichkeit, Sorglosigkeit, ease of operation: Bedien~* ‖ convenience * *Annehmlichkeit, Bequemlichkeit, Vorteil, Nutzen, Eignung, Angemessenheit* ‖ comfort * *Behaglichkeit, Wohlergehen, Bequemlichkeit, Gemütlichkeit* ‖ luxury * *Luxus, Wohlleben, (Hoch-) Genuss, Luxusartikel; Pracht* ‖ sophistication * *"Ausgefeiltheit", "Ausgeklügeltheit", Kultiviertheit*
komfortabel comfortable * *behaglich, gemütlich, reichlich, bequem* ‖ user-friendly * *→anwenderfreundlich, leicht zu bedienen* ‖ easy-to-handle * *leicht hantierbar/bedienbar* ‖ easy-to-use * *leicht bedienbar* ‖ convenient * *bequem, praktisch, (zweck-)dienlich, brauchbar* ‖ luxurious * *luxuriös* ‖ versatile * *vielseitig (verwendbar), wandelbar* ‖ sophisticated * *ausgeklügelt, ausgefeilt, raffiniert, hochentwickelt, anspruchsvoll, exquisit*
komisch comical ‖ comic ‖ funny * *lustig* ‖ strange * *seltsam* ‖ peculiar * *eigenartig, absonderlich, komisch*
Komitee committee
Kommanditgesellschaft limited partnership
Kommando command * *upon: auf*
Kommandodatei command file ‖ batch file
Kommandogabe command initiation
Kommandoregister instruction register
Kommandostufe auto-reversing stage * *zur Steuerung des Momentenwechsels bei Stromrichter* ‖ command stage * *zur Steuerung des Momenten-* *wechsels bei Stromrichter* ‖ sequential logic stage ‖ reversing logic module ‖ auto-reverse stage ‖ forward/reverse control ‖ switch-over logic ‖ auto-reversing module
Kommandozeile commandline * *Zeichensequenz z. Aufrufen e. Computerbefehls/-Programms (u.U. m. Übergabeparametern)*
kommen come ‖ be received * *be received from: ~ von (Signal, Nachricht)* ‖ approach * *herankommen* ‖ emerge * *auftauchen, heraus~, zum Vorschein ~, auftreten, entstehen, sich entwickeln*
kommende Meldung incoming message
Kommentar comment * *auch in (Quellsprache-) Programm*
kommentieren comment * *(up)on: etwas* ‖ annotate
kommerziell commercial * *geschäftlich, gewerblich, kommerziell, Handels-, Geschäfts-, handelsüblich*
Kommission commission * *Ausschuss, Arbeitsgruppe, Verkaufsprovision* ‖ consignment * *Sendung, Lieferung*
kommissionieren consign * *übersenden, verschicken, adressieren* ‖ bundle * *zusammenpacken/-bündeln* ‖ collate * *zusammenstellen* ‖ picking * *of ordered items* * *(Substantiv) im Auslieferungslager* ‖ batch orders * *zusammentragen von Bestellpositionen*
Kommissioniergerät order picking system * *z.B. in Hochregallager*
Kommissionierlager picking warehouse
Kommissionierung bundling * *gemeinsame Lieferung mehrerer Produkte* ‖ order-picking * *Zusammenstellung von Lieferungen im Auslieferungslager*
Kommissionsfertigung contract manufacturing * *Fertigung auf Bestellung* ‖ production on a just-in-time basis
Kommunikation communications * *Kommunikationsfunktionen/wesen* ‖ communication
Kommunikationsbaugruppe communication interface board * *Leiterkarte* ‖ communication board * *Leiterkarte* ‖ serial interface board * *Leiterkarte für serielle Kommunikation* ‖ communication module ‖ communications board
Kommunikationsbus bus ‖ data highway
kommunikationsfähig communications-capable ‖ able to communicate
Kommunikationsfähigkeit communications capability ‖ ability to communicate
Kommunikationsfehler communication error ‖ data transfer fault * *Datenübertragungsfehler*
Kommunikationsmodul communication module
Kommunikationsprotokoll communications protocol
Kommunikationsprozessor communications processor * *Abkürzg.: CP*
Kommunikationssubmodul communications submodule * *z.B. beim SIEMENS-Antriebsregelsystem SIMADYN D (R)*
Kommunikationsweg communications path
kommunizieren communicate
Kommutator commutator ‖ commutator/brush assembly * *inklusive Bürstenapparat (bei DC-Maschine)*
Kommutator-Anschlussfahne commutator riser
Kommutator-Lamellen commutator segments
Kommutatorlamelle commutator segment
Kommutatormotor commutator motor
kommutieren commutate
Kommutierfähigkeit commutating ability ‖ commutation ability
Kommutierung commutation * *Übergang d.*

Stromflusses, z.B. b. Stromrichter v. e. Zweig d. Schaltung i. d. Folgezweig
Kommutierungsblindleistung commutation reactive power || reactive power for commutation
Kommutierungsdauer commutation duration
Kommutierungsdrossel commutating reactor || commutating choke || commutation choke
Kommutierungseinbruch commutating dip || commutation notch
Kommutierungseinflüsse commutation effects
Kommutierungsfähigkeit commutation ability || commutating ability
Kommutierungsgrenzkurve black band
Kommutierungsinduktivität commutation inductance * *Gesamtinduktivität im Kommutierungskreis i. Reihe mit d. Kommutier.spanng.* || commutating inductance
Kommutierungskondensator commutating capacitor || commutation capacitor || commutating capacitance * *Kommutierungskapazität*
Kommutierungskreis commutation circuit
kommutierungsloser Stromrichter commutationless converter
Kommutierungsreaktanz commutation reactance * *induktiver Widerstand d. →Kommutierungsinduktivität bei der Grundwelle*
Kommutierungsspannung commutation voltage
Kommutierungsspannungsquelle commutation voltage source
Kommutierungsstrom commutation current
Kommutierungsverhalten commutation behaviour * *z.B. einer Gleichstrommaschine* || commutating properties
Kommutierungsverluste commutation losses
Kommutierungsvorgang commutation process
Kommutierungswinkel commutation angle
Kommutierungszahl commutation number * *Anzahl d.Kommutier.vorgänge zw.d.Stromrichter-Hauptzeigen während e.AC-Periode*
Kommutierungszeit commutating time || commutating period || commutator short-circuit period * *{el.Masch.}* || commutation interval
Kommutierverhalten commutation behaviour * *z.B. einer Gleichstrommaschine*
kompakt compact * *kompakt, fest, dicht, (zusammen)gedrängt, gedrungen* || tight * *gedrängt (auch Programmcode), dicht (gedrängt), eng, prall(voll), straff* || solid * *fest, stabil, massiv, derb, geschlossen, einheitlich*
Kompakt- compact || moulded case * *im geschlossenen Gehäuse (Leistungsschalter usw.)*
Kompakt-Fremdlüfter compact externally mounted fan * *extern montierter ~ bei Elektromotor*
Kompakt-Leistungsschalter moulded-case circuit breaker * *mit Isolierstoffgehäuse*
Kompaktbauform compact design || compact version
Kompaktbauweise compact design || compact type || compact type of construction
kompakter Aufbau compact design * *z.B. eines Geräts*
Kompaktgerät compact unit || chassis unit * *Chassis-/Einbaugerät auf offenem Tragblech ohne besondere Schutzart* || compact converter * *Kompaktstromrichter*
Kompaktheit compactness || compact design
Kompaktleistungsschalter moulded-case circuit braker * *in Isolierstoffgehäuse*
Komparator comparator
kompatibel compatible * *with: mit* || compliant * *zu einer Norm, Vorschrift usw.*
kompatibel mit compatible with
Kompatibilität compatibility * *with: mit*
Kompendium compendium * *Nachschlagewerk*
Kompensation compensation * *for: von* || correction * *Korrektur* || power factor compensation * *Blindleistungs-Kompensation*
Kompensationsanlage reactive-power compensation equipment * *zur Blindleistungskompensation* || power factor compensation equipment || reactive current compensation plant * *Blindstrom~* || Var compensator * *Blindleistungs-Kompensationseinrichtung* || reactive-power compensation system
Kompensationseinrichtung power factor compensation equipment || reactive current compensation plant || Var compensator * *Blindleistungs-Kompensationseinrichtung*
Kompensationskondensator power correction capacitor * *zur Blindleistungskompensation* || capacitor for power correction
Kompensationswicklung compensating winding || compensation winding || neutralizing winding || compound winding * *kompensiert Lastbhängigk. der Drehz.e. DC-Motors bei konst. Ankerspanng.*
kompensieren compensate * *for: etwas* || equalize * *siehe auch →ausgleichen*
kompensiert compensated * *auch DC-Motor, Tachogenerator*
kompensierter Motor compensated motor * *mit →Kompensationswicklung*
kompetent authorized * *befugt* || competent * *befugt, zuständig, befähigt (for: für), maßgeblich, zuverlässig* || qualified * *qualifiziert, befähigt, fähig* || responsible * *zuständig, verantwortlich (for: für)* || authoritative * *maßgeblich, maßgebend*
Kompetenz competence * *for: für* || authority * *Befugnis* || responsibility * *Zuständigkeit (for: für)* || technical knowledge * *siehe 'Fachwissen'* || competency * *Befähigung, Tauglichkeit (for: zu/für), Kompetenz, Zuständigkeit, Befugnis*
kompilieren compile * *{Computer} Quellspracheprogramm mit Compiler übersetzen (z.B. in →Maschinensprache)*
Kompilierer compiler * *Übersetzungsprogramm für Hochsprachprogramme*
komplementär complementary * *auch Signal (Differenzialsignal)*
komplett complete
Komplettanbieter full-liner || one stop shop * *(salopp)* || full-line supplier
Komplettanlage complete plant
Komplettgerät complete unit || self-contained unit * *autarkes, eigenständiges, selbstständig arbeitendes, in sich geschlossenes ~* || fully assembled unit || ready-to-run unit * *drei lauffähiges Gerät*
komplex complex * *vielschichtig* || intricate * *verwickelt; siehe auch →vielschichtig*
komplexe Größe complex variable * *{Mathem.}*
komplexe Zahl complex number
Komplexität complexity || complexedness
komplizieren complicate
kompliziert complicated || sophisticated * *hochentwickelt, hochgestochen, raffiniert, anspruchsvoll* || complex * *komplex, vielschichtig* || intricate * *Problem, Frage usw.; auch 'vielschichtig'* || tricky * *(salopp) verzwickt* || difficult * *schwierig* || hard * *schwierig, mühsam, anstrengend* || many-sided * *→vielschichtig, →facettenreich*
Komponente component * *Anteil, Teil* || component part * *Einzelteil* || element || part || member * *Glied* || constituent * *Bestandteil, Komponente*
Kompoundierung compounding
Kompoundwicklung compound winding * *Doppelschlusswicklung, kompensiert Lastabhängigkeit bei DC-Motor* || stabilizing series winding * *kompensiert Lastabhängigk.d.Drehzahl e.DC-Motors bei konst. Ankerspanng.*
Kompression compression
Kompressor compressor * *Verdichter*
Kompressorsatz compressor set * *Kombination*

mehrer Verdichter; häufig m.bedarfsweiser Zuschaltg. d. Einzelverdichter
komprimieren compress * *(auch figürl.)*
Komprimierprogramm data compression program
Komprimierung compression * *auch Datenkomprimierung* || data compression * *Datenkomprimierung* || encoding * *{Computer} von Sound- und Videodateien usw.*
Kompromiss compromise * *['kompromais] bad compromise: fauler ~* || trade-off * *make a trade-off: Kompromiss schließen; auch 'Absprache, Handel, Geschäft'* || tradeoff
kompromisslos uncompromising || uncompromisingly * *(Adv.)* || without compromise || thoroughgoing * *durch und durch, extrem, gründlich*
Kondensat condensate || condensed water * *Kondenswasser*
Kondensatbehälter condensate tank
Kondensationsturbine condensing turbine
Kondensator capacitor || condensor * *Verflüssiger in e. thermodynamisch. Kreislauf z.B. Wärmekraft/Kältemaschine*
Kondensatorbatterie capacitor bank
Kondensatorbremsung capacitor braking * *Gleichstrombremsung mit Kondensator als Stromlieferant (selten verwendet)*
Kondensatorerregung capacitor excitation
Kondensatorkaskade capacitor cascade
Kondensatormotor capacitor motor || single-phase capacitor motor * *m. Phasenschieberkondensator z.Erzeugg. e. Hilfsphase; Steinmetzschaltg.*
Kondensatorpufferung capacitor back-up
Kondensatorwickelmaschine capacitor winder
kondensieren condense
Kondenswasser condensed water || condensate * *Wasserfilm, d. sich b. feuchter Luft an Flächen bildet, der. Temp. unter d. Taupunkt liegt*
Kondenswasser-Ablaufloch condensate drain hole
Kondenswasserablauf drain hole * *~loch z.B. an tiefster Stelle e. Motors* || condensate drain hole * *~loch, ~öffnung*
Kondenswasserloch drain hole * *z.B. an der tiefsten Stelle e. Motors* || condensate drainhole * *z.B. an der tiefsten Stelle eines Motors* || condensation water drain hole * *z.B. a d.tiefsten Stelle eines Motors* || condensate drain hole
Konditionen conditions * *auch Verkaufs~*
Konditionierung conditioning
Konduktanz conductance * *Leitwert (1/R)*
konfektioniert ready-made
konfektionierte Leitung pre-assembled cable * *z.B. Stecklleitung*
konfektioniertes Kabel cable set || prefabricated cable || precut cable || pre-assembled cable * *z.B. Steckkabel*
Konferenz conference
Konfiguration configuration || environment * *Umgebung, äußere Bedingungen* || scenario * *→Konstellation*
Konfigurator configurator || modeller
konfigurierbar configurable * *freely configurable: frei ~* || arrangeable || programmable
Konfigurierbarkeit configurability
konfigurieren configure
Konflikt conflict * *Konflikt, Streit, Widerspruch, Zusammenstoß; come into conflict: in ~ geraten* || clash * *Streitigkeit, (feindlicher) Zusammenstoß, Zusammenprall, Widerstreit, Streitigkeit* || clashes * *Streitigkeiten* || conflict situation * *~situation* || collision * *Widerspruch, Gegensatz, Konflikt, Zusammenstoß*
Konfliktstoff matter of conflict || dynamite
konform in conformance * *with: zu/mit; siehe auch →gemäß* || conformable * *to: zu/mit* || in agreement with * *be in ageement with: ~ gehen mit* || in conformity * *with: zu* || conformant * *siehe auch →entsprechend*
Konformität conformance * *conformance declaration: Konformitätserklärung* || conformity * *in conformity with: konform zu*
Konformitätsbescheinigung conformance certification || certificate of conformance || certificate of conformity
Konformitätserklärung declaration of conformity * *z.B. Erklärung d.Herstellers, dass e.Produkt d. EG-Richtlinien entspricht* || EC Declaration of Conformity * *bezogen auf EG-Richtlinien*
Königswelle kingpin * *Königszapfen* || vertical shaft * *senkrecht angeordnete Welle*
Königszapfen central pivot * *zentrener Drehzapfen, z.B. für Kran, Bagger, Lastkraftwagen* || king post || king pin || kingpin || kingbolt * *Königsbolzen*
konisch tapered || conical
Konizität taper
Konjunktion conjunction * *Schaltalgebra*
konkav concave
konkret concrete * *konkret, greifbar, wirklich, dinglich, fest benannt* || actual * *wirklich, tatsächlich, eigentlich* || practical * *tatsächlich, praktisch, handgreiflich* || tangible * *fühlbar, handgreiflich* || definite * *bestimmt, fest, klar, deutlich, eindeutig, genau*
Konkurrenz competition * *stiff: starke ~* competitors * *Mitbewerb* || rivals * *Marktrivalen*
Konkurrenz- competitive
konkurrenzfähig competitive || able to compete
Konkurrenzfähigkeit competitive position || competitive power || competitiveness
Konkurrenzfirma rival firm || competitor || competition
Konkurrenzkampf competition * *cutthroat competition: mörderischer ~* || trade rivalry
konkurrenzlos without competition || unrivalled || unbeatable * *unschlagbar* || second-to-none * *unvergleichlich*
Konkurrenzprodukt competitive product
konkurrieren compete * *with: mit* || be in competition * *with: mit* || rival
konkurrierend competitive || competing || rival || vying * *wetteifernd; vying for the contract: um den Auftrag ~*
Konkurs bankruptcy * *['bängkraptsi] go bankrupt: Konkurs gehen* || insolvency
Konnektor connector
Konnektorbezeichner connector designator * *z.B. beim SIEMENS Regelsystem SIMADYN D (R)*
Können skill * *Fertigkeit* || powers || abilities * *Fähigkeiten* || prowess * *Tüchtigkeit* || efficiency * *Tüchtigkeit* || be able to * *(Verb)* || be capable of * *(Verb)* || be in the position to * *(Verb) zu etwas in der Lage sein* || be allowed to * *(Verb) dürfen* || be permitted to * *(Verb) dürfen* || can * *(Verb)*
Könner very able man || master || expert || wizzard * *(salopp) Zauberer* || crack * *(salopp)* || ace * *(salopp) Ass* || wiz * *(salopp) Kurzform für 'wizzard': Zauberer*
konsequent consequent || strict * *streng, durchgehend* || consistent * *folgerichtig, logisch, konsequent, gleichmäßig, stetig, übereinstimmend, durchgängig* || thorough-going * *gründlich* || comprehensive * *umfassend, vollständig* || consequential || rigorous * *(peinlich) genau, strikt* || consequentially * *(Adv.)* || single-minded * *zielbewusst, zielstrebig*
Konsequenz consequence * *Folge; take the consequences: die ~en tragen* || consistency * *Festigkeit, Folgerichtigkeit, Konsequenz, übereinstimmung* || conclusions * *Folgerungen; draw one's conclusi-*

konservieren 180

ons/act accordingly: die ~en ziehen || single-mindedness * *Zielstrebigkeit; Aufrichtigkeit*
konservieren preserve || preservation * *das Konservieren, die Konservierung* || conserve || tin * *in Büchsen* || can * *in Büchsen*
Konservierung preservation
konsistent consistent * *zusammengesetzt, zusammenpassend (z.B. Teildaten einer Informationsmenge)* || integral * *vollständig, unversehrt, zusammenpassend, z.B. zusammengehörige Daten aus einem Zyklus* || viscous * *Schmiermittel* || intact * *unversehrt* || whole * *ganz, unversehrt* || belonging together * *zusammengehörig*
Konsistenz integrity * *z.B. Zusammenpassen von Daten* || consistency * *Zusammensetzung*
Konsole bracket
Konsolfräsmaschine knee-type milling machine
Konsolidierung consolidation * *Konsolidierung (auch wirtschaftlich), (Be-) Festigung; Vereinigung/Zusammenschluss*
konstant constant * *stay constant: konstant bleiben* || consistent * *gleichmäßig, gleich bleibend, fest* || stable * *unveränderlich, stabil, dauerhaft, konstant*
Konstantanteil constant component
Konstantantrieb constant speed drive * *Festdrehzahlantrieb*
Konstantdrehzahlantrieb constant-speed drive
Konstante constant
Konstantfahrt running at constant speed
konstanthalten keep constant
Konstantlauf constant run || constant running
Konstantmoment-Antrieb constant torque drive || continuous torque drive
Konstantmomentantrieb constant-torque drive
Konstantspannungsgeber constant voltage generator || reference voltage generator * *Referenzspannungsgeber*
Konstantspannungsgenerator constant voltage generator * *Synchrongenerator m. eigenem Konstantspannungsgerät*
Konstantstromquelle impressed-current source || constant-current source || stabilized-current source || load-independent current source || constant current source
Konstantteil continuous section * *kontinuierlich produzierender Teil einer Papiermaschine (ab dem Stoffauflauf)*
Konstanz constancy * *Langzeitkonstanz (Einflussgrößen: Auflösung, Drift); siehe VDI/VDE 2185* || stability * *Kurzzeitkonstanz (Einflussgrößen: Last, Frequenz, Netzschwankungen)* || holding * *Lastausregelung* || repeatability * *Reproduzierbarkeit* || control stability * *Konstanz der Regelung* || long-term stability * *Langzeitkonstanz* || steady-state accuracy * *Genauigkeit im eingeschwungenen Zustand* || stability * *z.B. eines Kondensators*
Konstellation constellation || scenario
konstruieren construct || design * *entwerfen* || devise * *erfinden, ersinnen, ausdenken*
Konstrukteur design engineer || designer || mechanical designer * *Maschinenbau-Konstrukteur* || mechanical engineer * *Maschinenbau-Ingenieur* || designing engineer || technical designer || constructor * *Erbauer, Konstrukteur*
Konstruktion mechanical design || design || mechanical construction * *mechanische* || construction || structure * *Aufbau, Bauart* || engineering and design * *auch als Anteilungsbezeichnung*
Konstruktionsbüro mechanical engineering department || drawing office
Konstruktionselement structural element || constructional element || machine element * *Konstruktionselement, Maschinenelement* || constructional detail * *Konstruktionseinzelheit*

Konstruktionsfehler faulty design || constructional flaw || constructional defect
Konstruktionsgeschwindigkeit design speed * *Geschw. für die e. Maschine konstruiert ist (z.B. Papiermaschine)*
Konstruktionsmerkmal construction feature || constructional feature || feature of construction || design feature
Konstruktionsrichtlinien design rules
Konstruktionsteil structural part
Konstruktionszeichnung constructional drawing || workshop drawing || construction drawing || design drawing
konstruktive Ausführung mechanical design
konstruktive Gestaltung mechanical design
Konsultation consultation
konsultieren consult * *um Rat fragen* || take into consultation || ask * *fragen* || ask someone's advice * *jemanden um Rat fragen* || inquire * *nachfragen, erkundigen (for: nach/wegen)* || contact * *Kontakt aufnehmen/haben mit*
Konsum- consumer
Konsument consumer
Konsumqualität consumer quality * *im Gegensatz zur höherwertigen Industriequalität*
Kontakt contact * *auch elektrischer ~*
Kontakt aufnehmen contact
Kontakt aufnehmen mit contact
Kontakt-Abbrand contact erosion
Kontaktanordnung contact arrangement * *z.B. bei Relais, Taster usw.*
kontaktbehaftet with contacts
Kontaktbelastbarkeit contact rating || rated contact loading || contact load || switching capacity * *Schaltvermögen* || contact capacity || contact loading capability
Kontaktbelastung contact rating || contact load || switching capacity * *Schaltvermögen, Kontaktbelastbarkeit* || breaking capacity * *(Ab-) Schaltvermögen eines Leistungsschalters*
Kontaktbelegung connector assignment * *für Stecker* || contact assignment * *Relais, Schalter*
Kontaktbestückung contact arrangement * *von Relais, Schützen usw.*
Kontaktfläche contact face
kontaktieren connect
kontaktlos no-contact || non-contact || solid state * *in Halbleitertechnik ausgeführt (z.B. elektron. Relais)* || contactless
kontaktloser Schalter solid-state switch * *Halbleiterschalter*
kontaktloses Relais solid-state relay
Kontaktmittel electro-lubricant * *z.B. Kontaktspray*
Kontaktpartner contact partner
Kontaktplan ladder logic * *Darstellung einer Relais- oder SPS-Steuerung* || PLC Ladder programming language * *Programmiersprache für SPS* || ladder diagram * *auch Programmiersprache für SPS* || LAD * *Abkürzg. für 'LAdder Diagram': Relaisketten-Darstellung (SPS-Programmiersprache)*
Kontaktplan-Netzwerk ladder circuit * *(SPS)*
Kontaktprellen contact bounce
Kontaktsatz contact block * *für Relais/Schalter/ Taster*
Kontaktschraube contact screw || contact stud * *~ ohne Kopf* || terminal screw * *~ für Kabelanschluss/an einer Anschlussklemme*
Kontaktsicherheit contact reliability || contact stability * *high: große/hohe/gute* || reliable contacting
Kontaktwickler surface winder * *z.B. Druckwalzen/Doppeltragwalzenroller (m. Andruckrolle i. Gegens. z. Achswickler)* || contact winder

Kontaktwiderstand contact resistance * *z.B. eines Schaltkontakts oder Kommutators*
Kontaktzunge contact tab * *z.b. bei einer Flachsteckverbindung*
Kontermutter locknut || lock nut || check nut || counter nut
kontern check
Kontext context * *siehe auch →Zusammenhang*
Kontierung account * *Konto*
kontinuierlich continuous || stepless * *stufenlos* || continuously * *(Adv.)* || continual * *fortwährend, dauernd, immer wiederkehrend* || steplessly variable * *stufenlos verstellbar*
kontinuierlicher Regler continuous-action controller * *z.b. Analogregler*
Konto account
Kontrast contrast * *auch einer Anzeige/eines Bildschirms* || brilliance * *bei Bildschirm usw.*
Kontrollanlage checking system || inspection system
Kontrollbaustein control block * *Funktionsbaustein* || check block * *Funktionsbaustein*
Kontrollbescheinigung test certificate * *Prüfbescheinigung*
Kontrolle check * *Überprufung* || control * *Steuer-/Regelung; be in/keep the control of/have well in hand: unter ~ haben/halten* || checking * *Überprüfung, das Kontrollieren; for checking purposes: zur Kontrolle/für Kontrollzwecke* || inspection * *Inspektion, Durchsicht, Untersuchung, Prüfung* || recheck * *nochmalige ~*
Kontrolleinrichtung monitoring device
Kontrolleinrichtungen controls
Kontrolleuchte pilot lamp
Kontrolleur supervisor
kontrollierbar controllable
kontrollieren control || supervise * *siehe auch →inspizieren* || check * *überprüfen* || be in control of * *beherrschen* || keep track of * *nachverfolgen, 'auf der Spur bleiben'* || keep tab on * *(amerikan.) 'auf der Spur bleiben'* || verify * *die Richtigkeit ~, die Richtigkeit/Korrektheit nachweisen* || audit * *(Geschäfts-)Bücher ~* || inspect * *untersuchen, prüfen, nachsehen; inspizieren, be(auf)sichtigen*
kontrolliert in a controlled manner
Kontrollmessung check measurement
Kontrollwaage checkweigher * *[Verpack.technik]*
Kontur contour || outline || profile * *Profil, Seitenansicht, Kontur* || border * *Grenze; border line: Grenzlinie*
Konturfräsen contour milling
Konturschleifen contour sanding * *[Holz] mit Schmirgel*
Konus cone || taper sleeve * *hohler Konus* || tapered socket
Konvektion convection * *selbsttätiger Luft/Flüssigkeitsumlauf, Wärmeströmung (natural: freie)*
konvektionsgekühlt with free convection * *→selbstgekühlt* || free-convection
Konvektionskühlung convection cooling * *Kühlg. durch natürlichen Luftstrom (siehe auch →Luftselbstkühlung)*
konventionell conventional || traditional * *wie gewohnt* || classical * *klassisch, althergebracht* || classic * *klassisch*
konventioneller Speicher conventional memory * *von DOS ansprechbarer Speicherbereich 0 bis 640 K beim PC*
Konvergenz convergence
Konvergenzfehler misconvergence * *ungleiche Positionierung der Farben bei Bildschirm*
konvergieren converge * *zusammenlaufen, sich (einander) nähern*
Konverter converter || steel converter * *zur Stahlerzeugung*

Konvertierung conversion
Konvertierungsbaustein conversion block * *Funktionsbaustein*
Konvex convex
Konzentrator concentrator * *z.b. für Daten* || data concentrator * *für Daten*
konzentrieren concentrate * *auch 'sich ~' ((up)on: auf* || centre * *upon: auf* || center * *(amerikan.) upon: auf* || focus * *focus on: sich/seine Aufmerksamkeit/seine Kräfte konzentrieren auf* || give full concentration * *sich ~ (to: auf)*
konzentrisch concentric
Konzentrizität concentricity * *auch Fluchtfehler; bei Motor: siehe DIN 42955*
Konzept concept || idea || draft * *first/rough: Grob~, Grobentwurf* || scheme * *Schema, System, Plan, Entwurf* || model * *Muster, Vorbild, Vorlage, Modell, Bauweise* || arrangement * *Art der →Anlage*
konzeptionell conceptual || conceptive || in its outlines * *in der Anlage*
Konzession license * *(amerikan.) Lizenz, Verkaufsrecht* || privilege * *Privileg* || licence * *(brit.) Lizenz, Verkaufsrecht* || franchise * *(amerikan.) Verkaufsrecht* || concession * *Privileg; Entgegenkommen, Zugeständnis; make no conc. to a person: jdm. keine ~en machen*
Kooperation cooperation || teamwork || joint venture * *(vorübergehende) Zusammenarbeit mehrer Firmen bei einem Projekt*
Kooperationsvertrag cooperation agreement || collaboration agreement
Koordinate coordinate
Koordinaten coordinates
Koordinaten-Messmaschine three-dimensional measuring machine
Koordinatensystem system of coordinates || coordinate system
Koordinatentransformation coordinate transformation || transformation of coordinates
Koordinatenursprung coordinate origin * *z.B. Schnittpunkt von X- und Y-Achse*
Koordination coordination || co-ordination || harmonizing * *Harmonisieren, In-Einklang-Bringen*
koordinieren coordinate
Koordinierung coordination || co-ordination || harmonizing * *Harmonisieren, In-Einklang-Bringen*
Koordinierungsprozessor coordination processor
KOP LAD * *Abkürzung für 'LAdder Diagram': Kontaktplandarstellung für SPS* || ladder diagram
Kopf header * *z.B. eines Datensatzes/Menüs/Datentelegramms/Programms* || head || heading * *z.B. Kopfzeile(n)*
kopflastig top-heavy
Kopflastigkeit top-heaviness
Kopfleiste top bezel plate * *an Schaltschrank* || fascia plate * *an Schaltschrank*
Kopfstation head station || master station
Kopfstecker jumper header * *bei SPS*
Kopfsteuerung master controller || master PLC * *SPS, Programmierbare Steuerung* || master control
Kopfzeile header
Kopie copy
Kopie an c.c. || copy to
Kopierdrehmaschine copying lathe
kopieren copy
Kopierer copier
Kopierfräse copy router * *[Holz]*
kopierfräsen profile * *kopier-/fassonfräsen, fassonieren* || copy milling
Kopierfräsmaschine copying milling machine || profiler || duplicating milling machine || copying shaper * *[Holz]*
Kopierpapier copying paper
Koppelglied interface element || interface

Koppelkapazität coupling capacitance
Koppelkondensator coupling capacitor
Koppelmerker interprocessor communication flag * *bei SPS*
koppeln couple * *mechanisch ~, auch über Kommunikations-Schnittstelle* || connect * *verbinden, auch über serielle Schnittstelle* || interlink * *z.B. über serielle Schnittstelle ~* || link * *verbinden, auch über serielle Schnittstelle* || interconnect || network * *über ein Netzwerk ~*
Koppelpartner connected interface || connected slave * *bei Datenkopplung nach dem Master-Slave-Prinzip* || peer * *bei Datenkopplung ohne ausgeprägten Master* || connected station || linked unit || interface partner || communication partners * *(Plural)*
Koppelrelais interface relay
Koppelspeicher communications buffer * *bei Mehrprozessorkopplung (z.B. SIEMENS SIMA-DYN D (R))* || buffer memory || mailbox || common memory * *gemeinsamer Speicher* || coupling memory || mailbox memory * *zum Austausch von Daten, Nachrichten* || communications RAM || interface memory
Koppeltrieb driving mechanism * *z.B. bei Querschneider* || double drag-link gearing * *→Ungleichförmigkeitsgetriebe z. Antrieb d.→Messerpartie bei →Querschneider* || drag-link gearing * *→Ungleichförmigkeitsgetriebe*
Koppler coupler
Kopplung coupling * *auch Daten~ z.B.über serielle Schnittstelle* || interfacing || networking * *Vernetzung* || connection * *Verbindung* || link * *z.B. ~ über serielle Schnittstelle* || interprocessor communication * *siehe →Rechnerkopplung* || connection * *Verbindung, z.B. über Signale, Bus usw.*
Kops cop * *{Textil} Garn-/Faden~ (Spule mit aufgewickeltem Garn in der Spinnerei usw.)*
Korken kork
Korn grain || particle
Körner center punch * *zum →Ankörnen eines Bohrpunktes*
Korngröße particle size
kornorientiert grain-oriented * *z.B. Magnetblech*
Körnung punch * *An~ mit einem Körner(-schlag)* || granulation || graining || grain size * *Korngröße*
Körper body * *auch Raum/Fremdkörper*
Körperschall structure-born noise || sound conducted through solids || solid-born noise
Körperschluss fault to frame * *{el.Masch.} Isolierversagen (Masseschluss zum Gehäuse)*
Körperverletzung bodily injury || personal injury * *Personenschaden*
Korpus carcase
Korrektur correction
Korrekturfaktor correction factor
korrekturlesen revise * *revidieren; nochmals durchlesen, die zweite Korrektur lesen; Umbruch neu erstellen* || read the proofs * *den Vorabzug/die Druckfahnen ~* || correct the proofs * *den Vorabzug/die Druckfahnen ~* || proofread || proofreading * *(Substantiv)* || proof
Korrektursollwert corrective setpoint || correcting setpoint
Korrekturvektor correction vector
Korrekturwert corrective value
korrelieren correlate * *in Wechselbeziehung stehen (with: mit), voneinander abhängig sein*
Korrespondenz correspondence * *Übereinstimmung, Entsprechung, Briefwechsel*
korrespondieren correspond * *(to/with: mit)*
korrespondierend corresponding * *entsprechend*
korrigieren correct || alter * *ändern* || adjust * *anpassen* || revise * *revidieren, korrekturlesen,*

Druckerzeugnis/Dokument überarbeiten || modify * *modifizieren* || make corrections
korrodieren corrode
korrodierend corrosive
Korrosion corrosion
korrosionsbeständig non-corroding || corrosion-resistant
Korrosionsbeständigkeit resistance to corrosion
Korrosionsfestigkeit resistance to corrosion
Korrosionsgefahr danger of corrosion || corrosion risk
korrosionsgeschützt corrosion-proof || protected against corrosion
Korrosionsschutz protection against corrosion || anti-corrosion protection || corrosion protection || anticorrosion
Korund aluminium oxyde * *Schleifmittel*
kostbar valuable || precious
Kosten cost * *at the cost of: auf ~ von; fixed/running: fixe/l aufende; bear the cost: die ~ tragen* || costs * *Mehrzuhl* || expenses * *(Un)Kosten, Auslagen, Spesen; at the expense of: auf ~ von (auch figürl.)* || charges * *auch Gebühren; standing charges: laufende ~* || fees * *Gebühren* || expenditure * *Aufwand* || price * *Preis* || outlay * *(Geld-) Auslage; initial outlay: Anschaffungs~/einmalige ~* || cost * *(Verb)* || list for * *(Verb) einen Listenpreis von ... haben* || take * *(Verb) Mühe, Zeit usw. kosten* || require * *(Verb) benötigen, in Anspruch nehmen* || be * *(Verb) how much is it: was kostet das?* || sell for * *(Verb) für ... verkauft werden* || expense * *(Geld-) Ausgabe, Aufwand, Un~, Spesen; at great expense: mit großen ~*
Kosten sparen cut costs
Kostenbelastung cost burden
Kostendruck cost pressure || pressure to cut costs
Kosteneinsparung cost saving
kostenfrei at no charge || free of charge * *kostenlos*
kostengünstig cost-effective * *mit gutem Kosten/Nutzen-Verhältnis* || favourably-priced * *preisgünstig* || economical * *wirtschaftlich, sparsam* || worth the money * *preiswert/würdig* || low-priced * *preisgünstig* || budget-priced * *preisgünstig* || economy-priced * *preisgünstig* || good value * *(Adv.) be good value: preiswert sein* || low-cost * *mit niedrigem Preis* || at a reasonable price || well-priced || economic
kostenlos free-of-charge * *(Adj.)* || free || gratuitous * *gratis, unentgeltlich* || for nothing || without extra cost * *ohne zusätzliche Kosten* || without extra * *ohne zusätzliche Kosten* || at no charge || toll-free * *nicht gebührenpflichtig (z.B. Telefongespräch)* || free of charge * *(Adv.)*
kostenminimiert cost-minimized
kostenneutral not affecting the costs
kostenoptimal at minimum cost
kostenoptimiert cost-optimized
Kostenoptimierung cost optimization
kostenpflichtig invoiced * *berechnet, in Rechnung gestellt* || with costs || liable to pay costs
Kostenreduzierung cut in costs
Kostensenkung cost reduction || diminution of expenses || cut in expenses
kostensparend cost-saving
Kostenstelle cost center || accounting center * *Abrechnungsstelle*
Kostenträger cost unit
Kostenumlage cost allocation || cost distribution
Kostenvoranschlag cost estimate
Kostenvorteile cost benefits
Kostenziel cost target || cost aim || cost goal
kostspielig costly || expensive || extravagant * *aufwändig*
Kraft force * *~ [N], Stärke, Wucht; Einfluss, Wirkung, Nachdruck, (Rechts-)Gültigkeit; in*

force: in ~ || strength * *Stärke; gather strength: Kräfte sammeln* || power * *Macht; motive power: treibende ~* || energy * *Tat~, Energie* || worker * *Person, Arbeits~* || efficacy * *Wirksamkeit* || vigor * *(Körper-/Geistes-) Kraft, Vitalität, Energie, Nachdruck, Wirkung,* →*Härte* || might * *Macht, Gewalt, Stärke, Kraft; with all ones's might: mit aller ~*
Kraft-Wärme-Kopplungsanlage cogeneration plant
Kraft-Wärmekopplungsanlage combined-cycle plant * *{Kraftwerk} Kraftwerk mit (Fern-) Wärmeerzeugung*
Kraftangriff force application || torque application * *rotatorisch* || application of force
Kraftfahrzeug motor vehicle
Kraftfahrzeugelektronik automotive electronics
Kraftfahrzeugmotor vehicle engine
kräftig strong || robust || powerful * *tatkräftig, leistungsfähig, kraftvoll, mächtig* || rugged * →*robust, grob, stark, ungehobelt* || heavy * *Schlag usw.* || sturdy * →*stabil,* →*robust* || energetic * *tatkräftig*
Kraftliner kraft liner
Kraftlinie line of force * *magnetic: magnetische Kraftmaschine* prime mover * *wandelt Wasserkraft, Brennstoffe, Dampf, Windenergie usw. in mechan.Energie um* || primary mover
Kraftmesseinrichtung dynamometer
Kraftmessgerät dynamometer
Kraftpapier kraft paper
Kraftregelung force control * *z.B. Walzkraft*
Kraftschluss frictional connection || closed linkage || adhesion * *bei Rad-/Straße- bzw. Rad-/Schiene-System*
kraftschlüssig nonpositive || tensionally || frictionally || force-locking || friction-locked || adhesive * *bei Rad-Straße/Rad-Schiene-System* || press-fit || with an interference fit
Kraftsensor force sensor
Kraftübertragung power transmission || torque transmission * *rotatorisch* || mechanical transmission
Kraftübertragungselement power transmitting element || mechanical transmission element || power transmission element
Kraftverteilung force distribution
kraftvoll powerful
Kraftwerk power station || power plant
Kraftwerkleittechnik power station process control
Krampf rubbish * *(salopp) Müll, Abfall, Blödsinn, Quatsch, Schund, Ausschuss* || stuff and nonsense * *(figürl.) (salopp) Quatsch, Unsinn, Blödsinn*
Kran crane * *travelling crane: fahrbarer Kran; mobile crane: Mobilkran, auf Reifen/Ketten fahrend* || hoist * *auch Hebezeug, Flaschenzug, Winde, Lastenaufzug*
Kran-Kennlinie crane-torque characteristic * *{el. Masch.}*
Kranausleger crane jib
Kranbahn crane runway || crane track
Kranfahrwerk crane travel drive * *{Hebezeuge}* || crane travelling gear * *{Hebezeuge} bei Portalkran (im Gegensatz zum* →*Katzfahrwerk)* || crane travelling unit * *{Hebezeuge}*
Kranführer crane driver || crane operator
Kranhaken crane hook * *am Kran* || lifting eye * *Hebeöse an der zu hebenden Last (z.B. Motor, Schaltschrank)*
Kranhubwerk crane hoist || hoisting gear * *Hubwerk*
Kranschiene crane rail
Kransteuerung crane control

Kratzblech scraper plate
kratzen scratch * *scratchy noise: kratzendes Geräusch* || scrape * *schaben* || rabble * *Metall* || grate * *Geräusch* || rasp * *Geräusch*
Kratzenraumaschine cloth-raising machine * *{Textil} zum Aufrauen von Textilstoffen mit Kratzwalzen*
Kratzer scratch * *Kratzer, Schramme* || scraper * *Schaber*
Kratzscheibe scratching-type washer * *z.B. zur Herstellung einer sicheren Erdverbindung im Schaltschrank* || serrated washer * *Zackenscheibe, Fächerscheibe*
Kratzwalze raising roller * *{Textil} bei* →*Kratzenraumaschine*
Kräusel crimper * *{Textil}*
kräuseln crimp * *auch Chemiefasern*
Kreativität creativity
Kreditkarte credit card
Kreis circle * *auch Gruppe* || ring * *Ring* || circuit * *Strom~, Schalt~* || district * *Bezirk* || group * *Gruppe* || sphere * *Wirkungs~*
Kreiselpumpe centrifugal pump || fan pump || rotary pump || turbine pump || screw pump * →*Schraubenpumpe*
Kreiselverdichter centrifugal compressor
kreisen circle || move in a circle || spin around || revolve || rotate || circulate * *zirkulieren*
Kreisförderer continuous conveyor || circular conveyor
Kreisformfehler circulatory error * *{Werkz.-masch.}*
kreisförmig circular || round
kreisförmiges Drehfeld circular rotating field * *{el. Masch.}sin.förm. magn. Feld entlang d. Ständerbohrg. m. konstant. Amplitude*
Kreisfrequenz angular frequency * *[rad/sec]* || radian frequency
Kreisfunktion circular function || cyclic function
Kreislauf circulation || cycle || circle || round || cycling || circuit * *primary/secondary: innerer/äußerer ~ (bei Kühlsystem usw.)* || revolution || circular course
Kreislauf-Vakuumpumpe closed-circuit vacuum pump
kreislaufgekühlt closed-circuit cooled * *allg.* || closed-circuit ventilated * *geschlossener Kühlkreislauf mit Luftkühlung*
kreislaufgekühlt durch Eigenkühlung mittels Luft-Luft-Kühler closed-circuit self-ventilated with air-to-air heat exchanger * *Motor-Kühlart ICA01 A61 nach IEC34.6*
Kreislaufkühler heat exchanger * *Wärmetauscher* || closed-circuit cooler
Kreislaufkühlung closed-circuit cooling * *Motorkühlart m. geschloss. Primär-Kühlungskreislauf (Kühlart IC 81/85/86W)* || closed-circuit ventilation * *z.B. Motorkühlart mit geschlossenem Luft-Kühlkreislauf*
Kreismesser slitter * *z.B. zum Längsschneiden von Papier* || circular knife || cutting disk * *Tellermesser*
Kreissäge circular saw || circular saw bench || saw bench
Kreissägeblatt circular saw blade
Kreissägemotor circular saw motor * *in schlanker Bauform für große Schnitttiefe des Sägeblattes*
Kreisschere circle-cutting shear || disc cutting shear || circular shear
Kreisstrom circulating current
kreisstrombehaftet in circulating current mode || circulating current carrying || carrying circulating current
Kreisstromdrossel circulating-current reactor

kreisstromfrei in non circulating current mode || circulating-current-free || without circulating current
kreisstromfreie Gegenparallelschaltung circulating-current-free antiparallel bridge connection || double way bridge in non circulating current mode * *Gleichrichterbrücke*
kreisstromfreier Betrieb suppressed circulating current mode
kreisstromführend in circulating current mode || circulating current carrying || carrying circulating current
kreisstromführender Betrieb circulating current mode
Kreisverstärkung closed-loop gain || loop gain
Krempel card * *(Textil)* || teasel * *(Textil)* || rubbish * *(salopp)* Plunder, Kram || stuff * *(salopp)* Zeug
Krempelei cardroom * *(Textil)*
Krempelmaschine carding machine * *Textilmaschine z. Aufrauen/Kämmen v.Fasern/Wolle mit Kratzbändern* || crimper * *(Textil)* zum Aufrauen von Fasern
krempeln card * *(Textil)* kämmen/kardieren/Aufrauen von Fasern, Wolle, Vliesstoffen usw. || burr * *(Textil)*
Kreuz- cross- || four-way * *Vierweg-*
Kreuz-Wickelmaschine cross winder
kreuzen cross
kreuzförmig cruciform
Kreuzleger compensating stacker * *(z.B. in Druckerei) für Papierbögen*
Kreuzschaltung cross connection
Kreuzschienenverteiler crossbar interconnection || crossbar distributor || plugboard || crossbar switch
Kreuzschleifautomat cross-sanding machine * *(Holz)*
Kreuzschlitz-Schraubenzieher crosstip screwdriver
Kreuzschlitzschraube cross-recessed screw
Kreuzschlitzschraubenzieher crosstip screwdriver
Kreuztisch cross-sliding table * *(Werkz.masch.| Holz)*
Kriechbetrieb threading || running at crawl speed
kriechen crawl * *mit Schleichdrehzahl bzw. langsam fahren* || thread * *mit Einzieh/Schleichdrehzahl fahren* || creep * *schleichen, kriechen*
Kriechgeschwindigkeit thread speed * *Kriechgeschwindigkeit, z.B. zum Materialeinziehen* || crawl speed * *Schleichgeschwindigkeit/-drehzahl* || threading speed * *Einziehgeschwindigkeit*
Kriechsollwert crawl reference * *Sollwert für Kriech/Schleichgeschwindigkeit* || thread reference * *Sollwert für Einziehgeschwindigkeit/-Drehzahl*
Kriechstrecke creepage distance * *für Isolation (bei Stromrichtern: siehe IEC 664)*
Kriechstrom leakage current || creepage current || sneak current || surface leakage current
Kriechstromfestigkeit resistance to creepage currents || non-tracking quality
krimpen crimp
Krise crisis || critical stage * *kritische Phase* || critical situation * *Krisensituation*
Krisenmanagement crisis management
Kristall crystal
kristallin crystalline
Kriterien criteria
Kriterium criterion * *Merkmal*
Kritik criticism * *über/an: of* || censure * *Tadel, Verweis, Missbilligung* || review * *Besprechung*
Kritikpunkt criticism || criticized point || censure * *Tadel, Verweis, Missbilligung*
kritisch critical * *gewagt, riskant, bedenk-/gefährlich, genau/sorgfältig prüfend; be critic.of: etw. kritis.*
kritische Drehzahl critical speed * *(el. Masch.) Drehzahl, bei der mechan. Resonanz auftritt*

kritische Grenze critical limit
Kröpfung elbow || shoulder || bend || crimping || throw * *z.B. einer Kurbelwelle, eines Exzenters, einer Presse*
krumm crooked || curved * *gebogen* || hooked * *hakenförmig* || arched * *bogenförmig; siehe auch 'gewölbt'* || twisted * *verdreht, verbogen* || bended * *gebogen, gekrümmt*
krümmen crook || bend || twist || form a bend * *sich krümmen, eine Krümmung bilden* || form a curve * *sich krümmen, eine Krümmung bilden* || hook * *krümmen, biegen*
Krümmer elbow * *Kniestück, Krümmer, Winkel (-stück)* || bend * *Biegung, Krümmung, Windung, Kurve* || manifold * *Rohrverzweigung, Einlass~, Auspuff~*
Krümmung bend * *→Biegung* || curve * *Kurve* || curvature || warp * *Verziehen, Ver~, Verwerfung (von Holz usw.)*
Krümmungsradius bending radius || radius of curvature || duct radius * *eines Rohrs*
kubisch cubic || cubical
Kugelgewindespindel ball srew * *z.B. an Hubspindeltrieb*
Kugelgewindetrieb ball screw || ball screw drive
Kugellager ball bearing
Kugelmühle ball mill
Kugelpfanne ball socket
Kugelschreiber ball pen || ball-point pen || biro || ballpen || ball point pen
Kugelspindel ball screw
Kugelumlaufführung guidance system with circulating balls
Kugelventil ball valve
Kühl- cooling
Kühlart cooling method * *für Motoren: siehe VDE 0530 u. IEC 34 Teil 6; für Stromrichter: siehe VDE 558 Teil 1* || cooling type * *für el.Maschinen: siehe VDE 0530 u.IEC 34 T.6; für Stromrichter: siehe VDE 558 Teil 1* || method of cooling || type of cooling * *für elektr. Maschinen: siehe DIN IEC 34, Teil 6* || kind of cooling || cooling arrangement * *Kühl-Anordnung* || method of cooling * *z.B. für Motoren (siehe EN 60034-6)* || cooling principle * *Kühlprinzip/-verfahren*
Kühlbedingungen cooling conditions * *kennzeichn. Kühlmitteltemp.bereich; el. Maschin. VDE0530/IEC34; Stromricht. VDE558*
Kühlbett cooling bed || cooling table || cooling zone
kühlen cool * *auch 'abkühlen'*
Kühler cooler || heat exchanger * *Wärmetauscher* || radiator * *in Flüssigkeits-Kühlkreislauf*
Kühlfläche cooling surface area
Kühlflüssigkeit coolant || cooling fluid
Kühlkanal cooling duct * *für Kühlluft*
Kühlkörper heat sink
Kühlkreis cooling circuit * *Kreis(lauf), in dem d. Kühlmittel strömt, d.Verlustwärme aufnimmt u. abtransportiert*
Kühlkreislauf cooling circuit * *closed: geschlossener*
Kühlleistung cooling capacity
Kühlluft cooling air * *discharged: austretende* || warmed air * *~, nachdem sie ihre "Kühlarbeit" verrichtet hat (am ~-Austritt)*
Kühlluftaustritt cooling air outlet
Kühlluftbedarf cooling air flow requirement || cooling air requirement
Kühllufteintritt cooling air intake
Kühllüfter cooling fan
Kühlluftkanal cooling air duct
Kühlluftmenge cooling air rate
Kühlluftöffnung opening for the cooling air || ventilation opening || air inlet * *Lufteinlass*
Kühlluftstrom cooling air flow

Kühllufttemperatur cooling air temperature
Kühlmittel coolant || cooling medium * *Flüssigkeit (Wasser, Öl) oder Gas (Luft), mit deren Hilfe die Wärmeabfuhr erfolgt* || coolant agent || heat-transfer agent * *z.B. in einem Primär-Kühlkreislauf*
Kühlmittelpumpe cooling pump || coolant pump
Kühlmitteltemperatur coolant temperature || cooling medium temperature || cooling air temperature * *bei Luftkühlung* || heat transfer agent temperature * *z.B. in einem Primär-Kühlkreislauf*
Kühlofen annealing furnace * *für Glas* || annealing oven || cooling furnace * *z.b. in der Glasindustrie*
Kühlöl cooling oil
Kühlrippe cooling rib || heatsink fin * *eines Kühlkörpers* || cooling fin
Kühlschacht cooling chamber * *{Textil} hinter d. Spinnpumpe angeordnete Einrichtg. in d. Kunstfasererzeugung*
Kühlschlitz air slot * *zum Ein/Austritt von Kühlluft* || louvered cooling slot * *mit "Überdachungsstreifen"* || louver * *(amerikan.) Kiemenblech* || louvre * *(brit.)*
Kühlstrecke cooling section * *z.B. in Blechstraße* || cooling line
Kühlsystem cooling system * *Gesamtheit der Kühlkreise u.d. Kühlmittelströmung zur Wärmeabfuhr (VDE 0530 Teil 5)*
Kühltrommel cooling cylinder
Kühlturm cooling tower
Kühlung cooling || heat dissipation * *Wärmeabfuhr*
Kühlungsbedingungen cooling conditions
Kühlventilator cooling fan
Kühlwalze chill roll * *z.b. wassergekühlte Gießwalze in Folienmaschine hinter dem Extruder* || cooling cylinder * *z.b. für Zuckermasse bei d. Bonbonherstellung*
Kühlwasser cooling water
Kühlwassereintrittstemperatur cooling water inlet temperature
Kühlwassererwärmung cooling water temperature rise
Kühlwasserflansch cooling water flange
Kühlwassermenge cooling-water flow-rate
Kühlwasserpumpe cooling water pump
Kühlwassertemperatur cooling-water temperature
Kühlwerk cooler || cooling section * *Kühlstrecke, Kühlbett*
Kühlwirkung cooling effect || cooling efficiency
kulant fair * *Preis, Bedingungen usw.* || accommodating * *{Handelswesen} entgegenkommend, gefällig; anpassungsfähig* || obliging * *{Handelswesen} gefällig, zuvor-/entgegenkommend; verbindlich*
Kulanz fair dealing || good will || accomodation * *gütliche Einigung, Gefälligkeit* || goodwill
kümmern trouble * *s.bemühen (to do; zu tun),s. Mühe/Umstände machen; th. doesn't trouble me: d. kümm. mich nicht* || concern * *betreffen, angehen, interessieren, von Belang sein, beunruhigen* || regard * *angehen, betreffen* || attend * *to: sich um jmd. ~; beiwohnen, teilnehmen* || mind * *sich (be)~ um, (be)-achten, Acht geben auf, sorgen für, sehen nach* || look after * *aufpassen auf, sehen nach, sich ~ um, sorgen für* || take care of * *achten/Acht geben auf, sorgen für, ~ um, erledigen* || care about * *sich Gedanken machen um/über* || bother about * *sich Gedanken machen um, sich ~ um, sich plagen mit* || care for * *sorgen für, sich ~ um* || worry about * *sich ~ um* || pay attention to * *beachten* || take notice of * *beachten* || arrange * *arrangieren, regeln, in Ordnung bringen* || settle * *klären, regeln, erledigen, aus der Welt schaffen (Problem)*
kümmern um mind || take care of || worry about * *sich Gedanken machen über* || care about || bother about || look after

Kunde customer || client * *für Dienstleistung* || news * *Nachricht* || purchaser * *Käufer* || consumer * *Verbraucher (von Konsumartikeln)*
Kundenanforderungen customer's requirement
Kundenanfragen customer queries
Kundenanpassung customization || customizing * *Anpassung eines Produktes an Kundenbedürfnisse*
Kundenauftrag customer's order
Kundenberatung advisory service || customer advice || customer advisory service * *als Dienstleistung*
Kundenbesuch sales trip * *Geschäftsreise eines Vertriebsmannes zu Verkaufsgesprächen*
Kundenbesuche sales trips * *Geschäftsreisen zu Kunden zwecks Verkaufsgesprächen* || calls on customers
Kundendienst service department * *Abteilung* || customers service || service || after-sales service * *Kundenbetreuung nach Kauf e. Produkts (Service, Reparatur, Ersatzteilversorgung)* || client service
Kundenfang touting * *aufdringlich, durch üble Tricks*
Kundenklemmleiste user terminal block
Kundenkreis custom || customers || clientele
Kundennutzen customer benefit || advantage for the customer * *Vorteil für den Kunden* || customer benefits
kundenorientiert customer-oriented
Kundenorientierung client-conscious attitude * *als geistige Einstellung*
Kundenpersonal customer personnel
Kundenschulung customer training
kundenseitig on the customer side
Kundensignale external signals || plant signals
Kundenspezifikation custom-specification
kundenspezifisch custom-specific || custom-taylored * *auf Kundenbedürfnisse zugeschnitten* || custom || customer-specific || application-specific || customized || custom-design * *kundenspezifisch ausgeführt* || custom-built || customer-specified * *vom Kunden vorgegeben*
kundenspezifische Anpassung customization || custom-specific adaption
kundenspezifische Lösung customized solution || custom-taylored solution
kundenspezifische Projektierung customer-specific configuring
kundenspezifischer Schaltkreis custom-designed circuit || ASIC * *Abkürz. f. 'Application-Specific Integrated Circuit'*
Kundenvorteil customer benefit || customer advantage
Kundenwunsch customer requirements * *Kundenbedürfnisse; in accordance to: nach ~ (z.B. Spezialanfertigung)* || customer requirement || customer request * *on customer request: auf ~/Kundenanfrage*
Kundenzeitschrift house organ
Kündigung cancellation * *~ eines Vertrages* || resignation * *~ seitens des Arbeitsgebers; period of notice: ~sfrist* || notice to quit * *~ seitens des Arbeitsnehmers* || notice to leave * *~ seitens des Arbeitsnehmers* || notice of termination * *~ eines Vertrages* || notice of cancellation * *~ eines Vertrages* || anulment * *~ eines Vertrages usw.*
künftig future || next * *Jahr, Woche* || in times to come * *in künftigen Zeiten/Tagen* || prospective * *voraussichtlich* || potential * *eventuell später möglich, potenziell*
Kunst- man-made * *künstlich/synthetisch hergestellt, kein Naturprodukt, z.B. Kunstfasern* || artificial * *künstlich* || virtual * *scheinbar*
Kunstdruckpapier art paper

Kunstfaser man-made fibre || synthetic fiber
Kunstfasern man-made fibres || synthetic fibres
Kunstgriff knack || manipulation || trick
Kunstharz synthetic resin
künstlich artificial || synthetic * *synthetisch, künstlich hergestellt* || imitated * *nachgemacht, imitiert* || false * *unecht* || man-made * *künstlich hergestellt, synthetisch* || forced * *gezwungen* || artificially * *(Adv.)*
künstlich gealtert artificially aged
künstliche Alterung artificial ageing || artificial seasoning
künstliche Intelligenz artificial intelligence * *der menschlichen Intelligenz nachempfundene Computersoftware/Hardware* || AI * *Abk. für 'Artificial Intelligence'*
Kunststoff plastic || plastics || plastic material || synthetic material
Kunststoff-Folie plastic film || plastic foil
Kunststoff-Halbzeug plastic preform
Kunststoff-Lichtleiter plastic optocable
Kunststofffolie plastic film || plastic foil || cellophane * *Zellophan, Zellglas, Klarsichtverpackung(sfolie)*
Kunststofffolienkondensator film capacitor || foil capacitor
Kunststoffgranulat plastic granules || plastic pellets || plastic granulate
Kunststoffindustrie plastics industry
Kunststoffkondensator film capacitor * *metallisierter Kunststoff-oder Kunststofffolienkondensator*
Kunststoffmaschine plastics machine
Kunststoffriemen plastic belt
Kunststoffverarbeitung plastics processing
Kupfer copper
Kupferlackdraht enamelled copper wire || varnished copper wire
Kupferleiter copper conductor
Kupferleitung copper cable
Kupferlitze stranded copper
Kupferschiene copper bar
Kupferverluste copper losses * *Ohmsche Verluste in einer Wicklung, Stromwärmeverluste*
Kupferwicklung copper winding
Kupferzuschlag copper charge * *{el.Masch.} Zuschlag z.Motorpreis bei Überschreiten e. bestimmten Kupfer-Börsenpreises*
kuppeln couple || connect * *siehe auch →koppeln* || interlink * *über Signale* || operate the clutch * *die Kupplung betätigen*
Kupplung coupling * *nicht betriebsmäßig ein-/ausrückbare ~ (Faltenbalg~ usw.); el. Steck~, Kuppeln, Kopplung* || clutch * *betriebsmäßig ein-/ausrückbare ~ (z.B. Scheiben~ mit Reibbelag)*
Kupplungs-Brems-Kombination clutch/brake combination || combined clutch-brake unit
Kupplungs-Bremskombination clutch/brake unit || combined clutch-brake unit || clutch-brake unit || clutch/brake combined unit
Kupplungs-Druckplatte clutch pressure plate
Kupplungsabtrieb coupling output || clutch outdrive
Kupplungsantrieb coupled drive * *Motor treibt die Abeitsmaschine über Kupplung an (nicht über Riemen usw.)*
Kupplungsbelag clutch lining || clutch facing
Kupplungsdruckfeder release spring
Kupplungsgehäuse clutch housing
Kupplungshälfte coupling half
Kupplungshälften coupling halves
Kupplungslamelle clutch disc || clutch disk || clutch plate
Kupplungsring clutch ring
Kupplungsstern coupling star || claw-coupling half

Kupplungswelle clutch shaft
Kurbel crank
Kurbelantrieb crank drive
Kurbelgehäuse crankcase
Kurbelinduktor megger * *zur Isolationsprüfung/messung*
Kurbelradius crank radius
Kurbelwelle crank shaft || crankshaft
Kurierdienst courier service * *['kuriea söwis]*
Kurs training course * *Lehrgang* || course * *Lehrgang; refresher couse: Auffrischungs~; top-up course: Fortgeschrittenen~* || course of instruction * *Lehrgang* || practice * *Übung* || refresher course * *Auffrischungs~* || exchange rate * *Wechsel~* || rate of exchange * *Wechsel~* || course * *~ für Schiff und Politik* || route * *Route* || workshop * *~ mit praktischen Übungen*
Kursanmeldung course application * *Reservierung für e. Trainings-/Weiterbildungskurs usw.* || course reservation * *Resevierung für e. Trainings-/Weiterbildungskurs usw.* || course enrolment * *Kurseinschreibung; siehe auch →Anmeldung*
Kursbeginn start of course * *eines Aus/Weiterbildungskurses*
Kursbeschreibung course description * *eines Aus/Weiterbildungskurses*
Kursbüro course office * *einer Aus/Weiterbildungsstätte*
Kursgebühren course fees
Kursleiter instructor || chief instructor || head of course || teacher * *Lehrer* || tutor * *Studienleiter*
Kursplan schedule for courses * *Zeitplan, Kursprogramm*
Kursprogramm course program * *für Schulungskurse*
Kursteilnehmer student || participant in a course || course participant * *an Aus/Weiterbildungskurs*
Kurstitel title of course * *eines Schulungskurses*
Kursunterlagen course material * *Lehr/Lernunterlagen für einen Aus/Weiterbildungskurs* || training documentation * *Schulungsunterlagen*
Kurve curve * *[köw] polygon curve: aus Geradenstücken bestehende ~ (→Polygonzug)* || characteristic * *Kennlinie* || graph * *(statistisches) Kurvenbild* || bend * *Biegung, Kurve(auch e.Straße); sharp/hairpin/blind bend: scharfe/Haarnadel-/unüb.sichtl.~* || bow * *Bogen* || spline * *Kurvenlinie in d. Computergrafik* || turn * *Biegung, Kurve, Drehung, Kehre, Krümmung, Wendung, Umkehr* || corner * *(Straßen-) Ecke, Winkel, (auch Verb: um die Kurve biegen, e.Winkel bilden)* || cam * *(mechan.) Steuerkurve (eines Nockens, einer Kurvenscheibe usw.)* || curve of trajectory * *ballistische ~, Bahnkurve*
Kurvenanzeige curve display || trend display
Kurvendarstellung curve display * *am Bildschirm usw.* || graph || graphical representation || plot * *durch ein registrierendes Messgerät usw.* || trend diagram || curve diagram
Kurvenform curve shape || shape of the curve || wave form * *einer periodischen Größe* || waveform * *einer periodischen Größe*
Kurvenfräseinheit contour milling unit
Kurvenlineal spline
Kurvenlinie spline * *in der Computergrafik*
Kurvenschar family of curves || set of curves
Kurvenscheibe cam || cam disc || cam plate
Kurvenschreiber chart recorder * *auch 'Linienschreiber'* || plotter * *digital (punktweise) angesteuert*
Kurvenverlauf curve characteristic || curve shape
Kurvenzug characteristic * *Funktionskurve, Kennlinie* || curve trace * *von registrierendem Messgerät aufgezeichnet* || graph * *Kurvendiagramm, Kurvendarstellung* || curve path || polygon curve * *→Polygonzug*

kurz short * *räumlich und zeitlich* ~ || brief * *gedrängt, knapp; in brief:* ~ *gesagt* || concise * *gedrängt* || laconic * *treffend* || laconically * *(Adv.) treffend*
kurz gesagt in brief || briefly
Kurz- short || short form || shortened || brief || transient * *vorübergehend, bedingt durch Ein-schwing-/Ausgleichsvorgang*
Kurzangabe code * *Kodebuchstaben/-nummern* || short form designation * *Kurzbezeichnung*
Kurzangaben short infos
Kurzarbeit short-shift work || short time working
Kurzbeschreibung short description || brief description || brief outlines || outlines || product summary * *eines Produkts*
Kurzbetrieb short-time operation
Kurzbezeichnung short term || code * *Kode(nummer)* || short designation * *Kurzbezeichnung*
Kürze shortness * *(räumlich und zeitlich)* || short duration * *(zeitlich)* || brevity * *des Ausdrucks, auch zeitlich* || shortly * *in Kürze* || in near future * *in Kürze* || briefly * *in aller Kürze, kurz gesagt* || for short * *der Kürze halber*
kürzen cut * *z.B. Ausgaben, Gelder* || shorten || make shorter || abridge * *Buch usw.* || reduce * *{Mathem.} Bruch* || abbreviate * *(Zeit, Wörter, Rede usw.) abkürzen/kürzen.* || slash * *stark* ~, *zusammenstreichen, deutlich verringern (Gelder, Zeitaufwand, Gehalt usw.)*
kürzest shortest * *zeitlich und räumlich*
kürzestmöglich with the shortest possible length * *bezogen auf die Länge* || as short as possible
Kurzform short form * *(figürl.) in short form: in* ~
kurzfristig short-term || short-dated * *mit nahegelegenem Stichdatum* || short-period || of short duration * *kurz, nur kurz andauernd, von kurzer Dauer* || immediate * *sofortig* || within a short time * *(Adv.)* || at short notice * *(Adv.) z.B. liefern, absagen; available at short notice: kurzfristig lieferbar* || temporarily * *(Adv.) vorübergehend* || for a short period * *(Adv.) für kurze Zeit*
kurzgeschlossen short-circuited || shorted || shortened
Kurzhub- short-stroke
Kurzimpuls short pulse
Kurzkatalog abridged catalogue || short-form catalog
kürzlich recently * *(Adv)* || recent * *(Adj.)*
Kurzmotor short motor
kurzschließen short-circuit || short * *(salopp)* || jumper * *mit Hilfe einer Draht-/Klemmenbrücke*
Kurzschluss short-circuit || short
Kurzschluss-Scheinleistung locked-rotor apparent power * *{el.Masch.} Scheinleistg. e. Asynchr.mot. bei festgebremstem Läufer*
kurzschlussartig in the manner of a short-circuit
Kurzschlussausschaltung short-circuit breaking current * *eines Leistungsschalters*
Kurzschlussbremsung short-circuit braking * *Motorklemmen werden zum Bremsen kurzgeschlossen*
Kurzschlusseinschaltstrom short-circuit making current * *eines Leistungsschalters*
Kurzschlusserwärmung locked-rotor temperature rise * *{el.Masch.} Temp.anstieg d. Wicklungen e. Asynchr.masch. i.Stillstand*
Kurzschlussfall short-circuit condition || event of a short circuit
kurzschlussfest short-circuit proof || short-circuit resistant
kurzschlussfester Anschluss short-circuit-proof termination * *{el. Masch.} masch.seitiger Kurzschl. führt nicht z. KS i. Klemm.kasten*
Kurzschlussfestigkeit short-circuit strength || short-circuit capability || short-circuit capacity * *auch bei einer Sicherung*

Kurzschlussfortschaltung short-circuit clearing
Kurzschlusskäfig short-circuit cage
Kurzschlusskennlinie short-circuit characteristic
Kurzschlussläufer squirrel-cage rotor
Kurzschlussleistung fault power || short-circuit power || fault level
Kurzschlussleistungsverhältnis relative short-circuit power
Kurzschlussring short-circuit ring * *{el. Masch.} ringförmige Verbindung der Käfigstäbe bei Käfigläufer*
Kurzschlussscheinleistung locked-rotor apparent power * *{el.Masch.} Scheinleistung e.Motors bei festgebremst. Läufer*
Kurzschlussschutz short-circuit protection
kurzschlusssicher short-circuit proof || short-circuit protected
Kurzschlusssicherheit short-circuit withstandibility * *Kurzschlussfestigkeit* || short-circuit protection * *Kurzschlussschutz*
Kurzschlussspannung impedance voltage * *z.B.Drossel/Trafo (Eing.spann. bei d. am kurzgeschl. Ausg.d. Nennstrom fließt)* || percentage impedance * *z.B. bei Drossel/Trafo (siehe auch 'uk')* || impedance drop * *z.B. bei Drossel/Trafo (siehe auch 'uk')* || uk * referred voltage drop
Kurzschlussstrom short-circuit current || fault current * *Fehlerstrom* || fault level * *Pegel des Fehlerstroms*
Kurztaktverfahren short-cycle process
Kurzunterbrechung short interruption || transient interruption
Kurzversion short version
Kurzzeichen code || abbreviation * *Abkürzung* || symbol || short designation * *Kurzbezeichnung*
Kurzzeit- short-time || short-term
Kurzzeitbelastbarkeit short-time loading capacity
Kurzzeitbelastung short-time loading || short-time duty * *Betriebsart*
Kurzzeitbetrieb short-time operation || short-time duty * *{el.Masch.} Betriebsart S2* || short term operation || short-time duty type * *Betriebsart S2 nach VDE 0530/DIN 41756 für Motor/Stromrichter* || temporary duty || short time operation
kurzzeitig short-term || over short duration || instantaneous * *einen kurzen Moment lang* || short-time || momentary * *vorübergehend* || brief * *kurz(zeitig), gedrängt* || temporary * *zwischenzeitlich, vorübergehend* || temporarily * *(Adv.) zwischenzeitlich, vorübergehend* || briefly * *(Adv.) gedrängt, kurz* || for short periods of time * *(Adv.)*
kurzzeitige Überlastung short-term overload
kurzzeitiger Netzausfall momentary power loss
kurzzeitiger Spannungseinbruch voltage dip
kurzzeitiger Stromausfall voltage dip
Kurzzeitlebensdauer short-time life * *{el. Masch.} begrenzte Lebensdauer von Motoren mit extrem hohen Temperaturen*
Kurzzeitleistung short-time rating * *spezifizierte/zulässige* ~ *(bei d. Spezifikation der Überlastfähigkeit)*
Kurzzeitunterbrechung short-time interruption
Kusa-Schaltung stator-resistance starting circuit * *Abk.f. 'KUrzschluss-SAnftanlaufschaltg.' mit Serienwiderst.* || KUSA-connection * *Abk.f. 'KUrzschlus-SAnftanlaufschaltg.' für Asynchronmot. m. Serienwiderst. i Phase*
kW kW * *1 kW entspr. 1,3410 hp (Horsepowers) entspr. 1,3596 PS (Pferdestärken)* || hp * *horsepowers; 1 kW entspr. 1,3410 hp, 1,3596 PS (Pferdestärken)* || kilowatt * *['kilewot]*
KWh KWhr

L

l 1 * *Liter (1 l entspr. 0,2200 Engl.Gallonen entspr. 0,2642 US Gallonen)* || gal * *Gallone (1 l entspr. 0,2200 Engl.Gallonen entspr. 0,2642 US Gallonen)* || UK gal * *Engl.Gallonen; 1 l entspr. 0,2200 Engl.Gallonen* || US gal * *amerikan.Gallonen; 1 l entspr. 0,2642 US Gallonen* || cubic inch * *Kubikzoll; 1 l entspr 61,0237 cubic inch* || cubic foot * *Kubikfuß; 1 l entspr. 0,0353147 cubic foot* || UK pt * *Engl. Pint; 1 l entspr 1,7598 UK pt* || pt * *Engl. Pint; 1l entspr. 1,7598 pt*
Labor laboratory || lab
Labor-Netzgerät laboratory-type power supply || bench-type power supply
Laboratorium laboratory || lab
Laborausstattung laboratory equipment
Laboreinrichtung laboratory equipment
Labormessung lab measurement
Labormuster lab sample * *(salopp)* || laboratory sample * *siehe auch →Muster*
Labornetzgerät bench-type power supply || laboratory-type power supply
Labyrinthdichtung labyrinth seal
Lack paint || coating || varnish * *auf Ölbasis*
Lackdraht enamelled wire || lacquered wire || varnish-insulated wire || enameled wire
Lackieranlage laquering plant
Lackierbarkeit paintability
lackieren paint || varnish * *auch mit Isolierlack*
Lackiererei paint shop
Lackiergerät paint-application system
Lackierkabine spraying booth || laquer spray booth
Lackiermaschinen machines for applying laquers
Lackierofen baking oven
Lackierroboter spraying robot
Lackierstraße car spraying line * *für Autos*
Lackierung paint finish || painting
Lackierwalze varnishing roller * *(Druckerei)*
Lacktrockner laquer drier
ladbar loadable * *Software usw.*
Ladeeinheit loading unit * *z.B. Palette*
Ladefunktion load operation * *für ein (SPS-) Programm*
Ladegerät charger * *für Batterien*
Ladegleichrichter charging rectifier
Ladekondensator charging capacitor
Lademaschine charging machine * *Beschickungseinrichtung, siehe auch →Einschieber* || entry machine * *Eintragmaschine, z.B. für Ofen, Pressmaschine*
laden load * *auch Programm, Daten usw.* || download * *Programm oder Parameter in einen Rechner hinein* || charge * *Batterie, Kondensator usw.*
Laden schmeißen run the show
Ladentisch counter * *sell over/under the counter: über/unter dem Ladentisch verkaufen*
Ladespannung charging voltage * *bei Batterie/Akku*
Ladestrom charging current * *z.B. für Batterie, Kondensator*
Ladestromspitzen charging current peaks * *z.B. bei der Aufladung eines Kondensators*
Ladestufe charging level
Ladestufenumschaltung charge level selection
Ladung charge
Ladungsdichte charge density
Ladungspumpe charge pump * *Spannungswandler m.Chopper u.Kondensatoren (kann Spannungen hochsetzen u.invertieren)*
Ladungsträger charge carriers * *z.B. bei Halbleitern* || charge carrier * *(Singular) z.B. bei Halbleitern* || carrier * *bei Halbleitern*

Lage location * *Ort* || situation * *Situation* || position * *räumlich und figürlich; be in the position: in der ~ sein; auch bei ~regelung* || layer * *Ebene, Schicht, auch ~ einer Wicklung* || constellation * *Konstellation* || tier * *auch 'Reihe'; in tiers: ~nweise/in Reihen übereinander* || ply * *Faserstoff-Lage, Sperrholz-Schicht, Falte, (Garn-) Strähne; Hang, Neigung* || predicament * *(missliche) Lage; Kategorie*
Lageabweichung position error * *{Regel.techn. | NC-Steuerg}.* || positional variance * *Toleranz in einer mechan. Konstruktion*
Lageerfassung position sensing
Lagegeber position detector * *über Schaltkontakt* || position encoder * *absolute/incremental: absoluter/inkrementeller; rotary/linear: Dreh-/Linear-* || position transducer || position sensor
Lageistwert position value || position signal
Lagemesssystem position measuring system
Lagemessung position sensing * *Lageerfassung* || position measuring
Lagenwickelmaschine layer winding machine
Lageplan sketch map * *z.B. Landkartenskizze* || plan * *of: von* || layout plan * *(techn.) Anordnungsplan* || layout
Lager bearing * *Gleit-, Wälzlager* || stock * *Waren-, Ersatzteillager; supply from stock: ab Lager liefern; keep on stock: auf Lager haben* || storage * *Speicher* || yard * *~hof (im Freien; z.B. für Holz)* || warehouse * *für Waren, Produkte*
Lager mit Dauerschmierung prelubricated bearing
Lager-Temperaturüberwachung bearing-temperature monitoring
Lager-Thermometer bearing thermometer
Lagerabnutzung bearing wear
Lagerbehälter storage container
Lagerbelastung bearing load
Lagerbestand supplies || stock || store || stock inventory * *~saufnahme, festgestellter ~* || stock on hand || inventory
Lagerbock pedestal * *(Lager)Bock, Untergestell, Sockel, Postament, Säulenfuß, Grundlage (figürl.)* || bearing stand || bearing pedestal || bearing block || bearing bracket || pillow block * *Stehlager, Lagerbock*
Lagerbuchse bearing bush
Lagerbüchse bushing || bush || bearing bush || journal box
Lagerdeckel bearing cap * *siehe auch →Schlussdeckel (äußerer ~)* || bearing cover
Lagerdichtung bearing seal
Lagerdisposition warehouse disposition * *[wärhaus]*
Lageregelkreis position loop || position control loop
Lageregelung postion control
Lageregler position controller || motion controller * *Bahnkurvenregler*
Lagerfähigkeit storage life * *z.B. von Bauteilen* || shelf life * *z.B. von Bauteilen*
Lagerförderzeug stock conveyor
Lagerfutter bushing || bush
Lagergehäuse bearing housing || bearing box
Lagergeräusch bearing noise
Lagerhalle warehouse || storehouse || store
lagerhaltig available from stock
lagerhaltiges Produkt stock product * *ab Lager lieferbares/lagermäßiges Produkt*
Lagerhaltung inventory || parts inventory * *von Artikeln, Geräten usw.* || parts stocking * *von Artikeln, Geräten usw.* || stockkeeping || storekeeping || warehousing * *[wärhausing]* || stockage
Lagerist stockkeeper || stock clerk || stockman * *(amerikan.)*

Lagerkäfig bearing cage
Lagerkosten storage expenses || warehouse charges * *[wärhaus] an externe Unternehmen zu zahlende* || warehouse costs * *[wärhaus]*
Lagerlebensdauer bearing life * *L_10 in [h], die von 90% der Lager erreicht wird (DIN ISO 281)*
Lagerliste stock list || warehouse list
Lagerluft bearing clearance * *Lagerspiel* || bearing airgap || clearance * *Spiel* || internal bearing clearance * *Maß, um das sich d.Außenring gegen d.Innenring verschieben lässt*
lagermäßig available from the stock || from the stock || stocked || stock * *z.b. stock motor: ~er Motor* || ex-stock
lagern store * *speichern, auch in Waren-/Ersatzteillager* || keep on stock * *auf Lager halten* || support * *(unter)stützen, Maschine(nteil) ~/betten* || be warehoused * *auf Lager befindlich sein, auf Lager liegen* || be stored * *auf Lager befindlich* || mount in bearings * *in Wälz/Gleitlager ~* || pivot * drehbar ~, z.b. mittels Scharnier ~ || bed * *Maschine in Maschinenbettung ~* || seat * *Maschine in Maschinenbettung ~* || season * *altern, ab~ lassen (z.B. Holz)* || warehouse * *[wärhaus] aufbewahren, auf Lager halten* || stack * *auf Stapel ~*
Lagerölpumpe bearing oil pump || bearing-lubricating pump
Lagerplatz stockyard * *im Freien* || storage place || depot * *Vorratsplatz* || dump * *Stapelplatz* || storage area || storage location
Lagerreibung oil drag || bearing friction
Lagerschaden bearing-damage || bearing failure
Lagerschale bearing box
Lagerschild bearing shield || endshield * *geschlossenes ~ am Wellenende* || end shield * *geschlossenes ~ am Wellenende* || bearing bracket * *offenes ~* || end bracket * *offenes ~ am Wellenende* || bracket || end plate * *~ am Ende einer Welle (z.B. bei Motor)*
Lagerschildkühlluftöffnung endshield ventilation opening * *Motor*
Lagerschildnabe endshield hub * *führt die Welle, z.b. eines Motors*
Lagersilo storage silo
Lagerspiel axial clearance * *~ in Axialrichtung* || end play * *~ eines Wellenlagers in Axialrichtung* || radial clearance * *~ in Radialrichtung* || clearance * *(allg.)* || bearing play * *(allg.)*
Lagerstelle bearing arrangement * *Lagerung* || bearing
Lagerstrom bearing current * *el.Strom durch e. (Motor-) Lager (unerwünschter Effekt, kann z. Lagerschaden führen)*
Lagersystem storage system || stocking system || storage/retrieval sstem
Lagertemperatur storage temperature || bearing temperature * *bei Wälz-/Gleitlager (für [el. Masch.]: siehe VDE0530 Teil 1)*
Lagertyp type of bearing * *Gleit -/Wälzlager usw.*
Lagerüberwachung bearing monitoring
Lagerung storage * *Ein-/Lagerhaltung in einem Speicher/Lager* || bearing arrangement * *z.B. Gleit-/Wälzlager-Anordnung* || bearing system * *z.B. Gleit-/Wälzlager* || bearing * *Gleit-/Wälzlager* || warehousing * *[wärhaus] Ein~, i.Lager Aufbewahren, Auf-Lager-Halten* || bearing assembly * *zu einem Gleit/Wälzlager gehörende Teile* || seasoning * *Alterung, das Ablagern (z.B. von Holz)* || bearing application * *Verwendung/Montage e. mechan. Lagers* || mounting * *Befestigung* || bedding * *Bettung* || seating * *Sitz* || support * *Stützung, Aufhängung* || bearing design * *Ausführung/ Auslegung/konstruktive Gestaltung der ~*
Lagerungskosten storage expenses || warehouse charges * *[wärhaus ...] an dritte zu zahlende* || warehouse costs * *[wärhaus ...]*

Lagerungstemperatur storage temperature
Lagerungszeit period of storage
Lagerverriegelung bearing block * *Transportsicherung, z.B. bei [el.Masch.]*
Lagerverschleiß bearing wear
Lagerzapfen journal * *(Dreh-/Lager-/Wellen-) Zapfen; journal bearing: Achs-/Zapfenlager* || pivot pin || trunnion * *Schildzapfen*
Lagerzustand bearing condition * *eines Gleit/ Wälzlagers*
Lagerzustandserfassung bearing condition sensing * *eines Gleit/Wälzlagers*
Lagesollwert position setpoint || target position * *Zielposition* || position reference || position reference value
lahm lame * *(auch figürl.: Computer usw.)* || feeble * *(figürl.) kraftlos* || poor * *dürftig, mangelhaft, jämmerlich, schwach*
lahm legen deactivate * *→deaktivieren* || paralyze * *lähmen, paralysieren* || render ineffective * *ineffektiv machen* || neutralize
lähmen paralyze || cripple || hamstring * *(auch figürl.)* || tie up * *Verkehr usw.*
Laie amateur * *['ämete: r]; in this field: auf diesem Gebiet* || novice * *Neuling, Anfänger* || greenhorn * *(salopp) blutiger ~, siehe auch →Anfänger* || layman * *Nichtfachmann; technical: technischer ~* || non-professional * *'Nichtprofi'*
Lamelle segment * *eins Kommutators* || disc * *in Bremse, Kupplung; clutch disc: Kupplungslamelle* || disk * *Scheibe* || lamina * *Plättchen, Blättchen, dünne Schicht* || bar * *auch am Kommutator; Streifen, Strich* || lamella * *Plättchen, Blättchen* || blade * *am Kühler/Wärmetauscher (nicht 'grill'!)* || blade * *Blatt, Flügel, Schaufel, Klinge* || leaf * *Blatt, Flügel, dünne Folie, Blattfeder* || lamination * *Schichtung, blättrige Beschaffenheit* || plate * *auch in einer Kupplung/Bremse* || fin * *(Kühl-) Rippe* || rib * *an Kühler* || plate * *Scheibe einer Mehrscheiben-/Lamellenkupplung/-bremse*
Lamellenbremse multiple-disc brake * *Reibungsbremse mit mehreren Scheiben (abwechselnd eine ruhend, eine drehend)* || multiple-disk brake || multidisc brake || multi-plate brake
Lamellenkupplung multiple-disk clutch || multiple-disc clutch || multidisc clutch || multi-plate clutch
Lamellenpaket disc assembly * *einer Lamellenkupplung/-bremse* || plate assembly * *einer Mehrscheiben- bzw. Lamellenkupplung/-bremse* || plate stack * *einer Mehrscheiben- bzw. Lamellenkupplung/-bremse*
Lamellenpresse laminating press * *[Holz]*
Laminat laminate || laminated material || laminate material
Laminator laminator * *Einrichtg. z. Laminieren (übereinand. legen mehrerer Schichten) i. d. Folienherstellg. usw.*
laminieren laminate
Lampe lamp
Lampen-Kaltstrom lamp current in the cold state || warm-up lamp current * *erhöhter Lampenstrom während der Erwärmungsphase nach d. Einschalten*
Lampen-Warmstrom lamp current in the warm condition
Lampenfassung lamp holder || lamp socket
Lampenprüfung lamp test
Lampenwechsel lamp replacement
LAN LAN * *Abkürzg.f. 'Local Area Network': Nahbereichs-Computernetz (z.B. Ethernet)*
lancieren launch * *in Gang setzen, Starthilfe geben*
Länderkennzeichen country code * *(auch beim Zoll)*
länderspezifisch country-specific

Landesgesellschaft local company * *z.B. ausländische Niederlassung* || foreign branch * *Zweigniederlassung im Ausland* || foreign subsidiary * *Filiale/Tochtergesellschaft im Ausland*
landesspezifisch country-specific
Landessprache national language
Landmaschinen farm machinery
Landtransport land transport
Landwirtschaft farming || agriculture || agricultural industry
lang long * *räumlich und zeitlich* || for a long time * *zeitlich*
Länge length || duration * *zeitlich* || footage * *(Gesamt-) Länge/-Maß in Fuß (1 Fuß entspr. 30,48 cm*
Längenmaß linear size
Längenmesssystem length measuring system || linear measuring system
Längenzähler length counter || measuring counter
länger longer * *longer than: länger als (auch zeitlich); any longer: nicht länger/nicht mehr* || prolonged * *(zeitlich) verlängert, anhaltend; for a prolonged time. für ~e Zeit* || extended * *verlängert (zeitlich und räumlich)* || rather long * *ziemlich lang* || for some time * *für ~e/einige Zeit* || for a prolonged period * *über einen längeren Zeitraum*
langfristig long-term || long-range || over the long term
Langimpuls long pulse || extended-duration pulse
langjährige Erfahrung many years of experience
langjähriger Mitarbeiter long-serving employee
langlebig durable * *dauerhaft, haltbar, beständig* || long-lived || long-term * *lange geltend, langfristig*
Langlebigkeit durability * *Dauerhaftigkeit* || longevity || ruggedness * *Verschleißfestigkeit, Zerstörsicherheit, Robustheit*
Langloch oblong hole || elongated hole
Langlochbohrmaschine mortizing machine * *{Holz}* || slot-boring machine * *{Holz}*
Langlochfräsmaschine slot-mortizer * *{Holz} (amerikan.)* || slot mortiser * *{Holz} (brit.)*
Langmesser straight knife
längs along * *entlang* || alongside * *entlang; (of: an/zu)* || longitudinal * *der Länge nach* || vertical * *vertikal, senkrecht, lotrecht* || lengthwise * *der Länge nach*
Längs- longitudinal || material-directed * *in Materiallaufrichtung* || MD * *Abkürzg.f. 'Material-Directed': in Materiallaufrichtung*
langsam low-speed * *mit niedriger Geschwindigkeit* || slow || leisurely * *(Adv.) bedächtig, gemächlich* || unhurried * *ohne Eile* || tardy * *säumig* || sluggish * *träge* || heavy * *schwerfällig* || slow of comprehension * *geistig* || slow-speed * *mit langsamer Geschwindigkeit* || slowly * *(Adv.)* || slow-moving * *sich langsam bewegend*
langsam anwachsend creeping
langsam laufend low-speed
langsam wirkend slow-acting
Langsamkeit slowness || slow-motion * *Zeitlupentempo* || leisureness * *Gemächlichkeit, Bedächtigkeit* || slackness * *Unlust, Saumseligkeit, Trägheit, Nachlässigkeit*
langsamlaufend slow-running || slow-moving
Langsamläufer slow-moving machine || low-speed machine
Längsbohrung slot * *{Holz}*
Längsfeld longitudinal field
längsgeteilt slit * *slit strip: längsgeteiltes Blech-/Walzband*
Langsieb fourdrinier * *{Papier} bei einer Papiermaschine*
Langsiebmaschine fourdrinier machine * *Papiermaschine*

Langsiebpapiermaschine fourdrinier machine || fourdrinier paper machine
Längskraft longitudinal force || axial force * *Axialkraft*
Längsprofil longitudinal profile * *Längsverteilg. v.Flächengewicht, Feuchte, Temperatur,Dicke v.bandförm.Material* || MD profile * *Abkürzg.f. 'Material-Directed': (z.B. Fläch.gewichts-) Profil in Materiallaufrichtung*
Längsrecke material-directed orientation * *Längsreckeinrichtung in Folienmaschine* || MDO * *Abk.f. 'Material-Directed Orientation': Längsreckeinrichtung in Folienmaschine* || longitudinal orienter * *Längsstreckeinrichtung bei Kunststoff-Folienmaschine* || machine direction orienter * *{Kunststoff} bei Folienmaschine*
Längsregler in-phase regulator || stabilizing post regulator * *z.b. Spannungsregler mit Längstransistor b. stabilisierter Stromversorgg.*
Längsrichtung longitudinal direction || machine direction * *Maschinen(lauf)richtung* || material direction * *Materiallaufrichtung* || axial direction * *eines zylinderförmigen Teils (z.B. Motor)* || sense of length || MD direction * *Abk. für 'Machine Direction': Maschinenlaufrichtung*
Längsschneiden slitting
Längsschneider slitter * *siehe auch →Rollenschneider*
Längsschnitt longitudinal section * *in Konstruktionszeichnung* || cross section * *in Konstruktionszeichnung*
Längsteilanlage slitting line * *für Walzband, Blech usw.*
Längswelle line shaft || horizontal shaft
längswellenlos shaftless
Längswiderstand longitudinal resistance * *z.B. in einem Halbleiter*
langwierig time-consuming * *zeitaufwändig* || protracted * *in die Länge gezogen* || lengthy * *länglich* || unending * *endlos* || wearisome * *ermüdend* || long-drawn-out * *sich (schleppend) dahinziehend, sich in die Länge ziehend*
Langwort long word || double word * *z.B. 32 Bit langes Datenwort*
Langzeit- long-time
Langzeitdrift long-term drift
Langzeitgenauigkeit long-term accuracy
Langzeitkonstanz long-term stability
Langzeitlagerung long-time storage
Langzeitstabilität long-term stability
Langzeittest endurance test * *Dauertest, Belastungstest, Haltbarkeitstest* || long-duration test || life test * *Lebensdauerprüfung* || long-run test
Langzeitverlauf long-term trend * *von Messgrößen usw.* || long-term variation
Laplace-Transformation Laplace transform
Laplacescher Operator Laplacian
Laplacetransformation Laplace transform
Lappen rag * *z.B. Reinigungslappen* || lug * *→Lasche* || tab * *Streifen, Lappen (z.B. aus Blech), Zipfel, Nase an Maschinenteil*
läppen lap * *(passgenau) feinbearbeiten einer Fläche z.B. mit Werkzeugmaschine*
Läppmaschine lapping machine
Lärm noise
Lärmdämpfung noise suppression
Lärmminderung noise reduction
Lärmschutz noise protection || noise muffling * *Lärmminderung/-Dämpfung*
Lasche lug * *Ansatz(blech), Henkel, Ohr, Schlaufe, Zinke*, || strap * *Riemen (-Stück), Gurt, Band, Strippe, Steg, Bandeisen* || fishplate * *an Schienen* || butt strap * *Verbindungs~ an Stahlkonstruktion usw.* || tongue * *Zunge* || scarf * *La-

schung; scarf joint: Blattfuge/Falzverbindung/Verlaschung
Laserschneidmaschine laser-fusion cutting machine * Werkzeugmaschine
Last load * under load: unter Last || burden * (Trag-)Last, Bürde, Ladung || hook load * am Kran(haken)
Last abwerfen throw off the load || shed the load
lastabhängig load-dependent || load-sensitive
lastabhängige Feldschwächung load-dependent field weakening * ermöglicht bei Kran mit DC-Hubmotor schnelle Fahrt bei leichter Last
Lastabhängigkeit load impact effect || influence of load variation || effect of the load variations
Lastabwurf load shedding || load rejection
lastadaptiert load-adapted
lastadaptiv load-adapted
Lastanderung load change || variation in load * Lastschwankung
Lastanderungen load changes
Lastangriff load position * →Angriffspunkt der Last z.b. an einer Welle || point of load application * ~spunkt
Lastart character of the load
Lastaufnahmemittel load pick-up * {Hebezeuge}
Lastaufteilung load share || load sharing || load balance * zu gleichen Anteilen || load distribution * Lastverteilung || duty sharing * z.B. bei Kompressorsätzen
Lastausgleich load balance * zw. mehreren mechanisch o. über d. Material gekoppelten Antrieben z.B. Längswelle || load balancing * zw. mehreren mechanisch o. über das Material gekoppelte Antrieben || load compensation * Kompensation der Lasteinflüsse || load equalization || load share * Lastaufteilg., z.B. zw. mehren Motoren, d. starr od. über d.Material gekoppelt sind
Lastausgleichsregelung load share control || load equalization control || load sharing control || load balancing control * for drives coupled to a common load: für an dieselbe Last gekopp. Antriebe || load balance control || load distribution control * Lastverteilungsregelung zur Momentenaufteilung auf mehrere Motoren
Lastauto truck
Lastbedingungen load conditions
Lastenaufzug hoist || freight elevator || goods lift
Lastenheft specifications || specification * design specification
Lastfaktor load factor * Quotient aus Durchschnittslast und Spitzenlast
Lastführung load commutation * bei Stromrichterschaltung
lastgeführt load-commutated * Stromrichter, bei d.die Antriebsmaschine d.Kommutierungsspanng.-liefert (DIN 41750)
lästig troublesome * beschwerlich || burdensome * lästig, drückend || cumbersome * lästig, beschwerlich, unbequem || annoying * unangenehm, störend || uncomfortable * unbequem, unkomfortabel || inconvenient * unbequem || undesirable * unerwünscht || irksome * unangenehm || bothersome * lästig, unangenehm || tedious * langweilig, öde, ermüdend
Lastimpedanz load impedance
Lastkennlinie load characteristic || load torque characteristic * Verlauf des Lastmoments über der Drehzahl
lastkommutierter Stromrichter load-commutated converter
Lastkommutierung load commutation * für Stromrichterschaltung
Lastkreis load circuit
Lastmagnet lifting magnet * an Hebezeug
Lastmoment load torque * Drehmoment || load in-

ertia * Trägheitsmoment || counter torque * Gegenmoment
Lastnetzgerät external power supply * externe Kontaktstromversorgung (z.B. 24V) für (SPS-) Ein-/Ausgänge
Lastnetzteil external power supply * Kontaktstromversorgung (z.b. 24 V) für (SPS-) Ein-/Ausgänge
Lastpunkt load point
Lastrelais load relay
Lastschwankungen varying loads || fluctuations in the load || load fluctuations || variation in load || load changes
Lastseite load side || driven side * angetriebene Seite, Abtriebsseite
lastseitig load-side || at the load side
Lastspannung load voltage
Lastspannungsabfall load voltage drop
Lastspiel load cycle || operating sequence * "Programm" einer Arbeitsmaschine || duty cycle
Lastspitze peak load || peak overload * Über~
Lastspitzen shock load * (mechan.) || load transients
Laststoß load surge * kurzzeitige Überlast || sudden load variation || step change in load * Lastsprung || sudden loading || sudden load change || load thrust
Laststrom load current
Lasttrenner switch-disconnector || load-break switch || load disconnector || load interruptor || switch disconnecter
Lasttrennschalter switch-disconnector || load-break switch || load disconnector || load interruptor || switch disconnecter
Lastübernahme load transfer
lastunabhängig load-independent
Lastverhalten load characteristics || load response * einer Regelung
Lastverteiler load distribution plant * (Energieverteilung)
Lastverteilung load distribution * z.B. zwischen Antrieben, die starr oder über das Material gekoppelt sind
Lastvorsteuerung load precontrol || load feed-forward
Lastwagen truck
Lastwechsel load variation || load change || load transient * vorübergehender || changes of load
Lastwiderstand load resistance
Lastwinkel load angle || load angle * zw. Magnetisierungsstrom (d.parallel z. Fluss steht) u. dem Gesamtstrom
Lastwinkelregelung load angle control
Lastzustand load condition
Latch latch * Zwischenspeicher für Datenwort
Latenz latency * Verzögerung(szeit), Totzeit
Latenzzeit latency * Totzeit/Reaktionszeit/Wartezeit in Computer
Laterne cage * Käfig; z.B. zur Montage eines Impulsgebers am Motor || distance piece * Distanzstück
Latte lath || batten || strip board || slat * schmale ~
Laubholz hardwood
Laubsäge fretsaw
Lauf run * auch Durchlauf, Durchgang || running * smooth/swift: ruhiger/schneller Lauf; auch Maschine/Motor || operation * Betrieb || movement * Bewegung || motion * Bewegung || travel * Bewegung, Hub z.B.eines Kolbens || flow * Fluss || course * Kurs von Bahn/Schiff usw.; in course of time: im Laufe der Zeit || path * Weg || track * Spur || action * Gang, Funktionieren, Vorgang, Bewegung, Gangart, Handeln, Einwirkung * period * over the period of: im Laufe einer Zeitspanne || course * in the course of time: im Lauf der Zeit; in

Laufbahn the course of: *im Laufe einer Zeitspanne* || heat * *Durchlauf, Durchgang* || pass * *Durchlauf* || journey * *Weg (auch eines Produktes durch eine Maschine), Reise, Fahrt, Route, Gang*
Laufbahn race * *auch Kugel-/Rollenlaufbahn in Wälzlager*
Laufeigenschaften runability
laufen run * *smooth/swift: ruhig/schnell (auch Zeit); auch Softwareprogramm (on: auf (einem Computer))* || go * *Maschine* || work * *Maschine* || travel * *sich (hin- und her) bewegen, z.B.Kolben, Maschinenteil* || move * *sich bewegen* || pass * *sich hindurchbewegen, hindurch/hinüber-laufen/führen, hindurchgleiten lassen* || flow * *fließen* || circulate * *zirkulieren* || extend * *sich erstrecken (from...to: von...bis)* || elapse * *ablaufen von Zeit* || leak * *(aus)laufen (Gefäß)* || be in progress * *im Gange sein, gerade ablaufen, Fortschritte machen* || be in process * *im Gange sein, in Bearbeitung sein* || function * *Maschine*
laufend running * *im Laufen befindlich* || steady * *stetig, ständig, regelmäßig* || continuous * *ununterbrochen, (fort)laufend, unaufhörlich, kontinuierlich, progressiv* || current * *Ausgaben, Jahr, Produktion usw.; aktuell* || regular * *Geschäfte usw.; regelmäßig (z.B. Wartung); geregelt, geordnet, vorschriftsmäßig* || day-to-day * *tagtäglich, z.B. Geschäfte* || routine * *routinemäßig, z.B. Geschäft* || consecutive * *(fort)laufend, z.B. Nummerierung* || in circulation * *im Umlauf befindlich* || instant * *~en Monats, dringend* || inst * *Abkürzg. für 'instant': ~en Monats (the 10th inst: der 10. dieses Monats)* || up to date * *be up to date: auf dem ~en sein* || currently * *(Adv.) gegenwärtig, jetzig, aktuell, verbreitet, jetzt, zurzeit* || regularly * *(Adv.) regelmäßig, vorschriftsmäßig, geordnet, pünktlich, normal* || increasingly * *(Adv.) steigend* || continuously * *(Adv.) ununterbrochen, fortlaufend, kontinuierlich* || constantly * *(Adv.) unaufhörlich, stetig, regelmäßig, gleich bleibend, beharrlich, standhaft* || instantly * *(Adv.) sofortig, augenblicklich*
Läufer rotor * *{el.Masch.} drehender Teil einer elektr. Maschine* || armature * *Anker (einer Gleichstrommaschine)*
Läuferbleche rotor laminations
Läuferblechpaket rotor core || rotor lamination
Läuferblechung rotor lamination
Läuferdrehzahl rotor speed
Läuferdurchflutung rotor flux
Läuferfrequenz rotor frequency
läufergespeist rotor-fed
Läuferhaltevorrichtung rotor locking device * *Transportsicherung für Motor* || rotor shipping brace * *Transportsicherung für Motor* || shaft block * *Transportsicherung für Motor*
Läuferkäfig rotor-cage * *eines Käfigläufermotors*
Läuferkennzahl characteristic rotor resistance * *{el. Masch.} Rechengröße k [in Ohm] z. Bemessung d. Anlaßwiderstände*
Läuferklasse torque class * *kennzeichnet Momentenkennlinie von SIEMENS-Asynchronmotoren (besonders beim Anlauf)*
Läuferkörper spider || rotor body
läuferkritisch critical-rotor * *with thermally critical rotor* || thermally critical * *{el.Masch.} Maschine, bei der die Temperaturgrenze zuerst i. Läufer erreicht wird*
läuferkritischer Motor critical-rotor motor * *Läufer erreicht Grenztemperatur früher als Ständer* || motor with thermally critical rotor * *Läufer erreicht Grenztemperatur eher als Ständer*
Läufernabe rotor hub * *bei Elektromotor*
Läuferpaket rotor core
Läuferspannung rotor voltage
Läuferstab cage-bar * *bei Käfigläufer*

Läuferstillstandsspannung secondary voltage * *bei Schleifringläufermotor* || wound-rotor open-circuit voltage * *bei Schleifringläufermotor* || locked-rotor voltage || rotor standstill voltage * *{el. Masch.} an d. Schleifringen b.off. Läuferkreis i. Stillstand gemessen* || rotor standstill voltage * *z.B. b. Schleifringläufer, der im Stillstand d. höchste Läuferspanng. hat*
Läuferstrom rotor current
Läuferverluste rotor losses
Läuferwicklung rotor winding
Läuferwiderstand rotor resistance
Läuferzusatzverluste additional rotor losses
lauffähig ready-to-go * *z.B. Maschine* || ready to run * *z.B. Softwareprogramm* || runnable * *z.B. Programmcode auf einem Rechner* || executable * *z.B. Programmcode auf einem Rechner*
Lauffläche bearing surface * *in Wälz/Gleitlager* || contact face * *einer (Kohle)Bürste* || commutator end * *auf Kommutator* || tread * *eines Rades, Reifens usw.* || contact surface || sliding surface * *Gleitfläche* || tread * *eines Rades/(Gummi)Reifens; Spurweite*
Laufkatze trolley * *z.B. bei Kran (→Katzfahrwerk)* || crab * *Winde, Hebezeug, Laufkatze* || travelling crab || trolley hoist * *mit Hebezeug* || trolley block * *mit eingebautem Flaschenzug*
Laufprüfung running test * *z.B. für Motor nach DIN 51806*
Laufrad carrying wheel || impeller * *Gebläse/Lüfter/Pumpenrad, Flügelrad* || runner * *Turbine* || running wheel * *z.B. bei Lüfter* || fanwheel * *eines Lüfters*
Laufraddurchmesser carrying wheel diameter
Laufrichtung running direction || machine direction * *einer Maschine*
Laufring race * *in Wälzlager* || raceway * *in Kugellager*
Laufrolle cam roller
Laufruhe quiet running * *ruhiger Lauf* || smooth running * *Rundlauf, weicher/geschmeidiger Lauf* || quietness || smooth operation || smoothness of running || running smoothness * *geräusch- und schwingungsarme Bewegung mit gleichmäßiger Winkelgeschwindigkeit* || vibration-free operation
laufruhig smooth-running
Laufstreifen tread * *Lauffläche/Profil eines Reifens* || tire tread * *eines Reifens*
Laufwerk drive * *{Computer} z.B. Disketten-/Festplatten-/Band~*
Laufwerkschacht drive bay * *in Computer z.B. für Festplatten-/Disketten-/CD-ROM-Laufwerk*
Laufzeit run time * *e. Transportbandes; Zeit, während der ein Rechnerprogramm abläuft; Länge eines Films usw.* || operating time * *einer Maschine* || runtime * *Zeit, während der ein Computer ein Programm abarbeitet (during: zur)* || duration * *Dauer* || running time * *einer Maschine, eines Maschinenteils*
laufzeitoptimal run-time optimized * *(Adj.)* || with optimized execution time * *(Adv.)* || with optimum run time * *(Adv.)*
Lauge lye || caustic solution || liquor || steep || lixivium * *(chem.)* || lixiviant * *Laugungsmittel* || electrolyte solution * *Elektrolysebad*
Laugenkonzentration alkaline concentration
Laugenpumpe lye pump * *bei Wasch/Geschirrspülmaschine*
laut according to * *entsprechend, gemäß* || in accordance with * *entsprechend, gemäß* || in conformity with * *entsprechend, gemäß* || on the strength of * *kraft* || in pursuance of * *gemäß, zufolge, entsprechend* || by virtue of * *kraft* || under * *kraft* || as per * *kraft, gemäß (Gesetz, Vorschrift)* || loud * *(auch Adv.) laut vom Geräusch her* || loud-voiced * *Per-*

son || noisy * lärmend, geräuschvoll || boisterous * lärmend || audible * vernehmlich, hörbar || clear * klar, bestimmt || distinct * bestimmt, klar || sonorous * stark klingend || loudly * (Adv.) || aloud * (Adv.) laut, mit lauter Stimme || in a loud voice * (Adv.)mit lauter Stimme || openly * (Adv.) offen
lauten sound || run * Inhalt, Worte || read * Inhalt, Worte, Text; vor Doppelpunkt || say
Lautstärke sound intensity || loudness || loudness level * Geräuschpegel
Lawineneffekt avalanche phenomenon * z.B. bei Halbleitern (Durchbruch) || avalanche effect * bei Halbleitern
Leasing leasing
Leasingunternehmen leasing company
Lebensdauer lifetime * high: hohe || life * average: durchschnittliche; long: hohe || endurance || service life * Gebrauchsdauer, z.B. von Schmierfett, Kohlebürsten, Kontakten, Verschleißteilen || life expectancy * Lebenserwartung || durability * Langlebigkeit, Dauerhaftigkeit, Haltbarkeit || operating lifetime || operating life || lifespan || working life || life span || life expectancy * Lebenserwartung (für {el.Masch.}: siehe VDE 0530 Teil 1 und Beiblatt 1)
Lebensdauerberechnung life calculation || endurance calculation
lebensdauergeschmiert lubricated for life || lubricated for the service lifetime
Lebensdauerprüfung life test || endurance test
Lebensdauerschmierung lifetime lubrication
Lebenserwartung life expectancy * auch von techn. Geräten
lebensfähig viable * (auch figürl.)
Lebensgefahr danger to life || danger of life || danger of death
lebensgefährlich perilous || hazardous * hochgefährlich
lebenslange Garantie lifetime warranty
lebenslanges Lernen lifelong learning
Lebenslauf personal autobiography || curriculum vitae
Lebensmittel foodstuffs * (Plural) || food || foods * (Plural)
Lebensmittelindustrie foodstuff industry || food industry
lebenswichtig essential
Lebenszähler heartbeat counter
Lebenszyklus life cycle
Leck leakage || leak * Leck, Loch, undichte Stelle, Auslaufen, Durchsickern (auch figürl.); lecken, leck sein,streuen || leaking * (Adjektiv) || leaky * (Adjektiv)
Leck- leakage
Leckage leakage
Leckgefahr hazard of leaking
Leckstrom leakage current
Lecksuchgerät leak detector
LED-Anzeige LED display
LED-Anzeigen LED indicators
lediglich merely || exclusively || solely || purely
leer empty || unoccupied * unbelegt, unbesetzt || vacant * unbesetzt || evacuated * geräumt || blank * unausgefüllt, unbeschrieben, leer (Blatt, Raum, Frontplatte, Diskette usw.) || clean * glatt, unbeschrieben, glatt || void * bedeutungslos
Leer- empty * z.B. Flasche, Palette usw. || blank * Formular, Diskette, Speichermodul usw.
Leeranlaufschalthäufigkeit no-load starting frequency * {el.Masch.} →Leerschalthäufigkeit
Leeranlaufzeit no-load starting time * benötigt e.Motor ohne Last z. Hochlaufen auf Betriebsdrehzahl

Leeranstellung no-load screw-down * {Walzwerk} →Walzenanstellung während des "Leerweges"
Leerbefehl no operation || NOP * Abkürzung für 'No OPeration'
leeren empty || drain || void * ent~ || pour out * aus~ || clear * (weg-)räumen || evacuate * ent~, aus~ || deplete * ent~, ausräumen, z.B. Ladungsträger (of: von) || discharge * ent~, entladen, ausströmen lassen, herausbefördern
Leergehäuse empty housing || empty case
Leerlauf no-load operation || no-load condition * under: im/bei || idle run || idling || neutral gear * bei Schaltgetriebe || coast * im ~ fahren || no-load status || open-circuit * Schaltung ohne Last || open-circuit operation * Betrieb ohne Last || idle running || neutral * bei Schaltgetriebe || no-load
Leerlauf- no-load || open-circuit || at no load * (nachgetellt)
Leerlaufdrehmoment residual torque * Rest-Drehmoment einer Kupplung/Bremse in geöffnetem Zustand || idling torque * z.B. einer Kupplung oder Bremse || drag torque * z.B. einer Kupplung oder Bremse
Leerlaufdrehzahl no-load speed
leerlaufen run idle || be idling || coast * im Leerlauf fahren, trudeln, ungeführt/ohne treibendes Moment fahren || drain dry * Gefäß || windmill * mitgezogen werden (z.B. Lüfterantrieb) || turbine * mitgezogen werden (Pumpenantrieb)
leerlaufend running idle || coasting || idling || under no-load condition || in no-load operation || windmilling * (salopp) trudelnd, 'mitgezogen werdend' (z.B. von der Speisequelle getrennter Motor)
leerlauffest open-circuit proof * z.B. für Stromsignal 0-20mA || no-load proof || stable at no load * Antrieb
Leerlaufgleichspannung no-load DC voltage * ideal no-load DC voltage: →ideelle Gleichspannung || open-circuit DC voltage
Leerlaufkennlinie no-load characteristic * bei {el. Masch.}: Verlauf v. Leerlaufstrom/-verlusten i. Abhängigk. v.d.Spanng || open-circuit characteristic
Leerlaufmessung no-load measurement
Leerlaufmoment idling torque * z.B. einer Kupplung, Bremse || drag torque * z.B. einer Kupplung, Bremse || residual torque * Rest-Drehmoment einer Kupplung/Bremse in geöffnetem Zustand
Leerlaufspannung no-load voltage
Leerlaufstrom no-load current
Leerlaufverluste no-load losses * bei {el. Masch.}: Leistung, die ein Motor ohne Belastung a.d.Welle aufnimmt
Leerlaufverlustleistung no-load losses
Leerlaufversuch no-load test * z.B. zur Aufnahme der Kurve 'Leerlaufstrom als Funktion der Drehzahl' bei DC-Motor
Leermodul blank module
Leerschalthäufigkeit no-load starting frequency * {el.Masch.} zuläss. Anz. aufeinander folgender Anlaufvorgänge ohne Last
Leertaste space key || space bar
Leerumschalthäufigkeit no-load reversing frequency * {el. Masch.} zuläss. Anzahl aufeinanderfolg. Reversiervorgänge ohne Last
Leerzeichen blank || space
Leerzcile blank line
Legende legend * z.B. mit Bilderklarungen || key * z.B. auf Landkarte usw.
legiert alloyed || blended * Schmieröl || with additives * Schmieröl
Legierung alloy * light alloy: Leichtmetall~
Lehre gauge * Mess~, auch (Mess-) Normal/Eichmaß || template * Schablone || pattern * Muster, Vorlage, Modell || jig * Bohr~ || apprenticeship

Lehrgang * *eines Lehrlings* ‖ science * *Wissenschaft* ‖ theory * *Theorie* ‖ instruction * *Unterweisung* ‖ hint * *Wink* ‖ lesson * *Lektion,Warnung; let it be a lesson to you: lasse es Dir eine ~ sein* ‖ warning * *Warnung; let it be a warning to you: lasse Dir dies zur Warnung dienen*
Lehrgang course ‖ training course * *Schulungskurs* ‖ course of instruction * *Schulungskurs* ‖ practice * *Übung* ‖ refresher course * *Auffrischungskurs* ‖ workshop * *mit praktischen Übungen* ‖ seminar * *(Universitäts-) Seminar* ‖ course of study ‖ training seminar
Lehrgangsleiter instructor ‖ chief instructor ‖ head of a course
Lehrgangsteilnehmer student ‖ participant in a course
Lehrkräfte training personnel * *für berufliche Aus/Weiterbildung*
Lehrling apprentice ‖ novice * *Neuling* ‖ beginner * *Neuling* ‖ greenhorn * *Anfänger* ‖ newcomer * *Neuling*
Lehrmittel training aids
Lehrwerkstatt training workshop ‖ training shop ‖ apprentices' training shop
leicht simple * *einfach* ‖ light * *~ an Gewicht, ~ zu erledigen; (auch figürl.: Essen, Kleidg., Musik, Wein, Hand usw.)* ‖ light in weight * *~ anGewicht* ‖ in an easy way * *auf einfache Art* ‖ easy * *nicht schwierig, einfach* ‖ slight * *gering, z.B. Überdruck* ‖ light-weight * *~ an Gewicht* ‖ light-duty * *für geringe Beanspruchung, m.geringer Leistung, Schwachlast-; z.B. Schwachstrom-Relais* ‖ effortless * *mühelos* ‖ gentle * *sanft* ‖ trifling * *unbedeutend* ‖ petty * *unbedeutend (Fehler, Vergehen usw.)* ‖ minor * *unbedeutend* ‖ light-minded * *leichtfertig* ‖ with a light heart * *~en Herzens* ‖ light weight * *~ an Gewicht*
leicht bedienbar easy-to-use ‖ easy-to-operate
leicht erlernbar easy-to-learn ‖ easily assimilated
leicht gestrichenes Papier light weight coated paper ‖ LWC paper * *Light Weight Coated*
leicht verständlich easy-to-understand ‖ clear * *klar, einfach* ‖ easily-understood
Leichtbau light construction ‖ light-weight design ‖ lightweight construction * *{Bautechnik}*
Leichtbauplatte building board * *{Bautechnik}* ‖ lightweight building board * *{Bautechnik} für den Innenausbau*
leichte Erlernbarkeit ease of familiarization * *z.B. Bedienung einer Maschine/eines Geräts/eines Softwareprogramms*
leichte Handhabbarkeit operator convenience * *durch Bedienungsmann*
leichter Druck light pressure
leichtgängig smoothly moving ‖ easily moving ‖ easily going ‖ low-friction * *mit wenig Reibung* ‖ low-inertia * *mit kleiner Schwungmasse/Massenträgheit* ‖ smoothly running
Leichtgängigkeit ease of motion ‖ antifriction properties ‖ easy movement ‖ smooth action
leichtgewichtiges gestrichenes Papier lightweight coated paper ‖ LWC paper * *Abk.f. 'Light-Weight Coated'*
Leichtmetall light metal ‖ light alloy * *~legierung*
leiden suffer * *(from: an/unter)* ‖ suffer severely from: *schwer/stark ~ an/unter* ‖ be seriously affected * *(from: unter)* ‖ be adversely affected * *(ungünstig) beeinflusst sein (from: durch)*
leider unfortunately ‖ I'm afraid *'(in Ich-Satz, vorangestellt)
leihen lend * *an jdn. etwas aus~* ‖ lend out * *ver~* ‖ advance * *(amerikan.) Geld usw. vorstrecken* ‖ rent * *mieten, (Personal) anheuern; hire a thing from a person: sich etw. von jdm. aus~* ‖ rent * *(ver)mieten, (amerikan.) aus~, (amerikan.) sich etwas ~,* vermietet werden *(at/for: zu)* ‖ borrow * *borrow a thing from a person: sich etwas von jdm. aus~* ‖ loan * *(Geld usw.) ver~*
Leihgebühren rental fee ‖ lending fee ‖ rental costs * *Mietkosten*
Leihwagen rental car
Leim glue ‖ adhesive * *Klebstoff*
Leimauftrag glue application
Leimauftraggerät glue spreader * *{Holz}*
Leimauftragswalze glue application roller * *{Holz}*
Leimbinderpresse beam press * *{Holz}* ‖ gluelam beam press * *{Holz}*
leimen size * *Papier, Stoff usw. steifen* ‖ glue * *together: zusammen-* ‖ cement
Leimholz glued timber ‖ glued board
Leimpresse size press * *zur Papierbearbeitung* ‖ glue press * *{Holz}*
Leimung sizing * *Steifung von Papier*
Leinen linen * *['linin]*
leise quiet * *still* ‖ smooth * *weich, zügig, glatt, reibg.slos, geräuscharm, geschmeidig, z.B. Umrichter m.hoher Taktfrequ.* ‖ silent * *ruhig, geräuschlos* ‖ noiseless * *geräuschlos/arm* ‖ antinoise * *schallgedämpft* ‖ low * *kaum hörbar* ‖ soft * *(seiden)weich, sanft, gedämpft, sacht* ‖ faint * *matt, schwach, kraftlos (Stimme, Ton)* ‖ gentle * *sacht* ‖ slight * *kaum merkbar*
Leiste strap * *Lasche* ‖ strip * *Streifen, Holzleiste* ‖ rail * *Schiene; guide rail: Führungsschiene* ‖ ledge * *Sims, Leiste, vorstehender Rand, Felsbank, Lager/Ader in Bergwerk* ‖ bar * *am Bildschirm, z.B. toolbar: Werkzeug~, menu bar: Menü~* ‖ slat * *~ aus Holz, Metall usw., Rippe, Lamelle, Jalousie-Element*
leisten afford * *sich; auch: sich ~ können/die Mittel haben für* ‖ perform * *verrichten, Leistung bringen, Leistungsfähig sein; perform well: gute Leistung erbringen* ‖ be rated at * *(Nenn-) Leistung haben (in Zusammenhang mit Zahlenangabe)* ‖ carry out * *ausführen* ‖ execute * *ausführen* ‖ fulfil * *erfüllen* ‖ achieve * *vollbringen* ‖ accomplish * *vollbringen* ‖ supply * *liefern* ‖ provide * *liefern* ‖ render * *Dienst* ‖ make * *Zahlung usw. (payment)* ‖ effect * *Zahlung usw (payment)* ‖ be very efficient * *gute Arbeit ~*
Leistensäge rip saw * *{Holz} Längsteilsäge*
Leistung power ‖ performance * *~fähigkeit, ~svermögen; perform well: gute ~ erbringen* ‖ rating * *z.B. Nenn~ eines Stromrichters, Stellung im Produkt/~sspektrum* ‖ kW * *Kilowatt* ‖ horsepower * *Pferdestärken, amerikanisch; 1 kW entspr. 1,341 hp* ‖ HP * *Pferdestärken, amerikanisch* ‖ output * *Ausgangs~ z.B.eines Stromrichters oder Motors (Ausgangs~, Wellen~)* ‖ rate * *z.B. Motor~* ‖ output rating * *z.B. Ausgangsnenn~ eines Stromrichters* ‖ service * *Dienst~* ‖ power rating * *Nenn/Typenwert* ‖ output power * *abgegebene ~ (eines Motors, Stromrichters usw.)* ‖ input power * *aufgenommene ~* ‖ capacity * *Kapazität/Leistungsvermögen bezgl. Volumen, Durchsatz, Überlastbarkeit;* →*Kälte~ (DIN8928)* ‖ achievement * *Werk, Leistung, Errungenschaft, Großtat, Ausführung, Vollendung, das Vollbrachte*
Leistung reduzieren de-rate * *z.B. Ausgangsleistung i.techn Spezifikation bei gewissen Betriebsbedingungen* ‖ derate
Leistungs- power
Leistungsabgang power outlet ‖ power output
Leistungsabstufungen power stages
Leistungsanschluss power connection
leistungsarm low-power
Leistungsaufnahme power consumption ‖ power input ‖ power drain
Leistungsausbeute power yield

Leistungsbedarf power requirement || power demand * {el.Masch.} tatsächl.erforderl.
Antriebsleistung, muss < Motorbemess.leistg. sein
Leistungsbereich power range * elektr.leistung; high: oberer; lower: unterer || performance range * Leist.fähigkeit (upper/high-end: oberer), (medium/mid-range: mittlerer) || output range * z.b. einer Motoren- oder Stromrichterbaureihe || rating range * Bereich der Nennleistung(en), z.B. einer Stromrichter-Baureihe
Leistungsbilanz power budget * elektr.
Leistungsdaten performance data || ratings
Leistungsdichte power density
Leistungsdiode power diode
Leistungseinspeisung incoming powerline
Leistungselektronik power electronics
leistungsfähig powerful || efficient || high-performance || high-capacity * mit hohem Leistungsvermögen
leistungsfähig sein perform
leistungsfähiger sein als outperform
Leistungsfähigkeit performance || resources * zur Verfügung stehende Leistung || efficiency * Effizienz || potential * auch Wirkungsvermögen, innere Kraft || load capacity * Belastbarkeit || effectiveness * Effektivität, Wirksamkeit
Leistungsfaktor power factor * cos phi, Verhältnis Wirkleistung/Scheinleistung; unity power factor: ~ von eins || cos phi
Leistungsfluss power flow * grafisches Schema zur Darstellung der Wirkleistung und der Verluste
Leistungsgewicht power-weight ratio || power-to-weight ratio
leistungsgleich having the same rating * z.b. Stromrichter || having the same output * z.B. Motor || of the same rating
Leistungshalbleiter solid state power component
Leistungsherabsetzung output derating * d. Nennleistg.z.B. b. Motor/Umrichter aufgr.erschwerter (Umgeb.-)Bedingungen || derating
Leistungskabel power cable
Leistungskennzeichen rating code designation * für Thyristor-/Diodensatz nach DIN 41752
Leistungsklasse performance level * Leistungsfähigkeit || output class * Stromrichter, Motoren
Leistungskondensator power capacitor
Leistungsmerkmal performance feature
Leistungsmerkmale performance characteristics
Leistungsmesser wattmeter
Leistungsmessgerät power meter
Leistungsmessung power measurement || active-power measurement * Wirk~
Leistungsminderung derating * z.B. e. Geräts bei Überschreiten e. bestimmten Temperatur oder Aufstellungshöhe || de-rating * Reduzierg. d. Nennleistung auf Grund erschwerter Bedingungen
Leistungsmittelwert mean value of the power
Leistungspegel power level
Leistungsprüfstand efficiency-testing machine * zum Ermitteln der Leistungsfähigkeit/des Wirkungsgrades || power-measuring test-bench || dynamometer test stand
Leistungspufferung power back-up
Leistungsreduzierung derating * Herabsetzung z.B. auf Grund der Aufstellungshöhe, der höheren Temperatur usw.
Leistungsregelung closed-loop power control || power control || capacity modulation * für Verdichter und →Verdichtersätze (z.B. in Kälteanlage)
Leistungsreserve power reserves || power reserve || output margin * bei Motor/Umrichter
Leistungsreserven power reserves || output reserves * z.B. Stromrichter, Motor
Leistungsschalter circuit-breaker * (siehe z.B. DIN VDE 0660) || power switch || circuit breaker * siehe auch →Hauptschalter
Leistungsschild rating plate * für Motor siehe IEC 34-1 und VDE 0530 || name plate * Typenschild
Leistungsschub performance boost * Schub bezüglich der Leistungsfähigkeit || boost in performance
Leistungsschütz power contactor
Leistungsspektrum power spectrum
leistungsstark powerful || high-performance
Leistungssteigerung increase in efficiency || performance boost * (salopp) Steigerung der Leistungsfähigkeit || output rise * Steigerung der Ausgangsleistung, z.B. bei Motor, Umrichter usw. || power boost || output increase * z.B. eines Stromrichters || enhancement * Steigerung der Leistungsfähigkeit/des Funktionsumfangs || performance rise * Steigerung der Leistungsfähigkeit
Leistungssteller power controller unit || power amplifier * Leistungsverstärker || power actuator * Stellglied
Leistungsstellglied power output element || power controller * Steller || power actuator || power controlling element * z.B. Stromrichter
Leistungsteil power section || power unit || power set
Leistungsträger top performer * z.b. Mitarbeiter, Arbeitskraft (~ bezüglich d. Leistungsfähigkeit/Arbeitsleistung) || value performer * Produkt mit gutem Preis-/Leistungsverhältnis
Leistungstransformator power transformer * siehe IEC 76
Leistungstransistor power transistor
Leistungsumfang performance scope
leistungsunabhängig irrespective of the power rating * unabhängig von der Nennleistung
Leistungsverbindungen power connections
Leistungsverdrahtung power cabling
Leistungsvergleich benchmarking * z.B. von Computern, Firmen usw. || comparison of performance * Vergleich der Leistungsfähigkeit
Leistungsverhältnis power ratio
Leistungsverstärker power amplifier
Leistungverbindung power connection
Leit- master || supervisory * aufsichtsführend, Leitsystem || superimposed * übergeordnet, überlagert || coordinating
Leit-Folge-Antrieb master-slave drive
Leit-Folge-Betrieb master-slave operation || master-follower operation * bei Mehrmotorenantrieb
Leit-Folge-Steuerung master-slave control
Leit-Folge-Umschaltung master-slave changeover
Leit-Folgeantrieb master-slave drive
Leit-Folgebetrieb master-slave operation || master-follower operation
Leitachse master axis
Leitantrieb master drive || main drive * Hauptantrieb * speed master * gibt die Geschwindigkeit einer Verarbeitungs-/Produktionsstraße vor || pilot drive
Leitblech air guide * für (Kühl)Luft; siehe auch 'Luftleitblech' || guide plate * z.B. Tischleitblech bei Blechverarbeitung || baffle plate * Ablenk-/Prallplatte, Schlingerwand (in Tank usw.) || baffle
Leitdauer current flow time
Leitebene production control level * Produktions-Leitebene || plant control level || coordinating control level || factory control level
leiten lead * führen (auch im Team, Personal, eine Sitzung) || guide * führen, ge~ || conduct * führen (auch Strom), durchführen || steer * steuern, lenken, dirigieren || pilot * steuern, (durch)lotsen (auch figürl.), führen, lenken, leiten || convey * (be-) fördern, übermitteln, versenden, übertragen, senden || pass * (hindurch)führen/leiten, weiterreich./leiten, (Nachricht be)befördern, durchlass., dirigieren || route * Verkehr/Signalverkehr ~ (over:

leitend

über) || channel * *auf dem Dienstweg* || direct * *dirigieren, lenken, führen, adressieren(to: an)* || head * *anführen* || manage * *z.b. Betrieb, Projekt* ~ || control * *beaufsichtigen* || preside * *vorsitzen, over a meeting: eine Versammlung* ~ || be in the chair * *eine Versammlung* ~ || pipe * *durch ein Rohr* ~ || run * *betreiben, (Betrieb) führen/leiten, (Maschine/Anlage) laufen lassen* || chair * *Vorsitz führen, z.B. bei einer Sitzung*
leitend conductive * *(auch Strom-~)* || senior * *Angestellter, Ingenieur usw.* || chief * *z.b. Ingenieur* || managing || leading
leitender Angestellter senior executive
Leiter leader * *Führer* || conductor * *auch Stromleiter, (An-) Führer* || head * *einer Abteilung* || chief * *einer Abteilung* || president * *Direktor* || manager * *Verwalter, Geschäftsführer, Organisator* || works manager * *Werksleiter* || superintendent * *auch Vorsteher, Direktor* || technical director * *technischer Leiter* || principal * *Chef, Direktor, Vorsteher* || boss * *(salopp) Chef* || ladder * *zum Hochsteigen* || pair of steps * *Stehleiter* || steps * *Stehleiter* || lead * *Stromverbindung, Leitung, Leiter*
Leiter des Finanz- und Rechnungswesens chief financial officer
Leiterbahn printed circuit track || PC track || printed conductor || track || conductor track || conductor path || trace
Leiterende end of conductor
Leiterimpedanz conductor impedance
Leiterisolierung conductor insulation
Leiterkarte PC board || PCB * *Abkürzg. für 'Printed Circuit Board'* || printed circuit board || card || board
Leiterplatte board || pc-board || printed circuit board || PCB || card * *bestückt* || circuit board || printed board
Leiterplattenformat PC board format
Leiterplattentransformator board-mounted transformer
Leiterquerschnitt conductor cross-section || size of conductor || conductor size || cable cross-section
Leiterspannung line-to-line voltage * *Spannung zwischen 2 Hauptleitern, z.B. L1, L2, L3* || phase-to-phase voltage || supply voltage || line voltage || phase voltage * *z.B. eines Transformators*
Leitfaden guide || manual * *Handbuch* || primer * *Fibel, (Anfänger-) Lehrbuch, Leitfaden* || guideline * *Richtlinie*
leitfähig conductive * *(elektr.)*
leitfähige Unterlage conductive surface * *zur Hantierung von Elektronikbauteilen*
Leitfähigkeit conductivity
Leitfähigkeitsmessgerät conductivity meter
Leitfrequenz master frequency
Leitgruppe master group * *Antriebsgruppe bei Mehrmotorenverbund, die die Maschinengeschwindigkeit bestimmt* || master section * *Antriebsgruppe, die für die Anlagengeschwindigkt. maßgebl. ist (z.B. b. Papiermasch.)*
Leitlack conductive paint * *elektrisch leitender Lack*
Leitlinie guideline
Leitrad guide vane * *bei Ventilator*
Leitrechner host computer || host || supervisory computer || master computer
Leitregler master controller || superimposed controller * *überlagerter Regler* || host controller
Leitsollwert master reference || master reference value || master setpoint || line speed reference * *Linien-/Bahngeschwindigkeitssollwert für durchlaufende Warenbahn*
Leitspannung master reference voltage || line speed setpoint * *Bahngeschwindigkeits-Sollwert* || line speed reference * *Bahngeschwindigkeitssollwert* ||

line speed * *Bahngeschwindigkeit* || master reference * →*Leitsollwert, Bahngeschwindigkeitssollwert (bei Papiermaschine usw.)* || line speed reference voltage
Leitstand control station || control room || control console * *z.B. für Druckmaschine*
Leitstelle central coordination || coordination team || control point || coordinating department
Leitsteuerung coordinating control || master control || master controller || master PLC * *Leit-SPS, Kopfsteuerung*
Leitsystem host system || host || host computer * *Leitrechner* || higher-level system || supervisory system || overall control system
Leittechnik instrumentation and control
Leitung cable || wire || lead || line * *Übertragungs~, Telefon~, Druckluft~* || pipe * *Rohrleitung* || leading * *Führung, Leitung, Management* || management * *Management* || guidance * *Führung, Management* || leadership * *Leitung, Führung, Führerschaft* || tube * *Rohr(leitung), Schlauch* || direction || duct * *für (Kühl)Luft*
Leitungsabschirmung cable screen || cable shield
Leitungsabschluss cable termination * *z.B. durch Abschlusswiderstand (z.B. mit Wellenwiderst. 120 Ohm)* || line terminator * ~*glied, ~widerstand*
Leitungsanschluss cable connection || conduit connection * *z.B. in Motorklemmenkasten*
Leitungsbedämpfung cable damping * *z.B. für Leitung Umrichter-Motor zur Vermeidung von Resonanzen*
Leitungsbruch cable break || cable breakage * *auch 'Bruch(stelle), Bruchschaden'* || interrupted cable || wire break
Leitungsdurchmesser cable diameter || lead diameter
Leitungseinführung cable entry * *z.B. in Klemmenkasten*
Leitungsempfänger line receiver
Leitungsführung cable routing
leitungsgebunden cable-guided * *z.B. Störungen* || cable-fed * *leitungsgeführt, z.B. eingekoppelte Störungen* || cable-based * *z.B. Störungen* || cable-born * *z.B. Störabstrahlung von d. Leitung nach außen*
leitungsgebundene Störabstrahlung cable-born electromagnetic radiation
leitungsgebundene Störungen cable-fed disturbances || cable-fed noise || conducted noise || cable-born interference * *EMV-Test: siehe IEC 801-4*
leitungsgeführt cable-fed * *z.B.Störungen* || cable-guided * *leitungsgebunden, z.B.Störungen*
leitungsgeführte Störungen cable-fed disturbances
Leitungskanal cable duct || wiring duct || cable conduit || conduit
Leitungskapazität cable capacitance * *zwischen den Adern oder zwischen Ader und Schirm* || capacitance per unit length * →*Kapazitätsbelag*
Leitungskreis managing team || steering committee
Leitungslänge cable length || wiring distance || length of cable
Leitungsquerschnitt cable cross-section || wire gauge * *AWG: American Wire Gauge: amerikanische Maßeinheit für Leitungsquerschnitt* || lead cross section || conductor cross-section || AWG * *Abkürzg- für 'American Wire Gauge': amerikan. Maßeinheit für Leit.querschnitt* || wire size
Leitungsreste cable remains || remainders of cable materials
Leitungsrohr conduit pipe || cable conduit pipe * *für Kabel/Leitungen* || pipe * *water pipe: ~ für Wasser; gas pipe: ~ für Gas*
Leitungsschirm cable screen || cable shield
Leitungsschirmung cable shield

Leitungsschutz cable protection
Leitungsstecker cable connector
Leitungstreiber line driver
Leitungstreiber-Empfänger transceiver || line driver/receiver
Leitungsunterbrechung interrupted cable || cable breakage
Leitungsverlegung routing of cables * *legen, führen; Art der Leitungsverlegung* || cable installation || cable laying || cable routing
Leitungsverschraubung cable gland * *z.B. PG-Verschraubung bei Leitungs/Kabeleinführung* || cable entry screw gland * *bei einer Kabeleinführung*
Leitungsverstärker line amplifier || repeater * *inmitten der Leitung zum Erreichen einer größeren Reichweite*
Leitungswiderstand cable resistance || line resistance || resistance per unit length * *Widerstandsbelag, Widerstand pro Längeneinheit*
Leitverhalten conductive characteristic * *~ eines Stromleiters/der Leitfähigkeit*
Leitwalze guide roll || guide roller
Leitwarte control master station || control room * *Raum*
Leitwert conductance * *Ohmscher ~ (1/R)* || permeance * *magnetischer ~* || recommended value * *empfohlener Wert, Richtwert* || conductance value || master value * *Leitsollwert (z.B. Maschinen-/Liniengeschwindigkeits-Sollwert)*
Leitzentrale main control room
LEM-Wandler LEM transformer * *potenzialfreier Strom-Messwandler der Schweizer Fa. LEM (R)*
Lenkung steering * *eines Fahrzeugs*
Leonard-Schwungradumformer Ward-Leonard-Ilgner converter
Leonardsatz Ward-Leonard set || Ward-Leonard converter
Leonardumformer Ward-Leonard converter || Ward-Leonard set
lernen learn || study * *studieren* || practise * *üben* || get familiar * *vertraut werden (with: mit)* || familiarize oneself * *sich vertraut machen (with: mit)* || acquaint oneself * *sich vertraut machen mit (with: mit)* || pick up * *aufschnappen* || acquire * *erlernen, meistern* || master * *meistern* || serve one's apprenticeship * *seine Lehre/gewerbliche Ausbildung machen (with: bei)* || be apprenticed * *seine Lehre/gewerbliche Ausbildung machen (to: bei)*
Lernender trainee * *bei beruflicher Aus/Weiterbildung; siehe auch 'Lehrling'* || student * *Student, Kursteilnehmer* || course participant * *Kursteilnehmer (bei Aus/Weiterbildungskurs)*
Lernphase training phase || learning process
Lernziel training objectives || course objectives * *Kursziel* || objective || learning objective
Lesart system in reading
lesbar legible * *leserlich, ablesbar* || readable * *lesenswert, leserlich, lesbar*
Lese-Schreib-Speicher RAM memory
lesen read
Lesezugriff read access * *z.B. auf Speicher*
Lesezyklus read cycle
letzgenannt last-named
letzt last * *last-minute: im ~en Augenblick; next to last/last but one: vorl~; third from last: dritt~* || final * *endgültig, schließlich, definitiv, End-, Schluss-* || ultimate * *äußerst, aller~, endgültig, schließlich, Höchst-, Grenz-* || extreme * *äußerst* || lowest * *unterst, niedrigst* || bottom * *unterst* || latest * *neuest* || latter * *das ~ere* || in the end * *am Ende* || last but not least * *zu guter ~* || utmost * *do one's utmost: sein ~es hergeben/bestes tun; to the utmost: bis zum ~en*

letztendlich ultimately || in the last analysis * *letzten Endes, bei abschließender Betrachtung*
letzter last * *at the very last: als Letztes* || final * *endgültig* || ultimate * *äußerst, allerletzt, schließlich, endgültig*
letzterer latter
Letzteres latter || the last || the last one
Leuchtbalken bar graph * *→Balkenanzeige*
Leuchtdiode LED * *Abkürzg. für 'light emitting diode'* || light emitting diode || light-emitting diode
Leuchtdrucktaster illuminated pushbutton
Leuchte lamp || light || luminaire * *~nkörper, Lampe*
leuchten light * *be/er~, Licht machen* || shine * *scheinen, leuchten, strahlen* || flash * *auf~, blinken, aufblitzen* || switch on * *Licht einschalten* || lighten * *er~, erhellen* || be switched on * *eingeschaltet sein* || be lit * *erleuchtet sein, z.B. Lampe, LED, Meldeleuchte* || twinkle * *blinken* || be illuminated * *z.B. Lampe, LED* || illuminate * *sich erhellen*
leuchtend lit * *zum Leuchten gebracht, z.B. Meldeleuchte, LED*
Leuchtmelder indicator lamp || indicator light
Leuchtstofflampe fluorescent lamp
Leuchttaster illuminated pushbutton || lighted pushbutton
Leuchttisch luminous table
Lichtbogen arc * *heiße Gassäule (Plasma) aus elekr. geladenen Teilchen, die einen hohen Strom führt*
Lichtbogenofen electric arc furnace
Lichtbogenschweißen arc welding
lichtbogensicherer Klemmenkasten arcing-proof terminal box * *(el.Masch.)*
Lichtbogenspannung arc voltage * *z.B. bei Leistungsschalter, Sicherung usw.*
lichte Weite clear dimension
lichtempfindlich light-sensitive || light-sensing * *Sensor usw.*
Lichtleiter optical fibre || glass-fibre optocable * *Glasfaser-Lichtleiter* || plastic optocable * *Plastik-Lichtleiter*
Lichtleiteranschluss optical fibre interface
Lichtschranke light barrier
Lichtsensor light sensor
Lichtsignal light signal || visual signal
Lichtstrom luminous flux
Lichttechnik lighting technology || lighting engineering || illumination
Lichtvorhang optical sensor * *lichtempfindlicher Sensor* || light curtain
Lichtwellenleiter fiber-optic cable * *(amerikan.)* || fibre optic * *(brit.) vorangestellt in Wortkombination* || fiber optic cable * *Lichtwellenleiterkabel* || optical waveguide || opto-cable || FO cable * *Abk. f. 'Fiber-Optic' cable* || optical fibre * *(brit.)* || fiberoptic link * *~-Datenverbindung*
Lichtwellenleiterkabel fibre optic cable
Lieferadresse shipping address || address of delivery
Lieferant supplier || contractor * *auf Vertragsbasis* || distributor * *Verteiler, Zwischenhändler* || vendor * *Verkäufer* || subcontractor * *Unterlieferant* || provider
Lieferart term of shipment * *z.B. Unterscheidung Luft/Seefracht*
lieferbar available * *erhältlich, verfügbar; available on short lead times: kurzfristig ~* || deliverable || suppliable || in supply * *be in plentiful/short supply: in hohen Stückzahlen/schlecht ~* || available for delivery
Lieferbarkeit availability
Lieferbedingungen conditions of delivery * *siehe auch →allgemeine ~* || conditions for sale and deli-

Lieferdatum 198

very * *Verkaufs- und ~* || terms of delivery || conditions of sale * *Verkaufsbedingungen*
Lieferdatum date of delivery
Liefereinsatz availability * *become available in June 1999: ~ beginnt im Juni 1999* || start of shipping || readiness for delivery || commencement of delivery
Liefereinstellung discontinuation of delivery
Lieferengpass difficulties in delivery || delivery bottleneck || delivery problem || delivery shortage
Lieferer supplier * *Lieferant*
Lieferfirma supplier || vendor * *Verkäufer*
Lieferfreigabe release for delivery || release for general availability
Lieferfrist delivery time || delivery deadline * *maximale ~ aus Kundensicht* || delivery schedule * *planmäßige ~, ~ im Standardfall* || term of delivery
Liefergalette feeder godet * *{Textil} Fadenrolle zum Anfördern der Fäden*
Liefergeschwindigkeit delivery speed
Lieferlogistik supply logistics || delivery logistic
liefern deliver || ship * *ausliefern; ship back: zurück~* || supply * *versorgen mit; auch Ausgangssignal/Strom/Spannung/Leistung/Energie ~* || furnish * *with: jdn. mit etwas be~/ausstatten/ausrüsten; Beweise/Informationen ~/beschaffen* || provide * *be~/versehen (with: mit); z. Verfügg./bereitstellen,auch Strom/Spanng./Leistg./Energie* || supply with * *supply someone with: jdm. etwas (zu)~*
Lieferort place of delivery || available from * *kann geliefert werden von* || delivery location || supplied from * *lieferbar von* || supply location
Lieferpapiere delivery paperwork || shipping documents
Lieferproblem delivery problem || difficulties in delivery || delivery shortage || delivery bottleneck * *Lieferengpass*
Lieferprogramm delivery programme || product range * *range of products* * *lieferbare Produkte*
Lieferqualität delivery quality
Lieferquelle source of supply
Lieferschein delivery note || dispatch note || delivery slip || delivery receipt * *[... ri'ßiet] Lieferpapiere* || bill of lading * *Frachtbrief* || shipping note
Lieferschwierigkeiten delivery bottleneck
Liefertermin shipping date || delivery date || date of delivery || day of delivery || term of delivery
Liefertermine delivery schedules
Lieferumfang scope of supply * *be in the scope of supply: im ~ enthalten sein* || scope of delivery || extent of supply
Lieferung shipment * *Versand; auch: gelieferte Waren* || delivery * *An~; payable/cash on delivery: zahlbar bei ~* || supply || consignment * *Sendung* || lot * *Partie* || forwarding * *Beförderung*
Lieferverzug delay in delivery
Lieferwerk supplying factory || contractor * *Lieferant* || feeder unit * *Zuführeinrichtung* || delivery works || supplying works || suppliers
Lieferzeit delivery time || waiting period || shipping cycle
Lieferzentrum dispatch center
liegen lie * *the voltage lies between 8V and 25V: die Spannung liegt zwischen 8 und 25 V* || be lying || repose * *ruhen, verweilen, schlafen, beruhen (on: auf)* || rest * *ruhen* || be placed * *gelegen sein* || be situated * *gelegen sein* || be stationed * *stationiert sein* || be located * *gelegen sein* || reside * *liegen/ruhen bei, (inne)wohnen, untergebracht sein (in: in), ansässig sein*
liegen über exceed * *überschreiten*
liegend horizontal

LIFO LIFO * *Last In First Out, z.B.Stapelspeicher: d. zuletzt eingespeicherte wird zuerst wiederausgegeben*
Lift lift * *Aufzug* || elevator * *Aufzug*
LIM LIM * *Abk. f. 'Linear Induction Motor': linearer Induktionsmotor*
Lineal ruler || linear encoder * *→Linearmaßstab zur Positionserfassung bei e. translatorischer Bewegung*
linear linear || linearly * *(Adv.) degressive: abnehmend; progressive/increasing: ansteigend*
linear abnehmend linearly degressive
linear ansteigend linearly progressive || linearly increasing
linear interpoliert linearly interpolated
Linearantrieb linear actuator
Linearbewegung linear motion
lineare Bewegung linear motion
lineare Interpolation linear interpolation
lineare Programmierung linear programming * *ohne Sprünge*
lineares Programm linear program * *ohne Sprünge/Verzweigungen*
linearisieren linearize
Linearisierung linearization || linearizing || correction of nonlinearity
Linearisierungsstufe linearization block
Linearität linearity
Linearitätsfehler linearity error || nonlinearity
Linearmaßstab linear encoder * *(meist optischer) Encoder zur Lageerfassung bei Linearachse (z.B. Glaslineal)*
Linearmotor linear motor * *z.B. mit linear angeordnetem "Käfigläufer"*
Linearvorschub linear feed
Lineritätsfehler linearity error || nonlinearity
Linguistik linguistic
Linie line * *dashed: gestrichelte; dotted: punktierte; thin: dünne; heavy: dicke* || trend * *Tendenz* || course * *(politischer) Kurs*
Liniendruck nip pressure * *Anpressdruck an einer Klemmstelle (z.B. bei einer Papaierbahn)*
Liniengeschwindigkeit line speed
Linienintegral line integral
Linienraster grid pattern
Linienschreiber chart recorder || continuous line recorder
Liniensollwert line reference * *Geschwindigkeitssollwert für Produktions-/Bearbeitungslinie* || master line reference || line-speed reference || master reference * *Leitsollwert*
Linienstrom current loop * *Stromschleife, z.B. bei 20 mA/TTY-Schnittstelle* || line current
linke Seite lefthand side
Linker linker * *zum Zusammenbinden von Softwaremodulen ("Befriedigung" der gegenseitigen Adressreferenzen)*
linkes lefthand * *(Adj.)*
links counter clockwise * *in ~drehrichtung* || ccw * *in ~drehrichtung* || on the left || left hand side * *~seitig* || to the left * *nach ~; to the left of: ~ von* || left * *turn left: nach ~ drehen/abbiegen; left of: ~ von* || leftmost * *das sich am weitesten ~ befindende* || lefthanded * *~seitig* || left turn * *(Adv.) nach ~, ~herum* || lefthand * *~liegend, ~angeordnet* || at the left * *auf der linken Seite* || on the left side
links außen at the extreme left
links herum counterclockwise * *entgegen dem Uhrzeigersinn* || ccw * *Abkürzg. für 'conterclockwise': entgegen d. Uhrzeigersinn*
links oben at the top left || at the upper left || upper-left * *(Adjektiv)*
links unten at the bottom left
Linksanschlag left stop * *z.B. eines Potis*

linksbündig left-justified * z.B. in der Daten-/Textverarveitung
Linksdrehfeld counterclockwise phase sequence * entgegen d. Uhrzeigersinn || ccw rotating field * CounterClockWise: entgegen d.uhrzeigersinn
Linksdrehrichtung reverse direction of rotation || counter-clockwise direction of rotation
linksgängig left-hand * z.b. Schraube, Gewinde
linksgängige Schraube left-handed screw
Linksgewinde left-hand thread || left-handed thread || reverse thread * reverse threaded: mit ~ versehen sein
Linkslauf reverse running || counter-clockwise running || anti-clockwise rotation || counter-clockwise rotation || counter-clockwise rotating
linksschieben shift left
linksseitig lefthand
linse lens
Linsenkopfschraube fillister-head srew || dome-tipped srew || raised-head screw
Lippe lip
Lippendichtung lip-type seal
Liquidität liquidity
Liste list || listing * Computer-Ausdruck (z.B. Quellspracheprogramm) || register * amtliche || catalog * Katalog || catalogue * Katalog || schedule * Verzeichnis, Aufstellung, Tabelle, (Arbeits/Stunden)Plan || specification * detaillierte Aufstellung || roll * Gehalts~, Lohn~, Steuer~ || table * Tabelle, tabellarische Aufstellung
Listenform list form || listed form || statement-list notation * [SPS|Computer] Anweisungsliste als Quellsprachenotation
listenmäßig standard || catalogued || scheduled * in Liste/Katalog aufgeführt/eingetragen; festgelegt, geplant || catalog-listed || standard catalog || catalog
listenorientiert list-oriented
Listenpreis list price
Listenstruktur list structure * interne Darstellg.e.-Anwenderprogramms: Adreslisten d.abzuarb. Programm-u.Datenmodule
Listing listing * Ausdruck von einem Drucker (z.B. eines Quellspracheprogramms)
Liter litre * (brit.) || litres * (Plural) || liter * (amerikan.)
Literatur literature
Literaturhinweise bibliography * Literaturnachweis/-verzeichnis || bibliographical data * Literaturangaben
Lithiumbatterie lithium battery
lithiumverseift lithium-soaped * grease: Schmierfett
Lithiumzelle lithium cell * Batterie-Knopfzelle
Litzendraht stranded wire
Litzenleiter stranded conductor || stranded wire * Litzendraht
lizensieren license
Lizenz licence * under: in; grant licence for: ~ erteilen für || license * (amerikan.) under: in
Lizenzbau manufacture under licence || licensed construction || licensed production
Lizenzbedingungen license conditions
Lizenzfertigung licensed production
Lizenzgeber licenser || licensor
Lizenzgebühr licence fee || royalty || royalty fees * auch 'Patentgebühr'
Lizenznehmer licensee
lizenzpflichtig liable to payment of royalties
Lizenzvereinbarung licence agreement || license agreement * (amerikan.) || licence contract * Lizenzvertrag
Lizenzvertrag licence contract || licence agreement * Lizenzvereinbarung
LKW truck * trucking: LKW-Transport; light/heavy-duty truck: leichter/schwerer LKW || lorry * (brit.) || truck-trailer unit * Lastzug, LKW mit Anhänger || goods vehicle
Loch hole
Lochabstand hole clearance
lochen punch * auch Papier(bögen) || perforate || pierce * pierce holes into: Löcher in etwas ~; durchbohren/-dringen/-stechen
Locher puncher * auch im Büro || perforator * auch im Büro
Lochkreisdurchmesser pitch circle diameter * Kreis, auf dem Bohrungen für die Befestigung von Masachinenteilen liegen || PCD * Abk. für 'Pitch Circle Diameter' || diameter of the bolt pitch circle || bolthole circle diameter
Lochmaß bore size
Lochstanze hole punch
Lochung hole || punched hole * gestanztes Loch || perforation || boring || punching
Lochwalze peforated roll || rectifier roll * in Papiermaschine
Lochweite mesh size * eines Siebes
lockerer Sitz loose fit
lockern loosen * z.B. Schraube || slacken || ease * Passung, Sitz || work loose * sich lockern || become slack * sich lockern
logarithmisch logarithmic * graded logarithmically. ~ gestuft; logarithmic chararteristic; ~er Verlauf || logarithmic-scale * in ~em Maßstab
Logik logic || logic circuit * Logikschaltung || logic processing * Logikverarbeitung || interlock * Verriegelung
Logikanalysator logic analyzer
Logikfamilie logic family * Schaltkreisfamilie, z.B. TTL, CMOS
Logikgatter logic gate
Logikschaltung logic circuit
Logikverarbeitung logic processing
Logikzelle logic cell
logisch logical || logic * auch boolsch || logically * (Adv) ~erweise
logisch verknüpft logically combined
logische Gleichung logic equation
logische Verknüpfung logic operation
logischer Kanal logic channel
logischer Pegel logic level
logischer Zustand logic condition || logical state
Logistik logistic || logistics
logistisch logistical
Logo logo * Sinnbild
Lohn wages || salary * Gehalt || payment * Bezahlung || compensation * Entgelt || reward * Entgelt, (gerechter) Lohn, Belohnung, Vergütung
Löhne wages
lohnend profitable * profitabel, Gewinn bringend || paying * einträglich, rentabel || advantageous * vorteilhaft || worthwhile * der Mühe wert || lucrative * Gewinn bringend || rewarding * Gewinn bringend, auch figürlich || winning * gewinnend (auch figürlich), siegreich, einnehmend
Lohnstückkosten manpower cost per piece
lokalisieren localize || isolate * isolieren, absondern, abschneiden
Los batch * Fertigungs~, Partie || lot * Fertigungs~; lot code: ~nummer
lösbar detachable * abnehmbar || removable * abnehmbar || soluble * Problem, Aufgabe, Gleichung; Löslichkeit von Substanzen in Flüssigkeit || resolvable * Gleichung || solvable * Gleichung || dissoluble * [Chemie]
Losbrechen breakaway
Losbrechmoment break-away torque * Lastmoment im Stillstand, das größer ist als nach Beginn der Bewegung || initial starting torque || breakaway torque || static friction torque

Losbrechreibung break-away friction || starting friction || oil seal friction * augrund der anfänglichen "Klebewirkung" des Schmieröls || breakaway friction || static friction
Lösch-I2t-Wert interrupting I2t value * für Sicherung
Löschdauer turn-off time * eines Thyristors
Löschdiode arc-suppression diode * z.Beschaltg. v. Gleichstromschütz/Relaisspule (geg. Überspannng. b. Abschalt.)
Löscheinrichtung UV eraser * zum Löschen von EPROM-Modulen
löschen reset * den Inhalt eines Speichers ~, ein Signal ~ || erase * z.B. Daten in EPROM-Speicherbaustein ~ || cancel || go out * er~, z.B. Lampe || be turned off * gelöscht werden, z.B. Thyristor || turn off * z.B. Thyristor || turn-off * (Substantiv) das Löschen || extinguish * aus~, z.B. Lichtbogen, Feuer || extinction * (Subst.) Löschung, z.B. Lichtbogen, Feuer, Thyristor || arcing * (Subst.) Löschung, z.B. Lichtbogen, z.B. in Sicherung/Schütz/Leistungsschalter || unload * entladen || discharge * entladen, z.B. Schiffsladung || land * Waren || clear * z.B. Flipflop, Anzeigebit in Rechner ~ || switch off * z.B. Thyristor ~ || undo * rückgängig machen || quench * Funken, Asche, Koks usw. ~ || land * Schiffsladung ~
Löschgerät eraser * z.B.für EPROMs
Löschimpuls reset pulse * Richt/Rücksetzimpuls
Löschkondensator commutation capacitor * unterstützt den Kommutierungs-/Löschvorgang in e. Thyristor-Stromrichter
Löschkreis commutating circuit * ermöglicht die →Kommutierung in einer Stromrichterschaltung
Löschlampe eraser * für EPROM-Speicherbausteine/-module || UV eraser * UV-Löschlampe für EPROM-Speicher
Löschmittel fuse filler * in Sicherung, z.B. Sand || arc quenching medium * z.B. in Sicherung || arc extinguishing medium * z.B. in Sicherung
Löschtaste delete key || clear button || delete button || reset button
Löschthyristor commutation thyristor * Thyristor in einem Hilfs/Löschzweig, der den Kommutiervorgang unterstützt
Löschung extinction * z.B. ~ eines Thyristor, Lichtbogen, Feuer || turn-off * z.B. ~ eines Thyristors || cancellation || deletion || discharging * ~ einer Schiffsladung || quenching * ~ von Funken, Halbleiterventilen usw.
Löschwinkel turn-off angle || extinction angle * {Stromrichter} Winkel zw. Kommutier.ende u. Nulldurchg. d.Komm.spanng. IEC 146-1-1
Löschzweig turn-off arm
Lose dead travel * toter Gang,Spiel in einem Maschinenteil || backlash * ~ im Gewinde/Getriebe || slack * Durchhängen einer Warenbahn || loose * (Adj.) nicht fest; supplied loose: lose mitgeliefert || flying * (Adv) beweglich, fliegend (gelagert/angeordnet); flying leads: lose Anschlussleitungen
lösen loosen * z.B. Schraube, Befestigung, Knoten; loosening: das Lösen (einer Schraube usw.) || detach * abnehmen, abtrennen, abbauen, herausnehmen || release * Bremse, Kupplung, Griff, Befestigungsschraube usw. ~ || disengage * Kupplung ~ || disconnect * elektr. Verbindung || dissolve * Vertrag, sich ~ in einem Lösungsmittel || free * frei oneself from: sich befreien von || solve * Problem, Aufgabe, (Differenzial-) Gleichung ; overcome a problem: ein Problem ~ || terminate * Vertrag ~ || sever * abtrennen, Beziehungen ~ || unscrew * Schraube, Mutter ~ (abschrauben auch 'undo') || settle * Problem ~ || open * öffnen || resolve * Konflikt ~ || answer * Poblem ~, Frage beantworten || dispose * Schwierigkeiten, Konflikt ~ || break off

* Beziehung ~ || get loose * lose werden, sich ~ || come open * aufgehen, sich ~
Losgröße batch size || batch quantity || batch
Loslager floating bearing * i. Gegensatz z. Festlager || non-locating bearing
Losnummer lot number * eines Fertigungsloses || lot No
Lösung solution * Problem~, Realisierung, chemische ~ || loosening * Losmachen, z.B. Schraube, Knoten || detachment * Abtrennung, Abbau || answer * ~ eines Problems/einer Aufgabe, Gegenmittel, Gegenmaßnahme || break-up * (Auf)~ einer Beziehung || separation * ~ einer Beziehung
Lösungsmittel solvent
lösungsmittelarm low-solvent
lösungsmittelfrei solvent-free
loswerden get rid of
Lot solder
Lötbad soldering bath || solder bath
Lötbrücke solder jumper || soldering jumper || solder bridge
Lötdraht solder wire
löten solder * weich~ || soft-solder * weich- || brazc * hart~ || hard-solder * hart~ || dip-braze * tauch~
Lötfahne solder tag || solder lug || soldering tag
Lötfett soldering paste
lötfrei solderless * z.B. Verbindungstechnik
Lötkolben soldering iron || soldering gun * mit Pistolengriff
Lötmasse soldering compound || soldering paste
Lötöse solder lug || soldering lug
Lötpaste soldering paste || solder paste
Lötpistole soldering gun
Lötseite solder side * einer Leiterkarte
Lötspitze copper bit || red nose * (salopp)
Lötstelle solder joint * dry/faulty solder joint: kalte Lötstelle || soldered joint
Lötstift solder pin || solder post || soldering pin
Lötstoplack solder resist
Lötstopplack solder resist
Lötstützpunkt solder pin || solder-post * Pfosten || soldering post * Pfosten || solder tag || soldering terminal || solder lug * →Lötfahne
Lötverbindung soldered joint || soldered connection || solder connection || solder joint || brazed joint * hartgelötete ~
Lötverfahren soldering method || soldering technique
Lötwärmebeständigkeit resistance to soldering heat
Lötzinn tin solder || solder || lead-tin solder || plumber's solder || tinman's solder
Low-aktiv Low active
LOW-Signal anlegen de-energize * an Klemmen, Eingang
LRoS LRoS * Abkürzung für 'Lloyds Register of Shipping': Schiffs-Klassifizierungsgesellschaft
LRS LRS * Abkürzg. für 'Lloyd's Register of Shipping': britische Schiffsklassifikationsgesellschaft
LSB LSB * Abkürzg. für 'Least Significant Bit': niederwertigstes Bit
LSD LSD * Abkürzg. für 'Least Significant Digit': niederwertigstes Zeichen z.B. einer Ziffenanzeige
Lückadaption discontinuous current adaption
Lückbereich intermittent DC area || discontinuous-current area || discontinuous-current range || non-continuous area
Lückbetrieb discontinuous operation * {Stromrichter} bei DC-Stromrichter || discontinuous current state || pulsating-current operation || intermittent DC operation || pulsating DC operation || discontinuous conduction mode || discontinuous mode
Lückdauer non-continuous current period
Lücke gap * bridge the gap: Lücke schließen ||

blank * *leere Stelle* || interval * *Zwischenraum* || break * *Unterbrechung* || omission * *Auslassung* || void * *(figürl.) freier Raum, Leere, Lücke, Gefühl der Leere; fill the void: die ~ schließen*
lücken be pulsating * *z.B. Strom im netzgeführten Stromrichter* || be pulsating * *z.b. Strom im netzgeführten Stromrichter (→Lückstrom)*
Lückenbüßer stopgap
lückender Strom discontinuous current || intermittent current || pulsating current
lückenhaft incomplete || fragmentary || sketchy * *z.B. Kenntnisse* || patchy * *zusammengestoppelt, Flickwerk; (auch Kenntnisse, Wissen usw.)* || not accurately * *(Adv.)*
lückenlos continuous * *ununterbrochen, fortlaufend, zusammenhängend, fortwährend, (aus)dauernd* || without a gap || uninterrupted || complete * *vollständig* || unbroken * *unbeeinträchtigt, unvermindert, ungebrochen, gesamt, heil*
Lückgrenze crossover point from discontinuous to continuous current * *Aussteuerg. b. d. d. Strom nicht mehr lückt* || limit between pulsating and non-pulsating operation || starting point of non-continuous current
Lückstrom discontinuous current || intermittent current || pulsating current
Lückstromadaption discontinuous current adaption
Lückstromanpassung discontinuous current gain adaption * *P-Verstärkungs-Anhebung für Stromregler i. Lückbetrieb*
Luenberg-Beobachter Luenberg observer * *(Regel.techn.)*
Luft air
Luft- und Kriechstrecke air and creepage distances * *kürzester Abstand in der Luft u. auf Isolierfläche zw. leitenden Teilen*
Luft- und Raumfahrt aerospace
Luft-Innenkreislauf inner air cooling * *beim Motor*
Luft-Luft-Kühler air-to-air cooler || air to air heat exchanger * *mit Wärmetauscher* || air-air cooler
Luft-Luft-Wärmetauscher air-air heat exchanger
Luft-Selbstkühlung convection cooling
Luft-Wasser-Kühler air-to-water heat exchanger * *mit Wärmetauscher* || air-to-water cooler || air-water cooler
Luft-Wasser-Wärmetauscher air-to-water heat exchanger
Luftabschluss exclusion of air * *z.B. durch Schließen der Luftklappe in Heizungssystem*
Luftabstand air clearance || clearance in air
Luftauslass exhaust || air outlet || exhaust vent || vent
Luftauslassöffnung air outlet || air exhaust opening
Luftaustritt air outlet * *Auslassöffnung/-stutzen*
Luftaustrittsöffnung air outlet opening
Luftbedarf air consumption * *eines Druckluftsystems*
Luftbefeuchter air humidifier
Luftblase air bubble
luftdicht air-tight || hermetic || hermetically sealed * *luftdicht verschlossen*
Luftdrossel air-core reactor * *ohne Eisenkern* || air reactor * *kernlose Drossel (ohne Eisen- oder Ferritkern)*
Luftdruck air pressure || barometric pressure
Luftdruckmesser air pressure gauge
Luftdurchlässigkeit air permeance
Luftdurchsatz air flow rate
Lufteinführung air inlet || air inlet adapter * *Anschlussstück für den →Lufteinlass z.B. in e. Pneumatiksystem*
Lufteinlass air inlet || air inlet adapter * *Anschlußstück für den ~*
Lufteinlassöffnung air intake opening

Lufteintritt air inlet * *Einlassöffnung/-stutzen* || air intake
Lufteintritts-Jalousie air-intake louver * *als Kiemenblech ausgeführte Lufteintrittsabdeckung (z.B. für Motorbelüftung)* || air-intake louvre || air intake shutter
Lufteintrittsöffnung air intake opening
Lufteintrittstemperatur cooling air temperature * *für Kühlluft*
lüften release * *Bremse öffnen* || ventilate || air || aerate * *be~* || bleed * *Luft/Dampf ausströmen lassen, z.B. Hydraulik, Bremsleitung* || lift * *heben* || raise * *(an)heben* || disengage * *Bremse ~; auch 'gelüftet werden'*
Lüfter fan || blower * *Gebläse* || cooling fan * *zur Kühlung* || fan unit * *bei Motor: axially/radially/separately mounted: an-/aufgebauter/getrennt montierter ~*
Lüfter-Anschlussklemmen fan connecting terminals
Lüfter-Einschub slide-in fan unit
Lüfteraggregat fan unit
Lüfterantrieb fan drive
Lüfterbaugruppe fan module || fan assembly || fan subassembly
Lüftereinheit fan assembly
Lüftereinschub slide-in fan unit || withdrawable fan || fan slide-in unit
Lüfterflügel fan propeller || vane || fan blade
lüftergekühlt fan-cooled
Lüftergeräusch noise of the fan || fan noise || windage noise * *Luftrauschen*
Lüfterhaube fan cowl || fan shroud || fan cover
Lüfterkennlinie fan characteristic || square-law characteristic * *quadratische Kennlinie M~n²*
Lüftermotor fan motor || blower motor
Lüfterrad fan impeller || fan wheel || fan rotor
Lüfterüberwachung fan monitor * *als Einrichtung* || fan monitoring * *als Funktion*
Lüfterwirkung fan effect
Luftfahnenrelais air-vane relay * *zur Lüfter- oder Luftstromüberwachung*
Luftfahrt aviation
Luftfeuchte humidity * *relative: relative; absolute: absolute*
Luftfeuchtigkeit air humidity || humidity * *relative: relative* || atmospheric humidity || R.H. * *Abk. f. 'Relative Humidity': relative Luftfeuchtigkeit*
Luftfilter air filter
Luftfiltermatte air filter mat
Luftfracht air freight
Luftfrachtbrief airwaybill
Luftführung ventilation circuit * *Kreislauf* || air guide * *mechan. Element(e), z.B. Leitblech(e)* || air guiding * *Funktion* || air duct * *Luftkanal/-weg*
luftgekühlt air-cooled || ventilated
Luftgeschwindigkeit air velocity
luftisoliert air-insulated
Luftkanal air duct || air passage || ventilating duct
Luftkisseneffekt airfoil effect * *Bildung einer Gleitschicht aus Luft (z.B. bei Festplatte)*
Luftklappe air flap || air damper * *z.B. in Heizungssystem* || air supply damper * *z.B. in Heizungssystem*
Luftkühlung air-cooling || ventilation
Luftleitblech air baffle plate || air baffle || air guide || cooling baffle
Luftleitrad air guide wheel
Luftmenge air flow rate * *Luftdurchsatz (eines Kühl- oder Belüftungssystems)*
Luftöffnung air opening || air inlet * *Lufteinlassöffnung*
Luftreibung windage loss * *Reibung auf Grund des Luftwiderstandes*
Luftreinhaltung air pollution control

Luftrichtung air-flow direction || air direction
Luftschall airborne noise
Luftschlitz air slot
Luftschütz air-break contactor
luftselbstgekühlt natural-air cooled * *bei Leistungshalbleitern. Kühlung über freie Konvektion, kein Lüfter* || convection-cooled * *Kühlung (z.B. v. Leistungshalbleitern) über freie Luftkonvektion (kein Lüfter)* || natural-convection cooled
Luftselbstkühlung natural-air cooling * *von Leistungshalbleitern* || natural-convection cooling * *von Leistungshalbleitern* || natural air circulation cooling * *von Leistungshalbleitern* || convection cooling * *von Leistungshalbleitern*
Luftspalt gap * *z.B. bei elektr. Maschine (zwischen Läufer und Ständer)*, Luftspaltfeld air-gap field * *{el.Masch.} magn. Feld im Luftspalt e. drehenden el. Maschine*
Luftspaltfluss air gap flux
Luftspaltleistung air-gap power * *{el.Masch.} vom Ständer über d. Luftspalt auf Läufer übertragene Leistung*
Luftspaltmoment air-gap torque
Luftspaltüberwachung air-gap monitoring * *{el. Masch.} durch Messspulen i.d. Ständernuten, erkennt ungleichmäß. Luftspalt*
Luftspaltwicklung air gap winding
Luftspule air-core reactor || air-core inductor
Luftstabilisiert air-stabilized * *z.B. Motorläufer mit Permanentmagnet*
Luftstrecke air distance * *für Isolation* || air clearance * *für Isolation* || clearance || clearance distance * *für Isolation (bei Stromrichtern: siehe IEC 664)*
Luftstrom air flow || air stream || air convection * *Konvektion auf Grund von Temperaturunterschieden* || air flow rate * *Luftmenge pro Zeiteinheit* || airflow
Luftstromgeschwindigkeit air flow velocity
Luftstromüberwachung air flow monitoring * *als Funktion, z.B. zur Überwachung eines Lüfters* || air-flow monitor * *als Einrichtung/Gerät, z.B. zur Überwachung eines Kühlsystems*
Luftströmungswächter air-flow monitor
Luftstromwächter air pressure switch * *z.B. zur Lüfterüberwachung* || airflow monitor || air-flow monitor || air flow monitor
Lufttechnik air handling || air-handling systems
Lüftung release * *Bremse* || ventilation || aeration
Lüftungsanlage ventilation system * *in der Klimatechnik*
Lüftungsdach venting roof
Lüftungskanal ventilation duct
Lüftungskanäle ventilation ductwork
Lüftungsschlitz venting slot || ventilation slot || air slot
Luftverdichter air compressor
Luftverluste air leakage * *in Druckluftsystem*
Luftverschmutzung air pollution
Lüftweg release travel * *bei einer Bremse*
Luftwiderstand windage || air drag || drag of air || air resistance || drag || aerodynamic resistance
Luftwiderstandsverluste windage losses || air drag losses
Lüftzeit release time * *einer Bremse* || operation time * *einer Bremse mit Ruhestromprinzip*
Luftzirkulation air circulation
Luftzuführung air inlet * *Lufteinlass* || air supply * *Luftzufuhr, Luftversorgung* || air supply duct * *Zuführungskanal/weg*
Luminizenzdiode light-emitting diode || LED
Lumpen rags
Lunker shrinkhole
Lupe magnifier
Lupeneffekt zoom effect

Lupenfunktion detail function * *z.B. in Blockschaltbild* || zooming
Lüsterklemme connector strip || porcelain insulator || flying terminal block * *für "fliegende Verdrahtung"* || insulating terminal
LWC-Papier LWC paper * *Abkürzg. für 'Light-Weight-Coated' paper: leicht gestrichenes Papier*
LWL fibre optic link * →*Lichtwellenleiter* || shaftless * *längswellenlos*

M

m m * *Meter; 1 m entspr. 3,2808 ft (feet) entspr. 1,0936 yd (yards) entspr. 39.3701 in (inch)* || yd * *yard; 1 m entspr. 1,0936 yd* || ft * *feet; 1 m entspr. 3,2808 ft*
M-Leiter reference conductor || reference bus * *Schiene*
M-Potenzial zero reference voltage M || ground || GND * *Abkürzg. f. 'GrouND'*
M-Schiene reference bus
mäanderförmig meander-shaped || in concertina arrangement * *wie eine Ziehharmonika*
Machinensatz machine set * *z.B. Motor und Generator bei rotierendem Umformer*
Madenschraube grubscrew * *siehe auch* →*Feststellschraube* || grub screw
Magazin magazine || depot || store * *Speicher*
magazinieren warhouse * *in einem Lager einlagern*
Magnesium magnesium
Magnet magnet || solenoid * *Betätigungsspule bei Ventil, Relais usw.; Hubmagnet* || coil * *Betätigungsspule*
Magnet-Futter magnetic chuck * *z.B. zur Werkzeug-Aufnahme bei Werkzeugmaschine*
Magnetanker magnet armature || armature || keeper
Magnetantrieb solenoid-operator * *z.B.~ für Relais* || solenoid actuator * *elektromagnet. Stellantrieb* || solenoid operated drive * *z.B. ~ für Relais*
Magnetband magnetic tape
Magnetbandlaufwerk magnetic-tape drive || streamer * *zur Datensicherung/Datenarchivierung*
Magnetblech magnetic steel sheet || magnetic sheet steel
Magnetfeld magnetic field
magnetisch magnetic
magnetische Abstoßung magnetic repulsion
magnetische Anziehung magnetic attraction
magnetische Durchflutung ampere turns || magnetomotive force
magnetische Feldstärke magnetic field strength * *Maß für d.Kraft, mit d. e. magnet. Feld gedachte Elemetarmagnete ausrichten*
magnetische Flussdichte magnetic flux density
magnetische Geräusche magnetic noise
magnetische Kupplung magnetic clutch
magnetische Leitfähigkeit permeability
magnetische Sättigung magnetic saturation
magnetische Unsymmetrie magnetic unbalance
magnetischer Fluss magnetic flux
magnetischer Kreis magnetic circuit
magnetischer Leitwert permeance || magnetic permeance
magnetischer Widerstand reluctance
magnetisches Feld magnetic field || electro-magnetic field * *elektromagnet. Feld*
magnetisches Geräusch magnetic noise
Magnetisierung magnetizing || magnetization

Magnetisierungskennlinie magnetization curve * stellt den Zusammenhang zw. magn. Feldstärke und magnet. Flussdichte dar
Magnetisierungsleistung magnetization power
Magnetisierungsstrom magnetizing current || exciting current || magnetization current
Magnetismus magnetism
Magnetkörper coil body * Spulenkörper || magnet body || solenoid assembly * z.B. in einer Magnetbremse/Magnetventil
Magnetkraft magnetic force
Magnetkupplung magnetic clutch || magnetic coupling
Magnetlager magnetic bearing * berührungsloses Lager, z.b. für schnellaufende (Kompressor-) Antriebe || magnetic-suspension bearing * berührungsfrei mit magnetischen Führungskräften arbeitendes Lager
magneto-optisch magneto-optical
Magnetpulver-Bremse magnetic particle brake
Magnetpulver-Kupplung magnetic particle clutch
Magnetpulverkupplung magnetic particle coupling
Magnetschwebebahn magnetic levitation railway
Magnetspannfutter magnetic chuck
Magnetspannplatte magnetic clamping plate * Haltevorrichtung z.B. für Werkzeugmaschine
Magnetspule solenoid coil * einer Betätigungseinrichtung; bei Relais, Magnetventil, elektromagnet. Bremse usw.
Magnetteil solenoid component * z.B. bei einer elektromagnet. Bremse
Magnetton magnetic tone
Magnetventil solenoid valve || solenoid
Magnetzylinder magnetic cylinder
Mahlanlage refiner * zur Zellstofferzeugung || milling plant || pulverizing equipment || pulverizer mill || grinding plant || crushing mill
Mahlarbeit grinding energy
Mahlbarkeit beatability
mahlen grind || mill || pulverize * zu Pulver || crush * zerquetschen (z.B. Steine, Feldfrüchte usw.) || beat * Papier || refine * Zellstoff || spin * z.B. Räder im Schlamm
Mahlgrad degree of beating || grinding fineness || freeness * {Papier} bei Zellstoff || fineness of grinding
Mahnung reminder * Mahnbrief, Erinnerung || admonition * Ermahnung || exhortation * Ermahnung || dunning letter * Mahnschreiben
Mailbox mailbox * {Computer}
Maische mash
Makro macro * Zusammenfassung mehrerer Befehle zu einem Kurzbefehl
Makroprogramm macro program
mal multiplied by * multipliziert mit || times * ein vielfaches (Multiplikation); three times that of: dreimal so viel wie || once * einmal; once upon a time: es war einmal || time * (Subst.) for the first/ last time: zum ersten/letzten Mal; this time: dieses ~; two times: zwei- || by * bei Multiplikation
man one
Manager manager
Managerin manager || manageress
manchmal at times || now and then * hin und wieder
Mandat authorization || power || mandate * politisches
Mangel nuisance * →Missstand || shortage * Knappheit (of: an) || deficiency * of: an; Fehlen (von), Ausfall, ~haftigkeit, Schwäche, Unzulänglichkeit || defect * Defekt, Fehler, Mangel, Unvollkommenheit, Schwäche, Gebrechen || lack * Fehlen (of: von); (auch Verb: '~ haben an; nicht haben') || absence * Fehlen (of: von) || scarcity * Knappheit ||

shortcoming * →Fehler || drawback * →Nachteil || deficiency * Defizit (of: an) || insufficiency * Unzulänglichkeit, ~haftigkeit, Untauglichkeit || blemish * Fehler, Mangel, Makel, Schönheitsfehler; verunstalten, schaden, beflecken (figürl.)
Mangel an lack of
Mangel haben be short * of: an
mangelhaft unsatisfactory * unbefriedigend || poor * ungenügend, unzureichend, (auch Zeugnisnote) || inferior * minderwertig || defective * defekt, mangelhaft, unvollkommen, unvollständig, schadhaft || faulty * fehlerhaft, schadhaft, schlecht || imperfect * unvollkommen || incomplete * unvollständig || inadequate * unzulänglich, ungenügend, mangelhaft, unangemessen || deficient * →unzureichend, mangelhaft, ungenügend
Mängelliste complaint list || fault list || list of defects
Mängelrüge complaint || complaint about quality || deficiency claim || notification of defect || notice of defects
Mangelwalze nip roll * zum Durchfördern band-/ plattenförmigen Materials
Manipulation manipulation
manipulieren manipulate * Person, Information, technisches Gerät || handle * techn. Gerät
mannigfaltig manifold * siehe auch →vielfältig
Mannjahr man-year || man year
Mannmonat man-month
Mannschaft team || workforce * Belegschaft, Mitarbeiterschaft || crew || gang * Gruppe, Trupp, Abteilung, (Arbeits-) Kolonne
Mannstunde man-hour
Mannstunden working hours
Mannwoche man-week
Manometer manometer * Druckmesser
Manöver maneuver * (amerikan.) || manoeuvre * (engl.) || trick * trickreiches ~ || stratagem * (figürlich)
manövrieren maneuver * (amerikan.; auch figürl.) * z.B. Rolle beim Einlegen in Wickler ~ || manoeuvre * (engl.; auch figürl.) || practise tactical evolutions * taktisch ~
Manövrierpotentiometer manoeuvering potentiometer * (brit.) zum feinfühligen Vor-/Rückwärtsbewegen d. Wickelachse || maneuvering potentiometer (amerikan.)
Manövrierpoti maneuvering potentiometer * z.feinfühligen Positionieren d. Rolle bei Auf-/Abwickler aus d.Stillstand
Manschette sleeve || collar * Kragen || packing ring || U-seal * →Dichtmanschette
Mantel sheath * eines Kabels || jacket * Ummantelung, Hülle, Umwicklung
manuell manual || manually * (Adv.) manually operated: mit Handbedienung
manuelle Eingabe manual entry
Mappe folder || portfolio || brief-case * Aktentasche/ Mappe
Marine marine * Handelsmarine || navy * Kriegsmarine || naval forces * Seestreitkräfte
Marine- naval
Marine-Ausführung marine design
Marineschiff naval ship
Marke mark * Kennzeichen || index * (An)Zeiger, Zeichen || product label * eines ~nartikels || brand * eines ~nartikels || designation * Kennzeichnung || trade mark * Warenzeichen, Handels~; registered trade mark: eingetragenes Warenzeichen
Markenname brand name || brand label * Markenzeichen
Marketing marketing * Absatzpolitik,"zu Markte bringen", Erkunden/Wecken v. Kaufinteresse u. planvoller Verkauf
Markiereinrichtung marking device

markieren mark || highlight * *hervorheben, z.B. Passagen in einem Text* || indicate * *anzeigen, angeben, bezeichnen, zeigen, hinweisen auf*
Markierung mark * *auch Zeichen* || marking || designation * *Kennzeichnung* || index * *(An) Zeiger, Zeichen* || reference mark * *Einstell/Eichmarke*
Markt market * *in/on the market: auf dem ~; put on the market: auf den ~ bringen* || market place * *regionaler/nationaler ~*
Markt an sich reißen monopolize the market
Markt beherrschen control the market || hold the market || command the market || monopolize the market
Marktakzeptanz market acceptance
Marktanalyse market analysis || market survey
Marktanforderung market demand || market requirement
Marktanteil market share
Marktaufteilung division of the market
Marktbedürfnisse market requirements
marktbeherrschend controlling the market || monopolizing the market
marktbeherrschende Stellung dominant market position
Marktbeherrschung monopoly * *Monopol* || market control || market domination
Marktchance chance in the market place
Marktchancen market opportunities
Marktdurchdringung market penetration
Markteinführung launch || roll out || introduction into the market || market introduction || launching into the market
Markteintritt market entry
Marktentwicklung market development || market trend || trend of the market || market tendency
Markterfolg market success || marketing success
marktfähig marketable || saleable
Marktfähigkeit marketability
Marktforderung market requirement || market need
Marktforscher market researcher
Marktforschung market research
Marktführer market leader * *lead the market: den Markt anführen* || dominant company || main player || brand leader
marktgerecht marketable * *marktfähig, vermarktbar* || according to the market requirements * *den Marktbedürfnissen entsprechend* || in line with the real market conditions || tailored to the market requirements || tailored to the needs of the market || market-oriented * *am Markt orientiert* || market-adapted
Marktlage condition of market || market situation || market conditions
Marktlücke market gap * *fill: füllen* || gap in the market || opening
marktnah close to the market
Marktnähe proximity to the market || nearness to the market
Marktnische market niche
marktorientiert market-oriented
Marktpolitik market policy
marktpolitisch of market policy || relating to market policy
Marktposition market position
Marktpräsenz market presence
Marktpreis market price || current price || market rate || market price level * *Marktpreis-Niveau* || market pricing level || ASP * *Abkürzg. für 'Average Selling Price'*
marktpreiskonform in conformance with market pricing levels || in-line with market prices
Marktpreisverfall drop in market prices

marktreif market-ready
Marktsättigung market saturation
marktschreierisch puffing || obstrusive * *aufdringlich* || noisy * *schreierisch, grell, aufdringlich, krakeelend* || loud * *schreiend, auffallend, grell*
Marktschwerpunkt market focus * *Brennpunkt*
Marktsegment market segment
Marktstellung market position
Marktstruktur market structure
Marktstudie market study || market analysis
marktüblich usual in the market
Marktuntersuchung market research || market survey || market analysis * *Marktanalyse* || market investigation
Marktvolumen size of the market
Marktvorhersage market forecast
Marktvorsprung competitive edge * *gegenüber dem Mitbewerb; on: gegenüber*
Marktvorteil competitive edge * *Wettbewerbsvorsprung*
Marktwert market value || marketable value
Marktwirtschaft market economy * *siehe auch →freie Marktwirtschaft*
marktwirtschaftlich free-enterprise * *den Gesetzen der freien Marktwirtschaft gehorchend* || market-economy * *die Marktwirtschaft betreffend*
Marktzutritt entry into the market
Maß dimension * *Ausdehnung, Ab~ (z.B. Länge, Breite, Höhe)* || measure * *~einheit, ~stab, Aus~; be a measure of: ein ~ sein für; in a great m.: in großem ~* || proportion * *(Größen)Verhältnis* || rate * *Verhältnis* || extent * *Ausdehnung, Aus~* || size * *Größe* || quantity * *Menge* || volume * *Raummenge, Volumen* || gauge * *Eich~* || standard * *Eich~* || weights and measures * *~e und Gewichte* || moderation * *Mäßigung* || dose * *dosis* || degree * *Grad, Aus~; in a high degree: in hohem ~e* || scale * *~stab; on a large scale: in großem ~e*
Maßabweichung dimensional deviation || deviation in dimension || off-size condition * *als Zustand/Tatbestand* || tolerance * *zulässige ~*
Maßänderung change of dimension
Maßbild dimension drawing || dimensional drawing || dimension diagram || outline drawing * *Umrisszeichnung*
Maßblatt dimension drawing || dimension sheet
Masche mesh * *in einem Netzwerk*
Maschenspannung mesh voltage || interconnected voltage
Maschenstrom mesh current * *in einem Netzwerk*
Maschine machine || engine * *Verbrennungsmaschine*
maschinell mechanical * *machining: ~e Bearbeitung*
Maschinen machines || machinery
Maschinenausrüstung machinery
Maschinenbau mechanical engineering * *general: allgemeiner* || mechanics || machine building || machine industry * *~industrie* || machinery * *Maschinenausrüstung* || mechanical equipment manufacturers * *als Branche* || mechanical engineering industry * *~industrie* || machinery construction || industrial machinery * *als Industriezweig* || mechanical equipment * *Maschinenausrüstung*
Maschinenbauer mechanical engineer * *Maschinenbauingenieur/Techniker* || machine builder * *Herstellerfirma* || mechanical engineering firm * *Maschinenbaufirma* || machine constructor * *Erbauer, Konstrukteur* || machine manufacturer * *Maschinenhersteller* || machinery manufacturer * *Maschinenhersteller (als Unternehmen oder Industriezweig)* || machinery maker || machine supplier * *Maschinenlieferant*
Maschinenbefestigung machine mounting

Maschinenbett machine bed ‖ bedplate * *Fundamentplaate* ‖ bottom plate * *Fundamentplatte* ‖ bed ‖ bench
Maschinenbreite width of machine ‖ machine width
Maschineneinrichter machine setter ‖ machinist ‖ toolsetter ‖ setter
Maschinenelement machine element
Maschinenfluss motor flux * *bei Motor*
Maschinenführer machine operator ‖ machineman ‖ engineer
Maschinenführung motor commutation * *Motortaktung eines Stromrichters (Fremdführung durch d. Motor)* ‖ load commutation * *Lastführung eines Antriebsstromrichters* ‖ machine commutation * *Kommutierungssspannung e. Stromrichters wird von der el. Maschine geliefert*
Maschinenfußschraube holding-down bolt * *z.B. zur Motorbefestigung im Maschinenfundament* ‖ rag bolt * *Steinanker/Bartbolzen z.B. zur Motorbefestigung im Maschinenfundament*
maschinengeführt machine-commutated * *Stromrichter* ‖ motor commutated * *Stromrichter*
Maschinengeschwindigkeit machine velocity ‖ machine speed ‖ line speed * *Bahngeschwindigkeit*
Maschinengestell machine rack
maschinengetaktet voltage-clocked * *Umrichtertaktfrequ. stellt sich abhäng.v. d.Maschinenspanng. ein (Synchronmasch.)* ‖ machine-commutated * *Maschine gibt die Kommutierungszeitpunkte vor (→lastgeführter Stromrichter)* ‖ motor-commutated * *Motor gibt die Kommutierungszeitpunkte vor (→lastgeführter Stromrichter)*
Maschinenglättwerk inline calender ‖ calender stack * *mehrstufiges ~*
Maschinenhaus machine house
Maschinenhersteller machine manufacturer * *siehe auch →Maschinenbauer* ‖ machine supplier * *Maschinenlieferant*
Maschinenkalander inline calender * *Glättwerk, in Papiermaschine integriert*
Maschinenkommutierung motor commutation * *Lastkommutierg.e. (Thyristor)Stromrichters durch d.Läuferlage b.Synchronmotor*
Maschinenkörper machine frame
Maschinenlaufrichtung machine direction
maschinenlesbar machine readable
maschinenlesbare Fabrikatebezeichnung machine-readable product designation * *MLFB* ‖ machine-readable code
Maschinenmesser machine knife
Maschinenmodell machine model * *regelungstechnisches Modell z.B. eines Motors* ‖ motor model * *softwaremäßige Nachbildung eines Motors*
Maschinenpark machinery
Maschinenpersonal operating personnel ‖ engineering personnel
Maschinenprogramm machine code program * *[Computer] Programm im von der CPU ausführbaren ausführbaren Binärcode*
Maschinenraum maschine room ‖ engine room
Maschinenrichtlinie machinery directive * *der EG (89/106/EWG und 91/368/EWG in d.EG-Amtsblättern 89/L183/9 u. 91/L198/16)*
Maschinenrichtung machine direction * *Maschinenlaufrichtung*
Maschinenrolle jumbo roll * *Papiertambour*
Maschinensatz machine set ‖ generator set * *Motor-Generatorsatz*
Maschinenschraubstock machine vice
Maschinenschutz electrical machine protective relaying ‖ machine protection
Maschinenschwingungen machine vibrations
maschinenseitig on the motor side * *motorseitig, z.B. Stromrichter* ‖ load side * *z.B. zum Motor hin angeordneter Stromrichter* ‖ motor-side * *motorseitig (z.B. ~er Stromrichter)* ‖ load-side * *von Stromrichter aus gesehen*
Maschinensollwert machine speed setpoint * *Geschwindigkeitssollwert*
Maschinensprache machine language * *[Computer]* ‖ assembler language * *Assemblersprache, Mnemotechnische Maschinensprache* ‖ machine code * *Maschinencode* ‖ object language * *Objektsprache*
Maschinensteuerung machine control ‖ machine control system
Maschinenstrom motor current * *Motorstrom*
Maschinenstuhlung machine frame ‖ machine bed ‖ bedplate * *Bettung*
Maschinenteil machine element ‖ machine part ‖ machine component ‖ machine member ‖ machine section ‖ aggregate
Maschinenverluste machine losses * *z.B. durch Reibung*
Maschinenwicklung machine-made winding * *{el. Masch.} maschinell in das Blechpaket eingebrachte Wicklung*
Maße dimensions
Maßeinheit unit of measure ‖ unit
maßgebend decisive * *entscheidend, ausschlaggebend, bestimmend* ‖ authoritative * *autoritativ, maßgeblich, amtlich, gebieterisch* ‖ competent * *z.B. Behörde* ‖ relevant * *bestimmend; z.B. Bestimmung, Vorschrift* ‖ determining * *bestimmend* ‖ governing * *bestimmend* ‖ influential * *einflussreich, z.B. Kreise* ‖ leading * *führend, z.B. Kreise* ‖ applicable * *anwendbar; to: für* ‖ substantial * *beträchtlich* ‖ important * *wichtig*
maßgebend sein für determine * *bestimmen*
maßgebende Größe decisive factor * *entscheidende Größe* ‖ key factor * *wichtigste Einflussgröße*
maßgeblich decisive * *entscheidend, ausschlaggebend* ‖ authoritative * *maßgebend/geblich, autoritativ, amtlich, gebieterisch* ‖ competent * *z.B. Behörde* ‖ relevant * *bestimmend* ‖ governing * *bestimmend* ‖ influential * *einflussreich, z.B. Kreise* ‖ leading * *führend, z.B.Kreise* ‖ authentic * *Text* ‖ applicable * *anwendbar; to: für* ‖ substantial * *beträchtlich* ‖ important * *wichtig* ‖ determining * *bestimmend*
maßgebliche Größe decisive factor * *entscheidende Größe* ‖ key factor * *wichtigste Einflussgröße*
Maßgenauigkeit dimensional accuracy ‖ dimensional stability * *Maßhaltigkeit*
maßgeschneidert taylored ‖ made-to-measure * *nach Maß angefertigt* ‖ custom-specific * *→kundenspezifisch* ‖ tailor-made ‖ customized * *an die Kundenbedürfnisse angepasst* ‖ custom made * *für einen Kunden speziell angefertigt*
maßhaltig dimensionally correct ‖ true to size ‖ dimensionally true ‖ true to gauge ‖ dimensionally stable * *form/maßbeständig* ‖ to correct dimensions * *(Adv.) z.B. cut to correct dimensions: ~ geschnitten*
Maßhaltigkeit accuracy to size ‖ dimensional accuracy * *Maßgenauigkeit* ‖ dimensional stability * *Maß/Formbeständigkeit*
mäßig moderate * *gemäßigt; in: in* ‖ middling * *(salopp) mittel~* ‖ poor * *→dürftig*
Maske mask * *z.B. ~ auf dem Bildschirm, bei der Chip-Herstellung, Bit~* ‖ screen mask * *Bildschirm~* ‖ screen form * *~/Formular auf dem Bildschirm*
Maskenaufbau mask design * *[Computer] am Bildschirm*
maskieren mask * *Impulse ausblenden, Daten ausmaskieren usw.* ‖ mask out
Maßnahme measure * *take: ergreifen* ‖ action * *take: ergreifen*

Maßnahmen ergreifen take measures
Maßnahmen gegen Witterungseinflüsse weatherproofing measures
Maßpfeil dimension-line arrow
Maßreihe dimension series
Masse ground * *single-ended: auf ~ bezogen (z.B. Analogsignal; im Gegensatz zum Differenzialsignal)* || GND * *Kurzform für 'ground'* || 0V || earth * *Erde* || mass * *schwere ~ [kg]* || chassis * *Gehäuse~ (elektr,)* || frame * *Motorgehäuse-~ (elektr.)* || signal common * *Gemeinsame Bezugsspannung, "Wurzel"*
Masse-bezogener Analogeingang single-ended analog input * *im Gegensatz zum Differenzeingang* || analog input referred to ground
massebezogen single-ended * *Signal (im Gegensatz zum Differenzialsignal)* || referred to ground
Massegut bulk material
Massenkräfte inertia forces
Massenproduktion mass production * *siehe auch →Großserienfertigung*
Massenspeicher bulk storage || mass storage || bulk storage device || mass memory
Massenträgheit mass moment of inertia || mass inertia
Massenträgheitsmoment mass moment of inertia || moment of inertia * *J[kg m^2] Summe d. Produkte d.Einzelmassen u.d.Quadrat d.Entferng. z. Drehachse*
Masseschluss fault to frame * *{el. Masch.}*
Masseteilchen mass particle || particles of mass || mass particles
Masseverbindung ground connection
massiv solid * *fest, kräftig, stabil, gediegen, aus Vollmaterial* || massive * *massig, wuchtig, massiv, schwer* || heavy * *(auch figürl.) schwer, stark, kräftig, drückend* || powerful * *(figürlich) kraftvoll, mit Macht, wirksam, wuchtig* || rugged * *robust, stark, stabil*
Massivholz solid wood
Massivholzbearbeitung solid wood working
Massivholzplatte solid wood panel
Massivholzverarbeitung solid woodworking
Massivholzverleimmaschine laminating press
Massivläufer solid rotor * *ungeblechter ~ (→Turboläufer), hat eine höhere Grenzdrehzahl als e. geblechter Läufer*
Maßskizze dimension sketch
Maßstab scale * *on a reduced/larger scale: in verklein./vergrößertem ~; on a large scale: in großem ~*
Maßstäbe setzen set standards
maßstäblich to scale || true to scale * *maßstabsgerecht*
maßstabsgerecht true to scale
Massung grounding || earthing || earth connection
Mast pole
Maßtabelle dimension table
Master master
Master-Slave-Betrieb master-slave operation
Master-Slave-Prinzip master-slave principle || master-slave operation
Master-Slave-Steuerung master-slave control || master-slaving
Master-Slave-Verfahren master-slave technique
Master-Station master station * *z.B.bei Bussystem*
Masteranschaltung master interface * *für Bus*
Masterprogramm master program * *(SIEMENS SIMADYN D (R))*
Maßtoleranz dimension tolerance
Maßzeichnung dimension drawing || dimensional drawing
Material material || kind of material * *Art des Materials* || stuff || parts * *Teile (z.B. Ersatzteile)* || stock * *(Papier-/Faser-/Dick-) Stoff in der Prozeßtechnik; Füll-)Gut, Material, Brühe*

Materialdaten material data
Materialdurchlaufzeit material processing time
Materialeinsparung material saving
Materialeinzug material entry feed || material entry || material entry section
Materialermüdung fatigue || material fatigue
Materialfehler defect of material || flaw in the material || fault in the material || premature material fatigue * *vorzeitige Materialermüdung*
Materialfluss material flow || flow of material
Materialflussrichtung material-flow direction || material running direction * *bei durchlaufender Warenbahn*
Materialgeschwindigkeit material speed
Materialprüfung testing of material || inspection of material || materials testing || materials test
Materialqualität material quality
Materialstärke material thickness
Materialverfolgung material tracking
Materialverschleiß material wear
Materialwirtschaft materials management
Mathematik mathematics || math * *(salopp) Kurzform von 'mathematics'*
mathematisch mathematical || mathematically * *(Adv.)*
mathematische Fähigkeiten number skills || math skills
mathematische Gleichung mathematical equation
mathematisches Abbild mathematical image
mathematisches Modell mathematical model
Matrix matrix
Matrixdarstellung matrix notation
Matrixumrichter matrix converter * *Umrichter ohne Zwischenkreis mit Schaltermatrix zw. allen Netz- u Motoranschlüssen*
Matrize female die * *Stanzwerkzeug (Unterteil)* || bottom die * *Stanzwerkzeug (Unterteil)* || punching die * *Stanzmatrize (Unterteil)* || matrix * *Matrix*
Matrizen matrices
Mauerwerk mansonry
Maus mouse * *auch Computer~*
Mausklick mouse-click
Maustaste mouse key * *lefthand: linke; righthand: rechte*
Maustreiber mouse driver
maximal maximum * *maximum permissible: maximal zulässig* || max. || top * *Spitzen-* || peak * *Spitzen-* || maximum possible * *größtmöglicher*
Maximal- maximum
Maximaldrehzahl maximum speed || top speed || maximum rpm
Maximalfrequenz maximum frequency
Maximalgeschwindigkeit maximum speed || top speed
Maximalwert maximum value
Maximalwertauswahl maximum value selector || maximum selector
maximieren maximize
maximiert maximized * *for: bezüglich*
Maximum maximum
Maximumauswerter maximum selector
Mbaud Mbaud * *Megabits pro sec.*
MDF-Platte MDF board * *{Holz} Abk.f. 'Medium Density Fiber board': mittelharte Faserplatte*
Meall verarbeitend metalworking
Mechanik mechanics || mechanism * *Mechanismus, Triebwerk*
Mechaniker mechanic || mechanician || machinist * *→Schlosser, Maschinenschlosser/-einsteller*
Mechanikerdrehbank bench lathe
Mechanikerdrehmaschine precision bench lathe
mechanisch mechanical * *(Adj.)* || mechanically * *(Adv.)* || mechanically operated * *~ betätigt/angetrieben*
mechanisch betätigt mechanically operated

mechanisch entkoppelt mechanically decoupled
mechanisch gekoppelt mechanically coupled
mechanisch stabil mechanically rugged
mechanische Arbeit mechanical work
mechanische Ausführung mechanical design
mechanische Beanspruchung mechanical stressing
mechanische Bremsung mechanical braking * *mit Reibung arbeitende Bremse*
mechanische Festigkeit mechanical strength || mechanical stability
mechanische Grenzdrehzahl mechanical limiting speed || mechanical limit speed || mechanical speed limit
mechanische Größen mechanical quantities
mechanische Resonanz mechanical resonance
mechanische Schutzart mechanical degree of protection * *im Gegensatz z. Zündschutzart (Ex-Schutz),siehe z.B. IEC Publ.No.34*
mechanische Schwingungen mechanical vibrations * *für elektr. Maschinen: siehe DIN ISO 2373 (IEC34-14)*
mechanische Stöße mechanical shocks
mechanische Verriegelung mechanical interlock
mechanische Zeitkonstante mechanical time constant * *Zeit d.z.Erreichen v. 63,2% d. Enddrehzahl benötigt wird bei Nennbeding.*
mechanischer Aufbau mechanical design
mechanischer Verstellantrieb mechanical variable speed drive * *z.B. mit PIV-Getriebe oder Keilriemen-Verstellgetriebe*
mechanisches Stellglied mechanical actuator
Mechanismus mechanism
Meckerliste complaint list * *Mängelliste* || fault list * *Mängelliste* || list of defects * *Mängelliste*
Medien media * *Materialien*
Medienversorgung medium supply * *Herbeiförderung von Flüssigkeiten*
Medium medium * *Mittel, Träger, z.B. Speichermedium, Kühlmittel (Plural: 'media')*
medizinisch medical
medizinische Geräte medical devices
medizinische Technik medical engineering
Medizintechnik medical engineering
Meereshöhe sea level * *above sea level: über dem Meeresspiegel*
Meerwasser-Entsalzungsanlage desalination plant
Megabyte megabyte || meg * *(salopp)* || BMyte || MB
Megawatt megawatt
Megawattbereich megawatt range
Megneteil solenoid section * *bei einer elektromagnet. Bremse* || solenoid component * *z.B. bei einer elektromagnet.*
mehr more * *the more the better: je mehr desto besser; more than: ~ als; over/upwards of: ~ als (b.Zahlen)*
Mehr- multi || multiple || poly || additional * *Zusatz-*
Mehrachs- multi-axis
Mehrachs-Antrieb multi-axis drive || multi-shaft drive
Mehrachsantrieb multi-shaft drive || multi-axis drive
Mehrachsanwendung multi-axis application
Mehrachsbetrieb multiple axis operation
Mehrachsendrehzahlregler multi-axis speed controller
Mehrachsensteuerung coordinated motion control
Mehrachssteuerung multi-axis control
mehradrig multi-core || multi-wire
Mehraufwand extra costs || additional expense || additional efforts
Mehrbereichs- multi-range * *Spannung, Messbereich usw., auch -Messinstrument* || multigrade * *Schmieröl, Schmierfett usw.*
Mehrbereichsfett multigrade grease

Mehrbereichsöl multigrade oil
Mehrblock-Drahtziehmaschine multiblock wire-drawing machine * *{Draht}*
Mehrblockziehmaschine multiple block drawing machine * *{Draht} Drahtziehmaschine*
mehrdeutig ambiguous * *zweideutig*
Mehrdeutigkeit ambiguity
mehrdrähtig stranded * *(Kabel, Leitung) mit Litzenleiter (kein starrer Draht)* || multi-core || multi-conductor
mehrere several || some || a few || various * *verschiedene*
mehrfach multiple || multiplex || repeated * *wiederholt* || repeatedly * *(Adv.) wiederholt* || several times * *(Adv.) mehrere Male* || a multiple number of times * *(Adv.)* || a multiple number times * *(Adv.)* || manifold * *Mehr(fach)-, Mehrzweck-, vielfach, vielfältig, mannigfaltig*
mehrfach belegt multiple assigned * *z.B. Klemme*
Mehrfach- multi- || multiple || gang * *z.B. EPROM-Programmiergerät, Stanze, Fräser, Säge, Presse, Kondensator*
Mehrfach-Programmiergerät gang programmer * *zum Programmieren mehrer EPROMs usw. in einem Arbeitsgang*
Mehrfach-Stern-Dreieck-Anlauf multiple-star-delta starting
Mehrfaches multiple
Mehrfarbendruck color print * *(amerikan.)* || colour print * *(brit.)*
Mehrfunktions- multifunctional || multi-purpose * *Mehrzweck-*
Mehrheit majority * *great: überwiegende*
mehrkanalig multi-channel
mehrkanalige Steuerung multi-channel control
Mehrkosten additional cost || additional costs || additional expense || extra charges * *(Preis)Zuschlag* || extra costs
Mehrlagen- multilayer
Mehrlagen-Leiterplatte multilayer PCB
Mehrlagenleiterplatte multilayer printed board || multilayer PCB
mehrlagig multilayer || multi-layered
Mehrmotorenantrieb sectional drive * *auch: ein Antrieb im Mehrmotorenverbund* || multiple motor drive || multi-motor drive * *auch Betrieb von mehreren Motoren an einem Umrichter* || multi drive || parallel-motors drive * *mit mehreren Motoren an einem Stromrichter* || coordinated drive
Mehrmotorenantriebe sectional drives || multi-motor drives
Mehrmotorenbetrieb multi-motor operation
Mehrmotorensystem multi-motor system
Mehrphasenstrom polyphase current
Mehrphasensystem multi-phase system
mehrphasig multi-phase || polyphase
mehrpolig multi-pole
Mehrpreis extra price || extra cost * *Mehrkosten; at an extra cost: gegen Mehrpreis* || additional charge * *zusätzliche Berechnung* || additional price * *no: ohne; at: gegen* || price adder * *Preisaufschlag* || surcharge * →*Preisaufschlag*
Mehrprozessorsystem multiprocessor system
Mehrpunktwechselrichter multi-step inverter
Mehrquadrant- multi-quadrant
Mehrquadrantbetrieb multi-quadrant operation
Mehrquadranten- multi-quadrant
Mehrquadrantenantrieb multi-quadrant drive || multiple quadrant drive
Mehrquadrantenbetrieb multi-quadrant operation || multiple quadrant operation
Mehrscheiben- polydisc- || multisdisk-
Mehrscheibenbremse multiple-disk brake * *siehe auch →Lamellenbremse*
Mehrscheibenläufer multi-disc rotor

Mehrschichtbetrieb multi-shift operation
mehrschichtig multilayer || multilayered || many-sided * *(figürl.) vielschichtig*
Mehrspindel- multi-spindle * *z.B. Werkzeugmaschine* || multi-head * *z.B. Fräsmaschine* || gang * *z.b. Bohrmaschine* || multiple-spindle
Mehrspindeldrehmaschine multi-spindle lathe
mehrsprachig multi-lingual || multiple language
Mehrstufen- multi-stage
mehrstufig multi-stage || multi-speed * *(Schaltgetriebe) mit mehreren Übersetzungen* || multistage || multi-level * *z.B. Prozess, Vorgehen, Verfahren usw.*
mehrstufiger Stern-Dreieck-Anlauf multi-stage star-delta starting
mehrteilig multisectional || multipartite * *vielteilig, mehrteilig* || consisting of several parts || multi-segmented * *aus mehreren Segmenten bestehend, z.B. Schleifbürsten* || multi-fingered * *z.b. Kohlebürsten*
Mehrweg- re-usable
Mehrwegverpackung re-usable packaging * *(Verpack.technik)* || returnable packaging * *(Verpack.-technik)*
Mehrwert added value || value added || additional value
Mehrwertsteuer value-added tax || value added tax || VAT * *Abk. für 'Value-Added Tax'*
Mehrzahl majority * *Mehrheit, größerer Teil*
mehrzeilig multi-tier * *z.B. Baugruppenträger*
Meilenstein milestone || towering achievement * *in der technischen Entwicklung*
Meinung ändern change one's mind
Meißel chisel
meißeln chisel || carve
Meißelstemmmaschine chisel mortizer * *(Holz)*
meist most * *the most: am ~en (Adv.)* || most of * *most of the time: die ~e Zeit* || greatest * *größt* || the majority of * *die Mehrheit/Mehrzahl von* || the greater number * *die Mehrzahl* || the greater part * *das ~e, der größte Teil* || most of it * *das ~e* || the bulk of it * *der Großteil, die große Masse*
meistens usually * *gewöhnlich, normaler/gebräuchlicherweise* || generally * *gewöhnlich, im Allgemeinen* || as a rule * *in der Regel, im Normalfall, üblicherweise* || most * *am meisten* || most of all * *am meisten* || best known * *am meisten bekannt* || mostly * *meistenteils, meistens* || in most cases * *in den meisten Fällen* || for the most part * *meistenteils*
Meister master || registered master * *im Handwerk* || boss * *(salopp) Chef* || foreman * *Vorarbeiter* || shop foreman
meistern overcome * *Herr werden, lösen (z.B. Problem), überwinden* || master || control * *unter Kontrolle bringen, z.B. e. schwierige Lage* || cope with * *fertigwerden mit* || get over * *überwinden* || manage || succeed in * *(mit Gerundium) erfolgreich sein zu ...*
Meisterschalter master switch * *(Hebezeuge) beim Kran* || master controller * *(Hebezeuge) beim Kran* || joystick
Melde- signaling || signalling || message
Meldebaustein message block * *Funktionsbaustein* || signaling block * *Funktionsbaustein* || signaling block
Meldebürste signalling brush * *z.B. zur Bürstenverschleißmeldung in DC-Motor*
Meldedrucker event printer || message printer || logging printer || logger || message line printer || event recorder || log printer
Meldeeinrichtung signalling device || indicator module || annunciator || indicator
Meldefolgesystem message sequencing system || sequence message system

Meldefunktionsbaustein signaling function block
Meldegerät monitor || signaller
Meldekontakt message contact || signalling contact || signaling contact
Meldelampe signal lamp
Meldeleuchte pilot lamp || indicator lamp || signal lamp || indicator light
melden signal * *signalisieren, z.B. Zustand, Fehler* || report * *berichten, Bericht erstatten, Nachricht geben; dienstlich/amtlich melden/anzeigen* || inform * *informieren; of a thing: etwas* || notify * *amtlich/offiziell/schriftlich* ~ || signalize
Meldeprotokoll message printout || message listing || event log || event report
Meldepuffer message buffer || alarm buffer * *für Fehler- und Warnmeldungen*
Melder indicator || detector || monitor * *Wächter*
Melderelais signaling relay || signalling relay || alarm relay * *für Störmelde-/Überwachungszwecke*
Meldesteuerung signalling system
Meldesystem signalling system || event signalling system || reporting system || message system || signaling system
Meldetafel annunciator panel
Meldetext message text
Meldung message || report || signal * *Signal* || information * *Mitteilung* || notice * *Mitteilung, Ankündigung, Nachricht, Notiz* || news * *Neuigkeit* || signalling * *das Melden* || annunciation * *optische Meldung, z.B. an Leuchtmelder, Leuchttableau* || indication * *Anzeige*
Melkkuh cash cow * *(auch figürl.) sehr gute Einnahmequelle (z.B. Produkt oder Unternehmensbereich)* || gold mine * *(figürl.) sehr gute Einnahmequelle*
Membran membrane || diaphragm
Membrankupplung diaphragm coupling
Membranpumpe diaphragm pump
Menge amount * *Betrag, Summe, Höhe, Inhalt, Ergebnis, Wert* || quantity * *Anzahl, Menge, Betrag (Plural: quantities)* || q'ty. * *Abkürzg. für 'quantity':Anzahl* || aggregate * *auch Gesamt~, Anhäufung* || multitude * *Vielzahl* || heap * *Haufen* || batch * *Schub* || swarm * *Schwarm* || crowd * *Menschen~* || abundance * *Überfluss* || rate * *~ pro Zeiteinheit* || throughput * *Durchsatz (~ pro Zeiteinheit)*
Mengenflussregler mass flow controller
Mengengerüst volume scenario || volume proportion
Mengenrabatt quantity discount * *grant: gewähren*
Mengenregelung rate regulator * *Mengenregeler* || mass flow control * *Mengenstromregelung* || flow control * *Durchlussregelung*
Mengenstrom mass flow
Mensch- human || man
Mensch-Maschine-Dialog man-machine dialog
Mensch-Maschine-Kommunikation man-machine communications
Mensch-Maschine-Schnittstelle man-machine interface || HMI * *Abk. für "Human Machine Interface"* || human machine interface
Menü menu || menue
Menüführung menu prompting || menu guidance
menügeführt menu-prompted || menu-driven || menu-assisted || menu driven || with menu prompting
menügesteuert menu-prompted || menu-driven || menu-controlled || menu-assisted
Menüleiste menue bar * *(Computer) auf dem Bildschirm*
Menüpunkt menu item
Mercerisiermaschine mercerizing machine * *(Textil) zur Nachbehandlung ("Glänzendmachen") von Textilstoffen*

Merkblatt instruction card || instruction sheet || sheet of instructions || instruction leaflet
merken hold in memory * *im Speicher aufbewahren* || perceive * *wahrnehmen* || feel * *spüren, merken, fühlen* || bear in mind * *im Gedächtnis.behalten, nicht vergessen* || mark * *beachten, sich etwas ~, vormerken* || note * *beachten, notieren, aufschreiben* || notice * *wahrnehmen* || realize * *erkennen* || find out * *entdecken* || discover * *entdecken* || remember * *remember something: sich etwas ~* || make a mental note of * *sich etwas einprägen* || impress on someone's memory * *sich etwas einprägen* || stick in someone's mind * *sich einprägen*
Merker flag || marker || flag marker || relay equivalent * *bei Kontaktplandarstellung eines SPS-Programms* || internal relay * *bei Kontaktplandarstellung eines SPS-Programms*
Merker-Datum flag data * *bei SPS*
Merkerbereich flag area * *bei SPS* || flag address area * *bei SPS*
merklich considerable * *beträchtlich, erheblich, bedeutend* || noticeable * *wahrnehmbar, beachtlich, bemerkenswert* || distinct * *deutlich* || evident * *augenscheinlich, offensichtlich, klar* || visible * *sichtbar, offensichtlich, deutlich* || measurable * *messbar* || significant * *bedeutsam, bedeutend*
Merkmal feature * *Besonderheit, Eigenschaft; salient: herausragendes* || characteristic * *Besonderheit* || distinctive mark * *Kennzeichen* || property * *Eigenschaft* || mark * *Zeichen* || sign * *Zeichen*
Merkmale features
Mess- measuring || instrument
Mess- und Regelungstechnik instrumentation and control
Messanlage measuring equipment || scanner * *berührungslos arbeitende ~* || scanner system * *{Papier} z.B. zum berührungsfreien Messen von Feuchte, Flächengewicht, Dicke usw.*
Messaufnehmer sensor || transducer
Messband measuring strip * *zur Lage-/Längenerfassung* || tape measure
messbar measurable
Messbecher measuring cup
Messbereich measuring range || range
Messbereichs-Endwert full scale
Messbereichsendwert full scale
Messbereichsumschaltung measuring-range changeover || measuring changeover
Messbrücke measuring bridge || measurement bridge
Messbuchse measuring socket || test socket * *Prüfbuchse* || measuring jack
Messdaten measurement data || measuring data || test data || measured values * *Messwerte*
Messdatenerfassung data acquisition
Messe fair || exhibition * *Ausstellung* || trade fair * *Fachmesse, Industriemesse*
Messegelände fair ground || exhibition grounds || fair grounds || exhibition park * *Ausstellungsgelände*
Messeinrichtung measuring equipment || measuring unit
Messeinrichtungen instrumentation * *in der Anlagentechnik/Prozesstechnik*
messen measure || take the measurement * *ab~* || meter * *mit einem Messapparat, z.B. mit Stromzähler, auch dosieren, Menge ~* || gauge * *mit einer Lehre ~, z.B. (Draht-, Blech-) Dicke* || gage * *mit einer Lehre ~, z.B. (Draht-, Blech-) Dicke* || caliper * *mit einer Lehre ~* || time * *Zeit ~* || compete * *sich mit jdm. ~*
Messer knife * *slitting knife: Längsteil~* || blade * *Klinge* || slitter * *Kreis~, z.B. zum Längsschneiden von Papier, Blech usw.*
Messerhalter knife holder

Messerkontakt blade-type terminal || tab connector * →*Flachstecker*
Messerkopf milling cutter * *zum Fräsen* || cutter head
Messerleiste plug connector || male connector * *"männlicher" Steckverbinder* || male socket connector * *'männlicher' Steckverbinder*
Messermotor knife motor || knife-drum motor * *treibt Messertrommel(n) z.B. eines* →*Querschneiders an* || cutter motor
Messerpartie knife section * *z.B. bei* →*Querschneider* || knife rolls * *bei* →*Querschneider*
Messertrommel knife drum * *z.B. bei Querschneider*
Messerwalze knife drum * *z.B. bei Querschneider*
Messerwelle cutter block * *{Holz} in Holzfräse*
Messestand exhibit booth || booth || exhibition stand || stand || exhibition space * *Fläche* || exhibition booth
Messfehler measuring error || measurement error || error of measurement || incorrect measurement * *Fehlmessung*
Messfläche measuring surface * *zur Schalldruckmessung (entlang d. ~ (Kugel, Quader) sind Mikrofone angeordnet)* || measuring spot size * *z.B. bei (berührungsloser) Flächengewichts- und Feuchtemessung*
Messflächen-Schalldruckpegel measuring surface sound-pressure level * *gemessen entlang e.genormten Messfläche (z.B. Kugelfläche)*
Messflächenmaß measuring-surface * *Rechengröße zur Ermittlung der Schallleistung*
Messflächenschalldruckpegel measuring-surface sound-pressure level * *(mit Mikrofonen an e. Messfläche gemess.) DIN EN 21680 Tl.1*
Messfühler sensor || transducer * *Messgeber* || measuring probe * *Messsonde, Tastkopf*
Messgeber transducer * *[trans'dju: se]* || sensor * *auch Messfühler*
Messgenauigkeit measuring accuracy
Messgenerator signal generator
Messgerät meter || panel meter * *Einbau/Schalttafel-Instrument* || measuring meter * *Messinstrument* || measuring device * *allg.* || gauge * *für (Blech)Dicke, Breite, Stärke, Drahtdurchmesser, Druck usw* || measuring equipment * *Ausrüstung*
Messgeräte measuring equipment
Messgröße measured quantity || measured variable
Messing brass
Messinstrument meter || measuring instrument || panel meter * *Einbau/Schalttafel-Instrument*
Messklemme measuring terminal
Messkreis measuring circuit
Messlehre gauge
Messleitung measuring lead
Messmethode method of measuring || measuring method || technique of measurement * *Messtechnik, Messverfahren*
Messmikrofon measuring microphone * *zur Schallmessung*
Messnippel measuring nipple
Messnormal standard component || comparison standard || measuring reference
Messort measuring location || measuring site
Messplatz measuring station
Messprinzip measuring principle || method of measuring * *siehe* →*Messmethode*
Messpunkt test point || measuring point
Messrad measuring wheel * *Anlegerad zur Geschwindigkeitsmessung* || web-speed tachometer * →*Bahntacho*
Messreihe series of measurements
Messschieber sliding caliper * →*Schieblehre*
Messschreiber data recorder

Messschrieb strip chart * z.B. Papierstreifen v. Linienschreiber || record * Aufzeichnung z.b. von einem Messschreiber || record chart * aufgezeichnetes Messkurvenblatt || oscilloscopic trace * von einem Oszilloskop; make: aufnehmen/-zeichnen || oscilloscopic record * von einem Oszilloskop || plot
Messspannung measuring voltage * z.B. bei Isolationsprüfung || test voltage * →Prüfspannung
Messstelle measuring location || measuring point
Messstellenumschalter measuring-point selector
Messstrippe measuring lead
Messsystem measuring system
Messtechnik measurement || measurement engineering * als Fachrichtung || measurement technique || technique of measurement * i. einem konkreten Anwendungsfall, Messverfahren || measuring equipment * messtechnische Ausrüstung || measuring technology || instrumentation * Messausrüstung in der Prozess-/Verfahrenstechnik
Messumformer signal converter || transducer || measuring transducer || measured value transducer || measuring transmitter
Messung measurement || gauging * Abmessung (Länge), Eichung, Dickenmessung || surveying * Vermessung, Aufnahme, Prüfung, Begutachtung || reading * Ablesung, Ablesewert || measuring
Messunsicherheit measuring incertainty || measuring inaccuracy * Messungenauigkeit
Messverfahren measuring method || method of measuring || method of measurement || measuring technique
Messverstärker instrument amplifier
Messwandler transducer
Messwerk measuring element
Messwert measured value * gemessener Wert || reading * Ablesewert || indication * angezeigter Messwert || measurand * zu messender Wert || indicated value * angezeigter Wert || measuring sample * abgetasteter ~
Messwertanalyse measured value analysis
Messwertaufnehmer measuring sensor || transducer || sensor
Messwerterfassung measured-data acquisition || data acquisition || measuring data sensing || measured date sensing || measured value acquisition
Messwertschreiber data recorder
Messwertumformer measuring transducer || measured-value transducer
Messwiderstand resistance gauge * Widerstandsthermometer
Messzeit measuring time || measuring interval || measuring period
Metall metal
Metall-Suchgerät metal detector
Metallbalgkupplung metal bellows coupling
Metallband metal band || metal strip
Metallbearbeitung metalworking
Metallbearbeitungsmaschine metal working machine
metallbedampft metallized
Metallfaltenbalg metal bellow
Metallfilmwiderstand metal film resistor
Metallfolie metal foil
Metallgehäuse metal case || metallic enclosure
Metallgriffflasche metal lug * zum Ansatz des Ziehgriffs bei Sicherung
Metallindustrie metalworking industrie
metallisch metallic || metallical || metallically * (Adv.) z.B. metallically bright: ~ →blank
metallisch blank metallic bright || bright
metallisieren metallize
metallisiert metallized
metallisierter Kunststoffkondensator metallized film capacitor

Metallisierung metallization
Metallmantel metal sheath * z.B. einer Leitung
Metalloxid-Ableiter metal-oxide surge diverter || MOX arrester
Metalloxidableiter metal-oxid surge arrester * Überspannungsableiter || MOV surge suppressor * Abk. f. 'Metal Oxide Varistor'
Metallpapierkondensator metallized paper capacitor
Metallschere shear || plate shears || shears || snips * Hand'blechschere
Metallschichtwiderstand metal film resistor
Metallschlauch metal hose * flexible: biegsamer || flexible tube
Metallsuchgerät metal detctor * auch beim Papier-/Holzrecycling
Metallumformung metal forming * spanlose Metall-/Blechbearbeitung || metalworking * spanlose Metall-/Blechbearbeitung
metallurgisch metallurgical
Metallverarbeitung metal working
Metallveredelungsstraße metal finishing linc
Meter metre || meter * (amerikan.)
Meterware material sold by the metre || metered goods
Methode method || process || technique || system || way of doing || policy * Verfahrensweise
metrisch metric
metrisches Gewinde metric thread
Metrisches System metric system
mieten hire * siehe auch →leihen
Mietkauf hire-purchase
Mietwagen rental car
Migration migration
migrieren migrate * (langsam hinein-) wandern
Mikroampere microamp || microamps * (Plural)
Mikrobefehl micro operation * Teilbefehl, in den ein Maschinenbefehl in einer →CISC-CPU zerlegt wird || µop * (salopp) Teilbefehl, in den ein Maschinenbefehl in einer →CISC-CPU zerlegt wird
Mikrocomputer microcomputer
Mikroelektronik micro-electronics || microelectronics
Mikrofaser micro-fiber
Mikrometer micron
Mikroorganismen microorganisms
Mikroprogramm microprogram * Vorschrift für die sequenzielle Abarbeitung eines Maschinenfehls in →CISC-CPU
mikroprogrammgesteuert microprogrammed
mikroprogrammiert microcoded
Mikroprozessor microprocessor || CPU * Central Processing Unit eines Computers
Mikroprozessor-gesteuert microprocessor controlled
mikroprozessorgesteuert microprocessor-controlled
Mikroprozessorregelung microprocessor-based control
Mikroprozessortechnik microprocessor technology
Mikroreibung micro friction
Mikroschalter micro switch
Mikroskop microscope
militärisch military
Milliarde billion * (amerikan.) || milliard * (brit.) || thousand millions
Milliarden billion
Millimeter millimeter * (amerikan.) || millimetre * (brit.)
millimetergenau with millimeter accuracy
MIME MIME * 'Multi-Purpose Internet Mail Extensions': Konvertiert 8Bit Code in 7Bit →ASCII Code (f.E-Mails)

min-1 RPM * Abkürzg.f. 'Revolutions Per Minute': Umdrehungen pro Minute
Minderheit minority
Minderung decrease * Abnahme, Verringerung || diminution * Ver~, Verringerg., Verkleinerg., Abnahme, Nachlassen || reduction * Reduzierung || derating * Herabsetzen, z.B. der Leistg. eines Stromrichters bei größerer Aufstellhöhe || abatement * Abnehmen, Nachlassen, Linderg., Bekämpfung, (Preis)Nachlaß || depreciation * Herabsetzg., Geringschätzg., Wert~, Abschreibg., Ent/Abwertung
minderwertig low-quality * Ware, Produkt usw. || inferior * Qualität usw. || poor * Qualität usw. || of inferior quality * Arbeit, Produkt usw. || low-grade * Ware usw.
Mindest- minimum
Mindestabnahmemenge minimum quantity
Mindestanforderung minimum requirements * regarding: an
Mindestanforderungen minimum requirements
Mindestdrehzahl minimum speed
mindestens at least || at the last || at the very last || no less than * vor Zahlen, in einer Vorschrift || not under * vor Zahlen, in einer Vorschrift || at a minimum of || minimum
Mindestfrequenz minimum frequency
Mindestmaß minimum dimension * Längenmaß, Ausdehnung || minimum
Mine refill * zum Nachfüllen (z.B. für Kugelschreiber) || cartridge * Patrone, auch für Kugelschreiber || lead * für Bleistift || mine * Bergwerk; Sprengkörper
Mineralfaser mineral fiber
Mini-Diskette mini floppy disk * 5,25"-Diskette || minidiskette * 5,25"-Diskette
Miniatur- miniature- || micro- || mini-
Miniaturisierung miniaturization
minimal minimum || min. || insignificant * (figürl.) geringfügig, unbedeutend || trifling * (figürl.) unbedeutend, geringfügig, trivial
Minimal- minimum
Minimaldrehzahl minimum speed
Minimalfrequenz minimum frequency
Minimalkonfiguration minimum configuration
Minimalwert minimum value
Minimalwertauswahl minimum value selector || minimum selector
minimieren minimize || keep to a minimum * →gering halten
minimiert minimized * for: bezüglich
Minimierung minimization
Minimum minimum * cut to a minimum: auf ein ~ reduzieren/beschränken
Minimumauswahl minimum value selector || minimum selector
Minimumauswerter minimum selector
Minoritäts- minority
Minoritätsladungsträger minority carrier * bei Halbleitern
minus minus * ['maines]
Minuspol negative pole || negative terminal * Anschluss an Batterie, Kondensator
Minuszeichen negative sign || minus symbol
Minute minute
Misch- mixed || composite || mixing || compound
Mischanlage mixing plant * z.B. in der Gummi-/Reifenindustrie || blending system * besonders für Lebens-/Genussmittel || mixing installation
mischen merge * z.B. Daten(sätze) einsortieren || mix || mingle * sich ver~, verschmelzen, sich vereinigen/verbinden || combine * kombinieren || compound * Bestandteile ~ || intermix * ver~ || jumble * zusammenwerfen, in Unordnung bringen, durcheinanderwürfeln, (wahllos) ver~ || agitate * (um)rühren, schütteln, hin und her

bewegen || blend * bei Nahrungs- und Genussmitteln
Mischer mixer
Mischkalkulation composite calculation || mixed calculation
Mischkammer mixing chamber
Mischkonzern conglomerate
Mischpreis composite price
Mischpumpe fan pump * z.B. für Zellstoff
Mischreibung mixed friction
Mischung mixture || blend * ~ verschiedener Sorten bei Nahrungs- und Genussmitteln || combination * (chem.) || alloy * Legierung || adulteration * Verfälschung || amalgam * (auch figürl.)
Mischung, Gemenge, Verschmelzung
Mischungsverhältnis mixing ratio || proportion of ingredients
Mischventil mixer valve || mixing valve
Missbrauch misuse * falsche/unkorrekte An/Verwendung || improper use * unkorrekte An/Verwendung || abuse * übermäßige Beanspruchung, Misshandlung, falscher Gebrauch
Misserfolg failure * be unsuccessful: keinen →Erfolg haben || fiasco || flop * (salopp)
Missstand nuisance || bad state of affairs || grievance || defect * Mangel
Missverständnis misunderstanding || dissension * Meinungsverschiedenheit, leichter Streit, Zwietracht || difference * Meinungsverschiedenheit, Differenz || disagreement * Meinungsverschiedenheit || misinterpretation * Fehlinterpretation
mit with || having * habend/besitzend || as * as the overlap increases: mit steigender Überlappung || in the company of * in Begleitung von || full of * voll von || by * mit der Bahn, Post, Hauspost; mit Gewalt usw. || in * mit einer Währung bezahlen, mit Kugelschreiber schreiben || having a * ein ... besitzend
mit anderen Worten in other words
mit Blick auf die Antriebsseite facing the drive end || when looking at the drive end
mit dem Quadrat des with the square of the * vary with the square of the voltage: sich mit d. Quadrat d. Spannung ändern
mit dem Uhrzeigersinn clockwise || cw
mit dem Ziel aimed at
mit der dritten Potenz von with the cube of
mit der Zeit in the course of time
mit einem Mal all at once * alles auf einmal, plötzlich || suddenly * plötzlich
mit freundlichen Grüßen Regards, || Sincerely yours, || Best Regards,
mit Gewalt by force
mit Hilfe von with the aid of * einer Sache || with the help of * einer Person || by means of * mittels
mit hoher Geschwindigkeit at high speed
mit Leichtigkeit with great ease
mit LOW Signal ansteuern de-energize * z.B. eine Klemme
mit Nachdruck emphatically
mit nur einer Drehrichtung unidirectional
mit Rat und Tat zur Seite stehen provide advice and assistance
mit Rücksicht auf with respect to
mit sich bringen imply || involve
mit Störung abschalten fault trip
mit vorzüglicher Hochachtung Sincerely yours, || Regards, || Best Regards,
Mit- co- * z.B. co-operation: ~arbeit, co-worker: ~arbeiter || fellow- * z.B. fellow passenger: ~reisender || joint * z.B. joint owner: ~besitzer, joint liability: ~haftung || included || inclusive || join * (Verb)
Mitarbeiter colleague || staff * (Plural) || employee * Arbeitnehmer, Angestellter || assistant * Helfer,

Mitarbeiterschaft

Assistent, Zuarbeiter, Stellvertreter || co-worker || work-force * *(Plural) Belegschaft, Mitarbeiterschaft* || collaborator
Mitarbeiterschaft workforce || staff || work force
Mitbestimmung worker participation * *der Mitarbeiterschaft*
Mitbewerb competitors * *Konkurrenten* || rivals * *Marktgegner* || competition * *Konkurrenz*
Mitbewerber competitors * *Konkurrenz* || competition * *Mitbewerb* || rivals * *(Markt)Gegner* || contender * *Konkurrent, (Mit)bewerber, Mitanbieter*
Mitbewerbs- of the competition || competitive
Mitbewerbsprodukt competitive product
Mitgliedschaft membership || affiliation * *Zugehörigkeit, Angliederg., Zusammenschluss, Mitgliedschaft; Aufnahme (als Mitglied)*
Mithilfe assistance || aid || co-operation || cooperation
Mithöranlage monitoring equipment || snooping device * *hört passiv am Speicher (z.B. auch Cache-Speicher) mit*
Mitkomponente positive-sequence component * *symmetrischer Anteil am Dreiphasensystem* || positive component
Mitkopplung positive feedback
mitlaufen auf ride * *z.B. Tastrolle auf Warenbahn, Druckwalze auf Wickel*
Mitnehmer driver || dog || cam || driver pin * *~stift, ~bolzen* || carrier bolt * *~bolzen* || catch * *Fangeinrichtung/-behälter* || driving element * *antreibendes Teil* || tailstock * *Reitstock bei Drehbank* || driver disc * *~scheibe* || engaging piece
Mitnehmerbolzen drive pin || driving pin || carrier bolt
Mitnehmerflansch driving flange * *an Getriebeausgang usw.*
Mitnehmernut driving slot
Mitnehmerscheibe driver plate || driving plate
mitreißen sweep away || carry away * *(auch figürl.: begeistern)* || electrify * *(figürl.) jdn. begeistern* || drag along
Mitsystem positive-sequence system * *siehe →Mitkomponente*
Mittagessen lunch * *Mittagessen, zweites Frühstück; das ~ einnehmen* || dinner * *after dinner: nach Tisch; be at dinner: beim ~ sein* || early dinner
Mitte centre || center * *(amerikan.)* || middle || mid * *mittler, Mittel-; mid of may: Mitte Mai; in the mid 19th century: in der ~ d.19 Jahrhunderts* || mid point * *→Mittelpunkt* || center line * *Mittellinie*
mitteilen inform * *a person of: jmd. etwas mitt.* || make known * *bekannt machen* || tell about || put out * *herausgeben, z.B. offizielle Information einer Firma* || notify * *(amtlich/offiziell) mitteilen/bekannt geben*
Mitteilung information || notice || bulletin * *offizielle ~* || message * *Nachricht* || report * *Bericht*
Mittel medium * *Substanz* || average * *Mittel/Durchschnitts-Wert/Größe* || agent * *Wirkstoff* || compound * *Masse, Mischung; sealing compound: Dicht~* || middle * *(Adj.)* || central * *(Adj.) zentral* || average * *(Adj.) durchschnittlich* || medium * *(Adj.) durchschnittlich* || mean * *(Adj.) im statistischen ~; (Subst.)* || *(Hilfs-)Mittel, Werkzeug, Geld~* || middling * *(Adj.) mittelmäßig* || subordinate * *(Adj.) im Rang* || midrange * *(Adj.) im mittleren Bereich liegend* || medium range * *(Adj.) im mittleren Bereich liegend* || medium-sized * *(Adj.) von mittlerer Größe* || mean * *statistisches/arithmetisches ~* || mid-range * *in der Mitte rangierend, ~klasse-* || resources * *Mittel, Geld~, Reichtümer, Resourcen, Hilfsquelle(n), Hilfs~* || facilities * *Einrichtungen, Anlagen, Möglichkeiten, Gelegenheiten* || means * *(Plural) (Hilfs-/Geld-) Mittel, ~ und Wege, means to an end: ~ zum Zweck* || tool * *Hilfs~, Werkzeug (auch figürl.)*
Mittel- midrange || medium-
Mittelabgriff center tap * *(amerikan.)* || centre tap * *(brit.)*
Mittelanzapfung centre tap * *(brit.)* || center tap * *(amerikan.)*
mittelbar indirect || indirectly * *(Adv.)*
Mitteldruck- medium-pressure
Mittelfrequenzgenerator medium-frequency generator
mittelfristig in the medium-term * *(Adv.)* || medium term * *(Adj.)*
Mittelgang center aisle * *in der Mitte liegender Durchgang*
mittelgroß medium-sized || mid-sized
Mittelklasse- mid-range * *z.B. Produkt*
Mittellage centre position * *z.B. Tänzerwalze* || mid position
Mittelleiter neutral conductor * *im Drehstromsystem* || neutral * *im Drehstromsystem* || N * *im Drehstromsystem* || middle conductor * *bei Gleichstrom* || M conductor * *bei Gleichstrom* || protective neutral conductor * *mit Schutzleiterfunktion*
Mittellinie centre line
Mittelpunkt midpoint || centre * *Zentrum* || center * *(amerikan.) Zentrum* || central point * *zentraler Punkt* || focus * *Brennpunkt (auch figürl.: der Tätigkeit, des Interesses)* || heart * *(auch figürl.) Herz, Inneres, Mitte, Wesentliches, Kern (einer Sache/einer Stadt)* || neutral point * *(elektr.) eines Drehstromsystems* || centre point || center point * *(amerikan.)* || core * *Kern* || basis * *Grundlage* || focal point * *~ der Betrachtungen, des Interesses usw.*
Mittelpunktleiter neutral || neutral conductor || N
Mittelpunktschaltung midpoint connection * *Stromrichterschaltg.bei d.e. DC-Anschluss gleich dem Wechselstr.sternpkt. ist*
Mittelpunktsleiter neutral || N || neutral conductor
mittels by means of || by way of || through || with the help of || with
Mittelspannung medium voltage || MV * *Abkürzung f. 'Medium Voltage'* || medium-high voltage
Mittelspannungsschalter medium voltage circuit breaker || MV breaker * *Abkürzg.f. 'Medium Voltage': Mittelspannungs-*
mittelständische Firma small company * *kleine Firma* || medium-sized company * *mittelgroße Firma*
mittelständische Firmen small and medium-sized companies
mittelständischer Betrieb medium-sized company
mittelständisches Unternehmen medium-sized company
Mittelstellung mid position * *z.B. Tänzer* || centre position * *(brit.)* || center position * *(amerikan.)* || intermediate position
mittelträge medium time-lag * *Sicherung* || normal-blow * *Sicherung*
Mittelwert mean value * *arithmetischer ~* || average value * *Durchschnittswert* || average * *Durchschnitt* || arithmetic mean value * *arithmetischer ~* || RMS value * *Effektivwert (d.h. quadratischer ~)* || root mean square * *Effektivwert (d.h. quadratischer ~)* || median * *50%-Wert, Mittellinie/-wert, die Mitte bildend/einnehmend, Mittel-, mittlerer*
Mittelwertbildung mean value generation * *sliding-type/shifting-mode: gleitende ~* || averaging || average calculation * *durch Berechnung*
Mittenabstand centre-to-centre distance * *z.B. zwischen Bohrungen* || distance between centres
Mittenanschlag center stop || stop face * *an einer (Schraub)Klemme* || center dog

Mittenanzapfung centre tap ‖ center tap * *(amerikan.)* ‖ central tapping
Mittenfrequenz centre frequency * *für Bandpass/sperre* ‖ center frequency * *(amerikan.) eines Bandpasses, einer Bandsperre*
Mittigkeit centricity ‖ concentricity * *Konzentrizität*
mittlere Abweichung mean deviation ‖ standard deviation ‖ mean error * *mittlerer Fehler*
mittlerer mean * *arithm. Mittelwert* ‖ middle * *in der Mitte liegend* ‖ central * *in der Mitte liegend* ‖ intermediate * *dazwischenliegend* ‖ average * *durchschnittlich* ‖ medium-sized * *von ~ Größe* ‖ midrange * *z.B. mittlere Entfernung* ‖ medium * *durchschnittlich, mittelmäßig* ‖ central * *zentral* ‖ middling * *mittelmäßig* ‖ medium-performance * *von ~ Leistungsfähigkeit*
mittlerer Anlassstrom average starting current * *[el.Masch.] SQRT(Anl.spitzenstrom x Schaltstrom), kennzeichn.*→*Anlassschwere*
mittlerer Leistungsbereich mid-range ‖ medium performance level * *bezüglich der Leistungsfähigkeit/Rechenleistung*
MKL-Kondensator metallized-plastic capacitor * *Wickelkondensator aus metallisierter Kunststoffu.Lackfolie*
MKV-Kondensator metallized-dielectric capacitor * *mit metalliertem Papier-Kunststoff-Dielektrikum, verlustarm* ‖ low-loss metallized-dielectric capacitor * *m.metallisiert.Papier-Kunststoff-Dielektrikum, verlustarm*
MLFB MLFB * *Abkürzg. für 'Maschinenlesbare FabrikateBezeichnung' (SIEMENS-Bestellnummer)* ‖ Order No. ‖ machine-readable product designation ‖ machine-readable product code
mnemotechnisch mnemonic * *auch menemotechnische Notation eines Maschinenbefehls in einer* →*Assemblersprache* ‖ mnemotechnical
Möbelindustrie furniture industry
mobil mobile * *im Gegens. zu ortsfest*
mobiler Einsatz mobile applications
Mode mode * *Modus, Betriebsart* ‖ fashion ‖ style
Modell model * *auch mathematisches ~, Simulations~ usw.* ‖ type * *Typ, Bauart* ‖ design * *Ausführung, Bauart* ‖ prototype * *Prototyp* ‖ pattern * *(Gieß-) Muster* ‖ mould * *Form* ‖ type of construction * *Bauart* ‖ mock-up * *~ in natürlicher Größe, Attrappe, Nachbildung* ‖ imitation * *Nachbildung* ‖ reproduction * *Nachbildung* ‖ dummy * *Attrappe* ‖ facsimile * *genaue Nachbildung* ‖ replica * *genaue Nachbildung* ‖ image * *Abbild, Verkörperung* ‖ test model * *Versuchs~*
Modellbau pattern making
Modellbauer pattern maker
Modellbeschreibung model description
Modellbildung modeling
Modellparameter model parameters * *für Rechenmodell*
Modellreihe range of models ‖ model line
Modem modem * *Abk. f. 'Modulator/Demodulator': Gerät z.Übertragg. digitaler Nachrichten über Telefonleitg.*
modern modern ‖ advanced * *fortschrittlich* ‖ progressive * *fortschrittlich* ‖ up-to-date * *auf dem Stand der Technik* ‖ fashionable * *modisch* ‖ trendy * *im Trend* ‖ streamlined * *modernisiert, fortschrittlich, rationell* ‖ state-of-the-art * *d. neuest. Stand d. Technik entsprechend, auf d. neuest. Entwicklungsstand befindl.*
modernisieren modernize ‖ bring up to date ‖ remodel * *vom äußeren Aussehen her* ‖ refurbish
Modernisierung modernisation ‖ modernization * *(amerikan.)* ‖ retro-fitting * *nachträgliches Einbauen neuer Komponenten*
Modifikation modification

Modifizierte Lagerlebensdauer modified bearing life * *Wälzlagerlebensdauer bei bestimmtem Werkstoff. u. Betriebsbedingungen*
Modul module ‖ submodule * *Submodul, z.B. Speichermodul, "Huckepack-Kärtchen" usw.*
modular modular ‖ based on a modular concept
modular aufgebaut of modular design
modular erweiterbar extendable by modular design
modulare Bauweise modular construction ‖ modular design ‖ modular layout
modularer Aufbau modular design
modulares Umrichtersystem modular converter system
Modularität modular design ‖ modular concept ‖ modularity ‖ modular principle ‖ level of modularity * *Ausmaß der ~; with a high level of modularity: hochmodular*
Modularkonzept modular design ‖ modular nature of the design
Modulation modulation
Modulationsart type of modulation ‖ modulation mode
Modulationsfrequenz modulation frequency
Modulationsverfahren mode of modulation ‖ method of modulation ‖ type of modulation ‖ modulation method
Modulbibliothek module library
modulieren modulate
Modulparameterplan module parameter diagram * *(SIEMENS SIMADYN D (R))* ‖ board parameter diagram * *(SIEMENS SIMADYN D (R))*
Modus mode
möglich possible * *as fast as possible: so schnell wie ~; make possible: ~ machen* ‖ attainable * *erreichbar, erzielbar* ‖ practicable * *durchführbar* ‖ feasible * *aus/durchführbar* ‖ likely * *wahrscheinlich* ‖ eventual * *eventuell* ‖ potential * *potenziell* ‖ all sorts of * *alles ~e; all sorts of things: alles ~e* ‖ everything * *try everything: alles ~e tun* ‖ best * *do one's best: sein ~stes tun* ‖ manage to do * *~ machen* ‖ as possible * *(Adv.) as soon as possible: ~ bald* ‖ as ever possible * *(Adv.) sofern/wenn ~*
möglicherweise potential * *potenziell* ‖ if possible ‖ possibly ‖ perhaps * *vielleicht*
Möglichkeit possibility * *possibility of doing: ~ etwas zu tun* ‖ practicability * *Ausführbarkeit* ‖ feasibility * *Durchführbarkeit* ‖ chance * *Gelegenheit* ‖ opportunity * *gute Gelegenheit* ‖ alternative * *andere, zweite ~* ‖ potentiality * *Entwicklungs~* ‖ facility * →*Einrichtung, Vorteil, Gelegenheit; for: für,zu* ‖ capability * *Fähigkeit* ‖ potential * *Potenzial, Entwicklungs~(en)/-Reserven, Kraftvorrat* ‖ option * *(zusätzliche) Wahl~, Option*
möglichst possible * *the smallest possible: kleinstmöglich* ‖ as soon as possible: *~ bald*
Molkerei-Ausführung dairy-proof design * *Sonderausführung für den Einsatz in der Nahrungsmittelindustrie*
Molybdän molybdenum
Molybdän-Disulfid molybdenum disulphide * *MoS2 (Schmiermittel, z.B.* →*Molykote)*
Molykote Molycote ‖ MoS2 * →*Molybden-Disulfid (Schmiermittel)*
Moment moment * *Augenblick* ‖ instant * *Augenblick* ‖ torque * *[to: k] Drehmoment* ‖ inertia * *Trägheitsmoment, Schwungmoment* ‖ motive * *Beweggrund* ‖ factor * *Faktor* ‖ element * *Faktor, wesentlicher Umstand* ‖ momentum * *physikal. Impuls, "Moment einer Kraft", Triebkraft, Stosskraft, Wucht, Schwung, Fahrt* ‖ impetus * *(figürl.) Antrieb, (Auf)Schwung, Triebkraft* ‖ main point * *Hauptmoment* ‖ main factor * *Hauptmoment*

momentan momentary * *(Adj.) momentan, augenblicklich, vorübergehend, flüchtig, jed. Augenblick geschehend/mögl.* ‖ instantaneous * *(Adj.) augenblicklich (z.B. Wert, Geschwindigkeit), sofort* ‖ present * *(Adj.) gegenwärtig* ‖ actual * *(Adj.) aktuell, gegenwärtig, jetzig* ‖ temporary * *vorübergehend, zeitweise, provisorisch, zeit-/einstweilig* ‖ transient * *(elektr., mechan.) z.B. Stoss, Einbruch, Erhöhung* ‖ at the moment * *(Adv.) im Moment, zurzeit* ‖ at present * *(Adv.) im Augenblick, zurzeit* ‖ for the present * *(Adv.) zurzeit* ‖ for the present time * *(Adv.) zurzeit, im Moment* ‖ for the time being * *(Adv.) zum augenblicklichen Zeitpunkt, zurzeit* ‖ at the present time * *(Adv.) gegenwärtig* ‖ momentarily * *(Adv.) für e. Augenblick, kurz, vorübergehend, flüchtig* ‖ temporarily * *(Adv.) vorläufig, temporär, behelfsmäßig, provisorisch* ‖ instantaneously * *(Adv.) (auch physikal.) augenblicklich, sofort, unverzüglich, auf der Stelle* ‖ momentary * *(Adj.) vorübergehend, flüchtig* ‖ current * *laufend, gegenwärtig, jetzig, aktuell* ‖ currently * *(Adv.) jetzt, zurzeit*
Momentangeschwindigkeit instantaneous speed
Momentanwert instantaneous value
Momentanwertmessung instantaneous-value measurement * *z.B. schnelle Analog-/Digitalwandlung (Stufenumsetzer oder Flash)*
Momentenanregelzeit torque rise time
Momentenbegrenzung torque limiting
Momentenbelastbarkeit torque loadability * *für el. Motor: siehe z.B. VDE 0530 Teil 1, Beiblatt 2 und Teil 2*
momentenbildend torque-generating ‖ torque-producing
momentenfreie Pause period of no torque ‖ zero-torque interval ‖ torque-free interval ‖ dead time
Momentengenauigkeit torque accuracy
Momentengrenze torque limit * *along the torque limit: an der ~*
Momentenistwert torque actual value
Momentenkennlinie speed-torque characteristic * *{el. Masch.} Verlauf des Drehmoments über der Drehzahl*
Momentenklasse torque class * *bezeichnet mit 'KLxx' bei Motor*
momentenlose Pause zero-torque interval
Momentenregler torque controller
Momentenrichtung torque direction ‖ direction of torque
Momentenschale torque shell * *(z.B. beim SIEMENS-Regelsystem SIMADYN D (R))* ‖ torque block * *z.B. Stromregler, Kommandostufe und Steuersatz* ‖ converter block * *Softwarebaustein zur Leistungsteil-/Stromrichteransteuerung*
Momentenschwankungen torque fluctuations ‖ pulsation torque * *harte Schläge verursachend* ‖ torque ripple * *Momentenwelligkeit*
Momentensollwert torque setpoint ‖ torque reference
Momentenstoss torque surge
Momentenstütze torque arm ‖ torque bracket
Momentenüberlast overtorque
Momentenumkehr torque direction reversal ‖ torque reversal
Momentenumschaltung torque changeover ‖ torque-direction changeover * *Umschaltung der Momentenrichtung*
Momentenverlauf torque characteristic
Momentenwaage dynamometer * *z.B. Pendelmaschine* ‖ torque meter * *Drehmomenten-Messeinrichtung*
Momentenwechsel torque reversal ‖ change of torque direction * *Wechsel der Momentenrichtung* ‖ torque change * *Momentenänderung* ‖ torque direction change * *Wechsel der Momentenrichtung*
Momentenwelligkeit torque ripple
momentfreie Pause period of no torque ‖ zero-torque interval ‖ torque-free interval ‖ dead time
Monatsbericht monthly report ‖ monthly statement
Monitor screen * *Bildschirm* ‖ monitor * *Bildschirm, Überwachungseinrichtung* ‖ VDU * *Abk. f. 'Video Display Unit': Bildschirmanzeigegerät* ‖ CRT
Monitorprogramm monitor program * *Hilfsprogramm zum Testen von Anwenderprogrammen* ‖ monitoring program ‖ debugger * *Testhilfeprogramm zum Austesten von Softwareprogrammen*
Monopol monopoly
monostabil monostable
monoton monotone * *stetig fallend oder steigend (ohne Wendepunkte)* ‖ monotonic ‖ monotonous
Monotonität monotony * *stetiges Steigen oder Fallen ohne Wendepunkte* ‖ monoticity ‖ monotonousness
Montage installation * *Ein-/Aufbau (zu einer größeren Einheit); Industrieanlage, Maschine, Maschinenteil* ‖ assembly * *Zusammenbau, auch Leiterplattenbestückung* ‖ mounting * *Anbringung, Bestückung, Einbau, Befestigung auf e. Unterlage, auch Leiterplattenbestückung* ‖ setting up * *Aufstellen* ‖ fitting * *Anbau* ‖ erection * *Großanlage, Fabrik* ‖ assemblage * *das Zusammensetzen/-bauen*
Montageausschnitt mounting cutout
Montageband assembly-line ‖ assembly belt
Montagebohrung mounting hole
Montagebolzen mounting bolt
Montagebügel mounting bracket
Montagefläche mounting surface
Montagefolge sequence of assembling ‖ sequence of installation
Montagefüße mounting feet
Montagehalle assembly shop
Montagehilfe assembly aid
Montagehöhe installation height above sea level * *~ über d. Meeresspiegel*
Montageingenieur field engineer ‖ commissioning engineer * *Inbetriebnahmeingenieur* ‖ erection engineer
Montagelochung mounting hole
Montagemaße mounting dimensions ‖ mounting arrangements
Montageöffnung access hole * *siehe auch →Wartungsöffnung* ‖ mounting opening ‖ service opening * *→Wartungsöffnung*
Montageplan installation schedule * *Zeitplan* ‖ erection schedule * *Zeitplan* ‖ installation diagram * *techn. Zeichnung mit Darstellung des Auf- oder Einbaus* ‖ assembly drawing * *Montagezeichnung*
Montageplatte switch board * *z.B. für Mess- u. Bediengeräte, Schalttafel* ‖ mounting plate ‖ supporting plate * *Trägerplatte*
Montagepresse assembly press
Montagerahmen mounting frame
Montageschiene mounting rail * *z.B. Hutschiene* ‖ DIN rail * *Aufschnappmontageschiene C-/mini-C/G-förmig n. DIN EN50022/50045/50035 (35/15/32mm)* ‖ C-type mounting rail according to DIN EN 50022 * *C-Aufschnappinstallationsschiene 35 mm breit* ‖ mini C-type mounting rail according to DIN EN 50045 * *mini C-Aufschnappinstall.-schiene, 15mm breit* ‖ G-type mounting rail according to DIN EN 50035 * *asymmetr. G-Aufschnappinstall.schiene 32 mm breit*
Montageschriften mounting documentation ‖ installation instructions ‖ periodical installation guides * *regelmäßig erscheinende*
Montagesystem assembly system

Montagetechnik assembling systems
Montageüberwachung assembly supervision
Montagevorrichtung assembly equipment
Montagezeichnung installation diagram * *z.B. zur Montage e.Stromrichters i.schaltschrank* || assembly drawing || installation drawing
Montanindustrie coal and steel industry || coal and iron mining industry
Monteur electrician * *Elektriker* || mechanic * *Mechaniker, Schlosser* || assembly man * *in d.Fertigung* || engineer * *geschulter, erfahrener* ~ || fitter * *(mechan.) Monteur* || erector * *~ für Anlagenbaustelle* || maintenance man * *~ für Wartung, Instandhaltung* || plumber * *~ für sanitäre Anlagen* || technician * *service technician: Kundendienst~*
montierbar mountable
montieren mount * *bestücken, anbringen* || fit * *aufstellen, anbringen, einsetzen, montieren* || set up * *aufstellen* || assemble * *zusammenbauen* || install * *einrichten; auch Industrieanlage* || adjust * *einstellen* || fit into * *einbauen/setzen* || erect * *aufstellen, aufbauen, montieren, errichten*
montiert fixed * *on: an* || mounted
MOS MOS * *Abk. f. 'Metal-Oxyde Semiconductor'*
MOS-Transistor MOS transistor || metal-oxide semiconductor
Mosaikwarte mosaic-type control board * *Bedien-/Beobachtungssystem in der Verfahrenstechnik* || mosaic panel
motivieren motivate
Motor motor * *(elektr.)* || engine * *Verbrennungsmotor, Maschine, nicht-elektrischer Motor*
Motor für allgemeine Zwecke general-purpose motor * *{el.Masch.} siehe VDE 0530, Beiblatt 1*
Motor für Betrieb am Umrichter motor for operation with converter || inverter-duty motor
Motor in Fußbauform foot-mounting motor * *mit Fußbefestigung*
Motor mit Feldwicklung wound field motor * *Gegensatz: PM motor: permanenterregter Motor*
Motor-Baugröße motor size || frame size * *Baugröße, Achshöhe* || motor frame size
Motor-Nennimpedanz rated motor impedance
Motor-Nennstrom rated motor current
Motor-Wirkungsgrad motor efficiency
Motorabzweig motor feeder
Motoranbau motor mount * *Anbau/Montage (von Impulsgeber, Bremse usw.) an Motor*
Motoranbauten motor attachments * *z.B. Tacho, Bremse, Lüfter, Kupplung, Schutzdach*
Motoranlasser motor starter
Motoranschluss motor connection || motor supply lead * *Leitung* || motor supply cable * *Anschlusskabel*
Motoranschlusskabel motor feeder cable || motor supply cable
Motorausnutzung motor utilization
Motorbemessung motor rating * *{el. Masch.} elektromagnetische, thermische und mechanische Auslegung eines Motors*
Motorbetrieb motor operation
Motorbleche motor lamination
Motordaten motor specifications || motor data
Motordrehzahl motor speed || motor rpm
Motordrehzahl-Messinstrument motor speed meter || motor speedometer || speedometer
Motordrücker motor thruster * *f.Bremse/Kupplung/Schaltgetriebe m. groß. Stellkraft u.klein.Hub* || centrifugal thrustor * *centrifugal brake operator* * *elektromechan. oder hydraul. Betätigungselement für Bremse* || thrustor * * thrustor
Motoren großer Leistung high-rating motors
Motoren kleiner Leistung low-rating motors

Motorenbau motor building
Motorenbleche motor laminations
Motorfangschaltung flying-restart circuit * *siehe auch →Fangschaltung*
Motorgehäuse motor housing || motor frame
Motorgeräusch motor noise * *bei umrichtergespeistem AC-Käfigl.motor: siehe VDE 0530 Teil 1, Beiblatt 2*
Motorgröße motor frame size * *Baugröße, entspr. der Achshöhe* || frame size * *Baugröße, entspr. der Achshöhe*
Motoridentifikation motor identification * *durch Selbstoptimierungsvorgang* || motor identification run * *automat. Aufnahme der Motor-Kenngrößen durch Stromrichter* || motor identification routine * *automat. Ermittlung der Motor-Kenngrößen durch Antriebs-Stromrichter*
motorisch when motoring || motoring
motorisch angetrieben motorized
motorischer Betrieb motoring operation || motor operation
motorisches Moment torque when motoring
motorisches Sollwertpotentiometer motorized setpoint potentiometer
Motorkabel motor feeder cable * *Anschlusskabel* || motor supply cable * *Anschlusskabel*
Motorkaltleiter motor PTC resistor || motor PTC thermistor
Motorkatalog motor catalog
Motorkatze powered trolley * *{Hebezeuge}*
Motorkondensator motor capacitor * *z.B. für Einphasen-Wechselstrommotor*
Motorkreis motor circuit
Motorleistung motor rating || motor output || motor power || motor capacity || motorpower
Motorleitung motor lead * *Zuleitung zum Motor* || motor supply cable || motor feeder cable
Motorlüfter motor fan || motor blower
Motormoment motor torque * *das vom Motor an der Welle abgegebene Drehmoment*
Motornenndrehzahl rated motor speed
Motornennfrequenz rated motor frequency
Motornennleistung rated motor power || rated motor output || motor output rating || motor rating
Motornennmoment rated motor torque
Motornennschlupf rated motor slip * *((Synchrondrehzahl-Motornenndrehzahl)/Synchrondrehzahl) x 100%* || rated slip * *ca. 10% bei kleinen bis ca 0,3% bei großen Asynchronmotoren*
Motornennspannung rated motor voltage
Motornennstrom rated motor current || motor nominal current || motor current rating || rated motor amps
Motorola-Format Motorola format * *Datenformat: höherwertig.Byte/Wort auf niedriger, niederwertig.auf höherer Adresse* || big endian format * *Datenformat: höherwertig.Byte/Wort auf niedriger, niederwert.auf höherer Adesse*
Motorpotentiometer motorized potentiometer || motor operated potentiometer || motorpot ||
MOP * *Abkürzg. für 'motor operated potentiometer'* || motorpotentiometer || motor driven potentiometer
Motorpoti motorized potentiometer || motorpot ||
MOP * *Abkürzg.für 'MOtorized Potentiometer'*
Motorprüfstand engine test stand * *für Verbrennungsmotoren* || engine test bed * *für Verbrennungsmotoren*
Motorregelung motor control
motorschonend with low motor stressing
Motorschutz motor protection * *Schutzeinrichtg. gegen thermische Überlastung e. Motors*
Motorschutzrelais motor protecting relay || motor protection relay || thermal relay || motor protective relay || overload relay * *Überlast-Schutzrelais*

Motorschutzschalter motor circuit-breaker * *siehe z.B. DIN VDE 0660* || motor protecting switch * *(CEE 19)* || motor protector || motor protection circuit breaker || protective motor starter || starting circuit-breaker
motorseitig on the motor side || motor-side || load side * *von einem Stromrichter aus gesehen* || motor side
Motorsockel motor base
Motorstarter motor starter
Motorsteuergerät motor control unit * *z.B. Sanftanlaufgerät* || motor controller * *z.B. Sanftanlaufgerät*
Motorüberwachung motor monitoring
Motorverdichter motor-compressor * *Kompressor mit eingebautem Motor; hermetic: in e. gemeinsamen gasdichten Gehäuse*
Motorwelle motor shaft
Motorwicklung motor winding
Motorzuleitung motor lead || motor supply lead || incoming motor cable * *aus d. Sicht des Motorklemmenkastens* || motor supply cable || motor feeder cable || motor feeder
Motto slogan * *Schlagwort, Werbespruch, Slogan* || maxim * *Maxime, Grundsatz* || motto * *Motto, Wahlspruch, Sinnspruch* || catchphrase
MOV MOV * *Abkürzg. f. 'Metal-Oxide Varistor': Überspannungsschutzelement*
MP neutral * *neutrale Phase*
MPP MPP * *Abkürzung für 'Maximum Power Point': Betriebspunkt maximaler Leistung bei Solaranlage*
MPP-Regler MPPT * *Abk. f.'Maximum Power Point Tracker': regelt auf d. Punkt max. Leistung b.Solaranlage* || MPP tracker
MSB MSB * *Abkürzg.für 'Most Significant Bit': höchstwertiges Bit*
MSD MSD * *Abkürzg. für 'Most Significant Digit': Höchstwertiges Zeichen z.B. einer Ziffernanzeige*
MTBF MTBF * *MeanTime Between Failure: mittlere Ausfallzeit, Maß für die Zuverlässigkeit einer Einrichtung*
Muffe muff || sleeve * *Muffe, Buchse, Manschette, Kabelverbindung* || bushing * *Spannhülse, Muffe, Durchführungshülse* || bush || box * *Kabelverbindungs~* || junction box * *Kabelverbindungs~* || splice box * *Kabelverbindungs~* || pipe coupling * *Rohrverbindung* || pipe union * *Rohrverbindung* || coupling * *Rohrverbindung* || union * *Rohrverbindung*
Mühe trouble * *Mühe, Plage, Last, Belästigung, Verdruss, Scherereien, Unannehmlichkeiten* || pains * *Mühe, Bemühungen, Quälerei* || labour * *Arbeit* || effort * *Anstrengung* || difficulty * *Schwierigkeit* || waste of time * *vergebliche* ~ || waste of energy * *vergebliche* ~ || hassle * *(salopp) Mühe; (auch handgreifliche) Auseinandersetzung, Krach*
mühelos easy || without trouble || trouble-free || easily * *(Adv.)* || hassle-free
Mühle mill || crusher * *Brechmaschine, z.B. für Steine* || grinder * *Mahl/Quetschwerk, Schleifer* || pulverizer * *Zerkleinerer, Pulverisiermühle, Mahlanlage* || grinding mill * *Mahlwerk, Schleif/Reibmühle* || defibrizer * *[Holz] in der Spanplattenherstellung* || disintegrator * *Schlagmühle (z.B. in der Spanplattenherstellung)*
Mühlenluftgebläse mill blower
mühsam laborious * *mit viel Arbeit* || strenuous * *anstrengend* || toilsome * *anstrengend* || arduous * *anstrengend* || difficult * *schwierig* || hard * *schwierig* || tough * *schwierig* || tiring * *ermüdend* || irksome * *ermüdend* || troublesome * *mühevoll* || laboriously * *(Adv.)* || with an effort * *(Adv.)* || with difficulty * *(Adv.)*

Mulde cavity * *Höhlung* || depression * *Vertiefung* || hollow * *Hohlraum* || tray * *Bottich* || trough * *Bottich, Trog, Mulde, Wanne, Rinne*
Multi-Microcomputersystem multi-microprocessor system
Multicast- multicast * *an mehrere (Bus)Teilnehmer adressiert (z.B. Telegramm)*
Multicomputing multicomputing * *Mehrprozessortechnik*
Multifunktions- multi-function || multifunction || multifunctional || multi-purpose * *Mehrzweck-*
Multilayer-Leiterplatte multilayer PCB
Multimasterbetrieb multi-master operation * *bei Bussystem* || flying-master operation * *mit wechselndem Master bei Bussystem*
multimasterfähig multi-master capable
Multimasterfähigkeit multi-master capability * *z.B. Bussystem*
Multimeter multimeter || multi-range instrument
Multiplexbetrieb multiplex operation
Multiplexer multiplexer * *['maltiplexe]* || MUX * *Abk. f. 'MUltipleXer'*
Multiplikation multiplication
multiplizieren multiply * *by/with: mit; together: miteinander*
multiplizierend multiplying
Multiplizierer multiplier
Multiprozessorbetrieb multiprocessor operation || multiprocessor mode * *Betriebsart*
Multiprozessorfähigkeit multiprocessor capability
mündlich oral || orally * *(Adv.)*
Muss- mandatory
Mussfunktion mandatory function
Muster sample * *Probe, (Stück-/Typen-/Vorlageusw.) Muster* || pattern * *Impuls~, Bit~, Vorlage, Modell, Design, Motiv, Vorbild* || prototype * *Prototyp* || model * *Vorbild, (verkleinerte) Nachbildung, Arbeitsmodell, Vorlage, Bauweise, Bau~, Type* || type * *Bautype* || specimen * *Probestück* || standard * *Richtschnur* || example * *Vorbild, Beispiel* || paragon * *Vorbild, Muster, Ausbund* || design * *Entwurf, Zeichnung, Design, registered design: Gebrauchs~* || test model * *Versuchsmodell*
mustergültig exemplary || ideal || perfect || excellent || model * *als Beispiel/Muster dienend* || a model of * *musterhaft*
Mutter nut * *für Schraubverbindung*
Muttergesellschaft parent company || holding company * *Dachgesellschaft*

N

n-leitend n-conductive * *Halbleiterschicht*
N-Leiter N conductor || neutral conductor || neutral
N-Wellenende N-schaft end * *Non-drive side, Nicht-Antriebsseite, A-Seite eines Motors*
Nabe hub || nave || driving collar
Nabengehäuse hub housing
Nabennut hub keyway
nach according to * *entsprechend/gemäß einer Norm, Vorschrift, Beschreibung, Priorität, Reihenfolge* || in accordance with * *entsprechend einer Norm, Vorschrift; gemäß (eines Prinzips usw.)* || in conformity with * *entsprechend, gemäß (z.B. einer Vorschrift, Norm)* || after * *hinter (bezügl. Reihenfolge, Zeit usw.); bezüglich* || to * *zu einem Punkt hin, in Richtung; to the right:* ~ *rechts* || for * *zu einem Punkt hin (in Verbindung mit 'depart, leave, set out, bound, train')* || towards * *in Richtung; towards the top* ~ *oben; towards the*

bottom: ~ *unten zu* || bound for * *bestimmt* ~ || *subsequent* * *in der zeitlichen Reihenfolge (~-)folgend; subs. to: später als/~, in Anschluss an* || following * *in der zeitlichen Reihenfolge* || on * *on arrival:* ~ *(der) Ankunft; on receipt:* ~ *Erhalt* || by * *one by one: einer* ~ *dem anderen; little by little:* ~ *und* ~ || behind * *hinter* || about * *bezüglich* || by * *differenzieren (dv/dt); i.Zusammenhang mit Maßeinheit: sell by the weight:* ~ *Gewicht verkaufen*
nach Aufwand according to the effective expenses * *invoice: verrechnen* || depending on the actual time required * *beim Verrechnen von Dienstleistungen*
nach außen outside || towards the outside
nach der Begrenzung after limiting
nach hinten backwards * *zurück, rückwärts* || back || rearwards * *nach hinten zu* || to the rear * *zur Rückseite hin*
nach innen inside || towards the inside
nach links to the left
nach Maß taylored to suit
nach oben towards the top || upwards || upstairs * *in e. Haus*
nach rechts to the right
nach unten downwards || downstairs * *in Haus* || towards the bottom * *nach unten zu*
nach vorne to the front * *zur Vorderseite* || forward * *vorwärts* || ahead * *vorwärts* || towards the front * *nach vorne zu*
Nach- post- || re-
Nachbar- adjacent * *anstoßend, benachbart* || neighboring || neighbouring || adjoining * *Zimmer, Haus usw.*
Nachbau construction under license * *(amerikan.) Lizenzfertigung* || construction under licence * *(brit.) Lizenzfertigung* || copying || reproduction || imitation
Nachbaufertigung licensed production * *(amerikan.) Lizenzfertigung* || post production * ~ *von ausgelaufenen Geräten für den (Ersatzbedarf usw.)* || post-production phase * ~ *nach Fertigungseinstellung eines Produkts* || licenced production * *(brit.) Lizenzfertigung*
nachbearbeiten dress || refinish || remachine * *[Werkz.masch.]* || correct * *korrigieren (siehe auch* →*überarbeiten)* || post-machine * *[Werkz.-masch.]*
Nachbearbeitung finishing || secondary processing
Nachbehandlung further treatment || post-treatment || finishing
Nachbestellung further orders || repeat-order || second order || follow-up order * *Folgeauftrag* || reordering * *das Nachbestellen*
nachbilden simulate * *simulieren, nachbilden* || reproduce * *kopieren, wiedererzeugen, nachbilden* || imitate * *nachmachen/-ahmen, imitieren, kopieren* || emulate * *emulieren* || duplicate * *duplizieren, identisch nachbauen/-bilden* || copy * *kopieren* || model * *durch ein Modell* ~ *(auch ein mathematisches)* || counterfeit * *nachmachen, fälschen, heucheln, vorgeben; nachgemacht, unecht, gefälscht* || mimic * *nachahmen*
Nachbildung simulation * *Simulation, Nachahmung* || emulation * *Emulation* || imitation * *Nachbildung, Nachahmung, Imitation, das Nachgeahmte, Kopie* || copy * *Kopie* || reproduction * *Reproduktion, Nachbildung, Wiedererzeugung, Fortpflanzung, Wiedergabe* || replica * *genaue* ~ || facsimile * *genaue* ~ || mock-up * *Atrappe, Modell in natürlicher Größe* || dummy * *Atrappe* || modelling * *durch ein (mathemat.) Modell*
nacheilen lag * *behind: hinter; by: um* || run after
nacheilend lagging

Nacheilung lag * *der Zeitabstand* || lagging * *das Nacheilen (behind: hinter)*
nacheinander one after another || successively * *sukzessive* || consecutively * *aufeinander folgend* || by turns * *bei jedem Umlauf einmal* || in succession
nachfolgend downstream * *im Signalfluss* || subsequently * *(Adv.) z.B. in einer Beschreibung* || subsequent * *(Adj.)*
nachfolgend aufgeführt subsequently listed * *in Aufzählung, Liste*
Nachfolgeprodukt follow-up product || replacement product || successor product
Nachfolger succesor * *im Amt* || follower
Nachfolgetyp replacement type
Nachformierung reforming * *eines Kondensators nach langer Lagerung* || reforming process * *langames Aufladen e. Kondensators nach langer Lagerung*
Nachfrage demand * *low: geringe; high/great/heavy for hohe* ~ *nach (z.B. e.Produkt); keep up with: nachkommen* || inquiry * *das ~n, Rückfrage, Erkundigung* || rush * *rush for: starke* ~ *nach* || query * *Rückfrage, zweifelnde Frage*
nachfragen consult * *um Rat fragen* || inquire * *(nach)fragen nach, z.B. Lieferangebot, sich erkundigen* || enquire || query * *(aus/be)fragen, zweifelnd fragen, beanstanden, in Frage stellen, in Zweifel ziehen* || contact * *sich wenden an, Kontakt aufnehmen mit*
Nachführbetrieb tracking mode || follow-up operation || follow-up mode
nachführen track * *e. Größe e.anderen (to: auf); z.B. e. Prozessmodell, e. Hochlaufgeber auf d. Istwert usw.* || correct * *correct to the setpoint: auf den Sollwert* ~; *correct the setpoint: den Sollwert* ~ || keep track of * *(nach-)*→*verfolgen*
Nachführung tracking * *eine Größe einer anderen* || tracking function * *als Funktion* || tracker * *als Einrichtung/(Funktions-) Baustein* || matching * *Anpassung*
nachfüllen refill * *wieder füllen, nach-/auffüllen* || fill up || replenish * *(wieder) auffüllen, ergänzen (with: mit)* || top up * *Isolier-/Kühl-/Schmiermittel/Öl usw. auf den regulären Füllstand* ~
nachgeschaltet series-connected * *in Reihe geschaltet* || downstream * *im Leistungs-/Signalfluss* ~ || subsequent * *nachfolgend* || subsequently connected || connected in series || backend * *[Computer] z.B. ~es Softwareprogramm*
Nachgeschichte after-event history * *Inhalt eines Tracespeichers nach dem Triggerzeitpunkt* || post-mortem history * *nach einer Fehlerabschaltung aufgezeichnete Messdaten* || post-event history
nachgiebig elastic * *(auch figürl.) elastisch, federnd, dehnbar, biegsam, geschmeidig, anpasungsfähig* || pliable * *(auch figürl.) biegsam, geschmeidig, nachgiebig, fügsam, leicht zu beeinflussen* || yielding * *nachgebend, dehnbar, biegsam; (figürl.) gefügig, nachgiebig, zum Nachgeben bereit* || compliant * *willfährig, nachgiebig, zum Nachgeben bereit* || complaisant * *entgegenkommend, gefällig, höflich* || forbearing * *nachsichtig, geduldig* || indulgent * *nachsichtig (to: gegen, towards: gegenüber)* || flexible * *(auch figürl.) flexibel (Kupplung usw.), biegsam, beweglich, geschmeidig*
Nachgiebigkeit flexibility * *(auch figürl.)* || yieldingness * *Dehn-/Biegsamkeit; (figürl.) Nachgiebigkeit, Gefügigkeit* || complaisance * *(figürl.) Willfährigkeit* || elasticity * *Elastizität (auch figürl.), Feder-/Spannkraft* || compliance * *Willfährigkeit* || resilience * *elastische* ~, *z.B. einer Feder* || indulgence * *Nachsicht, Duldung, Toleranz; Gefälligkeit*

Nachkommastelle place after the decimal point || digit after the decimal point * *auf einer (Ziffern-) Anzeige*
Nachkommastellen decimal places
nachlassen pay out * *z.B. Papierbahn entspannen* || let out * *z.B. bei Papierbahn* || slack payout * *z.B. bei Papierbahn* || loosen * *lockern* || slacken * *lockern* || let go * *lockern* || make a reduction * *~ von Preisen usw.* || allow a discount * *~ von Preisn usw.* || diminish * *sich vermindern* || decrease * *sich vermindern, abnehmen* || soften * *milder werden* || weaken * *schwächer werden* || deteriorate * *schlechter werden* || cease * *aufhören* || slow down * *~ des Tempos* || drop * *~ von Preisen usw.* || fall off * *~ der Verkäufe usw.* || relax * *~ von Tätigkeit, Anspannung* || abate * *~ von Schwäche, Schmerz, Frost usw.*
Nachlassen-Taste slack-pay-out button * *z.B. bei Papiermaschine*
Nachlauf over-travel * *meist unerwünschtes Weiterlaufen e. Antriebs nach Erreichen der Ziel-/Halteposition* || track alignment * *Versatz zw. Radauflagepunkt und Lenkachse eines Fahrzeugrades* || last runnings * *beim Leeren eines Flüssigkeitsbehälters* || overtravel * *~en eines Antriebs über das Ziel hinaus* || overshoot || running on * *"Nachdieseln" eines Verbrennungsmotors* || dieseling * *"Nachdieseln" eines Verbrennungsmotors*
nachlaufen run after * *hinterherlaufen* || follow * *folgen* || trail * *hinter sich herziehen, (auf der Spur) verfolgen, (nach-) schleppen* || go behind * *hinterherlaufen* || lag behind * *mit (zeitlichem) Versatz hinterherlaufen* || overtravel * *über die Zielposition hinaus ~* || coast * *im Leerlauf/Freilauf fahren (ohne Antriebsdrehmoment)* || overshoot * *überschwingen, überschießen*
nachmessen check * *überprüfen* || re-measure * *erneut messen*
Nachoptimierung post-optimization
nachprüfen check * *kontrollieren* || verify * *verifizieren, auf die Richtigkeit hin prüfen, die Richtigkeit feststellen* || make sure * *sicherstellen* || investigate * *untersuchen* || inspect * *inspizieren, untersuchen* || check-up
Nachprüfung verification || verification of operation * *Funktionsprüfung* || check * *Kontrolle, Überprüfung, Nachprüfung* || check up * *Überprüfung, Kontrolle* || inspection || test * *Prüfung* || examination * *(nähere) Untersuchung, Prüfung, Besichtigung, Durchsicht, Revision, Examen, Verhör*
nachrechnen re-calculate
nachregeln re-adjust || correct
Nachricht message * *Meldung, Botschaft* || news * *Kunde, Neuigkeit(en), Zeitungsnachricht; good/bad: gute/schlechte* || information * *Mitteilung* || notice * *Mitteilung* || note * *Botschaft*
Nachrichtentechnik information technology || telecommunications technology
Nachrichtenübertragung communication || telecommunication || information transmission
nachrüstbar can be retrofitted || upgradable * *hochrüstbar* || refitable
nachrüsten retrofit * *nachträglich einbauen* || upgrade * *hochrüsten (Hard- oder Software)*
Nachrüstsatz upgrade set * *Hochrüstsatz* || retrofit assembly || add-on kit || retrofit set
nachschalten connect in series * *in Reihe schalten* || connect in outgoing circuit * *im Ausgangkreis* || connect on load side * *lastseitig* || connect downstream * *im Signal-/Leistungsfluss hinterherschalten*
nachschlagen consult * *z.B. in Buch/Betriebsanleitung ~* || refer to * *z.B. in Buch/Betriebsanleitung/Nachschlagewerk ~* || look up * *z.B. in Tabelle/Liste/Buch/Betriebsanleitung*

Nachschlagewerk reference book * *auch zu einem technische Hard- oder Softwareprodukt* || work of reference * *auch zu einem technischen Hard- oder Softwareprodukt* || encyclopaedia * *Enzyklopädie, Lexikon* || book of reference
nachschmierbar regreasable
Nachschmiereinrichtung relubricating device * *z.B. für Wälz-/Gleitlager (bestehend aus Schmiernippel und Rohrleitung)* || regreasing device * *z.B. für Elektromotor* || equipment for regreasing * *z.B. für Wälz-/Gleitlager* || relubrication device || relubricating device
nachschmieren regrease || relubricate || regreasing * *(Subst.)* || relubrication * *(Subst.)*
Nachschmierfrist relubrication interval || lubrication interval
Nachschmierung regreasing
Nachschnittsäge resaw * *[Holz]*
Nachspann trailer * *z.B. eines Datentelegramms, Datensatzes*
nachspannen re-tighten * *Riemen, Kette usw.* || restretch * *Riemen, Kette usw.*
Nachspeise dessert * *[di'söt]*
nächst next * *in der Reihenfolge* || nearest * *nächstgelegen*
nachstellbar adjustable || re-adjustable
nachstellen readjust * *Lager usw.*
Nachstellsicherung adjustment lock
Nachstellzeit integral-action time || integral time || reset time || integral time constant
nächster next * *in zeitlicher Reihenfolge* || following * *~ in der Reihenfolge* || nearest * *~ bezüglich Entfernung, Beziehung; am nächsten gelegen* || shortest * *kürzest* || next after * *(Präposition)* || next to * *(Präposition)* || close to * *(Präposition)*
nächstgelegen nearest
Nachtabsenkung night setback * *bei Heizung usw.*
Nachteil disadvantage * *big: großer; severe: schwerer; to the disadvantage of: zum ~ von* || drawback * *→Mangel, Beeinträchtigung, Schattenseite, Hindernis, 'Haken an der Sache'* || shortcoming * *Unzulänglichkeit, Fehler, Mangel* || detriment * *Schaden, Nachteil* || handicap * *(auch figürlich) ein Hinterherhinken bewirkend; be handicapped: im Nachteil/behindert sein* || prejudice * *Schaden, Nachteil*
nachteilig disadvantageous * *to: für* || adverse * *ungünstig; affect adversely: ~ beeinflussen* || unfavourable * *ungünstig* || harmful * *schädlich* || detrimental || bad
nachteilige Auswirkungen negative effects
Nachtisch dessert * *[di'söt]*
Nachtrag supplement * *z.B. zu einem Katalog* || addendum || appendix * *Anhang*
nachträglich additional * *ergänzend* || supplementary * *ergänzend* || subsequent * *später* || belated * *verspätet*
nachträglich einbauen retrofit
nachtriggern retrigger
Nachverarbeitung subsequent processing || finishing
nachverfolgen track || keep track of * *sich dauernd auf d.laufenden halten über*
Nachwalzgerüst temper mill * *Dressiergerüst in Walzwerk zur Nachbehandlung*
Nachweis proof || evidence || voucher * *['wautscher] Beleg* || record * *Unterlage* || certificate * *Zeugnis*
nachweisbar verifiable || provable || traceable || detectable || evident * *offenkundig*
nachweisen prove * *beweisen* || point out || show || establish * *feststellen, nachweisen*
Nachwirkung after-effect || consequences * *Folgen*
Nachwuchsförderung training for the next genera-

tion || encouragement of young talent || training of new recruits
Nachwuchskraft trainee || junior engineer * *Ingenieur*
nachwuchten rebalance || re-balance
Nachwuchtung re-balancing
Nadelfliesmaschine needling machine * *{Textil}*
nadelförmig needle-shaped
Nadelholz coniferous || softwood * *Weichholz* || coniferous wood
Nadelimpuls spike
Nadelkäfig needle cage * *bei Nadellager*
Nadellager needle bearing || needle roller bearing
Nadelloch pinhole * *kleines Loch, z.B. in Papier*
NAFTA NAFTA * *Abkürzg.f. 'North American Free Trade Agreement': Wirtsch.gemeinsch. zw. USA, Kanada u.Mexiko*
Nagel nail || pin
Nagelgerät nailer
nah close * *dicht (to: an/am/bei)* || near
Nahansicht close-up view
Nahaufnahme close-up * *Foto, Film*
Nahbus local bus
nahe close * *(to: an)*
nahe liegend obvious * *(figürl.)* || nearby * *in der Nähe* || near at hand * *in der Nähe verfügbar*
nahe Null close to zero
Nähe vicinity * *Umgebung* || proximity * *auch '(An-) Näherungs-'* || nearness || surroundings * *Umgebung* || neighbourhood * *Umgebung, Nachbarschaft*
nähen saw
näher in detail * *im Detail; go into detail: ~ ausführen; details: ~es* || nearer || closer || shorter * *(Weg usw.)* || more specific * *mehr ins Einzelne gehend, spezieller, genauer* || more detailed * *detaillierter* || more precise * *genauer* || further * *further particulars: ~e Einzelheiten*
Näheres particular || further information * *weitere Angaben/Informationen*
näherkommen get close * *get close to: sich an ... annähern* || approach * *→nähern*
nähern approach to * *sich etw. ~, sich an~ an* || near * *nähern, sich ~* || come nearer * *sich ~* || draw nearer * *sich ~* || get close to * *sich etwas ~, an~, näherkommen* || close in * *bedrohlich näherkommen* || approach * *sich einer Person ~*
Näherung approximation
Näherungsformel approximation formula * *siehe auch →Faustformel*
Näherungsfunktion approximation function * *siehe auch →Faustformel*
Näherungsrechnung approximate calculation
Näherungsschalter proximity switch
Näherungsverfahren approximation method || approximation procedure
näherungsweise by approximation || approximate * *ungefähr, angenähert, annähernd* || approximately * *(Adv.)*
Näherungswert rough value || estimated value * *Schätzwert* || approximate value || approximation value
nahezu nearly * *beinahe* || almost * *fast* || next to * *next to impossible: ~ unmöglich* || practically * *praktisch* || virtually * *eigentlich, praktisch, im Grunde genommen* || quasi * *gleichsam, gewissermaßen, sozusagen* || de-facto * *~ wirklich*
Nähmaschine sewing machine
Nahrungsmittel food || foods * *(Plural)* || foodstuff
Nahrungsmittelindustrie food industry || food-processing industry || foodstuff industry
Naht seam || weld * *z.B. Schweißnaht* || joint * *Verbindungsstelle*
nahtlos seamless * *(auch figürl.)*

Nahtschweißmaschine seam-welding machine
Nahtstelle interface || boundary position || interface * *Schnittstelle* || boundary position * *Grenzstelle, Rand* || control interface * *~ zum Prozess für Steuer- u. Regelsignale*
Nahverkehr local traffic || suburbian traffic || mass transit traffic || short-distance traffic || junction traffic || local service || short-distance communication * *Nahverkehrsverbindung*
Nahverkehrs-Triebwagen suburban rail car || suburban traction vehicle || rail-car for local transportation || short-haul rail-car
Nahverkehrsfahrzeug light rail vehicle * *leichtes Schienenfahrzeug*
Nahverkehrssystem suburban transportation system * *siehe auch →Personennahverkehr* || short-haul transport system
Name name || identifier * *Bezeichner* || label * *{Computer} Sprungmarke usw.* || designation * *Bezeichnung* || reputation * *Ruf, Ansehen* || term * *Fachausdruck*
Namen names || terminology * *Terminologie, Fachausdrücke*
nämlich namely || that is || that is to say || i.e. * *Abkürzung für 'id est': das heißt*
NAMUR NAMUR * *Normen-Arbeitsgemeinschaft für Mess- u.Regeltechnik i.d. chem.Industrie (deut.Normengremium)* || Standards Working Group for Instrumentation and Control in the Chemical Industry || German Standards Committee for Measurement and Control * *deutsches Normengremium in d. chem. Industrie*
NAMUR-Empfehlung NAMUR recommendation
narrensicher fool-proof
Nase nose * *Nase, Vorsprung, Schnabel, Mündung* || lug * *Ansatz, Lasche, Lappen (aus Blech usw.), Öse, Henkel, Zinke, Fahne (z. Schraub-/Lötverbindg.)* || tab * *Lappen, Zipfel, Schildchen, Nase, Öse, Fahne, Flachstift, Vorsprung* || projection * *vorspringender Teil, Vorsprung*
nass wet * *dripping wet: tropf~; become/get wet: ~ werden; wet: (Verb) ~machen* || damp * *feucht, dunstig* || moist * *feucht* || humid || soaked * *tropf~* || liquid
Nassbehandlung wet-treatment
Nassfestigkeit wet strength * *von Papier*
Nassi-Schneidermann-Diagramm Nassi-Schneidermann chart * *{Computer}* || NS chart * *{Computer}* || structured chart * *{Computer}* Struktogramm || structogram * *{Computer} Struktogramm*
Nasslauf wet running * *Kupplung, Bremse*
Nassläufer wet-rotor * *bei Pumpenantrieb: Läufer läuft in der zu fördernden Flüssigkeit*
Nassläufermotor wet-rotor motor * *bei Pumpenantrieb: Läufer dreht sich in der zu fördernden Flüssigkeit*
Nasspartie wet section * *vorderer Teil einer Papiermaschine* || wet end * *vorderer Teil einer Papiermaschine*
Nasspresse wet press * *zur Papierentwässerung in Papiermaschine*
national national || domestic * *heimisch, Binnen-*
Naturfaser natural fiber
Naturfasern natural fibers * *(amerikan.)* || natural fibres * *(brit.)*
Naturkautschuk natural rubber || pure india-rubber
natürlich natural || normal || genuine * *echt* || native * *angeboren* || unaffected * *ungekünstelt* || simple * *einfach* || real * *real size: natürliche Größe* || obviously * *(Adv.) auf der Hand liegend, offensichtlich, einleuchtend, augenfällig, klar*
natürliche Kommutierung natural commutation
natürliche Zahl natural number

natürlicher Zündzeitpunkt natural firing point || natural trigger instant
NC-gesteuert NC-controlled || numerically controlled
NC-Programmiersystem NC programming system
NC-Sprache NC programming language * *Programmiersprache für Bewegungssteuerungen, z.B. nach DIN 66025*
NC-Steuerung NC control * *Abk.f. 'Numeric Control': numerische Positioniersteuerg. (für Werkzeugmaschinen usw.)* || numeric control
Nebelschmierung mist lubrication * *durch Ölnebel*
neben by || by the side of || beside || alongside of || side by side with || next to * *unmittelbar ~* || near to * *dicht ~* || close by * *dicht ~* || against * *verglichen mit* || compared with * *verglichen mit* || apart from * *nebst* || besides to * *nebst* || in addition to * *zusätzlich zu* || amongst * *(mitten) unter, inmitten; amongst other things:* ~ *anderen Dingen* || among * *siehe →amongst* || adjacent to * *angrenzend/-stoßend an; benachbart zu; Nachbar-*
Neben- side- * *z.B. side letter: schriftliche ~vereinbarung* || auxiliary * *Hilfs-* || secondary * *zweitrangig, untergeordnet, sekundär; z.b. secondary aspect: ~aspekt* || supplementary * *Zusatz-, Hilfs-, Ersatz-* || additional * *zusätzlich* || by- * *by-product: ~produkt* || ancillary * *untergeordnet (to: zu), Hilfs-, Neben-*
Nebenantrieb slave drive || helper drive * *Hilfsantrieb* || auxiliary drive * *Hilfsantrieb* || speed follower * *Folgeantrieb, fährt mit einer vom Leitantrieb vorgegebenen Geschwindigkeit*
Nebenaspekt side aspect || secondary aspect || side issue
nebenbei gesagt by the way
Nebenbetriebe auxiliary plants || ancillary plants
Nebeneffekt side-effect || secondary phenomenon * *Begleiterscheinung* || attendant symptom * *Begleiterscheinung* || accompaniment * *Begleiterscheinung* || unintentional effect * *ungewollter/unbeabsichtigter Effekt*
nebeneinander side by side || next to one another || adjacent * →*benachbart*
Nebeneinrichtung auxiliary equipment
Nebenerscheinung secondary effect || side effect || secondary phenomenon
Nebenkosten incidental expenses || incidentals || extras || secondary costs
Nebenprodukt by-product
Nebenpunkt side issue || side aspect * *Nebenaspekt* || secondary aspect * *Nebenaspekt*
Nebenregister sidelay register * *(Druckerei)*
Nebenschluss shunt circuit || parallel connection * *Parallelschaltung*
Nebenschlusskennlinie shunt characteristic * *(el. Masch.)*
Nebenschlussmaschine shunt-wound machine
Nebenschlussmotor shunt-wound motor || shunt motor
Nebenschlusswicklung shunt winding || parallel winding
nebenstehend adjacent || marginal * *am Rande* || opposite * *in Dokument nebenan abgebildet*
Nebenstelle extension * *Telefon, Hausapparat*
Nebenstromkreis supplementary circuit
Nebenwiderstand shunt resistor || shunt || parallel resistor * *parallelgeschalteter Widerstand*
NEC NEC * *Abkürzg. für 'National Electrical Code': amerikanische Errichtungsbestimmungen*
negativ negative
negativ geladen negatively charged
negative Flanke falling edge * *fallende Flanke* || trailing edge * *Rückflanke* || negative edge || negative-going edge

negative Logik negative logic * *niedriger Spannungspegel entpr. logisch "1"*
negieren invert * *logisch*
Negierer sign inverter * *Vorzeichenumkehrer*
neigbar inclinable || canting
neigen zu tend to
Neigung inclination * *(auch figürl.: Hang, Vorliebe)* || slope * *geneigte Fläche* || incline * *geneigte Fläche* || gradient || dip * *z.B. ~ eines Schiffs, ~ einer Straße* || tilt * *Kipplage* || tilting * *Kipplage* || propensity * *(figürl.) Hang, Vorliebe (to/for: zu)* || tendency * *Tendenz (towards: zu)* || trend * *Tendenz (towards: zu)* || disposition * *(figürl.) Veranlagung*
Neigungswinkel angle of inclination
NEMA NEMA * *Abk.für 'National Electrical Manufacturers Association' Normengremium in den USA ähnlich VDE*
Nenn- nominal || rated * *z.B. rated current: Nennstrom*
Nenn-Anschlussquerschnitt rated connection cross-section || rated connecting capacity || rated wire range || rated conductor size || rated wire gauge
Nennanschlussspannung nominal input voltage || nominal AC voltage * *Stromrichter* || rated supply voltage * *z.B. für Motor, Umrichter*
Nennantriebsleistung rated input power * *eines Getriebes*
Nennbedingungen nominal rating conditions
Nennbelastbarkeit load rating
Nennbereich nominal range
Nennbetrieb type of operation * *Betriebsart z.B. für Motor nach VDE 0530 oder Stromrichter nach DIN 41756* || operation under rated conditions * *unter Nennbedingungen* || operating at the nominal working point * *am Nenn-Arbeitspunkt*
Nennbetriebsart type of operation * *Betriebsart z.B. für Motor nach VDE 0530 oder Stromrichter nach DIN 41756*
Nenndaten nominal data || rating data || ratings || nominal values || rated values || set of rated values || nominal rating
Nenndauerstrom rated continuous current || continuous current rating
Nenndrehmoment nominal torque || rated torque || break-down torque * *Kippmoment z.B. bei Asynchronmotor* || rated load torque * *Nenn-Lastmoment* || torque rating
Nenndrehzahl nominal speed || rated speed
Nenndurchmesser nominal diameter * *z.B. bei Ventilator*
nennen call * *bezeichnen* || term * *mit einem (Fach-) Ausdruck be~* || name * *mit einem Namen/Ausdruck ~* || designate * *bezeichnen* || mention * *erwähnen* || quote * *anführen* || style * *betiteln, benennen, bezeichnen, anreden mit/als* || nominate * *Kandidaten ~*
nennenswert significant || considerable || worth mentioning
Nenner denominator * *(mathematisch u. figürl.) reduce to a common denominator: auf einen ~ bringen*
Nennfrequenz rated frequency
Nenngeschwindigkeit nominal speed || nominal line speed * *Bahn-,Bandgeschwindigkeit* || nominal velocity || rated speed
Nenngleichspannung rated DC voltage
Nenngleichstrom rated DC current || rated direct current
Nennisolationsspannung rated insulation voltage
Nennisolationswechselspannung rated AC insulation voltage
Nennlast rated load * *under: bei*
Nennleistung rated power || rated output * *z.B. e.*

Stromrichters; Nennwellenleistung e.Motors (siehe DIN 42673, IEC72 u.VDE 0530) || rating || output rating * *z.B. Ausgangs~ eines Stromrichters* || nominal rating * *z.b. ~ eines (Stromrichter)Geräts* || power rating
Nennleistung haben   be rated * *Nennleistung von ... haben*
Nennleistungsfaktor   rated power factor
Nennmoment   rated torque || nominal torque
Nennpunkt   operating point * *Betriebspunkt* || rated point || nominal working point
Nennschlupf   rated slip * *b. kleinem bzw. großem Asynchronmotor ca. 10% bzw. 0,3% (s. auch →Motornennschlupf)* || rated motor slip * *((Synchrondrehzahl-Motornenndrehz.)/Synchrondrehzahl) x 100%*
Nennspannung   rated voltage || nominal voltage
Nennspannungen für elektrische Netze   nominal voltages of supply systems * *siehe IEC 38 (einphasig 120/240V, 3-phasig 400/480/690/1000V)*
Nennstrom   rated current || nominal current || current rating * *spezifizierter Nennstrom*
Nennwechselstrom   rated AC current || rated alternating current
Nennwert   nominal value * *gerundeter Zahlenwert einer Größe, mit dem e. Einrichtung/Gerät gekennzeichnet wird* || rated value || rating
netto   net || clear || pure * *rein*
Netto-   net
Nettodaten   net data * *z.B. Nutzdaten eines Telegramms (ohne Telegrammrahmen)*
Nettogewicht   net weight
Nettopreis   net price
Netz   line * *~~leitung, Speiseleitung, Strom~ (aus der Sicht des Verbrauchers)* || system * *Strom~, Energieversorgungssystem* || supply * *Versorgung, Einspeisung* || mains * *Strom~* || net || network * *Kommunikations~, Vertriebs~, Service~; Stromversorgungs~; (public: öffentliches)* || power-light system * *Hausstromversorgung* || power system * *Stromversorgungssystem* || supply system * *Versorgungs/Speisesystem; on the supply system: am ~* || distribution system * *Verteilungssystem, Stromsystem, Verteilungs~* || supply-system * *Versorgungs/Speisesystem; on the supply-system: am ~* || supply line * *Versorgungs~ vom Verbraucher aus gesehen* || supply network || line supply * *Stromversorgungs~*
Netz-   line || line-side * *netzseitig* || supply || mains
Netz-Aus   power off
Netz-Ein   power on
Netz-Grundschwingungsleistungsfaktor   line-side fundamental power factor
Netzanschluss   power connection || supply connection || connection to supply system || system connection || mains connection
Netzanschlussspannung   line supply voltage * *feed from a 400V line supply voltage: an einer ~ von 400V betreiben* || supply voltage
Netzausfall   input power failure * *die Eingangsspannung eines Geräts betreffend* || power failure || power loss || supply failure * *aus d.Sicht eines Verbrauchers z.B. Motor/Stromrichter* || mains failure || power fail * *Stromversorgungsausfall, z.B. eines Rechners* || line-voltage failure || mains interruption || supply interruption
Netzausfallüberbrückung   power loss ride-through * *Motor bleibt nicht stehen und läuft nach Netzwiederkehr weiter*
Netzbedingungen   line supply conditions * *poor: schlechte*
Netzbelastung   mains loading || line supply stressing * *z.B. durch Oberwellen*
Netzbetrieb   direct on-line operation * *eines Motors ohne Umrichter direkt am Netz* || on-line operation * *Betrieb am Stromnetz oder Kommunikationsnetz*

netzbetrieben   direct-on-line * *z.B.Motor i. Ggs. zum Umrichterbetrieb* || mains-operated
Netzbrumm   system hum
Netzdrossel   line reactor || line-side reactor || mains choke
Netzeinbruch   voltage dip * *kurzzeitiger ~* || supply dip * *kurzzeitiger ~, aus d. Sicht eines Verbrauchers/Motors/Stromrichters* || system voltage dip * *kurzzeitiger ~* || supply voltage dip * *kurzzeitiger ~*
Netzeinbruchüberbrückung   supply dip buffering || supply dip ride-through || supply voltage dip buffering || system voltage dip buffering
Netzeinspeisung   supply connection || commercial power supply input * *vom öffentlichen Netz* || supply system feeder * *zu Verbrauchern* || input to network * *Einspeisung ins Netz hinein* || supply feeder || supply feed || system infeed * *ins (Stromversorgungs-) Netz* || mains feeder
Netzfehler   supply failure
Netzfilter   line filter || AC line filter || mains filter || RFI filter * *Abk.f. 'Radio-Frequency Interference suppression filter': Funkentstörfilter*
netzfrequent   at system frequency || at line frequency
Netzfrequenz   line frequency || supply frequency || system frequency || power frequency || power line frequency || mainline frequency || mains frequency
Netzfrequenzanpassung   supply frequency adaption * *z.B. bei Steuersatz e. netzgeführten Stromrichters* || automatic adaption to the system frequency * *z.B. b. Steuersatz e.netzgeführten Stromrichters* || automatic line frequency tracking * *automat. Netzfrequ.nachführg. b.Steuersatz e.netzgef. Stromricht.* || adaption to the line frequency * *Netzfrequ.nachführg. d. Steuersatzsynchronisation e. netzgef. Stromr.* || line frequency tracking
Netzfrequenznachführung   line frequency tracking * *z.B. bei netzgeführten Stromrichter* || supply frequency tracking * *z.B. bei netzgeführtem Stromrichter* || adaption to the line frequency * *z.B. automat. Netzfrequenznachführung bei netzgeführtem Stromrichter*
Netzführung   line commutation * *für Stromrichterschaltung (z.B. bei Thyristorschaltung)*
netzgeführt   line-commutated * *Stromrichter bei d. das Netz die Kommutierungsspannung liefert* || line-commutating
netzgeführter Stromrichter   line-commutated converter * *siehe DIN EN 60146, VDE 0588 und IEC 146-1-1*
Netzgerät   power supply unit || power pack || mains power supply || power supply
netzgespeist   mains-operated
Netzgleichrichter   mains rectifier || line-side rectifier * *netzseitiger ~* || power rectifier
Netzhalbwelle   supply half wave
Netzimpedanz   system impedance || system supply impedance || line supply impedance || line impedance
Netzkabel   mains lead * *z.B. bei Kleingerät, Haushaltgerät* || power cord * *flexible Netzanschlussleitung, z.B. für Küchengerät*
Netzklemme   line terminal
Netzklemmen   line-side terminals * *z.B. e. Stromrichters*
Netzknoten   network node * *im Nachrichtennetz*
Netzkommutierungsdrossel   line commutating reactor
Netzkurzausfall   brief supply failure
Netzkurzschlussleistung   system fault power || system fault level
Netzleistungsfaktor   line power factor || system power factor

Netzleittechnik power systems control || power system management
Netzleitung line-supply cable || power cord * z.B. für mobiles Gerät || systems cable * z.B. Motor-/Stromrichterzuleitung || supply cable || mains lead
Netzmittel wetting agent * *Benetzungsmittel*
Netzmittelpunkt neutral point || supply-system neutral point
Netznulldurchgang supply voltage zero crossing || supply zero crossover
Netzreaktanz line reactance
Netzrückschalteinrichtung static bypass switch * überbrückt e. unterbrech.freie Stromversorgg. (→USV) nach Netzwiederkehr
Netzrückspeise-Einheit regenerative feedback unit
Netzrückspeiseeinheit regenerative feedback unit * zur Rückspeisung v. Bremsenergie ins Netz
Netzrückspeisung regenerative feedback * z.B. von Bremsenergie || regenerative feedback into the supply || regeneration back to the mains || regeneration of energy back into the mains || regeneration into the system
Netzrückwirkung system reaction * *Beeinflussung der Netzspannung durch Stromrichter/Motor* || system perturbation * *Beeinflussg. d. Netzspanng. durch Stromrichter/Motor* || harmonic effect on the system current || reaction on system || line reaction || line perturbation * *durch el. Masch.: siehe z.B. DIN VDE 0875, IEC 22G/21/CDV und IEC 1800-3*
Netzrückwirkungen harmonic effects on the supply * *Oberwellen-behaftete Belastung des Netzes durch Stromrichter* || noise injected back into the line * *Oberwell.-behaftete Belastg.d.Netzes durch taktenden Stromrichter* || harmonic noise injected back into the line * *Oberwellen-behaftete Netzbelastg. z.B. durch Stromrichter* || system perturbations || harmonic effects on the system || phase effects on the system || noise fedback into the supply system || line noise || line disturbances fed back into the distribution system || harmonics fed back into the supply || mains interference || line harmonics || system noise || noise feedback || harmonic disturbance on the mains supply || mains pollution * *(salopp)*
Netzschalter mains power switch || power switch || line switch || mains breaker * *netzseitiger Leistungsschalter*
netzschonend with low line supply stressing
Netzschutz network protective relaying || line protection || power system protection
Netzschütz line contactor
Netzschützansteuerung line contactor control || line contactor energizing
Netzschwankungen supply fluctuations || line supply fluctuations
Netzseite line side * *at/on the: auf der*
netzseitig line-side || at the line side || on the line side
netzseitiger Gleichrichter line-side rectifier * z.B. bei Frequenzumrichter
Netzsicherung line fuse || line-side fuse || mains-side fuse
Netzsimulator mains simulator
Netzspannung line voltage || supply voltage * *Eingangsspannung für Stromrichter* || mains voltage || phase to phase voltage * *Drehstrom* || system voltage
Netzspannungbereich mains voltage range
Netzspannungsabweichung deviation from rated system voltage * *vom Nennwert* || variation from rated system voltage * *vom Nennwert*
Netzspannungsausfall mains failure || power failure || line voltage failure
Netzspannungsbereich supply voltage range

Netzspannungseinbruch supply-system voltage dip || power line sag
Netzspannungsschwankungen supply voltage fluctuations || power supply variations * *bezogen auf die Netz-Versorgungsspannungen* || line voltage fluctuations || fluctuations in supply voltage || supply fluctuations * *bezogen auf die Netz-Versorgungsspannung*
Netzspannungsspitzen line voltage spikes
Netzspannungstoleranz line voltage tolerance
Netzspannungsüberwachung line voltage monitoring || line tolerance monitoring
Netzspannungswiederkehr mains restoration || power restoration
Netzstecker mains plug
Netzstörung line disturbances * *Einbrüche, Verzerrungen, Spitzen* || supply fault * *aus der Sicht eines Verbrauchers/Motors/Stromrichters* || line-voltage disturbance || line-voltage failure * *→Netzspannungsausfall*
Netzstrom line current || system current
Netzsymmetrie supply symmetry
Netzteil power supply || power pack
Netzthyristor line-frequency thyristor * *kann mit Schaltfrequenzen im Bereich der Netzfrequenz betrieben werden*
Netztransformator power transformer || line transformer || mains transformer || power-supply transformer
Netzüberwachung supply monitoring || line voltage monitoring * *Netzspannungsüberwachung*
Netzumschaltung system transfer * *{el.Masch.} Umschaltg. e. laufenden AC-Motors auf e.anders Netz (z.B. im Kraftwerk)*
Netzunterbrechung instantaneous power failure * *kurzzeitige* || line interruption
Netzverbund network interconnection * *Datennetz* || internetworking
Netzversorgung AC input power supply * *eines Geräts* || mains supply
Netzwerk network
Netzwerkanalyse network analysis
netzwerkfähig networkable
Netzwerkkarte network card * *z.B. Netzwerks-Anschaltungsbaugruppe in einem PC*
Netzwerkknoten network node
Netzwiederkehr system recovery || power restoration || resumption of power supply
Netzzuleitung supply feeder * *Einspeisung* || supply lead * *Leitung, Anschluss (-leitung)* || supply cable
neu new || fresh * *frisch, er~t: fresh start: ~er Anfang; with fresh energy: mit ~en Kräften* || brand-new * *ganz ~* || different * *bisher unbekannt* || renewed * *aufgefrischt, frisch, er~ert* || more * *weitere* || further * *weitere* || another * *ein ~er/ weiterer* || recent * *neu, jung, frisch, modern; kürzlich/vor kurzem/unlängst (geschehen/entstanden)* || modern * *modern* || latest * *~estes* || novel * *~artig* || additional * *zusätzlich, weiterer* || re- * als Vorsilbe (er~t, wieder-); auch 'um-'
neu definieren redefine
neu eingeben re-enter
neu laden bootstrap * *Programm ~*
Neu- re-
Neuauflage new edition * *~ eines Druckerzeugnisses*
neuentwickelt newly-developed
Neuentwicklung new development
neuer more recent * *(Komparativ) z.B. Softwareversion* || latter * *(Komparativ) später, neuer, jünger* || later * *zeitlich weniger lange zurückliegend (z.B. Softwareversion)*
neuerdings lately || of late || recently
neuest newest || most current * *am aktuellsten,*

höchst aktuell || latest || most recent * *jüngst, am kürzesten zurückliegend, frischest*
neueste Version latest version
Neuheit novelty * *auch einer Erfindung* || newness || originality * *Ursprünglichkeit*
Neuinbetriebnahme new start-up
Neuling newcomer || greenhorn * *(salopp) Grünschnabel* || beginner * →*Anfänger* || non-professional * *Nicht-Profi* || novice * →*Anfänger, Neuling*
neuronales Netz neural network * *d. menschl. Gehirn nachempfundene Computerarchitektur m. verteilten Knoten (Neuronen)* || neural net
Neustart new start || restart * *Rechner, Motor usw.* || cold restart * *Erstanlauf* || reset
neustarten re-start || re-boot * *einen Computer* ~
neutral neutral || impartial * *unparteiisch, gerecht, unvoreingenommen, unbefangen*
neutrale Lage neutral position
neutralisieren neutralize
Neutralleiter neutral conductor || neutral
neuzeitlich modern || state-of-the-art * *dem neuesten technischen Stand entsprechend* || advanced * *fortschrittlich*
Neuzustand new condition
Newton-Meter Newton-Meter
NF NF * *Abkürzg. für 'Norme Francaise': französische Norm*
NFPA NFPA * *Abkürzg. fur 'National Fire Protection Association': amerikan. Brandverhütungsgesellschaft*
NH-Sicherung low-voltage high-breaking-capacity fuse * *Niederspannungs-Hochleistungssicherung* || low-voltage high-rupturing-capacity fuse * *Niederspannungs-Hochleistungssicherung* || low-voltage high-rupturing-capacity fuse link * *Sicherungseinsatz*
NH-Sicherungseinsatz low-voltage high-rupturing-capacity fuse link * *Niederspannungs-Hochleistungssicherung*
Nibbelmaschine nibbling machine * *zur Blechschneidbearbeitung (z.B. die TRUMATIC (R) der Fa. Trumpf)*
nibbeln nibble * *Blechbearbeitung*
Nibble-codiert nibble-coded * *je 4 bits bilden eine Informationseinheit*
Nichols-Diagramm Nichols diagram
nicht abgesichert non-fused * *z.B. mit Schmelzsicherung*
nicht angetrieben non-driven || idling * *leerlaufend, z.B. Führungsrolle, Leitwalze* || running idle * *leerlaufend*
nicht anlaufen fail to start * *Motor, Maschine, Rechner*
nicht auf Lager out of stock
nicht ausreichend insufficient || poor * *siehe auch* →*ungenügend*
nicht beachten disregard || ignore * *ignorieren* || take no notice of || reject * →*ablehnen*, →*verwerfen* || refuse * *zurückweisen* || pass over * *übergehen (to: zu)*
nicht befolgen disregard || ignore
nicht belegt not connected * *nicht angeschlossen* || not assigned * *nicht (einer Funktion) zugeordnet* || NC * *Abkürzung für 'Not Connected': nicht angeschlossen (z.B. an Schaltkreis, Stecker)* || unoccupied * *z.b. Einbauplatz für Zusatzbaugruppe*
nicht durchgeschaltet not switched through || disabled * *gesperrt (z.B. Signal)* || not connected || not completed * *nicht geschlossen (Stromkreis)* || no signal flow * *Zustand eines Strompfades in Kontaktplandarstellung eines SPS-Programms*
nicht erlaubt unpermissible || unlegal * *nicht gesetzmäßig, illegal* || invalid * *ungültig, nichtig*
nicht fasernd non-linting * *nicht fusselnd, z.B. Lappen*

nicht funkender Lüfter non-sparking fan
nicht größer sein als not exceed
nicht in Anschlag bringen leave out of account
nicht interpretierbarer Befehl illegal operation || unused opcode * *nicht ausführbarer Maschinenbefehl*
nicht leitend non-conductive
nicht lückender Strom continuous current
nicht maßhaltig off gauge * *z.B. Draht, Blech* || not true to size
nicht maßstäblich not to scale
nicht maßstabsgetreu not to scale
nicht mehr no longer || not any longer
nicht mehr als no more than
nicht mehr gebräuchlich no longer used || outdated || obsolete
nicht nach meinem Geschmack not to my taste || not my cup of tea * *(salopp)*
nicht periodisch non-periodic
nicht rostend stainless * *Stahl* || rustless || rustproof || non-corroding
nicht sichtbar non-visible
nicht steuerbar non-controllable
nicht umlaufend non-rotating || stationary
nicht unterbrechender Kontakt bridging contact * *Wechs.kontakt dess. Kontakte während d. Schaltens kurz geschloss.sind* || make-before-break contact * *Wechs.kont. dess. Kontakte währ. d. Schaltens beide kurz geschloss. sind*
nicht vorhanden non-existent || unimplemented * *nicht implementiert/realisiert*
nicht zerstörende Werkstoffprüfung non-destructive material testing
nicht zugänglich inaccessible
nicht zusammenhängend discontiguous * *mit Lükken*
nicht- non- || dis-
Nicht-Antriebsseite non-drive end * *{el. Masch.}* || NDE * *{el. Masch.} Abk. für 'Non-Drive End'*
nicht-generatorische Last non-regenerative load
nicht-kondensierend non-condensing
nichtautorisiert unauthorized
Nichtbeachtung non-observance * *Nichtbefolgung, z.B. einer Vorschrift*
Nichteisen- non-ferrous
Nichteisenmetall non-ferrous metal
nichtelektrisch non-electrical
nichtelektrische Größen non-electrical quantities || non-electrical data
nichtfasernd non-linting * *z.B. Putzlappen*
nichtflüchtig nonvolatile * *non-volatile* || retentive
nichtflüchtig abspeichern lock in memory || store in a non-volatile mode
nichtflüchtige Speicherung bei Spannungseinbruch memory retention in case of power loss
nichtflüchtiger Speicher non-volatile memory || NOVRAM * *Abkürzung für 'nonvolatile random access memory'* || NVRAM * *Abkürzung für 'nonvolatile random access memory'* || EEPROM * *elektrisch löschbarer Nur-Lese-Speicher*
nichtinvertierend non-inverting
Nichtleiter insulator || dielectric
nichtleitfähig non-conductive
nichtlinear nonlinear || non-linear * *highly: stark*
Nichtlinearität nonlinearity * *non-linearity* || linearity error * *Linearitätsfehler*
Nichtlückbereich continuous-current area
nichtlückender Bereich continuous-current area || non-pulsating current area
nichtlückender Betrieb continuous operation || continuous current state || non-pulsating-current operation
nichtlückender Strom continuous current || non-pulsating current
nichtperiodischer Betrieb non-periodic duty

nichtrostender Stahl stainless steel
nichtsdestoweniger nevertheless
nichtsinusförmig non-sinusoidal
nichtspeichernd non-latching * *Signal ohne Selbsthaltung bei Relaissteuerung, Kontaktplan usw.* || momentary * *Signal ohne Selbsthaltung bei Relaissteuerung, Kontaktplan usw.* || not maintained * *Signal ohne Selbsthaltung bei Relaissteuerung, Kontaktplan usw.*
nichtstationär non-stationary
nichtzyklisch non-cyclic
Nickel nickel * *chrome-nickel steel: ~chromstahl*
Nickwerk luffing gear * *(Hebezeuge) zum Auf-/Abwärtsschwenken eines Kranauslegers*
nie never || at no time || hardly ever * *fast nie* || under no circumstances * *unter keinen Umständen* || on no account * *keinesfalls* || in no case * *keinesfalls* || not at all * *keineswegs* || by no means * *keineswegs* || anything but * *alles andere als*
Niederdruck- low-pressure
Niederdrucksammelschiene low-pressure steam mains * *zur Dampfversorgung*
Niedergang decay
Niederhalter pressure pad || press pad || jack || holding-down element || holding-down clamp
Niederlande Netherlands
Niederlassung branch office || branch * *Filiale* || representative || district office || place of business || administration centre || office
niederlegen lay down * *auch in einer Schrift*
niederohmig low-resistance || having a low Ohm value
niederprior of lower priority || minor * *untergeordnet, rangniedriger* || lower-ranking * *niedriger im Rang (auch Person)*
Niederschlag deposit * *Ablagerung* || sediment * *Ablagerung* || rainfall * *in Form von Regen*
Niederspannung low voltage * *alle Spannungen unter 1000V* || LV * *Abkürzung für 'Low Voltage'*
Niederspannungs- low-voltage
Niederspannungs-Schaltgeräte low-voltage switchgear || low-voltage switchgear and control-gear
Niederspannungsanlage low-voltage system
Niederspannungsmotor low voltage motor || LV motor
Niederspannungsnetz secondary distribution system
Niederspannungsrichtlinie Low-Voltage Directive * *der EG: 73/23/EWG, EG-Amtsblatt 73/L77/L29 u. 93/68/EWG "CE-Kennzeichng."*
Niederspannungsschalter low voltage circuit breaker || LV breaker * *Abkürzg.f. 'Low Voltage':* Niederspannungs-
Niederspannungsschaltgerät low-voltage switchgear
Niederspannungsseite secondary * *eines "Abwärtstrafos"*
Niederspannungsverteilung low-voltage distribution
Niedervolt- low-voltage
niederwertig less significant * *Bit, Dekade usw.* || lower-order * *Ziffer usw.* || low order * *Ziffer usw* || least significant * *niederwertigst, z.B. Bit, Dekade usw.*
niederwertigst least significant * *Bit, Dekade, Byte, Wort usw*
niederwertigste Ziffer least significant digit * *z.B. an Ziffernanzeige* || LSD * *Abkürzung für 'Least Significant Digit': niederwertigste Ziffer*
niederwertigstes Bit least significant bit || LSB * *Least Significant Bit*
niederwertigstes Zeichen least significant digit * *einer Ziffernanzeige* || LSD * *Least Significant Digit; einer Ziffernanzeige*

niedrig low
niedrige Aussteuerung high delay angle * *großer Steuerwinkel bei Phasenanschnittsteuerung* || low control setting || reduced output voltage || reduced output voltage ratio || low modulation * *in der Mess-/Nachrichtentechnik usw.*
niedrige Drehzahl low speed
niedrige Drehzahlen low speeds
niedrige Geräuschentwicklung low noise
niedrige Leistung low power
niedriger Ordnung of low order * *z.B. Oberschwingungen*
Niedrigpreis- economy || low price || low cost || low-budget
niedrigst lowest
Niet rivet
Niete rivet || flop * *(figürl.; auch 'failure')*
nieten rivet || stake || clinch * *ver~, sicher befestigen*
Nipco-Walze Nipco roll * *(Papier)*
Nippel nipple
Nische niche * *[nitsch] (auch figürl.: Marktnische, zugewiesener Platz usw.); fill: füllen/besetzen* || recess * *Vertiefung, Aussparung, Einschnitt, (versteckter) Winkel, Nische* || chamber * *Kammer*
Nitrolack cellulose paint
Niveau level * *(auch figürl.: at a low/high level: auf niedrigem/hohem ~)* || standard * *(figürl.) not up to standard: unter dem ~*
Niveaukasten flow box * *Stoffauflaufkasten bei Papiermaschine* || head box * *(amerikan.) Stoffauflaufkasten bei Papiermaschine*
Niveauschalter level switch
Niveauwächter level switch
nivellieren level || grade * *(Gelände) planieren, einebnen*
Nivellierung levelling * *Ausgleich von Höhenunterschieden*
Nm Nm * *Newtonmeter; 1 Nm entspr. 0,101972 kpm entspr 0,7376 lbf ft(pound-force feet) enspr. 8,851 lbf in*
nobel noble || elegant * *fein* || stylish * *fein "gestylt"* || generous * *großzügig, freigiebig* || plush * *luxuriös, plüschig* || posh * *(salopp) piekfein, todschick, fesch* || luxurious * *luxuriös* || swank * *(salopp) elegant, schick, protzig*
noch still
noch immer still
noch nicht not yet
noch nie never before
Nocke cam || lifter * *auch Hebeeinrichtung, Stößel*
Nocken cam || lifter * *auch Hebeeinrichtung, Stößel* || spline * *Keil, (Pass-) Feder; längliches, dünnes Stück* || lug * *Zinke, Ansatz, Lasche, "Ohr"*
Nockenkupplung cam clutch
Nockenschalter cam switch
Nockenschaltwerk cam-contactor group || cam group || cam-operated switchgroup
Nockensteuerwerk cam controller
Nockenwelle camshaft
Nomenklatur terms * *Fachausdrücke*
nominell nominal
Non-sparking-Ausführung non-sparking design * *Zündschutzart Typ 'N' nach BS 5000 Pt.16 (British Standard)*
Noppe nap || burl
NOR-Stufe NOR gate * *invertierte ODER-Verknüpfung*
Nordpol north pole * *auch bei einem Magnet*
Norm standard || rule * *Regel* || norm * *Arbeitsnorm*
Norm- standard || normative * *normativ, normende Kraft habend*
Norm-Asynchronmotor standard induction motor || standard asynchronous motor

normal   normal ‖ standard * *Abmessungen usw.* ‖ usual * *gebräuchlich* ‖ regular * *üblich* ‖ ordinary * *gewöhnlich* ‖ conventional * *herkömmlich* ‖ comparison standard * *(Substantiv)* →*Messnormal*
Normalausführung   standard design * *z.B. Motor* ‖ regular design * *übliche Bauart* ‖ basic design * *Grundtyp*
normale Anforderungen   normal requirements ‖ normal operating conditions * *normale Betriebsbedingungen*
normalerregt   standard excitation * *Motor*
normalerweise   normally ‖ as standard * *in der Standardausführung* ‖ usually * *gebräuchlicherweise* ‖ commonly * *gewöhnlich, (all)gemein, gemeinsam* ‖ generally * *im Allgemeinen, gewöhnlich, meistens* ‖ generally speaking * *im Allgemeinen, im Großen und Ganzen*
Normalform   canonical form
Normalklima   standard atmospheric conditions * *nach IEC*
normativ   normative
Normblatt   standard sheet
normen   standardize
Normenbehörde   institute of regulations ‖ regulatory body
Normengremium   institute of regulations
Normentwurf   draft standard ‖ specification draft ‖ draft
normgerecht   conforming to standards
Normgröße   standard size
normieren   scale * *auf eine Bezugs/Nenngröße* ‖ normalize * *normalisieren* ‖ standardize * *normen*
Normierfaktor   normalization factor ‖ scaling factor * *Gewichtungsfaktor*
normiert   normalized ‖ scaled * *auf eine Bezugsgröße bezogen (%, 10 V, 20 mA usw.)*
Normierung   normalization ‖ scaling ‖ scaling factor * *(→PROFIBUS-Profil)*
Normierungsfaktor   normalization factor
Normierungsparameter   normalization parameter
Normmotor   standard motor * *meist Käfigläufer-Asynchronmotor i. genormt. Baugröße nach IEC72, DIN 42673 und 42677* ‖ general-purpose motor * *normal üblicher Motor*
Normreihe   standard series ‖ standard range
Normspannung   standard voltage ‖ rated voltage * *siehe z.B. DIN 40030 für DC-Motoren*
Normspannungen   standard voltages * *Netzspannungen nach DIN IEC 38, z.B. 230 V, 400 V, 690 V, 1000 V usw.*
Normteil   standard part ‖ standard commercially available part * *im freien Handel erhältliches Teil* ‖ standardized element
Normung   standardization
Normvorschrift   standard specification ‖ standard
Norwegen   Norway
Not Aus   emergency stop * *Anlage wird schnellstmöglichst spannungslos gemacht (u.trudelt aus)* ‖ E-stop * *Anlage wird schnellstmöglichst spannungslos gemacht (u.trudelt aus)* ‖ emergency OFF
Not Halt   emergency stop
Not-   emergency ‖ stand-by * *Reserve*
Not-Aus   emergency stop ‖ E-stop ‖ emergency OFF
NOT-AUS-Knopf   all off button ‖ emergency button
NOT-AUS-Schalter   emergency stop switch ‖ emergency stop circuit braker * *Leistungsschalter*
NOT-AUS-Taster   emergency-stop pushbutton
Not-Halt   emergency stop
Not-Stop   emergency stop
Notabschaltung   emergency stop ‖ emergency shutdown * *Maschine, Atomkraftwerk usw.*
Notation   notation * *Schreibweise, Bezeichnung, (chem.) Formelzeichen; Notierung, Aufzeichnung*
Notbetrieb   emergency operation
Notbremsung   emergency braking

Notendschalter   emergency limit switch
Notfall   case of emergency ‖ emergency ‖ emergency situation
notfalls   if need may be * *(Adv.)* ‖ if need arises * *(Adv.)* ‖ in an emergency * *(Adv.)* ‖ if necessary * *(Adv.)* ‖ if need be * *(Adv.)* ‖ in the last resort * *(Adv.)*
Nothalt   emergency stop
notieren   make a note * *schriftliche Notiz machen, aufschreiben* ‖ document * *dokumentieren*
nötig   necessary ‖ needed * *benötigt* ‖ required * *benötigt* ‖ requisite * *eforderlich, notwendig* ‖ indicated * *angezeigt* ‖ recommended * *empfohlen*
nötig haben   require
nötig machen   necessitate
nötigenfalls   in case of need * *(Adv.)* siehe auch →*notfalls*
Notlaufeigenschaften   safety running features
Notlösung   expedient ‖ workaround * *~, die das eigentliche Problem umgeht* ‖ provisional solution * *behelfsmäßige Lösung* ‖ stopgap * *Notbehelf, Lückenbüßer, Überbrückung(slösung)*
Notschaltung   emergency engagement * *bei einer Kupplung*
Notsteuerung   emergency control
Notstromaggregat   emergency generating set ‖ stand-by generating set
Notstromversorgung   emergency power supply ‖ stand-by power supply ‖ emergency power supply system
notwendig   necessary * *necessitate: ~ machen* ‖ requisite * *erforderlich, ~* ‖ needful ‖ needed * *benötigt* ‖ required * *benötigt* ‖ urgent * *dringlich* ‖ essential * *wesentlich* ‖ indispensable * *unerläßlich* ‖ imperative * *unabdinglich*
notwendig machen   necessitate ‖ call for
notwendige Bedingung   necessary condition
notwendigerweise   necessarily ‖ of necessity
Notwendigkeit   necessity ‖ must ‖ urgency * *Dringlichkeit* ‖ requirement * *~ einer Sache* ‖ need
NOVRAM   NOVRAM * *Abk. f. 'NOn Volatile Random Access Memory': nichtflüchtiger Schreib/Lesespeicher*
NRZ   NRZ * *Abk. f. 'Non Return to Zero': Logikpegel mit Low-Signal ungleich Null*
Null   zero ‖ null ‖ o * *(salopp, gesprochen)* ‖ null and void * *Null und nichtig*
null und nichtig   null and void ‖ declare null and void: für ~ erklären
Nullabgleich   balance ‖ zero balancing ‖ offset adjustment * *Nullpunktabgleich/-korrektur* ‖ zero offset * *Nullpunktkorrektur (bei NC)* ‖ zero balance ‖ zeo adjust
Nulldurchgang   pass through zero ‖ zero passage * *z.B. einer Funktion/Kurve/Spannung* ‖ zero crossover ‖ zero crossing ‖ passage through zero ‖ voltage zero * *Nulldurchgang der Spannung*
Nullimpuls   index signal * *eines Winkelschritt-/Inkrementalgebers (auch 'Nullmarke')* ‖ marker pulse * *eines Winkelschritt-/Inkrementalgebers* ‖ zero pulse ‖ zero-marking pulse
Nullkomponente   zero component ‖ homopolar component * *Komponente mit 3 gleichphasigen Größen in e. unsymmetr. Drehstromsystem*
Nullkraftsockel   zero insertion force socket * *für ICs* ‖ ZIF socket * *Abk. f. 'Zero Insertion Force': Sockel ohne Steckkraft für ICs*
Nullage   neutral position ‖ unloaded position * *ohne Belastung* ‖ home position * *Nullpunkt der Weg-Koordinate bei Positionierregelung*
Nullleiter   neutral ‖ N ‖ directly earthed conductor ‖ neutral conductor
Nullmarke   zero mark ‖ 0 mark ‖ index signal * *Impulsgeber-Nullimpuls* ‖ marker pulse * *Nullimpuls eines Inkrementalgebers*

Nullmodem-Kabel null modem cable * *RS232-Kabel mit gekreuzten Transmit- u. Receive-Adern*
Nullpunkt zero point || zero || origin * *eines Koordinatensystems* || neutral point * *Sternpunkt* || star point * *Sternpunkt* || zero reference point * *Bezugspunkt, Koordinaten-Nullpunkt (z.B. bei NC)* || zero mark * *Nullmarke*
Nullpunkt-Mitte Instrument zero center meter
Nullpunktabgleich offset compensation || zero balancing || zero adjust
Nullpunktfehler offset error || zero error
Nullpunktsabgleich offset compensation || zero balancing
Nullpunktsfehler offset error || zero error
Nullpunktunterdrückung zero suppression || offset error suppression
Nullpunktverschiebung bias adjustment * *per Einstellung* || offset error * *Nullpunktfehler z.B. durch Drifteffekte* || zero offset || zero shift || zero displacement || drift of offset error * *Drift des Nullpunktfehlers*
nullsetzen set to zero || zero
nullspannungsgesichert non-volatile * *nichtflüchtig bei Abschalten der Versorgungsspannung (Speicher)* || retentive * *retentive flag: Haftmerker bei SPS*
Nullstelle pole
Nullstellung neutral position || unloaded position * *ohne Belastung* || zero position
Nullsystem zero-sequence system * *bei Drehstromsystem*
Nullung zeroing || reducing to zero || neutralization || earthing via neutral conductor * *MP dient gleichzeitig als Schutzleiter*
Nullwerden becoming zero || falling to zero * *Auf-Null-Abfallen* || decay to zero * *(auch Verb) Auf-Null-Absinken* || fall to zero * *(Verb)*
numerisch numeric || numerical || numerically * *(Adv.)*
numerisch gesteuert numerically controlled || NC-controlled
numerische Größen numerical quantities || numeric quantities
numerische Steuerung numeric controls || numerical control || NC * *Abk. für 'Numeric Control'* || numerical control system
numerischer Umschalter numeric switch * *Funktionsbaustein*
numerischer Wert numeric value
Nummer number || No * *Abkürzung*
nummerieren number * *consecutively: fortlaufend*
Nummerierung numbering
Nummern numbers || Nos * *Abkürzung für 'numbers'*
Nummernliste geänderter Parameterwerte parameter-change-flags array * *(→PROFIBUS-Profil)*
Nummernsystem coding system * *z.B. für Bestellnummern*
nur only || alone * *allein* || exclusively * *(Adv.) allein* || solely * *lediglich* || nothing but * *nichts als* || just * *bloß* || merely * *bloß* || except * *ausgenommen* || but * *ausgenommen* || simply * *einfach*
nur für internen Gebrauch for internal use only
nur in einer Richtung unidirectional
Nur-Lese-Speicher read-only memory || ROM
NURB NURB * *{Werkz.masch.} Non-Uniform Rational B-Spline: Darstellg.e. Kurvenstücks durch Polynom 3. Ordng.*
Nürnberg Nuremberg
Nut keyway * *für Passfeder z.B. auf Motorwelle* || notch || slot * *z.B. im Motorläufer* || groove * *Rinne, Nut, Vertiefg., Hohlkehle, Rille, Riefe, Nut b. Federverbindung i.Holzzkonstrukt.* || score * *Kerbe, Rille (siehe auch "Rille")*
Nut- slotting || grooving
Nut-und-Feder-Verbindung tongue and groove jointing * *{Holz}* || groove and feather joint * *{Holz}*
Nutauskleidung slot lining * *z.B. Motornut mit Isolierstoff*
Nute notch || groove * *Rinne, Nut, Vertiefung, Rille, Riefe*
nuten notch || slots * *(Subst.) z.B. eines Motors*
Nuten des Blechpakets slots of the core * *{el.-Masch.} zur Aufnahme der Wicklungen*
Nutenfräsmaschine groove cutter * *{Holz}*
Nutenstein T-nut * *in Nut eingreifende Befestigungsmutter (z.B. in Montageschienensystem)*
Nutform slot shape * *z.B. des Läufers einer el. Maschine* || slot form
Nutfräsmaschine groove-cutting machine * *{Holz}*
Nutisolierung slot insulation * *{el.Masch.} Flächenisolierung in den Nuten zw. Blechpaket und den Leitern*
Nutkeil key * *Passfeder, z.B. für Motorwelle* || spline * *Passfeder* || slot wedge * *{el. Masch.} Nutverschlusskeil: Halbrund- o.Trapezprofilstab z. Verschluss d. Wicklgssnuten*
Nutmesser scorer * *Kerbmesser, z.B. in der Buchbinderei* || cutter bar || keyway broach * *{Werkz.-masch.}*
Nutmutter T-nut * *für Schienenbefestigungssystem (siehe auch →Nutstein, →T-Nut, →Hammerschraube)*
Nutrasten cogging torque * *Drehmomentenwelligkeit eines Motors auf Grund der Polfühligkeit*
Nutstein T-nut * *in →T-Nut eingreifende Befestigungsmutter*
Nutthermometer slot thermometer * *{el. Masch.} PT100-Widerstandsthermometer, das in die Nuten eingebaut ist*
Nutverschlußkeil slot-termination wedge * *z.B. für Wicklungsnut in →Hochgeschwindigkeitsläufer eines Motors*
Nutwalze grooved roller || groove roll
Nutwand slot side
Nutwerkzeug groove-cutting tool * *{Holz}*
nutzbar useful || utilizable || effective || profitable * *Gewinn bringend* || available * *verfügbar* || productive * *Gewinn bringend*
nutzbar machen utilize || turn to account || take advantage of || harness * *Naturkräfte usw.* || make usable
Nutzbremsung regenerative braking * *mit Rückspeisg. d. Bremsenergie ins Netz bzw. Energierückgewinnung*
Nutzdaten useful data * *z.B. bei einem seriellen Telegramm* || net data * *Nettodaten* || user data
Nutzdatenlänge net data length
nutzen employ * *verwenden* || use || make use of || utilize * *(aus)nutzen* || put to account || exploit * *ausnützen* || avail oneself of * *Gelegenheit ~* || benefit * *(Substantiv) Vorteil* || gain * *(Substantiv) Gewinn* || advantage * *(Substantiv) Vorteil* || returns * *(Substantiv) Ertrag* || utility * *(Substantiv) Nutzen (to: für), Nützlichkeit, Nützliches, nützliche Einrichtung* || take advantage of * *vorteilhaft ~* || benefit from * *vorteilhaft nutzen, ~/Vorteile ziehen aus* || take advantage of * *vorteilhaft ~; take full advantage of: die Vorteile voll aus~* || value * *(Substantiv) Gegenwert, Gegenleistung, Wert; Gewicht (eines Wortes/einer Meinung usw.)* || harness * *nutzbarmachen (Naturkräfte usw.)* || repeat factor * *(Subst.) Anz.d.Teilstücke je Fläche (bei Leiterplattenfertigung) od. je Längeneinheit*
Nutzen ziehen aus benefit from
nützen be of use * *for: zu; a person: jemandem* || be useful || be of advantage * *vorteilhaft sein; to a person: jemandem* || be of benefit * *vorteilhaft sein; to a person: jemandem* || benefit * *jemandem nützlich sein; Vorteil haben (by: von, from:*

durch), Nutzen ziehen (aus) || help * *helfen* || use * *(be)nutzen* || make use of * *Gebrauch machen von* || utilize * *nutzen* || put to account * *nutzen* || exploit * *ausnützen* || avail * *avail oneself of: etwas ausnutzen (Gelegenheit usw.)* || seize * *(Gelegenheit usw.) ergreifen*
Nutzlast payload
Nutzleistung useful output * *nutzbare Ausgangsleistung; z.B. eines Motors, Stromrichters* || useful power || effective output * *Ausgangs-Effektivleistung* || useful horsepower || useful output power || duty * *Nutz-/Wirkleistg., das zu Leistende, Auslastg., die abgebbare/ausnutzbare Leistg.*
nützlich useful || beneficial || of use || serviceable * *dienlich* || helpful * *hilfreich* || advantageous * *vorteilhaft* || profitable * *Gewinn bringend, profitabel, vorteilhaft* || conducive * *[kondju: siw] dienlich, förderlich, fördernd, zielführend (to: für)*
Nutzsignal useful signal
Nutzung utilization
Nutzungsdauer service life
Nyquist-Kriterium Nyquist criterion

O

ö umlauted o || o with Umlaut || oe
O-Ring O-seal * *Dichtring* || O-ring * *Gummidichtring*
ob whether * *wether...or not: ob..oder nicht* || if * as if: *als ob* || as to whether
ob er wohl kommt I wonder if he will come
oben on top * *oben(auf); from top to bottom: von ~ nach unten; towards the top: nach ~* || above * *auch Bild, 'das ~ näher Erläuterte' usw. in einem Schriftsatz; from above: von ~* || overhead * *~liegend, ~gesteuert, Hoch-, Frei-* || at the top * *an der Spitze* || up * *in die Höhe, nach ~, hinauf, herauf, aufwärts, empor* || upstairs * *~ im Haus* || on the surface * *auf der Oberfläche* || this side up * *auf Versandkiste* || top * *ober(st)es Ende, Oberteil, Spitze* || upper * *ober, höher; at upper left: links ~* || upward * *aufwärts (Tendenz)* || upwards * *aufwärts, nach ~; with the shaft end pointing upwards: mit dem Wellenende nach ~* || top-mounted * *~ montiert*
oben erwähnt above mentioned || aforementioned || mentioned above * *(nachgestellt)*
oben genannt above-mentioned || afore mentioned || above-said || aforesaid || above
oben links at upper left || at the upper left || at the top left
oben montiert top-mounted
oben rechts at upper right || at the upper right || at the top right
obengenannt above mentioned || aforementioned
obenliegend top-mounted * *oben montiert* || overhead * *Frei-, Ober-, Hoch-, obengesteuert, obenliegend (z.B. Nockenwelle)* || located at the top
ober above || upper || higher || superior || senior * *(Substantiv) Chef-*
Ober- chief * *Chef-* || senior * *Chef-, Vorgesetzter, (rang)höher* || top * *obenliegend, oben angesiedelt (auch figürl.)* || upper * *oberes* || overhead * *obenliegend*
Oberbegriff generic term
Oberdeck-Motor deck-marine motor * *für Einsatz auf Schiffen (Schutzart IP56)*
Oberdeckaufstellung deck installation * *[el. Masch.] Aufstellung von Motoren auf Schiffsdeck*

obere Grenze upper limit
oberer Führungskreis senior management
oberer Grenzwert upper limit value
oberer Leistungsbereich upper performance range * *bezügl. der Leistungsfähigkeit* || high-rating range * *bezüglich der Nennausgangsleistung (Stromrichter, Motor usw.)* || upper performance level * *bezüglich der Leistungsfähigkeit* || high-end || high performance level
oberer Teil upper part
oberes upper
Oberfeld harmonic field * *Oberwelle eines (Magnet)Feldes*
Oberfelddrehmoment harmonic torque * *[el. Masch.] Beeinflussg. d. Moment.kennlin. durch unsymmetr.Wicklg. (Sattelmoment)*
Oberfläche surface || face || area * *Fläche(ninhalt)* || human interface * *Mensch/Maschinen-Schnittstelle, Bedien(er)oberfläche* || user interface * *Anwenderoberfläche einer Software/eines Rechnersystems* || GUI * *Abkürzg.für 'Graphic User Interface': grafische Betriebssystem/Programmoberfläche* || surface pattern * *Gestaltung-/Aussehen der Oberfläche* || top surface || surface area
Oberflächenbearbeitung surface finishing * *Nach-/Feinbearbeiten der Oberfläche* || surface machining * *mit Werkzeugmaschine* || facing * *Bearbeitung von Stirnflächen (mit Werkz maschine)* || surface treatment || surface finish * *End/Nachbearbeitung, Veredelung*
oberflächenbehandelt surface-treated
Oberflächenbehandlung surface treatment
oberflächenbelüftet totally enclosed self-ventilated * *→eigenbelüftet, Luftstrom ü.d. Motoroberfläche (z.B. IP54)* || totally enclosed forced-ventilated * *→fremdbelüftet, Luftstrom ü.d. Motoroberfläche* || totally enclosed fan-cooled * *Fremdlüfter bläst üb. Motoroberfläche; Motorkühlart IC0041 n. DIN IEC 34.6* || TEFC * *'Totally Enclosed Fan-Cooled': oberflächengekühlt mit Lüfter (Motor-Kühlart)* || totally enclosed non-ventilated * *lüfterlose Wärmeabfuhr über die Oberfläche des Motorgehäuses* || surface-cooled
Oberflächenbeschichtung surface coating
oberflächengekühlt totally-enclosed self-ventilated * *→eigenbelüftet, Luftstrom ü.d. Oberfläche d. Maschine(z.B.IP 54)* || totally enclosed forced-ventilated * *→fremdbelüftet, Luftstrom über die Maschinenoberfläche* || totally enclosed fan-cooled * *Fremdlüfter bläst üb.Motoroberfläche; Kühlart IC0041 nach DIN IEC 34.6* || TEFC * *'Totally-Enclosed Fan-Cooled': oberfläch. gekühlt m. Fremdlüfter (Motorkühlart)* || totally enclosed non-ventilated * *Wärmeabfuhr ohne Lüfter über die Oberfläche des Motorgehäuses* || surface-cooled
oberflächengeleimtes Papier surface-sized paper
Oberflächenkühlung surface cooling * *[el. Masch.] Ableitg. der Wärme über d. geschloss. äußere Oberfläche, VDE 0530 T.6* || surface ventilation * *z.B. über Kühlrippen bei Motor*
Oberflächenleimung surface sizing * *von Papier*
Oberflächenmontage surface mounting
oberflächenmontierbares Bauelement surface-mount device || SMD * *Abk. f. 'Surface-Mount Device'*
Oberflächenqualität surface quality * *z.B. v. Blech, Papier usw.*
Oberflächenschutz surface protection
Oberflächentechnik surface treatment || surface engineering
Oberflächenveredelung surface finish
Oberflächenwickler surface winder * *Druckwalzenprinzip: Zug wird durch Anpresswalze erzeugt (i. Ggs. z. Achswickler)* || reeler || reel drum winder || contact winder

Oberflächenzustand finish
oberflächlich superficial * *überschläglich* || rough * *ungefähr, grob*
Oberfräse overhead router * *{Holz}* || router * *{Holz}* || routing machine * *{Holz}*
Oberfräsmaschine router * *{Holz}*
Obergrenze upper limit
oberhalb above * *auch oberhalb eines Werts* || beyond * *jenseits, z.B. einer Grenze*
Oberkante top edge
Oberklasse- high-end * *z.B. Produkt*
Obermenge aggregate
Obermotor upper motor * *Walzwerk* || top-roll motor * *Walzwerk*
Oberschwingung harmonic * 5th harmonic: 5.Harmonische; harmonic of 5th order: Oberwelle 5.Ordung/d.Ordnungszahl 5 || harmonic oscillation
Oberschwingungen harmonics * *period.Wechselgrößen höherer Frequenz, d.der (Netz)Grundschwingung überlagert sind* || harmonic content * *Oberschwingungsanteil/gehalt*
Oberschwingungen höherer Ordnung higher order harmonics
Oberschwingungen niedriger Ordnungszahlen low-order harmonics
Oberschwingungsanteil harmonic content || ripple content
Oberschwingungsanteile harmonic components || harmonic content
oberschwingungsarm with low harmonic content || with a low harmonic content || low-harmonic
Oberschwingungsblindleistung harmonics reactive power
Oberschwingungsgehalt harmonic content * *ident.m. Klirrfaktor: Effektivwert d.Oberschwing.gen/Gesamteff.wert(DIN 40110)* || harmonic distortion || distortion factor * *Klirrfaktor* || harmonic factor * *Klirrfaktor*
oberschwingungshaltig containing harmonics
Oberschwingungskompensation harmonic suppression
oberschwingungsminimiert minimized for harmonics
Oberschwingungsspektrum harmonics spectrum
Oberschwingungsströme harmonic currents
Oberschwingungsverluste harmonic losses
Oberseil head rope * *{Aufzüge | Seilbahn}* || top cable * *{Aufzüge}* || upper rope * *{Aufzüge}* || top rope * *{Aufzüge}*
Oberseite top side || top face * *obere Fläche (einer Platte, eines Brettes usw.)*
oberstes Gebot sein have top priority || be imperative
Oberteil upper part || top part || top
Obertrum carrying run * *{z.B. Bergwerk} eines Bandförderers*
Oberwalze top roll || tongue roll || top roller * *auch z.B. in der Flachglas-Industrie*
Oberwelle harmonic || overtone * *(salopp)*
Oberwellen harmonics
Oberwellenanteil harmonic content || ripple rate
Oberwellengehalt harmonic content
Oberwellenunterdrückung harmonic suppression
Oberwelligkeit harmonic content
obig above * *z.B. Kapitel, Darlegung* || above mentioned * *oben genannt* || mentioned above * *(nachgestellt) oben genannt*
Objekt object
Objektcode object code * *Ausgangsprodukt eines Hochsprache-Übersetzers/-Compilers*
objektorientiert object oriented * *z.B. Programmiersprache (Visual C usw.) für grafische. Betriebssystem*
objektorientierte Programmierung object-oriented programming * *(z.B. für MS Windows; Gegensatz: prozedurale Programmierung)* || OOP * *{Computer} Abkürzg. für 'Object-Oriented Programming'*
Objektprogramm object program * *{Computer} (übersetztes) Programm in Maschinensprache*
Objektverzeichnis object dictionary * *{Network} beim →PROFIBUS* || object library * *{Network} beim →PROFIBUS*
obliegen be in the responsibility of
obligatorisch obligatory * *on: für* || mandatory * *obligatorisch, verbindlich, zwangsweise, Muss-; mandatory regulation: Mussvorschrift* || compulsory * *zwangsmäßig, ~, Zwangs-, bindend, Pflicht- (compulsory subject: Pflichtfach)* || inevitable * *unvermeidlich, unumgänglich, zwangsläufig* || indispensable * *unerlässlich* || stipulated * *zur Bedingung gemacht, (vertraglich) vereinbart*
obwohl although || though * *obwohl, obgleich, obschon, (je)doch; even though: wenn auch/selbst wenn; as though: als ob* || whilst * *wenn auch, obwohl* || while * *wenn auch, obwohl*
ODER OR
ODER-Glied OR gate
ODER-verknüpft ORed || OR'd
ODER-Verknüpfung OR operation
ODP ODP * *Abkürzung für 'Open Drip-Proof': offen, tropfwassergeschützt (Schutzart IP22 nach IEC und DIN)* || open drip-proof * *Schutzart IP22 nach IEC und DIN ('offen, tropfwassergeschützt')*
OE-Spinnmaschine OE spinning machine * *{Textil} im Gegensatz zur →Ringspinnmaschine* || open-end spinning machine * *{Textil} im Gegensatz zur →Ringspinnmaschine* || open-end rotor spinning machine * *{Textil}*
OEM OEM * *Original Equipment Manufacturer: stellt Maschinen o. Komponenten her u. liefert an →Endabnehmer*
Ofen stove * *Ofen, Herd, Brenn-/Trockenofen* || furnace * *z.B. Hochofen, Glühofen, Schmelzofen* || kiln * *Brenn/Trockenofen, z.B.Zementofen* || heater * *Heizkörper* || oven * *Backofen*
Ofen-Lademaschine furnace charging machine * *{Walzwerk}*
Ofen-Rollgang furnace roller table * *{Walzwerk}*
Ofenrollgang furnace roller table * *{Walzwerk}*
ofentrocken bone dry * *"knochentrocken"*
offen open || frankly * *(Adv.) ehrlich gesagt* || vacant * *Stelle* || non-proprietary * *nicht firmeneigen (z.B. Bussystem)* || open-circuit * *Kontakt*
offene Ausführung open design || open version
offene Bauart open type
offene Handelsgesellschaft general partnership
offene Kommunikation open communication * *allgemein zugängliche/firmenübergreifende Kommunikation (-shilfsmittel)*
offene Maschine open machine * *innengekühlte, durchgebelüftete Maschine (Schutzart IP23)* || open-type machine * *{el.Masch.} z.B. Motor mit →Innenkühlung*
offener Kollektor open collector * *Transistorschaltung*
offener Kontakt open contact
offenes Protokoll non-proprietary protocol * *nichtherstellerspezifisches Protokoll* || open protocol * *z.B. Busprotokoll zur seriellen Kommunikation*
offenlegen make open * *z.B. Protokoll*
offenliegend accessible * *auch von spannungsführenden Teile* || open * *offen (-gelegt) z.B. technische Lösung für andere Firmen* || exposed * *unbedeckt, offen; exponiert, gefährdet* || unprotected * *ungeschützt*
offensichtlich obvious || evident || manifest || visible || visibly * *(Adv.)*
öffentliches Netz public distribution system || public supply * *Energieversorgungsnetz* || public network

öffentliches Niederspannungsnetz public secondary distribution system
Öffentlichkeitsarbeit PR * *Abk. für 'Public Relations'* || public relations
Offline-Betrieb offline operation
öffnen open || release * *freigeben, (Bremse, Kupplung usw.)* *öffnen* || de-energize * *eine Klemme mit LOW-Signal ansteuern* || unlock * *Schloß, Sperre* ~
Öffner break contact || NC contact * *Abkürzg. für 'Normally Closed contact'* || normally closed contact || B-contact * *Abk. für 'break contact'* || form B contact
Öffnerkontakt break contact || normally closed contact || NC contact * *Abkürzung für 'Normally Closed contact'* || B-contact || form B contact || form-B contact
Öffnung opening || aperture || hole * *Loch* || gap * *Lücke* || slot * *Schlitz* || inlet * *Einlass* || outlet * *Auslass* || passage * *Durchlass* || vent * *Lüftung* || mouth * *Mündung (einer Flasche, eines Tunnels usw.)* || orifice * *Mündung* || cutout * *Ausschnitt* || lift * *{Masch.bau} eines Ventils*
Öffnungszeiten opening hours
Offset-Abgleich offset adjustment || offset compensation * *Nullpunktkorrektur* || zero balancing
Offset-Fehler offset error
Offset-Rollendruckmaschine web-fed offset press * *{Druckerei}*
Offsetdruck offset printing
Offsetdruckmaschine offset printing machine || offset printing press || offset press
Offsetpapier offset printing paper
Offsetpresse smoothing press * *zur Papierverarbeitung*
offshore offshore * *auf dem Meer (z.b. Ölplattform zum Abbau v. Bodenschätzen)*
oft gestellte Fragen FAQ * *Abk.f. 'Frequently Asked Questions'*
Ohmsch ohmic * *purely ohmic: rein ~; ohmic voltage drop: ~er Spannungsabfall* || resistive
Ohmsche Last resistive load || ohmic load
Ohmsche Verluste ohmic losses
Ohmscher Anteil resistive component || ohmic component
Ohmscher Spannungsabfall resistive voltage drop
Ohmscher Widerstand ohmic resistance
Ohmsches Gesetz Ohm's law
ohne without || w/o * *Abkürz. für 'without'* || -less * *(Vorsilbe)*
ohne Bedeutung of no account || of no significance || irrelevant || don't care * *in Liste/Tabelle*
ohne Betauung non-condensing
ohne Fehlerabschaltung trip-free
ohne Fleiß kein Preis no pains, no gains
ohne Gewähr without liability
ohne Vorzeichen unsigned
Öko- eco-
Öko-Audit Eco Audit * *Bestandteil des Umwelt-Managementsystems nach EN ISO 14.001*
Ökologie ecology
ökologisch ecological || environmental * *Umwelt-*
Oktavanalyse octave band analysis * *Darstellung d.unbeweerteten Schalldruckpegels in Frequenzbändern v. Oktavbreite*
Öl oil
Ölablassschraube oil drain plug
Ölanlasser oil-cooled starter * *Motoranlasser mit ölgekühlten Anlasswiderständen*
Ölbad oil bath
Ölbehälter oil tank
ölbeständig oil resistant
Ölbeständigkeit oil resistance
Öldampf oil vapor || lubricating vapour
Öldicht oil-tight || oiltight

öldichte Lagerung oil-proof bearing * *Lagerg.mit einseitig. Dichtung geg. druckloses Öl (z.B. mit →Radialwellendichtring)*
Öldruck oil pressure
Öleinführung oil inlet
Öleinfüllschraube oil filling plug
Öler lubricator
ölfrei non-lube * *z.b. Pneumatikzylinder, Kompressor*
ölfreie Vakuumpumpe oilfree vacuum pump
Ölfüllung oil charge * *z.B. in Motor, Getriebe, Verdichter* || oil filling
ölgefüllter Kondensator oil-type capacitor
ölgekühlt oil-cooled * *z.b. Transformator, Gleitlager usw.*
ölgeschmiert oil-lubricated
ölgeschwängerte Luft oil-laden air
Ölkanal oil duct || oil passage || oilway
Ölkapselung oil sealing * *Abdichtung gegen Öl* || oil immersion * *Schutzart 'o' nach EN50015 für Ex-geschützte el. Betriebsmittel*
Ölkessel oil-fired boiler
Ölkohle oil carbon
Ölkühler oil cooler
Ölleitung oil-lead || oil-feed * *oil pressure feed: Öldruckleitung* || oil supply pipe * *zur Schmierölversorgung* || lube-oil pipe * *zur Schmierölversorgung*
Ölmangel oil shortage || insufficient oil supply * *unzureichende Ölversorgung* || poor lubrication * *schlechte Schmierung*
Ölmenge oil quantity
Ölnebel oil mist || oil vapour || oil spray * *Sprühöl*
Ölpumpe oil pump || lubrication pump || oil-feed pump || hydraulic pump * *in Hydrauliksystem*
Ölschmierung oil lubrication
Ölstand oil level
Ölstandsanzeige oil level gauge * *~r* || oil level indication
Öltauchschmierung oil bath lubrication
Öltransformator oil-immersed transformer * *mit Öl als Kühlmittel*
Ölumlaufpumpe oil circulating pump
Ölumlaufschmierung forced oil lubrication || pressure lubrication
Ölversorgung oil supply || oil feed * *Ölzufuhr*
Online-Betrieb online operation
Online-Hilfe on-line-help
onshore onshore * *auf dem Lande (z.B. Abbau v. Bodenschätzen)*
Opazität opacity
Open-Emitter open emitter * *Transistorschaltung mit offenem Emitter*
Open-End-Spinnmaschine open-end spinning machine * *{Textil}* || OE spinning machine * *{Textil}*
Open-Kollektor open collector * *Transistorschaltung mit offenem Kollektor*
Operand operand || parameter * *Parameter (bei SPS)*
Operandenbereich operand area * *bei SPS*
Operandenkennzeichen operand identifier
Operation operation
Operationsausführung operation execution
Operationscode-Decoder opcode decoder * *Funktionsblock in der CPU eines Computers*
Operationsübersicht summary of operations * *Übersicht über SPS-Befehlssatz* || overview of operations * *Übersicht über SPS-Befehlssatz*
Operationsverstärker operational amplifier || Op Amp * *Abkürzg.f. 'Operational Amplifier'* || OpAmp * *Abkürzg.f. 'Operational Amplifier'* || OA * *Abkürzg. f. 'Operational Amplifier'*
Operationsvorrat operation set * *Befehlsvorrat einer SPS* || instruction set * *Befehlssatz eines Computers*

Operationszeit operation execution time * *Ausführungszeit einer (SPS-) Befehls* || instruction execution time * *Ausführungszeit eines Rechnerbefehls* || statement execution time * *Ausführungszeit eines (SPS-) Befehls*
operativ operational
opfern sacrifice * *(auch figürl.) Opfer bringen, mit Verlust verkaufen; Verzicht, Opfer, Aufopferung, Verlust*
Optik optics || appearance * *(figürl.) äußeres Aussehen, Erscheinung* || aspect * *(figürl.) Aussehen, Ansicht, Erscheinung, Gestalt, Gesichtspunkt, Blickwinkel*
optimal optimum || optimal || optimally * *(Adv.)* || most favourable * *günstigst* || best possible || highest possible || best || in the best possible way * *(Adv.)* || perfect
optimieren optimize || tune * *fein anpassen, eine Regelung* ~ || fine-tune * *feinoptimieren* || streamline * *stromlinienförmig machen* || establish optimum conditions
optimierender Compiler optimizing compiler
optimiert optimized * *for: bezüglich*
optimierte Pulsmuster optimized pulse patterns * *bei Pulsumrichter*
Optimierung optimization * *(amerikan.) siehe DIN 19236* || tuning * *z.B. eines Reglers* || optimizing || optimisation * *(brit.)* || streamlining * *z.B. eines Prozesses, einer Organisation*
Optimierungslauf optimization run || selftuning || self tuning routine || auto-tune || optimization routine
Optimierungsrichtlinien optimization guidelines
optimistisch optimistic || optimistically * *(Adv.)*
Optimum optimum
Option option
optional optional || option- || supplementary || additional
optionell optional * *als Option* || optionally * *(Adv.)*
Optionsbaugruppe option module || option board * *Leiterkarte*
optisch optical || visual * *Signal usw.*
optische Industrie optical industry
optische Kontrolle visual check || visual inspection
Optoelektronik optoelectronics || optoelectronics and electro-optics
optoelektronisch optoelectronic
Optokoppler opto-coupler || opto-isolator || optocoupler || photocoupler || optical coupler || optical isolator
ordentlich neat || tidy || in good order || well kept || orderly || regular * *regelrecht* || respectable * *achtbar, beachtlich* || decent * *anständig* || real * *wirklich*
Ordinalzahl ordinate number
Ordinate ordinate || axis of y * *Y-Achse* || Y-coordinate
ordnen put in order || settle * *abschließend* || sort * *sortieren* || align * *in Reih' und Glied bringen, ausrichten*
Ordner file * *siehe auch 'Aktenordner'* || binder * *auch für für Loseblattsammlung*
Ordnung putting in order * *das In-~-Bringen* || order * *ordentlicher Zustand, Reihenfolge; second/third order: zweiter/dritter* ~ || arrangement * *An~* || system * *System* || pattern * *Anlage*
ordnungsgemäß proper || orderly || regular || in due order || lawful * *gesetzmäßig* || duely * *(Adv.)* || correct
Ordnungsliebe love of order
Ordnungszahl ordinal number || ordinal index || harmonic number * ~ *einer Oberschwingung* || harmonic order * ~ *einer Oberschwingung* || order * ~ *einer Oberschwingung* || modal number * *z.B.* ~ *von Oberwellen* || order of harmonics * ~ *von*

Oberschwingungen || ordinal || order number * ~ *einer Oberschwingung*
Ordnungszahlen orders * *z.B. v. Oberschwingungen*
ORGALIME ORGALIME * *Abk. f. 'ORGanisme de Liason des Industries MEtalliques Européen': Maschinenbau/Metallverein*
Organisation organization || organizational structure * *Organisationsstruktur (z.B. eines Unternehmens)*
Organisationsbaustein organization block * *[SPS]* || executive module * *[SPS]*
Organisationseinheit organizational unit
Organisationsplan organization chart
organisatorisch organizational || organizing || managing || executive * *Verwaltungs-, geschäftsführend/-leitend*
organisch organic * *(Chemie)* || organically * *(Adv.) (Chemie)* || harmonic * *harmonisch*
organisieren organize || manage || arrange
orientieren orient * *to: an*
orientiert oriented * *ausgerichtet (auch figürl.), orientiert, informiert* || informed * *informiert, benachrichtigt, in Kenntnis gesetzt* || instructed * *informiert, unterrichtet, belehrt, unterwiesen*
Orientierung orientation * *auch von Fasern, Gewebe, Material* || attitude * *(Ein)Stellung (to/towards: to), Standpunkt, Verhalten, Haltung, Stellungnahme* || guidance * *for your guidance: zu Ihrer* ~
Original-Zubehör genuine accessories
Ort location || city * *Stadt* || place * *Platz, Ortschaft* || size || spot * *Fleck* || point * *Stelle* || locality * *Örtlichkeit* || town * *Stadt*
örtlich local
örtlich verteilt decentralized || distributed
ortsfest stationary || at a stationary position || at a permanent position
Ortskennzeichen location designation || location assignment || location identifier || higher level designation * *Anlagenkennzeichen*
Ortskurve circle diagram || locus diagram || locus
Ortszeit local time
Ortung position finding || orientation
OSB-Platte OSB board * *[Holz] Abk.f. 'Oriented Structural Board'*
Öse eye || eylet * *kleinere*
Osmose osmosis
Ossannakreis Heyland diagram * *Ortskurve der Ströme der Drehstrom-Induktionsmaschine* || Ossanna's circle diagram * *Stromortskurve*
Österreich Austria
Osteuropa eastern Europe
Oszillator oscillator
oszillieren oscillate
Oszillograf oscilloscope || scope || DSO * *Abkürzung für 'Digital Storage Oscilloscope': Digitalspeicher-Oszilloskop*
oszillografieren make an oscilloscope trace || make an oscilloscopic trace
Oszillogramm oscillogram || oscillographic curve || oscillographic record || trace * *Aufzeichnung eines Oszillokop-Bildes* || oscilloscopic trace || oscillogram trace || oscilloscope trace
Oszillograph oscilloscope || scope * *(salopp)* || DSO * *Abkürzung für 'Digital Storage Oscilloscope': Digitalspeicher-Oszilloskop*
Oszilloskop oscilloscope || scope || DSO * *Abkürzg. für 'Digital Storage Oscilloscope': Digitalspeicher-Oszilloskop*
OTP OTP * *Abk.f. 'One-Time Programmable': nur einmal programmierbar, nicht löschbar (Festwertspeicher)*
oval oval
Ovaldrehmaschine oval turning lathe

Overhead-Folie overhead transparency || overhead chart || chart || overhead slide || slide * *Dia* || overhead || transparency
Overheadfolie overhead transparency || overhead slide || overhead chart * *mit grafischer/tabellarischer Darstellung* || transparency || overhead || chart
Oxid oxide
oxidieren oxidize || oxygenize * *mit Sauerstoff verbinden*
Oxyd oxide
Ozonbegasungsanlage ozon injector * *zur Wasserbehandlung*

P

P-Anteil P component || proportional term
p-leitend p-conductive * *Halbleiterschicht*
P-Regler P-controller || proportional-action controller
P-schaltend switching to P potential * *geg. pos. Versorg.panng. schaltend (z B.Open Emitter-Digitalausg.stufe)*
P-Verstärkung P gain || proportional gain || proportional factor || gain
Paar pair || some * *(Adj.) ein paar, einige* || a few * *(Adj.) ein paar, einige* || a couple of * *(Adv.)* || twosome * *aus zwei Teilen (bestehend); (auch figürl.)*
Paarung mating || matching || combination
paarweise in pairs
paarweise verdrillt twisted in pairs * *Kabel* || twisted pair * *Kabel* || twisted pairs * *(Plural) Kabel*
paarweise verseilt twisted pair
packen pack * *ein/ver~* || pack up * *ein~* || wrap up * *ein~, einwickeln* || seize * *fassen* || grip * *fassen, greifen, packen (auch figürlich)* || clutch * *fassen, greifen* || lay hold of * *fassen* || tackle * *(an)packen, in Angriff nehmen, fertig werden mit* || grasp * *greifen*
Packer archiver
Packmittel packaging means * *z.B. für Verpakkungsmaschine* || packaging material
Packpapier packing paper || wrapping paper
Packpresse packpress machine * *Verpackungsmaschine*
Packprogramm packer program || archiver program || archiver * *['a: rkeiwer]* || compression program || packer
Packstoff packaging material * *z.B. für Verpackungsmaschine*
Packungsdichte packaging density || density * *high density: (mit) hohe(r) Packungsdichte*
Paket parcel || package * *großes ~; siehe auch →Paket; (auch figürl.: Daten-/Aufgabenpaket usw.)* || packet * *kleines ~* || kit * *(Bau-/Werkzeug-) Satz, Ausstattung, (Werkzeug-) Kasten*
Paketieranlage bundling system
Palette pallet * *Stapel~, Transport~* || palette * *Brett zum Mischen von Farben; Farbmodell/ Farbauswahlfunktion in der Computergrafik*
Palettentransportanlage pallet conveyor
Palettieranlage palletizing system * *(amerikan.)* || palletising equipment * *(brit.)*
Palettierer palletizing machine * *{Verpack.technik}* →*Palettiermaschine* || pallet loading machine * *{Verpack.technik}* →*Bepalettiermaschine* || palletiser * *(amerikan.) {Verpack.technik}* || palletiser * *(brit.) {Verpack.technik}*
Palettiermaschine palletizer * *Verpackungsmaschine* || palletizing machine * *Verpackungsmaschine* || pallet loading machine * *Bepalettiermaschine, Palettenbelademaschine* || palleting machine
Palettierroboter palletizing robot * *{Verpack.technik}*
Palettierung palletizing * *Aufstapeln von Teilen auf Palette*
PAM-Schaltung PAM-connection * *für polumschaltbaren Motor (Pol-Amplituden-Modulation)*
PAM-Wicklung PAM-winding * *Abk. f. 'Pole Amplitude Modulation': Polamplit.modulat. b. polumschaltb.Mot m. 1 Wicklg.* || Rawcliff winding * *nur 1 Wicklg.i.polumschaltb.Motor; Stromumkehr i.einz. Spulengruppen ändert Drehz.*
Panne breakdown || fault || trouble || failure || mishap * *(figürl.) unglücklich verlaufene Begebenheit, Unglück, Unfall, Panne* || blunder * *Schnitzer, grober Fehler*
Panzerung armor || armor plating || cladding
Papier paper * *auch Dokument,Urkunde, Wert~; identity papers: Ausweis~; filigreed paper: ~ mit Wasserzeich.* || document * *Dokument* || securities * *(Plural) Wert~e (z.B. Aktien, Anleihen)* || stationary * *~waren*
Papierabriss web breakage || paper break
Papierbahn paper web
Papierfabrik paper mill
Papierfaserstoff fibrous paper * *siehe auch* →*Zellstoff*
Papiergewicht paperweight || mass per unit area * *Flächengewicht* || basis weight * *Flächengewicht*
Papierherstellung paper manufacture
Papierindustrie paper-making industry || pulp and paper industry * *Papier- und Zellstoffindustrie*
Papierkorb waste-basket || waste-paper basket
Papierleitwalze paper guide roll
Papiermaschine paper machine || paper making machine
Papiersorte paper grade
Papierverarbeitungsmaschine paper processing machine
Papierveredelungsanlage paper finishing line
Papiervorschub paper feed * *bei Linienschreiber, Drucker usw.*
Papierzug paper tension || web tension
Papierzwischenlage separation paper * *z.B. in der Blech-/Gummiverarbeitung, auch in Kondensator*
Pappe pasteboard || millboard * *starke Pappe* || cardboard * *Kartonpappe; siehe auch* →*Karton* || board
Pappkarton cardboard box
Pappschachtel carton
parabelförmig parabolical
Parallelbetrieb von Motoren parallel running motors * *{el.Masch.}mehrere Mot. arbeit. auf denselb. Beweg.vorgang e. Arb.maschine*
parallel parallel || in parallel * *(Adv.)* || across * *connect across: ~schalten zu*
parallel schalten connect in parallel * *with: zu* || connect across || parallel || shunt || shunt across * *zu etwas ~*
Parallel-Hubspindeltrieb parallel jackscrew || parallel actuator
Parallelanschluss line connection in parallel * *{el. Masch.}*
Parallelbetrieb parallel operation || parallel working
Parallelbus parallel bus
Parallelbuskopplung parallel bus interface
parallele Rahmenkopplung parallel subrack interfacing
parallele Schnittstelle parallel link
parallelgeschaltet connected in parallel || paralleled || connected across || parallel connected

Parallelität

Parallelität parallelism
Parallelkondensator parallel capacitor || shunt capacitor
Parallelschaltung parallel connection || parallel configuration * *parallele Anordnung, auch von Stromrichtern, Motoren usw.* || parallel circuit || connecting in parallel * *das Parallelschalten* || paralleling || shunt arrangement || paralleling * *das Parallelschalten*
Parallelschnittstelle parallel link || parallel interface || parallel coupling
Parallelverarbeitung multi-tasking * *in einem Rechner* || parallel processing * *mehrerer Programme quasi-gleichzeitig in einem Rechner* || multi-computing * *einer Aufgabe auf mehreren Rechnern gleichzeitig*
Parallelverbindung parallel connection
Parallelwiderstand parallel resistor || shunt resistor
Parameter parameter * *[pe'rämite]* || operand * *einer (SPS-) Anweisung*
Parameterart type of parameter
Parameterausdruck parameter print-out * *auf einen Drucker*
Parameterbedieneinheit parameterization unit || programmer * *tragbare ~*
Parameterbeschreibung parameter description * *(auch →PROFIBUS-Profil)* || description of the parameter contents
Parametereinstellung parameter adjustment || parameter setting
Parameterempfindlichkeit parameter sensitivity
Parameteridentifikation parameter identification * *{Regel.techn.}*
Parameterkennung parameter identifier * *(→PROFIBUS-Profil)*
Parameternummer parameter number
Parameterpeicher parameter memory || parameter storage
Parametersatz parameter set
Parameterschnittstelle parameter interface || parameterization interface * *Parametrierschnittstelle*
Parameterschwankung parameter variation
Parametertyp data type || type of data || parameter type || type of parameter
Parameterwert parameter value
Parametrier-Software parameterizing software || parameterization software
parametrierbar programmable || can be parametrized * *kann parametriert werden*
Parametrierbaugruppe parameter assignment module * *bei SPS*
Parametriereinheit parameterization unit || parameterizing unit || parameterization panel
Parametriereinrichtung parameterizing device || parameterization device || parameter assignment device || parameter assignment unit
parametrieren parameterize || program || assign parameters * *bei SPS*
Parametrierhoheit parameter change rights * *(→PROFIBUS-Profil)*
Parametrierprogramm parameterizing program
parametriert parameterized || programmed
Parametrierung parameter setting || parameterization || programming || parameterizing || parameterization work * *Tätigkeit des Parametrierens z.B. durch e. Techniker* || parameter assignment
Parametrierwerkzeug parameterizing tool
parasitär parasitic
parasitärer Effekt parasitic effect || parasitic phenomenon || parasitic
Parität parity * *Datensicherung* || parity bit * *Paritätsbit*
Paritätsbit parity bit

Paritätserzeugung parity generation
Paritätsfehler parity error || incorrect parity bit
Paritätskontrolle parity check
Paritätsprüfung parity check || parity checking
Parity-Bit parity bit * *Datensicherungsbit, ergänzt ein Datenwort/-Byte auf e. gerad- oder ungeradzahlige Zahl*
partiell partial
Partikel particle
Partition partition * *logisches Laufwerk (z.B. C: , D: usw. auf Festplatte)*
Partner partner || interlocutor * *Gesprächspartner* || party * *Vertragspartner, Vertragspartei* || contact person * *Ansprechpartner* || companion * *Gefährte, Kamerad, Gesellschafter, Gegenstück*
Partnerschaft partnership
passen fit * *passen zu,auf* || be suitable * *passend geeignet sein* || be convenient * *geeignet, günstig sein; to/for: für* || tally * *zusammenpassen* || harmonize * *zusammenpassen* || agree together * *zusammenpassen* || match * *mehrere Dinge zueinander* || suit * *anpassen (to: an), passen zu, sich eignen zu/für, entsprechen, kleiden* || be suited * *geeignet sein* || be fit * *passend/geeignet/fähig/tauglich sein* || be proper * *richtig/passend/geeignet/angemessen/ordnungsgemäß/zweckmäßig sein* || make sense * *Sinn machen*
passen zu suit * *match* || be suitable for * *passend, geeignet sein für* || be convenient for * *geeignet/ günstig sein für*
passend appropriate || proper * *richtig, passend, geeignet, angemessen, ordnungsgemäß, zweckmäßig* || matching * *angepasst* || applicable * *anwendbar (to: auf), geeignet, zutreffend* || suitable * *suitable for use with: geeignet für die Verwendung bei/mit* || fit || suited || convenient * *(gut) geeignet, günstig (to/for: für/zu), bequem, günstig passend, gelegen* || fitting || opportune * *Zeit, Gelegenheit usw.* || corresponding * *entsprechend* || appropriately * *(Adv.)*
Passfeder key * *Nutkeil (DIN 748 Teil 1 u.3, DIN ISO 8821, VDE 0530 Teil 1)* || spline * *Nutkeil* || feather key * *Nutkeil* || featherkey * *half featherkey: halber ~ (Motor-Auswuchtverfahren mit z. Hälfte einbezogenem Federkeil)*
Passfedernut keyway
passieren pass || pass over || pass through || clear * *einen bestimmten (Kontroll-/Etappen-) Punkt ~* || happen * *sich ereignen* || occur * *sich ereignen* || take place * *sich ereignen* || come to pass * *sich ereignen* || passage * *(Substantiv)*
passiv passive
Passiva liabilities
passive Bauelemente passive components
passivieren passivate
passiviert passivated
Passivierung passivation * *z.B. Schutzauftrag aus Glas bei Hybridbauteil*
Passivität passivity
Passmarke pass mark
Passsitz snug fit
Passstift dowel pin || alignment pin
Passstiftloch dowel hole
Passung fit * *system of fits: Passungssystem (z.B. ISO)* || tolerance on fit * *Passungstoleranz* || fitting clearance * *Passungspiel* || fit size * *Passmaß* || tolerances
Passungsrost mated-surface rusting * *rostrote Verfärbung an d. Berührungsstelle von zwei Stahlteilen*
Passwort password
Paste paste * *breiige Masse, Paste, Brei, Kleister, Klebstoff, Glasmasse, Druck~ (bei Siebdruck)*
Patent patent * *Erfindungsurkunde (for: auf); apply for: anmelden; infringe on: verletzen; pending: angemeldet*

Patenterteilung granting of a patent
patentierbar patentable
patentieren patent * *take out a patent for a thing: sich etw. patentieren lassen* || protect by patent
patentiert patented
Patentverletzung patent infringement
Patina patina || oxide film * *Oxidschicht*
Patrone cartridge || refill * *zum Nachfüllen*
pauschal global || lump-sum * *-Preis, -Summe usw. z.B. lump-sum price: Pauschalpreis* || overall * *über alles, Gesamt-* || blanket * *allgemein, Blanko- (z.B. Erlaubnis-/Genehmigung); umfassend, Gesamt-, gemeinsam*
pauschal verrechnen invoice at a lump-sum price
Pauschale lump sum price * *Pauschalpreis*
Pauschalpreis lump-sum price
Pauschbetrag lump-sum * *Pauschalbetrag*
Pause pause || stop || interval * *['interwel]* || break * *z.b. ~ bei Besprechung; coffee break: Kafee~* || rest * *Ruhe~, auch einer Maschine* || deenergized interval * *z.b. bei Betrieb eines Motors* || machine at rest and de-energized * *[el.Masch.]* Stillstand mit stromlosen Wicklungen
Pausenzeit duration of pause * *Länge der ~* || stop interval || pause interval || no-load interval * *Zeitdauer ohne Belastung* || rest period || reduced-load interval * *Zeitdauer mit abgesenkter Belastung*
PC PC || personal computer
PE-Schiene PE rail * *Abk. für 'Protective Earth'*
Pedal pedal * *work the pedals: in die ~e treten* || foot lever * *Fußhebel*
Peer-to-Peer-Netzwerk peer to peer network * *Netzwerk mit gleichberechtigten PC's(jeder ist gleichzeitig Server u.Client)*
Peer-to-Peer-Schnittstelle Peer-to-Peer link * *Verbindg. zw. gleichberecht. Partnern, z.B. Stromrichtern (o. Master)*
Peer-to-Peer-Verbindung Peer-to-Peer link * *seriell. Verbindg. zw. gleichberecht. Partnern, z.B. Stromrichtern (o. Master)*
peforieren perforate
Pegel level * *auch Spannungspegel, logischer Pegel*
Pegelanpassung level adaption || signal level adaption * *für ein Signal* || level shifting
Pegelumschaltung level changeover
Pegelumsetzer level converter
Pegelwandler level shifter * *z.B. zwischen verschiedenen Logikfamilien wie TTL, CMOS* || level converter
Pegelwandlung level conversion || level changeover
Peigneur doffer * *[Textil] Abnahmewalze in →Krempelmaschine*
Pendant counterpart * *Gegenstück* || equivalent || matching piece * *zugehöriges Teil*
Pendel pendulum
Pendelarm rocker rarm || dancer arm * *einer Tänzerwalze/Tastrolle/Regelschwinge/Pendelwalze*
Pendeldämpfung pendulum control * *z.B. bei Kran* || oscillation suppression || oscillation damping || power stabilization * *im Stromversorgungsnetz* || grab oscillation damping * *[Hebezeuge] bei Greiferkran*
Pendelgenerator swinging-frame generator * *zur Drehmomentmessung* || cradle dynamometer || swinging-frame dynamometer || pendulum generator * *zur Drehmomentmessung* || swinging stator dynamometer
Pendelmaschine cradle dynamometer * *zur Drehmomentvorgabe/messung, z.B. für Prüfstand* || swinging-frame dynamometer * *zur Drehmomentvorgabe/messung* || pendulum generator * *zur Drehmomentmessung* || swinging-stator dynamometer * *zur Drehmomentmessung* || swinging frame generator * *zur Drehmomentmessung* || swinging-

frame machine * *el.Masch. mit ausschwingbar gelagertem Ständer zur Drehmomentmessung*
Pendelmoment oscillating torque * *[el. Masch.]* führt zu Drehschwingungen/Drehzahlschwankungen (b. kleiner Drehzahl)
pendeln oscillate * *oszillieren, schwingen, (hin und her) schwanken* || swing * *(hin und her) schwingen/schwenken/baumeln/pendeln lassen* || shuttle * *sich hin- und herbewegen; shuttle between: ~ zwischen (Fahrzeug)* || commute * *wenden, auswechseln, umtauschen* || reciproce * *sich hin- und herbewegen (z.B. Kolben)* || hunting * *(Substantiv) Schwingen/Pendeln einer instabilen Regelung* || gear meshing * *(Substantiv) um Schaltgetriebe-Zahnräder beim Schalten in Eingriff zu bringen* || oscillating * *(Substantiv) Schwingen* || oscillating motion * *(Substantiv) Hin- u. Herbewegung z. Erleichtern d.Gangwechsels bei Schaltgetriebe*
Pendelschleifmaschine swinging-frame grinding machine
Pendelung hunting * *auch Reglerschwingen bei Instabilität* || oscillation * *z.B. bei Synchronmotor; kleine Drehzahlvariationen z. leichten Schalten v. Schaltgetriebe*
Pendelwalze dancer roll * *Tänzerwalze/-rolle* || dancer * *Tänzer*
perfekt perfect || impeccable * *untadelig, einwandfrei* || excellent * *hervorragend, ausgezeichnet* || accomplished * *vollbracht, ausgeführt, erreicht*
perfektionieren perfect * *of technical perfection/technically perfect: technisch perfektioniert*
Perfektionismus perfectionism || perfectibilism
Perfektionist perfectionist
perforieren perforate
Periode period || cycle
Periodendauer period || time of oscillation || cycle time || cycle duration
periodisch periodical || periodically * *(Adv.)* || periodic || cyclic || repetitive * *~ auftretend/wiederkehrend* || recurrent * *wiederkehrend, sich wiederholend, ~ auftretend*
periodischer Aussetzbetrieb intermittent periodic duty * *[el.Masch.]* →Betriebsart S3
periodischer Aussetzbetrieb mit Einfluss des Anlaufvorgangs intermittent duty with starting * *[el.Masch.]* →Betriebsart S4
periodischer Aussetzbetrieb mit elektrischer Bremsung periodic intermittent duty with electrical braking * *[el.Masch]* →Betriebsart S5
periodischer Betrieb periodic duty
periodischer Verlauf periodic nature * *Art, Beschaffenheit*
Peripherie periphery || peripherals * *I/O-Baugruppen für SPS, (Prozess-)Rechner* || perimeter * *Umkreis (z.B. einer Stadt) Peripherie* || I/O * *Ein-/Ausgabefunktionen* || peripherals * *I/O-Baugruppen/Geräte für SPS/(Prozess-)Rechner*
Peripherie- peripheral || I/O- * *[SPS]*
Peripherie-Speicher-Umschaltung memory-I/O select * *Adressumschaltung bei SPS*
Peripheriebaugruppe I/O-module * *Ein/Ausgabebaugruppe bei Rechner/SPS* || peripheral board * *Leiterkarte* || peripheral module
Peripheriebaugruppen I/O modules
Peripheriebaustein peripheral * *z.B. eines Mikroprozessors* || interface chip * *Peripherieschaltkreis für Mikroprozessor (z.B. Interruptcontroller, USART)*
Peripheriebereich I/O area * *bei SPS*
Peripheriebyte I/O byte * *bei SPS* || peripheral byte * *bei SPS*
Peripheriegerät peripheral device * *eines Computers* || peripheral * *eines Computers* || peripheral unit

Peripheriegeräte peripherals * z.B. Drucker, Bildschirm bei SPS
Peripheriemodul peripheral module || interface module
Peripheriesignal peripheral signal
Peripherspeicher I/O memory || peripheral memory
Perle bead * Perlkorn, (Tau-/Schweiß-/Schmelz-) Perle, (Schaum-) Bläschen, Tröpfchen
permanent permanent || perpetual || constant || continually * (Adv.) →dauernd, fortwährend
permanenterregt permanent-field || with brushless excitation * mit bürstenloser Erregung || permanently-excited || permanent-magnet excited
permanenterregte Maschine permanent-field machine || permanently excited machine || machine with brushless excitation * ohne Bürsten
permanenterregter Gleichstrommotor permanent-magnet DC motor || PDMC motor * Abk. für 'Permanent-Magnet DC motor'
permanenterregter Läufer permanent-field rotor
permanenterregter Motor PM motor * PermanentMagnet-erregt.Motor; Geg.satz: wound field motor: Mot. m.Feldwicklg. || permanent-field motor || permanent-field synchronous motor * z.B. SIMOSYN (R) Motor von SIEMENS
permanenterregter Synchronmotor permanentmagnet synchronous motor || permanent-field synchronous motor * z.B. Siemosyn-Motor von SIEMENS
Permanenterregung permanent-field excitation
Permanentmagnet permanent-magnet
Permanentmagnetbremse permanent-magnet brake
Permanentmagnetmotor permanent magnet motor
Permanentspeicher permanent memory * nichtflüchtiger Speicher (z.B. EEPROM, NOVRAM, Flash-EPROM usw.) || non-volatile memory * nichtflüchtiger Speicher
Permanentverarbeitung permanent processing
permeabel permeable
Permeabilität permeability
Personal personnel || staff || employees * Angestellte || attendants * Bedienstete, Bedienungsmannschaft || servants * Beamte, Angestellte (öffentlicher Dienst); Dienstboten || workforce * Mitarbeiterschaft, Belegschaft
Personalabbau job-cutting || staff reduction
Personalausstattung work force
Personalchef personnel head || staff manager
Personalkosten manpower costs
Personalwesen personnel management || human resources * auch als Abteilungsbezeichnung
Personenbeförderung passenger transport || passenger service || passenger traffic || conveyance of passengers * auch bei Aufzügen
Personengefährdung risk of personnel injury || risk of injury or death || hazards to persons
Personengesellschaft partnership
Personennahverkehr suburban public transportation || short-distance passenger traffic || local traffic || short-haul passenger service
Personenschaden personal injury or death * Verletzung oder Tod || bodily injury * Körperverletzung
Personenschutz personal protection
Personenschutzeinrichtung personal protection gear
Personensicherheit personnel safety || operator safety
Personenverkehr passenger traffic || passenger transportation * (amerikan.)
Perspektive perspective * (auch figürl.: 'Ausblick') || prospect * (figürl.) Aussicht, Ausblick, Vor(aus)-schau, Zukunftsaussicht || view * (auch figürl.) Ansicht, Aussicht, Ausblick (of/over: auf), Überblick (of: über)
pessimistisch pessimistic || pessimistically * (Adv.)
PET PET * Abk. für 'PolyEthyleneterephThalate': Polyethylentheraftalat (für Kunststoffflaschen usw.)
petrochemisch petrochemical
Petroleum petroleum * (allg.) || paraffin * für Heizu. Leuchtzwecke || kerosene * (amerikan.) für Heiz- und Leuchtzwecke
Pfad path || section * Teil || track * Spur, Weg, Route, Fährte || branch * Zweig (auch in Programm)
Pfanne pan || copper * zur Würzegewinnung in einer Brauerei
pfeifen screech * kreischen (z.B. Lager), gellend/ durchdringend schreien || squeal * kreischen, grell/schrill schreien, quieken
Pfeil arrow
pfeilverzahnt herringbone-scewed * Zahnrad
Pferdestärke horse power || hp
pfiffig clever * geschickt, raffiniert, klug, gescheit, begabt || cunning * listig, schlau, schmitzt, geschickt, klug || artful * schlau, listig, verschlagen || sly * schlau, verschlagen, listig || slick * flott, raffiniert, geschickt || tricky * raffiniert, verschlagen, durchtrieben, trickreich
Pflege care || maintenance * →Wartung; preventive maintenance: vorbeugende Wartung
pflegen keep well || keep in good order || keep in good service || keep in good repair || conserve * erhalten || service || tend * Pflegehandlung ausüben
Pflicht duty || obligation * Verpflichtung || responsibility * Verantwortung
Pflichtenheft functional specification || final specification || target specification
Pfund Pound * brit. Währung || pound * brit. Gewichtseinheit (1 pound entspr. 453,6 g)
Pfusch bungle || bungling || botching || bad job || scamped work
pfuschen bungle || botch || scamp
PG steel conduit thread * verschraubte Kabeldurch/einführung (Abk.f.'Panzerrohr-Gewinde'-Verschraubung) || programming unit * Programmiergerät für SPS || programmer * Programmiergerät für SPS
PG-Anschaltung programming unit interface
PG-Schnittstelle programmer port * Anschluss (-stecker) z.B. an einer SPS zum Anschluss eines Programmiergeräts
PG-Verschraubung PG gland * Abk. f. 'PanzerrohrGewinde'-Verschraubg. (Kabeldurchfuhrg. m. 'Quetschgummi-Dichtg.') || threaded cable entry hole * Kabeldurchführung mit Verschraubung || heavy-gauge conduit gland || screwed conduit entry || cable entry gland || steel conduit thread * verschraubte Kabelein/-durchführung || cable gland * Leitungs/Kabelverschraubung || conduit fitting || conduit adapter || threaded cable entry * für abgedichtete Kabeleinführung || compression gland * siehe auch DIN 46320
pH-Wert pH-value
Phänomen phenomenon
phänomenal phenomenal
pharmazeutische Industrie pharmaceutical industry
Phase phase || period * Zeitraum || stage * Stufe
Phasenabweichung phase displacement || phase difference
Phasenanschnitt phase angle variation || phase angle control
Phasenanschnittsteuerung phase angle control * z.B. für netzgeführten Stromrichter, Thyristorsteller usw. || phase control
Phasenausfall phase loss || phase failure

Phasenausfallerkennung phase loss detection
Phasenausfallschutz phase-failure protection
Phasenausfallüberwachung phase failure monitoring
Phasenbaustein modular phase assembly
Phasenbruch phase failure || phase loss
Phasendrehung phase displacement * *Phasenverschiebung*
Phaseneinbruch phase dip * *kurzzeitiger*
Phasenfolge phase sequence
Phasenfolgelöschung interphase commutation || phase commutation || phase sequence commutation * *Kommutierungsart b. I-Umrichter: Speisung je 2er Wicklungsstränge i. Folge* || phase-sequence commutation * *Thyristorlöschung jew. durch Zündg. d. Folgethyristors ü. Löschkondensator*
Phasenfolgeüberwachung phase sequence monitoring
Phasengang phase response
Phasengeschwindigkeit phase velocity
phasengetrennt phase-segregated || phase-separated
phasengleich in phase
Phasengleichheit the same phase sequence * *z.B. zwischen Elektronik- und Leistungteilanschluss e. Stromrichters* || in-phase condition || phase coincidence || correct phasing * *phasenrichtiger Anschluss* || the same phase relationship
Phasenkennlinie phase characteristic || phase response * →*Phasengang*
Phasenkurzschluss phase-to-phase short-circuit || phase short-circuit
Phasenlage phase position || phase angle * *Phasenwinkel* || phase relation || phase relationship
Phasennacheilung phase lag
Phasenreserve phase margin * *{Regel.techn.} Phasenrand, Phasenreserve*
Phasenschieben phase shifting
Phasenschieber phase shifter || phase-shift network
Phasenschluss phase to phase short circuit || phase short-circuit || inter-phase short circuit
Phasenschnittpunkt phase intersection point
Phasenspannung phase-to-neutral voltage * *zwischen L1, L2, L3 und N* || phase voltage * *Strangspannung* || star voltage * *Sternspannung* || line-to-neutral voltage || phase-to-ground voltage
phasenstarr phase-locked
Phasenstrom phase current
Phasensymmetrie phase symmetry
Phasentrenner inter-phase insulation * *{el. Masch.} Isolierstoff zw. Spulengruppen u.Wicklungssträngen i.'Wickelkopf*
Phasenüberwachung phase sequence monitoring
Phasenunsymmetrie phase asymmetry || phase imbalance || phase unbalance || phase unsymmetry
Phasenvergleich phase comparison
Phasenverschiebung phase shifting || phase lag * *Phasen-Nacheilung* || lag * *negative ~, Phasennacheilung* || phase displacement || phase shift || phase lead * *Phasenvoreilung* || quadrature * *~ um 90 °* || phase delay * *Phasen-Nacheilung*
Phasenverschiebungswinkel phase angle || phase displacement angle || phase difference || phase displacement
phasenverschoben phase displaced || with a phase-angle displacement || out of phase * *by 90 degrees: um 90 Grad* || phase-shifted || phase-delayed * *verzögert* || phase-shifted * *with respect to: gegenüber*
Phasenverzögerung phase lag
Phasenvoreilung phase lead
Phasenwandler phase converter
Phasenwicklung phase winding * *z.B. eines Trafos, einer Drossel*

Phasenwinkel phase angle || phase delay || phase displacement angle || phase displacement
phosphatieren phosphate
photovoltaisch photovoltaic * *z.B. Solar-*
Physik physics || natural philosophy || physical science || hardware design * *hardwaremäßige Ausführ., z.B. serielle Schnittstelle*
physikalisch physical
physikalische Dimension engineering unit
physikalische Einheit engineering unit
physikalische Gesetze laws of physics
physikalische Grundlagen physical fundamental principles
physikalischer Kanal physical channel
PI-Regler PI-controller * *[pie 'ai ...]* || proportional plus integral controller || controller with PI characteristics || proportional plus integral-action controller
PID Regler PID controller
PID-Regler PID controller * *Regler mit Proportional-, Integral- u.Differenzialanteil*
Piepser beeper
Piezo- piezo
piezoelektrisch piezo-electric
Piktogramm pictograph || pictorial marking
Pilotanlage pilot plant || pilot equipment || test installation || prototype plant || pilot installation
Pilotprojekt pilot project
Pilotversion pilot version || trial version
Pilzrasttaster mit Drehentriegelung mushroom button with pushlock and twist-release
Pilztaster mushroom pushbutton * *z.B. Not-Aus Rasttaster*
Pin pin * *z.B. IC-/Steckerbeinchen*
Pinbelegung pinout * *eines IC* || pinning * *eines IC* || pinning diagram * *IC-Anschlussbild* || pin assignment * *z.B. eines Steckers/ICs*
Pinbezeichnung pin designation * *z.B. bei einem IC*
Pinsel brush || paint-brush
Pionierleistung pioneering feat * *Großtat, Kunst/ Meisterstück* || pioneering performance || pioneering achievement || pioneering accomplishment || pioneering work
Pipeline pipeline * *Rohrleitung für Öl, Gas usw.*
Pistole pistol * *z.B. Wire-Wrap-~, Schrumpf-~, Heißluft-~* || gun * *z.B. Löt-~, Wire-Wrap-~*
PIV-Getriebe PIV speed variator * *Markenname eines mechanisch stufenlos verstellbaren Getriebes*
Pixel pixel * *{Computer} Bildpunkt am Monitor*
Pixelabstand dot pitch * *bei einem Momitor*
Pixelfrequenz pixel frequency * *{Computer} Pixelanzahl x Bildwiederholrate x 1.3 bei Monitor* || pixel rate
Pixeltakt dot clock * *bei Monitor*
PKW parameter data * *(→PROFIBUS-Profil; Parameter/Kennung/Wert)*
PKW-Bedienhoheit parameter change rights * *(→PROFIBUS-Profil)*
placieren place * *z.B.Produkt am Markt; (auf)stellen/setz.; (sich) postieren, unterbring., (Bestellg.) aufg.*
Plakat poster || banner
Plan plan || intention * *Absicht* || project * *Vorhaben* || scheme * *Schema, Anlage, System,* →*Entwurf* || map * *Karte* || diagram * *grafische Darstellung* || blueprint * *Blaupause* || draft * *(Grob-)Entwurf* || chart * *Tabelle, Schaubild, grafische Darstellung* || layout * *Anlage, Anordnung* || schedule * *Zeitplan* || plain * *(Adj.) glatt* || level * *(Adj.) eben, waagerecht* || horizontal * *(Adj.) horizontal, waagerecht* || drawing * *Zeichnung* || plane * *(Adj) auch 'flach, eben'* || even * *(Adj.) eben, flach, gerade, waagerecht, in gleicher Höhe* || flat * *(Adj.) flach, eben, platt, umgelegt*

plandrehen face ‖ face down
Plandrehmaschine facing lathe ‖ front operated lathe
planen plan ‖ schedule * *zeitlich/terminlich ~ (auch Personaleinsatz)* ‖ project * *entwerfen, projektieren* ‖ map out * *voraus~, ausarbeiten, (seine) Zeit einteilen* ‖ propose * *vorhaben*
Planetengetriebe planetary gear ‖ planet gear ‖ planetary gearing ‖ planetary gear unit
Planetenrad planet wheel * *bei Planetengetriebe* ‖ planet pinion * *bei Planetengetriebe* ‖ satellite * *bei Planetengetriebe* ‖ epicyclic gear * *bei Planetengetriebe* ‖ satellite gear ‖ planet gear
Planetenträger planet carrier * *in einem →Planetengetriebe* ‖ sattelite carrier ‖ pinion cage ‖ epicyclic unit
Planetenverstellantrieb planetary variable speed drive
Planfläche plane surface ‖ face ‖ end face * *am Wellenende*
planieren level * *eben machen, nivellieren, einebnen, planieren* ‖ plane ‖ grade * *Gelände ~* ‖ planish * *Metall ~, glätten, (ab)schlichten, (Holz) abhobeln, (Metall) glatt hämmern/polieren* ‖ size * *{Buchbinderei}*
Planlage flatness ‖ facing
Planlauf axial eccentricity ‖ axial runout * *{el. Masch.} Fluchtfehler des Wellenendes bezogen auf den Flansch (DIN 42955, IEC 72-1)*
Planlauffehler axial eccentricity * *z.B.Rundlauffehl. e. Welle, Flanschzentrierg., Planlauf zw. Wellenende u.Flansch*
Planlauftoleranz axial eccentricity tolerance * *z.B. zw. Wellenende u. Flanschfläche e. Motors* ‖ axial eccentricity ‖ shaft-flange squareness * *zwischen Wellenende und Flanschfläche*
planparallel plane-parallel
Planscheibe faceplate ‖ face plate
planschneiden guillotine
Plansiebmaschine gyratory screener * *{Holz} für Holzschnitzel*
Planung planning * *be in the planning/under planning: in ~ sein; planning phase: ~sphase* ‖ plan ‖ layout * *Anlage, Anordnung* ‖ budgeting * *Ausgaben~, Haushalts~, Finanz~* ‖ diagram * *Zeichnung* ‖ drawing * *Zeichnung* ‖ scheduling * *zeitliche ~, Termin~*
Planungsaufgabe planning task
Planungsphase planning stage ‖ planning phase
Planungsunterlagen project documents ‖ project manual * *Handbuch*
Plasmaschneidmaschine plasma-jet cutting machine * *Werkzeugmaschine*
Plastikbeutel plastic bag
Plastikfolie plastic film ‖ plastic foil
plastisch plastic ‖ three-dimensional * *dreidimensional (bildliche Darstellung usw.)*
Platin platinum
Platine board * *Leiterkarte* ‖ PC board * *gedruckte Leiterkarte* ‖ card * *Flachbaugruppe, Karte* ‖ card module * *Flachbaugruppe* ‖ print * *Leiterkarte* ‖ PCB * *Abk. für 'Printed Circuit Board': gedruckte Leiterkarte*
Plättchen chip ‖ small plate ‖ lamina ‖ lamella ‖ splitting ‖ flake * *Flocke, dünne Schicht, Schuppe, Fetzen, Splitter*
Platte plate * *(Glas/Metall/Druck)Platte, Tafel, Scheibe, Grobblech, Blechtafel, (Batterie)Elektrode* ‖ panel * *(Schalt)Tafel* ‖ sheet * *Metall, Glas usw.* ‖ lamina * *dünne Platte, Lamelle* ‖ disk * *runde Scheibe* ‖ disc * *runde Scheibe* ‖ board * *Brett, Tafel, Planke*
Plattenaufteilanlage panel sizing plant * *{Holz} Zuschneidemaschine für Holzplatten* ‖ panel sizing system * *{Holz} Zuschneidemaschine* ‖ sizer * *{Holz} Zuschneidemaschine*
Plattenaufteilsäge panel dividing saw * *{Holz}* ‖ panel sizing saw * *{Holz}* ‖ panel beam saw * *{Holz}* ‖ panel cutting saw * *{Holz}* ‖ panel saw * *{Holz}* ‖ panel-sizing circular saw * *Plattenaufteil-Kreissäge*
Plattenband platform conveyor * *~förderer* ‖ slat * *Förder-/Vorschubeinrichtung*
Plattenbandförderer platform conveyor ‖ slat conveyor
Plattenbandvorschub slat feed * *Plattenbandförder/-Vorschubeinrichtung*
Plattenkondensator plate capacitor
Plattenwärmetauscher plate heat exchanger
Plattform platform * *(auch figürlich: Hard-/Software/Entwicklungsplattform)*
Platz position * *Stellung, Lage* ‖ place * *be in/out of place: am/nicht am ~ sein* ‖ spot ‖ point * *(amerikan.) Stelle* ‖ room * *Raum; make room/way for: ~ machen für* ‖ locality * *Örtlichkeit* ‖ site * *Lage (Bau)Platz* ‖ seat * *Sitz~* ‖ station * *~ zum Bedienen, Reparieren usw.* ‖ yard * *Hof(raum), (Bau-/Stapel-) Platz (im Freien)* ‖ place of work * *→Arbeitsplatz*
Platz sparend space-saving ‖ room-saving * *raumsparend*
Platzausnutzung space utilization ‖ room utilization * *Raumausnutzung*
Platzbedarf space requirements ‖ space request ‖ floor space required * *Bodenfläche z.B. in Raum, Gebäude* ‖ footprint * *Grundfläche* ‖ space requirement
Platzbedarf reduzieren downsize
Platzeinsparung space saving
Platzersparnis space saving
Platzhalter dummy
platzieren place * *z.B. Produkt am Markt; (auf)stell./setz.; (s.)postieren; unterbring.; (Bestellg.) aufgeb.* ‖ position * *anbringen, in die richtige Lage bringen, lokalisieren*
Platzmangel lack of space ‖ restricted place
Platznutzung space utilization ‖ room utilization * *Raumausnutzung*
Platzproblem space problem
Platzreduzierung downsizing
plausibel plausible * *make plausible: ~ machen; siehe auch →einleuchtend*
plausibel sein be plausible ‖ make sense * *Sinn machen*
Plausibilität plausibility
Plausibilitätsabfrage plausibility check
Plausibilitätskontrolle plausibility check
PLD PLD * *Abk. f. 'Programmable Logic Device': programmierbarer Logikbaustein* ‖ programmable logic device
Pleuellager rod bearing
Pleuelstange connecting rod ‖ piston rod * *Kolbenstange*
PLL-Schaltung PLL * *Phase-Locked Loop, phasenstarre Regelschleife: gleiche Phasenlage b. Aus-/Eing.pulsen*
plotten plot
Plotter plotter
plötzlich spontaneous * *spontan* ‖ sudden * *(Adj.)* ‖ suddenly * *(Adv.)* ‖ of a sudden * *(Adv.)*
Plungerpumpe plunger pump * *Tauchpumpe*
plus-minus plus or minus
Pluspol positive pole ‖ positive terminal * *Anschluss an Batterie, Kondensator*
Pluspunkt plus ‖ credit point ‖ plus point
Pluszeichen positive sign ‖ plus symbol
Pneumatik pneumatic * *[nju'mätik]*
Pneumatikkolben pneumatic piston * *[nju'mätik 'pisten]*

Pneumatikventil pneumatic valve
Pneumatikzylinder pneumatic cylinder * *[nju'mätik...]* || air cylinder
pneumatisch pneumatic * *[nju'mätik]* || pneumatically operated * *mit Druckluft betrieben* || pneumatically * *(Adv.) [nju: 'mätikälli]*
pneumatisch betätigt pneumatically operated * *[nju'mätikälli...]* || pneumatically actuated
Pofessor professor || lecturer * *Dozent* || university lecturer
Pol pole * *auch eines Magneten* || terminal * *(Batterie-) Anschluss*
Pol-Amplituden-Modulation pole-amplitude modulation * *[el.Masch.]* PAM-Schaltg. für Schleifringläufer m. Stromumkehr in 1 Wicklg
Polarität polarity
Polen Poland || pole * *(Verb)* || polarize * *(Verb)*
Polform pole shape * *z.B. einer el. Maschine*
polieren polish || burnish || buff * *glanzschleifen, polieren, schwabbeln*
Poliermaschine polishing machine || polisher
Polierpaste polishing paste
Politik policy
Polklemme pole terminal
Polpaar pole pair || pair of poles
Polpaarzahl number of pole pairs || pole pair number
Polrad rotor || magnet wheel || inductor
Polradlage rotor position || rotor angle
Polradlagegeber rotor position sensor || rotor position encoder
Polradwinkel rotor displacement angle || electrical rotor angle * *elektrischer* || rotor angle
Polschuh pole shoe || pole head
Polster pad * *Polster, Kissen (als Schutz gegen Stöße), Wulst, Bausch; oil pad: Schmierkissen* || padding * *Polsterung, Wattierung* || bolster * *Polster, Polsterung, Kissen, weiche Unterlage* || cushion * *Kissen* || upholstery
Polstermittel padding elements * *beim Verpacken* || padding materials * *[Verpack.techn.]*
polstern bolster * *(auf)polstern, (weich) abfedern*
Polteilung pole pitch * *[el. Masch.]* Abstand der Polmitten zweier benachbarter ungleichnamiger Pole
polumschaltbar pole changing * *n-speed pole-changing: n-fach ~*
polumschaltbarer Motor pole-changing motor || multispeed motor || change-pole motor || pole-changing multispeed motor || change-speed motor
Polumschalter pole-changing switch * *für polumschaltbaren Motor (z.B. in Dahlanderschaltung)*
Polumschaltung pole-changing * *Umschaltg. d.Ständerwicklg. e.AC-Motors z.Drehzahländerung (Dahlander/PAM-Schaltg.)*
Polumschaltung als Anlassverfahren starting by pole-changing * *[el.Masch.] therm. schonendes Anfahren b. polumschaltbarem Motor*
Polung poling || polarity || polarization * *Polarisation*
Polvorgabe pole assignment * *[Regel.techn.]*
Polyamid polyamide
Polyäthylen polyethylene
Polycarbonat polycarbonate
Polyester polyester
Polygon polygon * *aus Geradenstücken bestehende Kurve* || traverse
Polygonbaustein polygon curve block * *Funktionsbaustein* || polyline function block * *Funktionsbaustein*
Polygonzug polygon curve * *aus Geradenstücken bestehende Kurve* || polygon || polyline || polygon-based interpolation block * *Funktionsbaustein* || polygonal course
polykristallin polycristalline

Polymer polymer
Polynom polynominal
Polynom- polynominal || polynom
Polynomform polynom notation
Polyplanetengetriebe polyplanetary gear
Polypropylen polypropylene || HOSTALEN * *Markenname (als Isolierfolie usw.)* || TRESPAPHAN * *Markenname (z.B. als Isolierfolie usw.)*
Polystyrol polystyrene
Polyurethan polyurethane
Polzahl number of poles || pole count
Polzahlverhältnis ratio of the number of poles * *z.b. bei Maschinensatz* || pole-count ratio
Poperoller drum reel-up || single-drum surface winder || pope roller || reeler || reel || reel drum * *Andruckwalze, Stützwalze* || reel-up || reel-drum winder || reeling drum || pope reel winder * *Oberflächenwickler (d.h. kein Achswickler; meist mit einer Andruck-/Reibwalze)* || pope reel
Poren pores
Porosität porosity
portabel portable
Portabilität portability * *Übertragbarkeit von Programmen von einem Rechnertyp auf den anderen*
Portal- gantry * *z.B. ~kran, ~förderer, ~wickler, ~Werkzeugmaschine* || gantry-type * *z.B. ~kran, ~fräswerk*
Portalbauweise gantry type of construction
Portalfahrwerk gantry traversing unit * *eines Portalkrans*
Portalkran gantry crane || traveling gantry
Portalroboter gantry-type robot || gantry robot || portal robot
Portalsäge gantry saw * *[Holz]* || portal saw * *[Holz]* || overhead saw * *[Holz]*
portierbar portable * *übertragbar (z.B. Software von einer Hardwarebasis auf eine andere)*
Portierbarkeit portability * *z.B. von Software von e. Hardwareplattform/Betriebssystem auf e. andere(s)*
portieren port * *z.B. Softwareprogramm auf einer Zielhardware auf eine andere*
Portierung port * *[Computer] Software auf eine andere Hardware- oder Betriebssystemplattform*
Portugal Portugal
portugiesisch Portuguese || of Portugal
Position position * *Lage, auch bei [NC-Steuerung]* || rank * *Rang* || status * *Stand* || item * *~ in Liste/Aufstellung/Bestellformular/Lieferschein* || station * *geographische ~*
Position erreicht destination reached * *[NC-Steuerung]*
Positionier- positioning
Positionier-Regelung positioning control loop
Positionier-Regler position controller
Positionier-Rollgang positioning roller table || positioning roller
Positionierachse positioning axis
Positionierantrieb positioning drive || positioner * *als Stellantrieb z.B. für Proportionalventil*
Positionieraufgabe positioning task
Positionierbaugruppe position control module || positioning module || positioning board * *Leiterkarte* || positioning controller || positioning PCB * *Leiterkarte*
Positionierbereich positioning range
Positionierdatensatz positioning set || positioning block || positioning record
Positionieren positioning || inching * *z.B. Material(rolle) per Hand durch Tippen, Läufer einer elektr. Maschine*
Positionierfehler positioning error
Positioniergenauigkeit positioning accuracy
Positionierrechner position calculator || position controller

Positionierregler position controller || motion controller
Positionierrollgang positioning roller table || positioning roller
Positioniersteuerung positioning control || positioning controller || routing controller || point-to-point control
Positioniertisch positioning table
Positioniertoleranz positioning tolerance
Positionierung positioning || orientation || seek * *Schreib-/Lesekopf in Computer-Laufwerk* || inching * *eines Maschinenteils/einer Maschinenwelle z.B. im Tippbetrieb*
Positionierzeit positioning time
Positionserfassung position sensing
Positionsfehler position error || position deviation || positioning error * *Positionierfehler*
Positionsistwert actual position
Positionsregelung positioning control
Positionsschalter position switch
Positionssollwert position setpoint * *siehe auch →Lagesollwert* || position reference value || target position
Positionszähler position counter
positiv positive * *zustimmend, Plus- (positive pole: Pluspol), unumstößlich, eindeutig, zustimmend, bejahend*
positiv geladen positively charged
positive Flanke rising edge * *aufsteigende Flanke* || leading edge * *vordere Flanke* || positive edge || positive-going edge
positive Logik positive logic * *hoher Spannungspegel entspricht logisch "1"*
Postanschrift postal address
Postfach post-office box || P.O.B * *post-office box* || P.O. box
Postformingsäge postforming saw * *(Holz)*
Postleitzahl area code || zip code || post code
Postprozessor postprocessor * *(Computer)* || post processor * *(Computer)*
Potentiometer potentiometer * *[po'tenschemiete]* || pot || variable resistor
Potenz power * *Kraft, Stärke; auch mathematisch: power of: zur Potenz von* || square * *zweite Potenz* || cube * *dritte Potenz*
Potenzial potential
Potenzialausgleich potential equilization * *z.B. zwischen weit entfernten Anlagenteilen* || potential compensation || equipotential bonding * *über ~sleitung usw.* || potential bonding * *über ~sleitung usw.*
Potenzialausgleichsleitung potential compensation cable || potential equilization cable || equipotential bonding conductor || equi-potential bonding conductor
potenzialbehaftet non-floating || non-isolated || referred to a common signal ground * *auf d. gleiche Masse-Bezugspotenzial bezogen*
Potenzialdifferenz potential difference || difference of potential || potential gradient * *el. Spannungsanstieg/-unterschied zwischen zwei Punkten*
potenzialfrei isolated || floating || opto-coupled * *über Optokoppler gekoppelt* || opto-isolated * *über Optokoppler gekoppelt* || transformer-isolated * *durch Übertrager potenzialgetrennt* || potential-free || voltageless * *ohne Spannung* || with galvanic isolation || electrically isolated
potenzialfreie Stromistwerterfassung isolated current sensing
potenzialgebunden non-isolated * *nicht potenzialfrei* || non-floating
potenzialgetrennt electrically isolated || isolated || floating || with galvanic isolation || opto-isolated * *über Optokoppler*

Potenzialschwelle potential threshold || potential barrier * *bei Halbleitern*
potenzialtrennend isolating || voltage-isolating
Potenzialtrennung electrical isolation || optical isolation * *über Optokoppler/Lichtleiter* || isolation || insulation * *(weniger gebräuchlich)* || galvanic isolation || potential isolation || opto-isolation * *über Optokoppler*
potenziell potential || eventual * *eventuell* || contingent * *eventuell, →möglich; contingent fee: Erfolgshonorar* || possible * *möglich*
potenzielle Energie potential energy || hidden energy
potenzieller Kunde potential customer
Poti pot || potentiometer
PPP PPP * *'Point-To-Point Protocol': Datenübertr.protokoll im INTERNET über →MODEM u. Wählleitung (→SLIP)*
prädestiniert predestined
prädikativ predictive
prädiktiv predictive || looking ahead
prädiktiver Regler predictive controller * *"vorausschauender" Regler*
Präferenz preference
präferenzberechtigt preference eligible * *(SPS)*
Präferenzberechtigung preference authorization * *(beim Zoll)*
Präfix prefix
Prägeanlage embossing plant
Prägemaschine embossing machine
prägen emboss || stamp * *auf~, stanzen, pressen, stempeln* || impress * *(figürl.) on: ins Gedächtnis usw. (ein)~* || engrave * *(figürl.) on: ins Gedächtnis usw. ~* || form * *(auch figürl.)* || mint * *Münzen, Wörter, Begriffe usw. ~* || coin * *Metall, Münzen, Wörter, Begriffe usw. ~* || give distiction to * *(figürl.)*
praktikabel practicable || practical * *praktisch* || serviceable * *zweckdienlich, brauchbar, verwendbar, nützlich* || conducive * *(kondju: siw) (zweck)-dienlich, förderlich, zielführend*
Praktikant worker student * *Werkstudent* || laboratory student * *in Entwicklungsabteilung* || assistant * *Hilfskraft* || probationer * *Angestellter auf Probe, Probekandidat* || trainee
Praktiker practitioner || practical man || old hand * *(salopp)* || practitian
Praktikum practical course || laboratory sessions || practical training || practical training course
praktisch practical * *praktisch, tatsächlich, durchführbar; experience: Erfahrung, example: Beispiel* || useful * *zweckmäßig, nützlich* || serviceable * *zweckdienlich, brauchbar, verwendbar* || handy * *handlich, gut brauchbar, nützlich, praktisch veranlagt/begabt* || easy-to use * *leicht bedienbar/anwendbar (z.B. ein Gerät)* || practically * *(Adv.) so gut wie, in der Praxis* || virtually * *(Adv.) so gut wie, nahezu, fast gänzlich* || to all practical purposes * *(Adv.) so gut wie* || as good as * *(Adv.) so gut wie* || practical-minded * *praktisch veranlagt/denkend* || practised * *geübt* || matter-of-fact * *nüchtern* || down-to-earth * *nüchtern* || applied * *angewandt* || in practice * *(Adv.) i.d.Praxis/Anwendung; determined in actual practice: durch ~e Versuche ermittelt* || practicable * *durchführbar* || feasible * *durchführbar* || expedient * *tunlich, ratsam, zweckmäßig, praktisch*
praktische Übungen practical exercises || practical training
praktizieren practice
Prallbrecher rebound crusher
Prämisse premise * *Prämisse, Voraussetzung; das Oben Erwähnte* || overall aim * *übergeordnetes Ziel; with the overall aim to reduce costs: unter d. ~ der Kostensparung*

Präparationswalze preparation roll * *{Textil}* z.B. in der Chemiefaserherstellung
Präsentation presentation
Pratze pad * *{el.Masch.}* zur Motorbefestigung bei Bauformen IM B30, IM V30 u.IM V31 (3/4-Punktaufhängg.) || claw * *Klaue, (Greif-) Haken* || strap * *Bügel*
Pratzenanbau point fixing * *{el. Masch.}* Dreipunktbefestigung e. fuß-/flanschlosen Motors an 3 Pratzen am Gehäuse || three-point fixing * *{el. Masch}* →Dreipunktaufhängung || pad-mounting * *{el. Masch.}* von Motoren ohne Anbaufüße/Flansch; Bauformen IM B 30, IM V30 und IM V31
Praxis practice * *in practice: in der ~; put into praxis: in die ~ umsetzen* || experience * *Erfahrung* || usage * *Brauch* || exercise * *Übung* || real life * *in real life: im wirklichen/praktischen Leben*
praxiserprobt field proven || tried-and-tested
praxisgerecht application-oriented || practicable || practice-oriented
praxisnah close to practice
praxisorientiert practically oriented
präzise precise || exact * *genau, exakt* || punctual * *pünktlich* || accurate * *genau* || high-precision * *hoch~*
Präzision precision || accuracy * *Genauigkeit* || exactness * *Genauigkeit, Exaktheit*
Präzisions- precision
Präzisionsgetriebe precision gear unit
Präzisionsschleifmaschine precision grinding machine
Präzisionswerkzeug precision tool
Preis price * *at a lower price: zu einem niedrigeren ~* || pricing * *~stellung, ~gestaltung, for: für* || award * *Belohnung, Auszeichnung, Preis, Prämie, gutes Testurteil, z.B. in Zeitschrift* || premium * *Bonus, Prämie, Belohnung; be at a premium: hoch im Kurs stehen/sehr gesucht sein*
Preis-Leistungsverhältnis price-performance ratio
Preisangebot quotation
Preisanpassung price adaption || price revision
Preisaufschlag price adder || surcharge
Preisbildung generation of prices * *siehe auch 'Preisgestaltung'* || pricing || definition of prices || price generation
Preisblatt pricing sheet
Preisdisziplin pricing discipline
Preisdruck price pressure || pricing pressure * *high: großer*
Preisentwicklung price development
Preisgefüge price structure || price framework
Preisgestaltung pricing policy || pricing structure || price framework * *Preisgefüge* || pricing * *Preisstellung, Preiskalkulation* || price structure * *Preisgefüge*
preisgünstig favourably priced || low-cost || cheap * *billig* || economy || inexpensive * *nicht teuer* || at a reasonable price * *(Adv.)* || low-budget * *billig* || worth the money * *preiswert* || good value * *be good value: preiswürdig sein* || low-priced || budged-priced || economy-priced || favourably-priced * *more favourably priced: ~er* || affordable * *erschwinglich* || price-competitive * *~ verglichen mit Mitbewerbsprodukten* || reasonably priced
preisgünstiger more favourably priced
preisgünstigst most lowest price
Preishürde price barrier || price hurdle
Preiskampf price war * *throat-cutting: existenzgefährdender*
Preiskrieg price war
preislich in terms of price || regarding the price
Preisliste price list || pricelist
Preisnachlass discount || rebate * *Rabatt* || price reduction || price deduction || reduction in price || reduction of prices || allowance * *auf Grund von Mängeln*
Preisniveau price level
Preisreduzierung price reduction * *by: um*
Preisrückgang fall in prices || decline in prices || price drop || recession in prices || price recession
Preisschere price scissors || price gap
Preisschlager price sensation
Preisschraube price spiral
Preisschwankung price fluctuation
Preissegment price segment * *upper: oberes; lower: unteres*
preissenkend price reducing
Preissenkung price reduction || price cut || price cutting || reduction of prices || reduction in prices || reduced prices * *gesenkte Preise*
Preissteigerung price increase || rise in prices || rising prices
Preisstellung pricing * *for: für*
Preisstopp price stop || price freeze
Preisstruktur pricing structure || price structure || price framework
Preissturz sudden fall in prices || slump in prices
Preisstützung price support || price supporting || subsidization * *Subventionierung*
preistreibend price-raising
Preistreiberei forcing up of prices
Preisüberhöhung excessive prices
Preisunterbietung underselling
Preisunterschied price difference
Preisvereinbarung pricing agreement || pricing || price agreement
Preisverfall decay of prices || price erosion || price drop
preiswert worth the money * *seinen Preis wert* || low-priced * *mit niedrigem Preis* || favourably priced * *preisgünstig* || favourable in price || of good value || economy-priced || inexpensive || cheap * *billig* || low-budget * *"für den kleinen Geldbeutel", billig* || be good value * *preiswert sein* || budget-priced
Preiswürdigkeit good value || moderate price || cheapness
Preisziel price aim
Preiszuschlag additional charge
prellen bounce * *von Kontakten* || rebound * *zurückprallen/schnellen*
Prellschlag jarring blow
Press- press * *siehe auch* →Anpress-, →Quetsch-
Presse press * *Pressmaschine; Druck~; die Druckmedien, Zeitung; auch Trocken~ bei Papiermaschine* || press machine * *Pressmaschine* || journalism * *Journalismus* || wet press * *{Papier}* Nass~ in der Papierindustrie || clamp * *{Holz}* Formkasten~, Leim~, Schraubzwinge
Pressekonferenz press conference
Pressemitteilung press release
pressen press || compress * *zusammendrücken/pressen* || squeeze * *(zusammen-) drücken, (her-)aus~, (her-)ausquetschen (auch figürl.)* || extrude * *strang~* || force * *zwängen* || strain * *(durch ein Sieb) seihen*
Pressenantrieb press drive
Pressenpartie press section * *in Papiermaschine*
Pressenschleifer pocket grinder * *zum Holzschleifen*
Pressensicherheitsventil press safety valve
Pressereferent press agent
Presseur impression cylinder * *{Druckerei}*
Presskraft press force || power of press || contact pressure * →Anpresskraft
Pressluft compressed air
Pressmasse molding compound
Presspassung pressure fit || interference fit
Pressspanplatte pressboard * *{Holz}*

Pressung pressure || pressing || compression || shot * *Schub einer Pressmaschine*
Presswalze press roll * *z.B. in Papiermaschine*
Presszange crimping tool * *Quetschwerkzeug z.B. für Kabelschuhe, Aderendhülsen* || jointing clamp || pressing tool
prima first-class * *erstklassig* || first-rate * *erstklassig, großartig, ausgezeichnet* || great * *(salopp) groß(artig), überragend, famos, herrlich* || tremendous * *(salopp) kollossal, toll, gewaltig* || marvellous * *(salopp) fabelhaft, wunderbar* || fantastic * *(salopp) fantastisch* || swell * *(salopp) totschick, piekfein, stinkvornehm, feudal* || mashing * *(salopp) toll, sagenhaft, umwerfend* || groovy * *(salopp) klasse, toll* || fantastically * *(Adv., salopp) fantastisch* || marvellously *·(Adv., salopp) wunderbar, fabelhaft* || wizard * *erstklassig, prima* || top-notch * *(salopp) prima, erstklassig*
primär primary * *auch Trafowicklung*
Primäranschlüsse primaries * *z.B. eines Trafos*
Primärenergie primary energy
primärgetaktetes Schaltnetzteil primary-switched power supply * *auf der Netzseite (vor dem Übertrager) getaktetes Schaltnetzteil* || switched-mode power supply * *Schaltnetzteil*
Primärspannung primary voltage
Primärwicklung primary winding || primary
Print printed-circuit board || PC board || PCB
Printklemme PC board terminal
Prinzip principle * *on: aus; in: im; work on the principle of: nach dem ~ des ... arbeiten*
prinzipiell on principle * *aus Prinzip* || basical * *grundlegend, die Grundlage bildend, Grund-* || basically * *(Adv.) grundsätzlich, im Grunde*
Prinzipschaltbild schematic diagram || block diagram * *Blockschaltbild*
Prinzipschaltplan schematic diagram || block diagram * *Blockschaltbild* || single-line diagram * *in einpoliger Darstellung*
priorisieren prioritize * *Prioritäten zuteilen, z.B. Interrupts*
priorisiert prioritized * *z.B. Interrupts*
Priorisierung priority assignment * *z.B. von Interrupts, Buszugriffen usw.* || priority scheduling * *z.B. von Interrupts, Buszugriffen usw.*
Priorität priority * *low: niedrige; high: hohe: top: höchste*
Prioritätenfolge priority sequence
Prioritätenreihenfolge priority sequence
Prioritätsebene priority level
Prisma prism
prismatisch prismatic
pro per
pro Jahr per annum || p.a. * *Abk. für 'Per Annum'*
pro Tag per diem || p.d. * *Abk. für 'Per Diem'*
Probe sample * *Muster* || pattern * *Muster* || specimen * *Prüfstück* || proof * *Beweis (mathematischer)* || probation * *Bewährungsprobe* || check * *Überprüfung* || trial * *Erprobung* || test * *Erprobung* || experiment * *Versuch*
Probe- sample * *Muster* || tentative * *versuchsweise* || test * *Versuchs-*
Probelauf trial run || test run
Probelieferung trial shipment
Problem problem || basic issue * *Grundproblem* || fault * *Fehler* || defect * *Defekt* || malfunction * *Fehlfunktion* || trouble * *Schwierigkeiten, Schereien, Störung, Defekt, Unannehmlichkeiten* || issue * *Kernpunkt, Sachverhalt, Streitfrage*
Problembeschreibung problem description
problemlos without problems || easy * *leicht,einfach* || nearly maintenance-free * *wartungsarm* || reliable * *zuverlässig* || user-friendly * *anwenderfreundlich* || problem-free || trouble-free * *auch mü-*

helos || without trouble * *auch mühelos* || simple * *einfach, leicht* || without any difficulties || fussfree * *(salopp) ohne viel Getue*
Problemlösung problem solver
Produkt product * *Erzeugnis* || manufacture * *Erzeugnis, Fabrikat*
Produktabkündigung announcement that the product will be discontinued || product discontinuation * *Einstellung der Fertigung/Lieferung eines Produkts* || announcement of product discontinuation || announcement of discontinuation || notification of delivery-discontinuation
Produktankündigung product announcement
Produktauswahlliste buyers guide * *Einkaufsleitfaden für den Kunden*
Produktbeschreibung product description
Produktbetreuer product manager
Produktbetreuung product support || product management
Produktbezeichnung product designation
Produkterprobung product testing
Produktfamilie product family
Produktgruppe product group
Produkthaftung product liability
Produkthaftungsgesetz product liability law * *in Deutschland: Paragraph 823 BGB*
Produktidee product idea
Produktinformation product information || literature on the products * *(Plural) schriftliche* || folder * *Faltblatt* || brochure * *Broschüre* || pamphlet * *Druckschrift, Broschüre, Merkblatt* || prospectus * *Werbeprospect* || illustrated folder * *Faltblatt mit Bildern*
Produktinformationen literature on the products * *schriftliche*
Produktionsablauf production process
Produktionsanlage production plant || manufacturing facility || manufacturing line
Produktionsausfall production outage || loss of production
Produktionsbeginn beginning of production || start of production
Produktionseinstellung end of production
Produktionsführung production control
Produktionsgeschwindigkeit production speed || line speed * *einer kontinuierlich arbeitenden Produktionslinie* || production rate
Produktionskapazität production capacity
Produktionskette production sequence
Produktionsleitsystem production control system
Produktionsleittechnik production control || production management || manufacturing control || CIM * *Abkürzg.f. 'Computer-Integrated Manufacturing': rechnergestützte Fertigung* || production control technology || production control systems
Produktionslinie production line || manufacturing line
Produktionsmittel production equipment || means of production || capital equipment
Produktionsprogramm product range || manufacturing program
Produktionsprozess production process || manufacturing process
Produktionsstätte production facility || production facilities
Produktionssteigerung production increase || increase in production
Produktionsstufe production stage
Produktionstechnik production technology || manufacturing technology
Produktionsunterbrechung production stoppage || downtime || production break
Produktionswert je Beschäftigten output per employee

Produktivität productivity || production efficiency
Produktivitätserhöhung productivity increase
Produktivitätssteigerung productivity improvement
Produktmerkmal product feature || product highlight * *herausragendes*
Produktpalette sales program * *verkaufbare Produkte* || line of products * *Produktlinie/familie* || line of merchandise || programme * *Programm lieferbarer Produkte* || product line || product range || product programme * *Produktprogramm* || production programme || production range || assortment of products || production mix
Produktplanung product planning
Produktprogramm product range || product programme
Produktreife product maturity || degree of maturity
Produktreihe product line || product series || product range || product family * *Produktfamilie*
Produktschrift product brochure * *Werbebroschüre, Werbeblatt, Farbfaltblatt* || folder * *Faltblatt* || brochure * *Broschüre* || leaflet * *Faltblatt* || pamphlet * *Druckschrift, Broschüre, Merkblatt* || illustrated folder * *bebildertes Faltblatt* || prospectus * *Werbeprospekt* || product documentation
Produktsicherheit product safety
Produktspektrum product line * *Produktlinie* || product programme * *Produktprogramm* || product spectrum
Produktsteckbrief product profile
Produktübersicht product overview || buyers guide * *Einkaufsleitfaden für den Kunden*
Produktverantwortlicher product manager || person responsible for the product
Produzent manufacturer * *Hersteller* || producer * *z.b. in der Film/Werbeindustrie*
produzieren manufacture
PROFIBUS PROFIBUS * *Abkürzg. für 'PROcess Field BUS': firmenübergreifender Feldbus nach EN 50170,*
PROFIBUS-Profil 'Drehzahlveränderbare Antriebe' PROFIBUS profile 'Variable-Speed Drives' * *DIN 19245(PROFIBUS) u.VDI/VDE-Richtl.3689 (Profil)*
Profil profile * *(auch figürlich u.Anwenderprofil für Bussystem (z.B.PROFIBUS, DRIVECOM))* || contour || section * *z.B. von stangenförmigem Material (Stahl, Rohr usw.)* || shape * *Form, Gestalt, Umriss* || extruded section * *Stranggussprofil*
Profil entwickeln develop one's profile
Profil-Walzmaschine roller forming machine
Profildraht angular-shaped wire * *mit eckigen Profil* || rectangular wire * *mit rechteckförmigem Profil* || streamlined wire * *mit stromlinienförmigem Profil*
Profilfräsen profile milling || profile shaping * *[Holz]*
profilieren shape || forming of a section * *(Substantiv)* || distinguish oneself * *(figürl.: sich ~)* || acquire a strong image * *(figürl.: sich ~, e. guten Eindruck/gute Figur machen)* || mould * *[Holz]*
profiliert profiled * *z.B. Draht, Stahl, Blech, Holz*
Profilmaschine profiling machine
Profilschiene profile rail || mounting rail * *(Hutprofil-) Montageschiene, siehe auch →Installationsschiene* || DIN rail * *C- bzw. G-Aufschnapp-Montageschiene n. DIN EN50022 (35 mm breit) bzw. DIN EN50035 (32 mm)*
Profilschleifmaschine profile grinder * *[Werkz. masch.]* || profile sander * *[Holz]*
Profilschliff profile sanding * *[Holz]*
Profilschneiden profile cutting
Profilummantelungsanlage profile-wrapping plant * *[Holz] in der Möbelindustrie*
Profilwalzen profile rolling

Profit profit || earnings * *Geschäftsergebnis, Gewinn*
profitieren take benefit * *from: von/an* || reap * *profitieren von, ernten*
Proforma- proforma
Proforma-Rechnung proforma-invoice * *(für den Zoll)*
Prognose prognosis || forecast
Programm program * *Software~, Produkt~ usw.* || programme * *Fertigungs~, Verkaufs~, Schulungs~, Kurs~ usw.* || software * *Software~*
Programmabbruch program abort
Programmablauf program flow || program execution * *Programmausführung, Programmabarbeitung* || program run || program sequence * *z.B. in einer sequenziellen Steuerung*
Programmablaufplan program flowchart
Programmabschnitt program section
Programmabsturz program crash || program breakdown || hanging program * *"aufgehängtes" Programm*
Programmarchiv program library
Programmarchivierung storing in program library
Programmausführung program execution
Programmbaustein program block * *bei SPS* || program module * *bei SPS*
Programmbearbeitung program execution * *Ausführung*
Programmbibliothek program library
Programmcode program code
Programmeingabe program input * *z.B. SPS, CNC*
Programmelement program element * *bei SPS (bei Kontaktplandarstellung)*
Programmfehler software bug || bug
Programmieradapter programming adapter || programming adaptor * *z.b. zum "Schießen" von EPROMs/EPROM-Moduln/PLDs usw.*
programmierbar programmable
programmierbare Steuerung programmable controller || PLC * *Abk.f. 'Programmable Logic Controller': Speicher~*
programmierbarer Logikbaustein programmable logic device || PLD * *Abk. f. 'Programmable Logic Device'*
programmieren program || write a software program
programmieren im Kontaktplan ladder programming * *(SPS)*
Programmierer programmer || software developer * *Software-Entwickler* || software engineer * *Ingenieur* || software specialist * *Spezialist* || coder * *Hifskraft, Kodierer* || programming engineer || software designer || hacker * *(salopp) Trickprogrammierer, Programmierexperte; ~, der Sicherheitssysteme "knackt"*
Programmiergerät programming unit * *[SPS]* || programmer * *[SPS]* || EPROM burner * *für EPROMs* || programming device
Programmierkenntnisse programming knowledge || programming skills || computer programming expertise * *[expe'ties]* || knowledge of programming
Programmierplatz programming console * *für SPS* || programming station
Programmierschnittstelle programming port * *(für SPS)* || programming interface
Programmiersoftware programming software
Programmiersprache programming language
Programmierung programming
Programmkopf program header || program heading
Programmorganisation program organization
Programmpaket software package
Programmspeicher program memory || program storage
Programmspeicherplatz program capacity * *betrifft dessen Größe (large: groß)*
Programmteil program section

Programmtest program testing || program debugging
Programmübersicht program overview * *über Liefer/Verkaufs/Herstell/Softwareprogramm*
Programmunterbrechung interrupt
Programmzweig path || branch
progressiv progressive
Progressivität progression
Projekt project || activity * *Aktivität, Betätigung, Tätigkeit, Rührigkeit;* activities: *Unternehmungen*
Projektbearbeitung project processing
Projekteur designer || engineer || application engineer || planning engineer || configuring engineer * *(SIEMENS SIMADYN D (R))* || planner || project engineer || project planning engineer
projektierbar configurable * *(z.B. beim SIEMENS Antriebsregelsystem SIMADYN D (R));* freely: *frei*
Projektierbarkeit configurability * *Konfigurierbarkeit (z.B. SIEMENS-Regelsystem SIMADYN D (R))* || ability to be freely configured * *freie* ~ || ease of configuration * *Einfachheit/Komfort der* ~
projektieren engineer || design || configure * *soft-/hardwaremäßig konfigurieren/~ (z.B. SIEMENS-Regelsystem SIMADYN D (R))* || plan * *planen* || project * *entwerfen, planen* || lay out * *auslegen* || dimension * *auslegen, bemessen*
Projektierung design || engineering || planning || planning and design || application engineering || project management * *Abwicklung eines Projekts* || configuring * *Konfigurieren, Konfigurierung, (z.B. SIEMENS-Regelsystem SIMADYN D (R))* || project planning || configured system * *(SIEMENS SIMADYN D (R)) projektiertes (Regel)System* || system configuration * *einer Anlage* || engineering/ design
Projektierungsabteilung application department || application engineering department
Projektierungsanleitung planning guide || configuring instructions || engineering instructions
Projektierungsaufwand engineering expenditure
Projektierungsbeispiel configuring example || planning example || application example * *Anwendungsbeispiel*
Projektierungsblatt planning sheet
Projektierungsdaten engineering data || engineering information
Projektierungsfehler configuring error * *z.B. beim SIEMENS Regelsystem SIMADYN D (R)* || misengineering || incorrect dimensioning || layout error || incorrect engineering || incorrect configuring * *z.B. beim SIEMENS Regelsystem SIMADYN D (R)*
Projektierungsgerät configuring unit * *z.B. für Regelsystem SIMADYN D (R) von SIEMENS*
Projektierungshandbuch engineering manual || engineering guide || application book * *Applikations-(Hand-) Buch* || planning guide || configuring handbook || application manual * *Applikationshandbuch* || design selection manual * *zur richtigen Auswahl von techn. Produkten (z.B. Konstruktionselementen)*
Projektierungshilfe engineering aid || engineering guide * *gedruckte* ~
Projektierungshilfen configuring tools || configuring forms * *Formblätter* || engineering aids
Projektierungshinweis engineering information
Projektierungshinweise configuring instructions || application hints || engineering information
Projektierungskosten engineering costs
Projektierungsphase planning stage || configuring phase || planning phase
Projektierungsplan configuring diagram * *(SIEMENS SIMADYN D (R))*
Projektierungssoftware configuring software

Projektierungssprache configuring language * *z.B. 'STRUC' (R) beim Regelsystem SIMADYN D (R) von SIEMENS* || design language * *z.B. 'STRUC' (R) beim Regelsystem SIMADYN D (R)von SIEMENS*
Projektierungssystem configuring system
Projektierungsunterlagen engineering documentation * *Dokumentation, Applikationshinweise usw.* || engineering documents || project manual * *Projekthandbuch* || application manual * *Handbuch mit Anwendungshinweisen*
Projektierungsvorgang configuring * *z.B. beim SIEMENS Regelsystem SIMADYN D (R)*
Projektionsmethode projection method * *bei technischen Zeichnungen*
Projektleiter project leader || project manager || project director
Projektleitung project management
Projektmanagement project management
projizieren project
Promille per thousand || per mill || per mille || tenth of one percent
Prommer EPROM burner * *(salopp) Programmiergerät für EPROM-Speicher* || prommer * *(salopp) Programmiergerät für EPROMs usw.*
Promotion graduation * *Erteilung eines akadem. Grades* || degree day * ~*sfeier* || graduation exercises * *(amerikan.)* || commencement || commencement day * ~*sfeier* || promotion * *(Verkaufs-) Förderung*
promovieren take one's doctor's degree || graduate || confer a doctor's degree upon * *jemanden* ~, *jemandem den Doktortitel zuerkennen*
Propellermotor propulsion motor * *zum Antrieb eines Schiffes*
proportional proportional * *to/with: zu;* inversely/ directly: *umgekehrt/direkt;* be ~/in proportion to: ~ *sein zu* || proportionally * *(Adv.)* change proportionally with: *sich* ~ *ändern mit*
proportional zum Quadrat der in proportion with the square of the
Proportionalanteil proportional term
Proportionalbeiwert proportional-action coefficient || proportional coefficient || proportional gain * *Proportionalverstärkung*
Proportionalbereich proportional band
Proportionalglied proportional element || P-element
Proportionalität proportionality
Proportionalregler proportional-action controller || P controller || step-by-step controller * *z.B. zur Temperaturregelung*
Proportionalsteuerung proportional control
Proportionalventil proportional valve * *mit stetig veränderbarer Öffnung (z.B. für Pneumatik, Hydraulik usw.)*
Proportionalverstärkung proportional gain || gain || P-gain || proportional factor
Prospekt folder * *Faltblatt* || illustrated folder * *Faltblatt mit Bildern* || brochure * *Broschüre* || catalog * *Katalog* || catalogue * *Katalog* || list * *Liste* || product information * *Produktinformation* || pamphlet * *Druckschrift, Broschüre, Merkblatt* || prospectus * *Werbeprospekt* || leaflet * *Werbeblatt, Faltblatt* || flyer * *(amerikan.) Werbezettel, Prospekt* || flier * *(amerikan.) Werbezettel, Prospekt*
Prospekte brochures || sales literature
Protokoll protocol * *z.B. ~ für se(rielle) Datenübertragung* || report * *Bericht* || minutes of meeting * *Besprechungsbericht, Verhandlungs-/~;* enter in the minutes: *zu* ~ *nehmen* || operation log * *Betriebs~ z.B. für eine Produktionsanlage, Maschine* || log * *Betriebs~, Logbuch* || minutes * *Verhandlungs-/Besprechungs~;* enter in the minutes:

ins ~ aufnehmen || record * Verhandlung~; record in protocol: zu ~ nehmen || proceedings * Sitzungs-/Tätigkeitsbericht(e)
Protokollführer keeper of the minutes * bei Besprechungen, Sitzungen, Verhandlungen usw.
protokollieren log || record
Protokollierung logging || reporting || recording * Aufzeichung || data listing
Protokollumsetzer gateway * für serielle Kommunikation || data bridge || protocol converter
Prototyp prototype
Provision commission || brokerage * eines Maklers/ Vermittlers
Provisionsbasis commission basis * on a: auf
provisorisch provisional * vorläufig, einstweilen, behelfsmäßig || temporary * vorübergehend || tentative * vorläufig, versuchsweise, probehalber || make-shift * behelfsmäßig
Prozedur procedure * auch Rechner-(Unter)Programm || protocol * Kommunikationsprotokoll
prozedural procedural * z.B. Programmiersprache (Gegensatz: →objektorientierte Programmiersprache) || procedure-oriented
Prozent per cent || percentage * Prozentanteil, Prozentsatz || per unit * der Nenngröße || % || percent * (amerikan.) || rate of interest * Prozent Kapitalzins || p.u. * Abkürzung für 'per unit' (1% entspricht 0,01 p.u.)
Prozentanteil percentage
Prozentsatz percentage
prozentual per unit * Anteil des Nennwerts || percental * percentage: prozentualer Anteil || proportional || in per cent * (Adv.) || in terms of percentage * (Adv.)
prozentualer Spannungsabfall percentage voltage drop || % voltage drop || percental voltage drop
Prozentwert percentage value
Prozess process * Vorgang; Produktions~, Herstellungs~ || trial * Strafverfahren || lawsuit * Rechtsstreit || litigation * Rechtsstreit || action * Klage
Prozessabbild process image * z.B. bei SPS || process image table * gespeichertes Abbild der Prozess-Ein/Ausgänge bei SPS || process I/O image * Abbild des Zustandes der Ein/Ausgänge bei einer SPS
Prozessabbild der Ausgänge process output image * bei SPS
Prozessabbild der Eingänge process input image * bei SPS
Prozessalarm process interrupt
Prozessautomation process control || process automation
Prozessautomatisierung process automation || process control * Prozessleitung/-leitsystem
Prozessautomatisierungssystem process automation system
Prozessbedientastatur process communication keyboard * bei SPS
Prozessbeobachtung process monitoring || process visualization
Prozessdampf process steam
Prozessdaten process data
Prozessdatenführung process data control
Prozessführung process control
Prozessgröße process variable || process quantity
Prozessidentifizierung process identification
Prozessinterrupt process interrupt
Prozesskette process chain
Prozessleitebene process control level || master-process control level || process management level
Prozessleitsystem process control system || process control || host system * überlagertes Rechnersystem
Prozessleittechnik process control technology || process systems || process control and instrumentation technology

Prozessmesstechnik process instrumentation || instrumentation
Prozessmodell process model
prozessnahe Peripherie field devices * Sensoren, Aktuatoren usw. bei SPS
Prozessor processor * z.b. Mikroprozessor || CPU * Abk. für 'Central Processing Unit': Zentraleinheit eines Rechners
Prozessor-Nummer processor number * bei Mehrprozessorsystem
Prozessorabsturz processor crash * {Computer}
Prozessorauslastung processor loading || processor utilization
Prozessorfamilie family of processors
Prozessorleistung processor power || computing power * Rechenleistung
Prozessormodul processor module || processor board * als Leiterkarte || processor PCB * als Leiterkarte
Prozessormodule processor board * Leiterkarte, Flachbaugruppe
Prozessorprogramm processor program * z.B. beim Antriebsregelsystem SIMADYN D (R) der Fa. SIEMENS
Prozessperipherie process peripherals * bei SPS
Prozessrechner process control computer * Rechner m. schnell. Echtzeitbetriebssystem u.Digital/ Analog E/A-Funktionen || process computer
Prozessregelkreis process control loop
Prozesssignal process signal
Prozesssignalformer I/O conditioner module || I/O module
Prozesssteuerung process control
Prozesssteuerungssystem process control system
Prozessüberwachung process monitoring
Prozessunterbrechung process downtime
Prozessvisualisierung process visualization * (amerikan.) || process visualisation * (brit.)
Prozesswärme process heat
Prozesszustandskontrolle process status report * bei SPS
Prüfabteilung test department
Prüfadapter test adapter
Prüfanleitung instructions for testing
Prüfanzeige diagnostic display * bei SPS || diagnostic light * bei SPS
Prüfaufbau test-bed assembly
Prüfautomat automatic tester || ATE * Abkürzg.f. 'Automatic/Automated Test Equipment': automatische Testeinrichtung || ATE equipment
Prüfbarkeit testabilty
Prüfbedingung test condition
Prüfbericht test report
Prüfbescheinigung test certificate || certificate of inspection || inspection document * Bescheinigung, die der Besteller mit der Lieferung erhält (EN 10204)
Prüfbescheinigung für explosionsgeschützte Betriebsmittel test certificate for explosion-protected apparatuses * VDE 0171/02.61, DIN EN 50014,16,18-21
Prüfbit check bit
Prüfbuchse test socket || measuring socket * Messbuchse || test jack
Prüfdauer test duration * z.B. bei Isolationsprüfung
Prüfdrehmoment test torque
Prüfeinrichtung test equipment
prüfen test || check * über~ || inspect * inspizieren, abnehmen || investigate * untersuchen, erforschen, ermitteln, nachforschen || look into * nachsehen, inspizieren || analyze * analysieren || control * beaufsichtigen, überwachen, nach~ || audit * amtlich ~, einer Revision unterziehen || review * nach~, über~ (Entwicklungsvorhaben, Entscheidung,

Prüfergebnis

Schriftstück), inspizieren, revidieren || overhaul * überholen || verify * auf Richtigkeit ~ || try * erproben || consider * erwägen || study * (sorgfältig) untersuchen, erforschen, ~d ansehen, lesen || weigh * sorgsam ab-/erwägen || examine * auf Schule, Hochschule usw. || certify * bescheinigen, beglaubigen, bestätigen, bezeugen || scrutinize * genau prüfen/untersuchen/ansehen/studieren || gauge * (Ab)Maße kontrollieren, z.B. mit Lehre
Prüfergebnis testing result || test result
Prüffeld test department * Abteilung || test bay || testing station || test laboratory || test lab || testing department
Prüffeldleiter chief test engineer || head of test department
Prüfgerät testing device || test apparatus || test equipment || testing appliance
Prüfgeräte testing equipment
Prüfgeschwindigkeit testing rate
Prüfimpuls test pulse
Prüfklasse category || quality class
Prüfling test specimen || specimen || test object || test sample || test piece
Prüfmaschine testing machine
Prüfmuster test pattern
Prüfplan test schedule
Prüfplanung test planning || inspection planning
Prüfreihenfolge test sequence
Prüfschärfe test severity
Prüfschein test certificate
Prüfschild test label
Prüfspannung test voltage * für Isolationsprüfungen; für Niederspannungsmotoren: siehe VDE 0530
Prüfstand test stand || test rig || testbed || test bed || test bench
Prüfstecker test plug
Prüfstelle der Europäischen Union certification authority of EU
Prüfsumme checksum
Prüftechnik test engineering
Prüfüberdruck test overpressure
Prüfung test || check * Über~ || exam * z.B. ~ an Schule/Hochschule || examination * genaue Untersuchung, ~/Examen an Schule/Hochschule || analysis * Analyse || investigation * Untersuchung, Ermittlung, Nachforschung || consideration * Erwägung, Überlegung || studies * (wissenschaftliche) Untersuchung, Erforschung; genaue prüfende Betrachtung || verification * Nach~ (auf Richtigkeit) || checking * (Nach)Prüfung, Kontrolle || checkup * Nach~ || inspection * Inspektion, Über~, Kontrolle || service * Kundendienst, Wartung || audit * Revision (Buch, Geschäftsbücher usw.), Rechnungs~ || review * Revision, Über~ (Buch, Gerichtsurteil, Entscheidung, Entwicklungsvorhaben usw.) || trial * Erprobung || scrutiny * genaue Untersuchung, prüfender Blick || gauging * Maßkontrolle, ~ mit Lehre/Dickenmessgerät
Prüfung der Isolierung high-voltage test
Prüfungsschein test certificate
* →Prüfbescheinigung
Prüfverfahren test method || test procedure || testing method || method of testing || testing procedure
Prüfvorschrift test specification
Prüfzeichen mark of conformity * von einer Normen/Überwachungsbehörde (z.B. VDE, TÜV, UL) || testing mark || test mark
Prüfzeit test duration * z.B. bei Isolationsprüfung
Prüfzertifikat test certificate
Prüfzeugnis test certificate
Prüfzubehör accessories for testing
Prüfzwecke test purposes * for: für
PS PS * Pferdestärken; 1 PS entspr. 0,73550 kW

entspr. 0,9863 hp || DIN PS * deutsche PS; 1 PS entspr. 0,73550 kW entspr. 0,9863 hp || hp * horsepower; 1 PS entspr. 0,73550 kW entspr. 0,9863 hp
PT1-Glied first order filter element * Filterglied erster Ordnung || filter element || PT1 element || PT1-function || PT1-block * Funktionsbaustein || low pass filter * Tiefpassfilter || first-order delay element * Verzögerungsglied erster Ordnung
PT100-Messfühler PT100 temperatur sensor * Widerstandsthermometer aus Platindraht (PT), d. bei 0° e.Wid. v. 100 Ohm hat || PT 100 gauge
PTB PTB * Physikalisch-Techn. Bundesanstalt, amtl. deut. Behörde für d. Mess- und Eichwesen i. Braunschweig
PTC PTC * Positive Temperature Coefficient, Widerstand dessen Ohmwert sich mit steigend. Temperatur erhöht
PTC-Widerstand PTC resistor || PTC thermistor * als Temperaturfühler verwendet
Publikation publication
Puffer buffer || bumper * (mechan.) zum Abfangen/ federn eines (An)Stoßes
Pufferbatterie backup battery
Pufferbetrieb floating operation * Dauerladungsbetrieb für Batterie || trickle charge * Erhaltungsladung für Batterie
Puffereinheit battery backup unit * mit Pufferbatterie(n)
puffern buffer || cushion * (sanft) abfedern || back-up * unterstützen, den Rücken decken, (mit Batterie/Kondensator) puffern
Pufferspannung backup voltage || battery backup * Batteriepufferung bei Versorgungsspannungsausfall/-abschaltung
Pufferspeicher buffer storage || buffer
Pufferüberlauf buffer overflow
Pufferung battery backup * Ersatzstrom von Batterie bei Spannungsausfall/-Abschaltung || standby supply * bei Netzspannungsausfall || buffering * Zwischenspeicherung von Daten || backup * mit Batterie, Kondensatoren, Schwungrad usw.
Pufferungszeit backup time * Zeitdauer, die eine Pufferbatterie überbrücken kann
Pufferverwaltung buffer control
Pufferzeit buffer time
Pulper pulper * {Papier} Breimühle ("Holländer") zur Zellstofferzeugung
Puls-Pausenverhältnis pulse duty cycle * Tastverhältnis || mark-to-space ratio
Pulsation pulsation || beat || throb * Pochen, Klopfen, Hämmern, (Puls-) Schlag, Erregung
Pulsauswertung pulse evaluation
Pulsbreite pulse width
Pulsbreitenmodulation pulse width modulation || PWM || pulse duration modulation * Pulsdauermodulation
Pulsbreitenmodulator pulse width modulator
pulsbreitenmoduliert pulse-width modulated
pulsen pulsate
Pulsfolgefrequenz pulse rate
pulsförmig pulse || pulse-shaped || in pulse form || rectangular * rechteckförmig
Pulsformung pulse shaping
Pulsfrequenz pulse frequency || chopper frequency || PWM frequency * bei Pulsweitenmodulation || carrier frequency * bei Pulsweitenmodulation
Pulsgeber pulse encoder || pulse tachometer * Impulstachometer || pulse tach * Impulstacho || incremental encoder * Inkrementalgeber
Pulsieren pulsation
pulsierend pulsating
Pulsmuster pulse pattern * optimized: optimiertes (bei Pulsumrichter) || method of PWM modulation * bei pulsbreitenmoduliertem Umrichter

Pulsraster pulse pattern
Pulssteuerung pulse control
Pulstacho pulse tacho
Pulsumrichter pulse converter ‖ PWM converter * *mit Pulsbreitenmodulation* ‖ PWM inverter * *Wechselrichter mit Pulsbreitenmodulation* ‖ pulse-controlled converter * *mit konstanter Zwischenkreisspannung*
Pulsverfahren pulse technique * *für Pulsumrichter* ‖ pulse-pattern * *Pulsmuster, z.B. für Pulsumrichter* ‖ method of PWM modulation * *bei pulsbreitenmoduliertem Umrichter*
Pulswählverfahren pulse dialing * *Telefonwählverfahr.: für jede Ziffer wird e. entsprechende Zahl v. Pulsen übermittelt*
Pulswechselrichter pulse inverter ‖ PWM inverter * *mit Pulsbreitenmodulation* ‖ pulse-width modulated inverter
Pulsweitenmodulation pulse-width modulation ‖ PWM modulation
Pulswiderstand pulsed resistor * *z.B. Bremswiderstand für Frequenzumrichter* ‖ pulse resistor ‖ braking resistor * *Bremswiderstand* ‖ brake resistor * *Bremswiderstand* ‖ pulsed braking resistor
Pulszahl pulse number * *Anz.d.Kommutierungen in e.Stromrichter je Periode* ‖ number of pulses per revolution * *Impulsgeber/Tacho* ‖ PPR count * *Abkürzung für 'Pulses Per Revolution': Impulszahl pro Umdrehg.* ‖ pulse count * *z.b. Impulsgeber* ‖ number of increments * *bei (Inkremental)Impulsgeber* ‖ number of increments per revolution * *bei (Inkremental-) Impulsgeber* ‖ commutation number * *Kommutierungszahl p: Anzahl d. nicht gleichz. Kommutierungsvorgänge je AC-Periode*
Pult desk ‖ console ‖ control desk * *Bedien~*
Pulteinbau desk mounting ‖ desk installation
Pulver powder
pulverbeschichtet powder-coated
pulverförmig powder form
Pulvermetallurgie powder metallurgy
Pumpe pump
pumpen pump ‖ hunt * *instabiler Regler*
Pumpenantrieb pump drive
Pumpenkennlinie pump characteristic * *Drehmoment bzw. Leistg. über der Drehzahl (bei Kreiselpumpe: M ~ n^2, P ~ n^3)*
Pumpenlaufrad impeller ‖ impeller-wheel
Pumpenrad impeller * *Laufrad*
Pumpspeicherkraftwerk pumped-storage power station
Pumpspeicherwerksmaschine pumped-storage machine
Pumpstation booster station * *zur Druckerhöhung bei Rohrleitungen/Pipelines*
Pumpwerk pumping station
Punkt point ‖ item * *der Tagesordnung* ‖ topic * *Einzelthema* ‖ dot * *Satzzeichen, Tüpfelchen* ‖ spot * *Fleck, Punkt, Platz*
Punkt der maximalen Leistung maximum power point ‖ MPP * *Maximum Power Point, z.B. bei Solar/Windkraftwerk*
Punkt-zu-Punkt Verbindung point-to-point connection
Punkt-zu-Punkt- point-to-point
Punkt-zu-Punkt-Kopplung point-to-point connection
Punkt-zu-Punkt-Verbindung point-to-point connection
Punkte points
Punktentladung point discharge
punktförmig punctate ‖ punctiform ‖ point-like * *point of light: ~ Lichtquelle* ‖ dot-like ‖ crater-shaped * *Vertiefung, Grübchen*
punktgenau to-pinpoint accuracy * *(Adv.)*
punktieren dot-line * *Linie* ‖ punctuate

punktiert dotted * *Linie*
pünktlich on time ‖ punctual ‖ prompt * *unverzüglich, prompt, sofortig* ‖ accurate * *genau*
Punktschweißen spot welding
Punktschweißmaschine butt-welding machine
Punktsteuerung point-to-point control * *(Numer. Steuerg.)* ‖ point-to-point positioning * *(Numer. Steuerg.)*
punktuell at a certain point ‖ at certain points ‖ point-focal * *in der Optik*
putzen clean
Putztrommel tumbler ‖ rolling barrel ‖ polishing drum
Putzwalze cleaning roller * *{Textil} z.b. bei Kratzenraumaschine* ‖ fancy roller * *{Textil} Trommel~ bei Kratzenraumaschine*
Putzwolle cotton waste ‖ waste wool ‖ cleaning waste
PVC-isolierte Leitung PVC-insulated cable
Pyrolyse pyrolysis
PZD process data * *(→PROFIBUS-Profil; Abkürzg. für 'ProZessDaten')*

Q

Quader cube
quaderförmig cubic
Quadrant quadrant * *['kwodrent]*
Quadrat square * *increase/vary with the square of: mit dem ~ der ... anwachsen/ändern* ‖ sq * *Abkürzung für 'Square': Quadrat (sqmm: ~millimeter, mm² usw.)*
quadratisch square * *Fläche u. mathematisch* ‖ quadratic * *mathematisch* ‖ square-law * *~e Gesetzmäßigkeit, Kennlinie (z.B. →Lastkennlinie bei Lüfter-/Pumpenantrieb)* ‖ quadrangular * *viereckig, vierseitig* ‖ by the square * *(Adv.) z.B. mathem. Gesetzmäßigkeit*
quadratisch abhängig sein von vary with the square of
quadratisch mit der Spannung abnehmen decrease with the square of the voltage * *z.B. Drehmoment bei Asynchronmaschine* ‖ decrease as the square of the voltage
quadratische Drehmoment-Drehzahl-Kennlinie - square torque/speed characteristic * *z.B. Lastkennlinie e. Ström.maschine (Pumpe, Lüfter usw.)* ‖ square torque/speed relationship
quadratische Drehmomentenkennlinie square-law load torque characteristic * *z.B. bei Strömungsmaschinen (Pumpen/Lüfter)*
quadratische Drehzahl-Drehmoment-Kennlinie - square speed/torque relationship ‖ square speed/torque characteristic
quadratische Kennlinie square-law characteristic ‖ squared characteristic
quadratische Momentenkennlinie square-law load torque characteristic * *z.B. Lastkennlinie von Strömungsmasch. (Pumpen, Lüftern ...)*
quadratischer Momentenverlauf square-law torque * *quadratische Lastkennlinie, z.B. bei Lüfterantrieb*
quadratisches Gegenmoment square-law counter-torque * *z.B. beim Antrieb einer Strömungsmaschine (Pumpe, Lüfter)* ‖ square-law load torque * *Lastmoment*
quadratisches Lastmoment square-law load torque * *z.B. bei Strömungsmaschine (Pumpen, Lüfter)* ‖ square-law counter-torque * *Gegenmoment*

Quadratwurzel square root
quadrieren square || squaring * *(Substantiv)*
Quadrierer squaring element
quadriert squared
Qualifikation qualification
qualifiziert qualified * *geeignet, befähigt (for: für/ zu), berechtigt* || highly-trained * *gut ausgebildet/ geschult* || eligible * *geeignet, akzeptabel, befähigt (for: zu), berechtigt, in Frage kommend, erwünscht* || competent * *kompetent, fachkundig, (leistungs)fähig, tüchtig, zuständig, befugt, geschäftsfähig* || trained * *geschult, geübt, ausgebildet*
Qualität quality * *poor/low: schlechte, high: gute (siehe ISO 9000 ... 9004 (EN 29000 ... 29004))* || grade * →*Qualitätsstufe, Handelsklasse (von Standardwaren); high-grade: von höchster Qualität* || performance * *Leistungsfähigkeit* || type * *Sorte, Bauart* || kind * *Art* || characteristic * *Eigenschaft, charakteristisches Merkmal, Eigentümlichkeit, Kennzeichen* || durability * *Haltbarkeit, Langlebigkeit* || ruggedness * *Robustheit, Zerstörsicherheit, Verschleißfestigkeit, Stabilität* || rigidity * *Stand/Formfestigkeit, Stabilität*
qualitativ hochwertig high-quality
Qualitätsaudit quality audit
Qualitätsbeauftragter quality officer || person responsible for quality
qualitätsbewusst quality-conscious
Qualitätseinbuße quality loss * *Qualitätsverlust* || deterioration of quality * *Qualitätsbeeinträchtigung/-verschlechterung/-minderung*
Qualitätskontrolle quality control * *"Qualitätslenkung/-steuerung"* || quality tests || quality inspection * *Qualitätsprüfung*
Qualitätsleitsystem quality control system
Qualitätsleittechnik quality control technology
Qualitätsmanagement quality management * *siehe ISO 9000 ... 9004 (EN 2900 ... 29004)*
Qualitätsmangel quality defect
Qualitätsmerkmal quality feature
Qualitätssicherung quality assurance * *siehe ISO 9000...9004 (EN 29000...29004)* || quality management * *siehe ISO 9000...9004 (EN 29000...29004)* || quality control * *(für die Antriebstechnik relevant: DIN ISO 90001)*
Qualitätssicherungsbestimmungen quality control regulations || quality regulations
Qualitätssicherungsrichtlinien quality control regulations
Qualitätssicherungssystem quality assurance system || quality management system
Qualitätsstand quality status || quality standard
Qualitätsstufe quality grade
Qualitätsüberwachung quality control || quality surveillance
Qualitätsverbesserung improvement in quality
Qualitätsverlust quality loss
Qualitätszertifikat quality certificate
Quantisierung quantization
Quantisierungsfehler quantization error * *z.B. auf Grund der beschränkten Auflösung eines Analog/ Digital-Wandlers*
Quarz crystal || Xtal || quartz crystal || quartz
Quarzgenerator crystal pulse generator || quartz generator
Quarzoszillator crystal oscillator
quarzstabilisiert crystal-controlled
quasikontinuierlich quasi-continuous * *z.B. digitaler Regler mit hoher Abtastrate*
quasikontinuierlicher Regler quasi-continuous-action controller * *Digitalregler mit schneller Abtastung*

quasistetig quasi-continuous * *z.B. digitaler Regler mit hoher Abtastrate* || virtually-continuous
Quecksilber mercury
Quecksilberdampf-Gleichrichter mercury-arc rectifier
Quecksilberdampf-Stromrichter mercury-arc rectifier
Quecksilberdampfgleichrichter mercury-arc rectifier
Quellcode source code
Quelle source * *auch ~ eines Signals; Quellcode eines Programms*
Quellenauswahl source selection
quellenfreies Feld zero-divergence field
Quellhinweis source reference * *für Signal/Variable in Stromlaufplan/Blockschaltbild/ Programmlisting*
Quellsignal source signal
Quellsprache source language || source * *Quellspracheprogramm(listing)*
Quellsprache-Code source code || source * *(salopp)*
Quellsprachecode source code
Quellspracheebene source level
quer cross * *quer(liegend/verlaufend), schräg, sich überschneidend* || transverse * *quer(verlaufend), z.B. Kraft; schräg, diagonal; (to: zu)* || diagonal * *diagonal, schräg* || lateral * *seitlich* || horizontal * *waagerecht* || transversal * *querverlaufend (to: zu), Quer-, schräg, diagonal* || crosswise * *der Breite nach*
Quer- transversal || cross- || cross-directed * *in ~richtung* || transversal-directed * *in ~richtung* || CD * *Abkürzg.f. 'Cross-Directed': quer zur Materiallaufrichtung wirkend*
Querantrieb transverse drive
Querausschleuser side-shuttle * *bei Fördersystem*
Querband transverse belt * *in Fördersystem*
Querbeanspruchung transverse load || transverse stress || lateral stress
Querbewegung traverse movement || lateral movement || transverse travel || transverse motion
quergerichtet transverse
Querkraft cantilever force * *Radialkraft, greift senkrecht zur Drehachse (z.B. Motorwelle) an* || lateral force || transverse force || transverse load * *Balastung in Querrichtung* || radial force * *Radialkraft (z.B. auf eine Motorwelle wirkend)* || shearing force
Querkraftdiagramm radial capacity diagram * *{el.Masch.} Verlauf d.zulässigen Querkraft über d. Länge des Wellenstummels*
querliegend transverse || transverse * *querverlaufend*
Querprofil transverse profile * *Querverteilg z.B. v.Flächengewicht, Feuchte,Dicke, Temperat. v.bandförm.Material* || CD profile * *Abk.f. 'Cross-Directed': (z.B. Fläch.gewichts-) Profil quer zur Materiallaufrichtung*
querprüfen cross-check
Querrecke transversal-directed orientation * *Breitstreck-Einrichtung in Folienmaschine* || TDO * *Abk.f. 'Transversal Directed Orientation': Breitstreckeinrichtung in Folienmaschine* || transverse orienter * *Breitstreckeinrichtung bei Kunststoff-Folienmaschine*
Querrichtung tranversal direction || cross direction || CD direction * *Abk.für 'Cross Direction'*
Quersäge cross-cut saw * *{Holz}*
Querschlepper cross transfer * *z.B. in Walzwerk*
Querschneidemaschine guillotine * *mit Schlagmesser arbeitend (für Blech, Holzfurnier usw.)*
Querschneiden cutting * *z.B. von Papier (im Gegensatz zum Längs- oder Rollenschneiden)*
Querschneider cross cutter * *Bogenschneider für*

Papier, Blech || cut-off knife * *Bogenschneider für Papier* || sheeter * *Bogenschneider für Papier* || sheeting machine
Querschnitt cross-section * *auch für Kabel, Konstruktionszeichnung* || wire gauge * *für Kabel* || cutting to length * *das Quer-Schneiden (mit Messer/Schere)* || transverse section * *in technischer Zeichnung* || cross-sectional area * *Querschnittsfläche*
Querschnittsabnahme cross-sectional reduction * *bei Drahtziehmaschine*
Querstift cross pin
Querstrom cross flow
Querstromlüfter transverse-flow fan
Quersumme checksum || cross-check sum
Querteilanlage cutting-to-length line * *z.B. für Walzband, Blech usw.*
Quertransport cross transfer * *in Transferstraße*
Querverbindung cross connection
Querverkehr direct data transmission between substations * *[Network] z.B. zw. Slaves b.Master-Slave Feldbussystem*
Querverweis cross reference
Querverweisliste cross-reference list || cross reference || cross reference list
Quetsch- presser || squeezing || pinch || crush * *zerquetschend/ brechend* || compression * *siehe auch →Anpress-*
quetschen crimp * *pressen, z.B. Kabelschuh, Aderendhülse auf Leitungsende* || squeeze * *(zus.-) drücken, (aus-) pressen, (aus-/heraus-) quetschen, zwängen (into: in), erpressen* || pinch * *kneifen* || crush * *zerquetschen* || mash * *zerquetschen* || squash * *(zu Brei) zerquetschen, zus.drücken, flach-/breitschlagen, zerdrückt/-quetscht werden* || jam * *(ein-)zwängen/klemmen/keilen, zusammendrücken, quetschen (auch Körperteil b. Unfall)* || bruise * *Körperteil ~; Quetschung, Quetschwunde*
Quetschkontakt crimp contact || crimp connector * *Quetschverbinder*
Quetschverbinder crimp connector || pressure connector
Quetschwalze squeegee * *['skwiedjie] (Gummi-) Quetschwalze* || squeegee roll
Quetschzange crimping tool * *zum Anquetschen von Kabelschuhen und Kabel-Quetschverbindern*
quietschen squeak || squeal || screech * *kreischen (z.b. Bremse), gellend schreien*
quittieren acknowledge * *z.B. Fehler, Telegramm* || accept || clear * *Fehler usw.*
Quittiersignal acknowledge signal
quittiert acknowledged
Quittiertaste accept button || acknowledge pushbutton || O.K. key
Quittierung acknowledgement || acceptance || acknowledging * *als Tätigkeit oder Vorgang*
Quittierzeit acknowledgement time
Quittung acknowledgement * *Signal, Telegramm* || receipt * *[ri'βiet] Empfangsbescheinig.. Qittung (on/against: gegen); sales receipt: Kassenzettel* || discharge || voucher * *['wautscher] Beleg, Gutschein* || answer * *(figürl.)* || revenge * *(figürl.) Revanche* || quittance * *[Wirtschaft] ; auch Vergeltung, Bezahlung/Erlassen (einer Schuld)*
Quittungssignal acknowledge signal
Quittungsverzug no acknowledgement * *NAK bei serieller Kopplung* || acknowledgement timeout
Quotient quotient || ratio * *Verhältnis*

R

Rabatt discount * *→Preisnachlass (on: auf)* || rebate || abatement || price reduction * *→Preisnachlass* || bonus * *Bonus*
Rabattlinie discount line || dicount schedule * *Rabattstaffel*
Rabattstaffel discount schedule
Rad wheel || gear * *Zahnrad*
Radantrieb wheel drive * *bei Fahrzeug; wheel-driven: mit ~*
Räderwerk gear train
radial radial || radially * *(Adv.)*
radial bedrahtet radial-lead * *z.B. Kondensator, dessen beiden Anschlussdrähte nach unten herausschauen* || radially leaded || radial lead || with radial leads || with unidirectional leads * *nach IEC 286* || single-ended
Radialbewegung radial movement
Radialdichtung radial sealing ring || rotary shaft seal * *Wellendichtring* || radial shaft seal
Radialgebläse centrifugal blower || radial fan
Radialgleitlager radial slide bearing
Radialkraft radial force || radial load * *radiale Belastung* || cantilever force * *→Querkraft* || lateral force * *→Querkraft*
Radiallast radial load
Radiallüfter radial fan || radial-flow fan
Radialnuten radial slots * *z.B. an Kupplungs-/Bremslamelle*
Radialspiel radial play || radial clearance || gland clearance * *bei Schraubenverdichter*
Radialventilator centrifugal fan
Radialverdichter centrifugal compressor
Radialwellendichtring radial shaft seal ring * *siehe DIN 3760* || rotary shaft seal
Radian radian * *Winkelmaß/Bogenmaß; 1 rad entspr. 360 Grad/2 x Pi entspr. 57 Grad 17 min*
Radiant radian * *siehe →Radian*
Radien radii
radikal radical || extreme * *extrem*
radioaktive Strahlung radioactive radiation
Radius radius
radizieren extract the root * *(mathemat.)*
Radizierer square-root extractor || square root function
Radlader wheeled loader || rubber-tyred shovel || rubber-tired front-end loader || wheel-mounted front-end loader
Radnabe hub || nave || wheel nave
Radsatz wheel set * *[Bahn]* || wheel-frame * *[Bahn]* || Radgestell || bogie * *[Bahn] →Drehgestell*
Raffinerie refinery || finery
Raffineur refiner * *Fein-Mahlwerk für Holzstoff/ Hackschnitzel zur Zellstoffherstellung (→Refiner)*
raffiniert clever * *klug, gescheit, geschickt, pfiffig, begabt* || ingenious * *genial* || subtle * *schlau* || artful * *schlau, listig, verschlagen* || sly * *schlau, verschlagen, listig* || cunning * *pfiffig, listig, schlau, verschmitzt, geschickt, klug* || exquisite * *Aufmachung, Stil* || refined * *verfeinert, z.B. Zucker, Erdöl*
ragen project * *project beyond: hinaus~ über*
Rahmen frame * *auch →Zeichen-~, Daten-~, Telegramm-~* || chassis * *tragende Konstruktion, Fahrgestell, Tragblech* || framework * *(auch figürl.) Gerüst, Gerippe, Gestell, Gefüge, System; within th.framework of: i.~ von* || scope * *(Geltungs-) Bereich*
Rahmenabkommen skeleton agreement || letter of intent * *schriftl. Absichtserklärung*

Rahmenbedingungen boundary conditions
Rahmenkopplung subrack interfacing || cardrack interfacing || rack coupling || subrack coupling * *zwischen Elektronik-Baugruppenträgern* || subrack link || subrack expansion || rack link * *Datenkopplung zwischen Baugruppenträgern; serial: serielle; parallel: parallele*
Rahmenkopplungsmodul subrack interface board * *als Leiterkarte ausgeführte Anschaltung zur Datenkopplung zw. Baugr.trägern* || rack-interfacing module * *zur Datenkopplung zwischen Baugruppenträgern*
Rahmenprofil frame profile
RAID RAID * *'Redundant Array of Inexpensive Disks": Speicher aus mehreren jeweils redundanten Festplatten*
Rakel doctor blade * *Abstreifmesser, Streichmesser, (Druck)Rakel, Kratzeisen* || scraper bar * *Schaber, Kratzmesser, Ziehklinge* || wiper * *Abstreifer*
Rakelmesser doctor blade * *z.B. in Druckmaschine (siehe auch →Rakel)*
RAL RAL * *Deut.Normeninstitut f. Gütesicherg. u. Kennzeichng., früher ReichsAusschuss für Lieferbedingungen*
RAL-Farbe RAL color * *Farbton, der durch eine 4-stellige (internat. anerkannte) RAL-Nummer gekennzeichnet ist*
RAM RAM * *Abkürz. für 'Random Access Memory': Schreib/Lesespeicher*
Rampe ramp || slope * *Neigung, Schräge, Gefälle, Steigung, Anstieg, Rampe*
Rampenfunktion ramp function
Rand edge * *Kante, Ecke, Saum, Grenze; at the lower edge: am unteren ~* || brink * *(meist figürl.) on the brink of: am ~e des/von* || rim * *~ eines Rings, einer Scheibe, einer Vertiefung; Absatz* || margin * *~ einer Druckseite, eines Dokuments/Buches* || border * *Grenze* || periphery * *Umkreis, Peripherie* || fringe * *Einfassung, Um~ung, Franse; (figürl.): äußerer ~, Grenze* || brim * *~ eines Gefäßes, (Hut)Krempe* || collar * *Kragen*
Randbedingung boundary condition || limit condition
Rändelmaschine knurling machine * *Werkzeugmaschine*
Rändelmutter knurled nut
rändeln knurl
Rändelschraube finger screw * *mit der Hand zu betätigen* || knurled-head screw * *mit gerändeltem Kopf*
Randspritze edge cutter * *schneidet in Papiermaschine die Papierrandstreifen mit Wasserstrahl ab*
Randstreifen waste edges * *z.B. bei Papiermaschine* || edge trim * *abgeschnittene Randstreifen z.B. bei Kunstof-Folienmaschine* || edge || trimmings * *(Plural) abgeschnittene Stücke; z.B. von einer Papierbahn, einem Papierbogen*
Randstreifenabschneider edge cutter * *z.B. bei Papier(verarbeitungs)maschine*
Randstreifenschneider edge cutter
Rangfolge order || sequence || ranking
Rangier- marshalling * *bei Leitungen, Kontakten, Klemmen usw.*
Rangierdatenbaustein interface data block * *bei SPS*
rangieren route * *Signale* || allocate * *Signale ver-/zuteilen* || patch || reconnect * *umverbinden* || manoeuvre * *manövrieren* || rank * *im Rang; before: vor; with mit/unter* || be classed * *with: rangieren mit/unter* || marshal * *Leitungen usw.* || shunt * *{Bahnen}* || switch * *{Bahnen} (akerikan.)* : switcher: Rangierlokomotive
Rangierliste interface list * *bei Operationsbaustein einer SPS* || assignment list || allocation table
Rangierung jumpering * *z.B. von Leitungen über Rangierverteiler.* || wiring block * *zur Leitungsrangierung z.B. Rangierklemmen bei d.Leitungseinführung i. Schaltschrank* || marshalling
Rangierverteiler wiring block * *zur Leitungsverteilung z.B. bei der Leitungseinführung i. Schaltschrank* || cabinet wiring block || terminal board * *Klemmenbrett* || marshalling rack || multi-point connector * *Anschlussverteiler, Klemmenvervielfacher*
rapid rapid || rapidly * *(Adv.)*
rapide rapid || rapidly * *(Adv.)*
Rapport repeat size * *Abstand, in dem sich ein Muster wiederholt (bei Tapeten, Bodenbelägen, Textildruck)*
rasch quick || speedy * *zügig* || prompt * *sofortig* || snappy * *(salopp) flott, forsch, fix, schwungvoll, schmissig*
Raste notch * *Kerbe, Einschnitt, Raste (zum Einklinken)* || stop * *Anschlag, Sperre, Hemmung, Klappe* || detention point * *Rück/Festhaltepunkt* || lock * *Verschluss, Sperrvorrichtung* || clip * *Klammer, Klemme, Spange, Halter*
rastend locking type * *z.B. Taster* || latching type * *z.B. Taster* || push-lock * *"einrastend, wenn man es drückt", z.B. Rasttaster* || stay-put
Raster grid
Rasterfeld matrix board * *z.B. "Spielwiese" auf Leiterkarte z. nachträglichen Einlöten v. Bauelementen* || connection matrix * *"Spielwiese" auf Leiterkarten mit freien Bestückungsplätzen*
Rastermaß grid dimension || grid unit || lead spacing * *Abstand von Bauelemente-Anschlüssen* || grid
Rasterteilung basic grid dimension * *Grund-Rastermaß*
Rasterung grid || grid of modules
Rasterwalze anilox roller * *{Druckerei}*
Rastmoment cogging torque * *ruckartiges Drehmoment (z.B. Nutrasten auf Grund der Polfühligkeit eines Motors)*
Rastnase clip * *Klammer, Klemme, Spange, Halter*
Rasttaster pushlock-button || latched button * *latched emergency stop button: Not-Aus-Rasttaster* || lock button
Rasttaster mit Drehentriegelung pushlock-twist-release button * *z.B. Not-Aus Pilztaster*
Rastung latching || notching || notch * *Rastnase* || catch * *Sperrhaken, Schließhaken, Schnäpper, Sicherung, Verschluss* || detent * *Sperrhaken, Sperrklinke; Sperre, Auslösung*
Rate rate * *Ereignisse/Wert pro Zeiteinheit* || ratio * *Verhältniszahl*
raten advise * *jemandem* || guess * *erraten*
Ratgeber adviser || counsellor || counselor
Ratiomaßnahmen rationalization measures
Rationalisierung rationalization * *rationalization measures: ~smaßnahmen* || rationalizing process * *~sprozess*
rationell rational || efficient * *wirtschaftlich* || economical * *sparsam*
ratsam advisable
Rattermarken chatter marks
rattern rattle || chatter || clatter || roar || oscillate * *schwingen* || vibrate * *vibieren, pulsieren, zittern, beben*
rau rough || rugged || coarse * *grob* || rude * *grob* || harsh * *streng, hart, rau (auch Umgebungsbedingungen usw.), scharf, derb*
Rauchgas flue gas
Rauchgasentschwefelung flue gas desulpherization || flue gas desulphurization
raue Industriebedingungen harsh industrial environments
raue Umgebungsbedingungen harsh environments || tough environmental conditions || harsh ambient conditions || hostile environments || tough environments

rauen rough up * *z.B. Textilstoff* ~ || nap * *noppen, rauhen, z.B. von Textilstoff* ~
rauer Betrieb rough service || rough service conditions * *raue Betriebsbedingungen* || harsh environments * *raue Betriebs-/Umgebungsbedingungen*
rauer Industrieeinsatz rough industrial use || rugged industrial application
Rauheit roughness
Rauhigkeit roughness * *Bendtsen roughness:* ~ *nach Bendtsen*
Raum space * *Fläche, Platz, Zwischen*~, *Stelle, Lücke, Zeit*~, *Abstand, Dimension im Gegensatz zu 'Zeit'* || room * *Zimmer* || volume * ~*inhalt* || capacity * *Fassungsvermögen*
Raum- spatial || space || spacial
Raumausnutzung room utilization || space utilization * *Platzausnutzung*
Raumbedarf space requirement
Räumeisen tapping bar
Räumer reamer * *Reib-/Räumahle/Räumnadel-/platte zur Metallbearbeitung* || thickener * *Absetzbehälter zur Wasseraufbereitung* || scraper * *zur Wasseraufbereitung*
Raumladung space charge
räumlich of space * *an Raum* || relating to space * *auf den Raum/Platz bezogen* || three-dimensional * dreidimensional || spatial * *im Gegensatz zu 'zeitlich'* || cubic * *(mathemat.)* || solid * *(geometr.)* || spacial * *räumlich, Raum-* || in space || stereoscopic * *dreidimensional (in der Optik)* || relating to space
räumlich anordnen space * *auch räumlich einteilen, mit/in Zwischenräumen anordnen* || arrange || group
räumlich beengt close-quartered || cramped for space
räumlich getrennt separated || separate || spatially separated || physically segregated
räumliche Trennung spatial separation
räumliche Verteilung spatial distribution
Räummaschine broaching machine * *Werkz. masch.; internal/surface/turning broaching machine: Innen-/Außen-/Dreh-* || die-slotting machine
Räumnadel broach * *push/draw broach:* ~ *zum Stoßen/Ziehen*
Raumnutzung room utilization || space utilization * *Platz-/Raumausnutzung*
Raumpolygon space polygon
raumsparend space-saving || room-saving
Raumtemperatur normal room temperature || ambient temperature
Raumvektor space vector
Räumwerkzeug broaching tool
Raumzeiger space vector || sinor
Raumzeigerdiagramm space vector diagram || sinor diagram
Raumzeigermodulation space vector modulation
Raupe caterpillar || crawler * *Raupenschlepper* || bulldozer * *Planierraupe* || bead * *[Schweißtechnik]*
Raupenantrieb crawler drive * *z.B. bei Bagger* || track drive * *z.B. bei Bagger* || track-laying drive * *z.B. beim Bagger*
Raupenschlepper crawler tractor
Rauschen noise * *pink/random: rosa*
Rauschpegel noise level
Rawcliff-Wicklung Rawcliff winding * *für polumschaltbaren Motor ähnlich Dahlanderschaltung (siehe →PAM-Wicklung)* || PAM winding * *für polumschaltbaren Motor ähnlich Dahlanderschaltung (siehe →PAM-Wicklung')*
Raytracing raytracing * *Verfahren z. Berechnung v. 3D-Bildern unter Berücksichtigg. des Lichteinfalls (Schatten)*
RC-Beschaltung RC circuit || RC element * snubber circuit * *zum Überspannungsschutz, z.B. TSE-*

Beschaltung bei Thyristorsatz || RC snubber * *als* →*TSE-Beschaltung einer Thyristorschaltung*
RC-Glied RC element || R-C network
RC-Löschkombination RC damping element * *z.B. zur Beschaltung einer Schützspule z. Vermeidung von Spannungsspitzen* || RC surge suppression element
RDC RDC * *{Papier} Recommended Drive Capacity: empfohl. Antriebsleistg. (im stationär. Betrieb nach TAPPI)*
reagieren react * *on: auf* || respond * *to: auf*
Reaktanz reactance * *Blindwiderstand, induktiver Widerstand*
Reaktion reaction * *to: auf* || response * *Ansprechen, Antwort, Erwiderung; to: auf*
Reaktionsgeschwindigkeit speed of response
Reaktionsmoment torque reaction
Reaktionszeit response time || reaction time || latency * *Latenz-/Totzeit im Rechnersystem z.B. bei der Reaktion auf Interrupts*
reaktivieren reactivate || recall * *Person* ~ *(to service: zum Dienst)*
realisierbar realizable || practicable * →*durchführbar* || feasible || workable || viable * *machbar*
realisieren realize || implement * *ausführen, durchführen* || solve * *lösen (Problem, Aufgabe)*
realisiert realized
Realisierung realization || implementation * *auch Software*~ || solution * *Problemlösung* || arrangement * *Anordnung, Ausgestaltung* || design * *Entwurf, Gestaltung, Bauart, Ausführung, Konstruktion* || shaping * *Ausgestaltung, Ausformung, Ausprägung*
realistisch realistic
Realität reality
Realteil real component * *{Mathem.} bei komplexen Zahlen*
Rechen rake * *(Substantiv und Verb)* || grid * *in Kläranlage* || ratchet || squeegee || fork * *Harke*
Rechenaufgabe computation task * *für Rechner* || numeric task * *für Rechner* || arithmetic task * *für Computer*
Rechenbaustein arithmetic function block * *Funktionsbaustein* || numeric function block * *Funktionsbaustein*
Rechenfunktion arithmetic function || numeric function
Rechengeschwindigkeit processing speed * *Verarbeitungsgeschwindigkeit, z.B. eines Rechners* || computing speed * *Rechner* || execution speed * *für ein bestimmtes Programm/Befehl(sfolge)*
Rechenlauf computer run || computing procedure
Rechenleistung computing power || CPU performance * *der Zentraleinheit eines Rechners* || computational power || computer performance * *eines Rechners* || computing capacity || computational performance || computer performance || processor performance || processing capacity || processing power
Rechenoperation arithmetic operation || computing operation || calculation operation
Rechenschaltung computational circuit
Rechenschieber slide rule || calculating rule
Rechenverfahren computing method || computing procedure || calculation method || method of calculation || calculation technique
Rechenvorschrift process of calculation * *Rechengang* || algorithm * *Algorithmus* || calculation instruction || calculation procedure
Rechenwerk arithmetic logic unit || ALU * *Abkürzg. f. 'Arithmetic Logic Unit': Ausführungseinheit für arithmet. u.logische Befehle* || CPU * *Abkürzg. f. 'Central Processing Unit': Zentraleinheit*

Rechenwert useful value || calculated value * *berechneter Wert*
Rechenzeit run time * *während des Ablaufs eines Programms* || computation time * *für Programmablauf benötigte Zeit* || calculating time || computing time
rechenzeitoptimiert high response * *mit kurzer Reaktionszeit (bei Echtzeitanwendungen)*
Rechenzentrum computer centre || computer center * *(amerikan.)*
Rechenzyklus computational cycle
rechnen calculate * *(aus-) rechnen; miscalculate: falsch ~* || reckon * *(be-/er-) rechnen; reckon in: ein~; reckon over: nach~.; reckon with: ~ mit* || work out * *aus~* || sum up * *zusammen~* || estimate * veranschlagen || value * *bewerten* || charge * *mit Geldsumme belasten* || expect * *erwarten* || reckon with * *mit etwas Zukünftigem* ~ || face * *face a thing: mit etwas (Unangenehmem) ~ müssen* || be good at figures * *gut ~ können* || computation * *(Substantiv)*
Rechner computer
Rechnerauslastung computer loading || processor loading * *Prozessorauslastung* || processor utilization * *Prozessorauslastung* || computer utilization
rechnergesteuert computer-controlled
rechnergestützt computer-aided * *z.B. Entwurf, Konstruktion, Fertigungssteuerung* || computer-supported || computerized
rechnerisch mathematical || arithmetic || mathematically * *(Adv.)* || arithmetically * *(Adv.)* || by the way of calculation * *(Adv.)* || determined by calculation * *(Adv.)* || calculated * *berechnet*
Rechnerkopplung computer networking * *Vernetzung* || communication between computers || computer coupling || interlinking of computers || interprocessor communication || computer interfacing || computer link * *Schnittstelle zur Rechnerkopplung* || computer interface * *Schnittstelle zur Rechnerkopplung*
Rechnerleistung computing power
Rechnung invoice * *Waren~* || account * *Waren~, Be~, In-~-Stellen, (Ausgaben)Ab~* || bill * *Rechnung (-sschein/-szettel)* || calculation * *Be~* || tally * *{Wirtschaft} Rechnung, Ab~, Gegen~, Kontogegenbuch (eines Kunden); Zählstrich, Coupon*
Rechnung tragen take special account of * *siehe auch →berücksichtigen*
Rechnungsbetrag invoiced amount * *in Rechnung gestellter Betrag* || amount billed * *for: für*
Rechnungsprüfung auditing || examination of the accounts || accounting * *Rechnungswesen*
Rechnungswesen accounting || accountancy
rechte Seite righthand side || right hand || right hand side
Rechteck rectangle
Rechteck- rectangular || square * *quadratisch* || square-wave * *Impulse, Modulation usw.*
Rechteckbauweise rectangular design
rechteckförmig rectangular || rectangular-shaped
Rechteckgenerator square-wave generator
rechteckig rectangular
rechteckige Klammern square brackets
Rechteckimpuls rectangular pulse
Rechteckimpuls squarewave pulses * *mit Tastverhältnis 1: 1* || rectangular pulses || quadrature pulses * 2 um 90° versetzte Impulsketten m. Tastverhältnis 1: 1 (z.B. e.Inkrementalgebers)
Rechteckimpulsform rectangular pulse shape
Rechtecksignal rectangular signal || square-wave signal * *normalerweise mit Tastverhältnis 1:1*
rechter Winkel right angle * *be at right angles to each other: um 90 Grad gegeneinander versetzt sein*
rechtes right hand * *(Adj.)*

rechtfertigen justify || warrant || defend * *verteidigen*
Rechtfertigung justification
rechtlich legal || juridical
rechtliche Maßnahme legal action
rechts clockwise * *im Uhrzeigersinn (Drehrichtung usw.)* || cw * *in ~drehrichtung* || on the right || right hand side * *~seitig; to the right of: ~ von* || to the right * *nach ~; to the right of: ~ von* || right * *right from: ~ von; turn right: nach ~ drehen/abbiegen; on the right: ~* || rightmost * *das am weitesten ~ befindliche* || righthand * *zur Rechten (stehend usw.), ~ angegliedert, ~gängig, ~läufig, Rechts-* || at the right * *auf der rechten Seite* || on the right side
rechts außen at the extreme right
rechts herum clockwise * *mit dem Uhrzeigersinn* || cw * *Abkürzung für 'clockwise': im Uhrzeigersinn*
rechts oben at the top right || at the upper right
rechts unten at the bottom right
Rechts- legal * *rechtlich* || in law * *(nachgestellt) rechtlich, laut Gesetz* || right hand * *rechts befindlich* || clockwise * *im Urzeigersinn* || right-
Rechtsanschlag right stop * *z.B. eines Potis*
Rechtsberater legal advisor
rechtsbündig right-justified * *z.B. in der Daten-/Textverarbeitung*
Rechtsdrehfeld clockwise phase sequence || cw rotating field * *Abk. f. 'ClockWise': im Uhrzeigersinn* || clockwise rotating field
Rechtsdrehrichtung forward direction of rotation * *Vorwärtsdrehrichtung* || clockwise direction of rotation * *im Uhrzeigersinn* || cw rotation * *im Uhrzeigersinn*
rechtsgängig right-hand * *z.B. Schraube, Gewinde*
Rechtsgewinde right-hand thread || right-handed thread
Rechtslauf forward running * *Vorwärtslauf* || clockwise running || clockwise rotating || clockwise rotation
rechtsschieben shift right
rechtsseitig righthand
rechtsverbindlich binding * *(up)on: für* || legal
rechtswidrig illegal || unlawful || illicit
Rechtswidrigkeit illegality
rechtwinklig rectangular || right-angled || at right angles
Rechtwinkligkeit squareness || perpendicularity
rechtzeitig in time || on time || timely || well-timed || punctually * *(Adv.) pünktlich* || prompt * *pünktlich* || in plenty of time || in good time
recken stretch
Recycling recycling
Recycling- recycled * *aus Altmaterial gewonnen, z.B. ~Papier*
Redakteur editor
Redaktion editorial office
redaktionelle Bearbeitung editing
Redesign redesign || re-design
Redewendung phrase * *set/stock phrase: (fest)stehende Redensart/-wendung* || expression || idiom * *(Sprachbesonderheit)*
redigieren edit
Reduktion reduction || benefication * *Erz* || cut * *von Preisen, Kosten usw.* || reducing || derating * *~ der Leistungsdaten bei erschwerten Umgebungsbedingungen*
Reduktionsfaktor derating factor * *z.B. für verminderte Leistung e. Stromrichters bei größerer Aufstellhöhe* || reduction factor * *Verhältn. Motornennstrom/Stromtragfähigkeit d.i. Klemmkasten anschließb. Leitgn.*
Reduktionsgetriebe reduction gear || speed reducer * *siehe auch 'Reduziergetriebe'*
redundant redundant
Redundanz redundancy * *one-out-of-three redundancy: 1-aus-3-~*

Reduzier- reducing
reduzieren reduce * *by: um* || diminish * *vermindern, verringern* || lower * *absenken* || de-rate
* *z.B. Leistungsspezifikation bei Stromrichter in großer Aufstellhöhe ü.d. Meeresspiegel* || derate
* *siehe 'de-rate'* || cut * *Preise, Kosten, Ausgaben, Zeitbedarf, Platzbedarf usw.* || cut down * *Preise, Kosten, Ausgaben, Platz-/Zeitbedarf usw.* || go down * *sich reduzieren*
Reduziergetriebe reduction gear || speed reducer || gear reducer || reduction gearing || step-down gearing
Reduzierstück reducing adapter || reducer || adapter sleeve * *Reduzierhülse* || bore reducer * *zur Querschnittsverminderung einer Bohrung*
reduziert reduced
Reduzierung reduction || derating * *Herabsetzg., z.B. v.Kenndaten bei besond. Betriebsbedinggen. (z.B. große Aufstellhöhe)* || diminution * *Verminderung, Verringerung, Verkleinerung, Abnahme* || decrease * *Verringerung, Verminderung, Abnahme* || lessening * *Verringerung, Abnahme* || cut * *Kürzung, Streichung (Ausgaben, Kosten, Preise, Text usw.); cut in costs: Kosten~* || de-rating * *von Spezifikationen bei erschwerten Betriebsbedingungen (Temperatur/Taktfrequenz usw.)*
Reduzierventil pressure reducing valve
reentrant reentrant * *wiedereintrittsfähig, Fähigkeit eines Programms, sich selbst unterbrechen zu können*
Referent speaker * *Vortragender, Redner* || reporter || instructor * *bei einer Schulungsveranstaltung*
Referentenleitfaden presentation guidelines || presenter's guidelines
Referenz reference
Referenzdiode reference diode
Referenzfahrt homing * *Wegnullpunkt anfahren bei Positioniersteuerung (mit Inkrementalgeber)* || going to home position * *Anfahren des Referenzpunktes bei Posititioniersteuerung* || approach to reference point * *Anfahren d.Referenzpunktes bei Positioniersteuerung* || search for reference * *Anfahren d.Referenzpunktes bei Positioniersteuerung* || homing procedure || reference-point approach
Referenzfrequenz reference frequency
Referenzieren homing * *→Referenzfahrt, Anfahren des Wegnullpunktes bei einer Positioniersteuerung*
Referenzierlogik homing logic * *Steuerung d. →Referenzfahrt: Anfahren d. Nullpunktes bei e. Positioniersteuerung*
Referenzimpuls reference position pulse * *für Lagereregelung* || marker pulse * *Nullimpuls eines Dreh-Impulsgebers* || reference marker pulse * *Nullimpuls eines Dreh-Impulsgebers*
Referenzpunkt reference point || reference position * *bei Lageregelung* || home position * *bei Lageregelung*
Referenzpunktfahren search for reference * *bei Positioniersteuerung* || go to home position * *bei Positioniersteuerung/Numer. Steuerung* || approach to reference point * *bei Positioniersteuerung/Numer. Steuerung*
Referenzpunktfahrt homing procedure
Referenzspannung reference voltage * *genaue Spannung, ein Spannungsnormal darstellend* || reference potential * *Bezugsspannung für Signal (Masse)* || voltage reference * *Referenzspannungsquelle, auch als Bauelement*
Referenzspannungsquelle voltage reference * *auch als Bauelement* || reference voltage source || reference supply * *z.B. zum Anschluss eines Sollwertpotentiometers*
Refiner refiner * *zum Feinmahlen v.Holz-/Zellstoff; conical refiner: Kegel~; disc refiner: Scheiben~*

reflektieren reflect || refract * *[Optik] brechen, ablenken*
reflektierend reflective
Reflektion reflection
Reflexion reflection
reflexionsarm low-reflection
Reflowlöten reflow soldering * *Aufschmelzlöten z.B. von →SMD-Bauelementen*
Regal shelf || stack
Regalbediengerät stacker crane * *für Hochregallager* || order picker * *für (Hochregal-) Lager* || stacker truck * *Förderzeug für Hochregallager*
Regalförderer stacker crane
Regalförderzeug stacker crane * *Bediengerät für Hochregallager* || high-bay-rack conveyor * *siehe →Regalbediengerät* || order picker * *in (Hochregal-) Lagersystem* || stacker truck * *für Hochregallager* || high-bay racking vehicle || racking vehicle
Regallager rack store
Regel rule * *(auch mathem., physikal.)* rule of thumb: Daumen~ || standard * *Norm* || guiding principle * *Richtlinie* || principle * *Prinzip*
Regel- und Steueraufgaben closed-loop and open-loop control tasks
Regel- und Steuergeräte electronic control equipment * *elektronische ~*
Regelabweichung control deviation * *permanent: bleibende* || error signal * *als Signal* || setpoint-actual value difference * *Soll-Ist-Differenz* || control error || error || system deviation || offset
Regelalgorithmus control loop algorithm || control law || control algorithm
Regelantrieb variable-speed drive * *geregelter/ drehzahlveränderbarer Antrieb*
Regelart control method || control type || mode of control
Regelaufgabe closed-loop control task
regelbar controllable || adjustable || variable * *infinitely: stufenlos*
regelbarer Antrieb variable-speed drive || adjustable drive || adjustable-speed drive
Regelbarkeit controllability
Regelbaugruppe loop controller module || closed-loop control electronics board * *Leiterkarte* || control board * *Leiterkarte*
Regelbaustein control loop block * *Funktionsbaustein*
Regelbereich control range || range of adjustment
Regeldifferenz system deviation || error signal * *siehe auch →Regelabweichung* || system deviation * *→Regelabweichung* || control error
Regeldynamik dynamic response || response || control dynamics || dynamic performance
Regeleigenschaften response || response characteristics || closed-loop performance
Regeleinrichtung closed-loop control device || closed-loop control || closed-loop control system || closed-loop control equipment || system control equipment || controlling system || controlling equipment
Regelelektronik closed-loop control electronics || control electronics
Regelfall normal case || general case * *generally speaking: im ~*
Regelfunktion closed-loop control function || control
Regelfunktionen closed-loop control functions || controls
Regelgenauigkeit control accuracy * *Abweichung der Regelgröße vom eingestellten (Soll)Wert* || regulation accuracy * *Abweichung d. Regelgröße v. eingestellten (Soll)Wert* || control precision
Regelgerät controller || closed-loop controller || control unit || control equipment || control || control system * *Regelsystem*

Regelgetriebe

Regelgetriebe variable speed gearing || speed variator * *stufenlos regelbares ~*
Regelgröße controlled variable * *die Größe, die ein Regler dem Sollwert nachführen soll* || control variable
Regelgüte control performance * *Regeleigenschaften* || control quality || performance goodness
Regelkonstanz control constancy || closed-loop control stability
Regelkonzept control concept || loop-contol scheme || closed-loop control model * *Regelmodell*
Regelkreis control loop || closed-loop control circuit || control system || control circuit || automatic control loop
regelmäßig regular * *auch zeitlich* || on a regular basis || periodical * *(zeitlich)* || regulated * geordnet || regularly * *(Adv.)* || periodic * *in ~en Zeitabständen* || at regular intervals * *in ~en Abständen*
Regelmäßigkeit regularity
regeln control * *steuern, regeln, regulieren, verwalten* || regulate * *regeln, regulieren* || adjust * *regulieren, anpassen, angleichen, einstellen* || govern * *bestimmen, lenken, leiten* || arrange * *arrangieren, ordnen, in Ordnung bringen* || settle * *eine Vereinbarung treffen, sich vergleichen, einrichten* || direct * *lenken, leiten, führen, bestimmen* || put in order * *ordnen*
Regelparameter controller characteristic parameter
Regelprinzip closed-loop control principle
Regelreserve overshoot capability * *{Regel. techn.} über d.Nennwert hinausgehender Bereich des Istwerts/Stellgröße* || overmodulation capacity
Regelschleife control loop || closed-loop control circuit * *Regelkreis*
Regelschleifring regulator slipring * *{el. Masch.} mit dauernd aufliegenden Bürsten zur Drehz.regelg. durch Vorwiderst.*
Regelschleifringläufer-Motor regulator-slipring motor * *{el. Masch.}* || variable-speed slipring motor * *{el. Masch.}*
Regelschwinge dancer * *siehe →Tänzer(-walze), z.B. in der Gummi-/Reifenindustrie*
Regelschwingungen hunting * *Pendeln auf Grund von Instabilität* || oscillation of the controlled variable * *der Regelgröße*
Regelstrecke controlled system
Regelstruktur closed-loop control structure || control structure || closed-loop structure
Regelsystem closed-loop control system || closed-loop control || control system || control
Regeltechnik control engineering || controls * *Regelsysteme*
Regelteil closed-loop control section || closed-loop control unit || regulating section
Regeltrafo variable transformer * *siehe auch →Stelltrafo*
Regeltransformator variable transformer * *siehe auch →Stelltransformator*
Regelung closed-loop control || control || control system || loop control || control loop || feedback control || automatic control || servo-control * *~ mit Stellantrieb* || feedback control * *mit Rückführung arbeitend* || regulatory control || ruling * *~ durch Vereinbarung/Vorschrift usw.* || rule * *~Regel, Normalfall, Richtschnur, Grundsatz, Spielregel, Richtlinie, Verhaltensmaßregel* || regulation * *~Bestimmung, →Vorschrift, Verfügung*
Regelungs- und Steuerungstechnik control systems * *siehe DIN 19226*
Regelungsabtastzeit control sampling time
Regelungsart control method || control type || mode of control
Regelungsausführung control version
Regelungsbaugruppe loop controller module || control board * *Leiterkarte*

Regelungsbaustein control loop block * *bei SPS* || closed-loop control function block || control block * *Funktionsbaustein*
Regelungseigenschaften closed-loop performance
Regelungsprozessor loop processor * *bei SPS*
Regelungssystem closed-loop control system || control system || control
Regelungstechnik closed-loop controls || control engineering * *als Fachrichtung* || automatic control theory
Regelungstechniker control engineer || control technician
Regelungstheorie control theory
Regelungsverfahren control method
Regelventil servo valve || control valve || governor valve || positioning valve * *Stellventil*
Regelverfahren control method
Regelverhalten closed-loop response || response || closed-loop performance * *Regelgüte* || response characteristic || control characteristic || dynamic performance * *Regeldynamik* || control action || control behaviour || control performance * *Qualität/Güte/Leistungsvermögen der Regelung* || control response
Regelverstärker control amplifier
Regelverstärkung controller gain
Regelvorgang control process
Regelwalze controll roller
Regelwirkung correcting effect
regenerativ regenerative
regenerieren regenerate
Region region
regional regional || local || district
Register register * *Wortspeicher, Eintragungsbuch, (Inhalts-) Verzeichnis, Registriervorrichtg., Druckmarken~* || index * *(Einstell-, Teil-) Marke, ~, (An)Zeiger, Index, Inhalts-/Stichwortverzeichnis, Kennziffer*
Registerfehler register error * *{Druckerei}*
Registerhaltigkeit register accuracy * *{Druckerei}*
Registermarke register mark * *{Druckerei}*
Registerregelung register control * *{Druckerei} sorgt für passgenaues Übereinanderdrucken der Farbwerke/Passermarken*
Registerwalze table roll * *in Papiermaschine/Papierverarbeitungsmaschine*
Registriereinrichtung recording unit * *zum Aufzeichnen von Messwerten/Betriebsdaten*
registrieren register * *z.B. Betriebs- oder Fehlerdaten* || log * *Messwerte ~*
Registriergerät recording instrument * *z.B. Linienschreiber*
Registrierung record * *Aufzeichnung von Messwerten, Betriebsdaten usw.* || recording * *Aufzeichnung von Messwerten, Betriebsdaten usw.* || entry * *Eintragung* || registration || logging * *von Betriebsdaten* || printing out * *auf Drucker*
Regler regulator || controller || closed-loop controller || loop controller || loop-controller || control * *Regelung*
Regler sperren inhibit the controller
Regleranpassung controller adaption
Reglerausgang controller output || regulator output
Reglereingang controller input
Reglerentwurf controller design
Reglerfreigabe controller enable * *als Kommando* || controller enabling || regulator enable
Reglermatrix controller matrix
Regleroptimierung tuning of the regulator || controller optimization || optimizing the controller
Reglerparameter controller characteristic parameter || controller parameter

Reglerschwingen hunting * *siehe auch 'Instabilität'*
Reglerselbsteinstellung controller auto-tuning || controller self-setting
Reglersperre controller inhibit * *inhibit the controller: ~ setzen* || controller disable
Reglersperre setzen inhibit the controller
Reglerstruktur controller structure
Reglerteil regulating section * *siehe auch* →*Regelteil*
Reglerübersteuerung controller overshoot || controller overmodulation
Reglerüberwachung controller monitoring
Reglerverhalten control action
Regressforderung claim of damages
regulieren regulate || adjust * *einstellen* || set * *einstellen, setzen* || control * *steuern* || govern * *regieren, beherrschen, lenken, regeln, bestimmen, steuern*
Regulierung regulation || adjustment * *Einstellung*
Regulierventil control valve
Reib- friction
Reibahle reamer || reaming bit * *zum Einsetzen in Werkzeughalter* || broach
Reibarbeit friction energy || work of friction || friction energy absorption || friction losses * *Reibungsverluste* || friction work
Reibbelag friction lining || friction pad
reiben rub
Reibfestigkeit abrasion resistance * *Abriebfestigkeit*
Reibfläche friction face * *z.B. bei Bremse* || friction surface
Reibflächenverschleiß wear of the friction face * *bei Bremse/Kupplg.; specific: spezif. (Verschleißvolumen pro Reibarbeit)*
Reibkraft friction force
Reibkupplung friction clutch
Reibmaterial friction material
Reibmoment friction torque || frictional torque
Reibpaarung friction combination || friction-material mating
Reibrad friction wheel
Reibscheibe friction disk
Reibschluss frictional locking || friction contact || friction locking || close contact
reibschlüssig frictionally engaged || frictionally locked || non-positive
Reibung friction
Reibungsarbeit friction energy || work of friction || friction energy absorption || friction losses * *Reibverluste*
reibungsarm low-friction
Reibungskennlinie friction characteristic * *Reibmoment z.B. in Abhängigkeit von der Drehzahl*
Reibungskoeffizient coefficient of friction
Reibungskompensation friction compensation
Reibungskraft friction force
reibungslos without any problems * *problemlos* || frictionless * *ohne Reibung* || smooth * *glatt, reibungslos (auch figürl.), zügig, flüssig, geschmeidig* || without a hitch * *(Adv.) reibungslos, glatt, ungestört, ungehindert, ruckfrei*
Reibungsmoment friction torque
Reibungsverluste friction losses || friction energy absorpion || frictional losses || frictional heat * *Reibungswärme*
Reibungswärme frictional heat
Reibungszahl coefficient of friction || friction coefficient
Reibverhalten frictional behaviour || friction characteristic
Reibwalze friction drum || friction roll || friction roller
Reibwerkstoff friction material

Reibwert coefficient of friction
Reibzahl coefficient of friction
reichen stretch * *stretch from ... to: ~/sich erstrecken von ... bis* || extend * *sich erstrecken; to: bis; from ... to: von ... bis* || come up to * *hinauf~ bis* || go up to * *hinauf~ bis* || go down to * *hinab~ bis* || suffice * *genügen, aus~d/hin~d sein* || do * *genügen, aus~d sein; that will do!: das reicht!* || last out * *genügen, aus~d sein, (aus)reichen* || hold out * *andauern, sich halten, aus-/durchhalten, standhalten, genügen* || last * *dauern, währen, (sich) halten, (aus)reichen, genügen; durch-/aushalten, bestehen* || reach * *to: bis* || be sufficient * *aus~, ~d sein*
reichlich ample * *weit, groß, geräumig, genügend, ausführlich* || abundant * *reichlich (vorhanden), reich an, versehen (mit), abundant in/with: i. Überfluss besitzend* || copious * *~, ausgiebig, ergiebig, reich, umfassend, produktiv, viel schaffend; weitschweifig* || plentiful * *reich(lich), im Überfluss (vorhanden)* || plenty of * *(vor Substantiv)* || rather * *(Adv.) ziemlich* || amply * *(Adv.)* || awfully * *(salopp; Adv.) furchtbar, äußerst, sehr, riesig* || abundantly * *(Adv.) reichlich, völlig* || awful * *(salopp) furchtbar, riesig, kolossal*
Reichweite reach * *within reach/near at hand: in ~; medium reach: mittlere ~; out of reach: außer ~* || radius of action || range * *bei Funkübertragg. usw..; long/medium range: große/mittl. ~; outrange: an ~ übertreffen* || radius * *Umkreis, Wirkgs.-/Einflussbereich; rad. of action: Aktionsradius, Fahrbereich {Verkehr}* || operating span || radius of influence || reach of supply * *~ von Lagerbeständen (stocks are exhausted: die Vorräte sind erschöpft)* || cruising radius * *{Verkehr} ~ von Fahrzeugen (mit einer Tankfüllung)* || cruising range * *{Verkehr} ~ von Fahrzeugen* || flying range * *{Verkehr} ~ von Flugzeugen*
Reife maturity * *(auch figürl; →Produktreife) degree of maturity: ~grad; mature: zur ~ kommen/bringen* || ripeness * *(auch figürl.) siehe auch* →*Ausgereiftheit*
Reifen tire * *(amerikan.) Rad-/Auto-/Gummireifen* || tyre * *(brit.) Rad-/Auto-/Gummireifen* || ring || hoop * *z.B. Fassreifen*
Reihe series * *auch Bau~, a series of: eine ~ von; connected in series: in ~ geschaltet* || bank * *auch ~nanordnung* || range * *Kollektion, Angebot, Auswahl, Bereich* || line * *Linie, Bau~* || row * *(Häuser/Sitz usw.) Reihe, Zeile* || set * *Gruppe* || queue * *Schlange* || progression * *mathematische ~* || tier * *Reihe, Lage, Zeile (z.B. eines Baugruppenträgers); in tiers: in ~n/Lagen übereinander* || sequence * *(Aufeinander-) Folge, ~nfolge, Reihe, Serie; in sequence: der ~ nach*
Reihen- serial || inline
Reihenbohrmaschine gang drilling machine
Reihenfolge sequence * *alphabetic(al): alphabetische; inverse: umgekehrte* || succession || order * *alphabetic(al): alphabetische; rising/ascending: aufsteigende* || arrangement * *Einteilung, (An-) Ordnung*
Reiheninduktivität series inductance
Reihenklemme modular terminal || line-up terminal || modular terminal block
Reihenklemmenblock modular terminal block || terminal strip * *Klemmleiste*
Reihenkondensator series capacitor
Reihenschaltung series connection || series circuit
Reihenschluss-Kennlinie series characteristic * *{el.Masch.} Momentenkennlinie*
Reihenschlussmaschine series-wound machine * *Feldwicklung in Reihe mit Ankerwicklung*
Reihenschlussmotor series-wound motor * *Feldwicklung in Reihe mit Ankerwicklung*
Reihenschlußwicklung series winding

Reihenwiderstand series resistor ‖ series resistance
rein pure * *pur, unvermischt, bloß; pure P controller: reiner P-regler* ‖ purely * *(Adv.)* ‖ clean * *sauber* ‖ real * *eigentlich, tatsächlich, wirklich, wahr* ‖ net * *Gewinn* ‖ plain * *Wahrheit* ‖ mere * *bloß, lediglich (z.B. Formalität)* ‖ quite * *(Adv.) gänzlich* ‖ downright * *(Adv.) gänzlich, völlig, absolut, ausgesprochen, wirklich* ‖ sheer * *→bloß, rein, nicht als, by sheer force: mit bloßer/nackter Gewalt*
rein Ohmsch purely ohmic ‖ purely resistive
Reinheit purity ‖ pureness ‖ cleanness ‖ cleanliness ‖ clearness ‖ neatness ‖ tidiness ‖ fidelity * *Radio*
reinigen clean ‖ cleanse * *(auch figürl.)* ‖ tidy up * *putzen* ‖ clarify * *Flüssigkeit* ‖ purify * *Luft* ‖ clear * *säubern, reinigen*
Reinigung cleaning ‖ cleansing ‖ purification * *auch von Abwässern usw.* ‖ purge
Reinigungsmaschine cleaning machine ‖ purifying machine * *zum →Abscheiden von Fremdmaterial* ‖ washer * *zum Waschen*
Reinigungsmittel cleaning agent ‖ cleaner ‖ cleansing agent ‖ purifying agent ‖ detergent
Reinraum clean room
Reinwasser pure water ‖ purified water
Reißbrett drawing board
Reisebericht trip report
Reisebüro travel agency
reißen tear * *tear off: for~; tear out of: aus etwas heraus~; tear in two: entzwei~* ‖ rupture ‖ tug * *zerren (an), ziehen an* ‖ pull * *ziehen* ‖ jerk * *ruckartig ~* ‖ yank * *(amerikan.) ruckartig ~* ‖ snatch * *wegschnappen* ‖ drag * *fort~* ‖ drag along * *fort~* ‖ seize upon * *an sich ~* ‖ lay hold on * *an sich ~* ‖ usurp * *an sich ~, usurpieren; z.B. Macht, Markt* ‖ monopolize * *beherrschen, an sich ~, z.B. Gespräch, Markt* ‖ tear the lead * *die Führung an sich ~* ‖ pull down * *zu Boden ~* ‖ break * *brechen; z.B. Papierbahn, Draht, Kabel* ‖ snap * *brechen* ‖ burst * *bersten* ‖ get torn * *(Stoff)* ‖ split * *zer~, zer-/aufspalten, zerteilen, schlitzen; sich aufspalten/trennen/teilen*
Reiseweg route
Reißfestigkeit breaking strength ‖ tensile strength ‖ tearing strength
Reißnadel scriber
Reißnaht pressure-relief joint * *[el./Masch.] Sollbruchstelle in einem lichbogensicheren Klemmenkasten*
Reißwolf willow ‖ willowing machine
Reitstock tailstock * *bei Drehmaschine* ‖ headstock * *bei Drehmaschine*
Reklamation claim ‖ reclamation ‖ complaint * *Mängelrüge* ‖ protest * *Einspruch* ‖ objection * *Ein/Widerspruch*
Reklamationen claims
Reklame advertising
Rekombination recombination * *Neutralisierung von Ladungsträgern in Halbleiter*
rekonstruieren reconstruct ‖ retrieve * *wiederherstellen*
Rekonstruktion reconstruction
Rekursionsformel recurrence formula
rekursiv recursive
rekursiver Algorithmus recursive algorithm
Relais relay ‖ relays * *Mehrzahl*
Relais-Kontaktbelastung relay rating ‖ relay contact rating
Relaisausgang relay output ‖ relay logic output
Relaisflattern relay oscillation ‖ relay chatter ‖ relay rattling
Relaiskontakt relay contact
Relaisspule relay coil ‖ solenoid * *Betätigungsspule*

Relaissymbolik relay symbology * *bei Kontaktplandarstellung eines SPS-Programms*
Relation ratio * *Verhältnisfaktor* ‖ relation * *Verhältnis, Beziehung* ‖ relationship
relational relational * *z.B. Datenbank mit Querbeziehungen zwischen den Datensätzen*
Relationsfaktor ratio factor ‖ ratio
relativ relative * *(Adj.) to: zu* ‖ comparatively * *(Adv.) to: zu* ‖ relatively * *(Adv.)* ‖ in per unit * *bezogen auf Bezugsgröße, welche "1" oder "100%" entspricht* ‖ per unit * *bezogen auf Bezugsgröße, welche "1" oder "100%" entspricht*
Relativbewegung relative motion
Relativdrehzahl relative speed
relative Einschaltdauer cyclic duration factor * *(ED) prozentuales Verhältn. Lastzeit/Spieldauer, siehe VDE 0530 Teil 1* ‖ duty ratio
relative Feuchtigkeit relative humidity
relative Geschwindigkeitsdifferenz speed differential * *(v2-v1)/v1) x 100% zw. 2 Antrieben bei durchlaufender Warenbahn* ‖ draw
relative Luftfeuchtigkeit relative humidity ‖ relative air humidity ‖ R.H. * *Abk. f.'Relative Humidity'*
relevant relevant * *erheblich, zutreffend (to: für)* ‖ pertinent * *erheblich, zutreffend* ‖ important * *wichtig* ‖ significant * *wichtig, bedeutsam* ‖ applicable * *anwendbar, passend, geeignet, zutreffend (to: auf, für)*
Relevanz relevance * *Erheblichkeit* ‖ pertinence * *Erheblichkeit* ‖ importance * *Wichtigkeit* ‖ significance * *Wichtigkeit, Bedeutung* ‖ relevancy
Reluktanzmotor reluctance motor * *Drehstrom-Synchronmotor mit Käfigläufer (ohne besondere Erregung)*
remanent retentive * *Merker/Kontakt bei Kontaktplan/Relaissteuerung* ‖ latching * *Merker/Kontakt bei Kontaktplan/Relaissteuerung* ‖ maintained * *Merker/Kontakt bei Kontaktplan/Relaissteuerung* ‖ remanent * *magnet.* ‖ residual * *Rest-; z.B. remanent field: Restfeld/remanentes Feld*
Remanenz remanence ‖ residual magnetism * *Restmagnetismus* ‖ retentivity
Remanenzrelais remanence relay
Remanenzschaltung remanent circuit
Remanenzspannung remanence voltage
Rendite return on investment ‖ effective interest * *effektiver Gewinn, Effektivzins* ‖ yield * *→Gewinn, →Ertrag*
reparabel reparable
Reparatur repair * *under: in; have repaired: in ~ geben/reparieren lassen* ‖ overhaul * *Überholung* ‖ reconditioning * *Überholung*
Reparatur ausführen carry out the repair
reparaturbedürftig in need of repair ‖ defective * *defekt*
Reparaturbericht repair report ‖ report of repair
Reparaturdienst repair service
reparaturfähig reparable
Reparaturkosten costs of repair
Reparaturpauschale lump sum repair price * *Pauschalpreis für Reparatur*
Reparaturplatz repair station
Reparaturpreis repair price
Reparaturschalter repair switch
Reparaturstelle repair department * *Abteilung* ‖ repared spot ‖ repair facility
Reparaturwerkstatt repair shop ‖ service station
reparieren repair * *have repaired: reparieren lassen/in Reparatur geben* ‖ mend ‖ fix ‖ vamp * *flikken, reparieren, herrichten, zurechtschustern, "aufmotzen"*
Repeater repeater * *Leitungsverstärker zum Erreichen größerer Entfernungen bei Datenübertragung*
repräsentativ representative

repräsentieren represent
reproduzierbar reproducible || repeatable * *wiederholbar*
Reproduzierbarkeit reproducibility || repeatability * *Wiederholbarkeit*
reproduzieren reproduce || repeat * *wiederholen*
Reserve for future use * *für (zu)künftige Verwendung vorgesehen* || reserve || reserve capacity * *Leistungs/Kapazitätsreserve* || stand-by * auch als Vorsilbe; z.B. Umrichter, Maschine || spare * *(Adj.) Ersatz-* || margin * →*Spielraum; leave a margin: Spielraum lassen; margin of safety: Sicherheitsfaktor* || hot-standby * *zum sofortigen Einschalten/Übernehmen bereit, wenn das Hauptsystem ausfällt*
Reserve- standby || stand-by * z.B. Umrichter, Maschine || spare * z.b. Ersatzteil
Reserveumrichter standby converter * siehe auch →*Synchronisierzusatz* || spare converter * als Ersatzteil
reservieren reserve * *Zimmer, Flug, Restaurant-Tisch; auch '~ lassen'* || book * *~ lassen, buchen* || make a reservation for * *~ lassen*
reserviert reserved
Reset-Buchse reset socket || reset jack
Reset-Taster reset button
Resolver resolver * *Drehmelder als Lage/Drehzahlgeber mit sinusförmiger Erregg. u sin/cos-Signalen am Ausgang*
Resolver-Digital-Umsetzer R/D converter || resolver/digital converter
Resonanz resonance
Resonanzdämpfung resonance damping || resonance dampening
Resonanzdrehzahl resonance speed || resonant speed
Resonanzfaktor resonance factor * *relative Amplitudenerhöhung bei Resonanz*
Resonanzfrequenz resonant frequency || resonance frequency
Resonanzfrequenzüberhöhung resonance sharpness
Resonanzkreis resonance circuit
Resonanzkreisumrichter resonant-pole inverter
Resonanzpunkt resonance point
Resonanzschwingung resonant oscillation
Resonanzumrichter resonant-pole inverter || resonant converter
Rest rest || remainder * *auch mathematisch; Restbestand* || remnant * *Rest, Überbleibsel, letzter/kläglicher Rest, Spur, (Stoff- usw.) Rest*
Rest- residual * z.B. *Feld, Magnetisierung* || remanence * *restlich, übrig(bleibend); z.B. Spannung, Feld* || leakage * *Leck-* || vestige * -*Überrest, -Überbleibsel, -Spur*
Restfehler residual error || permanent error * *bleibender Fehler* || truncation error * *Rundungsfehler, ~ bei Rechenalgorithmus (Reihenentwicklung usw.)*
Restfeld residual field || remanent field
Restinduktivität residual inductance
restlich remaining
Restmagnetismus residual magnetism
Restspannung remanent voltage * *Remanenz* || residual voltage * *führt bei Relais/Schütz gerade noch nicht zum Anziehen*
Reststrom residual current * *bei Relais/Schütz: führt gerade noch nicht z. Anziehen* || leakage current * →*Leckstrom*
Restunwucht residual unbalance
Restweg remaining distance * *{NC-Steuerung}*
Restwelligkeit residual ripple
Resultat result || outcome || effect * *Wirkung*
resultieren result * *in: in; from (the fact that): aus*
resultierend resultant * *sich ergebend*
Resümee summary || resumé

Rèsumee summary || resumé
Retarder retarder * *(hydraul.) Strömungsbremse für LKWs, Schienenfahrzeuge usw.; Gleisbremse*
Retention retention * *Zurück(be)haltung*
Retourenbegleitschein returned products form * *zur Rücklieferung defekter Geräte*
retten save || retrieve * *wiedergewinnen/-erlangen* || recover * *(Güter, Geld, Stoff usw.) wiedergewinnen/-erlangen; (Verluste) wiedergutmachen/ersetzen* || backup * *{Computer} eine Sicherheitskopie anlegen*
reversibel reversible
Reversier- reversing * z.B. *Walzwerk*
Reversierantrieb reversing drive || reversible drive
Reversieren reversal || reverse * *(Verb)* || reversing
Reversiermotor reversing motor
Reversierzeit reversing time * z.B. *in Walzgerüst*
Revision inspection * *(techn.) Inspektion einer Maschine, Anlage usw.* || review * *Überprüfung (Buch, Gerichtsurteil, Entscheidung, Entwicklungsvorhaben usw.)* || audit * *{Handelswesen} Rechenschaftslegung, Überprüfung (der Geschäftsbücher), Rechnungsprüfung* || revision * *Revision, Durchsicht, Überarbeitung, Korrektur, verbesserte Ausgabe/Auflage, Ausgabestand* || maintenance * *Wartung* || revival * *~ eines Vertrages* || appeal * *{Rechtswesen} lodge an appeal: ~ einlegen* || rehearing * *{Rechtswesen} Wiederaufnahme* || auditing || overhaul * *technische Überholung*
Revisionsbetrieb maintenance operation || inspection mode
revolutionieren revolutionize
Revolver- turret * z.B. *turret lathe: ~drehbank*
Revolverbohrmaschine turret-head drilling machine
Revolverdrehmaschine turret lathe || capstan lathe
Revolverkopf turret head * z.B. *an Drehmaschine* || turret butt * →*Drehkreuz bei Wickler mit fliegendem Rollenwechsel*
Revolverkopf-Ankleber turret butt splice * *für fliegenden Rollenwechsel bei Auf/Abwickler*
Rezept recipe || formula * *Formel*
Rezeptsteuerung recipe control
Rezeptur recipe || formula
Rezeptursteuerung recipe control
Rezeptverwaltung recipe management || recipe handling || recipe administration
Rezession recession * *Wirtschaftskrise*
reziprok reciprocal || reciprocally * *(Adv.)*
Reziprokwert reciprocal value
RG-Erregung rotating rectifier excitation * '*Rotierender (Dioden)Gleichrichter',i.Rotor integriert* (→*Stromr.motor)* || rotating-rectifier excitation * *bürstenlose Erregg. über i.Rotor integrierten (Dioden)gleichrichter*
RG-Motor converter motor * →*Stromrichtermotor*
RGB stacker crane * *Abk. f. 'ReGalBediengerät'*
richten straighten * *geradebiegen durch Richtmaschine/Richtpresse usw.* || level * *gerade-/planvon (Metall-) Werkstücken, Blech usw.*
richten an forward to * z.B. *eine Anfrage* || address to * z.B. *Brief, Nachricht, Telegramm*
richtig right || correct || exact || proper * *gehörig, passend* || suitable * *geeignet* || appropriate * *angemessen* || fair * *gerecht* || accurate * *genau*
richtige Lage correct position || trim * z.B. *einer Tänzerrolle*
richtige Stellung correct position || trim
Richtimpuls initializing pulse || reset pulse * *Rücksetzimpuls* || power-on pulse * *wird beim Einschalten d. Stromversorgung erzeugt* || power-on reset || initialising pulse
Richtlinie guideline || guiding rule || guiding principle || general instructions * *Anweisung(en)* || directive * *Direktive, (An-) Weisung, Vorschrift*

Richtmaschine leveller * *rollt über mehrere Walzen gebogenes/welliges Blech plan* || straightening machine * *zum Planbiegen von Blech, Geradebiegen von Draht, Rohren usw.* || stretcher leveller * *für Feinblech* || levelling machine || tension-levelling machine * *Streckricht-Maschine* || tension leveller * *Streckricht-Maschine* || straightener
Richtplatte aligning plate * *zum Ausrichten* || straightening plate * *z.B. zum Glattrichten von Blech* || levelling plate
Richtschnur guiding principle || guidance
Richtung direction * *direction of rotation: Drehrichtung* || way * *Weg* || route * *(Fahr)Weg* || course * *Kurs* || trend * *Entwicklung* || tendency * *Tendenz* || orientation * *Einstellung* || policy * *Politik*
richtungsabhängig direction-dependent || depending on the direction of rotation * *drehrichtungsabhängig*
Richtungspfeil direction arrow
Richtungswechsel directional change || directional reversal * *Richtungsumkehr*
Richtwert guide value || rough value * *überschlägiger Wert* || recommended value * *empfohlener Wert* || empirical value * *Erfahrungswert* || pragmatical value * *plausibler Richtwert* || practically proven value * *praxisgerechter Wert* || orientative value * *(nur) zur Orientierung gedachter Wert* || average figure
Riefe groove || flute * *Rille, Riefe, Hohlkehle, (Span-) Nut*
Riefenbildung brinelling * *z.B. in Wälzlager* || pitting * *Lochfraß, Grübchenerosion* || threading * *an Kommutator* || ribbing || scoring || grooving * *z.B. an einem Kommutator*
Riegel bolt || key-bolt * *am Schloss* || latch * *Klinke, Schnäpper, Schnappriegel, Schnappschloss* || catch * *Sperr-/Schließhaken, Falle, Schnäpper, Verschluss, mechan.Sicherung*
Riemen belt * *z.B. V-belt: Keil~, toothed belt: Zahn~*
Riemenantrieb belt drive
Riemenförderer belt conveyor
Riemenrad belt pulley
Riemenscheibe belt pulley || pulley || pulley wheel
Riemenscheibendurchmesser pulley diameter * *bei Riementrieb*
Riemenspannrolle jockey pulley
Riemenspannung belt tension
Riementrieb belt drive
Riemenverstellgetriebe variable-speed belt drive
riesig giant * *gigantisch* || gigantic * *gigantisch* || colossal || enormous * *enorm* || huge || tremendous || immense || jumbo * *→übergroß*
Rille groove || flute || chamfer || score * *Kerbe, Rille* || race * *Lauf-~/Gleitbahn z.B. für Kugellager*
Rillenkugellager deep-groove ball bearing || deep-groove roller bearing
RINA RINA * *Abkürzg.f. 'Registro Italiano Navale': italienische Schiffsbauvorschriften*
Rinde bark * *{Holz}*
Ring ring || washer * *Unterlegscheibe, Dichtring* || collar * *Kragen* || coil * *aufgewickeltes Material, Bund (Draht, Kabel, Blech, Seil usw.)* || bundle * *Bund, Paket, Bündel, Ballen*
Ring- ringed || circular || ringlike || torodial * *z.B. Trafo, Stellwiderstand usw.* || annular
Ring-Spinnmaschine ring spinning machine * *Textilmaschine*
Ringbandkern torodial core * *z.B. bei Trafo*
Ringbolzen eye bolt * *Hebeöse, Kranöse*
Ringbus ring bus
ringförmig circular || toroidal || ringlike || annular
Ringkern-Transformator toroidal transformer || ring-wound transformer

Ringkernstromwandler toroidal-core current transformer
Ringkerntrafo ring wound transformer || toroidal transformer
Ringöse ring lug * *z.b. Hebeöse am Motor* || lifting eye * *z.b. Hebeöse an Motor, Schaltschrank* || lifting eybolt * *Hebeöse*
Ringpuffer circular buffer || ring buffer
Ringschelle ring clamp || ring clip
Ringschraube eye bolt * *Hebeöse, Kranöse* || eyebolt * *Schraube mit Hebeöse*
Ringspeicher circular buffer * *Ringpuffer* || ring buffer * *Ringpuffer*
Ringspindel ring spindle * *{Textil} in Ringspinnmaschine zur Garnerzeugung*
Ringspinnmaschine ring spinning machine * *{Textil} (i.Gegensatz zur →Open-End-Spinnmaschine)*
Ringventilator ring fan
Ringverdichter side-channel compressor
Ringverteiler pulse distributor * *verteilt d. Zündimpulse auf d. einz. Thyristoren e. Stromrichters*
Ringwickelmaschine toroidal coil winder
Ringzähler ring counter
Rinne groove * *Rille* || channel * *Rille* || trough * *Trog, Mulde, Wanne, Rinne, Kanal, Wellental, Tiefdruck* || gully * *Wasserablauf~* || sewer * *Wasserablauf~* || chute * *Rutsche* || furrow * *Furche*
Rippe rib * *auch Kühlrippe eines Motors/Kühlkörpers* || fin * *dünne Kühl/Heizrippe*
Rippengehäuse ribbed housing * *z.B. Motorgehäuse mit Kühlrippen*
rippengekühlt rib-cooled * *z.B. Motor* || fin-cooled * *z.B. Motor* || with rib cooling
rippengekühlter Motor rib frame motor || rib-cooled motor || fin-cooled motor
Rippenkühlkörper ribbed heat sink || finned heat sink
Rippenkühlung rib cooling * *Kühlung über Kühlrippen (z.B. Motor)*
Rippenrohr finned tube * *in einem Kühlsystem*
Rippenrohrkühler finned-tube radiator * *flüssigkeitsdurchströmter Wärmetauscher*
RISC RISC * *'Reduced Instruction Set Computer' mit einfachen aber schnellen u.kurzen Befehlen (→CISC)*
RISC-Prozessor RISC processor * *Reduced Instruction Set Computer: Zentraleinht. mit wenigen aber schnellen Befehlen*
Risiko risk * *take: eingehen, incur: übernehmen, at one's own risk: auf eigenes ~* || hazard * *Gefahr, Wagnis*
risikolos risk-free
Riss cleft || chink || crack * *Sprung* || scratch * *Schramme, Ritz* || flaw * *Sprung* || cracking * *~bildung* || fissure || breakage * *~ einer Material-/Papierbahn, eines Fadens/Drahts usw.* || break * *(Ab-/Zer-/Durch-) Brechen/Reißen; Bruch, ~ einer Papierbahn/eines Fadens usw.*
Rissbildung cracking
Ritzel pinion || cog || cog wheel || trunnion
Ritzelantrieb pinion drive
Roboter robot
Roboterarm robot arm
Robotersteuerung robot control || robot controller
Robotertechnik robotics || robot technology
robust rugged * *stark, stabil* || rigid * *standfest, formfest, stabil* || robust * *kräftig, stark, stabil (auch Parameterunempfindlichkeit e.Regelung),* widerstandsfähig || heavy-duty * *für hohe Beanspruchung/raue Umgebungsbedingungen; Schwerlast-* || tough * *stark, robust, widerstandsfähig* || sturdy * *→stabil, kräftig, standhaft (auch figürl.), massiv, fest*
Robustbauform heavy-duty type || rugged design ||

ruggedized model || A-version * *SPS* || construction suited for harsh environments
Robustbauweise ruggedized design
robuste Regelung robust control
Robustheit strength * *Stärke, auch einer mechanischen Konstruktion* || stability * *Stabilität* || ruggedness * *Zerstörsicherheit, Verschleißfestigkeit* || robustness * *Stabilität (auch e. Regelung), Stärke, Festigkeit* || sturdiness * *Kräftigkeit, Robustheit, Standhaftigkeit* || durability * →*Haltbarkeit, Dauerhaftigkeit, Beständigkeit, Lebensdauer* || rugged design * *robuste Konstruktion* || ruggedized design * *robuste Konstruktion*
roh raw * *unbearbeitet, wie aus der Natur gewonnen* || in native state * *im ursprünglichen Zustand* || rough * *rau, grob, roh (z.B. Entwurf), ungehobelt, unbearbeitet, im Rohzustand* || coarse * *grob* || unfinished * *nicht veredelt, nicht fertigbearbeitet* || unmachined * *nicht (fertig)bearbeitet (Werkstück auf Werkz.maschine)*
Roh- raw || base * *z.B. base paper: ~papier*
Rohdaten raw data
Rohkohle raw coal
Rohling blank * *unbearbeitetes Werkstück*
Rohöl crude oil
Rohr tube ± *Röhre; seamless; nahtlos* || pipe * *Rohrleitung, Leitungs~* || duct * *Röhre, Leitung, (Kabel/Kühlluft-) Kanal; duct away: durch Röhre abführen* || tubing * *~material, Ver~ung, ~anlage* || piping * *~material* || conduit * *(Kabel)Kanal* || tunnel * *Tunnel* || spout * *Einfüll~, Schütt~*
Rohr- tube || pipe || tubular * *röhrenförmig*
Rohrabschneider pipe cutter || tube cutter
Rohranschluss pipe adapter * *Anschluss (-stutzen)/Zwischenstück* || pipe connection || pipe ventilated * *pipe ventilated machine: Maschine mit Rohranschluss für Kühlmittel* || line pipe * *Leitungsrohr* || connecting of pipes * *das Anschließen von Rohren* || duct tube * *Anschlussrohr (für Ventilator)* || duct connection * *~ für Kühlluft* || pipe coupling
Rohrbiegeanlage pipe bending system
Rohrbiegemaschine tube-bending machine
Rohrbruch pipe burst
Rohrdurchführung pipe penetration
Rohrdurchmesser pipe diameter || pipe size * *Rohrquerschnitt* || tube diameter || line size * *bei Druckluftleitung usw.*
Röhre tube || pipe * *Leitungsrohr, Rohrleitung* || duct * *(Kabel/Leitungs/Luft)Kanal* || conduit * *(Kabel)Kanal* || spout * *Einfüll-/Schüttröhre, (Gieß-)Tülle, Abfluss-/Speirohr* || tunnel * *Tunnel* || valve * *Elektronenröhre* || cathode ray tube * *Kathodenstrahlröhre* || neon tube * *Neonröhre* || shaft * *Schacht*
röhrenförmig tubular || tubiform
Röhrenkühlung tube cooling * *[el. Masch.] Kühlart mit geschloss. Primär-Röhrenkühlkreislauf (IC511/611/516)*
Röhrenwärmetauscher pipe heat exchanger
rohrförmig tubular
Rohrkrümmer pipe bend || elbow
Rohrleitung pipe line * *Fernleitung* || pipeline * *Fernleitung* || tubing * *Röhrenmaterial, Rohr, Röhrenanlage, Rohrleitung(ssystem)* || piping * *Rohrleitung(snetz), Rohrverlegung* || ducting * *Rohrleitungen/Rohrleitungsnetz für Luft* || conduit * *Rohrleitung, Röhre, Kanal, Isolierrohr (für el. Leitung)* || duct * *Röhre, (Rohr-) Leitung, (Kabel-)Kanal*
Rohrleitungen pipework || piping
Rohrleitungsnetz pipeline network || piping
Rohrmantel jacket
Rohrmuffe pipe socket
Rohrnetz pipe system * *open: offenes* || mains * *für Gas, Wasser usw. aus der Sicht des Verbrauchers*

Rohrquerschnitt pipe size
Rohrrichtmaschine tube straightening machine
Rohrschelle pipe clamp || pipe clip * *meist aus Kunststoff*
Rohrschleuder centrifugal-cleaner * *Wirbelsichter/Hydrozyklon zur Reinigung, z.B. in der Zellstoffherstellung*
Rohrstutzen pipe socket || duct adapter * *für Kühlluft-Rohr usw.*
Rohrsystem piping system || piping * *Verrohrung*
Rohrwalzmaschine tube-forming machine
Rohrwalzwerk tube mill
Rohrzange pipe wrench
Rohrziehstein tube drawing die
Rohsignal raw signal
Rohsignalgeber sine-cosine encoder * *Drehzahl-/Lagegeber mit phasenverschobenen sin-Signalen* || raw-signal encoder
Rohstoff raw material
Rohstoffrückgewinnung recycling of raw materials
Rohwasser raw water || untreated water
Rohwert non-linearized value * *nichtlinearer Messwert (z.B. von Temperatursensor)* || raw signal * *Rohsignal*
Rollbalken rollbar * *am Bildschirm* || scrollbar * *auf dem Bildschirm*
Rollbandfeder coil spring ± *z.B. zum Niederhalten einer Kohlebürste*
Rollbandfederträger coil-spring holder * *z.B. z. Bürsten-Niederhalten in Gleichstrommaschine*
Rolle roll * *Rolle, Walze; live roll: angetriebene ~* || pulley * *Zug~ für Seil, Riemen, Kran, Flaschenzug, Sägeband* || roller * *Walze* || cylinder * *Hohlwalze* || coil * *Draht, Kabel, Seil* || reel * *Spule, Haspel, Rolle (mit Hülse)* || bolt * *bolt of cloth: Stoff~* || calender * *~ z. Pressen von Papier, Stoff usw.* || part * *(figürl.) (An-) Teil; play a part: eine ~ spielen* || role * *(figürl.) play a (significant/decisive) role: eine (wesentliche/entscheidende) ~ spielen* || castor * *Lauf~, Lenk~, ~ unter Möbeln* || caster * *Lauf~, Lenk~, ~ unter Möbeln* || sheave * *Scheibe, Rolle*
Rolle spielen be significant * *eine wesentliche ~*
rollen roll * *auch Bildschirmanzeige (up: hoch; down: hinunter/rückwärts)* || wheel * *auf Rädern* || calender * *Tuch/Papier mit Pressrolle* || roll up * (sich) aufrollen, z.B. Papier* || curl up * *sich aufrollen, z.B. Papier* || scroll * *(Bildschirm)Anzeige, die Ausschnitt e.größeren Bildes ist, up: hoch; down: hinunter*
Rollenbahn roller conveyor * *[Fördertechnik] Rollenförderer (live: angetriebene); buffer roller conveyor: Stau~*
Rollenbahnförderer roller conveyor
Rollenbock roller trestle * *z.B. für Langholz im Sägewerk*
rollende Reibung rolling friction || sliding friction
Rollendrucklager roller thrust bearing
Rollendruckmaschine web-fed printing press * *[Druckerei] im Gegensatz zur* →*Bogendruckmaschine; verarbeitet Papierrollen* || web-fed printing machine
Rollendruckpapier paper for web printing
Rollendurchmesser roll diameter
Rollenförderer roller conveyor
rollengelagert mounted on roller bearings
Rollenhebel roller lever
Rollenkette roller chain
Rollenlager roller bearing
Rollenoffsetmaschine web offset printing press * *[Druckerei]*
Rollenschneider roll slitter || slitter rewinder || slitter * *z.B. Längsschneider für Papier, Folie usw.* || rewinder * *Umroller, Wiederaufroller* || reel cut-

ting machine || slitterwinder || slitter-winder || reel cutter
Rollenschneidmaschine roll slitter || reel cutting machine || slitter rewinder || slitter winder || slitter-winder || slitterwinder || roll cutter || reel cutter
Rollenständer roll stand * *Abwickler* || unwind stand * *am Anfang einer Rotationsdruckmaschine*
Rollentisch roller table * *{z. B. Holz} Zuführeinrichtung*
Rollenträger unwind stand * *Abwickler*
Rollenwechsel splice * *Schneidevorgang bei (Nonstop-)Wickler* || roll change || changing of the rolls || roll exchange * *Austausch einer (abgenutzten) Rolle/Walze* || roll transfer * *bei Wickler* || reel change
Rollenwechselsteuerung splice control * *z. B. in Papierbearbeitungmaschine* || reel-change control * *z. B. zum Tambourwechsel bei Papiermaschine*
Rollenwechsler flying splicer * *Auf/Abwickler mit fliegendem Rollenwechsel* || non stop winder * *Wickler mit fliegendem Rollenwechsel* || splicer * *Auf-/Abroller mit fliegendem Rollenwechsel* || reel splicer * *Auf-/Abwickler mit fliegendem Rollenwechsel* || paster * *Rollenwechsel- u. -Anklebeeinrichtung; flying paster: ~ mit fliegendem Rollenwechsel*
Roller reeler * *Wickler für Papier (meist Umfangswickler)* || reel-up * *Wickler für Papier (meist Umfangswickler)*
Rollfeder coil spring
Rollgang roller table || roller conveyor * *Transportrollgang* || rolling table || positioning roller * *Positionierrollgang* || working roller table * *Arbeitsrollgang*
Rollgangsmotor roller-table motor * *{el. Masch.} Antriebsmotor für Rollgänge in Walzwerken* || roller-table drive motor
Rollreibung rolling friction * *z.B. eines Laufrades am Fahrzeug*
Rollring rolling ring || roller cage * *Lagerkäfig*
Rollspindel roller screw
Rollstange mandrel * *Wickeldorn* || tambour * *für Papier*
Rolltreppe escalator || moving staircase
röntgen X-ray || radiograph
Röntgenbild X-ray photograph || radiograph || X-ograph
Rost rust * *Eisenoxid* || grating * *Gitter* || grill * *Rost* || grate * *aus Metallstäben (z.B. in Ofen)*
Rostbildung rust formation || rusting
rosten rust || get rusty || oxidize * *oxidieren* || corrode * *korrodieren*
Rostfraß corrosion * *Korrosion* || rust attack
rostfrei stainless * *Stahl* || rustproof || rustless
rostgeschützt rustproof || rust-resistant || rust protected
rostig rusty || corroded * *korrodiert*
Rostschutz anti-rust || rust-proofing
Rostschutzfarbe rustproof coating || anti-corrosive paint || anti-rust coating || rust-preventing agent
rot red
rot glühend red hot
Rotation rotation || revolution || rotary movement || rotary press printing * *Rotationsdruck*
Rotations- rotary
Rotationsdruck rotary press printing * *{Druckerei}*
Rotationsdruckmaschine rotary printing machine * *{Druckerei}* || rotary printing press * *{Druckerei}* || web-fed printing machine * *{Druckerei}* || web-fed printing press * *{Druckerei}*
Rotationsenergie energy of rotation || kinetic energy of rotation
Rotationsschneider rotary cutter

rotationssymmetrisch axially symmetrical
Rotationsverdichter rotary compressor
rotatorisch rotary
Rotglut red heat
rotieren rotate || revolve
rotierend rotary || rotating
rotierende Masse rotating masses
rotierende Massen rotating masses
rotierende Teile rotating parts
rotierender Frequenzumformer rotating frequency converter * *zum Erreichen extrem hoher Antriebsdrehzahlen bei Asynchronmotor*
rotierender Umformer rotating converter
Rotor rotor || armature * *Anker*
Rotorfluss rotor flux
Rotorlage rotor position || rotor angle
Rotorlagegeber rotor position transmitter || rotor position sensor || rotor shaft angle encoder
Rotorstillstandsspannung locked rotor voltage
Rotorwelle rotor shaft
Rotorzeitkonstante rotor time constant
Routine routine * *Software-Unterprogramm, gleich bleibendes Verfahren, gewohnheitsmäßiger Gang*
routinemäßig routine * *(Adj.) siehe auch →turnusmäßig* || routinely * *(Adv.)* || regular
Routinewartung regular service procedure
RS232 RS232 * *'Recommended Standard': empfohlene Norm f.serielle Computerschnittstelle m. ±12V-Signalen (V24)*
RS485 RS485 * *"Recommended Standard": Norm f.serielle Daten-Schnittstelle m.5V-Differenzsignal, IEC 1158-2*
Rubrik rubric || column * *Spalte* || class * *Klasse, Gruppe* || category * *Kategorie, Klasse* || menue item * *{Computer} z.b. in Softwareprogramm, Mailbox usw.*
Ruck jerk * *erste Ableitung der Beschleunigung* || shock * *Stoß* || bump * *heftiger Stoß/Bums* || hitch * *Ruck, Stockung, ruckweise Bewegung* || yank * *(salopp)*
Rück- rear * *Hinter-, hinten angeordnet, hintenliegend* || re- * *wieder (zurück)* || reverse * *umgekehrt, in umgekehrter Richtung* || back || trailing || retro- * *zurück, rückwärtsgerichtet* || de- * *um-, ent-, weg-, ver-*
Rückansicht rear view
ruckartig jerky || by jerks || cogging * *unrund/ruckhaft laufend* || of a sudden * *(Adv.) plötzlich*
Ruckbegrenzung jerk limiting * *z.B. durch Hochlaufgeber mit Verrundung* || jerk control || jerk reduction || jerk smoothing || rate-of-change limiting || jerk limitation
Rückdokumentation reverse documentation || back documentation * *z.B. durch Recompiler, Disassembler* || graphic retranslation * *grafische ~* || back documentation
rückdokumentieren document reversely
Rückerstattung restitution || refund * *Geld* || refunding * *Geld* || reimbursement * *Auslagen, Kosten*
Rückfahrt return journey || return trip || return travel * *auch eines Maschinenteils/Krans usw.* || way back * *on the way back: auf der ~/dem Rückweg*
Rückfluss backward flow || backflow * *(amerikan.)* || recovery flow * *von Energie usw.* || reflux * *von Kapital usw.* || return * *auch von Kapital, Investitionen usw.* || back flow
Rückfrage query || question * *Frage* || inquiry * *Nachfrage, Erkundigung, Anfrage, Nachforschung* || check back * *zur Kontrolle*
Rückfragen queries || questions * *Fragen* || inquiries * *Nachfragen, Erkundigungen, Anfragen, Nachforschungen* || further inquiries || check back * *(Verb) zur Kontrolle* || consult * *(Verb) um Rat fragen* || enquire * *(Verb) sich erkundigen (nach), (er)fragen; for: nach; about: über; of someone:*

bei jdm. || inquire * *(Verb) fragen/s.erkundigen nach, (er)fragen; for: nach; about: über; of someone: bei jdm.*
ruckfrei jerkless * *erste Ableitung der Beschleunigung ist gleich Null* || bumpless * *stoßfrei* || smooth * *weich* || without cogging || jerk-free || judder-free * *frei von Vibrieren (z.B. Aufzug)*
ruckfreie Umschaltung bumpless switchover * *stoßfrei*
ruckfreier Übergang smooth transition || bumpless switchover || jerkfree changeover
Rückführkondensator feedback capacitor
Rückführung feedback || feedback loop || feedback path * *Zweig, Pfad*
Rückgang decay * *Schwund, Zerfall, Verfall, Schwäche* || decrease * *Abnahme, Verringerung, Verminderung* || falling-off * *Nachlassen, Abnahme, Abfall (Produktion, Umsatz usw.)* || decline * *Abnahme, Niedergang, Verschlechterung, Sinken (Preise usw.),* Verfall || recession * *Rezession, (Konjunktur)Rückgang* || regression * *Rückbildung, Rückentwicklung, Rückschritt* || downward movement * *Abwärtsbewegung* || return * *Rückweg* || retrogression * *Rückwärtsgehen* || downturn * *des Geschäfts, der Wirtschaft* || reduction * *Reduzierung*
rückgängig machen undo * *z.B. Computereingabe* || cancel * *z.B. Auftrag* || annul * *z.B. Vertrag* || break off * *abbrechen*
rückgewinnen recover || regenerate || regain * *wiedergewinnen (z.B. Information)* || retrieve * *wiederfinden, wiederbekommen, (sich etwas) zurückholen, (Fehler usw.) wiedergutmachen*
Rückgewinnung recovery || retrieval || recycling * *of raw materials: von Rohstoffen*
ruckhaft jerky || cogging * *unrund/ruckhaft laufend*
Rückholfeder return spring
Rückkopplung feedback || feedback loop * *Rückkopplungsschleife/zweig*
Rückkopplungskreis feedback loop
Rückkühlanlage re-cooling system * *z.B. bei wassergekühltem Stromrichter*
Rücklauf reverse running * *Lauf in Rückwärtsrichtung* || running in reverse direction * *Lauf in umgekehrte Richtung* || ccw operation * *conter-clockwise: entgegen dem Uhrzeigersinn, Linkslauf* || return motion || reverse movement * *auch "reversing, reverse operation"* || reverse action || backward movement || ramping-down * *along a ramp: entlang einer Hochlaufgeber-Rampe* || braking * *Bremsen* || deceleration * *Verzögerung* || slowing down * *Verringerung der Geschwindigkeit* || return * *Umkehr* || reverse stroke * *(Kolben-) Hub in Rückrichtung* || rewind * *Zurückspulen/Wickeln* || return flow * *Flüssigkeit* || recirculation * *Flüssigkeit, Gas* || flyback * *Zeilenrücklauf des Strahls auf Monitor, Bildschirm* || carriage return * *Zeilenrücklauf bei Drucker, Schreibmaschine* || back driving * *Zurückgedrückt-Werden bei Wegbleiben des Antriebsmoments*
Rücklaufeinrichtung reverse conveyor * *[Fördertechnik]*
Rücklauframpe deceleration ramp
Rücklaufsperre backstop || reverse running prevention || rollback lock || sprag clutch * *Sperr-/Spreizzahnkupplung, verhindert Rücklaufen des Antriebs (ähnl. Freilauf)* || backstopping clutch * *Kupplung, die ein Rücklaufen des Antriebs verhindert* || back stop || pawl stop * *durch Sperrklinke/Spreizhaken realisiert* || reverse blocking || anti-reversing device * *als Einrichtung* || rollback-stop * *mechanische* ~ || non-reverse ratchet * *Maschinenlement, das m. Kraft- oder Formschluss e. Drehen i.e. Richtg. verhindert*
Rücklaufzeit ramp-down time || deceleration time ||

decrease rate || decel time * *(salopp) Abk. f. 'deceleration time'*
Rückleitung return line * *für ein Signal* || return conductor || incoming conductor
rückmelden signal back || acknowledge * *Empfang bestätigen/quittieren* || report back || check back * *zur Kontrolle/Bestätigung* ~
Rückmeldesignal checkback signal || acknowledge * *Quittung, Bestätigung*
Rückmeldung acknowledge * *Quittungssignal, Qittungstelegramm* || acceptance signal * *Annahme* || reporting back * ~ *bei Rückkehr* || reply * *Antwort* || checkback * *i. Sinne einer Bestätigung* || checkback signal * *Rückmeldesignal (z.B. von Hauptschütz-Hilfskontakt)* || acknowledgement * *Quittung* || acknowledgement signal * *Quittungssignal* || check-back message
Rückruf recall * *z.b. am Telefon*
rückrufen call back * *am Telefon*
Rucksack patch * *behelfsmäßige Progammänderung durch Ansprung einer provisorischen Befehlssequenz* || binary patch * *[Computer]* || rucksack
Rückschlagventil non-return valve || check valve || nonreturn valve || back-pressure valve
Rückschritt step back || setback || regression || retrogression
Rückschwingthyristor ring-back thyristor
Rückschwingzweig ring-back arm
Rückseite back * *Hinterseite* || reverse * *andere, umgekehrte Seite* || reverse side * *andere, umgekehrte Seite* || rear panel * *Platte/Wand* || reverse side * *of this sheet: dieses (Papier)Blattes* || rear * *on/at the rear: auf der* ~ || backing * *rückwärtige Schicht/Verstärkung, Futter, Stützung*
rückseitig rear-side || rear
Rücksendung return || redelivery
rücksetzen reset || cancel * *Löschen, Wegnehmen eines Kommandos usw.* || unlatch * *ein in Selbsthaltung befindliches Relais, z.B. in Kontaktplandarstellg. bei SPS*
Rücksetzimpuls reset pulse
Rücksicht respect * *with respect to: hinsichtlich* || regard * *with regard to: mit* ~ *auf* || consideration * *in consideration of/to: mit* ~ *auf, in Anbetracht von, hinsichtlich*
Rückspeisediode reverse diode
Rückspeiseeinheit feedback unit || regenerating unit || energy-recovery unit
rückspeisefähig regenerative * *generatorischer Betrieb möglich*
rückspeisen feed back || regenerate * *z.B. (Brems-)Energie ins Netz* || recover * *wiedergewinnen, z.B. Bremsenergie*
Rückspeisung regeneration * *z.B. von Bremsenergie ins Netz* || regenerative feedback * *z.B. von Bremsenergie ins Netz* || feeding back || regenerative braking * *von Bremsenergie ins Netz* || feedback
Rücksprache consultation * *bei Fachmann; consultation with: nach* ~ *mit; consult s.o.: m. jdm.* ~ *nehmen*
Rücksprung return * *in e. Programm*
Rückstand residue * *Rest* || sediment * *Bodensatz* || backlog * *Liefer-/Arbeits~*
Rückstände arrears * *Schulden* || outstanding debts * *Sculden*
Rückstau back-pressure || hold-up * *im Verkehr* || bank-up * *im Verkehr*
Rückstaudruck back pressure
Rückstellfeder return spring || restoring spring
Rückstellknopf reset button || resetting button
Rückstellkraft reaction force || restoring force || reestablishing force || resiliency * *durch Federwirkung* || elastic force * *elastische* || return pressure * *Rückdruck, resetting force*

Rückstrom reverse current
Rücktransformation inverse transform
Rücktritt vom Vertrag withdrawal from the contract || cancellation of order * *Zurücknahme einer Bestellung*
rückübersetzen disassemble * *von Assembler-/Maschinenspracheprogrammen* || recompile * *von Hochsprachprogrammen* || decompile * *von Hochsprachprogrammen*
Rückübersetzer disassembler * *für Assembler-/Maschinensprachprogramme* || recompiler * *für Hochsprachprogramme* || decompiler * *für Hochsprachprogramme*
Rückübersetzung recompilation * *von Hochsprachprogrammen* || disassembling * *von Maschinensprache-/Assemblerprogrammen* || decompilation * *von Hochsprachprogrammen* || reverse translation
Rückwand back panel || rear-panel * *z.B. eines Schaltschranks, Geräts usw.* || back wall * *(Holz| Bautechnik)*
Rückwandbus backplane bus
Rückwandleiterplatte backplane
Rückwandplatine backplane || backplane assembly
Rückwandstecker rear panel connector
Rückwandverdrahtung backplane wiring * *eines Baugruppenträgers* || rear panel wiring * *z.B. eines Baugruppenträgers*
Rückware returned product || returned good * *z.B. zur Reparatur/Gutschrift rückgesandtes Teil* || product returned
rückwärtig rear * *auf der Rückseite (befindlich)*
rückwärts reverse * *Drehbewegung* || backward || ccw * *counter-clockwise: entgegen dem Uhrzeigersinn* || rev * *Abkürzung für 'reverse'* || back * *zurück*
rückwärts rollen roll down * *auf dem Bildschirm*
Rückwärtsdrehmoment reverse torque
Rückwärtsdrehrichtung reverse direction of rotation || reverse rotation || ccw speed direction * *Linksdrehrichtung* || reverse speed direction
Rückwärtsimpulse down pulses
Rückwärtskompatibilität backward-compatibility
Rückwärtsrichtung reverse direction
Rückwärtszähler down counter
Rückwasser back water || white water * *Siebwasser in Papiermaschine*
rückwirken retroact
Rückwirkung reaction || retroaction || repercussion * *Auswirkung* || feedback * *Rückkopplung* || effect || consequence * *Auswirkung* || retroactive effect * *zurückwirkender Effekt*
rückwirkungsfrei non-interacting || retroaction-free || reactive-free
Ruhekontakt normally closed contact || NC contact * *Abkürzung für 'Normally Closed contact'* || break contact * *Öffnerkontakt* || B contact * *Abkürzg. f. 'Break contact': Öffnerkontakt* || form B contact
Ruhelage neutral position * *z.B. einer Tänzerwalze*
Ruhepause rest period * *auch in einem Lastspiel*
Ruhestellung neutral position || unloaded position * *ohne Belastung* || normal position * *auch eines Relais* || inoperative position || idle position || zero position || home position || position of rest * *z.B. bei einem Schütz*
Ruhestrom closed-circuit current * *z.B. Bremse, Magnetventil* || electro-release * *'elektrische Lüftung' bei Ruhestrombremse (zum Lüften Strom erforderlich)* || quiescent current * *[kwai'esnt] bei Halbleitern* || bias current * *z.B. bei Operationsverstärker; input bias current: Eingangs~*
Ruhestrombremse electro-release brake * *Bremse, zu deren Lüftung elektr.Strom erforderlich ist* || fail-safe brake * *zum Lüften Strom erforderlich →Bremse fällt bei Stromausfall ein*

Ruhestromhaltebremse fail-safe holding brake
Ruhestromprinzip closed-circuit current principle * *z.B. Bremse: zum Lüften Strom erforderlich* || fail-safe principle * *z.B. b. Bremse: b. Stromausfall Bremse geschlossen; according to the...: nach d.* || closed-circuit principle * *elektromechan. Gerät geht in sicheren Zustand wenn kein Steuerstrom fließt*
ruhig smooth * *weich, zügig, reibungslos, geschmeidig, geräuscharm z.B.Umrichter m. hoher Taktfrequ.* || silent * *ruhig, geräuschlos* || quiet * *still* || cool * *gelassen* || cool-headed * *gelassen* || nerveless * *nervenstark* || even-tempered * *leidenschaftslos* || leisurely * *gemächlich (auch Adv.)* || at rest * *ausruhend* || anti-noise * *schallgedämpft* || low * *kaum hörbar* || soft * *(seiden)weich, sanft, gedämpft*
ruhiger Lauf smooth running * *z.B. eines Motors* || quiet running * *z.B. Motor* || silent running
rühren stir || agitate || mix * *mischen*
Rührer mixer
Rührwerk mixer || stirrer || agitator
rund in round figures * *(Adv.) in Zusammenhang mit Zahl* || about * *ungefähr* || approximately * *(Adv.) in Zusamm.hang mit Zahl; ungefähr (approx.: Abkürz.f. 'approximately')* || round * *abgerundet (auch Zahl), rund geformt* || smooth * *ruckfrei, geschmeidig, weich (z.B. Lauf einer Maschine/eines Motors usw.)* || true * *(ruckfreier) Lauf eines Motors ohne Rundlauffehler/Momentenwelligkeit* || circular * *(kreis-) rund, kreisförmig, Rund-, Kreis-, Ring-* || cylindrical * *zylinderförmig*
rund um die Uhr around the clock || 24 hours a day
Rund- rotary * *drehbar, sich drehend* || circular || round * *z.B. round iron: ~eisen*
Rundachse rotary axis
Rundeisen round iron || iron rod || round steel
runden round off * *z.B. gebrochene Zahlen auf/abrunden; to: auf* || round up * *aufrunden, to: auf* || round down * *abrunden, to: auf*
runderneuern rejuvenate * *(figürl.)* || retreat * *(Reifen)*
Rundfeuer heavy flashover || flashover || heavy brush flashover * →*Bürstenrundfeuer*
Rundgang tour
Rundholz logs || round timber || timber
Rundholzförderer log conveyor || timber conveyor
Rundholzplatz roundwood yard * *(Holz) in Sägewerk, Papier-/Zellstofffabrik usw.* || timber yard * *(Holz) im Sägewerk, Papier-/Zellstofffabrik* || round log merchandizing plant
Rundkabel round cable
Rundkopfschraube round-head srew
Rundlauf smooth running * *ruhiger/guter/weicher Lauf (mit kleiner Drehmomentwelligkeit)* || true * *running* || concentricity * *Konzentrität, ~fehler* || axial eccentricity * *Planlauf-/~fehler* || smooth running characteristics * *~eigenschaften* || smooth operation * *geschmeidiger Lauf, z.B. eines Motors* || shaft run-out * *(el. Masch) Komzentrizität, Wellenendes bezogen auf d.Flansch (DIN 42955, IEC 72-1)*
Rundlaufabweichung concentricity error || eccentricity || radial run-out || radial runout
Rundlaufeigenschaft smooth running characteristics
Rundlaufeigenschaften smooth running properties || smooth running characteristics
Rundlauffehler concentricity error || eccentricity || radial runout || radial eccentricity
Rundlaufgenauigkeit concentricity * *Konzentrizität, z.B. von Wellenende und Flansch bei Motor* || rotational accuracy
Rundlaufgüte smooth-running properties ||

true running characteristics * *Gleichmäßigkeit der Drehbewegung, besonders bei kleinen Drehzahlen*
Rundlauftoleranz radial eccentricity tolerance * *erlaubter Konzentrizitätsfehler z.b. bei Motorwelle (siehe DIN 42955)*
Rundlaufverhalten smooth running characteristics || true running
Rundleiter round core * *{Kabel}*
Rundleitung round cable
Rundmotor round-frame motor
Rundschälmaschine rotary cutting machine * *{Holz}* || wood peeling machine * für Baumstämme || debarker * *{Holz} Entrindungsmaschine*
Rundschälmesser peeling knife * *{Holz} z.B. zur Furniererzeugung*
Rundschleifmaschine cylindrical grinding machine * *{Werkz.masch.}* external: *Außen~*, internal: *Innen~* || rotary grinding machine * *{Werkz.masch.}*
Rundschneiden round corner cutting
Rundschreiben circular letter || memo
Rundstab round bar || rod || rounds || round rod
Rundstabläufer round-bar rotor * *{el. Masch.}*
Rundstecker circular connector
Rundsteckverbinder circular connector
Rundsteueranlage ripple control system
Rundsteuerempfänger ripple control receiver
Rundsteuersender ripple control transmitter
Rundsteuertelegramm ripple control pulse train || ripple telecontrol signal
Rundsteuerung power line carrier control || centralized ripple control || ripple control system * *Rundsteueranlage* || ripple control
Rundstrickmaschine circular knitting machine
Rundtisch rotary table * *Drehtisch, z.B. bei Förder-/Verpackungseinrichtung, Werkzeugmaschine*
Rundtisch-Fräsmaschine circular-table milling machine || rotary-table milling machine
Rundtisch-Wickler rotary table winder
Rundung rounding * *auch mathemath.* || rounding off * *(Zahl) Auf/Abrunden (nach oben oder unten)* || rounding down * *(Zahl) Abrunden (nach unten)* || rounding up * *(Zahl) Aufrunden (nach oben)* || roundness * *abgerundete Stelle* || curve * *abgerundete Stelle* || swelling * *Wölbung, Ausbauchung, Beule* || fillet * *Kehlnaht, Hohlkehle; (Fuß-) Ausrundung*
Rundungsfehler rounding off error || truncation error * *bei Rechnerprogramm*
Ruß carbon black || soot || lampblack
Rush-Effekt inrush || inrush effect
Rush-Moment inrush torque * *Momentenspitze, d.beim Einschalten u.Drehrichtungsumkehr e. Motors auftritt* || peak inrush torque * *Momentenspitze z.B. bei Motorstart u. Drehrichtungsumkehr* || torque inrush * *Momentenüberhöhung bei Start u. Drehrichtungsumkehr eines Motors* || rush torque * *Stoßmoment*
Rush-Strom inrush current * *Stromspitze beim Aufbau eines Feldes (z.B. bei Motor, Trafo usw.)* || peak inrush current * *Stromspitze b. Aufbau eines Feldes (z.B. Motor, Trafo usw.)* || current inrush * *Stromspitze beim Aufbau e. Feldes, z.B. bei Motor, Trafo usw.* || rush current * *Stoßstrom*
Rushstrom inrush current * *Stromspitze z.B. beim Einschalten induktiver Lasten*
Rüstzeit set-up time || setting time || equipping time || make-ready time * *z.B. bei Druckmaschine* || setting-up time
Rutenwebstuhl pile-wire loom * *{Textil} z.B. zur Teppichherstellung*
Rutschdrehmoment slipping torque * *einer Kupplung/Bremse*
Rutschdrehzahl slipping speed * *einer Kupplung*

Rutsche chute * *Rutsche; Rutsch-/Gleitbahn; Schüttröhre/-rinne; Abwurfschacht* || shoot * *Rutsche, Rutschbahn, Kipprinne* || slide
rutschen slip * *rutschen (auch Kupplung, Bremse), (aus)gleiten* || slide * *gleiten, (aus)rutschen* || glide * *(leicht) gleiten, dahingleiten*
rutschig slippery
Rutschkraft slip force
Rutschkupplung slipping clutch || friction clutch || safety slipping clutch * *Sicherheits-~ zur Drehmomentbegrenzung* || slipper clutch || torque-overload clutch * *rückt bei Überschreiten d. Nenndrehmoments aus (dient als Überlastungsschutz)*
Rutschmoment slipping torque * *bei einer Kupplung/Bremse*
Rutschzeit slipping time * *z.B. einer Kupplung, Bremse* || slipping period * *einer Kupplung/Bremse*
Rüttelbeanspruchung vibration stress
Rüttelbelastung vibration stressing
rüttelfest vibration-resistant || vibration-safe * *rüttelsicher*
Rüttelfestigkeit vibration resistance || vibrostability
rütteln vibrate || shake || rock
rüttelsicher vibration-safe || vibration resistant * *rüttelfest*
Rütteltisch jogging table || jolting table || vibrating table || vibration table * *auch zur Rüttelprüfung*
Rütteltopf vibration feeder * *{Verpack.techn.}*
Rüttler vibrator || jolter * *stoßender ~* || jar-ram * *z.B. zur Bodenfestigung* || jolt || ramming machine * *Stoßramme* || jarring machine * *schwirrend rüttelnd* || vibrating machine * *vibrierend* || jolt-ramming machine * *Stoßramme*
Rüttlermotor vibration motor

S

S-Bahn suburban railway || city-railway || metropolitan railway * *in einer Millionenstadt/Hauptstadt*
S-förmig S-shaped * *z.B. Hochlaufgeber-Rampe (mit Verrundung)*
S-Kurve S-curve * *abgerundete Hochlaufgeber-Rampe*
S-Motor S-motor * *energiesparender SIEMENS-Sondermotor mit hohem Wirkungsgrad*
S-Rolle S-roll * *Walzen mit großer Umschlingung, erzwingen konstante Geschwind.* || bridle roll * *z.B. i.Blechstraße: Rolle mit großer Umschlingung, erzwingt konstante Geschwind.*
S-Walze S-roll * *1 von 2 S-förmig angeordneten Walzen m.großer Umschlingung z.Fördern bandförmig. Materials* || bridle roll * *z.B. in Walzstraße: Rollen mit großer Umschlingg., erzwingen konstante Geschwind.* || pull roll * *Zugwalze*
S1-Betrieb continuous operation * *→Betriebsart "Dauerbetrieb"*
Sachbearbeiter official in charge || responsible person || contact person * *Ansprechpartner*
Sachgebiet field || subject
sachgemäß proper || appropriate || adequate || suitable || pertinent || correct || professionally * *(Adv.)*
sachgerecht proper || appropriate || pertinent || correct
Sachkenntnis expert knowledge || special knowledge || experience * *Erfahrung* || expertise * *[expe'ties] Expertenwissen, Fachwissen* || know-how * *das "Gewusst-Wie"*
sachlich real || realistic * *nüchtern* || unbiased * *unparteiisch* || impartial * *unparteiisch* || objective * *objektiv, unparteiisch* || functional * *zweckbetont*

Sachnummer 262

(*z.B. Design*) || factual * *sachbezogen* || practical-minded * *praktisch denken* || dispassional * *leidenschaftslos* || technical * *(rein) technisch* || unbiased * *unparteiisch* (*mit einem oder zwei "s" richtig*)
Sachnummer item No. * *z.B. in Werks-Lagerliste* || subject index
Sachschaden property damage || material damage
Sachverhalt facts || facts of the case || circumstances
Sachvermögen tangible property
Sachverstand expertise * *[expe'ties]* || know-how * *Gewusst-wie* || competence * *Kompetenz, Befähigung, Fähigkeit, Tüchtigkeit* || competency * *Kompetenz, Befähigung*
Sachverständiger expert || specialist || authority * *in/on: für*
Sack sack * *auch 'Beutel, Tüte'*
Sackbohrung blind hole || tapped blind hole * *Gewindesackloch*
Sackloch blind hole || tapped blind hole * *Gewinde-Sackloch* || hollow shaft bore * *in einer Welle* || blind hollow bore
Säge saw
Sägeblatt saw blade || saw web || saw-blade
Sägemaschine sawing machine
Sägemehl sawdust * *[Holz]*
sägen saw
Sägenfeilmaschine saw filer * *[Holz] zum Schärfen von Sägen*
Sägewagen saw carriage * *[Holz]*
Sägewerk sawmill
Sägezahn saw-tooth
Sägezahngenerator sawtooth generator * *erzeugt sägezahnförmige Dreieckimpulse*
Sägezahnspannung saw-tooth voltage
Saitenschneider snippers || cutting pliers || diagonal cutter
salzhaltige luft salt-laden air || salt air
Salzsäure hydrochloric acid || muriatic acid
Samarium-Kobalt Samarium-Cobalt || SmCo
Samarium-Kobalt-Magnet Samarium-Cobalt magnet
Sammel- group * *z.B. Warnung, Meldung* || general * *z.B. Katalog* || collective || multiple
Sammelalarm group alarm * *[SPS|Computer]* || combined alarm * *[SPS|COMPUTER]*
Sammelbehälter storage tank * *für Flüssigkeiten*
Sammelerder earthing bus
Sammelkatalog general catalog
Sammelmeldung group signal
sammeln collect || gather * *(an)~, anhäufen, (Personen) ver~; anziehen, gewinnen, erwerben, (auf)-lesen, ernten* || heap up * *an-/aufhäufen* || hoard * *horten, sammeln. anhäufen, hamstern; hoard up: aufhäufen/an~* || concentrate * , *sich (an)sammeln, (sich) konzentrieren; verdichten, vereinigen* || amass * *in Massen ~*
Sammelpackung group package
Sammelpunkt central point
Sammelschiene bus bar || bus-bar || bus || steam mains * *zur Dampfversorgung*
Sammelsignal group signal
Sammelstörmeldung group fault signal || collective fault signal
Sammlung compilation * *Zusammenstellung, Zusammengetragenes* || collection * *Gesammeltes, Zusammengestelltes, Auswahl*
samt plus || including
sämtliche all || all together || complete * *vollständig* || whole * *ganz* || entire * *ingesamt* || all of them || in a body * *zusammen, wie ein Mann*
Sandkapselung powder filling * *Schutzart 'q' nach EN50017 für Ex-geschützte el. Betriebsmittel*
Sandpapier sandpaper || abrasive paper || emery paper * *Schmirgelpapier*

Sandstrahlen sand blasting
Sandstrahlgebläse sandblast unit || sand-blasting machine || sandblaster || sand-blasting blower || sandblasting machine || sandblast blower
Sandstrahlmaschine sandblasting machine
Sanduhr hourglass * *auch auf dem Bildschirm*
sanft smooth * *Bewegung, Lauf eines Maschine/eines Motors usw.* || gentle || soft
sanftanlassen soft-start * *Motor*
Sanftanlasser soft starter
Sanftanlauf soft starting * *z.B.für Asynchronmotor* || softstart || smooth starting * *sanfter Anlauf* || softtorque starting * *[el.Masch.] Anlauf v.Käfigläufermotoren mit Anlasswiderstand o. Sanftanl.gerät*
Sanftanlauf-Schaltung soft-starting circuit
Sanftanlaufgerät reduced voltage starter * *für Drehstrommotoren* || soft starting unit || soft starter * *[el. Masch.] Leistgs.steller, d.durch Spannungsverringerg. e.Sanftanlauf ermöglicht*
Sanftauslauf soft braking * *z.B. mit Hilfe eines Dreh-/Wechselstromstellers* || soft-stop || soft stopping * *eines Motors z.B. mit Drehstromsteller*
sanfter Anlauf smooth starting
Satellitenrad satellite gear * *bei Planetengetriebe*
Satellitenrollspindel planetary-roller screw
Satellitenzahnrad satellite gear * *bei Planetengetriebe*
Satinierkalander web-glazing calender * *[Papier] zum Zusammenpressen und glatt/glänzend machen von Papier* || supercalender * *[Papier] Presse zur Herstellung hochkalandrierten Glanz-Papiers*
satt anliegend tight-fitting || snug || resting snugly * *against: an*
Satt- saturated
Sattdampfkessel saturated steam boiler
Sattellager bracket-mounted bearing || cradle-type bearing
Sattelmoment pull-up torque * *[el.Masch.] bei Käfigläufermotor (siehe VDE 0530 Teil 12)*
sättigbar saturable * *induktives Bauteil*
sättigen saturate
Sättigung saturation
Sättigungsdrossel saturable-core reactor
Satz set * *~ von Teilen* || bank * *~ von Bauteilen* || kit * *Bau~, Werkzeug~* || record * data record: Daten~ (auch Verfahr~ bei NC-Steuerung)* || block * *z.B. Daten~* || sentence * *~ aus Worten* || phrase * *Redewendung, Ausdruck, Phrase, kurzer ~* || theorem * *Lehr~, Theorem* || proposition * *Lehr~* || thesis * *These, Behauptung, Streit~* || tenet * *Grund~, Lehr~* || maxim * *Leit~, Maxime* || principle * *Prinzip* || batch * *Schub* || nest * *ineinanderpassende Gegenstände* || sediment * *Boden~* || jump * *Sprung* || law * *Lehr~, naturwissenschaftliches Gesetz* || type setting * *[Druckerei]* || block * →*[NC-Steuerung] Verfahrdaten~ bei NC (numerischer Positioniersteuerung)*
Satz fahren block mode * *bei Numeriksteuerung, Positioniersteuerung*
Satz von Fourier law of Fourier
Satzausblendung block skip * *[NC-Steuerung] Überlesen/Unterdrücken eines Verfahrsatzes*
Satzwechsel block transition * *[NC-Steuerung]*
sauber clean || cleanly * *(Adv.)* || neat * *ordentlich, aufgeräumt, klar, übersichtlich, geschickt (Arbeit, Äußeres, Handschrift usw.)* || tidy * *reinlich, ordentlich* || pretty * *hübsch* || fine * *fein* || nice * *hübsch, nett* || slick * *(salopp) flott, raffiniert, geschickt, pfiffig* || pure * *rein, nicht verunreigt* || unpolluted * *nicht verunreinigt*
Sauberkeit cleanliness
säubern clean || cleanse || tidy * *Zimmer usw. in Ordnung bringen, säubern, richten; tidy up: aufräumen* || clean up || clear * *freimachen (of: von)*
Sauerstoff oxygen

Saug- suction
Saugdrossel interphase transformer * *mit Mittelpunktanzapfung zur Stromssymmetrierung für 12-Puls-Stromrichter* || interbridge reactor * *{Stromrichter} zur Stromsymmetrierg. bei 12-pulsiger Stromrichterschaltung* || saturable reactor * *{Stromrichter}* || balance coil * *{Elektr.}*
Saugdrosselschaltung interphase transformer connection
Saugfähigkeit absorptive strength || absorptive capacity || absorbency
Saugfilter suction filter * *(mechan.)*
Saugfuß vacuum cup
Saughöhe capillary rise * *durch Kapillarwirkung* || suction lift * *einer Pumpe*
Saugkasten suction box * *in Siebpartie einer Papiermaschine*
Saugkreis harmonic absorber * *Filterkreis zur Oberschwingungskompensation* || acceptor circuit || series resonant circuit * *Serienresonanzkreis*
Saugleitung suction line * *z.B. eines Kompressors*
Saugnapf suction cup
Saugschaltung harmonic absorber * *Filterkreis zur Oberschwingungskompensation*
Saugwalze suction roll * *z.B. in der Siebpartie einer Papiermaschine* || suction couch roll * *Siebsaugwalze in der Siebpartie e. Papiermaschine*
Saugzuggebläse induced-draft fan * *in der Kraftwerkstechnik*
Saugzuglüfter induced-draft fan
Säule pillar * *(auch figürlich)* || column
Säulen- column-type
Säulenbohrmaschine pillar-type drilling machine || column-type drilling machine
Saumschere trimming shear * *zum seitlichen Abschneiden des Walzbandes in Blechstraße*
Säure acid * *(chem.)* || sourness
säurebeständig acid-resistant
Säurekonzentration acid concentration
Säurespritzer acid spills
schaben scrape || grate || rasp || scratch * *kratzen* || abrade * *abschaben* || rub * *abrubbeln* || shave * *dünne Späne abschälen, z.B. mit Schabeisen/Werkzeugmaschine*
Schaber scraper
Schablone template * *Bohr/Schneid/Gußschablone* || jig * *Bohrschablone* || stencil * *Zeichen/Malschablone* || model * *Modell, Muster* || pattern * *Modell, Muster*
Schablonendrehmaschine copying lathe * *{Holz}*
Schacht shaft || pit * *im Bergwerk* || well * *für Licht usw.* || vertical raceway * *senkrechter (Kabel)-Schacht* || vertical trunking * *senkrechter (Kabel)-Schacht* || elevator shaft * *→Aufzugsschacht*
Schachtel box || case || carton * *Papp~* || cardboard box * *Papp~* || folding-box * *Falt~*
schachteln nest * *Unterprogramme, Interrupts usw.* || interleave * *Blechpaket, Wicklung*
Schachteltiefe nesting depth * *z.B. von Unterprogrammen, Interrupts* || nesting level * *Schachtelungsebene* || level of program nesting * *Schachtelungsebene von Unterprogrammen*
Schachtelung nesting * *z.B. von Unterprogrammen, Interrupts* || interleaving * *Blechpaket, Wicklung*
Schachtelungstiefe nesting depth || nesting level
Schachtförderanlage mine hoist || mine winder * *Förderaufzug* || mining hoist
Schachtförderer mine winder
Schachtlüftermotor duct-fan motor * *{el. Masch.}*
Schaden breakdown * *zum Maschinenstillstand führender ~* || damage * *Beschädigung* || loss * *Verlust, auch finanziell* || injury * *Beschädigung (besonders von Personen), Schaden (to: jemandes)* || detriment * *Schaden, Nachteil* || prejudice * *Nachteil, Schaden, Beeinträchtigung*

Schaden erleiden incur damage || suffer injury || be damaged || sustain injury || get defective * *defekt werden*
Schadenanalyse breakdown analysis
Schadenersatz damages * *Geldsumme* || compensation || indemnity || indemnification || payment of damages * *Schadenersatzzahlung;* liable to payment of damages: *schadenersatzpflichtig*
Schadenersatz leisten pay damages
Schadenersatz verlangen claim damages
Schadenersatzanspruch claim for damages || liability claim || liability claims * *(Plural)* || claims * *(Plural)*
Schadenersatzforderung claim of damages
schadenersatzpflichtig liable for damages || liable to payment of damages
Schadenersatzzahlung payment of damages
Schadensbericht damage report
Schadensersatzanspruch claim for damages || liability claims * *(Plural)* || liability claim || claims * *(Plural)*
Schadensersatzansprüche claims
Schadensersatzzahlung payment of damages
Schadensverhütung damage prevention
schadhaft damaged * *beschädigt* || defective * *defekt, kaputt* || faulty * *fehlerhaft* || leaking * *Rohr, Gefäß mit Leck*
schädigen damage || impair || affect * *auch Rechte, Ruf usw.* || hurt * *verletzen* || prejudice * *benachteiligen* || weaken * *schwächen*
schädigende Einwirkung harmful effects
Schädigung damage * *to: von etwas* || impairment * *of: von etwas* || detriment * *to: von etwas* || weakening * *→Schwächung, →Abschwächung*
schädlich harmful * *nachteilig* || injurious * *für eine Person; nachteilig (to: für), beleidigend, verletzend* || noxious * *gesundheitsschädlich* || detrimental * *nachteilig* || prejudicial * *nachteilig* || disadvantageous * *nachteilig* || bad * *schlecht* || parasitic * *parasitär, störend* || dangerous * *gefährlich, gefahrvoll, bedenklich*
Schadstoff aggressive substance || harmful substance || pollutant * *in Luft, Flüssen usw.* || contaminant
Schadstoffe aggressive substances
Schaft shank * *einer Schraube, eines Werkzeugs, eines Schlüssels usw.* || shaft * *einer Säule usw.* || stem * *(Ventil-) Schaft, Stiel, Stängel, Stamm, Zapfen*
Schaftfräser shank cutter * *{Werkz.masch.}*
Schale shell || tray * *flache Schale, Tablett, Ablagekasten, Einsatz (i. Koffer usw.)*
Schalengreifer shell-type grab * *{Hebezeuge} bei Greiferkran*
Schalenkupplung muff coupling
Schall sound
Schall-Leistungspegel sound power level * *z.B. für el. Maschinen: siehe DIN VDE 0530 und DIN 45635* || noise power level
Schall-Reflexion sound reflection
Schallausbreitung sound propagation
Schalldämmelement sound-damping element || sound-proofing element
Schalldämmhaube noise-insulating cover || noise-reducing cover
schalldämpfend sound-absorbing || sound-deadening || sound-insulating
Schalldämpfer sound absorber || silencer || muffler * *z.B. für Verbrennungsmotor, Kompressor usw.*
Schalldämpfung sound absorption || sound insulation || sound attenuation || silencing || muffling * *(amerikan.)* || sound dampening || noise dampening
schalldicht sound-proof

Schalldruck sound pressure * *Effektivwert des Schallwechseldrucks in [N/m²]*
Schalldruckpegel sound pressure level * *[dB (A)] Messung: DIN 45635, EN 50144; Motorgeräusch: DIN VDE 0530, DIN 21680* || sound power level * *Schallleistungspegel, siehe z.B. EN 50144*
Schalleistungspegel sound power level * *ergibt s. aus Schalldruckpegel u.Messfläche; für El.motoren: siehe DIN VDE 0530*
schallgedämpft noise-reduced || sound-proofed || sound-insulated || sound-absorbing * *schalldämpfend*
schallisoliert sound-proof
Schallisolierung silencing * *als Funktion* || noise reduction * *Geräuschreduzierung* || noise insulation
Schallleistungspegel sound pressure level * *siehe z.B. EN 50144* || sound power level * *siehe z.B. EN 50144* || noise power level
Schallmessung noise measurement * *siehe z.B. DIN IEC 651* || noise level measurement * *Schallpegelmessung*
Schallpegel sound level || noise level * *Geräuschpegel*
Schallpegelmesser sound level meter || phonometer
Schallquelle sound source
Schallreflexion sound reflection
Schallschutz noise protection || sound protection
Schallschutzeinrichtung noise-reduction equipment
Schallübertragung sound transmission
Schälmaschine peeling machine * *z.B, für Baumstämme* || rotary cutting machine * *Rundschälmaschine* || debarker * *[Holz] Entrindungsmaschine* || veneer cutting machine * *[Holz] Rundschälmaschine zur Erzeugung von Furnier* || peeling lathe * *[Holz] zum Erzeugen von Furnier*
Schälmesser peeling knife * *[Holz]*
Schalt- switching || switch-mode * *z.B. -Netzteil*
Schalt- und Wendegetriebe gear-change and reversing gearbox
Schaltalgebra boolean algebra || logic algebra
Schaltanlage switchplant || switching station || switchgear system
Schaltanlagenbau switchgear manufacture
Schaltarbeit switched energy * *bei Kupplung, Bremse*
Schaltart winding connection * *Schaltgruppe z.B. Stern/Dreieck eines Motors/Trafos*
schaltbares Drehmoment engagement torque * *für Kupplung, Bremse (ist niedriger als das →übertragbare Drehmoment)*
Schaltbefehl switching signal || switching command
Schaltbetrieb intermittent periodic duty * *siehe auch 'Aussetzbetrieb' (Betriebsart für Motor/Stromrichter)* || intermittent service * *Aussetzbetrieb* || intermittent operation * *Aussetzbetrieb* || periodic duty * *sich wiederholendes Lastspiel, siehe auch 'Aussetzbetrieb'* || switching mode * *bei Wechselstromsteller* || intermittent duty
Schaltbild circuit diagram
Schaltbuch circuit manual * *auch für eine softwaremäßige Verdrahtung, z.B. in einem Regelsystem* || wiring manual
Schaltbügel terminal link * *Leiterstück zur Überbrückg.v. Klemmen z.B. KLemmer z.B.* || terminal jumper * *Klemmenbrücke* || terminal clip * *→Klemmbügel* || terminal link * *Klemmenbrücke z.B. in (Motor)Klemmenkasten z.Herstellung d. gewünschten Schaltgruppe*
Schaltcharakteristik switching characteristic || tripping characteristic * *z.B. eines Auslösegeräts/*

Überwachungsrelais || switching performance * *Schaltverhalten*
Schaltdiode switching diode
schalten switch || change * *Getriebe* || shift * *Getriebe* || connect * *Verbindg. herstell.; in series/parallel/star/delta: in Reihe/parallel-/Stern-/Dreieck*
schalten auf switch-through to * *ein Signal führen auf* || feed to * *ein Signal führen auf*
Schalter switch || circuit breaker * *Leistungsschalter*
Schalter-Sicherungseinheit switch-fuse unit
Schalteranbau built-on contactor * *[el.Masch.] Anbau des Motorschützes direkt an den Motor (statt Klemmenkasten)*
Schaltergehäuse switch enclosure
Schalterkonstante circuit breaker constant * *eines Leistungsschalters*
Schalterstellung switch position || breaker position * *eines Leistungsschalters*
schaltfest surge-proof * *z.B. Kondensator mit hoher Spitzenstrombelastbarkeit* || resistant to switching transients
Schaltfestigkeit dielectric strength * *von Kondensatoren (siehe auch →schaltfest)*
Schaltfläche button * *in einem Dialogfeld auf Computermonitor (zum Anklicken mit der Maus)*
Schaltflanke switching edge * *(extremely steep: extrem steil)*
Schaltfolge operating sequence || switching sequence
Schaltfrequenz switching frequency * *z.B. für PWM-Ansteuerung eines Leistungsteils* || carrier frequency * *z.B. für Pulsumrichter* || operating frequency * *von Lastspielen/Schaltspielen, siehe EN50032 und DIN IEC147-1D*
Schaltgabel striker fork * *z.B. für mechan. betätigte Kupplung* || striker clutch || shift fork || shift claw
Schaltgenauigkeit operating accuracy * *z.B. einer Kupplung/Bremse* || engagement accuracy * *einer Kupplung/Bremse*
Schaltgerät switchgear || switch device
Schaltgeräte switchgear || controlgear || switchgear and controlgear || switchgear units
Schaltgeräusch switching noise * *einer Bremse/Kupplung*
Schaltgerüst switchgear frame
Schaltgetriebe gear-change gearbox * *siehe auch →Wechselgetriebe* || change-speed gear || control gear || multispeed gearbox || multiple-speed gearbox || multi-speed gearbox || change-over gear
Schaltgruppe winding connection * *Verschaltung der Wicklungen, z.B. Stern/Dreieck bei Motor/Trafo* || vector group * *z.B. eines Transformators* || transformer connection * *eines Transformators*
Schalthandlung switching operation || switching action
Schalthäufigkeit operating frequency * *bei einem Last/Schalt/Betätigungsspiel* || switching rate * *~ von Kontakten* || number of startings * *bei Motor; higher: erhöht* || frequency of operation || engagement frequency * *einer Kupplung/Bremse* || starting frequency * *[el.Masch.] Anzahl period. wiederkehrender Schalthandlungen (Motorstarts je h)*
Schalthaus switchgear house * *Schaltanlage* || electrical equipment house
Schalthebel switching lever * *bei elektr. Schalter* || contact lever * *bei elektr. Schalter* || gear-shift lever * *für Schaltgetriebe* || control lever * *Bedienungshebel*
Schalthysterese switching hysteresis
Schaltknebel rocker * *an Handschalter*
Schaltkreis electric circuit || electronic circuit * *elektronischer ~* || integrated circuit * *integrier-*

ter ~ || IC * integrierter || chip * (salopp) integrierter ~ || ASIC * kundenspezifischer integrierter ~
Schaltkreisfamilie logic family * z.b. TTL, CMOS
Schaltkreistechnik circuit technology
Schaltleistung switching capacity || breaking capacity * Aus~, Ab~ || rupturing capacity * Aus~, Ab~(auch einer Sicherung) || making capacity * Ein~ || making and breaking capacity * Ein-/Aus~ || contact rating * ~ eines (Relais-/Schütz-) Kontakts
Schaltnetz network
Schaltnetzteil switch-mode power supply || switched-mode power supply
Schaltpegel switching level || logic level * einer Logikstufe
Schaltplan circuit diagram
Schaltpult control desk
Schaltpunkt switching point || operating point || switching instant * Zeitpunkt || operating temperature * Schalt-Temperatur, z.B. eines Thermoschalters
Schaltraum switch room || control room || switchgear room || electrical equipment room
Schaltregler switching controller || switching regulator * z.R in Schaltnetzteil
Schaltring actuator ring * z.D. für mechan. betätigte Kupplung
Schaltschema basic circuit diagram || wiring diagram * Verdrahtungsplan || block diagram * Blockschaltbild || connection diagram * Anschlussplan
Schaltschrank cabinet || cubicle || control cabinet * Steuerschrank || switchgear cabinet || switchgear cubicle
Schaltschrankbau control-cabinet manufacture
Schaltschrankbauer control-cabinet builder
Schaltschranktür cubicle door
Schaltschütz contactor
Schaltspannung switching voltage || voltage switched || breaking voltage * Ab~
Schaltspiel switching cycle || operating cycle * Lastspiel, Betätigungsspiel, siehe auch EN50032 und DIN IEC 147-1D || switching operation || break operation * Abschaltung || operation * bei Relais usw.
Schaltspitze switching transient
Schaltspitzen switching spikes
Schaltstellung open-close condition * eines Schalters, Relais, Schützes (z.B. opt. angezeigte) || switch position
Schaltstrom switching current
Schaltstück contact member * Schütz, Leistungsschalter usw. || contact piece * Schütz, Leistungsschalter || contact * Kontakt
Schalttafel control board || switch board || control panel * Steuertafel || switchboard
Schalttafel-Messinstrument panel-mounting measuring instrument || panel meter
Schalttafelauschnitt panel cutout * z.B. zum Einbau eines Schalttafelinstruments
Schalttafeleinbau panel mounting || flush panel mounting * mit der Schalttafel bündig abschließende Montage
Schalttafelinstrument panel meter
Schalttafelmontage wall mounting || panel-mount * für Schalttafelmontage || panel-mounting * →Durchsteck-Montage z.B. bei/von Messgeräten
Schaltung circuit * elektrische/elektronische ~ || connection * ~ mit Leistungshalbleitern, Leistungsteil, Trafo usw. || switching * Schaltvorgang, das Schalten || circuit arrangement * Ver~, ~sanordnung || circuit configuration * z.B. Brücken~ || circuitry * Teil~, Be~ || configuration * Anordnung, Konfiguration || circuit connection * ~ für

Stromrichter (siehe DIN V 41761), Schaltgruppe für Trafo usw. || engagement * bei Kupplung, Bremse || switching action * →Schalthandlung || switching operation * Schaltvorgang (z.B. einer mechan. Kupplung; innerhalb eines Lastspiels usw.)
Schaltungsänderung circuit modification
Schaltungsaufbau circuit design
Schaltungsauslegung circuit dimensioning || circuit design
Schaltungsentwurf circuit design
Schaltungsfaktor factor of connection * {el. Masch.} →Schalthäufigkeit || duty factor
Schaltungskonzept circuit concept
Schaltungstechnik circuit technique || circuit engineering
Schaltungsunterlagen circuit diagrams || schematic diagrams || wiring diagrams * Verdrahtungspläne || wiring manual * Schaltbuch
Schaltungswinkel circuit angle * z.B. bei Transformator
Schaltungszustand switching condition || circuit condition
Schaltverhalten switching characteristic || switching characteristics
Schaltverluste switching losses || switching losses * z.B bei Thyristor, Transistor
Schaltvermögen switching capacity || breaking capacity * z.B. ~ eines Leistungsschalters || interrupting capacity * Aus~ eines Leistungsschalters || contact rating * ~ eines (Relais-/Schütz-) Kontakts || switching capability
Schaltverzögerung propagation delay time * beim Durchschalten von Signalen z.B. durch Logikgatter || propagation delay
Schaltverzugszeit switching delay time * z.B. eines Leistungsschalters
Schaltvorgang switching sequence * Schaltfolge || gear-shift operation * (mechan.) bei Schaltgetriebe || switching process || switching transient * (elektr.) auf Energieversorgungsleitung/-netz || switching operation
Schaltwarte control room || control station || control centre
Schaltwerk sequential control * Ablaufsteuerung || contact mechanism * bei elektromechan. Schalter || switch group * bei elektromechan. Schalter || sequencing control * Ablaufsteuerung || sequence control * Ablaufsteuerung || sequence processor * elektronisches Ablaufsteuerwerk || cam-contactor group * Nockenschaltwerk
Schaltwippe rocker * an Handschalter
Schaltzahl engagement frequency * →Schalthäufigkeit einer Kupplung/Bremse
Schaltzeichen graphic symbol || logic symbol * für binäres Schaltelement || graphical symbol
Schaltzeit switching time
Schaltzustand switch status || switch position || switching state || switch state
Schaltzustandsanzeige switch status indicator * z.B. an Näherungsschalter, Schütz
Schalung shutter * {Bautechnik} || form * {Bautechnik} || sheathing * {Bautechnik}
Schalungsplatte shuttering panel * {Holz} für Betonbau
Schar family * von Kurven || set * von Kurven || group * z.B. von Kurven, Fäden
schären card * {Textil} →aufrauen (von Wolle usw.)
scharf critical * Überwachungseinrichtung || abrupt * abrupt || exact * sorgfältig, genau || precise * genau || salient * hervorspringend || strict * streng || active * aktiv, z.B. Überwachungseinrichtung || crisp * Bildschirm || sharp * Messer, Zähne, Kurve, (auch figürl.: rau, herb) || sharp-pointed * spitz ||

pointed * *spitz* || acute * *scharf, spitz, spitzwinklig, heftig, stechend* || stringent * *streng, zwingend, bindend (Gesetz, (An)Forderung, Regel)*
scharf kalkuliert closely calculated
Schärfe sharpness * *auch einer Bildschirm-Wiedergabe* || crispness * *eines Bildschirms*
Schärfegrad severity * *z.b. beim Funkstörgrad, bei Schwingfestigkeitsprüfung* || severity class || severity level || test severity * *Prüfschärfe*
schärfen sharpen || edge * *Kanten* || arm * *scharf machen, z.B. Gewehr, Triggerung eines Oszilloskops* || burr || resharpen * *nach~*
scharfkantig sharp-edged
scharfmachen arm * *z.b. Triggerung eines Oszilloskops* || activate || instigate * *(figürl.) an/aufreizen, anstiften, aufhetzen (to: zu; to do: zu tun)*
Schärfmaschine sharpener || sharpening machine
Schärgeschwindigkeit warping speed * *{Textil} in Schärmaschine*
Schärmaschine warping frame * *{Textil}* || sectional warper * *{Textil}* || warping machine * *{Textil}*
Scharnier hinge * *Scharnier, Gelenk, Angel* || hinged joint * *Scharnier, Gelenk* || hinge joint * *Scharnier, Gelenk*
Schatten-RAM shadow RAM * *(schneller) Speicher, in dem eine Kopie der Daten gehalten wird*
Schattenspeicher shadow memory || shadow RAM * *z.B. in →NOVRAM*
schätzen estimate * *abschätzen* || like * *(gern) mögen* || appreciate * *würdigen, hochschätzen, zu schätzen wissen*
Schätzung estimate * *superficial/rough estimate: überschlagsmäßige ~* || valuation || estimation
Schaubild graph || chart || diagram || schematic diagram * *Übersichtsdarstellung*
Schaufel shovel * *allg.; shovel dredger: ~bagger* || scoop * *zum Schöpfen, auch bei ~bagger* || blade * *bei Turbine, Lüfter usw.* || bucket * *~ bei ~bagger; Turbine usw.; Eimer* || paddle * *z.B. bei Mixer*
Schaufelrad bucket-wheel * *Bagger*
Schaufelradantrieb bucket-wheel drive * *Bagger*
Schaufelradbagger bucket-wheel excavator
Schauglas sight-glass * *z.B. zur Ölstandskontrolle*
schaukeln sway * *sich wiegen, schwanken*
Schaum foam
Schaumbildung foaming
schäumen foam
Schaumstoff foamed plastics || foamed material
Scheibchen lamella || chip || slice
Scheibe disk || disc * *weniger gebräuchlich als "disk"* || washer * *→Unterleg~* || slice || plate || pane * *Fenster~* || dial * *Wähl~* || lamella * *Blättchen, →Lamelle* || wheel * *Schleif~ usw.* || gasket * *→Dichtungs~* || pulley * *→Riemen~* || sheave * *→Seil~* || cake * *Wachs~, Teig~, (Spanplatten-) Kuchen usw.*
Scheibenabzug pull-off capstan * *konstant angetriebene Scheibe in Kabelmaschine, an die d. Kabel angepresst wird* || capstan * *große konstant angetriebene Scheibe in Kabelmaschine, an die d. Kabel angepreßt wird*
Scheibenbremse disk brake || disc brake
Scheibenfilter disc filter || polydisc filter * *Mehrscheibenfilter*
scheibenförmig discoidal || disk-shaped || disc-shaped
Scheibengehäuse disc case * *z.B. e. Thyristors*
Scheibenlagerschild disk-type endshield * *z.B. bei Motor*
Scheibenläufer disc rotor
Scheibenläufermotor disc-rotor motor || discoidal motor || disc-armature motor || pancake motor * *(salopp)*
Scheibenrefiner disc refiner * *{Papier} zur Zellstofferzeugung*

Scheibenthyristor disk thyristor || disk-type thyristor || disc thyristor || press-pack thyristor cell
Scheibenwischer wind-screen wiper || windshield wiper
Schein- apparent * *z.b. Leistung, Strom*
scheinbar apparent * *anscheinend* || seeming * *anscheinend* || false * *vorgeblich* || fictitious * *vorgeblich, erfunden, unecht, angenommen, fiktiv* || seemingly * *(Adv.) scheinbar; es scheint, dass* || ostensible * *scheinbar, angeblich, vorgeblich* || virtual * *virtuell* || it seems that * *(Adv.) anscheinend; es scheint, dass*
Scheinleistung apparent power * *Eff.wert d. Wechselspannung x Eff.wert d. Wechselstroms in [VA] oder [kVA]* || complex power
Scheinleistungsaufnahme apparent-power consumption
Scheinstrom apparent current
Scheinwiderstand impedance * *Eff.wert der Spannung geteilt den Eff.wert des Stroms*
Scheitel- peak * *z.B. -Wert, -Spannung* || crest * *z.B. -Wert, -Spannung*
Scheitelfaktor crest factor * *Maximalwert/Effektivwert eines periodischen Signals (1.41 bei sinusförmigem Signal)* || peak factor
Scheitelpunkt vertex * *(auch mathemat.); (Plural: vertices)* || apex * *Scheitelpunkt, →Spitze, Höhepunkt (figürl.)*
Scheitelspannung crest voltage || peak voltage
Scheitelwert amplitude || peak value || crest value
Schelle clip * *zum Befestigen* || saddle * *mit zwei Befestigungslappen für Rohr* || cleat * *auch Isolierschelle* || clamp * *Klammer, (Erdungs-) Schelle, Krampe, Zwinge, Klemmschraube*
Schema scheme || diagram || pattern * *Muster, Anordnung* || model || arrangement * *Anordnung* || system * *Anordnung, System*
schematisch schematic * *[ski: 'mätik]* || schematically * *(Adv.)* || systematic * *systematisch* || diagrammatic * *grafisch, diagrammatisch, schematisch, nach Schema* || according to rule * *nach Schema, entsprechend einer Regel/Vorschrift*
schematische Darstellung schematic diagram * *siehe auch →Übersichtsschaltbild* || schematic * *[ski: 'mätik]* || schematic representation || skeleton diagram
schematisieren schematize || standardize
Schenkelfeder leg spring
Schenkelpol salient pole * *ausgeprägter, hervorstehender Pol z.B. am Läufer e.Synchronmaschine (Ggs.: →Vollpol-)*
Scherbeanspruchung shear stress
Scherbelastung shear stress
Scherben cullet * *Bruchglas* || broken pieces * *go to pieces: in ~ gehen* || fragments || debris * *['debrie] Trümmer, (Gesteins-) Schutt*
Schere shears * *→Metall-/→Blech~, auch als Maschine; flying shears: →fliegende ~* || scissors * *allg.* || wire-cutter * *Draht~* || plate shears * *für Grobblech* || guillotine * *Schlag~* || shear * *→Metall-/→Blech~* || clipper * *Abschneider, Schermaschine*
Scherfestigkeit shear strength * *[in N/mm^2]*
Scherkraft shearing force
Scherquerschnitt shear section * *einer (Blech-) Schere [in mm^2]*
Scherspannung shearing stress || shear stress || shearing force || tangential stress || shearing resilience || shearing strain
scheuern scour * *scheuern, schrubben, polieren* || scrub * *schrubben, scheuern, reinigen* || chafe * *(sich) durch-/wundreiben, scheuern (against: an), verschleißen (z.B. Kabelisolierung)* || rub * *abreiben, reiben an, (ab)scheuern, (ab)schaben, abschleifen*

Schicht layer * *Ebene* || level * *Ebene* || film * *dünner Überzug* || skin * *Haut, äußerer Überzug* || coat * *Beschichtung, Auflage* || shift * *Arbeitsschicht*
Schicht- sandwich * *aus mehreren ~en bestehend* || compound * *aus Verbundmaterial bestehend* || shift * *~arbeit*
Schichtarbeit shift work
Schichtbetrieb shift operation
Schichtbürste sandwich brush * *Kohlebürste für Motor/Generator*
Schichtensystem stack * *System der 7...8 ISO-Schichten bei Feldbus*
Schichtholz laminated wood
Schichtkern laminated core || core stack
Schichtkondensator stacked-film capacitor * *Kunststoff-~, kein Keramik-Viel~* || multilayer capacitor * *(Keramik-) Vielschichtkondensator*
Schichtprotokoll shift report * *z.B. auf Meldedrucker ausgegebenes Protokoll einer Arbeitsschicht*
Schichtstoffplatte HPL board * *[Holz] Abk. für 'High-Pressure Laminated board': Dekorative ~ (DKS-Platte)*
Schichtwechsel change of shift || shift turnover
schichtweise in layers * *in Schichten (angeordnet)* || laminated * *geschichtet* || in shifts * *bei der Arbeit*
schicken ship * *Ware versenden/(aus)liefern* || send * *Brief, Post, Botschaft, Person (to: nach/zu; for: nach)* || dispatch * *ver/absenden, expedieren* || forward * *weiterleiten* || communicate * *übermitteln* || transmit * *übermitteln* || mail * *per Post, E-Mail usw.*
Schiebebalken rollbar * *am Bildschirm* || scrollbar * *am Bildschirm*
Schiebemuffe sliding sleeve
schieben shift * *verschieben* || push * *stoßen, schieben (into: in), stecken, drängen* || shove * *(beiseite) schieben/stoßen* || thrust * *vorwärtsstoßen* || slide * *gleiten lassen* || slip * *gleiten lassen* || telescope * *ineinander*
Schiebeoperation shift operation
Schieber pusher || slide || slide valve * *Steuerschieber, Schieberventil* || valve * *Ventil* || gate valve * *Ventil* || wrangler * *Betrüger* || profiteer * *Profithai* || black-marketeer * *Schwarzmarkthändler* || damper * *Schieber, (Zug-) Klappe*
Schieberegister shift register
Schiebeschalter slide switch
Schiebesitz sliding fit * *[Masch.bau] Passung, die ein Verschieben ermöglicht*
Schiebetisch sliding table * *[Werkz.masch.]*
Schiebetür sliding door
Schiebeverriegelung shift latch * *z.B. an (Sub-D) Stecker zur Arretierung u. Zugentlastung*
Schieblehre sliding caliper || caliper || slide gauge
schief skew * *schief, abgeschrägt, verdreht* || skewed * *abgeschrägt,verdreht* || oblique * *schief, schräg; oblique-angled: schiefwinklig* || sloping * *abfallend, geneigt, ansteigend* || inclined * *abfallend, geneigt, aufgelegt; inclined plane: schiefe Ebene* || slanting * *schräg* || lopsided * *nach einer Seite hängend, unsymmetrisch* || cock-eyed * *(salopp) nach einer Seite hängend, schielend, doof, betrunken* || crooked * *krumm, gekrümmt, gebeugt, unehrlich, bucklig, verwachsen* || distorted * *verdreht, verzogen, verzerrt, verrenkt, veformt, entstellt* || aslant * *(Adv.) schräg, quer, querüber, quer durch* || awry * *(Adv.) schief, krumm, ganz schief, verkehrt, scheel* || lop-sided * *nach einer Seite hängend, unsymmetrisch*
Schieflage awkward position * *(figürlich) Krise, unangenehme Lage* || oblique position * *Schräglage*
Schieflast asymmetrical load || unbalanced load
schiefwinklig oblique-angled || tilted

Schiene rail * *guide rail: Führungs~* || bar * *z.B. Strom~*
Schienenfahrzeug rail vehicle || railroad vehicle || rolling stock * *(Plural) Schienenfahrzeugpark* || LRV * *Abk. für 'Light-Rail Vehicle': leichtes/Kurzstrecken-~* || HRV * *Abk. für 'Heavy Rail Vehicle': schweres/Langstrecken-~* || railed vehicle
Schienenführung rail guide
schienengebunden rail-guided
Schienenprofil rail section
Schiff ship || vessel * *allg. Wasserfahrzeug*
Schiffbau marine industry || marine engineering * *als Fachrichtung*
Schiffbauvorschriften ship-building standards
Schiffs- marine
Schiffsantrieb marine drive * *z.B. Wellenantrieb* || propulsion || ship's propulsion || ship's propulsion system || marine propulsion drive * *Propellerantrieb*
Schiffsdeck ship deck
Schiffsdiesel marine Diesel engine
Schiffshebewerk ship-lifting device
Schiffsklassifikationsgesellschaft marine classification society
Schiffsmotor ship motor
Schiffsturbine marine turbine
Schiffswendegetriebe marine reversing gear
Schild shield * *Schutzschild, Lagerschild usw.* || plate * *z.B. Namen/Firmen/Tür/Typenschild* || signpost * *Wegweiser* || label * *Etikett* || signboard * *Firmen/Aushängeschild* || fascia * *Aushängeschild* || sign * *(Schrift-) Zeichen, (Aushänge-) Schild*
Schildlager bracket bearing
Schimmel mould || mold
Schimmelbildung fungi attack * *Schimmelpilzbefall*
schimmelig mouldy || moldy || musty || mildewy * *['mildju: i] modrig, schimm(e)lig, brandig*
Schirm shield * *Abschirmung, mechan. Schild* || screen * *Abschirmung, Bildschirm*
Schirmanschluss screening connection || shield connection
Schirmung screening * *z.B. einer Leitung* || shielding
Schirmwicklung shield winding * *bei Transformator*
Schlaffseilregelung slack-rope control * *bei Greiferkran* || slackening control * *bei Greiferkran* || slackness control * *bei Greiferkran*
Schlag hit || shock * *z.B. elektrischer; get a shock: einen ~ bekommen* || knock || stroke * *Hieb, Hub* || impact * *Aufprall, Auf/Einschlag, Schlagbelastung, heftige Einwirkung, Schlagfestigkeit* || out of round * *Unrundheit* || wobble * *Unrundheit* || tap * *leichter ~, Klaps* || blow * *Stoß, Streich, jarring blow: Prell~; at a blow: auf einen ~;strike a blow: e. ~ führ.* || bang * *schallender ~, Knall, Krach; Energie, Schwung*
schlagartig abrupt || sudden || prompt || of a sudden * *(Adv.) plötzlich* || all of a sudden * *(Adv.)* || abruptly * *(Adv.)* || from one minute to the other * *(Adv.)* || like a blow * *(Adv.)* || with a bang * *(Adv.)*
Schlagbohren percussion drilling
Schlagbohrer impact drill || percussion drill
Schlagbohrmaschine percussion drill || impact drill || hammer drill
schlagen beat * *auch wiederholt ~; übertreffen* || strike || knock * *hart* || hit || drive * *treiben (into: hinein), z.B. Nagel, Dorn* || crash * *krachend ~* || wobble * *unrund laufen, wackeln* || pulsate * *rythmisch stoßend* || tap * *leicht ~, klopfen/pochen (an, gegen), beklopfen, klopfen mit*

schlagfest impact-resistant || impact-proof
Schlagfestigkeit impact-resistance
Schlaglänge lay length * *beim Verseilen von Kabeln* || length of lay * *beim Verseilen von Kabeln* || length of twist * *beim Verseilen von Kabeln*
Schlagmühle beater mill * *z.B. für Steine* || crushing mill || hammer mill
Schlagschere guillotine
Schlagseite list * *(auch Verb: ~ haben); heeel over: ~ bekommen*
Schlagwetter firedamp
schlagwettergefährdeter Bereich area susceptible to firedamp || hazardous area
schlagwettergeschützt firedamp-proof || E EX I-protected * *Schutzart nach EN0514* || flameproof || explosion-proof * *explosionsgeschützt* || mine-type * *für Untertage-Einsatz* || flame proof
schlagwettergeschützter Motor firedamp motor
Schlagwetterschutz protection against firedamp || type of protection E EX I * *Ex-Schutzart nach EN50014* || protection against explosion
Schlagwort keyword * *Stichwort* || buzzword * *das in aller Munde ist* || catchphrase
Schlagwortregister index || key word index * *Stichwortverzeichnis*
Schlagwortverzeichnis keyword index || subject index * *Sachverzeichnis* || key word index || index
Schlamm mud || mire || slime * *schleimiger ~* || sludge * *Schlamm, Matsch, feuchter Bodensatz, Klärschlamm; Treibeis*
Schlammpumpe slush pump * *z.B. im Bergbau*
schlank low-profile * *v. d. Baugröße her* || slim * *(auch Verb: schlank machen)* || slim-line * *in schmaler/flacher Baugröße* || slender * *slenderize: schlank machen* || book-size * *in schmaler Bauform ("schmal wie ein Buch")* || slimline
schlanke Bauform low-profile type of construction
Schlauch hose * *zum Spritzen usw.* || flexible pipe || flexible tube || tube * *Rohr (-leitung), Schlauch, Luftschlauch, Röhre*
Schlauchanschluss hose connection * *Verbindung*
Schlauchbeutel tubular bag * *für Verpackungszwecke*
Schlauchbeutel-Füllmaschine tubular-bag filling machine
Schlauchleitung hose pipe * *Rohrleitung*
Schlaufe loop || runner || noose || sling || slack * *Lose, Durchhang*
schlecht bad * *(gelegentlich auch als Adv.)* worse: *~er; worst: ~est* || poor * *armselig,wertlos, unzureich. (Geschäft, Qualität, Schwingdämpfg., Aussicht.,Entschuldigg.)* || inferior * *Qualität, Ware* || low-quality * *von niedriger Qualität* || spoiled * *verdorben, ruiniert, kaputtgegangen, vernichtet* || mis- * *als Vorsilbe: ~es management: ~mismanagement* || badly * *(Adv.)* || worthless * *wertlos* || wretched * *erbärmlich, miserabel, schlecht, dürftig, unangenehm, ekelhaft, scheußlich* || wicked * *boshaft, verworfen* || evil * *böse, übel, boshaft, unglücklich* || ill * *(Adj. und Adv.) unbefriedigend, fehlerhaft, schlecht, böse, übel, ungünstig,* || scrap * *Schrott-* || damaged * *beschädigt* || waste * *Abfall-, unbrauchbar, abgängig, verschlissen* || lousy * *(salopp) mies, lausig*
schlechter worse * *(Komparativ); get/grow worse: schlechter werden*
Schleichdrehzahl thread speed * *Einziehdrehzahl zum Material-Einziehen* || crawl speed || jogging speed * *Tipp-Drehzahl* || threading speed || inching speed * *Tippdrehzahl*
schleichen thread * *mit Einzug/Schleichdrehzahl fahren* || crawl || creep * *kriechen*
Schleichgang crawl speed || creep feed || creep speed || thread speed * *Einziehgeschwindigkeit* || creep feedrate * *{Werkzeugmaschine}* || crawling
Schleichgangantrieb crawl speed drive * *Asynchronmotor mit Drehfeldmagnet auf 1 Welle z. Erreich. v. Schleichdrehzahlen*
Schleif- grinding || abrasive || sanding * *{Holz} mit Schmirgel arbeitend*
Schleifautomat automatic grinder * *{Werkz. masch.}* || automatic grinding machine * *{Werkz. masch.}* || automatic sander * *{Holz} mit Schmirgel arbeitend*
Schleifband grinding belt || abrasive belt || sanding beld * *{Holz} mit Sandpapier*
Schleifbock double-ended grinding machine * *Doppelschleifer* || wheel stand || grinding head
Schleifbürstmaschine brush grinding machine * *zur Reinigung in Blechstraße*
Schleife loop * *z.B. Regelschleife, Programmschleife*
schleifen grind || rub * *(ab)schleifen (with emery: abschmirgeln), (ab)feilen* || rub down * *anschleifen, z.B. vor einem Farbauftrag* || sand * *schmirgeln (z.B. Holz ~ mit Sandpapier/Bandschleifer usw.)*
Schleifenverstärkung over-all loop gain * *eines Regelkreises*
Schleifenzähler loop counter * *{Computer} z.B. Index in einer DO...FOR Programmschleife*
Schleifer grinder * *Mahlwerk, Schleifmaschine* || slider * *Mittelabgriff am Potentiometer, Schiebetrafo usw* || wiper * *an Potentiometer, Schiebetrafo usw.* || pick-off * *Abgriff, z.B. an einem Potentiometer* || sliding contact * *Schleifkontakt an Potentiometer usw.*
Schleiferabgriff slider pick-off * *z.B. an Potentiometer, Stelltrafo*
Schleiferei groundwood mill * *{Papier} Holz~ zur Zellstoffgewinnung* || grinding mill * *{Papier} in der Zellstoffindustrie* || grinding plant * *{Papier} in der Zellstofferzeugung*
Schleiferregelung grinder control * *für Holzschleifer i. d. Zellstoffherstellung*
Schleifertrog grinderpit * *bei Holzschleifer zur Zellstofferzeugung*
Schleifkontakt sliding contact || wiper || sliding-action contact || wipe contact || wiper contact
Schleifleitung daisy chain * *durchgeschleifte Steuer-/Datenleitung (1 kommende u. 1 gehende Leitg. je Teilnehmer)*
Schleifmaschine grinding machine || grinder || sander * *{Holz} mit Sandpapier arbeitend*
Schleifmittel abrasive products || abrasives || grinding abrasives
Schleifpapier sanding paper
Schleifpaste abrasive paste
Schleifring slip ring * *zur Stromzuführung auf sich drehende Maschinenteile (z.B. Läufer e. Asynchronmaschine)* || collector ring
Schleifringkupplung slipring clutch * *elektromagnet. Kupplung mit magnet. durchfluteten Kupplungsscheiben/-lamellen*
Schleifringläufer slipring rotor || wound rotor * *eine Wicklung tragender Läufer (kein Käfigläufer)*
Schleifringläufermotor slip ring motor * *z.B. Asynchronmot.m. Schleifkontakt z.Einschleif. e. ext. Anlasswiderstands* || wound-rotor motor * *Motor mit Läuferwicklung (i.Gegensatz zum Käfigläufer)*
schleifringlos without sliprings
Schleifringmotor slip ring motor
Schleifringübertrager slip-ring joint * *m.Schleifkontakten z.Übertragg. v. Strom/Signalen auf drehend. Maschinenteil*
Schleifscheibe grinding disc || grinding wheel || ab-

rasive disc || sanding disc * *{Holz}* mit Schmirgelpapier
Schleifstaub sanding dust * *{Holz}*
Schleifstein grindstone
Schleifwasser grinding solution
Schleifwerkzeug grinding tool
Schleifzylinder sanding drum * *{Holz}*
Schleim slime
Schleppabstand following error * *Soll-Ist-Abweichung/→Schleppfehler der Lage bei einer Lageregelung* || tracking error
schleppen drag || lug || haul * *auch mit Zugmaschine* || trail * *hinter sich herziehen* || carry * *(schwer) tragen* || tow * *(ab-) schleppen, ins Schlepptau nehmen, am Haken haben*
Schleppfähigkeit trailing capability * *eines Schleppkabels*
Schleppfehler tracking error * *z.B. bei einer Lageregelung* || following error * *z.b. statische Lage-Soll/Istabweichung bei einer Lageregelung* || contouring error * *beim "Konturfahren" mit einer NC (Numerische Mehrachs-Bewegungssteuerung)* || trajectory error * *Abweichung von der vorgegebenen Bahnkurve bei Positionier-/Bahn-/NC-Steuerung*
Schleppkabel trailing cable
Schleppleitung trailing cable
Schleuderdrehzahl overspeed test speed * *für Schleuderprüfung eines Motors, normalerweise mit 1,2-mal max. Drehzahl*
schleudern fling * *schleudern, werfen, aufreißen/zuschlagen (Tür usw.)*; eilen, stürzen || hurl * *schleudern, werfen; hurl oneself on: sich stürzen auf* || throw * *werfen, schleudern, zuwerfen, (Hebel, Schalter) umlegen, throw out: auskuppeln/-rücken* || veer * *sich drehen (auch Fahrzeug), umspringen, umschwenken (figürl.), wenden, drehen, schwenken* || sling * *mit einer Schleuder(vorrichtung) ~, verspritzen, katapultieren* || catapult * *mit einer Schleuder(vorrichtung) ~* || centrifuge * *mit einer Schleudermaschine/Zentrifuge ~* || spin-dry * *Wäsche ~* || swing * *(hin- u. her-) schwingen, s. drehen, (herum-) schwenken, schlenkern, durchdrehen, anwerfen* || skid * *sich drehen (Auto usw.)* || sideslip * *ins Schleudern kommen (Fahrzeug usw.)* || undersell * *zu billigen Preisen ver~* || sell below cost * *unter Preis ver~* || dump * *Exportgüter zu billigen Preisen ver~* || cure * *Zucker zentrifugieren* || strain * *Honig usw. ~; auch 'extract'* || spin * *herumwirbeln, schnell drehen, schleudern. trudeln lassen, spinnen* || side-slip * *Fahrzeug usw.*
Schleuderprüfung overspeed test * *z.B. mit 120% der max Drehzahl eines Motors (nach VDE 0530)*
Schleuderscheibe grease slinger * *z.B. bei Schmiersystem* || greasing slinger * *Fettschleuderscheibe/Fettmengenregler bei Schmiersystem* || slinger
Schleusenspannung threshold voltage * *(Diode, Transistor)*
Schließdruck engagement pressure * *einer Kupplung/Bremse*
schließen close * *auch Bremse, Kontakt, Datei usw. ~* || make * *Kontakt* || conclude * *Brief/Rede ~; auch 'logisch folgern'* || shut * *(ver)schließen, zumachen; auch figürl.: shut one's eyes to: seine Augen ver~ vor* || lock * *mit Schlüssel ~* || ed * *beenden* || finish * *beenden* || terminate * *beenden, auch Softwareprogramm* || deduce * *folgern* || infer * *folgern* || contract * *einen Vertrag ~* || bridge * *bridge the gap: die Lücke ~* || complete * *ab~, fertigstellen* || block up * *blockieren* || bolt * *mit Riegel ~* || seal * *hermetisch ver~, versiegeln, durch Kleben/Schweißen/Vergießen ~*
Schließer make contact || NO contact * *Abkürzung*

für 'Normally Open contact' || normally open contact || A-contact || form A contact
Schließerkontakt make contact || normally open contact || NO contact * *Abkürzung für 'Normally Open contact'* || A-contact || form A contact
schließlich final || last * *(Adj.)* || ultimate || conclusive * *abschließend* || eventual || finally * *(Adv.)* || ultimately * *(Adv.) zuletzt* || in the end * *(Adv.) zu guter Letzt* || at last * *(Adv.) endlich* || in the long run * *(Adv.) auf die Dauer, auf lange Sicht* || after all * *(Adv.) schießlich doch, schießlich und endlich* || all things considered * *im Grunde* || in the last analysis * *im Grunde* || after all * *(salopp) schießlich und endlich*
Schließwerk grab-closing gear * *{Hebezeuge} bei Greiferkran*
schlimmstenfalls at worst
Schlinge loop * *auch in Blechstraße* || noose * *zusammenziehbare*
Schlingengrube looping pit * *{Walzwerk} Warenspeicher in Walz/Blechstraße*
Schlingenheber looper * *{Walzwerk}* || loop lifter * *{Walzwerk}*
Schlingenkanal looping floor || sloping loop channel || loop channel
Schlingenleitblech release apron * *{Walzwerk}*
Schlingenregelung strip loop control * *{Walzwer-Sicherstellung e. definierten Durchhangs des Blechs i.d. Schlingengrube*
Schlingenspanner looper
Schlingenturm looping tower * *{Walzwerk}* || vertical accumulator * *senkrecht. Warenspeicher (für Kabel, Walzband) z.Überbrück.d. Rollenwechsels*
Schlingenwagen loop car * *{Walzwerk} dient als Warenspeicher bei Walzstraße*
Schlitten shuffle table * *Schlepptisch* || sliding carriage || cradle || slide * *gleitendes Teil, Führung, Schieber*
Schlitz slit * *z.B. zum Entweichen von Kühlluft* || slot || rift * *Spalt* || cleft * *Spalt* || crack * *Riss* || aperture * *Öffnung* || slotted hole
Schlitzbreite slit width || slot width
Schlitzmaschine mortize-cutting machine * *{Holz}*
Schlitzscheibe slotted disc || code disc * *z.B. im Gray-Code (einschrittiger Code) kodiert bei →Absolutwertgeber*
Schlitzschraube slotted-head screw
Schlitzschraubendreher flat-bladed screwdriver
Schlitzschraubenzieher flat-bladed screwdriver
Schlitzzahl pulse count * *Anzahl der Schlitze in der Schlitzscheibe eines (Inkremental-) Impulsgebers* || number of pulses per revolution * *Pulszahl pro Umdrehung bei Impulsgeber* || pulse number per revolution * *~ eines Impulsgebers* || number of increments * *~ eines (Inkremental-) Impulsgebers* || PPR count * *Abk.f. 'number of Pulses Per Revolution': Anzahl Impulse pro Umdrehung* || number of increments per revolution * *~ eines Impulsgebers* || number of slots per revolution * *~ eines optischen Drehimpulsgebers*
Schloss lock
Schlosser mechanic * *→Mechaniker* || machinist * *Maschinen~/-einsteller* || fitter * *Maschinen~* || toolmaker * *Werkzeugmacher*
Schlupf slip * *bei Asynchronmaschine: (Synchrondrehz. - Läuferdrehz.)/Synchrondrehz.; auch eines Riemens* || creep * *eines Riemens* || slippage * *Durchrutschen*
Schlupfbewegung slipping movement
Schlupfdrehzahl slip speed
schlupfen slip
schlüpfen slip
Schlupferfassung slip detecting || slip sensing
schlupffrei non-slip || slip-free || free from slip

Schlupffrequenz slip frequency * *Produkt Netzfrequenz x Schlupf*
Schlupfkompensation slip compensation
Schlupfkupplung slip clutch || slipping clutch
Schlupfläufer high-resistance rotor * *Läufer mit erhöhtem Schlupf; siehe →Widerstandsläufer* || high-resistance squirrel-cage rotor || slip rotor
Schlupfleistung slip power
Schlupfleistungsrückgewinnung slip-power recovery
Schlupfmoment slip torque || torque causing slipping
Schlupfnennfrequenz rated slip frequency
Schlupfregelung slip control || slip regulation
Schlupfregler slip regulator
Schlupfreibung slip friction
Schlupfspannung slip-frequency voltage
Schlupfsteuerung slip control * *{el. Masch.}*
schlupfsynchron slip-synchronous
Schlupfverluste slip losses || slip loss
Schluss end * *Ende* || inference * *(Schluss-) Folgerung, (Rück-) Schluß;* make inferences: *Schlüsse ziehen* || close * *(Ab-) Schluss* || conclusion * *Abschluss, Folgerung* || deduction * *→Folgerung, →Ableitung, →Herleitung* || stop * *Beendigung; stop it: ~ damit!*
Schluss- final
Schlussbemerkung conclusive comment || final remark || last remark
Schlussdeckel end cap || outer bearing cap * *äußerer Lagerdeckel* || bearing end cap * *äußerer Lagerdeckel*
Schlüssel key * *(auch figürl.)* || cipher * *Chiffriercode* || keyword * *Schlüsselwort* || spanner * *Schraubenschlüssel* || wrench * *Schraubenschlüssel* || code * *Code(Wort)*
Schlüssel- key
schlüsselfertig turnkey
schlüsselfertige Erstellung turnkey erection * *z.B. einer Fabrik*
Schlüsselfläche spanner surface * *zum Ansetzen eines Schraubenschlüssels* || flat face * *z.B. einer Mutter*
Schlüsselkunde key account customer || key account
Schlüsselparameter key parameter
Schlüsselposition key role || key position
Schlüsselposition einnehmen play a key role
Schlüsselschalter key switch || keylock switch || key-operated switch || key-actuated switch
Schlüsselstellung key role * *Schlüsselrolle*
Schlüsseltechnologie key technology
Schlüsselweite wrench opening * *eines Schraubenschlüssels* || spanner gap * *eines Schraubenschlüssels* || spanner opening * *eines Schraubenschlüssels* || width across flats * *eines Schraubenschlüssels* || width over flats * *eines Schraubenschlüssels*
Schlüsselwort password * *dient als Zugriffsperre für Unbefugte*
schmal narrow || thin * *dünn* || slender * *schlank* || slim * *schlank, flach, dünn* || small * *gering* || low-profile * *schlanke von der Baugröße her* || slimline * *in schlanker Bauform* || poor * *ungenügend* || book-size * *in schlanker Bauform, "wie ein Buch"* || narrow-profile
Schmalbandanalyse small-band analysis * *bei der Geräuschmessung*
schmalbandig narrow-band * *z.B. Frequenzband, Filter, Bandsperre*
schmaler narrower * *(Komparativ)*
Schmelz-I2t-Wert melting I2t value * *einer Sicherung*
Schmelz-I²t-Wert melting I²t value * *einer Sicherung*

Schmelze smelt || molten material || molten mass || smeltery * *{Metallurgie} Schmelzanlage*
Schmelzeinsatz fusing element * *Schmelzleiter in Sicherung* || fuse link * *Sicherungseinsatz (Sicherung ohne Unterteil)*
schmelzen melt * *melt down: einschmelzen* || fuse * *(ver-) schmelzen (auch figürl.), vermischen, durchbrennen (Sicherung usw.)*
Schmelzkleber hot-melt glue
Schmelzklebstoff hot-melt adhesive
Schmelzleiter fuse element * *in Sicherung* || fusable element * *in Sicherung* || fusing conductor * *in Sicherung* || fusing element * *in Sicherung* || fuse-element || fusible element
Schmelzofen melting furnace
Schmelzperlen beads of molten material
Schmelzsicherung fuse || fusible cut-out || safety fuse || fusible link || fusible cutout
Schmelzstraße smelting line * *im Hüttenwerk* || melting line * *im Hüttenwerk*
Schmelztiegel melting crucible * *{Metallurgie}* || melting pot * *(auch figürl.)* || melting tank
Schmelzzeit melting time * *(Sicherung)*
Schmerzgrenze threshold of pain
Schmerzschwelle pain threshold || threshold of pain
Schmiede forge || smith's shop
Schmiedemaschine forging machine
schmieden forge
Schmiedepresse forging press
Schmiedestahl forged steel
Schmiedestück forging
Schmiedeteil forging
schmiegen bevel * *{Holz}*
Schmiegenlehre beveled square * *{Holz} Winkelmaß-Lehre für Schrägschnitte*
Schmier- lubricating
Schmierbüchse oil cup * *für Öl* || oiler * *für Öl* || grease cup * *für Fett* || grease box * *für Fett* || Stauffer lubricator * *Staufferbüchse*
Schmiereinrichtung equipment for greasing * *z.B. für Wälzlager*
schmieren lubricate || grease * *mit Fett* || oil * *mit Öl*
Schmierfähigkeit lubricity || lubricating property * *Schmiereigenschaft*
Schmierfett lubricant || grease || lubricating grease
Schmierfilm lubricant film || oil film * *Ölfilm* || lubricating film || lubrication film
Schmierfrist lubrication interval || relubrication cycle
Schmiergeld palm-oil || bribe-money || slush-money
schmierig greasy * *fettig* || oily * *ölig* || sticky * *klebrig* || dirty * *schmutzig* || smarmy * *(salopp) schmeichlerisch, ölig, kriecherisch*
Schmierkanal lubrication duct
Schmierkreislauf oil circulation * *Ölumlauf*
Schmiermerker scratch flag * *bei SPS*
Schmiermittel lubricant || dressing agent * *{el.-Masch.} für Kommutator* || commutator agent * *{el.Masch.} für Kommutator*
Schmiermittel-Ausschwitzen lubricant exudation
Schmiermittelgebrauchsdauer lubricant life * *DIN51818/51852, DIN ISO281 Beibl.1, GfT Arb.bl. 2.4.1, Ges.f.Tribologie*
Schmiermittelmenge lubricant quantity
Schmiernippel grease nipple || lubrication nipple
Schmieröl lubricating oil || lube oil
Schmierpapier scrap paper
Schmierpumpe lubricating pump || oil pump * *Ölpumpe* || grease gun * *Abschmierpumpe, Fettpresse* || grease pump * *Fettpumpe*
Schmierschild lubricant plate * *mit Angaben zur Schmierung (Schmierstofftypen, Schmierfristen usw.)* || lubrication plate * *mit Angaben zur richtigen Schmierung*
Schmierspalt lubricating gap

Schmierstelle lubrication point
Schmierstoff lubricant
Schmierstoffmenge lubricant quantity
Schmierung lubrication * *DIN 51818/51852, DIN ISO 281 Beibl.1, GfT Arb.bl. 2.4.1 (Gesellsch. für Tribologie)* || greasing * *mit Schmierfett* || oiling * *mit Öl*
Schmiervorrichtung lubricator
Schmirgel emery
Schmirgelleinen abrasive cloth || emery cloth
Schmirgelpapier emery paper || abrasive paper
Schmitt-Trigger Schmitt trigger element * *Schwellwertschalter mit Hysterese*
Schmutz dirt || filth || smut * *(meist figürl.)* || mud * *Schlamm* || contamination * →*Verunreinigung,* →*Verschmutzung*
Schmutzablagerung dirt deposit
Schmutzeffekt adverse effect * *schädlicher/nachteiliger Effekt* || side effect * *Nebeneffekt (siehe auch dort)* || unintentional effect * *ungewollter/unbeabsichtigter Effekt*
schmutzfrei free of dirt
schmutzig dirty * *(auch figürl.: 'unanständig', 'unsauber')* || filthy * *verdreckt* || muddy * *schlammig* || soiled * *beschmutzt* || shabby * *schäbig* || messy * *unordentlich, unsauber, schmutzig* || contaminated * *verunreinigt*
Schnappdeckel snap lid || snap-fitting cover
schnappen snap * *z.B. auf eine Installationsschiene* || catch * *Schloss* || click * *einrasten/schnappen* || engage * *einrasten, in Eingriff kommen*
Schnappfeder catch spring
Schnappverschluss quick-release catch || snap-in fastener || snap lock
Schnecke worm * *Schneckenrad* || screw conveyor * *Schneckenförderer, Förderschnecke* || screw * *z.B. bei Extruder, Schneckengetriebe* || scroll * *Schnecke, Spirale, Schnörkel; z.B. ~ bei (Schrauben-) Kompressor* || auger * *~gang, ~bohrer*
Schnecken-Stirnradgetriebe worm-spur gear unit
Schneckendosierer auger-type dosing unit * *{Verpack.techn.}* || screw-type proportioning system || screw-type metering system
Schneckenförderer screw conveyor || spiral conveyor || helical conveyor || worm conveyor || auger-type feeder system * *{Verpack.techn.} siehe auch* →*Schneckendosierer*
schneckenförmig helical || spiral
Schneckenfüllmaschine metering-screw filling machine * *in der Verpackungstechnik*
Schneckengetriebe worm gear || worm gearing || screw gear || spiral gear || double-enveloping worm gear || worm drive * *Schnecken(an)trieb* || worm reducer * *Schnecken-Untersetzungsgetriebe*
Schneckenpumpe screw pump
Schneckenrad worm wheel || screw gear
Schneckenradgetriebe worm gear || worm gearbox || screw gear || spiral gear
Schneidabfälle shavings
Schneidbrenner blowpipe || cutting torch || flame cutter || oxygen cutter
Schneideisen tapping die * *Gewindebohrwerkzeug* || diestock || screw die || threading die * *zum Gewindeschneiden*
Schneidemaschine cutting machine || sizing machine * *{Holz} Zu~ für Holzplatten*
schneiden cut * *auch Kurve ~* || chop * *hacken* || shred * *schnitzeln* || clip * *mit Zange, Schere, Maschine (ab)schneiden/-hauen* || carve * *meißeln, schnitzen* || engrave * *ein~, stechen, gravieren, eingraben, einprägen* || cross * *sich ~ (z.B. zwei Kurven {Mathem.} oder zwei Straßen)* || slit * *der Länge nach ~* || rip * *(zer-) schlitzen, längs~ (z.B. Holzfurnier), der Länge nach auf~; rip up: aufschlitzen*

Schneidgerät cutting device * *z.B. für Lichtleiter*
Schneidkopf cutting head || cutter head
Schneidkraft cutting force * *bei Werkzeugmaschine*
Schneidmaschine slitter * →*Längsschneider* || cutter * →*Querschneider* || guillotine cutter * →*Schlagschere (für Blech usw.)*
Schneidöl cutting oil
Schneidpartie slitter unit * *Quer- oder Längsschneider für bandförmiges Material* || knife section * →*Messerpartie (z.B. bei Querschneider)*
Schneidplatte tip cutter * *{Werkz.masch.} hartmetallbestücktes Schneidwerkzeug*
Schneidwasser cutting water * *{Werkz.masch.}*
schnell quick || fast || rapid || rapidly * *(Adv.)* || speedy * *zügig* || prompt || immediate * *unverzüglich, unmittelbar* || racy * *(salopp) rasant* || quick-acting * *z.B. Sicherung* || high-speed * *Hochgeschwindigkeits-* || swift * *rasch, zügig, flüchtig (Bekanntschaft, Zeit), geschwind, eilig, flink, geschickt; Haspel*
schnell tippen progressive jog * *Kommando für Antrieb*
schnell wachsend quickly growing
Schnell- rapid || express * *Versand, Reparatur usw.* || quick-acting * *schnell arbeitend/auslösend, z.B. Schalter, Sicherung* || quick-action * *schnell arbeitend/auslösend, z.B. Schalter, Sicherung* || turbo- || high-speed
Schnellabhebung quick lifting * *quick llfting of the roll: Walzen~*
Schnellabschaltung quick breaking * *Stromkreis* || rapid shutdown || quick stopping * *Motor, Maschine* || emergency shutdown * *Notabschaltung*
Schnelladung boost charge || quick charge
schnellansprechend fast-response * *mit kurzer Reaktionszeit* || quick-acting * *z.B. Sicherung* || instantaneous-tripping * *Schutz/Überwachungseinrichtung* || quick-release * *Schutz/Überwachungseinrichtung* || high-speed response * *mit kürzester Reaktionszeit* || fast-operation
schnellaufend high-speed || fast-running
Schnellauslöser instantaneous release || quick-acting tripping device || instantaneous trip unit
Schnellauslösung instantaneous tripping
Schnellauswahl quick selection
Schnellauswahltabelle quick-selection table
Schnellbremsung rapid braking || quick stopping || emergency braking * *Notbremsung*
schnelle Antriebsregelungen highly-dynamic drive controls
schnelle Erledigung early settlement || immediate attention || prompt handling
Schnellentladung fast discharge * *z.B. von Zwischenkreiskondensatoren beim Abschalten eines Umrichters*
schneller faster || quicker || more rapid || more quickly * *(Adv.)*
schnellerregt fast excitation * *Motor*
Schnellerregung rapid excitation * *z.B. bei elektromagnet. Bremse/Kupplung mit erhöhter Spannung und Vorwiderstand*
Schnellgang rapid advance || rapid traverse || high gear
Schnellhalt quick stop || fast stop || hard stop || emergency stop * *Not-Aus* || E-stop * *Abkürzg. f. 'Emergency stop': Not-Aus/-Halt (Elektrikal Aus)*
Schnellhaltzeit quick-stopping time
Schnellkupplung quick-release coupling || quick coupling
Schnellmitteilung fast info
Schnellreparatur express repair
schnellschaltend fast-switching * *auch Schalttransistor* || fast-acting * *Sicherung*
Schnellschalter quick-break switch || high-speed circuit breaker * *Leistungsschalter*

Schnellspanneinrichtung quick-action fixture || quick-action clamping device
Schnellspannsystem fast-action clamping system || quick-release clamping system
schnellstmöglich at full speed || at high speed || as soon as possible * *baldmöglichst* || at top speed
Schnelltrennkupplung quick-release coupling
Schnellübersicht fast overview || quick-reference manual * *Kurzanleitung*
Schnellverschluss quick-release lock || snap lock * *Schnappverschluss* || quick-release catch
Schnitt cut || cutting || sectional view * *Schnittbild-Ansicht in Zeichnung* || section * *Schnitt(Bild), Profil*
Schnittbandkern cut strip-wound core * *z.B. eines Transformators* || C-core * *z.B. bei Trafo*
Schnittbild sectional view || section
Schnittgenauigkeit cut accuracy * *z.B. bei Querschneider, fliegender Schere usw.*
Schnitthöhe cutting height * *z.B. bei (Holz-) Kreissäge*
Schnittholz sawn timber || cut timber || converted timber || timber * *(brit.)* || lumber * *(amerikan.)*
Schnittholztrockner converted timber drier
Schnittiefe cutting depth || depth of cut || cut-deepness * *z.B. einer Säge*
Schnittkraft cutting force * *bei Werkzeugmaschine*
Schnittkraftregelung cutting-force control
Schnittlänge cut length * *z.B. bei Querschneider*
Schnittpunkt intersection point * *z.B. von Kurven oder Phasenspannungen* || break-even point * *z.B. Punkt, bei dem man in die Gewinnzone kommt* || concurrence || intersection || point of intersection
Schnittqualität quality of cut
Schnittstelle interface * *Daten-, Signal-, logische, serielle Schnittstelle.* || port * *(Signal-) Anschluss, z.B. serielle Schnittstelle* || link * *Verbindung, z.B.serielle*
Schnittstellenbaugruppe interface board * *(Leiterkarte) auch serielle Schnittstellenbaugruppe* || serial interface board * *serielle Schnittstellenbaugruppe (Leiterkarte)* || communication board * *(Leiterkarte) Kommunikationsbaugruppe* || interface module || interface card * *Leiterkarte* || communications board * *Kommunikationsbaugruppe (Leiterkarte)*
Schnittstellenkarte interface board || interface card || communication board
Schnittstellensubmodul interface submodule
Schnittstellentester interface tester
Schnittstellenumsetzer interface converter * *z.B. für serielle Schnittstelle*
Schnittwinkel cutting angle
Schnitzel chips * *z.B. aus Holz-/Kunststoffmaterial*
Schnur cord * *auch elektr. Anschluss~; flexible cord: flexible Anschluss~* || string * *Bindfaden* || thread * *Faden, Zwirn, Garn* || twine * *Bindfaden* || line * *Leine* || braid * *Litze* || flex * *(salopp)* Abkürzg. für 'flexible cord': flexible Anschlussschnur
Schockbeanspruchung shock stressing
Schockfestigkeit shock resistance || shockproofness * *Festigkeit gegen wiederholten mechan. Stoß*
schonen take care of * *pfleglich behandeln* || preserve * *erhalten* || save * *Kräfte, Vorrat usw. ~* || husband * *Kräfte, Vorrat usw.~* || treat a person with indulgence * *jemanden ~d behandeln* || respect * *Eigentum, Rechte usw.~* || save one's strength * *seine Kräfte sparen* || be kind to * *Material(schonung durch eine Maschine)* || protect * *schützen, erhalten*
schonend with low stressing || low-stressing || careful || gentle || tender * *pfleglich, zart, weich, empfindlich, heikel, bedacht (of: auf)*

Schonschicht underliner * *z.b. bei der Papier- oder Gummiverarbeitung*
Schonung low stressing * *geringe Beanspruchung* || careful treatment * *pflegliche Behandlung* || good care * *gute Behandlung* || protection * *Erhaltung, Schutz* || preservation * *Erhaltung*
Schonzeit hold-off interval * *Sicherheitszeit, die e. Stromrichter b. Umkommutieren gelassen wird* || hold-off time * *Sicherheitszeit, die e. Stromrichter b. Umkommutieren zum Ausgehen gelassen wird* || recovery time
Schopfschere cropping shear * *Querteilschere z.Abschneiden d.Brammen- oder Stahlbandenden i.Walzwerk/Blechstraße*
Schornstein chimney || smokestack * *Fabrik~* || funnel || stack || flue * *Abzugsrohr, →Kamin*
Schott bulkhead * *z.B. Trennwand in Schiff*
Schottky-Diode Schottky diode || Schottky barrier diode
Schottung compartmentalization * *Unterbringung in einzelnen Abteile* || barriers * *Schutzwände* || partitions * *durch Trennwände abbgeschottete Teile*
schraffieren cross-hatch || hatch
Schraffur hatching || hatched area
schräg skew * *schief, abge~t, verdreht,* || skewed * *abge~t,verdreht* || oblique * *schief, schräg* || slanting || sloping * *abfallend* || inclined * *abfallend, ~ angeordnet* || diagonal * *~ verlaufend* || transversal * *quer hindurchgehend* || bevel * *abge~t* || bevelled * *abge~t* || chamfered * *an/abgefast, abge~t*
Schrägbett-Drehmaschine inclined-bed lathe
Schräge chamfer * *Schrägkante* || bevel || slant || obliquity || slope * *schräge Fläche, Gefälle* || incline * *schräge Fläche*
Schrägförderband inclined belt conveyor || inclined conveyor * *{z.B. im Bergbau}*
Schrägförderer inclined conveyor
Schrägkugellager angular-contact ball bearing
Schräglage sloping position || skewing * *das 'Sich-Schräg-Legen'* || inclined position
Schrägschlitzanker skewed armature * *zur Verminderung der Nutwelligkeit (u.damit d. Momentenwelligkeit) eines Motors*
Schrägschnitt bevel cut * *{Holz}*
schrägstellen incline * *neigen* || tilt * *kippen*
Schrägstirnrad helical gear
Schrägstrich slash * *Zeichen auf Bildschirm, Drucker* || diagonal stroke || oblique stroke || diagonal line || backslash * *umgekehrter ~* || stroke
schrägverzahnt helical toothed || helical geared || helical || spiral toothed
Schrägwicklung skew-winding * *z.B eines Motor-Ankers*
Schrank cabinet * *Schaltschrank* || cubicle * *Schaltschrank* || enclosure
Schrankausführung cubicle design * *z.B. Aufbau eines Geräts als kompletter Schaltschrank*
Schrankbelüftung cubicle ventilation * *für Schaltschrank*
Schrankboden cubicle flooring section * *Bodenbereich/-Blech eines Schaltschranks*
Schrankdisposition cubicle layout
schränken dress * *{Holz} Säge scharf machen* || set * *{Holz} Säge scharf machen durch "Verbiegen" der Zähne*
Schrankgerät cubicle unit || cubicle mounted unit || cabinet unit
Schrankgerüst cubicle frame
Schranklüfter cubicle fan
Schrankmaschine saw setting machine * *{Holz} zum Schärfen von Sägen* || wavy setting machine * *{holz} Wellen~ zum Scharfmachen von Sägen* || setting machine * *{Holz} zum Schärfen von Sägen*
Schrankmontage cubicle mounting * *for: für*

Schranksystem cubicle system * *z.B. SIEMENS Schaltschranksystem 8MF*
Schranktür cubicle door
Schrankwand cabinet wall * *eines Schaltschranks*
Schraub-Steck-Klemme plug-and-screw terminal || pluggable screw terminal
Schraub-Verschließmaschine threaded closure machine * *in der Verpackungsindustrie* || screw-type sealing machine * *in der Lebensmittel-/Verpackungsindustrie*
Schraubanschluss screw terminal * *z.B. zum Herstellen der Kontaktierung an einem Bauteil*
Schraubantrieb nut running drive * *mit Kraftübertragung über Schraubspindel*
Schraubbefestigung screw fixing || bolt-on mounting || screw fitting || bolted connection * *Schraubverbindung*
Schraubdeckel screw cap
Schraube screw * *wood screw: Holz~; auch Schiffs~.; countersunk screw: eingelassene ~/ Senk~* || bolt * *dicke ~, ~ mit (teilweise gewindelosem) Schaft, Schraubbolzen* || propeller * *Schiffs~, Flugzeug~* || air-screw * *Flugzeugpropeller* || stud * *~ ohne Kopf, Stift~, Stehbolzen*
schrauben screw || tighten the screw * *fest~, fester an~* || loosen the screw * *los/ab~* || twist * *drehen* || spiral * *up/down: hoch/herunter~ (Preise usw.), sich spiralförmig nach oben/unten bewegen* || turn up * *in die Höhe ~*
Schraubenantrieb propulsion drive * *für Schiff*
Schraubenbolzen screw bolt
Schraubendreher screw driver || flat-bladed screwdriver * *Schlitz~ (für Schlitzschrauben)* || cross-tip screwdriver * *Kreuzschlitz~*
Schraubenfeder helical spring || spiral spring
schraubenförmig helical || spiral
Schraubenkompressor screw compressor
Schraubenkopf screw head || bolt head
Schraubenloch screw hole
Schraubenpumpe screw pump || axial pump || rotary pump * *→Kreiselpumpe*
Schraubenschlüssel wrench || spanner
Schraubensicherungsblech tab washer * *mit hochgezogenen Laschen*
Schraubensicherungselement screw-locking device || bolt locking element
Schraubensicherungslack screw locking varnish || screw-lock varnish
Schraubensicherungsring screw locking washer || screw retaining ring || lock washer
Schraubenverdichter screw compressor
Schraubenzieher screw driver || flat-bladed screwdriver * *Schlitz~ (für Schlitzschrauben)* || cross-tipped screwdriver * *Kreuzschlitz~ (für Kreuzschlitzschrauben)*
Schrauber screwing machine * *automatic screwing machine: Schraubautomat* || powered screw driver * *für Handbetrieb* || screw-driving equipment || screwdriving unit
Schraubklemme screw terminal || screw-type terminal
Schraubmaschine screw-setting machine * *z.B. in der Holzindustrie* || powered screw driver * *Hand~*
Schraubsteck-Klemme plug-and-screw terminal || pluggable screw terminal
Schraubsteckklemme pluggable screw terminal || plug-and-screw terminal || removable screw-terminal * *removable terminal* * *abziehbare Klemme* || quick-disconnect screw terminal || removable screw terminal
Schraubsteckklemmenblock quick-disconnect screw terminal block || quick-disconnect screw terminal strip
Schraubstock vice || vise * *(amerikan.)*
Schraubthyristor screw-in thyristor || stud-type thyristor || stud-casing thyristor || stud-mounting thyristor
Schraubverbindung bolted joint || screwed connection || screw connection || screwed joint || bolted connection || screwed pipe joint * *von Rohren* || screw union
Schraubverriegelung screw interlocking * *z.B. für (Sub-D) Stecker*
Schraubverschluss screw cap * *Schraubdeckel* || screw closure
Schraubwerkzeug screwdriver tool || screwdriver bit * *Schraubeinsatz*
Schraubzwinge screw clamp
Schreib-Lesespeicher RAM memory * *Abk. f. 'Random-Access Memory': Speicher mit wahlfreien Zugriff* || write-read memory
schreiben write * *auch ein Programm, be~ einer Speicherzelle usw.* || make a signal record * *einen Messschrieb aufzeichnen, z.B. mit einem Linienschreiber* || make an oscilloscope trace * *Messschrieb aufnehmen mit e. Oszilloskop*
Schreiber recorder * *Aufzeichnungsgerät, registrierendes Messgerät* || chart recorder * *Linien~*
schreibgeschützt write-protected || read-only
Schreibmarke cursor
Schreibpapier writing paper
Schreibschutz write protection * *z.B. für Diskette, Parameter*
Schreibweise notation
Schreibzugriff write access * *z.B. auf Speicher*
Schreibzyklus write cycle
Schreinerei joinery * *[Holz] siehe auch →Tischlerei*
Schrieb record * *von einem Messschreiber* || oscilloscopic record * *von einem Oszilloskop* || record chart * *aufgenommenes Kurvenblatt* || oscilloscopic trace * *von einem Oszilloskop*
schriftlich in writing || written || in letter * *brieflich* || in black and white * *schwarz auf weiß*
Schriftzeichen character || letter
Schritt step * *step by step: ~ für ~; take small steps: kleine ~e machen; in steps of: in ~en von* || pace * *keep pace: ~ halten; pace-maker: ~macher* || move * *(Schach-) Zug, Schritt, Maßnahme; make the first move: den ersten ~ machen*
Schritt für Schritt step by step
Schritt halten keep pace || keep step * *with: mit*
Schrittbaustein sequence block * *bei SPS*
Schritte voraus steps ahead
Schrittelement sequence step * *in einer Ablaufkette bei SPS* || drum sequencer step * *in einer Ablaufkette bei SPS*
Schrittfortschaltung step transition * *bei Ablaufsteuerung/-kette* || transition * *bei Ablaufsteuerung/-kette*
schritthalten keep pace * *with: mit*
Schrittmacher pace-maker || pacer * *set the pace for: ~ sein für*
Schrittmaß incremental dimension * *Kettenmaß (relative/inkrementelle Position) b. Numerik-/Positioniersteuerung*
Schrittmaß fahren incremental mode * *Betriebsart bei Positioniersteuerung (relative Lage anfahren, "Fahren Um")*
Schrittmerker step flag
Schrittmotor stepper motor || stepping motor
Schrittsteuerung sequential control
schrittweise stepwise || step by step * *Schritt für Schritt*
schrittweise erhöhen step up
schrittweise reduzieren step down
Schrittweises Durchschalten stepping through the functions * *bei SPS* || stepping through
Schrittweite step range || increment || step width || step size

Schrotmühle crushing mill * z.B. in der Malzherstellung
Schrott scrap * auch Eisen/Stahlschrott
Schrumpf- shrinking || heat-shrink * bei Hitze schrumpfend, z.B. Schlauch, Folie
schrumpfen shrink || contract * (sich) zusammenziehen/verengen, schrumpfen
Schrumpffolie shrink film || shrinking film || shrinkable film
Schrumpffolienverpackung shrink film wrapping || shrink wrapping
Schrumpfmaß degree of shrinkage
Schrumpfring shrunk-on ring
Schrumpfscheibe shrink disk * zur Wellenverbindung || shrink disc * zur Welle-Naben-Verbindung || shrink-pressure disc
Schrumpfschlauch heat-shrinkable tube || shrinkable tube || shrink sleeving || shrink tube || shrink tubing || heat-shrink tube
schrumpfschlauchisoliert shrunk-sleeve-insulated
Schrumpfung shrinkage || shrinking
Schub transverse force * Querkraft || shear * Scherkraft || thrust * ~ in Axialrichtung, Schiebekraft || batch * ~ von Ereignissen/Aufträgen usw., || push * Stoß || shove * Stoß || sudden rise * plötzlicher Anstieg (Preise, Kosten usw.) || impetus * (figürl.) Impuls, Schwung; z.B.: impetus to/of development: Entwicklungs~ || boost * Erhöhung der Leistungsfähigkeit
Schubgliederband push belt * geschobene Kette zur Kraftübertragg., stufenlos verstellbares Übersetz.verhältnis
Schubkraft thrust force || thrust || shearing force * Querschub, Scherkraft || shear * Querschub, Scherkraft
Schubkurbelantrieb crank drive
Schublade drawer
Schubrohr extension tube * herausgeschobenes Teil bei Spindeltrieb
Schubstange connecting rod || drive rod || side rod * bei Kurbelantrieb || torque rod * bei Kurbelantrieb || push rod || thrust rod || extension rod * herausgeschobenes Teil bei Spindeltrieb
Schukosteckdose socket outlet with earthing contact || grounding-type outlet || grounding outlet
Schukostecker plug with earthing contact || grounding-type plug
Schuld haben be responsible * verantwortlich sein || be to blame * for a thing: an etwas
Schulung training || schooling * Unterricht, Ausbildung || instruction * Belehrung, Ausbildung || practice * Übung || education * Bildung (Schule, Hochschule) || drill * strenge Schulung || indoctrination * Unterweisung, Belehrung, Schulung || coaching * Einarbeitung durch einen erfahrenen Kollegen
Schulungskurs training course || course * Kurs || course of instruction || practice * Übung || refresher course * Auffrischungskurs || workshop * ~ mit praktischen Übungen
Schulungsleiter trainer || instructor * auch Lehrer, Dozent || teacher * Lehrer
Schulungsprogramm training program
Schulungsseminar training course || training seminar
Schulungsunterlagen training documentation || course material * Kursunterlagen || training material || teaching aids * Lehrmittel
Schuss shot || weft * (Textil) beim Wirkvorgang in der Weberei || woof * (Textil) in der Weberei; warp and woof: Kette und ~ || rush * rasende Bewegung || rapid movement * rasende Bewegung || shooting * Emporschießen || working order * get into working order: in ~ bringen || good order * be in good order/running smoothly: gut in ~ sein

Schüttelfestigkeit vibration strength
schütteln shake || agitate || vibrate
Schüttgut bulk goods || bulk || bulk material || free-flowing products * (Verpack.techn.) || free-flowing material
Schüttkegel discharge pile * beim Entladen v. Schüttgut über eine Schüttrinne || material pile
Schüttrinne discharge chute * zum Entladen
Schüttstrommesser material flow meter * →Durchflussmesser für Schüttgut
Schutz protection * against: gegen, from: vor, legal: rechtlicher || coverage * Abdeckung, z.B. durch eine Versicherung, Gewährleistung || defence * Schutz, Verteidigung, Abwehr || defense * (amerikan.) Schutz, Verteidigung, Abwehr || care * Fürsorge * custody * Obhut || screen * Abschirmung || shield * Abschirmung (auch figürl.) || cover * Deckung, Abdeckung || insulation * gegen Wärme, Spannung, Lärm usw. || shelter * Zuflucht, seek/take: suchen || refuge * Zuflucht || guard * ~einrichtung/-Element/-Abdeckung z.B. an Maschinenteil || protective relaying * ~ durch Überwachungs- u. ~einrichtungen, z.B. Netz~, Maschinen~ || safeguard * ~vorrichtung, Vorsichtsmaßnahme (against: gegen), Sicherheitsklausel; sichern, schützen || protector * als Bauteil/Vorrichtung; edge protector: Kanten~
Schütz contactor * air-break contactor: Luft~; vacuum contactor: Vakuum~
Schutz- protective
Schutzabdeckung protective cover
Schutzabschaltung protective trip * auf Grund eines Fehlers, einer Grenzwertüberschreitung usw. || protective release
Schützansteuerung contactor control || contactor energizing
Schutzanstrich protective coating
Schutzart type of protection * (siehe auch 'IP') kennz.Schutz geg. Fremdkörper (IEC34-5/VDE0530/EN60034 Tl.5) || degree of protection * nur i.Zusammenhang m. Schutzarttyp verwenden, z.B. 'degree of protection IP22' || protection || type of enclosure || protection type || rating of enclosure || enclosure || protection mode || sealing * Abdichtung || mechanical degree of protection * im Gegensatz z.Zündschutzart (Ex-Schutz); siehe IEC-Publ. No.34 || degree of protection by enclosure * Gehäuse-Schutzart el. Maschinen, s. DIN IEC 34/VDE 0530 je Tl. 5 || degree of protection provided by enclosure * Gehäuse-Schutzart || protection class * Schutzklasse (siehe DIN 40050 und DIN VDE 0106 Teil 1) || enclosure protection
Schutzausrüstung safety equipment * persönliche
Schutzbeschaltung suppressor circuit * z.B. Überspannungs-Schutzbeschaltung, TSE-Beschaltung bei Thyristorsatz || suppressor network || RC circuit * RC-Beschaltung || snubber circuit * gegen Überspannung(sspitzen), z.B. TSE-Beschaltung bei Thyristorsatz || surge suppressor * Überspannungs-Schutzbeschaltung, Überspannungsableiter || protection circuit || protective circuit
Schutzbeschichtung protective coating
Schutzblech guard plate
Schutzbrille safety glasses || safety goggles * allseits geschlossene || goggles
Schutzdach canopy * z.B. für senkrecht eingebauten Motor, damit keine Fremdkörper i.d.Lüfter fallen || protective roof || shelter * auch 'Schuppen, Schutz, Unterstand, Anbau, Schuppen' || protection shield * z.B über senkrecht eingebauten Motor, damit keine Fremdkörper i.d.Lüfter fallen || protective top cover * z.B. über Motor b.Senkrechtmontage, damit keine Fremdkörper i.d.Lüfter fallen || protective cover

Schutzeinrichtung protective device || protection device || protection equipment || protection component * *Bauteil, Gerät* || protective equipment
Schutzelement protector || protective element
schützen protect * against: *gegen, from: v or, protect oneself from: sich ~ vor* || guard * against: *gegen, guard oneself against: sich ~ gegen* || defend * *verteidigen, schützen; against: gegen, from: vor* || secure * *sichern* || keep * bewahren, *from: vor* || shelter * *gegen Wetter usw. ~; from: vor* || cover * *(ab/be)decken* || shield * *abschirmen (auch figürl.)* || screen * *abschirmen* || preserve * *erhalten* || watch over * *bewachen*
schützend protecting
Schutzerde protective earth || safety ground
Schutzerdung protective earth
Schutzfunktion protective function || protection function
Schutzfunktionen protective features
Schutzgas protective gas || inert gas
Schutzgaskontakt reed contact
Schutzgaskontaktrelais reed relay
Schutzgasrelais reed relay
Schutzgebühr license fee * *z.B. für ein Druckerzeugnis* || licensing fee
Schutzgehäuse protective casing || protective housing
Schutzgeländer guard rail
Schutzgerät protecting device
Schutzgitter protective grid || barrier grid || guard grille || protective grille
Schutzgrad degree of protection * *Umfang d.Schutzes durch d.'Schutzart IP, VDE 0530 T.5: Schutzart durch Gehäuse*
Schutzhaube protective cover || protection hood || protective hood || cover || fan hood * *für Lüfter* || apron
Schutzhelm hard hat
Schutzkappe protection cover || protective cap
Schutzklasse protection class * *z.B. für elektr. Antriebe in Haushaltsgeräten nach VDE 0730 o. Explosionsschutzkl.* || safety class
Schutzkleidung protective clothing
Schutzkontakt earthing contact || ground contact || earth contact
Schützkontakt contactor contact
Schutzlack protective painting || resist
Schutzleiter protective earth conductor * *siehe z.B. VDE 0100 Teil 200* || protective conductor || protective earth * *'PE', Schutzerde* || PE conductor * *Abkürzung für 'Protective Earth': Schutzerde* || earth connection || ground connection || PE * *Abkürzung für 'Protective Earth': Schutzerde* || protective ground || equipment grounding conductor * *~ eines Geräts* || protective ground connector
Schutzleiteranschluss protective earth connection || protective earth terminal * *Klemme, Schraube* || safety earth connection || safety earth terminal * *Klemme, Schraube* || PE terminal * *Abk. f. 'Protective Earth'* || ground terminal || protective conductor connection
Schutzleiterklemme protective conductor terminal * *Erdungsklemme*
Schutzleiterpotential protective ground potential
Schutzleiterverbindung protective connection
Schutzmantel protective sheath * *für Kabel*
Schutzmaßnahme protective measure
Schutzrelais protective relay
Schutzring protecting ring || guard ring
Schutzrohr protecting tube || protection tube || cover tube || tubular jacket * *tubusförmige Ummantelung* || protection pipe
Schutzschalter protective circuit breaker || tripping relay || earth-leakage circuit breaker * *Fehlerstromschutzschalter, FI-Schutzschalter* || motor circuit-breaker * *→Motorschutzschalter* || protective motor starter * *→Motorschutzschalter*
Schutzschild protective shield
Schützspule contactor coil || contactor solenoid
Schützsteuerung contactor control || contactor control circuits
Schutzverpackung protective package
Schutzverriegelung protective interlocking || protective logic || protective interlock
Schutzvorrichtung safety device
Schutzzeichen protection mark
Schutzzünden protective firing
Schutzzündung protective firing * *von Thyristoren bei Überschreiten der max. zulässigen Sperrspannung*
Schwabbelmaschine buffing machine * *zum Hochglanzpolieren*
schwabbeln buff * *(hochglanz-) polieren*
Schwabbelscheibe buffing wheel * *zum Hochlanzpolieren* || buff || cloth wheel * *mit Textilbelag zum Hochglanzpolieren* || mop || dolly
schwach weak || poor * *schlecht, unzureichend* || feeble * *schwach, schwächlich, kraftlos, undeutlich, unbedeutend* || faint * *matt, kraftlos, schwach* || weary * *müde, matt, erschöpft (with: von/vor)* || light * *Schwach- (z.B. Last), leicht, leicht beladen* || flimsy * *dürftig, durchsichtig, schwach, fadenscheinig, lose, (hauch)dünn, zart*
Schwäche weakness || weak point * *Schwachpunkt, schwache Seite* || feebleness || faintness || exhaustion * *Erschöpfung*
schwächen weaken * *(auch figürlich)* || lessen * *vermindern* || diminish * *vermindern* || enfeeble * *entkräften* || impair || undermine * *unterminieren, z.B. Gesundheit* || debilitate * *entkräften*
Schwächephase weak phase * *wirtschaftliche*
schwaches Netz compliant supply * *nachgiebiges Netz mit hoher Netzimpedanz* || weak supply || unreliable supply system * *unsicheres Netz (mit häufigen Spannungseinbrüchen/Unterbrechungen)* || weak network
Schwachlast light load || light duty || partial load * *Teillast* || low-load conditions || small load
Schwachlastprüfung light load test
Schwachnetz compliant supply * *nachgiebig, mit hoher Netzimpedanz* || weak supply || unreliable supply * *unsicheres Netz (mit häufigen Spannungseinbrüchen, Unterbrechungen)* || weak supply network
Schwachpunkt weak point
Schwachstelle weak point
Schwachstellenanalyse weakest point analysis
Schwachstrom light current || low current || low voltage * *mit niedriger Spannung (arbeitend)*
Schwächung weakening * *(siehe auch 'Abschwächung')*
Schwallöten flow soldering || flow-soldering
schwallwassergeschützt splash-proof
schwanken vary * *variieren, sich ändern; auch Preise, Kurse* || oscillate * *schwingen* || wave * *schwanken, sich wiegen, sich hin und herbewegen; (to and fro: hin und her)* || swing * *schwingen, sich wiegen* || rock * *erbeben, sich wiegen, schaukeln* || wobble * *wackeln* || shake * *schütteln, erbeben* || falter * *zaudern, zögern* || waver * *zaudern, unschlüssig sein* || alternate * *abwechseln* || fluctuate * *~ (auch Preise, Kurse usw.), sich (ständig) verändern, fluktuieren; unschlüssig sein* || shillyshally * *unentschlossen sein, zaudern* || back and fill * *(amerikan.) unentschlossen sein, zaudern* || deviate * *abweichen, ablenken* || stagger * *wanken, schwanken, taumeln, torkeln* || reel * *wanken, taumeln* || yaw * *→gieren (Schiff, Flugzeug, Kran); abweichen (figürl.; from: von)*

schwankend varying * *Wert, Größe, Signal, Anzahl* || hesitant * *zögernd* || undecided * *unentschlossen* || unstable * *veränderlich, instabil*
schwankendes Netz unstable mains || fluctuating supply
Schwankung fluctuation || variation || deviation * *Abweichung* || oscillation * *Schwingung* || waving * *Schwenken, Hin- und Herbewegung* || rocking * *Schaukeln, Wiegen*
Schwankungen fluctuations || variations
Schwanz tail * *auch Stromschwanz, Rücken eines Impulses usw.*
schwarz black
Schwarz-Weiß-Monitor monochrome monitor
Schwärzung blackening
Schwarzweiß-Bildschirm monochrome monitor
Schwebe- suspension
Schwebedüse flotation nozzle * *z.B. zur Bildung e.Luftkissens unter einer Warenbahn*
schweben be suspended || hang || hang in the air || be poised || hover * *auf Luftkissen, über einer Stelle* || levitate * *frei schweben (lassen), z.B. Magnetschwebebahn* || float * *durch die Luft, in einer Flüssigkeit* || glide * *(durch die Luft) gleiten*
schwebend floating * *z.B. Bezugspotenzial zu einem Signal*
schwebendes Bezugspotenzial floating ground
Schwebesichter air sifter * *z.B. zum Sortieren von Holzschnitzeln* || suspension sifter * *Sortieranlage (z.B. für Holzspäne)*
Schwebestoffe suspended matters || floating particles * *Schwebeteilchen*
Schwebetrockner flotation dryer * *z.B. in einer Folienbeschichtungsanlage*
Schwebung beat
Schwebungsfrequenz beat frequency
Schweden Sweden
schwedisch Swedish
Schwefel sulfur
Schwefeldyoxid sulphur dioxide * *['salfur 'daioxaid] SO2* || SO2
Schwefelkohlenstoff carbon disulphide
Schwefelwasserstoff hydrogen sulphide * *['haidridjin 'salfaid] H2S* || H2S
Schweißbacke sealing bar * *{Verpack.techn.} z.B. zum Verschließen von Kunststoff-Schlauchbeuteln*
Schweißbrenner welding torch
schweißen weld
Schweißerei welding shop || welding plant
Schweißgerät welding equipment
Schweißmaschine welding machine
Schweißnaht welding seam || weld
Schweißtakt welding cycle
Schweißtechnik welding technology
Schweißtrafo welding transformer
Schweißung welding || weld || welding operation
Schweißzange welding gun
Schweiz Switzerland
schwelen smoulder || burn slowly || burn by a slow fire
Schwelle threshold || step * *Stufe* || barrier * *Barriere* || sill * *Türschwelle, Schwellbalken* || sleeper * *{Bahnen} Eisenbahn~*
Schwellenspannung threshold voltage
Schwellspannung threshold voltage
Schwellwert threshold value || threshold
Schwenk- swivel * *z.B. Achse, Hebel, Arm* || slewing || swiveling || revolving || hinged * *angelenkt, mit Scharnier versehen* || tilting * *z.B. Fräser* || swinging || pivoting || orientable * *im Winkel einstellbar* || angular * *Winkel-* || folding * *faltbar* || rocking * *hin- und herschwenkend/-schaukelnd* || rotating * *drehbar* || swing * *swing-out: ausschwenkbar* || turning * *sich drehend* || sluable || oscillating * *hin- und herschwingend* || swivelling

Schwenk-Neigefuß tilt/swivel stand * *für Computer-Monitor*
Schwenkarm swivel arm || slewing arm || swivelling arm || swinging lever
schwenkbar swivelling || pivoted || revolving * *drehbar* || slewing * *z.B. Kran* || sluable * *z.B. Kran* || hinged * *angelenkt, z.B. über Scharnier* || pivoted * *drehbar gelagert* || swing-out * *ausschwenkbar* || rotatable * *drehbar* || traversable * *(aus-) schwenkbar* || swivel-mounted * *schwenkbar gelagert* || which can be pivoted || tilting || slewable
Schwenkbiegemaschine folding machine * *zur Blechbearbeitung*
Schwenkbiegmaschine folding machine
schwenken traverse * *(seitwärts) schwenken, (sich) drehen, durchqueren* || rotate * *um eine Achse drehen* || slew round * *herumschwenken/drehen, im Kreis herumschwenken, baumeln lassen* || swing * *schwingen, schwenken, ein/ausschwenken lassen* || swivel * *auf einem Zapfen schwenken, drehen* || pivot * *(ein)schwenken, sich drehen (upon/on: um), drehbar lagern* || displace * *z.B. Phasenwinkel zur Steuersatzsynchronisierung mittels eines →Schwenktrafos* || slew * *herumdrehen, (herum)~* || sway * *schwenken, (sich) neigen, schaukeln, (sich) wiegen* || slue * *herumdrehen/-schwenken* || cant || tilt
Schwenkfräse tilting-spindle moulder * *{Holz}*
Schwenkfräsmaschine tilting-spindle shaper * *{Holz} zum Erzeugen von Profilen*
Schwenkhebel pivoted lever
Schwenkkran slewing crane
Schwenkrahmen swivel frame || swing-frame
Schwenktaster twist switch || rotating key || momentary-contact rotary switch
Schwenktrafo phase-displacement transformer * *z.Phasendrehung z.B.z. Erzeugung d. Synchron.-spanng. e. Steuersatzes*
Schwenktür hinged door
Schwenkvorrichtung swiveling mechanism || swinging device || swing-out mechanism * *Ausschwenkmechanismus*
Schwenkwerk slewing gear * *{Hebezeuge} bei Kran, Bagger* || slewing drive * *{Hebezeuge} bei Kran, Bagger* || swivel unit * *{Hebezeuge}* || swivelling gear * *{Hebezeuge}*
Schwenkwerksantrieb slewing drive * *Kran* || pivoting drive
Schwenkwinkel angle of traverse || swiveling angle || angle of displacement * *bei →Schwenktrafo*
schwer heavy * *Gewicht* || hard * *schwierig* || severe * *ernst, hoch (Kosten, Steuer, Gebühr, Strafe)* || heavy duty * *für Schwerlast- oder rauen Betrieb* || fatal * *fatal (z.B. Fehler), den ordnungsgemäßen Ablauf verhindernd, tödlich, verhängnisvoll* || difficult * *→schwierig*
schwer zugänglich hardly accessible || difficult to access
Schweranlauf heavy-duty starting || heavy starting || high-inertia starting || starting against high-inertia load || high starting duty || starting tough loads
schwere Belastung heavy load || tough load || heavy duty
Schwergut heavy load
Schwerkraft gravity || force of gravity
Schwerkraft- gravity-type
Schwerkraftförderer gravity conveyor
Schwerkrafttank gravity tank
Schwerlast heavy duty || heavy load
Schwerlast- heavy duty || heavy-duty
Schwerlastanlauf heavy-duty starting || heavy starting || high-inertia starting || starting against high-inertia load || starting tough loads

Schwerlastbetrieb heavy-duty operation || rough service
Schwerlasttransporter heavy duty lorry
Schwerpunkt focus * (figürl.) || Haugtaugenmerk, Schwerpkt. (d. Interesses, d.Bemühungen, d. Arbeit;) Brennpkt. || centre of masses * Massen~ || centre of gravity * (brit.) || center of gravity * (amerikan.) || main field * (figürlich) Haupt-Betätigungs/Anwendungsfeld || crucial point * (figürlich) || focal point * (figürlich) || emphasis * Nachdruck || priority * Vorrangigkeit || point of main effort || focus on * (Verb) ~ setzen auf, sein Hauptaugenmerk setzen auf || center of area * Flächen~
Schwerpunktanwendungen main applications
Schwerpunktmethode center of gravity method
schwerste Bedingungen stringent operating conditions
schwerzugängliche Stellen tight corners || awkward positions || hard-to-access positions
schwierig difficult * auch 'schwer zu behandeln' (Personen) || hard * hart, schwer, mühsam, anstrengend; work/job: Arbeit || tough * unangenehm, zäh, übel, 'bös', 'ein saures Stück Arbeit darstellend' || troublesome * mühsahm, lästig, beschwerlich || complicated * kompliziert || intricate * verwickelt || tricky * (salopp) verzwickt || precarious * misslich || trying * kritisch, unangenehm, nervtötend || critical * kritisch || problematical * problematisch || irksome * lästig, beschwerlich || onerous * lästig, beschwerlich || awkward * unangenehm || particular * eigen, wählerisch || exacting * anspruchsvoll (z.B. Kunde) || puzzling * question: Frage || arduous * schwierig, anstrengend, mühsam; task/job: Aufgabe || sticky * schwierig, heikel, kritisch; zäh, klebrig; schwül, stickig
Schwierigkeit difficulty || crisis * Krise || obstacle * Hindernis; put obstacles in a person's way: jdm. ~en machen || problem * Problem || dilemma * schwierige Lage || trouble * Schwierigkeiten; give trouble: einer Person ~en machen || bottleneck * →Engpass
schwimmend floating * auch elektr. Potenzial
Schwimmer float
schwinden dwindle * abnehmen || grow less * abnehmen || fall off * abnehmen || shrink * schrumpfen || contract * sich zusammenziehen, z.b. erhitztes Metall in der Abkühlungsphase || fade * z.B. Ton, Licht, Farbe || fade away * z.B. Ton, Licht, Farbe || disappear * ver~ || vanish * ver~
Schwingamplitude vibration amplitude * einer mechanischen Schwingung
Schwingbandförderer vibrator conveyor
schwingen vibrate * mechanisch || oscillate * periodisch || swing * sich wiegen, pendeln || sway * schwanken, schaukeln (sich) wiegen || rock
Schwingfestigkeit vibration resistance || vibratory stressability || anti-vibration || vibrostability || immunity to vibration
Schwingförderer vibrating conveyor
Schwingfrequenz vibration frequency * ~ einer mech. Schwingung
Schwinggeschwindigkeit vibration speed || oscillating velocity * bei Schwingung-/Rütteltest [mm/s] || vibration velocity * Augenblickswert d. Geschw. e. mech. Schwingg., DIN ISO3945, DIN5483, VDE0530 T.1
Schwinghebel rocker
Schwingkreis resonant circuit || oscillating circuit
Schwingkreis-Wechselrichter parallel-tuned inverter
Schwingkreiskondensator resonant circuit capacitor
Schwingkreiswechselrichter resonant circuit inverter
Schwingmeißel oscillating chisel * {Holz}

Schwingmetall rubber-metal vibration damper
Schwingquarz quartz crystal || Xtal * (salopp) || quarz crystal unit * als Komplettbauteil (im Gehäuse, mit Beschaltung usw.)
Schwingschleifer orbital sander * {Holz}
Schwingstärke vibration severity * mechan.; für el. Maschinen: siehe DIN ISO 2373, IEC34 Tl.14, DIN45660 u. VDI2065 || vibration severity grade * ~stufe, gemess. Eff.wert d. →Schwingeschwindigkeit, DIN ISO3945
Schwingstärkestufe vibration severity grade * Klassifiz.stufe für →Schwingstärke; el. Masch.: VDE0530 T.14, VDI 2056 || vibration level * für Elektromotoren: siehe z.B. DIN ISO 2373
schwingsteif vibro-stable
Schwingung vibration * mechanische ~ || oscillation * (elektr.) periodische; forced: erzwungene; free: freie; damped: gedämpfte || pulsation * Pulsieren || harmonic * (höhere) harmonische ~ || subharmonic * subharmonische ~ || wave * (elektr./ magn.) Welle
Schwingungen vibrations * Mechanische || oscillations * periodische || harmonics * (höhere) harmonische || subharmonics * subharmonische
Schwingungsamplitude vibration amplitude * Schwingwegamplitude einer mechanischen Schwingung || deflection amplitude * e. mechan. Schwingung (max.Durchbiegg. der in Schwingg. versetzten Teile)
Schwingungsanregung vibration excitation * Anregung einer mechan. Schwingung || oscillation excitation * Anregung einer elektr. Schwingung || excitation of vibrations * (mechanisch)
schwingungsarm low-vibration
Schwingungsbeanspruchung vibration stressing * mechanische || vibrational load
Schwingungsdämpfer vibration damper * gegen mechan. Schwingungen, z.B. zur Motoraufstellung/-hängung || shock absorber * Stoßdämpfer || vibration absorber * z.B. zur Aufhängung von Maschinen || antivibration mountings * z.b. zur Aufstellung eines Eletromotors
Schwingungsdämpfung vibration damping * mechan. || vibration control * mechan.; über Regel-/ Steuereinrichtung
Schwingungsdauer period of oscillation
Schwingungseigenschaften vibration characteristics * von mechan. Schwingungen
schwingungsfrei non-oscillating || non-oscillatory || non-vibrating * (mechan.)
Schwingungsgehalt pulsation factor || harmonic content * von Oberwellen
schwingungsisoliert vibration-damped
Schwingungsmesser vibration meter || vibrometer || instrument for measuring vibrations
Schwingungsmessgerät vibration meter || vibrometer || instrument for measuring vibrations
Schwingungsmessung vibration measurement || vibration measuring
Schwingungspaketsteuerung multi-cycle control * AC-Steller-Steuerverfahren m.Durchschaltg. ganz. Netzvollschwing.pakete || full-wave control * Vollwellensteuerung || multicycle control
Schwingungsstärke vibration severity * mechan.
schwingungssteif vibro-stable
schwingungstechnische Auslegung vibration-related design || vibration-related dimensioning
Schwingungsverhalten vibrational behavior * mechanisches || vibration characteristics * mechanisches
Schwingwegamplitude vibration amplitude * einer mechan. Schwingung || deflection amplitude * einer mechan. Schwingung (max.Durchbiegung des in Schwingung versetzten Teils)

Schwitzwasser condensate * *Kondenswasser* || condensation water
Schwitzwasserbildung deposit of condensate * *Ablagerung v. Schwitzwasser, Betauung* || deposit of moisture * *Feuchtigkeitsablagerung*
Schwitzwasserloch condensate drain hole * *Kondenswasser-Ablaufloch (z.B. bei {el. Masch.})*
Schwund decay * *Rückgang* || loss * *Verlust* || dwindling || shrinkage * *Schrumpfung* || leakage * *durch Aussickern* || fading * *Bremse, Kupplung, Radio*
Schwung swing * *Schwung, Schwingen, Schwingung, ~weite, Ausschlag (Pendel); in full swing: in vollem Gang* || headway * *Geschwindigkeit* || rise * *Auf~* || impetus * *(figürl.)* Antrieb, Tatkraft || energy * *Energie, Wucht* || vitality * *Vitalität* || drive * *Wucht* || verve * *Schmiss* || batch * *Menge, Schub* || thrust * *Schub(kraft)* || kick * *(salopp) Elan, (Stoß) Kraft, Energie, Nervenkitzel*
Schwungkraft centrifugal force * *Zentrifugalkraft, Fliehkraft* || momentum * *(auch figürl.) siehe auch →Schwung*
Schwungmasse rotating masses * *rotierende Masse* || inertia * *Trägheitsmoment* || rotating mass
Schwungmassenantrieb high-inertia drive * *Antrieb mit hoher Schwungmasse* || inertia drive
schwungmassenarm low-inertia
Schwungmoment moment of inertia * *Massenträgheitsmoment GD^2 in $[kpm^2]$; SI-Einheit: J in $[kgm^2]$* || inertia * *Massenträgheit* || flywheel moment * *auch "GD^2" genannt; Umrechnung: Trägh.mom. J entspr. $(GD^2)/4$; J $[kgm^2]$, GD^2 $[kpm^2]$* || flywheel efect || load flywheel effect * *als Motorlast beim Beschleunigen/Verzögern*
Schwungmomentenkompensation inertia compensation || acceleration compensation || moment of inertia compensation
Schwungmomentkompensation inertia compensation || moment of inertia compensation || acceleration compensation
Schwungrad flywheel || fly wheel
sebststperrend self-locking
Sechseckbauweise hexangular design || hexangular-frame design * *z.B. eines Elektromotors*
sechseckig hexagonal
Sechskantmutter hexagon nut || hex nut || hexagonal nut
Sechskantschraube hexagon-head bolt * *dickere Schraube, Schraube mit (teilw. gewindelosem) Schaft* || hexagon-head screw * *dünnere Schraube*
Sechsphasenmotor six-phase motor * *{el. Masch.} AC-Motor mit zwei Drehstromwicklungen in insgesamt 6 Wicklungssträngen*
Sechspuls- six-pulse
sechspulsig six-pulse
sechspulsige Brückenschaltung six-pulse bridge connection
See- marine || maritime * *-Recht, -Versicherung* || sea || navy * *Kriegsmarine-* || shipping * *Seefahrts-* || naval * *Marine-* || nautical * *nautisch, seemännisch*
Seefracht sea freight
Seekabel marine cable
seemäßig verpackt packed for sea transport
seemäßige Verpackung seaworthy package || seaworthy packaging
Seetransport sea transport
seewasserbeständig seawater-resistant
seewasserfest sea water-proof
Segment segment * *~ eines Kreises, eines Kommutators, einer Sieben~-Anzeige, Speicher~ usw.* || rung * *Strompfad in der Kontaktplandarstellung eines SPS-Programms*
Segmentanzeige segment display
Segmentierung segmentation
sehr very || strongly * *(Adv.) nachdrücklich, kräftig, stark* || particularly * *besonders* || very much * *(Adv.)* || highly * *(Adv.) höchst, äußerst* || greatly * *(Adv.) höchst, äußerst* || extremly * *(Adv.) extrem, höchst, äußerst* || a lot * *(salopp) sehr viel* || awfully * *(salopp; Adv.) siehe auch →reichlich* || pretty * *(salopp) ganz schön, beträchtlich, ziemlich, einigermaßen*
Sehr geehrte Dear
Sehr geehrte Herren Gentlemen:
Sehr geehrter Dear
Sehr geehrter Herr Dear Mr.
sehr gut excellent || perfect || well done * *(anerkennend)* || fine * *fein, vorzüglich* || superior * *hervorragend* || outstanding * *hervorragend*
Seil rope * *Tau, Strick, Strang; auch bei Aufzug, ~bahn, Kran; one-strand rope: eintrümmiges ~* || cable * *Tau, Draht~; auch bei ~bahn* || cord * *Schnur* || line * *Schnur, Leine, Tau* || wire rope * *Draht~* || pumping line * *für Bohrpumpenantrieb*
Seilantrieb rope drive || cable drive
Seilaufhängung rope suspension * *{Aufzüge}*
Seilbahn aerial railway || aerial ropeway || ropeway || cable railway || cable car || cableway
Seildehnung cable stretching * *{Aufzüge | Hebezeuge}* || rope stretching * *{Aufzüge}* || elongation of cable * *{Aufzüge}*
Seileinlaufkorrektur drum way correction * *{z.B. Aufzüge}*
Seilfutter drum lining * *{z.B. Aufzüge}*
Seilscheibe head wheel * *z.B. bei Aufzüge* || cable pulley || rope pulley || pulley || grooved roller || snatch block * *kleine ~, z.B. an Flaschenzug* || sheave * *Rillenscheibe, Leitrolle, Auflaufhaspel*
Seilschwingungen rope vibrations * *unerwünschter Effekt bei Aufzug*
Seiltrommel cable drum * *für Drahtseil* || rope drum * *auch für Drahtseil*
Seilwinde cable winch
seinerseits on his part * *(Adv.)* || for his part * *(Adv.)* || in his turn * *(Adv.)*
seit since * *von .. an; siehe auch →ab; since then/ from that time: seitdem* || for * *während; for some days: seit einigen Tagen*
Seite side * *on the side: auf der ~; aside: zur ~, bei~, abseits, seitwärts* || page * *~ in Schriftstück/ auf dem Bildschirm* || member * *~ einer Gleichung* || party * *Partei* || aspect * *~ einer Angelegenheit* || flank * *Flanke* || Pg. * *Abkürzung für 'Page'* || hand * *on the one/other hand: auf d.einen/anderen ~; on the right/left hand: auf d. recht./link. ~*
Seiten- side || lateral * *seitlich, Quer-, Seiten-, Neben-, von der Seite*
Seitenansicht side view || profile view * *siehe auch →Seitenriss*
Seitendeckel side cover
Seitenführung side guide * *z.B. Bandführungs-Einrichtung im Walzwerk* || side guiding * *Funktion bei durchlaufendem Bandmaterial* || guide plate * *Leitblech* || side ledge * *in Form einer (Anschlag-) Leiste*
Seitenkanal-Vakuumpumpe side-channel vacuum pump
Seitennummer page number
Seitenprofil side section
Seitenrahmen side frame
Seitenriss profile view * *in techn. Zeichnung* || side elevation * *in techn. Zeichnung* || profile section * *in techn. Zeichnung*
Seitenteil side panel * *seitliche Wand/Blech/Platte/Holz* || side wall * *Seitenwand, auch eines Gerätes*
Seitenwand side-panel * *z.B. eines Schaltschrankes* || side-wall || side plate
seitlich to the side || side || lateral || at the side * *(Adv.)* || on the side * *(Adv.)* || laterally * *(Adv.)*
seitlich links on the left hand side || on the left
seitlich rechts on the right-hand side || on the right

seitwärts sideways || sidewards || sideward || aside * beiseite, zur Seite, abseits, seitwärts; aside from: abgesehen von || lateral * seitlich, ~ befindlich || at the side * (Adv.) seitlich
Sekretärin secretary
Sektor sector * auch auf der Festplatte
Sektormotor sector motor * mit im Kreissektor angeordnetem "Käfigläufer" z.B. bei Pressen u. Steinbrechern
sekundär secondary
Sekundärdeite secondary * z.b. eines Transformators
sekundärgetaktetes Schaltnetzteil secondary switched power supply * hinter dem Übertrager getaktetes Schaltnetzteil
Sekundärseite secondary * eines Trafos
sekundärseitig secondary || secondary-side
Sekundärspannung secondary voltage
Sekundärwicklung secondary winding * Transformator || secondary
Sekundärwicklungen secondary windings || secondaries
Sekunde second
Sekundenbereich seconds range * in the : im ~ || vicinity of seconds * in the: im ~
selbst self || in person || personally || by oneself * ohne fremde Hilfe || alone * ohne fremde Hilfe || unaided * ohne fremde Hilfe || one's own * eigen || even * (Adv.) sogar || without assistance * ohne fremde Hilfe || even * auch nur || volontary * von ~, freiwillig || automatically * von ~, automatisch || in oneself * ohne fremde Hilfe || I myself * ich ~ || of itself * automatisch
Selbst- self- || auto-
Selbstabgleich autotuning || self-tuning || self calibration * eines Messsystems || automatic tuning * z.b. eines Reglers || auto-calibration * einer Messeinrichtung
Selbstanlauf self-starting
selbstanlaufend self-starting * Motor usw.
selbstanpassend self-adaption || self adapting || auto-adapting
selbstansaugend self-priming * Pumpe || self-aspirating * Verbrennungsmotor usw.
selbsttätig automatic * (korrekte Schreibweise ist → "selbsttätig") || auto-
selbstbelüftet non-ventilated * siehe →selbstgekühlt || surface-cooled || free-convection || with natural air cooling * mit Konvektionskühlung, z.B. Leistungsteil mit Kühlkörper ohne Lüfter || natural-convection || self-ventilated
Selbstbelüftung natural air cooling * Luftselbstkühlung
selbstbilanzierender Bereich profit center
Selbstdiagnose self diagnosis || auto diagnosis || self-diagnostic
selbstdichtend self-sealing
selbstdokumentierend self-documenting
selbsteinstellend sef-tuning * z.B. Regler || self-adapting * selbstanpassend
Selbsteinstellung auto-tuning * z.B. Regler || self-setting * z.B. Regler
Selbstentladung self-discharge
selbsterklärend self-explanatory || self-evident
Selbsterregung self-excitation
Selbsterregungsdrehzahl built-up speed * critical: kritische
selbstfahrend automated guided * z.B. Transportfahrzeug || self-moving || self-propelled
Selbstführung self commutation * für Stromrichterschaltung
selbstgeführt self-commutating || self-commutated
selbstgeführter Stromrichter self-commutated converter * benöt. kein. ext. Kommutier.spanng. (Ggs.: last-/netzgef.;DIN41750 Tl.1)

selbstgekühlt non-ventilated * z.b. Motor ohne Lüfter, Wärmeabfuhr durch Abstrahlung an der Gehäuseoberfläche || surface-cooled * oberflächengekühlt, z.B. Motor ohne Lüfter (f. Motor wenig gebräuchlicher Ausdruck) || free-convection * →luftselbstgekühlt (Kühlart für Leistungshalbleiter) || self-cooling || with free convection * konvektionsgekühlt (Leistungshalbleiter/-teil)
selbstgesteuert self-controlled * z.B. Stromrichter
selbstgetakteter Stromrichter self-clocked converter
Selbsthaltekontakt lock-type contact || seal-in contact || self-holding contact || remanent contact || maintained contact || latching contact
Selbsthalteschaltung remanent circuit || seal-in circuit
Selbsthaltung locking * remain locked in: in ~ gehen/bleiben || sealing in || sealing home * be sealed home: in ~ gehen || latching || sealing || maintained function * maintained command: Befehl mit ~ (Befehlsflanke speichert die Funktion)
selbstheilend self-recuperating * z.b. Kondensator nach Spannungüberbeanspruchung/Durchschlag || self-healing * Kondensator nach Spannungsüberbeanspruchung/Durchschlag || self-restoring * Isolierung usw.
Selbsthellung self-healing * z.b. von Kondensatoren nach einem Durchschlag des Dielektrikums
selbsthemmend self-locking * z.b. Gewinde, Getriebe (kein Rücklauf mögl. bei Wegbleiben o. Umkehr d. Antriebsmoments) || irreversible * nicht rückwärtslaufend (z.B. Getriebe)
Selbstinduktion self-induction || self-inductance
Selbstinduktivität self-inductance || coefficient of self-induction
Selbstklebe-Etikette adhesive label
Selbstklebeetikett adhesive label
selbstklebend self-adhesive || self-adherent || adhesive || gummed * gummiert || self-adhering
Selbstkostenpreis prime cost || cost price || net cost price || first cost || short price || cost to manufacture
selbstkühlend selfcooling * z.B. Motor || nonventilated * unbelüftet, z.B. Motor
Selbstkühlung self-cooling || natural cooling * ohne Lüfter || natural convection * durch natürliche Konvektion (ohne Lüfter) || natural-air cooling * →Luftselbstkühlung von Leistungshalbleitern nur mit Kühlkörper, ohne Lüfter || convection cooling * Konvektionskühlung
selbstlernend self-learning
Selbstoptimierend self-tuning || self-optimizing
Selbstoptimierung self-optimization || selftuning || autotuning || auto-tune
Selbstprüfung operator inspection * Werkstückprüfung durch den Maschinenbediener selbst (z.B. bei e. Werkz.masch.)
selbstreinigend self-cleaning
selbstschmierend self-lubricating
selbstschneidende Schraube self-tapping screw
selbstsichernd self-locking * z.B. Schraube, Mutter
selbstsperrend self-locking
selbstständig self-reliant * selbstbewusst, selbstsicher || independent * unabhängig, auch beruflich || self-contained * autark, eigenständig, unabhängig, in sich geschlossen; z.B. Gerät, Maschine || self-supporting || separate * getrennt, eigenständig || established * beruflich, geschäftlich || unaided * ohne Beistand/Hilfe || freelance * freiberuflich (tätig), unabhängig * without assistance * ohne Hilfe/Unterstützung || responsible * verantwortlich || independently * (Adv.) act independently/on one's own initiative: ~ handeln
selbstständig sein run one's own business

selbsttätig self-acting || automatic || auto- || automatically * *(Adv.)* || independent
Selbsttest selftest * *power-on selftest: b.Einschalten d.Versorgungsspanng. autom. durchgeführter* ~ || automatic check || self-testing || self test || selftest
selbsttragend self-supporting
Selbstüberwachung self-monitoring || self-supervision || self-diagnostics * *Selbstdiagnose*
selbstverlöschend self-extinguishing * *z.B. Brennbarkeitsstufe* || auto-extinguishing * *bei Feuer*
selbstverständlich self-evident || that goes without saying * *(Adv., nachgestellt) das ist* ~ || obvious || a matter of course * *Selbstverständlichkeit* || of course * *(Adv.)* || naturally * *(Adv.) natürlich* || it stands to reason that * *es ist* ~, *dass* || taken for granted * *take a thing for granted: etwas für* ~ *halten*
Selbstverständlichkeit matter of course || foregone conclusion || truism * *Binsenwahrheit* || self-evident fact
selbstzentrierend self-centering || self-nesting * *z.B. bei Steckverbindung*
selektieren select
selektiert selected
selektiv selective
Selektivität selectivity * *bei Schutzeinrichtung/system, Radio usw.* || discrimination * *Unterscheidung(svermögen); overcurrent discrimination: Selektivität e. Sicherung* || grading * *Stufung, Staffelung; time grading: Zeitselektivität*
Selen selenium
selten rare || infrequent || little used || seldom
Seltene-Erden-Magnet rare-earth magnet * *z.B. aus →Samarium-Kobalt (SmCo) für d. Läufer e. permanenterregten E-motors*
seltener more rarely * *(Adv.)*
Seltenerdmagnet rare-earth magnet
Semantik semantic * *auch einer Programmiersprache*
Semester semester
Semikolon semicolon
Seminar seminar * *Universität* || tutorial || course * *Lehrgang* || training course * *Lehrgang* || workshop * *Schulungskurs mit praktischen Übungen*
Sendeauftrag transmit request * *Sendeanforderung* || transmit task
Sendebaustein transmit block * *Funktionsbaustein* || transmitter block * *Funktionsbaustein* || transmitter
Sendebereitschaft ready to send * *als Signalname (z.B. "RTS" bei RS232-Schnittstelle)*
Sendedaten transmit data
Sendekanal transmit channel || transmission channel
senden transmit * *Nachricht/Signal über Funk, Draht, serielle Schnittstelle usw.* ~ || send * *Geld, Brief, Nachricht usw.* ~ || dispatch * *ver*~ || broadcast * *Nachricht an mehrere Empfänger gleichzeitig* ~
Sender transmitter || sender * *Ab*~ || broadcasting station * *Rundfunk~*
Sender-Empfänger transceiver * *Abk. f. 'TRANSmitter/reCEIVER': Zusammenfassung von Sende- u. Empfangsbaustein(en)*
Senderichtung transmit direction
Sendezeichen transmit character
Sendung shipment * *Absendung, Auslieferung von Gütern* || sending * *Absendung* || transmission * *Nachrichtenübermittlung/Übertragung* || transmittal * *Nachrichtenübermittlung* || broadcast * *Rundfunk/Fernsehsendung* || consignment * *[Wirtschaft] (Fracht-) Sendung, Lieferung, Konsignation, Komission(sware), Versand*
Senkbremsschaltung dynamic lowering circuit

* →*Gegenstrombremsung bei Kran mit Schleifringläufermotor*
Senkbremsung dynamic lowering * *el. Bremsg. zur Geschw.begrenzg. b.Hebezeug über Vorwid. bei Schleifringläufermotor*
Senke sink * *data sink: Daten~* || drain * ~ *für Flüssigkeit, Ladungsträger, Elektronen usw.*
senken reduce * *reduzieren (by: um)* || let down * *herunterlassen* || lower * *niedriger machen, herunterlassen (z.B. Last am Kran)* || dip * *neigen* || cut * *Ausgaben, Kosten, Preise, (Energie-) Verbrauch* || give way * *sich senken, nachgeben, z.B. Boden, Gebäude usw.* || settle * *sic ~ (Fundament usw.)* || sink * *sich ~, sinken* || drop * *sich ~, abfallen*
Senker countersink * *(kegelförmiges) Werkzeug zum Anfasen u. Entgraten von Bohrlöchern* || counterbore * *Werkzeug zum Entgraten/Anfasen von Bohrlochern*
Senkerodieren spark erosion
Senkkopfschraube countersunk head screw
senkrecht vertical || perpendicular * *(geometr.) einen rechten Winkel bildend, lotrecht (to: zu)* || vertically * *(Adv.) vertically above each other: ~/rechtwinklig aufeinander* || right-angeled * *→rechtwinklig*
Senkrechte vertical position || vertical plane || perpendicular position
senkrechte Bauform vertical type || vertical-shaft type * *Motor, Getriebe usw.* || book-size design * *schmal wie ein Buch*
Senkrechtförderer vertical conveyor
Senkschraube counter-sunk screw || countersunk screw
Senktiefe sinking depth
Senkung reduction * *Reduzierung (Preise usw.)* || cut * *Kosten, Preise* || lowering * *Preise usw.* || subsidence * *(Erd-/Fundament-) Senkung, Absinken, Nachlassen, Abflauen (auch figürl.)* || set * *Fundament usw.* || sag * *Mauer, Decke usw* || incline * *Neigung* || slope * *Neigung* || dip * *Neigung, Senkung, Gefälle, Neigungswinkel* || sinking * *(Ver)Sinken* || reducing * *Reduzierung*
sensibel sensitive
Sensor sensor || sensing device || sensing element
Sensorik sensors
SEP SPS * *Abk. f. 'Standard Packaging Slot': Standard-Einbauplatz*
separat separate || special || individual * *eigen*
Separator separator * *Trennanlage (z.B. mit Zentrifuge)*
Sequenz sequence * *Aufeinanderfolge, Reihenfolge, Reihe, Serie; Escape sequence: Steuersequenz (für Drucker)* || sequel * *(Aufeinander-) Folge, Folgeerscheinung, (Aus)Wirkung, Konsequenz; in the sequel: in d. Folge*
sequenziell sequential
SERCOS SERCOS * *Abk.f. 'SErial Realtime COmmunication System': schnell. Kommunikat.system für Numerikantriebe*
Serie series || range * *Bau-/Produktreihe* || programme * *(Verkaufs/Produkt-) Programm*
seriell serial
serielle Datenübertragung serial data transmission
serielle Kommunikation serial communications
serielle Kopplung serial link || serial interface || serial communication
serielle Kopplung zwischen gleichberechtigten Partnern peer-to-peer connection
serielle Schnittstelle serial interface || serial port * *Anschluss* || serial link * *serielle Verbindung* || serial communication * *Kommunikation, Datenaustausch* || serial connection * *Verbindung* || comm port * *(salopp) an einem Computer usw.*
serielle Übertragung serial transmission

serielle Verbindung serial link || serial communication * *Kommunikation*
serielle Vernetzung serial networking
Serienausführung standard design || standard type || series-produced type
Serieneinsatz series availability * *Verfügbarkeit in Serienstückzahlen* || start of volume production * *Beginn der Serienproduktion* || start of series production
Serienfertigung series production || mass production * *Fertigung in Großserie* || high-volume production * *Fertigung in hoher Stückzahl* || serial production
Serienherstellung series production || series manufacturing || mass production * *Großserienfertigung* || large-scale production * *Großserienherstellung*
Serieninduktivität series inductance
serienmäßig standard * *(Adj.)* || as standard * *(Adv.)* || series-produced || standard-design * *in der serienmäßigen Ausführung*
Seriennummer serial No.
Serienprodukt standard product || series-produced type
Serienproduktion series production * *Serienherstellung* || volume production * *siehe auch* →*Großserienfertigung*
Serienprüfung serial test || batch testing
Serienschaltung series connection || connecting in series * *das In-Serie-Schalten*
Serienstart start of series production * *siehe auch* →*Serieneinsatz*
Serienwiderstand series resistor * *als Bauteil* || series resistance * *als physikal. Größe, z.B. im Ersatzschaltbild*
Server server * *Zentralcomputer i.e. 'Client-Server Netzwerk', verwaltet d. gemeinsame (Dokumenten)Datenbank*
Service service || customer support * *Kundenunterstützung* || product support * *Kundenunterstützung für ein Produkt* || servicing * *das 'Service-Leisten', -leistung, Erbringen einer -leistung, Warten*
Service durchführen service
Service Factor Service Factor * *zuläss. Überlast/ Bemess.leistg. e.durchzugsbelüft. Elektromotors (NEMA MG1-14.36)* || SF * *Abk. für* →*Service Factor*
Service-Abteilung Field Services * *Außendienst* || Service Centre || Service Center * *(amerikan.)* || service department
Service-Ingenieur field engineer || service engineer
Service-Netz service network
Service-Schnittstelle service interface
Service-Techniker service engineer || service technician
Serviceabteilung service department || field service * *Außendienst* || Field Services * *Außendienst (Abteilungsbezeichnung)* || Service Centre * *Servicezentrum (Abteilungsbezeichnung)* || Service Center * *Servicezentrum (Abteilungsbezeichnung)*
Serviceaufwand maintenance costs * *Wartungsaufwand* || service effort
Serviceeinsatz service assignment * *von Servicepersonal* || service mission * *von Servicepersonal* || service call
Servicefahrzeug service vehicle
Servicefaktor service factor * *Verhältn. zulässige Motordauerleistung/Motornennleistung* || overload factor
Servicefall case of maintainance || when service work becomes necessary * *im ~*
servicefreundlich easy to service || easy to maintain * *wartungsfreundlich* || easy-to-service || service-friendly
Servicefreundlichkeit service-friendliness || ease of service

Servicefunktion servicing function
Servicefunktionen service functions
Servicegerät service unit || service device
Servicehilfe service support || service aid * *Hilfsmittel* || service instructions * *Service-/Wartungsanleitung*
Serviceintervall service interval
Servicenetz service network
Serviceöffnung service opening || service cover * *Wartungsklappe* || servicing cover * *Wartungsklappe/-Deckel*
Servicepersonal service personnel || service staff
Serviceprogramm servicing program
Servicestelle service location
Servicesystem service network * *Servicenetz*
Servicezwecke service purposes
Servoantrieb servo drive || actuator * *Stellantrieb*
Servomotor servo motor * *hochdynamischer Motor für Positionieraufgaben/Bewegungssteuerung* || synchronous 'servo' motor
Servoregler servo controller || servo amplifier * →*Servoverstärker*
Servotechnik servo technique
Servoventil servo valve || control valve * *Regelventil*
Servoverstärker servo amplifier * *Leistungs- und Regelteil z.Ansteuerung eines (kleinen)* →*Servomotors* || servo driver
setzen set * *z.B. Ausgang/Signal/Parameter/Flipflop/Merker ~* || place * *(auf-)stellen, setzen, legen, placieren, postieren, jdn. ernennen/in e.Amt ein~* || put || stack * *stapeln* || apply * *anbringen, auflegen, anwenden* || fix * *befestigen, fest~ (auch Termin), festmachen, anheften/bringen, ein~, unterbringen* || substitute * *anstelle von ... ~; er~* || affix * *befestigen, anbringen, anheften, ankleben, beilegen, hinzufügen* || settle * *(Erdreich, Fundament, Flüssigkeit, Bodensatz)* || clarify * *(Flüssigkeit)* || sit * *sich (auch zu Tisch) ~*
Setzimpuls set pulse
Setzwert preset value * *fest vorbesetzter Wert, z.B. Festsollwert* || setting value * *zu setzender Wert (z.B. für Integrator, Speicher)* || initial value * *Anfangs-/Urladewert*
SEV SEV * *Abk.f. 'Electrotechnical Institute of Switzerland': Schweizer. Elektrot. Verein (Normengremium)*
SF6-isoliert SF6-insulated
Shunt shunt
Shuntwandler shunt converter || DC transducer || shunt transducer * *Stromwandler zur (potenzialfreien) Strommessung mit Shunt-Widerstand*
SI-Einheiten International system of units
sich aufhängen lock up * *Softwareprogramm* || hang * *sich aufgehängt haben (Softwareprogramm)*
sich auszeichnen durch stand out for || be superior by || be characterized by * *herausragende Merkmale haben*
sich beschäftigen mit be involved with || occupy oneself with || be engaged in || work at || be busy doing something || be responsible for * *zuständig sein für*
sich beziehen auf refer to || apply to * *Anwendung finden bei, zutreffen auf*
sich drehen revolve || rotate
sich einfügen fit in * *Person und Sache; well: gut* || adapt oneself to * *Person*
sich ergeben arise * *auftreten, aufkommen (z.B. Schwierigkeiten, Probleme, Chance)* || emerge * *auftreten* || ensue * *auftreten* || result from * *aus* || follow from * *~ aus* || be a consequence of * *sich (logisch) ergeben aus*
sich ergeben aus result from
sich Gedanken machen über wonder about ||

sich halten an

worry about * *sorgend* || bother about * *sich sorgen/bemühen um*
sich halten an adhere to * *z.B. Betriebsanleitung, Vorschrift, Vertrag* || observe || follow || abide by * *festhalten/sich halten an, treubleiben* || act in conformity with || comply with || keep to || stick to
sich häufend cumulative
sich herausstellen als turn out to be
sich hin und herbewegen reciprocate * *z.B. Kolben*
sich hinwegsetzen über override
sich kümmern um mind || take care of || worry about * *sich Gedanken machen über* || care about || bother about
sich steigernd cumulative
sich summierend cumulative || aggregating
sich verändernd changing
sich verhalten behave
sich vertraut machen mit acquaint oneself with || get familiar with * *vertraut werden mit* || familiarize oneself with || get use to the idea * ~ *mit Gedanken/Idee/Vorschlag*
sich verziehen become distorted || warp * *[Holz]*
sich vorstellen imagine * *geistig* || introduce oneself * *sich jemandem* || make oneself known * *sich bekannt machen* || have an interview * *sich als Bewerber für eine Stelle ~, ein Bewerbungsgespräch führen*
sich widerspiegeln in be reflected by
sich widersprechend contradictory * *→widersprüchlich* || inconsistent
sich zusammensetzen aus be composed of || be formed out of
sicher safe * *from: vor; be on the safe side: auf der ~en Seite sein/~gehen* || reliable * *verlässlich, zuverlässig* || immune * *gefeit; from: gegen* || proof * *geschützt (from: gegen/vor; against: vor)* || firm * *fest* || certain * *gewiss* || sure * *gewiss* || definite * *bestimmt* || protected * *geschützt* || secure * *geschützt (from: vor), ge~t, fest, gewiss* || reliably * (Adv.) *zuverlässig* || confident * *selbst~, mit (Selbst-) Vertrauen*
sicher öffnender Kontakt positively driven opening contact * *zwangsgeführter (Relais-) Kontakt z.B. für Sicherheitsfunktion*
sichere Schiene secure bus * *(USV)*
sichere Trennung safe isolation || protective separation * *siehe DIN 0100, Tl. 410 u.DIN VDE 0106, Tl. 101* || safety separation || electrical separation || reliable isolation
Sicherheit security * *from/against: vor; auch [Wirtschaft]: security for payment* || safety * *Personen~, ~ von Wertpapieren; for safety reasons: ~shalber* || surety * *Gewissheit* || certainty * *Gewissheit; with certainty: mit ~* || reliability * *Zuverlässigkeit* || confidence * *Vertrauen* || assurance * *Garantie, Sicherheit (-sgefühl), Zuversicht, Selbst~, ~ im Auftreten* || cover * *Deckung* || protection * *Schutz* || trustworthiness * *Vertrauen(-swürdigkeit)*
Sicherheits- safety || protective * *Schutz-* || back-up * *zur Reserve (z.B. Sicherungsdiskette)*
Sicherheits-Rutschkupplung safety slipping clutch * *zur Drehmomentenbegrenzung* || safety-slip clutch
Sicherheitsabschaltung safety shutdown * *Antrieb* || emergency stop * *Antrieb* || safety disconnection * *Stromkreis* || emergency shutdown
Sicherheitsabstand safety clearance || safety distance || guardband * *z.B. bei Timing-Spezifikationen* || safety margin * *wertmäßig*
Sicherheitsanforderung safety requirements
Sicherheitsbeauftragter safety supervisor || safety officer
Sicherheitsbestimmungen safety regulations || safety requirements

Sicherheitsbremse fail-safe brake * *Bremse, die bei Stromausfall automatisch einfällt (Ruhestromprinzip)*
Sicherheitscode security code
Sicherheitseinrichtung safety device || safety equipment
Sicherheitseinrichtungen safety equipment
Sicherheitsendschalter safety limit-switch
Sicherheitsfaktor safety factor || reserve factor
sicherheitsgeprüft safety-tested
sicherheitsgerichtet safety-oriented || fail-safe
Sicherheitsglas safety glass
Sicherheitsgründe safety reasons * *for: aus* || precautional measures * *Vorsichtsmaßnahmen* || safety-related purposes * *for: aus*
sicherheitshalber to make sure * *um sicherzugehen* || to be on the safe side * *um sicherzugehen* || for reasons of safety * *aus Gründen der Sicherheit* || for safety reasons
Sicherheitshinweise safety instructions || safety precautions * *Vorsichtsmaßregeln* || safety notes || notes on safety || information on safety
Sicherheitskupplung safety coupling * *beginnt bei Überschreiten eines bestimmten Drehmoments zu rutschen* || torque-limiting clutch * *zur Drehmomentbegrenzung* || torque limiter * *zur Drehmomentbegrenzung* || safety slipping clutch * *→Schlupfkupplung* || slipping clutch * *→Schlupfkupplung* || overload slipping clutch * *Überlast verhindernde Rutschkupplung* || fail-safe coupling * *Reibkupplg. o. ~ mit Sollbruchstelle, öffn. bei Überschreiten eines Drehmoments*
Sicherheitsmaßregel safety precaution
Sicherheitsnorm safety standard * *für elektr. Antriebssysteme: z.B. EN 50178, EN 60204-1*
sicherheitsrelevant safety-related || concerned with safety
Sicherheitsventil safety valve * *siehe auch →Überdruckventil*
Sicherheitsverriegelung safety interlock
Sicherheitsvorschrift safety regulation
Sicherheitsvorschriften safety regulations
Sicherheitswartezeit base-block time * *bei Momentenwechsel eines DC oder AC Stromrichters*
sichern secure * ~ *(against: gegen; on/by: durch)*, *Sicherheit bieten, befestigen, (fest ver-)schließen* || safeguard * *sichern (auch Daten), schützen, (Interessen) wahrnehmen* || make a backup * *[Computer] Daten* ~ || make safe || lock || block || protect
sicherstellen secure * ~ *(auch i.Handelswesen), sichern (on/by: durch; z.B.Hypothek), Sicherheit bieten* || guarantee * *garantieren* || ensure * *für etw. sorgen, garantieren (that: dass; someone being: dass jemd. ... ist)*, ~ || confirm * *(Nachricht, Wahrheit, Auftrag) bestätigen, (Entschluss) bekräftigen, (Macht) festigen* || assure * *z.B. Einhaltung einer Vorschrift* ~ || give security || provide * *Vorsorge treffen, sich sichern (against: vor/gegen); bereit-/ zur Verfügung stellen*
Sicherstellung securing || guarantee || guaranty || cover * *durch Deckung* || assurance * *Garantie, Ver-/Zusicherung, Sicherstellung*
Sicherung fuse * *the fuse has blown: die ~ ist durchgebrannt/gefallen/hat ausgelöst* || fuse link * *~seinsatz ohne Unterteil* || lock * *(mechan.)* || catch * *(mechan.) Verschluss, Sicherung* || securing * *das Sichern, das Sicher-Machen (auch Datenübertragung, Transportlast usw.)* || locking * *das Sichern (to prevent: gegen), (against: gegen)*
Sicherungs-Lasttrenner fuse-switch disconnector || fused interrupter || fuse-disconnector || fuse disconnecting switch || fuse-switch disconnecter || fused load disconnect || fused load disconnect switch
Sicherungsansprechstrom fusing current

Sicherungsansprechzeit   melting time
Sicherungsaufsteckgriff   fuse puller
Sicherungsautomat   automatic circuit breaker ||
   MCB * *Abkürz. für 'Miniature Circuit Breaker'* ||
   miniature circuit breaker || automatic cut-out
Sicherungsblech   tab washer * *mit hochgebogenen
   Lappen zur Schraubensicherung*
Sicherungseinsatz   fuse link || fuse unit || fuse plug
   cartridge * *Sicherungs-Steckpatrone* || fuse cutout ||
   catridge fuse link
Sicherungseinsteckkontakte   fuse clips * *z.B.
   Lyrakontakte*
Sicherungsfall   blowing of fuse * *Durchbrennen
   e.Sicherung* || blowing of fuses * *Plural; Durchbrennen von Sicherungen* || fuse blowing || fuse
   rupturing || fuse blow || fuse blown
Sicherungsfassung   fuse mounting || fuse holder
   * *Sicherungshalter, Sicherungsunterteil*
Sicherungsgrenzstrom   minimum fusing current
Sicherungsgriff   fuse puller
Sicherungshalter   fuse holder || fuse carrier
Sicherungskennlinie   fuse time-current characteristic || charasteristic of fuse
Sicherungskontakt   fuse contact
Sicherungskopie   back-up copy
Sicherungskurzschlussvermögen   fuse interrupting rating
Sicherungslasttrenner   fused switch disconnector ||
   fuse switch disconnecter || fused load-disconnect ||
   fused interruptor || fuse disconnector || fuse disconnecting switch || fused load disconnector || fused
   load disconnect switch || fuse switch disconnector
   || fused disconnect
Sicherungsleiste   fuse block
sicherungslos   fuseless
Sicherungsnennstrom   rated fusing current || fuse
   rating
Sicherungspatrone   cartridge fuse
Sicherungsring   circlip * *Sprengring* || lock ring ||
   retaining ring
Sicherungsscheibe   lock washer || retaining washer
   || circlip * *Sprengring*
Sicherungsschmelzzeit   fuse clearing time || fuse
   acting time || fuse melting time
Sicherungsschraube   lock screw || securing screw
Sicherungsstift   lock pin
Sicherungsstromstärke   fuse current rating
Sicherungsträger   fuse holder || fuse base || fuse carrier
Sicherungstrenner   fuse-disconnector || fuse
   disconnecting switch || fuse switch disconnector
   * *Sicherungs-Lasttrenner* || fused interrupter
   * *Sicherungs-Lasttrenner* || fused disconnector
Sicherungstrennschalter   fused disconnect switch
Sicherungsüberwachung   fuse monitoring
Sicherungsunterteil   fuse holder
sichtbar   visible || noticeable * *wahrnehmbar* || perceptible * *wahrnehmbar* || conspicuous * *auffällig,
   auffallend* || evident * *offensichtlich, deutlich* ~ ||
   obvious * *offensichtlich* || appear * ~ *werden,
   erscheinen*
sichtbar machen   show || visualize
sichtbar werden   appear || show || become manifest
   * *(figürlich)*
Sichtbarkeit   visibility || readability * *(Ab)Lesbarkeit*
Sichter   separator * *trennt Stoffe z.B. durch Sieben,
   Zentrifugieren usw.* || sifter * *Siebeeinrichtung* ||
   grader * *Sortiereinrichtg., z.B. z. Trennen v.
   Schnitzeln unterschiedl.Größe i.d. Spanplattenherstellg.* || elutriator * *Sortiereinrichtung*
Sichtfenster   sight-glass * *z.B. zur Ölstandskontrolle* || inspection window * *für Servicezwecke*
Sichtgerät   CRT * *cathode-ray tube: Bildschirm-Monitor* || VDU * *video display unit* || video terminal || screen * *Monitor, z.B.für Personal Computer*

Sichtkontakt   visual contact
Sichtkontrolle   visual check
Sichtmelder   visual indicator
Sichtprüfung   visual check || visual inspection || optical inspection
Sichtweise   perspective
Sicke   bead * *wulstförmiger Rand* || corrugation
   * *Falte, Furche, Riefe, Welle* || crimp * *Kräuselung, Welligkeit* || stiffening corrugation * *zur Versteifung*
Sickmaschine   beading machine || seam-rolling
   machine
Sieb   wire * ~ *bei Papiermaschine; wire section:
   ~partie; wire side: ~seite* || sieve * *Durchwurf,
   Rätter, Sieb* || filter * ~ *für Flüssiges und elektr.
   Filterglied (z.B. Tief-/Hoch/Bandpass).* || strainer
   * ~ *für Flüssiges* || eliminator * *(elektr.)* || riddle
   * ~ *für Sand usw.* || screen * *Gitter~ (z.B. für Sand,
   Zellstoff usw.); auch beim* →*Siebdruck*
   machine wire * *{Papier} ~ bei Papiermaschine*
Siebdrossel   filter reactor
Siebdruck   screen printing * *Druckverfahren* || silk
   screening || silk screen * *auf Leiterkarte usw. aufgedruckter* ~
Siebdruckrahmen   silk-screen frame * *{Druckerei}*
sieben   filter || smooth * *glätten* || screen * *durch ein
   mechanisches Sieb (durch)sieben*
Siebensegment-Anzeige   seven-segment display
Sieberei   filtering plant * *z.B. in Zellstoff-Fabrik*
Siebfilzpresse   fabric press * *in Papiermaschine*
Siebglied   filter element || smoothing element
   * *Glättungsglied*
Siebgruppe   wire section * *bei Papiermaschine*
Siebkondensator   filter capacitor
Sieblaufregler   wire guide * *in Papiermaschine*
Siebleitwalze   wire roll * *in Papiermaschine*
Siebmaschine   screening machine * *{Holz} für zur
   Trennung der Holzschnitzel in der Spanplattenherstellung*
Siebpartie   wire section * *Entwässerungspartie in
   Papiermaschine, hinter dem Stoffauflauf angeordnet* || wet end * *nasses Ende, Entwässerungspartie
   einer Papiermaschine* || wire part * *in Papiermaschine*
Siebregulierwalze   wire-guide roll * *in Papiermaschine*
Siebsaugwalze   suction couch roll * *dient zur Entwässerung in Siebpartie einer Papiermaschine* ||
   wire suction roll * *{Papier} zum Entwässern des
   Papierbreis in d. Siebpartie e. Papiermaschine*
Siebsystem   screening system * *zum Trennen von
   Partikeln (z.B. zerkleinertes Holz)*
Siebung   filtering || smoothing
Siebzeit   filtering time || filtering time constant
   * *Siebzeitkonstante*
Siedebadkühlung   evaporation bath cooling * *z.B.
   bei Bahnumrichter* || evaporative bath cooling
   * *z.B. bei Bahn-Stromrichter*
Siedekühlung   vapour cooling
Siedepunkt   boiling point
siehe   refer to * *z.B.Verweis auf anderes Kapitel* ||
   see || confer * *vergleiche*
siehe auch   also refer to * *Verweis auf (Stelle in)
   Publikation* || see also
siehe oben   see above || refer above || same as above
   * *wie oben (angegeben)* || as mentioned above
siehe unten   see below
Siemosyn-Motor   Siemosyn-motor * *(R) permanenterregter Synchronmotor von SIEMENS* || permanent-field synchronous motor * *z.B. Siemosyn-Motor(R) von SIEMENS*
Signal   signal * *audible: akustisches, visible: sichtbares/optisches*
Signal ausgeben auf   direct a signal to * *z.B. Analogausgabe*

Signalanpassung

Signalanpassung signal adaption || signal conditioning * *Signalaufbereitung* || signal conditioning circuit * *Schaltung* || signal matching
Signalanschluss signal connection
Signalaufbereitung signal conditioning || signal preparing
Signalausgabe signal output
Signalaustausch signal interchange || signal transfer || signal exchange
Signalauswertung signal evaluation || signal interpolation * *Interpolation*
Signalbandkabel ribbon cable || signal ribbon cable
Signalbereich signal range
Signalbezeichner signal designator
Signalbildung signal generation
Signaleingabe signal input
Signalerfassung signal acquisition
signalerzeugend signal generating
Signalflanke signal edge * *negative/positive going: fallende/steigende*
Signalfluss signal flow || information flow
Signalflussplan signal flow chart
Signalgeber sensor || sensing element
signalisieren signal
Signalisierung signalling || indication * *Anzeige*
Signallampe indicator light || pilot lamp || signal lamp
Signallaufzeit propagation delay time || signal propagation delay || propagation delay
Signalleitung signal cable || signal lead || field cable * *zur (externen) Anlagenverdrahtung* || signal line
Signalmasse signal ground
Signalmuster signal pattern
Signalpegel signal level
Signalpfad signal path
Signalprozessor DSP * *Digital Signal Prozessor, mit großen Wortlängen und umfangreichen Arithmetikfunktionen* || DSP processor || Digital Signal Processor
Signalumformung signal conversion || signal transformation || signal conditioning * *Signalaufbereitung*
Signalumsetzer signal conditioner * *zur Signalaufbereitung* || signal converter || signal transducer * *besonders 'Messumformer'*
Signalverarbeitung signal processing
Signalverfolgung signal tracking || signal tracing
Signalverlauf signal sequence * *(Ab)Folge* || signal waveform * *Wellen/Kurvenform (meist für periodisches Signal)* || signal variation * *Änderung*
Signalverteiler signal dispatcher
signalvorverarbeitende Baugruppe intelligent I/O module
Signalvorverarbeitung signal pre-processing
Signalwandler signal converter || signal transducer * *z.B. Messwandler*
Signalwechsel signal change
Signalzustand signal status * *(Plural: signal statuses)*
Signalzustände signal statuses
Signalzustandsanzeige signal status display
Signet logo * *Sinnbild, Logo*
Silikatglas silicate glass
Silikon silicone
Silikonfett silicone grease
silikonfrei silicone-free * *z.B. Öl, Kabelisolierung usw.*
silikonisoliert silicone insulated
Silizium silicon
Siliziumdiode silicon diode
Siliziumgleichrichter silicon rectifier
Siliziumhalbleiter silicon semiconductor
Siliziumkarbid silicon carbide
Siliziumkristall silicon crystal

Siliziumplättchen die * *zur Erzeugung integrierter Schaltungen*
Siliziumscheibe silicon wafer || silicon slice || wafer
Silo silo * *ensilage: in ein ~ einlagern* || storage bin || grain elevator * *für Getreide*
Silumin siluminium * *verschleißfeste, korrosionsbeständige Alulegierung. mit ca. 12% Siliziumanteil*
SIMM SIMM * *[Computer] Single-Inline Memory Module': Speichermodul mit einreihigen Anschlusspins*
Simmerring spring-loaded sealing ring * *Gummiring m.innenliegender Spiralfeder, dichtet durch Anpressen an Welle* || radial sealing ring * *Radialdichtring* || sealing ring || oil seal ring || spring-energized seal * *Gummidichtring mit innenliegender Spiralfeder, dichtet Wellenaustritt* || radial shaft sealing ring || radial shaft seal
Simulation simulation
Simulationsbaugruppe simulation module
Simulationsbetrieb simulated operation
Simulationsprogramm simulation program || simulator program
simulieren simulate
Singularität singularity
sinken sink || drop * *Preise, Kurse, Messwerte* || fall * *Preise, Kurse* || go down * *Preise, Kurse, Messwerte* || decrease * *sich vermindern* || abate * *sich vermindern* || diminish * *sich vermindern* || decay * *verfallen* || decline * *verfallen*
Sinn reason * *Grund* || meaning * *Bedeutung* || sense * *Sinn, Verstand, Vernunft,Bedeutg., Richtg., Gefühl; in a narrow/broad sense: i.eng./weiterem~e* || mind * *Verstand, Meinung* || faculty * *Anlage, Geisteskraft* || taste * *Vorliebe, for: für* || liking * *Vorliebe, for: für* || inclination * *Neigung* || disposition * *Neigung* || flair * *Feingefühl, feines Gespür* || instinct * *Instinkt* || soul * *Seele* || heart * *Seele, Herz* || interpretation * *Auslegung* || construction * *Auslegung* || idea * *Gedanke* || basic idea * *Grundgedanke* || direction * *Richtung* || essence * *Wesen, Geist, das Wesentliche, der Kern, essence and purpose: ~ und Zweck*
Sinnbild symbol
sinngemäß analogous || corresponding || equivalent || accordingly * *(Adv.)* || analogously * *(Adv.)*
sinnvoll reasonable * *vernünftig, angemessen* || plausible * *plausibel* || practical * *praktikabel, praxisbewährt* || fraught with meaning * *bedeutungsvoll, bedeutungsschwer, bedeutungsschwanger, von Bedeutung* || meaningful * *von Bedeutung* || suggestive * *gehaltvoll, anregend* || wise * *klug* || sensible * *vernünftig; make sense: ~ sein/Sinn machen* || ingenious * *wohl ersonnen* || efficient * *zweckvoll* || good policy * *~e Verfahrensweise* || realistic * *realistisch, wirklichkeitsnah, sachlich* || useful * *zweckdienlich* || appropriate * *zweckdienlich, passend, angemessen* || intelligent * *klug* || practically * *(Adv.)*
sinnvoll sein make sense * *Sinn machen*
Sinter- sintered || sintering || sinter || carbide * *Sintermetall-* || powder
Sinterbelag sintered lining * *Reibbelag bei Kupplung, Bremse*
Sinterbronze sinter bronze
Sinterbronzelager porous-bronze bearing
Sinterlager sintered bearing * *Gleitlager* || sintered sleeve bearing
Sintermaterial sintered material
Sintermetall sintered metal
sintern sinter * *Erz, Stahl usw.* || bake || clinker
sinus sine
Sinus-Cosinus-Geber sine-cosine encoder * *Drehzahl-/Lagegeber mit phasenverschobenen Sinussignalen* || sinewave encoder * *Drehzahl-/Lagegeber mit sin- und cos-Signalen* || raw-signal encoder

* *Rohsignalgeber (gibt keine Rechteckimpulse, sondern sin-/cos-Signale aus)*
Sinusausgangsfilter sinusoidal output filter * *siehe auch 'Sinusfilter'*
sinusbewertet sinusoidal-weighted || sinusodially weighted || sinusodially evaluated * *z.B. Modulationsverfahren für Umrichter*
Sinusfilter sinusoidal filter * *aufwändiges Filt. am Pulsumrichterausg., mindert Motor-Spanngs.beanspr. u. Erwärmg.* || sine-wave filter || sinewavefilter * *zur Erzeugung sinusförmiger Spannungen am Ausgang eines Umrichters*
sinusförmig sinusoidal || according to a sine function || sine-wave || sine
sinusförmige Größe sinusoidal quantity
Sinusmodulation sine modulation
sinusmoduliert sinusoidally modulated || sine-modulated
Sinusperiode sine period || sine cycle || sine wave * *Sinusschwingung, Sinuswelle*
Sinusschwingung sine wave || sinusoidal oscillation * *sinusförmige Schwingung*
Sinuswelle sinusoidal wave || sine wave
Sinuswellenform sinusoidal waveform
Sirup sirup * *z.B. Zuckerrübensirup*
SITOR-Satz SITOR set * *(R) SIEMENS-Thyristorsatz*
Sitz fit * *Passung; tight: strammer* || seat * *~gelegenheit, Ventil~, mechan. ~fläche, Niederlassung; registered seat: Geschäfts~* || chair * *Stuhl* || place * *Ort, Platz; place of residence: Wohn~; place of business: Geschäfts~* || headquarters * *(Firmen-)Haupt~*
Sitzung meeting * *Besprechung* || conference * *Konferenz* || sitting * *einzelne Sitzung, z.B. beim Arzt* || hearing * *offizielle Anhörung*
Skala scale || dial
Skalar scalar * *(auch Adj.)*
skalare Größe scalar quantity || scalar variable
skalares Produkt scalar product
Skalarfeld scalar field
Skale scale || dial
Skalenendwert full scale
Skalenteil scale division
skalierbar scalable * *in Stufen ausbaubar/vergrößerbar*
Skalierung scaling || scaling information * *Skalierdaten, Skalierwerte*
Skilift ski-lift
Skin-Effekt skin effect * *Stromverdrängungseffekt*
Skizze sketch
skizzieren sketch * *(auch figürl.)* || make a sketch of || rough-draw || delineate * *sizzieren, zeichnerisch (in Strichzeichnung) darstellen, entwerfen* || make a rough sketch of * *etwas in groben Umrissen darstellen*
Slave slave
Slave-Anschaltung slave interface * *für Bus*
SLIP SLIP * *'Serial-Line Internet Protocol': DÜ-Protokoll für Internet-Verbindgn. über →Modem (jetzt →PPP)*
Slotblech slot bracket * *für Personalcomputer-Einbaukarte*
Slowakei Slovacia
SmCo SmCo * *Samarium-Kobalt* || Samarium-Cobalt * *Seltene-Erden-Permanentmagnetmaterial; z.B. für Motorläufer*
SMD SMT * *Abkürzg. f. 'Surface-Mount Technology': Oberflächenmontagetechnik* || SMD * *Abkürzg. f. 'Surface-Mount Device': oberflächenmontierbares Bauteil*
SMD-Kondensator SMD-type capacitor
SMD-Technik SMT * *Surface Mount Technology: Oberflächenmontage*
SMD-Widerstand SMD-type resistor

so like this * *solch, derartig; (Adv.) auf diese Weise* || like that * *derartig, (Adv.) auf diese Art/Weise* || this way * *(Adv.) auf diese Weise* || that way * *(Adv.) auf diese Weise* || so * *als, wie, folglich, also, auf diese Art und Weise; so that: sodass* || as * *als, wie; as ... as: in solchem Maße/so ... wie* || such a * *so ein* || thus * *dadurch, deshalb, daher* || what * *was für ein* || by this means * *(Adv.) auf diese Weise* || in this way * *(Adv.) auf diese Weise* || in such a way that * *so/in der Weise, dass* || that * *(Adv., salopp) solch, so (sehr), dermaßen; not that big: nicht so groß* || as follows * *folgendermaßen* || therefore * *folglich* || this is why * *deshalb, aus diesem Grunde*
so genannt so-called || pretended * *angeblich* || self-styled * *angeblich/vorgeblich*
so schnell wie möglich as fast as possible
so weit möglich as far as possible || within the bounds of possibility || if at all possible * *wenn irgend möglich*
sobald as soon as * *as soon as possible: sobald möglich* || the moment || at your convenience * *sobald es Ihnen möglich ist/wann es Ihnen beliebt*
sobald wie möglich as soon as possible || a.s.a.p. * *Abkürzg. f. 'as soon as possible'*
Sockel base * *Unterbau, Unterlage, Fundament, Fuß, Sohle, Basis, Grundlage, Stützpunkt, Hauptbestandteil* || socle * *~ eines Bauwerks* || pedestal * *Untergestell, (Lager)Bock, Säulenfuß, ~ eines Bauwerks* || socket * *Lampen~, IC-~, Steck~, Rohransatz, Muffe, Fassung* || pedestal * *Untergestell, (Lager)Bock, Säulenfuß, ~ eines Bauwerks*
Sodahaus soda shop * *z.B. in Papierfabrik*
sodass so that
Soffitte festoon * *z.B. Glühbirne zum Einsetzen in Meldeleuchte* || tubular lamp || double-capped tubular lamp * *Soffittenlampe*
Soffittenzieher festoon extractor * *zum Ausziehen einer Soffittenlampe aus Meldeleuchte usw.*
sofort at once || immediately * *(Adv.) effective immediately: mit ~iger Wirkung/ab ~ gültig* || immediate * *(Adj.) ~ig* || directly * *(Adv.)* || instantly * *(Adv.)* || forthwith * *sofort, umgehend, unverzüglich* || on the spot || straight away || right away || promptly * *prompt, pünktlich* || punctually * *pünktlich* || on time * *pünktlich*
sofortig immediate || prompt || instantaneous || punctual * *pünktlich* || on time * *pünktlich*
Software software
Software-Entwickler software designer || software developer
Software-Haus software manufacturer || software company
Software-Modul software module || software block * *Softwarebaustein*
Software-Version software version || software release
Software-Weiche software switch
Softwareänderung software modification || software change
Softwarebaustein software module || software function block * *Funktionsbaustein* || software block * *Funktionsbaustein*
Softwarebearbeiter software designer * *Entwickler*
Softwareentwickler software designer || software developer
Softwareentwicklung software development
Softwarehaus software manufacturer || software company
Softwarelösung software solution * *complete software solution: komplette ~*
softwaremäßig by software * *über/mit Hilfe der Software* || software || from the software side * *softwareseitig*

Softwaremodul software module || software block
* *Softwarebaustein*
Softwarepaket software package
Softwarestand software version || software release
Softwaretausch software upgrade * *auf eine neuere Version* || replacing of software * *Austausch*
Softwareversion software version || release
Softwareweiche software switch
Softwarewerkzeug software tool
söhlig horizontal * *(Bergbau)* || bottomed * *(Bergbau)*
Sohlplatte soleplate * *z.B.für das Aufstellen einer elektr. Maschine*
solch such || such as this * *(nachgestellt)* || such as that * *(nachgestellt)* || such a || such like this * *(nachgestellt)* || as such * *als solcher*
solcher of this nature * *a voltage notch of this nature: ein ~ Spannungseinbruch* || such * *such: solch einer* || of such a kind * *~ Art* || of this sort * *~ Art* || in such a manner * *~art, ~weise* || suchlike * *~lei* || in such a way * *~weise*
soll is to be * *z.B. is to be avoided: soll vermieden werden* || ref * *Kurzzeichen für 'reference': Sollwert* || set * *Kurzzeichen für 'setpoint': Sollwert* || target * *(Substantiv) (Produktions-, Liefer-) Ziel* || quota * *(Subst.) (Produktions-, Liefer-) Ziel; fixed: festgesetztes; production qu.: Produktionsziel* || debit * *(Substantiv; kaufmänn.) debit-side: auf d. Sollseite*
Soll- set
Soll-Ist-Abweichung setpoint-actual value difference || error signal * *als Signal* || system deviation || control error || error
Soll-Ist-Differenz setpoint-actual value difference || error * *Soll-Ist-Abweichung* || error signal * *Soll-Ist-Abweichung; siehe auch* →*Regelabweichung* || system deviation * →*Regelabweichung*
Soll-Ist-Überwachung setpoint-actual value monitoring
Soll-Istwert-Abweichung setpoint-actual value deviation || control error || system deviation
Soll-Istwert-Überwachung setpoint-actual value monitoring
Soll-Zustand specified condition || condition as specified || desirable condition * *gewünschter Zustand*
Sollbruchstelle rupture joint || preset breaking point || predetermined breaking point
sollen shall * *im Sine von Gebot, Pflicht* || ought to * *im Sinne von Verpflichtung (sittliche)* || be to * *auf Anweisung Dritter ~; auf Grund schicksalhafter Bestimmung ~* || have to * *müssen* || be obliged to * *müssen, verpflichtet sein zu* || must * *müssen* || should * *Möglichkeit* || would * *Vermutung, Wahrscheinlichkeit* || be said to * *angeblich ~* || be supposed to * *angeblich ~* || be believed to * *angeblich ~* || be reported to * *angeblich ~ (wie berichtet wird)* || be intended for * *bestimmt sein für* || be meant for * *bestimmt sein für* || be destined to * *schicksalhaft bestimmt sein für* || be fated to * *schicksalhaft bestimmt sein für*
Sollwert setpoint || reference value || reference || preset value * *Fest~* || setpoint value || set value || desired value * *gewünschter Wert*
Sollwertabminderung setpoint reduction
Sollwertanzeige setpoint indication || setpoint indicator * *Sollwert-Anzeigegerät*
Sollwertaufbereitung setpoint conditioning || setpoint generation * *Sollwerterzeugung* || setpoint preparation || setpoint processing
Sollwertbegrenzung setpoint limiting
Sollwertbereich setting range
Sollwertbildung setpoint generation
Sollwerteingang setpoint input || reference input

Sollwerteinsteller setpoint adjuster || setpoint setting device || motorized potentiometer * *motorischer*
Sollwerteinstellung setpoint adjustment || setpoint setting
Sollwertfreigabe setpoint enabling
Sollwertführung reference control
Sollwertgeber setpoint adjuster * *z.b. zur Drehzahlsollwertvorgabe* || setpoint source * *Sollwertquelle* || setpoint generator || setpoint transmitter
Sollwertglättung setpoint filtering || smoothing of reference value || averaging of setpoint value * *(gleitende) Mittelwertbildung* || setpoint smoothimg
Sollwertgrenze setpoint limit
Sollwertintegrator setpoint integrator * *Hochlaufgeber* || reference integrator * *Hochlaufgeber*
Sollwertkanal setpoint channel
Sollwertkaskade setpoint cascade || reference cascade || speed cascade * *für Geschwindigkeits-Sollwerte* || reference chain * *Sollwertkette* || reference value cascade
Sollwertkette setpoint cascade || setpoint chain
Sollwertpolarität setpoint polarity
Sollwertpotentiometer setpoint potentiometer
Sollwertquelle setpoint source || reference value source
Sollwertrechner setpoint computer
Sollwertsiebung setpoint filtering
Sollwertspeicher reference value memory
Sollwertstaffel setpoint cascade || setpoint link
Sollwertsteller setpoint adjuster || setpoint setting device || motorized potentiometer * *motorischer*
Sollwertsteuerung setpoint control
Sollwertverteilung setpoint distribution
Sollwertvorgabe setpoint setting || reference setting || setpoint input || setpoint entering || setpoint value setting || set-value input || setpoint issuing
Sollzustand desired state || specified condition || scheduled condition || desirable condition
somit so || thus || consequently * *folglich* || therewith * *daraufhin, demzufolge, darum, damit* || thereupon * *daraufhin, demzufolge, darauf, hierauf, danach, darum* || hence * *folglich, daher, deshalb; hieraus, daraus; hence it follows that: daraus folgt, dass*
Sonde probe
Sonder- special || particular * *besonders* || extra || customized * *auf Kundenwunsch (entwickelt, hergestellt, zugeschnitten usw.)* || special-purpose || custom-built * *nach Kundenvorgabe gebaut* || non-standard
Sonderanfertigung special design || special construction || customized product * *kundenspezifisch zugeschnittenes/entwickeltes Produkt* || special production
Sonderanstrich special coating
Sonderanwendung special application
Sonderausführung special design || special version || special construction * *Sonderkonstruktion*
Sonderbauart special design || special-purpose design || customized design * →*kundenspezifische Ausführung*
Sonderdruck reprint * *z.B. eines Zeitschriftenartikels* || off-print * *z.B. eines Zeitschriftenartikels* || separate print * *z.B. eines Zeitschriftenartikels* || special excerpt * *z.B. eines Zeitschriftenartikels*
Sonderfall special case || exceptional case || exception || exceptional condition * →*Ausnahmezustand*
Sonderfertigung specialized manufacturing
Sonderfett special grease
Sonderläufer special rotor * *[el. Masch.]*
Sonderlegierung special alloy
Sondermaschinen special machines || special-purpose machines
Sondermaßnahme special measure
Sondermaßnahmen special measures

Sondermotor special motor ‖ customized motor * *kundenspezifisch konstruierter Motor*
Sonderoption special option
Sonderpreis special price ‖ preferential price * *Vorzugspreis*
Sonderpreisvereinbarung special pricing agreement
Sonderstahl special steel
Sondertreiber special driver * *Softwaretreiber z.B. für Kommunikationsbaugruppe*
Sonderwunsch special requirement ‖ special wish
Sonderzubehör accessories
Sonderzweck special purpose
Sonneneinstrahlung solar radiation ‖ sun radiation ‖ insolation
Sonnenrad sun wheel * *bei →Planetengetriebe* ‖ sun gear * *in einem →Planetengetriebe*
sonst otherwise * *ansonsten* ‖ besides * *außerdem* ‖ in other respects * *im Übrigen* ‖ as a rule * *für gewöhnlich* ‖ usually * *für gewöhnlich* ‖ normally * *normalerweise* ‖ at any other time * *zu einer anderen Zeit* ‖ formerly * *ehemals*
sonstig other ‖ miscellaneous * *vermischt*
sorgen für care for ‖ provide for ‖ take care of * *dafür sorgen, dass* ‖ ensure * *sicherstellen (that: dass etwas getan wird)* ‖ provide * *beschaffen* ‖ maintain * *aufrechterhalten, z.B. eine konstante Drehzahl* ‖ assure * *sicherstellen, bürgen für, sichern* ‖ make for ‖ cater for * *(auch figürl.)* befriedigen, *sorgen für* ‖ facilitate * *ermöglichen, erleichtern, fördern; Einrichtung(en)/Resourcen bereitstellen für* ‖ make provision * *→Vorsorge treffen*
Sorgfalt care ‖ attention * *Aufmerksamkeit* ‖ exactness * *Genauigkeit* ‖ accuracy * *Genauigkeit* ‖ circumspection * *Umsicht* ‖ conscientiousness * *Gewissenhaftigkeit* ‖ scrupulousness * *Gewissenhaftigkeit* ‖ pains * *Mühe, Bemühung; take pains: sich Mühe geben/anstrengen* ‖ carefulness
sorgfältig careful * *achtsam, umsichtig, vorsichtig, bedacht, sorgsam, gründlich, genau, sparsam* ‖ attentive * *achtsam, aufmerksam* ‖ exact * *genau* ‖ conscientious * *gewissenhaft* ‖ scrupulous * *(über-) gewissenhaft, peinlich genau, ängstlich, vorsichtig* ‖ painstaking * *gewissenhaft, eifrig, rührig* ‖ cautious * *vorsichtig* ‖ carefully * *(Adv.)* ‖ with care * *(Adv.)* ‖ proper * *ordnungsgemäß, genau, passend, geeignet* ‖ diligent * *sorgfältig, gewissenhaft* ‖ meticulous * *peinlich genau*
Sorgfaltspflicht obligation to exercise due care * *(juristisch)* ‖ responsibility * *Verantwortung*
Sorte type ‖ kind ‖ sort ‖ variety * *Abart* ‖ species ‖ quality * *Qualität* ‖ grade * *Qualitätsstufe von Standardprodukten, z.B. von Papier, Öl usw.* ‖ brand * *Marke* ‖ design * *Bauform* ‖ type of construction * *Bauform*
Sortenspeicher grade memory * *z.B. bei der Papierherstellung*
Sortenwechsel grade change * *z.B. bei Papiermaschine* ‖ type change
Sortieranlage sorting plant ‖ grading plant * *zum Sortieren nach unterschiedlichen Produktqualitäten (z.B. bei Holz)* ‖ classifying plant ‖ sorting system ‖ sorting equipment
Sortiereinrichtung sorting device ‖ sorting equipment ‖ grader * *[Holz] für Holzspäne in der Spanplattenherstellung*
sortieren sort ‖ merge * *'(hin)einsortieren', z.B. Daten an die richtige Stelle in e. Datenbank* ‖ classify
Sortierkreisel distribution loop * *in (Hochregal-) Lager* ‖ distribution circle * *in (Hochregal-) Lager* ‖ carousel-type distribution system * *in (Hochregal-) Lager*
Sortiermaschine sorting machine
Sortierroller paper inspection machine * *wickelt Papier in beide Richtungen hin u.her z.Aussortieren v. Schadstellen*
Sortierung sorting ‖ sreening * *~ unterschiedl. großer Partikel, z.B.d.Holzspäne durch Siebung i.d. Spanplattenindustrie* ‖ assortment ‖ sizing ‖ grading ‖ classification
Sortiment line * *Produktpalette* ‖ variety * *of: an* ‖ assortment * *of: an* ‖ assortment kit * *Teilesatz*
sowieso in any case ‖ anyhow ‖ anyway ‖ as it is
sowohl als auch both ... and * *both in x and in y: sowohl in x als auch in y* ‖ as well ... as ‖ not only ... but also
Spachtel filler * *~masse* ‖ pore filler * *[Holz] ~masse* ‖ spatula * *~werkzeug* ‖ scraper * *Malerwerkzeug* ‖ smoother * *Schmierkelle (make smooth: spachteln)* ‖ surfacer * *[Holz] Spachtelmasse*
Spalt gap * *Lücke, Luftspalt usw.* ‖ slit * *Schlitz* ‖ chink * *Ritze* ‖ fissure * *Riss, Sprung* ‖ crevice * *Riss, Spalte* ‖ crack * *Riss, Sprung, Spalte, Schlitz* ‖ cleft * *Riss* ‖ rift * *Spalt(e), Ritze, Sprung, Riss*
Spaltdichtung groove seal ‖ grease-packed groove * *zur Abdichtung von Gehäusefugen (z.B. in ölgefülltem Gleit-/Wälzlager)* ‖ jointing gasket * *in Form einer Dichtungszwischenlage* ‖ jointing compound * *Fugen-Dichtmasse*
Spalte column * *in Text/Tabelle, auf Bildschirm (siehe auch 'Spalt')*
spalten divide ‖ split ‖ cleave * *[Holz]* ‖ decompose * *[Chemie]*
Spaltkreissäge circular resaw * *[Holz]*
Spaltpolmotor shaded-pole motor * *1-phas. AC-Klein-Induktionsmotor m.Käfigläufer, Kurzschlussring erzeugt Hilfsphase* ‖ split-pole motor
Spaltrohrmotor split-cage motor * *Nassläufer-Pumpenmotor mit Schutzrohr im Luftspalt zwischen Läufer u. Stator* ‖ canned motor * *el. Pumpenmotor, Ständer mit Blechpaket u. Wicklung durch Rohr v. Fördermedium getrennt*
Spaltsäge splitting saw * *[Holz]* ‖ resaw * *[Holz]* ‖ cleaving saw * *[Holz]*
Span chip ‖ shavings ‖ chips ‖ cuttings * *Metallspäne* ‖ facings * *Frässpäne* ‖ borings * *Bohrspäne* ‖ drillings * *Bohrspäne* ‖ filings * *Feilspäne* ‖ sliver * *Splitter, Span, Stückchen* ‖ splinter * *Splitter*
spanabhebend cutting ‖ metal-cutting
spanabhebende Bearbeitung cutting ‖ machining
Späne chippings ‖ chips ‖ shavings ‖ cuttings * *Metallspäne* ‖ facings * *Fräßspäne* ‖ borings * *Bohrspäne* ‖ drillings * *Bohrspäne* ‖ filings * *Feilspäne* ‖ particles * *[Holz] Ausgangsmaterial bei d. Spanplattenherstellung*
Späneabsauganlage chip extraction system * *[Holz]* ‖ chip extractor * *[Holz]* ‖ chip-exhaust plant * *[Holz]*
spanen chip * *[Werkz.masch.\ Holz]*
spanend cutting * *Bearbeitung, Werkzeug usw.*
spanend bearbeiten machine
spanende Bearbeitung cutting ‖ machining
spanende Fertigung cutting machining
Spänesichter chip sifter * *[Holz] zum Trennen der Späne in der Spanplattenherstellung*
Spanien Spain
spanisch Spanish
spanlos forming * *Werkstückbearbeitung (Gegensatz: →zerspanend)* ‖ chipless
spanlose Bearbeitung no-cutting working ‖ non-cutting working
spanlose Fertigung non-cutting working ‖ non-cutting machining
spanlose Umformung chipless forming ‖ chipless shaping
Spann- clamping ‖ tightening ‖ stretching ‖ clamp ‖ bracing ‖ tensioning

Spannbügel clamp || latch fastener || clamping brakket || clip
Spanndorn expanding mandrel * z.B. bei Werkzeugmaschine, Wickler, Haspel usw.
Spanne margin * z.b. Vertriebsspanne/marge || span || short distance * räumlich || short space of time * zeitlich
Spanneinrichtung clamping device * →Spannvorrichtung
Spannelement tensioning element * für Kette, Gurt, Riemen, Seil, Sägeblatt usw. || clamping element * z.B. zur Wellen-Nabeverbindung || locking device * auch für Wellen-Nabeverbindung || locking element || clamping device
spannen tighten * straffen, z.B. Riemen || stretch * strecken, spannen, straff ziehen, dehnen || bend * Feder, Bogen || grip * Werkstück || clamp * Werkstück auf~/ein~ || chuck * Werkstück auf~/ein~ || tension * dehnen, strecken, spannen (z.B. Riemen, Feder, Kette) || be too tight * zu eng sein, zu fest sitzen || charge * Feder || load * Feder || wind * (Spiral)Feder ~ || tensioning * (Substantiv) z.B. ~ eines Riemens
Spanner tensioner * für Kette, Gurt, Sägeblatt usw.
Spannfeder tension spring
Spannfutter chuck * bei Werkzeugmaschine
Spanngewicht rope-tightening weight * (kleines) Seil-Spanngewicht bei Aufzug || counterweight * →Gegengewicht, z.B. großes ~ bei Aufzug
Spannhülse clamping sleeve || bushing * Futter (stück) Buchse, Büchse || adapter sleeve || split taper sleeve * konische Hülse mit Schlitzen || locking bush
Spannkappe tesioning cap
Spannkraft tension load * elastische ~ || elastical force * elastische ~ || tensional force * elastische ~ || contact pressure * Anpressdruck (z.B. bei einer Spannvorrichtung). || clamping power * {Werz.-masch.} bei einer Spannvorrichtung || elasticity * (auch figürl.) || energy * (figürl.) Schwung, Spannkraft, Lebenskraft, Auftrieb || buoyancy * (figürl.) Schwung, Spann-/Lebenskraft; {Physik} Schwimm-, Tragkraft, Auftrieb
Spannplatte fixing plate * Spannvorrichtung · z.B. für Werkzeugmaschine || gripping plate || clamping plate * z.B. zum Aufspannen von Werkstücken
Spannpratze clamping claw || clamping shoe
Spannrahmen stentering machine * {Textil} Verstreckwerk mit Wärmebehandlung für Textilstoffe || drawing frame * {Textil} Verstreckwerk || textile stenter * {Textil} Verstreckwerk || stenter * {Textil} für Tuch || clamp frame * {Holz}
Spannring clamping ring
Spannrolle jockey pulley * z.B. bei Riementrieb || tightening pulley * z.B. an einem Riementrieb || tightener * für Riementrieb || idler pulley * für Riementrieb usw. || tension roller * für Riementrieb || tension pulley * für Riementrieb
Spannsatz clamping assembly * z.B. zur Welle-/ Nabebefestigung (bei Welle ohne Federkeil) || locking assembly z.B. zur Wellen-Naben-Verbindung || clamping kit || collet-type mounting clamp * zur Welle-Nabe-Verbindung über Klemmhülse (z.B. für Impulsgeberanbau)
Spannschiene slide rail * zum Spannen eines Riementriebs z.B. durch Verschieben des Motors
Spannschraube clamping bolt || clamp bolt || clamping screw * kleine ~ || tightening screw
Spannung voltage * (elektr.) || tension * (mechan.) Bahnspannung, Zug einer Warenbahn || strain * (mechan.) (verformende) Spann., Verdehng., Zerrg., Belastg.,Beanspruchg., Inanspruchnahme || voltage level * (elektr.) Spannungshöhe || stress * (mechan.) Beanspruchung, Spannung., Dehnung, Belastung
Spannungs-Frequenz-Kennlinie V/F characteristic || voltage-frequency characteristic
Spannungs-Frequenz-Umsetzer voltage-to-frequency converter || VFC * Voltage-To-Frequency Converter
Spannungs-Frequenz-Wandler voltage-to-frequency converter || VFC * Abkürz. f. 'Voltage-to-Frequency Converter' || voltage-frequency converter
Spannungs-Frequenz-Wandlung voltage-frequency conversion
Spannungs-Oberschwingungsfaktor harmonic voltage factor * kennzeichnet d. Abweichg. e. Wechselspanng. von d. Sinusform || HVC * Abkürz. für 'Harmonic Voltage Factor'
Spannungs-Unsymmetrie voltage asymmetry
Spannungs-Zeit-Fläche voltage-time-area
Spannungs-Zwischenkreis DC bus
Spannungsabfall voltage drop || line drop * entlang einer Leitung || voltage failure || voltage sag * 'Absacken' der Spannung || voltage failure * irregulärer Spannungsausfall/-einbruch || potential drop || voltage regulation * z.B. einer Spannungsquelle bei Laständerung || regulation * z.B. einer Spannungsquelle bei Laständerung
spannungsabhängig voltage-dependent
Spannungsabschaltung power shutdown * der Stromversorgung
Spannungsabsenkung voltage reduction
Spannungsänderung voltage variation
Spannungsanhebung voltage boost * z.B. zur Erhöhung d. Anfahrmoments e. Motors, um d. Losbrechreibg. zu überwinden || boost
Spannungsanschnittsteuerung voltage phase control
Spannungsanstieg voltage rise || rate of voltage rise * Anstiegsgeschwindigkeit (du/dt)
Spannungsanzapfung voltage tap * an einem Trafo usw.
Spannungsausfall voltage failure || power loss * Ausfall der Stromversorgung || supply failure || power failure * Ausfall der Versorgungsspannung || power outage
Spannungsbeanspruchung voltage stressing * bei umrichtergespeisten Drehstrommotoren: siehe VDE 0530 Teil 1, Beiblatt 2
Spannungsbegrenzung voltage limitation
Spannungsbereich voltage range
Spannungseinbruch voltage dip * kurzzeitiger || power loss * Ausfall der Stromversorgung || power failure * Stromausfall || voltage drop || voltage sag * 'Einsacken' der Spannung über längere zeit || power interruption * Unterbrechung der Stromversorgung
Spannungserfassung voltage sensing
Spannungserhöhung voltage rise || temporary overvoltage * vorübergehende
spannungsfest of high dielectric strength || surge-proof * robust bezüglich Spannungsspitzen (z.B. bei Kondensatoren) || voltage-proof
Spannungsfestigkeit dielectrical strength * Isolations-/Durchschlagsfestigkeit e. Isoliersystems, z.B. VDE 0530 T,15 || electric strength || voltage withstand capability || dielectric strength * z.B. eines Kondensator, eines Isoliersystems (el. Masch.: VDE0530 T.15) || voltage endurance
Spannungsfolger voltage follower * Impedanzwandler mit Verstärkung 1 (mit Operationsverstärker(n)) || buffer amplifier
spannungsfrei de-energized || dead * dead state; ~er Zustand || disconnected from the supply * vom Netz getrennt || off-circuit
spannungsfreie Pause zero-voltage interval

Spannungsfreiheit isolation from supply * *Trennung vom Netz; verify: feststellen* ‖ loss of voltage * *Spannungslosigkeit*
Spannungsfreischaltung voltage disconnection ‖ voltage disconnect
spannungsführende Teile components under voltage ‖ parts under voltage ‖ live parts * *Teile, die Personen gefährden können* ‖ components at hazardous voltage level ‖ life components
Spannungsgefälle potential gradient
Spannungskennziffer figure denoting voltage * *Bestellnummer-Ergänzung bei SIEMENS-Motor* ‖ voltage code * *Bestellnummer-Ergänzung bei SIEMENS-Motor*
Spannungsklasse voltage class
Spannungskonstante back-EMF constant * *Verhältnis EMK zu Drehzahl [V/upm] bei Elektromotor (1/→Drehzahlkonstante)*
Spannungskonstanthalter voltage stabilizer
spannungslos de-energized ‖ dead ‖ isolated * *Stromkreis* ‖ disconnected * *Stromkreis* ‖ off-circuit
spannungslose Pause zero-voltage interval
spannungsloser Zustand no-voltage condition * *in: in*
Spannungslücke zero-voltage interval
Spannungsmesser voltmeter
Spannungsmodell voltage simulation model * *Rechenmodell, z.B. für Drehstrommaschine*
Spannungspegel voltage level
Spannungspfad voltage path
spannungsproportional proportional to the voltage
Spannungsquelle voltage source
Spannungsregler voltage regulator
Spannungsreserve voltage reserve
Spannungsrückkehr voltage recovery
Spannungsschwankung voltage fluctuation ‖ voltage variation
Spannungsschwankungen voltage fluctuations
Spannungsspitze voltage peak ‖ voltage spike * *Nadelimpuls, z.B. Störimpuls* ‖ overvoltage spike * *Überspannungsspitze* ‖ overvoltage transient * *kurzzeitige Überspannung*
Spannungsspitzen voltage peaks ‖ voltage spikes * *Nadelimpulse*
Spannungsstabilisierung voltage stabilization
Spannungsständer tensioning stand * *[Textil] z.B. in Faserverstreckanlage*
Spannungssteilheit voltage gradient ‖ slew rate * *Anstiegsgeschwindigkeit der Spannung, z.B bei Operationsverstärker* ‖ voltage rate-of-rise
Spannungssteuerung voltage control
Spannungsstoß voltage surge ‖ surge
Spannungsteiler voltage divider ‖ attenuator * *Dämpfungs/Abschwächglied, z.B. für Messgerät, Gleichstromtacho-Auswertung*
Spannungsteiler-Widerstand voltage-dividing resistor
Spannungstoleranz voltage tolerance
Spannungsüberhöhung voltage increase ‖ voltage rise
Spannungsüberwachung voltage monitoring
spannungsumrüstbarer Motor voltage-conversion motor * *[el.Masch.] z.B. Bergwerksmotor, d. sich von 500 auf 1000 V umrüsten läßt*
spannungsumschaltbarer Motor multi-voltage Motor * *[el.Masch.]*
Spannungsunsymmetrie voltage unsymmetry ‖ voltage unbalance
Spannungsunterdrücker voltage suppressor
Spannungswandler voltage transformer ‖ DC-DC converter * *Gleichspannungswandler* ‖ voltage transducer * *Messwandler* ‖ voltage converter

Spannungswelligkeit voltage ripple
Spannungswiederkehr voltage restoration ‖ voltage recovery * *nach Spannungseinbruch*
Spannungszeiger voltage phasor
Spannungszwischenkreis voltage source DC-link ‖ DC-link ‖ DC-bus
Spannungszwischenkreis-Schiene DC bus
Spannungszwischenkreis-Umrichter voltage-source DC link converter ‖ voltage source converter ‖ DC link converter
Spannungszwischenkreisumrichter voltage source DC-link converter
Spannverbindung clamping connection ‖ claming fixture * *als Bau-/Konstruktionselement*
Spannvorrichtung jig * *Auf-/Einspannvorrichtung* ‖ chuck * *für Werkstücke* ‖ chucking appliance * *für Werkstücke* ‖ stretching device * *zum Stramm-/Langziehen* ‖ clamping device ‖ clamping ‖ fixture ‖ holding device ‖ gripping tool * *Greifwerkzeug* ‖ tensioning unit * *für Riemen, Kette, Seil usw.* ‖ clamp ‖ chucking device * *[Werkz.masch.]*
Spannwalze stretch roll ‖ tension roll
Spannweite opening * *eines Schraubenschlüssels* ‖ span * *eines Schraubenschlüssels*
Spannwelle expanding shaft
Spannzange collet
Spannzangenfutter collet chuck * *z.B. bei Werkzeugmaschine*
Spannzeug chucking tool ‖ fixture ‖ clamping device * →*Spannvorrichtung*
Spanplatte chipboard * *[Holz]* ‖ pressboard * *[Holz]* ‖ chip board * *[Holz]* ‖ particle board * *[Holz]* ‖ chipboard panel * *[Holz]*
Spanplattenpresse chipboard press * *[Holz]*
Spantiefe depth of cut
sparen save * *Geld, Kräfte, Mühe, Zeit usw.* ~ ‖ cut down expenses * *Ausgaben senken, sich einschränken* ‖ economize * *sparsam umgehen mit* ‖ spare * *Kosten, Mühe* ~ ‖ tighten the belt * *(salopp) den Gürtel enger schnallen*
sparsam saving ‖ economical * *of: mit* ‖ economizing
Sparschaltung economy circuit ‖ autotransformer * *Spartransformator*
Spartrafo autotransformer
Spartransformator autotransformer
später later ‖ subsequent * *nachfolgend* ‖ future * *(zu)künftig*
spätere Verwendung future use
Spediteur forwarding agent ‖ shipping agent ‖ forwarding agency * *Speditionsfirma*
Spedition freight agency ‖ forwarding agency ‖ transport company ‖ freight forwarding company ‖ shipping company
Speiche spoke * *z.B. eines Rades*
Speicher memory * *Datenspeicher* ‖ accumulator * *Zwischenspeicher, Warenspeicher zur Überbrückung d. Rollenwechselzeit bei Auf/Abwickler*
Speicherabzug memory dump
Speicheradresse memory address
Speicherausbau memory configuration * *Konfiguration* ‖ memory extension * *Vergrößerung* ‖ memory expansion * *Vergrößerung* ‖ memory space * *Speicherplatz/-raum* ‖ memory capacity * *Speicherkapazität/-vermögen*
Speicherbank memory bank * *[Computer]*
Speicherbaugruppe memory module ‖ memory submodule * *Speichermodul/-Subbaugruppe*
Speicherbaustein memory chip * *Integrierter Speicherschaltkreis* ‖ Integrierter Speicherschaltkreis ‖ memory module * *Speichermodul (evtl. auf kleinem Leiterkärtchen (z.B. →SIMM))*

Speicherbedarf memory requirements ‖ required memory capacity ‖ memory capacity
Speicherbelegung memory allocation ‖ memory assignment
Speicherbelegungsliste memory map
Speicherbelegungsplan memory map
Speicherfresser memory hog * *(salopp)* "Speicher-Vielfraß", Programm, das viel Speicherplatz beansprucht
Speicherglied storage element ‖ flipflop ‖ flip-flop ‖ memory cell * *Speicherzelle* ‖ latch * *mehrere parallel angeordnete Flipflops mit gemeinsamem Übergabe-Taktsignal*
Speichergröße memory size
Speicherinhalt contents of data memory ‖ memory contents
Speicherinhalte memory contents
Speicherkapazität memory capacity
Speicherkarte memory card
Speicherkondensator buffer capacitor ‖ storage capacitor ‖ reservoir capacitor
Speichermodul memory module ‖ memory pack ‖ memory submodul * *Speichersubmodul, Speicherkärtchen (z.B. →SIMM)*
speichern store * *abspeichern* ‖ buffer * *zwischenspeichern* ‖ accumulate * *ansammeln; auch in Ergebnisregister o. Warenspeicher für durchlaufende Warenbahnen* ‖ memorize * *im Speicher (auch spannungsausfallsicher) aufbewahren*
speichernd rücksetzen unlatch * *z.B. Relais*
speichernd setzen latch * *z.B. Relais*
speichernde Funktionen latching/unlatching functions * *bei SPS*
Speicherplatz memory location * *Speicherzelle(n) für bestimmte Daten* ‖ memory space * *zur Verfügung stehender Speicher* ‖ memory capacity * *zur Verfügung stehender ~, Speicherkapazität, Speicherausbau*
Speicherplatzbedarf memory requirements * *{Computer}* ‖ memory capacity * *{Computer}* ‖ memory capacity requirements * *{Computer}*
speicherprogrammierbar programmable * SPS
Speicherprogrammierbare Steuerung programmable controller ‖ PLC
speicherresidentes Programm resident program ‖ TSR program * *Terminate a.Stay Resident program, bleibt nach Beendig. i. Hauptspeicher* ‖ terminate and stay resident program
Speichersubmodul memory submodule ‖ memory sub-module
Speichertiefe memory depth * *eines Schiebespeichers (z.B. Trace-/Messwert-/FIFO-Speichers usw.*
Speicherung storage ‖ storing ‖ accumulation * *Ansammlung* ‖ capture * *eines Zählerstandes, Messwertes im laufenden Betrieb*
Speicherverwaltung memory management
Speicherverwaltungsprogramm memory manager ‖ memory administration programm
Speicherwert contents of data register * SPS
Speicherzelle memory cell ‖ memory location ‖ storage location
Speicherzugriffsfreigabe memory select * *(MEM-SEL) bei SPS*
Speisefrequenz supply frequency
Speiseleistung infeed power
Speiseleitung supply line ‖ feeder
speisen supply ‖ feed ‖ power * *mit Energie versorgen*
Speisequelle feed source ‖ power supply
Speisespannung supply voltage
Speisewalze feed roll ‖ feed roller * *z.B. bei einer Krempelmaschine {Textil}*
Speisewasser feed water
Speisewasserpumpe feedwater pump
Speisung supply ‖ feed

spektakulär spectacular
Spektrum spectrum ‖ spectra * *(Plural)*
Spektrumanalysator spectrum analyzer
Sperre inhibit * *als Kommando* ‖ disable * *als Kommando* ‖ blocking * *Versperren* ‖ suppression * *Unterdrückung* ‖ lock ‖ locking device * *Sperrvorrichtung* ‖ interlock * *Verriegelung* ‖ block * *Blockade, Versperren* ‖ lock out ‖ obstacle * *Hindernis* ‖ blockade * *Blockade, Sperre, Hindernis*
sperren disable ‖ block * *auch Diode, Thyristor* ‖ lock * *(mechan.)* ‖ be turned off * *z.B. Thyristor/Transistor* ‖ arrest * *feststellen, sperren, arretieren* ‖ inhibit ‖ ban * *verbieten, Verbot auferlegen, mit Verbot belegen* ‖ be blocked * *sich im gesperrten Zustand befinden, z.B. Halbleiter*
sperrend blocking * *z.B. Diode*
Sperrfähigkeit blocking ability * *z.B. Thyristor* ‖ blocking capacity * *eines Halbleiterbauelements*
Sperrfilter band-stop filter * *Bandsperre* ‖ rejector filter
Sperrholz plywood
Sperrichtung inverse direction * *bei Halbleitern*
sperrig bulky
Sperrklinke pawl * *Sperrhaken, Sperrklinke, Klaue* ‖ sprag * *Spreiz-/Sperrklinke, Bremsklotz*
Sperrkondensator blocking capacitor
Sperrkreis rejector circuit * *siehe auch 'Bandsperre'*
Sperrschaltung interlocking circuit
Sperrschicht barrier junction ‖ junction * *in Halbleiter; junction temperature: →Sperrschichttemperatur*
Sperrschichttemperatur barrier junction temperature ‖ junction temperature
Sperrspannung reverse voltage * *z.B. v. Diode, Transistor* ‖ off-state voltage * *z.B. Thyristor/Diode* ‖ blocking voltage * *z.B. Diode, Transistor, Thyristor* ‖ inverse voltage * *z.B. Diode, Transistor*
Sperrstrom reverse current * *(Diode, Thyristor usw.)* ‖ blocking-state current * *(Thyristor, Diode)* ‖ off-state current * *z.B. bei Diode, Thyristor, Transistor*
Sperrung blocking * *z.B. Thyristor, Regler, Verkehr, Konto* ‖ stoppage ‖ obstruction * *locking device * Vorrichtung* ‖ trip gear * *Vorrichtung*
Sperrverzugszeit reverse recovery time * *z.B. einer Diode*
Sperrzahnschraube self-locking screw * *flatheaded: mit Flachkopf*
Spesen expenses * *Ausgaben* ‖ charges * *an Kunden verrechnete Gebühren*
Spesenabrechnung expense voucher ‖ expense statement
Spezial- special ‖ special-purpose * *für einen speziellen Verwendungszweck* ‖ non-standard * customized * *kundenspezifisch (angepasst)* ‖ speciality
Spezialauslegung special design
Spezialgebiet specialist area ‖ special branch ‖ special subject
Spezialisierung specialization
Spezialist specialist ‖ specialist engineer ‖ expert ‖ professional * *Profi*
Spezialität speciality
Spezialkenntnisse special knowledge * *siehe auch →Fachwissen*
Speziallösung dedicated solution * *zugeschnitten auf ein Problem/eine Anwendung* ‖ customized solution * *auf einen Kunden zugeschnitten*
Spezialwerkzeug special tool
Spezialwissen special knowledge
speziell special ‖ separate * *gesondert* ‖ certain * *bestimmt, gewiss* ‖ especially * *(Adv.)* ‖ particular * *besonders* ‖ individual * *gesondert, eigen* ‖ specific * *spezifisch, speziell, bestimmt, typisch, kennzeichnend, eigen(tümlich), sonder, genau* ‖ specifi-

cally * *(Adv.) speziell, besonders, definitiv* || speciality * *Spezial-, besonders*
Spezifikation   specification || spec * *Kurzform für 'specification'*
Spezifikum   special feature || speciality * *Spezialität* || specialty * *(amerikan.) Spezialität* || special property || detail
spezifisch   specific
spezifische Wärme   specific heat
spezifisches Gewicht   specific weight || specific gravity || relative density || density
spezifisches Volumen   specific volume
spezifizieren   specify
spezifiziert   specified
Sphäroguss   ductile iron || spheroidal graphite iron
Spiegel   mirror
Spiegelbild   mirror image
spiegelbildlich   mirror-inverted || mirror-image || mirrored
Spiegelglas   plate glass
spiegeln   mirror * *(auch figürlich) auch 'wider~'* || reflect * *reflektieren (auch figürl.)* || be mirrored * *sich ~* || be reflected * *sich ~*
spiegeln an   reflect on * *z.B. den Anforderungen* || mirror in
Spiegelplatte   mirror disk * *redundante Festplatte (2 Platten mit gleichen Daten; siehe auch →RAID)*
Spiegelplattensystem   RAID * *'Redundant Array of Inexpensive Disks'. Speicher mit redundanten Festplatten* || hot-swap disk array * *~ mit unter Spannung wechselbaren Festplatten*
Spiegelsymmetrie   mirror symmetry
spiegelsymmetrisch   mirror symmetric
Spiel   slack * *Lose/Sack z.B. in einer Warenbahn* || cycle * *Last~* || dead travel * *toter Gang, Lose in einem Maschinenteil* || backlash * *Lose in Getriebe/Gewinde* || slackness * *toter Gang* || clearance * *freier Abstand, z.B. Ventil~, Lager~* || play * *come into play: ins ~ kommen*
spielarm   low-backlash || nearly backlash-free * *nahezu spielfrei*
Spieldauer   time for one cycle * *eines Lastspiels* || duty cycle time * *eines Lastspiels* || cycle duration
spielfrei   zero-backlash || free from backlash || with zero backlash * *z.B. Getriebe* || backlash-free
Spielraum   room to move || free play * *(figürl.)* || elbow-room * *Bewegungsfreiheit* || leeway * *(figürl.)* || margin * *Frist, Spanne* || clearance * *räumlich* || play * *Spiel, siehe auch 'Spiel'* || room of move || scope * *Spanne; leave/allow someone scope: jdm. ~ lassen* || margin * *Spanne* || latitude * *in der Auslegung*
Spielzeug   toy || gadget * *(amerikan.) Apparat, Gerät, Vorrichtung, techn.Spielerei, Apparätchen* || gimmick * *(salopp) technische Spielerei, Apparätchen, Spielkram, unnötige Apparatur*
Spindel   spindle * *z.B. ~ bei Werkzeugmaschine* || screw * *Schraub~, Gewindestange, Hub~* || mandrel * *Spanndorn, Haupt~* || beam * *~ einer Drehmaschine* || lead screw || bobbin * *{Textil} Spule*
Spindel-Mutter-Getriebe   spindle-nut gear || lead screw drive * *Leitspindelantrieb*
Spindelantrieb   spindle drive || mandrel drive * *Haupt~*
Spindeldrehzahl   spindle speed * *{Werkz.masch.| Textil}*
spindelgetrieben   operated through spindle gearbox
Spindelhals   spindle collar * *z.B. einer Bohrmaschine*
Spindelkasten   headstock
Spindelmesssystem   spindle measuring system * *{Werkz.masch.}*
Spindelorientierung   spindle orientation
Spindelpositionierung   spindle positioning ||

spindle orientation * *Ausrichtung* || nut setting * *bei Hub/Gewinde/Positionierspindel*
Spindelpotentiometer   helical potentiometer || lead-screw potentiometer || spindle-actuated potentiometer || helipot * *(salopp) Markenname der Firma Beckman* || multiturn potentiometer
Spindelsteigung   pitch of screw || screw lead
Spinn-Streck-Spulmaschine   spinning/drawing/winding machine * *zur Erzeugung von Chemiefasern*
Spinndüse   spinning nozzle || spinneret
spinnen   spin * *{Textil}Fäden, Garn usw; spin wool into yarn: Wolle zu Garn spinnen* || cocoon * *(sich) ein~, einhüllen (figürl.) (Gerät usw.) einmotten/außer Betrieb setzen*
Spinnereimaschine   spinning machine
Spinnextruder   spinning extruder * *{Textil} zur Chemiefaserherstellung*
Spinnfaden   spinning thread || spinning fiber * *Spinnfaser*
Spinngeschwindigkeit   spinning speed * *{Textil} bei Spinnmaschine*
Spinnmaschine   spinning machine * *Textilmaschine*
Spinnmotor   spinning frame motor
Spinnpumpe   spinning pump
Spinnschacht   spinning duct * *{Textil} bei der Chemiefaserherstellung*
Spinnstelle   spinning section * *{Textil} in Chemiefasermaschine* || spinning unit * *z.Chemiefaserherstellg., bestehend aus Spinnpumpe, Kühlschacht, Verstreckwerk,Wickler*
Spinnwebverfahren   cobwebbing || cocooning
Spiralbohrer   twist drill
Spirale   spiral || helical curve * *spiralförmige Kurve*
Spiralfeder   helical spring || spiral spring || coil spring
spiralförmig   helical || spiral
spiralgerillt   spiral grooved
Spiralrillen   spiral grooves * *z.B. an Kupplungs-/Bremslamelle*
Spiritus   spirit
spitz   pointed || peaked || acute * *(Winkel usw.)* || tapered * *~ zulaufend; taper (off): ~ zulaufen/sich verjüngen*
spitz zulaufen   taper || taper off * *auch 'sich verjüngen'*
spitz zulaufend   tapered
Spitzbohrer   spade drill
Spitze   peak * *(auch elektr.)* || spike * *Nadelimpuls* || end * *äußerstes Ende, Stirnseite* || inrush * *Strom/Momentenspitze beim Einschalten und Drehrichtungswechsel* || point || tip * *(mechan.)* || apex * *Spitze (auch eines Kegels), Gipfel, Scheitelpunkt, Höhepunkt (figürl.)* || top * *(figürl.) einer Hierachie/Reihenfolge usw.*
Spitze-Spitze   peak-to-peak
Spitze-Spitze-Spannung   peak-to-peak voltage
Spitze-zu-Spitze-Wert   peak-to-peak value
Spitzen-   high end * *an der Spitze stehend* || front end * *an der Spitze stehend* || high performance * *Hochleistungs-* || super * *überragend, höchst* || superlative * *überragend, höchst* || ultimate * *ultimativ* || peak * *kurzzeitig vorkommender Maximalwert* || maximum * *Maximal-* || transient * *schlagartige, vorübergehende Erhöhung* || top || highest || first-class || leading * *führend* || top-notch * *(salopp) spitzenmäßig, →prima*
Spitzenbedarf   peak demand
Spitzenbelastung   peak load
Spitzenbremsleistung   peak brake power
Spitzenlast   peak load
Spitzenlastzeit   peak hours
Spitzenleistung   peak power * *[kW]* || peak output * *z.B. Stromrichter, Motor* || peak performance

* höchste Leistungsfähigkeit || maximum output
* z.b. eines Motors, eines Stromrichters || top performance * beste Leistungsfähigkeit/Ausführung/ Durchführung/Verrichtung/Hervorbringung
Spitzenprodukt front end product || top-quality product || high-end product
Spitzenspannung peak voltage || surge voltage * Höhe von Spannungsspitzen, z.B. nach IEC || crest voltage * →Scheitelspannung
Spitzensperrspannung peak inverse voltage * eines Halbleiters
Spitzenstrom peak current
Spitzentemperatur peak temperature
Spitzenwert peak value
spitzer Winkel acute angle
Spitzsenker countersink
spitzwinklig acute-angled
Spitzzahnkette ripping-tooth chain * {Holz} für Kettensäge
Spitzzahnung pointed teeth * {Holz} bei Säge
Spitzzange needle-nosed pliers || pointed pliers
Spleiß splice * Rollenwechsel Klebe/Schneidvorgang bei fliegender Auf/Abwicklung
Spleißstelle splice
Spline spline * Kurvendarstellung im CAD- und Numerikbereich durch Polynom 3. Ordnung
Splint split pin * geschlitzter ~ || cotter || cotter pin || split pin * Spreizsplint || alburnum * {Holz} Splint, Splintholz || sap * {Holz} Splint(holz)
Splintloch split-pin hole
Splitter splinter || shiver || fragment * Bruchstück || chip * Span || shivers * (Plural) || edges ~ {Holz} Spreißel
splittern splinter || split
Spontanmeldung parameter change report * (→PROFIBUS-Profil) || spontaneous message * (→PROFIBUS-Profil)
sporadisch sporadic || sporadically * (Adv.) || now and then * hin und wieder || at times * gelegentlich, manchmal || occasional * gelegentlich || random * wahllos/von Fall zu Fall auftretend (z.B. Fehler), Zufalls-
Sprache language * Fremdsprache, Programmiersprache || voice * Stimme,Ton || foreign language * Fremdsprache
Sprachebene language level
Sprachmittel language resources * {Computer} || syntax element * {Computer} Syntaxelement
Sprachraum language subset * bei SPS
Spraydose spray tin
Sprecher speaker || talker || spokesman * offizieller ~ einer Firma, Behörde usw.
Spreize spreader * Spreize, Abstandshülse, "Auseinanderhalter", Zerstäuber, Spritzdüse
spreizen spread || straddle * Beine/sich ~, mit gespreizten Beinen stehen/gehen; straddle carrier; Containertransporter
sprengen blow up || blast || force open * gewaltsam öffnen || break open * aufbrechen || sprinkle * z.B. Wasser ~ || spray * z.B. Wasser ~, mit dem Schlauch ~ || be out of proportion * den Rahmen ~
Sprengring circlip || snap ring || retaining ring
springen jump || branch * verzweigen
spritzen splash * be~ (mit Wasser usw.) || spray * (ver-)sprühen, zerstäuben, bespritzen, be~, spritzlackieren, sprengen || squirt * (hervor-/heraus-/be-) spritzen, hervorsprudeln; auch Substantiv: Strahl, Spritze || syringe * (aus-/ab-/be-/ein-) spritzen || sprinkle * besprengen || injection-mould * thermoplastischen Kunststoff ~ || spout * heraus~ || dash * (salopp) eilen || flit * (salopp) eilen || spurt * (heraus)spritzen (Wasser, Flüssigkeit) || splatter * (be-/umher-) spritzen; beschmutzen, sprenkeln; planschen || water * wässern || hose * mit Schlauch ~

Spritzer splash || speck * Spenkel, Tupfer, Fleckchen, Pünktchen
Spritzgießmaschine injection-moulding machine * für Kunststoff
Spritzgießwerkzeug injection molding tool * für Kunststoff-Spritzgießmaschine
Spritzguss die cast * (Nichteisen-) Metall-Druckguss; die-cast aluminium alloy: Aluminium-~ || injection moulding * z.b. Kunststoff~ || injection diecasting || injection molding
Spritzgussform die-casting die * für Metall-Spritzguss || die mould * für Metallspritzguss || die mold * für Metallspritzguss
Spritzgussmasse molding compound
Spritzgussteil injection-molded part * aus Kunststoff || die-cast part * aus Metall || injection-moulded part * aus Kunststoff
Spritzpistole spray gun
Spritzroboter spraying robot * zum Lackieren
Spritzstand spraying booth * Lackierkabine
Spritzwasser splashing water || spraying water || drip water * Tropfwasser
Spritzwasser-geschützt splash-proof || drip-proof * Tropfwasser-geschützt
spritzwassergeschützt splash-proof || drip-proof * tropfwassergeschützt
Spritzwasserschutz protection against splashing water
spröde brittle
Sprödigkeit brittleness
Sprühdüse spray nozzle
sprühen spray
Sprühentladung corona discharge || partial discharge
Sprühölschmierung flood lubrication || spray oil lubrication
Sprühwasser spraying water || spray water
Sprung jump * auch in Programm || step * ~ eines Signals || step change * ~artige Änderung || sudden change * ~artige/plötzliche Änderung || stepchange signal * Stimulus für Regleroptimierung (zur Aufnahme der ~antwort) || sudden variation * ~artige Änderung || skip * Überspringen (Seiten, Programmbefehle usw.) || crack * Riss || flaw * Bruch, Riss, Materialfehler, fehlerhafte Stelle || throw * ~ in einer Wicklung || branch * Programmverzweigung || bound * Satz, 'Hüpfer' || step function * ~funktion
Sprungadresse jump address || branch address
Sprungantwort step response || step-forced response || transient response || response to step changes
sprungartig step || by leaps and bounds * sprungweise, sprunghaft || sudden
sprungartige Änderung step change * in: einer Größe || sudden change || sudden variation
Sprungbefehl jump instruction
Sprungfunktion step function
Sprungfunktionsgenerator step function generator * z.B. Funktionsbaustein zur Regleroptimierung
sprunghaft with a step function || jerky * ruckhaft (auch Markt usw.) || spasmodic * sprunghaft (auch Markt), krampfartig || by leaps and bounds * (Adv.) rise by leaps and bounds: ~ steigen || dramatically * (Adv.) dramatisch, z.B. Preisanstieg
Sprungleiste multiple branch
Sprungmarke jump label
Sprungoperation jump operation
Sprungziel jump destination * {Computer} in Softwareprogramm
SPS PLC * programmable logic controller: Speicherprogrammierbare Steuerung (EN 61131-3, IEC 1131-3) || programmable controller
Spulantrieb coil drive

Spule choke * *(elektr.) Drossel (eher kleinere)* || reactor * *(elektr.) Drossel (eher größere)* || inductor * *(elektr.) Drossel* || coil * *Haspel; Betätigungs~ (z.b. Relais, Schütz), ~ einer (Motor)Wicklung* || solenoid * *Betätigungs~ (z.b. Magnetventil)* || bobbin * *Garn-/Spinn~* || spool * *Garn-/Spinn-/ Draht-/Faden~, Rolle, Haspel* || reel * *Webspule, Rad; ~ bei Auf/Abwickler, Webstuhl, Film, Tonband* || drum * *Trommel* || inductance * *(elektr.) Induktivität*
Spule mit einer Windung single-turn coil
spülen flush * *z.B. Getriebekasten* || wash || swill || rinse
Spulenabzug bobbin drawing device * *{Textil} in der Spinnerei*
Spulendraht magnet wire
Spulengatter creel * *{Textil} z.b. am Anfang einer Schärmaschine* || warping creel * *{Textil} in der Weberei*
Spulengruppe coil group * *z.b. in einer Motorwicklung*
Spulenkopf coil end * *{el.Masch.} aus dem Bleckpaket herausragender Teil einer Spule*
Spulenseite coil side * *{el.Masch.} in der Nut eines Blechpakets liegender Teil einer Spule*
Spulenweite coil pitch * *{el.masch.} Ausdehnung einer Spule i.Verhältn. z. Umfang der Ständerbohrg.*
Spulenwickelmaschine coil winding machine * *für Magnetspulen*
Spulenwicklung coil winding
Spuler coiler * *~ für Blechband, Draht usw.* || pay-off * *Ab~/Abwickler, z.B. für Blechband, Kabel* || take-up * *Auf~/Aufwickler, z.B. für Blechband, Kabel* || winder * *Wickler* || spooler * *~ bei Drahtziehmaschine*
Spulkopf winding head || coil end * *z.b. bei Aufwickler (für Fäden, Draht usw.)*
Spulmaschine coil machine
Spülmaschine dish washer * *Geschirrspülmaschine* || rinse machine
Spülöl flushing oil
Spülpumpe mud pump * *zum Ausspülen von (Öl-) Bohrlöchern*
Spur track * *auch eines Impulsgebers*
spürbar significant || considerable * *beträchtlich* || sensible * *fühlbar* || marked * *deutlich* || distinct * *deutlich*
Spurkennbit track-ID bit * *bei SPS*
Spurversatz track offset * *Impulsgeber* || track displacement * *Impulsgeber*
squentiell sequential
SR-Motor SR motor * *Abkürzg. für 'Switched Reluctance motor'* || switched reluctance motor
SRAM SRAM * *Static Random Access Memory, statischer Schreib/Lesespeicher, braucht keine Auffrischzyklen*
ssignalflußanzeige power-flow indication * *Anzeige des durchgeschalteten Leistungsstrangs* || powerflow readout * *Anzeige des durchgeschalteten Leistungsstrangs*
ssignalformer signal conditioner || I/O module * *Ein/Ausgangsmodul* || process interface module * *Prozeßsignalformer (z.B. binäre/analoge Ein/Ausgabebaugruppen)*
Staatsunternehmen state company
Stab bar * *Gitterstab, Metallstab* || rod * *Stange* || team * *Stab von Experten usw.* || staff * *Mitarbeiterstab*
stabil stable * *auch für Regelung* || solid * *fest, robust* || rugged * *stark, stabil, →robust, grob, ungehobelt* || strong * *stark, robust* || steady * *gleichmäßig, stetig, ohne Abweichung* || sturdy * *kräftig, stabil, standhaft, fest, →massiv*
stabiler Bereich stable region

stabiler Betrieb stable operation
Stabilisator stabilizer
stabilisieren stabilize
stabilisiert stabilized * *z.B. Stromversorgung*
Stabilisierung stabilization || stabilizing || stabilizer * *stabilisierendes (Konstruktions-) Element* || steadying
Stabilität stability || rubustness * *auch Regelung* || ruggedness * *der Bauart usw.*
Stabilitätsgrenze stability limit || limit of stability
Stabilitätskriterium stability criterion * *(Plural: 'criteria')*
Stabilitätsprobleme stability problems
Stabisolierung bar insulation * *{el. Masch.} Isolierung e. stabförmigen Leiters*
Stabkerndrossel bar-core reactor
Stabschwingungen bar vibrations * *{el. Masch.} Schwingungen v. losen Stäben e. Käfigläufers mit doppelter Netzfrequenz*
Stabwicklung bar winding * *{el. Masch.} Wicklung, deren Leiter aus Stäben besteht*
Stack stack * *Stapelspeicher (zum Ablegen von Rücksprungadressen und Registerinhalten); LIFO-Speicher*
Stadium stage * *early: frühes (siehe auch →Stufe)* || phase * *siehe auch →Phase*
Stadtbahnwagen light-rail vehicle
Stadtrundfahrt city tour || tour of the city
Stadtwerke municipal authorities
Staffel group * *Antriebe* || sequence * *(Reihen)Folge; z.B. Antriebe* || graduation * *~ung, Abstufung, Einteilung; z.b. Preise, Steuern, Löhne* || grading * *Stufung, ~ung; z.b. Qualität, Güte, Rang Sorte, Klasse* || grouping * *Aufteilung in Gruppen* || classification * *Aufteilung in Klassen*
staffeln grade * *(ab)stufen, staffeln; z.b. Qualität, Güte, Rang, Sorte, Klasse* || graduate * *abstufen, einteilen; z.B. Preise, Steuern, Löhne* || differenciate * *differenzieren, Unterschiede machen; z.B. Preise, Steuern, Löhne* || stagger * *versetzt anordnen; z.B. Kohlebürsten, Arbeitszeit* || time-grade * *zeitlich ~; z.b. bei Schutzeinrichtungen* || divide into sections * *in Abschnitte, Gruppen usw. einteilen* || group * *in Gruppen einteilen* || classify * *in Klassen einteilen*
Stagnation stagnation || stagnancy || sluggishness * *geschäftliche, wirtschaftliche ~*
stagnieren stagnate * *z.B. Geschäft, Markt* || be stagnant || be at a standstill || be sluggish
stagnierend stagnant * *z.B. Geschäft, Markt* || stagnated * *z.B. Geschäft, Markt*
Stahl steel * *(Bau)Stahl; siehe z.B. DIN 17006, DIN 17100 (z.B. ST37 mit Zugfestigkeit 370 N/mm²)*
Stahlband strip-steel || steel band || steel strip
Stahlbandförderer steel belt conveyor * *{Fördertechnik}*
Stahlbandmaß steel measuring tape || steel tape measure
Stahlblech sheet steel * *dünnes Blech, Feinblech* || sheet metal * *(dünnes) Metallblech, Feinblech* || steel plate * *dickes Blech, Grobblech*
Stahlcord steel cord * *{Reifen}*
Stahlfeder steel spring
Stahlfundament steel base * *aus Stahlteilen geschweißtes Maschinenfundament*
Stahlgehäuse steel frame * *{el.Masch.}*
Stahlguss cast steel || cast steel * *Gusswerkstoff aus niedrig legiert. Eisen; z.B. GS 38 m. Zugfestigkt. 380 N/mm²; DIN 1681*
Stahlnagel steel pin || masonry pin * *zum Einschlagen in Mauerwerk, z.B. "IMPU"-Nagel zur Leitungsverlegung im Putz*
Stahlseil wire rope
Stahlstift steel pin

Stahlwerk steel works || steelmaking plant || steelworks
stammen originate * *seinen Ursprung haben (in: in); entstehen, entspringen, von jdm. ausgehen (from: von)* || be descended * *abstammen, herkommen (from: von)* || be derived * *abgeleitet sein* || stem * *(amerikan) (from: von)*
Stammhaus head office * *Leitungsbüro, koordinierende Abteilung, Stammsitz* || corporate headquarters * *Firmen-Hauptsitz* || headoffice
Stammhausvertrieb head office sales/marketing || headoffice sales/marketing
Stammverzeichnis root directory * *Hauptverzeichnis eines Comuterlaufwerks (z.B. einer Festplattenpartition)* || main directory
Stand booth * *Messe~* || stand * *Messe~, Gestell, ~ort, Platz; Ausrüstung* || position * *(auch figürl.)* || state * *Zus~* || situation * *Lage* || level * *Niveau* || rate * *Markt-Kurs/-Preis* || status * *(Zu)Stand, (Gültigkeits- usw.) Datum* || footing * *sichere Stellung; secure footing: sicherer ~; gain a footing: festen Fuß fassen*
Stand der Technik state of the art * *the present: der gegenwärtige*
standard standard || off-the-shelf * *handelsüblich*
Standard- standard || common || commercial * *handelsüblich*
Standard-Einbauplatz standard plug-in station * *(bei SPS) Rastermaß für Steckbaugruppen-Breite*
Standardabweichung standard deviation * *(Wahrscheinlichkeitsrechnung)*
Standardausführung standard design * *Standardbauart* || standard type * *Standardmodell* || standard version * *Normalversion* || basic design * *Grundtyp*
Standardbaustein standard block * *Funktionsbaustein*
Standardbedienfeld standard operator panel
Standardeinbauplatz standard plug-in station * *SEP bei SPS (Maß für die Breite einer Steckbaugruppe)* || SPS * *Abk. für 'Standard Packaging Slot'*
Standarderweiterung standard option
standardisieren standardize
standardisiert standardized || standardised
Standardisierung standardization
Standardkomponente standard component
Standardlösung standard solution
Standardmaße standard dimensions
standardmäßig in standard * *(Adv.)* || as standard * *(Adv.) als Standard* || normally * *(Adv.) normalerweise* || standard * *(Adj.)* || as standard * *im Normalfall* || routine * *routinemäßig* || routinely * *(Adv.) routinemäßig*
Standardmodell standard type || standard design
Standardperipherie standard peripherals * *bei SPS*
Standardprojektierung standard configured package * *(z.B. mit SIEMENS-Regelsystem SIMADYN D (R) erstellt)* || standard configured system * *z.B. bei freiprojektierbarem Regelsystem SIMADYN D (R) von SIEMENS* || standard function package * *(z.B. erstellt mit SIEMENS-Regelsystem SIMADYN D (R))* || standard configuring
Standardtyp standard type || standard design * *Standardbauart* || standard type of construction * *standardmäßige Bauform*
Standdienst booth duty * *auf einer Messe/Ausstellung* || stand duty * *auf einer Messe/Ausstellung*
Ständer Stator * *Stator e. elektr. Maschine* || yoke * *(Magnet-) Joch* || post * *Pfosten, Pfosten, Ständer, Stab* || column * *Säule, Pfeiler* || pillar * *Pfeiler, Ständer, z.B. bei Bohrmaschine* || base * *Sockel,*

Unterteil, Unterbau, Fundament (auch einer Maschine), Fuß
Ständer- pedestal-type * *{Werkz.masch.}* || column-type * *{Werkz.masch.}*
Ständeranlasser line resistance starter * *{el. Masch.} 3-phasiger Vorwiderstand zum Anlassen e. Käfigläufermotors*
Ständerbleche stator laminations
Ständerblechpaket stator core
Ständerbohrmaschine column-type drilling machine
Ständerfrequenz stator frequency
Ständergehäuse stator frame * *{el. Masch.}* || stator housing * *{el. Masch.}*
ständergespeist stator-fed
Ständerjoch stator yoke
ständerkritisch thermally-critical-stator * *{el. Masch.} zuläss. Grenzübertemperatur wird zuerst im Ständer erreicht* || stator-critical * *{el. Masch.}*
ständerkritischer Motor stator-critical motor * *Ständer thermisch kritischer als der Läufer*
Ständerpaket stator core
Ständerspannung stator voltage
Ständerstrom stator current
Ständerwicklung stator winding * *{el. Masch.} Wicklung im feststehenden Teil einer el. Masch.*
Ständerwiderstand stator resistance
Standgerät floor-standing unit
standhalten stand * *aushalten; a test: einer Prüfung ~* || sustain * *einem Angriff ~* || withstand * *aushalten* || resist * *einer Person oder Sache ~*
ständig continuous * *laufend, ohne Unterbrechung* || steady * *stetig, gleich bleibend/mäßig, unveränderlich* || permanent * *fortdauernd, bleibend, permanent, dauerhaft* || constant * *fortwährend, gleich bleibend* || fixed * *fest (Einkommen usw.)* || regular * *regelmäßig (Einkommen usw.)* || established * *fest begründet, eingeführt, feststehend, z.B. Praxis, Regel* || permanently * *(Adv.)* || constantly * *(Adv.)* || continuously * *(Adv.)* || for ever * *(Adv.)* || continual * *fortwährend, dauernd; immer wiederkehrend, (sehr) häufig, oft wiederholt* || continually * *(Adv.) fortwährend, dauernd, immer wieder*
Standort location || position || site
Standpunkt point of view * *from this point of view: von diesem ~ aus betrachtet* || view * *take the view that: auf dem ~ stehen, dass* || viewpoint
Standpunkt vertreten take the view * *that: dass* || take the line || hold * *that: dass*
Standriefen brinelling * *Riefenbildung, z.B. in Wälzlager*
Standverteiler floor-mounting distribution board
Standzeit life || useful life || tool life * *Werkzeug* || stability time * *Lager, Fett usw.* || down-time * *Maschinen/Anlagen-Stillstandszeit* || service life * *eines Verschleißteils, z.B. Kohlebürsten, Lager* || lifetime || working life
Stange bar || rod || pole * *Pfosten, Pfahl, Mast, (Telegrafen-, Zelt-) stange, Stab, (Wagen-) Deichsel*
Stangenmagazin stick magazine * *z.B. für Bauteile der Leiterplattenbestückung*
Stangenziehstein rod drawing die
Stanniol tinfoil
Stanzabfälle punchings
Stanzautomat automatic blanking press
Stanzblech punching sheet
Stanze pinch || punching press * *Stanzpresse* || punching tool || punching machine || stamping machine
stanzen punch || stamp || blank * *auschneiden* || perforate * *lochen*
Stanzerei punching shop
Stanzmaschine punching machine || punching press || stamping machine

Stanzmatrize   punching die
Stanzpresse   punching press
Stanzstempel   punch || die
Stanzteil   punching || stamping || punched part
Stanzwerkzeug   punching die * *Stanzmatrize (Unterteil)* || punch * *Stanzstempel (Oberteil)* || blanking die || stamping tool || press tool
Stapel   pile || stack || staple * *(auch Fadenlänge/-qualität)*
Stapelanlage   stacking system || stacker
Stapelband   stack conveyor * *Transportsystem*
Stapeldatei   batch file * *Kommandodatei (Textdatei, die eine Folge von Betriebssystem-Kommandos enthält)*
Stapelfaktor   stacking factor * *{el. Masch.} Verhältnis d. reinen Eisenlänge e. Blechpakets zu dessen Gesamtlänge*
stapeln   stack || pile
Stapelspeicher   stack || LIFO stack || LIFO memory * *Abk. f. 'Last-In-First-Out': Die neueste Information wird als Erstes ausgelesen*
Stapelzeiger   stack pointer * *Adresse des aktuellen Elements eines Stapelspeichers (stack)*
Stapler   stacker * *z.B. Einrichtung zum Ablegen von Papierbögen auf Stapel* || layboy * *Papierbogenableger* || fork-lift truck * *Gabel~* || piler * *Vorrichtung zum Aufeinanderstapeln von Teilen* || piling device
stark   strong || robust * *kräftig* || powerful * *mächtig, leistungsfähig* || intense * *intensiv* || violent * *heftig* || bad * *schlimm* || large * *beträchtlich* || high-powered * *z.B. Motor* || great * *Nachfrage usw.* || widely * *(Adv.) weit, stark (verstreut, streuend, abweichend, bekannt, verschieden usw.)* || stiff * *(figürl.) z.B. Wind, Dosis, Gegner, Wettbewerb, Konkurrenz* || highly * *(Adv., figürl.) highly nonlinear: stark nichtlinear* || markedly * *(Adv., figürl.) deutlich, merklich, auffällig, ausgesprochen; increase markedly: ~ ansteig.* || severe * *schwer, schlimm (z.B. Verlust, Verschleiß, Fehlfunktion, Krankheit, Wetter)* || drastically * *(Adv.) drastisch*
stark ansteigen   be suddenly increasing * *z.B. plötzliches Ansteig. einer Spannung* || increase markedly
stark beeinflusst   largely influenced
stark erweitert   significantly expanded
Stärke   strength * *(auch figürlich: starke Seite(n) einer Person/eines Unternehmens)* || thickness * *Dicke* || power || force || gauge * *Dicke von Draht, Blech, einer (Gehäuse-/Rohr-) Wandung usw.* || vigour * *Tatkraft* || energy * *Tatkraft* || strong point * *starke Seite* || forte * *starke Seite* || intensity * *Intensität* || starch * *{Chemie} Stärke(pulver), Stärkemehl/-Kleister*
Starkstrom   power current || high-voltage current || heavy current
Starkstromanlage   high voltage installation || electrical power installation
Starkstromkabel   power cable
Starkstromleitung   power cable || power line || power transmission line
Starkstromtechnik   power engineering
starr   rigid * *(auch figürl.)* || stiff * *steif* || inflexible * *unflexibel, unelastisch, nicht biegsam, starr; (auch figürl.; siehe auch 'bürokratisch')*
starr gekoppelt   rigidly coupled
starr gekuppelt   rigidly coupled
starre Kupplung   rigid coupling
starres Netz   stiff system
Starrfräse   rigid-spindle moulder * *{Holz} im Gegensatz zur →Schwenkfräse* || rigid shaper * *{Holz}*
Starrfräsmaschine   rigid-spindle shaper * *{Holz} im Gegensatz zur →Schwenkfräsmaschine*
Startbedingungen   starting conditions

startbereit   ready to run || ready to go
Startbit   start bit * *kennzeichnet den Beginn eines Zeichens bei serieller Übertragung (normalerweise 0-Pegel)*
starten   start || take-off * *Flugzeug* || launch * *beginnen, lancieren, in Gang setzen, unternehmen, Starthilfe geben, Computerprogramm* ~ || boot * *Computer neu~/hochlaufen lassen*
Starthilfe   initial impulse * *(figürlich)* || starting aid * *z.B. für Motor* || take-off assistance * *z.b. Flugzeug*
Startwert   starting value || base value || initial value || initialization value
Startzeichen   start character * *kennzeichnet z.b. das Ende eines Datentelegramms (häufig STX entspr. 02hex)*
Statik   droop * *Weichmachen e. Reglers durch P-Rückführung d. Reglerausgangssignals* || droop function || elastic feedback * *nachgebende Rückführung (negative Rückkopplung des Reglerausgangs auf d. Eingang)* || statics * *{Bautechnik}* || static feedback * *nachgebende Reglerrückführung; siehe auch →Kennlinienneigung*
Statik-Aufschaltung   droop * *Weichmachen e. Regelung durch P-Rückführung d. Reglerausgangssignals*
Statikaufschaltung   droop * *Weichmachen einer Regelung durch P-Rückführung d. Reglerausgangssignals*
Station   station
stationär   steady-state * *z.B. im eingeschwungenen Zustand/Beharrungszustand* || stationary * *auch ortsfest* || steady * *gleich bleibend* || constant * *gleich bleibend*
stationärer Betrieb   steady-state operation || stationary operation || steady-state condition * *als Betriebszustand*
stationärer Einsatz   stationary application
stationärer Fehler   steady-state error || permanent error * *bleibender Fehler* || steady-state deviation * *Abweichung im stationären Zustand*
stationärer Verlauf   steady-state variation
stationärer Wert   steady-state value
stationärer Zustand   steady state || steady-state condition
statisch   static || steady-state * *im stationären/eingeschwungenen Zustand* || statical
statische Elektrizität   static electricity
statische Genauigkeit   static accuracy || steady-state accuracy || stationary precision
statische Verluste   fixed losses || static losses
statischer Fehler   static error * *Fehler im eingeschwungenen Zustand*
statistisch   statistical || statistically * *(Adv.)*
statistische Auswertung   statistical evaluation || statistical analysis
statistische Berechnung   statistical evaluation
Stativ   tripod || stand * *Ständer*
Stator   stator
Statorgehäuse   stator frame * *eines Motors*
Statorpaket   stator lamination || stationary core
stattfinden   take place || happen || come-off || be held * *abgehalten werden*
Status   status || condition * *Lage*
Statusabfrage   status check || status scan * *Test-/Diagnosefunktion bei SPS*
Statusanzeige   status display || status indication
Statusbearbeitung   status processing * *bei SPS*
Statusbit   status bit
Statusinformation   status information
Statusmeldung   status report
Statuswort   status word
Staub   dust || powder * *Puder*
Staubablagerung   dust deposit || dust accumulation || deposits of dust

Staubabsauganlage dust extraction system
Staubabsauggerät dust extractor || dust exhauster || dust collector * *Staubabscheider* || dust aspirator
Staubabsaugungsanlage dust extraction system || dust exhaust plant
staubdicht dust-proof || dust-tight
staubdichte Ausführung dust-tight design * *z.B. in Schutzart IP65 (EN 60529)*
Staubentzündung dust-ignition
Staubexplosionsschutz protection against dust-ignition * *Schutz gegen Zündung explosionsfähiger Stäube (DIN VDE 0165)*
Staubfilter dust filter
staubfrei dust-free
staubig dusty
Staubkorn dust particle
Staubsauger vacuum cleaner
Staubschutz dust protection || protection against dust
Stauchapparat swaging unit * *(Holz) zum Schärfen von Sägen* || swager * *(Holz) zum Schärfen von Sägen*
stauchen upset * hot/cold upsetting: *Heiß-/Kaltstauchen* || compress * *zusammendrücken, zusammenpressen, komprimieren* || swage * *(Holz) Schärfmethode für Sägen* || pressure-forge * *in der Schmiedetechnik* || roll on edge * *(Walzwerk)* || buckle * *(Walzwerk)*
Stauchmaschine swaging machine * *(Holz) zum Schärfen von Sägen*
Stauchpresse sizing press * *(Walzwerk)*
Staudruck back-pressure
Staufferbüchse Stauffer lubricator || grease cup
Staufferfett cup grease
Stauförderer accumulating conveyor * *(Fördertechnik) mit Produktespeicher-Funktion*
Staukettenförderer buffer chain conveyor * *(Fördertechnik) mit Produktespeicher-Funktion*
Stechbeitel ripping chisel || firmer chisel || chisel
Stechzirkel dividers
Steck- plug in || plug-in
Steckanschluss plug connection || plug-in connection || fast-on connector * *(Flach)Steckverbinder* || pluggable connection
steckbar plug-in || plug-in type
steckbar sein can be plugged in
Steckbaugruppe plug-in module
Steckbrief profile * *z.B. eines Produkts*
Steckbrücke plug-in jumper || jumper
Steckdose wall socket || socket outlet || wall outlet || wall plug || plug socket || receptacle || outlet * *Netzsteckdose* || socket
stecken insert * *z.B. eine Steckbaugruppe oder ein IC in seinen Einbauplatz* ~ || plug * *z.B. einen Stecker/eine Steckbaugruppe* ~ || be hidden * *verborgen sein* || stick fast * *festsitzen* || be stuck * *festsitzen* || slide in * *einschieben* || insertion * *(Substantiv) z.B. einer Steckbaugruppe oder eines ICs*
Stecken unter Spannung live insertion || hot plugging
Stecker connector * *male/plug connector: Stift~; female connector: Buchsen~* || plug || plug connector || power plug * *für Starkstrom*
Steckeranschluss plug connection || plug-in connection * *Anschluss mittels Stecker*
Steckerbelegung connector assignment || connector pin assignment
Steckerbezeichnung cennector designation * *z.B. "X6"*
Steckerbuchse female plug-connector element
Steckergehäuse connector housing || connector receptacle || plug-connector housing || plug housing
Steckernetzteil plug-in power supply

Steckerstift male plug-connector element || connector pin
Steckhülse quick-connect terminal * *Leitungsanschluss-Element* || push-on contact * *"weiblicher" Steckkontakt zum Aufstecken* || FASTON tab receptacle * *FASTON Steckhülse* || tab receptacle * *Gegenstück für Flachstecker/Messerkontakt*
Steckkabel plug-in cable
Steckkartenhalter plug-in card holder * →*Einzelkartenhalterung*
Steckklemme plug-in terminal * *siehe auch 'Schraubsteckklemme'* || plug-type terminal
Steckkupplung plug-in connector
Steckleitung plug-in cable
Steckmodul plug-in module || plug-in board * *Leiterkarte, Flachbaugruppe* || plug-in PCB * *Leiterkarte*
Steckplatz slot * *Einschub-Platz z.B. für Steckbaugruppe* || plug-in location || plug-in station || module plug-in location || receptacle * *Aufnahmeöffnung/schacht, z.B. für Speichersubmodul bei SPS* || expansion slot * *Erweiterungs~ für Leiterkarten usw.*
Steckplatzabdeckung slot cover * *für Baugruppenträger zum Abdecken eines nicht bestückten Steckplatzes*
stechplatzunabhängig random module insertion * *Steckbaugruppe kann in e. beliebigen Einbauplatz eingeschoben werden*
Steckschlüssel socket wrench || socket spanner
Stecksockel plug socket || plug-in socket
Steckverbinder plug connector || connector || plug-in connector
Steckverbindung plug-in connection || plug connection
Steckzyklen mating cycles * *eines Steckers* || insertion/withdrawal cycles * *einer Steckverbindung*
Stehbolzen stud * *Stehbolzen, Stift, Zapfen; mittels Schraubenbolzen sichern* || stay-bolt || standoff * *als Abstandshalter*
stehen stand * *(auch figürl.: stand for: ~ für)* || be written * *geschrieben sein/stehen* || be listed * *in Liste, Tabelle, Aufzählung* ~ || stand still * *still~* || have stopped * *~geblieben sein* || point to * *zeigen auf (z.B. Zeigerinstrument auf einen Wert)* || stop * *anhalten* || leave * *unverändert lassen* || be * *sein, s. befinden; be under the direction of: unter jds. Führung ~; be at issue: z.Debatte ~* || appear * *in Liste, auf Bildschirm usw.* ~ || be located * *sich befinden* || rank * *im Rang ~* || be laid down * *festgelegt sein, z.B. in Gesetz, Vorschrift, Norm* || be faced with * *vor etw. unangenehmem* ~ || point * *auf Mssßinstrument ~/angezeigt sein (to: auf e. Wert)*
stehen bleiben stall * *(Motor) kippen (AC-Motor), absterben, stecken bleiben, abgewürgt werden* || come to a standstill || come to a stop || stop || halt || come to a halt || remain unchanged * *unverändert bleiben* || remain untouched * *unberührt ~* || leave off * *beim Reden, Lesen usw.* ~ || be left * *nicht verändert werden, übrig bleiben*
stehende Schwingung standing vibration * *(mechan.)*
stehende Welle standing wave || stationary wave
Stehlager pedestal bearing * *auf Lagerbock/Untergestell angeordnet*
Stehwelle upright shaft || vertical shaft * *siehe auch* →*Königswelle*
steif rigid || inflexible * *unbiegsam* || fixed * *fest* || firm * *fest* || thick * *dickflüssig* || stiff * *(auch figürlich) zäh, dick, stark, formell, steif, starr* || wooden * *(figürlich) hölzern* || formal * *förmlich* || starchy * *gezwungen* || awkward * *linkisch* || clumsy * *linkisch*
Steifigkeit rigidity || stiffness
steigen go up * *auch Preise, Kurse usw.* || ascend ||

mount ‖ climb * *klimmen, auf/hochsteigen* ‖ rise * *Zahlenwert, Wasserspiegel, Temperatur, Preise, Kurse (to: bis); in die Luft* ~ ‖ increase * *zunehmen* ‖ move upward ‖ advance ‖ be increased * *an~ (by: um)* ‖ be increasing * *im ~ begriffen sein* ‖ escalate * *eskalieren, steigen, in die Höhe gehen (auch Preise usw.)*
steigend increasing ‖ rising * *z.B. Flanke, Kennlinie*
steigern increase ‖ boost * *(salopp)* ‖ step up * *Produktion usw. steigern/→ankurbeln*
Steigerung increase * *Vergrößerung, Erhöhung, Zunahme, (An)Wachsen, Zuwachs, Wachstum, Vermehrung* ‖ raising * *Erhöhung, (Ver)Stärkung, Vergrößerung, Vermehrung* ‖ rise * *(Auf/An)Steigen, Aufstieg (auch figürl.), Aufschwung, Zuwachs, Zunehmen* ‖ enhancement * *Steigerung (Leistungsfähigkeit, Funktionsumfang), Erhöhung, Vermehrung* ‖ intensification * *Verstärkung, Intensivieren* ‖ heightening * *Erhöhung, Vergrößerung (figürl.), Stärkung, Anhebung* ‖ boost * *Ankurbelg., Erhöh. (z.B. Druck, Spanng., Drehmoment), Verstärkg., Förderg., Auftrieb* ‖ raise * *Erhöhung, Steigerung (auch Straße), Aufbesserung (z.B. Gehalt)* ‖ aggravation * *Erschwerung, Verschlimmerung, Verschärfung* ‖ comparison * *(linguistisch)* ‖ gradation * *Abstufung* ‖ climax * *auch 'Gipfel, Höhepunkt, Krisis'* ‖ gain * *→Zunahme, (Wert-/Leistungs-) →Zuwachs*
Steigung pitch * *Gewinde~ (achsparaller Abstand der Gewindegänge), Gewindeteilung* ‖ rise * *Anstieg* ‖ gradient * *z.B. einer Kurve, eines Messwertes, einer Straße* ‖ ascent * *Anstieg, z.B. einer Kurve* ‖ slope * *Neigung* ‖ inclination * *Neigung* ‖ lead angle * *Steigungswinkel eines Gewindes; siehe auch →Gewindesteigung* ‖ lead * *Gewinde~ in Axialrichtung* ‖ grade * *Steigung, Neigung, Gefälle, Niveau*
steil steep
steile Schaltflanken steep switching edges
Steilgewinde steep-lead-angle thread
Steilheit rate of rise * *z.B. einer Impulsflanke* ‖ slew rate * *Anstiegsgeschwindigkeit, z.B. einer Impulsflanke (z.B. in [mV/μs])* ‖ rate of change * *z.B. einer Signalflanke* ‖ rate ‖ steepness
Steilimpuls steep-fronted pulse * *mit steiler Anstiegsflanke, z.B.zur Thyristoransteuerung*
Steinbohrer masonry drill * *zum Bohren in Mauerwerk*
Steinbrecher stone crusher ‖ rock crusher
Steinkohle hard coal
Steinmetz-Schaltung Steinmetz connection * *Einphasenbetrieb eines Drehstrommotors mit Betriebskondensator*
Steinmühle rock crusher ‖ stone crusher
Steinschraube rag bolt * *z.B. zur Motorbefestigung (auch "Steinanker, Bartbolzen")*
Stellantrieb servo drive * *(schneller) Positionierantrieb* ‖ actuator * *z.B. für Ventil, Schieber, Klappe* ‖ positioning drive ‖ positioner * *Stellgerät m.Positionsregelg. für proportional wirkend. Stellglied (z.B. Stellventil)* ‖ solenoid actuator * *Magnetantrieb, z.B. für Ventil* ‖ motor actuator * *motorisch betriebener Stellantrieb* ‖ actuating drive
stellbar adjustable ‖ variable
Stellbereich control range ‖ speed control range * *der Drehzahl* ‖ speed range * *der Drehzahl*
Stellbereich bis Null control range down to zero
Stelle position ‖ location * *Ort, Platz* ‖ digit * *Stelle einer Zahl/Ziffernanzeige; Zeichen in e. Textanzeige* ‖ job * *Arbeits~* ‖ spot * *Punkt, Fleck; hot spot: heiße ~; thick spot: Verdickung* ‖ place * *→Platz, →Ort; auch Dezimal~, ~ in einer Bestellnummer usw.* ‖ stand * *Standort* ‖ employment * *Arbeits~* ‖

authority * *~ bei einer Behörde* ‖ figure * *~ einer Zahl* ‖ point * *Punkt*
stellen put ‖ place ‖ set * *auch Schalter* ‖ position * *auch Schalter* ‖ pose * *(Frage) ~, (Behauptung) auf~, (Anspruch) erheben, hin~/ausgeben (as: als)*
Steller power controller * *Leistungs~* ‖ actuator * *Stellglied* ‖ amplifier * *Verstärker* ‖ positioner * *Stellgerät*
Stellgerät actuator ‖ final controlling element ‖ positioner * *mit Lage/Positionsverstellung* ‖ actuator control unit ‖ final control element ‖ regulating unit
Stellglied final controlling element ‖ actuator ‖ positioning element * *zur Lageeinstellung (z.b. Stellventilantrieb)* ‖ final control element ‖ controlling element
Stellgröße correcting variable * *Reglerausgangsgröße, d. verstellt wird, um d. Regelgröße d. Sollwert nachzuführen* ‖ regulating variable ‖ manipulated variable ‖ actuating variable
Stellkraft actuating force
Stellmotor servo motor ‖ motor actuator ‖ positioning motor ‖ actuating motor
Stellorgan control element ‖ regulating element ‖ final controlling element * *→Stellglied*
Stellsystem actuator system
Stelltrafo variable-ratio transformer ‖ variable transformer * *rotary transformer* ‖ *Drehtrafo* ‖ slide transformer
Stelltransformator variable-ratio transformer ‖ variable transformer ‖ rotary transformer * *Drehtrafo* ‖ slide transformer
Stellung setting * *eines Schalters, Parameters, Sollwerts, Potentiometers usw.* ‖ position * *örtliche/taktische Lage, berufliche Position (of: als), Rang* ‖ employment * *berufliche Anstellung* ‖ job * *berufliche Anstellung* ‖ standing * *Ansehen* ‖ legal position * *Rechtsstellung* ‖ arrangement * *Anordnung*
Stellungnahme response * *Einstellung, Haltung, Standpunkt (to(wards): zu)* ‖ comment * *Erklärung (on: zu)* ‖ answer * *Beantwortung* ‖ return letter * *Antwortschreiben*
Stellventil positioning valve ‖ servo valve ‖ control valve * *Regelventil* ‖ governor valve * *Regelventil* ‖ regulating valve
Stellwert value of correcting variable * *Wert der Stellgröße* ‖ value of manipulated variable * *Wert der Stellgröße*
Stellwiderstand adjustable resistor
Stellzeit positioning time
Stemmmaschine mortizer * *{Holz} (amerikan.)* ‖ mortizing machine * *{Holz} mit Meißeln arbeitend* ‖ mortiser * *{Holz} (brit.}*
Stempel stamp ‖ punch * *zum Stanzen*
stempeln stamp * *z.B. auf Typenschild ~* ‖ punch * *auch Angabe auf Typenschild ~* ‖ engrave * *eingravieren, auch Typenschildangaben*
Stern star * *Schaltgruppe z.B. für Trafo/Motor; in star connection: in ~ geschaltet* ‖ wye * *elektr. Schaltgruppe z.B. für Trafo/Motor; wye-connected: in ~ geschaltet*
Stern-Doppelstern-Anlauf star-double star starting * *mit Dahlander bzw. PAM-Schaltung (siehe dort)*
Stern-Dreieck-Anlauf starting * *Sternschaltung reduziert Anlaufstrom (und -moment) auf 1/(Wurzel 3)*
Stern-Dreieck-Schaltung star-delta connection
Stern-Dreieckschalter star-delta switch * *zum Starten eines Drehstrommotors mit verringertem Anlaufstrom*
Stern-Dreieckschaltung star-delta connection ‖ Y-delta connection ‖ wye-delta connection
Sternchen asterisk * *Zeichen auf Tastatur/Monitor/Drucker*

Sternpunkt neutral point || star point || neutral
Sternpunkterdung neutral earthing || neutral grounding
Sternpunktkasten neutral box * *{el.Masch.} z.B. zur Bildung des Sternpunktes in Hochspannungsmaschine*
Sternpunktschiene neutral busbar
Sternpunktspannung voltage to neutral
Sternpunktwiderstand star point resistor
Sternschaltung star connection * *z.B. Motor/Trafo; star-connected: in ~* || Y-connection * *z.B. Motor/Trafo* || wye connection
Sternspannung line-to-neutral voltage * *zwischen Phase und Mittelpunktsleiter* || star voltage * *zw. Phase u. Mittelpunktsleiter* || phase-to-neutral voltage * *zw. Phase und Mittelpunktsleiter* || voltage to neutral
Sternverteiler hub * *für Rechner-Netzwerk ('hub': Nabe, von der die Speichen ausgehen)*
stetig steady * *gleichmäßig, unerschütterlich* || continual * *fortwährend, dauernd, immer wieder(kehrend)* || continuous * *ununterbrochen, (fort)laufend, kontinuierlich* || constant * *konstant, beständig, gleich bleibend, unveränderlich*
stetiger Übergang continuous transition
Stetigschleifer continuous grinder * *{Papier} zur Zellstofferzeugung*
stets at all times || always || constantly * *ständig* || continuously * *ständig*
Steuer tax || duty * *Abgabe, Gebühr, Zoll* || steering gear * *{Schifffahrt} Ruder*
Steuer- und Regelaufgabe control task
Steuer- und Regelfunktionen open-loop and closed-loop control functions
Steueradresse internal control address
Steueranschluss gate electrode * *z.B. e. Thyristors* || gate terminal * *z.B. eines Thyristors* || gate * *z.B. eines Thyristors*
Steueranschlüsse control connections
Steueranweisung emdedded command * *bei SPS* || control * *für Compiler/Assembler usw.* || control statement || control * *definiert Ausführungsoptionen, z.B. für Compiler, Assembler, Linker usw.* || compiler switch * *Compilerschalter, aktiviert/deaktiviert Übersetzungsoptionen*
Steueraufgabe control task
Steuerausgang control output
steuerbar controllable * *z.B. Ventil, Stromrichter, Gleichrichter*
steuerbare Stromrichter-Reihenschaltung boost-and-buck converter connection * *2 i.Reihe geschaltete Stromrichter m. getrennter Ansteuerg.*
steuerbarer Feldgleichrichter contollable field rectifier || controllable field converter
steuerbarer Gleichrichter controllable rectifier
steuerbares Ventil controllable valve
Steuerbarkeit controllability
Steuerbaustein control function block * *Funktionsbaustein* || control block * *z.B. Funktionsbaustein in SPS* || control
Steuerbefehl control command || internal control command * *interner Steuerbefehl bei SPS*
Steuerbehörde tax authority || board of assessment
Steuerbelastung tax burden || incidence of taxation
Steuerberater tax consultant || tax advisor || tax expert || tax preparer
Steuerbereich control range
Steuerbescheid tax notice
Steuerbit control bit
Steuerblindleistung control reactive power * *z.B. auf Grund der Phasenanschnittsteuerung eines Stromrichters* || reactive power due to the phase-angle control * *~ auf Grund der Phasenanschnittsteuerung* || phase-control reactive power * *{Stromrichter} auf Grund der Phasenanschnittsteuerung*
Steuerbühne control pulpit * *Steuerkanzel z.B. im Walzwerk*
Steuerbus control bus * *beinhaltet z.B. Read-/Write-Signale bei Mikroprozessorsystem*
Steuerebene control level || individual control level * *Einzel~*
Steuereingang control input || control logic input * *binärer*
Steuereinrichtung control equipment || trigger equipment * *An~ für Stromrichter (erzeugt die Ansteuer-/Zünd-/Gateimpulse)*
Steuerelektronik control electronics || electronic control || electronic control unit * *als Baueinheit/Gerät*
steuerfrei non-taxable
Steuerfunktion control function || force on-off facility * *Zwangssteuerungsfunktion für Ein-/Ausgänge einer SPS (Testfunktion)*
Steuergerät control unit || control
Steuergröße open-loop controlled variable
Steuerimpuls firing pulse * *Zündimpuls z.B. für Thyristor; siehe auch DIN 41750 Teil 7* || gate control pulse * *An~ z.B. für Thyristor* || control pulse * *auch An~ für Thyristor* || gating pulse * *z.B. zur Ansteuerung eines Thyristors* || trigger pulse
Steuerkabel control cable
Steuerkarte control board * *Leiterkarte*
Steuerkennlinie control characteristic * *z.B. b. Phasenanschnittsteuerg.: Ausgangsspannung als Fkt. d. Aussteuerwinkels*
Steuerklemme control terminal
Steuerklemmen control terminals
Steuerklemmleiste control terminal strip
Steuerkreis control circuit
Steuerleistung control power || triggering power * *z.B. zur Ansteuerung eines Thyristors* || driving power
Steuerleitung control lead * *Leitung, über die Steuer- u. Meldesignale übertragen werden* || control cable || control line * *Steuersignal in einem Prozessorbus* || low-current cable
Steuerlogik control logic
Steuermatrix control matrix
steuern control || vary * *durch einen Bereich (hin-)durchsteuern, variieren* || fire * *zünden z.B. Thyristor* || tax * *{Wirtschaft} (Substantiv) after tax: nach Steuern* || force * *{SPS} zwangssteuern von Ein-/Ausgängen bei SPS (Testfunktion)* || open-loop control * *im Gegensatz zum 'regeln'* || forcing * *(Substantiv) {SPS} zwangssteuern von Ein-Ausgängen (Testfunktion)*
steuernde Wirkung control action
Steuerorgan control element
steuerpflichtig subject to taxation
Steuerprogramm control program
Steuerpult control desk
Steuerraum control room
Steuerrechner control computer
Steuersatz gating unit * *Zünd/Steuerimpulserzeugung* || trigger set * *Zündimpulserzeugung* || gating circuit * *Zünd-/Steuerimpulserzeugung* || firing circuit * *Zündimpulserzeugung besonders für Thyristorsatz* || modulator * *(Pulsweiten-) Modulator bei Pulsumrichter* || gating module * *Zünd-/Steuerimpulserzeugung für ein Stromrichter-Leistungsteil* || tax rate * *{Wirtschaft}* || rate of assessment * *{Wirtschaft}*
Steuerschalter control switch
Steuerschrank control cubicle || control cabinet
Steuersignal control signal || logic signal || control command * *Steuerkommando/-befehl* || command * *Steuerbefehl*

Steuersignalfehler control signal error * *bei SPS*
Steuerspannung control voltage * *z.B. Eingangssignal eines Steuersatzes ("Vergleichsspannung für Sägezahn")* || control supply voltage * *Versorgungsspannung für Steuerkreise/Steuer elektronik*
Steuerstand control room * *Warte, Leitraum* || control station || control cabin * *Steuerkabine, z.B. an Kran*
Steuerstrom control current || driving current || gating current * *z.B. für Thyristor*
Steuerstromkreis control signal loop
Steuertafel control board || switchboard || control switchboard || control panel || supervision panel
Steuerteil control unit || control circuit || control section || open-loop control section
Steuertrafo control circuit transformer || control transformer
Steuertransformator control circuit transformer || control transformer
Steuerung control || open-loop control || control system * *System* || control unit * *Gerät* || management * *auch Verwaltung* || control setting * *Setzen des Steuerwinkels e. Thyristorschaltung* || controller * *(Steuer-) Gerät* || PLC * *Abkürzung für 'Programmable Logic Controller': speicherprogrammierbare ~ (PLC)* || gating * *An~ eines Thyristor-/ Transistor-Leistungsteils* || logic control
Steuerungsablauf control sequence
Steuerungsaufgabe control task || control function
Steuerungsebene control level
Steuerungsfunktionen open-loop control functions
Steuerungskabel control cable
Steuerungsprogramm control program
Steuerungsprozessor boolean processor * *bei SPS*
Steuerungssystem PLC system * *SPS-System*
Steuerungstechnik control engineering
Steuerungstechniker control technician
Steuerventil control valve
Steuerwerk sequence control || sequencing control || state machine * *elektronisches Zustandsfolge-Steuerwerk* || processor
Steuerwinkel firing angle * *für Thyristor-Gate-Ansteuerung* || gating angle || delay angle * *z.B. bei Phasenanschnittsteuerung* || gate angle || trigger delay angle
Steuerwinkelvorsteuerung firing angle precontrol || gating angle feed-forward
Steuerwort control word * *(auch →PROFIBUS-Profil)*
Steuerzeichen control character * *z.B. in einem Datentelegramm*
Stich prick * *mit Nadel usw.* || stitch * *mit Nähnadel usw.* || pass * *[Walzwerk]*
Stichleitung radial line || spur line * *z.B. im Netz* || dead-end line || tap line || stub-end feeder * *im Versorgungsnetz* || single feeder || spur || line tab || branch * *Abzweig* || branch pipe * *Rohrleitung* || stub line
Stichprobe random sample * *Zufallsstichprobe* || sample
Stichprobenentnahme sampling
Stichprobenkontrolle random test || sampling test || test on selected samples || sampling inspection
Stichprobenprüfung random test || sampling test || sampling inspection || spot checking
Stichsäge jig saw || keyhole saw * *[Holz]* || pad saw
Stichwort keyword
Stichwortverzeichnis keyword index || index of keywords || index
Stickstoff nitrogen
Stift pin || bolt * *Bolzen* || tack * *(Heft)Zwecke* || pencil * *Zeichen~, Schreib~* || apprentice * *Lehrling* || youngster * *(salopp) Lehrjunge, junger Bursche, Junge, Knirps*

Stiftdübel pin plug * *zum Einschlagen von Stahlstiften in harte Wände*
Stiftleiste male connector * *"männlicher" Steckverbinder* || plug connector
Stiftschraube stud bolt || tap bolt || setscrew * *→Feststellschraube, →Madenschraube* || stud
Stiftstecker male connector * *"männlicher" Stecker* || plug connector
Stillsetzautomatik automatic shut-down || automatic stop control
Stillsetzbremse shut-down brake
stillsetzen stop || ramp down to zero * *entlang einer Rampe ~* || ramp back to zero * *entlang einer Rampe auf Drehzahl Null fahren (lassen)* || brake * *Abbremsen* || shut down || bring to a standstill || arrest * *feststellen, sperren, arretieren* || halt * *Halt machen* || shutdown
Stillsetzung shutdown || stopping
Stillstand standstill * *at: im; come to a standstill: zum ~ kommen* || stall || zero speed * *Drehzahl Null* || shutdown * *~ einer Machine/Anlage* || rest * *Ruhe, Ruhepause, Pause* || idleness * *~ einer Maschine* || stagnation * *~ von Geschäften, der Entwicklung usw.* || deadlock * *(figürl.) völliger ~, Sackgasse, toter Punkt, Auf-d.-Stelle-Treten (Verhandlgen.)* || paralyse * *Lahmlegung, Lähmung; be paralysed: lahmgelegt/unwirksam/zum ~ gekommen sein*
Stillstand mit stromlosen Wicklungen rest and de-energized * *[el. Masch.] Stillstand ohne Zufuhr el. o. mechan. Energie,"Pause"*
Stillstands- standstill
Stillstandsbremse standstill brake || holding brake * *Haltebremse*
Stillstandserregung standstill excitation * *z.B. reduz. Feldstrom b. Gleichstrommotor, um Überhitzung zu vermeiden* || standstill field
Stillstandsfeld standstill field * *z.B. bei DC-Motor: Absenkung d. Feldstroms i. Stillstand, um Überhitzg. zu vermeiden* || standstill excitation
Stillstandsheizung anti-condensing heating * *z.B. bei Elektromotor zur Vermeidung von Kondensatbildung* || standstill heating * *z.B. bei Getriebe für möglichst geringe Reibung bei Neuanlauf* || space heater || anti-condensation heater || anti-condensation heating * *z.B. in Motor zur Vermeidung von Kondenswasserbildung* || space heating
Stillstandslogik standstill logic
Stillstandsmoment static torque
Stillstandsmotor static-torque motor
Stillstandsspannung standstill voltage
Stillstandsüberwachung zero-speed monitoring * *als Funktion* || zero speed monitor * *als Gerät/ Einrichtung* || zero-speed switch * *Stillstandswächter* || zero-speed relay * *Stillstandsrelais* || standstill monitoring
Stillstandswächter zero-speed monitor * *löst bei Drehzahl Null aus* || zero-speed switch * *löst bei Drehzahl Null aus* || zero-speed relay * *Stillstandsrelais*
Stillstandszeit downtime * *einer Maschine, Anlage (auch wegen Störung)* || rest period
Stillstandszug stall tension * *meist kleiner als Betriebszug, z.B. um b.DC-Motor den Kommutator zu schonen*
stillstehen stand still || come to a standstill * *zum Stillstand kommen* || stop * *anhalten* || be paralyzed * *lahmgelegt sein, z.B. Betrieb, Verkehr*
stillstehend stationary || at a standstill || out of operation * *außer Betrieb* || blocked * *blockiert*
stimulieren stimulate
Stimulus stimulus
Stirnbohrung end hole * *[Holz]*
Stirndeckel front-side cover

Stirnfläche face || end face || front end
Stirnrad spur gear
Stirnrad-Schneckengetriebe helical worm gear
Stirnradgetriebe helical gear || helical gearbox || spur gear || spur gearing
Stirnradgetriebemotor helical-geared motor
Stirnradschneckengetriebe helical worm gearbox
Stirnseite front face || face * *(auch "stirnseitig")* || front || front side || end || front end
stirnseitig montiert face-mounted
Stirnwand end wall
Stockwerk storey || floor || story * *(amerikan.)*
Stoff material * *Material* || agent * *Mittel, Wirkstoff* || medium * *Mittel* || textile * *Textil* || substance * *Substanz* || compound * *Masse, zusammengesetzter Werkstoff* || stock * *Dick-/Dünn-/Grundstoff z.B. in d. Zellstoff-/Papierherstellung*
Stoffaufbereitung stock preparation * *Herstellg.d. Papier-Grundmaterials z.weiteren Verarbeitg. i.d. Papiermaschine* || stock preparation and equipment * *{Papier} Vorstufe der Papiererzeugung*
Stoffauflauf headbox * *{Papier} (amerikan.)* ~*kasten bei Papiermaschine vor der Siebpartie* || pulp arrical * *{Papier} Das Auflaugen des Stoffes auf das Sieb bei e. Papiermaschine* || flow box * ~*kasten bei Papiermaschine* || breast box
Stoffauflaufkasten headbox * *(amerikan) bei Papiermaschine; siehe auch* →*Stoffauflauf* || flow box
Stoffauflaufpumpe headbox feed pump * *{Papier} am Beginn einer Papiermaschine*
Stoffauflaufregelung headbox control * *{Papier}- regelt auf Sieb auflauf. Stoffmenge mit Stoffdruck u. Lippenverstellg.*
Stofflöser pulper * *für Zellstoff*
Stoffpumpe stock pump * *zum Fördern von Dick-/ Dünn-/Zellstoff in der Zellstoffaufbereitung*
Stoffstrom material flow
Stoffventil stock valve * *in Zellstoff-/Papierherstellung*
Stolperstein . stumbling-block * *(figürl.) Stolperstein, Stein des Anstoßes,* →*Hindernis (to: für)*
Stopfbüchse gland || stuffing box
Stopfbuchsenverschraubung cable gland || compression gland
Stopfbuchsverschraubung compression gland * *bei Kabeleinführung, siehe* →*PG-Verschraubung und DIN 46320* || screw gland || packed gland
Stopfen plug * *Stöpsel, Dübel, Zapfen, Propfen, Verschlussschraube* || stuffing * *Stopfbuchse, Füllung, Füllmaterial* || stopper * *Stöpsel, Pfropfen (z.B. für Flasche), Absperrvorrichtung* || filler plug * *Blindstopfen, z.B. für Kabeleinführungsloch* || blanking plug * *Blindstopfen*
Stopp stop
Stoppbit stop bit * *kennzeichnet das Ende eines Zeichens bei serieller Übertragung (meist logisch "1")*
stoppen stop
Stopper stopper
Stör- noise- * *Geräusch, elektromagnetische Störung* || interference- * *(elektromagnetische) Störung* || parasitic * *parasitär*
Störabschaltung fault trip || shut down after fault || tripping || trip || shut down on fault
Störabstand noise margin || signal-to-noise ratio || permissible noise level * *zulässiger Störpegel*
Störabstrahlung noise radiation || noise emission || electromagnetic radiation
Störanalyse fault analysis
störanfällig noise sensitive * *anfällig gegen elektromagnetische Störungen* || susceptible to interference * *anfällig gegen elektromagnetische Störungen* || trouble-prone || susceptible to trouble || prone to breakdown || unreliable * *unzuverlässig*

Störanfälligkeit susceptibility to interference * *gegen elektromagnetischen Störungen* || noise sensitiveness * *gegen elektromagnetische Störungen* || trouble proneness * *Fehleranfälligkeit* || susceptibility to failure * *Fehleranfälligkeit* || susceptibility to breakdown * *zum Ausfallen neigend* || noise sensitivity * *gegen elektromagnetische Störungen* || susceptibility to faults * *Ausfall-Anfälligkeit* || unreliability * *Unzuverlässigkeit*
Störanzeige fault indication || fault signalling
Störaussendung RFI-Emission * *Radio Frequency Interference, elektromagnetische Störabstrahlung* || electromagnetic emission * *elektromagnetische ~* || noise radiation * *Stör(ab)strahlung*
Störbeeinflussung electrical interference * *[inte'fi: rens]* || noise interference * *[inte'fi: rens] elektromagnetische*
Störeinkopplung coupled-in noise || noise injection
Störeinstrahlung interference radiation || disturbing irridiation
Störeinstreuung injection of interference
störempfindlich noise sensitive * *gegen elektromagnetische Störungen* || susceptible to interference * *gegen elektromagnetische Störungen* || trouble-prone * *fehleranfällig* || susceptible to faults * *fehleranfällig* || prone to breakdown * *zum Ausfall neigend* || susceptible to trouble * *fehleranfällig* || unreliable * *unzuverlässig*
Störempfindlichkeit susceptibility to interference * *gegen elektromagnetische Störungen* || noise sensitivity * *gegen elektromagnetische Störungen* || noise sensitiveness * *gegen elektromagnetische Störungen* || trouble proneness * *Fehleranfälligkeit* || susceptibility to breakdown * *zum Ausfallen neigend* || unreliability * *Unzuverlässigkeit*
stören disturb * *durcheinander bringen, belästigen, behindern, in Unordng.bringen* || interfere * *behindern, belästigen, dazwischenkommen, sich einmischen, with: bei Ablauf/Verrichtung* || trouble * *belästigen, beunruhigen* || bother * *belästigen* || annoy * *belästigen, ärgern, schikanieren* || irritate * *ärgern* || vex * *ärgern* || upset * *durcheinander bringen* || disarrange * *durcheinander bringen* || interrupt * *unterbrechen* || jam * *sperren, hemmen, blockieren, klemmen, verstopfen, Radiosender stören* || be intruding * *sich einmischen/eindrängen* || meddle * *sich einmischen* || be in the way * *im Weg sein* || be inconvenient * *unangenehm sein* || be awkward * *unangenehm sein* || be intrusive * *aufdringlich sein* || intrude * *unangenehm bemerkbar machen,aufdrängen,sich einmischen/eindrängen*
störend disturbing || troublesome * *unangenehm* || inconvenient * *unbequem* || awkward * *peinlich* || parasitic * *parasitär, schädlich*
Störerfassung fault sensing || fault recording * *Aufzeichnung*
Störfeldstärke interference field strength
Störfestigkeit noise immunity || interference immunity * *(siehe z.B. EN61000-4)* || interference resistance || electromagnetic immunity * *elektromagnet. ~* || immunity level * *Grad der ~*
Störfestigkeitsklasse noise immunity class * *siehe IEC 801-4/5*
Störfolgespeicher fault sequence memory * *(→PROFIBUS-Profil)*
Störgenerator noise generator || interference generator * *noise simulator * Störsimulator* || EMI generator * *Abkürzg. für 'ElectroMagnetic Interference': elektromagn. Störungen* || EMI simulator * *Störsimulator*
Störgröße disturbance variable * *in einem Regelkreis* || disturbance || disturbance effect * *Störeffekt in e. Regelkreis* || interference variable
Störgrößen disturbances || disturbance variables

Störgrößenbeobachter disturbance observer
Störgrößenunterdrückung disturbance suppression * *bei einer Regelung* || disturbance rejection * *bei einer Regelung* || disturbance suppression control * *bei einer Regelung*
Störimpuls noise pulse || interfering pulse
Störleistung interference power
Störlichtbogen arcing fault
störlichtbogenfest resistant to arcing faults
Störmelderelais fault signalling relay
Störmeldesignal fault signal
Störmeldung fault message || alarm || error message * *Fehlermeldung* || fault alarm || fault signal || fault indication * *Störungsanzeige* || fault signalling * *Störmeldefunktion*
stornieren cancel * *z.B. Auftrag* || withdraw
Störpegel noise level || interference level || background level * *akustischer ~*
Störquittierung fault acknowledgement
Störschreiber line fault recorder * *zum Aufzeichnen von Netzstörungen* || line fault documenter * *zum Aufzeichnen von Netzstörungen*
Störschutz interference suppression || interference protection || EMI suppression
Störschutzdrossel interference suppression choke || suppression choke || EMI suppression choke
Störschutzkondensator interference suppresion capacitor
störsicher noise-immune * *gegen elektromagnetische Störungen* || noise insensitive * *gegen elektromagnetische Störungen* || interference immune * *gegen elektromagnetische Störungen* || noise-free * *keine elektromagnetische Störungen aussendend/abstrahlend* || reliable * *betriebssicher, zuverlässig* || trip-free * *ohne Störabschaltungen* || interference-proof * *gegen elektromagnetische Störungen*
Störsicherheit noise immunity * *gegen elektromagnetische Störungen* || reliability * *Betriebssicherheit, Zuverlässigkeit* || noise insensitivity * *gegen elektromagnetische Störungen* || interference immunity
Störsignal noise signal || noise
Störsignalunterdrückung disturbance suppression * *bei einer Regelung* || disturbance rejection * *bei einer Regelung* || disturbance suppression control * *bei einer Regelung*
Störsimulator EMI simulator || noise simulator || fault simulator || noise generator * *Störgenerator* || EMI generator * *Störgenerator*
Störspannung interference voltage || noise voltage || parasitic voltage
Störspannungen noise voltages || interference voltages
Störspannungsfestikeit noise immunity || interference immunity
Störspannungsprüfung radio interference voltage test || RIV test * *Abk. für 'Radio Interference Voltage'*
Störspeicher fault memory || diagnostic memory * *Diagnosespeicher*
Störspitze noise voltage spike || interference voltage spike
Störstrahlfestigkeit immunity to radiated noise
Störstrahlung interfering radiation || radiation noise || radiated noise * *Störabstrahlung (siehe z.B. EN55011)* || radiated emissions
Störstrahlungsfestigkeit immunity to radiated noise
Störstrom noise current || interference current || parasitic current
Störübertragungsfunktion disturbance transfer function * *{Regel.techn.}*
störunempfindlich noise insensitive * *gegen elektromagnetische Störungen* || noise resistant * *gegen elektromagnetische Störungen* || insusceptible to interference * *gegen elektromagnetische Störungen* || trouble-free || trip-free * *ohne Störabschaltungen* || robust * *robust*
Störunempfindlichkeit noise insensitivity * *gegen elektromagnetische Störungen* || insensibility to interference * *gegen elektromagnetische Störungen* || noise immunity * *gegen elektromagnetische Störungen* || robustness * *Robustheit*
Störung fault * *Fehler, Mangel; fault condition occurs: Störfall tritt auf* || malfunction * *Fehlfunktion* || noise * *elektromagnetische ~, Rauschen* || interference * *[inter'fi: rens] Einmischung, elektromagnetische ~* || EMI problem * *EMV-Problem, elektromagnetische Störbeeinflussung* || disturbing * *das Stören* || disturbance * *auch Unruhe, Brumm, Beeinträchtigung, Behinderung, z.b. Kommutierungseinbruch* || break-down * *Maschinenschaden, Zusammenbruch, Versagen, Scheitern, Panne* || inconvenience * *Unbequemlichkeit, Unannehmlichkeit, Schwierigkeit,* || annoyance * *Ärgernis* || irritation * *Ärgernis, Ärger, Reizung* || interruption * *Unterbrechung* || hitch * *Stockung* || obstruction * *Behinderung* || disarrangement * *Unordnung* || disorder * *Unordnung* || trouble * *auch 'Unannehmlichkeiten, Schwierigkeiten, Sch erereien, Plage, Belästigung'* || jam * *Hemmung, Stockung, Klemmung, Radio~* || intrusion * *Einmischung, Eindringen, Zu-/Aufdringlichkeit, Belästigung* || problem * *Problem*
Störungsabstrahlung noise radiation || noise emission
störungsanfällig susceptible to faults
störungsbedingt fault related
Störungsbehebung fault rectification
Störungsbeseitigung troubleshooting || repairing of faults || fault finding || fault fixing || fault rectification
Störungseinkopplung coupled-in noise
störungsfrei fault-free || disturbance-free * *(auch figürl.: ohne Beeinträchtigung/Behinderung)* || trouble-free || faultless
Störungskanal fault channel
Störungsmeldung fault message || fault notification * *schriftliche* || fault signalling || fault indication || fault signal || fault signaling * *(kann mit einem oder zwei "l" geschrieben werden)*
Störungsquittierung fault acknowledgement
Störungsstrom fault current * *z.B. bei Kurzschluss, Wechselrichterkippen usw.*
Störungssuche fault finding || fault location || troubleshooting
Störungsunterdrückung EMI suppression * *Abkürzg. f. 'Electro-Magnetic Interference': elektromagnetische ~* || noise rejection || interference suppression || radio interference suppression * *Funkentstörung*
Störungsursache cause of the fault || fault cause || cause of malfunction
Störunterdrückung EMI suppression * *Abk. f. 'ElectroMagnetic Interference': elektromagnetische Störung* || noise rejection || radio interference suppression * *Funkentstörung* || RFI suppression * *Funkentstörung* || disturbance suppression * *Störgrößenunterdrückung bei einer Regelung*
Störverhalten response to disturbances * *Reaktion eines Regelkreises bei Einprägung von Störgrößen* || response to process-side and mains-side changes * *Reaktion e.Regelg.auf Störgrößen u.Netzschwankungen*
Stoß surge * *plötzlicher Anstieg, Spannungs~* || impulse * *elektr. Impuls, (An)Stoß, Antrieb* || step change * *stufenförmiger Sprung (z.B. als Stimulus z.Aufzeichnung d. Regler-Sprungantwort)* || shock * *mechanischer* || bump * *heftiger ~, Bums* || im-

stoßartig pact * *Aufprall, Schlag, Stoß, Schlagfestigkeit* || jerk * *Ruck* || joint * *Fuge* || butt joint * *(Stoß)Fuge zw. aneinandergereihten Teilen* || junction * *Verbindung* || sudden change * *schlagartige Änderung; sudden change in torque: Drehmomenten~*
stoßartig abrupt * *plötzlich, aprupt; schroff, kurz angebunden, jäh, steil* || pulse || shock-type || sporadic * *in Stößen/Schüben/unregelmäßig auftretend* || intermittent * *in Stößen/mit Unterbrechungen auftretend* || sporadically * *(Adv.)* || jerky * *ruckhaft*
Stoßbeanspruchung shock stressing * *mechanische ~* || impact stressing * *mechanische Schlagbeanspruchung* || jerk stressing * *durch ruckhafte Lasten* || surge stressing * *durch Spannungs-/Stromstöße* || impulse stressing * *Impulsbelastung* || impact load * *Schlag/Stoßbeanspruchung*
Stoßbelastung shock stressing || impulse loading
stoßdämpfend shock absorbing
Stoßdämpfer shock absorber
Stoßdrehmoment shock overload torque || peak torque
Stößel ram || plunger || rammer || driver rod * *siehe auch →Schubstange* || pushing rod * *siehe auch →Schubstange*
stoßen inject a step-change signal * *sprungförm. Signal anlegen für Regleroptimierg. (Sprungantw. aufnehmen)* || impulse-test * *Prüfen mit Stoßspannung* || push * *stoßen, schieben, treiben, drängen* || shove * *schieben (z.B. beiseite)* || thrust * *heftig ~, schubsen* || kick * *mit Fuß ~* || punch * *(aus)stanzen, stempeln, durchschlagen, lochen* || knock * *schlagen* || strike * *schlagen* || nudge * *(leise/heimlich) an~, stupsen, pochen* || ram * *rammen* || drive * *treiben* || slot * *stanzen, (Nuten) stoßen* || pulverize * *pulverisieren* || border on * *angrenzen* || about on * *angrenzen* || adjoin * *an~, angrenzen* || touch * *berühren* || butt * *zusammen-/aneinander-/-fügen (against: an-/gegen(einander)), aneinandergefügt sein* || bump * *(heftig) stoßen, (an)prallen, bumsen (gegen), zusammen~ (mit)*
stoßen auf encounter * *begegnen, treffen auf*
Stoßfaktor load factor * *→Lastfaktor (Verhältnis Durchschnitts- zu Spitzenlast)* || shock factor * *kennzeichnet d.Zahl d.Lastzyklen je Zeiteinheit (maßgebl. für Getriebeauslegung usw.)*
stoßfest shock-proof * *(mechan.)* || surge-proof * *(elektr.) gegen Stoßspannungen*
Stoßfestigkeit shock resistance * *(mechan.)* || impact resistance * *(mechan.)* || anti-shock * *(mechan.)* || impulse strength * *elektrische ~* || surge stressability || shock tolerance
stoßfrei bumpless * *ohne (heftigen) Stoß, Bums* || joltfree * *ohne Rütteln, Holpern* || smooth * *weich, sanft* || jerk-free * *ohne Ruck* || bumplessly * *(Adv.)*
stoßfreie Umschaltung bumpless changeover || jolt-free changeover
stoßfreier Übergang continuous transition || bumpless transition
Stoßfrequenz shock frequency || excitation frequency * *Schwingungs-anregende Frequenz*
Stoßgrad cyclic irregularity * *Ungleichförmigkeitsgrad einer Drehbewegung*
Stoßimpulsmessung shock pulse measurement * *z.B. z. Verschleißmessg. v.Wälzlagern durch Ermittlg. d. Körperschallspektr.*
Stoßkondensator pulse capacitor
Stoßlast sudden load || load surge * *Laststoß* || shock load
Stoßmaschine shaping machine * *{Werkz.masch.}* || shaper * *{Werkz.maschine}* || slotting machine * *{Werkz.masch.}* || slotter * *{Werkz.masch.} slotting tool: Stoßmeißel*
Stoßmoment transient torque || peak transient torque

Stoßprüfung shock test * *für explosionsgeschützte Geräte (DIN EN 50014)*
Stoßspannung impulse voltage || voltage surge
Stoßspannungsfestigkeit impulse withstand voltage || surge withstand capability || impulse strength
Stoßspannungsprüfung impulse test || impulse voltage test
Stoßstelle transition point || butt joint * *Stoßverbindung, Stoßnaht* || butt point || butt * *Berührungsstelle von Bauteilen, Stoß*
Stoßstrom surge current || peak current || impulse current || transient current
stoßweise pulsating * *(auch figürlich)* || intermittently * *mit Unterbrechungen, zeitweilig aussetzend* || sporadically * *sporadisch auftretend* || by jerks * *ruck/stoß/sprungweise* || by fits and starts * *stoß/ruckweise* || in waves * *wellenförmig*
straff tight * *Riemen, Terminplan usw.*
straff halten keep tight
straff spannen tighten || stretch
Straffung streamlining * *einer Organisation, von Abläufen usw.* || tightening * *(mechan.)* || making compact * *(figürl.)* || rendering compact * *(figürl.)*
Strahl beam || jet || ray || straight line * *{Mathem.}* || squirt * *(Wasser-) Strahl, Spritze* || radius * *{Mathem.}*
Strahlenbelastung dose rate * *[rd/h]*
Strahlenschutz radiation protection
Strahlung radiation || rays * *Strahlen*
Strahlungseinflüsse radiation influences
Strahlungswärme radiated heat
Strahlwasser water jets * *protection against water jets: Schutz gegen ~ (siehe →IP65)* || jet-water || hose-water
Strahlwasserschutz protection against water jets * *z.B. Schutzart IP55, IP56*
stramm tight * *straff; fit tightly: ~ sitzen*
Strang strand * *aus Metall, eines Seils/Taus* || phase * *Phase eines Wechselstromsystems* || track * *Schienenstrang* || train * *Kette, Folge; drive train: Antriebs~*
Strangdrossel phase reactor * *in d. wechselstromseitigen Zuleitg. z. Stromrichter*
Stranggebäckmaschine rout-out buiscuit machine
stranggepresst extruded
Stranggießanlage continuous casting plant || continuous casting line || continuous caster || continuous casting mill
Strangguss continuous casting || extrusion casting || extruded section * *Stranggussprofil*
Stranggussanlage continuous casting plant || continuous casting line || continuous caster || continuous casting machine
Stranggussprofil extruded section
Stranggussschere extrusion shear
Strangpresse extrusion press || extruding press
strangpressen extrude || extrusion molding * *(Substantiv)*
Strangpressprofil extruded section
Strangsicherung a.c. side fuse || AC side fuse * *in d. wechselstromseitigen Zuleitung z. einem Stromrichter* || AC side fuse link * *Sicherungseinsatz ohne Unterteil* || AC line fuse || line fuse * *netzseitige Sicherung*
Strangspannung phase voltage || phase-to-neutral voltage * *Phasenspannung zw. L1, L2, L3 und jeweils N* || phase-to-ground voltage
Strangstrom phase current
Strangzahl number of phases
Strangziehmaschine continuous drawing machine
strapazierfähig rugged || heavy-duty
Straße road || line * *Verabeitungs-/Produktions~*
Straßenbahn tram || street car * *(amerikan.) Straßenbahntriebwagen, evtl. mit Anhänger*
Straßenbau roadway construction

Straßentransport   road transport || trucking
Strategie   strategy || policy * *Taktik, Verfahrensweise*
strategisch   strategic || strategical
Strebe   strut * *Strebe, Stütze, Spreize* || brace * *Stütze, Strebe, Klammer, Versteifung, (mechan.)* **Anker** || prop * *Stütze, Stempel, Stützbalken, Stützpfahl, Stelze* || stay * *Stütze, Strebe, Verspannung, (mechan.)* **Anker, Korsett** || support * *Stütze, Träger, Ständer, Strebe, Absteifung, Bettung, Stativ* || crossbeam * *Quer~* || traverse * *Quer~* || bar * *Stange, Stab, Querbalken/-stange, Schranke, Sperre* || bracket * *Träger, Halter, Konsole, Stützbalken, Winkelstütze*
streben   strive * *after/towards: nach* || struggle * *for: nach* || aim * *to: nach* || endeavour * *sich anstrengen* || tend * *zu (einer Richtung) hin ~, (to wards): nach*
Strecke   distance * *Entfernung; distance covered: zurückgelegte ~* || system * *Regel~* || route * *Fahrstrecke, →Weg* || stage * *Teil~* || span * *Spanne* || course * *Bahn, Kurs* || straight line * *(geometr.)* || section * *Abschnitt*
strecken   stretch || extend * *(aus)dehnen* || lengthen * *in die Länge ~* || draw * *ziehen* || straighten * *geradebiegen, (auf)richten*
Streckenparameter   system parameters || parameters of the controlled system
Streckensteuerung   point-to-point control * *Positioniersteuerung*
Streckfaktor   stretching factor * *bei der Chemiefasern usw.* || stretching ratio * *bei Chemiefasern usw.*
Streckfestigkeit   tensible strength || yield strength
Streckgrenze   yield point * *Fließ-/Streck-/Nachgiebigkeitsgrenze* || elastic limit
Streckmaschine   drawing machine * *für Textilfasern*
Streckricht-Maschine   tension-levelling machine || tension leveller
Streckrichter   leveller * *Streckrichtmaschine z. Geraderichten von Blech über viele dünne Walzen*
Streckrichtmaschine   tension-levelling machine || tension leveller
Streckwerk   drawing frame || draw stand || streching unit || stretching machine || drafting device * *{Textil | Draht} in Spinnmaschine; zur Durchmesser-Verringerung bei Drahtziehmaschin*
Streichanlage   coating plant
streichen   delete * *löschen* || cancel * *löschen, anullieren* || paint * *anstreichen*
Streichleiste   foil * *bei Streichmaschine* || blade * *bei Streichmaschine*
Streichmaschine   coating machine * *Beschichtungsmaschine für Papier oder Folien* || coater * *zur Beschichtung von Papier oder Folien* || blade coater * *{Papier}*
Streichmesser   blade * *bei Streichmaschine* || scraper * *bei Streich-/Papierbeschichtungsmaschine usw.* || doctor blade * *Rakel* || spreading knife
Streichpaste   coating paste * *{Papier} zum Streichen des Papiers in der →Streichmaschine*
Streifen   stripe || streak * *Streif(en) Strich, (Licht-) Streifen, Strahl* || strip * *kurzes/schmales Stück; Papierstreifen, Filmstreifen* || tape * *Band* || shred * *Schnipsel* || braid * *Litzenbandleiter* || touch * *(Verb; auch figürl. (z.B. ein Thema)) berühren* || graze * *(Verb) streifen an* || brush * *(Verb) brush against: streifen an* || glide over * *(Verb) über etwas hingleiten* || skim over * *(Verb) über etwas hingleiten* || strip off * *(Verb) abstreifen*
Streifenhobelmesser   thin planing knife * *{Holz}*
Streik   strike
streng   rigid * *unnachgiebig, ohne Ausnahme* || strict * *scharf, bestimmt, genau; z.B. Reihenfolge, An-* wendg. e.Prinzips || stringent * *zwingend, bindend (z.B. Maßnahme, Regel, Vorschrift, (An)Forderung)* || severe || rigorous || harsh * *hart* || severely * *(Adv.)* || strictly * *(Adv.) strictly confidential: vertraulich; adhere strictly to: sich ~ an etw. halten*
streng zeitzyklisch   in a rigidly cyclic time frame
streng zyklisch   rigidly cyclic
Stretchfolie   stretch film * *{Verpack.techn.}*
Streu-   leakage * *z.B. Streureaktanz* || stray
Streublatt   leaflet * *Werbeblatt* || flier * →*Prospekt* || flyer * *(amerikan.)* →*Prospekt*
streuen   deviate * *abweichen; significantly: stark* || spread * *statistisch~* || leak * *magnetisch ~* || disperse || scatter * *statistisch~, physikalisch~*
Streufeld   leakage field || stray field || extraneous field
Streufluss   leakage flux * *{el. Masch.} nicht moment.bildender Teil d. magn. Flusses, d. sich auf Nebenwegen schließt* || stray flux
Streuinduktivität   leakage inductance * *Induktivität d.* →*Streuflusses*
Streureaktanz   leakage reactance * *Streublindwiderstand auf Grund der Induktionswirkung der* →*Streuflüsse*
Streuung   leakage * *Verlusteffekt einer elektromagnetische Größe (Fluss, Durchflutung)...* || strewing * *Aus-/Bestreuen, z.B. Sand, Werbematerial* || deviation * *Abweichung* || dispersion * *statistische ~* || spread * *statistische ~* || scattering * *statistische ~, physikal. ~* || leakage * *magnetische ~* || stray losses * *magnet, Streuverluste* || standard deviation * *(statistische) Standardabweichung* || scatter * *statistische ~*
Strich   line * *Linie* || dash * *Gedankenstrich* || stroke * *Pinsel-/Feder-/Kennzeichnungs~*
stricheln   dash-line
Strichgewicht   coated weight * *{Papier} Flächengewicht des Auftrages durch eine* →*Streichmaschine*
Strichpunkt   semicolon
strichpunktiert   chain-dotted || dot-and-dash
Strichzahl   pulse number per revolution * *eines Impuls-/Winkelschrittgebers* || PPR count * *Abkürzung für 'Pulses Per Revolution': Impulse pro Umdrehung* || number of pulses per revolution * *number of increments per revolution * eines Inkrementalgebers/Impulsgebers* || number of slots per revolution * *eines optischen Drehimpulsgebers*
stricken   knit
Strickmaschine   knitting machine
strikt   rigorous || rigorous * *peinlich genau, strikt* || strictly * *(Adv.)*
String   string * *{Computer} Zeichenkette, Folge von Textzeichen*
Stroboskop   stroboscope
Stroboskopeffekt   stroboscopic effect || persistance of vision || strobe effect
Strom   current * *elektrischer* || amps * *Stromstärke* || flow * *von Material/Luft/Verkehr* || rate * *Menge pro Zeiteinheit*
Strom-Null-Erkennung   current zero detection
Strom-Spannungs-Ebene   current/voltage chart * *Koordinatensystem*
Strom-Spannungs-Kennlinie   voltage-current characteristic
Stromabbau   current suppression || current decay * *Abklingen des Stroms*
Stromabfall   current decay
stromabgebendes   outgoing * *zu löschendes {Ventil in Stromrichterschaltung}*
stromabhängig   current-dependent
Stromabnehmer   current collector * *{Bahnen}* || bow collector * *{Bahnen} ~bügel*
Stromänderungsgeschwindigkeit   current rate of rise || speed of current alternation

Stromanregelzeit current rise time
Stromanstieg current rise
Stromaufbauzeit current build-up time
Stromaufnahme current consumption || current input || power consumption || power input || current draw * z.b. eines Geräts/Bauteils || current drain
Stromaufteilung current sharing || current distribution || current division
Stromausfall power failure || power loss
Strombahn current path * z.B. in Auslösegerät, Leistungsschalter usw
Strombedarf current demand || current draw * z.B. Stromaufnahme eines Geräts/Bauteils || power demand * bezüglich der Stromversorgung || current requirement
strombegrenzt current limited
Strombegrenzung current limiting || current limit * Stromgrenze || current limiter * als Einrichtung, Modul, Funktionsbaustein, Schaltung
Strombegrenzungsregelung current limiting control
Strombegrenzungsregler current limiting controller
Strombelag electric loading || specific loading
Strombelastbarkeit current carrying capacity * von Leitungen: siehe VDE 0113/EN 60204, jeweils Teil 1 || current loading capability || current carrying || current capability || current handling capability || contact rating * spezifizierte Nenn-~ eines (Relais, Schütz-, Schalter-) Kontakts || carrying capacity
Strombelastung current loading
Stromdichte current density * Verteilung des el. Stroms auf den Leiterquerschnitt in [A/mm²]
stromdurchflossen current-carrying
stromdurchflossener Leiter current-carrying conductor
Stromdurchgang conductive continuity || continuity
strömen flow * fließen, strömen || stream * allg., auch Personen || gush * quellen, schießen || pour * Regen usw.
Stromerzeugung power generation
Stromfaktor current factor
Stromflußrichtung direction of the current || direction of the current flow || direction of the flow of current
Stromfluss current flow || current flowing
Stromflussüberwachung conduction monitor
Stromflusswinkel current flow angle || forward flow angle
Stromflusszeit conductive interval
stromführend conducting * z.B. (Stromrichter-) Ventil || current-carrying * auch (Stromrichter-) Ventil || current-energized
Stromführungsdauer conduction period || current conduction duration
Stromführungszeit current flow period || conducting interval
stromgeregelt current controlled
Stromgrenze current limit * current limit acceleration: Hochlauf an der ~
Stromgrenze erreicht current limit reached
Stromistwert current actual value || actual current || act. current || current feedback || current signal
Stromistwertanzeige current actual value display || current feedback indication
Stromistwerterfassung current actual value sensing || current sensing || actual current measuring circuit * Messschaltung || actual-current sensing || actual-current sensing circuit * als Schaltung/Einrichtung
Stromkraft electrodynamic force
Stromkreis electric circuit || circuit
Stromlaufplan circuit diagram * z.B. für Leiterkarte || schematic diagram || schematic * auch 'schematische Darstellung, Übersicht, Schema'
stromlinienförmig streamlined
stromlos de-energized || dead || zero-current || at zero current || current-free
stromlose Pause no-current interval * wird z. Sicherheit von Kommandostufe bei Momentenumkehr eingefügt || zero-current interval * z.b. durch Kommandostufe erzeugt || period of no current * z.B. durch Kommandostufe erzeugt || zero current flow interval * wird z. Sicherheit v. Kommandostufe eingefügt bei Momentenwechsel
stromloser Zustand zero-current condition || current-free state
Stromlücken intermittent current condition || pulsating current
Strommesser ammeter
Stromnulldurchgang zero current || current zero || zero crossing of current
Stromnullerfassung current zero sensing || current zero detection
Stromnullmeldung current zero signalling || zero current signal
Strompfad ladder diagram line * in Relaissteuerungs-/Kontaktplan (KOP) || rung * in Relaissteuerungs-/Kontaktplan (KOP)
Stromquelle current source || impressed-current source * eingeprägten Strom liefernd
Stromregelung current control * secondary: unterlagert || closed-loop current control
Stromregler current controller || current regulator
Stromreglerausgang current loop output || current controller output
Stromrichter converter * siehe DIN EN 60146, IEC Publ.146 u.411, DIN 41750/51/52, DIN 57558 und VDE 0558 || inverter * Wechselrichter (mit Wechselspannungsausgang) || drive * (amerikan.) Antriebsstromrichter; aber: im engl. bedeutet "drive" Antrieb || static converter * (veraltet) nicht rotierender Stromrichter (i. Gegens. z.B. zum Leonard-Umformer) || convertor * (veraltet) || converters * (Plural) || convertors * (Plural, veraltet) || drive converter * Antriebsstromrichter || power converter
Stromrichter mit Gleichspannungsausgang DC converter || DC drive * (nur amerikan.) für Motoransteuerung
Stromrichteranlage converter plant
Stromrichteranschaltung drive converter interface * zu einem Antriebsstromrichter
Stromrichterantrieb converter-fed drive
Stromrichteraufwand converter cost
Stromrichterbetrieb converter supply * ~ eines Motors am Stromrichter/Umrichter (nicht direkt am Netz) || converter operation || use with thyristor converters * Betrieb z.B. eines (Gleichstrom-) Motors am Thyristor-Stromrichter
Stromrichtereinheit converter unit
Stromrichtergerät converter unit || servo-amplifier * zur Ansteuerung eines Servomotors bzw.eines Positionierantriebs || drive converter unit * Antriebsstromrichter || power converter unit || converter equipment
stromrichtergespeist converter fed
stromrichtergespeister Antrieb converter-fed drive
Stromrichterkaskade converter cascade * subsynchronous: untersynchrone
Stromrichtermotor converter motor * Sychronmot.1-250MW m.Ständerspeisg. durch →lastkommut. I-Umrichter u. →RG-Erregung
stromrichternah converter-related || converter-specific
Stromrichtersatz converter set || converter assembly

Stromrichterschaltung converter connection || converter circuit configuration * *Anordnung* || converter circuit || converter circuit configuration
Stromrichterschaltungen converter circuits * *DIN 41761* || converter connections || types of converter connections * *Arten von Stromrichterschaltungen*
stromrichterspezifisch converter-specific || converter-related
Stromrichtertechnik converter technology
Stromrichtertrafo converter transformer
Stromrichtertransformator converter transformer
Stromrichterventil converter valve
Stromrichterzweig valve arm || converter leg || converter arm
Stromrichtgrad conversion factor * $(U\_dc. \times I\_dc/$ *Grundschwingungsleistung auf der AC-Seite) bei DC-Stromrichter*
Stromrichtung current direction || direction of current || direction of current flow
Stromrichtungsumkehr current reversal || converter reversal * *in einem Stromrichter*
Stromschiene bus-bar || conductor bar || busbar || power rail * *in der Kontaktplandarstellung eines SPS-Programms*
Stromschienensystem busbar system || busbar trunking system
Stromschleife current loop
Stromsenke current sink
Stromsensor current sensor
Stromsignal active current signal * *z.B. 0-20 mA* || current loop signal * *z.B. 0-20mA; siehe auch 'eingeprägter Strom'*
Stromsollwert current setpoint || current reference
stromsparend energy-saving
Stromspitze current peak || current spike * *Nadelimpuls*
Stromspitzen current peaks
Stromstärke amperage || current intensity || current
Stromsteilheit current gradient
Stromstoß current surge || current rush * *inrush current surge:* →*Anfahrstromstoß*
Stromstoßrelais current-impulse relay
Stromtragfähigkeit ampacity * *(salopp) Kurzform für "amps capacity"* || current carrying capacity * *(von Leitungen: siehe VDE 0113/EN 60204, jeweils Teil 1)* || current loading capability || current carrying || amps capacity || current carrying capability
Stromträgheit current inertia
Stromübergang current transfer
Stromüberlastung excess current
Stromübernahme current takeover
Strömungsbremse hydrodynamic brake || retarder * *hydraul. ~/Retarder für LKWs, Schienenfahrzeuge usw.*
Strömungskupplung hydraulic coupling
Strömungsmaschinen fans and pumps || fans, pumps and compressors || fans and centrifugal pumps and compressors
Strömungsmaschinen-Antriebe pump and fan drives
Strömungsmaschinenantrieb fan/pump drive || fan/pump/compressor drive
Strömungsmaschinenantriebe fan and pump drives * *Lüfter- und Pumpenantriebe*
Strömungsmesssonde flow measuring probe * *z.B. in der Belüftungstechnik*
Strömungswächter flow monitor
Strömungswiderstand resistance to flow || drag * *siehe auch* →*Luftwiderstand* || aerodynamic resistance * →*Luftwiderstand*
Stromverbrauch power consumption || electricity consumption * *am Stromnetz*
Stromverdrängung current displacement * *z.B. Skin-Effekt, Heaviside-Effekt*

stromverdrängungsarm with low current displacement || anti-skin-effect
Stromverdrängungseffekt proximity effect || skin effect * *Verdrängung des Stroms ins Leiteräußere auf Grund der Wirbelströme* || Heaviside effect || deep bar effect
Stromverdrängungsläufer deep-bar squirrel-cage rotor * *{el. Masch.}* || eddy current cage rotor * *{el. Masch.}*
Stromverdrängungsläufermotor deep-bar squirrel-cage motor
Stromverdrängungsverluste skin-effect losses
Stromversorgung power supply || power supply module * *als Funktionsmodul* || electric supply
Stromversorgungseinheit power supply unit
Stromversorgungsunternehmen power supply company || power utility || power company
Stromverstärkung current amplification * *z.B. eines (bipolaren) Transistors ("Beta")*
Stromvorsteuerung current pre-control || current feed forward
Stromwandler current transducer || currrent transformer || CT * *Abkürzung für 'Current Transducer'* || current sensor
Stromwandlerübersetzung current transformer ratio * *Übersetzungsverhältnis*
Stromwärmeverluste ohmic losses
Stromwelligkeit current ripple
Stromwender commutator * *Kommutator*
Stromwendung commutation * *Kommutierung*
Stromzähler energy meter || electricity meter
Stromzange current probe
Stromzwischenkreis current source DC-link
Stromzwischenkreis-Umrichter current source inverter || current fed converter || current-source DC link converter
Stromzwischenkreisumrichter current source DC-link converter
Struktogramm structured chart || structogram || Nassi-Schneidermann chart * *Nassi-Schneidermann-Diagramm* || NS chart * *Nassi-Schneidermann Diagramm*
Struktur structure || pattern * *Gestalt(ung), Anlage, Muster, Schema, Gesetzmäßigkeit(en), (wirtsch./soziale) Struktur* || texture * *Textur, Maserung, Struktur, Gefüge, Oberflächenbeschaffenheit*
Struktur- structural
Strukturbild structure diagram
strukturbildorientiert function-diagram oriented * *z.B. SIEMENS Antriebsregelsystem SIMADYN D (R)* || structural-diagram oriented || structure diagram oriented
strukturieren structure || configure * *konfigurieren*
strukturiert structured
strukturierte Programmierung structured programming * *z.B.nach Dijkstra, Nassi Schneidermann (z.B. direkte Sprünge verboten)*
Strukturierung structuring || orgaization * *Organisation*
Strukturproblem structural problem
Strukturwandel structural change || state of flux
Stück item || *Zahl, ~ Ware, Artikel, Teil* || piece * *(Einzahl)* || pieces * *(Mehrzahl)* || pcs * *Abkürzg. für 'Pieces'.* || quantity * *~zahl* || qty. * *(Abkürzung) ~zahl*
Stückholz lump wood * *z.B Brennholz*
Stückliste parts list
Stückpreis price per item || unit price || price per piece
Stückprüfung routine test * *jedes gefertigten Exemplars (i.Gegensatz zur Typprüfung); für Motor: siehe VDE 0530* || 100% inspection * *vollständige Qualitätsprüfung an allen Einheiten eines Fertigungsloses*
Stückzahl volume || quantity * *for a quantity of:*

Stückzahlen

bei einer ~ von; *(high-) volume quatities: hohe ~en* || number of pieces * *z.B. ~ bei einer Bestellung* || unit volume * *Gesamt~/verkaufte ~ eines Produkts* || number of items
Stückzahlen quantities
Student student || undergraduate
Studie analysis || study * *of: über* || survey * →*Untersuchung*
Stufe stage * *(elektronische) Schalt~, Entwicklungsstadium, Ausbau~* || step || increment * *Inkrement* || grade * *Grad, Klasse, Rang, Qualitäts~, Art, Gattung, Sorte, Dienstgrad* || phase * *Phase, auch Entwicklungsphase* || degree * *Grad, Rang* || level * *Niveau* || standard * *Niveau* || rank * *Rang* || gate * *logisches Gatter, z.B. UND, ODER*
stufen step || grade * *staffeln*
stufenförmig in steps * *(Adj. und Adv.)* || in the form of steps || terraced * *Gelände usw.* || in stages * *(figürl.) Prozess, Entwicklung usw.* || in tiers * *(Adv.) in Reihen/Lagen übereinanderliegend* || arranged in tiers
Stufenhöhe step range * *Schrittweite* || step height
stufenlos stepless || without steps || infinitely variable * *stufenlos einstell/regelbar* || infinitely adjustable * *stufenlos einstell/regelbar* || continuous * *ununterbrochen, fortlaufend, kontinuierlich* || continual * *stetig, fortwährend, dauernd*
stufenlos vergrößern zoom in
stufenlos verstellbar infinitely adjustable || infinitely variable || continuously changeable * *z.B. Getriebe* || continuously variable
stufenlos verstellbares Getriebe continuously changeable gearbox || speed variator
stufenlose Maßstabsveränderung zooming
Stufensprung step range * *Schrittweite* || resolution * *Auflösung* || step change * *Stimulationssignal für Regleroptimierung (Aufnahme der Sprungantwort)*
Stufentransformator stepping transformer
stufenweise step-by-step * *(Adj.)* || step by step * *(Adv.)* || stepwise || gradual * *~ fortschreitend, allmählich* || in steps || stepped
stufenweise erhöhen step up
stufenweise Näherung successive approximation * *z.B. schnelles Verfahren für A/D-Umsetzer*
stufenweise reduzieren step down
Stufenzahl number of starter steps * *{el. Masch.}* *Anzahl d. abschaltbaren Teilwiderstände eines Motoranlassers*
Stufung step range * *Schrittweite* || grading
stukturieren structure
stülpen turn * *inside out: das Innere nach außen; upside down: das Obere nach unten* || put over * *auf/überstülpen* || dish * *wölben*
Stumpfschweißmaschine butt welding machine
Stunde der Wahrheit moment of truth || hour of decision * *Stunde der Entscheidung* || showdown * *Aufdecken der Karten, spannender Höhepunkt, entscheidende Kraftprobe*
Stundensatz hourly rate || rate per hour
Sturz fall * *sudden fall: unerwarteter ~; have a bad fall: einen schweren ~ tun* || tumble * *Fall, Sturz, Purzelbaum, Salto* || crash * *lauter/spektakulärer/aufsehenerregender ~* || smash * *lauter ~* || plunge * *~ ins Wasser* || camber * *Rad~* || lintel * *Fenster~, Tür~* || drop * *Wetter~, Temperatur~* || collapse * *~ von Kursen Preisen usw.* || ruin * *Untergang, Ruin*
Stütz- supporting
Stütze support || support leg * *Stützbein* || prop * *Pfahl, Stempel, Strebe, Stützbalken*
Stutzen muff * *auch ~Muffe, ~Flanschstück* || duct * *inlet duct: Einlass~; discharge duct: Auslass~* || connecting piece * *z.B. ~ für Rohrverbindung (siehe auch →Gewindestutzen)* || nipple * *kleiner ~* || clip * *(Verb) abschneiden* || cut short * *(Verb) beschneiden* || trim * *(Verb)*

stützen support || buffer * *Versorgungsspannung bei Spannungseinbruch ~* || be based on * *abgestützt sein, sich ~ (on: auf)* || hold up * *halten, stützen, aufrechterhalten, hochheben* || subsidize * *subventionieren* || sustain * *stützen, tragen, (auf recht)erhalten in Gang halten, (Interesse) wachhalten, unterhalten*
Stützisolator supporting insulator * *für eine Schraubklemme*
Stützkondensator supporting capacitor * *z.Stützen d. Stromversorgg. auf Elektron.leiterkarte, 'schluckt' Schaltspitzen* || back-up capacitor * *z. Stützen d. Stromversorgg. auf Elektron.leiterkarte, 'schluckt' Schaltspitzen* || backup capacitor * *verteilte ~en puffern Spannungseinbrüche b. Schaltvorgängen i. Elektronikschaltungen* || buffer capacitor || supporting capacitor
Stützpunkt point * *eines Polygonzuges* || base * *z.B. Handelsstützpunkt* || interpolation point * *eines Polygonzuges* || support point * *eines Polygonzuges* || support value * *eines Polygonzuges/Kennlinienbausteins*
Stützung buffering * *z.B. durch Kondensator, Batterie, kinetische Energie* || supporting * *auch von Preisen usw. (siehe auch →Subventionierung)*
Stützwalze recl drum * *für Oberflächenwickler, z.B. bei Rollenschneidmaschine* || backing drum || backing roll || supporting drum
Stützwert support value * *Stützpunkt einer aus Geradenstücken entstehenden Kennlinie (Polygonzug)* || interpolation point * *Stützpunkt eines Polygonzuges*
Sub D Stecker Sub D connector || Subminiature D connector
SUB-D Stecker SUB-D connector || D-type connector
Sub-D-Stecker Sub-D connector * *auch "Cannon-Stecker", Grundfläche hat die Form eines "D"*
Sub-Leiterkarte daughter board || sub-board
Subindex parameter index * *(→PROFIBUS-Profil)*
Subminiatur D-Stecker Sub D connector || Subminiature D connector
Submodul submodule || daughter board * *Leiterkarte* || sub-module
Substanz substance * *(auch figürlich: Grundlage)* || medium || agent * *Wirkstoff* || essence * *innere Natur, Wesen, das Wesentliche, der Kern (der Sache)*
substanziell substantial
substituieren substitute * *for: durch* || replace * *ersetzen; by: durch*
Substitution substitution || replacement * *Ersetzen, Ersatz*
Substitutionsbefehl substitution operation * *bei SPS*
Substrat substrate
subtrahieren subtract
Subtrahierer subtractor
Subtraktion subtraction
subtropisches Klima subtropical climate
Subvention subsidies || subvention
subventionieren subsidize || support * *stützen (auch Preis, Währung)*
Subventionierung subsidization || subsidies || supporting * *Stützung (auch von Preisen, einer Währung usw.)*
Suchbegriff search key || key word || keyword
Suche search * *in search/quest of: auf der ~ nach* || hunt * *dringende ~ (for: nach)* || look-out * *be on the look-out for: auf der ~ nach ... sein/Ausschau halten nach* || exploration * *Suche (nach Bodenschätzen), Erforschung, Untersuchung*
suchen search * *for: nach* || speed-search * *(Subst.) Suchen d. aktuell. Motordrehz. nach Spanngs.einbruch z. Zuschalt. e. Umrichters* || seek || trace

* *aufspüren (Fehler, Ursache, Spur usw.)* || desire * *wünschen, haben wollen* || want * *wünschen, haben wollen* || look for || hunt for * *eilig, hastig ~* || rummage for * *wühlend, kramend ~* || look up * *nachschlagen (in Liste, Buch, Naaachschlagewerk, Tabelle im Programmspeicher)* || find * *finden* || locate * *ausfindig machen*
Suchen-Fangen speed-search * *Umrichter schaltet nach Netzspannungseinbruch wieder auf d. laufenden Motor auf* || flycatcher * *Umrichter schaltet nach Netzspannungseinbruch wieder auf d. laufenden Motor auf* || auto speed search
Suchlauf search || search function
Suchvorgang search operation || search sequence * *aufeinander folgende Suchvorgänge z.B. zum Fangen e. Motors nach Spannungseinbruch* || lookup operation * *in einer Tabelle (z.B. durch Rechner)*
Sudhaus brewing room
Südpol south pole * *auch bei einem Magnet*
Sulfat sulphate * *{Chemie} Salz der Schwefelsäure*
Sulfid sulphide * *{Chemie} Salz des Schwefelwasserstoffs*
Sulfit sulphite * *{Chemie} Salz der schwefligen Säure*
Summationspunkt summing point
Summe sum || total * *Gesamtsumme* || totality * *(figürl.)* || amount * *Betrag* || deficit * *fehlende ~, Fehlergebnis* || addition * *Addition* || accumulated number * *aufsummierte Anzahl* || amount * *(Geld)-Betrag, Wert, Höhe*
summen hum * *brummen* || buzz * *hochfrequent*
Summen- group * *Sammel-, gemeinsamer* || total * *Gesamt-, Über-Alles-* || overall * *Gesamt-, Über-Alles-* || summary * *(in) gedrängte(r) Übersicht, summarisch, Übersichts-, abgekürzt, Schnell-*
Summenimpulse summing pulses || group pulses
Summenlöschung common turn-off * *Löschung mehrerer Thyristoren über e. gemeinsamen Löschzweig*
Summensignal group signal * *z.B. Sammel(fehler)meldung*
Summenstörmeldung group fault signal
Summenwarnung group alarm
Summer buzzer
summieren sum || sum up * *auf~* || totalize * *auf~* || total up * *(sich) auf~ (to: auf)* || accumulate * *sich auf~, sich ansammeln, sich aufakkumulieren*
Summierer sum block * *Funktionsbaustein* || summing element || adder * *Addierer* || summator
Summierglied summing element || adder * *Addierer*
Summierpunkt summing point || summator || summing junction
Summierstelle summing point
Summierung summation
Summierverstärker summing amplifier
super super * *(salopp) großartig* || first-class * *(salopp) erstklassig* || first-rate * *(salopp) großartig, ausgezeichnet* || great * *(salopp) großartig, famos, herrlich, überragend, prima, toll* || fabulous * *(salopp) fabelhaft*
superflink super fast * *Sicherung* || very quick acting * *Sicherung* || high-speed || super-rapid || super-fast acting * *Sicherung* || fast-acting * *Sicherung*
superflinke Sicherung high-speed fuse || very quick acting fuse || fast 'semiconductor' type fuse * *~ zum Schutz von Halbleitern* || SCR type high speeed fuse * *~ zum Thyristorschutz*
Superkalander supercalender * *Mehrstufig angeordnete Glättzylinder zum Erzeugen hochglatten (satinierten) Papiers*
superkalandrieren supercalender * *satinieren, hochfein glätten von Papier durch Glättzylinder*

superleitend superconductive
supraleitend superconducting || superconductive
Supraleiter superconductor
Suspension suspension * *Flüssigkeit mit ungelösten, gleichmäßig verteilten (Schweb-) Stoffen/Partikeln*
Süßwarenmaschine convectionary machine || candy machine
Suszeptibilität susceptibility || magnetizability
Symbiose symbiosis
Symbol symbol
Symbolik symbology * *auch die '(Bedeutung der) definierten Symbole'* || symbolism
symbolisch symbolic || symbolical
symbolische Adressierung symbolical addressing * *{Computer} durch e. mnemotechn. Namen, nicht durch eine physikal. Adresse*
symbolischer Name symbolic name
Symmetrie symmetry * *(auch von Spannungen und Strömen)*
symmetrieren balance || symmetrize || neutralize || align * *→abgleichen*
Symmetrierung balancing || symmetry * *Symmetrie*
Symmetrierwiderstand balancing resistor || balacing resistance
symmetrisch symmetric || symmetrical || balanced
symmetrisch geteilte Wendepolwicklung symmetrically split interpole winding
symmetrische Komponenten symmetric components * *Zerlegung e.unsymmetr. Drehsromsystems in →Mit-, Gegen und Nullkomponente*
symmetrisches Optimum symmetrical optimum * *Regleroptimierung,sodass Regelfläche gleich null ist (hohes Überschwingen)*
Symposium symposium
synchron synchronous || in synchronism * *with: mit; run in synchronism: ~ laufen* || in step * *im Gleichschritt, gleich schnell (laufend)*
Synchron-Frequenzumformer synchronous frequency converter * *rotierender*
Synchrondrehzahl synchronous speed * *{el. Masch.} Drehz. d. Drehfelds bei. AC-Motor [upm]: 120 x Netzfrequ./Polzahl*
synchrone Drehzahl synchronous speed * *e. AC-Motors in [upm]: (120-mal Netzfrequenz/Polzahl)*
synchroner Lauf synchronous running
Synchronimpuls synchronizing pulse
Synchronisation synchronization
Synchronisier- synchronizing
Synchronisierbaugruppe synchronization module * *z.B. für Steuersatz e. netzgeführten Stromrichters* || synchronizing board * *Leiterkarte zur Synchronisierung e.fremdgeführten (Thyristor-)Stromrichters*
synchronisieren synchronize
Synchronisierimpuls synchronizing pulse
Synchronisiermarke sync mark * *Signal, das den Referenzpunkt bei Winkel-/Weg-/Gleichlaufregelung vorgibt* || synchronizing mark
Synchronisierspannung synchronizing voltage || synchronising voltage || sync voltage * *Kurzform für 'synchronizing voltage'* || synchronization voltage
Synchronisierstufe synchronization block || synchronization circuit
synchronisiert synchronized
synchronisierter Asynchronmotor synchronized asynchronous motor * *nach erfolgt. Hochl. Umschaltg. in Synchronbetr. (Läufer an Gleichsp.)* || synchronized induction motor * *{el. Masch.}*
Synchronisiertrafo synchronizing transformer * *erzeugt Synchronisiersignal, z.B. für netzgeführten Stromrichter*
Synchronisiertransformator synchronizing trans-

Synchronisierung

former * erzeugt Synchronisiersignal z.B. für netzgeführten Stromrichter
Synchronisierung synchronisation || synchronization
Synchronisierungsmoment synchronization torque * bei Synchronmaschine
Synchronisierzusatz synchronizing option * z.B. z. Synchronisierg. Umrichter-Umr. (Reserveumr.)/ Umr.-Netz (Hochfahrumr.)
Synchronismus synchronism || synchronous running * synchroner Lauf
Synchronlauf synchronous running || synchronism || synchronous operation
Synchronmaschine synchronous machine * el. AC-Masch. mit Erregg. durch Dauermagnete oder DC-Feldwicklg. i.Läufer
Synchronmotor synchronous motor
Synergie synergy || synergism || synergistic effect * Synergieeffekt || synergic effect * Synergieeffekt || synergetic effect * Synergieeffekt
Synergieeffekt synergistic effect || synergetic effect || synergic effect
Syntax syntax * auch einer Programmiersprache (Schreibweise von Anweisungen, Befehlen)
Syntaxfehler syntax error
Syntaxprüfung syntax check * z.B. ob ein Quellspracheprogramm den Syntaxregeln einer Programmiersprache entspricht
Synthese synthesis
Synthetiköl synthetic oil
synthetisch synthetic || man-made * künstlich hergestellt
synthetisches öl synthetic oil
System system * systematize: in ein ~ bringen/~atisieren * plan * Plan || method * Methode || doctrine * Lehre || structure * Struktur, Gefüge || framework * Gefüge
System-Software operating system || system software
Systemabsturz system crash * auch ~ auf Grund eines Softwarefehlers
Systemanbieter system supplier || OEM * Abk. für 'Original Equipment Manufacturer': Systemhersteller, z.B. Maschinenbauer
Systemarchitektur system architecture || systems architecture
Systematik systematics || system * Aufbau || taxonomy * systematische Einordnung (z.B. Botanik, Zoologie) || classification * Einteilung in Klassen/Gruppen || framework * Gefüge
Systematiker systematic person
systematisch systematical || methodical || systematically * (Adv.) || methodically * (Adv.)
systematisieren systemize || systematize || classify * in Klassen/Gruppen einteilen
Systematisierung systematization
Systemaufruf system call * (Computer)
systembedingt inherent in the system || due to the system || system-inherent
Systembibliothek system library
Systembus system bus
Systemdaten system data
Systemdruck system pressure * in Hydrauliksystem
Systementwickler system developer || system designer
Systemfehler system error
systemgerecht system-compatible
Systemhersteller OEM * Original System Manufacturer, stellt s. Systeme z.T.aus Zulieferteilen zusammen || system vendor * Systemanbieter
Systemintegration system integration
Systemintegrator system integrator * integriert für →OEM o. →Endabnehmer Systeme aus Kompo-

nenten verschied. Hersteller || systems integrator || system supplier * Systemanbieter
Systemkomponente system component
Systemkonfiguration system configuration
Systemleistung system performance
Systemlösung system solution
Systemmanagement system management
Systemmatrix system matrix
Systemmerkmale system features
Systemparameter system parameter
Systemprogramm system program
Systemprüfung system checkout
systemseitig system-side
Systemsoftware systems software || operating system * Betriebssystem || system cernel * (Betriebs-) Systemkern
Systemspezialist system specialist
Systemsteuerung system control
Systemtakt system clock || system tick || clock tick
Systemtest system test || system checkout
Systemtransferdaten system transfer data
systemübergreifend system-overlapping
Systemumgebung system environment
Systemzeit real time clock * Echtzeituhr
Szenarium scenario

T

t t * metrische Tonne (entspr. 1000 kg) || ton * in Großbritannien: 1 Brit. ton entspr. 1016 kg; in USA: 1 US-ton entspr. 907,2 kg || sh tn * short ton, US-ton: 1 sh tn entspr. 907,2 kg; 1000 kg entspr. 1,1023 short tons || cwt * hundredweight; 1000 kg entspr. 19,6841 cwt; 1 cwt enspr. ca. 1 Zentner || metric ton * metrische Tonne; 1 metric ton entspr. 1000 kg entspr. 0,9842 ton || tonne * metrische Tonne (entspr. 1000 kg)
T-Glied T-element * z.B. Tiefpass aus zwei Widerständen und einem Kondensator
T-Nut T-Slot * z.B. ~ für variable Schraubbefestigung mit "Hammerschrauben"
T-Profil T section
t6-Zeit t6-time * (el.Masch.) zuläss. Einschaltdauer e. AC-Motors mit blockiert. Läufer u. 6-fach. Bemess.strom
tabellarisch tabular || tabulated || in tabular form * (Adv.) || tabularly * (Adv.)
Tabelle table || chart * als Schaubild || look-up table * "Nachschlage-/Zuordnungstabelle" in Computer (von Adressen, Koordinatenwerten usw.)
Tabellen-Arbeitsblatt spreadsheet * (Computer) in einer Tabellenkalkulations-Software
Tabellenblatt spread-sheet
Tabellenform tabular form * in a tabular form: in ~
Tabellenheft pocket guide * Kurzübersicht/Kurzbetriebsanleitung im Format DIN A5 oder kleiner || pocket reference
Tabellenkalkulationsprogramm spread-sheet program
Tabellenwert table value
Tabulator tab * (salopp) Taste, die die Schreibmarke um mehrere Zeichenabstände weiterbewegt || tabulator * Taste an Computer-Tastatur, die die Schreibmarke um mehrere Leerzeichen weiterbewegt
Tacho tacho * ['täko] || tach * [täk] || tachometer * ['täkomiete] || speed sensor || pulse tach * Impulstacho || tachogenerator
Tachoanbau tachometer mounting

Tachoanpassung tachometer adaption
Tachoausfall tachometer failure || tacho loss * *Ausfall des Tachosignals z.B. durch Leitungsbruch*
Tachobruch tacho loss * *Wegbleiben des Tachosignals z.B.durch Leitungsbruch* || tachometer failure * *Tacho-Fehler* || tachometer interruption * *Unterbrechung des Tachosgnals/der Tacholeitung* || tacho break-down
Tachodynamo tacho-generator * *Gleichstrom- (manchmal auch Drehstrom-) Generator zur Drehzahlmessung*
Tachofehler tacho loss * *z.B. durch Drahtbruch der Tacholeitung* || speed feedback loss * *Wegbleiben des Drehzahlistwertsignals*
Tachogeber tacho-generator
Tachogenerator tachometer generator || tacho-generator || tachogenerator || tacho || speed sensor * *Drehzahlgeber*
Tachokupplung tachometer coupling
Tacholäufer tachometer rotor || tacho rotor
Tachomaschine tacho-generator || tachometer generator || tachogenerator
Tachometer tachometer * *['täkomete]* || tacho * *['täko]* || tach * *[täk]* || pulse tach * *Impulsgeber* || speed sensor || web tach * *Bahn~* || surface tachometer * *Bahn~, Anlege~, Messrad* || portable tachometer * *Hand~* || manual tachometer * *Hand~*
Tachometerdynamo tachogenerator
Tachometerkupplung tachometer coupling
Tachospannung tachometer voltage
Tachoständer tachometer stator
Tachostörung tachometer fault || tacho loss * *Ausfall des Tachosignals* || tachometer interruption * *Tachobruch, Unterbrechung in der Tacholeitung*
Tachoüberwachung tacho loss monitoring || tacho failure monitoring || tachometer interruption monitoring
Tacker staple gun
Tafel board * *für Anschläge und zum Schreiben* || panel * *Holz~, Blechplatte, auch zur Wandverkleidung* || plate * *Platte, z.B. Blech~* || chart * *grafische Darstellung* || blackboard * *Wand~*
Tafelblech plate * *Grobblech, dickes Blech*
Tafelglas sheet glass
Tafelschere guillotine shear * *Schlagschere für Blechtafeln*
Tagebau opencast working * *[Bergbau]* || opencast mining * *[Bergwerk]* || surface mining * *[Bergwerk]* || open-pit mine * *[Bergbau] ~-Grube, ~-Bergwerk*
Tages- daily
Tagesordnung agenda
täglich daily || everyday || a day * *(Adv) twice a day: zweimal ~* || day-to-day * *Einsatz, Gebrauch usw.* || per day * *pro Tag* || diurnal * *täglich (wiederkehrend), Tages-, Tag-, tageszeitlich; nur bei Tag auftretend*
tagsüber during daylight hours
Taiwan Taiwan
Takt clock || clock pulse * *~impuls* || cycle * *(bei Verbrennungsmotor, Fertigungslinie, Schweißgerät)* || phase || step * *out of step: außer ~* || rythm * *Rythmus* || tactfulness * *~gefühl* || tick * *~ eines Rechners* || interval * *at 10 minute intervals/every 10 minutes: im 10-Minuten-~*
takten clock
Taktfrequenz clock frequency || clock rate || switching frequency * *für Leistungsteil* || carrier frequency * *z.B. für PWM-Modulation, auch Trägerfrequenz* || pulse frequency * *Pulsfrequenz, z.B. einer Thyristor-Stromrichterschaltg.* || clock speed * *z.B. eines Mikroprozessors, einer getakteten Schaltung* || chopper frequency || switch frequency * *für Leistungsteil* || clock speed * *[Computer]* || chopper frequency * *eines Gleichstromstellers*

Taktgeber clock generator || clock pulse generator || clock
Taktgenerator clock generator
Taktik tactics || policy * *Verfahrensweise, Taktik*
taktisch tactical
Taktpresse fixed-cycle press * *[Holz] im Gegensatz zur kontinuierlichen Presse (für Spanplatten)* || cycle-controlled press * *[Holz]* || cycle press
Taktraster clock grid
Taktsignal clock signal
Taktung pulsing * *bei Stromrichter, Leistungsimpulse erzeugend* || clocking * *Taktsignale erzeugend/verarbeitend*
Taktversorgung clock supply
Taktzahl cycle frequency * *Anzahl Takte je Zeiteinheit (bei Verpackungsmaschine, Presse usw.)* || cycle rate
Taktzeit cycle time || machining period * *[Werkz. masch.]* || clock time || interval * *at 10 minute intervals: im 10-Minuten-Takt* || station time * *[Verkehr] bei Zug, Bus usw.*
Talfahrt downhill operation
Tambour jumbo reel * *[Papier]* || jumbo roll * *Wickeldorn für/mit Papier, Maschinenrolle* || tambour * *Wickeldorn für/mit Papier* || reeling drum || reel * *Maschinenrolle für/mit Papier* || paper reel * *[Papier]* || jumbo * *Wickeldorn für/mit Papier* || spool
Tambourstarter tambour starter * *[Papier]*
Tambourwechsel reeling drum exchange * *Wechsel volle gegen leere Papierrollle z.B. am Ende einer Papiermaschine* || reel change * *Wechsel der Papierrolle in Papiermaschine an der Aufrollung*
Tandem tandem
Tandem- tandem
Tandem-Motor tandem motor * *[el.Masch.] 2 Motoren i. 1 Gehäuse hintereinand. eingebaut, kleiner Durchm. u.Trägheit*
tangens delta loss tangent * *tangens zw. Wirk- u-Blindstrom, Maß für die Verluste im Dielektrikum eines el. Feldes*
Tangente tangent
tangential tangential
Tangentialkraft tangential force
tangieren touch || be tangent to || affect * *beeinflussen*
Tantal tantalum
Tantalkondensator tantalum capacitor
Tänzer dancer * *zur Zugregelung bei Wickelantrieben usw.* || dancer roll
Tänzerabstützung dancer loading * *z.B. durch Pneumatikzylinder einstellbare Tänzerbelastung (verändert den Bahnzug)*
Tänzerarm dancer arm
Tänzerbelastung dancer loading * *z.B. Abstützung durch Pneumatikzylinder mit einstellbarer Kraft zur Zugeinstellung*
Tänzerlage dancer position || dancer arm position
Tänzerlageregelung closed-loop dancer roll position control || dancer position control
Tänzerlageregler dancer position controller || dancer controller
Tänzerlagesignal dancer position signal || dancer feedback signal
Tänzerlagesollwert dancer position setpoint || dancer position reference
Tänzerpotentiometer dancer roll potentiometer * *erfasst Tänzerlage,häufig berührungslos arbeitend, z.B. Feldplattenpoti*
Tänzerpoti dancer potentiometer * *zur Erfassung des Ausschlags e. Tänzerwalze (Istwert für Tänzerlageregelung)*
Tänzerregelung dancer control * *(direkte) Zugregelung mit Tänzerwalze*

Tänzerregler dancer controller || dancer position controller * *Tänzerlageregler*
Tänzerrolle dancer roll || dancer
Tänzerwalze dancer roll || dancer
Taragewicht tare weight
Tariersatz balancing set * *zum Auswuchten*
Tarif tariff * *(mit 'ff')* || rate * *freight rate: Frachttarif* || duty || charge * *Gebühr* || scale * *für Löhne, Steuern usw.*
Tarifvertrag wage agreement || collective agreement * *(amerikan.)*
Taschenbürstenhalter box-type brush holder
Taschenlampe flash-light * *(amerikan.)* || pocket lamp || electric torch * *Stablampe* || torch * *Stablampe* || hand torch || lantern * *~ mit großem Reflektor, Laterne*
Taschenmesser pocket knife
Taschenrechner pocket calculator || electronic calculator
Tastatur keyboard || keypad * *kleines Tastenfeld*
Tastbetrieb inching * *Tippbetrieb eines Antriebs* || jogging * *Tippbetrieb eines Antriebs*
Taste key
Taste betätigen hit the key || press the key || type in * *über Tastatur eingeben*
Tastenbelegung key assignment
Tastenbetätigung keystroke
Tastenblock keypad
Tastenfeld keypad * *kleines ~* || keyboard * *Tastatur*
Tastenfolge keystroke sequence || key sequence
Taster pushbutton * *Druckknopf* || button || key * *Taste* || momentary switch || momentary-on type switch * *mit Schließerkontakt* || momentary-off type switch * *mit Öffnerkontakt* || momentary-contact switch
Tastkopf probe * *z.B. für Oszilloskop* || probing head || sensing probe * *z.B. für Näherungsschalter* || scanning head * *Abtastkopf (z.B. für Duckmarken)*
Tastlehre caliper
Tastperiode sampling period || sampling interval * *Abtastintervall*
Tastrolle dancer roll * *{Draht} zur Zugerfassung und -Regelung bei Drahtziehmaschine*
Tastverhältnis duty factor || pulse duty factor * *~ einer Impulskette* || duty cycle || mark-to-space ratio * *Puls-Pausenverhältnis eines impulsförmigen Signals* || pulse duty factor || mark-space ratio * *Puls-Pausenverhältnis eines impulsförmigen Signals*
Tastzirkel caliper * *Tastlehre zur Ermittlung von Innenmaßen*
Tatbestand fact
Tätigkeit activity || action || function * *Funktion* || occupation * *Beschäftigung; Besitz(-nahme)*
Tatsache matter of fact || fact
tatsächlich actual || real * *wirklich* || actually * *(Adv.)* || really * *(Adv.) wirklich* || factual || based on fact || in fact * *(Adv.)* || in reality * *(Adv.)* || the fact is that * *(einleitend)* || in real terms * *(Adv.)* eigentlich, in Wirklichkeit || in actual fact * *(Adv.)* in Wahrheit, in Wirklichkeit* || effective * *wirksam, in Kraft*
tatsächlicher Betrieb real service
Tau dew * *~ durch Kondensierung; siehe auch →Betauung* || rope || cable || hawser * *{Schifffahrt}*
Tauchdauer dwell time * *Verweildauer im Tauchbad*
tauchen dip * *auch in Löt-/Galvanisierbad usw.* || plunge * *plötzlich/kräftig ~* || douse * *plötzlich/ kräftig ~* || dowse * *plötzlich/kräftig ~* || dive * *tief ~*
tauchlöten dip-solder || dip-soldering * *(Substantiv)*
Tauchmotorpumpe submersible motor-pump

Tauchpumpe immersion pump || submersible pump
Tauchpumpenmotor submersible-pump motor
Tauchschmierung splash lubrication
Tauchspule plunger coil
Tauchumhüllung dip coating
taumeln reel * *wanken, taumeln* || stagger * *Schwanken, wanken, ~, torkeln, unsicher werden (figürl.), ins Wanken bringen, verblüffen* || tumble * *schleudern, Purzelbäume schlagen, durcheinanderwerfen, durchwühlen, umstürzen*
Taumelsiebmaschine tumbler screening machine * *{Holz} zur Trennung von Holzschnitzeln unterschiedlicher Größe*
Taupunkt dew point * *Temperatur, unterhalb derer die Betauung beginnt*
Taupunktkurve dew point curve * *Abhängigkeit der Taupunkt-Temp. von der absoluten Luftfeuchte*
Tausch replacement * *Ersetzen, Austausch (z.B. bei Reparatur/eines defekten Geräts)* || exchange || replacing * *das Austauschen, Ersetzen (z.B. von Verschleiß-/Ersatzteilen)*
tauschen exchange || replace * *aus~, z.b. im Reparaturfall* || swap * *miteinander ver~, aus~, z.B. zwei Bytes in e. Datenwort/Briefmarken*
Tauschgerät replacement unit
tausend thousand || thsnd. * *(Abkürzung)*
Tausenderstelle thousands digit
tausendfach thousand times || thousand fold
Taxi taxi || cab
TCP-IP TCP-IP * *Transmission-Control Protocol/ Internet Protocol: Datenübertr.protokollfamilie f. Internet..*
Teamarbeit teamwork * *by: in*
Teamgeist team spirit
Technik engineering * *angewandte* || technology * *Wissenschaft, Technologie* || technique * *Verfahren* || systems * *Systeme*
Techniker technician || engineer * *Ingenieur* || specialist * *Spezialist*
Technikgeschichte history of technology
technisch technical * *for technical reasons: aus ~en Gründen* || engineering
technische Anforderungen technical requirements || technical demands
technische Angaben technical specifications || specifications || technical data * *technische Daten* || technical information
technische Beschreibung technical description || reference manual * *technisches Referenzhandbuch* || instructions * *Betriebsanleitung*
technische Daten technical specifications || technical data || specification data || specifications
technische Dienste engineering services
technische Dokumentation technical documentation * *siehe DIN-Entwürfe 8418, 66055 und VDI-Richtlinie 4500*
technische Erläuterungen general technical information * *allgemeine ~*
Technische Hochschule university of technology || college of technology || technical university
technische Merkmale technical features
technische Pionierleistung pioneering technical performance || pioneering technical achievement || pioneering technical feat * *Großtat, Kunst/ Meisterstück* || pioneering technical accomplishment
Technische Redaktion Technical Editorial Office * *Abteilungsbezeichnung*
technische Textilien technical textiles
technische Voraussetzungen technical requirements
technische Zeichnung engineering drawing || blueprint * *(Blau)Pause* || drawing
technischer Fortschritt technological advance

Technischer Zeichner draughtsman || draftsman || designer
Technologe technologist
Technologie technology || high level control * *Überlagerte Regelung und Steuerung*
technologieabhängig technology-dependent || process-dependent
Technologiebaugruppe technology board * *Leiterkarte* || technology module || process-oriented PCB * *Leiterkarte* || process-oriented board * *Leiterkarte* || process-related module
Technologiefunktionen technology functions
Technologieparameter technology parameter
Technologieregler technology controller || process regulator || process-oriented controller || process controller || application controller
Technologieschema process flowchart
Technologiezusatz technological supplementary unit
technologisch technological || process-oriented || technology || process-related * *z.B. Kenngrößen* || process-specific
technologische Funktion technological function || process-oriented function
technologische Funktionen technology functions || process-oriented functions || application-specific functions
technologische Regelungen technological closed-loop controls
Teer tar
TEFC TEFC * *Abkürzg.f. 'Totally Enclosed Fan Cooled': vollk. geschlossen u.lüftergekühlt (Schutzart IP44)*
Teflon Teflon
Teigkneter dough kneader
Teigwalze dough roller * *zur Brot-/Gebäckherstellung*
Teil fraction * *Bruch~* || part * *Bau~, Artikel* || item * *Artikel* || piece * *Stück Ware* || spare part * *Ersatz~* || portion * *An~, Bruch~* || component * *~ einer zusammengesetzten Sache; Bestand~, An~* || section * *auch ~abschnitt, ~element* || particle * *Partikel, Körper, ~chen* || foreign body * *Fremdkörper* || element * *Element, Grundbestand~*
Teil- partial * *teilweise; An/Bruchteil* || component * *Bestandteil eines zusammengesetzten Ganzen; z.B. Teilwechselrichter je Phase* || element * *Element, Glied* || part- || sub- * *Unter-, untergeordnet* || reduced * *reduziert* || section * *Teilabschnitt, (Teil-) Element*
Teilabschnitt segment * *auch ~ eines Computernetzwerks*
Teilansicht partial view
Teilapparat dividing head
Teilaufgabe sub task
Teildrehzahlen part-speed * *under/during part-speed operation: bei Teildrehzahlen*
Teile des Nennwerts per unit
teilen share * *(sich) teilen (with: mit); teilhaben (lassen) (an: an)* || divide * *auch auf~, sich ~; auch Frequenz, Meinung* || split * *aufspalten* || dismember * *zerstückeln* || distribute * *ver/aus~* || portion out * *aus~ (in Portionen)* || separate * *absondern/trennen* || partition off * *absondern* || take part in * *teilhaben/nehmen an* || part * *sich ~* || branch out * *(Straße, Programmfluss usw.)* || fork * *sich ~/gabeln/spalten, z.B. Straße* || go halves * *sich in 2 Hälften ~* || be divisible by * *teilbar sein durch* || segregate * *trennen, (sich) absondern, ausschneiden, (sich) abspalten, s. chem. abscheiden, abgesondert*
Teilentladung partial discharge * *Mikrolichtbögen im Dielektrikum bei Überschreiten d. Durchschlagfestigkeit*

Teilenummer part number
Teileprogramm part program * *(NC-Steuerung)*
Teiler divider * *z.B. Frequenzteiler, Spannungsteiler* || denominator * *(mathemat.) Nenner* || divisor * *(mathemat.) Nenner, Divisor* || attenuator * *Spannungsteiler zur Abschwächung von Spannungssignalen*
Teilfunktion partial function || subfunction * *Unterfunktion*
teilgesteuerte Stromrichterschaltung non-uniform converter connection * *siehe IEC 146-1-1 u. EN 60146-1-1; z.B. für 1Q-DC-Stromrichter*
Teilhaber shareholder * *z.B. Aktionär* || partner * *an einer Firma* || associate * *an einer Firma* || participator * *Teilnehmer (in: an)* || joint proprietor * *Mitinhaber*
Teilkreis pitch circle * *z.B. bei Zahnrad*
Teilkreisdurchmesser pitch circle diameter * *z.B. eines Zahnrades* || pitch diameter * *z.B. bei Zahnrad*
Teillast partial load * *under partial load condition: bei ~* || part load || light load * *Schwachlast*
Teillastbereich partial load range
Teillastbetrieb part-load operation || part-load condition || operation at light loads || partial-load operation
Teillieferung partial delivery || part-delivery || installment * *eines Sammelwerkes (Buch, Nachschlagewerk, Katalog usw.)* || part shipment
Teillösung partial solution
Teilnahme participation * *in: an* || cooperation * *Mitarbeit* || attendance * *an einer Versammlung, Besprechung usw.*
teilnehmen participate * *in: an* || take part * *in: an* || join in * *zusammen mit anderen ~* || be present at * *anwesend sein* || attend at * *anwesend sein* || cooperate * *mitwirken (in: an), zusammenarbeiten (with: mit, to: zu e.Zweck, in: an), beitragen (to: zu)* || contribute * *beitragen* || cooperate
Teilnehmer participant * *an Kurs, Besprechung usw.* || node * *Knoten eines Datennetzes bzw. Kommunikationsbusses* || participator * *in: an* || partner * *Partner, Teilhaber, 'Kompagnon', Gesellschafter* || sharer * *Teilhaber* || member * *Mitglied* || student * *an Lehrgang, Kurs, Seminar* || competitor * *Mitbewerber (z.B. an Ausschreibung, Wettkampf)* || subscriber * *Telefonanschluss; ~ am Nachrichtennetz* || party * *(Vertrags)Partei, Teilnehmer/haber, Beteiligter, Fraktion* || those present * *(Plural) die Anwesenden* || station * *z.B. Gerät an einem seriellen Bus* || master station * *Master-Teilnehmer (z.B. an seriellem Bus)* || slave station * *Slave-Teilnehmer (z.B. an seriellem Bus)* || attending persons * *Anwesende*
Teilnehmeradresse node address * *bei Bussystem (z.B. PROFIBUS)* || slave address * *Slave-Teilnehmeradresse bei Master/Slave-Bussystem* || station address * *z.B. bei Bussystem*
Teilnehmernummer station number * *bei Bussystem*
Teilschere dividing shear * *Quer~ in Walzstraße/Blechbehandlungsanlage* || cross-cutting shears * *Quer~*
Teilspannungen reduced voltages * *reduzierte Spannungen*
Teilstrom component current
Teilstromrichter half-converter * *z.B. Brückenhälfte, Teilbrücke* || component converter || converter section || partial converter
Teilung division || separation * *Trennung* || dismemberment * *Zergliederung* || pitch * *(Zahnrad-, Kommutator-, Nuten-, Lochteilung usw.)* || spacing * *Abstandsteilung* || graduation * *Skalenteilung* || module width * *Breiten-Raster von Einschubmo-*

dulen || partition * *(Auf-, Ver-)Teilung, Trennung, Absonderung*
teilweise   partial * *partiell* || partially * *(Adv.)* || partly * *(Adv.) zum Teil* || in part * *(Adv.) teilweise, zum Teil* || in parts * *teilweise, zum Teil, in Teilen, Teillieferungen* || to some extent * *in gewissem Ausmaß* || in some cases * *manchmal, in einigen Fällen*
Teilwicklungsanlauf   part-winding starting * *{el. Masch.} Anlassen e. Käfigläufermot. mit nur einem Teil d. Ständerwicklg.*
Teilzeichnung   part drawing
Teilzeit-   part-time
Tel   Pho. || Tel.
Telefon   telephone || phone * *answer the phone: den ~hörer abnehmen; bo on the phone: am ~ sein*
telefonisch   by phone
Telefonverkauf   tele-sale
Telegramm   telegram * *auch Datentelegramm* || message * *Nachricht, Meldung, Sendung*
Telegramm an alle   broadcast telegram
Telegramm-Ausfallzeit   telegram failure time || permissible interval of time between two incoming data telegrams || telegram timeout * *Ansprechen der Telegramm-Zeitüberwachung*
Telegrammausfall   telegram failure
Telegrammausfallüberwachung   telegram failure monitoring || telegram timeout monitoring * *Zeitüberwachung der Telegrammpausen*
Telegrammausfallzeit   telegram failure time
Telegrammbearbeitung   processing of telegrams
Telegrammfolge   telegram sequence
Telegrammkopf   telegram header
Telegrammrahmen   telegram frame * *Aufbau eines Datentelegramms: Plätze für Nettozeichen plus Vorspann plus Nachspann*
Telegrammüberwachungszeit   telegram failure monitoring time
Telegrammverkehr   telegram data transfer * *auf Bus, serieller Verbindung usw.*
Telekommunikation   telecommunication
Telemetrie   telemetry * *drahtlose Messdatenübertragung* || telemetering * *drahtlose Messwertübertragung*
Teleskop   telescope
Teleskop-   telescopic
Teleskopieren   telescoping * *ineinander verschieben; z.B. beim Aufwickler: innere Lagen werden herausgepresst*
Tellerfeder   cup spring
Tellermesser   cutting disk * *siehe auch →Kreismesser*
Tellerschleifmaschine   disk sanding machine * *{Holz}*
Temperatur   temperature
Temperatur an der heißesten Stelle   hot spot temperature
Temperaturabhängigkeit   temperature dependence
Temperaturabnahme   temperature decrease || temperature drop * *Temperaturabfall*
Temperaturänderung   temperature change || variation of temperature
Temperaturanstieg   temperature rise || rise of temperature
Temperaturauswertung   temperature evaluation
Temperaturbeanspruchung   thermal stress || thermal cycling * *durch wechselnde Temperaturen/ Temperaturzyklen*
Temperaturbeiwert   temperature coefficient
Temperaturbereich   temperature range
Temperaturbeständigkeit   heat resistance || thermostability || thermal stability || resistance to heat
Temperaturdifferenz   difference in temperature
Temperatureinfluss   influence of ambient temperature * *Einfluss d. Umgebungstemperatur* || temperature dependency * *Temperatureinfluss*
Temperaturerfassung   temperature sensing
Temperaturerhöhung   temperature increase * *siehe auch →Übertemperatur*
Temperaturfühler   temperature sensor || thermistor * *Heiß- oder Kaltleiter (NTC oder PTC)* || RTD * *Abk.f. 'Resistance Temperature Detector': Widerstands-~ (NTC od. PTC)* || thermal sensor
Temperaturgang   temperature sensitivity || response to temperature changes || temperature coefficient * *Temperaturkoeffizient* || temperature stability * *Unempfindlichkeit gegenüber Temperaturänderungen/-schwankungen* || temperature excursion || temperature response || function of temperature
Temperaturgeber   temperature sensor || temperature transmitter
Temperaturgefälle   temperature gradient
Temperaturgleichgewicht   temperature equilibrium
Temperaturklasse   temperature class * *Zündgruppe T1..6 nach EN50014 bis -20, kennzeichn. jew. Gase ähnl. Zündtemperatur*
Temperaturkoeffizient   temperature coefficient
Temperaturkompensation   temperature compensation
temperaturkompensiert   temperature-compensated
Temperaturmesser   thermometer
Temperaturmessgerät   thermometer
Temperaturregelung   temperature closed-loop control || closed-loop temperature control
Temperaturregler   temperature regulator || temperature loop controller || temperature controller || thermostat
Temperaturschutz   thermal protection
Temperaturschwankungen   temperature fluctuations || fluctuations of temperature
Temperatursensor   thermal sensor || thermistor * *→Heiß- oder →Kaltleiter (PTC/NTC Widerstand)* || thermocouple * *→Thermoelement* || RTD * *Abk.f. 'Resistance Temperature Detector': temp.abh. Widerstand (Heiß-/Kaltleiter)* || thermal sensing device || temperature sensor
Temperatursturz   sudden temperature drop || thermal shock
Temperaturüberlauf   excessive temperature rise * *{el. masch.} Differenz d. Wicklungs- z.gemess. Temp. nach Motorabschaltg.*
Temperaturüberwachung   temperature monitoring
Temperaturverteilung   temperature distribution
Temperaturwächter   overheating protector || thermal protector * *Baueinheit von Temperaturfühler und Stromkreisunterbrecher* || thermal release * *Auslösegerät zum Übertemperaturschutz* || temperature monitor
Temperaturwechsel   change of temperature
Temperguss   malleable cast iron
temperieren   temper || anneal * *Metall* || set the temperature * *die Temperatur (richtig) einstellen*
tempern   temper * *Wärmebehandlung/Glühen zum Ändern des Gefüges*
Tempo   speed * *Geschwindigkeit* || pace * *Gangart; set/increase the pace: das ~ angeben/steigern* || rate * *Grad der Geschwindigkeit*
temporär   temporary
Tendenz   trend || tendency * *downward/upward tendency: ~ nach unten/oben*
tendieren   tend
TENV   TENV * *Abkürz.f. 'Totally Enclosed Non Ventilated': vollkomm. geschloss. o. Lüfter (z.B.Schutzart IP54)*
Teppichgarn   carpet yarn
Teppichherstellung   carpet manufacture

Teppichwebmaschine   carpet-weaving machine
Term   term * *in einer Formel*
Termin   appointed time * *festgesetzte Zeit* || appointment * *Verabredung, Zusammenkunft* || appointed day * *festgelegter Tag* || date * *Zeitpunkt, Datum; final date: äußerster ~; target date: Ziel~* || target date * *Zieldatum* || deadline * *äußerster Termin* || date of completion * *Fertigstellungs~* || term * *Frist* || time-limit * *Frist* || scheduled date * *geplanter Zeitpunkt*
Termin anberaumen   appoint a date || fix a date || schedule a meeting * *für Besprechung*
Terminal   terminal
Termine   schedules * *in Zeitplan* || dates
termingemäß   to schedule || in due time || on the due date || on time * *pünktlich*
termingerecht   to schedule || in due time || on the due date || on time * *pünktlich* || according to schedule || timely || in time
Terminologie   terminology || nomenclature
Terminplan   schedule || time schedule || schedules
Terminplanung   scheduling || date scheduling
Terminzusage   schedule promise || promised date || schedule confirmation
Termitenschutz   termite protection * *relevant z.b. bei Langzeitlagerung von el.motoren*
Terpentin   turpentine * *['törpentain]*
Terzanalyse   third-band analysis * *schmalbandige Messung des Geräuschspektrums*
Tesafilm   sellotape * *Klebeband aus durchsichtiger dünner Kunststoff- oder Zellstofffolie* || cellotape * *durchsichtiges Klebeband*
Test   test || debugging * *von Software*
Testbaugruppe   test module
Testbaustein   test block * *Funktionsbaustein*
Testbericht   test report
Testbetrieb   test mode || test operation
Testbuchse   test socket || measuring socket * *Messbuchse* || test jacket || test jack
testen   test
Testergebnis   test result || testing result
Testfeld   testing panel * *z.B. an Mikrocomputersystem*
Testhilfe   testing aid
Testhilfeprogramm   debugger * *zum Testen von u. zur Fehlerfindung in Softwareprogrammen*
Testlauf   test run
Testliner   kraft faced liner
Testmodul   test adapter
Testroutine   test routine
Testsignal   test signal
teuer   expensive || costly * *kostspielig, kostenaufwändig* || valuable * *wertvoll* || pricy * *(salopp)* || high-cost * *Produkt* || at a high price * *(Adv.)* || how much is it * *wie teuer ist es* ?
Text   text || message * *Melde-/Aufforderungstext auf einem Computerbildschirm*
Textanzeigegerät   text display unit
Texte   texts
Texteingabe   text entry * *in Rechner* || manual text input * *von Hand in einen Rechner*
Textil   textile
Textilanwendung   textile application
textiler Stoff   textile
Textilfabrik   textile mill
Textilherstellung   textile manufacture
Textilien   textiles
Textilindustrie   textile industry
Textilmaschine   textile machine
Textilmotor   textile motor * *[el. Masch.] m. Schutz geg. Faserablagerg. (niedr. Kühlluftgeschw., kein Fettaustritt)*
Textilveredelung   textile finishing
Texturiermaschine   texturizing machine * *[Textil] raut glatte Chemiefasern auf u.macht sie bauschig*

wie Naturfasern || texturing machine * *[Textil] raut glatte Chemiefasern auf und macht sie bauschig wie Naturfasern*
Textverarbeitung   text processing || text processor * *~sprogramm* || word processing
Textverarbeitungsprogramm   word processor || text processor || text editor * *einfaches (z.B. z. Eingabe v. Quellprogrammen) ohne Formatierungsmöglichkt.*
Textverarbeitungssystem   word processor || text processor
Textzeile   text line || row
Thema   subject * *Bericht, Brief, Besprechung, Tagung, Vortrag* || topic * *auch Gegenstand, Besprechungs~; current topics: aktuelle ~n* || theme * *Stoff; ~ e. wissenschaftlichen oder Examensarbeit; Musik~* || issue * *Kernpunkt, wesentlicher Punkt, Streit-/Diskussionspunkt*
theoretisch   theoretical || in theory * *(Adv.)* || theoretically * *(Adv.)*
theoretische Untersuchung   theoretical analysis
Theorie   theory
thermisch   thermal || thermally * *(Adv.)*
thermisch ausgenutzt   thermally utilized
thermisch verzögerter Überlastschutz   thermally delayed overload protection
thermische Belastbarkeit   thermal loadability * *z.B. eines Isoliersystems im Motor*
thermische Belastung   thermal stress
thermische Ersatzzeitkonstante   equivalent thermal time constant * *beschreibt d. Temperaturänderg. nach sprunghafter Belastungsänderg.*
thermische Klassifizierung   thermal classification * *Kennzeichn. d. →Wärmeklasse (früher: Isolierstoffklasse) e. Isoliersystems*
thermische Motorzeitkonstante   thermal motor time constant
thermische Reserve   thermal reserve
thermische Stabilität   thermal stability
thermische Überlastbarkeit   thermal overload capacity
thermische Überlastung   thermal overload
thermische Zeitkonstante   thermal rise time constant || thermal time constant * *Zeit,nach d. e. Körper bei e. Sprung d. Umgeb.temp. 63% d. Endtemp. erreicht*
thermischer Beharrungszustand   thermal steady state
thermischer Motorschutz   thermal motor protection * *[el. Masch.] Maschinenschutzeinrichtung z.Vermeidg. unzuläss. hoher Temp.*
thermischer Schutz   thermal protection * *siehe IEC 34 Teil 11 für in Motor eingebaute Schutzelemente*
thermischer Überlastschutz   thermal overload protection
thermischer Widerstand   thermal resistance
thermisches Abbild   thermal replica || thermal model
thermisches Gleichgewicht   thermal equilibrium
Thermistor   thermistor || PTC
Thermoauslöser   thermal trip unit * *thermisch verzögerter (Motorschutz-) Schalter*
thermodynamisch   thermodynamic
Thermoelement   thermocouple * *Temperaturmessgeber bestehend aus 2 verschieden miteinand. leitend verbunden.Metallen*
Thermofühler   thermo sensor || temperature sensor
Thermokontakt   thermal bimetal contact || thermal contact || thermistor contact
thermomechanisch   thermo-mechanical
Thermometerverfahren   thermometer method * *zur Ermittlg d.Temperatur oder Übertemperatur an d. Oberfläche m.Thermometer*

Thermopaar thermocouple * →*Thermoelement zur Temperaturmessung*
Thermoplast thermoplastic || thermoplastic material || thermoplastic resin
Thermoschalter thermo switch || thermal switch || thermostat || thermoswitch || thermal circuit breaker || thermostatic cut-out
Thermospannung thermal EMF * *Abk.f. 'Electro-Motive Force'* || thermoeletric voltage
Thermostat thermostat || thermal switch
Thomson-Messbrücke Thomson measuring bridge
Thyristor thyristor * *['theiriste] siehe DIN 41786/87* || SCR * *silicon controlled rectifier* || silicon controlled rectifier
Thyristor-Stromrichter thyristor converter
Thyristor-Umkehrsteller thyristor reversing controller
Thyristorbaustein thyristor module || thyristor block
Thyristorblock thyristor module
Thyristorcheck thyristor check
thyristorgesteuerter Gleichstromantrieb thyristor fed DC drive
Thyristormodul thyristor module
Thyristorsatz thyristor set || thyristor assembly || thyristor stack
Thyristorsätze thyristor sets || thyristor assemblies
Thyristorschrank thyristor cubicle
Thyristorschutz thyristor protection
Thyristorsicherung SCR type high speed fuse * *silicon-controlled rectifier: zum Thyristorschutz* || fast 'semiconductor' type fuse * *Halbleitersicherung* || fuse for protection of thyristors || high speed SCR fuse
Thyristorsteller thyristor power controller || thyristor controller || thyristor AC controller * *Wechselstromsteller*
Thyristortablette thyristor wafer
tief deep * *(auch figürl.: Erkenntnisse, Wissen, Datenspeicher usw.)* || profound * *Wissen, Kenntnisse* || low * *niedrig* || in-depth * *in die Tiefe gehend (z.B. Kenntnisse)*
Tiefdruck platen-printing || rotogravure printing * *Rollentiefdruck* || rotary photogravure * *Rollentiefdruck*
Tiefdruck-Rollendruckmaschine rotogravure press * *{Druckerei}*
Tiefdruckpapier gravure paper || copper plate paper
Tiefdruckzylinder gravure cylinder * *{Druckerei}*
Tiefe depth * *['däbth] (auch figürlich) in-depth: in die ~ gehend* || profoundness * *(figürlich) Wissen, Kenntnisse, Erfahrung* || profundity * *(figürlich)* ~ *von Wissen, Kenntnissen, Erfahrung* || deep * *Abgrund* || abyss * *Abgrund*
Tiefenschnittsäge rip-cut saw * *{Holz}*
tiefer lower
Tiefer-Taste decrease button || lower key || lower button
Tiefer-Taster lower button * *z.B. für Motorpotentiometer*
tiefgehend in-depth
tiefgezogen deep-drawn
tiefkühlen freeze
Tieflaufzeit ramp-down time * *eines Hochlaufgebers*
Tiefnutläufer high bar rotor * *{el. Masch.} bei Käfigläufermotor (Stromverdrängungsläufer)*
Tiefpass low pass filter || low-pass filter * *first-order low-pass filter: ~ erster Ordnung*
Tiefpassfilter low pass filter
Tiefseemotor deep-sea motor * *{el. Masch.} zum Betrieb in großen Meerestiefen, z.B. zur Gas-/Ölförderung*
Tieftemperaturfett low-temperature grease
tiefziehen deep-draw || thermoforming * *von Kunst-*

stoff (z.B. für Verpackungszwecke, Yogourtbecher, Armaturentafeln)
Tiegel platen * *{Druckerei} Drucktiegel/-Platte für Bogendruck* || crucible * *Schmelz-/Stahl~; Feuerprobe (figürl.)*
Tierhaare animal hairs * *{Textil}*
Timing timing
Timing einhalten observe the timing || keep within the timing || maintain the timing * *beibehalten, aufrechterhalten*
Timing-Diagramm timing chart || timing diagram
Tinte ink
Tintenschreiber ink recorder
Tintenstrahldrucker ink-jet printer * *{Computer}*
Tintenstrahlschreiber ink-jet recorder
Tipp hint || tip || pointer * *Fingerzeig, Tipp*
Tippbetrieb jogging || inching || inching mode || inching operation
Tippdrehzahl jog speed || inching speed
tippen inch * *kurzzeitige Aufschaltung eines (niedrigen) Drehzahl/Frequenzsollwerts über Steuerkommando* || jog * *kuzzeitige Aufschaltung eines (niedrigen) Drehzahl/Frequenzsollwerts über Steuerkommando* || inching * *(Substantiv)* || jogging * *(Substantiv)*
tippen rückwärts inch reverse || jog reverse || jog backward
Tippen schnell progressive jog
tippen vorwärts inch forward || jog forward
Tippgeschwindigkeit jog speed || inching speed
Tippschaltung jog control || inch control
Tippsollwert inching setpoint || inch reference || jog reference || jogging setpoint
Tisch desk || table || platform * *Plattform* || bench * *Werkbank, Werk~, Experimentier~*
Tischbohrmaschine bench-type drilling machine
Tischdrehmaschine bench lathe || bench-type lathe
Tischfräse spindle moulder * *{Holz}*
Tischfräsmaschine single-spindle moulder * *{Holz}*
Tischgerät desktop model || bench-type * *für den Labortisch*
Tischkreissäge circular saw bench * *{Holz}*
Tischleitblech table-guide plate * *z.B. bei Richtmaschine*
Tischler joiner || cabinet maker * *Möbel~, Kunst~* || carpenter * *Zimmermann, Tischler*
Tischlerei joinery * *Handwerkszweig* || joiner's workshop * ~*werkstatt* || cabinet shop * *Möbel~, Kunst~*
Tischlereimaschine joiner's machine
Tischlerkreissäge joiner's circular saw
Tischlerplatte blockboard * *{Holz}* || battenboard * *{Holz}* || coreboard * *{Holz}* || laminboard * *{Holz} mit Stäbchenzwischenlage*
Tischmesssystem table measuring system * *(in der Werkzeugmaschinentechnik)*
Tischverstellung table adjustment * *{Werkz.masch.}*
Tischvorlage handout * *bei Besprechung verteilte/ausgehändigte Unterlage*
Tissuemaschine tissue machine * *{Papier} zur Herstellung von Hygienepapier*
Titer titre size * *{Textil} Fadendicke*
TN-Netz TN system * *geerdetes Netz mit Schutzleiter-Verbindung zwischen den Sternpunkten (VDE 0100 Teil 410)* || grounded system with protective conductor || grounded supply || grounded system
Tochtergesellschaft subsidiary || subsidiary company
Toilette men's room * *Herren~* || ladies' room * *Damen~* || gentlemen's room * *Herren~* || lavatory || washroom || toilet * *(besonders i. Amerika)*
Token token * *Sendeberechtigung bei einer seriellen Kommunikation*
Token Passing token passing * *Weitergabe der Sendeberechtigung z.B. von einem Busteilnehmer zu e. anderen*

Token-Weitergabe token passing * *Weitergabe der Sendeberechtigung bei einem (seriellen) Bussystem*
Tokenweitergabe token-passing * *Weitergabe der Zugriffsberechtigung von Teilnehmer zu Teilnehmer bei Bussystem*
Toleranz tolerance * *out of: außerhalb der; close: enge* || limit of error * *check for specified limits: auf Einhaltung d.toleranzen prüfen* || maximum error || deviation * *Abweichung* || allowance * *zulässige Abweichung, Spielraum* || allowable variation * *zulässige Abweichung* || permissible variation * *zulässige Abweichung* || correct clearance * *zulässiges Spiel* || toleration * *Duldung, Tolerierung* || tolerant attitude * *tolerante(r) Standpunkt, Geisteshaltung, Einstellung* || permissible variation * *erlaubte Schwankungsbreite*
Toleranzausgleich tolerance compensation * *z.B. bei Werkzeugmaschine* || compensation of tolerances
Toleranzband tolerance band || tolerance range * *Toleranzbereich* || error band || tolerance window * *Toleranzfenster*
Toleranzbereich tolerance range || tolerance band * *Toleranzband*
Toleranzfeld tolarance zone || tolerance band * *Toleranzband*
Toleranzgrenze tolerance limit
Toleranzring tolerance ring
tolerierbar tolerable
tolerieren tolerate
toll great * *(salopp) groß(artig), prima, famos, herrlich, überragend* || marvellous * *(salopp) fabelhaft, wunderbar* || fantastic * *(salopp) fantastisch* || terrific * *(salopp) fantastisch, gewaltig* || fabulous * *(salopp) fabelhaft* || smashing * *(salopp) sagenhaft, umwerfend, prima* || a hell of a * *(salopp) verdammt gut*
Tonfrequenz audio frequency
Tonnage tonnage
Tonne tun || barrel || cask || ton * *Gewichtseinheit; siehe auch 't (1 Brit. ton entspr. 1016 kg; 1 US-ton entspr. 907,2 kg)* || metric ton * *Gewichtseinheit "metrische Tonne", siehe auch 't (1 metric ton entspr. 1000 kg)* || tonne * *Gewichtseinheit "metrische Tonne" (entspricht 100 kg)*
Tonwahlverfahren touch-tone dialing
Tonwählverfahren touch-tone dialing * *Telefonwählverfahren (→Frequenzwählverfahren, →Pulswählverfahren)* || tone dialing * *Telefonwählverfahren: für jede Ziffer wird e.entsprechende Tonfrequenz übertragen*
Topfbauform cup type
Topfgehäuse cup housing * *Gehäusebauform z.B. einer Kupplung*
Topfmotor canned motor * *B-seitiges Lagerschild bildet mit d.Gehäuse eine Einheit → hohe Schutzart (z.B.IP67)*
Topologie topology * *Beziehg. einzeln. Komponenten bezügl.d. räuml./örtlich. Anordnung oder d. Datenflusses*
Tor gate * *(auch figürl.)* || portal * *Portal, Tor (auch figürl.), (Haupt-) Eingang* || gateway * *(auch figürl.) Einfahrt, ~weg*
Torsion torsion
Torsionsschwingung torsional vibration
Torsionsschwingungen torsional vibrations
torsionssteif torsionally rigid
Torsionssteifigkeit torsional stiffness || torsion resistance
Torsteuerung gate control * *z.B. bei der Verarbeitung von Impulsen*
Torx-Schraube TORX screw
Totalausfall total failure || catastrophic failure * *plötzlich auftretender schwerer Totalausfall* || in-

operative * *totaler Funktionsfehler* || total loss || dead loss
Totband dead band || dead zone * *Totzone* || neutral zone
totlegen disable * *sperren* || block * *sperren* || kill * *zum Absterben bringen, unterdrücken* || suppress * *unterdrücken* || eliminate * *totmachen, eliminieren*
Totpunkt dead center * *z.B. bei Kolbenmaschine, Exzenterpresse usw.; top: oberer, bottom: unterer*
Totzeit dead-time || dead time || latency * *Latenz-/Verzugszeit z.B. bei der Reaktion auf ein Ereignis in einem Rechner* || delay * *Verzögerungszeit*
Totzeitglied delay block * *Funktionsbaustein* || delay element * *Verzögerungsglied*
Totzone dead zone || dead band * *Totband*
Toxität toxity
Trace trace * *(meist triggerbarer) Speicher zum Aufzeichnen von Signalen über eine gewisse Zeit*
Trace-Speicher trace memory || trace buffer
Traceaufzeichnung trace recording
Tracepuffer trace buffer
Tracespeicher trace memory || trace buffer
Trafo-Nennspannung rated transformer voltage
Trafoeinspeisung transformer incoming feeder
Trafoleistung transformer capacity
Tragarm supporting arm
Tragbalken spider arm
tragbar portable * *in der Hand ~ vom Gewicht/Volumen her, z.B. Gerät, Rechner* || bearable * *(figürl.) (er)tragbar* || acceptable * *annehmbar* || reasonable * *zumutbar* || within reason * *im Rahmen des Tragbaren* || mobile * *nicht ortsfest*
Tragbild contact pattern * *bei Zahnradflanken, Lagerlauffläche usw.* || appearance of bearing surface * *eines Gleit-/Wälzlagers*
Tragblech support plate || mounting plate || supporting plate || chassis * *Grundplatte; Fahrgestell*
träge time-lag * *Sicherung* || slow-acting * *Sicherung* || delayed || sluggish * *träge, langsam, schwerfällig*
Tragegriff lifting handle || carrier handle
tragen carry * *auch 'stützen, Lasten ~'* || take * *mitnehmen* || convey * *(be)fördern* || transport * *befördern* || lift * *heben* || support * *stützen* || uphold * *stützen* || hold * *stützen* || bear * *hervorbringen; Verantwortung, Folgen, Verlust, Kosten usw. tragen* || yield * *ein-/hervorbringen (Ernte, Gewinn, Früchte, Zinsen, Resultat)* || produce * *hervorbringen (Gewinn, Zinsen, Früchte)* || include * *einschließen* || comprise * *enthalten, beinhalten* || incorporate * *eingebaut haben*
tragendes Teil supporting element
Träger girder * *Tragbalken, Stahl~* || carrier * *z.B. bei Modulation, Ladungsträger* || support * *Stütze, Ständer, Strebe, Absteifung, Bettung, Stativ* || holder * *Halter, Haltevorrichtung* || bracket * *Wandkonsole* || mount
Trägerband carrier tape * *z.B. zur Gurtung von Bauteilen*
Trägerfrequenz carrier frequency
Trägermodul supporting module || support module || motherboard * *Trägerleiterkarte, Hauptplatine*
Trägerschicht substrate * *Trägermaterial, Unter-/Grundlage, Unterschicht, Träger, Medium, Grundschicht, Substrat* || substratum * *Trägermaterial*
Tragfähigkeit load-carrying capacity * *auch mechan. ~, z.B. von Wälz-/Gleitlagern, Zahnrädern* || load rating || carrying capacity * *z.B. bei Förder-/Hebezeug* || safe load * *~ von Seil, Hebezeug, Brücke* || load capacity || maximum payload * *maximale Nutzlast* || lifting capacity * *bei Hebezeug, Kran* || tonnage * *~ eines Schiff* || buoyancy * *Schwimmkraft, Auftrieb* || bearing strength * *~ eines Gleit-/Wälzlagers* || lifting power * *bei Hebe-*

zeug, Kran || breaking strength * Trag-/Bruchfestigkeit
Traggriff handle || lifting handle
Trägheit inertia * *Massenträgheit* || laziness * *Faulheit* || indolence * *Faulheit* || inactivity * *Untätigkeit* || slowness * *Langsamkeit (auch geistige)* || sluggishness * *Langsamkeit, Trägheit, Schwerfälligkeit*
trägheitsarm low-inertia
Trägheitsfaktor factor of inertia * *(el. Masch.)* Verhältnis Gesamtträgheitsmoment zu Motorträgheitsmoment
trägheitslos inertialess || instantaneous * *trägheitslos/unverzögert ansprechend*
Trägheitsmoment inertia * *[kg x m²]* || moment of inertia
Trägheitsmoment-Kompensation inertia compensation
Trägheitsmoment-Vorsteuerung inertia compensation || inertia precontrol
Trägheitsradius radius of gyration
Traglager thrust bearing
Tragöse lifting eye * *z.B. an el.Masch. (ab 30 kg Gewicht) zum Einhängen des Kran-Lasthakens*
Tragrohr bracing tube || suspension pipe || pylon bracing
Tragrolle bearing pulley || bogie wheel
Tragscheibe support disk
Tragschiene mounting rail * *siehe auch →Installationsschiene* || supporting rail || DIN rail * *C- bzw.G-Aufschnappinstall.schiene n. DIN EN50022 (35 mm breit) bzw. DIN EN 0035 (32mm)*
Tragsicherheit loading ratio * *Kenngröße für Wälzlager, bestimmend für dessen Lebensdauer*
Tragwalze drum * *(Papier) z.B. bei →Doppeltragwalzenroller, →Poperoller* || transportation roller
Tragwalze 1 back drum * *eines Doppeltragwalzenrollers* || rear drum * *eines Doppeltragwalzenrollers*
Tragwalze 2 front drum * *eines Doppeltragwalzenrollers*
Tragwalzenroller drum winder || drum weel-up
Traktion traction
Traktionsantrieb traction drive
Tränk-Isoliermittel impregnation insulation material
tränken impregnate * *z.B. Wicklungsdraht mit Isolier-Tränkharz* || soak * *sich vollsaugen, durchtränkt werden, durchtränken, aufsaugen, imprägnieren*
Tränkharz impregnating resin * *z.B. zur Wicklungsisolierung in Trafo, Motor* || insulating enamel
Tränkmittel impregnating material || impregnation compound || impregnant || impregnating agent
Tränktechnik impregnating technique
Tränkung impregnation * *Füllung einer Wicklung mit Isolierharz oder -Lack o. anschließender Aushärtung*
Transaktion transaction
Transduktor transductor * *Transduktor (-drossel), (vormagnetisierte) Regeldrossel*
Transfer transfer
Transferfunktion transfer operation * *bei SPS*
transferieren transfer
Transferlinie transfer line * *Transferstraße*
Transferstraße transfer line
Transformation transformation
Transformationsmatrix transformation matrix
Transformationsverhältnis transformation ratio
Transformator transformer
Transformator mit belastbarem Sternpunkt transformer having a loadable neutral
Transformatorleistung transformer capacity
Transformatorschaltung transformer connection

transformieren transform * *(auch mathem. u. physikal.)* || step down * *heruntertransformieren* || step up * *hochtransformieren*
Transformierte transform
transient transient * *vorübergehend/in einem Übergang auftretend*
Transientenrecorder transient recorder
Transistor transistor * *(Abk.f. TRANSfer resistOR') transistorized: transistorisiert, mit ~en ausgestattet*
Transistorausgang transistor output
Transistorpulsumrichter transistor PWM converter * *mit pulsweitenmodulierter Ansteuerung*
Transistorschalter transistor switch
Transistorsteller transistor chopper * *[Stromrichter] z.B. für Servo-DC-Antriebe*
Transmission transmission
Transnormmotor non-standard motor * *AC Motor, in d.Leistung über d.Normreihe (DIN 42669..81) liegend (über ca.200kW)* || trans-standard motor * *AC Motor, in der Leistung über die Normreihe liegend (über ca.200kW)*
transparent transparent || diaphanous
transparent machen provide a clear insight to || gain a clear insight * *sich ~*
Transparentfolie transparency * *für Overhead-Projektion*
Transparenz transparency
Transport carriage * *von Frachtgut* || conveyance * *Beförderung, Fördern, Übersendung, Zufuhr* || transport || transportation * *(amerikan.)* || shipment || haulage * *mit Eisenbahn, Lastwagen usw.* || transfer * *von Daten*
Transportbahn transport track * *in Transferstraße, z.B. Rollenbahn*
Transportband conveyor belt
Transportbehälter transport container
Transporteinheit transport unit || transportable unit
Transporteinrichtung transport equipment
Transportfahrzeug transport vehicle || transportation vehicle
Transportkette conveyor chain || carrier chain
Transportöse lifting eyebolt
Transportpapiere transport documents * *siehe auch →Begleitpapiere*
Transportplatte transport plate * *z.B. in Transferstraße* || pallet * *→Palette*
Transportriemen transport belt
Transportrolle pinch roll * *z.B. in Walzwerk/ Stranggießanlage* || drive roller
Transportrollgang transport roller table || roller table || roller conveyor || roller transporter
Transportschaden transport damage
Transportsicherung shipping brace || bearing block * *in Form einer Lagerverriegelung zur Vermeidung v.Lageschäden*
Transportsystem transport system * *auch bei serieller Kommunikation*
Transporttarif transport tarif
Transportweg route || transport path
Transvektor-Regelung TRANSVEKTOR control * *Vektorregelung (Warenzeichen der Fa. SIEMENS)*
Transvektorregelung TRANSVEKTOR control * *Vektorregelung (Warenzeichen d. Fa. SIEMENS)*
transversal transversal
Transversalflussmotor motor with transversal flux
trapezförmig trapezoidal
Trapezfunktion trapezium function
Trapezgewinde acme thread * *z.B. bei Spindeltrieb, Hubspindel* || trapezoidal thread
Trapezgewindespindel acme-screw spindle * *z.B. bei Spindeltrieb*
Träufelwicklung fed-in winding * *Wicklungsart bei Motor: Drähte werden von Hand in die Nut*

"eingeträufelt" || mush winding * Wicklungsart bei Motor: Drähte werden von Hand in die Nut "eingeträufelt" || random-wound winding
Traverse cross-arm || beam || crossbeam
traversieren traverse * sich hin- und herbewegen
Traversiergeschwindigkeit traversing speed || scanning speed * z.b. bei berührungsloser Messanlage in der Papierindustrie
treffen meet * sich ~ || encounter * auf etwas ~ || hit * schlagen, stoßen (auf), passen, finden, erreichen || strike * (auf/an/ein)schlagen, zufällig ~/entdecken || hit hard * empfindlich ~
Treffer hit
Treffpunkt meeting point || meeting place || meeting location * z.B. für eine Besprechung
treiben drive * mechan. u. elektrisch (Leistungsverstärker usw.) || drive * (Substantiv) || cause to flow * zum Fließen veranlassen, z.b.Strom (through: durch) || emboss * Metall || carry on * Geschäfte, Handel ~
treibend driving || input * Maschinenteil
treibende Kraft driving force || moving force
Treiber driver * z.B. Leitungs~, Software~, Device~ für Computer-Peripherie || buffer * Vertärkerstufe für Datensignale/-leitungen (meist mit Speicherwirkung) || handler * Software-Hantierungsbaustein || amplifier * Verstärker || device driver * {Computer} Peripheriegeräte~ || pinch roll * {Walzwerk} Band-Transportrolle
Treiberfähigkeit driving capacity * z.B. bei Leistungstreiber/Digitalausgang usw. || fan out * eines Logikschaltkreises
Treibersoftware driver software
Treiberspalt pinch roll gap * Walzwerk
Treiberstufe driver circuit
Treibhülse driving sleeve * z.b. zum Aufdrücken eines Lagers auf eine Welle
Treibriemen driving belt || transmission belt
Treibrolle pinch roll * {im Walzwerk}
Treibscheibe driving pulley * Riemen-/Seilscheibe (auch bei Aufzügen)
Treibstoffverbrauch fuel consumption
Treibwalze drive roll || driving roll || pinch roll * {Walzwerk}
Trend trend || mainstream * allgemeiner/langanhaltender ~ || tendency * Tendenz (to: zu/für)
Trendanalyse trend analysis || historical trending
trennen disconnect * Leitung, Stromkreise, Nachrichtenverbindung ~ || isolate * isolieren (auch Stromkreise), absondern (from: von), chem. rein darstellen, abschließen || interrupt * unterbrechen, (Stromkreise) trennen || separate * (ab)trennen (from: von), (ab)spalten, absondern, unterscheiden zwischen, zentrifugieren || divide * teilen || segregate * ab~, isolieren, absondern, sich abspalten/absondern || dissolve * auflösen
Trenner disconnector || isolating switch
Trennfläche joint face || plane of separation
Trennfuge parting line || separating line || plane of separation || slit * →Fuge, →Schlitz || isolating joint * z.B. in Motorläufer
Trennkondensator blocking capacitor
Trennkreissäge circular resaw * {Holz} || resaw * {Holz} || circular rip saw * {Holz}
Trennlinie separating line
Trennsäge resaw * {Holz} || cross-cut saw * {Holz}
Trennschalter circuit breaker * zum Abschalten eines Starkstromkreises || isolation switch || disconnecting switch || disconnector
Trennschleifmaschine cutting-off machine || cutting-off grinder
Trennstelle isolation point || ~ mit galvanischer Trennung

Trenntrafo isolation transformer || isolating transformer
Trennung separation * auch von Stoffen/Bestandteilen || disconnection * Lösen einer (auch elektr.) Verbindung || isolation * Isolierung, Absonderung, galvanische ~, Abschließung || segregation * Absonderung || division * Teilung
Trennverstärker isolating amplifier || signal isolator * zur potenzialgetrennten Signalübertragung || isolation amplifier
Trennwandler isolating transformer * z.B. Trenntransformator, potenzialfreier Strom/Spannungswandler || isolating transducer
Trennzeichen separator || separating character
treppenförmig stair-step || stepped || terraced || in the form of stairs
Treppenfunktion stepped characteristic || staircase function
Treppenstufe staircase step
Triac Triac * Zweirichtungs-Thyristortriode, DIN 41787 Tl.2 || bi-directional triode thyristor
Trichter funnel * pour through a funnel: durch ~ gießen; funnel-shaped: ~förmig || hopper * Einfüll~ || infeed hopper * Einfüll~
trichterförmig funnel-shaped
Trick trick || artifice * List, Kunstgriff, List
Triebwagen rail car * {Bahn} || traction vehicle || prime mover * Straßenbahn~ || motor carriage * Straßenbahn~ || rail-car * {Bahn}
Trigger-Ereignis trigger event
Triggerbaustein trigger block * Funktionsbaustein
Triggerbedingung trigger condition * bei Logikanalysator, Störspeicher/schreiber, Registriergerät
Triggerung triggering
Triggerzeitpunkt trigger time instant || trigger instant
trigonometrische Funktion trigonometrical function
trimmen trim * i.d. richtige Lage bringen, z.B. e.Tänzer durch Tänzerlageregler, der die Drehz. korrigiert
Trimmer trimmer
Trimmkondensator trimming capacitor || trimmer capacitor
Trimmpotentiometer trim potentiometer || trim pot || trimmer
Trimmpoti trim pot
Trinkgeld tip || tips
Trittgrenzen stability limits * Aussteuergrenzen für netzgeführten Stromrichter || commutation limits * Aussteuerwinkelgrenzen bei netzgeführtem Stromrichter || firing limits * Aussteuergrenzen für netzgeführten Stromrichter
trocken dry
Trocken-Luftfilter dry-type air filter
trockene Wärme dry heat
trockenes Ende dry end * einer Papiermaschine
Trockengehalt dry content
Trockengruppe dryer section * bei Papiermaschine
Trockenkupplung dry-running clutch
trockenlaufend dry running * z.B. Kupplung
Trockenmittel desiccant || desiccation agent * desiccation bag: ~beutel || drying agent
Trockenofen drying oven || baking oven || drier furnace || seasoning kiln
Trockenpartie dryer section * bei Papiermaschine || drying section * in Papiermaschine
Trockentrommel drying drum
Trockenzylinder dryer cylinder || Yankee cylinder * großer Trocken-/Glättzylinder bei Papiermaschine (bis 6 m Durchmesser) || drier cylinder || drying cylinder || drying drum * z.B. in der Papierindustrie || drying roll
trocknen dry || dry up || desiccate * (aus-) trocknen/

dörren ‖ season * ~ *durch Lagerung* ‖ dehydrate * ~ *durch Wasserentzug* ‖ drain * *trockenlegen (Land usw.)*
Trockner dryer ‖ drying machine ‖ desiccator ‖ drier ‖ dehumidifier
Trocknung drying
Trocknungsanlage drying plant ‖ drying system
Trocknungseinrichtung drying equipment
Trocknungstrommel drying drum
Trommel drum
Trommelbelag drum lining * *z.b. bei Antriebstrommel eines Förderbandes*
Trommelhacker drum chipper * *(Holz) zur Herstellung von Holzschnitzeln* ‖ drum hog * *(Holz) zur Erzeugung von Holzschnitzeln*
Trommellüfter drum-type fan ‖ centrifugal fan * *Walzenlüfter, Fliehkraftlüfter*
Trommelmotor external rotor motor * *Außenläufermotor* ‖ drum-integrated motor * *in Antriebstrommel integriert (z.B. für Förderband)* ‖ motor-pulley * *Außenläufer-Motor z.b. für Bandförderer*
Trommelofen rotary furnace ‖ drum-type oven ‖ revolving-cylinder roaster * *Röst-/Abschwelofen*
Trommelputzwalze fancy roller * *(Textil) bei Kratzenraumaschine*
tropenfest tropicalized ‖ tropics-proof
tropenfeste Isolierung tropicalized insulation * *z.B. einer Motorwicklung* ‖ tropics-proof insulaton
Tropenschutz tropic proofing
Tropenschutzverpackung tropicalized package ‖ climate package * *Klimaverpackung*
Tropfen drop * *einer Flüssigkeit; drop-shaped: ~förmig* ‖ gob * *Schleimklumpen, Auswurf, flüssiger Glas~ (zur Hohlglasherstellung)*
Tropfenverteiler scoop * *in der Hohlglasherstellung*
tropffrei drip-free
Tropfwasser dripping water ‖ drip water
tropfwassergeschützt drip-proof * *Schutzarten IP21 und IP22* ‖ DP * *Abkürzg. für 'Drip-Proof'; Schutzarten IP21 und IP22* ‖ DPBV * *'Drip Proof Blower Vented': ~e Motorschutzart mit Anbaulüfter* ‖ DPFG * *'Drip Proof Fully Guarded': ~e Mot.-schutzart mit Einbau-Fremdlüfter*
Tropfwasserschutz drip-water protection
tropisch feucht tropically humid
tropische Klimata tropical climates
tropisches Klima tropical climate
trotz despite ‖ in spite of ‖ in the face of * *gegen Widerstände* ‖ in the teeth of * *gegen Widerstände* ‖ for all that * *~ alledem* ‖ for all these efforts * *~ aller Bemühungen*
trotz Schwierigkeiten despite difficulties
trotzdem nevertheless * *(Adv)* ‖ in spite of this * *(Adv)* ‖ all the same * *(Adv)* ‖ though * *(Adv., nachgestellt)* ‖ although * *(Bindewort)* ‖ even though * *(Bindewort)* ‖ notwithstanding that * *(Bindewort)*
trüb dull * *(auch figürl.: glanzlos, unklar)* ‖ muddy * *schlammig, trüb(e) (auch Licht usw.)*, ‖ turbid * *dick(flüssig), trübe, schlammig, verschwommen (figürl.)* ‖ cloudy * *Flüssigkeit usw.* ‖ hazy * *dunstig, diesig* ‖ thick * *dick(flüssig), neblig, trübe (auch Wetter)* ‖ bleak * *(figürl.) trostlos, trüb, düster (Aussichten usw.)* ‖ dim * *(halb)dunkel, düster, trübe, undeutlich, verschwommen, schwach, blass, matt (Farbe usw.)*
Trübung muddiness ‖ turbidness ‖ turbidity ‖ cloudiness ‖ dimness ‖ haze * *Dunst, leichter Nebel*
trudeln coast ‖ coast freely ‖ windmill * *(salopp) mitgezogen werden, ohne Antriebsmoment freilaufen (z.B. Motor)* ‖ spin * *z.B. freilaufender Motor*
Trum end of rope * *eines Seils* ‖ end * *eines Seils/eines Riemens; slack trum: schlaffes ~; driving end: straffes ~* ‖ strand * *einer Kette, Eines Riemens; siehe auch* →*Obertrum*
Trumm end * *eines Seils, eines Riemens; siehe auch* →*Trum* ‖ strand * *einer Kette/eines Riemens; siehe auch* →*Trum*
Tschechische Republik Czech Republic
TSE-Beschaltung snubber network * *RC-Beschaltg. geg. d. 'TrägerSpeicherEffekt' b. Abkommutier. e. Thyristors* ‖ surge suppression network * *RC-Beschaltg. geg. d. 'TrägerSpeicherEffekt' b. Abkommutier. e. Thyristors* ‖ surge suppression circuit * *RC-Beschaltg. geg. d. 'TrägerSpeicherEffekt' b.Abkommutieren e.thyristors* ‖ snubber circuitry ‖ snubber circuit ‖ RC snubber circuit ‖ snubber * *(salopp)* ‖ thyristor snubbers * *Schutzbeschaltung für Thyristoren* ‖ suppressor circuit
TSE-Beschaltungselemente surge-suppressor components * *verhindern den TSE- (Träger-Speicher-) -Effekt bei e. Thyristorschaltg.*
TSE-Schutzbeschaltung snubber network * *RC-Beschaltg. gegen d. 'TrägerSpeicherEffekt' b. Abkommutieren e. Thyristors* ‖ surge suppression network ‖ snubber circuit ‖ RC snubber circuit
TSR-Programm TSR program * *Abk.f. "Terminate and Stay Resident": verbleibt auch nach Beendigung im Speicher*
TT-Netz TT system * *auf Einspeise- u.Verbraucherseite geerdeter Sternpunkt, jedoch ohne Schutzleit. dazwisch.* ‖ grounded system without protective conductor * *(VDE 0100 Teil 410)* ‖ grounded supply ‖ grounded system
TTL-Logik TTL logic
TTL-Signal TTL signal
Tube tube ‖ collapsible tube
Tubenhütchen tube cap
Tubus tube ‖ tubus ‖ tubulature ‖ tubular jacket * *tubusförmige Umhüllung/Ummantelung*
tubusförmig tubular
tüchtig able * *auch fähig* ‖ fit ‖ capable * *tüchtig* ‖ competent ‖ qualified * *qualifiziert, geeignet, befähigt (for: für)* ‖ efficient * *leistungsfähig, gewandt* ‖ clever * *geschickt, gewitzt* ‖ skilful * *geschickt* ‖ proficient * *erfahren, geübt* ‖ experienced * *erfahren* ‖ excellent * *vorzüglich*
Tülle socket ‖ spout * *Gießröhre* ‖ sleeve * *Manschette, Muffe* ‖ grommet * *für Kabeldurchführungen usw.*
Tunnel tunnel ‖ tube * *U-Bahn-Tunnel* ‖ subway * *Unterführung* ‖ gallery * *im Bergwerk* ‖ duct * *im Maschinenbau*
Tunnelbelüftung tunnel ventilation
Tunneldiode tunnel diode
Tunnelentlüftung stale air extraction for road tunnels * *Absaugung der Abluft bei Straßentunnels*
Tür door * *auch eines Geräts, eines Schaltschrankes*
Turbine turbine
Turbinenregelung turbine control
Turbogeneratorsatz turbine-generator set
Turbokompressor turbo compressor
Turbokühlung turbo-cooling
Turbokupplung turbo coupling
Turboläufer turbo-rotor * *Hochgeschwindigkeitsläufer, häufig als ungeblechter* →*Massivläufer ausgeführt* ‖ high-speed rotor * *Läufer eines schnellaufenden Motors*
Turbosatz turboset * *Turbinen-/Generatorsatz zur Erzeugung elektr. Energie*
Turboventilator turbo fan
Turboverdichter turbo-compressor ‖ turbocompressor
Turbulenz turbulence
Türkenklopfwalze turks head roll

türkis turquoise
turnusmäßig regular || in rotation || regularly recurring || routine * *rountinemäßig*
TÜV TÜV * *authorized inspection agency: Technischer Überwachungsverein* || authorized inspection agency * *Technischer Überwachungsverein* || Technical Inspectorate * *Techn. Überwachungsverein*
Typ type || model * *einer Produktreihe* || design * *Bauart* || type of construction * *Bauform, Bauart*
Typbezeichnung type designation || product designation
Typenbezeichnung type designation * *bei el. Masch. z.B. entspr. d. Baugrößenbezeichng. nach IEC 72* || product designation || type code * *Kurz~, z.b. in Bestellnummer*
Typenblatt data sheet * *Datenblatt*
Typenprüfung type test
Typenrad daisy-wheel
Typenraddrucker daisy-wheel printer
Typenreduzierung reduced number of types
Typenschild nameplate || rating plate || name plate || rate plate * *Leistungsschild* || type plate
Typenschildangaben rating plate indications || rating plate data
Typenschilddaten rating plate data || rating plate indications
Typenschlüssel type code || type key * *in Bestellnummer usw.*
Typenspektrum type spectrum || product range * *Produktpalette/-Reihe*
Typenübersicht type summary
Typenvielfalt variety of different types || number of different types || abundance of designs || number of types * *lower: geringere*
typgeprüft type-tested
typisch typical || representative * *kennzeichnend (of: für), repräsentativ (Auswahl, Querschnitt i. Sinne d. Statistik)*
typische Merkmale typical features || characteristic features || prominent features * *hervorstechende Merkmale* || salient features * *herausragende Merkmale* || typical properties * *Eigenschaften* || essential properties * *kennzeichende Eigenschaften* || pecularities * *Eigentümlichkeiten*
typischer Wert representative value
Typprüfung prototype test * *~ der Erstausführung; für Motor: siehe VDE 0530* || type test || factory type test * *im Herstellerwerk*

U

U V * *voltage: Spannung*
ü umlauted u || u with umlaut || ue
U-Bahn underground railway || underground train || subway || tube
U-Bügel U-shaped bracket || U-shaped washer * *Unterlegscheibe* || U-bolt
U-f-Kennlinie V/F characteristic || V/F pattern || voltage-frequency characteristic || voltage-frequency control characteristic
U-f-Wandler voltage to frequency converter || VCO * *Abkürzg. für 'Voltage Controlled Oscillator': spannungsgesteuerter Oszillator* || voltage controlled oscillator * *spannungsgesteuerter Oszillator* || analog to frequency controller || V/F-converter
U-Profil U section
UART UART * *'Universal Asynchronous Receiver/Transmitter': parallel<>seriell-Wandler f. asynchr.Schnittst.*
üben exercise * *Lehrstoff/Geist ~; jdn. ausbilden* ||

practise * *ein~, sich ~ in, jdn. schulen* || train * *schulen, ausbilden, jdm. etwas beibringen* || drill * *streng schulen, drillen*
über via * *mit Hilfe von, mittels, auf dem Wege ~* || above * *oberhalb von, auch für Wert, Temperatur, Spannung* || through * *durch, auch 'mittels'* || by means of * *mittels* || by way of * *auf die Art und Weise* || over * *auch betrags/zahlen/zeitmäßig; over the complete range: ~ den gesamten Bereic* || higher than * *höher als* || more than * *mehr als* || exceeding * *mehr als (amtlich, techn.; Zahlenwert), wertmäßig über ... hinaus* || across * *quer, hin~, dr~ hinweg* || on account of * *wegen* || during * *während* || concerning * *betreffend* || versus * *z.B. Kurve/Funktion über der X-Achse* || on * *~ etwas (Abhandlung, Vortrag usw.)* || about * *sprechen, nachdenken* || of * *sprechen* || with * *deal with: handeln ~/von (Abhandlung, Vortrag, Buch, Film usw.)* || beyond * *~ etwas hinaus* || all over * *(Adv.) ~ und ~, ganz (und gar), ~all*
über alles overall * *gesamt* || total * *gesamt* || entire * *gesamt* || in total * *(Adv.)*
über dem Meeresspiegel above sea level
über den gesamten Bereich over the complete range
über den Ladentisch off-the-shelf * *aus dem Verkaufsregal*
über Kreuz cross-wise || in diametrically opposite sequence * *z.B. beim Anziehen von Schrauben* || of a different opinion * *(figürl.) unterschiedlicher Meinung* || at a variance * *(figürl.) zerstritten, im Widerstreit/Widerspruch* || diagonally opposite
über längere Zeit over a longer time period
über Tage above ground
Über- over || super
über-alles overall
überarbeiten modify * *modifizieren* || review * *Schriftstück, Druckerzeugnis usw. ~* || revise * *Gerät, Dokument usw. ~* || go over again * *Manuskript ~* || touch up || retouch || work oneself to the bones * *sich ~/überanstrengen* || redesign * *umkonstruieren, anders auslegen, "umentwickeln", umbauen* || re-edit * *Schriftstück ~/neu bearbeiten/ neu herausgeben*
überarbeitet revised * *z.B. Buch, Software, Druckschrift*
Überarbeitung modification * *Modifizierung (to/of: von)* || revision * *~, Durchsicht, Korrektur, verbesserte Ausgabe a. Dokuments, Buchs, Softwareprogramms* || overwork * *zu viel Arbeit* || redesign * *Umkonstruktion, Entwicklungsänderung, Umdimensionierung* || re-editing * *Neubearbeitung eines Druckerzeugnisses usw.*
Überbeanspruchung overstressing || overloading || overworking * *(figürl.)*
überbelasten overload
überbieten beat * *z.B.Gegner* || outperform * *bezüglich der Leistung(sfähigkeit) ~*
Überblick overview * *in Beschreibung, Vortrag usw.* || survey || summary * *Zusammenfassung, Übersicht, Abriss, Inhaltsangabe* || synopsis * *Zusammenfassung, Übersicht, Abriss*
Überblick geben outline || give a survey * *comprehensive: umfassenden* || provide an overview || review || summarize
überbrücken jumper * *mit Draht, z.B.Klemmen-/ Draht-/Steck-/Lötbrücke* || shunt || shunt out || shortcircuit * *kurzschließen* || bypass * *umgehen, umleiten, vermeiden* || link-up * *verbinden* || bridge * *(auch figürl.)* || bridge over * *(auch figürl.)* || short out * *elektr. kurzschließen* || short || overcome * *→überwinden*
Überbrückung bypass || ride-through * *eines Netzspannungseinbruchs bei Antrieb,ohne Abzuschal-

Überbrückungskondensator 320

ten || bridge over || jumpering * *Steckbrücke, Klemmen usw*
Überbrückungskondensator bypass capacitor * *Ableit(quer)kondensator*
Überbrückungsrelais bypass relay * *z.B. bei USV, Motor-Sanftanlaufgerät (überbrückt dieses nach erfolgtem Anlauf)* || short-circuiting relay * *z.B. bei Sanftanlaufgerät*
Überbrückungsschütz bypass contactor * *bei Umrichter-Zwischenkreis-Vorladeschaltung, Sanftanlaufgerät, USV usw.* || short-circuiting contactor * *z.B. f. Zwischenkreis-Ladewiderstand e.Umrichters, USV, Sanftanlaufgerät*
Überbrückungszeit buffer time * *Pufferungszeit* || ride-through interval * *Netzausfall/Einbruchzeit, die überbrückt werden muss/kann*
überdacht roofed-over || roofed-in || sheltered
Überdeckung overlay || overlapping * *Überlappung*
Überdeckungsbreite contact width * *Kohlebürstenabmessung in tangentialer Richtung (Breite der abgerundeten Kante)*
Überdeckungsgrad contact ratio
Überdeckungssperre overlay inhibit * *am Bildschirm*
überdimensionieren oversize || overdimension || overrate || dimension generously * *großzügig dimensionieren*
überdimensioniert oversized || generously dimensioned * *großzügig dimensioniert* || overdimensioned || overrated
Überdimensionierung oversizing || overdimensioning
überdrehen skim * *z.B. Kommutator auf Drehmaschine ~* || overspeed * *mit Überdrehzahl fahren*
Überdrehzahl overspeed
Überdrehzahlauslöser overspeed trip || overspeed relay || overspeed monitor || overspeed release
Überdrehzahlschutz overspeed protection
Überdruck pressure || overpressure * *übermäßiger Druck; pressurized: unter ~ stehend* || pressure difference || positive pressure || excess pressure * *übermäßiger Druck* || above atmospheric pressure || gauge pressure * *überatmosphärischer Druck* || pressure above atmospheric * *überatmosphärischer Druck*
Überdruckkapselung pressurized enclosure * *Schutzart'p' n. EN50016 f.Ex-geschützte el. Betriebsmittel; Zündschutzgasfüllg.*
Überdruckkapselung mit ständiger Durchspülung open-circuit pressurized enclosure * *Zünd/Explosionsschutzart* || pressurized enclosure with leakage compensation * *Zünd/Exschutzart mit Ausgleich d.Schutzgasverluste*
Überdruckprüfung overpressure test
Überdruckventil relief valve || pressure-relief valve || blowoff valve || excess-pressure valve || relief pressure valve || governor relief valve
überdurchschnittlich above average || outstanding || above-average
übereinander above one another || one upon the other || stacked * *~gestapelt* || on top of each other
Übereinkunft agreement || arrangement || understanding || settlement * *Vergleich, Schlichtungsergebnis* || compromise * *Kompromiss* || stipulation * *vertragliche ~*
übereinstimmen coincide * *örtlich/zeitlich zusammenfallen/treffen, sich decken/genau entsprechen (with: mit)* || agree * *with a person on: mit jemandem ~ über* || share opinion * *share his opinion: mit ihm ~* || correspond * *entsprechen (einer Sache usw.), in Einklang stehen (to: mit)* || harmonize * *zusammenpassen, im Einklang sein, harmonieren* || be in agreement * *in Übereinstimmung/Eintracht/Einigkeit sein* || tally * *übereinstimmen, ent-*

sprechen, aufgehen, stimmen (with: mit)* || square * *übereinstimmen, passen* || meet * *~ mit/entsprechen/gerecht werden (Anforderung, Vorschrift, Norm usw.)* || match up * *with: mit* || be in phase * *zeitlich ~* || be in line * *with: mit* || match * *~ mit*
übereinstimmend in accordance * *(Adv.) with: mit* * *z.B. Norm, Vorschrift* || in conformity * *(Adv.) with: mit* || corresponding || conformable || concurring * *z.B. Meinung, Erfahrung* || consistent * *folgerichtig* || unanimous * *einstimmig* || identical * *identisch*
übereinstimmend mit in accordance with || in compliance with * *gemäß*
Übereinstimmung accordance * *in accord. with: in ~ mit; according to: gemäß; be i. accord. with: übereinst. mit* || agreement * *Übereinst., Einklang, Abkommen, Vereinbarg., Vertrag; in agreem. with: in ~ mit* || correspondence * *auch 'Angemesenheit, Entsprechung'; (with: mit; between: zwischen)* || conformity * *auch 'Gleichförmigkeit, Ähnlichk.'; in conformity with: in ~ mit/gemäß* || conformance * *in conformance with: gemäß, in ~ mit* || concurrence * *Einverständnis, Zustimmung, Gleichzeitigkeit, Zusammentreffen* || harmony * *Eintracht, Einklang, Harmonie* || accord * *Einigkeit, Zustimmung, Übereinkommen, Vergleich* || unison * *Einklang, Gleichklang, Übereinstimmung; in unison with: in Einklang mit* || consistency * *Vereinbarkeit, Folgerichtigkeit, Konsequenz, Zusammengehörigkeit, Einklang* || coincidence * *Zusammentreffen (räumlich oder zeitlich), Übereinstimmung, Gleichzeitigkeit* || equality * *Gleichheit* || equivalence * *Gleichwertigkeit* || compliance * *Befolgung, Erfüllung, Einhaltung; in compliance with: gemäß (Vorschrift, Norm usw.)*
Übererhitzung overheating || excessive temperature rise || superheating
übererregt overexcited || over-excited * *z.B. Synchronmaschine*
Übererregung overexcitation * *z.B. z. beschleunigten Momentenrichtungswechsel durch Feldumkehr* || over-excitation
Übererwärmung overheating || excessive temperature rise || superheating
überfällig overdue * *verspätet*
Überflurförderer elevated conveyor
Überfluss abundance * *of: an/von* || plenty || profusion || wealth * *Reichtum, Fülle (of: an/von)* || surplus * *Überschuss* || excess * *Übermaß, Überfluss, Unmäßigkeit* || affluence * *Fülle, Überfluss, Reichtum, Wohlstand* || excessive supply * *Überangebot, Überversorgung* || glut * *Fülle, Überfluss, Überangebot; a glut in the market: e. Überschwemmg./Sättigung d. Marktes* || plethora * *Überfülle, Übermaß, Zuviel (of: an)*
überflüssig superfluous * *unnötig; render superfluous: ~ machen* || unnecessary * *unnötig* || useless * *nutzlos, unnötig* || undesired * *unerwünscht, nicht gewünscht* || surplus * *überschüssig* || excess * *überschüssig* || excessive * *übermäßig* || undue * *übermäßig, übertrieben, unangemessen, unzulässig*
überfluten flood * *auch überfließen*
Überflutung flooding || submersion * *Untertauchung* || flood || inundation || submerging * *das Versenken*
Überfurnierpresse veneering press * *{Holz}*
Übergabe transfer || handing-over || presentation * *von Dokumenten* || delivery * *Aus-/Ablieferung*
Übergabeleistung transfer power
Übergabepuffer transfer buffer || mailbox * *→Koppelspeicher*
Übergabestelle transfer position
Übergabetaste ENTER key || ENTER || ENTER button || entry key

Übergabetrafo substation transformer * *in der Energieverteilung*
Übergang transition || transfer * *Wechsel, Übertragung, Transfer* || passage || crossing || change || changeover || junction * *in einem Halbleiter, z.B. pn-junction: pn-~* || move * *Aufbruch, (Fort-) Bewegung, Schritt, Entwicklung, Maßnahme, Fortschreiten*
Übergangs- transitional || provisional * *provisorisch, vorläufig*
Übergangsdrehzahl transition speed
Übergangsfrequenz cross-over frequency
Übergangsfunktion response characteristic * *in der Regelungstechnik* || response || step-forced response * *Sprungantwort* || transient function || transient response * *Übergangsverhalten* || transfer function
Übergangsphase transition phase
Übergangspunkt cross-over point
Übergangsstecker plug connector || adapter plug * *Adapterstecker*
Übergangsverhalten transient response
Übergangswiderstand contact resistance * *eines Schaltkontaktes, Schleifkörpers oder Kommutators* || transfer resistance || passive resistance || transition resistance
Übergangszeit transition period || transition time || interim period * *Zwischenzeit* || intervening period * *dazwischenliegende Zeitspanne* || meantime * *Zwischenzeit; in the meantime: in der Zwischenzeit/derweil*
übergeben deliver up || give up || hand over * *aushändigen; auch (Fabrikations)Anlage, Aufgabe usw.* übergeben || present * *überreichen, bringen, einreichen, schenken* || transfer * *übertragen, beschaffen, wechseln* || changeover * *hinübergehen, hinüberwechseln* || render * *z.B. Gewinn, Festung*
übergehen pass over * *to: zu; auch 'übersehen', →'nicht beachten' (in silence: stillschweigend)* || change over * *hinüberwechseln* || switch over * *to: zu etwas* || pass on * *weitergehen (to: zu)* || overlook * *übersehen, nicht beachten* || ignore * *ignorieren, →nicht beachten* || omit * *auslassen* || skip * *überspringen* || leave out * *jemanden ~* || neglect * *jemanden ~; →nicht beachten, außer Acht lassen, missachten* || pass into * *~ in* || change into * *~ in* || merge * *von Farbtönen/Firma (into: in)* || fade * *von Farbtönen usw.* || pass to * *von Besitz*
übergeordnet higher-level * *z.B. Automatisierungssystem* || higher-ranking * *im Rang* || master || supervisory * *aufsichtsführend, Leit-(System)* || generic * *allgemein, generell, generisch; z.B. Begriff, Norm usw.* || superordinate
übergeordnete Norm generic standard * *Grundnorm*
übergeordnete Regelung master control || higher level loop control
übergeordneter Rechner higher level computer
übergeordnetes Automatisierungssystem higher-level automation system
Übergeschwindigkeit overspeed
übergreifend overlapping || superimposed * *darüberliegend/gelegt, überlagert* || higher-level * *übergeordnet* || supervisory * *aufsichtsführend, leitend* || generic * *allgemein (z.B. Norm), generell, generisch, Gattungs-*
übergroß jumbo * *(salopp)* || oversized || huge * *→riesig*
überhäufen overwhelm * *(with: mit)* || swamp * *(with: mit (Arbeit, Vorwürfen, Anrufen, Aufträgen usw.))* || clutter * *vollstopfen, überhäufen, anfüllen*
überhitzen overheat || superheat

Überhitzung overheating || overheat || excessive temperature rise || superheating
Überhöhung increase || cant * *einer Kurve usw* || superelevation * *von Schienen usw.* || overshoot * *überschwingen*
überholen pass || overtake * *Fahrzeug im Verkehr ~, Mitbewerber ~* || outdistance * *übertreffen* || outperform * *an Leistung(sfähigkeit) übertreffen* || outrun * *übertreffen* || outstrip * *übertreffen* || outpace * *übertreffen* || overhaul * *nachsehen, ausbessern, instand setzen, technisch überprüfen, wieder i.Ordnung bringen* || recondition * *techn. ~* || service * *technisch nachsehen, warten, überholen* || pass by * *vorbeigehen, vorübergehen (auch zeitlich)* || overrun * *überlaufen, z.B. Softwarezyklus* || check up || refurbish * *renovieren, aufpolieren, überholen (z.B. Maschine)*
Überholung overhaul * *Überholung, gründliche Überprüfung; complete overhaul: ~ von Grund auf* || check-up * *Überprüfung, Überholung* || refurbishing * *z.B. ~ einer Maschine* || reconditioning
Überholungsarbeiten overhaul work
überkritisch over-critical || above critical
überlackieren overpaint
überlagern superimpose * *on: etwas/über etwas* || overlay * *überdecken* || mask * *verhüllen, maskieren* || conceal * *verdecken* || superpose
überlagert superimposed * *on: über etwas; auch Wechselspannungsanteil* || outer * *Regelkreis* || higher-level * *übergeordnet* || master * *Leit-* || major || superordinated || superposed
überlagerte Regelung outer loop control || higher-level loop control || outer control loop || superimposed controls || high-level controls || higher-level closed-loop control
überlagerte Wechselspannung superimposed alternating voltage || ripple voltage * *Spannungs(rest)welligkeit*
überlagerter Regelkreis higher-level control loop || outer control loop
überlagerter Regler higher-level controller || master controller || superimposed controller || outer controller || supervisory controller
überlagerter Wechselstrom superimposed alternating current || ripple current * *Strom(rest)welligkeit*
Überlagerung superposition || superimposition || ripple * *Welligkeit (z.B. auf einer Gleichspannung)* || interference * *Störung*
Überlagerungsverfahren superposition method * *zur Ermittlung von Temperaturen*
überlappen overlap || take overlap || lap
überlappend with overlap * *auch zeitlich* || overlapping * *auch zeitlich*
Überlappung overlap * *auch zeitlich* || overlapping
Überlappungswinkel overlap angle || angle of overlap
Überlappungszeit overlap time || overlap interval || overlap period
überlassen let * *let a person have a thing: jemandem etwas überlassen* || leave * *leave a thing to a person: jemandem etwas anheimstellen* || leave to * *abtreten* || entrust * *anvertrauen; to: an* || be up to * *it is up to me to do: es ist mir überlassen/meine Sache zu tun*
Überlast overload || overload condition * *Überlast-Zustand/-Betrieb* || overtorque * *zu hohes Drehmoment*
überlastbar overload-capable || overloadable
Überlastbarkeit overload capability * *dynamic: dynamische; für E-Motoren nach EN 60034 (2 min mit 1,5-fach. Strom)* || overload capacity * *wert/betragsmäßig; el.Maschinen nach VDE0530 für 2 min m. 1,5fach.Strom belastb.* || overload factor

überlasten 322

* *Überlastbarkeitsfaktor (max. Leistung/Strom zu Nennleistung/Strom)* || overload rating * *spezifizierte Überlastfähigkeit* || permissible overload * *zugelassene Überlast (short-time: kurzzeitige)*
überlasten overload * *by: um* || overcharge || overburden * *(figürlich)*
überlastet overloaded
überlastfähig overload-capable || overloadable
Überlastfähigkeit overload capacity || overload capability || overload margin * *Bereich der ~*
Überlastfaktor overload factor * *Verhältnis 'zulässige Motordauerleistung/Motornennleistung'* || service factor * *Service/Betriebsfaktor*
Überlastmöglichkeit overload capability
Überlastrelais overload relay
Überlastschutz overload protection
Überlastung overloading || overload * *Überlast*
Überlauf overflow || carry * *Übertragsbit/-merker bei Rechenoperation, z.B. Addition, Multiplikation* || spillage * *Flüssigkeit*
überlegen consider * *bedenken* || reflect on * *bedenken* || think over * *bedenken* || reconsider * *noch einmal ~* || superior * *(Adj.) funktionell überragend, i. Vorteil; (to: gegenüber/verglichen mit; in: an/bezügl.)* || in superior style * *(Adv.)* || outperform * *an Leistungsfähigkeit ~ sein im Vergleich zu*
Überlegenheit superiority || preponderance * *Übergewicht*
Überlegung consideration * *Betrachtung; upon mature consideration: nach reiflicher ~* || reflection * *Nachdenken* || thought * *Gedanke; on second thoughts: bei näherer ~*
überleiten pass over
Überleitung transition * *Übergang* || transfer * *Übertragung, Übergang, Transfer, Wechsel; to: zu* || passing over
Übermagnetisierung overmagnetization
Übermaß oversize * *bezüglich des Abmaßes* || excess * *Überfluss, Übermaß, Unmäßigkeit*
übermäßig excessive || immoderate || undue * *unnötig*
übermorgen the day after tomorrow
Übernahme takeover || taking over || acceptance * *Abnahme, Annahme, Empfang; acceptance test: Abnahmeprüfung* || undertaking * *einer Aufgabe, Verpflichtung, Garantie* || assumption * *Annahme, Übernahme, Aneignung* || adoption * *(auch figürlich; Idee, Lösungsweg usw.)* Annahme, Übernahme, Aneignung || taking charge of * *Verwaltg., Aufsicht, Obhut, Verantwortg., Leitg., Besitz, Posten* || take possession of * *Besitz ergreifen von, in Besitz nehmen* || succession to * *Amt, Erbschaft usw.* || entering upon * *Amt*
Übernahme-Taste set key || enter key || transfer key || entry key
Übernahmetaste entry key * *z.B. an SPS-Programmiergerät*
übernehmen take over || receive * *empfangen* || undertake * *Arbeit, Verantwortung ~* || take * *Führung, Befehl, Risiko ~* || accept * *akzeptieren, annehmen, Pflicht/Ware/Gewährleistung/Garantie ~* || take possession of * *Besitz ~* || adopt * *Verfahren ~* || enter upon * *Amt ~* || succeed to * *in ein Amt nachfolgen* || overextend * *sich ~* || carry * *z.B. Strom/Stromfluss ~* || assume * *Pflicht, Gefahr, Schuld, Strom ~* || buy out * *aufkaufen (Firma usw.)* || take care of * *sorgen für*
überproportional over-proportional
überprüfen check * *for: bezüglich* || examine || review * *Druckschrift überarbeiten, Ablauf eines Projektes ~* || verify * *verifizieren, auf Richtigkeit/Echtheit untersuchen, die (fehlerfrei) Funktion ~* || investigate * *untersuchen* || inspect * *nachschauen, besichtigen, inspizieren* || re-check * *nochmalig ~* || check-up * *→nachprüfen*

überprüft checked
Überprüfung check * *make a check: eine ~ durchführen* || examination || inspection * *Inspektion* || checking * *das Überprüfen* || analysis * *Analyse* || investigation * *Untersuchung, Ermittlung, Nachforschung* || consideration * *Erwägung, Überlegung* || studies * *(wissenschaftliche) Untersuchung/Erforschung; genaue, prüfende Betrachtung* || verification * *Nachprüfung (auf Richtigkeit)* || checking * *(Nach)Prüfung, Kontrolle* || check up * *Nachprüfung* || inspection * *Inspektion, Überprüfung* || service * *Kundendienst, Wartung* || audit * *Revision (Geschäftsbücher usw.), Rechnungsprüfung* || auditing * *Rechnungsprüfung* || review * *Revision, ~ (Entwickl.vorhaben, Entscheidg., Gerichtsurteil), (Buch-) Rezension* || trial * *Erprobung* || scrutiny * *genaue Untersuchung, prüfender Blick*
überragen protrude over * *herausragen/hervorstehen über* || project over * *hervorstehen über*
überragend superior || outstanding || brilliant || paramount * *(figürl.)*
überschaubar clear * *Lage, Problem* || easy to grasp * *leicht fassbar; z.B. Problem* || visible at a glance
Überschlag flashover * *durch Überspannung* || arc over * *durch Überspannung* || spark-over
überschlägig rough * *z.B. Schätzung, Rechnung, Wert* || estimated * *geschätzt* || superficial * *oberflächlich*
überschläglich superficial * *oberflächlich, flüchtig* || rough * *grob; estimation: Berechnung/Abschätzung* || estimated * *geschätzt*
überschlagsmäßig superficial * *oberflächlich, flüchtig* || rough * *grob; estimation: Berechnung/Schätzung* || estimated * *geschätzt*
Überschlagsspannung flashover voltage || arc-over voltage
überschneiden overlap * *überlappen* || intersect * *zwei Linien*
überschreiben overwrite * *z.B. Speicherinhalt, Parametersatz* || rewrite * *z.B. einer Speicherzelle, Texteingabe auf Bildschirm*
überschreiten exceed * *Maß/Termin/Wert/Grenze usw. ~ (by: um); x times the: um das x-fache von* || overstep * *Maß, Grenzlinie ~* || go beyond * *Maß ~* || fail to meet * *Termin ~* || cross * *überqueren* || pass over * *überqueren* || go across * *überqueren* || transgress * *übertreten, z.B. Gesetz, Vorschrift* || infringe * *Gesetz/Vorschrift übertreten* || cross * *kreuzen*
Überschreitung exceeding || overrange * *des Werte-/Zahlen-/Messbereichs* || transgression || infringement * *Verletzung, Bruch, Übertretung, Verstoß von/gegen Gesetz/Vorschrift* || crossing * *Kreuzen, Kreuzung, Durch/Überquerung (z.B. einer Grenze/Linie)*
Überschrift title || headline * *Kopf/Schlagzeile* || heading || caption * *Überschrift, Titel, (Bild-) Unterschrift; Rubrik*
Überschuss surplus || excess || balance * *Saldo* || margin * *(Geld)Differenz* || profit * *Gewinn*
überschüssig excess || surplus * *überschüssig, Überschuss-, Mehr-; surplus weight: Mehr-/Übergewicht*
überschütten overwhelm * *überhäufen, überwältigen, überschütten, erdrücken*
überschwemmen inundate * *(auch figürl.)* || flood || swamp || deluge * *(figürl.)*
überschwingen overshoot * *(auch Substantiv)*
Überschwinger overshoot
überschwingungsfrei without overshoot
Überschwingweite overshoot * *z.B. einer Sprungantwort* || maximum overshoot || overshoot amplitude

Übersee overseas || oversea || world (without Europe) * *von Europa aus gesehen*
übersetzen translate * *into: in* || compile * *Hochspracheprogramm mit Hilfe eines Compilers* || interpret * *verdolmetschen; auch Quellprogramm während des Programmablaufs (z.B. BASIC)* || assemble * *Assembler/Maschinenspracheprogramm m.Hilfe eines Assemblers*
Übersetzer translator || interpreter * *für Fremdsprachen (bes. mündlich), Dolmetscher; online-Übersetzer für Hochspracheprogr.* || compiler * *für Hochspracheprogramme* || assembler * *für Assembler/Maschinenspracheprogramme* || converter * *Umsetzer, Umsetzprogramm; Konverter*
Übersetzung ratio * *(~s-) Verhältnis, z.B. Getriebe/Trafo/Stromwandlerübersetzung* || gear ratio * →*Getriebe~ ('i' entspr. Getriebe-Eingangs- zu Ausgangsdrehzahl)* || transformation ratio * *Übersetzungsverhältnis von Trafo, Strom-/Spannungswandler usw.* || compilation * *durch Compiler (Hochspracheprogramm in Maschinen/Objektspracheprogramm)* || assembling * *~ eines Assemblerprogramms ("Mnemomic-Sprache") in Maschinensprache/Objektcode* || code generation * *Erzeugung von Programmcode* || translation * *z.B. in eine andere Landessprache; auch 'Übertragung, Auslegung'* || interpretation * *Auslegung, Deutung, (mündliche) Wiedergabe, Sprachübersetzung, Interpretation* || reduction ratio * *~sverhältnis eines Untersetzungsgetriebes*
Übersetzungsgetriebe transmission gearing || gearbox * *Getriebe allg.* || step-up gearing * *Hochsetzgetriebe* || step-up gear * *Hochsetzgetriebe*
Übersetzungslauf compilation * *für Hochsprache-Programm* || compiler pass * *einer von mehreren Durchläufen eines Hochsprache-Übersetzers* || translation run
Übersetzungsverhältnis gear ratio * *bei Getriebe* || speed reduction ratio * *~ eines Untersetzungsgetriebes* || transmission ratio * *Getriebe ("i" entspr. Getriebe-Eingangs- zu Ausgangsdrehzahl)* || speed ratio * *Getriebe* || transformation ratio * *Trafo, Strom/Spannungswandler usw.* || voltage ratio * *Trafo* || ratio * *Trafo; one-to-one ratio: Übersetzungsverhältnis 1: 1* || reduction ratio * *~ eines Untersetzungsgetriebes*
Übersicht overview * *of: über; Überblick; provide an overview: eine ~ geben; fast overview: Schnell~* || survey * *auch Überblick, Gutachten, Begutachtung, (Prüfungs-)Bericht* || view * *Überblick, Ausblick* || review * *Besprechung, Rückblick, Kritik, kritische Besprechung, auch 'eine ~ geben über'* || outline * *auch Abriss, Umriss, Ausblick, Darstellung in groben Zügen* || summary * *Zusammenfassung, Abriss, Inhaltsangabe, tabellarische ~ (of: über)* || summary outline * *Kurz-Zusammenfassung* || synopsis * *Zusammenfassung, Abriss, vergleichende Zusammenschau (meist in einem Buch)* || control * *Kontrolle* || list * *Aufstellung* || table * *(tabellarische) Aufstellung* || chart * *Schaubild, Diagramm, grafische Darstellung, Kurvenblatt, Karte, Tabelle* || general view
übersichtlich clear * *klar dargestellt* || clearly arranged * *~ angeordnet* || easy to survey * *leicht überschaubar* || lucid * *klar, deutlich in Stil, licht im Geiste* || open * *Gelände* || transparent * *klar (im Stil), durchsichtig, offen, ehrlich* || well organized || straightforward * *unkompliziert, einfach*
Übersichtlichkeit transparency || clearness || lucidity * *Klarheit, Deutlichkeit* || clarity * *Klarheit* || distinctness * *Klarheit, Deutlichkeit*
Übersichtsdarstellung overview representation * *z.B. eines SPS-Programms*
Übersichtsdiagramm overview diagram

Übersichtsplan block diagram || schematic diagram || survey diagram || layout plan || general plan
Übersichtsschaltbild basic circuit configuration || block diagram * *Blockschaltbild* || schematic diagram || single-line diagram * *in einpoliger Darstellung* || survey diagram
Übersichtsschaltplan schematic circuit diagram || schematic diagram || block diagram
überspannen overtight * *z.B. Riemen, Kette*
Überspannung overvoltage || overvoltage condition * *Auftreten/Vorhandensein einer ~* || voltage surge * *kurzzeitige ~, ~sspitze* || excess voltage
Überspannungs-Schutzbeschaltung suppressor circuit || snubber circuit
Überspannungsableiter voltage surge suppressor || voltage surge diverter || varistor || surge suppressor || surge protector || surge diverter || over-voltage arrester || transient suppressor || surge arrester || lightning arrester
Überspannungsabschaltung overvoltage trip * *Fehlerabschaltung auf Grund unzulässig hoher Spannung*
Überspannungsauslöser overvoltage tripping unit || overvoltage trip || overvoltage release
Überspannungsfestigkeit overvoltage strength * *siehe z.B. DIN VDE 0160* || surge strength * *Festigkeit gegen Spannungsspitzen* || dielectric strength * *z.B. bei Kondensator* || impulse strength * *Festigkeit gegenüber Stoßspannungen* || overvoltage handling capacity
Überspannungskategorie overvoltage category * *kennzeichnet Isolations-/Überspanngs.festigkeit nach VDE0110* || surge strength class * *kennzeichnet Überspannungsfestigkeit nach VDE 0110 Teil 2*
Überspannungsschutz surge suppressor * *Überspannungsableiter* || overvoltage protection * *allg.* || Überspannungsverhinderung, Schutz gegen Zerstörung || transient protection * *gegen Spannungsspitzen* || surge protector * *Überspannungsschutzelement* || surge suppresion * *gegen Spannungsspitzen*
Überspannungsschutzdiode surge absorber diode
Überspannungsschutzkondensator snubber capacitor || surge-protection capacitor
Überspannungsspitze overvoltage surge
Übersprechen crosstalk * *Störeinflüsse v. benachbarten Informationen (auf Leitungen, magn. Datenträger, Display)*
überspringen skip * *Schritt/Rechnerbefehl(e) usw. weglassen; skip over something: etwas ~/übergehen* || flash across * *Funken* || leap over || overleap || spark over * *Funken*
überstehen overcome * *(Schwierigkeiten usw.)* || überwinden, bestehen || surmount * *überwinden* || endure * *ertragen* || get over * *erfolgreich ~, überwinden* || survive * *überleben* || jut out * *(über etwas) hinausragen, hervorstehen* || project * *vorspringen, vorstehen, (her)vorragen (over: über), (her)vortreten (lassen)* || protrude * *herausstehen, vorstehen, →hervorragen, hervortreten (lassen), herausstrecken*
übersteigen exceed * *überschreiten (Wert/Größe)* || go beyond * *surpass* || Erwartungen usw. ~ || be too much for * *Kräfte, Verstand, Fähigkeiten usw. ~* || climb over || surmount
übersteuern overmodulate || override * *über ein anderes Signal hinweg ~* || override * *über ein anderes Signal hinweg ~* || drive into saturation * *in die Sättigung hineinsteuern/-treiben* || overcharge * *overload* || *z.B. Messgerät ~* || overmodulation * *z.B. ~ eines Regelkreises* || bias * *auch ~sollwert, elektrische Vorspannung* || overloading * *z.B. ~ eines Messgeräts* || saturation * *Sättigung* || override

Übersteuerungssollwert bias * *zur Übersteueung eines Reglers*
Übersteuerungswert bias
Überstrom overcurrent || overcurrent capacity * *Überlastbarkeit, Überlastfähigkeit* || current overload * *Überlaststrom, Stromüberlastung*
Überstromabschaltung overcurrent trip * *Störabschaltung* || overcurrent cut-off
Überstromauslöser overcurrent releasing device * *zum thermischen Motorschutz* || overcurrent tripping unit || overcurrent release || overcurrent trip || excess current switch
Überstromauslösung overcurrent tripping || overcurrent trip
Überstrombegrenzung overcurrent limiting
Überstrombelastbarkeit overcurrent capability || overcurrent capacity
Überstrombelastung current overload
Überstromrelais overcurrent relay
Überstromschutz overcurrent protection
Überstunde overtime hour
Überstunden overtime work * *work overtime: ~ machen*
Überstundenzuschlag overtime charges
übersynchron oversynchronous || supersynchronous || above-synchronism * *z.B. Betrieb e. Asynchronmotors in der Feldschwächung*
übersynchrone Bremsung oversynchronous braking
Übertemperatur over-temperature || temperature rise * *Temp.erhöhg. gegenüb. Vergleichspkt. (Umgebgs./Kühlmitteltemp.); f. Motor: VDE0530* || overtemperature || excessive temperature rise * *übermäßige Temperaturerhöhung* || overheat * *Überhitzung*
Übertemperatur bei Anlauf starting-up temperature rise * *{el.Masch.}*
Übertrag carry * *einer Addition, eines Saldos, usw* || borrow * *bei einer Subtraktion*
übertragbar transferable || transmittable * *z.B. Drehmoment, Datenmenge* || portable * *Programm auf einen andern Rechner*
übertragbares Drehmoment transmittable torque * *z.B. über eine Kupplung ~ (ist kleiner als das →schaltbare Drehmoment)* || static torque * *z.B. über eine Kupplung (ist kleiner als das →schaltbare Drehmoment)* || static clutch torque * *statisch ~ einer Kupplung (gilt nicht für Schaltvorgänge)*
übertragbares Moment transmittable torque
Übertragbarkeit portability * *z.B. von Software auf eine andere Hardware- oder Betriebssystemplattform*
übertragen transfer * *auch elektr. Energie, Telegramm ~* || transferred * *(passiv)* || download * *z.B. Arbeitsdaten/Parameter zu einem Gerät ~* || delegate * *Aufgabe, Vollmacht; abordnen, delegieren, bevollmächtigen* || transmit * *Daten, Nachrichten, Signale, Kraft, Drehmoment, Kraft ~* || communicate * *Stimmung, Panik usw. ~* || pass over * *überleiten, überführen, übertragen,* || translate * *Grundsätze/Bewegung usw. ~ (into: in; to: auf); transl. into action: in die Tat umsetzen*
Übertrager transformer * *Transformator f.Signale; pulse transformer: Impuls~ (z.B. z. Ansteuerung v.Thyristoren)* || pulse transformer * *Impuls~ (für Thyristoren usw.)*
Übertragsbit carry bit * *Überlaufbit z.B. bei Addition, Subtraktion* || carry flag * *Übertrags-Merkerbit in Recheneinheit eines Rechners* || borrow bit * *Unterlaufbit z.B. bei Subtraktion*
Übertragung transmission * *z.B. von Daten, Zündimpulsen, Kraft, Drehmoment, Bewegung* || transmittal * *auch Kraft, Drehmoment* || transfer || transferring

Übertragungselement transmitting element || transmission element || torque-transmission component * *zur Drehmomentübertragung, z.B. von el.Motor auf die Arbeitsmasch.*
Übertragungsentfernung transmission distance || transfer distance
Übertragungsfaktor transfer ratio
Übertragungsfehler transmission error || communication error * *Kommunikationsfehler* || transfer error
Übertragungsfrequenz transmission frequency
Übertragungsfunktion transfer function || response characteristic * *z.b. eines Regelkreises (siehe auch 'Sprungantwort')*
Übertragungsgeschwindigkeit transmission rate * *z.B. bei serieller Kommunikation* || transmission speed || data-transfer rate * *Daten~*
Übertragungsglied transfer element || transmission element * *mechanical: mechanisches*
Übertragungskennlinie transfer characteristic
Übertragungsleistung signal power
Übertragungsleitung transmission line
Übertragungsmedium transmission medium * *z.B. Bus(system)*
Übertragungsprotokoll data transfer protocol || transmission protocol * *zur Datenübertragung*
Übertragungsrate transmission rate * *z.B. Baudrate bei serieller Datenübertragung* || baud rate * *Baudrate [bits/sec]* || data-transfer rate * *Daten~*
Übertragungsrichtung transmission direction
Übertragungssicherheit transmission reliability || transmission security || transmission data integrity
Übertragungssignal transfer signal
Übertragungsstrecke transmission line || data transmission line * *zur Datenübertragung* || transmission path || transmission distance * *Entfernung/ Länge der Übertragungsstrecke* || transmission length * *Länge der Übertragungsstrecke*
Übertragungsverfahren transmission method || transmission mode
Übertragungsverhalten dynamic behaviour * *siehe DIN 19229* || response characteristic || dynamic response || response
Übertragungszeit transmission time
übertreffen exceed || beat * *schlagen* || surpass * *überholen* || outrun * *in der Leistung ~* || outperform * *in der Leistungsfähigkeit ~* || excel * *übertreffen, sich auszeichnen, hervorragen (in/at: in/ bei)*
übertrieben excessive || magnified || extreme
überwachen monitor * *auf richtige Funktion ~ durch "Mithören"* || supervise * *beaufsichtigen, superintend * *beaufsichtigen* || control * *kontrollieren* || inspect * *inspizieren, kontrollieren* || watch over * *aufpassen auf* || survey * *genau betrachten, inspizieren, prüfen, begutachten* || check * *überprüfen* || test * *prüfen* || keep track of * *nachverfolgen, kontrollieren*
Überwachung monitoring || monitor * *als Vorrichtung* || monitor module * *als Baugruppe/Gerät* || supervision * *Aufsicht* || check * *Überprüfung* || surveillance * *Beaufsichtigung, genaue Betrachtung/ Prüfung* || observation * *~ durch einen Menschen, Beobachtung*
Überwachungsbaugruppe monitoring module || watchdog module * *überwacht den zyklischen Programmablauf e. Rechners/SPS*
Überwachungseinrichtung monitor
Überwachungsfunktion monitoring function
Überwachungsgerät monitoring equipment
Überwachungsrelais watchdog relay * *mit eingebauter Überwachungseinrichtung* || monitoring relay
Überwachungsschaltung monitoring circuit
Überwachungssignal monitoring signal

Überwachungssystem monitoring system || supervisory system
Überwachungsteil alarm section * *eines Geräts*
Überwachungszeit monitoring time * *für Zeitüberwachung* || check time
Überweisung transfer
überwiegen predominate * *vorherrschen* || prevail * *over: (gegen)über* || preponderate || have overweight
überwiegend prevalent || prevailing || predominant || preponderant || vast * *Mehrheit* || overwhelming * *Mehrheit* || majority * *überwiegender Teil; in the majority of cases: in der ~en Zahl der Fälle* || predominant * *über-/vorwiegend, vorherrschend*
überwinden overcome * *(auch figürl.) Problem, Schwierigkeit, Hemmungen, Widerstand, Reibung, Lastmoment usw.* ~ || get over * *Schwierigkeiten usw.* ~ || overpower || surmount * *Schwierigkeiten usw.* ~ || conquer * besiegen || cope with * *bewältigen, meistern, fertig werden mit, gewachsen sein* || tackle * *(Aufgabe, Problem) lösen, fertig werden mit, etwas "packen"*
Überwurfmutter screw cap || cap nut || box nut || flare nut
überzeugen convince * *of: von* || persuade * *of: von, that: dass; auch 'überreden'*
überzeugend convincing * *advantages: Vorteile; arguments: Argumente* || persuasive * *überredend, überzeugend*
Überzug coat || coating || plating * *Plattierung, Metallüberzug* || cover || film * *hauchdünner* || lining * *Verkleidung; protective: Schutz-*
Überzugmaschine covering machine * *z.B. in der Verpackungstechnik*
üblich usual || customary || conventional * *herkömmlich* || common * *gewöhnlich* || ordinary * *gewöhnlich* || standard * *normal*
übrig left * ~ *gelassen/-geblieben; have something left: etwas ~haben* || over * *übrig, über; have something over: etwas ~ haben* || remaining * *~geblieben* || residual * *Rest-* || odd * *überschüssig, überzählig* || superfluous * *überflüssig* || others * *the others: die ~en, die anderen* || rest * *for the rest: im ~en; the rest: die ~en* || otherwise * *~ens* || by the way * *beiläufig, ~ens* || further * *zusätzlich* || additional * *zusätzlich* || residuary * *Rest-* || left over * ~ *gelassen* || rest * *der Rest, das ~e; the rest of: das restliche/~e; for the rest: im ~en*
übrigens by the way || incidentally * *nebenbei bemerkt, übrigens; beiläufig, nebenbei, zufällig*
Übung practice || exercise || training || drill || practice * *'practice makes perfect': '~ macht den Meister'*
Übungen exercises * *practical: praktische*
Uhrzeigersinn clockwise direction * *clockwise: im Uhrzeigersinn; counter/anticlockwise: entgegen d. Uhrzeigersinn*
Uhrzeit time of day
uk uk * *Kurzschlusspann. z.B.Drossel (Eing.spann. bei d. bei kurzgeschloss. Ausgang der Nennstrom fließt)* || impedance voltage * *siehe auch →Kurzschlusspannung (bei Drossel, Trafo)* || percent impedance * *siehe auch →Kurzschlusspannung (bei Drossel, Trafo)* || impedance drop * *siehe auch →Kurzschlusspannung (bei Drossel, Trafo)*
UL UL * *Abkürzg. für 'Underwriters Laboratory': USA-Normengremium der Feuerversicherer*
UL approbiert UL approved || UL recognized || UL listed
Ultraschall ultrasonics || sonar * *nach dem Echoprinzip funktionierend*
Ultraschall- ultrasonic
Ultraschallgeber ultrasonic transducer

Ultraschallmotor ultrasonic motor
um about * *(Adj. u. Adv.)auch 'ungefähr' und zeitlich (about the time: um die Zeit herum)* || near * *zeitlich ungefähr* || towards * *zeitlich ungefähr* || at * *zeitlich genau* || around * *zeitlich/örtlich/wertmäßig 'um herum'* || for * *Lohn, Preis* || by * *Maß, Wert; increase by a half: um die Hälfte erhöhen; by a great deal: um ein bedeutendes Stück* || all the * *umso; all the better: umso besser; all the more/less: umso mehr/weniger* || for the sake of * *um einer Sache willen* || in order to * *um zu* || through * *rotate through 90 Degrees:* ~ *90 Grad drehen; shifted through 90 Degrees:* ~ *90 Grad versetzt*
um 60 Grad verschoben at 60 degrees to each other
um 90 Grad versetzt displaced by 90 degrees || quadrature * *z.B. Rechteck-Impulssignale eines Impulsgebers* || phase-shifted by 90 degrees * *z.B. Ausgangssignale eines inkrementellen Impulsgebers*
um 90 Grad versetzt sein be 90 degrees apart || be at right angles to each other * *Teile gegeneinander*
Um- re- * *anders herum, neu, anders, wieder, zurück* || modification of the
umarbeiten work over || remodel * *gänzlich* ~ || modify * *modifizieren* || improve * *verbessern* || re-edit * *Schriftsatz* ~ || revise * *Buch/Schriftstück* ~ || rewrite * *Schriftstück/Softwareprogramm* ~
Umarbeitung working over || modification *Modifikation* || remodelling || revision || readaption || rewriting * *Schriftsatz, Softwareprogramm*
umarrangieren re-arrange
Umbau reconstruction || rebuilding || alterations * *Änderungen* || modifications * *(verbessernde) Änderungen z.B. an einem Gerät* || remodeling * *an e. Gerät* || reorganization * *(figürlich) Um/Neuorganisation* || conversion * *zu einem neuen Zweck* || re-arrangement * *Anders-Anordnen* || reconditioning * *z.B. einer Maschine* || retrofitting * *Einbau/Nachrüstung neuer Teile* || revamp * *(salopp) Renovieren, Herausputzen, "Aufmotzen"* || resetting * *Umstellung, (Neu-)Einstellung, Umrüstung, Nachjustierung, Rückstellung*
umbauen reconstruct * *neu konstruieren, umbauen, wiederaufbauen/-herstellen, umformen, rekonstruieren* || redesign * *umentwickeln* || rebuild * *umbauen, wiederaufbauen, wiederherstellen (figürl.)* || remodel * *umbilden, umbauen, umformen, umgestalten* || alter * *teilweise* ~*, Teilbereiche* ~ || modify * *modifizieren, ändern, verbessern* || convert * *zu einem neuen Zweck* ~ *(into: zu)* || reorganize * *(figürl.) anders organisieren/anordnen, umorganisieren* || retrofit * *neue Teile einbauen/nachrüsten* || build round * *um etwas her~* || revamp * *(salopp) neu herausputzen, aufpolieren, 'aufmotzen', herrichten, renovieren* || re-set * *(Maschine, Anlage usw.) umrüsten, neu einrichten*
Umbausatz retrofit set
umbenennen rename || re-designate
umbiegen turn * *abbiegen; up/down: nach oben/unten* || bend || bend over
Umbruch justification * *(Druckerei) line justification: Zeile~* || paging * *(Druckerei) Seiten~* || make up * *(Druckerei) ~ eines Druckerzeugnisses* || radical change * *radikale Änderung* || upheaval * *politischer/wirtschaftlicher/gesellschaftlicher ~*
Umcodierung code conversion
Umdrehung revolution * *[rewol'u: schen]* || turn || turning round || rotation || rotary motion * *Drehbewegung*
Umdrehungen revolutions || turns * *Drehungen* || RPM * *Abkürzung für 'Revolutions Per Minute': ~ pro Minute*

Umdrehungen pro Minute revolutions per minute || RPM * Abkürz. f. 'Revolutions Per Minute' || revolutions
Umfang scope * Anwendbarkeits/Geltungsbereich, Inhalt, beinhaltete Funktionen || range * Bereich, Reichweite || circumference * ~ eines Kreises || extent * Ausdehnung, Ausmaß || size * Größe (auch figürlich) || round * be 10 inches round: einen ~ von 10 inches haben || radius * Umkreis, Wirkungs/Einflussbereich || volume * Rauminhalt || scale * large-scale: in großem ~ || entirety * das Ganze, Gesamtheit, Ganzheit; in its entirety: in vollem ~ || perimeter * (geometr.) Umkreis
umfangreich comprehensive * umfassend || extensive * ausgedehnt || voluminous * körperlich || sizeable * ziemlich groß, ansehnlich, beträchtlich || sizable * siehe 'sizeable' || wide-ranging * weit reichend, weitgehend
Umfangs- peripheral || peripheric * peripheric holes: ~bohrungen
Umfangsbohrungen peripheric holes
Umfangsgeschwindigkeit peripheral speed || circumferential velocity || circumferential speed || peripheral velocity || surface metre per minute * in m/min || surface speed * Oberflächengeschwindigkeit || surface velocity * Oberflächengeschwindigkeit
Umfangskraft peripheral force || tangential force * Tangentialkraft
Umfangswickler surface winder || reeler || reel drum * Stütz-/Tragwalze || surface-driven winder || contact winder
Umfangswinkel angle of circumference
umfassen comprise * in sich schließen, einschließen,enthalten || cover * abdecken || include * beinhalten || encompass * einschließen, beinhalten; umfassen, umgeben, umringen, || embrace * umschließen, umgeben, umklammern, einschließen (auch figürl.), umfassen
umfassend comprehensive * ausführlich || global * mit weitem Geltungsbereich || extensive * umfangreich, ausgedehnt || exhaustive * erschöpfend || in-depth * in die Tiefe gehend; in-depth knowledge: ~e Kenntnisse
Umfeld environment
umformen convert || transform
Umformer converter || static converter * nicht rotierender (i. Gegens. zum Leonhard-Umformer) || transducer * Messumformer || inverter * Wechselrichter || convertor * (veraltete Schreibweise) || rotary converter * rotierender ~ (Maschine(nsatz) z. Änderung der Frequenz, Spannung oder Phasenzahl)
Umformung conversion || forming * von Werkstoffen
Umfrage inquiry * all round: generelle ~ an alle; make inquiries: ~ durchführen || survey
umgeben surround * umgeben, umringen (auch figürl.) || encompass * umfassen, umgeben, umringen, einschließen || environ * umgeben, umringen (with: mit)
umgebend ambient * auch 'Umgebungs-(z.B. Temperatur), umkreisend, Neben-(z.B. Geräusch)' || surrounding
Umgebung environs || environment * auch Programmierumgebung, Betriebssystem-Umgebung || surroundings * das Umgebende, der umgebende Raum z.B. einer Stadt, einer Maschine || outside * das Außenliegende, die Außenwelt
Umgebungs- ambient * z.B. Temperatur, Geräusch || environment * Umwelt-, Milieu || environmental * Umwelt-, Milieu
Umgebungsbedingungen ambient conditions * tough: raue || environmental conditions || environments || working conditions * →Betriebsbedingungen
Umgebungsluft ambient air

Umgebungstemperatur ambient temperature * bei el. Masch. normalerweise max. 40 Grad C. nach VDE 0530 T.1 || ambient air temperatur
Umgebungsvariable environment variable
umgehen by-pass * Signal/Energie/Verkehrsfluss/Hochlaufgeber ~, vorbeistoßen an || go round * um etwas her~ || circulate * die Runde machen || deal with * mit etwas ~, behandeln, sich befassen mit || avoid * vermeiden || circumvent * ein Gesetz/eine Vorschrift ~ || bypass || handle * hantieren (with/of: mit)
umgehend as a matter of priority * hochprior || speedy * zügig || punctual * pünktlich || prompt * pünktlich
Umgehung by-pass || bypassing || bypass || by-passing
umgekehrt converse || the other way round || reverse * z.b. Richtung || vice versa * dasselbe ~ || v.v. * Kurzform für 'vice versa' || conversely * (Adv.) || by the same token * (Adv.) genauso, mit gleichem Recht || inverted * auch 'mit ~em Vorzeichen, hängend (montiert)' || opposite * entgegengesetzt, gegenteilig, widersprechend; the opposite holds good for: d. ~e gilt für || contrary * (meist figürlich) entgegengesetzt || inverse * in inverse proportion to: im ~en Verhältn. zu; inversely proportional: ~ proportional || just the other way round * gerade anders herum || quite the contrary * genau das Gegenteil || in the opposite direction * in ~e Richtung || reciprocal * reziprok, umgekehrt, wechsel-/gegenseitig
umgekehrt proportional inversely proportional * to: zu || inverse in proportion * to: zu || inversely in proportion * (Adv.) to: zu || reciprocal in proportion || in inverse proportion * to: zu
umgekehrte Richtung reverse direction
umgekehrter Schrägstrich back-slash
umgeschaltet changed-over
umgestalten redesign || modify || remodel || rearrange * anders/neu anordnen || reform * verbessern, reformieren || reorganize * neu/umorganisieren || alter || recast || transform
umhüllen sheathe * ummanteln, überziehen, armieren, z.B. Kabel || cover || wrap up * in: mit || envelope * with/in: mit || encase || mantle || coat
Umhüllende envelope
Umhüllung encapsulation || coating || envelope || jacket || casing || sheathing || covering || box
umkämpft highly-competitive * Markt usw.
Umkehr reversal * z.B. Drehrichtungs-~, Drehmomenten-~, Drehrichtungs~ || turning back || return * (auch figürl.) to: zu || inversion * z.B. Vorzeichen-~ || change * Änderung
Umkehr-Walzmotor reversing rolling motor
Umkehrantrieb reversing drive || four-quadrant drive * Vierquadrantantrieb || reversible drive
umkehrbar reversible * z.B. Drehrichtung
umkehren reverse || return || turn back || inverse * Vorzeichen
Umkehrgetriebe reversing gearbox
Umkehrmotor reversible motor * in der Drehrichtung
Umkehrpunkt reversal point
Umkehrschaltung reversible connection * 4-Quadrant Stromrichter
Umkehrstromrichter reversible converter || double converter * Doppelstromrichter mit 2 Stromrichterhälften, z.B. Gegenparallelschaltung || regenerative converter * rückspeisefähiger Stromrichter || dual-way converter || double-way converter || two-way converter
Umkehrverstärker inverting amplifier || reversing amplifier
umklappen toggle * zwischen zwei Zuständen || turn down * nach unten || fold back * nach hinten falten/knicken

umklemmen reconnect || change-over the terminal connections * *die Klemmenverdrahtung ändern*
Umkodierung code conversion || code converting
umladen recharge * *z.B. Kondensator*
Umladestrom charge/discharge current * *z.B. bei Kondensator, langer Leitung (auf Grund d. Leitungskapazität) usw.* || re-charging current * *z.B. bei einer langen Leitung auftretend*
Umlauf circulation || revolution * *(Um)Drehung* || turn
Umlauf- circulating
umlaufen rotate || circulate * *z.B. Flüssigkeit, Dokument* || revolve
umlaufend rotating || rotary || circulating || orbital || revolving
Umlaufintegral circulation
Umlaufliste poll list * *z.B. zum Aufruf der Slave-Teilnehmer vom Master bei Bussystem* || polling list * *z.b. bei Bussystem* || circulating list
Umlaufpuffer circular buffer || ring buffer || cyclic buffer
Umlaufpumpe circulating pump
Umlaufschmierung forced-circulation oil lubrication
Umlaufspeicher circulating buffer || circulating stack || ring buffer
Umlaufvermögen floating capital * *Vermögensteile, d.nur kurz i.e.Firma vorh.sind: Roh-/Fertigstoffe, Außenstände usw.* || current assets * *{Wirtschaft}*
Umlaufzeit bus cycle time * *eines Bussystems (wird benötigt, um jeden Teilnehmer einmal anzusprechen)*
umlegen turn down * *Gegenstand von der Vertikalen in die Horizontale* || apportion * *(Kosten, Gelder usw.)* umlegen/→zuteilen
Umleimer lipping * *{Holz}* || edge band * *{Holz}* || overlapping edge band * *{Holz}* um *die Kante herumgewickelter* ~
umlenken turn round * *herum-/zurücklenken* || turn back * *zurücklenken* || turn aside * *zur Seite lenken* || deflect * *ablenken, abbiegen (from: von), umlenken*
Umlenkrolle deflection roll || guide roller || deflector roll || guide pulley * *→Führungsrolle* || deviating roller
Umlenkung deflection
Umluft circulating air || recirculated air
Umluftfilter circulation filter
Umluftofen forced-air oven
Umlufttrockner circulating-air drier
Ummagnetisierung magnetic reversal || reversal of magnetization
ummanteln sheathe * *einhüllen, armieren, z.B. Kabel*
Ummantelung sheathing * *z.B. von Kabel* || wrapping || jacketing
Ummantelungs-Etikettiermaschine wrap-around labelling machine
Umnormierung renormalization || rescaling
Umorganisation reorganization || re-organization
umorganisieren re-organize || restructure * *umstrukturieren*
umpolen reverse the polarity
Umpolung polarity reversal
umrangieren reconnect * *Signale, Leitungen* || rearrange * *anders anordnen*
umrechnen convert * *into: in*
Umrechnung conversion * *Umwandlung* || transformation * *einer Gleichung* || recalculation * *z.B. einer Gleichung*
Umrechnungsfaktor conversion factor
Umrechnungsindex unit of measure * *(→PROFIBUS-Profil)*
Umrechnungstabelle conversion table
Umreifungsmaschine strapping machine * *Verpackungsmaschine z.Umreifen von Kartons/Paletten mit Kunststoff- o.Stahlband* || bundling machine * *{Verpack.technik} zum Verpacken von Paletten usw.*
umrichten convert
Umrichter converter || inverter * *Wechselrichter* || convertor * *(veraltete Schreibweise)* || frequency converter * *Frequenz~*
Umrichter im Motor motor-integrated inverter
Umrichter-Störung converter fault
Umrichterauslastung inverter load
umrichtergespeist converter-fed || converter powered * *Antriebssystem usw.* || inverter-fed
Umrichterleistung converter output * *Ausgangsleistung* || converter rating
Umrichternennstrom rated converter current
Umrichterscheinleistung apparent converter output || converter apparent power
Umrichterspeisung converter supply * *Speisung e. AC-Maschine durch e. Umrichter, nicht direkt vom Netz*
Umriss outline || contour
Umrisszeichnung outline drawing
umrollen rewind * *z.B. eine Papierbahn*
Umroller rewinder || rereeler || roll slitter * *→Rollenschneider* || slitter rewinder * *→Rollenschneidmaschine* || re-reeler
umrühren stir || stir up
Umrührer stirring unit || stirrer
umrüsten re-equip * *z.B. Werkzeugmaschine*
Umrüstzeit re-equipping time * *z.B. für eine (Textil-)Maschine* || change-over time || setting time
Umsatz sales volume * *turnover* || volume || sales || sales turnover
Umschaltautomatik automatic changeover circuit
umschaltbar switchable || switch-selectable * *mit Schalter wählbar* || with ratio selection * *Trafo* || reversible * *z.B. Motor mit umschaltbarer Drehrichtung* || jumper-selectable * *über (Steck-) Brücken* || selectable * *wählbar* || changeable * *änderbar*
Umschaltdrehzahl changeover speed
umschalten change over || switch over || flick * *an einem Handschalter* || toggle * *hin- und herschalten* || shift * *hinüberschieben* || throw over || shift * *Getriebe* || reconnect * *umverbinden/klemmen* || switch || change-over switching * *(Substantiv)*
Umschalter selector switch * *Mehrfach-Umschalter* || change-over switch || selector || multiplexer
Umschaltkontakt changeover contact || C contact || double-throw contact || DTC * *Abkürzg. f. 'Double-Throw Contact'; 4PDTC: 4-poliger Umschaltkontakt* || form 'C' contact || form C contact || two-way contact || SPDT contact * *Abk. für 'Single-Pole Double-Throw contact: einpoliger* ~ || form-C contact
Umschaltpause dead interval || dead interval on reversing * *bei Momentenumkehr (in Kommandostufe)*
Umschaltschwelle changeover threshold
Umschaltstrom change-over current
Umschaltung changeover || switchover || selection * *Aus/Anwahl z.B. über Schalter* || reversal * *Umkehr* || select * *(toggling* * *Hin- und Herschalten zwischen zwei Zuständen*
Umschaltvorgang changeover sequence || switch-over sequence || reversing process * *Drehrichtungs/Momentenrichtungsumkehr*
Umschaltzeit switch-over time || changeover time
umschließen enclose * *enthalten, beinhalten* || wrap around * *umhüllen, umwickeln*
umschließend wrapping around * *z.B. ein Kabel durch eine Schelle* || enclosing * *beinhaltend*
umschlingen entangle || loop around || wind around || loop || wrap around || clasp || embrace

Umschlingung entangling || hug || embrace || angle of contact * ~swinkel (bei Seilscheibe, Zugwalze usw.)
Umschlingungswinkel looping angle || angle of grip || angle of contact * für Riemen, Seilscheibe, Zugwalze usw.
umschlüsseln convert
Umschnürungsmaschine lacing machine * in der Verpackungstechnik
Umschwingdrossel ring-around reactor * unterstützt d. Kommutierg. in e. selbstgeführten Stromrichter
Umschwingen polarity reversal * Wechsel der Polarität || ring-around * eines Thyristors
Umschwingkreis ring-around arm * sorgt für d. Kommutierung i.e. selbstgeführten Stromrichter || ring-around circuit
Umschwingthyristor ring-around thyristor
umseitig overleaf * see overleaf : siehe ~ || on the reverse page * auf der Rückseite || on the next page * auf der nächsten Seite || on the reverse
umsetzen reposition * z.B. Bauteil an eine andere Position || convert * umwandeln || realize * realisieren || translate * →übertragen; translate ideas into action: Gedanken in die Tat ~ || rearrange * neu anordnen || put in effect * zur Wirksamkeit bringen || implement * realisieren, in die Tat ~ || reverse * {Bahnen} Zug/Waggon in die Gegenrichtung umrangieren
Umsetzer converter || adaptor || conditioner * Signalumsetzer || transducer * (Mess)Umformer || encoder * Kodierer, Verschlüsselungseinrichtung
Umsetzprinzip conversion principle * z.B. für A/D-Wandler
Umsetzung conversion * Umwandlung || transposition || transformation || realization * Realisierung || implementation * Realisierung, z.B. softwaremäßig
umso größer je the greater the
umspulen rewind
Umstand circumstance * in/under all circumstances: unter allen Umständen || fact * Tatsache, Tatbestand || detail * Einzelheit || condition * Umstände || state of affairs * Lage der Dinge || factors * Einflussfaktoren; favourable factors: günstige Umstände || particular || particularity * besonderer
umständlich complicated * verwickelt, kompliziert || troublesome * unbequem || fussy * (übertrieben) umständlich, geschäftig, aufgeregt, kleinlich, affektiert || circumstantial * Schilderung usw. || long-winded * langatmig || minute * sehr genau || detailed * in allen Einzelheiten || elaborate * komliziert, umständlich; sorgfältig
umsteigen step up * (figürl.) to: auf eine bessere Lösung usw. || change * bei einer Fahrt mit Verkehrsmittel (to: nach)
umstellen rearrange
Umstellung changeover
Umsteuern reversing
Umsteuervorgang changeover || reversing process * Momentenumkehr bei 4Q-Stromrichter
Umsteuerzeit reversal time || reversing time * zur Drehzahl- oder Momentenumkehr benötigte Zeit
umstrukturieren restructure || reorganize * umorganisieren
Umstürzen push-over
umverdrahten re-wire || modify the cabling || re-wire
Umverdrahtung rewiring || modification of the cabling
Umverpackung final packing || final wrapping
Umwälz- circulating * z.B. Pumpe, Ventilator
umwälzen circulate
Umwälzpumpe circulating pump
Umwälzung circulation * z.B. von Luft

umwandeln convert * auch Energie, Signale
Umwandlung conversion
Umwandlungsoperation conversion operation
Umwelt environment || world around us || global environment * global betrachtet
Umwelt- environmental || eco- * Öko-
Umweltbeauftragter environmental protection delegate || environmental officer
Umweltbedingungen environmental conditions
Umweltbeeinflussung environmental influence || impact on the environment
umweltbelastend environmentally detrimental
Umweltbewusstsein environmental awareness
Umwelteinflüsse environmental factors || environmental influences
umweltfreundlich environmentally friendly || eco-friendly || environmentally safe || environmentally green * (salopp) || green * (salopp) || environment-friendly || with minimum environmental impact * (Adv.) || environmentally compatible
Umweltklasse environmental class * nach DIN IEC 721-3-3
Umweltprüfverfahren environmental testing procedure
Umweltschutz environmental protection || environment protection
Umweltschutzbeauftragter environmental protection delegate || environmental officer
Umweltschutztechnik environmental protection engineering || environmemtal protection technology
Umwelttechnik environmental technology
Umweltverschmutzung environmental pollution || pollution || earth's pollution
umweltverträglich environmentally acceptable || eco-friendly * umweltfreundlich || environmentally green * (salopp) || green * (salopp) || environmentally safe || environmental compatible
umwickeln rewind * von einer Rolle auf die andere || wrap around || envelop * einwickeln, einschlagen, einhüllen || wind on || wind over || tape * mit Band ~/binden/heften
umziehen move * to: nach || relocate || remove * to: nach || change one's residence || change one's clothes * Kleidung
Umzug move || relocation
un- in- * nicht || un- * nicht || dis- * hinweg- || not * nicht || non- * nicht || -less * ohne || im- * nicht
unabdingbar essential * unbedingt erforderlich, wesentlich, (lebens)wichtig || indispensable * unerlässlich, unentbehrlich || indispensible * unerlässlich, unentbehrlich || imperative * unumgänglich, zwingend, dringend (nötig), unbedingt erforderlich || unalterable * unabänderlich, unveränderlich
unabhängig independent * of: von; as to whether ...or: davon ob...oder; of each other: voneinander || insensitive * unempfindlich || regardless * regardless of whether...or not: unabhängig davon, ob...oder nicht || self-contained * (in sich) geschlossen, selbstständig (arbeitend), autark, || absolute * unbeschränkt, unumschränkt (herrschend) || freelance * freiberuflich tätig || irrespective * irrespective of: ohne Rücksicht auf; irrespective of whether: ungeachtet dessen, ob
unabhängig davon ob regardless of whether * or not: oder nicht || regardless whether * or not: oder nicht || irrespective of whether * or not: oder nicht || irrespective whether * or not: oder nicht || regardless of the fact that || irrespective of the fact that || independent of the fact that || independent as to whether * or not: oder nicht
unabhängig von irrespective of * ~ Tatsachen/Umstand; ohne Rücksicht auf || regardless of * ~ Tatsachen/Umstand; ohne Rücksicht auf || independent of
unabhängig von der Drehrichtung regardless of

the direction of rotation || independent of the speed direction
unabhängig voneinander einstellbar independently adjustable * *z.B. Hoch- und Rücklaufzeit e. Hochlaufgebers*
Unabhängigkeit independence
unabsichtlich inadvertent * *auch 'versehentlich'* || unintentional * *unbeabsichtigt* || undesigned * *unbeabsichtigt* || involuntary * *unfreiwillig, unwillkürlich* || accidental * *zufällig* || erroneously * *(Adv.) versehentlich, irrtümlicherweise* || inadvertently * *(Adv.) ohne Absicht* || by a mistake * *(Adv.) versehentlich, irrtümlicherweise, fehlerhafterweise*
unachtsam inattentive || absent-minded || careless * *nachlässig* || negligent * *nachlässig* || inadvertent * *unachtsam, nachlässig, unabsichtlich, versehentlich*
unangemessen excessive * *zu hoch* || unsuitable || inadequate * *auch 'unzulänglich'* || incongruous * *aus der Proportion* || improper * *unschicklich*
unangenehm ugly * *garstig, widerwärtig* || unpleasant * *unangenehm, unerfreulich* || annoying * *ärgerlich, lästig, verdrießlich* || troublesome * *verdrießlich, lästig, beschwerlich* || irksome * *lästig, beschwerlich, ärgerlich, verdrießlich, beschwerlich* || unwelcome * *unwillkommen* || distasteful * *zuwider* || hateful * *zuwider, verhasst*
Unannehmlichkeit unpleasantness || difficulty || inconvenience * *Unbequemlichkeit, Lästigkeit, Schwierigkeit* || drawback * *Nachteil, Hindernis, 'Haken', Schatten/Kehrseite* || trouble * *~en* || disadvantage * *Nachteil*
unaufhaltsam unstoppable || uncheckable || relentless * *anhaltend, unbarmherzig, schonungslos, hart*
unbeabsichtigt unintentional || inadvertent * *auch 'versehentlich'* || undesigned || involuntary * *unfreiwillig, unwillkürlich* || accidental * *zufällig, unabsichtigt* || erroneously * *(Adv.) versehentlich, irrtümlicherweise* || inadvertently * *ohne Absicht* || by a mistake * *versehentlich, irrtümlicherweise, fehlerhafterweise*
unbearbeitet unmachined * *(durch Werkzeugmaschine)* || unfinished * *ohne Endbehandlung*
unbeaufsichtigt unattended || uncontrolled || without supervision || not looked after
unbedeutend insignificant || trifling || trivial
unbedingt unconditional * *unbedingt, bedingungslos (auch Programmsprung), uneingeschränkt, vorbehaltlos* || absolute * *absolut, unbeschränkt, unumschränkt, uneingeschränkt* || implicit * *absolut, vorbehaltlos, bedingungslos; z.B. Glaube, Anhänger, Vertrauen* || unquestioning * *bedingungslos, blind; z.B. Anhänger, Glaube, Vertrauen* || unreserved * *uneingeschränkt (z.B. Billigung), vorbehaltlos, rückhaltlos, völlig, offen, freimütig* || unconditionally * *(Adv.) bedingungslos* || absolutely * *(Adv.) bedingungslos* || by all means * *(Adv.) unter allen Umständen, mit allen Mitteln* || badly * *(Adv.) dringend; need badly: unbedingt/dringend brauchen* || urgently * *(Adv.) dringend*
unbedingt erforderlich imperative || essential
unbedingter Sprung unconditional jump * *im Programm*
unbeeinflusst unaffected * *unbeeinflusst/beeindruckt; by: von* || uninfluenced
unbefriedigend unsatisfactory || poor * *siehe auch 'ungenügend'*
Unbefugter unauthorized person
unbegrenzt unlimited || boundless || indefinitely * *(Adv.) unbeschränkt, auf unbestimmte Zeit* || absolute * *(Voll)Macht, Befugnis* || infinitely * *(Adv.)*

unendlich, ungeheuer, unbegrenzt; infinitely expandable: ~ erweiterbar
Unbehagen discomfort * *Unbehagen, (körperliche) Beschwerde* || uneasiness * *körperl./geistiges ~, (innere) Unruhe, Unbehaglichkeit*
unbehindert non-obstructed
unbekannt unknown || unfamiliar * *nicht vertraut*
unbekannter Befehl unused opcode * *{Computer} vom Befehlsdecoder als unbekannt zurückgewiesener (Maschinen-) Befehl* || non-recognizable command
unbelastet unloaded || off-load || idling || no-load
unbelüftet non-ventilated * *z.B. Motor ohne Lüfter (Wärmeabstrahlung über die Gehäuseoberfläche)* || surface-cooled * *oberflächengekühlt (wenig gebräuchlicher Ausdruck für Motorkühlart)* || free-convection * *→luftselbstgekühlt (bei Leistungshalbleitern)*
unbemannt unmanned || automated-guided * *selbstlenkend, fahrerlos*
unbemerkt unnoticed * *by: von*
unberechtigt unauthorized || unqualified * *unqualifiziert, unberechtigt, nicht approbiert; ungeeignet, untauglich, unbefähigt*
unberührt not involved * *(figürl.) nicht einbezogen* || untouched || virgin * *in der Natur* || not to be affected * *by: von Gesetz, Regel usw.* || not to fall within the scope of * *von Vorschrift. Gesetz usw. ~ bleiben*
unbesäumt unedged * *{Holz}*
unbeschädigt undamaged
unbeschaltet undamped * *Schutz/Relaisspule mit RC-Glied oder Diode zum Überspannungsschutz* || not connected * *Anschluss, Klemme* || open-circuit
unbestimmt indefinite || indeterminate * *undeutlich* || vague * *vage* || uncertain * *unsicher*
unbestritten uncontested || undisputed || without doubt * *(Adv.)* || irrefutable * *unwiderlegbar, nicht zu widerlegen* || unchallenged || indisputably * *(Adv.)*
unbestückt bare * *z.B. Leiterkarte*
unbestückte Leiterplatte bare board || bare PCB
unbewegliches Vermögen immovable property * *in Gebäuden, Grundstücken usw. steckendes Vermögen* || immovables * *in Gebäuden, Grundstücken usw. steckendes Vermögen*
unbezahlten Urlaub nehmen take unpayed time off work
unbrauchbar useless || of no use || unserviceable * *nicht betriebsfähig, z.B. Maschine* || waste * *Ausschuss/Abfall* || unworkable * *Plan usw.* || impracticable * *nicht praktikabel/durchfürbar (Plan usw.)*
unbrennbar non-inflammable || incombustible
unbürokratisch unbureaucratic || unbureaucratically * *(Adv.)* || without any difficulties * *(Adv.) problemlos* || smooth * *reibungslos* || simply * *(Adv.)* || straightforward * *unkompliziert, einfach, geradlinig*
UND AND
UND-Gatter AND gate
UND-verknüpft ANDed || AND'd
UND-Verknüpfung AND operation || AND logic
undefiniert undefined
undicht leaky || leaking * *leak: undicht sein* || tight || not tight || not watertight * *nicht wasserdicht* || not airtight * *luftdurchlässig, nicht luftdicht* || porous * *porös* || not waterproof * *nicht wasserdicht*
undurchlässig tight || impermeable
uneben uneven || not flat
uneingeschränkt unrestricted * *ungehindert* || unlimited * *unbegrenzt* || uncontrolled || full || unqualified * *unbedingt, uneingeschränkt, ausgesprochen*
UNEL UNEL * *ital. Norm entspr. DIN*

unempfindlich unaffected * *unbeeinflusst, unbeeindruckt (by: durch; to: gegen)* || insensible * *to: gegen* || insensitive * *to: gegen (Druck, Wettereinflüsse usw.)* || rigid * *robust* || impervious * *unempfindlich (to: gegen), unzugänglich (to: für), undurchlässig (to: für), taub*
Unempfindlichkeit insensibility || insensitivity || insensitiveness || ruggedness * *Robustheit*
Unempfindlichkeitsbereich dead band
unendlich unlimited || infinite || endless * *endlos*
unentbehrlich essential * *unbedingt erforderlich, (lebens)wichtig* || indispensable || imperative * *unumgänglich, zwingend, unbedingt erforderlich* || indispensible
unerfahren unexperienced || novice * *Anfänger-, Neuling-*
unerlässlich essential * *unbedingt erforderlich, (lebens)wichtig* || indispensable || imperative * *unumgänglich, zwingend (erforderlich)* || a must * *this feature is a must: diese Eigenschaft ist unerlässlich* || indispensible
unersetzbar irreplaceable
unerwartet unexpected || unforeseen * *unvorhergesehen, unerwartet* || surprise * *überraschend, Überraschungs-* || unanticipated || unexpectedly * *(Adv.)* || all of a sudden * *(Adv.) ganz plötzlich, auf einmal* || surprising * *überraschend*
unerwünscht undesirable || non-desirable * *nicht wünschenswert* || unwelcome || undesirably * *(Adv.)* || parasitic * *parasitär, schädlich, störend* || unwanted * *ungewünscht* || undesired
unfachgemäß inexpert || not correct || improper
Unfall accident
Unfallgefahr danger of accident || safety hazard
Unfallschutz prevention of accidents || accident precautions
Unfallverhütung accident prevention || prevention of accidents
Unfallverhütungsvorschrift safety rule || safety rules * *(Plural)* || safety standards * *(Plural)* || accident prevention regulation || safety regulation
Unfallverhütungsvorschriften accident prevention regulations || safety regulations * *Sicherheitsvorschriften* || accident prevention legislation * *gesetzliche ~*
unflexibel inflexible * *(auch figürl.; siehe auch →bürokratisch)*
unformatiert unformatted * *z.B. Text ohne Angaben von Schriftarten, Absatzgestaltung usw.*
Ungarn Hungaria
ungeachtet irrespective of || regardless of || despite * *trotz* || in spite of * *trotz*
ungedämpft undampened * *oscillation: Schwingung* || continuous * *Welle usw.*
ungedämpfte Schwingung undamped oscillation
ungeeignet not suitable || unfit * *for: zu* || unqualified * *untauglich, ungeeignet, unbefähigt (auch Person)* || inopportune * *z.B. Moment, Augenblick, Situation* || unsuitable
ungeerdet ungrounded || unearthed
ungeerdetes Netz ungrounded system || unearthed system || isolated supply system * *operation on/ from: Betrieb am* || ungrounded supply network * *aus der Sicht eines Verbrauchers (z.B. Motor, Stromrichter usw.)* || ungrounded supply * *z.B. Netzanschluss für Stromrichter/Motor (from: an)* || IT system * *siehe →IT-Netz*
ungefähr approximately || approx. || appr. || roughly * *→etwa, annähernd* || about || in the neighbourhood of || around
ungefährlicher Fehler non-fatal fault || non-fatal error
ungeglättet unsmoothed
ungehindert unrestricted || unlimited * *unbegrenzt*
ungekapselt non-encapsulated || unenclosed

ungeleimtes Papier waterleaf paper
ungelernter Arbeiter unskilled worker
ungenau inaccurate || inexact || vague * *(figürl.) vage* || hazy * *(figürl.) verschwommen, nebulös, vage* || incorrect * *unrichtig, ungenau, irrig, falsch*
Ungenauigkeit inaccuracy
ungenügend insufficient * *unzulänglich, untauglich, mangelhaft, unzureichend* || poor * *schwach, schlecht (siehe auch 'mangelhaft')* || inadequate * *unzulänglich, unangemessen* || under- * *(Vorsilbe) z.B. underpaid: unterbezahlt* || unsatisfactory * *unbefriedigend* || imperfect * *unvollkommen*
ungeordnet unarranged || unsettled || disorderly * *verwirrt, unordentlich, ordnungswidrig; disorder: ~e Verhältnisse* || neglected * *ungepflegt, vernachlässigt*
ungeplant unplanned || undesired * *unerwünscht*
ungepolt bipolar || non-polar * *z.B. Kondensator* || non-polarized * *Kondensator, Relais usw.*
ungepolter Kondensator non-polarized capacitor
ungerade odd
ungerade Zahl odd number
ungeradzahlig odd-numbered || odd || uneven
ungeradzahlige Oberschwingungen odd-order harmonics
ungeradzahlige Oberwelle odd-order harmonic
ungeregelt uncontrolled || unregulated * *z.B. Stromversorgung* || open-loop controlled * *gesteuert*
ungeregelte Stromversorgung unregulated power supply
ungeregelter Antrieb constant-speed drive || fixed-speed drive * *Festdrehzahlantrieb* || uncontrolled drive
ungeregeltes Netzteil unregulated power supply
ungesättigt unsaturated
ungeschirmt unscreened * *z.B. Kabel/Leitung* || unshielded * *z.B. Kabel/Leitung*
ungesichert unsecured || unprotected * *ungeschützt* || unguarded || non-backed-up * *Daten usw.* || exposed * *→offenliegend*
ungesteuert uncontrolled * *z.B. Brückenschaltung, Ventil, Gleichrichter* || non-controllable * *nicht steuerbar, z.B.Brückenschaltung aus Dioden* || non-controlled * *z.B. Gleichrichter aus Dioden (nicht mit Thyristoren)*
ungesteuerter Gleichrichter uncontrolled rectifier
ungesteuertes Ventil uncontrolled valve
ungeteilt unsplit || undivided * *(auch figürl.)* || solid
ungewöhnlich unusual * *unüblich* || abnormal * *unnormal* || uncommon * *it's uncommon for someone to do: es es für jemanden ungewohnt ... zu tun* || novel * *neuartig* || exceptional * *außergewöhnlich, ungewöhnlich (gut)* || odd * *sonderbar, seltsam, kurios, merkwürdig*
ungewollt unintentional || involuntary * *unfreiwillig* || unwanted * *→unerwünscht* || by mistake * *→versehentlich* || unintentionally * *(Adv.)*
ungewünscht unwanted * *siehe auch 'unerwünscht'*
unglaublich incredible || unbelievable
ungleich unequal || different || unlike * *unähnlich* || dissimilar || varying * *schwankend*
ungleiche Belastung unbalanced load
ungleichförmig irregular * *unregelmäßig; auch (Dreh)bewegung* || not uniform || unequal || non-uniform
Ungleichförmigkeit cyclic irregularity * *einer Drehbewegung, Ungleichförmigkeitsgrad* || irregularity || inequality || diversion * *Abweichung* || variation * *Schwankung, Abweichung* || uneveness || difference in shape * *Ungleichheit in der Form/im Aussehen* || rotational irregularity * *einer Drehbewegung*
Ungleichförmigkeitsgetriebe double drag-link gear * *z.B. zum Antrieb der →Messerpartie bei*

→*Querschneider* || drag-link gearing * *z.B. zum Antrieb der* →*Messerpartie bei* →*Querschneider*
Ungleichförmigkeitsgrad degree of irregularity || cyclic irregularity * *Stossgrad einer Drehbewegung*
ungleichmäßig non-uniform || uneven || unsymmetrical || disproportionate || unbalanced || irregular * *unregelmäßig*
unglücklicherweise unfortunately
ungültig invalid * *invalidate: für ~ erklären; be no longer valid: ~ werden* || unvalid || void * *(rechts)unwirksam, ungültig, nichtig, frei, unbesetzt* || null and void * *null und nichtig; declare null and void: für ~ erklären* || inoperative * *unwirksam, ungültig, nicht in Kraft (Gesetz usw.)* || illegal * *entgegen der Vorschrift, ungesetzlich*
ungültig machen cancel
ungünstig unfavourable * *auch nachteilig (z.B. Bedingungen), nicht dienlich* || disadvantageous * *nachteilig* || adverse * *nachteilig; affect adversely: ~ beeinflussen* || untoward * *widrig, ungünstig, unglücklich (Umstand usw.)*, widerspenstig, ungefügig* || unfavorable * *(amerikan.)* || onerous * *lästig, beschwerlich*
ungünstigst least favourable
unhörbar inaudible || non-audible || imperceptible * *nicht wahrnembar*
UNI UNI * *Abkürzg. für 'Ente Nazionale Italiano di Unificazione': italienisches Normenbüro*
universal universal || all-purpose * *Allzweck-* || multi-purpose * *Vielzweck-*
Universal- universal || general-purpose * *Allzweck* || all-purpose * *Allzweck* || multi-purpose || multi-range * *Mehrbereichs-, z.B. Messgerät* || multi-
Universal-Messgerät multimeter
Universalbauform universal type of contruction
Universaldiode universal diode
Universalmotor AC-DC motor || universal motor * *mit Kommutator für Gleich- und Wechselstromanschluss in Hauptschlussschaltg.*
Universalverstärker universal amplifier
Universalverstärkerbaugruppe OpAmp board || general-purpose amplifier board
universell universal || allround || all-purpose
Universität university || collegiate * *Universitäts-, akademisch*
unklar ambiguous * *zwei/mehrdeutig* || indistinct * *undeutlich* || vague * *vage* || not clear || obscure * *undeutlich* || muddy * *trüb* || misty * *neblig* || muddled * *verworren* || woolly * *verschwommen, wolkig (z.B. Gedanken)* || fuzzy * *Gedanken* || be in the dark * *im Unklaren sein* || leave guessing * *im Unklaren lassen* || inexact * *ungenau* || inaccurate * *ungenau* || wooly * *siehe 'woolly'*
Unklarheit ambiguity * *Zwei-/Mehrdeutigkeit* || vagueness * *Verschwommenheit, Unbestimmtheit, Undeutlichkeit, Unverständlichkeit* || obscurity * *Undeutlichkeit, Unverständlichkeit* || want of clearness || open points * *offene Fragen* || uncertainty * *Unsicherheit* || unclarity
unkompensiert non compensated * *z.B. Motor, Tachometermaschine* || uncompensated * *z.B. Motorwicklung*
unkompensierter Motor uncompensated motor * *ohne* →*Kompensationswicklung*
unkompliziert uncomplicated || simple * *einfach* || straight-forward * *(Personen u. Sachen)* →*einfach, direkt, offen,gerade(her)aus, freimütig, aufrichtig*
unkontrollierbar uncontrollable * *nicht unter Kontrolle zu halten, nicht regelbar* || unverifiable * *nicht nachprüfbar*
unkontrolliert uncontrolled || unchecked * *ungeprüft*

Unkosten costs || expenses || charges || overhead expenses * *allgemeine Kosten, (All)Gemeinkosten* || overheads * *allgemeine Kosten, (All)Gemeinkosten*
unlauterer Wettbewerb unfair competition
unlegiert unalloyed
unlegierter Stahl unalloyed steel || carbon steel * *Kohlenstoffstahl*
unlöslich insoluble * *durch Lösungsmittel* || permanent * *unlösbar*
unmerklich imperceptible
unmittelbar direct || immediate
unmittelbar hintereinander consecutively
unmittelbar vorher immediately beforehand
unnormal abnormal
unnormiert non-normalized
unnötig unnecessary
unpassend unsuited || unsuitable || inappropriate * *unangebracht* || out of place * *unangebracht, deplatziert* || improper * *unschicklich* || unseasonable * *zur Unzeit* || untimely * *zur Unzeit*
unpassierbar impassable
unprogrammiert blank * *z.B. EPROM-Speicher*
unqualifiziert unqualified
unregelmäßig irregular || erratic * *ungleichmäßig, regel-/ziellos, unstet, unberechenbar, sprunghaft*
Unregelmäßigkeit irregularity
unruhiger Lauf unsmooth running || irregular running * *unruhiger Lauf* || uneven running || unsteady running
unrund out of round || eccentric || untrue || out of true || noncircular * *nicht kreisförmig* || cornered * *eckig* || distorted * *Drehfeld usw.*
unrund laufen run out of true || run untruly || cog * *ruckhafter Lauf bei kleinen Drehzahlen* || run unsmoothly
unrunder Lauf unsmooth running || cogging * *ruckhafter Lauf bei kleinen Drehzahlen* || jerky running * *ruckhafter Lauf* || untrue running
Unrundheit out-of-round || untrue running * *unrunder Lauf*
unsachgemäß incorrect || unprofessional || careless * *unvorsichtig* || carelessly * *(Adv.) unvorsichtig* || improper * *improper usage/handling: ~e Behandlung*
unscharf blurred || fuzzy * *['fassi] unscharf, verschwommen, kraus, flockig, flaumig, faserig, fusselig* || purely defined || out of focus || hazy * *(figürl.) verschwommen, nebelhaft, nebulös, dunstig*
unscharfe Mengen fuzzy sets
Unschärfe lack of definition || insufficient focus || unsharpness || focus-shift * *Fehlfokussierung, Unschärfe (auch auf einem Bildschirm)* || uncertainty * *(figürl.)* || diffusiveness * *Verschwommenheit*
unschlagbar unbeatable
Unsicherheit uncertainty * *Ungewissheit, Unverherseharkeit* || insecurity || unsteadiness * *Schwanken*
unsichtbar non-visible || invisible * *to: für; siehe auch* →*verborgen* || imperceptible * *nicht wahrnehmbar*
unstabilisiert non-stabilized * *z.B. Versorgungsspannung/Netzteil*
unstetig discontinuous || unsteady
Unstetigkeit unsteadiness || inconstancy
Unsumme enormous sum
Unsymmetrie unbalance || unsymmetry || dissymmetry || imbalance
unsymmetrisch asymmetric || asymmetrical || unsymmetric
unsystematisch unsystematic || unmethodical
untätig idle || inactive
Untätigkeitsschleife idle loop || endless loop * *Endlosschleife* || unproductive loop

unten below * *see below: siehe ~; as below: wie ~ näher bezeichnet; far below: weit ~* || beneath * *unterhalb* || downstairs * *~ im Haus, ~ an der Treppe* || downwards * *nach ~; facing downwards: nach ~ zeigend; levelled/pointed downwards: nach ~ gerichtet* || at the foot of * *am Fuße von* || at the bottom of * *~in Gefäß/im Wasser/auf Buchseite; bottom up: d. Untere nach oben gedreht, kieloben* || down there * *da ~* || bottom * *from the bottom: von ~; from top to bottom: von ~ bis/nach oben; towards the bottom: nach ~* || right up from below * *von ~ an* || underside * *from the underside: von ~* || down there * *da ~; down on the left: links ~* || down at the bottom * *am Boden, auf dem Grund* || from below * *von ~, von ~ herauf* || down * *deep down: weit ~; deeper/further down: weiter ~; from down there: von hier ~ (aus)*
unten erwähnt mentioned below * *(nachgestellt)* || below * *(amerikan.) (Adj. nachgestellt) (Adv.) please find below: unten stehend finden Sie* || undermentioned || hereinafter * *(Adv.)* || hereinbelow * *(Adv.)*
unten links at the bottom left
unten rechts at the bottom right
unten stehend mentioned below * *(nachgestellt)* || below * *(Adj. (nachgestellt) und Adv.) please find below: unten stehend finden Sie* || as mentioned below || hereinbelow * *(Adv.)* || hereinafter * *(Adv.)* || undermentioned
untenliegend located at the bottom || mounted underneath * *unterhalb montiert* || bottom-mounted * *unten montiert*
unter under * *auch ~ einem Namen* || below * *auch ~halb einer Zahl/eines Wertes* || beneath || underneath || among * *zwischen, mitten ~, bei; among other things: ~ anderem* || amongst * *zwischen, mitten ~, bei* || in the midst of * *mitten ~* || low * *niedrig* || lower * *niedrig(er)* || inferior * *~geordnet, geringer(wertig)* || minor * *klein(er), unbedeutend(er), geringfügig, ~geordnet (Rang)* || during * *während* || from under * *~ hervor* || amidst * *mitten ~* || less * *less than: weniger/niedriger als* || bottom * *untenliegend; bottom half: ~e Hälfte*
unter allen Umständen in any case || at all events || by all means * *auf alle Fälle, mit allen Mitteln* || in all circumstances
unter anderem among other things || including || such as || inter alia
unter Auftrag nehmen contract
unter Belastung under load || on load
unter Berücksichtigung von taking into account || in consideration of || with regard to || considering that || under consideration of || with due regard to * *unter gebührender Berücksichtigung von*
unter den Tisch fallen lassen discard * *ausscheiden, (Gewohnh.,Vorurteil) ablegen; aufgeben, entlassen* || neglect * *→vernachlässigen*
unter der Annahme assuming
unter der Bedingung provided * *that: dass (kann auch weggelassen werden)* || on condition * *that: dass* || under the condition
unter der Bedingung dass provided || provided that || on condition that || under the condition that
unter der Voraussetzung assuming * *angenommen; that: dass* || on condition * *that: daß* || on the understanding * *that: dass*
unter diesen Umständen under these circumstances || as matters stand || in these circumstances
unter keinen Umständen under no circumstances || on no account || not on any account
unter keiner Bedingung on no account
unter Last on load || loaded * *belastet* || under load
unter Spannung under voltage || live * *live removal: Ziehen ~* || hot * *hot unplugging: Ziehen ~* || with the power on

unter Tage underground
unter Umständen possibly * *möglicherweise* || it is possible that * *es ist möglich, dass* || perhaps * *vielleicht* || if necessary * *falls erforderlich* || if need be * *notfalls, nötigenfalls* || under certain circumstances
unter Vernachlässigung von neglecting || leaving out of account * *außer Betracht lassend*
Unter- sub- * *untergeordnet* || lower || bottom
Unterbegriff derivative term * *abgeleiteter Begriff* || subsumption * *(gemeinsame) Einordnung, Zusammenfasssung (under: unter)*
unterbieten undercut * *Preis ~* || dump * *auf dem Weltmarkt ~* || undersell * *die Konkurrenz ~* || lower * *Rekord ~*
unterbrechen interrupt || suspend * *vorübergehend anhalten* || stop * *anhalten* || break * *auch Leiterbahn* || cut short || disconnect * *Leitung, Stromkreis* || stop || hold up * *aufhalten, hindern* || isolate * *(potenzialmäßig) trennen, z.b. Stromkreise* || discontinue * *unterbrechen, →einstellen, aussetzen, aufhören (to do: etwas zu tun)*
unterbrechender Kontakt break-before-make contact * *Wechselkontakt, der unterbrechend (ohne Überlappung) schaltet* || non-bridging contact * *Wechselkontakt, der ohne Überbrückung arbeitet*
Unterbrechung disconnection * *einer (Leitungs)Verbindung* || interruption || break || suspension * *Aussetzung, Hemmung, vorübergehende Aufhebung, Aufschieben* || stopover * *~ der Fahrt* || open circuit * *Stromkreis-~* || disruption * *Unterbrechung, Zerreißung, Spaltung; Zerrüttung, Zerfall*
Unterbrechungsanzeigewort interrupt condition code word * *bei SPS*
unterbrechungsfrei uninterrupted * *z.B. Stromversorgung, Produktion, Betrieb* || uninterruptible * *nicht unterbrechbar, z.B. Stromversorgung* || interruption-free
unterbrechungsfreie Stromversorgung uninterruptible power supply * *UPS * Abkürzg. für 'Uninterruptible Power Supply'* || uninterrupted power supply || uninterruptable power supply
unterbrechungsfreier Betrieb uninterrupted duty
Unterbrechungszeit interruption time
unterbringen accommodate * *(räumlich) auch 'beherbergen', auch von Komponenten/Geräten* || place * *(auch figürlich; z.B. Person, Aufträge usw. ~)* || house * *beherbergen* || store * *lagern* || install * *installieren, einbauen, anbringen, montieren, aufstellen* || fit * *into: in* || incorporate * *einbauen, (in sich) aufnehmen, enthalten*
unterbrochen interrupted || disconnected * *Stromkreis* || discontinued
Unterdeck-Aufstellung under-deck installation * *z.B. v. El.motoren auf Schiffen unt.Deck (Temp.-/Feuchte-/Rüttelbelastg.!)*
unterdimensionieren undersize || underdimension
Unterdruck low air pressure * *Luftdruck* || negative air pressure * *Luftdruck* || partial vacuum || vacuum || negative pressure
unterdrücken suppress || oppress * *bedrücken*
unterdrückt suppressed
unterdrückter Nullpunkt suppressed zero
Unterdrückung suppression || rejection
unterdurchschnittlich sub-average || below normal || below-average
untere Grenze lower limit
untereinander one beneath the other * *(räumlich)* || one with another * *miteinander* || among one another || mutually * *gegen-/wechselseitig* || with respect to each other * *aufeinander bezogen, →gegeneinander*
unterer Grenzwert lower limit value
unterer Leistungsbereich low-end performance

range * *bezüglich der Leistungsfähigkeit* || lowrating range * *bezüglich der (Aus-/Eingangs-) Leistung* || low-output range * *mit kleiner Ausgangsleistung,z.b. Stromrichter, Motor* || fractional-hp range * *im Leistungsbereich unter 1 PS (z.B. Motor)* || low-end
unterer Strich bottom line
Unterfräsmaschine spindle moulder * *{Holz}* || table moulding machine * *{Holz}*
Unterfunktion subfunction * siehe auch →*Teilfunktion*
untergebracht accommodated * *z.B. in einem Gehäuse* || housed || located * *örtlich/räumlich* || placed || installed || fitted
untergeordnet subordinate || secondary
Untergliederung subdivision
Untergrenze lower limit
Untergruppe sub-group
unterhalb below || not exceeding * *nicht überschreitend (Wert, Zahl usw.)* || underneath * *unter* || beneath
Unterhaltung maintenance * *Wartung, Instandhaltung* || servicing * *Wartung* || upkeep * *Instandhaltung* || conversation * *Gespräch* || talk * *Gespräch* || entertainment * *Vergnügen*
Unterhaltungselektronik entertainment electronics || consumer electronics * *Konsumelektronik*
Unterhandlung negotiation * siehe auch →*Verhandlung*
Unterkante bottom edge || lower edge
Unterkette secondary sequencer * *bei SPS*
unterkritisch sub-critical || subcritical
Unterlage document * *schriftliche* || base * *Auflage, Unterbau, Fundament* || paper * *Papier, Dokument, schriftliche Ausarbeitung*
Unterlagen material * *schriftliche* ~ || documents * *schriftliche* ~ || documentation
Unterlagenverzeichnis document listing
unterlagert subordinate || secondary * *untergeordnet* || subordinated * ~ *angeordnet, z.b. Regelkreis (to: einer Einrichtung)* || inner * *z.B. inner current control loop: ~e Stromregelung* || cascaded * *hintereinander geschaltet/angeordnet*
unterlagerte Stromregelung inner current control loop || secondary current control
unterlagerter Regelkreis minor control loop || inner control loop || secondary control loop || subordinate control loop || subordinated control circuit
unterlagerter Regler subcontroller || subordinated controller || inner control loop * *unterlagerte Regelung* || subordinate controller
Unterlauf borrow * *Unterlaufbit/Merker bei Rechenoperationen. z.B. Subtraktion* || underflow * *bei Rechenoperation*
Unterlegscheibe washer
Unterlieferant subcontractor
unterliegen undergo * *ausgesetzt sein* || be governed by * *Vorschrift, Gesetz* || be liable to * *unterworfen sein, z.B. Gebühr, Zoll, Steuer, Gewährleistung*
Untermaß undersize || minus allowance * *zulässiges*
untermauern confirm || back up * *mit Rückendeckung versehen* || bolster up * *unterstützen/-mauern*, *unterfüttern* || corroborate * *bekräftigen, bestätigen, erhärten*
Untermenge subset || sub-quantity
Untermenü sub menu
Untermotor lower motor * *Walzwerk* || bottom-roll motor * *Walzwerk*
untermotorisiert under-powered
Unternehmen enterprise || corporation * *großes* || company * *Firma* || firm * *Firma*
Unternehmensberater management consultant
Unternehmensberatung consulting company
Unternehmensbereich company group || division

Unternehmenserfolg company success
Unternehmensgruppe group of companies
Unternehmenskultur corporate culture
Unternehmensleitsatz corporate mission guideline
Unternehmensprofil company profile
Unternehmergeist entrepreneurial spirit || spirit of enterprise || entrepreneurship * *Unternehmertum*
Unterölmotor oil-immersed motor * *in (Hydraulik-)Öl betriebener Motor, z.b. Hydraulikpumpenmotor*
Unterprogramm subroutine || procedure
Unterricht teaching || training || drill * *scharfer*
unterschätzen undervalue || underestimate * *Fähigkeiten usw.* || underrate
unterscheidbar distinguishable || discernible
unterscheiden distinguish * *between: zwischen* || make a distinction * *between: zwischen* || differentiate * *Unterschied machen/hervorheben/differenzieren; between: zwischen* || differ * *sich unterscheiden; from: von* || discriminate * *(scharf)* ~, *e.Unterschied machen, unterschiedl. behandeln; d. against: benachteiligen*
Unterscheidung differentiation || distinction || discrimination || difference * *Unterschied*
Unterscheidungskriterium differentiating criterion || distinctive mark * *Unterscheidungsmerkmal* || distinctive feature * *Unterscheidungsmerkmal*
Unterscheidungsmerkmal differentiating criterion || distinctive feature || distinctive mark || distinctive characteristic
Unterschied difference * *between: zwischen* || distinction * *Unterscheidg.(smerkmal), Unterschied; make a distinction between: d.* ~ *machen zwischen* || variation * *Veränderung, Abweichung, Schwankung, Wechsel* || mismatch * *Nicht-Übereinstimmung, Nicht-Zusammenpassen*
unterschiedlich different * *significantly different: stark* ~ || differing * *nicht übereinstimmend, schwankend* || variable * *schwankend* || varying * *schwankend* || uneven * *schwankend* || varied * *abwechslungsreich, mannigfaltig, verschieden (artig), varriiert* || diverse * *verschieden(e), ungleich, mannigfaltig* || diversified * *verschieden (artig), mannigfaltig, abwechslungsreich*
Unterschnitt undercut * *Unterschnitt, Unterhöhlung, {Werkz.masch.}* →*Freistich*
unterschreiten fall below || be less than
Unterschrift signature
unterschwingen undershoot * *(auch Substantiv)*
Unterschwingung undershoot
Unterseil tail rope * *{Aufzüge | Seilbahn}* || bottom cable * *{Aufzüge}* || lower rope * *{Aufzüge}* || bottom rope * *{Aufzüge}*
Unterseite bottom side || underside || bottom face * *untere Fläche*
untersetzen step down * *Trafo, Getriebe* || gear down * *Getriebe*
Untersetzer scaler * *z.B. Frequenzteiler für Impulsketten*
untersetzt reduction * *Getriebe* || geared down * *Getriebe*
Untersetzung reduction * *v. Getriebe* || gear reduction * *v. Getriebe* || stepping down * *Getriebe, Trafo usw.* || reduction ratio * *Untersetzungsverhältnis* || speed reduction
Untersetzungsgetriebe reduction gear * *zur Drehzahlreduzierung* || speed reducer || gear reducer || step-down gear || reduction gearbox
Untersetzungsverhältnis reduction ratio * *z.B. eines Getriebes*
Unterspannung undervoltage || under-voltage
Unterspannungsabschaltung undervoltage trip * *Fehlerabschaltung auf Grund unzulässig kleiner Spannung*
Unterspannungsauslöser undervoltage release

Unterspannungserfassung undervoltage sensing
Unterspannungsschutz undervoltage protection
Unterspannungsüberwachung undervoltage monitoring || undervoltage monitor * *als Schaltung/ Einrichtung* || undervoltage monitor module * *als Baugruppe/Gerät* || undervoltage monitoring module * *als Baugruppe/Gerät*
Unterstation substation
unterstreichen underscore * *(auch figürlich: betonen)* || highlight * *heraus/hervorheben, betonen, herausstreichen, Schlaglicht setzen* || emphasize * *(figürl.) betonen, hervorheben, unterstreichen* || underline * *Text mit einer Linie ~*
Unterstrich underscore * *Zeichen '_' auf Tastatur/ Bildschirm/Drucker*
unterstützen support || back up || assist * *helfen* || aid * *helfen* || second * *beistimmen, sekundieren* || carry * *Antrag* || back || promote * *fördern, unterstützen, (im Rang) befördern, werben für, organisieren*
unterstützende Maßnahmen supportive measures
Unterstützung support || assistance * *Hilfe, Beistand* || aid * *Hilfe, auch finanzielle ~* || subsidy * *staatliche Gelder* || benefit payment * *~saufwendung* || benefits * *~sleistungen* || promotion * *Förderung, Befürwortung, Beförderung* || back-up service
untersuchen examine || prüfen || check * *überprüfen* || overhaul * *überholen* || inquire * *untersuchen, erforschen* || inspect * *untersuchen, prüfen, nachsehen, besichtigen, inspizieren* || scrutinize * *genau ~* || test * *for: auf* || explore * *erforschen* || investigate * *technisch/wissenschaftlich/juristisch ~* || analyze * *analysieren* || lab-examine * *im Labor ~* || go over * *(gründlich) überprüfen/untersuchen, z.B. Maschine*
Untersuchung examination || investigation || scrutiny * *genaue* || inquiry || test || analysis * *Analyse* || treatise * *Abhandlung* || survey * *Übersicht* || check * *Überprüfung* || exploration * *Erforschung, Untersuchung* || survey * *Prüfung, (Ab-) Schätzung, Begutachtung, Umfrage, Überblick, Aufnahme (v.Tatbeständen)*
untersynchron subsynchronous * *converter cascade: Stromrichterkaskade*
untersynchrone Stromrichterkaskade subsynchronous converter cascade * *I-Umrichter f. Schleifr.-läufer-Async.mot. (Ständ.am Netz; 1..20 MW)*
Untertage underground mining
Untertagemotor mining motor * *Bergbaumotor*
Untertagetechnik underground mining
untertauchen submerge * *untersinken, untertauchen, überschwemmen, unter Wasser setzen, übertönen* || immerse * *(ein)tauchen, versenken, (sich) vertiefen (figürl.), verwickeln (figürl.; in: in)* || dip * *eintauchen* || submersion * *(Subst.) Ein-/Untertauchen, Überschwemmung*
Unterteil lower part || holder * *Halter; z.B. fuse holder: Sicherungsunterteil* || lower section
unterteilen subdivide || break down * *statistisch ~* || classify * *in Gruppen/Klassen ~* || sectionalize * *in Abschnitte (räumlich) ~* || split * *aufspalten, sich aufspalten* || group * *in Gruppen ~*
unterteilt divided * *into: in* || subdivided * *into: in* || broken down * *statistisch* || classified * *in Klassen* || grouped * *in Gruppen*
Unterteilung subdivision
Unterwalze bottom roll || lower carrier roll
Unterwasser-Motor underwater motor
unterwerfen submit to * *einer Prüfung, einem Schiedsgericht usw.; auch 'sich ~'* || subdue || be subject to * *(einer Sache/Vorgehensweise) unterworfen sein*
Unterwerk substation
unterziehen undergo * *sich unterziehen, unterzogen werden*

unübertroffen unsurpassed || unmatched || unexcelled || unequalled
unüberwindbar unsurmountable * *Probleme, Schwierigkeit* || invincible * *Barriere, Schwierigkeit, Widerstände, Festung usw.*
unüblich uncommon
unumgänglich imperative * *unbedingt erforderlich* || essential * *unbedingt erforderlich, (lebens)wichtig* || indispensible * *unerlässlich/entbehrlich*
ununterbrochener Betrieb continuous duty * *Motor- u. Stromrichter'betriebsart S6 bis S8, siehe auch →Durchlaufbetrieb* || continuous operation duty type * *als Betriebsart* || continuous operation
ununterbrochener Betrieb mit Anlauf und elektrischer Bremsung continuous-operation periodic duty type with starting and electric braking * *Mot. Betr.art S7 VDE0530*
ununterbrochener Betrieb mit periodischer Drehzahländerung continuous-operation duty type with related load/speed changes * *Motorbetriebsart S8 nach VDE0530*
unveränderlich unchangeable || invariable || consistent * *fest* || stable * *stabil* || fixed * *fest* || unalterable * *unabänderlich, unveränderlich*
unverändert unchanged * *remain unchanged: unverändert bleiben* || unaltered || the same as before || the same * *keep the same: unverändert beibehalten*
unverändert beibehalten keep the same
unverbindlich non-binding * *nicht bindend* || non obligatory * *ohne Verpflichtung, nicht (rechts)verbindlich* || informal * *zwanglos, nur zur Information* || non-committal * *neutral, zurückhaltend, nichts sagend (Stellungnahme usw.)* || tentatively * *(Adv.) versuchsweise, probehalber* || tentative * *Versuchs-, Probe-, vorläufig, provisorisch*
unvereinbar opposing * *to: mit* || incompatible * *with: mit*
unvergleichlich unparalleled || incomparable || peerless || unique * *einzigartig*
unvergossen unpotted * *z.B. Baugruppe*
unverlierbar captive * *z.B. Schraube* || never-lost
unverlierbare Mutter captive nut || clip nut * *mit Montage-/Halteklammer*
unverlierbare Schraube captive screw
unvermeidbar unavoidable || inevitable || unfailing || inevitably * *(Adv.)*
unvermeidlich inevitable || unavoidable || unfailing || inevitably * *(Adv.)*
unvermindert unimpaired * *(Adj.) ungeschwächt, ungeschmälert, unvermindert, unverhindert* || undiminished * *(Adj.) nicht reduziert* || with undiminished violence * *(Adv.) mit unverminderter Kraft/ Gewalt*
unverpackt unpacked || loose * *lose*
unverständlich unintelligible || incomprehensible || inconceivable || obscure * *z.B.(Beweg-) Gründe* || complicated * *verwickelt, kompliziert* || cryptic * *rätselhaft, verborgen* || fussy * *unverständlich, kleinlich, heikel, verschwommen* || troublesome * *~unbequem* || elaborate * *~kompliziert, →umständlich,; sorgfältig*
unverträglich incompatible
Unverträglichkeit incompatibility
unverwechselbar non-interchangeable || keyed * *mit Kodierung versehen (z.B. Stecker)* || noninterchangeable
unverzögert non-delayed || undelayed || immediate * *unverzüglich* || instant * *unverzüglich* || without delay * *(Adv.)* || at once * *(Adv.) unverzüglich* || instantaneous * *unverzüglich, sofortig, augenblicklich (z.B. Ansprechen e. Schutzeinrichtung)* || highspeed * *Hochgeschwindigkeits-* || quick-acting * *schnell arbeitend (z.B. Schalter)*
unvollkommen imperfect || defective * *mangelhaft,*

unvollkommen, unvollständig || wanting * *fehlend, mangelnd; be in wanting of: z. wünschen übr. lassen bezügl./nicht gerecht werden*
Unvollkommenheit imperfection
unvollständig incomplete
unvorhergesehen unforeseen
Unvorhergesehenes unforeseen
unvorschriftsmäßig improper * *unsachgemäß* || irregular || contrary to the regulations
unvorsichtig incautious || inconsiderate * *unüberlegt* || imprudent * *unklug* || rash * *übereilt, unüberlegt* || careless * *sorglos, nicht sorgfältig*
Unvorsichtigkeit incautiousness || carelessness || imprudence * *Unklugheit*
unwahrscheinlich unlikely || improbable
unweigerlich inevitable * *unvermeidlich, zwangsläufig* || inevitably * *(Adv.)*
unwesentlich negligible * *vernachlässigbar* || insignificant * *geringfügig* || inessential * *nebensächlich* || unimportant * *unwichtig*
unwirksam ineffective || not effective * *nicht wirksam, nicht in Kraft*
Unwirksamkeit inefficiency || inefficacy || inoperativeness || inactivity * *{Chemie}* || futility * *Vergeblichkeit, Zweck-/Nutz-/Wertlosigkeit; Nichtigkeit*
unwirtschaftlich uneconomic || unprofitable
Unwucht unbalance * *nicht ausgeglichene Fliehkräfte (z.B. U in [gmm] entspr. u (Unwuchtmasse [g]) x r [mm])* || unbalanced state
Unwuchtmessgerät unbalance measuring equipment
unzerstörbar indestructible || non-destructive
unzugänglich inaccessible || unapproachable
Unzulänglichkeit shortcoming * *auch 'Fehler, Mangel, Pflichtversäumnis, Fehlbetrag'*
unzulässig inadmissible || undue || inacceptable * *untragbar, unannehmbar* || unlawful * *ungesetzmäßig* || invalid * *ungültig* || unrecognized * *nicht anerkannt*
unzureichend insufficient || poor * *armselig, schlecht, mager, mangelhaft, jämmerlich* || inadequate * *unzulänglich, ungenügend, →mangelhaft, unangemessen*
unzuverlässig unreliable || undependable || uncertain * *unsicher* || unsafe * *unsicher* || shaky * *Methode usw.* || treacherous * *trügerisch* || nonreliably * *(Adv.)*
Unzuverlässigkeit unreliability || undependability || uncertainty || unthrustworthiness || treacherousness
unzweideutig unambiguous
upm rpm * *revolutions per minute [rewol'u: schen ...] : Umdrehungen pro Minute* || r.p.m. || rev./min || revolutions per minute || revolutions/ minute
Urheberrecht copyright
URL URL * *{Network} Abk. für 'Universe Resource Locator': Adresse im World Wide Web des Internet*
urladen initial loading * *z.B. Programm ~* || bootstrap * *Programme usw.* ~ || bootstrap loading * *Programm ~* || establish the factory default condition * *Parametersatz ~* || preset the factory values * *Parameter, Datensätze usw. ~* || initial program loading * *Programm ~*
Urladewert factory setting * *eines Parameters* || initial value * *eines Parameters* || default setting * *eines Parameters*
Urladezustand initialized status || factory setting * *Werkseinstellung z.B. von Parametern* || factory default condition * *Werkseinstellung z.B. von Parametern* || 'as delivered' condition * *Auslieferungszustand*

Urlaub holiday * *take a holiday: ~ nehmen; be on holiday: in ~ sein*
Urlöschen overall reset * *z.b. bei SPS* || general reset * *bei SPS* || ERASE PROGRAM * *Programm (im RAM-Speicher) löschen bei SPS*
Ursache cause * *the cause lies in: die ~ liegt/ist zu suchen bei/in* || reason * *Grund* || occasion * *Anlass* || motive * *Beweggrund*
ursächlich causal || causative
Ursprung origin * *originate in/from: seinen ~ haben in; auch ~ eines Koordinatensystems* || source * *Quelle, Ursprung, Herkunft (auch einer Nachricht)*
ursprünglich initial * *anfänglich* || original || in the beginning * *(Adv.)* || at first * *(Adv.)*
Ursprungsdokument original document
Ursprungserzeugnis original manufacturer's equipment
Ursprungszeugnis certificate of origin
USA U.S.A.
USA und Canada North America
USART USART * *Universal Synchronous/Asychronous Receiver/Transmitter: Schaltg. f. sychr./ asynchr. Schnittst.*
USV UPS * *Abk.f. 'Uninterruptible Power Supply': Unterbrechungsfreie Stromversorgung* || uninterruptible power supply * *Unterbrechungsfreie Stromversorgung* || uninterruptable power supply || uninterrupted power supply
USV-Anlage UPS system * *Abk. f. 'Uninterruptible Power Supply': unterbrechungsfreie Stromversorgung*
usw etc || and so on || so on and so forth * *und so weiter und so fort*
UTE UTE * *Abkürzg. für 'Union Technique de l'Electricite': französ. elektrotechn. Vereinigung*
UV-löschbar UV-erasable * *z.B. Speicherbausteine*
UV-Löscheinrichtung UV eraser
UV-Löschgerät UV eraser * *zum Löschen von EPROMS*

V

V-Ring V-ring
V-Schaltung V-connection * *Messschaltung zur Stromistwerterfassung mit 2 Stromwandlern und Diodenbrücke*
V24-Schnittstelle RS232 link || RS232 port * *Anschluss* || RS232 interface
VAC VAC * *Volt Wechselspannung*
Vakuum vacuum
Vakuum-Fluoreszenz-Anzeige vacuum-fluorescent display * *selbstleuchtende Ziffernanzeige*
Vakuum-Leistungsschalter vacuum circuit breaker
Vakuumbeschichtung vacuum deposition
Vakuumhebegerät vacuum lifting device || vacuum lifting equipment
Vakuumpumpe vacuum pump
Vakuumröhre vacuum bottle || vacuum tube
Vakuumschalter vacuum circuit breaker * *Hochspannungsschalter* || VCB * *Abkürzg.f. 'Vacuum Circuit Breaker': Hochspannungs-~*
Vakuumschütz vacuum contactor
validierbar validatable
variabel variable || variably * *(Adv.)*
Variable variable
Variante variant || variety * *Abart, Spielart* || alternative * *Alternative* || version * *Version* || kind

Varianz 336

* *Art, Sorte, Gattung* || solution * *Lösung* || way * *Art und Weise, (Lösungs)Weg* || type * *Typ, Bauform* || design * *Bauart* || type of construction * *Bauweise* || way of realization * *Art u. Weise d. Realisierung/Verwirklichg./Aus-/Durchführung* || way of implementation * *Art und Weise der Aus/ Durchführung* || answer * *Lösung; Antwort auf ein Problem (to: auf)* || form * *Form, Gestalt, Fasson, Art u. Weise, Schema, System* || shape * *(auch figürlich) Form, Gestalt, Umriss*
Varianz variance
variieren vary
Varistor varistor * *spannungsabhängiger Widerstand, z.B. zum Überspannungsschutz* || VDR * *Abkürzg. für 'Voltage Dependent Resistor': spannungsabhängiger Widerstand* || voltage dependent resistor || MOV * *Abkürzung für 'Metal-Oxide Varistor': Metalloxid-Varistor*
Vaseline vaseline || petrolatum
VDC VDC * *Volt Gleichspannung*
VDE VDE * *Abkürzg. für 'Verein Deutscher Elektrotechniker': Union of German Electrical Engineers* || Union of German Technical Engineers * *Verein Deutscher Elektrotechniker*
VDE-Bestimmung VDE regulation * *elektrotechn. Sicherheitsbestimmung des →VDE (für el. Masch.: VDE 0560, IEC 34)* || VDE-regulation
VDE-Vorschrift VDE regulation
VDMA VDMA * *'Verband Deutscher Maschinenu. Anlagenbau': German Machinery Plant Manufacturer's Assoc.*
VDW VDW * *Abk.f.'Verein Deutscher Werkzeugmaschinenfabriken e.V.': German Machine Tool Builders Associat.*
Vektor vector || phasor
Vektor-Regelung vector control || TRANSVEKTOR control * *(R) Variante des Erfinders dieses Verfahrens, der Firma SIEMENS*
Vektordiagramm vector diagram
Vektordreher vector rotator
Vektordrehung vector rotation || vector circulation
Vektorfeld vector field
vektorgeregelt vector controlled
vektoriell vector || vectorial
vektorielle Größe vector quantity || vector variable
vektorielles Produkt vector product
Vektormodulation vector modulation
Vektormultiplikation vectorial multiplication
Vektorregelung vector control || TRANSVEKTOR control * *(R) Variante der Firma SIEMENS, des Erfinders dieses Verfahrens*
Ventil valve * *auch Halbleiter~, Thyristor~; safety-valve: Sicherheits~*
Ventilansteuerung valve control
Ventilation ventilation
Ventilator fan || air fan || blower * *Lüfter* || ventilator || motorized impeller || ventilating fan
Ventilatorantrieb fan drive
Ventilatorbaugruppe fan module || fan assembly || fan unit
Ventilatorflügel vane * *with adjustable pitch: einstellbarer ~* || fan blade
Ventilatorkennlinie fan characteristic * *Lastkennlinie (M als f(n)) eines Lüfters (normalerweise M~n²)*
Ventilatorlaufrad fan propeller
Ventilatorrad impeller
Ventilatorschaufel vane
Ventilatortrockner fan dryer
Ventilatorüberwachung fan monitor * *als Einrichtung* || fan monitoring * *als Funktion*
Ventilelement valve device
Ventilinsel valve block * *zusammenfasste Einheit aus mehreren (Pneumatik-) Ventilen*
Ventilrichtung valve direction
ventilseitig valve-side

Ventilverschluss valve seal * *z.B. für Verpackung*
Ventilzweig valve arm
ver- inter- * *unter/miteinander* || mis- * *falsch, fälschlich(er weise)* || dis- * *auseinander*
Verallgemeinerung generalization
veralten become obsolete || become antiquated || go out of fashion || go out of date
veraltet obsolete || out of date || outdated || outmoded * *altmodisch* || antiquated * *antiquiert* || old-fashioned * *altmodisch*
veränderbar variable * *variabel* || changeable * *änderbar*
veränderbarer Widerstand variable resistor
veränderlich variable || varying || changeable
verändern vary * *variieren, abwechseln, sich verändern* || change * *ändern* || alter || modify
verändert modified * *modifiziert*
Veränderung variation * *Variierung, Variation; in: in/von; to: an* || change || fluctuation * *Schwankung(en)*
verankern anchor * *(auch figürl.)* || stay * *absteifen, verankern, ab-/verspannen; hemmen, aufhalten* || fix * *to foundation: im Fundament ~* || brace * *verstreben, befestigen, verklammern, verankern, versteifen*
veranlassen induce * *veranlass., bewegen, überreden, verursachen, bewirken, herbeiführen, führen zu,auslösen* || cause * *veranlassen, verursachen, bewirken, hervorrufen, herbeiführen* || bring * *dazu bringen/bewegen (to: zu)*
Veranlassung reason * *Grund* || occasion * *Anlass, Veranlassung, Gelegenheit* || cause * *Ursache, Grund* || suggestion * *Anregung, Vorschlag; at my suggestion: auf meine ~* || instance * *Ansuchen, (dringende) Bitte; at the instance of: auf ~ von* || for further action * *zur (weiteren) ~* || background * *Hintergrund, Beweggrund*
veranschaulichen illustrate || be illustrative of
veranschlagen estimate * *(ab)schätzen* || value * *bewerten, ansetzen*
Veranstalter organizer * *z.B. einer Messe, Tagung usw.* || promoter
Veranstaltung event
Veranstaltungsort place of event || location of course * *eines Kurses*
verantwortlich responsible * *for: für* || answerable * *for: für* || liable * *haftbar, verpflichtet, for: für* || accounting * *accounting for: ~ sein für, Rechnung/Rechenschaft ableg. über; erklärend, begründend* || accountable * *verantwortlich, rechenschaftspflichtig; erklärlich*
Verantwortlichkeit responsibility
Verantwortung responsibility * *carry: tragen; under sole responsibility: unter alleiniger ~*
Verantwortung übernehmen take the responsibility || accept the responsibility || assume the responsibility
verarbeiten execute * *Daten, Programm, Eingangssignale* || handle * *handhaben, umgehen mit, abwickeln* || manufacture || process || convert * *into: zu* || treat * *behandeln* || machine * *maschinell ~* || run * *Daten in e. Computerprogramm ~, eine durchlaufende Warenbahn in e. Maschine ~*
verarbeitet processed || finished * *endver-/bearbeitet*
Verarbeitung processing * *Programm, Daten, Signale, Nachrichten; Herstellungsprozess* || manufacture || treatment || workmanship * *qualitative Ausführung* || execution * *Ausführung, z.B. eines Rechnerprogramms*
Verarbeitungsanlage processing unit * *z.B. in einer Fabrik*
Verarbeitungsgeschwindigkeit processing speed * *z.B. Rechner*

Verarbeitungsgröße processed variable * *in der Software* || processed quantity * *in der Software*
Verarbeitungsleistung processing power || computing power * *Rechenleistung* || performance * *Leistungsfähigkeit* || computer performance * *eines Rechners*
Verarbeitungsmaschine processing machine || process machine
Verarbeitungsstraße processing line || manufacturing line * *Herstellungsstraße*
Verarbeitungstiefe processing depth
Verarbeitungszeit processing time * *in einem Computer*
verbacken bonded * *verklebt, ausgehärtet* || baked * *siehe auch →sintern* || caked together * *miteinander ~* || sticked together * *miteinander verklebt, aneinander festsitzend*
Verband association * *z.B. berufsständige Organisation* || committee * *Komitee, Ausschuss, Kommission* || federation * *Vereinigung* || union * *Vereinigung, Verbindung, Bund* || formation * *Formation, Kampfverband* || trade association * *Fachvereinigung von Unternehmen* || dressing * *in der Medizin* || bandage * *in der Medizin* || board * *Interessenvereinigung* || lattice * *(mathem.)*
verbessern improve * *besser machen* || enhance * *in der Leistungsfähigkeit/im Funktionsumfang* ~ || correct * *berichtigen* || modify * *umgestalten* || revise * *revidieren, z.B. Buch, Beschreibung, Software* || refine * *verfeinern "ausklügeln", veredeln, raffinieren* || streamline * *modernisieren, wirkungsvoller/zügiger/zweckmäßiger/reibungsloser gestalten*
verbessert improved || enhanced * *in Leistungsfähigkeit oder Funktionsumfang ~* || corrected * *berichtigt*
Verbesserung improvement || correction * *Korrektur* || advancement * *Fortschrit, Wachstum, Vorankommen* || enhancement * *Erhöhung der Leistungsfähigkeit/der Funktionalität*
Verbesserungsvorschlag suggestion of improvement
verbiegen twist * *auch 'sich verbiegen'* || bend || distort || warp * *[Holz]* || deform
Verbiegung deflection * *Ablenkung, Abbiegung* || twisting * *Verdrehung* || buckling * *Krümmen, Verziehen, Wölben* || warping * *Verziehen, Verwerfen, Krümmen, z.B. Holz*
verbinden connect * *to: mit; via: durch/über; by means of/mit Hilfe von* || interconnect * *miteinander ~* || link * *together: miteinander, z.B. Ventile in einer Stromrichterschaltung* || incorporate * *vereinigen, zusammenschließen, einverleiben, eingliedern* || associate * *(gedanklich) assoziieren* || interlink * *miteinander ~ (z.B. über Bussystem)* || interface * *z.B. Rechnersysteme miteinander ~* || internetwork * *über ein (Computer-) Netzwerk ~, vernetzen* || software * *per Software "verdrahten"* || combine * *kombinieren, verbinden, (in sich) vereinigen* || join * *verbinden, vereinigen, zusammenfügen (to/onto: mit)*
Verbinder interconnect * *z.B. Steck~* || connector
verbindlich binding * *upon/for: für* || obligatory * *bindend (rechts)verbindlich, obligatorisch* || compulsory * *bindend, Pflicht-, obligatorisch, Zwangs-; make compulsory: für ~ erklären* || mandatory * *~ vorgeschrieben* || certified * *garantiert, bescheinigt, beglaubigt*
verbindlich vorgeschrieben mandatory
Verbindlichkeiten accounts payable
Verbindung connection * *(auch figürl.) in connection with: in ~ mit; elektr. Anschluss* || interconnection || conjunction * *logisch; in conjunction with: in ~/Zusammenhang mit* || implication * *enge logische ~* || junction * *mechan. ~/Anschluss, ~spunkt, Zusammentreffen, Berührung, Knotenpunkt* || compound * *(chem.)* || joint * *(mechan.)* || link * *~sglied, (serielle/parallele) Daten-/Signal~*
Verbindung herstellen make the connection || establish the connection
Verbindungs- connecting || junction || trunk * *in der Telekommunikation* || coupling || transmission * *zur Kraft-/Datenübertragung* || contact || communication || combining || fixing * *Befestigungs-* || joint || transfer
Verbindungselement fastener * *Befestigungselement (z.b. Schraube, Mutter, Niete, Bolzen)* || connecting element * *z.B. in einer mechan. Konstruktion/einem Rohrleitungssystem* || transmission element * *zur Kraft-/Datenübertragung* || connection link || fastening element * *Befestigungselement*
Verbindungskabel connecting cable
Verbindungslasche connecting lug || connecting plate || link
Verbindungsleitung connecting cable || connecting wire || lead wire || connecting pipe * *Rohrleitung*
Verbindungslinie connecting line
Verbindungsmann contact man || liaison man
Verbindungsnetz interconnecting network
Verbindungsperson contact person || liaison person || liaison man * *Verbindungsmann*
verbindungsprogrammiert hardwired * *Steuerung (im Gegensatz zur SPS)* || hard-wired
Verbindungsstelle connection point || connecting point
Verbindungstester continuity tester * *Durchgangsprüfer*
Verbindungsweg route
verbleiben be left || remain
verbleibend remaining || sustained
verborgen concealed * *verborgen, unübersichtlich, versteckt* || hidden * *verborgen, geheim* || secret * *geheim* || latent * *(physikalisch) schlafend, schlummernd, verborgen, versteckt* || invisible * *unsichtbar* || covert * *versteckt, verborgen, verschleiert, heimlich; Obdach, Schutz, Versteck, Dickicht*
verboten prohibited || not allowed * *nicht erlaubt*
Verbrauch consumption
verbrauchen consume || wear up * *abnutzen* || exhaust * *erschöpfen* || run down * *z.B. Batterie* || spend * *Geld, Zeit usw.; (on: für; in doing: bei)* || draw * *herausholen, entnehmen, ziehen; auch Strom, Wasser, Leistung usw.* || dissipate * *(Verlustleistung)*
Verbraucher load * *Last, energieaufnehmender Teil e. Stromkreises* || consumer * *Konsument, Energie~ (auch Gerät, Bauelement)* || user * *Anwender, Benutzer*
Verbraucherabzweig load branch
verbreitet popular * *beliebt; with: bei* || widespread || spreaded || wide-spreaded * *(weit) verbreitet* || propagated * *Lehre, Meinung usw.* || prevalent * *(vor)herrschend, überwiegend, weit ~* || be in vogue * *Mode sein* || come into vogue * *Mode werden* || widely held * *(Ansicht usw.)* || common * *(allgemein) üblich*
Verbrennungsanlage incinerator * *zur Einäscherung/Verbrennung z.B. von Müll*
Verbrennungsmotor internal-combustion engine || combustion engine
Verbund network * *Netz(werk)* || interconnection * *z.B. ~ von Stromnetzen; interconnected operation: ~betrieb* || alliance * *Verbündung, Allianz* || compound * *~maschine, ~material, ~werkstoff* || co-operation * *Zusammenarbeit* || pool * *Interessen/Arbeitsgemeinschaft, Ring* || shared forces * *gemeinsame Kräfte* || other SIEMENS sales divisions * *SIEMENS-interne ~vertriebspartner in an-*

Verbund- 338

deren Unternehm.bereichen || composite * *Verbund-, Misch-, Zusammensetzg. (v. Materialien usw.); composite construction: −bauweise* || integrated operation * *~ von Firmen usw.*
Verbund- sandwich * Mehrschicht- || compound * *z.B. Maschine, Verbundmaterial, Materialmischung* || composite * *zusammengesetzt* || multiple- * Mehrfach- || laminated * *aus mehreren Schichten bestehend, z.B. Glas, Kunststoff, Folie, Holz* || grid * *z.b. -Kraftwerk, -Stromnetz*
Verbund-Kompressorsatz multiple compressor set || compound compressor set * *häufig mit bedarfsweiser Zu- und Wegschaltung einzelner Kompressoren*
Verbundbetrieb combined operation * *z.B. mehrere Motoren an einem Umrichter* || interconnected system operation * *von Stromnetzen* || interconnected operation
Verbunddokument container * *[Computer] Datei, die eingebette (OLE-) Objekte enthält*
verbunden mit associated with * *einhergehend mit*
verbunden sein mit be associated with * *einhergehen mit* || involve * *mit sich bringen, Folgen haben, nötig machen* || be bound up with * *eng ~* || be combined with
Verbundmaterial composite material || compound material || compound
Verbundvertrieb inter-group sales * *inter-group sales department: ; ~ als Abteilung* || business with in-house partners
Verbundwerkstoff composite || compound
verchromt chromium-plated
verdampfen vaporize || evaporate
Verdampfer vaporizer || evaporator * *z.B. in der Kältetechnik*
Verdampfungstemperatur evaporating temperature
Verdampfungsziffer evaporation coefficient
verdecken cover up * *zudecken, verdecken* || cover * *bedecken, abdecken, verdecken, verbergen, verhüllen* || hide * *verstecken* || conceal * *verbergen, verstecken, verdecken, verheimlichen* || mask * *maskieren, abdecken, verheimlichen* || cloak * *bemänteln* || obscure * *verbergen*
verdeckt covered * *bedeckt, abgedeckt, verdeckt, verborgen, verhüllt* || covered up * *zugedeckt, verdeckt* || hidden * *verborgen, versteckt* || concealed * *verdeckt, verheimlicht, versteckt* || masked * *(aus-) maskiert* || obscured * *verborgen* || obscure * *verborgen, unbekannt* || covert * *heimlich, versteckt (auch Weg usw.), verborgen, verschleiert; geschützt*
verdeutlichen make clear || make plain || elucidate * *aufhellen, aufklären, erklären, erläutern* || illustrate * *durch Beispiele, Bilder usw.*
verdichten compress
Verdichter compressor
Verdichterkennlinie compressor characteristic * *Lastkennlinie M als f(n) eines Kompressors (M~n^2 bei Zentrifugalkompr.)*
Verdichtersatz compressor set * *Kombination mehrerer Verdichter; häufig m.bedarfsweiser Zuschaltung d.Einzelverdichter*
Verdickung thickening || thick spot * *dicke Stelle*
verdienen earn || make || deserve * *Lob, Tadel usw.*
Verdienstspanne profit margin || sales margin
verdoppeln double * *auch 'sich verdoppeln'*
verdrahten wire-up || wire || connect * *verbinden* || connect-up * *"hinverdrahten", anschließen, auflegen, z.B. auf eine Klemme*
Verdrahtung wiring || cabling
Verdrahtungsänderung cabling changing * *in der Leitungsverlegung* || rewiring || wiring modification
Verdrahtungsaufwand cost of cabling * *Kosten* || amount of wiring || cabling costs || wiring costs || wiring overhead

Verdrahtungskosten wiring costs * *siehe auch →Verdrahtungsaufwand*
Verdrahtungsplan wiring diagram || connection diagram
verdrängen push away || thrust aside || displace * *(figürl. und physikal.)* || supersede * *als Nachfolger ~* || oust * *aus Stellung, Amt ~* || crush * *(ver)drängen, (unter)drücken, (zer)quetschen; (auch figürl.: Konkurrenten usw. ~)* || repress * *(psycholog.)*
Verdrängung displacement * *volumenmäßig* || supersession * *(figürl.) Ablösung, Ersatz (durch Nachfolger)* || repression * *(psychologisch)* || pushing away * *mit Gewalt* || thrusting aside * *mit Gewalt*
Verdreh- torsional || of twist * *(nachgestellt)* || twisting || torsion
Verdrehbarkeit ability to rotate
verdrehen twist * *against: gegenüber* || distort || wrench || turn
Verdrehschutz torsion lock * *z.B. für Kabelschuhe in Motorklemmenkasten*
verdrehsicher protected against twisting || cannot be rotated || keyed * *verpolungssicher, z.B. Stecker mit unsymmetrischem Aufbau*
Verdrehspiel torsional backlash
Verdrehsteifigkeit torsional rigidity * *siehe auch →Drehsteifigkeit* || torsional stiffness
Verdrehung torsion || twisting || twist || rotational twist
Verdrehwinkel torsional angle * *bei Torsionsbeanspruchung* || angle of twist || torsion angle
verdreifachen triple
verdrillen twist
verdrillt twisted * *z.B. Kabel; in pairs: paarweise*
verdrillte Zweidrahtleitung twisted-pair cable
Verdünner thinner
verdunsten evaporate
Verdunstung evaporation
veredeln finish * *Güter, Rohstoffe, z.B. Metall* || refine * *verfeinern* || process * *Rohstoffe* || convert * *umformen, umwandeln, veredeln, weiterverarbeiten*
Veredelung finishing * *Güter, Rohstoffe, z.B. Metall, Textil* || refinement * *Verfeinerung*
Veredelungsmaschine finishing machine
vereinbaren agree upon || arrange || stipulate * *festsetzen, vereinbaren, ausbedingen, zur Bedingung machen (for: etwas)* || negotiate * *vertraglich ~*
Vereinbarung agreement * *Abmachung; make an agreement: e. ~ treffen* || arrangement * *Vereinbarg., Verabredg., Ab-/Übereinkommen, Schlichtung, as arranged: wie vereinbart* || clause * *Klausel* || understanding * *Vereinbarung, Übereinkunft* || appointment * *Verabredung* || condition * *Klausel, Vorbehalt, Bestimmung*
vereinen combine * *kombinieren, verbinden (auch Eigenschaften, Vorteile usw.)* || pool * *(Kräfte, Resourcen, Kapital usw.) vereinen/gemeinsam nutzen* || unite * *verbinden, (sich) vereinigen, sich anschließen, in sich vereinigen (Eigenschaften usw.)* || join * *verbinden, vereinigen, zusammmenfügen, sich anschließen/zusammentun* || amalgate * *fusionieren, →vereinigen* || mesh * *verschmelzen (mit), aufgehen lassen (in), fusionieren; be meshed in: aufgehen in*
vereinfachen simplify || make easier || reduce * *Formel* || ease * *leichter machen, erleichtern, entlasten, befreien*
vereinfacht simplified || reduced * *Formel*
Vereinfachung simplification * *to simplify matters: zur ~*
vereinheitlichen standardize || unify || harmonize * *harmonisieren*

Vereinheitlichung unification || standardization || harmonization * *Harmonisierung*
vereinigen incorporate * *in: in; vereinigen, in sich schließen* || combine * *verbinden* || integrate * *zusammenschließen; within: in* || unite || join || pool * *Kräfte, Resourcen usw.* ~ || coordinate * *gleichschalten, koordinieren* || associate * *(sich) vergesellschaften* || amalgate * *fusionieren* || consolidate * *fusionieren* || merge * *fusionieren; into: zu* || assemble * *versammeln* || reconcile * *in Einklang bringen*
Vereinigte Staaten U.S.
Vereinigungsmenge aggregate
vereinzeln separate * *absondern, abtrennen, ausscheiden* || destacking * *(Substantiv) z.B. bei einer Verpackungsmaschine* || individualize * *einzeln behandeln*
Vereinzelung separate feeding * *{Verpack.techn.}*
verengen narrow || reduce || contract
verfahrbar travelling || mobile || moving || sliding * *verschiebbar, gleitend*
Verfahrbereich travel range * *bei Positionierantrieb*
Verfahrdatensatz positioning data set || traversing record || travelling data block || block * *{NC-Steuerung}*
Verfahren procedure * *Vorgehensweise, Ablauf, Prozedur, Verfahrensweise* || method * *Methode* || mode * *Art und Weise* || process || technique || system
Verfahrenstechnik process technology || process engineering
verfahrenstechnisch process-related || process-oriented
verfahrenstechnischer Regler process controller * *für Temperatur, Durchfluss, Druck, Feuchte, Flächengewicht usw.*
Verfahrensweise procedure || method || policy * *Richtlinie, Schema* || system * *Schema, System* || process
Verfahrlänge travel length * *z.B. eines Positionierantriebs*
Verfahrprogramm traversing program * *{NC-Steuerung}*
Verfahrrichtung traversing direction * *bei Positionierantrieb* || travelling direction || direction of motion
Verfahrsatz positioning data set || traversing record || travelling data block || block * *bei NC (numerischer Positioniersteuerung)*
Verfahrstrecke travel path || travel distance || travelling distance
Verfahrweg travel distance || travel range * *Verfahrbereich (von ... bis)* || travelling distance * *z.B. bei einer Positioniersteuerung* || travelled distance * *ver-/gefahrener/zurückgelegter Weg*
Verfahrzeit move time * *bei Positioniersteuerung* || traversing time || duration of the traversing process || travelling period
Verfall decay || drop * *z.B. von Preisen*
verfälschen falsify || invalidate * *ungültig machen* || corrupt * *z.B. Daten, Signale* || interfere * *störend einwirken auf* || deform * *verzerren* || distort * *verzerren* || mix up * *vermischen, durcheinandermischen, (völlig) durcheinander bringen*
verfälscht invalid * *ungültig* || corrupt * *verdorben, unecht, verfälscht, z.B. Daten* || distorted * *verzerrt* || deformed * *verzerrt*
Verfärbung discoloration || discolouration
verfeinern refine * *(auch figürl.: Methode, Arbeit, Stil, Geschmack usw.)* || sophisticate * *(figürl.)* || subtilize * *(figürl.)* || perfect * *vervollkommnen, perfektionieren*
Verfestigung strain hardening * *Aushärtung* || consolidation * *Festwerden, Festigung, Konsolidie-*
rung (auch wirtschaftl.), Verdichtung || bonding * *Abbindung, z.B. einer Klebestelle* || hardening * *Härtung, Härten*
Verflechtung implication || interweaving || interlocking || interlacement || involvement * *Verstrickung*
Verflüssiger condenser * *z.B. im Kühlkreislauf, in der Kältetechnik*
Verflüssigung condensation || liquefaction || liquefying
Verflüssigungstemperatur condensing temperature * *z.B. in einem Kühlkreislauf*
Verfolgbarkeit traceability * *z.B. eines Produkts in der Vertriebskette zum Kunden*
verfolgen pursue * *Ziel, Zweck, Plan, Weg, Kurs* ~ || follow * *einen Vorgang usw.* ~ || keep track * *auf der Spur bleiben, nach~ (of: etwas), weiter~* || track * *nach~* || trace * *eine Spur* ~
Verfolgung tracking * *Nachverfolgung eines Signals/Exemplars/(Verwaltungs-) Vorgangs/von Material* || tracking back * *Zurückverfolgung* || pursuit * *Verfolgung (-sjagd)* || persecution * *drangsalierende Verfolgung* || carrying on * *z.B. ein Geschäft, einen Plan* || pushing forward * *das Vorantreiben*
Verfolgungssystem tracking system * *z.B.für Klebe-/Schweiß-/Ausschussstellen b. Rotationsdruck, Blechverarbeit.straße*
verformbar workable * *(Metall)* || deformable * *(Metall)* || mouldable * *(Kunststoff)* || thermoplastic * *warm verformbar*
verformen torm * *formen* || deform * *deformieren, verzerren, verunstalten* || work * *bearbeiten* || shape * *in e.Form bringen, bearbeiten* || distort * *verzerren, verwinden* || deform * *deformieren, verzerren*
verformt distorted * *z.B.Kurvenform e.Spannung*
Verformung deforming * *Deformierung* || deformation || working * *non-cutting working: spanlose* ~ || distortion * *unerwünschte* ~*/Verdrehung/Verwerfung*
Verformungsmaschine forming machine
verfügbar available * *freely: frei* || disposable || at hand * *vefügbar, zur Hand, bei der Hand* || existing * *existierend*
Verfügbarkeit availability || disposability || up-time * *~sdauer, Dauer ungestörten Betriebs* || operating availability * *relative ~ eines Geräts/einer Anlage/ einer Maschine* || plant uptime * *~ einer Anlage* || reliability * *Zuverlässigkeit (high level of: hohe)*
Verfügbarkeitsdauer up-time || uptimes
verfügen über possess * *besitzen, im Besitz haben, beherrschen* || dispose of * *gebieten über, verwenden* || have at one's disposal * *zur Verfügung haben* || control || be provided with * *ausgestattet sein mit* || be equipped with * *ausgestattet sein mit* || have * *ausgestattet sein mit* || make use of * *verwenden*
vergeben place * *place with: Auftrag usw.* ~ *an* || confer * *übertragen* || let slip * *Chance* ~ || miss * *Chance* ~ || allocate * *anweisen, zuteilen (z.B. Speicherplatz, Namen usw.)* || assign * *zuordnen, zuweisen, zuteilen (auch Speicherplatz, Namen usw.)*
vergehen expire * *Zeit, Frist*
vergießen pot * *z.B. Baugruppe mit Gießharz* || encapsulate || seal
Vergleich comparison * *in comparision to: im* ~ *zu* || analogy * *Analogie, Entsprechung* || settlement * *Einigung; out-of-court settlement: außergerichtlicher* ~
vergleichbar comparable * *to: mit* || analogous * *analog, entsprechend (to/with: mit)*
vergleiche refer to * *siehe auch* || compare || cp. * *Abkürzung für 'compare'* || confer * *Hinweis in Schriftstück; confer page 23: ~ Seite 23* || cf. * *Abkürzg. für 'confer'*

vergleichen compare * *(with: mit) (to: mit (gleichstellend))* || collate * *Texte usw.* ~ *(with: mit)* || liken * *bildhaft* ~ *(to: mit)* || check against * *zur Kontrolle* ~ *mit*
Vergleicher comparator
Vergleichmäßigung smoothing || making even
Vergleichs- reference || comparative
Vergleichsbaustein compare block * *Funktionsbaustein* || comparator || comparison block * *Funktionsbaustein*
Vergleichsfunktion comparison operation * *Computerbefehl* || relational operation * *Computerbefehl* || compare function
Vergleichsoperation comparison operation
Vergleichsoperator relational operator || comparison operator
Vergleichstest benchmark test || benchmark || benchmarking * *Durchführung eines* ~*s*
vergleichsweise comparatively
Vergleichswert comparison value
verglichen compared * *with: mit*
vergoldet gold-plated
vergossen encapsulated * *(ein)gekapselt, z.B. Wicklung* || potted * *z.B. Baugruppe*
vergriffen out of stock
vergrößern increase * *erhöhen* || enlarge * *größer machen; auch einen Wert/eine Zahl* || zoom * *Maßstab stufenlos verändern* || expand * *(sich) ausdehnen* || extend || widen * *verbreitern* || augment * *(sich) vermehren/vergrößern, zunehmen* || add to || boost * *(salopp) 'aufblasen', ankurbeln, Auftrieb geben, steigern*
vergrößert dargestellt shown on a larger scale
Verguss sealing || encapsulation
Vergussmasse sealing compound || moulding compound || casting compound || filling compound * *z.B. für Kabelmuffe* || potting material
Vergussmaterial potting material * *siehe auch* →*Vergussmasse*
vergüten anneal * *glühen (Metall)* || temper * *Wärmebehandlung zur Änderung d. Gefüges o. d. Oberflächeneigensch. von festen Werkstoffen* || heat-treat * *mit Wärme behandeln* || refund * *Ausgaben/Auslagen* ~ || reimburse * *Ausgaben* ~ || compensate for * *Ausgaben/Auslagen* ~
Vergütungsstahl heat-treated steel || heat-treatable steel * *vor der Vergütung* || tempering steel
Verhalten behavior || attitude * *Haltung* || characteristics || behaviour || characteristic * *Merkmal, Eigentümlichkeit, Eigenschaft* || capability * *Vermögen, Fähigkeit* || performance * *Leistungsfähigkeit, Betriebsverhalten*
Verhältnis ratio * *of...to: von...zu* || proportion || conditions * *Umstände* || relation * *Beziehung*
verhältnismäßig relative || comparative || proportional || relatively * *(Adv.)* || comparatively * *(Adv.)* || comparatively speaking * *(Adv.)*
Verhältnisregelung ratio control * *auch bei Winkelgleichlaufregelung mit 'Impulsübersetzung'*
verhandeln negotiate || treat * *for: wegen* || confer * *konferieren* || hear * *Zivilrecht* || try * *Strafrecht* || talk * *besprechen*
Verhandlung transaction || discussion * *Diskussion, Gespräch* || conference * *Konferenz, Besprechung* || talks * *Gespräche* || negotiation || hearing * *Anhörung, öffentliche* ~, *zivilrechtliche* ~ || trial * *strafrechtliche* ~ || meeting * *Besprechung*
Verhandlungen transactions || negotiations
verharren remain * *in: bei/in; z.B. in einer bestimmten Stellung* || hold still * *(in einer bestimmten Stellung) stillhalten* || persist * *in: bei/auf (z.B. einer Meinung)* || persevere * *beharren (in), ausdauern, aushalten (bei), fortfahren (mit), festhalten (an)* || stick * *to: bei* || abide * *by: bei/auf (z.B. einer Meinung)* || adhere * *to: bei*

verhindern prevent || protect against * *schützen gegen* || stop * *aufhalten* || avoid * *vermeiden*
verjüngen taper * *sich* ~, *spitz zulaufen, konisch sein* || taper off * *sich* ~
verjüngt tapered * *konisch, spitz zulaufend*
Verjüngung taper * *Konizität, z.B. abnehmende Wickelhärte bei Aufwickler* || tapering
Verkabelung cabling || wiring || cable installation
Verkabelungsaufwand cabling costs
verkanten tilt * *schrägstellen, neigen, kippen* || set on edge
Verkapselmaschine capsuling machine * *Verpakkungsmaschine*
verkapseln capsule || can * *"eindosen"* || seal curling * *(Substantiv) verschließen durch Bördelung (Dose, Becherkondensator usw.)* || case * *umkleiden, verkleiden, in ein Gehäuse stecken*
verkaufen sell
Verkäufer vendor * *(Firma)* || seller || shop assistant * *Ladengehilfe* || clerk * *(amerikan.)* || salesman * *Vertriebsmann* || salesclerk * *(amerikan.)* || retailer * *Einzelhändler, Wiederverkäufer*
Verkaufs- und Lieferbedingungen conditions of sale and delivery || conditions of sale/delivery
Verkaufsbedingungen conditions of sale * *siehe auch* →*Allgemeine Lieferbedingungen* || terms of sale
Verkaufsbüro sales agency || agency
Verkaufsgebiet sales territory
Verkaufspreis sales price || selling price || retail price * *Endverbraucher-/Einzelhandelspreis, Kleinmengen-Preis für Wiederverkäufer* || resale price * *Wieder*~ || market value * *Marktwert*
Verkaufsvereinbarung sales agreement
Verkehr traffic || transport * *von Gütern und Personen* || transportation * *(amerikan.) von Gütern und Personen* || communication * *(Kommunikations-, Verkehrs-, Brief-) Verbindung* || service * *Verkehrsdienst; withdraw from service: aus dem Verkehr ziehen* || commerce * *Handelsverkehr* || trade * *Handelsverkehr* || intercourse * *persönlicher Verkehr*
verkehren communicate * *kommunizieren; with: mit* || run * *Verkehrsmittel*
Verkehrsbetrieb transport company * *siehe auch* →*Verkehrsunternehmen*
Verkehrsleitsystem traffic-routing system
Verkehrssystem traffic system || transport system
Verkehrstechnik traffic engineering || traffic systems || transport systems || science of transport * *als Wissenschafts-/Forschungsgebiet*
Verkehrsunternehmen transport company || transportation company || transport authority * *Staatliches/städtisches* ~
verkehrt wrong || incorrect * *unrichtig, nicht korrekt* || false * *falsch* || inverted * *umgekehrt* || reversed * *verkehrt herum*
verketten concatenate || interlink || chain up || chain || queue * *eine Schlange bilden*
verkettete Spannung phase-to-phase voltage || line-to-line voltage * *zwischen den Hauptleitern, z.B. L1-L2-L3* || interlinked voltage || mesh voltage * *Maschenspannung* || interconnected voltage * *Maschenspannung*
Verkettung interlinkage * *siehe auch* →*Verknüpfung* || interlinking || enchainment * *(figürlich)* || concatenation * *(figürlich)* || linking || interlocking * *Ineinanderhaken, Verzahnung, Inandergreifen* || interconnection * *gegenseitige Verbindung* || chaining || sequencing * *das Aufeinanderfolgen-Lassen*
verkitten cement || putty
verklagen sue * *gerichtlich belangen, verklagen (for: auf)* || sue out: *(Gerichtsbeschluss usw.) erwirken* || take legal action
verkleben paste over || paste up || plaster over || ce-

ment * *kitten* || stick together * *kitten* || glue * *mit Leim* || bond * *Metall*
Verkleidung covering || cover sheets * ~ *aus Blech* || lining
verkleinern make smaller || reduce * *reduzieren; auch Maßstab/Zeichnung* || reduce in size * *in den Abmaßen* ~ || scale down * *Zeichnung/Maßstab* || decrease * *verringern, vermindern* || diminish * *vermindern* || grow smaller * *kleiner werden* || cut back * *Produktion, Umsatz, Ausgaben usw.* ~; *auch 'sich verkleinern'* || minimize * *auf ein Minimum* ~ || size down * *Fläche, (Bau-) Volumen verringern (z.B. durch Neukonstruktion)* || shrink * *schrumpfen; z.B. Halbleiterstrukturen durch verbesserten Herstellprozess* || lessen * *vermindern* || depreciate * *herabsetzen; z.B. Preis, Wert* || downsize * *Fläche/Bauvolumen verringern (z.B. durch Neukonstruktion)* || reduce the size || cut the size || dwarf * *verkümmern lassen, verkleinern, klein erscheinen lassen*
verkleinert dargestellt shown on a reduced scale
verkleinerter Maßstab reduced scale
Verkleinerung miniaturization || down-sizing || shrinking * *Schrumpfen, z.B. von Halbleiterstrukturen durch verbesserten Herstellprozess* || reduction || reduction in size || diminution || cutback
Verklemmung locking-up * *auch softwaremäßige* ~ || jamming || freezing * *softwaremäßige* ~ || hanging * *Sich-Aufhängen einer Software*
Verknappung shortage
verknüpfen concatenate * *auch 'verketten, kaskadieren'* || combine * *kombinieren; logically combined: logisch verknüpft* || gate * *Signale* ~ || link together * *(figürl.) miteinander* ~, *z.B. Ideen* || string together * *(figürl.) miteinander* ~ || interlink * *(figürl.) miteinander* ~ || associate * *closely associated: eng miteinander verknüpft sein (with: mit)* || connect * *verbinden* || knot * *verknoten* || tie * *zusammenbinden* || tie together * *zusammenknüpfen* || attach * *to: mit* || link || software * *per Software "verdrahten"* || chain * *verketten (to: mit)* || knit * *(figürl. u. konkret) zusammenfügen,verbinden, vereinigen; knit up: (sich eng) verbinden*
verknüpft combined * *logically: logisch*
Verknüpfung interlocking * *Verriegelung, vor allem steuerungsmäßig* || connection * *Verbindung, z.B. von Signalen; Zusammenhang, Beziehung, Anschluss* || association * *Zusammenhang, Beziehung, Verbindung, Vereinigung* || interlinking || connexion * *Verbindung, z.B. von Signalen; Zusammenhang, Beziehung, Anschluss* || gating * *z.B. von Signalen; logic: logische* || logic operation * *logische/binäre Funktion* || linking * *Verkettung, Verkopplung,z.B. v. Signalzuständen; Zusammenknüpfen* || knotting together * *Verbindung, Verknotung, Zusammenknüpfung* || concurrence * *Zusammentreffen, Mitwirkung, Gleichzeitigkeit* || logic * *logische, binäre* || binary logic * *binäre* || nexus * *Zusammenhang, Verknüpfung*
Verknüpfungsergebnis boolean result * *einer logischen Operation* || calculation result * *accumulator data* || result of logic operation * *RLO: VKE bei SPS* || RLO * *Abkürzg.f. 'Result of Logic Operation'*
Verknüpfungsfunktion logic function || binary logic function || gate function * *Logikgatterfunktion (z.B. UND, ODER usw.)*
Verknüpfungsglied logic element || combinative element
verknüpfungsintensiv logic-intensive * *z.B. SPS-Programm*
Verknüpfungssteuerung logic control
Verknüpfungstiefe logic nesting depth
verkomplizieren complicate
verkörpern present * *darstellen* || represent * *darstellen, verkörpern* || incarnate || embody || personify * *personifizieren, versinnbildlichen* || materialize * *verwirklichen, e.Sache stoffliche Form geben, Gestalt annehmen, sich* ~, *zustandekommen*
verkraften cope with * *fertigwerden/es aufnehmen mit, gewachsen sein, bewältigen* || bear * *ertragen* || stand * *ertragen, aushalten* || deal with * *bewältigen* || handle * *bewältigen*
verkupfern copper-plate
verkürzen short || shorten * *auch 'sich ~'* || clip * *beschneiden* || abridge * *abkürzen* || curtail * *beschränken* || cut down * *herab/zurechtstutzen* || become shorter * *sich* ~ || foreshorten * *(geometr.) perspektivisch* ~ || cut short * *plötzlich beenden, abkürzen; es kurz machen, ins Wort fallen* || bypass * *umgehen, übergehen*
Verkürzung shortening * *(zeitl. und räumlich)*
Verladung shipment
Verlag publishing house || press || publishers || publisher
verlagern shift * *auch 'sich verlagern'* || displace || dislocate || remove * *to: nach* || transfer
Verlagerung displacement * *auch 'Ausrichtungsfehler'* || shifting || transfer || removal || basic change * *(figürl.)* || shift * *(figürl.)* || misalignment * *Fluchtfehler, Fehler in der Ausrichtung* || movement * *Bewegung*
verlangen require * *benötigen, erfordern* || demand * *fordern, begehren (of/from: von, that: dass, to do: zu tun), bedürf.,erford. (z.B.Sorgfalt)* || claim * *Anspruch erheben auf, beanspruchen* || desire * *wünschen* || charge * *(Geldbetrag) berechnen* || insist on * *bestehen auf* || call for * *er/einfordern* || ask for * *verlangen nach, bitten um*
verlängern lengthen || elongate || stretch * *strecken* || prolong * *zeitlich* ~, *z.B. Lebensdauer, Frist* || extend * *vergrößern, erweitern, ausfahren; auch figürl.: Patent, Kredit, Zeitintervall* ~ || renew * *Vertrag* ~
Verlängerungs- extension
Verlängerungskabel extension cable
Verlängerungsleitung extension cable
verlangsamen slow down || decelerate * *verzögern*
verlassen rely * *vertrauen (on: jdm./etwas)* || exit * *herausgehen* || exitted * *herausgegangen sein,* ~ *haben* || leave || quit * *gänzlich* ~ || forsake * *im Stich lassen* || desert * *treulos* ~, *desertieren* || rely on * *sich* ~ *auf* || depend on * *sich* ~ *auf* || count on * *sich* ~ *auf, zählen auf* || abandon * *(auf immer)* ~, *(völlig) aufgeben, verzichten auf, preisgeben, überlassen, im Stich lassen* || terminate * *beenden*
Verlauf variation * *z.B.* ~ *einer Funktion* || curve * ~ *einer Funktion/Kurve; Verlauf von Strom/Spannung usw.* || characteristic * ~ *einer Kennlinie/Kurve usw.* || lapse * ~/*Ablauf von Zeit* || course * ~ *von Zeit, eines Vorgangs, Grenze, Linie, Kurve; in the course of: im* ~ *von* || progress * *Fortschreiten,* ~ *eines Vorgangs* || development * *Entwicklung, Entstehen, Bildung, Wachstum, Erschließung,Ausbau,Umgestaltung* || trend * *Tendenz* || sequel * *(Aufeinander-)Folge, weiterer* ~ || run * *Ablauf* || turn * ~ *einer Sache* || end * *Ausgang* || sequence * ~ *von Zeit* || shape * ~ *eines Impulses/Signals, einer Kurve* || characteristic * ~ *einer Kennlinie/Kurve/(mathem.) Funktion* || waveform * ~ *eines periodischen Signals* || profile * ~ *Profil,Kontur, Seitenansicht, Längs-/Querschnitt; i.Profil/Quer-/Längsschnitt darstellen*
verlaufen pass * *Zeit* || elapse * *Zeit* || proceed * *Vorgang* || run * *Grenze, Weg, Leiterbahn, Farben usw.* || lose one's way * *sich* ~ || disperse * *sich* ~ *(Menge usw.)* || take a course * *einen Verlauf nehmen* || cut untrue * *abweichen einer Säge* || drift out of true * *z.B. Bohrer beim Bohrvorgang*

verlegen lay * *Kabel* ‖ install * *Kabel, Rohre* ‖ run * *Kabel* ‖ transfer * *versetzen* ‖ move * *bewegen, versetzen* ‖ shift * *verschieben, versetzen; (auch figürlich: Termin usw.)* ‖ relocate * *versetzen (an einen anderen Ort/Platz)* ‖ route * *leiten, auch verlegen von Kabel/Leitungen* ‖ defer * *zeitlich verschieben* ‖ put off * *zeitlich verschieben (to: auf)* ‖ postpone * *zeitlich verschieben* ‖ adjourn * *vertagen* ‖ reschedule * *Termin* ‖ traverse * *in Querrichtung hin- und herbewegen, z.b. Kabel dei der Aufwicklung* ‖ misplace * *an die falsche Stelle, unauffindbar*
Verlegung routing * *Art und Weise der Leitungs~* ‖ laying of cables * *von Kabeln/Leitungen* ‖ removal ‖ shifting ‖ installation * *z.B. ~ von Kabeln/Leitungen* ‖ traversing * *das Hin- und Herbewegen in Querrichtung, z.B. von Kabel, Draht, Faden bei d. Aufwicklung* ‖ traversing unit * *Einrichtung zum Hin- u. Herbewegen von Kabel beim Aufwickeln*
Verleih hire service
verleihen lend ‖ rent * *z.B. Auto* ‖ hire out * *gegen Miete* ‖ impart * *geben, gewähren, verleihen, erteilen, mitteilen (auch physikal.: Kraft, Schwung usw.)*
Verleimmaschine gluing machine * *{Holz}*
Verleimung gluing * *{Holz}* ‖ pasting * *{Papier}*
verletzen violate * *Regel, erlaubte Grenze, Begrenzung, Vorschrift, Gesetz ~* ‖ infringe * *Regel, Begrenzung usw. ~* ‖ damage * *beschädigen* ‖ injure * *verwunden* ‖ hurt * *verwunden (auch figürlich: Gefühle usw. ~)* ‖ offend against * *Vorschrift, Anstand ~* ‖ sacrifice * *Gesetz, Vorschrift ~* ‖ infringe on * *z.B. Patent ~*
Verletzungsgefahr danger of injury
Verlitzmaschine twisting machine * *für Kabel*
verloren gehen be lost ‖ disappear * *verschwinden* ‖ get lost ‖ miscarry * *z.B. Brief*
verlöschen extinguish * *auch Feuer, Lichtbogen usw.*
verlöten solder * *Weichlöten* ‖ braze * *hartlöten* ‖ hard-solder * *hartlöten*
Verlust loss * *auch finanzieller ~, ~ von Daten/Parametereinstellungen usw.* ‖ leakage * *Streuung e. elektromagnetischen Größe* ‖ damage * *Schaden* ‖ waste * *Abfall, Abgang* ‖ wastage * *Material~, Verschleiß, Abfall, Abgang* ‖ red figures * *rote Zahlen; run into red: in die ~zone gelangen*
verlustarm low-loss ‖ with low losses * *(Adv.)* ‖ low power * *mit niedriger (Verlust)Leistung*
verlustbehaftet high-loss ‖ low-efficiency ‖ inefficient ‖ with associated losses
Verlustbremsung non-regenerative braking
Verluste losses * *Geld, Reibung, Energie* ‖ loss
Verlustfaktor loss-factor ‖ dissipation factor ‖ power-loss factor ‖ loss factor * *→tangens delta eines Dielektrikums (kennzeichnet Verluste e. el. Feldes)*
verlustfrei loss-free ‖ no-loss ‖ loss-less ‖ low-loss * *verlustarm*
Verlustgeschäft losing business ‖ losing bargain ‖ lossmaker * *verlustbringendes Produkt/Geschäft*
Verlustleistung power loss ‖ power loss dissipation * *als Wärme abgegebene ~* ‖ power dissipation ‖ power losses ‖ losses * *z.B. ~ eines Motors* ‖ heat loss * *in Wärme umgesetzte ~* ‖ heat loss * *in Wärme umgesetzte ~; auch 'Wärmeverluste' bei e. thermodynamischen Prozess* ‖ dissipation * *Abführung der ~, abgeführte/abzuführende ~*
Verlustmeldung report of loss
Verlusttrennung loss segregation * *Aufteilung der Verluste auf die einzelnen Anteile bei d. Wirkungsgrad-Ermittlung*
Verlustwärme heat losses ‖ power loss * *→Verlustleistung* ‖ thermal losses

Verlustwinkel loss angle
Verlustziffer loss index * *Summe Wirbelstrom- u. Hystereseverluste (1,3.. 4 W/kg) d. (ungestanzten) Magnetblechs*
vermarktbar marketable
vermarkten market
vermascht meshed * *netzartig, maschig; network: Netz* ‖ complex * *komplex, vielschichtig* ‖ intricate * *verzweigt, verschlungen, verwickelt*
Vermaschung meshing ‖ mixing up ‖ association * *→Verknüfung* ‖ intermeshing
vermehren augment * *vergrößern, (sich) vermehren, zunehmen, ergänzen* ‖ increase * *auch 'sich vermehren' (by: um)* ‖ multiply * *an Zahl ~/vergrößern* ‖ propagate * *fortpflanzen* ‖ add to * *beitragen zu*
vermeidbar avoidable ‖ evitable ‖ preventable * *abwendbar, "verhütbar"*
vermeiden avoid ‖ prevent * *verhindern* ‖ eliminate * *eliminieren, beseitigen, ausschließen, entfernen, tilgen* ‖ get out of the way * *aus dem Wege gehen* ‖ steer clear of * *umschiffen* ‖ dodge * *aus dem Wege gehen, umgehen, sich drücken vor, Ausflüchte machen*
Vermerk note ‖ notice ‖ entry * *Eintrag* ‖ endorsement * *Vermerk, Zusatz, Eintrag (auf Urkunden usw.)*
vermerken note down ‖ record ‖ enter * *eintragen* ‖ make an entry of * *eintragen*
vermessen take the measurement * *abmessen* ‖ measure
vermieten let * *to let: zu ~ (z.B. Haus); siehe auch 'leihen'* ‖ rent * *(besonders i. Amerikan.)* ‖ hire * *on hire: zu ~* ‖ hire out ‖ lease
vermindern diminish ‖ reduce * *reduzieren* ‖ lessen ‖ lower * *senken (Kosten, Temperaturen, Preise usw.)* ‖ alleviate * *(ver-)mindern, erleichtern, mildern, lindern*
Verminderung decrease ‖ diminution ‖ lessening ‖ reduction ‖ cut
vermitteln get across * *schlichtend ~* ‖ give * *Vorstellung, Eindruck, Bild ~* ‖ offer * *Vorstellung, Eindruck, Bild ~* ‖ impart * *(z.B. Wissen) vermitteln, unterrichten* ‖ intervene * *between: zwischen, z.B. bei einem Streit ~, intervenieren* ‖ arrange * *zustandebringen* ‖ procure * *be-/verschaffen* ‖ mediate * *sich als Vermittler betätigen (between: zw.), e. Bindeglieg bilden, dazwischen liegen* ‖ act as a mediator * *als Vermittler fungieren (in: bei)*
vermittels by means of ‖ through ‖ with the help of
Vermittler middleman * *z.B. Zwischenhändler* ‖ intermediary * *Mittelsmann* ‖ go-between * *Schlichter* ‖ liaison man * *Verbindungsmann* ‖ agent * *von Geschäftskontakten*
Vermittlungstechnik communications switching * *in der Telekommunikation*
Vermögenswerte assets
vermutlich presumably * *(Adv.)* ‖ I suppose * *(Adv.) ich vermute* ‖ presumable ‖ supposed ‖ probable * *wahrscheinlich* ‖ likely * *wahrscheinlich*
vernachlässigbar negligible ‖ which can be neglected
vernachlässigen neglect * *nicht beachten, z.B. einens Nebeneffekt ~* ‖ fail in one's duty * *seine Pflicht ~*
Vernadelungsmaschine needling machine * *{Textil}*
Vernetzbarkeit multi-drop capability * *z.B. über Bussystem* ‖ capability to be networked * *z.B. über Kommunikations-Netzwerk/Bus*
vernetzen network * *together: miteinander* ‖ inter-network
vernetzt internetworked * *z.B. durch Bussystem oder Peer-to-Peer Netz ~* ‖ interconnected in a bus

configuration * *durch einen Kommunikations-Bus ~* || networked || crosslinked * *[Chemie] (Kunststoffe)*
Vernetzung networking * *Kommunikation* || networked connection || connectivity * *als Funktion* || internetworking || networked operation
vernichten destroy * *zerstören* || exterminate * *auslöschen, ausrotten* || erase * *ausradieren, auslöschen, (aus)tilgen (from: aus)* || dissipate * *energy: Energie* || annihilate * *vernichten, zunichtemachen, aufheben, aufreiben*
vernickeln nickel-plate
vernickelt nickel-plated
vernieten rivet || clinch
vernünftig reasonable * *vernunftsgemäß, angemessen* || rational * *vernunftsmäßig, rational, vernunftbegabt* || judicious * *→wohlüberlegt*
verodern OR
verodert OR'd * *logische Funktion* || ORed * *logische Funktion*
veröffentlichen publish * *Buch, Aufsatz/Artikel* || make public
Veröffentlichung publication * *auch Schrift, Buch, Aufsatz, Broschüre* || announcement * *Bekanntmachung, Ankündigung*
Verölung fouling by oil
Verordnung enactment
verpacken pack || pack up || package * *Einzelstück; besonders 'maschinell verpacken'* || wrap up * *einwickeln*
verpackt packed
Verpackung packing * *die Hülle, das Hüllmaterial* || packaging * *die ~, das Verpacken; seaworthy packaging: seegemäße ~* || packing material * *~smaterial* || wrapping * *~ aus Papier, Karton oder Folie*
Verpackungs- packaging
Verpackungs-Recycling packaging recycling
Verpackungsart mode of dispatch * *Angabe in Bestellformular*
Verpackungsband packaging strap * *streifen-/reifenförmig* || packaging thread * *Schnur, Kordel, Bindfaden*
Verpackungsboden packing base
Verpackungseinheit unit pack || packing unit
Verpackungsgewicht packed weight * *Gewicht inklusive Verpackeung*
Verpackungskarton packing box || package carton
Verpackungsmaschine packaging machine || wrapping machine * *Einwickelmaschine (z.B. für Folienverpackung)*
Verpackungsmaschinen paging machinery || packaging equipment
Verpackungsmaterial packing material
Verpackungstechnik packaging technology
verpflichten oblige * *durch Umstände, Stellung, Vertrag, Umstände, Gesetz ~* || obligate * *vertraglich ~* || engage * *vertraglich, oneself to do a thing: sich zu etwas ~* || bind * *oneself: sich; oneself to do a thing: sich zu etwas ~* || sign * *on: zu Arbeitsleistungen usw. ~* || commit * *oneself to do a thing: sich zu etwas ~* || undertake * *sich ~ (to do: zu tun), sich verbürgen (that: dass), Risiko/Verantwortg. übernehmen* || agree * *to: einwilligen in, z.B. in eine Vertragsbestimmung* || be liable * *gesetzlich verpflichtet sein* || be bound * *by law/contract: gesetzlich/vertraglich verpflichtet sein*
Verpflichtung commitment * *übernommene/eingegangene ~* || obligation * *undertake an obl.: eine ~ eingehen; be under obl.: unter ~ sein* || liability * *gesetzliche/vertragliche ~; assume/incur liabilities: ~en eingehen* || duty * *Pflicht* || engagement * *übernommene ~, enter into an eng.: eine ~ eingehen* || bond * *~, Bürgschaft, Kaution, Vertrag, Schuldschein, Obligation, Wertpapier, Zollpapiere*

Verpolschutz protection against false polarity
Verpolung false polarity * *falsche Polung* || incorrect polarity * *falsche Polung* || polarity reversal * *Umpolung* || reverse polarity
Verpolungsschutz protection against polarity reversal
verpolungssicher keyed * *Stecker mit Codierung* || polarized * *unsymmetrisch (Stecker)* || protected against polarity reversal * *z.B. Digitaleingang, Stromversorgungsanschluss*
verrauscht noise-corrupted || noisy
verrechnen charge * *belasten* || invoice * *buchungstechnisch; in Rechnung stellen, fakturieren* || balance * *ausgleichen* || compensate * *gegeneinander* || miscalculate * *sich* || pass to account * *verbuchen* || set off against * *gegeneinander aufrechnen* || compensate * *gegeneinander aufrechnen* || clear * *im Verrechnungsverkehr* || make a mistake * *einen Fehler machen* || be out in one's reckoning * *sich verrechnet haben, sich ~* || be mistaken * *sich verrechnet/geirrt haben*
Verrechnung invoicing || account * *Berechnung; only for account/not negotiable: nur zur ~ (auf Scheck)* || charging * *Belastung* || clearing * *gegenseitige ~ (im ~sverkehr)*
Verrechnungssatz rate * *z.B. Stundensatz für Servicepersonal* || hourly rate * *Stundensatz z.B. für Personaleinsatz*
verriegeln interlock * *logisch, steuerungsmäßig, mechanisch usw. ~* || lock out * *gegen Wiedereinschalten ~* || block * *blockieren*
verriegelt interlocked
Verriegelung interlocking || interlock * *logische, steuerungstechnische, mechanische usw. ~* || locking * *Sperre* || interlocking function * *~sfunktion*
Verriegelungshebel locking lever * *z.B. an Flachbandkabelstecker*
Verriegelungsschiene locking bar * *an e. Baugruppenträger als 'Herausziehsperre' für Baugruppen*
Verriegelungssteuerung interlock control
verringern diminish || reduce || decrease || lessen || cut down * *Ausgaben, Preise usw.*
Verringerung reduction || decrease || diminution || cut
verrippt ribbed * *z.B. mit Kühlrippen versehen (Motor, Kühlkörper usw.)*
Verrippung ribbing || finning
verript finned
Verrohrung piping
verrundet smoothed || rounded || curved
verrundete Hochlaufgeberrampe rounded ramp || smoothed ramp * *S ramp* * *in Form eines 'S' abgerundete Rampe*
Verrundung smoothing * *z.B. einer Rampe* || rounding-off * *z.B. Hochlaufgeber* || rounding || S-curve rounding * *z.e.Hochlaufgeber-Rampe, sodass e. S-förmige Kurve entsteht* || rounding function * *bei Hochlaufgeber*
Verrundungszeit rounding time * *z.B. bei einem Hochlaufgeber mit Verrundung*
versagen fail || fail to work || fail to act || break down || failure * *(Substantiv; auch figürl.: einer Person)*
versammeln convene * *(sich) versammeln, z.B. zu e. Besprechung; einberufen * *(Personen) versammeln; sich ~; (Dinge) (an)sammeln/anhäufen* || convoke * *einberufen* || meet * *sich ~*
Versand shipping || dispatch * *Absendung, Expedition* || delivery * *Auslieferung* || shipment
Versandanschrift shipping address || forwarding address
Versandanzeige dispatch note
Versandhaus mail-order house || mail-order company

Versandkosten freight charge || shipping costs
Versandpackung shipping package
Versandpalette dispatch pallet
Versandpapiere shipping documents
Versandrohr mailing tube
Versandschaden shipping damage
Versandspesen shipping costs || freight charge
Versatz displacement || offset * z.B. Software: Basisadresse, Spuren eines Impulsgebers; time offset: zeitlicher ~ || misalignment * Versatzfehler, Fehlausrichtung/Justierung
Versatzwinkel displacement angle || angle of displacement
verschalten assign * zuweisen, zuteilen || interconnect || wire up * verdrahten
verschandeln disfigure || spoil || ruin || litter * mit Abfall, Dreck
verschiebbar sliding || movable || adjustable * einstellbar || displaceable || slidable || moveable || displacable || slideable || that can be shifted
Verschiebeläufermotor sliding-rotor motor * (Hebezeug)Motor m. integr. Bremse, Lüftg.durch Verschiebg. d.konisch. Läufers
verschieben shift * auch 'sich ~' || delay * zeitlich verzögern || move || displace * versetzen, verrükken, verlagern, verschieben; auch Phasenwinkel || shunt * Eisenbahn || disarrange * in Unordnung bringen || defer * zeitlich ~ || put off * zeitlich ~ || postpone * zeitlich ~, Termin ~ || adjourn * vertagen || sell underhand * unter der Hand verkaufen, z.B. auf d. Schwarzmarkt ~ || get out of place * sich ~ || offset * bezüglich Phase/Spannung ~; auch 'ausgleichen' || telescope * ineinander ~ || slide * gleiten lassen || slip * gleiten lassen || reschedule * (Termin) neu festsetzen
Verschiebewagen shuttle car * z.B. zur Verteilg. d. Güter zu/von Regalbediengerät(en) i. Hochregallager
Verschiebung shift || shifting || displacement * z.B. aus Sollposition || rescheduling * Termin~, Neufestlegung eines Termins
Verschiebungsfaktor displacement factor * auch cos phi1: Verhältn. Grundschw.wirkleistg./Grundschw.scheinleistg.
Verschiebungsfluss displacement flux || dielectric flux
Verschiebungsstrom displacement current * bei Halbleitern
Verschiebungswinkel displacement angle
verschieden various * diverse || different * unterschiedlich; from: von || several * mehrere || miscellaneous * vermischte || distinct * ver/unterschieden, getrennt, abgesondert, unverkennbar || dissimilar * unähnlich || unlike * unähnlich || varied * wechselnd
verschiedene miscellaneous * diverse || various * diverse || a number of * einen Anzahl von, mehrere || various * →diverse
Verschiedenes miscellaneous
verschiedenst widely differing * z.B. widely differing applications: ~e Anwendungen || varied * verschiedenartig, mannigfaltig || heterogeneous * verschiedenartig, heterogen
Verschienung conductor bars || busbars || bar conductors
verschlechtern make worse * become worse: schlechter werden || impair || get worse * sich ~, schlechter werden || deteriorate * auch 'sich ~, schlechter werden' || debase * in der Qualität ~ || fall off * sich ~ (in: an/in)
verschleifen smooth * z.B. Signale in Form eines weichen Übergangs ~ || round * verrunden, z.B. Signale
Verschleiß wear * Abnutzung (high: großer/hoher) || fatigue * Ermüdung || abrasion * mechanischer Abrieb || ageing * Alterung || erosion * Abnutzung, Schwund, Zerfressung; z.B. Kontakte || wear-out || wear and tear * Abnutzung || corrosion * Korrosion, Anfressung || wastage * Verbrauch
Verschleißanzeige wear indication * z.b. für Brems-/Kupplungbelag || wear indicator * Verschleißanzeiger, z.B. für Kupplungs-/Bremsbelag
Verschleißanzeiger wear indicator * z.B. für Bremsbelag, Kohlebürste usw.
verschleißarm low-wear || low-abrasive
Verschleißausfall wearout failure
Verschleißausgleich wear compensation * z.b. an einer Kupplung/Bremse
verschleißbehaftet subject to wear || subject to abrasion * sich abreibend || wearing out || wearing
verschleißen wear || wear out * abnutzen, sich verschleißen || age * altern || fatigue * ermüden, altern || abrade * abreiben, abnutzen || become worn || wear down * sich 'abrubbeln'
verschleißfest wear-resistant || hardwearing || abrasion-proof * ~ gegen Abrieb || hard wearing
Verschleißfestigkeit resistance to wear || abrasion resistance * gegen Abrieb
verschleißfrei free of wear || wearless || wear-resistant * verschleißfest || abrasion-proof * abriebfest || wearfree || non-wearing
Verschleißteil wearing part || consumable * Verbrauchsartikel || part subject to wear || worn-out part * verschlissenes Teil
Verschleißverhalten wear characteristics
verschließbar lockable
verschließen shut * auch figürl.: shut one's eyes to: die Augen ~ vor || close || lock up * mit Schlüssel ~ || block * blockieren || bolt * mit Riegel usw. ~ || seal * hermetisch ~, z.B. durch Kleben, Schweißen, Vergießen
Verschließmaschine closing machine * (Verpack.techn.) || closure machine * (Verpack.techn.) || sealing machine * (Verpack.techn.) z.B. für Flaschen, Tuben, zur Verpackung von Lebensmitteln usw. || capping machine * (Verpack.techn.) zum Verschließen von Flaschen, Tuben usw. mit Verschlussdeckeln
Verschließmittel sealing materials * (Verpack.techn.)
Verschliff smoothening * weicher Übergang
verschlossen closed || shut || locked || locked up || plugged * durch Stöpsel/Blindstopfen ~ (z.B. Kabeleinführung) || reserved * (figürlich) reserviert
Verschluss latch * zum Verriegeln || fastener * ~mittel, Befestigungseinrichtung || fastening * ~mittel, das Verschließen || lock * Schloss; keep under lock (and key): unter ~ halten; lock nut: ~mutter || catch * Schnapp~ || stopper * ~ einer Flasche usw. || plug * Stöpsel || seal * Dichtung || shutter * an (Foto)Kamera; shutter release: Verschlussauslösung || closure * ~(vorrichtung)
Verschlüsselung conversion * durch einen Analog-/Digitalumsetzer usw. || encryption * von Daten zur Geheimhaltung
Verschlüsselungszeit conversion time * z.B. eines Analog/Digital-Umsetzers
Verschlusskappe cap closure * in der Verpakkungstechnik
Verschlussschraube plug * z.B. für Kabeleinführung, Ölablass usw. || locking screw || closure plug || drain plug * Ablassschraube || screw plug
Verschlussstopfen plug * z.B. Blindstopfen für Kabeleinführung in Klemmenkasten || gland * Kabelverschraubung, z.B. an Motorklemmenkasten || drain plug * (Flüssigkeits-) Ablassstopfen || dummy plug * Blindstopfen, z.B. bei Kabeleinführung
verschmutzen soil || pollute * Wasser, Umwelt usw.

|| get dirty * *schmutzig werden* || contaminate * *verunreinigen* || foul * *z.B. durch Öl*
verschmutzt   soiled || polluted * *Wasser, Umwelt usw.* || fouled * *verölt, beschmutzt, befleckt, verstopft, verölt* || dirty * *schmutzig* || contaminated * *verunreinigt, besudelt, vergiftet, verseucht (auch Umwelt, Atmosphäre)*
Verschmutzung   pollution || ingress of dirt * *Eintreten/-dringen von Schmutz* || contamination * →*Verunreinigung* || dirt contamination || fouling * *Verschmutzen, Verstopfung, Verölung, Versperrung, Befleckung*
Verschmutzungsgrad   pollution degree || degree of pollution * *siehe z.B. DIN 0110, Teil 1* || amount of fouling * *Ausmaß d. Verscmutzung (durch Öl, Ruß usw.), Befleckg., Verstopfg., Verdreckg.*
verschoben   shifted || out of phase * *periodische Signale gegeneinander; by 90 ° (vorangestellt): um 90° gegeneinander* || displaced * *versetzt, verrückt, verschoben, verlagert; displaced by 90°: um 90° verschoben*
verschobener Nullpunkt   live zero * *z.B. 4-20 mA Stromsignal mit Drahtbrucherkennungsmöglichkeit* || life zero
verschrauben   screw * *siehe auch* →*festschrauben (to: mit)* || screw on || screw together * *zusammenschrauben* || bolt down * *festschrauben* || screw home * *festschrauben (am Einbauplatz)* || screw in place * *am vorgesehenen Einbauplatz* →*festschrauben*
verschraubt   screwed || bolted * *mit Schraubenbolzen*
Verschraubung   gland * *~ einer Kabeldurchführung (siehe auch* →*PG-Verschraubung)* || screwed gland * *~ einer Kabeldurchführung* || screwed joint || threaded joint * *~ ohne Mutter* || bolted joint || union * *Rohrverbindung* || screw connection || screw fitting * *z.B. ~ für Rohrverbindung, Kabeldurchführung*
verschrotten   scrap
Verschrottung   scrapping || disposal * *Entsorgung*
Verschweißen   welding || contact welding * *von Kontakten* || fusion welding * *Kontaktverschleiß*
verschwinden   disappear || vanish || dissolve * *sich auflösen* || fade away * *hinwegschwinden* || disappearance * *(Substantiv) das Verschwinden*
verschwindend klein   infinitely small || infinitesimal || microscopic * *winzig* || minute * *winzig* || negligibly small
versehen   provide * *with: mit* || equip * *with: mit* || furnish || ausstatten; *with: mit etwas* || supply * *versorgen; with: mit* || perform * *Pflichten ~* || discharge * *Pflichten ~* || hold * *Amt ~* || act as * *Amt ~; fungieren* || administer * *Amt ~* || fill * *Amt/Dienst ~* || do the work of * *Dienst ~* || oversight * *(Substantiv)* || mistake * *(Substantiv)* || inadvertance * *(Substantiv)*
versehenlich   erraneously
versehentlich   inadvertent * *unbeabsichtigt* || by a mistake * *irrtümlicherweise* || through oversight * *auf Grund eines Flüchtigkeitsfehlers* || erroneously * *irrtümlicherweise* || unintentional * *unabsichtlich* || undesigned * *unbeabsichtigt* || involuntary * *unfreiwillig* || accidental * *zufällig* || by mistake || inadvertantly * *ohne Absicht*
Verseilanlage   stranding line * *z.B. für Kabel* || twisting line * *z.B. für Kabel* || twister * *z.B. für Kabel*
verseilen   strand * *Seil, Kabel, Litzen eines Litzenleiters usw.*
Verseilkorb   twisting cage * *bei (Kabel-/Draht-) Verseilmaschine* || stranding cage * *bei (Kabel-/ Draht) Verseilmaschine*
versenden   ship * *Fracht* || deliver * *liefern* || mail

* *per Post* || forward * *befördern, schicken, verladen, weiterbefördern (Brief)* || send || consign
versenken   countersink * *z.B. Schraubenkopf* || sink
versetzen   stagger * *versetzt anordnen, staffeln, auf Lücke setzen* || displace * *displaced by 90 degrees: um 90 Grad versetzt* || transplant * *transplantieren, z.b. Baum* || transpose * *mit etwas vertauschen, umstellen, umsetzen* || transfer * *Mitarbeiter an einen anderen Ort/Platz* || put into * *in eine Lage/einen Zustand* || place into * *in eine Lage/einen Zustand* || set vibrating * *in Schwingungen*
versetzt   staggered * *~ angeordnet, staffeln, auf Lücke setzen* || displaced * *displaced by 90 degrees: um 90 Grad ~* || transplanted * *(z.B. Baum)* || transposed * *mit etwas vertauscht, umgesetzt, umgestellt* || transferred * *(z.B. Mitarbeiter an einen anderen Ort)* || put into * *in eine Lage/einen Zustand ~* || placed into * *in eine Lage/einen Zustand ~* || set vibrating * *in Schwingungen ~* || spaced * *auf Abstand gesetzt, spaced at 60°: um 60° ~* || offset * *offset by 60°: um 60° ~; against/with respect to each other: gegeneinander*
versetzt anordnen   stagger
versichern   assure * *sichern, sicherstellen, ver-/zusichern (z.B. Leben), beruhigen, überzeugen* || protest * *beteuern* || make sure of * *sich e. Sache ~* || insure * *Eigentum ~* || affirm * *beteuern* || declare * *behaupten* || confirm * *bestätigen*
Versicherung   insurance * *auch 'Versicherungspolice/prämie'* || affirmation * *Behauptung, Versicherung, Bestätigung, Bejahung, Beteuerung* || assurance * *Ver/Zusicherung, Garantie* || protestation * *Beteuerung* || guarantee * *Garantie, Gewähr*
versiegeln   seal
versiert   versed * *well versed: gut bewandert* || experienced * *erfahren* || expert * *in: in*
versilbert   silver-plated || silver-coated
Version   version * *auch eines Softwareprogramms* || release * *Ausgabestand; auch eines Softwareprogramms* || edition * *Ausgabe/Auflage eines Druckerzeugnisses* || variant * *Variante* || variety * *Abart* || kind * *Art, Sorte, Gattung* || solution * *Lösung* || answer * *Lösung(sweg)/Antwort auf ein Problem; to: auf* || way * *Art und Weise* || type * *Typ, Art, Bauform* || design * *Bauart* || type of construction * *Bauweise* || way of realization * *Art und Weise der Realisierung/Ausführung/Durchführung* || way of implementation * *Art und Weise der Aus/ Durchführung*
Versionsanpassung   version matching || version adaption
Versionsnummer   version-number * *z.B. einer Software*
versorgen   provide * *with: mit* || supply * *z.B. mit Spannung/Strom; with: mit* || furnish * *ausstatten; with: mit* || support * *unterhalten, unterstützen* || maintain * *unterhalten*
Versorgung   supply * *Lieferung, Bereitstellung, Beschaffung, Speisung (with: mit)* || provision * *Beschaffung, Bereitstellung, Vorkehrung* || providing * *Ausstattung, Vorsorge, Bereitstellung* || parameter assignment * *SPS* || feeding * *Speisung, Zuführung, Zuleitung, Zufuhr*
Versorgungsleitung   supply cable || supply line || feeder
Versorgungsnetz   public supply system
Versorgungsquelle   supply source
Versorgungsspannung   supply voltage
Versorgungsspannungsschwankungen   fluctuations in supply voltage
Versorgungsunternehmen   utility company || power utility * *Stromversorgungsunternehmen* || power company * *Stromversorgungsunternehmen* || utility
verspannen   brace * *versteifen, verstreben, befesti-*

Verspannung

gen, verklammern, verankern || guy * mit Tau sichern, verspannen || preload * →vorspannen || distort * (sich) →verziehen || stay * ab-/verspannen, →verankern
Verspannung bracing * Verstrebung, Verklammerung || rigging * mit Tauen || locked-up stress * (psycholog.) innere ~ || clamping force * Klemmkraft, Verklemmung || distortion * Verdrehung, Verziehung, Verwindung || rig * ~ aus (Draht)Seilen, Takelage
verspannungsfrei free from distortion || distortion-free
versplinten cotter
verstanden understood
Verständigung communication || information * Benachrichtigung || notification * Benachrichtigung
verständlich clear * klar, leicht verständlich || understandable || intelligible || distinct * deutlich || popular * allgemein verständlich || within everybody's grasp * allgemein verständlich; difficult to grasp: schwer verständlich || easy-to-understand * leicht verständlich || comprehensible * begreiflich, (leicht) verständlich
Verständnis appreciation * richtige Beurteilg., Einsicht, krit.Würdigg., (günstige) Kritik, Kennenlernen (of: für) || understanding * Verstehen, Einsicht; show understanding for a person: jdm, Verständnis entgegenbringen || comprehension * Verstehen || insight * Einsicht, Verständnis, Einblick || sympathy * Mitgefühl || appreciate * Verständnis haben für || understand * Verständnis haben für
verstärken amplify || boost * Spannungs/Strom/Drehmoment in bestimmten Arbeitsbereichen anheben || reinforce * mechanisch ~/fester machen || strengthen * (mechan. u. figürl.) fester machen, (be-/ver-) stärken (auch zahlenmäß.), stärker werden || increase * Druck usw. || intensify * intensivieren
Verstärker amplifier || repeater * Zwischenverstärker für Kommunikationsleitung || booster * Hochsetzschaltung || driver * z.B. Leitungstreiber || buffer * Puffer z.B. für Datenleitung
verstärkt reinforced * mechanisch stärker/kräftiger gemacht; auch 'strengthened'; siehe auch →verstärken || amplified * elektrisch ~ || boosted * angehoben, erhöht, gesteigert, angekurbelt || intensified * intensiviert || increased * gesteigert; increase one's efforts: ~e Anstrengungen machen || increasing * ansteigend || more energetically * (Adv.) energischer || more dynamically * (Adv.) energisher, tatkräftiger || with more power * (Adv.) mit mehr Energie/Tatkraft || more strongly * (Adv.) || more intensely * (Adv.) || with added force * (Adv.) mit mehr Energie/Tatkraft || with increased efforts * (Adv.) mit ~en/erhöhten Anstrengungen || fortified * Gewebe; nylon fortified: mit Nylon ~; (auch 'heavy-duty' für Schwerlastbetrieb) || aggressively * (Adv.) aggressiv, "bissig", more aggressively: mit ~em Einsatz/"Biss" || stiffened * versteift; high-capacity: mit ~er Tragfähigkeit (z.B. Lager) || with more energy * (Adv.) mit mehr Energie/Tatkraft || with special emphasis * (Adv.) mit besonderem Nachdruck || with focussed energy * mit konzentrierten Kräften || enhanced * gesteigert (auch i.d. Leistungsfähigkeit), erhöht, vergrößert, angehoben; z.B. Kühlung
verstärkte Ausführung heavy-duty design || ruggedized design || reinforced design
verstärkte Kühlung forced cooling * z.B. von Leistungshalbleitern mit Lüfter oder flüssigen Kühlmitteln || forced air cooling * →verstärkte Luftkühlung, z.B. von Leistungshalbleitern mit Lüfter
verstärkte Lagerung reinforced bearing design
verstärkte Luftkühlung forced air cooling * Kühlung von Leistungshalbleitern usw. mit Lüfter || forced ventilation * Kühlung von Leistungshalbleitern usw. mit Lüfter || assisted air cooling * Luftkühlung mit Lüfter(n), z.B. bei Leistungshalbleitern || enhanced ventilation * Kühlung z.b. von Leistungshalbleitern mit Lüfter || enhanced air cooling * Kühlung v.Leistungshalbleitern usw. mit Lüfter || forced-air cooling * z.b. von Leistungshalbleitern mit Lüfter
verstärkter Stern-Dreieck-Anlauf combined star-delta starting * [el. Masch.] Kombination v. Stern-Dreieck- u. Wicklungsumschaltung
Verstärkung amplification || boost * Anhebung von Spannung/Drehmoment/Druck usw. (z.B. in e. bestimmten Arbeitsbereich) || gain * Verstärkungsfaktor || gain factor * Verstärkungsfaktor || reinforcement * mechanisch
Verstärkungsadaption gain adaption
Verstärkungsfaktor gain factor || gain || proportional gain * P-Verstärkung
Verstärkungsfehler gain error
verstecken hide
verstehen understand || understanding * (Subst.) Verstehen, Verständnis, Verständigg., Vereinbarg., Übereinkunft, Abmachung
versteifen strut || prop || brace || stiffen || reinforce * versteifen
Versteifung strut * Strebe || prop * Stütze || stay * Stütze, Strebe, Verspannung, Verankerung || bracing * Verstärkung, Abstützung || brace * Stütze, Strebe || stiffening element || bracket * Stütze, Träger || reinforcement * Verstärkung, Armierung || bracing * ~ durch Strebe, Stütze, Anker, Klammer, Gurt usw.
verstellbar adjustable || variable
verstellen adjust || vary * variieren, auch verändern, wechseln || shift || obstruct * versperren || block * versperren
Verstellgeschwindigkeit setting speed || speed of adjustment || positioning speed * Positioniergeschwindigkeit
Verstellgetriebe speed variator * stufenlos verstellbares (z.B. →PIV-Getriebe) || variator || variable-speed gearing
Verstellpropellerantrieb adjustable propeller drive
Verstellung variation || adjustment * Einstellung || misadjustment * fehlerhafte/ungewollte ~ || setting * durch Bedienpersonal, Stellantrieb usw. || displacement * Verschiebung, Verlagerung
verstiften pin * pinned fitting: Verstiftung || dowel * verdübeln || connect with pins || fasten with pins || secure with pins * mit Stiften sichern
verstopfen block * versperren || choke up * versperren (Rohr usw.) || clog * versperren, sich zusetzen || obstruct * hemmen || jam * Straße usw. || stop * abdichten (Loch usw.) || plug * abdichten (Loch usw.) || plug up * abdichten (Loch usw.) || bung * abdichten (Loch usw.)
verstopft clogged up * z.B.Filter
Verstopfung obstruction || clogging || stopping || block || blockage || fouling * durch Verschmutzung
Verstoß offence || violation * Zuwiderhandlung (of: gegen) || contravention * Zuwiderhandlung || mistake * Fehler || infringement * (Rechts-, Vertrags-, Patent-) Verletzung, Bruch, Übertretung, Verstoß; (up)on: gegen || offense * (amerikan.)
verstoßen offend * against: gegen || violate * ~ gegen || contravene * ~ gegen
verstoßen gegen violate || offend against
verstreben brace || strut
Verstrebung strut || strutting || brace
Verstreckanlage stretching system * z.B. für Textilfasern || fiber stretching plant * (amerikan.) Fa-

ser~ || fibre stretching line * *Faserverstreckstraße* || stretching line
verstrecken draw || stretch
Verstreckmaschine drawing frame * *Spannrahmen, z.B. für Textilfasern* || fibre stretching plant * *Faserverstreckanlage* || fiber stretching plant * *(amerikan.) Faserverstreckanlage*
Verstreckung stretching || stretching-up * *z.B. von Textilfäden*
Verstreckverhältnis stretching ratio
Verstreckwerk drawing frame * *{Textil} Spannrahmen (Textilmaschine)* || stentering machine * *{Textil} Spannrahmen* || fiber stretching plant * *{Textil} (amerikan.) Faserverstreckanlage* || fibre stretching line * *{Textil} Faserverstreckstraße* || drawing machine * *{Textil} für (Kunst-) Fasern* || stretching unit
verstreichen elapse * *Zeit, Frist* || pass away * *Zeit* || slip by * *verrinnen (Zeit)* || expire * *Frist* || spread * *Farbe/Salbe/Auftrag* ~
Verstümmelung garbling * *bei Datenübertragung usw.*
Versuch experiment || test * *conduct a test: e. Versuch durchführen* || attempt || effort * *Anstrengung* || try-out * *Bemühung* || trial * *Versuch (of: mit), Erprobung, Probe, Prüfung* || try * *(salopp)* || endeavour * *Bemühung* || endeavor * *Bemühung*
versuchen attempt || endeavour * *sich bemühen* || make an effort * *sich bemühen* || make an attempt * *den Versuch machen*
Versuchs- test || experimental
Versuchsanordnung test arrangement || experimental set-up
Versuchskaninchen guinea pig * *['gini pig] (figürl.), eigentlich 'Meerschweinchen'*
versuchsweise tentatively || by way of an experiment || on trial || by way of trial
vertagen adjourn || reschedule
vertauschen exchange || twist * *drehen* || swap * *vertauschen, z.B. Bytes in einem Wort* || mix up * *mischen* || interchange * *(miteinander) vertauschen, auswechseln, austauschen, abwechseln lassen* || reverse * *Phasen*
vertauschungssicher keyed * *z.B. Stecker*
verteilen distribute * *ver/aus/zuteilen, ausgeben; among: auf/unter/an; auch statistisch* || hand out * *austeilen, aushändigen, z.B. Besprechungs-/Schulungsunterlagen* || apportion * *zuteilen* || allot * *zuteilen* || allocate * *zuteilen; auch Speicherplatz durch Übersetzungs/Bindeprogramm)* || share * *unter sich teilen* || spread * *Farbe, Auftrag; (auch figürl.) over: über e. Zeitraum; evenly: gleichmäßig* || disperse * *sich verteilen/ver-/ausbreiten, z.B. Geschwulst, Nebel, Mienschlag, zerstreuen* || be distributed * *sich verteilen (among: unter); verteilt angeordnet werden* || divide * *(ein)teilen* || cast * *Rollen (z.B. im Theater)* || deal out * *austeilen* || scatter * *verstreuen, bestreuen (with: mit), verbreiten, ~; well-scattered: gleichmäßig verteilt*
Verteiler distributor * *auch Großhändler, Signal~, Energie~, Luft~; ignition distributor: Zünd~* || copies to * *in Brief, Bericht usw.* || distribution list * *für Informationsdienst* || mailing list * *für den Versand, z.B. von Informationsmaterial* || distribution list * *Anschriftenliste z. Verteilen schriftlicher Informationen*
Verteilerabzweig distribution branch
Verteilerstation distribution station * *z.B. elektr. Energie, Erdgas usw.* || distribution substation * *Unterstation*
verteilt distributed * *auch statistisch ~, auch 'örtlich/dezentral/~ angeordnet'* || decentralized * *dezentral angeordnet*
verteilt sein be spread * *über Raum oder Zeit* ~ || be distributed * *auch statistisch* ~

verteilte Intelligenz distributed intelligence
Verteilung distribution system * *elektrische* || distribution * *Ver/Zuteilung von Briefen/Waren usw; auch figürl.: z.B. statistisch/räuml./zeitl.* || apportionment * *gleichmäßige/gerechte Ver/Zu/Einteilung* || allotment * *Ver/Zuteilung* || dissemination * *Ausstreuung, Ver/Ausbreitung* || allocation * *Zu/Verteilung, Zu/Anweisung; z.b. Kosten* || dispersion * *Zerstreuung, Verbreitung, Verteilung (von Nebel, Teilchen, Stoffen usw.)* || power distribution * *zur elektr. Energieversorgung*
Verteilungsnetz distribution network * *Elektrizitäts~ (public: öffentliches)*
Vertiefung deepening * *(auch figürlich: in e. Materie tiefer Eindringen)* || hollow * *Höhlung* || cavity * *Aushöhlung, Hohlraum* || recess * *Aussparung, Einschnitt, Ausbuchtung, Einsenkung* || absorption * *(figürlich) auch 'vertieft sein'* || impression * *Einprägung, Einpressung* || groove * *Riefe* || depression || pit * *Grube, Grübchen, Narbe*
vertikal vertical || vertically * *(Adv.)*
Vertikalfrequenz vertical frequency * *{Computer} bei einem Monitor*
Vertrag contract * *draft: entwerfen, ausarbeiten; under contract: unter* ~ || agreement || convention * *Abkommen*
Vertrag schließen make an agreement || enter into an agreement || agree upon a contract || arrange a contract
vertraglich verpflichten contract * *to/for: zu; auch 'sich* ~*; contr. for someth.: sich etw. ausbedingen* || bind * *oneself: sich; someone to something: jdm. durch Vertrag zu etwas* || commit * *oneself: sich*
verträglich compatible * *kompatibel, auch elektromagnetisch* || consistent * *übereinstimmend, vereinbar, im Einklang (with: mit)* || sociable * *im menschlichen Umgang* || good-natured * *gutmütig*
vertragliche Regelung contractual agreement
Verträglichkeit compatibility * *Kompatibilität* || sociability * *im menschlichen Umgang*
Verträglichkeitspegel compatibility level
Vertrags- contractual
Vertragsbedingungen contract conditions || terms of contract || provisions of the agreement
Vertragsentwurf draft contract
Vertragshändler authorized dealer || appointed dealer || distributor * *Generalvertreter, Großhändler*
Vertrauen trust * *(auch Verb) in: auf* || confidence
vertraulich confidential || in strict confidence * *streng* ~ || strictly confidential * *streng* ~ || secret * *geheim; top secret: streng geheim*
vertraut machen mit get acquainted with || get to grips with * *in den Griff bekommen, sich die Handgriffe aneignen* || familiarize with * *auch sich* ~
vertraut sein mit ~ be familiar with || be acquainted with || feel at home with || be at home in * *sich (in einem Fachgebiet) auskennen* || be well versed in || be conversant with
vertreiben sell * *verkaufen* || distribute * *z.B. durch einen Zwischenhändler* || market * *vermarkten* || drive away * *hinwegtreiben*
vertretbar reasonable * *annehmbar, angemessen, tragbar, vernünftig* || justifiable * *defendable*
vertreten represent * *jemanden/etwas* ~ *(Firma, Partei usw.)* || replace * *ersetzen* || act for * *im Amt* ~
Vertreter substitute * *Stell~, Ersatzmann* || assistant * *Assistent* || deputy * *amtlicher Stell~* || representative * *Repräsentant, auch Handels~; local: örtlicher* || agent * *Handels~* || sales representative * *Verkäufer* || champion * *Verfechter*
Vertretung representation * *Repräsentanz* || agency * *Industrie~, Handels~* || agent * *Vertreter* || sales office * *Vertriebsbüro* || importer * *Importeur* || re-

Vertrieb

presentative * z.B. Industrie~, Handels~, Verkaufsniederlassung, Vertriebsbeauftragter || sales agency * Handels~, Industrie~
Vertrieb sale || sales/marketing || marketing || sales department * ~sabteilung || sales network * ~snetz || distribution network * ~snetz (mehr im Sinne von Verteilung, Logistik, Lieferung) || distribution * Vertrieb, Absatz, Handel
vertrieblich marketing || sales || sales/marketing
vertriebliche Zielsetzung sales/marketing goals
Vertriebsabteilung sales department || sales and marketing department
Vertriebsbüro sales office
Vertriebsergebnis sales profit || sales profit margin * prozentuales ~
Vertriebsfreigabe sales release * z.B. eines neuentwickelten Produkts || release for sale || release for launching into the market
Vertriebsgesellschaft sales company || distributing company || trading company * Handelsgesellschaft || marketing corporation
Vertriebshilfsmittel sales/marketing resources
Vertriebsingenieur sales engineer
Vertriebskanal sales channel || channel of distribution || distribution channel
Vertriebskosten distribution costs || sales expense * at great sales expense: mit hohen ~
Vertriebsleiter sales manager || head of sales department || sales and marketing manager
Vertriebsmann sales manager * leitender || cannvasser * im Außendienst, (Handels-) Vertreter || sales engineer * Vertriebsingenieur || salesman
Vertriebsnetz sales network || distribution network * (mehr im Sinne von Verteilen, Logistik, Lieferung)
Vertriebsniederlassung sales office || sales representative * Handelsvertretung, Handelsvertreter
Vertriebsorganisation sales organization || sales network * Vertriebsnetz
Vertriebspersonal sales personnel
Vertriebsregion sales region
Vertriebsspanne sales margin || profit margin
Vertriebssystem sales network * Vertriebsnetz || distribution network * Vertriebsnetz inkl. Auslieferungslager, Großhändler und Einzelhändler
Vertriebsvereinbarung sales agreement
Vertriebsweg sales channel || channel of distribution || distribution channel
verunden AND
verundet ANDed
Verundung ANDing
verunreinigen contaminate || pollute || soil * beschmutzen, besudeln, verunreinigen, beflecken; schmutzig werden
verunreinigt contaminated
Verunreinigung contamination || pollution || dirt * Schmutz || impurity * Unreinheit
verursachen cause || produce || create || entail * mit sich bringen, zur Folge haben, nach sich ziehen, erfordern
verursacht caused
vervielfachen multiply * auch 'sich ~' || reproduce * nachbilden || duplicate * duplizieren || make copies * kopieren
Vervielfacher grouping block * für Kontakte/Anschlüsse
Vervielfältigung duplication || copying || multiplication * (allg.) || duplicating || reproduction
vervierfachen quadruple * z.B. Impuls-/Inkrementalgebersignale, Produktion usw. ~
vervollständigen complete
verwalten administer * auch 'als Verwalter fungieren' || manage || conduct * führen || control * überwachen, steuern || supervise * überwachen || hold * (amerikan.) ein Amt ~/innehaben || schedule

* (zeitlichen) Ablauf von Programmen ~ || handle * hantieren, umgehen mit, abwickeln
Verwaltung management || administration
Verwandschaft relationship * between: zwischen; to: zu/mit
verwandt related * to: mit || analogous * entsprechend, analog (to: zu, mit) || similar * ähnlich (to: zu, mit) || associated * (eng) verbunden, verwandt (with: mit)
verwechseln change by mistake || interchange by mistake * versehentlich vertauschen || exchange || confound * durcheinander bringen (with: mit) || confuse * confuse with: verwechseln/durcheinander bringen mit || mix up * durcheinander bringen (with: mit) || take the wrong one || mistake a person for another * jemanden mit jemandem ~
Verwechslung mistake || confusion || mix-up || interchanging by mistake * versehentliche Vertauschung
verweilen stay || linger || dwell * on: bei
Verweilzeit dwell time || retention time
Verweis reference * Hinweis, Referenz, Verweisung (to: auf) || reprimand * Tadel || reproof * Vorwurf, Tadel
verweisen refer * refer to: (sich) beziehen auf/~ auf/~ an || make reference * make reference to: erwähnen/eingehen auf || reference
verweisen auf refer to
verwendbar applicable || usable
verwenden use || apply || employ * einsetzen || utilize * ausnutzen, nutzbar machen || spend * aufwenden || expend * aufwenden
verwendet employed * eingesetzt || used || applied
Verwendung application * Anwendung, Anwendbarkeit || use * Verwendung(szweck), Verwendbarkeit, Zweck, Nutzen, Nützlichkeit || employment * Einsatz, Gebrauch, Anwendung || utilization * Nutzbarmachung, (Aus)Nutzung, (Nutz)Anwendung || expenditure * Verbrauch, Ausgabe || versatility * vielseitige Verwendung/Verwendbarkeit, flexible Einsetzbarkeit || duty * (Be)Nutzung, Aufgabe, Amt, Dienst || intercession * Fürsprache
verwerfen reject * ablehnen || drop * fallen lassen || give up * aufgeben || discard * ablegen,aufgeben (Gewohnheit, Vorurteil, Kleider), ausscheiden/-schalten/-rangieren || dismiss * fallen lassen, aufgeben (z.B. Thema); ab-/zurückweisen (z.B. Vorschlag); as: als...abtun || overrule * Antrag/Entscheidung ~ || turn down * Vorschlag usw. ~ || refuse * ablehnen, zurückweisen, abweisen || neglect * übergehen, außer acht lassen, übersehen, missachten || warp * (Holz) sich ~
Verwiegung weighing
Verwindung torsion
verwindungssteif torsionally-rigid
verwirren entangle * (konkret) Garn, Fäden usw.; (figürl.) in Schwierigkeiten usw. verwirren/verstricken || confound * in Unordnung bringen, verwirren (someone: jemanden), vermengen || bewilder * verwirren, irremachen, irreführen (someone: jdn.) || embarrass * verwirren, erschweren, komplizieren, in Verlegenheit bringen || confuse * verworren/undeutlich machen, in Unordnung bringen, vermengen, verwirren || get entangled * sich ~, hängen bleiben, verwirrt werden
verwirrend confusing
Verwirrung entanglement * (konkret) || confusion || disorder * (von Sachen) || mix-up * (von Sachen) || distraction * Zerstreuung, Zerstreutheit, Ablenkung, Verwirrung, Bestürzung, Wahnsinn
verwoben interwoven || meshed * (figürl.) verzahnt closely: eng
verzahnen hob * Zahnrad ~ mit Werkzeugmaschine

Verzahnmaschine gear cutting machine * *{Werkz. masch.}*
verzahnt meshed * *(figürl.)* closely: *eng* || toothed * *externally: außen, internally: innen* || geared
Verzahnung gearing || toothing || tooth system || gear teeth || gear-tooth system || tooth construction || interlocking * *(figürl.)* || meshing * *(mechan. u. figürl.)* *Ineinander-Greifen*
Verzahnungsbeanspruchung gearing stress
Verzahnungsmaschine gear-cutting machine * *{Werkz.masch.}* || gear-shaping machine * *{Werkz.masch.}*
Verzeichnis listing * *Aufzählung, Auflistung* || directory * *z.B. ~ von Dateien, Telefonnummern* || list * *Liste* || catalogue * *Katalog* || table * *Tabelle* || index * *meist alphabetisch/alphanumerisch sortiertes ~* || catalog * *Katalog*
Verzeichnung image distortion * *{Optik\Computer}* *z.B. auf dem Bildschirm, bei der Bildverarbeitung usw.*
verzerrt distorted * *auch 'verzogen'*
Verzerrung distortion * *eines Signals* || shift * *Verschiebung (z.B. Farbverschiebung)*
verzerrungsarm low-distortion
Verzerrungsfaktor deformation factor * *Verhältnis →Leistungsfaktor/→Grundschwingungsleist.faktor (cos Phi/cos Phi_1)*
Verzerrungsleistung distortion power || harmonic power || distortive power
Verzicht renunciation * *of·auf* || resignation || disclaimer
verzichten dispense * *with: auf; ohne etwas auskommen* || do without * *ohne etwas auskommen; I can do without it: ich kann darauf ~* || renounce * *~ auf* || resign * *~ auf* || relinquish || forego || disclaim * *~ auf, ablehnen* || waive * *~ auf, auskommen ohne, sich e. Rechtes/Vorteils begeben*
verziehen distort * *verzerren, verdrehe, verziehen, be distorted: sich ~, verformen* || warp * *sich ~ (Holz usw.)*
verzinken zinc-coat || zinc-plate || zinc || galvanize || finger-joint * *{Holz}*
verzinkt zinc plated
verzinnt tin-coated
verzogen distorted || warped * *{Holz}* || moved away * *umgezogen*
verzögern decelerate * *['di: celereit]* || ramp-down * *entlang einer Rampe ~* || slow-down * *langsamer machen, Geschwindigkeit/Drehzahl verringern* || decel * *Kurzform für 'decelerate'* || delay || retard * *~, verlangsamen, bremsen, hemmen, nachstellen (Zündg.), verspäten, auf-/zurückhalten* || be delayed * *sich ~* || lag * *sich ~, zurückbleiben, nachhinken*
verzögert delayed
Verzögerung deceleration * *['dieβelereischen] negative Beschleunigung* || delay * *~szeit, zeitliche ~* || time-lag * *Hinterherhinken* || lag || retardation * *Verzögerung, Verlangsamung, Verspätung, Aufschub*
Verzögerungsbaustein delay block * *Funktionsbaustein*
verzögerungsfrei instantaneous || without time delay
Verzögerungsglied delay element * *first-order delay element: ~ erster Ordnung*
Verzögerungsglied erster Ordnung first-order delay element * *PT1-Glied* || first-order filter element * *PT1-Glied*
Verzögerungsmoment deceleration torque || retardation torque
Verzögerungspunkt deceleration point || deceleration-initiation point * *→Bremseinsatzpunkt*

Verzögerungsrampe slow-down ramp || deceleration ramp
Verzögerungsrelais time-delay relay || time-lag relay
Verzögerungsweg deceleration distance
Verzögerungszeit delay || delay time || time delay * *Zeitverzögerung*
verzollen pay duty on
Verzug delay * *→Verzögerung*
Verzugszeit delay || latency * *Latenzzeit/Totzeit bei der Reaktion auf ein Ereignis in einem Rechner* || delay time
verzweigen branch
Verzweigung branch * *auch im Programm: multiple branch: Mehrfach~* || branching * *auch ~ im Programmlauf*
Verzweigungspunkt branch-off point
VESA VESA * *Abk.f. 'Video Electronics Standards Association': Normengremium für Bildschirmstandards*
Vetriebsabteilung sales department || sales/marketing department
VGA-Monitor VGA monitor || VGA screen
VHDL VHDL * *Abk.f. 'Very Highly integrated circuits Description Language': Beschreibungssprache für ASICs*
Vibration vibration || pulsation * *pochend, schlagend* || judder * *in Aufzug, Flugzeug usw.* || shaking * *Schütteln*
Vibrationsfestigkeit vibration resistance
Vibrationsmotor vibration motor * *{el. Masch.} Motor erzeugt gewollte Schwingbewegung durch eingebaute Unwuchtscheibe*
Vibrationsschleifer vibratory sander * *{Holz}*
Vibrationssiebmaschine vibration screening machine * *{Holz} z.B. zum Trennen von Holzspänen in der Spanplattenherstellung*
Vibrationswalze vibration roller
vibrieren vibrate || judder * *(Substantiv) im Flugzeug, Aufzug usw.; Verwackeln*
Videograbber videograbber * *{Computer} →Framegrabber mit einer Digitalisierrate von 25 Bildern (Frames) pro sec*
Videorecorder VCR * *Abkürzung für 'Video Cassette Recorder'*
viel much * *much better: ~ besser; too much: zu ~* || a lot of || lots of || plenty of * *reichlich* || a lot * *(Adv.) thanks a lot: ~en Dank* || a great many * *sehr ~e*
Viel Erfolg Good Luck
vieladrig multi-core
viele many || a lot * *thanks a lot: ~n Dank* || a lot of || lots of || plenty of || a great many * *sehr ~*
vielfach multiple * *(auch Substantiv: das Vielfache)* || in many cases * *(Adv.)* || frequently * *(Adv.)* || widely * *(Adv.)* || many times over * *um ein Vielfaches*
Vielfach-Messgerät multimeter
Vielfach-Messinstrument multimeter || multi-range instrument
Vielfaches multiple * *as a multiple of/as multiples of: als ~ von* || many times more * *um ein ~*
Vielfachinstrument multimeter || multi-range instrument
Vielfachmessinstrument multimeter || multi-range instrument
Vielfachpolumschaltung multi-pole switching
Vielfalt diversity * *Mannigfaltigkeit, Vielgestaltigkeit, Abwechslung, Verschiedenheit* || variety * *Mannigfaltigkeit, Auswahl, Anzahl, Vielseitigkeit; a wide variety of: eine große ~ an* || multitude * *Mannigfaltigkeit, Vielzahl; a multitude of: eine Vielzahl von* || multiplicity * *Vielfalt, Vielzahl, Menge, Mehrzahl, Mehrfachheit, Mehrwertigkeit*
vielfältig manifold * *mannigfaltig, ~, vielfach,*

Vielkeilwelle 350

Mehr(fach)-, Mehrzweck-, Sammelleitung, Rohrverzweigg. || multifarious * *mannigfältig* || diverse * *mannigfaltig, verschieden, ungleich* || various * *verschieden(artig), mehrere, verschiedene* || varied * *verschieden(artig), mannigfaltig, abwechslungsreich, bunt*
Vielkeilwelle spline shaft
vielmehr rather || on the contrary * *im Gegenteil* || in fact * *in Wirklichkeit*
Vielschicht- multilayer
vielschichtig complex * *komplex* || many-sided * *facettenreich* || many-layered || intricate * *verzweigt, verschlungen, verwickelt, knifflig, schwierig, verworren* || stratified * *geschichtet, schichtförmig* || tricky * *(salopp) verzwickt, kompliziert, heikel* || multi-faceted * *facettenreich*
Vielschichtkondensator multilayer capacitor * siehe auch →Schichtkondensator
vielseitig versatile * *['wörsetail]* || flexible * *flexibel; flexibel use: ~er Einsatz* || all-round || multipurpose * *~ verwendbar*
vielseitig einsetzbar versatile in application || meeting manifold requirements
Vielseitigkeit versatility
Vielzahl multitude * *a multitude of: eine ~ von/an* || multiplicity || diversity * *Vielfalt, Verschiedenheit, Mannigfaltigkeit; a wide diversity of: e. große ~ an* || manifoldness * *Vielfalt* || variety * *Vielfalt, Mannigfaltigkeit, Anzahl, Auswahl; a wide variety of: eine große Vielfalt an*
Vierdrahtbetrieb four-wire operation
Vierdrahtleitung four-wire cable * *z.B. serielle Datenverbindung nach RS422*
Vierdrahtsystem 4-wire system * *z.B. für serielle Datenübertragg. nach RS422 oder Stromversorgg. mit Fühlerleitg.*
Viereckbauweise square-frame * *z.B. bei Elektromotor* || square type of construction * *z.B. bei Elektromotor* || square-frame design * *z.B. eines Elektromotors*
viereckig square || quadrangular || rectangular
vierfach quad
Vierfach- quadruple || quad
Vierkant square
Vierkantschraube square-head bolt
Vierpol four-terminal network || two-terminal-pair network || quadripole || two-port
vierpolig four-pole
Vierpunktaufhängung four-point fixing * *z.B. bei Motor mit →Pratzenanbau*
Vierquadrant four-quadrant
Vierquadrant- four-quadrant || 4-quadrant || regenerative * *rückspeisefähig* || four quadrant
Vierquadrantbetrieb four-quadrant operation || 4-quadrant operation
Vierquadrantenbetrieb four-quadrant operation
Viertaktmotor four-cycle engine || four-stroke engine
viertel quarter
viertletzt fourth from last
Vierwegeventil four-way valve
VIK VIK * *Verband d. Industriellen Energie- u. Kraftwirtsch., legt techn. Anforderungen für AC-Motoren fest*
violett purple
Viren viruses * *auch Computer~*
Virenbefall virus infection * *auch durch Computerviren*
virenfrei virus-free * *auch "frei von Computerviren"*
virtuell virtual * *(techn/physikal.) nur scheinb. vorhanden; (umgangssprachl.) faktisch, praktisch, eigentlich* || apparent * *scheinbar* || seeming * *scheinbar* || pretended * *vorgetäuscht, scheinbar* || ostensible * *scheinbar, vorgetäuscht* || virtually * *(Adv.)*

Virus virus * *auch Computer~*
Vision vision * *Seherblick, Weitblick, Phantasie, Vision, Traum-/Wunschbild* || enlightment * *Erleuchtung, Vision*
visionär visionary || enlightened * *erleuchtet, seherisch*
Visitenkarte visiting card || calling card * *(amerikan.)* || business card
Viskosereibung viscous friction * *Gleitreibung*
Viskosimeter viscometer * *zum Messen der Zähigkeit von Flüssigkeiten* || viscosimeter
Viskosität viscosity || viscosity grade * *Viskositätsstufe (von Öl usw.)*
Viskositätsklasse viscosity class * *bei Schmieröl usw.*
visualisieren visualize
Visualisierung visualization || display * *Anzeige* || visibility * *die "Sichtbarmachung" als Funktion*
Visualisierungssystem visualization system || visualization || process display sytem
Vitrine showcase * *für Werbe-/Ausstellungs-/Verkaufszwecke*
Vize- vice-
Vlies fleece || fibre
Vliese non-wovens * *{Textil}*
Vliesstoff non-woven material || fleece material
VLSI VLSI * *Abkürzg. für 'Very Large Scale Integration': hochintegriert*
VOB VOB * *Abk. für 'VerfügungsanOrdnung für Bauleistungen': German Construction Contract Procedures*
Vokabel word || term * *(Fach-) Ausdruck* || technical term * *Fachausdruck*
Volant fancy roller * *{Textil} Trommelputzwalze zum Reinigen der Kratztrommel bei Kratzenraumaschine*
Volkswirtschaft national economy * *Wirtschaft eines Landes* || political economy || economics * *Volkswirtschaftslehre*
Volkswirtschaftslehre economics || political economy
voll full * *(auch figürl.) vollständig, gefüllt* || filled * *gefüllt* || solid * *massiv, aus Vollmaterial* || whole * *ganz (z.B. Betrag)* || complete * *komplett, vollständig* || entire * *gesamt* || fully * *(Adv.)* || in full * *(Adv.)* || entirely * *(Adv.)vollständig* || entire * *(Adj.) gesamt* || totally * *(Adv.) vollkommen* || round * *prall* || tanked * *(salopp) betrunken*
voll digital fully digital
voll geblecht fully-laminated
voll geschlossen totally enclosed * *Gehäuse*
voll kompatibel fully compatible * *with: mit*
voll programmierbar fully programmable
Voll- full * *z.B. vollständig* || solid * *z.B. aus Vollmaterial, ganz mit Material ausgefüllt* || complete * *vollständig* || entire * *gesamt* || fully * *(Adv.) vor Adjektiv*
Vollanbieter full liner * *Anbieter mit vollständigem Produktspektrum* || broad liner * *Anbieter mit breitem Produktspektrum* || one stop shop * *bietet alles aus einer Hand*
Vollast full load * *at: bei* || rated load * *Nennlast* || full-load duty * *Betrieb bei ~, d.h. mit der höchsten zugelassenen Leistung am Bemessungspunkt*
Vollaussteuerung zero delay angle setting * *z.B. bei Phasenanschnittsteuerung; at: bei* || minimum delay angle setting * *z.B. bei Phasenanschnittsteuerung* || zero delay firing angle setting * *bei Phasenanschnittsteuerung* || maximum control setting * *allg.* || unity delay-angle setting * *{Stromrichter}*
vollautomatisch fully-automatic
Vollblechung full lamination
Vollblocksteuerung full block control
volldigital fully digital || all digital

volldigitale Plattenbelichtung computer-to-plate
 * *[Druckerei] im grafischen Gewerbe, in der Druckvorstufe*
Volldraht solid wire * *im Gegensatz zum Litzenleiter*
vollduplex full duplex
volle Geschwindigkeit full speed || top speed
volle Rolle full roll * *bei Wickler*
vollelektronisch all-electronic
voller Federkeil full key
voller Wert full value
Vollgas full throttle * *bei Verbrennungsmotor* || full speed
Vollgattersäge multi-blade frame saw * *[Holz]*
vollgeblecht fully laminated || fully-laminated
vollgeschlossen totally enclosed * *hohe Schutzart*
vollgesteuert fully controlled * *z.b. Thyristorbrücke*
vollgesteuerte Drehstrombrückenschaltung fully-controlled three-phase bridge connection * *z.b. B6C-Schaltung für 1Q-DC-Stromrichter*
vollgesteuerte Einphasen-Brückenschaltung fully controlled single-phase bridge connection * *B2C-Schaltung für 1Q-DC-Umrichter*
vollgesteuerte Gegenparallel-Drehstrombrückenschaltung fully controlled anti-parallel three-phase bridge connecti * *(B6)A(B6)C-Schaltung f. 4Q-DC-Stromricht.*
vollgesteuerte Gegenparallelschaltung fully controlled inverse parallel bridge connection * *(B6)A(B6)C Schaltung*
vollgesteuerte Stromrichterschaltung uniform converter connection * *siehe IEC 146-1-1 und EN 60146-1-1*
vollgrafisch fully-graphic
Vollgummi solid rubber
völlig full * *ganz* || entire * *ganz* || complete * *vollständig* || total * *vollständig* || thorough * →*gründlich* || perfect * *vollkommen* || absolute * *z.b. Gewissheit* || quite * *(Adv.)* || totally * *(Adv.)* || radically * *(Adv.)* →*grundlegend, von Grund auf, radikal* || completely * *(Adv.) almost completely: fast* ~
völlig geschlossen totally enclosed * *Gehäuse, z.B. einer elektr. Maschine*
Vollimprägnierung complete immersion
Vollkeilwuchtung full-key balancing * *Federkeil wird beim Auswuchten des Motorsläufers voll berücksichtigt*
vollkommen perfect * *perfekt,makellos* || full * *völlig, vollständig* || absolute * *Macht, Befugnisse, Recht usw.*
vollkommen geschlossen totally enclosed * *Schutzart* || TEFC = 'Totally Enclosed Fan Cooled': ~e *Motorschutzart IP44 mit Eigenbelüftung* || TENV = 'Totally Enclosed Non Vented': *~e Motorschutzart IP54 ohne Kühlung* || TEPV = 'Tot. Enclosed Pipe Vented': *~e Motorschutzart mit Kühlung über Rohranschluss* || TEUC = 'Totally Enclosed Unit Cooled': *~e Motorschutzart mit Wärmetauscher* || TEAO = 'Tot. Enclosed Air-Over': *~e Motorschutzart mit Gehäuse-Außenbelüftung* || closed-circulation * *~es Kühlsystem, ~er Kühlkreislauf* || closed-loop * *~es Kühlsystem, ~er Kühlkreislauf*
vollkommen unterschiedlich distinctly different
vollkompatibel fully compatible * *with: mit*
vollkompensiert fully compensated
Vollpolläufer solid-pole rotor || non-salient pole rotor || cylindrical rotor
Vollpolmaschine cylindrical-rotor machine * *i. Gegensatz zur* →*Schenkelpol-Maschine (kein 'gezackter' Läufer)*
Vollschwingung full-wave
Vollschwingungssteuerung multi-cycle control * *bei Thyristorsteller: es wird e. ganze Anz. v. Netzschwingungen durchgelass.* || multicycle control || full-wave control * *bei Wechsel/Drehstromsteller* || full-wave voltage controlled operation * *Leistungssteller-Betriebsart (lässt ganze Netzschwing. durch)*
vollständig complete || entire || total * *gesamt* || integral || whole || thorough * *vollkommen, völlig, durch und durch, gänzlich, gründlich*
Vollständigkeit completeness || entirity || totality || integrity
vollstopfen cram * *vollstopfen, an-/überfüllen (with: mit), hineinstopfen/-zwängen* || stuff * *(voll-) stopfen; stuff up: ver-/zustopfen/mit Fett imprägnieren/abschmieren* || crowd * *hineinstopfen/-pressen (with: mit); überfüllen, vollstopfen, zusammendrängen*
Volltreffer haben hit it right
Vollwelle solid shaft * *z.B. eines Getriebes*
Vollwellengleichrichtung full-wave rectification
Vollwellensteuerung multi-cycle control * *AC-Steller-Steuerverfahren mit Durchschaltung ganzer Netzvollschwingungspakete* || multicycle control || full-wave control * *bei Wechsel/Drehstromsteller*
vollwertig of full value || full || up to standard
Vollzylinder solid cylinder
Volt Spitze-Spitze Vp-p || Vpp
Voltmeter voltmeter || volts meter
Volumen volume * *(auch figürlich), Gesamtbetrag, Stückzahl; low: kleines/niedriges* || dimensions * *Abmessungen* || size * *Größe* || capacity * *Inhalt, Fassungsvermögen* || turnover * *Umsatz* || account * *Wert(igkeit), Wichtigkeit, Gewinn, Bedeutung (z.B. e. Kunden)*
Volumendurchfluss-Messgerät volume flow meter
Volumenintegral volume integral
Volumenregelung volume control * *z.B.mechan. "Hubraumänderg." bei Schraubenkompressor zur Variation d. Kälteleistung*
Volumenstrom volume flow || volume flow rate * *durchströmendes Volumen pro Zeiteinheit, z.B. bei Lüfter [Kubikmeter/h]* || flow
volumetrisch volumetric
vom Ausland from abroad
vom Netz trennen disconnect from mains || isolate from the supply system || interrupt the supply feeder
von Anfang an right from the start
von außen from the outside || from outside || from without || without * *außerhalb* || on the outside * *außerhalb*
von ausschlaggebender Bedeutung of the essence
von Bedeutung sein be of importance || be relevant * *in diesem Zusammenhang ~; zu: für* || matter || be important || be of significance || be significant
von der Last her gesehen seen from the load
von Hand manually || by hand * *handwerklich, per Handarbeit*
von Haus aus inherently * *(z.B. auf Grund des Funktionsprinzips) innewohnend* || by nature || originally || by birth
von hinten from the rear || from behind
von innen from the inside
von oben nach unten from the top to the bottom || from bottom to top || top down
von unten from the bottom || from underneath || bottom up * *von unten nach oben, das Untere nach oben gekehrt, kieloben* || from below
von unten nach oben from the bottom to the top || from bottom to top || bottom up * *das untere nach oben gekehrt* || from below to the top
von vorn from the front
von Vorteil beneficial
von Zeit zu Zeit from time to time * *siehe auch* →*gelegentlich*

vor before * *zeitlich, räumlich, im Signalfluss usw.*
~neliegend || in front of * *räumlich* || prior to * *früher als* || previous to * *früher als, zeitlich* ~ || ago * ~ *soundso langer Zeit, nachgestellt* || ahead of * *~weg* || in the presence of * *in Gegenwart von* || opposite * *gegenüber* || from * *schützen, verstecken, warnen* || against * *schützen, verstecken, warnen* || beforehand * *zu~, im ~aus* || upstream of * *im Signal-/Daten-/Energie-/Kühlmedien-/Materialfluss ~gelagert*
vor allem first and foremost || especially * *besonders* || particular * *besonders*
vor dem Hochlaufgeber before the ramp-function generator || before the ramp
vor der Begrenzung before limiting
vor Ort local || on site * *an Ort und Stelle, auf der Baustelle/Anlage/beim Kunden* || at site * *auf der Anlage/Baustelle* || on the shop floor * *direkt an/bei den Produktionseinrichtungen* || at the customer's facilities * *beim Kunden* || on-the-spot
Vor Rückwärtszähler up/down counter
vor Steuern pre-tax * *ohne Abzug der Steuern*
Vor- pre- || preliminary * *vorläufig, vorbereitend* || basic * *grundlegend* || rough * *Grob-* || draft- * *(Vor)Entwurf/Norm* || upstream * *im Signal-/Energiefluss vorgeschaltet* || series * *in Serie geschaltet*
Vor-Ort- local || remote
Vor-Ort-Betrieb local operation || local mode
Vor-Ort-Sollwert local reference
Vor-Ort-Vertrieb local sales office || direct field sales office * *Abteilung, Geschäftsstelle* || field sales services * *~s-Organisation*
Vor-Rückwärtsauswertung up-down interpreter * *Funktionsbaustein* || forward-reverse evaluation || cw-ccw detection
Vor-Rückwärtszähler up-down counter
Vorab- preliminary || advance
Vorabinformation early information || early info
Vorabzug initial pay-out * *z.B. bei Folienmaschine*
voraltern season
Vorarbeiten preliminary work || preparatory work * *vorbereitende Arbeiten* || general preparations || premachine * *(Verb) mit (Werkzeug-) Maschine vorarbeiten*
voraus in front || ahead || in advance * *of: von/gegenüber; in advance: im* ~ || advance * *Voraus-*
voraus- pre- || advance || in advance * *(nachgestellt)* || in front || ahead || fore-
vorausberechnen predetermine * *vorherbestimmen* || pre-calculate || forecast * *voraussagen/sehen* || predict * *vorher/voraussagen* || precalculate
Vorausberechnung precalculation || prediction * *Vorhersage*
vorausgehen preceed * *vorangehen (zeitlich früher)*
vorausgesetzt provided * *that: dass (kann auch weggelassen werden)* || assuming * *angenommen* || on condition * *that: vorausgesetzt, dass* || assumed * *it is assumed that: ~/angenommen dass*
vorausgesetzt dass provided that || on condition that || provided
vorausplanen plan in advance || schedule * *planen, festlegen*
vorausschauend foresighted || predictive * *prädiktiv, z.B. Regler* || with preview action * *z.B. Regler* || looking ahead || look-ahead * *Algorithmus*
voraussetzen suppose * *als Notwendigkeit/gegeben/mögl. voraussetz./annehmen; s.vorstellen, vermuten, glauben* || presuppose * *voraussetzen, zur Voraussetzung/als Bedingung haben, erfordern* || require * *erfordern, benötigen, verlangen,fordern, wünschen* || assume * *annehmen, unterstellen, sich anmaßen* || presume * *annehmen, vermuten, schließen* || take for granted * *als bekannt/gesichert;*

stillschweigend voraussetzen || provide * *provided that: vorausgesetzt, dass; sicherstell., vorsehen/-schreiben, Vorsorge treffen* || pect * erwarten || demand * *erfordern*
Voraussetzung premise || qualification * *an Wissen/Können/Ausbildung* || prerequisite * *Vorbedingung, (erste) Voraussetzung* || precondition * *Vorbedingung* || presupposition || supposition || assumption * *Annahme, Vermutung; on the assumption that: in der Annahme, dass* || requirement * *Bedingung, (An)Forderg., Erfordernis; meet the requirements: die ~gen erfüllen* || basic requirement * *Grundbedingung, wesentliche Erfordernis* || condition * *Bedingg., Voraussetzg.; on condition that/under the condition that: vorausgesetzt, dass* || provision * *Vorsorge, Vorkehrung, Maßnahme; provided that: vorausgesetzt, dass* || presuppose * *(Verb) zur ~ haben* || imperative * *zwingende ~*
voraussichtlich prospective * *(Adj.) in Aussicht stehend, (zu)künftig* || probable * *(Adj.) vermutlich* || presumable * *(Adj.) vermutlich, mutmaßlich, wahrscheinlich* || expected * *(Adj.) erwartet, erhofft, womit gerechnet wird* || estimated * *(Adj.) geschätzt, schätzungsweise* || likely * *(Adj.) wahrscheinlich, voraussichtlich* || probably * *(Adv.) wahrscheinlich* || be likely to * *(Adv.) wahrscheinlich* || be expected to * *(Adv.) erwartet werden, zu* || it is planned that * *(Adv.) es ist geplant, dass* || as it stands now * *wie die Dinge jetzt liegen, aus heutiger/jetziger Sicht* || scheduled * *(zeitlich ein)geplant*
Vorauswahl preselection || pre-selection
Vorauszahlung payment in advance
Vorband rough strip * *Grundmaterial im Walzwerk*
Vorbedingung precondition || prerequisite || basic requirement * *Grundbedingung, Grundvoraussetzung*
Vorbehandlung pretreatment
vorbei along * *entlang, längs* || by * *(nahe) bei/an, neben, via, an ... entlang/~* || past * *(räumlich und zeitlich)* || over * *(zeitlich)* || gone * *vergangen* || missed * *gefehlt, verfehlt*
vorbeifahren travel past || drive past
vorbeilaufen run past || drive past
Vorbelastung preloading || prestressing * *vorangehende Beanspruchung*
Vorbelegung preassignment || pre-assignment || default * *Vorbesetzung von Parametern, Variablen usw.*
vorbereiten prepare * *for: auf/zu; being prepared: in Vorbereitung befindlich*
vorbereitend preparatory * *preparatory to: als Vorbereitung zu/für*
vorbereitet prepared || pre-engineered * *konstruktiv vorbereitet, vorprojektiert* || pre-configured * *vorkonfiguriert, vorprojektiert*
Vorbereitung preparation * *in preparation: in ~; preparatory to: als ~ zu/für*
Vorbereitungsmaschine preparing machine * *z.B. in d. Textilmasch./Spinnerei*
Vorbereitungszeit preparatory time
vorbesetzen default * *Speicherzelle, Parameter usw. auf einen Initialisierungswert* || pre-assign * *vorbelegen, vorläufige Zuordnung herstellen*
Vorbesetzung default * *Parameter, Speicherzelle, Wert usw.* || presetting || preset value * *vorbesetzter Wert* || preset
Vorbesetzungswert default value * *Speicherzelle, Parameter usw.*
vorbeugende Wartung preventive maintenance
vorbohren predrill || pre-drill || rough-drill || prebore
Vorbrecher precrusher * *{Holz} z.B. in der Spanplattenherstellung*

vordefiniert predetermined * *z.B. Software-Regelbaustein*
vorder front || fore || anterior * *vorder, vorhergehend, früher (to: als)* || forward
Vorder- front
Vorderansicht front view
vordere Tragwalze front drum * *Tragwalze 2 eines Doppeltragwalzenrollers*
vorderer Bereich front area || towards the front * *im vorderen Bereich*
Vordergrund foreground || prime importance * *höchste Bedeutung/Wichtigkeit (of: von)*
Vordergrundspeicher foreground memory
Vordergrundverarbeitung foreground task * *Rechnerprogramme, die mit hoher Priorität laufen*
Vorderkante leading edge * *z.b. eines Impulses* || front edge
Vorderseite front side || front || face || front face
vorderseitiger Anschluss front connection
vorderst foremost * *am weitesten vorne gelegen* || first || front-
Vordruck form * *Formular* || blank * *(amerikan.)*
Vordruckregler inlet pressure controller * *für eine Turbine*
voreilen lead * *by: um; the voltage leads the current: die Spannung eilt dem Strom vor* || be advanced
voreilend leading
voreilender Kontakt leading contact
Voreilung lead || leading || forward slip * *{Walzwerk}*
Voreilwinkel lead angle || phase lead * *Phasen-Voreilwinkel* || angle of lead || advance angle * *{Stromrichter} siehe IEC 146-1-1*
voreingestellt preset || pre-set * *z.b. Einstellparameter im Herstellerwerk*
voreinstellen preset
Voreinstellung presetting || default * *vorbesetzter Wert für Parameter, Speicherzelle usw.* || initial setting * *ürsprünglich gesetzter Wert* || set-up * *auch Vorgabe der Hardwareparameter für das BIOS-System eines PCs* || factory setting * *Werkseinstellung z.B. v. Parametern* || preset option * *bei SPS-Programmiergerät* || preset || pre-setting
Voreinstellwert default value * *für Speicherzelle, Parameter usw.* || default
Vorendschalter leading limit switch * *bei Positioniersteuerung, Aufzug usw.* || preliminary contact * *bei Positioniersteuerung, Aufzug usw.*
Vorentwurf rough draft
vorerst first of all * *zuallererst* || for the present * *vorderhand* || temporary * *zeitweilig, vorübergehend* || for the time being * *vorderhand*
vorfabriziert ready-made || canned * *'eingedost' ('Schwarzer Kasten')*
Vorfertigung parts manufacture || parts production || preproduction || pre-fabrication
Vorführ- demonstration
vorführen demonstrate || present * *präsentieren*
Vorführkoffer demonstration case || presentation case
Vorführraum show-room
Vorführung demonstration || presentation * *Präsentation*
Vorführwagen demonstration vehicle
Vorgabe setting * *z.B. eines Sollwerts* || specification * *Aufgabenbeschreibung*
Vorgang event * *Ereignis* || procedure * *Vorgehensweise, Ablauf* || process * *Ablauf, z.B. Herstellungs-, Produktions-* || operation * *Arbeits(vor)gang, Prozess, Betätigung, Wirken, Tätigkeit* || action * *Aktion, Handlung* || sequence * *(Ab-) Folge* || cycle * *in (regelmäßigen) Abständen sich wiederholender ~*
Vorgänger predecessor

Vorgängermodell previous model
Vorgangsweise procedure
vorgeben enter * *Signal z.B. Sollwertsignal ~* || specify || establish * *bilden, einrichten, herstellen (Verbindung), etablieren; to: jdm./etwas* || set * *z.B. einen Sollwert von Hand ~* || apply * *anlegen, z.b. eine Spannung/einen Sollwert ~* || input * *z.b. ein Signal anlegen*
vorgebohrt pre-drilled || predrilled || prebored || rough-drilled
vorgefertigt prefabricated
vorgegeben given || predetermined * *vorherbestimmt, festgesetzt, genau definiert* || preset * *vorbesetzt, fest* || fixed * *fest* || specified || set
vorgehen proceed * *verfahren; by stage: schrittweise; according to/as described in: gemäß (z.B. Betriebsanltg.)* || procedure * *(Substantiv) Vorgehens-/Verfahrensweise* || course of action * *(Substantiv) Vorgehensweise* || have priority * *Vorrang haben, höherprior sein (gegenüber)* || come first * *Vorrang haben* || act * *handeln; in conformity with: gemäß/in Übereinstimmung mit* || go ahead * *voran/vorausgehen* || advance || go forward || be more important * *wichtiger sein (than: als)* || approach * *(figürl.) herangehen (an e. Sache/Aufgabe/Problem); (auch Subst.: Herangehensweise)*
Vorgehensweise procedure || action * *Handlung, Aktion* || kind of doing || proceeding || course of action
Vorgelege transmission gear * *vor dem 'eigentlichen' Getriebe angeordnetes Zwischengetriebe* || back bear || intermediate gear * *vor dem 'eigentlichen' Getriebe angeordnetes Zwischengetriebe* || connecting gear
vorgenannt already mentioned || mentioned above * *→oben erwähnt, vorgenannt* || foregoing || abovementioned * *→obengenannt*
Vorgerüst roughing stand * *zum Grobwalzen im Walzwerk*
vorgeschaltet series-connected * *in Reihe geschaltet* || upstream * *im Signal/Leistungsfluss ~* || lineside * *auf der Netzseite* || in incoming circuit || connected upstream * *im Signal-/Leistungsfluss ~* || frontend * *{Computer} z.B. ~es Programm (einer Datenbank usw.)*
Vorgeschichte history * *z.B. eines Fehlers, eines Messwertspeichers vor dem Triggerzeitpunkt* || preevent history
vorgeschrieben statutory * *(gesetzlich) vorgeschrieben, →gesetzlich, satzungsgemäß* || specified * *spezifiziert* || prescribed * *angeordnet* || mandatory * *obligatorisch, →verbindlich, zwangsweise, Muss- (Vorschrift, Bedingung usw.)*
vorgesehen provided * *for a thing: etwas; provided that: ~, dass* || planned * *geplant* || scheduled * *(zeitlich) geplant* || assigned * *for: für einen Zweck~* || intended * *beabsichtigt, geplant (something: etwas; to do/doing: zu tun)* || conceived * *vorgesehen, bestimmt, ersonnen, geplant (for: für)* || earmarked * *for: für einen Zweck ~*
vorgesehen sein für be provided for * *zur Verfügung gestellt sein für* || be intended for * *bestimmt sein für* || be assigned for * *bestimmt sein für (einen Zweck)* || be dedicated to * *bestimmt/gewidmet sein für*
vorgespannt preloaded || prestressed || biased * *mit einer elektr. Spannung*
vorgespanntes Lager preloaded bearing * *zur Vermeidung von Axialspiel*
vorgewickelt prewound * *z.B. fertig gewickelte Spule zum Einziehen in einen Motor*
vorhaben intend * *beabsichtigen, vorhaben, planen, bezwecken; to do/doing: etwas zu tun* || mean * *beabsichtigen, vorhaben, entschlossen sein; mean to do: etwas zu tun gedenken* || have in mind

Vorhalt

* etwas im Sinn haben, beabsichtigen; have a mind to do sth.: Lust haben etw.zu tun || propose
* beabsichtigen, sich vornehmen || plan * planen || have something on * etwas geplant haben, z.B. 'für den Abend' || be going to * planen; im Begriff sein, etwas zu tun || intention * (Substantiv) Absicht, Vorhaben, Vorsatz, Plan, Zweck, Ziel || intent * (Substantiv) Absicht, Vorsatz, Zweck || project * (Substantiv) Projekt || plan * (Substantiv) Plan
Vorhalt derivative component * {Regel.techn} D-Anteil, Differenzierer
Vorhaltezeit derivative-action time * eines Differenzialgliedes || rate time
Vorhaltzeit derivative-action time * für Differenzierer || derivative time * für Differenzierer || differentiation time * für Differenzierer || derivative gain * eines Differenziergliedes || derivative time constant * eines Differenziergliedes
vorhanden present || at hand || available * verfügbar, zur Verfügung gestellt || in stock * auf Lager || existing * bestehend, existierend || existent * bestehend, existierend || on stock * auf Lager || provided * zur Verfügung gestellt || included * beinhaltet
Vorhandensein presence * auch 'Anwesenheit'
Vorheizung preheating
vorher previous * (Adj.) vorherig || before * (Adj.) || previously * (Adv.) || in advance * (Adv.) im voraus || before * (Adv.) voraus || before-hand * (Adv.) zuvor, voraus, im Voraus || beforehand
vorher- pre-
vorherbestimmbar deterministic || determinable * bestimmbar
vorherbestimmen predetermine || precalculate * vorausberechnen || forecast * voraussagen, vorhersehen || pre-determine
vorherbestimmt predetermined
vorhergehend preceding || previous || foregoing
vorherig previous || former * vorig, ehemalig || preceding || foregoing || past * vergangen, ehemalig || prior * früher, älter (to: als); prior patent: älteres Patent; prior use: Vorbenutzung
vorherrschen predominate || prevail || be predominant
vorherrschend prevalent || prevailing || predominant
Vorhersage forecast || prediction || prognosis
vorhersagen predict || forecast
vorhersehbar foreseeable || predictable * voraussagbar
vorhersehen foresee || anticipate * voraussehen, vorausahnen, vorausempfinden || foreknow * vorherwissen || forecast * vorhersagen || predict * vorhersagen
Vorinbetriebnahme pre-start-up
Vorkehrung provision || precaution || measure
Vorkenntnisse technical knowledge * Fachwissen || experience * Erfahrung || previous commercial training || basic knowledge * Grundkenntnisse || previous knowledge
Vorkommastelle place preceding the decimal point || digit preceding the decimal point * auf einer (Ziffern-) Anzeige
vorkommen occur
vorkonfektioniert prefabricated * z.B. Kabel || precut * Kabel || preassembled * z.B. Kabel
vorkonfektioniertes Kabel cable set || precut cable || prefabricated cable || preassembled cable set
Vorkontakt leading contact || early contact || preliminary contact
Vorkopf pre-header * bei Funktions-/Datenbaustein einer SPS
Vorladeeinrichtung pre-charging device * z.B. z. Aufladen des Zwischenkreises bei Umrichter || pre-charging unit * z.B. z. Begrenzen d. Einschaltstroms von Umrichter-Zwischenkreiskondensatoren || precharge circuit * zum sanften Aufladen der Zwischenkreiskondensatoren eines Umrichters
Vorladekreis pre-charging circuit * zur Aufladg. d. Zwischenkreiskondensatoren bei Umrichter
vorladen pre-charge
Vorladeschaltung precharge circuit * z.B. zum langsamen Aufladen d. Z.kreiskondensatoren b. Spannungszwischenkreisumr. || pre-charging circuit
Vorladeschütz pre-charging contactor
Vorladung pre-charging || precharging
Vorlage printed form * gedruckte
Vorlauf advance || preparation time * Vorbereitungszeit || preparatory period * Vorbereitungszeit
vorläufig preliminary || tentative * versuchsweise || temporary || provisional || interim || for the present * vorerst, vorderhand || for the time being * vorerst, vorderhand
Vorlaufzeit set-up time
vorletzt next to last || last but one || preultimate
vorletzter last but one || preultimate || next to last
vorliegen be in hand * Antrag, Auftrag, Bestellung usw. || be there * vorhanden sein || exist * vorhanden sein, existieren || be submitted * vorgelegt/beantragt sein || be under consideration * behandelt werden || be present * vorhanden sein
vormachen demonstrate * demonstrieren || put before * Brett usw. || humbug * (salopp) jemandem etwas vormachen || fool * jemandem/sich etwas vormachen
vormagnetisieren premagnetize
Vormaterial rough material * z.B. im Walzwerk
Vormontage preassembly
vormontieren preassemble
vormontiert preassembled || prefabricated
vorn in front * at the lower front: ~e unten; towards the front: nach ~ || before || ahead || at the head * an der Spitze (of: von) || right in front * ganz ~ || at the beginning * am Anfang || forward * nach ~ || from the front * von ~ || from before * von ~ || begin at the beginning * von ~ anfangen || anew * von neuem anfangen || make a new start * von neuem anfangen || from front to back * von ~ bis hinten || from first to last * von ~ nach hinten, vom ersten zum letzten
vornehmen undertake * (Aufgabe) übernehmen/beginnen, (Sache) in die Hand nehmen, unternehmen || take in hand * beginnen, in die Hand nehmen || deal with * behandeln || make * Änderungen, Anpassungen usw. ~ || make up one's mind to do something * sich etwas ~ || implement * realisieren || occupy oneself with * sich beschäftigen mit || busy oneself with * sich beschäftigen mit || effect * ausführen, erledigen, vollziehen; abschließen (Versicherung usw.) || implement * realisieren, z.B. Preisänderungen
Vornorm tentative standard || draft standard * Normentwurf || pre-standard
Vorort-Betrieb local control * Vorort-Steuerung || local operation || local mode * Betriebsart
vorprojektieren preconfigure
vorprojektiert pre-engineered || pre-configured || predesigned
Vorprüfung precheck || preliminary test
Vorrang priority * Vorrang, Vordringlichkeit (over: gegenüber, top: höchsten) || precedence * take precedence of: ~ haben vor || pre-eminence || of highest priority || of top priority * to be processed with top priority: ~ zu bearbeiten
Vorrat stock || supply || store || provision * of: an || reserve * Reserve || tankage || ~ an flüssigen Gü-

tern || set * *z.B.* ~ *an Funktionsbausteinen, Rechnerbefehlen usw.* || repertoire * *Repertoire, Bestand* || repertory * *Repertoire, Bestand* || collection * *(An-) Sammlung, Kollektion, Auswahl, Anhäufung* || assortment * *Zusammenstellung, Sortiment, Auswahl, Sammlung, Mischung* || choice * *Auswahl* || variety * *Anzahl, Reihe, Vielfalt, Auswahl* || buffer * *Puffer*
vorrätig available || on hand || in stock * *out of stock: nicht mehr vorrätig; keep in stock: vorrätig halten* || on stock || stocked * *auf Lager liegend, vorrätig gehalten*
Vorreißer licker-in * *{Textil} bei* →*Krempelmaschine*
Vorreiter pace-setter
Vorrichtung device * *Gerät* || equipment * *Ausrüstung* || fixture * *z. Montieren* || mechanism || jig * *zur Werkzeugführung z.B. bei der Metallbearbeitung* || fitting device * *zur Montage* || chuck * *Vorrichtung z. Montieren, Spannvorrichtung* || appliance
Vorroller rereeler || rewinder * *Umroller, Wiederaufwickler* || re-reeler
Vorschalt- series || line-side * *netzseitig angeordnet* || ballast * *z.B. für Leuchte; auch 'Vorschaltgerät' für Leuchtstoffröhre* || upstream * *im Signal-/Energiefluss vorgeschaltet*
Vorschaltdrossel series reactor ~ || starting reactor * *als Anlaufhilfe für Motorstart* || series inductor * *in Reihe geschaltete* ~ || series inductance * *Vorschaltinduktivität*
vorschalten connect in series * *in Reihe schalten* || connect in incoming circuit || connect on line side * *auf der Netzseite* ~, *z.B. Netzdrossel*
Vorschaltgerät ballast * *für Leuchtstofflampen* || lighting ballast * *für Leuchtstofflampen*
Vorschaltinduktivität series inductance * *siehe auch* →*Vorschaltdrossel* || series reactor * *Vorschalt-Leistungsdrossel*
Vorschalttrafo series transformer * *siehe auch 'Vorschalttransformator'* || step-down transformer * *zur Spannungsreduzierung* || infeed transformer * *Einspeisetrafo*
Vorschalttransformator series transformer || step-down transformer * *zur Spannungsreduzierung* || line-side transformer * *netzseitiger Transformator* || isolating transformer * *Isoliertrafo* || intermediate transformer * *Zwischentransformator* || matching transformer * *Anpasstranformator* || infeed transformer * *Einspeisetransformator*
Vorschlag proposal || suggestion || recommendation * *Empfehlung*
vorschlagen propose || suggest || recommend * *empfehlen* || slate * *(vorläufig) aufstellen, (für eine Posten usw.)* ~; *be slated for: vorgesehen sein für*
vorschreiben prescribe || lay down * *einen Grundsatz festlegen/aufstellen* || set * *Bedingungen* ~ || stipulate * *festsetzen, abmachen, vertraglich festlegen* || specify * *spezifizieren, festlegen* || dictate * *diktieren, vorschreiben, gebieten, auferlegen, befehlen*
Vorschrift regulation || instruction * *Anweisung; to instructions: nach* ~ || direction * *Anweisung* || order * *Befehl* || rule * *Regel, Dienst*~ || prescription || precaution * *Vorsichtsmaßregel* || specification * *Spezifikation, Festlegung* || standard * *Norm* || provision * *(gesetzl.) Bestimmg.,Vorschrift; come within th.prov.s of th.law: u.d. gesetzl. Best.fal.*
Vorschriftenstelle institute of regulations * *z.B. Normengremium*
Vorschub feed * *Werkzeugmaschine* || feed function || feed rate * *Vorschubgeschwindigkeit* || feed speed * *Vorschubgeschwindigkeit*
Vorschubachse feed axis * *{Werkz.masch.} (Plural: 'axes')*

Vorschubantrieb feed drive * *Werkzeugmaschine*
Vorschubapparat automatic feeder * *{Werkz.-masch.| Holz}* || feeding unit
Vorschubeinrichtung feeding device
Vorschubgeschwindigkeit feed rate * *{Werkz.-masch.}* || feed speed || feed velocity
Vorschubtisch feeding table * *{Holz}*
vorsehen provide for * ~ *für, sorgen für; the law provides that: das Gesetz sieht vor, dass* || plan * *planen* || schedule * *(zeitlich) planen* || assign * *for: für einen Zweck* || intend * *beabsichtigen, wollen; intend for: bestimmen für/zu* || conceive * *planen, er-/ausdenken, ersinnen; conceived for: vorgesehen/bestimmt für* || take care * *sich* ~ || be careful * *vorsichtig sein* || earmark * *for: für einen Zweck* ~ || be on one's guard against * *sich* ~/*auf der Hut sein vor* || look out for * *sich* ~ *vor* || be the program * *auf dem Programm stehen; what is the program today: was ist für heute vorgesehen* || be on the programme * *auf dem Programm stehen* || envision * *ins Auge fassen, gedenken (doing: zu tun), für möglich halten* || envisage * *ins Auge fassen, gedenken (doing: zu tun), für möglich halten* || apply * *ver-/anwenden, anbringen, gebrauchen*
Vorsicht caution || care * *Behutsamkeit* || circumspection * *Umsicht*
Vorsicht bitte caution! || danger! * *Gefahr* || take care! * *vorsichtig behandeln* || with care! * *Aufschrift auf Kiste usw.*
vorsichtig careful * *behutsam* || cautious || conservative * *konservativ (Schätzung usw.)*
Vorsichtsmaßnahme precautionary measure || precaution
Vorsichtsmaßregel precaution || safety precaution * *Sicherheitshinweis/Vorschrift/Maßregel/Vorkehrung*
Vorsitzender chairman || president || chairwoman * *(weiblich)*
Vorsorge precautions * *take precautions:* ~ *treffen* || provisions * *Vorkehrung/-sorge, Maßnahme; make provisions:* ~ *treffen; provide against:* ~ *treffen gegen* || providence * *Vorsorge, (weise) Voraussicht*
Vorsorge treffen take precautions || make provision || provide * *against: gegen*
vorsorgliche Wartung preventive maintenance
Vorspann header * *z.B. eines Datentelegramms, Datensatzes*
vorspannen bias * *(elektr.) z.B. durch e. additiv wirkende elektr. Spannung/Signal* || preload * *(mechan.) z.B. eine Feder, ein Wälz-/Gleitlager in Axialrichtung zur Spielvermeidung* || pretension * *(mechan.)* || prestress * *(mechan.)*
Vorspannfaktor prestressing factor * *z.B. Verhältn. treibende Umfangskraft/Vorspannquerkraft bei Riementrieb*
Vorspannung prestressing * *z.B. bei Riementrieb* || initial stress * *z.B. bei Riementrieb* || bias * *(elektrisch)* || preloading * *z.B. durch Feder zur Vermeidung von Axialspiel i. e. Lager* || bias voltage * *(elektrisch)* || pre-tensioning
Vorspannung geben bias
Vorsprung lead * *Abstand (of: vor)* || advantage * *Überlegenheit, Vorteil (of: gegenüber); have the adv.of s.o.: jdm. gegenüber i.Vorteil s.* || leadership * *Führung* || ledge * *das Hervorspringende (Teil)* || shoulder * *Schulter, Vorprung, (mechan.) Ansatz z.B. an Maschinenteil* || edge * *(salopp) Vorteil, Vorsprung; market edge: Markt*~; *(on: gegenüber)* || overhang * *Überhang* || projection * *vorspringender Teil, Überhang,* →*Ausladung* || ledge * *Fels*~ *usw.* || margin * *zeit-/wertmäßiger* ~
Vorstand board of directors || executive board || managing committee || executive committee || directorate || chairman of the board * ~*svorsitzender*

Vorstandsmitglied head * *Einzelperson* || principal * *Einzelperson* || member of the executive board * *~smitglied* || member of the board * *~smitglied*
Vorstandsmitglied member of the executive board || member of the board
Vorstandsvorsitzender chairman of the board || Chief Executive Officer * *(Großschreibg. i. Verbindung mit Namen)* || CEO * *Abk. f. 'Chief Executive Officer'* || president
vorstehend projecting || protruding || prominent || jutting out || salient * *(her)vorspringend/-stehend* || above mentioned * *oben erwähnt*
vorstehend genannt previously mentioned * *siehe auch →obengenannt, →obig*
vorstellen introduce * *einführen, (her-) einbringen, bekannt machen; someone to somebody: jemandem eine Person* || represent * *darstellen, präsentieren* || visualize * *(sich) ein Bild machen, (sich) bildlich vorstellen, visualisieren* || imagine * *sich vorstellen, sich ein Bild machen* || present * *präsentieren* || envisage * *sich (geistig) vorstellen, (Ziel) ins Auge fassen* || have an interview * *sich vorstellen, ein Bewerbungsgespräch führen* || unveil * *erstmalig präsentieren, enthüllen, den Schleier fallen lassen*
Vorstellung introduction * *Einführung* || presentation * *Präsentation* || interview * *Bewerbungsgespräch (with: bei)* || vision * *Vision* || explanation * *Erläuterung*
Vorstellungsgespräch interview
Vorstellungsgespräch führen conduct an interview
Vorstellungstermin appointment of interview * *für Bewerbungsgespräch (with: bei)*
Vorsteueranteil pre-control component
vorsteuern pre-control * *bekannte Führungs- oder Störgröße direkt auf d. Stellgröße e. Regelkreises aufschalten* || feed-forward
Vorsteuersignal precontrol signal || feed forward signal
Vorsteuerung pre-control || precontrol || feed forward || feed-forward control
Vorsteuerwert precontrol value || feedforward signal
Vorsteuerwinkel precontrol angle
Vorstudie pre-project study * *zu einem Projekt*
Vorstufe first step || preliminary stage || input stage * *(elektr.) Eingangsstufe*
Vortaste shift key || function select key * *Zweitfunktionstaste*
Vorteil advantage * *over: gegenüber* || benefit * *Nutzen* || profit * *Gewinn* || edge * *Vorsprung, Vorteil; have the edge on s.o.: einen ~ gegenüber jdm. haben* || payoff * *Investitionsrückfluss, Profit, das 'Sich-Auszahlen'*
Vorteiler prescaler * *für Frequenzen*
vorteilhaft advantageous || profitable * *to: für* || lucrative || favourable * *günstig* || beneficial * *günstig* || advantageously * *(Adv.)* || to the best advantage * *(Adv.) aufs vorteilhafteste* || of advantage * *be of advantage: ~ sein* || convenient * *geeignet, →zweckmäßig, bequem, passend, zweckdienlich*
Vortrafo isolating transformer * *Trenntrafo* || step-down transformer * *zur Spannungsreduzierung* || series transformer * *siehe auch →Vortransformator* || infeed transformer * *Einspeisetrafo*
Vortrag report * *Bericht, Referat* || elocution * *Vortragsweise, Sprechtechnik, Redekunst* || lecture * *Vorlesung* || address * *Ansprache, Rede; deliver an address: eine Ansprache/Rede halten* || discourse * *ausführliche Rede* || speech * *Rede; make a speech: eine Rede halten* || talk * *Gespräch*
Vortragender speaker * *Redner* || instructor * *bei einer Schulungsveranstaltung*

Vortransformator line-side transformer * *netzseitiger Trafo* || isolating transformer * *Isoliertrafo* || step-down transformer * *zur Spannungsreduzierung* || series transformer * *in Reihe geschalteter Trafo* || intermediate transformer * *Zwischentransformator* || matching transformer * *Anpasstransformator* || infeed transformer * *Einspeisetransformator*
vortrocknen pre-dry
vorübergehend in the meantime * *zwischenzeitlich* || meanwhile * *zwischenzeitlich* || preliminary * *vorläufig* || provisional * *provisorisch* || temporary * *vorübergehend, zwischenzeitlich* || tentative * *versuchsweise* || passing * *auch flüchtig, vergehend* || transient * *zeitlich; flüchtig; transient deviation: ~e Abweichung* || temporarily * *(Adv.)* || transitory || for the present * *vorderhand*
vorübergehende Abweichung transient deviation || transient
Voruntersuchung pre-project study * *zu einem Projekt* || preliminary examination
vorurteilsfrei open-minded
vorverarbeiten pre-process || preprocess
Vorverarbeitung preprocessing || pre-processing * *auch von Daten*
vorverdichten supercharge
vorverdrahten prewire
Vorverdrahtung basic wiring
Vorverstärker preamplifier
Vorverstärkung preamplifying
Vorvertrag letter of intent || agreement of understanding
Vorwahl preselection || country code * *Ländervorwahlnummer am Telefon* || area code * *Vorwahlnummer am Telefon* || presetting * *Vorbesetzung* || pre-selection
vorwählbar pre-selectable
vorwählen preselect || predial * *Vorwahlnummer am Telefon* || select
Vorwahlnummer area code * *im Inland* || country code * *für Ausland* || call prefix || STD code * *(engl.) Abkürzg. für 'subscriber trunk dialing'*
Vorwahlschalter selection switch
Vorwahlzähler preset counter
vorwalzen break down * *{Walzwerk} Blech ~* || bloom * *{Walzwerk} Walzblöcke ~*
vorwärmen preheat
Vorwärmofen preheating furnace * *z.B. bei Blechstraße*
Vorwarnsignal prewarning signal || early warning signal
Vorwarnung pre-warning || pre-alarm || early warning
vorwärts forward || cw * *im Uhrzeigersinn* || fwd * *Abkürzung für 'forward'* || ahead * *vorwärts, voran, voraus*
vorwärts rollen roll up * *auf dem Bildschirm*
Vorwärts-Rückwärts-Auswerter up-down interpreter * *Funktionsbaustein*
Vorwärts-Rückwärtsauswertung up-down interpreter * *Funktionsbaustein* || forward-reverse evaluator || cw-ccw detector * *Rechts-Linkserkennung*
vorwärtsbewegen travel forward * *sich ~* || move forward
Vorwärtsdrehrichtung forward rotation || clockwise rotation * *Rechtslauf (im Uhrzeigersinn)* || cw. rotation * *Rechtslauf, Abkürzung für clockwise: im Uhrzeigersinn* || forward speed direction
Vorwärtsimpulse up pulses
vorwickeln prewind * *z.B. Spule vor dem Einziehen in einen Motor bei Motorenfertigung*
Vorwiderstand series resistor
vorwiegend predominant || predominantly * *(Adv.)* || preponderant || mainly * *(Adv.) hauptsächlich* ||

mostly * *(Adv.) meist(ens)* || largely * *(Adv.)* zum größten Teil
Vorwort preface * *meist vom Autor selbst*
foreword * *meist von jemand anderem als dem Autor* introduction * *Einleitung*
Vorzeichen sign || polarity * *eines (Analog-) Signals*
Vorzeichenauswertung polarity evaluation
vorzeichenbehaftet with sign || signed
vorzeichenrichtig with the correct sign
Vorzeichenumkehr polarity reversal || sign reversal || sign inversion || change of sign || polarity inverting
Vorzeichenumschaltung polarity reversal * *einer Spannung, eines Signals* || sign reversal
Vorzeigeobjekt prestigious object
vorzeitig premature * *frühzeitig, verfrüht, voreilig, vorschnell, übereilt* || precocious * *frühzeitig, vor der (normalen) Zeit, frühzeitig entwickelt, frühreif*
Vorziehpartie drawing section * *z.B. bei* →*Querschneider* || draw rolls * *z.B. bei* →*Querschneider*
Vorzug preference * *Bevorzugung (above/over/to: vor), Vorliebe* || virtue * *Tugend, gute Eigenschaft, Wirksamkeit* || highlight * *herausragendes (positives) Merkmal* || pilot train * *{Bahnen}*
vorzüglich excellent || superior || first-rate * *erstklassig* || first-class * *erstklassig* || fantastic * *(salopp) fantastisch*
Vorzugs- preferable * *vorzuziehend* || preferred * *bevorzugt* || standard * *Standard* || superior *± hervorragend, überlegen (auch zahlenmäßig), bevorrechtigt, erlesen* || preferential * *Vorzugs-, bevorzugt, bevorrechtigt* || recommended * *empfohlen* || preference
Vorzugslage preferred state * *eines Speichers (Flip Flop usw.)* || preferable operating position * *räumliche Anordnung eines Bauteils, Einbaulage*
Vorzugspreis preferential price || special price
Vorzugsrichtung preferred orientation
Vorzugstyp preferred type
Vorzugstypen standard types
vorzugsweise preferably || by preference || chiefly || mostly || preferentially
vorzuziehen preferable
VRAM VRAM * *Video-RAM (Speicher auf Grafikkarte, wird v.Grafik-Chip beschrieben und von Video-DAC gelesen)*
vulkanisieren vulcanize || cure
Vulkanisiermittel vulcanizing agent
VVIDD VVIDD * *'Video Interface for Digital Displays': Norm der* →*VESA für dig.Bildschirmsteuerung (LCD...)*

W

W watt * *Einzahl* || watts * *Mehrzahl*
Waage balance || pair of scales || scales || scale || weigher * *{Verpack.techn} Abwiegeeinrichtung*
waagerecht horizontal
Waagerechte horizontal plane || horizontal position || horizontal plane * *Horizontalebene*
Wabe honeycomb * *honeycomb radiator: Wabenkühler*
Waben- honeycomb || dual-lateral * *Spule*
wabenförmig honeycomb || dual-lateral * *Spule*
Wachs wax
wachsen grow || expand * *sich ausdehnen* || develop * *sich entwickeln* || increase * *ansteigen*
Wachstum growth

Wachstumsrate growth rate
Wächter monitor * *Überwachungseinrichtung*
Wackelkontakt loose contact || loose connection || poor terminal connection * *an (Klemmen)Anschluss* || bad connection || intermittent electrical contact
wackeln shake || rock * *schwanken* || reel * *taumeln*
Waffelmuster waffle pattern * *z.B. an Kupplungs-/ Bremslamelle*
Wägeeinrichtung weigher || weighing machine
Wagen carriage * *(Transport)Wagen, Karren, Fahrgestell, Schlitten* || car || coach * *{Bahnen}* || waggon * *{Bahnen} Güter~* || cart * *Karren* || trolley * *(zweirädriger) Wagen, Laufkatze, Förder~, Draisine*
wägen weigh
Wägesystem weighing system
Wagnis risk || venture
Wagniskapital venture capital || risk capital
Wahl selection * *Auswahl* || choice || alternative * *Alternative* || option * *freie Wahl* || election * *(politisch)* || selecting * *das Auswählen* || choosing
Wahlausgang select output * *z.B. analoge/digitale/ binäre Ausgangsklemme* || assignable output * *frei mit einer Funktion belegbar* || general-purpose output * *Mehrzweckausgang* || programmable output
wählbar selectable || optional || programmable * *programmierbar, parametrierbar* || user programmable * *vom Anwender programmierbar* || multiple-choice || can be selected * *kann gewählt werden* || pre-selectable *+ vor··*
Wahleingang select input * *z.B. binäre/digitale/ analoge Eingangsklemme* || assignable input * *frei mit einer Funktion belegbar* || general-purpose input * *Mehrzweckeingang* || programmable input
wählen select || choose || take one's choice * *seine Wahl treffen* || dial * *am Telefon* || opt * *sich entscheiden, optieren (for: für, between: zwischen)* || opt for * *sich entscheiden/optieren für*
wahlfrei random * *beliebig (z.B. Möglichkeit, auf alle Speicherplätze belieb. m. gleicher Geschw. zuzugreifen)* || freely selectable * *frei wählbar*
wahlfreier Zugriff random access * *mit gleicher Zugriffszeit auf alle Daten (to: auf)*
Wahlklemme select terminal * *programmable terminal* || terminal with selectable function || general purpose terminal || assignable terminal
Wahlschalter selection switch || selector || select switch || selector switch
Wahlsignal select signal || selectable signal || programmable signal
wahlweise alternative || optional || selective || optionally * *(Adv.)*
wahr true || real * *echt* || veritable * *echt* || genuine * *echt* || proper * *eigentlich, richtig* || sincere * *aufrichtig*
während during * *(zeitlich); during the configuration phase: ~ der Projektierung* || while * *Gegensatz* || whereas * *Gegensatz* || whilst * *~ demgegenüber*
während des Betriebes during operation || on-line * *ohne abzuschalten* || during normal operation * *während des normalen Betriebes* || on the fly * *ohne anzuhalten* || online || hot * →*unter Spannung (Stecken von Baugruppen usw.)*
Wahrheitstabelle truth table
wahrnehmbar noticeable || perceptible || visible * *sichtbar* || audible * *hörbar*
wahrnehmen perceive * *empfinden, wahrnehmen, (be)merken, spüren, verstehen, erkennen, begreifen* || notice * *bemerken, beobachten, wahrnehmen, achten auf, mit Aufmerksamkeit behandeln* || observe * *bemerken, wahrnehmen (auch Termin), beobachten* || become aware of * *etw. gewahr werden* || make

use of * *Gelegenheit* ~ || avail oneself of * *Gelegenheit* ~ || look after * *Interesse* ~ || protect * *Interesse* ~ || safeguard * *sicherstellen von Interessen usw.* || exercise the fuction of * *ein Amt/eine Funktion* ~
Wahrnehmung perception * *(sinnliche, geistige) Wahrnehmung, Empfindung* || observation * *Beobachtung(svermögen), Wahrnehmung, Überwachung* || safeguarding * ~ *der Interessen usw.* || acting on behalf * *of a person: der Interessen e. Person*
Wahrnehmungsschwelle perception threshold
wahrscheinlich probably * *(Adv.)* || likely || probable * *(Adj.)*
Wahrscheinlichkeit probability || likelihood
Wahrscheinlichkeitsverteilung probability distribution || cumulative distribution
wahrscheinlichst most likely
Währung currency
Walkarbeit churning work * *z.B. in Lager, Reifen* || churning || churning loss * *Verluste auf Grund der Walkarbeit*
walken churn * *(durch-) schütteln, aufwühlen, buttern, sich heftig bewegen, schäumen* || full || felt * *Textilien/Filz* ~
Walkleistung churning power
Walzantrieb mill-stand drive
Walzblech rolled plate
Walzdraht wire rod || rolled wire
Walzdruck roll pressure || pressure of rolling
Walze roll || drum * *Trommel* || cylinder * *hohle Walze* || roller * *(Stoff-, Garn-)Rolle, Zylinder, Druckwalze (in Druckmaschine), Lauf/Gleit/Führungsrolle* || barrel * *Rolle, Walze, Trommel, Zylinder, Rohr* || reel * *Haspel, Winde, Rolle*
Walzeisen rolled iron
walzen roll || grind * *zermahlen*
wälzen roll * *auch 'sich* ~*'* || pore over * *Bücher* ~ || turn over * *auch Gedanken usw.* ~
Walzenanstellung screw-down operation * *{Walzwerk}* || roll adjustment * *{Walzwerk}* || screwdown * *{Walzwerk}* || screwing down * *{Walzwerk}*
Walzenbezug roller covering * *{z.B. in Druckerei}*
Walzendrehmaschine roll turning lathe
walzenförmig cylindrical
Walzenheizung roll heating * *z.B. bei Papiermaschine, Kalander usw.*
Walzenlüfter centrifugal fan * *Fliehkraftlüfter* || drum-type fan * *Trommellüfter*
Walzenmantel roll shell * *z.B. bei Siebsaugwalze einer Papiermaschine*
Walzenschleifmaschine roll grinding machine
Walzenschnellabhebung quick lifting of the roll
Walzenverschiebung roll shifting * *Walzwerk*
Walzenverschleiß roll wear
Walzenverstellvorrichtung roller adjustment drive
Walzenvorschub roll feed
Walzenwechsel roll changing || roll exchange * *Austausch*
wälzfräsen hob
Wälzfräser hob
Walzgerüst milling stand * *(Walzwerk)* || mill stand * *(Walzwerk)* || rolling stand * *(Walzwerk)* || stand of rolls * *(Walzwerk)* || roll stand
Walzgerüst-Antrieb millstand drive * *(Walzwerk)*
Walzgerüst-Motor mill-stand motor
Walzgerüstmotor mill-stand motor
Walzgeschwindigkeit rolling speed
Walzkante rolled edge
Walzknüppel billet
Walzkraft roll force || rolling load
Wälzlager rolling-contact bearing || rolling bearing || rolling-elememt bearing || anti-friction bearing * *DIN 611 T.1 3, 625, 628 T.1, DIN5412 T.1, DIN ISO76, 281, el.Masch. DIN 42966*

Wälzlagerstahl bearing steel
Walzmotor mill motor || rolling mill motor || rolling motor || mill-stand motor
Walzrichtung rolling direction
Walzspalt rolling gap || nip
Walzspaltregelung rolling gap control
Walzstahl rolled steel
Walzstraße rolling mill line || rolling line
Walzwerk rolling mill || steel mill * *Stahlwalzwerk*
Walzwerk-Hauptantrieb rolling-mill main drive
Walzwerkmotor rolling mill motor || mill motor
Walzwerksantrieb rolling mill drive
Walzwerksmotor rolling mill motor || mill motor || millstand motor * *Walzgerüst-Motor*
WAN WAN * *Abkürzg.f. 'Wide Area Network': Weitbereichs-Computernetz*
Wand plate * *Vorder~, Rück~* || side * *Seiten~* || panel || wall
wandbefestigt wall mounted
Wandbefestigung wall mounting * *eines Geräts*
Wanddicke wall thickness || gauge * *eines Rohrs/ Behälters*
Wanddurchführungs-Montage through-hole mounting
Wanderfeld travelling field
Wanderfeldmotor travelling field motor * *z.B. Linear/Sektormotor mit linear/im Kreissektor angeordn. "Käfigläufer"*
wandern creep * *kriechen* || travel || migrate * *wegwandern, langsam hinein~ (auch techn.), ziehen, migrieren* || drift * *(ab)treiben, ziehen, stömen, ab~, auseinanderleben*
Wandgehäuse wall-mounted casing || wall-mount housing
Wandhalter wall bracket
Wandler transformer || transducer || instrument transformer * *Mess~* || converter
Wandlungszeit conversion time * *z.B. eines Analog/Digital-Umsetzers*
Wandmontage wall mounting * *eines Geräts; wall-mount: für* ~
Wandstärke wall thickness || thickness of pipe wall * *eines Rohres* || thickness of wall || section || thickness
Wandtafel blackboard
Wandverteiler wall-mounted distribution board
Wanne bowl * *Napf, Schale, Schüssel, Becken, Wölbung* || tank * *Flüssigkeitsbehälter*
Ward-Leonard-Umformer Ward-Leonard set
Ware ware * *Ware(n), Artikel, Erzeugnis(se)* || article * *Artikel* || article of commerce * *Handelsartikel* || commodity * *Handels/Gebrauchsartikel, Waren* || product * *Erzeugnis* || material * *z.B. Rohstoff, Grundstoff/material, Textilstoff* || goods * *(Transport)Güter, Waren, Gegenstände, Frachtgut* || manufactured goods * *Fertigwaren, Fabrikwaren, Erzeugnisse* || merchandise * *Ware(n), Handelgüter*
Waren goods || merchandise || commodities || articles || products * *Erzeugnisse*
Warenbahn material web || web * *vor allem durchlaufende Papier/Textil/Kunststoffbahn* || material * *Material* || line
Warenbahngeschwindigkeit line speed || web speed * *besonders Papier, Textil, Kunststofffolie* || material speed * *Materialgeschwindigkeit*
Warenbahnspeicher accumulator || web accumulator || strip accumulator * *für Walzband* || looper
Warenbahnzug web tension * *Zugkraft*
Warengeschwindigkeit line speed * *Bahn/Bandgeschwindigkeit* || web speed * *besonders einer Papier-/Textil/Kunststofffolienbahn* || material speed * *Materialgeschwindigkeit (z.B. bei einer Produktionsanlage)*
Warenrücksendung returned good * *zurückge-

sandte Ware || redelivery of goods * *das Zurücksenden*
Warenspeicher accumulator * *für durchlauf. Warenbahn zum Zwischenpuffern des Materials bei Rollenwechsel* || storage || web storage * *für durchlauf. Papier/Textil/Kunststoffbahn, z.B. z. Überbrückg. d. Rollenwechselzeit* || strip storage * *für durchlaufendes Blechband (z.B. zur Überbrückung der Coil-Wechselzeit)* || looper * *für durchlaufende Warenbahn zur Überbrückung von Rollenwechselzeiten*
Warenzeichen trade-mark * *registered: eingetragenes*
Warenzug web tension * *Zugkraft*
warm warm
Warmbandstraße hot-strip mill * *Walzwerk*
Warmbreitband hot wide-strip * *Walzwerk*
Wärme heat * *absorb: aufnehmen; dissipate: abgeben; conduct: leiten; carry off: abführen; dry: trockne*
Wärme- thermal
Wärmeabfuhr heat discharge || heat removal || heat dissipation
Wärmeabführung heat discharge || heat removal || heat dissipation
Wärmeabgabe heat dissipation * *Verlustwärme*
Wärmeableitung heat dissipation || heat conduction * *Wärmeleitung* || leading-off of the heat * *Wärmeabführung* || heat transfer * *Wärmeübertragung*
Wärmeabstrahlung heat emission || heat radiation
Wärmeaufnahmefähigkeit thermal absorption capacity
Wärmeausdehnung thermal expansion
Wärmeausdehnungskoeffizient coefficient of thermal expansion
Wärmeausgleich temperature compensation
Wärmeaustauscher heat exchanger
Wärmebehandlung heat treatment || annealing * *durch Glühen (meist für Metall)*
Wärmebelastbarkeit thermal capacity || permissible thermal load || thermal rating * *spezifizierte Nenn-~, z.B. einer Kupplung/Bremse*
Wärmebelastung thermal load * *high: hohe* || thermal stress * *Wärmebeanspruchung* || thermal loading
wärmebeständig heat-resistant
Wärmebeständigkeit heat stability
wärmedämmend heat-insulating
Wärmedämmung thermal insulation || heat insulation
Wärmedehnung thermal expansion * *Längendehnung bei Erwärmung (Eisen 0.1%/100K; Kupfer 0.15 %/100 K; Alu 0.2 %/100 K)*
Wärmeentwicklung development of heat
Wärmeerzeuger heater
Wärmegefälle heat drop
Wärmegrenzwert thermal limiting value
Wärmehaushalt thermal economy
wärmeisoliert thermally insulated
Wärmekapazität thermal capacity
Wärmekennwert rated thermal capacity * *spezifizierte Wärmebelastbarkeit einer Kupplung, Bremse*
Wärmeklasse temperature rise class * *kennzeichnet max. zuläss. Übertemperaturen (el. Masch.: IEC 43, VDE 0530)* || thermal classification || temperature class || thermal raise class || temperature rise * *Kurzform für 'temperature rise class'* || thermal class * *kennzeichnet max. zuläss. Temp. i. el. Gerät nach IEC 85 (z.B. Y, A, E, B, F, H)*
Wärmekraftwerk thermal power station
wärmeleitend heat-conducting
Wärmeleitfähigkeit thermal conductivity
Wärmeleitpaste thermolube || thermo-lubricant || heat tranfer compound || heat-conducting paste
Wärmeleitung thermal conduction

Wärmemenge heat quantity || amount of heat
Wärmenest hot spot
Wärmepumpe heat pump
warmer Betriebszustand warm operating condition
Wärmerohr heat pipe
Wärmerückgewinnung heat recovery
Wärmespeicher heat storage * *heat-storage capacity: ~vermögen*
Wärmestau build-up of heat
Wärmestrahlung heat radiation
Wärmestrom heat flow
Wärmetauscher heat exchanger * *Einrichtung z. Übertragung v.Wärme von einem primären auf ein sekundäres Kühlmittel*
Wärmetransport heat transport
Wärmeübergang heat transfer
Wärmewiderstand thermal resistance || thermal impedance
Warmpresse warm press || forging press
Warmstart restart || jump start || flying restart * *z.B. durch Fangschaltung bei Umrichtergespeistem Drehstrommotor* || warm restart * *eines Rechners oder Antriebs*
Warmwalzen hot rolling
Warmwalzwerk hot rolling mill
Warmwiderstand resistance in the hot state || resistance when hot * *z.B. einer Lampe (ist höher als der Kaltwiderstand)*
warnen warn * *of/against: vor; warn someone against doing: jdn. davor ~, etw. zu tun* || caution * *against: vor*
warnend cautionary || warning
Warnhinweis warning information || warning instruction
Warnhinweise safety information * *in Betriebsanleitung*
Warnleuchte warning light || alarm lamp
Warnmeldung alarm signal || alarm message || alarm
Warnrelais alarm relay
Warnschild warning sign || warning plate || warning label * *Warn-Aufkleber*
Warnung alarm || warning || caution || admonition * *Ermahnung, Verweis* || warning signal * *Warnsignal*
Warnungsfolgespeicher alarm sequence memory * *(→PROFIBUS-Profil)*
Warnungskanal warning channel
Warnungsmeldung warning signal || warning message || warning
Warnzeichen warning sign || danger signal || warning symbol
Wartbarkeit maintainability
Warte central control room * *zentraler Steuerraum* || control room * *Leitraum* || control station || switchboard gallery || control master station * *→Leitwarte*
warten service * *Wartungsdienste leisten, z.B. an Maschine* || maintain * *z.B. Maschine ~* || inspect * *inspizieren* || overhaul * *überholen* || wait * *for: auf* || insert a delay * *eine Wartezeit einfügen* || carry out the maintenance * *Wartungsarbeiten durchführen*
warten bis man dran ist wait for one's cycle to come around
Wartentechnik control-room systems
Wärter attendant || operator
Warteschlange queue * *[Computer] [kju:] z.B. für Rechnerbefehle/Tasks*
Warteschleife idle loop * *z.B.in Programmm*
Wartezeit delay time * *Verzögerungszeit* || delay * *Verzögerung* || waiting period * *Wartefrist* || wait time || latency
Wartung maintenance * *z.B. an Maschine, einem Gerät; preventive: vorbeugende* || servicing * *z.B.*

wartungsanfällig 360

an Maschine || maintenance routine * *laufende* || service || attendance * *Pflege, Wartung* || upkeep * *Instandhaltung, Unterhalt* || checking * *Nachprüfung, Kontrolle* || inspection * *Inspektion, Überprüfung* || checkup * *Nachprüfung* || overhaul * *Überholung*
wartungsanfällig high-maintenance
Wartungsanleitung maintenance guide || maintenance instructions || maintenance manual * *Handbuch* || service instructions
Wartungsarbeiten maintenance work
wartungsarm nearly maintenance-free * *nahezu wartungsfrei* || low-maintenance
Wartungsaufwand maintenance requirements * *Bedarf, Erfordernisse; low: geringer* || maintenance costs
Wartungsdienst service department
wartungsfrei maintenance-free || no-maintenance
Wartungsfreiheit maintenance freedom || freedom from maintenance || ease of maintenance * *Wartungsfreundlichkeit* || zero maintenance
wartungsfreundlich easy to maintain || nearly maintenance-free * *nahezu wartungsfrei* || easy to service
Wartungsfreundlichkeit ease of maintenance || minimum maintenance requirements || ease of service || ease of serviceability || easy serviceability || service capability * *Wartbarkeit*
wartungsintensiv high-maintenance
Wartungsintervall maintenance period || service interval || maintenance interval
Wartungskosten maintenance costs
Wartungsöffnung service opening || service cover * *Wartungsklappe* || maintenance opening || access hole
Wartungspersonal maintenance personnel || maintenance staff
Wartungsplan maintenance schedule
Wartungstechniker maintenance engineer
Wartungsvertrag service contract
was anbetrifft as far as
Wäsche washing
waschen wash
Wäscherei washing plant
Wäschetrockner clothes drier || electric drier || clothes dryer * *(kann mit "i" oder "y" geschrieben werden)* || clothes-horse || tumbler
Waschmaschine washing machine || washer * *für Nicht-Textilien, z.B. für Behälter in der Getränkeindustrie* || cleansing machine * *Reinigungsmaschine*
Wasser-Wasser-Rückkühlanlage water-to-water heat exchanger * *bei einem geschlossenen Wasserkühlkreislauf*
Wasserablass water outlet
Wasserablauf water drain || water outlet * *Wasserauslass/-austritt*
Wasserablaufbohrung water drain hole
wasserabweisend water-repellent
Wasseranschluss water connection || water supply * *Wasserversorgung*
wasseranziehend hygroscopic
Wasseraufbereitung water treatment
Wasseraufbereitungsanlage water preparing plant
Wasserbau hydraulic engineering
wasserdicht waterproof || watertight || impermeable to water * *wasserundurchlässig, für Wasser undurchdringlich*
Wassereinlass water inlet * *water inlet temperature: ~temp.; b. wassergekühlt. Wärmetauscher f. El.motor 5...25 °C*
Wasserfüllung water filling
wassergekühlt water-cooled || watercooled
Wasserhahn water cock

Wasserkraftgenerator hydroelectric generator || waterwheel generator
Wasserkraftwerk hydroelectric power station || hydro-electric power plant
Wasserkühlung water cooling || water-cooling system
Wasserleitung water pipe
Wasserschlauch water hose
Wasserschutz water protection
Wasserstandsregelung water level closed-loop control
Wasserstoff hydrogen
Wasserstrahl water jet
Wasserstrahlschneiden water jet cutting * *{Werkz.masch.}*
Wasserversorgung water supply || water service
Wasserwaage spirit level || water-gauge || carpenter's level
Wasserwerk water works || water utility * *Wasserversorgungsunternehmen* || water supply installation || waterworks || water pumping station || water supply plant || waterwork
Watchdog watchdog * *Zeitüberwachung eines zyklischen Vorgangs (Zeitglied, muss zyklisch getriggert werden)*
Watt watt
weben weave
Webmaschine weaving machine
Webstuhl weaver's loom || loom || weaving machine * *Webmaschine* || textile loom || weaving loom
Webstuhlmotor loom motor
Webwarenfabrik weaving mill
Wechsel change * *auch ~ von Geldsorten* || changeover * *Hinüber ~* || turnover * *Hinüber~* || exchange * *Tausch* || succession * *Aufeinanderfolge* || rotation * *regelmäßiger Austausch* || fluctuation * *Schwankung* || changing || replacement * *Austausch z.B. e. defekten Teils*
Wechselanteil alternating component || AC component
Wechselbeanspruchung alternating stress || cyclic load || cyclic stressing || cycling
Wechselbeziehung correlation
Wechselfeld alternating field
Wechselfluss alternating flux
Wechselgetriebe changeable gearbox || multi-speed gearbox || gear-change gearbox || change-speed gear
Wechselgröße alternating quantity * *physikal.Größe mit periodisch wiederkehrendem Zeitverlauf, z.B. Strom/Spann.*
Wechselkontakt change-over contact || C contact || double-throw contact || DTC * *Abkürzg. f. 'Double-Throw Contact'; 4PDTW: 4-poliger Wechselkontakt* || form C contact || two-way contact
Wechselkurs exchange rate
Wechsellast varying load || changing load || alternating load || cyclic load * *zyklisch sich wiederholendes Lastspiel* || alternating stress * *(mechanisch)* || pulsed load
Wechsellastfestigkeit stability against pulsed load * *z.B. einer Sicherung* || stability against alternating load * *z.B. einer Sicherung*
wechseln change || change-over * *hinüber~* || alternate * *ab~, ab~ lassen* || vary * *variieren, (sich) verändern, verstellen* || exchange * *(miteinander) austauschen* || shift * *Platz/Lage/Szene ~* || reverse * *umkehren*
wechselnd changing || varying * *(sich) verändernd* || alternating * *(sich) ab~* || changeable * *veränderlich, veränderbar*
wechselnde Belastung varying load
Wechselpuffer alternating buffer || multi-buffer
Wechselpuffersystem alternating buffer system
Wechselrad change gear * *zur Änderung der Übersetzung bei → Wechselradgetriebe* || interchangable

gear || loose change gear || pick-off gear wheel || changeable gear wheel || interchangeable gear wheel * *zur Änderung der Übersetzung bei* →*Wechselradgetriebe*
Wechselradgetriebe change-gear-type gearbox * *Getriebe mit durch Zahnradwechsel änderbarer Übersetzung* || gearbox with interchangeable gear wheels * ~ *mit durch (manuellen) Zahnradwechsel änderbar.Übersetzg.* || gearbox with pick-off gear wheels || gearbox with interchangeable gear wheels || change gears
wechselrichten invert
Wechselrichter inverter || invertor * *(veraltet)* || inverter module * *Wechselrichtereinheit*
Wechselrichterbetrieb inverter operation * *Umwandlung von Gleich- in Wechselstrom* || inverter range * *bei Umkehrstromrichter* || regenerative mode * *Rückspeisebetrieb (ins Netz zurück) eines Stromrichters*
Wechselrichtereinheit inverter unit * *bei modularem Umrichtersystem* || inverter module
Wechselrichterkippen inverter shoot-through * *Überstrom i.Thyristorschaltg.: neues Ventil gezündet u. altes noch stromführ.* || shoot-through || inverter commutation failure * *Thyristorzerstörg.- falls alt. Ventil noch nicht stromlos u. neues gezünd.* || converter conduction-through
Wechselrichterschieben shifting the firing pulses to the inverter range * *{Stromrichter}*
Wechselrichtertrittgrenze inverter stability limit * *max. Steuerwink., darf z.Vermeidg. d. Wechselr.- kippens nicht überschrit. werd.* || maximum firing angle * *max. Steuerwinkel, darüber besteht Gefahr des Wechselrichterkippens* || inverter limit || inversion limit || maximum gating angle
Wechselschalter reversing switch
wechselseitig mutual || reciprocal
Wechselspannung AC voltage || alternating voltage || a.c. voltage
Wechselspannungsanteil AC voltage component || ripple content * *Welligkeits-Anteil*
Wechselspannungsquelle AC source
Wechselspiel interplay || interaction * *Wechselwirkung/-beziehungen (between: zwischen)*
Wechselstrom alternating current || AC current || a.c. current || AC
Wechselstrom-Direktumrichter direct AC converter || cycloconverter
Wechselstromanteil AC component
Wechselstromantrieb AC drive
Wechselstromasynchronmaschine single-phase induction machine
Wechselstrombelastbarkeit ripple current capability * *z.B. eines Kondensators*
Wechselstromgrößen AC quantities * *siehe auch DIN 40110*
Wechselstrommotor AC motor || a.c. motor || single-phase AC motor * *einphasiger ~*
Wechselstromnetz AC line || AC system || a.c. line
Wechselstromschütz AC contactor
Wechselstromseite AC side * *on the: auf der*
wechselstromseitige Zuleitung AC feeder
Wechselstromsteller AC controller * *wandelt fest.- Wechselspanng. (z.B. Netz) i.e. Wechselspanng. m. variabl. Effektivwert* || single-phase AC controller * *einphasig* || AC power controller * *erzeugt aus e.festen e.variable Wechselspannung, siehe DIN 57875, VDE 0558 Tl.4* || AC power converter || direct AC voltage converter || a.c. power controller * *mit Phasenanschnitt- oder Vollschwing.steuerg. für Heizungszwecke o. Softstart*
Wechselstromsystem AC system
Wechselstromtacho AC tacho || AC tach || AC tachogenerator

Wechselstromtachogenerator AC tachogenerator || AC tacho || AC tach
Wechselstromzwischenkreis AC link
Wechselwirkung interaction || reciprocal action * *gegenseitige* || interplay * *auch 'Wechselspiel'*
Wechselzeit change-over time * *z.B. für den Materialwechsel in einer Verarbeitungsmaschine benötigte Zeit*
Wechsler changeover contact || C contact || double-throw contact || DTC * *Double-throw contact; 4PCTC: 4-poliger Wechslerkontakt* || form C contact || two-way contact
Wechselkontakt change-over contact || C contact || form C contact || double-throw contact || DTC * *Abk.f. 'Double-Throw Contact'; 4PDTW: 4-poliger Wechselkontakt* || two-way contact
Weckalarm timer interrupt * *time interrupt*
Wecker timer * *Zeitglied in der Digital-/Rechnertechnik* || interrupt timer * *Interrupt-erzeugendes Zeitglied* || alarm clock
Weckfehler collision of two timer interrupts
weder noch neither nor
Weg path * *(auch figürl.) Pfad, Signal-, Leistung-; position: Lage* || way * *Fuß/Fahr~, Art und Weise, Richtung; stand in the way: im ~ stehen; midway: in der Mitte des ~es* || distance * *Strecke* || travel * *~ eines Machinenteils* || direction * *Richtung* || manner * →*Art und Weise, Methode* || method * *Methode, Art und Weise* || course * *Bahn, ~ zum Ziel* || road * *Straße; road to success: ~ zum Erfolg* || channel * *Kanal (Vertriebs, diplomatischer usw.); through the channel of: auf dem ... ~e* || track * *Spur; be on the right track: auf d. richtigen Spur/d. richtigen ~ sein* || away * *(Adv.) fort* || off * *(Adv.) fort* || gone * *(Adv.) verloren, gegangen sein* || lost * *(Adv.) verloren* || procedure * *Verfahrensweise* || route * *Reise~, Fahrtroute/- Weg,* →*Strecke* || passage * *Durchgang* || walk * *Gang* || steps * *by legal steps: auf gerichtlichem ~e*
Wegaufnehmer travel transducer || position transducer * *Lage/Positionsgeber*
wegen because of || on account of * *wegen, um ... willen; on his account: seinet~* || by reason of * *auf Grund; for this reason: aus diesem Grund, deshalb* || owing to * *infolge, wegen, dank; be owing to: zu rückzuführen sein auf/... zuzuschreiben sein* || as a result of * *auf Grund von, als Ergebnis von* || for the sake of * *um ... willen* || regarding * *betreffend* || due to * *auf Grund, verursacht durch, ... zuzuschreiben/zurückzuführen sein*
Wegendschalter limit switch
Wegerfassung position sensing || position detection || position decoder * *Decodierer* || position measurement
wegfliegen jump apart * *davonspringen (z.B. schlecht befestigtes Maschinenteil)* || lift off * *mit Flugzeug*
wegführen feed away || direct away
Weggeber position encoder || position transducer
Wegistwert position actual value || position feedback
weglassen omit || leave out || suppress * *unterdrücken* || drop * *auslassen, fallen lassen, aufgeben, nicht bearbeiten*
weglaufen drift * *z.B. Spannung, Messwert*
Wegmessgeber position encoder
wegnehmen withdraw * *ein Kommando/Signal ~* || take away || remove || capture * *ein/auffangen (auch in Zwischenspeicher), an sich reißen* || take up * *Raum, Zeit* || occupy * *Raum, Zeit usw* || detract * *entziehen, ~, herabsetzen, beeinträchtigen (from: etwas/e.Eigenschaft), schmälern*
Wegregelung closed-loop position control || position control

Wegregler position controller
wegschalten disconnect * *z.B. Stromkreis oder e. Motor e. Gruppenantriebs von e. Umrichter*
Wegsollwert position reference value || position reference || position setpoint
weich smooth * *auch Regelcharakteristik, Kennlinie, Laufeigenschaften eines Antriebs* || soft
Weiche points * *work the points: die ~ stellen* || switch * *work/shift the switches: die ~ stellen (auch figürl.)* || course * *(figürlich) Kurs, Richtung; set the course: die ~ stellen* || separating filter * *z.B. Impuls~* || software switch * *Software~* || T-switch * *Daten~, z.B. für parallele/serielle Datenleitungen* || embranchment * *Verzweigung* || branching * *Verzweigung* || distribution * *Verteilung, Verzweigung* || branch * *Verzweigung* || point of diversion * *Umleitungspunkt, Ableitstelle, Umlenkungsposition, Weiche*
Weicheiseninstrument soft-iron instrument * *misst Effektivwert*
Weichenstellung setting the course * *Vorgabe des Kurses (for: auf/in Richtung)*
weicher Übergang smooth transition
weiches Anfahren smooth start || soft start
weiches Anlaufen smooth starting
weiches Netz non-rigid supply system * *Netz mit hoher Kurzschlussleistung (z.B. über 1%)*
Weichheit softness
Weichlöten soldering
Weichmacher plasticizer || plasticiser || softening agent
weichmagnetisch soft magnetic
Weichschaum soft foam
Weise manner || way || mode || fashion || style
Weiße whiteness
Weißgrad whiteness * *z.B. von Papier*
Weißlauge white liquor
weit wide * *weit, ausgedehnt, breit, ~gehend, umfassend, reich (Erfahrung usw.), ~ offen* || far * *weit, fern, entfernt, ~ draußen, ~aus; far and wide: ~ entfernt; so far: bisher/bis jetzt* || vast * *weit, ausgedehnt, unermesslich, riesig*
weit reichend wide-ranging
weit verbreitet widespread || widely held * *Ansicht* || wide-spreaded || popular * *beliebt* || propagated * *Lehre, Meinung usw.* || prevalent * *auch 'vorherrschend, überwiegend'* || widely distributed || widely used * *häufig verwendet*
Weitbereichs- wide-range
Weitbereichswicklung varying-voltage winding * *für e. Bemessungsspannungsbereich ausgelegte Wicklg. e. Drehstrommotors*
weiter further * *(figürl.)* || more distant * *entfernter* || another * *another 5 times: ~e 5 Male* || farther * *weiter (entfernt)*
weiter Bereich wide range || broad range
weiter Sitz loose fit
Weiterbildung retraining || further education || continued education || advanced education || advanced training || technical training * *technische*
weitere more * *mehr* || other * *zusätzliche, andere* || additional * *zusätzliche* || further * *(besonders figürlich)*
weitere Angaben further information || further details
weitere Einzelheiten further details
weitere Informationen further information
Weiterentwicklung advance || further development || advancement
weiterer further
weiterführender Kurs secondary course * *Aus-/Weiterbildungskurs* || advanced training course
Weitergabe passing-on || transmission * *auch von Daten* || forwarding * *Weiterbefördern, z.B. von Briefen*

Weitergabe der Sendeberechtigung token passing
weitergeben pass on * *to: an* || transmit || forward * *z.B. Brief* || turn over * *die Sache an: the matter over to* || pass-on || transfer * *transferieren, übertragen, übergeben* || relay * *(allg.) weitergben; mit Relais steuern*
weitergehend further || continuing * *fortgesetzt* || more extensive * *umfassender* || more intensive * *stärker, heftiger, intensiver, verstärkt*
weiterhin further on || in future * *zukünftig* || further more * *ferner* || moreover * *ferner* || furthermore * *ferner, darüber hinaus*
weiterkommen get ahead * *auch beruflich*
weiterleiten forward * *Brief usw. (to: an)* || transmit * *übertragen (auch Kraft), übersenden, übermitteln, (ver)senden* || refer * *Antrag, Fall, Vorgang (to: an)* || transfer || route * *Aufträge auf d. Dienstweg, Material/Transportgüter, Datenpakete in Rechnernetz ~*
weiterreichen pass || pass on || forward * *z.B. Brief*
Weiterschaltbedingung step criterion * *(Plural: criteria) bei Ablaufsteuerung* || step enabling condition || transition condition
weiterschalten advance || commutate * *kommutieren*
Weiterverarbeitung further processing || converting * *Umformung, Veredelung* || subsequent processing
weiterzählen continue counting
weitestgehend essential * *wesentlich, eigentlich, wichtig* || extensive * *umfassend, ausgedehnt, extensiv* || far-reaching || wide * *weitgehend, z.B. Vollmacht* || essentially * *(Adv.) im Wesentlichen, eigentlich, wichtig, in der Hauptsache*
weitgehend extensive * *umfassend, ausgedehnt, extensiv* || far-reaching || weit reichend || full * *Verständnis usw.* || wide * *Vollmacht, Befugnisse usw.* || essential * *wesentlich, eigentlich, wichtig* || essentially * *(Adv.) im Wesentlichen, eigentlich, in der Hauptsache* || largely * *(Adv.)* || within wide bounds * *(Adv.) in weiten Grenzen* || predominant * *überwiegend, vorwiegend, vorherrschend, überlegen*
Weitspannungs- with a wide voltage range
Well- corrugated * *z.B. Pappe, Blech, Schlauch*
Welle shaft * *(mechan.) Maschinenteil* || wave * *elektrische ~/Schwingung; Wasser-, Angreifer usw.; standing: stehende*
Welle mit Abtriebsbuchse sleeved shaft
Welle-Naben-Verbindung shaft/hub connection
Welle-Welle-Verbindung coupling of shafts
Wellenabdichtung shaft seal
Wellenachse shaft axis
Wellenantrieb propulsion drive * *für Schiff*
Wellenbauch antinode
Wellenbelastung shaft loading || shaft load
Wellenberg wave crest
Wellenbohrung shaft bore * *z.B. in einem Lagerschild*
Wellenbruch shaft rupture
Wellenbund shaft shoulder
Wellendehnung shaft expansion * *Längsdehnung einer Welle auf Grund von Erwärmung*
Wellendichtring shaft seal ring * *siehe auch →Simmerring* || shaft seal || radial shaft seal ring || radial shaft sealing ring * *Radial-Wellendichtring* || radial shaft seal
Wellendichtung shaft seal || shaft sealing || shaft packing
Wellendrehzahl rotational speed of the shaft
Wellendurchbiegung shaft deflection || shaft sag
Wellendurchmesser shaft diameter
Wellendurchtritt shaft exit * *Motor*
Wellenende shaft end * *z.B. einer Motorwelle (free: freies)* || shaft extension * *für elektr. Maschine: siehe DIN 748/IEC 74, Teil 3 u.IEC 72*

Wellenende nach oben shaft extension pointing upwards
Wellenform waveform
wellenförmig wavy ‖ corrugated * *gewellt, gefurcht, gerieft; z.B. Blech* ‖ undulatory * *wellenförmig, Wellen-*
Wellengenerator shaft generator * *im Schiff*
Wellenhöhe shaft height
Wellenlage shaft position * *horizontal: horizontal; vertical: vertikal*
Wellenlänge wavelength
Wellenlast shaft load
Wellenleistung shaft power ‖ shaft output * *z.B. eines Motors*
Wellenlöten wave soldering
Wellenmitte center line of shaft
Wellenmoment shaft torque
Wellenmutter shaft nut ‖ collar nut
Wellennut shaft keyway
wellenparallel parallel-shaft
Wellenprofil flute * *z.B. in Wellpappe*
Wellenschulter shaft shoulder
Wellenschutz shaft-end guard * *Kappe als Transportschutz des Wellenendes*
Wellenspannung shaft voltage * *{el.Masch.}* unerwünschte Spannung zw. Welle u. Motorgehäuse, erzeugt →*Lagerstrom*
Wellenstrang shafting ‖ shaft assembly ‖ shaft train
Wellenstrom shaft current * *elektr. Strom durch (Motor-) Welle (unerwünschter Effekt)*
Wellenstummel shaft end ‖ shaft stub ‖ shaft extension
Wellentest shaft test * *{el.Masch.} Festigkeitsprüfung des Wellenmaterials*
Wellenumdrehungen shaft revolutions
Wellenversatz shaft offset ‖ radial misalignment * *paralleler Fluchtfehler* ‖ shaft misalignment * *Fehlausrichtung der Welle(n)*
Wellenwiderstand wave resistance * *z.B. einer Leitung* ‖ surge impedance ‖ characteristic resistance ‖ characteristic impedance ‖ iterative impedance
Wellenzapfen shaft extension * *Wellenende* ‖ shaft journal ‖ shaft end * *Wellenende*
Welligkeit ripple * *z.B. von Spannung/Strom/Drehmoment* ‖ ripple content * *Effektivwert d.Wechselspannungsanteils e.Gleichspannung; DIN 41750 Tl.4 u.41755 Tl.1* ‖ RMS ripple factor * *Effektivwert des Wechselanteils* ‖ waviness * *z.B. von Papier* ‖ percentage ripple * *Anteil der Welligkeit in Prozent* ‖ corrugation * *Wellen, Runzeln, Furchen, Riefen, Falten (z.B. in Papier, Pappe, Stahl usw.)*
Wellpappe corrugated board ‖ corrugated cardboard ‖ cellular board
Welt world
weltbekannt known throughout the world
Weltfirma global enterprise
Weltmarkt world market ‖ global market
Weltmarktanteil worldwide market share
Weltmarktführer global market leader
Weltmarktniveau world class level
weltweit worldwide ‖ throughout the world ‖ global ‖ around the world ‖ globally * *(Adv.)*
weltweit tätiger Anbieter global player * *(salopp)*
weltweit tätiges Unternehmen worldwide company ‖ globally operating company ‖ global player * *(salopp)*
Wendefeld commutating field * *Motor* ‖ compole field * *z.B. in Gleichstrommaschine*
Wendegetriebe reversing gearbox * *z.B. im Schiffbau* ‖ reversing gear * *z.B. im Schiffbau*
Wendelpotentiometer helical potentiometer * →*Spindelpotentiometer* ‖ helipot * *(salopp) Markenname der Fa. Beckman*
wenden turn * *about/round: um-* ‖ reverse ‖ turn over

* *Buchseite; please turn over/P.T.O./pto: bitte wenden* ‖ pto * *Abk. für 'please turn over': bitte (Buchseite) wenden* ‖ address * *ansprechen, sich wenden an* ‖ consult * *sich (Rat suchend) wenden an*
Wendeplattenwerkzeug turnblade tool * *{Werkz.masch.! Holz}*
Wendepol commutating pole ‖ interpole ‖ compole ‖ reversing pole
Wendepole commutating poles
Wendepolkern interpole core
Wendepolwicklung commutating winding ‖ interpole winding ‖ compole winding
Wendepunkt turning point
Wender turner
Wendeschalter reversing switch * *zur Motor-Drehrichtungsumkehr durch Tausch von Ständer-/Anker- oder Feldleitungen*
Wendeschneidplatte reversible tip cutter * *{Werkz.masch.}*
Wendeschütz reversing contactor * *in Ständer-, Anker- oder Feldkreis zur Drehrichtungsumkehr eines Motors*
Wendestange turner bar * *{Druckerei} im Falzapparat*
Wendestarter reversing starter
Wendestation turnover station * *{Werkz.maschine|Holz}*
Wendevorrichtung turning unit * *board turning unit: ~ für Holzplatten* ‖ turnover unit * *für Holzplatten usw.*
Wendewalze stripper roll * *{Textil}* ‖ turning cylinder
Wendewickler non-stop rewinder
wenig little ‖ few
wenig größer just larger
weniger less ‖ fewer ‖ minus * *{Mathem.}* ‖ less-than * *(Adv.) z.B. less-than-optimal: ~ optimal*
wenn if ‖ where * *in Fällen,in denen* ‖ when
wenn nicht unless ‖ if not ‖ except if ‖ except when
wenn nötig if required
wenn überhaupt if at all
Werbeagentur advertising agency
Werbeaktion advertising campaign ‖ promotion campaign
Werbeanzeige advertisement
Werbeartikel publicity article ‖ promotion article ‖ promotion item
Werbeblatt publicity leaflet * *Faltblatt,* →*Prospekt* ‖ flyer * *(amerikan.) Reklamezettel,* →*Prospekt* ‖ flier * *(amerikan.) Reklamezettel,* →*Prospekt* ‖ advertising leaflet ‖ folder * *Faltprospekt, Faltblatt, Broschüre*
Werbeetat promotional budget ‖ advertising budget
Werbefachmann advertising specialist ‖ promotion expert * *(amerikan.)* ‖ adman * *(salopp)*
Werbefeldzug advertising campaign ‖ promotion campaign
Werbegeschenk advertising gift ‖ publicity gift
Werbekampagne advertising campaign
Werbematerial advertising material ‖ promotional material ‖ *zur Verkaufsförderung* ‖ promotion matter ‖ advertising publications * *Druckerzeugnisse*
Werbemittel advertizing material * *(amerikan.) siehe auch 'Werbematerial'* ‖ advertizing publications * *(amerikan.) Druckerzeugnisse* ‖ promotional items ‖ advertising material * *(brit.)* ‖ advertising publications * *(brit.)* ‖ sales promotion articles ‖ advertising media
werben advertise * *Werbung machen, z.B. über Inserate* ‖ advertize * *Reklame machen, inserieren* ‖ make propaganda
Werbeplakat advertisement poster ‖ poster
Werbeprospekt publicity leaflet ‖ flyer * *(amerikan.)* ‖ flier * *(amerikan.)* ‖ advertising leaflet * *Faltblatt, Prospekt*

Werbeschrift publicity leaflet * *Faltblatt,* →*Prospekt* || flyer * *(amerikan.) Reklamezettel, Prospekt* || flier * *(amerikan.) Reklamezettel, Prospekt* || advertizing leaflet * *Faltblatt, Prospekt* || advertizing brochure * *(amerikan.)* || sales brochure || advertising brochure
Werbeschriften promotional literature
Werbespruch slogan || advertisement slogan
Werbetext advertisement text
Werbetexter copywriter
Werbeunterlagen advertizing material * *(amerikan.)* || promotional material * *verkaufsfördernde Unterlagen* || advertizing documentation || advertising material * *(brit.)*
werbewirksam having publicity appeal || eye-catching * *optisch, das Auge auf sich ziehend* || promotionally effective
Werbewirksamkeit publicity appeal || pull || eye appeal * *optisch*
werblich advertising || promotional || for publicity purposes * *für Werbezwecke*
Werbung advertising * *Werbung, Reklame* || publicity * *Reklame, Werbung* || propaganda * *Reklame, Werbung* || advertisement * *über Anzeigen* || sales promotion * *Verkaufsförderung* || ad * *Abkürzg. für advertising* || public relations * *Werbe-/Presse-/Öffentlichkeitsarbeit*
Werk factory * *Fabrik; ex factory: ab ~* || production facility * *Produktionsanlage,-stätte* || works * *Fabrik; ex works: ab ~* || industrial plant * *Fabrikanlage* || plant * *Fabrikanlage* || workmanship * *kunstvolles Stück Arbeit* || work * *Arbeitsergebnis, Aufgabe, (Fabrikations)Betrieb* || movement * *z.B. ~ einer Uhr*
Werkbank workbench || bench
Werkhalle workshop hall || workshop || manufacturing shop
Werkmeister foreman || supervisor
Werknormblatt works standard sheet
Werknummer job number || works order number
Werksarzt company doctor
Werksbescheinigung certificate of compliance with order * →*Prüfbescheinigung, die der Lieferant dem Besteller übergibt*
Werksbesichtigung factory tour
Werkseinstellung factory setting * *z.B. von Einstellparametern; reset to factory setting: ~ herstellen* || initialized status * *Urladezustand* || factory default * *werksseitig eingestellter Wert z.B. eines Parameters* || default || factory adjustment
werksgeschult factory-trained
Werkshalle manufacturing shop || workshop || shop || shed * *Schuppen, kleine Werkhalle* || workshop hall
Werksmitteilung factory info || factory memo
werksmontiert factory-fitted
Werkstatt workshop || shop || craftshop * *eines Handwerksbetriebes*
Werkstattleiter shop foreman
Werkstoff material
Werkstoffprüfung materials testing
Werkstofftechnik materials engineering
Werkstück workpiece * *z.B. bei der Metallverarbeitung* || working part || work || production part || part
Werkstückansauggerät vacuum cup clamping device * *{Werkz.masch.| Holz]*
Werkstückmessung workpiece measuring
Werkstückrückführung return conveyor
Werkstudent working student || part-time student
Werktag business day
Werktage working days || workdays
werktags weekdays || on weekdays || on workdays || on working days
Werkzeug tool * *auch Software~; tool kit: ~atz;* power tool: *Elektro~ (auch Heimwerkergerät usw.)* || mould * *(Guss)Form*
Werkzeugbau toolmaking || tool manufacture
Werkzeugbauer tool engineer || toolmaker * *Werkzeugmacher*
Werkzeugbruch tool breakage
Werkzeugeinsatz bit
Werkzeughalter tool holder || bit holder * *für Werkzeugeinsatz*
Werkzeugkasten toolbox || tool kit * *Satz von Werkzeugen*
Werkzeugkorrektur tool compensation * *{NC-Steuerung} kompensiert Werkzeugverschleiß* || tool offset * *{NC-Steuerung} kompensiert Werkzeuglänge (Nullpunktkorrektur)*
Werkzeugleiste tool bar * *auf dem Bildschirm in einer grafischen Bedienoberfläche*
Werkzeugmaschine machine tool || tool machine
Werkzeugmaschinenbau machine tools manufacturing || machine tool builders * *~er*
Werkzeugmaschinenbauer machine-tool manufacturer
Werkzeugmessgerät tool measuring instrument
Werkzeugsatz tool kit
Werkzeugschleifmaschine tool grinding machine || tool sharpening machine * *zum Schärfen* || tool grinder
Werkzeugstahl tool steel
Werkzeugtasche tool bag || tool kit * *Satz*
Werkzeugwechsel tool change || tool changeover
Werkzeugwechseleinrichtung tool changer
Werkzeugwechselsystem tool-changing system * *{Werkz.masch.}* || tool change system * *{Werkz.masch.}*
Werkzeugwechsler tool changer * *{Werkz.masch.}*
Werkzeugzurichtung tool dressing * *{Werkz.masch.}* schärfen, geradeschleifen (Schleifscheibe usw.)
Wert value || equivalent * *Gegen~* || asset * *Vermögens~* || valence * *~igkeit, Verhältniszahl, mitwirkende Kraft, Faktor* || coefficient * *Koeffizient* || factor * *Factor* || data * *Datum* || worth * *innerer/kultureller ~*
Wert auf die Analogausgabe ausgeben direct a value to the analog output
Wert legen auf lay significance on || attach importance to * *great/significant: großen*
Wertangabe value indication * *z.B. bei Meldesystem* || value information || declaration of value * *Angabe des Geldwerts* || declared value * *Angabe des Geldwertes* || numerical date * *Zahlenangabe* || numerical data * *Zahlenangaben*
Wertebereich value range
Wertepuffer value buffer
Werterhaltung value retention || preservation of value
Wertespeicher register area * *{SPS}* || time and data register area * *{SPS}*
Wertigkeit valence || weighting * *Gewichtung; (Umsatz-) ~ eines Kunden* || value * *Wert* || account * *Wert, Wichtigkeit, Bedeutung, z.B. eines Kunden* || weight || significance || annual purchasing volume * *Jahresbedarf eines Kunden* || evaluation * *Abschätzung, Bewertung, Beurteilung, Auswertung*
wertlos worthless * *(auch figürlich) worthless stuff: ~es Zeug* || valueless || no good || useless * *nutzlos* || of no use * *nutzlos*
wertmäßig quantitative * *['kwontitetiw] quatitativ, Mengen-* || in terms of figures * *zahlenmäßig* || ad valorem * *(Adj. und Adv.) kaufmännisch* || quantified * *quantifiziert, quantitativ bestimmt*
Wertminderung depreciation || decrease in value || decline in value
Wertpapier banknote paper * *Papiersorte*

Wertpapierdruckmaschine security printing machine || banknote printing machine * *für Banknoten*
Wertschöpfung value-adding
wertvoll valuable * *(auch figürl.) be of value to s.o.: jdm. wertvoll sein* || precious
wesentlich essential * *wichtig (auch Substantiv: ~er Punkt)* || intrinsic * *innewohnend, inner, wahr, eigentlich, wirklich, Eigen-* || substantial * *in der Substanz, auch 'beträchtlich, substanziell, grundlegend'* || significant * *deutlich* || largely * *(Adv.)* || considerable * *(Adj.)* beträchtlich, erheblich, bedeutend || considerably * *(Adv.)* beträchtlich, bedeutend || mainly * *hauptsächlich* || basic * *grundlegend, Grund-, die Grundlage bildend* || essentially * *(Adv.)* || material * *von Belang, sachlich wichtig (to: für)* || vital * *unerlässlich* || fundamental * *grundlegend* || major * *hauptsächlich* || basically * *(Adv.) im~en, im Grunde, grundsätzlich* || salient * *hearusragend, hervorstechend, ins Auge springend; salient point: springender Punkt* || principal * *erst, hauptsächlich, Haupt-; Stamm-*
wesentliche Merkmale essential features
wesentlicher Punkt essential || essential point || most important point * *wichtigster Punkt* || vital point || main issue * *Kernpunkt* || basic aspect * *wesentlicher Gesichtspunkt* || basic item * *der Tagesordnung usw.*
Wettbewerb competition * *Konkurrenz; stiff: stark; fierce: heftig*
Wettbewerber competitor || rival firm * *Konkurrenzfirma*
Wettbewerbs- of the competition * *die Konkurrenz/Mitbewerber betreffend* || competitive
wettbewerbsfähig competitive || able to compete
Wettbewerbsfähigkeit competitive strength || competitiveness || competitivity || competitive edge * *Wettbewerbsvorsprung*
Wettbewerbskraft competitive strength
Wettbewerbslage competitive situation
Wettbewerbsvorsprung competitive edge
Wettbewerbsvorteil competitive advantage || competitive edge
wetterfest weather resistant
wettergeschützt weatherproof
wettergeschützte Ausführung weather protected design * *z.B. Schutzart IPW24 für Motor*
Wetterschutz weather protection
wettmachen make up for || make good * *Versäumnis, Verlust*
Wheatstone-Messbrücke Wheatstone measuring bridge
wichtig important * *to: für* || essential * *wesentlich, →Kern-* || vital * *unerlässlich* || momentous * *sehr wichtig* || weighty * *gewichtig; z.B. Gründe, Argumente* || key * *Schlüssel-*
Wichtigkeit importance * *of prime importance: von größter* || significance * *Wichtigkeit; Bedeutung, (tieferer) Sinn* || seriousness * *Ernst* || relevance * *Erheblichkeit* || pertinence * *Erheblichkeit*
wichtigst most important || main * *hauptsächlich* || top-priority * *von höchster Priorität, hochprior*
Wickel coil * *Bund* || wound material * *aufgewickeltes Material* || winding || cop * *{Textil} Garn-/Fadenwickel, Kops (in der Spinnerei usw.)* || beam * *{Textil} Baum, aufgewickelter Textilstoff*
Wickelantrieb winder drive || winding drive || coil drive * *Spulerantrieb, Haspelantrieb (für Blech, Draht usw.)*
Wickelautomat automatic winding machine
Wickelbremse tension brake * *z. Aufbringen d.Zugmoments bei Abwickler (magnet./mechan./hydraul./pneumat.Bremse)*
Wickeldorn mandrel || tambour * *für/mit Papier* || mandril

Wickeldurchmesser winding diameter || coil diameter
Wickelgeschwindigkeit winding speed
Wickelgut coiled product * *z.B. Spule, Drossel, Transformator, aufgewickeltes Blech* || coiled component * *z.B. Spule, Trafo* || wound material * *von Auf-/Abwickler gewickeltes Material*
Wickelgüter coiled products * *z.B. Spulen, Drosseln, Trafos, aufgewickeltes Blech* || wound goods || coiled components || chokes and transformers * *Drosseln und Trafos*
Wickelhärte core hardness * *im Wickelkern erhöhte Wickelhärte* || taper tension * *mit steigendem Durchmesser abnehmender Zug (zur Schonung d. Innenlagen bei Aufwickler)* || winding characteristic * *Wickelhärtenkennlinie (Zugkraftverlauf) abhängig vom Durchmesser* || winding pressure || roll tightness || winding density
Wickelhärtenkennlinie taper tension curve * *sorgt für abnehmende Wickelhärte beim Aufwickler* || taper tension characteristic * *sorgt für abnehmende Wickelhärte beim Aufwickler* || winding characteristic * *m. d.Durchmess. abnehm. ~ verhindert Beschädigg. d. Innenlagen bei Aufwickler* || roll's tension characteristic * *mit d. Durchm. abnehm.~ verhind. Teleskopieren u. Quetschg. b. Aufwickler*
Wickelhärtensteuerung taper tension control * *z.B. abnehmende Wickelhärte beim Aufwickler; taper: Verjüngung* || taper control || slip core control * *Zugreduzierg. b. steig. Durchm.; verhind. Beschädigg. d. Innenlagen b. Aufwickler* || winding density control
Wickelhülse core || paper core * *zum Aufwickeln von Papier* || winding sleeve * *z.B. Papphülse zum Wickeln von Papier, Folie, Textilien usw.*
Wickelkern mandrel * *Wickeldorn* || core
Wickelkondensator wound capacitor
Wickelkopf winding overhang * *{el. Masch.} aus d. Blechpaket herausragender Teil e. Wicklung, z.B. eines Motors* || overhang * *z.B. eines Motors* || end turns || end winding overhang
Wickelkopfabstützung coil-end bracing * *{el. Masch.} mechan. Versteifung d. →Wickelkopfs z.B. bei Hochspannungsmotoren* || end-winding support || overhang bracket
Wickelkopfbandage end-turn banding * *{el. Masch.} Bandagierung d.Wickelkopfs mit Isolierbändern b. Niederspannungsmotor*
Wickelkopfisolierung winding-head insulation * *Isolierg. nebeneinanderliegend. Spulen → Wickelkopf durch →Phasentrenner*
Wickelmaschine winding machine || wrapping machine * *{Verpack.techn.} zum Umwickeln mit Verpackungsmaterial (z.B. Folie, Papier)* || wrapper * *{Verpack.techn.}* || wraparounder * *{Verpack.techn.} Umwickelmaschine* || reeling machine
wickeln wind * *auch auf~* || coil * *Metall, Draht, Kabel, Seil ~* || unwind * *ab~* || rewind * *wiederauf~ (z.B. beim Rollenschneider)* || overwind * *~ von oben* || underwind * *~ von unten*
wickeln von oben overwind
wickeln von unten underwind
Wickelpappe millboard * *geformte Pappe, Graupappe, Zellstoffpappe* || press-rolled board
Wickelpistole wire-wrap gun * *zum Herstellen von Wire-Wrap Verbindungen*
Wickelrechner winding calculator * *berechnet d.Durchmesser b. Wickler u. gibt das d.Sollzug entspr. Drehmoment vor*
Wickelrolle wrapping roll * *Haspelhilfsantrieb im Walzwerk*
Wickelstation winder station || wind || winding station || wind stand
Wickeltechnik winding technology

Wickelverhältnis maximum diameter ratio * *Verhältnis Kern/Max. Durchmesser bei Auf/Abwickler* || maximum buildup * *Durchmesserverhältnis bei Aufwickler* || maximum builddown * *Durchmesserverhältnis beim Abwickler* || diameter ratio * *Durchmesserverhältnis (max. Durchmesser/ Kerndurchmesser)*
Wickelwelle winding shaft
Wickler winder || centre winder * *Achswickler* || surface winder * *Oberflächenwickler, z.B. Stützwalzenroller, Doppeltragwalzenroller* || center winder * *(amerikan.) Achswickler* || armature winder * *Achswickler mit DC-Antrieb und Durchmesseranpassung i.ankerkreis* || CEMF winder * *Achswickler mit DC-Antrieb und Durchmesseranpassung i.feldkreis*
Wickler mit fliegendem Rollenwechsel nonstop winder || winder with flying splice || splicer
Wicklerantrieb winder drive
Wicklung winding * *eines Motors/Trafos usw.; tapped: unterteilte ~ mit Anzapfungen*
Wicklungsanfang start of winding || lead of winding
Wicklungsende end of winding || winding end
Wicklungserneuerung winding renewal
Wicklungserwärmung winding temperature rise
Wicklungsfaktor winding factor * *Reduzierung d. in verteilten Spulen induz. Spanng gegenüber konzentrierter Anordng.*
Wicklungsisolation winding insulation
Wicklungsnut winding slot
Wicklungsprüfung high-voltage test || separate-source voltage-withstand test * *mit Fremdspannung*
Wicklungsstrang winding circuit || winding phase * *eines Drehstromsystems*
Wicklungssystem winding system * *z.B. einer elektr. Maschine*
Wicklungsteil winding section
Wicklungstemperatur winding temperature
Wicklungsverluste winding losses
Wicklungswiderstand winding resistance
Wicklungszuleitung winding lead * *z.B. einer Motorwicklung*
wider Erwarten against all expectations
Widerruf withdrawal || cancellation * *Löschung, Stornierung*
widerrufen cancel * *widerrufen, aufheben, annullieren (auch Auftrag), rückgängig machen, absagen, streichen* || withdraw * *(Auftrag, Aussage usw.) ~, zurückziehen, zurücktreten, (Geld, Personal) abziehen* || undo * *Computereingaben usw.* →*rückgängig machen*
widersetzen act in opposition to * *z.B. sich einer Kraft ~* || oppose * *sich ~, angehen gegen, bekämpfen* || resist * *widersetzen, Widerstand leisten, sich streuuben gegen* || set one's facts against || struggle against || disobey * *sich einem Befehl/einem Gesetz ~*
widerspiegeln reflect * *(auch figürl.)* || be reflected * *sich ~(in: in)*
widersprechend contradictory || conflicting * *widerstrebend/-sprechend (Forderungen, Gesetze, Gefühle usw.)*
Widerspruch contradiction * *i. contradiction to: im ~ zu; contradiction in terms: ~ in sich selbst* || opposition * *Wider-/Einspruch z.B. gegen Patentanspruch* || inconsistency * *innerer, logischer ~; be inconsistent with: im ~ stehen zu* || discrepancy * *Widerspruch, Zwiespalt, Unstimmigkeit* || antithesis * *Antithese, Gegensatz, Widerspruch* || conflict * *Konflikt, Widerstreit, Widerspruch, Gegensatz, Zusammenstoß, Kampf*
widersprüchlich contradictory * *without contradiction/uncontradicted: widerspruchslos* || incon-
sistent * *be inconsistent to: im Widerspruch stehen zu* || in contradiction * *to: zu*
Widerstand resistor * *Bauteil* || resistance * *elektr. Größe; (auch figürl.); low/high: klein/groß* || opposition * *Opposition, Wider-/Einspruch* || air drag * *Luft~* || strength * *~skraft eines Materials* || impedance * *Schein~ (bestehend aus* →*Wirk- und* →*Blindwiderstand)*
Widerstandsanlasser rheostatic starter || resistor starter
Widerstandsbelag resistance per unit length * *Widerstand pro Längeneinheit bei Kabel, Leitung*
Widerstandsbremsung dynamic braking * *Gleich- und Drehstrommotor* || rheostatic braking * *z.B. bei Drehstrom- und 1Q-Gleichstromantrieb*
Widerstandsdraht resistance wire
widerstandsfähig robust || rugged
Widerstandsfähigkeit resistance * *to: gegen* || strength || stability || robustness
Widerstandsheizung resistance heating
Widerstandslast resistive load
Widerstandsläufer high-resistance rotor * *Käfigläufer m.Stäben erhöhten Widerstandes* →*weichere Drehmomentkennlinie* || high-resistance squirrel-cage rotor * *Käfigläufer* || slip rotor || increased resistance rotor
Widerstandsmoment load torque * →*Lastmoment*
Widerstandspaste resistor paste * *in der Hybridtechnik*
Widerstandsschweißen resistance welding
Widerstandssteuerung resistance speed-control * *{el. Masch.} Drehz.veränderung e. Schleifringläufermot. mit Vorwiderständen*
Widerstandsthermometer resistance thermometer * *el. Thermometer, nutzt d.Temperaturabhängigkeit d. Leiterwiderstandes* || thermistor * *Heiß/ Kaltleiter-Temperaturfühler (mit NTC-, PTC-Widerstand)* || resistor temperature detector
Widerstandsverfahren resistance method * *zur Ermittlg. d. Wicklungstemp. e. el. Masch. durch Messg. d. Wickl.widerstands*
Widerstandswert resistor value || resistance
widerstehen withstand || resist
widerstrebend reluctant * *widerwillig, widerstrebend, zögernd; be reluctant to do: s.sträuben, etw. zu tun*
widrig adverse * *ungünstig* || untoward * *widrig, ungünstig, unglücklich (Umstände usw.), ungefügig, widerspenstig* || contrary * *widerspenstig, eigensinnig, gegensätzlich/-teilig*
wie bisher as in the past || as before || as until now
wie erwartet up to expectations
wie gewohnt as usual
wie möglich as possible * *as soon as possible: sobald ~*
wie oben as mentioned above || same as above || as above
wie oft how often * *wie häufig* || how many times * *wie viel mal* || the number of times * *so viel wie*
wie schon erwähnt as already mentioned * *siehe auch* →*oben erwähnt*
wieder anfangen resume
wieder aufnehmen resume
wieder- re- * *Vorsilbe für Substantiv/Verb*
wiederanfangen resume
Wiederanlauf restart * *Motor, Rechner usw.; flying restart: ~ einer laufenden Maschine nach Netzausfall* || re-start || warm restart * *Warmstart* || re-initialization * *Programm* || initialization * *Initialisierungsprogramm* || restarting || reset * *Rücksetzen*
wiederaufbereiten recycle || reprocess
Wiederaufbereitung recycling

wiederaufladbar rechargeable
Wiederaufnahme resumption * *z.B. einer Tätigkeit, von Zahlungen usw.*
Wiederaufnahme des Betriebes resumption of service
Wiederaufschalten restart || jump start
wiedereinbauen reinstall
wiedereinführen re-import * *(aus dem Ausland)*
Wiedereinschaltautomatik automatic restart * *siehe auch →Suchen, →Fangschaltung* || automatic restart feature || automatic restart on the fly * *Aufschalten e. Stromrichters auf lauf. Motor nach Netzspannungsausfall*
Wiedereinschalten re-start * *Neustart* || restart * *(Subst. und Verb)* || re-closure * *z.B. Stromkreis nach Wartungsarbeiten* || reconnection * *(auch Stromkreis)*
Wiedereinschalten bei Restspannung restarting at residual voltage * *{el. Masch.} Einschalten e. nach Abschaltg. noch laufenden Motors*
Wiedereinschaltsperre re-start lockout * *für Antrieb* || reclosing lockout * *für einen freigeschalteten Stromkreis z.B. bei Wartungsarbeiten* || restart disabling * *für Antrieb*
Wiedereinschaltverzögerung restart time delay * *z.B. zur Vermeidung zu hoher Schalthäufigkeit*
wiedereintreten re-enter
wiedererlangen recover || regain
Wiedererlangung recovery
wiedergewinnen recover || recycle || regenerate || reclaim * *aus Altmaterial ~, regenerieren; zurückbringen/-führen (from: von/aus; to: zu)*
Wiedergewinnung regeneration || recycling * *dem Zyklus wieder zuführen* || recoveiy * *Rückgewinnung*
wiederherstellen restore || repair || *reparieren* || re-establish * *z.B. einen Zustand, eine Datenverbindung*
Wiederherstellung retrieval || restoration || re-establishing * *einer Verbindung usw.* || recovery * *Rückgewinnung, Erholung* || restitution * *~ eines Rechts, des alten Zustandes usw.*
Wiederhol- repetitive
Wiederholbarkeit repeatability
wiederholen repeat || say again * *verbal* || sum up * *zusammenfassen* || happen again * *nochmal geschehen* || take back * *zurückholen*
Wiederholfaktor iteration factor * *Anzahl d. Wiederholungen einer Befehlssequenz bei SPS/Rechner* || repeat factor
Wiederholgenauigkeit reproducibility || repeatability
wiederholt repeated || repeatedly * *(Adv.)* || repetitive * *→wiederkehrend* || recurrent * *wiederkehrend, sich wiederholend, periodisch auftretend*
Wiederholung repetition || repeat
Wiederholungszahl repeat number
Wiederholzyklus repetitive cycle
Wiederinbetriebnahme recommissioning
Wiederkehr return * *auch von Spannung/Leistung/Versorgung* || recurrence * *periodische* || recovery * *z.B. Spannung nach Spannungseinbruch* || restoration * *z.B. Spannung nach Spannungseinbruch*
Wiederkehr der Spannung voltage recovery
wiederkehren recover * *z.B. Spannung nach einem Spannungseinbruch* || recur * *sich wiederholen* || come back || return || re-occur * *wieder auftreten, wieder vorkommen*
wiederkehrend recurring * *in der Reihenfolge ~* || periodical || recurrent || periodical * *periodisch* || repetitive
wiederkehrende Spannung recovery voltage
Wiederverkäufer reseller || retailer
wiederverwenden re-use || re-employ

Wiederverwendung reuse
wiederverwerten recycle
Wiederverwertung recycling
wievielmal how often || how many times || the number of times
willkürlich arbitrary * *zufällig* || any * *irgendein* || random * *beliebig, wahllos, zufällig (statistisch gesehen)*
Winde windlass * *Winde, Förderhaspel, Ankerspill* || winch * *Winde, Haspel, Kurbel* || capstan * *auf Schiff* || hoist * *Kranwinde, Hebewinde* || crab * *Winde, Hebezeug, Laufkatze*
Windeinfluß windage
Windenantrieb winch drive
Winderhitzer air preheater * *zur Stahlerzeugung*
Windgenerator wind-driven generator
Windkanal wind tunnel
Windrad impeller * *auch Lüfterrad*
Windung turn * *einer Spule* || coil * *Blech, Draht, Kabel, Seil* || worm * *Gewinde* || thread * *Gewindegang* || layer * *Lage*
Windungsisolierung interturn insulation * *Isolierung zw. benachbarten Windungen z.B. eines Elektromotors*
Windungsschluss interturn short-circuit
Windungszahl number of turns || number of turns per unit length * *pro Längeneinheit*
Winkel angle * *['ängel] (geometr.) auch Zündwinkel, Phasenwinkel; at an angle of 45°: im Winkel von 45°* || L-bracket * *Haltewinkel, Blechwinkel (in 'L'-Form)* || bracket * *Haltewinkel* || angle bracket * *Halte~, Blech~*
Winkel- angular
Winkelabweichung angular misalignment * *Fluchtfehler* || angular deflection || angular distance * *Winkeldifferenz*
Winkelauflösung angular resolution
Winkelbeschleunigung angular acceleration
Winkelfunktion trigonometrical function
winkelgenauer Gleichlauf angular synchronism * *siehe auch 'Winkelgleichlaufregelung'* || angular synchronous control * *winkelgenaue Gleichlaufregelung*
Winkelgenauigkeit angular accuracy || angle exactness * *{Werkz.masch.| Holz} z.B. bei der Bearbeitung von Holzplatten usw.*
Winkelgeschwindigkeit angular velocity * *[rad/sec]*
winkelgetreuer Gleichlauf exact synchronism || accurate synchronism || angular synchronism
Winkelgetriebe mitre gear * *z.B. Kegelradgetriebe* || right angle gear * *Ein- und Ausgangswelle bilden e. rechten Winkel*
Winkelgleichlauf angular-locked synchronism || angular synchronism || exact synchronism || operation in perfect synchronism || angular synchronisation || 'electronic shaft' type of operation * *"elektronische Welle"* || absolute synchronism || angular-locked synchronizing control * *~regelung* || accurate synchronism
Winkelgleichlaufregelung angular synchronous control || angular-locked synchronizing control || synchronizing control * *Gleichlaufregelung*
Winkelgrad angular degree * *in angular degrees/ in angular measure: in ~n*
Winkelhebel angle lever
Winkelkante angular edge * *{Holz}*
Winkelkodierer angular encoder || absolute shaft-angle encoder
Winkellage angular position || shaft position * *einer Welle*
Winkelminute angular minute
Winkelnachgiebigkeit angular compliance * *z.B. einer elastischen Kupplung*
Winkelposition angular position

Winkelpositionsgeber deflection pickup || absolute shaft encoder * *rotierender Absolutwertgeber*
Winkelregelung angular control || angular regulation
Winkelregler phase-angle controller * *z.B. bei USV*
Winkelschiebeeinrichtung angular shifting device
Winkelschleifer angle sander || grinder || sander
Winkelsekunde angular second
Winkelstellung angular position
winkelsynchron angle-synchronized || in angular-locked synchronism
Winkelübergabe angular transfer * *{Fördertechnik}* || angle transfer * *{Fördertechnik}*
Winkelverschiebung angular displacement
winklig angular || full of corners * *verwinkelt*
Wippe rocker
wir wären dankbar für we would appreciate
Wirbelfeld curl field
Wirbelstrom eddy-current * *wirbelförm.Wechselstrom aufgr. d.Induktion v.Wechselfeldern in massiven metall.Körpern*
Wirbelstrombremse eddy-current brake * *el. Masch. zur Erzeugung e. reibungs- u. verschleißfreien Bremswirkung*
Wirbelstromkupplung eddy-current clutch * *el. steuerbare Kupplg. mit berührungslos. Kraftübertragg. nach d. Indukt.prinzip*
Wirbelstromläufer deep-bar cage rotor * *{el.-Masch.} →Stromverdrängungsläufer* || deep-bar squirrel-cage rotor
Wirbelstromverluste eddy-current losses
Wirewrap-Stift wirewrap post
Wirkdurchmesser pitch-circle diameter * *Teilkreisdurchmesser, z.B. eines Zahnrades*
wirken take effect * *wirksam werden, (Ein-) Wirkung/Einfluss haben; have an effect on: (ein)wirken auf* || be in force * *wirksam sein* || behave * *sich verhalten* || act * *eingreifen, arbeiten, fungieren, agieren, dienen (as: als)* || influence * *beeinflussen* || affect * *beeinflussen* || operate * *arbeiten, in Betrieb sein, funktionieren, laufen (on: auf)* || knit * *Textilstoff ~/stricken; verknüpfen, zusammenfügen, verbinden, vereinigen, fest verbinden*
wirken auf operate on || act on
Wirkleistung active power * *nutzbarer Leistungsanteil b. Wechselstromsystem: P entspr. SQRT(3) x U x I x cos (phi)* || true power || true output || effective power
wirklich actual || true * *echt, wahr, wahrhaftig* || genuine * *echt, tatsächlich* || substantial * *wesentlich* || visible * *sichtbar, festgestellt* || effective * *wirksam, gültig, wirkend, in Kraft* || really * *(Adv.)* || actually * *(Adv.)* || truly * *(Adv.)* || in fact * *(Adv.) in Wirklichkeit*
Wirklichkeit reality || actuality || truth || real life || hard facts * *raue*
Wirkmaschine knitting machine * *{Textil}* || warp knitting machine * *{Textil}*
Wirkrichtung effective direction || positive direction * *Wirkungsrichtung*
wirksam effective * *wirkend, gültig, in Kraft; become effective/into effect: wirksam werden* || efficient * *effizient (z.B. Person, Arbeit)*
Wirksamkeit effectiveness || impressiveness * *Wirksamkeit, Nachhaltigkeit, das Eindrucksvolle* || efficacy || efficiency
Wirkstrom active current * *derj. Anteil e. Wechselstroms, d. mit d. Spannung phasengleich ist (I_eff x cos phi)*
Wirkstromregelung active current closed-loop control || active current control
Wirkung effect * *with effect from: mit ~ von* || operation * *Tätigkeit* || action * *Tätigkeit* || consequence * *Folge* || result * *Ergebnis* || impression * *Eindruck* || impact * *Aus~, (heftige) Ein~, Auftreffen* || reaction * *(Gegen)Reaktion*
Wirkungsgrad efficiency * *Verhältn."eta": abgegeb. z.zugeführt.Wirkleistg. (typ.DC-Stromrichter 99%,DC-Motor 94%)* || power efficiency * *%uales Verhältnis Ausgangs-/Eingangsleistung; high-efficiency: mit hohem ~*
Wirkungsrichtung direction of force || line of action * *Angriffsrichtung (einer Kraft usw.)*
wirkungsvoll effective || powerful * *höchst wirksam*
Wirkungsweise mode of action || mode of operation || method of functioning || working || mechanism || operating principle * *Funktionsprinzip*
Wirkwiderstand effective resistance || real resistance || active resistance || nonreactive resistance || ohmic resistance * *Ohmscher Widerstand*
Wirtschaft economy || economic system * *~ssystem eines Gemeinwesens* || trade and industry * *gewerbliche ~* || economics * *Welt~, ~swissenschaft, Volks~, Volks~slehre* || free competitive system * *freie (Markt-) ~* || public house * *Wirtshaus* || pub * *Kneipe* || doings * *Treiben* || goings-on * *Treiben* || housekeeping * *Haushaltung*
wirtschaftlich economic || economical * *wirtschaftlich, haushälterisch, sparsam* || financial * *finanziell* || cost-effective * *kostengünstig, zu günstigen/niedrigen Kosten* || efficient * *leistungsfähig, rationell* || profitable * *Ertrag bringend* || commercial * *geschäftlich* || business || paying * *ertragreich, sich auszahlend* || economically viable * *~ machbar/realisierbar* || economically * *(Adv.)*
wirtschaftlich durchführbar econimically feasible
Wirtschaftlichkeit cost-effectiveness * *Kostengünstigkeit (Anschaffungs- und Betriebskosten eines Investitionsgutes)* || profitability * *Rentabilität* || economy of operation * *eines Geräts/einer Maschine z.B. im Unterhalt auf Grund von Wartungsfreiheit* || efficiency * *Effizienz, Preis/Leistungsverhältnis* || good management * *wirtschaftliches Handeln* || economic operation * *Betrieb mit geringen Betriebskosten* || economic efficiency || economic quality || operational economics * *z.B. einer Maschine/Anlage* || economy || economic viability || operational efficiency || economy in operation * *z.B. einer Maschine/Anlage* || economic characteristics || cost-efficiency
Wirtschafts- economic
Wirtschaftsprüfer auditor
Wirtschaftswachstum economic growth
Wischer spurious pulse * *Wischimpuls* || passing contact * *Wischkontakt* || wiper * *Scheibenwischer an Fahrzeug* || transient pulse * *Kurzzeitimpuls* || flicker || momentary pulse
Wischermotor wiper motor * *Scheiben~*
Wischimpuls momentary pulse || transition-sensitive impulse || impulse contact || momentary impulse || spurious pulse || flicker
Wischkontakt passing contact || wiper contact
Wischrelais impulse relay
Wissen knowledge * *extensive: umfangreiches; well-grounded: fundiert* || know-how * *Erfahrungsschatz* || technical knowledge * *praktische Erfahrung; wide: groß; extensive: umfangreich* || information * *~sgut* || scholarship * *Gelehrsamkeit* || wealth of technical knowledge * *Erfahrungsschatz* || know * *(Verb)* || have knowledge of * *(Verb)* || be informed of * *(Verb)* || expertise * *[expe'ties] Experten~, Fach-/Sachkenntnis* || understanding * *Verstehen, Intelligenz, Verständnis, Verstand, Vereinbarung, Verständigung*
Wissenschaft science || science and research * *~ und Forschung*

Wissenschaftler scientist || researcher * *Forscher*
wissenschaftlich scientific
Witterungseinflüsse influence of the weather || weather factors * *weatherproofing measures: Maßnahmen gegen ~*
Wobbelamplitude sweep amplitude || wobble amplitude * *für Wobbelgenerator bei Textil-Changierantrieb*
Wobbelbandbreite sweep bandwidth || sweep width
Wobbeleinrichtung wobbulation equipment * *{Textil} für Changierantrieb beim Aufwickeln von Fäden* || wobbling unit * *{Textil}*
Wobbelfrequenz sweep frequency * *z.B. bei Funktionsgenerator* || wobbling frequency || wobbulation frequency
Wobbelgenerator sweep generator || sweep frequency generator || wobble generator * *sorgt f. guten Spulenaufbau bei Textil-Changierantrieb durch period.Zusatzsollwert* || wobbling unit * *erzeugt period.Zusatzsollw.(z.B.Dreieck) f. Textil-Changierantrieb> gut. Spulenaufbau* || wobbulation generator * *für Changierantrieb i. Textilmaschine z.Fadenaufwickeln mit gutem Spulenaufbau*
Wobbelung wobbulation
wobei whereby * *wodurch, wobei (gleichzeitig)* || being * *the thyristor being blocked; ~ der Thyristor gesperrt ist* || with being * *with the thyristor being blocked: ~ der Thyristor gesperrt ist* || where || at which || in doing so || in the course of which || through which * *wodurch*
wohlüberlegt judicious * *vernünftig, klug, wohlüberlegt* || well-considered * *deliberate*
Wohngebiet living area || housing estate * *Wohnsiedlung* || residential settlement * *Wohnsiedlung* || residential area || residential district
Wölbung vault || arch || dome * *Kuppel* || curvature * *gewölbte Form* || camber * *leichte ~, Krümmung* || buckling || crossfall * *z.B. einer Straße* || turtleback * *(schildkrötenförmig) gewölbte Fläche*
Wolfram tungsten
Wolframkarbid tungsten carbide * *Hartmetall*
Wolfszahnung peg teeth * *{Holz} bei Säge*
Wolle wool * *Schurwolle* || cotton * *Baumwolle*
Wolllappen woolen rag
womöglich wherever possible * *wo immer es möglich ist* || if possible * *wenn möglich* || possibly * *wenn möglich*
Workstation workstation * *leistungsfähiger Rechner mit hoher Grafikleistung, oft vernetzt, z.B.für* →*CAD und* →*DTP*
Wort word * *auch Datenwort* || term * *Ausdruck*
Wörterbuch dictionary * *of: über/für* || vocabulary * *Vokabelverzeichnis, Wörterbuch, Vokabular, Wortschatz*
Wortlaut wording || text || tenor * *genauer* || run as follows * *folgenden ~ haben*
Wortprozessor byte processor * *Prozessor für Byteverarbeitung bei SPS* || word processor * *bei SPS*
Wuchtart balancing type
wuchten balance
Wuchtgüte balance quality * *{el.Masch.} Bahngeschw. d. Schwerpunkts e. Rotors mit Restunwucht (VDI 2056 u.2060)*
Wuchtung balancing * *Auswuchtung*
Wulst bead * *auch am Autoreifen* || beading || bulge * *Bauchung* || hump * *Buckel*
Wulstmaschine embossment machine || curling machine || beading machine * →*Sickmaschine*
Wunsch wish * *with the best wishes: mit den besten Wünschen* || desire * *as desired: (je)nach ~* || request * *Bitte; on request: auf ~; by popular request: auf allgemeinen ~* || ambition * *hochgestecktes Ziel* || is there anything else I can do for you * *haben Sie noch eine ~?*

Wunschdenken wishful thinking
wünschen wish || desire * *leave much to be desired: viel zu ~ übrig lassen* || want * *wollen, wünschen* || request * *bitten, ersuchen um; request sth. from someone: jdn. um etwas ersuchen* || long for * *sehnend ~*
wünschenswert desirable
wunschgemäß requested * *angefordert* || as requested * *(Adv.)* || as desired * *(Adv.)* || according to one's wishes * *(Adv.)*
Würfel cube
würgen throttle || choke || strangle || bunch * *beim Verseilen von Kabeln*
Wurzel root * *auch mathematisch; square/cubic root: zweite/dritte Wurzel* || common connection * *gemeinsamer Anschluss, z.B. bei Digital-Ein/Ausgängen* || common * *gemeinsamer Anschluss bei Relais-Wechslerkontakt, Binärein/Ausgang*
Wurzelortskurve root locus diagram
Wurzelung grouping || connecting to common potential * *bei Digital-Ein-/Ausgängen*
WWW WWW * *'World-Wide Web': multimediafähiger Dienst innerhalb des INTERNET-Datennetzes (*→*HTML,* →*HTTP)*

X

x-Ablenkung x-deflection
X-Achse axis of x || abcissa * *Abzisse* || x-axis
x-y-Positionsgeber joystick controller
X-Y-Schreiber X-Y recorder || X-Y plotter

Y

Y-Achse axis of y || ordinate * *Ordinate* || y-axis
Yankee Zylinder yankee dryer * *großer Trocken- und Glättzylinder in Papiermaschine* || yankee cylinder
Yankee-Zylinder Yankee cylinder

Z

z Hd Att. * *zu Händen*
Z-Lager Z-bearing * *mit Wellenabdichtungsfunktion* || Z-type bearing * *mit Wellenabdichtungsfunktion* || deep-groove bearing with sideplate * *mit Wellenabdichtungsfunktion*
Z-Transformation Z-transform
Zackenscheibe serrated washer
zäh tough * *zäh, hart, widerstandsfähig, robust, hartnäckig; toughen: zäher machen* || tenacious * *tenacity: Zähfestigkeit/Zähigkeit* || viscous * *Flüssigkeit* || ropy * *Flüssigkeit* || ductile * *Metall* || thickly liquid * *zähflüssig*
zähflüssig high-viscous
Zähigkeit toughness || tenacity * *auch 'Zähfestigkeit', (auch figürl.: 'Verbissenheit'))* || ropiness * ~ *von Flüssigkeiten* || viscosity * ~ *von Flüssigkeiten* || ductility * ~ *bei Metall* || doggedness * *verbissene ~*

Zahl number * *auch Anzahl (even: gerade, odd: ungerade)* || figure * *Zahlenangabe, Ziffer* || numeral * *Ziffer, auch 'Betrag, Wert'* || digit * *Stelle, Ziffer*
Zähl-Füllmaschine counting filling machine * *{Verpack.tech.}*
zahlbar payable * *zu zahlen*
Zählbaugruppe count module
Zählbereich count range || counting range
Zahlen figures
zählen count || integrate * *aufintegrieren*
Zahlen- numerical || of numbers || of figures
Zahlendarstellung number representation || numerical notation
Zahleneingabe numerical input
zahlenmäßig in terms of figures * *siehe auch →wertmäßig*
Zahlenschalter thumb wheel switch * *Daumenradschalter* || rotary switch * *Drehschalter* || thumbwheel switch * *Daumenradschalter* || numeric switch || BCD switch * *im BCD-Code arbeitend*
Zahlensystem number system
Zahlenwert numerical value || numeric value
Zahlenwertgleichung numerical value equation
Zähler counter || numerator * *(mathem.) das über dem Bruchstrich stehende* || electricity meter * *→Stromzähler* || timer * *Zeit~*
Zählerbaugruppe counter module || count module
Zählerbereich count range
Zählerinhalt counter content || count
Zählfrequenz count rate
Zählgerät counting instrument
zahlreich numerous || in great number * *(Adv.)* || great many || in large numbers * *(Adv.)*
zahlreiche a large number of
Zahlung payment * *offer as payment: in ~ geben* || settlement * *~ einer Schuld* || clearance * *~ einer Schuld* || disbursement * *~ von Unkosten*
Zahlungsbedingungen terms of payment
Zahlungseingang receipt of payment
zahlungsfähig solvent
Zählwert count || count value || counter content * *Zählerinhalt*
Zählwort counter word
Zahn tooth * *auch an Zahnrad*
Zahndicke tooth thickness * *z.B. bei Zahnrad*
Zahneingriff tooth engagement * *bei Zahnrädern usw.* || tooth meshing * *bei Zahnrädern usw.*
Zähnezahl number of teeth * *bei Zahnrad und Zahnriemen-Scheibe*
Zahnflanke tooth flank * *z.B. an einem Zahnrad* || tooth profile
Zahnflankenspiel backlash of teeth || tooth-flank backlash
Zahnform tooth shape * *z.B. an einem Zahnrad*
Zahnkranz ring gear
Zahnkranzbohrfutter ring gear chuck * *für Bohrmaschine*
Zahnkupplung tooth clutch || tooth-clutch
Zahnprofil tooth profile * *z.B. eines Zahnrades*
Zahnrad gear wheel || cog-wheel * *kleines Zahnrad, Ritzel* || pinion * *Ritzel, Antriebs(kegel)rad*
Zahnradantrieb cog-wheel drive
Zahnradbahn rack-and-pinion railway
Zahnradpumpe gear-type pump || geared pump
Zahnradtrieb gear drive
Zahnriemen toothed belt
Zahnriemenscheibe toothed-belt pulley
Zahnritzel pinion
Zahnschiene cog rail
Zahnspitze tooth tip * *bei Zahnrad usw.* || tip of tooth * *bei Zahnrad usw.*
Zahnspitzenspiel tooth tip clearance * *bei Zahnrad*
Zahnstange toothed rack || rack || spur rack || gear rack || cog rack

Zahnstangenantrieb rack-and-pinion drive
Zahnstangengetriebe rack-and-pinion gear
Zange pliers || pair of pliers || tongs * *große* || pulling dog * *z.b. als Ziehwerkzeug beim Drahtziehen*
Zapfen stud * *Stift, Zapfen, Stehbolzen, Pfosten* || pin * *Bolzen, Stift, Dorn; pin with thread/theaded pin: Gewinde~* || pivot * *Drehzapfen eines Scharnieres/Gelenks* || gudgeon * *Zapfen, Bolzen* || journal * *Dreh-/Lager-/Wellenzapfen; journal box: Lagerbüchse; journal bearing: ~lager* || spigot * *Zapfen, (Leitungs-/Fass-) Hahn, Muffenverbindung* || tenon * *{Holz} (auch Verb: 'verzapfen')*
Zapfenfräse tenoning machine * *{Holz} im Zimmereihandwerk*
Zapfenfräsmaschine tenoner * *{Holz}* || tenon-cutting machine * *{Holz}*
Zapfenlager journal bearing || pivot bearing || trunnion seat || spindle bearing
Zapfenschneidmaschine tenoner * *{Holz}* || tenoning machine * *{Holz}* || tenon cutting machine * *{Holz}*
Zapfenschneidscheibe tenoning disc * *{Holz}*
Zapfwellenantrieb power take-off * *z.B. an Traktor* || power-take-off shaft * *Zapfwelle, z.B. an Traktor*
zappeln jitter * *z.B. Impulse*
Zarge frame * *{Holz}*
zB e.g. * *example given: zum Beispiel*
Zehnerstelle tens digit
Zehnertastatur numeric keypad || numeric keyboard || numerical keypad
Zeichen character * *z.B. abdruckbares ~, Text~, ASCII-~, ANSI-~* || reference * *Akten~ z.B.in einem Brief, our reference: unser ~* || sign * *Symbol, Wink, Kenn~, Hinweis~* || token * *Anzeichen, Merkmal, Beweis, (umlaufende) Zugriffsberechtigg. für Bus-Kommunikation* || symbol * *Symbol; auch Plus~, Minus~, Frage~ usw.* || mark * *Marke, Markierung, Einstellmarke, Kennung, Kerbe, Merk~* || indication * *Anzeichen, Kenn~, Hinweis, Andeutung, Symptom, Anzeige, Angabe* || signal * *Signal* || omen * *vorausahnenlassendes ~, An~* || index * *of: ~ für/von, Fingerzeig, Hinweis* || digit * *z.B. BCD-/Hexa-~/Ziffer an Siebensegment-Anzeige o.Zahlenschalter* || byte * *in 8 Bit verschlüsseltes ~ (z.B. im ASCII-code)* || characteristic * *Kenn~, Eigentümlichkeit, charakteristisches Merkmal*
Zeichenbrett drawing board
Zeichenfolge string * *von Text-/ASCII-Zeichen usw.* || character string || text string * *Text* || character sequence
Zeichenkette character string || character chain
zeichenorientiert character-oriented * *z.B. serielles Übertragungsverfahren*
Zeichenrahmen character frame * *z.B. 11-bit frame: 11 Bit Zeichenr. (1 Startbit, 8 Datenbits, 1 Paritätsb., 1 Stopbit)* || frame
Zeichenrahmenfehler framing error * *bei serieller Datenübertragung*
Zeichensatz character set || code page * *Codeseite: länderspezif.ASCII-Zeichentabelle z.Bildschirmdarstellung i. PC b.MS-DOS (R)*
zeichnen draw || draft * *skizzieren, entwerfen* || sketch * *(auch figürl.) skizzieren* || outline * *(auch figürl.) skizzieren, umreißen* || mark * *be-/kenn~* || plot * *einen Plan anfertigen von, etwas in e. Plan ein~, etwas planen/entwerfen*
Zeichner draftsman
Zeichnung diagram || graph || figure || blueprint * *Pause* || chart * *Schaubild, Diagramm*
Zeichnungsnummer drawing No
Zeichnungsverzeichnis list of drawings || list of documents * *(Zeichnungs-) Unterlagen-Verzeichnis*
zeigen show * *auch darstellen (Bild/Tabelle/An-*

hang) || point at * *deuten auf* || point out * *deuten auf, darlegen* || indicate * *an~, ~ auf (z.B. Pfeil); the values are indicated i.the table: die Tabelle zeigt d.Werte an* || exhibit * *zur Schau stellen (auf Messe, in Ausstellungsraum usw.), aufweisen, zeigen, an d.Tag legen* || display * *zur Schau stellen, zur Anzeige bringen* || demonstrate * *demonstrieren* || prove * *sich ~, herausstellen* || appear * *erscheinen* || turn up * *plötzlich erscheinen* || showcase * *zur Schau stellen, ausstellen (auf Messe, in Vitrine, in Ausstellungsraum usw.)* || be illustrated by * *durch eine Abbildung, Beispiel usw. gezeigt werden* || illustrate * *durch Bild, Beispiel usw. ~* || manifest * *sich ~/offenbaren/kundtun; manifestieren; be-/erweisen; erscheinen* || illustrate * *bildhaft ~, an einem Beispiel ~; ... is illustrated in Fig. 3: ... zeigt Bild 3* || reveal * *(figürl.) offenbaren, enthüllen, zeigen, erkennen lassen, aufdecken, verraten; Fensterrahmen* || point * *gerichtet sein; to the rear: nach hinten; upwards: nach oben*
Zeiger indicator * *~ eines Messgeräts* || pointer * *~ eines Messgeräts, Adress~ auf eine Variablenadresse, indirekte Adresse* || needle * *Nadel, z.B. Messnadel eines Messinstruments* || index * *auch eines Messgeräts, (Einstell)Marke/Strich* || vector * *Vektor, Raum~* || phasor * *Vektor*
Zeigerausschlag pointer deflection * *~ eines Messgerät*
Zeigerbild vector diagram
Zeigerdiagramm vector diagram
Zeigerinstrument pointer instrument
Zeile line || row * *Reihe (z.B. in Tabelle, Baugruppenträger)* || tier * *Reihe, Lage, Zeile in Baugruppenträger*
Zeilenabstand line space
Zeilenbreite line width
Zeilenhonorar lineage * *für Übersetzer, Datenerfasser*
Zeit time * *only one at a time: nur einer zur ~, jeweils nur einer; save time: ~ sparen* || period * *Zeitraum/Spanne/Dauer, Frist, Periode* || space * *~raum* || space of time * *~raum* || duration * *~dauer* || stage * *Stadium* || phase * *Phase, Stadium* || epoch * *~alter* || era * *~alter* || age * *~alter* || delay * *~abstand, ~verzögerung* || interval * *~intervall, ~abstand* || lag * *Rückstand, Verzögerung, Nacheilung, negative Phasenverschiebg., ~abstand* || instant * *(genauer) ~punkt, Augenblick* || period of time * *~spanne*
Zeit-Füllmaschine filling machine by time * *in der Verpackungstechnik*
zeitabhängig related to time * *zeitbezogen* || time-dependent || time-variant || as a function of time || time-dependant
Zeitablauf time lapse
Zeitablenkung time sweep * *(Oszilloskop)* || time base * *(Oszilloskop) Zeitbasis*
Zeitabschnitt time segment || period || interval || time slot * *Zeitscheibe für Softwareprogramme*
Zeitabstand interval || lag * *Nacheilung; z.B.jet lag: Zeitverschiebung zw.verschiedenen Ländern* || time interval
Zeitachse time base || time axis
Zeitangabe time specification
Zeitangaben time information
Zeitauflösung time resolution
Zeitaufwand expenditure of time || time spent * *aufgewendete Zeit* || sacrifice of time * *Zeitverlust, Zeiteinbuße, "Zeitopfer"* || a great deal of time * *take up a great deal of time: mit großem ~ verbunden sein*
zeitaufwändig time-consuming
Zeitbereich interval * *Zeitspanne* || time range * *time domain * *{Regel.techn.}*
Zeitdauer period || duration

Zeiteinheit unit of time
Zeiteinstellung timing
Zeitersparnis saving of time
Zeitfunktion function of time * *Funktion (über) der Zeit*
Zeitgeber timer || clock unit * *Taktgeber* || real-time clock * *Echtzeituhr* || time generator * *für Uhrzeit*
zeitgemäß seasonable || opportune || timely || up to date || modern * *modern* || current * *aktuell* || state-of-the-art
zeitgerecht in time * *rechtzeitig, mit der Zeit, im (richtigen) Takt* || on time * *(Adv.) pünktlich, termingerecht, rechtzeitig* || according to schedule * *(Adv.) termingerecht, wie geplant* || in due time * *rechtzeitig, termingerecht*
zeitgesteuert time controlled
Zeitgewinn saving of time
Zeitglied timer || monoflop * *Impulsbildner, Zeitstufe* || time element
Zeitgründe time reasons * *for: aus*
Zeitkonstante time constant || T.C.
zeitkritisch critical from a time perspective || time-critical || requiring fast response * *{SPS | Computer} kurze Reaktionszeiten erfordernd* || critical with respect to time || where time is crucial * *(nachgestellt) wo (Rechen-, Reaktions-) Zeit ausschlaggebend ist*
zeitlich begrenzt for a limited time || time-limited || limited for a specific time
zeitlich versetztes Einschalten serial start-up * *z.B. bei Verbund-→Verdichtersatz zur Begrenzung des Anlaufstroms*
zeitlicher Ablauf timing
zeitlicher Verlauf variation with time * *Änderung mit der Zeit* || time characteristic || characteristics over time || trend * *Tendenz* || characteristic
zeitliches Verhalten variation in time || time response
Zeitlimit time limit
Zeitmarke time tag || time stamp * *→Zeitstempel, z.B. bei Diagnose-/Tracespeicher*
Zeitplan time schedule * *tight: eng/knapp* || schedule
Zeitplanung scheduling
zeitproportional proportional to time || time-proportional
Zeitpunkt moment || instant || time || point of time || timing * *gewählter ~, Wahl des ~s* || time instant
zeitraffende Vorlagerung burn-in * *durch zyklische Wärmebehandlung*
Zeitrampe time slope
Zeitraster time base
zeitraubend time consuming
Zeitraum period || interval * *['interwel] Zeitabstand* || space of time || space || time period
Zeitrelais timing relay || time lag relay || time-delay relay || TDR || *Abk. für 'Time-Delay Relay'* || timer relay
Zeitschalter time switch
Zeitscheibe time slot * *Rechnerzyklus* || time slice
Zeitscheibenverwalter time slot manager || time slot scheduler
Zeitschrift magazine * *[mege'sien]* || journal || periodical * *regelmäßig erscheinende ~*
Zeitschriftenaufsatz article || paper * *Abhandlung, Fachaufsatz, Referat*
Zeitselektivität time grading * *z.B. von Schutz-/Überwachungseinrichtungen*
Zeitspanne period of time * *in: nach e. Zeitspanne* || time interval || space of time || span || interval * *Intervall* || duration * *Dauer* || period || space
zeitsparend time-saving
Zeitsparmaßnahme time saver

Zeitstempel

Zeitstempel time stamp * z.B. Echtzeitgabe für Daten in einem Messwert-/Diagnose-/Tracespeicher
Zeitstrahl time beam * Zeitachse für Zeit-/Projektplan
Zeitstufe monoflop * monostable multivibrator: monostabiler Multivibrator || timer || delay || one-shot * Impulsbildner, Monoflop, monostabiler Multivibrator
zeitsynchron in synchronism * with: mit
Zeittaktverteiler clock distributor
Zeitüberschneidung time overlap
Zeitüberschreitung time out * Überschreitung einer Überwachungszeit || violation of the time limit
Zeitüberwachung time monitoring || time check || timeout monitoring || time supervision || watchdog timer * Zeitglied zur Überwachung d. zyklischen Programmablaufes eines Rechners
Zeitungsdruckpapier newsprint
Zeitungspapier newsprint
zeitunkritisch not critical from a time perspective || not requiring fast response * keine schnelle Reaktion erfordernd || not time-critical
Zeitverhalten time response * bezüglich der Reaktionszeit
Zeitverkürzung reduction of time
Zeitverlauf variation with time * zeitlicher Verlauf
Zeitverlust loss of time
Zeitverschiebung time gap * Lücke || jet lag * bei Langstreckenflug || time shift
Zeitverschwendung waste of time
zeitverzögert time-delayed || time-lag * relay: Relais
Zeitverzögerung time delay || lag
zeitweise for a time || from time to time * von Zeit zu Zeit || at times * manchmal || occasional * gelegentlich || only intermittently * (Adv.) mit Unterbrechungen, nur zeitweise
Zeitwert time value || time * bei SPS
Zeitwertspeicher time register * bei SPS
Zell- cellular
Zelle cell
Zellen- cellular
Zellenfilter cellular filter
Zellstoff pulp * Ausgangsmaterial zur Papiererzeugung; Zellstoff, (Zellstoff-) Brei || cellulose * Zellulose, Zellstoff || chemical pulp * Ausgangsmaterial für Papierherstellung; bleached: gebleichter || dissolving pulp * Chemie-Faserzellstoff, Ausgangsmaterial für Papierherstellung || mechanical pulp * Holzstoff (Zellstoffrohprodukt) || semi-chemical pulp * Halbzellstoff || soda pulp * Natron-/Sulfatzellstoff || sulphite pulp * Sulfitzellstoff || wood pulp * aus Holz gewonnener ~
Zellstoffbleiche pulp bleaching
Zellstofffabrik pulp factory || pulp mill
Zellstoffkarton pulp board
Zellstoffkocher pulp digester
Zellstofftrockenmaschine pulp drying machine
Zellulose cellulose
Zellwolle staple fiber || rayon staole
Zement cement
Zementdrehofen rotary cement kiln
Zementierungspumpe cement pump
Zementindustrie cement industry
Zementmühle cement mill
Zenerbarriere Zener barrier
Zenerdiode Zener diode
Zenerspannung Zener voltage
zentral central || centralized * an einer Stelle zusammmengefasst || centrally * (Adv.) || centrally coordinated * mit zentraler Koordination (z.B. Zentraleinkauf)
Zentral-Baugruppenträger controller rack * bei SPS

Zentralantrieb center drive * (amerikan.) || centre drive * (brit.)
Zentralbaugruppe central processing unit * bei SPS || CPU * bei SPS
Zentralbaugruppenträger controller rack * bei SPS
Zentralbaustein central function block * Funktionsbaustein || central block * Funktionsbaustein
Zentrale headquarters * Firmen~ || control centre * ~ Warte, Leitstelle || control room * Warte || main station * Hauptstation || gear centre distance * Achsabstand in einem Getriebe || head office * Firmen~ || central office * Firmen~ || central station * Zentralstation || power house * Schalthaus || power station * Kraftwerk || control station * Steuer~ auch auf Schiff || central switchboard * ~ Steuertafel || main control room * Leit~ || central control room * zentraler Steuerraum, Leitwarte
zentrale Baugruppe central controller module * bei SPS
zentrale Einspeiseeinheit central rectifying unit || common rectifier unit * gemeinsame
Zentraleinheit CPU * Central Processing Unit; eines Computers || central processing unit
Zentralelektronik central control electronic || central electronics
zentraler Aufbau centralized configuration
Zentralerweiterungsgerät controller expansion unit * bei SPS
Zentralgerät central controller * (SPS)
Zentralprozessor central processor
Zentralrechner host computer || host
Zentralschmierung central lubrication
Zentralwickler center winder * (amerikan.) siehe auch →Achswickler || centre winder * (brit.) || core winder || axial winder
Zentrier- centering
Zentrierbohrer centering drill
Zentrierbohrung center hole * (amerikan.) || tapped center hole * (amerikan.) als Gewindebohrung ausgeführt || centering bore || center bore || centre hole * (brit.) || centering hole
Zentrierdurchmesser centering diameter
zentrieren center * (amerikan.) || centre * (brit.) || centering * (Substantiv)
Zentrierfläche spigot face
Zentriergewinde tapped centre hole * z.B. am Wellenende eines Motors
Zentrierlager centering bearing
Zentrierrand centering spigot || centering shoulder
Zentrierstift centering pin || pilot pin
Zentrierung centering
Zentrifugalkraft centrifugal force
Zentrifuge centrifuge
zentrifugieren centrifuge || spin * zentrifugieren, schnell drehen, trudeln lassen, spinnen (Fäden) || centrifugalize
zentrisch centric || concentric || concentrically * (Adv.) || centrical
Zentrizität centricity || concentricity * Konzentrizität
Zentrumswickler centre winder * Achswickler || center winder * (amerikan.) Achswickler || core winder
zerbrechlich fragile || breakable
zerbrochen broken
Zerfall decay
Zerfaserer defiberizer * (Holz) || defibrator * (Holz | Papier)
zerfasern reduce to fibers || pulp wood grinding * (Papier) (Subst.) || defibrating * (Holz) (Subst.) || Holzzerfaserung
Zerfaserungsmaschine defiberizer * (Holz) in der Spanplattenherstellung
zerfressen corrode * zer-/anfressen, korrodieren ||

erode * zer-/an-/wegfressen; siehe auch 'fressen' ||
pit * an-/zerfressen, Grübchen/Löcher/Narben bilden || eat away * zerfressen, nagen an, ein Loch fressen || attack * angreifen, befallen, aggressiv einwirken
Zerhacker chopper
Zerkleinerer chopper * z.B. zum Zerhacken von Abfallholz || schredder * für Akten, Lumpen, Abfälle usw. || reducer * {Holz}
zerkleinern comminute * zerkleinern, zerstückeln, zerreiben || reduce to small pieces || chop * {Holz} || chop up * {Holz} || crush * {Steine} || grind * zermahlen || pulverize * zu Pulver zermahlen || reduce * {Holz} || disintegrate * aufspalten/-lösen/-schließen, ~; (auch intransitiv: zerfallen, sich zersetzen)
zerlegbar capable of being disassembled || able to be disassembled || divisible * teilbar (auch {Mathem.}) || detachable * abnehmbar || collapsible * (zusammen-) klappbar, Klapp-, Falt-
zerlegen disassemble * auseinanderbauen || dismantle || demount || split up * (figürlich, mathemat.) in Bestandteile aufspalten || break down || separate * in seine physikalische Bestandteile ~ || fraction * in seine physikalische/chemische Bestandteile ~
Zerreißarbeit tensile energy absorption
Zerreißprüfmaschine tension testing machine
Zerreißprüfung tension test || rupture test
Zerrung strain
zerspanen machine-cut
zerspanend cutting * Bearbeitung, Werkzeug usw. (Gegensatz: →spanlos)
zerspanend bearbeiten machine
zerspanende Bearbeitung cutting || machining
Zerspaner chipper * {Holz} in der Spanplattenherstellung || flaker * {Holz} in der Spanplattenherstellung || disintegrator * {Holz}
Zerspanung chipping
zerstören destroy || demolish * auch 'niederreißen' || ruin || wreck || devastate * verwüsten || blast * sprengen, zugrunderichten, vernichten, (salopp) 'wegpusten'
zerstörende Prüfung destructive test * von Werkstoffen usw.
zerstörende Werkstoffprüfung destructive material testing
zerstörsicher indestructible * unzerstörbar || nondestroyable || destruction protected || protected against damaging
Zerstörsicherheit safety against damage || surge immunity * gegen Überspannungen geschützt
Zerstörung destruction || demolition || ruin || damage * Beschädigung || wreckage
zerstörungsfrei non-destructive * z.B. Materialprüfung
Zerstörungsprüfung destruction test || destructive testing
Zertifikat certificate
zertifizieren certify
zertifiziert certified || certificated
Zertifizierung certification
Zeug stuff || material || gear * Ausrüstung, Zubehör
Zeugnis letter of recommendation * ~ vom Arbeitgeber || certificate * Zertifikat, Prüfungszeugnis || report * Schul~ || diploma * Diplom || reference * Führungs~ für Angestellte || testimonial * Führungs~
Ziegelpresse brick press || brick extruder * kontinuierliche Strang-~
ziehen draw * auch Strom aus Stromversorgung(snetz) ~ || withdraw * z.B. eine Steckbaugruppe ~ || pull || unplug * Stecker/Steckbaugruppe ~ || extract * herausziehen || disconnect * Stecker ~ || drag * →schleppen; auch 'mit der (Computer-) Maus ~'

Ziehen unter Spannung live removal || hot unplugging
Ziehgriff pulling handle * an Steckbaugruppe || pulling grip * an Steckbaugruppe || extracting handle
Ziehhilfe extractor || pulling lever * Hebel z.B. an Flachbandleitung-Stecker || pulling aid
Ziehklinge scraper * {Holz}
Ziehmaschine drawing machine
Ziehmatritze drawing die * {Werkz.masch.} z.B. in der Blechdosenherstellung
Ziehöl drawing oil * zum Drahtziehen
Ziehpresse extrusion press
Ziehscheibe drawing disk * bei Drahtziehmaschine || capstan * b. Drahtziehmaschine: Rolle m. Mehrfach-Drahtumschlingung, zieht Draht durch d. Ziehstein
Ziehstein drawing die * bei Drahtziehmaschine || die
Ziehtrommel capstan * {Draht | Kabel}
Ziel aim * Richtung, Zweck, Absicht; with the aim of/aimed at: mit dem ~ || target * Leistungs~, Produktions~, Soll, ~scheibe || destination * ~ des Signalflusses, der Reise usw. || end * Zweck, Absicht, Nutzen || object * Gegenstand, Absicht, Zweck || objective * taktisches/strateg./unternehmerisches ~, ~setzg., Operations-/Angriffs-/Lern~ || winning post * ~ im Sport || goal * reach/attain one's goal: sein ~ erreichen, auch Ziel/Tor im Sport || purpose * Zweck
Ziel- target || goal
zielen aim * at: auf || drive * at: auf || take aim * zielen, als/aufs Ziel nehmen (at: auf) || target * zielen auf
zielführend conducive * {kondju: siw} || purposeful
zielgenau accurately taking the aim
zielgerichtet purposeful || purposive * zweckdienlich, zweckmäßig || systematical * systematisch || single-minded * nur dies eine im Sinn habend
Zielgruppe target group * für Werbung, Marketing, Schulung, Dokumentation usw.
Zielhafen port of destination
Zielhardware target hardware * auf die eine Software letztendlich ablaufen soll
Zielhinweis destination reference * für Signale/Daten i. Stromlaufplan/Blockschaltbild/Programmlisting
Zielmarkt target market
Zielposition target position
Zielsetzung objective || target || goal || aims
Zielsteueranlage target tracking system * {Fördertechnik\ Hebezeuge}
Zielsystem target system * Hard-/Softwareystem, auf dem d. Software ablaufen soll (siehe auch →Ablaufumgebung)
Zieltermin target date
ziemlich rather || pretty || fairly
Ziffer digit || figure || numeral
Zifferanzeige numeric display
Zifferntastatur numeric keypad || numeric keyboard
Zifferntasten numeric keys || numeric keypad * Zehner-/Zifferntastenfeld || digital keyboard * Zifferntastatur
Zimmerbestellung room reservation
Zimmereimaschine carpentry machine * {Holz}
Zink zinc
Zinkdruckguss zinc diecast || die-cast zinc
Zinke prong || tine || tooth * z.B. eines Kamms
Zinkenfräsmaschine dovetailing machine * {Holz}
Zinkenverbindung finger jointing * {Holz} || pin joint * {Holz}
Zinnlot tin solder
Zins interest * Kapital~; in full measure: mit ~ und ~es~; yield interest: ~en tragen || rate of interest * Prozent Kapital~

Zinsen interest * *Kapitalzins*
Zinslast interest charges
Zinssatz interest rate
zirka about ‖ approximately ‖ approx. * *Abkürzung für 'approximately'* ‖ appr. * *Abkürzung für 'approximately'*
Zirkel pair of dividers ‖ pair of compasses ‖ compass
Zirkulation circulation
Zitat quotation
zittern tremble ‖ shake ‖ quiver ‖ vibrate * *vibrieren* ‖ dither * *zittern, bibbern, zaudern*
ZN local office * *(SIEMENS-) Zweigniederlassung* ‖ local sales office * *(SIEMENS-) Zweigniederlassung* ‖ regional office * *(SIEMENS-)* →*Zweigniederlassung*
zögern hesitate
Zoll duty * *Gebühr, Abgabe, Steuer* ‖ customs * *auch ~behörde; pay customs: ~ bezahlen* ‖ inch * *Längenmaß (1 Zoll entspr. 1 inch entspr. 0,393701 cm)* ‖ toll * *Straßen-/Wege-/Brücken~, Standgeld, Wegegeld*
Zollabfertigung custom clearing
Zollager bonded warehouse
Zollbehörde customs authority ‖ customs ‖ customs office * *Zollamt* ‖ customhouse * *Zollamt*
Zollformalitäten custom formalities
zollfrei duty-free
Zollgewinde inch thread ‖ thread measured in inches
zollpflichtig dutiable ‖ subject to duty
Zollstock yard-stick ‖ folding rule ‖ foot-rule ‖ inch rule
Zollsystem imperial system * *Maßeinheitensystem*
Zollverschluss customs seal * *in bond: unter ~*
Zone zone
Zoneneinteilung zone-classification * *of hazardous areas: in explosionsgefährdeten Bereichen (DIN VDE 0165)*
zoomen zoom
Zopf- top ‖ twisting ‖ plait * *Zopf, Flechte, (Haar-/Stroh-, Textil-) Geflecht, Falte: (ver)flechten (Verb)*
zu regarding * *betreffend* ‖ too * *too small: zu klein* ‖ towards * *Richtung* ‖ closed * *geschlossen* ‖ close * *Befehl/Kommando*
zu einem späteren Zeitpunkt at a later date
zu erwarten expected
zu Händen att.
zu Händen von att: ‖ for the attention of
zu jedem Zeitpunkt at any point in time
zu Null werden be reduced to zero ‖ fall to zero * *auf Null zurückfallen*
Zu- und Gegenschaltung reversing connection ‖ boost-and-buck connection * *z.B. b.Thyristorsatz z. Erreichung e. hohen Spannspanng. üb. Spezialtrafo*
Zubehör accessories ‖ attachment * *Zusatzgerät* ‖ option * *wahlweise erhältliches* ‖ add-on * *Zusatzeinrichtung* ‖ extras ‖ optional extras * *wahlweise erhältliches*
Zubehörteile accesories ‖ fittings * *Montagezubehör* ‖ spare parts * *Ersatz-/Reserveteile*
Zucker sugar
Zuckerfabrik sugar mill ‖ sugar factory
Zuckerindustrie sugar industry
Zuckerkristalle sugar crystals
Zuckermühle sugar mill
Zuckerzentrifuge sugar centrifuge
zueinander to one another ‖ mutually * *gegenseitig* ‖ together * *aneinander* ‖ to each other ‖ with respect to each other * *aufeinander bezogen, in Bezug ~* ‖ one upon the other * *gegenseitig,* →*aufeinander*
zuerkannt bekommen be awarded * *Auszeichnung, Auftrag, Zertifikat usw.*

zuerst first * *als Erster* ‖ first of all * *vor allem (Übrigen)* ‖ in the first place * *vor allem (Übrigen), erstens* ‖ above all * *vor allem (Übrigen)* ‖ at first * *anfangs* ‖ at the beginning * *zu Beginn, anfangs* ‖ initially * *anfangs, zunächst* ‖ for the first time * *zum ersten Mal* ‖ first * *als Erster, vor allem Übrigen, zum ersten Mal* ‖ for the moment * *vorerst, vorläufig*
zufällig accidental ‖ incidental * *nebenbei* ‖ random * *(statistisch) wahllos* ‖ by accident
Zufallsausfall random failure
Zufallsgenerator random-number generator
Zufallsrauschen random noise ‖ pink noise * *rosa Rauschen*
zufrieden contented * *(with: mit)* ‖ content * *zufrieden; rest/be content with: sich ~geben mit; bereit, willens* ‖ satisfied * *~gestellt, befriedigt* ‖ pleased * *zufrieden (with: mit), erfreut (at: über), angenehm berührt* ‖ gratified * *erfreut, befriedigt, angenehm berührt; be gratified: sich freuen*
zufrieden stellend satisfactory ‖ satisfactorily * *(Adv.)*
Zufriedenheit contentment ‖ satisfaction * *to my greatest satifaction: zu meiner größten ~* ‖ contentedness * *Zufriedenheit, Genügsamkeit*
Zufuhr supply ‖ conveying * *Zuförderung* ‖ delivery ‖ feeding ‖ entry * *Eintrag, z.B. in Ofen, Pressmaschine* ‖ arrivals
Zuführeinrichtung feeding equipment ‖ loading equipment * *Beladungs-/Verladungseinrichtung* ‖ feed mechanism ‖ applicator * *Anlege-/Aufbringeinrichtung* ‖ charging device * *Beschickungs-/Fülleinrichtung* ‖ feeder ‖ feeding mechanism
zuführen route * *Leitung, Signal ~* ‖ carry ‖ carry up ‖ convey ‖ lead in * *Leitung, Draht, Kabel ~* ‖ feed * *to: zu etwas; auch Material, Versorgungsgüter, Waren ~* ‖ deliver * *auch Material ~* ‖ take in * *hereinlassen (Flüssigkeit, Gas usw.)* ‖ supply * *versorgen (mit)* ‖ convey * *fördern (to: zu)* ‖ lead * *leiten, führen, lenken, bringen (to: nach, zu)* ‖ bring * *(her)bringen, herbeischaffen* ‖ lead in * *Draht ~*
Zuführsystem feeding system * *Antransport/Zuführung von Teilen/Material in eine Maschine*
Zuführung infeed * *z.B. von Material in eine Maschine* ‖ feeder * *Zuführ-/Zufördereinrichtung* ‖ feeding * *das Zuführen, Herbeifördern* ‖ feeding in * *das Hineinführen* ‖ feed * *Maschinenteil; pressure feed: mit Druck arbeitende ~* ‖ supply * *Versorgung* ‖ conveyance * *(An-) Förderung* ‖ delivery * *Anlieferung, Anfördern* ‖ lead * *Draht, Leitung* ‖ approach * *Annäherungsweg, (Heran-) Nahen, Annäherung, Nahekommen, Zugang/-fahrt* ‖ intake * *Einlass (-Öffnung), (Neu-) Aufnahme, Zustrom, aufgenommene Menge, Ein-/Absaugen*
Zug tension * *Zugkraft* ‖ train * *Eisenbahnzug* ‖ draw * *Verstreckung, das Ziehen* ‖ pull * *Ziehen, Zerren, Zug, Schubkraft, Anzug, Ruck, Anziehungskraft* ‖ traction * *~kraft, ~leistung {Bahn}, Ziehen, Bodenhaftung, Griffigkeit (Reifen), Fortbewegung* ‖ stress * *Beanspruchung* ‖ course * *Lauf, Bahn, Weg, Gang, Ablauf, Verlauf; in the course of: im ~e des*
Zug- tensional * *Spann(ungs)-, Dehn(ungs)-*
Zug- und Schubraupe belt caterpillar * *Kabelförder mit mechan. Klemmung durch Gurt in Kabelmaschine* ‖ caterpillar
Zugang access * *(auch figürl.) Zugriff, Zugangsmöglichkeit; unobstructed: ungehindert* ‖ approach * *Weg* ‖ gateway * *Tor* ‖ entry * *Tür* ‖ receipts * *Einnahmen*
zugängig accessible * *to: für* ‖ open * *to: für; (auch figürl. z.B. für Gründe)*
zugänglich accessible * *to: für; easily/free: leicht/frei; front-accessible: von vorn ~* ‖ open * *to: für; (auch figürl., z.B. für Gründe)*

zugänglich machen disclose * *offenlegen, bekannt geben; aufdecken* || make accessible * *zugreifbar machen* || make available * *verfügbar machen* || throw open to the public * *der Öffentlickeit* ~
Zugänglichkeit accessibility
Zugangsberechtigung access right || right of access
Zugbeanspruchung tensile stress || tensile load || tensional stress * *tension-free: ohne* ~
zugehörig associated || related || suitable * *dazu passend* || corresponding * *entsprechend, zugeordnet* || belonging to * *a person: zu jemandem* || appertaining * *zugehörig, gebührend, zugestanden* || pertinent * *zur Sache gehörig, sachdienlich, (zu)gehörig, (to: zu), angemessen, passend* || matching * *in Form, Farbe, Größe, Typ usw.* ~/*passend/angepasst* || affiliated * *angegliedert, angeschlossen, verknüpft, verbunden, Tochter-, Zweig-, Filial-* || accompanying * *begleitend* || inherent * *innewohnend, zugehörend, eigen* || constituent * *einen (Bestand-) Teil bildend, constituent part: wesentlicher Bestandteil*
Zugehörigkeitsfunktion fuction of membership
zugelassen authorized * *amtlich* ~ || listed * *bei einer Normenbehörde/Zulasssungsbehörde* ~ || approved * *bei einer Normenbehörde* ~ || licensed * *(amerikan.) lizensiert (auch Auto usw.)* || registered * *amtlich eingetragen/registriert* || certified * *offiziell zugelassen/bestätigt/bescheinigt/beglaubigt* || licenced * *siehe 'licensed'* || permitted * *erlaubt, gestattet*
Zugentlastung strain relief || pull relief
Zugentlastungsstutzen strain relief bushing * *Durchführungshülse mit Kabel-Klemmvorrichtung zur Zugentlastung*
zugeordnet associated * *with: zu* || allocated * *z.B. Speicherplatz* || corresponding * *bezogen auf* || assigned * *festgesetzt, angegeben, zugeschrieben/-wiesen, zugeteilt, übertragen (Aufgabe usw.)*
zugeschnitten tailored * *to: z.b. auf die Kundenbedürfnisse* ~ || fit * *to: an* || adapted || accommodated || adjusted * *to (meet): auf eine Norm, einen Zweck* ~ || dedicated * *speziell auf einen Verwendungszweck/eine Anwendung* ~ || definite-purpose * *auf einen bestimmten Anwendungszweck* ~
Zugfeder tension spring
Zugfederklemme cage-clamp terminal * *schraubenlose Leitungsanschlussklemme*
Zugfestigkeit tensile strength || breaking strength * *Zug-/Reißfestigkeit (z.B. in [N/mm²])* || breaking strain
zügig brisk * *flink, flott, lebhaft* || speedy || swift || smooth * *reibungslos* || easy * *reibungslos* || uninterrupted * *ohne Unterbrechung*
Zugistwert tension actual value
Zugkraft tension force || tension * *Zug* || web force * *einer Warenbahn (Papier, Textil, Kunststoff)* || tensile force || traction power * *[Bahnen] eines (Schienen-) Fahrzeugs* || tractive power * *[Bahnen]*
Zugkraftmessgeber tension transducer || tension dynamometer || load cell * *Zugmessdose*
Zugkraftmessgerät tension meter
Zugkraftregler tension controller
Zugkraftsteuerung tension control
zugleich simultaneously * *(Adv.) gleichzeitig* || at the same time || together
Zugmagnet pull magnet
Zugmessdose load cell || tension transducer || force transducer * *Zugkraft-Messdose*
Zugmesseinrichtung tension transducer || force transducer * *(Zug-) Kraftaufnehmer* || load cell * *Zugmessdose* || tensiometer || tension sensing device
Zugmesswalze tension measuring roll
Zugreduzierung taper tension * *abnehmende Wickelhärte b. steigend. Durchmesser e.Aufwicklers zur Innenlagenschonung*
Zugregelung closed-loop tension control || tension control || strip-tensioning control * *Bandzugregelung in Walzstraße*
Zugregler tension controller || tension regulator
zugreifbar accessible * *siehe auch →zugänglich, →zugänglich machen*
zugreifen access * ~; *access a memory location: auf eine Speicherzelle* ~ || have access * *Zugriff haben; to: auf*
zugreifen auf access || have access to
Zugriff access * *z.b. auf Speicher (to: zu/auf/bei)*
Zugriff freigeben permit access
Zugriffsberechtigung access right || right of access || access authorization || change rights * *zum Ändern/Schreiben, z.b. von Parametern*
Zugriffschutz access protection * *z.b.gegen unbefugte Parameteränderung*
Zugriffskonflikt access collision * *z.b. bei gleichzeitigem Speicherzugriff von 2 Teilnehmern*
Zugriffspriorität access priority
Zugriffsrecht access authorization * *z.b. auf Speicher, Bus, Datenpuffer usw.*
Zugriffsrechte access rights * *(PROFIBUS-Profil)*
Zugriffsschlüssel security access code * *z.b. Codewort zur Vermeidung unberechtigten Zugriffs*
Zugriffsstufe access level || access stage
Zugriffsverfahren access technique || access mechanism || access method
Zugriffszeit access time * *auf Speicher*
zugrunde legen take as a basis * *for: für*
zugrunde liegen be the basis
zugrundelegen take as a basis || apply * *anwenden (to: zu), passen, zutreffen, gelten, z.B. Vorschrift, Betriebsanleitung usw.* || use as basis
zugrundeliegend underlying
Zugschwankungen tension variations || tension fluctuations
Zugsollwertpotentiometer tension setting potentiometer
Zugspannung web tension * *Zugkraft einer Warenbahn* || draw * *z.B. ~ einer Papierbahn (siehe →relative Geschwindigkeitsdifferenz)* || tension * *Zug* || tensile load || tensile stress * *→Zugbeanspruchung*
zugunsten in favour of || in favor of || for the benefit of || to the credit of
Zugwalze pull roll || tension roll || tension roller
Zugwerk tensioner * *z.B. S-Walzengruppe in Folienmaschine* || bridle * *(Walzwerk) Zugwalzenpaar*
zukünftig future * *intended* * *geplant, beabsichtigt* || in future * *(Adv.)* || for the future * *(Adv.)* || prospective * *Person* || in times to come * *in künftigen Zeiten/Tagen, in der Zukunft*
zukunftsorientiert future-oriented || progressive * *→fortschrittlich* || forward-looking
Zukunftsperspektive future prospect
zukunftssicher compatible with the future || future-oriented * *→zukunftsorientiert* || resistant to obsolence * *nicht veraltend* || future-proof
zukunftsweisend pointing the way to the future || trend-setting || pioneering * *→bahnbrechend* || future-oriented || forward-looking
Zulage bonus
zulassen admit * ~ *(amtlich) anerkennen/zulassen/ gewähren/gestatten* || approve * *billigen, gutheißen, genehmigen, anerkennen* || authorize * *ermächtigen, bevollmächtigen, berechtigen, genehmigen, billigen, genehmigen* || license * *behördlich* || tolerate * *geschehenlassen* || suffer * *geschehenlassen, dulden* || allow * *erlauben, gestatten*
zulässig permissible || admissible || allowable * *zulässig, zulässig, rechtmäßig* || authorized * *befugt,*

zulässige Abweichung 376

zulässig, autorisiert, bevollmächtigt || approved * *offiziell anerkannt, zugelassen* || tolerable * *tolerierbar, zulässig, erträglich, leidlich* || acceptable * *annehmbar, tragbar (to: für), willkommen* || valid * →*gültig* || permitted * *zugelassen, erlaubt*
zulässige Abweichung   tolerance
zulässiger Bereich   permissible range
Zulassung   approval * *z.B. von einer Norm/Prüfbehörde* || certification * *z.B. von einer Prüfbehörde/ Zulassungsstelle* || registration || licence || license * *(amerikan.)* || admission * *zu Prüfung/Studium/ Amt usw.*
Zulassungsprüfung   qualification test * *z.B. bei einem Kunden zur Zulassung eines Lieferanten*
Zuleitung   feeder * *z.B. für Motor* || supply cable * *z.B. (Netzanschluss)Leitung für Motor, Stromrichter* || feeder cable || feed line || leads || feed wire || feeding pipe * *Rohrleitung* || feed pipe * *Rohrleitung* || supply * *Rohrleitung* || supply line || incoming cable * *elektr. Kabel/Leitung*
Zuleitungsinduktivität   feeder inductance
Zuleitungskabel   feeder cable
Zuleitungswiderstand   feeder resistance
zuletzt   finally || in the end || at last * *schließlich* || ultimately * *schließlich* || at last * *als Letzter* || most recently * *(Adv.) neuerdings, zeitlich am wenigsten zurückliegend, jüngst, frischest, neuest*
Zulieferant   subcontractor * *Unterlieferant* || supplier
Zulieferer   subcontractor * *Unterlieferant* || supplier
Zulieferfirma   supplier || subcontractor
Zulieferindustrie   subcontracting industry || ancillary industry || subcontracting industries
zuliefern   supply
Zulieferteil   subcontracting part || part delivered by a subcontractor || part delivered by an external manufacturer
Zulieferung   supply || subcontracting parts * *Zulieferteile*
Zuluft   combustion air * *für Verbrennung in Kraftwerk/Ofen/Verbrennungsmotor* || cooling air * *Kühlluft* || fresh air * *für Kühlung* || intake air * *angesaugte (Kühl)Luft* || supply air
Zuluftanlage   air supply system
Zulufttemperatur   air intake temperature || cooling air temperature * *von Kühlluft*
zum Antrieb von Ventilatoren   for driving fans
zum Beispiel   for example || for instance || e.g. * *Abkürzung f. 'example given'*
zum ersten Mal   for the first time
zum größten Teil   largely || mostly || for the most part
zum Teil   partly || in part || to some extent * *zu einem gewissen Ausmaß*
zum Vertrieb freigeben   release for sale || launch into the market
zumutbar   reasonable || bearable * *(er)tragbar* || acceptable * *annehmbar* || within reason * *im Rahmen des Zumutbaren* || fair * *gerecht, angemessen, anständig*
zunächst   initially * *anfänglich* || first of all * *vor allem* || above all * *vor allem* || to begin with * *erstens* || in the first instance * *im ersten Moment* || for the present * *vorläufig* || for the time being * *vorläufig, vorerst* || tentatively * *versuchsweise* || next to * *am nächsten gelegen* || first * *erstens, (zu)erst, in erster Linie, vor allem* || at first * *zuerst, anfangs* || first of all * *zu aller erst, vor allen Dingen* || for one thing * *erstens* || in the first place * *erstens, vor allem andern* || for the moment * *vorerst, vorläufig*
Zunahme   increase || gain * *Steigerung, Zuwachs,* →*Erhöhung*
Zünddurchschlag   ignition breakdown || transmission of internal ignition || spark ignition || transmission of igniting flame

Zünddurchschlagsfähigkeit   ignition breakdown capacity * *Fähigkt. brennbar. Gase, durch engen Spalt hindurchzuzünden;* →*Ex.gruppe*
Zünddurchschlagsvermögen   spark ignition capacity * *bestimmt die Explosionsgruppe z.B. von Motoren*
zünden   fire * *z.B. einen Thyristor* || turn on * *z.B. Thyristor* || firing * *(Substantiv) eines Thyristors* || ignite * *z.B. Gase, Motor* || electrify * *Zuhörerschaft*
zündend   be fired * *Thyristor* || stirring * *erregend, aufwühlend, mitreißend, bewegt, bewegend; z.B. Ereignis, Rede, Vision* || electrifying * *Idee, Vision, Rede usw.*
Zündenergie   gating pulse energy * *Zündimpuls-Energie (zum Zünden von Thyristoren)* || firing pulse energy * *z. Zünden von Thyristoren*
zündfähig   triggerable * *z.B. Thyristor*
Zündgefahren   possible explosion hazards
Zündgruppe   ignition group * *Explosionsgruppe nach alter Norm VDE0170/0171* || explosion group
Zündimpuls   firing pulse * *z.B.für Thyristor; siehe auch DIN 41750 Teil 7* || trigger pulse || gate control pulse || gating pulse
Zündimpulsübertrager   trigger pulse transformer
Zündkennlinie   control characteristic * *z.B. bei Gasentladungsröhre* || ignition characteristic
Zündleitung   firing cable * *für Thyristor(en)*
Zündquelle   ignition source
Zündschutzart   flameproof enclosure || type of protection * *siehe EN 50014 und 50016 bis 50021* || degree of protection
Zündschutzart druckfeste Kapselung   flameproof enclosure protection || degree of protection flameproof enclosure
Zündschutzart EEx d   explosion-proof EEx d type of protection * *Explosionsschutzart für Motor*
Zündschutzart EEx e   EEx e increased safety type of protection * *Explosionsschutzart für Motor*
Zündschutzgas   pressurizing gas * *zum Explosionsschutz z.B. von Motoren*
Zündsperrbaugruppe   pulse suppression module * *z.B. für Thyristor-Leistungsteil* || firing inhibit board * *für Thyristorsatz*
Zündstrom   gate trigger current * *(Stromrichter)* || igniting current
Zündtemperatur   ignition temperature * *z.B. von Gasen, Stäuben*
Zündübertrager   pulse transformer * *(Stromrichter)*
Zündung   firing * *z.B. ~ eines Thyristors* || ignition * *z.B. von Gasen/des Benzin-/Luftgemisches im Motor usw.*
Zündwinkel   firing angle * *z.B. bei Thyristor (siehe auch 'Steuerwinkel')* || gating angle * *z.B. bei Thyristor* || ignition angle * *bei Benzinmotor*
Zündwinkelvorsteuerung   firing angle precontrol * *zur Verbesserung der Dynamik bei DC-Stromrichter*
Zündzeitpunkt   firing point * *für Thyristor* || firing time * *für Thyristor* || triggering instant * *für Thyristor*
zunehmen   increase * *increase in number: an Zahl ~* || grow * *wachsen; grow worse: schlechter werden,* grow stronger: *stärker werden* || grow larger * *anwachsen, größer werden* || grow bigger * *anwachsen, umfangreicher werden* || rise || get longer * *(die Tage, wenn es länger hell bleibt) ;* get worse: *schlechter werden*
zunehmend   increasing || growing
zuordnen   assign * *(z.B. Aufgabe/Arbeit) zuweisen (to: zu); zuschreiben (to: etwas/jdm.)* || adjoin * *beiordnen/fügen* || coordinate * *beiordnen, an die Seite stellen* || appoint * *appoint a person as assistant to: jmd. e. Person als Mitarbeiter/Helfer ~* ||

class with * *einordnen/einstufen bei, gleichstellen mit* || relate to * *in Beziehung setzen/bringen zu;* i.*Zusammenhang bringen mit; verbinden mit* || allocate * *zuteilen, zuordnen, zuweisen (to: an; auch Speicherplatz), den Platz bestimmen für*
Zuordnung assignment * *An/Zuweisung, Bestimmung, Festsetzung; Aufgaben/Arbeits/Postenzuweisung* || matching * *passende* ~ || coordination * *Beiordnung, z.B. von Personen* || allocation * *z.B. von Speicherplatz* || classification * *Einteilung; Klassifizierung*
Zuordnungsliste assignment list
Zuordnungstabelle assignment table * *z.B. Umrichter zu Motor* || coordination table || allocation table || matching table * *z.B. Umrichter zu Motor*
zur Folge haben result in || lead to || mean * *bedeuten*
zur Hand handy * *have something handy: etwas zur Hand haben* || ready to hand || within reach * *in Reichweite, greifbar*
zur Sicherheit for safety reasons || to be on the safe side * *um sicherzugehen*
zur Verfügung gestellt provided
zur Verfügung stehen be available * *to: für* || be at hand
zur Verfügung stellen supply * *bereitstellen, versorgen, ausstatten, liefern* || provide * *bereitstellen; versehen, versorgen, ausstatten(with: mit); sorgen für,liefern* || make available || pass to * *an jdn./etwas weitergeben*
zur weiteren Veranlassung for further action
zurück back || backward || backwards || reverse * *umgekehrt* || behind * *hinten; lag behind: im Rückstand sein* || behindhanded * *im Rückstand*
zurück- re- || away * *(nachgestellt) aus dem Weg, weg; z.b. fold back: ~/wegklappen*
zurückerstatten restore || return || refund * *Ausgaben, Kosten, Geld* || repay * *Ausgaben, Kosten*
zurückfedern spring back
zurückführen feed back || derive * *ab/herleiten; from: von* || reduce * *to: auf eine Regel usw.* ~ || track back * *to: auf eine Ursache usw.* ~
zurückgeben give back || return || restore * *zurückerstatten/bringen/geben/stellen* || surrender * *übergeben, ausliefern, aushändigen; to: an* || render * *zurückgeben/erstatten, z.B. Gewinn*
zurückgehen go back || return * *wiederkehren* || fall back * *zurückfallen, ins Hintertreffen gelangen (to: auf)* || originate * *seinen Ursprung haben (in/from: von einer Sache/Person)* || decrease * *sich verringern* || diminish * *sich verringern* || fall off * ~ *des Geschäfts usw.* || decline * ~ *des Geschäfts usw.* || drop * *abfallen (z.B. Spannung, Leistung)* || be decreased * *verringert werden* || decay * *abnehmen, schwinden, verfallen*
zurückgespeist fed back
zurückgreifen auf resort to * *auch 'Zuflucht nehmen zu'* || fall back on * *auf Notlösung* || fall back upon * *auf Notlösung* || have recourse to || refer to * *sich wenden an, befragen (Uhr, (Wörter-)Buch usw.)*
zurückhalten hold back || keep back || retain * *zurückbehalten, einbehalten, beibehalten, (Eigenschaft/Posten/i.Gedächtnis) behalten* || withhold * *vorenthalten* || keep to oneself * *bei sich behalten* || delay * *verzögern* || retard * *verzögern* || restrain * ~, *Einhalt gebieten, abhalten (to do: von); zügeln, unterdrücken (Gefühl), beschränken* || check * *drosseln, bremsen, sperren* || retention * *(Substantiv) Zurückhaltung, Einbehaltung, Beibehaltung, Festhalten, Merken, Bewahrung*
zurückhinken lag * *zeitlich (behind: hinter)*
zurücklaufen run reversely || move in reverse direction || act reversely * *in Gegenrichtung wirken* || return * *umkehren, zurücklaufen* || run back ||

ramp-down * *entlang einer (Hochlaufgeber-) Rampe* ~ || brake * *bremsen* || decelerate * *verzögern* || slow down * *Geschwindigkeit/Drehzahl reduzieren* || flow back * *Flüssigkeit* || recirculate * *Flüssigkeit in Kreislauf* || back out * *rückwärts herausfahren* || rewind * *zurückwickeln/spulen* || recoil * *zurückwickeln/spulen (Blech, Draht, Kabel, Seil)* || fly back * *Zeilenrücklauf des Strahls bei Monitor/Bildschirm*
zurückliefern recover * *zurückgewinnen* || deliver back || return * *z.B. Vertriebsergebnis/Profit/ Bremsleistung ans Netz* ~ || regenerate * *z.B. Bremsenergie ins Netz* ~
zurückmelden signal back || report back || check back * *zur Bestätigung* ~
zurücknehmen reduce * *reduzieren* || take back || withdraw * *Vorwürfe, Urteil, Kritik usw.* ~ || retract * *Vorwürfe, Urteil, Kritik usw.* ~ || eat one's words * *(salopp) Vorwürfe, Kritik usw.* || revoke * *widerrufen* || go back from * *Zusage usw.* ~ || throttle back * *drosseln (Gas beim Motor usw.)*
zurückschalten switch back || return || fall back * *automatisch in einen Zustand geringerer Leistung(sfähigkeit)* ~ *(z.B. Baudrate)*
zurücksenden send back || return
Zurücksendung redelivery || return
zurücksetzen reset
zurückspeisen refeed || feed back * *z.B. Bremsenergie ins Netz* ~ || feed back regeneratively * *z.B. Bremsenergie ins Netz* ~ || regenerate * *z.B. Bremsenergie ins Netz* ~ || return
zurückspeisend regenerative * *Bremsenergie ins Netz*
zurückverfolgen trace back
Zurückverfolgung back-tracking
zurückweisen reject * *ablehnen, abweisen* || refuse * *ablehnen; refuse to accept: Annahme verweigern* || decline || repulse * *Angriff*
Zurückweisung rejection || refusal || dismissal
zurückwirken retroact
zurückziehen withdraw
zurückziehend fold-back * *z.B. Kennlinie einer Stromversorgung (Spannung wird bei Überstrom zurückgezogen)* || retracting * *z.B. Kennlinie*
zurückzuführen auf due to || traceable to || to be explained by
zurzeit at present || at the moment || for the time being || at the present || presently * *momentan, gegenwärtig* || at the present time
Zusage promise || word || assent * *Einwilligung* || undertaking * *Verpflichtung* || acceptance * *Annahme* || approval * *Billigung* || confirmation * *Bestätigung*
zusammen together || along with * ~ *mit* || in company with * ~ *mit* || in conjunction with * ~ *mit* || at the same time * *(Adv.) gleichzeitig* || simultaneous * *gleichzeitig* || contemporaneous * *gleichzeitig* || coincident * ~*fallend* || paired * *paired with: gepaart mit*
Zusammen- co- * *gemeinsam mit*
Zusammenarbeit cooperation * *close: enge* || teamwork * *Teamarbeit* || joint operation * *z.B. von Abteilungen, Firmen*
zusammenarbeiten co-operate || work together || collaborate * *mitarbeiten, zusammenarbeiten, behilflich sein* || cooperate || interact
Zusammenbau assembly || assembling
zusammenbauen assemble * *into: zu* || combine * *kombinieren, (miteinander) verbinden; z.B. Motor- und Anbaugetriebe* || make
zusammenbrechen break down || collapse || smash * *(salopp) zusammenkrachen, Bruch machen, zu Schrott fahren*
Zusammenbruch breakdown || collapse * *völliger* ~ || smash * *Vernichtung, Ruin (auch finanzieller)*

|| crack-up * Zusammenbruch, Bruchlandung || crash * Absturz, Zusammenstoß/bruch, schwerer Unfall, Ruin

zusammendrücken compress || press together || squeeze together || sqeeze * (zusammen)drücken, auspressen/-quetschen, schröpfen

zusammenfassen summarize * gedrängt darstellen, z.B. einen Textes || outline * Übersicht/Überblick geben || sum up * noch einmal ~ || recapitulate * noch einmal ~; rekapitulieren || comprise * in sich fassen || comprehend * in sich fassen || embrace * in sich fassen || collect * sammeln; auch Gedanken || combine * miteinander verbinden || concentrate * miteinander verbinden zu höherer Dichte/Kompaktheit; auch Kräfte || pool * Resourcen, Material bündeln/gemeinsam benutzen || integrate * integrieren || condense * Schriftwerk ~ || compile * zusammentragen (Material, Unterlagen usw.) || unite * miteinander verbinden || consolidate * vereinigen, zusammenlegen/schließen, verdichten || merge * verschmelzen (in: mit), aufgehen lassen in, einverleiben, fusionieren || group * in Gruppen ~ || organize * in ein System bringen; organize into groups: in Gruppen ~

zusammenfassend summing up

Zusammenfassung summary || abstract * Abriss, Übersicht, Inhaltsangabe, Auszug || overview * Übersicht || outline * Abriss, Übersicht, Überblick || resumé * Resümee, Zusammenfassung

zusammenführen bring together || let flow together * zusammenfließen lassen || agglomerate * zusammenballen || concentrate * bündeln || join together * zusammenfügen

zusammengefasst compiled * z.B. Angaben gesammelter Daten in einer Tabelle || summarized * kurz dargestellt || summed up * kurz dargestellt

zusammengehören belong together || be correlated * (figürl.) || form pairs * je zwei Sachen || be fellows * mehrere Sachen

zusammengehörig belonging together || belonging to one another || homogeneous * gleichartig, homogen || related * verwandt, in Beziehung/Zusammenhang miteinander stehend || consistent * übereinstimmd., vereinb.,i.Einklang (with: mit), zusamm.gehörig(Teile e.Datensatzes)

zusammengesetzt composed * of: aus || compound * zusammengesetzt, Verbund-, aus Verbundmaterial bestehend || consisting * of: bestehend aus

Zusammenhang coherence || connexion * Beziehung, Verbindung; siehe →connection || connection * Beziehung; in this connection: in diesem ~ || Verbindung, Verwandtschaft || correlation * Wechselbeziehung, gegenseitige Abhängigkeit, Entsprechung || relation * Beziehung, Verhältnis, kausaler ~ || continuity * enge Verbindung, ununterbrochene Folge, Fortlaufendes || context * Kontext, inhaltlicher ~, ~ von Textstellen, Umgebung, Milieu || association * ~ von Ideen, Beziehung, Verknüpfung, (Gedanken-)Verbindung || relationship * Beziehung, Verhältnis, Rechtsverhältn., Verwandtschaft; between...and: zwischen...und || implication * enge Verbindung, Verflechtung, Begleiterscheinung, Folge || conjunction * Verbindung; in conjunction with: in Verbindung mit

zusammenhängend coherent * auch Gedanke, Rede || continuous * fortlaufend || connected * in Beziehung stehend || related * verwandt, in enger Beziehung stehend || allied * verwandt, in enger Beziehung stehend || interdependent * voneinander abhängend || associated * miteinander verbunden, in Zusammenhang stehend || contiguous * angrenzend, berührend, benachbart, lückenlos ~; discontinuous: nicht ~/mit Lücken || joined * zusammen-

gefasst, vereinigt || integrated * zusammenhängend, einheitlich, gleichmäßig, integriert

zusammenpassen fit together || match * nach Form, Farbe, Funktion usw. || be well matched * gut ~ || go well together * gut ~ || harmonize * harmonisieren, gut ~ || agree || fit || fit into one another

zusammenpressen compress || press together || squeeze together || condense * verdichten

Zusammenschalten interconnection || synchronizing * zum Synchronlauf

Zusammenschaltung interconnection

zusammenschließen incorporate * vereinigen, verbinden, einbauen, einverleiben || link together || join-in * auch 'sich ~' (closely: eng) || unite * (sich) vereinigen || integrate into * zu einem Ganzen ~ || consolidate * zu einem Ganzen ~ || join forces * sich ~ || combine * gemeinschaftliche Sache machen || form an alliance * zu einem Bündnis ~ || merge * (sich) zusammenschließen (Firmen usw.)

Zusammenschluss merging * von Firmen || merger * von Firmen

zusammensetzen put together || compose * zu einem Ganzen; be composed of: sich ~ aus || compound * aus Teilkomponenten ~ (z.B. Verbundmaterial) || assemble * zusammenbauen, montieren, Mannschaft zusammenstellen || sit together * sich (z.B. zu einer Besprechung) ~ || get together * zusammenkommen, z.B. zu einer Besprechung || make up * bilden, ~; be made up of: bestehen/sich ~ aus; zusammenstellen (Schriftstück)

Zusammensetzung compound * Verbund-, Mischung, Masse, ~ einer Substanz, chem. Verbindung || ingredients * Bestandteile || structure * Struktur, Gefüge || composition * (gestalterische) Anordnung, Aufbau, Struktur, Beschaffenheit, Synthese || make-up * das Bilden einer Mannschaft, eines Teams usw.

Zusammenspiel interaction * Wechselwirkung || interplay * of forces: der Kräfte; auch 'Wechselwirkung' || teamwork * gemeinsame Zusammenarbeit im Team || cooperation * Zusammenarbeit, siehe auch →Zusammenwirken || team-work * von Personen in einer Gruppe

zusammenstellen arrange * (an)ordnen || group * nach Gruppen ~ || classify * nach Klassen ~ || assort * nach Sorten ~ || match * nach zusammenpassender Ausführung, Größe, Farbe usw. || make up * Liste, Unterlagen usw. ~ || assemble * zusammensetzen, montieren, bereitstellen, (Mannschaft, Tatsachen) zusammenstellen || compile * zusammentragen, z.B. Besprechungsunterlagen, Liste, Ersatzteilsortiment usw. || combine * zusammenfassen, vereinigen || place together * an einen gemeinsamen Ort ~/vereinigen || compose * Menü ~ || make out * Liste, Rechnung usw. ~ || form * Mannschaft, Team, Arbeitsgruppe usw. ~ || set up * Mannschaft usw. aufstellen/etablieren || tabulate * in einer Tabelle ~

Zusammenstellung compilation || combination || arrangement || grouping || table * Tabelle || summarizing sheet * Übersichts-Blatt || synopsis * Zusammenfassung, Übersicht, Abriss, (vergleichende) Zusammenschau || comparison table * (vergleichende) tabellarische ~ || survey * Übersicht, Überblick (of: über)

zusammentreffen coincide * zeitlich ~, ~ von Umständen || meet || concur * zeitlich ~, ~ von Umständen || meeting * (Substantiv) || coincidence * (Substantiv) zeitlich~, ~ von Umständen || concurrence * (Substantiv) zeitlich~, ~ von Umständen || encounter * (Verb,Subst.) jdm./e.Sache beggenen, auf Widerstände stoßen, jdm. entgegentreten

zusammenwirken co-operate || collaborate || work together || share forces * Kräfte vereinigen || com-

bine * *sich verbinden* || co-operation * *(Substantiv)* || combined action * *(Substantiv)* || joint operation * *(Substantiv)* || concurrence * *(Substantiv)* ~ *von Umständen* || joint venture * *(Substantiv)* ~ *von Unternehmen bei einem bestimmten Projekt* || interworking * *(Substantiv)* || cooperate || interact * *aufeinander wirken, sich gegenseitig beeinflussen* || interplay * *(Substantiv) Zusammenspiel, Wechselwirkung* || interaction * *(Substantiv) Wechselwirkung*
zusammenziehen contract * *(sich) zusammenziehen, verengern, schrumpfen, kleiner werden* || draw together * *(auch figürl.)* || concentrate * *konzentrieren*
Zusatz suffix * *z.B. an Bestellnummer* || addendum * *zu einem Schriftstück* || addition || appendix * *Anhang* || supplement * *Ergänzung* || add-on * *Zubehör (-Teil)* || option * *wählbarer ~, (gegen Aufpreis) zusätzlich lieferbarer ~*
Zusatz- additional * *z.B. additional information: Zusatzinformation* || option || supplementary || helper * *Hilfs-* || add-on * *Zubehör* || side * *-Nutzen, -Abkommen, -Effekt usw.* || auxiliary * *Hilfs-* || complementary * *Ergänzungs-, Begleit-* || added * *vermehrt, erhöht, zusätzlich*
Zusatzangaben additional information || additional data || supplementary information
Zusatzaufwand additional efforts || additional costs * *Zusatzkosten; incur additional costs: Zusatzkosten auf sich nehmen (müssen)* || added cost
Zusatzausstattung additional equipment
Zusatzbaugruppe supplementary module || additional module || option module || option board * *Leiterplatte* || supplementary board * *Leiterplatte* || additional board * *Leiterplatte* || add-on card * *Leiterkarte*
Zusatzblatt supplement sheet
Zusatzbremse auxiliary brake * *z.B. für Sicherheitszwecke*
Zusätze additives * *z.B. Hilfsstoffe*
Zusatzeinrichtung auxiliary equipment || auxiliary device
Zusatzfunktion supplementary function || additional function || special function * *Spezial-/Sonderfunktion*
Zusatzgerät supplementary unit
Zusatzgeräusche additional noise
Zusatzkosten additional costs * *incur additional costs: ~ auf sich nehmen (müssen)* || added cost
zusätzlich in addition * *(Adv.)* || additonal * *(Adj.)* || extra || further || added || supplementary || auxiliary * *Hilfs-*
zusätzliche Merkmale additional features || supplementary features
Zusatzmaßnahmen additional measures || supplementary measures
Zusatznutzen side benefit || added value || goody * *(salopp) 'Bonbon', Annehmlichkeit*
Zusatzsollwert supplementary setpoint * *Hilfssollwert* || additional setpoint || trim signal * *z.B. als Stellsignal von einem überlagerten Regler* || supplementary reference value
Zusatzstoff additive
Zusatzstoffe additional ingredients || aggregates
Zusatzvereinbarung side letter * *zu einem Vertrag*
Zusatzverluste additional losses || stray losses * *(magnet.) Streuverluste*
Zusatzverlustmoment stray loss torque * *eines Motors auf Grund der Streuverluste (VDE 0530, Teil 2)*
Zusatzwert supplementary value || additional value
zuschalten switch in || switching-in * *(Substantiv)* || switch-in * *(Substantiv)* || connect to the system * *ans Netz* || enable * *freigeben* || connect-in || bring on line * *Maschinenteil zur laufenden Arbeitsmaschine hin~*

Zuschaltung switch-in
zuschicken mail * *per Post/E-Mail* || send * *to: jemandem* || forward * *weiterleiten; to: an*
Zuschlag surcharge * *Aufschlag, (Preis-) Zuschlag* || extra charge * *Preis~, zusätzliche Gebühr, Berechnung* || additional charge * *zusätzliche Gebühr, Berechnung* || addition || extra price * *Preisaufschlag, "Zusatzpreis"* || award * *b.Ausschreibg., Auftragsvergabe; of contract/order: Vergabe eines Liefer/Bestellauftrags* || allowance * *(Sicherheits-/ Gehalts-) Zuschlag, Quote, (Geld-) Zuwendung, Spielraum, zuläss. Abweichung* || acceptance of tender * *~ bei einer Ausschreibung; obtain the contract: den ~ erhalten*
Zuschlag bekommen be awarded * *bei Auftragsvergabe; be awarded th.contract: d.Liefervertrag zuerteilt bekommen*
zuschneiden taylor * *auch ein Produkt to: auf eine Anwendung/auf Kundenbedürfnisse* || customize * *auf Kundenanforderungen/-bedürfnisse* || cut to length * *{Holz}*
Zuschnitt cutting out || cut-out * *das zugeschnittene Teil* || cutting to size
Zuschnittoptimierung size-cutting optimization * *{Holz}*
zusetzen clog * *→verstopfen* || lose money * *Geld einbüßen* || be a loser * *Geld einbüßen (by: bei)* || press hard * *press a person hard; jdm. zusetzen* || importune * *jdn. →belästigen (with: mit)* || ply with * *jdn. mit Fragen usw. →belästigen* || urge * *drängend, mahnend* || obstruct * *→verstopfen* || add * *→hinzufügen* || adulterate * *durch Zusätze verfälschen, verschlechtern*
zusichern assure * *assure a person of a thing: jdm. etwas zusichern* || guarantee * *guarantee a thing to a person: jdm. etwas zusichern* || promise * *versprechen; promise a thing to a person: jdm. etw. versprechen*
Zustand state * *z.B. Betriebs~; steady: stationärer; solid: fester* || status * *~ eines Signals/Binärein/ ausgangs (Plural: statuses)* || level * *Pegel eines logischen/Binär-Signals* || condition * *Zustand, Verfassung, Beschaffenheit, Umstände, Lage; fault cond.: Fehler~* || situation * *Lage* || position * *Lage* || circumstances * *Umstände* || phase * *Phase* || status * *rechtlicher, politischer ~* || conditions * *Verhältnisse*
Zustände statuses
zustandebringen bring about * *by: mittels/durch*
Zustandekommen realization * *Realisierung, Verwirklichung, Aus/Durchführung* || accomplishment * *Vollendung, Ausführung, Erfüllung, Bildung* || taking place * *Stattfinden*
zuständig responsible * *verantwortlich; z.B. Abteilung, Bearbeiter* || competent * *befugt* || proper * *maßgebend, passend, richtig* || appropriate * *passend, richtig* || serving * *~ Vertriebsniederlassung usw.*
Zustandsanzeige status indication || monitor
Zustandsbeobachter state observer
Zustandsbeschreibung status description
Zustandsbit status bit
Zustandsfolgediagramm state transition diagram || state graph || state-sequence diagram
Zustandsgröße state variable
Zustandsmeldung state signal || status signal || state information || status bit * *Zustandsbit* || status message
Zustandsraum state space
Zustandsregelung state control
Zustandsregister status register
Zustandsregler state controller
Zustandsschätzung state estimation
Zustandssteuerwerk sequence control module || state machine * *{Computer}* || sequencer
Zustandswort status word

zuteilen allocate * *auch Speicherplatz, Zugriffsrechte usw.* ~ || apportion * *(Anteil, Aufgabe usw.)* ~, *(gleichmäßig/gerecht)* ein-/verteilen, *(Lob)* erteilen || distribute * verteilen || attach * *eine Person jemandem* ~ || delegate * *powers: Befugnisse* ~ || grant * genehmigen || allow * genehmigen || issue * ausgeben || allot * aus-/verteilen, bewilligen, abtreten, *(für e. Zweck usw.)* bestimmen
Zuteilung assignment * *An/Zuordnung, Bestimmung, Festsetzung, Übertragung* || allocation * *Zu/Verteilung, An/Zuteilung, Bewilligung* || allotment * *Ver/Zuteilung, Anteil, das Zugeteilte* || distribution * *Ver/Aus/Einteilung*
zutreffen apply * *to/for: auf/bei/für; gelten, anwendb. sein, Anwendg. finden, passen,gelten (Regel,Gesetz)* || be true * *of/for: bei/für/auf* || be correct || be the case || be right * *richtig sein* || fit * *passen (Beschreibung usw.)*
zutreffend applicable * anwendbar, geeignet || right || true || correct
zuverlässig reliable || dependable * verlässlich || reliably * *(Adv.)*
Zuverlässigkeit reliability * *siehe auch →Betriebssicherheit* || dependability * *auch Verlässlichkeit, Vertrauen, Verlass*
zuvor beforehand
Zuwachs increase * *in: an/von* || increment || accretion * *Zunahme, Zuwachs, Anwachsen, Hinzugekommenes, Hinzufügung, Wertzuwachs* || buildup * *Aufbau, starke Zunahme* || augmention * *of: an/von* || growth * *(An)Wachsen (in: von)* || gain * *(Zu)Gewinn, z.B. von Kapital*
zuweisen assign || allocate * *zuteilen, zuweisen, z.B. Speicherplatz* || allot || apportion || grant * genehmigen || allow * genehmigen || issue * ausgeben || distribute * verteilen || attach * *jemandem eine Person* ~ || delegate * *powers: Befugnisse*
Zuweisung assignment || allocation * *Speicherplatz usw.*
ZVEI ZVEI * *Abkürzg.f. 'Zentralverband der Elektrotechnischen Industrie'*
Zwangs- forced
Zwangsabschaltung trip * *auf Grund eines Fehlers* || forcibly actuated shutdown
zwangsbelüftet forced-air cooled || forced-ventilated
Zwangsbelüftung forced ventilation || forced cooling
zwangsgeführt positively driven || positive-action || positively-opening * *Öffnerkontakt, z.B. an einem Überwachung-/Schutzrelais*
zwangsgekühlt forced-cooled
zwangsgeschmiert force-lubricated
Zwangskommutierung forced commutation
zwangsläufig inevitable || necessary || guided * *Bewegung e. Maschinenteils* || geared * *Bewegung e. Maschinenteils* || with necessity * *(Adv.)* || inevitably * *(Adv.)* || automatically * *(Adv.)* || positive * *Zwangs-, zwangsläufig, formschlüssig; bestimmt, definitiv, ausdrückl., fest, tatsächl.*
zwangsmäßig tun müssen be bound to * *dazu bestimmt sein*
zwangssteuern force * *durch künstliche Einprägung von Steuerbefehlen für Inbetriebnahme/Testzwecke* || forcing * *(Substantiv) Festvorgabe von SPS-Ein-/Ausgängen für Testzwecke* || permanent forcing * *(Substantiv) Festvorgabe von SPS-Ein-/Ausgängen zum Test* || sustained forcing * *(Substantiv) Festvorgabe von SPS-Ein-/Ausgangswerten zum Test*
Zwangssteuern beenden force release || force disable
zwangszünden force-fire * *→Schutzzündung von Thyristoren bei Erreichen der max. zulässigen Sperrspannung*

Zweck purpose || reason * *Grund* || object * *Ziel* || aim * *Ziel* || end * *Ziel; to this end: zu diesem* ~ || intent * *Absicht* || intended use * *Bestimmung* || application * *Verwendung* || function
Zwecke purposes
zweckmäßig expedient || appropriate || well-directed || suitable || practical || proper || functional || advisable * ratsam || purposive * *zweckmäßig, zweckdienlich* || making sense * *make sense: sinnvoll sein/Sinn machen* || convenient * *geeignet, bequem, passend, zweckdienlich*
zwei- bi- || two- || twin || double
zweiadrig two-core
zweideutig ambiguous
Zweidraht-Verbindung two-wire connection
Zweidrahtbetrieb two-wire operation
Zweidrahtkabel two-wire cable * *z.B. für serielle Verbindung nach RS485* || twisted pair cable * verdrillte Zweidrahtleitung || two-conductor cable * *z.B. für ein serielles Datenübertragungsverfahren nach RS485*
Zweidrahtsystem two wire system * *für serielle Datenübertragung*
zweifach polumschaltbar two-speed pole-changing
Zweifach- twin- || double * *Doppel-* || two- || tandem * *in Reihen/Kaskadenanordnung, hintereinander angeordnet* || bi-
Zweifachstromrichter twin converter || double converter * *z.B. gegenparallele 4 Q Drehstrombrückenschaltung*
zweifellos definitely * *gewiss, zweifellos, entschieden* || undoubted || doubtless || undoubtedly * *(Adv.)*
Zweiflächen- twin-face * *z.B. Bremse, Kupplung*
Zweiflächenbremse twin-face brake
Zweig path || section * *Teil* || arm * *z.B. einer Stromrichter/Brückenschaltung* || branch || leg * *z.B. einer Stromrichterschaltung*
Zweiganggetriebe two-speed gearbox
Zweigbetrieb satellite plant
Zweigbüro sub-office || branch office || subsidiary * *Filiale, Tochtergesellschaft*
Zweigdrossel arm reactor * *in e. Zweig e. Stromrichterschaltung. (nicht i.d. Wechselstromzuleitung)*
Zweigniederlassung branch || branch establishment || subsidiary * *Filiale, Tochtergesellschaft* || foreign branch * *im Ausland* || branch office * *Zweigbüro* || sub-office * *Zweigbüro* || regional office
Zweigsicherung arm fuse * *in e. Zweig e. Stromrichterschaltung (nicht i.d. Wechselstromzuleitung)* || arm fuse link * *Sicherungseinsatz ohne Unterteil* || thyristor branch fuse * *in Thyristor-Brücke* || branch fuse || leg fuse
Zweigstelle' branch * *~ einer Firma; siehe auch →Zweigbüro und →Zweigniederlassung*
Zweileiteranschluss two-wire connection
Zweileiterbetrieb two-phase operation * *am Drehstromnetz*
Zweilochmutter nut with two holes
Zweilochmutterndreher pin-type face spanner
Zweilochmutterschlüssel socket screw wrench
zweimal twice * *twice as large: ~ so groß*
zweiphasig two-phase || 2-phase || two-cannel * *Impulsgeber, Inkrementalgeber*
Zweipol two-terminal circuit || two-terminal network
zweipolig two-pole
Zweipuls- double-pulse
zweipulsig 2-pulse || double-pulse
Zweipunktregelung two-step control || bang-bang control || flip-flop control || on-off control
Zweipunktregler two-state controller || two-step action controller || on-off controller || two-position action controller || bang-bang controller || two-point controller || single-circuit controller * *z.B. zur Tem-*

peraturregelung || two-level controller * z.B. bei Füllstandsregelung
Zweiquadrant- two-quadrant || unidirectional * *nur in einer Drehrichtung arbeitend*
Zweiquadrantbetrieb two-quadrant operation
Zweirichtungs- bidirectional
Zweisäulenpresse two-column press
Zweischichtwicklung double-layer winding * *Motorwicklgs.art: jede Nut enthält Drähte zweier Spulen →wenig Oberwellen*
zweiseitig two-sided || bilateral * *Vertrag usw.* || double-action * ~ *arbeitend* || two-way * *Zweiweg-* || duplex * *z.B. in d. Druckerei* || bipartite * *Verhandlungen, Verwaltung usw.; doppelt ausgefertigt (Dokumente)* || at both ends * *auf beiden Seiten* || on both sides * *auf beiden Seiten* || two-side || double-ended || double-end
zweiseitig gelagert double-bearing
zweiseitig saugend double-inlet * *Ventilator*
Zweiseitigkeit two-sidedness
zweisprachig bilingual
Zweiständerpresse straight-sided press
zweistufig 2-stage || double reduction * *bei Untersetzungsgetriebe* || double-stage || two-step || two-stage * *auch bei Getriebe*
zweit second || 2nd * *Abkürzung für 'second'* || two * *(nachgestellt) z.B. step two: zweiter Schritt*
Zweitaktmotor two-stroke engine || two-cycle engine
zweite Wahl second grade || low grade
zweite Wurzel square root
zweiteilig bipartite || two-piece || split * *in (zwei) Teile aufgespalten* || two-part
zweitens secondly || in the second place
zweiter Ordnung second-order || 2nd order
zweites Wellenende second shaft end || second shaft extension || non-driving shaft end * *auf Nicht-Antriebsseite (B-Seite) eines Motors* || rear shaft * *hinteres Wellenende*
Zweitfunktionstaste shift key || function key
Zweithersteller second source
Zweitimpuls secondary pulse * *z.B bei Steuersatz für Thyristorsatz*
zweitletzter last but one || next to the last
Zweitrommelwinde two-drum hoist
Zweiweg- double-way
Zweiweggleichrichter double-way rectifier
Zweiweggleichrichtung double-way rectification
Zweiwegschaltung double-way connection * *z.B. für Stromrichter*
Zweiwicklungstrafo two-winding transformer
zweizeilig double-line * *Anzeige, (Aus)Druck* || double-tier * *Baugruppenträger* || double-row * *Baugruppenträger* || two-tier
Zwillingsantrieb twin drive * *z.B. Walzmotor-Paar*
Zwillingsbürste split brush * *Kohlebürste für Motor/Generator*
zwingend imperative * *zwingend erforderlich*
zwingend erforderlich imperative
Zwirn thread || twisted thread || sewing cotton || twine * *Spinnereigarn* || twisted yarn * *Garn*
zwirnen twist * *[Textil] Fäden usw*
zwischen between * *zwischen zweien* || among * *zwischen/unter mehreren* || amongst * *(mitten) unter, zwischen, inmitten*
Zwischen- intermediate * *zeitlich, räumlich* || interim * *zeitlich (z.B. Bilanz, Bericht)* || adapter * *zur Anpassung dienend* || semi- * *Halb-* || inter- * *untereinander* || auxiliary * *Hilfs-* || indirect * *indirekt* || provisional * *vorläufig, provisorisch, behelfsmäßig* || temporary * *vorübergehend*
Zwischenablage clipboard * *[Computer] z.B. in WINDOWS*
Zwischenbescheid interim reply

Zwischenergebnis intermediate result
Zwischenflansch intermediate flange || joining flange || adapter flange || adaptor flange
Zwischenfrequenz intermediate frequency
Zwischengehäuse intermediate housing || adapter housing
zwischengeschaltet interposed
Zwischenglied intermediate part * *Maschinenelement; z.B. in einer elastischen Kupplung* || connecting link || intermediate element
Zwischenklemme intermediate terminal
Zwischenkreis DC link || intermediate circuit * *weniger gebräuchlich* || DC bus * *Zwischenkreis-Schiene* || link || d.c. intermediate circuit * *(weniger gebräuchlich) Gleichspannungszwischenkreis*
Zwischenkreisankopplung DC link coupling
Zwischenkreisdrossel DC-link reactor || link circuit reactor
Zwischenkreiskondensator DC link capacitor || dc-link capacitor || link-circuit capacitor
Zwischenkreisschiene DC link bus || common DC link bus * *gemeinsame*
Zwischenkreissicherung DC link fuse * *bei Umrichter*
Zwischenkreisspannung DC link voltage || link voltage || intermediate voltage || intermediate DC voltage
Zwischenkreisstrom DC link current
Zwischenkreistaktung DC-link pulsing
Zwischenkreisumrichter DC-link converter || DC link converter || indirect AC converter || link converter || intermediate-circuit converter
Zwischenkühlmittel intermediate cooling medium
Zwischenlage separating layer * *~ zum Trennen aufeinander liegender Lagen* || intermediate layer || interlayer || separator * *z.B. Wicklungs~ (aus Hostaphan usw.)* || spacing layer * *~ zum Abstand-Halten* || spacer * *~ zum Abstand-Halten* || interlining * *Zwischenbelag, Einlagestoff/-material* || intermediary || insulation layer * *Isolations~ bei Trafos, Drosseln usw.*
Zwischenlagerung intermediate storage
Zwischenlösung interim solution * *siehe auch →Notlösung* || provisional solution * *behelfsmäßige Lösung*
Zwischenplatte intermediate plate
Zwischenraum spacing || gap * *Lücke* || space * *Leerzeichen zwischen Buchstaben* || clearance * *lichter/freier Raum, Zwischenraum, Freilegung, Räumung*
Zwischenraumtaste space bar
zwischenschalten connect in between * *as intermediate element: als Zwischenglied*
Zwischenschritt intermediate step
Zwischenspeicher latch * *~ für ein Datenwort* || cache memory * *~ zur Entkopplung zweier verschieden schneller Partner, z.B. CPU und Hauptspeicher* || buffer memory || buffer storage || temporary storage || scratch pad memory * *schneller "Notizblock"-~* || buffer || register * *~ für Datenwort* || capture-register * *Auffangspeicher für Zählerstände usw. (für 'Momentaufnahmen')*
zwischenspeichern store temporarily || latch * *in Datenregister(n)*
Zwischenspeicherung intermediate storage || data buffering * *von Daten* || temporary storage
Zwischenstufe interstage
Zwischentrafo intermediate transformer * *siehe auch →Zwischentransformator u. →Vorschalttransformator*
Zwischentransformator intermediate transformer * *siehe auch →Vorschalttransformator* || matching transformer * *Anpasstransformator*
Zwischenwert intermediate value
Zwischenzeit time in between || meantime * *in the*

meantime: in der Zwischenzeit || interim period || intervening period * siehe auch →*Übergangszeit* || interval * *Zeitabstand, Zwischenzeit, Zeitperiode, Pause*
zwischenzeitlich   temporary
zwölfpulsig   twelve-pulse || 12-pulse
zwölfpulsige Stromrichterschaltung   twelve-pulse converter configuration
zyklisch   cyclic || cyclically * *(Adv.)* || in a cyclic time frame * *Programmabarbeitung*
zyklisch gesteuert   cyclically controlled
zyklische Bearbeitung   cyclic processing
zyklische Dienste   cyclic services * z.B. bei einem *Feldbussystem*
zyklische Folge   cyclic succession
zyklische Übertragung   cyclic transmission * z.B. über einen seriellen Bus
zyklische Verarbeitung   cyclic processing || cyclic operation * *Arbeitsweise*
zyklischer Aufruf   cyclic call
Zykloide   cycloid
Zykloiden-Schneckengetriebe   cycloidal worm gear
zykloidisch   cycloidal
Zyklus   cycle
Zyklusbetrieb   cycle operation * z.B. →*Betriebsart S5*

Zyklusüberlauf   cycle-overrun || cycle-timeout || exceeding of sample time || exceeding of scan time
Zyklusüberwachung   sampling time monitoring * *erkennt Überlauf eines Softwarezyklus*
Zykluszeit   cycle time || scan time * *Abtastzeit z.B. bei SPS/Regelsystem* || sampling time * *Abtastzeit bei Regelsystem*
Zykluszeitüberschreitung   timeout || exceeding of scan time
Zykluszeitüberwachung   scan time monitoring || scan time monitor * *als Einrichtung/Gerät* || watchdog timer * *Überwachungs-Zeitglied, das anspricht, wenn es nicht zyklisch getaktet wird*
Zylinder   cylinder * *['silinder]*
zylinderförmig   cylindrical
Zylinderkopfschraube   cylinder-head screw || cheese-head screw
Zylinderrollenlager   cylindrical-roller bearing
Zylinderschleifmaschine   cylinder grinding machine * *z.B. für Tiefdruckzylinder [Druckerei]*
Zylinderschraube   cylindrical screw || socket screw
zylindrisch   cylindrical
µ   µ || micro || micron * *Dickenmaß 1/1000mm*
µop   Mikrobefehl * *(salopp) Teilbefehl, in den ein Maschinenbefehl in einer* →*CISC-CPU zerlegt wird*

Wörterbuch Antriebstechnik

Dictionary of Drives

Teil 2
Englisch – Deutsch
English – German

% Prozent
% voltage drop prozentualer Spannungsabfall
'as delivered' condition jungfräulicher Zustand
* *Lieferzustand* || Urladezustand * *Auslieferungszustand*
'as supplied' condition Auslieferungszustand
'electronic shaft' type of operation Winkelgleichlauf * *"elektronische Welle"*
'experience' value Erfahrungswert
(B6)A(B6)C connection B6-A-B6-C-Schaltung
* *vollgesteuerte Drehstrombrücke i.Antiparallelschaltung für 4Q-DC-Stromrichter*
-ability -barkeit
-able -fähig * *in der Lage ... zu werden; z.B. expendable: erweiterbar*
-action -wirkend * *z.B. double-action: doppelt~*
-based -basiert
-capable -fähig
-clad -ummantelt
-coated -beschichtet
-colored -farben
-core -polig * *Leitung, Kabel*
-count -Zahl
-defining -bestimmend
-dependant -abhängig
-dependent -abhängig
-determining -bestimmend
-digit -stellig * *z.B. Ziffernanzeige, Bestellnummer*
-er -Maschine * *z.B. coater: Streich~*
-fashioned -artig * *ausgeführt, geformt*
-fast -beständig * *fest, beständig, haltbar* || -fest * *fest, beständig, haltbar*
-fed -gespeist
-figure number -stellig * *Zahl*
-fold -fach
-free -frei * *ohne*
-hood -schaft * *z.B. neighbourhood:Nachbarschaft*
-impregnated -geschwängert
-independent -unabhängig
-internal -intern
-laden -geschwängert
-language -sprachig
-lead -adrig * *z.B. 3-lead cord: 3-adrige (Geräteanschluss-) Leitung*
-less -frei * *-los, ohne* || *-los* || *ohne* * *(Vorsilbe)* ||
un- * *ohne*
-like -ähnlich || -artig * *ähnlich* || -förmig
-line -zeilig * *z.B.alphanumerisches Display*
-ment -ung
-ness -heit || -keit * *z.B. friendliness: Freundlich~* ||
-schaft * *z.B. readiness:Bereitschaft*
-number -Zahl
-oriented -orientiert
-pin -polig * *Stecker, IC*
-plated -beschichtet * *mit Metall-/Hartmetallschicht*
-ply -lagig * *Textilstoff, Karton, Sperrholz usw.; z.B. five-ply carton: 5-lagiger Karton*
-pole -polig * *bei elektr. Maschine, Resolver usw.*
-powered -betrieben * *mit Energie versorgt durch ...* || -gespeist * *z.B. mit Versorgungsspannung*
-proof -beständig || -dicht || -fest || -geschützt || -sicher * *geschützt*
-proportional -proportional
-protected -beständig * *geschützt vor* || -dicht * *geschützt vor* || -fest * *geschützt* || -geschützt || -sicher * *geschützt*
-raising -treibend * *ein Ansteigen verursachend*
-readiness -bereitschaft
-related -bedingt * *mit ... in Zusammenhang stehend* || -bezogen || -relevant || -spezifisch * *-nah, in Zusammenhang stehend mit*
-resistance -Festigkeit
-resistant -fest * *beständig, widerstandsfähig gegen* || -sicher * *beständig/widerstandsfähig gegen*

-rise -Anstieg
-safe -sicher * *gegen Gefahr*
-saturated -geschwängert
-selection switch -Stellungsschalter * *n-fach-Umschalter*
-sensitive -abhängig * *-empfindlich* || -empfindlich
-shaped -förmig * *Profil, Umriss, Querschnitt, äußere Gestalt; z.B. e. Impulses/Profileisens*
-ship -schaft * *z.B. leadership:Führerschaft, partnership:Partnerschaft*
-side -seitig
-specific -spezifisch
-speed -tourig * *z.b. bei Getriebe, Motor*
-speed gearbox -ganggetriebe
-stage -stufig
-strength -Festigkeit
-stroke -takt * *z.B. 4-stroke engine: Viertaktmotor*
-supported -gestützt
-terminal circuit -Pol * *z.B. 3-terminal circuit: Dreipol*
-tested -geprüft
-tier -reihig * *in ... Reihen/Lagen* || -zeilig * *-reihig, z.B. Baugruppenträger*
-tight -fest * *gegen mechan.Einflüsse z.B.Staub,Öl,Wasser*
-times -fach * *z.B. ten times: zehnfach; 1.5 times:1,5-fach*
-track -spurig * *Impulsgeber, Impulstacho; single-track: ein~*
-turn -gängig * *z.B. 20-turn potentiometer: 20-gängiges Potentiometer*
-turn potentiometer -Gang-Potentiometer * *z.B. 10-turn potentiometer:10-Gang-Potentiometer*
-type -artig * *Sorte, Modell, Ausführung* || -Ausführung || -förmig
-type of construction -Ausführung
-variant -abhängig
-wise -weise
0 mark Nullmarke
0V Masse
1-to-1 connection Eins-Zu-Eins-Verbindung
100% inspection Stückprüfung * *vollständige Qualitätsprüfung an allen Einheiten eines Fertigungsloses*
12-pulse 12-pulsig || zwölfpulsig
19-inch housing 19-Zoll-Gehäuse
19-inch plug-in unit 19-Zoll-Einschub
19-inch rack 19-Zoll-Gestell
19-inch slide-in 19-Zoll-Einschub
1st erst * *Abkürzung für 'first'* || erstes
1st order erster Ordnung * *(vorangestellt)*
2 power of 14 2 hoch 14
2 wire connection 2-Draht-Verbindung
2-phase zweiphasig
2-pulse zweipulsig
2-stage zweistufig
20mA current loop 20mA-Schnittstelle
24 hours a day rund um die Uhr
2nd zweit * *Abkürzung für 'second'*
2nd order zweiter Ordnung
3-ph. 3 AC * *3-phasige Wechselgröße*
3-phase Drehstrom- || dreiphasig
3-phase AC controller Drehstromsteller
3-phase AC voltage controller Drehstromsteller
3-phase bridge connection Drehstrombrückenschaltung
3-phase current Drehstrom
3-phase line Drehstromnetz
3-phase motor Drehstrommotor * *dreiphasig*
3-position switch Dreistellungsschalter
3-pulse dreipulsig
3-stage dreistufig
3-step inverter Dreipunktwechselrichter
3-terminal terminal box dreiklemmiger Klemmenkasten * *bei explosionsgeschütztem Motor*

3rd dritt * *Abkürzung für 'third'*
4 quad 4Q
4 quadrant 4Q
4-quadrant Vierquadrant-
4-quadrant operation Vierquadrantbetrieb
4-wire connection 4-Drahtverbindung
4-wire operation 4-Draht Betrieb
4-wire system Vierdrahtsystem * *z.B. für serielle Datenübertragg. nach RS422 oder Stromversorgg. mit Fühlerleitg.*
6-pulse 6-pulsig
7-segment display 7-Segment-Anzeige || 7-Segmentanzeige

A

a bit etwas * *(Adv.)*
a certain something etwas * *ein gewisses Etwas*
A contact Arbeitskontakt
a couple of Paar * *(Adv.)*
a day täglich * *(Adv) twice a day: zweimal* ~
a few mehrere || Paar * *(Adj.) ein paar, einige*
a function of -abhängig * *(vorangestellt)*
a great deal of time Zeitaufwand * *take up a great deal of time: mit großem* ~ *verbunden sein*
a great many viel * *sehr* ~*e* || viele * *sehr* ~
a hell of a toll * *(salopp) verdammt gut*
a large number of zahlreiche
a little etwas * *(Adv.)*
a little bit geringfügig * *(Adv.)*
a little less than knapp * *(Adv.) vor Zahlen*
a lot sehr * *(salopp) sehr viel* || viel * *(Adv.) thanks a lot: ~en Dank* || viele * *thanks a lot: ~n Dank*
a lot of viel || viele
a matter of course selbstverständlich * *Selbstverständlichkeit*
a model of mustergültig * *musterhaft*
a multiple number of times mehrfach * *(Adv.)*
a multiple number times mehrfach * *(Adv.)*
a must unerlässlich * *this feature is a must: diese Eigenschaft ist unerlässlich*
a number of verschiedene * *einen Anzahl von, mehrere*
a series of eine Reihe von
a total of insgesamt * *in Zusammenhang mit Zahl*
a wealth of eine Fülle von * *ideas:Einfälle/Ideen; knowledge:Wissen*
a with umlaut ä
A-B A-B * *Abkürzung des Firmennamens 'Allen Bradley'* || Allen Bradley * *Kurzform*
A-contact Schließer || Schließerkontakt
A-to-D converter Analog-Digital-Umsetzer || Analog-Digital-Wandler
A-version Robustbauform * *SPS*
A-weighting filter A-Filter * *zur Schallmessung mit Gewichtung entsprechend d.menschl. Ohr*
a.c. current Wechselstrom
a.c. line Wechselstromnetz
a.c. motor Drehstrommotor || Wechselstrommotor
a.c. power controller Wechselstromsteller * *mit Phasenanschnitt- oder Vollschwing.steuerg. für Heizungszwecke o.Softstart*
a.c. side fuse Strangsicherung
a.c. voltage Wechselspannung
a.c.-d.c. motor Allstrommotor
a.c.-d.c. relay Allstrom Relais || Allstromrelais
a.s.a.p. sobald wie möglich * *Abkürzg. f. 'as soon as possible'*
A/D conversion Analog-Digital-Umsetzung

A/D converter Analog-Digital-Umsetzer || Analog-Digital-Wandler
abandon verlassen * *(auf immer)* ~, *(völlig) aufgeben, verzichten auf, preisgeben, überlassen, im Stich lassen*
abate fallen * *(figürl.) nachlassen* || nachlassen * ~ *von Schwäche, Schmerz, Frost usw.* || sinken * *sich vermindern*
abatement Bonus * *Rabatt* || Minderung * *Abnehmen, Nachlassen, Linderg., Bekämpfung, (Preis)-Nachlaß* || Rabatt
abbreviate abkürzen * *(Zeit, Wörter, Rede usw.) abkürzen/kürzen* || kürzen * *(Zeit, Wörter, Rede usw.) abkürzen/kürzen.*
abbreviation Abkürzung || Kennzeichen * *Abkürzung* || Kurzzeichen * *Abkürzung*
abcissa Abzisse * *X-Achse* || X-Achse * *Abzisse*
abduct ableiten * *Hitze*
abide verharren * *by:bei/auf (z.B. einer Meinung)*
abide by befolgen * *festhalten an, z.B. Grundsatz* || *sich halten an* * *festhalten/sich halten an, treubleiben*
abilities Können * *Fähigkeiten*
ability Fähigkeit
ability to be freely configured Projektierbarkeit * *freie* ~
ability to be integrated Integrationsfähigkeit
ability to communicate Kommunikationsfähigkeit
ability to rotate Verdrehbarkeit
able -bar || fähig * *to: zu* || tüchtig * *auch fähig*
able to be -bar
able to be activated aktivierbar * *möglich, aktiviert zu werden*
able to be disassembled zerlegbar
able to be flange-mounted anflanschbar
able to be swung-out ausklappbar
able to communicate kommunikationsfähig
able to compete konkurrenzfähig || wettbewerbsfähig
able to operate perfectly funktionsfähig || funktionstüchtig
abnormal anormal || ungewöhnlich * *unnormal* || unnormal
abnormal condition irregulärer Zustand
abolish aufheben * *abschaffen* || beseitigen * *abschaffen*
abondon abgehen * *von einem Vorhaben/einer Lösung*
abort abbrechen * *Programm/Vorgang/Ablauf z.B. auf Grund eines Fehlers* ~ || Abbruch * *eines Programms* || beenden * *abbrechen*
aborted abgebrochen * *vorzeitig abgebrochen (z.B. Ablauf e. Programms, z.B. wegen eines Fehlers*
about annähernd * *(Adv.)* || danach * *ask about: danach fragen* || etwa * *ungefähr* || gegen * *ungefähr* || herum * *um herum (in Verbindung mit Zahl)* || nach * *bezüglich* || rund * *ungefähr* || über * *sprechen, nachdenken* || um * *(Adj. u.Adv.) auch 'ungefähr' und zeitlich (about the time: um die Zeit herum)* || ungefähr || zirka
about that darüber * *in dieser Hinsicht*
about to dabei * *im Begriffe; do a thing:etwas zu tun*
above ab * *oberhalb (auch figürlich)* || oben * *auch Bild, 'das* ~ *näher Erläuterte' usw. in einem Schriftsatz; from above: von* ~ || oben genannt || ober || oberhalb * *auch oberhalb eines Werts* || obig * *z.B. Kapitel, Darlegung* || über * *oberhalb von, auch für Wert, Temperatur, Spannung*
above all in erster Linie || zuerst * *vor allem (Übrigen)* || zunächst * *vor allem*
above atmospheric pressure Überdruck
above average überdurchschnittlich
above base speed operation Feldschwächbetrieb

* *Betrieb oberhalb der Ablösedrehzahl (bei der d.
Feldschwächung einsetzt)*
above critical   überkritisch
above ground   über Tage
above it   darüber * *(auch figürl.: over and above:
darüber hinaus)*
above mentioned   oben erwähnt || obengenannt ||
obig * *oben genannt* || vorstehend * *oben erwähnt*
above one another   übereinander
above sea level   über dem Meeresspiegel
above that   darüber hinaus * *bei mehr/größerem*
above-average   überdurchschnittlich
above-mentioned   oben genannt || vorgenannt
* →*obengenannt*
above-said   oben genannt
above-synchronism   übersynchron * *z.B. Betrieb e.
Asynchronmotors in der Feldschwächung*
abrade   abschaben * *abreiben, abschürfen, abscheuern, aufscheuern* || schaben * *abschaben* ||
verschleißen * *abreiben, abnutzen*
abraded particles   Abrieb * *abgeriebenes Material*
abrasion   Abnutzung * *Abrieb* || Abrieb || Verschleiß
* *mechanischer Abrieb*
abrasion resistance   Abriebfestigkeit || Reibfestigkeit * *Abriebfestigkeit* || Verschleißfestigkeit * *gegen Abrieb*
abrasion-proof   abriebfest || verschleißfest * *~ gegen Abrieb* || verschleißfrei * *abriebfest*
abrasive   Schleif-
abrasive belt   Schleifband
abrasive cloth   Schmirgelleinen
abrasive disc   Schleifscheibe
abrasive paper   Sandpapier || Schmirgelpapier
abrasive paste   Schleifpaste
abrasive products   Schleifmittel
abrasive-band grinding machine   Bandschleifmaschine
abrasive-belt grinding machine   Bandschleifmaschine
abrasives   Schleifmittel
abridge   kürzen * *Buch usw.* || verkürzen * *abkürzen*
abridged catalogue   Kurzkatalog
abroad   Ausland * *ins Ausland, im Ausland; from
abroad: aus dem Ausland; trip abroad: Auslandsreise*
abrupt   scharf * *abrupt* || schlagartig || stoßartig
* *plötzlich, aprupt; schroff, kurz angebunden, jäh,
steil*
abruptly   blitzschnell * *(Adv.) plötzlich* || schlagartig * *(Adv.)*
ABS   ABS * *'American Bureau of Shipping': amer.
Schiffsklassifikationsgesellsch., erlässt Prüfvorschriften*
absence   Abwesenheit || Freiheit * *Nicht-Vorhandensein, Abwesenheit* || Mangel * *Fehlen (of: von)*
absence of   -Freiheit * *Fehlen, Abwesenheit von*
absent   abwesend
absent-minded   unachtsam
absolute   absolut || unabhängig * *unbeschränkt, unumschränkt (herrschend)* || unbedingt * *absolut, unbeschränkt, unumschränkt, uneingeschränkt* || unbegrenzt * *(Voll)Macht, Befugnis* || völlig * *z.B. Gewissheit* || vollkommen * *Macht, Befugnisse, Recht usw.*
absolute encoder   Absolutwertgeber * *erzeugt e.absolute Winkel-/Lageinformation z.B. i. Gray-Code
(einschrittiger Code)*
absolute error   absolute Abweichung || absoluter
Fehler
absolute measuring system   Bezugsmaßsystem
* *Fahren auf absolute (immer auf d.gleichen Wegnullpunkt bezogene) Position*
absolute position encoder   Absolutwertgeber * *Lagegeber*
absolute shaft encoder   Winkelpositionsgeber * *rotierender Absolutwertgeber*

absolute shaft-angle encoder   Winkelkodierer
absolute synchronism   absoluter Gleichlauf * *z.B.
→Winkelgleichlauf* || Winkelgleichlauf
absolute value   Absolutwert || Betrag * *Absolutwert*
absolute-value function   Absolutwertbildner || Absolutwertbildung
absolute-value generation   Betragsbildung
absolute-value generator   Absolutwertbildner ||
Absolutwertbildung || Betragsbildner
absolute-value module   Absolutwertbildner
absolutely   unbedingt * *(Adv.) bedingungslos*
absorb   absorbieren || aufnehmen * *absorbieren, in
sich ~, "schlucken", aufsaugen, z.b. Wärme, Stöße*
|| Beschlag * *mit ~ belegen, ganz in Anspruch nehmen* || dämmen * *absorbieren* || dämpfen * *Stöße,
Schall, Schwingungen absorbieren/dämpfen*
absorb sound   dämpfen * *Geräusche ~*
absorbed   aufgenommen * *z.B. Wärmeaufnahme/
Leistungsaufnahme eines Motors*
absorbency   Saugfähigkeit
absorption   Absorption || Aufnahme * *Aufsaugung,
auch Wärme* || Vertiefung * *(figürlich) auch 'vertieft sein'*
absorptive capacity   Saugfähigkeit
absorptive strength   Saugfähigkeit
abstract   abstrakt || Zusammenfassung * *Abriss,
Übersicht, Inhaltsangabe, Auszug*
abundance   Fülle || Menge * *Überfluss* || Überfluss
* *of: an/von*
abundance of designs   Typenvielfalt
abundance of orders in hand   Auftragspolster
abundant   reichlich * *reichlich (vorhanden), reich
an, versehen (mit), abundant in/with· i. Überfluss
besitzend*
abundantly   reichlich * *(Adv.) reichlich, völlig*
abuse   Missbrauch * *übermäßige Beanspruchung,
Misshandlung, falscher Gebrauch*
abut   aneinander stoßen
abut on   stoßen * *angrenzen*
abyss   Tiefe * *Abgrund*
AC   Drehstrom * *auch 'Wechselstrom'* || Drehstrom-
* *auch 'Wechselstrom-'* || Wechselstrom
AC commutator machine   Drehstrom-Kommutatormaschine
AC commutator motor   Drehstrom-Kommutatormotor || Drehstrom-Nebenschlussmotor
AC commutatormotor   Drehstromkommutatormotor
AC component   Wechselanteil || Wechselstromanteil
AC contactor   Wechselstromschütz
AC controller   Wechselstromsteller * *wandelt fest.-
Wechselspanng.(z.B.Netz) i.e. Wechselspanng. m.
variabl. Effektivwert*
AC current   Wechselstrom
AC drive   Drehstromantrieb || Wechselstromantrieb
AC feeder   wechselstromseitige Zuleitung
AC input power supply   Netzversorgung * *eines
Geräts*
AC line   Wechselstromnetz
AC line filter   Netzfilter
AC line fuse   Strangsicherung
AC link   Wechselstromzwischenkreis
AC machine   Drehfeldmaschine
AC motor   Drehstrommotor || Wechselstrommotor
AC power controller   Wechselstromsteller * *erzeugt aus e.festen e.variable Wechselspannung,
siehe DIN 57875, VDE 0558 Tl.4*
AC power converter   Wechselstromsteller
AC quantities   Wechselstromgrößen * *siehe auch
DIN 40110*
AC servo motor   Brushless DC-Motor * *(salopp)* ||
bürstenloser Gleichstrommotor * *(salopp) AC Synchronmot. m."elektron.Kommutator" (Blockstromsteuerg.)*

AC side

AC side Wechselstromseite * *on the: auf der*
AC side fuse Strangsicherung * *in d. wechselstromseitigen Zuleitung z. einem Stromrichter*
AC side fuse link Strangsicherung * Sicherungseinsatz ohne Unterteil
AC source Wechselspannungsquelle
AC system Wechselstromnetz || Wechselstromsystem
AC tach Wechselstromtacho || Wechselstromtachogenerator
AC tacho Wechselstromtacho || Wechselstromtachogenerator
AC tachogenerator Wechselstromtacho || Wechselstromtachogenerator
AC voltage Wechselspannung
AC voltage component Wechselspannungsanteil
AC winder Drehstromwickler
AC-DC Allstrom
AC-DC motor Allstrommotor * *mit Kommutator, z. Anschluss an Gleich- und Wechselstrom (Hauptschlussschaltung)* || Universalmotor
AC-DC relais Allstromrelais
AC-DC relay Allstrom Relais
AC-induction motor Asynchronmotor
academy Fachhochschule
accel beschleunigen * *Kurzform für 'accelerate'*
accel time Hochlaufzeit * *(salopp) Abk. f. 'acceleration time'*
accelerate beschleunigen || hochfahren * *beschleunigen* || hochlaufen * *beschleunigen*
accelerated beschleunigt
accelerating ansteigend * *Geschwindigkeit, Drehzahl*
accelerating current Beschleunigungsstrom
accelerating torque Anzugsmoment * *Beschleunigungsmoment* || Beschleunigungsmoment
acceleration Beschleunigung * $[m/s^2]$ || Hochlauf * *Beschleunigung; current limit acceleration: ~ an der Stromgrenze*
acceleration belt Beschleunigungsband * *bei Bandförderer*
acceleration compensation Beschleunigungsaufschaltung || Beschleunigungsausgleich || Beschleunigungskompensation || Beschleunigungsvorsteuerung * →*Beschleunigungsausgleich* || Schwungmomentenkompensation || Schwungmomentkompensation
acceleration compensator Beschleunigungsaufschaltung * →*Beschleunigungsausgleich (als Einrichtg., Schaltg., Funktionsbaustein)* || Beschleunigungsausgleich * *als Schaltung, Funktionsbaustein, Einrichtung*
acceleration distance Beschleunigungsweg
acceleration due to gravity Fallbeschleunigung
acceleration feed forward Beschleunigungsaufschaltung * *Beschleunigungsvorsteuerung*
acceleration figure Beschleunigungszahl * *[el. Masch.] Walzwerk] bei Rollgangsmotor*
acceleration force Beschleunigungskraft
acceleration power Beschleunigungsleistung
acceleration process Beschleunigungsvorgang
acceleration ramp Hochlauframpe
acceleration rate Beschleunigung * *wertmäßig*
acceleration sensor Beschleunigungsaufnehmer
acceleration time Anlaufzeit * *(mechan.) Hochlaufzeit* || Beschleunigungszeit || Hochlaufzeit * *auch mechanische Hochlaufzeit e. Arbeitsmaschine bei konstantem Antriebsmoment*
acceleration torque Beschleunigungsmoment
acceleration transducer Beschleunigungsaufnehmer
accelerometer Beschleunigungsaufnehmer || Beschleunigungsmesser
accentuate anheben * *anheben (z.B. Frequenzkurve), hervorheben, akzentuieren, betonen* || hervor-

heben * *hervorheben, Wertlegen auf, betonen, akzentuieren*
accept abnehmen * *(offiziell) anerkennen, annehmen, akzeptieren* || anerkennen * *als gültig* ~ || annehmen || aufnehmen * *annehmen; auch eine Baugruppe/ein IC in e. Einbauplatz* || billigen * *akzeptieren, anerkennen, gelten lassen, bejahen, hin/annehmen, sich mit ... abfinden* || empfangen * *annehmen* || hereinnehmen * *entgegen/annehmen, z.B. Auftrag* || hinnehmen * *annehmen* || quittieren || übernehmen * *akzeptieren, annehmen, Pflicht/Ware/Gewährleistung/Garantie* ~
accept button Quittiertaste
accept the responsibility Verantwortung übernehmen
acceptable akzeptabel * →*tragbar; (to: für)* || annehmbar * *to:für* || tragbar * *annehmbar* || zulässig * *annehmbar, tragbar (to: für), willkommen* || zumutbar * *annehmbar*
acceptance Abnahme * *z.B. Abnahmeprüfung eines Produktes durch Kunden oder Prüfinstitut* || Akquisition * *Hereinnahme von Aufträgen* || Akzeptanz * *enjoy a high level of acceptance:eine hohe Akzeptanz haben* || Annahme * *Empfang; auch figürlich; Akzeptanz* || Genehmigung * *Billigung (of:von/des)* || Quittierung || Übernahme * *Abnahme, Annahme, Empfang; acceptance test: Abnahmeprüfung* || Zusage * *Annahme*
acceptance certification Bescheinigung * *über eine Abnahmeprüfung z.b. durch Prüfbehörde (z.B. TÜV, PTB)*
acceptance of orders Bestellannahme
acceptance of tender Zuschlag * ~ *bei einer Ausschreibung; obtain the contract: den* ~ *erhalten*
acceptance procedure Annahmeverfahren
acceptance report Abnahmebericht
acceptance signal Rückmeldung * *Annahme*
acceptance test Abnahme * *Abnahmeprüfung durch Kunden bzw. Prüfinstitut/Kontrollbehörde (z.B. TÜV, PTB)* || Abnahmeprüfung || Annahmeprüfung || Eingangsprüfung * *Abnahmeprüfung bestellter Waren*
acceptance test report Abnahmeprüfprotokoll
acceptance-test certificate Abnahmeprüfzeugnis
acceptation Bedeutung * *gebräuchlicher Sinn (eines Wortes)*
acceptor Akzeptor
acceptor circuit Saugkreis
accesories Zubehörteile
access Ansprechen * *zugreifen auf* || eingreifen * *zugreifen auf* || Zugang * *(auch figürl.) Zugriff, Zugangsmöglichkeit; unobstructed: ungehindert* || zugreifen * ~; *access a memory location: auf eine Speicherzelle* ~ || zugreifen auf || Zugriff * *z.B. auf Speicher (to:zu/auf/bei)*
access authorization Zugriffsberechtigung || Zugriffsrecht * *z.B. auf Speicher, Bus, Datenpuffer usw.*
access collision Zugriffskonflikt * *z.B. bei gleichzeitigem Speicherzugriff von 2 Teilnehmern*
access hole Montageöffnung * *siehe auch* →*Wartungsöffnung* || Wartungsöffnung
access level Zugriffstufe
access mechanism Zugriffsverfahren
access method Zugriffsverfahren
access priority Zugriffspriorität
access protection Zugriffsschutz * *z.B.gegen unbefugte Parameteränderung*
access right Zugangsberechtigung || Zugriffsberechtigung
access rights Zugriffsrechte * *(PROFIBUS-Profil)*
access stage Zugriffsstufe
access technique Zugriffsverfahren
access time Zugriffszeit * *auf Speicher*
accessibility Zugänglichkeit

accessible frei herausgeführt * *with accessible cable ends: mit ~n Leitungen* || offenliegend * *auch von spannungsführenden Teile* || zugängig * *to: für* || zugänglich * *to: für; easily/free: leicht/frei; front-accessible: von vorn ~* || zugreifbar * siehe auch →*zugänglich,* →*zugänglich machen*
accessories Sonderzubehör || Zubehör
accessories for testing Prüfzubehör
accident Unfall
accident precautions Unfallschutz
accident prevention Unfallverhütung
accident prevention legislation Unfallverhütungsvorschriften * *gesetzliche ~*
accident prevention regulation Unfallverhütungsvorschrift
accident prevention regulations Unfallverhütungsvorschriften
accidental unabsichtlich * *zufällig* || unbeabsichtigt * *zufällig, unbeabsichtigt* || versehentlich * *zufällig* || zufällig
accommodate aufnehmen * *räumlich ~ (z.B. Zusatzbaugruppe), beherbergen* || entgegenkommen * *z.B. jemanden bei einer Verhandlung* || unterbringen * *(räumlich) auch 'beherbergen', auch von Komponenten/Geräten*
accommodated untergebracht * *z.b. in einem Gehäuse* || zugeschnitten
accommodating kulant * *{Handelswesen} entgegenkommend, gefällig; anpassungsfähig*
accommodation Anpassung * *mehrere Dinge aneinander* || Aufnahme * *Beherbergung, Unterkunft, Platz für...* || Einbau * *Unterbringung*
accomodate fassen * *unterbringen/aufnehmen können*
accomodation Kulanz * *gütliche Einigung, Gefälligkeit*
accompaniment Begleiterscheinung || Nebeneffekt * *Begleiterscheinung*
accompanying zugehörig * *begleitend*
accompanying documents Begleitpapiere
accompanying paperwork Begleitpapiere * *z.B. Versandpapiere*
accompanying the einhergehend mit
accomplish bewerkstelligen || erfüllen * *Aufgabe* || leisten * *vollbringen*
accomplished perfekt * *vollbracht, ausgeführt, erreicht*
accomplishment Zustandekomen * *Vollendung, Ausführung, Erfüllung, Bildung*
accord Abkommen || gewähren * *bewilligen* || Übereinstimmung * *Einigkeit, Zustimmung, Übereinkommen, Vergleich*
accordance Übereinstimmung * *in accord. with: in ~ mit; according to: gemäß; be i. accord. with: übereinst. mit*
according as je nachdem
according to entsprechend * *(Adv.) gemäß (auch 'accordingly')* || gemäß || je nach || je nachdem || laut * *entsprechend, gemäß* || nach * *entsprechend/gemäß einer Norm, Vorschrift, Beschreibung, Priorität, Reihenfolge*
according to a sine function sinusförmig
according to an exponential function exponentiell * *entspr. e. Exponentialfunktion*
according to directions auftragsgemäß * *anweisungsgemäß*
according to instructions auftragsgemäß * *den Anweisungen entsprechend*
according to it danach * *gemäß*
according to one's wishes wunschgemäß * *(Adv.)*
according to requirement bedarfsgerecht
according to rule schematisch * *nach Schema, entsprechend einer Regel/Vorschrift*
according to schedule fristgerecht * termingerecht || termingerecht || zeitgerecht * *(Adv.) termingerecht, wie geplant*
according to the effective expenses nach Aufwand * *invoice: verrechnen*
according to the market requirements marktgerecht * *den Marktbedürfnissen entsprechend*
accordingly danach * *entsprechend* || infolgedessen * *folglich, demgemäß, entsprechend* || sinngemäß * *(Adv.)*
account Abrechnung * *Rechnung* || Anschlag * *Betracht; take into (leave out of) account:(nicht) in ~ bringen* || Kontierung * *Konto* || Konto || Rechnung * *Waren~, Be~, In~-~-Stellen, (Ausgaben)Ab~* || Verrechnung * *Berechnung; only for account/not negotiable: nur zur ~ (auf Scheck)* || Volumen * *Wert(igkeit), Wichtigkeit, Gewinn, Bedeutung (z.B. e. Kunden)* || Wertigkeit * *Wert, Wichtigkeit, Bedeutung, z.B. eines Kunden*
accountable verantwortlich * *verantwortlich, rechenschaftspflichtig; erklärlich*
accountancy Buchhaltung * *Buchhaltung* || Buchführung * *Buchhaltung* || Rechnungswesen
accounting Buchführung * *Rechnungswesen* || Rechnungsprüfung * *Rechnungswesen* || Rechnungswesen * *verantwortlich * accounting for: ~ sein für, Rechnung/Rechenschaft ableg. über; erklärend,begründend*
accounting center Kostenstelle * *Abrechnungsstelle*
accounts payable Verbindlichkeiten
accounts receivable Außenstände
accredit akkreditieren * *offiziel/als berechtigt anerkennen/bestätigen/beglaubigen*
accretion Zuwachs * *Zunahme, Zuwachs, Anwachsen, Hinzugekommenes, Hinzufügung, Wertzuwachs*
accrual Ansammlung
accrue anfallen * *Gewinn, Zinsen usw.* || eintreten * *Versicherungs/Gewährleistungsfall* || entstehen * *Kosten usw. (from:durch)*
accumulate akkumulieren || speichern * *ansammeln; auch in Ergebnisregister o. Warenspeicher für durchlaufende Warenbahnen* || summieren * *sich auf~, sich ansammeln, sich aufakkumulieren*
accumulated number Summe * *aufsummierte Anzahl*
accumulating conveyor Stauförderer * *{Fördertechnik} mit Produktespeicher-Funktion*
accumulation Ablagerung || Anhäufung || Ansammlung || Haufen * *Ansammlung, Häufung* || Häufung * *(auch figürl.)* || Speicherung * *Ansammlung*
accumulator Akkumulator * *Ergebnisregister in Computer-Rechenwerk* || Speicher * *Zwischenspeicher, Warenspeicher zur Überbrückung d. Rollenwechselzeit bei Auf/Abwickler* || Warenbahnspeicher || Warenspeicher * *für durchlauf. Warenbahn zum Zwischenpuffern des Materials bei Rollenwechsel*
accumulator data Verknüpfungsergebnis
accuracy -Genauigkeit || Genauigkeit * *steady state accuracy: im eingeschwungenen Zustand* || Präzision * *Genauigkeit* || Sorgfalt * *Genauigkeit*
accuracy of indication Anzeigegenauigkeit
accuracy to size Maßhaltigkeit
accurate exakt * *genau* || genau * *with centimeter accuracy: zentimeter~* || genau || pünktlich * *genau* || richtig * *genau*
accurate synchronism winkelgetreuer Gleichlauf || Winkelgleichlauf
accurate synchronism control Gleichlaufregelung * *Winkelgleichlauf-Regelung*
accurately taking the aim zielgenau
ace Könner * *(salopp) Ass*
acetylene welding Autogenschweißen
achievable erreichbar || erzielbar

achieve erreichen * hingelangen, erzielen || erzielen * zustandebringen, ausführen, (Ziel) erreichen, (Erfolg) erzielen || leisten * vollbringen
achievement Erfolg || erreichen * (Subst.) Erreichung eines Ziels usw. || Leistung * Werk, Leistung, Errungenschaft, Großtat, Ausführung, Vollendung, das Vollbrachte
acid Säure * (chem.)
acid concentration Säurekonzentration
acid spills Säurespritzer
acid-resistant säurebeständig
acknowledge anerkennen || bestätigen * den Empfang bestätigen, quittieren || quittieren * z.B. Fehler, Telegramm || rückmelden * Empfang bestätigen/quittieren || Rückmeldesignal * Quittung, Bestätigung || Rückmeldung * Quittungssignal, Qittungstelegramm
acknowledge pushbutton Quittiertaste
acknowledge signal Quittiersignal || Quittungssignal
acknowledged quittiert
acknowledgement Anerkennung || Bestätigung || Empfangsbestätigung * Qittierung, auch für Datentelegramm || Quittierung || Quittung * Signal, Telegramm || Rückmeldung * Quittung
acknowledgement signal Rückmeldung * Quittungssignal
acknowledgement time Quittierzeit
acknowledgement timeout Quittungsverzug
acknowledging Quittierung * als Tätigkeit oder Vorgang
acme thread Trapezgewinde * z.b. bei Spindeltrieb, Hubspindel
acme-screw spindle Trapezgewindespindel * z.B. bei Spindeltrieb
acoustic akustisch
acoustic coupler Akustikkoppler * zur Datenübertragung über Telefon
acquaint informieren * with: über; mitteilen (with a thing: etwas, that: dass)
acquaint oneself lernen * sich vertraut machen mit (with: mit)
acquaint oneself with sich vertraut machen mit
acquainted bekannt * be acquainted with: mit etwas/jmd. bekannt sein
acquire anschaffen || bekommen * erwerben || erarbeiten * z.B. Wissensstoff, Problemlösung ~ || erfassen * z.B. Istwerte, Messwerte, (Analog-) Signale || erlangen || erreichen * erlangen || erschließen * erwerben, erlangen, erreichen, gewinnen || erwerben * erwerben, erlangen, gewinnen, bekommen, (er-) lernen, erfassen (z.B. Messwerte) || gewinnen * erlangen, erreichen || kaufen * erwerben || lernen * erlernen, meistern
acquire a strong image profilieren * (figürl.: sich ~, e. guten Eindruck/gute Figur machen)
acquiring Erwerb
acquisition -Erfassung * ~ von Messdaten || Akquisition * (käuflicher) Erwerb, (An-) Kauf, Erwerbung, Erfassen, Aneignung || Beschaffung || Erfassung * z.B. von Messwerten || Erwerb || Kauf * Erwerb, (An)Kauf, (Neu)Anschaffung, das Erworbene
acquisition system Erfassungssystem * für Messwerte
acquisitional argumentation booklet Argumentenheft
acronym Abkürzung * (kryptisches) Kürzel
across parallel * connect across: ~schalten zu || über * quer, hin~, dr~ hinweg
act agieren * as:als || angreifen * z.B. eine Kraft an einem Punkt || arbeiten * handeln, tätig sein, wirken, (eingreifen || betätigen * sich betätigen (as: als) || dienen * fungieren, arbeiten, tätig sein, agieren, wirken, eingreifen (as:als) || eingreifen * on:

in/auf || einwirken * (up)on: auf; through:durch; wirksam werden, be-/hinwirken(for: auf), verursach., betätigen || fungieren * as:als || handeln || Handlung || ist * Kurzform für Istwert || vorgehen * handeln; in conformity with: gemäß/in Übereinstimmung mit || wirken * eingreifen, arbeiten, fungieren, agieren, dienen (as:als)
act as arbeiten * fungieren/agieren/amtieren/dienen als || dienen als * fungieren/agieren/arbeiten/ tätig sein/eingreifen als || versehen * Amt ~; fungieren
act as a mediator vermitteln * als Vermittler fungieren (in:bei)
act for vertreten * im Amt ~
act in conformity with sich halten an
act in opposition to entgegenwirken * z.B. einer Kraft || widersetzen * z.B. sich einer Kraft ~
act on wirken auf
act reversely zurücklaufen * in Gegenrichtung wirken
act upon bearbeiten * mit etw. befasst sein || beaufschlagen
act. current Stromistwert
act. value Istwert
acting on behalf Wahrnehmung * of a person: der Interessen e. Person
action -Wirkung * Funktionieren, (Ein-)Wirkung, Wirksamkeit, Einfluss, Handlung, Tätigkeit, Verrichtung || Aktion * Handeln, Handlung, Tat, Tätigkeit, Gang, Funktionsweise; enter into action:in Aktion treten || Einfluss * (Ein-)Wirkung, Wirksamkeit, Einfluss, Vorgang, Prozess, Handeln, Tätigkeit || eingreifen * (Substantiv) figürlich || Einsatz * be in action: im ~ sein; be brought into action: zum ~ kommen || Einwirkung || Gang * Wirkungsweise || Handlung * das Handeln || Lauf * Gang, Funktionieren, Vorgang, Bewegung, Gangart, Handeln, Einwirkung || Maßnahme * take:ergreifen || Prozess * Klage || Tätigkeit || Vorgang * Aktion, Handlung || Vorgehensweise * Handlung,Aktion || Wirkung * Tätigkeit
activate aktivieren || anregen || ansteuern || anstoßen * aktivieren || auslösen * aktivieren (auch Stör/ Warnmeldung) || einschalten * aktivieren || scharfmachen
activated aktiviert
activation Aktivierung || Ansteuerung
active aktiv || scharf * aktiv, z.B. Überwachungseinrichtung
active carbon filter Aktivkohlefilter * z.B. in der Wasser-/Abwasseraufbereitung
active current Wirkstrom * derj. Anteil e. Wechselstroms, d. mit d. Spannung phasengleich ist (I_eff x cos phi)
active current closed-loop control Wirkstromregelung
active current control Wirkstromregelung
active current signal Stromsignal * z.B. 0–20 mA
active field Klickfeld * auf dem Bildschirm
active filter aktives Filter
active front end Ein-Rückspeiseeinheit * selbstgeführt. Umkehrstromrichter z. Zwischenkreispeis. (z.B. m. Transistoren)
active parts Aktivteile * {el. Masch.}
active power Wirkleistung * nutzbarer Leistungsanteil b. Wechselstromsystem: P entspr. SQRT(3) x U x I x cos (phi)
active resistance Wirkwiderstand
active state aktiver Zustand
active-power measurement Leistungsmessung * Wirk~
activity Aktion * Tätigkeit, Betätigung, Unternehmung, Rührigkeit, Wirksamkeit, Betrieb || Aktivität || Fach * Tätigkeit, Betätigung(sfeld) || Projekt

* *Aktivität, Betätigung, Tätigkeit, Rührigkeit; activities: Unternehmungen* || Tätigkeit
acts of God   höhere Gewalt
actual   aktuell * *wirklich, tatsächlich* || *betriebsmäßig* * *tatsächlich auftretend* || *eigentlich* * *wirklich, tatsächlich* || Ist- * *z.B. Istwert, Istfrequenz, Istdrehzahl* || *jetzig* * *aktuell* || *konkret* * *wirklich, tatsächlich, eigentlich* || *momentan* * *(Adj.) aktuell, gegenwärtig, jetzig* || *tatsächlich* || *wirklich*
actual condition   Istzustand
actual current   Stromistwert
actual current measuring circuit   Stromistwerterfassung * *Messschaltung*
actual frequency   Frequenzistwert
actual parameter   Aktualoperand * *bei SPS*
actual position   Positionsistwert
actual speed   Drehzahlistwert
actual speed sensing   Drehzahlistwerterfassung
actual speed value   Drehzahlistwert
actual speed value acquisition   Drehzahlistwerterfassung
actual value   aktueller Wert || Istwert * *['äktjuel 'wälju]*
actual value acquisition   Istwerterfassung
actual value channel   Istwertkanal
actual value conditioning   Istweraufbereitung || Istwertaufbereitung
actual value encoder   Istwertgeber * *digitaler Drehzahl~*
actual value evaluator   Istwerterfassung * *als Einrichtung*
actual value filtering   Istwertglättung || Istwertsiebung
actual value information   Istwertinformation
actual value matching circuit   Istwertanpassung * *hardwaremäßig*
actual value of speed   Drehzahlistwert
actual value polarity   Istwertpolarität
actual value polarity reversal   Istwertumpolung
actual value sensing   Istwertabfrage || Istwerterfassung
actual value sensor   Istwertgeber
actual value smoothing   Istwertglättung || Istwertsiebung * *Istwertglättung*
actual value transducer   Istwertgeber
actual value transmitter   Istwertgeber
actual-current sensing   Stromistwerterfassung
actual-current sensing circuit   Stromistwerterfassung * *als Schaltung/Einrichtung*
actual-speed signal adapter   Drehzahlistwertanpassung * *als Einrichtung/Gerät*
actual-speed value   Drehzahlistwert
actual-value adjustment   Istwertabgleich
actual-value cable   Istwertleitung * *[Regel.techn]*.
actual-value estimation   Istwertschätzung
actual-value input   Istwerteingang
actual-value matching   Istwertabgleich || Istwertanpassung
actuality   Wirklichkeit
actualize   aktualisieren
actually   eigentlich * *(Adv.) tatsächlich* || *tatsächlich* * *(Adv.)* || *wirklich* * *(Adv.)*
actuate   ansteuern * *betätigen, z.B. Magnetventil* || *antreiben* * *betätigen, auslösen, anstoßen* || *betätigen* * *auch Taste, Bremse, Kupplung* || *drücken* * *betätigen (z.B. eine Taste)* || *in Betrieb setzen* * *in Gang bringen* || *inbetriebnehmen* * *in Gang bringen* || *inbetriebsetzen* * *in Gang bringen*
actuated   betätigt * *hydraulically: hydraulisch; pneumatically: pneumatisch; electromagnetically: elektromagnet.*
actuating drive   Stellantrieb
actuating element   Bedienelement * *an einem Stellglied/Aktutor* || Bedienungselement * *an einem Stellglied/Aktuator*

actuating force   Stellkraft
actuating magnet   Betätigungsmagnet
actuating mechanism   Betätigungsmechanismus
actuating motor   Stellmotor
actuating shaft   Antriebswelle * *Schalter, Relais*
actuating solenoid   Betätigungsmagnet
actuating system   Antriebssystem * *~ für Schalter, Relais*
actuating unit   Betätigungseinheit
actuating variable   Stellgröße
actuation   Betätigung
actuation method   Betätigungsart * *z.B. bei Kupplungen*
actuation unit   Betätigungseinheit
actuator   Aktor || Aktuator * *Stellglied* || Antrieb * *Stell~* || Bedienelement * *Stell-/Betätigungsglied, Aktuator* || Bedienungselement * *Aktuator, Stellglied, Betätigungselement* || Servoantrieb * *Stellantrieb* || Stellantrieb * *z.B. für Ventil, Schieber, Klappe* || Steller * *Stellglied* || Stellgerät || Stellglied
actuator control unit   Stellgerät
actuator ring   Schaltring * *z.B. für mechan. betätigte Kupplung*
actuator stem   Antriebsspindel * *eines Ventils*
actuator system   Stellsystem
acute   aktuell * *Problem* || *scharf* * *scharf, spitz, spitzwinklig, heftig, stechend* || *spitz* * *(Winkel usw.)*
acute angle   spitzer Winkel
acute-angled   spitzwinklig
acyclic   azyklisch
acyclical   azyklisch
acyclically   azyklisch * *(Adv.)*
Ad   Anzeige * *(salopp) Abkürzung für 'advertisement': Werbe~* || Inserat * *(salopp) Abkürzg. für 'advertisement'* || Werbung * *Abkürzg. für advertising*
AD converter   ADU
ad valorem   wertmäßig * *(Adj. und Adv.) kaufmännisch*
adapt   adaptieren || anpassen * *auch 'sich ~' (to:an); adapt oneself to: sich ~ an*
adapt oneself to   sich einfügen * *Person*
adaptability   Änderbarkeit * *Anpassbarkeit* || Anpassbarkeit || Anpassungsfähigkeit
adaptable   anpassbar || anpassfähig * *anpassungsfähig*
adapted   adaptiert || angepasst || zugeschnitten
adapter   Adapter || Anschaltung * *[Computer] für Monitor, Netzwerk usw.* || Einsatz * *(An-) Passstück* || Zwischen- * *zur Anpassung dienend*
adapter board   Adapterbaugruppe * *Leiterkarte*
adapter cable   Adapterkabel
adapter casing   Adaptionskapsel
adapter connector   Adapterstecker
adapter flange   Zwischenflansch
adapter housing   Zwischengehäuse
adapter module   Adapterbaugruppe || Adaptionskapsel
adapter plate   Adapterplatte
adapter plug   Übergangsstecker * *Adapterstecker*
adapter sleeve   Reduzierstück * *Reduzierhülse* || Spannhülse
adapting circuit   Anpassschaltung
adaption   -Anpassung || Adaption || Anpassung * *(to: an)*
adaption function   Adaption
adaption module   Anpassbaugruppe * *siehe auch →Anpassmodul*
adaption to the line frequency   Netzfrequenzanpassung * *Netzfrequ.nachführg. d. Steuersatzsynchronisation e. netzgef.Stromr.* || Netzfrequenznachführung * *z.B. automat. Netzfrequenznachführung bei netzgeführtem Stromrichter*
adaptive control   adaptive Regelung
adaptive controller   Adaptivregler

adaptive regulator adaptiver Regler
adaptor Adapter || Anpasselektronik || Anpassstufe || Umsetzer
adaptor board Anpassbaugruppe * Leiterkarte
adaptor cable Adapterkabel
adaptor connector Adapterstecker
adaptor flange Zwischenflansch
adaptor module Anpassbaugruppe * Baustein, Modul
adaptor plug Adapterstecker
ADC ADU * Abkürz. f. 'Analog to Digital Converter': Analog/Digital-Umsetzer || Analog-Digital-Umsetzer || Analog-Digital-Wandler
add addieren || anbauen * Gebäude || anschließen * anfügen || ansetzen * anfügen || beifügen || beipacken * hinzufügen || ergänzen * hinzufügen || erweitern * erweitern um, hinzufügen || hinzufügen || zusetzen * →hinzufügen
add to vergrößern || vermehren * beitragen zu
add-on Zubehör * Zusatzeinrichtung || Zusatz * Zubehör (-Teil) || Zusatz- * Zubehör
add-on card Zusatzbaugruppe * Leiterkarte
add-on kit Nachrüstsatz
added hoch * vermehrt, erhöht, zusätzlich || Zusatz- * vermehrt, erhöht, zusätzlich || zusätzlich
added cost Zusatzaufwand || Zusatzkosten
added value Mehrwert || Zusatznutzen
addendum Nachtrag || Zusatz * zu einem Schriftstück
adder Addierer || Additionsglied || Summierer * Addierer || Summierglied * Addierer
adding additiv
addition Summe * Addition || Zusatz || Zuschlag
additional extra * zusätzlich || Mehr- * Zusatz- || nachträglich * ergänzend || Neben- * zusätzlich || neu * zusätzlich, weiterer || optional || übrig * zusätzlich || weitere * zusätzliche || Zusatz- * z.B. additional information: Zusatzinformation
additional board Zusatzbaugruppe * Leiterplatte
additional charge Aufschlag * zusätzliche Berechnung, Verrechnung, Gebühr || Mehrpreis * zusätzliche Berechnung || Preiszuschlag || Zuschlag * zusätzliche Gebühr, Berechnung
additional cost Mehrkosten
additional costs Mehrkosten || Zusatzaufwand * Zusatzkosten; incur additional costs: Zusatzkosten auf sich nehmen (müssen) || Zusatzkosten * incur additional costs: ~ auf sich nehmen (müssen)
additional data Zusatzangaben
additional efforts Mehraufwand || Zusatzaufwand
additional equipment Zusatzausstattung
additional expense Mehraufwand || Mehrkosten
additional features zusätzliche Merkmale
additional function Zusatzfunktion
additional information Zusatzangaben
additional ingredients Zusatzstoffe
additional losses Zusatzverluste
additional measures Zusatzmaßnahmen
additional module Zusatzbaugruppe
additional noise Zusatzgeräusche
additional price Aufpreis || Mehrpreis * no:ohne; at:gegen
additional reference value Hilfssollwert
additional rotor losses Läuferzusatzverluste
additional setpoint Zusatzsollwert
additional terminal modules Klemmenerweiterung
additional value Mehrwert || Zusatzwert
additive additiv * (Adjektiv), Zusatz (Substantiv) || Hilfsstoff || Zusatzstoff
additives Hilfsstoffe || Zusätze * z.B. Hilfsstoffe
additonal zusätzlich * (Adj.)
address Adresse || adressieren || Anschrift || Vortrag * Ansprache, Rede; deliver an address: eine Ansprache/Rede Halten || wenden * ansprechen, sich wenden an
address bus Adressbus
address decoding Adressdekodierung
address displacement Adressversatz
address error Adressfehler
address line Adressleitung
address map Adressliste * {Computer}
address of delivery Lieferadresse
address range Adressiervolumen
address space Adressiervolumen || Adressraum
address to richten an * z.B. Brief, Nachricht, Telegramm
addressable adressierbar
addressee Adressat || Empfänger * Adressat, z.B. eines Briefes
addressing Adressierung
addressing error Adressierfehler
addressing range Adressiervolumen
adequate ausreichend * passend, hinreichend, entsprechend, genügend (to:dem) || gut * angemessen || sachgemäß
adhere aneinander hängen * aneinandergehängt sein || haften * (an-) kleben; to: an || halten * adhere to: sich an etwas ~/fest~ an/treubleiben || verharren * to:bei
adhere to aufrechterhalten * festhalten an, bleiben bei, (Urteil usw.) bestätigen, etw./jdm. treu bleiben || befolgen * Grundsatz~, Regel ~ || beibehalten * Grundsatz beibehalten, sich halten an || einhalten * sich halten an, einhalten (z.B. an eine Regel) || festhalten * (figürl.) an Meinung, Grundsatz, Vorgehen usw. ~ || kleben * kleben bleiben || sich halten an * z.B. Betriebsanleitung, Vorschrift, Vertrag
adherence Befolgung * to:von || Einhaltung * to:von
adhesion Adhäsion || Haftung * Adhäsion || Kraftschluss * bei Rad-/Straße- bzw. Rad-/Schiene-System
adhesion lining Adhäsionsbelag * z.B. bei einem Treibriemen, z.B. aus Chromleder
adhesive Kleber || Klebstoff || kraftschlüssig * bei Rad-Straße/Rad-Schiene-System || Leim * Klebstoff || selbstklebend
adhesive application Beleimung * {Holz} Aufbringung v.Klebstoff, z.B. z.Furnier- oder Kantenanleimung
adhesive applicator Beleimmaschine * {Holz}
adhesive label Aufkleber || Selbstklebe-Etikette || Selbstklebeetikett
adhesive power Adhäsionskraft
adhesive strip Klebeband || Klebestreifen
adhesive symbol Haftbild
adhesive tape Klebeband
adhesiveness Adhäsionskraft
adjacent angrenzend * to:an || anschließend * räumlich || benachbart * to: zu/an || daneben * angrenzend/-liegend/-stoßend; benachbart || Nachbar- * anstoßend, benachbart || nebeneinander * →benachbart || nebenstehend
adjacent to benachbart * neben * angrenzend/-stoßend an; benachbart zu; Nachbar-
adjoin beilegen * z.B. in eine Verpackung || stoßen * an~, angrenzen || zuordnen * beiordnen/fügen
adjoining angrenzend * to:an || benachbart * angrenzend, aneinander stoßend || Nachbar- * Zimmer, Haus usw.
adjourn aufschieben * zeitlich || aussetzen * vertagen || verlegen * vertagen || verschieben * vertagen || vertagen
adjust Abgleich * Kurzform für Beschreibung e. Bedienelements || abgleichen * anpassen, einstellen, justieren || anpassen * einem Zweck ~; adjust oneself to: sich ~ an || ausregeln || ausrichten * (auch

figürl.: Benehmen usw.) || einstellen * *justieren; (auch figürl.: adjust oneself to: sich einstellen auf)* || justieren || korrigieren * *anpassen* || montieren * einstellen || regeln * *regulieren, anpassen, angleichen, einstellen* || regulieren * einstellen || verstellen
adjustable einstellbar || nachstellbar || regelbar || stellbar || verschiebbar * *einstellbar* || verstellbar
adjustable drive regelbarer Antrieb
adjustable factor Anpassfaktor
adjustable propeller drive Verstellpropellerantrieb
adjustable resistor Stellwiderstand
adjustable speed drive Antrieb mit Drehzahleinstellung || drehzahlveränderbarer Antrieb
adjustable-speed drehzahlgeregelt || drehzahlstellbar || drehzahlveränderbar
adjustable-speed drive geregelter Antrieb || regelbarer Antrieb
adjusted bemessen * *(Adj.)* || eingestellt || zugeschnitten * *to (meet): auf eine Norm, einen Zweck* ~
adjusting Einstell- * *z.B. ~schraube, ~knopf, ~hebel, ~markierung*
adjusting device Einstellgerät || Justiereinrichtung
adjusting gauge Einstelllehre
adjusting instructions Einstellanweisung
adjusting resistor Abgleichwiderstand
adjustment -Anpassung || Abgleich || Anpassung || Einstell- * *Einstell-, Regulier-* || Einstellung * ~ *von Parametern* || Justage || Regulierung * *Einstellung* || Verstellung * *Einstellung*
adjustment factor Anpassfaktor
adjustment instructions Einstellhinweise || Einstellvorschrift
adjustment knob Einstellknopf
adjustment lock Nachstellsicherung
adjustment range Einstellbereich
adjustments Einstellarbeiten
adman Werbefachmann * *(salopp)*
administer abwickeln * *verwalten, handhaben, durchführen* || versehen * *Amt* ~ || verwalten * *auch 'als Verwalter fungieren'*
administration Abwicklung * *verwaltungsmäßige* ~ || Bearbeitung * *Handhabung, Durchführung, Verabreichung* || Verwaltung
administration centre Niederlassung
administration of inquiries Angebotsbearbeitung
administration of orders Auftragsbearbeitung
administrative administrativ
administrator Administrator * *auch Verwalter eines Computernetzes*
admirably suited bestens geeignet
admissible gültig * *zulässig* || zulässig
admission Aufnahme * *Zulassung* || Eintritt * *zu einer Veranstaltung usw.* || Zulassung * *zu Prüfung/Studium/Amt usw.*
admit beaufschlagen * *zuführen, einlassen* || gewähren * *zulassen, erlauben* || zulassen * *(amtlich) anerkennen/zulassen/gewähren/gestatten*
admit of gelten lassen
admit to aufnehmen * *in einen Verein* ~
admittance Admittanz
admitting different interpretations interpretationsfähig
admonition Mahnung * *Ermahnung* || Warnung * *Ermahnung, Verweis*
adopt annehmen * *Antrag, Haltung* || einführen * *Maßnahmen usw. übernehmen, annehmen, einführen, ergreifen* || übernehmen * *Verfahren* ~
adoption Annahme * *Übernahme* || Übernahme * *(auch figürlich; Idee, Lösungsweg usw.) Annahme, Übernahme, Aneignung*
adulterate zusetzen * *durch Zusätze verfälschen, verschlechtern*

adulteration Mischung * *Verfälschung*
advance anziehen * *von Preisen usw.* || Aufschlag * *Erhöhung, Hochsetzung (Preis, Kurs usw.)* || erhöhen * *Preise usw.* ~ || fortschreiten || Fortschritt || Fortschritte machen || leihen * *(amerikan.) Geld usw. vorstrecken* || steigen || Vorab- || voraus * *Voraus-* || voraus- || vorgehen || Vorlauf || Weiterentwicklung || weiterschalten
advance angle Voreilwinkel * *[Stromrichter] siehe IEC 146-1-1*
advance to a higher position aufrücken * *im Rang* ~
advanced erhöht * *Preise usw.* || fortgeschritten || fortschrittlich || hochentwickelt || modern * *fortschrittlich* || neuzeitlich * *fortschrittlich*
advanced education Weiterbildung
advanced training Weiterbildung
advanced training course Fortgeschrittenenkurs || weiterführender Kurs
advancement Aufstieg * *sozialer* ~ || Beförderung * *im Rang* || fortschreiten * *(Substantiv)* || Fortschritt || Verbesserung * *Fortschrit, Wachstum, Vorankommen* || Weiterentwicklung
advantage nutzen * *(Substantiv) Vorteil* || Vorsprung * *Überlegenheit, Vorteil (of: gegenüber); have the adv. of s.o.: jdm. gegenüber i.Vorteil s.* || Vorteil * *over: gegenüber*
advantage for the customer Kundennutzen * *Vorteil für den Kunden*
advantageous günstig * *vorteilhaft* || gut * *vorteilhaft* || lohnend * *vorteilhaft* || nützlich * *vorteilhaft* || vorteilhaft
advantageously vorteilhaft * *(Adv.)*
advent erscheinen * *(Substantiv) das Erscheinen, das Kommen, die Ankunft*
adverse nachteilig * *ungünstig; affect adversely:* ~ *beeinflussen* || ungünstig * *nachteilig; affect adversely:* ~ *beeinflussen* || widrig * *ungünstig*
adverse effect Schmutzeffekt * *schädlicher/nachteiliger Effekt*
advert Anzeige * *(salopp) Werbeanzeige*
advertise bekanntmachen * *durch Werbung, in der Zeitung* || werben * *Werbung machen, z.B. über Inserate*
advertisement Anzeige * *Inserat, Werbeanzeige* || Bekanntgabe * *in den Medien* || Bekanntmachung * *in den Medien* || Inserat || Werbeanzeige || Werbung * *über Anzeigen*
advertisement poster Werbeplakat
advertisement slogan Werbespruch
advertisement text Werbetext
advertiser Inserent
advertising Reklame || werblich || Werbung * *Werbung, Reklame*
advertising agency Werbeagentur
advertising brochure Werbeschrift
advertising budget Werbeetat
advertising campaign Werbeaktion || Werbefeldzug
advertising gift Werbegeschenk
advertising leaflet Faltblatt * *Werbe-Faltblatt* || Werbeblatt || Werbeprospekt * *Faltblatt, Prospekt* || Werbeschrift * *Faltblatt, Prospekt*
advertising material Werbematerial || Werbemittel * *(brit.)* || Werbeunterlagen * *(brit.)*
advertising media Werbemittel
advertising publications Werbematerial * *Druckerzeugnisse* || Werbemittel * *(brit.)*
advertising specialist Werbefachmann
advertize werben * *Reklame machen, inserieren*
advertizer Inserent * *(amerikan.)*
advertizing brochure Werbeschrift * *(amerikan.)*
advertizing campaign Werbekampagne
advertizing documentation Werbeunterlagen
advertizing material Werbemittel * *(amerikan.)*

siehe auch 'Werbematerial' || Werbeunterlagen * (amerikan.)
advertizing publications Werbemittel * (amerikan.) Druckerzeugnisse
advice Beratung * Rat, Raterteilung || Hinweis * Rat, Richtlinie
advisable angebracht * ratsam || ratsam || zweckmäßig * ratsam
advise beraten || informieren * of: über; benachrichtigen (of: von, that: dass), beraten || raten * jemandem
adviser Ratgeber
advising Beratung * Beratung, Raterteilung, (An-) Empfehlung
advisor Berater
advisory board Beirat
advisory council Beirat
advisory service Beratung * als Dienstleistung, Dienststelle || Kundenberatung
advocate fördern * befürworten, eintreten für
ae ä
AEE AEE * 'Asociación Electrotécnica Espanola': Spanischer Elektrotechn.Verband (ähnlich VDE)
aerate belüften || lüften * be~
aerating plant Belüftungsanlage
aeration Belüftung || Lüftung
aerator Belüfter * z.B. bei Wasseraufbereitungsanlage
aerial Antenne
aerial railway Seilbahn
aerial ropeway Seilbahn
aerodynamic aerodynamisch
aerodynamic resistance Luftwiderstand || Strömungswiderstand * →Luftwiderstand
aerospace Luft- und Raumfahrt
affair Angelegenheit || Ereignis * Angelegenheit || Fall * Angelegenheit || Gegenstand * Angelegenheit || Geschäft * Angelegenheit
affect auswirken * sich auswirken auf, beeinflussen || beeinflussen * einwirken auf; oft negativ || beinträchtigen || berühren * betreffen, (ein-) wirken auf, beeinflussen, beeinträchtigen, bewegen || betreffen || einwirken * angreifen, einwirken auf, beeinflussen, beeinträchtigen || schädigen * auch Rechte, Ruf usw. || tangieren * beeinflussen || wirken * beeinflussen
affect adversely beeinträchtigen
affected betroffen * auch 'beeinflusst, beeinträchtigt'
affected with -behaftet * (vorangestellt)
affiliated zugehörig * angegliedert, angeschlossen, verknüpft, verbunden, Tochter-, Zweig-, Filial-
affiliation Mitgliedschaft * Zugehörigkeit, Angliederg., Zusammenschluss, Mitgliedschaft; Aufnahme (als Mitglied)
affirm versichern * beteuern
affirmation Aussage * Behauptung, Bestätigung, Beteuerung || Versicherung * Behauptung, Versicherung, Bestätigung, Bejahung, Beteuerung
affix anbringen * befestigen, anbringen, anheften, anklebеn, beilegen, hinzufügen (to:an) || anhängen || befestigen * befestigen, anbringen, anheften, ankleben || hinzufügen * beiheften/-kleben, beilegen || setzen * befestigen, anbringen, anheften, ankleben, beilegen, hinzufügen
afflicted betroffen * belastet, behaftet, geplagt, leidend
affluence Überfluss * Fülle, Überfluss, Reichtum, Wohlstand
afford bieten * gewähren, liefern || gewähren * gewähren, spenden || leisten * sich; auch: sich ~ können/die Mittel haben für
affordable erschwinglich || preisgünstig * erschwinglich
AFNOR AFNOR * Association Francaise de Normalisation': französischer Normenverband ähnlich DIN/VDE
afore mentioned oben genannt
aforementioned oben erwähnt || obengenannt
aforesaid oben genannt
after hinter * in der (Reihen/Rang) Folge, z.B. des Signalflusses || nach * hinter (bezügl. Reihenfolge, Zeit usw.); bezüglich
after all endlich * (Adv.) endlich doch || schließlich * (Adv.) schießlich doch, schießlich und endlich || schließlich * (salopp) schließlich und endlich
after limiting nach der Begrenzung
after ramp hinter dem Hochlaufgeber
after that danach
after the ramp-function generator hinter dem Hochlaufgeber
after-effect Folge * Nachwirkung, ~erscheinung || Folgeerscheinung || Nachwirkung
after-event history Nachgeschichte * Inhalt eines Tracespeichers nach dem Triggerzeitpunkt
after-sales service Kundendienst * Kundenbetreuung nach Kauf e.Produkts(Service, Reparatur, Ersatzteilversorgung)
aftermath Folge * ernste ~
afterwards danach * später
again erneut * (Adv.)
against an || entgegen * Gegensatz, Richtung || gegen * ~sätzlich || neben * verglichen mit || vor * schützen, verstecken, warnen
against all expectations wider Erwarten
against each other gegeneinander
against it dagegen * (Adv.)
against one another aufeinander * gegeneinander || gegeneinander
age altern * Maschinenteil || verschleißen * altern || Zeit * ~alter
aged alt * bejahrt, gealtert (auch technisch)
ageing Alterung || Verschleiß * Alterung
ageing effects Alterungserscheinungen
ageing phenomenon Alterungserscheinungen
ageing symptoms Alterungserscheinungen
agency Geschäftsstelle || Verkaufsbüro || Vertretung * Industrie~, Handels~
agenda Tagesordnung
agent -Mittel * Wirkstoff; z.B.Kühlmittel || Mittel * Wirkstoff || Stoff * Mittel, Wirkstoff || Substanz * Wirkstoff || Vermittler * von Geschäftskontakten || Vertreter * Handels~ || Vertretung * Vertreter
agglomerate zusammenführen * zusammenballen
agglomeration Ansammlung * Anhäufung, Zusammenballung
aggravate erschweren * erschweren, verschlimmern, verschärfen, verstärken
aggravating gravierend * (auch jurist.) erschwerend
aggravation Steigerung * Erschwerung, Verschlimmerung, Verschärfung
aggregate Aggregat || Anhäufung * Anhäufung, (Gesamt)Menge, Summe || betragen * mehrere Posten || Gesamt- * angehäuftes, angesammeltes, vereinigtes || Gesamtmenge || Inbegriff * Summe, Anhäufung, Zusammenfassung || Maschinenteil || Menge * auch Gesamt~, Anhäufung || Obermenge || Vereinigungsmenge
aggregate figure Gesamtanzahl
aggregates Zusatzstoffe
aggregating sich summierend
aggregation Anhäufung
aggressive aggressiv
aggressive gases aggressive Gase
aggressive substance Schadstoff
aggressive substances aggressive Medien || Schadstoffe
aggressively verstärkt * (Adv.) aggressiv, "bissig", more aggressively: mit ~em Einsatz/"Biss"

agile beweglich * *(figürl.) behände, wendig, beweglich, agil*
aging Alterung
aging effects Alterungserscheinungen
aging phenomenon Alterungserscheinungen
aging resistance Alterungsbeständigkeit
aging symptoms Alterungserscheinungen
agitate mischen * *(um)rühren, schütteln, hin und her bewegen* || rühren || schütteln
agitator Rührwerk
ago vor * ~ *soundso langer Zeit, nachgestellt*
agree absprechen * *verabreden* || ausmachen * *vereinbaren* || einverstanden sein * *to:mit* || übereinstimmen * *with a person on: mit jemandem ~ über* || verpflichten * *to: einwilligen in, z.B. in eine Vertragsbestimmung* || zusammenpassen
agree to billigen || einlassen * *sich auf Vorschlag usw.* ~ || genehmigen * *gutheißen*
agree together passen * *zusammenpassen*
agree upon ausmachen * *vereinbaren, verabreden, sich einigen auf/über* || vereinbaren
agree upon a contract Vertrag schließen
agreeable angenehm * *gefällig, liebenswürdig, von angenehmem Wesen* || günstig * *befriedigend*
agreement Abkommen * *Vereinbarung (make/enter into:treffen)* || Abmachung || Festlegung * *Abkommen, Vereinbarung, Übereinkunft, Vertrag* || Übereinkunft || Übereinstimmung * *Übereinst., Einklang, Abkommen, Vereinbar., Vertrag; in agreem. with: in ~ mit* || Vereinbarung * *Abmachung; make an agreement: e. ~ treffen* || Vertrag
agreement of understanding Absichtserklärung || Vorvertrag
agricultural industry Landwirtschaft
agriculture Landwirtschaft
AGV führerloses Fahrzeug * *Abkürzg. für 'Automated-Guided Vehicle'*
Ah Ah * *Amperestunde*
ahead nach vorne * *vorwärts* || voraus || voraus- || vorn || vorwärts * *vorwärts, voran, voraus*
ahead of vor * *~weg*
ahead of time in voraus
AI künstliche Intelligenz * *Abk. für 'Artificial Intelligence'*
aid Hilfe * *with the aid of:mit Hilfe von* || Hilfsmittel * *Hilfe, Hilfsmittel, Hilfsgerät, Mittel* || Mithilfe || unterstützen * *helfen* || Unterstützung * *Hilfe, auch finanzielle ~*
aided -gestützt
aids Arbeitsmittel * →*Hilfsmittel* || Hilfsmittel * *(Plural)*
aim streben * *to: nach* || Ziel * *Richtung, Zweck, Absicht; with the aim of/aimed at: mit dem ~* || zielen * *at:auf* || Zweck * *Ziel*
aim at anstreben || betreiben * *hinarbeiten auf*
aim for anstreben * *beabsichtigen, im Sinne haben, hinzielen auf*
aim of development Entwicklungsziel
aimed at mit dem Ziel
aims Zielsetzung
air Luft || lüften
air and creepage distances Luft- und Kriechstrecke * *kürzester Abstand in der Luft u. auf Isolierfläche zw. leitenden Teilen*
air baffle Luftleitblech
air baffle plate Luftleitblech
air brake Druckluftbremse
air bubble Luftblase
air circulation Luftzirkulation
air clearance Luftabstand || Luftstrecke * *für Isolation*
air compressor Luftverdichter
air conditioner Klimaanlage * *Gerät* || Klimagerät
air conditioning Klimaanlage || Klimatechnik || Klimatisierung

air conditioning and refrigeration Klima- und Kältetechnik
air conditioning system Klimaanlage
air consumption Luftbedarf * *eines Druckluftsystems*
air convection Luftstrom * *Konvektion auf Grund von Temperaturunterschieden*
air cylinder Pneumatikzylinder
air damper Luftklappe * *z.B. in Heizungssystem*
air direction Luftrichtung
air distance Luftstrecke * *für Isolation*
air drag Luftwiderstand || Widerstand * *Luft~*
air drag losses Luftwiderstandsverluste
air draught Fahrtwind
air duct Luftführung * *Luftkanal/-weg* || Luftkanal
air ejector Gasstrahler
air exhaust opening Luftauslassöffnung
air fan Ventilator
air filter Luftfilter
air filter mat Luftfiltermatte
air flap Luftklappe
air flow Luftstrom
air flow monitor Luftstromwächter
air flow monitoring Luftstromüberwachung * *als Funktion, z.B. zur Überwachung eines Lüfters*
air flow rate Luftdurchsatz || Luftmenge * *Luftdurchsatz (eines Kühl- oder Belüftungssystems)* || Luftstrom * *Luftmenge pro Zeiteinheit*
air flow velocity Luftstromgeschwindigkeit
air freight Luftfracht
air gap Luftspalt
air gap flux Luftspaltfluss
air gap winding Luftspaltwicklung
air guide Leitblech * *für (Kühl)Luft; siehe auch 'Luftleitblech'* || Luftführung * *mechan. Element(e), z.B. Leitblech(e)* || Luftleitblech
air guide wheel Luftleitrad
air guiding Luftführung * *Funktion*
air handling Lufttechnik
air humidifier Luftbefeuchter
air humidity Luftfeuchtigkeit
air inlet Kühlluftöffnung * *Lufteinlass* || Lufteinführung || Lufteinlass || Lufteintritt * *Einlassöffnung/-stutzen* || Luftöffnung * *Lufteinlassöffnung* || Luftzuführung * *Lufteinlass*
air inlet adapter Lufteinführung * *Anschlussstück für den →Lufteinlass z.B. in e. Pneumatiksystem* || Lufteinlass * *Anschlußstück für den ~*
air inlet end Ansaugseite * *[el. Masch.]*
air intake Lufteintritt
air intake opening Lufteinlassöffnung || Lufteintrittsöffnung
air intake shutter Lufteintritts-Jalousie
air intake temperature Zulufttemperatur
air leakage Luftverluste * *in Druckluftsystem*
air opening Luftöffnung
air outlet Luftauslass || Luftauslassöffnung || Luftaustritt * *Auslassöffnung/-stutzen*
air outlet opening Luftaustrittsöffnung
air passage Luftkanal
air permeance Luftdurchlässigkeit
air pollution Luftverschmutzung
air pollution control Luftreinhaltung
air preheater Winderhitzer * *zur Stahlerzeugung*
air pressure Luftdruck
air pressure gauge Luftdruckmesser
air pressure switch Luftstromwächter * *z.B. zur Lüfterüberwachung*
air reactor Luftdrossel * *kernlose Drossel (ohne Eisen- oder Ferritkern)*
air resistance Luftwiderstand
air sifter Schwebesichter * *z.B. zum Sortieren von Holzschnitzeln*
air slot Kühlschlitz * *zum Ein/Austritt von Kühlluft* || Luftschlitz || Lüftungsschlitz

air spacing Hohlraum
air stream Luftstrom
air supply Luftzuführung * *Luftzufuhr, Luftversorgung*
air supply damper Luftklappe * *z.B. in Heizungssystem*
air supply duct Luftzuführung * *Zuführungskanal/weg*
air supply system Zuluftanlage
air to air heat exchanger Luft-Luft-Kühler * *mit Wärmetauscher*
air velocity Luftgeschwindigkeit
air vent Entlüftung
air-air cooler Luft-Luft-Kühler
air-air heat exchanger Luft-Luft-Wärmetauscher
air-break contactor Luftschütz
air-condition klimatisieren
air-cooled luftgekühlt
air-cooling Luftkühlung
air-core inductor Luftspule
air-core reactor Luftdrossel * *ohne Eisenkern* || Luftspule
air-flow direction Luftrichtung
air-flow monitor Luftstromüberwachung * *als Einrichtung/Gerät, z.B. zur Überwachung eines Kühlsystems* || Luftströmungswächter || Luftstromwächter
air-gap field Luftspaltfeld * *[el. Masch.] magn. Feld im Luftspalt e. drehenden el. Maschine*
air-gap monitoring Luftspaltüberwachung * *[el. Masch.] durch Messspulen i.d. Ständernuten, erkennt ungleichmäß. Luftspalt*
air-gap power Luftspaltleistung * *[el. Masch.] vom Ständer über d. Luftspalt auf Läufer übertragene Leistung*
air-gap reactance Hauptreaktanz
air-gap torque Luftspaltmoment
air-handling systems Lufttechnik
air-inlet end Ansaugseite * *[el. Masch.] (für Kühlluft)*
air-insulated luftisoliert
air-intake louver Lufteintritts-Jalousie * *als Kiemenblech ausgeführte Lufteintrittsabdeckung (z.B. für Motorbelüftung)*
air-intake louvre Lufteintritts-Jalousie
air-operated Druckluft-
air-pressure Druckluft-
air-screw Schraube * *Flugzeugpropeller*
air-stabilized Luftstabilisiert * *z.B. Motorläufer mit Permanentmagnet*
air-tight luftdicht
air-to-air cooler Luft-Luft-Kühler
air-to-water cooler Luft-Wasser-Kühler
air-to-water heat exchanger Luft-Wasser-Kühler * *mit Wärmetauscher* || Luft-Wasser-Wärmetauscher
air-vane relay Luftfahnenrelais * *zur Lüfter- oder Luftstromüberwachung*
air-water cooler Luft-Wasser-Kühler
airborne noise Luftschall
airflow Luftstrom
airflow monitor Luftstromwächter
airfoil Flügel * *eines Flugzeugpropellers*
airfoil effect Luftkisseneffekt * *Bildung einer Gleitschicht aus Luft (z.B. bei Festplatte)*
airplane Flugzeug
airport Flughafen
airwaybill Luftfrachtbrief
aisle Flügel * *eines Gebäudes* || Gang * *Durchgang (zwischen Bänken, Tischen usw.); Schneise (figürl.)*
alarm Alarm * *auch Programmunterbrechung in einem Computer* || Störmeldung || Warnmeldung || Warnung

alarm buffer Meldepuffer * *für Fehler- und Warnmeldungen*
alarm clock Wecker
alarm code Fehlernummer * *Warnungsnummer*
alarm event Alarmereignis
alarm horn Alarmhupe
alarm lamp Warnleuchte
alarm message Alarmmeldung || Warnmeldung
alarm processing Alarmbearbeitung * *Verarbeitung von Fehler- und Warnungsmeldungen, z.B. durch ein Leitsystem* || Alarmverarbeitung
alarm relay Melderelais * *für Störmelde-/Überwachungszwecke* || Warnrelais
alarm section Überwachungsteil * *eines Geräts*
alarm sequence memory Warnungsfolgespeicher * *(→PROFIBUS-Profil)*
alarm signal Alarmmeldung || Warnmeldung
alburnum Splint * *[Holz] Splint, Splintholz*
alertness Aufmerksamkeit * *Wachsamkeit*
algorithm Algorithmus || Rechenvorschrift * *Algorithmus*
align abgleichen * *abgleichen/-fluchten, ausgleichen/-fluchten, symmetrieren, zum Fluchten bringen* || anreihen || ausrichten * *ausfluchten, z.B. Motor, Objekte an einer Linie* || fluchten * *in e. gerade Linie bringen, ausrichten, ausfluchten* || ordnen * *in Reih' und Glied bringen, ausrichten* || symmetrieren * *→abgleichen*
aligned ausgerichtet * *zum Fluchten gebracht*
aligned position Fluchtstellung
aligner Justiereinrichtung * *zum mechanischen Ausrichten*
aligning plate Richtplatte * *zum Ausrichten*
alignment Abgleich * *Anpassung, Ausrichtung* || Ausrichtung * *z.B. eines Motors* || Deckung * *Fluchtung; bring into alignment: zur ~ bringen* || Fluchtstellung * *Ausrichtung (z.B. einer Welle)* || Justage * *Ausrichtung, Ausfluchtung; Ausrichten (in einer geraden Linie usw.)*
alignment coupling Ausgleichskupplung
alignment pin Passstift
alike ähnlich
alkaline concentration Laugenkonzentration
alkyd resin Alkydharz
all alle * *ganz* || ganz * *(Adv.) sehr; not at all: ~ und gar nicht* || sämtliche
all at once auf einmal * *gleichzeitig* || gleichzeitig * *mit einem Mal* || *alles auf einmal, plötzlich*
all digital volldigital
all in one alles in einem
all kinds of aller Art * *(vorangestellt) all kinds of machines: Maschinen aller Art*
all of a sudden blitzschnell * *(Adv.) plötzlich* || schlagartig * *(Adv.)* || unerwartet * *(Adv.) ganz plötzlich, auf einmal*
all of them sämtliche
all off button NOT-AUS-Knopf
all over ganz * *über das ~e (hinweg)* || über * *(Adv.) ~ und ~, ganz (und gar), ~all*
all possible means alle mögliche * *try all possible means:alles Mögliche versuchen (auch 'try everything')*
all sorts of alle mögliche || aller Art * *(vorangestellt)* || möglich * *alles ~e; all sorts of things: alles ~e*
all sorts of things alles mögliche
all the um * *umso; all the better:umso besser; all the more/less:umso mehr/weniger*
all the same trotzdem * *(Adv)*
all things considered schließlich * *im Grunde*
all together sämtliche
all-climate-proof klimabeständig
all-electronic vollelektronisch
all-in-one Allzweck-
all-pole disconnection allpoliges Abschalten

all-pole switching allpoliges Schalten
all-purpose Allzweck- * *universal* || universal * *Allzweck-* || Universal- * *Allzweck* || universell
all-round vielseitig
all-wheel drive Allradantrieb
Allen Bradley A-B
alleviate vermindern * *(ver-)mindern, erleichtern, mildern, lindern*
alliance Arbeitsgemeinschaft * *Bund, Bündnis, (Interessen-) Gemeinschaft, Allianz* || Verbund * *Verbündung, Allianz*
allied zusammenhängend * *verwandt, in enger Beziehung stehend*
allocate allokieren * *(Speicherplatz) zuweisen/-teilen (to: an), anweisen, d.Platz bestimmen für* || allozieren * *(Speicherplatz) zuteilen* || anweisen * *z.B. Speicherplatz* || rangieren * *Signale ver-/zuteilen* || vergeben * *anweisen, zuteilen (z.B. Speicherplatz, Namen usw.)* || verteilen * *zuteilen; auch Speicherplatz durch Übersetzungs/Bindeprogramm)* || zuordnen * *zuteilen, zuordnen, zuweisen (to: an; auch Speicherplatz), den Platz bestimmen für* || zuteilen * *auch Speicherplatz, Zugriffsrechte usw.* ~ || zuweisen * *zuteilen, zuweisen, z.B. Speicherplatz*
allocated zugeordnet * *z.b. Speicherplatz*
allocation Belegung * →*Zuordnung* || Verteilung * *Zu/Verteilung, Zu/Anweisung; z.B. Kosten* || Zuordnung * *z.b. von Speicherplatz* || Zuteilung * *Zu/Verteilung, An/Zuteilung, Bewilligung* || Zuweisung * *Speicherplatz usw.*
allocation table Rangierliste || Zuordnungstabelle
allot verteilen * *zuteilen* || zuteilen * *aus-/verteilen, bewilligen, abtreten, (für e. Zweck usw.) bestimmen* || zuweisen
allotment Aufteilung * *Ver/Zuteilung, Anteil, das Zugeteilte* || Verteilung * *Ver/Zuteilung* || Zuteilung * *Ver/Zuteilung, Anteil, das Zugeteilte*
allow erlauben || ermöglichen * *gewähren, möglich machen, ermöglichen (of:etwas), zulassen* || gelten lassen || gestatten || gewähren * *bewilligen, einräumen* || zulassen * *erlauben, gestatten* || zuteilen * *genehmigen* || zuweisen * *genehmigen*
allow a discount nachlassen * ~ *von Preisn usw.*
allow for berücksichtigen * *auch 'in Betracht ziehen', z.B. e. Umstand* || einschließen * *berücksichtigen, in Betracht ziehen, einberechnen*
allow to Gelegenheit geben
allow to pass durchlassen
allowable zulässig * *erlaubt, zulässig, rechtmäßig*
allowable load Belastbarkeit * *zulässige/maximale* ~
allowable variation Toleranz * *zulässige Abweichung*
allowance Abzug * *Preis~/-Nachlass, Rabatt; Vergütg., (zahlen-/wertmäßige) Berücksichtigung (for:von)* || Bonus * *Rabatt* || Erlaubnis * *make allowance for:berücksichtigen* || Preisnachlass * *auf Grund von Mängeln* || Toleranz * *zulässige Abweichung, Spielraum* || Zuschlag * *(Sicherheits-/Gehalts-) Zuschlag, Quote, (Geld-) Zuwendung, Spielraum, zuläss. Abweichung*
allowed erlaubt
alloy Legierung * *light alloy: Leichtmetall~* || Mischung * *Legierung*
alloyed legiert
allround universell
allrounder Generalist
allude to hinweisen auf * *anspielen auf*
almost annähernd * *fast* || etwa * *beinahe, fast* || gleichsam * *beinahe* || nahezu * *fast*
alone allein || nur * *allein* || selbst * *ohne fremde Hilfe*
along an * *entlang (an)* || entlang * *Weg, Straße (vorangestellt)* || längs * *entlang* || vorbei * *entlang, längs*

along general lines in groben Zügen
along the torque limit an der Momentengrenze
along with zusammen * ~ *mit*
alongside längs * *entlang; (of: an/zu)*
alongside of neben
alotting Aufteilung * *in Teile, Austeilung, Verteilung (auch durch Los)*
aloud laut * *(Adv.) laut, mit lauter Stimme*
alphabetic alfabetisch || alphabetisch
alphabetic order alphabetische Reihenfolge
alphabetic sequence alphabetische Reihenfolge
alphabetically alphabetisch * *(Adv.)*
alphanumeric alphanumerisch
already bereits
already mentioned vorgenannt
also auch * *gleichfalls* || ebenfalls
also refer to siehe auch * *Verweis auf (Stelle in) Publikation*
alter ändern || korrigieren * *ändern* || umbauen * *teilweise* ~, *Teilbereiche* ~ || umgestalten || verändern
alterable änderbar
alteration Änderung * *(Ab/Um)Änderung, Veränderung, Umbau*
alterations Umbau * *Änderungen*
alterations reserved Änderungen vorbehalten
altered geändert
alternate ablösen * *sich* ~ *(in:bei)* || abwechseln || schwanken * *abwechseln* || wechseln * *ab~, ab~ lassen*
alternately abwechselnd * *(Adv.)*
alternating abwechselnd || wechselnd * *(sich) ab~*
alternating buffer Wechselpuffer
alternating buffer system Wechselpuffersystem
alternating component Wechselanteil
alternating current Wechselstrom
alternating field Wechselfeld
alternating flux Wechselfluss
alternating load Wechsellast
alternating quantity Wechselgröße * *physikal.-Größe mit periodisch wiederkehrendem Zeitverlauf, z.B. Strom/Spanng.*
alternating stress Wechselbeanspruchung || Wechsellast * *(mechanisch)*
alternating voltage Wechselspannung
alternative alternativ || Alternative * *to: zu/für; have no alternative: keine Alternative haben* || Ausweg * *(figürl.) Ausweichmöglichkeit, Alternative* || Möglichkeit * *andere, zweite* ~ || Variante * *Alternative* || Wahl * *Alternative* || wahlweise
alternator Generator * *für Drehstrom/Wechselstrom*
although auch * *obwohl, wenn auch* || obwohl || trotzdem * *(Bindewort)*
altitude Betriebshöhe * *Höhe * *geographisch, above sea level: über dem Meeresspiegel*
altitude above sea level Höhe über dem Meeresspiegel
altitude of site Aufstellhöhe || Aufstellungshöhe
altogether insgesamt * ~ *gänzlich, ganz und gar, im Ganzen genommen, völlig*
ALU ALU * *Abkürzg. für 'Arithmethic and Logic Unit': Rechenwerk eines Computers* || Rechenwerk * *Abkürzg. f. 'Arithmetic Logic Unit': Ausführungseinheit für arithmet. u.logische Befehle*
aluminium Aluminium
aluminium alloy Aluminiumlegierung
aluminium foil Aluminiumfolie
aluminium foil capacitor Aluminium-Folienkondensator
aluminium housing Aluminiumgehäuse
aluminium oxyde Korund * *Schleifmittel*
aluminum Aluminium * *(amerikan.)*
aluminum alloy Aluminiumlegierung * *(amerikan.)*

always

always grundsätzlich * *(Adv.) immer* || stets
always the same gleich bleibend
amalgam Mischung * *(auch figürl.)* Mischung, Gemenge, Verschmelzung
amalgamate fusionieren
amalgamation Fusion
amalgate vereinen * *fusionieren,* →*vereinigen* || vereinigen * *fusionieren*
amass sammeln * *in Massen* ~
amateur Laie * *['ämete:r]; in this field: auf diesem Gebiet*
ambient umgebend * *auch 'Umgebungs-(z.B. Temperatur), umkreisend, Neben-(z.B. Geräusch)'* || Umgebungs- * *z.B. Temperatur, Geräusch*
ambient air Außenluft * *Umgebungsluft* || Umgebungsluft
ambient air temperatur Umgebungstemperatur
ambient conditions Umgebungsbedingungen * *tough: raue*
ambient temperature Raumtemperatur || Umgebungstemperatur * *bei el.Masch. normalerweise max. 40 Grad C. nach VDE 0530 T.1*
ambiguity Mehrdeutigkeit || Unklarheit * *Zwei-/Mehrdeutigkeit*
ambiguous interpretationsfähig * →*mehrdeutig* || mehrdeutig * *zweideutig* || unklar * *zwei/mehrdeutig* || zweideutig
ambition Wunsch * *hochgestecktes Ziel*
ambitious anspruchsvoll * *von Sachen* || eifrig * *strebsam, ehrgeizig*
amend ergänzen * *(Gesetz, Vorschrift, Norm usw.) abändern, ergänzen; verbessern, berichtigen*
amendment Änderung * *Berichtigung, Verbesserung, Neufassung, Nachtrag, Zusatzartikel; (für Norm, Gesetz usw.)* || Ergänzung * *zu Gesetz, Norm, Vorschrift*
amidst unter * *mitten* ~
ammeter Amperemeter || Strommesser
among neben * *siehe* →*amongst* || unter * *zwischen, mitten* ~, *bei; among other things:* ~ *anderem* || *zwischen* * *zwischen/unter mehreren*
among one another untereinander
among other things unter anderem
among them darunter * *unter einer Anzahl von Dingen*
amongst neben * *(mitten) unter, inmitten; amongst other things:* ~ *anderen Dingen* || unter * *zwischen, mitten* ~, *bei* || *zwischen* * *(mitten) unter, zwischen, inmitten*
amongst others außerdem * →*unter anderem*
amorphous amorph
amortization Amortisation || Amortisierung
amortize amortisieren
amount Ausmaß * *Betrag, Summe, Höhe, (Geld)-Wert, Inhalt, Ergebnis* || Betrag * *Menge, Summe, Höhe, Geldbetrag, quantitative Größe* || Größe * *Betrag, Summe, Höhe, Ergebnis, Wert, Bedeutung* || Höhe * *Betrag, Summe, Höhe, Ergebnis, Wert* || Menge * *Betrag, Summe, Höhe, Inhalt, Ergebnis, Wert* || Summe * *Betrag* || Summe * *(Geld)-Betrag, Wert, Höhe*
amount billed Rechnungsbetrag * *for:für*
amount of fouling Verschmutzungsgrad * *Ausmaß d. Verschmutzung (durch Öl, Ruß usw.), Befleckg., Verstopfg., Verdreckg.*
amount of heat Wärmemenge
amount of information Informationsmenge
amount of wiring Verdrahtungsaufwand
amount to ausmachen * *(zahlen-/wertmäßig) betragen* || betragen * *wert/zahlenmäßig* ~ || entsprechen * →*betragen*
amounts of data Datenmenge
amp Ampere * *1 Ampere*
AMP blade-type terminal AMP-Steckverbinder * *Flachsteckverbinder, z.B. von der Fa. AMP*

ampacity Stromtragfähigkeit * *(salopp) Kurzform für "amps capacity"*
amperage Stromstärke
ampere turns magnetische Durchflutung
ampere-hour Amperestunde
ampere-turns Amperewindungen || Durchflutung
ampersand Kaufmannsund * *Zeichen " auf Tastatur/Bildschirm/Drucker*
ample genügend * *reich(lich), (vollauf) genügend, weit, groß, geräumig, weitläufig, ausführlich, umfassend* || groß * *weit (-läufig/-räumig), geräumig, reichlich (bemessen) ausführlich* || reichlich * *weit, groß, geräumig, genügend, ausführlich*
amplification Verstärkung
amplified verstärkt * *elektrisch* ~
amplifier Steller * *Verstärker* || Treiber * *Verstärker* || Verstärker
amplify aufschaukeln * →*verstärken* || verstärken
amplitude Amplitude || Betrag * *Amplitude* || Höhe * *e. elektr.Größe, z.B. Spannung Strom, Leistung* || Scheitelwert
amplitude response Amplitudengang
amplitude spectrum Amplitudenspektrum
apply reichlich * *(Adv.)*
amply dimensioned großzügig dimensioniert * *mit reichlich Platz ausgestattet (z.B. Klemmenkasten)*
amply-sized geräumig
amps Ampere || Strom * *Stromstärke*
amps capacity Stromtragfähigkeit
amps meter Amperemeter
an abundance of eine Fülle von
analog analog
analog based in Analogtechnik
analog closed-loop control analoge Regelung
analog control Analogregelung
analog input Analogeingabe || Analogeingang
analog input module Analog-Eingabebaugruppe || Analogeingabebaugruppe
analog input referred to ground Masse-bezogener Analogeingang
analog meter analoges Messinstrument || Analoginstrument || Analogmessgerät
analog multiplexer Analogmultiplexer
analog output Analogausgabe || Analogausgang
analog output channel Analogausgabe * *Ausgabekanal*
analog output module Analog-Ausgabebaugruppe
analog select input analoger Wahleingang
analog select output analoger Wahlausgang
analog setpoint Analogsollwert
analog signal Analogsignal
analog switch Analogschalter
analog tachometer Analogtacho
analog technology Analogtechnik * *in: in der*
analog to digital converter Analog-Digitalwandler
analog to frequency controller U-f-Wandler
analog value Analoggröße || Analogwert
analog variable Analoggröße
analog-digital converter Analog-Digital-Umsetzer || Analog-Digital-Wandler
analog-digital convertor Analog-Digital-Umsetzer * *(veraltet)*
analog-to-digital conversion Analog-Digital-Umsetzung || Digitalisierung * *Analog-Digital-Wandlung*
analog-to-digital converter Analog-Digital-Umsetzer || Analog-Digital-Wandler
analogous ähnlich * *to/with:zu* || gleichartig * *analog, ähnlich* || sinngemäß || vergleichbar * *analog, entsprechend (to/with:mit)* || verwandt * *entsprechend, analog (to: zu, mit)*
analogous to ähnlich * *entsprechend, analog/parallel zu* || analog zu || entsprechend * *sinngemäß*
analogously analog * *(Adv.)* || sinngemäß * *(Adv.)*
analogue analog

analogy Ähnlichkeit * *Analogie, Ähnlichkeit, Übereinstimmung* || Entsprechung * *Übereinstimmung, Analogie, Ähnlichkeit* || Vergleich * *Analogie, Entsprechung*
analysis Abschätzung * *Analyse* || Analyse || Aufgliederung || Auswertung * *Analyse* || Prüfung * *Analyse* || Studie || Überprüfung * *Analyse* || Untersuchung * *Analyse*
analysis of feasibility Durchführbarkeitsstudie
analysis program Auswerteprogramm
analytical process Analyse
analyze analysieren || aufgliedern || auswerten || prüfen * *analysieren* || untersuchen * *analysieren*
analyzer Analysator || Analysegerät
anchor Anker * *Befestigungselement* || verankern * *(auch figürl.)*
anchor winch Ankerwinde
anchor winder Ankerwinde
anchor winding gear Ankerwinde
anchoring screw Feststellschraube
ancient alt * *geschichtlich alt*
ancillary Hilfs- * *untergeordnet (to: zu); Hilfs-, Neben-* || Neben- * *untergeordnet (to: zu), Hilfs-, Neben-*
ancillary industry Zulieferindustrie
ancillary plants Nebenbetriebe
AND UND || verunden
and above ab * *ab einschließlich (Reihenfolge; nachgestellt)*
AND gate UND-Gatter
AND logic UND-Verknüpfung
AND operation UND-Verknüpfung
and so on usw
AND'd UND-verknüpft
ANDed UND-verknüpft || verundet
ANDing Verundung
anew erneut * *(Adv.)* || vorn * *von neuem anfangen*
angle Ecke * *spitzer Punkt* || Gesichtspunkt * *(amerikan.) Blickwinkel* || Knick * *Winkel* || Knie * *Winkel* || Winkel * *['ængel] (geometr.) auch Zündwinkel, Phasenwinkel; at an angle of 45°: im Winkel von 45°*
angle bracket Winkel * *Halte~, Blech~*
angle exactness Winkelgenauigkeit * *{Werkz.-masch.\ Holz} z.B. bei der Bearbeitung von Holzplatten usw.*
angle lever Winkelhebel
angle of circumference Umfangswinkel
angle of contact Umschlingung * *~swinkel (bei Seilscheibe, Zugwalze usw.)* || Umschlingungswinkel * *für Riemen, Seilscheibe, Zugwalze usw.*
angle of displacement Schwenkwinkel * *bei →Schwenktrafo* || Versatzwinkel
angle of grip Umschlingungswinkel
angle of inclination Neigungswinkel
angle of lead Voreilwinkel
angle of overlap Überlappungswinkel
angle of rotation Drehwinkel
angle of traverse Schwenkwinkel
angle of twist Verdrehwinkel
angle sander Winkelschleifer
angle transfer Winkelübergabe * *{Fördertechnik}*
angle-synchronized winkelsynchron
angled -winklig
angular eckig || gewinkelt || Schwenk- * *Winkel-* || Winkel- || winklig
angular acceleration Winkelbeschleunigung
angular accuracy Winkelgenauigkeit
angular compliance Winkelnachgiebigkeit * *z.B. einer elastischen Kupplung*
angular control Winkelregelung
angular deflection Winkelabweichung
angular degree Winkelgrad * *in angular degrees/ in angular measure: in ~n*
angular displacement Winkelverschiebung

angular distance Winkelabweichung * *Winkeldifferenz*
angular edge Winkelkante * *{Holz}*
angular encoder Winkelkodierer
angular flexibility Drehelastizität || Elastizität * *Dreh~, z.B. einer Kupplung*
angular frequency Kreisfrequenz * *[rad/sec]*
angular minute Winkelminute
angular misalignment Beugungswinkel * *{Masch.-bau} Fluchtfehler bei der Ausrichtung von Wellen usw.* || Winkelabweichung * *Fluchtfehler*
angular momentum Drehimpuls
angular position Winkellage || Winkelposition || Winkelstellung
angular regulation Winkelregelung
angular resolution Winkelauflösung
angular second Winkelsekunde
angular shifting device Winkelschiebeeinrichtung
angular synchronisation Winkelgleichlauf
angular synchronism winkelgenauer Gleichlauf * *siehe auch 'Winkelgleichlaufregelung'* || winkelgetreuer Gleichlauf || Winkelgleichlauf
angular synchronous control winkelgenauer Gleichlauf * *winkelgenaue Gleichlaufregelung* || Winkelgleichlaufregelung
angular transfer Winkelübergabe * *{Fördertechnik}*
angular velocity Drehgeschwindigkeit * *Winkelgeschwindigkeit* || Winkelgeschwindigkeit * *[rad/sec]*
angular-contact ball bearing Schrägkugellager
angular-locked synchronism Winkelgleichlauf
angular-locked synchronizing control Winkelgleichlauf * *~regelung* || Winkelgleichlaufregelung
angular-shaped wire Profildraht * *mit eckigem Profil*
anilox roller Rasterwalze * *{Druckerei}*
animal hairs Tierhaare * *{Textil}*
animal-powered mechanism Göpel * *Vorrichtung zum Antrieb landwirtsch. Maschinen, Pumpen usw. durch Zugtiere*
anneal anlassen * *glühen* || glühen * *Metallveredelung* || temperieren * *Metall* || vergüten * *glühen (Metall)*
annealer Glühofen
annealing Glühbehandlung * *zur Vergütung von Metall* || Wärmebehandlung * *durch Glühen (meist für Metall)*
annealing furnace Glühe * *Glühofen für Blechbehandlung im Walzwerk* || Glühofen * *z.B. in Walzwerk/Bandbehandlungsanlage* || Kühlofen * *für Glas*
annealing oven Kühlofen
annealing plant Glühanlage * *z.B. für Blech*
annex Anbau * *an Gebäude* || anbauen * *Gebäude (to: an)* || Anhang * *auch einer Beschreibung* || beifügen
annexed hierbei * *beigefügt, beifolgend*
annihilate vernichten * *vernichten, zunichtemachen, aufheben, aufreiben*
annotate kommentieren
announce ankündigen || anmelden || bekannt geben || bekanntmachen * *verkünden, ankündigen*
announcement Ankündigung || Anmeldung || Anzeige * *Ankündigung* || Bekanntgabe * *Verkündigung* || Bekanntmachung * *Verkündigung* || Hinweis * *Ankündigung, of: von* || Veröffentlichung * *Bekanntmachung, Ankündigung*
announcement of discontinuation Abkündigung * *dass ein Produkt nicht mehr gefertigt/geliefert wird* || Produktabkündigung
announcement of product discontinuation Produktabkündigung
announcement that a product will be discontinued Abkündigung eines Produktes

announcement that the product will be discontinued Produktabkündigung
annoy belästigen || stören * belästigen, ärgern, schikanieren
annoyance Ärgernis * Störung, Belästigung, Plage(geist), Ärger(nis) || Belästigung || Störung * Ärgernis
annoying lästig * unangenehm, störend || unangenehm * ärgerlich, lästig, verdrießlich
annual Jahres- * jährlich, pro Jahr || jährlich
annual average Jahresdurchschnitt
annual demand Jahresbedarf
annual holidays Jahresurlaub
annual purchasing volume Jahresbedarf * jährl. Einkaufsvolumen eines Kunden || Wertigkeit * Jahresbedarf eines Kunden
annual report Geschäftsbericht * jährlicher ~ || Jahresbericht
annual requirements Jahresbedarf
annual sales Jahresumsatz
annual sales volume Jahresumsatz || Jahresvolumen * z.B. Bestellvolumen eines Kunden
annual shut period Betriebsferien
annual turnover Jahresumsatz
annually jährlich * (Adv.)
annul annullieren || auflösen * Vertrag usw. ~ || rückgängig machen * z.B. Vertrag
annular Ring- || ringförmig
annulment Annullierung
annunciate anzeigen * z.B. Fehlermeldung ~ || aufstellen * verkünden, z.b. Lehrsatz, Axiom, Richtlinie
annunciation Anzeige * Anzeige (z.b. von Fehler, Zustand), Verkündigung || Meldung * optische Meldung, z.B. an Leuchtmelder, Leuchttableau
annunciator Anzeige * ~gerät, z.b. Störleuchte || Anzeigeelement || Anzeiger || Anzeigetableau || Meldeeinrichtung
annunciator board Anzeigetableau || Anzeigetafel
annunciator panel Anzeigetableau || Anzeigetafel || Meldetafel
anode Anode
anode terminal Anode * Anodenanschluss
anode-cathode voltage Anoden-Kathoden-Spannung
anodize eloxieren
anodized eloxiert
anomalous anormal
anomaly Abweichung
another anders * be another story: ~ (gelagert) sein || neu * ein ~er/weiterer || weiter * another 5 times: ~e 5 Male
ANSI ANSI * 'American National Standards Institute': nationales USA-Normeninstitut (Ähnlich DIN)
answer Antwort * to: auf || antworten * to: auf || beantworten * with: mit || Gegenmaßnahme * Gegenmittel, (Problem-) →Lösung || lösen * Poblem ~, Frage beantworten || Lösung * ~ eines Problems/einer Aufgabe, Gegenmittel, Gegenmaßnahme || Quittung * (figürl.) || Stellungnahme * Beantwortung || Variante * Lösung; Antwort auf ein Problem (to: auf) || Version * Lösung(sweg)/Antwort auf ein Problem; to:auf
answerable verantwortlich * for:für
answering machine Anrufbeantworter
antagonism Gegensatz * Widerspruch (between: zwischen)
antenna Antenne * (amerikan.) ferrite-bar antenna: Ferrit~
anterior vorder * vorder, vorhergehend, früher (to: als)
anti- -Schutz * (vorangestellt) || Gegen- * entgegen, z.B. anti-clockwise:entgegen d. Uhrzeigersinn
anti-aliasing Kantenglättung * Glättung von Trep-

peneffekten bei Computergrafik, z.B. durch Interpolieren
anti-clockwise entgegen dem Uhrzeigersinn
anti-clockwise rotation Linkslauf
anti-condensation heater Stillstandsheizung
anti-condensation heating Stillstandsheizung * z.b. in Motor zur Vermeidung von Kondenswasserbildung
anti-condensing heating Stillstandsheizung * z.B. bei Elektromotor zur Vermeidung von Kondensatbildung
anti-corrosion protection Korrosionsschutz
anti-corrosive paint Rostschutzfarbe
anti-friction bearing Wälzlager * DIN 611 T.1 3, 625, 628 T.1, DIN5412 T.1, DIN ISO76, 281, el. Masch. DIN 42966
anti-frost Frostschutz
anti-kink sleeve Knickschutztülle
anti-noise leise * schallgedämpft || ruhig * schallgedämpft
anti-noise measures Entstörmaßnahmen
anti-parallel antiparallel * anti-parallel connected: ~ geschaltet || gegenparallel
anti-parallel connection Antiparallelschaltung || Gegenparallelschaltung
anti-reversing device Rücklaufsperre * als Einrichtung
anti-rotation drehsteif
anti-rust Rostschutz
anti-rust coating Rostschutzfarbe
anti-shock Stoßfestigkeit * (mechan.)
anti-skin-effect stromverdrängungsarm
anti-stall protection Kippschutz
anti-static antistatisch
anti-static clothes Antistatikbekleidung
anti-vibration Schwingfestigkeit
anti-vibration compound Antidröhnmittel
anticipate erwarten * erwarten, erhoffen, voraussehen, vorausahnen, vorausempfinden || vorhersehen * voraussehen, vorausahnen, vorausempfinden
anticorrosion Korrosionsschutz
anticorrosive Antikorrosiv
antiferromagnetic antiferromagnetisch
antifreezing agent Frostschutz * Frostschutzmittel
antifreezing protection Frostschutz
antifriction properties Leichtgängigkeit
antifriction property Gleiteigenschaften * Reibarmut, →Leichtgängigkeit
antimagnetic antimagnetisch
antinode Wellenbauch
antiparallel bridge connection Gegenparallelschaltung * z.B. Gleichrichterbrücke
antiparallel connection Antiparallelschaltung
antiquated veraltet * antiquiert
antistatic antistatisch
antistatic agent Antistatikmittel
antithesis Gegensatz * Gegensatz, Widerspruch || Gegenteil * Gegensatz, Widerspruch || Widerspruch * Antithese, Gegensatz, Widerspruch
antitype Gegenstück * Gegenbild, Vorbild
antivibration mountings Schwingungsdämpfer * z.B. zur Aufstellung eines Eletromotors
anulment Kündigung * ~ eines Vertrages usw.
anvil Amboss
anxious eifrig * be very anxious to: eifrig bestrebt sein, zu
any beliebig * z.B. ein Beliebiger || etwaig * irgendwas || jedes * ~ Beliebige; without any doubt: ohne jeden Zweifel; at any time:z u jeder Zeit || jedes beliebige || willkürlich * irgendein
any given beliebig * jedes beliebig herausgegriffene Ding
any possible etwaig
anyhow eigentlich * (Adv.) what do you want anyhow: was willst Du eigentlich || sowieso

anything but nie * *alles andere als*
anyway sowieso
apart Abstand * *125 mm apart: im ~ von 125 mm* ‖ entfernt * *voneinander ~* ‖ gesondert * *einzeln, für sich; apart from: (ab)~ von; siehe auch →trennen* ‖ im Abstand von * *(nachgestellt) 125 mm apart: im Abstand von 125 mm*
apart from abgesehen von ‖ außer * *abgesehen von* ‖ neben * *nebst*
aperiodic aperiodisch ‖ azyklisch * *nicht periodisch*
aperiodical aperiodisch
aperture Ausbruch * →*Öffnung* ‖ Ausschnitt * *Öffnung* ‖ Durchbruch * *Öffnung* ‖ Öffnung ‖ Schlitz * *Öffnung*
apex Scheitelpunkt * *Scheitelpunkt,* →*Spitze, Höhepunkt (figürl.)* ‖ Spitze * *Spitze (auch eines Kegels), Gipfel, Scheitelpunkt, Höhepunkt (figürl.)*
API Betriebssystemschnittstelle * *[Computer] Application Program Interface: Schnittst. z. Anwenderprogr.*
apologize entschuldigen * *oneself: sich*
apology Entschuldigung * *Abbitte*
apothecary Apotheke * *(amerikan.)*
app Anwendung * *(salopp) Kurzform für 'application' (Computer-Anwendungsprogramm)*
apparatus Apparat * *(auch figürlich)* ‖ Einrichtung * *Vorrichtung* ‖ Gerät * *Apparat, Gerät, Vorrichtung, Apparatur, Maschinerie*
apparatuses Apparatur * *Apparate*
apparent Schein- * *z.B. Leistung, Strom* ‖ scheinbar * *anscheinend* ‖ virtuell * *scheinbar*
apparent converter output Umrichterscheinleistung
apparent current Scheinstrom
apparent power Scheinleistung * *Eff.wert d. Wechselspannung x Eff.wert d. Wechselstroms in [VA] oder [kVA]*
apparent-power consumption Scheinleistungsaufnahme
appeal Revision * *[Rechtswesen] lodge an appeal: ~ einlegen*
appealing ansprechend * *gefallend* ‖ eindrucksvoll
appear erscheinen * *auch auf einer Anzeige, einem Bildschirm, vor Gericht* ‖ herausbilden * *sich herausbilden* ‖ herauskommen * *erscheinen* ‖ herausstellen * *sich ~ (to be: als)* ‖ sichtbar * *~ werden, erscheinen* ‖ sichtbar werden ‖ stehen * *in Liste, auf Bildschirm usw. ~* ‖ zeigen * *erscheinen*
appearance Anstrich * *Aussehen* ‖ auftreten * *(Substantiv) Erscheinen* ‖ erscheinen * *(Substantiv)* ‖ Erscheinung * *äußere* ‖ Form * *(äußeres) Erscheinung(sbild); (to) keep up appearances: die Form wahren/der Form halber* ‖ Optik * *(figürl.) äußeres Aussehen, Erscheinung*
appearance of bearing surface Tragbild * *eines Gleit-/Wälzlagers*
append beifügen ‖ hinzufügen * *einen Nachtrag ~, eine Anlage ~*
appendices Anhänge
appendix Anhang * *eines Buches,einer Beschreibung, eines Berichts* ‖ Anhängsel * *auch Anhang, Ansatzteil* ‖ Ergänzung * *Anhang* ‖ Nachtrag * *Anhang* ‖ Zusatz * *Anhang*
appertaining zugehörig * *zugehörig, gebührend, zugestanden*
appliance Apparat * *z.B. Haushaltsgerät* ‖ Einrichtung * *Vorrichtung* ‖ Gerät * *z.B. Haushalts-~* ‖ Haushaltsgerät ‖ Hilfsmittel * *(Hilfs)Mittel, Gerät, Vorrichtung, Apparat* ‖ Vorrichtung
appliance for domestic purposes Haushaltsgerät * *VDE 0700*
applicability Anwendbarkeit ‖ Anwendungsfall * *Anwendungsmöglichkeit* ‖ Eignung * *Eignung, Anwendbarkeit (to: für, auf, zu)*

applicable anwendbar * *to: auf; auch 'passend, zutreffend'* ‖ einsetzbar * *anwendbar* ‖ geeignet * *anwendbar* ‖ gültig * *anwendbar, zutreffend (to: auf)* ‖ jeweilig * *anwendbar, passend, geeignet* ‖ maßgebend * *anwendbar; to:für* ‖ maßgeblich * *anwendbar; to: für* ‖ passend * *anwendbar (to: auf), geeignet, zutreffend* ‖ relevant * *anwendbar, passend, geeignet, zutreffend (to: auf,für)* ‖ verwendbar ‖ zutreffend * *anwendbar, geeignet*
applicant Bewerber * *for:für/um*
application Anmeldung * *Patent, Trainingskurs usw.* ‖ Antrag * *Gesuch* ‖ Anwendbarkeit * *Anwendungsmöglichkeit* ‖ Anwenderprogramm * *Anwendung * Einsatzfall, Einsatzmöglichkeit* ‖ Anwendungsfall * *Anwendungszweck* ‖ Applikation ‖ Auftrag * *von Farbe, Leim, Klebstoff usw.* ‖ Bewerbung * *for: um* ‖ Branche * *Anwendungsbereich* ‖ einfallen * *(Substantiv) einer Bremse* ‖ Einsatz * *Anwendung* ‖ Verwendung * *Anwendung, Anwendbarkeit* ‖ Zweck * *Verwendung*
application area Anwendungsgebiet ‖ Einsatzbereich ‖ Einsatzgebiet
application areas Einsatzgebiete
application book Projektierungshandbuch * *Applikations-(Hand-) Buch*
application class Betriebsklasse * *für Sicherung (siehe DIN VDE 0636)*
application conditions Einsatzbedingungen
application controller Technologieregler
application department Projektierungsabteilung
application document Applikationsschrift
application engineer Applikationsingenieur ‖ Entwickler * *Projekteur, Anwendungsentwickler* ‖ Projekteur
application engineering Projektierung
application engineering department Projektierungsabteilung
application example Anwendungsbeispiel ‖ Projektierungsbeispiel * *Anwendungsbeispiel*
application guide Anwendungsrichtlinien ‖ Applikationshandbuch
application guidelines Anwendungshinweise ‖ Applikationshinweise
application hints Anwendungshinweise * *Tipps/ Ratschläge für die Anwendung* ‖ Einsatzhinweise * *Tipps/Ratschläge für die Anwendung* ‖ Projektierungshinweise
application in domestic premises Haushaltsanwendung
application information Anwendungshinweise ‖ Einsatzhinweise * *siehe auch 'Anwendungshinweise'*
application instructions Anwendungshinweise ‖ Einsatzhinweise
application layer Anwenderschicht * *Schicht 7 im ISO 7-Schichten-Kommunikationsmodell*
application macros Anwendungsmakros
application manual Applikationshandbuch ‖ Projektierungshandbuch * *Applikationshandbuch* ‖ Projektierungsunterlagen * *Handbuch mit Anwendungshinweisen*
application note Applikationshinweis ‖ Applikationsschrift
application notes Anwendungshinweise ‖ Applikationshinweise
application of force Kraftangriff
application oriented anwendungsorientiert
application possibilities Anwendungsmöglichkeiten
application potential Anwendungsmöglichkeiten
application program Anwenderprogramm ‖ Anwendungsprogramm
application program interface Betriebssystemschnittstelle * *[Computer] zum Anwenderprogramm*

application related

application related -spezifisch * *für eine Anwendung*
application roller Auftragswalze
application scope Anwendungsbereich
application spectrum Anwendungsmöglichkeiten * *Anwendungsspektrum* || Anwendungsspektrum
application time Einfallzeit * *einer Bremse*
application-oriented anwenderorientiert || anwendungsorientiert || branchenorientiert * *anwendungsorientiert* || praxisgerecht
application-related anwendungsbezogen
application-specific anwenderspezifisch || anwendungsspezifisch || kundenspezifisch
application-specific functions technologische Funktionen
application-specific interface Betriebssystemschnittstelle * *(Computer)*
applications Anwendungsbereich || Anwendungsmöglichkeiten
applications engineering Anwendungstechnik
applicator Zuführeinrichtung * *Anlege-/Aufbringeinrichtung*
applicator roll Auftragswalze * *bei Streich-/Beschichtungsmaschine*
applied eingefallen * *z.B. Bremse* || praktisch * *angewandt* || verwendet
applied economics Betriebswirtschaft
applied force Anpresskraft * *z.b. bei Bremse, Kupplung usw.*
applied sciences angewandte Wissenschaften
applied voltage angelegte Spannung
apply anbringen * *to: an, auf* || anlegen * *a voltage across/to: eine Spannung zwischen/an; apply a standard: einen Maßstab* ~ || ansetzen * *anlegen, ansetzen (Hebel, Meißel usw.) (to:an)* || anwendbar sein * *to:Anwendung finden bei; passen/zutreffen auf* || anwenden * *to:auf; auch Gesetz, Prinzip* || Anwendung finden * *to:bei* || anziehen * *Bremse* ~ || aufbringen * *anwenden, anbringen, bestreichen usw.* || auflegen * *Beschichtung/Pflaster usw.* ~ || auftragen * *aufbringen* || beaufschlagen || auflegen/-tragen, anbringen (to: an/auf), zur Anwendung bringen || bewerben * *for a job:sich um eine Stelle; to a person:bei jdm.* || einsetzen * *verwenden, anwenden (to:bei/auf), in eine Formel einsetzen* || gelten * *Anwendung finden (to: bei); passen, zutreffen (to: für/auf/bei); z.B. Vorschrift, Norm* || handhaben * *Methode usw.* || setzen * *anbringen, auflegen, anwenden* || verwenden || vorgeben * *anlegen, z.B. eine Spannung/einen Sollwert* ~ || vorsehen * *ver-/anwenden, anbringen, gebrauchen* || zugrundelegen * *anwenden (to:zu), passen, zutreffen, gelten, z.B. Vorschrift, Betriebsanleitung usw.*
apply for anmelden * *Patent, Konkurs*
apply to anwendbar sein auf || Anwendung finden bei || beziehen * ~ *auf* || beziehen auf || einspeisen || gelten für * *sich anwenden lassen auf* || sich beziehen auf * *Anwendung finden bei, zutreffen auf*
appoint ansetzen * *Termin* || beauftragen * *berufen* || ernennen * *zum Amtsinhaber; he was appointed chairman: er wurde zum Vorsitzenden ernannt* || zuordnen * *appoint a person as assistant to: jmd. e. Person als Mitarbeiter/Helfer* ~
appoint a date Termin anberaumen
appointed day Termin * *festgelegter Tag*
appointed dealer Vertragshändler
appointed time Frist * *festgesetzte Zeit* || Termin * *festgesetzte Zeit*
appointment Festlegung * *besonders für Termin, Verabredung, Zusammenkunft* || Termin * *Verabredung, Zusammenkunft* || Vereinbarung * *Verabredung*
appointment as a university lecturer Habilitation
appointment of interview Vorstellungstermin * *für Bewerbungsgespräch (with: bei)*

apportion umlegen * *(Kosten, Gelder usw.) umlegen/→zuteilen* || verteilen * *zuteilen* || zuteilen * *(Anteil, Aufgabe usw.)* ~, *(gleichmäßig/gerecht) ein-/verteilen, (Lob) erteilen* || zuweisen
apportionment Verteilung * *gleichmäßige/gerechte Ver/Zu/Einteilung*
appr. ca || ungefähr || zirka * *Abkürzung für 'approximately'*
appraisal Abschätzung * *(Ab-) Schätzung, Taxierung, Schätzung, Schätzwert, Bewertung* || Bewertung * *(Ab)Schätzung, Taxierung, Schätzungswert, Bewertung*
appraise bewerten * *abschätzen*
appreciable fühlbar * *beträchtlich*
appreciate anerkennen * *lobend* ~ || schätzen * *würdigen, hochschätzen, zu schätzen wissen* || Verständnis * *Verständnis haben für*
appreciation Verständnis * *richtige Beurteil., Einsicht, krit.Würdigg., (günstige) Kritik, Kennenlernen (of:für)*
apprentice Auszubildender * *Lehrling* || Lehrling || Stift * *Lehrling*
apprentices' training shop Lehrwerkstatt
apprenticeship Lehre * *eines Lehrlings*
approach anfahren * *annähern, sich nähern, nahekommen, (fast) erreichen* || angreifen * *(figürlich) Aufgabe usw.* || annähern || Annäherung || Ansatz * *logisches Herangehen (to:an)* || einfahren * *sich nähern, heranfahren (z.B. Aufzug, Zug), ein-/anfliegen, nahekommen (auch figürl.)* || Einfahrt * *Annäherung, (Heran-)Nahen, Anmarsch, Nahekommen (auch figürl.); auch bei Zug, Aufzug usw.* || Einstellung * *"Herangehen", Haltung (to: gegenüber/zu), Ansatz* || kommen * *herankommen* || näherherkommen * *→nähern* || nähern * *sich einer Person* — || vorgehen * *(figürl.) herangehen (an e. Sache/Aufgabe/Problem); (auch Subst.: Herangehensweise)* || Zuführung * *Annäherungsweg, (Heran-) Nahen, Annäherung, Nahekommen, Zugang/-fahrt* || Zugang * *Weg*
approach to nähern * *sich etw.* ~, *sich an*~ *an*
approach to reference point Referenzfahrt * *Anfahren d.Referenzpunktes bei Positioniersteuerung* || Referenzpunktfahren * *bei Positioniersteuerung/Numer. Steuerung*
appropriate angebracht * *passend, gut* ~; *inappropriate/out of place/ill-timed: un*~ || angemessen * *diesbezüglich* || passend || entsprechend * *passend (auch "adequate": →angemessen); (Adv.: appropriately)* || passend || richtig * *angemessen* || sachgemäß || sachgerecht || sinnvoll * *zweckdienlich, passend, angemessen* || zuständig * *passend, richtig* || zweckmäßig
appropriately passend * *(Adv.)*
approval Anerkennung * *z.B. durch eine Normenbehörde* || Approbation * *z.B. Zulassung/Anerkennung bei einer Normenbehörde* || Freigabe * *Genehmigung, Anerkennung* || Genehmigung * *Billigung, Erlaubnis (of:von/des)* || Zulassung * *z.B. von einer Norm/Prüfbehörde* || Zusage * *Billigung*
approval procedure Genehmigungsverfahren
approval test Freigabeuntersuchung * *Abnahmeprüfung*
approve abnehmen * *genehmigen, anerkennen, bestätigen* || anerkennen * *billigend/offiziell* ~ || billigen || einverstanden sein * *of:mit* || genehmigen * *billigen, gutheißen, billigen, gutheißen, genehmigen, anerkennen* || zulassen * *billigen, gutheißen, genehmigen, anerkennen*
approve of billigen
approved approbiert * *z.B. bei Normenbehörde UL, CSA usw.* || zugelassen * *bei einer Normenbehörde* ~ || zulässig * *offiziell anerkannt, zugelassen*
approx. ca || etwa * *ungefähr (Abkürzg. für approx-*

imate(ly)) || ungefähr || zirka * *Abkürzung für 'approximately'*
approximate angenähert * *annähernd* || annähern || annähernd * *siehe auch* →*näherungsweise* || näherungsweise * *ungefähr, angenähert, annähernd*
approximate calculation Näherungsrechnung
approximate value Anhaltswert || Näherungswert
approximately ca || etwa * *ungefähr* || *näherungsweise* * *(Adv.)* || rund * *(Adv.) in Zusamm.hang mit Zahl; ungefähr (approx.: Abkürzg. f. 'approximately')* || ungefähr || zirka
approximation Annäherung * *z.b. mathematisch* || Approximation || Näherung
approximation formula Faustformel * *Näherungsformel* || Näherungsformel * *siehe auch* →*Faustformel*
approximation function Näherungsfunktion * *siehe auch* →*Faustformel*
approximation method Näherungsverfahren
approximation procedure Näherungsverfahren
approximation value Näherungswert
apron Schutzhaube
apt fähig * *passend, geeignet (to: für/zu)*
aptitude Eignung * *(einer Person) Befähigung, Fähigkeit, Tauglichkeit, Begabung (for: für, zu)*
aptitude test Eignungstest * *für Bewerber, Personal*
arbiter Arbitrierungseinheit * *organisiert die Zusteilung d. Buszugriffsrechte bei Multiprozessorsystem*
arbitrary beliebig * *willkürlich* || willkürlich * *zufällig*
arbitration Arbitrierung * *Verwaltung u. Zuteilung der Bus-Zugriffsrechte bei Multiprozessorsystem*
arbitration unit Arbitrierungseinheit * *organisiert Zusteilg. d. Buszugriffsrechte bei Multiprozessorsystem*
arbor Dorn * *Aufsteck~, Aufnahme~, Aufsteckhalter; z.B. zum Aufstecken eines Schleif-/Fräswerkzeugs*
arc Bogen * *auch Licht~* || Lichtbogen * *heiße Gassäule (Plasma) aus elekr. geladenen Teilchen, die einen hohen Strom führt*
arc discharge Bogenentladung
arc extinguishing medium Löschmittel * *z.B. in Sicherung*
arc measure Bogenmaß
arc over Überschlag * *durch Überspannung*
arc quenching medium Löschmittel * *z.B. in Sicherung*
arc voltage Lichtbogenspannung * *z.B. bei Leistungsschalter, Sicherung usw.*
arc welding Lichtbogenschweißen
arc-free funkenfrei * *ohne Lichtbogen*
arc-free commutation funkenfreie Kommutierung * *z.B. bei Gleichstrommotor*
arc-over voltage Durchschlagsspannung * *Überschlagsspannung* || Überspannung
arc-suppression diode Löschdiode * *z. Beschaltg. v. Gleichstromschütz/Relaisspule (geg. Überspannng. b. Abschalt.)*
arch Biegung * *Wölbung* || Wölbung
arch file Aktenordner
arched gewölbt || hohl * *gewölbt* || krumm * *bogenförmig; siehe auch 'gewölbt'*
architecture Architektur * *(auch figürl.) auch eines Rechners*
archive Archiv * *['a:kaiv] zur Ablage/Sicherung von Software, Akten* || archivieren * *['a:kaiv] Software, Akten, usw.*
archiver Packer || Packprogramm * *['a:rkeiwer]*
archiver program Packprogramm
archiving Archivierung
archiving floppy disk Archivdiskette
archrival Erzrivale * *auch auf dem Markt*

arcing löschen * *(Subst.) Löschung, z.B. Lichtbogen, z.B. in Sicherung/Schütz/Leistungsschalter*
arcing fault Störlichtbogen
arcing-proof terminal box lichtbogensicherer Klemmenkasten * *{el.Masch.}*
ardent heftig * *lebhaft*
arduous mühsam * *anstrengend* || schwierig * *schwierig, anstrengend, mühsam; task/job: Aufgabe*
area Bereich * *(örtlich und figürl.)* || Fläche * *(begrenzte) Fläche, ~ninhalt, Ober/Grund~* || Oberfläche * *Fläche(ninhalt)*
area code Postleitzahl || Vorwahl * *Vorwahlnummer am Telefon* || Vorwahlnummer * *im Inland*
area of application Anwendungsbereich || Anwendungsgebiet
area of industry Industriebereich
area of research Forschungsgebiet
area susceptible to firedamp schlagwettergefährdeter Bereich
argument Argument || Grund * *Beweis~, arguments for and against: Gründe für u. wider*
argumentation booklet Argumentenheft
argumentation brochure Argumentenheft || Argumentenheft
arguments booklet Argumentenheft
arise auftauchen * *(figürl.) Frage, Problem usw.* || auftreten * *erscheinen, entstehen; z.B. Nebeneffekt, Verluste in e. Motor* || eintreten * *Fall, Notwendigkeit, Umstände* || entstehen * *erscheinen, auftreten, hervorgehen, herrühren, stammen (from/out of: von)* || ergeben * *entstehen, erscheinen, auftreten* || sich ergeben * *auftreten, aufkom men (z.B. Schwierigkeiten, Probleme, Chance)*
arise from hervorgehen aus * *stammen von*
arithmetic arithmetisch || rechnerisch
arithmetic average arithmetischer Mittelwert
arithmetic average value arithmetischer Mittelwert
arithmetic block Arithmetikbaustein
arithmetic function Arithmetische Funktion || Rechenfunktion
arithmetic function block Rechenbaustein * *Funktionsbaustein*
arithmetic functions Arithmetikfunktionen
arithmetic instruction arithmetischer Befehl * *Rechner, SPS*
arithmetic logic unit Rechenwerk
arithmetic mean arithmetischer Mittelwert || arithmetisches Mittel
arithmetic mean value arithmetischer Mittelwert || Mittelwert * *arithmetischer ~*
arithmetic operation Arithmetische Funktion || Rechenoperation
arithmetic task Rechenaufgabe * *für Computer*
arithmetically rechnerisch * *(Adv.)*
arm Hebel * *Arm (auch Maschinenteil)* || schärfen * *scharf machen, z.B. Gewehr, Triggerung eines Oszilloskops* || scharfmachen * *z.B. Triggerung eines Oszilloskops* || Zweig * *z.B. einer Stromrichter/Brückenschaltung*
arm fuse Zweigsicherung * *in e. Zweig e. Stromrichterschaltung (nicht i.d. Wechselstromzuleitung)*
arm fuse link Zweigsicherung * *Sicherungseinsatz ohne Unterteil*
arm reactor Zweigdrossel * *in e. Zweig e. Stromrichterschaltg. (nicht i.d. Wechselstromzuleitung)*
arm voltage Ankerspannung * *(Kurzform)*
armature Anker * *['a:matje] an elektr. Maschine, Relais, Bremse; Magnetanker* || Ankerplatte * *bei Bremse* || Läufer * *Anker (einer Gleichstrommaschine)* || Magnetanker || Rotor * *Anker*
armature ampere conductors Ankerstrombelag
armature ampere-turns Ankerdurchflutung

armature band Ankerbandage
armature choke Ankerdrossel
armature circuit Ankerkreis
armature circuit inductance Ankerkreisinduktivität
armature circuit resistance Ankerkreiswiderstand
armature control range Ankerstellbereich || Ankersteuerbereich
armature core loss Ankereisenverlust
armature current Ankerstrom * *['a:matje...]*
armature current actual value Ankerstromistwert
armature current controller Ankerstromregler
armature current limits Ankerstromgrenzen
armature current regulator Ankerstromregler
armature current setpoint Ankerstromsollwert
armature direct-axis field Ankerlängsfeld
armature disk Ankerscheibe * *bei einer Bremse*
armature field Ankerfeld
armature flux Ankerdurchflutung
armature flux density Ankerflussdichte
armature gating unit Ankersteuersatz
armature inductance Ankerinduktivität || Ankerkreisinduktivität
armature induction Ankerflussdichte || Ankerinduktion
armature iron Ankereisen
armature lamination Ankerblech
armature leakage flux Ankerstreufluss
armature leakage inductance Ankerstreuinduktivität
armature leakage reactance Ankerstreureaktanz
armature plate Ankerscheibe * *bei einer (elektromagnet.) Bremse/Kupplung*
armature quadrature-axis field Ankerquerfeld
armature reaction Ankerrückwirkung
armature reactor Ankerdrossel
armature resistance Ankerkreiswiderstand || Ankerwiderstand
armature reversal Ankerkreisumschaltung * *Momentenumkehr b.1Q-Stromrichter über Schütze i. Ankerkreis*
armature ring Ankerscheibe * *einer elektromagnet. Bremse/Kupplung*
armature spider Ankerstern
armature time constant Ankerzeitkonstante
armature voltage Ankerspannung
armature voltage actual value Ankerspannungsistwert
armature voltage controller Ankerspannungsregler
armature voltage feedback Ankerspannungsistwert
armature voltage sensing Ankerspannungserfassung
armature winder Ankergesteuerter Wickler * *DC-Auf/Abwickler mit Durchmesseranpassg. i. Ankerkreis* || *Wickler * Achswickler mit DC-Antrieb und Durchmesseranpassung i.Ankerkreis*
armature winding Ankerwicklung
armature winding overhang Ankerwickelkopf
armature-circuit converter Ankerstromrichter
armature-circuit time constant Ankerkreiszeitkonstante * *typ. 20 msec bei Gleichstrommaschinen*
armature-reaction reactance Hauptreaktanz
armor Panzerung
armor plating Panzerung
armoring yarn Armierungsgarn * *{Kabel}*
armour Bewehrung * *z.B. für Kabel*
armoured cable bewehrtes Kabel
around herum || um * *zeitlich/örtlich/wertmäßig 'um herum'* || ungefähr
around the clock rund um die Uhr
around the world weltweit
arrange ablegen * *anordnen* || absprechen * *verabreden, arrangieren* || anordnen || ausmachen * *vereinbaren* || einordnen * *anordnen* || einrichten * *anordnen, aufstellen, einteilen* || einteilen * *anordnen; in:in Gruppen* || gliedern * *anordnen* || kümmern * *arrangieren, regeln, in Ordnung bringen* || organisieren || räumlich anordnen || regeln * *arrangieren, ordnen, in Ordnung bringen* || vereinbaren || vermitteln * *zustandebringen* || zusammenstellen * *(an)ordnen*
arrange a contract Vertrag schließen
arrange in a series anreihen * *in (einer) Reihe anordnen*
arrange in proper order einordnen * *in der richtigen Ordnung/Reihenfolge* ~
arrange matters disponieren
arrange side by side aneinander reihen || anreihen
arrangeable konfigurierbar
arranged gegliedert * *angeordnet*
arranged in -förmig * *(vorangestellt) angeordnet in/als*
arranged in tiers stufenförmig
arrangement Abmachung || Anlage * *Art der ~*, →Konzept || Anordnung || Aufbau * *Anordnung* || Disposition * *Anordnung* || Einrichtung || Einteilung * *Anordnung* || Festlegung * *Vereinbar., Verabredg. (make: treffen), Einteilg., Ab/Übereinkommen, Auf/Zusammenstellg.* || Konzept * *Art der* →*Anlage* || Ordnung * *An~* || Realisierung * *Anordnung, Ausgestaltung* || Reihenfolge * *Einteilung, (An-) Ordnung* || Schema * *Anordnung* || Stellung * *Anordnung* || Übereinkunft || Vereinbarung * *Vereinbarg., Verabredg., Ab-/Übereinkommen, Schlichtung, as arranged:wie vereinbart* || Zusammenstellung
arrangement as to size Größenordnung * *Ordnung/Einteilung nach der Größe*
arrangement drawing Anordnungsplan
array Feld * *{Computer} Daten~, mehrdimensionale Variable*
arrears Rückstände * *Schulden*
arrest ableiten * *aufhalten, hemmen* || ablenken * *aufhalten, hemmen, ableiten, (Aufmerksamkeit usw.) fesseln/festhalten* || arretieren || aussetzen * *an/aufhalten, hemmen, hindern* || sperren * *feststellen, sperren, arretieren* || stillsetzen * *feststellen, sperren, arretieren*
arrester Ableiter
arresting element Arretierung
arrival Anfahrt * *Ankunft* || Ankunft * *upon arrival.: bei Ankunft* || Einfahrt || erreichen * *(Subst.)* →*Ankunft, Eintreffen, Gelangen (auch figürl.; at: zu); Erreichen e. Wertes usw.*
arrivals Zufuhr
arrive anfahren * *ankommen (at: bei)* || ankommen || anreisen * *ankommen* || einfahren || einlaufen * *{Bahnen}* || eintreffen * *ankommen* || erfolgen * *sich einstellen*
arrow Pfeil
ART Art * *{el.Masch.} Abkürzg. für 'Allowable Running-up Time: zulässige Anlaufzeit*
art paper Kunstdruckpapier
artful pfiffig * *schlau, listig, verschlagen* || raffiniert * *schlau, listig, verschlagen*
article Artikel * *Aufsatz/Bericht z.B. in Zeitschrift* || Aufsatz * *in Zeitschrift* || Ware * *Artikel* || Zeitschriftaufsatz
article of commerce Ware * *Handelsartikel*
articles Waren
articulated Gelenk-* *z.B. bei Stadtbahn-~wagen*
articulated-arm-type robot Knickarmroboter
articulation Gelenk * *Gelenkverbindung*
artifice Trick * *List, Kunstgriff, List*
artificial Kunst- * *künstlich* || künstlich
artificial ageing künstliche Alterung
artificial intelligence künstliche Intelligenz * *der*

404

menschlichen Intelligenz nachempfundene Computersoftware/Hardware
artificial seasoning künstliche Alterung
artificially künstlich * *(Adv.)*
artificially aged künstlich gealtert
artisan Handwerker * *(Kunst)Handwerker, Mechaniker*
artist Grafiker * *advertising artist: Werbe~*
artworking Entflechtung
AS AS * *Australian Standard (Australische Norm ähnlich DIN und VDE)* || dadurch * *weil* || mit * *as the overlap increases:mit steigender Überlappung* || so * *als, wie; as ... as: in solchem Maße/so ... wie*
as a consequence of infolge
as a function of in Abhängigkeit von * *(mathem.)*
as a function of time zeitabhängig
as a general principle grundsätzlich
as a matter of fact eigentlich * *(Adv.)* tatsächlich
as a matter of priority umgehend * *hochprior*
as a result dabei * *dadurch* || infolgedessen
as a result of auf Grund * *als Ergebnis von* || auf Grund * *als Ergebnis von, ~ von* || auf Grund von || durch * *als Ergebnis von* || gemäß * *zufolge* || wegen * *auf Grund von, als Ergebnis von*
as a result of the design bauartbedingt
as a rule in der Regel || meistens * *in der Regel, im Normalfall, üblicherweise* || sonst * *für gewöhnlich*
as above wie oben
as against gegenüber * *im Vergleich zu*
as already mentioned wie schon erwähnt * *siehe auch →oben erwähnt*
as before wie bisher
as compared to im Vergleich zu
as compared with bezogen auf * *verglichen mit*
as customary in gewohnter Weise
as delivered Auslieferungszustand * *im Auslieferungszustand*
as desired beliebig * *(Adv.) wie es beliebt, nach Belieben* || wunschgemäß * *(Adv.)*
as easy as ABC kinderleicht * *(salopp)*
as ever possible möglich * *(Adv.) sofern/wenn ~*
as far as bezüglich * *so weit ... betrifft, betreffend* || was anbetrifft
as far as concerned hinsichtlich * *as far as the performance is concerned: ~ der Leistungsfähigkeit*
as far as possible so weit möglich
as fast as possible so schnell wie möglich
as follows folgendermaßen || so * *folgendermaßen*
as good as praktisch * *(Adv.) so gut wie*
as in the past bisher * *wie bisher* || wie bisher
as it is sowieso
as it stands now voraussichtlich * *wie die Dinge jetzt liegen, aus heutiger/jetziger Sicht*
as it was alt * *wie üblich/letztes Mal/immer; everything remains as it was: es bleibt alles beim Alten*
as it were gewissermaßen
as many as you like beliebig viele
as many times as required jederzeit * *beliebig oft, sooft wie gewünscht/erforderlich*
as matters stand unter diesen Umständen
as mentioned above siehe oben || wie oben
as mentioned below unten stehend
as of ab * *~ einem bestimmten Datum; as of 1996: ~ 1996*
as often as required beliebig oft
as opposed to anstatt || gegenüber * *verglichen mit, im Vergleich zu, im Gegensatz zu* || im Gegensatz zu
as ordered auftragsgemäß
as per entsprechend * *laut, gemäß* || laut * *kraft, gemäß (Gesetz, Vorschrift)*
as per order auftragsgemäß
as possible möglich * *(Adv.) as soon as possible: ~ bald* || möglichst * *as soon as possible: ~ bald* || wie möglich * *as soon as possible: sobald ~*
as requested wunschgemäß * *(Adv.)*
as shipped Anlieferungszustand * *im ~* || Auslieferungszustand * *im Auslieferungszustand*
as short as possible kürzestmöglich
as soon as sobald * *as soon as possible:sobald möglich*
as soon as possible schnellstmöglich * *baldmöglichst* || sobald wie möglich
as standard normalerweise * *in der Standardausführung* || serienmäßig * *(Adv.)* || standardmäßig * *(Adv.) als Standard* || standardmäßig * *im Normalfall*
as such solch * *als solcher*
as the case may be je nachdem
as to whether ob
as until now wie bisher
as usual in gewohnter Weise * *wie gewöhnlich, gewohntermaßen* || wie gewohnt
as well auch * *(nachgestellt) gleichfalls* || ebenfalls
as well ... as sowohl als auch
as-delivered condition Anlieferungszustand * *siehe auch →Auslieferungszustand* || Auslieferungszustand
asbestos Asbest
asbestos-free asbestfrei
ascend ansteigen * *auf/empor/hinaufsteigen, (schräg) in die Höhe gehen* || aufsteigen * *Rang, Zahlen, Kennlinie, Töne, Ballon usw.* || bewegen * *nach oben* || steigen
ascending ansteigend * *Rang, Töne, Zahlen, Kennlinie usw.* || aufsteigend * *Rang, Töne, Zahlen, Kennlinie, Töne usw.*
ascending branch ansteigender Ast
ascending speed ramp ansteigende Hochlaufgeber-Rampe * *für Drehzahlsollwert* || ansteigende Rampe * *e.Drehzahlhochlaufgebers*
ascension Aufstieg * *(amerikan.)*
ascent Anstieg * *Aufstieg, Steigung, Gefälle, Auffahrt, Rampe* || Aufstieg || Steigung * *Anstieg, z.B. einer Kurve*
ascertain ermitteln * *feststellen* || feststellen * *ermitteln*
ascertainment Bestimmung * *Ermittlung, Feststellung* || Ermittlung * *Feststellung*
ASCII ASCII * *Abk.f.'American Standard Code for Information Interchange': Code zur Textdarstellg, ISO 646*
ASCII character ASCII-Zeichen
aseismic erdbebensicher
ash Asche * *auch als Bestandteil von Papier* || Füllstoff * *{Papier} bei der ~messung erfasste Bestandteile*
ash measurement Aschemessung * *zur Messung des Füllstoffgehalts bei Papier*
ash tray Aschenbecher
ASIC ASIC * *Abk. f. 'Application-Specific Integrated Circuit': kundenspezifischer Integrierter Schaltkreis* || kundenspezifischer Schaltkreis * *Abkürzg. f. 'Application-Specific Integrated Circuit'* || Schaltkreis * *kundenspezifischer integrierter ~*
aside seitwärts * *beiseite, zur Seite, abseits, seitwärts; aside from: abgesehen von*
aside from abgesehen von || außer * *(amerikan.)* abgesehen von
ask bitten * *for:um* || konsultieren * *fragen*
ask for anfragen || verlangen * *verlangen nach, bitten um*
ask someone's advice konsultieren * *jemanden um Rat fragen*
ask to come bestellen * *kommen lassen*
aslant schief * *(Adv.) schräg, quer, querüber, quer durch*
ASP Marktpreis * *Abkürzg. für 'Average Selling Price'*
aspect Aspekt || Blickwinkel * *Gesichtspunkt* || Ge-

aspirant

sichtspunkt || Optik * (figürl.) Aussehen, Ansicht, Erscheinung, Gestalt, Gesichtspunkt, Blickwinkel || Seite * ~ einer Angelegenheit
aspirant Bewerber * to:für/um
aspire to anstreben * er-/anstreben, streben/trachten nach
assail angreifen * (auch figürl.) feindlich
assemblage Montage * das Zusammensetzen/-bauen
assemble anfertigen * zusammenbauen/-setzen, montieren || aufbauen * zusammensetzen, montieren || aufstellen * Maschine ~ || einbauen * zusammensetzen/bauen, montieren || fertigen * zusammenbauen, montieren || montieren * zusammenbauen || übersetzen * Assembler/Maschinenspracheprogramm m.Hilfe eines Assemblers || vereinigen * versammeln || zusammenbauen * into: zu || zusammensetzen * zusammenbauen, montieren, Mannschaft zusammenstellen || zusammenstellen * zusammensetzen, montieren, bereitstellen, (Mannschaft, Tatsachen) zusammenstellen
assembled board bestückte Leiterplatte
assembler Assembler * Übersetzungsprogramm für Maschinenspracheprogramme; →Assemblersprache || Assemblierer * Übersetzungsprogramm für Maschinenspracheprogramme || Übersetzer * für Assembler/Maschinenspracheprogramme
assembler language Maschinensprache * Assemblersprache, Mnemotechnische Maschinensprache
assembling Bestückung * Bestücken, Einbau, Zusammenbau || Übersetzung * ~ eines Assemblerprogramms ("Mnemomic-Sprache") in Maschinensprache/Objektcode || Zusammenbau
assembling systems Montagetechnik
assembly -Paket || -Satz * Baugruppe, zusammengesetzte Teile || Aufbau * Zusammensetzung, Zusammenbau || Baugruppe * aus Einzelteilen zusammengebaute ~ || Montage * Zusammenbau, auch Leiterplattenbestückung || Zusammenbau
assembly aid Montagehilfe
assembly belt Montageband
assembly code Assemblercode
assembly drawing Montageplan * Montagezeichnung || Montagezeichnung
assembly equipment Montagevorrichtung
assembly line Band * →Fließband || Fließband * come off the assembly line: vom Band laufen
assembly man Monteur * in d.Fertigung
assembly press Montagepresse
assembly set Bausatz
assembly shop Montagehalle
assembly supervision Montageüberwachung
assembly system Montagesystem
assembly-line Montageband
assent Genehmigung * Billigung (to:von/des) || Zusage * Einwilligung
assent to billigen * zustimmen, billigen, beipflichten, genehmigen || genehmigen * gutheißen
assert behaupten * behaupten, erklären, (Anspruch, Recht) geltend machen, sich behaupten/durchsetzen
assertion Aussage * Behauptung, Erklärung, Geltendmachung
assess bemessen * abschätzl. einschätzen, veranlagen, bewerten || beurteilen * (ein)schätzen (z.B. Schaden), bewerten, veranschlagen, (z.B. Steuer) festsetzen || bewerten * abschätzen || feststellen * einschätzen, bewerten, veranschlagen, festsetzen, schätzen (besonders Schaden)
assessing Abschätzung * Einschätzen, Bewertung, Beurteilung
assessment Abschätzung * (steuerliche usw.) Einschätzen, Veranlagung || Aufnahme * ~ eines Schadens || Beurteilung * Einschätzung, Bewertung || Bewertung * (Wert)Einschätzung, Bewer-

tung (auch figürl.), (Schadens)Feststellung || Festlegung * Festsetzung einer Zahlung/Steuer, Bewertung
asset Wert * Vermögens~
assets Aktiva || Guthaben || Vermögenswerte
assign abordnen || belegen * zuweisen, z.B. Speicherplatz ~, Klemme mit einer Funktion ~ || entsenden * abordnen || führen * zuweisen, zuteilen, to: an || vergeben * zuordnen, zuweisen, zuteilen (auch Speicherplatz, Namen usw.) || verschalten * zuweisen, zuteilen || vorsehen * for: für einen Zweck || zuordnen * (z.B. Aufgabe/Arbeit) zuweisen (to: zu); zuschreiben (to:etwas/jdm.) || zuweisen
assign parameters parametrieren * bei SPS
assign to einteilen * Arbeit, Aufgaben ~
assignable input Wahleingang * frei mit einer Funktion belegbar
assignable output Wahlausgang * frei mit einer Funktion belegbar
assignable parameter Formaloperand
assignable terminal Wahlklemme
assigned vorgesehen * for: für einen Zweck~ || zugeordnet * festgesetzt, angegeben, zugeschrieben/-wiesen, zugeteilt, übertragen (Aufgabe usw.)
assignment Anweisung * Zahlung || Belegung * Zuordnung z.B. für Stecker-Kontakte, IC-Pins, Ein/Ausgänge usw. || Einsatz * Anweisung, Arbeitsauftrag, taktischer Auftrag || Einweisung * Arbeits/Aufgabenzuteilung, Anweisung || Kennzeichen * Zuordnung, Zuweisung || Zuordnung * An/Zuweisung, Bestimmung, Festsetzung; Aufgaben/Arbeits/Postenzuweisung || Zuteilung * An/Zuordnung, Bestimmung, Festsetzung, Übertragung || Zuweisung
assignment list Rangierliste || Zuordnungsliste
assignment of cable cores Kabelbelegung * Zuordnung der Kabeladern zu (Signal-) Funktionen
assignment of conductors Kabelbelegung
assignment table Zuordnungstabelle * z.B. Umrichter zu Motor
assist helfen * assistieren, Beistand leisten; assist us in doing: ~ Sie uns bei || unterstützen * helfen
assistance Förderung * Unterstützung, Mithilfe || Hilfe * Beistand, Unterstützung, Hilfestellung, Assistenz || Mithilfe || Unterstützung * Hilfe, Beistand
assistant Assistent || Assistent * Assistent(in), Helfer, Mitarbeiter, Gehilfe, Angestellte(r), Verkäufer, Hilfs- || Hilfs- || Mitarbeiter * Helfer, Assistent, Zuarbeiter, Stellvertreter || Praktikant * Hilfskraft || Vertreter * Assistent
assisted air cooling verstärkte Luftkühlung * Luftkühlung mit Lüfter(n), z.B. bei Leistungshalbleitern
assmbly language Assemblersprache * Maschinensprache in mnemotechn. Notation (1 Statement entspr. 1 Maschinenbefehl)
associate Gesellschafter || Teilhaber * an einer Firma || verbinden * vereinigen, verknüpfen, verbinden, (gedanklich) assoziieren || vereinigen * (sich) vergesellschaften || verknüpfen * closely associated: eng miteinander verknüpft sein (with: mit)
associated betreffend * zugeordnet || dazugehörig * zugeordnet || verwandt * (eng) verbunden, verwandt (with: mit) || zugehörig || zugeordnet * with: zu || zusammenhängend * miteinander verbunden, in Zusammenhang stehend
associated company Beteiligungsgesellschaft
associated with verbunden mit * einhergehend mit
association Assoziation * Zusammenhang, Gedankenverbindung/verknüpfung, Vereinigung, Zusammenschluss || Verband * z.B. berufsständige Organisation || Verknüpfung * Zusammenhang, Beziehung, Verbindung, Vereinigung || Vermaschung * →Verknüfung || Zusammenhang * ~ von Ideen, Beziehung, Verknüpfung, (Gedanken-)Verbindung

associative assoziativ * z.B. (Cache-) Speicher
associative memory assoziativer Speicher
assort zusammenstellen * nach Sorten ~
assortment Sortierung || Sortiment * of: an || Vorrat * Zusammenstellung, Sortiment, Auswahl, Sammlung, Mischung
assortment kit Sortiment * Teilesatz
assortment of products Produktpalette
assume annehmen * Gestalt, vermuten, auch Wert annehmen || übernehmen * Pflicht, Gefahr, Schuld, Strom ~ || voraussetzen * annehmen, unterstellen, sich anmaßen
assume liability haften * Haftung übernehmen
assume the responsibility Verantwortung übernehmen
assumed vorausgesetzt * it is assumed that: ~/angenommen dass
assuming unter der Annahme || unter der Voraussetzung * angenommen; that: dass || vorausgesetzt * angenommen
assumption Annahme * Vermutung, Voraussetzung || Aufnahme * ~ einer Tätigkeit || Übernahme * Annahme, Übernahme, Aneignung || Voraussetzung * Annahme, Vermutung; on the assumption that: in der Annahme, dass
assurance Aussage * Ver-/Zusicherung, Garantie || Sicherheit * Garantie, Sicherherheit (-sgefühl), Zuversicht, Selbst~, ~ im Auftreten || Sicherstellung * Garantie, Ver-/Zusicherung, Sicherstellung || Versicherung * Ver-/Zusicherung, Garantie
assure behaupten * versichern || sicherstellen * z.B. Einhaltung einer Vorschrift ~ || sorgen für * sicherstellen, bürgen für, sichern || versichern * sichern, sicherstellen, ver-/zusichern (z.B. Leben), beruhigen, überzeugen || zusichern * assure a person of a thing:jdm. etwas zusichern
asterisk Sternchen * Zeichen auf Tastatur/Monitor/ Drucker
astonishing außerordentlich * erstaunlich
asychronous freilaufend * asynchron
asymmetric asymmetrisch || unsymmetrisch
asymmetrical asymmetrisch || unsymmetrisch
asymmetrical DIN rail G-Schiene * asymmetrische Aufschnapp-Install.-Schiene (32mm breit)
asymmetrical load Schieflast
asymptotical asymptotisch
asymtotically asymptotisch * (Adv.)
asynchronous asynchron
asynchronous frequency converter Asynchron-Frequenzumformer * rotierender
asynchronous generator Asynchrongenerator
asynchronous induction machine Asynchronmaschine
asynchronous motor Asynchronmotor
asynchronously asynchron * (Adv.)
at am * at the input: am Eingang || an || auf || bei * z.B. Spannung, Drehzahl, Betriebsbedingung; at high frequencies: bei hohen Frequenzen || bei * in Verbindung mit Institutionen und Personen || um * zeitlich genau
  at 60 degrees each other um 60 Grad verschoben
  at a certain point punktuell
  at a distance of im Abstand von
  at a glance auf einen Blick
  at a high price teuer * (Adv.)
  at a later date zu einem späteren Zeitpunkt
  at a minimum of mindestens
  at a mouse click auf Knopfdruck * mit einem Mausklick
  at a permanent position ortsfest
  at a premium entscheidend * be at a premium: ausschlaggebend/~ sein
  at a reasonable price kostengünstig || preisgünstig * (Adv.)
  at a standstill stillstehend
  at a stationary position ortsfest
  at a time je * immer; three at a time: je drei
  at a variance über Kreuz * (figürl.) zerstritten, im Widerstreit/Widerspruch
  at all costs Gewalt * um jeden Preis
  at all events auf alle Fälle || unter allen Umständen
  at all times jederzeit || stets
  at any other time sonst * zu einer anderen Zeit
  at any point in time zu jedem Zeitpunkt
  at any time jederzeit
  at best bestenfalls || günstig * im ~sten Falle || höchstens * im besten Falle; b. Zahlenangaben nachgestellt
  at both ends beidseitig * an beiden Enden, z.B. Kabel, Welle || doppelseitig * beidseitig || zweiseitig * auf beiden Seiten
  at certain points punktuell
  at each other aneinander
  at first anfänglich * (Adv.) || anfangs || erst * anfangs || erstens * anfangs, zuerst || ursprünglich * (Adv.) || zuerst * anfangs || zunächst * zuerst, anfangs
  at frequency Frequenz erreicht
  at full speed schnellstmöglich
  at hand verfügbar * vefügbar, zur Hand, bei der Hand || vorhanden
  at high speed mit hoher Geschwindigkeit || schnellstmöglich
  at high speeds bei hohen Geschwindigkeiten
  at home erfahren * bewandert, vertraut (in: in)
  at last endlich * (Adv.) || schließlich * (Adv.) endlich || zuletzt * schließlich || zuletzt * als Letzter
  at least mindestens
  at length ausführlich
  at line frequency netzfrequent
  at little expense aufwandsarm * (Adv.) mit geringen Kosten verbunden
  at low speeds bei niedrigen Geschwindigkeiten
  at minimum cost kostenoptimal
  at most höchstens
  at no charge kostenfrei || kostenlos
  at no load Leerlauf- * (nachgetellt)
  at no time nie
  at once auf einmal * zugleich || sofort || unverzögert * (Adv.) unverzüglich
  at one blow gleichzeitig * (Adv.)
  at one end einseitig * auf einer Seite
  at one end only einseitig * nur auf einer Seite
  at one side einseitig * auf einer Seite
  at operating temperature betriebswarm
  at pleasure beliebig * (Adv.)
  at present im Augenblick || im Moment * gegenwärtig || momentan * (Adv.) im Augenblick, zurzeit || zurzeit
  at rated-load operating temperature betriebswarmer Zustand
  at reduced output voltage herabgesteuert * z.B. Stromrichter
  at regular intervals regelmäßig * in ~en Abständen
  at rest ruhig * ausruhend
  at right angles rechtwinklig
  at short notice kurzfristig * (Adv.) z.B. liefern, absagen; available at short notice: kurzfristig lieferbar
  at site anlagenseitig * auf der Baustelle || vor Ort * auf der Anlage/Baustelle
  at speed Drehzahl erreicht
  at system frequency netzfrequent
  at that auch * (nachgestellt) übrigens
  at the back hinten
  at the back of hinter
  at the beginning vorn * am Anfang || zuerst * zu Beginn, anfangs

at the bottom left links unten || unten links
at the bottom of unten * ~in Gefäß/im Wasser/auf Buchseite; bottom up: d.untere nach oben gedreht, kieloben
at the bottom right rechts unten || unten rechts
at the customer's beim Kunden
at the customer's facilities vor Ort * beim Kunden
at the drive end antriebsseitig
at the end hinten * am Ende
at the extreme left links außen
at the extreme right rechts außen
at the foot of unten * am Fuße von
at the head vorn * an der Spitze (of: von)
at the installation anlagenseitig
at the last mindestens
at the left links * auf der linken Seite
at the line side netzseitig
at the load side lastseitig
at the moment augenblicklich * (Adv.). || im Augenblick || im Moment || momentan * (Adv.) im Moment, zurzeit || zurzeit
at the most best * (Adv.) im ~en Falle || höchstens * b. Zahlenangaben nachgestellt
at the present zurzeit
at the present time derzeit || gegenwärtig * siehe auch 'momentan' || momentan * (Adv.) gegenwärtig || zurzeit
at the right rechts * auf der rechten Seite
at the same time dabei * gleichzeitig || daneben * gleichzeitig || gleichzeitig * (Adv.) || hierbei * gleichzeitig || zugleich || zusammen * (Adv.) gleichzeitig
at the side seitlich * (Adv.) || seitwärts * (Adv.) seitlich
at the top oben * an der Spitze
at the top left links oben || oben links
at the top right oben rechts || rechts oben
at the upper left links oben || oben links
at the upper right oben rechts || rechts oben
at the very best bestenfalls
at the very last mindestens
at this hierbei
at times gelegentlich * (Adv.) || manchmal || sporadisch * gelegentlich, manchmal || zeitweise * manchmal
at top speed schnellstmöglich
at upper left oben links
at upper right oben rechts
at which wobei
at worst schlimmstenfalls
at your convenience sobald * sobald es Ihnen möglich ist/wann es Ihnen beliebt
at your leisure gelegentlich * (Adv.) wann es Ihnen passt
at zero current stromlos
ATE Prüfautomat * Abkürzg.f. 'Automatic/Automated Test Equipment': automatische Testeinrichtung
ATE equipment Prüfautomat
ATM Geldautomat * Abk. für 'Automated Teller Machine': Bankautomat
atmosphere Atmosphäre
atmosphere of departure Aufbruchstimmung * 'Abschieds/Abreisestimmung'
atmospheric humidity Luftfeuchtigkeit
Att. z Hd * zu Händen || zu Händen
att: zu Händen von
attach anbauen * Maschinenteil || anbringen || anhängen || anschlagen * Hebezeug || befestigen * to:an || beifügen || hinzufügen * beilegen, beifügen, anheften, befestigen, verbinden, verknüpfen, angliedern || verknüpfen * to: mit || zuteilen * eine Person jemandem ~ || zuweisen * jemandem eine Person ~
attach importance to Wert legen auf * great/significant: großen

attaché case Aktenkoffer
attached Anlage * in der Anlage (z.B. eines Briefes) || befestigt || hierbei * angeschlossen, zugeteilt, dazugehörig
attachement Befestigung * Befestigung, Anbringung, Verbindung, Anschluss
attaching pad Anbauflansch
attachment Anbau * Anbringung, Befestigung || Anlage * zu einem Handbuch, Bericht, Brief || Zubehör * Zusatzgerät
attachments Anbauten
attack angreifen * (auch figürlich) feindlich || zerfressen * angreifen, befallen, aggressiv einwirken
attain annehmen * erreichen; auch 'Wert erreichen' || erlangen * erreichen, erzielen || erreichen * erlangen, erzielen, z.B. Maximalwert erreichen || erzielen * erlangen, erreichen
attainable erreichbar * erzielbar || erzielbar || möglich * erreichbar, erzielbar
attempt Anstrengung * Versuch, Bemühung || Versuch || versuchen
attend bedienen * sich kümmern/sorgen um, Acht geben/achten auf, Dienst tun || betreuen || kümmern * to: sich um jmd. ~; beiwohnen, teilnehmen
attend at teilnehmen * anwesend sein
attendance Anwesenheit * at:bei || Beteiligung * Teilnehmerzahl || Teilnahme * an einer Versammlung, Besprechung usw. || Wartung * Pflege, Wartung
attendance mistake Bedienungsfehler
attendant Arbeiter * ~ an der Maschine || Bedienungsmann || Wärter
attendant symptom Begleiterscheinung || Nebeneffekt * Begleiterscheinung
attendants Personal * Bedienstete, Bedienungsmannschaft
attending persons Gesprächspartner * Anwesende || Teilnehmer * Anwesende
attention Achtung || Aufmerksamkeit * pay attention to: jmd./e.Sache Aufmerksamkeit schenken || Beachtung * pay attention to: etwas ~ schenken || Sorgfalt * Aufmerksamkeit
attentive sorgfältig * achtsam, aufmerksam
attentiveness Aufmerksamkeit
attenuate abschwächen || dämpfen * (elektr.) abschwächen || herunterleiten * z.B. Spannung(ssignal) über einen Spannungsteiler ~
attenuation Abschwächung * Bedämpfung * Dämpfung, Abschwächung || Dämpfung * Abschwächg. (z.B. Signal auf Leitg./an Messgerät.), Dämpfg. elektr./opt./mech. Schwingg.
attenuator Abschwächer || Dämpfungsglied * (elektr.) Abschwächer || Spannungsteiler * Dämpfungs/Abschwächglied, z.B. für Messgerät, Gleichstromtacho-Auswertung || Teiler * Spannungsteiler zur Abschwächung von Spannungssignalen
attest bescheinigen
attitude Einstellung * mental: geistige ~, personal: persönliche ~ || Orientierung * (Ein)Stellung (to/towards: to), Standpunkt, Verhalten, Haltung, Stellungnahme || Stellungnahme * Einstellung, Haltung, Standpunkt (to(wards):zu) || Verhalten * Haltung
attract anziehen * Magnet/Kapital ~
attraction Anziehung * (auch figürl.) magnetic: magnetisch; electrostatic: elektrostatisch || Anziehungskraft * (figürl.)
attractive ansprechend * attraktiv || attraktiv || interessant * attraktiv
attractive force Anziehungskraft * z.B. magnetische
attractive power Anziehungskraft * (physikal.)
attribute Attribut * (Computer) (Zusatz-) Kennzeichen einer Variablen/einer Datei usw.

attune to abstimmen * z.B.
audible deutlich * hörbar || hörbar * inaudible: nicht hörbar || laut * vernehmlich, hörbar || wahrnehmbar * hörbar
audible alarm akustische Meldung * Warnsignal
audible noise Geräusch * hörbares ~
audible range Hörbereich
audible signal akustische Meldung
audio frequency Tonfrequenz
audit Audit || kontrollieren * (Geschäfts-)Bücher ~ || prüfen * amtlich ~, einer Revision unterziehen || Prüfung * Revision (Buch, Geschäftsbücher usw.), Rechnungs~ || Revision * (Handelswesen) Rechenschaftslegung, Überprüfung (der Geschäftsbücher), Rechnungsprüfung || Überprüfung * Revision (Geschäftsbücher usw.), Rechnungsprüfung
auditing Rechnungsprüfung || Revision || Überprüfung * Rechnungsprüfung
auditor Wirtschaftsprüfer
auditorium Hörsaal
auger Schnecke * ~gang, ~bohrer
auger doser Dosiermaschine * mit Dosierschnecke
auger-type dosing unit Schneckendosierer * (Verpack.techn.)
auger-type feeder system Schneckenförderer * (Verpack.techn.) siehe auch →Schneckendosierer
augment ergänzen * (sich) vermehren, vergrößern, zunehmen, ergänzen || erhöhen * steigern (to: auf, by:um) || vergrößern * (sich) vermehren/vergrößern, zunehmen || vermehren * vergrößern, (sich) vermehren, zunehmen, ergänzen
augmention Zuwachs * of: an/von
Austria Österreich
authentic maßgebend * Text || maßgeblich * Text
authentic record Beleg
authenticate bescheinigen * amtlich ~
authenticity Glaubwürdigkeit
authoritative kompetent * maßgeblich, maßgebend || maßgebend * autoritativ, maßgeblich, amtlich, gebieterisch || maßgeblich * maßgebend/geblich, autoritativ, amtlich, gebieterisch
authoritative body Gremium
authority Befugnis || Behörde || Dienststelle * bei einer Behörde || Gewalt * z.B. gesetzgebende, || Instanz * Dienststelle || Kompetenz * Befugnis || Sachverständiger * in/on:für || Stelle * ~ bei einer Behörde
authorization Berechtigung * Ermächtigung, Vollmacht, Genehmigung (to: zu) || Genehmigung * Ermächtigung; authorize a person to:jdm. zu etw. ermächtigen || Mandat
authorize beauftragen * ermächtigen || berechtigen || billigen * gutheißen, billigen, genehmigen || genehmigen * gutheißen, billigen, rechtfertigen, ermächtigen, bevollmächtigen, berechtigen || zulassen * ermächtigen, bevollmächtigen, berechtigen, gutheißen, billigen, genehmigen
authorized autorisiert * siehe auch 'befugt' || befugt * to:zu || kompetent * befugt || zugelassen * amtlich ~ || zulässig * befugt, zulässig, autorisiert, bevollmächtigt
authorized dealer Vertragshändler
authorized inspection agency TÜV * Technischer Überwachungsverein
auto diagnosis Selbstdiagnose
auto restart automatischer Wiederanlauf * z.B. nach einem Netzspannungseinbruch
auto speed search Suchen-Fangen
auto tracking automatische Nachführung
auto- Automatik- || automatisch * (als Vorsilbe) || Selbst- || selbsttätig || selbsttätig
auto-adapting selbstanpassend
auto-calibration Selbstabgleich * einer Messeinrichtung

auto-extinguishing selbstverlöschend * bei Feuer
auto-restart Fangen || Fangschaltung
auto-reverse stage Kommandostufe
auto-reversing module Kommandostufe
auto-reversing stage Kommandostufe * zur Steuerung des Momentenwechsels bei Stromrichter
auto-routing automatische Entflechtung * einer Leiterkarte
auto-tune Optimierungslauf || Selbstoptimierung
auto-tuning Selbsteinstellung * z.B. Regler
autogenous autogen * z.b. Schweißverfahren
autogenous welding Autogenschweißen
automat Automat * automatic lathe: Drehautomat; siehe auch →Sicherungsautomat
automate automatisieren
automated automatisch * automatisiert || automatisiert
automated guided fahrerlos * Fahrzeug || selbstfahrend * z.B. Transportfahrzeug
automated guided vehicle system fahrerloses Transportsystem
automated teller machine Geldautomat * Bankautomat
automated-guided unbemannt * selbstlenkend, fahrerlos
automated-guided vehicle führerloses Fahrzeug
automatic Automatik- || automatisch * (Adj.) || selbsttätig * (korrekte Schreibweise ist → "selbsttätig") || selbsttätig
automatic adaption to the system frequency automatische Netzfrequenzanpassung * z.B. bei Steuersatz e. netzgeführten Stromrichters || Netzfrequenzanpassung * z.B. b. Steuersatz e. netzgeführten Stromrichters
automatic blanking press Stanzautomat
automatic changeover circuit Umschaltautomatik
automatic check Selbsttest
automatic circuit breaker Sicherungsautomat
automatic component insertion equipment Bestückungsautomat
automatic control Automatik || Regelung
automatic control loop Regelkreis
automatic control theory Regelungstechnik
automatic cut-out Sicherungsautomat
automatic drilling machine Bohrautomat * (Werkz.masch.)
automatic edger Besäumautomat * (Holz)
automatic feeder Bogenanleger * (Druckerei) automatic ~ für Bogendruckmaschine || Vorschubapparat * (Werkz.masch.\ Holz)
automatic grinder Schleifautomat * (Werkz.masch.)
automatic grinding machine Schleifautomat * (Werkz.masch.)
automatic image recognition Bilderkennung * automat. Bilderkennung
automatic imaging Bilderkennung
automatic lathe Drehautomat
automatic line frequency tracking automatische Netzfrequenzanpassung * aut. Netzfrequ.nachführg. b.Steuersatz e. netzgef. Stromrichters || Netzfrequenzanpassung * aut. Netzfrequ.-nachführg. b.Steuersatz e. netzgef. Stromricht.
automatic machine Automat
automatic mode Automatikbetrieb * ~sart
automatic operation Automatik || Automatikbetrieb
automatic ordering beleglose Bestellung
automatic pattern recognition Bilderkennung * automatische Mustererkennung
automatic pick-and-place machine Bestückungsautomat
automatic picture recognition Bilderkennung
automatic restart Wiedereinschaltautomatik * siehe auch →Suchen, →Fangschaltung

automatic restart feature Wiedereinschaltautomatik
automatic restart on the fly Wiedereinschaltautomatik * *Aufschalten e. Stromrichters auf lauf. Motor nach Netzspannungsausfall*
automatic sander Schleifautomat * *{Holz} mit Schmirgel arbeitend*
automatic shut-down Stillsetzautomatik
automatic source matching automatische Anpassung an die Netzspannung
automatic start up Anfahrautomatik
automatic start-stop control Ein-Ausschaltautomatik
automatic starting control Anfahrautomatik
automatic stop control Stillsetzautomatik
automatic supply voltage matching automatische Anpassung an die Netzspannung
automatic switch-on and switch-off circuit Ein-Ausschaltautomatik
automatic switch-on and switch-off control Ein-Ausschaltautomatik
automatic tester Prüfautomat
automatic transmission Automatik * *automatisches Getriebe*
automatic tuning Selbstabgleich * *z.B. eines Reglers*
automatic warm restart automatischer Wiederanlauf
automatic winding machine Wickelautomat
automatic wire winder Drahtwickelautomat
automatically automatisch * *(Adv.)* || selbst * von ~, automatisch || selbsttätig * *(Adv.)* || zwangsläufig * *(Adv.)*
automation Automatisierung || Automatisierungstechnik
automation concept Automatisierungskonzept
automation control system Automatisierungssystem
automation engineering Automatisierungstechnik * *als Fachrichtung*
automation environment Automatisierungslandschaft
automation interface Automatisierungsschnittstelle
automation level Automatisierungsebene * *lower/medium/higher:untere/mittlere/obere*
automation scenario Automatisierungslandschaft
automation setpoint Automatiksollwert || Automatisierungssollwert
automation solution Automatisierungslösung
automation system Automatisierung * *Automatisierungssystem* || Automatisierungssystem
automation task Automatisierungsaufgabe
automation technology Automatisierungstechnik
automation unit Automatisierungsgerät
automatism Automatik
automatization Automatisierungstechnik
automotive electronics Automobilelektronik || Kraftfahrzeugelektronik
automotive industry Autoindustrie || Automobilbau || Automobilindustrie
autonomous abgeschlossen * autark || autark * *für sich allein lebens-/betriebsfähig* || eigenständig
autotransformer Sparschaltung * *Spartransformator* || Spartrafo || Spartransformator
autotransformer starter Anlasstransformator * *{el.Masch.} als Spartrafo ausgelegte Starthilfe für Motor*
autotuning Selbstabgleich || Selbstoptimierung
auxiliaries Hilfsbetriebe
auxiliary Hilfs- || Hilfsmittel * *Hilfs-/Behelfs-/Ausweich- (Mittel)* || Neben- * Hilfs- || Zusatz- * Hilfs- || zusätzlich * Hilfs- || Zwischen- * Hilfs-
auxiliary arm Hilfszweig * *z.B. eines Stromrichters*

auxiliary brake Zusatzbremse * *z.b. für Sicherheitszwecke*
auxiliary circuit Hilfsstromkreis
auxiliary contact Hilfskontakt * *z.B. an Schütz, Relais, Leistungsschalter*
auxiliary device Hilfseinrichtung || Hilfsmittel * *Gerät, Einrichtung* || Zusatzeinrichtung
auxiliary drive Nebenantrieb * *Hilfsantrieb*
auxiliary energy Hilfsenergie
auxiliary equipment Hilfsmittel || Hilfsstoff || Nebeneinrichtung || Zusatzeinrichtung
auxiliary flag Hilfsmerker * *bei SPS*
auxiliary function Hilfsfunktion
auxiliary machine Hilfsmaschine
auxiliary material Hilfsmittel * *i. weitesten Sinne* || Hilfsstoff * *siehe auch →Hilfsmittel*
auxiliary materials Arbeitsmittel
auxiliary motor Hilfsmotor
auxiliary phase Hilfsphase * *z.b. bei Einphasen-/ →Kondensatormotor*
auxiliary plants Nebenbetriebe
auxiliary power Hilfsenergie
auxiliary power supply Hilfsstromversorgung
auxiliary program Hilfsprogramm
auxiliary starting winding Hilfswicklung * *z.B. zur Anlaufhilfe von Einphasenmotoren*
auxiliary supply Hilfsspannung * *Versorgungsspannung*
auxiliary terminal Hilfsklemme * *z.b. für Temperaturgeber und Tacho in Motorklemmenkasten*
auxiliary thyristor Hilfsthyristor
auxiliary voltage Hilfsspannung
auxiliary winder Hilfswickler * *z.b. zum Auf-/Abwickeln des Trägermaterials*
auxiliary winding Anlaufwicklung * *Hilfswicklung* || Hilfswicklung
av arithmetischer Mittelwert * *Abkürzg. für 'Average'*
avail nützen * *avail oneself of:etwas ausnutzen (Gelegenheit usw.)*
avail oneself of nutzen * *Gelegenheit ~* || wahrnehmen * *Gelegenheit ~*
availability Lieferbarkeit || Liefereinsatz * *become available in June 1999: ~ beginnt im Juni 1999* || Verfügbarkeit
available erhältlich || erreichbar * *verfügbar* || greifbar * *verfügbar* || lieferbar * *erhältlich, verfügbar; available on short lead times: kurzfristig ~* || nutzbar * *verfügbar* || verfügbar * *freely: frei* || vorhanden * *verfügbar, zur Verfügung gestellt* || vorrätig
available for delivery lieferbar
available from Lieferort * *kann geliefert werden von*
available from stock lagerhaltig
available from the stock ab Lager lieferbar || lagermäßig
available off the shelf ab Lager * *ab Lager lieferbar* || ab Lager lieferbar
available on the market handelsüblich
available soon in Kürze lieferbar
avalanche current Durchbruchstrom * *z.B. bei Halbleiter (Thyristor usw.)*
avalanche effect Lawineneffekt * *bei Halbleitern*
avalanche phenomenon Lawineneffekt * *z.B. bei Halbleitern (Durchbruch)*
average arithmetischer Mittelwert || Durchschnitt * *on (an) average:im Durchschnitt* || durchschnittlich || Mittel * *Mittel/Durchschnitts-Wert/Größe* || Mittel * *(Adj.) durchschnittlich* || Mittelwert * *Durchschnitt* || mittlerer * *durchschnittlich*
average calculation Mittelwertbildung * *durch Berechnung*
average figure Richtwert
average starting current mittlerer Anlassstrom

* *{el. Masch.}* SQRT(Anl.spitzenstrom x Schaltstrom), kennzeichn.→Anlassschwere
average value Mittelwert * *Durchschnittswert*
averaged gemittelt * *over 10 sec:über 10 sec*
averaging Mittelwertbildung
averaging of setpoint value Sollwertglättung * *(gleitende) Mittelwertbildung*
avert abwehren * *Unheil* || abwenden * *Gefahr/Unheil* ~
averting Ablenkung * *das Abwenden/-wehren*
aviation Flugwesen || Luftfahrt
avoid abbiegen * *(figürl.) Konflikt, Nachteil usw.* verhindern || umgehen * *vermeiden* || verhindern * *vermeiden* || vermeiden
avoidable vermeidbar
award Auftragserteilung * *bei einer Ausschreibung* || Belohnung * *Verleihung, Zuerkennung* || erteilen * *z.B. (einen Lieferauftrag) vergeben, gewähren, zuteilen, zuerkennen, zusprechen* || Preis * *Belohnung, Auszeichnung, Preis, Prämie, gutes Testurteil, z.B. in Zeitschrift* || Zuschlag * *b.Ausschreibg., Auftragsvergabe; of contract/order: Vergabe eines Liefer/Bestellauftrags*
award of contract Auftragserteilung * *bei Ausschreibung/gegen Konkurrenz*
award-winning ausgezeichnet * *mit einem Preis/ einer Auszeichnung* ~
awarded ausgezeichnet * *mit einer Auszeichnung/ einem Preis* ~
away ab * *nicht (mehr) vorhanden (from: von), weg* || abwesend || davon * *hinweg/-fort* || Weg * *(Adv.) fort* || zurück- * *(nachgestellt) aus dem Weg, weg; z.B. fold back: ~/wegklappen*
awful reichlich * *(salopp) furchtbar, riesig, kolossal*
awfully reichlich * *(salopp; Adv.) furchtbar, äußerst, sehr, riesig* || sehr * *(salopp; Adv.) siehe auch →reichlich*
AWG Leitungsquerschnitt * *Abkürzg- für 'American Wire Gauge': amerikan.Maßeinheit für Leit.-querschnitt*
awkward schwierig * *unangenehm* || steif * *linkisch* || störend * *peinlich*
awkward position Schieflage * *(figürlich) Krise, unangenehme Lage*
awkward positions schwerzugängliche Stellen
awning Decke * *Plane*
awry schief * *(Adv.) schief, krumm, ganz schief, verkehrt, scheel*
axes Achsen * *(Plural von 'axis') auch bei Werkzeugmaschine*
axial axial
axial blower Axialgebläse
axial clearance Axialspiel * *freier Abstand* || Lagerspiel * ~ *in Axialrichtung*
axial direction Längsrichtung * *eines zylinderförmigen Teils (z.B. Motor)*
axial eccentricity Planlauf || Planlauffehler * *z.B.Rundlauffehl. e.Welle, Flanschzentrierg., Planlauf zw. Wellenende u.Flansch* || Planlauftoleranz || Rundlauf * *Planlauf/-/fehler*
axial eccentricity tolerance Planlauftoleranz * *z.B. zw. Wellenende u. Flanschfläche e. Motors*
axial fan Axiallüfter || Axialventilator
axial flux Axialfluss
axial force Axialkraft || Axialschub || Längskraft * *Axialkraft*
axial load Axialbelastung || Axialkraft * *axiale Belastung* || Axiallast
axial movement Axialbewegung
axial play Axialspiel
axial pressure Axialdruck
axial pump Schraubenpumpe
axial runout Planlauf * *{el. Masch.} Fluchtfehler des Wellenendes bezogen auf den Flansch (DIN 42955, IEC 72-1)*

axial sealing ring Axialdichtring * *z.b. Simmerring*
axial thrust Axiallast
axial winder Achswickler || Zentralwickler
axial-flow fan Axiallüfter || Axialventilator
axial-lead axial bedrahtet
axially axial * *(Adv.)*
axially parallel achsparallel
axially symmetrical rotationssymmetriesch
axiom Grundsatz * *unbestreitbarer* ~
axis Achse * *geometrisch*
axis module Achsmodul * *bei Modular- oder Servoumrichter*
axis of rotation Drehachse
axis of x Abzisse * *X-Achse* || X-Achse
axis of y Ordinate * *Y-Achse* || Y-Achse
axle Achse
axle bearing Achslager
axle drive Achsantrieb
axle journal Achsschenkel * *{Auto}*
axle neck Achsschenkel * *{Auto}*
AZNIE AZNIE * *Associazione Nazionale Industre Elettrotecniche: ital. Verband d. Elektroindustrie (wie ZVEI)*

B

B contact Ruhekontakt * *Abkürzg. f. 'Break contact': Öffnerkontakt*
B type of construction B-Bauform * *Motor*
B-contact Öffner * *Abk. für 'break contact'* || Öffnerkontakt
B.Sc. Diplomingenieur * *(brit.) Abk. f. 'Bachelor of Science'*
B2C connection B2C-Schaltung * *vollgesteuerte Einphasen-Gegenparallel-Brückenschaltg. für 4Q-DC-Stromrichter*
B2H connection B2H-Schaltung * *halbgesteuerte Einphasen-Brückenschaltung für 1Q-DC-Stromrichter*
B6C connection B6C-Schaltung * *vollgesteuerte Drehstrombrückenschaltung für 1Q-DC-Stromrichter*
Bachelor of Engineering Diplomingenieur
back Antriebsseite * *Rückseite, hintere Seite* || Gegen- * *in Gegen-/Rückwärtsrichtung* || Hinter- || nach hinten || Rück- || Rückseite * *Hinterseite* || rückwärts * *zurück* || unterstützen || zurück
back and fill schwanken * *(amerikan.) unentschlossen sein, zaudern*
back and forth hin und her
back axle Hinterachse
back bear Vorgelege
back cloth Drucktuch * *{Druckerei}*
back documentation Rückdokumentation * *z.B. durch Recompiler, Disassembler* || Rückdokumentation
back driving Rücklauf * *Zurückgedrückt-Werden bei Wegblieben des Antriebsmoments*
back drum hintere Tragwalze * *Tragwalze 1 eines Doppeltragwalzenrollers* || Tragwalze 1 * *eines Doppeltragwalzenrollers*
back EMF Gegen-EMK || Gegenspannung * *EMK eines Motors (EMF: ElectroMagnetic Force)*
back flow Rückfluss
back of hinter * *(salopp)*
back out zurücklaufen * *rückwärts herausfahren*
back panel Rückwand
back pressure Rückstaudruck
back stop Rücklaufsperre

back to back gegenüberliegend * *Rücken an Rücken*
back to front falsch herum * *die Rückseite nach vorn*
back up fördern * *den Rücken stärken* || untermauern * *mit Rückendeckung versehen* || unterstützen
back voltage Gegenspannung
back wall Rückwand * *{Holz\ Bautechnik}*
back water Rückwasser
back-EMF Gegenspannung
back-EMF constant Spannungskonstante * *Verhältnis EMK zu Drehzahl [V/upm] bei Elektromotor (1/→Drehzahlkonstante)*
back-pressure Gegendruck || Rückstau || Staudruck
back-pressure valve Rückschlagventil
back-slash umgekehrter Schrägstrich
back-to-back gegeneinander * *Rücken an Rücken, aneinander*
back-to-back connection Antiparallelschaltung
back-tracking Zurückverfolgung
back-up puffern * *unterstützen, den Rücken decken, (mit Batterie/Kondensator) puffern* || Sicherheits- * *zur Reserve (z.B. Sicherungsdiskette)*
back-up capacitor Stützkondensator * *z. Stützen d. Stromversorgg. auf Elektron.leiterkarte, 'schluckt' Schaltspitzen*
back-up copy Sicherungskopie
back-up service Unterstützung
back-voltage Gegenspannung
backbone Bus-Hauptstrang * *von dem die Vor-Ort-Kommunikationsstränge über Stichleitungen abzweigen* || Haupt-Busstrang * *in Computer-Netzwerk* || Hauptnervenstrang * *(figürl.) Haupt-Busstrang, von d.neben-/Vor-Ort-Busse/Stichleitungen abzweigen*
backend nachgeschaltet * *{Computer} z.B. ~es Softwareprogramm*
backer Förderer * *Gönner*
backflow Rückfluss * *(amerikan.)*
background Hintergrund * *auch langsame Programmablaufebene (Programme mit niedriger Priorität)* || Veranlassung * *Hintergrund, Beweggrund*
background level Störpegel * *akustischer ~*
background memory Hintergrundspeicher
background noise Fremdgeräusch
background process Hintergrundprozess
background processing Hintergrundverarbeitung * *Rechnerprogramme, die mit niedriger Priorität ablaufen (as: in)*
background task Hintergrundverarbeitung * *Rechnerproramm, das nicht auf externe Ereignisse reagieren muss*
backing Rückseite * *rückwärtige Schicht/Verstärkung, Futter, Stützung*
backing drum Stützwalze
backing roll Gegenwalze * *z.B. bei Streichmaschine, Beschichtungsmaschine* || Stützwalze
backlash Flankenspiel * *im Getriebe zw. Zahnflanken* || Gang * *toter Gewinde-* || Lose * *~ im Gewinde/Getriebe* || Spiel * *Lose in Getriebe/Gewinde*
backlash of teeth Zahnflankenspiel
backlash-free spielfrei
backlight Hintergrundbeleuchtung * *z.B. bei einer LCD-Anzeige*
backlit hintergrundbeleuchtet * *z.B. LCD-Anzeige*
backlog Rückstand * *Liefer-/Arbeits~*
backlog of unfilled orders Auftragspolster * *Auftragsüberhang* || Auftragsrückstand
backplane Busplatine || Rückwandleiterplatte || Rückwandplatine
backplane assembly Rückwandplatine
backplane bus Rückwandbus
backplane connector Basisstecker * *auf e. Leiterkarte zur Rückwandplatine hin angeordnet*

backplane wiring Rückwandverdrahtung * *eines Baugruppenträgers*
backslash Schrägstrich * *umgekehrter ~*
backstop Rücklaufsperre
backstopping clutch Rücklaufsperre * *Kupplung, die ein Rücklaufen des Antriebs verhindert*
backup Pufferung * *mit Batterie, Kondensatoren, Schwungrad usw.* || retten * *{Computer} eine Sicherheitskopie anlegen*
backup battery Pufferbatterie
backup capacitor Stützkondensator * *verteilte ~en puffern Spannungseinbrüche b. Schaltvorgängen i. Elektronikschaltungen*
backup time Pufferungszeit * *Zeitdauer, die eine Pufferbatterie überbrücken kann*
backup voltage Pufferspannung
backward rückwärts || zurück
backward flow Rückfluss
backward movement Rücklauf
backward-compatibility Rückwärtskompatibilität
backwards hinten * *nach hinten* || nach hinten * *zurück, rückwärts* || zurück
bad groß * *(Fehler)* || nachteilig || schädlich * *schlecht* || schlecht * *(gelegentlich auch als Adv.) worse: ~er; worst: ~est* || stark * *schlimm*
bad connection Wackelkontakt
bad job Pfusch
bad state of affairs Missstand
badly dringend * *(Adv.) want badly: ~ brauchen* || schlecht * *(Adv.)* || unbedingt * *(Adv.) dringend; need badly: unbedingt/dringend brauchen*
baffle Leitblech
baffle plate Leitblech * *Ablenk-/Prallplatte, Schlingerwand (in Tank usw.)*
bag Beutel
bag filling machine Beutelmaschine * *{Verpack.-techn.} Beutelfüllmaschine*
baggy bauschig
bags packaging machine Beutelmaschine * *{Verpack. techn.}*
bail Bügel * *Henkel, Bügel, (Hand-) Griff; Schranke; Bürge, Kaution; (Wasser) schöpfen*
bake sintern
baked gesintert || verbacken * *siehe auch →sintern*
bakery machine Bäckereimaschine
baking oven Lackierofen || Trockenofen
balacing resistance Symmetriewiderstand
balance Abgleich * *(Null)Abgleich, Ausgleich, Kompensation* || abgleichen * *austarieren, (auch Konten ~)* || abstimmen * *austarieren* || ausbalancieren * *ins Gleichgewicht bringen, ausbalancieren, auswuchten, Konto ausgleichen* || Ausgleichs- || auswuchten || Bilanz || bilanzieren * *{Wirtschaft} Konto ausgleichen, aufrechnen, saldieren, abschließen* || Gleichgewicht * *unbalance: aus dem ~ bringen; loose one's: verlieren; balance: ins ~ bringen* || Guthaben * *my balance stands at 5 $: mein ~ beträgt 5 $; balance standing for my favour: mein ~* || Nullabgleich || symmetrieren || Überschuss * *Saldo* || verrechnen * *ausgleichen* || Waage || wuchten
balance coil Saugdrossel * *{Elektr.}*
balance dynamically dynamisch auswuchten
balance out auswuchten
balance quality Wuchtgüte * *{el.Masch.} Bahngeschw. d. Schwerpunkts e. Rotors mit Restunwucht (VDI 2056 u.2060)*
balance sheet Bilanz * *Aufstellung*
balance weight Gegengewicht
balanced abgeglichen || ausgeglichen || symmetrisch
balancing Abgleich * *Austarieren* || Ausgleich || Ausgleichs- || Auswuchtung || Symmetrierung || Wuchtung * *Auswuchtung*
balancing machine Auswuchtmaschine
balancing resistor Symmetriewiderstand

balancing ring Auswuchtring
balancing set Tariersatz * *zum Auswuchten*
balancing type Wuchtart
bale Ballen
bale press Ballenpresse * *(Verpack.techn.)*
baling press Ballenpresse * *(Verpack.techn.)*
baling ress Bündelpresse * *Ballenpresse*
ball bearing Kugellager
ball mill Kugelmühle
ball pen Kugelschreiber
ball point pen Kugelschreiber
ball screw Kugelgewindetrieb || Kugelspindel
ball screw drive Kugelgewindetrieb
ball socket Kugelpfanne
ball srew Kugelgewindespindel * *z.B. an Hubspindeltrieb*
ball valve Kugelventil
ball-point pen Kugelschreiber
ballast Vorschalt- * *z.b. für Leuchte; auch 'Vorschaltgerät' für Leuchtstoffröhre* || Vorschaltgerät * *für Leuchtstofflampen*
ballast circuit Ballastschaltung
balloon Ballon
ballpen Kugelschreiber
balustrade Geländer
ban sperren * *verbieten, Verbot auferlegen, mit Verbot belegen*
banana plug Bananenstecker * *z.B. 4mm-Stecker*
banana roll Breitstreckwalze * *(salopp) Bananenförmige Walze*
band Band * *Frequenzband, Metallband* || Beschlag * *an Kisten* || Fenster * *Band (z.B. Toleranzband)*
band annealing line Bandglühanlage
band annealing plant Bandglühanlage
band brake Bandbremse
band conveyor Bandförderer || Förderband
band iron Bandsäge
band saw Bandsäge
band saw blade Bandsägeblatt
band-conveyor Förderband
band-pass Bandpass
band-pass filter Bandpassfilter
band-sawing machine Bandsägemaschine
band-shaped bandförmig
band-stop filter Bandsperre || Sperrfilter * *Bandsperre*
bandage Bandage || Verband * *in der Medizin*
banderoling machine Banderoliermaschine * *(Verpack.technik)*
banding Bandage * *Wicklung* || Bandagierung * *Wicklung*
banding machine Banderoliermaschine * *(Verpack.technik)*
bandwidth Bandbreite * *z.b. eines Frequenzbandes/Bandpasses/Datenübertragungskanals*
bang Schlag * *schallender ~, Knall, Krach; Energie, Schwung*
bang-bang control Zweipunktregelung
bang-bang controller Zweipunktregler
banisters Geländer
bank Bank * *Geldinstitut; Speicherbank, Kondensatorbank, Reihenanordnung* || Batterie * *Reihe von Geräten/Bauelementen, z.B. von Kondensatoren* || Reihe * *auch ~nanordnung* || Satz * *~ von Bauteilen*
bank-up Rückstau * *im Verkehr*
banknote paper Wertpapier * *Papiersorte*
banknote printing machine Wertpapierdruckmaschine * *für Banknoten*
bankruptcy Konkurs * *['bӕŋkrʌptsi] go bankrupt:Konkurs gehen*
banner Plakat
bar Balken * *auf Anzeige, Messgerät, Bildschirm usw.* || Lamelle * *auch am Kommutator; Streifen,* Strich || Leiste * *am Bildschirm, z.b. toolbar: Werkzeug~, menu bar: Menü~* || Schiene * *z.B. Strom~* || Stab * *Gitterstab, Metallstab* || Stange || Strebe * *Stange, Stab, Querbalken/-stange, Schranke, Sperre*
bar conductors Verschienung
bar diagram Balkendiagramm
bar display Balkenanzeige
bar graph Balkenanzeige || Leuchtbalken * →*Balkenanzeige*
bar insulation Stabisolierung * *(el. Masch.) Isolierung e. stabförmigen Leiters*
bar vibrations Stabschwingungen * *(el. Masch.) Schwingungen v. losen Stäben e. Käfigläufers mit doppelter Netzfrequenz*
bar winding Stabwicklung * *(el.Masch.) Wicklung, deren Leiter aus Stäben besteht*
bar-core reactor Stabkerndrossel
bar-type current transformer Durchsteckstromwandler * *mit integriertem Durchsteck-Primärleiter*
bar-type transformer Durchsteckwandler * *mit integriertem Durchsteck-Primärleiter*
bare abisolieren || blank * *nicht isoliert (z.B. Draht), nicht abgedeckt, bloß* || bloß * *unbedeckt* || knapp * *knapp, kaum hinreichend, arm (of: an); a bare mile: e.~ Meile; bare majority: ~ Mehrheit* || unbestückt * *z.B. Leiterkarte*
bare board unbestückte Leiterplatte
bare PCB unbestückte Leiterplatte
bareled ballig * *fassförmig*
barelled ballig * *fassförmig*
barely kaum * *nur gerade* || knapp * *(Adv.) mit ~er Not; barely sufficient: ~ ausreichend*
bargain Geschäft * *(~s-) Abkommen, Handel, Geschäft, vorteilhaftes ~, günstiger Kauf* || Handel * *Geschäft(sabkommen), vorteilhaftes Geschäft* || handeln * *feilschen (for: um)* || Kauf * *guter ~, "Schnäppchen"*
bargraph Balken * *zur Messwertanzeige auf Messinstrument/Bildschirm* || Balkenanzeige || Balkendiagramm
bargraph display Balkenanzeige
bark Rinde * *(Holz)*
barker Entrindungsmaschine * *(Holz)* || Holzschälmaschine * *Entrindungsmaschine*
barm Hefe
barometric pressure Luftdruck
barrel einfüllen * *in Fass* || Fass * *(auch als Verb: 'in ein Fass füllen')* || füllen * *in Fässer ~* || Tonne || Walze * *Rolle, Walze, Trommel, Zylinder, Rohr*
barrier abschranken || Abschrankung || Grenze * *Barriere* || Hindernis * *Barriere* || Schwelle * *Barriere*
barrier grid Schutzgitter
barrier junction Sperrschicht
barrier junction temperature Sperrschichttemperatur
barriers Schottung * *Schutzwände*
basal surface Grundfläche
base Ansatz * *logischer* || aufbauen * *gründen, stützen, sich verlassen (on:auf)* || Auflage * *Unterlage* || basieren * *(up)on:auf* || Basis * *auch Grund, Grundlage, Fundament, e.Transistors, Sockel, Unterbau, Ausgangspunkt* || bauen * *Hoffnung, Urteil usw.; (upon:auf)* || Fundament || Fuß * *Maschinenteil* || Grund- || Grundfläche || Grundlage || Rohr- * *z.B. base paper: ~papier* || Sockel * *Unterbau, Unterlage, Fundament, Fuß, Sohle, Basis, Grundlage, Stützpunkt, Hauptbestandteil* || Ständer * *Sockel, Unterteil, Unterbau, Fundament (auch einer Maschine), Fuß* || Stützpunkt * *z.B. Handelsstützpunkt* || Unterlage * *Auflage, Unterbau, Fundament*

base board Grundbaugruppe * Leiterkarte || Grundleiterplatte
base connector Basisstecker
base frame Grundrahmen * z.B. für die Aufstellung einer elektr. Maschine
base frequency Eckfrequenz * z.B. e. U/f-Kennlinie beim Pulsumrichter, bei dieser Frequ. wird die max. Spanng. erreicht || Grundfrequenz || Grundwelle
base frequency sine wave Grundwelle
base load Grundlast * wird bei Definition einer Überlastbarkeit zugrundegelegt
base load current Grundlaststrom * wird bei der Definition einer Überlastbarkeit zugrundegelegt
base load duty Grundlastbetrieb
base load time Grundlastdauer
base material Basismaterial
base mounting Fußanbau
base plate Fußplatte * z.B. an Getriebemotor zur Befestigung
base sampling time Grundabtastzeit * oft als "T0" bezeichnet
base speed Ablösedrehzahl * oberhalb derer die Feldschwächung beginnt || Grunddrehzahl * Drehz. bei Nennspannung/Nennfrequenz, oberhalb derer die Feldschwächung beginnt
base speed control range Grunddrehzahlbereich
base speed range Grunddrehzahlbereich * Drehzahlen unterhalb des Feldschwächbereichs
base unit Grundgerät
base unit operator panel Grundgerätebedienfeld
base value Bezugsgröße || Startwert
base-block time Sicherheitswartezeit * bei Momentenwechsel eines DC oder AC Stromrichters
base-emitter circuit Basis-Emitter-Strecke
based -gestützt
based on auf der Basis von
based on a modular concept modular
based on fact tatsächlich
baseplate Grundplatte
BASIC BASIC * Beginner's All-purpose Symbolic Instruction Code, einfache höhere Programmiersprache || basisch * (chem.) || eigentlich * grundlegend, die Grundlage bildend || entscheidend * grundlegend, Grund-, die Grundlage bildend || Grund- || grundlegend || Haupt- * grundlegend, die Grundlage bildend || hauptsächlich * grundlegend, die Grundlage bildend || Vor- * grundlegend || wesentlich * grundlegend, Grund-, die Grundlage bildend
basic aspect wesentlicher Punkt * wesentlicher Gesichtspunkt
basic axis Grundachse * eines Roboters
basic block Grundbaustein
basic board Grundbaugruppe * Leiterkarte
basic building block Grundbaustein
basic change Verlagerung * (figürl.)
basic circuit Grundschaltung
basic circuit configuration Übersichtsschaltbild
basic circuit diagram Schaltschema
basic complement Grundbestückung
basic concept grundlegendes Konzept
basic concepts Grundbegriffe
basic configuration Grundausbau
basic converter Grundgerät * Stromrichter (aus Sicht einer Zusatzbaugruppe)
basic converter connection Grundschaltung * (Stromrichter)
basic converter operator control panel Grundgerätebedienfeld * eines Stromrichters
basic criterion Grundlage * grundlegendes Kriterium (Plural: 'criteria')
basic design Grundausführung || Grundtyp * z.B. eines Geräts || Normalausführung * Grundtyp || Standardausführung * Grundtyp
basic function Grundfunktion

basic grid Grundraster
basic grid dimension Rasterteilung * Grund-Rastermaß
basic idea Sinn * Grundgedanke
basic industry Grundstoffindustrie * z.B. Stahl-, Zellstoff-, Papierindustrie
BASIC interpreter BASIC-Interpreter
basic issue Kernpunkt || Problem * Grundproblem
basic item wesentlicher Punkt * der Tagesordnung usw.
basic knowledge Grundkenntnisse || Grundwissen || Vorkenntnisse * Grundkenntnisse
basic load Grundlast * 'Sockelbetrag' der dauernd erlaubten Belastung bei einer Überlastbarkeits-Spezifikation
basic logic funktion Grundfunktion * logische
basic module Grundbaugruppe || Grundbaustein
basic operation Grundoperation
basic price Grundpreis
basic principle Grundprinzip || Grundregel
basic principles Grundlagen
basic profile Bezugsprofil * einer Verzahnung
basic requirement Voraussetzung * Grundbedingung, wesentliche Erfordernis || Vorbedingung * Grundbedingung, Grundvoraussetzung
basic requirements allgemeine Anforderungen
basic sampling time Grundabtastzeit
basic science Grundlagenwissenschaft
basic screen mask Grundmaske * am Bildschirm
basic speed Ablösedrehzahl * oberhalb derer die Feldschwächung beginnt
basic terms Grundbegriffe
basic type Grundausführung
basic type of construction Grundbauform
basic understanding Grundwissen
basic unit Grundgerät
basic wiring Vorverdrahtung
basic-reserve changeover Grund-Reserve-Umschaltung
basical prinzipiell * grundlegend, die Grundlage bildend, Grund-
basically eigentlich * (Adv.) im Grunde grundsätzlich || grundsätzlich || im Prinzip * grundsätzlich, im Grunde || prinzipiell * (Adv.) grundsätzlich, im Grunde || wesentlich * (Adv.) im~en, im Grunde, grundsätzlich
basics Grundlagen * z.B. Wissensgrundlagen eines Fachgebiets
basin Behälter * für Flüssigkeiten || Gefäß * Bassin, Schüssel
basis Basis * auch Grundlage, Fundament, Basis eines Transistors; on the basis of: auf d. ~ von || Grundfläche || Grundlage * on the basis of: auf ~ von || Mittelpunkt * Grundlage
basis material Ausgangsmaterial || Basismaterial
basis weight Flächengewicht * Papier || Papiergewicht * Flächengewicht
batch Charge * auch Fertigungslos || diskontinuierlich * Chargen-; batch process: Chargenprozess/ ~er (Produktions-) Prozess || Fertigungslos || Los * Fertigungs~, Partie || Losgröße || Menge * Schub || Satz * Schub || Schub * ~ von Ereignissen/Aufträgen usw., || Schwung * Menge, Schub
batch file Batchdatei * Textdatei, die eine Folge von Betriebssystem-Kommandos enthält || Kommandodatei || Stapeldatei * Kommandodatei (Textdatei, die eine Folge von Betriebssystem-Kommandos enthält)
batch orders kommissionieren * zusammentragen von Bestellpositionen
batch process Chargenprozess * →diskontinuierlicher Prozess, z.B. in der Zement-, Zucker-, chemischen Industrie || diskontinuierlicher Prozess
batch quantity Losgröße

batch size Losgröße
batch testing Serienprüfung
bate beizen * *Häute* ~
bath soldering Badlöten
bathtub curve Badewannenkurve * *Kurve d. Ausfallwahrscheinlichkt. über d.Gebrauchsdauer (hohe Früh- u. Spätausfälle)*
batten Latte
battenboard Tischlerplatte * *[Holz]*
batteries Batterien
battery Batterie
battery back-up Batteriepufferung
battery backed batteriegepuffert
battery backup Pufferspannung * *Batteriepufferung bei Versorgungsspannungsausfall/-abschaltung* || Pufferung * *Ersatzstrom von Batterie bei Spannungsausfall/-Abschaltung*
battery backup unit Puffereinheit * *mit Pufferbatterie(n)*
battery box Batteriekästchen * *z.B. zur Vorgabe v. Sollwerten/Sollwertsprüngen für d. Antriebsregler-Optimierung*
battery buffer Batteriepufferung
battery charger Batterieladegerät
battery life Batterie-Lebensdauer || Batterielebensdauer
battery low Batterieausfall * Meldung '(Puffer-) Batterie entladen' (z.B. bei SPS)
battery operation Batteriebetrieb
battery-backed Batterie-gepuffert || batteriegestützt * *z.b. nichtflüchtiger Datenspeicher* || gepuffert * *batteriegepuffert*
Baud Baud * *['bod] bits pro sec.*
baud rate Baudrate * *Maßeinheit bits pro sec, Übertragungsgeschwindigkeit bei einer seriellen Kommunikation* || Übertragungsrate * *Baudrate [bits/sec]*
bay Fach * *Fach (in Regal, Hochregallager), Abteil-(ung), Nische, Laufwerks-Einbauschacht in Computer* || Feld * ~ *in einer (Freiluft-) Schaltanlage*
bayonet Bajonett
bayonet catch Bajonettverschluss
bayonet joint Bajonettverschluss
bayonet socket Bajonettfassung
BBS BBS * *Abk.f. 'Bulletin Board System': Mailbox-System, z.B. Support-Mailbox mit Zugriff über Modem*
BCC Blockprüfzeichen * *Abk. f. 'Block Check Character': Datensicherungszeichen b. seriell. Übertragg.* || Blockprüfzeichen * *Abkürzung für "Block-Check Character"*
BCD BCD * *Abkürzg.f. 'Binary-Coded Decimal': binär codierte Dezimal- (Zahl).*
BCD switch Zahlenschalter * *im BCD-Code arbeitend*
be betragen * *zahlen/wertmäßig ~* || darstellen * *sein* || herrschen * *bestehen* || *sein * (Verb) how much is it: was kostet das?* || stehen * *sein, s. befinden; be under the direction of: unter jds. Führung ~; be at issue: z.Debatte* ~
be 90 degrees apart um 90 Grad versetzt sein
be a consequence of sich ergeben * *sich (logisch) ergeben aus*
be a loser zusetzen * *Geld einbüßen (by: bei)*
be a match for gewachsen sein
be a matter of handeln * *sich um etwas* ~
be a parallel to ähneln * *entsprechen* || gleichen * *entsprechen*
be a question of handeln * *sich um etwas* ~
be a spin-off abfallen * *als (zufälliges) Nebenprodukt*
be able to Können * *(Verb)*
be absent fehlen * *abwesend/nicht vorhanden sein*
be acquainted with vertraut sein mit

be adequate ausreichen * *angemessen sein* || genügen * *angemessen/ausreichend sein*
be advanced voreilen
be adversely affected leiden * *(ungünstig) beeinflusst sein (from: durch)*
be agreeable einverstanden sein * *to: mit*
be aligned fluchten * *in Linie gebracht/ausgerichtet sein* || fluchtend * *in Linie gebracht/ausgerichtet sein*
be allotted to entfallen * *auf jemanden/etwas (anteilmäßig)* ~
be allowed to dürfen || Können * *(Verb) dürfen*
be among gehören * *zählen zu*
be analogous to ähneln * *entsprechen* || gleichen * *entsprechen*
be anxious to bedacht sein auf
be applicable anwendbar sein * *to:auf*
be applicable to gelten für * *anwendbar sein für*
be applied Anliegen * *(Verb) angelegt sein, z.B. eine Spannung* || anstehen * *angelegt sein, z.B. Spannung* || einfallen * *Bremse*
be apprenticed lernen * *seine Lehre/gewerbliche Ausbildung machen (to: bei)*
be as per entsprechen * *wie bei ... sein*
be assigned for vorgesehen sein für * *bestimmt sein für (einen Zweck)*
be associated with gehören * ~ *zu, zusammen~ zu, zugeordnet sein zu* || verbunden sein mit * *einhergehen mit*
be at a premium an erster Stelle stehen * *in der Priorität*
be at a standstill stagnieren
be at hand zur Verfügung stehen
be at home in vertraut sein mit * *sich (in einem Fachgebiet) auskennen*
be at right angles to each other um 90 Grad versetzt sein * *Teile gegeneinander*
be available Anliegen * *(Verb) z.B. Versorgungsspannung* || anstehen * *Fehler/Warnsignal* || existieren * *vorhanden/verfügbar sein* || zur Verfügung stehen * *to: für*
be awarded bekommen * *Preis, Auszeichnung, Auftrag* || erhalten * *Auftrag/Auszeichnung erteilt/verliehen bekommen; z.B. 'he was awarded the prize'* || zuerkannt bekommen * *Auszeichnung, Auftrag, Zertifikat usw.* || Zuschlag bekommen * *bei Auftragsvergabe; be awarded th. contract: d. Liefervertrag zuerteilt bekommen*
be awarded the order Auftrag erteilt bekommen
be awkward stören * *unangenehm sein*
be based basieren * *beruhen, sich stützen ((up)on: auf)* || beruhen * *basieren (on:auf), zur Grundlage haben*
be based on basieren auf || beruhen auf || stützen * *abgestützt sein, sich ~ (on:auf)*
be believed to sollen * *angeblich* ~
be blocked sperren * *sich im gesperrten Zustand befinden, z.B. Halbleiter*
be blown fallen * *Sicherung*
be bound verpflichten * *by law/contract: gesetzlich/vertraglich verpflichtet sein*
be bound to dazu bestimmt sein * *etwas tun zu müssen* || zwangsmäßig tun müssen * *dazu bestimmt sein*
be bound up with verbunden sein mit * *eng* ~
be busy doing something sich beschäftigen mit
be capable in der Lage sein * *fähig/im Stande sein (of: zu)*
be capable of Können * *(Verb)*
be careful vorsehen * *vorsichtig sein*
be careful to bedacht sein auf
be caused entstehen * *verursacht sein (by: durch)*
be characterized by sich auszeichnen durch * *herausragende Merkmale haben*

be classed rangieren * *with: rangieren mit/unter*
be combined with verbunden sein mit
be committed engagiert sein * *to: bei, an*
be comparable to ähneln * *vergleichbar sein mit* || gleichen * *vergleichbar sein mit*
be composed bestehen * *zusammengesetzt sein (of: aus)*
be composed of sich zusammensetzen aus
be concerned beteiligen * *beteiligt/betroffen sein (in: an)* || betreffen * *betroffen sein* || handeln * *sich ~*
be concerned about bedacht sein auf
be concerned with erstrecken * *(figürlich) sich ~ auf*
be conducive to dienen * *dienlich/förderlich sein*
be considered as gelten * *betrachtet werden als* || gelten als
be convenient passen * *geeignet, günstig sein; to/for:für*
be convenient for passen zu * *geeignet/günstig sein für*
be conversant with vertraut sein mit
be correct zutreffen
be correlated zusammengehören * *(figürl.)*
be covered of fallen * *in eine Kategorie ~*
be critical of beanstanden * *etwas ~/kritisieren*
be damaged Schaden erleiden
be de-energized abfallen * *Relais*
be decisive ausschlaggebend sein
be decreased zurückgehen * *verringert werden*
be dedicated to Aufgabe haben || vorgesehen sein für * *bestimmt/gewidmet sein für*
be defensive abwehren * *abwehrend/in der Defensive sein*
be delayed verzögern * *sich ~*
be demanding hohe Ansprüche stellen
be dependent abhängen * *finanziell usw.*
be derived stammen * *abgeleitet sein*
be descended stammen * *abstammen, herkommen (from: von)*
be destined for dazu bestimmt sein * *bestimmt/vorgesehen sein für*
be destined to sollen * *schicksalhaft bestimmt sein für*
be determined entschlossen sein * *to do:etwas zu tun*
be discontinued auslaufen * *Produktion eines Modells/einer Type; discontinue: ~ lassen*
be disposed to geneigt * *etwas begünstigen, zu etw. ~ sein*
be distributed verteilen * *sich verteilen (among:-unter); verteilt angeordnet werden* || verteilt sein * *auch statistisch ~*
be divisible by teilen * *teilbar sein durch*
be due to beruhen auf * *zurückzuführen sein auf* || entstehen * *zurückzuführen sein auf*
be effective gelten * *Gesetz*
be endangered gefährdet sein
be endowed with besitzen * *Talent, Gaben usw. ~*
be engaged befassen * *(figürl.) be engaged in: sich befassen mit* || engagiert sein * *beschäftigt sein/arbeiten (in:mit/an), nicht abkömml.s., in Anspruch genomm.s.*
be engaged in betreiben * *Beruf, Geschäft usw. ~* || sich beschäftigen mit
be enough ausreichen
be equal to betragen || gewachsen sein || gleich sein || gleichen
be equipped with besitzen * *ausgestattet sein mit*, verfügen über * *haben * ausgestatet sein mit* || verfügen über * *ausgestattet sein mit*
be equivalent to entsprechen * *gleichkommen* || gleichkommen
be exacting hohe Ansprüche stellen * *streng/genau/anspruchsvoll sein (z.B. Anwendung, Kunde)*

be executed ablaufen * *Programm* || erfolgen * *aus-/durchgeführt werden*
be expected to voraussichtlich * *(Adv.) erwartet werden, zu*
be faced with stehen * *vor etw. unangenehmem ~*
be falling fallen * *Luftdruck usw.*
be familiar auskennen * *with:sich auskennen mit*
be familiar with vertraut sein mit
be fated to sollen * *schicksalhaft bestimmt sein für*
be fellows zusammengehören * *mehrere Sachen*
be fired gehen * *(salopp) 'rausgeschmissen werden'* || zünden * *Thyristor*
be fit passen * *passend/geeignet/fähig/tauglich sein*
be fixed haften * *befestigt sein; (auch figürl.: Gedanken usw.)*
be formed out of sich zusammensetzen aus
be founded basieren * *sich gründen, beruhen (on: auf)*
be founded on beruhen auf * *sich gründen auf*
be frozen in place festsitzen * *klemmen (z.B. Passsitz)*
be fulfilled eintreffen * *sich erfüllen*
be functional funktionieren * *funktionsfähig sein*
be given erhalten * *gegeben/verliehen bekommen*
be going to to vorhaben * *planen; im Begriff sein, etwas zu tun*
be good at figures rechnen * *gut ~ können*
be good value preiswert * *preiswert sein*
be governed by unterliegen * *Vorschrift, Gesetz*
be headquartered Hauptsitz haben
be held stattfinden * *abgehalten werden*
be held responsible haften * *zur Haftung gezogen werden*
be hidden stecken * *verborgen sein*
be identical with identisch sein mit
be idling leerlaufen
be illuminated leuchten * *z.B. Lampe, LED*
be illustrated by zeigen * *durch eine Abbildung, Beispiel usw. gezeigt werden*
be illustrative of veranschaulichen
be imperative oberstes Gebot sein
be important von Bedeutung sein
be in agreement übereinstimmen * *in Übereinstimmung/Eintracht/Einigkeit sein*
be in competition konkurrieren * *with:mit*
be in control of kontrollieren * *beherrschen*
be in danger gefährdet sein
be in existence existieren
be in favour of für etwas sein
be in force gelten * *in Kraft sein, z.B. Vorschrift* || wirken * *wirksam sein*
be in hand vorliegen * *Antrag, Auftrag, Bestellung usw.*
be in line fluchten * *in einer Linie/Achse ausgerichtet sein* || übereinstimmen * *with: mit*
be in need of benötigen || brauchen * *nötig haben, →benötigen*
be in operation gelten
be in phase in Phase sein || übereinstimmen * *zeitlich ~*
be in possession of besitzen
be in process laufen * *im Gange sein, in Bearbeitung sein*
be in process of development entstehen * *in Entwicklung befindlich*
be in progress ablaufen * *im Gange sein, fortschreiten, in Entwicklung befindlich sein* || laufen * *im Gange sein, gerade ablaufen, Fortschritte machen*
be in the chair leiten * *eine Versammlung ~*
be in the dark unklar * *im Unklaren sein*
be in the making entstehen * *im Entstehen begriffen sein*
be in the position in der Lage sein * *to do something: ~ etwas zu tun*

be in the position to Können * *(Verb) zu etwas in der Lage sein*
be in the responsibility of obliegen
be in the way stören * *im Weg sein*
be in vogue verbreitet * *Mode sein*
be in working order funktionieren * *funktionsfähig sein*
be inapplicable entfallen * *nicht in Frage kommen*
be inclusive of einschließen * *im Preis usw. eingeschlossen sein*
be inconvenient stören * *unangenehm sein*
be incorporated eingebaut sein
be increased ansteigen || steigen * *an~ (by: um)*
be increasing ansteigen * *im Ansteigen begriffen sein* || steigen * *im ~ begriffen sein*
be incurred anfallen * *Kosten, finanzielle Verluste; losses incurred: angefallene Verluste* || auftreten * *Kosten, finanzielle Verluste usw.* || entstehen * *Kosten, Nachteile usw. (by:durch)*
be informed erfahren * *(Verb)*
be informed of Wissen * *(Verb)*
be intended for dazu bestimmt sein * *bestimmt/vorgesehen sein für* || sollen * *bestimmt sein für* || vorgesehen sein für * *bestimmt sein für*
be intruding stören * *sich einmischen/eindrängen*
be intrusive stören * *aufdringlich sein*
be involved in beteiligen * *verwickelt/involviert sein in*
be involved with sich beschäftigen mit
be irregular aussetzen * *öfter aussetzen*
be kind to schonen * *Material(schonung durch eine Maschine)*
be laid down stehen * *festgelegt sein, z.B. in Gesetz, Vorschrift, Norm*
be left stehen bleiben * *nicht verändert werden, übrig bleiben* || verbleiben
be less than unterschreiten
be liable haften * *bürgen; for: für; without limitation: unbeschränkt; personally: persönlich* || verpflichten * *gesetzlich verpflichtet sein*
be liable to unterliegen * *unterworfen sein, z.B. Gebühr, Zoll, Steuer, Gewährleistung*
be like ähneln * *ähnlich sein* || gleichen * *ähnlich sein*
be likely to voraussichtlich * *(Adv.) wahrscheinlich*
be listed stehen * *in Liste, Tabelle, Aufzählung ~*
be lit aufleuchten || leuchten * *erleuchtet sein, z.B. Lampe, LED, Meldeleuchte*
be located liegen * *gelegen sein* || stehen * *sich befinden*
be looked upon as gelten als * *betrachtet werden als*
be lost verloren gehen
be lying liegen
be made erfolgen * *payment must be made: Zahlung muss ~*
be made of bestehen aus * *hergestellt sein aus (Material)*
be made up of bestehen aus * *sich →zusammensetzen aus, zusammengesetzt sein aus*
be meant for dazu bestimmt sein * *bestimmt/vorgesehen sein für* || sollen * *bestimmt sein für*
be mirrored spiegeln * *sich ~*
be missing abgehen * *fehlen* || fehlen
be mistaken verrechnen * *sich verrechnet/geirrt haben*
be more important vorgehen * *wichtiger sein (than: als)*
be moving fahren * *in Fahrt sein*
be necessary benötigen * *notwendig sein*
be O.K. funktionieren * *funktionsfähig sein, in Ordnung sein*
be obliged to sollen * *müssen, verpflichtet sein zu*
be obtained einstellen * *sich einstellen* || ergeben * *sich ergeben*

be of betragen * *zahlen/wertmäßig ~*
be of advantage nützen * *vorteilhaft sein; to a person: jemandem*
be of benefit nützen * *vorteilhaft sein; to a person: jemandem*
be of consequence ausmachen * *it is of no consequence: das macht nichts aus*
be of importance von Bedeutung sein
be of significance von Bedeutung sein
be of the opinion that Ansicht vertreten
be of the same -gleich
be of use nützen * *for: zu; a person: jemandem*
be of value gelten
be on one's guard against vorsehen * *sich ~/auf der Hut sein vor*
be on show ausgestellt sein * *auf Messe, Ausstellung usw.*
be on the payrol beschäftigt sein * *angestellt sein, auf der Gehaltsliste stehen*
be on the programme vorsehen * *auf dem Programm stehen*
be on the right track auf dem richtigen Weg sein
be on the rise ansteigen * *im Anstieg begriffen sein*
be operated arbeiten * *betrieben werden*
be operational in Betrieb sein
be opposite gegenüberliegen
be out in one's reckoning verrechnen * *sich verrechnet haben, sich ~*
be out of place aus dem Rahmen fallen
be out of proportion sprengen * *den Rahmen ~*
be owing to beruhen auf * *zurückführbar sein auf*
be owned by gehören
be paralyzed stillstehen * *lahmgelegt sein, z.B. Betrieb, Verkehr*
be permeable to durchlassen * *(physikal.) durchlässig sein für*
be permitted to dürfen || Können * *(Verb) dürfen*
be pervious to durchlassen * *(physikal.) durchlässig sein für*
be phased out auslaufen * *allmählich ~; phase out: ~ lassen*
be placed liegen * *gelegen sein*
be plausible plausibel sein
be poised schweben
be practicable anwendbar sein * *ausführbar sein*
be precipitated abstürzen * *(figürlich) hinein/hinabgestürzt sein (into/in)*
be predominant dominieren || vorherrschen
be present Anliegen * *(Verb) vorhanden sein, z.B. Signal, Warnung, Meldung* || anstehen * *Spannung, Signal* || existieren * *vorhanden sein* || vorliegen * *vorhanden sein*
be present at teilnehmen * *anwesend sein*
be pressing drängen * *dringend sein*
be promoted aufrücken * *im Rang ~* || aufsteigen * *beruflich be/gefördert werden*
be proper passen * *richtig/passend/geeignet/angemessen/ordnungsgemäß/zweckmäßig sein*
be provided erfolgen * *zur Verfügung gestellt werden (by: von)*
be provided for vorgesehen sein für * *zur Verfügung gestellt sein für*
be provided with besitzen * *ausgestattet sein mit, verfügen über* || haben * *ausgestattet sein mit* || verfügen über * *ausgestattet sein mit*
be published erscheinen * *Druckerzeugnis*
be pulsating lücken * *z.B. Strom im netzgeführten Stromrichter* || lücken * *z.B. Strom im netzgeführten Stromrichter (→Lückstrom)*
be put on einfallen * *Bremse*
be rated Nennleistung haben * *Nennleistung von ... haben*
be rated at leisten * *(Nenn-) Leistung haben (in Zusammenhang mit Zahlenangabe)*
be ready in Bereitschaft stehen

be realized erfolgen * *realisiert/durchgeführt werden*
be received kommen * *be received from: ~ von (Signal, Nachricht)*
be reduced to zero zu Null werden
be reflected spiegeln * *sich ~* || widerspiegeln * *sich ~(in: in)*
be reflected by sich widerspiegeln in
be released Ansprechen * *(Verb) Überwachungseinrichtung*
be relevant anwendbar sein * *einschlägig/relevant sein* || von Bedeutung sein * *in diesem Zusammenhang ~; to: für*
be reported to sollen * *angeblich ~ (wie berichtet wird)*
be responsible haften * *verantwortlich sein, bürgen* || Schuld haben * *verantwortlich sein*
be responsible for Aufgabe haben * *verantwortlich sein für* || sich beschäftigen mit * *zuständig sein für*
be right zutreffen * *richtig sein*
be right for gelten für * *richtig sein für*
be rising ansteigen * *~d sein*
be said to sollen * *angeblich ~*
be satisfied einverstanden sein * *zufrieden sein (with: mit)*
be seriously affected leiden * *(from: unter)*
be settled erledigen * *sich erledigen*
be short Mangel haben * *of:an*
be significant Rolle spielen * *eine wesentliche ~* || von Bedeutung sein
be similar to ähneln * *ähnlich sein* || gleichen * *ähnlich sein*
be situated liegen * *gelegen sein*
be sluggish stagnieren
be sorry bedauern
be spread verteilt sein * *über Raum oder Zeit ~*
be stagnant stagnieren
be stalled festsitzen * *bestgefahren sein, steckengeblieben sein*
be stationed liegen * *stationiert sein*
be stored lagern * *auf Lager befindlich*
be stuck festsitzen * *festgefahren sein, steckengeblieben sein* || stecken * *festsitzen*
be subject abhängen * *to:von einer Zustimmung, Vorschrift usw.abhängen* || abhängig * *to:von einer Zustimmung, Vorschrift usw. abhängig sein*
be subject of ausgesetzt sein * *z.B. einer Belastung/Gefahr*
be subject to erfahren * *(einem Effekt) ausgesetzt sein* || unterwerfen * *(einer Sache/Vorgehensweise) unterworfen sein*
be subject to potential injury gefährdet sein * *Verletzungsgefahr*
be submitted vorliegen * *vorgelegt/beantragt sein*
be suddenly increasing stark ansteigen * *z.B. plötzliches Ansteig. einer Spannung*
be sufficient ausreichen || genügen * *aus-/hinreichend sein* || reichen * *aus~, ~d sein*
be suitable anwendbar sein * *geeignet sein* || passen * *passend geeignet sein*
be suitable for passen zu * *passend, geeignet sein für*
be suited passen * *geeignet sein*
be superior by sich auszeichnen durch
be supposed to sollen * *angeblich ~*
be suspended schweben
be switched off ausgehen * *ausgeschaltet werden, z.B. Lampe*
be switched on leuchten * *eingeschaltet sein*
be tangent to tangieren
be tender of bedacht sein auf
be tentatively scheduled angedacht sein * *unverbindlich eingeplant sein*
be the basis zugrunde liegen

be the case zutreffen
be the leader an erster Stelle stehen * *führend sein, z.b. am Markt*
be the main player an erster Stelle stehen * *Marktführer sein*
be the program vorsehen * *auf dem Programm stehen; what is the program today: was ist für heute vorgesehen*
be there vorliegen * *vorhanden sein*
be to sollen * *auf Anweisung Dritter ~; auf Grund schicksalhafter Bestimmung ~*
be to blame Schuld haben * *for a thing: an etwas*
be told erfahren * *(Verb)*
be too much for übersteigen * *Kräfte, Verstand, Fähigkeiten usw. ~*
be too tight spannen * *zu eng sein, zu fest sitzen*
be true gelten * *Formel, techn. Angabe usw. (for: für)* || zutreffen * *of/for: bei/für/auf*
be true for gelten für * *Formel, Technische Angabe usw.*
be turned off löschen * *gelöscht werden, z.b. Thyristor* || sperren * *z.B. Thyristor/Transistor*
be under consideration vorliegen * *behandelt werden*
be unsuited to the occasion aus dem Rahmen fallen * *Benehmen*
be up anstehen * *for decision:zur Entscheidung*
be up to gewachsen sein || überlassen * *it is up to me to do: es ist mir überlassen/meine Sache zu tun*
be urgent drängen * *dringend sein*
be used dienen * *in: verwendet werden für*
be useful nützen
be valid gelten * *gültig sein; become valid: gültig werden; remained valid unchanged: unverändert ~*
be very efficient leisten * *gute Arbeit ~*
be vogue herrschen * *in Mode sein*
be warehoused lagern * *auf Lager befindlich sein, auf Lager liegen*
be well matched zusammenpassen * *gut ~*
be well versed in vertraut sein mit
be winner gewinnen * *siegen*
be written geschehen * *geschrieben sein/stehen*
bead bördeln * *mit Wulst/mit Perlen versehen* || Perle * *Perlkorn, (Tau-/Schweiß-/Schmelz-) Perle, (Schaum-) Bläschen, Tröpfchen* || Raupe * *[Schweißtechnik]* || Sicke * *wulstförmiger Rand* || Wulst * *auch am Autoreifen*
beading Wulst
beading machine Sickmaschine || Wulstmaschine * *→Sickmaschine*
beads of molten material Schmelzperlen
beaker Becher * *großer Becher, Becherglas (chem.)*
beam Baum * *[Textil] Wickel, aufgewickelter Textilstoff* || Holm * *Balken, Tragbalken, Holm* || Spindel * *~ einer Drehmaschine* || Strahl || Traverse || Wickel * *[Textil] Baum, aufgewickelter Textilstoff*
beam planer Balkenhobel * *[Holz]*
beam press Leimbinderpresse * *[Holz]*
bear tragen * *hervorbringen; Verantwortung, Folgen, Verlust, Kosten usw. tragen* || verkraften * *ertragen*
bear in mind achten auf * *daran denken, nicht vergessen* || beachten * *berücksichtigen, nicht vergessen* || berücksichtigen * *beachten, nicht vergessen* || merken * *im Gedächtnis behalten, nicht vergessen*
bearable tragbar * *(figürl.) (er)tragbar* || zumutbar * *(er)tragbar*
bearing Lager * *Gleit-, Wälzlager* || Lagerstelle || Lagerung * *Gleit-/Wälzlager*
bearing airgap Lagerluft
bearing application Lagerung * *Verwendung/Montage e. mechan. Lagers*
bearing area Auflagefläche

bearing arrangement Lagerstelle * *Lagerung* || *Lagerung* * *z.b. Gleit-/Wälzlager-Anordnung*
bearing assembly Lagerung * *zu einem Gleit/ Wälzlager gehörende Teile*
bearing block Lagerbock || Lagerverriegelung * *Transportsicherung, z.B. bei [el. Masch.]* || Transportsicherung * *in Form einer Lagerverriegelung zur Vermeidung v. Lageschäden*
bearing box Lagergehäuse || Lagerschale
bearing bracket Lagerbock || Lagerschild * *offenes ~*
bearing bush Lagerbuchse || Lagerbüchse
bearing cage Lagerkäfig
bearing cap Lagerdeckel * *siehe auch →Schlussdeckel (äußerer ~)*
bearing capacity of floor Bodenbelastbarkeit
bearing clearance Lagerluft * *Lagerspiel*
bearing condition Lagerzustand * *eines Gleit/ Wälzlagers*
bearing condition sensing Lagerzustandserfassung * *eines Gleit/Wälzlagers*
bearing cover Lagerdeckel
bearing current Lagerstrom * *el. Strom durch e. (Motor-) Lager (unerwünschter Effekt, kann z.Lagerschaden führen)*
bearing design Lagerung * *Ausführung/Auslegung/ konstruktive Gestaltung der ~*
bearing end cap Schlussdeckel * *äußerer Lagerdeckel*
bearing extraction Abziehen des Lagers
bearing extractor Abzieher * *für Kugellager* || Abziehvorrichtung * *z.B. für Kugellager*
bearing face Auflagefläche
bearing failure Lagerschaden
bearing friction Lagerreibung
bearing housing Lagergehäuse
bearing life Lagerlebensdauer * *L_10 in [h], die von 90% der Lager erreicht wird (DIN ISO 281)*
bearing load Lagerbelastung
bearing monitoring Lagerüberwachung
bearing noise Lagergeräusch
bearing oil pump Lagerölpumpe
bearing pedestal Lagerbock
bearing play Lagerspiel * *(allg.)*
bearing pressure Anpressdruck || Anpresskraft
bearing pulley Tragrolle
bearing remove Abziehen des Lagers * *Entfernen des Lagers von der Welle mit Abziehwerkzeug*
bearing seal Lagerdichtung
bearing shield Lagerschild
bearing stand Lagerbock
bearing steel Wälzlagerstahl
bearing strength Tragfähigkeit * *~ eines Gleit-/ Wälzlagers*
bearing surface Auflagefläche || Lauffläche * *in Wälz/Gleitlager*
bearing system Lagerung * *z.B. Gleit-/Wälzlager*
bearing temperature Lagertemperatur * *bei Wälz-/Gleitlager (für [el. Masch.]: siehe VDE0530 Teil 1)*
bearing thermometer Lager-Thermometer
bearing wear Lagerabnutzung || Lagerverschleiß
bearing-damage Lagerschaden
bearing-lubricating pump Lagerölpumpe
bearing-temperature monitoring Lager-Temperaturüberwachung
beat klopfen * *schlagen* || mahlen * *Papier* || Pulsation || schlagen * *auch wiederholt ~; übertreffen* || Schwebung || überbieten * *z.B. Gegner* || übertreffen * *schlagen*
beat back abwehren * *Angriff usw.*
beat frequency Schwebungsfrequenz
beatability Mahlbarkeit
beater Holländer * *[Papier] zum Feinmahlen der Faserstoffe bei der Zellstoff-/Papierherstellung*

beater mill Schlagmühle * *z.B. für Steine*
because dadurch * *weil*
because of auf Grund * *because of the fact that: ~ der Tatsache, dass* || wegen
become antiquated veralten
become apparent herausstellen * *sich ~*
become aware of wahrnehmen * *etw. gewahr werden*
become distorted sich verziehen
become established Fuß fassen
become familiar with einarbeiten * *sich in etwas ~*
become functional in Funktion treten
become increasingly sigificant an Bedeutung gewinnen
become inoperative ausfallen
become instable Kippen * *(Verb) ~ einer el. Maschine*
become manifest sichtbar werden * *(figürlich)*
become obsolete veralten
become operational in Betrieb gehen
become shorter verkürzen * *sich ~*
become slack lockern * *sich lockern*
become worn verschleißen
becoming zero Nullwerden
bed lagern * *Maschine in Maschinenbettung ~* || Maschinenbett
bed in einschleifen * *Kohlebürsten*
bed-type milling machine Bettfräsmaschine
bedding Auflagefläche || Lagerung * *Bettung*
bedplate Maschinenbett * *Fundamentplatte* || Maschinenstuhlung * *Bettung*
beeper Piepser
before vor * *zeitlich, räumlich, im Signalfluss usw. ~neliegend* || vorher * *(Adj.)* || vorher * *(Adv.) voraus* || vorn
before limiting vor der Begrenzung
before the ramp vor dem Hochlaufgeber
before the ramp-function generator vor dem Hochlaufgeber
before-hand vorher * *(Adv.) zuvor, voraus, im Voraus*
beforehand im Voraus || vor * *zu~, im ~aus* || vorher || zuvor
begin beginnen || einsetzen * *anfangen*
begin at the beginning vorn * *von ~ anfangen*
begin operation in Betrieb gehen
begin to do ansetzen * *zu einer Handlung*
beginner Anfänger || Einsteiger * *Anfänger* || Lehrling * *Neuling* || Neuling * *→Anfänger*
beginning Anfang || Beginn || Einsatz * *Anfang, Beginn; (initiation): Einleitung, Beginn, Aufnahme)* || Entstehung * *Anfang, Beginn, Ursprung*
beginning of production Produktionsbeginn
beginning of work Inbetriebsetzung * *Aufnahme des Betriebes einer (Fabrik)Anlage*
behave betragen * *(sich) benehmen (towards: gegenüber)* || sich verhalten || wirken * *sich verhalten*
behavior auftreten * *(Substantiv) Verhalten, Benehmen* || Verhalten
behaviour Verhalten
behind hinten * *from behind: von ~* || hinter || nach * *hinter* || zurück * *hinten; lag behind:im Rückstand sein*
behindhanded zurück * *im Rückstand*
beige beige
being wobei * *the thyristor being blocked: ~ der Thyristor gesperrt ist*
being free of -Freiheit
being in existence existent * *vorhanden*
being prepared in Vorbereitung || in Vorbereitung befindlich
belated nachträglich * *verspätet*
Belgium Belgien
bell-shaped glockenförmig

bell-shaped rotor Glockenanker * z.B. für Mikromotor (System "Faulhaber")
bellow Faltenbalg
bellow-type coupling Faltenbalgkupplung
bellows Faltenbalg
bellows coupling Faltenbalgkupplung
belong gehören * to: zu
belong together zusammengehören
belonging to zugehörig * a person: zu jemandem
belonging to it dazugehörig
belonging to one another zusammengehörig
belonging together konsistent * zusammengehörig || zusammengehörig
below darunter || unten * see below: siehe ~; as below: wie ~ näher bezeichnet; far below: weit ~ || unten erwähnt * (amerikan.) (Adj. nachgestellt) (Adv.) please find below: unten stehend finden Sie || unten stehend * (Adj. (nachgestellt) und Adv.) please find below: unten stehend finden Sie || unter * auch ~halb einer Zahl/eines Wertes || unterhalb
below normal unterdurchschnittlich
below-average unterdurchschnittlich
belt Band * Gurt, (Förder-) Band || Gurt || Riemen * z.B. V-belt: Keil~, toothed belt: Zahn~
belt caterpillar Zug- und Schubraupe * Kabelförder mit mechan. Klemmung durch Gurt in Kabelmaschine
belt conveyor Bandförderer || Fließband * Förderband || Förderband || Gurtbandförderer || Gurtförderer || Riemenförderer
belt conveyor scale Bandwaage || Bandwaage * in Förderband integrierte Waage
belt drive Bandantrieb * z.B. Antrieb für Förderband; Gurtantrieb || Gurtantrieb || Riemenantrieb || Riementrieb
belt feed Bandvorschub * bei Fördersystem
belt press Bandfilterpresse * {Nahrungsmittel} z.B. zur Saftherstellung
belt pulley Riemenrad || Riemenscheibe
belt return Bandrücklauf * eines Transportbandes
belt sander Bandschleifmaschine * {Holz}
belt sanding machine Bandschleifmaschine * {Holz}
belt saw Bandsäge
belt scale Bandwaage
belt stretcher Gurtspanner
belt take-off Bandabzug
belt tension Riemenspannung
belt tensioner Gurtspanner
belt weigher Bandwaage
bench Maschinenbett || Tisch * Werkbank, Werk~, Experimentier~ || Werkbank
bench lathe Mechanikerdrehbank || Tischdrehmaschine
bench-type Tischgerät * für den Labortisch
bench-type drilling machine Tischbohrmaschine
bench-type lathe Tischdrehmaschine
bench-type power supply Labor-Netzgerät || Labornetzgerät
benchmark Benchmark * Vergleichstest || bewerten * bezüglich der Leistungsfähigkeit/Messgrößen im Vergleich zu anderen ~ || Vergleichstest
benchmark test Vergleichstest
benchmarking Bewertung * der Leistungsfähigkeit (im Vergleich) || Leistungsvergleich * z.B. von Computern, Firmen usw. || Vergleichstest * Durchführung eines ~s
bend abbiegen * krümmen || abwinkeln * down: nach unten; upwards: nach oben || biegen || Biegung * siehe auch →Krümmung || Durchbiegung * Biegung, Krümmung || Knick * Biegung, Krümmung; auch ~ in Kennlinie usw. || Kröpfung || krümmen || Krümmer * Biegung, Krümmung, Windung, Kurve || Krümmung * →Biegung || Kurve * Biegung, Kurve(auch e.Straße); sharp/

hairpin/blind bend: scharfe/Haarnadel-/unüb.-sichtl.~ || spannen * Feder, Bogen || umbiegen || verbiegen
bend off abbiegen
bend over umbiegen
bend radius Biegeradius
bended gekrümmt || krumm * gebogen, gekrümmt
bender Biegemaschine
bending Biegung || Durchbiegung
bending characteristic abgeknickte Kennlinie
bending device Abkantvorrichtung
bending force Biegekraft
bending machine Biegemaschine
bending moment Biegemoment
bending press Abkantpresse * zum Biegen von Blech (-profilen) usw.
bending radius Biegeradius || Krümmungsradius
bending resistance Biegefestigkeit || Biegesteifigkeit || Biegewiderstand
bending strength Biegefestigkeit
bending torque Biegemoment
bending vibrations Biegeschwingungen
beneath unten * unterhalb || unter || unterhalb
beneath it darunter
benefication Reduktion * Erz
beneficial förderlich * heilsam || günstig * vorteilhaft || nützlich || von Vorteil || vorteilhaft * günstig
benefit nutzen * (Substantiv) Vorteil || nützen * jemandem nützlich sein; Vorteil haben (by: von, from: durch), Nutzen ziehen (aus) || Vorteil * Nutzen
benefit from nutzen * vorteilhaft nutzen, ~/Vorteile ziehen aus || Nutzen ziehen aus
benefit payment Unterstützung * ~saufwendung
benefits Unterstützung * ~sleistungen
bent gebogen * backwards: nach hinten
BERO BERO * Markenzeichen für einen 'BErührungslosen Näherungsschalter mit Rückgekoppeltem Oszillator'
Beryllium Beryllium
beside daneben * vorbei; beside the mark: am Ziel vorbei || neben
beside it daneben * ~liegend
beside the abgesehen von
besides außerdem * außerdem, ferner, überdies || dabei * überdies || daneben * außerdem || sonst * außerdem
besides to neben * nebst
best best * (Adj. und Adv.) best: am ~en; at best: im ~en Falle; in the best (possible) manner: ~ens || größtmöglich || möglich * do one's best: sein ~stes tun || optimal
best known meistens * am meisten bekannt
best possible bestmöglich || optimal
Best Regards Hochachtungsvoll
Best Regards, mit freundlichen Grüßen || mit vorzüglicher Hochachtung
better besser * all the better: umso besser
better half größerer Teil * besserer Teil von zweien
between zwischen * zwischen zweien
bevel abgeschrägt || abschrägen || Anfasung || schmiegen * {Holz} || schräg * abge~t || Schräge
bevel cut Schrägschnitt * {Holz}
bevel gear Kegelrad || Kegelradgetriebe
bevel gear unit Kegelradgetriebe
bevel square Schmiegenlehre * {Holz} Winkelmaß-Lehre für Schrägschnitte
bevel-gear wheel Kegelrad
beveled abgeschrägt * {Kante}
bevelled abgeschrägt * {Kante} || schräg * abge~t
bevelling Abschrägung
beverage Getränk
beverage industry Getränkeindustrie

beware bewusst * *be beware of: sich einer Sache ~ sein*
bewilder verwirren * *verwirren, irremachen, irreführen (someone: jdn.)*
beyond außer * *außer, abgesehen von, über...hinaus, jenseits* || jenseits || oberhalb * *jenseits, z.b. einer Grenze* || über * *~ etwas hinaus*
beyond it darüber hinaus
bezel Frontblende * *Frontblende (z.B. f. Messinstrument), Skalenfenster (f. Messinstrument, Radio), Guckloch*
bi- zwei- || Zweifach-
bi-directional bidirektional
bi-directional triode thyristor Triac
bias Übersteuerung * *auch ~ssollwert, elektrische Vorspannung* || Übersteuerungssollwert * *zur Übersteuerung eines Reglers* || Übersteuerungswert || vorspannen * *(elektr.) z.B. durch e. additiv wirkende elektr. Spannung/Signal* || Vorspannung * *(elektrisch)* || Vorspannung geben
bias adjustment Nullpunktverschiebung * *per Einstellung*
bias current Ruhestrom * *z.B. bei Operationsverstärker; input bias current: Eingangs~*
bias voltage Vorspannung * *(elektrisch)*
biased einseitig * *parteiisch* || vorgespannt * *mit einer elektr. Spannung*
bibliographical data Literaturhinweise * *Literaturangaben*
bibliography Bibliographie || Literaturhinweise * *Literaturnachweis/-verzeichnis*
bichromated bichromatisiert
bid Angebot * *~ bei öffentlicher Ausschreibung; Gebot (bei Versteigerung), Bewerbung (for: um), Bemühung* || Gebot * *Angebot (z.B. bei Versteigerung)*
bidder Bewerber * *bei einer Ausschreibung*
bidirectional bidirektional || drehrichtungsunabhängig * *in beiden Drehrichtungen arbeitend/wirkend* || Zweirichtungs-
bidirectional fan drehrichtungsunabhängiger Lüfter
bifilar winding bifilare Wicklung
big dick * *massig* || groß * *umfangreich; grow big: ~ werden*
big company Großunternehmen
big endian format Motorola-Format * *Datenformat: höherwertig. Byte/Wort auf niedriger, niederwert. auf höherer Adesse*
big player Großanbieter
biggest größt
bilateral gegenseitig * *zweiseitig* || zweiseitig * *Vertrag usw.*
bilateral drive doppelseitiger Antrieb
bilding lumber Bauholz
bilingual zweisprachig
bill berechnen * *jdm. einen Geldbetrag ~* || Rechnung * *Rechnung (-sschein/-szettel)*
bill of lading Frachtbrief * *(amerikan.)* || Lieferschein * *Frachtbrief*
billet Knüppel * *Metall/Walzknüppel (Stahl/Walzwerk)* || Walzknüppel
billet mill Knüppelwalzwerk
billion Milliarde * *(amerikan.)* || Milliarden
bimetal Bimetall
bimetal relay Bimetallrelais
bimetal release Bimetallschalter * *{el. Masch.}*
bimetal switch Bimetallschalter * *z.B. zum thermischen Motorschutz*
bimetallic Bimetall-
bimetallic release Bimetallauslöser
bimetallic trip Bimetallauslöser || Bimetallschalter * *Bimetallauslöser*

bimetallic-element switch Bimetallschalter * *z.B. zum thermischen Motorschutz*
bin Behälter * *für Schüttgut* || Bunker * *(großer) Behälter, (Vorrats-) Bunker, Fülltrichter,Lagerfach, Kasten, Tasche*
binary binär
binary code Binärcode || Dual-Code || Dualcode
binary function binäre Funktion || Binärfunktion
binary input Binäreingabe || Binäreingang || Digitaleingang * *Binäreingang*
binary logic Verknüpfung * *binäre*
binary logic function Verknüpfungsfunktion
binary number Dualzahl
binary observer binärer Beobachter * *{Regel.-techn.}*
binary operation binäre Funktion * *SPS-Befehl* || Binärfunktion * *SPS-Befehl*
binary output Binärausgabe || Binärausgang || Digitalausgang * *Binärausgang*
binary patch Balkon || Rucksack * *{Computer}*
binary quantity Binärgröße
binary scaler Binäruntersetzer
binary signal Binärsignal
binary statement Binäranweisung * *z.B. für SPS*
binary value Binärgröße
binary variable Binärgröße
binary vector Binärvektor
bind binden * *(an-/um-/fest)binden, Bücher (ein)~, binden {Chemie}, verpflichten (figürl.)* || verpflichten * *oneself: sich; oneself to do a thing: sich zu etwas ~* || vertraglich verpflichten * *oneself: sich; someone to something: jdm. durch Vertrag zu etwas*
binder Aktenordner || Ordner * *auch für für Loseblattsammlung*
binding bindend || fest * *verbindlich, z.B. Angebot* || rechtsverbindlich * *(up)on:für* || verbindlich * *upon/for:für*
binding agent Bindemittel * *{Chemie}*
binding department Buchbinderei * *als Abteilung*
binding machine Bindemaschine * *Walzwerk*
bio-degradable biologisch abbaubar
biological biologisch
BIOS BIOS * *'Basic Input Output System': D. Betriebssystem unterlagerte →Firmware z.Ansprechen d. Hardware*
bipartite zweiseitig * *Verhandlungen, Verwaltung usw.; doppelt ausgefertigt (Dokumente)* || zweiteilig
bipolar bipolar || ungepolt
biro Kugelschreiber
birth Entstehung * *Geburt, Ursprung, Entstehung*
bistable bistabil
bistable amplifier Kippverstärker
bistable relay bistabiles Relais || Haftrelais * *bistabiles Relais*
bit Bit * *Abkürzg. für 'Binary Digit' (Informationseinheit)* || Einsatz * *zum Einsetzen in Werkzeughalter (z.B. Bohr-/Schraub~)* || Werkzeugeinsatz
bit assignment Bitbelegung
bit by bit bitweise
bit for bit bitweise
bit holder Werkzeughalter * *für Werkzeugeinsatz*
bit level Bitebene
bit pattern Bitmuster
bit processor Bitprozessor * *Spezialprozessor zur schnellen Bitverarbeitung in SPS*
bit sequence Bitmuster * *Bitfolge*
bit string Bitfolge
bite angreifen * *Säge usw.* || fassen * *Werkzeug usw. ~*
bitumen Bitumen
bitwise bitweise
black schwarz
black band Kommutierungsgrenzkurve

black-marketeer Schieber * *Schwarzmarkthändler*
blackboard Tafel * *Wand~* || Wandtafel
blackening Schwärzung
blade Abstreifer * *doctor blade:* →*Rakel* || Blatt * *Klinge* || Flügel * *z.B. eines Lüfters* || Lamelle * *Blatt, Flügel, Schaufel, Klinge* || Messer * *Klinge* || Schaufel * *bei Turbine, Lüfter usw.* || Streichleiste * *bei Streichmaschine* || Streichmesser * *bei Streichmaschine*
blade coater Streichmaschine * *{Papier}*
blade contact Faston-Anschluss * *Messerkontakt*
blade-type terminal Flachsteckkontakt || Messerkontakt
blameless einwandfrei * *tadellos*
blank Formblatt || Formular || jungfräulich * *(figürl.)* leer, unbeschrieben || leer * *unausgefüllt, unbeschrieben, leer (Blatt, Raum, Frontplatte, Diskette usw.)* || Leer- * *Formular, Diskette, Speichermodul usw.* || Leerzeichen || Lücke * *leere Stelle* || Rohling * *unbearbeitetes Werkstück* || stanzen * *auschneiden* || unprogrammiert * *z.B. EPROM-Speicher* || Vordruck * *(amerikan.)*
blank form Formblatt
blank line Leerzeile
blank module Leermodul
blank out ausstanzen
blanket Decke * *Stoffdecke* || Gummituch * *{Druckerei} auch Stoffdecke, Filzunterlage* || pauschal * *allgemein, Blanko-* (*z.b. Erlaubnis-/Genehmigung); umfassend, Gesamt-, gemeinsam*
blanking die Stanzwerkzeug
blanking plug Blindstopfen || Stopfen * *Blindstopfen*
blast sprengen || zerstören * *sprengen, zugrunderichten, vernichten, (salopp) 'wegpusten'*
blast furnace Hochofen
blatant eklatant * *brüllend, marktschreierisch, lärmend, eklatant (Beispiel, Unsinn, Lüge usw.)*
blazing eklatant * *schreiend, auffallend, offenkundig, eklatant, (salopp) verteufelt*
bleach bleichen
bleach plant Bleicherei * *z.B. für Papier*
bleached gebleicht * *bleached chemical pulp: ~er Zellstoff*
bleacher Bleicher * *{Papier} in der Zellstoff-/Papierindustrie*
bleaching Bleiche * *das Bleichen, z.B. von Papier*
bleaching apparatus Bleicher * *{Papier} in der Zellstoff-/Papierindustrie*
bleaching plant Bleicherei * *z.B. in der Textilindustrie*
bleak trüb * *(figürl.) trostlos, trüb, düster (Aussichten usw.)*
bleed entlüften * *Bremse, Hydraulikanlage* || lüften * *Luft/Dampf ausströmen lassen, z.B. Hydraulik, Bremsleitung*
bleed resistor Ableitwiderstand || Bremswiderstand
bleeder Ableitwiderstand || Bremswiderstand
bleeding Entlüftung * *einer hydraulischen Bremse*
blemish Fehler * *Makel* || Mangel * *Fehler, Mangel, Makel, Schönheitsfehler; verunstalten, schaden, beflecken (figürl.)*
blend mischen * *bei Nahrungs- und Genussmitteln* || Mischung * *~ verschiedener Sorten bei Nahrungs- und Genussmitteln*
blended legiert * *Schmieröl*
blending system Mischanlage * *besonders für Lebens-/Genussmittel*
blind hole Sackbohrung || Sackloch
blind hollow bore Sackloch
blink blinken * *Leuchtdiode, Meldeleuchte, Anzeige, Cursor*
blinking blinkend
blinking signal Blinksignal

blister Blase * *an der Oberfläche*
blister machine Blistermaschine * *{Verpack.-techn.}*
blister package Blisterverpackung
blistering Blasenbildung
block abwehren || arretieren || Baustein * *Software-Funktionsbaustein* || Block || blockieren || deaktivieren * *sperren* || Flaschenzug || Funktionsbaustein || Glied * *Funktionsbaustein* || Satz * *z.B. Daten~* || Satz * →*{NC-Steuerung} Verfahrdaten~ bei NC (numerischer Positioniersteuerung)* || sichern || Sperre * *Blockade, Versperren* || sperren * *auch Diode, Thyristor* || totlegen * *sperren* || Verfahrdatensatz * *{NC-Steuerung}* || Verfahrsatz * *bei NC (numerischer Positioniersteuerung)* || verriegeln * *blockieren* || verschließen * *blockieren* || verstellen * *versperren*
block and tackle Flaschenzug
block call Bausteinaufruf
block check character Blockprüfungszeichen * *Datensicherungszeichen, z.B. bei serieller Datenübertragung*
block circuit diagram Blockschaltbild * *Schaltbildübersicht*
block current Blockstrom * *rechteckmodulierter Strom z. Ansteuerung* →*bürstenloser Servomotoren (SIEMENS 1FT5)*
block design Blockbauform * *z.B. bei SPS*
block diagram Blockschaltbild || Prinzipschaltbild * *Blockschaltbild* || Prinzipschaltplan * *Blockschaltbild* || Schaltschema * *Blockschaltbild* || Übersichtsplan || Übersichtsschaltbild * *Blockschaltbild* || Übersichtsschaltplan
block end Bausteinende * *Ende eines Funktionsbausteins bei SPS*
block mode Satz fahren * *bei Numeriksteuerung, Positioniersteuerung*
block package Bausteinpaket * *Paket aus Software-/Funktionsbausteinen*
block parameter Bausteinparameter
block skip Satzausblendung * *{NC-Steuerung} Überlesen/Unterdrücken eines Verfahrsatzes*
block transition Satzwechsel * *{NC-Steuerung}*
block up schließen * *blockieren*
block-check character Blockprüfzeichen * *sorgt für die Datensicherung bei einer Datenübertragung (z.B. Quersumme)*
blockade Sperre * *Blockade, Sperre, Hindernis*
blockage Blockierung || Verstopfung
blockboard Tischlerplatte * *{Holz}*
blocked blockiert || gesperrt || stillstehend * *blockiert*
blocking -Sperre || Arretierung || blockieren * *(Substantiv)* || Sperre * *Versperren* || sperrend * *z.B. Diode* || Sperrung * *z.B. Thyristor, Regler, Verkehr, Konto*
blocking ability Sperrfähigkeit * *z.B. Thyristor*
blocking capacitor Sperrkondensator || Trennkondensator
blocking capacity Sperrfähigkeit * *eines Halbleiterbauelements*
blocking protection Blockierschutz
blocking voltage Sperrspannung * *z.B. Diode, Transistor, Thyristor*
blocking-state current Sperrstrom * *(Thyristor, Diode)*
bloom vorwalzen * *{Walzwerk} Walzblöcke ~*
blotted fleckig * *mit Schmierflecken*
blow Ansprechen * *(Verb) Sicherung* || Blase * *im Inneren* || blasen || durchbrennen * *Sicherung* || Erschütterung * *Schlag* || fallen * *Sicherung* || Schlag * *Stoß, Streich, jarring blow: Prell~; at a blow: auf einen ~; strike a blow: e. ~ führ.*
blow off ablassen * *Dampf ~*
blow out ausblasen * *mit Druckluft usw.*

blow up aufblasen || sprengen
blower Gebläse || Lüfter * *Gebläse* || Ventilator
 * *Lüfter*
blower drive Gebläseantrieb
blower motor Lüftermotor
blowing machine Blasmaschine * *z.b. für Glas*
blowing of fuse Sicherungsfall * *Durchbrennen e. Sicherung*
blowing of fuses Sicherungsfall * *Plural; Durchbrennen von Sicherungen*
blowing roll Blaswalze
blowoff valve Überdruckventil
blowpipe Brenner * *Schweißbrenner* || Schneidbrenner
blue blau
blue-print Blaupause
blueprint Plan * *Blaupause* || technische Zeichnung * *(Blau)Pause* || Zeichnung * *Pause*
blunder Panne * *Schnitzer, grober Fehler*
bluntly frei * *(Adv.) unverblümt*
blurred unscharf
BMyte Megabyte
BNC BNC * *Abkürzg. für 'BajoNet Connector': Bajonett-Steckverbindung für 50 Ohm Koax-Leitung usw.*
board Baugruppe * *Leiterkarte* || Brett || einsteigen * *an Bord eines Schiffes* || Flachbaugruppe || Gremium || Holzplatte || Karte * *Leiterkarte* || Karton * *~papier* || Leiterkarte || Leiterplatte || Pappe || Platine * *Leiterkarte* || Platte * *Brett, Tafel, Planke* || Tafel * *für Anschläge und zum Schreiben* || Verband * *Interessenvereinigung*
board gluing line Brettverleimanlage * *[Holz]*
board gluing machine Bretterverleimmaschine
board identification Baugruppen Identifikation * *(automatische) Baugruppen-Erkennung*
board machine Kartonmaschine
board of assessment Steuerbehörde
board of directors Vorstand
board parameter diagram Modulparameterplan * *(SIEMENS SIMADYN D (R))*
board-mounted transformer Leiterplattentransformator
boast aufweisen * *etwas aufzuweisen haben*
bobbin Spindel * *[Textil]* Spule || Spule * *Garn-/Spinn~*
bobbin drawing device Spulenabzug * *[Textil] in der Spinnerei*
BOD BOD * *Break-Over Diode: Diode zum Thyristor-Schutzzünden bei Überschreiten d. max. zuläss. Sperrspanng.*
Bode diagram Bode-Diagramm || Bodediagramm
bodily injury Körperverletzung || Personenschaden * *Körperverletzung*
body Behörde || Gegenstand * *(Fest)Körper, Masse, Substanz* || Gehäuse * *äußere Gestalt* || Körper * *auch Raum/Fremdkörper*
body conformity Anpassung * *an die Oberfläche eines Körpers*
bodying Eindickung * *von Öl usw.*
bogie Drehgestell * *[Bahn]* || Radsatz * *[Bahn]* →*Drehgestell*
bogie wheel Tragrolle
boiler Kessel || Kocher
boiler feed pump Kesselspeisepumpe * *für Kraftwerk*
boiling point Siedepunkt
boisterous laut * *lärmend*
bold Fett * *(Adj., typografisch) in bold print: fettgedruckt*
bold print Fettdruck
bolster Polster * *Polster, Polsterung, Kissen, weiche Unterlage* || polstern * *(auf)polstern, (weich) abfedern*

bolster up untermauern * *unterstützen/-mauern, unterfüttern*
bolt anschrauben || Bolzen || Dorn * *Bolzen* || festschrauben || Riegel || Rolle * *bolt of cloth: Stoff~* || schließen * *mit Riegel ~* || Schraube * *dicke ~, ~ mit (teilweise gewindelosem) Schaft, Schraubbolzen* || Stift * *Bolzen* || verschließen * *mit Riegel usw. ~*
bolt down festschrauben * *z.B. Deckel* || verschrauben * *festschrauben*
bolt head Schraubenkopf
bolt into place festschrauben * *Maschinenteil an seinen Einbauplatz, Schaltgerät an Montageplatte*
bolt locking element Schraubensicherungselement
bolt on aufschrauben * *anschrauben*
bolt screw Bolzenschraube
bolt tightening scheme Anziehschema * *für Schrauben*
bolt to anschrauben
bolt-on Anschraub- || anschrauben
bolt-on fixing Anschraubbefestigung
bolt-on mounting Schraubbefestigung
bolted angeschraubt || verschraubt * *mit Schraubenbolzen*
bolted connection Schraubbefestigung * *Schraubverbindung* || Schraubverbindung
bolted joint Schraubverbindung || Verschraubung
bolted on angeschraubt
bolthole circle diameter Lochkreisdurchmesser
bon appétit guten Appetit * *("guten Appetit" in England u. USA ungebräuchlich).*
bond aushärten * *abbinden* || binden * *ab~ (Lack, Harz, Klebestelle usw.)* || härten * *(chem.) (ab-) binden, aushärten (z.B. Klebstoff)* || kleben * *Metall kleben, chem./techn. binden* || verkleben * *Metall* || Verpflichtung * *~, Bürgschaft, Kaution, Vertrag, Schuldschein, Obligation, Wertpapier, Zollverschluss*
bond note Begleitpapiere * *zollamtliche ~*
bonded verbacken * *verklebt, ausgehärtet*
bonded warehouse Zollager
bonding Aushärtung * *(chem.) von Klebstoff usw.* || Bondierung * *Befestigung der Anschlussdrähte eines Halbleiterchips auf dem Trägermaterial* || Verfestigung * *Abbindung, z.B. einer Klebestelle*
bonding agent Bindemittel
bone dry ofentrocken * *"knochentrocken"*
bonus Bonus || Rabatt * *Bonus* || Zulage
book anmelden * *buchen* || bestellen * *buchen (Platz, Zimmer usw.)* || buchen || eintragen || hereinnehmen * *(ver)buchen (z.B. Auftrag)* || reservieren * *~ lassen, buchen*
book of reference Nachschlagewerk
book-binding Buchbinderei
book-size flach * *flach/schmal bauend wie ein Buch* || schlank * *in schmaler Bauform ("schmal wie ein Buch")* || schmal * *in schlanker Bauform, "wie ein Buch"*
book-size design senkrechte Bauform * *schmal wie ein Buch*
bookbindery Buchbinderei * *Buchbinderwerkstatt/-Firma*
bookbinding Buchbinderei
booking Anmeldung
bookkeeper Buchhalter
bookkeeping Buchführung
booklet Broschüre || Heft * *Büchlein, Broschüre*
bookprinting Buchdruckerei
boolean binär * *boolsch* || boolsch
boolean algebra Schaltalgebra
boolean processor Steuerungsprozessor * *bei SPS*
boolean result Verknüpfungsergebnis * *einer logischen Operation*
boolean variable boolsche Größe
boost -Anhebung || -Steigerung * *(salopp; nachgestellt)* || anheben * *Strom/Spannung/Drehmoment*

durch eine besondere Maßnahme || Anhebung * *z.B. Spannung/Leistung/Drehmoment in einem bestimmten Arbeitsbereich* || ankurbeln * *(figürl.)* ankurbeln, in die Höhe treiben, Auftrieb geben, anpreisen* || erhöhen * *Spannung/Strom/Drehmoment in einem bestimmten Arbeitspunkt/durch eine besondere Maßnahme* || Erhöhung * *Leistungs~, Spannungs~, Druck~* || hochtransformieren * *anheben, verstärken (z.B. Druck, Spannung)* || Schub * *Erhöhung der Leistungsfähigkeit* || Spannungsanhebung* || steigern * *(salopp)* || Steigerung * *Ankurbelg., Erhöhg. (z.B. Druck, Spanng., Drehmoment), Verstärkg., Förderg., Auftrieb* || vergrößern * *(salopp) 'aufblasen'*, ankurbeln, Auftrieb geben, steigern* || verstärken * *Spannungs/Strom/Drehmoment in bestimmten Arbeitsbereichen anheben* || Verstärkung * *Anhebung von Spannung/Drehmoment/Druck usw. (z.B. in e.bestimmten Arbeitsbereich)*
boost charge Schnelladung
boost in performance Leistungsschub
boost-and-buck connection Zu- und Gegenschaltung * *z.B. b.Thyristorsatz z. Erreichung e. hohen Schaltspanng. üb.Spezialtrafo*
boost-and-buck converter connection steuerbare Stromrichter-Reihenschaltung * *2 i.Reihe geschaltete Stromrichter m. getrennter Ansteuerg.*
boosted erhöht * *(salopp) aufgeblasen, z.B. Leistungsvermögen, Wirkung usw.* || verstärkt * *angehoben, erhöht, gesteigert, angekurbelt*
booster Verstärker * *Hochsetzschaltung*
booster station Pumpstation * *zur Druckerhöhung bei Rohrleitungen/Pipelines*
boot hochfahren * *einen Computer* || starten * *Computer neu~/hochlaufen lassen*
booth Ausstellungsstand || Messestand || Stand * *Messe~*
booth duty Standdienst * *auf einer Messe/Ausstellung*
bootstrap neu laden * *Programm ~* || urladen * *Programme usw. ~*
bootstrap loading urladen * *Programm ~*
border bördeln * *einfassen, (um-)säumen* || Grenze * *Landes~; border on: grenzen an (auch figürl.)* || Kontur * *Grenze; border line: Grenzlinie* || Rand * *Grenze*
border line Grenze * *Grenzlinie*
border on stoßen * *angrenzen*
bore ausbohren || ausdrehen * *z.B. mit Drehbank, Ausbohrmaschine* || bohren * *ausbohren, mit Bohrmeißel ~, mit Holzbohrer ~* || Bohrung
bore diameter Bohrungsdurchmesser
bore out ausbohren
bore reducer Reduzierstück * *zur Querschnittsverminderung einer Bohrung*
bore size Lochmaß
borehole Bohrloch * *durch Aufbohrung; auch in Holz und in der Erde (zur Öl-/Wasser-/Gasgewinnung)*
borehole pump Bohrlochpumpe
boring Lochung
boring and milling center Bohr- und Fräszentrum * *{Werkz.masch.}*
boring and moulding center Bohr- und Fräszentrum * *{Holz}*
boring bit Bohrer * *(Aus)Bohreinsatz, z.B. für Holz*
boring head Bohrkopf * *{Holz}*
boring jig Bohrfutter
boring machine Ausbohrmaschine || Bohrwerk
boring mill Bohrwerk
boring socket Bohrfutter
boring tool Aufbohrwerkzeug
borings Span * *Bohrspäne* || Späne * *Bohrspäne*
borrow leihen * *borrow a thing from a person: sich etwas von jdm. aus~* || Übertrag * *bei einer Subtraktion* || Unterlauf * *Unterlaufbit/Merker bei Rechenoperationen. z.B. Subtraktion*
borrow bit Übertragsbit * *Unterlaufbit z.b. bei Subtraktion*
borrow from entnehmen * *aus einem Buch usw. ~*
borrowed capital Fremdkapital
boss Chef * *salopp* || Leiter * *(salopp) Chef* || Meister * *(salopp) Chef*
botch pfuschen
botching Pfusch
both ... and sowohl als auch * *both in x and in y:sowohl in x als auch in y*
bother Ärgernis * *Last, Plage, Mühe, Ärger, Schererei, Aufregung, Getue* || stören * *belästigen*
bother about kümmern * *sich Gedanken machen um, sich ~ um, sich plagen mit* || kümmern um || sich Gedanken machen über * *sich sorgen/bemühen um* || sich kümmern um
bothersome lästig * *lästig, unangenehm*
bottle einfüllen * *in Flasche* || füllen * *auf/in Flaschen ~*
bottle box Flaschenkasten
bottle cleaning machine Flaschenreinigungsmaschine
bottle crate Flaschenkasten
bottle filler Flaschenabfüllmaschine
bottle filling machine Flaschenabfüllanlage
bottle rinsing machine Flaschenreinigungsmaschine
bottle sorting plant Flaschensortieranlage
bottle-filling machine Flaschenabfüllmaschine
bottle-making machine Hohlglasmaschine * *zur Herstellung von Glasflaschen*
bottleneck Engpass * *(auch figürlich) delivery bottleneck: Liefer~* || Engstelle * *Flaschenhals* || Flaschenhals * *(auch figürlich: Engstelle, kritischer Weg, den Durchsatz bestimmendes Element)* || Schwierigkeit * *→Engpass*
bottling Abfüllung * *von Flaschen*
bottling line Flaschenabfüllanlage
bottling machine Abfüllmaschine * *zur Flaschenfüllung, z.B. in d. Getränkeindustrie* || Getränkeabfüllmaschine
bottling plant Abfüllanlage * *{Verpack.techn.}* || Flaschenabfüllanlage
bottling process Füllung * *Vorgang des Flaschen-Abfüllens*
bottom Boden * *eines Gefäßes, Schrankes, Gehäuses usw.* || Grund * *~ von Gefäßen* || letzt * *unterst* || unten * *from the bottom: von ~; from top to bottom: von ~ bis/nach oben; towards the bottom:-nach ~* || unter * *untenliegend; bottom half: ~e Hälfte* || Unter-
bottom cable Unterseil * *{Aufzüge}*
bottom die Matrize * *Stanzwerkzeug (Unterteil)*
bottom edge Unterkante
bottom face Unterseite * *untere Fläche*
bottom line unterer Strich
bottom plate Bodenblech || Maschinenbett * *Fundamentplatte*
bottom roll Unterwalze
bottom rope Unterseil * *{Aufzüge}*
bottom side Unterseite
bottom up von unten * *von unten nach oben, das Untere nach oben gekehrt, kieloben* || von unten nach oben * *das untere nach oben gekehrt*
bottom-mounted untenliegend * *unten montiert*
bottom-roll motor Untermotor * *Walzwerk*
bottomed söhlig * *{Bergbau}*
bounce prellen * *von Kontakten*
bound bestimmt * *for: zu/für; be bound to: dazu ~ sein, etwas tun zu müssen* || Grenze * *Grenze, Schranke, Bereich; within wide bounds:in weiten ~n; within b.:in ~en halten* || Sprung * *Satz, 'Hüpfer'*

bound for nach * *bestimmt* ~
boundary Grenze
boundary condition Grenzbedingung || Randbedingung
boundary conditions Rahmenbedingungen
boundary load Grenzbelastung * *z.B. bei einem Walzgerüst*
boundary position Nahtstelle || Nahtstelle * *Grenzstelle, Rand*
boundary value Grenzwert
boundless unbegrenzt
bow Bügel * *federnder Bügel (z.B. für Brille); bow collector: Stromabnehmer*~ || Kurve * *Bogen*
bow collector Stromabnehmer * *[Bahnen] ~bügel*
bowl Wanne * *Napf, Schale, Schüssel, Becken, Wölbung*
box Behälter * *Kasten* || Gehäuse || Kapsel || Kassette || Kasten * *auch auf dem Bildschirm* || Kiste || Muffe * *Kabelverbindungs~* || Schachtel || Umhüllung
box board Karton * *~agenpappe* || Kartonage * *~npappe*
box nut Überwurfmutter
box pallet Boxpalette
box spanner Inbusschlüssel
box-type brush holder Taschenbürstenhalter
boxboard Faltschachtelkarton * *Kartonagenpappe*
bps Baud * *bits pro sec*
brace Anker * *Befestigungselement* || Holm || Klammer * *geschweifte* ~ || Strebe * *Stütze, Strebe, Klammer, Versteifung, (mechan.) Anker* || verankern * *verstreben, befestigen, verklammern, verankern, versteifen* || verspannen * *versteifen, verstreben, befestigen, verklammern, verankern* || versteifen || Versteifung * *Stütze, Strebe* || verstreben || Verstrebung
bracing Spann- || Verspannung * *Verstrebung, Verklammerung* || Versteifung * *Verstärkung, Abstützung* || Versteifung * ~ *durch Strebe, Stütze, Anker, Klammer, Gurt usw.*
bracing tube Tragrohr
bracket Bügel || Halter * *(Wand)Konsole, Winkelstütze, Stützbalken, offenes Lagerschild* || Halterung * *Halter, Konsole, Träger, Winkelstütze, Wandarm* || Klammer * *Bauteil und Schriftzeichen (meist eckige ~); bracketed: in ~n (gesetzt)* || Klemmbügel * *(eckige) Klammer* || Konsole || Lagerschild || Strebe * *Träger, Halter, Konsole, Stützbalken, Winkelstütze* || Träger * *Wandkonsole* || Versteifung * *Stütze, Träger* || Winkel * *Haltewinkel*
bracket bearing Schildlager
bracket depth Klammertiefe
bracket level Klammerebene
bracket-mounted bearing Sattellager
brackets Klammern * *auch als Schriftzeichen*
brackish water Brackwasser
braid Geflecht * *Litze, ~ aus feindrähtigen Drähten* || Schnur * *Litze* || Streifen * *Litzenbandleiter*
braided screen Geflechtschirm * *bei abgeschirmtem Kabel*
braiding machine Flechtmaschine * *[Textil]*
brain wave Idee * *Geistesblitz, guter Einfall*
brake abbremsen || Abkantpresse * *zum Biegen von Blech ("brake" und "break" ist beides richtig)* || Bremse || bremsen || herunterbremsen || stillsetzen * *Abbremsen* || zurücklaufen * *bremsen*
brake caliper Bremssattel * *an Scheibenbremse*
brake chopper Bremschopper
brake closing time Bremsenschließzeit
brake coil Bremsspule
brake cone Bremskonus
brake contactor Bremsschütz
brake control Bremsensteuerung
brake control circuit Bremsstufe
brake delay time Einfallzeit * *einer Bremse*

brake disc Bremslamelle || Bremsscheibe
brake disk Bremslamelle || Bremsscheibe
brake down herunterbremsen
brake fan Bremslüfter
brake for static duty Haltebremse
brake generator Bremsgenerator * *z.B. Abwickler*
brake hub Bremsnabe
brake lining Bremsbelag
brake monitor Bremswächter
brake motor Bremsmotor * *Motor mit Anbau-/Einbaubremse*
brake motor with fail-to-safe brake Federdruck-Bremsmotor * *mit Anbaubremse nach d. Ruhestromprinzip (bremst, wenn kein Strom)*
brake plate Bremslamelle || Bremsscheibe
brake release Bremsenöffnung
brake resistance Bremswiderstand
brake resistor Bremswiderstand || Pulswiderstand * *Bremswiderstand*
brake shoe Bremsbacke
brake-application point Bremseinsatzpunkt
braked gebremst
brakemotor Bremsmotor * *Motor mit Anbau-/Einbaubremse*
braking Abbremsung || bremsend || Bremsung || generatorisch * *im Bremsbetrieb* || Rücklauf * *Bremsen*
braking angle Bremswinkel * *bis zum Stillstand noch zurückgelegter Winkel ab Einschalten einer Bremse*
braking by armature short-circuiting Ankerkurzschlussbremsung
braking by reversal Gegenstrombremsung
braking characteristic Bremskennlinie
braking control Bremsensteuerung
braking copper Bremschopper
braking current Bremsstrom
braking device Bremsgerät
braking distance Bremsweg
braking effect Bremsung * *Bremswirkung* || Bremswirkung
braking energy Bremsarbeit || Bremsenergie
braking equipment Bremseinrichtung
braking facility Bremseinrichtung
braking gear Bremseinrichtung
braking interval Bremszeit
braking module Bremseinheit || Bremseinrichtung || Bremsmodul
braking operation Bremsbetrieb || Bremsung * *Bremsvorgang* || Bremsvorgang
braking power Bremsleistung
braking pressure Bremsdruck
braking relay Bremsrelais
braking resistor Bremswiderstand || Pulswiderstand * *Bremswiderstand*
braking rheostat Bremswiderstand
braking time Bremszeit || Bremszeit * *[el. Masch.] Dauer einer Bremsung vom Ausschalten d. Motors bis zum Stillstand*
braking torque Bremsmoment
braking unit Bremseinheit || Bremseinrichtung || Bremsgerät
braking voltage Bremsspannung
branch Ableger * *eines Unternehmens/einer Firma* || Abzweig || Abzweigung || Anspruch || Ast * *einer Kurve; ascending: aufsteigender; descending: abfallender* || Bereich * *Branche* || Branche || Fach * *Branche* || Niederlassung * *Filiale* || Pfad * *Zweig (auch in Programm)* || Programmzweig * *verzweigen* || Sprung * *Programmverzweigung* || Stichleitung * *Abzweig* || verzweigen || Verzweigung * *auch im Programm: multiple branch: Mehrfach~* || Weiche * *Verzweigung* || Zweig || Zweigniederlassung || Zweigstelle → * ~ *einer Firma; siehe auch* →*Zweigbüro und* →*Zweigniederlassung*

branch address Sprungadresse
branch box Abzweigdose
branch circuit Abzweig
branch establishment Zweigniederlassung
branch fuse Zweigsicherung
branch of business Geschäftszweig
branch of industry Branche || Industriebereich || Industriezweig
branch off abbiegen * *Straße usw.* || abführen * *abzweigen* || abgehen * *Weg usw.*
branch office Außenbüro || Niederlassung || Zweigbüro || Zweigniederlassung * *Zweigbüro*
branch out teilen * *(Straße, Programmfluss usw.)*
branch pipe Stichleitung * *Rohrleitung*
branch trade Fachhandel
branch-off point Verzweigungspunkt
branch-oriented branchenorientiert
branch-specific branchenbedingt
branching Abzweigung || Verzweigung * *auch ~ im Programmlauf* || Weiche * *Verzweigung*
branching box Abzweigdose
branching off Abzweigung
brand Fabrikat * *Herstellermarke* || Marke * *eines ~nartikels* || Sorte * *Marke*
brand label Handelsmarke || Markenname * *Markenzeichen*
brand leader Marktführer
brand name Markenname
brand-new fabrikneu || neu * *ganz ~*
brass Messing
braze hartlöten || löten * *hart~* || verlöten * *hartlöten*
brazed joint Lötverbindung * *hartgelötete ~*
brazing hartlöten * *(Substantiv)*
break abreißen || abschalten * *z.b. durch Relaiskontakt ~* || ausschalten || bersten * *Eis, Glas usw* || brechen * *auch Steine/Vertrag/Rekord/Versprechung/Widerstand/Knochen/Kanten; break down: zusammen~* || durchreißen || knacken * *(figürl.) Software, Zugangscode usw.* || Knick * *Abbruch, Wechsel, Umschwung, Sturz (z.B. Preise, Wetter, Wirtschaft, Geschäft),* || knicken * *brechen* || Lücke * *Unterbrechung* || Pause * *z.B. ~ bei Besprechung; coffee break: Kafee~* || reißen * *brechen; z.B. Papierbahn, Draht, Kabel* || Riss * *(Ab-/Zer-/Durch-) Brechen/Reißen; Bruch, ~ einer Papierbahn/eines Fadens usw.* || unterbrechen * *auch Leiterbahn* || Unterbrechung
break away anziehen * *Motor*
break contact Öffner || Öffnerkontakt || Ruhekontakt * *Öffnerkontakt*
break down aufgliedern * *"herunterbrechen", aufgliedern, aufschlüsseln, analysieren* || aufreißen * *auch Schmierfilm* || aufschlüsseln * *statistisch aufgliedern/schlüsseln* || ausfallen || Durchlegieren * *bei Halbleiterbauelement; siehe auch* →Durchschlag || Durchschlag * *Isolationsversagen (z.B. bei Überspannung)* || unterteilen * *statistisch ~* || versagen || vorwalzen * *[Walzwerk] Blech ~* || zerlegen || zusammenbrechen
break in einbrechen || eindringen * *(hin)einbrechen* || einfahren * *(amerikan.) neues Auto usw.*
break in two durchbrechen * *entzweibrechen*
break into einbrechen * *gewaltsam ~*
break off abbrechen * *Beziehung usw. ~* || Abbruch || abreißen || aussetzen * *unterbrechen* || herausbrechen || lösen * *Beziehung ~* || rückgängig machen * *abbrechen*
break open sprengen * *aufbrechen*
break operation Schaltspiel * *Abschaltung*
break out durchbrechen * *zum Vorschein kommen* || entstehen * *Feuer usw.*
break through durchbrechen * *(figürl. und gegenständl.)*
break time Abschaltzeit

break up aufbrechen * *(auch figürl.)* || aufgliedern || aufspalten
break-away friction Losbrechreibung
break-away torque Losbrechmoment * *Lastmoment im Stillstand, das größer ist als nach Beginn der Bewegung*
break-before-make contact unterbrechender Kontakt * *Wechselkontakt, der unterbrechend (ohne Überlappung) schaltet*
break-down Störung * *Maschinenschaden, Zusammenbruch, Versagen, Scheitern, Panne*
break-down torque Nenndrehmoment * *Kippmoment z.B. bei Asynchronmotor*
break-even point Schnittpunkt * *z.B. Punkt, bei dem man in die Gewinnzone kommt*
break-over diode BOD * *zum Schutzzünden v. Thyristoren bei Überschreiten der max. zulässigen Sperrspannung*
break-up Lösung * *(Auf)~ einer Beziehung*
breakable zerbrechlich
breakage Abriss * *Draht, Faden, Materialbahn usw.* || Bruch * *Draht, Material; auch 'Bruchstelle, Bruchschaden'* || Riss * *~ einer Material-/Papierbahn, eines Fadens/Drahts usw.*
breakage-proof bruchfest
breakaway Losbrechen
breakaway friction Losbrechreibung
breakaway starting current Anzugsstrom * *Motor*
breakaway torque Anzugsdrehmoment * *Motor* || Anzugsmoment * *Motor* || Losbrechmoment
breakaway voltage Anzugsspannung * *Motor*
breakdown -Ausfall * *Zusammenbruch* || Aufgliederung * *Aufschlüsselung/gliederung, Analyse* || Betriebsstörung || Durchlegieren * *bei Halbleiterbauelement* || Durchschlag * *Isolationsversagen* || Fehler * *Zusammenbruch; zum Maschinenstillstand führender ~* || Kippen * *z.B. ~ eines Asynchronmotor* || Panne || Schaden * *zum Maschinenstillstand führender ~* || Zusammenbruch
breakdown analysis Schadenanalyse
breakdown current Durchbruchstrom
breakdown strength Durchschlagfestigkeit * *Widerstandsfähigkeit gegen Isolierversagen*
breakdown torque Kippmoment * *bei Synchron- u. Asynchronmotor*
breakdown voltage Durchbruchspannung || Durchschlagspannung
breaker plant Brechanlage * *z.B. für Steine, Koks*
breaker position Schalterstellung * *eines Leistungsschalters*
breaking Brechung
breaking capacity Abschaltvermögen * *eines Leistungsschalters, einer Sicherung usw.* || Ausschaltvermögen * *z.B. Sicherung, Leistungsschalter* || Kontaktbelastung * *(Ab-) Schaltvermögen eines Leistungsschalters* || Schaltleistung * *Aus~, Ab~* || Schaltvermögen * *z.B. ~ eines Leistungsschalters*
breaking current Ausschaltstrom * *eines Leistungsschalters*
breaking resistant bruchfest
breaking strain Zugfestigkeit
breaking strength Reißfestigkeit || Tragfähigkeit * *Trag-/Bruchfestigkeit* || Zugfestigkeit * *Zug-/Reißfestigkeit (z.B. in [N/mm^2])*
breaking voltage Schaltspannung * *Ab~*
breaking-down Abbau * *Niederreißen, Abbrechen, Zerlegen, Auflösen*
breaking-in period Einarbeitungszeit
breakpoint Haltepunkt * *[Computer] zum Programmtest* || Knickpunkt * *~ einer Kurve, Kennlinie usw.*
breakproof bruchfest
breakthrough Durchbruch * *(auch figürl.)* || Durchzündung * *(Thyristor)*

breaktime Ausschaltzeit * *eines Leistungsschalters*
breast box Stoffauflauf
breathe atmen
breather hole Entlüftungsbohrung
breed Art * *(Menschen-) Schlag, Art*
brevity Kürze * *des Ausdrucks, auch zeitlich*
brewery Brauerei
brewing room Sudhaus
bribe-money Schmiergeld
brick extruder Ziegelpresse * *kontinuierliche Strang-~*
brick press Ziegelpresse
bridge Brücke * *auch Dioden/Transistor/Thyristorbrückenschaltung* || schließen * *bridge the gap: die Lücke ~* || überbrücken * *(auch figürl.)*
bridge circuit Brückenschaltung * *Messbrücke*
bridge configuration Brückenschaltung * *Anordnung*
bridge connection Brückenschaltung * *Stromrichter*
bridge exciter Feldgleichrichter * *Brückengleichrichter*
bridge over überbrücken * *(auch figürl.)* || Überbrückung
bridge rectifier Brückengleichrichter
bridging contact nicht unterbrechender Kontakt * *Wechs.kontakt dess. Kontakte während d.Schaltens kurz geschloss.sind*
bridle Zugwerk * *{Walzwerk} Zugwalzenpaar*
bridle roll S-Rolle * *z.B. i. Blechstraße: Rolle mit großer Umschlingung, erzwingt konstante Geschwind.* || S-Walze * *z.b. in Walzstraße: Rollen mit großer Umschlingg., erzwingen konstante Geschwind.*
brief informieren * *eine Kurzinformation/Einweisung geben* || kurz * *gedrängt, knapp; in brief: ~ gesagt* || Kurz- || kurzzeitig * *kurz(zeitig), gedrängt*
brief description Kurzbeschreibung
brief outlines Kurzbeschreibung
brief supply failure Netzkurzausfall
brief-bag Aktentasche
brief-case Aktentasche || Mappe * *Aktentasche/Mappe*
briefcase Aktenkoffer * *Aktentasche*
briefing Einweisung * *(genaue) Anweisung, Instruktion, Einsatzbesprechung, Befehlsausgabe*
briefly kurz gesagt || Kürze * *in aller Kürze, kurz gesagt* || kurzzeitig * *(Adv.) gedrängt, kurz*
bright blank * *(metallically) bright: metallisch blank* || metallisch blank
brightener Aufheller
brightness Helligkeit
brightness control Helligkeitssteuerung
brilliance Kontrast * *bei Bildschirm usw.*
brilliant brillant || eklatant * *ausgezeichnet, glänzend* || überragend
brim Kante * *Rand (auch eines Gefäßes), Krempe* || Rand * *~ eines Gefäßes, (Hut)Krempe*
brinelling Riefenbildung * *z.B. in Wälzlager* || Standriefen * *Riefenbildung, z.B. in Wälzlager*
bring veranlassen * *dazu bringen/bewegen (to: zu)* || zuführen * *(her)bringen, herbeischaffen*
bring about bewerkstelligen * *by: mittels/durch* || zustandebringen * *by: mittels/durch*
bring down to a round figure abrunden * *eine Zahl nach unten abrunden*
bring forward einbringen * *auch Vorlage, Vorschlag*
bring in einbringen * *auch Vorlage, Vorschlag* || einfahren * *Ernte ~*
bring into action aktivieren * *ins Spiel bringen, einsetzen*
bring off bewerkstelligen
bring on line zuschalten * *Maschinenteil zur laufenden Arbeitsmaschine hin~*

bring onto load belasten * *auf die Last aufschalten (z.B. Motor)*
bring out herausführen * *z.B. Signale aus einem Gerät (to: terminals: auf Klemmen ~)*
bring to a close erledigen * *beenden*
bring to a standstill stillsetzen
bring to a stop anhalten || herunterfahren * *stillsetzen*
bring to an end beenden
bring together zusammenführen
bring up to a round figure abrunden * *eine Zahl nach oben aufrunden* || aufrunden * *eine Zahl nach oben*
bring up to date modernisieren
bring up-to-date aktualisieren * *auf den neuesten Stand bringen*
brink Rand * *(meist figürl.) on the brink of: am ~e des/von*
briquetting machine Brikettiermaschine
briquetting press Brikettpresse
brisk zügig * *flink, flott, lebhaft*
brittle spröde
brittleness Sprödigkeit
broach ausräumen * *{Werkz.masch.} z.b. mit Räumnadel/Räumeisen* || Räumnadel * *push/draw broach: ~ zum Stoßen/Ziehen* || Reibahle
broaching machine Räummaschine * *Werkz.masch.; internal/surface/turning broaching machine: Innen-/Außen-/Dreh~*
broaching tool Räumwerkzeug
broad breit
broad liner Vollanbieter * *Anbieter mit breitem Produktspektrum*
broad range weiter Bereich
broad-band breitbandig
broadband breitbandig
broadcast Broadcast * *{Network} {Telegramm} an alle* || senden * *Nachricht an mehrere Empfänger gleichzeitig ~* || Sendung * *Rundfunk/Fernsehsendung*
broadcast telegram Telegramm an alle
broadcasting station Sender * *Rundfunk~*
broaden erweitern * *z.B. Wissen*
brochure Broschüre || Druckschrift * *Broschüre* || Heft * *Broschüre* || Produktinformation * *Broschüre* || Produktschrift * *Broschüre* || Prospekt * *Broschüre*
brochures Prospekte
broke Ausschuss * *z.B. bei der Papierherstellung*
broken defekt || kaputt * *entzwei, zerbrochen* || zerbrochen
broken down ausgefallen * *auf Grund eines Fehlers/einer Störung* || unterteilt * *statistisch*
broken pieces Scherben * *go to pieces: in ~ gehen*
brokerage Honorar * *Provision eines Maklers usw.* || Provision * *eines Maklers/Vermittlers*
bronze Bronze
bronze sleeve bearing Bronzegleitlager
broom Besen
brought to a close abgeschlossen * *zum Abschluss gebracht*
brown braun || brünieren * *Metall*
brown-out Durchschlag * *Isolationsversagen*
browse anschauen * *durchschauen/-stöbern v. Datenbeständen, Dateien, Verzeichnissen, Datennetz-Inhalten usw.* || blättern * *{Computer} durch Datenbestand/Bildschirm-/Internetseiten ~*
browser Browser * *Programm zur Anzeige verschiedenster Daten/Dateien in einer Schnell-/Inhaltsübersicht*
bruise quetschen * *Körperteil ~; Quetschung, Quetschwunde*
brush Bürste * *auch Kohlebürste in Motor/Generator* || Kohlebürste || Pinsel || Streifen * *(Verb) brush against: streifen an*

brush access Bürstenzugänglichkeit * *external: von außen, easy: leichte*
brush arm Bürstenhebel
brush assembly Bürstenapparat * *einer DC-Maschine*
brush flashover Bürstenfeuer || Bürstenrundfeuer
brush gear Bürsteneinrichtung * *an Gleichstrommaschine usw.*
brush grinding machine Schleifbürstmaschine * *zur Reinigung in Blechstraße*
brush guide Bürstenführung * *z.b. bei Gleichstrommaschine*
brush holder Bürstenhalter * *für Kohlebürsten in Motor/Generator* || Kohlebürstenhalter
brush length Bürstenlänge
brush life Bürstenlebensdauer || Bürstenstandzeit
brush lifting and short-circuiting device Bürstenabhebevorrichtung * *{el.Masch.} Kurzschließ- und ~ für Schleifringläufermotor*
brush lifting device Bürstenabhebevorrichtung * *{el. Masch.}*
brush lifting gear Bürstenhebevorrichtung
brush machine Bürstmaschine * *z.B. zur Reinigung in Blechstraße*
brush mechanism Bürstenapparat
brush monitoring Bürstenüberwachung * *z.b. der Restlänge der Kohlebürsten in Gleichstrommotor*
brush monitoring device Bürstenüberwachung * *als Gerät/Einrichtung*
brush raising device Bürstenhebevorrichtung
brush rocker Bürstenbrücke * *für Kohlebürsten z.b. in Gleichstrommaschine*
brush roller Bürstenwalze || Bürstwalze * *z.B. in Bandbehandlungsanlage zum Reinigen des Walzbandes*
brush service life Bürstenlebensdauer || Bürstenstandzeit
brush set Bürstensatz * *z.b. in Gleichstrommaschine*
brush sparking Bürstenfeuer * *Funkenbildung am Kommutator*
brush spindle Bürstenhalterbolzen || Bürstenlineal
brush spring Bürstenfeder * *für Kohlebürste*
brush unit Bürstenapparat
brush wear Bürstenabrieb * *Bürstenverschleiß* || Bürstenverschleiß || Kohlebürstenabrieb * *Bürstenverschleiß*
brush-holder arm Bürstenlineal * *z.B. im Gleichstrommaschine*
brush-holder stud Bürstenhalterbolzen * *z.B. bei Gleichstrommaschine* || Bürstenlineal
brush-holderarm Bürstenhalterbolzen * *z.B. bei Gleichstrommaschine*
brush-replacement Bürstenwechsel
brush-rocker gear Bürstenverstellapparat * *bei Drehstromkommutatormotor* || Bürstenverstelleinrichtung * *bei Drehstromkommutatormotor*
brushgear Bürstenapparat * *z.B. einer Gleichstrommaschine*
brushless bürstenlos
brushless d.c. motor bürstenloser Gleichstrommotor * *AC Synchronmotor mit "elektron. Kommutator" (Blockstromansteuerung)*
brushless DC motor Brushless DC-Motor * *permanenterregter Synchron-Servomotor mit Block- oder Sinusmodulation* || bürstenloser Gleichstrommotor * *AC Synchronmotor mit "elektron. Kommutator" (Blockstromansteuerung)*
brushwear monitoring Bürstenüberwachung * *Überwachung d.Bürsten e. Gleichstrommaschine auf Verschleiß*
BS BS * *Abk. für 'British Standard': britische Norm*
bubble Blase
bubble formation Blasenbildung
bucket Baggereimer || Becher * *bei Bagger* || Grei-

fer * *~ eines Baggers/Krans* || Schaufel * *~ bei ~bagger; Turbine usw.; Eimer*
bucket conveyor Becherwerk * *Förderer*
bucket-wheel Schaufelrad * *Bagger*
bucket-wheel drive Schaufelradantrieb * *Bagger*
bucket-wheel excavator Schaufelradbagger
buckle knicken * *sich krümmen, verbiegen* || stauchen * *{Walzwerk}*
buckling Verbiegung * *Krümmen, Verziehen, Wölben* || Wölbung
budged-priced preisgünstig
budget Budget || einteilen * *(Geld)Ausgaben ~* || Einteilung * *~ der Finanzen* || Etat
budget-priced kostengünstig * *preisgünstig* || preiswert
budgeting Planung * *Ausgaben~, Haushalts~, Finanz~*
buff polieren * *glanzschleifen, polieren, schwabbeln* || schwabbeln * *(hochglanz-) polieren* || Schwabbelscheibe
buffer entkoppeln * *puffern durch Treiber- oder Speicherstufe* || Puffer || puffern || Pufferspeicher || speichern * *zwischenspeichern* || stützen * *Versorgungsspannung bei Spannungseinbruch ~* || Treiber * *Vertärkerstufe für Datensignale/-leitungen (meist mit Speicherwirkung)* || Verstärker * *Puffer z.B. für Datenleitung* || Vorrat * *Puffer* || Zwischenspeicher
buffer amplifier Spannungsfolger
buffer capacitor Speicherkondensator || Stützkondensator
buffer chain conveyor Staukettenförderer * *{Fördertechnik} mit Produktespeicher-Funktion*
buffer control Pufferverwaltung
buffer memory Koppelspeicher || Zwischenspeicher
buffer overflow Pufferüberlauf
buffer storage Pufferspeicher || Zwischenspeicher
buffer time Pufferzeit || Überbrückungszeit * *Pufferungszeit*
buffered gepuffert
buffering Pufferung * *Zwischenspeicherung von Daten* || Stützung * *z.B. durch Kondensator, Batterie, kinetische Energie*
buffing machine Schwabbelmaschine * *zum Hochglanzpolieren*
buffing wheel Schwabbelscheibe * *zum Hochlanzpolieren*
bug Fehler * *~ in Softwareprogramm* || Programmfehler
bug fixing Fehlerbereinigung * *in Software*
bugfix Fehlerbehebung * *~ in (bereits ausgelieferter) Software* || Fehlerbeseitigung * *in (bereits ausgelieferter) Software*
buggy fehlerhaft * *(salopp) Software*
build bauen || erstellen * *Gebäude ~*
build in einbauen
build into einbauen * *in ... hinein*
build round umbauen * *um etwas her~*
build up aufbauen * *auch Strom, Spannung, Feld, Druck, Wickeldurchmesser, Kundenstamm, Theorie usw.* || aufschaukeln * *anschwellen*
build-down Abbau * *z.B des Durchmessers beim Abwickler*
build-in position Einbauplatz
build-up Aufbau * *z.B. Spannung, Wickel beim Aufwickler, Wärme in e. Wärmenest, Magnetfeld* || aufbauen * *auch Feld, Druck, Wickeldurchmesser, Kundenstamm, Theorie usw.* || Erhöhung * *Aufbau, z.B. Temperatur in e. Wärmenest, Durchmesser bei e. Aufwickler*
build-up of heat Wärmestau
build-up time Anstiegszeit || Aufbauzeit * *z.B. für Motor-Feld*
builder's saw bench Baukreissäge * *{Holz}*

building Aufbau || Bau * ~ *eines Gebäudes* || Gebäude
building and civil engineering Bauwesen * *Hoch- und Tiefbau*
building automation Gebäudeautomatisierung
building block Baustein
building board Leichtbauplatte * *{Bautechnik}*
building construction Bauwesen * *Hochbau*
building crane Baukran
building in Einbau
building material Baustoff
building material industry Baustoffindustrie
building site Baustelle * *Haus~, Anlagen~*
building systems Gebäudetechnik
building-up Aufbau * *z.B. Spannung, Durchmesser beim Aufwickler*
buildup Aufbau * *siehe 'build-up'* || Zuwachs * *Aufbau, starke Zunahme*
built-in eingebaut || integriert * *eingebaut* || intern * *eingebaut*
built-in accessories Einbauten * *eingebaute Zubehörteile*
built-in components Einbauten
built-in motor Einbaumotor * *"nackter" Motor, nur aus Ständer- und Läuferpaket bestehend*
built-in unit Einbaugerät
built-on Anbau- || angebaut * *onto: an* || aufgebaut * *(oben) angebaut*
built-on accessories Anbauten
built-on circuit breaker Anbauschalter * *z.B.am Motor angebauter EIN/AUS, Polum- oder Stern/ Dreieckschalter*
built-on contactor Anbauschalter * *{el. Masch.} direkt an den Motor angebautes Motorschütz* || Schalteranbau * *{el. Masch.} Anbau des Motorschützes direkt an den Motor (statt Klemmenkasten)*
built-on gearbox Anbaugetriebe
built-on switch Anbauschalter
built-up Aufbau * *das Aufgebaute, z.B. Spannung, Durchmesser beim Aufwickler*
built-up speed Selbsterregungsdrehzahl * *critical:- kritische*
built-up time Anschwingzeit || Einschwingzeit * *z.b. zum Aufbau der Versorgungsspannung*
bulge Wulst * *Bauchung*
bulk Größe * *Rauminhalt* || Schüttgut
bulk goods Schüttgut
bulk material Massegut || Schüttgut
bulk storage Massenspeicher
bulk storage device Massenspeicher
bulkhead Schott * *z.B. Trennwand in Schiff*
bulky bauschig * *massig, sperrig; (sehr) umfangreich* || dick * *massig* || groß * *(sehr) umfangreich, massig, sperrig: bulky goods: Sperrgut* || sperrig
bulldozer Raupe * *Planierraupe*
bulletin Informationsdienst * *Folge eines ~es* || Informationsschrift * *kleines Informationsblatt aus aktuellem Anlass* || Mitteilung * *offizielle ~*
bulletin board Brett * *Anschlagbrett, Diskussionsforum (in Mailbox usw.)*
bump anstoßen * *(heftig) anstoßen/-prallen* || Ruck * *heftiger Stoß/Bums* || Stoß * *heftiger ~, Bums* || stoßen * *(heftig) stoßen, (an)prallen, bumsen (gegen), zusammen~ (mit)*
bumper Puffer * *(mechan.) zum Abfangen/federn eines (An)Stoßes*
bumpless ruckfrei * *stoßfrei* || stoßfrei * *ohne (heftigen) Stoß, Bums*
bumpless changeover stoßfreie Umschaltung
bumpless switchover ruckfreie Umschaltung * *stoßfrei* || ruckfreier Übergang
bumpless transition stoßfreier Übergang
bumplessly stoßfrei * *(Adv.)*
bunch bündeln * *zusammen~, zusammenfassen (zu Gruppe, Haufen usw., z.b. beim Verseilen von Kabeln)* || würgen * *beim Verseilen von Kabeln*
bundle Ballen || binden * *bündeln* || Bund * *Bündel, Bund, Paket, Ballen* || Bündel || bündeln * *(zusammen-) bündeln, zusammenknüpfen, zusammenfassen (zu Gruppe, Haufen), kombinieren* || kommissionieren * *zusammenpacken/-bündeln* || Ring * *Bund, Paket, Bündel, Ballen*
bundle press Bündelpresse
bundle with beilegen * *zusammenpacken mit*
bundling Kommissionierung * *gemeinsame Lieferung mehrerer Produkte*
bundling machine Bündelmaschine || Umreifungsmaschine * *{Verpack.technik} zum Verpacken von Paletten usw.*
bundling press Bündelpresse
bundling system Paketieranlage
bung verstopfen * *abdichten (Loch usw.)*
bungle Pfusch || pfuschen
bungling Pfusch
bunker Bunker * *Behälter z.B. für Kohle*
buojancy Auftrieb * *eines Körpers in einer Flüssigkeit, e.Flugzeuges, Tragkraft*
buoyancy Spannkraft * *{figürl.} Schwung, Spann-/ Lebenskraft; {Physik} Schwimm-, Tragkraft, Auftrieb* || Tragfähigkeit * *Schwimmkraft, Auftrieb*
burden belasten * *mit (Trag)Last/Bürde ~; burden someone with: jdn. mit etwas ~* || Belastung * *Traglast, Tragfähigkeit, (auch finanzielle) Bürde, Ladung* || Belastung * *financial: finanzielle ~* || Bürde || Last * *(Trag-)Last, Bürde, Ladung*
burden resistor Bürde * *Bürdenwiderstand* || Bürdenwiderstand
burdensome lästig * *lästig, drückend*
bureaucratic bürokratisch
bureaucratically bürokratisch * *(Adv.)*
burglary Einbruch * *krimineller ~, ~sdiebstahl (besonders b. Dunkelheit)*
burl Noppe
burn anbrennen || brennen
burn away abbrennen
burn by a slow fire schwelen
burn down abbrennen
burn in Erwärmungsprüfung * *Hitzebehandlung zur Voralterung und Erfassung von Frühausfällen*
burn mark Brandstelle * *z.B. an Kommutator*
burn out durchbrennen * *z.B. Wicklung*
burn slowly schwelen
burn through durchbrennen * *z.B. Wicklung*
burn-in Burn-In || zeitraffende Voralterung * *durch zyklische Wärmebehandlung*
burn-in oven Einbrennofen
burn-off Abbrand
burner Brenner
burnish brünieren * *polieren* || polieren
burr Grat || krempeln * *{Textil}* || schärfen
burst aufreißen * *bersten* || bersten * *brechen* * *bersten, (zer)platzen, (auf-/zer-) springen* || Bruch * *eines Rohrs, Tanks usw.* || Impulsgruppe * *von Nadelimpulsen* || Impulskette * *Paket von Nadelimpulsen* || Impulspaket * *~ von Nadelimpulsen* || reißen * *bersten*
burst pulses Kettenimpulse * *z.B. zur Thyristor-Ansteuerung*
bursting protection Berstschutz
bursting strength Berstfestigkeit
bus Bus || Kommunikationsbus || Sammelschiene
bus access Buszugriff
bus access inhibit Buszugriffsperre
bus activity Busaktivität
bus arbiter Busverwalter || Buszuteiler
bus arbitration Busarbitrierung * *Verwaltung und Zuteilung der Zugriffsrechte auf einen (Multiprozessor-) Bus* || Busverwaltung * *kollisionsfreie*

bus bar

Buszuteilung z.B. bei einem Multiprozessorbus || Buszuteilungslogik
bus bar   Sammelschiene
bus cable   Buskabel || Busleitung
bus circulating list   Busumlaufliste
bus circulating time   Busumlaufzeit * *wird insgesamt benötigt, alle Busteilnehmer einmal anzusprechen*
bus connection   Busanschluss
bus connector   Busanschluss * *Busanschlussstecker* || Busanschlussstecker
bus converter   Busumsetzer
bus coupler   Buskoppler
bus coupling module   Buskoppelbaugruppe
bus cycle time   Busumlaufzeit || Umlaufzeit * *eines Bussystems (wird benötigt, um jeden Teilnehmer einmal anzusprechen)*
bus enable   Busfreigabe
bus error   Busfehler
bus fault   Busfehler
bus interconnection   Busverbindung
bus interface   Busanschaltung || busfähige Schnittstelle || Buskopplung
bus interface board   Busanschaltung * *Leiterkarte*
bus interface module   Busanschaltung * *Baugruppe*
bus jumper   Busbrücke
bus line   Busleitung * *logisch gesehen*
bus linkage   Busverbindung
bus management   Busverwaltung
bus node   Busteilnehmer * *Knoten*
bus operation   Busbetrieb
bus PCB   Busplatine * *Bus-Leiterplatte*
bus polling list   Busumlaufliste
bus protocol   Busprotokoll
bus station   Busteilnehmer
bus system   Bussystem
bus terminating connector   Busabschlussstecker
bus terminating module   Busabschlußmodul
bus terminating possibility   Busabschlußmöglichkeit
bus terminating resistor   Busabschlusswiderstand * *zum Abschluss der Busleitung mit dem Wellenwiderstand (z.b. 120 Ohm)*
bus termination   Busabschluss * *z.b. mit Wellenwiderstand der Busleitung (z.B. 120 Ohm)* || Busabschlussstecker * *Busabschluss*
bus termination resistor   Busabschlusswiderstand
bus terminator   Busabschlussstecker * *Busabschluss-Element*
bus transceiver   Buskoppler * *Sende-/Empfangsschaltung*
bus transmission   Busübertragung
bus-bar   Sammelschiene || Stromschiene
bus-capable   busfähig * *Gerät*
bus-capable interface   busfähige Schnittstelle
bus-coupled   busgekoppelt
bus-coupling module   Busanschaltungsmodul
bus-interfacing module   Busanschaltungsmodul || Buskoppelbaugruppe
bus-type   busförmig
bus-type connection   Buskopplung * *busförmige Verbindung*
busable   busfähig
busable serial link   busfähige Schnittstelle * *serielle*
busbar   Stromschiene
busbar system   Stromschienensystem
busbar trunking system   Stromschienensystem
busbars   Verschienung
buses   Busse * *auch Kommunikationsbusse*
bush   Buchse * *Maschinenteil: Büchse, Buchse* || Hülse * *Buchse, Büchse, Lagerfutter* || Lagerbüchse || Lagerfutter || Muffe
bushing   Buchse * *Muffe, Spannhülse, Durchführungshülse* || Durchführung * *Leitungs-Durchführungshülse, Muffe, Spannhülse* || Durchführungshülse * *z.B. für Kabel* || Hülse * *Muffe, Spann~, Durchführungs~ (für Kabel)* || Kabeldurchführung * *~shülse* || Lagerbüchse || Lagerfutter || Muffe * *Spannhülse, Muffe, Durchführungshülse* || Spannhülse * *Futter(stück)* Buchse, Büchse
bushing-type capacitor   Durchführungskondensator
business   Angelegenheit || Betrieb * *Unternehmen, Firma* || Fach * *Geschäft(stätigkeit), Aufgabe, Angelegenheit* || Geschäft * *(allg.)* || Geschäftsbeziehung * *do business with: in ~ stehen mit* || Handel * *Geschäft* || wirtschaftlich
business acumen   Geschäftssinn
business administration   Betriebswirtschaft || kaufmännische Leitung * *auch als Abteilungsbezeichnung*
business area   Geschäftsbereich
business card   Visitenkarte
business connection   Geschäftsverbindung
business connections   Geschäftsbeziehungen
business contact   Geschäftsverbindung
business contract   Geschäftsabschluss
business day   Werktag
business economist   Betriebswirt
business forms press   Formulardruckmaschine
business group   Geschäftsbereich * *Abteilung*
business interest   Geschäftsinteresse
business life   Geschäftsleben
business offer   Angebot
business outlook   Geschäftslage * *Geschäftsaussichten*
business partner   Geschäftspartner
business premises   Geschäftsräume
business relation   Geschäftsverbindung
business relations   Geschäftsbeziehung || Geschäftsbeziehungen
business report   Geschäftsbericht
business risk   Geschäftsrisiko
business secret   Geschäftsgeheimnis
business sense   Geschäftssinn
business situation   Geschäftslage
business space   Geschäftsräume
business transaction   Geschäftsabschluss
business trip   Dienstreise * *Geschäftsreise* || Geschäftsreise
business unit   Geschäftsbereich || Geschäftseinheit
business vacation   Betriebsferien
business with in-house partners   Verbundvertrieb
businessman   Geschäftsmann
busy   besetzt * *(amerikan.) Telefon*
busy oneself with   vornehmen * *sich beschäftigen mit*
but   allein * *aber* || außer * *außer, als, ohne dass* || bloß * *aber* || erst * *bloß* || jedoch || nur * *ausgenommen*
but then   dagegen * *indessen*
butt   aneinander fügen || aneinander reihen * *(stumpf) aneinander stoßen (lassen)* || anreihen * *auf Stoß anordnen, aneinander stoßen/fügen* || anstoßen * *aneinander stoßen, (stumpf) aneinander fügen, (an)grenzen, auf Stoß aneinandergefügt sein* || Bütte || stoßen * *zusammen-/aneinander~/fügen (against: an-/gegen(einander)), aneinandergefügt sein* || Stoßstelle * *Berührungsstelle von Bauteilen, Stoß*
butt joint   Stoß * *(Stoß)Fuge zw. aneinandergereihten Teilen* || Stoßstelle * *Stoßverbindung, Stoßnaht*
butt mounting   Anreihmontage
butt strap   Lasche * *Verbindungs~ an Stahlkonstruktion usw.*
butt together   aneinander reihen
butt welding machine   Stumpfschweißmaschine
butt-join   aneinander fügen * *auf Stoß*

butt-jointed aneinander stoßend
butt-mount aneinander reihen * *aneinander stoßend befestigen/montieren* || anreihen * *angereiht/ auf Stoß befestigen/montieren*
butt-mountable anreihbar
butt-welding machine Punktschweißmaschine
buttable aneinanderreihbar
butted aufeinander folgend * *aneinandergereiht* || benachbart * *(stumpf) aneinander gefügt*
button Druckknopf || Klickfeld * *auf dem Bildschirm* || Knopf || Schaltfläche * *in einem Dialogfeld auf Computermonitor (zum Anklicken mit der Maus)* || Taster
button cell Knopfzelle * Batterie
buy anschaffen || beziehen * *Ware* ~ *(from:von)* || erkaufen * *(auch figürl.)* || Kauf * *good buy: guter* ~ || kaufen
buy out aufkaufen * *Firma usw.* || übernehmen * *aufkaufen (Firma usw.)*
buy up aufkaufen * *Firma usw.*
buyer Abnehmer * Käufer || Besteller || Käufer
buyers guide Produktauswahlliste * *Einkaufsleitfaden für den Kunden* || Produktübersicht * *Einkaufsleitfaden für den Kunden*
buying Erwerb || Kauf * *das* ~*en*
buying decision Kaufentscheidung
buzz summen * hochfrequent
buzzer Summer
buzzword Schlagwort * *das in aller Munde ist*
by -weise * *z.B. bit by bit: bit~, step by step: stufen~* || an || anhand * *durch* || aus || dadurch * *indem* || durch * *z.B. by pressing the key: durch Drücken der Taste* || für * *step by step: Schritt für Schritt, piece by piece: Stück für Stück, day by day:Tag für Tag* || gegen * *(amerikan.) zeitlich* || indem || je * *three by three: je drei* || mal * *bei Multiplikation* || mit * *mit der Bahn, Post, Hauspost; mit Gewalt usw.* || nach * *one by one: einer ~ dem anderen; little by little: ~ und ~* || nach * *differenzieren (dv/dt); i. Zusammenhang mit Maßeinheit: sell by the weight: ~ Gewicht verkaufen* || neben || um * *Maß, Wert; increase by a half:um die Hälfte erhöhen; by a great deal:um ein bedeutend entlang/~*
by a mistake irrtümlich * *(Adv.)* ~*erweise, versehentlich* || unabsichtlich * *(Adv.) versehentlich, irrtümlicherweise, fehlerhafterweise* || unbeabsichtigt * *versehentlich, irrtümlicherweise, fehlerhafterweise* || versehentlich * *irrtümlicherweise*
by accident zufällig
by all means auf alle Fälle || unbedingt * *(Adv.) unter allen Umständen, mit allen Mitteln* || unter allen Umständen * *auf alle Fälle, mit allen Mitteln*
by analogy analog * *(Adv.)*
by approximation näherungsweise
by birth von Haus aus
by degrees allmählich * *(Adv.)*
by experiments empirisch * *durch Versuche*
by fits and starts stoßweise * *stoß/ruckweise*
by force gewaltsam || mit Gewalt
by habit in gewohnter Weise * *gewohnheitsmäßig*
by hand von Hand * *handwerklich, per Handarbeit*
by hardware hardwaremäßig * *(Adv.) durch/mittels Hardware*
by heart auswendig * *know by heart: ~ wissen/können; learn by heart: ~ lernen*
by hook or crook Gewalt * *mit aller Gewalt*
by it dadurch * *auf solche Weise*
by jerks ruckartig || stoßweise * *ruck/stoß/sprungweise*
by leaps and bounds sprungartig * *sprungweise, sprunghaft* || sprunghaft * *(Adv.) rise by leaps and bounds:* ~ *steigen*
by means of anhand * *mittels* || durch * *mittels* || mit Hilfe von * *mittels* || mittels || über * *mittels* || vermittels

by memory auswendig * *aus dem Gedächtnis*
by mistake irrtümlich || ungewollt * →*versehentlich* || versehentlich
by nature von Haus aus
by no means auf keinen Fall || ganz * *(Adv.)* ~ *und gar nicht, auf keinen Fall* || keinesfalls * *als Antwort* || nie * *keineswegs*
by oneself allein * *ohne Hilfe, automatisch* || selbst * *ohne fremde Hilfe*
by order of im Auftrag von
by phone telefonisch
by preference bevorzugt * *(Adv.) mit Vorliebe, begünstigt* || vorzugsweise
by reason of wegen * *auf Grund; for this reason: aus diesem Grund, deshalb*
by rights eigentlich * *von Rechts wegen*
by sectors abschnittweise
by software softwaremäßig * *über/mit Hilfe der Software*
by steps abschnittweise * →*schrittweise*
by substitution ersatzweise
by tests empirisch * *durch Versuche*
by that means dadurch * *auf diesem Weg, hierdurch*
by the end of Ende * *z.B. by the end of September: Ende September*
by the example of anhand * *am Beispiel von*
by the same token umgekehrt * *(Adv.) genauso, mit gleichem Recht*
by the side of neben
by the square quadratisch * *(Adv.) z.B. mathem. Gesetzmäßigkeit*
by the way nebenbei gesagt || übrig * *beiläufig,* ~*ens* || übrigens
by the way of durch * *mittels*
by the way of calculation rechnerisch * *(Adv.)*
by the way of substitution ersatzweise
by this means so * *(Adv.) auf diese Weise*
by turns abwechselnd * *(Adv.)* || nacheinander * *bei jedem Umlauf einmal*
by virtue of auf Grund || kraft *(Gesetz, Vollmacht usw.), auf Grund von, vermöge* || laut * *kraft*
by way of mittels || über * *auf die Art und Weise*
by way of an experiment experimentell || versuchsweise
by way of experiments empirisch * *durch Versuche*
by way of trial versuchsweise
by- Neben- * *by-product: ~produkt*
by-pass umgehen * *Signal/Energie/Verkehrsfluss/Hochlaufgeber ~, vorbeistoßen an* || Umgehung
by-passing Umgehung
by-product Nebenprodukt
bye Auf Wiedersehen
bye bye Auf Wiedersehen
bypass abkürzen * *einen Umweg* ~ || überbrücken * *ohne zusammen, umleiten, vermeiden* || Überbrückung || umgehen || Umgehung || verkürzen * *umgehen, übergehen*
bypass arm Freilaufzweig
bypass capacitor Ableitkondensator * *Querkondensator* || Überbrückungskondensator * *Ableit(-quer)kondensator*
bypass contactor Überbrückungsschütz * *bei Umrichter-Zwischenkreis-Vorladeschaltung, Sanftanlaufgerät, USV usw.*
bypass diode Freilaufdiode
bypass relay Überbrückungsrelais * *z.B. bei USV, Motor-Sanftanlaufgerät (überbrückt dieses nach erfolgtem Anlauf)*
bypass- Ableit-
bypassing Ausschleusung * *am Hauptstrom vorbei* || Umgehung
byte Byte || Zeichen * *in 8 Bit verschlüsseltes* ~ *(z.B. im ASCII-code)*

byte processor Wortprozessor * *Prozessor für Byteverarbeitung bei SPS*
bytes Byte * *(Plural)*

C

C contact Umschaltkontakt || Wechselkontakt || Wechsler || Wechslerkontakt
C-axis C-Achse * *{Werkz.masch.}* Hauptspindel bei Werkzeugmaschine
C-core Schnittbandkern * *z.B. bei Trafo*
C-type mounting rail C-Schiene * *Aufschnappmontageschiene nach DIN EN 500222 (35 mm breit)*
C-type mounting rail according to DIN EN 50022 Hutprofilschiene * *C-Schiene zur Aufschnappmontage (35 mm breit)* || Hutschiene * *C-Schiene zur Aufschnappmontage, 35 mm breit* || Installationsschiene * *zur Aufschnappmontage, 35 mm breit* || Montageschiene * *C-Aufschnappinstallationsschiene 35 mm breit*
c.c. cc * *Kopie an* || Durchschlag an * *Abk. f. 'Carbon 'Copy': Schreibmaschinendurchschlag/Fotokopie an* || Kopie an
c.p.s. Hertz * *Cycles Per Second* || Hz * *Cycles Per Second*
cab Taxi
cabin Kabine
cabinet Schaltschrank || Schrank * *Schaltschrank*
cabinet maker Tischler * *Möbel~, Kunst~*
cabinet shop Tischlerei * *Möbel~, Kunst~*
cabinet unit Schrankgerät
cabinet wall Schrankwand * *eines Schaltschranks*
cabinet wiring block Rangierverteiler
cable Kabel || Leitung || Seil * *Tau, Draht~; auch bei ~bahn* || Tau
cable and hose drag chain Energieführungskette * *(für Kabel und Schläuche)* || Energiekette * *für Kabel und Schläuche*
cable bearer Kabelpritsche
cable box Anschlusskasten || Klemmenkasten
cable break Leitungsbruch
cable breakage Drahtbruch * *Kabel/Leitungsbruch* || Kabelbruch || Leitungsbruch * *auch 'Bruch(stelle), Bruchschaden'* || Leitungsunterbrechung
cable bushing plate Kabeldurchführungsplatte
cable capacitance Kabelkapazität * *zwischen den Adern oder zwischen Ader und Schirm* || Leitungskapazität * *zwischen den Adern oder zwischen Ader und Schirm*
cable car Seilbahn
cable channel Kabelkanal
cable clamp Kabelklemme * *siehe auch →Kabelschelle* || Kabelschelle || Klemmschelle * *zur Kabelbefestigung*
cable clamping Kabelbefestigung * *durch Schellen*
cable clamping device Kabelabfang-Einrichtung * *zur Zugentlastung und Schirmauflegung*
cable cleat Kabelklemme * *Kabel-Klemmschelle aus Isoliermaterial* || Kabelschelle * *Kabel-Klemmschelle aus Isoliermaterial*
cable clip Kabelschelle * *meist aus Kunststoff*
cable conduit Kabelkanal || Leitungskanal
cable conduit pipe Leitungsrohr * *für Kabel/Leitungen*
cable connection Kabelanschluss || Kabelverbindung || Leitungsanschluss
cable connector Anschlussstecker || Leitungsstecker
cable core Kabelader

cable cross-section Kabelquerschnitt || Leiterquerschnitt || Leitungsquerschnitt
cable damping Leitungsbedämpfung * *z.B. für Leitung Umrichter-Motor zur Vermeidung von Resonanzen*
cable diameter Leitungsdurchmesser
cable drag chain Energieführungskette * *(für Kabel)* || Energiekette * *für (Schlepp-)Kabel bei Werkzeugmaschine usw.*
cable drive Seilantrieb
cable drum Kabeltrommel || Seiltrommel * *für Drahtseil*
cable duct Kabelführung * *z.B. Kabelkanal* || Kabelkanal || Leitungskanal
cable entry Kabeldurchführung * *Kabeleinführung* || Kabeleinführung * *z.B. in Klemmenkasten (für el. Masch.: siehe z.B. DIN 42925)* || Leitungseinführung * *z.B. in Klemmenkasten*
cable entry gland PG-Verschraubung
cable entry plate Einführungsplatte * *zur Kabeleinführung in Klemmenkasten*
cable entry screw gland Leitungsverschraubung * *bei einer Kabeleinführung*
cable entry-hole Kabeleinführungsöffnung * *(z.B. in Klemmenkastenwand)*
cable exit Kabelausführung
cable feeder Kabelzuführung
cable form Kabelbaum
cable gauge Kabelquerschnitt
cable gland Antichronstutzen * *für PG-Verschraubung* || Kabelstutzen || Leitungsverschraubung * *z.B. PG-Verschraubung bei Leitungs/Kabeleinführung* || PG-Verschraubung * *Leitungs/Kabelverschraubung* || Stopfbuchsenverschraubung
cable harness Kabelbaum
cable housing Kabelgehäuse
cable installation Kabelverlegung || Leitungsverlegung || Verkabelung
cable laying Kabellegung || Kabelverlegung || Leitungsverlegung
cable length Kabellänge || Leitungslänge
cable lug Kabelschuh
cable outlet direction Kabelabgangsrichtung * *z.B. bei Motorklemmenkasten*
cable plug connector Kabelstecker
cable protection Leitungsschutz
cable pulley Seilscheibe
cable rack Kabelpritsche
cable railway Seilbahn
cable reel Kabeltrommel
cable remains Kabelreste * *siehe auch →Leitungsreste* || Leitungsreste
cable resistance Leitungswiderstand
cable route Kabeltrasse
cable routing Kabelführung || Kabelverlegung || Leitungsführung || Leitungsverlegung
cable run Kabeltrasse
cable running Kabelzuführung
cable screen Kabelabschirmung || Kabelschirm || Leitungsabschirmung || Leitungsschirm
cable set Anschlusskabel * *konfektionierter Kabelsatz* || konfektioniertes Kabel || vorkonfektioniertes Kabel
cable sheathing Kabelumhüllung || Kabelumwicklung * *Armierung*
cable sheathing line Kabelummantelungsanlage
cable shelf Kabelpritsche
cable shield Kabelabschirmung || Kabelschirm || Leitungsabschirmung || Leitungsschirm || Leitungsschirmung
cable shoe Kabelschuh
cable size Kabelquerschnitt
cable sleeve Kabeltülle * *→Knickschutz bei Kabelausführung*
cable strand Kabelader

cable stranding device Kabelverseilvorrichtung
cable stranding line Kabelverseilanlage
cable stranding machine Kabelverseilmaschine
cable stranding plant Kabel-Verseilanlage
cable stretching Seildehnung * *{Aufzüge | Hebezeuge}*
cable support Kabelpritsche
cable termination Kabelanschluss || Leitungsabschluss * *z.B. durch Abschlusswiderstand (z.B. mit Wellenwiderst. 120 Ohm)*
cable tie Kabelbinder
cable ties Kabelbinder * *(Plural)*
cable twister Kabelverseilvorrichtung
cable twisting machine Kabelverseilmaschine
cable twisting plant Kabel-Verseilanlage || Kabelverseilanlage
cable winch Seilwinde
cable-based leitungsgebunden * *z.B. Störungen*
cable-born leitungsgebunden * *z.B. Störabstrahlung von d.* Leitung nach außen
cable-born electromagnetic radiation leitungsgebundene Störabstrahlung
cable-born interference leitungsgebundene Störungen * *EMV-Test: siehe IEC 801-4*
cable-clamping bar Kabelabfangschiene
cable-fed leitungsgebunden * *leitungsgeführt, z.B. eingekoppelte Störungen* || leitungsgeführt * *z.B. Störungen*
cable-fed disturbances leitungsgebundene Störungen || leitungsgeführte Störungen
cable-fed noise leitungsgebundene Störungen
cable-guided leitungsgebunden * *z.B. Störungen* || leitungsgeführt * *leitungsgebunden, z.B. Störungen*
cableway Seilbahn
cabling Verdrahtung || Verkabelung
cabling changing Verdrahtungsänderung * *in der Leitungsverlegung*
cabling costs Verdrahtungsaufwand || Verkabelungsaufwand
cache memory Cachespeicher * *schnell. Pufferspeicher z. Zwischenspeichern v.Befehlen u.Daten f. CPU-Schnellzugriff* || Zwischenspeicher * *~ zur Entkopplung zweier verschieden schneller Partner, z.B. CPU und Hauptspeicher*
CAD CAD * *Abkürzg.für 'Computer Aided Design': Rechnergestützte Konstruktion/Entwicklung*
CAD workstation CAD-Arbeitsplatz
cadmium plated cadmiert
CAE CAE * *Abkürzg. für 'Computer-Aided Engineering': rechnergetütztes Konstruieren*
cage Abstandshalter * *in einem Wälzlager (Kugelkäfig)* || Baugruppenträger || Kabine * *{Aufzüge}* || Käfig * *z.B. für Käfigläufermotor oder Kugelkäfig Wälzlager* || Laterne * *Käfig; z.B. zur Montage eines Impulsgebers am Motor*
cage bar Käfigstab * *in einem Motorläufer*
cage clamp terminal Federzugklemme * *Käfigzugfederklemme (mit Befestigung ohne Schraube)* || Käfigzugfederklemme * *schraubenlose Kabelanschlussklemme*
cage motor Käfigläufermotor
cage resistance Käfigwiderstand
cage rotor Käfigläufer
cage-bar Läuferstab * *bei Käfigläufer*
cage-clamp terminal Zugfederklemme * *schraubenlose Leitungsanschlussklemme*
cake Scheibe * *Wachs~, Teig~, (Spanplatten-) Kuchen usw.*
caked together verbacken * *miteinander ~*
calculable berechenbar
calculate ausrechnen || bemessen * *berechen, auslegen (for:nach)* || berechnen || errechnen || rechnen * *(aus-) rechnen; miscalculate: falsch ~*
calculated berechnet || rechnerisch * *berechnet*

calculated value Rechenwert * *berechneter Wert*
calculating rule Rechenschieber
calculating time Rechenzeit
calculation Berechnung || Kalkulation || Rechnung * *Be~*
calculation documents Berechnungsunterlagen
calculation example Berechnungsbeispiel
calculation factor Berechnungsfaktor
calculation instruction Rechenvorschrift
calculation method Rechenverfahren
calculation operation Rechenoperation
calculation procedure Rechenvorschrift
calculation result Verknüpfungsergebnis
calculation technique Rechenverfahren
calculus Kalkül
calendar Kalender
calendar week Kalenderwoche
calender Glättwerk * *Kalander, glättet Papier, Textil usw.* || Glättzylinder/Dampfzylinder || Glättzylinder * *Kalander zur Glättung/Satinierung von Papier und Textilstoffen* || Kalander * *beheizter Glättzylinder zur Glättung/Satinierung v.Papier oder Textilstoff durch Pressen* || kalandrieren * *(z.B. Papier/Tuch/Stoff) pressen/glätten/satinieren mit Hilfe von Kalanderwalzen* || Rolle * *~ z. Pressen von Papier, Stoff usw.* || rollen * *Tuch/Papier mit Pressrolle*
calender roll Kalanderwalze
calender stack Kalander * *Gruppe senkrecht übereinander angeordneter Kalanderwalzen* || Maschinenglättwerk * *mehrstufiges ~*
calenders Kalander * *(Plural) siehe auch* →Maschinenkalander, →Superkalander
calendize kalandrieren * *mit Hilfe eines* →Kalanders glätten (z.B. Papier, Textil)
calibrate abgleichen * *eichen, z.B. Messinstrument* || eichen || einmessen || kalibrieren
calibrated geeicht * *Messgerät*
calibration Abgleich * *Eichung, Kalibrierung, Abgleich einer Messeinrichtung* || Eichung
calibration device Eichgerät || Kalibriereinrichtung
calibration factor Eichfaktor
calibrator Eichgerät || Justiereinrichtung * *zum Eichen von Messgeräten* || Kalibrator
caliometric kaliometrisch
caliper Dicke * *von Papier* || Endmaß * *Tastvorrichtung z.B. zur Messung von Innendurchmessern* || messen * *mit einer Lehre ~* || Schieblehre || Tastlehre || Tastzirkel * *Tastlehre zur Ermittlung von Innenmaßen*
caliper gauge Endmaß * *Tast-/Rachenlehre*
caliper measurement Dickenmessung * *z.B. für Papier*
caliper-type brush Doppelschenkelbürste * *zur Stromübertragung auf bewegte Maschinenteile*
call Abruf * *on call: auf ~~* || abrufen || Anruf || anrufen * *call about someth.: wegen etwas anrufen* || Aufforderung || Aufruf * *z.B. eines Unterprogramms* || aufrufen * *z.B. Unterprogramm* || benennen * *nennen, bezeichnen als, heißen (after: nach), halten für* || einberufen * *eine Sitzung, Versammlung, Besprechung usw. ~; for eight o'clock: für 8 Uhr* || Forderung * *for:nach* || nennen * *bezeichnen*
call attention to hinweisen auf * *call a person's attention to: jdn. ~ auf*
call back rückrufen * *am Telefon*
call for anfordern || Anforderung * *Anforderung von, z.B. von Tagungsbeiträgen (call for papers)* || anwählen || beanspruchen * *verlangen, abrufen, (er)fordern* || bedingen * *erforden, verlangen* || (er)fordern || fordern || holen * *ab~* || notwendig machen || verlangen * *er/einfordern*
call for host control Führung gefordert * *(→PROFIBUS-Profil)*

call for tenders Ausschreibung * *für Lieferangebote*
call off abbrechen
call on customers Dienstreise * *Kundenbesuch (at: an einem Ort)*
call prefix Vorwahlnummer
call together einberufen * *Teilnehmer (zu einer Versammlung) zusammenrufen*
call up abrufen * *z.B. Diagnosedaten* || anrufen
call-up abrufen * *z.B. einzelne Lieferungen bei Rahmenvertrag* || aufrufen * *auch 'abrufen'; z.B. Hilfetext, Diagnosedaten*
called genannt * *mit e.Ausdruck bezeichnet*
caller Anrufer * *am Telefon*
calling card Visitenkarte * *(amerikan.)*
calls on customers Kundenbesuche
caloric value Heizwert
calorimetric kalorimetrisch
calorimetric measuring principle kalorimetrisches Messprinzip
calotte Kalotte
CAM assoziativer Speicher * *Abk.f. 'Content-Addressable Memory':Speicher, dess.Inhalt als Adresse dient* || CAM * *Abkürzg.für 'Computer Aided Manufacturing': rechnergestützte Fertigung* || CAM * *Abkürzg.für 'Content-Addressable Memory': assoziativer Speicher (siehe dort)* || Kurve * *(mechan.) Steuerkurve (eines Nockens, einer Kurvenscheibe usw.)* || Kurvenscheibe || Mitnehmer || Nocke || Nocken
cam clutch Nockenkupplung
cam controller Nockensteuerwerk
cam disc Kurvenscheibe
cam group Nockenschaltwerk
cam plate Kurvenscheibe
cam roller Laufrolle
cam switch Nockenschalter
cam-contactor group Nockenschaltwerk || Schaltwerk * *Nockenschaltwerk*
cam-operated switchgroup Nockenschaltwerk
camber Sturz * *Rad*~ || Wölbung * *leichte ~, Krümmung*
cambered ballig || gewölbt
campaign Aktion * *z.B. Werbeaktion, Kampagne, Feldzug* || Kampagne * *(Werbe-) Feldzug*
camshaft Nockenwelle
CAN CAN * *Abk.f. 'Controller Area Network': seriell. Bussystem n. ISO DIS 11898 n.11519-1 (urspr. f. Auto)* || Dose || Können * *(Verb)* || konservieren * *in Büchsen* || verkapseln * *"eindosen"*
can be -bar
can be activated aktivierbar * *kann aktiviert werden*
can be adjusted einstellbar * *einstellbar sein*
can be applied anwendbar sein
can be calculated ergeben * *can be calculated from: ergibt sich rechnerisch/physikal.aus*
can be combined kombinierbar * *freely: frei*
can be continuously loaded dauernd belastbar * *with: mit*
can be decontaminated dekontaminierbar
can be dispensed with entfallen * *~ können, weggelassen werden können, verzichtbar/entbehrlich sein*
can be freely parameterized frei parametrierbar sein
can be loaded belastbar
can be obtained ergeben * *kann man erhalten/erlangen/bekommen*
can be parameterized parametrierbar * *kann parametriert werden*
can be plugged in steckbar sein
can be retrofitted nachrüstbar
can be selected wählbar * *kann gewählt werden*

can be separately adjusted getrennt einstellbar * *kann getrennt eingestellt werden*
can be set einstellbar * *einstellbar sein*
can be turned drehbar * *drehbar sein*
can coiler Kannenablage * *{Textil} zur Ablage von Fäden*
can filling machine Dosenfüllmaschine
can take a knock or two einstecken * *etwas ~ können*
can-making machine Blechdosen-Herstellungsmaschine
cancel abbestellen || annullieren * *Auftrag* || aufheben * *each other: einander; auch Erlass/Verbot ~* || auflösen * *Vertrag usw.* ~ || löschen || rückgängig machen * *z.B. Auftrag* || rücksetzen * *Löschen, Wegnehmen eines Kommandos usw.* || stornieren * *z.B. Auftrag* || streichen * *löschen, anullieren* || ungültig machen || widerrufen * *widerrufen, aufheben, annullieren (auch Auftrag), rückgängig machen, absagen, streichen*
cancel and not replace ersatzlos streichen
canceled abgebrochen * *(kann mit 'l' oder mit 'll' geschrieben werden)*
cancellation Absage || Annullierung || Aufhebung || Kündigung * *~ eines Vertrages* || Löschung || Widerruf * *Löschung, Stornierung*
cancellation of order Rücktritt vom Vertrag * *Zurücknahme einer Bestellung*
cancelled abgebrochen * *gestrichen, annulliert, widerrufen, ungültig gemacht, storniert* || gelöscht * *ungültig/rückgängig gemacht*
candidate Bewerber
candidature Bewerbung * *for: um*
candy machine Süßwarenmaschine
canister Kanister
canned gekapselt * *'eingedost'* || vorfabriziert * *'eingedost' ('Schwarzer Kasten')*
canned motor Spaltrohrmotor * *el.Pumpenmotor, Ständer mit Blechpaket u.Wicklung durch Rohr v.Fördermedium getrennt* || Topfmotor * *B-seitiges Lagerschild bildet mit d.Gehäuse eine Einheit –> hohe Schutzart (z.B.IP67)*
cannibalize ausschlachten * *z.B. Fahrzeug* || kannibalisieren * *aus-/abschlachten (auch Auto usw.)*
cannot be rotated verdrehsicher
cannvasser Vertriebsmann * *im Außendienst, (Handels-) Vertreter*
canonical form Normalform
CANOpen CANOpen * *[Network] erweitertes →CAN Busprotokoll mit Anwenderprofilen (z.B. DSP-402 für el.Antriebe)*
canopy Schutzdach * *z.B. für senkrecht eingebauten Motor, damit keine Fremdkörper i.d. Lüfter fallen*
cant schwenken || Überhöhung * *einer Kurve usw*
canteen Kantine
cantilever force Querkraft * *Radialkraft, greift senkrecht zur Drehachse (z.B. Motorwelle) an* || Radialkraft * *→Querkraft*
cantilevered construction freitragende Konstruktion
canting neigbar
canvass akquirieren * *Kunden bereisen, Aufträge hereinholen*
canvasser Akquisiteur * *Vertriebsmann im Außendienst, (Handels-) Vertreter*
canvassing Akquisition * *Kundenwerbung*
caoutchouc Kautschuk
cap Deckel * *Kappe* || Hülse * *Kappe; slip-on cap: (Kugelschreiber-/Füllhalter-) Aufsteck-* || Kalotte || Kappe || Kapsel * *(Verschluss-) Kappe, Deckel* || kapseln * *mit Deckel versehen*
cap closure Verschlusskappe * *in der Verpakkungstechnik*
cap nut Hutmutter || Überwurfmutter

capability -Fähigkeit || Fähigkeit || Kapazität * *Fähigkeit (of something:zu etwas), Vermögen, Tauglichkeit* || Möglichkeit * *Fähigkeit* || Verhalten * *Vermögen, Fähigkeit*
capability of connecting Anschlussmöglichkeit für
capability of development Ausbaubarkeit * *Entwicklungsfähigkeit, z.B. eines Marktes*
capability to be networked Vernetzbarkeit * *z.B. über Kommunikations-Netzwerk/Bus*
capable fähig * *to/of: zu etwas; auch 'in der Lage'* || gut * *fähig, tüchtig* || tüchtig * *tüchtig*
capable in business geschäftstüchtig
capable of being -bar * *z.b. capable of being switched off: abschaltbar*
capable of being disassembled zerlegbar
capable of being switched off abschaltbar
capacitance Kapazität * *elektr. Kapazität*
capacitance per unit length Kabelkapazität * →*Kapazitätsbelag* || Kapazitätsbelag * *Kabel, Leitung* || Kapazitätsbelag * *Kapazität pro Längeneinheit einer Leitung/eines Kabels (z.b.100 pF/m)* || Leitungskapazität * →*Kapazitätsbelag*
capacitive kapazitiv
capacitive load kapazitive Last
capacitive reactance kapazitiver Blindwiderstand
capacitor Kapazität * *Kondensator* || Kondensator
capacitor back-up Kondensatorpufferung
capacitor bank Kondensatorbatterie
capacitor braking Kondensatorbremsung * *Gleichstrombremsung mit Kondensator als Stromlieferant (selten verwendet)*
capacitor cascade Kondensatorkaskade
capacitor excitation Kondensatorerregung
capacitor for power correction Kompensationskondensator
capacitor motor Kondensatormotor
capacitor winder Kondensatorwickelmaschine
capacity -Fähigkeit * *wert/betragsmäßig, z.B. Überlast* || Belastbarkeit * *z.b. thermal capacity: Wärme~* || Kapazität * *(Fassungs)Vermögen, (Leistungs)Fähigkeit, (Nutz)Leistung, Höchstmaß* || Leistung * *Kapazität/Leistungsvermögen bezgl. Volumen, Durchsatz, Überlastbarkeit;* →*Kälte~ (DIN8928)* || Raum * *Fassungsvermögen* || Volumen * *Inhalt, Fassungsvermögen*
capacity modulation Leistungsregelung * *für Verdichter und* →*Verdichtersätze (z.B. in Kälteanlage)*
caper Bocksprünge machen * *Kapriolen/Luftspünge machen*
capillary effect Kapillarwirkung
capillary fissure Haarriss
capillary rise Saughöhe * *durch Kapillarwirkung*
capital groß * *capital letter: ~buchstabe* || Kapital
capital cost Investitionskosten
capital equipment Produktionsmittel
capital from outside sources Fremdkapital
capital goods Investitionen * *Investitionsgüter* || Investitionsgüter
capital investments Investitionen
capital letter Großbuchstabe
capital resources Eigenkapital
capital returns Kapitalrückfluss
capital stock Eigenkapital
capital stock and reserve Eigenkapital
capital tied up Kapitalbindung
capital-intensive kapitalintensiv
capping machine Verschließmaschine * *{Verpack.techn.} zum Verschließen von Flaschen, Tuben usw. mit Verschlussdeckeln*
capstan Ankerwinde || Scheibenabzug * *große konstant angetriebene Scheibe in Kabelmaschine, an die d.Kabel angepreßt wird* || Winde * *auf Schiff* || Ziehscheibe * *b.Drahtziehmaschine: Rolle m. Mehrfach-Drahtumschlingung, zieht Draht durch d. Ziehstein* || Ziehtrommel * *{Draht \ Kabel}*

capstan drive Bandantrieb * *für Magnetbandgerät*
capstan lathe Revolverdrehmaschine
capsule Hülse * *Kapsel* || Kapsel || verkapseln
capsuling machine Verkapselmaschine * *Verpakkungsmaschine*
caption Bildunterschrift || Überschrift * *Überschrift, Titel, (Bild-) Unterschrift; Rubrik*
captive unverlierbar * *z.b. Schraube*
captive nut unverlierbare Mutter
captive screw unverlierbare Schraube
capture auffangen * *einen Messwert, Zählerstand usw. in einen Zwischenspeicher ~* || erfassen * *auffangen, z.B. in Tracespeicher* || Erfassung * *~ eines Messwertes oder Zählerstandes im laufenden Betrieb in einen Zwischenspeicher* || Speicherung * *eines Zählerstandes, Messwertes im laufenden Betrieb* || wegnehmen * *ein/auffangen (auch in Zwischenspeicher), an sich reißen*
capture-register Zwischenspeicher * *Auffangspeicher für Zählerstände usw. (für 'Momentaufnahmen')*
car Kabine * *{Aufzüge} Fahrkorb* || Wagen
car manufacturing Automobilbau
car rental Autoverleih
car spraying line Lackierstraße * *für Autos*
carbide Hartmetall * *cemented-carbide: gesintertes* ~ || Sinter- * *Sintermetall-*
carbide metal Hartmetall
carbide tip Hartmetallplättchen * *{Werkz.masch.} an Werkzeug* || Hartmetallschneidplatte
carbide tips Hartmetallbesatz * *{Werkz.masch.} bei (Schneid-) Werkzeug*
carbide-tipped hartmetallbestückt * *{Werkz.masch.}*
carbide-tipped drill Hartmetallbohrer
carbide-tipped tool Hartmetallwerkzeug * *Werkzeug mit (Schneid-/Bohr-/Fräs-) Einsatz aus Hartmetall*
carbon black Druckerschwärze || Druckfarbe * *{Druckerei} Druckerschwärze* || Ruß
carbon brush Kohlebürste
carbon brush holder Kohlebürstenhalter
carbon content Kohlenstoffgehalt
carbon dioxide Kohlendioxid || Kohlendioxyd * *$CO2$* || Kohlensäure
carbon disulphide Schwefelkohlenstoff
carbon dust Bürstenabrieb * *Abriebpulver, Grafitstaub* || Kohlebürstenabrieb
carbon dust deposits Bürstenabrieb * *abesetzter Kohlebürstenstaub (am Kommutator einer Gleichstrommaschine usw.)*
carbon paper Kohlepapier
carbon steel unlegierter Stahl * *Kohlenstoffstahl*
carbonic acid gas Kohlensäure
carborundum block Carborundstein * *zum Schleifen*
carboy Ballon * *für Chemikalien*
carcase Korpus
card aufrauen * *{Textil} Wolle ~* || Baugruppe * *Leiterkarte* || Flachbaugruppe || kardieren * *{Textil} krempeln/kämmen/kardieren von Wolle, Fasern, Vliesstoffen usw.* || Karte * *auch Leiterkarte* || Krempel * *{Textil}* || krempeln * *{Textil} kämmen/kardieren/Aufrauen von Fasern, Wolle, Vliesstoffen usw.* || Leiterkarte || Leiterplatte * *bestückt* || Platine * *Flachbaugruppe, Karte* || schären * *{Textil}* → *aufrauen (von Wolle usw.)*
card cage Baugruppenträger
card frame Baugruppenträger * *für Leiterkarten*
card index Kartei
card module Flachbaugruppe || Platine * *Flachbaugruppe*
card rack Baugruppenträger * *für Flachbaugruppen/Leiterkarten*
cardan joint Kardangelenk

cardan shaft Gelenkwelle || Kardanwelle
cardboard Karton * ~papier || Pappe * Kartonpappe; siehe auch →Karton
cardboard box Karton * Schachtel || Pappkarton || Schachtel * Papp~
cardboard factory Kartonfabrik
cardboard machine Kartonmaschine
cardbox Kartei * Karteikasten
cardinal point Angelpunkt * (figürlich)
carding machine Kardenmaschine * {Textil} zum Aufrauen von Fasern || Krempelmaschine * Textilmaschine z. Aufrauen/Kämmen v.Fasern/Wolle mit Kratzbändern
carding plant Karde * {Textil}
cardrack interfacing Rahmenkopplung
cardroom Krempelei * {Textil}
care Betreuung * of/for:von/für || Pflege || Schutz * Fürsorge || Sorgfalt || Vorsicht * Behutsamkeit
care about kümmern * sich Gedanken machen um/ über || kümmern um || sich kümmern um
care for betreuen || kümmern * sorgen für, sich ~ um || sorgen für
careful behutsam || gewissenhaft * →sorgfältig || schonend || sorgfältig * achtsam, umsichtig, vorsichtig, bedacht, sorgsam, gründlich, genau, sparsam || vorsichtig * behutsam
careful treatment Schonung * pflegliche Behandlung
carefully sorgfältig * (Adv.)
carefully directed gezielt
carefulness Sorgfalt
careless fahrlässig || unachtsam * nachlässig || unsachgemäß * unvorsichtig || unvorsichtig * sorglos, nicht sorgfältig
carelessly unsachgemäß * (Adv.) unvorsichtig
carelessness Fahrlässigkeit || Unvorsichtigkeit
cargo Fracht || Frachtgüter
carousel distributor Drehverteiler * in Förder-/ Dosiersystem
carousel winder Drehkreuzwickler
carousel-type distribution system Sortierkreisel * in (Hochregal-) Lager
carpenter Tischler * Zimmermann, Tischler
carpenter's level Wasserwaage
carpentry machine Zimmereimaschine * {Holz}
carpet manufacture Teppichherstellung
carpet yarn Teppichgarn
carpet-weaving machine Teppichwebmaschine
carriage Beförderung * von Frachtgut || Transport * von Frachtgut || Wagen * (Transport)Wagen, Karren, Fahrgestell, Schlitten
carriage return Rücklauf * Zeilenrücklauf bei Drucker, Schreibmaschine
carrier Halter * Träger eines mechan. Bauteils || Halterung * Träger || Klaue * Nase, Mitnehmer || Ladungsträger * bei Halbleitern || Träger * z.B. bei Modulation, Ladungsträger
carrier bolt Mitnehmer * ~bolzen || Mitnehmerbolzen
carrier chain Transportkette
carrier frequency Pulsfrequenz * bei Pulsweitenmodulation || Schaltfrequenz * z.B. für Pulsumrichter || Taktfrequenz * z.B. für PWM-Modulation, auch Trägerfrequenz || Trägerfrequenz
carrier handle Tragegriff
carrier tape Trägerband * z.B. zur Gurtung von Bauteilen
carrousel Drehkreuz * Karussell
carrousel table Drehtisch
carry bewegen * befördern,fördern || durchflossen werden von * führen, z.B. Strom || führen * z.B. Signal, (gefährliche) Spannung, Strom ~ || schleppen * (schwer) tragen || tragen * auch 'stützen', Lasten ~ '|| Überlauf * Übertragsbit/-merker bei Rechenoperation, z.B. Addition, Multiplikation || übernehmen * z.b. Strom/Stromfluss ~ || Übertrag * einer Addition, eines Saldos, usw || unterstützen * Antrag || zuführen
carry away mitreißen * (auch figürl.: begeistern)
carry bit Übertragsbit * Überlaufbit z.B. bei Addition, Subtraktion
carry flag Übertragsbit * Übertrags-Merkerbit in Recheneinheit eines Rechners
carry off abführen * eine Sache/Hitze ~ || abtragen
carry on betreiben * (Geschäft usw.) betreiben; fortführen, fortsetzen || treiben * Geschäfte, Handel ~
carry one's point durchsetzen * seine Meinung ~; with:bei
carry out ausführen * z.b. a work:eine Arbeit, a repair:eine Reparatur || durchführen * ausführen, z.b. Reparatur, Anlagenprojektierung || erfüllen * Pflicht ~ || erledigen * aus-/durchführen, erfüllen || herausführen || leisten * ausführen
carry out the maintenance warten * Wartungsarbeiten durchführen
carry out the repair Reparatur ausführen
carry through durchsetzen * durchhelfen/bringen, durchführen || erledigen * durchführen
carry up anfahren * Material antransportieren || zuführen
carrying capacity Strombelastbarkeit || Tragfähigkeit * z.b. bei Förder-/Hebezeug
carrying circulating current kreisstrombehaftet || kreisstromführend
carrying off Abfuhr * einer Sache, von Wärme usw
carrying on Verfolgung * z.B. ein Geschäft, einen Plan
carrying out Abwicklung * ~ eines Auftrags
carrying run Obertrum * {z.B. Bergwerk} eines Bandförderers
carrying wheel Laufrad
carrying wheel diameter Laufraddurchmesser
carrying-out Durchführung * Aus/Durchführung
cart Wagen * Karren
Cartesian kartesisch * z.B. Koordinatensystem
carton Karton * ~pappe, Schachtel || Pappschachtel || Schachtel * Papp~
cartridge Einsatz * Patrone, (Filter-) Einsatz || Kartusche * z.B. für Tinte || Kassette * Patrone, →Einsatz || Mine * Patrone, auch für Kugelschreiber || Patrone
cartridge fuse Sicherungspatrone
cartridge-type brush Köcherbürste * zur Stromübertragung auf bewegte Maschinenteile
carve meißeln * schneiden * meißeln, schnitzen
cascade aneinander hängen * kaskadieren, hintereinanderschalten || hintereinander anordnen || hintereinander schalten * kaskadieren, in Kaskade schalten || Kaskade * connected in cascade: in Kaskade geschaltet, kaskadiert || kaskadieren
cascade arrangement Kaskadenschaltung
cascade circuit Kaskadenschaltung
cascade connection Kaskadenschaltung
cascade control Kaskadenregelung * Regelungseinrichtung mit mehreren einander unterlagerten Regelkreisen || kaskadierte Regelung * Regeleinrichtung mit mehreren einander unterlagerten Regelkreisen
cascade element Kettenelement * in Ablaufkette
cascaded gestaffelt * kaskadiert || kaskadiert || unterlagert * hintereinander geschaltet/angeordnet
cascaded control kaskadierte Regelung
cascading Kaskadierung
cascode Kaskode * cascode amplifier: ~verstärker
case Behälter * Kasten || CASE * {Computer} Abkürzg. für 'Computer-Aided Software Engineering' || Fall * be never the case: nie der ~ sein; in that/ either/no case: in diesem/auf jeden/keinen ~ ||

Gehäuse ‖ Hülse * Einfassung, Mantel, Gehäuse, Behälter, Kapsel, Schutzhülle ‖ Kapsel ‖ Kassette * Kasten, Kästchen, Koffer, Kapsel, Schachtel, Behälter, Gehäuse, Hülle ‖ Kiste * Kasten, Kiste, Behälter, Schachtel ‖ Schachtel ‖ verkapseln * umkleiden, verkleiden, in ein Gehäuse stecken
case hardened   einsatzgehärtet
case interior   Gehäuseinneres
case of emergency   Notfall
case of failure   Fehlerfall * in: im
case of maintainance   Servicefall
case temperature   Gehäusetemperatur * z.B. eines Getriebes, Halbleiterbauelements usw.
caseharden   härten * einsatzhärten
casement   Flügel * eines Fensters
cash   Bahrzahlung
cash cow   Goldgrube * (figürl.) Melkkuh, sehr gute Einnahmequelle (z.b. Produkt oder Unternehmensbereich) ‖ Melkkuh * (auch figürl.) sehr gute Einnahmequelle (z.b. Produkt oder Unternehmensbereich)
cash dispenser   Geldautomat
cash payment   Bahrzahlung
cash register   Kasse
casing   Gehäuse ‖ Umhüllung
cask   Fass * (auch als Verb: 'in ein Fass füllen') ‖ Tonne
casket   Kästchen
cassette   Kassette * [ke'sset] z.b. Video-/Audio-/ Magnetband-/Pneumatikventilkassette
cast   gegossen ‖ Gießen * (Verb) Metall, Gussstücke ‖ Guss- ‖ verteilen * Rollen (z.B. im Theater)
cast aluminium   Aluminium-Druckguss
cast bronze   Gussbronze
cast coating   Gussstreichverfahren * z.B. zur Papierbeschichtung
cast glass   Gussglas
cast housing   Gussgehäuse
cast in   eingießen * z.B. in Kunstharz
cast integrally   angegossen * z.b. Aufstellfüße eines Motors
cast iron   Grauguss * kohlenstoffhaliges Gusseisen (Kohlenst. liegt als Lamellengrafit vor; DIN 1691) ‖ Gusseisen
cast piece   Gussteil
cast resin   Gießharz
cast steel   Stahlguss ‖ Stahlguss * Gusswerkstoff aus niedrig legiert. Eisen (z.B. GS 38 m. Zugfestigkt. 380 N/mm²; DIN 1681
cast-iron   Grauguss
cast-iron housing   Graugussgehäuse * z.B. für Elektromotor
cast-resin   Gießharz
cast-resin insulated   gießharzisoliert
caster   Gießmaschine ‖ Rolle * Lauf~, Lenk~, ~ unter Möbeln
casting   Gießen * z.B. Gusseisen ‖ Guss * z.B. Herstellung von Gusseisen ‖ Gussteil
casting compound   Vergussmasse
casting epoxy   Epoxidharz * Gießharz
casting machine   Gießmaschine
casting plant   Gießanlage * Stahl/Walzwerk ‖ Gießerei
casting process   Gießverfahren
castor   Rolle * Lauf~, Lenk~, ~ unter Möbeln
casual   gelegentlich * unregelmäßig, ~; zufällig, unerwartet; ungewohnt, ungezwungen, zwanglos, beliebig
catalog   Katalog ‖ katalogmäßig * siehe auch →listenmäßig ‖ Liste * Katalog ‖ listenmäßig ‖ Prospekt * Katalog ‖ Verzeichnis * Katalog
catalog No   Bestellnummer * Artikelnummer im Katalog
catalog-listed   listenmäßig
catalogue   Katalog ‖ Liste * Katalog ‖ Prospekt * Katalog ‖ Verzeichnis * Katalog

catalogued   listenmäßig
catalyst   Katalysator
catapult   schleudern * mit einer Schleuder(vorrichtung) ~
catastrophic failure   Totalausfall * plötzlich auftretender schwerer Totalausfall
catch   einklinken ‖ einrasten ‖ erfassen * greifen (auch figürl.) ‖ Fangen * (Verb) ‖ fassen * fangen ‖ Klaue * Nase ‖ Klinke * Sperr~ ‖ Mitnehmer * Fangeinrichtung/-behälter ‖ Rastung * Sperrhaken, Schließhaken, Schnäpper, Sicherung, Verschluss ‖ Riegel * Sperr-/Schließhaken, Falle, Schnäpper, Verschluss, mechan.Sicherung ‖ schnappen * Schloss ‖ Sicherung * (mechan.) Verschluss, Sicherung ‖ Verschluss * Schnapp~
catch fire   anbrennen
catch hold of   erfassen * ergreifen ‖ greifen
catch spring   Schnappfeder
catch up   auffangen ‖ aufholen * Vorsprung, Verspätung usw. ~ ‖ aufschließen * zur Spitzengruppe ~ ‖ einholen * with: jemanden
catching type   einrastend
catchphrase   Motto ‖ Schlagwort
catchy   einprägsam * z.B. Motto, Slogan
categorize   einordnen * in Kategorie ~ (as: unter)
category   Art * Gattung, Klasse ‖ Gruppe * Kategorie, Gattung, Klasse; fall in the category: zur ~ gehören ‖ Kategorie * (Begriffs-) Klasse, Gattung; fall in the category: in die Kategorie fallen ‖ Klasse * Kategorie ‖ Prüfklasse ‖ Rubrik * Kategorie, Klasse
category voltage   Dauergrenzspannung * Kategoriespannung, z.b. nach IEC ‖ Kategoriespannung * Dauergrenzspannung, z.B. nach IEC
catenary control   Durchhangregelung * z.B. b. Kabelummantelungsanlage; catenary: "e. mathem. Seil-/Kettenlinie folgend".
cater for   sorgen für * (auch figürl.) befriedigen, sorgen für
caterpillar   Raupe ‖ Zug- und Schubraupe
cathode   Kathode
cathode ray   Elektronenstrahl * Kathodenstrahl
cathode ray tube   Röhre * Kathodenstrahlröhre
cathode-ray tube   Bildröhre * Kathodenstrahlröhre
catridge fuse link   Sicherungseinsatz
causal   ursächlich
causative   ursächlich
cause   auslösen * verursachen ‖ bedingen * verursachen ‖ bewirken * veranlassen,verursachen ‖ Grund * Ursache ‖ Ursache * the cause lies in: die ~ liegt/ist zu suchen bei/in ‖ veranlassen * veranlassen, verursachen, bewirken, hervorrufen, herbeiführen ‖ Veranlassung * Ursache, Grund ‖ verursachen
cause of malfunction   Störungsursache
cause of the fault   Störungsursache
cause to flow   treiben * zum Fließen veranlassen, z.B. Strom (through:durch)
caused   bedingt * verursacht; by:durch ‖ verursacht
caustic   Beize
caustic solution   Lauge
causticizing   Kaustizierung * z.B. in der Zellstoff-Herstellung
caustisizer   Kaustifizierer * [Papier] in der Zellstoffindustrie
caution   Achtung * als Hinweis z.B. auf Schild/in Betriebsanleitung ‖ Vorsicht ‖ warnen * against: vor ‖ Warnung
caution!   Vorsicht bitte
cautionary   warnend
cautious   behutsam ‖ sorgfältig * vorsichtig ‖ vorsichtig
cavity   Grube * Höhlung ‖ Hohlraum ‖ Mulde * Höhlung ‖ Vertiefung * Aushöhlung, Hohlraum
ccw   entgegen dem Uhrzeigersinn ‖ links * in

ccw operation 438

~drehrichtung || links herum * Abkürzg. für 'conterclockwise': entgegen d. Uhrzeigersinn || rückwärts * counter-clockwise: entgegen dem Uhrzeigersinn
ccw operation Rücklauf * conter-clockwise: entgegen dem Uhrzeigersinn, Linkslauf
ccw rotating field Linksdrehfeld * CounterClockWise: entgegen d.Uhrzeigersinn
ccw speed direction Rückwärtsdrehrichtung * Linksdrehrichtung
CD Quer- * Abkürzg.f. 'Cross-Directed': quer zur Materiallaufrichtung wirkend
CD direction Querrichtung * Abk.für 'Cross Direction'
CD profile Querprofil * Abk.f. 'Cross-Directed': (z.B. Fläch.gewichts-) Profil quer zur Materiallaufrichtung
CDF ED * Abk. f. 'Cyclic Duration Factor': relative Einschaltdauer (e. el. Maschine: s. VDE 0530 Teil 1)
CE CE * französ. Abk. für 'Communautéen Europeéen': Europäische Gemeinschaft
CE mark CE-Kennzeichen
CE marking CE-Kennzeichnung
CE-marking of explosion protected eletrical motors CE-Kennzeichnung explosionsgeschützter Elektromotoren
cease aufhören * zu Ende gehen; cease to/from: aufhören zu (mit Infinitiv)/mit || nachlassen * aufhören
cease production Fertigung einstellen
ceaseless andauernd * unaufhörlich
CEB CEB * Abk. für Comité Électrotechnique Belge: Belgisches Electrotechn. Komitee (ähnl. d. deutschen DKE)
CEF CEF * Abk. für Comité Électrotechnique Francais: französ. Eletektrotechn. Komitee (ähnl. dem DKE)
CEI CEI * Abkürzg. für 'Comitato Electrotechnico Italieno': italienisches Normengremium || CEI * französ. Abk.für Comité Électrotechnique International: Internat. Elektrotechn. Kommission (IEC)
ceiling Decke * eines Raums
cell Zelle
cell phone Handy
cell- Insel-
cellophane Kunststofffolie * Zellophan, Zellglas, Klarsichtverpackung(sfolie)
cellotape Tesafilm * durchsichtiges Klebeband
cellular Zell- || Zellen-
cellular board Wellpappe
cellular filter Zellenfilter
cellular phone Handy
cellulose Cellulose || Zellstoff * Zellulose, Zellstoff || Zellulose
cellulose paint Nitrolack
CEMA CEMA * Canadian Electrical Manufacturers Association: Verband d. canad, Elektroindustrie (ähnl. ZVEI)
cement kleben * kitten || Klebstoff * Kitt || leimen || verkitten * verkleben * kitten || Zement
cement industry Zementindustrie
cement mill Zementmühle
cement pump Zementierungspumpe
cementing agent Bindemittel
CEMEP CEMEP * Abk. für. 'Comité des Constructeurs d'Machines Electriques et d'Electronique de Puissance'
CEMF EMK * Abkürzung für 'counter electromagnetic force': Gegen-EMK
CEMF crossover voltage Ablösespannung * oberhalb dieser Spanng.beginnt d. Feldschwächereich b. DC-Motor
CEMF winder Wickler * Achswickler mit DC-Antrieb und Durchmesseranpassung i.Feldkreis

CEN CEN * Abkürzg. f. 'Comité Européen de Coordination des Normes': Europäisch. Komitee für Normung
CENELEC CENELEC * European Commitee for Electrotechnical Standardization (europäisches Normengremium)
cennector designation Steckerbezeichnung * z.B. "X6"
censure Kritik * Tadel, Verweis, Missbilligung || Kritikpunkt * Tadel, Verweis, Missbilligung
center Abteilung * Zentrum || konzentrieren * (amerikan.) upon: auf || Mitte * (amerikan.) || Mittelpunkt * (amerikan.) Zentrum || zentrieren * (amerikan.)
center aisle Mittelgang * in der Mitte liegender Durchgang
center bore Zentrierbohrung
center distance Achsabstand
center dog Mittenanschlag
center drive Achsantrieb * (amerikan.) bei Wickler || Zentralantrieb * (amerikan.)
center frequency Mittenfrequenz * (amerikan.) eines Bandpasses, einer Bandsperre
center hole Zentrierbohrung * (amerikan.)
center line Mitte * Mittellinie
center line of rotation Drehachse
center line of shaft Wellenmitte
center of area Flächenschwerpunkt || Schwerpunkt * Flächen~
center of gravity Schwerpunkt * (amerikan.)
center of gravity method Schwerpunktmethode
center of rotation Drehpunkt
center point Mittelpunkt * (amerikan.)
center position Mittelstellung * (amerikan.)
center punch Körner * zum →Ankörnen eines Bohrpunktes
center stop Mittenanschlag
center tap Mittelabgriff * (amerikan.) || Mittelanzapfung * (amerikan.) || Mittenanzapfung * (amerikan.)
center winder Achswickler * (amerikan.) || Wickler * (amerikan.) Achswickler || Zentralwickler * (amerikan.) siehe auch →Achswickler || Zentrumswickler * (amerikan.) Achswickler
center-piece Herzstück * (amerikan.)
centering Zentrier- || zentrieren * (Substantiv) || Zentrierung
centering bearing Zentrierlager
centering bore Zentrierbohrung
centering diameter Zentrierdurchmesser
centering drill Zentrierbohrer
centering hole Zentrierbohrung
centering pin Zentrierstift
centering shoulder Zentrierrand
centering spigot Zentrierrand
central Mittel * (Adj.) zentral || mittlerer * in der Mitte liegend || mittlerer * zentral || zentral
central block Zentralbaustein * Funktionsbaustein
central control electronic Zentralelektronik
central control room Warte * zentraler Steuerraum || Zentrale * zentraler Steuerraum, Leitwarte
central controller Zentralgerät * {SPS}
central controller module zentrale Baugruppe * bei SPS
central coordination Leitstelle
central electronics Zentralelektronik
central function block Zentralbaustein * Funktionsbaustein
central lubrication Zentralschmierung
central office Zentrale * Firmen~
central pivot Königszapfen * zentraler Drehzapfen, z.B. für Kran, Bagger, Lastkraftwagen
central point Mittelpunkt * zentraler Punkt || Sammelpunkt

central processing unit   CPU || Zentralbaugruppe * *bei SPS* || Zentraleinheit
central processor   Zentralprozessor
central rectifying unit   zentrale Einspeiseeinheit
central station   Zentrale * *Zentralstation*
central switchboard   Zentrale * ~ *Steuertafel*
central tapping   Mittenanzapfung
centralized   zentral * *an einer Stelle zusammmengefasst*
centralized configuration   zentraler Aufbau
centralized ripple control   Rundsteuerung
centrally   zentral * *(Adv.)*
centrally coordinated   zentral * *mit zentraler Koordination (z.B. Zentraleinkauf)*
centre   Kern * *z.B. Stadt~, Zentrum* || konzentrieren * *upon: auf* || Mitte || Mittelpunkt * *Zentrum* || zentrieren * *(brit.)*
centre drive   Achsantrieb * *(brit.)* || Zentralantrieb * *(brit.)*
centre frequency   Mittenfrequenz * *für Bandpass/sperre*
centre hole   Zentrierbohrung * *(brit.)*
centre line   Mittellinie
centre of gravity   Schwerpunkt * *(brit.)*
centre of masses   Schwerpunkt * *Massen~*
centre point   Mittelpunkt
centre position   Mittellage * *z.B. Tänzerwalze* || Mittelstellung * *(brit.)*
centre tap   Mittelabgriff * *(brit.)* || Mittenanzapfung * *(brit.)* || Mittenanzapfung
centre winder   Achswickler * *(brit.)* || Wickler * *Achswickler* || Zentralwickler * *(brit.)* || Zentrumswickler * *Achswickler*
centre-piece   Herzstück
centre-to-centre distance   Mittenabstand * *z.B. zwischen Bohrungen*
centric   zentrisch
centrical   zentrisch
centricity   Mittigkeit || Zentrizität
centrifugal blower   Radialgebläse
centrifugal brake operator   Motordrücker * *elektromechan. oder hydraul. Betätigungselement für Bremse*
centrifugal clutch   Fliehkraftkupplung
centrifugal compressor   Kreiselverdichter || Radialverdichter
centrifugal contactor   Fliehkraftschalter
centrifugal fan   Radialventilator || Trommellüfter * *Walzenlüfter, Fliehkraftlüfter* || Walzenlüfter * *Fliehkraftlüfter*
centrifugal force   Fliehkraft || Schwungkraft * *Zentrifugalkraft, Fliehkraft* || Zentrifugalkraft
centrifugal lift   Fliehkraftabhebung * *z.B. bei Klemmzahn-→Rücklaufsperre/-→Freilauf*
centrifugal overspeed switch   Fliehkraftschalter * *Überdrehzahlschalter*
centrifugal pump   Kreiselpumpe
centrifugal starting switch   Anwerfschalter
centrifugal switch   Fliehkraftregler || Fliehkraftschalter * *centrifugal starting switch: ~ z. Abschalten e. Anlas-Hilfswicklung bei AC-Motor*
centrifugal thrustor   Motordrücker
centrifugal-cleaner   Rohrschleuder * *Wirbelsichter/Hydrozyklon zur Reinigung, z.B. in der Zellstoffherstellung*
centrifugalize   zentrifugieren
centrifuge   schleudern * *mit einer Schleudermaschine/Zentrifuge* ~ || Zentrifuge || zentrifugieren
CEO   Vorstandsvorsitzender * *Abk. f. 'Chief Executive Officer'*
CEP   CEP * *Abk.für 'Comissao Electrotécnica Portoguêse' portugiesische Elektrotechn.Kommission (wie DKE)*
ceramic   Keramik || keramisch
ceramic capacitor   Keramikkondensator

ceremony   Form * *Förmlichkeit*
certain   bestimmt * *gewiss; up to a certain limit: bis zu einer ~en Grenze* || bestimmte || bestimmter * *gewisser* || gewisser || sicher * *gewiss* || spezziell * *bestimmt, gewiss*
certainty   Sicherheit * *Gewissheit; with certainty: mit ~*
certificate   Approbation || Bescheinigung * *Dokument, (Prüf)Zeugnis, Prüfbescheinung einer Abnahmestelle (z.B. TÜV, PTB)* || Gutachten * *Dokument* || Nachweis * *Zeugnis* || Zertifikat || Zeugnis * *Zertifikat, Prüfungszeugnis*
certificate of compliance with order   Werksbescheinigung * *→Prüfbescheinigung, die der Lieferant dem Besteller übergibt*
certificate of conformance   Konformitätsbescheinigung
certificate of conformity   Konformitätsbescheinigung
certificate of inspection   Prüfbescheinigung
certificate of origin   Ursprungszeugnis
certificated   zertifiziert
certification   Anerkennung * *Zertifikation* || Approbation * *Zertifizierung/Approbation* || Bescheinigung * *Vorgang des Bescheinigens* || Zertifizierung || Zulassung * *z.b. von einer Prüfbehörde/Zulassungsstelle*
certification authority of EU   Prüfstelle der Europäischen Union
certified   verbindlich * *garantiert, bescheinigt, beglaubigt* || zertifiziert || zugelassen * *offiziell zugelassen/bestätigt/bescheinigt/beglaubigt*
certify   bescheinigen * *z.B. Erfüllung einer Norm ~* || bestätigen * *bescheinigen* || prüfen * *bescheinigen, beglaubigen, bestätigen, bezeugen* || zertifizieren
CES   CES * *Abk.für 'Comité Électrotechnique Suisse': Schweizer. Elektrotechn. Komitee (ähnlich DKE)*
cf.   vergleiche * *Abkürzg. für 'confer'*
CFC   CFC * *'Continuous Function Chart': Grafische Beschreib.sprache für kontinuierl. Steuer-/Regelaufgaben*
chafe   aufscheuern * *(sich) durch-/wundreiben, scheuern (against: an), verschleißen (z.B. Kabelisolierung)* || scheuern * *(sich) durch-/wundreiben, scheuern (against: an), verschleißen (z.B. Kabelisolierung)*
chain   aneinander hängen || hintereinander schalten * *verketten* || Kette || verketten || verknüpfen * *verketten (to:mit)*
chain conveyor   Förderkette || Kettenförderer
chain drive   Kettenantrieb
chain grinder   Kettenschleifer * *Stetigschleifer z.B. bei der Zellstoffherstellung*
chain guard   Kettenschutz
chain link   Kettenglied
chain mortizer   Kettenfräse * *{Holz}*
chain mortizing machine   Kettenfräse * *{Holz}*
chain saw   Kettensäge
chain steel   Kettenstahl
chain stretcher   Kettenspanner
chain tensioner   Kettenspanner
chain tightener   Kettenspanner
chain transfer   Kettenschlepper * *z.B. in Walzwerk*
chain up   verketten
chain-dotted   strichpunktiert
chain-driven   kettengetrieben
chain-sprocket   Kettenrad
chaining   Kettung || Verkettung
chair   leiten * *Vorsitz führen, z.B. bei einer Sitzung* || Sitz * *Stuhl*
chairman   Aufsichtsratvorsitzender || Vorsitzender
chairman of the board   Aufsichtsratvorsitzender || Vorstand * *~svorsitzender* || Vorstandsvorsitzender

chairman of the supervisory board Aufsichtsratsvorsitzender || Aufsichtsratvorsitzender
chairwoman Vorsitzender * *(weiblich)*
challenge Aufgabe * *schwierige/lockende Aufgabe, Herausforderung* || Herausforderung * *auch technische*
chamber Kammer * *auch mechan. Gehäusekammer* || Nische * *Kammer*
chamber of commerce Handelskammer
Chamber of Industry and Commerce Industrie- und Handelskammer
chamfer abschrägen || anfasen || Anfasung || Kehle * *{Bautechnik}* || Rille || Schräge * *Schrägkante*
chamfered abgeschrägt * *(Kante)* || schräg * *an/abgefast, abge~t*
champion Vertreter * *Verfechter*
chance Möglichkeit * *Gelegenheit*
chance in the market place Marktchance
chanelling in Einschleusung
change Abweichung * *Änderung* || ändern || Änderung * *in: in einem Ab/Verlauf; of: von einer Größe* || auswechseln * *Verschleißteil (Reifen, Batterie, Öl)* || erneuern * *Verschleißteil, Öl usw. (aus)wecheln* || schalten * *Getriebe* || Übergang || Umkehr * *Änderung* || umsteigen * *bei einer Fahrt mit Verkehrsmittel (to: nach)* || verändern * *ändern* || Veränderung || Wechsel * *auch ~ von Geldsorten* || wechseln
change by mistake verwechseln
change gear Wechselrad * *zur Änderung der Übersetzung bei →Wechselradgetriebe*
change gears Wechselradgetriebe
change in -Änderung
change in frequency Frequenzänderung
change into übergehen * *~ in*
change note Änderungsmitteilung
change of -Änderung || -Wechsel
change of dimension Maßänderung
change of operating mode Betriebsartenänderung || Betriebsartenwechsel * *Betriebsartenänderung*
change of shift Schichtwechsel
change of sign Vorzeichenumkehr
change of temperature Temperaturwechsel
change of torque direction Momentenwechsel * *Wechsel der Momentenrichtung*
change one's clothes umziehen * *Kleidung*
change one's mind Meinung ändern
change one's residence umziehen
change over übergehen * *hinüberwechseln* || umschalten
change rate Änderungsgeschwindigkeit || Anstiegsgeschwindigkeit * *Änderungsgeschwindigkeit*
change rate limiter Anstiegsbegrenzer
change rights Zugriffsberechtigung * *zum Ändern/Schreiben, z.B. von Parametern*
change-gear-type gearbox Wechselradgetriebe * *Getriebe mit durch Zahnradwechsel änderbarer Übersetzung*
change-log file Änderungsdatei
change-over wechseln * *hinüber~*
change-over contact Wechselkontakt || Wechslerkontakt
change-over current Umschaltstrom
change-over gear Schaltgetriebe
change-over point Ablösepunkt
change-over switch Umschalter
change-over switching umschalten * *(Substantiv)*
change-over the terminal connections umklemmen * *die Klemmenverdrahtung ändern*
change-over time Umrüstzeit || Wechselzeit * *z.B. für den Materialwechsel in einer Verarbeitungsmaschine benötigte Zeit*
change-pole motor polumschaltbarer Motor
change-speed gear Schaltgetriebe || Wechselgetriebe

change-speed motor polumschaltbarer Motor
changeability Änderbarkeit
changeable änderbar || umschaltbar * *änderbar* || veränderbar * *änderbar* || veränderlich || wechselnd * *veränderlich, veränderbar*
changeable gear wheel Wechselrad
changeable gearbox Wechselgetriebe
changed-over umgeschaltet
changeover Übergang || übergeben * *hinübergehen, hinüberwechseln* || Umschaltung || Umstellung || Umsteuervorgang || Wechsel * *Hinüber ~*
changeover contact Umschaltkontakt || Wechsler
changeover control ablösende Regelung
changeover sequence Umschaltvorgang
changeover speed Umschaltdrehzahl
changeover threshold Umschaltschwelle
changeover time Umschaltzeit
changes of -Änderung
changes of load Lastwechsel
changing Ersatz * *Auswechslung (besonders eines Verschleißteils, z.B. Batterie, Reifen, Bremsbelag)* || sich verändernd || Wechsel || wechselnd
changing load Wechsellast
changing of the rolls Rollenwechsel
changing-load operation Betrieb mit wechselnder Last
channel -Kanal || führen * *durch einen Kanal/ein Ventil usw. hindurchleiten/-lenken* || Kanal || Kehle || leiten * *auf dem Dienstweg* || Rinne * *Rille* || Weg * *Kanal (Vertriebs, diplomatischer usw.); through the channel of: auf dem ... ~e*
channel control Kanalverwaltung * *für Datenübertragungskanal*
channel of distribution Vertriebskanal || Vertriebsweg
chaotic chaotisch
chapter Kapitel * *in einem Buch*
character Art * *Wesen, Natur, Eigenart* || Beschaffenheit * *Wesen, Natur, (Eigen-) Art, Merkmal, Kennzeichen, Eigenart* || Buchstabe * *Zeichen* || Schriftzeichen || Zeichen * *z.B. abdruckbare ~, Text~, ASCII-~, ANSI-~*
character chain Zeichenkette
character frame Zeichenrahmen * *z.B. 11-bit frame:11 Bit Zeichenr.(1 Startbit, 8 Datenbits, 1 Paritätsb.,1 Stopbit)*
character of the load Lastart
character sequence Zeichenfolge
character set Zeichensatz
character string Zeichenfolge || Zeichenkette
character-oriented zeichenorientiert * *z.B. serielles Übertragungsverfahren*
characteristic Abhängigkeit * *Kennlinie, math.-Funktion* || Ausprägung * *Eigentümlichkeit, Kennzeichen, Eigenschaft, charakteristisches Merkmal* || Beschaffenheit * *charakteristisches Merkmal, Eigenschaft, Eigenschaft, Eigentümlichkeit, Kennzeichen* || charakteristisch * *feature/property: Eigenschaft* || Eigenart || Eigenheit || Eigenschaft || Funktion * *Kennlinie* || Kennlinie * *falling/rising:-fallende/(an)steigende ~* || Kennwert || Kurve * *Kennlinie* || Kurvenzug * *Funktionskurve, Kennlinie* || Merkmal * *Besonderheit* || Qualität * *Eigenschaft, charakteristisches Merkmal, Eigentümlichkeit, Kennzeichen* || Verhalten * *Merkmal, Eigentümlichkeit, Eigenschaft* || Verlauf * *~ einer Kennlinie/Kurve usw.* || Verlauf * *~ einer Kennlinie/ Kurve(mathem.) Funktion* || Zeichen * *Kenn~, Eigentümlichkeit, charakteristisches Merkmal* || zeitlicher Verlauf
characteristic curve Kennlinie
characteristic data Kenndaten
characteristic data set Kenndatensatz * *z.B. wählbarer umschaltbarer Parametersatz im Stromrichter*

characteristic droop Kennlinienneigung * *siehe* →*Statik*
characteristic equation charakteristische Gleichung
characteristic feature Kennzeichen * *Haupt/ Schlüsseleigenschaft*
characteristic features typische Merkmale
characteristic impedance Wellenwiderstand
characteristic line Kennlinie
characteristic number Eigenwert
characteristic resistance Wellenwiderstand
characteristic rotor resistance Läuferkennzahl * *{el. Masch.} Rechengröße k [in Ohm] z. Bemessung d. Anlaßwiderstände*
characteristic setting Kennlinieneinstellung
characteristic type Kennlinientyp
characteristic value Kenngröße * *Kennwert* || Kennwert
characteristics Ausprägung || Betriebswerte || Eigenschaften || Verhalten
characteristics of special note besondere Merkmale
characteristics over time zeitlicher Verlauf
characterize beschreiben || kennzeichnen * *charakterisieren*
characterized gekennzeichnet * *charakterisiert; by: durch*
charasteristic of fuse Sicherungskennlinie
charge aufladen * *z.B. Kondensator, Batterie, Akku* || Auftrag * *Befehl, Pflicht* || aufziehen * *Feder* ~ || beauftragen || befüllen * *Behälter, Zentrifuge usw.* ~ || Belastung * *finanzielle* ~ *(besonders Steuern, Zinsen, Abgaben) auch e. Kontos, Grundstücks* || berechnen * *in Rechnung stellen, Preis berechnen, (Konto) belasten* || Berechnung * *in Rechnung stellen (Preis, Kosten)* || einfüllen * *(Behälter usw.)* befüllen; Metall (-guss), Granulat usw. einfüllen || eintragen * *beschicken (z.B. Ofen)* || fordern * *Preis* ~ || fördern * *in einen Behälter hinein*~ || füllen * *beladen, aufladen* || Füllung * *Beschickung, Befüllung, Ladung, Einsatz (metall.)* || Gebühr || laden * *Batterie, Kondensator usw.* || Ladung || rechnen * *mit Geldsumme belasten* || spannen * *Feder* || Tarif * *Gebühr*
charge a fee Gebühr erheben
charge carrier Ladungsträger * *(Singular) z.B. bei Halbleitern*
charge carriers Ladungsträger * *z.B. bei Halbleitern*
charge density Ladungsdichte
charge level selection Ladestufenumschaltung
charge pump Ladungspumpe * *Spannungswandler m.Chopper u.Kondensatoren (kann Spannungen hochsetzen u.invertieren)*
charge/discharge current Umladestrom * *z.B. bei Kondensator, langer Leitung (auf Grund d. Leitungskapazität) usw.*
charged aufgeladen * *z.B. Kondensator, Batterie, Akku* || berechnet * *in Rechnung gestellt, belastet*
charger Einschieber * *Beschickungseinrichtung* || Füller * *Befülleinrichtung* || Ladegerät * *für Batterien*
charges Gebühren || Kosten * *auch Gebühren; standing charges: laufende* ~ || Spesen * *an Kunden verrechnete Gebühren* || Unkosten
charges deducted abzüglich * *abzüglich der Spesen*
charging Beschickung || Einschleusung || Eintrag * *z.B. in Ofen, Presse* || Füllung || Verrechnung * *Belastung*
charging capacitor Ladekondensator
charging current Ladestrom * *z.B. für Batterie, Kondensator*
charging current peaks Ladestromspitzen * *z.B. bei der Aufladung eines Kondensators*

charging device Zuführeinrichtung * *Beschickungs-/Fülleinrichtung*
charging equipment Beschickungsanlage
charging level Ladestufe
charging machine Belademaschine || Eintragmaschine * *Beschickungs-/Lademaschine, z.B. für Ofen, Presse* || Lademaschine * *Beschickungseinrichtung, siehe auch* →*Einschieber*
charging rectifier Ladegleichrichter
charging system Beladeeinrichtung || Beschickungssystem
charging unit Befüllungsanlage || Beschickungseinrichtung * *z.B. für Ofen; siehe auch* →*Einschieber,* →*Lademaschine* || Einschieber * *Beschickungseinrichtung, z.B. für Ofen*
charging voltage Ladespannung * *bei Batterie/ Akku*
chart Abbildung * *Diagramm, Tabelle, Schaubild, Kurvenblatt, grafische Darstellung* || Bild * *Diagramm, Schaubild, Tabelle, Kurve(nblatt), Karte, Skizze* || darstellen * *grafisch* ~, *mit Diagrammen* ~ || Diagramm * *grafische Darstellung, Kurvenblatt, Schaubild* || Ebene * *im Koordinatensystem, z.B. Strom/Spannungsebene* || Grafik * *Diagramm, grafische Darstellung, Schaubild, Kurvendarstellung* || grafisch darstellen * *in Schaubild, Diagramm* || grafische Darstellung || Overhead-Folie || Overheadfolie || Plan * *Tabelle, Schaubild, grafische Darstellung* || Schaubild || Tabelle * *als Schaubild* || Tafel * *grafische Darstellung* || Übersicht * *Schaubild, Diagramm, grafische Darstellung, Kurvenblatt, Karte, Tabelle* || Zeichnung * *Schaubild, Diagramm*
chart recorder Kurvenschreiber * *auch 'Linienschreiber'* || Linienschreiber || Schreiber * *Linienschreiber*
chassis Gehäuse * *tragende Konstruktion, Tragblech* || Gehäuse * *Gestell (-Rahmen), Grundplatte* || Grundplatte * *Tragblech, tragende Konstruktion* || Masse * *Gehäuse~ (elektr.)* || Rahmen * *tragende Konstruktion, Fahrgestell* || Tragblech * *Grundplatte; Fahrgestell*
chassis earth Gehäuse-Erde
chassis module Einbaueinheit * *auf offenem Tragblech montiert, ohne besonderer Schutzart, für Schrankenbau*
chassis unit Chassis-Gerät * *Baueinheit auf offenem Tragblech ohne besondere Schutzart* || Einbaugerät * *auf offenem Tragblech montiert, m. niedriger Schutzart (meist IP20), für Schrankenbau* || Kompaktgerät * *Chassis-/Einbaugerät auf offenem Tragblech ohne besondere Schutzart*
chassis- gehäuselos * *z.B. offenes Gerät, nur auf Tragblech montiert*
chatter flattern * *klappern, z.B. Relais* || rattern
chatter marks Rattermarken
chattering Instabilität * *Reglerschwingen, Rattern (bei Antriebsregelung)*
cheap billig || günstig * *billig* || preisgünstig * *billig* || preiswert * *billig*
cheapness Preiswürdigkeit
check Abfrage * *Überprüfung,Prüfung* || abfragen * *überprüfen; against a limit: auf eine Grenze* ~ || durchmessen * *überprüfen* || kontern || Kontrolle * *Überprüfung* || kontrollieren * *überprüfen* || nachmessen * *überprüfen* || nachprüfen * *kontrollieren* || Nachprüfung * *Kontrolle, Überprüfung, Nachprüfung* || Probe * *Überprüfung* || prüfen * *über~* || Prüfung * *Über~* || überprüfen * *for: bezüglich* || Überprüfung * *make a check: eine* ~ *durchführen* || überwachen * *überprüfen* || Überwachung * *Überprüfung* || untersuchen * *überprüfen* || Untersuchung * *Überprüfung* || zurückhalten * *drosseln, bremsen, sperren*
check against vergleichen * *zur Kontrolle* ~ *mit*
check against a limit auf eine Grenze abfragen

check back Rückfrage * *zur Kontrolle* || Rückfragen * *(Verb) zur Kontrolle* || rückmelden * *zur Kontrolle/Bestätigung* ~ || zurückmelden * *zur Bestätigung* ~
check bit Prüfbit
check block Kontrollbaustein * *Funktionsbaustein*
check measurement Kontrollmessung
check nut Gegenmutter || Kontermutter
check off abhaken || anstreichen * *abhaken*
check time Überwachungszeit
check up Nachprüfung * *Überprüfung, Kontrolle* || überholen || Überprüfung * *Nachprüfung*
check valve Absperrventil * →*Rückschlagventil* || Rückschlagventil
check-back message Rückmeldung
check-up Inspektion || nachprüfen || Überholung * *Überprüfung, Überholung* || überprüfen * →*nachprüfen*
checkback Rückmeldung * *i. Sinne einer Bestätigung*
checkback signal Rückmeldesignal || Rückmeldung * *Rückmeldesignal (z.b. von Hauptschütz-Hilfskontakt)*
checked abgesichert * *überprüft* || geprüft * *überprüft* || überprüft
checking Kontrolle * *Überprüfung, das Kontrollieren; for checking purposes: zur Kontrolle/für Kontrollzwecke* || Prüfung * *(Nach)Prüfung, Kontrolle* || Überprüfung * *das Überprüfen* || Überprüfung * *(Nach)Prüfung, Kontrolle* || Wartung * *Nachprüfung, Kontrolle*
checking system Kontrollanlage
checklist Checkliste
checksum Prüfsumme || Quersumme
checkup Prüfung * *Nach~* || Wartung * *Nachprüfung*
checkweigher Kontrollwaage * *{Verpack.technik}*
cheese-head screw Zylinderkopfschraube
chemical Chemikalie || chemisch
chemical industry chemische Industrie
chemical processed pulp Chemiefaserzellstoff * *{Papier}*
chemical pulp Zellstoff * *Ausgangsmaterial für Papierherstellung; bleached: gebleichter*
chemical reaction chemische Reaktion
chemical recovery Chemikalienrückgewinnung
chemically chemisch * *(Adv.)*
chemically aggressive chemisch aggressiv
chemicals Chemikalien
chemicals preparation Chemikalienaufbereitung
chemist's shop Apotheke
chemistry Chemie
chest Bütte * *z.B. in der Zellstofferzeugung*
chest level control Büttenregelung * *Füllstandsregelung für Bütte in der Zellstoffaufbereitung*
chief Chef * *Oberhaupt, Vorgesetzter* || Haupt- * *Ober-, Höchst-, Haupt-, in chief: leitender; chief part: ~rolle* || leitend * *z.B. Ingenieur* || Leiter * *einer Abteilung* || Ober- * *Chef-*
chief designer Chef * *Konstrukteur, Entwickler*
chief editor Chefredakteur
chief engineer Chef * *Ingenieur*
chief executive Geschäftsführer
Chief Executive Officer Vorstandsvorsitzender * *(Großschreibg. i.Verbindung mit Namen)*
chief financial officer Leiter des Finanz- und Rechnungswesens
chief instructor Kursleiter || Lehrgangsleiter
chief test engineer Prüffeldleiter
chiefly bevorzugt * *(Adv.) hauptsächlich* || größtenteils || vorzugsweise
child's play kinderleicht * *it's mere child's play to them: es ist ein Kinderspiel für sie* || Kinderspiel * *(auch figürl.) it's mere child's play to you: das ist ein ~ für dich*

Chile mill Kollergang * *Walzen-Mahlwerk*
chill abkühlen * *schlagartig* ~ || Kapsel * *Gusskapsel*
chill roll Gießwalze * *wassergekühlte Walze in der Kunststoff-Folienherstellung (hinter Extruder angeordnet)* || Kühlwalze * *z.B. wassergekühlte Gießwalze in Folienmaschine hinter dem Extruder*
chimney Kamin * →*Schornstein* || Schornstein
china clay Kaolin
chink Riss || Spalt * *Ritze*
chip Baustein * *integrierter Baustein* || IC || integrierte Schaltung * *(salopp) Halbleiter-Chip* || integrierter Schaltkreis || Plättchen || Schaltkreis * *(salopp) integrierter* ~ || Scheibchen || Span || spanen * *{Werkz.masch.\ Holz}* || Splitter * *Span*
chip board Spanplatte * *{Holz}*
chip extraction system Späneabsauganlage * *{Holz}*
chip extractor Späneabsauganlage * *{Holz}*
chip sifter Spänesichter * *{Holz} zum Trennen der Späne in der Spanplattenherstellung*
chip-card Chip-Karte
chip-exhaust plant Späneabsauganlage * *{Holz}*
chip-type capacitor Chip-Kondensator * *z.B. für SMD-Bestückung*
chip-type resistor Chip-Widerstand * *z.B. für SMD-Bestückung*
chipboard Graupappe || Spanplatte * *{Holz}*
chipboard panel Spanplatte * *{Holz}*
chipboard press Spanplattenpresse * *{Holz}*
chipless spanlos
chipless forming spanlose Umformung
chipless shaping spanlose Umformung
chipper Zerspaner * *{Holz} in der Spanplattenherstellung*
chipper chain Hobelzahnkette * *{Holz} für Kettensäge*
chipping Zerspanung
chipping plant Hackerei * *in Zellstoff-Fabrik*
chippings Späne
chips Hackschnitzel * *{Holz}* || Schnitzel * *z.B. aus Holz-/Kunststoffmaterial* || Span || Späne
chisel Meißel || meißeln || Stechbeitel
chisel mortizer Meißelstemmmaschine * *{Holz}*
chlorine Chlor * *chlorine-free: ~frei (z.B. ungebleichtes oder mit anderen Stoffen gebleichtes Papier)*
choice Alternative * *Wahlmöglichkeit; have no choice: keine Alternative haben* || Auswahl * *(Aus)-Wahl* || Vorrat * *Auswahl* || Wahl
choices Auswahlmöglichkeiten
choke abwürgen * *Motor* || Drossel * *kleine* || Drosselspule * *kleinere* || Spule * *(elektr.) Drossel (eher kleinere)* || verstopfen * *versperren (Rohr usw.)*
choke coil Drossel * *Drosselspule* || Drosselspule
choke up verstopfen * *versperren (Rohr usw.)*
choked beengt * *zurückgedrängt, abgewürgt, zusammengedrängt*
chokes and transformers Wickelgüter * *Drosseln und Trafos*
choose auswählen || wählen
choosing Wahl
chop schneiden * *hacken* || zerkleinern * *{Holz}*
chop by half halbieren * *Zeitaufwand, Kosten usw.* ~
chop up zerkleinern * *{Holz}*
chopper Chopper || Zerhacker || Zerkleinerer * *z.B. zum Zerhacken von Abfallholz*
chopper converter Chopper-Umrichter * *mit nachgeschaltetem Gleichspannungssteller im Zwischenkreis*
chopper frequency Pulsfrequenz || Taktfrequenz * *eines Gleichstromstellers*
chosen gewählt
chromated chromatisiert

chrome Chrom * *chromic: Chrom-*
chromium Chrom
chromium-plated verchromt
chronological chronologisch
chronological order chronologische Reihenfolge
chuck Bohrfutter || Futter * *für Drehbank, Bohrmaschine usw.* || spannen * *Werkstück auf~/ein~* || Spannfutter * *bei Werkzeugmaschine* || Spannvorrichtung * *für Werkstücke* || Vorrichtung * *Vorrichtung z. Montieren, Spannvorrichtung*
chuck key Bohrfutterschlüssel
chucking appliance Spannvorrichtung * *für Werkstücke*
chucking device Spannvorrichtung * *{Werkz.-masch.}*
chucking tool Spannzeug
churn walken * *(durch-) schütteln, aufwühlen, buttern, sich heftig bewegen, schäumen*
churning Walkarbeit
churning loss Walkarbeit * *Verluste auf Grund der Walkarbeit*
churning power Walkleistung
churning work Walkarbeit * *z.B. in Lager, Reifen*
chute Gleitbahn * *Rutschbahn, Gleitbahn, Rutsche* || Rinne * *Rutsche* || Rutsche * *Rutsche; Rutsch-/Gleitbahn; Schüttröhre/-rinne; Abwurfschacht*
CIGRE CIGRE * *Conference Internationale des Grandes Réseaux Electriques: Intern. Hochspanngs.normenkonferenz*
CIM CIM * *Abkürzg. für: 'Computer Integrated Manufacturing': rechnerbasierte Fertigung* || Produktionsleittechnik * *Abkürzg. f. 'Computer-Integrated Manufacturing': rechnergestützte Fertigung*
cipher Schlüssel * *Chiffriercode*
circle Kreis * *auch Gruppe* || kreisen || Kreislauf
circle diagram Ortskurve
circle-cutting shear Kreisschere
circlip Sicherungsring * *Sprengring* || Sicherungsscheibe * *Sprengring* || Sprengring
circuit Beschaltung || Kreis * *Strom~, Schalt~* || Kreislauf * *primary/secondary: innerer/äußerer ~ (bei Kühlsystem usw.)* || Schaltung * *elektrische/elektronische ~* || Stromkreis
circuit angle Schaltungswinkel * *z.B. bei Transformator*
circuit arrangement Schaltung * *Ver~, ~sanordnung*
circuit board Leiterplatte
circuit breaker Leistungsschalter * *siehe auch →Hauptschalter* || Schalter * *Leistungsschalter* || Trennschalter * *zum Abschalten eines Starkstromkreises*
circuit breaker constant Schalterkonstante * *eines Leistungsschalters*
circuit concept Schaltungskonzept
circuit condition Schaltungszustand
circuit configuration Schaltung * *z.B. Brücken~*
circuit connection Schaltung * *~ für Stromrichter (siehe DIN V 41761), Schaltgruppe für Trafo usw.*
circuit design Schaltungsaufbau || Schaltungsauslegung || Schaltungsentwurf
circuit diagram Schaltbild || Schaltplan || Stromlaufplan * *z.B. für Leiterkarte*
circuit diagrams Schaltungsunterlagen
circuit dimensioning Schaltungsauslegung
circuit engineering Schaltungstechnik
circuit manual Schaltbuch * *auch für eine softwaremäßige Verdrahtung, z.B. in einem Regelsystem*
circuit modification Schaltungsänderung
circuit technique Schaltungstechnik
circuit technology Schaltkreistechnik
circuit-breaker Leistungsschalter * *(siehe z.B. DIN VDE 0660)*

circuitry Beschaltung || Schaltung * *Teil~, Be~*
circular kreisförmig || Ring- || ringförmig || rund * *(kreis-) rund, kreisförmig, Rund-, Kreis-, Ring-* || Rund-
circular buffer Ringpuffer || Ringspeicher * *Ringpuffer* || Umlaufpuffer
circular connector Rundstecker || Rundsteckverbinder
circular conveyor Kreisförderer
circular course Kreislauf
circular function Kreisfunktion
circular knife Kreismesser
circular knitting machine Rundstrickmaschine
circular letter Rundschreiben
circular measure Bogenmaß * *Maßeinheit rad (1 rad entspr. 360 Grad/2 Pi entspr. 57 Grad 17 min)*
circular resaw Spaltkreissäge * *{Holz}* || Trennkreissäge * *{Holz}*
circular rip saw Trennkreissäge * *{Holz}*
circular rotating field kreisförmiges Drehfeld * *{el.Masch.}sin.förm. magn.Feld entlang d.Ständerbohrg. m.konstant.Amplitude*
circular saw Kreissäge
circular saw bench Kreissäge || Tischkreissäge * *{Holz}*
circular saw blade Kreissägeblatt
circular saw motor Kreissägemotor * *in schlanker Bauform für große Schnitttiefe des Sägeblattes*
circular shear Kreisschere
circular-table milling machine Rundtisch-Fräsmaschine
circulate ausbreiten || kreisen * *zirkulieren* || laufen * *zirkulieren* || umgehen * *die Runde machen* || umlaufen * *z.B. Flüssigkeit, Dokument* || umwälzen
circulating Umlauf- || umlaufend || Umwälz- * *z.B. Pumpe, Ventilator*
circulating air Umluft
circulating buffer Umlaufspeicher
circulating current Kreisstrom
circulating current carrying kreisstrombehaftet || kreisstromführend
circulating current mode kreisstromführender Betrieb
circulating list Umlaufliste
circulating pump Umlaufpumpe || Umwälzpumpe
circulating stack Umlaufspeicher
circulating-air drier Umlufttrockner
circulating-current reactor Kreisstromdrossel
circulating-current-free kreisstromfrei
circulating-current-free antiparallel bridge connection kreisstromfreie Gegenparallelschaltung
circulation Ausbreitung || Kreislauf || Umlauf || Umlaufintegral || Umwälzung * *z.B. von Luft* || Zirkulation
circulation filter Umluftfilter
circulatory error Kreisformfehler * *{Werkz.-masch.}*
circumference Umfang * *~ eines Kreises*
circumferential speed Umfangsgeschwindigkeit
circumferential velocity Umfangsgeschwindigkeit
circumspection Sorgfalt * *Umsicht* || Vorsicht * *Umsicht*
circumstance Umstand * *in/under all circumstances: unter allen Umständen*
circumstances Sachverhalt || Zustand * *Umstände*
circumstantial umständlich * *Schilderung usw.*
circumvent umgehen * *ein Gesetz/eine Vorschrift ~*
CISC CISC * *'Complex Instruction Set Computer': Computer m.komplexem Befehlssatz (mächtige Befehle; →RISC)*
city Ort * *Stadt*
city tour Stadtrundfahrt
city-railway S-Bahn
civil engineering Bautechnik * *Hoch- und Tiefbau* || Bauwesen * *Tief- und Hochbau*

civil servant Beamter * *Berufsgruppe*
cladding Panzerung
claim Anspruch * *auch 'für s. in ~ nehmen'; auch Patent~; (to: auf); for damages: auf Schadeners.* || beanspruchen * *ein Recht ~* || behaupten * geltend machen || fordern * *rechtlich ~/beanspruchen* || Forderung * *Anpruch (for:auf)* || Reklamation || verlangen * *Anspruch erheben auf, beanspruchen*
claim damages Schadenersatz verlangen
claim for damages Schadenersatzanspruch || Schadensersatzanspruch
claim of damages Regressforderung || Schadenersatzforderung
claims Reklamationen || Schadenersatzanspruch * *(Plural)* || Schadensersatzanspruch * *(Plural)* || Schadensersatzansprüche
clam Greifer * *~ eines Baggers*
claming fixture Spannverbindung * *als Bau-/Konstruktionselement*
clamp abfangen * *z.B Kabel mit Hilfe einer Schelle festklemmen* || arretieren || befestigen * *mit Klammer(n)* || Bügel * *Klammer, Schelle* || Halter * Festklemmer || Halterung * *Klemme, Klammer* || Klammer * *Klammer, Krampe, Klemmschraube, Zwinge, Erdungsschelle* || Klemme * *(mechan.)* || klemmen || Presse * *(Holz) Formkasten~, Leim~, Schraubzwinge* || Schelle * *Klammer, (Erdungs-) Schelle, Krampe, Zwinge, Klemmschraube* || Spann- || Spannbügel || spannen * *Werkstück auf~/ein~* || Spannvorrichtung
clamp bolt Spannschraube
clamp collar Klemmring
clamp frame Spannrahmen * *(Holz)*
clamp strap Klemmbügel
clamp tight festspannen * *z.B. Werkstück in Schraubstock/Spannvorrichtung*
clamp-on anschellen * *Befestigung mittels einer Schelle*
clamping Anpress- * *(Druckerei)* || Befestigung * *mit Klammer(n)* || Klammerung || Klemmung || Spann- || Spannvorrichtung
clamping assembly Spannsatz * *z.B. zur Welle-/Nabebefestigung (bei Welle ohne Federkeil)*
clamping bolt Spannschraube
clamping bracket Spannbügel
clamping claw Spannpratze
clamping connection Spannverbindung
clamping device Arretierung || Spanneinrichtung * *→Spannvorrichtung* || Spannelement || Spannvorrichtung || Spannzeug * *→Spannvorrichtung*
clamping diode Klemmdiode
clamping element Spannelement * *z.B. zur Wellen-Nabeverbindung*
clamping force Verspannung * *Klemmkraft, Verklemmung*
clamping hub Klemmnabe
clamping kit Spannsatz
clamping plate Spannplatte * *z.B. zum Aufspannen von Werkstücken*
clamping power Spannkraft * *(Werz.masch.) bei einer Spannvorrichtung*
clamping ring Druckring * *übt axialen Druck auf ein e. Lager aus (siehe →angestelltes Lager)* || Klemmring || Spannring
clamping saddle Klemmbügel * *an einer Schraubklemme*
clamping screw Klemmschraube || Spannschraube * *kleine ~*
clamping shoe Spannpratze
clamping sleeve Spannhülse
clamping strap Klemmlasche
clamping washer Klemmscheibe
clamshell Greifer * *~ eines Baggers*
clandestine geheim * *heimlich, verborgen, verstohlen, unerlaubt*

clap klappen
clarification Aufgabenklärung * *(Ab)Klärung* || Klärung
clarify klären || klarstellen || reinigen * *Flüssigkeit* || setzen * *(Flüssigkeit)*
clarity Übersichtlichkeit * *Klarheit*
clash kollidieren * *(figürl.)* || Kollision * *von Interessen* || Konflikt * *Streitigkeit, (feindlicher) Zusammenstoß, Zusammenprall, Widerstreit, Streitigkeit*
clashes Konflikt * *Streitigkeiten*
clasp Beschlag * *Klammer, Haken, Schnalle, Spange, Schließe, Schloss; Umklammerung* || umschlingen
class -Klasse || Art * *Klasse, Gruppe* || Baureihe || bezeichnen * *klassifizieren* || einordnen * *in Klasse(n) ~ (with/as:unter)* || Grad * *z.B. Klasse, Funkstörgrad* || Gruppe * *Klasse* || kennzeichnen * *klassifizieren, z.B. in Schärfegrade beim Funkstörgrad* || Klasse * *auch 'Typdeklaration' in der objektorientierten Programmierung (z.B. bei C)* || klassifizieren * *z.B. in Schärfegrade beim Funkstörgrad* || Rubrik * *Klasse, Gruppe*
class B insulation Isolierstoffklasse B * *Isolierung entsprechend Isolierstoffklasse B*
class of climate Klimagruppe * *Klassifizierung v. natürl. Umwelteinflüssen (Temp., Luftfeuchte... nach IEC721-2-1)*
class of inflammable gases and vapours Explosionsgruppe * *kennzeichn. Gase ähnlich. Zünddurchschlagvermögens (EN50014)*
class of insulation system Isolationsklasse || Isolierstoffklasse
class of rating Bemessungsklasse * *(el. Masch.)*
class with zuordnen * *einordnen/einstufen bei, gleichstellen mit*
classic klassisch * *klassisch, althergebracht, anerkannt* || konventionell * *klassisch*
classical klassisch * *konventionell, üblich* || konventionell * *klassisch, althergebracht*
classification -Klasse || Abstufung * *Aufteilung in Klassen* || Aufteilung * *in Klassen* || Einteilung * *z.B. in Klassen* || Gruppierung * *Aufteilung in Klassen* || Klasse * *Klassifizierung, z.B.in einer Norm* || Klassifizierung || Sortierung || Staffel * *Aufteilung in Klassen* || Systematik * *Einteilung in Klassen/Gruppen* || Zuordnung * *Einteilung; Klassifizierung*
classified abgestuft * *in Klassen eingeteilt* || gegliedert * *in Gruppen/Klassen eingeteilt* || geheim * *classified matter; ~e Dienstsache* || unterteilt * *in Klassen*
classify abstufen * *in Klassen einteilen* || einordnen * *in Klasse(n) ~ (as:unter)* || einteilen * *in Klassen ~* || gliedern * *in Gruppen/Klassen aufteilen* || klassifizieren || sortieren || staffeln * *in Klassen einteilen* || systematisieren * *in Klassen/Gruppen einteilen* || unterteilen * *in Gruppen/Klassen ~* || zusammenstellen * *nach Klassen ~*
classifying plant Sortieranlage
clatter rattern
clause Absatz * *kurzer Abschnitt in einem Text* || Vereinbarung * *Klausel*
claw Greifer * *Klaue, Kralle; siehe auch "Klaue"* || Klaue * *Kralle* || Pratze * *Klaue, (Greif-) Haken*
claw clutch Klauenkupplung
claw coupling Klauenkupplung
claw-coupling half Kupplungsstern
clean aufgeräumt * *ordentlich, sauber* || blank * *sauber* || leer * *glatt, unbeschrieben, glatt* || putzen * *sauber* || reinigen * *sauber* || säubern
clean room Reinraum
clean up säubern
cleaner Reinigungsmittel
cleaning Reinigung

cleaning agent Reinigungsmittel
cleaning machine Reinigungsmaschine
cleaning roller Ausputzwalze * *{Textil} z.b.* bei →*Kratzenraumaschine* || Bürstwalze * *{Textil}* bei →*Kratzenraumaschine* || Putzwalze * *{Textil} z.B. bei Kratzenraumaschine*
cleaning waste Putzwolle
cleanliness Reinheit || Sauberkeit
cleanly sauber * *(Adv.)*
cleanness Reinheit
cleanse reinigen * *(auch figürl.)* || säubern
cleansing Reinigung
cleansing agent Reinigungsmittel
cleansing machine Waschmaschine * *Reinigungsmaschine*
clear abbauen * *eine Verbindung* ~ || abtrennen * *a connection: eine Nachrichtenverbindung* || anschaulich * →*deutlich, übersichtlich,* →*klar, eindeutig* || ausleeren || ausräumen || deutlich * *klar* || durchsichtig * *Farbe* || eindeutig * *klar, deutlich; clearly identified:eindeutig gekennzeichnet* || einfach * *klar* || einleuchtend || frei * *Straße, von Schulden usw.; swing clearly:frei schwingen* || klar || klären || laut * *klar, bestimmt* || leeren * *(weg-) räumen* || leicht verständlich * *klar, einfach* || löschen * *z.B. Flipflop, Anzeigebit in Rechner* ~ || netto || passieren * *einen bestimmten (Kontroll-/Etappen-) Punkt* ~ || quittieren * *Fehler usw.*
clear away aufräumen * *wegräumen, wegschaffen* || beseitigen * *Hindernisse, Fehler usw.* ~
clear button Löschtaste
clear dimension lichte Weite
clear of -frei * *(vorangestellt) von Verunreinigungen usw.*
clear off aufräumen * *ausräumen, reinigen; beseitigen, loswerden, erledigen* || ausräumen * *Waren usw.* ~
clear the line auflegen * *Telefon* ~
clear the port auslaufen * *{Schifffahrt} aus dem Hafen* ~
clear up ausräumen * *(figürl.) Missverständnis usw.* ~ || klären
clearance Abstand * *lichter Raum, Zwischenraum, Spiel(raum)* || Bereinigung * *Räumung, Beseitigung, Leerung* || Freiraum * *unobstructed: unbehindert; z.b.erforderlich unter- u.oberhalb e. Stromrichtergeräts* || Lagerluft * *Spiel* || Lagerspiel * *(allg.)* || Luftstrecke || Spiel * *freier Abstand, z.B. Ventil~, Lager~* || Spielraum * *räumlich* || Zahlung * ~ *einer Schuld* || Zwischenraum * *lichter/freier Raum, Zwischenraum, Freilegung, Räumung*
clearance distance Luftstrecke * *für Isolation (bei Stromrichtern: siehe IEC 664)*
clearance in air Luftabstand
clearing Abrechnung * *Abrechnungs/Bankverkehr* || Verrechnung * *gegenseitige* ~ *(im ~sverkehr)*
clearing capacity Ausschaltvermögen
clearing time Abschaltzeit
clearing up Klärung
clearly arranged übersichtlich * ~ *angeordnet*
clearly structured klar strukturiert
clearness Reinheit || Übersichtlichkeit
cleat befestigen * *mit Klammer(n)* || Schelle * *auch Isolierschelle*
cleave aufspalten || spalten * *{Holz}*
cleaving saw Spaltsäge * *{Holz}*
cleft Riss || Schlitz * *Spalt* || Spalt * *Riss*
clerk Angestellter * *Büro~* || Verkäufer * *(amerikan.)*
clever ausgeklügelt || fähig * *gewitzt, gescheit* || findig || geschäftstüchtig * *gewieft* || geschickt * *gewand, geschickt, raffiniert, klug, begabt* || pfiffig * *geschickt, raffiniert, klug, gescheit, begabt* || raffiniert * *klug, gescheit, geschickt, pfiffig, begabt* || tüchtig * *geschickt, gewitzt*

cleverly geschickt * *(Adv.) geschickt ausgedacht, genial*
cliché Druckplatte * *{Druckerei} Druckstock, Bildstock, Klischee*
click knacken * *metallisch* || schnappen * *einrasten/schnappen*
click into place einrasten
click-on anklicken * *mit der Computermaus*
client Abnehmer || Auftraggeber * *Kunde* || Client * *(Personal)Computer, d. als Anwenderstation i.e. 'Client-Server-Netzwerk' dient (s. 'Server')* || Kunde * *für Dienstleistung*
client service Kundendienst
client-conscious attitude Kundenorientierung * *als geistige Einstellung*
clientele Kundenkreis
climate Klima
climate classification Klimagruppe * *siehe z.B. IEC-Publikation 721-2-1*
climate group Klimagruppe
climate package Klimaverpackung * *siehe auch* →*Tropenschutzverpackung* || Tropenschutzverpackung * *Klimaverpackung*
climate-proof klimabeständig || klimafest
climatic klimatisch
climatic categorie Klimakategorie * *z.B. nach IEC*
climatic category Klimaklasse * *z.B. nach IEC*
climatic conditions klimatische Bedingungen
climatic sequence Klimafolge
climatic zone Klimazone
climax Steigerung * *auch 'Gipfel, Höhepunkt, Krisis'*
climb aufsteigen || klimmen, auch Flugzeug || steigen * *klimmen, auf/hochsteigen*
climb over übersteigen
clinch einklinken || einrasten || nieten * *ver~, sicher befestigen* || vernieten
cling haften * *(an-) kleben; to: an*
clinging eng * ~ *anliegend*
clinic roll cutter Fehlrollenschneider * *{Papier}*
clinker sintern
clinkered gesintert
clip abgraten * *Gussnaht* || abschneiden || ausschneiden || beschränken * *beschneiden* || Clip || entgraten * *Gussnaht* || Halter * *Festhalteklammer, Festklemmer* || Klaue * *Klammer, Klemme, Halter, Spange* || Klemmbügel || Klemmnase || Raste * *Klammer, Klemme, Spange, Halter* || Rastnase * *Klammer, Klemme, Spange, Halter* || Schelle * *zum Befestigen* || schneiden * *mit Zange, Schere, Maschine (ab)schneiden/-hauen* || Spannbügel || Stutzen * *(Verb)* abschneiden || verkürzen * *beschneiden*
clip nut unverlierbare Mutter * *mit Montage-/Halteklammer*
clip on aufstecken * *rastend*
clip-on Ansteck- || Schnapp- || Aufsteck- * *z.B. clip-on heatsink: ~-Kühlkörper*
clip-on fixing Aufschnappmontage
clipboard Zwischenablage * *{Computer} z.B. in WINDOWS*
clipper Schere * *Abschneider, Schermaschine*
cloak verdecken * *bemänteln*
clock Takt || takten || Uhrtakt
clock distributor Zeittaktverteiler
clock frequency Taktfrequenz
clock generator Taktgeber || Taktgenerator
clock grid Taktraster
clock pulse Takt * ~-*impuls*
clock pulse generator Taktgeber
clock rate Taktfrequenz
clock signal Taktsignal
clock speed Taktfrequenz * *z.B. eines Mikroprozessors, einer getakteten Schaltung* || Taktfrequenz * *{Computer}*

clock supply Taktversorgung
clock tick Systemtakt
clock time Taktzeit
clock unit Zeitgeber * *Taktgeber*
clocked getaktet
clocking Taktung * *Taktsignale erzeugend/verarbeitend*
clockwise im Uhrzeigersinn || mit dem Uhrzeigersinn || rechts * *im Uhrzeigersinn (Drehrichtung usw.)* || rechts herum * *mit dem Uhrzeigersinn* || Rechts- * *im Urzeigersinn*
clockwise direction Uhrzeigersinn * *clockwise: im Uhrzeigersinn; counter/anticlockwise: entgegen d. Uhrzeigersinn*
clockwise direction of rotation Rechtsdrehrichtung * *im Uhrzeigersinn*
clockwise phase sequence Rechtsdrehfeld
clockwise rotating Rechtslauf
clockwise rotating field Rechtsdrehfeld
clockwise rotation Rechtslauf || Vorwärtsdrehrichtung * *Rechtslauf (im Uhrzeigersinn)*
clockwise running Rechtslauf
clog verstopfen * *versperren, sich zusetzen* || zusetzen * →*verstopfen*
clogged up verstopft * *z.B.Filter*
clogging Verstopfung
close abschließen * *zum Ende bringen* || beenden * *(ab)schließen* || beschließen * *abschließen* || Beschluss * *Ende, Beendigung, Abschluss* || einschalten * *Kontakt schließen* || eng * *dicht, nah; close-tolerance: ~ toleriert* || nah * *dicht (to:an/am/bei)* || nahe * *(to:an)* || schließen * *auch Bremse, Kontakt, Datei usw.* ~ || Schluss * *(Ab-) Schluss* || verschließen || zu * *Befehl/Kommando*
close by dabei * *nahebei* || neben * *dicht ~*
close by it daneben * *dicht ~*
close contact Reibschluss
close in nähern * *bedrohlich näherkommen*
close to bei * *(räumlich)* || nächster * *(Präposition)*
close to practice praxisnah
close to the market marktnah
close to zero nahe Null
close together ähnlich * *dicht beieinander*
close tolerances enge Toleranzen
close up aufrücken || aufschließen * *zum Vordermann ~, einholen*
close version geschlossene Ausführung
close-quartered räumlich beengt
close-up Nahaufnahme * *Foto, Film*
close-up view Nahansicht
closed geschlossen * *auch Schaltkontakt* || verschlossen || zu * *geschlossen*
closed circuit geschlossener Kreislauf
closed circuit of the heat exchanger geschlossener Kühlmittelkreislauf
closed circulation geschlossener Kreislauf
closed contact geschlossener Kontakt
closed control loop geschlossener Regelkreis
closed coolant circuit geschlossener Kühlmittelkreislauf * *siehe auch →Kühlmittel*
closed linkage Kraftschluss
closed loop tension control direkte Zugregelung * *mit Zugerfassung durch Tänzer oder Zugmessdose*
closed loop tension regulation direkte Zugregelung
closed machine geschlossene Maschine
closed on all sides allseitig geschlossen
closed-circuit geschlossen * *Kreislauf-*
closed-circuit cooled kreislaufgekühlt * *allg.*
closed-circuit cooler Kreislaufkühler
closed-circuit cooling Kreislaufkühlung * *Motorkühlart m. geschloss. Primär-Kühlungskreislauf (Kühlart IC 81/85/86U)*

closed-circuit current Ruhestrom * *z.B. Bremse, Magnetventil*
closed-circuit current principle Ruhestromprinzip * *z.B. Bremse: zum Lüften Strom erforderlich*
closed-circuit principle Ruhestromprinzip * *elektromechan. Gerät geht in sicheren Zustand wenn kein Steuerstrom fließt*
closed-circuit self-ventilated with air-to-air heat exchanger kreislaufgekühlt durch Eigenkühlung mittels Luft-Luft-Kühler * *Motor-Kühlart ICA01 A61 nach IEC34.6*
closed-circuit vacuum pump Kreislauf-Vakuumpumpe
closed-circuit ventilated kreislaufgekühlt * *geschlossener Kühlkreislauf mit Luftkühlung*
closed-circuit ventilation Kreislaufkühlung * *z.B. Motorkühlart mit geschlossenem Luft-Kühlkreislauf*
closed-circulation vollkommen geschlossen * *~es Kühlsystem, ~er Kühlkreislauf*
closed-loop vollkommen geschlossen * *~es Kühlsystem, ~er Kühlkreislauf*
closed-loop and open-loop control tasks Regel- und Steueraufgaben
closed-loop armature current control Ankerstromregelung
closed-loop control Regeleinrichtung || Regelsystem || Regelung
closed-loop control circuit geschlossener Regelkreis || Regelkreis || Regelschleife * *Regelkreis*
closed-loop control device Regeleinrichtung
closed-loop control electronics Regelelektronik
closed-loop control electronics board Regelbaugruppe * *Leiterkarte*
closed-loop control equipment Regeleinrichtung
closed-loop control function Regelfunktion
closed-loop control function block Regelungsbaustein
closed-loop control functions Regelfunktionen
closed-loop control model Regelkonzept * *Regelmodell*
closed-loop control principle Regelprinzip
closed-loop control section Regelteil
closed-loop control stability Regelkonstanz
closed-loop control structure Regelstruktur
closed-loop control system Regelsystem || Regelungssystem
closed-loop control task Regelaufgabe
closed-loop control unit Regelteil
closed-loop controlled geregelt
closed-loop controlled field geregeltes Feld * *mit Feldstromrückführung*
closed-loop controller Regelgerät || Regler
closed-loop controls Regelungstechnik
closed-loop current control Stromregelung
closed-loop dancer roll position control Tänzerageregelung
closed-loop drive control system Antriebsregelungssystem
closed-loop field current control Feldstromregelung
closed-loop frequency control Frequenzregelung
closed-loop gain Kreisverstärkung
closed-loop performance Regeleigenschaften || Regelungseigenschaften || Regelverhalten * *Regelgüte*
closed-loop position control Wegregelung
closed-loop power control Leistungsregelung
closed-loop response Regelverhalten
closed-loop speed control Drehzahlregelung
closed-loop structure Regelstruktur
closed-loop system Regeleinrichtung
closed-loop temperature control Temperaturregelung
closed-loop tension control Zugregelung
closed-loop-controlled drive geregelter Antrieb

closely calculated scharf kalkuliert
closely meshed eng verzahnt * *(figürl.)*
closely related eng zusammenhängend
closeness Dichtheit * *Dichte, Festigkeit, Knappheit, Verschlossenheit, Enge*
closer näher
closing Abschluss * *Beendigung*
closing delay Einschaltverzögerung * *Schalter, Kontakt*
closing machine Verschließmaschine * *{Verpack.-techn.}*
closing shape Formschluss
closing time Ansprechzeit * *eines Schließerkontakts*
closure Verschluss * *~(vorrichtung)*
closure machine Verschließmaschine * *{Verpack.-techn.}*
closure plug Verschlussschraube
cloth Gewebe * *Tuch, Stoff, Leinwand {Buchbinderei}, Lappen*
cloth wheel Schwabbelscheibe * *mit Textilbelag zum Hochglanzpolieren*
cloth-raising machine Kratzenraumaschine * *{Textil} zum Aufrauen von Textilstoffen mit Kratzwalzen*
clothes drier Wäschetrockner
clothes dryer Wäschetrockner * *(kann mit "i" oder "y" geschrieben werden)*
clothes-horse Wäschetrockner
clothing industry Bekleidungsindustrie
cloudiness Trübung
cloudy trüb * *Flüssigkeit usw.*
clumsy steif * *linkisch*
cluster Cluster * *kleinste logische Einheit auf der Festplatte (1 Festplatte hat unter MS-DOS 65536 Cluster)* || Haufen * *Ansammlung, Haufen, Menge, Schwarm, Gruppe*
clutch erfassen * *(fest) packen; (mechanisch)* || Griff * *klammernder ~ (at:nach)* || Kupplung * *betriebsmäßig ein-/ausrückbare ~ (z.B. Scheiben~ mit Reibbelag)* || packen * *fassen, greifen*
clutch disc Kupplungslamelle
clutch disk Kupplungslamelle
clutch facing Kupplungsbelag
clutch housing Kupplungsgehäuse
clutch lining Kupplungsbelag
clutch outdrive Kupplungsabtrieb
clutch plate Kupplungslamelle
clutch pressure plate Kupplungs-Druckplatte
clutch ring Kupplungsring
clutch shaft Kupplungswelle
clutch-brake unit Kupplungs-Bremskombination
clutch/brake unit Kupplungs-Bremskombination
clutch/brake combination Kupplungs-Brems-Kombination
clutch/brake combined unit Kupplungs-Bremskombination
clutter überhäufen * *vollstopfen, überhäufen, anfüllen*
cm cm * *Zentimeter; 1 cm entspr. 0,39370l in (inch)*
CMR Gleichtaktunterdrückung * *Abk. für 'Common-Mode Rejection'*
CNC CNC * *{Werkz.masch} Abk. für "Computer Numerical Controller": Computergestützte numerische Steuerung*
CNC control CNC-Steuerung * *Abk. f. 'Computerized Numeric Control': numerische Steuerung (Bewegungssteuerung)*
CNC interpreter CNC-Interpreter * *interpretiert Programme für Bewegungssteuerungen, z.B. nach DIN 66025*
CNC-controlled CNC-gesteuert
co- Mit- * *z.b. co-operation: ~arbeit, co-worker: ~arbeiter* || Zusammen- * *gemeinsam mit*

co-operate entgegenkommen * *mitwirken, beitragen, helfen; to:an/bei/zu* || zusammenarbeiten || zusammenwirken
co-operation Mithilfe || Verbund * *Zusammenarbeit* || zusammenwirken * *(Substantiv)*
co-operative entgegenkommend * *(figürl.)* || hilfsbereit
co-ordination Koordination || Koordinierung
co-worker Kollege * *Mitarbeiter* || Mitarbeiter
Co. Fa * *Company: Firma*
coach Ausbilder * *Trainer, einarbeitender Kollege, Einpauker* || einarbeiten * *jdm. einarbeiten, ausbilden, trainieren, instruieren (meist durch älteren Kollegen)* || Wagen * *{Bahnen}*
coaching Einarbeitung * *von einem erfahrenen Kollegen* || Einweisung * *jdm. einarbeiten, trainieren, ausbilden, instruieren (meist durch älteren Kollegen)* || Schulung * *Einarbeitung durch einen erfahrenen Kollegen*
coal Kohle
coal and iron mining industry Montanindustrie
coal and steel industry Montanindustrie
coal power station Kohlekraftwerk
coarse grob * *auch für Einstell/Justierelement, Schleifmittel* || rau * *grob* || roh * *grob*
coarse adjust Grobabgleich
coarse adjustment Grobabgleich || Grobeinstellung || Grobjustierung
coarse chips Hackschnitzel * *{Holz}*
coarse control Grobregelung
coarse pulse Grobimpuls * *z.B. zur Positionserfassung*
coarse setting Grobabgleich || Grobeinstellung * *z.B. an einem Potentiometer*
coarse tuning Grobabgleich
coast anhalten * *austrudeln, im Leerlauf fahren* || auslaufen * *→trudeln* || im Leerlauf fahren || Leerlauf * *im ~ fahren* || leerlaufen * *im Leerlauf fahren, trudeln, ungeführt/ohne treibendes Moment fahren* || nachlaufen * *im Leerlauf/Freilauf fahren (ohne Antriebsdrehmoment)* || trudeln
coast down auslaufen * *austrudeln (Maschine, Motor)* || austrudeln
coast freely trudeln
coast stop aus * *Kommando z.spannungsfrei machen u. Austrudeln eines Antriebs (AUS2 im →PROFIBUS-Profil)* || AUS2 * *Steuerbit im →PROFIBUS-Profil: Motor mögl.schnell spannungslos mach., Antrieb trudelt aus*
coast to rest austrudeln
coast to standstill auslaufen * *bis zum Stillstand austrudeln (Maschine, Motor)*
coast to stop austrudeln * *bis zum Stillstand*
coaster Fahrgeschäft * *Achterbahn usw.* || Karussel * *Fahrgeschäft (Achterbahn usw.)*
coasting Freilauf * *das Freilaufen, Trudeln, ungeführter/nicht angetriebener Lauf* || freilaufend || leerlaufend
coat anstreichen || Anstrich || Auflage * *Anstrich, Beschichtung* || auftragen * *beschichten. eine Schicht auftragen* || Belag * *Überzug* || belegen * *mit (Schutz-) Überzug usw. ~* || beschichten || Schicht * *Beschichtung, Auflage* || Überzug || umhüllen
coat of paint Farbanstrich
coat on aufbringen * *Farbe, Auftrag, Beschichtung*
coated beschichtet || gestrichen * *Papier*
coated paper gestrichenes Papier
coated weight Strichgewicht * *{Papier} Flächengewicht des Auftrages durch eine →Streichmaschine*
coater Auftragswerk * *bei Streichmaschine* || Streichmaschine * *zur Beschichtung von Papier oder Folien*
coating Anstrich || Belag || Überzug, Auftrag || Be-

coating machine 448

schichtung || Farbe * *Lack, Beschichtung* || *Lack* || *Überzug* || *Umhüllung*
coating machine   Beschichtungsmaschine || Streichmaschine * *Beschichtungsmaschine für Papier oder Folien*
coating paste   Streichpaste * *(Papier) zum Streichen des Papiers in der →Streichmaschine*
coating plant   Beschichtungsanlage || Streichanlage
coating roller   Auftragswalze * *in (Papier-) Streichmaschine, Folienbeschichtungsmaschine usw.*
coax cable   Koaxialkabel || Koaxialleitung
coaxial   koaxial * *Kabel, Buchse usw.*
coaxial cable   Koaxialkabel || Koaxialleitung
cobald-samarium   Kobald-Samarium * *permanentmagnet. Material, z.B. für Elektromotor*
cobwebbing   Spinnwebverfahren
cock   Absperrhahn
cock-eyed   schief * *(salopp) nach einer Seite hängend, schielend, doof, betrunken*
cockpit   Kabine * *Führerraum eines Fahrzeugs*
cocoon   spinnen * *(sich) ein~, einhüllen (figürl.) (Gerät usw.) einmotten/außer Betrieb setzen*
cocooning   Spinnwebverfahren
codable   codierbar
code   Abkürzung * *Kode(nummer); siehe auch 'Kurzbezeichnung'* || Code * *1-out-of-8 code: 1-aus-8 Code* || Gesetz * *~buch, Vorschriftenbuch* || Kennbuchstabe || Kennung || Kennzeichen * *Kurzbezeichnung* || Kennziffer || kodieren * *Programmcode erstellen* || Kurzangabe * *Kodebuchstaben/-nummern* || Kurzbezeichnung * *Kode(nummer)* || Kurzzeichen || Schlüssel * *Code(Wort)*
code conversion   Umcodierung || Umkodierung
code converting   Umkodierung
code designation   Kennzeichen * *Kurzbezeichnung*
code digit   Kennziffer
code disc   Schlitzscheibe * *z.B. im Gray-Code (einschrittiger Code) kodiert bei →Absolutwertgeber*
code generation   Übersetzung * *Erzeugung von Programmcode*
code generator   Codegenerator * *erzeugt z.B. Hochsprache- oder Assemblerquellcode z.B. aus einer graf. Darstellung*
code letter   Kennbuchstabe
code No.   Kennziffer
code number   Identnummer || Kennzahl || Kennziffer
code of identification   Kennzeichnung * *~ durch Kennbuchstaben, Ziffern, Balken-/Farbcode usw.*
code page   Zeichensatz * *Codeseite: länderspezif. ASCII-Zeichentabelle z. Bildschirmdarstellung i. PC b. MS-DOS (R)*
code system   Kennzeichnungssystem * *mit Kurzbezeichnungen*
CODEC   CODEC * *COmpressor/DECompressor: Einrichtg. z. Echtzeit-Komprimieren/Dekomprimieren v. Videosignalen*
coded   kodiert
coder   Kodierer * *Programmierer, der Programmcode erstellt* || Programmierer * *Hifskraft, Kodierer*
coding   Codier- || Codierung || Kodierung
coding element   Kodierelement * *z.B. bei vertauschungssicherem Stecker*
coding pin   Codierzapfen * *an einem unverwechselbaren Steckverbinder*
coding plug   Codierstecker
coding switch   Codierschalter
coding system   Nummernsystem * *z.B. für Bestellnummern*
coefficient   -Grad || Beiwert * *Koeffizient* || Kennziffer * *Koeffizient, Verhältniszahl* || Koeffizient * *z.B. coefficient of friction: Reibzahl* || Wert * *Koeffizient*
coefficient notation   Koeffizientenschreibweise

coefficient of elasticity   Elastizitätskoeffizient
coefficient of friction   Reibungskoeffizient || Reibungszahl || Reibwert || Reibzahl
coefficient of self-induction   Selbstinduktivität
coefficient of thermal expansion   Wärmeausdehnungskoeffizient
coeffidient of   -Zahl * *Koeffizient, Konstante*
coercive   koerzitiv || Koerzitäts- * *~feldstärke, ~kraft usw*
coercivity   Koerzivitäts-
coffee break   Kaffeepause
cog   Ritzel || unrund laufen * *ruckhafter Lauf bei kleinen Drehzahlen*
cog rack   Zahnstange
cog rail   Zahnschiene
cog wheel   Kettenrad * *Ritzel* || Ritzel
cog-wheel   Zahnrad * *kleines Zahnrad, Ritzel*
cog-wheel drive   Zahnradantrieb
cogeneration plant   Kraft-Wärme-Kopplungsanlage
cogged   gezahnt
cogging   ruckartig * *unrund/ruckhaft laufend* || ruckhaft * *unrund/ruckhaft laufend* || unrunder Lauf * *ruckhafter Lauf bei kleinen Drehzahlen*
cogging torque   Nutrasten * *Drehmomentenwelligkeit eines Motors auf Grund der Polfühligkeit* || Rastmoment * *ruckartiges Drehmoment (z.B. Nutrasten auf Grund der Polfühligkeit eines Motors)*
coherence   Kohärenz * *Zusammengehörigkeit in Ort und/oder Zeit* || Zusammenhang
coherent   zusammenhängend * *auch Gedanken, Rede*
cohesion   Kohäsion
cohesive force   Kohäsionskraft
cohesiveness   Kohäsionskraft
coil   aufwickeln || Ausgang * *in Kontaktplandarstellung eines SPS-Programms* || Bund * *aufgewickeltes Material (Draht, Blech, Seil usw.)*, Blechwickel, Spule || Magnet * *Betätigungsspule* || Ring * *aufgewickeltes Material, Bund (Draht, Kabel, Blech, Seil usw.)* || Rolle * *Draht, Kabel, Seil* || Spule * *Haspel; Betätigungs~ (z.B. Relais, Schütz), ~ einer (Motor)Wicklung* || Wickel * *Bund* || wickeln * *Metall, Draht, Kabel, Seil ~* || Windung * *Blech, Draht, Kabel, Seil*
coil body   Magnetkörper * *Spulenkörper*
coil car   Bundwagen
coil conveyor   Bundabtransport * *Fördereinrichtung im Walzwerk*
coil diameter   Wickeldurchmesser
coil drive   Spulantrieb || Wickelantrieb * *Spulerantrieb, Haspelantrieb (für Blech, Draht usw.)*
coil end   Spulenkopf * *(el.Masch.) aus dem Bleckpaket herausragender Teil einer Spule* || Spulkopf * *z.B. bei Aufwickler (für Fäden, Draht usw.)*
coil group   Spulengruppe * *z.B. in einer Motorwicklung*
coil machine   Spulmaschine
coil pitch   Spulenweite * *(el.masch.) Ausdehnung einer Spule i. Verhältn. z. Umfang der Ständerbohrg.*
coil removal   Bundabtransport * *Funktion im Walzwerk*
coil side   Spulenseite * *(el. Masch.) in der Nut eines Blechpakets liegender Teil einer Spule*
coil spring   Rollbandfeder * *z.B. zum Niederhalten einer Kohlebürste* || Rollfeder || Spiralfeder
coil up   aufwickeln * *Blech, Draht, Kabel usw.*
coil winding   Spulenwicklung
coil winding machine   Spulenwickelmaschine * *für Magnetspulen*
coil-end bracing   Wickelkopfabstützung * *(el. Masch.) mechan. Versteifung d. →Wickelkopfs z.B. bei Hochspannungsmotoren*

coil-spring holder Rollbandfederträger * z.B. z. Bürsten-Niederhalten in Gleichstrommaschine
coiled component Wickelgut * z.b. Spule, Trafo
coiled components Wickelgüter
coiled product Wickelgut * z.b. Spule, Drossel, Transformator, aufgewickeltes Blech
coiled products Wickelgüter * z.b. Spulen, Drosseln, Trafos, aufgewickeltes Blech
coiler Aufhaspel * für Blech, Draht, Kabel || Aufwickelhaspel || Aufwickler * Aufhaspel für Blech, Draht, Kabel || Aufwicklung * Aufhaspel || Haspel || Spuler * ~ für Blechband, Draht usw.
coiler drive Haspelantrieb
coiler unit Haspelanlage
coin prägen * Metall, Münzen, Wörter, Begriffe usw. ~
coincide entsprechen * übereinstimmen, sich decken (mit), etwas genau ~ || übereinstimmen * örtlich/zeitlich zusammenfallen/treffen, sich decken/ genau entsprechen (with:mit) || zusammentreffen * zeitlich ~, ~ von Umständen
coincidence Gleichzeitigkeit || Kollision * zeitliche || Übereinstimmung * Zusammentreffen (räumlich oder zeitlich), Übereinstimmung, Gleichzeitigkeit || zusammentreffen * (Substantiv) zeitlich~, ~ von Umständen
coincidence factor Gleichzeitigkeitsfaktor
coincident gleichzeitig * zusammenfallend || zusammen * ~fallend
coke Koks
coke drum Kokstrommel
coke furnace Koksofen
coke oven Koksofen
cold kalt || Kälte
cold extruded part Kaltflusspressteil
cold press Kaltpresse
cold production Kälteerzeugung
cold resistance Kaltwiderstand * z.B. Lampe (ist wesentlich niedriger als der Warmwiderstand)
cold restart Erstanlauf || Kaltstart * kompletter Neuanlauf (eines Computers usw.) || Neustart * Erstanlauf
cold rolled kaltgewalzt * z.B. Blechband
cold rolling Kaltwalzen
cold rolling mill Kaltwalzwerk
cold start Kaltstart
cold-drawn steel Blankstahl * {Draht}
collaborate zusammenarbeiten * mitarbeiten, zusammenarbeiten, behilflich sein || zusammenwirken
collaboration agreement Kooperationsvertrag
collaborator Mitarbeiter
collapse brechen * zusammen~, kollabieren || einbrechen * kollabieren, zusammenbrechen || Sturz * ~ von Kursen Preisen usw. || zusammenbrechen || Zusammenbruch * völliger ~
collapsible Falt- * faltbar, zusammenlegbar || faltbar || Klapp- * zusammenklappbar || klappbar * zusammenklappbar || zerlegbar * (zusammen-) klappbar, Klapp-, Falt-
collapsible tube Tube
collar Bund * z.B. einer Welle || Manschette * Kragen || Rand * Kragen || Ring * Kragen
collar nut Wellenmutter
collate kommissionieren * zusammenstellen || vergleichen * Texte usw. ~ (with: mit)
colleague Kollege || Mitarbeiter
collect abholen || auffangen || eintreiben * Schulden, Steuern ~ || erarbeiten * zusammentragen || erfassen * ein-/aufsammeln || sammeln || zusammenfassen * sammeln; auch Gedanken
collection abholen * (Substantiv) || Ansammlung || Sammlung * Gesammeltes, Zusammengestelltes, Auswahl || Vorrat * (An-) Sammlung, Kollektion, Auswahl, Anhäufung

collection of overhead slides Foliensatz
collective gemeinsam * kollektiv, zusammengenommen || Sammel-
collective agreement Tarifvertrag * (amerikan.)
collective fault signal Sammelstörmeldung
collector Kollektor * Anschluss eines Transistors
collector connection Kollektorschaltung
collector ring Schleifring
college Fachhochschule
college of technology Technische Hochschule
collegiate Universität * Universitäts-, akademisch
collet Klemmring * Klemmhülse, Klemmring, Kragen, Konushülse, Spannhülse || Spannzange
collet chuck Spannzangenfutter * z.b. bei Werkzeugmaschine
collet-type mounting clamp Spannsatz * zur Welle-Nabe-Verbindung über Klemmhülse (z.b. für Impulsgeberanbau)
collide kollidieren * kollidieren, zusammenstoßen, im Widerspruch stehen (with: mit/zu)
collision Kollision * auch von Interessen, Interrupts, Datenzugriffen || Konflikt * Widerspruch, Gegensatz, Konflikt, Zusammenstoß
collision of two timer interrupts Weckfehler
collision-free kollisionsfrei * z.b. Speicherzugriff von 2 Teilnehmern
colloquium Kolloquium
colloquy Gespräch * gelehrtes ~, Kolloquium
colon Doppelpunkt
color Farbe || färben
color code Farbkennzeichnung
color coding system Farbkennzeichnung * Farbkennzeichnungssystem
color depth Farbtiefe * bei Monitor; z.B. ~ 8 bit - > 256 Farben darstellbar (16/24Bit: High/True Color)
color graphics Farbgrafik
color monitor Farbbildschirm || Farbmonitor
color print Mehrfarbendruck * (amerikan.)
color structuring Farbgebung * z.B. Verwendung v. Farben am Bildschirm
color VDU Farbbildschirm * Abk. f. 'Video Display Unit': Prozessvisualisierungsstation
colossal riesig
colour färben
colour graphics Farbgrafik
colour monitor Farbmonitor
colour print Mehrfarbendruck * (brit.)
coloured gefärbt
column Rubrik * Spalte || Säule || Spalte * in Text/ Tabelle, auf Bildschirm (siehe auch 'Spalt') || Ständer * Säule, Pfeiler
column-type Säulen- || Ständer- * {Werkz.masch.}
column-type drilling machine Säulenbohrmaschine || Ständerbohrmaschine
comb Kamm
combination Kombination || Mischung || Paarung || Zusammenstellung
combination pliers Kombizange
combination possibilities Kombinationsmöglichkeiten
combination starter Anlassschützkombination
combinative element Verknüpfungsglied
combine kombinieren || mischen * kombinieren || verbinden, kombinieren, verbinden, (in sich) vereinigen || vereinen * kombinieren, verbinden (auch Eigenschaften, Vorteile usw.) || vereinigen * verbinden || verknüpfen * kombinieren; logically combined: logisch verknüpft || zusammenbauen * kombinieren, (miteinander) verbinden; z.B. Motor und Anbaugetriebe || zusammenfassen * miteinander verbinden, zusammenschließen * gemeinschaftliche Sache machen || zusammenstellen * zusammenfassen, vereinen, vereinigen, zusammenwirken * sich verbinden

combined gemeinsam * *kombiniert, zusammengenommen* || gemischt * *kombiniert* || kombiniert * *with: mit* || verknüpft * *logically: logisch*
combined action zusammenwirken * *(Substantiv)*
combined alarm Sammelalarm * *{SPS | COMPUTER}*
combined clutch-brake unit Kupplungs-Brems-Kombination || Kupplungs-Bremskombination
combined generating and heating plant Heizkraftwerk
combined operation Anlagenverbund || Verbundbetrieb * *z.B. mehrere Motoren an einem Umrichter*
combined star-delta starting gemischter Stern-Dreieck-Anlauf * *{el.Masch.} Kombination v. Stern-Dreieck- und Wicklungsumschaltung* || verstärkter Stern-Dreieck-Anlauf * *{el.Masch.} Kombination v. Stern-Dreieck- u. Wicklungsumschaltung*
combined with in Verbindung mit * *in Kombination mit*
combined-cycle plant Kraft-Wärmekopplungsanlage * *{Kraftwerk} Kraftwerk mit (Fern-) Wärmeerzeugung*
combining Verbindungs-
combustable entzündbar
combustibility Brennbarkeit
combustible brennbar
combustion air Zuluft * *für Verbrennung in Kraftwerk/Ofen/Verbrennungsmotor*
combustion chamber Feuerraum * *in Heizkessel usw.*
combustion engine Verbrennungsmotor
come erfolgen * *sich einstellen* || kommen
come about eintreten * *geschehen, passieren, entstehen, umspringen (Wind, Markt)*
come across finden * *entdecken, stoßen auf*
come back wiederkehren
come first vorgehen * *Vorrang haben*
come for abholen || holen * *ab~*
come from hervorgehen aus * *stammen von*
come fully up to expectation Erwartungen voll erfüllen
come in einfahren || einlaufen || eintreffen || eintreten * *hinein-/hereinkommen*
come into action eingreifen * *taktisch*
come into being entstehen
come into effect Ansprechen * *(Verb) wirksam werden*
come into force in Kraft treten
come into vogue verbreitet * *Mode werden*
come nearer nähern * *sich ~*
come off abgehen * *sich loslösen* || ablösen * *sich ~ (z.B. Farbe)*
come off winner hervorgehen aus * *als Sieger ~*
come on-line anlaufen * *einer Produktionslinie, z.B. nach der Montage* || in Betrieb gehen
come open lösen * *aufgehen, sich ~*
come out herauskommen * *auch 'ruchbar werden'*
come out winner hervorgehen aus * *als Sieger ~*
come to betragen * *zahlen/wertmäßig ~* || erzielen * *Übereinkunft usw. ~*
come to a dead stop abreißen * *ein Prozess, eine Beziehung usw.*
come to a halt stehen bleiben
come to a standstill stehen bleiben || stillstehen * *zum Stillstand kommen*
come to a stop anhalten || stehen bleiben
come to an end ablaufen * *sein Ende nehmen*
come to know erfahren * *(Verb)*
come to meet entgegenkommen * *treffen*
come to mind einfallen * *Gedanke*
come to pass passieren * *sich ereignen*
come true eintreffen * *sich erfüllen*
come under fallen * *unter Gesetz, in eine Kategorie*

come up aufsteigen * *Gewitter, Unheil* || herauskommen * *hervor-/heraufkommen, Mode werden* || hervorkommen
come up to reichen * *hinauf~ bis*
come up to expectations allen Ansprüchen gerecht werden * *die Erwartungen erfüllen*
come-off stattfinden
COMEL COMEL * *'Comité des Coord. d. Contr. d. Machines Tournantes Elect. d. Marché Commun.': El.masch.verband*
comensating weight Ausgleichsgewicht
comfort Komfort * *Behaglichkeit, Wohlergehen, Bequemlichkeit, Gemütlichkeit*
comfortable angenehm * *komfortabel* || komfortabel * *behaglich, gemütlich, reichlich, bequem*
comic komisch
comical komisch
coming into being Entstehung * *das Entstehen, das Erschaffen-Werden*
coming the other way entgegenkommend * *z.B. Fahrzeug, Verkehr*
comm port serielle Schnittstelle * *(salopp) an einem Computer usw.*
command Anweisung * *Befehl, Kommando* || Befehl * *Steuerbefehl/Kommando* || Führung * *~ einer Regelung/Steuerung* || Kommando * *upon:auf* || Steuersignal * *Steuerbefehl*
command code Befehlscode * *Bediengerät*
command disable Befehl sperren * *bei SPS*
command enable Befehl freigeben * *bei SPS*
command execution time Befehlsausführungszeit * *{SPS}*
command file Kommandodatei
command initiation Kommandogabe
command input Binäreingang * *für Steuerbefehl* || Digitaleingang * *für Steuerbefehl*
command output Befehlsausgabe * *bei SPS*
command set Befehlssatz * *Bediengerät*
command stage Kommandostufe * *zur Steuerung des Momentenwechsels bei Stromrichter*
command the market Markt beherrschen
command variable Führungsgröße * *Sollwert in e. Regelkreis*
commandline Kommandozeile * *Zeichensequenz z. Aufrufen e. Computerbefehls/-Programms (u.U. m. Übergabeparametern)*
commence beginnen
commencement Anfang * *→Beginn* || Beginn * *Anfang, Beginn, (Tag der) Feier* || Promotion
commencement day Promotion * *~sfeier*
commencement of delivery Liefereinsatz
commencement of operation Inbetriebnahme * *Einweihung, Produktionsbeginn*
commencement of operations Inbetriebsetzung * *Aufnahme des Betriebes einer (Fabrik)anlage*
commensurate angemessen * *in Einklang stehend (with/to: mit), angemessen, entsprechend*
commensurate with entsprechend * *im Verhältnis stehend zu, Einklang stehend mit, angemessen*
comment Anmerkung * *Kommentar* || Bemerkung * *Kommentar; on: über* || Erklärung * *Kommentar* || Kommentar * *auch in (Quellsprache-) Programm* || kommentieren * *(up)on: etwas* || Stellungnahme * *Erklärung (on: zu)*
comment on beurteilen * *kommentieren, Meinung abgeben*
commerce Geschäft * *Handel* || Handel * *in großem Maßstab* || Verkehr * *Handelsverkehr*
commercial handelsüblich || kaufmännisch || kommerziell * *geschäftlich, gebräuchlich, kommerziell, Handels-, Geschäfts-, handelsüblich* || Standard- * *handelsüblich* || wirtschaftlich * *geschäftlich*
commercial power supply input Netzeinspeisung * *vom öffentlichen Netz*
commercial vessel Handelsschiff

comminute zerkleinern * *zerkleinern, zerstückeln, zerreiben*
commission Auftrag * *Beauftragung* || Honorar * *Provision* || inbetriebnehmen || inbetriebsetzen || Kommission * *Ausschuss, Arbeitsgruppe, Verkaufsprovision* || Provision
commission basis Provisionsbasis * *on a:auf*
commissioning IBN * *Inbetriebnahme* || IBS * *Inbetriebnahme* || Inbetriebnahme || Inbetriebsetzung
commissioning engineer Inbetriebnahme-Ingenieur || Inbetriebnahmeingenieur || Inbetriebnehmer || Montageingenieur * *Inbetriebnahmeingenieur*
commissioning expenses Inbetriebnahmeaufwand
commissioning personnel Inbetriebnahme-Personal
commissioning time Inbetriebnahmezeit
commit engagiert sein * *commit oneself to: ~ bei* || verpflichten * *oneself to do a thing: sich zu etwas ~* || vertraglich verpflichten * *oneself: sich*
commit to paper festhalten * *schriftlich ~*
commitment Verpflichtung * *übernommene/eingegangene ~*
commitment to take delivery Abnahmeverpflichtung * *of: von Lieferungen*
committee Komitee || Verband * *Komitee, Ausschuss, Kommission*
commodities Waren
commodity Ware * *Handels/Gebrauchsartikel, Waren*
common allgemein * *üblich* || gebräuchlich * *gewöhnlich, üblich* || gemeinsam * *to: mehren Dingen/Personen; common to all: allen ~; common denominator: ~er. Nenner* || Standard- || *üblich * gewöhnlich* || verbreitet * *(allgmein) üblich* || Wurzel * *gemeinsamer Anschluss bei Relais-Wechslerkontakt, Binärein/Ausgang*
common booth Gemeinschaftsstand * *gemeinsamer (Messe-) Stand*
common collector circuit Kollektorschaltung
common connection Wurzel * *gemeinsamer Anschluss, z.b. bei Digital-Ein/Ausgängen*
common DC bus gemeinsamer Zwischenkreis
common DC link bus gemeinsamer Zwischenkreis || Zwischenkreisschiene * *gemeinsame*
common DC-bus Gleichstromschiene * *gemeinsame*
common emitter circuit Emitterschaltung
common ground Bezugsspannung * *gemeinsame Masse*
common interest Gemeinsamkeit
common memory Koppelspeicher * *gemeinsamer Speicher*
common mode Gleichtakt
common mode rejection Gleichtaktunterdrückung
common rectifier unit zentrale Einspeiseeinheit * *gemeinsame*
common sense gesunder Menschenverstand
common shaft gemeinsame Welle
common supply Gruppenspeisung * *z.B. von parallelgeschalteten Motoren an einem Umrichter*
common turn-off Summenlöschung * *Löschung mehrerer Thyristoren über e. gemeinsamen Löschzweig*
common-mode control Gleichtaktaussteuerung
common-mode rejection Gleichtaktunterdrückung
commonly im Allgemeinen * *üblicherweise* || normalerweise * *gewöhnlich, (all)gemein, gemeinsam*
commonly encountered gebräuchlich * *häufig anzutreffen*
commonly used gängig * *→weit verbreitet* || gebräuchlich
commonness Gemeinsamkeit
communicate kommunizieren || schicken * *übermitteln* || übertragen * *Stimmung, Panik usw. ~* || verkehren * *kommunizieren; with: mit*

communication Fernmeldewesen || Kommunikation || Nachrichtenübertragung || Verbindungs- || Verkehr * *(Kommunikations-, Verkehrs-, Brief-) Verbindung* || Verständigung
communication between computers Rechnerkopplung
communication board Kommunikationsbaugruppe * *Leiterkarte* || Schnittstellenbaugruppe * *(Leiterkarte) Kommunikationsbaugruppe* || Schnittstellenkarte
communication cable Anschlusskabel * *Kommunikationsleitung, z.B. für serielle Schnittstelle* || Anschlussleitung * *für (serielle) Kommunikationsverbindung*
communication error Datenübertragungsfehler * *Kommunikationsfehler* || Kommunikationsfehler || Übertragungsfehler * *Kommunikationsfehler*
communication interface board Kommunikationsbaugruppe * *Leiterkarte*
communication module Kommunikationsbaugruppe || Kommunikationsmodul
communication partners Koppelpartner * *(Plural)*
communications Kommunikation * *Kommunikationsfunktionen/wesen*
communications board Kommunikationsbaugruppe || Schnittstellenbaugruppe * *Kommunikationsbaugruppe (Leiterkarte)*
communications buffer Koppelspeicher * *bei Mehrprozessorkopplung (z.B. SIEMENS SIMADYN D (R))*
communications capability Kommunikationsfähigkeit
communications path Kommunikationsweg
communications processor Kommunikationsprozessor * *Abkürzg.: CP*
communications protocol Kommunikationsprotokoll
communications RAM Koppelspeicher
communications submodule Kommunikationssubmodul * *z.B. beim SIEMENS-Antriebsregelsystem SIMADYN D (R)*
communications switching Vermittlungstechnik * *in der Telekommunikation*
communications-capable kommunikationsfähig
communique Bekanntgabe * *(offizielle) Verlautbarung* || Bekanntmachung * *(offizielle) Verlautbarung*
community Gemeinsamkeit
commutate kommutieren || weiterschalten * *kommutieren*
commutating ability Kommutierfähigkeit || Kommutierungsfähigkeit
commutating capacitance Kommutierungskondensator * *Kommutierungskapazität*
commutating capacitor Kommutierungskondensator
commutating choke Kommutierungsdrossel
commutating circuit Löschkreis * *ermöglicht die →Kommutierung in einer Stromrichterschaltung*
commutating dip Kommutierungseinbruch
commutating field Wendefeld * *Motor*
commutating inductance Kommutierungsinduktivität
commutating period Kommutierungszeit
commutating pole Wendepol
commutating poles Wendepole
commutating properties Kommutierungsverhalten
commutating reactor Kommutierungsdrossel
commutating time Kommutierungszeit
commutating winding Wendepolwicklung
commutation Kommutierung * *Übergang d. Stromflusses, z.B. b. Stromrichter v. e. Zweig z. Schaltung i.d. Folgezweig* || Stromwendung * *Kommutierung*

commutation ability Kommutierfähigkeit || Kommutierungsfähigkeit
commutation angle Kommutierungswinkel
commutation behaviour Kommutierungsverhalten * z.B. einer Gleichstrommaschine || Kommutierverhalten * z.B. einer Gleichstrommaschine
commutation capacitor Kommutierungskondensator || Löschkondensator * *unterstützt den Kommutierungs/Löschvorgang in e. Thyristor-Stromrichter*
commutation choke Kommutierungsdrossel
commutation circuit Kommutierungskreis
commutation current Kommutierungsstrom
commutation duration Kommutierungsdauer
commutation effects Kommutierungseinfluüsse
commutation inductance Kommutierungsinduktivität * *Gesamtinduktivität im Kommutierungskreis i. Reihe mit d. Kommutier.spanng.*
commutation interval Kommutierungszeit
commutation limits Trittgrenzen * *Aussteuerwinkelgrenzen bei netzgeführtem Stromrichter*
commutation losses Kommutierungsverluste
commutation notch Kommutierungseinbruch
commutation number Kommutierungszahl * *Anzahl d.Kommutier.vorgänge zw. d. Stromrichter-Hauptzeigen während e. AC-Periode* || Pulszahl * *Kommutierungszahl p: Anzahl d.nicht gleichz. Kommutierungsvorgänge je AC-Periode*
commutation process Kommutierungsvorgang
commutation reactance Kommutierungsreaktanz * *induktiver Widerstand d. →Kommutierungsinduktivität bei der Grundwelle*
commutation reactive power Kommutierungsblindleistung
commutation thyristor Löschthyristor * *Thyristor in einem Hilfs/Löschzweig, der den Kommutiervorgang unterstützt*
commutation voltage Kommutierungsspannung
commutation voltage source Kommutierungsspannungsquelle
commutationless converter kommutierungsloser Stromrichter
commutator Kollektor * *z.B. bei Gleichstrommaschine* || Kommutator || Stromwender * *Kommutator*
commutator agent Schmiermittel * *{el. Masch.} für Kommutator*
commutator end Lauffläche * *auf Kommutator*
commutator motor Kommutatormotor
commutator riser Kommutator-Anschlussfahne
commutator segment Kommutatorlamelle
commutator segments Kommutator-Lamellen
commutator short-circuit period Kommutierungszeit * *{el. Masch.}*
commutator/brush assembly Kommutator * *inklusive Bürstenapparat (bei DC-Maschine)*
commute pendeln * *wenden, auswechseln, umtauschen*
compact fest * *von ~er Beschaffenheit* || geschlossen * *(figürlich) z.B. Stil* || klein * *mit hocher Dichte; kompakt* || Klein- || kompakt * *kompakt, fest, dicht, (zusammen)gedrängt, gedrungen* || Kompakt-
compact converter Kompaktgerät * *Kompaktstromrichter*
compact design Kompaktbauform || Kompaktbauweise || kompakter Aufbau * *z.B. eines Geräts* || Kompaktheit
compact externally mounted fan Kompakt-Fremdlüfter * *extern montierter ~ bei Elektromotor*
compact PCB Kleinleiterplatte
compact type Kompaktbauweise
compact type of construction Kompaktbauweise
compact unit Kompaktgerät

compact version Kompaktbauform
compact whole geschlossen * *(Substantiv) ~es Ganzes*
compactness Dichtheit * *Kompaktheit, Festigkeit, Gedrängtheit* || Festigkeit * *Kompaktheit, Gedrängtheit* || Kleinheit * *Kompaktheit* || Kompaktheit
companion Gegenstück * *zugehöriges ~* || Gesellschafter || Kollege * *Kamerad, Gefährte* || Partner * *Gefährte, Kamerad, Gesellschafter, Gegenstück*
company Firma || Unternehmen * *Firma*
company division Geschäftsbereich
company doctor Werksarzt
company founder Firmengründer
company group Unternehmensbereich
company neutral firmenneutral * *z.B. Bussystem*
company owned firmeneigen * *der Firma gehörend*
company policy Firmenpolitik
company premises Firmengelände
company profile Firmenprofil || Unternehmensprofil
company success Unternehmenserfolg
company's betrieblich * *innerbetrieblich*
company-neutral firmenübergreifend * *firmenneutral*
company-specific firmeneigen * *firmenspezifisch* || firmenspezifisch
comparable vergleichbar * *to:mit*
comparative Vergleichs- || verhältnismäßig
comparatively relativ * *(Adv.) to: zu* || vergleichsweise || verhältnismäßig * *(Adv.)*
comparatively speaking verhältnismäßig * *(Adv.)*
comparator Grenzwertmelder * *Komparator* || Komparator || Vergleicher || Vergleichsbaustein
compare vergleiche || vergleichen * *(with: mit) (to: mit (gleichstellend))*
compare badly with abfallen * *~ gegenüber, im Vergleich zu ... ~*
compare block Vergleichsbaustein * *Funktionsbaustein*
compare function Vergleichsfunktion
compared verglichen * *with: mit*
compared to gegenüber * *im Vergleich zu* || im Vergleich zu
compared with gegen * *verglichen mit* || gegenüber * *im Vergleich zu* || neben * *verglichen mit*
comparison Gegenüberstellung * *Vergleich* || Steigerung * *(linguistisch)* || Vergleich * *in comparision to: im ~ zu*
comparison block Vergleichsbaustein * *Funktionsbaustein*
comparison of performance Leistungsvergleich * *Vergleich der Leistungsfähigkeit*
comparison operation Vergleichsfunktion * *Computerbefehl* || Vergleichsoperation
comparison operator Vergleichsoperator
comparison standard Messnormal || normal * *(Substantiv)* →Messnormal
comparison table Zusammenstellung * *(vergleichende) tabellarische ~*
comparison value Vergleichswert
compartment Fach * *Abteil, Ablage~* || Kabine * *(Zug-) Abteil*
compartmentalization Schottung * *Unterbringung in einzelnen Abteile*
compass Zirkel
compatibility Kompatibilität * *with:mit* || Verträglichkeit * *Kompatibilität*
compatibility level Verträglichkeitspegel
compatible kompatibel * *with: mit* || verträglich * *kompatibel, auch elektromagnetisch*
compatible with kompatibel mit
compatible with the future zukunftssicher

compendium Kompendium * Nachschlagewerk
compensate abgleichen * kompensieren || aufheben * aufwiegen, ausgleichen, kompensieren || ausgleichen * kompensieren || ausregeln * for: etwas || kompensieren * for:etwas || verrechnen * gegeneinander || verrechnen * gegeneinander aufrechnen
compensate for ausgleichen * etwas ~ || vergüten * Ausgaben/Auslagen ~
compensated -kompensiert || kompensiert * auch DC-Motor, Tachogenerator
compensated motor kompensierter Motor * mit →Kompensationswicklung
compensating Ausgleichs-
compensating current Ausgleichsstrom
compensating stacker Kreuzleger * {z.B. in Druckerei} für Papierbögen
compensating winding Kompensationswicklung
compensation Abgleich || Ausgleich || Kompensation * for: von || Lohn * Entgelt || Schadenersatz
compensation of tolerances Toleranzausgleich
compensation winding Kompensationswicklung
compensative ausgleichend
compensatory ausgleichend
compete konkurrieren * with:mit || messen * sich mit jdm. ~
competence Befugnis * Zuständigkeit || Berechtigung * Zuständigkeit || Entscheidungsbefugnis || Kompetenz * for: für || Sachverstand * Kompetenz, Befähigung, Fähigkeit, Tüchtigkeit
competency Kompetenz * Befähigung, Tauglichkeit (for: zu/für), Kompetenz, Zuständigkeit, Befugnis || Sachverstand * Kompetenz, Befähigung
competent befugt * zuständig; for a thing: für etwas;to do a thing:etwas zu tun || fachkundig || fachmännisch * Arbeit || fähig + tuchtig, wissend || kompetent * befugt, zuständig, befähigt (for: für), maßgeblich, zuverlässig || maßgebend * z.B. Behörde || maßgeblich * z.B. Behörde || qualifiziert * kompetent, fachkundig, (leistungs)fähig, tüchtig, zuständig, befugt, geschäftsfähig || tüchtig || zuständig * befugt
competing konkurrierend
competition Konkurrenz * stiff: starke || Konkurrenzfirma || Konkurrenzkampf * cutthroat competition: mörderischer ~ || Mitbewerb * Konkurrenz || Mitbewerber * Mitbewerb || Wettbewerb * Konkurrenz; stiff: stark; fierce: heftig
competitive Konkurrenz- || konkurrenzfähig || konkurrierend || Mitbewerbs- || Wettbewerbs- || wettbewerbsfähig
competitive advantage Wettbewerbsvorteil
competitive edge Marktvorsprung * gegenüber dem Mitbewerb; on: gegenüber || Marktvorteil * Wettbewerbsvorsprung || Wettbewerbsfähigkeit * Wettbewerbsvorsprung || Wettbewerbsvorsprung || Wettbewerbsvorteil
competitive position Konkurrenzfähigkeit
competitive power Konkurrenzfähigkeit
competitive procurement procedure Ausschreibung * (meist staatliches) Ausschreibungsverfahren für Beschaffung
competitive product Konkurrenzprodukt || Mitbewerbsprodukt
competitive situation Wettbewerbslage
competitive strength Wettbewerbsfähigkeit || Wettbewerbskraft
competitiveness Konkurrenzfähigkeit || Wettbewerbsfähigkeit
competitivity Wettbewerbsfähigkeit
competitor Bewerber * Mitbewerber bei einer Ausschreibung || Konkurrenzfirma || Teilnehmer * Mitbewerber (z.B. an Ausschreibung, Wettkampf) || Wettbewerber

competitors Konkurrenz * Mitbewerb || Mitbewerb * Konkurrenten || Mitbewerber * Konkurrenz
compilation Ausarbeitung * Zusammenstellung, Zusammentragung von (schriftl.) Material || Sammlung * Zusammenstellung, Zusammengetragenes || Übersetzung * durch Compiler (Hochspracheprogramm in Maschinen/Objektspracheprogramm) || Übersetzungslauf * für Hochsprache-Programm || Zusammenstellung
compile aufstellen * zusammentragen, z.b. tabellarisch || erarbeiten * zusammentragen || kompilieren * {Computer} Quellspracheprogramm mit Compiler übersetzen (z.B. in →Maschinensprache) || übersetzen * Hochspracheprogramm mit Hilfe eines Compilers || zusammenfassen * zusammentragen (Material, Unterlagen usw.) || zusammenstellen * zusammentragen, z.b. Besprechungsunterlagen, Liste, Ersatzteilsortiment usw.
compiled zusammengefasst * z.b. Angaben gesammelter Daten in einer Tabelle
compiler Compiler * Übersetzungsprogramm für Hochspracheprogramme || Kompilierer * Übersetzungsprogramm für Hochspracheprogramme || Übersetzer * für Hochspracheprogramme
compiler pass Übersetzungslauf * einer von mehreren Durchläufen eines Hochsprache-Übersetzers
compiler switch Steueranweisung * Compilerschalter, aktiviert/deaktiviert Übersetzungsoptionen
complain beanstanden * of: etwas ~; reklamieren; sich beklagen/beschweren (of/about: über; that: dass)
complaint Beanstandung || Mängelrüge || Reklamation * Mängelrüge
complaint about quality Mängelrüge
complaint list Mängelliste || Meckerliste * Mängelliste
complaisance Nachgiebigkeit * (figürl.) Willfährigkeit
complaisant nachgiebig * entgegenkommend, gefällig, höflich
complement Bestückung * eines Baugruppenträgers, z.B. bei SPS || Ergänzung * das Ergänzte, Vervollständigung, Ergänzung; full complement:- volle Anzahl/Menge || invertiertes Signal * Binärsignal, z.B. Impulsgeber-Signal
complementary ergänzend * ergänzend, sich ergänzend, Ergänzungs-; Komplementär- || invers * komplementär || invertiert * Impulse || komplementär * auch Signal (Differenzialsignal) || Zusatz- * Ergänzungs-, Begleit-
complete abrunden * vollenden || abschließen * beenden, komplettieren, vervollständigen || ausbauen * fertigstellen || beenden * abschließen, beenden, vollenden, fertigstellen, erledigen || ergänzen || erweitern * vollständig machen || fertigstellen || ganz * vollständig || gesamt * komplett || komplett || lückenlos * vollständig || sämtliche * vollständig || schließen * ab~, fertigstellen || vervollständigen || voll * komplett, vollständig || Voll- * vollständig || völlig * vollständig || vollständig
complete immersion Vollimprägnierung
complete plant Komplettanlage
complete the circuit durchschalten * Stromkreis schließen
complete unit Komplettgerät
completed abgeschlossen * beendet || beendet * abgeschlossen || durchgeschaltet * completed circuit: durchgeschalteter/geschlossener Stromkreis
completely völlig * (Adv.) almost completely: fast ~
completely accurate einwandfrei * genau
completely hardened durchgehärtet
completeness Vollständigkeit

completion Abschluss * *Fertigstellung* ‖ Ausbau * *Fertigstellung* ‖ Beendigung * *Fertigstellung* ‖ Durchführung * *Fertigstellung, Vollendung, Abschluss, Vervollständigung, Erfüllung* ‖ Ergänzung * *Vervollständigung, Vollendung, Fertigstellung, Abschluss* ‖ Fertigstellung
complex facettenreich * *komplex* ‖ komplex * *vielschichtig* ‖ kompliziert * *komplex, vielschichtig* ‖ vermascht * *komplex, vielschichtig* ‖ vielschichtig * *komplex*
complex component Imaginärteil
complex number komplexe Zahl
complex power Scheinleistung
complex variable komplexe Größe * *[Mathem.]*
complex variable domain Bildbereich
complexedness Komplexität
complexity Aufwand * *Kompliziertheit* ‖ Komplexität
compliance Befolgung * *with:von* ‖ Einhaltung * *Erfüllung; with:von* ‖ Erfüllung ‖ Nachgiebigkeit * *Willfährigkeit* ‖ Übereinstimmung * *Befolgung, Erfüllung, Einhaltung; in compliance with:gemäß (Vorschrift, Norm usw.)*
compliant kompatibel * *zu einer Norm, Vorschrift usw.* ‖ nachgiebig * *willfährig, nachgiebig, zum Nachgeben bereit*
compliant supply schwaches Netz * *nachgiebiges Netz mit hoher Netzimpedanz* ‖ Schwachnetz * *nachgiebig, mit hoher Netzimpedanz*
complicate erschweren * *(ver-) komplizieren, verwickelt machen, erschweren* ‖ komplizieren ‖ verkomplizieren
complicated aufwändig * *kompliziert* ‖ kompliziert ‖ schwierig * *kompliziert* ‖ umständlich * *verwickelt, kompliziert* ‖ unverständlich * *verwickelt, kompliziert*
comply with befolgen * *Vorschrift ~* ‖ einhalten * *Vorschriften, Norm* ‖ entsprechen * *einer Vorschrift, Norm, Regel, Bedingung, Bitte ~* ‖ erfüllen * *erfüllen, entsprechen (z.B. Norm, Vorschrift, Bedingung, Bitte, Regel)* ‖ sich halten an
compole Wendepol
compole field Wendefeld * *z.B. in Gleichstrommaschine*
compole winding Wendepolwicklung
component -Anteil * *einen Teil bildend, Bestandteil* ‖ -Glied ‖ Anteil * *Komponente, z.B. integral component: Integral~, direct component: Gleich~* ‖ Bauelement ‖ Baugruppe * *Komponente, Bauteil* ‖ Bauteil ‖ Bestandteil ‖ Einzelteil * *Komponente, Bauteil* ‖ Geräteteil ‖ Glied ‖ Komponente * *Anteil, Teil* ‖ Teil * *~ einer zusammengesetzten Sache; Bestand~, An~* ‖ Teil- * *Bestandteil eines zusammengesetzten Ganzen;z.B.Teilwechselrichter je Phase*
component converter Teilstromrichter
component current Teilstrom
component inspection Bauteilprüfung
component mounting diagram Bestückungsplan * *z.B. für Leiterkarte*
component mounting position Bestückungsplatz * *für Bauelemente auf Leiterkarte* ‖ Einbauplatz * *z.B. für Einbauplatz auf Leiterkarte*
component part Einzelteil ‖ Komponente * *Einzelteil*
component set Bausatz ‖ Bestückung * *Satz von zum Einbau vorgesehenen Teilen*
component side Bauteilseite * *einer Leiterkarte* ‖ Bestückungsseite * *einer Leiterkarte*
components at hazardous voltage level berührungsgefährdete Teile ‖ spannungsführende Teile
components inspection Bauteilprüfung
components under voltage spannungsführende Teile
compose ausarbeiten * *schriftlich* ‖ zusammensetzen * *zu einem Ganzen; be composed of: sich ~ aus* ‖ zusammenstellen * *Menü ~*
composed zusammengesetzt * *of:aus*
composite gemischt * *(aus verschiedenen Elementen) zusammengesetzt, gemischt, vielfältig, Misch-; Mischung* ‖ Misch- ‖ Verbund * *Verbund-, Misch-, Zusammensetzg. (v. Materialien usw.); composite construction: ~bauweise* ‖ Verbund- * *zusammengesetzt* ‖ Verbundwerkstoff
composite calculation Mischkalkulation
composite material Verbundmaterial
composite price Mischpreis
composition Ausarbeitung * *Abfassung, Entwurf, Aufsatz, (Schrift-) Werk* ‖ Beschaffenheit * *Wesen, Natur, Anlage; (chemische) Zusammensetzung/ Verbindung* ‖ Mischung * *(chem.)* ‖ Zusammensetzung * *(gestalterische) Anordnung, Aufbau, Struktur, Beschaffenheit, Synthese*
compound -Mittel * *Masse, Mischung; z.B. Dichtmittel* ‖ Misch- ‖ mischen * *Bestandteile ~* ‖ Mittel * *Masse, Mischung; sealing compound: Dicht~* ‖ Schicht- * *aus Verbundmaterial bestehend* ‖ Stoff * *Masse, zusammengesetzter Werkstoff* ‖ Verbindung * *(chem.)* ‖ Verbund * *~maschine, ~material, ~werkstoff* ‖ Verbund- * *z.B. Maschine, Verbundmaterial, Materialmischung* ‖ Verbundmaterial ‖ Verbundwerkstoff ‖ zusammengesetzt * *zusammengesetzt, Verbund-, aus Verbundmaterial bestehend* ‖ zusammensetzen * *aus Teilkomponenten ~ (z.B. Verbundmaterial)* ‖ Zusammensetzung * *Verbund-, Mischung, Masse, ~ einer Substanz, chem. Verbindung*
compound compressor set Verbund-Kompressorsatz * *häufig mit bedarfsweiser Zu- und Wegschaltung einzelner Kompressoren*
compound material Verbundmaterial
compound winding Compoundwicklung * *[el. Masch.]* ‖ Hilfs-Reihenschlusswicklung * *kompensiert Lastabhängigk. d.Drehz. e.DC-Motors bei konst. Ankerspanng.* ‖ Hilfsreihenschlusswicklung * *in DC-Motor* ‖ Kompensationswicklung * *kompensiert Lastbhängigk. der Drehz.e. DC-Motors bei konst. Ankerspanng.* ‖ Kompoundwicklung * *Doppelschlusswicklung, kompensiert Lastabhängigkeit bei DC-Motor*
compounding Kompoundierung
comprehend aufnehmen * *geistig erfassen, verstehen, begreifen, umfassen, einschließen* ‖ erfassen * *geistig* ‖ zusammenfassen * *in sich fassen*
comprehensible verständlich * *begreiflich, (leicht) verständlich*
comprehension Verständnis * *Verstehen*
comprehensive ausführlich * *umfassend* ‖ gründlich * *umfassend* ‖ intensiv * *umfassend* ‖ konsequent * *umfassend, vollständig* ‖ umfangreich * *umfassend* ‖ umfassend * *ausführlich*
compress komprimieren * *(auch figürl.)* ‖ pressen * *zusammendrücken/-pressen* ‖ stauchen * *zusammendrücken, zusammenpressen, komprimieren* ‖ verdichten ‖ zusammendrücken ‖ zusammenpressen
compressed air Druckluft ‖ Pressluft
compressed-air Druckluft-
compressed-air brake Druckluftbremse
compressed-air clutch Druckluftkupplung
compressed-air connection Druckluftanschluss
compressed-air container Druckluftbehälter
compressed-air cylinder Druckluftzylinder
compressed-air filter Druckluftfilter
compressed-air hose Druckluftschlauch
compressed-air line Druckleitung * *für Druckluft* ‖ Druckluftleitung
compressed-air pipe Druckleitung * *für Druckluft* ‖ Druckluftleitung

compressed-air reservoir Druckluftbehälter * Zwischenbehälter/Ausgleichsbehälter für Preßluft
compressed-air system Druckluftanlage
compressed-air tank Druckluftbehälter
compressing spring Druckfeder
compression Druck * Zusammendrücken/-pressen, Verdichtung, Druck, Kompression || Kompression || Komprimierung * auch Datenkomprimierung || Pressung || Quetsch- * siehe auch →Anpress-
compression gland PG-Verschraubung * siehe auch DIN 46320 || Stopfbuchsenverschraubung || Stopfbuchsverschraubung * bei Kabeleinführung, siehe →PG-Verschraubung und DIN 46320
compression program Packprogramm
compression spring Druckfeder * z.B. zur Lagervorspannung (siehe →vorgespannt (-es Lager))
compression stress Druckbeanspruchung
compressive pre-tensioning Druckvorspannung
compressive strength Druckfestigkeit || Druckspannung
compressive stress Druckbeanspruchung
compressor Kompressor * Verdichter || Verdichter
compressor characteristic Verdichterkennlinie * Lastkennlinie M als f(n) eines Kompressors (M~n^2 bei Zentrifugalkompr.)
compressor set Kompressorsatz * Kombination mehrer Verdichter; häufig m. bedarfsweiser Zuschaltg. d. Einzelverdichter || Verdichtersatz * Kombination mehrer Verdichter; häufig m. bedarfsweiser Zuschaltung d. Einzelverdichter
comprise beinhalten || bestehen * of: aus || bestehen aus * enthalten, umfassen || einschließen * umfassen, enthalten || enthalten * einschließen, umfassen, beinhalten || tragen * enthalten, beinhalten || umfassen * in sich schließen, einschließen, enthalten || zusammenfassen * in sich fassen
comprising einschließlich * beinhaltend
compromise Kompromiss * ['kompromais] bad compromise: fauler ~ || Übereinkunft * Kompromiss
compulsary circumstances höhere Gewalt
compulsory obligatorisch * zwangsmäßig, ~, Zwangs-, bindend, Pflicht- (compulsory subject: Pflichtfach) || verbindlich * bindend, Pflicht-, obligatorisch, Zwangs-; make compulsory: für ~ erklären
computation Berechnung || rechnen * (Substantiv)
computation functions Arithmetikfunktionen
computation task Rechenaufgabe * für Rechner
computation time Rechenzeit * für Programmablauf benötigte Zeit
computational circuit Rechenschaltung
computational cycle Rechenzyklus
computational performance Rechenleistung
computational power Rechenleistung
compute ausrechnen || berechnen || errechnen
computer EDV || Rechner
computer center Rechenzentrum * (amerikan.)
computer centre Rechenzentrum
computer coupling Rechnerkopplung
computer engineering Datenverarbeitung * als Fachrichtung
computer from other manufacturers Fremdrechner
computer interface Rechnerkopplung * Schnittstelle zur Rechnerkopplung
computer interfacing Rechnerkopplung
computer link Rechnerkopplung * Schnittstelle zur Rechnerkopplung
computer loading Rechnerauslastung
computer networking Rechnerkopplung * Vernetzung
computer performance Rechenleistung * eines Rechners || Rechenleistung || Verarbeitungsleistung * eines Rechners

computer programming expertise Programmierkenntnisse * [expe'ties]
computer run Rechenlauf
computer science Datenverarbeitung * als Wissenschaftszweig || Informatik * Computertechnik/-wissenschaft
computer utilization Rechnerauslastung
computer-aided rechnergestützt * z.B. Entwurf, Konstruktion, Fertigungssteuerung
computer-controlled rechnergesteuert
computer-supported rechnergestützt
computer-to-cylinder technology automatische Belichtung * {Druckerei} bei Tiefdruck
computer-to-plate volldigitale Plattenbelichtung * {Druckerei} im grafischen Gewerbe, in der Druckvorstufe
computer-to-plate technology automatische Belichtung * {Druckerei} bei Offsetdruck
computerized automatisch * rechnergesteuert || rechnergestützt
computerized ordering beleglose Bestellung
computing Berechnung * das Berechnen || Datenverarbeitung * Rechnertechnik (als Fachgebiet) || EDV
computing capacity Rechenleistung
computing method Rechenverfahren
computing operation Rechenoperation
computing power Prozessorleistung * Rechenleistung || Rechenleistung || Rechnerleistung || Verarbeitungsleistung * Rechenleistung
computing procedure Rechenlauf || Rechenverfahren
computing speed Rechengeschwindigkeit * Rechner
computing time Rechenzeit
con Zusammen-
concatenate kaskadieren * auch 'verketten,verknüpfen' || verketten || verknüpfen * auch 'verketten, kaskadieren'
concatenated connection Kaskadenschaltung
concatenation Kaskadierung * Schaltung in Kaskade, Verkettung || Verkettung * (figürlich)
concatenation connection Kaskadenschaltung
concave hohl * vertieft || konkav
conceal überlagern * verdecken || verdecken * verbergen, verstecken, verdecken, verheimlichen
concealed geheim * verborgen || verborgen * verborgen, unübersichtlich, versteckt || verdeckt * verheimlicht, verheimlicht, versteckt
concede gewähren * einräumen
conceive ausdenken * erdenken, ersinnen, planen, in Worten ausdrücken || gestalten * er/ausdenken, ersinnen, planen, in Worten ausdrücken || vorsehen * planen, er-/ausdenken, ersinnen * vorgesehen für: vorgesehen/bestimmt für
conceived vorgesehen * vorgesehen, bestimmt, ersonnen, geplant (für: für)
concentrate konzentrieren * auch 'sich ~' ((up)on: auf || sammeln * , sich (an)sammeln, (sich) konzentrieren; verdichten, vereinigen || zusammenfassen * miteinander verbinden zu höherer Dichte/Kompaktheit; auch Kräfte || zusammenführen * bündeln || zusammenziehen * konzentrieren
concentration Anreicherung || Ansammlung || Dichte * {Chemie} || Eindampfung || Eindickung
concentrator Konzentrator * z.B. für Daten
concentric koaxial || konzentrisch || zentrisch
concentrically zentrisch * (Adv.)
concentricity Koaxialität * auch Fluchtfehler; für Motor: siehe DIN 42955, IEC 72-1 || Konzentrizität * auch Fluchtfehler; für Motor: siehe DIN 42955 || Mittigkeit * Konzentrizität || Rundlauf * Konzentrizität, ~fehler || Rundlaufgenauigkeit * Konzentrizität, z.B. von Wellenende und Flansch bei Motor || Zentrizität * Konzentrizität

concentricity error Rundlaufabweichung || Rundlauffehler
concept Gedanke * *Begriff* || Konzept
conception Begriff * *Konzeption* || Idee * *Vorstellung*
conceptive konzeptionell
conceptual konzeptionell
concern Angelegenheit * *Sache (that's his concern: d. ist seine ~), Geschäft, Bedenken, Interesse, Beziehung* || Bedeutung * *Wichtigkeit, ~; Unruhe, Sorge, Bedenken (at/about/for: um); be a concern: von ~ sein* || betreffen * *angehen, interessieren, v. Belang sein für, beunruhigen; c. oneself with: sich befass. mit* || Betrieb * *Konzern* || Beziehung * *Geschäft, Unternehmen; Beziehung; have no concern with: nichts zu tun haben mit* || erstrecken * *(figürlich) sich ~ auf, betreffen* || Geschäft * *Konzern* || Interesse * *~ (for:für; in:an), Unruhe, Sorge, Bedenken (at/about/for: um/wegen), Wichtigkeit* || kümmern * *betreffen, angehen, interessieren, von Belang sein, beunruhigen*
concern oneself with befassen * *sich befassen mit*
concerned betroffen * *berührt, beteiligt, verwickelt*
concerned with safety sicherheitsrelevant
concerning betreffend || bezüglich * *betreffend* || über * *betreffend*
concession Konzession * *Privileg; Entgegenkommen, Zugeständnis; make no conc. to a person: jdm. keine ~en machen*
concise bündig * *Stil, Rede* || kurz * *gedrängt*
conclude abschließen * *abschließend beenden* || beschließen * *beschließen, beendigen, abschließen* || folgern * *schließen * Brief/Rede ~; auch 'logisch folgern'*
concluding abschließend
conclusion Beschluss * *Beendigung, Abschluss, Folgerung* || Folgerung * *draw a conclusion: eine ~ ziehen* || Schluss * *Abschluss, Folgerung*
conclusions Fazit * *Folgerungen* || Konsequenz * *Folgerungen; draw one's conclusions/act accordingly: die ~en ziehen*
conclusive bündig * *überzeugend* || entscheidend * *schlüssig* || schließlich * *abschließend*
conclusive comment Schlussbemerkung
concomitance Gleichzeitigkeit
concomitant Begleiterscheinung
concomitant phenomenon Begleiterscheinung
concrete anschaulich * *konkret, greifbar, wirklich, dinglich* || Beton || greifbar * *(figürl.) konkret (siehe auch dort)* || konkret * *konkret, greifbar, wirklich, dinglich, fest benannt*
concrete mixer Betonmischer
concur zusammentreffen * *zeitlich ~, ~ von Umständen*
concurrence Gleichzeitigkeit || Schnittpunkt || Übereinstimmung * *Einverständnis, Zustimmung, Gleichzeitigkeit, Zusammentreffen* || Verknüpfung * *Zusammentreffen, Mitwirkung, Gleichzeitigkeit* || zusammentreffen * *(Substantiv) zeitlich~, ~ von Umständen* || zusammenwirken * *(Substantiv) ~ von Umständen*
concurrent gleichzeitig * *gemeinsam ablaufend, z.B.Programme*
concurring übereinstimmend * *z.B. Meinung, Erfahrung*
concussion Erschütterung * *Erschütterung, Aufschlag, Stoß*
condensate Kondensat || Kondenswasser * *Wasserfilm, d. sich b. feuchter Luft an Flächen bildet, der. Temp. unter d. Taupunkt liegt* || Schwatzwasser * *Kondenswasser*
condensate drain hole Kondenswasser-Ablaufloch || Kondenswasserablauf || *~loch, ~öffnung* || Kondenswasserloch || Schwatzwasserloch * *Kondenswasser-Ablaufloch (z.B. bei {el.Masch.})*

condensate drainhole Kondenswasserloch * *z.B. an der tiefsten Stelle eines Motors*
condensate tank Kondensatbehälter
condensation Verflüssigung
condensation water Schwitzwasser
condensation water drain hole Kondenswasserloch * *z.B. a d.tiefsten Stelle eines Motors*
condense kondensieren || zusammenfassen * *Schriftwerk ~* || zusammenpressen * *verdichten*
condensed water Kondensat * *Kondenswasser* || Kondenswasser
condenser Verflüssiger * *z.B. im Kühlkreislauf, in der Kältetechnik*
condensing temperature Verflüssigungstemperatur * *z.B. in einem Kühlkreislauf*
condensing turbine Kondensationsturbine
condensor Kondensator * *Verflüssiger in e. thermodynamisch. Kreislauf z.B. Wärmekraft/Kältemaschine*
condition aufbereiten * *konditionieren, i.e. bestimmten Zustand bringen; z.B. Signale* || Bedingung * *on condition that: unter der Bedingung, dass; fulfill/comply with:erfüllen* || Betriebszustand || Form * *(z.B. körperliche) Verfassung/Leistungsfähigkeit* || Status * *Lage* || Umstand * *Umstände* || Vereinbarung * *Klausel, Vorbehalt, Bestimmung* || Voraussetzung * *Beding., Voraussetzg.; on condition that/under the condition that: vorausgesetzt, dass* || Zustand * *Zustand, Verfassung, Beschaffenheit, Umstände, Lage; fault cond.: Fehler~*
condition as specified Soll-Zustand
condition code Anzeige * *bei SPS*
condition of market Marktlage
condition of operation Betriebszustand * *Betriebsbedingungen (auch an einem bestimmten Arbeitspunkt)*
condition when supplied Auslieferungszustand
conditional abhängig * *bedingt (durch), abhängig (von); make something conditional on: etwas abh. machen von* || bedingt * *abh. von e. Bedingung, z.B. Programmsprung (on: durch), Funktionsbaustein usw.*
conditional branch bedingter Sprung * *im Programm*
conditional jump bedingter Sprung * *im Programm*
conditioner Umsetzer * *Signalumsetzer*
conditioning -Anpassung * *Aufbereitung* || Anpassung * *Signal~* || Aufbereitung * *Signale* || Konditionierung
conditions Anforderungen * *(Betriebs-) Bedingungen* || Beanspruchung * *(Betriebs-) Beanspruchung* || Bedingungen * *Verhältnisse* || Konditionen * *auch Verkaufs~* || Verhältnis * *Umstände* || Zustand * *Verhältnisse*
conditions for sale and delivery Lieferbedingungen * *Verkaufs- und ~*
conditions of delivery Lieferbedingungen * *siehe auch →allgemeine ~*
conditions of sale Lieferbedingungen * *Verkaufsbedingungen * siehe auch →Allgemeine Lieferbedingungen*
conditions of sale and delivery Verkaufs- und Lieferbedingungen
conditions of sale/delivery Verkaufs- und Lieferbedingungen
conditions of use Gebrauchsbedingungen
condone billigen * *stillschweigend billigen, verzeihen, (Fehltritt usw.) entschuldigen*
conducive förderlich * *[kondju:siw] dienlich, förderlich (to: für)* || gut * *[kondju:siw] dienlich, förderlich, nützlich, ersprießlich, zielführend (to: für)* || nützlich * *[kondju:siw] dienlich, förderlich, fördernd, zielführend (to: für)* || praktikabel * *[kond-*

ju:siw] (zweck)dienlich, förderlich, zielführend ||
zielführend * *[kondju:siw]*
conduct   durchführen * *führen/leiten, z.B.*
Kurs, Vorlesung, Tagung || Führung * *Führung, Leitung, Verwaltung, Handhabung, Haltung, Verhalten, Betragen* || halten * *z.b. einen Kurs, eine Vorlesung ~* || leiten * *führen (auch Strom), durchführen* || verwalten * *führen*
conduct an interview   Vorstellungsgespräch führen
conduct outwards   abführen * *nach außen ~ (z.b. Kühlluft)*
conductance   Konduktanz * *Leitwert (1/R)* || Leitwert * *Ohmscher ~ (1/R)*
conductance value   Leitwert
conducted noise   leitungsgebundene Störungen
conducting   durchlässig * *(Strom-) leitend* || Führung * *Leitungs-, das Leiten (auch von Strom)* || stromführend * *z.B. (Stromrichter-) Ventil*
conducting interval   Stromführungszeit
conducting state   Durchgängigkeit * *stromdurchlässiger/-führender Zustand* || Durchlasszustand
conducting-state voltage drop   Durchlassspannung * *z.b. einer Diode*
conduction   Führung * *Leitung (von Strom), (Zu-) Führung, Übertragung*
conduction interval   Einschaltdauer * *Stromflusszeit*
conduction monitor   Stromflussüberwachung
conduction period   Stromführungsdauer
conductive   durchgängig * *(den Strom) leitend* || durchlässig * *~ für Strom* || leitend * *(auch Strom-~)* || leitfähig * *(elektr.)*
conductive characteristic   Leitverhalten * *~ eines Stromleiters/der Leitfähigkeit*
conductive coating   elektrisch leitende Beschichtung || elektrisch leitender Anstrich
conductive continuity   Stromdurchgang
conductive interval   Stromflusszeit
conductive losses   Durchgangsverluste * *z.b. bei Halbleiter*
conductive paint   Leitlack * *elektrisch leitender Lack*
conductive surface   leitfähige Unterlage * *zur Hantierung von Elektronikbauteilen*
conductivity   Leitfähigkeit
conductivity meter   Leitfähigkeitsmessgerät
conductor   Ableiter * *Leiter* || Ader || Draht * *Leiter* || Leiter * *auch Stromleiter, (An-) Führer*
conductor bar   Stromschiene
conductor bars   Verschienung
conductor breakage   Drahtbruch
conductor breakage detection   Drahtbrucherkennung
conductor cross-section   Leiterquerschnitt || Leitungsquerschnitt
conductor end sleeve   Aderendhülse
conductor impedance   Leiterimpedanz
conductor insulation   Leiterisolierung
conductor path   Leiterbahn
conductor rail   Fahrleitung * *als Schiene ausgelegte ~*
conductor size   Anschlussquerschnitt || Leiterquerschnitt
conductor track   Leiterbahn
conduit   Kabelkanal || Leitungskanal || Rohr * *(Kabel)Kanal* || Röhre * *(Kabel)Kanal* || Rohrleitung * *Rohrleitung, Röhre, Kanal, Isolierrohr (für el.-Leitung)*
conduit adapter   PG-Verschraubung
conduit box   Klemmenkasten
conduit connection   Leitungsanschluss * *z.B. in Motorklemmenkasten*
conduit fitting   PG-Verschraubung
conduit mounting plate   Anschlussplatte
conduit pipe   Leitungsrohr

cone   Kegel || Konus
confer   siehe * *vergleiche* || vergeben * *übertragen* || vergleiche * *Hinweis in Schriftstück; confer page 23: ~ Seite 23* || verhandeln * *konferieren*
confer a doctor's degree upon   promovieren * *jemanden ~, jemandem den Doktortitel zuerkennen*
conference   Beratung * *Besprechung* || Besprechung * *Konferenz* || Gespräch * *Konferenz* || Konferenz || Sitzung * *Konferenz* || Verhandlung * *Konferenz, Besprechung*
conferring of contract   Auftragserteilung * *das "Zuerteilt-Bekommen" eines Auftrags*
confidence   Sicherheit * *Vertrauen* || Vertrauen
confident   sicher * *selbst~, mit (Selbst-) Vertrauen*
confidential   geheim * *vertraulich* || intern * *vertraulich* || vertraulich
confidential report   Beurteilung * *in Personalakte*
configurability   Konfigurierbarkeit || Projektierbarkeit * *Konfigurierbarkeit (z.B. SIEMENS-Regelsystem SIMADYN D (R))*
configurable   konfigurierbar * *freely configurable: frei ~* || projektierbar * *(z.B. beim SIEMENS Antriebsregelsystem SIMADYN D (R)); freely: frei*
configuration   Anordnung * *Konfiguration, Gestalt(ung), Struktur* || Aufbau * *Anordnung, Konfiguration* || Gestaltung * *Konfiguration, Anordnung* || Konfiguration || Schaltung * *Anordnung, Konfiguration*
configurator   Konfigurator
configure   aufbauen * *(auf)bauen, gestalten, konfigurieren* || auslegen * *anordnen, konfigurieren, gestalten, strukturieren, bauen* || konfigurieren || projektieren * *soft-/hardwaremäßig konfigurieren/~ (z.B. SIEMENS-Regelsystem SIMADYN D (R))* || strukturieren * *konfigurieren*
configured system   Projektierung * *(SIEMENS SIMADYN D (R)) projektiertes (Regel)System*
configuring   Projektierung * *Konfigurieren, Konfigurierung, (z.B. SIEMENS-Regelsystem SIMADYN D (R))* || Projektierungsvorgang * *z.B. beim SIEMENS Regelsystem SIMADYN D (R)*
configuring diagram   Projektierungsplan * *(SIEMENS SIMADYN D (R))*
configuring engineer   Projekteur * *(SIEMENS SIMADYN D (R))*
configuring error   Projektierungsfehler * *z.B. beim SIEMENS Regelsystem SIMADYN D (R)*
configuring example   Projektierungsbeispiel
configuring forms   Projektierungshilfen * *Formblätter*
configuring handbook   Projektierungshandbuch
configuring instructions   Projektierungsanleitung || Projektierungshinweise
configuring language   Projektierungssprache * *z.B. 'STRUC' (R) beim Regelsystem SIMADYN D (R) von SIEMENS*
configuring phase   Projektierungsphase
configuring software   Projektierungssoftware
configuring system   Projektierungssystem
configuring tools   Projektierungshilfen
configuring unit   Projektierungsgerät * *z.B. für Regelsystem SIMADYN D (R) von SIEMENS*
confine   beschränken * *confine oneself to: sich beschränken auf* || einschränken * *beengen, begrenzen, einschränken, beschränken*
confined   beengt || beschränkt
confirm   bestätigen || sicherstellen * *(Nachricht, Wahrheit, Auftrag) bestätigen, (Entschluss) bekräftigen, (Macht) festigen* || untermauern || versichern * *bestätigen*
confirmation   Bestätigung || Zusage * *Bestätigung*
confirmation of enrolment   Anmeldebestätigung * *für Vorlesung, Kurs, Tagung usw.*
confirmation of order   Auftragsbestätigung

confirmation of receipt   Empfangsbestätigung * *für Ware, Brief, Zahlung*
conflict   kollidieren * *(figürl.)* || Kollision || Konflikt * *Konflikt, Streit, Widerspruch, Zusammenstoß; come into conflict:in ~ geraten* || Widerspruch * *Konflikt, Widerstreit, Widerspruch, Gegensatz, Zusammenstoß, Kampf*
conflict situation   Konflikt * *~situation*
conflicting   widersprechend * *widerstrebend/-sprechend (Forderungen, Gesetze, Gefühle usw.)*
conform to   entsprechen * *einer Norm/Vorschrift ~* || erfüllen * *einer Norm/Vorschrift usw. entsprechen* || genügen * *einer Vorschrift ~; siehe auch →entsprechen*
conformable   konform * *to: zu/mit* || übereinstimmend
conformance   Konformität * *conformance declaration: Konformitätserklärung* || Übereinstimmung * *in conformance with: gemäß, in ~ mit*
conformance certification   Konformitätsbescheinigung
conformance test   Abnahmeprüfung * *Normprüfung eines Produkts; siehe auch →Abnahme* || Freigabeuntersuchung * *zur Sicherstellung der Übereinstimmung mit einer Norm/Vorschrift*
conformant   konform * *siehe auch →entsprechend*
conforming to   entsprechend * *z.B. einer Norm/Vorschrift (auch "in line with": →in Übereinstimmung mit)* || gemäß * *(eine Norm, Vorschrift usw.) erfüllend*
conforming to standards   normgerecht
conformity   Konformität * *in conformity with: konform zu* || Übereinstimmung * *auch 'Gleichförmigkeit, Ähnlichk.'; in conformity with: in ~ mit/gemäß*
confound   verwechseln * *durcheinander bringen (with: mit)* || verwirren * *in Unordnung bringen, verwirren (someone: jemanden), vermengen*
confronted   gegenüber * *be confronted with: sich gegenübersehen, konfrontiert sein mit*
confuse   verwechseln * *confuse with: verwechseln/durcheinander bringen mit* || verwirren * *verworren/undeutlich machen, in Unordnung bringen, vermengen, verwirren*
confusing   verwirrend
confusion   durcheinander * *(Subst.)* || Verwechslung || Verwirrung
conglomerate   Mischkonzern
congruent   deckungsgleich * *(geometr.)*
conical   kegelförmig * *konisch* || kegelig * *konisch* || konisch
conical refiner   Kegelmühle * *→Refiner mit konischen Mahlwalzen in der Zellstoffindustrie (mahlt Holzstoff)*
conical wheel   Kegelrad
conical-roller bearing   Kegelrollenlager
conical-roller mill   Kegelmühle
coniferes   Holz * *Nadelhölzer*
coniferous   Holz * *Nadel~* || Nadelholz
coniferous wood   Nadelholz
coniform   kegelförmig
conjunction   Konjunktion * *Schaltalgebra* || Verbindung * *logisch; in conjunction with: in ~/Zusammenhang mit* || Zusammenhang * *Verbindung; in conjunction with: in Verbindung mit*
connect   anklemmen * *Draht (to a terminal: an eine Klemme)* || ankoppeln || anschalten * *mit Draht* || anschließen * *verbinden, anschließen, auflegen* || befestigen * *miteinander verbinden* || durchschalten || durchstellen * *jemanden am Telefon weitervermitteln* || führen * *ein Signal ankoppeln, schalten, verbinden* || kontaktieren || koppeln * *verbinden, auch über serielle Schnittstelle* || kuppeln * *siehe auch →koppeln* || schalten * *Verbindg. herstell.; in series/parallel/star/delta:in Reihe/paral-* lel-/Stern/Dreieck || verbinden * *to:mit; via: durch/über; by means of/mit Hilfe von* || verdrahten * *verbinden* || verknüpfen * *verbinden*
connect across   parallel schalten
connect back to back   gegeneinanderschalten
connect downstream   nachschalten * *im Signal-/Leistungsfluss hinterherschalten*
connect in a bus-type configuration   busförmig verbinden
connect in between   zwischenschalten * *as intermediate element: als Zwischenglied*
connect in cascade   kaskadieren
connect in incoming circuit   vorschalten
connect in outgoing circuit   nachschalten * *im Ausgangkreis*
connect in parallel   parallel schalten * *with: zu*
connect in series   hintereinander schalten * *in Reihe schalten (with: zu)* || in Reihe schalten * *with: zu* || in Serie schalten * *with:zu* || nachschalten * *in Reihe schalten* || vorschalten * *in Reihe schalten*
connect in tandem   kaskadieren
connect on line side   vorschalten * *auf der Netzseite ~, z.B. Netzdrossel*
connect on load side   nachschalten * *lastseitig*
connect through   durchschalten || durchverbinden
connect to earth   erden
connect to terminal   herausführen * *auf Klemme*
connect to the system   zuschalten * *ans Netz*
connect with   beziehen * *~ auf*
connect with pins   verstiften
connect-in   zuschalten
connect-up   anschließen || auflegen * *"hinverdrahten/verbinden", z.B. Leitungen an Klemme* || hinverdrahten || verdrahten * *"hinverdrahten", anschließen, auflegen, z.B. auf eine Klemme*
connected   aufgelegt * *z.B. ein Signal auf eine Klemme* || durchgeschaltet || zusammenhängend * *in Beziehung stehend*
connected across   parallelgeschaltet
connected in delta   in Dreieck geschaltet
connected in parallel   parallelgeschaltet
connected in series   hintereinandergeschaltet * *(elektr.)* || in Reihe geschaltet * *with: zu* || nachgeschaltet
connected in star   in Stern geschaltet
connected interface   Koppelpartner
connected load   Anschlussleistung
connected slave   Koppelpartner * *bei Datenkopplung nach dem Master-Slave-Prinzip*
connected station   Koppelpartner
connected to common potential   gewurzelt * *mit gemeins. Bezugspotenzial (z.B. Digital-Ein-/Ausgänge)*
connected upstream   vorgeschaltet * *im Signal-/Leistungsfluss ~*
connected via a serial link   gekoppelt * *über serielle Schnittstelle ~*
connected with   in Zusammenhang mit * *be connected with: ~ stehen mit; have no connexion with: nicht ~ stehen mit*
connecting   Anschluss- || Verbindungs-
connecting adapter   Anschlussstück * *für Schlauch, Rohr usw.*
connecting bar   Anschlussschiene
connecting cable   Anschlusskabel || Anschlussleitung || Verbindungskabel || Verbindungsleitung
connecting dimensions   Anschlussmaße
connecting element   Befestigungselement * *→Verbindungselement* || Verbindungselement * *z.B. in einer mechan. Konstruktion/einem Rohrleitungssystem*
connecting example   Anschlussbeispiel
connecting gear   Vorgelege
connecting in parallel   Parallelschaltung * *das Parallelschalten*

connecting in series Serienschaltung * *das In-Serie-Schalten*
connecting lead Anschluss * *~leitung* || Anschlussdraht || Anschlussleitung
connecting line Anschlussleitung * *Kommunikations-/Telefon~* || Verbindungslinie
connecting link Zwischenglied
connecting lug Anschlussfahne || Anschlusslasche || Verbindungslasche
connecting of pipes Rohranschluss * *das Anschließen von Rohren*
connecting panel Anschlussfeld
connecting passage Durchgang * *für Flüssigkeiten, Gase*
connecting piece Anschlussstück || Anschlussstutzen * *Rohr* || Stutzen * *z.B. ~ für Rohrverbindung (siehe auch →Gewindestutzen)*
connecting pipe Verbindungsleitung * *Rohrleitung*
connecting plate Verbindungslasche
connecting plug Anschlussstecker
connecting point Anschlusspunkt || Verbindungsstelle
connecting rod Kolbenstange * *Verbindungs-/Pleuelstange* || Pleuelstange || Schubstange
connecting surface Anschlussfläche * *(mechanisch)*
connecting tag Anschlussöse * *z.B. an Lötfahne*
connecting terminal Anschlussklemme
connecting to common potential Wurzelung * *bei Digital-Ein-/Ausgängen*
connecting unit Anschlusseinheit
connecting wire Verbindungsleitung
connecting-up Anbindung * →*Verbindung* || anschließen * *(Substantiv)* || Anschluss * *das Anschließen, der Vorgang des Anschließens*
connecting-up designation Anschlussbezeichnung
connecting-up diagram Anschlussplan
connecting-up instructions Anschlusshinweise
connection Anbindung * *Verbindung; siehe auch* →*Verknüpfung; (to: an)* || Ankopplung * *to:an* || Anschluss * *Verbindung, Leitungs~, auch Netz~* || Beziehung * *with: zu Personen, Firmen* || Kopplung * *Verbindung* || Kopplung * *Verbindung, z.B. über Signale, Bus usw.* || Schaltung * *~ mit Leistungshalbleitern, Leistungsteil, Trafo usw.* || Verbindung * *(auch figürl.) in connection with: in ~ mit; elektr. Anschluss* || Verknüpfung * *Verbindung, z.B. von Signalen; Zusammenhang, Beziehung, Anschluss* || Zusammenhang * *Beziehung; in this connection: in diesem ~; Verbindung, Verwandtschaft*
connection circuit Anschlussschaltung
connection cross-section Anschlussquerschnitt
connection designation Anschlussbezeichnung
connection diagram Anschlussbild || Anschlussplan || Anschlussschaltbild || Anschlussschema || Schaltschema * *Anschlussplan* || Verdrahtungsplan
connection link Verbindungselement
connection matrix Rasterfeld * *"Spielwiese" auf Leiterkarten mit freien Bestückungsplätzen*
connection plate Knotenblech
connection point Anschlusspunkt || Anschlussstelle || Verbindungsstelle
connection system Anschlusstechnik
connection to supply system Netzanschluss
connectivity Vernetzung * *als Funktion*
connector Anschluss * *~vorrichtung, Stecker* || Konnektor || Stecker * *male/plug connector: Stift~; female connector: Buchsen~* || Steckverbinder || Verbinder
connector assignment Kontaktbelegung * *für Stecker* || Steckerbelegung
connector designator Konnektorbezeichner * *z.B. beim SIEMENS Regelsystem SIMADYN D (R)*

connector housing Steckergehäuse
connector pin Steckerstift
connector pin assignment Steckerbelegung
connector receptacle Steckergehäuse
connector sleeve Aderendhülse
connector strip Anschlussleiste || Lüsterklemme
connexion Verknüpfung * *Verbindung, z.B. von Signalen; Zusammenhang, Beziehung, Anschluss* || Zusammenhang * *Beziehung, Verbindung; siehe* →*connection*
connotation Assoziation * *Nebenbedeutung, 2. Bezeichnung, Begriffsinhalt*
conquer überwinden * *besiegen*
conscientious gewissenhaft || sorgfältig * *gewissenhaft*
conscientiousness Sorgfalt * *Gewissenhaftigkeit*
conscious bewusst * *be conscious of: sich einer Sache ~ sein*
consecutive aufeinander folgend || fortlaufend * *aufeinander folgend* || laufend * *(fort)laufend, z.b. Nummerierung*
consecutive effect Folgeerscheinung
consecutive fault Folgefehler
consecutive numbers fortlaufende Nummern
consecutive symptom Folgeerscheinung * *Symptom*
consecutively nacheinander * *aufeinander folgend* || unmittelbar hintereinander
consent einverstanden sein * *to:mit*
consent to billigen || genehmigen * *gutheißen (of:etwas/jemanden)*
consequence -Wirkung * *Folge* || Auswirkung * *Rückwirkung, Folge* || Bedeutung * *Folge, Resultat, Wirkung, Wichtigkeit, Bedeutung; of no consequence: ohne ~* || Erfolg * *Folge* || Folge * *Wirkung, logische ~, ~erscheinung* || Folgeerscheinung || Konsequenz * *Folge; take the consequences: die ~en tragen* || Rückwirkung * *Auswirkung* || Wirkung * *Folge*
consequences Nachwirkung * *Folgen*
consequent konsequent
consequential konsequent
consequentially konsequent * *(Adv.)*
consequently also * *(Adv.)* || daher * *folglich* || folglich || infolgedessen || somit * *folglich*
conservation Erhaltung * *Bewahrung, Instandhaltung, Schutz; Haltbarmachung, Konservierung*
conservative vorsichtig * *konservativ (Schätzung usw.)*
conserve erhalten * *erhalten, bewahren, beibehalten* || konservieren || pflegen * *erhalten*
consider beachten * *berücksichtigen* || bearbeiten * *bedenken, überlegen* || befassen * *sich prüfend befassen mit* || berücksichtigen * *beachten, i. Betracht ziehen, bedenken, erwägen, genehmigen* || betrachten * *as:als* || erfassen * *berücksichtigen* || finden * *dafürhalten* || prüfen * *erwägen* || überlegen * *bedenken*
considerable beachtlich || bedeutend * *(Adj.)* || beträchtlich * *(Adj.) auch 'erheblich, ansehnlich, bedeutend'* || deutlich * *beträchtlich* || erheblich * *beträchtlich, bedeutend* || fühlbar * *beträchtlich* || merklich * *beträchtlich, erheblich, bedeutend* || nennenswert || spürbar * *beträchtlich* || wesentlich * *(Adj.) beträchtlich, erheblich, bedeutend*
considerably bedeutend * *(Adv.)* || beträchtlich * *(Adv.)* || erheblich * *(Adv.) beträchtlich, bedeutend* || wesentlich * *(Adv.) beträchtlich, bedeutend*
consideration Beachtung * *Berücksichtigung; give due consideration to sth.: etwas gebührend* →*berücksichtigen* || Beratschlagung * *(of:über)* || Berücksichtigung * *be considered: ~ finden; in consideration of: unter ~ von* || Erfassung || Prüfung * *Erwägung, Überlegung* || Rücksicht * *in consideration of/to: mit ~ auf, in Anbe-*

tracht von, hinsichtlich || Überlegung * Betrachtung; upon mature consideration: nach reiflicher ~ || Überprüfung * Erwägung, Überlegung
considered as parts of a cohesive whole ganzheitlich betrachtet
considering angesichts || für * in Anbetracht von || gegenüber * angesichts
considering that unter Berücksichtigung von
consign kommissionieren * übersenden, verschicken, adressieren || versenden
consignment Kommission * Sendung, Lieferung || Lieferung * Sendung || Sendung * (Wirtschaft) (Fracht-) Sendung, Lieferung, Konsignation, Komission(sware), Versand
consignment note Frachtbrief
consist bestehen * of: aus
consist of bestehen aus
consistency Dichte * Konsistenz, Beschaffenheit, →Dichtheit, Dicke, Übereinstimmung, Folgerichtigkeit || Konsequenz * Festigkeit, Folgerichtigkeit, Konsequenz, übereinstimmung || Konsistenz * Zusammensetzung || Übereinstimmung * Vereinbarkeit, Folgerichtigkeit, Konsequenz, Zusammengehörigkeit, Einklang
consistent durchgängig * konsequent, folgerichtig, logisch, übereinstimmend, stetig, gleichmäßig || geschlossen * (figürlich) Arbeit, Leistung usw. || gleich bleibend * gleichmäßig, stetig, ausgeglich., ständig, fest, unveränderl., stabil (Preise) || gleichmäßig * gleichmäßig, stetig, übereinstimmend, →homogen, ebenmäßig im Gefüge || homogen * →gleichmäßig || konsequent * folgerichtig, logisch, konsequent, gleichmäßig, stetig, übereinstimmend, durchgängig || konsistent * zusammengesetzt, zusammenpassend (z.B. Teildaten einer Informationsmenge) || konstant * gleichmäßig, gleich bleibend, fest || übereinstimmend * folgerichtig, unveränderlich * fest || verträglich * übereinstimmend, vereinbar, im Einklang (with: mit) || zusammengehörig * übereinstimmd., vereinb., i. Einklang (with: mit), zusamm.gehörig(Teile e. Datensatzes)
consisting zusammengesetzt * of: bestehend aus
consisting of several parts mehrteilig
console Pult
consolidate ausbauen * (ver)festigen || fusionieren * (amerikan.) (with: mit) || vereinigen * fusionieren || zusammenfassen * vereinigen, zusammenlegen/ schließen, verdichten || zusammenschließen * zu einem Ganzen ~
consolidation Fusion * (amerikan.) || Konsolidierung * Konsolidierung (auch wirtschaftlich), (Be-)Festigung; Vereinigung/Zusammenschluss || Verfestigung * Festwerden, Festigung, Konsolidierung (auch wirtschaftl.), Verdichtung
conspicuous sichtbar * auffällig, auffallend
constancy Genauigkeit * Konstanz || Konstanz * Langzeitkonstanz (Einflussgrößen: Auflösung, Drift); siehe VDI/VDE 2185
constant -Zahl * -Konstante || beständig * (an)dauernd, gleich bleibend || fest * standhaft || gleich bleibend * unveränderlich, konstant || gleichmäßig * konstant, gleich bleibend || konstant * stay constant: konstant bleiben || Konstante || permanent || ständig * fortwährend, gleich bleibend || stationär * gleich bleibend || stetig * konstant, beständig, gleich bleibend, unveränderlich
constant component Gleichanteil || Konstantanteil
constant current source Konstantstromquelle
constant run Konstantlauf
constant running Konstantlauf
constant speed drive Festdrehzahlantrieb || Konstantantrieb * Festdrehzahlantrieb
constant torque drive Konstantmoment-Antrieb
constant voltage generator Konstantspannungsgeber || Konstantspannungsgenerator * Synchrongenerator m.eigenem Konstantspannungsgerät
constant-current source Konstantstromquelle
constant-power range Feldschwächbereich * Drehzahl steigt u. max. Drehmoment sinkt –> annähernd const. Leistung
constant-speed drive Konstantdrehzahlantrieb || ungeregelter Antrieb
constant-torque drive Konstantmomentantrieb
constantly laufend * (Adv.) unaufhörlich, stetig, regelmäßig, gleich bleibend, beharrlich, standhaft || ständig * (Adv.) || stets * ständig
constellation Konstellation || Lage * Konstellation
constituent Anteil * (Bestand-) Teil, Komponente || Bestandteil * Teil, Bestandteil, (physikal.) Komponente, Auftrag-/Vollmachtgeber || Komponente * Bestandteil, Komponente || zugehörig * einen (Bestand-) Teil bildend, constituent part: wesentlicher Bestandteil
constituent part Bestandteil * (wesentlicher) Bestandteil
constitute bilden * ausmachen, bilden, ernennen, einsetzen, in Kraft setzen, einsetzen || ernennen
constituted beschaffen * gebildet, gestaltet, geartet
constraint Beschränkung
constrict einschnüren * zusammenziehen, zusammenschnüren, einengen, zusammenpressen
constriction Einschnürung * mechan.
construable interpretationsfähig
construct aufbauen * gestalten, formen, entwerfen, konstruieren || bauen || erstellen * Gebäude ~ || konstruieren
construction Aufbau * Bauweise, Konstruktion || Ausführung * Aufbau, Konstruktion || Bau * ~ von Maschinen usw. || Bau- || Konstruktion || Sinn * Auslegung
construction circular saw Baukreissäge * (Holz)
construction component Bauelement * (mechanisches/maschinenbautechnisches) Konstruktionselement
construction drawing Konstruktionszeichnung
construction engineering Bautechnik * Hochbau
construction feature Konstruktionsmerkmal
construction industry Bauindustrie
construction kit Bausatz
construction length Baulänge * z.B. eines Motors
construction machinery Baumaschinen
construction material Baustoff
construction site Baustelle
construction size Baugröße
construction suited for harsh environments Robustbauform
construction under licence Nachbau * (brit.) Lizenzfertigung
construction under license Nachbau * (amerikan.) Lizenfertigung
construction volume Bauvolumen
construction-site crane Baukran
constructional defect Konstruktionsfehler
constructional detail Konstruktionselement * Konstruktionseinzelheit
constructional drawing Konstruktionszeichnung
constructional element Konstruktionselement
constructional feature Konstruktionsmerkmal
constructional flaw Konstruktionsfehler
constructor Konstrukteur * Erbauer, Konstrukteur
consult anfragen * um Rat fragen || konsultieren * um Rat fragen || nachfragen * um Rat fragen || nachschlagen * z.B. in Buch/Betriebsanleitung ~ || Rückfragen * (Verb) um Rat fragen || wenden * sich (Rat suchend) wenden an
consultancy Beratung || Beratungsfirma
consultancy service Beratung * (Industrie-) Beratung als Dienstleistung
consultant Berater || Beratungsfirma || Fachberater

* *von externem Unternehmen* || Freiberufler
* *freier Berater* || freier Mitarbeiter * *Berater, Spezialist*
consultant company Ingenieurbüro
consultant firm Ingenieurbüro * *Beraterfirma*
consultation Beratung * *Besprechung mit e. Fachmann (with a person: mit jemandem)* || Konsultation || Rücksprache * *bei Fachmann; on consultation with: nach ~ mit; consult s.o.: m. jdm. ~ nehmen*
consulting Beratung * *Industrieberatung(sfirma)*
consulting company Unternehmensberatung
consulting engineer's office Ingenieurbüro
consumable Verschleißteil * *Verbrauchsartikel*
consumable load Betriebslast
consume verbrauchen
consumed aufgenommen * *verbraucht, z.B. Leistung*
consumer Abnehmer * *Verbraucher* || Konsum- || Konsument || Kunde * *Verbraucher (von Konsumartikeln)* || Verbraucher * *Konsument, Energie~ (auch Gerät, Bauelement)*
consumer electronics Unterhaltungselektronik * *Konsumelektronik*
consumer quality Konsumqualität * *im Gegensatz zur höherwertigen Industriequalität*
consumption Aufnahme * *Verbrauch* || Verbrauch
contact Anpress- || Ansprechen * *(Verb) jemanden* || berühren * *z.B. mit der Hand, von Maschinenteilen untereinander usw.* || Berührung || konsultieren * *Kontakt aufnehmen/haben mit* || Kontakt * *auch elektrischer ~* || Kontakt aufnehmen || Kontakt aufnehmen mit || nachfragen * *sich wenden an, Kontakt aufnehmen mit* || Schaltstück * *Kontakt* || Verbindungs-
contact area Auflagefläche
contact arrangement Kontaktanordnung * *z.B. bei Relais, Taster usw.* || Kontaktbestückung * *von Relais, Schützen usw.*
contact assignment Kontaktbelegung * *Relais, Schalter*
contact block Kontaktsatz * *für Relais/Schalter/Taster*
contact bounce Kontaktprellen
contact capacity Kontaktbelastbarkeit
contact erosion Abbrand * *~ von Kontakten* || Kontakt-Abbrand
contact face Kontaktfläche || Lauffläche * *einer (Kohle)Bürste*
contact lever Schalthebel * *bei elektr. Schalter*
contact load Kontaktbelastbarkeit || Kontaktbelastung
contact loading capability Kontaktbelastbarkeit
contact man Verbindungsmann
contact mechanism Schaltwerk * *bei elektromechan. Schalter*
contact member Schaltstück * *Schütz, Leistungsschalter usw.*
contact partner Ansprechpartner || Kontaktpartner
contact pattern Tragbild * *bei Zahnradflanken, Lauffläche usw.*
contact person Ansprechpartner || Bearbeiter * *Kontaktperson, z.B.in Vertriebsniederlassung* || Gesprächspartner * *Ansprechpartner* || Partner * *Ansprechpartner* || Sachbearbeiter * *Ansprechpartner* || Verbindungsperson
contact piece Schaltstück * *Schütz, Leistungsschalter*
contact pressure Anpressdruck || Anpresskraft * *Anpressdruck* || Flächenpressung * *siehe auch →Spannkraft* || Presskraft * *→Anpresskraft* || Spannkraft * *Anpressdruck (z.B. bei einer Spannvorrichtung)*
contact rating Kontaktbelastbarkeit || Kontaktbelastung || Schaltleistung * *~ eines (Relais-/Schütz-) Kontakts* || Schaltvermögen * *~ eines (Relais-/ Schütz-) Kontakts* || Strombelastbarkeit * *spezifizierte Nenn-~ eines (Relais, Schütz-, Schalter-) Kontakts*
contact ratio Überdeckungsgrad
contact reliability Kontaktsicherheit
contact resistance Kontaktwiderstand * *z.b. eines Schaltkontakts oder Kommutators* || Übergangswiderstand * *eines Schaltkontaktes, Schleifkörpers oder Kommutators*
contact roller Führungsrolle
contact screw Kontaktschraube
contact stability Kontaktsicherheit * *high: große/ hohe/gute*
contact stud Kontaktschraube * *~ ohne Kopf*
contact surface Lauffläche
contact tab Kontaktzunge * *z.b. bei einer Flachsteckverbindung*
contact welding Verschweißen * *von Kontakten*
contact width Überdeckungsbreite * *Kohlebürstenabmessung in tangentialer Richtung (Breite der abgerundeten Kante)*
contact winder Kontaktwickler || Oberflächenwickler || Umfangswickler
contact wire Fahrdraht || Fahrleitung * *(Bahn)*
contact-free berührungslos
contactless berührungsfrei || berührungslos || kontaktlos
contactor Schaltschütz || Schütz * *air-break contactor: Luft~; vacuum contactor: Vakuum~*
contactor coil Schützspule
contactor contact Schützkontakt
contactor control Schützansteuerung || Schützsteuerung
contactor control circuits Schützsteuerung
contactor energizing Schützansteuerung
contactor relay Hilfsschütz
contactor solenoid Schützspule
contain aufnehmen * *enthalten* || enthalten || halten * *enthalten*
container Behälter || Container || Gefäß || Verbunddokument * *(Computer) Datei, die eingebette (OLE-) Objekte enthält*
container crane Containerkran
containing harmonics oberschwingungshaltig
contaminant Schadstoff
contaminate verschmutzen * *verunreinigen* || verunreinigen
contaminated schmutzig * *verunreinigt* || verschmutzt * *verunreinigt, besudelt, vergiftet, verseucht (auch Umwelt, Atmosphäre)* || verunreinigt
contaminated water Brackwasser * *verschmutztes Wasser*
contamination Schmutz * *→Verunreinigung, →Verschmutzung* || Verschmutzung * *→Verunreinigung* || Verunreinigung
contemplate andenken * *ins Auge fassen, erwägen, nachdenken (über), überdenken, erwarten, rechnen mit* || betrachten * *sinnend ~*
contemporaneous zusammen * *gleichzeitig*
contemporary heutig * *zeitgenössisch*
contender Mitbewerber * *Konkurrent, (Mit)bewerber, Mitanbieter*
content Anteil * *Inhalt, Gehalt, z.B. an Oberwellen* || Gehalt * *Inhalt* || Inhalt || zufrieden * *zufrieden; rest/be content with: sich ~geben mit; bereit, willens*
content-addressable memory assoziativer Speicher
contented zufrieden * *(with: mit)*
contentedness Zufriedenheit * *Zufriedenheit, Genügsamkeit*
contention Argument * *Argument, Behauptung, Streitpunkt*
contentment Zufriedenheit

contents Inhaltsverzeichnis
contents of data memory Speicherinhalt
contents of data register Speicherwert * *SPS*
contestable einwandfrei * *unanfechtbar*
context Kontext * *siehe auch* →*Zusammenhang* ||
 Zusammenhang * *Kontext, inhaltlicher ~, ~ von Textstellen, Umgebung, Milieu*
contiguous angrenzend || zusammenhängend * *angrenzend, berührend, benachbart, lückenlos ~; discontinuous: nicht ~/mit Lücken*
contingent abhängig * *(up)on:von Umständen* || eventuell * *eventuell, möglich, zufällig, ungewiss, gelegentlich* || potenziell * *eventuell,* →*möglich; contingent fee: Erfolgshonorar*
contingent on abhängig von * *von Umständen*
continual beständig * *ausdauernd* || dauernd * *immer wiederkehrend, (sehr) häufig, oft, wiederholt, fortwährend, dauernd* || kontinuierlich * *fortwährend, dauernd, immer wiederkehrend* || ständig * *fortwährend, dauernd; immer wiederkehrend, (sehr) häufig, oft wiederholt* || stetig * *fortwährend, dauernd, immer wieder(kehrend)* || stufenlos * *stetig, fortwährend, dauernd*
continually dauernd * *(Adv.) fortwährend, dauernd, immer wieder, ständig* || permanent * *(Adv.)* →*dauernd, fortwährend* || ständig * *(Adv.) fortwährend, dauernd, immer wieder*
continually running durchlaufend * *ohne Unterbrechung/fortwährend ~; immer wiederkehrend*
continuation Fortsetzung
continue fortführen * *fortsetzen* || fortsetzen
continue counting weiterzählen
continued education Weiterbildung
continuing weitergehend * *fortgesetzt*
continuity Durchgang * *Stromdurchgang, geschlossene Verbindung* || Durchgängigkeit * *ununterbrochener Zusammenhang, Kontinuität, zusammenhängendes Ganzes* || Gleichmäßigkeit || Stromdurchgang || Zusammenhang * *enge Verbindung, ununterbrochene Folge, Fortlaufendes*
continuity tester Durchgangsprüfer || Verbindungstester * *Durchgangsprüfer*
continuos running duty Betriebsart S1 * *[el. Masch.]* Dauerbetrieb
continuous andauernd * *ununterbrochen, (fort)laufend, zusammenhängend, unaufhörlich, andauernd, fortwährend* || Dauer- || dauernd * *auch ständig* || durchgehend || Durchlauf- || durchlaufend * *gleich bleibend* * *ununterbrochen, fortlaufend* || kontinuierlich || laufend * *ununterbrochen, (fort)laufend, unaufhörlich, kontinuierlich, progressiv* || lückenlos * *ununterbrochen, fortlaufend, zusammenhängend, fortwährend, (aus)dauernd* || ständig * *laufend, ohne Unterbrechung* || stetig * *ununterbrochen,(fort)laufend, kontinuierlich* || stufenlos * *ununterbrochen, fortlaufend, kontinuierlich* || ungedämpft * *Welle usw.* || zusammenhängend * *fortlaufend*
continuous brake power Dauerbremsleistung
continuous braking power Dauerbremsleistung
continuous caster Stranggießanlage || Stranggussanlage
continuous casting Strangguss
continuous casting line Stranggießanlage || Stranggussanlage
continuous casting machine Stranggussanlage
continuous casting mill Stranggießanlage
continuous casting plant Stranggießanlage || Stranggussanlage
continuous conveyor Kreisförderer
continuous current Dauerstrom || nicht lückender Strom || nichtlückender Strom
continuous current rating Nenndauerstrom
continuous current state nichtlückender Betrieb
continuous DC current Dauergleichstrom

continuous DC forward current Dauergleichstrom * *einer Diode*
continuous DC on-state current Dauergleichstrom * *eines Thyristors*
continuous direct current Dauergleichstrom
continuous drawing machine Strangziehmaschine
continuous duty Dauerbelastung * *das dauernd zu leistende (Betriebsart)* || Dauerbetrieb * *Belastung, Betriebsart* || ununterbrochener Betrieb * *Motor- u. Stromrichter*→*betriebsart S6 bis S8, siehe auch* →*Durchlaufbetrieb*
continuous duty cycle Dauerbetrieb
continuous duty type Dauerbetrieb * *Betriebsart S1 nach VDE 0530/DIN 41756 für Motoren/Stromrichter*
continuous furnace Durchlaufofen
continuous grinder Stetigschleifer * *[Papier] zur Zellstofferzeugung*
continuous light Dauerlicht * *z.B. einer Diagnose-Leuchtdiode*
continuous line recorder Linienschreiber
continuous load Dauerbelastung * *Dauerlast* || Dauerlast
continuous load current Dauerlaststrom
continuous load limit Dauerlastgrenze
continuous loading Dauerbelastung || Dauerlast
continuous loading capability Dauerbelastbarkeit
continuous loop Dauerschleife || Endlosschleife * *Dauerschleife*
continuous lubrication Dauerschmierung
continuous moisture metering equipment Durchlauf-Feuchtemessanlage
continuous operation Dauerbetrieb * *für Motor: siehe EN 60034-1* || Durchlaufbetrieb * *bei [el. Masch.]:* →*Betriebsarten S6 bis S8* || durchlaufender Betrieb || nichtlückender Betrieb || S1-Betrieb * *Betriebsart "Dauerbetrieb"* || ununterbrochener Betrieb
continuous operation duty type ununterbrochener Betrieb * *als Betriebsart*
continuous operation with short-time loading Durchlaufbetrieb mit Kurzzeitbelastung
continuous output Dauerleistung * *z.B. ~ eines Motors/Stromrichters*
continuous output current Dauer-Ausgangsstrom
continuous oven Durchlaufofen
continuous power Dauerleistung * *maximum permissible: maximal erlaubte*
continuous press Durchlaufpresse * *z.B. für Spanplatten (im Gegensatz zur* →*Taktpresse)*
continuous pulse Dauerimpuls
continuous rated current Dauernennstrom
continuous rating Dauerleistung * *Nenn~, z.B. eines Motors*
continuous resistance annealer Durchlaufwiderstandsglühofen * *[Draht]*
continuous running Dauerbetrieb || Durchlaufbetrieb
continuous running duty type Dauerbetrieb * *[el. Masch.] Betriebsart S1*
continuous section Konstantteil * *kontinuierlich produzierender Teil einer Papiermaschine (ab dem Stoffauflauf)*
continuous signal Dauersignal
continuous stressing Dauerbelastung * *Dauerbeanspruchung*
continuous test Dauerversuch
continuous torque drive Konstantmoment-Antrieb
continuous transition stetiger Übergang || stoßfreier Übergang
continuous web durchlaufende Warenbahn
continuous-action controller kontinuierlicher Regler * *z.B. Analogregler*
continuous-current area Nichtlückbereich || nichtlückender Bereich

continuous-flow dryer Durchlauftrockner
continuous-operation duty type with related load/ speed changes ununterbrochener Betrieb mit periodischer Drehzahländerung * *Motorbetriebsart S8 nach VDE0530*
continuous-operation periodic duty Betriebsart S6 * *[el. Masch] Ununterbrochener periodischer Betrieb* || Durchlaufbetrieb mit Aussetzbelastung
continuous-operation periodic duty type Durchlaufbetrieb mit Aussetzbelastung * *Motorbetriebsart S6 nach VDE0530/IEC34-1*
continuous-operation periodic duty with electric braking Betriebsart S7 * *[el. Masch.]*
continuous-operation periodic duty with related load/speed changes Betriebsart S8 * *[el. Masch.]*
continuous-path control Bahnsteuerung * *[NC-Steuerung]*
continuous-running duty Dauerbetrieb * *Betriebsart S1 für Motoren/Stromrichter nach VDE 0530*
continuously dauernd * *(Adv.)* || kontinuierlich * *(Adv.)* || laufend * *(Adv.)* ununterbrochen, fortlaufend, kontinuierlich || ständig * *(Adv.)* || stets * *ständig*
continuously changeable stufenlos verstellbar * *z.B. Getriebe*
continuously changeable gearbox stufenlos verstellbares Getriebe
continuously running durchlaufend
continuously variable stufenlos verstellbar
contollable field rectifier steuerbarer Feldgleichrichter
contour Kontur || Profil || Umriss
contour control Bahnsteuerung * *[NC-Steuerung]*
contour milling Konturfräsen
contour milling unit Kurvenfräseinheit
contour sanding Konturschleifen * *[Holz] mit Schmirgel*
contoured gebogen * *Fläche*
contouring control Bahnsteuerung * *[NC-Steuerung]* Bewegung entlang einer Kontur Steuern
contouring error Schleppfehler * *beim "Konturfahren" mit einer NC (Numerische Mehrachs-Bewegungssteuerung)*
contra-flow Gegenstrom * *von Luft oder einer Flüssigkeit*
contra-rotating gegenläufig * *in Gegenrichtung rotierend*
contract Auftrag * *Liefervereinbarung, öffentlicher Auftrag, Vertrag* || beauftragen * *vertraglich* || schließen * *einen Vertrag* ~ || schrumpfen * *(sich) zusammenziehen/verengen, schrumpfen* || schwinden * *sich zusammenziehen, z.B. erhitztes Metall in der Abkühlungsphase* || unter Auftrag nehmen || verengen || Vertrag * *draft: entwerfen, ausarbeiten; under contract: unter* ~ || vertraglich verpflichten * *to/for: zu; auch 'sich ~'; contr. for someth.: sich etw. ausbedingen* || zusammenziehen * *(sich) zusammenziehen, verengern, schrumpfen, kleiner werden*
contract administration Auftragsabwicklung || Bestellabwicklung
contract award Auftragserhalt * *gegen Konkurrenz* || Auftragserteilung * *Auftrags-Gewährung/Vergabe (meist gegen Konkurrenzangebote)*
contract conditions Vertragsbedingungen
contract ID Auftragskennzeichen
contract manufacturing Auftragsfertigung || Kommissionsfertigung * *Fertigung auf Bestellung*
contract No Auftragsnummer
contract number Auftragskennzeichen
contract ref. Auftragskennzeichen
contract ref. No. Auftragskennzeichen
contract reference Auftragskennzeichen
contract reference number Auftragskennzeichen

contract secured Geschäftsabschluss
contract value Auftragswert
contract volume Auftragswert
contraction Einschnürung * *magn.*
contractor Auftragnehmer || Lieferant * *auf Vertragsbasis* || Lieferwerk * *Lieferant*
contractual Vertrags-
contractual agreement vertragliche Regelung
contradiction Widerspruch * *i. contradiction to: im ~ zu; contradiction in terms: ~ in sich selbst*
contradictory sich widersprechend * *→widersprüchlich* || widersprechend || widersprüchlich * *without contradiction/uncontradicted: widerspruchslos*
contrary entgegengesetzt * *(figürlich)* || Gegensatz * *of: zu/von* || gegensätzlich || Gegenteil * *(to:von)-quite on the contrary to:ganz im Gegensatz zu* || gegenteilig || umgekehrt * *(meist figürlich) entgegengesetzt* || widrig * *widerspenstig, eigensinnig, gegensätzlich/-teilig*
contrary to entgegen * *Gegensatz; contrary to all expectations: ~ allen Erwartungen* || gegenüber * *im Gegensatz zu* || im Gegensatz zu
contrary to the regulations unvorschriftsmäßig
contrast Gegensatz * *to:zu; in contrast to/with:im Gegensatz zu* || Kontrast * *auch einer Anzeige/eines Bildschirms*
contrast with abstechen * *~ gegen(-über)*
contravene verstoßen * *~ gegen*
contravention Verstoß * *Zuwiderhandlung*
contribute beitragen * *to:zu* || beteiligen * *Beitrag leisten, sich beteiligen (to: zu/an)* || teilnehmen * *beitragen*
contribution Anteil * *Beitrag, Mitwirkung; make a contribution: einen Beitrag liefern* || Beitrag * *(allg.) auch zu einem Projekt, Geldbeitrag usw.* || Beteiligung * *(to: an/von)*
contrive ausdenken || bewerkstelligen || einfädeln * *(figürl.) zustande bringen, ermöglichen, es einrichten (to: zu)*
control Abwicklung * *Management* || ansteuern || Ansteuerung || aussteuern * *z.B.Thyristor bei Phasenanschnittsteuerung; to give the voltage x: auf die Spannung x* || bedienen * *steuern, führen, lenken, beaufsichtigen, überwachen* || Bedienung || Bedienungs- || beeinflussen * *beherrschen* || Beeinflussung * *Steuerung* || Beherrschung * *auch des Marktes; self-control: Selbst~* || betätigen * *steuern* || Einfluss * *Beherrschung* || Führung * *~ eines Unternehmens/Prozesses/Antriebs* || Gewalt * *Herrschaft (of:über); beyond one's control: höhere Gewalt* || Kontrolle * *Steuer-/Regelung; be in/keep the control of/have well in hand: unter* ~ *haben/halten* || kontrollieren || leiten * *beaufsichtigen* || meistern * *unter Kontrolle bringen, z.B. e. schwierige Lage* || prüfen * *beaufsichtigen, überwachen, nach~* || Regeleinrichtung
control accuracy Regelgenauigkeit * *Abweichung der Regelgröße vom eingestellten (Soll)Wert*
control action Regelverhalten || Reglerverhalten || steuernde Wirkung
control algorithm Regelalgorithmus
control amplifier Regelverstärker
control and operation Bedienung
control behaviour Regelverhalten
control bit Steuerbit
control block Kontrollbaustein * *Funktionsbaustein* || Regelungsbaustein * *Funktionsbaustein* || Steuerbaustein * *z.B. Funktionsbaustein in SPS*
control board Regelbaugruppe * *Leiterkarte* || Regelungsbaugruppe * *Leiterkarte* || Schalttafel || Steuerkarte * *Leiterkarte* || Steuertafel
control bus Steuerbus * *beinhaltet z.B. Read-/Write-Signale bei Mikroprozessorsystem*

control buttons Bedientasten * *Druckknöpfe*
control cabin Steuerstand * *Steuerkabine, z.B. an Kran*
control cabinet Schaltschrank * *Steuerschrank* || Steuerschrank
control cable Steuerkabel || Steuerleitung || Steuerungskabel
control card Elektronikbaugruppe
control centre Schaltwarte || Zentrale * ~ *Warte, Leitstelle*
control character Steuerzeichen * *z.B. in einem Datentelegramm*
control characteristic Regelverhalten || Steuerkennlinie * *z.B. b. Phasenanschnittsteuer.: Ausgangsspannung als Fkt. d. Aussteuerwinkels* || Zündkennlinie * *z.B. bei Gasentladungsröhre*
control circuit Regelkreis || Steuerkreis || Steuerteil
control circuit transformer Steuertrafo || Steuertransformator
control command Steuerbefehl || Steuersignal * *Steuerkommando/-befehl*
control computer Steuerrechner
control concept Regelkonzept
control connections Steueranschlüsse
control console Bedienpult || Leitstand * *z.B. für Druckmaschine*
control constancy Regelkonstanz
control cubicle Steuerschrank
control current Steuerstrom
control desk Bedienpult || Bedienungspult * *siehe auch →Bedienpult* || Pult * *Bedien~* || Schaltpult || Steuerpult
control deviation Regelabweichung * *permanent: bleibende*
control device Bedienungselement
control dynamics Regeldynamik
control electronics Regelelektronik || Steuerelektronik
control element Bedienelement || Bedienungselement || Stellorgan || Steuerorgan
control elements Bedienungselemente
control engineer Regelungstechniker
control engineering Regeltechnik || Regelungstechnik * *als Fachrichtung* || Steuerungstechnik
control equipment Regelgerät || Steuereinrichtung
control error Regelabweichung || Regeldifferenz || Soll-Ist-Abweichung || Soll-Istwert-Abweichung
control factor Aussteuergrad || Aussteuerungsgrad
control function Steuerfunktion || Steuerungsaufgabe
control function block Steuerbaustein * *Funktionsbaustein*
control gear Schaltgetriebe
control input Steuereingang
control interface Nahtstelle * *~ zum Prozeß für Steuer- u. Regelsignale*
control keys Bedientasten * *Steuertaster*
control knob Drehknopf
control law Regelalgorithmus
control lead Steuerleitung * *Leitung, über die Steuer- u. Meldesignale übertragen werden*
control level Steuerebene || Steuerungsebene
control lever Bedienungshebel || Schaltebel * *Bedienungshebel*
control line Steuerleitung * *Steuersignal in einem Prozessorbus*
control location Bedienort
control logic Ansteuerlogik || Steuerlogik
control logic input Steuereingang * *binärer*
control loop geschlossener Regelkreis || Regelkreis || Regelschleife || Regelung
control loop algorithm Regelalgorithmus
control loop block Regelbaustein * *Funktionsbaustein* || Regelungsbaustein * *bei SPS*

control master station Leitwarte || Warte * *→Leitwarte*
control matrix Steuermatrix
control method Regelart || Regelungsart || Regelungsverfahren || Regelverfahren
control module Ansteuerbaugruppe * *Steuerbaugruppe*
control output Steuerausgang
control panel Bedienfeld || Schalttafel * *Steuertafel* || Steuertafel
control performance Führungsverhalten * *Regelverhalten* || Regelgüte * *Regeleigenschaften* || Regelverhalten * *Qualität/Güte/Leistungsvermögen der Regelung*
control point Leitstelle
control power Ansteuerleistung || Steuerleistung
control precision Regelgenauigkeit
control problem Automatisierungsaufgabe
control process Regelvorgang
control program Steuerprogramm || Steuerungsprogramm
control pulpit Steuerbühne * *Steuerkanzel z.B. im Walzwerk*
control pulse Ansteuerimpuls || Steuerimpuls * *auch An~ für Thyristor*
control quality Regelgüte
control range Aussteuerbarkeit * *Aussteuerbereich* || Aussteuerbereich || Regelbereich || Stellbereich || Steuerbereich
control range down to zero Stellbereich bis Null
control reactive power Steuerblindleistung * *z.B. auf Grund der Phasenanschnittsteuerung eines Stromrichters*
control relay Hilfsschütz
control requested Führung gefordert * *(→PROFIBUS-Profil)*
control response Regelverhalten
control room Leitstand || Leitwarte * *Raum* || Schaltraum || Schaltwarte || Steuerraum || Steuerstand * *Warte, Leitraum* || Warte * *Leitraum* || Zentrale * *Warte*
control sampling time Regelungsabtastzeit
control section Steuerteil
control sequence Ablaufkette || Steuerungsablauf
control setting Aussteuergrad || Aussteuerung * *Aussteuergrad* || Steuerung * *Setzen des Steuerwinkels e. Thyristorschaltung*
control signal Steuersignal
control signal error Steuersignalfehler * *bei SPS*
control signal loop Steuerstromkreis
control stability Konstanz * *Konstanz der Regelung*
control statement Steueranweisung
control station Leitstand || Schaltwarte || Steuerstand || Warte || Zentrale * *Steuer~ auch auf Schiff*
control structure Regelstruktur
control supply voltage Steuerspannung * *Versorgungsspannung für Steuerkreise/Steuerelektronik*
control switch Steuerschalter
control switchboard Steuertafel
control system Regelgerät * *Regelsystem* || Regelkreis || Regelsystem || Regelung || Regelungssystem || Steuerung * *System*
control system flowchart Ablaufplan * */SPS/* || Funktionsplan * *beim SIEMENS Regelsystem SIMADYN D (R)*
control system function chart Funktionsplan * */SPS/* || FUP
control systems Regelungs- und Steuerungstechnik * *siehe DIN 19226*
control task Steuer- und Regelaufgabe || Steueraufgabe || Steuerungsaufgabe
control technician Regelungstechniker || Steuerungstechniker
control terminal Bedienterminal || Steuerklemme

control terminal strip Steuerklemmleiste
control terminals Steuerklemmen
control the market Markt beherrschen
control theory Regelungstheorie
control transformer Steuertrafo || Steuertransformator
control type Regelart || Regelungsart
control unit Elektronikteil * *Elektronikeinheit* || Regelgerät || Steuergerät || Steuerteil || Steuerung * *Gerät*
control valve Regelventil || Regulierventil || Servoventil * *Regelventil* || Stellventil * *Regelventil* || Steuerventil
control variable Regelgröße
control version Regelungsausführung
control voltage Steuerspannung * *z.B. Eingangssignal eines Steuersatzes ("Vergleichsspannung für Sägezahn")*
control word Steuerwort * *(auch →PROFIBUS-Profil)*
control-cabinet builder Schaltschrankbauer
control-cabinet manufacture Schaltschrankbau
control-room systems Wartentechnik
controlgear Schaltgeräte
controll roller Regelwalze
controllability Regelbarkeit || Steuerbarkeit
controllable gesteuert * *steuerbar, z.B. Gleichrichterbrücke* || handhabbar * *kontrollierbar, beherrschbar* || kontrollierbar || regelbar || steuerbar * *z.B. Ventil, Stromrichter, Gleichrichter*
controllable field converter steuerbarer Feldgleichrichter
controllable rectifier steuerbarer Gleichrichter
controllable valve steuerbares Ventil
controlled geführt * *gesteuert* || geregelt || gesteuert || gezielt
controlled bridge exciter Feldgleichrichter * *gesteuerter*
controlled field geregeltes Feld
controlled field-current supply geregelte Feldstromversorgung
controlled rectifier gesteuerter Gleichrichter
controlled system Regelstrecke
controlled valve geteuertes Ventil
controlled variable Regelgröße * *die Größe, die ein Regler dem Sollwert nachführen soll*
controller Regelgerät || Regler || Steuerung * *(Steuer-) Gerät*
controller adaption Regleranpassung
controller auto-tuning Reglerselbsteinstellung
controller characteristic parameter Regelparameter || Reglerparameter
controller design Reglerentwurf
controller disable Reglersperre
controller enable Reglerfreigabe * *als Kommando*
controller enabling Reglerfreigabe
controller expansion unit Zentralerweiterungsgerät * *bei SPS*
controller gain Regelverstärkung
controller inhibit Reglersperre * *inhibit the controller: ~ setzen*
controller input Reglereingang
controller matrix Reglermatrix
controller monitoring Reglerüberwachung
controller optimization Regleroptimierung
controller output -Reglerausgang || Reglerausgang
controller overmodulation Reglerübersteuerung
controller overshoot Reglerübersteuerung
controller parameter Reglerparameter
controller rack Zentral-Baugruppenträger * *bei SPS* || Zentralbaugruppenträger * *bei SPS*
controller self-setting Reglerselbsteinstellung
controller structure Reglerstruktur
controller with constricting output limiting signals

ablösender Regler * *Regler mit ablösender Begrenzung*
controller with dynamically acting limit signals ablösender Regler * *Regler m. ablös. Begrenzg. (Grenze dynam. geführt)*
controller with override characteristic ablösender Regler
controller with PI characteristics PI-Regler
controller with transfer characteristic ablösender Regler * *z.B. n-Regler mit Steuerung d. Ausgangsbegrenzg (Moment).*
controlling Führung * *der ~sgröße in einem Regelkreis* || Führungs- * *die Steuerung ausübend*
controlling authority Aufsichtsbehörde * *z.B. TÜV*
controlling element Stellglied
controlling equipment Regeleinrichtung
controlling system Regeleinrichtung
controlling the market marktbeherrschend
controls Bedienungselemente || Kontrolleinrichtungen || Regelfunktionen || Regeltechnik * *Regelsysteme*
convection Konvektion * *selbsttätiger Luft/Flüssigkeitsumlauf, Wärmeströmung (natural: freie)*
convection cooling Konvektionskühlung * *Kühlg. durch natürlichen Luftstrom (siehe auch →Luftselbstkühlung)* || Luft-Selbstkühlung || Luftselbstkühlung * *von Leistungshalbleitern* || Selbstkühlung * *Konvektionskühlung*
convection-cooled luftselbstgekühlt * *Kühlung (z.B. v. Leistungshalbleitern) über freie Luftkonvektion (kein Lüfter)*
convectionary machine Süßwarenmaschine
convene einberufen * *Versammlung usw. ~* || versammeln * *(sich) versammeln, z.B. zu e. Besprechung; einberufen*
convene a conference Besprechung einberufen
convene a meeting Besprechung einberufen
convenience Annehmlichkeit * *Bequemlichkeit, Komfort, Vorteil, Nutzen, Nützlichkeit* || Komfort * *Annehmlichkeit, Bequemlichkeit, Vorteil, Nutzen, Eignung, Angemessenheit*
convenient geeignet * *gut geeignet, günstig* || komfortabel * *bequem, praktisch, (zweck-)dienlich, brauchbar* || passend * *(gut) geeignet, günstig (to/for: für/zu), bequem, günstig passend, gelegen* || vorteilhaft * *geeignet, →zweckmäßig, bequem, passend, zweckdienlich * *zweckmäßig * *geeignet, bequem, passend, zweckdienlich*
convention Vertrag * *Abkommen*
conventional handelsüblich * *herkömmlich* || herkömmlich * *(auch eine techn. Konstruktion); in the conventional sense: im ~n Sinn* || klassisch * *konventionell, herkömmlich* || konventionell || normal * *herkömmlich* || üblich * *herkömmlich*
conventional agreement Abmachung * *vertragliche*
conventional memory konventioneller Speicher * *von DOS ansprechbarer Speicherbereich 0 bis 640 K beim PC*
converge konvergieren * *zusammenlaufen, sich (einander) nähern*
convergence Konvergenz
conversation Gespräch || Unterhaltung * *Gespräch*
converse umgekehrt
conversely umgekehrt * *(Adv.)*
conversion Konvertierung || Umbau * *zu einem neuen Zweck* || Umformung || Umrechnung * *Umwandlung* || Umsetzung * *Umwandlung* || Umwandlung || Verschlüsselung * *durch einen Analog-/Digitalumsetzer usw.*
conversion block Konvertierungsbaustein * *Funktionsbaustein*
conversion factor Stromrichtgrad * *(U_dc. x I_dc/ Grundschwingungsleistung auf der AC-Seite) bei DC-Stromrichter* || Umrechnungsfaktor

conversion operation Umwandlungsoperation
conversion principle Umsetzprinzip * z.B. für A/ D-Wandler
conversion table Umrechnungstabelle
conversion time Verschlüsselungszeit * z.B. eines Analog/Digital-Umsetzers || Wandlungszeit * z.B. eines Analog/Digital-Umsetzers
convert umbauen * zu einem neuen Zweck ~ (into: zu) || umformen || umrechnen * into:in || umrichten || umschlüsseln || umsetzen * umwandeln || umwandeln * auch Energie, Signale || verarbeiten * into: zu || veredeln * umformen, umwandeln, veredeln, weiterverarbeiten
converted timber Schnittholz
converted timber drier Schnittholztrockner
converter Adapter || Konverter || Stromrichter * siehe DIN EN 60146, IEC Publ. 146 u. 411, DIN 41750/51/52, DIN 57558 und VDE 0558 || Übersetzer * Umsetzer, Umsetzprogramm; Konverter || Umformer || Umrichter || Umsetzer || Wandler
converter apparent power Umrichterscheinleistung
converter arm Stromrichterzweig
converter assembly Stromrichtersatz
converter block Momentenschale * Softwarebaustein zur Leistungsteil-/Stromrichteransteuerung
converter cascade Stromrichterkaskade * subsynchronous:untersynchrone
converter circuit Stromrichterschaltung
converter circuit configuration Stromrichterschaltung * Anordnung
converter circuits Stromrichterschaltungen * DIN 41761
converter conduction-through Wechselrichterkippen
converter configuration Stromrichterschaltung
converter connection Stromrichterschaltung
converter connections Stromrichterschaltungen
converter cost Stromrichteraufwand
converter door Gerätetür * eines Stromrichters
converter duty Betrieb am Umrichter * Motor
converter equipment Stromrichtergerät
converter fan Gerätelüfter * eines Stromrichters
converter fault Umrichter-Störung
converter fed stromrichtergespeist
converter leg Stromrichterzweig
converter limit Gleichrichtertrittgrenze
converter motor RG-Motor * →Stromrichtermotor || Stromrichtermotor * Sychronmot. 1-250MW m. Ständerspeisg. durch →lastkommut. I-Umrichter u. →RG-Erregung
converter operation Betrieb am Umrichter * Motor || Stromrichterbetrieb
converter output Geräteleistung * eines Stromrichters || Umrichterleistung * Ausgangsleistung
converter plant Stromrichteranlage
converter powered umrichtergespeist * Antriebssystem usw.
converter rating Umrichterleistung
converter ratings Gerätenenndaten * eines Stromrichters
converter replacement Gerätetausch * von Stromrichtergeräten
converter response Gerätereaktion * (→PROFIBUS-Profil)
converter reversal Stromrichtungsumkehr * in einem Stromrichter
converter section Brückenhälfte * eines Gleichrichters || Teilstromrichter
converter set Stromrichtersatz
converter supply Stromrichterbetrieb * ~ eines Motors am Stromrichter/Umrichter (nicht direkt am Netz) || Umrichterspeisung * Speisung e. AC-Maschine durch e. Umrichter, nicht direkt vom Netz

converter technology Stromrichtertechnik
converter transformer Stromrichtertrafo || Stromrichtertransformator
converter unit Stromrichtereinheit || Stromrichtergerät || Stromrichtersatz
converter valve Stromrichterventil
converter-fed umrichtergespeist
converter-fed drive Stromrichterantrieb || stromrichtergespeister Antrieb
converter-fed operation Betrieb am Umrichter * eines Motors
converter-related stromrichternah || stromrichterspezifisch
converter-specific stromrichternah || stromrichterspezifisch
converters Stromrichter * (Plural)
converting Weiterverarbeitung * Umformung, Veredelung
convertor Stromrichter * (veraltet) || Umformer * (veraltete Schreibweise) || Umrichter * (veraltete Schreibweise)
convertors Stromrichter * (Plural, veraltet)
convex ballig || gewölbt || Konvex
convey bewegen * fördern || fördern * Material transportieren, z.B. durch ein Förderzeug || leiten * (be-) fördern, übermitteln, versenden, übertragen, senden || tragen * (be)fördern || zuführen || zuführen * fördern (to: zu)
conveyance Beförderung * Materialtransport, Transport, Beförderung, Spedition || Förderung * Transport, Beförderung, Überbringung, Zufuhr || Transport * Beförderung, Fördern, Übersendung, Zufuhr || Zuführung * (An-) Förderung
conveyance of passengers Personenbeförderung * auch bei Aufzügen
conveying Zufuhr * Zuförderung
conveying capacity Förderleistung * eines Förderzeugs
conveying rate Förderleistung * eines Fördersystems
conveying system Fördersystem * z.B. zum Gütertransport
conveyor -Förderer || Förderanlage || Förderband * Förderzeug || Fördereinrichtung || Förderer * Förderzeug, Materialförderer || Fördermaschine * Förderer || Förderzeug
conveyor belt Fließband * Förderband || Förderband || Fördermaschine * Förderband || Transportband
conveyor belt scales Bandwaage
conveyor capacity Förderkapazität * eines Fördersystems
conveyor chain Transportkette
conveyor control Bänderregelung * für Transportband
conveyor drive Förderantrieb
conveyor dryer Durchlauftrockner
conveyor equipment Fördereinrichtung
conveyor system Förderanlage || Fördereinrichtung * Transport/Fördersystem || Fördersystem
conveyor technology Fördertechnik
conveyor-type weigher Bandwaage
convince überzeugen * of:von
convincing überzeugend * advantages: Vorteile; arguments: Argumente
convoke einberufen * z.B. Sitzun ~g || versammeln * einberufen
cook up ausdenken * (salopp) auskochen
cooker Kocher
cool kühlen * auch 'abkühlen' || ruhig * gelassen
cool-headed ruhig * gelassen
cool-off abkühlen
coolant Kühlflüssigkeit || Kühlmittel
coolant agent Kühlmittel
coolant pump Kühlmittelpumpe

coolant temperature Kühlmitteltemperatur
cooler Kühler || Kühlwerk
cooling Abkühlung || Entwärmung || Kühl- || Kühlung
cooling agent Kältemittel * für Kälteanlage
cooling air Kühlluft * discharged: austretende || Zuluft * Kühlluft
cooling air duct Kühlluftkanal
cooling air flow Kühlluftstrom
cooling air flow requirement Kühlluftbedarf
cooling air intake Kühllufteintritt
cooling air outlet Kühlluftaustritt
cooling air rate Kühlluftmenge
cooling air requirement Kühlluftbedarf
cooling air temperature Kühllufttemperatur || Kühlmitteltemperatur * bei Luftkühlung || Lufteintrittstemperatur * für Kühlluft || Zulufttemperatur * von Kühlluft
cooling arrangement Kühlart * Kühl-Anordnung
cooling baffle Luftleitblech
cooling bed Kühlbett
cooling capacity Kälteleistung * eines Kompressors/einer Kältemaschine (DIN 8928, ISO 9309) || Kühlleistung
cooling chamber Kühlschacht * [Textil] hinter d. Spinnpumpe angeordnete Einrichtg. in d. Kunstfasererzeugung
cooling circuit Kühlkreis * Kreis(lauf), in dem d. Kühlmittel strömt, d.Verlustwärme aufnimmt u. abtransportiert || Kühlkreislauf * closed: geschlossener
cooling conditions Kühlbedingungen * kennzeichn. Kühlmitteltemp.bereich; el. Maschin. VDE0530/ IEC34; Stromricht. VDE558 || Kühlungsbedingungen
cooling cylinder Kühltrommel || Kühlwalze * z.B. für Zuckermasse bei d. Bonbonherstellung
cooling down Abkühlung * z.B. eines Motors nach dem Abschalten
cooling duct Kühlkanal * für Kühlluft
cooling effect Kühlwirkung
cooling efficiency Kühlwirkung
cooling fan Kühllüfter || Kühlventilator || Lüfter * zur Kühlung
cooling fin Kühlrippe
cooling fluid Kühlflüssigkeit
cooling furnace Kühlofen * z.B. in der Glasindustrie
cooling line Kühlstrecke
cooling medium Kühlmittel * Flüssigkeit (Wasser, Öl) oder Gas (Luft), mit deren Hilfe die Wärmeabfuhr erfolgt
cooling medium temperature Kühlmitteltemperatur
cooling method Kühlart * für Motoren: siehe VDE 0530 u. IEC 34 Teil 6; für Stromrichter: siehe VDE 558 Teil 1
cooling oil Kühlöl
cooling principle Kühlart * Kühlprinzip/-verfahren
cooling pump Kühlmittelpumpe
cooling rib Kühlrippe
cooling section Kühlstrecke * z.B. in Blechstraße || Kühlwerk * Kühlstrecke, Kühlbett
cooling surface area Kühlfläche
cooling system Kühlsystem * Gesamtheit der Kühlkreise u.d. Kühlmittelströmung zur Wärmeabfuhr (VDE 0530 Teil 5)
cooling table Kühlbett
cooling tower Kühlturm
cooling type Kühlart * für el. Maschinen: siehe VDE 0530 u. IEC 34 T.6; für Stromrichter: siehe VDE 558 Teil 1
cooling water Kühlwasser
cooling water flange Kühlwasserflansch
cooling water inlet temperature Kühlwassereintrittstemperatur

cooling water pump Kühlwasserpumpe
cooling water temperature rise Kühlwassererwärmung
cooling zone Kühlbett
cooling-down curve Abkühlungskurve
cooling-off period Abkühlphase
cooling-water flow-rate Kühlwassermenge
cooling-water temperature Kühlwassertemperatur
cooperate beteiligen * sich helfend beteiligen (in: an) || teilnehmen * mitwirken (in: an), zusammenarbeiten (with: mit, to: zu e.Zweck, in: an), beitragen (to: zu) || teilnehmen || zusammenarbeiten || zusammenwirken
cooperation Beteiligung * Mitwirkung || Kooperation || Mithilfe || Teilnahme * Mitarbeit || Zusammenarbeit * close: enge || Zusammenspiel * Zusammenarbeit, siehe auch →Zusammenwirken
cooperation agreement Kooperationsvertrag
cooperative Genossenschaft
coordinate absprechen * koordinieren || abstimmen * koordinieren, aufeinander abstimmen, richtig anordnen || harmonisieren * koordinieren, aufeinander abstimmen, richtig anordnen || Koordinate || koordinieren || vereinigen * gleichschalten, koordinieren || zuordnen * beiordnen, an die Seite stellen
coordinate origin Koordinatenursprung * z.B. Schnittpunkt von X- und Y-Achse
coordinate system Koordinatensystem
coordinate transformation Koordinatentransformation
coordinated drive Mehrmotorenantrieb
coordinated motion control Mehrachsensteuerung
coordinates Koordinaten
coordinating Disposition * Koordinierung || Leitung
coordinating control Leitsteuerung
coordinating control level Leitebene
coordinating department Leitstelle
coordination Abstimmung * Koordierung || Anordnung * richtige An/Bei/Zuordnung || Koordination || Koordinierung || Zuordnung * Beiordnung, z.B. von Personen
coordination of insulation Isolationskoordinaten
coordination processor Koordinierungsprozessor
coordination table Zuordnungstabelle
coordination team Leitstelle
cop Kops * [Textil] Garn-/Faden~ (Spule mit aufgewickeltem Garn in der Spinnerei usw.) || Wickel * [Textil] Garn-/Fadenwickel, Kops (in der Spinnerei usw.)
cope with beherrschen * fertigwerden mit || fertig werden mit || fertigwerden * mit etwas ~ || gewachsen sein * fertigwerden mit || im Griff haben * bewältigen, meistern, fertig werden mit, gewachsen sein || meistern * fertigwerden mit || überwinden * bewältigen, meistern, fertig werden mit, gewachsen sein || verkraften * fertigwerden/es aufnehmen mit, gewachsen sein, bewältigen
copier Fotokopierer || Kopierer
copies to Verteiler * in Brief, Bericht usw.
copious reichlich * ~, ausgiebig, ergiebig, reich, umfassend, produktiv, viel schaffend; weitschweifig
copper Kupfer || Pfanne * zur Würzegewinnung in einer Brauerei
copper bar Kupferschiene
copper bit Lötspitze
copper cable Kupferleitung
copper charge Kupferzuschlag * [el. Masch.] Zuschlag z. Motorpreis bei Überschreiten e. bestimmten Kupfer-Börsenpreises
copper conductor Kupferleiter
copper losses Kupferverluste * Ohmsche Verluste in einer Wicklung, Stromwärmeverluste
copper plate paper Tiefdruckpapier

copper winding Kupferwicklung
copper-plate verkupfern
copy Abbild || abbilden || Ausgabe * Exemplar, z.B. eines Buches || Duplikat * Kopie || Durchschlag * Kopie; c.c.: Durchschlag an ("carbon copy") || Exemplar * Exemplar/Kopie eines Druckerzeugnisses/Buchs/Dokuments || Kopie || kopieren || nachbilden * kopieren || Nachbildung * Kopie
copy milling kopierfräsen
copy router Kopierfräse * {Holz}
copy to cc * Kopie an || Kopie an
copy-book Heft
copying Nachbau || Vervielfältigung
copying lathe Kopierdrehmaschine || Schablonendrehmaschine * {Holz}
copying milling machine Kopierfräsmaschine
copying paper Kopierpapier
copying shaper Kopierfräsmaschine * {Holz}
copyright Urheberrecht
copywriter Werbetexter
cord Anschlussleitung * flexible Netz~ z.B. für Kleingeräte || Anschlussschnur || binden * mit Stricken ~ || Schnur * auch elektr. Anschluss~; flexible cord: flexible Anschluss~ || Seil * Schnur
cording Bund * {Buchbinderei}
cordless batteriebetrieben * ohne Netzanschluss (z.B. Elektrowerkzeug)
core -adrig * Leitung, Kabel || -Paket * Blechpaket bei Trafo, Motor, Drossel usw || Ader * ~ einer mehradrigen Leitung/Kabel || Blechpaket || Hülse * Wickelkern; z.b. paper core: ~ zum Aufwickeln von Papier || Kern * das Innere; auch Spulen/ Transformator/Wickel~; (auch figürl.: get to the core: z. Kern kommen) || Kern- * Inneres || Mittelpunkt * Kern || Wickelhülse || Wickelkern
core assembly Blechpaket
core burning Eisenbrand * {el.Masch.} Zerstörung der Blechisolation und Verschmelzen der Bleche
core competence Kernkompetenz
core cross-section Aderquerschnitt
core diameter Hülsendurchmesser * z.B. einer Papphülse bei Wickler
core group Kerngruppe
core hardness Wickelhärte * im Wickelkern erhöhte Wickelhärte
core identification Aderkennzeichnung
core loss Eisenverlust
core pair Aderpaar
core stack Schichtkern
core strength Kernkompetenz
core team Kerngruppe * von Personen
core type Kerntyp * bevorzugter Typ im Produktspektrum
core winder Achswickler || Zentralwickler || Zentrumswickler
core-and-winding-asssembly Aktivteile * Teile einer el. Masch., in denen sich die Energieumwandlung vollzieht
coreboard Tischlerplatte * {Holz}
coreless kernlos * z.B. Induktionsofen, Motor usw.
corner Ecke * (auch figürl.) || Kante * Ecke || Kurve * (Straßen-) Ecke, Winkel, (auch Verb: um die Kurve biegen, e. Winkel bilden)
corner frequency Eckfrequenz * z.B. ~ einer U/f-Kennlinie (bei dieser Frequenz wird die max. Spannung erreicht)
corner point Eckpunkt || Knickpunkt * Stützpunkt einer Kurve, eines Polygonzuges usw.
cornered eckig || unrund * eckig
corona discharge Glimmentladung * Teilentladung; Mikrolichtbögen i. Dielektrikum b. Überschreit. d. Durchschl.festigk. || Sprühentladung
corona inception voltage Glimmeinsatzspannung
corona shielding Glimmschutz * {el. Masch.}

Schutz vor →Glimmentladg. an durch Feldstärkespitzen gefährd. Stellen
corporate culture Unternehmenskultur
corporate headquarters Stammhaus * Firmen-Hauptsitz
corporate mission guideline Unternehmensleitsatz
corporate orientation Ausrichtung * ~ eines Unternehmens
corporation Unternehmen * großes
correct ausregeln || einwandfrei || fehlerfrei || gut * richtig, korrekt || in Ordnung || korrigieren || nachbearbeiten * korrigieren (siehe auch →überarbeiten) || nachführen * correct to the setpoint: auf den Sollwert ~; correct the setpoint: den Sollwert ~ || nachregeln || ordnungsgemäß || richtig || sachgemäß || sachgerecht || verbessern * berichtigen || zutreffend
correct clearance Toleranz * zulässiges Spiel
correct functioning Funktionsfähigkeit
correct phasing Phasengleichheit * phasenrichtiger Anschluss
correct position richtige Lage || richtige Stellung
correct the proofs korrekturlesen * den Vorabzug/ die Druckfahnen ~
corrected fehlerbereinigt * korrigiert || verbessert * berichtigt
correcting effect Regelwirkung
correcting setpoint Hilfssollwert * Korrektursollwert || Korrektursollwert
correcting variable Stellgröße * Reglerausgangsgröße, d.verstellt wird, um d.Regelgröße d.Sollwert nachzuführen
correction Berichtigung || Kompensation * Korrektur || Korrektur || Verbesserung * Korrektur
correction factor Korrekturfaktor
correction of nonlinearity Linearisierung
correction time Ausregelzeit
correction vector Korrekturvektor
corrective maintenance Instandsetzung
corrective measure Abhilfemaßnahme
corrective setpoint Korrektursollwert
corrective value Korrekturwert
correlate korrelieren * in Wechselbeziehung stehen (with:mit), voneinander abhängig sein
correlation Wechselbeziehung || Zusammenhang * Wechselbeziehung, gegenseitige Abhängigkeit, Entsprechung
correspond korrespondieren * (to/with: mit) || übereinstimmen * entsprechen (einer Sache usw.), in Einklang stehen (to: mit)
correspond to ähneln * entsprechen || entsprechen || gleichen * entsprechen
correspondence Korrespondenz * Übereinstimmung, Entsprechung; Briefwechsel || Übereinstimmung * auch 'Angemessenheit, Entsprechung'; (with: mit; between: zwischen)
corresponding entsprechend || jeweilig * entsprechend, gemäß || korrespondierend * entsprechend || passend * entsprechend || sinngemäß || übereinstimmend || zugehörig * entsprechend, zugeordnet
corresponding to ähnlich * entsprechend || analog zu * entsprechend || bezogen auf || entsprechend * in Bezug stehend mit || zugeordnet * bezogen auf
correspondingly jeweilig * (Adv.) entsprechend, demgemäß
corresponds to entspricht
corridor Gang * ~ im Haus
corroborate untermauern * bekräftigen, bestätigen, erhärten
corrode anfressen * zer/anfressen, angreifen, korrodieren, rosten, s.einfressen || angreifen * korrodieren, chemisch zersetzen || fressen * zer~, an~, korrodieren, wegätzen, angreifen || korrodieren || rosten * korrodieren || zerfressen * zer-/anfressen, korrodieren

corroded rostig * *korrodiert*
corrosion Korrosion || Rostfraß * *Korrosion* || Verschleiß * *Korrosion, Anfressung*
corrosion protection Korrosionsschutz
corrosion risk Korrosionsgefahr
corrosion-proof korrosionsgeschützt
corrosion-resistant korrosionsbeständig
corrosion-resistant design ätz- und säurefeste Ausführung
corrosive aggressiv * *chem. zersetzend, zerfressend, ätzend, angreifend; z.B. Gase, Atmosphäre* || korrodierend
corrosive atmosphere aggressive Atmosphäre
corrosive gases aggressive Gase * *Korrosion verursachend*
corrugated geriffelt * *gewellt* || gewellt || Well- * *z.B. Pappe, Blech, Schlauch* || wellenförmig * *gewellt, gefurcht, gerieft; z.B. Blech*
corrugated board Wellpappe
corrugated cardboard Wellpappe
corrugation Sicke * *Falte, Furche, Riefe, Welle* || Welligkeit * *Wellen, Runzeln, Furchen, Riefen, Falten (z.B. in Papier, Pappe, Stahl usw.)*
corrupt verfälschen * *z.B. Daten, Signale* || verfälscht * *verdorben, unecht, verfälscht, z.B. Daten*
corrupted inkonsistent * *verfälscht, z.B. Text, Dokumentation, Daten*
cos phi cos phi * *Leistungsfaktor (Verhältnis Wirk-/Scheinleistung)* || Leistungsfaktor
cosine cosinus
cost Ausgabe * *Unkosten* || Kosten * *at the cost of: auf ~ von; fixed/running: fixe/l aufende; bear the cost: die ~ tragen* || Kosten * *(Verb)*
cost aim Kostenziel
cost allocation Kostenumlage
cost benefits Kostenvorteile
cost burden Kostenbelastung
cost center Kostenstelle
cost distribution Kostenumlage
cost estimate Kostenvoranschlag
cost goal Kostenziel
cost of cabling Verdrahtungsaufwand * *Kosten*
cost optimization Kostenoptimierung
cost pressure Kostendruck
cost price Selbstkostenpreis
cost reduction Kostensenkung
cost saving Kosteneinsparung
cost savings Einsparungen
cost target Kostenziel
cost to manufacture Selbstkostenpreis
cost unit Kostenträger
cost-effective kostengünstig * *mit gutem Kosten/Nutzen-Verhältnis* || wirtschaftlich * *kostengünstig, zu günstigen/niedrigen Kosten*
cost-effectiveness Wirtschaftlichkeit * *Kostengünstigkeit (Anschaffungs- und Betriebskosten eines Investitionsgutes)*
cost-efficiency Wirtschaftlichkeit
cost-minimized kostenminimiert
cost-optimized kostenoptimiert
cost-saving kostensparend
costly aufwändig * *kostspielig, teuer* || kostspielig || teuer * *kostspielig, kostenaufwändig*
costomer benefit Kundenvorteil
costs Aufwand * *Kosten* || Kosten * *Mehrzahl* || Unkosten
costs incurred entstandene Kosten
costs of repair Reparaturkosten
cosy angenehm * *behaglich*
cotter Splint || versplinten
cotter pin Splint
cotton Garn * *Baumwollgarn* || Wolle * *Baumwolle*
cotton waste Putzwolle
cotton-spinning plant Baumwollspinnerei

couch press Gautschpresse * *zur Papier-Entwässerung*
couch roll Gautschpresse * *zur Papier-Entwässerung*
coucher Gautscher * *in der Papierindustrie*
counsel beraten || Beratung * *Rat (to a person:jemanden beraten), Raterteilung*
counsellor Ratgeber
counselor Ratgeber
count ankommen auf * *zählen, wichtig sein* || Anzahl || bauen * *sich verlassen, zählen; (on:auf)* || gelten * *zählen* || zählen || Zählerinhalt || Zählwert
count down abwärtszählen
count module Zählbaugruppe || Zählerbaugruppe
count on verlassen * *sich ~ auf, zählen auf*
count range Zählbereich || Zählerbereich
count rate Zählfrequenz
count up aufwärtszählen
count value Zählwert
counter Gegen- * *z.B. Gegenmaßnahme, Gegenmoment/spannung* || Ladentisch * *sell over/under the counter:über/unter dem Ladentisch verkaufen* || Zähler
counter clockwise links * *in ~drehrichtung*
counter content Zählerinhalt || Zählwert * *Zählerinhalt*
counter EMF Gegen-EMK || Gegenspannung * *Gegen-EMK*
counter module Zählerbaugruppe
counter nut Kontermutter
counter torque Gegendrehmoment || Gegenmoment * *Moment, das die Arbeitsmaschine d.Motor entgegensetzt (Last/Beschleunigungsmoment)* || Lastmoment * *Gegenmoment*
counter word Zählwort
counter- Ausgleichs-
counter-act Gegenmaßnahmen ergreifen
counter-clockwise direction of rotation Linksdrehrichtung
counter-clockwise rotating Linkslauf
counter-clockwise rotation Linkslauf
counter-clockwise running Linkslauf
counter-EMF EMK
counter-excitation Gegenerregung * *z.B. bei elektromagnet. Kupplung/Bremse zum Erreichen schneller Ansprechzeiten*
counter-measure Abhilfe * *Abhilfe-/Gegenmaßnahme (possible: mögliche)* || Abhilfemaßnahme * *possible: mögliche* || Abhilfemaßnahme
counter-pressure Gegendruck
counter-rotating gegenläufig * *in Gegenrichtung rotierend*
counter-sunk screw Senkschraube
counter-weight Ausgleichsgewicht
counteract entgegenwirken * *to:gegen*
counterbalance auswuchten || Gegengewicht * *(auch Verb: ein ~ bilden, ausgleichen, aufwiegen, die Waage halten)*
counterbalance weight Ausgleichsgewicht
counterbore ansenken * *z.B. einer Bohrung mit Spitzsenker* || Senker * *Werkzeug zum Entgraten/Anfasen von Bohrlochern*
counterclockwise entgegen dem Uhrzeigersinn || links herum * *entgegen dem Uhrzeigersinn*
counterclockwise phase sequence Linksdrehfeld * *entgegen d. Uhrzeigersinn*
countercurrent braking Gegenstrombremsung
counterelectrode Gegenelektrode
counterfeit nachbilden * *nachmachen, fälschen, heucheln, vorgeben; nachgemacht, unecht, gefälscht*
countermand abbestellen
countermeasure Abhilfemaßnahme * *Gegenmaßnahme* || Gegenmaßnahme

counterpart Entsprechung * *Pendant* || Gegenstück || Pendant * *Gegenstück*
counterpoise Gegengewicht * *(auch figürl.) Gleich-/Gegengewicht (auch Verb: als ~ wirken zu, ausgleichen)*
counterrotating gegenläufig * *in Gegenrichtung rotierend*
counterrotation Gegenläufigkeit
countersink Senker * *(kegelförmiges) Werkzeug zum Anfasen u. Entgraten von Bohrlöchern* || Spitzsenker || versenken * *z.B. Schraubenkopf*
countersunk head screw Senkkopfschraube
countersunk screw Senkschraube
counterweight Gegengewicht * *z.B. bei Aufzug* || Spanngewicht * →*Gegengewicht, z.B. großes ~ bei Aufzug*
counting filling machine Zähl-Füllmaschine * *[Verpack.tech.]*
counting instrument Zählgerät
counting range Zählbereich
country code Länderkennzeichen * *(auch beim Zoll)* || Vorwahl * *Ländervorwahlnummer am Telefon* || Vorwahlnummer * *für Ausland*
country-specific länderspezifisch || landesspezifisch
couple ankoppeln * *to:an; with: koppeln mit/ankoppeln an* || befestigen * *aneinanderkoppeln* || koppeln * *mechanisch ~, auch über Kommunikations-Schnittstelle* || kuppeln
couple into einkoppeln
coupled gekoppelt
coupled drive Kupplungsantrieb * *Motor treibt die Abeitsmaschine über Kupplung an (nicht über Riemen usw.)*
coupled-in noise Störeinkopplung || Störungseinkopplung
coupler Koppler
coupling Ankopplung || Einführung * *~ eines Rohrs* || Einkopplung || Kopplung * *auch Daten~ z.B. über serielle Schnittstelle* || Kupplung * *nicht betriebsmäßig ein-/ausrückbare ~ (Faltenbalg~ usw.); el. Steck~, Kuppeln, Kopplung* || Muffe * *Rohrverbindung* || Verbindungs-
coupling capacitance Koppelkapazität
coupling capacitor Koppelkondensator
coupling end Antriebsseite * *Kupplungsseite*
coupling half Kupplungshälfte
coupling halves Kupplungshälften
coupling memory Koppelspeicher
coupling of shafts Welle-Welle-Verbindung
coupling output Kupplungsabtrieb
coupling star Kupplungsstern
courier service Kurierdienst * *['kuriea söwis]*
course Gang * *~iner Mahlzeit; Bahn, Lauf, Verlauf; course of business: Geschäftsverlauf* || Kurs * *Lehrgang; refresher couse: Auffrischungs~; top-up course: Fortgeschrittenen~* || Kurs * *~ für Schiff und Politik* || Lauf * *Kurs von Bahn/Schiff usw.; in course of time: im Laufe der Zeit* || Lauf * *in the course of time:im Lauf der Zeit; in the course of: im Laufe einer Zeitspanne* || Lehrgang || Linie * *(politischer) Kurs* || Richtung * *Kurs* || Schulungskurs * *Kurs* || Seminar * *Lehrgang* || Strecke * *Bahn, Kurs* || Verlauf * *~ von Zeit, eines Vorgangs, Grenze, Linie, Kurve; in the course of: im ~ von* || Weg * *Bahn, ~ zum Ziel* || Weiche * *(figürlich) Kurs, Richtung; set the course: die ~ stellen* || Zug * *Lauf, Bahn, Weg, Gang, Ablauf, Verlauf; in the course of: im ~e des*
course application Kursanmeldung * *Reservierung für e. Trainings-/Weiterbildungskurs usw.*
course description Kursbeschreibung * *eines Aus/Weiterbildungskurses*
course enrolment Kursanmeldung * *Kurseinschreibung; siehe auch* →*Anmeldung*

course fees Kursgebühren
course material Kursunterlagen * *Lehr/Lernunterlagen für einen Aus/Weiterbildungskurs* || Schulungsunterlagen * *Kursunterlagen*
course objectives Lernziel * *Kursziel*
course of action vorgehen * *(Substantiv) Vorgehensweise* || Vorgehensweise
course of instruction Kurs * *Lehrgang* || Lehrgang * *Schulungskurs* || Schulungskurs
course of study Lehrgang
course office Kursbüro * *einer Aus/Weiterbildungsstätte*
course participant Kursteilnehmer * *an Aus/Weiterbildungskurs* || Lernender * *Kursteilnehmer (bei Aus/Weiterbildungskurs)*
course program Kursprogramm * *für Schulungskurse*
course reservation Kursanmeldung * *Resevierung für e. Trainings-/Weiterbildungskurs usw.*
courteous höflich
courtesy Aufmerksamkeit * *Höflichkeit*
cousin Gegenstück
cover Abdeckblech || abdecken * *auch figürlich, z.B. einen Markt, einen Leistungs/Anforderungsbereich, ein Risiko* || Abdeckplatte || Abdeckung || auslegen * *auskleiden* || bedecken || Bezug * *Überzug* || Deckel * *Abdeckung, auch Buch~; side/top cover: Seiten-/obenliegender ~* || decken * *auch figürlich* || Deckung * *~ der Kosten usw.; Sicherheit* || einbeziehen * *abdecken* || erstrecken * *sich ~ über, z.B. Garantie, Versicherungsschutz* || gelten * *abdecken, einschließen,* || Haube * *Abdeckung; top cover: obere Abdeckhaube/Deckel* || Schutz * *Deckung, Abdeckung* || schützen * *(ab/be)decken* || Schutzhaube || Sicherheit * *Deckung* || Sicherstellung * *durch Deckung* || Überzug
cover note Anschreiben
cover of terminal box Klemmenkastendeckel
cover plate Abdeckplatte || Deckscheibe * *z.B. zur Abdichtung bei Wälzlager*
cover screw Deckelschraube
cover sheet Abdeckblech || Deckblatt * *z.B.eines Berichts/einer Telefax-Nachricht/eines Dokuments*
cover sheets Verkleidung * *~ aus Blech*
cover tube Schutzrohr
cover up verdecken * *zudecken, verdecken*
cover washer Deckscheibe * *z.B. bei Wälzlager*
coverage Schutz * *Abdeckung, z.B. durch eine Versicherung, Gewährleistung*
covered abgedeckt * *(auch figürlich: durch Versicherung, Garantie usw.)* || verdeckt * *bedeckt, abgedeckt, verdeckt, verborgen, verhüllt*
covered up verdeckt * *zugedeckt, verdeckt*
covering Abdeckung || Beschichtung || Bezug * *Überzug* || Deckung || Hülle || Umhüllung || Verkleidung
covering machine Überzugmaschine * *z.B. in der Verpackungstechnik*
covert verborgen * *versteckt, verborgen, verschleiert, heimlich; Obdach, Schutz, Versteck, Dickicht* || verdeckt * *heimlich, versteckt (auch Weg usw.), verborgen, verschleiert; geschützt*
cowl Haube * *z.B. für Lüfter; Motorhaube*
cp. vergleiche * *Abkürzung für 'compare'*
cps Hertz * *Cycles Per Second* || Hz * *Cycles Per Second*
CPU CPU * *Abkürzung für 'Central Processing Unit': Zentraleinheit eines Rechners* || Mikroprozessor * *Central Processing Unit eines Computers* || Prozessor * *Abk. für 'Central Processing Unit': Zentraleinheit eines Rechners* || Rechenwerk * *Abkürzg. f. 'Central Processing Unit': Zentraleinheit eines Rechners* || Zentralbaugruppe * *bei SPS* || Zentraleinheit * *Central Processing Unit; eines Computers*
CPU board CPU-Baugruppe * *Flachbaugruppe*

CPU card CPU-Baugruppe * *Flachbaugruppe*
CPU module CPU-Baugruppe * *Zentraleinheit eines Rechners/einer SPS*
CPU performance Rechenleistung * *der Zentraleinheit eines Rechners*
crab Laufkatze * *Winde, Hebezeug, Laufkatze* || Winde * *Winde, Hebezeug, Laufkatze*
crab drive Katzfahrwerk * *[Hebezeuge]*
crack aufreißen * *reißen* || bersten * *einen Riss bekommen* || brechen * *bersten, platzen* || Bruch * *Riss, Sprung* || knacken * *(auch figürl.: harte Nuss, Software usw.)* || knicken * *brechen, zerspringen, ausreißen, einbrechen, (zer-) spalten* || Könner * *(salopp)* || Riss * *Sprung* || Schlitz * *Riss* || Spalt * *Riss, Sprung, Spalte, Schlitz* || Sprung * *Riss*
crack off absprengen
crack-up Zusammenbruch * *Zusammenbruch, Bruchlandung*
cracking Riss * *~bildung* || Rissbildung
cradle Schlitten
cradle dynamometer Pendelgenerator || Pendelmaschine * *zur Drehmomentvorgabe/messung, z.B. für Prüfstand*
cradle relay Kammrelais
cradle-type bearing Sattellager
craft Handwerk
craft business Handwerksbetrieb
craftshop Handwerksbetrieb || Werkstatt * *eines Handwerksbetriebes*
craftsman Handwerker
craftsman's business Handwerksbetrieb
craftsman's establishment Handwerksbetrieb
craftsman's workshop Handwerksbetrieb
craftsmen's trade Handwerk * *Berufsstand; craftsmen enterprises: ~sbetriebe*
cram füllen * *vollstopfen* || vollstopfen * *vollstopfen, an-/überfüllen (with: mit), hineinstopfen/-zwängen*
cramp einschnüren * *einzwängen* || Klammer * *Krampe, Klammer; Schraubzwinge; Fessel, Einengung*
cramped beengt * *eng, beengt*
cramped for space räumlich beengt
cramped space enger Raum
crane Hebezeug * *Kran* || Kran * *travelling crane:-fahrbarer Kran; mobile crane:Mobilkran, auf Reifen/Ketten fahrend*
crane control Kransteuerung
crane driver Kranführer
crane hoist Hubwerk * *~ eines Krans* || Kranhubwerk
crane hook Kranhaken * *am Kran*
crane jib Kranausleger
crane long-travel gear Kranfahrwerk * *[Hebezeuge] bei Portalkran (im Gegensatz zum →Katzfahrwerk)*
crane operator Kranführer
crane rail Kranschiene
crane runway Kranbahn
crane track Kranbahn
crane travel drive Kranfahrwerk * *[Hebezeuge]*
crane travelling gear Kranfahrwerk * *[Hebezeuge]*
crane travelling unit Kranfahrwerk * *[Hebezeuge]*
crane-torque characteristic Kran-Kennlinie * *[el. Masch.]*
crane-type motor Hebezeugmotor * *z.B. AC-Schleifringläufermotor bemessen für period. Aussetzbetrieb*
cranes and hoisting systems Hebezeuge
crank Antriebskurbel || Hebel * *Kurbel* || Kurbel
crank drive Kurbelantrieb || Schubkurbelantrieb
crank radius Kurbelradius
crank shaft Kurbelwelle
crank up ankurbeln * *(konkret u. figürl.)*

crank-handle Handkurbel
crankcase Kurbelgehäuse
crankshaft Kurbelwelle
crash -Sturz || Absturz * *Programm, Computer, Flugzeug usw.; system crash: System~ (eines Comuters usw.)* || abstürzen * *Programm, Flugzeug usw.* || Kollision * *Zusammenstoß* || schlagen * *krachend ~* || Sturz * *lauter/spektakulärer/aufsehenerregender ~* || Zusammenbruch * *Absturz, Zusammenstoß/bruch, schwerer Unfall, Ruin*
crate Kasten * *Lattenkiste, (Getränkeflaschen-) Kasten* || Kiste * *aus Holzlatten*
crate loading machine Kastenstapler * *in der Getränkeindustrie*
crate unloading machine Kastenentstapler * *in der Getränkeindustrie*
crater-shaped punktförmig * *Vertiefung, Grübchen*
crawl kriechen * *mit Schleichdrehzahl bzw. langsam fahren* || schleichen
crawl reference Kriechsollwert * *Sollwert für Kriech/Schleichgeschwindigkeit*
crawl speed Einziehgeschwindigkeit * *Kriechgeschwindigkeit* || Kriechgeschwindigkeit * *Schleichgeschwindigkeit/-drehzahl* || Schleichdrehzahl || Schleichgang
crawl speed drive Schleichgangantrieb * *Asynchronmotor mit Drehfeldmagnet auf 1 Welle z. Er-reich. v. Schleichdrehzahlen*
crawler Raupe * *Raupenschlepper*
crawler drive Fahrantrieb * *eines Baggers* || Raupenantrieb * *z.B. bei Bagger*
crawler tractor Raupenschlepper
crawling Schleichgang
CRC CRC * *Abk.f. 'Cyclic Redundancy Check': Datensicherungsverfahren mit zykl.berechneten Prüfmustern*
cream best * *(Subst., salopp) the cream: das Beste*
crease brechen * *Papier* || Falte
crease-freedom Faltenfreiheit * *z.B. beim Aufwickler (kann z.B. durch abnehmende Wickelhärte erreicht werden)*
create anlegen * *erzeugen, schaffen* || erstellen * *erschaffen, erzeugen, z.B. Dokumentation* || erzeugen * *(er)schaffen, hervorbringen, erzeugen; verursachen, ein-/errichten, ernennen (to:zu)* || herstellen * *(er)schaffen* || verursachen
creativity Kreativität
credibility Glaubwürdigkeit
credit Guthaben || gutschreiben || Gutschrift * *for credit: zur ~*
credit card Kreditkarte
credit point Pluspunkt
crediting Erstattung * *z.B. Gewährleistungsansprüche*
creel Spulengatter * *[Textil] z.B. am Anfang einer Schärmaschine*
creep kriechen * *schleichen, kriechen* || schleichen * *kriechen* || Schlupf * *eines Riemens* || wandern * *kriechen*
creep feed Schleichgang
creep feedrate Schleichgang * *[Werkzeugmaschine]*
creep speed Schleichgang
creepage current Kriechstrom
creepage distance Kriechstrecke * *für Isolation (bei Stromrichtern: siehe IEC 664)*
creeping langsam anwachsend
crêpe gekreppt * *Papier usw.*
crest Scheitel- * *z.B. -Wert, -Spannung*
crest factor Amplitudenfaktor || Scheitelfaktor * *Maximalwert/Effektivwert eines periodischen Signals (1.41 bei sinusförmigem Signal)*
crest value Scheitelwert
crest voltage Scheitelspannung || Spitzenspannung * *→Scheitelspannung*

crevice Spalt * Riss, Spalte
crew Mannschaft
crimp anquetschen * z.B. Kontakt an Kabel usw. || kräuseln * auch Chemiefasern || krimpen || quetschen * pressen, z.b. Kabelschuh, Aderendhülse auf Leitungsende || Sicke * Kräuselung, Welligkeit
crimp connector Crimpverbinder || Kabelschuh * Quetschverbinder || Kerbverbinder || Quetschkontakt * Quetschverbinder || Quetschverbinder
crimp contact Crimpkontakt || Quetschkontakt
crimp on anquetschen * z.b. Kontakt an Kabel usw.
crimp snap-in contact Crimpkontakt * zum Einrasten z.b. in Steckergehäuse
crimp snap-on connection Crimpanschluss
crimp snap-on connector Crimpanschluss * Crimpanschlussverbinder
crimp termination Crimpanschluss
crimp-on lug Kabelschuh * Flach-Quetschverbinder
crimped gekreppt
crimper Kräusel * {Textil} || Krempelmaschine * {Textil} zum Aufrauen von Fasern
crimping Kröpfung
crimping tool Crimpzange * zum Anquetschen von Kabelschuhen und Kabel-Quetschverbindern || Kerbzange * Quetsch-/Presszange für Quetschkontakte, Kabelschuhe usw. || Presszange * Quetschwerkzeug z.B. für Kabelschuhe, Aderendhülsen || Quetschzange * zum Anquetschen von Kabelschuhen und Kabel-Quetschverbindern
cripple lähmen
crisis Krise || Schwierigkeit * Krise
crisis management Krisenmanagement
crisp scharf * Bildschirm
crispness Schärfe * eines Bildschirms
criteria Kriterien
criterion Gesichtspunkt * Merkmal, Kriterium || Kriterium * Merkmal
critical entscheidend * Augenblick || gefährlich * kritisch, ernst || heikel || knapp * critical items: ~ Waren || kritisch * gewagt, riskant, bedenk-/gefährlich, genau/sorgfältig prüfend; be critic.of: etw. kritis. || scharf * Überwachungseinrichtung || schwierig * kritisch
critical frequency Grenzfrequenz * kritische Frequenz
critical frequency lockout band Frequenzausblendband || Frequenzausklammerungsband
critical frequency rejection Ausblendfrequenz * ausgeblendet.Frequ.sollw. f. Frequ.umrichter z. Vermeidg. mechan. Resonanz
critical from a time perspective zeitkritisch
critical limit kritische Grenze
critical load Grenzbelastung
critical situation Krise * Krisensituation
critical speed Grenzdrehzahl * bei Gleichstrommotor z.B.durch Kommutator begrenzt || kritische Drehzahl * {el. Masch.} Drehzahl, bei der mechan. Resonanz auftritt
critical speed rejection Ausblendband * Ausblendung v. Drehzahlsollwertbereichen zur Vermeidung mechan. Resonanzen || Frequenzausblendung || Frequenzausklammerung || Frequenzbandausklammerung
critical speed rejection band Frequenz-Ausblendband * siehe 'Frequenzausblendband' || Frequenzausblendband * Ausblendung von (mechan.) Resonanzfrequenzen im Sollwert || Frequenzausklammerungsband
critical stage Krise * kritische Phase
critical with respect to time zeitkritisch
critical-rotor läuferkritisch
critical-rotor motor läuferkritischer Motor * Läufer erreicht Grenztemperatur früher als Ständer

criticism Kritik * über/an: of || Kritikpunkt
criticize aussetzen * kritisieren, etwas auszusetzen haben || beurteilen * fachmännisch
criticized point Kritikpunkt
crook krümmen
crooked gekrümmt || krumm || schief * krumm, gekrümmt, gebeugt, unehrlich, bucklig, verwachsen
cropping shear Schopfschere * Querteilschere z.Abschneiden d.Brammen- oder Stahlbandenden i. Walzwerk/Blechstraße
cross kreuzen || quer * quer(liegend/verlaufend), schräg, sich überschneidend || schneiden * sich ~ (z.B. zwei Kurven {Mathem.} oder zwei Straßen) || überschreiten * überqueren || überschreiten * kreuzen
cross connection Kreuzschaltung || Querverbindung
cross cutter Querschneider * Bogenschneider für Papier, Blech
cross direction Querrichtung
cross flow Querstrom
cross pin Querstift
cross reference Querverweis || Querverweisliste
cross reference list Querverweisliste
cross section Längsschnitt * in Konstruktionszeichnung
cross section of the core stack Blechschnitt * Querschnitt eines (Motorläufer-/-stator-) Blechpakets
cross transfer Ausschleusung * seitliche ~ || Querschlepper * z.B. in Walzwerk || Quertransport * in Transferstraße
cross winder Kreuz-Wickelmaschine
cross- Kreuz- || Quer-
cross-arm Traverse
cross-beam Holm * Querträger/-balken
cross-check querprüfen
cross-check sum Quersumme
cross-cut kappen * {Holz} (an den Enden) quersägen
cross-cut saw Ablängsäge * {Holz} || Kappsäge * {Holz} || Quersäge * {Holz} || Trennsäge * {Holz}
cross-cutting and trenching machine Abbundanlage * {Holz}
cross-cutting saw Kappsäge * {Holz}
cross-cutting shears Teilschere * Quer~
cross-directed Quer- * in ~richtung
cross-hatch schraffieren
cross-over frequency Übergangsfrequenz
cross-over point Übergangpunkt
cross-recessed screw Kreuzschlitzschraube
cross-reference list Querverweisliste
cross-sanding machine Kreuzschleifautomat * {Holz}
cross-section Querschnitt * auch für Kabel, Konstruktionszeichnung
cross-sectional area Querschnitt * Querschnittsfläche
cross-sectional reduction Querschnittsabnahme * bei Drahtziehmaschine
cross-sliding table Kreuztisch * {Werkz.masch.| Holz}
cross-tip screwdriver Schraubendreher * Kreuzschlitz~
cross-tipped screwdriver Schraubenzieher * Kreuzschlitz~ (für Kreuzschlitzschrauben)
cross-travel gear Katzfahrwerk * {Hebezeuge}
cross-traversing gear Katzfahrwerk * {Hebezeuge}
cross-wise über Kreuz
crossbar distributor Kreuzschienenverteiler
crossbar interconnection Kreuzschienenverteiler
crossbar switch Kreuzschienenverteiler
crossbeam Strebe * Quer~ || Traverse
crosscut saw Ablängsäge * {Holz}

crossfall Wölbung * z.B. einer Straße
crossing Übergang ‖ Überschreitung * Kreuzen, Kreuzung, Durch/Überquerung (z.B. einer Grenze/Linie)
crosslinked vernetzt * [Chemie] (Kunststoffe)
crossover point Ablösepunkt
crossover point from discontinuous to continuous current Lückgrenze * Aussteuerg. b. d. d. Strom nicht mehr lückt
crosstalk Übersprechen * Störeinflüsse v. benachbarten Informationen (auf Leitungen, magn.Datenträger, Display)
crosstip screwdriver Kreuzschlitz-Schraubenzieher ‖ Kreuzschlitzschraubenzieher
crosswise quer * der Breite nach
crowd drängen * sich ~ ‖ Haufen * Menge, Ansammlung, Haufen, Häufung, Fülle, Schwarm ‖ Menge * Menschen~ ‖ vollstopfen * hineinstopfen/-pressen (with: mit); überfüllen, vollstopfen, zusammendrängen
crowned ballig
crowned with success erfolgreich * von großem Erfolg gekrönt
CRT Bildröhre * Abkürzg. f. 'Cathode-Ray tube': Kathodenstrahlröhre ‖ Monitor ‖ Sichtgerät * cathode-ray tube: Bildschirm-Monitor
CRT unit Bildschirmgerät * Abk. für 'Cathode-Ray Tube':Kathodenstrahl-Bildröhre
CRT-based programmer Bildschirmprogrammiergerät * für SPS
crucial ausschlaggebend * entscheidend, kritisch; crucial point: springender Punkt ‖ entscheidend * kritisch; crucial point: ~der Punkt
crucial point Hauptsache * springender Punkt ‖ Schwerpunkt * (figürlich)
crucible Tiegel * Schmelz-/Stahl~; Feuerprobe (figürl.)
cruciform kreuzförmig
crude oil Erdöl * Rohöl ‖ Rohöl
cruising radius Reichweite * [Verkehr] ~ von Fahrzeugen (mit einer Tankfüllung)
cruising range Reichweite * [Verkehr] ~ von Fahrzeugen
crush brechen * mahlen ‖ mahlen * zerquetschen (z.B. Steine, Feldfrüchte usw.) ‖ Quetsch- * zerquetschend/-brechend ‖ quetschen * zerquetschen ‖ verdrängen * (ver)drängen, (unter)drücken, (zer)quetschen; (auch figürl.: Konkurrenten usw. ~) ‖ zerkleinern * [Steine]
crusher Mühle * Brechmaschine, z.B. für Steine
crushing mill Mahlanlage ‖ Schlagmühle ‖ Schrotmühle * z.B. in der Malzherstellung
crushing plant Brechanlage * z.B. für Erz
cryptic geheim * verschlüsselt, verborgen, schwer entzifferbar/entschlüsselbar ‖ unverständlich * rätselhaft, verborgen
crystal Kristall ‖ Quarz
crystal oscillator Quarzoszillator
crystal pulse generator Quarzgenerator
crystal-controlled quarzstabilisiert
crystalline kristallin
CSA CSA * Abkürzg. für 'Canadian Standard Association': kanadisches Normengremium
CSF Funktionsplan * [SPS] Control System Function chart: FUP (mit Grafiksymbolen; DIN 40700, IEC 117-5) ‖ FUP * 'Control System Function chart': Funktionsplan für SPS (siehe DIN 40700, IEC117-17)
CT Stromwandler * Abkürzung für 'Current Transducer'
cube dritte Potenz * increase (proportionally) with the cube of:mit der dritten Potenz von ... anwachsen ‖ Potenz * dritte Potenz ‖ Quader ‖ Würfel

cubic kubisch ‖ quaderförmig ‖ räumlich * (mathemat.)
cubic contents Größe * eines Gefäßes
cubic foot 1 * Kubikfuß; 1 l entspr. 0,0353147 cubic foot
cubic inch 1 * Kubikzoll; 1 l entspr 61,0237 cubic inch
cubic root dritte Wurzel
cubical kubisch
cubicle Kabine * abgeteilter Raum ‖ Schaltschrank ‖ Schrank * Schaltschrank
cubicle design Schrankausführung * z.B. Aufbau eines Geräts als kompletter Schaltschrank
cubicle door Schaltschranktür ‖ Schranktür
cubicle fan Schranklüfter
cubicle flooring section Schrankboden * Bodenbereich/-Blech eines Schaltschranks
cubicle frame Schrankgerüst
cubicle layout Schrankdisposition
cubicle mounted unit Schrankgerät
cubicle mounting Schrankmontage * for: für
cubicle system Schranksystem * z.B. SIEMENS Schaltschranksystem 8MF
cubicle unit Schrankgerät
cubicle ventilation Schrankbelüftung * für Schaltschrank
cullet Scherben * Bruchglas
culpable negligence grobe Fahrlässigkeit
cultivate ausbauen * pflegen
cumbersome lästig * lästig, beschwerlich, unbequem
cumulative sich häufend ‖ sich steigernd ‖ sich summierend
cumulative distribution Wahrscheinlichkeitsverteilung
cunning pfiffig * listig, schlau, verschmitzt, geschickt, klug ‖ raffiniert * pfiffig, listig, schlau, verschmitzt, geschickt, klug
cup Becher ‖ Buchse * Fettbuchse ‖ Kalotte
cup grease Staufferfett
cup housing Topfgehäuse * Gehäusebauform z.B. einer Kupplung
cup packing Dichtmanschette
cup spring Federscheibe * Tellerfeder (z.B. zur axialen Vorspannung eines Lagers) ‖ Tellerfeder
cup type Topfbauform
cure aushärten ‖ beseitigen * Fehler ~ ‖ härten * aus~ (mit Wärme, Wartezeit usw.; z.B. Gummi, Kunststoff) ‖ schleudern * Zucker zentrifugieren ‖ vulkanisieren
cured abgebunden * verfestigt (z.B. Klebeverbindung), ausgehärtet
curing Aushärtung
curing equipment Härtungseinrichtung * zum Aushärten
curl bördeln * kräuseln; seal-curling: verschließen durch Bördeln (z.B. Kondensator, Dose ...)
curl field Wirbelfeld
curl up rollen * sich aufrollen, z.B. Papier
curling machine Wulstmaschine
currency Währung
current aktuell ‖ gebräuchlich ‖ geläufig * allgemein bekannt ‖ jetzig ‖ laufend ‖ jeweilig * aktuell ‖ laufend * Ausgaben, Jahr, Produktion usw.; aktuell ‖ momentan * laufend, gegenwärtig, jetzig, aktuell ‖ Strom * elektrischer ‖ Stromstärke ‖ zeitgemäß * aktuell
current actual value Stromistwert
current actual value display Stromistwertanzeige
current actual value sensing Stromistwerterfassung
current amplification Stromverstärkung * z.B. eines (bipolaren) Transistors ("Beta")
current assets Umlaufvermögen * [Wirtschaft]
current build-up time Stromaufbauzeit

current capability Strombelastbarkeit
current carrying bestromt || Strombelastbarkeit || Stromtragfähigkeit
current carrying capability Stromtragfähigkeit
current carrying capacity Strombelastbarkeit * von Leitungen: siehe VDE 0113/EN 60204, jeweils Teil 1 || Stromtragfähigkeit * (von Leitungen: siehe VDE 0113/EN 60204, jeweils Teil 1)
current collector Abnehmer * Stromabnehmer || Stromabnehmer * (Bahnen)
current conduction duration Stromführungsdauer
current consumption Stromaufnahme
current control Stromregelung * secondary: unterlagert
current controlled stromgeregelt
current controller Stromregler
current controller output Stromreglerausgang
current conveying bestromt
current DC-link converter I-Umrichter * →Stromzwischenkreisumrichter
current decay Stromabbau * Abklingen des Stroms || Stromabfall
current demand Strombedarf
current density Stromdichte * Verteilung des el. Stroms auf den Leiterquerschnitt in $[A/mm^2]$
current direction Stromrichtung
current displacement Stromverdrängung * z.B. Skin-Effekt, Heaviside-Effekt
current distribution Stromaufteilung
current division Stromaufteilung
current drain Stromaufnahme
current draw Stromaufnahme * z.B. eines Geräts/Bauteils || Strombedarf * z.B. Stromaufnahme eines Geräts/Bauteils
current factor Stromfaktor
current fed converter Stromzwischenkreis-Umrichter
current fed inverter I-Umrichter * Stromzwischenkreisumrichter
current feed forward Stromvorsteuerung
current feedback Stromistwert
current feedback indication Stromistwertanzeige
current flow Stromfluss
current flow angle Stromflusswinkel
current flow period Stromführungszeit
current flow time Leitdauer
current flowing Stromfluss
current gradient Stromsteilheit
current handling capability Strombelastbarkeit
current inertia Stromträgheit
current input Stromaufnahme
current inrush Rush-Strom * Stromspitze beim Aufbau e.Feldes, z.B. bei Motor, Trafo usw.
current intensity Stromstärke
current limit Strombegrenzung * Stromgrenze || Stromgrenze * current limit acceleration: Hochlauf an der ~
current limit reached Stromgrenze erreicht
current limited strombegrenzt
current limiter Strombegrenzung * als Einrichtung, Modul, Funktionsbaustein, Schaltung
current limiting Strombegrenzung
current limiting control Strombegrenzungsregelung
current limiting controller Strombegrenzungsregler
current loading Strombelastung
current loading capability Strombelastbarkeit || Stromtragfähigkeit
current loop Linienstrom * Stromschleife, z.B. bei 20 mA/TTY-Schnittstelle || Stromschleife
current loop output Stromreglerausgang

current loop signal Stromsignal * z.B. 0-20 mA; siehe auch 'eingeprägter Strom'
current overload Überstrom * Überlaststrom, Stromüberlastung || Überstrombelastung
current path Strombahn * z.B. in Auslösegerät, Leistungsschalter usw
current peak Stromspitze
current peaks Stromspitzen
current pre-control Stromvorsteuerung
current price Marktpreis
current probe Stromzange
current rate of rise Stromänderungsgeschwindigkeit
current rating Nennstrom * spezifizierter Nennstrom
current reference Stromsollwert
current regulator Stromregler
current requirement Strombedarf
current reversal Stromrichtungsumkehr
current ripple Stromwelligkeit
current rise Stromanstieg
current rise time Stromanregelzeit
current rush Stromstoß * inrush current surge: →Anfahrstromstoß
current sensing Stromistwerterfassung
current sensor Stromsensor || Stromwandler
current setpoint Stromsollwert
current sharing Stromaufteilung
current signal Stromistwert
current sink Stromsenke
current source Stromquelle
current source DC-link Stromzwischenkreis
current source DC-link converter Stromzwischenkreisumrichter
current source inverter I-Umrichter || Stromzwischenkreis-Umrichter
current spike Stromspitze * Nadelimpuls
current state Istzustand
current suppression Stromabbau
current surge Stromstoß
current takeover Stromübernahme
current transducer Stromwandler
current transfer Stromübergang
current transformer ratio Stromwandlerübersetzung * Übersetzungsverhältnis
current zero Stromnulldurchgang
current zero detection Strom-Null-Erkennung || Stromnullerfassung
current zero sensing Stromnullerfassung
current zero signalling Stromnullmeldung
current-carrying durchströmt * elektr. Strom führend || stromdurchflossen || stromführend * auch (Stromrichter-) Ventil
current-carrying conductor stromdurchflossener Leiter
current-dependent stromabhängig
current-energized stromführend
current-free stromlos
current-free state stromloser Zustand
current-impulse relay Stromstoßrelais
current-source DC link converter I-Umrichter || Stromzwischenkreis-Umrichter
current/voltage chart Strom-Spannungs-Ebene * Koordinatensystem
currently gerade * (Adv.) momentan || im Moment || laufend * (Adv.) gegenwärtig, jetzig, aktuell, verbreitet, jetzt, zurzeit || momentan * (Adv.) jetzt, zurzeit
curriculum vitae Lebenslauf
currrent transformer Stromwandler
cursor Schreibmarke
curt bündig * schroff
curtail verkürzen * beschränken
curtain coating process Gießverfahren * zum Auftragen von Leim, Klebstoffen usw. auf eine Fläche

curvature Biegung * →*Krümmung* || Krümmung || Wölbung * *gewölbte Form*
curve Biegung * *Kurve, Wölbung* || Kennlinie * *Kurve* || Krümmung * *Kurve* || Kurve * *[köw] polygon curve: aus Geradenstücken bestehende ~ (→Polygonzug)* || Rundung * *abgerundete Stelle* || Verlauf * ~ *einer Funktion/Kurve; Verlauf von Strom/Spannung usw.*
curve characteristic Kurvenverlauf
curve diagram Kurvendarstellung
curve display Kurvenanzeige || Kurvendarstellung * *am Bildschirm usw.*
curve of trajectory Bahnkurve || Kurve * *ballistische ~, Bahnkurve*
curve path Kurvenzug
curve shape Kurvenform || Kurvenverlauf
curve trace Kurvenzug * *von registrierendem Messgerät aufgezeichnet*
curved abgerundet || ballig || gebogen * *(ab)gerundet* || gekrümmt || geschwungen || krumm * *gebogen* || verrundet
curved roll Breitstreckwalze * *mit gebogener Achse zur Vermeidg.v. Faltenbildg. b. Papier(verarbeitungs)maschine*
curved-tooth coupling Bogenzahnkupplung
cushion auffangen * *abfedern/puffern, z.B. Stoß* || dämpfen * *(mechanisch) puffern, abfedern, weich betten, (Stoß, Fall) dämpfen* || Polster * *Kissen* || puffern * *(sanft) abfedern*
cushioning Dämpfung * *Abfederung, weiche Aufhängung* || Federung
custody Schutz ≙ *Obhut*
custom Kundenkreis || kundenspezifisch
custom clearing Zollabfertigung
custom formalities Zollformalitäten
custom made maßgeschneidert * *für einen Kunden speziell angefertigt*
custom-built kundenspezifisch || Sonder- * *nach Kundenvorgabe gebaut*
custom-design kundenspezifisch * *kundenspezifisch ausgeführt*
custom-designed circuit kundenspezifischer Schaltkreis
custom-specific kundenspezifisch || maßgeschneidert * →*kundenspezifisch*
custom-specific adaption kundenspezifische Anpassung
custom-specific production Einzelfertigung * *kundenspezifische Fertigung*
custom-specification Kundenspezifikation
custom-taylored kundenspezifisch * *auf Kundenbedürfnisse zugeschnitten*
custom-taylored solution kundenspezifische Lösung
customary gebräuchlich * *herkömmlich* || herkömmlich * *gebräuchlich, herkömmlich, üblich, Gewohnheits-* || üblich
customer Abnehmer * *Kunde* || Anwender * *Kunde* || Auftraggeber * *Kunde* || Besteller || Käufer * *Kunde* || Kunde
customer advantage Kundenvorteil
customer advice Kundenberatung
customer advisory service Kundenberatung * *als Dienstleistung*
customer benefit Kundennutzen
customer benefits Kundennutzen
customer personnel Kundenpersonal
customer queries Kundenanfragen
customer request Kundenwunsch * *on customer request: auf ~/Kundenanfrage*
customer requirement Kundenwunsch
customer requirements Kundenwunsch * *Kundenbedürfnisse; in accordance to: nach ~ (z.B. Spezialanfertigung)*
customer support Service * *Kundenunterstützung*

customer training Kundenschulung
customer's order Kundenauftrag
customer's requirement Kundenanforderungen
customer-oriented kundenorientiert
customer-specific kundenspezifisch
customer-specific configuring kundenspezifische Projektierung
customer-specified kundenspezifisch * *vom Kunden vorgegeben*
customers Kundenkreis
customers service Kundendienst
customhouse Zollbehörde * *Zollamt*
customisation Anpassung * *(brit.) ~ an Kundenanforderungen*
customizable anpassbar * *an Kundenwünsche* || anpassfähig * *an Kundenanforderungen* || anpassungsfähig * *an Kundenbedürfnisse*
customization Anpassung * *(amerikan.) ~ an Kundenbedürfnisse/-spezifikationen* || Kundenanpassung || kundenspezifische Anpassung
customize anpassen an Kundenwünsche || zuschneiden * *auf Kundenanforderungen/-bedürfnisse*
customized anwenderspezifisch * *auf einen Kunden zugeschnitten* || kundenspezifisch || maßgeschneidert * *an die Kundenbedürfnisse angepasst* || Sonder- * *auf Kundenwunsch (entwickelt, hergestellt, zugeschnitten usw.)* || Spezial- * *kundenspezifisch (angepasst)*
customized design Sonderbauart * →*kundenspezifische Ausführung*
customized motor Sondermotor * *kundenspezifisch konstruierter Motor*
customized product Sonderanfertigung * *kundenspezifisch zugeschnittenes/entwickeltes Produkt*
customized solution kundenspezifische Lösung || Speziallösung * *auf einen Kunden zugeschnitten*
customizing Kundenanpassung * *Anpassung eines Produktes an Kundenbedürfnisse*
customs Zoll * *auch ~behörde; pay customs: ~ bezahlen* || Zollbehörde
customs authority Zollbehörde
customs office Zollbehörde * *Zollamt*
customs permit Begleitpapiere * *zollamtliche ~*
customs seal Zollverschluss * *in bond: unter ~*
cut Abbau * *Löhne, Preise* || abschneiden || Abschnitt * *abgeschnittenes Stück* || ausschneiden * ~ *und in die Zwischenablage einfügen (z.B. unter WINDOWS)* || Ausschnitt || beschränken * *beschneiden* || Einschnitt || fräsen || herabsetzen * *kürzen* || Herabsetzung * *Kürzung* || kappen * *Tau usw.* || Kerbe * *Einschnitt* || kürzen * *z.B. Ausgaben, Gelder* || Reduktion * *von Preisen, Kosten usw.* || reduzieren * *Preise, Kosten, Ausgaben, Zeitbedarf, Platzbedarf usw.* || Reduzierung * *Kürzung, Streichung (Ausgaben, Kosten, Preise, Text usw.); cut in costs: Kosten~* || schneiden * *auch Kurve* || Schnitt || senken * *Ausgaben, Kosten, Preise, (Energie-) Verbrauch* || Senkung * *Kosten, Preise*
cut accuracy Schnittgenauigkeit * *z.B. bei Querschneider, fliegender Schere usw.*
cut back verkleinern * *Produktion, Umsatz, Ausgaben usw. ~; auch 'sich verkleinern'*
cut capers Bocksprünge machen * *Kapriolen/Lüftsprünge machen*
cut costs Kosten sparen
cut down Abbau * *Personal* || abbauen * *Personal, Kosten usw. ~* || kürzen || reduzieren * *Preise, Kosten, Ausgaben, Platz-/Zeitbedarf usw.* || verkürzen * *herab/zurechtstutzen* || verringern * *Ausgaben, Preise usw.*
cut down expenses sparen * *Ausgaben senken, sich einschränken*
cut in costs Kostenreduzierung
cut in expenses Kostensenkung
cut in half halbieren

cut it fine knapp * *(Adv.)* ~ *berechnen (Preis)*
cut length Schnittlänge * *z.B. bei Querschneider*
cut off abschalten || Abschaltung || abschneiden || abstechen * *mit Drehbank* ~ || ausschalten
cut out aufhören || ausschalten || ausschalten * *(sich) ausschalten, aussetzen, ausschneiden* || ausschneiden
cut short abbrechen * *plötzlich* ~ || Stutzen * *(Verb) beschneiden* || unterbrechen || verkürzen * *plötzlich beenden, abkürzen, es kurz machen, ins Wort fallen*
cut strip-wound core Schnittbandkern * *z.B. eines Transformators*
cut the size verkleinern
cut timber Schnittholz
cut to length ablängen || zuschneiden * *{Holz}*
cut untrue verlaufen * *abweichen einer Säge*
cut width beschnittene Breite * *{Papier}*
cut-deepness Schnittiefe * *z.B. einer Säge*
cut-in speed Ablösedrehzahl * *oberhalb derer die Feldschwächung beginnt*
cut-off current Durchlassstrom * *~ einer Sicherung/eines Leistungsschalters im ausgeschalteten Zustand*
cut-off frequency Grenzfrequenz
cut-off knife Querschneider * *Bogenschneider für Papier*
cut-off machine Abschneidemaschine * →*Querteilanlage z.B. für Blech*
cut-off saw Kappsäge * *{Holz}*
cut-off speed Abschaltdrehzahl
cut-out Ausschnitt * *z.B. in einem Blech* || Aussparung * *Ausschnitt* || Zuschnitt * *das zugeschnittene Teil*
cut-out frequency Ablösefrequenz * *Frequenz bei der die Spannungsanhebung b.e. Frequenzumrichter f/U-Kennlinie endet*
cut-out point Ablösepunkt
cut-out speed Ablösedrehzahl * *oberhalb derer die Feldschwächung beginnt*
cutback Verkleinerung
cutoff frequency Eckfrequenz || Knickfrequenz
cutout Ausbruch * →*Ausschnitt, z.B. in Blechkonstruktion* || Ausschnitt || Aussparung * *Ausschnitt* || Einschnitt * *Aussparung, Ausschnitt* || Öffnung * *Ausschnitt*
cutt-off abtrennen
cutter Schneidmaschine * →*Querschneider*
cutter arbor Fräserdorn * *{Werkz.masch.}*
cutter bar Führungsschiene * *{Holz} für Kettensäge* || Nutmesser
cutter block Fräskopf || Messerwelle * *{Holz} in Holzfräse*
cutter head Messerkopf || Schneidkopf
cutter motor Messermotor
cutter set Fräsersatz * *{Holz}*
cutterhead Fräskopf
cutting Querschneiden * *z.B. von Papier (im Gegensatz zum Längs- oder Rollenschneiden)* || Schnitt || spanabhebend || spanabhebende Bearbeitung || spanend * *Bearbeitung, Werkzeug usw.* || spanende Bearbeitung || zerspanend * *Bearbeitung, Werkzeug usw. (Gegensatz:* →*spanlos)* || zerspanende Bearbeitung
cutting angle Schnittwinkel
cutting depth Schnittiefe
cutting device Schneidgerät * *z.B. für Lichtleiter*
cutting disk Kreismesser * *Tellermesser* || Tellermesser * *siehe auch* →*Kreismesser*
cutting force Schneidkraft * *bei Werkzeugmaschine* || Schnittkraft * *bei Werkzeugmaschine*
cutting head Schneidkopf
cutting height Schnitthöhe * *z.B. bei (Holz-) Kreissäge*
cutting machine Abschneidmaschine || Schneidemaschine

cutting machining spanende Fertigung
cutting oil Schneidöl
cutting out Zuschnitt
cutting pliers Saitenschneider
cutting to length Querschnitt * *das Quer-Schneiden (mit Messer/Schere)*
cutting to size Zuschnitt
cutting tool Drehstahl * *für* →*Drehbank* || Hobelstahl
cutting torch Schneidbrenner
cutting water Schneidwasser * *{Werkz.masch.}*
cutting-force control Schnittkraftregelung
cutting-off grinder Trennschleifmaschine
cutting-off machine Abstechmaschine * *Werkzeugmaschine* || Trennschleifmaschine
cutting-to-length line Querteilanlage * *z.B. für Walzband, Blech usw.*
cuttings Span * *Metallspäne* || Späne * *Metallspäne*
cw im Uhrzeigersinn * *Abkürzg. für 'clockwise'* || mit dem Uhrzeigersinn || rechts * *in ~drehrichtung* || rechts herum * *Abkürzung für 'clockwise':im Uhrzeigersinn* || vorwärts * *im Uhrzeigersinn*
cw rotating field Rechtsdrehfeld * *Abk. f. 'Clock-Wise': im Uhrzeigersinn*
cw rotation Rechtsdrehrichtung * *im Uhrzeigersinn*
cw-ccw detection Vor-Rückwärtsauswertung
cw-ccw detector Vorwärts-Rückwärtsauswertung * *Rechts-Linkserkennung*
cw. rotation Vorwärtsdrehrichtung * *Rechtslauf, Abkürzung für clockwise: im Uhrzeigersinn*
cwt t * *hundredweight; 1000 kg entspr. 19,6841 cwt; 1 cwt enspr. ca. 1 Zentner*
cycle Durchlauf * *Zyklus* || Kreislauf || Periode || Spiel * *Last~* || Takt * *(bei Verbrennungsmotor, Fertigungslinie, Schweißgerät)* || Vorgang * *in (regelmäßigen) Abständen sich wiederholender ~* || Zyklus
cycle duration Periodendauer || Spieldauer
cycle frequency Taktzahl * *Anzahl Takte je Zeiteinheit (bei Verpackungsmaschine, Presse usw.)*
cycle of operation Arbeitsgang * *Maschine* || Betriebszyklus
cycle operation Zyklusbetrieb * *z.B.* →*Betriebsart S5*
cycle press Taktpresse
cycle rate Taktzahl
cycle time Periodendauer || Taktzeit || Zykluszeit
cycle-controlled press Taktpresse * *{Holz}*
cycle-overrun Zyklusüberlauf
cycle-timeout Zyklusüberlauf
cycles per second Hertz || Hz
cyclic periodisch || zyklisch
cyclic buffer Umlaufpuffer
cyclic call zyklischer Aufruf
cyclic duration factor ED * *relative* →*Einschaltdauer* * *{el.Masch.} relative ~, z.B. eines Motors bei Aussetzbetrieb* || relative Einschaltdauer * *(ED) prozentuales Verhältn. Lastzeit/Spieldauer, siehe VDE 0530 Teil 1*
cyclic function Kreisfunktion
cyclic irregularity Stoßgrad * *Ungleichförmigkeitsgrad einer Drehbewegung* || Ungleichförmigkeit * *einer Drehbewegung, Ungleichförmigkeitsgrad* || Ungleichförmigkeitsgrad * *Stossgrad einer Drehbewegung*
cyclic load Wechselbeanspruchung || Wechsellast * *zyklisch sich wiederholendes Lastspiel*
cyclic operation zyklische Verarbeitung * *Arbeitsweise*
cyclic processing zyklische Bearbeitung || zyklische Verarbeitung
cyclic services zyklische Dienste * *z.B. bei einem Feldbussystem*
cyclic stressing Wechselbeanspruchung

cyclic succession zyklische Folge
cyclic transmission zyklische Übertragung * z.B. über einen seriellen Bus
cyclically zyklisch * (Adv.)
cyclically controlled zyklisch gesteuert
cycling Kreislauf || Wechselbeanspruchung
cyclo-converter Direktumrichter * setzt ohne Zw.kreis e. AC-Eingangs- i.AC-Ausg.spanng. mit max 0,5-facher Frequ.um
cycloconverter Hüllkurvenumrichter * Direktumrichter || Wechselstrom-Direktumrichter
cycloid Zykloide
cycloidal zykloidisch
cycloidal worm gear Zykloiden-Schneckengetriebe
cylinder Rolle * Hohlwalze || Walze * hohle Walze || Zylinder * ['silinder]
cylinder grinding machine Zylinderschleifmaschine * z.B. für Tiefdruckzylinder (Druckerei)
cylinder-head screw Zylinderkopfschraube
cylindrical rund * zylinderförmig || walzenförmig || zylinderförmig || zylindrisch
cylindrical grinding machine Rundschleifmaschine * (Werkz.masch.)external: Außen~, internal: Innen~
cylindrical rotor Vollpolläufer
cylindrical screw Zylinderschraube
cylindrical-roller bearing Zylinderrollenlager
cylindrical-rotor machine Vollpolmaschine * i. Gegensatz zur →Schenkelpol-Maschine (kein 'gezackter' Läufer)
cyrogen Frigen
Czech Republic Tschechische Republik

D

D end A-Seite * Abk. f. 'Drive end': Antriebsseite eines Motors
DE F-und-E * Abkürzg. für 'Development and Research': Entwicklung und Forschung
D-component D-Anteil
D-connected in Dreieck geschaltet
D-end Antriebsseite * Abkürzung für drive end || antriebsseitig || AS * Abkürzg. für 'Drive end': Antriebsseite
D-shaft end D-Wellenende * Drive side, Antriebsseite, A-Seite eines Motors
D-type conector Cannon-Stecker * mit "D"-förmigem Querschnitt
D-type connector D-Sub Stecker * mit "D"-förmigem Querschnitt || SUB-D Stecker
d.c. Gleichstrom
d.c. chopper Gleichstromsteller
d.c. drive Gleichstromantrieb
d.c. intermediate circuit Gleichspannungszwischenkreis || Zwischenkreis * (weniger gebräuchlich) Gleichspannungszwischenkreis
d.c. link Gleichstromzwischenkreis
D.C. machine Gleichstrommaschine
d.c. motor Gleichstrommotor
d.c. side fuse Gleichstromsicherung
d.c. voltage Gleichspannung
d.c.-a.c. Allstrom
DA converter DAU
DAC DAU * Abk. für 'Digital to Analog Converter' || Digital-Analog-Wandler
daemon Hintergrundprozess * vom Betriebssystem in Auftrag gegebener ~ bei UNIX
Dahlander-connection Dahlander-Schaltung * für polumschalbaren Motor, ermöglicht 2 Drehzahlen

daily Tages- || täglich
dairy-proof design Molkerei-Ausführung * Sonderausführung für den Einsatz in der Nahrungsmittelindustrie
daisy chain Kettenschaltung * z.B. v. Kommunikationspartnern (i.Gegensatz. z. stern-/busförmigen Kommunikation) || Schleifleitung * durchgeschleifte Steuer-/Datenleitung (1 kommende u. 1 gehende Leitg.je Teilnehmer)
daisy-chain aneinander hängen * mit Hilfe v.Signalen, die jew.von einem Element zum nächsten durchgeschleift sind
daisy-wheel Typenrad
daisy-wheel printer Typenraddrucker
dam Damm
damage beschädigen || Beschädigung * incur damage:beschädigt werden || Schaden * Beschädigung || schädigen || Schädigung * to: von etwas || verletzen * beschädigen || Verlust * Schaden || Zerstörung * Beschädigung
damage prevention Schadensverhütung
damage report Schadensbericht
damaged beschädigt || defekt || schadhaft * beschädigt || schlecht * beschädigt
damages Schadenersatz * Geldsumme
damp anfeuchten || dämpfen * (elektrisch, mechanisch, akustisch); Schwingung, auch Stimmung ~ || feucht * feucht, dunstig (z.B. Luft) || nass * feucht, dunstig
damp heat feuchte Wärme
damped beschaltet * Relais/Schützspule mit RC-Glied oder Diode zum Überspannungsschutz || gedämpft * z.B. elektr./mechan./akust. Schwingung; (siehe auch 'dämpfen')
damped oscillation gedämpfte Schwingung
dampen dämpfen * (elektrisch, mechanisch, akustisch) auch Stimmung; dampen down: abdämpfen
dampening Bedämpfung
damper Abkühlung * (figürlich) || Dämpfungselement * mechanisch, elektrisch und akustisch || Dämpfungsglied * (mechan., elektr. und akustisch) || Drossel * strömungstechnische ~, z.B. in Rohrleitung || Schieber * Schieber, (Zug-) Klappe
damper weight Ausgleichsgewicht * "Wuchtgewicht" zur Vermeidung unrunden Laufs
damper winding Dämpferwicklung
damping Bedämpfung || Beruhigung * Dämpfung || Dämpfung * (elektr., mechan., akust.) abschwächen, drosseln, auch Schwingungen
damping cage Dämpferkäfig * im Läufer eines AC-Motors
damping circuit Dämpfungskreis
damping coefficient Dämpfungsgrad
damping effect Dämpfung * Dämpfungseffekt
damping roller Feuchtwalze * (Druckerei)
damping unit Feuchtwerk * (Druckerei)
dancer Pendelwalze * Tänzer || Regelschwinge * siehe →Tänzer (-walze), z.B. in der Gummi-/Reifenindustrie || Tänzer * zur Zugregelung bei Wickelantrieben usw. || Tänzerrolle || Tänzerwalze
dancer arm Pendelarm * einer Tänzerwalze/Tastrolle/Regelschwinge/Pendelwalze || Tänzerarm
dancer arm position Tänzerlage
dancer control Tänzerregelung * (direkte) Zugregelung mit Tänzerwalze
dancer controller Tänzerlageregler || Tänzerregler
dancer feedback signal Tänzerlagesignal
dancer loading Tänzerabstützung * z.B. durch Pneumatikzylinder einstellbare Tänzerbelastung (veränderd den Bahnzug) || Tänzerbelastung * z.B. Abstützung durch Pneumatikzylinder mit einstellbarer Kraft zur Zugeinstellung
dancer position Tänzerlage
dancer position control Tänzerlageregelung

dancer position controller Tänzerlageregler || Tänzerregler * *Tänzerlageregler*
dancer position reference Tänzerlagesollwert
dancer position setpoint Tänzerlagesollwert
dancer position signal Tänzerlagesignal
dancer potentiometer Tänzerpoti * *zur Erfassung des Ausschlags e. Tänzerwalze (Istwert für Tänzerlageregelung)*
dancer roll Pendelwalze * *Tänzerwalze/-rolle* || Tänzer || Tänzerrolle || Tänzerwalze || Tastrolle * *[Draht] zur Zugerfassung und -Regelung bei Drahtziehmaschine*
dancer roll potentiometer Tänzerpotentiometer * *erfasst Tänzerlage,häufig berührungslos arbeitend, z.B. Feldplattenpoti*
dandy roll Egoutteur * *bei Papierherstellung*
danger Gefahr
danger area Gefahrenbereich
danger of accident Unfallgefahr
danger of corrosion Korrosionsgefahr
danger of death Lebensgefahr
danger of injury Verletzungsgefahr
danger of life Lebensgefahr
danger signal Warnzeichen
danger to life Lebensgefahr
danger! Vorsicht bitte * *Gefahr*
dangerous gefährlich || schädlich * *gefährlich, gefahrvoll, bedenklich*
dangerous substance Gefahrstoff
dank feucht * *nasskalt, (unangenehm) feucht*
dark dunkel
Darlington stage Darlingtonstufe * *Transistorschaltung (aus zwei "hintereinandergeschalteten" Transistoren)*
dash Bindestrich * *Gedankenstrich* || Gedankenstrich || spritzen * *(salopp) eilen* || Strich * *Gedankenstrich*
dash-line stricheln
dash-lined gestrichelt
dashed line gestrichelte Linie
data Angaben || Daten * *(Plural, nie mit End-s verwenden: "datas" ist falsch!)* || Informationen * *Daten* || Wert * *Datum*
data acquisition Datenerfassung * *von Messdaten* || Messdatenerfassung || Messwerterfassung
data area Datenbereich || Datensatz * *Datenbereich*
data array Datenbereich * *Datenfeld*
data backup Datensicherung
data base Datenbasis
data bit Datenbit
data block Datenbaustein * *z.B. bei SPS* || Datenbereich || Datensatz * *Datenblock*
data bridge Busumsetzer * *Protokollumsetzer* || Protokollumsetzer
data buffer Datenpuffer
data buffering Zwischenspeicherung * *von Daten*
data bus Datenbus
data carrier Datenträger * *Diskette, Magnetband usw.; mobile data carrier: mobiler ~ (z.B. in Materialflussystem)*
data communication Datenaustausch || Datenübermittlung || Datenübertragung || Datenverkehr
data communications Datenaustausch
data compression Komprimierung * *Datenkomprimierung*
data compression program Komprimierprogramm
data concentrator Konzentrator * *für Daten*
data consistency Datenkonsistenz
data contents Dateninhalte
data entry Dateneingabe
data exchange Datenaustausch || Datenverkehr
data flow Datenfluss || Datenstrom * *Datenfluss*
data flowchart Datenflussplan

data form Datenblatt * *Formular zum Eintragen individueller Werte (Rechenblatt)*
data handling block Hantierungsbaustein * *bei SPS*
data highway Bus || Datenautobahn || Kommunikationsbus
data integrity Datenkonsistenz * *keine Durchmischung von Teildaten aus verschiedenen Zyklen in einem Datensatz* || Datensicherheit * *Unverfälschtheit der Daten*
data interchange Datenaustausch || Datenverkehr
data item Datum * *(Singular von 'Daten')*
data line Datenleitung
data listing Protokollierung
data loss Datenverlust
data management Datenhaltung * *Verwaltung*
data media Datenträger * *(Plural)*
data medium Datenträger
data memory Datenspeicher
data module Datenbaustein * *z.B. bei SPS*
data packet Datenpaket
data processing Datenverarbeitung || DV * *Datenverarbeitung* || Informationsverarbeitung
data processing system Datenverarbeitungssystem || DV-System
data protection Datensicherung
data quantities Datenmengen
data record Datenbereich * *Datensatz* || Datensatz
data recorder Messschreiber || Messwertschreiber
data recording Datenerfassung
data register Datenspeicher * *Datenregister bei SPS*
data representation Datendarstellung
data retention without battery batterielose Pufferung * *nichtflüchtige Datenspeicherung ohne Batterie (z.B. in NOVRAM)*
data save Datensicherung
data save routine Datenrettroutine
data security Datensicherheit || Datensicherung
data segment Datenbereich * *Datensegment (im Gegensatz zum Programmcode-Segment)*
data set Datensatz
data sheet Datenblatt || Typenblatt * *Datenblatt*
data sheet information Datenblattangabe
data storage Datenhaltung * *Speicherung*
data storage area Datenbereich * *im Speicher einer SPS*
data storage medium Datenträger
data stream Datenstrom
data traffic Datenverkehr
data transfer Datenaustausch * *Datenübertragung* || Datentransfer || Datenübertragung || Datenverkehr
data transfer control Datenübertragungssteuerung
data transfer fault Kommunikationsfehler * *Datenübertragungsfehler*
data transfer program Datentransferprogramm
data transfer protocol Übertragungsprotokoll
data transmission Datenaustausch * *Übertragung* || Datenübermittlung || Datenübertragung || Datenverkehr
data transmission control Datenübertragungssteuerung
data transmission error Datenübertragungsfehler
data transmission line Übertragungsstrecke * *zur Datenübertragung*
data type Datentyp || Parametertyp
data width Datenbreite
data word Datenwort * *meist 16-bit-Information*
data-save routine Daten-Rettroutine
data-transfer rate Datenübertragungsgeschwindigkeit || Datenübertragungsrate || Übertragungsgeschwindigkeit * *Daten~* || Übertragungsrate * *Daten~*
database Datenbank

date Datum * *Tag/Monat/Jahr* || Termin * *Zeitpunkt, Datum; final date: äußerster ~; target date: Ziel~*
date of completion Frist * *Fertigstellungs/Erledigungstermin* || Termin * *Fertigstellungs~*
date of delivery Lieferdatum || Liefertermin
date of manufacture Baujahr
date scheduling Terminplanung
date-stamping system Datiereinrichtung * *z.B. in der Nahrungsmittel- u. Verpackungsindustrie*
dates Termine
datum Daten * *(Singular von 'data') ein Einzeldatum* || Datum * *(Singular von 'data') ein Einzeldatum*
daughter board Sub-Leiterkarte || Submodul * *Leiterkarte*
daughtercard Huckepackbaugruppe * *Leiterkarte*
day of delivery Liefertermin
day-to-day laufend * *tagtäglich, z.B. Geschäfte* || täglich * *Einsatz, Gebrauch usw.*
DC Gleichstrom
DC bus Gleichspannungs-Zwischenkreis || Gleichspannungszwischenkreis-Schiene || Spannungs-Zwischenkreis || Spannungszwischenkreis-Schiene || Zwischenkreis * *Zwischenkreis-Schiene*
DC busbar Gleichspannungsschiene * *Sammelschiene* || Gleichstromschiene
DC chopper Gleichstromsteller * *erzeugt aus konst. Gleichsp. e. variable Gleichsp., DIN 57558/ VDE 0558 jew.Tl.3*
DC chopper converter Gleichstromsteller
DC chopper regulator Gleichstromsteller
DC circuit Gleichstromkreis
DC common bus gemeinsamer Zwischenkreis
DC component Gleichanteil * *bei Strom/Spannung/Leistung* || Gleichstromanteil
DC conversion Gleichrichtung
DC converter Stromrichter mit Gleichspannungsausgang
DC current Gleichstrom
DC drive Gleichstromantrieb || Stromrichter mit Gleichspannungsausgang * *(nur amerikan.) für Motoransteuerung*
DC excitation Gleichstromerregung * *z.B. Motor, Magnetbremse*
DC fuse Gleichstromsicherung
DC injection braking Gleichstrombremsung * *durch Gleichstrom-Einprägung in vom Netz getrennte Ständerwicklg. e.AC-Motors*
DC intermediate circuit Gleichspannungszwischenkreis
DC link Gleichspannungszwischenkreis || Gleichstromzwischenkreis || Zwischenkreis
DC link bus Zwischenkreisschiene
DC link capacitor Zwischenkreiskondensator
DC link circuit Gleichstromzwischenkreis
DC link converter Spannungszwischenkreis-Umrichter || Zwischenkreisumrichter
DC link coupling Zwischenkreisankopplung
DC link current Zwischenkreisstrom
DC link fuse Zwischenkreissicherung * *bei Umrichter*
DC link voltage Zwischenkreisspannung
DC machine Gleichstrommaschine
DC motor Gleichstrommotor
DC permanent magnet motor Gleichstrom-Permanentmagnetmotor
DC rectifier unit Einspeiseeinheit
DC shunt-wound motor Gleichstrom-Nebenschlussmotor || Gleichstromnebenschlussmaschine
DC side Gleichstromseite * *on the: auf der*
DC side fuse link Gleichstromsicherung * *Sicherungseinsatz*
DC supply Gleichspannungsquelle * *Versorgungspannung* || Gleichspannungsversorgung
DC supply network Gleichspannungsnetz
DC supply system Gleichstromnetz
DC system Gleichstromsystem
DC tach Gleichstromtacho || Gleichstromtachogenerator
DC tacho Gleichstromtacho || Gleichstromtachogenerator
DC tachogenerator Gleichstromtachogenerator
DC tachometer Gleichstromtacho
DC traction supply system Gleichstrom-Fahrleitung
DC traction system Gleichstrombahn
DC transducer Shuntwandler
DC voltage Gleichspannung
DC voltage auxiliary bus Gleichspannungshilfsschiene
DC voltage bus Gleichspannungsschiene
DC voltage component Gleichspannungsanteil
DC-bus Gleichstromschiene * *on:an* || Spannungszwischenkreis
DC-DC converter Gleichspannungswandler || Gleichspannungwandler || Spannungswandler * *Gleichspannungswandler*
DC-link Spannungszwischenkreis
dc-link capacitor Zwischenkreiskondensator
DC-link converter Zwischenkreisumrichter
DC-link pulsing Zwischenkreistaktung
DC-link reactor Zwischenkreisdrossel
dc-to-dc converter Gleichspannungswandler
DCS DCS * *abkürzg. f. 'Digital Control System': Digital-Regelsystem, vorzugsw. f. verfahrenstechn. Prozesse*
DDC digitale Regelung * *Abkürzg. für 'Direct Digital Control'*
DDE dynamischer Datenaustausch * *Abkürzg. f. 'Dynamic Data Exchange' (siehe dort)*
de Ent- * *Vorsilbe für Substantiv*
de- Rück- * *um-, ent-, weg-, ver-*
de-aerate entlüften * *(chem.)*
de-aerating Entlüftung
de-aeration Entlüftung
de-clutch auskuppeln * *eine Kupplung*
de-coil abwickeln * *Draht/Blech ~*
de-energize abschalten * *mit Low-Signal ansteuern* || freischalten || LOW-Signal anlegen * *an Klemmen, Eingang* || mit LOW Signal ansteuern * *z.B. eine Klemme* || öffnen * *eine Klemme mit LOW-Signal ansteuern*
de-energized abgefallen * *Relais, Schütz* || spannungsfrei || spannungslos || stromlos
de-facto nahezu * *~ wirklich*
de-ionized entionisiert * *z.B. Wasser (für Kühlkreislauf, zur Erzielung einer geringen Leitfähigkeit)*
de-rate Leistung reduzieren * *z.B. Ausgangsleistung i.techn Spezifikation bei gewissen Betriebsbedingungen* || reduzieren * *z.B. Leistungsspezifikation bei Stromrichter in großer Aufstellhöhe ü.d. Meeresspiegel*
de-rating Leistungsminderung * *Reduzierg. d. Nennleistung auf Grund erschwerter Bedingungen* || Reduzierung * *von Spezifikationen bei erschwerten Betriebsbedingungen (Temperatur/Taktfrequenz usw.)*
deactivate abschalten * *deaktivieren, z.B. eine Überwachung* || deaktivieren || lahm legen * *→deaktivieren*
dead spannungsfrei * *dead state: ~er Zustand* || spannungslos || stromlos
dead band Hysterese * *Totband/zone* || Totband || Totzone * *Totband* || Unempfindlichkeitsbereich
dead center Totpunkt * *z.B. bei Kolbenmaschine, Exzenterpresse usw.; top: oberer, bottom: unterer*
dead interval Umschaltpause
dead interval on reversing Umschaltpause * *bei Momentenumkehr (in Kommandostufe)*

dead loss Totalausfall
dead stop Anschlag * ~ *einer Messeinrichtung*
dead time momentenfreie Pause || momentfreie Pause || Totzeit
dead travel Gang * *toter ~/Spiel/Lose bei Maschinenteilen* || Lose * *toter Gang, Spiel in einem Maschinenteil* || Spiel * *toter Gang, Lose in einem Maschinenteil*
dead wood Holz * *abgelagertes Holz, totes ~*
dead zone Totband * Totzone || Totzone
dead-end line Stichleitung
dead-time Totzeit
deadbeat aperiodisch * *auch 'aperiodisch gedämpft'*
deaden dämpfen * *(akustisch)*
deadline äußerster Termin || Frist * *maximal akzeptierte/äußerste* ~ || Termin * *äußerster Termin*
deadlock Absturz * *"Auf-Der-Stelle-Treten" z.B. eines Softwareprogramms* || Stillstand * *(figürl.) völliger* ~, *Sackgasse, toter Punkt, Auf-d.-Stelle-Treten (Verhandlgen.)*
deaerate entlüften
deaeration Entlüftung
deal Abschluss * ~ *eines Geschäfts* || Geschäft * *Handel* || handeln * *Handel treiben (in: mit Waren); deal with:* ~ *von/über*
deal is off Geschäft ist geplatzt
deal out verteilen * *austeilen*
deal with abhandeln || bearbeiten * *behandeln* || befassen * *(figürl.) sich befassen mit* || befassen * *sich befassen mit, behandeln (z.B. in Aufsatz, Abhandlung)* || behandeln * *auch Thema* || umgehen * *mit etwas ~, behandeln, sich befassen mit* || verkraften * *bewältigen* || vornehmen * *behandeln*
dealer Fachhändler || Händler
Dear Sehr geehrte || Sehr geehrter
Dear Mr. Sehr geehrter Herr
debarker Entrindungsmaschine * *{Holz}* || Holzschälmaschine * *Entrindungsmaschine* || Rundschälmaschine * ~ *{Holz} Entrindungsmaschine* || Schälmaschine * *{Holz} Entrindungsmaschine*
debarking Entrindung
debarking drum Entrindungstrommel * *{Holz}*
debase verschlechtern * *in der Qualität* ~
debilitate schwächen * *entkräften*
debit abbuchen || belasten * *debit an account: ein Konto* ~ || Belastung * ~ *eines Kontos* || Berechnung * *Belastung (eines Kontos)* || soll * *(Substantiv; kaufmänn.) debit-side: auf d. Sollseite*
debouncing Entprellen || Entprellung
debris Scherben * *['debrie] Trümmer, (Gesteins-) Schutt*
debt claim Forderung * *Schuldforderung*
debug Bearbeitungskontrolle * *bei der Abarbeitung eines SPS-Programms* || beheben * *Softwarefehler* ~
debug function Ablaufkontrolle * *{SPS}*
debug function block Diagnosebaustein * *Funktionsbaustein*
debugged fehlerbereinigt * *Software*
debugger Monitorprogramm * *Testhilfeprogramm zum Austesten von Softwareprogrammen* || Testhilfeprogramm * *zum Testen von u. zur Fehlerfindung in Softwareprogrammen*
debugging Ablaufkontrolle * *{SPS}* || Fehlerbehebung * *beim Test (noch nicht ausgelieferter) Software* || Fehlerbereinigung * *in Software* || Fehlerbeseitigung * *Software* || Fehlersuche * *in Software* || Test * *von Software*
deburr abgraten || entgraten
decade Dekade || dekadisch
decanter Dekanter * *Zentrifuge zur Trennung von Substanzen z.B. Fest- u.Flüssigstoffe i. d. Getränkeindustrie*
decay Abbau * *Rückgang, Schwund* || Abfall * *Ver-*

fall, Schwächerwerden || abklingen * *schwinden, abnehmen, schwach werden, (herab)sinken, zerfallen, verfallen, zugrundegehen* || abnehmen * *schwinden, sinken, verfallen* || abnehmen * *schwinden, abnehmen, (ab)sinken, schwach werden, ver-/zerfallen* || Fall * *Niedergang* || Niedergang || Rückgang * *Schwund, Zerfall, Verfall, Schwäche* || Schwund * *Rückgang* || sinken * *verfallen* || Verfall || Zerfall || zurückgehen * *abnehmen, schwinden, verfallen*
decay of prices Preisverfall
decay time Abklingzeit
decay to zero Nullwerden * *(auch Verb) Auf-Null-Absinken*
decaying abklingend * *z.b. Schwingungen*
decel verzögern * *Kurzform für 'decelerate'*
decel time Rücklaufzeit * *(salopp) Abk.f. 'deceleration time'*
decelerate bremsen * *verzögern* || herunterfahren * *verzögern* || herunterlaufen * *verzögern* || verlangsamen * *verzögern* || verzögern * *['di:celereit]* || zurücklaufen * *verzögern*
decelerating abnehmend * *Geschwindigkeit, Drehzahl*
deceleration Abbremsung * *Verzögerung* || Rücklauf * *Verzögerung* || Verzögerung * *['dießelereischen] negative Beschleunigung*
deceleration distance Verzögerungsweg
deceleration point Bremseinsatzpunkt || Verzögerungspunkt
deceleration ramp Rücklauframpe || Verzögerungsrampe
deceleration time Anhaltezeit * *Rücklaufzeit* || Bremszeit * *Zeit, während der verzögert wird* || Rücklaufzeit
deceleration torque Bremsmoment * *Verzögerungsmoment* || Verzögerungsmoment
deceleration-initiation point Bremseinsatzpunkt || Verzögerungspunkt * *→Bremseinsatzpunkt*
decent ordentlich * *anständig*
decentral peripheral device dezentrales Peripheriegerät
decentralized dezentral || örtlich verteilt || verteilt * *dezentral angeordnet*
decentralized peripherals dezentrale Peripherie * *ferngesteuerte Klemmenleiste, z.B. SIMATIC (R) ET100, ET200 von SIEMENS*
decibel Dezibel
decide beschließen * *sich entschließen, entscheiden* || entscheiden * *on: über; in favo(u)r of: für/zugunsten; against (doing): gegen/nicht zu tun*
decidedly entschieden * *(Adv.) entschieden, fraglos, bestimmt, deutlich*
decimal dezimal || Dezimalzahl
decimal number Dezimalzahl
decimal places Nachkommastellen
decimal point Dezimalpunkt
decimal system Dezimalsystem
decise ausdenken
decision Beschluss || Entscheidung * *on: über; of: des; take a decision: eine ~ treffen*
decision maker Entscheider || Entscheidungsträger
decision-making Entscheidungsprozess
decision-making power Entscheidungsspielraum * *Entscheidungsgewalt*
decision-making process Entscheidungsprozess
decisive ausschlaggebend * *entscheidend, maßgebend, endgültig; be decisive in: verantwortlich für* || entscheidend * *z.B. Kriterium, Eigenschaft* || maßgebend * *entscheidend, ausschlaggebend, bestimmend* || maßgeblich * *entscheidend, ausschlaggebend*
decisive factor Ausschlag * *ausschlaggebender Punkt* || maßgebende Größe * *bestimmende Größe* || maßgebliche Größe * *entscheidende Größe*

deck Deck * Schiffsdeck || Etage * auf Schiff
deck installation Oberdeckaufstellung * {el.-Masch.} Aufstellung von Motoren auf Schiffsdeck
deck-marine motor Oberdeck-Motor * für Einsatz auf Schiffen (Schutzart IP56)
deckwater-proof deckwassergeschützt * z.B. Motor-Schutzart für Einsatz auf Schiffen (Schutzart IP56)
declaration Aussage * Erklärung, Aussage || Deklaration || Erklärung * Aussage, Feststellung (auch rechtsverbindliche)
declaration of conformity Konformitätserklärung * z.B. Erklärung d.Herstellers, dass e.Produkt d. EG-Richtlinien entspricht
declaration of value Wertangabe * Angabe des Geldwerts
declare angeben * erklären || behaupten * erklären || erklären * aussprechen, kundtun || feststellen * erklären || versichern * behaupten
declare null and void annullieren * für null und nichtig erklären || aufheben * für ungültig erklären
declared value Wertangabe * Angabe des Geldwertes
decline abfallen * abnehmen, nachlassen, zurückgehen, sich verschlechtern, schwächer werden, verfallen || ablehnen || abweisen * zurückweisen || Einbruch * Nieder-/Rückgang, Abnahme, Verschlechterung || fallen * (figürl.)nachlassen, sich neigen/senken, abnehmen, nachlassen (auch Preise), zurückgehen || Geschäftsrückgang || Rückgang * Abnahme, Niedergang, Verschlechterung, Sinken (Preise usw.), Verfall || sinken * verfallen || zurückgehen * ~ des Geschäfts usw. || zurückweisen
decline in prices Preisrückgang
decline in value Wertminderung
decline with thanks danken * ablehnend
declutch auskuppeln * eine Kupplung || ausrücken * auskuppeln, Kupplung ~ || ausrücken * Kupplung
declutter entwirren * Ungeordnetes übersichtlich gestalten (z.B. Bildschirmaufbau/-darstellung)
decode dekodieren
decode switch Codierschalter
decoder Dekodierer
decoiler Abwickelhaspel
decompilation Rückübersetzung * von Hochspracheprogrammen
decompile rückübersetzen * von Hochspracheprogrammen
decompiler Rückübersetzer * für Hochspracheprogramme
decompose auflösen * zerlegen, zerspalten, vereinzeln || spalten * {Chemie}
decomposed aufgelöst * zerlegt, auseinandergespaltet, auseinandergelegt, vereinzelt
decompress dekomprimieren * auch Daten || entpacken * Daten ~
decompression Druckabfall || Druckverminderung
decontaminable dekontaminierbar
decontamination Entgiftung
decorating paper Dekorpapier
decorative textiles Dekotextilien
decouple abkoppeln || auskuppeln || entkoppeln
decoupling Entkopplung
decoupling diode Entkopplungsdiode * z.B. bei Stromversorgung: schaltet d. höhere Spannung z. Verbraucher durch
decrease abfallen * sich vermindern (wert-/zahlenmäßig) || abfallen * Geschwindigkeit || Abnahme * Verringerung || abnehmen * reduzieren, kleiner werden || Abschwächung || Absenkung * Herabminderung, (in:von) || absinken * sich vermindern || erniedrigen * kleiner machen || herabsetzen * vermindern, verringern, reduzieren; abnehmen, sich verringern || Minderung * Abnahme, Verringerung || nachlassen * sich verringern, abnehmen

|| Reduzierung * Verringerung, Verminderung, Abnahme || Rückgang * Abnahme, Verringerung, Verminderung || sinken * sich vermindern || verkleinern * verringern, vermindern || Verminderung || verringern || Verringerung || zurückgehen * sich verringern
decrease as the square of the voltage quadratisch mit der Spannung abnehmen
decrease button Tiefer-Taste
decrease in value Wertminderung
decrease rate Rücklaufzeit
decrease with the square of the voltage quadratisch mit der Spannung abnehmen * z.B. Drehmoment bei Asynchronmaschine
decreasing abnehmend
decree Entscheidung
dedicated bestimmt * vorgesehen, gewidmet (für einen ~en Zweck) || eigen * speziell zugeordnet, zugeeignet, gewidmet || eigen * speziell zugeordnet/ zugeschnitten; zugeeignet, gewidmet || eigener * speziell zugeordnet, zugeeignet, gewidmet || fest zugeordnet * zugeschnitten, speziell zugeordnet || zugeschnitten * speziell auf einen Verwendungszweck/eine Anwendung ~
dedicated solution Speziallösung * zugeschnitten auf ein Problem/eine Anwendung
dedication Einsatz * Hingabe, Arbeits~
deduce ableiten * folgern; schließen (from:aus); herleiten z.B. einer Formel || ableiten * ab-/herleiten (from: von); folgern, schließen (from: aus); vermuten || folgern || schließen * folgern
deducting abzüglich
deduction Ableitung * (Schluss-) Folgerung, →Herleitung, (logischer) Schluss || Abzug * Geld, Preisnachlass; charges deducted: nach ~ der Kosten || Folgerung || Schluss * →Folgerung, →Ableitung, →Herleitung
deem halten * halten, erachten für, betrachten als, denken, meinen
deenergized abgefallen * Relais, Schütz
deenergized interval Pause * z.B. bei Betrieb eines Motors
deep tief * (auch figürl.: Erkenntnisse, Wissen, Datenspeicher usw.) || Tiefe * Abgrund
deep bar effect Stromverdrängungseffekt
deep-bar cage rotor Wirbelstromläufer * {el. Masch.} →Stromverdrängungsläufer
deep-bar squirrel-cage motor Stromverdrängungsläufermotor
deep-bar squirrel-cage rotor Stromverdrängungsläufer * {el.Masch.} || Wirbelstromläufer
deep-draw tiefziehen
deep-drawn tiefgezogen
deep-groove ball bearing Rillenkugellager
deep-groove bearing with sideplate Z-Lager * mit Wellenabdichtungsfunktion
deep-groove roller bearing Rillenkugellager
deep-sea motor Tiefseemotor * {el. Masch.} zum Betrieb in großen Meerestiefen, z.B. zur Gas-/Ölförderung
deepen erhöhen * Eindruck ~
deepening Vertiefung * (auch figürlich: in e. Materie tiefer Eindringen)
default initialisieren * {Computer} auf den Vorbesetzungswert setzen (Variable, Parameter usw.) || Vorbelegung * Vorbesetzung von Parametern, Variablen usw. || vorbesetzen * Speicherzelle, Parameter usw. auf einen Initialisierungswert ~ || Vorbesetzung * Parameter, Speicherzelle, Wert usw. || Voreinstellung * vorbesetzter Wert für Parameter, Speicherzelle usw. || Voreinstellwert || Werkseinstellung
default setting Urladewert * eines Parameters
default value Defaultwert * Vorbesetzungswert ||

defect

Vorbesetzungswert * *Speicherzelle, Parameter usw.* || Voreinstellwert * *für Speicherzelle, Parameter usw.*
defect defekt * *(Substantiv) in: an* || Fehler * *Mangel, Defekt, Ausfall, hidden defect: versteckter ~* || Mangel * *Defekt, Fehler, Mangel, Unvollkommenheit, Schwäche, Gebrechen* || Missstand * *Mangel* || Problem * *Defekt*
defect of material Materialfehler
defective defekt || fehlerhaft * *defekt, kaputt, ausgefallen* || kaputt * *defekt* || mangelhaft * *defekt, mangelhaft, unvollkommen, unvollständig, schadhaft* || reparaturbedürftig * *defekt* || schadhaft * *defekt, kaputt* || unvollkommen * *mangelhaft, unvollkommen, unvollständig*
defence Schutz * *Schutz, Verteidigung, Abwehr*
defence of thesis Diplomarbeit
defend rechtfertigen * *verteidigen* || schützen * *verteidigen, schützen; against: gegen, from: vor*
defendable vertretbar
defense Schutz * *(amerikan.) Schutz, Verteidigung, Abwehr*
defer aufschieben * *zeitlich* || aussetzen * *aufschieben* || verlegen * *zeitlich verschieben* || verschieben * *zeitlich ~*
defiberizer Zerfaserer * *{Holz}* || Zerfaserungsmaschine * *{Holz} in der Spanplattenherstellung*
defibrating zerfasern * *{Holz} (Subst.) Holzzerfaserung*
defibrator Zerfaserer * *{Holz | Papier}*
defibrizer Mühle * *{Holz} in der Spanplattenherstellung*
deficiency Mangel * *of:an; Fehlen (von), Ausfall, ~haftigkeit, Schwäche, Unzulänglichkeit* || Mangel * *Defizit (of: an)*
deficiency claim Mängelrüge
deficient mangelhaft * *→unzureichend, mangelhaft, ungenügend*
deficit Summe * *fehlende ~, Fehlergebnis*
define angeben * *by: durch (Definition)* || bestimmen || definieren || erklären * *definieren* || festlegen * *definieren, genau bestimmen, kennzeichnen*
defined definiert
definite deutlich * *bestimmt, fest, klar, eindeutig, genau* || eindeutig || endgültig * *eindeutig, genau, definitiv, bestimmt, fest* || genau * *klar, deutlich, eindeutig, definitiv, bestimmt ,endgültig, fest* || klar * *eindeutig, genau, deutlich, fest, bestimmt* || konkret * *eindeutig, fest, klar, deutlich, eindeutig, genau* || sicher * *bestimmt*
definite-purpose zugeschnitten * *auf einen bestimmten Anwendungszweck ~*
definitely entschieden * *(Adv.)* || zweifellos * *gewiss, zweifellos, entschieden*
definition Definition || Erklärung * *Begriffsbestimmung*
definition of prices Preisbildung
definitions of terms Begriffsdefinitionen
definitive abschließend
deflate entleeren * *Luft usw. ablassen, aus-/entleeren*
deflation Entleerung * *Ablassen von Luft; {Wirtschaft} Deflation*
deflect abbiegen * *(sich) biegen, krümmen, neigen* || ablenken * *Licht, Feld, Strahlen usw.* || auslenken || umlenken * *ablenken, abbiegen (from: von), umlenken*
deflection Ableit- || Ablenkung * *Verbiegung, Abweichung, Ausschlag, Ablenkung (auch eines Magnetfeldes)* || Abweichung * *Verbiegung, Ablenkung, Ausschlag* || Auslenkung || Ausschlag * *z.B. eines Zeigers* || Biegung * *Biegung, Krümmung, Faltung* || Durchbiegung * *Abbiegung, Ablenkung, Durchbiegung (einer Antriebswelle usw.); Aus-

schlag* || Umlenkung || Verbiegung * *Ablenkung, Abbiegung*
deflection amplitude Schwingungsamplitude * *e. mechan. Schwingung (max. Durchbiegg. der in Schwingg. versetzten Teile)* || Schwingwegamplitude * *einer mechan. Schwingung (max.Durchbiegung des in Schwingung versetzten Teils)*
deflection axis Beugungsachse
deflection pickup Winkelpositionsgeber
deflection roll Umlenkrolle
deflection turn entry Einschleusung * *durch Umlenkung*
deflection turn exit Ausschleusung * *durch Umlenken*
deflector Abschieber * *{Verpack.techn.}*
deflector roll Umlenkrolle
deflexion Ablenkung || Durchbiegung * *Ablenkung, Abbiegung*
deform deformieren || verbiegen || verfälschen * *verzerren* || verformen * *deformieren, verzerren, verunstalten* || verformen * *deformieren, verzerren*
deformable verformbar * *(Metall)*
deformation Verformung
deformation factor Verzerrungsfaktor * *Verhältnis →Leistungsfaktor/→Grundschwingungsleist.-faktor (cos Phi/cos Phi_1)*
deformed verfälscht * *verzerrt*
deforming Verformung * *Deformierung*
defrost Abtauung
defuzzification Defuzzierung * *bei Fuzzy-Regelung*
degradation Änderung * *Verminderung, Schwächung, Verschlechterung*
degrade herabsetzen * *(im Rang) herabsetzen, vermindern, heruntersetzen, absinken, (sich) verschlechtern*
degrease entfetten
degree Grad * *Winkelgrad, Grad elektr.* || Maß * *Grad, Aus~; in a high degree: in hohem ~e* || Stufe * *Grad, Rang*
degree day Promotion * *~sfeier*
degree of -Grad * *vorangestellt*
degree of accuracy Genauigkeitsgrad
degree of automation Automatisierungsgrad
degree of beating Mahlgrad
degree of expansion Ausbaugrad
degree of extension Ausbaugrad
degree of fineness Feinheitsgrad
degree of freedom Freiheitsgrad
degree of integration Integrationsgrad
degree of irregularity Ungleichförmigkeitsgrad
degree of maturity Produktreife
degree of perfection Ausgereiftheit
degree of pollution Verschmutzungsgrad * *siehe z.B. DIN 0110, Teil 1*
degree of protection Schutzart * *nur i. Zusammenhang m. Schutzarttyp verwenden, z.B. 'degree of protection IP22'* || Schutzgrad * *Umfang d. Schutzes durch d. →Schutzart IP, VDE 0530 T.5: Schutzart durch Gehäuse* || Zündschutzart
degree of protection by enclosure Schutzart * *Gehäuse-Schutzart el. Maschinen, s. DIN IEC 34/ VDE 0530 je Tl. 5*
degree of protection flameproof enclosure Zündschutzart druckfeste Kapselung
degree of protection IP21 IP21 * *Schutzart geg. Fingerberührg., Fremdkörper >12 mm u. senkrecht. Tropfwass.*
degree of protection IP22 IP22 * *Schutzart geg. Fingerberührg., Fremdkörper > 12 mm u. Tropfwasser bis 15°* || IP23 * *Schutzart geg. Fingerberührg., Fremdkörper >12 mm u. Sprühwasser bis 60°*

degree of protection IP44   IP44 * *Schutzart geg. Berühr.m.Werkzeug, Fremdkörper >1mm und Spritzwasser*
degree of protection IP54   IP54 * *Schutzart mit vollständ Berührschutz, geg. Staubablagerg. u. Spritzwasser*
degree of protection IP55   IP55 * *Schutzart mit vollständ. Berührschutz, geg.Staubablagerg. u. Strahlwasser*
degree of protection IP56   IP56 * *Schutzart m.vollständ. Berührschutz, geg.Staubablagerg. u. schwere Seen*
degree of protection IP65   IP65 * *Schutzart mit vollständ. Berühr- u.Staubschutz u. gegen Strahlwasser*
degree of protection IP67   IP67 * *Schutzart mit vollständ. Berühr und Staubschutz u. für Unterwasserbetrieb*
degree of protection provided by enclosure - *Schutzart* * *Gehäuse-Schutzart*
degree of radio interference   Funkstörgrad
degree of shrinkage   Schrumpfmaß
degree of synchronism   Gleichlaufgüte
degree thesis work   Diplomarbeit
degrees   Grad * *(Plural) Winkelgrad, Grad elektr.*
degressive   abnehmend
dehumidifier   Entfeuchter || Trockner
dehydrate   entwässern * *Wasser entziehen* || trocknen * *~ durch Wasserentzug*
deionized water   Feinwasser * *z.B. im Kühlwasserkreislauf e. Stromrichters/Motors*
DEK   DEK * *Abk.f. Dansk Elektroteknisk Komite': Dänisches Elektrotechn. Komitee (ähnlich DKE)*
delay   aufschieben * *verzögern* || Totzeit * *Verzögerungszeit* || verschieben * *zeitlich verzögern* || verzögern || Verzögerung * *~szeit, zeitliche ~* || Verzögerungszeit || Verzug * *→Verzögerung* || Verzugszeit || Wartezeit * *Verzögerung* || Zeit * *~abstand, ~verzögerung* || Zeitstufe || zurückhalten * *verzögern*
delay angle   Steuerwinkel * *z.B. bei Phasenanschnittsteuerung*
delay angle setting   Ansteuerung * *Einstellg. d. Steuerwinkels für Phasenanschnittsteuerg.*
delay block   Totzeitglied * *Funktionsbaustein* || Verzögerungsbaustein * *Funktionsbaustein*
delay element   Totzeitglied * *Verzögerungsglied* || Verzögerungsglied * *first-order delay element: ~ erster Ordnung*
delay in delivery   Lieferverzug
delay time   Verzögerungszeit || Verzugszeit || Wartezeit * *Verzögerungszeit*
delayed   träge || verzögert
delegate   abordnen || Beauftragter * *Abgeordneter, Bevollmächtigter, Delegierter* || entsenden * *abordnen; auch als Vertreter ~* || übertragen * *Aufgabe, Vollmacht; abordnen, delegieren, bevollmächtigen* || zuteilen * *powers: Befugnisse ~* || zuweisen * *powers: Befugnisse*
delete   ausfügen * *auf PG-Tastatur für SPS-Programmierung* || streichen * *löschen*
delete button   Löschtaste
delete key   Löschtaste
deleted   gelöscht
deletion   Löschung
deliberate   bewusst * *überlegt, wohlwogen, bewusst, absichtlich, vorsätzlich; besonnen, vorsichtig* || wohlüberlegt
deliberation   Beratung * *Beratschlagung (on:über)* || Besprechung * *Beratung*
delicate   anfällig * *empfindlich, zart, zerbrechlich, fein, filigran, heikel, schwierig* || dünn * *zart* || empfindlich * *zart, fein, zerbrechlich, leicht, dünn, fein, genau, heikel, schwierig* || feinfühlig * *zartfühlend* || filigran || heikel * *Angelegenheit usw.*

delimitation   Abgrenzung
delimiter   Endezeichen
delimiting character   Endezeichen
delineate   skizzieren * *sizzieren, zeichnerisch (in Strichzeichnung) darstellen, entwerfen*
deliver   abgeben * *auch Drehmoment, Leistung, Strom usw. ~* || anliefern || ausrichten * *Botschaft usw. ~* || fördern * *z.B. Gase, Flüssigkeit durch Pumpe ~* || halten * *Rede ~* || liefern || versenden * *liefern* || zuführen * *auch Material ~*
deliver back   zurückliefern
deliver up   aushändigen || übergeben
deliverable   lieferbar
delivered   abgegeben || geliefert
delivered condition   Auslieferungszustand
delivery   Abgabe || Anlieferung || Förderleistung * *einer Pumpe* || Förderung * *(Be-)Förderung (durch Pumpe usw.), (Zu-)Leitung, Zuführung, Aus-/Übergabe, Lieferung* || Lieferung * *An~; payable/cash on delivery: zahlbar bei ~* || Übergabe * *Aus-/Ablieferung* || Versand * *Auslieferung* || Zufuhr || Zuführung * *Anlieferung, Anfördern*
delivery bottleneck   Lieferengpass || Lieferproblem * *Lieferengpass* || Lieferschwierigkeiten
delivery date   Ausliefertermin || Auslieferungsdatum || Auslieferungstermin || Liefertermin
delivery deadline   Lieferfrist * *maximale ~ aus Kundensicht*
delivery head   Förderhöhe * *einer Pumpe*
delivery location   Lieferort
delivery logistic   Lieferlogistik
delivery note   Lieferschein
delivery paperwork   Lieferpapiere
delivery pipe   Druckleitung * *Pumpenausgangsleitung, Zuführleitung*
delivery problem   Lieferengpass || Lieferproblem
delivery programme   Lieferprogramm
delivery quality   Lieferqualität
delivery rate   Durchflussgeschwindigkeit * *Förderrate einer Pumpe* || Förderkapazität * *(pumpe, Lüfter usw.)* || Förderleistung * *einer Pumpe/eines Ventilators; siehe auch →Förderkapazität*
delivery rate per hour   Durchflussgeschwindigkeit * *Förderrate einer Pumpe, z.B. in Kubikmeter pro h*
delivery receipt   Lieferschein * *[... ri'ßiet] Lieferpapiere*
delivery schedule   Lieferfrist * *planmäßige ~, ~ im Standardfall*
delivery schedules   Liefertermine
delivery shortage   Lieferengpass || Lieferproblem
delivery slip   Lieferschein
delivery speed   Liefergeschwindigkeit
delivery time   Lieferfrist || Lieferzeit
delivery works   Lieferwerk
delta   Dreieck * *Schaltgruppe z.B. für Trafo/Stromrichter/Motor; connected in delta: in ~ geschaltet* || Dreieck- * *Schaltgruppe, Schaltung*
delta connection   Dreieckschaltung * *z.B. Trafo, Motor; delta-connected: in ~*
delta function   Dreieckfunktion
delta modulation   Dreieckmodulation
delta voltage   Dreieckspannung
delta-connected   in Dreieck geschaltet
delta-I control   Delta-I-Regelung
deluge   überschwemmen * *(figürl.)*
demagnetization   Magnetisierung || Demagnetisierung || Entmagnetisierung
demagnetize   abmagnetisieren || entmagnetisieren
demand   anfordern || Anforderung * *place higher/ strigent demands on:höhere/strenge Ansprüche stellen an* || Anspruch * *Forderung (for: nach); make high demands:hohe Ansprüche stellen* || Beanspruchung * *Anspruch (on: an), Anforderung* || Bedarf * *auch an Stromversorgung; meet the de-*

demanded 484

mand for: dem ~/der Nachfrage nach ... nachkommen || erfordern * verlangen, fordern, bedürfen || fordern || Forderung * for:nach; on: an || Nachfrage * low: geringe; high/great/heavy for hohe ~ nach (z.B. e. Produkt); keep up with: nachkommen || verlangen * fordern, begehren (of/from:von, that: dass, to do: zu tun), bedürf., erford. (z.B. Sorgfalt) || voraussetzen * erfordern
demanded angefordert * z.B. Momentenrichtung || gefordert
demanded torque direction angeforderte Momentenrichtung
demanding anspruchsvoll * technische Anwendung/Anforderung; geistig ~; highly demanding: höchst-/sehr ~
demands Anforderungen * make high demands on: hohe Ansprüche stellen an; increased: erhöhte || Ansprüche * rising: steigende; place demands on: ~ stellen an
demarcation Abgrenzung
demijohn Ballon * Korbflasche
deminishing degressiv
demolish zerstören * auch 'niederreißen'
demolition Zerstörung
demonstrate beweisen * demonstrieren || erklären * veranschaulichen, zeigen || illustrieren * (figürl.) || vorführen || vormachen * demonstrieren || zeigen * demonstrieren
demonstration Vorführ- || Vorführung
demonstration case Vorführkoffer
demonstration vehicle Vorführwagen
demount zerlegen
demountable abnehmbar
demur to beanstanden * Forderung usw. ~
Denmark Dänemark
denominate benennen * (be)nennen, bezeichnen
denomination Benennung
denominator Divisor * Nenner, das unter dem Bruchstrich stehende || Nenner * (mathematisch u. figürl.) reduce to a common denominator: auf einen ~ bringen || Teiler * (mathemat.) Nenner
denote kennzeichnen * bezeichnen, kennzeichnen, anzeigen, andeuten, bedeuten
dense dicht
density Dichte * high-/low-density mit geringer/hoher ~ || Dichtheit * →Dichte, Dichtheit, Gedrängtheit, Enge || Packungsdichte * high density: (mit) hohe(r) Packungsdichte || spezifisches Gewicht
depacker Auspackmaschine
depalletizer Entpalettiermaschine
depart abgehen * Zug usw. || abreisen
department Abteilung || Dienststelle || Fach * Fach, Gebiet, Geschäftsbereich, Abteilung
department name Abteilungsbezeichnung
departmental chief Abteilungsleiter
departmental classification Aufgliederung * eines Unternehmens
departmental order number BZ-Nummer
departure Abreise * for: nach; on my departure: bei meiner ~ || Abweichung * from: von einer Regel || Abzug
depend abhängen * (up)on:von
depend on beruhen auf * abhängen von || verlassen * sich ~ auf
dependability Zuverlässigkeit * auch Verlässlichkeit, Vertrauen, Verlass
dependable zuverlässig * verlässlich
dependance Funktion * in dependance of: als ~ von (z.B. bei Kennlinie: auf X-Achse aufgetragene Variable
dependant abhängig * siehe 'dependent'
dependence Abhängigkeit
dependent abhängig * on/upon:von || bedingt * abhängig
dependent on abhängig von || in Abhängigkeit von

dependent on whether je nachdem ob * abhängig davon ob
dependent upon abhängig von
depending abhängig * (up)on: von/davon; on whether...or not: davon ob...oder nicht
depending as to whether abhängig davon ob
depending on abhängig von || je nach || je nachdem * abhängig davon; depending on how you do it: je nachdem, wie Du es machst
depending on the actual time required nach Aufwand * beim Verrechnen von Dienstleistungen
depending on the direction of rotation drehrichtungsabhängig || richtungsabhängig * drehrichtungsabhängig
depending on whether abhängig davon ob * or not: oder nicht
depict abbilden || beschreiben * darstellen, schildern, (anschaulich) ~, darstellen, (ab)malen || darstellen * (bildlich) darstellen, veranschaulichen, zeichnen, (ab)malen, schildern
deplete ausleeren * (auch figürlich) entleeren, entblößen, erschöpfen || entleeren || erschöpfen * entleeren || leeren * ent~, ausräumen, z.B. Ladungsträger (of:von)
depletion Entleerung * (Ent)Leerung, Erschöpfung, Entblößung
deployment Einsatz * von Arbeitskräften
deposit abgeben * Gepäck usw. deponieren || Ablagerung * das Abgelagerte; dust: Staub; moisture: Feuchtigkeit; dirt: Schmutz || ablegen || absetzen * ablegen, sich ~ || abstellen * ablegen || Anzahlung || hinterlegen * deponieren, hinterlegen, übergeben || Niederschlag * Ablagerung
deposit of condensate Schwitzwasserbildung * Ablagerung v. Schwitzwasser, Betauung
deposit of moisture Schwitzwasserbildung * Feuchtigkeitsablagerung
deposition Ablage * das Ablegen; Aufbewahrungsort, Magazin || Ablagerung || Aussage * schriftliche (rechtlich verwertbare) Erklärung || Beschichtung * Schicht, Belag, Ablagerung, Niederschlag, (Boden)satz, Aufdampfung
deposits of dust Staubablagerung
depot Ablage * Lagerstelle, Ablegestelle || Lagerplatz * Vorratsplatz || Magazin
depreciate verkleinern * herabsetzen; z.B. Preis, Wert
depreciation Abschreibung || Herabsetzung * Verächtlichmachung || Minderung * Herabsetzg., Geringschätzg., Wert~, Abschreibg., Ent/Abwertung || Wertminderung
depress betätigen * Taste || drücken * niederdrücken, auch Taste, Druckknopf
depression Abfall || Abnahme * Fallen (z.B. von Preisen), ~, Schwäche, Schwächung, Flaute, Depression, Wirtschaftskrise || Anschlag * einer Taste || Mulde * Vertiefung || Vertiefung
depriciate herabsetzen * i. Preis/Wert ~; Abschreibungen machen von; verächtlich machen, herabwürdigen
dept Abt. * Abk. für 'department'; Abteilung
Dept. Abteilung || Dienststelle
depth Tiefe * ['däbth] (auch figürlich) in-depth: in die ~ gehend
depth of cut Schnittiefe || Spantiefe
depth of immersion Eintauchtiefe * in Flüssigkeit; z.B. Getrieberad in Ölbad
depth of thread Gewindetiefe
depth of throat Ausladung * ~ z.B. einer Blechschere, Überhang
deputy Vertreter * amtlicher Stell~
derate herabsetzen * absenken, heruntersetzen; z.B. e. Spezifikation (by: um) bei bestimmt. Betriebsbedingn. || Leistung reduzieren || reduzieren * siehe 'de-rate'

derated loading Gesamtbelastbarkeit * *verminderte*
derating Herabsetzung * *Absenkung; z.b.* ~ *einer Spezifikation bei bestimmten Umgebungsbedingungen* || Leistungsherabsetzung || Leistungsminderung * *z.B. e. Geräts bei Überschreiten e. bestimmten Temperatur oder Aufstellungshöhe* || Leistungsreduzierung * *Herabsetzung z.B. auf Grund der Aufstellungshöhe, der höheren Temperatur usw.* || Minderung * *Herabsetzen, z.B. der Leistg. eines Stromrichters bei größerer Aufstellhöhe* || Reduktion * ~ *der Leistungsdaten bei erschwerten Umgebungsbedingungen* || Reduzierung * *Herabsetzg., z.B. v.Kenndaten bei besond. Betriebsbedinggen.(z.B. große Aufstellhöhe)*
derating factor Reduktionsfaktor * *z.B. für verminderte Leistung e. Stromrichters bei größerer Aufstellhöhe*
derivate Derivat
derivation Ableitung * *mathematisch* || Herleitung
derivative Ableitung * *[Mathem.] das Abgeleitete/ das Differenzierte (z.B Beschleunigung: dv/dt)* || Anstiegsgeschwindigkeit * *(math.) erste Ableitung* || Differenzialquotient * *of the third order: dritter Ordnung*
derivative action D-Verhalten
derivative characteristic D-Verhalten
derivative component D-Anteil || Vorhalt * *[Regel.techn] D-Anteil, Differenzierer*
derivative element Differenzialglied || Differenzierer * *Differenzierglied liefert am Ausg. die 1.Ableit. (nach d. Zeit) der Eingangsgröße* || Differenzierglied
derivative gain Vorhaltzeit *+ eines Differenziergliedes*
derivative regulator Differenzialregler
derivative term D-Anteil || Unterbegriff * *abgeleiteter Begriff*
derivative time Vorhaltzeit * *für Differenzierer*
derivative time constant Vorhaltzeit * *eines Differenziergliedes*
derivative-action element Differenzialglied || Differenzierglied
derivative-action regulator Differenzialregler
derivative-action time Differenzierzeit || Vorhaltezeit * *eines Differenzialglieds* || Vorhaltzeit * *für Differenzierer*
derive ableiten * *herleiten (auch Gleichung, Formel), zurückführen (from:von), mathematisch differenzieren* || differenzieren * *erste Ableitung bilden (math.)* || entnehmen * *ab/herleiten* || gewinnen * *durch chem. Prozess ~* || herleiten || zurückführen * *ab/herleiten; from: von*
derust entrosten
desalination Entsalzung
desalination plant Entsalzungsanlage || Meerwasser-Entsalzungsanlage
desaturation Entsättigung
descale entzundern * *von Stahl*
descaling entzundern * *(Substantiv)*
descaling machine Entzundungsmaschine * *[Walzwerk]*
descend abwärts fahren * *[Aufzüge]* || bewegen * *nach unten* || einfahren * *[Bergwerk]*
descend steeply abstürzen * *abschüssig sein, steil abfallen*
descending abfallend * *in descending order: in ~er Reihenfolge*
descending branch absteigender Ast
describe beschreiben * *schildern, (Kreis, Bahn usw.) ~; indescribable/beyond all description: nicht zu ~* || darstellen * *beschreiben*
description Angabe * *Bezeichnung, Beschreibung* || Beschreibung * *Erläuterung, Beschreibung*
description of the parameter contents Parameterbeschreibung

descriptive beschreibend
descriptive information Beschreibung
desert verlassen * *treulos ~, desertieren*
deserve verdienen * *Lob, Tadel usw.*
desiccant Trockenmittel
desiccate trocknen * *(aus-) trocknen/dörren*
desiccation agent Trockenmittel * *desiccation bag: ~beutel*
desiccator Trockner
design -Ausführung || anlegen * *planen* || anordnen || Anordnung || Aufbau * *Konstruktion* || Aufbau * *Konstruktion, konstruktiver/schaltungstechnischer ~* || ausführen * *konstruktiv ~/auslegen* || Ausführung * *Konstruktion, Auslegung, Bauart* || auslegen * *entwerfen, auch bemessen* || Auslegung || Bauart || bauen * *entwerfen* || Bauform || Bauweise || bemessen * *(for: nach)* || Bemessung || Design * *Konstruktion, Auslegung, Entwicklung* || dimensionieren * *eine Konstruktion, for: nach einem Zielkriterium* || Einrichtung * *Bauart* || entwerfen
design armature current Bemessungs-Ankerstrom * *eines Stromrichters*
design armature voltage Bemessungs-Ankerspannung * *eines Stromrichters*
design calculation Berechnung * *zur Auslegung/ Dimensionierung e. Konstruktion/Schaltung usw.*
design criteria Auslegungskriterien
design drawing Konstruktionszeichnung
design engineer Entwickler * *Konstrukteur, Entwerfer* || Entwicklungsingenieur * *siehe auch 'Entwickler'* || Konstrukteur
design feature Konstruktionsmerkmal
design field current Bemessungsfeldstrom * *eines Stromrichters*
design language Projektierungssprache * *z.B. 'STRUC'(R) beim Regelsystem SIMADYN D (R)von SIEMENS*
design of the frame Gehäuseausführung * *Motor*
design patent Gebrauchsmuster
design rating Bauleistung * *z.B. eines Transformators*
design review Entwurfsprüfung * *Qualitätsprüfung an einem (Konstruktions-) Entwurf (DIN 55350 T.17)*
design rules Konstruktionsrichtlinien
design selection manual Projektierungshandbuch * *zur richtigen Auswahl von techn. Produkten (z.B. Konstruktionselementen)*
design specification Lastenheft
design speed Konstruktionsgeschwindigkeit * *Geschw. für die e.Maschine konstruiert ist (z.B. Papiermaschine)*
design suitable for use in explosive atmospheres explosionsgeschützte Ausführung
design target Entwicklungsziel
design tool Entwicklungshilfsmittel || Entwicklungswerkzeug
design tools Entwicklungshilfsmittel * *(Plural)*
design voltage Bemessungsspannung
designate benennen * *bezeichnen, (be)nennen; kennzeichnen* || bezeichnen * *mit Namen, Markierung (auch Klemme, Kabel usw.)* || kennzeichnen * *benennen, (be)nennen, kennzeichnen; berufen, ernennen, ausersehen* || nennen * *bezeichnen*
designated autorisiert * *ernannt, bestimmt*
designation Belegung * *Bezeichnung, Kennzeichnung, Bestimmung, z.B. von Klemmen* || Benennung * *Bezeichnung* || Bezeichnung * *Name, Kennzeichnung* || Kennzeichen || Kennzeichnung * *Bezeichnung* || Marke * *Kennzeichnung* || Markierung * *Kennzeichnung* || Name * *Bezeichnung*
designation system Kennzeichnungssystem
designator Bezeichner

designer Entwickler * *Entwerfer, Konstrukteur* || Konstrukteur || Projekteur || Technischer Zeichner
designing engineer Konstrukteur
desintegrate aufschließen * *(Chemie)*
desirable wünschenswert
desirable condition Soll-Zustand * *gewünschter Zustand* || Sollzustand
desire suchen * *wünschen, haben wollen* || verlangen * *wünschen* || Wunsch * *as desired:(je)nach ~* || wünschen * *leave much to be desired:viel zu ~ übrig lassen*
desired gewünscht * *erwünscht*
desired function gewünschte Funktion
desired speed Drehzahlsollwert * *gewünschte Drehzahl*
desired state Sollzustand
desired value Sollwert * *gewünschter Wert*
desist from Abstand nehmen von
desk Pult || Tisch
desk installation Pulteinbau
desk mounting Pulteinbau
desktop model Tischgerät
despite trotz || ungeachtet * *trotz*
despite difficulties trotz Schwierigkeiten
dessert Nachspeise * *[di'söt]* || Nachtisch * *[di'söt]*
destacking Abstapelung || vereinzeln * *(Substantiv) z.B. bei einer Verpackungsmaschine*
destination Ziel * *~ des Signalflusses, der Reise usw.*
destination reached Position erreicht * *(NC-Steuerung)*
destination reference Zielhinweis * *für Signale/Daten i.Stromlaufplan/Blockschaltbild/Programmlisting*
destined bestimmt * *vorgesehen; be destined for: ~ sein für*
destroy vernichten * *zerstören* || zerstören
destruction Zerstörung
destruction protected zerstörsicher
destruction test Zerstörungsprüfung
destructive material testing zerstörende Werkstoffprüfung
destructive test zerstörende Prüfung * *von Werkstoffen usw.*
destructive testing Zerstörungsprüfung
desulpherization Entschwefelung
detach ablösen * *lösen, abtrennen/montieren* || abnehmen * *lösen* || abziehen * *entfernen, z.B. ein Steckkabel* || ausbauen * *abbauen (z.B. Maschinenteil)* || demontieren * *abbauen/-montieren* || lösen * *abnehmen, abtrennen, abbauen, herausnehmen* || trennen * *ab-/loslösen*
detachable abnehmbar || lösbar * *abnehmbar* || zerlegbar * *abnehmbar*
detached einzeln * *abgetrennt*
detachement Ablösung * *(Ab)Lösung, Abtrennung, Abbau*
detachment Lösung * *Abtrennung, Abbau*
detail Detail || einteilen * *Arbeit ~* || Einzelheit * *(on: über), with full details: mit allen ~en; go into details: sich mit ~en befassen* || einzeln * *(Substantiv) Einzelheit; in detail: im Einzelnen* || Spezifikum || Umstand * *Einzelheit*
detail display Detaildarstellung * *am Bildschirm*
detail function Lupenfunktion * *z.B. in Blockschaltbild*
detail representation Detaildarstellung
detailed ausführlich || detailliert || eingehend * *detailliert, in Einzelne gehend* || genau * *ins Einzelne gehend* || im Detail || umständlich * *in allen Einzelheiten*
detailed drawing Detaildarstellung * *Detailzeichnung*
detailed information Detailinformationen
detailed representation Feindarstellung

details Angaben || Einzelheiten * *go into details: sich mit ~ befassen* || Information * *nähere Einzelheiten*
detect erfassen || erkennen * *entdecken, detektieren; z.B. Fehler* || feststellen * *herausfinden, ermitteln, entdecken, wahrnehmen, aufdecken, enthüllen*
detectable nachweisbar
detection Erkennung
detector -Melder || Geber || Melder
detent Rastung * *Sperrhaken, Sperrklinke; Sperre, Auslösung*
detention point Raste * *Rück/Festhaltepunkt*
detergent Reinigungsmittel
deteriorate nachlassen * *schlechter werden* || verschlechtern * *auch 'sich ~, schlechter werden'*
deterioration of quality Qualitätseinbuße * *Qualitätsbeeinträchtigung/-verschlechterung/-minderung*
determinable vorherbestimmbar * *bestimmbar*
determinant beherrschend * *bestimmend, entscheidend; entscheidender Faktor* || Determinante * *(Mathem.)* || entscheidend * *bestimmend, entscheidend; ~der Faktor* || Faktor * *bestimmender ~*
determinate eindeutig * *~ bestimmt (Fehlerzustand usw.), bestimmt, fest(gesetzt), entschieden*
determination Berechnung * *Bestimmung* || Entscheidung || Ermittlung * *Bestimmung* || Festlegung * *Festsetzung, Entschluss, Beschluss, Bestimmung* || Feststellung * *Ermittlung, Feststellung, Bestimmung, Festsetzung*
determinative ausschlaggebend * *entscheidend*
determine begrenzen * *festlegen* || beschließen * *bestimmen, entscheiden, sich entschließen, festsetzen* || bestimmen * *festlegen, ermitteln* || entscheiden || ermitteln * *bestimmen,festsetzen* || festlegen * *bestimmen* || feststellen * *bestimmen, festlegen, ermitteln, herausfinden* || maßgebend sein für * *bestimmen*
determined bestimmt * *festgelegt* || festgelegt * *bestimmt*
determined by calculation rechnerisch * *(Adv.)*
determined by law gesetzlich vorgeschrieben
determining ausschlaggebend * *determining factor: entscheidender Umstand/Faktor* || bestimmend || ermitteln * *(Substantiv)* || Ermittlung || maßgebend * *bestimmend* || maßgeblich * *bestimmt*
deterministic deterministisch * *vorausbestimmbar/-berechenbar* || vorherbestimmbar
detonate bersten * *explodieren, detonieren*
detract wegnehmen * *entziehen, ~, herabsetzen, beeinträchtigen (from: etwas/e.Eigenschaft), schmälern*
detriment Nachteil * *Schaden, Nachteil* || Schaden * *Schaden, Nachteil* || Schädigung * *to: von etwas*
detrimental nachteilig || schädlich * *nachteilig*
Deutschmark DM * *Einzahl*
Deutschmarks DM * *Mehrzahl*
devastate zerstören * *verwüsten*
develop ausbauen * *entwickeln* || ausbilden * *entfalten, bilden, wachsen* || entstehen * *(sich) entwickeln, z. Vorschein kommen, auftreten, s. zeigen* || entwickeln * *auch 'sich ~'* || erschließen * *entwickeln, nutzbar machen, (sich) entfalten; auch "Baugelände usw. ~"* || herausbilden * *(sich) entwickeln, entstehen, sich zeigen, bekanntwerden, erweitern, ausdehnen* || wachsen * *sich entwickeln*
develop one's profile Profil entwickeln
developed as far as it can go ausgereizt * *eine techn. Konstruktion*
developer Entwickler
developing country Entwicklungsland
developing trend Entwicklungstrend
development Ausbau * *Erschließung, Ausbau* || Ausbildung * *Entfaltung, Entstehen, Bildung, Wachstum* || Bildung * *Entstehung, Entfaltung* || entstehen * *(Substantiv) Bildung, Entstehung, Ent-

faltung || Entwicklung * *eines Produktes; be in/under development: in ~ sein* || Entwicklungsabteilung || Erschließung || Herleitung * →*Ableitung einer Formel usw.* || Verlauf * *Entwicklung, Entfaltung, Entstehen, Bildung, Wachstum, Erschließung, Ausbau, Umgestaltung*
development aim   Entwicklungsziel
development costs   Entwicklungsaufwand * *Entwicklungskosten* || Entwicklungskosten
development department   Entwicklungsabteilung
development engineer   Entwickler
development environment   Entwicklungsumgebung * *die Gesamtheit aller Entwicklungswerkzeuge für eine Softwareentwicklung*
development group   Entwicklungsgruppe
development jump   Entwicklungssprung
development of heat   Wärmeentwicklung
development platform   Entwicklungsplattform
development programme   Entwicklungsvorhaben
development project   Entwicklungsvorhaben
development step   Ausbaustufe * *Entwicklungsschritt*
development system   Entwicklungssystem * *z.B. für Mikroprozessor-Software*
development target   Entwicklungsziel
development tool   Entwicklungshilfsmittel || Entwicklungswerkzeug
development trend   Entwicklungstendenz || Entwicklungstrend
deviate   abweichen * *from:von; by:um* || schwanken * *abweichen, ablenken* || streuen * *abweichen; significantly: stark*
deviating   abweichend
deviating roller   Umlenkrolle
deviation   Abweichung * *transient deviation: vorübergehende ~, frequency deviation: Frequenz~* || Schwankung * *Abweichung* || Streuung * *Abweichung* || Toleranz * *Abweichung*
deviation from rated system voltage   Netzspannungsabweichung * *vom Nennwert*
deviation in dimension   Maßabweichung
device   -Mittel * *Vorrichtung* || Apparat * *Gerät, Vorrichtung* || Apparatur || Bauelement || Bauteil || Betriebsmittel || Einrichtung * *Gerät, Vorrichtung* || Gerät * *electrical: elektrisches* || Hilfsmittel * *Vorrichtung, Gerät, Behelf* || Vorrichtung * *Gerät*
device condition code   Geräteanzeige * *z.B. bei SPS*
device driver   Gerätetreiber * *Softwaretreiber für Rechner-Peripheriegeräte* || Treiber * *{Computer} Peripheriegeräte~*
device link   Gerätekoppling || Gerätekopplung
device network   Feldbus * *z.Anschlus v. Feldgeräten (Sensoren, Aktuatoren, Remote-I/O, dezentrale Intelligenz)*
device parts list   Apparate-Stückliste
device protection   Geräteschutz
device type   Gerätetyp
devise   ausdenken || entwickeln * *ausdenken, ersinnen, erfinden, konstruieren, auch Software* || konstruieren * *erfinden, ersinnen, ausdenken*
dew   Betauung * *Tau* || Tau * *~ durch Kondensierung; siehe auch* →*Betauung*
dew point   Taupunkt * *Temperatur, unterhalb derer die Betauung beginnt*
dew point curve   Taupunktskurve * *Abhängigkeit der Taupunkt-Temp. von der absoluten Luftfeuchte*
dewater   entwässern
dewatering   Entwässerung
DF   Klirrfaktor * *Abkürzung für 'Distortion Factor'*
diagnose   diagnostizieren || erkennen * *diagnostizieren, z.B. Fehler*
diagnose a fault   Fehler erkennen || Fehler finden
diagnosis   Diagnose * *[daieg'nousis]*

diagnosis connector   Diagnosestecker
diagnosis function block   Diagnosebaustein * *Funktionsbaustein*
diagnosis unit   Diagnosegerät
diagnostic aid   Diagnosehilfe
diagnostic aid system   Diagnoseeinrichtung
diagnostic aids   Diagnosemittel
diagnostic block   Diagnosebaustein * *Funktionsbaustein*
diagnostic display   Diagnoseanzeige || Prüfanzeige * *bei SPS*
diagnostic equipment   Diagnoseeinrichtung
diagnostic function   Diagnosefunktion
diagnostic help   Diagnosehilfe
diagnostic indicator   Diagnoseanzeige
diagnostic LED   Diagnaose-LED
diagnostic light   Prüfanzeige * *bei SPS*
diagnostic memory   Diagnosespeicher || Störspeicher * *Diagnosespeicher*
diagnostic message   Diagnosemeldung
diagnostic plug   Diagnosestecker
diagnostic signal   Diagnosemeldung
diagnostic support   Diagnosehilfe * *Unterstützung*
diagnostic system   Diagnosesystem
diagnostic tool   Diagnosehilfe * *Diagnosehilfsmittel/Werkzeug* || Diagnosemittel
diagnostic unit   Diagnosegerät
diagnostics   Diagnose
diagnostics purposes   Diagnosezwecke
diagnostics technology   Diagnosetechnik
diagonal   diagonal || Diagonale || quer * *diagonal, schräg* || schräg * *~ verlaufend*
diagonal cutter   Saitenschneider
diagonal line   Schrägstrich
diagonal stroke   Schrägstrich
diagonally opposite   über Kreuz
diagram   Abbildung || Bild * *Diagramm, grafische Darstellung* || Diagramm || grafische Darstellung || Kennlinie * *Kurvenbild* || Plan * *grafische Darstellung* || Planung * *Zeichnung* || Schaubild || Schema || Zeichnung
diagrammatic   schematisch * *grafisch, diagrammatisch, schematisch, nach Schema*
dial   Scheibe * *Wähl~* || Skala || Skale || wählen * *am Telefon*
dialog   Dialog * *(amerikan.)* || Gespräch * *Zwiesprache*
dialog box   Eingabefeld * *Dialog-Feld auf einem Bildschirm*
dialogue   Dialog
diameter   Durchmesser
diameter buildup   Durchmesseraufbau * *Durchmessererhöhung bei Aufwickler während des Wickelvorgangs*
diameter calculator   Durchmesserrechner * *bei Auf/Abwickler z.B. nach der Formel [Durchm. entspr. const x V/n)]*
diameter of root circle   Fußkreisdurchmesser * *eines Zahnrades*
diameter of the bolt pitch circle   Lochkreisdurchmesser
diameter preset value   Durchmesser-Setzwert || Durchmessersetzwert
diameter range   Durchmesserbereich * *bei Auf-/Abwickler*
diameter ratio   Durchmesserbereich * →*Durchmesserverhältniss (bei Auf-/Abwickler)* || *Durchmesserverhältniss * *beim Wickler* || Wickelverhältnis * *Durchmesserverhältnis (max. Durchmesser/Kerndurchmesser)*
diameter sensing   Durchmessererfassung
diameter sensor   Durchmesser-Sensor * *z.B. mit Ultraschall für Auf/Abwickler*
diameter-dependent   Durchmesser-abhängig
diamond   Diamant

diamond tool Diamantwerkzeug
diamond-tipped diamantbestückt * *{Werkz.masch.\ Holz}* Werkzeug usw.
diaphanous durchlässig * ~ *für Licht* || transparent
diaphragm Membran
diaphragm coupling Membrankupplung
diaphragm pump Membranpumpe
dicharged cooling air austretende Kühlluft
dicount schedule Rabattlinie * *Rabattstaffel*
dictate vorschreiben * *diktieren, vorschreiben, gebieten, auferlegen, befehlen*
dictionary Wörterbuch * *of: über/für*
die Form * *Spritzguss~, Strangpress~* || Gesenk || Siliziumplättchen * *zur Erzeugung integrierter Schaltungen* || Stanzstempel || Ziehstein
die away abklingen * *schwächer werden, nachlassen, sich verlieren*
die cast Guss- * →*Druckguss-* || Spritzguss * *(Nichteisen-) Metall-Druckguss; die-cast aluminium alloy: Aluminium-~*
die mold Spritzgussform * *für Metallspritzguss*
die mould Spritzgussform * *für Metallspritzguss*
die out abklingen
die-cast Druckguss
die-cast aluminium Aluminium-Druckguss
die-cast aluminium alloy Aluminium-Spritzguss
die-cast metal Druckguss * *Druckgussmetall/-material*
die-cast part Spritzgussteil * *aus Metall*
die-cast rotor Druckgussläufer * *{el. Masch.}*
die-cast zinc Zinkdruckguss
die-casting die Druckgießwerkzeug || Spritzgussform * *für Metall-Spritzguss*
die-casting machine Druckgießmaschine
die-sinking machine Gesenkfräsmaschine
die-slotting machine Räummaschine
die-stock Kluppe * *{Werkz.masch.}* an Drehbank
dielectric Dielektrikum * *allg.Ausdruck f. e. festen, flüssigen, gasförmigen Isolierstoff o. Vakuum als Isolierst.* || dielektrisch || Nichtleiter
dielectric constant Dielektrizitätskonstante
dielectric flux Verschiebungsfluss
dielectric strength Durchschlagfestigkeit || Isolationsfestigkeit || Schaltfestigkeit * *von Kondensatoren (siehe auch →schaltfest)* || Spannungsfestigkeit * *z.B. eines Kondensator, eines Isoliersystems (el. Masch.: VDE0530 T.15)* || Überspannungsfestigkeit * *z.B. bei Kondensator*
dielectric stress dielektrische Beanspruchung * *Beanspruchung eines Isolierstoffs (Dielektrikum) durch ein elektr. Feld*
dielectrical strength Spannungsfestigkeit * *Isolations-/Durchschlagsfestigkeit e. Isoliersystems, z.B. VDE 0530 T.15*
Diesel generator Dieselaggregat || Dieselgenerator
Diesel power station Dieselkraftwerk
diesel-electric dieselelektrisch
diesel-electric drive dieselelektrischer Antrieb
diesel-engine operation Dieselbetrieb
Diesel-generator set Dieselaggregat || Dieselgenerator
diesel-generator supply Dieselsstromversorgung
diesel-hydraulic dieselhydraulisch
dieseling Nachlauf * *"Nachdieseln" eines Verbrennungsmotors*
diestock Schneideisen
differ abweichen * *from (one another): von (einander)* || unterscheiden * *sich unterscheiden; from: von*
difference Abstand * *Unterschied* || Abweichung || Differenz * *between:zwischen* || Missverständnis * *Meinungsverschiedenheit, Differenz* || Unterscheidung * *Unterschied* || Unterschied * *between: zwischen*
difference in shape Ungleichförmigkeit * *Ungleichheit in der Form/im Aussehen*

difference in temperature Temperaturdifferenz
difference of potential Potenzialdifferenz
differenciate staffeln * *differenzieren, Unterschiede machen; z.B. Preise, Steuern, Löhne*
different ander || anders * *(Adj.) verschieden* || neu * *bisher unbekannt* || ungleich || unterschiedlich * *significantly different: stark ~* || verschieden * *unterschiedlich; from: von*
differential Differenzial * *differential gear: ~getriebe; differential equation: ~gleichung*
differential amplifier Differenzverstärker
differential coefficient Differenzialquotient
differential compound winding Anticompoundwicklung
differential driver Differenzialtreiber
differential equation Differenzialgleichung * *solve: lösen*
differential equation system Differenzialgleichungssystem
differential gear Differenzialgetriebe
differential input Differenzeingang * *z.B. eines Verstärkers (nicht massebezogen)*
differential line driver Differenzial-Leitungstreiber || Differenzialleitungstreiber || Differenzialtreiber * *Leitungstreiber*
differential line receiver Differenzempfänger * *Leitungsempfänger* || Differenzial-Leitungsempfänger || Differenzialempfänger * *Leitungsempfänger* || Differenzialleitungsempfänger
differential pressure Differenzdruck
differential pressure transmitter Differenzdruckgeber
differential receiver Differenzempfänger || Differenzialempfänger
differential selsyn Ausgleichswelle * *Ausführungsform d. elektr.Welle mit Asynchron-Schleifringläufermotoren*
differential signal Differenzialsignal * *z.B. für Analogeingangsklemme, RS485 (Gegensatz: single ended signal)* || Differenzsignal * *z.B. für Analogeingangsklemme, Verstärker, RS482-Signal*
differential term D-Anteil
differential voltage Differenzspannung * *z.B. an Differenzverstärker-Eingang*
differentiate differenzieren * *unterscheiden, einen Unterschied machen* || unterscheiden * *Unterschied machen/hervorheben/differenzieren; between: zwischen*
differentiating criterion Unterscheidungskriterium || Unterscheidungsmerkmal
differentiation Unterscheidung
differentiation of cases Fallunterscheidung
differentiation time Differenzierzeit || Vorhaltzeit * *für Differenzierer*
differentiation time constant Differenzier-Zeitkonstante
differentiator Differenzierer * *~ als Funktionsbaustein* || Differenzierglied
differently anders * *(Adv.) verschieden*
differing unterschiedlich * *nicht übereinstimmend, schwankend*
difficult aufwändig * *schwierig* || kompliziert * *schwierig* || mühsam * *schwierig* || schwer * →*schwierig* || schwierig * *auch 'schwer zu behandeln' (Personen)*
difficult to access schwer zugänglich
difficulties in delivery Lieferengpass || Lieferproblem
difficulty Hindernis * *Schwierigkeit* || Mühe * *Schwierigkeit* || Schwierigkeit || Unannehmlichkeit
diffraction Ablenkung
diffraction angle Beugungswinkel * *{Optik}*
diffuse breit * *weitschweifig, langatmig, diffus* || diffundieren

diffusion Diffusion
diffusion current Diffusionsstrom
diffusiveness Unschärfe * *Verschwommenheit*
dig graben
digester Kocher * *z.B. für Zellstoff*
digit Stelle * *Stelle einer Zahl/Ziffernanzeige; Zeichen in e. Textanzeige* || Zahl * *Stelle, Ziffer* || Zeichen * *z.B. BCD-/Hexa-~/Ziffer an Siebensegment-Anzeige o. Zahlenschalter* || Ziffer
digit after the decimal point Nachkommastelle * *auf einer (Ziffern-) Anzeige*
digit preceding the decimal point Vorkommastelle * *auf einer (Ziffern-) Anzeige*
digital digital
digital control Abtastregelung || digitale Regelung || digitales Regelsystem * *Einrichtung*
digital control system digitales Regelsystem
digital controller Digitalregler
digital display unit Digitalanzeigegerät
digital function digitale Funktion
digital input Bináreingang || Digitaleingang
digital input module Digital-Eingabebaugruppe
digital keyboard Zifferntasten * *Zifferntastatur*
digital operation digitale Funktion
digital output Bináraugang || Digitalausgang
digital output module Digital-Ausgabebaugruppe
digital position decoding module digitale Wegerfassung * *als Baustein*
digital position sensing digitale Wegerfassung
digital readout system Digitalanzeigegerät
digital regulator Digitalregler
digital signal processing digitale Signalverarbeitung
Digital Signal Processor Signalprozessor
digital speed actual value evaluation digitale Drehzahlistwerterfassung
digital speed evaluation digitale Drehzahlistwerterfassung
digital speed measurement digitale Drehzahlistwerterfassung
digital speed sensing digitale Drehzahlistwerterfassung
digital storage oscilloscope Digitalspeicher-Oszilloskop
digital tachometer Digitaltacho
digital technique Digitaltechnik
digital technology Digitaltechnik
digital to analog converter Digital-Analogwandler
digital-analog converter Digital-Analog-Wandler
digital-analog convertor Digital-Analogwandler * *(veraltet)*
digitalization Digitalisierung
digitalize digitalisieren
digitalized digitalisiert
digitally digital * *(Adv.)*
digitization Digitalisierung
digitize digitalisieren * *digitizer: Digitalisiereinrichtung; digitizing tablet: Digitalisiertablett*
digitized digitalisiert
digitizer Digitalisiereinrichtung
digitizing Digitalisierung
dignified fein * *gediegen, edel*
digress abgehen * *from: von einem Thema ~/abschweifen*
dilatation Dehnung * *z.B. Wärme~*
dilemma Schwierigkeit * *schwierige Lage*
diligent sorgfältig * *sorgfältig, gewissenhaft*
diluted dünn * *verdünnt*
dim trüb * *(halb)dunkel, düster, trübe, undeutlich, verschwommen, schwach, blass, matt (Farbe usw.)*
dimension Abmessung || auslegen * *dimensionieren* || bemaßen || bemessen * *dimensionieren (for: nach)* || Dimension * *auch physikalische ~* || dimensionieren * *bemessen, auch einer Schaltung* || Maß

* *Ausdehnung, Ab~ (z.B. Länge, Breite, Höhe)* || projektieren * *auslegen, bemessen*
dimension diagram Maßbild
dimension drawing Maßbild || Maßblatt || Maßzeichnung
dimension generously überdimensionieren * *großzügig dimensionieren*
dimension information Dimensionsangabe
dimension series Maßreihe
dimension sheet Maßblatt
dimension sketch Maßskizze
dimension table Maßtabelle
dimension tolerance Maßtoleranz
dimension with unspecified tolerance Freimaß * *Maßangabe ohne Toleranz*
dimension-line arrow Maßpfeil
dimensional accuracy Maßgenauigkeit || Maßhaltigkeit * *Maßgenauigkeit*
dimensional deviation Maßabweichung
dimensional drawing Maßbild || Maßzeichnung
dimensional stability Maßgenauigkeit * *Maßhaltigkeit* || Maßhaltigkeit * *Maß/Formbeständigkeit*
dimensionally correct maßhaltig
dimensionally stable formbeständig || maßhaltig * *form/maßbeständig*
dimensionally true maßhaltig
dimensioned ausgelegt * *dimensioniert* || bemessen * *(Adj.)* || dimensioniert
dimensioning Auslegung * *Dimensionierung* || Bemessung * *Dimensionierung* || Dimensionierung * *Bemessung, auch einer Schaltung*
dimensioning criterium Auslegungskriterium
dimensions Abmaße * *WxHxD:BxHxT (Breite x Höhe x Tiefe)* || Abmessungen * *of smaller dimensions: mit kleineren ~* || Ausmaß * *Abmessungen* || Größe * *Abmessungen* || Maße || Volumen * *Abmessungen*
diminish abnehmen * *vermindern, verringern, verkleinern, herabsetzen* || abschwächen || absenken * *reduzieren* || beeinträchtigen * *vermindern, herabsetzen, (ab)schwächen* || herabsetzen * *vermindern, verringern, verkleinern* || nachlassen * *sich vermindern* || reduzieren * *vermindern, verringern* || schwächen * *vermindern* || sinken * *sich vermindern* || verkleinern * *vermindern* || vermindern || verringern || zurückgehen * *sich verringern*
diminution Absenkung * *Verminderung, Verringerung, Verkleinerung* || Minderung * *Ver~, Verringerg., Verkleinerg., Abnahme, Nachlassen* || Reduzierung * *Verminderung, Verringerung, Verkleinerung, Abnahme* || Verkleinerung || Verminderung || Verringerung
diminution of expenses Kostensenkung
dimness Trübung
DIN DIN * *Abkürzg. für 'Deutsche Industrienorm bzw. Deut. Institut für Normg.': German Industry Standard(s)*
DIN climatic category Anwendungsklasse * *Klimaklasse nach DIN*
DIN EN DIN EN * *Deutsche Fassung einer Europäischen Norm*
DIN IEC DIN IEC * *Deutsche Fassung einer IEC-Norm*
DIN ISO DIN ISO * *Deutsche Fassung einer internationalen ISO-Norm*
DIN mounting rail DIN-Schiene * *Aufschnapp-Installationsschiene* || →C-Schiene oder →G-Schiene || Hutschiene * *Normschiene zur Aufschnappmontage z.B. nach DIN EN 50022 (35 mm breit)*
DIN PS PS * *deutsche PS; 1 PS entspr. 0,73550 kW entspr. 0,9863 hp*
DIN rail Befestigungsschiene * *DIN-Montageschiene (Hutschiene)* || Hutschiene * *z.Aufschnappmontage: C-Schiene DIN EN50022 (35mm), G-*

Schiene DIN EN 50035 (32 mm breit) ‖ Installationsschiene * C- bzw. G- Montageschiene n. DIN EN50022 (35 mm breit) bzw. DIN EN50035 (32mm) ‖ Montageschiene * Aufschnappmontageschiene C-/mini-C/G-förmig n. DIN EN50022/ 50045/50035 (35/15/32mm) ‖ Profilschiene * C-bzw. G-Aufschnapp-Montageschiene n. DIN EN50022 (35 mm breit) bzw. DIN EN50035 (32 mm) ‖ Tragschiene * C-bzw.G-Aufschnappinstall.schiene n. DIN EN50022 (35 mm breit) bzw. DIN EN 0035 (32 mm)
DIN VDE DIN VDE * *VDE-Bestimmung, die gleichzeitig eine Deutsche Elektrotechnische Norm ist*
dinner Abendessen * *Hauptmahlzeit, Mittagessen, Abendessen* ‖ Mittagessen * *after dinner: nach Tisch; be at dinner: beim ~ sein*
diode Diode
diode bridge Diodenbrücke
diode converter Diodengleichrichter
diode rectifier Diodengleichrichter
diode set Diodensatz
diode stack Diodensatz
dip -Einbruch * *kurzzeitiger* ‖ Abfall * *kurzzeitig, z.B. Spannung* ‖ beizen * *Metall* ~ ‖ Einbruch * *kurzzeitiger, z.B. Spannungseinbruch* ‖ Einsattelung * *in einer Kurve/Kennlinie* ‖ Neigung * *z.B. ~ eines Schiffs, ~ einer Straße* ‖ senken * *neigen* ‖ Senkung * *Neigung, Senkung, Gefälle, Neigungswinkel* ‖ tauchen * *auch in Löt-/Galvanisierbad usw.* ‖ *untertauchen* * *eintauchen*
dip coating Tauchumhüllung
dip in eintauchen
dip into eintauchen * *~ in*
DIP switch Codierschalter * *Dual-In-Line-Schalter (auf Leiterkarte)*
dip-braze löten * *tauch~*
dip-solder tauchlöten
dip-soldering tauchlöten * *(Substantiv)*
DIP-type switch DIL-Schalter * *Abkürzg. für 'Dual In Line package':im DIL-Gehäuse*
diploma Abschlusszeugnis ‖ Zeugnis * Diplom
diploma thesis Diplomarbeit
diploma'd engineer Diplomingenieur
Dirac pulse Dirac-Funktion ‖ Dirac-Impuls * *"unendlich hoher" Nadelimpuls, Sprungantwort eines Differenzierers* ‖ Einheitsstoß
dircharge point Entladeort * *für Kran usw.*
direct anordnen * *lenken, steuern, leiten, jdn. anweisen* ‖ ausgeben * *leiten, führen; z.B.auf eine Analogausgabe* ‖ beauftragen * *direkt* ‖ führen * *ein Signal ~* ‖ gerade * *direkt* ‖ leiten * *dirigieren, lenken, führen, adressieren(to:an)* ‖ regeln * *lenken, leiten, führen, bestimmen* ‖ unmittelbar
direct a signal to Signal ausgeben auf * *z.B. Analogausgabe*
direct a value to the analog output Wert auf die Analogausgabe ausgeben
direct AC converter Hüllkurvenumrichter ‖ Wechselstrom-Direktumrichter
direct AC voltage converter Wechselstromsteller
direct access Direktzugriff
direct approach Direkteinfahrt * *[Aufzüge]*
direct away wegführen
direct component Gleichanteil
direct connection to the line direkter Netzanschluss * *z.B. ohne Trafo oder Stromrichter*
direct cooling direkte Kühlung * *ohne Sekundärkühlkreislauf*
direct cooling of the conductors direkte Leiterkühlung * *z.B. Direktkühlung der Wicklungsleiter bei Großmotor*
direct current Gleichstrom
direct data transmission between substations -

Querverkehr * *[Network] z.B. zw. Slaves b. Master-Slave Feldbussystem*
direct DC converter Gleichstromsteller
direct digital control digitale Regelung
direct drive Direktantrieb
direct driven getriebelos * *direkt angetrieben*
direct electrical connection galvanische Verbindung
direct field sales office Vor-Ort-Vertrieb * *Abteilung, Geschäftsstelle*
direct line direkter Draht
Direct Memory Access direkter Speicherzugriff * *Peripherelement trennt CPU v.Speicher u.greift auf diesen direkt zu*
direct on-line operation Betrieb am Netz * *Betrieb eines Motor ohne Stromrichter direkt am Netz* ‖ Netzbetrieb * *eines Motors ohne Umrichter direkt am Netz*
direct online supply Betrieb am Netz
direct-cooled-conductor winding direkt-leitergekühlte Wicklung * *[el.Masch.] Kühlmittel umfließt direkt die Wicklungsleiter*
direct-current voltage Gleichspannung
direct-on-line direkt am Netz * *z.B. Betrieb eines Motors ohne Stromrichter/Sanftanlaufhilfe* ‖ netzbetrieben * *z.B.Motor i.Ggs. zum Umrichterbetrieb*
direct-on-line starting Direkteinschaltung * *Direktstart e. AC-Motors am Netz ohne Anlasswiderst. oder Anlasshilfe* ‖ direktes Einschalten * *[el. Masch.] e. Motors direkt am Netz ohne Anlasshilfe/Umrichter/Softstarter* ‖ Direktstart * *eines Motors direkt am Netz ohne Umrichter/Sanftanlaufgerät*
directed gezielt
directed to specific objectives gezielt * *Maßnahmen usw.*
direction Anordnung * *Anweisung* ‖ Anweisung * *Weisung, Anordnung, Vorschrift, Richtlinie, Belehrung* ‖ Führung * *Leitung* ‖ Hinweis * *(An-) Weisung, Anleitung, Anordnung, Belehrung* ‖ Leitung ‖ Richtung * *direction of rotation: Drehrichtung* ‖ Sinn * *Richtung* ‖ Vorschrift * *Anweisung* ‖ Weg * *Richtung*
direction arrow Drehrichtungspfeil ‖ Richtungspfeil
direction of current Stromrichtung
direction of current flow Stromrichtung
direction of force Angriffsrichtung * *einer Kraft* ‖ Wirkungsrichtung
direction of motion Fahrtrichtung * *Bewegungsrichtung* ‖ Verfahrrichtung
direction of phase rotation Drehfeld * *Richtung des Drehfeldes*
direction of rotation Drehrichtung * *clockwise: rechts; counterclockwise:links; forward: vorwärts; reverse: rückwärts* ‖ Drehsinn * *für elektr. Maschinen: siehe DIN VDE 0530 Teil 8 (IEC 34-8)*
direction of the current Stromflußrichtung
direction of the current flow Stromflußrichtung
direction of the flow of current Stromflußrichtung
direction of torque Drehmomentenrichtung ‖ Drehmomentrichtung ‖ Momentenrichtung
direction-dependent richtungsabhängig
direction-of-rotation switch Drehrichtungsumschalter
directional change Richtungswechsel
directional reversal Richtungswechsel * *Richtungsumkehr*
directions for use Gebrauchsanleitung * *siehe auch 'Betriebsanleitung'*
directive Richtlinie * *Direktive, (An-) Weisung, Vorschrift*
directly direkt * *(Adv.)* ‖ sofort * *(Adv.)*

directly coupled direkt gekoppelt * *z.B. ohne Getriebe/Riementrieb*
directly earthed conductor Nullleiter
directly fed direkt gespeist
directly proportional direkt proportional
directly soldered-in direkt eingelötet * *z.B. ohne Sockel*
director Direktor
directorate Vorstand
directory Katalog * *Verzeichnis* || Verzeichnis * *z.B. ~ von Dateien, Telefonnummern*
dirt Schmutz || Verunreinigung * *Schmutz*
dirt contamination Verschmutzung
dirt deposit Schmutzablagerung
dirty schmierig * *schmutzig* || schmutzig * *(auch figürl.: 'unanständig', 'unsauber')* || verschmutzt * *schmutzig*
dis- nicht- || un- * hinweg- || ver- * auseinander
disable -Sperre * *als Steuerkommando* || abschalten * *Freigabe für Ein/Ausgang (z.B. e. SPS)/Überwachung usw.* ~ || ausschalten * sperren, deaktivieren || blockieren * *sperren* || deaktivieren * *sperren* || Sperre * *als Kommando* || sperren || totlegen * *sperren*
disable interrupt Alarm sperren
disabled gesperrt || nicht durchgeschaltet * gesperrt *(z.B. Signal)*
disabled persons Behinderte
disabling -Sperre
disadvantage Fehler * *Nachteil* || Nachteil * *big: großer; severe: schwerer; to the disadvantage of: zum ~ von* || Unannehmlichkeit * *Nachteil*
disadvantageous nachteilig * *to: für* || schädlich * *nachteilig* || ungünstig * *nachteilig*
disagreement Missverständnis * Meinungsverschiedenheit
disallow absprechen * *Ansprüche (z.B. auf Schadenersatz)* negativ bescheiden
disappear schwinden * *ver*~ || verloren gehen * *verschwinden* || verschwinden
disappearance verschwinden * *(Substantiv) das Verschwinden*
disarrange stören * *durcheinander bringen* || verschieben * *in Unordnung bringen*
disarrangement Störung * Unordnung
disassemble abbauen * *auseinanderlegen, demontieren, zerlegen* || ausbauen * *auseinanderbauen* || auseinanderbauen || auseinandernehmen || demontieren || rückübersetzen * *von Assembler-/Maschinenspracheprogrammen* || zerlegen * *auseinanderbauen*
disassembler Rückübersetzer * *für Assembler-/Maschinenspracheprogramme*
disassembling Rückübersetzung * *von Maschinensprache-/Assemblerprogrammen*
disassembly Abbau || Demontage
disburden entlasten * *(meist figürl.) von einer Bürde befreien, entlasten (of/from:von)* || erleichtern * *von einer Last befreien*
disbursement Zahlung * ~ *von Unkosten*
disc Lamelle * *in Bremse, Kupplung; clutch disc: Kupplungslamelle* || Platte * *runde Scheibe* || Scheibe * *weniger gebräuchlich als "disk"*
disc assembly Lamellenpaket * *einer Lamellenkupplung/-bremse*
disc brake Scheibenbremse
disc case Scheibengehäuse * *z.B. e. Thyristors*
disc cutting shear Kreisschere
disc filter Scheibenfilter
disc refiner Scheibenrefiner * *[Papier] zur Zellstofferzeugung*
disc rotor Scheibenläufer
disc thyristor Scheibenthyristor
disc-armature motor Scheibenläufermotor
disc-rotor motor Scheibenläufermotor

disc-shaped scheibenförmig
discard unter den Tisch fallen lassen * *ausscheiden, (Gewohnh.,Vorurteil) ablegen; aufgeben, entlassen* || verwerfen * *ablegen,aufgeben (Gewohnheit, Vorurteil, Kleider), ausscheiden/-schalten/-rangieren*
discern erkennen * wahrnehmen
discernible unterscheidbar
discharge Abfuhr * *Leerung eines Behälters* || abführen * *(Behälter) leeren* || Ableit- || ableiten * *Strom, Ladung* || ablösen * *Schuld* ~ || Auslauf * *von/für Flüssigkeit* || ausräumen * *entleeren* || ausströmen * *ausströmen (lassen), z.B. Wasser; ausstoßen* || Austrag * *Herausförderung aus einem Behälter, einem Ofen usw.* || austragen * *entladen; Material/Teile aus Behälter/Ofen/Fertigungslinie austragen* || austreten * *ausströmen (lassen), z.B. Kühlluft* || entladen * *Kondensator (auch "sich* ~ *"), Behälter; oneself: sich (Person b.Hantierung m. el. Bauteilen)* || entlasten * *Vorstand* ~ || Entlastung || entsorgen * *entladen, austragen* || fördern * *aus einem Behälter heraus*~ || leeren * *ent*~, *entladen, ausströmen lassen, herausbefördern* || löschen * *entladen, z.B. Schiffsladung* || Quittung || versehen * *Pflichten* ~
discharge chute Schüttrinne * *zum Entladen*
discharge current Entladestrom
discharge device Ablass * *Entleerungsvorrichtung* || Ablauf * *Ablassvorrichtung z.B. an einem Behälter*
discharge inception voltage Glimmeinsatzspannung
discharge line Druckleitung * *am Ausgang eines Kompressors*
discharge pile Schüttkegel * *beim Entladen v. Schüttgut über eine Schüttrinne*
discharge pressure controller Enddruckregler * *für eine Turbine*
discharge pump Austragspumpe
discharge resistor Ableitwiderstand
discharge system Abnahmesystem
discharge table Ablauftisch * *zur Materialabfuhr*
discharge time Entleerungszeit
discharged air Abluft
discharger Entladeeinrichtung
discharging Ausschleusung || Löschung * ~ *einer Schiffsladung*
discharging device Entladeeinrichtung * *z.B. für Chargenprozess*
discharging machine Abräumer * *z.B. in der Verpackungstechnik* || Austragmaschine * *Entlademaschine z.B. für Ofen*
discipline Disziplin * *Wissenschaftszweig, Selbstdisziplin, Zucht*
disciplined geordnet
disclaim verzichten * ~ *auf, ablehnen*
disclaimer Ausschluss * *Haftungs-/Gewährleistungs~(erklärung)* || Verzicht
disclose erschließen * *aufdecken, enthüllen, offenbaren* || zugänglich machen * *offenlegen, bekannt geben; aufdecken*
disclosure Bekanntgabe * *Verlautbarung* || Bekanntmachung * *Verlautbarung*
discoidal scheibenförmig
discoidal motor Scheibenläufermotor
discoloration Verfärbung
discolouration Verfärbung
discomfort Unbehagen * *Unbehagen, (körperliche) Beschwerde*
disconnect abbauen * *eine Verbindung* ~ || abklemmen || abschalten * *Stromkreis/Verbindung unterbrechen* || abtrennen || auftrennen * *(elektr.) Stromkreis, Leitung* || auskuppeln * *trennen, lösen, auskuppeln/-rücken, ab/ausschalten* || ausschalten * *Stromkreis* ~ || freischalten * *Stromkreis* ~ || Frei-

schaltung || lösen * elektr. Verbindung ~ || trennen * Leitung, Stromkreise, Nachrichtenverbindung ~ || unterbrechen * Leitung, Stromkreis || wegschalten * z.B. Stromkreis oder e. Motor e. Gruppenantriebs von e. Umrichter || ziehen * Stecker ~
disconnect from mains   vom Netz trennen
disconnected   spannungslos * Stromkreis || unterbrochen * Stromkreis
disconnected from the supply   ausgeschaltet * vom Netz getrennt; when disconnected from the supply: im ~en Zustand || spannungsfrei * vom Netz getrennt
disconnectible   abschaltbar
disconnecting switch   Trennschalter
disconnecting test   Ausschaltprüfung * z.B. für Sicherungen
disconnection   Abschaltung * eines Stromkreises || Abtrennung || Freischaltung || Trennung * Lösen einer (auch elektr.) Verbindung || Unterbrechung * einer (Leitungs)Verbindung
disconnection on faults   Fehlerabschaltung * eines Stromkreises
disconnector   Trenner || Trennschalter
discontiguous   nicht zusammenhängend * mit Lücken
discontinuated   eingestellt * z.B. Fertigung
discontinuation   Einstellung * Aufgabe; z.B. product discontinuation: Einstell. d.Fertigung/Lieferung e.Produkts
discontinuation of delivery   Liefereinstellung
discontinuation of production   Fertigungseinstellung
discontinue   abbestellen * Abonnement usw. || aufhören * abbrechen || aussetzen * mit (Tätigkeit usw.) aussetzen || beenden * aufhören mit, aufgeben || einstellen * beenden/abbrechen, z.B. Fertigung eines Produktes || unterbrechen * unterbrechen, →einstellen, aussetzen, aufhören (to do: etwas zu tun)
discontinue the production   Fertigung einstellen
discontinued   abgekündigt * nicht mehr gefertigtes Produktes || unterbrochen
discontinued product   auslaufendes Produkt
discontinuous   diskontinuierlich || unstetig
discontinuous conduction mode   Lückbetrieb
discontinuous current   lückender Strom || Lückstrom
discontinuous current adaption   Lückadaption || Lückstromadaption
discontinuous current gain adaption   Lückstromanpassung * P-Verstärkungs-Anhebung für Stromregler i. Lückbetrieb
discontinuous current state   Lückbetrieb
discontinuous mode   Lückbetrieb
discontinuous operation   Lückbetrieb * {Stromrichter} bei DC-Stromrichter
discontinuous-action controller   diskontinuierlicher Regler * z.B. Digitalregler mit relativ langsamer Abtastung
discontinuous-current area   Lückbereich
discontinuous-current range   Lückbereich
discontinuously operating   diskontinuierlich
discount   Abzug * Skonto, Preisnachlass, Rabatt, Abschlag (on:auf) || Bonus * Rabatt (on:auf) || Preisnachlass || Rabatt * →Preisnachlass (on:auf)
discount line   Rabattlinie
discount schedule   Rabattstaffel
discourse   Vortrag * ausführliche Rede
discover   erkennen * entdecken || ermitteln * entdecken, ausfindig machen || finden * entdecken || herausbekommen * herausfinden, entdecken || merken * entdecken
discovery   Ermittlung * Fund, Entdeckung, Enthüllung
discrepancy   Abweichung * z.B. in einem Vertrags-
ablauf || Widerspruch * Widerspruch, Zwiespalt, Unstimmigkeit
discrete   diskret * einzeln, einzeln/getrennt aufgebaut; z.B. Vorgang, Regelung, Bauelement || einzeln * getrennt, getrennt betrachtetes Teil, eins von mehreren Teilen || einzeln aufgebaut * z.B. Schaltung aus Einzelbauteilen || gesondert * einzeln || getrennt * einzeln
discrete semiconductor   Einzelhalbleiter
discriminate   aufschlüsseln * (scharf) unterscheiden, einen Unterschied machen || unterscheiden * (scharf) ~, e.Unterschied machen, unterschiedl. behandeln; d. against: benachteiligen
discrimination   Selektivität * Unterscheidung(svermögen); overcurrent discrimination: Selektivität e. Sicherung || Unterscheidung
discuss   beraten * etwas beratschlagen || beurteilen * besprechen, diskutieren || diskutieren || durchsprechen
discussion   Beratung * Diskussion || Besprechung * Gespräch || Diskussion || Durchsprache || Gespräch * Diskussion || Verhandlung * Diskussion, Gespräch
disengage   auskuppeln || ausrücken * Kupplung || ausschalten * Kupplung trennen || lösen * Kupplung ~ || lüften * Bremse ~; auch 'gelüftet werden'
disengage the clutch   auskuppeln * eine Kupplung
disengaged   ausgerastet
disengagement   Abschaltung * einer Kupplung || ausrücken * (Substantiv) Kupplung
disengagement force   Ausrückkraft * zum Öffnen einer Kupplung
disengagement time   Ausschaltzeit * einer Kupplung
disengagement torque   Ausrückmoment * einer (Sicherheits-) Kupplung
disengagement travel   Ausrückweg * einer Kupplung
disentangle   auflösen * entwirren || entwirren * auch 'sich ~'; befreien, auf-/loslösen
disfigure   verschandeln
dish   stülpen * wölben
dish up   auffahren * Speisen, Getränke usw. ~ lassen
dish washer   Spülmaschine * Geschirrspülmaschine
disintegrate   aufreißen * Schmierfilm || aufspalten * {Chemie} || zerkleinern * aufspalten/-lösen/-schließen, ~; (auch intransitiv: zerfallen, sich zersetzen)
disintegrator   Mühle * Schlagmühle (z.B. in der Spanplattenherstellung) || Zerspaner * {Holz}
disjoin   trennen * zusammengehöriges ~
disjunction   Disjunktion * Schaltalgebra
disk   Diskette || Lamelle * Scheibe || Platte * runde Scheibe || Scheibe
disk brake   Scheibenbremse
disk sanding machine   Tellerschleifmaschine * {Holz}
disk thyristor   Scheibenthyristor
disk-shaped   scheibenförmig
disk-type endshield   Scheibenlagerschild * z.B. bei Motor
disk-type thyristor   Scheibenthyristor
diskette   Diskette
diskette drive   Diskettenlaufwerk
dislocate   verlagern
dismantle   abbauen * demontieren || abreißen * Werksanlage, Maschine || ausbauen * de/abmontieren, auseinandernehmen, zerlegen, niederreißen || demontieren * auseinanderbauen || zerlegen
dismantlement   Abbruch
dismantling   Abbau * z.B. einer Maschine || Demontage * ganzer Werksanlagen
dismember   teilen * zerstückeln

dismemberment Teilung * Zergliederung
dismiss verwerfen * fallen lassen, aufgeben (z.B. Thema); ab-/zurückweisen (z.B. Vorschlag); as:als...abtun
dismissal Entlassung || Zurückweisung
dismount abbauen * abmontieren, ausbauen, auseinandernehmen || auseinandernehmen
dismountable demontierbar
dismounting Ausbau * Abmontieren, Herausbauen
disobey widersetzen * sich einem Befehl/einem Gesetz ~
disorder Störung * Unordnung || Verwirrung * (von Sachen)
disorderly ungeordnet * verwirrt, unordentlich, ordnungswidrig; disorder: ~e Verhältnisse
disparage herabsetzen * verächtlich machen, in Verruf bringen
dispassional sachlich * leidenschaftslos
dispatch abschicken * ab/versenden || ausliefern * (schnell) ver/absenden || expedieren || Förderung * Absendung, Versand, Abfertigung, Beförderung, schnelle Erledigung || schicken * ver/absenden, expedieren || senden * ver~ || Versand * Absendung, Expedition
dispatch center Lieferzentrum
dispatch note Lieferschein || Versandanzeige
dispatch pallet Versandpalette
dispel ausräumen * (figürl.) Bedenken usw. ~
dispense verzichten * with: auf; ohne etwas auskommen
disperse streuen || verlaufen * sich ~ (Menge usw.) || verteilen * sich verteilen/ver-/ausbreiten, z.B. Geschwulst, Nebel, Niederschlag, zerstreuen
dispersing Entsorgung * (Weiter-) Verteilung von Daten z.B. bei SPS
dispersion Streuung * statistische ~ || Verteilung * Zerstreuung, Verbreitung, Verteilung (von Nebel, Teilchen, Stoffen usw.)
displacable verschiebbar
displace schwenken * z.B. Phasenwinkel zur Steuersatzsynchronisierung mittels eines →Schwenktrafos || verdrängen * (figürl. und physikal.) || verlagern || verschieben * versetzen, verrücken, verlagern, verschieben; auch Phasenwinkel || versetzen * displaced by 90 degrees: um 90 Grad versetzt
displaceable verschiebbar
displaced verschoben * versetzt, verrückt, verschoben, verlagert; displaced by 90°: um 90° verschoben || versetzt * displaced by 90 degrees: um 90 Grad ~
displaced by 90 degrees um 90 Grad versetzt
displacement Auslenkung || Hubraum * z.B. Verbrennungsmotor, Kompressor || Verdrängung * volumenmäßig || Verlagerung * auch 'Ausrichtungsfehler' || Versatz || Verschiebung * z.B. aus Sollposition || Verstellung * Verschiebung, Verlagerung
displacement angle Versatzwinkel || Verschiebungswinkel
displacement current Verschiebungsstrom * bei Halbleitern
displacement factor Verschiebungsfaktor * auch cos phi1: Verhältn. Grundschw.wirkleistg./Grundschw.scheinleistg.
displacement flux Verschiebungsfluss
displacement power factor Verschiebungsfaktor
display Abbildung * an Bildschirm, Anzeigeeinrichtung || Anzeige * z.B. auf LED~ oder Bildschirm || Anzeige- || anzeigen || auf Display ausgeben || aufblenden * anzeigen || ausgeben * anzeigen || Ausstellung * von Waren || Bildschirm * Anzeige || Visualisierung * Anzeige || zeigen * zur Schau stellen, zur Anzeige bringen
display building Bildaufbau * manuelle Erstellung eines Bildschirmbildes

display buildup Bildaufbau
display construction Bildaufbau * manuelle Erstellung eines Bildschirm-Bildes
display control Anzeigeelement
display data Anzeigedaten
display element Anzeigeelement
display limits Anzeigebereich * (Anzeigegrenzen)
display mode Anzeigemodus
display model Attrappe * für Ausstellungszwecke
display monitor Bildschirm
display panel Anzeigeeinheit * Anzeigefeld || Anzeigefeld
display parameter Anzeigeparameter || Beobachtungsparameter * Anzeigeparameter
display range Anzeigebereich
display tube Bildröhre
display unit Anzeigeeinheit || Anzeigegerät
displayable anzeigbar
displayed angezeigt || ausgegeben * angezeigt
displayed value Anzeigewert
disposability Verfügbarkeit
disposable Einweg- || verfügbar
disposal Beseitigung || Disposition * Verfügung || Entsorgung * z.B. von Schrott || Verschrottung * Entsorgung
dispose disponieren * over:über etwas/jdn. || entsorgen * of:etwas || lösen * Schwierigkeiten, Konflikt ~
dispose of beseitigen * Abfalls usw. ~ || einteilen * Zeit ~ || erledigen * aus der Welt schaffen || haben * verfügen über || verfügen über * gebieten über, verwenden
disposition Ansatz * Anlage (figürl.) || Aufbau * Anlage, Anordnung || Disposition || Einrichtung * Anordnung || Neigung * (figürl.) Veranlagung || Sinn * Neigung
disposition diagram Anordnungsplan || Dispositionsplan
disproportionate ungleichmäßig
disregard außer Acht lassen || nicht beachten || nicht befolgen
disruption Unterbrechung * Unterbrechung, Zerreißung, Spaltung; Zerrüttung, Zerfall
dissemination Verteilung * Ausstreuung, Ver/Ausbreitung
dissension Missverständnis * Meinungsverschiedenheit, leichter Streit, Zwietracht
dissertation Dissertation || Doktorarbeit
dissimilar ungleich || verschieden * unähnlich
dissipate abführen * z.B. Verlustwärme; auch '(sich) auflösen/zerstreuen/verflüchtigen' || abgeben * z.B. heat: (Verlust)Wärme ~ || abklingen * sich zerstreuen/auflösen/verflüchtigen || ableiten * Wärme || ableiten * Hitze || entwickeln * (Verlust-) Wärme ~ || verbrauchen * (Verlustleistung) || vernichten * energy:Energie
dissipation Abfuhr * of heat: Wärmeabgabe || Abführung * von Verlustwärme usw. || Verlustleistung * Abführung der ~, abgeführte/abzuführende ~
dissipation factor Verlustfaktor
dissoluble lösbar * {Chemie}
dissolution Auflösung * Firma, Stoff in einer Flüssigkeit usw.
dissolve aufheben * auflösen (Organisation usw.) || auflösen * Verein, Ehe, Versammlung, Stoffe in Flüssigkeit usw. || lösen * Vertrag, sich ~ in einem Lösungsmittel || trennen * auflösen || verschwinden * sich auflösen
dissolving pulp Chemiefaserzellstoff * {Papier} || Zellstoff * Chemie-Faserzellstoff, Ausgangsmaterial für Papierherstellung
dissymmetry Unsymmetrie
distance Abstand || Entfernung || Strecke * Entfernung; distance covered: zurückgelegte ~ || Weg * Strecke

distance between -Abstand * *(vorangestellt)*
distance between achses Achsabstand
distance between centres Mittenabstand
distance between signal edges Flankenabstand * *von (Impuls-) Signalen*
distance link Distanzstück
distance measurement transducer Abstandsmessgeber * *z.B. nach d. Ultraschallprinzip*
distance piece Distanzstück || Laterne * *Distanzstück*
distant entfernt
distant-reading thermometer Fernthermometer
distasteful unangenehm * *zuwider*
distinct ausgeprägt * *entschieden, klar, deutlich, merklich, unverkennbar, eigen, selbstständig* || deutlich * *klar, ausgeprägt, entschieden, merklich, spürbar* || fühlbar * *deutlich* || klar * *ausgeprägt, deutlich, entschieden, merklich, spürbar* || laut * *bestimmt, klar* || merklich * *deutlich* || spürbar * *deutlich* || verschieden * *ver/unterschieden, getrennt, abgesondert, unverkennbar* || verständlich * *deutlich*
distinction Unterscheidung || Unterschied * *Unterscheidg.(smerkmal), Unterschied; make a distinction between: e. ~ machen zwischen*
distinction between cases Fallunterscheidung
distinction of cases Fallunterscheidung
distinctive characteristic Unterscheidungsmerkmal
distinctive feature Unterscheidungskriterium * *Unterscheidungsmerkmal* || Unterscheidungsmerkmal
distinctive mark Merkmal * *Kennzeichen* || Unterscheidungskriterium * *Unterscheidungsmerkmal* || Unterscheidungsmerkmal
distinctly different vollkommen unterschiedlich
distinctness Übersichtlichkeit * *Klarheit, Deutlichkeit*
distinguish erkennen * *wahrnehmen* || herausstellen * *erkennbar machen* || unterscheiden * *between: zwischen*
distinguish oneself profilieren * *(figürl.: sich ~)*
distinguishable unterscheidbar
distinguished charakteristisch * *heraus-/hervorragend, bemerkenswert, charakteristisch*
distort verbiegen || verdrehen || verfälschen * *verzerren* || verformen * *verzerren, verwinden* || verspannen * *(sich)* →*verziehen* || verziehen * *verzerren, verdrehe, verziehen, be distorted: sich ~, verformen*
distorted schief * *verdreht, verzogen, verzerrt, verrenkt, veformt, entstellt* || unrund * *Drehfeld usw.* || verfälscht * *verzerrt* || verformt * *z.B.Kurvenform e.Spannung* || verzerrt * *auch 'verzogen'* || verzogen
distortion Klirrfaktor * *(salopp)* || Verformung * *unerwünschte ~/Verdrehung/Verwerfung* || Verspannung * *Verdrehung, Verziehung, Verwindung* || Verzerrung * *eines Signals*
distortion factor Klirrfaktor * *Verhältn. Effekt.-wert d. Oberschwingungen/Gesamteff.wert (DIN40110)* || Oberschwingungsgehalt * *Klirrfaktor*
distortion power Verzerrungsleistung
distortion-free verspannungsfrei
distortive power Verzerrungsleistung
distraction Ablenkung || Verwirrung * *Zerstreuung, Zerstreutheit, Ablenkung, Verwirrung, Bestürzung, Wahnsinn*
distribute aufteilen * *verteilen* || einteilen * *verteilen* || teilen * *ver/aus~* || verteilen * *ver/aus/zuteilen, ausgeben; among:auf/unter/an; auch statistisch* || vertreiben * *z.B. durch einen Zwischenhändler* || zuteilen * *verteilen* || zuweisen * *verteilen*
distributed dezentral * *verteilt angeordnet* || örtlich verteilt || verteilt * *auch statistisch ~, auch 'örtlich/dezentral/~ angeordnet'*
distributed configuration dezentraler Aufbau
distributed control dezentrale Steuerung
distributed intelligence verteilte Intelligenz
distributed peripherals dezentrale Peripherie * *ferngesteuerte/intelligente Klemmenleiste, z.B. ET 100/200 von SIEMENS*
distributed periphery dezentrale Peripherie * *ferngesteuerte Klemmenleiste, z.B. SIMATIC (R) ET100, ET200 von SIEMENS*
distributed power station Blockkraftwerk
distributing company Vertriebsgesellschaft
distributing warehouse Auslieferungslager
distribution Absatz * *das Absetzen auf dem Markt* || Abzweig || Aufteilung * *Verteilung* || Einteilung * *Verteilung* || Verteilung * *Ver/Zuteilung von Briefen/Waren usw; auch figürl.:z.B. statistisch/räuml./zeitl.* || Vertrieb * *Vertrieb, Absatz, Handel* || Weiche * *Verteilung, Verzweigung* || Zuteilung * *Ver/Aus/Einteilung*
distribution box Abzweigdose
distribution branch Verteilerabzweig
distribution channel Vertriebskanal || Vertriebsweg
distribution circle Sortierkreisel * *in (Hochregal-) Lager*
distribution costs Vertriebskosten
distribution list Adressliste * *für die Verteilung eines Informationsdienstes* || Verteiler * *für Informationsdienst* || Verteiler * *Anschriftenliste z. Verteilen schriftlicher Informationen*
distribution loop Sortierkreisel * *in (Hochregal-) Lager*
distribution network Verteilungsnetz * *Elektrizitäts~ (public: öffentliches)* || Vertrieb * *~snetz (mehr im Sinne von Verteilung, Logistik, Lieferung)* || Vertriebsnetz * *(mehr im Sinne von Verteilen, Logistik, Lieferung)* || Vertriebssystem * *Vertriebsnetz inkl. Auslieferungslager, Großhändler und Einzelhändler*
distribution of electricity Energieverteilung * *von elektr. Energie*
distribution of functions Funktionsaufteilung * *over: zwischen*
distribution station Verteilerstation * *z.B. elektr. Energie, Erdgas usw.*
distribution substation Verteilerstation * *Unterstation*
distribution system Netz * *Verteilungssystem, Stromsystem, Verteilungs~* || Verteilung * *elektrische*
distributor Distributor || Großhändler * *Vertragshändler* || Lieferant * *Verteiler, Zwischenhändler* || Verteiler * *auch Großhändler, Signal~, Energie~, Luft~; ignition distributor: Zünd~* || Vertragshändler * *Generalvertreter, Großhändler*
district Kreis * *Bezirk* || regional
district heating Fernwärme
district heating plant Fernwärmeanlage
district heating power plant Heizkraftwerk
district heating power station Heizkraftwerk
district office Niederlassung
district power station Blockkraftwerk
disturb in Unordnung bringen || stören * *durcheinander bringen, belästigen, behindern, in Unordng.bringen*
disturbance Störgröße || Störung * *auch Unruhe, Brumm, Beeinträchtigung, Behinderung, s.a. Kommutierungseinbruch*
disturbance effect Störgröße * *Störeffekt in e. Regelkreis*
disturbance observer Störgrößenbeobachter
disturbance rejection Störgrößenunterdrückung * *bei einer Regelung* || Störsignalunterdrückung

* *bei einer Regelung* ‖ Störunterdrückung * *Störgrößenunterdrückung bei einer Regelung*
disturbance suppression Störgrößenunterdrückung * *bei einer Regelung* ‖ Störsignalunterdrückung * *bei einer Regelung*
disturbance suppression control Störgrößenunterdrückung * *bei einer Regelung* ‖ Störsignalunterdrückung * *bei einer Regelung*
disturbance transfer function Störübertragungsfunktion * *{Regel.techn.}*
disturbance variable Störgröße * *in einem Regelkreis*
disturbance variables Störgrößen
disturbance-free störungsfrei * *(auch figürl.: ohne Beeinträchtigung/Behinderung)*
disturbances Störgrößen
disturbed gestört * *durcheinander/in Unordnung gebracht*
disturbing störend ‖ Störung * *das Stören*
disturbing irridiation Störeinstrahlung
disturbing voltage Fremdspannung * *Störspannung*
dither zittern * *zittern, bibbern, zaudern*
dithering Dithering * *{Computer} Erzeugung von Zwischenfarben durch über-/nebeneinanderlegen von Farbpunkten*
ditto dito
diurnal täglich * *täglich (wiederkehrend), Tages-, Tag-, tageszeitlich; nur bei Tag auftretend*
dive eintauchen * *(intransitiv, aktiv) into:in* ‖ tauchen * *tief ~*
dive in eintauchen
divergence Divergenz
diverse unterschiedlich * *verschieden(e), ungleich, mannigfaltig* ‖ vielfältig * *mannigfaltig, verschieden, ungleich*
diversification Diversifikation
diversified unterschiedlich * *verschieden(artig), mannigfaltig, abwechslungsreich*
diversion Ablenkung ‖ Ausschleusung * *Umleitung, Ableitung* ‖ Ungleichförmigkeit * *Abweichung*
diversity Vielfalt * *Mannigfaltigkeit, Vielgestaltigkeit, Abwechslung, Verschiedenheit* ‖ Vielzahl * *Vielfalt, Verschiedenheit, Mannigfaltigkeit; a wide diversity of: e. große ~ an*
divert ableiten * *z.B. Spannungsspitzen; auch ablenken,abwenden; from:von, to:nach* ‖ ablenken * *from:von, to:nach* ‖ abwenden * *from: von*
diverter Ableiter
divide aufspalten * *(ein)teilen (in(to): in)* ‖ aufteilen ‖ dividieren * *by:durch* ‖ einteilen * *in/into:in* ‖ gliedern * *(ein-) teilen* ‖ spalten ‖ teilen * *auch auf~, sich ~; auch Pause, Meinung* ‖ trennen * *teilen* ‖ verteilen * *(ein)teilen*
divide into halves halbieren * *divide into equal halves: in gleich große Hälften (auf)teilen*
divide into sections staffeln * *in Abschnitte, Gruppen usw. einteilen*
divide up aufspalten * *zerteilen, zerlegen, zerstückeln, spalten* ‖ aufteilen
divided gegliedert * *eingeteilt (into: in)* ‖ geteilt ‖ unterteilt * *into:in*
dividend Dividend
divider Dividierer ‖ Teiler * *z.B. Frequenzteiler, Spannungsteiler*
dividers Stechzirkel
dividing head Teilapparat
dividing shear Teilschere * *Quer~ in Walzstraße/Blechbehandlungsanlage*
divisible zerlegbar * *teilbar (auch {Mathem.})*
division Abteilung * *Unternehmensbereich* ‖ Aufteilung * *Teilung (into:in), Gliederung, Verteilung* ‖ Bereich * *eines Unternehmens* ‖ Division ‖ Einteilung ‖ Fach * *Abteil* ‖ Geschäftsgebiet * *Abteilung* ‖ Teilung ‖ Trennung * *Teilung* ‖ Unternehmensbereich
division director Bereichsleiter
division head Abteilungsleiter * *Leiter eines Unternehmensbereiches*
division of labour Arbeitsteilung
division of the market Marktaufteilung
divisional gemeinsam * *~ innerhalb einer Abteilung, eines Unternehmensbereichs*
divisor Divisor ‖ Teiler * *(mathemat.) Nenner, Divisor*
DIY Heimwerker- * *Abkürzg. für 'Do It Yourself'* ‖ Hobby- * *Abkürzg. für 'Do It Yourself'*
DKE DKE * *Abk.f. 'Deutsche Elektrotechn. Kommission' im DIN und VDE* ‖ DKE * *Abk. für 'Deutsche Elektrotechnische Kommission im DIN und VDE': Normengremium aus VDE u. DIN*
DLL DLL * *'Dynamic Link Library': Unterprogrammbibliothek, die erst bei Programmaufruf "gelinkt" wird*
DMA direkter Speicherzugriff * *Abk.f. 'Direct Memory Access'*
DMK DM * *Deutsche Mark*
DNV DNV * *Abk. für 'Det Norske Veritas': Schiffsklassifizierungsgesellschaft*
do machen * *that will hardly do: das wird kaum ~* ‖ dito * *Abkürzung für 'ditto':dito* ‖ durchführen * *tun* ‖ reichen * *genügen, aus~d sein; that will do!: das reicht!*
do away with aufräumen * *(figürl.) mit etwas aufräumen*
do business Geschäfte machen ‖ Geschäfte tätigen
do business with Geschäftsbeziehungen haben mit ‖ Geschäftsverbindung haben mit ‖ in Geschäftsbeziehung stehen mit
do gymnastics Klimmzüge machen * *(figürl., salopp)*
Do It Yourself Heimwerker-
do mental acrobatics Klimmzüge machen * *(figürl., salopp) geistige ~*
do the job funktionieren * *Aufgabe/Funktion erfüllen*
do the work of versehen * *Dienst ~*
do well gehen * *bei Geschäften, Absatz usw.*
do without verzichten * *ohne etwas auskommen; I can do without it: ich kann darauf ~*
do-it-yourself Hobby-
do-it-yourself store Heimwerkermarkt
dock anlegen ‖ anlegen * *Schiff*
docket Etikett * *Adresszettel, Etikett, Vermerk; Liste; Zollquittung, Lieferschein*
doctor Arzt
doctor blade Rakel * *Abstreifmesser, Streichmesser, (Druck)Rakel, Kratzeisen* ‖ Rakelmesser * *z.B. in Druckmaschine (siehe auch →Rakel)* ‖ Streichmesser * *Rakel*
doctoral thesis Doktorarbeit
doctrine System * *Lehre*
document Akte ‖ Beleg ‖ Unterlage ‖ dokumentieren ‖ notieren * *dokumentieren* ‖ Papier * *Dokument* ‖ Unterlage * *schriftliche*
document case Aktentasche * *Aktenmappe*
document listing Unterlagenverzeichnis
document reversely rückdokumentieren
documentation Dokumentation * *bezüglich techn. Dokumentation siehe DIN-Entwurf 8418, 66055 u. VDI-Richtl. 4500* ‖ Druckschrift ‖ Unterlagen
documentation package Dokumentiersatz
documents Unterlagen * *schriftliche ~*
dodge vermeiden * *aus dem Wege gehen, umgehen, sich drücken vor, Ausflüchte machen*
doffer Abnehmer * *{Textil} Abnahmewalze in Krempel-/Kardenmaschine* ‖ Doffer * *{Textil} Abnehmerwalze in bei Krempel-/Kardenmaschine* ‖

Peigneur * *[Textil]* Abnahmewalze in →*Krempelmaschine*
dog Klaue * *Mitnehmer, Nase* || Mitnehmer
dog clutch Klauenkupplung
doggedness Zähigkeit * *verbissen* ~
doings Wirtschaft * *Treiben*
DOL direkt am Netz * *Abkürzg. für 'Direct On Line'* || direktes Einschalten * *[el.Masch.]Direct-On-Line Starting: Start e.AC-Motors ohne Umrichter/Anlasser*
dolly Schwabbelscheibe
domain Bereich * *Gebiet, Sphäre, Reich, Domäne* || Domäne * *Sphäre, →Gebiet, →Bereich, Reich; übergeordnete Adresskennung im WWW* || Gebiet * *Domäne, →Bereich, Sphäre, Reich*
dome Wölbung * *Kuppel*
dome-tipped srew Linsenkopfschraube
domed gewölbt
domestic Heim- || inländisch || national * *heimisch, Binnen-*
domestic appliance Hausgerät || Haushaltsgerät
domestic application Haushaltsanwendung
domestic market Binnenmarkt || Inlandsmarkt
domestic supply Haushaltsnetz
domestic supply network Haushaltsnetz
domestic textile industry Heimtextilindustrie
dominant beherrschend * *(be)herrschend, vorherrschend, entscheidend; emporragend, weithin sichtbar* || dominant * *factor: Umstand, Faktor*
dominant company Marktführer
dominant market position marktbeherrschende Stellung
dominate dominieren || herrschen * *dominieren, be~*
dominating beherrschend * *(be)herrschend, dominierend, emporragend über* || dominant * *dominierend*
domination Beherrschung
dominion Gewalt * *Herrschaft (over:über)*
don't care ohne Bedeutung * *in Liste/Tabelle*
don't mention it bitte * *als Antwort auf "Danke!"*
done fertig * *abgeschlossen*
door Tür * *auch eines Geräts, eines Schaltschrankes*
dope dotieren * *Halbleiter*
doping Dotierung * *Halbleiter*
DOS level DOS-Ebene
dosage Dosierung
dose Dosieren * *(Verb)* || Maß * *dosis*
dose rate Strahlenbelastung * *[rd/h]*
dosing Dosieren || Dosierung
dosing auger Dosierschnecke
dosing equipment Dosiereinrichtung || Dosiermaschine
dosing plant Dosieranlage * *z.B. in der Getränkeindustrie*
dosing pump Dosierpumpe
dosing scales Dosierwaage
dot Punkt * *Satzzeichen, Tüpfelchen*
dot clock Pixeltakt * *bei Monitor*
dot pitch Pixelabstand * *bei einem Momitor*
dot-and-dash strichpunktiert
dot-like punktförmig
dot-line punktieren * *Linie*
dotted gepunktet * *line:Linie* || punktiert * *Linie*
double doppel || doppelt * *(Adj.u.Adv.)* || Duo- || verdoppeln * *auch 'sich verdoppeln'* || zwei- || Zweifach- * *Doppel-*
double bearing Doppellager
double brake Doppelbremse
double break contact Doppelkontakt * *z.B. bei Schütz*
double converter Doppelstromrichter || Umkehrstromrichter * *Doppelstromrichter mit 2 Stromrichterhälften, z.B. Gegenparallelschaltung* ||

Zweifachstromrichter * *z.B. gegenparallele 4 Q Drehstrombrückenschaltung*
double drag-link gear Ungleichförmigkeitsgetriebe * *z.B. zum Antrieb der →Messerpartie bei →Querschneider*
double drag-link gearing Koppeltrieb * *→Ungleichförmigkeitsgetriebe z. Antrieb d.→Messerpartie bei →Querschneider*
double explosion-safety Doppelschutzart * *Kombinat.d. Zündschutzarten d- Druckfeste Kapselg. u. e- Erhöhte Sicherheit*
double fan Doppellüfter
double fault Doppelfehler
double mitre cutting saw Doppelgehrungssäge * *[Holz]*
double mitre saw Doppelgehrungssäge * *[Holz]*
double pulse Doppelimpuls
double reduction zweistufig * *bei Untersetzungsgetriebe*
double shunt-wound motor Doppelschlussmotor * *DC-Motor*
double sided doppelseitig
double tachometer Doppeltacho * *z.B. zur Gewährleistung einer gewissen Sicherheitsstufe*
double the doppelt * *das Doppelte von, doppelt so viel wie*
double way gegenparallel
double way bridge connection Gegenparallelschaltung * *z.B. Gleichrichterbrücke*
double way bridge in non circulating current mode kreisstromfreie Gegenparallelschaltung * *Gleichrichterbrücke*
double winder Doppelwickler * *z.B für fliegenden Rollenwechsel*
double word Doppelwort * *z.B. Datenwort mit einer Länge von 32 Bit* || Langwort * *z.B. 32 Bit langes Datenwort*
double- Doppel-
double-action zweiseitig * *~ arbeitend*
double-action press doppelt wirkende Presse
double-bearing zweiseitig gelagert
double-buffer Doppelpuffer * *[Computer] zur Datenverwaltung/-übergabe*
double-capped tubular lamp Soffitte * *Soffittenlampe*
double-circuit controller Dreipunktregler * *z.B. Temperaturregler*
double-click doppelklicken * *mit Computermaus (on: auf)*
double-crank drawing press Breitziehpresse
double-deck squirrel-cage motor Käfigläufermotor mit Anlauf- und Betriebswicklung
double-delta Doppeldreieck * *z.B. Schaltung für polumschaltbaren Motor*
double-disc Doppelscheiben-
double-disk Doppelscheiben-
double-end zweiseitig
double-ended zweiseitig
double-ended drive doppelseitiger Antrieb
double-ended grinding machine Schleifbock * *Doppelschleifer*
double-enveloping worm gear Schneckengetriebe
double-Eurocard Doppeleuropakarte * *Leiterkarte T x H: 160 x 233.4 mm oder 220 x 233.4 mm*
double-Euroformat Doppeleuropaformat * *T x H: 160 x 233.4 mm oder 220 x 233.4 für Leiterkarte*
double-Europe format Doppeleuropaformat * *T x H: 160 x 233.4 mm oder 220 x 233.4 für Leiterkarte*
double-height doppelt hoch
double-height Eurocard format Doppel-Europaformat * *Leiterkartenformat, T x H: 160 x 233,4 mm oder 220 x 233,4 mm*
double-inlet zweiseitig saugend * *Ventilator*
double-layer winding Zweischichtwicklung * *Mo-*

torwicklgs.art: jede Nut enthält Drähte zweier Spulen –> wenig Oberwellen
double-leg brush Doppelschenkelbürste * *zur Stromübertragung auf bewegte Maschinenteile*
double-level terminal Doppelstockklemme
double-line zweizeilig * *Anzeige, (Aus)Druck*
double-moving contact Doppelschaltstück
double-precision doppelter Genauigkeit
double-pulse Zweipuls- || zweipulsig
double-row zweizeilig * *Baugruppenträger*
double-sealed bearing gedeckeltes Lager * *Rillenkugellager mit eingebauten Lagerdichtungen*
double-size Eurocard Doppeleuropakarte * *Leiterkarte T x H: 160 x 233.4 mm oder 220 x 233.4 mm*
double-stage zweistufig
double-star Doppelstern * *z.B. Schaltung für polumschaltbaren Motor*
double-star connection Doppelsternschaltung
double-throw contact Umschaltkontakt || Wechselkontakt || Wechsler || Wechslerkontakt
double-tier zweizeilig * *Baugruppenträger*
double-tier transformer Doppelstocktrafo * *z.B. 3-Wickl.~ z.Speisg. e. 12puls-Stromricht., 2 um 30 versetzte Sek.wickl.*
double-way Zweiweg-
double-way connection Zweiwegschaltung * *z.B. für Stromrichter*
double-way converter Umkehrstromrichter
double-way rectification Zweiweggleichrichtung
double-way rectifier Zweiweggleichrichter
double-wound bifilar gewickelt
double-wound winding bifilare Wicklung
doubling machine Doubliermaschine * *zum Doublieren von Papier*
doubly doppelt * *(Adv.vor Adj.)*
doubtless zweifellos
dough kneader Teigkneter
dough roller Teigwalze * *zur Brot-/Gebäckherstellung*
douse tauchen * *plötzlich/kräftig ~*
dovetailing machine Zinkenfräsmaschine * *{Holz}*
dowel Dübel || eintreiben * *Dübel ~* || verstiften * *verdübeln*
dowel hole Dübelloch * *{Holz}* || Passstiftloch
dowel hole boring machine Dübellochbohrmaschine * *{Holz}*
dowel pin Passstift
dowel-inserting machine Dübeleintreibmaschine * *{Holz}*
doweldriving machine Dübeleintreibmaschine * *{Holz}*
dowelling gun Dübeleinschießpistole
down abwärts || herunter || hinab * *right down to 0 V: bis hinab zu 0 Volt* || hinunter || unten * *deep down: weit ~; deeper/further down: weiter ~; from down there: von hier ~ (aus)*
down at the bottom unten * *am Boden, auf dem Grund*
down counter Rückwärtszähler
down hole pump Bohrlochpumpe
down pulses Rückwärtsimpulse
down there herunter || hinab || hinunter || unten * *da ~* || unten * *da ~; down on the left: links ~*
down time Betriebsstörung * *Stillstandszeit*
down to bis * *bis herunter zu*
down-sizing Verkleinerung
down-time Ausfalldauer || Standzeit * *Maschinen/Anlagen-Stillstandszeit*
down-to-earth praktisch * *nüchtern*
downhill operation Talfahrt
download Download * *Transfer von Dateien zum eigenen Computer (z.B. über d. –>FTP-Protokoll im Internet)* || einlesen * *hinunterladen von Daten/Programmen/Parametern in ein Gerät hinein* || herunterladen * *von Dateien* || laden * *Programm oder Parameter in einen Rechner hinein* || übertragen * *z.b. Arbeitsdaten/Parameter zu einem Gerät ~*
downright rein * *(Adv.) gänzlich, völlig, absolut, ausgesprochen, wirklich*
downsize abspecken || Platzbedarf reduzieren || verkleinern * *Fläche/Bauvolumen verringern (z.b. durch Neukonstruktion)*
downsized klein * *ver~ert, in Baugröße/Volumen reduziert*
downsizing Abbau * *Belegschaft, Produktion usw.* || Platzreduzierung
downstairs nach unten * *in Haus* || unten * *~ im Haus, ~ an der Treppe*
downstream nachfolgend * *im Signalfluss* || nachgeschaltet * *im Leistungs-/Signalfluss ~*
downstream of hinter * *im Signal-/Daten-/Material-/Kühlmittelfluss weiter hinten liegend*
downtime Ausfallzeit * *z.B. einer Maschine, Anlage* || Betriebsunterbrechung * *Stillstandszeit einer Maschine/Anlage (auch auf Grund eines Fehlers)* || Produktionsunterbrechung || Stillstandszeit * *einer Maschine, Anlage (auch wegen Störung)*
downturn Abschwung * *economic: wirtschaftlicher* || Rückgang * *des Geschäfts, der Wirtschaft*
downward abwärts || herunter || hinab || hinunter
downward movement Rückgang * *Abwärtsbewegung*
downward pressure Abtrieb * *nach unten wirkender Druck*
downwards abwärts || herunter || hinab || hinunter || nach unten || unten * *nach ~; facing downwards: nach ~ zeigend; levelled/pointed downwards: nach ~ gerichtet*
downwards compatible abwärtskompatibel
dowse tauchen * *plötzlich/kräftig ~*
DP Datenverarbeitung * *Abk. f. 'Data Processing'* || Differenzdruck * *Abkürzg. für 'Differential Pressure'* || DP * *Abkürzg.für 'drip-proof': tropfwassergeschützt (Schutzart IP22)* || DP * *Abk. für 'Data Processing':Datenverarbeitung* || DV * *Abkürzg. für 'data processing': Datenverarbeitung* || EDV * *Abk. für 'Data Processing': (elektron.) Datenverarbeitung* || tropfwassergeschützt * *Abkürzg.für 'Drip-Proof'; Schutzarten IP21 und IP22*
DP system DV-System
DPBV tropfwassergeschützt * *'Drip Proof Blower Vented': ~e Motorschutzart mit Anbaulüfter*
DPFG tropfwassergeschützt * *'Drip Proof Fully Guarded': ~e Mot.schutzart mit Einbau-Fremdlüfter*
Dpt. Abt * *Abk. für 'department'; Abteilung*
draft Bild * *Skizze* || entwerfen * *Vertrag, Vorschrift, Norm usw. ~* || Entwurf * *fist/rough: Grobentwurf* || Grobentwurf || Konzept * *first/rough: Grob~, Grobentwurf* || Normentwurf || Plan * *(Grob-)Entwurf* || zeichnen * *skizzieren, entwerfen*
draft contract Vertragsentwurf
draft standard Normentwurf || Vornorm * *Normentwurf*
draft- Vor- * *(Vor)Entwurf/Norm*
drafting device Streckwerk * *{Textil| Draht} in Spinnmaschine; zur Durchmesser-Verringerung bei Drahtziehmaschin*
draftsman Technischer Zeichner || Zeichner
drag Belastung * *(figürlich) Hemmschuh* || Luftwiderstand || reißen * *fort~* || schleppen || Strömungswiderstand * *siehe auch –>Luftwiderstand* || ziehen * *–>schleppen; auch 'mit der (Computer-) Maus ~'*
drag along mitreißen || reißen * *fort~*
drag of air Luftwiderstand
drag torque Leerlaufdrehmoment * *z.B. einer Kupplung oder Bremse* || Leerlaufmoment * *z.B. einer Kupplung, Bremse*

drag-link gearing Koppeltrieb
* →*Ungleichförmigkeitsgetriebe* || Ungleichförmigkeitsgetriebe * *z.B. zum Antrieb der* →*Messerpartie bei* →*Querschneider*
drain Ablass * *z.b. für Flüssigkeiten* || ablassen * *Wasser, Öl* ~ || Ablauf * *Ablass(vorrichtung), z.b. für Flüssigkeiten* || Auslauf * *Ableitung, Abfluss, Entwässerung* || ausleeren * *Flüssigkeit leeren/ableiten/abführen* || Belastung * *finanzielle* ~ || entleeren * *Flüssigkeit* || entwässern * *ablaufen lassen* || leeren || Senke * ~ *für Flüssigkeit, Ladungsträger, Elektronen usw.* || trocknen * *trockenlegen (Land usw.)*
drain dry leerlaufen * *Gefäß*
drain hole Ablass * *Ablauf-/Ablassöffnung* || Ablauf * *Ablassöffnung* || Kondenswasserablauf * *~loch z.b. an tiefster Stelle e. Motors* || Kondenswasserloch * *z.b. an der tiefsten Stelle e. Motors*
drain off abfließen * *besonders Flüssigkeiten; auch 'abfließen lassen'* || abführen * *Wasser usw.* ~ || ablaufen * *auch Flüssigkeit* || ableiten * *z.b. Waser* || ausströmen * *ausströmen lassen, z.b. Wasser*
drain plug Ablassschraube || Verschlussschraube * *Ablassschraube* || Verschlussstopfen * *(Flüssigkeits-) Ablassstopfen*
drainage Drainage || Dränage || Entleerung * *Entleerung (v. Flüssigkeiten), Drainage, Abfluss, Ableitung, Entwässerung, Trockenlegung* || Entwässerung * *z.b. von Papier*
draining off Abfuhr * *Wasser usw.*
DRAM DRAM * *Dynamic Random Access Memory, dynamisch aufzufrischender Schreib/Lesespeicher*
dramatic drastisch
dramatically drastisch * *(Adv.)* || sprunghaft * *(Adv.) dramatisch, z.B. Preisanstieg*
drastic drastisch
drastical drastisch
drastically drastisch * *(Adv.)* || stark * *(Adv.) drastisch*
draughtsman Technischer Zeichner
draw abziehen || anziehen * *(allg.)* || beziehen * *Gelder, Gehalt, Einkommen, Energie aus dem Netz usw.* ~ || dehnen * *ziehen, verstrecken* || Drehzahlverhältnis * →*relative Geschwindigkeitsdifferenz, z.B. zw.mehreren Papiermaschinen-Antrieben* || empfangen * *Geldmittel usw.* ~ || entnehmen * *z.b. Strom aus dem Netz ~ (from:von/aus)* || Geschwindigkeitsrelation * →*relative Geschwindigkeitsdifferenz [(v2-v1)/v1] x 100% zw. 2 Antrieben* || relative Geschwindigkeitsdifferenz || strecken * *ziehen* || verbrauchen * *herausholen, entnehmen, ziehen; auch Strom, Wasser, Leistung usw.* || verstrecken || zeichnen || ziehen * *auch Strom aus Stromversorgung(snetz)* ~ || Zug * *Verstreckung, das Ziehen* || Zugspannung * *z.B.* ~ *einer Papierbahn (siehe* →*relative Geschwindigkeitsdifferenz)*
draw a person's attention to hinweisen * *jdn. auf etwas* ~
draw attention to aufmerksam machen auf || hinweisen auf * *draw a person's attention to: jdn.* ~ *auf*
draw from entnehmen * *Geld ~; aus einem Buch usw.* ~
draw in ansaugen * *Luft*
draw nearer nähern * *sich* ~
draw off abziehen
draw on ample resources aus dem Vollen schöpfen
draw rolls Vorziehpartie * *z.B. bei* →*Querschneider*
draw stand Streckwerk
draw together zusammenziehen * *(auch figürl.)*
draw up anfertigen * *schriftlich* || ausarbeiten * *ent-

werfen* || entwerfen * *Vertrag, Schriftsatz, Regelstruktur usw.* ~ || erstellen * *Dokumentation, Zeichnungen usw.* ~
draw work Hebewerk * *z.b. bei Bohrplattform*
draw-in unit Einlaufwerk * *in Faserverstreckanlage usw.*
draw-out section Einschub * *bei Leistungsschalter usw.*
drawback Fehler * *Beeinträchtigung, Nachteil, Hindernis, Haken* || Mangel * →*Nachteil* || Nachteil * →*Mangel, Beeinträchtigung, Schattenseite, Hindernis, 'Haken an der Sache'* || Unannehmlichkeit * *Nachteil, Hindernis, 'Haken', Schatten/ Kehrseite*
drawer Fach * *Schub~* || Schublade
drawing Abbildung * *Zeichnung* || Bild * *Zeichnung,Skizze* || Förderung * *von Bodenschätzen* || grafische Darstellung * *Zeichnung* || Plan * *Zeichnung* || Planung * *Zeichnung* || technische Zeichnung
drawing board Reißbrett || Zeichenbrett
drawing cage Förderkorb
drawing die Ziehmatritze * *(Werkz.masch.) z.B. in der Blechdosenherstellung* || Ziehstein * *bei Drahtziehmaschine*
drawing disk Ziehscheibe * *bei Drahtziehmaschine*
drawing engine Göpel
drawing frame Spannrahmen * *(Textil) Verstreckwerk* || Streckwerk || Verstreckmaschine * *Spannrahmen, z.B. für Textilfasern* || Verstreckwerk * *(Textil) Spannrahmen (Textilmaschine)*
drawing machine Streckmaschine * *für Textilfasern* || Verstreckwerk * *(Textil) für (Kunst-) Fasern* || Ziehmaschine
drawing No Zeichnungsnummer
drawing off Abzug * *das Abziehen*
drawing office Konstruktionsbüro
drawing oil Ziehöl * *zum Drahtziehen*
drawing section Vorziehpartie * *z.B. bei* →*Querschneider*
drawing-through ventilation Durchzugsbelüftung
dredge Bagger
dredger Bagger
dredger bucket Baggereimer * *Baggerschaufel (auch eines Schwimmbaggers)*
dregs Hefe * *Bodensatz*
dress abrichten * *Schleifscheibe egalisieren, Säge usw. scharfmachen* || aufbereiten * *Erz* || nachbearbeiten || schränken * *(Holz) Säge scharf machen*
dressing Aufbereitung * *Nachbearbeitung, Zurichtung, Aufbereitung (auch von Erz)* || Bearbeitung * *Zurichtung* || Verband * *in der Medizin*
dressing agent Schmiermittel * *(el.Masch.) für Kommutator*
dressing machine Egalisiermaschine * *(Holz) zum Pflegen/Schärfen von Sägen, Schleifscheiben usw.*
dressing tool Abrichtwerkzeug * *(Werkz.masch.) zum Scharfmachen von Sägen, Schleifscheiben usw.*
drier Trockner
drier cylinder Trockenzylinder
drier furnace Trockenofen
drift Änderung * *z.B. temperatur/langzeitabhängig* || Drift || wandern * *(ab)treiben, ziehen, stömen, ab~, auseinanderleben* || weglaufen * *z.B. Spannung, Messwert*
drift of offset error Nullpunktverschiebung * *Drift des Nullpunktfehlers*
drift out of true verlaufen * *z.B. Bohrer beim Bohrvorgang*
drift-free driftfrei
drill ausbilden * *streng schulen* || Ausbildung * *strenge Schulung* || bohren || Bohrer || Bohrung * *nach Bodenschätzen (Erdöl, Gas); lower a drill:*

eine ~ niederbringen || Schulung * strenge Schulung || üben * streng schulen, drillen || Übung || Unterricht * scharfer
drill bit Bohrer * Bohrereinsatz (s.B. Spiral-, Steinbohrer) || Bohrmeißel * z.B. bei Erdöl-Bohreinrichtung
drill chuck Bohrfutter
drill hole Bohrung
drill machine Bohrmaschine
drill milk Bohrmilch * {Werkz.masch.}
drill out ausbohren
drill pipe Bohrstange * für Erdölbohrung
drill stand Bohrständer
drill string Bohrgestänge * z.B. für Erdöl-/Erdgasbohrung
drill-gun Handbohrmaschine
drill-hole Bohrloch * Bohrung ins Volle, Vollbohrung; auch zur Erdölgewinnung || Bohrung * Bohrloch
drilling Bohrung
drilling drive Bohrantrieb
drilling jig Bohrfutter
drilling machine Bohrmaschine
drilling platform Bohrplattform
drilling template Bohrschablone
drillings Span * Bohrspäne || Späne * Bohrspäne
drip water Spritzwasser * Tropfwasser || Tropfwasser
drip-free tropffrei
drip-proof DP * tropfwassergeschützt (IP22) || Spritzwasser-geschützt * Tropfwasser-geschützt || spritzwassergeschützt * tropfwassergeschützt || tropfwassergeschützt * Schutzarten IP21 und IP22
drip-water protection Tropfwasserschutz
dripping water Tropfwasser
drive Aktion * Vorstoß, (Sammel-) Aktion, Kampagne, || ansteuern || antreiben * Maschine || Antrieb * Maschine, Motor-Stromrichterkombination; im amerikan. häufig nur d.Stromrichter gemeint || Antriebsstromrichter * (amerikan.) || fahren * selbstlenkend ~ || Fahrt * mit Wagen || klopfen * Nagel in die Wand usw. || Laufwerk * {Computer} z.B. Disketten-/Festplatten-/Band~ || schlagen * treiben (into: hinein), z.B. Nagel, Dorn || Schwung * Wucht || stoßen * treiben || Stromrichter * (amerikan.) Antriebsstromrichter; aber: im engl. bedeutet "drive" Antrieb || treiben * mechan. u. elektrisch (Leistungsverstärker usw.) || zielen * at:- auf
drive against auffahren * auf etwas ~, rammen
drive at set speed Drehzahl erreicht
drive at speed Drehzahl erreicht
drive away vertreiben * hinwegtreiben
drive bay Laufwerkschacht * in Computer z.B. für Festplatten-/Disketten-/CD-ROM-Laufwerk
drive concept Antriebkonzept || Antriebskonzept
drive configuration Antriebskonzept
drive connected to the bus Antrieb am Bus
drive control Antriebsregelung || Antriebssteuerung
drive control level Antriebssteuerungsebene
drive control system Antriebsregelungssystem
drive controller Antriebsstromrichter
drive converter Antriebsstromrichter || Stromrichter * Antriebsstromrichter
drive converter interface Stromrichteranschaltung * zu einem Antriebsstromrichter
drive converter unit Stromrichtergerät * Antriebsstromrichter
drive data Antriebsdaten
drive drum Antriebstrommel
drive element Antriebselement
drive end A-Seite * Antriebs-Seite eines Motors, an der die Arbeitsmaschine angekuppelt wird || Antriebsseite || AS * Antriebs-Seite eines Motors

drive engineering Antriebsprojektierung || Antriebstechnik
drive equipment Antriebsausrüstung
drive group Antriebsgruppe || Antriebsstaffel
drive healthy Antrieb Betriebsbereit * Steuersignal
drive in einfahren || eintreiben
drive into place einfahren
drive into saturation übersteuern * in die Sättigung hineinsteuern/-treiben
drive line Antriebsstrang
drive location Antriebsstelle
drive mechanics Antriebsmechanik
drive motor Antriebsmaschine || Antriebsmotor
drive noise Antriebsgeräusche
drive package Antriebspaket * z.B. Stromrichter und Motor
drive past vorbeifahren || vorbeilaufen
drive pin Mitnehmerbolzen
drive pinion Antriebsritzel
drive position Antriebsstelle
drive power Antriebsleistung
drive pulse Ansteuerimpuls * Treiberimpuls
drive ready Antrieb Betriebsbereit
drive ready-to-run Antrieb Betriebsbereit
drive regulation Antriebsregelung
drive response Gerätereaktion * Fernsteuermöglichktn. ü.d.serielle Automatis.-Schnittstelle e. Antriebsstromrichters
drive rod Schubstange
drive roll Treibwalze
drive roller Transportrolle
drive section Antriebsgruppe
drive sequence Antriebsstaffel
drive shaft Antriebswelle
drive shaft end AS-Wellenende * {el.Masch.} antriebsseitiges Wellenende (z.B. eines El.motors)
drive side A-Seite * Antriebs-Seite eines Motors || Antriebsseite * z.B. eines Motors || AS * Antriebs-Seite eines Motors
drive solution Antriebslösung
drive speed Antriebsdrehzahl * z.B. Drehzahl n2 einer Getriebeausgangswelle
drive spindle Antriebsspindel
drive station Antriebsstelle
drive system Antriebsausrüstung || Antriebssystem || Antriebsverbund
drive systems Antriebstechnik * Spektrum antriebstechnischer Produkte
drive task Antriebsaufgabe
drive technology Antriebstechnik
drive torque Antriebsdrehmoment || Antriebsmoment
drive train Antriebsstrang
drive unit Antriebseinheit
drive wheel Antriebsrad * z.B. eines Fahrwerks
drive-end antriebsseitig
drive-out Abtrieb
drive-ramp Hochlaufgeber-Rampe * für Drehzahlsollwert
drive-related antriebsnah * in engem Zusammenhang mit dem Antrieb stehend (z.B. Steuerung, Regelung)
drive-related control Antriebssteuerung * antriebsnahe Steuerung
drive-related open-loop control antriebsnahe Steuerung
drive-specific antriebsnah
drive-specific control antriebsnahe Steuerung
drive-specific open-loop control antriebsnahe Steuerung
drive-through mechanism Durchtrieb * sealed: abgedichteter; z.B. z.Antreiben e.in Vakuum befindl. Antriebsstelle
driven angetrieben || getrieben

driven end of shaft Abtrieb
driven machine Arbeitsmaschine * *angetriebene Maschine*
driven side Antriebsseite * *angetriebene Seite* || Lastseite * *angetriebene Seite, Abtriebsseite*
driver Ansteuerung || Klaue * *Mitnehmer, Nase* || Mitnehmer || Treiber * *z.B. Leitungs~, Software~, Device~ für Computer-Peripherie* || Verstärker * *z.B. Leitungstreiber*
driver circuit Treiberstufe
driver disc Mitnehmer * *~scheibe*
driver module Ansteuerbaugruppe * *Verstärkerbaugruppe*
driver pin Mitnehmer * *~stift, ~bolzen*
driver plate Mitnehmerscheibe
driver rod Stößel * *siehe auch →Schubstange*
driver software Treibersoftware
driver software module Hantierungsbaustein * *Treibersoftware-Baustein*
driverless fahrerlos * *z.B. Transportfahrzeug, Kran*
driverless vehicle führerloses Fahrzeug
drives Antriebstechnik
drives and controls Antriebs- und Regelungstechnik
driving Ansteuerung || treiben * *(Substantiv)* || treibend
driving belt Treibriemen
driving capacity Treiberfähigkeit * *z.B. bei Leistungstreiber/Digitalausgang usw.*
driving circuit Ansteuerung
driving collar Nabe
driving component Antriebselement
driving current Steuerstrom
driving direction Fahrtrichtung * *{Verkehr} facing the front/the engine: in ~*
driving drum Antriebstrommel
driving element Mitnehmer * *antreibendes Teil*
driving end Antriebsseite
driving flange Mitnehmerflansch * *an Getriebeausgang usw.*
driving force Antriebskraft || treibende Kraft
driving maschine Antriebsmaschine
driving mechanism Antriebsmechanik * *Antriebsmechanismus* || Koppeltrieb * *z.B. bei Querschneider*
driving motor Antriebsmaschine
driving pin Mitnehmerbolzen
driving pinion Antriebsritzel
driving plate Mitnehmerscheibe
driving power Ansteuerleistung || Antriebsleistung || Steuerleistung
driving pulley Antriebsscheibe || Treibscheibe * *Riemen-/Seilscheibe (auch bei Aufzügen)*
driving roll Treibwalze
driving shaft Antriebswelle
driving sheave Antriebsscheibe
driving side Antriebsseite
driving sleeve Treibhülse * *z.B. zum Aufdrücken eines Lagers auf eine Welle*
driving slot Mitnehmernut
driving speed Fahrgeschwindigkeit * *Fahrzeug*
driving torque Antriebsdrehmoment || Antriebsmoment
driving-point Eingangs-
droop Abfall * *Herabhängen; Statik eines Reglers* || Statik * *Weichmachen e.Reglers durch P-Rückführung d.Reglerausgangssignals* || Statik-Aufschaltung * *Weichmachen e. Regelung durch P-Rückführung d. Reglerausgangssignals* || Statikaufschaltung * *Weichmachen einer Regelung durch P-Rückführung d. Reglerausgangssignals*
droop function Statik
droopless driftfrei
drop Abfall * *z.B. Spannungsabfall an einem Widerstand, Druckabfall, Abfall e.Relaiskontakts* || abfallen * *Relais, Schütz, Druck, Spannung an einem Widerstand usw. (by:um)* || abgehen * *von einem Thema usw.* || absinken * *sinken, (herunter/ab)fallen, z.B. Spannung; schwächer werden* || Absturz * *Fallen, Sinken, Sturz, Rückgang; drop in prices: Preisrückgang* || abstürzen * *z.B. Last vom Kran* || abwerfen * *z.B. Behälter* ~ || einbrechen * *z.B. Spannung* || Fall * *Sturz; auch von Preisen* || fallen * *abfallen; auch Preise usw.; auch ~ lassen (Bemerkung, Gegenstand, Plan, Ansprüche)* || fallen lassen || nachlassen * *~ von Preisen usw.* || senken * *sich ~, abfallen* || sinken * *Preise, Kurse, Messwerte* || Sturz * *Wetter~, Temperatur~* || Tropfen * *einer Flüssigkeit; drop-shaped: ~förmig* || Verfall * *z.B. von Preisen* || verwerfen * *fallen lassen* || weglassen * *auslassen, fallen lassen, aufgeben, nicht bearbeiten* || zurückgehen * *abfallen (z.B. Spannung, Leistung)*
drop down absenken * *(figürl.) wert-/zahlenmäßig* ~
drop in market prices Marktpreisverfall
drop in pressure Druckabfall
drop in speed Drehzahleinbruch
drop out abfallen * *Relais, Schütz* || aussteigen
drop suddenly absacken
drop-out Abfall * *eines Relais-/Schützkontakts*
drop-out delayed abfallverzögert
drop-out time Abfallverzögerungszeit || Abfallzeit * *Schütz, Relais*
dropout Abfall * *eines Relais-/Schützkontakts*
dropout delay Abfallverzögerung
dropout delayed contactor abfallverzögertes Schütz
dropout time Abfallverzögerungszeit || Abfallzeit
dropped out abgefallen * *Relais, Schütz*
dropping Abwurf
drug store Apotheke
drum Spule * *Trommel* || Tragwalze * *{Papier} z.B. bei →Doppeltragwalzenroller, →Poperoller* || Trommel || Walze * *Trommel*
drum brake Backenbremse * *Trommelbremse*
drum chipper Trommelhacker * *{Holz} zur Herstellung von Holzschnitzeln*
drum controller Ablaufkette
drum hog Trommelhacker * *{Holz} zur Erzeugung von Holzschnitzeln*
drum lining Seilfutter * *{z.B. Aufzüge}* || Trommelbelag * *z.B. bei Antriebstrommel eines Förderbandes*
drum reel-up Poperoller
drum sequencer Ablaufkette
drum sequencer step Schrittelement * *in einer Ablaufkette bei SPS*
drum way correction Seileinlaufkorrektur * *{z.B. Aufzüge}*
drum weel-up Tragwalzenroller
drum winder Tragwalzenroller
drum-integrated motor Trommelmotor * *in Antriebstrommel integriert (z.B. für Förderband)*
drum-type fan Trommellüfter || Walzenlüfter * *Trommellüfter*
drum-type oven Trommelofen
dry trocken || trocknen
dry content Trockengehalt
dry end trockenes Ende * *einer Papiermaschine*
dry heat trockene Wärme
dry ice Kohlensäure * *feste ~, Trockeneis*
dry running trockenlaufend * *z.B. Kupplung*
dry up trocknen
dry-fluid coupling Fliehkraftkupplung
dry-running clutch Trockenkupplung
dry-type air filter Trocken-Luftfilter
dryer Trockner
dryer cylinder Trockenzylinder

dryer section Trockengruppe * *bei Papiermaschine* || Trockenpartie * *bei Papiermaschine*
drying Trocknung
drying agent Trockenmittel
drying cylinder Trockenzylinder
drying drum Trockentrommel || Trockenzylinder * *z.b. in der Papierindustrie* || Trocknungstrommel
drying equipment Trocknungseinrichtung
drying machine Trockner
drying oven Trockenofen
drying plant Trocknungsanlage
drying roll Trockenzylinder
drying section Trockenpartie * *in Papiermaschine*
drying system Trocknungsanlage
DSO Digitalspeicher-Oszilloskop * *Abkürzg. für 'Digital-Storage Oscilloscope'* || Oszillograf * *Abkürzung für 'Digital Storage Oscilloscope': Digitalspeicher-Oszilloskop* || Oszillograph * *Abkürzung für 'Digital Storage Oscilloscope':Digitalspeicher-Oszilloskop* || Oszilloskop * *Abkürzg. für 'Digital Storage Oscilloscope':Digitalspeicher-Oszilloskop*
DSP DSP * *Digitaler SignalProzessor* || Signalprozessor * *Digital SignalProzessor,mit großen Wortlängen und umfangreichen Arithmetikfunktionen*
DSP processor Signalprozessor
DSS1 DSS1 * *Abk.f. 'Digital Subscriber System No. 1': Europäisches Übertragungsprotokoll für →ISDN*
DT1-block DT1-Glied * *Funktionsbaustein*
DT1-function DT1-Glied
DTC Umschaltkontakt * *Abkürzg. f. 'Double-Throw Contact'; 4PDTC:4-poliger Umschaltkontakt* || Wechselkontakt * *Abkürzg. f. 'Double-Throw Contact'; 4PDTW:4-poliger Wechselkontakt* || Wechsler * *Double-throw contact; 4PCTC: 4-poliger Wechslerkontakt* || Wechselkontakt * *Abk.f. 'Double-Throw Contact'; 4PDTW:4-poliger Wechselkontakt*
DTP DTP * *{Computer} Abk. f. 'DeskTop Publishing': Layout von Druckerzeugnissen per Computer*
dual doppel || Doppel-
dual port RAM interface board Dual-Port-RAM-Anschaltung * *Leiterkarte, Flachbaugruppe*
dual take-up Doppelspuler * *Doppelaufwickler (meist mit fliegendem Spulenwechsel) z.B. in Kabelmaschine*
dual-lateral Waben- * *Spule* || wabenförmig * *Spule*
dual-port RAM interface Dual-Port-RAM-Anschaltung
dual-way converter Umkehrstromrichter
duct führen * *durch Kanal/Röhre* ~ || Führung * *Kanal, Weg, Gang, Röhre; z.B. für Kabel, Kühlluft* || Gang * *Röhre* || Kanal * *(Zu-/Ab-) Führungskanal, z.B. für Kabel, Luft; ventilation ductwork: Lüftungskanäle* || Leitung * *für (Kühl)Luft* || Rohr * *Röhre, Leitung, (Kabel/Kühlluft-) Kanal; duct away: durch Röhre abführen* || Röhre * *(Kabel/Leitungs/Luft)Kanal* || Rohrleitung * *Röhre, (Rohr-) Leitung, (Kabel-) Kanal* || Stutzen * *inlet duct: Einlass~; discharge duct: Auslass~* || Tunnel * *im Maschinenbau*
duct adapter Anschlussstutzen * *für Luft* || Rohrstutzen * *für Kühlluft-Rohr usw.*
duct connection Rohranschluss * ~ *für Kühlluft*
duct radius Krümmungsradius * *eines Rohrs*
duct tube Rohranschluss * *Anschlussrohr (für Ventilator)*
duct-fan motor Schachtlüftermotor * *{el.Masch.}*
duct-ventilated fremdbelüftet * *mit Kühlluftzufuhr/-abfuhr über Rohranschluss* || fremdbelüftet über Rohranschluss * *z.B. Motorkühlarten IC05/ 06/17/37 nach IEC34 Teil 6*

duct-ventilation Fremdbelüftung über Rohranschluss || Fremdkühlung * *Motorkühlart mit Luftkühlung über Rohranschluss*
ductile zäh * *Metall*
ductile iron Sphäroguss
ductility Zähigkeit * ~ *bei Metall*
ducting Rohrleitung * *Rohrleitungen/Rohrleitungsnetz für Luft*
due angemessen * *gebührend, angemessen, fällig, passend, vorschriftsmäßig, recht(zeitig), erwartet* || genau * *(Adv.) genau, gerade; due west: ~ westlich*
due to -bedingt || auf Grund * *verursacht durch; due to the fact that: ~ der Tatsache, dass* || auf Grund von || durch * *auf Grund* || infolge * *auf Grund* || wegen * *auf Grund, verursacht durch, ... zuzuschreiben/zurückzuführen auf* || zurückzuführen auf
due to the conditions in the particular trade - branchenbedingt
due to the system systembedingt
duely ordnungsgemäß * *(Adv.)*
dull trüb * *(auch figürl.: glanzlos, unklar)*
dummy Attrappe || Blind- * *z.B. dummy plug: Blindstopfen* || Modell * *Attrappe* || Nachbildung * *Atrappe* || Platzhalter
dummy parameter formaler Parameter || Formalparameter
dummy plug Blindstecker || Blindstopfen || Verschlussstopfen * *Blindstopfen, z.B. bei Kabeleinführung*
dump entladen * *Schüttgut* ~ || Halde * *Schutt-/ Müllhaufen, (Müll-) Abladeplatz, Stapel-/Lagerplatz (für Schüttgut)* || Lagerplatz * *Stapelplatz* || schleudern * *Exportgüter zu billigen Preisen ver~* || unterbieten * *auf dem Weltmarkt* ~
dunning letter Mahnung * *Mahnschreiben*
duo- Duo-
dupe Duplikat * *(salopp)*
duplex Doppel- * *zweifach* || zweiseitig * *z.B. in d. Druckerei*
duplex bearing Doppellager
duplicate doppelt * *(Adj.)* || Duplikat || duplizieren || nachbilden * *duplizieren, identisch nachbauen/-bilden* || vervielfachen * *duplizieren*
duplicating Vervielfältigung
duplicating milling machine Kopierfräsmaschine
duplication Vervielfältigung
durability Haltbarkeit || Langlebigkeit * *Dauerhaftigkeit* || Lebensdauer * *Langlebigkeit, Dauerhaftigkeit, Haltbarkeit* || Qualität * *Haltbarkeit, Langlebigkeit* || Robustheit * →*Haltbarkeit, Dauerhaftigkeit, Beständigkeit, Lebensdauer*
durable dauerhaft * *dauerhaft, haltbar, beständig* || fest * *dauerhaft; auch (Geschäfts)Beziehung* || langlebig * *dauerhaft, haltbar, beständig*
duration Dauer || Länge * *zeitlich* || Laufzeit * *Dauer* || Zeit * ~*dauer* || Zeitdauer || Zeitspanne * *Dauer*
duration of a trip Fahrzeit * *Fahrtdauer*
duration of acceleration Beschleunigungszeit
duration of operation Betriebsdauer
duration of pause Pausenzeit * *Länge der* ~
duration of the traversing process Verfahrzeit
during bei * *(zeitlich) während* || über * *während* || unter * *während* || während * *(zeitlich): during the configuration phase: ~ der Projektierung*
during daylight hours tagsüber
during decades jahrzehntelang
during normal operation während des Betriebes * *während des normalen Betriebes*
during operation während des Betriebes
dust Abrieb * *(abgeriebener) Staub, carbon dust: Kohlebürstenabrieb* || Staub
dust accumulation Staubablagerung

dust aspirator Staubabsauggerät
dust collector Staubabsauggerät * *Staubabscheider*
dust deposit Staubablagerung
dust exhaust plant Staubabsaugungsanlage
dust exhauster Entstauber * *Staub-Absauger* || Staubabsauggerät
dust extraction system Entstaubungseinrichtung || Staubabsauganlage || Staubabsaugungsanlage
dust extractor Entstauber || Entstaubungsgerät * *z.B. bei Holzbearbeitungsmaschine* || Staubabsauggerät
dust filter Staubfilter
dust particle Staubkorn
dust protection Staubschutz
dust removal Entstaubung
dust removal system Entstaubungsanlage
dust remover Entstauber
dust removing Entstaubung * *siehe auch →Staubabsauggerät*
dust separation system Entstaubungsanlage
dust-free staubfrei
dust-ignition Staubentzündung
dust-proof staubdicht
dust-tight staubdicht
dust-tight design staubdichte Ausführung * *z.B. in Schutzart IP65 (EN 60529)*
dusty staubig
dutiable zollpflichtig
duty Abgabe * *z.B.Steuer,Zoll* || Aufgabe * *auch Pflicht* || Beanspruchung * *das zu Leistende, Nutzleistung* || Belastung * *das zu leistende, Nutzleistung* || Betriebsart * *Kurzform für duty class* || Nutzleistung * *Nutz-/Wirkleistg., das zu Leistende, Auslastg., die abgebbare/ausnutzbare Leistg.* || Pflicht || Steuer * *Abgabe, Gebühr, Zoll* || Tarif || Verpflichtung * *Pflicht* || Verwendung * *(Be)Nutzung, Aufgabe, Amt, Dienst* || Zoll * *Gebühr, Abgabe, Steuer*
duty class Belastungsklasse * *Betriebsart für elektr. Maschinen (EN 60530) u. Maschinen (EN 60141, IEC 146-1-1)* || Betriebsart * *genormtes Lastspiel elektr.Maschinen lt.VDE 0530 (z.B.Kurzzeit/Dauerbetrieb)* || Betriebsklasse * *siehe →Betriebsart (nach VDE0530/IEC 34-1)*
duty cycle Betriebsart * *Lastspiel, Art des Lastspiels* || ED * *Abk. für →Einschaltdauer; relative Einschaltdauer* || Einschaltdauer * *relative ~ (→ED) in % des gesamten Lastzyklus* || Lastspiel || Tastverhältnis
duty cycle time Spieldauer * *eines Lastspiels*
duty factor Einschaltdauer * *relative ~* || Schaltungsfaktor || Tastverhältnis
duty ratio Einschaltdauer * *relative ~* || relative Einschaltdauer
duty sharing Lastaufteilung * *z.B. bei Kompressorsätzen*
duty type Betriebsart * *genormtes Lastspiel el. Maschinen n.VDE0530/IEC34-1 u. v. Stromrichtern nach DIN41756*
duty type S1 Betriebsart S1 * *{el.Masch.} Dauerbetrieb*
duty type S10 Betriebsart S10 * *{el.Masch.} Betrieb mit einzelnen konstanten Belastungen*
duty type S12 Betriebsart B12 * *{el.Masch.} wie →Betriebsart S3,jedoch Motor während Stillstandszeit unter Spannung*
duty type S2 Betriebsart S2 * *{el.Masch.} Kurzzeitbetrieb*
duty type S3 Betribart S3 * *{el.Masch.} periodischer Aussetzbetrieb*
duty type S4 Betriebsart S4 * *{el.Masch.} periodischer Aussetzbetrieb mit Einfluss des Anlaufvorgangs*
duty type S5 Betriebsart S5 * *{el.Masch.} Periodischer Aussetzbetrieb mit elektr.Bremsung*
duty type S6 Betriebsart S6 * *{el.Masch.} Ununterbrochener periodischer Betrieb*
duty type S7 Betriebsart S7 * *{el.Masch.} Ununterbrochener periodischer Betrieb mit elektrischer Bremsung*
duty type S8 Betriebsart S8 * *{el.Masch.} ununterbrochener periodischer Betrieb mit Last-/Drehzahländerungen*
duty type S9 Betriebsart S9 * *{el.Masch.} Betrieb mit nichtperiodischen Last- und Drehzahländerungen*
duty with discrete constant loads Betriebsart S10 * *{el.Masch.}*
duty with non-periodic load and speed variations Betriebsart S9 * *{el.Masch.}*
duty-free zollfrei
dv/dt dv-dt
dV/dt filter dV-dt-Filter * *am Umrichterausgang z. Schaltflanken-Verschliff (zur EMV u.Motorschonung)*
dv/dt-filter du-dt-Filter * *zur Spanngs.anstiegsbegrenzg. z.b. hinter Umrichter,schützt Motor vor Spannungspitzen*
dwarf verkleinern * *verkümmern lassen, verkleinern, klein erscheinen lassen*
dwell verweilen * *on:bei*
dwell time Tauchdauer * *Verweildauer im Tauchbad* || Verweilzeit
dwindle schwinden * *abnehmen*
dwindling Schwund
DWORD Doppelwort * *Abkürzung für 'double word'*
dye plant Färberei
dyeing Färberei
dyeing machine Färbereimaschine * *für Textilien*
dynamic dynamisch
dynamic behavior dynamisches Verhalten
dynamic behaviour Dynamik * *dynam. Verhalten* || dynamisches Verhalten || Übertragungsverhalten * *siehe DIN 19229*
dynamic braking dynamische Bremsung || Widerstandsbremsung * *Gleich- und Drehstrommotor*
dynamic characteristics dynamisches Verhalten
dynamic control deviation dynamische Regelabweichung
dynamic data exchange dynamischer Datenaustausch * *feste Verknüpfung e."Wirtsdatei" mit Elementen anderer Dateien*
dynamic losses dynamische Verluste
dynamic lowering Senkbremsung * *el.Bremsg. zur Geschw.begrenzg. b.Hebezeug über Vorwid. bei Schleifringläufermotor*
dynamic lowering circuit Senkbremsschaltung * *Gegenstrombremsung bei Kran mit Schleifringläufermotor*
dynamic overload capability dynamische Überlastbarkeit
dynamic performance Dynamik * *Leistungsfähigkeit bezügl. der (Regel-) Dynamik* || Regeldynamik || Regelverhalten * *Regeldynamik*
dynamic performance characteristic dynamische Eigenschaften
dynamic response Dynamik || Regeldynamik || Übertragungsverhalten
dynamic response characteristic dynamische Eigenschaften
dynamic speed-torque characteristic dynamische Momentenkennlinie * *{el.Masch.} M(n)-Kennl. e. AC-Motors b. schneller Drehz.änderg.*
dynamical loading dynamische Beanspruchung
dynamically in Betrieb * *dynamisch, (Änderung) während des Betriebes*
dynamics Dynamik
dynamite Konfliktstoff

dynamometer Belastungsmaschine * *mit Drehmomenten-Messeinrichtung (für Prüfstände usw.)* || Drehmoment-Messeinrichtung * *Pendelmaschine, Momentenwaage* || Drehmomentenwaage || Drehmomentmessgerät || Dynamometer * *Drehmomentvorgabe- und Messeinrichtung, z.B. Pendelmaschine, z.B. für Prüfstand* || Kraftmesseinrichtung || Kraftmessgerät || Momentenwaage * *z.B. Pendelmaschine*
dynamometer test stand Leistungsprüfstand

E

E beam Elektronenstrahl
E EX I-protected schlagwettergeschützt * *Schutzart nach EN0514*
e-function e-Funktion
E-stop aus * *Abk.f. 'Emergency stop':Nothalt (entspr. meist 'coast stop'; siehe dort)* || AUS2 * *Abk. für 'emergency stop' (Not Aus)* || Not Aus * *Anlage wird schnellstmöglichst spannungslos gemacht (u.trudelt aus)* || Not-Aus || Schnellhalt * *Abkürzg.f. 'Emergency stop': Not-Aus/-Halt (Elektrisch Aus)*
E.F.T.A. EFTA * *European Free Trade Association*
e.g. zB * *example given:zum Beispiel* || zum Beispiel * *Abkürzung f. 'example given'*
each einzeln * *each one: jeder ~e (aus einer Gruppe)* || je * *jedes* || jedes * *- Einzelne; each thing: ~ Ding*
each other gegenseitig * *einander; praise each other: sich ~ loben; neutralize each other: sich ~ aufheben*
each time jedes Mal
eager eifrig
eager to help hilfsbereit
earlier früher * *zeitiger*
early contact Vorkontakt
early dinner Mittagessen
early failure Frühausfall
early info Vorabinformation
early information Vorabinformation
early settlement schnelle Erledigung
early stage Anfang * *früher Zeitpunkt (eines Projekts usw.)* || Beginn * *früher Zeitpunkt (eines Projekts usw.)*
early warning Frühwarnung || Vorwarnung
early warning signal Vorwarnsignal
earmark vorsehen * *for: für einen Zweck ~*
earmarked vorgesehen * *for: für einen Zweck ~*
earn erwerben * *durch Arbeit, auch Achtung usw.* ~ || verdienen
earnings Einnahme * *(Plural) Gewinn, Ergebnis, Verdienst* || Ergebnis * *Geschäftsergebnis, Gewinn, Profit* || Ertrag * *Gewinn, (Geschäfts-) Ergebnis, Profit* || Gewinn * *Profit,Geschäftsergebnis* || Profit * *Geschäftsergebnis, Gewinn*
earth Erde * *auch elektr.; protective earth: Schutz~* || erden || Erdungsleitung || Masse * *Erde*
earth connection Erdung * *Erdverbindung* || Erdverbindung || Massung || Schutzleiter
earth connection screw Erdungsschraube
earth contact Schutzkontakt
earth current Erdstrom
earth fault Erdschluss
earth fault protection Erdschlussschutz
earth leakage Fehlerstrom * *~ gegen Erde*
earth leakage breaker Fehlerstromschalter * *siehe auch →FI-Schutzschalter*

earth leakage circuit-breaker FI Schutzschalter * *Sicherheitseinrichtg., schaltet bei Erdfehlerstrom den Stromkreis ab* || Fi-Schutzschalter * *Sicherheitseinrichtg., schaltet bei Erdfehlerstrom ab*
earth leakage current Erdstrom * *Fehlerstrom* || Fehlerstrom * *~ gegen Erde*
earth leakage relay Erdschlusswächter
earth terminal Erdanschluss || Erdungsanschluss || Erdungsanschlusspunkt
earth's pollution Umweltverschmutzung
earth-fault monitoring Erdschlussüberwachung
earth-fault resistance Erdschlussfestigkeit
earth-fault resistant erdschlussfest
earth-fault-proof erdschlussfest
earth-free erdfrei
earth-free network erdfreies Netz
earth-leakage circuit breaker Schutzschalter * *Fehlerstromschutzschalter, FI-Schutzschalter*
earth-leakage monitor Erdschlusswächter
earthed-neutral supply system geerdetes Netz
earthing Erdung || Massung
earthing accessories Erdungsgarnitur
earthing brush Erdungsbürste * *z.B. zum Erden eines Motorläufers*
earthing bus Sammelerder
earthing conductor Erdungsleitung
earthing connection Erdverbindung
earthing contact Schutzkontakt
earthing terminal Erdklemme || Erdungsanschluss || Erdungsklemme
earthing via neutral conductor Nullung * *MP dient gleichzeitig als Schutzleiter*
earthquake-proof erdbebensicher
ease Einfachheit * *Leichtigkeit, Lockerheit, Bequemlichkeit* || entspannen * *sich ~ (Lage usw.)* || erleichtern * *vereinfachen* || Komfort * *Leichtigkeit, Einfachheit, Bequemlichkeit, Sorglosigkeit, ease of operation: Bedien~* || lockern * *Passung, Sitz* || vereinfachen * *leichter machen, erleichtern, entlasten, befreien*
ease of configuration Projektierbarkeit * *Einfachheit/Komfort der ~*
ease of familiarization leichte Erlernbarkeit * *z.B. Bedienung einer Maschine/eines Geräts/eines Softwareprogramms*
ease of maintenance Wartungsfreiheit * *Wartungsfreundlichkeit* || Wartungsfreundlichkeit
ease of motion Leichtgängigkeit
ease of operation Anwenderfreundlichkeit * *Bedienkomfort * einfache Bedienung, leichte Bedienungsweise* || Bedienfreundlichkeit || Bedienungskomfort || bequeme Bedienung || einfache Bedienbarkeit * *bequeme Bedienung; siehe auch 'Bedienungsfreundlichkeit'* || Handhabbarkeit * *leichte Bedienung, Bedienungsfreundlichkeit*
ease of service Servicefreundlichkeit || Wartungsfreundlichkeit
ease of serviceability Wartungsfreundlichkeit
ease of use Anwenderfreundlichkeit || Benutzerfreundlichkeit
easily mühelos * *(Adv.)*
easily assimilated leicht erlernbar
easily going leichtgängig
easily moving leichtgängig
easily remembered einprägsam
easily-understood leicht verständlich
eastern Europe Osteuropa
easy einfach * *leicht* || geläufig * *fließend* || leicht * *nicht schwierig, einfach* || mühelos || problemlos * *leicht,einfach* || zügig * *reibungslos*
easy movement Leichtgängigkeit
easy operation einfache Bedienung * *siehe auch 'Bedienungsfreundlichkeit'*
easy serviceability Wartungsfreundlichkeit

easy to grasp überschaubar * *leicht fassbar; z.B. Problem*
easy to handle aufwandsarm * *leicht hantierbar*
easy to install einfach inbetriebzunehmen
easy to maintain anspruchslos * *wartungsfreundlich* || servicefreundlich * *wartungsfreundlich* || wartungsfreundlich
easy to read gut lesbar
easy to remember einprägsam
easy to service servicefreundlich || wartungsfreundlich
easy to survey übersichtlich * *leicht überschaubar*
easy to use bedienungsfreundlich
easy-to use praktisch * *leicht bedienbar/anwendbar (z.B. ein Gerät)*
easy-to-apply anwenderfreundlich * *leicht anwendbar*
easy-to-handle komfortabel * *leicht hantierbar/bedienbar*
easy-to-learn leicht erlernbar
easy-to-operate anwenderfreundlich * *einfach zu bedienen* || leicht bedienbar
easy-to-remove abnehmbar * *leicht ~/demontierbar* || demontierbar * *leicht ~; siehe auch* →abnehmbar
easy-to-service servicefreundlich
easy-to-understand leicht verständlich || verständlich * *leicht verständlich*
easy-to-use anwenderfreundlich * *anwendungsfreundlich* || bedienungsfreundlich || komfortabel * *leicht bedienbar* || leicht bedienbar
eat away zerfressen * *zerfressen, nagen an, ein Loch fressen*
eat one's words zurücknehmen * *(salopp) Vorwürfe, Kritik ~*
ebb abklingen * *(figürlich) verebben, versiegen, abnehmen, (dahin-) schwinden* || aufhören * *allmählich verebben*
EC EG * *Abkürzg.f. 'European Community':Europäische Gemeinschaft*
EC Declaration of Conformity EG-Konformitätserklärung * *Herstellererklärung der Übereinstimmung mit EG-Normen/Richtlinien* || Konformitätserklärung * *bezogen auf EG-Richtlinien*
EC Directive EG-Richtlinie
EC Low-Voltage Directive EG-Niederspannungsrichtlinie
ECC Guideline EG-Richtlinie * *siehe* →EG
eccentric ausgefallen * *exzentrisch* || exzentrisch || unrund
eccentric cam Exzenter * *auf Welle*
eccentric disk Exzenterscheibe
eccentric element Exzenter
eccentric gear reducer Exzentergetriebe * *Untersetzungs-~*
eccentric movement Exzenter
eccentric press Exzenterpresse * *mit Schwungrad zur Verringerung der Laststöße*
eccentric shaft Exzenterwelle
eccentricity Exzentrizität * *z.B. von Motorwelle u. Anbauflansch (DIN 42995)* || Rundlaufabweichung || Rundlauffehler
Eco Audit Öko-Audit * *Bestandteil des Umwelt-Managementsystems nach EN ISO 14.001*
eco- Öko- || Umwelt- * *Öko-*
eco-friendly umweltfreundlich || umweltverträglich * *umweltfreundlich*
ecological ökologisch
ecology Ökologie
econimically feasible wirtschaftlich durchführbar
econimize ausnutzen * *sparsam umgehen mit, (so gut wie möglich) ausnützen*
economic kostengünstig || wirtschaftlich || Wirtschafts-
economic characteristics Wirtschaftlichkeit

economic efficiency Wirtschaftlichkeit
economic growth Wirtschaftswachstum
economic operation Wirtschaftlichkeit * *Betrieb mit geringen Betriebskosten*
economic quality Wirtschaftlichkeit
economic system Wirtschaft * *~ssystem eines Gemeinwesens*
economic viability Wirtschaftlichkeit
economical aufwandsarm * →*kostengünstig* || kostengünstig * *wirtschaftlich, sparsam* || rationell * *sparsam* || sparsam * *of:mit* || wirtschaftlich * *wirtschaftlich, haushälterisch, sparsam*
economically wirtschaftlich * *(Adv.)*
economically viable wirtschaftlich * *~ machbar/ realisierbar*
economics Volkswirtschaft * *Volkswirtschaftslehre* || Volkswirtschaftslehre || Wirtschaft * *Welt~, ~swissenschaft, Volks~, Volks~slehre*
economize einsparen * *on: etwas* || sparen * *sparsam umgehen mit*
economizing Einsparung || sparsam
economy anspruchslos * *preisgünstig* || Niedrigpreis- || preisgünstig || Wirtschaft || Wirtschaftlichkeit
economy circuit Sparschaltung
economy in operation Wirtschaftlichkeit * *z.B. einer Maschine/Anlage*
economy of operation Wirtschaftlichkeit * *eines Geräts/einer Maschine z.B. im Unterhalt auf Grund von Wartungsfreiheit*
economy-priced kostengünstig * *preisgünstig* || preisgünstig || preiswert
eddy current cage rotor Stromverdrängungsläufer * *{el. Masch.}*
eddy-current Wirbelstrom * *wirbelförm.Wechselstrom aufgr. d.Induktion v.Wechselfeldern in massiven metall.Körpern*
eddy-current brake Wirbelstrombremse * *el.-Masch. zur Erzeugung e. reibungs- u. verschleißfreien Bremswirkung*
eddy-current clutch Wirbelstromkupplung * *el.-steuerbare Kupplg. mit berührungslos. Kraftübertragg. nach d. Indukt.prinzip*
eddy-current losses Wirbelstromverluste
edge besäumen * *{Holz} Kanten von Brettern/Platten gerade- bzw. glattschneiden* || bördeln * *durch Schmieden* || Ecke * *Kante* || Falz || Flanke * *eines Impulses* || Kante * *lower: Unter~; upper: Ober~-; sharp: scharfe ~* || Rand * *Kante, Ecke, Saum, Grenze; at the lower edge: am unteren ~* || Randstreifen || schärfen * *Kanten* || Vorsprung * *(salopp) Vorteil, Vorsprung; market edge: Markt~; (on: genüber)* || Vorteil * *Vorsprung, Vorteil; have the edge on s.o.: einen ~ gegenüber jdm. haben*
edge band Umleimer * *{Holz}*
edge change Flankenwechsel
edge clearance Flankenabstand * *Zwischenraum zwischen den (Impuls-) Flanken*
edge cutter Randspritze * *schneidet in Papiermaschine die Papierrandstreifen mit Wasserstrahl ab* || Randstreifenabschneider * *z.B. bei Papier(verarbeitungs)maschine* || Randstreifenschneider
edge detection Flankenauswertung * *Erkennung von Impulsflanken, z.B. one. Impulsgeber*
edge displacement Flankenabstand * *von (Impuls-) Signalen*
edge evaluation Flankenauswertung * *z.B. von Impulsgebersignalen*
edge guide Kantenführung * *Einrichtung*
edge guiding Kantenführung * *Funktion*
edge joint Eckverbindung * *{Holz}*
edge mill Kollergang * *Walzen-Mahlwerk* || Kollergang * *Kollermühle*
edge protection Kantenschutz * *als Funktion*
edge protector Kantenschutz * *als Bauteil*

edge trigger flag Flankenmerker * bei SPS
edge trim Randstreifen * abgeschnittene Randstreifen z.b. bei Kunstof-Folienmaschine
edge working Kantenbearbeitung * *{Holz}*
edge working machine Kantenbearbeitungsmaschine * *{Holz}*
edge-banding machine Kantenanleimmaschine * *{Holz}* in der Möbelindustrie
edge-gluing machine Kantenanleimmaschine * *{Holz}* bei der Möbelherstellung
edge-runner mill Kollergang * *Kollermühle*
edge-triggered flankengetriggert
edge-trimming tool Kantenbearbeitungswerkzeug * *{Holz}*
edged besäumt * *{Holz}*
edger Besäumer * *{Holz}* zum Kantenabschneiden bei Holzplatten || Besäumsäge * *{Holz}*
edging circular saw Besäumkreissäge * *{Holz}* zum Beschneiden der Brettkanten
edging saw Besäumsäge * *{Holz}*
edgings Splitter * *{Holz}* Spreißel
edifice of ideas Gedankengebäude
edifice of thoughts Gedankengebäude
edit bearbeiten * *Text, Artikel, Buch, Gafik usw. ~/ redigieren/druckfertig machen* || editieren || eingeben * *Text/Daten in einen Rechner* ~ || redigieren
editing redaktionelle Bearbeitung
edition Auflage * *eines Druckerzeugnisses* || Ausgabe * *Auflage einer Druckschrift* || Fassung * *Ausgabe eines Druckerzeugnisses* || Version * *Ausgabe/Auflage eines Druckerzeugnisses*
editor Herausgeber * *Herausgeber, Redakteur* || Redakteur
editor in chief Chefredakteur
editorial office Redaktion
EDP EDV * *Abk. für 'Electronic Data Processing'*
educate ausbilden * *Schule, Hochschule*
education Ausbildung * *Bildung (Schule, Hochschule, Universität)* || Bildung * *durch Schule/ Hochschule* || Schulung * *Bildung (Schule, Hochschule)*
education manager Ausbildungsleiter
EEC EG * *Abkürzg.f. 'European Economic Community':Europ. Wirtschaftsgemeinsch.: EWG*
EEMAC EEMAC * *Abkürzg. f. 'Electrical Electronic Manufacturers Association of Canada':Canad. Norm.gremium*
EEPROM EEPROM * *Electrically Erasable Programmable Read Only Memory, elektrisch löschbarer Nur-Lese-Speicher* || nichtflüchtiger Speicher * *elektrisch löschbarer Nur-Lese-Speicher*
EEx e *increased safety type of protection* Zündschutzart EEx e * *Explosionsschutzart für Motor*
effect -Wirkung * *Funktionieren (z.B. cooling effect: Kühl~), (Ein-)Wirkung, Wirksamkeit, Mechanismus* || ausrichten * *bewirken* || auswirken * *bewirken* || Auswirkung * *Wirkung* || bewerkstelligen || bewirken || Effekt * *Effekt, (Ein-) Wirkung, Einfluß, Erfolg, Folge* || Einfluss * *Wirkung, Ein/Auswirkung* || Einwirkung || Erfolg * *Wirkung* || Folge * *Wirkung* || leisten * *Zahlung usw (payment)*. || Resultat * *Wirkung* || Rückwirkung || vornehmen * *ausführen, erledigen, vollziehen; abschließen (Versicherung usw.)* || Wirkung * *with effect from: mit ~ von*
effect of the load variations Lastabhängigkeit
effective aktiv || effektiv * *wirksam, effektvoll; effectively required: ~ benötigt* || effizient * *wirsam, erfolgreich, wirkungsvoll, kräftig, effektvoll; wirklich, tatsächlich* || erfolgreich || förderlich * *wirksam* || gültig * *wirksam, in Kraft; from: ab* || nutzbar || tatsächlich * *wirksam, in Kraft* || wirksam * *wirksam, gültig, wirkend, in Kraft* || wirksam * *wirkend, gültig, in Kraft; become effective/into effect: wirksam werden* || wirkungsvoll

effective direction Wirkrichtung
effective immediately ab sofort * *mit sofortiger Wirksamkeit/Gültigkeit* || hiermit * *mit sofortiger Wirkung*
effective interest Rendite * *effektiver Gewinn, Effektivzins*
effective output Nutzleistung * *Ausgangs-Effektivleistung*
effective power Wirkleistung
effective resistance Wirkwiderstand
effectiveness Effektivität || Leistungsfähigkeit * *Effektivität, Wirksamkeit* || Wirksamkeit
efficacy Kraft * *Wirksamkeit*
efficiacy Wirksamkeit
efficiency Effekt * *Wirkungsgrad, Leistungsfähigkeit, Tauglichkeit, Tüchtigkeit* || Effizienz || Fachkönnen || Können * *Tüchtigkeit* || Leistungsfähigkeit * *Effizienz* || Wirksamkeit || Wirkungsgrad * *Verhältn."eta":abgegeb. z.zugeführt.Wirkleistg. (typ.DC-Stromrichter 99%,DC-Motor 94%)* || Wirtschaftlichkeit * *Effizienz, Preis/Leistungsverhältnis*
efficiency report Beurteilung * *Leistungsbeurteilung in Personalakte*
efficiency-testing machine Leistungsprüfstand * *zum Ermitteln der Leistungsfähigkeit/des Wirkungsgrades*
efficiency/power-factor value Betriebsgüte * *Produkt (Wirkungsgrad mal cos phi) eines Motors*
efficient effizient || fähig * *tüchtig* || funktionsfähig || funktionstüchtig || gut * *tüchtig, effizient* || leistungsfähig || rationell * *wirtschaftlich* || sinnvoll * *zweckvoll* || tüchtig * *leistungsfähig, gewandt* || wirksam * *effizient (z.B. Person, Arbeit)* || wirtschaftlich * *leistungsfähig, rationell*
efficient in business geschäftstüchtig
effort Anstrengung * *Bemühung* || Arbeit * *Mühe* || Aufwand * *Mühe, Anstrengung, Kraftanspannung* || Bemühung || Mühe * *Anstrengung* || Versuch * *Anstrengung*
effortless leicht * *mühelos*
efforts Einsatz * *Anstrengung*
egoutteur Egoutteur * *{Papier}* bei Papiermaschine hinter dem Sieb
EIA EIA * *Abkürzg. f. 'Electronic Industries Association': US-Normengremium*
eigen value Eigenwert
either auch * *[am. i:the; engl. aithe] (immer nachgestellt) auch i.Zusamm.hang m. Verneinung(not either)* || jedes * *~ von zweien*
either or entweder oder
eject ausstoßen || auswerfen * *auch 'hinauswerfen, ausstoßen'* || Auswurfhebel * *Auswerfer*
eject lever Auswurfhebel * *z.B. für Flachbandkabel-Steckverbinder*
ejecting Auswurf
ejection Auswurf
ejector Auswerfer || Ejektor
elaborate aufwändig * *ausführlich, kompliziert, sorgfältig ausgearbeitet/-geführt; (sorgfältig) ausarbeiten* || ausarbeiten * *sorgfältig ~; auch herausarbeiten (siehe auch →erarbeiten)* || erarbeiten * *sorgfältig aus-/durcharbeiten* || herausarbeiten * *sorgfältig; auch ausarbeiten* || umständlich * *komliziert, umständlich; sorgfältig* || unverständlich * *→kompliziert, →umständlich,; sorgfältig*
elaboration Ausarbeitung * *Ausarbeitung, genaue Darlegung, (Weiter-) Entwicklung, Vervollkommnung*
elapse ablaufen * *vergehen, verstreichen von Zeit* || laufen * *ablaufen von Zeit* || verlaufen * *Zeit* || verstreichen * *Zeit, Frist*
elapsed abgelaufen * *Zeit*
elapsed-hour meter Betriebsstundenzähler
elapsed-time counter Betriebsstundenzähler

elapsed-time meter Betriebsstundenzähler
elastic dehnbar || elastisch || federnd || nachgiebig
 * *(auch figürl.)* elastisch, federnd, dehnbar, biegsam, geschmeidig, anpassungsfähig
elastic band Gummiband
elastic constant Elastizität * ~skonstante
elastic feedback Statik * nachgebende Rückführung (negative Rückkopplung des Reglerausgangs auf d. Eingang)
elastic force Rückstellkraft * elastische
elastic limit Elastizitätsgrenze || Streckgrenze
elastic modulus E-Modul || Elastizitätsmodul
elastical force Spannkraft * elastische ~
elasticity Elastizität * *(auch figürlich)* || Nachgiebigkeit * *Elastizität (auch figürl.)*, Feder-/Spannkraft || Spannkraft * *(auch figürl.)*
elasticity constant Federkonstante * Elastizitätskonstante
elastomer Elastomer
elbow Knick * Biegung, Krümmung, Ecke, Knie, Kniestück, Krümmer, Winkel(stück) || Knie * Rohr~ || Kröpfung || Krümmer * Kniestück, Krümmer, Winkel (-stück) || Rohrkrümmer
elbow-pushing aggressiv * gegenüber Mitbewerbern mit 'Ellenbogen' vorhebend
elbow-room Spielraum * Bewegungsfreiheit
ELCB FI Schutzschalter * Earth Leakage Circuit-Breaker; Sicherh.einrichtg., schaltet b.Erdfehlerstrom ab || Fi-Schutzschalter * Earth Leakage Circuit-Breaker; Sicherh.einrichtg., schaltet b.Erdfehlerstrom ab
election Wahl * *(politisch)*
electric elektrisch * meist direkt mit Elektrizität behaftet, z.B. electric field:elektr.Feld
electric arc furnace Lichtbogenofen
electric brake elektrische Bremse
electric braking elektrische Bremsung
electric bulb Glühbirne || Glühlampe
electric chain hoist Flaschenzug * mit elektr. Antrieb
electric circuit Schaltkreis || Stromkreis
electric constant Durchlässigkeit * elektrische
electric drier Wäschetrockner
electric drive Antrieb * elektrischer ~ || Elektroantrieb
electric field elektrisches Feld
electric field intensity elektrische Feldstärke
electric field strength elektrische Feldstärke || Feldstärke * elektrische ~
electric filter Elektrofilter * z.B. zur Abgasreinigung
electric flux density elektrische Flussdichte
electric heat Elektrowärme
electric heating equipment Elektrowärmeanlage
electric line of force elektrische Feldlinie
electric loading Strombelag
electric motor Elektromotor
electric motor driven elektromotorisch betrieben
electric power line Freileitung
electric power utilization Energieanwendung * elektrische ~
electric shock elektrischer Schlag
electric strength Spannungsfestigkeit
electric supply Stromversorgung
electric torch Taschenlampe * Stablampe
electric vehicle Elektrofahrzeug
electric-motor actuated elektromotorisch * Stellantrieb
electric-motor driven elektromotorisch
electrical elektrisch * meist nicht direkt mit Elektrizität behaftet, z.B. electrical engineer:Elektroingenieur
electrical craftsmen Elektrohandwerk
electrical degree Grad * Grad elektrisch, z.B. Steuerwinkel bei Phasenanschnittsteuerung

electrical degrees elektrische Grade || Grad elektrisch
electrical design Elektrokonstruktion
electrical disturbance elektrische Störung
electrical drive elektrischer Antrieb || Elektroantrieb
electrical engineering Elektrokonstruktion * auch als Abteilungsbezeichnung || Elektrotechnik
electrical equipment elektrische Betriebsmittel
electrical equipment house E-Haus || Schalthaus
electrical equipment room E-Raum || Schaltraum
electrical industry Elektroindustrie
electrical installation Elektroinstallation
electrical interference Störbeeinflussung * [inte'-fi:rens]
electrical isolation galvanische Trennung || Potenzialtrennung
electrical life elektrische Lebensdauer
electrical loadability elektrische Festigkeit * z.B. eines Isoliersystems im Motor
electrical machine elektrische Maschine
electrical machine protective relaying Maschinenschutz
electrical machines elektrische Maschinen
electrical OFF AUS2 * 'Elektrisch AUS', Steuerbit im →PROFIBUS-Profil 'Drehzahlveränderbare Antriebe' || elektrisch AUS * entspr. 'AUS2' beim Profibusprofil 'Drehzahlveränderbare Antriebe'
electrical personnel Elektropersonal
electrical power installation Starkstromanlage
electrical quantities elektrische Größen
electrical rotor angle Polradwinkel * elektrischer
electrical science Elektrizitätslehre
electrical separation sichere Trennung
electrical sheet steel Dynamoblech
electrical simulated shaft elektrische Welle
electrical trade fair Elektrofachmesse
electrical trades Elektrohandwerk
electrically elektrisch * *(Adv.)*
electrically isolated potenzialfrei || potenzialgetrennt
electrician Elektriker || Monteur * Elektriker
electrician's screwdriver Elektrikerschraubendreher || Elektrikerschraubenzieher
electricity Elektrizität
electricity consumption Elektrizitätsverbrauch || Stromverbrauch * am Stromnetz
electricity meter Stromzähler || Zähler * →Stromzähler
electricity supply company Elektrizitäts-Versorgungsunternehmen || Elektrizitätsgesellschaft
electrify mitreißen * *(figürl.)* jdn. begeistern || zünden * Zuhörerschaft
electrifying zündend * Idee, Vision, Rede usw.
electro-chemical elektrochemisch
electro-erosion Elektroerosion * z.B. Verschleißeffekt bei elektr. Kontakten
electro-lubricant Kontaktmittel * z.B. Kontaktspray
electro-magnetic field elektromagnetisches Feld || magnetisches Feld * elektromagnet. Feld
electro-mechanical elektromechanisch
electro-release Ruhestrom * 'elektrische Lüftung' bei Ruhestrombremse (zum Lüften Strom erforderlich)
electro-release brake Ruhestrombremse * Bremse, zu deren Lüftung elektr.Strom erforderlich ist
electroacoustics Elektroakustik
electrode Elektrode
electrodynamic force Stromkraft
electroheat Elektrowärme
electroheating Elektrowärme
electrolysis Elektrolyse
electrolysis plant Elektrolyseanlage
electrolyte Elektrolyt

electrolyte solution Lauge * *Elektrolysebad*
electrolytic elektrolytisch
electrolytic capacitor Elektrolyt-Kondensator || Elektrolytkondensator || Elko
electromagnet Elektromagnet
electromagnetic elektromagnetisch
electromagnetic brake Elektromagnet-Bremse || elektromagnetische Bremse
electromagnetic clutch Elektomagnet-Kupplung || elektromagnetische Kupplung
electromagnetic compatibility elektromagnetische Verträglichkeit
electromagnetic emission Störaussendung * *elektromagnetische ~*
electromagnetic immunity Störfestigkeit * *elektromagnet. ~*
electromagnetic interference Elektromagnetische Störung
electromagnetic radiation Störabstrahlung
electromagnetic single-disc clutch Elektromagnet-Einscheibenkupplung
electromagnetically actuated elektromagnetisch betätigt
electromagnetism Elektromagnetismus
electromechanical elektromechanisch
electromotive elektromotorisch
electromotor Elektromotor
electron beam Elektronenstrahl
electronic elektronisch
electronic board Elektronikbaugruppe * *Leiterkarte*
electronic calculator Taschenrechner
electronic circuit elektronische Schaltung || Schaltkreis * *elektronischer ~*
electronic commutator elektronischer Kommutator
electronic control Steuerelektronik
electronic control equipment Regel- und Steuergeräte * *elektronische ~*
electronic control unit Steuerelektronik * *als Baueinheit/Gerät*
electronic gearbox elektronisches Getriebe
electronic gearing elektronisches Getriebe
electronic modul Elektronikbaugruppe * *Baustein, Modul*
electronic module Elektronikteil * *Baugruppe*
electronic motor Elektronikmotor * *regelbarer Kleinmotor 0,6 bis 60 W und 6 bis 60 V*
electronic ordering beleglose Bestellung
electronic ordering procedure beleglose Bestellabwicklung
electronic terminal elektronische Klemmenleiste
electronic terminal block elektronische Klemmenleiste
electronic terminal strip elektronische Klemmenleiste
electronically elektronisch * *(Adv.)*
electronics Elektronik
electronics board Elektronikbaugruppe * *Leiterkarte*
electronics ground Elektronik-Masse
electronics housing Elektronikgehäuse
electronics module Elektronikbaugruppe
electronics power supply Elektronik-Stromversorgung || Elektronikstromversorgung
electronics section Elektronikteil
electroplating Galvanik
electrostatic charging elektrostatische Aufladung
electrostatic sensitive devives elektrostatisch gefährdete Bauteile
electrotechnical elektrotechnisch
elegant nobel * *fein*
element -Anteil * *Grundbestandteil* || -Glied || Element || Glied * *active:aktives* || Komponente || Moment * *Faktor, wesentlicher Umstand* || Teil * Element, Grundbestand~ || Teil- * Element, Glied
elementar elementar
elements Grundlage * ~ *einer Wissenschaft usw.*
elevate erhöhen * *emporheben (in die Luft)* || heben * *hoch~, empor~, aufrichten, er~, befördern (auch Mitarbeiter)*
elevated erhöht * *emporgehoben* || hoch * *Lage*
elevated conveyor Überflurförderer
elevated platform Hubbühne
elevating capacity Förderleistung * *eines Vertikalförderers/Aufzugs*
elevating platform Hubtisch
elevating screw Gewindespindel * *Hubspindel* || Hubspindel
elevation Anhebung * *in die Höhe* || Aufriss * *Zeichnung*
elevator Aufzug * *(amerikan.; siehe EN 81))* || Fahrstuhl || Lift * *Aufzug*
elevator bank Aufzuggruppe
elevator cage Aufzugskabine
elevator closed-loop control Aufzugsregelung * *(amerikan.)*
elevator control Aufzugsregelung * *(amerikan.)* || Aufzugssteuerung || Aufzugsteuerung
elevator control cable Aufzugsteuerleitung
elevator control system Aufzugssteuerung || Aufzugsteuerung
elevator drive Aufzugsantrieb * *(amerikan.)*
elevator hoisting cable Aufzugseil || Aufzugsseil * *(amerikan.)*
elevator motor Aufzugmotor || Aufzugsmotor
elevator shaft Aufzugschacht || Schacht * →*Aufzugsschacht*
elevator/lift control Aufzugsregelung
eligibility Berechtigung * *Befähigung (for:zu)*
eligible qualifiziert * *geeignet, akzeptabel, befähigt (for:zu), berechtigt, in Frage kommend, erwünscht*
eliminate ausgliedern || ausschließen * *eliminieren* || beseitigen * *auch Fehler, Problem ~* || eliminieren || totlegen * *totmachen, eliminieren* || vermeiden * *eliminieren, beseitigen, ausschließen, entfernen, tilgen*
elimination Behebung || Beseitigung
eliminator Sieb * *(elektr.)*
ellipse Ellipse
elliptic elliptisch
elliptical elliptisch
elliptical field elliptisches Drehfeld * *[el.Masch.] läuft entlang d. Umfangs d.Ständerbohrg.mit lageabhäng. Amplitude*
eloctroflotation Elektroflotation
elocution Vortrag * *Vortragsweise, Sprechtechnik, Redekunst*
elongate ausdehnen || dehnen * *in die Länge ziehen* || verlängern
elongated hole Langloch
elongation Ausdehnung * *Streckung* || Dehnung * *verformende/bleibende ~*
elongation of cable Seildehnung * *{Aufzüge}*
eloxadize eloxieren
else anders * *z.B. somebody else: jemand ~*
elucidate verdeutlichen * *aufhellen, aufklären, erklären, erläutern*
elutriator Sichter * *Sortiereinrichtung*
embarrass verwirren * *verwirren, erschweren, komplizieren, in Verlegenheit bringen*
embed einbetten
embed in concrete einbetonieren
embedded eingebaut * *eingebettet* || eingebettet || eingebracht * *z.B. Wicklung in Motornut* || integriert * *eingebettet, eingebaut*
embedded controller Ein-Chip-Computer || Ein-Chip-Mikroprozessor
embodiment Ausprägung * *Verkörperung, Dar-*

embody

stellung, Inbegriff || Inbegriff * Verkörperung, Darstellung, Inbegriff
embody aufnehmen * eingliedern, in: in || verkörpern
emboss prägen || treiben * Metall
embossed geprägt
embossing machine Prägemaschine
embossing plant Prägeanlage
embossment machine Wulstmaschine
embrace umfassen * umschließen, umgeben, umklammern, einschließen (auch figürl.), umfassen || umschlingen || Umschlingung || zusammenfassen * in sich fassen
embranchment Weiche * Verzweigung
EMC elektromagnetische Verträglichkeit * Abkürzg.f. 'ElectroMagnetic Compatibility': →EMV || EMV * ElectroMagnetic Compatibility: elektromagnet. Verträglichkeit (z.B. EN 61000)
EMC behaviour EMV-Verhalten
EMC directive EMV-Richtlinie * z.B. der EG (EG-Richtlinien 89/336/EWG, 92/31/EWG, 93/68/EWG)
EMC measurement EMV-Messung
EMC measures EMV-Maßnahmen
EMCD EMV-Richtlinie * Abkürzg. für 'Elektro-Magnetic Compatibility Directive'
emdedded command Steueranweisung * bei SPS
emerge auftauchen || austreten * auftauchen, herauskommen, erscheinen, hervortreten, zum Vorschein kommen || entstehen * auftauchen, zum Vorschein kommen, herauskommen (from:aus) || herauskommen * erscheinen || kommen * auftauchen, heraus~, zum Vorschein ~, auftreten, entstehen, sich entwickeln || sich ergeben * auftreten
emerge a winner hervorgehen aus * als Sieger ~
emerge from hervorgehen aus * stammen von
emergence Entstehung * Heraus-/Hervorkommen, Auftreten, Erscheinen, Auftauchen, Entstehen
emergency dringend * Notfall, Notlage, Notruf || Not- || Notfall
emergency braking Notbremsung || Schnellbremsung * Notbremsung
emergency button NOT-AUS-Knopf
emergency control Notsteuerung
emergency engagement Notschaltung * bei einer Kupplung
emergency generating set Notstromaggregat
emergency limit switch Notendschalter
emergency OFF Not Aus || Not-Aus
emergency operation Notbetrieb
emergency power supply Notstromversorgung
emergency power supply system Notstromversorgung
emergency shutdown Notabschaltung * Maschine, Atomkraftwerk usw. || Schnellabschaltung * Notabschaltung || Sicherheitsabschaltung
emergency situation Notfall
emergency stop aus * Nothalt (entspricht meist 'coast stop', siehe dort) || AUS2 * Not Aus || Not Aus * Anlage wird schnellstmöglichst spannungslos gemacht (u.trudelt aus) || Not Halt || Not-Aus || Not-Halt || Not-Stop || Notabschaltung || Nothalt || Schnellhalt * Not-Aus || Sicherheitsabschaltung * Antrieb
emergency stop circuit braker NOT-AUS-Schalter * Leistungsschalter
emergency stop switch NOT-AUS-Schalter
emergency-stop pushbutton NOT-AUS-Taster
emery Schmirgel
emery cloth Schmirgelleinen
emery paper Sandpapier * Schmirgelpapier || Schmirgelpapier
EMF EMK * Abküzg. für 'Electro-Motive Force' oder 'ElectroMagnetic Force': elektromagnetische Kraft

EMF actual value EMK-Istwert
EMF closed-loop control EMK-Regelung * Abkürzg.f. 'ElectroMagnetic Force':elektromagnetische Kraft
EMF control EMK-Regelung
EMF controller EMK-Regler
EMF feed-forward EMK-Vorsteuerung
EMF measurement EMK-Messung
EMF pre-control EMK-Vorsteuerung
EMF reference value EMK-Sollwert
EMF sensing EMK-Erfassung
EMF setpoint EMK-Sollwert
EMI Elektromagnetische Störung * Abkürz. für 'electromagnetic interference' || EMV * Abk. f. 'ElectroMagnetic Interference': elektromagnetische Störung
EMI filter Entstörfilter * ElectroMagnetic Interference: elektromagnetische Störung
EMI generator Störgenerator * Abkürzg. für 'ElectroMagnetic Interference':elektromagn. Störungen || Störsimulator * Störgenerator
EMI noise Einstreuungen
EMI problem Störung * EMV-Problem, elektromagnetische Störbeeinflussung
EMI simulator Störgenerator * Störsimulator || Störsimulator
EMI suppression Entstörung * Unterdrückung v. 'ElectroMagnetic Interference': Störunterdrückung || Störschutz || Störungsunterdrückung * Abkürzg. f. 'Electro-Magnetic Interference': elektromagnetische ~ || Störunterdrückung * Abk.f. 'ElectroMagnetic Interference':elektromagnetische Störung
EMI suppression capacitor Entstörkondensator * zur Sicherstellung der →EMV
EMI suppression choke Entstördrossel * gegen elektromagn. Beeinflussung von außen u. nach innen || Störschutzdrossel
EMI suppression filter Entstörfilter * ElectroMagnetic Interference: elektromagnetische Störung
EMI suppression measures Entstörung * Entstörmaßnahmen
EMI-suppressing components Entstörmittel
EMI-suppression measures Entstörmaßnahmen
eminent bedeutend * hervorragend
emission Abstrahlung * Ausstrahlung, Ausströmung || Aussendung || Emission
emit abgeben * Wärme usw. ~ || abstrahlen || ausströmen
emitter Emitter
emitter connection Emitterschaltung
emitter follower Emitterfolger * Transistorschaltung
emitting sparks funken Sprühend
emphasis Gewicht * Betonung, Schwerpunkt, Nachdruck, Deutlichkeit; lay emphasis on: ~/Wert legen auf || Schwerpunkt * Nachdruck
emphasize herausstellen * (figürl.), z.B. Gedanken ~ || hervorheben * (nachdrücklich) betonen, hervorheben, unterstreichen || hinweisen auf * betonen || unterstreichen * (figürl.) betonen, hervorheben, unterstreichen
emphatically mit Nachdruck
empirical empirisch
empirical value Daumenwert * Erfahrungswert || Erfahrungswert || Richtwert * Erfahrungswert
employ anwenden * for; einsetzen || benutzen || einsetzen * verwenden, beschäftigen || nutzen * verwenden || verwenden * einsetzen
employed verwendet * eingesetzt
employee Angestellter || Arbeitnehmer || Mitarbeiter * Arbeitnehmer, Angestellter
employees Personal * Angestellte
employer Arbeitgeber || Auftraggeber * Arbeitgeber || Chef * Arbeitgeber

Employer's Liability Assurance Association Berufsgenossenschaft
employment Anstellung || Anwendung * *Verwendung, Beschäftigung, Einsatz* || Arbeit * *Berufstätigkeit* || Arbeitsplatz * *Anstellung, Arbeits/Beschäftigungsverhältnis* || Arbeitsstelle || Arbeitsverhältnis * *Anstellung,Stellung* || Gebrauch || Stelle * *Arbeits~* || Stellung * *berufliche Anstellung* || Verwendung * *Einsatz, Gebrauch, Anwendung*
empowered autorisiert * *bevollmächtigt, ermächtigt* || befugt * *bevollmächtigt, ermächtigt*
empowerment Berechtigung * *→Befugnis, Ermächtigung*
empty ausleeren || ausräumen || entleeren || hohl * *leer* || leer || Leer- * *z.B. Flasche, Palette usw.* || leeren
empty case Leergehäuse
empty housing Leergehäuse
empty space Hohlraum
emptying Entleerung
emulate emulieren * *nachbilden, simulieren, nachahmen (z.B. ein Computer-Zielsystem)* || nachbilden * *emulieren*
emulation Nachbildung * *Emulation*
emulator Emulator
emulsion Emulsion
EN EN * *Abkürzg. für 'Europanorm': European Standard (zuständiges elektrotechn. Normengremium: →CENELEC)* || Europäische Norm * *Abkürzg. für European Standard:Europäsche Norm* || Europanorm
en bloc geschlossen * *(Adv.) als Gesamtheit, alles in einem Rutsch*
enable aktivieren * *freigeben* || durchschalten * *(Signal) durchschalten/freigeben* || ermöglichen || Freigabe * *(Verb) als Kommando* || freigeben * *eine Funktion, ein Signal, einen Regler ~* || instand setzen * *in die Lage versetzen* || zuschalten * *freigeben*
enable interrupt Alarm freigeben
enable ramp-function generator Hochlaufgeberfreigabe
enable signal Betriebsfreigabe * *als Signal* || Freigabesignal
enabled durchgeschaltet * *(Signal) freigegeben* || fähig * *fähig gemacht, in die Lage versetzt (to: für/zu)* || freigegeben
enabling Freigabe * *das Freigeben*
enactment Verordnung
enamel emaillieren
enameled wire Lackdraht
enamelled copper wire Kupferlackdraht
enamelled wire Lackdraht
encapsulate eingießen * *(ein-) kapseln, umgießen* || einkapseln || einschließen * *einkapseln* || vergießen
encapsulated gekapselt * *auch 'vergossen'* || vergossen * *(ein)gekapselt, z.B. Wicklung*
encapsulated module Kapsel * *gekapselte Baugruppe*
encapsulation Umhüllung || Verguss
encase einschließen * *in Gehäuse ~* || umhüllen
encased geschlossen * *mit Gehäuse versehen*
enchainment Verkettung * *(figürlich)*
enclose beifügen * *einem Brief (with: in/bei)* || beipacken * *loose: lose* || einkapseln * *einschließen* || einschließen * *einschließen, einkapseln* || umschließen * *enthalten, beinhalten*
enclosed Beipack * *in der Anlage beiliegend, beigefügt, im ~* || gekapselt || geschlossen * *(durch Gehäuse) umschlossen, ab~* || hierbei * *ein/beigefügt*
enclosed self-ventilated innenbelüftet * *innengekühlt, →eigenbelüftet; Motorkühlart*
enclosed space geschlossener Raum * *wettergeschützter Bereich/Ort/Platz*
enclosed ventilated durchzugbelüftet * *durchzugsbelüftet, innengekühlt (el.Maschine mit Gehäuse-interner Lüftung)* || innengekühlt * *→durchzugsbelüftet (el. Maschine mit Kühlluftstrom durchs Gehäuse)*
enclosed ventilation Durchzugsbelüftung * *(el.-Masch.)*
enclosed version geschlossene Ausführung
enclosed-ventilated innenbelüftet
enclosing umschließend * *beinhaltend*
enclosure Anlage * *zu einem Brief, einer Sendung* || Gehäuse * *auch b. Motor* || Schrank || Schutzart
enclosure protection Schutzart
encode kodieren
encoder Geber * *digitaler Drehzahl/Lagegeber* || Kodierer || Umsetzer * *Kodierer, Verschlüsselungseinrichtung*
encoder feedback Geberrückführung * *Istwert von Drehzahl/Lagegeber wird zur Regelung verwendet* || Impulsgeberrückführung
encoder interface Geber-Interface * *Anschaltung für Drehzahl-/Lagegeber*
encoderless geberlos * *ohne Drehzahl-/Lagegeber*
encoding Komprimierung * *(Computer) von Sound- und Videodateien usw.*
encompass einschließen * *auch figürlich* || umfassen * *einschließen, beinhalten; umfassen, umgeben, umringen,* || umgeben * *umfassen, umgeben, umringen, einschließen*
encounter antreffen || begegnen * *jemanden ~,einer Sache ~* || stoßen auf * *begegnen, treffen auf* || treffen * *auf etwas ~* || zusammentreffen * *(Verb,Subst.) jdm./e.Sache begegnen, auf Widerstände stoßen, jdm. entgegentreten*
encountered anzutreffen
encourage anregen * *ermuntern*
encouragement Anregung * *Ermutigung* || Förderung * *Ermutigung*
encouragement of young talent Nachwuchsförderung
encouraging günstig * *ermutigend*
encryption Verschlüsselung * *von Daten zur Geheimhaltung*
encyclopaedia Nachschlagewerk * *Enzyklopädie, Lexikon*
end ablaufen * *Ausgang nehmen* || Abschluss || beenden || beschließen * *beenden, abschließen* || End- || Ende * *auch Wellenende, Ende einer Liste, einer Wicklung usw.* || enden || schließen * *beenden* || Schluss * *Ende* || Spitze * *äußerstes Ende, Stirnseite* || Stirnseite || Trum * *eines Seils/eines Riemens; slack trum: schlaffes ~; driving end: straffes ~* || Trumm * *eines Seils, eines Riemens; siehe auch →Trum* || Verlauf * *Ausgang* || Ziel * *Zweck, Absicht, Nutzen* || Zweck * *Ziel; to this end: zu diesem*
end bracket Lagerschild * *offenes ~ am Wellenende*
end cap Schlussdeckel
end character Endezeichen * *kennzeichnet z.B. das Ende eines Datentelegramms (häufig ETX entspr. 03hex)*
end customer Endkunde
end diameter Enddurchmesser * *z.B. bei Wickler*
end face Planfläche * *am Wellenende* || Stirnfläche
end finishing Endfertigung
end frequency Endfrequenz * *z.B.der f/U-Kennlinie eines Frequenzumrichters*
end hole Stirnbohrung * *(Holz)*
end in führen zu * *enden mit*
end lock Endarretierung
end module Bausteinende * *Befehl z. Beenden eines Funktionsbaustein bei einer SPS*
end of conductor Leiterende
end of production Fertigungseinstellung || Produktionseinstellung

end of rope Trum * *eines Seils*
end of winding Wicklungsende
end plate Lagerschild * ~ *am Ende einer Welle (z.B. bei Motor)*
end play Axialspiel * *z.B. an einem Wellenende* || Lagerspiel * ~ *eines Wellenlagers in Axialrichtung*
end point Endpunkt * *z.B. einer Kennlinie*
end point of the voltage boost Ablösefrequenz * *Frequenz,bei d.die Spannungsanhebung bei e. Frequenzumrichter endet*
end position Endlage || Endstellung
end product Endprodukt
end result Endergebnis
end shield Lagerschild * *geschlossenes ~ am Wellenende*
end sleeve Aderendhülse
end stop Anschlag * *End~*
end turns Wickelkopf
end user Endabnehmer * *z.B. einer Ware/eines Geräts* || Endanwender || Endkunde * *letztendlicher (eigentlicher) Anwender*
end value Endwert
end wall Außenwand * *Stirnwand* || Stirnwand
end winding Wickelkopf
end-float washer Ausgleichsscheibe * *zur Einstellung des axialen Lagerspiels* || Ausgleichsscheibe * *zur Einstellung des axialen Lagerspiels*
end-to-end cooling Durchzugsbelüftung
end-turn banding Wickelkopfbandage * *[el. Masch.] Bandagierung d.Wickelkopfs mit Isolierbändern b. Niederspannungsmotor*
end-winding support Wickelkopfabstützung
endanger gefährden
endeavor Anstrengung || Versuch * *Bemühung*
endeavour Anstrengung * *Bemühung, Anstrengung, Bestreben* || streben * *sich anstrengen* || Versuch * *Bemühung* || versuchen * *sich bemühen*
ended abgeschlossen * *beendet*
ending Ausgang * *Ende*
endless endlos || unendlich * *endlos*
endless loop Dauerschleife * *Endlosschleife* || Endlosschleife || Untätigkeitsschleife * *Endlosschleife*
endless saw blade Bandsägeblatt
endorsement Vermerk * *Vermerk, Zusatz, Eintrag (auf Urkunden usw.)*
endshield Lagerschild * *geschlossenes ~ am Wellenende*
endshield hub Lagerschildnabe * *führt die Welle, z.B. eines Motors*
endshield ventilation opening Lagerschildkühlluftöffnung * *Motor*
endstop Anschlag * *mechanischer End~, z.B. an Potentiometer*
endurance Lebensdauer
endurance calculation Lebensdauerberechnung
endurance test Dauerprüfung || Langzeittest * *Dauertest, Belastungstest, Haltbarkeitstest* || Lebensdauerprüfung
endure überstehen * *ertragen*
enduring bleibend * *(an-, fort-) dauernd, bleibend* || dauerhaft * *nachhaltig, bleibend*
energetic kräftig * *tatkräftig*
energization Erregung * *einer Federdruck-Magnetbremse*
energize an Spannung legen || anregen || ansteuern * *e. Ein/Ausgang mit High-Signal; eine Relaispule; de-energize: mit LOW-Signal* || bestromen || betätigen * *Relais, Kupplung usw.* || einschalten * *an Spannung legen, HIGH-Signal an eine Klemme anlegen* || erregen * *z.B. die Spule eines Magneten, einer Bremse* || HIGH-Signal anlegen * *Klemme*
energized angesteuert * *Klemme/Steuereingang mit HIGH-Signal* || angezogen * *Relais* || bestromt || eingeschaltet * *mit High-Signal angesteuert*

energized state eingeschalteter Zustand * *z.b. Klemme, Relais, Motor*
energizing Ansteuerung * *z.B. eines Steuereingangs mit HIGH-Signal*
energizing unit Ansteuerbaustein * *z.b. zur Schützansteuerung*
energy Arbeit * *auch als physikalische Größe in [Ws] oder [J]* || Energie || Kraft * *Tat~, Energie* || Schwung * *Energie, Wucht* || Spannkraft * *(figürl.) Schwung, Spannkraft, Lebenskraft, Auftrieb* || Stärke * *Tatkraft*
energy balance Energiebilanz
energy balancing Energieausgleich * *z.B. zw. mehreren Umrichtern am gemeinsamen Zwischenkreis*
energy buffering Energiepufferung
energy consumption Energieverbrauch
energy conversion Energieumwandlung * *z.b. von elektrischer in mechanische Energie*
energy demand Energiebedarf
energy distribution Energieverteilung
energy efficient energiesparend
energy equalization Energieausgleich * *z.B. durch gemeinsamen Zwischenkreis bei Umrichtern*
energy flow Energiefluss
energy index Energiekennziffer
energy loss Energieverlust
energy meter Stromzähler
energy of rotation Rotationsenergie
energy production Energieerzeugung
energy recovery Energierückspeisung
energy saving Energieeinsparung
energy source Energiequelle
energy storage Energiespeicher
energy transducer Energiewandler
energy transfer Energieaustausch
energy-recovery unit Rückspeiseeinheit
energy-saving energiesparend || stromsparend
enfeeble schwächen * *entkräften*
enforce durchsetzen * *erzwingen* || erzwingen * *(besonders gesetzlich)*
enforcement Durchführung * *Durchsetzung, Vollzug (Gesetz, Vorschrift)*
engage einfallen * *Bremse* || eingreifen * *in/with:in* || einklinken || einlegen * *Kupplung, Bremse, Gang usw.* || einrasten || einrasten * *in: an/in* || einrücken * *Kupplung usw.* || einschalten * *Kupplung, Bremse* || einstellen * *Arbeitskräfte usw. engagieren* || engagiert sein * *engage oneself: sich engagieren (in:bei)* || ineinander greifen || schnappen * *einrasten, in Eingriff kommen* || verpflichten * *vertraglich, oneself to do a thing: sich zu etwas ~*
engaged besetzt * *Telefon* || eingerastet
engagement eingreifen * *(Substantiv)* || Eingriff * *von Zahnrädern, Kupplung usw.* || einschalten * *(Substantiv) das Einschalten einer Kupplung/Bremse* || Schaltung * *bei Kupplung, Bremse* || Verpflichtung * *übernommene ~, enter into an eng.: eine ~ eingehen*
engagement accuracy Schaltgenauigkeit * *einer Kupplung/Bremse*
engagement delay Ansprechverzögerung * *einer Kupplung/Bremse*
engagement force Einrückkraft * *zum Schließen einer Kupplung*
engagement frequency Schalthäufigkeit * *einer Kupplung/Bremse* || Schaltzahl * →*Schalthäufigkeit einer Kupplung/Bremse*
engagement pressure Schließdruck * *einer Kupplung/Bremse*
engagement torque schaltbares Drehmoment * *für Kupplung, Bremse (ist niedriger als das →übertragbare Drehmoment)*
engaging dogs Greifer * *Greifhaken*
engaging piece Mitnehmer
engine Maschine * *Verbrennungsmaschine* || Motor

* *Verbrennungsmotor, Maschine, nicht-elektrischer Motor*
engine room Maschinenraum
engine test bed Motorprüfstand * *für Verbrennungsmotoren*
engine test stand Motorprüfstand * *für Verbrennungsmotoren*
engineer Bearbeiter * *Ingenieur, Techniker* || bewerkstelligen || Ingenieur || Maschinenführer || Monteur * *geschulter, erfahrener* ~ || Projekteur || projektieren || Techniker * *Ingenieur*
engineer's office Ingenieurbüro
engineering -Technik * *Fachrichtung* || Dimensionierung * *Entwurf, konstruktive/technische Auslegung* || Projektierung || Technik * *angewandte* || technisch
engineering aid Projektierungshilfe
engineering aids Projektierungshilfen
engineering and design Konstruktion * *auch als Anteilungsbezeichnung*
engineering consultant Ingenieurbüro
engineering costs Projektierungskosten
engineering data Projektierungsdaten
engineering dimension Dimension * *physikalische* ~
engineering documentation Projektierungsunterlagen * *Dokumentation, Applikationshinweise usw.*
engineering documents Projektierungsunterlagen
engineering drawing technische Zeichnung
engineering effort Entwicklungsaufwand
engineering expenditure Projektierungsaufwand
engineering expertise Ingenieurwissen
engineering firm Ingenieurbüro
engineering guide Projektierungshandbuch || Projektierungshilfe * *gedruckte* ~
engineering information Projektierungsdaten || Projektierungshinweis || Projektierungshinweise
engineering instructions Projektierungsanleitung
engineering manual Projektierungshandbuch
engineering of the drive system Antriebsprojektierung * *für eine Maschine/Anlage*
engineering office Ingenieurbüro
engineering operation Industriebetrieb
engineering personnel Maschinenpersonal
engineering sample Entwicklungsmuster
engineering services technische Dienste
engineering term Ausdruck * *technischer (Fach)-Ausdruck* || Fachausdruck * *technischer* ~ || Fachterminus * *technischer* ~
engineering terms Fachausdrücke * *technische* || Fachbegriffe * *technische* || Fachtermini * *technische* ~
engineering unit Einheit * *physikalische* ~ || physikalische Dimension || physikalische Einheit
engineering/design Projektierung
English englisch
English-language englischsprachig
engrave gravieren || prägen * *(figürl.) on: ins Gedächtnis usw.* ~ || schneiden * *ein*~, *stechen, gravieren, eingraben, einprägen* || stempeln * *eingravieren, auch Typenschildangaben*
engraved graviert
engraving Gravur
enhance ausbauen * *verbessern* || erhöhen * *Wirkung, Leistungsfähigkeit, Zuverlässigkeit usw.* ~ || erweitern * *verbessern* || verbessern * *in der Leistungsfähigkeit/im Funktionsumfang* ~
enhanced erhöht * *Wirkung usw.* || verbessert * *in Leistungsfähigkeit oder Funktionsumfang* ~ || verstärkt * *gesteigert (auch i.d. Leistungsfähigkeit), erhöht, vergrößert, angehoben; z.B. Kühlung*
enhanced air cooling verstärkte Luftkühlung * *Kühlung v.Leistungshalbleitern usw. mit Lüfter*
enhanced ventilation verstärkte Luftkühlung * *Kühlung z.B. von Leistungshalbleitern mit Lüfter*

enhancement Änderung * *Verbesserung* || Anreicherung || Fortschritt * *Verbesserungen* || Leistungssteigerung * *Steigerung der Leistungsfähigkeit/des Funktionsumfangs* || Steigerung * *Steigerung (Leistungsfähigkeit, Funktionsumfang), Erhöhung, Vermehrung* || Verbesserung * *Erhöhung der Leistungsfähigkeit/der Funktionalität*
enlarge ausbauen * *vergrößern* || ausdehnen || vergrößern * *größer machen; auch einen Wert/eine Zahl*
enlargement Erweiterung
enlightened visionär * *erleuchtet, seherisch*
enlightment Vision * *Erleuchtung, Vision*
enlist auflisten * *in Liste eintragen* || einsetzen * *jmd. heranziehen/gewinnen/engagieren/anheuern (in:für)*
enormous außerordentlich * *enorm, ungeheuer* || enorm || groß * *enorm, riesig* || höchst * *enorm* || riesig * *enorm*
enormous sum Unsumme
enough ausreichend * *genug* || genug || genügend
enquire nachfragen || Rückfragen * *(Verb) sich erkundigen (nach), (er)fragen; for:nach; about:-über; of someone: bei jdm.*
enrichment Anreicherung
enrol aufnehmen * *in einen Verein* ~ || eintragen * *(sich) eintragen/einschreiben (für Kurs, an Hochschule usw.), sich immatrikulieren*
enrollment Einschreibung
enrolment Anmeldung * *Einschreibung* || Aufnahme * *Einschreibung* || Einschreibung * *z.B. zu Vorlesung, Kurs, Tagung*
ensue sich ergeben * *auftreten*
ensure bewirken * *sicherstellen* || garantieren * *sicherstellen, gewährleisten, für, für etwas sorgen* || gewährleisten * *sicherstellen, Gewähr bieten für, garantieren, für etw. sorgen* || sicherstellen * *für etw. sorgen, garantieren (that: dass; someone being: dass jemd. ... ist),* ~ || sorgen für * *sicherstellen (that: dass etwas getan wird)*
entail führen zu * *mit sich bringen, zur Folge haben, nach sich ziehen, erfordern* || verursachen * *mit sich bringen, zur Folge haben, nach sich ziehen, erfordern*
entangle umschlingen || verwirren * *(konkret) Garn, Fäden usw.; (figürl.) in Schwierigkeiten usw. verwirren/verstricken*
entanglement Verwirrung * *(konkret)*
entangling Umschlingung
enter aufführen * *eintragen* || aufnehmen * *eintragen* || eintreten, eindringen in || einfahren || Eingabetaste || eingeben * *z.B. über Tastatur/Bediengerät/Klemme* ~ || einlaufen * *{Schifffahrt} eines Schiffes, einer Wahrenbahn* || einsteigen || eintragen * *buchen usw.; into: in* || eintreten * *hin*~ *(auch Dienst, Beruf), into:in Verhandlungen, als Teilhaber usw.* || Übergabetaste || vermerken * *eintragen* || vorgeben * *Signal z.B. Sollwertsignal* ~
enter a protest Einspruch einlegen * *against:gegen*
ENTER button Übergabetaste
enter by force eindringen * *mit Gewalt*
enter into an agreement Vertrag schließen
enter into the books buchen
ENTER key Eingabetaste || Übergabetaste || Übernahme-Taste
enter one's name for belegen * *Vorlesung usw.* ~
enter the license code freischalten * *Software* ~ *(z.B. durch Eingabe der Registriernummer)*
enter upon übernehmen * *Amt* ~
entered eingegeben
entered into aufgenommen * *Verhandlungen*
entering Eingabe * *z.B. über Tastatur*
entering upon Übernahme * *Amt*

enterprise Betrieb * *Unternehmen* || Geschäft * *Unternehmmen* || Unternehmen
entertainment Unterhaltung * *Vergnügen*
entertainment electronics Unterhaltungselektronik
enthusiastic eifrig
entire ganz * *vollständig* || gesamt || Gesamt- || sämtliche * *ingesamt* || über alles * *gesamt* || voll * *gesamt* || voll * *(Adj.) gesamt* || Voll- * *gesamt* || völlig * *ganz* || vollständig
entirely ganz * *(Adv.) vollständig, völlig* || voll * *(Adv.)vollständig*
entirety Umfang * *das Ganze, Gesamtheit, Ganzheit; in its entirety: in vollem ~*
entirity Vollständigkeit
entitled befugt
entitlement Berechtigung * *Anrecht, Anspruch*
entity Größe * *Einheit, Grundelement, (Zeichnungs-) Objekt, Gebilde*
entrance Einfahrt || Eintritt * *Eintritt (zu einer Veranstaltung), Einlass, Zutritt, Beginn*
entrance examination Aufnahmeprüfung
entrepreneurial spirit Unternehmergeist
entrepreneurship Unternehmergeist * *Unternehmertum*
entries Einträge
entruck einladen * *Lastwagen beladen*
entrust beauftragen || betrauen * *with:mit; auch 'anvertrauen'* || überlassen * *anvertrauen; to:an*
entry Aufnahme * *~ in eine Liste* || Aufzeichnung * *Eintrag* || Durchführung * *Einlass-/Einführungs(-stelle) z.b. von Leitungen* || Einführung * *z.B. Kabel~* || Eingabe * *per Bedienung, z.B. in einen Computer, einem Bedienfeld* || Einlauf * *z.b. einer Walzstraße, einer Warenbahn in e. Verarbeitungsmaschine* || Eintrag * *in Tabelle, Liste usw.; Buchung* || Eintritt || Einzug || Registrierung * *Eintragung* || Vermerk * *Eintrag* || Zufuhr * *Eintrag, z.B. in Ofen, Pressmaschine* || Zugang * *Tür*
entry feed Einzug * *fördert Material hinein*
entry field Eingabefeld * *auf einer LCD-/Bildschirmanzeige*
entry fitting Anschlussstutzen
entry into the market Marktzutritt
entry key Übergabetaste || Übernahme-Taste || Übernahmetaste * *z.B. an SPS-Programmiergerät*
entry machine Eintragmaschine * *z.B. für Ofen* || Lademaschine * *Eintragmaschine, z.B. für Ofen, Pressmaschine*
entry of strip Bandeinlauf * *im Walzwerk: Stelle, an der das Walzband einläuft*
entry plate Einführungsplatte * *z.b. zur Kabeleinführung in einen (Motor-) Klemmenkasten*
entry section Einlauf * *Einlaufteil z.B. bei Blechstraße* || Einlaufteil * *z.B. einer Walzstraße* || Einzug * *Materialeinzugsystem bei Verarbeitungsstraße*
entry-level Einsteiger-
entry-level solution Einstiegslösung
enumerate aufführen * *aufzählen*
enumerated aufgeführt * *aufgezählt*
envelop umwickeln * *einwickeln, einschlagen, einhüllen*
envelope Briefumschlag || Einhüllende || Hüllkurve * *Einhüllende* || umhüllen * *with/in:mit* || Umhüllende || Umhüllung
envelope curve Einhüllende * *Hüllkurve* || Hüllkurve
enveloping dimensions Hüllmaße * *Abmessung eines gedachten Quaders, den der Umriss e. Gegenstands kennzeichnet*
enviromentally green umweltverträglich * *(salopp)*
environ umgeben * *umgeben, umringen (with: mit)*

environmemtal protection technology Umweltschutztechnik
environment Konfiguration * *Umgebung, äußere Bedingungen* || Umfeld || Umgebung * *auch Programmierumgebung, Betriebssystem-Umgebung* || Umgebungs- * *Umwelt-, Milieu* || Umwelt
environment protection Umweltschutz
environment variable Umgebungsvariable
environment-friendly umweltfreundlich
environmental ökologisch * *Umwelt-* || Umgebungs- * *Umwelt-, Milieu* || Umwelt-
environmental awareness Umweltbewusstsein
environmental class Umweltklasse * *nach DIN IEC 721-3-3*
environmental compatible umweltverträglich
environmental conditions Betriebsbedingung * *(Plural) Umgebungsbedingungen* || Umgebungsbedingungen || Umweltbedingungen
environmental factors Umwelteinflüsse
environmental influence Umweltbeeinflussung
environmental influences Umwelteinflüsse
environmental officer Umweltbeauftragter || Umweltschutzbeauftragter
environmental pollution Umweltverschmutzung
environmental protection Umweltschutz
environmental protection delegate Umweltbeauftragter || Umweltschutzbeauftragter
environmental protection engineering Umweltschutztechnik
environmental technology Umwelttechnik
environmental testing procedure Umweltprüfverfahren
environmentally acceptable umweltverträglich
environmentally compatible umweltfreundlich
environmentally detrimental umweltbelastend
environmentally friendly umweltfreundlich
environmentally green umweltfreundlich * *(salopp)*
environmentally safe umweltfreundlich || umweltverträglich
environments Umgebungsbedingungen
environs Umgebung
envisage vorsehen * *ins Auge fassen, gedenken (doing: zu tun), für möglich halten* || vorstellen * *sich (geistig) vorstellen, (Ziel) ins Auge fassen*
envision vorsehen * *ins Auge fassen, gedenken (doing: zu tun), für möglich halten*
enyoy it guten Appetit
epicyclic gear Planetenrad * *bei Planetengetriebe*
epicyclic unit Planetenträger
epitaxial epitaxial
epoch Zeit * *~alter*
epoch-making bahnbrechend * *epochemachend*
epoxy Epoxid || Epoxidharz || Epoxyd
epoxy casting resin Epoxidharz * *Gießharz*
epoxy coated epoxidbeschichtet
epoxy laminated paper Hartpapier * *mit Epoxidharz verstärkt/getränkt*
epoxy resin Epoxidharz
epoxy-resin Epoxydharz
epoxy-resin paint Epoxydharzlack
EPROM burner EPROM-Programmiergerät || Programmiergerät * *für EPROMs* || Prommer * *(salopp) Programmiergerät für EPROM-Speicher*
EPROM eraser EPROM-Löschgerät
EPROM programmer EPROM-Programmiergerät
equableness Gleichmäßigkeit
equal entsprechen * *identisch sein mit* || gleich sein * *(Verb)* || gleichen || gleichförmig || identisch sein mit
equal to gleich * *identisch, mathematisch gleich mit (something equals ...: etwas ist ~ ... (in Gleichung)* || gleichwertig
equal to or greater than größer gleich
equal to or less than kleiner gleich

equal-priority gleichberechtigt * *mit gleicher Priorität*
equality Gleichheit || Übereinstimmung * *Gleichheit*
equalization Ausgleich * *z.B. energy equalization:* Energieausgleich
equalization effect Ausgleichsvorgang
equalization operation Ausgleichsvorgang
equalization process Ausgleichsvorgang
equalize abgleichen * *ab-/ausgleichen, egalisieren, entzerren, gleichmachen/-stellen, nivellieren* || ausgleichen * *ausgleichen, kompensieren, stabilisieren, egalisieren* || egalisieren * *siehe auch 'ausgleichen'* || kompensieren * *siehe auch 'ausgleichen'*
equalizing Ausgleichs-
equalizing charge Erhaltungsladung
equalizing current Ausgleichsstrom || Ausgleichstrom
equalizing ring Ausgleichsscheibe * *z.B. zum Ausgleichen des axialen Lagespiels*
equalizing rolling mill Egalisierwalzwerk * *{Walzwerk}*
equally gleichermaßen
equation Gleichung * *mathematical: mathematische; boolean: Boolsche/logische*
equation of motion Bewegungsgleichung
equi-potential bonding conductor Potenzialausgleichsleitung
equidistant äquidistant
equilibrium Beharrungszustand * *Gleichgewichtszustand* || Gleichgewicht * *~szustand*
equip ausrüsten || ausstatten || einrichten * *ausrüsten* || versehen * *with: mit*
equiphase gleichphasig
equipment Anlage || Apparatur || Ausrustung * *~/stattung, ~sgegenstände, techn. Material/Gerätschaft/Einrichtung, Rüstzeug* || Ausstattung || Betriebmittel || Einrichtung * *Ausrüstung* || Gerät * *→Ausrüstung (-sgegenstände), techn. Material/~schaft, Apparatur(en)* || Geräte || Vorrichtung * *Ausrüstung*
equipment circuit diagram Geräteschaltplan
equipment class Geräteklasse * *(→PROFIBUS-Profil)*
equipment designation Betriebsmittelkennzeichnung
equipment error Gerätefehler
equipment failure Geräteausfall
equipment fan Gerätelüfter
equipment for greasing Schmiereinrichtung * *z.B. für Wälzlager*
equipment for regreasing Nachschmiereinrichtung * *z.B. für Wälz-/Gleitlager*
equipment grounding conductor Schutzleiter * *~ eines Geräts*
equipment identification Betriebsmittelkennzeichnung
equipment manual Gerätehandbuch
equipment parts list Apparate-Stückliste
equipment safety law Gerätesicherheitsgesetz
equipotential Äquipotenzial-
equipotential bonding Potenzialausgleich * *über ~sleitung usw.*
equipotential bonding conductor Potenzialausgleichsleitung
equipped ausgerüstet * *with:ausgestattet mit* || ausgestattet * *with:mit* || bestückt * *ausgerüstet, ausgestattet (with: mit)*
equipping Bestückung * *das Bestücken (z.B. eines Baugruppenträgers), Ausrüsten*
equipping time Rüstzeit
equity Eigenkapital * *Wert nach Abzug aller Belastungen, reiner Wert* || Eigenkapital
equity capital Eigenkapital * *~ einer Gesellschaft;* Wert nach Abzug aller Belastungen, "reiner Wert" || Eigenkapital
equivalence Gleichheit * *Äquivalenz, Gleichwertigkeit* || Gleichwertigkeit || Übereinstimmung * *Gleichwertigkeit*
equivalent äquivalent || Entsprechung * *Gegenstück, Entsprechung, Äquivalent* || Ersatz- || Gegenstück || Gegenwert || gleichbedeutend * *to: mit* || gleichwertig * *to: mit* || Pendant || sinngemäß || Wert * *Gegen~*
equivalent circuit Ersatzschaltbild || Ersatzschaltung
equivalent circuit diagram Ersatzschaltbild
equivalent continuous rating gleichwertiger Dauerbetrieb * *{el.Masch.} Dauerbetrieb mit gleicher Erwärmung wie reale Betriebsart*
equivalent thermal time constant thermische Ersatzzeitkonstante * *beschreibt d.Temperaturänderg. nach sprunghafter Belastungsänder.*
equivalent to entsprechend * *gleichwertig*
era Zeit * *~alter*
erase löschen * *z.B. Daten in EPROM-Speicherbaustein ~* || vernichten * *ausradieren, auslöschen, (aus)tilgen (from:aus)*
ERASE PROGRAM Urlöschen * *Programm (im RAM-Speicher) löschen bei SPS*
eraser Löschgerät * *z.B.für EPROMs* || Löschlampe * *für EPROM-Speicherbausteine/-module*
erect aufbauen * *errichten, bauen, aufstellen,* montieren, einrichten || aufstellen * *Bauten ~* || bauen * *erricheten* || errichten || erstellen * *Gebäude ~* || montieren * *aufstellen, aufbauen, errichten*
erecting machine Aufrichtemaschine * *{Verpack.-techn.} z.B. für Kartons*
erection Aufbau || Bau * *~ eines Gebäudes, einer Anlage usw.* || Erstellung * *~ einer Fabrik, einer Anlage, eines Gebäudes usw.* || Montage * *Großanlage, Fabrik*
erection altitude Aufstellungshöhe * *above sea level:über dem Meeresspiegel*
erection engineer Montageingenieur
erection schedule Montageplan * *Zeitplan*
erection site Baustelle
erector Aufrichter * *{Verpack.techn.}* || Monteur * *~ für Anlagenbaustelle*
ergonomical ergonomisch
ergonomics Ergonomie
erode abtragen * *wegfressen, erodieren* || anfressen * *an/zer/wegfressen, verschleißen, abnützen; z.B. Kontakte* || fressen * *an-/zer-/weg~, erodieren,* || zerfressen * *zer-/an-/wegfressen; siehe auch 'fressen'*
erosion Abnutzung * *Verschleiß, Abnützung, Schwund, Zerfressen* || Verschleiß * *Abnutzung, Schwund, Zerfressung; z.B. Kontakte*
erraneously versehenlich
errata Druckfehler * *(Plural)*
erratic unregelmäßig * *ungleichmäßig, regel-/ziellos, unstet, unberechenbar, sprunghaft*
erratum Druckfehler * *(Singular)* || Fehler * *Druck~ (Plural: "errata")*
erroneous falsch * *irrig, unrichtig, irrtümlich, versehentlich* || Fehl- || fehlerhaft * *falsch, unrichtig, irrtümlich; z.B.Signalzustand,Daten* || irrtümlich
erroneous function Fehlfunktion
erroneously aus Versehen * *siehe auch →versehentlich* || falsch * *(Adv.) irrtümlicherweise, zu Unrecht, aus Versehen* || irrtümlich * *(Adv.) ~erweise* || irrtümlicherweise * *(Adv.) siehe auch →versehentlich* || unabsichtlich * *(Adv.) versehentlich, irrtümlich, irrtümlicherweise* || unbeabsichtigt * *(Adv.) versehentlich, irrtümlicherweise* || versehentlich * *irrtümlicherweise*
error Abweichung * *Regel~, mean error: mittlere*

error band

~ || Fehler * auch Mess~, Regelabweichung (Abweichung auch 'deviation'); typographic error: Druck~ || Regelabweichung || Soll-Ist-Abweichung || Soll-Ist-Differenz * *Soll-Ist-Abweichung*
error band Toleranzband
error cause Fehlerursache
error characteristic Fehlerbild
error check Fehlerüberprüfung
error code Fehlercode || Fehlerkennung || Fehlernummer
error condition Fehlerzustand
error correction Fehlerbereinigung || Fehlerkorrektur
error counter Fehlerzähler
error detection Fehlererkennung
error detector Fehlerdetektor * *erfasst Funktionsfehler*
error display Fehleranzeige
error elimination Fehlerbehebung || Fehlerbeseitigung
error fix Fehlerbehebung || Fehlerbeseitigung
error fixing Fehlerbeseitigung || Fehlersuche * *Fehlerbeseitigung*
error flag Fehleranzeige * *Fehlermerker*
error indication Fehleranzeige
error logging Fehleraufzeichnung
error management Fehlerbehandlung
error message Fehlermeldung || Störmeldung * *Fehlermeldung*
error of measurement Messfehler
error phenomenon Fehlerbild
error probability Fehlerwahrscheinlichkeit
error profile Fehlerbild
error recovery Fehlerkorrektur * *bei Datenübertragung/-Speicherung*
error signal Fehlermeldung || Regelabweichung * *als Signal* || Regeldifferenz * *siehe auch →Regelabweichung* || Soll-Ist-Abweichung * *als Signal* || Soll-Ist-Differenz * *Soll-Ist-Abweichung; siehe auch →Regelabweichung*
error statistics Fehlerstatistik
error status Fehlerzustand
error-free fehlerfrei
error-proof fehlersicher * *es können kaum Fehler auftreten*
escalate eskalieren || steigen * *eskalieren, steigen, in die Höhe gehen (auch Preise usw.)*
escalator Rolltreppe
escape ausströmen * *Gas, Dampf* || austreten * *entweichen; z.B.von Kühlluft*
ESD EGB * *Abk f. 'Electrostatically Sensitive Devices': elektrostatisch gefährdete Bauteile* || elektrostatisch gefährdete Bauteile * *Abkürzg. für 'Electrostatic Sensitive Devices': EGB* || ESD * *Abk. für 'ElectroStatic Discharge': Elektrostatische Entladung (bezüglich EMV: siehe EN61000-4-2*
especially insbesondere * *(Adv.)* || speziell * *(Adv.)* || vor allem * *besonders*
especially suited besonders geeignet
essence Inbegriff * *Substanz, Wesen, Geist, das Wesentliche, Kern; of the essence:von wesentlicher Bedeutg.* || Kern * *Wesen, das Wesentliche; in essence: im ~* || Sinn * *Wesen, Geist, das Wesentliche, der Kern, essence and purpose: ~ und Zweck* || Substanz * *innere Natur, Wesen, das Wesentliche, der Kern (der Sache)*
essential beträchtlich * *wesentlich* || eigentlich * *wesentlich* || entscheidend * *wesentlich, (lebens)wichtig, unentbehrlich* || erforderlich * *wichtig* || grundlegend || Hauptsache || hauptsächlich || Kern- * *wesentlich, Kernstück, Kernthema* || lebenswichtig || notwendig * *wesentlich* || unabdingbar * *unbedingt erforderlich, wesentlich, (lebens)wichtig* || unbedingt erforderlich || unentbehrlich * *unbedingt*

erforderlich, (lebens)wichtig || unerlässlich * *unbedingt erforderlich, (lebens)wichtig* || unumgänglich * *unbedingt erforderlich, (lebens)wichtig* || weitgehend * *wesentlich, eigentlich, wichtig* || weitgehend * *wesentlich, eigentlich, wichtig* || wesentlich * *wichtig (auch Substantiv: ~er Punkt)* || wesentlicher Punkt || wichtig * *wesentlich, →Kern*
essential features wesentliche Merkmale
essential function Grundfunktion
essential part Kernteil
essential point wesentlicher Punkt
essential properties typische Merkmale * *kennzeichende Eigenschaften*
essential quantity Kenngröße
essentially im Prinzip * *im Wesentlichen, eigentlich, im Prinzip* || im Wesentlichen || weitgehend * *(Adv.) im Wesentlichen, eigentlich, wichtig, in der Hauptsache* || weitgehend * *(Adv.) im Wesentlichen, eigentlich, in der Hauptsache* || wesentlich * *(Adv.)*
establish aufbauen * *z.B. Strom, Spannung, Feld, eine Organisation, eine Kommunikationsverbindung* || aufnehmen * *Beziehungen ~* || aufstellen * *System ~, Rekord ~* || bekanntmachen * *z.b. ein Produkt/eine Marke am Markt* || bilden || durchsetzen * *establish oneself: sich etablieren/durchsetzen, z.b. am Markt/eine neue Technik* || einführen * *Einrichtungen usw.; well established: gut eingeführt (Firma, Produkt)* || einrichten * *z.b. Firma, Geschäft, Niederlassung ~* || ermitteln * *feststellen, festsetzen, nachweisen, etablieren* || etablieren * *festlegen * festsetzen, einführen, etablieren, erlassen, durchsetzen* || feststellen * *festsetzen, nachweisen, begründen* || gründen * *z.b. eine Firma, Niederlassung; auch sich niederlassen* || herstellen * *errichten, durchsetzen, schaffen; z.b. (Betriebs-) Zustand, Leitungs-/Telefonverbindung* || nachweisen * *feststellen, nachweisen* || vorgeben * *bilden, einrichten, herstellen (Verbindung), etablieren; to: jdm./etwas*
establish a company Firma gründen
establish a foothold Fuß fassen * *(auch figürl.: Unternehmen usw.)*
establish optimum conditions optimieren
establish the connection Verbindung herstellen
establish the factory default condition urladen * *Parametersatz ~*
established feststehend * *feststehend (Tatsache, Brauch usw), planmäßig, fest begründet, unzweifelhaft.* || selbstständig * *beruflich, geschäftlich* || ständig * *fest begründet, eingeführt, feststehend, z.B. Praxis, Regel*
establishing Aufnahme * *~ von Beziehungen*
establishment Einführung * *~ von Einrichtungen* || Einrichtung * *Gründung; öffentliche/private Einrichtung/Institution* || Festlegung * *Festsetzung, Feststellung* || Feststellung * *Feststellung, Festsetzung* || Gründung * *Errichtung*
estate Grund * *Land, ~ u. Boden*
estimate abschätzen * *schätzen* || Abschätzung * *superficial estimate: überschlagsmäßige/grobe ~* || anschlagen * *(ver)anschlagen, schätzen* || bemessen * *abschätzen* || Berechnung * *(Ab)Schätzung* || beurteilen * *abschätzen* || bewerten * *abschätzen* || rechnen * *veranschlagen* || schätzen * *abschätzen* || Schätzung * *superficial/rough estimate: überschlagsmäßige ~* || veranschlagen * *(ab)schätzen*
estimated bewertet * *abgeschätzt* || überschlägig * *geschätzt* || überschläglich * *geschätzt* || überschlagsmäßig * *geschätzt* || voraussichtlich * *(Adj.) geschätzt, schätzungsweise*
estimated value Näherungswert * *Schätzwert*
estimation Abschätzung * *Bewertung * *(Ein)Schätzung, Meinung, Urteil* || Ermittlung * *durch Abschätzung* || Schätzung

etc usw
etch ätzen
etching acid Beize
etching solution Beize
ethylene Äthylen
ETSI ETSI * *Akürzg.f. 'European Telecommunications Standards Institute':europ.Norm.grem. f.telekommunikat*
etui Etui
EU EU * *Abkürzung für 'European Union': Europäische Union*
EU Directive EG-Richtlinie * *Abk. für 'Union Europan'*
EURO flange EURO-Flansch
Euro voltage Eurospannung * *genormte Netzspannungen, z.B. 230 V, 400 V, 690 V usw.*
Euro-card Europakarte * *Leiterkarte 100 x 160 mm*
Euro-supply voltages Euro-Netzspannungen * *siehe →Eurospannungen*
Euro-voltages Eurospannungen * *Netz-Normspannungen nach DIN IEC 38/5.87 (z.B. 230V, 400V, 690V usw.)*
Eurocard standard format Europaformat * *100 x 160 mm für Leiterkarte*
Euroformat Europaformat * *100 x 160 mm für Leiterkarte*
Europe Europa
Europe format Europaformat * *100 x 160 mm für Leiterkarte*
European europäisch
European Council Directive EG-Richtlinie
European Standard EN || Europäische Norm || Europanorm
European standard supply voltages Eurospannungen
evacuate abpumpen || ausräumen || entleeren || evakuieren * *auch Luft* || leeren * *ent~, aus~*
evacuate the air from entlüften
evacuated leer * *geräumt*
evacuation Abtransport * *Aus/Entleerung, Verlegung, Räumung, Umsiedlung* || Entleerung * *Aus-/Entleerung, Abtransport, Räumung*
evacuation of air Entlüftung
evaluate auswerten || bewerten * *abschätzen, berechnen, beurteilen, bewerten, auswerten* || ermitteln * *berechnen, abschätzen, auswerten*
evaluation Abschätzung * *auch Berechnung* || Auslegung * *Abschätzung* || Auswertung * *Abschätzung, Bewertung, Auswertung, Beurteilung* || Beurteilung * *Abschätzung, Bewertung, Beurteilung, Auswertung* || Bewertung * *Abschätzung, Bewertung, Beurteilung, Auswertung* || Einschätzung || Ermittlung * *durch Berechnung, Messung* || Wertigkeit * *Abschätzung, Bewertung, Beurteilung, Auswertung*
evaluation block Auswertebaustein * *Funktionsbaustein*
evaluation electronics Auswerteelektronik
evaluation factor Bewertungsfaktor
evaluation unit Auswertegerät * *z.B. für (Temperatur-) Messgeber, Näherungsschalter usw.*
evaluator Auswerter
evaporate verdampfen || verdunsten
evaporating temperature Verdampfungstemperatur
evaporation Eindampfung || Verdunstung
evaporation bath cooling Siedebadkühlung * *z.B. bei Bahnumrichter*
evaporation coefficient Verdampfungsziffer
evaporative bath cooling Siedebadkühlung * *z.B. bei Bahn-Stromrichter*
evaporator Eindampfanlage || Verdampfer * *z.B. in der Kältetechnik*
eveloping measuring method Hüllkurvenanalyse

* *zur Ermittlung des Körperschalls (z.B. zur Erkennung von Lagerschäden)*
even auch * *selbst, sogar* || eben * *flach, waagerecht* || gerade * *Zahl* || geradzahlig || gleich bleibend * *gleich/regelmäßig, ausgeglichen, glatt, ruhig, gerade* || gleichmäßig * *ausgeglichen* || Plan * *(Adj.) eben, flach, gerade, waagerecht, in gleicher Höhe* || selbst * *(Adv.) sogar* || selbst * *auch nur*
even if auch * *wenn auch*
even though trotzdem * *(Bindewort)*
even-numbered geradzahlig
even-tempered ruhig * *leidenschaftslos*
evener roll Egalisierwalze * *bei der Papierherstellung*
evening class Abendkurs
evening meal Abendessen
evenness Gleichmäßigkeit
event Ereignis || Fall * *Ereignis, Vor~, Begebenheit* || Veranstaltung || Vorgang * *Ereignis*
event counter Ereigniszähler
event log Meldeprotokoll
event of a short circuit Kurzschlussfall
event printer Meldedrucker
event recorder Drucker * *Meldedrucker* || Meldedrucker
event report Meldeprotokoll
event signalling system Meldesystem
event-controlled ereignisgesteuert
event-driven azyklisch * *ereignisgesteuert* || ereignisgesteuert
eventual etwaig * *eventuell, möglich* || eventuell || möglich * *eventuell* || potenziell * *eventuell* || schließlich
eventually eventuell * *(Adv.)*
ever je * *jemals*
ever before jemals zuvor
every alle * *every 4 sec:alle 4 Sekunden* || jedes * *~ insgesamt; every day: mit jedem Tage; everyone: jeder*
every time jedes Mal
everyday täglich
everything alle * *alles; everything new: alles neue; everything else but:alles andere als* || möglich * *try everything: alles ~e tun*
evidence Aussage * *give evidence: eine ~ machen* || Beweis || beweisen * *zeigen, beweisen* || Beweismittel || Nachweis
evident deutlich * *offensichtlich* || einleuchtend || klar * *augenscheinlich, offenbar* || merklich * *augenscheinlich, offensichtlich, klar* || nachweisbar * *offenkundig* || offensichtlich || sichtbar * *offensichtlich, deutlich ~*
evil schlecht * *böse, übel, boshaft, unglücklich*
evitable vermeidbar
evolution Entwicklung * *Entfaltung, Entwicklung* || Evolution
evolve entwickeln * *entfalten, herausbilden, hervorrufen, erzeugen,ausscheiden, (sich) ~/-falten, entstehen*
evolvent Evolvente
ex works ab Herstellerwerk || ab Werk
ex- alt * *ehemalig*
ex-stock ab Lager || lagermäßig
exact eigentlich * *genau* || exakt || genau || präzise * *genau, exakt* || richtig || scharf * *sorgfältig, genau* || sorgfältig * *genau*
exact synchronism winkelgetreuer Gleichlauf || Winkelgleichlauf
exacting anspruchsvoll * *streng, genau; z.B. Kunde; be exacting: hohe Anforderungen stellen* || schwierig * *anspruchsvoll (z.B. Kunde)*
exacting requirements hohe Anforderungen
exactly eigentlich * *(Adv.) genau* || genau * *(Adv.)* || gerade * *(Adv.)*

exactly as genauso
exactness Genauigkeit || Präzision * *Genauigkeit, Exaktheit* || Sorgfalt * *Genauigkeit*
exam Prüfung * *z.B. ~ an Schule/Hochschule*
examination Besichtigung * *prüfende ~* || Examen || Nachprüfung * *(nähere) Untersuchung, Prüfung, Besichtigung, Durchsicht, Revision, Examen, Verhör* || Prüfung * *genaue Untersuchung, ~/Examen an Schule/Hochschule* || Überprüfung || Untersuchung
examination of the accounts Rechnungsprüfung
examine befassen * *sich prüfend befassen mit* || betrachten * *genau, prüfend ~* || inspizieren * *prüfen* || prüfen * *auf Schule, Hochschule usw.* || überprüfen || untersuchen * *prüfen*
example Beispiel * *for example/e.g.: zum Beispiel; exemplified by/by the example of: am Beispiel von* || Muster * *Vorbild, Beispiel*
example for order Bestellbeispiel
example for wiring up Anschlussbeispiel
example of application Anwendungsbeispiel || Einsatzbeispiel
excavation Hohlraum * *Aushöhlung*
excavator Bagger * *z.B. Trocken-/Greifbagger*
excavator drive Baggerantrieb
exceed größer sein als || liegen über * *überschreiten* || überschreiten * *Maß/Termin/Wert/Grenze usw. ~ (by: um); x times the: um das x-fache von* || übersteigen * *überschreiten (Wert/Größe)* || übertreffen
exceeding größer * *wert/betragsmäßig größer sein als* || über * *mehr als (amtlich, techn.; Zahlenwert), wertmäßig über ... hinaus* || Überschreitung
exceeding of sample time Zyklusüberlauf
exceeding of scan time Zyklusüberlauf || Zykluszeitüberschreitung
excel übertreffen * *übertreffen, sich auszeichnen, hervorragen (in/at: in/bei)*
excellent ausgezeichnet * *hervorragend, vorzüglich* || exzellent || gut * *ausgezeichnet (auch 'perfect)* || herausragend * *ausgezeichnet, hervorragend* || hervorragend * *exzellent, vorzüglich, ausgezeichnet* || mustergültig || perfekt * *hervorragend, ausgezeichnet* || sehr gut || tüchtig * *vorzüglich* || vorzüglich
excellently hervorragend * *(Adv.)*
excelsior Holzwolle * *(amerikan.)*
excenter drive Exzenterantrieb
except außer * *ausgenommen; except that: außer dass* || bis auf * *außer* || nur * *ausgenommen*
except for abgesehen von
except if wenn nicht
except when wenn nicht
exception Ausnahme || Sonderfall
exceptional außerordentlich * *außergewöhnlich, Ausnahme-, Sonder-, ungewöhnlich (gut)* || besonders * *außergewöhnlich, Sonder-* || extrem * *außergewöhnlich, Ausnahme-, ungewöhnlich* || höchst * *ungewöhnlich* || ungewöhnlich * *außergewöhnlich, ungewöhnlich (gut)*
exceptional case Ausnahmefall || Sonderfall
exceptional condition Ausnahmezustand * *exception handler: Programm(teil) zum Reagieren auf ~/Sonderfälle* || Sonderfall || —Ausnahmezustand
exceptional situation Ausnahmesituation * *exception handling: Behandlung von ~en [Computer]*
exceptional status Ausnahmezustand
exceptionally außerordentlich * *(Adv.)* || besonders * *(Adv.) ausnahmsweise, außergewöhnlich*
excerpt Auszug * *aus einem Dokument*
excess Überfluss * *Übermaß, Überfluss, Unmäßigkeit* || überflüssig * *überschüssig* || Übermaß * *Überfluss, Übermaß, Unmäßigkeit* || Überschuss || überschüssig
excess current Stromüberlastung
excess current switch Überstromauslöser
excess pressure Überdruck * *übermäßiger Druck*

excess voltage Überspannung
excess-pressure valve Überdruckventil
excess-three code Drei-Exzess-Code
excess-torque capacity Drehmoment-Überlastbarkeit * *{el.Masch.} kurzzeitige Überlastbarkeit, ohne dass d.Motor stehenbleibt*
excessive extrem * *übermäßig* || überflüssig * *übermäßig* || übermäßig || übertrieben || unangemessen * *zu hoch*
excessive prices Preisüberhöhung
excessive supply Überfluss * *Überangebot, Überversorgung*
excessive temperature rise Temperaturüberlauf * *{el.masch.} Differenz d. Wicklungs- z.gemess. Temp. nach Motorabschaltg.* || Übererhitzung || Übererwärmung || Überhitzung || Übertemperatur * *übermäßige Temperaturerhöhung*
exchange Austausch * *das ~en (for:gegen)* || austauschen * *vertauschen* || auswechseln || Ersatz * *Auswechslung* || Tausch || tauschen || vertauschen || verwechseln || Wechsel * *Tausch* || wechseln * *(miteinander) austauschen*
exchange of energy Energieaustausch
exchange of experience Erfahrungsaustausch * *interdepartmental: zwischen Abteilungen eines Unternehmens*
exchange rate Kurs * *Wechsel~* || Wechselkurs
exchangeable austauschbar * *wechselbar* || auswechselbar
excitation Anregung * *einer Schwingung* || Erreger- || Erregung * *Motor, Bremse usw.* || Feld * *Erregung*
excitation control Erregerstromregelung
excitation control unit Erregergerät * *z.B. für Bremse*
excitation controller Feldregler
excitation converter Erregerstromrichter * *z.B. für Synchronmaschine*
excitation current Erregerstrom
excitation current control Erregerstromregelung
excitation frequency Stoßfrequenz * *Schwingungs-anregende Frequenz*
excitation losses Erregerverluste
excitation of vibrations Schwingungsanregung * *(mechanisch)*
excitation power Erregerleistung
excitation power supply Erregerversorgung
excitation rectifier Erregergleichrichter
excitation setpoint Feldstromsollwert
excitation strength Erregerdurchflutung
excitation supply Erregerversorgung
excitation voltage Erregerspannung
excitation winding Erregerwicklung
excite anfachen * *Schwingung* || anregen * *Schwingung, einspeisen eines Stimulations-Signals* || erregen * *z.B. elektr. Maschine*
excited erregt * *z.B. (Motor)Feld, Relaisspule, Bremse*
exciter Erregermaschine
exciter supply converter Erregerstromrichter
exciter winding Erregerwicklung
exciting current Magnetisierungsstrom
exciting field Feld * *Erreger~*
excitor Erregermaschine
exclude ausschließen
excluded ausgeschlossen || ausschließlich * *(nachgestellt) z.B. packing excluded: ausschließlich Verpackung*
exclusion Ausschluss
exclusion of air Luftabschluss * *z.B. durch Schließen der Luftklappe in Heizungssystem*
exclusive ausschließlich * *of:von* || einseitig * *ausschließlich*
exclusive of abgesehen von
exclusive OR Antivalenz

exclusively allein * *(Adv.) * ausschließlich* || ausschließlich * *(Adv.)* || lediglich || nur * *(Adv.)*
excursion Abweichung * *Abweichung; Abschweifung; Ausflug* || Ausflug || Fahrt * *Ausflug*
excuse entschuldigen || Entschuldigung * *it is unexcusable/there is no excuse for it: dafür gibt es keine ~*
excuse me Entschuldigung * *(Ausspruch) in Verbindung mit einer Bitte*
executable ablauffähig * *ausführbar, z.B. Programmcode auf einem Rechner* || ausführbar * *z.B. Softwareprogramm* || lauffähig * *z.B. Programmcode auf einem Rechner*
execute abarbeiten * *Programm* || ausführen * *Auftrag, Aufgabe, Computerbefehl/Programm/Prozedur* || durchführen * *aus/durchführen* || erledigen * *aus/durchführen* || leisten * *ausführen* || verarbeiten * *Daten, Programm, Eingangssignale*
execution Abarbeitung * *Ausführung, z.B. eines Programms* || Ausführung * *z.B. Programm, Befehl* || Durchführung * *Ausführung, z.B. eines Projektes* || Verarbeitung * *Ausführung, z.B. eines Rechnerprogramms*
execution speed Rechengeschwindigkeit * *für ein bestimmtes Programm/Befehl(sfolge)*
execution time Ausführungszeit * *für Rechnerbefehl/-programm* || Bearbeitungszeit * *Ausführungszeit eines (SPS-/Computer-) Programms/-Befehls*
executive geschäftsführend || Geschäftsführer || organisatorisch * *Verwaltungs-, geschäftsführend/leitend*
executive board Vorstand
executive case Aktenkoffer
executive committee Vorstand
executive management Geschäftsführung
executive module Organisationsbaustein * *[SPS]*
executive president Geschäftsführer
executives Führungsebene || Führungskräfte
exemplary beispielhaft || mustergültig
exemplified by am Beispiel von
exemplify belegen * *durch Beispiele* ~ || illustrieren * *(figürl.) durch Beispiele*
exempt frei * *from:von Steuer, Zoll, Gebühren usw.*
exemption Freiheit * *from: von Lasten usw.*
exercise Aufgabe * *Übungs-/Schulaufgabe* || Praxis * *Übung* || üben * *Lehrstoff/Geist ~; jdn. ausbilden* || Übung
exercise an influence on beeinflussen * *Einfluss ausüben auf*
exercise the fuction of wahrnehmen * *ein Amt/eine Funktion ~*
exercises Übungen * *practical:praktische*
exert ausüben * *gebrauchen, anwenden, z.B. Druck, Einfluss* || erzeugen * *ausüben*
exhaust abführen * *Gas/Dampf ~* || absaugen * *Gas* || Absauggerät || Abzug * *Lüftung* || angreifen * *erschöpfen* || aufbrauchen * *erschöpfen* || Auspuff * *exhaust box: Auspufftopf* || ausströmen * *Gas, Dampf* || ermüden * *erschöpfen* || Luftauslass || verbrauchen * *erschöpfen*
exhaust air Abluft * *ausgestoßene/ausströmende Luft, Abluft, Abgas*
exhaust air purification Abluftreinigung
exhaust box Auspufftopf
exhaust fan Absauggerät * *Absauggebläse*
exhaust gas Abgas
exhaust manifold Auspuffkrümmer
exhaust system Absauganlage
exhaust vent Luftauslass
exhauster Absauganlage || Absaugung * *Gasabsauger*
exhaustion Abbau * *Erschöpfung* || Anstrengung * *Erschöpfung* || Schwäche * *Erschöpfung*
exhaustive eingehend * *gründlich, erschöpfend* ||

erschöpfend * *z.b. Information, Auskunft* || umfassend * *erschöpfend*
exhibit aufweisen * *aufweisen, zeigen, an den Tag legen, entfalten, ausstellen (Waren usw.)* || zeigen * *zur Schau stellen (auf Messe, in Ausstellungsraum usw.), aufweisen, zeigen, an d.Tag legen*
exhibit booth Messestand
exhibition Ausstellung || Messe * *Ausstellung*
exhibition area Ausstellungsfläche * *auf einer Messe*
exhibition booth Ausstellungsstand || Messestand
exhibition grounds Messegelände
exhibition park Ausstellungsgelände || Messegelände * *Ausstellungsgelände*
exhibition room Ausstellungsraum
exhibition space Ausstellungsstand * *Fläche* || Messestand * *Fläche*
exhibition stand Ausstellungsstand || Messestand
exhibitor Aussteller * *auf Ausstellung/Messe*
exhibits ausgestellte Produkte * *auf Messe* || Exponate * *auf Messe,Ausstellung*
exhortation Mahnung * *Ermahnung*
exigency Dringlichkeit * *Dringlichkeit, dringlicher Fall, (dringendes) Erfordernis*
exist anstehen * *vorhanden sein; fault condition exists:Fehler steht an* || existieren || herrschen * *bestehen* || vorliegen * *vorhanden sein, existieren*
existence Existenz * *Vorhandensein, Leben, Lebensunterhalt*
existent existent || vorhanden * *bestehend, existierend*
existing existierend || jetzig * *bestehend* || verfügbar * *existierend* || vorhanden * *bestehend, existierend*
existing message anstehende Meldung
exit Ausgang * *eines Raums/Gebäudes/Programms* || Auslauf || aussteigen * *ein SPS-Programm verlassen* || Durchtritt * *z.B. einer Welle* || herausgehen || verlassen * *herausgehen*
exit feed Abzug * *dort wird Material herausgefördert*
exit machine Austragmaschine * *z.B. für Ofen*
exit opening Durchtrittsöffnung
exit roller table Auslaufrollgang * *Walzwerk*
exit section Abzug * *fördert (bandförmiges) Material aus einer Verarbeitungsstraße heraus* || Auslauf * *Auslaufteil z.B. bei Blechstraße* || Auslaufteil * *z.B. einer Walzstraße*
exitted verlassen * *herausgegangen sein, ~ haben*
EXOR Antivalenz * *Exklusiv-Oder*
expand ausbauen * *erweitern* || ausbreiten || ausdehnen * *erweitern, verlängern* || ausweiten * *erweitern* || dehnen * *ausdehnen, z.B. Impulse* || ergänzen * *erweitern* || erweitern * *vergrößern* * *(sich) ausdehnen* || wachsen * *sich ausdehnen*
expandability Erweiterbarkeit * *modular:modulare*
expandable erweiterbar
expandibility Ausbaufähigkeit
expanding mandrel Spanndorn * *z.B. bei Werkzeugmaschine, Wickler, Haspel usw.*
expanding shaft Spannwelle
expansion Ausdehnung || Dehnung * *Ausdehnung* || Erweiterung || Erweiterungs-
expansion board Erweiterungsbaugruppe
expansion capability Ausbaubarkeit
expansion loop Dehnungsschlaufe * *in (Halbleiterschutz-) Sicherung*
expansion rack Erweiterungs-Baugruppenträger
expansion slot Erweiterungssteckplatz * *Steckplatz* * *Erweiterungs- für Leiterkarten usw.*
expansion stage Ausbaustufe
expansion unit Erweiterungsgerät
expansion valve Expansionsventil * *zum Verdampfen, z.B. bei einer Kältemaschine*
expect erwarten || rechnen * *erwarten*

expect a great deal hohe Ansprüche stellen * *from:an jdm.*
expectancy Erwartung * *Erwartung (-shaltung), Hoffnung, Aussicht*
expectation Erwartung * *(Gegenstand der) Erwartung; come up to/exceed expectations: ~n erfüllen/ übertreffen*
expectation value Erwartungswert
expected erwartet ‖ voraussichtlich * *(Adj.) erwartet, erhofft, womit gerechnet wird* ‖ zu erwarten
expedient Ausweg * *Notbehelf* ‖ Hilfsmittel * *(Hilfs-)Mittel, (Not)Behelf, Ausweg* ‖ Notlösung ‖ praktisch * *tunlich, ratsam, zweckmäßig, praktisch* ‖ zweckmäßig
expedited vorrangig * *beschleunigt (z.B. auszuführen, voranzutreiben, abzusenden)*
expeditious beschleunigt * *schnell, zügig*
expel ausstoßen * *auch (Kühlluft) 'ausblasen'*
expend aufwenden ‖ verwenden * *aufwenden*
expendability Ausbaubarkeit * *Ausbaubarkeit*
expendable ausbaubar ‖ ausbaufähig
expenditure Aufwand * *~ an Geld, Zeit, Kraft; of:an; considerable: großer* ‖ Ausgabe * *Geldausgabe* ‖ Kosten * *Aufwand* ‖ Verwendung * *Verbrauch, Ausgabe*
expenditure of time Zeitaufwand
expenditures Ausgaben * *von Geld*
expense Aufwand * *an Geld; Auslagen, Kosten; invoice according to the effective expenses: nach ~ verrechn.* ‖ Ausgabe * *Geldausgabe* ‖ Kosten * *(Geld-) Ausgabe, Aufwand, Un~, Spesen; at great expense: mit großen ~*
expense statement Spesenabrechnung
expense voucher Spesenabrechnung
expenses Ausgaben * *von Geld* ‖ Kosten * *(Un)Kosten, Auslagen, Spesen; at the expense of: auf ~ von (auch figürl.)* ‖ Spesen * *Ausgaben* ‖ Unkosten
expensive aufwändig * *teuer* ‖ kostspielig ‖ teuer
experience erfahren * *(Verb) erleben* ‖ Erfahrung ‖ Fachkenntnisse * *Erfahrung* ‖ Fachwissen * *Erfahrung* ‖ Knowhow ‖ Praxis * *Erfahrung* ‖ Sachkenntnis * *Erfahrung* ‖ Vorkenntnisse * *Erfahrung*
experience gained in the field Felderfahrung
experienced erfahren ‖ tüchtig * *erfahren* ‖ versiert * *erfahren*
experienced in the trade branchenkundig
experienced person Erfahrungsträger
experienced personnel Erfahrungsträger * *(Plural)*
experiences gained in operation Betriebserfahrungen
experiment experimentieren * *with: mit* ‖ Probe * *Versuch* ‖ Versuch
experimental empirisch * *experimentell* ‖ experimentell ‖ Versuchs-
experimental set-up Versuchsanordnung
expert erfahren * *in:in* ‖ Facharbeiter ‖ Fachingenieur ‖ fachkundig ‖ Fachmann * *Experte (on:für); in this field:auf diesem Gebiet* ‖ fachmännisch ‖ Könner ‖ Sachverständiger ‖ Spezialist ‖ versiert * *in: in*
expert consultation Fachberatung
expert knowledge Expertenwissen ‖ Sachkenntnis
expert opinion Gutachten
expert system Expertensystem
expert worker Facharbeiter
expert's report Gutachten
expertise Erfahrung * *[expe'ties] Fachwissen, Expertenwissen* ‖ Expertenwissen * *[expe'ties]* ‖ Fachkenntnisse * *[expe'ties] Expertenwissen* ‖ Fachwissen * *[expe'ties] Expertenwissen* ‖ Kenntnis * *[expe'ties] Expertenwissen* ‖ Knowhow * *[expe'ties] Expertenwissen* ‖ Sachkenntnis * *[expe'ties] Expertenwissen, Fachwissen* ‖ Sachverstand * *[expe'ties]* ‖ Wissen * *[expe'ties] Experten~, Fach-/Sachkenntnis*
expertly fachmännisch * *(Adv.)*

expiration Ablauf * *einer Frist* ‖ Ablauf * *einer Frist, Zeit usw.*
expire ablaufen * *Frist, Zeit, Garantie usw.* ‖ vergehen * *Zeit, Frist* ‖ verstreichen * *Frist*
expired abgelaufen * *Frist, Überwachungszeit*
explain erläutern
explanation Ausführung * *Darlegung, Erklärung* ‖ Erklärung * *Erläuterung (of:für/zu)* ‖ Erläuterung ‖ Vorstellung * *Erläuterung*
explanatory erklärend
explanatory notes Erläuterungen
explane erklären * *erläutern*
explicit ausdrücklich ‖ explizit * *ausdrücklich, deutlich, bestimmt*
explode bersten * *explodieren*
exploded view explodierte Darstellung ‖ Explosionsdarstellung
exploit abbauen * *Bodenschätze ~* ‖ ausnutzen * *auswerten, (kommerziell) verwerten, ausbeuten (auch figürl.), ausnutzen, ausschlachten* ‖ erschließen * *nutzbar machen; auch "Bodenschätze usw. ~ "* ‖ nutzen * *ausnützen* ‖ nützen * *ausnützen*
exploitation Abbau * *Abbau/Gewinnung (von Bodenschätzen), Verwertung/Ausnutzung/-beutung (auch figürl.)*
exploration Suche * *Suche (nach Bodenschätzen), Erforschung, Untersuchung* ‖ Untersuchung * *Erforschung, Untersuchung*
explore untersuchen * *erforschen*
explosion class Explosionsklasse * *Explosionsgruppe nach alter Norm VDE0170/0171*
explosion group Explosionsgruppe * *kennzeichnet jew. Gase ähnlichen Zünddurchschlagvermögens, siehe EN50014 bis 50020* ‖ Zündgruppe
explosion proof druckfest * *(amerikan.) Explosionsschutzart*
explosion protected Ex-geschützt * *siehe EN50014* ‖ explosionsgeschützt * *siehe EN50014*
explosion protection Ex-Schutz ‖ Explosionsschutz
explosion-proof explosionsgeschützt ‖ schlagwettergeschützt * *explosionsgeschützt*
explosion-proof EEx d type of protection Zündschutzart EEx d * *Explosionsschutzart für Motor*
explosion-proof enclosure AD-PE * *Schutzart druckf.Kapselg.i.Italien: Antideflagranti Prove di Esplosione* ‖ druckfeste Kapselung * *(amerikan.) Zünd-/Explosionsschutzart*
explosion-protected design explosionsgeschützte Ausführung
explosion-protected motor explosionsgeschützter Motor
explosion-protected zone Ex-Schutzzone
explosion-protection directive Explosionsschutz-Richtlinie
explosions-protected explosionsgeschützt
explosive explosionsfähig
exponential exponentiell
exponential characteristic e-Funktion
export ausführen * *exportieren*
export authorization Ausfuhrgenehmigung
export licence Ausfuhrgenehmigung * *(brit.)*
export license Ausfuhrgenehmigung * *(amerikan.)*
export regulation Exportvorschrift
export share Exportanteil
export-minded exportorientiert
expose ausstellen * *zur Schau stellen, z.B. auf Messe*
expose to aussetzen * *Hitze, Wind, Wetter, Unbill, Gefahr usw.*
expose to danger gefährden
exposed offenliegend * *unbedeckt, offen; exponiert, gefährdet* ‖ ungesichert * *—>offenliegend*
exposition Ausstellung * *(amerikan.)*
exposure to radiation Bestrahlung

express ausdrücklich || Schnell- * *Versand, Reparatur usw.*
express repair Schnellreparatur
expression Ausdruck || Redewendung
expressly ausdrücklich * *(Adv.)*
exquisite raffiniert * *Aufmachung, Stil*
extend ausbauen * *vergrößern, erweitern* || ausbreiten || ausdehnen || ausweiten * *erweitern* || erstrecken * *auch 'sich ~' (to: bis zu; over: über); extend up to: sich ~ bis (zu)* || erweitern || laufen * *sich erstrecken (from...to:von...bis)* || reichen * *sich erstrecken; to: bis; from ... to: von ... bis* || strecken * *(aus)dehnen* || vergrößern || verlängern * *vergrößern, erweitern, ausfahren; auch figürl.: Patent, Kredit, Zeitintervall ~*
extendable ausbaufähig || erweiterungsfähig
extendable by modular design modular erweiterbar
extended länger * *verlängert (zeitlich und räumlich)*
extended area erweiterter Bereich * *z.B. Datenbereich bei SPS*
extended I/O area erweiterte Peripherie * *bei SPS*
extended range erweiterter Bereich
extended-duration pulse Langimpuls
extendible dehnbar
extensible ausbaufähig || ausdehnbar || ausziehbar * *verlängerbar* || erweiterungsfähig
extension Ansatz * *Ansatzstück* || Ausbau * *Erweiterung* || Dehnung * *permanent: bleibende* || Durchwahl * *Haus-Telefonapparat* || Erweiterung || Erweiterungs- || Hausapparat * *interne Telefonnummer* || Nebenstelle * *Telefon, Hausapparat* || Verlängerungs-
extension board Erweiterungsbaugruppe || Erweiterungskarte * *Leiterkarte*
extension cable Verlängerungskabel || Verlängerungsleitung
extension kit Anbausatz
extension rod Schubstange * *herausgeschobenes Teil bei Spindeltrieb*
extension slot Erweiterungssteckplatz * *für Leiterkarte*
extension stage Ausbaustufe
extension step Ausbaustufe
extension tube Schubrohr * *herausgeschobenes Teil bei Spindeltrieb*
extension unit Erweiterungsgerät
extensive ausführlich || besonders * *umfangreich, ausgedehnt; siehe auch 'umfassend'* || groß * *weit* || umfangreich * *ausgedehnt* || umfassend * *umfangreich, ausgedehnt* || weitestgehend * *umfassend, ausgedehnt, extensiv* || weitgehend * *umfassend, ausgedehnt, extensiv*
extent Ausdehnung * *Ausmaß* || Ausmaß * *to a large extent:in hohem Ausmaß* || Maß * *Ausdehnung, Aus~* || Umfang * *Ausdehnung, Ausmaß*
extent of data Datenmenge
extent of supply Lieferumfang
extent of work Arbeitsumgang || Aufwand * *Arbeits~*
exterior äußerer
exterminate vernichten * *auslöschen, ausrotten*
external Außen- || äußerer || äußeres * *außen befindlich* || außerhalb * *äußer(lich), Außen-, external to: ~ von* || extern || fremd * *extern* || Fremd-
external appearance Außenansicht * *äußeres Erscheinungsbild* || äußeres Erscheinungsbild
external blower Fremdlüfter
external circuitry Beschaltung
external commutation Fremdführung * *für Stromrichterschaltung*
external conditions äußere Umstände
external control system externe Steuerung
external fan Außenlüfter || Fremdlüfter

external forces äußere Umstände
external gearing Außenverzahnung
external influences äußere Einflüsse
external moment of inertia Fremdträgheitsmoment * *z.B. Trägheitsmoment der Last/der Arbeitsmaschine bei e. Motor*
external operator Fernbedienung * *als Gerät*
external power supply Lastnetzgerät * *externe Kontaktstromversorgung (z.B. 24V) für (SPS-) Ein-/Ausgänge* || Lastnetzteil * *Kontaktstromversorgung (z.B. 24 V) für (SPS-) Ein-/Ausgänge*
external rotor Außenläufer
external rotor motor Außenläufer-Motor || Trommelmotor * *Außenläufermotor*
external signals Anlagensignale || Kundensignale
external supply Fremdeinspeisung
external teeth Außenverzahnung
external thread Außengewinde
external to außerhalb der || außerhalb des
external ventilation Fremdbelüftung
external vibrations Fremdschwingungen * *(mechan.)* || immitierte Schwingungen * *von außen eingeleitete mechanische ~*
external view Außenansicht
external-rotor motor Außenläufermotor
externally außen * *außerhalb, nach außen hin, extern* || außerhalb * *(Adv.)* || extern * *(Adv.)* || Fremd- * *(Adv.)*
externally commutated fremdgeführt * *Stromrichter; netz- oder lastgeführt i.Ggs.zu selbstgeführt*
externally commutated converter fremdgeführter Stromrichter * *netz- oder lastgeführt i.Ggs.zu selbstgeführt*
externally driven fremdangetrieben
externally ventilated fremdbelüftet
externally-driven fan Fremdlüfter
externally-mounted fan Fremdlüfter
extinction löschen * *(Subst.) Löschung, z.B. Lichtbogen ,Feuer, Thyristor* || Löschung * *z.B. ~ eines Thyristor, Lichtbogen, Feuer*
extinction angle Löschwinkel * *[Stromrichter] Winkel zw. Kommutier.ende u.Nulldurchg. d.Komm.spanng. IEC 146-1-1*
extinction voltage Abreißspannung * *Löschspannung*
extinguish abbauen * *abschaffen, aufheben, (aus)löschen, ersticken* || auslöschen || löschen * *aus~, z.B. Lichtbogen, Feuer* || verlöschen * *auch Feuer, Lichtbogen usw.*
extra extra || Sonder- || zusätzlich
extra charge Aufpreis * *zusätzl. Verrechnung/Gebühr/Belastung* || Aufschlag * *zusätzliche Berechnung* || Zuschlag * *Preis~, zusätzliche Gebühr, Berechnung*
extra charges Mehrkosten * *(Preis)Zuschlag*
extra cost Aufpreis * *zusätzliche Kosten, zusätzlicher Preis* || Mehrpreis * *Mehrkosten; at an extra cost: gegen Mehrpreis*
extra costs Aufschlag * *supply at small extra costs: zu einem kleinen (Preis)Aufschlag liefern* || Mehraufwand || Mehrkosten
extra price Aufpreis || Aufschlag * *Zusatzpreis, Preisaufschlag* || Mehrpreis || Zuschlag * *Preisaufschlag, "Zusatzpreis"*
extra slot Erweiterungssteckplatz * *für Leiterkarte*
extract abziehen || ausblenden * *herausholen, (her)ausziehen (from:aus/von)* || auskoppeln * *z.B. Signal* || Auszug || erarbeiten * *herausarbeiten, ab/ herleiten* || extrahieren * *herausziehen* || gewinnen * *Bodenschätze, Stoff durch chemische Prozess usw. ~* || herauslocken || ziehen * *herausziehen*
extract the root radizieren * *(mathemat.)*
extracted air Abluft * *abgesaugte Luft*
extracting handle Ziehgriff
extracting screw Abdrückschraube

extracting tool Abziehvorrichtung
extraction Abfuhr || Absaugung || Entnahme || Förderung * *von Bodenschätzen*
extraction tool Abzieher * *Werkzeug* || Abziehwerkzeug
extractor Absauganlage || Absauggerät || Abzieher || Abziehwerkzeug || Ziehhilfe
extraneous fremd * *nicht dazugehörig*
extraneous field Streufeld
extraordinarily außerordentlich * *(Adv.)*
extraordinary außergewöhnlich
 * →*außerordentlich* || außerordentlich
 * →*ungewöhnlich*, →*außergewöhnlich* || höchst
 * *außerordentlich*
extrapolate extrapolieren
extrapolation Hochrechnung
extras Nebenkosten || Zubehör
extravagant kostspielig * *aufwändig*
extreme außerordentlich * →*extrem* || besonders * *sehr, extrem* || extrem || höchst * *äußerst, weitest, höchst, außergewöhnlich, radikal* || letzt * *äußerst* || *radikal* * *extrem* || übertrieben
extreme requirements höchste Anforderungen
extreme value selection Extremwertauswahl * *als Funktion*
extreme-value Extremwert
extreme-value selector Extremwertauswahl * *als Funktionsbaustein/Einrichtung*
extremely extrem * *(Adv.)*
extremely accurate hochgenau
extremely high level Höchstmaß * *an extremely high level of: ein ~ von*
extremity Grenze * *äußerstes Ende*
extremly sehr * *(Adv.) extrem, höchst, äußerst*
extremum Extremwert
extrude extrudieren || pressen * *strang~* || strangpressen
extruded extrudiert || Guss- * →*Strangguss-* || stranggepresst
extruded aluminium Alu-Profil || Aluminium-Strangguss || Aluminiumprofil
extruded aluminium alloy Aluminium-Strangguss * *aus Aluminium-Legierung*
extruded section Profil * *Stranggussprofil* || Strangguss * *Stranggussprofil* || Stranggussprofil || Strangpressprofil
extruder Extruder
extruding press Strangpresse
extrusion Extrusion || Extrusion * *'Strangpressen' von erhitztem Gummi, Kunststoff.., durch Schnekken-/Spalt-/Blasextruder*
extrusion casting Strangguss
extrusion coating Extrusionsbeschichtung * *z.B. zur Erzeugung kunststoffbeschichteten Papiers*
extrusion molding strangpressen * *(Substantiv)*
extrusion press Strangpresse || Ziehpresse
extrusion shear Stranggussschere
exudation ausschwitzen * *(Subst.) Ausschwitzen, Auschwitzung (von Schmiermittel usw.)*
exude ausschwitzen
eye Öse
eye appeal Werbewirksamkeit * *optisch*
eye bolt Ringbolzen * *Hebeöse, Kranöse* || Ringschraube * *Hebeöse, Kranöse*
eye catcher Blickfang
eye-catching attraktiv * *das Auge anziehend* || einprägsam * *optisch* || formschön * *das Auge auf sich ziehend, auffällig* || werbewirksam * *optisch, das Auge auf sich ziehend*
eyebolt Ringschraube * *Schraube mit Hebeöse*
eylet Öse * *kleinere*

F

f.b. ist * *Abkürzg.f. 'feedback'*
F/V-converter f-u-Wandler
fab Fabrik * *(salopp) Abk. f. 'FABrication plant': Produktionsanlage; nofab company: Firma ohne Fabrik*
fab new fabrikneu
fab-wide fabrikweit * *(salopp) über die ganze Fabrikationsanlage*
fabric press Siebfilzpresse * *in Papiermaschine*
fabricate anfertigen
fabrication Herstellung
fabulous super * *(salopp) fabelhaft* || toll * *(salopp) fabelhaft*
face abdrehen * *Stirnfläche mit einer Drehmaschine* || Fläche * *Stirnfläche, Hammer-/Ambossbahn usw.* || fräsen * *~ von Stirnflächen* || gegenüberliegen * *Oberfläche* || plandrehen || Planfläche || rechnen * *face a thing: mit etwas (Unangenehmem) ~ müssen* || Stirnfläche || Stirnseite * *(auch "stirnseitig")* || Vorderseite
face down plandrehen
face plate Planscheibe
face to face with gegenüber * *Personen*
face-mounted stirnseitig montiert
face-mounting motor Flachmotor * *[el.Masch.]*
faceplate Frontplatte || Planscheibe
facilitate erleichtern * *auch 'fördern'* || ermöglichen * *erleichtern, fördern* || fördern * *erleichtern* || sorgen für * *ermögliche, erleichtern, fördern; Einrichtung(en)/Resourcen bereitstellen für*
facilities -Werk || Mittel * *Einrichtungen, Anlagen, Möglichkeiten, Gelegenheiten*
facility Anlage * *Betriebs-* || Betrieb * *Einrichtung* || Einrichtung * *Ausrüstung, Institution, (Instrie-) Anlage, Werkstatt, Labor, (öffentliche) Einrichtung* || Möglichkeit * →*Einrichtung, Vorteil, Gelegenheit; for: für,zu*
facing Beschichtung * *Außenschicht, Belag, Überzug* || gegenüber || gegenüberliegend || Oberflächenbearbeitung * *Bearbeitung von Stirnflächen (mit Werkz.maschine)* || Planlage
facing lathe Plandrehmaschine
facing the drive end mit Blick auf die Antriebsseite
facings Span * *Frässpäne* || Späne * *Fräßspäne*
facsimile Modell * *genaue Nachbildung* || Nachbildung * *genaue ~*
fact Tatbestand || Tatsache || Umstand * *Tatsache, Tatbestand*
factor Einflussgröße * *(allg.)* || Faktor || Gesichtspunkt * *Faktor* || Größe * *Einflußgröße,Faktor* || Moment * *Faktor* || Wert * *Factor*
factor of connection Schaltungsfaktor * *[el.-Masch.]* || Schalthäufigkeit
factor of inertia Trägheitsfaktor * *[el.Masch.]* Verhältnis Gesamtträgheitsmoment zu Motorträgheitsmoment
factor which determines the -bestimmend
factors Umstand * *Einflussfaktoren; favourable factors: günstige Umstände*
factors which determine the speed drehzahlbestimmende Größen
factory -Werk || Betrieb * *Fabrik(anlage)* || Fabrik || Fertigungsbetrieb || Herstellerwerk || Werk * *Fabrik; ex factory: ab ~*
factory adjustment Werkseinstellung
factory automation Fabrikautomatisierung
factory bus Fabrikbus
factory control level Leitebene

factory default Werkseinstellung * *werksseitig eingestellter Wert z.B. eines Parameters*
factory default condition Urladezustand * *Werkseinstellung z.b. von Parametern*
factory environments Fabrikumgebung
factory equipment Betriebsmittel
factory floor Fabrikumgebung * *(salopp) at the factory floor: in der (rauen)* ~ || Feld * *(figürl.) in der untersten Ebene der Fertigung/Produktion, (harter) Praxiseinsatz* || Fertigungsbereich * *on: im*
factory info Werksmitteilung
factory inspector Gewerbeaufsichtsbeamter
factory memo Werksmitteilung
factory network Fabriknetz
factory setting jungfräulicher Zustand * *Werkseinstellung, z.B. von Parametern* || Urladewert * *eines Parameters* || Urladezustand * *Werkseinstellung z.B. von Parametern* || Voreinstellung * *Werkseinstellung z.B. v. Parametern* || Werkseinstellung * *z.b. von Einstellparametern; reset to factory setting: ~ herstellen*
factory tour Werksbesichtigung
factory type test Typprüfung * *im Herstellerwerk*
factory wide fabrikweit
factory-assembled anschlussfertig || fertig montiert
factory-fitted werksmontiert
factory-trained werksgeschult
facts Daten * *Tatsachen* || Sachverhalt
facts of the case Sachverhalt
factual sachlich * *sachbezogen* || tatsächlich
faculty Sinn * *Anlage, Geisteskraft*
fade schwinden * *z.B. Ton, Licht, Farbe* || übergehen * *von Farbtönen usw.*
fade away abklingen * *dahinschwinden, verschwinden, vergehen, zerrinnen* || schwinden * *z.B. Ton,Licht,Farbe* || verschwinden * *hinwegschwinden*
fading Abschwächung * *Schwund* || Schwund * *Bremse, Kupplung, Radio*
fail ausfallen * *auch Spannung, Gerät, Maschine, Netz* || aussetzen * *versagen; failure:das Versagen* || versagen
fail in one's duty vernachlässigen * *seine Pflicht ~*
fail to act versagen
fail to meet überschreiten * *Termin ~*
fail to start nicht anlaufen * *Motor, Maschine, Rechner*
fail to work versagen
fail-safe ausfallsicher || drahtbruchsicher || fehlersicher || fehlersicher * *Fehler können keine unsicheren Zustände herbeiführen* || funktionssicher || sicherheitsgerichtet
fail-safe brake Federdruckbremse * *nach dem Ruhestromprinzip arbeitend (schließt bei Spannungsausfall per Federdruck)* || Ruhestrombremse * *zum Lüften Strom erforderlich –> Bremse fällt bei Stromausfall ein* || Sicherheitsbremse * *Bremse, die bei Stromausfall automatisch einfällt (Ruhestromprinzip)*
fail-safe coupling Sicherheitskupplung * *Reibkupplg. o. ~ mit Sollbruchstelle, öffn. bei Überschreiten eines Drehmoments*
fail-safe holding brake Ruhestromhaltebremse
fail-safe principle Ruhestromprinzip * *z.B. b.Bremse: b. Stromausfall Bremse geschlossen; according to the...:nach d.*
failing daneben * *verpassend; fail to hit: am Ziel vorbeigehen/nicht treffen*
failure Ausfall * *Ausbleiben, Fehlen, Versagen, Zusammenbruch, Fehler,* || Betriebsstörung || Fehler * *Versagen, Fehlschlag, Zusammenbruch* || Fehlfunktion * *Fehler* || Misserfolg * *be unsuccessful: keinen →Erfolg haben* || Panne || versagen * *(Substantiv; auch figürl.: einer Person)*
failure phenomena Ausfallerscheinungen

failure prone fehleranfällig * *siehe auch 'störanfällig'*
failure rate Ausfallrate || Fehlerrate * *Ausfallrate*
faint leise * *matt, schwach, kraftlos (Stimme, Ton)* || schwach * *matt, kraftlos, schwach*
faintness Schwäche
fair akzeptabel || angemessen * *reell, z.B. Preis* || annehmbar * *ehrlich, fair* || Ausstellung * *Messe* || fair || kulant * *Preis, Bedingungen usw.* || Messe || richtig * *gerecht* || zumutbar * *gerecht, angemessen, anständig*
fair dealing Kulanz
fair ground Messegelände
fair grounds Ausstellungsgelände * *Messegelände* || Messegelände
fairground ride Fahrgeschäft
fairly einigermaßen * *ziemlich* || ziemlich
fairly exact annähernd
fall absinken * *fallen, z.B. Preise* || Absturz || Einbruch * *wirtschaftlicher ~* || Fall * *das ~en; auch von Preiseen; give a fall: zu ~ bringen* || fallen * *auch hin~; auch Preise/Bemerkung usw; fall within the scope of: in eine Kategorie ~* || sinken * *Preise, Kurse* || Sturz * *sudden fall: unerwarteter ~; have a bad fall: einen schweren ~ tun*
fall back zurückgehen * *zurückfallen, ins Hintertreffen gelangen (to:auf)* || zurückschalten * *automatisch in einen Zustand geringerer Leistung(sfähigkeit) ~ (z.B. Baudrate)*
fall back on zurückgreifen auf * *auf Notlösung*
fall back upon zurückgreifen auf * *auf Notlösung*
fall below unterschreiten
fall down abstürzen
fall in pressure Druckabfall
fall in prices Preisrückgang
fall off abfallen * *abnehmen, hinunterfallen* || nachlassen * *~ der Verkäufe usw.* || schwinden * *abnehmen* || verschlechtern * *sich ~ (in:an/in)* || zurückgehen * *~ des Geschäfts usw.*
fall out herausfallen
fall out of step außer Schritt fallen || außer Tritt fallen || Kippen * *(Verb) ~ einer el. Maschine*
fall out of synchronism außer Schritt fallen || außer Tritt fallen * *Synchronität verlieren* || Kippen * *(Verb) ~ einer el. Maschine*
fall to a person's share entfallen * *auf jemanden (anteilmäßig) entfallen*
fall to zero auf Null zurückfallen || Nullwerden * *(Verb)* || zu Null werden * *auf Null zurückfallen*
fall under gehören * *zählen zu*
falling abfallend || fallend
falling edge abfallende Flanke * *~ eines Impulses* || fallende Flanke * *eines Impulses* || hintere Flanke * *fallende Flanke* || negative Flanke * *fallende Flanke*
falling off of business Geschäftsrückgang
falling off of orders Auftragsrückgang
falling to zero Nullwerden * *Auf-Null-Abfallen*
falling-off Rückgang * *Nachlassen, Abnahme, Abfall (Produktion, Umsatz usw.)*
false falsch || irrtümlich || künstlich * *unecht* || scheinbar * *vorgeblich* || verkehrt * *falsch*
false measurement Fehlmessung
false polarity Verpolung * *falsche Polung*
falsify verfälschen
falter schwanken * *zaudern, zögern*
familiar bekannt * *be familiar with: mit etwas bekannt/vertraut sein*
familiarization period Einarbeitungsphase
familiarize einarbeiten * *someone with his new work: jmd. in seine neue Aufgabe; oneself: sich*
familiarize oneself lernen * *sich vertraut machen (with:mit)*
familiarize oneself with sich vertraut machen mit
familiarize with vertraut machen mit * *auch sich ~*

familiary geläufig * bekannt, vertraut
family Schar * von Kurven
family business Familienbetrieb
family of curves Kurvenschar
family of processors Prozessorfamilie
family-owned business Familienbetrieb
family-owned company Familienbetrieb * Unternehmen in Familienbesitz
family-run company Familienbetrieb
fan Fächer * siehe auch 'Fach' || Gebläse * Lüfter, Ventilator || Lüfter || Ventilator
fan and pump drives Strömungsmaschinenantriebe * Lüfter- und Pumpenantriebe
fan assembly Lüfterbaugruppe || Lüftereinheit || Ventilatorbaugruppe
fan blade Lüfterflügel || Ventilatorflügel
fan characteristic Lüfterkennlinie || Ventilatorkennlinie * Lastkennlinie (M als f(n)) eines Lüfters (normalerweise M~n²)
fan connecting terminals Lüfter-Anschlussklemmen
fan cover Lüfterhaube
fan cowl Lüfterhaube
fan drive Gebläseantrieb * Lüfterantrieb || Lüfterantrieb || Ventilatorantrieb
fan dryer Ventilatortrockner
fan effect Lüfterwirkung
fan heater Heizlüfter
fan hood Schutzhaube * für Lüfter
fan impeller Lüfterrad
fan louvre Kiemenblech * Lüfterjalousie, Abdeckblech mit Schlitzen für Be-/Entlüftungsöffnung
fan module Lüfterbaugruppe || Ventilatorbaugruppe
fan monitor Lüfterüberwachung * als Einrichtung || Ventilatorüberwachung * als Einrichtung
fan monitoring Lüfterüberwachung * als Funktion || Ventilatorüberwachung * als Funktion
fan motor Lüftermotor
fan noise Lüftergeräusch
fan out Treiberfähigkeit * eines Logikschaltkreises
fan pressure Druckdifferenz * bei Lüfter || Drukkerhöhung * bei Ventilator
fan propeller Lüfterflügel || Ventilatorlaufrad
fan pump Kreiselpumpe || Mischpumpe * z.B. für Zellstoff
fan rotor Lüfterrad
fan shroud Lüfterhaube
fan slide-in unit Lüftereinschub
fan subassembly Lüfterbaugruppe
fan unit Lüfter * bei Motor: axially/radially/separately mounted: an-/aufgebauter/getrennt montierter ~ || Lüfteraggregat || Ventilatorbaugruppe
fan wheel Lüfterrad
fan-cooled lüftergekühlt
fan/pump drive Strömungsmaschinenantrieb
fan/pump/compressor drive Strömungsmaschinenantrieb
fancy roller Putzwalze * {Textil} Trommel~ bei Kratzenraummaschine || Trommelputzwalze * {Textil} bei Kratzenraummaschine || Volant * {Textil} Trommelputzwalze zum Reinigen der Kratztrommel bei Kratzenraummaschine
fans and centrifugal pumps and compressors - Strömungsmaschinen
fans and pumps Strömungsmaschinen
fans, pumps and compressors Strömungsmaschinen
fantastic prima * (salopp) fantastisch || toll * (salopp) fantastisch || vorzüglich * (salopp) fantastisch
fantastically prima * (Adv., salopp) fantastisch
fanwheel Laufrad * eines Lüfters
FAQ FAQ * Abk. f.. 'Most frequently Asked Questions": die am häufigsten gestellten Fragen || Fragesammlung * Frequently Asked Questions: Kollektion häufig gestellter Fragen u. ihrer Anworten || oft gestellte Fragen * Abk.f. 'Frequently Asked Questions'
far entfernt || weit * weit, fern, entfernt, ~ draußen, ~aus; far and wide: ~ entfernt; so far: bisher/bis jetzt
far-away entfernt
far-reaching weitestgehend || weitgehend * weit reichend
Faraday's law Induktionsgesetz
farm machinery Landmaschinen
farming Landwirtschaft
farther weiter * weiter (entfernt)
fascia Firmenschild || Schild * Aushängeschild
fascia plate Kopfleiste * an Schaltschrank
fascinating interessant * faszinierend
fashion Art * Art u.Weise, Stil, Brauch, Sitte, || Art und Weise * Stil, Brauch, Sitte || Mode || Weise
fashionable aktuell * modisch || modern * modisch
fashioned -förmig * ausgeführt, geformt
fast flink || schnell
fast 'semiconductor' type fuse Halbleiter-Sicherung * zum Schutz v. Leistungshalbleitern || Halbleitersicherung * zum Schutz von Leistungshalbleitern || superflinke Sicherung * ~ zum Schutz von Halbleitern || Thyristorsicherung * Halbleitersicherung
fast discharge Schnellentladung * z.B. von Zwischenkreiskondensatoren beim Abschalten eines Umrichters
fast excitation schnellerregt * Motor
fast info Schnellmitteilung
fast overview Schnellübersicht
fast response Dynamik * hohe Dynamik || echtzeitfähig * mit kurzen Reaktionszeiten || hohe Dynamik
fast stop Schnellhalt
fast-acting flink * Sicherung || schnellschaltend * Sicherung || superflink * Sicherung
fast-action clamping system Schnellspannsystem
fast-on connector Flachsteckverbinder || Steckanschluss * (Flach)Steckverbinder
fast-operation schnellansprechend
fast-response hochdynamisch || schnellansprechend * mit kurzer Reaktionszeit
fast-running schnellaufend
fast-switching schnellschaltend * auch Schalttransistor
fasten anschlagen * befestigen || ansetzen * befestigen (to:an) || befestigen * fest machen
fasten with pins verstiften
fastener Befestigung * Verschluss, Befestigungsteil || Befestigungselement * Verschluss, Halter, Befestigungsvorrichtung, Sicherung || Halter * Befestigung, Verschluss, Halter || Verbindungselement * Befestigungselement (z.B. Schraube, Mutter, Niete, Bolzen) || Verschluss * ~mittel, Befestigungseinrichtung
fastening Befestigung || Verschluss * ~mittel, das Verschließen
fastening bolt Befestigungsschraube
fastening element Befestigungselement || Verbindungselement * Befestigungselement
fastening hole Befestigungsbohrung
fastening parts Befestigungsteile
fastening screw Befestigungsschraube
fastenings Befestigungsteile
faster schneller
faston connector Faston-Anschluss
FASTON tab Flachstecker * Faston-Steckzunge
FASTON tab connector Faston-Stecker || Flachstecker
FASTON tab receptacle Faston-Stecker * Flachsteckhülse, "weiblicher" Faston-Stecker || Flach-

stecker * *weiblicher Faston-Stecker, Flächsteckhülse* || Flachsteckhülse * *'Weiblicher' Faston-Stecker* || Steckhülse * *FASTON Steckhülse*
FASTON terminal Flachstecker * *Faston-Flachsteckeranschluss/Verbindung*
FAT Dateibelegungstabelle * *'File Allocation Table' (siehe →FAT) zur Dateibuchhaltung in Betriebssystem* || dick * *fett* || FAT * *'File Allocation Table': enthält für jeden →Cluster d.Festplatte e. Verweis auf d.Folgecluster*
fatal schwer * *fatal (z.B. Fehler), den ordnungsgemäßen Ablauf verhindernd, tödlich, verhängnisvoll*
fatigue altern * *ermüden* || Alterung * *[fe'tieg] Ermüdung* || ermüden * *von Material* || Ermüdung * *Materialermüdung* || Materialermüdung * Verschleiß * *Ermüdung* || verschleißen * *ermüden, altern*
fatigue crack Alterungsriss * *[fe'tieg ...] Ermüdungsriss*
fault defekt * *(Substantiv) in:an* || Fehler * *Fehlfunktion, Vergehen, Schuld, (Charakter) Schwäche; diagnose a fault: e. ~ erkennen* || Funktionsfehler || Panne || Problem * *Fehler* || Störung * *Fehler, Mangel; fault condition occurs: Störfall tritt auf*
fault acknowledge Fehlerquittierung * *als Kommando*
fault acknowledgement Fehlerquittierung || Störquittierung || Störungsquittierung
fault alarm Fehlermeldung || Störmeldung
fault alarm relais Fehlermelderelais
fault analysis Fehleranalyse || Fehleruntersuchung || Störanalyse
fault cause Fehlerursache || Störungsursache
fault channel Störungskanal
fault clear Fehlerquittierung * *Signal*
fault code Fehlerkennung || Fehlernummer || Fehlerschlüssel
fault condition Fehlerfall * *under:im* || Fehlerzustand
fault condition exists Fehler steht an
fault correction Fehlerbehebung || Fehlerbereinigung || Fehlerbeseitigung
fault current Fehlerstrom * *auch bei ~ Kurzschluss, Wechselrichterkippen, Isolationsversagen usw.* || Kurzschlussstrom * *Fehlerstrom* || Störungsstrom * *z.B. bei Kurzschluss, Wechselrichterkippen usw.*
fault description Fehlerbeschreibung
fault detector Fehlerdetektor * *erfasst Funktionsfehler*
fault diagnosis Fehlerdiagnose * *[daieg'nousis]* || Fehlersuche * *Fehlerdiagnose*
fault diagnostics memory Fehlerdiagnosespeicher
fault evaluation Fehlerauswertung
fault finding Fehlersuche || Störungsbeseitigung || Störungssuche
fault fix Fehlerbehebung * *in (bereits ausgelieferter) Software*
fault fixing Fehlerbehebung || Fehlerbereinigung || Fehlerbeseitigung || Störungsbeseitigung
fault frequency Fehlerhäufigkeit || Fehlerrate
fault handling Fehlerbehandlung
fault history Fehlerhistorie || Fehlervorgeschichte
fault identification Fehlerkennung || Fehlersuche * *"Dingfest-Machen" von Fehlern*
fault in the material Materialfehler
fault indication Fehleranzeige || Fehlerkennung || Störanzeige || Störmeldung * *Störungsanzeige* || Störungsmeldung
fault instant Fehleraugenblick
fault level Kurzschlussleistung || Kurzschlussstrom * *Pegel des Fehlerstroms*

fault list Fehlerliste || Mängelliste || Meckerliste * *Mängelliste*
fault localization Fehlersuche * *Lokalisierung, Findung, Ortung von Fehlern*
fault locating Fehlersuche
fault location Fehlerortung * *z.b. auf einer Leitung* || Störungssuche
fault log Fehlerprotokoll || Fehlerspeicher
fault management Fehlerbehandlung
fault memory Fehlerspeicher || Störspeicher
fault message Fehlermeldung || Störmeldung || Störungsmeldung
fault notification Störungsmeldung * *schriftliche*
fault power Kurzschlussleistung
fault profile Fehlerbild
fault rate Fehlerhäufigkeit || Fehlerrate
fault recording Störerfassung * *Aufzeichnung*
fault rectification Störungsbehebung || Störungsbeseitigung
fault related störungsbedingt
fault search Fehlersuche
fault sensing Fehlererfassung || Störerfassung
fault sequence memory Störfolgespeicher * *(→PROFIBUS-Profil)*
fault signal Störmeldesignal || Störmeldung || Störungsmeldung
fault signaling Störungsmeldung * *(kann mit einem oder zwei "l" geschrieben werden)*
fault signaling Störanzeige || Störmeldefunktion || Störungsmeldung
fault signalling relay Störmelderelais
fault simulator Störsimulator
fault storage Fehlerspeicher
fault suppression Fehlerausblendung || Fehlerunterdrückung
fault text Fehlertext
fault to frame Körperschluss * *{el.Masch.} Isolierversagen (Masseschluss zum Gehäuse)* || Masseschluss * *{el.Masch.}*
fault tolerance Fehlertoleranz
fault trip Ansprechen * *Fehlerauslösung, Fehlerabschaltung* || Fehlerabschaltung * *mit Störung abschalten* || Störabschaltung
fault-free fehlerfrei || störungsfrei
fault-tolerant fehlersicher || fehlertolerant || hochverfügbar * *fehlertolerant*
faulted gestört * *fehlerhaft geworden*
faultless einwandfrei || fehlerfrei || störungsfrei
faulty ausgefallen * *auf Grund eines Fehlers/einer Störung* || defekt || Fehl- * *siehe auch →fehlerhaft* || fehlerhaft || kaputt * *schadhaft, fehlerhaft* || mangelhaft * *fehlerhaft, schadhaft, schlecht* || schadhaft * *fehlerhaft*
faulty design Konstruktionsfehler
faulty operator control Bedienungsfehler
faulty program execution gestörter Programmablauf
faulty solder connection kalte Lötstelle
faulty solder joint kalte Lötstelle
faulty-reel cutter Fehlrollenschneider * *{Papier}*
faux pas Fehler * *Taktlosigkeit*
favorable günstig * *(amerikan.)*
favorable price günstiger Preis * *(amerikan.)*
favorably günstig * *(Adv., amerikan.)*
favored bevorzugt * *(amerikan.)*
favorite beliebt * *Lieblings-*
favourable günstig * *to: für; be favourable/favour to: ~ sein für* || gut * *günstig* || vorteilhaft * *günstig*
favourable in price preiswert
favourable price günstiger Preis
favourably günstig * *(Adv.)*
favourably priced günstig * *preis~* || preisgünstig || preiswert * *preisgünstig*
favourably-priced kostengünstig * *preisgünstig* || preisgünstig * *more favourably priced: ~er*

favoured bevorzugt
fax Fax
fax machine Faxgerät
FCC FCC * *Flux Current Control, Regelverfahren f.Asynchronmotor für optimalen Wirkungsgrad u.hohe Dynamik* || Fluss-Stromregelung * *Flux Current Control*
FDD Diskettenlaufwerk * *Abkürzung für 'Floppy Disk Drive'*
feasibility Anwendbarkeit * *Durchführbarkeit* || Ausführbarkeit || Durchführbarkeit || Möglichkeit * *Durchführbarkeit*
feasibility analysys Durchführbarkeitsstudie
feasibility study Durchführbarkeitsstudie
feasible durchführbar * *ausführbar, durchführbar, möglich* || möglich * *aus/durchführbar* || praktisch * *durchführbar* || realisierbar
feather key Feder * *Passfeder* || Feder * *Passfeder, Federkeil* || Passfeder * *Nutkeil*
featherkey Federkeil || Passfeder * *half featherkey: halber ~ (Motor-Auswuchtverfahren mit z.Hälfte einbezogenem Federkeil)*
feature aufweisen * *als Besonderheit/Merkmal haben, sich auszeichnen durch* || Besonderheit * *Merkmal* || Eigenschaft * *auch 'Eigenschaft haben/sich auszeichnen durch'* || herausstellen * *z.B. in der Werbung, Presse ~* || Merkmal * *Besonderheit, Eigenschaft; salient: herausragendes*
feature of construction Konstruktionsmerkmal
features Daten * *Eigenschaften* || Eigenschaften * *Merkmale* || Merkmale
fed geführt * *geleitet; fed out to terminals: auf Klemmen herausgeführt (z.B. Signal)* || gespeist
fed back zurückgespeist
fed-in winding Träufelwicklung * *Wicklungsart bei Motor: Drähte werden von Hand in die Nut "eingeträufelt"*
federation Verband * *Vereinigung*
fee Gebühr || Honorar * *eines Arztes usw.*
feeble lahm * *(figürl.) kraftlos* || schwach * *schwach, schwächlich, kraftlos, undeutlich, unbedeutend*
feebleness Schwäche
feed anlegen * *(Druckerei) Papierbögen usw.* || ansteuern * *z.B.Motor* || einspeisen || fördern * *zuführen* || führen * *ein Signal ~; to:auf/zu; through:hindurch, über* || speisen || Speisung || Vorschub * *Werkzeugmaschine* || zuführen * *to: zu etwas; auch Material, Versorgungsgüter, Waren ~* || Zuführung * *Maschinenteil; pressure feed: mit Druck arbeitende ~*
feed aside ausschleusen
feed away wegführen
feed axis Vorschubachse * *(Werkz.masch.) (Plural: 'axes')*
feed back rückspeisen || zurückführen || zurückspeisen * *z.B. Bremsenergie ins Netz ~*
feed back regeneratively zurückspeisen * *z.B. Bremsenergie ins Netz ~*
feed drive Vorschubantrieb * *Werkzeugmaschine*
feed forward Vorsteuerung
feed forward signal Vorsteuersignal
feed function Vorschub
feed hopper Einfülltrichter
feed in einführen * *zuführen* || einspeisen || einziehen
feed information informieren * *to somebody: jdm. Informationen zukommen lassen*
feed line Zuleitung
feed mechanism Zuführeinrichtung
feed opening Einfüllöffnung
feed out herausführen * *auch Signale, Leitungen usw.*
feed performance Förderleistung
feed pipe Zuleitung * *Rohrleitung*

feed rate Förderkapazität * *siehe auch →Förderleistung* || Förderleistung || Vorschub * *Vorschubgeschwindigkeit* || Vorschubgeschwindigkeit * *(Werkz.masch.)*
feed roll Speisewalze
feed roller Speisewalze * *z.b. bei einer Krempelmaschine (Textil)*
feed source Speisequelle
feed speed Vorschub * *Vorschubgeschwindigkeit* || Vorschubgeschwindigkeit
feed through durchführen * *hindurchführen* || Durchführung * *z.B. von Kabeln* || hindurchleiten
feed to schalten auf * *ein Signal führen auf*
feed tube Einfüllschacht
feed velocity Vorschubgeschwindigkeit
feed water Speisewasser
feed wire Zuleitung
feed-forward vorsteuern
feed-forward control Vorsteuerung
feed-through Durchführung * *z.B. Kabel~*
feed-through capacitor Durchführungskondensator
feed-wheel Galette
feedback Istwert || Rückführung || Rückkopplung || Rückspeisung || Rückwirkung * *Rückkopplung*
feedback cable Istwertleitung * *(Regel.techn.)*
feedback capacitor Rückführkondensator
feedback control Regelung || Regelung * *mit Rückführung arbeitend*
feedback loop Rückführung || Rückkopplung * *Rückkopplungsschleife/zweig* || Rückkopplungskreis
feedback path Rückführung * *Zweig, Pfad*
feedback signal processing Istwertaufbereitung
feedback unit Rückspeiseeinheit
feedback-signal input Istwerteingang
feedback-signal matching Istwertanpassung
feeder Abzweig * *einer Energieverteilungsanlage* || Anleger * *(Druckerei)* || Einschieber * *Beschickungseinrichtung, z.B. für Ofen* || Einspeisung * *Leistungs/Netzeinspeisung* || Speiseleitung || Versorgungsleitung || Zuführeinrichtung || Zuführung * *Zuführ-/Zufördereinrichtung* || Zuleitung * *z.B. für Motor*
feeder cable Zuleitung || Zuleitungskabel
feeder godet Liefergalette * *(Textil) Fadenrolle zum Anfördern der Fäden*
feeder inductance Zuleitungsinduktivität
feeder module Einspeiseeinheit || Einspeisemodul
feeder resistance Zuleitungswiderstand
feeder system Beschickung * *~sanlage*
feeder unit Einspeiseeinheit || Lieferwerk * *Zuführeinrichtung*
feedforward signal Vorsteuerwert
feeding Beschickung || Versorgung * *Speisung, Zuführung, Zuleitung, Zufuhr* || Zufuhr || Zuführung * *das Zuführen, Herbeifördern*
feeding back Rückspeisung
feeding device Vorschubeinrichtung
feeding equipment Zuführeinrichtung
feeding in Zuführung * *das Hineinführen*
feeding mechanism Zuführeinrichtung
feeding pipe Zuleitung * *Rohrleitung*
feeding section Einspeisung
feeding system Zuführsystem * *Antransport/Zuführung von Teilen/Material in eine Maschine*
feeding table Vorschubtisch * *(Holz)*
feeding unit Einzug || Vorschubapparat
feedwater pump Speisewasserpumpe
feel Griff * *"Anfasseigenschaften" von Stoff, Papier usw.* || merken * *spüren, merken, fühlen*
feel at home with vertraut sein mit
feeling Gefühl
fees Kosten * *Gebühren*
feet Füße * *auch Aufstell~ eines Motors*

feet recesses Fußnischen * z.B. zur E-Motor-Befestigung
fellow Gegenstück || Kollege * Gefährte, Kamerad
fellow- Mit- * z.b. fellow passenger: ~reisender
felt Filz * auch zur Führung noch nassen Papiers in der Papiermaschine || walken * Textilien/Filz ~
felt guide Filzlaufregler * Filzregulier-/Führungswalze in Papiermaschine
felt guide roll Filzleitwalze
felt roll Filzleitwalze * in der Entwässerungspartie einer Papiermaschine
felt sealing ring Filzring * zum Abdichten (z.B. eines Wälzlagers)
felt tip pen Filzschreiber
felt-guide roll Filzregulierwalze * in Papiermaschine
felt-tip Filzschreiber
felt-tipped pen Filzschreiber
female connector Buchsenleiste * "weiblicher" Steckverbinder || Buchsenstecker * "weiblicher" Stecker || Federleiste * 'weiblicher' Steckverbinder, Buchsenstecker
female die Matrize * Stanzwerkzeug (Unterteil)
female plug-connector element Steckerbuchse
female socket connector Federleiste * 'weiblicher' Steckverbinder, Buchsenstecker
female thread Innengewinde
fence abschranken || Abschrankung || Gitter * Zaun
fencing Absicherung * Einzäunung, Abschrankung
FEPROM FEPROM * Abkürzg. für 'Flash EPROM': elektrisch schnell löschbarer Festwertspeicher
fermentation Gärung
ferrite Ferrit
ferrite magnet Ferritmagnet
ferromagnetic ferromagnetisch
ferrule Aderendhülse
festoon Soffitte * z.B. Glühbirne zum Einsetzen in Meldeleuchte
festoon extractor Soffittenzieher * zum Ausziehen einer Soffittenlampe aus Meldeleuchte usw.
FET Feldeffekt-Transistor || Feldeffekttransistor
fetch abholen || abrufen * herbeiholen, ab-/hervorholen || erhalten * eine Sache ~ || holen
few wenig
fewer weniger
FFT Fourier-Transformation * Abk.f. 'Fast Fourier Transform': schnelle ~ z.B. mit Digitalrechner
FHP motor Kleinmotor * Abkürzg.f. 'Fractional-Horsepower motor': Motor bis 1 hp (entspr. 0,746 kW)
fiasco Misserfolg
fiber Faden * (amerikan.) Faser || Faser * (amerikan.)
fiber flocks Faserflocken * (amerikan.) {Textil}
fiber fly Faserflug * {Textil}
fiber grader Fasersichter * zum Trennen unterschiedlicher Faserpartikel
fiber line Faserlinie * z.B. in Zellstoff- oder Textilfabrik
fiber optic cable Lichtwellenleiter * Lichtwellenleiterkabel
fiber reinforced faserverstärkt * (amerikan.)
fiber roving Faserlunte * {Textil} (amerikan.) bei der Fadenherstellung
fiber stretching Fadenverstreckung
fiber stretching drive Faserverstreckantrieb * (amerikan.)
fiber stretching line Faserverstreckstraße * (amerikan.)
fiber stretching plant Faserverstreckanlage * (amerikan.) || Verstreckanlage * (amerikan.) Faser~ || Verstreckmaschine * (amerikan.) Faserverstreckanlage || Verstreckwerk * {Textil} (amerikan.) Faserverstreckanlage

fiber-optic cable Lichtwellenleiter * (amerikan.)
fiber-optic link Lichtwellenleiter * ~-Datenverbindung
fiberboard Faserplatte * {Holz} || Feinpappe * (amerikan.)
fibers Fasern * (amerikan.)
fibre Faden * Faser || Faser * (brit.) virgin fibre: Primärfaser (zur Papierherstellung) || Vlies
fibre flocks Faserflocken * (brit.) {Textil}
fibre optic Lichtwellenleiter * (brit.) vorangestellt in Wortkombination
fibre optic cable Lichtwellenleiterkabel
fibre optic link LWL * →Lichtwellenleiter
fibre roving Faserlunte * {Textil} (brit.) bei der Fadenherstellung
fibre stretching drive Faserverstreckantrieb
fibre stretching line Faserverstreckanlage || Faserverstreckstraße * (brit.) || Verstreckanlage * Faserverstreckstraße || Verstreckwerk * {Textil} Faserverstreckstraße
fibre stretching plant Verstreckmaschine * Faserverstreckanlage
fibre-optic Glasfaser * Lichtleiter-
fibre-optic cable Glasfaserkabel
fibre-optic device Glasfaser-Element
fibre-reinforced faserverstärkt * (brit.)
fibreboard Feinpappe * (brit.) || Hartpappe
fibres Fasern
fibrous material Fasermaterial * {Textil} || Faserstoff * siehe auch →Zellstoff (Papierfaserstoff)
fibrous paper Papierfaserstoff * siehe auch →Zellstoff
fictitious fiktiv || scheinbar * vorgeblich, erfunden, unecht, angenommen, fiktiv
fidelity Genauigkeit * Wiedergabetreue || Reinheit * Radio
field Anlage * praktischer Einsatz, Einsatzfeld, Praxis, Wirklichkeit, Branche || Bereich * z.B. Industrie~, Wirtschafts~, Branche, ~ im Markt, Anwendungs~ || Branche * Bereich, (Arbeits)gebiet, Fach || Erreger- || Erregung * Feld || Fach * (Betätigungs/Sach/Arbeits)Gebiet, Branche || Fachgebiet || Feld * el./magn. ~, Bereich, Sachgebiet; zero-divergence: quellenfreies; irrotational: wirbelfreies || Gebiet * Fach~, →Branche || Sachgebiet
field ampere turns Erregerdurchflutung
field area Feldbereich * untere Ebene der 'Automatisierungs-Pyramide'
field bus Feldbus
field cable Signalleitung * zur (externen) Anlagenverdrahtung
field characteristic Feldkennlinie
field characteristic plotting Feldkennlinienaufnahme
field choke Felddrossel
field circuit Erregerkreis || Feldkreis
field circuit time constant Erregerkreiszeitkonstante * ca 1 bis 4,5 sec bei Gleichstrommaschinen
field component Flusskomponente
field control Feldregelung
field control range Feldsteuerbereich * Feldschwächbereich eines AC-Asynchron- oder DC-Motors
field controller Feldregler
field converter Erregergleichrichter || Erregerstromrichter
field current Erregerstrom * Feldstrom
field current control Erregerstromregelung
field current controller Feldstromregler
field current limit Feldstromgrenze
field current limiting Feldstrombegrenzung
field current reduction Feldstromreduzierung * z.B. bei Stillstand, um eine Aufheizung des Motors zu vermeiden
field current setpoint Feldstromsollwert

field current supply Erregerstromversorgung
field density Feldliniendichte
field device Feldgerät * Endgerät an einem Feldbussystem
field devices prozessnahe Peripherie * Sensoren, Aktuatoren usw. bei SPS
field distribution Feldverteilung * z.B. in einem Motor
field economizing Feldstromreduzierung * DC-Motor beim Stillstand
field engineer Montageingenieur || Service-Ingenieur
field experience Betriebserfahrung
field gating unit Feldsteuersatz
field line Feldlinie
field loss Feldausfall || Feldkreisunterbrechung * Feldausfall
field losses Erregerverluste
field of activity Geschäftsfeld
field of application Anwendungsbereich || Anwendungsgebiet
field of industry Industriezweig
field of use Anwendungsbereich
field orientation Feldorientierung
field pattern Feldlinienbild
field pole Erregerpol
field power supply Erregerstromversorgung
field proven praxiserprobt
field rating Erregerleistung * z.B. Nennerregerleistung eines DC-Motors
field reactor Felddrossel
field rectifier Erregergleichrichter || Feldgleichrichter
field resistance Feldkreiswiderstand
field reversal Drehfeldumkehr || Erregerkreisumschaltung * Feldumkehr z.B. zur Momentenumkehr bei DC 1Q Antrieb || Feldkreisumschaltung * Feldumkehr z.B. zur Momentenumkehr bei 1Q Gleichstromantrieben || Feldumkehr || Feldumschaltung * Feldumkehr z.B.zur Momentenumkehr bei 1Q Gleichstromantrieb
field reversing Feldumkehr
field rotation reversal Drehfeldumkehr
field sales service Außendienst * von Vertriebspersonal
field sales services Vor-Ort-Vertrieb * ~s-Organisation
field service Außendienst * z.B. von Servicepersonal || Inbetriebnahme * Abteilungsbezeichnung || Inbetriebsetzung * als Tätigkeitsfeld/Abteilungsbezeichnung || Serviceabteilung * Außendienst
Field Services Service-Abteilung * Außendienst || Serviceabteilung * Außendienst (Abteilungsbezeichnung)
field strength Feldstärke
field supply Erregerstromversorgung || Erregerversorgung || Feldspeisegerät || Feldstromversorgung || Feldversorgung
field supply unit Feldgerät * zur Feldstromversorgung z.B. einer DC-Maschine || Feldspeisegerät
field test Felderprobung || Feldversuch
field trial Felderprobung || Feldversuch
field voltage Erregerspannung
field voltage supply Feldstromversorgung
field weakened speed range Feldschwächbereich * Drehzahlerhöhung durch Feldreduzierung
field weakening Feldschwächung
field weakening control range Feldschwächstellbereich
field weakening controller Feldschwächregler
field weakening operating range Feldschwächbereich
field weakening operation Feldschwächbetrieb
field weakening point Ablösepunkt * bei diesem (EMK-)Wert beginnt die Feldschwächung des Motors
field weakening speed Feldschwächdrehzahl * Drehzahl bei max. Feldschwächung eines AC- oder DC-Motors
field winding Erregerwicklung || Feldwicklung
field winding inductance Erregerwicklungsinduktivität
field winding resistance Erregerwicklungswiderstand
field-circuit rectifier Feldgleichrichter
field-circuit reversal Feldkreisumschaltung
field-current supply Feldstromversorgung * controlled:geregelte || Feldversorgung * controlled:-geregelte
field-effect transistor Feldeffekt-Transistor || Feldeffekttransistor
field-oriented feldorientiert * z.B. Vektorregelung
field-oriented closed-loop control feldorientierte Regelung * für Asynchronmotor
field-oriented control feldorientierte Regelung * für Asynchronmotor
field-oriented principle Feldorientierung
field-proven bewährt * im breiten (Feld)Einsatz bewährt, →felderprobt || felderprobt
field-weakening control Feldschwächregelung
field-weakening range Feldschwächbereich || Feldschwächebereich * siehe →Feldschwächbereich || Feldschwächstellbereich
field-weakening speed Feldschwächdrehzahl
fields of application Einsatzgebiete * z.B. eines Geräts
fierce heftig * wild, heftig (auch Konkurrenz usw.), hitzig, verbissen
FIFO FIFO * Abk. f. 'First In First Out': Schieberegister mit einer Breite von mehreren Bits
FIFO memory Fallregister * Abk.f.'First-In-First-Out': ältester Eintrag wird als Erster geles. (Schieberegister) || FIFO-Speicher * Abk. für "First In First Out": zuerst eingegebene Daten werden als Erstes ausgegeben
figure Abbildung * besonders als Bildunterschrift mit Zahl, z.B. Fig. 23: Abb. 23 || Bild * Abbildung (in Dokument) || darstellen * grafisch, rechnerisch ~ || Form * Figur, (geometr. Körper)Form, Gestalt, Aussehen, Erscheinung, Charakter || Stelle * ~ einer Zahl || Zahl * Zahlenangabe, Ziffer || Zeichnung || Ziffer
figure denoting type of construction Bauform-Kennziffer * Bestellnummer-Ergänzung bei SIEMENS Motor
figure denoting voltage Spannungskennziffer * Bestellnummer-Ergänzung bei SIEMENS-Motor
figure out ausrechnen * (amerikan.) || berechnen * (amerikan.) →ausrechnen, 'rauskriegen || ermitteln * (amerikan.) ausrechnen, herausbekommen || herausbekommen * (amerikan.) ermitteln, ausrechnen
figures Zahlen
filament Faden * Faden, Faser, Glüh-/Heizfaden || Faser * (allg.) feine Faser || Filament || Glühfaden
filament group Fadenschar * {Textil} bei der Chemiefaserherstellung
filament lamp Glühlampe
filament stretching Fadenverstreckung
filaments Fasern
file ablegen * Akten, Briefe || Akte * Aktenordner, abgelegte Akten; put on file: zu den Akten legen || Aktenordner || archivieren * Dokumente , Akten usw. || Datei || einordnen * Akten usw. ~ || Feile || feilen || Ordner * siehe auch 'Aktenordner'
file an objection Einspruch einlegen * against:gegen (schriftlich)
file card Karteikarte
file case Aktenkoffer

file of the changes Änderungsdatei
file off abfeilen
filing Ablage * *das Ablegen von Akten, Dokumenten usw.* || Archivierung * *von (Papier-) Dokumenten, Akten*
filings Span * *Feilspäne* || Späne * *Feilspäne*
fill füllen || versehen * *Amt/Dienst* ~
fill in einfüllen || eintragen * *ausfüllen; z.B. Angaben in einen Fragebogen/ein Formular* ~
fill into einfüllen
fill out ausfüllen * *auch Formular*
fill up auffüllen || besetzen * *offene Stelle* || nachfüllen
fill-up füllen * *auffüllen*
filled voll * *gefüllt*
filler Füller * *Füllmaterial* || Füllstoff || Füllstoffe || Spachtel * *~masse*
filler cap Einfüllstutzen
filler plug Blindstopfen * *z.B. für Kabeleinführungsloch* || Einfüllschraube || Stopfen * *Blindstopfen, z.b. für Kabeleinführungsloch*
filler screw Einfüllschraube
filler socket Einfüllstutzen
filler stub Einfüllstutzen
fillet Hohlkehle || Rundung * *Kehlnaht, Hohlkehle; (Fuß-) Ausrundung*
filling Abfüllung * *von Behältern, Flaschen usw.* || Befüllung || Füllung
filling compound Vergussmasse * *z.B. für Kabelmuffe*
filling funnel Einfüllschacht
filling hole Einfüllöffnung
filling level Füllstand
filling line Abfüllanlage * *(Verpack.techn) z.B. für Flaschen*
filling machine Abfüllmaschine * *(allg.)* || Füllmaschine * *(Verpack.techn.) (siehe →Dosier~, →Kolben~, →Durchfluss~, →Schlauchbeutel~)*
filling machine by time Zeit-Füllmaschine * *in der Verpackungstechnik*
filling plant Abfüllanlage * *(Verpack.techn.) z.B. für Flaschen*
filling plug Einfüllschraube
filling system Beladeeinrichtung
filling weigher Abfüllwaage * *(Verpack.technik)*
fillister-head srew Linsenkopfschraube
film Anstrich * *hauchdünn* || Belag * *Häutchen, Überzug* || Folie * *meist aus Kunststoff* || Schicht * *dünner Überzug* || Überzug * *hauchdünner*
film capacitor Kunststofffolienkondensator || Kunststoffkondensator || metallisierter Kunststoff- oder Kunststofffolienkondensator
film manufacturing line Folienmaschine
film printing press Foliendruckmaschine * *für Kunststofffolie*
film production Folienherstellung || Folienproduktion
film winding machine Folienwickler * *Auf-/Abwickler für Kunststofffolie*
filmy hauchdünn
filter durchlassen * *filtern* || Filter * *auch Glättungsglied, Staubfilter* || filtern || glätten * *filtern, sieben* || Sieb * ~ *für Flüssiges und elektr. Filterglied (z.B. Tief-/Hoch/Bandpass).* || sieben
filter capacitor Filterkondensator || Glättungskondensator || Siebkondensator
filter circuit Filterkreis
filter element Glättungsglied || PT1-Glied || Siebglied
filter for radio interference suppression Funkentstörfilter
filter mat Filtermatte
filter mounting Filteranbau
filter network Filterkreis
filter pad Filtermatte

filter press Filterpresse * *z.b. bei Wasseraufbereitungsanlage*
filter reactor Siebdrossel
filter time constant Filterzeitkonstante
filterbed Klärbecken
filtered geglättet * *durch Glättungsfilter* ~
filtering Filterung || Glättung * *Filterung über Glättungsfilter* || Siebung
filtering capacitor Filterkondensator || Glättungskondensator
filtering plant Sieberei * *z.b. in Zellstoff-Fabrik*
filtering time Siebzeit
filtering time constant Siebzeit * *Siebzeitkonstante*
filth Schmutz
filthy schmutzig * *verdreckt*
filtration Filterung * *(mechan.)* || Filtration * *Filterung*
filtration plant Filteranlage
fin Lamelle * *(Kühl-) Rippe* || Rippe * *dünne Kühl-/Heizrippe*
fin-cooled rippengekühlt * *z.b. Motor*
fin-cooled motor rippengekühlter Motor
final abschließend * *endgültig* || ausschlaggebend * *endgültig* || End- * *endgültig, Schluss-, letzt* || endgültig * *schließlich* || endlich * *(Adj.)* || entscheidend * *endgültig* || Fertig- || letzt * *endgültig, schließlich, definitiv, End-, Schluss-* || letzter * *endgültig* || schließlich || Schluss-
final assembly Endfertigung * *Endmontage* || Endmontage
final character Endezeichen
final coat Deckanstrich
final control element Aktuator * →*Stellglied* || Stellgerät || Stellglied
final controlling element Stellgerät || Stellglied || Stellorgan * →*Stellglied*
final customer Endabnehmer * *Endkunde* || Endkunde
final date äußerster Termin || Frist * *äußerste* ~
final discharge voltage Entladeschlussspannung * *bei Batterie/Akku*
final extension stage Endausbau * *letzte Ausbaustufe*
final inspection Abnahme * *Abnahmeprüfung* || Endkontrolle || Endprüfung
final packing Umverpackung
final product Endprodukt
final remark Schlussbemerkung
final result Endergebnis
final rounding Endverrundung
final rounding-off Endverrundung * *für Hochlaufgeber-Rampe*
final specification Pflichtenheft
final stage Endstufe * *letzte/abschließende Stufe*
final temperature Endtemperatur
final test Endprüfung
final treatment Endbehandlung * *z.B. bei Faserherstellung*
final user Endabnehmer
final value Endwert
final wrapping Umverpackung
finalize abschließen * *beenden, vollenden, endgültig erledigen, endgültige Form geben*
finally abschließend * *(Adv.)* || endlich * *(Adv.)* || schließlich * *(Adv.)* || zuletzt
finance Kapital
financial wirtschaftlich * *finanziell*
financial results Geschäftsbericht * *Ergebnisbericht*
financial transaction Geldgeschäft
find beschaffen * *z.B. Arbeit, Kapital* || finden || suchen * *finden*
find a fault Fehler finden

find fault with aussetzen * *etwas auszusetzen haben*
find out erkennen * *herausfinden* || ermitteln * *herausfinden* || feststellen * *herausfinden* || herausbekommen * *herausfinden* || merken * *entdecken*
finding Erkenntnis * *etwas Herausgefundenes*
findings Befund || Ermittlung * *Feststellungen* || Feststellung * *Tatsachenfeststellung, Befund, Fund, Entdeckung*
fine ausgezeichnet || dünn * *fein, zart* || fein || gut * *fein, prächtig* || sauber * *fein* || sehr gut * *fein, vorzüglich*
fine adjust Feinabgleich
fine adjustment Feinabgleich || Feinanpassung || Feineinstellung || Feinjustierung
fine dust Feinstaub
fine interpolation Feininterpolation * *(NC-Steuerung)*
fine mechanics Feinmechanik
fine paper Feinpapier
fine pulse Feinimpuls
fine setting Feinjustierung
fine thread Feingewinde
fine tuning Feinabgleich || Feineinstellung
fine-dust filter Feinstaubfilter
fine-finishing planer Feinputzhobel * *(Holz)*
fine-grained fein * *feinkörnig, z.B. Schleifmittel, Bimsstein* || feinkörnig * *z.B. Schleifmittel, Bimsstein*
fine-pore feinporig
fine-tune optimieren * *feinoptimieren*
finely stranded feindrähtig
fineness Feinheit * *z.B. von Fäden; finer points: die ~en (figürl.)*
fineness of grinding Mahlgrad
finery Raffinerie
fines Feinstoff
finger Hebel * *kurzer Hebel, Klinke, Sperrhaken, Finger usw.*
finger joint Keilzinkenverbindung * *(Holz)*
finger jointing Keilverzinkung * *(Holz)* || Keilzinkenverbindung * *(Holz)* || Zinkenverbindung * *(Holz)*
finger jointing gluing press Keilzinkenverleimpresse * *(Holz)*
finger jointing machine Keilzinkenmaschine * *(Holz)*
finger screw Rändelschraube * *mit der Hand zu betätigen*
finger-joint verzinken * *(Holz)*
finish ausarbeiten * *vollständig ~* || ausrüsten * *Papier, Tuch* || beenden || behandeln * *nach-/end~* || beschließen * *beenden* || erledigen * *beenden* || fertigbearbeiten || Oberflächenzustand || schließen * *beenden* || veredeln * *Güter, Rohstoffe, z.B. Metall*
finish-bore fertigbohren
finish-roll fertigwalzen
finished fertig * *abgeschlossen, beendet, ~ bearbeitet* || Fertig- || verarbeitet * *endver-/bearbeitet*
finished part Fertigteil
finished procuct Endprodukt * *Fertigprodukt*
finished product Fertigerzeugnis || Fertigprodukt
finishing Ausrüstung * *Weiterverarbeitung/Endbehandlung z.B. von Papier, Textilstoffen* || Endbehandlung * *z.B. bei Papier,Metall* || Fertig- || Nachbearbeitung || Nachbehandlung || Nachverarbeitung || Veredelung * *Güter, Rohstoffe, z.B. Metall, Textil*
finishing machine Ausrüstungsmaschine || Ausstattungsmaschine * *z.b. in der Verpackungstechnik* || Veredelungsmaschine
finishing mill Fertigstraße * *Walzwerk*
finishing stand Fertiggerüst * *Walzwerk*
finite endlich * *(Adj.) begrenzt, nicht unendlich*

finite-element method Finite-Elemente-Methode * *CAD-Berechnungsverfahren: Annäherung e. Kontur durch kleinste Teilflächen*
finite-elements method Finite-Elemente-Verfahren * *CAD-Berechnungsverfahren: Annäherung einer Kontur durch kleinste Flächen*
finned gerippt * *z.B. Kühlkörper, Motorgehäuse* || verript
finned heat sink Rippenkühlkörper
finned tube Rippenrohr * *in einem Kühlsystem*
finned-tube radiator Rippenrohrkühler * *flüssigkeitsdurchströmter Wärmetauscher*
finning Verrippung
FIP FIP * *Abkürz. für 'Field Instrumentation Protocol': französisches Feldbusprotokoll*
fire ansteuern * *Thyristor, Schneidmesser bei Rollenwechsel usw.* || auslösen * *z.b. knife:Messer beim fliegenden Rollenwechsel* || steuern * *zünden z.B.Thyristor* || zünden * *z.B. einen Thyristor*
fire hazard Brandgefahr
fire off abdrücken * *Gewehr usw.*
fire protection Brandschutz
fire risk Brandgefahr
fire-retardant feuerhemmend
fire-retarding feuerhemmend
fired angesteuert * *Thyristor* || ausgelöst * *z.B. Messer bei fliegendem Rollenwechsel* || gezündet * *z.B. Thyristor*
firedamp Schlagwetter
firedamp motor schlagwettergeschützter Motor
firedamp-proof schlagwettergeschützt
firewood Holz * *Brenn~*
firing Ansteuerung * *eines Leistungshalbleiters bei Phasenanschnittsteuerung* || Heizung || zünden * *(Substantiv) eines Thyristors* || Zündung * *z.B. ~ eines Thyristors*
firing angle Aussteuergrad * *Steuerwinkel besonders bei Phasenanschnittsteuerung* || Aussteuerung * *Aussteuerwinkel* || Aussteuerwinkel * *für netzgeführten Stromrichter (Phasenanschnittsteuerung)* || Steuerwinkel * *für Thyristor-Gate-Ansteuerung* || Zündwinkel * *z.B. bei Thyristor (siehe auch 'Steuerwinkel')*
firing angle precontrol Steuerwinkelvorsteuerung || Zündwinkelvorsteuerung * *zur Verbesserung der Dynamik bei DC-Stromrichter*
firing angle setting Aussteuerung * *Aussteuerwinkel bei Thyristorsatz*
firing cable Ansteuerleitung * *für Thyristor(en)* || Zündleitung * *für Thyristor(en)*
firing circuit Steuersatz * *Zündimpulserzeugung besonders für Thyristorsatz*
firing circuit module Ansteuerbaugruppe * *Steuersatz für Thyristor-Stromrichter*
firing inhibit board Zündsperrbaugruppe * *für Thyristorsatz*
firing limits Aussteuerbegrenzung * *Steuerwinkelbereich für netzgeführten Stromrichter* || Trittgrenzen * *Aussteuergrenzen für netzgeführten Stromrichter*
firing point Zündzeitpunkt * *für Thyristor*
firing pulse Ansteuerimpuls * *für Thyristor* || Steuerimpuls * *Zündimpuls z.B. für Thyristor; siehe auch DIN 41750 Teil 7* || Zündimpuls * *z.B.für Thyristor; siehe auch DIN 41750 Teil 7*
firing pulse blocking Impulssperre * *für Thyristorsatz*
firing pulse enable Impulsfreigabe * *für Thyristor(satz)*
firing pulse energy Zündenergie * *z. Zünden von Thyristoren*
firing pulse transformer Impulsübertrager * *zur Thyristoransteuerung*
firing time Zündzeitpunkt * *für Thyristor*
firm Betrieb * *Firma* || fest * *fest, stark, beständig;*

auch *'unerschütterlich',* auch *Markt, Kurse usw.* ||
Firma || Geschäft * *Firma* || hart * *fest* || sicher
* *fest* || steif * *fest* || Unternehmen * *Firma*
firmer chisel   Stechbeitel
firmness   Festigkeit * *Beständigkeit, Entschlossenheit*
firmware   Firmware * *in einem Festwertspeicher (z.B. EPROM) hinterlegte Software*
firmware release   Firmwarestand * *Ausgabestand einer in Festwertspeicher hinterlegten Software*
firmware version   Firmwarestand
first   Anfangs- || erst * *zu~, zuvor* || erstens || erstes || vorderst || zuerst * *als Erster* || zuerst * *als Erster, vor allem Übrigen, zum ersten Mal* || zunächst * *erstens, (zu)erst, in erster Linie, vor allem*
first aid   Erste Hilfe
first and foremost   vor allem
first commissioning   Erstinbetriebnahme
first cost   Anschaffungskosten || Investition * *einmalige Kosten, Anschaffungskosten* || Selbstkostenpreis
first cyclical run   Erstlauf * *eines zyklisch abgearbeiteten Programms*
first fault   Erstfehler
first fault memory   Erstfehlerspeicher
first harmonic   Grundschwingung
first harmonic field   Grundfeld * *(el.Masch.)*
first input bit scan   Erstabfrage * *bei SPS*
first message identification   Erstmeldungserkennung
first occurrence   erstmaliges Auftreten
first of all   erstens * *an erster Stelle, zunächst* || in erster Linie || vorerst * *zuallererst* || zuerst * *vor allem (Übrigen)* || zunächst * *vor allem* || zunächst * *zu aller erst, vor allen Dingen*
first order   erster Ordnung * *(vorangestellt) z.B. Oberwelle*
first order filter element   Glättungsglied * *PT1-Glied* || PT1-Glied * *Filterglied erster Ordnung*
first run   Erstlauf
first step   Vorstufe
first stuff   Halbzeug * *Papier usw.*
first-class   ausgezeichnet || erstklassig || prima * *erstklassig* || Spitzen- || super * *(salopp) erstklassig* || vorzüglich * *erstklassig*
first-ever   erstmalig
first-fault identification   Erstfehlererkennung
first-grade   erstklassig
first-order delay element   PT1-Glied * *Verzögerungsglied erster Ordnung* || Verzögerungsglied erster Ordnung * *PT1-Glied*
first-order filter element   Verzögerungsglied erster Ordnung * *PT1-Glied*
first-rate   prima * *erstklassig, großartig, ausgezeichnet* || super * *(salopp) großartig, ausgezeichnet* || vorzüglich * *erstklassig*
first-time user   Einsteiger || Erstanwender
first-up indicator   Erstwerterkennung * *Funktionsbaustein*
first-up signalling   Erstwertmeldung
firstly   erstens
fiscal report   Geschäftsbericht || Jahresbericht * *(Jahres-) Geschäftsbericht*
fiscal year   Geschäftsjahr
fishplate   Knotenblech || Lasche * *an Schienen*
fissure   Riss || Spalt * *Spalt, Riss, Sprung*
fit   anbauen || anbringen || anpassen * *on: an* || aufziehen * *z.B. Lager, Kupplung, Zahnrad, Riemenscheibe* ~ || einbauen * *into:in* || fähig * *passend, geeignet, fähig, tauglich, in (guter) Form; (for: für/zu)* || montieren * *auftellen, anbringen, einsetzen, montieren* || passen * *passen zu,auf* || passend || Passung * *system of fits: Passungssystem (z.B. ISO)* || Sitz * *Passung; tight: strammer* || tüchtig || unterbringen * *into:in* || zugeschnitten * *to: an* ||

zusammenpassen || zutreffen * *passen (Beschreibung usw.)*
fit in   einfügen * *Sache und Person; (well:gut)* || einpassen || sich einfügen * *Person und Sache; well:- gut*
fit into   einlassen * *in Holz/Metall* || einpassen || einsetzen * *einbauen, einmontieren* || montieren * *einbauen/setzen*
fit into one another   zusammenpassen
fit into place   einpassen
fit out   ausrüsten || ausstatten
fit size   Passung * *Passmaß*
fit tightly   festsitzen
fit together   zusammenpassen
fitment   Anbau
fitness   Eignung * *(auch einer Person) Eignung, Fähigkeit, Tauglichkeit, Zweckmäßigkeit, Angemessenheit*
fitted   ausgerüstet * *mit Zubehör, Lager, Beschlag usw.* || untergebracht
fitted out   ausgestattet
fitter   Einrichter * *Installateur, Maschinenschlosser* || Monteur * *(mechan.) Monteur* || Schlosser * *Maschinen*~
fitting   Armatur || Befestigung * *Zubehör(teil), Beschlag, Armatur (für Rohre, Kabel usw.)* || Einbau || Montage * *Anbau* || passend
fitting clearance   Passung * *Passungsspiel*
fitting device   Vorrichtung * *zur Montage*
fitting dimensions   Anschlussmaße
fitting out   Ausrüstung
fitting plate   Befestigungslasche
fitting sleeve   Einbauhülse
fitting tool   Aufdrückvorrichtung || Aufziehvorrichtung * *z.B. für Lager*
fitting-insertion machine   Beschlagsetzmaschine * *[Holz]*
fittings   Armatur || Befestigungsteile * *für Rohre usw.* || Einbauten || Einrichtung * *z.B. eines Ladens* || Zubehörteile * *Montagezubehör*
fix   anbringen * *befestigen* || anschlagen * *befestigen* || ansetzen * *Frist, Termin, Preis* || befestigen || beheben * *instand setzen, (Software-) Fehler* ~, *in Ordnung bringen* || Behebung * *Fehler ~, Problem* ~ || benennen * *einen Termin usw.* ~ || Fehlerbehebung || Fehlerbeseitigung || festlegen || instand setzen * *(amerikan.)* || reparieren || setzen * *befestigen, fest~ (auch Termin), festmachen, anheften/bringen, ein~, unterbringen* || verankern * *to foundation: im Fundament* ~
fix a date   Termin anberaumen
fix in place   festhalten * *in Lage/Stellung* ~
fixation   Festlegung * *Festsetzung*
fixed   fehlerbereinigt * *in Ordnung gebracht* || fest * *orts~, be~igt, starr, unveränderlich, ~gesetzt* || Fest- || festgelegt * *festgesetzt, unveränderlich* || feststehend * *ortsfest, befestigt, fest (auch figürl.: Preis, Grundsatz, Termin), laufend (Ausgaben)* || montiert * *on:an* || ständig * *fest (Einkommen usw.)* || steif * *fest* || unveränderlich * *fest* || vorgegeben * *fest*
fixed assets   Anlagevermögen * *unbewegl. (Immobilien), bewegliches(Maschinen..) u. immateriell.- Vermög. (Patente..)*
fixed capacitor   Festkondensator
fixed frequency   Festfrequenz * *fester Frequenzsollwert*
fixed losses   statische Verluste
fixed price   Festpreis
fixed resistor   Festwiderstand
fixed setpoint   Festsollwert
fixed speed   Festdrehzahl
fixed value   Festwert
fixed-command control   Festwertregelung

fixed-cycle press Taktpresse * *(Holz) im Gegensatz zur kontinuierlichen Presse (für Spanplatten)*
fixed-mounted fest eingebaut || fest montiert
fixed-point number Festpunktzahl
fixed-point value Festpunktwert
fixed-speed drive Festdrehzahl-Antrieb || Festdrehzahlantrieb || ungeregelter Antrieb * *Festdrehzahlantrieb*
fixing Befestigung || Befestigungs- || Behebung * ~ *von Fehlern* || Fehlerbehebung || Fehlerbereinigung * *das In-Ordnung-Bringen* || Festlegung * *auch Festsetzung, Termin~* || Verbindungs- * *Befestigungs-*
fixing accessories Befestigungselemente
fixing bolt Befestigungsschraube
fixing dimensions Anbaumaße * *z.B. für AC-Motoren: siehe IEC 72, DIN 42673 u. DIN 42677, jew. Blatt 1*
fixing element Befestigungselement * *Halteelement*
fixing flange Anbauflansch || Befestigungsflansch * *für Motor: siehe DIN 42984*
fixing hole Befestigungsbohrung
fixing lug Befestigungslasche
fixing material Befestigungselemente
fixing parts Befestigungsteile
fixing plate Spannplatte * *Spannvorrichtung z.B. für Werkzeugmaschine*
fixing rail Befestigungsschiene
fixture Befestigung * *(Befestigungs-/Aufspann-/Halte-) Vorrichtung, Befestigung, Spannzeug* || Beschlag * *Befestigg., fester Gegenstand, festes Zubehör, (Aufspann-/Halte-) Vorrichtg., Befestigung* || Gegenstand * *fester ~, Installationsteil, Inventarstück, zum Inventar gehörend (auch Mitarbeiter)* || Halterung || Spannvorrichtung || Spannzeug || Vorrichtung * *z. Montieren*
flag Kennbit || Merker
flag address area Merkerbereich * *bei SPS*
flag area Merkerbereich * *bei SPS*
flag data Merker-Datum * *bei SPS*
flag marker Merker
flair Sinn * *Feingefühl, feines Gespür*
flake Plättchen * *Flocke, dünne Schicht, Schuppe, Fetzen, Splitter*
flake off blättern * *abblättern*
flaker Zerspaner * *(Holz) in der Spanplattenherstellung*
flame cutter Schneidbrenner
flame proof schlagwettergeschützt
flame resistance Feuerbeständigkeit
flame resistant flammbeständig || flammwidrig
flame retardance Flammwidrigkeit
flame retardant Flammschutzmittel
flame-cut brennschneiden
flame-cutting machine Brennschneidmaschine
flame-proof gekapselt * →*explosionsgeschützt*
flame-proof enclosure druckfeste Kapselung * *Zündschutzart 'd' nach EN50018 für Ex-geschützte el. Betriebsmittel*
flame-retardant flammhemmend
flameproof druckfest * *(brit.)* →*explosionsgeschützt* || Ex-geschützt * *siehe EN50014* || explosionsgeschützt * *siehe EN50014* || schlagwettergeschützt
flameproof enclosure Zündschutzart
flameproof enclosure protection Zündschutzart druckfeste Kapselung
flameproof enclosure type d EEx d * *Schutzklasse EEx d (druckfeste Kapselung)*
flammability Brennbarkeit * *(amerikan.) Entflammbarkeit* || Entflammbarkeit * *(amerikan.)*
flammable brennbar * *entflammbar* || entflammbar * *(amerikan.)*
flange bördeln * *flanging press: Bördelpresse;*

double-flanged butt weld: Bördelschweißung || Flansch * *(bei Elektromotoren: siehe DIN 42948)*
flange accuracy Flanschgenauigkeit * *für Elektromotoren: siehe DIN 42955*
flange bearing end shield Flanschlagerschild * *z.B. bei Motor*
flange centering Flanschzentrierung
flange centering rim Flanschzentrierrand
flange design Flanschausführung
flange endshield Flanschlagerschild
flange face Flanschfläche * *z.B. an Motor-Wellenseite*
flange housing Flanschgehäuse * *Gehäusebauform z.B. einer Kupplung*
flange hub Flanschnabe
flange mounting Flanschmontage * *z.B. eines Motors (im Gegensatz zur Fußmontage)*
flange on anflanschen
flange pan Flanschwanne
flange shovel Flanschwanne * *schaufelförmige Flanschplatte z.B. zur Verbindung von Motorfuß u. Getriebe*
flange-mount anflanschen
flange-mounted angeflanscht
flange-mounted design Flanschausführung * *z.B. Motor mit Flanschbefestigung (i.Ggs.z. Fußausführung)*
flange-mounted motor Flanschmotor
flange-mounting design Flanschbauform
flanged geflanscht
flanged endshield Flanschlagerschild
flanged joint Flanschverbindung
flanged on angeflanscht
flanged type of construction Flanschausführung * *z.B. Motor mit Flansch- (statt Fuß-) Befestigung* || Flanschbauform * *z.B. Motorbauart mit Flanschbefestigung*
flanging machine Bördelmaschine
flank Flanke * *Flanke (auch von Getriebezähnen), Weiche (bei Tier usw.), Seite, Flügel (beim Militär usw.)* || Seite * *Flanke*
flap Klappe || klappen
flare nut Überwurfmutter
flash aufleuchten || blinken * *Leuchtdiode, Meldeleuchte, Anzeige, Cursor, Verkehrsampel* || leuchten * *auf~, blinken, aufblitzen*
flash across überspringen * *Funken*
flash memory Flash-Speicher * *nichtflüchtiger Flash-→EPROM-Speicher*
flash point Flammpunkt * *Temperatur, bei der eine Substanz (z.B. Luft-Lösungsmittelgemisch) zündfähig wird*
flash up aufleuchten
flash-light Taschenlampe * *(amerikan.)*
flashes Funken * *(Plural)*
flashing blinkend * *z.B. Anzeigelampe,LED-Anzeige*
flashing light Blinklicht
flashing signal Blinksignal
flashlight Blinklicht
flashover Funkenüberschlag || Rundfeuer || Überschlag * *durch Überspannung*
flashover current Durchbruchstrom
flashover voltage Durchbruchspannung || Durchschlagspannung * *Überschlagspannung* || Überschlagspannung
flat eben * *flach* || flach || flächig * *Plan * (Adj.) flach, eben, platt, umgelegt*
flat bar Flacheisen
flat belt Flachriemen
flat cable Flachleitung
flat connector Flachanschluss * *gelochtes ~blech zum Anschluss der Leitung, z.B. bei Trafo, Drossel usw.* || Flachanschluss * *Flach(steck)verbinder* ||

Flachstecker * *für Flachbandleitung* || Flachsteckverbinder
flat core   Flachleiter * *{Kabel}*
flat design   Flachbauform
flat face   Schlüsselfläche * *z.B. einer Mutter*
flat glass   Flachglas
flat grinding machine   Flachschleifmaschine * *{Werkz.masch.}*
flat iron   Flacheisen * *flat iron section: ~profil*
flat leather belt   Flachlederriemen
flat nose pliers   Flachzange
flat pliers   Flachzange
flat plug   Flachstecker
flat spot   Flachstelle
flat surface   ebene Fläche
flat type   Flachbauform
flat-bladed screwdriver   Schlitzschraubendreher || Schlitzschraubenzieher || Schraubendreher * *Schlitz~ (für Schlitzschrauben)* || Schraubenzieher * *Schlitz~ (für Schlitzschrauben)*
flat-frame motor   Flachmotor * *{el.Masch.}*
flat-head srew   Flachkopfschraube
flat-headed screw   Flachkopfschraube
flat-screen   Flachbildschirm
flatness   Planlage
flats   Flacheisen
flatscreen   Flachbildschirm
flatten   abflachen * *auch 'sich ~'* || ausbeulen
flattened   abgeflacht
flattened motor   Flachmotor * *{el.Masch.}*
flaw   defekt * *(Substantiv) fehlerhafte Stelle, Fehler, Defekt, Bruch, Riss, Sprung* || Riss * *Sprung* || Sprung * *Bruch, Riss, Materialfehler, fehlerhafte Stelle*
flaw detector   Fehlerdetektor * *erfasst Fehler im Material*
flaw in the material   Materialfehler
flawless   fehlerfrei
fleece   Vlies
fleece material   Vliesstoff
fleecing system   Fließanlage * *{Textil}*
fleet   Fuhrpark
flex   Schnur * *(salopp) Abkürzg. für 'flexible cord':* flexible Anschlussschnur
flexibility   Elastizität || Flexibilität || Nachgiebigkeit * *(auch figürl.)*
flexible   anpassungsfähig || beweglich * *elastisch, beweglich (auch figürl.)* || biegsam || dehnbar || elastisch || feindrähtig || flexibel || gleitend * *(figürl.)* || nachgiebig * *(auch figürl.) flexibel (Kupplung usw.), biegsam, geschmeidig* || vielseitig * *flexibel; flexible use: ~er Einsatz*
flexible conductor   feindrähtige Leitung || feindrähtiger Leiter
flexible cord   Anschlussschnur
flexible coupling   elastische Kupplung || flexible Kupplung
flexible pipe   Schlauch
flexible shaft   biegsame Welle
flexible stranded   feindrähtig
flexible tube   Metallschlauch || Schlauch
flexible working hours   Gleitzeit
flexible-conductor   feindrähtig
flexitime   Gleitzeit
flexographic printing machine   Flexodruckmaschine
flexographic roller   Flexodruckwalze * *{Druckerei}*
flexural stiffness   Biegesteifigkeit
flexural strength   Biegesteifigkeit
flexure   Biegung * *Krümmung* || Durchbiegung * *Biegung, Krümmung*
flick   umschalten * *an einem Handschalter*
flicker   flackern * *z.B. Licht, Lampe* || flimmern * *eines Bildschirms* || Wischer || Wischimpuls

flicker-free   flimmerfrei * *Bildschirm, Monitor*
flier   Faltblatt * *(amerikan.) Prospekt, Reklamezettel* || Prospekt * *(amerikan.) Werbezettel, Prospekt* || Streublatt * *→Prospekt* || Werbeblatt * *(amerikan.) Reklamezettel, →Prospekt* || Werbeprospekt * *(amerikan.)* || Werbeschrift * *(amerikan.) Reklamezettel, Prospekt*
flight No.   Flug Nr
flimsy   dünn * *Gewebe* || dürftig * *fadenscheinig, →schwach* || schwach * *dürftig, durchsichtig, schwach, fadenscheinig, lose, (hauch)dünn, zart*
fling   schleudern * *schleudern, werfen, aufreißen/ zuschlagen (Tür usw.); eilen, stürzen*
flip   blättern * *flip through: (ein Buch) durchblättern*
flip side of the coin   Kehrseite der Medaille
flip-flop   Speicherglied
flip-flop control   Zweipunktregelung
flipflop   Speicherglied
flit   spritzen * *(salopp) eilen*
float   schweben * *durch die Luft, in einer Flüssigkeit* || Schwimmer
float charge   Dauerladen * *Batterie*
float glass   Flachglas * *Glasschmelze wird über ein Bett aus flüss.Zinn (>600°C) geleitet u. härtet dort aus* || Floatglas * *siehe auch →Flachglas*
float limiting shim   Ausgleichscheibe * *zur Reduzierung des axialen Lagerspiels*
floatation   Flotation
floating   ausschwitzen * *(Subst.) ~ von Farbe* || erdfrei || Flotation || hochohmig * *ohne festes Potenzial* || potenzialfrei || potenzialgetrennt || schwebend * *z.B. Bezugspotenzial zu einem Signal* || schwimmend * *auch elektr. Potenzial*
floating bearing   Loslager * *l. Gegensatz z. Festlager*
floating capital   Umlaufvermögen * *Vermögensteile, d.nur kurz i.e.Firma vorh.sind:Roh-/Fertigstoffe, Außenstände usw.*
floating ground   schwebendes Bezugspotenzial
floating operation   Pufferbetrieb * *Dauerladungsbetrieb für Batterie*
floating particles   Schwebestoffe * *Schwebeteilchen*
floating point   Fließkomma || Gleitkomma || Gleitpunkt * *(mathem.) Gleitkomma, Fließkomma*
floating point arithmetic   Fließkommaarithmetik || Gleitpunktarithmetik
floating point arithmetic unit   Gleitpunkt-Arithmetikeinheit
floating supply system   erdfreies Netz
floating-point number   Gleitpunktzahl
floating-point unit   Gleitpunkteinheit
floating-type gear   Aufsteckgetriebe
flocculation   Flockung * *{Physik}*
flood   überfluten * *auch überfließen* || Überflutung || überschwemmen
flood lubrication   Sprühölschmierung
flooding   Überflutung
flooding, lightning etc.   höhere Gewalt * *(versicherungstechnisch) Überschwemmung, Unwetter usw.*
floor   Boden * *eines Raums* || Etage || Stockwerk
floor loading   Bodenbelastung
floor plan   Grundriss * *eines Raums, Gebäudes*
floor space   Aufstellfläche * *auf den (Fuß-) Boden*
floor space required   Platzbedarf * *Bodenfläche i.B. in Raum, Gebäude*
floor-mounting distribution board   Standverteiler
floor-standing unit   Standgerät
floorspace   Grundfläche * *Bodenfläche*
flop   Fehlschlag * *(salopp)* || Misserfolg * *(salopp)* || Niete * *(figürl.; auch 'failure')*
floppy   Diskette
floppy disk   Diskette
floppy disk drive   Diskettenlaufwerk

flossy bauschig * *flaumig, seidig*
flotation Flotation
flotation dryer Schwebetrockner * *z.B. in einer Folienbeschichtungsanlage*
flotation nozzle Schwebedüse * *z.B. zur Bildung e. Luftkissens unter einer Warenbahn*
flow Durchfluss || Durchflussmenge || fließen * *Strom, Programm* || Fluss * *von Signalen/Daten/Programm, Durchfluss, Fließen* || Lauf * *Fluss* || laufen * *fließen* || Strom * *von Material/Luft/Verkehr* || strömen * *fließen, strömen* || Volumenstrom
flow back zurücklaufen * *Flüssigkeit*
flow box Niveaukasten * *Stoffauflaufkasten bei Papiermaschine* || Stoffauflauf * *~kasten bei Papiermaschine* || Stoffauflaufkasten
flow chart Ablaufdiagramm * *Flussdiagramm* || Fließbild || Flussdiagramm
flow control Durchflussregelung || Durchflusssteuerung || Mengenregelung * *Durchlussregelung*
flow controller Durchflussregler
flow diagram Fließbild || Flussdiagramm
flow down ablaufen * *abfließen*
flow forth ausströmen * *hervorströmen*
flow measuring probe Strömungsmesssonde * *z.B. in der Belüftungstechnik*
flow medium Förderstrom
flow meter Durchflussgeber * *Durchflussmesser* || Durchflussmesser
flow monitor Durchflusswächter || Strömungswächter
flow of material Materialfluss
flow off abfließen || ablaufen * *abfließen*
flow rate Durchfluss * *Durchflussrate, z.B. in Kubikmeter pro h* || Durchflussgeschwindigkeit * *Förderrate z.B.in [Volumen/h]* || Durchsatz * *Durchflussrate* || Förderleistung * *z.B. eines Lüfters* || Fördermenge * *z.B. eines Lüfters*
flow rate control Durchflussregelung
flow rate controller Durchflussregler
flow rate regulator Durchflussregler
flow sensor Durchflussgeber || Durchflussmesser
flow soldering Schwalllöten
flow through durchfließen || Durchfluss || durchströmen
flow transducer Durchflussgeber * *Messumformer* || Durchflussmesser
flow transmitter Durchflussgeber || Durchflussmesser
flow velocity Durchflussgeschwindigkeit
flow-metering filling machine Durchflussmess-Füllmaschine * *in der Verpackungstechnik*
flow-soldering Einschwallen || Schwalllöten
flowing durchlaufend
flowing in Einfluss * *das Einfließen*
flowmeter Durchflussmesser
fluctuate bewegen * *schwanken, z.B. Preise* || schwanken * *~ (auch Preise, Kurse usw.), sich (ständig) verändern, fluktuieren; unschlüssig sein*
fluctuating supply schwankendes Netz
fluctuation Schwankung || Veränderung * *Schwankung(en)* || Wechsel * *Schwankung*
fluctuations Schwankungen
fluctuations in supply voltage Netzspannungsschwankungen || Versorgungsspannungsschwankungen
fluctuations in synchronism Gleichlaufschwankungen
fluctuations in the load Lastschwankungen
fluctuations of temperature Temperaturschwankungen
flue Kamin * *Abzug(rohr)* || Schornstein * *Abzugsrohr,* →*Kamin*
flue gas Rauchgas
flue gas desulpherization Rauchgasentschwefelung

flue gas desulphurization Rauchgasentschwefelung
flue gas exhauster Brandgasentlüfter * *zum Absaugen der Brandgase im Falle eines Gebäudebrands*
fluff Fussel
fluid Flüssigkeit
fluid coupling Flüssigkeitskupplung
fluid friction Flüssigkeitsreibung
fluid level Füllstand * *einer Flüssigkeit*
fluorescent lamp Leuchtstofflampe
flush bündig * *fluchtend/fluchtrecht (with:mit); be flush with: bündig abschließen mit* || spülen * *z.B. Getriebekasten*
flush out ausspülen
flush panel mounting Schalttafeleinbau * *mit der Schalttafel bündig abschließende Montage*
flush-mounted bündig * *bündig abschließend montiert*
flushing oil Spülöl
flute Hohlkehle || Kehle || Riefe * *Rille, Riefe, Hohlkehle, (Span-) Nut* || Rille || Wellenprofil * *z.B. in Wellpappe*
fluted geriffelt * *gerieft, gerillt*
flutter flattern
flux Fluss * *magnetischer* || Flussmittel
flux braking Flussbremsung
flux controller Flussregler
flux current control Fluss-Stromregelung
flux density Flussdichte * *magnetic: magnetische; electric: elektrische* || Induktion * *Feldliniendichte, Flussdichte (Dichte der magnet. Feldlinien)*
flux distribution characteristic Feldlinienverlauf
flux reduction Flussabsenkung
flux vector Flussvektor
fly fliegen
fly back zurücklaufen * *Zeilenrücklauf des Strahls bei Monitor/Bildschirm*
fly wheel Schwungrad
flyback Rücklauf * *Zeilenrücklauf des Strahls auf Monitor, Bildschirm*
flycatcher Fangschaltung || Suchen-Fangen * *Umrichter schaltet nach Netzspannungseinbruch wieder auf d.laufenden Motor auf*
flyer Faltblatt * *(amerikan.) Prospekt, Reklamezettel* || Prospekt * *(amerikan.) Werbezettel, Prospekt* || Streublatt * *(amerikan.)* →*Prospekt* || Werbeblatt * *(amerikan.) Reklamezettel,* →*Prospekt* || Werbeprospekt * *(amerikan.)* || Werbeschrift * *(amerikan.) Reklamezettel, Prospekt*
flying fliegend * *fliegend, flüchtig, kurz, beweglich; flying (re-) start: fliegender (Neu-) Start* || Lose * *(Adv) beweglich, fliegend (gelagert/angeordnet); flying leads: lose Anschlussleitungen*
flying changeover fliegende Umschaltung
flying master fliegender Master * *z.B. bei Multiprozessor- oder Feldbussystem*
flying range Reichweite * *[Verkehr] ~ von Flugzeugen*
flying restart Fangschaltung * *Einschalten e.Umrichters auf e.s. drehende Maschine(frequ./phasen/amplitud.gerecht)* || Warmstart * *z.B. durch Fangschaltung bei Umrichtergespeistem Drehstrommotor*
flying restart circuit Fangschaltung
flying restart function Fangschaltung * *als Funktion*
flying restart module Fangschaltung
flying saw fliegende Säge * *läuft während des Sägevorgangs mit dem Material mit*
flying shears fliegende Schere * *läuft während des Schneidvorgangs mit d. Material mit (Materialfluss stoppt nicht)*
flying splice fliegender Rollenwechsel
flying splicer Rollenwechsler * *Auf/Abwickler mit fliegendem Rollenwechsel*
flying start fliegender Start

flying terminal block Lüsterklemme * *für "fliegende Verdrahtung"*
flying-master operation Multimasterbetrieb * *mit wechselndem Master bei Bussystem*
flying-restart Fangen * *Aufschaltg. e. Frequenzumrichters auf laufenden Motor z.B. nach Spannungseinbruch*
flying-restart circuit Motorfangschaltung * *siehe auch →Fangschaltung*
flywheel Schwungrad
flywheel backup kinetische Pufferung
flywheel efect Schwungmoment
flywheel effect Fliehkraft
flywheel moment Schwungmoment * *auch "GD^2" genannt; Umrechnung: Trägh.mom. J entspr. $(GD^2)/4; J [kgm^2], GD^2 [kpm^2]$*
FM FM * *Frequenzmodulation*
FMS FMS * *Abkürzg.für 'Field bus Message Specification': Busprotokoll, b.* PROFIBUS *genormt (DIN 19245,T.1)*
FO cable Lichtwellenleiter * *Abk.f. 'Fiber-Optic' cable*
foam Schaum || schäumen
foam-up ausschäumen
foamed material Schaumstoff
foamed plastics Schaumstoff
foaming Schaumbildung
focal point Mittelpunkt * *~ der Betrachtungen, des Interesses usw.* || Schwerpunkt * *(figürlich)*
focal question Hauptsache * *zentrale Frage*
focus Brennpunkt || bündeln * *konzentrieren, vereinigen (sich) sammeln, fokussieren* || Gewicht * *Brenn-/Schwerpunkt (des Interesses, der Bemühungen, der Arbeit usw.)* || konzentrieren * *focus on: sich/seine Aufmerksamkeit/seine Kräfte konzentrieren auf* || Mittelpunkt * *Brennpunkt (auch figürl.: der Tätigkeit, des Interesses)* || Schwerpunkt * *(figürl.) Haugtaugenmerk, Schwerpkt. (d.Interesses, d.Bemühungen, d.Arbeit;) Brennpkt.*
focus on Schwerpunkt * *(Verb) ~ setzen auf, sein Hauptaugenmerk setzen auf*
focus-shift Unschärfe * *Fehlfokussierung, Unschärfe (auch auf einem Bildschirm)*
foil Folie * *meist aus Metall* || Streichleiste * *bei Streichmaschine*
foil capacitor Kunststofffolienkondensator
foil machine Folienmaschine
foil printing press Foliendruckmaschine * *für Metallfolie*
foil producing machine Folienmaschine
foil sealing machine Folieneinschweißmaschine || Folienschweißmaschine
foil stretching machine Folienreckanlage
foil-covered keyboard Folientastatur
foiling machine Foliermaschine * *[Verpack.techn]*
fold brechen * *Papier* || falten || Falz || falzen || klappen * *falten; fold away: weg-/zurückklappen* || Knick * *Faltung in Papier Blech usw.* || knicken * *Papier ~*
fold back umklappen * *nach hinten falten/knicken*
fold-back zurückziehend * *z.B. Kennlinie einer Stromversorgung (Spannung wird bei Überstrom zurückgezogen)*
fold-back characteristic abgeknickte Kennlinie * *zurückgebogende Kennlinie (bei Netzteil: Spannungsreduzierung b.Überstrom)*
fold-out Faltblatt
foldable faltbar
folded box Faltschachtel
folder Aktenordner * *Ringbuch, dünner Aktenordner* || Broschüre * *Werbeblatt, Faltblatt, Broschüre, Heft* || Faltblatt * *Faltprospekt, Faltblatt, Broschüre, Heft* || Falzapparat * *am Ende einer Druckmaschine zum Sammeln und Falten der Exemplare* || Mappe || Produktinformation * *Faltblatt* || Produktschrift * *Faltblatt* || Prospekt * *Faltblatt* || Werbeblatt * *Faltprospekt, Faltblatt, Broschüre*
folder shutter Falzklappe * *[Druckerei] im Falzapparat*
folding Falt- || klappbar || Schwenk- * *faltbar*
folding box Faltkiste
folding boxboard Faltschachtelkarton * *Faltschachtelkartonpappe*
folding endurance Falzwiderstand * *von Kartonpapier*
folding machine Falzapparat * *am Ende einer (Zeitungs-) Druckmaschine* || Falzmaschine || Schwenkbiegemaschine * *zur Blechbearbeitung* || Schwenkbiegmaschine
folding press Abkantbank * *zum Biegen/Abkanten von Blech* || Abkantpresse
folding rule Zollstock
folding-box Faltschachtel || Karton * *Faltschachtel* || Schachtel * *Falt~*
folding-type Klapp- * *faltbar*
foliage trees Holz * *Laub~*
follow anschließen * *einem Beispiel folgen, nachfolgen* || beachten * *befolgen* || befolgen * *Ratschläge ~* || einhalten || folgen * *auch logisch* || nachlaufen * *folgen* || sich halten an || verfolgen * *einen Vorgang usw. ~*
follow a rule gehen * *nach einer Regel ~*
follow from hervorgehen aus * *sich als Folge ergeben aus; from this follows that: daraus geht hervor, dass* || sich ergeben * *~ aus*
follow up dranbleiben * *seine Idee, Stoßrichtung usw. verfolgen; energetisch:mit Nachdruck*
follow-up mode Nachführbetrieb
follow-up operation Nachführbetrieb
follow-up order Anschlussauftrag || Folgeauftrag * *Anschlussauftrag* || Nachbestellung * *Folgeauftrag*
follow-up product Nachfolgeprodukt
follower Folgeantrieb * *z.B. bei Leit-Folgebetrieb, Lastausgleichsregelung usw.* || Nachfolger
following anschließend * *zeitlich* || Befolgung * *Befolgung, Beachtung, 'Sich-Richten-Nach'* || entsprechend * *(Adv.) entsprechend* || folgend || gemäß * *(Vorschrift, Betriebsanleitung usw.) beachtend/folgend* || im Anschluss an || nach * *in der zeitlichen Reihenfolge* || nächster * *~ in der Reihenfolge*
following drive Folgeantrieb
following error Schleppabstand * *Soll-Ist-Abweichung/→Schleppfehler der Lage bei einer Lageregelung* || Schleppfehler * *z.B. statische Lage-Soll/ Istabweichung bei einer Lageregelung*
food Lebensmittel || Nahrungsmittel
food industry Lebensmittelindustrie || Nahrungsmittelindustrie
food-processing industry Nahrungsmittelindustrie
foods Lebensmittel * *(Plural)* || Nahrungsmittel * *(Plural)*
foodstuff Nahrungsmittel
foodstuff industry Lebensmittelindustrie || Nahrungsmittelindustrie
foodstuffs Lebensmittel * *(Plural)*
fool vormachen * *jemandem/sich etwas vormachen*
fool-proof narrensicher
foolproof einfach * *kinderleicht, narrensicher, idiotensicher, ungefährlich, betriebssicher* || kinderleicht * *→narrensicher*
foot Fuß * *auch Befestigungs/Aufstellfuß eines Motors*
foot lever Pedal * *Fußhebel*
foot mounting Fußanbau || Fußmontage * *eines Motors (im Gegensatz zur Flanschmontage)*
foot mounting type of construction Fußausführung * *Motor mit Fußbefestigung (Ggs.: Flanschbefestigung)*
foot pedal Fußpedal

foot-mounted design Fußausführung * z.B. (Motor mit) Fußbefestigung
foot-mounted version Fußbauform * z.B. Motor/Getriebe mit Aufstellfüßen
foot-mounting Fußbefestigung * Motor
foot-mounting motor Motor in Fußbauform * mit Fußbefestigung
foot-mounting type Fußbauform * z.B. Motor mit Aufstellfüßen (→angegossen oder →angeschraubt)
foot-operated potentiometer Fußpedalpotentiometer
foot-rule Zollstock
footage Länge * (Gesamt-) Länge/-Maß in Fuß (1 Fuß entspr. 30,48 cm)
footer Fußzeile
footing Stand * sichere Stellung; secure footing: sicherer ~; gain a footing: festen Fuß fassen
footnote Fußnote
footprint Aufstellfläche * für die Montage benötigte Grundfläche || Grundfläche * benötigte Aufstell-/Einbaufläche (z.B. eines Geräts, Schaltschranks usw.) || Platzbedarf * Grundfläche
for -Gründen * z.B. for accuracy: aus Genauigkeitsgründen || -weise * z.B. bit for bit: Bit für Bit || bei * für, bezogen auf || für || gegen * Mittel ~, z.B. ein Leiden || nach * zu einem Punkt hin (in Verbindung mit 'depart, leave, set out, bound, train') || seit * während; for some days: seit einigen Tagen || um * Lohn, Preis
for a limited time zeitlich begrenzt
for a long time lang * zeitlich
for a prolonged period länger * über einen längeren Zeitraum
for a quantity of bei einer Stückzahl von
for a short period kurzfristig * (Adv.) für kurze Zeit
for a time zeitweise
for all that dabei * dennoch || trotz * ~ alledem
for all these efforts trotz * ~ aller Bemühungen
for driving fans zum Antrieb von Ventilatoren
for ever ständig * (Adv.)
for example etwa * zum Beispiel || zum Beispiel
for further action Veranlassung * zur (weiteren) ~ || zur weiteren Veranlassung
for further information Auskunft erteilt
for future use für spätere Verwendung || Reserve * für (zu)künftige Verwendung vorgesehen
for his part seinerseits * (Adv.)
for industrial purposes Industrie-
for instance zum Beispiel
for internal use only nur für internen Gebrauch
for nothing kostenlos
for one thing erstens || zunächst * erstens
for plant-specific reasons anlagenbedingt
for publicity purposes werblich * für Werbezwecke
for reasons -bedingt * for safety reasons: sicherheits~; for plant-specific reasons: anlagen~
for reasons concerning the aus Gründen der * den ... betreffend
for reasons of safety sicherheitshalber * aus Gründen der Sicherheit
for reasons of the aus Gründen der
for safety reasons sicherheitshalber || zur Sicherheit
for short Kürze * der Kürze halber
for short periods of time kurzzeitig * (Adv.)
for some time länger * für ~e/einige Zeit
for that purpose deshalb * für diesen Zweck
for that reason deshalb * aus diesem Grund, als Folge davon || deswegen * →daher
for the attention of zu Händen von
for the benefit of zugunsten

for the first time erstmalig || zuerst * zum ersten Mal || zum ersten Mal
for the future zukünftig * (Adv.)
for the moment zuerst * vorerst, vorläufig || zunächst * vorerst, vorläufig
for the most part größtenteils || meistens * meistenteils || zum größten Teil
for the present im Moment * gegenwärtig || momentan * (Adv.) zurzeit || vorerst * vorderhand || vorläufig * vorerst, vorderhand || vorübergehend * vorderhand || zunächst * vorläufig
for the present time momentan * (Adv.) zurzeit, im Moment
for the purpose of im Rahmen von * für Zweck/Ziel/Vorhaben
for the sake of auf Grund * um (einer Sache) willen || um * um einer Sache willen || wegen * um ... willen
for the time being bis auf weiteres * momentan || im Moment || momentan * (Adv.) zum augenblicklichen Zeitpunkt, zurzeit || vorerst * vorderhand || vorläufig * vorerst, vorderhand || zunächst * vorläufig, vorerst || zurzeit
for this account deshalb * deswegen
for use in hazardous locations explosionsgeschützt
for use later für spätere Verwendung
forbearing nachgiebig * nachsichtig, geduldig
force durchsetzen * zwingen || erzwingen || Gewalt * zwingende Kraft, Gewalttätigkeit; by force/forcibly: mit Gewalt; force majeure: höhere Gewalt || Kraft * ~ [N], Stärke, Wucht; Einfluss, Wirkung, Nachdruck, (Rechts-)Gültigkeit; in force: in ~ || pressen * zwängen || Stärke || steuern * {SPS} zwangssteuern von Ein-/Ausgängen bei SPS (Testfunktion) || zwangssteuern * durch künstliche Einprägung von Steuerbefehlen für Inbetriebnahme/Testzwecke
force application Kraftangriff
force away abdrücken
force control Kraftregelung * z.B. Walzkraft
force disable Zwangssteuern beenden
force distribution Kraftverteilung
force down Absturz * zum ~ bringen || anpressen * niederhalten
force majeure höhere Gewalt
force of acceleration Beschleunigungskraft
force of gravity Schwerkraft
force off abdrücken * z.B. Lager von Welle
force on-off facility Steuerfunktion * Zwangssteuerungsfunktion für Ein-/Ausgänge einer SPS (Testfunktion)
force one's way durchbrechen * mit Gewalt seinen Weg machen
force open sprengen * gewaltsam öffnen
force release Zwangssteuern beenden
force sensor Kraftsensor
force transducer Zugmessdose * Zugkraft-Messdose || Zugmesseinrichtung * (Zug-) Kraftaufnehmer
force-cooled fremdbelüftet || fremdgekühlt
force-cooling fremdgekühlt
force-fire zwangszünden * →Schutzzündung von Thyristoren bei Erreichen der max. zulässigen Sperrspannung
force-locking kraftschlüssig
force-lubricated zwangsgeschmiert
force-ventilated fremdbelüftet
forced erzwungen || künstlich * gezwungen || Zwangs-
forced air cooling forcierte Luftkühlung * z.B. bei Leistungshalbleitern || verstärkte Kühlung * →verstärkte Luftkühlung, z.B. von Leistungshalbleitern mit Lüfter || verstärkte Luftkühlung * Kühlung von Leistungshalbleitern usw. mit Lüfter

forced commutation Zwangskommutierung
forced cooling Fremdkühlung || verstärkte Kühlung * z.b. von Leistungshalbleitern mit Lüfter oder flüssigen Kühlmitteln || Zwangsbelüftung
forced oil lubrication Olumlaufschmierung
forced ventilation Fremdbelüftung * z.B. Motor || verstärkte Luftkühlung * Kühlung von Leistungshalbleitern usw. mit Lüfter || Zwangsbelüftung
forced-air cooled zwangsbelüftet
forced-air cooling Fremdbelüftung || Fremdkühlung * Fremdbelüftung über fremdangetriebenen Lüfter oder Rohranschluss || verstärkte Luftkühlung * z.b. von Leistungshalbleitern mit Lüfter
forced-air oven Umluftofen
forced-circulation oil lubrication Umlaufschmierung
forced-cooled fremdgekühlt || zwangsgekühlt
forced-ventilated fremdbelüftet * z.B. Motorkühlart mit getrennt angetriebenem Lüfter || fremdgekühlt * Motorkühlung durch getrennt angetriebenen Lüfter || zwangsbelüftet
forcibly actuated shutdown Zwangsabschaltung
forcing steuern * (Substantiv) [SPS] zwangssteuern von Ein-Ausgängen (Testfunktion) || zwangssteuern * (Substantiv) Festvorgabe von SPS-Ein-/Ausgängen für Testzwecke
forcing up of prices Preistreiberei
forcing-off screw Abdrückschraube * zum Abziehen eines im Presssitz montierten Maschinenteils
fore vorder
fore- voraus-
forecast Prognose || vorausberechnen * voraussagen/sehen || vorherbestimmen * voraussagen, vorhersehen || Vorhersage || vorhersagen || vorhersehen * vorhersagen
forego verzichten
foregoing vorgenannt || vorhergehend || vorherig
foregone conclusion Selbstverständlichkeit
foreground Vordergrund
foreground memory Vordergrundspeicher
foreground task Vordergrundverarbeitung * Rechnerprogramme, die mit hoher Priorität laufen
foreign ausländisch
foreign bodies Fremdkörper * (Plural)
foreign body Fremdkörper || Gegenstand * Fremdkörper || Teil * Fremdkörper
foreign branch Auslandsniederlassung || Landesgesellschaft * Zweigniederlassung im Ausland || Zweigniederlassung * im Ausland
foreign computer Fremdrechner
foreign country Ausland
foreign language Fremdsprache || Sprache * Fremdsprache
foreign matter Fremdkörper
foreign objects Fremdkörper * protection from ingress of foreign objects: Schutz gegen Eindringen von ~n
foreign particles Fremdkörper * (Plural)
foreign subsidiary Auslandsniederlassung || Landesgesellschaft * Filiale/Tochtergesellschaft im Ausland
foreknow vorhersehen * vorherwissen
foreman Meister * Vorarbeiter || Werkmeister
foremost vorderst * am weitesten vorne gelegen
foresee vorhersehen
foreseeable vorhersehbar
foreshorten verkürzen * (geometr.) perspektivisch ~
foresighted vorausschauend
forestry Forstwirtschaft
forestry industry Forstwirtschaft
foreword Vorwort * meist von jemand anderem als dem Autor
forge Schmiede || schmieden
forged geschmiedet

forged steel Schmiedestahl
forging Schmiedestück || Schmiedeteil
forging machine Schmiedemaschine
forging press Schmiedepresse || Warmpresse
fork Rechen * Harke || teilen * sich ~/gabeln/spalten, z.b. Straße
fork spanner Gabelschlüssel
fork wrench Gabelschlüssel
fork-lift truck Gabelstapler || Stapler * Gabel~
form ausbilden * formen, gestalten || ausmachen * einen Teil bilden/formen; ausbilden || Ausprägung * Form, Gestalt, Fasson, Manier, Art u.Weise, Schema, System || bilden || Form * Gestalt, Förmlichkeit; pro forma: der ~ halber; be off form: nicht in ~ sein || Formblatt || Formular || gestalten || prägen * (auch figürl.) || Schalung * [Bautechnik] || Variante * Form, Gestalt, Fasson, Art u. Weise, Schema, System || verformen * formen || Vordruck * Formular || zusammenstellen * Mannschaft, Team, Arbeitsgruppe usw. ~
form 'C' contact Umschaltkontakt
form a bend krümmen * sich krümmen, eine Krümmung bilden
form A contact Arbeitskontakt || Schließer || Schließerkontakt
form a curve krümmen * sich krümmen, eine Krümmung bilden
form an alliance zusammenschließen * zu einem Bündnis ~
form B contact Öffner || Öffnerkontakt || Ruhekontakt
form C contact Umschaltkontakt || Wechselkontakt || Wechsler || Wechslerkontakt
form factor Formfaktor * Effektiv-/Arithmet.-Gleichricht-Mittelwert (1 bei Gleichspanng.; 1,11 bei sin-Spanng)
form pairs zusammengehören * je zwei Sachen
form part of gehören * einen Teil bilden von
form-B contact Öffnerkontakt
form-C contact Umschaltkontakt
form-closed formschlüssig
form-fit formschlüssig
form-locking formschlüssig
form-wound coil Formspule * [el.Masch.]
formal formal || steif * förmlich
formal operand Formaloperand
formal parameter formaler Parameter || Formalparameter
formaldehyde Formaldehyd
formalities Formalitäten
formalize formalisieren
format Format * auch von Daten || formatieren || Größe * Format
format size Format
formation Aufbau * (Aus)Formung || Ausbildung * Formung, Gestaltung, Bildung || Entstehung * Bildung, Formung, Entstehung, Gestaltung || Formation || Gründung * auch ~ einer Firma || Verband * Formation, Kampfverband
formatted formatiert
former alt * ehemalig || bisherig || früher * ehemalig || vorherig * vorig, ehemalig
formerly sonst * ehemals
forming Bearbeitung * Formung, Gestaltung || Bildung * Formung || Formierung * z.b. langsame Aufladung eines (Elektrolyt-) Kondensators nach langer Lagerung || Formveränderung * bei Verarbeitungsmaschine || spanlos * Werkstückbearbeitung (Gegensatz: →zerspanend) || Umformung * von Werkstoffen
forming machine Formmaschine * z.B. für Pralinen || Verformungsmaschine
forming of a section profilieren * (Substantiv)
forming part of it dazugehörig

forming process Formierung * *Vorgang der ~, z.B. eines Elektrolytkondensators*
formless formlos
formula Formel * *using/according to the following formula: nach folgender ~* || Rezept * Formel || Rezeptur
formula for success Erfolgsrezept
formulae Formeln
formulate ausarbeiten * *schriftlich*
forsake verlassen * *im Stich lassen*
forte Stärke * *starke Seite*
forth hervor
forthwith sofort * *sofort, umgehend, unverzüglich*
fortified verstärkt * *Gewebe; nylon fortified: mit Nylon ~; (auch 'heavy-duty': für Schwerlastbetrieb)*
fortunately glücklicherweise
forward hervor || nach vorne * *vorwärts* || schicken * *weiterleiten* || versenden * *befördern, schicken, verladen, weiterbefördern (Brief)* || vorder || vorn * *nach ~* || vorwärts || weitergeben * *z.B. Brief* || weiterleiten * *Brief usw. (to:an)* || weiterreichen * *z.b. Brief* || zuschicken * *weiterleiten; to: an*
forward characteristic Durchlasskennlinie * *(Diode)*
forward converter Durchflusswandler * *bei Schaltnetzteil*
forward current Durchlassstrom * *bei Diode, Thyristor usw.*
forward direction Durchlassrichtung * *z .B. Thyristor/Diode* || Flussrichtung * *bei Halbleitern*
forward direction of rotation Rechtsdrehrichtung * *Vorwärtsdrehrichtung*
forward flow angle Stromflusswinkel
forward losses Durchlassverluste * *z.b. bei Diode/Transistor/Thyristor*
forward planning Disposition * *Vorausplanung*
forward rotation Vorwärtsdrehrichtung
forward running Rechtslauf * *Vorwärtslauf*
forward slip Voreilung * *(Walzwerk)*
forward speed direction Vorwärtsdrehrichtung
forward to richten an * *z.b. eine Anfrage*
forward voltage Durchlassspannung * *(Diode)*
forward voltage drop Durchlassspannung * *z.b. bei Diode/Thyristor/Transistor*
forward-looking fortschrittlich || zukunftsorientiert || zukunftsweisend
forward-reverse evaluation Vor-Rückwärtsauswertung
forward-reverse evaluator Vorwärts-Rückwärtsauswertung
forward/reverse control Drehrichtungssteuerung || Kommandostufe
forwarding Lieferung * *Beförderung* || Weitergabe * *Weiterbefördern, z.B. von Briefen*
forwarding address Versandanschrift
forwarding agency Spediteur * *Speditionsfirma* || Spedition
forwarding agent Spediteur
foul faul * *schmutzig, schmutzig, verdorben, stinkend, widerlich* || verschmutzen * *z.B. durch Öl*
fouled verschmutzt * *verölt, beschmutzt, befleckt, verstopft, verölt*
fouling Verschmutzung * *Verschmutzen, Verstopfung, Verölung, Versperrung, Befleckung* || Verstopfung * *durch Verschmutzung*
fouling by oil Verölung
found gründen * *z.B. eine Firma*
found a company Firma gründen
foundation Fundament * *eines Gebäudes, einer Maschine* || Grund * *Fundament* || Grundlage || Gründung * *~ einer Firma, Vereinigung usw.*
foundation block Fundamentklötzchen * *z.B. zur Motorbefestigung*
foundation load Fundamentbelastung * *für el.-*

Masch: siehe: DIN ISO 3945, VDI 2060 und VDE 0530 Teil 14
foundation vibrations Fundamentschwingungen
foundations Grundlagen * *Unterbau, Fundamente, Grundfesten, Stützen*
foundry Gießerei * *siehe auch →Hüttenwerk* || Hüttenwerk * *Gießerei, Hüttenwerk*
fountain Brunnen
four quadrant 4Q || Vierquadrant-
four-cycle engine Viertaktmotor
four-point fixing Vierpunktaufhängung * *z.b. bei Motor mit →Pratzenanbau*
four-pole vierpolig
four-quadrant Vierquadrant || Vierquadrant-
four-quadrant drive Umkehrantrieb * *Vierquadrantantrieb*
four-quadrant operation Vierquadrantbetrieb || Vierquadrantenbetrieb
four-stroke engine Viertaktmotor
four-terminal network Vierpol
four-way Kreuz- * *Vierweg-*
four-way valve Vierwegeventil
four-wheel drive Allradantrieb * *bei Vierrad-Fahrzeug*
four-wire cable Vierdrahtleitung * *z.b. serielle Datenverbindung nach RS422*
four-wire operation Vierdrahtbetrieb
fourdrinier Langsieb * *(Papier) bei einer Papiermaschine*
fourdrinier machine Langsiebmaschine * *Papiermaschine* || Langsiebpapiermaschine
fourdrinier paper machine Langsiebpapiermaschine
Fourier analysis Fourier-Analyse
Fourier series Fourier-Reihe
Fourier transform Fourier-Transformation || Fourier-Transformierte
Fourier transformation Fourier-Transformation
fourth from last viertletzt
FPGA FPGA * *'Field-Programmable Gate Array': On-Board/im Anwendersystem programmierbarer Logikbaustein*
fraction Anteil * *Bruchteil* || Bruch * *(mathem.)* || Bruchteil || Teil * *Bruch~* || zerlegen * *in seine physikalische/chemische Bestandteile ~*
fraction bar Bruchstrich * *(mathem.)*
fractional horse-power motor Kleinmotor * *unter 1 PS (0,74 kW) Leistung*
fractional horsepower kleiner Leistung * *fractional horse power motor: Motor ~*
fractional-hp motor Kleinmotor * *Motor bis 1 hp Leistung (entspr. 0,746 kW)*
fractional-hp range unter 1 PS * *im Leistungsbereich unter 1 PS (z.B. Motor)*
fractional-slot winding Bruchlochwicklung * *(el. Masch.) Nutenzahl je Pol/Nutenzahl je Wicklungsstrang ist gebroch.Zahl*
fractionize brechen
fracture brechen * *(zer)brechen* || Bruch
fragile empfindlich * *~zerbrechlich* || zerbrechlich
fragment Splitter * *Bruchstück*
fragmentary lückenhaft
fragments Scherben
frame Baugröße * *bei Elektromotor (Achshöhe)* || Gehäuse * *Motorgehäuse* || Gerüst * *z.B. Schalt~* || Gestell * *auch 'Gerüst'* || Masse * *Motorgehäuse-~ (elektr.)* || Rahmen * *auch 'Zeichen~, Daten~, Telegramm~'* || Zarge * *(Holz)* || Zeichenrahmen
frame profile Rahmenprofil
frame rate Bilderneuerungsfrequenz * *Häufigkeit d.Ausgabe bewegter Videobilder (flüssige Bewegg. ab 25 frames/s)* || Bildwiederholfrequenz * *softw.-mäßige Ausgabehäufigkt. v.Videobildern; flüssige Bewegg. ab 25 frames/s*

frame saw Gattersäge * *(Holz)* zersägt Baumstämme zu Brettern
frame size Achshöhe || Ah * *Baugröße/Achshöhe AH bei Motor* || Baugröße * e. Elektromotors *(ident.m. Achshöhe (z.B. BG225: Achshöhe 225mm; s.IEC Publ. 72 (1971))* || BG * *Motor-Baugröße, Achshöhe* || Gehäusegröße * *eines Motors* || Motor-Baugröße * *Baugröße, Achshöhe* || Motorgröße * *Baugröße, entspr. der Achshöhe*
framed crosscut saw Bügelsägemaschine
framegrabber Framegrabber * *(Computer) Einrichtung z. Echtzeit-Digitalisierung u.speicherung analoger Videosignale*
frameless gehäuselos * *bei Elektromotor: Blechpaket dient gleichzeitig als Gehäuse*
framework Gefüge || Rahmen * *(auch figürl.)* Gerüst,Gerippe, Gestell, Gefüge, System; *within th. framework of:i.~ von* || System * *Gefüge* || Systematik * *Gefüge*
framing error Zeichenrahmenfehler * *bei serieller Datenübertragung*
France Frankreich
franchise Konzession * *(amerikan.)* Verkaufsrecht
frank frei * *freimütig, unumwunden*
frankly offen * *(Adv.) ehrlich gesagt*
fraught with meaning sinnvoll * *bedeutungsvoll, bedeutungsschwer, bedeutungsschwanger, von Bedeutung*
free -los || befreien * *from:von* || frei * *from/of:von; free (of charge):unentgeltlich* || freizügig * *frei, zwanglos, ungezwungen, ungebunden, nach Belieben, unbeengt, unbelastet* || kostenlos || lösen * *free oneself from: sich befreien von*
free and easy frei * *ungezwungen*
free competitive system freie Marktwirtschaft || Wirtschaft * *freie (Markt-) ~*
free economy freie Marktwirtschaft
free enterprise freie Marktwirtschaft
free from backlash spielfrei
free from distortion verspannungsfrei
free from error fehlerfrei
free from slip schlupffrei
free from vibration erschütterungsfrei
free length freie Länge
free market economy freie Marktwirtschaft
free of -frei * *(vorangestellt)*
free of charge gratis || kostenfrei * *kostenlos* || kostenlos * *(Adv.)*
free of defects fehlerfrei * *Holz usw.*
free of dirt schmutzfrei
free of wear verschleißfrei
free play Spielraum * *(figürl.)*
free shaft end freies Wellenende * *z.B. ~ an der Nicht-Antriebsseite eines Motors (z.Anbau von Bremse/Handrad ...)*
free size Freimaß * *ohne Toleranzangabe*
free space Freiraum * *freier Raum*
free time Freizeit
free to move freizügig
free-convection konvektionsgekühlt || selbstbelüftet || selbstgekühlt * *→luftselbstgekühlt (Kühlart für Leistungshalbleiter)* || unbelüftet * *→luftselbstgekühlt (bei Leistungshalbleitern)*
free-enterprise marktwirtschaftlich * *den Gesetzen der freien Marktwirtschaft gehorchend*
free-flowing material Schüttgut
free-flowing products Schüttgut * *(Verpack.-techn.)*
free-of-charge kostenlos * *(Adj.)*
free-wheeling Freilauf * *im Freilauf fahrend* || freilaufend
free-wheeling arm Freilaufzweig
free-wheeling current Freilaufstrom
free-wheeling diode Freilaufdiode * *z.B. bei halbgesteuerter Thyristorbrücke*

free-wheeling path Freilaufzweig
free-wheeling valve Freilaufventil * *meist Diode in einem Freilaufzweig*
freedom Freiheit * *from: von; degree of freedom:* →Freiheitsgrad
freedom from maintenance Wartungsfreiheit
freedom in decision-making Entscheidungsspielraum
freedom of choice Entscheidungsspielraum * *Entscheidungsfreiheit*
freehand sketch Handskizze
freelance frei * *freiberuflich* || Freiberufler || freiberuflich * *work freelance:freiberufl. tätig sein* || freier Mitarbeiter * *Freiberufler* || selbstständig * *freiberuflich (tätig), unabhängig* || unabhängig * *freiberuflich tätig*
freely frei * *(Adv.)* || frei- || freizügig * *(Adv.)*
freely assignable frei belegbar * *z.B. Tasten, Ein/Ausgangsklemmen mit einer Funktion*
freely configurable frei projektierbar * *the control system can be freely configured:d.Regelsystem ist frei projektierb.* || frei verschaltbar * *frei konfigurierbar/projektierbar* || freiprojektierbar * *z.B. Regelsystem SIEMENS SIMADYN D (R)*
freely interconnectable frei verschaltbar
freely selectable wahlfrei * *frei wählbar*
freely-assignable freidefinierbar * *frei zuzuordnen (z.B. eine Funktion zu einer Klemme)*
freely-definable freidefinierbar
freely-programmable frei projektierbar || freiprogrammierbar * *z.B. Funktion einer Klemme*
freeness Mahlgrad * *(Papier) bei Zellstoff*
freewheel Freilauf
freewheel clutch Freilauf || Freilaufkupplung
freewheel diode Freilaufdiode
freeze einfrieren * *auch figürl.: Meßwert usw.* || festfressen * *sich festfressen* || festhalten * *einfrieren (Integrator, Messwerte usw.)* || fressen * *z.B. Kolben* || tiefkühlen
freezer Gefriergerät
freezer unit Gefriergerät
freezing Verklemmung * *softwaremäßige ~*
freight Fracht
freight agency Spedition
freight charge Versandkosten || Versandspesen
freight elevator Lastenaufzug
freight forwarding company Spedition
freight goods Frachtgüter
freight traffic Frachtverkehr
freight transportation Frachtverkehr * *(amerikan.)*
French französisch
frequencies Frequenzen
frequency Frequenz * *Schwingungen, Impulse* || Häufigkeit
frequency actual value Frequenzistwert
frequency arrival Frequenz erreicht * *Substantiv*
frequency at which field weakening starts Einsatzfrequenz der Feldschwächung
frequency avoidance Frequenzausblendung || Frequenzausklammerung
frequency avoidance zone Ausblendband * *zur Ausblendung (mechan.) Resonanzfrequenzen im Sollwert* || Frequenzausblendband || Frequenzausklammerungsband || Frequenzbandausklammerung
frequency band Frequenzband || Frequenzbereich
frequency band inhibit Frequenzbandausklammerung * *Ausblendband für Frequenzsollw. z. Vermeidg. mechan. Resonanzen*
frequency change Frequenzänderung
frequency changer Frequenzumformer * *(el. Masch.)*
frequency changer set Frequenzumformer * *(el. Masch.) rotierender Drehfeldumformer (Maschinensatz)*

frequency characteristic Frequenzkennlinie
frequency control Frequenzregelung
frequency control range Frequenzstellbereich
frequency converter Frequenzumformer * *auch rotierender Umformer (z. Erreichen extrem hoher Antriebdrehzahlen)* || Frequenzumrichter || Umrichter * *Frequenz~*
frequency counter Frequenzzähler
frequency distribution Häufigkeitsverteilung
frequency divider Frequenzteiler
frequency domain Frequenzbereich * *(Regel.-techn.)*
frequency fluctuations Frequenzschwankungen
frequency generator Frequenzgeber || Frequenzgenerator
frequency inhibit band Ausblendband * *Ausblendung von Frequenzsollwerten zur Vermeidung mechan. Resonanzen*
frequency input Frequenzeingang
frequency inverter Frequenzumformer || Frequenzumrichter
frequency jumping Ausblendband * *Ausblendung bestimmter Frequenzsollwerte zur Vermeidung mechan. Resonanzen* || Ausblendfrequenz * *ausgeblendeter Frequ.sollwert f.Frequenzumrichter z.Vermeidg.mechan.Resonanzen* || Frequenzausblendung * *Ausblendung kritischer (mechan.) Resonanzfrequenzen im Sollwert* || Frequenzausklammerung * *Ausblendung (mechan.) Resonanzfrequenzen im Sollwert* || Frequenzbandausklammerung
frequency jumping band Frequenzausblendband || Frequenzausklammerungsband
frequency meter Frequenzmesser
frequency modulation FM
frequency of operation Schalthäufigkeit
frequency output Frequenzausgang
frequency range Frequenzbereich
frequency response Frequenzgang || Frequenzkennlinie * →*Frequenzgang*
frequency scaler Frequenzteiler
frequency setting range Frequenzstellbereich
frequency signal Frequenzsignal
frequency skipping Frequenzausblendung || Frequenzausklammerung
frequency spectrum Frequenzspektrum
frequency suppression band Ausblendband
frequency thyristor Frequenzthyristor * *für über der Netzfrequ. liegende Schaltfrequenzen (für fremdgeführt.Schaltgn.)*
frequency to voltage converter Frequenz-Spannungswandler
frequency tracking Frequenznachführung
frequency variation Frequenzänderung
frequency-converter fed frequenzumrichtergespeist
frequency-converter powered frequenzumrichtergespeist
frequency-dependent frequenzabhängig
frequency-dependent variable frequenzabhängige Größe
frequency-related frequenzabhängig
frequency-voltage conversion Frequenz-Spannungs-Wandlung
frequent häufig
frequent ocurring Häufung * *Wiederholung, häufiges Vorkommen*
frequently häufig * *(Adv.)* || vielfach * *(Adv.)*
fresh erneut || jungfräulich * *frisch* || neu * *frisch, er~t: fresh start: ~er Anfang; with fresh energy: mit ~en Kräften*
fresh air Frischluft || Zuluft * *für Kühlung*
fresh oil Frischöl
fresh up frisch machen * *z.B.sich waschen u.umziehen*

fretsaw Laubsäge
friction Friktion * *Reibung* || Reib- || Reibung
friction bearing Gleitlager
friction characteristic Reibungskennlinie * *Reibmoment z.B. in Abhängigkeit von der Drehzahl* || Reibverhalten
friction clutch Reibkupplung || Rutschkupplung
friction coefficient Reibungszahl
friction combination Reibpaarung
friction compensation Reibungskompensation
friction contact Reibschluss
friction disk Reibscheibe
friction drive Friktionsantrieb
friction drum Reibwalze
friction energy Reibarbeit || Reibungsarbeit
friction energy absorpion Reibungsverluste
friction energy absorption Reibarbeit || Reibungsarbeit
friction face Reibfläche * *z.B. bei Bremse*
friction force Reibkraft || Reibungskraft
friction lining Reibbelag
friction locking Reibschluss
friction losses Reibarbeit * *Reibungsverluste* || Reibungsarbeit * *Reibverluste* || Reibungsverluste
friction material Reibmaterial || Reibwerkstoff
friction of rest Haftreibung
friction pad Reibbelag
friction roll Reibwalze
friction roller Friktionswalze || Reibwalze
friction surface Reibfläche
friction torque Reibmoment || Reibungsmoment
friction wheel Reibrad
friction work Reibarbeit
friction-drum motor Außenläufermotor * *für Förderband usw.*
friction-locked kraftschlüssig
friction-material mating Reibpaarung
frictional behaviour Reibverhalten
frictional connection Kraftschluss
frictional heat Reibungsverluste * *Reibungswärme* || Reibungswärme
frictional locking Reibschluss
frictional losses Reibungsverluste
frictional torque Reibmoment
frictionally kraftschlüssig
frictionally engaged reibschlüssig
frictionally locked reibschlüssig
frictionless reibungslos * *ohne Reibung*
frigen Frigen
fringe Rand * *Einfassung, Um~ung, Franse; (figürl.): äußerer ~, Grenze*
frisk Bocksprünge machen * *herumtollen*
frolic Bocksprünge machen * *herumtoben, (ausgelassen) herumtollen*
from ab * *zeitlich und örtlich; from this time onwards:ab diesem Zeitpunkt* || aus * *von e.Ort, aus Material; z.B.: from cast iron: ~ Grauguss; from plastic material: ~ Kunststoff* || vor * *schützen, verstecken, warnen*
from a block aus dem Vollen * *(Werkz.masch.) z.B. ~ drehen/fäsen*
from a full-liner alles aus einer Hand * *vom Komplettanbieter*
from abroad vom Ausland
from and above ab * *ab einschließlich*
from before vorn * *von ~*
from behind hervor * *hinter ... hervor* || von hinten
from below unten * *von ~, von ~ herauf* || von unten
from below to the top von unten nach oben
from bottom to top von unten nach oben
from first to last vorn * *von ~ nach hinten, vom ersten zum letzten*
from front to back vorn * *von ~ bis hinten*

from memory auswendig * *aus dem gedächtnis; memorize: auswendig lernen, sich einprägen*
from now on an * *(Adv.) von nun ~*
from one minute to the other schlagartig * *(Adv.)*
from outside von außen
from stock ab Lager
from the bottom von unten
from the bottom to the top von unten nach oben
from the front von vorn || vorn * *von ~*
from the hardware point of view hardwaremäßig * *(Adv.) aus dem Blickwinkel der Hardware*
from the hardware side hardwaremäßig * *(Adv.) hardwareseitig*
from the inside von innen
from the last -letzt * *z.B. fourth from the last: viert~*
from the outside von außen
from the perspective aus der Sicht * *from the price perspective:aus der Preis-Sicht* || hinsichtlich * *from the performance perspective: ~ der Leistungsfähigkeit* || in Hinblick auf * *aus der ... Perspektive*
from the point of view aus der Sicht
from the rear von hinten
from the software side softwaremäßig * *softwareseitig*
from the stock lagermäßig
from the top to the bottom von oben nach unten
from time to time gelegentlich * *(Adv.) von Zeit zu Zeit* || von Zeit zu Zeit * *siehe auch →gelegentlich* || zeitweise * *von Zeit zu Zeit*
from top to bottom von oben nach unten
from under hervor * *unter ... hervor* || unter * *~ hervor*
from underneath von unten
from version x onwards ab Version x
from without von außen
front Frontseite || frontseitig || Stirnseite || vorder || Vorder- || Vorderseite
front area vorderer Bereich
front connection vorderseitiger Anschluss
front connector Frontstecker
front cover Gehäusedeckel * *vorderer ~*
front drum Tragwalze 2 * *eines Doppeltragwalzenrollers* || vordere Tragwalze * *Tragwalze 2 eines Doppeltragwalzenrollers*
front edge Vorderkante
front end Einspeiseeinheit * *zur Speisung einer Gleichstrom-Zwischenkreisschiene vom Netz* || Hochleistungs- * *Spitzen-* || Spitzen- * *an der Spitze stehend* || Stirnfläche || Stirnseite
front end product Spitzenprodukt
front face Stirnseite || Vorderseite
front operated lathe Plandrehmaschine
front panel Frontplatte || Frontseite * *Frontplatte*
front panel cut-out Frontplattenauschnitt
front plug Frontstecker
front side Frontseite || Stirnseite || Vorderseite
front view Ansicht von vorn || Aufriss * *Vorderansicht* || Frontansicht || Vorderansicht
front- Front- || vorderst
front-end Bedienungsoberfläche * *{Computer} Software, die e. (komfortablere) ~ für e. anderes Programm erzeugt*
front-panel cut-out Frontplattenaussparung
front-panel element Frontplatten-Einbauelement * *z.B. Schalter, Potentiometer*
front-side cover Stirndeckel
frontend vorgeschaltet * *{Computer} z.B. ~es Programm (einer Datenbank usw.)*
frontier Grenze * *(Landes)~, Grenzbereich (figürl.); frontier town: Grenzstadt; new frontiers: neue Ziele*
frontpanel Frontplatte
frontplate Frontplatte

frost protection Frostschutz
frow upon beanstanden * *etwas ~*
frozen eingefroren * *auch Messwerte, Speicherinhalt usw.*
frugality Einfachheit * *Genügsamkeit, Einfachheit*
frustrum of cone Kegelstumpf
ft m * *feet; 1 m entspr. 3,2808 ft*
FTP FTP * *Abk.f. 'File-Transfer Protocol': Datenübertrag.protokoll z. Dateitransfer von/zu einem Rechner*
fuction of membership Zugehörigkeitsfunktion
fuel Brennstoff
fuel cell Brennstoffzelle
fuel consumption Treibstoffverbrauch
fulcrum point Drehpunkt
fulfil leisten * *erfüllen*
fulfill erfüllen * *Bedingung, Aufgabe ~, requirements: Anforderung, Spezifikation*
full ganz * *voll(er), voll(ständig), ganz, völlig, gänzlich; full of:erfüllt mit/reich an* || uneingeschränkt || voll * *(auch figürl.) vollständig, gefüllt* || Voll- * *z.B. vollständig* || völlig * *ganz* || vollkommen * *völlig, vollständig* || vollwertig || walken || weitgehend * *Verständnis usw.*
full block control Vollblocksteuerung
full duplex vollduplex
full key voller Federkeil
full lamination Vollblechung
full liner Vollanbieter * *Anbieter mit vollständigem Produktspektrum*
full load Vollast * *at: bei*
full of mit * *voll von*
full of corners winklig * *verwinkelt*
full roll volle Rolle * *bei Wickler*
full scale Endwert * *eines Meßbereichs* || Messbereichs-Endwert || Messbereichsendwert || Skalenendwert
full speed Hochdruck * *(figürl) work at full speed/ at full blast: mit ~ arbeiten* || volle Geschwindigkeit || Vollgas
full throttle Vollgas * *bei Verbrennungsmotor*
full value Endwert || voller Wert
full-key balancing Vollkeilwuchtung * *Federkeil wird beim Auswuchten des Motorsläufers voll berücksichtigt*
full-line supplier Komplettanbieter
full-liner Generalist * *Vollanbieter* || Komplettanbieter
full-load duty Vollast * *Betrieb bei ~, d.h. mit der höchsten zugelassenen Leistung am Bemessungspunkt*
full-range Endausschlag * *(Messgerät)* || Ganzbereichs-
full-scale Endausschlag * *(Messgerät)*
full-wave Vollschwingung
full-wave control Schwingungspaketsteuerung * *Vollwellensteuerung* || Vollschwingungssteuerung * *bei Wechsel/Drehstromsteller* || Vollwellensteuerung * *bei Wechsel/Drehstromsteller*
full-wave rectification Vollwellengleichrichtung
full-wave voltage controlled operation Vollschwingungssteuerung * *Leistungssteller-Betriebsart (lässt ganze Netzschwing.durch)*
fully voll * *(Adv.)* || Voll- * *(Adv.) vor Adjektiv*
fully assembled unit Komplettgerät
fully compatible voll kompatibel * *with: mit* || vollkompatibel * *with: mit*
fully compensated vollkompensiert
fully controlled vollgesteuert * *z.B. Thyristorbrücke*
fully controlled anti-parallel three-phase bridge connecti vollgesteuerte Gegenparallel-Drehstrombrückenschaltung * *(B6)A(B6)C-Schaltung f.4Q-DC-Stromricht.*
fully controlled inverse parallel bridge connec-

fully controlled single-phase antiparallel bridge connection 540

tion   vollgesteuerte Gegenparallelschaltung * *(B6)A(B6)C-Schaltung*
fully controlled single-phase antiparallel bridge connection   B2C-Schaltung
fully controlled single-phase bridge connection   vollgesteuerte Einphasen-Brückenschaltung * *B2C-Schaltung für 1Q-DC-Umrichter*
fully developed   ausgereift * *technische Lösung, Konstruktion usw.*
fully digital   voll digital || volldigital
fully functional   betriebsfähig
fully laminated   vollgeblecht
fully programmable   voll programmierbar
fully tightened   fest angezogen * *z.B. Schraubenverbindung*
fully-automatic   vollautomatisch
fully-controlled inverse-parallel three-phase bridge connection   B6-A-B6-C-Schaltung * *vollgest. gegenpar.Drehstrombrücke*
fully-controlled three-phase bridge connection   B6C-Schaltung || vollgesteuerte Drehstrombrückenschaltung * *z.B. B6C-Schaltung für 1Q-DC-Stromrichter*
fully-enclosed   geschlossen * *voll ~ (Gehäuse-Schutzart)*
fully-graphic   vollgrafisch
fully-laminated   voll geblecht || vollgeblecht
function   Anwendungszweck * *Funktion* || Aufgabe || fungieren * *as:als* || Funktion * *as a funct.of: als math. ~ von; variation of y as a function of x :Verlauf v. y als ~ v. x* || funktionieren * *correctly:richtig* || gehen * *bei Maschine/Mechanismus usw.* || laufen * *Maschine* || Tätigkeit * *Funktion* || Zweck
function block   Baustein * *Funktionsbaustein* || Funktionsbaustein
function block call   Bausteinaufruf
function block interconnection   Bausteinverbindung * *zwischen Funktionsbausteinen*
function block library   Baustein-Bibliothek * *mit Software-Funktionsbausteinen* || Funktionsbausteinbibliothek
function block mask   Bausteinmaske * *Bildschirmmaske für Funktionsbaustein; z.B. b.SIEMENS Regelsystem SIMADYN D (R)*
function block package   Bausteinpaket
function button   Fuktionstaste || Funktionstaste
function call   Funktionsaufruf
function chart   Funktionsplan * *(SPS)* '*FUP*' || FUP * *Funktionsplan für SPS*
function check   Funktionskontrolle
function checking   Funktionskontrolle
function class   Funktionsklasse
function curve   Kennlinie * *eine math. Funktion darstellend*
function description   Funktionsbeschreibung
function diagram   Funktionsplan || Funktionsschaltbild
function element   Funktionsglied
function expansion   Funktionserweiterung
function generator   Funktionsgeber || Funktionsgenerator || Kennliniengeber
function key   Fuktionstaste || Funktionstaste || Zweitfunktionstaste
function module   Funktionsbaustein
function of temperature   Temperaturgang
function of time   Zeitfunktion * *Funktion (über) der Zeit*
function package   Funktionspaket * *(SIEMENS SIMADYN D (R))*
function plan   Funktionsplan * *beim SIEMENS Regelsystem SIMADYN D (R)*
function range   Funktionsumfang
function scope   Funktionsumfang
function section   Funktionsabschnitt
function select key   Vortaste * *Zweitfunktionstaste*

function selection switch   Betriebsartenschalter
function stage   Funktionsstufe
function symbol   Funktionssymbol
function table   Funktionstabelle
function test   Funktionstest
function-diagram oriented   strukturbildorientiert * *z.B. SIEMENS Antriebsregelsystem SIMADYN D (R)*
function-oriented diagram   Funktionsschaltplan
function-related   funktionsbezogen
function-tested   funktionsgeprüft
functional   funktional || funktionell || Funktions- || funktionsfähig || funktionstüchtig || sachlich * *zweckbetont (z.B. Design)* || zweckmäßig
functional area   Funktionsbereich
functional block   Funktionsbaustein
functional characteristics   Funktionsumfang * *funktionelle Eigenschaften*
functional circuit diagram   Funktionsschaltbild
functional description   Funktionsbeschreibung
functional diversity   Funktionsvielfalt
functional element   Funktionselement
functional expansion   Funktionserweiterung
functional extent   Funktionsumfang
functional group   Funktionsgruppe
functional module   Funktionseinheit
functional scope   Funktionalität * *Funktionsumfang* || Funktionsbereich || Funktionsumfang
functional sequence   Funktionsablauf
functional specification   Pflichtenheft
functional symbol   Funktionssymbol
functional test   Funktionsprüfung
functional unit   Funktionseinheit
functional upgrade   Funktionserweiterung * *Hochrüstung*
functionality   Funktionalität || Funktionsumfang * *Funktionalität*
functionally identical   funktionsidentisch
functioning   funktionsfähig || Funktionsfähigkeit || funktionstüchtig
functioning correctly   funktionsfähig * *ordnungsgemäß/richtig funktionierend*
fundamental   elementar || grundlegend * *als Grundlage dienend, grundlegend, wesentlich* || grundsätzlich * *(Adj.)* || wesentlich * *grundlegend*
fundamental component   Grundschwingung * *~santeil* || Grundschwingungsanteil || Grundschwingungsgehalt * *Grundschwingungsanteil*
fundamental factor   Grundschwingungsgehalt * *Anteil*
fundamental field   Grundfeld * *{el.Masch.}* räuml. sin.förmig. Drehfeld mit gleicher Polzahl wie die Wicklung*
fundamental frequency   Grundschwingung * *Grundfrequenz*
fundamental frequency component   Grundschwingungsgehalt
fundamental frequency content   Grundschwingungsgehalt * *Verhältnis Effekt.wert der Grundschw./Gesamteff.wert (DIN40110)*
fundamental power   Grundschwingungsleistung
fundamental power factor   Grundschwingungsleistungsfaktor * *%-Verhältnis Grundschw.wirk-/Grundschw.scheinleistung (cos Phi_1)*
fundamental principle   Grundprinzip
fundamental principles   Grundlagen * *einer Wissenschaft/Technik/Fachrichtung*
fundamental rule   Grundregel
fundamental terms   Grundbegriffe
fundamental wave   Grundschwingung
fundamental-frequency active power   Grundschwingungswirkleistung
fundamental-frequency apparent power   Grundschwingungsscheinleistung

fundamental-frequency content Grundschwingungsanteil
fundamental-frequency current Grundschwingungsstrom
fundamental-frequency reactive power Grundschwingungsblindleistung
fundamentally grundsätzlich * *(Adv.)*
fundamentals Grundlage * ~ *einer Wissenschaft usw.* || Grundlagen * *Grundlagen, Grundprinzip, Grundbegriffe*
fungi attack Schimmelbildung * *Schimmelpilzbefall*
funnel Einfülltrichter || Schornstein || Trichter * *pour through a funnel:durch* ~ *gießen; funnelshaped: ~förmig*
funnel tube Einfülltrichter
funnel-shaped trichterförmig
funny komisch * *lustig*
furnace Ofen * *z.b. Hochofen, Glühofen, Schmelzofen*
furnace charging machine Ofen-Lademaschine * *(Walzwerk)*
furnace roller table Ofen-Rollgang * *(Walzwerk)* || Ofenrollgang * *(Walzwerk)*
furnish ausrüsten || ausstatten * *versehen, aurüsten, einrichten, möblieren* || beschaffen * *liefern* || gewähren * *liefern, beschaffen, bieten, ausstatten, ausrüsten, versehen* || liefern * *with: jdn. mit etwas be~/ausstatten/ausrüsten; Beweise/Informationen ~/beschaffen* || versehen * *ausstatten; with: mit etwas* || versorgen * *ausstatten; with:mit*
furnished ausgerüstet * *auch technisches Gerät* || ausgestattet * *auch technisches Gerät*
furnishings Einrichtung * *z.B. eines Hauses*
furniture Einrichtung * *z.b. eines Hauses*
furniture industry Möbelindustrie
furrow Rinne * *Furche*
further außerdem * *ferner, überdies, außerdem, weiter, entfernter* || darüberhinaus || ferner || näher * *further particulars: -e Einzelheiten* || neu * *weitere* || übrig * *zusätzlich* || weiter * *(figürl.)* || weitere * *(besonders figürlich)* || weiterer || weitergehend || zusätzlich
further details weitere Angaben || weitere Einzelheiten
further development Weiterentwicklung
further education Weiterbildung
further information Näheres * *weitere Angaben/Informationen* || weitere Angaben || weitere Informationen
further inquiries Rückfragen
further more weiterhin * *ferner*
further on weiterhin
further orders Nachbestellung
further processing Weiterverarbeitung
further treatment Nachbehandlung
furthermore außerdem * *ferner, überdies, außerdem, weiterhin* || darüberhinaus || ferner || weiterhin * *ferner, darüber hinaus*
fusable element Schmelzleiter * *in Sicherung*
fuse absichern * *mit Sicherung/Kurzschlussschutz* || durchbrennen * *Sicherung* || schmelzen * *(ver-) schmelzen (auch figürl.), vermischen, durchbrennen (Sicherung usw.)* || Schmelzsicherung || Sicherung * *the fuse has blown: die ~ ist durchgebrannt/gefallen/hat ausgelöst*
fuse acting time Sicherungsschmelzzeit
fuse base Sicherungssockel
fuse block Sicherungsleiste
fuse blow Sicherungsfall
fuse blowing Sicherungsfall
fuse blown Sicherungsfall
fuse carrier Sicherungshalter || Sicherungsträger
fuse clearing time Sicherungsschmelzzeit

fuse clips Sicherungseinsteckkontakte * *z.b. Lyrakontakte*
fuse contact Sicherungskontakt
fuse current rating Sicherungsstromstärke
fuse cutout Sicherungseinsatz
fuse disconnecting switch Sicherungs-Lasttrenner || Sicherungslasttrenner || Sicherungstrenner
fuse element Schmelzleiter * *in Sicherung*
fuse filler Löschmittel * *in Sicherung, z.b. Sand*
fuse for protection of semiconductors Halbleiter-Sicherung * *Halbleiterschutz* || Halbleiterschutz-Sicherung || Halbleitersicherung * *schützt Leistungshalbleiter*
fuse for protection of thyristors Thyristorsicherung
fuse for semiconductor protection Halbleiterschutz-Sicherung
fuse holder Sicherungsfassung * *Sicherungshalter, Sicherungsunterteil* || Sicherungshalter || Sicherungsträger || Sicherungsunterteil
fuse interrupting rating Sicherungskurzschlussvermögen
fuse link Schmelzeinsatz * *Sicherungseinsatz (Sicherung ohne Unterteil)* || Sicherung * *~seinsatz ohne Unterteil* || Sicherungseinsatz
fuse melting time Sicherungsschmelzzeit
fuse monitoring Sicherungsüberwachung
fuse mounting Sicherungsfassung
fuse plug cartridge Sicherungseinsatz * *Sicherungs-Steckpatrone*
fuse protection Absicherung * *mit Schmelzsicherung usw.*
fuse puller Sicherungsaufsteckgriff || Sicherungsgriff
fuse rating Sicherungsnennstrom
fuse rupturing Sicherungsfall
fuse switch disconnecter Sicherungslasttrenner
fuse switch disconnector Sicherungslasttrenner || Sicherungstrenner * *Sicherungs-Lasttrenner*
fuse time-current characteristic Sicherungskennlinie
fuse unit Sicherungseinsatz
fuse-disconnector Sicherungs-Lasttrenner || Sicherungslasttrenner || Sicherungstrenner
fuse-element Schmelzleiter
fuse-switch disconnecter Sicherungs-Lasttrenner
fuse-switch disconnector Sicherungs-Lasttrenner
fused abgesichert * *z.B. mit Schmelzsicherung*
fused disconnect Sicherungslasttrenner
fused disconnect switch Sicherungstrennschalter
fused disconnector Sicherungstrenner
fused interrupter Sicherungs-Lasttrenner || Sicherungstrenner * *Sicherungs-Lasttrenner*
fused interruptor Sicherungslasttrenner
fused link Absicherung
fused load disconnect Sicherungs-Lasttrenner
fused load disconnect switch Sicherungs-Lasttrenner || Sicherungslasttrenner
fused load disconnector Sicherungslasttrenner
fused load-disconnect Sicherungslasttrenner
fused switch disconnector Sicherungslasttrenner
fuseless sicherungslos
fuses Absicherung * *mit meheren Schmelzsicherungen*
fusible cut-out Schmelzsicherung
fusible cutout Schmelzsicherung
fusible element Schmelzleiter
fusible link Schmelzsicherung
fusing Absicherung * *z.B. mit Schmelzsicherung*
fusing conductor Schmelzleiter * *in Sicherung*
fusing current Sicherungsansprechstrom
fusing element Schmelzeinsatz * *Schmelzleiter in Sicherung* || Schmelzleiter * *in Sicherung*
fusion welding Verschweißen * *Kontaktverschleiß*
fuss-free problemlos * *(salopp) ohne viel Getue*

fussy anspruchsvoll * *['fassi] übertrieben anspruchsvoll, wählerisch, heikel* || umständlich * *(übertrieben) umständlich, geschäftig, aufgeregt, kleinlich, affektiert* || unverständlich * *unverständlich, kleinlich, heikel, verschwommen*
futility Unwirksamkeit * *Vergeblichkeit, Zweck-/Nutz-/Wertlosigkeit; Nichtigkeit*
future Folge * *Zukunft* || künftig || später * *(zu)künftig* || zukünftig
future order Folgeauftrag
future prospect Zukunftsperspektive
future use spätere Verwendung
future-oriented zukunftsorientiert || zukunftssicher * →*zukunftsorientiert* || zukunftsweisend
future-proof zukunftssicher
fuzz Fussel
fuzzification Fuzzifizierung * *bei Fuzzy-Regelung*
fuzzy unklar * *Gedanken* || unscharf * *['fassi] unscharf, verschwommen, kraus, flockig, flaumig, faserig, fusselig*
fuzzy control Fuzzy-Regelung
fuzzy sets unscharfe Mengen
fwd vorwärts * *Abkürzung für 'forward'*

G

g g * *Gramm; 1 g entspr. 0,0352740 oz*
G-type mounting rail according to DIN EN 50035 Installationsschiene * *asymmetr. G-Aufschnappmontageschiene, 32mm breit* || Montageschiene * *asymmetr. G-Aufschnappinstall.schiene 32 mm breit*
G-type mounting rail according to DIN EN 50045 G-Schiene * *asymmetr. Aufschnapp-Install.schiene (32mm breit)*
GaAs Galliumarsenid
gadget Gerät * *(amerikan.) Apparat, Gerät, Vorrichtung, techn.Spielerei,; (auch Verb: mit ~en ausstatten)* || Spielzeug * *(amerikan.) Apparat, Gerät, Vorrichtung, techn.Spielerei, Apparätchen*
gage messen * *mit einer Lehre ~, z.B. (Draht-, Blech-) Dicke*
gain Ausbeute || Erhöhung * *Zunahme, Steigerung, (Wert-/Leistungs-) Zuwachs* || erreichen * *gewinnen* || erwerben * *(sich) Reichtum, Achtung usw. ~* || Gewinn * *[Elektr.] Verstärkung* || gewinnen * *z.B. Erfahrung* || nutzen * *(Substantiv) Gewinn* || P-Verstärkung * *Proportionalverstärkung* || Steigerung * →*Zunahme, (Wert-/Leistungs-)* →*Zuwachs* || Verstärkung * *Verstärkungsfaktor* || Verstärkungsfaktor || Zunahme * *Steigerung, Zuwachs,* →*Erhöhung* || Zuwachs * *(Zu)Gewinn, z.B. von Kapital*
gain a clear insight transparent machen * *sich ~*
gain a foothold Fuß fassen
gain adaption Verstärkungsadaption
gain by working erarbeiten * *mühsam ~*
gain error Verstärkungsfehler
gain factor Verstärkung * *Verstärkungsfaktor* || Verstärkungsfaktor
gain in gewinnen * *(an Bedeutung, Ansehen usw.) ~; gain ground: an Boden ~*
gal 1 * *Gallone (1 l entspr. 0,2200 Engl.Gallonen entspr. 0,2642 US Gallonen)*
gallery Bühne * *Galerie, Laufgang, Empore; Stollen im Bergbau* || Tunnel * *im Bergwerk*
gallium arsenide Galliumarsenid
galvanic isolation galvanische Trennung || Potenzialtrennung

galvanically isolatet galvanisch getrennt
galvanization Galvanik
galvanize galvanisieren || verzinken
galvanized galvanisiert
galvanizing Galvanik
galvanizing shop Galvanik * *Galvanikwerkstatt/-Anlage*
gambol Bocksprünge machen * *herumtanzen, Luftsprünge machen*
gang aneinander reihen * *z.B.Klemmen* || anreihen || Mannschaft * *Gruppe, Trupp, Abteilung, (Arbeits-) Kolonne* || Mehrfach- * *z.B. EPROM-Programmiergerät, Stanze, Fräser, Säge, Presse, Kondensator* || Mehrspindel- * *z.B. Bohrmaschine*
gang drilling machine Reihenbohrmaschine
gang programmer Mehrfach-Programmiergerät * *zum Programmieren mehrer EPROMs usw. in einem Arbeitsgang*
gang saw Gattersäge
gangsaw Gattersäge * *{Holz}*
gangway Gang * *~ zwischen Sitzreihen*
gantry Brücke * *Portal (bei Kränen; Füll~ bei der Flaschenabfüllung usw.)* || Gerüst * *Gerüst, Stütze, Bock, (Signal-, Bedienungs-) Brücke, Portal-* || Portal- * *z.B. ~kran, ~förderer, ~wickler, ~Werkzeugmaschine*
gantry crane Brückenkran || Portalkran
gantry robot Portalroboter
gantry saw Portalsäge * *{Holz}*
gantry traversing unit Portalfahrwerk * *eines Portalkrans*
gantry type of construction Portalbauweise
gantry-type Portal- * *z.B. ~kran, ~fräswerk*
gantry-type robot Portalroboter
gap Abstand * *Lücke* || Lücke * *bridge the gap: Lücke schließen* || Luftspalt * *z.B. bei elektr. Maschine (zwischen Läufer und Ständer), Bremse* || Öffnung * *Lücke* || Spalt * *Lücke, Luftspalt usw.* || Zwischenraum * *Lücke*
gap in the market Marktlücke
garbled fehlerhaft * *verstümmelt (Daten, Telegramme usw.)*
garbling Verstümmelung * *bei Datenübertragung usw.*
gas Benzin * *(amerikan.)* || Gas
gas blower Gasgebläse
gas cutting machine Brennschneidmaschine
gas exhausting Gasabsaugung
gas extraction Gasabsaugung
gas welding Autogenschweißen
gas-ring compressor Gasringverdichter
gas-tight gasdicht
gas-turbine Gasturbine
gaseous gasförmig
gaseous discharge Gasentladung
gases Gase
gasket Dichtung * *Dichtungsring/-manschette/-zwischenlage* || Scheibe * →*Dichtungs~*
gasketed abgedichtet * *mit* →*Dichtung ~ (z.B. Schranktür mit Gummiwulst)*
gasoline engine Benzinmotor
gate ansteuern * *z.B. Transistor/Thyristor-Leistungsteil mit Ansteuer/Zündimpulsen* || Basis * *Steueranschluss bei Thyristor, Transistor usw.* || Durchgang * *eines Ventils* || Gatter * *auch logisches/elektronisches ~* || Steueranschluss * *z.B. eines Thyristors* || Stufe * *logisches Gatter, z.B. UND, ODER* || Tor * *(auch figürl.)* || verknüpfen * *Signale ~*
gate angle Steuerwinkel
gate control Ansteuerung * *Transistor/Thyristor* || Torsteuerung * *z.B. bei der Verarbeitung von Impulsen*
gate control pulse Steuerimpuls * *An~ z.B. für Thyristor* || Zündimpuls

gate driving circuit Ansteuerschaltung * für Transistor/Thyristor-Leistungsteil
gate electrode Steueranschluss * z.B. e. Thyristors
gate function Verknüpfungsfunktion * Logikgatterfunktion (z.B. UND, ODER usw.)
gate pulse Ansteuerimpuls * z.B. für Thyristor/Transistor
gate pulse amplifier Impulsverstärker * zur Ansteuerg.v.Leistungsthyristoren/Transistoren
gate pulse blocking Impulssperre * für Ansteuerimpulse
gate terminal Steueranschluss * z.b. eines Thyristors
gate trigger current Zündstrom * {Stromrichter}
gate turn-off thyristor Abschaltthyristor * Thyristor, der sich über einen Gate-Anschluss abschalten lässt || GTO * GTO-Thyristor || GTO-Thyristor
gate turn-off voltage Abschaltspannung * (Mindest-) Abschaltspannung e. Thyristors
gate valve Absperrventil || Schieber * Ventil
gate-turn-off abschaltbar * Thyristor (→GTO)
gateway Busumsetzer * Protokollumsetzer || Durchgang * (Eingangs)Tor, Zugang (auch figürl.), Einfahrt, Eingang || Protokollumsetzer * für serielle Kommunikation || Tor * (auch figürl.) Einfahrt, ~weg || Zugang * Tor
gather entnehmen * schließen (from:aus), folgern, (auf)lesen, sich zusammenreimen, sich denken || folgern * from:aus || sammeln * (an)~, anhäufen, (Personen) ver~; anziehen, gewinnen, erwerben, (auf)lesen, ernten || versammeln * (Personen) versammeln; sich ~; (Dinge) (an)sammeln/anhäufen
gather information informieren * sich informieren, Informationen sammeln
gating Ansteuerung * Thyristor, Leistungstransistor, Schütz || Aussteuerung * Ansteuerung, z.B. eines Thyristorsatzes || Steuerung * An~ eines Thyristor-/Transistor-Leistungsteils || Verknüpfung * z.b. von Signalen; logic:logische
gating angle Steuerwinkel || Zündwinkel * z.B. bei Thyristor
gating angle feed-forward Steuerwinkelvorsteuerung
gating board Ansteuerbaugruppe * Leiterkarte
gating cable Ansteuerleitung * für Leistungstransistoren/Thyristoren
gating circuit Ansteuerschaltung * für Transistor/Thyristor-Leistungsteil || Steuersatz * Zünd-/Steuerimpulserzeugung
gating current Steuerstrom * z.B. für Thyristor
gating electronics Ansteuerelektronik * z.B. für Transistor/Thyristor-Leistungsteil
gating logic Ansteuerlogik * z.B. für Transistor
gating module Ansteuerbaugruppe * zur Ansteuerung von Leistungstransistoren/Thyristoren || Steuersatz * Zünd-/Steuerimpulserzeugung für ein Stromrichter-Leistungsteil
gating pulse Steuerimpuls * z.B. zur Ansteuerung eines Thyristors || Zündimpuls
gating pulse energy Zündenergie * Zündimpuls-Energie (zum Zünden von Thyristoren)
gating ratio Aussteuergrad || Aussteuerung * Aussteuergrad
gating unit Ansteuerbaugruppe * zur Ansteuerung von Leistungshalbleitern || Ansteuereinrichtung * für Thyristorschaltung || Steuersatz * Zünd/Steuerimpulserzeugung
gauge abmessen * genau abmessen (hauptsächlich Längen/Dickenmaß) || Anzeiger * z.B. für Druck, Batterie-Restkapazität || Dicke * Stärke von Draht, Blech, einer Rohr/Gehäusewandung usw. || Dikkenmessanlage || Lehre * Mess~, auch (Mess-) Normal/Eichmaß || Maß * Eich~ || messen * mit einer Lehre ~, z.B. (Draht-, Blech-) Dicke || Messgerät * für (Blech)Dicke, Breite, Stärke, Drahtdurchmesser, Druck usw || Messlehre || prüfen * (Ab)Maße kontrollieren, z.B. mit Lehre || Stärke * Dicke von Draht, Blech, einer (Gehäuse-/Rohr-) Wandung usw. || Wanddicke * eines Rohrs/Behälters
gauge pressure Überdruck * überatmosphärischer Druck
gauging Messung * Abmessung (Länge), Eichung, Dickenmessung || Prüfung * Maßkontrolle, ~ mit Lehre/Dickenmessgerät
Gaussean distribution Gauß-Verteilung
Gaussean process Gauß-Verteilung
Gbaud Gbaud * Gigabits pro sec.
GDI GDI * Abk. f. 'Graphics Device Interface': Bindeglied zw. Betriebssystem u. Grafikkarte/Drucker usw.
gear -Ausrüstung || -Mittel * Ausrüstung || Ausrüstung * Werkzeug, Gerät, Zubehör; besonders in zusammengesetzten Wörtern || eingreifen * Getriebe, Maschinenteile; in(to):in || einstellen * (figürl.) einrichten, anpassen, einstellen (to/for: auf/an); gear with: passen zu || Gang * Getriebe (low/high: niedriger/hoher); shift into: wechseln in || Gerät * Ausrüstung || Getriebe * primary/secondary gear: ~ein-/ausgangsstufe || Rad * Zahnrad || Zeug * Ausrüstung, Zubehör
gear case Getriebegehäuse
gear centre distance Zentrale * Achsabstand in einem Getriebe
gear component set Getriebeeinbausatz
gear cutting machine Verzahnmaschine * {Werkz.masch.}
gear down untersetzen * Getriebe
gear drive Zahnradtrieb
gear frame Getriebegehäuse || Getriebekasten
gear housing Getriebegehäuse
gear into each other greifen * ineinander~
gear meshing pendeln * (Substantiv) um Schaltgetriebe-Zahnräder beim Schalten in Eingriff zu bringen
gear output Getriebeausgang
gear output speed Getriebeausgangsdrehzahl
gear rack Zahnstange
gear ratio Getriebeübersetzung * "i" entspr. Getriebe-Eingangs- zu Ausgangsdrehzahl || Getriebeverhältnis || Übersetzung * →Getriebe~ ('i' entspr. Getriebe-Eingangs- zu Ausgangsdrehzahl) || Übersetzungsverhältnis * bei Getriebe
gear reducer Reduziergetriebe || Untersetzungsgetriebe
gear reduction Getriebeuntersetzung || Untersetzung * v. Getriebe
gear reduction ratio Getriebeuntersetzung * ~sverhältnis
gear stage Getriebestufe
gear stiffness Getriebesteifigkeit
gear teeth Verzahnung
gear testing machine Getriebeprüfstand
gear train Räderwerk
gear transmission Getriebe
gear unit Getriebe
gear wheel Zahnrad
gear wheel testbed Getriebeprüfstand
gear-box Getriebe
gear-change and reversing gearbox Schalt- und Wendegetriebe
gear-change gearbox Schaltgetriebe * siehe auch →Wechselgetriebe || Wechselgetriebe
gear-cutting machine Verzahnungsmaschine * {Werkz.masch.}
gear-shaping machine Verzahnungsmaschine * {Werkz.masch.}
gear-shift lever Schalthebel * für Schaltgetriebe
gear-shift operation Schaltvorgang * (mechan.) bei Schaltgetriebe
gear-tooth system Verzahnung

gear-type pump Zahnradpumpe
gearbox Getriebe || Getriebekasten || Übersetzungsgetriebe * *Getriebe allg.*
gearbox motor Getriebemotor
gearbox oil Getriebeöl
gearbox output Getriebeausgang
gearbox output speed Getriebeausgangsdrehzahl
gearbox protection Getriebeschonung
gearbox protection ramp Getriebeschonung * *durch Hochlaufgeber vor oder hinter Momentenregler*
gearbox stage Getriebestufe
gearbox stage changeover Getriebestufenumschaltung
gearbox test bed Getriebeprüfstand
gearbox test stand Getriebeprüfstand
gearbox with interchangeable gear wheels Wechselradgetriebe * *~ mit durch (manuellen) Zahnradwechsel änderbar.Übersetzg.* || Wechselradgetriebe
gearbox with pick-off gear wheels Wechselradgetriebe
gearbox-side getriebeseitig
gearcase Getriebekasten
geared formschlüssig * *über Zahnräder verbunden* || verzahnt || zwangsläufig * *Bewegung e. Maschinenteils*
geared down untersetzt * *Getriebe*
geared motor Getriebemotor
geared pump Zahnradpumpe
geared up for export exportorientiert * *{Wirtschaft}*
gearhead Anbaugetriebe * *mit dem Motorflansch verschraubtes ~* || Aufsteckgetriebe * *mit Motorflansch verschraubtes Getriebe* || Getriebe * *an Getriebemotor*
gearhead motor Getriebemotor
gearing Getriebe || Verzahnung
gearing stress Verzahnungsbeanspruchung
gearless getriebelos
gearless drive Direktantrieb * *ohne Getriebe*
gearmotor Getriebemotor
gearmotor unit Getriebemotor
gearshift Getriebeumschaltung
general allgemein || Allgemeines || durchgängig * *allgemein, umfassend, gängig, allgemein üblich* || generell * *(Adj.)* || Gesamt- * *allgemein, umfassend, General-, →Haupt-* || Sammel- * *z.B. Katalog*
general case Regelfall * *generally speaking: im ~*
general catalog Sammelkatalog
general conditions of sale and delivery allgemeine Lieferbedingungen
general contractor Generalunternehmer
general data allgemeine Daten
general information allgemeine Hinweise || Allgemeines * *Kapitelüberschrift in Dokument*
general instructions Richtlinie * *Anweisung(en)*
general license Generallizenz
general manager Geschäftsführer
general mechanical engineering allgemeiner Maschinenbau
general note allgemeiner Hinweis
general operating test Funktionsprüfung
general partnership offene Handelsgesellschaft
general plan Übersichtsplan
general preparations Vorarbeiten
general purpose terminal Wahlklemme
general requirements allgemeine Anforderungen
general reset Urlöschen * *bei SPS*
general rule Faustregel
general specifications allgemeine Daten * *technische Daten*
general technical information technische Erläuterungen * *allgemeine ~*

general trend allgemeiner Trend
general use allgemeine Anwendung
general view allgemeine Übersicht * *of: über* || Übersicht
general view of allgemeine Übersicht über
general-purpose allgemein verwendbar || Allzweck- || für allgemeine Zwecke || Universal- * *Allzweck*
general-purpose amplifier board Universalverstärkerbaugruppe
general-purpose input Wahleingang * *Mehrzweckeingang*
general-purpose motor Motor für allgemeine Zwecke * *{el.Masch.} siehe VDE 0530, Beiblatt 1* || Normmotor * *normal üblicher Motor*
general-purpose output Wahlausgang * *Mehrzweckausgang*
generality Allgemeingültigkeit * *Allgemeingültigkeit, allgemeine Regeln*
generalization Verallgemeinerung
generally allgemein * *(Adv.) im Allgemeinen* || generell * *(Adv.)* || im Allgemeinen || in der Regel || meistens * *gewöhnlich, im Allgemeinen* || normalerweise * *im Allgemeinen, gewöhnlich, meistens*
generally speaking im Allgemeinen || in der Regel * *im Allgemeinen* || normalerweise * *im Allgemeinen, im Großen und Ganzen*
generally used gebräuchlich || geläufig * *→gebräuchlich*
generate anlegen * *erzeugen* || aufbauen * *aufbringen, erzeugen* || aufbringen * *erzeugen, z.B. Drehmoment* || bilden * *erzeugen* || erstellen * *erzeugen, herstellen, z.B. Dokumente, Unterlagen, Zeichnungen* || erzeugen || generieren || herstellen * *erzeugen, bilden, entwickeln, bewirken, verursachen, hervorrufen*
generated gebildet * *erzeugt, abgeleitet (from:aus)*
generating -erzeugend || generatorisch * *Energie erzeugend; z.B. Drehmoment, Betriebsart eines Stromrichters*
generating set Aggregat * *zur Stromerzeugung*
generation Bildung * *Erzeugung* || Erstellung * *Erzeugung, Erstellung (auch von Dokumenten)* || Erzeugung * *auch von Programmen, Dokumenten, Zeichnungen usw.* || Generation * *z.B. einer Gerätereihe, auch 'Erzeugung'* || Generierung * *z.B. eines (ablauffähigen) Softwarepakets z.B. durch Zusammenbinden mehrerer Module*
generation of electricity Energieerzeugung * *Stromerzeugung*
generation of prices Preisbildung * *siehe auch 'Preisgestaltung'*
generator Generator * *electric: elektrischer*
generator operation Generatorbetrieb
generator set Maschinensatz * *Motor-Generatorsatz*
generic allgemein * *oberbegrifflich; generic term: Oberbegriff; generic standard:übergreifende Norm* || generell * *(Adj.)* || übergeordnet * *allgemein, generell, generisch; z.B. Begriff, Norm usw.* || übergreifend * *allgemein (z.B. Norm), generell, generisch, Gattungs-*
generic standard Fachgrundnorm || Grundnorm || übergeordnete Norm * *Grundnorm*
generic term Gattungsname * *auch 'Oberbegriff'* || Oberbegriff
generically allgemein * *(Adv.) oberbegrifflich* || generell * *(Adv.)*
generous freizügig * *großzügig* || nobel * *großzügig, freigiebig*
generously dimensioned überdimensioniert * *großzügig dimensioniert*
generously sized großzügig dimensioniert
generously-dimensioned großzügig dimensioniert

genesis Entstehung * *Ursprung, Beginn, Entstehung, Werden*
gentle behutsam * *sachte* || leicht * *sanft* || leise * *sacht* || sanft || schonend
gentlemen's room Toilette * *Herren~*
Gentlemen: Sehr geehrte Herren
genuine echt || natürlich * *echt* || wahr * *echt* || wirklich * *echt, tatsächlich*
genuine accessories Original-Zubehör
geometric geometrisch
geometrical geometrisch
geometrical sum geometrische Summe
geometry Geometrie
German deutsch
German Industry Standard DIN
German Lloyd Germanischer Lloyd * *Schiffs-Klassifizierungs- und -Versicherungsgesellschaft*
German Standards Committee for Measurement and Control NAMUR * *deutsches Normengremium in d.chem. Industrie*
German-language deutschsprachig
germanium Germanium
germanium diode Germaniumdiode
Germany Deutschland
get bekommen || erhalten * *bekommen* || erreichen * *bekommen* || holen
get acquainted with vertraut machen mit
get across ausgleichen * *vermitteln* || vermitteln * *schlichtend ~*
get ahead weiterkommen * *auch beruflich*
get back herausbekommen * *Geld, Investition usw.* ~ || herausbekommen * *wiederbekommen*
get change herausbekommen * *Wechselgeld ~*
get close näherkommen * *get close to: sich an annähern*
get close to auffahren * *sicht an etwas annähern; get too close to: zu dicht auf etwas ~ (im Verkehr)* || nähern * *sich etwas ~, an~, näherkommen*
get defective Schaden erleiden * *defekt werden*
get dirty verschmutzen * *schmutzig werden*
get entangled verwirren * *sich ~, hängen bleiben, verwirrt werden*
get familiar lernen * *vertraut werden (with:mit)*
get familiar with einarbeiten * *sich in etwas ~* || sich vertraut machen mit * *vertraut werden mit*
get in einsteigen * *in ein Fahrzeug*
get in touch Ansprechen * *(Verb) jemanden*
get into einsteigen * *in ein Fahrzeug*
get longer zunehmen * *(die Tage, wenn es länger hell bleibt) ; get worse: schlechter werden*
get loose lösen * *lose werden, sich ~*
get lost verloren gehen
get open aufbringen * *öffnen, offen bekommen*
get out aussteigen * *auch aus einem Fahrzeug/Unternehmen/einer Firma (of:aus)* || herausbekommen || herauskommen * *entfliehen*
get out of commission ausfallen
get out of order ausfallen
get out of place verschieben * *sich ~*
get out of the way vermeiden * *aus dem Wege gehen*
get over fertigwerden * *mit Problem, Kummer usw.* ~ || meistern * *überwinden* || überstehen * *erfolgreich ~, überwinden* || überwinden * *Schwierigkeiten usw. ~*
get ready fertigwerden
get rid of beseitigen * *Problem, Gegner ~* || loswerden
get rusty rosten
get the knack in den Griff bekommen * *of a thing: etwas ~*
get to grips with vertraut machen mit * *in den Griff bekommen, sich die Handgriffe aneignen*
get together zusammensetzen * *zusammenkommen, z.B. zu einer Besprechung*

get torn durchreißen || reißen * *(Stoff)*
get use to the idea sich vertraut machen mit * *~ mit Gedanken/Idee/Vorschlag*
get wise to herausbekommen * *herausfinden*
get worse verschlechtern * *sich ~, schlechter werden*
getaway power Anzug * *Anzugskraft bei Fahrzeug*
GFCI Fi-Schutzschalter * *Abk.für 'Ground Fault Circuit Interrupter'*
giant gigantisch || riesig * *gigantisch*
gigantic riesig * *gigantisch*
gill Lamelle * *am Kühler/Wärmetauscher (nicht 'grill'!)*
gimmick Spielzeug * *(salopp) technische Spielerei, Apparätchen, Spielkram, unnötige Apparatur*
ginger up ankurbeln * *anfeuern, ankurbeln, scharfmachen; aufmöbeln*
girder Träger * *Tragbalken, Stahl~*
give angeben * *z.B. Grenzwerte in einer Norm, Kenndaten in einem Datenblatt ~* || aufführen * *z.b. Angabe in einer Tabelle* || bestellen * *Grüße usw.* || bieten * *schenken, geben, (dar)bieten, gewähren, liefern* || ermöglichen || gewähren * *geben, darbieten* || halten * *Vorlesung ~* || vermitteln * *Vorstellung, Eindruck, Bild ~*
give a description of beschreiben * *schildern*
give a survey Überblick geben * *comprehensive:umfassenden*
give an answer antworten
give an order Auftrag erteilen
give an order for bestellen
give back zurückgeben
give consideration to achten auf
give distiction to prägen * *(figürl.)*
give first aid Erste Hilfe leisten
give full concentration konzentrieren * *sich ~ (to: auf)*
give full play gewähren * *gewähren lassen*
give notice of bekanntmachen
give off abgeben * *z.B. heat: Verlustwärme ~*
give response Ansprechen * *(Verb) z.B.Messinstrument*
give security absichern * *sicherstellen* || sicherstellen
give short measure knapp * *(Adv.) ~ bemessen*
give up aufgeben * *übergeben, preisgeben, z.B. Geschäft* || aussteigen * *aufgeben* || übergeben * *verwerfen* * *aufgeben*
give way absinken * *z.B. Boden* || senken * *sich senken, nachgeben, z.B. Boden, Gebäude usw.*
given aufgeführt * *z.B. Angabe in einer Tabelle* || vorgegeben
GL GL * *Abkürzg. für 'Germanischer Lloyd': deutsche Schiffsklassifikationsgesellschaft*
gland Gewindestutzen * *z.b. zur Kabeldurchführung (bei →PG-Verschraubung)* || Stopfbüchse || Verschlussstopfen * *Kabelverschraubung, z.B. an Motorklemmenkasten* || Verschraubung * *~ einer Kabeldurchführung (siehe auch →PG-Verschraubung)*
gland clearance Radialspiel * *bei Schraubenverdichter*
glare protection Blendschutz
glass-fibre optocable Glasfaser-Lichtleiter
glass Glas
glass container production machine Hohlglasmaschine
glass epoxy Epoxydharz
glass industry Glasindustrie
glass silk Glasseide
glass-fiber reinforced glasfaserverstärkt * *(amerikan.)*
glass-fibre Glasfaser
glass-fibre optocable Lichtleiter * *Glasfaser-Lichtleiter*

glass-fibre reinforced glasfaserverstärkt * *(brit.)*
glass-fibre reinforced plastic glasfaserverstärkter Kunststoff
glassfibre Glasfaser
glassfibre cable Glasfaserkabel
glaze glätten * *Papier kalandrieren; (Papier/Leder) satinieren*
glazing cylinder Glättzylinder * *zur Satinierung von Papier, Textilstoffen, Leder usw.*
gleam blinken * *glänzen, leuchten, schimmern*
glide gleiten * *(leicht) gleiten, dahingleiten* || rutschen * *(leicht) gleiten, dahingleiten* || schweben * *(durch die Luft) gleiten*
glide over Streifen * *(Verb) über etwas hingleiten*
gliding gleitend
glim lamp Glimmlampe
global geräteübergreifend * *(PROFIBUS)* || Gesamt- * *mit weitem Geltungsbereich* || pauschal || umfassend * *mit weitem Geltungsbereich* || weltweit
global enterprise Weltfirma
global environment Umwelt * *global betrachtet*
global market Weltmarkt
global market leader Weltmarktführer
global player weltweit tätiger Anbieter * *(salopp)* || weltweit tätiges Unternehmen * *(salopp)*
globalization Globalisierung
globally weltweit * *(Adv.)*
globally operating company weltweit tätiges Unternehmen
gloss Glanz || Glätte * *Glanz, Politur*
glossary Glossar * *Liste von Fachausdrücken mit kurzen Erklärungen*
glow glühen || Glut
glow discharge Glimmentladung
glow lamp Glimmlampe
glue anleimen * *→leimen* || beleimen || kleben || Kleber || Klebstoff * *Leim; auch 'umgebende Logik (m.niedr.Integrationsgrad) bei e. Mikroprozessor-Schaltung'* || Leim || leimen * *together: zusammen-* || verkleben * *mit Leim*
glue application Leimauftrag
glue application roller Leimauftragswalze * *{Holz}*
glue applicator Beleimgerät * *{Holz}*
glue on ankleben || anleimen
glue penetration ausschwitzen * *{Holz} (Subst.) ~ von Leim*
glue press Leimpresse * *{Holz}*
glue spreader Leimauftraggerät * *{Holz}*
glue-applicating machine Beleimmaschine * *{Holz}*
glued board Leimholz
glued timber Leimholz
glued timber construction Holzleimbau
gluelam beam construction Holzleimbau
gluelam beam press Leimbinderpresse * *{Holz}*
gluelam construction Holzleimbau
gluing Beleimung || Verleimung * *{Holz}*
gluing machine Verleimmaschine * *{Holz}*
glut Überfluss * *Fülle, Überfluss, Überangebot; a glut in the market:e. Überschwemmg./Sättigung d. Marktes*
GMA GMA * *Gesellschaft für Mes- u.Automatisierungstechnik: Assoc. for Instrumentation Automation Engin.*
GND Bezugspotenzial * *Abkürzg. für 'GrouND'* || M-Potenzial * *Abkürzg. f. 'GrouND'* || Masse * *Kurzform für 'ground'*
GND fault Erdschluss
GNP Bruttosozialprodukt * *Abkürzung für 'Gross National Product'*
go fahren * *in einem (beliebigen) Fahrzeug* || gehen auf * *z.B. signal goes High/Low: Signal geht auf High/Low* || laufen * *Maschine*

go across überschreiten * *überqueren*
go ahead fortfahren || fortsetzen * *weitermachen* || vorgehen * *voran/vorausgehen*
go away abgehen
go back zurückgehen
go back from zurücknehmen * *Zusage usw. ~*
go back to zero auf Null zurückgehen
go behind nachlaufen * *hinterherlaufen*
go beyond hinausgehen über * *Zahl, Maß, Wert usw.* || überschreiten * *Maß* ~ || übersteigen
go big groß herauskommen
go by plane fliegen * *per Flugzeug reisen*
go down abnehmen || absinken * *z.B. Last am Kran* || abstürzen || fallen * *Preise usw.* || reduzieren * *sich reduzieren* || sinken * *Preise, Kurse, Messwerte*
go down to reichen * *hinab~ bis*
go for holen
go forward vorgehen
go halves teilen * *sich in 2 Hälften ~*
go into datail about beschreiben * *genau ~*
go into operation in Betrieb gehen
go off abgehen || erlöschen * *z.B. Lampe, Lichtbogen*
go on angehen * *Lampe, Licht, Anzeige* || aufleuchten * *angehen, eingeschaltet werden, z.B. Lampe/Leuchte*
go on stream in Betrieb gehen
go on with fortführen * *fortsetzen*
go online in Betrieb gehen * *Kraftwerk ans Netz*
go out ausgehen * *z.B. Licht, Lampe, LED, Thyristor* || erlöschen * *z.B. Lampe, Lichtbogen* || löschen * *er~, z.B. Lampe*
go out of date veralten
go out of fashion veralten
go over untersuchen * *(gründlich) überprüfen/untersuchen, z.B. Maschine*
go over again überarbeiten * *Manuskript ~*
go round umgehen * *um etwas her~*
go through erfahren * *(Verb) erleben, durchmachen*
go to home position Referenzpunktfahren * *bei Positioniersteuerung/Numer. Steuerung*
go up steigen * *auch Preise, Kurse usw.*
go up to reichen * *hinauf~ bis*
go well together zusammenpassen * *gut ~*
go-between Vermittler * *Schlichter*
goad anspitzen * *(salopp) goad someone (into action): jdn. ~/scharfmachen/anstacheln/-treiben*
goal Ziel * *reach/attain one's goal: sein ~ erreichen, auch Ziel/Tor im Sport* || Ziel- || Zielsetzung
gob Tropfen * *Schleimklumpen, Auswurf, flüssiger Glas~ (zur Hohlglasherstellung)*
godet Galette * *{Textil} Rolle für Fäden z.B. in Textilfaserstraße*
goggles Schutzbrille
going to dabei * *im Begriffe*
going to home position Referenzfahrt * *Anfahren des Referenzpunktes bei Posititioniersteuerung*
goings-on Wirtschaft * *Treiben*
gold contact Goldkontakt
gold mine Goldgrube * *(auch figürl.)* || Melkkuh * *(figürl.) sehr gute Einnahmequelle*
gold-plated vergoldet
gold-point contact Goldspitzkontakt
goldplated contact Goldkontakt
gone vorbei * *vergangen* || Weg * *(Adv.) verloren, gegangen sein*
good gut * *for:für; (Adv.: 'well')*
good bye Auf Wiedersehen
good care Schonung * *gute Behandlung*
Good Luck Viel Erfolg
good management Wirtschaftlichkeit * *wirtschaftliches Handeln*

good order Schuss * *be in good order/running smoothly: gut in ~ sein*
good policy sinnvoll * *~e Verfahrensweise*
good value kostengünstig * *(Adv.) be good value:- preiswert sein* || preisgünstig * *be good value: preiswürdig sein* || Preiswürdigkeit
good will Kulanz
good-natured verträglich * *gutmütig*
goods -Güter * *Waren, Güter, Fachtgut* || Artikel * *(Plural) (Transport)Güter, Waren, Gegenstände, Frachtgut* || Fracht * *Frachtgut* || Frachtgüter || Ware * *(Transport)Güter, Waren, Gegenstände, Frachtgut* || Waren
goods lift Lastenaufzug
goods traffic Frachtverkehr
goods vehicle LKW
goodwill Kulanz
goody Annehmlichkeit * *(salopp) Zusatznutzen, 'Bonbon'* || Zusatznutzen * *(salopp) 'Bonbon', Annehmlichkeit*
govern ausschlaggebend sein * *z.B. Auslegungskriterium* || bestimmen * *maßgeblich sein für* || herrschen * *regieren, lenken, regeln, steuern, bestimmend/maßgebend sein für, leiten* || regeln * *bestimmen, lenken, leiten* || regulieren * *regieren, beherrschen, lenken, regeln, bestimmen, steuern*
governing beherrschend * *bestimmend (Prinzip usw.), leitend* || maßgebend * *bestimmend* || maßgeblich * *bestimmend*
government Führung * *Herrschaft, Kontrolle, Leitung, Verwaltung, Regierung* || Gewalt * *Kontrolle, Leitung, Verwaltung, Regierung*
government body Behörde * *Regierungs-/staatliche Behörde*
governor relief valve Überdruckventil
governor valve Regelventil || Stellventil * *Regelventil*
grab greifen * *(hastig) ~ (auch mit Greiferkran), an sich reißen, fassen, packen, schnappen, einheimsen* || Greifer * *~ eines Krans/Baggers*
grab bucket Greifer * *~schaufel eines Krans/Baggers*
grab crane Greiferkran
grab hoisting gear Greiferhubwerk * *{Hebezeuge}*
grab oscillation damping Pendeldämpfung * *{Hebezeuge} bei Greiferkran*
grab-closing gear Schließwerk * *{Hebezeuge} bei Greiferkran*
grab-holding gear Haltewerk * *{Hebezeuge} bei Greiferkran*
grabbing operation Greifvorgang * *{Hebezeuge} bei Greiferkran/Bagger*
gradation Steigerung * *Abstufung*
grade abstufen || Anstieg * *Steigung, Gefälle, Neigung* || bewerten * *einstufen, klassifizieren* || einteilen * *in Klassen einteilen, abstufen* || Güte * *→Qualitätsstufe, Handelsklasse, →Qualität* || nivellieren * *(Gelände) planieren, einebnen* || planieren * *Gelände ~* || Qualität * *→Qualitätsstufe, Handelsklasse (von Standardwaren); high-grade: von höchster Qualität* || Sorte * *Qualitätsstufe von Standardprodukten, z.B. von Papier, Öl usw.* || staffeln * *(ab)stufen, staffeln; z.B. Qualität, Güte, Rang, Sorte, Klasse* || Steigung * *Steigung, Neigung, Gefälle, Niveau* || Stufe * *Grad, Klasse, Rang, Qualitäts~, Art, Gattung, Sorte, Dienstgrad* || stufen * *staffeln*
grade change Sortenwechsel * *z.B. bei Papiermaschine*
grade memory Sortenspeicher * *z.B. bei der Papierherstellung*
graded abgestuft || gestaffelt * *abgestuft; siehe 'staffeln'* || gestuft
grader Sichter * *Sortiereinrichtg.,z.B. z.Trennen v. Schnitzeln unterschiedl.Größe i.d. Spanplattenher-* stellg. || Sortiereinrichtung * *{Holz} für Holzspäne in der Spanplattenherstellung*
gradient Änderungsgeschwindigkeit * *Neigung, Steigung* || Anstieg * *Gradient, Steigung (auch physikal.), Neigung, Gefälle* || Gefälle || Gradient || Neigung || Steigung * *z.b. einer Kurve, eines Messwertes, einer Straße*
grading Abstufung || Selektivität * *Stufung, Staffelung; time grading: Zeitselektivität* || Sortierung || Staffel * *Stufung, ~ung; z.b. Qualität, Güte, Rang Sorte, Klasse* || Stufung
grading plant Sortieranlage * *zum Sortieren nach unterschiedlichen Produktqualitäten (z.B. bei Holz)*
grading system Güteklassensortiermaschine * *z.b. für Holz*
gradual allmählich || stufenweise * *~ fortschreitend, allmählich*
gradual change allmähliche Änderung
gradually allmählich * *(Adv.) nach und nach, allmählich*
graduate abstufen || einteilen * *in Grade* || promovieren || staffeln * *abstufen, einteilen; z.B. Preise, Steuern, Löhne*
graduated abgestuft || gestuft
graduated engineer Diplomingenieur * *→graduierter Ingenieur* || graduierter Ingenieur * *undergraduated engineer: noch nicht ~*
graduation Abstufung || Einteilung * *~ in Grade, Abstufung* || Promotion * *Erteilung eines akadem. Grades* || Staffel * *~ung, Abstufung, Einteilung; z.B. Preise, Steuern, Löhne* || Teilung * *Skaleneinteilung*
graduation exercises Promotion * *(amerikan.)*
grahics-oriented grafikorientiert
grain Faser * *Holzfaser; crosscut: quer zur ~ sägen* || Korn
grain elevator Silo * *für Getreide*
grain size Körnung * *Korngröße*
grain-oriented kornorientiert * *z.B. Magnetblech*
graining Körnung
grammage Flächengewicht * *Flächenmasse, z.B. von Papier [g/m^3]* || Flächenmasse * *z.B. von Papier [g/m^3]*
grammar Grammatik * *auch einer Programmiersprache*
grant bewilligen || garantieren * *bewähren, bewilligen, zugestehen* || genehmigen * *bewilligen* || Genehmigung * *Bewilligung* || gewähren * *bewilligen* || zuteilen * *genehmigen* || zuweisen * *genehmigen*
granting of a patent Patenterteilung
granular material Granulat
granular matter Granulat
granulate Granulat * *plastic granulate: Kunststoff~* || granulieren
granulated material Granulat
granulation Granulierung || Körnung
granules Granulat
graph Abbildung * *grafische Darstellung, Schaubild, Kurvenbild/blatt* || Bild * *grafische Darstellung, Schaubild, Kurve(ndarstellung), Kurve* || Diagramm * *['grä:f] Schaubild, grafische Darstellung, Kurvenbild/-blatt* || Grafik * *z.B. Kurvendarstellung* || grafische Darstellung || Kurve * *(statistisches) Kurvenbild* || Kurvendarstellung || Kurvenzug * *Kurvendiagramm, Kurvendarstellung* || Schaubild || Zeichnung
graphic anschaulich * *plastisch, lebendig, grafisch, zeichnerisch* || bildlich || grafisch * *(amerikan.)*
graphic adapter Grafikkarte
graphic configuring grafische Projektierung
graphic driver Grafiktreiber
graphic operator interface grafische Oberfläche
graphic presentation grafische Darstellung
graphic representation grafische Darstellung

graphic retranslation Rückdokumentation * *grafische ~*
graphic symbol Schaltzeichen
graphic user interface grafische Oberfläche * *für Rechnerbenutzer*
graphical grafisch * *(engl.)*
graphical configuring grafische Projektierung * *z.B. m.d.SIEMENS Projektiersprache STRUC G für d.Regelsystem SIMADYN D (R)*
graphical editor Grafikeditor
graphical representation grafische Darstellung || Kurvendarstellung
graphical symbol Grafiksymbol || Schaltzeichen
graphical user interface grafische Anwenderoberfläche * *(z.B. MS-WINDOWS (R))*
graphically grafisch * *(Adv.)*
graphics Grafik
graphics accelerator Grafikbeschleuniger * *Grafikkarte mit eigener "Bildaufbau-Intelligenz"*
graphics accelerator card Grafikbeschleuniger
graphics card Grafikkarte
graphics performance Grafikleistung * *z.B. Geschwindigkeit eines Computers b. Bildaufbau auf e. Monitor*
graphite Grafit
grapple claw Greifer * *~ eines Fördermechanismus*
grasp erfassen * *greifen (auch figürl.)* || fassen * *packen, fassen, (er-)greifen (at:nach), an sich reißen, zugreifen/-packen, streben, trachten* || fassen * *packen, greifen (Werkzeug usw.)* || greifen || Griff * *(Zu-) Griff, Macht,Gewalt, Verständnis; within one's grasp:in Reichweite/in jmdes. Gewalt* || packen * *greifen*
grate Gitter * *Rost* || kratzen * *Geräusch* || Rost * *aus Metallstäben (z.B. in Ofen)* || schaben
grated gitterförmig
grateful dankbar * *innerlich*
gratified zufrieden * *erfreut, befriedigt, angenehm berührt; be gratified: sich freuen*
grating Gitter || Rost * *Gitter*
gratuitous kostenlos * *gratis, unentgeltlich*
grave erheblich * *Schaden, Verlust* || gefährlich * *ernst, schwer*
gravel Kies
gravitational acceleration Fallbeschleunigung
gravity Schwerkraft
gravity conveyor Schwerkraftförderer
gravity tank Schwerkrafttank
gravity-type Schwerkraft-
gravure cylinder Tiefdruckzylinder * *(Druckerei)*
gravure paper Tiefdruckpapier
Gray code Gray-Code
Gray excess three code Drei-Exzess-Gray-Code
Gray-coded excess three code Drei-Exzess-Gray-Code
graze Streifen * *(Verb)* streifen an
grease abschmieren || einfetten * *mit Fett* || Fett || schmieren * *mit Fett* || Schmierfett
grease box Schmierbüchse * *für Fett*
grease chamber Fettvorratsraum * *in Schmiersystem, Lager usw.*
grease change interval Fettwechselfrist
grease consumption Fettverbrauch * *von Schmierfett, z.B. in einem Lager*
grease cup Schmierbüchse * *für Fett* || Staufferbüchse
grease gun Schmierpumpe * *Abschmierpumpe, Fettpresse*
grease life Fettstandzeit * *Schmiermittel-Gebrauchsdauer*
grease life time Fettgebrauchsdauer
grease lubrication Fettschmierung * *grease lubricated bearing: Lager mit ~*
grease nipple Schmiernippel

grease packing Fettfüllung * *z.B. eines Lagers*
grease pump Schmierpumpe * *Fettpumpe*
grease quantity Fettmenge
grease renewal Fetterneuerung
grease replacement Fetterneuerung
grease slinger Fettmengenregler * *Fettschleuderscheibe (zum Abschleudern überschüssigen Lagerfetts)* || Fettschleuderscheibe || Schleuderscheibe * *z.B. bei Schmiersystem*
grease stability time Fettgebrauchsdauer
grease-lubricated fettgeschmiert
grease-packed groove Spaltdichtung * *zur Abdichtung von Gehäusefugen (z.B. in ölgefülltem Gleit-/Wälzlager)*
greaseproofness Fettdichtigkeit
greasing Schmierung * *mit Schmierfett*
greasing slinger Schleuderscheibe * *Fettschleuderscheibe/Fettmengenregler bei Schmiersystem*
greasy schmierig * *fettig*
great groß || hoch * *groß* || prima * *(salopp) groß (artig), überragend, famos, herrlich* || stark * *Nachfrage usw.* || super * *(salopp) großartig, famos, herrlich, überragend, prima, toll* || toll * *(salopp) groß(artig), prima, famos, herrlich, überragend*
Great Britain Großbritannien
great many zahlreich
greater größer * *(Komparativ) than:als*
greater part größerer Teil * *of:von*
greater than größer als
greater than or equal to größer gleich
greatest größt || höchst * *größt* || meist * *größt*
greatest possible größtmöglich
greatly höchst * *(Adv.)* || sehr * *(Adv.)* höchst, äußerst
Greece Griechenland
green grün || umweltfreundlich * *(salopp)* || umweltverträglich * *(salopp)*
Green Supply and Delivery Conditions Grüne Lieferbedingungen
green wood Holz * *grünes ~*
greenhorn Anfänger * *(salopp) Grünschnabel* || Laie * *(salopp) blutiger ~, siehe auch →Anfänger* || Lehrling * *Anfänger* || Neuling * *(salopp) Grünschnabel*
grey grau
grey cast-iron Grauguss
grid Gitter || Rost * *auf Diagramm/Landkarte; Raster* || Raster || Rastermaß || Rasterung || Rechen * *in Kläranlage* || Verbund- * *z.B. -Kraftwerk, -Stromnetz*
grid dimension Rastermaß
grid of modules Rasterung
grid pattern Linienraster
grid unit Rastermaß
grid-box pallet Gitterboxpalette
grievance Missstand
grill Rost * *Rost*
grille Gitter * *an Tür, Fenster usw.*
grind mahlen || schleifen || walzen * *mahlen* || zerkleinern * *zermahlen*
grind off abschleifen
grinder Mühle * *Mahl/Quetschwerk, Schleifer* || Schleifer * *Mahlwerk, Schleifmaschine* || Schleifmaschine || Winkelschleifer
grinder control Schleiferregelung * *für Holzschleifer i. d. Zellstoffherstellung*
grinderpit Schleifertrog * *bei Holzschleifer zur Zellstofferzeugung*
grinding Schleif-
grinding abrasives Schleifmittel
grinding belt Schleifband
grinding disc Schleifscheibe
grinding energy Mahlarbeit
grinding fineness Mahlgrad
grinding head Schleifbock
grinding machine Schleifmaschine

grinding mill Mühle * *Mahlwerk, Schleif/Reibmühle* || Schleiferei * *{Papier}* in der Zellstoffindustrie
grinding plant Mahlanlage || Schleiferei * *{Papier}* in der Zellstofferzeugung
grinding solution Schleifwasser
grinding tool Schleifwerkzeug
grinding wheel Schleifscheibe
grindstone Schleifstein
grip anfassen || erfassen * *packen (auch figürl.)* || festhalten * *greifen, packen (firmly: fest/sicher)* || greifen * *packen, (er)greifen, festhalten, festklemmen* || Griff * *Zu~, Halte~* || Handgriff * *Griff zum Anfassen, auch Art des Zugreifens* || Klaue * *Greifer, Klemmer* || packen * *fassen, greifen, packen (auch figürlich)* || spannen * *Werkstück*
grip device Fangvorrichtung * *{Aufzüge}*
grip release Fangaufhebung * *{Aufzüge}*
gripper Greifer * *Greifer, Halter; gripping lever:- Spannhebel; gripping Tool: Spannwerkzeug; auch b. Roboter*
gripping device Fangvorrichtung * *{Aufzüge}* || Greifer * *Greifvorrichtung*
gripping plate Spannplatte
gripping tool Greifwerkzeug * *z.B. für Roboter* || Spannvorrichtung * *Greifwerkzeug*
grist Feinheit * *{Textil}* von Garn
grommet Tülle * *für Kabeldurchführungen usw.*
groove Einstich * *mit Drehbank erzeugte Rille* || falzen * *{Holz}* nuten, auskehlen || Fuge * →*Nut,* →*Falz,* →*Rille, Furche, Rinne,* →*Hohlkehle, Kerbe, Vertiefung, Aushöhlung* || fugen * *nuten, auskehlen* || Nut * *Rinne, Nut, Vertiefg., Hohlkehle, Rille, Riefe, Nut b. Nut/Federverbindung i.Holzkonstrukt.* || Nute * *Rinne, Nut, Vertiefung, Rille, Riefe* || Riefe || Rille || Rinne * *Rille* || Vertiefung * *Riefe*
groove and feather joint Nut-und-Feder-Verbindung * *{Holz}*
groove cutter Nutenfräsmaschine * *{Holz}*
groove roll Nutwalze
groove seal Spaltdichtung
groove-cutting machine Nutfräsmaschine * *{Holz}*
groove-cutting tool Nutwerkzeug * *{Holz}*
grooved roller Nutwalze || Seilscheibe
grooving Grübchenbildung * *Riefenbildung z.B. an einem Kommutator* || Nut- || Riefenbildung * *z.B. an einem Kommutator*
groovy prima * *(salopp) klasse, toll*
gross brutto
gross national product Bruttosozialprodukt
gross negligence grobe Fahrlässigkeit
gross weight Bruttogewicht
ground Bezugspotenzial * *Masse* || Boden * *(Unter)-Grund; gain ground: an ~/Bedeutung gewinnen* || Erde || erden || Fundament || geschliffen || Grund * *Boden, Erde, Fläche, Gebiet, Motiv, Beweg~, ~lage, on grounds of: auf ~ von* || M-Potenzial || Masse * *single-ended: auf ~ bezogen (z.B. Analogsignal; im Gegensatz zum Differenzialsignal)*
ground connection Masseverbindung || Schutzleiter
ground contact Schutzkontakt
ground conveyor Flurförderzeug * *driverless/automated-guided: fahrerlos*
ground fault Erdschluss
ground fault circuit interrupter Fi-Schutzschalter
ground fault current Fehlerstrom
ground fault monitor Erdschlusswächter
ground fault monitoring Erdschlussüberwachung
ground fault monitoring device Erdschlussüberwachung * *als Gerät*
ground fault proof erdschlusssicher
ground fault protected erdschlussfest
ground fault protection Erdschlussschutz || Erdschlusssicherheit

ground loop Erdschleife
ground terminal Bezugsklemme * *Massepotenzial* || Erdungsanschlusspunkt || Schutzleiteranschluss
ground vibrations Fundamentschwingungen
ground-controlled flurgesteuert * *{Hebezeug}*
ground-fault resistant erdschlussfest
ground-plan Grundriss * *z.b. eines Gebäudes*
grounded supply TN-Netz || TT-Netz
grounded supply network geerdetes Netz * *aus der Sicht eines Verbrauchers (z.B. Motor, Stromrichter usw.)*
grounded system TN-Netz || TT-Netz
grounded system with protective conductor TN-Netz
grounded system without protective conductor TT-Netz * *(VDE 0100 Teil 410)*
grounding Erdung || Massung
grounding brush Erdungsbürste * *z.B. zum Erden eines Motorläufers*
grounding make-proof switch Erdungsdraufschalter
grounding outlet Schukosteckdose
grounding terminal Erdanschluss * *Klemme* || Erdungsklemme
grounding-type outlet Schukosteckdose
grounding-type plug Schukostecker
groundwood mill Schleiferei * *{Papier} Holz~ zur Zellstoffgewinnung*
groundwood pulp Holzschliff
groundwork Grundlage
group abstufen * *in Gruppen einteilen* || Abteilung * *Gruppe von Abteilungen, Unternehmensbereich* || anordnen * *gruppieren* || einteilen * *in Gruppen ~* || gliedern * *gruppieren, in Gruppen aufteilen* || Gremium * *siehe auch* →*Arbeitsgruppe* || Gruppe || gruppieren * *siehe auch* →*einteilen* || Kreis * *Gruppe* || räumlich anordnen || Sammel- * *z.B. Warnung, Meldung* || Schar * *z.B. von Kurven, Fäden* || Staffel * *Antriebe* || staffeln * *in Gruppen einteilen* || Summen- * *Sammel-, gemeinsamer* || unterteilen * *in Gruppen* || zusammenfassen * *in Gruppen ~* || zusammenstellen * *nach Gruppen ~*
group alarm Sammelalarm * *{SPS\Computer}* || Summenwarnung
group control Gruppensteuerung
group converter Gruppenumrichter * *speist mehrere parallelgeschaltete Motoren*
group drive Gruppenantrieb
group fault signal Sammelstörmeldung || Summenstörmeldung
group of companies Firmengruppe || Unternehmensgruppe
group package Sammelpackung
group pulses Summenimpulse
group signal Sammelmeldung || Sammelsignal || Summensignal * *z.B. Sammel(fehler)meldung*
group supply Gruppeneinspeisung
grouped gegliedert * *gruppiert* || gestaffelt * *in Gruppen aufgeteilt* || gewurzelt * *mit gemeinsamem Bezugspotenzial (z.B. Digital-Ein-/Ausgänge)* || unterteilt * *in Gruppen*
grouping Abstufung * *Aufteilung in Gruppen* || Anordnung * *Gruppierung* || Aufteilung * *in Gruppen* || Einteilung * *~ in Gruppen, Klassen* || Gruppierung || Staffel * *Aufteilung in Gruppen* || Wurzelung || Zusammenstellung
grouping block Vervielfacher * *für Kontakte/Anschlüsse*
grow entstehen * *erwachsen; out of:aus* || wachsen || zunehmen * *wachsen; grow worse: schlechter werden; grow stronger: stärker werden*
grow bigger zunehmen * *anwachsen, umfangreicher werden*
grow hard härten * *hart werden*

grow larger zunehmen * anwachsen, größer werden
grow less schwinden * abnehmen
grow old altern * Lebewesen
grow smaller verkleinern * kleiner werden
grow up herausbilden * auf-/heranwachsen (into: zu), entstehen, sich einbürgern
grow warm erwärmen * sich erwärmen
growing zunehmend
growth Anstieg * Wachstum || Wachstum || Zuwachs * (An)Wachsen (in:von)
growth rate Wachstumsrate
grub screw Gewindestift * Madenschraube || Madenschraube
grubscrew Feststellschraube * Madenschraube || Gewindestift * Madenschraube || Madenschraube * siehe auch →Feststellschraube
GTO Abschaltthyristor * Abkürzung für 'Gate Turn-Off Thyristor' || GTO * Abkürzg. f. 'Gate Turn-Off thyristor':abschaltbarer Thyristor
GTO thyristor GTO-Thyristor
GTO-thyristor GTO * GTO-Thyristor
guarantee Garantie * Garantie, Gewähr, Bürgschaft || garantieren * auch 'sicherstellen, (sich ver) bürgen (für), Garantie leisten' || gewährleisten * garantieren, sicherstellen, (sich ver-) bürgen (für), Grantie leisten || Gewährleistung * Garantie, Gewähr, Bürgschaft || haften * garantieren, gewährleisten, verbürgen, sicherstellen, Garantie leisten || Haftung * Gewährleistung || sicherstellen * garantieren || Sicherstellung || Versicherung * Garantie, Gewähr || zusichern * guarantee a thing to a person:jdm. etwas zusichern
guarantee period Garantiezeit
guaranteed garantiert
guaranty Garantie || Gewährleistung || Sicherstellung
guard Abdeckung || Abschrankung || Schutz * ~einrichtung/-Element/-Abdeckung z.b. an Maschinenteil || schützen * against: gegen, guard oneself against: sich ~ gegen
guard against absichern
guard grille Schutzgitter
guard plate Schutzblech
guard rail Geländer * Schutzgeländer || Schutzgeländer
guard ring Schutzring
guardband Sicherheitsabstand * z.B. bei Timing-Spezifikationen
guarded against abgesichert * geschützt gegen
guarding Absicherung * Schutz
gudgeon Zapfen * Zapfen, Bolzen
gudgeon pin Kolbenbolzen
guess annehmen * vermuten || glauben * annehmen || raten || erraten
GUI grafische Anwenderoberfläche * Abk.f. 'Graphical User Interface' (z.B. WINDOWS (R)) || grafische Oberfläche * Abk.f. 'Graphic User Interface':graf. Benutzeroberfläche (z.B. MS-Windows) || Oberfläche * Abkürzg.für 'Graphic User Interface': grafische Betriebssystem/Programmoberfläche
guidance Anleitung * Belehrung, Richtschnur || Beratung * (Berufs-, Ehe- usw.) -Beratung || Einweisung * Anleitung, Belehrung, Beratung, Führung || Führung * Leitung, Führung (einem Ziele zu), Anleitung, Belehrung, Beratung, Richtschnur || Leitung * Führung, Management || Orientierung * for your guidance: zu Ihrer ~ || Richtschnur
guidance system with circulating balls Kugelumlaufführung
guide Beschreibung * Führer || führen * lenken, führen, leiten, geleiten || Führung * mechanische; auch '~ von Luft' usw. || Führungs- * (nachgestellt) z.B. Rolle, Schiene || Führungsleiste || Handbuch * Leitfaden || leiten * führen, ge~ || Leitfaden

guide arm Führungsleiste
guide bearing Führungslager
guide bolt Führungsbolzen
guide plate Leitblech * z.B. Tischleitblech bei Blechverarbeitung || Seitenführung * Leitblech
guide pulley Führungsrolle || Umlenkrolle * →Führungsrolle
guide rail Führungsleiste * Führungsschiene || Führungsschiene
guide roll Leitwalze
guide roller Führungsrolle || Leitwalze || Umlenkrolle
guide slide Führung * mechan. (Gleit-) Führung
guide through hindurchleiten
guide value Daumenwert * Richtwert, empfohlener Wert || Richtwert
guide vane Leitrad * bei Ventilator
guided geführt * auch durch eine mechan. Führung || zwangsläufig * Bewegung e. Maschinenteils
guideline Leitfaden * Richtlinie || Leitlinie || Richtlinie
guideline for setup Einstellhinweise
guidelines Hinweise * als Richtschnur dienende ~
guideway Führung * (mechan.) || Gleitbahn * zur Führung
guiding Führungs-
guiding principle Regel * Richtlinie || Richtlinie || Richtschnur
guiding rule Richtlinie
guild Innung
guillotine planschneiden || Querschneidemaschine * mit Schlagmesser arbeitend (für Blech, Holzfurnier usw.) || Schere * Schlag~ || Schlagschere
guillotine cutter Schneidmaschine * →Schlagschere (für Blech usw.)
guillotine shear Tafelschere * Schlagschere für Blechtafeln
guinea pig Versuchskaninchen * ['gini pig] (figürl.), eigentlich 'Meerschweinchen'
gullet Kehle * Schlund
gully Rinne * Wasserablauf~
gum anleimen * gummieren || beleimen * gummieren
gum elastic Kautschuk
gummed gummiert || selbstklebend * gummiert
gun Pistole * z.B. Löt~, Wire-Wrap~
gun metal Bronze
gush strömen * quellen, schießen
gusset plate Knotenblech
gut demontieren * ausweiden (auch techn. Anlage usw.)
guy verspannen * mit Tau sichern, verspannen
gypsum Gips
gypsum board Gipsplatte * (Bautechnik) für den Innenausbau
gypsum fiberboard Gipsfaserplatte * (Bautechnik) zum Innenausbau
gyratory screener Plansiebmaschine * (Holz) für Holzschnitzel

H

H-bridge B2-Schaltung * (Stromrichter)
h.v. distribution Hochspannungsverteilung
h.v. switch Hochspannungsschalter
H2S Schwefelwasserstoff
habitate habilitieren
habilitation Habilitation
habitually in gewohnter Weise * gewohnheitsmäßig

habour master Hafenmeister
hack knacken * *(salopp)* Software, Zugangscode usw.
hack sawing machine Bügelsägemaschine
hacker Programmierer * *(salopp) Trickprogrammierer, Programmierexperte; ~, der Sicherheitssysteme "knackt"*
hacksaw Bügelsägemaschine
hair drier Föhn
hair dryer Haartrockner
hairline Faden * *Haarlinie*
hairline crack Haarriss
half halb * *half an apple: ein ~er Apfel; halfway: auf ~em Wege; half as much: ~ so viel* || Hälfte * *half the amount: die ~ des Betrages; half as much: halb so viel; halfway: auf halbem Wege*
half bridge Brückenhälfte * *z.B. einer Stromrichter-Brückenschltung*
half duplex halbduplex
half key halber Federkeil
half wave Halbschwingung || Halbwelle
half- Halb- * *z.B. half-controlled bridge: halbgesteuerte Brücke*
half-controlled halbgesteuert * *z.B. Brückenschaltung*
half-controlled circuit halbgesteuerte Schaltung
half-controlled single-phase bridge connection B2H-Schaltung * *halbgesteuerte 1-Phasen-Brückenschaltung* || halbgesteuerte Einphasen-Brückenschaltung * *B2H-Schaltung für 1Q-DC-Stromrichter*
half-converter Teilstromrichter * *z.B. Brückenhälfte, Teilbrücke*
half-duplex halbduplex
half-finished product Halbprodukt
half-key Halbkeil * *zur Motorauswuchtung (Federkeil wird nur zur Hälfte mit einbezogen)*
half-key balancing Halbkeilwuchtung * *Federkeil wird beim Auswuchten des Motorläufers zur Hälfte berücksichtigt*
half-wave rectification Einweggleichrichtung
half-wave rectifier Einweggleichrichter
halfway auf halbem Weg
hall Halle
Hall effect Hall-Effekt
hall generator Hall-Generator
Hall sensor Hallsensor
halogen Halogen
halogen-free halogenfrei
halt anhalten || stehen bleiben || stillsetzen * *Halt machen*
halting place Haltestelle
halve halbieren
halves Hälften
hammer Hammer
hammer drill Schlagbohrmaschine
hammer mill Hammermühle * *für Steine, Erz usw.* || Schlagmühle
Hammig distance HD * *→Hammingdistanz, Maß für Fehlererkenn- u. korrigierbarkeit eines Codes*
Hamming distance Hamming-Distanz * *Maß HD für Datensicherheit: HD - 1 Bitfehler pro Datenblock können erkannt werd.* || Hammingdistanz * *Maß →HD für Datensicherheit b.serieller Übertragung (HD-1 Fehlbits sind erkennbar)*
Hamming metric Hamming-Distanz * *Maß für Datensicherheit: HD - 1 Bitfehler pro Datenblock können erkannt werd.* || Hammingdistanz * *→HD* || HD * *→Hammingdistanz*
hamper behindern * *(be-) hindern, hemmen, verstopfen*
hampered beengt
hamstring lähmen * *(auch figürl.)*
hand Hand * *handy: →zur Hand* || Seite * *on the one/other hand: auf d.einen/anderen ~; on the right/left hand: auf d. recht./link. ~*
hand crank Handkurbel
hand drilling machine Handbohrmaschine
hand lever Handhebel
hand out herausgeben * *aushändigen* || verteilen * *austeilen, aushändigen, z.B. Besprechungs/Schulungsunterlagen*
hand over aushändigen * *to a person:jemandem* || übergeben * *aushändigen; auch (Fabrikations)Anlage, Aufgabe usw. übergeben*
hand speed counter Handtacho * *digitaler*
hand tacho Handtacho
hand torch Taschenlampe
hand- Hand-
hand-drill Handbohrmaschine
hand-held Hand- * *~gehalten, ~gefüht (z.B. Werkzeug, Bediengerät, Computer)*
hand-held controller Handeingabegerät
hand-held device Handgerät
hand-held keybord unit Handeingabegerät * *mit Tasten*
hand-held operator Handbediengerät * *tragbares ~*
hand-held operator panel Handeingabegerät
hand-held programmer Handprogrammiergerät
hand-held programming unit Handprogrammiergerät
hand-held tools Handwerkzeuge
hand-held unit Handgerät
hand-made winding Handwicklung * *{el.Masch.} Wicklung, der. Spulen v.Hand i.d. Nuten eingebracht/geträufelt werd.*
hand-operated handbetätigt
hand-pallet truck Hubwagen * *handbedienter Paletten-Hub-/Rollwagen*
hand-rail Geländer * *Handgeländer, Handlauf*
handbook Fachbuch * *→Handbuch* || Handbuch
handheld terminal Handeingabegerät
handhold Griff * *Festhalte~*
handicap beeinträchtigen * *behindern* || behindern * *benachteiligen, behindern, belasten* || Belastung * *Erschwernis* || Nachteil * *(auch figürlich) ein Hinterherhinken bewirkend; be handicapped:im Nachteil/behindert sein*
handicraft Handwerk
handing out Ausgabe * *Aushändigung, z.B. von Besprechungsunterlagen*
handing-over Übergabe
handle abwickeln * *durchführen* || angreifen * *anfassen* || bearbeiten * *erledigen, hantieren, handhaben, abwickeln* || bedienen * *handhaben, hantieren mit, umgehen mit, führen, leiten* || befassen || behandeln * *auch in einem Buch* ~ || Griff * *Halte~, Stiel; auch "Anfasseigenschaften" von Papier, Textilstoff usw.* || Handgriff * *Griff zum Anfassen/Bedienen* || handhaben || Hebel * *Handgriff* || Heft * *Griff* || manipulieren * *techn. Gerät* || Traggriff || umgehen * *hantieren (with/of: mit)* || verarbeiten * *handhaben, umgehen mit, abwickeln* || verkraften * *bewältigen* || verwalten * *hantieren, umgehen mit, abwickeln*
handler Treiber * *Software-Hantierungsbaustein*
handling Bearbeitung * *~ eines Falles, Vorgangs usw.* || Bedienung * *Hantierung, Handhabung* || Behandlung * *Handhabung* || Handhabung * *ease/simplicity of use: einfache ~* || Handlung * *Handhabung* || Hantierung
handling device Handhabungsgerät
handling of orders Auftragsbearbeitung
handling of quotations Angebotsbearbeitung
handling system Handhabungssystem
handout Tischvorlage * *bei Besprechung verteilte/ausgehändigte Unterlage*
handwheel Handrad * *(auch bei Aufzug); elevat-*

ing: zur Höhenverstellung; *traversing:* zur Seitenverstellung
handy griffbereit || handlich || praktisch * *handlich, gut brauchbar, nützlich, praktisch veranlagt/begabt* || zur Hand * *have something handy: etwas zur Hand haben*
hang aufhängen * *auch Programm* || schweben || sich aufhängen * *sich aufgehängt haben (Softwareprogramm)*
hang in einhängen
hang in the air schweben
hang up aufhängen || auflegen * *Telefon* ~ || einhängen * *aufhängen*
hangar Halle * *(Flugzeug)~, Schuppen*
hanging Verklemmung * *Sich-Aufhängen einer Software*
hanging load hängende Last * *am Kran usw.*
hanging program Programmabsturz * *"aufgehängtes" Programm*
Hanover Exhibition Hannover Messe
Hanover Fair Hannover Messe
Hanover Industrial Fair Hannover-Messe Industrie
happen eintreffen * *geschehen* || eintreten * *sich ereignen, geschehen* || erfolgen * *sich ereignen* || passieren * *sich ereignen* || stattfinden
happen again wiederholen * *nochmal geschehen*
hard aufwändig * schwierig || dringend * *Bitte usw.* || fest * *hart von ~er Beschaffenheit* || hart * *(auch figürl.) (auch Adv.) auch 'schwierig'* || kompliziert * *schwierig, mühsam, anstrengend* || mühsam * *schwierig* || schwer * *schwierig* || schwierig * *hart, schwer, mühsam, anstrengend; work/job:-Arbeit*
hard alloy Hartmetall
hard by it daneben * *dicht* ~
hard chrome plated hartverchromt
hard chromium plated hartverchromt
hard coal Steinkohle
hard disk Festplatte
hard disk data compression program Festplattenkomprimierprogramm
hard disk drive Festplatte * *Festplattenlaufwerk* || Festplattenlaufwerk
hard drive Festplatte || Festplattenlaufwerk
hard facts Wirklichkeit * *raue*
hard hat Schutzhelm
hard metal Hartmetall
hard stop Schnellhalt
hard substances Feststoffe
hard to please anspruchsvoll * *schwer zufriedenzustellen*
hard up knapp * *(salopp)* ~ *an Geld*
hard wearing verschleißfest
hard work Einsatz * *harte Arbeit*
hard-solder löten * *hart*~ || verlöten * *hartlöten*
hard-to-access positions schwerzugängliche Stellen
hard-wired verbindungsprogrammiert
hardcopy Ausdruck * *durch einen Drucker (z.B. Bildschirminhalt, Parametersatz)* || Bild * *z.B.* ~ *von Oszillografenkamera, Computerausdruck d. Bildschirminhalts*
harden aushärten * *non-hardening: nicht ~d* || härten
hardened gehärtet
hardening Aushärtung * *z.B. von Isolier-Tränkharz, Kunstharz* || Verfestigung * *Härtung, Härten*
hardening steel Einsatzstahl
hardly kaum * *schwerlich, fast nicht, mühsam, schwer*
hardly accessible schwer zugänglich
hardly any kaum * *kaum irgendwelche*
hardly ever kaum jemals || nie * *fast nie*
hardness Härte * *mechan. Härte, ~ von Material*

hardware Beschlag * *(Holz) (eingelassener)* ~ *an Möbeln usw.* || Hardware || hardwaremäßig
hardware address Hardwareadresse
hardware change Hardwareänderung
hardware components Hardwarekomponenten
hardware configuration Hardwarekonfiguration * *(SPS)*
hardware design Hardwareentwicklung || Physik * *hardwaremäßige Ausführ., z.B. serielle Schnittstelle*
hardware designer Entwickler * *Hardwareentwickler*
hardware development Hardwareentwicklung
hardware fault Hardwarefehler
hardware modification Hardwareänderung
hardware module Baugruppe * *Hardware-~*
hardware platform Hardware-Plattform
hardware recessing machine Beschlageinlassmaschine * *(Holz) in der Möbelindustrie*
hardwearing verschleißfest
hardwired fest verdrahtet || festverdrahtet || hardwaremäßig * *hardwaremäßig/fest verdrahtet* || verbindungsprogrammiert * *Steuerung (im Gegensatz zur SPS)*
hardwood Holz * *Hart~, Laub~* || Laubholz
harmful nachteilig * *schädlich* || schädlich * *nachteilig*
harmful effects schädigende Einwirkung
harmful substance Schadstoff
harmonic Harmonische || Oberschwingung * *5th harmonic:5.Harmonische;harmonic of 5th order:Oberwelle 5.Ordung/d.Ordnungszahl 5* || Oberwelle || organisch * *harmonisch* || Schwingung * *(höhere) harmonische* ~
harmonic absorber Saugkreis * *Filterkreis zur Oberschwingungskompensation* || Saugschaltung * *Filterkreis zur Oberschwingungskompensation*
harmonic analysis Fourier-Analyse
harmonic compensation Blindleistungskompensation
harmonic components Oberschwingungsanteile
harmonic content Oberschwingungen * *Oberschwingungsanteil/gehalt* || Oberschwingungsanteil || Oberschwingungsanteile * *Oberschwingungsgehalt * ident.m. Klirrfaktor:Effektivwert d.Oberschwing.gen/Gesamteff.wert(DIN 40110)* || Oberwellenanteil || Oberwellengehalt || Oberwelligkeit || Schwingungsgehalt * *von Oberwellen*
harmonic currents Oberschwingungsströme
harmonic distortion Klirrfaktor || Oberschwingungsgehalt
harmonic disturbance on the mains supply Netzrückwirkungen
harmonic effect on the system current Netzrückwirkung
harmonic effects on the supply Netzrückwirkungen * *Oberwellen-behaftete Belastung des Netzes durch Stromrichter*
harmonic effects on the system Netzrückwirkungen
harmonic factor Formfaktor || Klirrfaktor || Oberschwingungsgehalt * *Klirrfaktor*
harmonic field Oberfeld * *Oberwelle eines (Magnet)Feldes*
harmonic losses Oberschwingungsverluste
harmonic noise injected back into the line Netzrückwirkungen * *Oberwellen-behaftete Netzbelastg.z.B. durch Stromrichter*
harmonic number Ordnungszahl * ~ *einer Oberschwingung*
harmonic order Ordnungszahl * ~ *einer Oberschwingung*
harmonic oscillation Oberschwingung
harmonic power Verzerrungsleistung

harmonic suppression Oberschwingungskompensation || Oberwellenunterdrückung
harmonic torque Oberfelddrehmoment * {el. Masch.] Beeinflussg.d. Moment.kennlin.durch unsymmetr.Wicklg. (Sattelmoment)
harmonic voltage factor Spannungs-Oberschwingungsfaktor * kennzeichnet d. Abweichg. e. Wechselspanng. von d. Sinusform
harmonics Oberschwingungen * period.wechselgrößen höherer Frequenz, d.der (Netz)Grundschwingung überlagert sind || Oberwellen || Schwingungen * (höhere) harmonische
harmonics fed back into the supply Netzrückwirkungen
harmonics reactive power Oberschwingungsblindleistung
harmonics spectrum Oberschwingungsspektrum
harmonization Abstimmung || Harmonisierung || Vereinheitlichung * Harmonisierung
harmonize abstimmen * in Einklang bringen, harmonisieren || harmonisieren || in Übereinstimmung bringen * harmonisieren || passen * zusammenpassen || übereinstimmen * zusammenpassen, im Einklang sein, harmonieren || vereinheitlichen * harmonisieren || zusammenpassen * harmonisieren, gut ~
harmonized abgestimmt * harmonisiert || einheitlich * ver~t
harmonizing Koordination * Harmonisieren, In-Einklang-Bringen || Koordinierung * Harmonisieren, In-Einklang-Bringen
harmony Übereinstimmung * Eintracht, Einklang, Harmonie
harness nutzbar machen * Naturkräfte usw. || nutzen * nutzbarmachen (Naturkräfte usw.)
harsh hart * streng (z.B. Test), grob, schroff || rau * streng, hart, rau (auch Umgebungsbedingungen usw.), scharf, derb || streng * hart
harsh ambient conditions raue Umgebungsbedingungen
harsh environments raue Umgebungsbedingungen || rauer Betrieb * raue Betriebs-/Umgebungsbedingungen
harsh industrial environments raue Industriebedingungen
harshness Härte * (figürl.) Strenge, Schärfe
hassle Mühe * (salopp) Mühe; (auch handgreifliche) Auseinandersetzung, Krach
hassle-free mühelos
hatch schraffieren
hatched area Schraffur
hatching Schraffur
hateful unangenehm * zuwider, verhasst
haul schleppen * auch mit Zugmaschine
haul off abführen * eine Sache ~
haulage Förderung * Ziehen, Schleppen, Transport, auch Fördern von Bodenschätzen || Transport * mit Eisenbahn, Lastwagen usw.
haulage capacity Förderleistung * max. Förderleistung z.B. einer Bandförderanlage {Bergwerk}
haulage system Förderanlage
hauling Förderung * Ziehen, Schleppen, Transport, auch Fördern von Bodenschätzen
hauling off Abfuhr * von Sachen
have aufweisen || besitzen * eine Eigenschaft ~ || besitzen * auch Eigenschaft, Talent usw. ~ || haben || verfügen über * ausgestattet sein mit
have a bearing on beeinflussen * einwirken auf
have a capacity of fassen * Fassungsvermögen von ... haben
have a look at betrachten
have a nice day Auf Wiedersehen * einen schönen Tag noch
have a nice evening Auf Wiedersehen * einen schönen Abend noch

have a share in beteiligen * geschäftlich beteiligt sein (in: an)
have access zugreifen * Zugriff haben; to: auf
have access to zugreifen auf
have an adverse affect on beeinträchtigen
have an effect beeinflussen * einen Einfluss ausüben; have an adverse effect on: nachteilig/schädlich ~
have an effect upon einwirken * Ein/Auswirkung haben auf
have an interest beteiligen * Geschäftsanteile haben (in: an)
have an interview bewerben * ein Bewerbungsgespräch führen || sich vorstellen * sich als Bewerber für eine Stelle ~, ein Bewerbungsgespräch führen || vorstellen * sich vorstellen, ein Bewerbungsgespräch führen
have at one's disposal verfügen über * zur Verfügung haben
have at ones's disposal haben * zur Verfügung haben
have business relations with Geschäftsbeziehungen haben mit
have done aufhören * (mit Gerundium)
have effect on einwirken * Ein/Auswirkung haben auf
have in mind vorhaben * etwas im Sinn haben, beabsichtigen; have a mind to do sth.: Lust haben etw.zu tun
have influence on beeinflussen
have jurisdiction befugt * over:zuständig sein für
have knowledge of Wissen * (Verb)
have on stock führen * Artikel im Einzel-/Großhandel ~
have overweight überwiegen
have plenty aus dem Vollen schöpfen
have priority vorgehen * Vorrang haben, höherprior sein (gegenüber)
have recourse to zurückgreifen auf
have reference to beziehen * sich ~ auf
have regard to berücksichtigen
have respect to berücksichtigen
have something on vorhaben * etwas geplant haben, z.B. 'für den Abend'
have stocked führen * Artikel im Einzel-/Großhandel ~
have stopped stehen * ~geblieben sein
have the feel im Griff haben * of a thing:etwas
have the knack im Griff haben * of a thing:etwas
have the right befugt * befugt sein; have no right to do so:dazu nicht befugt sein
have the right to dürfen
have the upperhand dominieren
have to sollen * müssen
have to accept Kauf * etwas in ~ nehmen müssen
have to pay a heavy price for erkaufen * (auch figürl.) etwas teuer ~ müssen
have top priority an erster Stelle stehen * höchste Priorität haben || oberstes Gebot sein
having mit * habend/besitzend
having a mit * ein ... besitzend
having a high dynamic response hochdynamisch
having a higher pulse number höherpulsig * Stromrichterschaltung
having a low Ohm value hochohmig || niederohmig
having a ripple geriffelt
having capability -fähig * z.B.: having alarm capabilty: alarm~
having equal rights gleichberechtigt
having interrupt capability interruptfähig
having publicity appeal werbewirksam
having the same -gleich || -gleich * z.B. having the same rating: leistungs~

having the same output leistungsgleich * z.B. Motor
having the same rating leistungsgleich * z.B. Stromrichter
hawser Tau * {Schifffahrt}
hazard Gefahr || Risiko * Gefahr, Wagnis
hazard of leaking Leckgefahr
hazardous explosionsgefährdet || gefährlich * riskant, gewagt, gefährlich || lebensgefährlich * hochgefährlich
hazardous area explosionsgefährdete Umgebung || explosionsgefährdeter Bereich || explosionsgefährdeter Raum || schlagwettergefährdeter Bereich
hazardous location explosionsgefährdete Umgebung || explosionsgefährdeter Ort
hazardous type explosionsgeschützt
hazardous voltage levels gefährliche Spannungen
hazardous-duty explosionsgeschützt
hazardous-duty design explosionsgeschützte Ausführung
hazardous-duty motor explosionsgeschützter Motor
hazardous-duty type explosionsgeschützte Ausführung
hazardous-location regulations Explosionsschutz-Vorschriften
hazards to persons Personengefährdung
haze Trübung * Dunst, leichter Nebel
hazy trüb * dunstig, diesig || ungenau * (figürl.) verschwommen, nebulös, vage || unscharf * (figürl.) verschwommen, nebelhaft, nebulös, dunstig
HD Hamming-Distanz * Abkürzg. f. 'Hamming Distance': Maß für Datensicherheit eine Codes/Protokolls || Hammingdistanz * Maß für Datensicherheit z.B. b.seriell.Übertragg. (HD-1 Falschbits erkennb.) || HD * →Hammingdistanz
HDD Festplatte * Abkürzg. für 'Hard Disk Drive': Festplattenlaufwerk || Festplattenlaufwerk * Abkürzg. für 'Hard Disk Drive'
HDF board HDF-Platte * {Holz} Abk. für 'High-Density Fiberboard': hochdichte Faserplatte
HDLC HDLC * Abk.f. 'High-Level Data Link Control': Datenübertragungsprotokoll mit Fehlerkorrektur
HDU Festplatte * Abkürzg. für 'Hard Disk Unit': Festplattenspeichereinheit
he who derjenige
head Büroleiter || Chef * Führer, Leiter, Vorstand, Direktor || Förderhöhe * einer Pumpe; head of water: Wassersäule || Gefälle * z.B. bei Wasserkraftanlage || Kopf || leiten * anführen || Leiter * einer Abteilung || Vorstand * Einzelperson
head box Niveaukasten * (amerikan.) Stoffauflaufkasten bei Papiermaschine
head manager Chef * "Obermanager"
head of a course Lehrgangsleiter
head of course Kursleiter
head of department Abteilungsleiter
head of sales department Vertriebsleiter
head of section Gruppenführer
head of test department Prüffeldleiter
head off abwehren * Unheil
head office Hauptsitz * Stammhaus, Hauptniederlassung || Hauptverwaltung || Stammhaus * Leitungsbüro, koordinierende Abteilung, Stammsitz || Zentrale * Firmen~
head office sales/marketing Stammhausvertrieb
head rope Oberseil * {Aufzüge | Seilbahn}
head station Kopfstation
head wheel Seilscheibe * z.B. bei Aufzug
headbox Stoffauflauf * {Papier} (amerikan.) ~kasten bei Papiermaschine vor der Siebpartie || Stoffauffaufkasten * (amerikan) bei Papiermaschine; siehe auch →Stoffauflauf
headbox control Stoffauflaufregelung * {Papier}-regelt auf Sieb auflauf. Stoffmenge mit Stoffdruck u. Lippenverstellg.
headbox feed pump Stoffauflaufpumpe * {Papier} am Beginn einer Papiermaschine
header Kopf * z.B. eines Datensatzes/Menüs/Datentelegramms/Programms || Kopfzeile || Vorspann * z.B. eines Datentelegramms, Datensatzes
heading Kopf * z.B. Kopfzeile(n) || Überschrift
headless screw Gewindestift
headline Überschrift * Kopf/Schlagzeile
headoffice Stammhaus
headoffice sales/marketing Stammhausvertrieb
headpiece Aufsatz * Oberteil
headquarter Hauptsitz
headquarters Hauptsitz || Sitz * (Firmen-) Haupt~ || Zentrale * Firmen~
headroom Freiraum * lichte Höhe, Platz für Erweiterungen
headstock Reitstock * bei Drehmaschine || Spindelkasten
headway Fortschritt * make headway:Fortschritte machen || Schwung * Geschwindigkeit
healthy betriebsbereit || förderlich * gesund (auch figürl.), heilsam, förderlich, fehlerfrei funktionierend || intakt
heap Ansammlung || Haufen || Menge * Haufen
heap up sammeln * an-/aufhäufen
heaping Häufung * Anhäufung; (auch figürl.: Überhäufung/Überschüttung)
hear verhandeln * Zivilrecht
hearing Sitzung * offizielle Anhörung || Verhandlung * Anhörung, öffentliche ~, zivilrechtliche ~
heart Kern * of the matter: ~ einer Sache || Mittelpunkt * (auch figürl.)Herz, Inneres, Mitte, Wesentliches, Kern (einer Sache/einer Stadt) || Sinn * Seele, Herz
heartbeat counter Lebenszähler
heat Durchgang * Runde, Lauf, Endrunde, Entscheidungskampf || erwärmen || Glut || Hitze || Lauf * Durchlauf, Durchgang || Wärme * absorb:aufnehmen; dissipate:abgeben; conduct:leiten; carry off:abführen; dry: trockne
heat build-up Erwärmung * Entwicklung höherer Temperaturen z.B. bei Motor/Getriebe i.Dauerbetrieb
heat conduction Wärmeableitung * Wärmeleitung
heat discharge Entwärmung || Wärmeabfuhr || Wärmeabführung
heat dissipation Entwärmung || Kühlung * Wärmeabfuhr || Wärmeabfuhr || Wärmeabführung || Wärmeabgabe * Verlustwärme || Wärmeableitung
heat drop Wärmegefälle
heat emission Wärmeabstrahlung
heat exchanger Kreislaufkühler * Wärmetauscher || Kühler * Wärmetauscher || Wärmeaustauscher || Wärmetauscher * Einrichtung z.Übertragung v.Wärme von einem primären auf ein sekundäres Kühlmittel
heat flow Wärmestrom
heat insulation Wärmedämmung
heat loss Verlustleistung * in Wärme umgesetzte ~ || Verlustleistung * in Wärme umgesetzte ~; auch 'Wärmeverluste' bei e. thermodynamischen Prozess
heat losses Erwärmungsverluste || Verlustwärme
heat pipe Wärmerohr
heat pump Wärmepumpe
heat quantity Wärmemenge
heat radiation Wärmeabstrahlung || Wärmestrahlung
heat recovery Wärmerückgewinnung
heat removal Wärmeabfuhr || Wärmeabführung
heat resistance Hitzebeständigkeit || Temperaturbeständigkeit
heat run Erwärmungsprüfung * Erwärmungslauf

heat sink Kühlkörper
heat stability Wärmebeständigkeit
heat storage Wärmespeicher * *heat-storage capacity: ~vermögen*
heat test Erwärmungsprüfung
heat to be dissipated abzuführende Verlustleistung
heat tranfer compound Wärmeleitpaste
heat transfer Wärmeableitung * *Wärmeübertragung* ‖ Wärmeübergang
heat transfer agent temperature Kühlmitteltemperatur * *z.B. in einem Primär-Kühlkreislauf*
heat transport Wärmetransport
heat treatment Wärmebehandlung
heat up aufheizen * *auch sich aufheizen* ‖ erwärmen * *auch 'sich erwärmen'*
heat-conducting wärmeleitend
heat-conducting paste Wärmeleitpaste
heat-insulating wärmedämmend
heat-resistant hitzebeständig ‖ wärmebeständig
heat-shrink Schrumpf- * *bei Hitze schrumpfend, z.B. Schlauch, Folie*
heat-shrink tube Schrumpfschlauch
heat-shrinkable tube Schrumpfschlauch
heat-transfer agent Kühlmittel * *z.B. in einem Primär-Kühlkreislauf*
heat-treat vergüten * *mit Wärme behandeln*
heat-treatable steel Vergütungsstahl * *vor der Vergütung*
heat-treated steel Vergütungsstahl
heatable beheizbar ‖ heizbar
heater Erhitzer ‖ Heizung * *Heizeinrichtung* ‖ Ofen * *Heizkörper* ‖ Wärmeerzeuger
heater control Heizungsregler
heating Erwärmung * *(Auf)Heizung, Erhitzung* ‖ Heizung * *auch 'Heizungsanlage'*
heating boiler Heizkessel
heating control Heizungsregelung
heating plate Heizplatte
heating power Heizwert
heating power station Heizkraftwerk
heating pump Heizungspumpe
heating rod Heizstab
heating up Erwärmung * *Aufheizen, Sich-Aufheizen*
heating zone Heizzone * *['hieting 'soun] z.B. bei Extruder, Kunststoffspritzmaschine*
heating-ventilation-air-conditioning Heizung-Klima-Lüftung
heatsink fin Kühlrippe * *eines Kühlkörpers*
Heaviside effect Stromverdrängungseffekt
heavy beträchtlich * *Kosten, Verluste* ‖ erheblich * *Kosten, Verluste* ‖ groß * *schwer (Maschine, Fahrzeug, Motor, Last, Verlust, Unwetter), stark, heftig, umfangreich* ‖ hoch * *Steuer, Gebühr, Zoll, Strafe, Unkosten* ‖ kräftig * *Schlag usw.* ‖ langsam * *schwerfällig* ‖ massiv * *(auch figürl.) schwer, stark, kräftig, drückend* ‖ schwer * *Gewicht*
heavy brush flashover Bürstenrundfeuer ‖ Rundfeuer * →*Bürstenrundfeuer*
heavy current Starkstrom
heavy demands hohe Ansprüche * *make heavy demands on: ~ stellen an*
heavy duty Hochleistungs- * *für Schwerlast- oder rauen Betrieb* ‖ schwer * *für Schwerlast- oder rauen Betrieb* ‖ schwere Belastung ‖ Schwerlast ‖ Schwerlast-
heavy duty lorry Schwerlasttransporter
heavy flashover Rundfeuer
heavy load schwere Belastung ‖ Schwergut ‖ Schwerlast
heavy machine Großmaschine
heavy plate Grobblech ‖ Groblech
heavy rail traffic Fernverkehr * *Bahn*
heavy rail vehicle Fernverkehrsfahrzeug * *Schienenfahrzeug*

heavy starting Schweranlauf ‖ Schwerlastanlauf
heavy-duty Groß- * *Schwerlast* ‖ robust * *für hohe Beanspruchung/raue Umgebungsbedingungen; Schwerlast-* ‖ Schwerlast- ‖ strapazierfähig
heavy-duty design verstärkte Ausführung
heavy-duty operation Schwerlastbetrieb
heavy-duty starting Schweranlauf ‖ Schwerlastanlauf
heavy-duty type Robustbauform
heavy-gauge dick * *Draht, Kabel, Blech*
heavy-gauge conduit gland PG-Verschraubung
heed berücksichtigen * *beachten, Acht geben auf*
height Höhe * *['hait]*
heightening Steigerung * *Erhöhung, Vergrößerung (figürl.), Stärkung, Anhebung*
helical schneckenförmig ‖ schrägverzahnt ‖ schraubenförmig ‖ spiralförmig
helical bevel gearbox Kegel-Stirnradgetriebe ‖ Kegelstirnradgetriebe
helical conveyor Schneckenförderer
helical curve Spirale * *spiralförmige Kurve*
helical gear Schrägstirnrad ‖ Stirnradgetriebe
helical gearbox Stirnradgetriebe
helical geared schrägverzahnt
helical potentiometer Spindelpotentiometer ‖ Wendelpotentiometer * →*Spindelpotentiometer*
helical spring Schraubenfeder ‖ Spiralfeder
helical toothed schrägverzahnt
helical worm gear Stirnrad-Schneckengetriebe
helical worm gearbox Stirnradschneckengetriebe
helical-geared motor Stirnradgetriebemotor
helicoil Gewindeeinsatz * *zur Reparatur eines ausgerissenen Gewindes*
helipot Spindelpotentiometer * *(salopp) Markenname der Firma Beckman* ‖ Wendelpotentiometer * *(salopp) Markenname der Fa. Beckman*
help helfen ‖ Hilfe ‖ nützen * *helfen*
help message Hilfetext
help oneself bedienen * *sich (selbst) bedienen*
help text Hilfetext
help texts Hilfetexte
helper Hilfs- * *z.B. Antrieb* ‖ Zusatz- * *Hilfs-*
helper drive Hilfsantrieb ‖ Nebenantrieb * *Hilfsantrieb*
helper motor Hilfsmotor
helpful nützlich * *hilfreich*
hemp Hanf
hence also * *folglich, daher, deshalb* ‖ daher * *folglich, deshalb, also* ‖ damit * *daher, deshalb, somit* ‖ deshalb * *so, daher, folglich* ‖ folglich * *daher, deshalb, also* ‖ somit * *folglich, daher, deshalb; hieraus, daraus; hence it follows that: daraus folgt, dass*
hence it follows daraus folgt * *that: dass*
here you are bitte * *nach e. erbetenen Handreichung,z.B.auf haben Sie Feuer?*
hereafter im Folgenden
hereby hierbei * *hierdurch, hiermit* ‖ hiermit * *hierdurch, hiermit*
hereinafter unten erwähnt * *(Adv.)* ‖ unten stehend * *(Adv.)*
hereinbelow unten erwähnt * *(Adv.)* ‖ unten stehend * *(Adv.)*
herewith hierbei * *hierdurch, hiermit* ‖ hiermit
hermetic hermetisch ‖ luftdicht
hermetically sealed hermetisch abgeschlossen ‖ hermetisch dicht ‖ hermetisch dicht ‖ luftdicht * *luftdicht verschlossen*
herringbone-scewed pfeilverzahnt * *Zahnrad*
Hertz Hertz ‖ Hz
hesitant schwankend * *zögernd*
hesitate zögern
heterogenous heterogen * *aus verschiedenartigsten Teilen zusammengesetzt* ‖ verschieden * *verschiedenartig, heterogen*

hex monitor Hex-Monitor * *Hilfsprogramm z. Testen v.Anwenderprogrammen mit hexadezimaler Anzeige d.Speicherinhalte*
hex number Hexazahl
hex nut Sechskantmutter
hexadecimal hexadezimal
hexadecimal number Hexazahl
hexadecimal system Hexadezimalsystem
hexagon 6-Kant-
hexagon nut Sechskantmutter
hexagon-head bolt Sechskantschraube * *dickere Schraube, Schraube mit (teilw. gewindelosem) Schaft*
hexagon-head screw Sechskantschraube * *dünnere Schraube*
hexagon-socket Innensechskant
hexagon-socket screw Inbusschraube || Innensechskantschraube
hexagon-socket spanner Inbusschlüssel
hexagonal sechseckig
hexagonal nut Sechskantmutter
hexangular design Sechseckbauweise
hexangular-frame design Sechseckbauweise * *z.B. eines Elektromotors*
Heyland diagram Heylandkreis * *Ortskurve der Ströme der Drehstrom-Induktionsmaschine* || Ossannakreis * *Ortskurve der Ströme der Drehstrom-Induktionsmaschine*
HF- HF- * *High Freqency, Hochfrequenz-*
hidden geheim * *versteckt* || intern * *verborgen* || verborgen * *verborgen, geheim* || verdeckt * *verborgen, versteckt*
hidden energy potenzielle Energie
hide verdecken * *verstecken* || verstecken
hierarchic hierarchisch
hierarchical hierarchisch * *hierarchical layered: ~ aufgebaut*
hierarchical layer Hierarchieebene
hierarchical level Hierarchieebene
hierarchy Hierarchie
hierarchy level Hierarchieebene
high groß * *zahlen-/wertmäßig hoch, z.B. Spannung, Strom, Qualität* || hoch * *auch wert/zahlenmäßig* || höchster Gang * *Getriebe*
high accuracy hohe Genauigkeit
High active High-aktiv
high bar rotor Tiefnutläufer * *[el.Masch.] bei Käfigläufermotor (Stromverdrängungsläufer)*
high bay racking system Hochregallager
high control setting hohe Aussteuerung
high degree of hoch * *ein hohes Maß an, hoher Grad an*
high degree of reliability hohes Maß an Zuverlässigkeit
high delay angle niedrige Aussteuerung * *großer Steuerwinkel bei Phasenanschnittsteuerung*
high dynamic performance hohe Dynamik
high dynamic response drive hochdynamischer Antrieb
high end Hochleistungs- * *Spitzen-* || Spitzen- * *an der Spitze stehend*
high energy Hochenergie-
high gear höchster Gang * *Getriebe* || Schnellgang
high impedance hochohmig
high level control Technologie * *Überlagerte Regelung und Steuerung*
high level language Hochsprache * *Programmiersprache* || höhere Programmiersprache
high modulation hohe Aussteuerung * *in der Mess-/Nachrichtentechnik usw.*
high output voltage ratio hohe Aussteuerung
high pass Hochpass
high performance Hochleistungs- * *mit hoher Leistungsfähigkeit* || hochwertig * *bezüglich der Leistungsfähigkeit* || Spitzen- * *Hochleistungs-*
high performance level oberer Leistungsbereich
high pressure Hochdruck * *auch figürl.: work at high pressure: mit ~arbeiten*
high quality hochwertig * *qualitativ hochstehend*
high rel requirements erhöhte Anforderungen * *Kurzform für 'high reliability ..'*
high requirements erhöhte Anforderungen
high response dynamisch * *mit hoher Dynamik* || hochdynamisch || rechenzeitoptimiert * *mit kurzer Reaktionszeit (bei Echtzeitanwendungen)*
high speed Hochgeschwindigkeit
high speed SCR fuse Thyristorsicherung
high starting duty Schweranlauf
high tech Hochleistungs- * *hochtechnologisch*
high value hochwertig
high voltage Hochspannung || Hochspannungs-
high voltage DC power transmission HGÜ * *Hochspannungs-Gleichstromübertrag. elektr. Stroms über Leitg.*
high voltage DC transmission HGÜ
high voltage installation Starkstromanlage
high-accuracy hochgenau
high-alloy hochlegiert
high-availability hochverfügbar
high-bandwidth breitbandig || Hochfrequenz- * *mit großer Bandbreite (bei Nachrichtenübertragung)*
high-bay racking vehicle Regalförderzeug
high-bay storage Hochregallager
high-bay warehouse Hochregallager
high-bay-rack conveyor Regalförderzeug * *siehe →Regalbediengerät*
high-brow anspruchsvoll * *hochgestochen, betont intellektuell*
high-capacity Hochleistungs- * *mit hoher Leistung/hohem Leistungs-/Fassungsvermögen* || leistungsfähig * *mit hohem Leistungsvermögen*
high-coercive hochkoerzitiv
high-cost teuer * *Produkt*
high-definition hochauflösend * *Bildschirm*
high-density hochdicht
high-dynamic hochdynamisch
high-efficiency Hochleistungs- * *mit hoh. Leistungsfähigkeit/Leistung/Wirkungsgrad/Funktionsgüte/Effizienz*
high-end Hochleistungs- * *der Spitzenklasse zuzuordnen* || hoher Leistungsbereich * *bezügl. der Leistungsfähigkeit* || oberer Leistungsbereich || Oberklasse- * *z.B. Produkt*
high-end product Spitzenprodukt
high-energy Hochenergie- || Hochleistungs-
high-energy battery Hochenergiebatterie
high-frequency hochfrequent || Hochfrequenz-
high-frequency interference Hochfrequenzstörung
high-frequency radiation hochfrequente Einstrahlung
high-grade hochwertig
high-inertia drive Schwungmassenantrieb * *Antrieb mit hoher Schwungmasse*
high-inertia starting Schweranlauf || Schwerlastanlauf
high-insulating hochisolierend
high-level controls überlagerte Regelung
high-load hochbelastet * *z.B. durch mechan./elektr. Last*
high-loss verlustbehaftet
high-maintenance wartungsanfällig || wartungsintensiv
high-MTBF hochverfügbar * *mit hoher MTBF ('Mean Time Between Failures':Durchschnittl. fehlerfreie Betriebszeit)*
high-output range hoher Leistungsbereich * *bezügl. der Ausgangsleistung, z.B. Motor/Stromrichter*
high-pass filter Hochpassfilter

high-performance hochperformant || leistungsfähig || leistungsstark
high-power Hochleistungs- * *mit hoher elektr./mechan.Leistung*
high-power drive Antrieb großer Leistung || Antrieb hoher Leistung || Großantrieb * *Antrieb mit hoher Leistung*
high-powered stark * *z.B. Motor*
high-precision hochgenau || präzise * *hoch~*
high-pressure Hochdruck-
high-pressure line Druckleitung * *Hochdruckleitung*
high-pressure steam mains Hochdrucksammelschiene * *zur Dampfversorgung*
high-pressure test Dichtigkeitsprüfung * *~ mit Überdruck*
high-quality gut * *von hoher Qualität (auch 'highgrade')* || qualitativ hochwertig
high-ranking hoch * *hochrangig (Person)*
high-rating groß * *(Adj.) z.B. Motor, Stromrichter, Pumpe mit hoher Leistung* || großer Leistung * *(Adj.) z.B. Motor, Stromrichter, Pumpe, Ventilator* || Hochleistungs- * *mit hoher Abgabeleistung/ hohem Strom/hoher Spannung, z.B. Stromrichter*
high-rating converter Hochleistungsstromrichter
high-rating drive Großantrieb
high-rating motors Motoren großer Leistung
high-rating range hoher Leistungsbereich * *bezüglich der Nennleistung, z.B. Motor/Stromrichter* || oberer Leistungsbereich * *bezüglich der Nennausgangsleistung (Stromrichter, Motor usw.)*
high-reliability requirements erhöhte Anforderungen
high-resistance hochohmig
high-resistance rotor Schlupfläufer * *Läufer mit erhöhtem Schlupf; siehe →Widerstandsläufer* || Widerstandsläufer * *Käfigläufer m.Stäben erhöhten Widerstandes →weichere Drehmomentkennlinie*
high-resistance squirrel-cage rotor Schlupfläufer || Widerstandsläufer * *Käfigläufer*
high-resistive hochohmig
high-resolution feinstufig * *mit hoher Auflösung* || hochauflösend
high-speed Hochgeschwindigkeits- || hochtourig || schnell * *Hochgeschwindigkeits-* || Schnell- || schnellaufend || superflink || unverzögert * *Hochgeschwindigkeits-*
high-speed circuit breaker Schnellschalter * *Leistungsschalter*
high-speed DC excitation Gleichstrom-Schnellerregung
high-speed fuse superflinke Sicherung
high-speed response schnellansprechend * *mit kürzester Reaktionszeit*
high-speed rotor Hochgeschwindigkeitsläufer * *siehe auch →Massivläufer, →Turboläufer* || Turboläufer * *Läufer eines schnelllaufenden Motors*
high-speed shaft Antriebswelle * *bei einem Untersetzungsgetriebe*
high-speed train Hochgeschwindigkeitszug
high-temperature grease Hochtemperaturfett
high-tensile hochwertig * *hochfest, z.B. Stahl*
high-threshold logic Hochpegellogik
high-viscosity hochviskos
high-viscous zähflüssig
high-voltage capacitor Hochspannungskondensator
high-voltage circuit breaker Hochspannungsschalter
high-voltage current Starkstrom
high-voltage distribution Hochspannungsverteilung
high-voltage motor Hochspannungsmotor * *ca. 1 bis 12 kV und 200 bis 16000 kW*

high-voltage rectifier Hochspannungsgleichrichter
high-voltage switch Hochspannungsschalter
high-voltage system Hochspannungsanlage
high-voltage test Hochspannungsprüfung || Isolationsprüfung || Prüfung der Isolierung || Wicklungsprüfung
high-volume production Serienfertigung * *Fertigung in hoher Stückzahl*
higher erhöht * *höher* || höher * *(auch Adv.)* || ober
higher level assignment Anlagenkennzeichen
higher level computer übergeordneter Rechner
higher level designation Anlagenkennzeichen || Ortskennzeichen * *Anlagenkennzeichen*
higher level loop control übergeordnete Regelung
higher order harmonics Oberschwingungen höherer Ordnung
higher than über * *höher als*
higher up höher * *(Adv.) weiter oben*
higher-level übergeordnet * *z.B. Automatisierungssystem* || übergreifend * *übergeordnet* || überlagert * *übergeordnet*
higher-level automation system übergeordnetes Automatisierungssystem
higher-level closed-loop control überlagerte Regelung
higher-level control loop überlagerter Regelkreis
higher-level control system externe Steuerung * *überlagertes Steuersystem*
higher-level controller überlagerter Regler
higher-level loop control überlagerte Regelung
higher-level system Leitsystem
higher-order höherer Ordnung * *z.B. Oberwelle*
higher-quality höherwertig * *von höherer Qualität*
higher-ranking höherprior * *im Rang höherstehend (Person)* || übergeordnet * *im Rang*
higher-rating größer * *mit höherer Leistung, z.B. Motor, Stromrichter, Pumpe*
highest größt || höchst || Spitzen-
highest degree Höchstmaß * *(figürl.) of: an*
highest level Höchstmaß * *(figürl.) of: an*
highest possible optimal
highest-ranking höchst * *am rang~en (Person)*
highest-rating größt * *mit der größten Leistung, z.B. Motor, Stromrichter, Pumpe*
highlight Besonderheit * *herausragendes Merkmal* || herausragendes Merkmal || herausstellen * *hervorheben, betonen, herausstreichen* || markieren * *hervorheben, z.B. Passagen in einem Text* || unterstreichen * *heraus/hervorheben, betonen, herausstreichen, Schlaglicht setzen* || Vorzug * *herausragendes (positives) Merkmal*
highlite hervorheben * *hervorheben, betonen, herausstreichen; auch in einem Dokument, z.B. durch Fettschrift*
highly hoch- * *z.B. highly-integrated: hochintegriert; highly-flexible: hochflexibel* || höchst * *(Adv.)* || sehr * *(Adv.) höchst, äußerst* || stark * *(Adv., figürl.) highly nonlinear: stark nichtlinear*
highly accurate hochgenau
highly complex hochkomplex
highly delicate hoch empfindlich * *zart, zerbrechlich*
highly developed hochentwickelt
highly dynamic response hochdynamische Regeleigenschaften
highly flexible hochelastisch || hochflexibel
highly integrated hochintegriert
highly permeable magnet materials hochpermeable Magnetmaterialien
highly proven erprobt * *wohl erprobt, gut bewährt*
highly sensitive hoch empfindlich * *feinfühlig, empfindsam, hochauflösend (Messgerät)*
highly sophisticated hochkomplex
highly-blocking hochsperrend
highly-competitive umkämpft * *Markt usw.*

highly-dynamic hochdynamisch
highly-dynamic drive controls schnelle Antriebsregelungen
highly-flexible hochflexibel
highly-polished chromium-plated hochglanzverchromt
highly-proven bewährt * *sehr bewährt*
highly-trained qualifiziert * *gut ausgebildet/geschult*
hinder behindern * *aufhalten, hindern, verhindern (from: zu tun), abhalten (from:von), im Wege sein*
hinge abhängen * *hinge (up)on: letztlich abhängen von* || anlenken * *mit Scharnier/Gelenk/Angel versehen* || Gelenk * *Scharnier* || Gelenk- || Scharnier * *Scharnier, Gelenk, Angel*
hinge joint Scharnier * *Scharnier, Gelenk*
hinged angelenkt * *mit Scharnier/Gelenk/Angel versehen, (zusammen)klappbar* || ausklappbar * *angelenkt, mit Scharnier versehen* || drehbar * *angelenkt, mit Scharnier versehen* || Gelenk- || Klapp- * *angelenkt, mit Scharnier versehen* || klappbar * *mit Scharnier/Gelenk/Angel versehen, (zusammen)klappbar* || Schwenk- * *angelenkt, mit Scharnier versehen* || schwenkbar * *angelenkt, z.B. über Scharnier*
hinged door Schwenktür
hinged joint Gelenk || Scharnier * *Scharnier, Gelenk*
hinged rack Klapprahmen * *Baugruppenträger für Flachbaugruppen*
hint Hinweis * *Rat, Wink, Tipp, Fingerzeig* || Lehre * *Wink* || Tipp
hint at andeuten * *einen Hinweis geben auf* || hinweisen * *anspielen auf* || hinweisen auf * *anspielen/hinweisen auf*
hire anstellen * *Personal* || leihen * *mieten, (Personal) anheuern; hire a thing from a person: sich etw. von jdm. aus~* || mieten * *siehe auch →leihen* || vermieten * *on hire: zu ~*
hire out verleihen * *gegen Miete* || vermieten
hire service Verleih
hire-purchase Mietkauf
hires hochauflösend * *(salopp) Abkürzg.f. 'high-resolution'*
historical fault chronology Fehlerhistorie || Fehlervorgeschichte
historical trending Trendanalyse
history Vorgeschichte * *z.B. eines Fehlers, eines Messwertspeichers vor dem Triggerzeitpunkt*
history of technology Technikgeschichte
hit betätigen * *Taste* || drücken * *eine Taste* || Schlag || schlagen || treffen * *schlagen, stoßen (auf), passen, finden, erreichen* || Treffer
hit hard treffen * *empfindlich ~*
hit it right Volltreffer haben
hit the key Taste betätigen
hitch Ruck * *Ruck, Stockung, ruckweise Bewegung* || Störung * *Stockung*
hitherto bisher || bislang
hitherto existing bisherig
hitting of a key Knopfdruck * *just hitting a key: auf Tastendruck/lediglich durch Betätigen einer Taste*
HLL Hochsprache * *Abkürzung für 'high level language' (Programmiersprache)* || höhere Programmiersprache * *Abkürzung für 'High-Level Language'*
HMI Mensch-Maschine-Schnittstelle * *Abk. für "Human Machine Interface"*
hoard sammeln * *horten, sammeln. anhäufen, hamstern; hoard up: aufhäufen/an~*
hob verzahnen * *Zahnrad ~ mit Werkzeugmaschine* || wälzfräsen || Wälzfräser
hobby Hobby-
hogged wood Hackschnitzel * *{Holz}*

hogger Hackmaschine * *{Holz} in der Spanplattenherstellung*
hogging knife Hackmesser
hoist Aufzug * *Lastenaufzug* || heben * *durch Winde/Aufzug/Kran* || Hebezeug * *electric hoist: elektrisch angetriebenes ~* || Hubwerk * *→Hebezeug, Winde, Kran, Flaschenzug, Lastenaufzug* || Kran * *auch Hebezeug, Flaschenzug, Winde, Lastenaufzug* || Lastenaufzug || Winde * *Kranwinde, Hebewinde*
hoist drive Hubwerk * *~santrieb z.B. eines Krans/Förderzeugs*
hoisting capacity Hubleistung * *eines Hebezeugs, einer Vertikalförderanlage usw.*
hoisting device Hebewerk || Hebezeug
hoisting drive Hubwerksantrieb * *bei Kran*
hoisting equipment Hebezeug || Hebezeuge
hoisting gear Hebezeug || Hubwerk * *Hubeinrichtung z.B. eines Krans* || Kranhubwerk * *Hubwerk*
hoisting height Hubhöhe * *{Hebezeuge}*
hoisting lug Hebeöse * *in Form einer Lasche*
hoisting motor Aufzugsmotor
hoisting rope Aufzugseil || Aufzugsseil
hoisting speed Hubgeschwindigkeit * *{Hebezeuge}*
hoisting winch Aufzugwinde
hoistway Aufzugschacht
hold anhalten * *(an)halten (z.B. Durchmesserrechner, SPS-Programm usw.)* || bleiben * *hold the line, please:bleiben Sie bitte am Telefon* || dranbleiben * *am Telefon* || durchführen * *Kurs/Seminar usw. ~* || enthalten * *fassen (Gefäß), enthalten, Platz haben für, in sich schließen* || fassen * *aufnehmen können, Fassungsvermögen von ... haben* || festhalten || gelten * *gelten; (sich) halten, festhalten, standhalten; hold to: festhalten an* || Griff * *Halt, Griff, Stütze; Gewalt, Macht (on/over/of: über); catch hold of: zu fassen kriegen* || halten * *fest-/auf-/zurück-/an-/stand-/ent-~, (Versammlung) ab~* || Standpunkt vertreten * *that:dass* || tragen * *stützen* || versehen * *Amt ~* || verwalten * *(amerikan.) ein Amt ~/innehaben*
hold a meeting Besprechung abhalten || einberufen * *eine Besprechung abhalten*
hold back zurückhalten
hold in memory merken * *im Speicher aufbewahren*
HOLD mode HALT * *Betriebsart bei SPS*
hold out halten * *stand~* || reichen * *andauern, sich halten, aus-/durchhalten, standhalten, genügen*
hold still verharren * *(in einer bestimmten Stellung) stillhalten*
hold that Ansicht vertreten * *meinen, der Ansicht sein*
hold the line dranbleiben * *am Telefon*
hold the market Markt beherrschen
hold tight festhalten * *packen*
hold time Haltezeit
hold up stützen * *halten, stützen, aufrechterhalten, hochheben* || unterbrechen * *aufhalten, hindern*
hold-off interval Freihaltezeit || Schonzeit * *Sicherheitszeit, die e. Stromrichter b. Umkommutieren gelassen wird*
hold-off time Schonzeit * *Sicherheitszeit, die e. Stromrichter b. Umkommutieren zum Ausgehen gelassen wird*
hold-up Rückstau * *im Verkehr*
holder Besitzer * *Inhaber, auch von Aktien, Kreditkarten, Rechten usw.* || Geschäftsinhaber || Halter * *auch Haltevorrichtung || Halterung * *Halter, Halterung, Fassung (für Lampe, Sicherung, Werkzeug usw.)* || Inhaber * *einer Firma, Lizenz, Genehmigung usw.* || Kassette * *Halter* || Träger * *Halter, Haltevorrichtung* || Unterteil * *Halter; z.B. fuse holder: Sicherungsunterteil*

holding Dachgesellschaft || Konstanz * *Lastausregelung*
holding brake Haltebremse * *fail-safe holding brake: Ruhetrom-~* || Stillstandsbremse * *Haltebremse*
holding company Dachgesellschaft || Muttergesellschaft * *Dachgesellschaft*
holding current Haltestrom * *benötigt ein Thyristor, um den Strom im Ein-Zustand aufrechtzuerhalten*
holding device Halterung || Spannvorrichtung
holding down Anpress- * *zum Niederhalten*
holding torque Haltemoment
holding-down bolt Fußschraube * *zum Festschrauben eines Maschinen-/Motorfußes* || Maschinenfußschraube * *z.b. zur Motorbefestigung im Maschinenfundament*
holding-down clamp Niederhalter
holding-down element Niederhalter
holdings Beteiligung * *(Plural) durch Aktienbesitz*
hole Bohrung || Loch || Lochung || Öffnung * *Loch*
hole clearance Lochabstand
hole diameter Bohrungsdurchmesser
hole in foot Fußloch * *z.B. in Motorbefestigungsfuß*
hole plug Blindstopfen * *z.B. für Kabeleinführungsloch*
hole punch Lochstanze
holed gelocht
holiday Feiertag || Urlaub * *take a holiday: ~ nehmen; be on holiday: in ~ sein*
hollow Grube * *Höhlung* || hohl || Hohlraum || Mulde * *Hohlraum* || Vertiefung * *Höhlung*
hollow cylinder Hohlzylinder
hollow glass Hohlglas
hollow groove Hohlkehle || Kehle * *→Hohlkehle*
hollow section Hohlprofil
hollow shaft Hohlwelle
hollow shaft bore Sackloch * *in einer Welle*
hollow space Hohlraum
hollow-glass machine Hohlglasmaschine
hollow-shaft design Hohlwellenausführung
hollow-shaft tacho Hohlwellentacho
home position Nulllage * *Nullpunkt der Weg-Koordinate bei Positionierregelung* || Referenzpunkt * *bei Lageregelung* || Ruhestellung
homegrown heimisch * *Pflanzen usw.*
homing Referenzfahrt * *Wegnullpunkt anfahren bei Positioniersteuerung (mit Inkrementalgeber)* || Referenzieren * *→Referenzfahrt, Anfahren des Wegnullpunktes bei einer Positioniersteuerung*
homing logic Referenzierlogik * *Steuerung d. →Referenzfahrt: Anfahren d. Nullpunktes bei e. Positioniersteuerung*
homing procedure Referenzfahrt || Referenzpunktfahrt
homogeneity Homogenität
homogeneous gleichartig || homogen * *[homo'd-schi:njes]* || zusammengehörig * *gleichartig, homogen*
homogeneousness Homogenität
homogenize homogenisieren
homokinetic homokinetisch
homopolar gleichpolig
homopolar component Nullkomponente * *Komponente mit 3 gleichphasigen Größen in e. unsymmetr. Drehstromsystem*
hone honen || honen
honeycomb Wabe * *honeycomb radiator: Wabenkühler* || Waben- || wabenförmig
honing machine Honmaschine * *Werkzeugmaschine*
honorarium Honorar
hood Dach || Haube

hooded louvre plate Kiemenblech * *zur Abdeckung einer Belüftungöffnung*
hook einhaken * *an-/ein-/fest-/zuhaken* || Greifer * *Haken* || Haken || krümmen * *krümmen, biegen*
hook in einhaken || einhängen * *into: in*
hook into einhaken
hook load Last * *am Kran(haken)*
hook pin Haken * *Hakenbolzen*
hook position Hakenstellung * *(Hebezeuge)*
hook up andocken || ankoppeln * *anschließen, "andocken", anhaken* || anschließen * *z.B. Signal/Leitung an Empfänger/Klemme/Stecker ~/andocken/anstecken/auflegen* || einhaken * *anhaken, (mit Haken) aufhängen, anschließen* || kombinieren * *ein Teil an ein anderes "andocken"*
hook wrench Hakenschlüssel
hook-spanner Hakenschlüssel
hooked krumm * *hakenförmig*
hooked spanner Hakenschlüssel
hooked up aufgelegt * *z.B. ein Signal an eine Klemme*
hooked wrench Hakenschlüssel
hooking-on arm Greifer * *~ eines Krans*
hoop Reifen * *z.b. Fassreifen*
hop Hopfen
hopper Bunker * *Schüttgutbehälter, Vorratsbehälter, Fahrzeug mit Schnellentlade-Bodentrichter* || Einfülltrichter || Trichter * *Einfüll~*
horizontal eben * *horizontal* || horizontal || liegend || Plan * *(Adj.) horizontal, waagerecht* || quer * *waagerecht* || söhlig * *(Bergbau)* || waagerecht
horizontal frequency Horizontalfrequenz * *(Computer) Anzahl Zeilen/sec b. Monitor (Anz.Zeilen x Bildwiederholrate x 1,1)*
horizontal plane Waagerechte || Waagerechte * *Horizontalebene*
horizontal position Waagerechte
horizontal projection Grundriss
horizontal shaft Längswelle
horn Hupe
horse power Pferdestärke
horse power output Ausgangsleistung * *in Pferdestärken*
horsepower Leistung * *Pferdestärken, amerikanisch; 1 kW entspr. 1,341 hp*
hose Schlauch * *zum Spritzen usw.* || spritzen * *mit Schlauch ~*
hose connection Schlauchanschluss * *Verbindung*
hose nipple Anschlussstück * *für Schlauch*
hose pipe Schlauchleitung * *Rohrleitung*
hose-water Strahlwasser
host Leitrechner || Leitsystem || Zentralrechner
host computer Leitrechner || Leitsystem || Leitrechner || Zentralrechner
host controller Leitregler
host system Leitsystem || Prozessleitsystem * *überlagertes Rechnersystem*
HOSTALEN Polypropylen * *Markenname (als Isolierfolie usw.)*
hostile environments raue Umgebungsbedingungen
hot fliegend * *unter Spannung; hot plugging: Ziehen/Stecken unt.Spannung, hot fixing: Reparatur u.Spanng.* || heiß * *(auch figürl.: hot topic: ~es/~ diskutiertes Thema)* || unter Spannung * *hot unplugging: Ziehen ~* || während des Betriebes * *→unter Spannung (Stecken von Baugruppen usw.)*
hot air Abluft * *ausströmende erwärmte (Kühl-) Luft*
hot galvanized feuerverzinkt
hot platen Heizplatte
hot plugging Stecken unter Spannung
hot press Heizpresse * *(Holz/ Gummi)*
hot rolling Warmwalzen

hot rolling mill Warmwalzwerk
hot spot Heißpunkt * *heiß(est)er Punkt i.Motorwicklung usw., auch Feld in Grafikelement z.Anklicken mit d.Maus* || Wärmenest
hot spot temperature Temperatur an der heißesten Stelle
hot swapping fliegender Wechsel * *z.b. fliegender Tausch von Baugruppen während des Betriebes/unter Spannung*
hot unplugging Ziehen unter Spannung
hot wide-strip Warmbreitband * *Walzwerk*
hot-air blower Heißluftgebläse
hot-dip galvanized feuerverzinkt
hot-dip galvanizing Feuerbeschichtung * *Galvanisierung, z.B. Verzinkung* || Feuerverzinkung
hot-galvanize feuerverzinken
hot-melt adhesive Schmelzklebstoff
hot-melt glue Schmelzkleber
hot-standby Reserve * *zum sofortigen Einschalten/Übernehmen bereit, wenn das Hauptsystem ausfällt*
hot-strip mill Warmbandstraße * *Walzwerk*
hot-swap disk array Spiegelplattensystem * *~ mit unter Spannung wechselbaren Festplatten*
hotel reservation Hotelreservierung
hottest spot heißester Punkt
hotwater boiler Heißwasserkessel
hour counter Betriebsstundenzähler
hour of decision Stunde der Wahrheit * *Stunde der Entscheidung*
hourglass Sanduhr * *auch auf dem Bildschirm*
hourly rate Stundensatz || Verrechnungssatz * *Stundensatz z.B. für Personaleinsatz*
hours run Betriebsstunden
hours-meter Betriebsstundenzähler
hours-run meter Betriebsstundenzähler
house einbauen * *befestigen, verzapfen/schrauben* || einschließen * *in Gehäuse* ~ || Gebäude || unterbringen * *beherbergen*
house organ Kundenzeitschrift
housebreaking Einbruch * *~sdiebstahl*
housed untergebracht
household appliance Haushaltsgerät
household appliances Haushaltsgeräte
household application Haushaltsanwendung * *Anwendung für den Privathaushalt*
household device Haushaltsgerät
household equipment Haushaltsgeräte
housekeeping Wirtschaft * *Haushaltung*
housing Gehäuse
housing construction Gehäuseausführung
housing design Gehäuseausführung * *konstruktive Ausführung des Gehäuses*
housing dimensions Gehäuseabmessungen
housing estate Wohngebiet * *Wohnsiedlung*
housing feet Gehäusefüße * *z.B. an Motor; cast: angegossen*
housing roof Gehäusedach
housing size Gehäusegröße
housing type Gehäuseform
housing version Gehäuseausführung
housing vibrations Gehäuseschwingungen * *{el. Masch.}*
housing width Gehäusebreite
hover schweben * *auf Luftkissen, über einer Stelle*
How are you Guten Tag
how far inwieweit
how many times wie oft * *wie viel mal* || wievielmal
how much is it teuer * *wie teuer ist es ?*
how often wie oft * *wie häufig* || wievielmal
however allein * *jedoch* || dagegen * *andererseits, jedoch* || jedoch
hp kW * *horsepowers; 1 kW entspr. 1,3410 hp entspr. 1,3596 PS (Pferdestärken)* || Leistung * *Pferdestärken, amerikanisch* || Pferdestärke || PS * *horsepower; 1 PS entspr. 0,73550 kW entspr. 0,9863 hp*
HPL board Schichtstoffplatte * *{Holz} Abk. für 'High-Pressure Laminated board': Dekorative ~ (DKS-Platte)*
HRV Schienenfahrzeug * *Abk. für 'Heavy Rail Vehicle': schweres/Langstrecken-~*
HSS HSS * *Abk. für 'High-Speed Steel': Hochleistungs-Schnell(schnitt)-Stahl*
HTL Hochpegellogik * *Abk. f. 'High-Threshold Logic': z.B. mit 24V-Signalen arbeitende Logikschaltungen* || HTL * *Abkürzg. für 'High-level Transistor Logic': hochpegelige Transistorlogik (mit 24V-Pegel)*
HTML HTML * *'HyperText Markup Language': Seitenbeschreib.-Sprache mit Hyperlinks, Grafik, Sound für* →WWW
HTTP HTTP * *'HyperText Transfer Protocol': Protokoll zur Übertragung von* →HTML-*Seiten im* →WWW *des Internet*
hub Nabe || Radnabe || Sternverteiler * *für Rechner-Netzwerk ('hub': Nabe, von der die Speichen ausgehen)*
hub housing Nabengehäuse
hub keyway Nabennut
hue Farbe * *Farbton*
hug Umschlingung
huge enorm * *sehr groß, riesig, ungeheuer, gewaltig* || groß * *riesig* || riesig || übergroß * →*riesig*
hull Hülse * *Hülse (besonders von Feldfrüchten), Schale, Rumpf*
hum Brumm * *Geräusch, elektr. Brummen* || Brummen * *Geräusch, elektr.Brummen* || summen * *brummen*
hum voltage Brummspannung
hum-free brummfrei
human Mensch-
human communication interface Bedienungsoberfläche
human interface Anwenderoberfläche || Bedienungsoberfläche || Oberfläche * *Mensch/Maschinen-Schnittstelle, Bedien(er)oberfläche*
human machine interface Mensch-Maschine-Schnittstelle
human resources Arbeitskraft * *menschliche* || Personalwesen * *auch als Abteilungsbezeichnung*
humbug vormachen * *(salopp) jemandem etwas vormachen*
humid feucht * *(physikal.) Luft* || nass
humidification Befeuchtung
humidifying Befeuchtung
humidity Feuchte * *absolute humidity: absolute ~ [g/m³], relative humidity: relative Luftfeuchte [%]* || Feuchtigkeit || Luftfeuchte * *relative: relative; absolute: absolute* || Luftfeuchtigkeit * *relative:relative*
humidity class Feuchteklasse
humidity classification Feuchteklasse * *siehe DIN 40040*
humidity measuring equipment Feuchtigkeitsmessgerät
humidity rating Feuchtebeanspruchung * *zugelassene, spezifizierte* || Feuchteklasse
humidity sensor Feuchtesensor
hump Wulst * *Buckel*
hundred fold hundertfach
hundred times hundertfach
hundreds digit Hundertersteile
hung aufgehängt * *z.B. Programm*
Hungaria Ungarn
hunt pumpen * *instabiler Regler* || Suche * *dringende ~ (for: nach)*
hunt for suchen * *eilig, hastig ~*
hunting Drehzahlschwankungen * *periodische*

Schwankungen der Drehzahl (Kenngröße: Ungleichförmigkeitsgrad) || Instabilität * Reglerschwingen/pendeln || pendeln * (Substantiv) Schwingen/Pendeln einer instabilen Regelung || Pendelung * auch Reglerschwingen bei Instabilität || Regelschwingungen * Pendeln auf Grund von Instabilität || Reglerschwingen * siehe auch 'Instabilität'
hurl schleudern * schleudern, werfen; hurl oneself on: sich stürzen auf
hurry antreiben * zur Eile || drängen * zur Eile ~
hurt schädigen * verletzen || verletzen * verwunden (auch figürlich: Gefühle usw. ~)
husband schonen * Kräfte, Vorrat usw.~
husk Hülse * Schale, (wertlose) Hülle (auch figürl.); Hülse, Schote (besonders von Feldfrüchten)
HV Hochspannung * Abkürz. für 'High Voltage' || Hochspannungs- * Abkürzung für 'High Voltage'
HV circuit breaker Hochspannungsschalter
HV distribution Hochspannungsverteilung
HV motor Hochspannungsmotor * ca. 1 bis 12 kV und 200 bis 16000 kW
HV switch Hochspannungsschalter
HVAC Heizung-Klima-Lüftung * Abkürzg.f. 'Heating, Ventilation and Air Conditioning'
HVC Spannungs-Oberschwingungsfaktor * Abkürzg. für 'Harmonic Voltage Factor'
HVDC transmission HGÜ * Abkürzg. für "High Voltage Direct Current"
hybrid hybrid
hybrid circuit Hybridschaltung
hybrid device Hybridbaustein * z.b. in Dick- oder Dünnschichttechnik auf Keramikträger
hybrid module Hybridbaustein * z.B. in Dick- oder Dünnschichttechnik auf Keramikträger || Hybridmodul
hybrid technology Hybridtechnik * z.B. Schaltung in Dickschichttechnik auf Keramikträger
hydraulic hydraulisch
hydraulic coupling Strömungskupplung
hydraulic cylinder Hydraulikzylinder
hydraulic engineering Hydrotechnik || Wasserbau
hydraulic fluid Hydraulikflüssigkeit || Hydrauliköl
hydraulic motor Hydraulikmotor || Hydromotor * axial piston type: in Axialkolbenbauweise
hydraulic press Hydraulikpresse
hydraulic pump Hydraulikpumpe * zur Druckerzeugung in Hydrauliksystem || Ölpumpe * in Hydrauliksystem
hydraulically hydraulisch * (Adv.)
hydraulically actuated hydraulisch betätigt
hydraulically operated hydraulisch betätigt
hydro-electric power plant Wasserkraftwerk
hydrochloric acid Salzsäure
hydrodynamic hydrodynamisch
hydrodynamic brake Strömungsbremse
hydroelectric generator Wasserkraftgenerator
hydroelectric power station Wasserkraftwerk
hydrogen Wasserstoff
hydrogen sulphide Schwefelwasserstoff * ['haidridjin 'salfaid] H2S
hydrostatic hydrostatisch
hygroscopic wasseranziehend
hyperbolic hyperbolisch
hyphen Bindestrich * auch 'mit Bindestrich versehen'; hyphenated: mit Bindestrich versehen
hypoid gear Hypoidgetriebe
hypothetical hypothetisch || ideell * hypothetisch, angenommen, mutmaßlich
hysteresis Hysterese
hysteresis brake Hysteresebremse
hysteresis clutch Hysteresekupplung
Hz Hertz || Hz

I

I component I-Anteil
I controller I-Regler || Integralregler
I guess ich glaube * (amerikan.)
I hope you like it guten Appetit
I myself selbst * ich ~
I suppose vermutlich * (Adv.) ich vermute
I think ich glaube
I wonder if he will come ob er wohl kommt
I'm afraid leider * (in Ich-Satz, vorangestellt)
I-element Integrierglied
I-square-t monitoring I-quadrat-t Überwachung
I-square-t-value I-quadrat-t-Wert
i.e. also * d.h. || dh * das heißt || nämlich * Abkürzung für 'id est':das heißt
I/O Peripherie * Ein-/Ausgabefunktionen
I/O area Peripheriebereich * bei SPS
I/O board Ein-Ausgabe-Baugruppe * Leiterkarte, Flachbaugruppe || Ein-Ausgabebaugruppe * Leiterkarte/Flachbaugruppe
I/O byte Peripheriebyte * bei SPS
I/O conditioner module Prozesssignalformer
I/O memory Peripheriespeicher
I/O module Ein-Ausgabe-Baugruppe || Ein-Ausgabebaugruppe || Ein-Ausgabemodul || Prozesssignalformer || ssignalformer * Ein/Ausgangsmodul
I/O modules Peripheriebaugruppen
I/O- Peripherie- * {SPS}
I/O-module Peripheriebaugruppe * Ein/Ausgabebaugruppe bei Rechner/SPS
I/U-system Ein-Ausgabe-System
I/Q/F reference list Belegungsplan * Abk. f.'Input/Output/Flag': Beleg.plan für Ein-/Ausgänge und Merker bei SPS
I2t-value I2t-Wert * "I-Quadrat t Wert",Grenzlastintegral, (thermische) Belastungsgrenze
IBM-compatible PC IBM-kompatibler PC
IC Baustein * Integrierter Baustein || IC * Abkürzg.f. 'Integr. Circuit':integr. Schaltg.' u. 'Internat. Cooling':Kühlsyst. nach DIN IEC 34 || integrierte Schaltung * Abkürzung für 'integrated circuit' || integrierter Schaltkreis || Schaltkreis * integrierter
IC socket IC-Sockel
ICE Emulations- und Testadapter * Abk. f. 'In-Circuit Emulator'
ice-breaker Eisbrecher
icing fan brechersicherer Lüfter * {Schifffart} Lüfter, d.e. plötzl.Blockade durch Brecherschlag verkraftet
icon Icon
ID Kennung * Abkürzg. für 'identifier':Kennung
ID bit Kennbit
idea Begriff * Vorstellung || Gedanke || Idee || Konzept || Sinn * Gedanke
ideal ideal * (auch Substantiv) for:zu/für || ideell || mustergültig
ideal AC content ideeller Wechselspannungsgehalt
ideal case Idealfall
ideal conditions Idealfall * under ideal conditions/ideally: im ~
ideal no-load DC voltage ideelle Gleichspannung * an DC-Stromrichterausgang ohne Ohmsche Spann.abfälle u.Schwachlast || ideelle Leerlaufgleichspannung
ideal no-load direct voltage ideelle Gleichspannung * ~ am DC-Stromrichter-Ausgang ohne Ohmsch. Spann.abfälle o. Schwachlast
ideal value ideeller Wert
idealized idealisiert || ideell * idealisiert
ideally suited bestens geeignet

identical identisch || übereinstimmend * *identisch*
identical copy Duplikat * *Kopie*
identification Erkennung * *Identifizierung* || Identifikation * *auch einer Regelstrecke/Baugruppe* || Identifizierung || Kennung || Kennzeichnung * ~ *zur Unterscheidung/Vertauschungssicherheit*
identification code Identnummer
identification letter Kennbuchstabe
identification number Kennziffer
identification symbol Kennzeichnung
identification system Identifikationssystem || Kennzeichnungssystem
identified gekennzeichnet
identifier Bezeichnung * *Kennzeichen, Kennung* || Identifizierung * *Bezeichner* || Kennung || Kennzeichen || Kennzeichnung * *Kennzeichen* || Kennziffer || Name * *Bezeichner*
identifier bit Kennbit
identify erkennen * *identifizieren (by:an/durch)* || ermitteln * *indentifizieren* || feststellen * *identifizieren* || identifizieren || kennzeichnen * *mit e. Erkennungsmerkmal versehen*
identity Identität
idiom Redewendung * *(Sprachbesonderheit)*
idle faul * *träge* || im Leerlauf fahren || untätig
idle current Blindstrom
idle loop Dauerschleife * *Untätigkeitsschleife* || Endlosschleife * *Untätigkeitsschleife* || Untätigkeitsschleife || Warteschleife * *z.B.in Programmm*
idle position Ruhestellung
idle run Leerlauf
idle running Leerlauf
idle wheel Führungsrolle
idleness Stillstand * ~ *einer Maschine*
idler Führungsrolle * *z.b. in Fördersystem*
idler pulley Spannrolle * *für Riementrieb usw.*
idling Leerlauf || leerlaufend || nicht angetrieben * *leerlaufend, z.B. Führungsrolle, Leitwalze* || unbelastet
idling torque Leerlaufdrehmoment * *z.B. einer Kupplung oder Bremse* || Leerlaufmoment * *z.B. einer Kupplung, Bremse*
IEC IEC * *Abkürzg. für 'International Electrotechnical Commission'; internationales Normengremium*
IEEE IEEE * *Abkürzg.f. 'Institute of Electrical and Electronic Engineers': amerikan. Ingenieursverband*
if ob * *as if:als ob* || wenn
if any falls vorhanden
if applicable gegebenenfalls * *falls zutreffend/anwendbar*
if at all wenn überhaupt
if at all possible so weit möglich * *wenn irgend möglich*
if available falls vorhanden * *falls verfügbar*
if existing eventuell * *falls vorhanden*
if necessary eventuell * *(Adv.) falls nötig* || falls erforderlich || gegebenenfalls * *falls notwendig/nötig* || notfalls * *(Adv.)* || unter Umständen * *falls erforderlich*
if need arises notfalls * *(Adv.)*
if need be eventuell * *(Adv.) falls nötig, notfalls* || falls erforderlich || gegebenenfalls * *falls nötig* || notfalls * *(Adv.)* || unter Umständen * *notfalls, nötigenfalls*
if need may be notfalls * *(Adv.)*
if not wenn nicht
if possible möglicherweise || womöglich * *wenn möglich*
if required bei Bedarf || falls erforderlich || gegebenenfalls * *falls erforderlich* || wenn nötig
IGBT IGBT * *Insulated Gate Bipolar Transistor, schneller Leistungstransistor (Schaltzeit 50...100ns)*

ignitability Entflammbarkeit * *Entzündbarkeit (Explosion verursachend)*
ignitable entzündbar * *eine Explosion verursachend*
ignite zünden * *z.b. Gase, Motor*
igniting current Zündstrom
ignition Zündung * *z.b. von Gasen/des Benzin-/Luftgemisches im Motor usw.*
ignition angle Zündwinkel * *bei Benzinmotor*
ignition breakdown Zünddurchschlag
ignition breakdown capacity Zünddurchschlagsfähigkeit * *Fähigkt. brennbar. Gase, durch engen Spalt hindurchzuzünden;* →*Ex.gruppe*
ignition characteristic Zündkennlinie
ignition group Zündgruppe * *Explosionsgruppe nach alter Norm VDE0170/0171*
ignition source Zündquelle
ignition temperature Zündtemperatur * *z.b. von Gasen, Stäuben*
ignore ignorieren || nicht beachten * *ignorieren* || nicht befolgen || übergehen * *ignorieren,* →*nicht beachten*
Ilgner flywheel equalizing set Ilgner-Umformer
Ilgner system Ilgner-Umformer
ill schlecht * *(Adj. und Adv.) unbefriedigend, fehlerhaft, schlecht, böse, übel, ungünstig,*
illegal rechtswidrig || ungültig * *entgegen der Vorschrift, ungesetzlich*
illegal operation nicht interpretierbarer Befehl
illegality Rechtswidrigkeit
illicit rechtswidrig
illuminate illustrieren || leuchten * *sich erhellen*
illuminated pushbutton Leuchtdrucktaster || Leuchttaster
illumination Lichttechnik
illustrate belegen * *durch Beispiele* ~ || darstellen * *illustrieren* || erklären * *veranschaulichen* || erläutern * *illustrieren anhand eines Beispiels/eines Bildes* || grafisch darstellen || illustrieren || veranschaulichen || verdeutlichen * *durch Beispiele, Bilder usw.* || zeigen * *durch Bild, Beispiel usw.* ~ || zeigen * *bildhaft* ~, *an einem Beispiel* ~; ... *is illustrated in Fig. 3: ... zeigt Bild 3*
illustrated dargestellt
illustrated folder Produktinformation * *Faltblatt mit Bildern* || Produktschrift * *bebildertes Faltblatt* || Prospekt * *Faltblatt mit Bildern*
illustration Abbildung || Bild * *Illustration, Abbildung* || Erklärung * *Veranschaulichung* || grafische Darstellung
IM IM * *{el.Masch.} 'International Mounting': Bauform und Aufstellung (VDE 0530 Teil7)*
im- un- * *nicht*
image Abbild || Abbildung * *Abbild* || Bild * *Abbild, optisches Bild* || Image || Modell * *Abbild, Verkörperung*
image distortion Verzeichnung * *{Optik|Computer} z.B. auf den Bildschirm, bei der Bildverarbeitung usw.*
image quality Bildqualität * *eines Computer-Monitors*
image refresh rate Bildwiederholfrequenz
image setter Belichter * *{grafisches Gewerbe}*
imaginary component Imaginärteil
imaginary part Imaginärteil
imagine ausdenken * *vorstellen* || sich vorstellen * *geistig* || vorstellen * *sich vorstellen, sich ein Bild machen*
imaging Bilderkennung || bildgebend
imbalance Unsymmetrie
imbed in einlassen
imitate nachbilden * *nachmachen/-ahmen, imitieren, kopieren*
imitated künstlich * *nachgemacht, imitiert*
imitation Modell * *Nachbildung* || Nachbau * Nach-

bildung * *Nachbildung, Nachahmung, Imitation, das Nachgeahmte, Kopie*
immaterial geistig * *unkörperlich*
immediate augenblicklich || kurzfristig * *sofortig* || schnell * *unverzüglich, unmittelbar* || sofort * *(Adj.)* ~*ig* || sofortig || unmittelbar || unverzögert * *unverzüglich*
immediate address direkte Adresse * *im Befehlscode steht direkt e.(konstanter)Wert, nicht d.Adresse einer Konstanten*
immediate attention schnelle Erledigung
immediately ab sofort || sofort * *(Adv.) effective immediately: mit ~iger Wirkung/ab ~ gültig*
immediately beforehand unmittelbar vorher
immense außerordentlich * *immens, ungeheuer* || groß * *immens* || riesig
immerse eintauchen * *in Flüssigkeit* || untertauchen * *(ein)tauchen, versenken, (sich) vertiefen (figürl.), verwickeln (figürl.; in: in)*
immersion eintauchen * *(Substantiv)*
immersion pump Tauchpumpe
immoderate übermäßig
immovable property unbewegliches Vermögen * *in Gebäuden, Grundstücken usw. steckendes Vermögen*
immovables unbewegliches Vermögen * *in Gebäuden, Grundstücken usw. steckendes Vermögen*
immune beständig * *geschützt, unempfänglich, gefeit (from/against:gegen)* || sicher * *gefeit; from: gegen*
immunity Beständigkeit || Festigkeit * *Immunität, Geschützt-/Sicher-/Unempfänglich-Sein, Unempfindlichkeit (against:gegen)*
immunity level Störfestigkeit * *Grad der ~*
immunity to radiated noise Störstrahlfestigkeit || Störstrahlungsfestigkeit
immunity to vibration Schwingfestigkeit
impact Anschlag * *Anprall* || Aufschlag * *Aufprall (on:auf)* || Auswirkung * *Beeinflussung* || Einfluss * *Einwirkung* || Gewalt * *Wucht* || Schlag * *Aufprall, Auf/Einschlag, Schlagbelastung, heftige Einwirkung, Schlagfestigkeit* || Stoß * *Aufprall, Schlag, Stoß, Schlagfestigkeit* || Wirkung * *Aus~, (heftige) Ein~, Auftreffen*
impact drill Schlagbohrer || Schlagbohrmaschine
impact load Stoßbeanspruchung * *Schlag/Stoßbeanspruchung*
impact on the environment Umweltbeeinflussung
impact resistance Stoßfestigkeit * *(mechan.)*
impact stressing Stoßbeanspruchung * *mechanische Schlagbeanspruchung*
impact-proof schlagfest
impact-resistance Schlagfestigkeit
impact-resistant schlagfest
impair beeinträchtigen || schädigen || schwächen || verschlechtern
impairment Schädigung * *of: von etwas*
impart verleihen * *geben, gewähren, verleihen, erteilen, mitteilen (auch physikal.: Kraft, Schwung usw.)* || vermitteln * *(z.B. Wissen) vermitteln, mitteilen*
impartial neutral * *unparteiisch, gerecht, unvoreingenommen, unbefangen* || sachlich * *unparteiisch*
impassable unpassierbar
impeccable perfekt * *untadelig, einwandfrei*
impedance Impedanz || Scheinwiderstand * *Eff.wert der Spannung geteilt durch den Eff.wert des Stroms* || Widerstand * *Schein~ (bestehend aus →Wirk- und →Blindwiderstand)*
impedance converter Impedanzwandler
impedance drop Kurzschlussspannung * *z.B. bei Drossel/Trafo (siehe auch 'uk')* || uk * *siehe auch →Kurzschlussspannung (bei Drossel, Trafo)*
impedance transformer Impedanzwandler

impedance voltage Kurzschlussspannung * *z.B.Drossel/Trafo (Eing.spann. bei d. am kurzgeschl. Ausg.d. Nennstrom fließt)* || uk * *siehe auch →Kurzschlussspannung (bei Drossel, Trafo)*
impede behindern * *(be-) hindern, verhindern, erschweren* || erschweren * *(be-) hindern, erschweren, verhindern*
impel antreiben * *(figürlich) treiben, zwingen, nötigen*
impeller Flügelrad * *z.B. eines Lüfters* || Laufrad * *Gebläse/Lüfter/Pumpenrad, Flügelrad* || Pumpenlaufrad * *Pumpenrad * Laufrad* || Ventilatorrad || Windrad * *auch Lüfterrad*
impeller-wheel Pumpenlaufrad
imperative dringend * *~/zwingend notwendig* || notwendig * *unabdinglich* || unabdingbar * *unumgänglich, zwingend, dringend (nötig), unbedingt erforderlich* || unbedingt erforderlich || unentbehrlich * *unumgänglich, zwingend, unbedingt erforderlich* || unerlässlich * *unumgänglich, zwingend (erforderlich)* || unumgänglich * *unbedingt erforderlich* || Voraussetzung * *zwingende ~* || zwingend erforderlich || zwingend erforderlich
imperceptible unhörbar * *nicht wahrnembar* || unmerklich || unsichtbar * *nicht wahrnehmbar*
imperfect defekt * *mangel/fehlerhaft* || mangelhaft * *unvollkommen* || ungenügend * *unvollkommen* || unvollkommen
imperfection Fehler * *Unvollkommenheit, Material~* || Unvollkommenheit
imperial system Zollsystem * *Maßeinheitensystem*
imperil gefährden
impermeable abgedichtet * *undurchlässig, undurchdringlich; impermeable to water: wasserdicht* || undurchlässig
impermeable to water wasserdicht * *wasserundurchlässig, für Wasser undurchdringlich*
impervious unempfindlich * *unempfindlich (to: gegen), unzugänglich (to: für), undurchlässig (to: für), taub*
impetuous heftig * *stürmisch, mit Macht*
impetus Antrieb * *innerer ~* || Dynamik * *(Auf-) Schwung (z.B. der Wirtschaft, einer Entwicklung(stendenz) usw.)* || Impuls * *Stoß/Tiebkraft z.B. für Wirtschaftswachstum* || Moment * *(figürl.) Antrieb, (Auf)Schwung, Triebkraft* || Schub * *(figürl.) Impuls, Schwung; z.B.: impetus to/of development: Entwicklungs~* || Schwung * *(figürl.) Antrieb, Tatkraft*
implement ausführen * *durchführen, vollenden, (Vertrag) erfüllen* || durchführen * *gestalten * realisieren* || implementieren || realisieren * *ausführen, durchführen* || umsetzen * *realisieren, in die Tat ~* || vornehmen * *realisieren * vornehmen * realisieren, z.B. Preisänderungen*
implementability Durchführbarkeit
implementation Ausführung * *~ eines Gesetzes/Befehls, Softwarerealisierung* || Durchführung * *Erfüllung, Ausführung, (Software) Realisierung* || Realisierung * *(Software) Realisierung* || Umsetzung * *Realisierung, z.B. softwaremäßig*
implication Auswirkung * *Konsequenz, Begleit-/Folgeerscheinung; by implication:als (natürliche) Folge* || Folgerung * *stillschweigende ~* || Verbindung * *enge logische ~* || Verflechtung || Zusammenhang * *enge Verbindung, Verflechtung, Begleiterscheinung, Folge*
implicit inbegriffen * *(mit/stillschweigend) inbegriffen, implizit* || unbedingt * *absolut, vorbehaltlos, bedingungslos; z.B. Glaube, Anhänger, Vertrauen*
imply bedeuten * *in sich schließen* || enthalten * *beinhalten* || mit sich bringen
import einführen * *Waren aus dem Ausland ~*
import license Einfuhrgenehmigung

importance Bedeutung * Wichtigkeit; gain in importance: an ~ gewinnen ‖ Relevanz * Wichtigkeit ‖ Wichtigkeit * of prime importance: von größter ~
important Achtung * vor Hinweis ‖ bedeutend ‖ beträchtlich * wesentlich, bedeutend, wichtig ‖ erheblich * wichtig ‖ hoch * wichtig, bedeutsam ‖ maßgebend * wichtig ‖ maßgeblich * wichtig ‖ relevant * wichtig ‖ wichtig * to: für
importation Einführung * Import aus d.Ausland
importer Vertretung * Importeur
importune zusetzen * jdn. →belästigen (with: mit)
impose ausüben * (Kraft, Druck usw.) ~ (upon/on: auf); beanspruchen; (Pflicht, Steuer usw.) auferlegen ‖ beanspruchen * auferlegen, aufbürden ((up)on:e. Ding, e. Person), beanspruchen, z.B. mit Oberwellen ‖ belasten * z.B. mit einer Plicht, Steuer, Oberwellen ~
impracticable unbrauchbar * nicht praktikabel/durchführbar (Plan usw.)
impregnant Imprägniermittel ‖ Tränkmittel
impregnate imprägnieren ‖ tränken * z.B. Wicklungsdraht mit Isolier-Tränkharz
impregnated imprägniert
impregnating agent Imprägniermittel * z.B. für Motorwicklung ‖ Imprägniermittel ‖ Tränkmittel
impregnating compound Imprägniermittel
impregnating material Imprägniermittel ‖ Tränkmittel
impregnating resin Imprägnierharz ‖ Tränkharz * z.B. zur Wicklungsisolierung in Trafo, Motor
impregnating technique Tränktechnik
impregnation Imprägnierung * auch einer Wicklung mit Isolierlack oder Tränkharz ‖ Tränkung * Füllung einer Wicklung mit Isolierharz oder -Lack u. anschließender Aushärtung
impregnation compound Tränkmittel
impregnation insulation material Tränk-Isoliermittel
impress aufprägen * (on:auf) ‖ einprägen * (auch figürl.: jemd. etwas einprägen, siehe auch 'merken') auch Strom, Spannung usw. ‖ prägen * (figürl.) on: ins Gedächtnis usw. (ein)~
impress on someone's memory merken * sich etwas einprägen
impressed eingeprägt * auch Strom/Spannung
impressed current eingeprägter Strom
impressed EMF eingeprägte EMK
impressed voltage eingeprägte Spannung
impressed-current source Konstantstromquelle ‖ Stromquelle * eingeprägten Strom liefernd
impression Vertiefung * Einprägung, Einpressung ‖ Wirkung * Eindruck
impression cylinder Druckzylinder * {Druckerei} ‖ Gegendruckzylinder * {Druckerei} ‖ Presseur * {Druckerei}
impressive eindrucksvoll ‖ einprägsam * eindrucksvoll
impressiveness Wirksamkeit * Wirksamkeit, Nachhaltigkeit, das Eindrucksvolle
imprint aufprägen ‖ einprägen
imprinting unit Eindruckwerk * {Druckerei}
improbable unwahrscheinlich
improper unangemessen * unschicklich ‖ unfachgemäß ‖ unpassend * unschicklich ‖ unsachgemäß * improper usage/handling: ~e Behandlung ‖ unvorschriftsmäßig * unsachgemäß
improper use Missbrauch * unkorrekte An/Verwendung
improve ausbauen * entwickeln, verbessern ‖ ertüchtigen * verbessern ‖ umarbeiten * verbessern ‖ verbessern * besser machen
improved besser * verbessert ‖ verbessert
improvement Fortschritt * Verbesserung ‖ Verbesserung

improvement in quality Qualitätsverbesserung
imprudence Unvorsichtigkeit * Unklugheit
imprudent unvorsichtig * unklug
impulse Anregung ‖ Anstoß * (figürl.) ‖ Antrieb * Impuls ‖ Stoß * elektr. Impuls, (An)Stoß, Antrieb
impulse contact Wischimpuls
impulse current Stoßstrom
impulse loading Stoßbelastung
impulse relay Wischrelais
impulse sound-power level Impuls-Schalldruckpegel
impulse strength Stoßfestigkeit * elektrische ~ ‖ Stoßspannungsfestigkeit ‖ Überspannungsfestigkeit * Festigkeit gegenüber Stoßspannungen
impulse stressing Stoßbeanspruchung * Impulsbelastung
impulse test Stoßspannungsprüfung
impulse voltage Stoßspannung
impulse voltage test Stoßspannungsprüfung
impulse withstand voltage Stoßspannungsfestigkeit
impulse-test stoßen * Prüfen mit Stoßspannung
impurity Verunreinigung * Unreinheit
in auf * in the market: auf dem Markt ‖ cm * inch; 1 cm entspr. 0,39370 l in ‖ mit * mit einer Währung bezahlen, mit Kugelschreiber schreiben
in a body gemeinsam * (Adv.) geschlossen * geschlossen * (Adv.) alle gemeinsam ‖ insgesamt * zusammen, geschlossen ‖ sämtliche * zusammen, wie ein Mann
in a broad sense im weitesten Sinne
in a bus-type configuration busförmig
in a controlled manner kontrolliert
in a cyclic time frame zyklisch * Programmabarbeitung
in a defined fashion definiert * (Adv.) auf ~e Art und Weise
in a flash blitzschnell * (Adv.) wie der Blitz
in a grid pattern gitterförmig * z.B. Halbleiterstruktur
in a loud voice laut * (Adv.)mit lauter Stimme
in a manner of speaking gewissermaßen
in a narrow sense im engeren Sinne
in a rigidly cyclic time frame streng zeitzyklisch
in a way gewissermaßen * in gewissem Maße
in accordance übereinstimmend * (Adv.) with:mit z.B. Norm, Vorschrift
in accordance with entsprechend * z.B. gemäß einer Norm/Vorschrift ‖ gemäß ‖ in Einklang mit ‖ in Übereinstimmung mit ‖ laut * entsprechend, gemäß ‖ nach * entsprechend einer Norm, Vorschrift; gemäß (eines Prinzips usw.) ‖ übereinstimmend mit
in actual fact tatsächlich * (Adv.) in Wahrheit, in Wirklichkeit
in addition darüberhinaus ‖ zusätzlich * (Adv.)
in addition to neben * zusätzlich zu
in addition to it darüber hinaus * (figürl.)
in advance im Voraus ‖ voraus * of: von/gegenüber; in advance: im ~ ‖ voraus- * (nachgestellt) ‖ vorher * (Adv.) im voraus
in agreement with gemäß ‖ in Übereinstimmung mit ‖ konform * be in ageement with: ~ gehen mit
in all circumstances unter allen Umständen
in an easy way einfach * (Adv.) auf einfache Art ‖ leicht * auf einfache Art
in an emergency notfalls * (Adv.)
in angular-locked synchronism winkelsynchron
in any case auf alle Fälle ‖ sowieso ‖ unter allen Umständen
in ascending order aufsteigend * (Adv.)
in black and white schriftlich * schwarz auf weiß
in bold print fettgedruckt
in brief kurz gesagt
in brief outlines in groben Zügen

in broad terms   in groben Zügen
in case of   bei * *im Falle von, bei Eintreten von*
in case of need   nötigenfalls * *(Adv.) siehe auch →notfalls*
in circulating current mode   kreisstrombehaftet || kreisstromführend
in circulation   laufend * *im Umlauf befindlich*
in company with   zusammen * ~ *mit*
in comparison with   im Vergleich zu
in compliance with   entsprechend * *(Adv.) gemäß, entsprechend (e. Vorschrift/Wunsches usw.)* || gemäß * *gemäß, befolgend* || übereinstimmend mit * *gemäß*
in concertina arrangement   mäanderförmig * *wie eine Ziehharmonika*
in conclusion   abschließend * *(Adv.)*
in concurrence   gemeinsam * *(Adv.) act in concurrence/conjointly/in concert with:* ~ *handeln mit*
in conformance   konform * *with: zu/mit; siehe auch →gemäß*
in conformance with   -konform * *(nachgestellt)*
in conformance with market pricing levels   - marktpreiskonform
in conformity   konform * *with: zu* || übereinstimmend * *(Adv.) with:mit*
in conformity with   gemäß || in Übereinstimmung mit || laut * *entsprechend, gemäß* || nach * *entsprechend, gemäß (z.B. einer Vorschrift, Norm)*
in confusion   durcheinander * *(Adj.)*
in conjunction with   in Verbindung mit || in Zusammenhang mit * *siehe auch →in Verbindung mit* || zusammen * ~ *mit*
in connection with   im Anschluss an * *in Zusammenhang mit, mit Bezug auf* || in Verbindung mit
in connexion with   im Anschluss an
In consequence of   gcmäß * *zufolge*
in consequence of this   daher
in consideration of   unter Berücksichtigung von
in contradiction   widersprüchlich * *to: zu*
in contrast to   im Gegensatz zu
in course of time   im Laufe der Zeit
in delta connection   in Dreieck geschaltet
in dependance of   als Funktion von * *in Abhängigkeit von, z.B. Wert auf der X-Achse einer Kennlinie*
in detail   ausführlich || detailliert * *im Einzelnen* || eingehend * *(Adv.) detailliert, im Einzelnen* || im Detail || im Einzelnen || näher * *im Detail; go into detail:* ~ *ausführen; details:* ~*es*
in diametrically opposite sequence   über Kreuz * *z.B. beim Anziehen von Schrauben*
in direct proportion to   direkt proportional zu
in doing so   dabei * *gleichzeitig* || hierbei * *dabei, gleichzeitig* || wobei
in doubt   im Zweifelsfall
in due order   ordnungsgemäß
in due time   termingemäß || termingerecht || zeitgerecht * *rechtzeitig, termingerecht*
in each case   jeweilig * *(Adv.)*
in exchange   gegen * *als Entgelt (for: für)*
in exchange for   ersatzweise * *im (Aus-) Tausch für* || für * *als Ersatz für*
in face of   angesichts * *konfrontiert mit*
in fact   tatsächlich * *(Adv.)* || vielmehr * *in Wirklichkeit* || wirklich * *(Adv.) in Wirklichkeit*
in favor of   zugunsten
in favour of   für * *zugunsten von* || zugunsten
in force   gültig * *in Kraft*
in front   voraus || voraus- || vorn * *at the lower front:* ~*e unten; towards the front: nach* ~
in front of   gegenüber || vor * *räumlich*
in full   ganz * *(Adv.) (voll)ständig, "in Gänze"; write in full: etw. ausschreiben* || voll * *(Adv.)*
in full action   in vollem Betrieb
in future   weiterhin * *zukünftig* || zukünftig * *(Adv.)*

in general   allgemein * *(Adv.)* || im Allgemeinen
in good order   ordentlich
in good time   rechtzeitig
in great number   zahlreich * *(Adv.)*
in hand   bestellt * *vom Lieferer hereingenommener Auftrag*
in harmony with   in Übereinstimmung mit
in his turn   seinerseits * *(Adv.)*
in ideas   gedanklich
in incoming circuit   vorgeschaltet
in inverse proportion   umgekehrt proportional * *to:zu*
in its entirety   in vollem Umfang
in its infancy   in den Kinderschuhen * *be still in its infancy: noch in den* ~ *stecken*
in its outlines   konzeptionell * *in der Anlage*
in keeping with   in Übereinstimmung mit
in large numbers   zahlreich * *(Adv.)*
in law   Rechts- * *(nachgestellt) rechtlich, laut Gesetz*
in layers   schichtweise * *in Schichten (angeordnet)*
in letter   schriftlich * *brieflich*
in like manner   gleichermaßen
in line   fluchtend * *in einer Linie/Achse ausgerichtet*
in line with   entlang * *z.B. entlang einer Rampe* || in Übereinstimmung mit * *(amerikan.) in Übereinstimmung/Einklang mit, einhergehend mit*
in line with the real market conditions   marktgerecht
in many cases   vielfach * *(Adv.)*
in most cases   meistens * *in den meisten Fällen*
in native state   roh * *im ursprünglichen Zustand*
in near future   in nicht zu ferner Zukunft * *in naher Zukunft* || Kürze * *in Kürze*
in need of repair   reparaturbedürftig
in no case   auf keinen Fall || keinesfalls * *nie * keinesfalls*
in no-load operation   leerlaufend
in non circulating current mode   kreisstromfrei
in one go   auf einmal
in one step   auf einmal * *in einem Schritt/Arbeitsgang* || in einem Schritt
in oneself   selbst * *ohne fremde Hilfe*
in operation   betriebsmäßig || in Betrieb
in opposite direction   gegenläufig
in opposition   entgegengesetzt * *in entgegengesetzter Anordnung* || gegenphasig || in entgegengesetzter Richtung
in opposition to   entgegen * *im Gegensatz zu* || im Gegensatz zu
in order   funktionsfähig * *in Ordnung*
in order to   um * *um zu*
in other respects   sonst * *im Übrigen*
in other words   in andern Worten || mit anderen Worten
in our times   heutzutage
in overhung position   fliegend angeordnet
in pairs   paarweise
in parallel   parallel * *(Adv.)*
in parantheses   in Klammern
in part   teilweise * *(Adv.) teilweise, zum Teil* || zum Teil
in particular   im Besonderen || insbesondere
in parts   teilweise * *teilweise, zum Teil, in Teilen, Teillieferungen*
in per cent   prozentual * *(Adv.)*
in per unit   relativ * *bezogen auf Bezugsgröße, welche "1" oder "100%" entspricht*
in person   selbst
in phase   gleichphasig || phasengleich
in plenty of time   rechtzeitig
in practice   praktisch * *(Adv.) i.d.Praxis/Anwendung; determined in actual practice:durch* ~*e Versuche ermittelt*

in praxis in der Praxis
in principle im Prinzip
in private praxis freiberuflich * *Arzt, Rechtsanwalt usw.*
in proportion as je nachdem
in proportion with the square of the proportional zum Quadrat der
in pulse form impulsförmig || pulsförmig
in pursuance of gemäß * *gemäß e. Norm, Vorschrift, Gesetz* || laut * *gemäß, zufolge, entsprechend*
in real life in der Praxis
in real terms tatsächlich * *(Adv.) eigentlich, in Wirklichkeit*
in reality tatsächlich * *(Adv.)*
in relation to abhängig von * *im Verhältnis zu*
in respect to an * *hinsichtlich*
in return gegen * *als Entgelt (for: für)*
in return for für * *(Gegenwert ausdrückend)*
in rotation turnusmäßig
in round figures abgerundet * *(Adv.) ganze Zahl* || rund * *(Adv.) in Zusammenhang mit Zahl*
in series hintereinander
in sheltered areas in geschützten Räumen
in shifts schichtweise * *bei der Arbeit*
in small steps feinstufig
in so doing dabei * *wenn/indem man das tut, hierdurch, gleichzeitig* || hierbei * *dabei, gleichzeitig*
in some cases teilweise * *manchmal, in einigen Fällen*
in space räumlich
in spite of trotz || ungeachtet * *trotz*
in spite of this trotzdem * *(Adv)*
in stages abschnittweise * →*stufenweise* || stufenförmig * *(figürl.) Prozess, Entwicklung usw.*
in standard standardmäßig * *(Adv.)*
in standby mode betriebsbereit * *in Bereitschaft stehend*
in star connection in Stern geschaltet
in step synchron * *im Gleichschritt, gleich schnell (laufend)*
in steps gestaffelt * *gestuft* || gestuft * *(Adv.)* || stufenförmig * *(Adj. und Adv.)* || stufenweise
in steps of in Schritten von
in stock auf Lager * *have stocked: ~ haben* || vorhanden * *auf Lager* || vorrätig * *out of stock:nicht mehr vorrätig; keep in stock:vorrätig halten*
in store auf Lager
in strict confidence geheim * *streng vertraulich* || vertraulich * *streng ~*
in succession hintereinander || nacheinander
in such a manner derartig || solcher * *~art, ~weise*
in such a way derartig || solcher * *~weise*
in such a way that so * *so/in der Weise, dass*
in superior style überlegen * *(Adv.)*
in supply lieferbar * *be in plentiful/short supply: in hohen Stückzahlen/schlecht ~*
in synchronism synchron * *with: mit; run in synchronism: ~ laufen* || zeitsynchron * *with:mit*
in tabular form tabellarisch * *(Adv.)*
in tandem arrangement hintereinander * *hintereinander angeordnet*
in terms of -lich * *z.B. in terms if price: preis~* || betreffend || bezüglich * *im Sinne von*
in terms of figures wertmäßig * *zahlenmäßig* || zahlenmäßig * *siehe auch* →*wertmäßig*
in terms of percentage prozentual * *(Adv.)*
in terms of price preislich
in that case eventuell * *(Adv.) gegebenenfalls*
in the beginning anfänglich * *(Adv.)* || anfangs || ursprünglich * *(Adv.)*
in the best possible manner bestmöglich * *(Adv.)*
in the best possible way optimal * *(Adv.)*
in the case of bei * *im Falle des* || im Falle von
in the company of mit * *in Begleitung von*

in the course of im Rahmen von * *während, des Verlaufes/der Abwicklung von* || im Zuge dessen
in the course of time mit der Zeit
in the course of which wobei
in the end letzt * *am Ende* || schließlich * *(Adv.) zu guter Letzt* || zuletzt
in the face of entgegen * *gegenüberstehend, angesichts* || gegen * *~sätzlich* || gegenüber * *angesichts* || trotz * *gegen Widerstände*
in the first instance zunächst * *im ersten Moment*
in the first place erstens || in erster Linie || zuerst * *vor allem (Übrigen), erstens* || zunächst * *erstens, vor allem andern*
in the following text im Folgenden * *in Schriftstück*
in the form of -förmig * *(vorangestellt)*
in the form of stairs treppenförmig
in the form of steps stufenförmig
in the last analysis letztendlich * *letzten Endes, bei abschließender Betrachtung* || schließlich * *im Grunde*
in the last resort notfalls * *(Adv.)*
in the long run auf lange Sicht * *auf die Dauer* || schließlich * *(Adv.) auf die Dauer, auf lange Sicht*
in the manner of a -artig
in the manner of a short-circuit kurzschlussartig
in the market auf dem Markt
in the meantime vorübergehend * *zwischenzeitlich*
in the medium-term mittelfristig * *(Adv.)*
in the midst of unter * *mitten ~*
in the neighbourhood of gegen * *ungefähr* || ungefähr
in the opposite direction in entgegengesetzter Richtung || umgekehrt * *in ~e Richtung*
in the opposite sense gegensinnig
in the presence of vor * *in Gegenwart von*
in the range im Bereich || im Bereich von
in the range of im Bereich von
in the range of seconds im Sekundenbereich
in the rear hinten * *am Ende, auf der Rückseite* || hinter * *am Ende*
in the same sense gleichsinnig
in the second place zweitens
in the seconds range im Sekundenbereich
in the simplest case im einfachsten Fall
in the teeth of trotz * *gegen Widerstände*
in the third place drittens
in the vicinity of seconds im Sekundenbereich
in the warm state betriebswarm
in the way of an * *hinsichtlich*
in the widest sense im weitesten Sinne
in the wrong order falsch herum * *in der falschen Reihenfolge*
in theory theoretisch * *(Adv.)*
in these circumstances unter diesen Umständen
in these days heutzutage
in this hierbei
in this connexion hierbei * *in diesem Zusammenhang*
in this manner dadurch * *auf diese Art und Weise (am Satzanfang)*
in this way dadurch * *auf diese Art und Weise (am Satzanfang)* || so * *(Adv.) auf diese Weise*
in tiers stufenförmig * *(Adv.) in Reihen/Lagen übereinanderliegend*
in time fristgerecht || rechtzeitig || termingerecht || zeitgerecht * *rechtzeitig, mit der Zeit, im (richtigen) Takt*
in times to come künftig * *in künftigen Zeiten/Tagen* || zukünftig * *in künftigen Zeiten/Tagen, in der Zukunft*
in total insgesamt || über alles * *(Adv.)*
in turn andererseits
in turns abwechselnd * *(Adv.)*

in use gebräuchlich * *(Adj.)* in *Gebrauch/Benutzung* || in Benutzung || in Gebrauch
in view of angesichts * *in view of the fact that: angesichts der Tatsache, dass* || gegenüber * *in Anbetracht von*
in waves stoßweise * *wellenförmig*
in what respect inwieweit * *in welcher Hinsicht*
in what way inwieweit * *in welcher Hinsicht*
in whole ganz * *(Adv.) vollständig; in whole or in part: ~ oder teilweise* || insgesamt * *(Adv.) als Gesamtes/Gesamtheit*
in whole or in part ganz oder teilweise
in working condition betriebsfähig
in working order funktionsfähig * *in betriebsfähigem Zustand* || funktionstüchtig || intakt * *siehe auch 'funktionsfähig'*
in writing schriftlich
in- un- * *nicht*
in-circuit emulator Emulations- und Testadapter || Emulator * *online-Emulator als Mikroprozessor-Entwicklungshilfsmittel*
in-depth gründlich * *tiefgehend* || tief * *in die Tiefe gehend (z.B. Kenntnisse)* || tiefgehend || umfassend * *in die Tiefe gehend; in-depth knowledge: ~e Kenntnisse*
in-house firmeneigen * *auch ~e technische Lösung* || firmenintern || firmenspezifisch || herstellerspezifisch
in-house production Eigenfertigung
in-house vehicle Flurförderzeug
in-line Geradeaus- || koaxial * *z.B. Armatur*
in-line with -konform
in-line with market prices marktpreiskonform
in-phase condition Phasengleichheit
In-phase regulator Längsregler
in-rush current surge Anfahrstromstoß * *beim Einschalten eines (induktiven) Verbrauchers (Motor, Drossel usw.)*
in/output data Ein-Ausgabedaten
inacceptable unzulässig * *untragbar, unannehmbar*
inaccessible nicht zugänglich || unzugänglich
inaccuracy Ungenauigkeit
inaccurate ungenau || unklar * *ungenau*
inactivation of capital Kapitalbindung
inactive untätig
inactivity Trägheit * *Untätigkeit* || Unwirksamkeit * *{Chemie}*
inadequate dürftig || mangelhaft * *unzulänglich, ungenügend, mangelhaft, unangemessen* || unangemessen * *auch 'unzulänglich'* || ungenügend * *unzulänglich, unangemessen* || unzureichend * *unzulänglich, ungenügend, →mangelhaft, unangemessen*
inadmissible unzulässig
inadvertance versehen * *(Substantiv)*
inadvertantly versehentlich * *ohne Absicht*
inadvertent irrtümlich * *unbeabsichtigt* || unabsichtlich * *auch 'versehentlich'* || unachtsam * *unachtsam, nachlässig, unabsichtlich, versehentlich* || unbeabsichtigt * *auch 'versehentlich'* || versehentlich * *unbeabsichtigt*
inadvertently irrtümlich * *(Adv.) unbeabsichtigt* || unabsichtlich * *(Adv.) ohne Absicht* || unbeabsichtigt * *ohne Absicht*
inappropriate unpassend * *unangebracht*
inattentive unachtsam
inaudible unhörbar
inboard innenliegend * *{Schifffahrt} innenbords*
Inc. AG * *Incorporated: Aktien/Kapitalgesellschaft*
incandescent bulb Glühbirne
incandescent lamp Glühlampe
incarnate verkörpern
incarnation Ausprägung * *Verkörperung, Inbegriff* || Inbegriff * *Verkörperung, Inbegriff (auch figürl.)*

incautious unvorsichtig
incautiousness Unvorsichtigkeit
incentive Anstoß * *Ansporn, Antrieb, (Leistungs-) Anreiz* || Antrieb * *Anreiz, Ansporn, Antrieb, Leistungsanreiz*
incentives of innovation Innovationsanstöße
inception Beginn * *Beginn, Anfang, Einsatz*
incessant andauernd * *unaufhörlich, unablässig, ständig*
inch tippen * *kurzzeitige Aufschaltung eines (niedrigen) Drehzahl/Frequenzsollwerts über Steuerkommando* || Zoll * *Längenmaß (1 Zoll entspr. 1 inch entspr. 0,393701 cm)*
inch control Tippschaltung
inch forward tippen vorwärts
inch reference Tippsollwert
inch reverse tippen rückwärts
inch rule Zollstock
inch thread Zollgewinde
inching Positionieren * *z.B. Material(rolle) per Hand durch Tippen, Läufer einer elektr. Maschine* || Positionierung * *eines Maschinenteils/einer Maschinenwelle z.B. im Tippbetrieb* || Tastbetrieb * *Tippbetrieb eines Antriebs* || Tippbetrieb || tippen * *(Substantiv)*
inching mode Tippbetrieb
inching operation Tippbetrieb
inching setpoint Tippsollwert
inching speed Schleichdrehzahl * *Tippdrehzahl* || Tippdrehzahl || Tippgeschwindigkeit
incidence of taxation Steuerbelastung
incident Ereignis * *Vorfall*
incidental zufällig * *nebenbei*
incidental expenses Nebenkosten
incidentals Nebenkosten
incidentially übrigens * *nebenbei bemerkt, übrigens; beiläufig, nebenbei, zufällig*
incinerator Verbrennungsanlage * *zur Einäscherung/Verbrennung z.B. von Müll*
inclinable neigbar
inclination Neigung * *(auch figürl.: Hang, Vorliebe)* || Sinn * *Neigung* || Steigung * *Neigung*
incline Gefälle || geneigt * *geneigt sein (auch figürl.)* || Neigung * *geneigte Fläche* || Schräge * *schräge Fläche* || schrägstellen * *neigen* || Senkung * *Neigung*
inclined geneigt * *schräg, abschüssig, (auch figürl.; to: zu); downward/upward: nach unten/oben* || schief * *abfallend, geneigt, aufgelegt; inclined plane: schiefe Ebene* || schräg * *abfallend, ~ angeordnet*
inclined belt conveyor Schrägförderband
inclined conveyor Förderberg * *{Bergbau} Schrägfördereinrichtung/Rollenbahn im Bergbau* || Förderbergbahn * *Schrägförderband* * *{z.B. im Bergbau}* || Schrägförderer
inclined position Schräglage
inclined-bed lathe Schrägbett-Drehmaschine
include aufnehmen * *eingliedern, into: in* || beinhalten * *einschließen* || beipacken || einbeziehen * *einschließen* || einschließen * *enthalten* * *ein-einschließen, beinhalten* || tragen * *einschließen* || umfassen * *beinhalten*
included eingeschlossen * *inbegriffen; siehe auch →einschließen* || enthalten * *inbegriffen* || inbegriffen || Mit- || vorhanden * *beinhaltet*
included loose Beipack * *lose im ~ beiliegend*
including darunter * *einschließlich* || einschließlich || samt || unter anderem
inclusion Aufnahme * *Einbeziehung (into:in/bei)* || Einbeziehung
inclusive Mit-
inclusive of einschließlich
incombustible unbrennbar
income Einkünfte

incomer section Einspeisung
incoming ankommend * *Telegramm, Strom, Signal* || Eingangs- || eingehend * *hereinkommend*
incoming cable Zuleitung * *elektr. Kabel/Leitung*
incoming conductor Rückleitung
incoming feeder Einspeisung * *Schalteinrichtung zur Netzeinspeisung*
incoming fuse Eingangssicherung
incoming inspection Eingangsprüfung
incoming line ankommende Leitung
incoming message kommende Meldung
incoming motor cable Motorzuleitung * *aus d. Sicht des Motorklemmenkastens*
incoming orders Auftragseingang
incoming powerline Einspeisung * *Strom/Energieversorgung* || Leistungseinspeisung
incoming rectifier Eingangsgleichrichter
incoming section Einspeisung
incoming supply Einspeisung
incoming supply cubicle Einspeiseschrank
incoming unit Einspeiseeinheit * *(allg.)*
incoming voltage Anschlussspannung
incomparable unvergleichlich
incompatibility Unverträglichkeit
incompatible unvereinbar * *with:mit* || unverträglich
incomplete lückenhaft || mangelhaft * *unvollständig* || unvollständig
incomprehensible unverständlich
inconceivable unverständlich
incongruous unangemessen * *aus der Proportion*
inconsiderate unvorsichtig * *unüberlegt*
inconsistency Widerspruch * *innerer, logischer ~; be inconsistent with: im ~ stehen zu*
inconsistent sich widersprechend || widersprüchlich * *be inconsistent to: im Widerspruch stehen zu*
inconstancy Unstetigkeit
inconvenience Störung * *Unbequemlichkeit, Unannehmlichkeit, Schwierigkeit,* || Unannehmlichkeit * *Unbequemlichkeit, Lästigkeit, Schwierigkeit*
inconvenient lästig * *unbequem* || störend * *unbequem*
incorporate aufnehmen * *eingliedern, in: in* || beinhalten * *eingebaut haben* || einarbeiten * *einfließen lassen, einbauen, hin~* || einbauen * *in:in* || einbeziehen * *einverleiben, vereinigen, verbinden, aufnehmen, eine Körperschaft bilden* || eingebaut haben || integrieren * *einbauen (in/into: in), (in sich) aufnehmen, einverleiben* || tragen * *eingebaut haben* || unterbringen * *einbauen, (in sich) aufnehmen, enthalten* || verbinden * *vereinigen, zusammenschließen, einverleiben, eingliedern* || vereinigen * *in:in; vereinigen, in sich schließen* || zusammenschließen * *vereinigen, verbinden, einbauen, einverleiben*
incorporated eingebaut
incorporated company AG * *eingetragene Gesellschaft, Aktiengesellschaft*
incorporation Aufnahme * *Eingliederung (into:in)* || Einbau * *Eingliederung, Einverleibung* || Einbeziehung * *into:in*
incorrect falsch * *unrichtig* || Fehl- || fehlerhaft * *nicht richtig* || gestört * *nicht richtig, nicht korrekt* || ungenau * *unrichtig, ungenau, irrig, falsch* || unsachgemäß || verkehrt * *unrichtig, nicht korrekt*
incorrect configuring Projektierungsfehler * *z.B. beim SIEMENS Regelsystem SIMADYN D (R)*
incorrect dimensioning Projektierungsfehler
incorrect engineering Projektierungsfehler
incorrect measurement Messfehler * *Fehlmessung*
incorrect operation Bedienungsfehler
incorrect operator command Fehlbedienung * *bei einer Kommandoeingabe*
incorrect operator control Fehlbedienung

incorrect parity bit Paritätsfehler
incorrect phase sequence falsches Drehfeld
incorrect polarity Falschpolung || Fehlpolung || Verpolung * *falsche Polung*
incorrect poling Falschpolung || Fehlpolung
incorrect program run gestörter Programmablauf
incorrect setting Einstellfehler
incorrectly falsch * *(Adv.) unrichtig*
incorrectly connected falsch angeschlossen
incorrectness Fehler * *Unrichtigkeit, ~haftigkeit, Ungeschicklichkeit*
increase Anhäufung * *Anstieg* || anheben * *erhöhen* || Anhebung || ansteigen * *zunehmen (by:um)* || Anstieg || aufsteigen * *zunehmen* || erhöhen * *auch 'sich ~'; (by:um; to:auf)* || Erhöhung || Häufung * *Ansteigen* || heraufsetzen || höher * *Kommando für Motorpotentiometer* || steigen * *zunehmen* || steigern || Steigerung * *Vergrößerung, Erhöhung, Zunahme, (An)Wachsen, Zuwachs, Wachstum, Vermehrung* || Überhöhung || vergrößern * *erhöhen* || vermehren * *auch 'sich vermehren' (by:um)* || verstärken * *Druck usw.* || wachsen * *ansteigen* || Zunahme
increase button Höher-Taste
increase in -Steigerung
increase in efficiency Leistungssteigerung
increase in production Produktionssteigerung
increase key Höher-Taster
increase markedly stark ansteigen
increase rate Hochlaufzeit
increased erhöht * *gesteigert* || verstärkt * *gesteigert; increase one's efforts: ~e Anstrengungen machen*
increased accuracy erhöhte Genauigkeit
increased output erhöhte Leistung * *Motor/Stromrichter*
increased resistance rotor Widerstandsläufer
increased safety erhöhte Sicherheit * *Schutzart 'e' nach EN50019 f.Ex-geschützte el.Betriebsmitt.(-keine Funkenbildg.)*
increased-safety type Ex e * *Zünd/Explosionsschutzart "Druckfeste Kapselung" nach EN 50019*
increasing ansteigend || aufsteigend * *zunehmend* || steigend || verstärkt * *ansteigend* || zunehmend
increasingly laufend * *(Adv.) steigend*
incredible unglaublich
increment Inkrement || Schrittweite || Stufe * *Inkrement* || Zuwachs
incremental inkrementell
incremental data set Kettensatz * *Satz mit Relativ-Lagesollwerten für Lageregelung/Positioniersteuerung*
incremental dimension Kettenmaß * *Schrittmaß/Positionsdifferenz bei Numerik-/Positioniersteuerung* || Schrittmaß * *Kettenmaß (relative/inkrementelle Position) b.Numerik-/Positioniersteuerung*
incremental encoder Drehimpulsgeber || Impulsgeber * *Inkrementalgeber* || Impulstacho * *Inkrementalgeber, Winkelschrittgeber* || Inkrementalgeber * *rotary: Dreh-* || Pulsgeber * *Inkrementalgeber*
incremental measuring system Kettenmaßsystem * *Fahren auf relative Positionen bei Positioniersteuerung ("Fahren um")*
incremental mode Schrittmaß fahren * *Betriebsart bei Positioniersteuerung (relative Lage anfahren, "Fahren Um")*
incremental pulse encoder Drehimpulsgeber
incur erleiden * *erleiden (damage: Schaden; losses: Verluste), auf sich laden, geraten in*
incur damage beschädigt werden || Schaden erleiden
indefinite unbestimmt
indefinitely unbegrenzt * *(Adv.) unbeschränkt, auf unbestimmte Zeit*
indemnification Schadenersatz

indemnity Schadenersatz
indent einrücken * *Absatz in einem Dokument, Text auf dem Bildschirm usw.*
independence Unabhängigkeit
independent autark * *unabhängig* || eigenständig * *unabhängig* || frei * *unabhängig* || selbstständig * *unabhängig, auch beruflich* || selbsttätig || unabhängig * *of:von; as to whether...or:davon ob...oder; of each other:voneinander*
independent as to whether unabhängig davon ob * *or not:oder nicht*
independent blower Fremdlüfter
independent of unabhängig von
independent of the direction of rotation drehrichtungsunabhängig * *z.B. Lüfter, der i. beiden Drehrichtungen arbeitet*
independent of the fact that unabhängig davon ob
independent of the speed direction unabhängig von der Drehrichtung
independently selbstständig * *(Adv.) act independently/on one's own initiative:* ~ *handeln*
independently adjustable unabhängig voneinander einstellbar * *z.B. Hoch- und Rücklaufzeit e. Hochlaufgebers*
indestructible unzerstörbar || zerstörsicher * *unzerstörbar*
indeterminate unbestimmt * *undeutlich*
index Index || indizieren || Inhaltsverzeichnis * *Stichwort-/~, Register; (An)Zeiger; (Hand)Zeichen, Index, Kennziffer* || Kennziffer * *(Plural: indices)* || Marke * *(An)Zeiger, Zeichen* || Markierung * *(An)Zeiger, Zeichen* || Register * *(Einstell-, Teil- Marke, ~, (An)Zeiger, Index, Inhalts-/Stichwortverzeichnis, Kennziffer* || Schlagwortregister || Schlagwortverzeichnis || Stichwortverzeichnis || Verzeichnis * *meist alphabetisch/alphanumerisch sortiertes ~* || Zeichen * *of: ~ für/von, Fingerzeig, Hinweis* || Zeiger * *auch eines Messgeräts, (Einstell)Marke/Strich*
index card Karteikarte
index file Kartei
index of keywords Stichwortverzeichnis
index signal Nullimpuls * *eines Winkelschritt-/Inkrementalgebers (auch 'Nullmarke')* || Nullmarke * *Impulsgeber-Nullimpuls*
indexed indiziert * *z.B. Variable, Parameter*
indexed parameter indizierter Parameter
indexing table Drehtisch * *Positioniertisch mit einstellbarer Winkellage*
India Indien
india-rubber Kautschuk
indicate andeuten * *kurz erwähnen* || angeben * *z.B.Richtung; e.Einzelheit in/durch e. Diagramm/Grafik/Spezifikation; Größe i.e.Dimension* || anzeigen || hinweisen || hinweisen auf * *hinweisen auf, angeben, bezeichnen* || markieren * *anzeigen, angeben, bezeichnen, zeigen, hinweisen auf* || zeigen * *an~, ~ auf (z.B. Pfeil); the values are indicated i.the table:die Tabelle zeigt d.Werte an*
indicated angebracht * *angezeigt, angezeigt* || angezeigt || nötig * *angezeigt*
indicated value Anzeigewert || Messwert * *angezeigter Wert*
indicating Anzeige- || anzeigend
indicating instrument Anzeigeinstrument
indicating lamp Anzeigelampe
indication Anzeige * *auch v. Messinstrument* || Hinweis * *Anhaltspunkt* || Kennzeichnung * *Angabe/Anzeige z.B. auf Typenschild* || Meldung * *Anzeige* || Messwert * *angezeigter Messwert* || Signalisierung * *Anzeige* || Zeichen * *Anzeichen, Kenn~, Hinweis, Andeutung, Symptom, Anzeige, Angabe*
indicator Anzeige * *~gerät, ~element (z.B. Lampe, LED, LCD-Anzeige)* || Anzeige- || Anzeigeinstrument* || Anzeiger || Meldeeinrichtung || Melder || Zeiger * *~ eines Messgeräts*
indicator lamp Anzeigelampe || Leuchtmelder || Meldeleuchte
indicator light Leuchtmelder || Meldeleuchte || Signallampe
indicator module Meldeeinrichtung
indirect indirekt * *(Adj.)* || mittelbar || Zwischen- * *indirekt*
indirect AC converter Zwischenkreisumrichter
indirect cooling indirekte Kühlung * *mit Sekundär-Kühlkreislauf*
indirect-cooled winding indirekt gekühlte Wicklung * *{el.Masch.} nicht →direkt-leitergekühlte Wicklung*
indirectly indirekt * *(Adv.)* || mittelbar * *(Adv.)*
indispensable notwendig * *unerläßlich* || obligatorisch * *unerlässlich* || unabdingbar * *unerlässlich, unentbehrlich* || unentbehrlich || unerlässlich
indispensible unabdingbar * *unerlässlich, unentbehrlich* || unentbehrlich || unerlässlich || unumgänglich * *unerlässlich/entbehrlich*
indistinct unklar * *undeutlich*
individual eigen || eigener || Einzel- * *einzeln, individuell, eigen(tümlich), bestimmmt, charakteristisch* || einzeln * *eigen, für sich allein, eigenständig* || einzelner || gesondert * *einzeln* || getrennt * *einzeln* || jeweilig * *einzeln* || separat * *eigen* || speziell * *gesondert, eigen*
individual control Einzelsteuerung * *bei SPS*
individual control level Einzelsteuerebene || Einzelsteuerungsebene || Steuerebene * *Einzel~*
individual control module Einzelsteuerungsglied * *bei SPS*
individual drive Einzelantrieb * *bei mehreren Antriebsstellen*
individual function Einzelfunktion
individual initiative Eigeninitiative
individual license Einzellizenz
individual operating unit Einzelbediengerät
individual part Einzelteil
individual supply Einzelspeisung
individual workplace Einzelarbeitsplatz
individualize vereinzeln * *einzeln behandeln*
individually einzeln * *jedes Teil für sich* || einzeln * *(Adv.)für sich/~ genommen*
individually assigned fest zugeordnet
individually-acting einzelnwirkend
indoctrination Ausbildung * *Unterweisung, Belehrung, Schulung* || Schulung * *Unterweisung, Belehrung, Schulung*
indolence Trägheit * *Faulheit*
indolent faul * *träge*
indoor innen * *im Innenraum (nicht im Freien)* || Innenraum * *Innenraum-; für die Aufstellung/Verwendung in geschlossenen Innenräumen*
indoor installation Innenraumaufstellung
indoor- Innenraum- * *nicht im Freien*
induce anfachen || anregen * *Schwingung* || bewegen * *jemanden/etwas zu etwas bewegen* || induzieren || veranlassen * *veranlass., bewegen, überreden, verursachen, bewirken, herbeiführen, führen zu,auslösen*
induced-draft fan Saugzuggebläse * *in der Kraftwerkstechnik* || Saugzuglüfter
inducement Antrieb * *Bewegrund*
inductance Induktivität * *Fähigkeit e. stromdurchflossenen Spule, magnet. Energie zu speichern* || Spule * *(elektr.) Induktivität*
inductance value Induktivität * *~swert*
induction Induktion * *elektromagn. ~: Erzeugung e. Spanng. in e. Leiter, d. sich i.e.magn. Wechselfeld befindet*
induction air Ansaugluft * *Verbrennungsmotor*
induction furnace Induktionsofen

induction generator Asynchrongenerator
induction hardening Induktionshärtung
induction heating Elektrowärme * *induktiv erzeugte*
induction machine Asynchronmaschine ‖ Induktionsmaschine
induction motor Asynchronmotor * *siehe auch* →*Käfigläufermotor* ‖ Induktionsmotor
induction regulator Drehtransformator * *ähnl. Asynchronmotor m. fixiertem Läufer f.stetige Spanngs.-u. Phasenänderg.*
induction voltage Induktionsspannung
inductive induktiv
inductive load induktive Belastung ‖ induktive Last
inductive reactance induktiver Blindwiderstand
inductive sensor induktiver Messaufnehmer
inductive voltage drop induktiver Spannungsabfall
inductively coupled induktiv gekoppelt
inductor Drossel ‖ induktives Bauelement ‖ Polrad ‖ Spule * *(elektr.) Drossel*
indulgence Nachgiebigkeit * *Nachsicht, Duldung, Toleranz; Gefälligkeit*
indulgent nachgiebig * *nachsichtig (to: gegen, towards: gegenüber)*
industial systems department Anlagenabteilung
industrial Industrie- ‖ industriell
industrial accident Arbeitsunfall
industrial and capital investment goods Investitionsgüter
industrial application Industrieanwendung
industrial area Industriegebiet
industrial automation Industrieautomatisierung
industrial data acquisition Betriebsdatenerfassung
industrial district Industriegebiet
industrial electronics Industrieelektronik
industrial engineering Anlagentechnik
industrial enterprise Industriebetrieb
industrial environments Industrieumgebung * *rugged: raue*
industrial equipment Industrieausrüstungen
industrial establishment Industriebetrieb
industrial estate Industriegebiet * *z.B. am Rande eines Ortes*
industrial fair Industriemesse
industrial furnace Industrieofen
industrial injuries Arbeitsunfall
industrial machinery Maschinenbau * *als Industriezweig*
industrial network Industrienetz * *Energieversorgungsnetz*
industrial oven Industrieofen
industrial PC Industrie-PC
industrial PC computer Industrie-PC
industrial plant Industrieanlage ‖ Werk * *Fabrikanlage*
industrial plant construction Industrieanlagenbau
industrial quality Industriequalität * *i. Ggs. zur kurzlebigen Konsumqualität*
industrial region Industriezone
industrial robot Industrieroboter
industrial safety Arbeitssicherheit
industrial safety equipment Arbeitssicherheitseinrichtung
industrial sector Branche * *in der Indutrie*
industrial standard Industriestandard
industrial systems Anlagentechnik
industrial systems engineering Anlagentechnik
industrial technology Anlagentechnik
industrial vacuum cleaner Industriestaubsauger
industrial zone Industriezone
industry Branche * *Industrie* ‖ Gebiet * *Industrie*, →*Branche* ‖ Industrie
industry standard Industriestandard

ineffective unwirksam
inefficacy Unwirksamkeit
inefficiency Unwirksamkeit
inefficient verlustbehaftet
inequality Ungleichförmigkeit
inermediate transformer Zwischentransformator * *siehe auch* →*Vorschalttransformator*
inert gas Schutzgas
inertia Moment * *Trägheitsmoment, Schwungmoment* ‖ Schwungmasse * *Trägheitsmoment* ‖ Schwungmoment * *Massenträgheit* ‖ Trägheit * *Massenträgheit* ‖ Trägheitsmoment * *[kg x m²]*
inertia compensation Beschleunigungsausgleich * *Trägheitsmoment-Kompensation: Vorsteuerg. d. Beschleun./Verzög.moments* ‖ Beschleunigungskompensation * *Vorsteuerg. d. Schwungmassen-abhäng. Beschleun./Verzögerungsmoments* ‖ Beschleunigungsvorsteuerung * *Trägheitsmoment-Vorsteuerung* ‖ Schwungmomentenkompensation ‖ Schwungmomentkompensation ‖ Trägheitsmoment-Kompensation ‖ Trägheitsmoment-Vorsteuerung
inertia drive Schwungmassenantrieb
inertia energy kinetische Energie
inertia forces Massenkräfte
inertia precontrol Beschleunigungsausgleich * *Trägheitsmoment-Vorsteuerung* ‖ Beschleunigungskompensation ‖ Beschleunigungsvorsteuerung ‖ Trägheitsmoment-Vorsteuerung
inertialess trägheitslos
inessential unwesentlich * *nebensächlich*
inevitable obligatorisch * *unvermeidlich, unumgänglich, zwangsläufig* ‖ unvermeidbar ‖ unvermeidlich ‖ unweigerlich * *unvermeidlich, zwangsläufig* ‖ zwangsläufig
inevitably unvermeidbar * *(Adv.)* ‖ unvermeidlich * *(Adv.)* ‖ unweigerlich * *(Adv.)* ‖ zwangsläufig * *(Adv.)*
inexact ungenau ‖ unklar * *ungenau*
inexpensive billig ‖ preisgünstig * *nicht teuer* ‖ preiswert
inexpert unfachgemäß
infancy Kinderschuhe * *be still in its infancy: noch in den Kinderschuhen stecken*
infeed Einlauf * *Zuführ-/Zufördereinrichtung* ‖ Einschleusung ‖ Einzug ‖ Zuführung * *z.B. von Material in eine Maschine*
infeed current Einspeisestrom
infeed hopper Einfülltrichter ‖ Trichter * *Einfüll~*
infeed module Einspeiseeinheit
infeed power Speiseleistung
infeed system Beschickungssystem
infeed transformer Einspeisetrafo ‖ Einspeisetransformator ‖ Vorschalttrafo * *Einspeisetrafo* ‖ Vorschalttransformator * *Einspeisetransformator* ‖ Vortrafo * *Einspeisetrafo* ‖ Vortransformator * *Einspeisetransformator*
infer folgern ‖ schließen * *folgern*
infer from entnehmen * *schließen/folgern/ableiten aus*
inference Folgerung * *['inferens] (Schluß-) ~, (Rück-) Schluss; make inferences: Schlüsse ziehen* ‖ Herleitung * *(Schluss-) Folgerung, (Rück-) Schluß* ‖ Schluss * *(Schluss-) Folgerung, (Rück-) Schluß; make inferences: Schlüsse ziehen*
inferior mangelhaft * *minderwertig* ‖ minderwertig * *Qualität usw.* ‖ schlecht * *Qualität, Ware* ‖ unter * *~geordnet, geringer(wertig)*
infiltrate eindringen * *hinein, auch 'eindringen lassen, einsickern, unterwandern'*
infiltration eindringen * *(Substantiv) auch 'Einsickern'*
infinite unendlich
infinitely unbegrenzt * *(Adv.) unendlich, ungeheuer, unbegrenzt; infinitely expandable: ~ erweiterbar*

infinitely adjustable stufenlos * *stufenlos einstell/ regelbar* || stufenlos verstellbar
infinitely small verschwindend klein
infinitely variable stufenlos * *stufenlos einstell/regelbar* || stufenlos verstellbar
infinitesimal verschwindend klein
inflammability Entflammbarkeit
inflammabilty Brennbarkeit * *Entflammbarkeit*
inflammable entflammbar || entzündbar * *highly:-leicht* || feuergefährlich
inflate aufblasen || füllen * *aufpumpen (Reifen usw.)*
inflected characteristic abknickende Kennlinie
inflexible starr * *unflexibel, unelastisch, nicht biegsam,starr; (auch figürl.;siehe auch 'bürokratisch')* || steif * *unbiegsam* || unflexibel * *(auch figürl.; siehe auch →bürokratisch)*
influence beeinflussen * *on:auf; with:bei* || Beeinflussung * *Einfluss* || Einfluss || eingreifen * *beeinflussen* || einwirken * *beeinflussen,einwirken auf* || Einwirkung * *Einfluss (of:von)* || wirken * *beeinflussen*
influence beyond one's control höhere Gewalt
influence of ambient temperature Temperatureinfluss * *Einfluss d. Umgebungstemperatur*
influence of load variation Lastabhängigkeit
influence of the weather Witterungseinflüsse
influenced beeinflusst
influencing variable Einflussgröße
influential maßgebend * *einflussreich, z.B. Kreise* || maßgeblich * *einflussreich, z.B.Kreise*
influx Einfluss * *das Einfließen*
info Information * *(salopp) Kurzform für 'Information'* || Informationsdienst
inform angeben * *informieren* || informieren * *of/on/about:über; oneself:sich* || melden * *informieren; of a thing:etwas* || mitteilen * *a person of: jmd. etwas mitt.*
informal unverbindlich * *zwanglos, nur zur Information*
informatics Informatik
information Angabe * *Information (on: über)* || Angaben || Auskunft || Aussage * *give information about: Aufschluss geben über* || Hinweis * *on/about:zu/über; relevant information: sachdienliche ~e* || Information * *(Singular und Plural) on/about: über; auch 'Auskunft'* || Informationen * *nie in d.Mehrzahl verwenden ("informations" ist falsch!)* || Informationsdienst || Meldung * *Mitteilung* || Mitteilung || Nachricht * *Mitteilung* || Verständigung * *Benachrichtigung* || Wissen * *~sgut*
information counter Informationsstand * *z.B. bei Messe*
information event Informationsveranstaltung
information flow Informationsfluss || Signalfluss
information material Informationsmaterial
information on commissioning Inbetriebnahmehinweise
information on safety Sicherheitshinweise
information processing Informationsverarbeitung
information procurement Informationsbeschaffung
information regarding Angaben zu
information science Informatik
information science Informationstechnik * *→Informatik*
information technology Datenverarbeitung || Informatik * *→Informationstechnik* || Informationstechnik || Nachrichtentechnik
information transmission Nachrichtenübertragung
information-rich informativ
informative aufschlussreich || aussagekräftig * *Informations-* || *informativ* || informativ
informative discussion Informationsgespräch
informatory informativ

informed orientiert * *informiert, benachrichtigt, in Kenntnis gesetzt*
infrared Infrarot
infrared drying Infrarot-Trocknung
infrastructure Infrastruktur
infrequent selten
infringe beeinträchtigen * *jemandes Rechte ~* || überschreiten * *Gesetz/Vorschrift übertreten* || verletzen * *Regel, Begrenzung usw. ~*
infringe on verletzen * *z.B. Patent ~*
infringement Überschreitung * *Verletzung, Bruch, Übertretung, Verstoß von/gegen Gesetz/Vorschrift* || Verstoß * *(Rechts-, Vertrags-, Patent-) Verletzung, Bruch, Übertretung, Verstoß; (up)on: gegen*
ingenious ausgeklügelt || fähig * *einfallsreich, genial, erfinderisch, klug* || findig || genial || geschickt * *geschickt ausgedacht* || hervorragend * *klug erdacht* || raffiniert * *genial* || sinnvoll * *wohl ersonnen*
ingoing shaft Eingangswelle * *bei Getriebe*
ingot Barren * *Metallbarren, auch 'Stange, Block'*
ingredients Zusammensetzung * *Bestandteile*
ingress eindringen * *(Substantiv) Eintritt, Eintreten, Zutritt, Zugang* || Eintritt * *z.B. von Fremdkörpern*
ingress of dirt Verschmutzung * *Eintreten/-dringen von Schmutz*
inherent eigen * *innewohnend, zugehörig, eigen* || Eigen- * *innewohnend, Selbst-* || innerer * *innewohnend, →zugehörig, angeboren, →eigen, eingewurzelt* || innewohnend * *in:dem* || zugehörig * *innewohnend, zugehörend, eigen*
inherent in -bedingt
inherent in the system systembedingt
inherent in the type of construction bauartbedingt
inherently von Haus aus * *(z.B. auf Grund des Funktionsprinzips) innewohnend*
inhibit Sperre * *als Kommando* || sperren
inhibit the controller Regler sperren || Reglersperre setzen
inhibited gesperrt * *z.B. Regler*
inhomogeneous inhomogen * *[inhomo'dschi:njes]*
inhouse hauseigen * *→firmeneigen, allein einer Firma zugeordnet/gehörend*
iniative Anstoß * *Iniative, Anstoß, Anregung, Unternehmungsgeist*
initial anfänglich * *(Adj.)* || Anfangs- || ursprünglich * *anfänglich*
initial address Anfangsadresse
initial condition Anfangsbedingung || Ausgangsstellung
initial cost einmalige Kosten * *nur zu Beginn einmal aufzubringende (Initialisierungs-) Kosten*
initial defects Kinderkrankheiten * *von technischen Geräten, Einrichtungen*
initial impulse Starthilfe * *(figürlich)*
initial loading urladen * *z.B. Programm ~*
initial operation Inbetriebnahme * *erstmaliger Betrieb, z.B. eines Kleingeräts*
initial outlay Anschaffungskosten
initial pay-out Vorabzug * *z.B. bei Folienmaschine*
initial position Anfangslage || Ausgangsstellung
initial program loading urladen * *Programm ~*
initial rounding Anfangsverrundung
initial rounding-off Anfangsverrundung * *für Hochlaufgeber-Rampe*
initial run Erstlauf
initial setting Grundeinstellung * *z.B.von Potentiometern/Parametern* || Voreinstellung * *ürsprünglich gesetzter Wert*
initial start Anlauf * *bei SPS*
initial start-up Erstinbetriebnahme || Inbetriebnahme * *erstmalige ~* || Inbetriebsetzung * *Erstinbetriebnahme*

initial starting torque Losbrechmoment
initial state Grundzustand || jungfräulicher Zustand
initial stress Vorspannung * z.b. bei Riementrieb
initial symmetrical short-circuit power Anfangskurzschlussleistung
initial value Anfangswert || Defaultwert || Setzwert * *Anfangs-/Urladewert* || Startwert || Urladewert * *eines Parameters*
initialisation Bedienungseintrag * *{SPS} (brit.)*
initialising pulse Richtimpuls
initialization Anlauf * *Anlaufprozedur für Computer* || Anlaufprogramm || Bedienungseintrag * *{SPS} (amerikan.)* || Hochlauf * *Rechner* || Initialisierung || Wiederanlauf * *Initialisierungsprogramm*
initialization error Initialisierungsfehler
initialization procedure Anlaufprogramm || Hochlauf * *Anlaufprozedur eines Rechners*
initialization value Startwert
initialize initialisieren
initialized status Urladezustand || Werkseinstellung * *Urladezustand*
initializing pulse Richtimpuls
initially am Anfang * *(Adv.)* || anfänglich * *(Adv.)* || zuerst * *anfangs, zunächst* || zunächst * *anfänglich*
initiate anfahren * *beginnen, einleiten, einführen, aufnehmen* || anregen || Anstoß * *(Verb) den ersten ~ geben* || anstoßen * *einleiten, initiieren* || aufrufen * *initiieren* || auslösen * *veranlassen, bewirken, initiieren* || einarbeiten * *someone into a job: jemanden in eine Aufgabe ~* || einführen * *Maßnahmen usw.* ~; *initiate someone into: jdn. einweihen in* || einleiten * *anstoßen, initieren, initieren* || *in die Wege leiten* || initieren || initiieren * *einleiten, anstoßen*
initiation Aufnahme * *Beginn*
initiative Antrieb * *Initiative* || Initiative * *take the initiative:d. Initiative ergreifen; on one's own initiative:aus eigen. Antrieb*
inject anlegen * *(Signal) einprägen, anlegen* || aufschalten * *hineinleiten, z.B. Signale (siehe auch →vorsteuern), Rechtecksprünge, Feldstrom* || einkoppeln || einprägen * *z.B. Strom* || einspeisen * *z.B. einen Sollwert, ein Stimulus-Signal*
inject a step-change signal stoßen * *sprungförm. Signal anlegen für Regleroptimierg. (Sprungantw. aufnehmen)*
injection Aufschaltung * *Einspeisung e. (Zusatz-Stimulations-/Kompensations-) Signals* || Einkopplung * *z.B. von (Stimulations-) Signalen* || Einspritzung
injection diecasting Spritzguss
injection molding Spritzguss
injection molding tool Spritzgießwerkzeug * *für Kunststoff-Spritzgießmaschine*
injection moulding Spritzguss * *z.B. Kunststoff~*
injection of interference Störeinstreuung
injection-molded part Spritzgussteil * *aus Kunststoff*
injection-mould spritzen * *thermoplastischen Kunststoff ~*
injection-moulded part Spritzgussteil * *aus Kunststoff*
injection-moulding machine Spritzgießmaschine * *für Kunststoff*
injure beeinträchtigen || verletzen * *verwunden*
injurious schädlich * *für eine Person; nachteilig (to:für), beleidigend, verletzend*
injury Schaden * *Beschädigung (besonders von Personen), Schaden (to:jemanden)*
ink Druckfarbe * *{Druckerei}* || Farbe * *Druck~* || Tinte
ink dosing equipment Färbeeinrichtung
ink pre-setting system Farbvoreinstellsystem * *{Druckerei}*
ink recorder Tintenschreiber

ink zone Farbzone * *{Druckerei}*
ink-jet printer Tintenstrahldrucker * *{Computer}*
ink-jet recorder Tintenstrahlschreiber
ink-zone presetting Farbzonenvoreinstellung * *{Druckerei}*
inking roller Auftragswalze * *{Druckerei}* || Farbwalze * *{Druckerei}*
inking unit Farbwerk * *{Druckerei}*
inlayed work Furnierung
inlaying Furnierung
inlet Durchführung * *Enlass(stelle), z.B. von Leitungen* || Eingang * *Einlassöffnung* || Einlass * *inlet valve: ~ventil* || Einlauf * *Einlass* || Eintritt * *Einlassöffnung* || Eintrittsöffnung * *z.B. für Kühlluft* || Öffnung * *Einlass*
inlet duct Einlasskanal * *z.B. für Kühlluft*
inlet pressure controller Vordruckregler * *für eine Turbine*
inlet side Ansaugseite * *Pumpe*
inlet temperature Eintrittstemperatur * *z.B. von Kühlluft/Wasser*
inline hintereinander * *entlang einer (gedachten) Linie angeordnet* || Reihen-
inline calender Maschinenglättwerk || Maschinenkalander * *Glättwerk, in Papiermaschine integriert*
inline installation Direkteinbau * *z.B. Magnetventil in e. Rohrleitung*
inline operation Anlagenverbund
inline pump Inline-Pumpe * *Kreiselpumpe mit integriertem Motor*
inner Innen- || innerer || unterlagert * *z.B. inner current control loop: ~e Stromregelung*
inner air cooling Luft-Innenkreislauf * *beim Motor*
inner control loop unterlagerter Regelkreis || unterlagerter Regler * *unterlagerte Regelung*
inner current control loop unterlagerte Stromregelung
inner parts Innenteile
inner pole excitation Innenpolerregung
innovate innovieren
innovated innoviert
innovation Innovation
innovation cycle Innovationszyklus
innovation strategy Innovationsstrategie
innovative fortschrittlich * *innovativ* || innovativ
innovative strength Innovationskraft
inoperative außer Betrieb || Totalausfall * *totaler Funktionsfehler* || ungültig * *unwirksam, ungültig, nicht in Kraft (Gesetz usw.)*
inoperative position Ruhestellung
inoperativeness Unwirksamkeit
inopportune ungeeignet * *z.B. Moment, Augenblick, Situation*
inorganic anorganisch * *{Chemie}*
input Anregung * *Signal* || Antriebsleistung * *Generator* || aufgenommene Leistung * *z.B. eines Motors* || aufgenommene Wirkleistung * *eines Motors* || Aufnahme * *eingespeiste/zugeführte Menge/Spannung/Leistung* || Aufnahmeleistung * *z.B. eines Motors* || Eingang * *Eingangs-* || eingangsseitig || eingeben || eingespeist || treibend * *Maschinenteil* || vorgeben * *z.B. ein Signal anlegen*
input board Eingangsbaugruppe * *Leiterkarte*
input byte Eingangsbyte
input check Eingabekontrolle
input circuit Eingangsschaltung
input clock frequency Eingangstaktfrequenz
input current Eingangsstrom
input device Eingabegerät
input end Antriebswelle * *z.B. eines Getriebes/einer Kupplung*
input error Eingabefehler
input filter against EMI Funkentstörfilter * *am*

Eingang e.Geräts geg. 'ElectroMagnetic Interference':el.magn. Störungen
input function Eingangsfunktion
input impedance Eingangsimpedanz || Eingangswiderstand * *Eingangsimpedanz*
input information Eingabeinformation
input level Eingangspegel
input mask Eingabemaske * *am Bildschirm*
input module Eingangsbaugruppe
input power Anschlussleistung * *z.B. eines Motors* || Eingangsleistung || Leistung * *aufgenommene* ~
input power failure Netzausfall * *die Eingangsspannung eines Geräts betreffend*
input pulse Ansteuerimpuls
input quantity Eingangsgröße * *zahlen-/wert-/pegelmäßig*
input resistance Eingangswiderstand
input sequence Eingabesequenz
input shaft Antriebswelle * *Getriebeeingangswelle*
input signal Eingangsgröße || Eingangssignal
input speed Antriebsdrehzahl * *eines Getriebes* || Eintriebsdrehzahl * *eines Getriebes*
input stage Eingangsstufe || Vorstufe * *(elektr.) Eingangsstufe*
input terminal Eingangsklemme
input to network Netzeinspeisung * *Einspeisung ins Netz hinein*
input value Eingangsgröße * *Eingangswert*
input variable Eingangsgröße
input voltage Anschlussspannung || Eingangsspannung * *to:für/an* || Eingangsspannung
input voltage range Eingangsspannungsbereich
input word Eingangswort
input-/output module Ein-Ausgabemodul
input-side eingangsseitig
input/output/flag reference list Belegungsplan * *Belegungsplan für Ein-/Ausgänge und Merker bei SPS*
input/output/memory map Belegungsplan * *Speicherbelegungsliste eines Rechners/einer SPS*
inquery Information * *(Ergebnis von) Nachforschungen*
inquire anfragen * *for: nach* || konsultieren * *nachfragen, erkundigen (for:nach/wegen)* || nachfragen * *(nach)fragen nach, z.B. Lieferangebot, sich erkundigen* || Rückfragen * *(Verb) fragen/s.erkundigen nach, (er)fragen; for:nach; about:über; of someone:bei jdm.* || untersuchen * *untersuchen, erforschen*
inquiries Rückfragen * *Nachfragen, Erkundigungen, Anfragen, Nachforschungen*
inquiry Abfrage * *Erkundigung, Nachforschung, Untersuchung, Auskunft-Einholen* || Anfrage * *z.B. nach e. Lieferangebot, auch 'Nachfrage, Erkundigung'; make:stellen* || Frage * *Erkundigung* || Nachfrage * *das ~n, Rückfrage, Erkundigung* || Rückfrage * *Nachfrage, Erkundigung, Anfrage, Nachforschung* || Umfrage * *all round: generelle ~ an alle; make inquiries: ~ durchführen* || Untersuchung
inrush Einschalt-Rush * *z.B. erhöhte Leistungsaufnahme b. Einschalten einer induktiven Last* || Einschaltleistung * *erhöhte Leistung im Einschaltaugenblick* || Einschaltstoß * *Stoßstrom/Spannung, z.B. beim Einschalten induktiver Last* || Rush-Effekt || Spitze * *Strom/Momentenspitze beim Einschalten und Drehrichtungswechsel*
inrush current Anzugsstrom * *erhöhter Einschaltstrom b.Schalten induktiver Lasten (Relais, Schütz, Magnetventil)* || Einschaltstoßstrom * *z.B. beim Einschalten induktiver Last* || Einschaltstrom * *hoher* ~; *z.B. beim Schalten induktiver Lasten* || Einschaltstromstoß * *beim Einschalten induktiver Lasten* || Rush-Strom * *Stromspitze beim Aufbau eines Feldes (z.B. bei Motor, Trafo usw.)* || Rushstrom * *Stromspitze z.B. beim Einschalten induktiver Lasten*

inrush effect Rush-Effekt
inrush torque Rush-Moment * *Momentenspitze, d. beim Einschalten u.Drehrichtungsumkehr e.Motors auftritt*
inrush-current limiting Einschaltstrombegrenzung * *begrenzt den Überstrom beim Einschalten induktiver Verbraucher*
inscribe gravieren
inscribe for belegen * *Vorlesung usw.* ~
insecurity Unsicherheit
insensibility Unempfindlichkeit
insensibility to interference Störunempfindlichkeit * *gegen elektromagnetische Störungen*
insensible unempfindlich * *to:gegen*
insensitive unabhängig * *unempfindlich* || unempfindlich * *to: gegen (Druck, Wettereinflüsse usw.)*
insensitiveness Unempfindlichkeit
insensitivity Unempfindlichkeit
insert aufnehmen * *Klausel/Passus* ~ || aufstecken * *z.B. Stecker/Steckbaugruppe* || einbauen * *einfügen, into:in* || einfügen || einlegen * *Werkstück, Batterie usw.* || Einsatz. * *Teil zum Einsetzen in etwas, Einsatzstück* || einschieben || einsetzen * *einfügen* || einstecken || eintragen * *einfügen* || eintreiben || hineinstecken || stecken * *z.B. eine Steckbaugruppe oder ein IC in seinen Einbauplatz* ~
insert a delay warten * *eine Wartezeit einfügen*
insert into einführen * *in eine Öffnung* ~
inserted aufgenommen * *Klausel*
inserted-blade milling cutter Fräskopf
insertion Bestückung * *von Bauelementen auf einer Leiterkarte* || Einbau * *das Einfügen* || einfügen * *(Substantiv) Einfügung, Einfügen* || Einfügung || Einführung * *Einsetzen eines Bauteils, Markt~* || Einfügen * *(Substantiv) z.B. einer Steckbaugruppe oder eines ICs*
insertion depth Einstecktiefe
insertion of components Bauelemente-Bestückung || Bauelemente-Bestückung
insertion/withdrawal cycles Steckzyklen * *einer Steckverbindung*
inside innen * *auch "nach (dr)innen"* || Innen- || innerhalb * *auf der Innenseite, im Inneren* || nach innen
inside diameter Innendurchmesser
inside parts Innenteile
inside room Innenraum * *im Gebäude*
inside the -intern * *(vorangestellt)*
insight Verständnis * *Einsicht, Verständnis, Einblick*
insightful aufschlussreich || einleuchtend * *aufschlußreich*
insignificant geringfügig * *bedeutungs/belanglos, unwichtig, unbedeutend, nichtssagend* || irrelevant * *bedeutungslos* || klein * *unbedeutend* || minimal * *(figürl.) geringfügig, unbedeutend* || unbedeutend || unwesentlich * *geringfügig*
insist on verlangen * *bestehen auf*
insolation Sonneneinstrahlung
insoluble unlöslich * *durch Lösungsmittel*
insolvency Konkurs
inspect abnehmen || betrachten * *genau* ~ || inspizieren || kontrollieren * *untersuchen, prüfen, nachsehen; inspizieren, be(auf)sichtigen* || nachprüfen * *inspizieren, überprüfen* || prüfen * *inspizieren, abnehmen* || überprüfen * *nachschauen, besichtigen, inspizieren* || überwachen * *inspizieren, kontrollieren* || untersuchen * *untersuchen, prüfen, nachsehen, besichtigen, inspizieren* || warten * *inspizieren*
inspection Besichtigung * *prüfende* ~ || Inspektion || Kontrolle * *Inspektion, Durchsicht, Untersuchung, Prüfung* || Nachprüfung || Prüfung * *Inspektion, Über-~, Kontrolle* || Revision * *(techn.) Inspektion einer Maschine, Anlage usw.* || Überprü-

fung * *Inspektion* || Überprüfung * *Inspektion, Überprüfung* || Wartung * *Inspektion, Überprüfung*
inspection certificate Abnahmeprüfzeugnis
inspection document Prüfbescheinigung * *Bescheinigung, die der Besteller mit der Lieferung erhält (EN 10204)*
inspection machine Inspektionsmaschine * *Sortierroller z.Hin- u. Herwickeln v. Papier usw. z. Aussortieren v.Schadstell.*
inspection mode Revisionsbetrieb
inspection of material Materialprüfung
inspection planning Prüfplanung
inspection report Abnahmeprüfprotokoll
inspection system Kontrollanlage
inspection window Sichtfenster * *für Servicezwecke*
inspector Abnahmebeauftragter
inspiration Idee * *Einfall, Eingebung, Erleuchtung, Anregung*
inst laufend * *Abkürzg. für 'instant': ~en Monats (the 10th inst: der 10. dieses Monats)*
instability Instabilität
install anlegen * einrichten || aufbauen * *intallieren, montieren, aufstellen, einbauen* || aufstellen * *Maschine ~; auch 'mount' (Motor, Maschine ~)* || einbauen * *installieren* || einrichten || in Betrieb setzen * *intallieren* || inbetriebnehmen * *installieren* || inbetriebsetzen * *installieren* || installieren * *siehe auch →einbauen* || montieren * *einrichten; auch Industrieanlage* || unterbringen * *installieren, einbauen, anbringen, montieren, aufstellen* || verlegen * *Kabel, Rohre*
installation Anlage * *Einrichtung, Anlage; Installierung, Errichtung, Einbau* || Apparatur * *Aufbau* * *Errichtung, Einbau, Intallierung* || Aufstellung * *Installation, Montage* || Aufstellung * *Installation, Montage; indoor: Innenraum-~; outdoor: ~ im Freien* || Einbau || Einrichtung * *Einbau, Anlage* || Installation * *auch eines Computer-Programms* || Montage * *Ein-/Aufbau (zu einer größeren Einheit); Industrieanlage, Maschine, Maschinenteil* || Verlegung * *z.B. ~ von Kabeln/Leitungen*
installation altitude Aufstellhöhe || Aufstellungshöhe * *above sea level:über dem Meeresspiegel* || Betriebshöhe * *above sea level:über dem Meeresspiegel*
installation configuration Anlagenkonfiguration
installation diagram Montageplan * *techn. Zeichnung mit Darstellung des Auf- oder Einbaus* || Montagezeichnung * *z.B. zur Montage e.Stromrichters i.Schaltschrank*
installation drawing Montagezeichnung
installation engineering Installationstechnik
installation guide Inbetriebnahmeanleitung
installation guideline Aufbaurichtlinie * *zum störsicheren Einbau u.Verdrahtung v. Steuer- u. Regelsystemen-/Geräten*
installation height above sea level Einbauhöhe * *über d. Meeresspiegel* || Montagehöhe * *~ über d. Meeresspiegel*
installation instructions Einbauanleitung || Einbauhinweise || Montageschriften
installation location Einbauort
installation material Installationsmaterial
installation position Einbaulage
installation recommendations Einbauhinweise * *Empfehlungen den Einbau betreffend*
installation reference-conditions Installations-Bezugsbedingungen * *bei Installation e. Motors zu beachtende ~, VDE 0530 Beiblatt 1*
installation schedule Montageplan * *Zeitplan*
installation side Anlagenseite
installation technology Installationstechnik * *als Fachrichtung oder Geschäftszweig*
installations Fabrik * *Fabrikationsanlagen*

installed untergebracht
installed load Anschlussleistung
installment Teillieferung * *eines Sammelwerkes (Buch, Nachschlagewerk, Katalog usw.)*
instance Fall * *Einzel~* || Fall * *(einzelner) Fall, Beispiel* || Instanz * *(auch in der Datenverarbeitg.: Exemplar eines mehrfach ablaufenden Programms)* || Veranlassung * *Ansuchen, (dringende) Bitte; at the instance of: auf ~ von*
instant Augenblick || laufend * *~en Monats, dringend* || Moment * *Augenblick* || unverzögert * *unverzüglich* || Zeit * *(genauer) ~punkt, Augenblick* || Zeitpunkt
instant of turn-on Einschaltmoment * *Augenblick des →Einschaltens, z.b. eines Thyristors*
instantaneous augenblicklich || kurzzeitig * *einen kurzen Moment lang* || momentan * *(Adj.) augenblicklich (z.b. Wert, Geschwindigkeit), sofort* || sofortig || trägheitslos * *trägheitslos/unverzögert ansprechend* || unverzögert * *unverzüglich, sofortig, augenblicklich (z.B. Ansprechen e. Schutzeinrichtung)* || verzögerungsfrei
instantaneous power Augenblickswert * *~ der Leistung*
instantaneous power failure Netzunterbrechung * *kurzzeitige*
instantaneous release Schnellauslöser
instantaneous speed Momentangeschwindigkeit
instantaneous status augenblicklicher Zustand * *z.B. der Zustandsgrößen eines Steuer/Regelsystems i. e. Moment*
instantaneous trip unit Schnellauslöser
instantaneous tripping Schnellauslösung
instantaneous value Augenblickswert || Momentanwert
instantaneous value of speed Drehzahlistwert * *Augenblickswert der Drehzahl*
instantaneous-tripping schnellansprechend * *Schutz/Überwachungseinrichtung*
instantaneous-value measurement Momentanwertmessung * *z.B. schnelle Analog-/Digitalwandlung (Stufenumsetzer oder Flash)*
instantaneously momentan * *(Adv.) (auch physikal.) augenblicklich, sofort, unverzüglich, auf der Stelle*
instantly laufend * *(Adv.) sofortig, augenblicklich* || sofort * *(Adv.)*
instead anstelle * *of:von*
instead of anstatt || für * *anstatt*
instigate scharfmachen * *(figürl.) an/aufreizen, anstiften, aufhetzen (to:zu; to do:zu tun)*
instinct Sinn * *Instinkt*
institute of regulations Normenbehörde || Normengremium || Vorschriftenstelle * *z.B. Normengremium*
institution Einrichtung * *öffentliche Einrichtung*
instruct anordnen * *befehlen* || ausbilden * *belehren* || informieren * *anweisen*
instructed orientiert * *informiert, unterrichtet, belehrt, unterwiesen*
instruction Anordnung * *Befehl* || Anweisung * *auch Computerbefehl, besonders für Assembler/ Maschinensprache* || Auftrag * *Weisung* || Ausbildung * *Belehrung, Ausbildung* || Befehl * *[Computer] (Assembler-/Maschinen) befehl* || Hinweis * *Anleitung (for:zu)* || Lehre * *Unterweisung* || Schulung * *Belehrung, Ausbildung* || Vorschrift * *Anweisung; to instructions: nach ~*
instruction book Anleitung * *Handbuch* || Benutzerhandbuch || Beschreibung * *Handbuch* || Handbuch
instruction card Merkblatt
instruction code Befehlscode * *Rechner*
instruction cycle Befehlszyklus
instruction execution Befehlsausführung

instruction execution time Befehlsausführungszeit * *[Computer]* || Operationszeit * *Ausführungszeit eines Rechnerbefehls*
instruction leaflet Merkblatt
instruction list Anweisungsliste
instruction manual Anleitung * *Betriebsanleitung* || Betriebsanleitung * *Handbuch*
instruction register Befehlsregister * *(Rechner)* || Kommandoregister
instruction set Befehlssatz * *Rechner* || Operationsvorrat * *Befehlssatz eines Computers*
instruction sheet Merkblatt
instructions Anleitung || Beschreibung * *Betriebsanleitung* || Betriebsanleitung || Gebrauchsanleitung || Gebrauchsanweisung * *siehe auch →Betriebsanleitung* || Handbuch || technische Beschreibung * *Betriebsanleitung*
instructions for testing Prüfanleitung
instructions for use Gebrauchsanleitung || Gebrauchsanweisung
instructive aufschlussreich || aussagekräftig
instructor Ausbilder * *auch Lehrer, Dozent* || Dozent * *Kursleiter* || Kursleiter || Lehrgangsleiter || Referent * *bei einer Schulungsveranstaltung* || Schulungsleiter * *auch Lehrer, Dozent* || Vortragender * *bei einer Schulungsveranstaltung*
instrument -Messer || -Messgerät || -Mittel * *Instrument, Werkzeug, Gerät, Apparat* || -Zeug * *zum Messen, Prüfen* || Apparat * *feinmechanischer ~, Messeinrichtung* || Gerät * *Mess~* || Instrument || Messinstrument
instrument amplifier Messverstärker
instrument for measuring -Messer * *(vorangestellt)* z.B. *instrument for measuring vibrations:-Schwingungsmesser* || -Messgerät * *(vorangestellt)* z.B. *instrument f.measuring vibrations:Schwingungsmessgerät*
instrument for measuring vibrations Schwingungsmesser || Schwingungsmessgerät
instrument range Anzeigebereich * *eines Messinstruments*
instrument transformer Wandler * *Mess~*
instrumentality -Mittel * *auch Vermittlung, Mitwirkung, Mithilfe*
instrumentation Instrumentierung * *Mess- und Regelausrüstungen für verfahrenstechnische Prozesse* || Messeinrichtungen * *in der Anlagentechnik/Prozesstechnik* || Messtechnik * *Messausrüstung in der Prozess-/Verfahrenstechnik* || Prozessmesstechnik
instrumentation and control Leittechnik || Mess- und Regelungstechnik
insufficiency Mangel * *Unzulänglichkeit, ~haftigkeit, Untauglichkeit*
insufficient nicht ausreichend || ungenügend * *unzulänglich, untauglich, mangelhaft, unzureichend* || unzureichend
insufficient focus Unschärfe
insufficient oil supply Ölmangel * *unzureichende Ölversorgung*
insulant Isolierstoff
insular solution Insellösung
insulate dämmen * *z.B. Schall, Wärme ~* || isolieren * *durch Isolierstoff*
insulated isoliert * *z.B. Kabel*
insulated bearing isoliertes Lager * *[el.Masch.]* zur Vermeidung von Lagerströmen, die zum Lagerschaden führen können
insulated sleeve Isolierstoffhülse
insulating bushing Durchführung * *isolierte Leitungsdurchführungshülse* || Durchführungsisolator || Kabeldurchführung * *isolierte ~shülse*
insulating cover Abdeckung * *Isolierung*
insulating enamel Isolierlack || Tränkharz
insulating glass Isolierglas

insulating layer Isolierschicht
insulating material Dämmstoff || Isolierstoff
insulating sheet Flächenisolierstoff * *z.B. als Wicklungsisolation für el.Maschine, Trafo, Drossel*
insulating sheeting material Flächenisolierstoff
insulating sleeve Isolierumhüllung * *buchsen-/schlauchförmige ~*
insulating sleeving Isolierschlauch
insulating system Isoliersystem * *z.B. eines Motors*
insulating tape Isolierband
insulating terminal Lüsterklemme
insulating varnish Isolierlack
insulating washer Isolierscheibe * *siehe auch →Glimmerplättchen*
insulation Isolation || Isolierung * *(für [el.Masch.]: siehe VDE 530, Teile 18-32)* || Potenzialtrennung * *(weniger gebräuchlich)* || Schutz * *gegen Wärme, Spannung, Lärm usw.*
insulation class Isolationsgruppe * *Isolationsklasse, Isolierstoffklasse* || Isolationsklasse || Isolierklasse || Isolierstoffklasse * *siehe VDE 0530 Teil 1/ DIN IEC 85; class F insulation: ~ F*
insulation class B Isolierstoffklasse B
insulation coordination Isolationskoordinaten * *z.B. Luft- und Kriechstrecken (für Stromrichter: siehe IEC 664)*
insulation failure Isolierungsfehler
insulation fault Isolationsfehler
insulation group Isolationsgruppe
insulation layer Isolierschicht || Zwischenlage * *Isolations~ bei Trafos, Drosseln usw.*
insulation monitor Isolationswächter
insulation monitoring Isolationsüberwachung
insulation rating Isolationsbemessung * *~sspannung z.B. einer Wicklung*
insulation resistance Isolationswiderstand * *z.B. von Kabel, Leitung, Wicklung (siehe z.B. IEC 79-15 für el.Masch.)*
insulation sheets Flächenisolierstoffe
insulation strength Isolationsfestigkeit || Isoliervermögen * *(eines →Isoliersystems, kann mit (Stoß)-Spannungsprüfung ermittelt werden)*
insulation system Isoliersystem * *z.B. für eine Motorwicklung*
insulation test Isolationsmessung * *→Isolationsprüfung* || Isolationsprüfung
insulation tester Isolationsmessgerät
insulation testing Isolationsmessung * *→Isolationsprüfung* || Isolationsprüfung
insulation voltage Isolationsspannung
insulator Isolator || Nichtleiter
insurance Versicherung * *auch 'Versicherungspolice/prämie'*
insure versichern * *Eigentum ~*
insusceptible to interference störunempfindlich * *gegen elektromagnetische Störungen*
intact intakt || konsistent * *unversehrt*
intake Aufnahme * *Zustrom, aufgenommene Menge, Ein/Ansaugen* || Einlass || Eintritt || Zufuhrung * *Einlass -(öffnung), (Neu-) Aufnahme, Zustrom, aufgenommene Menge, Ein-/Absaugen*
intake air Ansaugluft || Frischluft * *angesaugte Luft; siehe →Kühlluft* || Zuluft * *angesaugte (Kühl)Luft*
intake diameter Ansaugdurchmesser * *z.B. bei Lüfter*
intake equipment Beladeeinrichtung * *z.B. für Chargenprozess*
intake filter Ansaugfilter
intake flange Ansaugstutzen * *z.B. für (Kühl)Luft bei elektr.Maschine*
intake pressure Ansaugdruck
intake side Ansaugseite * *Lüfter*

intake stub Ansaugstutzen * *Pumpe*
intake valve Ansaugventil
integer Festpunktzahl mit Vorzeichen ‖ ganze Zahl ‖ ganzzahlig
integer multiple ganzzahliges Vielfaches
integer-frequency harmonic ganzzahlige Oberwelle
integral eingebaut * *integriert* ‖ ganzzahlig ‖ Integral * *[in'ti:grel] of:* von/über ‖ integriert * *eingebaut, beinhaltet, beinhaltend;* ganz, vollständig ‖ intern * *eingebaut, beinhaltet* ‖ konsistent * *vollständig, unversehrt, zusammenpassend, z.B.zusammengehörige Daten aus einem Zyklus* ‖ vollständig
integral action I-Verhalten
integral component I-Anteil ‖ Integralanteil * *[in'ti:grel]*
integral gain Integrationszeitkonstante ‖ Integrierzeit ‖ Integrierzeitkonstante
integral slot winding Ganzlochwicklung * *{el. Masch.} AC-Motorwicklg., mit ganzzahlig. Nutzahl je Pol u.Wicklungsstrang*
integral term I-Anteil ‖ Integralanteil
integral time Integrationszeitkonstante ‖ Integrationszeitkonstante ‖ Integrierzeit * *[in'ti:grel]* ‖ Integrierzeitkonstante ‖ Nachstellzeit
integral time constant Integrationszeit ‖ Integrationszeitkonstante ‖ Integrierzeit ‖ Integrierzeitkonstante ‖ Nachstellzeit
integral-action controller I-Regler ‖ Integralregler
integral-action time Integrationszeit * *[inti:grel]* ‖ Integrationszeitkonstante ‖ Integrierzeit * *[in'ti:grel]* ‖ Integrierzeitkonstante * *[in'ti:grel]* ‖ Nachstellzeit
integral-key shaft Keilwelle
integrally cast feet angegossene Füße * *z.B. am Motor*
integrate aufnehmen * *eingliedern, integrieren, within: in* ‖ beinhalten * *in sich tragen* ‖ einbauen * *eingliedern, integrieren* ‖ einbinden * *integrieren; into:in* ‖ integrieren * *of: von; over: über* ‖ vereinigen * *zusammenschließen; within:in* ‖ zählen * *aufintegrieren* ‖ zusammenfassen * *integrieren*
integrate into zusammenschließen * *zu einem Ganzen ~*
integrated -integriert ‖ eingebaut * *integriert* ‖ integriert ‖ intern * *eingebaut* ‖ zusammenhängend * *zusammenhängend, einheitlich, gleichmäßig, integriert*
integrated circuit integrierte Schaltung ‖ integrierter Schaltkreis ‖ Schaltkreis * *integrierter ~*
integrated operation Verbund * *~ von Firmen usw.*
integrating integrierend
integrating element Integrierglied
integrating measurement integrierende Messung * *z.B. bei A/D-Umsetzer nach d."Dual-Slope-Verfahren" (langsam aber störarm)*
integration Aufnahme * *Eingliederung (within:in)* ‖ Integration
integration time Integrierzeitkonstante
integrator Integrator ‖ Integrierer ‖ Integrierglied
integrator setting value Integratorsetzwert
integrity Konsistenz * *z.B. Zusammenpassen von Daten* ‖ Vollständigkeit
Intel format Intel-Format * *Datenformat: niederwertiges Byte/Wort auf niederer, höherwertig.auf höherer Adresse*
intellectual gedanklich ‖ geistig * *intellektuell*
intellectual property geistiges Eigentum
intelligence Intelligenz
intelligent intelligent ‖ sinnvoll * *klug*
intelligent I/O module IP ‖ signalvorverarbeitende Baugruppe

intelligent I/O-module intelligente Peripheriebaugruppe * *z.B. bei SPS*
intelligent IO-module IP * *(SPS) intelligente Peripheriebaugruppe*
intelligent peripheral intelligente Peripheriebaugruppe
intelligible deutlich * *verständlich* ‖ verständlich
intend beabsichtigen ‖ vorhaben * *beabsichtigen, vorhaben, planen, bezwecken; to do/doing: etwas zu tun* ‖ vorsehen * *beabsichtigen, wollen; intend for: bestimmen für/zu*
intended beabsichtigt ‖ bestimmt * *intended for: vorgesehen für* ‖ gewollt * *beabsichtigt* ‖ gewünscht * *beabsichtigt, bezweckt* ‖ gezackt * *eingekerbt* ‖ vorgesehen * *beabsichtigt, geplant (something: etwas; to do/doing: zu tun)* ‖ zukünftig * *geplant, beabsichtigt*
intended use Anwendungszweck * *Bestimmung* ‖ Zweck * *Bestimmung*
intense groß * *(Hitze)* ‖ heftig * *Schmerz usw.* ‖ stark * *intensiv*
intensification Steigerung * *Verstärkung, Intensivieren*
intensified erhöht * *verstärkt* ‖ verstärkt * *intensiviert*
intensify ausbauen * *intensivieren, z.B. Geschäft(sbeziehungen)* ‖ erhöhen * *verstärken, intensivieren* ‖ intensivieren ‖ verstärken * *intensivieren*
intensity Stärke * *Intensität*
intensive intensiv
intensive course Intensivkurs
intent vorhaben * *(Substantiv) Absicht, Vorsatz, Zweck* ‖ Zweck * *Absicht*
intention Gedanke * *Absicht* ‖ Plan * *Absicht* ‖ vorhaben * *(Substantiv) Absicht, Vorhaben, Vorsatz, Plan, Zweck, Ziel*
intentional bewusst * *absichtlich* ‖ gewollt
intentionally gewollt * *(Adv.)*
inter alia unter anderem
inter- ver- * *unter/miteinander* ‖ Zwischen- * *untereinander*
inter-group sales Verbundvertrieb * *inter-group sales department:; ~ als Abteilung*
inter-phase insulation Phasentrenner * *{el.-Masch.} Isolierstoff zw. Spulengruppen u.Wicklungssträngen i.→Wicklungskopf*
inter-phase short circuit Phasenschluss
interact zusammenarbeiten ‖ zusammenwirken * *aufeinander wirken, sich gegenseitig beeinflussen*
interaction Beeinflussung * *gegenseitige ~; between: zwischen* ‖ gegenseitige Beeinflussung * *Wechselspiel * Wechselwirkung/-beziehungen (between: zwischen)* ‖ Wechselwirkung ‖ Zusammenspiel * *Wechselwirkung* ‖ zusammenwirken * *(Substantiv) Wechselwirkung*
interactive dialogfähig ‖ dialoggesteuert ‖ interaktiv
interactive program Dialogprogramm
interbridge reactor Saugdrossel * *{Stromrichter} zur Stromsymmetrierg. bei 12-pulsiger Stromrichterschaltung*
intercede for eintreten * *für jdn. Fürsprache einlegen, intervenieren*
intercept auffangen * *Funkspruch, Brief usw.*
intercession Verwendung * *Fürsprache*
interchangable gear Wechselrad
interchange Austausch * *z.B. ~ von Signalen, Telegrammen usw.* ‖ austauschen ‖ auswechseln ‖ Ersatz * *Auswechslung* ‖ vertauschen * *(miteinander) vertauschen, auswechseln, austauschen, abwechseln lassen*
interchange by mistake verwechseln * *versehentlich vertauschen*
interchangeability Austauschbarkeit

interchangeable austauschbar * *untereinander* || auswechselbar * *one with another:miteinander* ·
interchangeable gear wheel Wechselrad * *zur Änderung der Übersetzung bei* →*Wechselradgetriebe*
interchanging by mistake Verwechslung * *versehentliche Vertauschung*
interconnect koppeln || verbinden * *miteinander* ~ || Verbinder * *z.B. Steck*~ || verschalten
interconnected in a bus configuration vernetzt * *durch einen Kommunikations-Bus* ~
interconnected operation Verbundbetrieb
interconnected system operation Verbundbetrieb * *von Stromnetzen*
interconnected voltage Maschenspannung || verkettete Spannung * *Maschenspannung*
interconnecting network Verbindungsnetz
interconnection Verbindung || Verbund * *z.B.* ~ *von Stromnetzen; interconnected operation: ~betrieb* || Verkettung * *gegenseitige Verbindung* || Zusammenschalten || Zusammenschaltung
intercourse Verkehr * *persönlicher Verkehr*
interdependence Abhängigkeit * *gegenseitige* ~
interdependent zusammenhängend * *voneinander abhängend*
interdisciplinary interdisziplinär
interest Anteil * *(Wirtschaft) Beteiligung* || Beteiligung * *Anteil, Kapitalbeteiligung* || Interesse * *auch (An)Teilnahme, Reiz, Wichtigkeit; be of (little) interest:v.(geringer) Bedeutg.sein* || Zins * *Kapital~; in full measure: mit* ~ *und* ~*es*~; *yield interest:* ~*en tragen* || Zinsen * *Kapitalzins*
interest charges Zinslast
interest in business Geschäftsinteresse
interest rate Zinssatz
interesting interessant || interessierend
interface ankoppeln * *verbinden, anschalten (to: an)* || Ankopplung * *to:an* || Anpass- || anschalten * *Schnittstelle* || Anschaltung || Interface || Koppelglied || Nahtstelle * *Schnittstelle* || Schnittstelle * *Daten-, Signal-, logische, serielle Schnittstelle.* || verbinden * *z.B. Rechnersysteme miteinander* ~
interface board Anschaltbaugruppe * *Leiterkarte zum Anschluss eines externen Gerätes, eines Busses usw.* || Anschaltungsbaugruppe * *Leiterkarte* || Schnittstellenbaugruppe * *(Leiterkarte) auch serielle Schnittstellenbaugruppe* || Schnittstellenkarte
interface card Anschaltung * *als Leiterkarte; z.B. network interface card: Netzwerk*~ || Schnittstellenbaugruppe * *Leiterkarte* || Schnittstellenkarte
interface chip Peripheriebaustein * *Peripherieschaltkreis für Mikroprozessor (z.B. Interruptcontroller, USART)*
interface circuit Anpasselektronik
interface converter Schnittstellenumsetzer * *z.B. für serielle Schnittstelle*
interface data area Bereich-Anschaltung * *(SPS) Datenbereich für Anschaltungen*
interface data block Rangierdatenbaustein * *bei SPS*
interface electronics Anpasselektronik
interface element Koppelglied
interface facilities Anschlussmöglichkeiten * *für Peripheriegeräte*
interface list Rangierliste * *bei Operationsbaustein einer SPS*
interface memory Koppelspeicher
interface modul Anpassmodul
interface module Anschaltbaugruppe * *z.B. für ein externes Gerät, einen Feldbus usw.* || Anschaltung * *als Baugruppe/Baustein* || Anschaltungsbaugruppe || Interfacemodul || Peripheriemodul || Schnittstellenbaugruppe
interface partner Koppelpartner

interface relay Koppelrelais
interface submodule Schnittstellensubmodul
interface tester Schnittstellentester
interface with bus capability busfähige Schnittstelle
interfacing Anbindung * *durch Signale, serielle Schnittstelle usw.* || Ankopplung * *Signalankopplung, Datenkopplung (to:an)* || Kopplung
interfere beeinflussen * *s.einmischen, (be)hindern, eingreifen, stören, beeinträchtigen* || eingreifen * *störend* ~ || stören * *behindern, belästigen, dazwischenkommen, sich einmischen, stören bei Ablauf/Verrichtung* || verfälschen * *störend einwirken auf*
interfered gestört * *behindert, belästigt; auch durch elektromagnet. Störung, Netzrückwirkung usw.*
interference Beeinflussung * *[inter'fierens] störende, schädliche* ~ *(Störung, Beeinträchtigung, Übersprechen)* || Einfluss * *Eingreifen (with:in), Beeinflussung, Beeinträchtigung, Störung,Überlagerung* || Einkopplung * *von Störungen* || Einstreuungen || gegenseitige Beeinflussung * *störende* || Interferenz || Störung * *[inter'fi:rens] Einmischung, elektromagnetische* ~ || Überlagerung * *Störung*
interference current Störstrom
interference field strength Störfeldstärke
interference filter Entstörfilter
interference fit Presspassung
interference generator Störgeneratoro 584
interference immune störsicher * *gegen elektromagnetische Störungen*
interference immunity Störfestigkeit * *(siehe z.B. EN61000-4)* || Störsicherheit || Störspannungsfestikeit
interference level Störpegel
interference power Störleistung
interference protection Störschutz
interference radiation Störeinstrahlung
interference resistance Störfestigkeit
interference suppresion capacitor Störschutzkondensator
interference suppression Entstörung * *Störunterdrückung* || Funkentstörung || Störschutz || Störungsunterdrückung
interference suppression choke Entstördrossel || Funkentstördrossel || Störschutzdrossel
interference suppression coil Entstördrossel
interference suppression filter Entstörfilter
interference suppression measures Entstörung * *Entstörmaßnahmen*
interference suppressor Entstörmittel
interference variable Störgröße
interference voltage Fremdspannung || Störspannung
interference voltage spike Störspitze
interference voltages Störspannungen
interference- Stör- * *(elektromagnetische) Störung*
interference-proof störsicher * *gegen elektromagnetische Störungen*
interference-supression capacitor Entstörkondensator
interfering pulse Störimpuls
interfering radiation Störstrahlung
interim vorläufig || Zwischen- * *zeitlich (z.B. Bilanz, Bericht)*
interim period Übergangszeit * *Zwischenzeit* || Zwischenzeit
interim reply Zwischenbescheid
interim solution Zwischenlösung * *siehe auch* →*Notlösung*
interior innen || innenliegend || Innenraum * *das Innere* || inner || Inneres || innerhalb * *innen gelegen* || intern * *innen, innengelegen*

interior construction   Innenausbau * *{Bautechnik}*
interior decoration   Innenausbau * *{Bautechnik}*
interior finishing   Innenausbau * *{Bautechnik}*
interior fittings   Innenausbau
interior parts   Innenteile
interior space   Innenraum * *der innere Raum*
interiors   Innenteile
interlacement   Verflechtung
interlayer   Zwischenlage
interleave   schachteln * *Blechpaket, Wicklung*
interleaving   Schachtelung * *Blechpaket, Wicklung*
interlining   Zwischenlage * *Zwischenbelag, Einlagestoff/-material*
interlink   koppeln * *z.B. über serielle Schnittstelle* ~ || kuppeln * *über Signale* || verbinden * *miteinander* ~ *(z.B. über Bussystem)* || verketten || verknüpfen * *(figürl.) miteinander* ~
interlinkage   Verkettung * *siehe auch* →*Verknüpfung*
interlinked voltage   verkettete Spannung
interlinking   Anbindung * →*Verknüpfung* || Verkettung || Verknüpfung
interlinking of computers   Rechnerkopplung
interlock   blockieren * *verriegeln* || Logik * *Verriegelung* || Sperre * *Verriegelung* || verriegeln * *logisch, steuerungsmäßig, mechanisch usw.* ~ || Verriegelung * *logische, steuerungstechnische, mechanische usw.* ~
interlock control   Verriegelungssteuerung
interlocked   verriegelt
interlocking   Verflechtung || Verkettung * *Ineinanderhaken, Verzahnung, Inandergreifen* || Verknüpfung * *Verriegelung, vor allem steuerungsmäßig* || Verriegelung || Verzahnung * *(figürl.)*
interlocking circuit   Sperrschaltung
interlocking function   Verriegelung * *~sfunktion*
interlocutor   Gesprächspartner || Partner * *Gesprächspartner*
intermediary   Einlage * *{Kabel | Draht}* || Vermittler * *Mittelsmann* || Zwischenlage
intermediate   dazwischenliegend * *(zeitlich u. räumlich)* || mittlerer * *dazwischenliegend* || Zwischen- * *zeitlich, räumlich*
intermediate belt   Beschleunigungsband * *bei Bandförderer*
intermediate circuit   Zwischenkreis * *weniger gebräuchlich*
intermediate cooling medium   Zwischenkühlmittel
intermediate DC voltage   Zwischenkreisspannung
intermediate element   Bindeglied || Zwischenglied
intermediate flange   Zwischenflansch
intermediate frequency   Zwischenfrequenz
intermediate gear   Vorgelege * *vor dem 'eigentlichen' Getriebe angeordnetes Zwischengetriebe*
intermediate housing   Zwischengehäuse
intermediate layer   Zwischenlage
intermediate part   Zwischenglied * *Maschinenelement; z.B. in einer elastischen Kupplung*
intermediate plate   Zwischenplatte
intermediate position   Mittelstellung
intermediate product   Halbprodukt * *Zwischenprodukt*
intermediate result   Zwischenergebnis
intermediate step   Zwischenschritt
intermediate storage   Zwischenlagerung || Zwischenspeicherung
intermediate terminal   Zwischenklemme
intermediate transformer   Vorschalttransformator * *Zwischentransformator* || Vortransformator * *Zwischentransformator* || Zwischentrafo * *siehe auch* →*Zwischentransformator u.* →*Vorschalttransformator*
intermediate value   Zwischenwert
intermediate voltage   Zwischenkreisspannung

intermediate-circuit converter   Zwischenkreisumrichter
intermeshing   Vermaschung
intermit   aussetzen * *unterbrechen*
intermittent   intermittierend || stoßartig * *in Stößen/ mit Unterbrechungen auftretend*
intermittent current   lückender Strom || Lückstrom
intermittent current condition   Stromlücken
intermittent DC area   Lückbereich
intermittent DC flow operation   Lückbetrieb
intermittent duty   Aussetzbetrieb || Schaltbetrieb
intermittent duty with starting   periodischer Aussetzbetrieb mit Einfluss des Anlaufvorgangs * *{el. Masch.}* →*Betriebsart S4*
intermittent electrical contact   Wackelkontakt
intermittent operation   Aussetzbetrieb || Schaltbetrieb * *Aussetzbetrieb*
intermittent periodic duty   Betriebart S3 * *{el. Masch.} periodischer Aussetzbetrieb* || periodischer Aussetzbetrieb * *{el.Masch.}* →*Betriebsart S3* || Schaltbetrieb * *siehe auch 'Aussetzbetrieb' (Betriebsart für Motor/Stromrichter)*
intermittent periodic duty type   Aussetzbetrieb * *Betriebsart S3 nach VDE0530/DIN41756 für Motor/Stromrichter*
intermittent periodic duty type with starting   Aussetzbetrieb mit Einfluss des Anlaufvorgangs * *Motor-Betriebsart S4 nach VDE0530/IEC34-1*
intermittent periodic duty type without starting   Aussetzbetrieb ohne Einfluss des Anlaufvorgangs * *Motorbetriebsart S3 nach VDE0530*
intermittent periodic duty with electric braking   - Betriebsart S5 * *{el.Masch.}*
intermittent periodic duty with starting   Betriebsart S4 * *{el.Masch.} period. Aussetzbetrieb mit Einfluss d. Anlaufs*
intermittent service   Schaltbetrieb * *Aussetzbetrieb*
intermittently   stoßweise * *mit Unterbrechungen, zeitweilig aussetzend*
intermix   mischen * *ver~*
internal   betrieblich * *intern* || innen || Innen- || innenliegend || inner || innerer || intern * *(Adj.)*
internal bearing clearance   Lagerluft * *Maß, um das sich d.Außenring gegen d.Innenring verschieben lässt*
internal consumption   Eigenverbrauch
internal control address   Steueradresse
internal control command   Steuerbefehl * *interner Steuerbefehl bei SPS*
internal fan   Innenlüfter
internal gear   Innenverzahnung
internal gearing   Innenverzahnung
internal hexagon   Innensechskant
internal impedance   Ausgangsimpedanz * *Innenwiderstand z.B. einer Spannungs- oder Signalquelle* || Innenwiderstand
internal lubrication   Innenölung
internal mail   Hauspost
internal relay   Merker * *bei Kontaktplandarstellung eines SPS-Programms*
internal resistance   Ausgangswiderstand * *z.B. einer Spannungs- oder Signalquelle* || Innenwiderstand
internal rotor   Innenläufer
internal source resistance   Ausgangswiderstand * *Quellwiderstand e. Spannungs-/Signalquelle*
internal teeth   Innenverzahnung
internal thread   Innengewinde
internal use   interner Gebrauch
internal view   Innenansicht
internal-combustion engine   Verbrennungsmotor
internal-rotor motor   Innenläufermotor
internally   intern * *(Adv.)*
internally cooled   innengekühlt

internally ventilated innengekühlt * elektr. Maschine
International system of units SI-Einheiten
internetwork verbinden * über ein (Computer-) Netzwerk ~, vernetzen || vernetzen
internetworked vernetzt * z.b. durch Bussystem oder Peer-to-Peer Netz ~
internetworking Netzverbund || Vernetzung
interoperability Interoperability * Betreibbarkeit von Geräten mehrerer Hersteller an einem Bussystem
interoperable gemeinsam betreibbar * z.b. Geräte verschiedener Hersteller an einem Feldbussystem
interpedendence Beeinflussung * gegenseitige Abhängigkeit
interphase commutation Phasenfolgelöschung
interphase transformer Saugdrossel * mit Mittelpunktanzapfung zur Stromssymmetrierung für 12-Puls-Stromrichter
interphase transformer connection Saugdrosselschaltung
interplay Wechselspiel || Wechselwirkung * auch 'Wechselspiel' || Zusammenspiel * of forces: der Kräfte; auch 'Wechselwirkung' || zusammenwirken * (Substantiv) Zusammenspiel, Wechselwirkung
interpolate interpolieren * linearly:linear
interpolated interpoliert
interpolation Interpolation * linear interpolation: Linear ~
interpolation algorithm Interpolationsalgorithmus
interpolation point Stützpunkt * eines Polygonzuges || Stützwert * Stützpunkt eines Polygonzuges
interpole Wendepol
interpole core Wendepolkern
interpole winding Wendepolwicklung
interpose intervenieren * vermittelnd eingreifen, "dazwischengehen"
interposed dazwischenliegend || zwischengeschaltet
interpret interpretieren || übersetzen * verdolmetschen; auch Quellprogramm während des Programmablaufs (z.B. BASIC)
interpretable interpretationsfähig
interpretation Auslegung * Interpretation || Auswertung || Erklärung * Deutung || Interpretation || Sinn * Auslegung || Übersetzung * Auslegung, Deutung, (mündliche) Wiedergabe, Sprachübersetzung, Interpretation
interprete auslegen * interpretieren, z.B. Norm, Vorschrift || auswerten || erklären * deuten
interpreter Auswerter || Dolmetscher || Interpreter * z.B. Hochspracheübersetzer, der während des Programmablaufs arbeitet (z.B. →BASIC) || Übersetzer * für Fremdsprachen (bes. mündlich), Dolmetscher; online-Übersetzer für Hochspracheprogr.
interprocessor communication Kopplung * siehe →Rechnerkopplung || Rechnerkopplung
interprocessor communication flag Koppelmerker * bei SPS
interpulse period Impulslücke || Impulspause
interrogate abfragen * be/ausfragen (auch einen Binärausgang), vernehmen
interrogation Abfrage * Frage, Befragung, Vernehmung
interrupt abbrechen * unterbrechen || Alarm * [SPS | Computer] || auftrennen * (elektr.) Stromkreis || ausschalten || aussetzen * unterbrechen || Interrupt || Programmunterbrechung || stören * unterbrechen || trennen * unterbrechen, (Stromkreise) trennen || unterbrechen
interrupt address Interruptadresse
interrupt condition code word Unterbrechungsanzeigewort * bei SPS
interrupt event Alarmereignis || Interruptereignis

interrupt generating interruptfähig * alarmbildend, einen Interrupt erzeugend (z.b. Digitaleingang)
interrupt handler Alarmprogramm || Interruptbearbeitung * ~sprogramm || Interruptprogramm
interrupt identification Interruptkennung
interrupt input Alarmeingang
interrupt latency Interrupt-Verzugszeit
interrupt level Interruptebene
interrupt list Alarmliste * z.b. bei SPS
interrupt procedure Alarmbearbeitung * durch Computer || Interruptbearbeitung
interrupt processing Alarmbearbeitung * durch Computer, SPS usw. || Alarmverarbeitung || Interruptbearbeitung
interrupt program Interruptbearbeitung || Interruptprogramm
interrupt request Interruptanforderung
interrupt response time Interrupt-Reaktionszeit || Interrupt-Verzugszeit
interrupt service routine Alarmbearbeitung * durch Computer || Alarmprogramm || Interruptbearbeitung * ~sprogramm || Interruptprogramm
interrupt task Interruptprogramm
interrupt the supply feeder vom Netz trennen
interrupt timer Wecker * Interrupt-erzeugendes Zeitglied
interrupt vektor Interruptadresse * indirekte Interruptadresse, Zeiger auf das Interruptprogramm
interrupt-capable alarmfähig * z.B. Digital-/Binäreingang || interruptfähig * z.B. Binär-/Digitaleingang
interrupt-controlled alarmgesteuert || interruptgesteuert
interrupt-driven alarmgesteuert || interruptgesteuert
interruptable abschaltbar
interrupted unterbrochen
interrupted cable Kabelbruch || Leitungsbruch || Leitungsunterbrechung
interruptible abschaltbar
interrupting capacity Ausschaltvermögen || Schaltvermögen * Aus~ eines Leistungsschalters
interrupting I2t value Lösch-I2t-Wert * für Sicherung
interrupting rating Ausschaltvermögen
interruption Abschaltung * Unterbrechung || Störung * Unterbrechung || Unterbrechung
interruption of operation Betriebsunterbrechung
interruption of service Betriebsstörung * Stockung/Unterbrechung des Betriebs || Betriebsunterbrechung
interruption time Abschaltzeit * ~ einer Sicherung/eines Leistungsschalters usw. || Unterbrechungszeit
interruption-free unterbrechungsfrei
intersect überschneiden * zwei Linien
intersection Schnittpunkt
intersection point Schnittpunkt * z.B. von Kurven oder Phasenspannungen
interstage Zwischenstufe
interturn insulation Windungsisolierung * Isolierung zw. benachbarten Windungen z.B. eines Elektromotors
interturn short-circuit Windungsschluss
interval Abstand * zeitlich; in short intervals: in kurzen Abständen || Frist * Zwischenzeit/raum || Intervall || Lücke * Zwischenraum || Pause * ['interwel] || Takt * at 10 minute intervals/every 10 minutes: im 10-Minuten-~ || Taktzeit * at 10 minute intervals: im 10-Minuten-Takt || Zeit * ~intervall, ~abstand || Zeitabschnitt || Zeitabstand || Zeitbereich * Zeitspanne || Zeitraum * ['interwel] Zeitabstand || Zeitspanne * Intervall || Zwischenzeit * Zeitabstand, Zwischenzeit, Zeitperiode, Pause

interval of zero torque drehmomentfreie Pause || drehmomentlose Pause
intervene dazwischenkommen || eingreifen * vermittelnd ~ (in:in) || eintreten * ein Ereignis; auch 'plötzlich ~, unerwartet dazwischenkommen, eingreifen' || intervenieren || vermitteln * between: zwischen, z.B. bei einem Streit ~, intervenieren
intervening period Übergangszeit * dazwischenliegende Zeitspanne || Zwischenzeit * siehe auch →Übergangszeit
intervention Abgleich * Einstellung des Eingriffs, das Abgleichen || Bewertung * Eingriff eines Signals (z.B. Zusatzsollwert) || eingreifen * (Substantiv) Eingriff; manual intervention:Eingreifen von Hand || Eingriff * auch:eines Signals/Sollwerts || Gewichtung * Eingriff
intervention factor Bewertungsfaktor * Eingriff-Faktor
interview Besprechung * Befragung || Bewerbung * Bewerbungs-/Vorstellungsgespräch || Vorstellung * Bewerbungsgespräch (with:bei) || Vorstellungsgespräch
interweaving Verflechtung
interworking zusammenwirken * (Substantiv)
interwoven verwoben
intimate andeuten * zu verstehen geben, andeuten/-kündigen, mitteilen || eng * innig
into the supply ins Netz
intricate facettenreich * verwickelt, verschlungen, vielschichtig, knifflig, schwierig || komplex * verwickelt; siehe auch →vielschichtig || kompliziert * Problem, Frage usw.; auch 'vielschichtig'|| schwierig * verwickelt || vermascht * verzweigt, verschlungen, verwickelt || vielschichtig * verzweigt, verschlungen, verwickelt, knifflig, schwierig, verworren
intrinsic eigen * innewohnend, inner, wahr, eigentlich, wirklich || Eigen- * innewohnend || eigentlich * innewohnend || wesentlich * innewohnend, inner, wahr, eigentlich, wirklich, Eigen-
intrinsic capacitance Eigenkapazität
intrinsic safety Eigensicherheit * Schutzart 'i' nach EN50020 für Ex-geschützte el. Betriebsmittel
intrinsically safe eigensicher * Ex-Schutzart 'i' nach EN50020
introduce aufbringen * Mode || bekanntmachen * einführen, vorstellen (auch eine Person) || einbringen * z.B. Vorschlag, Gesetz; auch 'Thema/ Frage anschneiden/aufwerfen/vorstellen'|| einführen * (auch figürl.) in eine Öffnung ~; well introduced: gut eingeführt (Person, Produkt) || vorstellen * einführen, (her-) einbringen, bekannt machen; someone to somebody: jemandem eine Person
introduce oneself sich vorstellen * sich jemandem
introduce to the marketplace am Markt einführen
introduction Anleitung * Einführung || Einführung * ~ in Wort oder Schrift; ~einer Person; ~ von Maßnahmen || Vorstellung * Einführung || Vorwort * Einleitung
introduction into the market Markteinführung
introductory training course Einführungskurs
intrude stören * unangenehm bemerkbar machen,aufdrängen,sich einmischen/eindrängen
intrusion Störung * Einmischung, Eindringen, Zu-/ Aufdringlichkeit, Belästigung
intuitive intuitiv
inundate überschwemmen * (auch figürl.)
inundation Überflutung
invade eindringen * in einen Markt/ein Land
invalid fehlerhaft * ungültig || nicht erlaubt * ungültig, nichtig || ungültig * invalidate: für ~ erklären; be no longer valid: ~ werden || unzulässig * ungültig || verfälscht * ungültig
invalidate verfälschen * ungültig machen

invariable gleich bleibend * unveränderbar/lich || gleichförmig * →unveränderlich || unveränderlich
invariant invariant
invasion eindringen * (Substantiv) Eindringen, Einfall, Angriff
invent ausdenken
invention Erfindung
inventory Bestandsaufnahme * take: machen || Lagerbestand || Lagerhaltung
inverse invers || umgekehrt * in inverse proportion to: im ~en Verhältn. zu; inversely proportional: ~ proportional || umkehren * Vorzeichen
inverse direction Sperrichtung * bei Halbleitern
inverse in proportion umgekehrt proportional * to:zu
inverse parallel gegenparallel
inverse sign entgegengesetztes Vorzeichen
inverse transform Rücktransformation
inverse voltage Sperrspannung * z.B. Diode, Transistor
inverse-parallel antiparallel
inverse-parallel connection Gegenparallelschaltung * 4-Quadranten DC-Doppelstromrichter in Brückenschaltung
inversely in proportion umgekehrt proportional * (Adv.) to:zu
inversely proportional umgekehrt proportional * to:zu
inversion Invertierung * für Binär- und Analogsignal (logische ~ bzw.Vorzeichenumkehr) || Umkehr * z.B. Vorzeichen~
inversion limit Wechselrichtertrittgrenze
invert invertieren || negieren * logisch || wechselrichten
inverted invers * invertiert || invertiert * Analogsignal (Vorzeichenumkehr) oder Digitalsignal || umgekehrt * auch 'mit ~em Vorzeichen, hängend (montiert)'|| verkehrt * umgekehrt
inverted commas Gänsefüßchen
inverted signal invertiertes Signal * Analog- und Digitalsignal
inverted-plan view Ansicht von unten
inverter Invertierer || Stromrichter * Wechselrichter (mit Wechselspannungsausgang) || Umformer * Wechselrichter || Umrichter * Wechselrichter || Wechselrichter
inverter commutation failure Wechselrichterkippen * Thyristorzerstör.falls alt.Ventil noch nicht stromlos u.neues gezünd.
inverter limit alpha w || alpha-W || Wechselrichtertrittgrenze
inverter load Umrichterauslastung
inverter module Wechselrichter * Wechselrichtereinheit || Wechselrichtereinheit
inverter operation Wechselrichterbetrieb * Umwandlung von Gleich- in Wechselstrom
inverter range Wechselrichterbetrieb * bei Umkehrstromrichter
inverter shoot-through Wechselrichterkippen * Überstrom i.Thyristorschaltg.: neues Ventil gezündet u.altes noch stromführ.
inverter stability limit alpha w * Wechselrichtertrittgrenze || alpha-W || Wechselrichtertrittgrenze * max.Steuerwink.,darf z.Vermeidg.d.Wechselr.-kippens nicht überschrit.werd.
inverter unit Wechselrichtereinheit * bei modularem Umrichtersystem
inverter-duty motor Motor für Betrieb am Umrichter
inverter-fed umrichtergespeist
inverting invertierend || Invertierung * für Binär- und Analogsignal
inverting amplifier Umkehrverstärker
invertor Wechselrichter * (veraltet)
invest anlegen * Kapital usw. || investieren

invested capital Anlagevermögen
investigate nachprüfen * *untersuchen* || prüfen * *untersuchen, erforschen, ermitteln, nachforschen* || überprüfen * *untersuchen* || untersuchen * *technisch/wissenschaftlich/juristisch* ~
investigation Ermittlung * *Untersuchung* || Prüfung * *Untersuchung, Ermittlung, Nachforschung* || Überprüfung * *Untersuchung, Ermittlung, Nachforschung* || Untersuchung
investment Anlage * *Kapitalanlage* || Beteiligung * *durch Kapitalanlage* || Investition || Investitionskosten
investment expenses Investitionskosten
investment goods Investitionsgüter
investment returns Kapitalrückfluss
investor Investor
invincible unüberwindbar * *Barriere, Schwierigkeit, Widerstände, Festung usw.*
invisible unsichtbar * *to:für; siehe auch* →*verborgen* || verborgen * *unsichtbar*
invitation Aufforderung * *Einladung* || Einladung * *auch zu einer Besprechung*
invitation to bid Ausschreibung * *für Angebote*
invite einladen * *jemanden zu Besuch, Besprechung usw.* ~
invoice belasten * *Konto* ~ || berechnen * *Kosten in Rechnung stellen, fakturieren* || fakturieren || Rechnung * *Waren*~ || verrechnen * *buchungstechnisch; in Rechnung stellen, fakturieren*
invoice at a lump-sum price pauschal verrechnen
invoiced berechnet * *in Rechnung gestellt* || kostenpflichtig * *berechnet, in Rechnung gestellt*
invoiced amount Rechnungsbetrag * *in Rechnung gestellter Betrag*
invoicing Abrechnung * *{Wirtschaft}* || Berechnung * *in Rechnung stellen* || Fakturierung || Verrechnung
invoicing department Abrechnungseinheit * *Dienststelle*
invoke aktivieren * *aufrufen (Programm usw.)* || aufrufen * *Programm usw.*
involuntary unabsichtlich * *unfreiwillig, unwillkürlich* || unbeabsichtigt * *unfreiwillig, unwillkürlich* || ungewollt * *unfreiwillig* || versehentlich * *unfreiwillig*
involute Evolvente * *z.B. Flankenform einer Verzahnung*
involve bedingen * *mit sich bringen* || einbeziehen || handeln * *~ um, angehen, betreffen, berühren, mit sich bringen* || mit sich bringen || verbunden sein mit * *mit sich bringen, Folgen haben, nötig machen*
involved beteiligt || betreffend * *betroffen* || betroffen * *beteiligt, verwickelt, berührt, involviert (in/with: in/an)* || einbezogen
involvement Beteiligung * *das "Verwickelt-/Beteiligt-/Betroffen-Sein"* || Verflechtung * *Verstrickung*
involving betreffend * *einbeziehend*
IO Ein-Ausgabe
IO-module Ein-Ausgabe-Baugruppe
ion exchange unit Ionenaustauscher * *z.B. zum* →*Entionisieren von Kühlwasser in einem geschlossenen Kühlkreislauf*
ion exchanger Ionentauscher * *z.B. zum* →*Entionisieren des Wassers i.e. Kühlwasserkreislauf (zur Leitfähik.reduz.)*
ionization Ionisierung
IP IP * *'International Protection': kennzeichnet Schutz geg.Berührg. Fremdkörper, Wasser, Staub (IEC 34)*
IPC Industrie-PC * *Abkürzg.f. 'Industrial PC'*
IPM IPM * *'Intelligent Power Module': hochintegrierter Leist.halbleiter, z.B. in Dick-/Dünnschichttech.*
IR compensation I-mal-R-Kompensation * *z.B.*

Komp. d. Ankerwiderstands bei tacholoser n-Regelg. || IxR Kompensation || IxR-Kompensation * *Ankerwiderstandskomp. für tacholosen Betrieb e. DC Motors*
irksome lästig * *unangenehm* || mühsam * *ermüdend* || schwierig * *lästig, beschwerlich* || unangenehm * *lästig, beschwerlich, ärgerlich, verdrießlich, beschwerlich*
iron bügeln || Eisen
iron and steel plant Hüttenwerk
iron core Eisenkern
iron losses Eisenverluste
iron rod Rundeisen
iron-core reactor Eisendrossel
iron-loss Eisenverlust
ironworks Hüttenwerk * *Stahlwerk*
IRQ Interruptanforderung * *Abk. f.'Interrupt Request'* || IRQ * *{Computer\ SPS} Abkürzg. für 'Interrupt ReQuest': Interrupt-Anforderung*
irrefutable unbestritten * *unwiderlegbar, nicht zu widerlegen*
irregular abweichend || irregulär || ungleichförmig * *unregelmäßig; auch (Dreh)bewegung* || ungleichmäßig * *unregelmäßig* || unregelmäßig || unvorschriftsmäßig
irregular running unruhiger Lauf
irregularity Ausreißer || irregulärer Zustand || Ungleichförmigkeit || Unregelmäßigkeit
irrelevant irrelevant * *to:für* || ohne Bedeutung
irreplaceable unersetzbar
irrespective unabhängig * *irrespective of:ohne Rücksicht auf; irrespective of whether:ungeachtet dessen, ob*
irrespective of -unabhängig * *(vorangestellt)* || unabhängig von * *~ Tatsachen/Umstand; ohne Rücksicht auf* || ungeachtet
irrespective of the fact that unabhängig davon ob
irrespective of the power rating leistungsunabhängig * *unabhängig von der Nennleistung*
irrespective of whether unabhängig davon ob * *or not: oder nicht*
irrespective whether unabhängig davon ob * *or not:oder nicht*
irreversible selbsthemmend * *nicht rückwärtslaufend (z.B. Getriebe)*
irridation Bestrahlung
irritate stören * *ärgern*
irritation Störung * *Ärgernis, Ärger, Reizung*
is beträgt
is eliminated entfällt * *wird überflüssig (gemacht)*
is to be soll * *z.B. is to be avoided: soll vermieden werden*
ISDN ISDN * *Abkürzg.für 'Integrated Services Digital Network': dienstintegrierendes digit. Nachrichtennetz*
island power system Inselnetz
ISO ISO * *Abkürz. f. 'International Organization for Standardization':Internationale Organis.für Normen*
ISO tolerance ISO-Passung
isolate abtrennen || fernhalten * *from: von; z.B. Störungen vom Netz* ~ || freischalten * *Stromkreis* ~; *siehe VDE 0100 Teil 2* || isolieren * *potenzialmäßig trennen, spannungs-/stromlos machen, (sicherheitsmäßig) abschalten* || lokalisieren * *isolieren, absondern, abscheiden* || trennen * *isolieren (auch Stromkreise), absondern (from: von), chem. rein darstellen, abschließen* || unterbrechen * *(potenzialmäßig) trennen, z.B. Stromkreise*
isolate faults Fehler lokalisieren
isolate from the supply freischalten * *vom Netz trennen*
isolate from the supply system vom Netz trennen
isolated abgeschlossen * *abgesondert, isoliert* || einzeln * *für sich allein* || galvanisch getrennt || iso-

liert * *potenzialgetrennt, (zur Sicherheit) abgeschaltet* || potenzialfrei || potenzialgetrennt || spannungslos * *Stromkreis*
isolated current sensing potenzialfreie Stromistwerterfassung
isolated fact Einzelheit * *abgesonderter/abgeschlossener Tatbestand*
isolated network isoliertes Netz
isolated operation Inselbetrieb * *z.B. Inselnetz, Dieselgenerator*
isolated power station Inselkraftwerk
isolated supply IT-Netz * *(VDE 0100 Teil 410)*
isolated supply system ungeerdetes Netz * *operation on/from: Betrieb am*
isolating potenzialtrennend
isolating amplifier Trennverstärker
isolating facility Freischaltmöglichkeit * *Auftrennbarkeit des Stromkreises*
isolating joint Trennfuge * *z.B. in Motorläufer*
isolating switch Trenner
isolating transducer Trennwandler
isolating transformer Trenntrafo || Trennwandler * *z.B. Trenntransformator, potenzialfreier Strom/Spannungswandler* || Vorschalttransformator * *Isoliertrafo* || Vortrafo * *Trenntrafo* || Vortransformator * *Isoliertrafo*
isolation galvanische Trennung || Isolation * *Potenzialtrennung, (Sicherheits)Abschaltung e. Stromkreises* || Potenzialtrennung || Trennung * *Isolierung, Absonderung, galvanische ~, Abschließung*
isolation amplifier Isolierverstärker * *Trennverstärker*
isolation coordinates Isolationskoordinaten
isolation from supply Spannungsfreiheit * *Trennung vom Netz; verify:feststellen*
isolation point Trennstelle * *~ mit galvanischer Trennung*
isolation resistance measurement Isolationsmessung
isolation switch Trennschalter
isolation transformer Trenntrafo
ISP ISP * *Abkürzg.f. 'Interoperable Systems Project':Feldbus-Normungsvorhaben*
issue abgeben * *a declaration: eine Erklärung ~* || abschicken * *z.B. ein Telegramm auf Bus* || Aspekt * *(Kern)Punkt, Sachverhalt, (Streit/Kern)Frage* || Ausgabe * *Herausgabe/Folge/Nummer, z.B. einer Zeitschrift* || ausgeben * *ein (Warn/Zustands) Signal* || ausströmen || ausströmen * *herauskommen, herausströmen, hervorbrechen, entspringen* || Exemplar * *Ausgabe (einer Zeitschrift usw.)* || Gegenstand * *Diskussionspunkt, Streitfrage* || Gesichtspunkt || herausgeben * *z.B. Material, Buch ~* || Kernpunkt || Problem * *Kernpunkt, Sachverhalt, Streitfrage* || Thema * *Kernpunkt, wesentlicher Punkt, Streit-/Diskussionspunkt* || zuteilen * *ausgeben* || zuweisen * *ausgeben*
IT Datenverarbeitung * *Abk. für 'Information Technology'* || EDV * *Abk. für 'Information Technology'* || Informationstechnik * *Abkürz. f. 'Information Technology'*
it depends on je nachdem
it is planned that voraussichtlich * *(Adv.) es ist geplant, dass*
it is possible that unter Umständen * *es ist möglich, dass*
it seems that scheinbar * *(Adv.) anscheinend; es scheint, dass*
it stands to reason that selbstverständlich * *es ist ~, dass*
IT system IT-Netz * *erdfreies Netz (Sternpunkte bei Einspeisung u.Verbraucher ungeerdet)* || ungeerdetes Netz * *siehe →IT-Netz*
it's all right bitte * *(salopp) 'schon gut/in Ordnung' als Antwort auf Entschuldigung*

Italian italienisch
Italy Italien
item Artikel * *Stück Ware* || Eintrag || Einzelheit * *Einzelgegenstand, Stück, Posten, Position, Artikel* || Einzelteil * *(Einzel)Gegenstand, Stück* || Gegenstand * *Punkt (der Tagesordnung usw.), (Einzel)Gegenstand, Stück, Posten, Artikel* || Gesprächsthema * *Einzelpunkt (von mehreren)* || Position * *~ in Liste/Aufstellung/Bestellformular/Lieferschein* || Punkt * *der Tagesordnung* || Stück * *Zahl, ~ Ware, Artikel, Teil* || Teil * *Artikel*
item No Bestellnummer * *Artikelnummer*
item No. Artikel-Nr || Sachnummer * *z.B. in Werks-Lagerliste*
itemize aufführen * *(amerikan.) im Einzelnen ~*
itemized aufgeführt * *(amerikan.) im Einzelnen ~*
iteration factor Wiederholfaktor * *Anzahl d.Wiederholungen einer Befehlssequenz bei SPS/Rechner*
iteration method Iterationsverfahren
iteration technique Iterationsverfahren * *Rechenmit Annäherung ans Endergebnis durch wiederholte gleiche Rechenschritte*
iterative impedance Wellenwiderstand
IxR compensation IxR Kompensation || IxR-Kompensation
I²T Value I²T-Wert * *"I-Quadrat t Wert",Grenzlastintegral, (thermische) Belastungsgrenze*

J

jack Bock || Buchse * *Steckdose, Buchse (z.B. für 4 oder 2mm Prüfstecker), Klinkenbuchse* || Klinke * *(elektr.)* || Klinkenstecker || Niederhalter
jack plug Klinkenstecker
jack screw Abdrückschraube * *zum Demontieren/ Abdrücken eines Maschinenteils (z.B. Nabe von einer Welle)*
jack shaft Antriebsspindel * *im Walzwerk*
jack up anheben * *mit Vorrichtung*
jacket Buchse * *(elektrisch)* || Hülse * *Ummantelung, Hülle; tubular jacket: tubusförmige Hülle* || Mantel * *Ummantelung, Hülle, Umwicklung* || Rohrmantel || Umhüllung
jacketing Ummantelung
jacking lug Ansetzstelle * *zum Heben*
jacking spindle Hubspindel
jackscrew Hubspindel
jagged gezackt
jam festfressen * *klemmen, hemmen, stocken, verstopfen, blockieren* || klemmen * *festsitzen, festklemmen* || quetschen * *(ein-)zwängen/klemmen/ keilen, zusammendrücken, quetschen (auch Körperteil b.Unfall)* || stören * *sperren, hemmen, blokkieren, klemmen, verstopfen, Radiosender stören* || Störung * *Hemmung, Stockung, Klemmung, Radio~* || verstopfen * *Straße usw.*
jamming Verklemmung
jar-ram Rüttler * *z.B. zur Bodenfestigung*
jarring blow Prellschlag
jarring machine Rüttler * *schwirrend rüttelnd*
jaw Klaue * *(Klemm-)Backe, Backen, Klaue; jaw clutch: Klauenkupplung*
jaw clutch Klauenkupplung
jaw coupling Klauenkupplung
jeopardize gefährden * *in Frage stellen*
jerk Bewegung * *ruckweise* || reißen * *ruckartig ~* || Ruck * *erste Ableitung der Beschleunigung* || Stoß * *Ruck*
jerk control Ruckbegrenzung

jerk limitation Ruckbegrenzung
jerk limiting Ruckbegrenzung * z.B. durch Hochlaufgeber mit Verrundung
jerk reduction Ruckbegrenzung
jerk smoothing Ruckbegrenzung
jerk stressing Stoßbeanspruchung * durch ruckhafte Lasten
jerk-free ruckfrei || stoßfrei * ohne Ruck
jerkfree changeover ruckfreier Übergang
jerkless ruckfrei * erste Ableitung der Beschleunigung ist gleich Null
jerky ruckartig || ruckhaft || sprunghaft * ruckhaft (auch Markt usw.) || stoßartig * ruckhaft
jerky running unrunder Lauf * ruckhafter Lauf
jet Strahl
jet lag Zeitverschiebung * bei Langstreckenflug
jet tube Düsenrohr
jet-water Strahlwasser
jig Lehre * Bohr~ || Schablone * Bohrschablone || Spannvorrichtung * Auf-/Einspannvorrichtung || Vorrichtung * zur Werkzeugführung z.B. bei der Metallbearbeitung
jig saw Stichsäge
jitter flattern * z.B. Impulse || zappeln * z.B. Impulse
job Anstellung || Arbeit * Berufstätigkeit, Arbeitsplatz, Aufgabe || Arbeitsplatz. * Stelle; change ones job: ~ wechseln; maintenance: Erhaltung; security: Sicherheit || Arbeitsstelle || Aufgabe || Handwerk * (figurllch) || Stelle * Arbeits~ || Stellung * berufliche Anstellung
job card Begleitkarte * z.B. für ein Produkt in der Fertigung
job number Werknummer
job planning Auftragsplanung || Auftragsvorbereitung
job table Busumlaufliste
job title Berufsbezeichnung
job-cutting Personalabbau
job-trained angelernt * Arbeiter
job-trained worker angelernter Arbeiter
joblessness Arbeitslosigkeit
jockey pulley Riemenspannrolle || Spannrolle * z.B. bei Riementrieb
jog tippen * kuzzeitige Aufschaltung eines (niedrigen) Drehzahl/Frequenzsollwerts über Steuerkommando
jog backward tippen rückwärts
jog control Tippschaltung
jog forward tippen vorwärts
jog reference Tippsollwert
jog reverse tippen rückwärts
jog speed Tippdrehzahl || Tippgeschwindigkeit
jogging Tastbetrieb * Tippbetrieb eines Antriebs || Tippbetrieb || tippen * (Substantiv)
jogging setpoint Tippsollwert
jogging speed Schleichdrehzahl * Tipp-Drehzahl
jogging table Rütteltisch
join aneinander fügen || anschließen * anfügen || ansetzen * anfügen || beifügen * to:an || eintreten * in Verein/Verband usw. ~ || Mit- * (Verb) || verbinden * verbinden, vereinigen, zusammenfügen (to/onto: mit) || vereinen * verbinden, vereinigen, zusammenfügen, sich anschließen/zusammentun || vereinigen || zusammenschließen * auch 'sich ~' (closely: eng)
join forces zusammenschließen * sich ~
join in teilnehmen * zusammen mit anderen ~
join in series hintereinander schalten
join to anlegen
join together aneinander reihen || zusammenführen * zusammenfügen
join up aufschließen * zum Verband
joined zusammenhängend * zusammengefasst, vereinigt

joiner Tischler
joiner's circular saw Tischlerkreissäge
joiner's machine Tischlereimaschine
joiner's workshop Tischlerei * ~werkstatt
joinery Schreinerei * {Holz} siehe auch →Tischlerei || Tischlerei * Handwerkszweig
joining Anschluss * (allg., mechanisch, auch figürlich)
joining flange Zwischenflansch
joint Eckpunkt * Verbindungspunkt || Fuge * Verbindung(s~) || fugen * (zusammen)fügen, fugen, verlaschen || Gelenk || Gelenk- || gemeinsam * joint/concerted action: ~e Aktion; joint property: ~es Eigentum || Mit- * z.B. joint owner: ~besitzer, joint liability: ~haftung || Naht * Verbindungsstelle || Stoß * Fuge || Verbindung * (mechan.) || Verbindungs-
joint coupling Anschlussstück
joint face Trennfläche
joint operation Zusammenarbeit * z.B. von Abteilungen, Firmen || zusammenwirken * (Substantiv)
joint proprietor Teilhaber * Mitinhaber
joint sealant Fugendichtungsmittel
joint venture Gelegenheitsgesellschaft * vorübergehende Zusammenarbeit mehrerer Firmen zu einem Geschäftszweck || Gemeinschaftsunternehmen || Kooperation * (vorübergehende) Zusammenarbeit mehrer Firmen bei einem Projekt || zusammenwirken * (Substantiv) ~ von Unternehmen bei einem bestimmten Projekt
joint-gluing line Fugenverleimanlage * {Holz}
jointing clamp Presszange
jointing compound Spaltdichtung * Fugen-Dichtmasse
jointing gasket Spaltdichtung * in Form einer Dichtungszwischenlage
jointing machine Fügemaschine * {Holz}
jointly gemeinsam * (Adv.)
jointly agree upon abstimmen * etwas gemeinsam vereinbaren
jolt Erschütterung || Rüttler
jolt-free changeover stoßfreie Umschaltung
jolt-ramming machine Rüttler * Stoßramme
jolter Rüttler * stoßender ~
joltfree stoßfrei * ohne Rütteln, Holpern
jolting table Rütteltisch
journal Fachzeitschrift || Lagerzapfen * (Dreh-/Lager-/Wellen-) Zapfen; journal bearing: Achs-/Zapfenlager || Zapfen * Dreh-/Lager-/Wellenzapfen; journal box: Lagerbüchse; journal bearing: ~lager || Zeitschrift
journal bearing Achslager * Zapfenlager || Gleitlager || Zapfenlager
journal box Lagerbüchse
journalism Presse * Journalismus
journey Fahrt * Reise || Lauf * Weg (auch eines Produktes durch eine Maschine), Reise, Fahrt, Route, Gang
journeyman Geselle * Handwerker
joystick Meistersteller
joystick controller x-y-Positionsgeber
judder Vibration * in Aufzug, Flugzeug usw. || vibrieren * (Substantiv) im Flugzeug, Aufzug usw.; Verwackeln
judder-free ruckfrei * frei von Vibrieren (z.B. Aufzug)
judge bemessen * bewerten (by:nach) || beurteilen * by:nach; of:über; misjudge:falsch beurteilen
judgement Beurteilung * auch 'Urteil, Ansicht'; in my judgement:meines Erachtens || Bewertung * Urteil, Beurteilung, Ansicht, Einschätzung || Entscheidung * gerichtliche ~
judgment Beurteilung * siehe 'judgement' || Bewertung * siehe 'judgement'

judicious vernünftig * →*wohlüberlegt* || wohlüberlegt * *vernünftig, klug, wohlüberlegt*
jug Kanne
jumble mischen * zusammenwerfen, in Unordnung bringen, durcheinanderwürfeln, (wahllos) ver~
jumbo riesig * →*übergroß* || Tambour * *Wickeldorn für/mit Papier* || übergroß * *(salopp)*
jumbo reel Tambour * *(Papier)*
jumbo roll Maschinenrolle * *Papiertambour* || Tambour * *Wickeldorn für/mit Papier, Maschinenrolle*
jump Satz * *Sprung* || springen || Sprung * *auch in Programm*
jump address Sprungadresse
jump apart wegfliegen * *davonspringen (z.B. schlecht befestigtes Maschinenteil)*
jump destination Sprungziel * *(Computer) in Softwareprogramm*
jump frequency Ausblendfrequenz * *ausgeblendeter Frequ.sollwert f.Frequenzumrichter zur Vermeidg.mechan.Resonanzen*
jump frequency band Frequenzausklammerungsband
jump instruction Sprungbefehl
jump label Sprungmarke
jump operation Sprungoperation
jump start Warmstart || Wiederaufschalten
jumper Brücke * *Steck-, Klemmenbrücke* || brücken * *über~, z.B. mit Löt-/Klemmen-/Drahtbrücke* || Drahtbrücke || kurzschließen * *mit Hilfe einer Draht-/Klemmenbrücke* || Steckbrücke || überbrücken * *mit Draht, z.B.Klemmen-/Draht-/Steck-/Lötbrücke*
jumper header Kopfstecker * *bei SPS*
jumper plug Brückenstecker
jumper-selectable umschaltbar * *über (Steck-)Brücken*
jumpered terminals Klemmenbrücke * *gebrückte Klemmen*
jumpering Rangierung * *z.B. von Leitungen über Rangierverteiler* || Überbrückung * *Steckbrücke, Klemmen usw*
junction Abzweigung || Sperrschicht * *in Halbleiter; junction temperature:* →*Sperrschichttemperatur* || Stoß * *Verbindung* || Übergang * *in einem Halbleiter, z.B. pn-junction: pn-~* || Verbindung * *mechan. ~/Anschluss, ~spunkt, Zusammentreffen, Berührung, Knotenpunkt* || Verbindungs-
junction and tapping box Anschluss- und Abzweigdose
junction box Abzweigdose || Anschlussdose || Muffe * *Kabelverbindungs~*
junction plate Knotenblech
junction temperature Sperrschichttemperatur
junction traffic Nahverkehr
junior engineer Nachwuchskraft * *Ingenieur*
junk Altmaterial
juridical rechtlich
jurisdiction Befugnis * *Zuständigkeit* || Entscheidungsbefugnis * *rechtlich*
just bloß * *nur* || erst * *bloß* || genau * *(Adv.) just as good: ~ so gut* || gerade * *(Adv.)* || knapp * *(Adv.) gerade eben* || nur * *bloß*
just as genauso
just hitting a key auf Knopfdruck * *siehe auch* →*Knopfdruck*
just larger wenig größer
just now augenblicklich * *(Adv.)*
just the other way round umgekehrt * *gerade anders herum*
just the same thing genau dasselbe
just under knapp * *(Adv.) vor Zahlen*
justifiable vertretbar
justification Berechtigung * *Rechtfertigung* || Rechtfertigung || Umbruch * *(Druckerei) line justificitation: Zeile~*
justified bündig * *ausgerichtet; right-/left-justified: rechts-/linksbündig*
justify ausrichten * *(Text usw.) am Rand ~; richten, justieren, richtigstellen; rechtfertigen, jdm. Recht geben* || entschuldigen * *rechtfertigen* || rechtfertigen
jut out herausragen * *vorspringen,* →*herausragen* || hervorstehen || überstehen * *(über etwas) hinausragen, hervorstehen*
jutting out herausragend * *z.B. Maschinenteil* || hervorragend * →*herausragend (z.B. Maschinenteil)* || vorstehend

K

kaolin Kaolin
Kbaud Kbaud * *Kilobits pro sec*
kbs Kbaud * *Kilobits pro sec* || Kbd * *Kilobaud, Kilobits pro sec*
keen eifrig
keep aufheben * *aufbewahren* || behalten || beibehalten || erfüllen * *Versprechen ~* || erhalten * *bewahren, beibehalten, aufrecht~* || halten * *(bei)be-/fest-/an-/zurück-/ver~; Versprechg. ~; bleiben; keep to/low: s. ~ an/niedrig ~* || schützen * *bewahren, from:vor*
keep an eye on beobachten
keep away fernhalten * *from/of: von; auch 'sich ~'*
keep back zurückhalten
keep clear freihalten * *z.B. Straße*
keep constant konstanthalten
keep down gering halten
keep free freihalten * *from: von*
keep in good condition erhalten * *in gutem Zustand halten*
keep in good order pflegen
keep in good repair erhalten * *in gutem Zustand halten* || pflegen
keep in good service pflegen
keep in position festhalten * *in Lage/Stellung ~*
keep in stock auf Lager halten
keep low gering halten
keep on fortführen * *fortsetzen*
keep on stock lagern * *auf Lager halten*
keep open freihalten * *z.B. Angebot*
keep pace Schritt halten || schritthalten * *with: mit*
keep ready bereitstellen * *in Bereitschaft halten*
keep short knapp * *~ halten*
keep step Schritt halten * *with: mit*
keep tab on kontrollieren * *(amerikan.) 'auf der Spur bleiben'*
keep the same unverändert beibehalten
keep tight straff halten
keep to erhalten * *term: Frist* || sich halten an
keep to a minimum gering halten * *so gering wie möglich halten* || minimieren * →*gering halten*
keep to oneself zurückhalten * *bei sich behalten*
keep track verfolgen * *auf der Spur bleiben, nach~ (of: etwas), weiter~*
keep track of kontrollieren * *nachverfolgen, 'auf der Spur bleiben'* || nachführen * *(nach-)* →*verfolgen* || nachverfolgen * *sich dauernd auf d.laufenden halten über* || überwachen * *nachverfolgen, kontrollieren*
keep up beibehalten
keep well pflegen
keep with einhalten * *Forderung usw.*

keep within einhalten * *limits:Grenzen; time limit:Frist; appointed time:Termin*
keep within the timing Timing einhalten
keeper Magnetanker
keeper of the minutes Protokollführer * *bei Besprechungen, Sitzungen, Verhandlungen usw.*
keg Fass * *kleines Fass, z.B. für Getränke*
kept in good condition gepflegt
kept in good order gepflegt
kept in good repair gepflegt
kept in good service gepflegt
kernel Kern * *Kern(punkt), das Innerste, das Wesen, Guss~, Getreidekorn*
kerosene Petroleum * *(amerikan.) für Heiz- und Leuchtzwecke*
key Druckknopf * *Drucktaste* || Drucktaste || Drucktaster * *Taste(r)* || Feder * →*Passfeder, Nutkeil* || Federkeil * *Nutkeil, Passfeder* || Haupt- * *Schlüssel-, entscheidend; z.B. Kunde, Rolle* || Keil * *auch* →*Federkeil* || Kern- * *wichtigst, Schlüssel-* || Legende * *z.B. auf Landkarte usw.* || Nutkeil * *Passfeder, z.B. für Motorwelle* || Passfeder * *Nutkeil (DIN 748 Teil 1 u.3, DIN ISO 8821, VDE 0530 Teil 1)* || Schlüssel * *(auch figürl.)* || Schlüssel- || Taste || Taster * *Taste* || wichtig * *Schlüssel-*
key account Hauptkunde * *Schlüsselkunde mit hohem Umsatz* || Schlüsselkunde
key account customer Schlüsselkunde
key assignment Tastenbelegung
key factor maßgebende Größe * *wichtigste Einflussgröße* || maßgebliche Größe * *wichtigste Einflussgröße*
key feature Haupteigenschaft || herausragendes Merkmal
key parameter Schlüsselparameter
key position Schlüsselposition
key role Schlüsselposition || Schlüsselstellung * *Schlüsselrolle*
key seat Keilnut
key sequence Tastenfolge
key switch Schlüsselschalter
key technology Schlüsseltechnologie
key word Suchbegriff
key word index Schlagwortregister * *Stichwortverzeichnis* || Schlagwortverzeichnis
key-account Großkunde
key-account customer Großkunde
key-actuated switch Schlüsselschalter
key-bolt Riegel * *am Schloss*
key-operated switch Schlüsselschalter
key-seating machine Keilnuten-Stoßmaschine
key-stroke Anschlag * *Tasten~*
keyboard Tastatur || Tastenfeld * *Tastatur*
keyboard unit Bedienfeld * *Tastenfeld* || Bediengerät * *mit Tasten*
keyed formschlüssig * *mit Federkeil (auf Welle) befestigt* || unverwechselbar * *mit Kodierung versehen (z.B. Stecker)* || verdrehsicher * *verpolungssicher, z.B. Stecker mit unsymmetrischem Aufbau* || verpolungssicher * *Stecker mit Codierung* || vertauschungssicher * *z.B. Stecker*
keyed-bar cage rotor Keilstabläufer * *{el.Masch.} Stromverdrängungsläufer bei Käfigläufermotor*
keyhole saw Stichsäge * *{Holz}*
keying Codierung * *von Steckern zur Vertauschungssicherheit* || Kodierung * *von Steckern zur Vertauschungssicherheit*
keying element Kodierelement * *z.B. bei vertauschungssicherem Stecker*
keying pin Codierzapfen * *an einem unverwechselbaren Steckverbinder*
keying plug Kodierstift * *z.B. für vertauschungssicheren Stecker*
keying slot Kodiernut * *an Stecker zur Gewährleistung der Vertauschungs-/Verpolungssicherheit*

keyless shaft glatte Welle * *Welle ohne Federkeil*
keylock switch Schlüsselschalter
keypad Tastatur * *kleines Tastenfeld* || Tastenblock || Tastenfeld * *kleines ~*
keyseating Keilnut
keystroke Tastenbetätigung
keystroke sequence Tastenfolge
keyway Keilnut || Nut * *für Passfeder z.B. auf Motorwelle* || Passfedernut
keyway broach Nutmesser * *{Werkz.masch.}*
keywayed genutet * *mit Paßfedernut versehen (z.B. Motorwelle)*
keyword Schlagwort * *Stichwort* || Schlüssel * *Schlüsselwort* || Stichwort || Suchbegriff
keyword index Schlagwortverzeichnis || Stichwortverzeichnis
kg kg * *Kilogramm; 1 kg entspr. 2,2046 lb (pounds)*
kick Schwung * *(salopp) Elan, (Stoß) Kraft, Energie, Nervenkitzel* || stoßen * *mit Fuß ~*
kill abstechen * *totmachen (auch figürl.)* || totlegen * *zum Absterben bringen, unterdrücken*
kiln Ofen * *Brenn/Trockenofen, z.B.Zementofen*
kilowatt Kilowatt * *['kilewot]* || kW * *['kilewot]*
kilowatt hour Kilowattstunde * *['kilewot 'auer]*
kilowatt hour meter Kilowattstundenzähler
kilowatt hours Kilowattstunden
kind Art * *Sorte, Wesen, Gattung, Natur* || Ausprägung * *Art, Sorte, Wesen, Gattung, Natur* || Qualität * *Art* || Sorte || Variante * *Art, Sorte, Gattung* || Version * *Art, Sorte, Gattung*
kind of cage Käfigform * *{el.Masch.} bei Käfigläufermotor (Rundstab-, Hochstab-, L-Stab-, Keilstabkäfig usw.)*
kind of cooling Kühlart
kind of doing Art und Weise * *des Handels, Vorgehensweise* || Vorgehensweise
kind of material Material * *Art des Materials*
kindly note bitte zu beachten
kinetic kinetisch
kinetic backup kinetische Pufferung
kinetic buffering kinetische Pufferung * *Überbrückg. v. Netzausfällen durch Ausnutzg. d. kinet. Energie e.Antriebs*
kinetic energy Bewegungsenergie || kinetische Energie
kinetic energy of rotation Rotationsenergie
kinetic moment of inertia kinetische Schwungenergie * *Schwungmoment*
kinetic ride through kinetische Pufferung * *Überbrückg.v. Netzausfällen durch Nutzung d. kinet.-Energie e.Antriebs*
kinetics Kinetik
king pin Königszapfen
king post Königszapfen
kingbolt Königszapfen * *Königsbolzen*
kingpin Herzstück * *(figürl.)* || Königswelle * *Königszapfen* || Königszapfen
kink Knick * *in Draht, Kennlinie usw.* || knicken * *knicken (z.B. Kabel), kräuseln*
kink protection Knickschutz
kink-protection sleeve Knickschutztülle
Kirchhoff's law Kirchhoffsches Gesetz
kit Paket * *(Bau-/Werkzeug-) Satz, Ausstattung, (Werkzeug-) Kasten* || Satz * *Bau~, Werkzeug~*
Kloss' equation Kloss'sche Gleichung * *{el.-Masch.} Formel für die Momentenkennlinie einer idealen Asynchronmaschine*
knack Kniff * *Dreh, Kniff, Apparätchen, "Dingsda"* || Kunstgriff
knead kneten
kneader Kneter || Knetmaschine
kneading line Knetstraße * *in der Backwarenindustrie*
kneading machine Knetmaschine || Knetwerk

knee Knick * *Knie (-Stück, Rohr-), Winkel, Ausbeulung; auch ~ in Kennlinie usw.* || Knie * *auch Rohr~; bring to its knees: in die ~ zwingen*
knee-type milling machine Konsolfräsmaschine
kneepoint Knickpunkt * *z.B. e. Kurve, Kennlinie, Polygonzuges usw.*
knife Messer * *slitting knife: Längsteil~*
knife drum Messertrommel * *z.B. bei Querschneider* || Messerwalze * *z.B. bei Querschneider*
knife holder Messerhalter
knife motor Messermotor
knife rolls Messerpartie * *bei →Querschneider*
knife section Messerpartie * *z.B. bei →Querschneider* || Schneidpartie * *→Messerpartie (z.B. bei Querschneider)*
knife-drum motor Messermotor * *treibt Messertrommel(n) z.B. eines →Querschneiders an*
knit stricken || verknüpfen * *(figürl. u. konkret) zusammenfügen, verbinden, vereinigen; knit up:(sich eng) verbinden* || wirken * *Textilstoff ~/stricken; verknüpfen, zusammenfügen, verbinden, vereinigen, fest verbinden*
knitting machine Strickmaschine || Wirkmaschine * *{Textil}*
knob Drehknopf || Griff * *Tür~* || Knopf * *auch Dreh/Druckknopf*
knob-operated switch Knebelschalter
knock anstoßen || klopfen || Schlag || schlagen * *hart* || stoßen * *schlagen*
knock-in nut Einpressmutter
knot verknüpfen * *verknoten*
knothole Astloch * *{Holz}*
knotting together Verknüpfung * *Verbindung, Verknotung, Zusammenknüpfung*
know erkennen * *make oneself known: sich zu ~ geben* || Wissen * *(Verb)*
know-how Erfahrung * *"gewußt wie"* || Expertenwissen * *"gewusst wie"* || Fachkenntnisse * *"gewußt wie"* || Fachwissen * *"gewusst wie"* || Kenntnis * *"Gewusst Wie"* || Knowhow || Sachkenntnis * *das "Gewusst-Wie"* || Sachverstand * *Gewusst-wie* || Wissen * *Erfahrungsschatz*
know-how development characteristic Erfahrungskurve
knowledge Erkenntnis * *Wissen* || Fachkönnen || Fachwissen * *Wissen* || Kenntnis || Wissen * *extensive: umfangreiches; well-grounded: fundiert*
knowledge level Kenntnisstand
knowledge of programming Programmierkenntnisse
knowledge of the trade Branchenkenntnis
knowlegeable erfahren * *(salopp) klug, kenntnisreich*
known bekannt * *to:jemandem* || bewusst * *bekannt*
known throughout the world weltbekannt
knurl rändeln
knurled gerändelt || geriffelt * *gerändelt*
knurled nut Rändelmutter
knurled-head screw Rändelschraube * *mit gerändeltem Kopf*
knurling machine Rändelmaschine * *Werkzeugmaschine*
kork Korken
kraft faced liner Testliner
kraft liner Kraftliner
kraft paper Kraftpapier
KUSA-connection Kusa-Schaltung * *Abk.f. 'KUrzschlus-SAnftanlaufschaltg.' für Asynchronmot. m. Serienwiderst. i Phase*
kW kW * *1 kW entspr. 1,3410 hp (Horsepowers) entspr. 1,3596 PS (Pferdestärken)* || Leistung * *Kilowatt*
KWhr KWh

L

l 1 * *Liter (1 l entspr. 0,2200 Engl.Gallonen entspr. 0,2642 US Gallonen)*
L-bracket Befestigungswinkel * *z.B. Blechwinkel* || Halter * *Haltewinkel* || Winkel * *Haltewinkel, Blechwinkel (in 'L'-Form)*
lab Labor || Laboratorium
lab measurement Labormessung
lab sample Labormuster * *(salopp)*
lab-examine untersuchen * *im Labor ~*
label Aufkleber || Beschriftung * *mittels Aufkleber* || Etikett * *adhesive label: Selbstklebe-Etikett* || etikettieren || Name * *{Computer} Sprungmarke usw.* || Schild * *Etikett*
label printer Etiketten-Druckmaschine
labeled gekennzeichnet * *beschriftet*
labeling Beschriftung * *(amerikan.) auf einem Etikett*
labeling machine Etikettiermaschine
labeling strip Beschriftungsstreifen * *z.B. eines Baugruppenträgers*
labelled beschriftet
labelling Beschriftung * *z.B. auf einem Etikett* || Kennzeichnung * *~ mit Schild; Beschriftung*
labelling bar Beschriftungsschiene * *z.B. bei einem Baugruppenträger*
labelling machine Etikettiermaschine
labelling strip Beschriftungsstreifen * *z.B. bei einer Flachbaugruppe/eines Baugruppenträgers*
lable benennen * *bezeichnen, mit Namen/Symbol/Schild usw. versehen*
labor-intensive arbeitsintensiv
labor-saving aufwandsarm * *Arbeit einsparend*
laboratory Labor || Laboratorium
laboratory equipment Laborausstattung || Laboreinrichtung
laboratory sample Labormuster * *siehe auch →Muster*
laboratory sessions Praktikum
laboratory student Praktikant * *in Entwicklungsabteilung*
laboratory-type power supply Labor-Netzgerät || Labornetzgerät
laborious mühsam * *mit viel Arbeit*
laboriously mühsam * *(Adv.)*
labour Mühe * *Arbeit*
labour-union Gewerkschaft
labourer Arbeiter
labyrinth seal Labyrinthdichtung
lacing machine Umschnürungsmaschine * *in der Verpackungstechnik*
lack haben * *nicht ~, Mangel ~ an* || Mangel * *Fehlen (of: von); auch Verb: '~ haben an; nicht haben')*
lack of Mangel an
lack of definition Unschärfe
lack of space Platzmangel
laconic kurz * *treffend*
laconically kurz * *(Adv.) treffend*
lacquered wire Lackdraht
LAD Kontaktplan * *Abkürzg. für 'LAdder Diagram': Relaisketten-Darstellung (SPS-Programmiersprache)* || KOP * *Abkürzung für 'LAdder Diagram': Kontaktplandarstellung für SPS*
ladder Leiter * *zum Hochsteige*
ladder circuit Kontaktplan-Netzwerk * *(SPS)*
ladder diagram Kontaktplan * *auch Programmiersprache für SPS* || KOP
ladder diagram line Strompfad * *in Relaissteuerungs-/Kontaktplan (KOP)*
ladder logic Kontaktplan * *Darstellung einer Relais- oder SPS-Steuerung*

ladder programming programmieren im Kontaktplan * *(SPS)*
ladies' room Toilette * *Damen~*
ladle Gießtiegel * *z.b. im Stahlwerk*
lag nacheilen * *behind:hinter; by:um* || Nacheilung * *der Zeitabstand* || Phasenverschiebung * *negative ~, Phasennacheilung* || verzögern * *sich ~, zurückbleiben, nachhinken* || Verzögerung || Zeit * *Rückstand, Verzögerung, Nacheilung, negative Phasenverschiebg., ~abstand* || Zeitabstand * *Nacheilung; z.B.jet lag:Zeitverschiebung zw.verschiedenen Ländern* || Zeitverzögerung || zurückhinken * *zeitlich (behind:hinter)*
lag behind nachlaufen * *mit (zeitlichem) Versatz hinterherlaufen*
lagging nacheilend || Nacheilung * *das Nacheilen (behind:hinter)*
laid out ausgelegt * *z.b. konstruktiv ausgelegt, dimensioniert* || dimensioniert * *ausgelegt*
lame faul * *lahm, mangelhaft, stockend; (auch figürl:lame excuse: ~e) Ausrede* || lahm * *(auch figürl.: Computer usw.)*
lamella Lamelle * *Plättchen, Blättchen* || Plättchen || Scheibchen || Scheibe * *Blättchen,* →*Lamelle*
lamina Lamelle * *Plättchen, Blättchen, dünne Schicht* || Plättchen || Platte * *dünne Platte, Lamelle*
laminate beschichten * *(eine von mehreren) Teilschichten auftragen* || kaschieren * *mit Schicht/ Plättchen belegen* || Laminat || laminieren
laminate material Laminat
laminated beschichtet || geblecht || schichtweise * *geschichtet* || Verbund- * *aus mehreren Schichten bestehend, z.b. Glas, Kunststoff, Folie, Holz*
laminated core Blechpaket || Schichtkern
laminated material Laminat
laminated paper Hartpapier
laminated wood Schichtholz
laminating machine Kaschiermaschine
laminating press Lamellenpresse * *[Holz]* || Massivholzverleimmaschine
lamination Beschichtung * *Schichtung, Auftrag (einer) von (mehreren) Teilschichten* || Blechpaket || Blechung || Kaschierung || Lamelle * *Schichtung, blättrige Beschaffenheit*
laminator Laminator * *Einrichtg. z. Laminieren (übereinand. legen mehrerer Schichten) i.d.Folienherstellg. usw.*
laminboard Tischlerplatte * *[Holz] mit Stäbchenzwischenlage*
lamp Lampe || Leuchte
lamp current in the cold state Lampen-Kaltstrom
lamp current in the warm condition Lampen-Warmstrom
lamp holder Lampenfassung
lamp replacement Lampenwechsel
lamp socket Lampenfassung
lamp test Lampenprüfung
lampblack Ruß
LAN LAN * *Abkürzg.f. 'Local Area Network':Nahbereichs-Computernetz (z.B. Ethernet)*
land Grund * *Land, ~ u. Boden* || löschen * *Waren* || löschen * *Schiffsladung ~*
land transport Landtransport
language Sprache * *Fremdsprache, Programmiersprache*
language level Sprachebene
language resources Sprachmittel * *[Computer]*
language subset Sprachraum * *bei SPS*
lantern Taschenlampe * *~ mit großem Reflektor, Laterne*
lap läppen * *(passgenau) feinbearbeiten einer Fläche z.B. mit Werkzeugmaschine* || überlappen
Laplace transform Laplace-Transformation || Laplacetransformation
Laplacian Laplacescher Operator

lapping machine Läppmaschine
lapse ablaufen * *z.B. Frist,Vertrag* || Verlauf * *~/ Ablauf von Zeit*
laquer drier Lacktrockner
laquer spray booth Lackierkabine
laquering plant Lackieranlage
large breit * *ausgedehnt* || dick * *groß, massig* || groß * *umfangreich, large motor: ~motor* || Groß- || stark * *beträchtlich*
large area of großflächig * *(Adj.)*
large batch production Großserienfertigung
large computer Großcomputer || Großrechner
large drive Großantrieb
large hp drive Großantrieb * *'mit vielen horsepowers'*
large motor Großmotor
large portion Großteil
large surface of großflächig * *(Adj.)*
large-capacity Groß- * *z.B. Ventilator, Pumpe, Zentrifuge* || Großraum- * *mit großem Fassungsvermögen*
large-scale groß * *in großem Maßstab/Stil, ~ angelegt* || in großem Umfang
large-scale enterprise Großunternehmen
large-scale production Großserienfertigung || Serienherstellung * *Großserienherstellung*
large-series production Großserienfertigung
largely größt * *(Adv.) zum größten Teil* || vorwiegend * *(Adv.) zum größten Teil* || weitgehend * *(Adv.)* || wesentlich * *(Adv.)* || zum größten Teil
largely influenced stark beeinflusst
largely perfected ausgereizt
largeness Größe * *Umfang, Format*
larger größer
larger half größerer Teil * *von zweien*
larger than größer als
largescale aufwändig * *ausgedehnt, in großem Maßstab*
largest größt
largest possible größtmöglich
laser-fusion cutting machine Laserschneidmaschine * *Werkzeugmaschine*
last ausreichen || dauern || letzt * *last-minute: im ~en Augenblick; next to last/last but one: vorl~; third from last: dritt~* || letzter * *at the very last: als Letztes* || reichen * *dauern, währen, (stets) halten, (aus)reichen, genügen; durch-/aushalten, bestehen* || schließlich * *(Adj.)*
last but not least letzt * *zu guter ~*
last but one vorletzt || vorletzter || zweitletzter
last out reichen * *genügen, aus~ sein, (aus)reichen*
last remark Schlussbemerkung
last resort Ausweg * *letzter Ausweg*
last runnings Nachlauf * *beim Leeren eines Flüssigkeitsbehälters*
last-named letztgenannt
lasting andauernd * *dauernd, anhaltend, beständig, nachhaltig; dauerhaft, haltbar* || beständig * *dauerhaft, lang andauernd* || bleibend * *ever-lasting: ewig ~* || dauerhaft * *langwährend, nachhaltig, bleibend* || fest * *dauerhaft, bleibend*
latch einklinken * *on the latch:eingeklinkt* || einrasten * *(sich) einklinken/einschnappen; on the latch: eingeklinkt* || einschnappen * *auch 'einrasten'* || Klinke * *Rast~* || Latch * *Zwischenspeicher für Datenwort* || Riegel * *Klinke, Schnäpper, Schnappriegel, Schnappschloss* || Speichergliod * *mehrere parallel angeordnete Flipflops mit gemeinsamem Übergabe-Taktsignal* || speichernd setzen * *z.B. Relais* || Verschluss * *zum Verriegeln* || Zwischenspeicher * *~ für ein Datenwort* || zwischenspeichern * *in Datenregister(n)*
latch fastener Spannbügel

latch output Ausgang setzen * *in Kontaktplandarstellung einer SPS*
latched button Rasttaster * *latched emergency stop button:Not-Aus-Rasttaster*
latching Rastung || remanent * *Merker/Kontakt bei Kontaktplan/Relaissteuerung* || Selbsthaltung
latching contact Selbsthaltekontakt
latching relay bistabiles Relais * *Haftrelais* || Haftrelais
latching type rastend * *z.B. Taster*
latching/unlatching functions speichernde Funktionen * *bei SPS*
lately neuerdings
latency Latenz * *Verzögerung(szeit), Totzeit* || Latenzzeit * *Totzeit/Reaktionszeit/Wartezeit in Computer* || Reaktionszeit * *Latenz-/Totzeit im Rechnersystem z.b. bei der Reaktion auf Interrupts* || Totzeit * *Latenz-/Verzugszeit z.B. bei der Reaktion auf ein Ereignis in einem Rechner* || Verzugszeit * *Latenzzeit/Totzeit bei der Reaktion auf ein Ereignis in einem Rechner* || Wartezeit
latent verborgen * *(physikalisch) schlafend, schlummernd, verborgen, versteckt*
later neuer * *zeitlich weniger lange zurückliegend (z.B. Softwareversion)* || später
later on danach * *später*
lateral quer * *seitlich* || Seiten- * *seitlich, Quer-, Seiten-, Neben-, von der Seite* || seitlich || seitwärts * *seitlich, ~ befindlich*
lateral force Querkraft || Radialkraft * →*Querkraft*
lateral movement Querbewegung
lateral stress Querbeanspruchung
laterally seitlich * *(Adv.)*
latest letzt * *neuest* || neu * ~*estes* || neuest
latest version neueste Version
lath Latte
lathe Drehbank || Drehmaschine * *{Werkz.masch.}-Drehbank*
lathe chuck Drehfutter
lathe tool Drehling * *Werkzeug für Drehmaschine* || Drehstahl * *für* →*Drehbank*
latitude Freiheit * *Spielraum* || Spielraum * *in der Auslegung*
latter letzt * *das* ~*ere* || letzterer || Letzteres || neuer * *(Komparativ) später, neuer, jünger*
lattice Gitter || Verband * *(mathem.)*
latticed gitterförmig
launch auf den Markt bringen * *z.B. ein Produkt* || einführen * *Mode, neues Produkt usw.* ~ || in Gang setzen * *(figürlich) lancieren, Starthilfe geben, unternehmen, beginnen, vom Stapel lassen* || lancieren * *in Gang setzen, Starthilfe geben* || Markteinführung || starten * *beginnen, lancieren, in Gang setzen, unternehmen, Starthilfe geben, Computerprogramm* ~
launch into the market zum Vertrieb freigeben
launching into the market Markteinführung
lavatory Toilette
law Algorithmus * *Bildungsgesetz, Rechenvorschrift* || Gesetz * *auch Natur~, naturwissenschaftliches* ~; *enact: erlassen* || Satz * *Lehr~, naturwissenschaftliches Gesetz*
law of Fourier Satz von Fourier
law of induction Induktionsgesetz
lawful ordnungsgemäß * *gesetzmäßig*
laws of physics physikalische Gesetze
lawsuit Prozess * *Rechtsstreit*
lawyer Anwalt
lay auslegen * *Kabel* || verlegen * *Kabel*
lay alongside anlegen * *Schiff {of: an}*
lay claim to in Anspruch nehmen * *Anspruch erheben auf*
lay down ablegen || aufstellen * *Grundsatz* ~ || darlegen * *in einer Schrift* || festlegen * *Grundsatz/ Regel* ~ *z.B. in einer Norm/Vorschrift* || niederle-

gen * *auch in einer Schrift* || vorschreiben * *einen Grundsatz festlegen/aufstellen*
lay hold of packen * *fassen*
lay hold on reißen * *an sich* ~
lay in anlegen * *Vorrat*
lay in seperate ducts getrennt verlegen * *z.B. Motor- und Signalleitungen in unterschiedlichen Kabelkanälen/-pritschen*
lay length Schlaglänge * *beim Verseilen von Kabeln*
lay out anlegen * *auslegen* || auslegen * *konstruktiv* ~, *dimensionieren* || dimensionieren * *auslegen, entwerfen* || entwerfen * *auslegen* || projektieren * *auslegen*
lay significance on Wert legen auf
layboy Bogenableger * *für Papierbögen* || Bogenstapler * *für Papierbögen* || Stapler * *Papierbogenableger*
layer -lagig * *z.B. Leiterplatte, Wicklung* || Auflage * *Schicht* || Belag * *Schicht* || Ebene * *Schicht* || Lage * *Ebene, Schicht, auch* ~ *einer Wicklung* || Schicht * *Ebene* || Windung * *Lage*
layer winding machine Lagenwickelmaschine
laying down Festlegung * *Festsetzen einer Regel/Bedingg.; i.e. Vertrag niederlegen*
laying of cables Verlegung * *von Kabeln/Leitungen*
layman Laie * *Nichtfachmann; technical: technischer* ~
layout Anordnung * *räumliche, konstruktive* ~ || Auslegung || Disposition || Entwicklung || Lageplan || Plan * *Anlage, Anordnung* || Planung * *Anlage, Anordnung*
layout diagram Anordnungsplan * *(auch bei SIMADYN D (R) von SIEMENS)* || Grundriss * *Anordnungsplan*
layout drawing Dispositionszeichnung * *mit Darstellung einer Baugruppenträger/Schrankbelegung*
layout error Projektierungsfehler
layout of drive equipment Antriebsprojektierung * *Auslegung der Antriebsausrüstung*
layout plan Anlagenschema || Lageplan * *(techn.) Anordnungsplan* || Übersichtsplan
laziness Trägheit * *Faulheit*
lazy faul * *träge*
lb kg * *pounds; 1 kg entspr. 2,2046 lb*
lb (av) kg * *pounds; 1 kg entspr. 2,2046 lb (av)*
lb (tr) kg * *pounds; 1 kg entspr. 2,6792 lb (tr)*
LCD Flüssigkeitskristall-Anzeige * *Abkürz. für 'liquid crystal display'* || Flüssigkeitskristallanzeige * *Abkürz. für 'Liquid Crystal Display'*
LDC Entwicklungsland * *Less Developed Country*
lead Anfang * *z.B. einer Wicklung* || anführen * *vorne stehen* || Anschluss * *Zuleitung* || Blei || Draht * *Leitung* || Führungs- * *Führung, Leitung, Spitze; Vorsprung* || Gewindesteigung * *in axialer Richtung (Höhe einer vollen Umdrehung)* || leiten * *führen (auch in einem Team, Personal, eine Sitzung)* || Leiter * *Stromverbindung, Leitung, Leiter* || Leitung || Mine * *für Bleistift* || Steigung * *Gewinde~ in Axialrichtung* || vorlaufen * *by: um; the voltage leads the current: die Spannung eilt dem Strom vor* || Voreilung * *Vorsprung* * *Abstand (of: vor)* || zuführen * *leiten, führen, lenken, bringen (to: nach, zu)* || Zuführung * *Draht, Leitung*
lead angle Gewindesteigung * *Steigungswinkel* || Steigung * *Steigungswinkel eines Gewindes; siehe auch* →*Gewindesteigung* || Voreilwinkel
lead away abführen || fortführen * *weg bewegen* || fortleiten * *fortführen, hinwegführen*
lead cross section Leitungsquerschnitt
lead diameter Leitungsdurchmesser
lead in einführen * *eine elektr. Leitung* ~ || zuführen * *Leitung, Draht, Kabel* ~ || zuführen * *Draht* ~
lead of winding Wicklungsanfang

lead off abführen || ansetzen * *den Anfang machen*
lead screw Spindel
lead screw drive Spindel-Mutter-Getriebe * *Leitspindelantrieb*
lead spacing Rastermaß * *Abstand von Bauelemente-Anschlüssen*
lead to führen zu * ~ *nach/zu* || zur Folge haben
lead wire Anschlussdraht || Verbindungsleitung
lead-acid battery Bleiakkumulator || Bleibatterie
lead-in Einführung * ~ *von Leitung, Kabel*
lead-screw potentiometer Spindelpotentiometer
lead-tin solder Lötzinn
lead-wire Anschlussleitung * *einer Wicklung*
leaded bedrahtet * *mit Anschlussdraht versehen*
leader Leiter * *Führer*
leadership Führung * *Leitung* || Leitung * *Leitung, Führung, Führerschaft* || Vorsprung * *Führung*
leadership position führende Position * *z.B. eines Herstellers am Markt*
leading führend || Führungs- || leitend || Leitung * *Führung, Leitung, Management* || maßgebend * *führend, z.B. Kreise* || maßgeblich * *führend, z.b.Kreise* || Spitzen- * *führend* || voreilend || Voreilung
leading contact voreilender Kontakt || Vorkontakt
leading edge ansteigende Flanke * *Vorderflanke eines Impulses* || positive Flanke * *vordere Flanke* || Vorderkante * *z.B. eines Impulses*
leading feature Haupteigenschaft
leading limit switch Vorendschalter * *bei Positioniersteuerung, Aufzug usw.*
leading off Abfuhr || Abführung
leading turn Anfangswindung
leading zero digits führende Nullen
leading-edge führend * *z.B. Firma, Produkt*
leading-off of the heat Wärmeableitung * *Wärmeabführung*
leads Zuleitung
leaf Blatt * *in Buch; Folie, Blatt einer Blattfeder* || Flügel * *einer Tür* || Lamelle * *Blatt, Flügel, dünne Folie, Blattfeder*
leaf spring Blattfeder
leaf-wood Holz * *Laub~*
leaflet Faltblatt || Produktschrift * *Faltblatt* || Prospekt * *Werbeblatt, Faltblatt* || Streublatt * *Werbeblatt*
leak durchlassen * *Flüssigkeit ~* || laufen * *(aus)laufen (Gefäß)* || Leck * *Leck, Loch, undichte Stelle, Auslaufen, Durchsickern (auch figürl.); lecken, leck sein,streuen* || streuen * *magnetisch ~*
leak detector Lecksuchgerät
leak off abfließen * *durch ein Leck* || ableiten * *Strom*
leak test Dichtigkeitsprüfung * *leak-tested: auf Dichtigkeit geprüft*
leak-proof auslaufsicher
leakage Ableit- || Ableitung * *Streuung, Auslaufen, Lecken, Schwund, Entweichen* || Leck || Leck- || Leckage || Rest- * *Leck-* || Schwund * *durch Aussickern* || Streu- * *z.B. Streureaktanz* || Streuung * *Verlusteffekt einer elektromagnetischen Größe (Fluss, Durchflutung)...* || Streuung * *magnetische ~* || Verlust * *Streuung e. elektromagnetischen Größe*
leakage current Fehlerstrom || Kriechstrom || Leckstrom || Reststrom * *→Leckstrom*
leakage field Streufeld
leakage flux Streufluss * *{el.Masch.} nicht moment.bildender Teil d.magn.Flusses, d.sich auf Nebenwegen schließt*
leakage inductance Streuinduktivität * *Induktivität d. →Streuflusses*
leakage reactance Streureaktanz * *Streublindwiderstand auf Grund der Induktionswirkung der →Streuflüsse*

leakiness Durchlässigkeit
leaking Leck * *(Adjektiv)* || schadhaft * *Rohr, Gefäß mit Leck* || undicht * *leak:undicht sein*
leaky durchlässig * *leck, ~ für Flüssigkeiten* || Leck * *(Adjektiv)* || undicht
lean dünn * *mager*
leap over überspringen
learn erfahren * *(Verb)* || lernen
learning objective Lernziel
learning process Lernphase
learning time Einarbeitungszeit
lease vermieten
leasing Leasing
leasing company Leasingunternehmen
least favourable ungünstigst
least significant niederwertig * *niederwertigst, z.B. Bit, Dekade usw.* || niederwertigst * *Bit, Dekade, Byte, Wort usw*
least significant bit niederwertigstes Bit
least significant digit niederwertigste Ziffer * *z.B. an Ziffernanzeige* || niederwertigstes Zeichen * *einer Ziffernanzeige*
leave abgehen * *Zug, Post usw.* || abreisen || gehen * *weggehen (for: nach), aus dem Dienst ~* || stehen * *unverändert lassen* || überlassen * *leave a thing to a person:jemandem etwas anheimstellen* || verlassen
leave alone gewähren * *sich selbst überlassen, allein/in Ruhe lassen*
leave guessing unklar * *im Unklaren lassen*
leave off aufhören || stehen bleiben * *beim Reden, Lesen usw. ~*
leave out übergehen * *jemanden ~* || weglassen
leave out of account außer Acht lassen || nicht in Anschlag bringen
leave to überlassen * *abtreten an*
leaven Hefe
leaving abgehend * *Zug usw.*
leaving out abgesehen von
leaving out of account unter Vernachlässigung von * *außer Betracht lassend*
leaving-certificate Abschlusszeugnis
lecture Vortrag * *Vorlesung*
lecturer Dozent || Pofessor * *Dozent*
LED Leuchtdiode * *Abkürzg. für 'light emitting diode'* || Luminizenzdiode
LED display LED-Anzeige
LED indicators LED-Anzeigen
ledge Absatz * *vorstehender Rand* || Kante * *Leiste, vorstehender ~, Sims, Anschlag~* || Leiste * *Sims, Leiste, vorstehender Rand, Felsbank, Lager/Ader in Bergwerk* || Vorsprung * *das Hervorspringende (Teil)* || Vorsprung * *Fels~ usw.*
leeway Spielraum * *(figürl.)*
left links * *turn left: nach ~ drehen/abbiegen; left of: ~ von* || übrig * *~ gelassen/-geblieben; have something left: etwas ~haben*
left hand side links * *~seitig*
left over übrig * *~ gelassen*
left stop Linksanschlag * *z.B. eines Potis*
left turn links * *(Adv.) nach ~, ~herum*
left-hand linksgängig * *z.B. Schraube, Gewinde*
left-hand thread Linksgewinde
left-handed screw linksgängige Schraube
left-handed thread Linksgewinde
left-justified linksbündig * *z.B. in der Daten-/Textverarbeitung*
lefthand linkes * *(Adj.)* || links * *~liegend, ~ angeordnet* || linksseitig
lefthand side linke Seite
lefthanded links * *~seitig*
leftmost links * *das sich am weitesten ~ befindende*
leg Zweig * *z.B. einer Stromrichterschaltung*
leg fuse Zweigsicherung
leg spring Schenkelfeder

legal erlaubt * *per Gesetz ~*, *den Vorschriften entsprechend* || gültig * *zulässig* || rechtlich || Rechts- * *rechtlich* || rechtsverbindlich
legal action rechtliche Maßnahme
legal advisor Rechtsberater
legal domicile Gerichtsstand * *Gerichtsort*
legal force Gültigkeit * *~ e. Gesetzes*
legal issue gesetzliche Bestimmung
legal position Stellung * *Rechtsstellung*
legal regulation gesetzliche Vorschrift
legality Gültigkeit * *Zulässigkeit*
legalize bestätigen * *amtlich*
legend Beschriftung * *z.B. an einem Messinstrument* || Bildunterschrift * *Legende, Bildinschrift* || Legende * *z.B. mit Bilderklärungen*
legend plate Beschriftungsschild * *z.B.für Bedienelement*
legible lesbar * *leserlich, ablesbar*
legislation Gesetzgebung
legislative provision gesetzliche Vorschrift
leisure time Freizeit
leisurely langsam * *(Adv.) bedächtig, gemächlich* || ruhig * *gemächlich (auch Adv.)*
leisureness Langsamkeit * *Gemächlichkeit, Bedächtigkeit*
LEM transformer LEM-Wandler * *potenzialfreier Strom-Messwandler der Schweizer Fa. LEM (R)*
lend leihen * *an jdn. etwas aus~* || verleihen
lend an ear to anhören * *jdn. anhören; jdm. zuhören*
lend out leihen * *ver~*
lending fee Leihgebühren
length Länge
length counter Längenzähler
length measuring system Längenmesssystem
length of brake path Bremsweg
length of cable Leitungslänge
length of jib Ausladung * *eines Drehkrans*
length of lay Schlaglänge * *beim Verseilen von Kabeln*
length of thread Gewindelänge
length of time Dauer * *Zeitlänge*
length of twist Schlaglänge * *beim Verseilen von Kabeln*
lengthen strecken * *in die Länge ~* || verlängern
lengthwise längs * *der Länge nach*
lengthy langwierig * *länglich*
lens linse
less -los || abzüglich || darunter * *weniger* || kleiner * *(Komparativ) zahlenmäßig; than:als* || unter * *less than: weniger/niedriger als* || weniger
less developed country Entwicklungsland
less significant niederwertig * *Bit, Dekade usw.*
less-than weniger * *(Adv.) z.B. less-than-optimal: ~ optimal*
lessen schwächen * *vermindern* || verkleinern * *vermindern* || vermindern || verringern
lessening Reduzierung * *Verringerung, Abnahme* || Verminderung
lesson Lehre * *Lektion, Warnung; let it be a lesson to you: lasse es Dir eine ~ sein*
let überlassen * *let a person have a thing:jemandem etwas überlassen* || vermieten * *to let: zu ~ (z.B. Haus); siehe auch 'leihen'*
let alone gewähren * *allein/in Ruhe lassen, sich selbst überlassen*
let alone that abgesehen davon dass
let down senken * *herunterlassen*
let flow together zusammenführen * *zusammenfließen lassen*
let go nachlassen * *lockern*
let go out ausschalten * *Lampe, Licht, LED-Anzeige ~*
let have his way gewähren * *gewähren lassen*
let in einlassen

let into einlassen * *in Holz/Metall (z.B. Möbelbeschlag)*
let into concrete einbetonieren
let off ablassen
let out nachlassen * *z.B. bei Papierbahn*
let pass durchlassen || durchreichen * *durchlassen* || gelten lassen
let slip vergeben * *Chance ~*
let through durchlassen
let us assume that angenommen * *(Adv.)*
letter Buchstabe || Schriftzeichen
letter of intent Absichtserklärung * *schriftliche* || Rahmenabkommen * *schriftl. Absichtserklärung* || Vorvertrag
letter of recommendation Empfehlungsschreiben || Zeugnis * *~ vom Arbeitgeber*
letter of thanks Dankschreiben
letter press Druckerpresse || Hochdruck * *(Druckerei) Druckverfahren*
letter symbol Kennbuchstabe
lettering Beschriftung * *z.B. Werbeaufschrift*
lettering strip Beschriftungsstreifen
letterpress printing Buchdruck
level abflachen * *auch 'sich ~'* || Ausmaß * *(figürl.) Niveau, Stufe, Stand, Ebene* || eben * *waagerecht* || Ebene * *Grad, Stufe, Klasse* || Füllstand || Grad * *z.B. Funkstörgrad, Schärfegrad* || Klasse * *Ebene* || Niveau * *(auch figürl.: at a low/high level: auf niedrigem/hohem ~)* || nivellieren || Pegel * *auch Spannungspegel, logischer Pegel* || Plan * *(Adj.) eben, waagerecht* || planieren * *eben machen, nivellieren, einebnen, planieren* || richten * *gerade-/plan~ von (Metall-) Werkstücken, Blech usw.* || Schicht * *Ebene* || Stand * *Niveau* || Stufe * *Niveau* || Zustand * *Pegel eines logischen/Binär-Signals*
level adaption Pegelanpassung
level changeover Pegelumschaltung || Pegelwandlung
level conversion Pegelwandlung
level converter Pegelumsetzer || Pegelwandler
level filling machine Höhen-Füllmaschine
level gauge Füllstandsmesser || Füllstandsmessgerät
level indicator Füllstandsanzeiger
level meter Füllstandsmesser
level monitoring Füllstandskontrolle
level monitoring system Füllstandskontrolle * *als Einrichtung*
level of automation Automatisierungsebene || Automatisierungsgrad
level of modularity Modularität * *Ausmaß der ~; with a high level of modularity: hochmodular*
level of program nesting Schachteltiefe * *Schachtelungsebene von Unterprogrammen*
level sensor Füllstandssensor
level shifter Pegelwandler * *z.B. zwischen verschiedenen Logikfamilien wie TTL, CMOS*
level shifting Pegelanpassung
level switch Niveauschalter || Niveauwächter
level up ausgleichen * *aufs gleiche Niveau heben*
leveller Richtmaschine * *rollt über mehrere Walzen gebogenes/welliges Blech plan* || Streckrichter * *Streckrichtmaschine z. Geraderichten von Blech über viele dünne Walzen*
levelling Nivellierung * *Ausgleich von Höhenunterschieden*
levelling machine Richtmaschine
levelling plate Richtplatte
lever Griff * *Hebel* || Hebel || hebeln * *hebeln, stemmen; lever out (of): aus~/herausstemmen aus* || Hebelwirkung * *Hebelübersetzung, Hebelkraft/-wirkung; Einfluss (figürl.)*
lever arm Hebelarm
lever closure Hebelverschluss * *in der Verpakkungstechnik*
lever handle Handhebel

leverage Einfluss * *Mittel/Hebel zur Einflussnahme, Einfluss, Hebelwirkung* || Hebelwirkung * *Hebelkraft/-Wirkung, Hebelanordnung/-Anwendung*
levitate schweben * *frei schweben (lassen), z.B. Magnetschwebebahn*
liabilities Passiva
liability Haftung * *z.b. für Schäden; limited:beschränkte; personal:persönliche; exemption from: Ausschluss* || Verpflichtung * *gesetzliche/vertragliche ~; assume/incur liabilities: ~en eingehen*
liability claim Schadenersatzanspruch || Schadensersatzanspruch
liability claims Schadenersatzanspruch * *(Plural)* || Schadensersatzanspruch * *(Plural)*
liable verantwortlich * *haftbar, verpflichtet, for: für*
liable for damages schadenersatzpflichtig
liable to -pflichtig * *(vorangestellt) unterliegen, z.B. Schadenersatz, Haftung usw.*
liable to pay costs kostenpflichtig
liable to payment of damages schadenersatzpflichtig
liable to payment of royalties lizenzpflichtig
liaison man Verbindungsmann || Verbindungsperson * *Verbindungsmann* || Vermittler * *Verbindungsmann*
liaison person Verbindungsperson
liberty Freiheit * *politische/persönliche ~; take the liberty of· sich die ~ nehmen zu*
library Bibliothek * *auch von Softwaremodulen (Plural: libraries)* || Bücherei
licence Berechtigung * *(brit.) offizielle ~* || Genehmigung * *(behördliche/offizielle) Zulassung* || Konzession * *(brit.) Lizenz, Verkaufsrecht* || Lizenz * *under: in; grant licence for: ~ erteilen für* || Zulassung
licence agreement Lizenzvereinbarung || Lizenzvertrag * *Lizenzvereinbarung*
licence contract Lizenzvereinbarung * *Lizenzvertrag* || Lizenzvertrag
licence fee Lizenzgebühr
licenced zugelassen * *siehe 'licensed'*
licenced production Nachbaufertigung * *(brit.) Lizenzfertigung*
licencing procedure Genehmigungsverfahren
license Berechtigung * *(amerikan.) von Behörde zugesprochene/offizielle ~* || genehmigen * *amtlich ~, lizensieren* || Genehmigung * *(amerikan.) offizielle Zulassung* || Konzession * *(amerikan.) Lizenz, Verkaufsrecht* || lizensieren || Lizenz * *(amerikan.) under: in* || zulassen * *behördlich* || Zulassung * *(amerikan.)*
license agreement Lizenzvereinbarung * *(amerikan.)*
license conditions Lizenzbedingungen
license fee Schutzgebühr * *z.B. für ein Druckerzeugnis*
license to export Ausfuhrgenehmigung
license to import Einfuhrgenehmigung
licensed autorisiert || zugelassen * *(amerikan.) lizensiert (auch Auto usw.)*
licensed construction Lizenzbau
licensed production Lizenzbau || Lizenzfertigung || Nachbaufertigung * *(amerikan.) Lizenzfertigung*
licensee Lizenznehmer
licenser Lizenzgeber
licensing fee Schutzgebühr
licensing procedure Genehmigungsverfahren * *(amerikan.)*
licensor Lizenzgeber
licker-in Briseur * *{Textil} bei →Krempelmaschine* || Vorreißer * *{Textil} bei →Krempelmaschine*
lid Deckel * *hinged lid: Klapp~ (bei Behälter usw.), lid catch: ~verschluss*

lie aufliegen * *ruhen, lasten, liegen, gelegen sein, sich befinden (upon: auf)* || liegen * *the voltage lies between 8V and 25V:die Spannung liegt zwischen 8 und 25 V*
lie opposite gegenüberliegen
life Gebrauchsdauer || Haltbarkeit * *Lebensdauer* || Lebensdauer * *average: durchschnittliche; long: hohe* || Standzeit
life calculation Lebensdauerberechnung
life components spannungsführende Teile
life cycle Lebenszyklus
life expectancy Lebensdauer * *Lebenserwartung* || Lebensdauer * *Lebenserwartung (für {el.Masch.}): siehe VDE 0530 Teil 1 und Beiblatt 1)* || Lebenserwartung * *auch von techn. Geräten*
life span Lebensdauer
life test Langzeittest * *Lebensdauerprüfung* || Lebensdauerprüfung
life zero verschobener Nullpunkt
lifelong learning lebenslanges Lernen
lifespan Lebensdauer
lifetime Gesamtlebensdauer || Lebensdauer * *high:-hohe* || Standzeit
lifetime lubrication Lebensdauerschmierung
lifetime warranty lebenslange Garantie
LIFO LIFO * *Last In First Out, z.B.Stapelspeicher: d.zuletzt eingespeicherte wird zuerst wiederausgegeben*
LIFO memory Stapelspeicher * *Abk. f. 'Last-In-First-Out':Die neueste Information wird als Erstes ausgelesen*
LIFO stack Stapelspeicher
lift anheben * *heben* || Auftrieb || Aufzug * *(siehe EN 81)* || erhöhen * *in die Luft heben; auch Preise usw. ~* || Fahrstuhl || heben || Hub * *~ e. Ventils* || Lift * *Aufzug* || lüften * *heben* || Öffnung * *{Masch.bau} eines Ventils* || tragen * *heben*
lift capacity Förderleistung * *{Aufzüge}*
lift car Aufzugskabine
lift control Aufzugsregelung || Aufzugssteuerung || Aufzugsteuerung
lift drive Aufzugsantrieb || Hubantrieb
lift element Hubelement
lift hoisting rope Aufzugseil
lift machine Aufzugmotor
lift motor Aufzugmotor
lift off abdrücken * *z.B. Lager von Welle* || abheben || abreisen * *mit Flugzeug* || wegfliegen * *mit Flugzeug*
lift rope Aufzugseil
lift shaft Aufzugschacht
lift table Hubtisch
lift transfer Ausschleusung * *durch Hubbewegung*
lift trolley Hubwagen
lift up aufheben * *emporheben* || erhöhen * *(amerikan.) Preise usw. ~*
lift winch Aufzugwinde
lift-truck Hubwagen
lifted erhöht * *angehoben (in die Luft)*
lifter Nocke * *auch Hebeeinrichtung, Stößel* || Nocken * *auch Hebeeinrichtung, Stößel*
lifting appliance Hebezeug
lifting capacity Hubleistung * *eines Hubwerks* || Tragfähigkeit * *bei Hebezeug, Kran*
lifting drive Hubantrieb || Hubwerksantrieb * *bei Kran*
lifting element Hubelement
lifting equipment Hebezeug
lifting eyebolt Ringöse * *Hebeöse*
lifting eye Hebeöse * *an Motor, Schaltschrank* || Kranhaken * *Hebeöse an der zu hebenden Last (z.B. Motor, Schaltschrank)* || Ringöse * *z.B. Hebeöse an Motor, Schaltschrank* || Tragöse * *z.B. an el. Masch. (ab 30 kg Gewicht) zum Einhängen des Kran-Lasthakens*

lifting eyebolt Anhängevorrichtung * *beim Krahn* || Hebeöse * *z.B. an Motor, Schaltschrank* || Transportöse
lifting force Hubkraft
lifting gear Hebezeug || Hubwerk * ~ *eines Krans/Förderzeugs*
lifting gear and cranes Hebezeuge
lifting handle Tragegriff || Traggriff
lifting height Hubhöhe
lifting lug Ansetzstelle * *zum Heben* || Hebeöse * *in Form einer Lasche*
lifting magnet Hubmagnet || Lastmagnet * *an Hebezeug*
lifting pin Anhängevorrichtung * *beim Krahn*
lifting platform Hebebühne || Hubbühne * *siehe auch* →*Hebebühne*
lifting power Tragfähigkeit * *bei Hebezeug, Kran*
lifting screw Abdrückschraube
lifting solenoid Hubmagnet
lifting speed Hubgeschwindigkeit
lifting spindle Hubspindel
lifting table Hubtisch
lifting trolley Hubwagen
lifting unit Hebezeug * →*Hubwerk* || Hubwerk
light dünnflüssig * *Öl* || einschalten * *(auf)leuchten lassen, z.B. LED, Meldeleuchte; lit: eingeschaltet* || gering * *gering (z.b. Strom, Gewicht), leicht, Schwach-, leicht beladen* || leicht * ~ *an Gewicht, ~ zu erledigen; (auch figürl.: Essen, Kleidg., Musik, Wein, Hand usw.)* || Leuchte || leuchten * *be/er~, Licht machen* || schwach * *Schwach- (z.B. Last), leicht, leicht beladen*
light alloy Leichtmetall * ~*legierung*
light barrier Lichtschranke
light construction Leichtbau
light current Schwachstrom
light curtain Lichtvorhang
light duty Schwachlast
light emitting diode Leuchtdiode
light in weight leicht * ~ *anGewicht*
light load Schwachlast || Teillast * *Schwachlast*
light load test Schwachlastprüfung
light metal Leichtmetall
light pressure leichter Druck
light rail vehicle Nahverkehrsfahrzeug * *leichtes Schienenfahrzeug*
light sensor Lichtsensor
light sheet Feinblech
light signal Lichtsignal
light up aufleuchten
light weight leicht * ~ *an Gewicht*
light weight coated paper leicht gestrichenes Papier
light-duty leicht * *für geringe Beanspruchung, m.geringer Leistung, Schwachlast-; z.B. Schwachstrom-Relais*
light-emitting diode Leuchtdiode || Luminizenzdiode
light-gauge dünn * *Draht, Kabel, Blech*
light-minded leicht * *leichtfertig*
light-rail vehicle Stadtbahnwagen
light-sensing lichtempfindlich * *Sensor usw.*
light-sensitive lichtempfindlich
light-weight leicht * ~ *an Gewicht*
light-weight coated paper leichtgewichtiges gestrichenes Papier
light-weight design Leichtbau
lighted pushbutton Leuchttaster
lighten erleichtern * *eine Bürde* ~ || leuchten * *er~, erhellen*
lighting Beleuchtung
lighting ballast Vorschaltgerät * *für Leuchtstofflampen*
lighting engineering Lichttechnik

lighting technology Beleuchtungstechnik || Lichttechnik
lightning Blitz || blitzschnell * *(Adj.)*
lightning arrester Überspannungsableiter
lightning conductor Blitzableiter
lightning stroke Blitzschlag
lightning withstand voltage Blitzstoßspannung
lightning-fast blitzschnell
lightweight building board Leichtbauplatte * *[Bautechnik] für den Innenausbau*
lightweight construction Leichtbau * *[Bautechnik]*
lignite coal Braunkohle
lignite open-cast mining Braunkohlentagebau * *Abbauverfahren*
lignite open-pit mine Braunkohlentagebau * *Bergwerk*
like ähnlich || schätzen * *(gern) mögen*
like a blow schlagartig * *(Adv.)*
like a shot blitzschnell * *(Adv.) wie aus der Pistole geschossen*
like that so * *derartig, (Adv.) auf diese Art/Weise*
like this so * *solch, derartig; (Adv.) auf diese Weise*
likelihood Wahrscheinlichkeit
likeliness Ähnlichkeit
likely möglich * *wahrscheinlich* || vermutlich * *wahrscheinlich* || voraussichtlich * *(Adj.) wahrscheinlich, voraussichtlich* || wahrscheinlich
liken vergleichen * *bildhaft* ~ *(to: mit)*
likewise auch * *eben/gleichfalls desgleichen, ebenso* || ebenfalls || gleichermaßen
liking Sinn * *Vorliebe, for: für*
LIM LIM * *Abk. f. 'Linear Induction Motor': linearer Induktionsmotor*
lime kiln Kalkofen
limit begrenzen * *to: auf (einen Wert)* ~ || Begrenzung * *Grenze* || beschränken || End- * *Grenz-* || Grenze * *auch 'Begrenzung'; upper/lower/utmost: obere/untere/äußerste; (at: an)* || Grenzwert
limit between pulsating and non-pulsating operation Lückgrenze
limit class Grenzwertklasse * *siehe VDE 0875, EN 55011; DIN VDE 0871 bezügl. Funkentstörung*
limit condition Randbedingung
limit controller Begrenzungsregler
limit current Grenzstrom
limit curve Grenzkurve
limit detector Grenzwertmelder
limit indication Grenzwertmeldung
limit monitor Grenzwertmelder
limit monitoring Grenzwertüberwachung
limit of error Toleranz * *check for specified limits: auf Einhaltung d.Toleranzen prüfen*
limit of stability Stabilitätsgrenze
limit position Endlage
limit rating Grenzleistung
limit signal Grenzwertmeldung
limit signal transmitter Grenzsignalgeber * *z.B. Endschalter*
limit speed Grenzdrehzahl
limit stop Anschlag * *End~ z.B. ~ einer Messeinrichtung*
limit switch Endschalter || Wegendschalter
limit value Grenzwert
limit value indicator Grenzwertmelder
limit value monitor Grenzwertmelder
limit violation Grenzwertüberschreitung
limitation Begrenzung || Beschränkung || Einschränkung
limitations Einschränkungen
limited bedingt * *beschränkt* || begrenzt || beschränkt || eingeschränkt || knapp * *beschränkt; my time is limited: meine Zeit ist ~ bemessen*

limited company Gesellschaft mit beschränkter Haftung
limited for a specific time zeitlich begrenzt
limited partnership Kommanditgesellschaft
limited room enger Raum * beengte Raumverhältnisse
limiter Begrenzer
limiter module Begrenzer
limiting Begrenzung * before/after the limiting:- vor/nach der Begrenzung || Grenze * Begrenzung
limiting characteristic Grenzkennlinie || Grenzkurve
limiting continuous current Grenzdauerstrom
limiting control Begrenzungsregelung
limiting diode Begrenzungsdiode
limiting frequency Grenzfrequenz * z.B. von AC-Motor
limiting load Grenzbelastung
limiting speed Grenzdrehzahl * mechanical: mechanische
limiting temperature Grenztemperatur
limiting temperature rise Grenzübertemperatur
limiting value Grenzwert
line Anlage * (Fertigungs-, Walz-, Verarbeitungs-, Behandlungs-) Straße/Linie || Anschluss * Telefon~ || ausfüttern * (aus) füttern, auskleiden, ausschlagen, (auf der Innenseite) überziehen || Baureihe || belegen * Kupplung, Bremse usw. mit Belag ~/auskleiden || Branche * Fach, Gebiet, Tätigkeitsfeld, Sparte, Geschäftszweig || Fach * Gebiet, Tätigkeitsfeld, Sparte, Fach, Berufszweig || kaschieren || Leitung * Übertragungs~, Telefon~, Druckluft~ || Linie * dashed: gestrichelte; dotted: punktierte; thin: dünne; heavy: dicke || Netz * ~~leitung, Speiseleitung, Strom~ (aus der Sicht des Verbrauchers) || Netz- || Reihe * Linie, Bau~ || Schnur * Leine || Seil * Schnur, Leine, Tau || Sortiment * Produktpalette || Straße * Verabeitungs-/Produktions~ || Strich * Linie || Warenbahn || Zeile
line amplifier Leitungsverstärker
line breaker Hauptschalter
line circuit breaker Hauptschalter * Leistungsschalter in Netzeinspeisung
line commutating reactor Netzkommutierungsdrossel
line commutation Netzführung * für Stromrichterschaltung (z.B. bei Thyristorschaltung)
line connection in parallel Parallelanschluss * {el.Masch.}
line contactor Netzschütz
line contactor control Netzschützansteuerung
line contactor energizing Netzschützansteuerung
line current Linienstrom || Netzstrom
line disturbances Netzstörung * Einbrüche, Verzerrungen, Spitzen
line disturbances fed back into the distribution system Netzrückwirkungen
line drive Bandantrieb * z.B. in Blechstraße, Bandbehandlungsanlage, Walzwerk
line driver Leitungstreiber
line driver/receiver Leitungstreiber-Empfänger
line drop Spannungsabfall * entlang einer Leitung
line fault documenter Störschreiber * zum Aufzeichnen von Netzstörungen
line fault recorder Störschreiber * zum Aufzeichnen von Netzstörungen
line filter Netzfilter
line frequency Netzfrequenz
line frequency tracking Netzfrequenzanpassung || Netzfrequenznachführung * z.B. bei netzgeführtem Stromrichter
line fuse Netzsicherung || Strangsicherung * netzseitige Sicherung
line harmonics Netzrückwirkungen

line impedance Netzimpedanz
line integral Linienintegral
line interruption Netzunterbrechung
line noise Netzrückwirkungen
line of action Angriffsrichtung * auch einer Kraft || Wirkungsrichtung * Angriffsrichtung (einer Kraft usw.)
line of business Geschäftszweig
line of force Feldlinie * magnetic: magnetische, electric: elektrische || Kraftlinie * magnetic:magnetische
line of merchandise Fertigungsprogramm * lieferbare Produkte || Produktpalette
line of products Fertigungsprogramm || Produktpalette * Produktlinie/familie
line of training Ausbildungsweg * z.b. in beruflicher Aus- und Weiterbildung
line perturbation Netzrückwirkung * durch el.-Masch.: siehe z.b. DIN VDE 0875, IEC 22G/21/CDV und IEC 1800-3
line pipe Rohranschluss * Leitungsrohr
line power factor Netzleistungsfaktor
line printer Drucker * Zeilendrucker
line protection Netzschutz
line reactance Netzreaktanz
line reaction Netzrückwirkung
line reactor Netzdrossel
line receiver Leitungsempfänger
line reference Liniensollwert * Geschwindigkeitssollwert für Produktions-/Bearbeitungslinie
line resistance Leitungswiderstand
line resistance starter Ständeranlasser * {el.-Masch.} 3-phasiger Vorwiderstand zum Anlassen e. Käfigläufermotors
line shaft Längswelle
line side Netzseite * at/on the:auf der
line size Rohrdurchmesser * bei Druckluftleitung usw.
line space Zeilenabstand
line speed Bahngeschwindigkeit || Leitspannung * Bahngeschwindigkeit || Liniengeschwindigkeit || Maschinengeschwindigkeit * Bahngeschwindigkeit || Produktionsgeschwindigkeit * einer kontinuierlich arbeitenden Produktionslinie || Warengeschwindigkeit || Warengeschwindigkeit * Bahn/Bandgeschwindigkeit
line speed reference Leitsollwert * Linien-/Bahngeschwindigkeitssollwert für durchlaufende Warenbahn || Leitspannung * Bahngeschwindigkeitssollwert
line speed reference voltage Leitspannung
line speed setpoint Leitspannung * Bahngeschwindigkeits-Sollwert
line supply Netz * Stromversorgungs~
line supply conditions Netzbedingungen * poor: schlechte
line supply fluctuations Netzschwankungen
line supply impedance Netzimpedanz
line supply stressing Netzbelastung * z.B. durch Oberwellen
line supply voltage Netzanschlussspannung * feed from a 400V line supply voltage: an einer ~ von 400V betreiben
line switch Hauptschalter || Netzschalter
line tab Stichleitung
line terminal Netzklemme
line terminator Leitungsabschluss * ~glied, ~widerstand
line tolerance monitoring Netzspannungsüberwachung
line transformer Netztransformator
line up aneinander reihen || anreihen
line voltage Anschlussspannung * Netzspannung || Leiterspannung || Netzspannung
line voltage failure Netzspannungsausfall

line voltage fluctuations Netzspannungsschwankungen
line voltage monitoring Netzspannungsüberwachung || Netzüberwachung * Netzspannungsüberwachung
line voltage spikes Netzspannungsspitzen
line voltage tolerance Netzspannungstoleranz
line width Zeilenbreite
line-communtated converter netzgeführter Stromrichter * siehe DIN EN 60146, VDE 0588 und IEC 146-1-1
line-commutated netzgeführt * Stromrichter bei d.das Netz die Kommutierungsspannung liefert
line-commutating netzgeführt
line-frequency thyristor Netzthyristor * kann mit Schaltfrequenzen im Bereich der Netzfrequenz betrieben werden
line-side Netz- * netzseitig || netzseitig || vorgeschaltet * auf der Netzseite || Vorschalt- * netzseitig angeordnet
line-side fundamental power factor Netz-Grundschwingungsleistungsfaktor
line-side fuse Eingangssicherung * netzseitig angeordnete ~, →Netzsicherung || Netzsicherung
line-side reactor Netzdrossel
line-side rectifier Eingangsgleichrichter * netzseitiger ~ || Netzgleichrichter * netzseitiger ~ || netzseitiger Gleichrichter * z.B. bei Frequenzumrichter
line-side rectifier unit Einspeiseeinheit
line-side terminals Netzklemmen * z.B. e. Stromrichters
line-side transformer Vorschalttransformator * netzseitiger Transformator || Vortransformator * netzseitiger Trafo
line-speed reference Liniensollwert
line-supply cable Netzleitung
line-to-line voltage Leiterspannung * Spannung zwischen 2 Hauptleitern, z.B. L1, L2, L3 || verkettete Spannung * zwischen den Hauptleitern, z.B. L1-L2-L3
line-to-neutral voltage Phasenspannung || Sternspannung * zwischen Phase und Mittelpunktsleiter
line-up terminal Reihenklemme
line-voltage disturbance Netzstörung
line-voltage failure Netzausfall || Netzstörung * →Netzspannungsausfall
lineage Zeilenhonorar * für Übersetzer, Datenerfasser
linear linear
linear actuator Hubspindeltrieb || Linearantrieb
linear encoder Lineal * →Linearmaßstab zur Positionserfassung bei e. translatorischer Bewegung || Linearmaßstab * (meist optischer) Encoder zur Lageerfassung bei Linearachse (z.B. Glaslineal)
linear equation Geradengleichung
linear feed Linearvorschub
linear interpolation lineare Interpolation
linear measuring system Längenmesssystem
linear motion Linearbewegung || lineare Bewegung
linear motor Linearmotor * z.B. mit linear angeordnetem "Käfigläufer"
linear movement geradlinige Bewegung
linear program lineares Programm * ohne Sprünge/Verzweigungen
linear programming lineare Programmierung * ohne Sprünge
linear size Längenmaß
linear slope gerade * (Substantiv) gerade Linie
linear temperature rise adiabatische Erwärmung * Temperaturanstieg über d.Zeit b.reiner Wärmespeicherg. (keine Wärmeleitg.)
linearity Linearität
linearity error Linearitätsfehler || Linearitätsfehler || Nichtlinearität * Linearitätsfehler

linearization Linearisierung
linearization block Linearisierungsstufe
linearize linearisieren
linearized value Fertigwert * linearisierter Messwert, z.B. von Temperaturgeber
linearizing Linearisierung
linearly linear * (Adv.) degressive:abnehmend; progressive/increasing:ansteigend
linearly degressive linear abnehmend
linearly increasing linear ansteigend
linearly interpolated linear interpoliert
linearly progressive linear ansteigend
lined kaschiert
linen Leinen * ['linin]
liner Buchse * Zylinderbuchse
linger verweilen
linguistic Linguistik
lining Auflage * Futter || Auskleidung || Belag * z.B. Reib~, Brems~, Kupplungs~ || Beschichtung * Belag (Bremse, Kupplung usw.), Auskleidung, Überguss, Ausfütterung * Futter * Ausfütterung, Futter, Auskleidung, Verkleidung || Kaschierung || Überzug * Verkleidung; protective:Schutz- || Verkleidung
lining material Auskleidung * Auskleidungsmaterial
link ankoppeln * to:an; z.B. über serielle Schnittstelle || Bindeglied || binden * z.B. Softwaremodule zusammen~ ("Befriedigung" der externen Adressbeziehungen) || Brücke * Verbindungs(stück) || Gelenk || Glied * Verbindungsglied || koppeln * verbinden, auch über serielle Schnittstelle || Kopplung * z.B. ~ über serielle Schnittstelle || Schnittstelle * Verbindung, z.B.serielle || verbinden * together: miteinander, z.B. Ventile in einer Stromrichterschaltung || Verbindung * ~sglied, (serielle/parallele) Daten-/Signal~ || Verbindungslasche || verknüpfen || Zwischenkreis
link circuit reactor Zwischenkreisdrossel
link converter Zwischenkreisumrichter
link into einbinden
link to anbinden
link together aneinander fügen * miteinander verbinden, verknüpfen, verketten || verknüpfen * (figürl.) miteinander ~, z.B. Ideen || zusammenschließen
link voltage Zwischenkreisspannung
link-circuit capacitor Zwischenkreiskondensator
link-up überbrücken * verbinden
linkage Anbindung
linked unit Koppelpartner
linked-up gekoppelt
linker Binder * zum Zusammenbinden von Softwaremodulen ("Befriedigung" der gegenseitigen Adressreferenzen) || Linker * zum Zusammenbinden von Softwaremoduln ("Befriedigung" der gegenseitigen Adressreferenzen)
linking Verkettung || Verknüpfung * Verkettung, Verkopplung z.B. v.Signalzuständen; Zusammenknüpfen
lint Fussel
lintel Sturz * Fenster~, Tür~
lip Lippe
lip-type seal Lippendichtung
lipping Umleimer * [Holz]
liquefaction Verflüssigung
liquefying Verflüssigung
liquid flüssig || Flüssigkeit || nass
liquid cooler Flüssigkeitskühler
liquid cooling Flüssigkeitskühlung
liquid crystal display Flüssigkeitskristall-Anzeige
liquid gas Flüssiggas
liquid level Flüssigkeitsstand || Füllstand * einer Flüssigkeit

liquid level closed-loop control Füllstandsregelung * *für Flüssigkeitsbehälter*
liquid level indicator Füllstandsanzeiger * *für Flüssigkeiten*
liquid level sensor Füllstandssensor * *für Flüssigkeiten*
liquid manometer Druckmessgerät
liquid pump Flüssigkeitspumpe
liquid resistor Flüssigkeitswiderstand * *mit Flüssigkeit als Widerstandsmedium, z.B.Wasser* || Flüssigkeitswiderstand * *mit Flüssigkeit als Widerstandsmedium, z.b.Wasser*
liquid separator Flüssigkeitsabscheider
liquid-cooled flüssigkeitsgekühlt
liquid-crystal display Flüssigkeitskristallanzeige
liquid-resistor starter Flüssigkeitsanlasser * *Flüssigk.-Anlasswiderst. f.Großmotor z.B.mit Sodalauge u.Verstellelektroden*
liquid-ring vacuum pump Flüssigkeitsring-Vakuumpumpe
liquidate auflösen * *Firma* ~ || beseitigen * *liquidieren, ausschalten*
liquidation Amortisation || Beseitigung * *(figürl.)*
liquidity Liquidität
liquor Lauge
list aufführen * *in einer Liste* || auflisten * *z.B. in Aufzählung/Tabelle/Liste* ~ || aufnehmen * *eintragen* || Aufstellung || buchen || eintragen || Katalog * *Liste* || Liste || Prospekt * *Liste* || Schlagseite * *(auch Verb: ~ haben); heeel over: ~ bekommen* || Übersicht * *Aufstellung* || Verzeichnis * *Liste*
list for Kosten * *(Verb) einen Listenpreis von ... haben*
list form Listenform
list of contents Inhaltsübersicht * *~ in Listenform* || Inhaltsverzeichnis
list of defects Mängelliste || Meckerliste * *Mängelliste*
list of documents Zeichnungsverzeichnis * *(Zeichnungs-) Unterlagen-Verzeichnis*
list of drawings Zeichnungsverzeichnis
list of exhibitors Ausstellerverzeichnis * *bei einer Messe/Ausstellung*
list of spare parts Ersatzteilliste
list of suppliers Bezugsquellenverzeichnis
list price Listenpreis
list structure Listenstruktur * *interne Darstellg.e.-Anwenderprogramms: Adreslisten d.abzuarb. Programm-u.Datenmodule*
list-oriented listenorientiert
listed approbiert * *z.B. UL listed:beim UL-Normengremium approbiert* || aufgeführt * *in einer Liste* || zugelassen * *bei einer Normenbehörde/Zulassungsbehörde* ~
listed form Listenform
listen to anhören
listing Approbation * *z. B.UL* || Aufnahme * *~ in eine Liste* || Aufstellung || Ausdruck * *Programm* || Erfassung || Liste * *Computer-Ausdruck (z.B. Quellspracheprogramm)* || Listing * *Ausdruck von einem Drucker (z.B. eines Quellspracheprogramms)* || Verzeichnis * *Aufzählung, Auflistung*
lit eingeschaltet * *Leuchtdiode, Meldeleuchte* || leuchtend * *zum Leuchten gebracht, z.B. Meldeleuchte, LED*
liter Liter * *(amerikan.)*
literature Informationen * *schriftliche* || Literatur
literature on the products Produktinformation * *(Plural) schriftliche* || Produktinformationen * *schriftliche*
lithium battery Lithiumbatterie
lithium cell Lithiumzelle * *Batterie-Knopfzelle*
lithium-soaped lithiumverseift * *grease: Schmierfett*
litigation Prozess * *Rechtsstreit*

litre Liter * *(brit.)*
litres Liter * *(Plural)*
litter Abfall * *herumliegender Abfall* || verschandeln * *mit Abfall, Dreck*
little gering || geringfügig || klein || wenig
little box Kästchen
little by little allmählich * *(Adv.).*
little case Etui * *siehe auch* →Kästchen
little endian format Intel-Format * *Datenformat: niederwertig.Byte/Wort auf niederer, höherwert.-auf höher.Adresse*
little used selten
littleness Kleinheit
live bestromt * *Draht oder Leiter* || existieren * *leben* || unter Spannung * *live removal: Ziehen* ~
live insertion Stecken unter Spannung
live parts spannungsführende Teile * *Teile, die Personen gefährden können*
live removal Ziehen unter Spannung
live steam Frischdampf
live zero verschobener Nullpunkt * *z.B. 4-20 mA Stromsignal mit Drahtbrucherkennungsmöglichkeit*
livering Eindickung * *von Farbe usw,*
living area Wohngebiet
lixiviant Lauge * *Laugungsmittel*
lixivium Lauge * *(chem.)*
load Auslastung * *Belastung* || beaufschlagen * *(elektr.) belasten* || belasten || Belastung * *Last; (maximum) permissible: zulässige; per unit load: pro Flächeneinheit* || Belastungs- || Bürde * *Last,- Belastung* || eintragen * *beschicken (z.B. Ofen)* || Fracht || füllen * *beladen* || laden * *auch Programm, Daten usw.* || Last * *under load: unter Last* || spannen * *Feder* || Verbraucher * *Last, energieaufnehmender Teil e. Stromkreises*
load angle Belastungswinkel || Lastwinkel || Lastwinkel * *zw. Magnetisierungsstrom (d.parallel z. Fluss steht) u. dem Gesamtstrom*
load angle control Lastwinkelregelung
load balance Lastaufteilung * *zu gleichen Anteilen* || Lastausgleich * *zw. mehreren mechanisch o. über d. Material gekoppelten Antrieben z.B. Längswelle*
load balance control Lastausgleichsregelung
load balancing Lastausgleich * *zw. mehreren mechanisch o. über das Material gekoppelte Antrieben*
load balancing control Lastausgleichsregelung * *for drives coupled to a common load:für an dieselbe Last gekopp. Antriebe*
load branch Verbraucherabzweig
load capability Beanspruchbarkeit * *Belastbarkeit* || Belastbarkeit * *z.B. ~ eines Analog/Digital-Ausgangs*
load capacity Belastbarkeit * *Belastungsvermögen (zahlen/wertmäßig)* || Leistungsfähigkeit * *Belastbarkeit* || Tragfähigkeit
load cell Zugkraftmessgeber * *Zugmessdose* || Zugmessdose || Zugmesseinrichtung * *Zugmessdose*
load change Laständerung || Lastwechsel
load changes Belastungsänderungen || Laständerungen || Lastschwankungen
load characteristic Belastungskennlinie || Lastkennlinie
load characteristics Lastverhalten
load circuit Lastkreis
load commutation Lastführung * *bei Stromrichterschaltung* || Lastkommutierung * *für Stromrichterschaltung* || Maschinenführung * *Lastführung eines Antriebsstromrichters*
load compensation Lastausgleich * *Kompensation der Lasteinflüsse*
load condition Belastungszustand || Lastzustand
load conditions Lastbedingungen

load current Laststrom
load cycle Lastspiel
load cycle characteristic Belastungsverlauf * zeitlicher Verlauf eines Lastzyklus, eines Lastdrehmoments usw.
load disconnector Lasttrenner || Lasttrennschalter
load distribution Lastaufteilung * Lastverteilung || Lastverteilung * z.B. zwischen Antrieben, die starr oder über das Material gekoppelt sind
load distribution control Lastausgleichsregelung * Lastverteilungsregelung zur Momentenaufteilung auf mehrere Motoren
load distribution plant Lastverteiler * (Energieverteilung)
load equilization Lastausgleich
load equilization control Lastausgleichsregelung
load factor Belastungsfaktor * (el.Masch.) kennzeichnet die erlaubte Schalthäufigkeit e. Motors || ED || Lastfaktor * Quotient aus Durchschnittslast und Spitzenlast || Stoßfaktor * →Lastfaktor (Verhältnis Durchschnitts- zu Spitzenlast)
load feed-forward Lastvorsteuerung
load fluctuations Belastungsänderungen * Lastschwankungen || Lastschwankungen
load flywheel effect Schwungmoment * als Motorlast beim Beschleunigen/Verzögern
load impact effect Lastabhängigkeit
load impedance Lastimpedanz
load in einladen * Transportgüter
load inertia Lastmoment * Trägheitsmoment
load interruptor Lasttrenner || Lasttrennschalter
load machine Belastungsmaschine
load operation Ladefunktion * für ein (SPS-) Programm
load pick-up Lastaufnahmemittel * (Hebezeuge)
load point Lastpunkt
load position Lastangriff * →Angriffspunkt der Last z.B. an einer Welle
load precontrol Lastvorsteuerung
load range Belastungsbereich
load rating Belastbarkeit * spezifizierte~ , zugelassene ~ || Nennbelastbarkeit || Tragfähigkeit
load rejection Lastabwurf
load relay Lastrelais
load resistance Lastwiderstand
load resistor Belastungswiderstand || Bürdenwiderstand * Lastwiderstand
load response Lastverhalten * einer Regelung
load rheostat Bremswiderstand
load share Lastaufteilung || Lastausgleich * Lastaufteilg., z.B. zw. mehren Motoren, d. starr od.- über d.Material gekoppelt sind
load share control Lastausgleichsregelung
load sharing Lastaufteilung
load sharing control Lastausgleichsregelung
load shedding Lastabwurf
load side Lastseite || maschinenseitig * z.B. zum Motor hin angeordneter Stromrichter || motorseitig * von einem Stromrichter aus gesehen
load surge Laststoß * kurzzeitige Überlast || Stoßlast * Laststoß
load test Belastungstest * z.B. ~ für el. Maschine, Stromrichter
load thrust Laststoß
load time ratio Einschaltdauer * relative ~
load torque Gegenmoment * Lastmoment || Lastmoment * Drehmoment || Widerstandsmoment * →Lastmoment
load torque characteristic Gegenmomentverlauf * ~ über der Drehzahl || Lastkennlinie * Verlauf des Lastmoments über der Drehzahl
load transfer Lastübernahme
load transient Lastwechsel * vorübergehender
load transients Lastspitzen

load unit Bürdeneinheit * Belastungseinheit aus Widerständen zur Strommessung
load variation Lastwechsel
load voltage Lastspannung
load voltage drop Lastspannungsabfall
load-adapted lastadaptiert || lastadaptiv
load-break switch Lasttrenner || Lasttrennschalter
load-carrying capacity Tragfähigkeit * auch mechan. ~, z.B. von Wälz-/Gleitlagern, Zahnrädern
load-commutated lastgeführt * Stromrichter, bei d.die Antriebsmaschine d.Kommutierungsspanng.- liefert (DIN 41750)
load-commutated converter lastkommutierter Stromrichter
load-dependent lastabhängig
load-dependent field weakening lastabhängige Feldschwächung * ermöglicht bei Kran mit DC-Hubmotor schnelle Fahrt bei leichter Last
load-independent belastungsunabhängig || eingeprägt * z.B. Strom/Spannung || lastunabhängig
load-independent current eingeprägter Strom
load-independent current source Konstantstromquelle
load-independent voltage eingeprägte Spannung
load-sensitive lastabhängig
load-side lastseitig || maschinenseitig * von Stromrichter aus gesehen
loadability Belastbarkeit
loadable belastbar || ladbar * Software usw.
loaded belastet || unter Last * belastet
loaded with -behaftet * (vorangestellt) mit Schulden usw. ~
loading Auslastung || Beanspruchung * Belastung || Belastung
loading capability Belastbarkeit
loading equipment Beladeeinrichtung || Beschikkungseinrichtung || Zuführeinrichtung * Beladungs-/Verladungseinrichtung
loading facility Beladeeinrichtung
loading limit Belastungsgrenze
loading machine Belastungsmaschine
loading ratio Tragsicherheit * Kenngröße für Wälzlager, bestimmend für dessen Lebensdauer
loading system Beladeeinrichtung
loading unit Ladeeinheit * z.B. Palette
loan leihen * (Geld usw.) ver~
local dezentral || Insel- || örtlich || regional || vor Ort || Vor-Ort-
local bus Nahbus
local company Landesgesellschaft * z.B. ausländische Niederlassung
local control Vorort-Betrieb * Vorort-Steuerung
local control level Handebene
local dealer Fachhändler * am Orte
local mode Vor-Ort-Betrieb || Vorort-Betrieb * Betriebsart
local office ZN * (SIEMENS-) Zweigniederlassung
local operation Vor-Ort-Betrieb || Vorort-Betrieb
local reference Vor-Ort-Sollwert
local representative Bezirksvertreter * örtlicher Vertreter
local sales office Vor-Ort-Vertrieb || ZN * (SIEMENS-) Zweigniederlassung
local service Nahverkehr
local time Ortszeit
local traffic Nahverkehr || Personennahverkehr
locality Ort * Örtlichkeit || Platz * Örtlichkeit
localize lokalisieren
locate ablegen || anordnen * örtlich || ausmachen * orten || ermitteln * Aufenthaltsort || feststellen * Ort, Lage, Fehler lokalisieren || finden * auffinden, lokalisieren, ausfindig machen, "dingfest machen" || suchen * ausfindig machen
locate a fault Fehler lokalisieren

located befindlich * örtlich || untergebracht * örtlich/räumlich
located at the bottom untenliegend
located at the top obenliegend
locating bearing Festlager
location Anordnung * räumliche || Bereich * Stelle, Platz || Lage * Ort || Ort || Standort || Stelle * Ort, Platz
location assignment Ortskennzeichen
location bearing Festlager || Führungslager * →Festlager
location designation Anlagenkennzeichen * Ortskennzeichen || Ortskennzeichen
location diagram Anordnungsplan || Dispositionsplan * →Anordnungsplan
location identifier Ortskennzeichen
location of business Geschäftssitz
location of course Veranstaltungsort * eines Kurses
location of terminal box Klemmenkastenlage * z.b. eines Motorklemmenkastens (siehe z.B. DIN 42673 Teil 1)
lock -Sperre * Verriegelung || abschließen * ver-/zusperren || arretieren || Arretierung || feststellen * verriegeln || Raste * Verschluss, Sperrvorrichtung || schließen * mit Schlüssel ~ || Schloss || sichern || Sicherung * (mechan.) || Sperre || sperren * (mechan.) || Verschluss * Schloss; keep under lock (and key): unter ~ halten; lock nut: ~mutter
lock button Rasttaster
lock home einrasten
lock in einschließen * mit Schlüssel, Riegel usw. ~
lock in memory nichtflüchtig abspeichern
lock in place arretieren
lock nut Befestigungsmutter || Gegenmutter || Kontermutter
lock of rotational direction Drehrichtungssperre
lock out ausschließen || Sperre || verriegeln * gegen Wiedereinschalten ~
lock pin Sicherungsstift
lock ring Sicherungsring
lock screw Sicherungsschraube
lock the rotor festbremsen * Motor
lock up aufhängen * →sich aufhängen (Softwareprogramm) || einschließen * mit Schlüssel, Riegel usw. ~ || sich aufhängen * Softwareprogramm || verschließen * mit Schlüssel ~
lock washer Schraubensicherungsring || Sicherungsscheibe
lock-bolt Feststellschraube
lock-out band Ausblendband
lock-out function Einschaltsperre
lock-type contact Selbsthaltekontakt
lockable abschließbar || verschließbar
locked abgeschlossen * abgesperrt || blockiert * versperrt || festgebremst || geschlossen * verschlossen, verriegelt, versperrt || gesperrt || verschlossen
locked bearing Festlager
locked rotor blockierter Läufer || festgebremster Läufer
locked rotor voltage Rotorstillstandsspannung
locked up verschlossen
locked-motor torque Anzugsmoment * Motor
locked-rotor festgebremst * Motor
locked-rotor apparent power Einschaltscheinleistung * Kurzschluss-Scheinleistg. e. Asynchronmotors bei festgebremst. Läufer || Kurzschluss-Scheinleistung * {el.Masch.} Scheinleistg. e. Asynchr.mot. bei festgebremstem Läufer || Kurzschlussscheinleistung * {el.Masch.} Scheinleistung e. Motors bei festgebremst. Läufer
locked-rotor current Anlassstrom * wird vom Motor beim Anlaufen aufgenommen || Anlaufstrom * wird z.B. von AC-Motor beim Anlaufen benötigt || Anzugsstrom * ~ eines Motors im Startmoment

locked-rotor protection Blockierschutz * für Motor
locked-rotor temperature rise Kurzschlusserwärmung * {el.Masch.} Temp.anstieg d. Wicklungen e. Asynchr.masch. i.Stillstand
locked-rotor torque Anzugsmoment * Anlaufmoment e. Motors aus d. Stillstand heraus
locked-rotor voltage Läuferstillstandsspannung
locked-up stress Verspannung * (psycholog.) innere ~
locking Arretierung || Feststellung * mechan. Arretierung || Formschluss || Selbsthaltung * remain locked in: in ~ gehen/bleiben || Sicherung * das Sichern (to prevent: gegen), (against: gegen) || Verriegelung * Sperre
locking assembly z.b. zur Wellen-Naben-Verbindung Spannsatz
locking bar Verriegelungsschiene * an e. Baugruppenträger als 'Herausziehsperre' für Baugruppen
locking bush Spannhülse
locking button Feststellknopf
locking device Feststellvorrichtung || Spannelement * auch für Wellen-Nabeverbindung || Sperre * Sperrvorrichtung || Sperrung * Vorrichtung
locking element Spannelement
locking lever Verriegelungshebel * z.B. an Flachbandkabelstecker
locking ring Klemmring
locking screw Feststellschraube || Verschlussschraube
locking sleeve Klemmhülse
locking type rastend * z.B. Taster
locking-up Verklemmung * auch softwaremäßige
locknut Kontermutter
lockout -Schutz
lockup Absturz * ~/Sich-Verklemmen eines Softwareprogramms
locus Ortskurve
locus diagram Ortskurve
log aufzeichnen || Aufzeichnung * aufgezeichnete Betriebsdaten usw. || Holz * Rundholz, unbehauener ~klotz, gefällter Baumstamm || Protokoll * Betriebs~, Logbuch || protokollieren || registrieren * Messwerte ~
log band saw Blockbandsäge * {Holz} Holzsäge mit umlaufendem Sägeblatt
log conveyor Rundholzförderer
log frame saw Gattersäge * {Holz} für Holzstämme
log off abmelden * im Computernetzwerk, Steuerfunktion
log on anmelden * als Steuerfunktion ; im Computernetz ~
log planer Blockhobelmaschine * {Holz}
log printer Meldedrucker
log-in einloggen * Starten einer Arbeitssitzung in einem Computer-Netzwerk
log-log doppellogarithmisch
log-off ausloggen * Arbeitssitzung in einem Computer-Netzwerk beenden
log-on einloggen * in Computer-Netzwerk
logarithmic logarithmisch * graded logarithmically: ~ gestuft; logarithmic characteristic: ~er Verlauf
logarithmic-scale logarithmisch * in ~em Maßstab
logger Meldedrucker
logging Aufzeichnung * von Betriebsdaten || Protokollierung || Registrierung * von Betriebsdaten
logging printer Meldedrucker
logic Logik || logisch * auch boolsch || Verknüpfung * logische, binäre
logic algebra Schaltalgebra
logic analyzer Logikanalysator
logic cell Logikzelle

logic circuit Logik * Logikschaltung || Logikschaltung
logic condition logischer Zustand
logic control Steuerung || Verknüpfungssteuerung
logic element Verknüpfungsglied
logic equation logische Gleichung
logic family Logikfamilie * Schaltkreisfamilie, z.B. TTL, CMOS || Schaltkreisfamilie * z.b. TTL, CMOS
logic function binäre Funktion || Binärfunktion || Verknüpfungsfunktion
logic gate Logikgatter
logic input Binäreingang || Digitaleingang
logic level logischer Pegel || Schaltpegel * einer Logikstufe
logic nesting depth Verknüpfungstiefe
logic operation logische Verknüpfung || Verknüpfung * logische/binäre Funktion
logic output Digitalausgang * Binärausgang
logic processing Logik * Logikverarbeitung || Logikverarbeitung
logic signal Steuersignal
logic symbol Schaltzeichen * für binäres Schaltelement
logic variable Binärgröße
logic-intensive verknüpfungsintensiv * z.B. SPS-Programm
logical logisch
logical channel logischer Kanal
logical state logischer Zustand
logically also * (Adv.) logischerweise || logisch * (Adv) ~erweise
logically combined logisch verknüpft
logistic Logistik
logistical logistisch
logistics Logistik
logo Logo * Sinnbild || Signet * Sinnbild, Logo
logon procedure Anmeldeprozedur * (Computer) für Betriebssystem, Netzwerk, Mailbox usw.
logs Rundholz
long lang * räumlich und zeitlich
long for wünschen * sehnend ~
long life hohe Standzeit
long pulse Langimpuls
long word Langwort
long-distance Fern- * im Verkehrswesen, in der Telekommunikation usw.
long-distance traffic Fernverkehr
long-drawn-out langwierig * sich (schleppend) dahinziehend, sich in die Länge ziehend
long-duration test Langzeittest
long-established alt * firm:alt eingeführte Firma || alteingesessen
long-haul Fern- * (Bahnen)
long-haul locomotive Fernbahn-Lokomotive
long-lived langlebig
long-minded breit * weitschweifig
long-range auf lange Sicht || langfristig
long-run test Langzeittest
long-serving employee langjähriger Mitarbeiter
long-standing alt * seit langem bestehend (Freudschaft, Beziehungen usw.)
long-term auf lange Sicht || Dauer- || dauerhaft * (zeitlich) || langfristig || langlebig * lange geltend, langfristig
long-term accuracy Langzeitgenauigkeit
long-term drift Langzeitdrift
long-term endurance Dauerfestigkeit
long-term lubrication Dauerschmierung * Langzeitschmierung
long-term stability Konstanz * Langzeitkonstanz || Langzeitkonstante || Langzeitstabilität
long-term trend Langzeitverlauf * von Messgrößen usw.
long-term variation Langzeitverlauf

long-time Langzeit-
long-time storage Langzeitlagerung
long-time test Dauertest || Dauerversuch
longer länger * longer than: länger als (auch zeitlich); any longer: nicht länger/nicht mehr
longeron Holm * Längs-/Rumpfholm usw.
longevity Langlebigkeit
longitudinal längs * der Länge nach || Längs-
longitudinal direction Längsrichtung
longitudinal field Längsfeld
longitudinal force Längskraft
longitudinal orienter Längsrecke * Längsstreckeinrichtung bei Kunststoff-Folienmaschine
longitudinal profile Längsprofil * Längsverteilg. v.Flächengewicht, Feuchte, Temperatur,Dicke v.bandförm.Material
longitudinal resistance Längswiderstand * z.B. in einem Halbleiter
longitudinal section Längsschnitt * in Konstruktionszeichnung
longitudinal stress Dehnung * Längsspannung/-beanspruchung
longwinded umständlich * langatmig
look after kümmern * aufpassen auf, sehen nach, sich ~ um, sorgen für || kümmern um || wahrnehmen * Interesse ~
look at anschauen * (auch figürl.) || ansehen || betrachten
look for suchen
look into prüfen * nachsehen, inspizieren
look like ähneln
look out for vorsehen * sich ~ vor
look up nachschlagen * z.B. in Tabelle/Liste/Buch/Betriebsanleitung || suchen * nachschlagen (in Liste, Buch, Naaachschlagewerk, Tabelle im Programmspeicher)
look upon betrachten * as: als
look-ahead vorausschauend * Algorithmus
look-out Suche * be on the look-out for: auf der ~ nach ... sein/Ausschau halten nach
look-up operation Suchvorgang * in einer Tabelle (z.B. durch Rechner)
look-up table Tabelle * "Nachschlage-/Zuordnungstabelle" in Computer (von Adressen, Koordinatenwerten usw.)
looking after Betreuung
looking ahead prädiktiv || vorausschauend
loom aufsteigen * drohendes Ereignis || Webstuhl
loom motor Webstuhlmotor
loop Schlaufe || Schleife * z.B. Regelschleife, Programmschleife || Schlinge * auch in Blechstraße || umschlingen
loop around umschlingen
loop car Schlingenwagen * (Walzwerk) dient als Warenspeicher bei Walzstraße
loop channel Schlingenkanal
loop control Regelung
loop controller Regler
loop controller module Regelbaugruppe || Regelungsbaugruppe
loop counter Schleifenzähler * (Computer) z.B. Index in einer DO...FOR Programmschleife
loop gain Kreisverstärkung
loop into einschleifen * loop into a circuit: in einen Stromkreis einschleifen
loop lifter Schlingenheber * (Walzwerk)
loop output -Reglerausgang
loop processor Regelungsprozessor * bei SPS
loop through durchschleifen
loop-contol scheme Regelkonzept
loop-controller Regler
loop-in einschleifen * z.B. Signal, Leitung
looper Bandspeicher * für Metallband || Schlingenheber * (Walzwerk) || Schlingenspanner || Warenbahnspeicher || Warenspeicher * für durchlaufende

Warenbahn zur Überbrückung von Rollenwechselzeiten
loophole Ausweg * *Ausflucht, Ausweichmöglichkeit*
looping angle Umschlingungswinkel
looping floor Schlingenkanal
looping pit Schlingengrube * *{Walzwerk} Warenspeicher in Walz/Blechstraße*
looping tower Schlingenturm * *{Walzwerk}*
loose Lose * *(Adj.) nicht fest; supplied loose: lose mitgeliefert* || unverpackt * *lose*
loose change gear Wechselrad
loose connection Wackelkontakt
loose contact Wackelkontakt
loose fit lockerer Sitz || weiter Sitz
loosen lockern * *z.B. Schraube* || lösen * *z.B. Schraube, Befestigung, Knoten; loosening: das Lösen (einer Schraube usw.)* || nachlassen * *lockern*
loosen the screw schrauben * *los/ab~*
loosening Lösung * *Losmachen, z.B. Schraube, Knoten*
lop-sided schief * *nach einer Seite hängend, unsymmetrisch*
lopsided schief * *nach einer Seite hängend, unsymmetrisch*
lorry LKW * *(brit.)*
lose money zusetzen * *Geld einbüßen*
lose one's way verlaufen * *sich ~*
losing bargain Verlustgeschäft
losing business Verlustgeschäft
loss -Ausfall * *Fehlen von Signal/Spannung/Strom/Phase* || Ausfall * *Wegbleiben e. Signals, Spannung oder Strom* || Bruch * *Wegbleiben eines Signals, z.B. Tachobruch* || Dämpfung * *von Energien* || Schaden * *Verlust, auch finanziell* || Schwund * *Verlust* || Verlust * *auch finanzieller ~, ~ von Daten/Parametereinstellungen usw.* || Verluste
loss angle Verlustwinkel
loss factor Verlustfaktor * *→tangens delta eines Dielektrikums (kennzeichnet Verluste e. el. Feldes)*
loss index Verlustziffer * *Summe Wirbelstrom- u. Hystereseverluste (1,3.. 4 W/kg) d. (ungestanzten) Magnetblechs*
loss of data Datenverlust
loss of production Produktionsausfall
loss of time Zeitverlust
loss of voltage Spannungsfreiheit * *Spannungslosigkeit*
loss segregation Verlusttrennung * *Aufteilung der Verluste auf die einzelnen Anteile bei d. Wirkungsgrad-Ermittlung*
loss tangent tangens delta * *tangens zw. Wirk- u. Blindstrom, Maß für die Verluste im Dielektrikum eines el. Feldes*
loss-factor Verlustfaktor
loss-free verlustfrei
loss-less verlustfrei
loss-summation method Einzelverlustverfahren * *zur Ermittlung der Verluste und des Wirkungsgrads*
losses Verluste * *Geld, Reibung, Energie* || Verlustleistung * *z.B. ~ eines Motors*
losses due to forward current Durchlassverluste * *bei Diode, Thyristor usw.*
lossmaker Verlustgeschäft * *verlustbringendes Produkt/Geschäft*
lost Weg * *(Adv.) verloren*
lost property office Fundbüro
lot Fertigungslos || Lieferung * *Partie* || Los * *Fertigungs~; lot code: ~nummer*
lot No Losnummer
lot number Losnummer * *eines Fertigungsloses*
lots of viel || viele

loud laut * *(auch Adv.) laut vom Geräusch her* || marktschreierisch * *schreiend, auffallend, grell*
loud-voiced laut * *Person*
loudly laut * *(Adv.)*
loudness Lautstärke
loudness level Lautstärke * *Geräuschpegel*
lousy schlecht * *(salopp) mies, lausig*
louver Jalousie * *auch an Luftein-/auslassöffnung* || Kühlschlitz * *(amerikan.) Kiemenblech*
louvered cooling slot Kühlschlitz * *mit "Überdachungsstreifen"*
louvered cover Jalousie * *mit "überdachten" Kühlschlitzen versehene Abdeckung (einer Luftein-/Auslassöffnung)* || Kiemenblech * *Abdeckblech mit Schlitzen, z.B. für Be-/Entlüftungsöffnung*
louvre Kühlschlitz * *(brit.)*
love of order Ordnungsliebe
low gering * *niedrig* || klein * *niedrig bezügl. Zahlenwert/Höhe, z.B. Spannung, Leistung, Reaktanz* || leise * *kaum hörbar* || niedrig || ruhig * *kaum hörbar* || tief * *niedrig* || unter * *niedrig*
Low active Low-aktiv
low air pressure Unterdruck * *Luftdruck*
low control setting niedrige Aussteuerung
low cost Niedrigpreis-
low current Schwachstrom
low delay angle setting hohe Aussteuerung * *kleiner Steuerwinkel bei Phasenanschnittsteuerung*
low end anspruchslos * *am unteren Ende des Typenspektrums*
low gearbox stressing Getriebeschonung * *z.B. durch Hochlaufgeber vor oder hinter dem Momentenregler*
low grade zweite Wahl
low induction induktionsarm
low modulation niedrige Aussteuerung * *in der Mess-/Nachrichtentechnik usw.*
low noise niedrige Geräuschentwicklung
low order niederwertig * *Ziffer usw*
low pass filter Filter * *Tiefpassfilter* || PT1-Glied * *Tiefpassfilter* || Tiefpass || Tiefpassfilter
low performance anspruchslos * *mit niedriger Leistungsfähigkeit*
low power kleiner Leistung || niedrige Leistung || verlustarm * *mit niedriger (Verlust)Leistung*
low price günstiger Preis || Niedrigpreis-
low speed niedrige Drehzahl
low speeds kleine Drehzahlen || niedrige Drehzahlen
low stressing Schonung * *geringe Beanspruchung*
low voltage Niederspannung * *alle Spannungen unter 1000V* || Schwachstrom * *mit niedriger Spannung (arbeitend)*
low voltage circuit breaker Niederspannungsschalter
low voltage motor Niederspannungsmotor ·
low- -arm * *vorangestellt; z.B.: low-harmonics: oberschwingungsarm; low-maintenance: wartungsarm*
low-abrasive verschleißarm
low-backlash spielarm
low-budget Niedrigpreis- || preisgünstig * *billig* || preiswert * *"für den kleinen Geldbeutel", billig*
low-cost billig * *siehe auch →preisgünstig* || kostengünstig * *mit niedrigem Preis* || preisgünstig
low-current cable Steuerleitung
low-distortion verzerrungsarm
low-efficiency verlustbehaftet
low-end unterer Leistungsbereich
low-end performance range unterer Leistungsbereich * *bezüglich der Leistungsfähigkeit*
low-friction leichtgängig * *mit wenig Reibung* || reibungsarm
low-grade minderwertig * *Ware usw.*
low-harmonic oberschwingungsarm

low-impedance induktivitätsarm
low-inductance induktivitätsarm
low-inertia leichtgängig * *mit kleiner Schwungmasse/Massenträgheit* || schwungmassenarm || trägheitsarm
low-level Kleinsignal-
low-load conditions Schwachlast
low-loss verlustarm || verlustfrei * *verlustarm*
low-loss metallized-dielectric capacitor MKV-Kondensator * *m.metallisiert.Papier-Kunststoff-Dielektrikum, verlustarm*
low-maintenance wartungsarm
low-noise geräuscharm
low-order harmonics Oberschwingungen niedriger Ordnungszahlen
low-output range unterer Leistungsbereich * *mit kleiner Ausgangsleistung,z.B. Stromrichter, Motor*
low-pass filter Tiefpass * *first-order low-pass filter: ~ erster Ordnung*
low-power energiesparend || leistungsarm
low-pressure Niederdruck-
low-pressure steam mains Niederdrucksammelschiene * *zur Dampfversorgung*
low-priced billig || kostengünstig * *preisgünstig* || preisgünstig || preiswert * *mit niedrigem Preis*
low-profile dünn * *in flacher Bauart* || flach * *flach bauend* || Flach- || schlank * *v. d. Baugröße her* || schmal * *schlanke von der Baugröße her*
low-profile type of construction Flachbauform || schlanke Bauform
low-quality minderwertig * *Ware, Produkt usw.* || schlecht * *von niedriger Qualität*
low-quality goods Ausschuss * *schlechte Ware(n)*
low-rating klein * *mit ~er Leistung, z.B. Motor, Stromrichter, Pumpe*
low-rating motors Motoren kleiner Leistung
low-rating range unterer Leistungsbereich * *bezüglich der (Aus-/Eingangs-) Leistung*
low-reflection reflexionsarm
low-resistance niederohmig
low-solvent lösungsmittelarm
low-speed langsam * *mit niedriger Geschwindigkeit* || langsam laufend
low-speed machine Langsamläufer
low-stressing schonend
low-temperature grease Tieftemperaturfett
low-vibration schwingungsarm
low-viscosity dünnflüssig
low-voltage Niederspannungs- || Niedervolt-
Low-Voltage Directive Niederspannungsrichtlinie * *der EG: 73/23/EWG, EG-Amtsblatt 73/L77/L29 u. 93/68/EWG "CE-Kennzeichng."*
low-voltage distribution Niederspannungsverteilung
low-voltage high-breaking-capacity fuse NH-Sicherung * *Niederspannungs-Hochleistungssicherung*
low-voltage high-rupturing-capacity fuse NH-Sicherung * *Niederspannungs-Hochleistungssicherung*
low-voltage high-rupturing-capacity fuse link NH-Sicherung * *Sicherungseinsatz* || NH-Sicherungseinsatz * *Niederspannungs-Hochleistungssicherung*
low-voltage switchgear Niederspannungs-Schaltgeräte || Niederspannungsschaltgerät
low-voltage switchgear and controlgear Niederspannungs-Schaltgeräte
low-voltage system Niederspannungsanlage
low-voltage version Kleinspannungsausführung
low-wear verschleißarm
lower erniedrigen * *kleiner machen, reduzieren* || herabsetzen * *erniedrigen, verringern* || kleiner * *(Komparativ) niedriger* || reduzieren * *absenken* || senken * *niedriger machen, herunterlassen (z.B.*

Last am Kran) || tiefer || unter * *niedrig(er)* || Unter- || unterbieten * *Rekord* ~ || vermindern * *senken (Kosten, Temperaturen, Preise usw.)*
lower button Tiefer-Taste || Tiefer-Taster * *z.B. für Motorpotentiometer*
lower carrier roll Unterwalze
lower edge Unterkante
lower key Tiefer-Taste
lower limit untere Grenze || Untergrenze
lower limit value unterer Grenzwert
lower motor Untermotor * *Walzwerk*
lower part Unterteil
lower rope Unterseil * *{Aufzüge}*
lower section Unterteil
lower-order niederwertig * *Ziffer usw.*
lower-ranking niederprior * *niedriger im Rang (auch Person)*
lower-rating kleiner * *mit kleiner Leistung, z.B. Motor, Stromrichter, Pumpe*
lowering Absenkung * *Niedriger-Machen* || Herabsetzung * *Verringerung, Erniedrigung* || Senkung * *Preise usw.*
lowest kleinst || letzt * *unterst, niedrigst* || niedrigst
lowest-rating kleinst * *mit ~er Leistung, z.B. Motor, Stromrichter, Pumpe*
LRoS LRoS * *Abkürzung für 'Lloyds Register of Shipping': Schiffs-Klassifizierungsgesellschaft*
LRS LRS * *Abkürzg. für 'Lloyd's Register of Shipping': britische Schiffsklassifikationsgesellschaft*
LRV Schienenfahrzeug * *Abk. für 'Light-Rail Vehicle': leichtes/Kurzstrecken-~*
LSB LSB * *Abkürzg. für 'Least Significant Bit':niederwertigstes Bit* || niederwertigstes Bit * *Least Significant Bit*
LSD LSD * *Abkürzg. für 'Least Significant Digit':-niederwertigstes Zeichen z.B. einer Ziffernanzeige* || niederwertigste Ziffer * *Abkürzung für 'Least Significant Digit':niederwertigste Ziffer* || niederwertigstes Zeichen * *Least Significant Digit; einer Ziffernanzeige*
LSI hochintegriert * *Abkürzg.für 'Large Scale Integration': (IC) mit hohem Integrationsgrad*
lube oil Schmieröl
lube-oil pipe Ölleitung * *zur Schmierölversorgung*
lubricant Gleitmittel * *Schmiermittel* || Schmierfett || Schmiermittel || Schmierstoff
lubricant exudation Schmiermittel-Ausschwitzen
lubricant film Schmierfilm
lubricant life Schmiermittelgebrauchsdauer * *DIN51818/51852, DIN ISO281 Beibl.1, GfT Arb.bl. 2.4.1, Ges.f.Tribologie*
lubricant plate Schmierschild * *mit Angaben zur Schmierung (Schmierstofftypen, Schmierfristen usw.)*
lubricant quantity Schmiermittelmenge || Schmierstoffmenge
lubricate abschmieren || schmieren
lubricated for life lebensdauergeschmiert
lubricated for the service lifetime lebensdauergeschmiert
lubricating Schmier-
lubricating film Schmierfilm
lubricating gap Schmierspalt
lubricating grease Schmierfett
lubricating oil Schmieröl
lubricating property Schmierfähigkeit * *Schmiereigenschaft*
lubricating pump Schmierpumpe
lubricating vapour Öldampf
lubrication Schmierung * *DIN 51818/51852, DIN ISO 281 Beibl.1, GfT Arb.bl. 2.4.1 (Gesellsch. für Tribologie)*
lubrication duct Schmierkanal
lubrication film Schmierfilm

lubrication interval Fettgebrauchsdauer || Nachschmierfrist || Schmierfrist
lubrication nipple Schmiernippel
lubrication plate Schmierschild * mit Angaben zur richtigen Schmierung
lubrication point Schmierstelle
lubrication pump Ölpumpe
lubricator Öler || Schmiervorrichtung
lubricity Schmierfähigkeit
lucid übersichtlich * *klar, deutlich in Stil, licht im Geiste*
lucidity Übersichtlichkeit * *Klarheit, Deutlichkeit*
lucrative gut * *lukrativ; business:Geschäft* || lohnend * *Gewinn bringend* || vorteilhaft
Luenberg observer Luenberg-Beobachter * *(Regel.techn.)*
luffing gear Nickwerk * *(Hebezeuge) zum Auf-/ Abwärtsschwenken eines Kranauslegers*
lug Lappen * →*Lasche* || Lasche * *Ansatz(blech), Henkel, Ohr, Schlaufe, Zinke,* || Nase * *Ansatz, Lasche, Lappen (aus Blech usw.), Öse, Henkel, Zinke, Fahne (z. Schraub-/Lötverbindg.)* || Nocken * *Zinke, Ansatz, Lasche, "Ohr"* || schleppen
lug connector Kabelschuh * *Flach-/Messer~*
lumber Bauholz * *(amerikan.)* || Holz * *(amerikan.) Nutz~* || Schnittholz * *(amerikan.)*
lumlnaire Leuchte * *~nkörper, Lampe*
luminous flux Lichtstrom
luminous table Leuchttisch
lump sum price Pauschale * *Pauschalpreis*
lump sum repair price Reparaturpauschale * *Pauschalpreis für Reparatur*
lump wood Stückholz * *z.B Brennholz*
lump-sum pauschal * *-Preis, -Summe usw. z.B. lump-sum price: Pauschalpreis* || Pauschbetrag * *Pauschalbetrag*
lump-sum price Pauschalpreis
lunch Mittagessen * *Mittagessen, zweites Frühstück; das ~ einnehmen*
luxurious komfortabel * *luxuriös* || nobel * *luxuriös*
luxury Komfort * *Luxus, Wohlleben, (Hoch-) Genuss, Luxusartikel; Pracht*
LV Niederspannung * *Abkürzung für 'Low Voltage'*
LV breaker Niederspannungsschalter * *Abkürzg.f. 'Low Voltage': Niederspannungs-*
LV motor Niederspannungsmotor
LWC paper gestrichenes Papier * *Abk. f. 'Light-Weight-Coated paper': leicht ~* || leicht gestrichenes Papier * *Light Weight Coated* || leichtgewichtiges gestrichenes Papier * *Abk.f. 'Light-Weight Coated'* || LWC-Papier * *Abkürzg. für 'Light-Weight-Coated' paper: leicht gestrichenes Papier*
lye Lauge
lye pump Laugenpumpe * *bei Wasch/Geschirrspülmaschine*
lytic capacitor Elektrolytkondensator * *Kurzform* || Elko * *Kurzform*

M

m m * *Meter; 1 m entspr. 3,2808 ft (feet) entspr. 1,0936 yd (yards) entspr. 39.3701 in (inch)*
M conductor Mittelleiter * *bei Gleichstrom*
M. G. cylinder Glättzylinder * *Abk. für 'Machine Glazing cylinder'*
machine Apparat * *Maschine* || bearbeiten * *Werkstück(e) maschinell ~ (meist spanabhebend)* || Maschine || spanend bearbeiten || verarbeiten * *maschinell ~* || zerspanend bearbeiten

machine at rest and de-energized Pause * *(el.-Masch.) Stillstand mit stromlosen Wicklungen*
machine bed Maschinenbett || Maschinenstuhlung
machine builder Maschinenbauer * *Herstellerfirma*
machine building Maschinenbau
machine code Maschinensprache * *Maschinencode*
machine code program Maschinenprogramm * *(Computer) Programm im von der CPU ausführbaren ausführbaren Binärcode*
machine commutation Maschinenführung * *Kommutierungssspannung e. Stromrichters wird von der el. Maschine geliefert*
machine component Maschinenteil
machine constructor Maschinenbauer * *Erbauer, Konstrukteur*
machine control Maschinensteuerung
machine control system Maschinensteuerung
machine direction Längsrichtung * *Maschinen(-lauf)richtung* || Laufrichtung * *einer Maschine* || Maschinenlaufrichtung || Maschinenrichtung * *Maschinenlaufrichtung*
machine direction orienter Längsrecke * *(Kunststoff) bei Folienmaschine*
machine element Einzelteil * *Maschinenteil, Konstruktionselement* || Konstruktionselement * *Konstruktionselement, Maschinenelement* || Maschinenelement || Maschinenteil
machine frame Maschinenkörper || Maschinenstuhlung
machine glazing cylinder Glättzylinder
machine house Maschinenhaus
machine industry Maschinenbau * *~industrie*
machine knife Maschinenmesser
machine language Maschinensprache * *(Computer)*
machine losses Maschinenverluste * *z.B.durch Reibung*
machine manufacturer Maschinenbauer * *Maschinenhersteller* || Maschinenhersteller * *siehe auch →Maschinenbauer*
machine member Maschinenteil
machine model Maschinenmodell * *regelungstechnisches Modell z.B. eines Motors*
machine mounting Maschinenbefestigung
machine operator Maschinenführer
machine part Maschinenteil
machine protection Maschinenschutz
machine rack Maschinengestell
machine readable maschinenlesbar
machine section Maschinenteil
machine set Machinensatz * *z.B. Motor und Generator bei rotierendem Umformer* || Maschinensatz
machine setter Maschineneinrichter
machine speed Maschinengeschwindigkeit
machine speed setpoint Maschinensollwert * *Geschwindigkeitssollwert*
machine supplier Maschinenbauer * *Maschinenlieferant* || Maschinenhersteller * *Maschinenlieferant*
machine tool Werkzeugmaschine
machine tool builders Werkzeugmaschinenbau * *~er*
machine tools manufacturing Werkzeugmaschinenbau
machine velocity Maschinengeschwindigkeit
machine vibrations Maschinenschwingungen
machine vice Maschinenschraubstock
machine vision analysis Bilderkennung
machine width Maschinenbreite
machine wire Sieb * *(Papier) ~ bei Papiermaschine*
machine with brushless excitation permanenterregte Maschine * *ohne Bürsten*
machine-coated gestrichen * *Papier*

machine-commutated maschinengeführt * Stromrichter || maschinengetaktet * Maschine gibt die Kommutierungszeitpunkte vor (→lastgeführter Stromrichter)
machine-cut zerspanen
machine-made winding Maschinenwicklung * [el. Masch.] maschinell in das Blechpaket eingebrachte Wicklung
machine-readable code maschinenlesbare Fabrikatebezeichnung
machine-readable product code MLFB
machine-readable product designation maschinenlesbare Fabrikatebezeichnung * MLFB || MLFB
machine-tool manufacturer Werkzeugmaschinenbauer
machineman Maschinenführer
machinery Maschinen || Maschinenausrüstung || Maschinenbau * Maschinenausrüstung || Maschinenpark
machinery construction Maschinenbau
machinery directive Maschinenrichtlinie * der EG (89/106/EWG und 91/368/EWG in d.EG-Amtsblättern 89/L183/9 u. 91/L198/16)
machinery maker Maschinenbauer
machinery manufacturer Maschinenbauer * Maschinenhersteller (als Unternehmen oder Industriezweig)
machines Maschinen
machines for applying laquers Lackiermaschinen
machining Bearbeitung * zerspanende ~ (durch Werkzeugmaschine) || spanabhebende Bearbeitung || spanende Bearbeitung || zerspanende Bearbeitung
machining center Bearbeitungszentrum * [Werkz.-masch.](amerikan.)
machining centre Bearbeitungszentrum * (brit)
machining period Taktzeit * [Werkz.masch.]
machining station Bearbeitungsstation * [Werkz.-masch.] mit spanabhebender Bearbeitung
machining steel Automatenstahl
machinist Maschineneinrichter || Mechaniker * →Schlosser, Maschinenschlosser/-einsteller || Schlosser * Maschinen~/-einsteller
macro Makro * Zusammenfassung mehrerer Befehle zu einem Kurzbefehl
macro program Makroprogramm
made up gebildet * zusammengesetzt (of:aus)
made-to-measure maßgeschneidert * nach Maß angefertigt
magazine Fachzeitschrift * populärwissenschaftliche ~ || Magazin || Zeitschrift * [mege'sien]
magnesium Magnesium
magnet Magnet
magnet armature Magnetanker
magnet body Magnetkörper
magnet system Antrieb * Relais~, Schütz~
magnet wheel Polrad
magnet wire Spulendraht
magnetic magnetisch
magnetic attraction magnetische Anziehung
magnetic bearing Magnetlager * berührungsloses Lager, z.B. für schnellaufende (Kompressor-) Antriebe
magnetic chuck Magnet-Futter * z.B. zur Werkzeug-Aufnahme bei Werkzeugmaschine || Magnetspannfutter
magnetic circuit magnetischer Kreis
magnetic clamping plate Magnetspannplatte * Haltevorrichtung z.B. für Werkzeugmaschine
magnetic clutch magnetische Kupplung || Magnetkupplung
magnetic coupling Magnetkupplung
magnetic cylinder Magnetzylinder
magnetic field Magnetfeld || magnetisches Feld
magnetic field strength magnetische Feldstärke * Maß für d.Kraft, mit d. e. magnet.Feld gedachte Elemetarmagnete ausrichtet
magnetic flux Durchflutung || magnetischer Fluss
magnetic flux density magnetische Flussdichte
magnetic force Magnetkraft
magnetic levitation railway Magnetschwebebahn
magnetic noise magnetische Geräusche || magnetisches Geräusch
magnetic particle brake Magnetpulver-Bremse
magnetic particle clutch Magnetpulver-Kupplung
magnetic particle coupling Magnetpulverkupplung
magnetic permeance magnetischer Leitwert
magnetic repulsion magnetische Abstoßung
magnetic reversal Ummagnetisierung
magnetic saturation magnetische Sättigung
magnetic sheet steel Dynamoblech || Elektroblech * [el.Masch.] Ausgangsmaterial der Bleche für Läufer und Ständer || Magnetblech
magnetic steel sheet Magnetblech
magnetic tape Magnetband
magnetic tone Magneton
magnetic unbalance magnetische Unsymmetrie
magnetic-suspension bearing Magnetlager * berührungsfrei mit magnetischen Führungskräften arbeitendes Lager
magnetic-tape drive Magnetbandlaufwerk
magnetically operated brake Arbeitsstrombremse * Bremsen nur mit Strom (d.h. nicht bei Spannungsausfall) möglich
magnetism Magnetismus
magnetizability Suszeptibilität
magnetization Magnetisierung
magnetization current Magnetisierungsstrom
magnetization curve Magnetisierungskennlinie * stellt den Zusammenhang zw. magn.Feldstärke und magnet. Flussdichte dar
magnetization power Magnetisierungsleistung
magnetize aufmagnetisieren
magnetized durchflutet * magnetisch ~
magnetizing Magnetisierung
magnetizing current Magnetisierungsstrom
magnetizing inductance Hauptinduktivität
magnetizing reactance Hauptreaktanz * [el.Masch.] Blindwiderst. e. Wicklungsstrangs mit Berücksicht. d. →Hauptfelds
magneto-optical magneto-optisch
magneto-resistive potentiometer Feldplattenpotentiometer || Feldplattenpoti
magnetomotive force Durchflutung * Summe aller Ströme (Ampere-Windungen), die ein bestimmtes Magnetfeld erzeugen || magnetische Durchflutung
magnified übertrieben
magnifier Lupe
magnitude Amplitude || Betrag * Größe, Höhe, Betrag (auch eines Vektors) || Größe * Betrag, Ausmaß, Bedeutung || Höhe * auch e.Stromes/Spannung/Impulses/Drehzahl/usw.
mail schicken * per Post, E-Mail usw. || versenden * per Post || zuschicken * per Post/E-Mail
mail-order company Versandhaus
mail-order house Versandhaus
mailbox Koppelspeicher || Mailbox * [Computer] || Übergabepuffer * →Koppelspeicher
mailbox memory Koppelspeicher * zum Austausch von Daten, Nachrichten
mailing list Adressliste * für Postversand || Verteiler * für den Versand, z.B. von Informationsmaterial
mailing tube Versandrohr
main Haupt- || hauptsächlich || wichtig * hauptsächlich
main actual value Hauptistwert
main application Anwendungsschwerpunkt || Einsatzschwerpunkt || Hauptanwendung

main applications Hauptanwendungen || Schwerpunktanwendungen
main arm Hauptzweig * *{Stromrichter} eines Leistungsteils*
main board Grundplatine || Hauptplatine * *Leiterkarte, auch für PC*
main breaker Hauptschalter * *Leistungsschalter*
main circuit Hauptkreis || Hauptstromkreis
main circuit board Hauptplatine * *Leiterkarte*
main circuit breaker Hauptschalter * *Leistungsschalter*
main components Hauptkomponenten
main conductor Hauptleiter * *z.B. L1, L2, L3 bei Drehstromnetz*
main contact Hauptkontakt
main contactor Hauptschütz
main contactor control Hauptschützansteuerung
main control room Leitzentrale || Zentrale * *Leit~*
main converter Ankerstromrichter || Betriebsumrichter * *Übernimmt nach d.Hochfahren; siehe auch →Synchronisierzusatz, →Hochfahrumrichter*
main criticism Hauptkritikpunkt
main directory Stammverzeichnis
main distribution Hauptverteilung
main drive Hauptantrieb || Leitantrieb * *Hauptantrieb*
main factor Moment * *Hauptmoment*
main field Hauptfeld * *{el.Masch.} magn.Feld, das Ständer- u.Läuferwicklg. einer Asynchr.masch. gemeinsam haben* || Schwerpunkt * *(figürlich) Haupt-Betätigungs/Anwendungsfeld*
main field of application Hauptanwendungsgebiet
main inductance Hauptinduktivität
main inspection Hauptinspektion
main issue Hauptaspekt || Hauptsache * *Hauptaspekt, Kernpunkt* || Kern * *der wichtigste Punkt* || wesentlicher Punkt * *Kernpunkt*
main line traffic Fernverkehr
main memory Arbeitsspeicher * *Hauptspeicher, z.B. eines PC* || Hauptspeicher * *z.B. eines PC*
main menu Hauptmenü
main player Marktführer
main point Hauptsache || Moment * *Hauptmoment*
main pole Hauptpol
main pole core Hauptpolkern
main portion Hauptanteil
main program Hauptprogramm
main property Haupteigenschaft
main provider Hauptanbieter
main reference Hauptsollwert
main reference value Hauptsollwert
main setpoint Hauptsollwert
main shaft Hauptspindel * *{Werkz.masch.}*
main shaft drive Hauptspindelantrieb * *Werkzeugmaschine*
main spindle Hauptspindel * *{Werkz.masch.}*
main spindle axis C-Achse * *{Werkz.masch.} Hauptspindel*
main spindle drive Hauptspindelantrieb * *Werkzeugmaschine*
main spindle motor Hauptspindelmotor
main station Zentrale * *Hauptstation*
main switch Hauptschalter
main transformer Haupttransformator
main-line Fern- * *{Bahn}*
main-shaft mounted fan Eigenlüfter * *in elektr. Maschine*
mainframe Großcomputer || Großrechner
mainframe computer Großcomputer || Großrechner
mainline frequency Netzfrequenz
mainly hauptsächlich * *(Adv.)* || im Wesentlichen * *hauptsächlich* || vorwiegend * *(Adv.) hauptsächlich* || wesentlich * *hauptsächlich*

mains Netz * *Strom~* || Netz- || Rohrnetz * *für Gas, Wasser usw. aus der Sicht des Verbrauchers*
mains breaker Netzschalter * *netzseitiger Leistungsschalter*
mains choke Netzdrossel
mains connection Netzanschluss
mains failure Netzausfall || Netzspannungsausfall
mains feeder Netzeinspeisung
mains filter Netzfilter
mains frequency Netzfrequenz
mains fuse Anschlusssicherung * *Netzsicherung*
mains interference Netzrückwirkungen
mains interruption Netzausfall
mains isolating switch Hauptschalter * *in Netzzuleitung*
mains lead Netzkabel * *z.B. bei Kleingerät, Haushaltgerät* || Netzleitung
mains loading Netzbelastung
mains plug Netzstecker
mains pollution Netzrückwirkungen * *(salopp)*
mains power supply Netzgerät
mains power switch Netzschalter
mains rectifier Netzgleichrichter
mains restoration Netzspannungswiederkehr
mains simulator Netzsimulator
mains supply Netzversorgung
mains switch Hauptschalter * *in Netzeinspeisung*
mains transformer Netztransformator
mains voltage Netzspannung
mains voltage range Netzspannungsbereich
mains-operated netzbetrieben || netzgespeist
mains-side fuse Netzsicherung
mainstream allgemeiner Trend || Trend * *allgemeiner/langanhaltender ~*
maintain aufrechterhalten * *(einen Zustand aufrecht-) erhalten, beibehalten, (be-) wahren* || beibehalten || erhalten * *(aufrecht)erhalten, beibehalten, unterstützen, (Grenzen, Spezifikation) beachten* || erhalten * *aufrecht~* || sorgen für * *aufrechterhalten, z.B. eine konstante Drehzahl* || versorgen * *unterhalten* || warten * *z.B. Maschine ~*
maintain the timing Timing einhalten * *beibehalten, aufrechterhalten*
maintainability Wartbarkeit
maintained remanent * *Merker/Kontakt bei Kontaktplan/Relaissteuerung*
maintained contact Selbsthaltekontakt
maintained function Selbsthaltung * *maintained command: Befehl mit ~ (Befehlsflanke speichert die Funktion)*
maintained signal Dauersignal
maintenance Aufrechterhaltung || Erhaltung * *~ von Maschinen, Frieden usw.* || Instandhaltung * *siehe z.B. DIN 31051* || Pflege * *→Wartung; preventive maintenance: vorbeugende Wartung* || Revision * *Wartung* || Unterhaltung * *Wartung, Instandhaltung* || Wartung * *z.B. an Maschine, einem Gerät; preventive: vorbeugende*
maintenance costs Serviceaufwand * *Wartungsaufwand* || Wartungsaufwand || Wartungskosten
maintenance engineer Wartungstechniker
maintenance freedom Wartungsfreiheit
maintenance guide Wartungsanleitung
maintenance instructions Wartungsanleitung
maintenance interval Wartungsintervall
maintenance man Monteur * *~ für Wartung, Instandhaltung*
maintenance manual Wartungsanleitung * *Handbuch*
maintenance opening Wartungsöffnung
maintenance operation Revisionsbetrieb
maintenance period Wartungsintervall
maintenance personnel Wartungspersonal
maintenance requirements Wartungsaufwand * *Bedarf, Erfordernisse; low: geringer*

maintenance routine Wartung * *laufende*
maintenance schedule Wartungsplan
maintenance staff Wartungspersonal
maintenance work Wartungsarbeiten
maintenance-free wartungsfrei
major bedeutend || gängig * *hauptsächlich* || Haupt- * *bedeutend, größer, hauptsächlich; z.B. Hauptfach, Hauptachse, Hauptanbieter* || hauptsächlich * *größer, bedeutend, hauptsächlich* || höherprior * *höherrangig* || überlagert || wesentlich * *hauptsächlich*
major account Hauptkunde * *~ mit hohem Umsatz*
major customer Hauptkunde
major factor entscheidender Faktor
major player Hauptanbieter * *(salopp)*
major supplier Hauptanbieter
majority größerer Teil * *einer größeren Menge* || Mehrheit * *great: überwiegende* || Mehrzahl * *Mehrheit, größerer Teil* || überwiegend * *überwiegender Teil; in the majority of cases: in der ~en Zahl der Fälle*
make anfertigen || aufstellen * *an assertion: eine Behauptung ~* || bauen * *herstellen* || bewerkstelligen || durchführen * *bilden, formen, bilden, machen* || erwerben * *ein Vermögen, Einkommen ~* || erzielen * *Gewinn usw. ~* || Fabrikat * *Marke, Typ, Bauart, Machart, Fabrikat; Anfertigung, Herstellung* || fertigen || herstellen || leisten * *Zahlung usw. (payment)* || schließen * *Kontakt* || verdienen || vornehmen * *Änderungen, Anpassungen usw. ~* || zusammenbauen
make a backup sichern * *[Computer] Daten ~*
make a clean sweep of aufräumen * *mit etwas aufräumen*
make a distinction differenzieren * *unterscheiden, einen Unterschied machen* || unterscheiden * *between: zwischen*
make a mental note of merken * *sich etwas einprägen*
make a mistake verrechnen * *einen Fehler machen*
make a new start vorn * *von neuem anfangen*
make a note notieren * *schriftliche Notiz machen, aufschreiben*
make a note of festhalten * *Gedanken usw. schriftlich ~*
make a reduction nachlassen * *~ von Preisen usw.*
make a reservation for reservieren * *~ lassen*
make a rough sketch of skizzieren * *etwas in groben Umrissen darstellen*
make a signal record schreiben * *einen Messschrieb aufzeichnen, z.B. mit einem Linienschreiber*
make a sketch of skizzieren
make accessible erschließen || zugänglich machen * *zugreifbar machen*
make agree in Übereinstimmung bringen * *with:-mit*
make allowance for berücksichtigen * *berücksichtigen, bedenken, im Budget reservieren*
make an agreement Vertrag schließen
make an appointment with bestellen * *eine Verabredung treffen mit*
make an attempt versuchen * *den Versuch machen*
make an effort versuchen * *sich bemühen*
make an entry of vermerken * *eintragen*
make an oscilloscope trace oszillografieren
make an oscilloscopic trace oszillografieren || schreiben * *Messschrieb aufnehmen mit e. Oszilloskop*
make arrangements disponieren * *Vorkehrungen treffen*
make available beschaffen * *sich beschaffen, verfügbar machen* || erstellen || zugänglich machen * *verfügbar machen* || zur Verfügung stellen

make clear verdeutlichen
make contact aufliegen * *z.B. Kohlebürsten auf d. Kommutator einer Gleichstrommaschine* || Schließer || Schließerkontakt
make contakt Arbeitskontakt * *Schließerkontakt*
make copies vervielfachen * *kopieren*
make corrections korrigieren
make deals Geschäfte machen || Geschäfte tätigen
make dispositions disponieren
make easier erleichtern * *gegenüber vorher ~* || vereinfachen
make easy erleichtern * *einfach machen* || fördern * *erleichtern*
make feasable ermöglichen
make for sorgen für
make good wettmachen * *Versäumnis, Verlust*
make headway Fortschritte machen
make heavy demands hohe Ansprüche stellen * *on:an*
make highly unlikely ausschließen * *weitgehend unwahrscheinlich machen*
make it a condition ausmachen * *ausbedingen, zur Bedingung machen*
make known bekanntmachen || mitteilen * *bekannt machen*
make mention of hinweisen auf * *erwähnen*
make one's own erarbeiten * *sich einen Wissensstoff/e. Thema ~/aneignen*
make oneself known sich vorstellen * *sich bekannt machen*
make open offenlegen * *z.B. Busprotokoll*
make out aufstellen * *Liste, Rechnung usw. ~* || ausmachen * *sichten, feststellen* || herausbekommen * *Sinn usw. ~* || zusammenstellen * *Liste, Rechnung usw. ~*
make perfect abrunden * *vollkommen machen*
make plain verdeutlichen
make possible ermöglichen
make propaganda werben
make provision sorgen für * *→Vorsorge treffen* || Vorsorge treffen
make public bekanntmachen * *öffentlich* || herausstellen * *an die Öffentlichkeit bringen* || veröffentlichen
make redundant abbauen * *Personal ~*
make reference verweisen * *make reference to: erwähnen/eingehen auf*
make reference to Bezug nehmen auf
make safe sichern
make sense passen * *Sinn machen* || plausibel sein * *Sinn machen* || sinnvoll sein * *Sinn machen*
make shorter kürzen
make smaller verkleinern
make stand out hervorheben
make sure nachprüfen * *sicherstellen*
make sure of versichern * *sich e. Sache ~*
make the connection Verbindung herstellen
make the most of ausschlachten * *ausnutzen*
make tight abdichten
make up aufstellen * *Bilanz ~, Rechnung ~* || ausmachen * *einen Teil bilden* || Umbruch * *[Druckerei] ~ eines Druckerzeugnisses* || zusammensetzen * *bilden, ~; be made up of: bestehen/sich ~ aus; zusammensetzen (Schriftstück)* || zusammenstellen * *Liste, Unterlagen usw. ~*
make up for aufholen * *Zeit, Verspätung usw. ~* || einarbeiten * *by extra work: einen Rückstand durch Zusatzarbeit* || wettmachen
make up for lost time aufholen * *Verspätung ~*
make up one's mind beschließen * *sich ~; to do: etwas zu tun*
make up one's mind to do something vornehmen * *sich etwas ~*
make usable nutzbar machen

make use beanspruchen * *Gebrauch machen (of:- von)*
make use of anwenden * *Gebrauch machen von* || nutzen || nützen * *Gebrauch machen von* || verfügen über * *verwenden* || wahrnehmen * *Gelegenheit ~*
make use of flexible working hours gleiten * *Gleitzeit in Anspruch nehmen*
make worse verschlechtern * *become worse: schlechter werden*
make-before-break contact nicht unterbrechender Kontakt * *Wechs.kont. dess.Kontakte währ. d. Schaltens beide kurz geschloss.sind*
make-current principle of operation Arbeitsstromprinzip * *b. el.mechan. Gerät (z.B. magn. Bremse schließt bei Stromfluss)*
make-proof earthing einschaltfeste Erdung * *Personen-Schutzeinrichtung/-Maßnahme für Servicearbeiten an Hochspannung*
make-proof earthing switch Erdungsdraufschalter
make-ready time Rüstzeit * *z.B. bei Druckmaschine*
make-shift provisorisch * *behelfsmäßig*
make-up bilden || Zusammensetzung * *das Bilden einer Mannschaft, eines Teams usw.*
maker Hersteller
makeshift behelfsmäßig
making Herstellung
making and breaking capacity Schaltleistung * *Ein-/Aus~*
making capacity Einschaltleistung * *Schalter, Relais usw.* || Schaltleistung * *Ein~*
making compact Straffung * *(figürl.)*
making current Einschaltstrom * *~ eines Leistungschalters*
making even Vergleichmäßigung
making sense zweckmäßig * *make sense: sinnvoll sein/Sinn machen*
mal- Fehl-
male connector Messerleiste * *"männlicher" Steckverbinder* || Stiftleiste * *"männlicher" Steckverbinder* || Stiftstecker * *"männlicher" Stecker*
male plug-connector element Steckerstift
male socket connector Messerleiste * *'männlicher' Steckverbinder*
malfunction Fehlfunktion || Funktionsfehler * *Fehlfunktion, Funktionsstörung* || Funktionsstörung || Problem * *Fehlfunktion* || Störung * *Fehlfunktion*
malleable cast iron Temperguss
maloperation Fehlbedienung
man besetzen * *durch Bedienungspersonal* || Mensch-
man year Mannjahr
man-hour Arbeitsstunde || Mannstunde
man-machine communications Mensch-Maschine-Kommunikation
man-machine dialog Bedienen und Beobachten || Mensch-Maschine-Dialog
man-machine interface Bedieneroberfläche * *siehe auch →Bedienungsoberfläche* || Bedienoberfläche || Bedienungsoberfläche || Mensch-Maschine-Schnittstelle
man-made Kunst- * *künstlich/synthetisch hergestellt, kein Naturprodukt, z.B. Kunstfasern* || künstlich * *künstlich hergestellt, synthetisch* || synthetisch * *künstlich hergestellt*
man-made fiber Chemiefaser * *(amerikan.)*
man-made fiber industry Chemiefaserindustrie
man-made fibre Chemiefaser * *(brit.)* || Kunstfaser
man-made fibres Kunstfasern
man-month Mannmonat
man-week Mannwoche
man-year Mannjahr
manage abwickeln * *verwalten, durchführen* || behandeln * *lenken, meistern* || betreiben * *Unternehmen, Verkehrslinie usw. ~* || bewerkstelligen || durchführen * *organisieren* || fertigwerden * *mit jdm./etwas ~* || handhaben || leiten * *z.B. Betrieb, Projekt ~* || meistern || organisieren || verwalten
manage to do möglich * *~ machen*
manageable handbabar || hantierbar
manageableness Handhabbarkeit
management Abwicklung || Bearbeitung * *Verwaltung, Handhabung, Behandlung* || Behandlung * *Verwaltung, Handhabung, Behandlung* || Betrieb * *Betrieb, Leitung, Verwaltung* || Führung * *Leitung, Verwaltung* || Führungsebene || Führungskräfte || Leitung * *Management* || Steuerung * *auch Verwaltung* || Verwaltung
management behaviour Führungsverhalten * *~ eines Vorgesetzten/einer Führungskraft*
management consultant Unternehmensberater
management level Führungsebene
management style Führungsverhalten * *~ einer Führungskraft*
manager Chef * *Verwalter* || Leiter * *Verwalter, Geschäftsführer, Organisator* || Manager || Managerin
manageress Managerin
managerial class Führungsebene
managerial economics Betriebswirtschaft
managing Führungs- || leitend || organisatorisch
managing body Führungsebene
managing committee Vorstand
managing director Direktor * *Geschäftsführer* || Geschäftsführer
managing team Leitungskreis
mandate Mandat * *politisches*
mandatory Muss- || obligatorisch * *obligatorisch, verbindlich, zwangsweise, Muss-; mandatory regulation: Mussvorschrift* || verbindlich * *~ vorgeschrieben* || verbindlich vorgeschrieben * *vorgeschrieben* || obligatorisch, →verbindlich, zwangsweise, Muss- (Vorschrift, Bedingung usw.)
mandatory function Mussfunktion
mandrel Dorn * *Wickeldorn, →Spanndorn, Dreh~, Hauptspindel* || Rollstange * *Wickeldorn* || Spindel * *Spanndorn, Haupt~* || Wickeldorn || Wickelkern * *Wickeldorn*
mandrel drive Dornantrieb * *für Wickeldorn z.B. in Walzwerk, Blechverarbeitungsstraße* || Spindelantrieb * *Haupt~*
mandril Dorn * *Wickel~, Spann~, Dreh~* || Wickeldorn
maneuver Manöver * *(amerikan.)* || manövrieren * *(amerikan.; auch figürl.) z.B. Rolle beim Einlegen in Wickler ~*
maneuvering potentiometer Manövrierpoti * *z. feinfühligen Positionieren d.Rolle bei Auf-/Abwickler aus d.Stillstand*
maneuvering potentiometer (amerikan.) Manövrierpotentiometer
manifest offensichtlich || zeigen * *sich ~/offenbaren/kundtun; manifestieren; be-/erweisen; erscheinen*
manifold Krümmer * *Rohrverzweigung, Einlass~, Auspuff~* || mannigfaltig * *siehe auch →vielfältig* || mehrfach * *Mehr(fach)-, Mehrzweck-, vielfach, vielfältig, mannigfältig* || vielfältig * *mannigfaltig, ~, vielfach, Mehr(fach)-, Mehrzweck-, Sammelleitung, Rohrverzweigg.*
manifoldness Vielzahl * *Vielfalt*
manipulate bedienen * *handhaben, geschickt behandeln* || behandeln * *betätigen* || betätigen * *bedienen* || manipulieren * *Person, Information, technisches Gerät*
manipulated variable Stellgröße
manipulation Bedienung || Handgriff * *(meist Plural) Bedienung* || Handhabung * *auch durch Robo-*

manipulator 606

ter, Transfereinrichtung usw. || Kunstgriff || Manipulation
manipulator Handhabungsgerät
manned besetzt * bemannt, mit Personal ~
manner Art * Art u.Weise, auch 'Verhalten, Betragen, Auftreten' || Art und Weise || Form * Art und Weise || Weg * →Art und Weise, Methode || Weise
manoeuvering potentiometer Manövrierpotentiometer * (brit.) zum feinfühligen Vor-/Rückwärtsbewegen d. Wickelachse
manoeuvre Manöver * (engl.) || manövrieren * (engl.; auch figürl.) || rangieren * manövrieren
manometer Manometer * Druckmesser
manpower Arbeitskraft || Arbeitsleistung
manpower cost per piece Lohnstückkosten
manpower costs Personalkosten
mansonry Mauerwerk
mantle umhüllen
manual Anleitung * Handbuch || Beschreibung * Handbuch || Betriebsanleitung * Handbuch || Hand- || Handbuch || händisch || Leitfaden * Handbuch || manuell
manual bypass Handumgehung * handbetätigter Überbrückungsschalter für unterbrechungsfreie Stromversorgung
manual control Handeingabegerät
manual control level Handebene
manual entry manuelle Eingabe
manual intervention Eingreifen von Hand || Handeingriff
manual mode Handbetrieb
manual optimizing Handoptimierung
manual override Handumgehung
manual release Handlüftung * einer Bremse (z.B. mittels Handhebel bei Aufzugs-/Hebezeugantrieb)
manual release device Handlüfteinrichtung * zum Lösen einer Bemse von Hand (z.B. über einen Hebel)
manual setpoint Hand-Sollwert || Handsollwert
manual tachometer Tachometer * Hand~
manual text input Texteingabe * von Hand in einen Rechner
manual workmanship Handarbeit
manual-automatic mode selector switch Hand-Automatik-Umschalter
manually Hand * von ~ || händisch * (Adv.) || manuell * (Adv.) manually operated: mit Handbedienung || von Hand
manually actuated Hand-
manually actuated potentiometer Handpotentiometer
manually operated Hand- || handbetätigt
manufacture anfertigen || Bau * Herstellung, Erzeugung || bauen * herstellen || Betrieb * Herstellungsgang || Erzeugung || Fabrikat * Erzeugnis || fertigen || Fertigung * Erzeugung, Fertigung, Herstellung, Fabrikation, Industrie (-zweig) || herstellen * produzieren, herstellen, erzeugen || Herstellung || Produkt * Erzeugnis, Fabrikat || produzieren || verarbeiten || Verarbeitung
manufacture under licence Lizenzbau
manufactured hergestellt * from: aus (...-Material)
manufactured by a subcontractor fremdgefertigt * durch einen Unterlieferanten
manufactured from aus * hergestellt aus (... Material)
manufactured goods Ware * Fertigwaren, Fabrikwaren, Erzeugnisse
manufacturer Fabrikant || Hersteller * on the manufacturer's premises: beim ~ || Produzent * Hersteller
manufacturer's declaration Herstellererklärung
manufacturer's designation Herstellerzeichen
manufacturer's specification Herstellerangabe

manufacturer-neutral firmenneutral * z.B. offenes Bussystem || firmenübergreifend * z.B. offenes Bussystem
manufacturer-specific herstellerspezifisch
manufacturing Bau * das ~en, z.B. einer Maschine || Fertigung || Herstellung
manufacturing area Fertigungsbereich
manufacturing automation Fabrikautomatisierung || Fertigungsautomatisierung
manufacturing by a subcontractor Fremdfertigung * von Unterlieferant gefertigt
manufacturing control Fertigungssteuerung || Produktionsleittechnik
manufacturing costs Herstellkosten
manufacturing data recording Betriebsdatenerfassung
manufacturing defect Fertigungsfehler || Herstellfehler
manufacturing deficiency Fertigungsfehler
manufacturing district Industriegebiet * Fabrikgebiet
manufacturing engineering Fertigungsplanung
manufacturing equipment Betriebsmittel || Fertigungseinrichtungen
manufacturing facilities Fertigung * Fertigungseinrichtung(en) || Fertigungsanlagen || Fertigungseinrichtungen
manufacturing facility Fertigungsanlage || Produktionsanlage
manufacturing line Fertigungsstraße || Fließband || Produktionsanlage || Produktionslinie || Verarbeitungsstraße * Herstellungsstraße
manufacturing method Fertigungsmethode
manufacturing plant Betrieb * Fabrikanlage || Fabrik * Fabrikationsanlage || Fertigungsbetrieb || Industriebetrieb
manufacturing process Herstellungsprozess || Produktionsprozess
manufacturing program Fertigungsprogramm || Produktionsprogramm
manufacturing shop Fertigung * Fertigungsabteilung || Halle * Werkshalle || Werkhalle || Werkshalle
manufacturing system Fertigungssystem * flexible: flexibles
manufacturing technology Fertigungstechnik || Fertigungstechnologie || Produktionstechnik
manufacturing tolerance Herstellungstoleranz
many viele
many thanks danke * dankeschön!
many times over vielfach * um ein Vielfaches || Vielfaches * um ein ~
many years of experience langjährige Erfahrung
many-layered facettenreich * vielschichtig || vielschichtig
many-sided facettenreich || kompliziert * →vielschichtig, →facettenreich || mehrschichtig * (figürl.) vielschichtig || vielschichtig * facettenreich
map abbilden * map an input: einen Eingang ~ || Karte * Landkarte || Plan * Karte
map out einteilen * (im Einzelnen voraus)planen, Zeit ~, mittels einer Karte/Grafik ~ || planen * voraus~, ausarbeiten, (seine) Zeit einteilen
march out ausrücken * Feuerwehr usw.
margin Abstand * vom Rand e.Druckseite; Vorsprung (z.B. zu einem Mitbewerber auf dem Markt); (Handels)Spanne || Grenze * Toleran~, Spielraum || Rand * ~ einer Druckseite, eines Dokuments/Buches || Reserve * →Spielraum; leave a margin: Spielraum lassen; margin of safety: Sicherheitsfaktor || Spanne * z.B. Vertriebsspanne/ marge || Spielraum * Frist, Spanne || Spanne * Spanne || Überschuss * (Geld)Differenz || Vorsprung * zeit-/wertmäßiger ~

marginal nebenstehend * *am Rande*
marine Marine * *Handelsmarine* || Schiffs- || See-
marine cable Seekabel
marine classification society Schiffsklassifikationsgesellschaft
marine design Marine-Ausführung
marine Diesel engine Schiffsdiesel
marine drive Schiffsantrieb * *z.B. Wellenantrieb*
marine engineering Schiffbau * *als Fachrichtung*
marine industry Schiffbau
marine propulsion drive Schiffsantrieb * *Propellerantrieb*
marine reversing gear Schiffswendegetriebe
marine turbine Schiffsturbine
maritime See- * -*Recht, -Versicherung*
mark anstreichen * *Fehler, Textstelle* ~ || anzeichnen || Beschriftung * *Markierung* || bezeichnen * *mit Markierung, Marke* || kennzeichnen * *markieren, bezeichnen, kennzeichnen* || Kennzeichnung * *Kennzeichen, Markierung* || Marke * *Kennzeichen* || markieren || Markierung * *auch Zeichen* || merken * *beachten, sich etwas ~, vormerken* || Merkmal * *Zeichen* || Zeichen * *Marke, Markierung, Einstellmarke, Kennung, Kerbe, Merk~* || zeichnen * *be-/kenn~*
mark of conformity Prüfzeichen * *von einer Normen/Überwachungsbehörde (z.B. VDE, TÜV, UL)*
mark of distinction Kennzeichnung * *zur Unterscheidung dienende ~*
mark out anreißen * *markieren* || festlegen * *bestimmen, festsetzen, ausersehen(for:für,zu), bezeichnen, markieren*
mark up erhöhen * *Preise usw.* ~ || heraufsetzen * *z.B. Preise*
mark-space ratio Tastverhältnis * *Puls-Pausenverhältnis eines impulsförmigen Signals*
mark-to-space ratio Puls-Pausenverhältnis || Tastverhältnis * *Puls-Pausenverhältnis eines impulsförmigen Signals*
marked ausgeprägt * *markant, deutlich, merklich,* || deutlich * *merklich, auffällig, markant, ausgeprägt* || fühlbar * *deutlich* || gekennzeichnet * *markiert* || spürbar * *deutlich*
markedly stark * *(Adv., figürl.) deutlich, merklich, auffällig, ausgesprochen; increase markedly:~ ansteig.*
marker Merker
marker pulse Nullimpuls * *eines Winkelschritt-/Inkremetalgebers* || Nullmarke * *Nullimpuls eines Inkremetalgebers* || Referenzimpuls * *Nullimpuls eines Dreh-Impulsgebers*
market Markt * *in/on the market: auf dem ~; put on the market: auf den ~ bringen* || vermarkten || vertreiben * *vermarkten*
market acceptance Marktakzeptanz
market analysis Marktanalyse || Marktstudie || Marktuntersuchung * *Marktanalyse*
market conditions Marktlage
market control Marktbeherrschung
market demand Marktanforderung
market development Marktentwicklung
market domination Marktbeherrschung
market economy Marktwirtschaft * *siehe auch →freie Marktwirtschaft*
market entry Markteintritt
market focus Marktschwerpunkt * *Brennpunkt*
market forecast Marktvorhersage
market gap Marktlücke * *fill:füllen*
market introduction Markteinführung
market investigation Marktuntersuchung
market leader Marktführer * *lead the market: den Markt anführen*
market need Marktforderung
market niche Marktnische
market opportunities Marktchancen

market penetration Marktdurchdringung
market place Markt * *regionaler/nationaler* ~
market policy Marktpolitik
market position Marktposition || Marktstellung
market presence Marktpräsenz
market price Marktpreis
market price level Marktpreis * *Marktpreis-Niveau*
market pricing level Marktpreis
market rate Marktpreis
market report Geschäftsbericht * *~ über die Marktlage*
market requirement Marktanforderung || Marktforderung
market requirements Marktbedürfnisse
market research Marktforschung || Marktuntersuchung
market researcher Marktforscher
market saturation Marktsättigung
market sector Branche * *Marktsektor*
market segment Marktsegment
market share Marktanteil
market situation Marktlage
market structure Marktstruktur
market study Marktstudie
market success Markterfolg
market survey Marktanalyse || Marktuntersuchung
market tendency Marktentwicklung
market trend Marktentwicklung
market value Marktwert || Verkaufspreis * *Marktwert*
market-adapted marktgerecht
market-economy marktwirtschaftlich * *die Marktwirtschaft betreffend*
market-oriented marktgerecht * *am Markt orientiert* || marktorientiert
market-ready marktreif
marketability Marktfähigkeit
marketable marktfähig || marktgerecht * *marktfähig, vermarktbar* || vermarktbar
marketable value Marktwert
marketing Marketing * *Absatzpolitik,"zu Markte bringen", Erkunden/Wecken v. Marktinteresse u. planvoller Verkauf* || Vertrieb || vertrieblich
marketing corporation Vertriebsgesellschaft
marketing success Markterfolg
marking Beschriftung * *z.B. von Steckern, Kabeln, Klemmen usw.* || Bezeichnung * *Kennzeichnung, z.B. Anschlussbezeichnung an einer Klemme* || Kennzeichnung * *Markierung, Kennzeichnung (auch von Anschlussklemmen usw.)* || Markierung
marking device Markiereinrichtung
marking label Beschriftungsstreifen
marshal rangieren * *Leitungen usw.*
marshalling Rangier- * *bei Leitungen, Kontakten, Klemmen usw.* || Rangierung
marshalling rack Rangierverteiler
marvellous prima * *(salopp) fabelhaft, wunderbar* || toll * *(salopp) fabelhaft, wunderbar*
marvellously prima * *(Adv., salopp) wunderbar, fabelhaft*
maschine room Maschinenraum
mash Maische || quetschen * *zerquetschen*
mask ausblenden * *maskieren, z.B. Impulse, Daten* || Maske * *z.B. ~ auf dem Bildschirm, bei der Chip-Herstellung, Bit~* || maskieren * *Impulse ausblenden, Daten ausmaskieren usw.* || überlagern * *verhüllen, maskieren* || verdecken * *maskieren, abdecken*
mask design Maskenaufbau * *[Computer] am Bildschirm*
mask out ausblenden * *ausmaskieren; z.B. Impulse, Daten* || maskieren
masked verdeckt * *(aus-) maskiert*

masonry drill Steinbohrer * *zum Bohren in Mauerwerk*
masonry pin Stahlnagel * *zum Einschlagen in Mauerwerk, z.B. "IMPU"-Nagel zur Leitungsverlegung im Putz*
mass Haufen * *Häufung* || Masse * *schwere ~ [kg]*
mass flow Mengenstrom
mass flow control Mengenregelung * *Mengenstromregelung*
mass flow controller Durchflussregler * *Mengenstromregler* || Mengenflussregler
mass inertia Massenträgheit
mass memory Massenspeicher
mass moment of inertia Massenträgheit || Massenträgheitsmoment
mass particle Masseteilchen
mass particles Masseteilchen
mass per unit area Flächengewicht || Flächenmasse || Papiergewicht * *Flächengewicht*
mass production Großserienfertigung || Massenproduktion * *siehe auch →Großserienfertigung* || Serienfertigung * *Fertigung in Großserie* || Serienherstellung * *Großserienfertigung*
mass storage Massenspeicher
mass transit traffic Nahverkehr
massive massiv * *massig, wuchtig, massiv, schwer*
master beherrschen * *meistern* || Busteilnehmer * *aktiver* || Führungs- * *Leit-* || im Griff haben * *meistern* || in den Griff bekommen * *meistern* || Könner || Leit- || lernen * *meistern* || Master || Meister || meistern || übergeordnet || überlagert * *Leit-*
master axis Leitachse
master board Grundplatine
master computer Leitrechner
master control Fernsteuerung * *durch ein überlagertes System* || Kopfsteuerung || Leitsteuerung || übergeordnete Regelung
master controller Führungsregler || Kopfsteuerung || Leitregler || Leitsteuerung || Meisterschalter * *[Hebezeuge] beim Kran* || überlagerter Regler
master drive Leitantrieb
master frequency Leitfrequenz
master group Leitgruppe * *Antriebsgruppe bei Mehrmotorenverbund, die die Maschinengeschwindigkeit bestimmt*
master interface Masteranschaltung * *für Bus*
master line reference Liniensollwert
master PLC Kopfsteuerung * *SPS, Programmierbare Steuerung* || Leitsteuerung * *Leit-SPS, Kopfsteuerung*
master program Masterprogramm * *(SIEMENS SIMADYN D (R))*
master reference Leitsollwert || Leitspannung * *→Leitsollwert, Bahngeschwindigkeitssollwert (bei Papiermaschine usw.)* || Liniensollwert * *Leitsollwert*
master reference value Leitsollwert
master reference voltage Leitspannung
master section Leitgruppe * *Antriebsgruppe, die für die Anlagengeschwindigkt. maßgebl. ist (z.B. b. Papiermasch.)*
master setpoint Leitsollwert
master station Busteilnehmer * *aktive Station, Leitstation* || Kopfstation || Master-Station * *z.B.bei Bussystem* || Teilnehmer * *Master-Teilnehmer (z.B. an seriellem Bus)*
master switch Hauptschalter || Meisterschalter * *[Hebezeuge] beim Kran*
master value Leitwert * *Leitsollwert (z.B. Maschinen-/Liniengeschwindigkeits-Sollwert)*
master-follower operation Leit-Folge-Betrieb * *bei Mehrmotorenantrieb* || Leit-Folgebetrieb
master-process control level Prozessleitebene
master-slave changeover Leit-Folge-Umschaltung
master-slave control Leit-Folge-Steuerung || Master-Slave-Steuerung
master-slave drive Leit-Folge-Antrieb || Leit-Folgeantrieb
master-slave operation Leit-Folge-Betrieb || Leit-Folgebetrieb || Master-Slave-Betrieb || Master-Slave-Prinzip
master-slave principle Master-Slave-Prinzip
master-slave technique Master-Slave-Verfahren
master-slaving Master-Slave-Steuerung
mastery Beherrschung * *z.B. Prozess, Technik, Sprache*
match abgleichen * *mehre Teile aneinander anpassen* || abstimmen * *anpassen (to: an/auf)* || anpassen * *mehrere (meist 2) Dinge aneinander; to: an; closely to: weitgehend an; auch 'sich ~'* || entsprechen * *einer Anforderung, einem zweiten Teil ~* || passen * *mehrere Dinge zueinander* || passen zu || übereinstimmen * *~ mit* || zusammenpassen * *nach Form, Farbe, Funktion usw.* || zusammenstellen * *nach zusammenpassender Ausführung, Größe, Farbe usw. ~*
match the requirements Anforderung erfüllen
match up übereinstimmen * *with: mit*
matched angepasst
matching -Anpassung * *mehrere Dinge aneinander* || Abgleich * *~ von mehreren (meistens 2) Dingen aneinander* || Anpass- || Anpassung * *mehrere Dinge aneinander* || Nachführung * *Anpassung* || Paarung || passend * *angepasst* || zugehörig * *in Form, Farbe, Größe, Typ usw. ~/passend/angepasst* || Zuordnung || passende ~
matching amplifier Anpassverstärker
matching circuit Anpassschaltung || Anpassungsschaltung
matching module Anpassmodul
matching piece Gegenstück * *zugehöriges ~* || Pendant * *zugehöriges Teil*
matching table Zuordnungstabelle * *z.B. Umrichter zu Motor*
matching transformer Anpasstrafo || Anpasstransformator || Anpassungstrafo || Vorschalttransformator * *Anpasstranformator* || Vortransformator * *Anpasstransformator* || Zwischentransformator * *Anpasstransformator*
matching unit Anpassstufe
mate kombinieren * *zwei Dinge passend miteinander ~*
mated-surface rusting Passungsrost * *rostrote Verfärbung an d. Berührungsstelle von zwei Stahlteilen*
material -Mittel * *Material* || -Zeug || ausschlaggebend * *von Belang, sachlich wichtig (to:für)* || gut * *(Substantiv) Material (auch 'goods': Güter, Waren, Gegenstände, Frachtgut)* || Material * *Material* || Unterlagen * *schriftliche ~* || Ware * *z.B. Rohstoff, Grundstoff/material, Textilstoff* || Warenbahn * *Material* || Werkstoff || wesentlich * *von Belang, sachlich wichtig (to:für)* || Zeug
material damage Sachschaden
material data Materialdaten
material direction Längsrichtung * *Materiallaufrichtung*
material entry Materialeinzug
material entry feed Materialeinzug
material entry section Materialeinzug
material fatigue Materialermüdung
material flow Materialfluss || Stoffstrom
material flow meter Schüttstrommesser * *→Durchflussmesser für Schüttgut*
material for construction Baustoff
material pile Schüttkegel
material processing time Materialdurchlaufzeit
material quality Materialqualität
material running direction Materialflussrichtung * *bei durchlaufender Warenbahn*

material saving Materialeinsparung
material sold by the metre Meterware
material speed Bahngeschwindigkeit * *Material* ||
Materialgeschwindigkeit || Warenbahngeschwindigkeit * *Materialgeschwindigkeit* || Warengeschwindigkeit * *Materialgeschwindigkeit (z.B. bei einer Produktionsanlage)*
material strength Festigkeit * ~ *eines Werkstoffs*
material thickness Materialstärke
material tracking Materialverfolgung
material wear Materialverschleiß
material web Warenbahn
material-directed Längs- * *in Materiallaufrichtung*
material-directed orientation Längsrecke * *Längsreckeinrichtung in Folienmaschine*
material-flow direction Materialflussrichtung
materialize verkörpern * *verwirklichen, e.Sache stoffliche Form geben, Gestalt annehmen, sich ~, zustandekommen*
materials engineering Werkstofftechnik
materials management Materialwirtschaft
materials test Materialprüfung
materials testing Materialprüfung || Werkstoffprüfung
math Mathematik * *(salopp) Kurzform von 'mathematics'*
math functions Arithmetikfunktionen
math skills mathematische Fähigkeiten
mathematical mathematisch || rechnerisch
mathematical equation mathematische Gleichung
mathematical image mathematisches Abbild
mathematical model mathematisches Modell
mathematically mathematisch * *(Adv.)* || rechnerisch * *(Adv.)*
mathematics Arithmetikfunktionen || Mathematik
mating Paarung
mating connector Gegenstück * ~ *für Stecker/Buchse*
mating cycles Steckzyklen * *eines Steckers*
mating piece Gegenstück
matrices Matrizen
matrix Matrix || Matrize * *Matrix*
matrix board Rasterfeld * *z.B. "Spielwiese" auf Leiterkarte z. nachträglichen Einlöten v. Bauelementen*
matrix converter Matrixumrichter * *Umrichter ohne Zwischenkreis mit Schaltermatrix zw. allen Netz- u Motoranschlüssen*
matrix notation Matrixdarstellung
matter Angelegenheit || ausmachen * *that does not matter: das macht nichts aus; it matters a great deal: es macht viel aus* || Fall * *Angelegenheit* || Gegenstand * *Angelegenheit* || gelten * *wichtig sein* || von Bedeutung sein
matter of conflict Konfliktstoff
matter of course Selbstverständlichkeit
matter of fact Tatsache
matter-of-fact praktisch * *nüchtern*
mature ausgereift * *reif, voll entwickelt, reichlich erwogen; fällig, zahlbar, zur Reife bringen*
matured ausgereift * *(aus)gereift; abgelagert*
maturing ausgereift
maturity Reife * *(auch figürl;* →*Produktreife) degree of maturity: ~grad; mature: zur ~ kommen/bringen*
maverick Ausreißer
max. maximal
maxim Grundsatz * *Maxime, Lebensregel, Leitsatz* || Motto * *Maxime, Grundsatz* || Satz * *Leit~, Maxime*
maximize maximieren
maximized maximiert * *for: bezüglich*
maximum äußeres * *maximales* || größt * *wert-/zahlenmäßig* || größtmöglich || höchst * *zahlen-/*

wertmäßig am ~en, maximal || maximal * *maximum permissible:maximal zulässig* || Maximal- || Maximum || Spitzen- * *Maximal-*
maximum builddown Wickelverhältnis * *Durchmesserverhältnis beim Abwickler*
maximum buildup Wickelverhältnis * *Durchmesserverhältnis bei Aufwickler*
maximum control setting Vollaussteuerung * *allg.*
maximum delay angle alpha w || alpha-W
maximum diameter ratio Wickelverhältnis * *Verhältnis Kern/Max. Durchmesser bei Auf/Abwickler*
maximum dimension Höchstmaß * *größte Abmessung*
maximum error Toleranz
maximum firing angle alpha w * *Wechselrichtertrittgrenze* || alpha-W || Wechselrichtertrittgrenze * *max. Steuerwinkel, darüber besteht Gefahr des Wechselrichterkippens*
maximum frequency Grenzfrequenz || Maximalfrequenz
maximum gating angle Wechselrichtertrittgrenze
maximum limit stress Grenzbeanspruchung
maximum load Belastbarkeit * *maximale ~*
maximum mean forward current Dauergrenzstrom * *Diode*
maximum mean on-state current Dauergrenzstrom * *Thyristor*
maximum operating frequency Grenzfrequenz
maximum output Spitzenleistung * *z.B. eines Motors, eines Stromrichters*
maximum overshoot Überschwingweite
maximum payload Tragfähigkeit * *maximale Nutzlast*
maximum permissible höchstzulässig
maximum permissible temperature Grenztemperatur
maximum permissible temperature rise Grenzübertemperatur * *max. zulässige Temperaturerhöhung (z.B. VDE0530 Teile 1 u. 22)*
maximum possible maximal * *größtmöglicher*
maximum power point Punkt der maximalen Leistung
maximum rpm Maximaldrehzahl
maximum selector Maximalwertauswahl || Maximumauswerter
maximum speed Höchstgeschwindigkeit || Maximaldrehzahl || Maximalgeschwindigkeit
maximum temperature rise Grenzübertemperatur * *z.B für Motor*
maximum value Endwert * *Maximalwert* || Maximalwert
maximum value selector Extremwertauswahl * *Maximalwertauswahl-Baustein* || Maximalwertauswahl
maximum width of gap Grenzspaltweite * *oberhalb besteht Zündgefahr; Kriterium für Einstufg. i.Explosionsklasse*
MB Megabyte
Mbaud Mbaud * *Megabits pro sec.*
MCB Sicherungsautomat * *Abkürz. für 'Miniature Circuit Breaker'*
MD Längs- * *Abkürzg.f. 'Material-Directed': in Materiallaufrichtung*
MD direction Längsrichtung * *Abk. für 'Machine Direction': Maschinenlaufrichtung*
MD profile Längsprofil * *Abkürzg.f. 'Material-Directed': (z.B. Fläch.gewichts-) Profil in Materiallaufrichtung*
MDF board Faserplatte * *{Holz} Abkürzg. für 'Medium Density Fiber board': mittelharte ~* || MDF-Platte * *{Holz} Abk.f. 'Medium Density Fiber board': mittelharte Faserplatte*
MDO Längsrecke * *Abk.f. 'Material-Directed Orientation': Längsreckeinrichtung in Folienmaschine*

meager   dürftig * *(amerikan.) spärlich, mager* || knapp * *(amerikan.) dürftig, kärglich, mager*
meagre   dürftig * *spärlich, mager* || knapp * *dürftig, kärglich, mager*
mean   bedeuten || Durchschnitt || durchschnittlich || Mittel * *(Adj.) im statistischen ~; (Subst.) (Hilfs-) Mittel, Werkzeug, Geld~* || Mittel * *statistisches/ arithmetisches ~* || mittlerer * *arithm. Mittelwert* || vorhaben * *beabsichtigen, vorhaben, entschlossen sein; mean to do: etwas zu tun gedenken* || zur Folge haben * *bedeuten*
mean deviation   mittlere Abweichung
mean error   mittlere Abweichung * *mittlerer Fehler*
mean value   Mittelwert * *arithmetischer ~*
mean value generation   Mittelwertbildung * *sliding-type/shifting-mode: gleitende ~*
mean value of the power   Leistungsmittelwert
meander-shaped   mäanderförmig
meaning   Bedeutung || Sinn * *Bedeutung*
meaningful   sinnvoll * *von Bedeutung*
means   -Mittel * *Hilfsmittel* || Hilfsmittel * *(Plural) Mittel, Werkzeug, Wege, (Geld)Mittel* || Mittel * *(Plural) (Hilfs-/Geld-) Mittel, ~ und Wege, means to an end: ~ zum Zweck*
means of production   Produktionsmittel
meantime   Übergangszeit * *Zwischenzeit; in the meantime: in der Zwischenzeit/derweil* || Zwischenzeit * *in the meantime: in der Zwischenzeit*
meanwhile   vorübergehend * *zwischenzeitlich*
measurable   merklich * *messbar* || messbar
measurand   Messwert * *zu messender Wert*
measure   abmessen || Aktion * *Maßnahme* || bemessen * *bewerten (by:nach)* || erfassen * *messen* || Maß * *~einheit, ~stab, Aus~; be a measure of: ein ~ sein für; in a great m.: in großem ~e* || Maßnahme * *take:ergreifen* || messen || vermessen || Vorkehrung
measure off   abmessen
measure out   Dosieren * *(Verb)*
measure up to   gewachsen sein
measured   bemessen * *(Adj.)* || gemessen
measured against   gemessen an
measured date sensing   Messwerterfassung
measured quantity   Messgröße
measured value   Messwert * *gemessener Wert*
measured value acquisition   Messwerterfassung
measured value analysis   Messwertanalyse
measured value transducer   Messumformer
measured values   Messdaten * *Messwerte*
measured variable   Messgröße
measured-data acquisition   Messwerterfassung
measured-value transducer   Messwertumformer
measurement   Erfassung * *Messung* || Messtechnik || Messung
measurement bridge   Messbrücke
measurement data   Messdaten
measurement dimensions   Abmaße
measurement engineering   Messtechnik * *als Fachrichtung*
measurement error   Messfehler
measurement technique   Messtechnik
measuring   Mess- || Messung
measuring accuracy   Messgenauigkeit
measuring bridge   Messbrücke
measuring changeover   Messbereichsumschaltung
measuring circuit   Messkreis
measuring counter   Längenzähler
measuring cup   Messbecher
measuring data   Messdaten
measuring data sensing   Messwerterfassung
measuring device   Messgerät * *allg.*
measuring element   Messwerk
measuring equipment   Messanlage || Messeinrichtung || Messgerät * *Ausrüstung* || Messgeräte || Messtechnik * *messtechnische Ausrüstung*

measuring error   Messfehler
measuring inaccuracy   Messunsicherheit * *Messungenauigkeit*
measuring incertainty   Messunsicherheit
measuring instrument   Instrument * *Mess~* || Messinstrument
measuring interval   Messzeit
measuring jack   Messbuchse
measuring lead   Messleitung || Messstrippe
measuring location   Messort || Messstelle
measuring meter   Messgerät * *Messinstrument*
measuring method   Messmethode || Messverfahren
measuring microphone   Messmikrofon * *zur Schallmessung*
measuring nipple   Messnippel
measuring of the temperature rise   Erwärmungsmessung
measuring period   Messzeit
measuring point   Messpunkt || Messstelle
measuring principle   Messprinzip
measuring probe   Messfühler * *Messsonde, Tastkopf*
measuring range   Messbereich
measuring reference   Messnormal
measuring sample   Messwert * *abgetasteter ~*
measuring sensor   Messwertaufnehmer
measuring site   Messort
measuring socket   Buchse * *Messbuchse* || Messbuchse || Prüfbuchse * *Messbuchse* || Testbuchse * *Messbuchse*
measuring spot size   Messfläche * *z.B. bei (berührungsloser) Flächengewichts- und Feuchtemessung*
measuring station   Messplatz
measuring strip   Messband * *zur Lage-/Längenerfassung*
measuring surface   Messfläche * *zur Schalldruckmessung (entlang d. ~ (Kugel, Quader) sind Mikrofone angeordnet)*
measuring surface sound-pressure level   Messflächen-Schalldruckpegel * *gemessen entlang e.genormten Messfläche (z.B. Kugelfläche)*
measuring system   Messsystem
measuring tape   Bandmaß
measuring technique   Messverfahren
measuring technology   Messtechnik
measuring terminal   Messklemme
measuring time   Messzeit
measuring transducer   Messumformer || Messwertumformer
measuring transmitter   Messumformer
measuring unit   Messeinrichtung
measuring voltage   Messspannung * *z.B. bei Isolationsprüfung*
measuring wheel   Messrad * *Anlegerad zur Geschwindigkeitsmessung*
measuring-point selector   Messstellenumschalter
measuring-range changeover   Messbereichsumschaltung
measuring-surface level   Messflächenmaß * *Rechengröße zur Ermittlung der Schallleistung*
measuring-surface sound-pressure level   Messflächenschalldruckpegel * *(mit Mikrofonen an e. Messfläche gemess.) DIN EN 21680 Tl.1*
mechanic   Handwerker * *Mechaniker* || Mechaniker || Monteur * *Mechaniker, Schlosser* || Schlosser * *→Mechaniker*
mechanical   maschinell * *machining: ~e Bearbeitung* || mechanisch * *(Adj.)*
mechanical actuator   mechanisches Stellglied
mechanical braking   mechanische Bremsung * *mit Reibung arbeitende Bremse*
mechanical construction   Konstruktion * *mechanische*
mechanical degree of protection   mechanische

Schutzart * im Gegensatz z. Zündschutzart (Ex-Schutz), siehe z.B. IEC Publ.No.34 || Schutzart * im Gegensatz z.Zündschutzart (Ex-Schutz); siehe IEC-Publ. No.34
mechanical design   Konstruktion || konstruktive Ausführung || konstruktive Gestaltung || mechanische Ausführung || mechanischer Aufbau
mechanical designer   Konstrukteur * Maschinenbau-Konstrukteur
mechanical engineer   Konstrukteur * Maschinenbau-Ingenieur || Maschinenbauer * Maschinenbauingenieur/Techniker
mechanical engineering   Maschinenbau * general: allgemeiner
mechanical engineering department   Konstruktionsbüro
mechanical engineering firm   Maschinenbauer * Maschinenbaufirma
mechanical engineering industry   Maschinenbau * ~industrie
mechanical equipment   Maschinenbau * Maschinenausrüstung
mechanical equipment manufacturers   Maschinenbau * als Branche
mechanical input power   Antriebsleistung * eines Generators
mechanical interlock   mechanische Verriegelung
mechanical limit speed   mechanische Grenzdrehzahl
mechanical limiting speed   mechanische Grenzdrehzahl
mechanical outfit   Apparatur
mechanical pulp   Holzschliff || Holzstoff * {Papier} Zellstoffrohprodukt || Zellstoff * Holzstoff (Zellstoffrohprodukt)
mechanical quantities   mechanische Größen
mechanical rack system   Aufbausystem * mit Baugruppenträgern und Steckbaugruppen
mechanical resonance   mechanische Resonanz
mechanical shocks   mechanische Stöße
mechanical speed limit   mechanische Grenzdrehzahl
mechanical stability   mechanische Festigkeit
mechanical strength   Festigkeit * mechanische ~ || mechanische Festigkeit
mechanical stressing   mechanische Beanspruchung
mechanical time constant   mechanische Zeitkonstante * Zeit d.z.Erreichen v. 63,2% d. Enddrehzahl benötigt wird bei Nennbeding.
mechanical transmission   Kraftübertragung
mechanical transmission element   Kraftübertragungselement
mechanical treatment   Bearbeitung * mechanische ~
mechanical variable speed drive   mechanischer Verstellantrieb * z.B. mit PIV-Getriebe oder Keilriemen-Verstellgetriebe
mechanical vibrations   mechanische Schwingungen * für elektr. Maschinen: siehe DIN ISO 2373 (IEC34-14)
mechanical woodpulp   Holzschliff
mechanical work   mechanische Arbeit
mechanically   in gewohnter Weise * gewohnheitsmäßig || mechanisch * (Adv.)
mechanically coupled   mechanisch gekoppelt
mechanically decoupled   mechanisch entkoppelt
mechanically operated   mechanisch * ~ betätigt/angetrieben || mechanisch betätigt
mechanically rugged   mechanisch stabil
mechanician   Mechaniker
mechanics   Maschinenbau || Mechanik
mechanism   Apparat * Mechanismus || Einrichtung * Vorichtung || Mechanik * Mechanismus, Triebwerk || Mechanismus || Vorrichtung || Wirkungsweise

meddle   stören * sich einmischen
media   Medien * Materialien
median   Mittelwert * 50%-Wert, Mittellinie/-wert, die Mitte bildend/einnehmend, Mittel-, mittlerer
mediate   vermitteln * sich als Vermittler betätigen (between: zw.), e.Bindeglied bilden, dazwischen liegen
medical   medizinisch
medical devices   medizinische Geräte
medical engineering   medizinische Technik || Medizintechnik
medical man   Arzt * (salopp)
medium   -Mittel * Medium (Plural: media) || durchschnittlich * Größe, Preis, Qualität usw. || Medium * Mittel, Träger, z.B. Speichermedium, Kühlmittel (Plural: 'media') || Mittel * Substanz || Mittel * (Adj.) durchschnittlich || mittlerer * durchschnittlich, mittelmäßig || Stoff * Mittel || Substanz
medium flow   Förderstrom
medium performance level   mittlerer Leistungsbereich * bezüglich der Leistungsfähigkeit/Rechenleistung
medium range   Mittel * (Adj.) im mittleren Bereich liegend
medium supply   Medienversorgung * Herbeiförderung von Flüssigkeiten
medium term   mittelfristig * (Adj.)
medium time-lag   mittelträge * Sicherung
medium voltage   Mittelspannung
medium voltage circuit breaker   Mittelspannungsschalter
medium-   Mittel-
medium-frequency generator   Mittelfrequenzgenerator
medium-high voltage   Mittelspannung
medium-performance   mittlerer * von ~ Leistungsfähigkeit
medium-pressure   Mitteldruck-
medium-sized   Mittel * (Adj.) von mittlerer Größe || mittelgroß || mittlerer * von ~ Größe
medium-sized company   mittelständische Firma * mittelgroße Firma || mittelständischer Betrieb || mittelständisches Unternehmen
medley   Gemisch * (figürl.)
meet   aufbringen * Kosten || einhalten * Vorschriften, Norm || entgegenkommen * treffen || entsprechen * einer Bitte, einer Vorschrift * || erfüllen * Bedingung, Erwartung, Norm, Vorschrift, Anforderung, Bedürfnis ~ || treffen * sich ~ || übereinstimmen * ~ mit/entsprechen/gerecht werden (Anforderung, Vorschrift, Norm usw.) || versammeln * sich ~ || zusammentreffen
meet all requirements   allen Ansprüchen gerecht werden
meet the requirements   Anforderung erfüllen
meet with   begegnen || finden * antreffen
meeting   Besprechung || Gespräch * Besprechung || Sitzung * Besprechung || Verhandlung * Besprechung || zusammentreffen * (Substantiv)
meeting location   Treffpunkt * z.B. für eine Besprechung
meeting manifold requirements   vielseitig einsetzbar
meeting minutes   Besprechungsbericht
meeting place   Treffpunkt
meeting point   Treffpunkt
meeting report   Gesprächsprotokoll
meeting request   Einladung * zu einer Besprechung || Gesprächseinladung
meg   Megabyte * (salopp)
megabyte   Megabyte
megawatt   Megawatt
megawatt range   Megawattbereich
megger   Kurbelinduktor * zur Isolationsprüfung/messung

melt durchbrennen * *schmelzen (Sicherung usw.)* ||
schmelzen * *melt down: einschmelzen*
melting crucible Schmelztiegel * *[Metallurgie]*
melting furnace Schmelzofen
melting I2t value Schmelz-I2t-Wert * *einer Sicherung*
melting I²t value Schmelz-I²t-Wert * *einer Sicherung*
melting line Schmelzstraße * *im Hüttenwerk*
melting pot Schmelztiegel * *(auch figürl.)*
melting tank Schmelztiegel
melting time Schmelzzeit * *(Sicherung)* || Sicherungsansprechzeit
member Einzelteil * *Glied* || Komponente * *Glied* || Seite * ~ *einer Gleichung* || Teilnehmer * *Mitglied*
member of the board Aufsichtsratsmitglied || Vorstand * ~*smitglied* || Vorstandsmitglied
member of the executive board Vorstand * ~*smitglied* || Vorstandsmitglied
member of the supervisory board Aufsichtsratmitglied || Aufsichtsratsmitglied
membership Mitgliedschaft
membrane Membran
membrane keyboard Folientastatur * *sealed: hermetisch dicht*
membrane keypad Folientastatur * *kleines Tastenfeld*
membrane panel Folientastatur
membrane-type keyboard Folientastatur
memo Aktennotiz || Aktenvermerk || Rundschreiben
memorandum Aktennotiz || Besprechungsbericht
memorize abspeichern * *in Speicher festhalten, aufbewahren* || einprägen * *→auswendig lernen, sich ~* || speichern * *im Speicher (auch spannungsauffallsicher) aufbewahren*
memory Speicher * *Datenspeicher*
memory address Speicheradresse
memory address space Adressraum
memory administration programm Speicherverwaltungsprogramm
memory allocation Speicherbelegung
memory assignment Speicherbelegung
memory bank Speicherbank * *[Computer]*
memory capacity Speicherausbau * *Speicherkapazität/-vermögen* || Speicherbedarf || Speicherkapazität || Speicherplatz * *zur Verfügung stehender ~, Speicherkapazität, Speicherausbau* || Speicherplatzbedarf * *[Computer]*
memory capacity requirements Speicherplatzbedarf * *[Computer]*
memory card Speicherkarte
memory cell Speicherglied * *Speicherzelle* || Speicherzelle
memory chip Speicherbaustein * *Integrierter Speicherschaltkreis*
memory configuration Speicherausbau * *Konfiguration*
memory contents Speicherinhalt || Speicherinhalte
memory depth Speichertiefe * *eines Schiebespeichers (z.B. Trace-/Messwert-/FIFO-Speichers usw.*
memory dump Speicherabzug
memory expansion Speicherausbau * *Vergrößerung*
memory extension Speicherausbau * *Vergrößerung*
memory hog Speicherfresser * *(salopp) "Speicher-Vielfraß", Programm, das viel Speicherplatz beansprucht*
memory IC Speicherbaustein * *Integrierter Speicherschaltkreis*
memory location Speicherplatz * *Speicherzelle(n) für bestimmte Daten* || Speicherzelle
memory loss Datenverlust * *in einem Datenspeicher*
memory management Speicherverwaltung

memory manager Speicherverwaltungsprogramm
memory map Adressliste * *[Computer]* || Belegungstabelle * *Speicherbelegungsliste bei Rechner/SPS* || Speicherbelegungsliste || Speicherbelegungsplan
memory module Speicherbaugruppe || Speicherbaustein * *Speichermodul (evtl. auf kleinem Leiterkärtchen (z.B. →SIMM))* || Speichermodul
memory pack Speichermodul
memory requirements Speicherbedarf || Speicherplatzbedarf * *[Computer]*
memory retention in case of power loss nichtflüchtige Speicherung bei Spannungseinbruch
memory select Speicherzugriffsfreigabe * *(MEMSEL) bei SPS*
memory size Speichergröße
memory space Adressraum || Speicherausbau * *Speicherplatz/-raum* || Speicherplatz * *zur Verfügung stehender Speicher*
memory sub-module Speichersubmodul
memory submodul Speichermodul * *Speichersubmodul, Speicherkärtchen (z.B. →SIMM)*
memory submodule Speicherbaugruppe * *Speichermodul/-Subbaugruppe* || Speichersubmodul
memory-I/O select Peripherie-Speicher-Umschaltung * *Adressumschaltung bei SPS*
men's room Toilette * *Herren~*
mend instand setzen * *ausbessern, flicken, reparieren, richten* || reparieren
mental gedanklich || geistig
mention anführen * *erwähnen* || erwähnen || nennen * *erwähnen*
mentioned erwähnt * *above: oben; below: unten; earlier: bereits/früher* || genannt * *erwähnt*
mentioned above oben erwähnt * *(nachgestellt)* || obig * *(nachgestellt) oben genannt* || vorgenannt * *→oben erwähnt, vorgenannt*
mentioned below unten erwähnt * *(nachgestellt)* || unten stehend * *(nachgestellt)*
menu Menü
menu driven interaktiv * *menügesteuert* || menügeführt
menu guidance Menüführung
menu item Menüpunkt
menu prompting Menüführung
menu-assisted menügeführt || menügesteuert
menu-controlled menügesteuert
menu-driven menügeführt || menügesteuert
menu-prompted interaktiv * *menügesteuert* || menügeführt || menügesteuert
menue Menü
menue bar Menüleiste * *[Computer] auf dem Bildschirm*
menue driven dialoggesteuert
menue item Rubrik * *[Computer] z.B. in Softwareprogramm, Mailbox usw.*
mercerizing machine Mercerisiermaschine * *[Textil] zur Nachbehandlung ("Glänzendmachen") von Textilstoffen*
merchandise Ware * *Ware(n), Handelgüter* || Waren
merchant Händler * *(Groß)Händler*
mercury Quecksilber
mercury-arc rectifier Quecksilberdampf-Gleichrichter || Quecksilberdampf-Stromrichter || Quecksilberdampfgleichrichter
mere rein * *bloß, lediglich (z.B. Formalität)*
merely allein * *(Adv.) nur, bloß* || bloß * *lediglich* || lediglich || nur * *bloß*
merge fusionieren * *auch Firmen* || mischen * *z.B. Daten(sätze) einsortieren* || sortieren * *'(hin)einsortieren', z.B. Daten an die richtige Stelle in e. Datenbank* || übergehen * *von Farbtönen/Firma (into: in)* || vereinigen * *fusionieren; into:zu* || zusammenfassen * *verschmelzen (in:mit), aufgehen*

lassen in, einverleiben, fusionieren || zusammenschließen * (sich) zusammenschließen (Firmen usw.)
merger Fusion * auch von Firmen || Zusammenschluss * von Firmen
merging Fusion * auch von Firmen || Zusammenschluss * von Firmen
merry-go-round Fahrgeschäft * Karussel
merry-goround Karussel
mesh eingreifen * Maschinenteile ; in(to):in; with:ineinander || Geflecht * z.B. aus Draht; fine: feines; coarse: grobes || greifen * ineinander~ || ineinander greifen || Masche * in einem Netzwerk || vereinen * verschmelzen (mit), aufgehen lassen (in), fusionieren; be meshed in: aufgehen in
mesh current Maschenstrom * in einem Netzwerk
mesh size Lochweite * eines Siebes
mesh voltage Maschenspannung || verkettete Spannung * Maschenspannung
meshed vermascht * netzartig, maschig; network:- Netz || verwoben * (figürl.) verzahnt closely:eng || verzahnt * (figürl.) closely: eng
meshing Eingriff * von Zahnrädern usw. || Vermaschung || Verzahnung * (mechan. u. figürl.) Ineinander-Greifen
message Auftrag * Botschaft || Aussage * Mitteilung, Bescheid, Botschaft || Melde- || Meldung || Mitteilung * Nachricht || Nachricht * Meldung, Botschaft || Telegramm * Nachricht, Meldung, Sendung || Text * Melde-/Aufforderungstext auf einem Computerbildschirm
message block Meldebaustein * Funktionsbaustein
message buffer Meldepuffer
message contact Meldekontakt
message line printer Meldedrucker
message listing Meldeprotokoll
message printer Meldedrucker
message printout Meldeprotokoll
message sequencing system Meldefolgesystem
message system Meldesystem
message text Meldetext
messy schmutzig * unordentlich, unsauber, schmutzig
met erfüllt * z.B. Vorschrift, Voraussetzung
metal Metall
metal band Metallband
metal bellow Metallfaltenbalg
metal bellows coupling Metallbalgkupplung
metal case Metallgehäuse
metal detctor Metallsuchgerät * auch beim Papier-/Holzrecycling
metal detector Metall-Suchgerät
metal film resistor Metallfilmwiderstand || Metallschichtwiderstand
metal finishing line Metallveredelungsstraße
metal fitting Beschlag
metal foil Metallfolie
metal forming Metallumformung * spanlose Metall-/Blechbearbeitung
metal hose Metallschlauch * flexible: biegsamer
metal lug Metallgrifflasche * zum Ansatz des Ziehgriffs bei Sicherung
metal plate working Blechbearbeitung * Grob~
metal sheath Metallmantel * z.B. einer Leitung
metal sheet Feinblech * thin metal sheet: Feinstblech
metal strip Metallband
metal working Metallverarbeitung
metal working machine Metallbearbeitungsmaschine
metal-cased capacitor Becherkondensator
metal-cutting spanabhebend
metal-oxid surge arrester Metalloxidableiter * Überspannungsableiter
metal-oxide semiconductor MOS-Transistor
metal-oxide surge diverter Metalloxid-Ableiter
metallic metallisch
metallic bright metallisch blank
metallic enclosure Metallgehäuse
metallic isolation galvanische Trennung
metallical metallisch
metallically metallisch * (Adv.) z.b. metallically bright: ~ →blank
metallically separated galvanisch getrennt
metallization Metallisierung
metallize metallisieren
metallized metallbedampft || metallisiert
metallized film capacitor metallisierter Kunststoffkondensator
metallized paper capacitor Metallpapierkondensator
metallized-dielectric capacitor MKV-Kondensator * mit metalliertem Papier-Kunststoff-Dielektrikum, verlustarm
metallized-plastic capacitor MKL-Kondensator * Wickelkondensator aus metallisierter Kunststoff- u. Lackfolie
metallurgical metallurgisch
metallurgical plant Hüttenwerk
metallurgy Hüttentechnik || Hüttenwesen
metalworking Meall verarbeitend || Metallbearbeitung || Metallumformung * spanlose Metall-/Blechbearbeitung
metalworking industrie Metallindustrie
meter -Messer || -Messgerät || Anzeigeinstrument * Messinstrument || Dosieren * (Verb) || erfassen || messen * mit einem Messapparat, z.B. mit Stromzähler, auch dosieren, Menge ~ || Messgerät || Mcssinstrument || Meter * (amerikan.)
metered goods Meterware
metering Dosierung
metering accuracy Dosiergenauigkeit
metering machine Dosiermaschine
metering pump Dosierpumpe
metering screw Dosierschnecke * in der Verpakkungstechnik || Füllschnecke * (Verpack.techn.)
metering system Dosiereinrichtung
metering unit Dosiereinrichtung
metering-piston filling machine Kolbenfüllmaschine * in der Verpackungstechnik
metering-screw filling machine Schneckenfüllmaschine * in der Verpackungstechnik
method Art * Methode || Art und Weise * Verfahren, Methode || Ausführung * Methode, Verfahren || Methode || System * Methode || Verfahren * Methode || Verfahrensweise || Weg * Methode, Art und Weise
method of calculation Rechenverfahren
method of cooling Kühlart || Kühlart * z.B. für Motoren (siehe EN 60034-6)
method of data representation Datendarstellung * Art d. Datendarstellung
method of functioning Funktionsprinzip || Wirkungsweise
method of measurement Messverfahren
method of measuring Messmethode || Messprinzip * siehe →Messmethode || Messverfahren
method of modulation Modulationsverfahren
method of PWM modulation Pulsmuster * bei pulsbreitenmoduliertem Umrichter || Pulsverfahren * bei pulsbreitenmoduliertem Umrichter
method of representation Darstellungsart
method of testing Prüfverfahren
methodical systematisch
methodically systematisch * (Adv.)
meticulous sorgfältig * peinlich genau
métier Handwerk * (figürlich)
metre Meter
metric metrisch
metric system Metrisches System

metric thread metrisches Gewinde
metric ton t * *metrische Tonne; 1 metric ton entspr. 1000 kg entspr. 0,9842 ton* || Tonne * Gewichtseinheit "metrische Tonne", siehe auch →t *(1 metric ton entspr. 1000 kg)*
metropolitan railway S-Bahn * *in einer Millionenstadt/Hauptstadt*
mica Glimmer * *z.B. bei Kondensator, Isolierung (zwischen Kommutatorlamellen)*
mica flake Glimmerblättchen || Glimmerplättchen || Glimmerscheibe
mica lamella Glimmerblättchen || Glimmerplättchen || Glimmerscheibe
mica lamina Glimmerblättchen * *zur Isolierung Leistungshalbleiter* →*Kühlkörper/Leiterkarte* || Glimmerplättchen * *zur Isolierung Leistunghalbleiter* →*Kühlkörper/Leiterkarte* || Glimmerscheibe * *zur Isolierung Leistungshalbleiter* →*Kühlkörper/Leiterkarte*
mica segment Glimmerzwischenlage * *zur Isolierung zwischen den Kommutatorlamellen*
mica separator Glimmerzwischenlage * *zur Isolierung zwischen den Kommutatorlamellen*
mica splitting Glimmerblättchen || Glimmerplättchen || Glimmerscheibe
micro Klein- || μ
micro friction Mikroreibung
micro operation Mikrobefehl * *Teilbefehl, in den ein Maschinenbefehl in einer* →*CISC-CPU zerlegt wird*
micro switch Mikroschalter
micro wave engineering Höchstfrequenztechnik
micro- Kleinst- || Miniatur-
micro-converter Kleinstumrichter
micro-electronics Mikroelektronik
micro-fiber Mikrofaser
micro-filter Feinstaubfilter
micro-wave technology Höchstfrequenztechnik
microamp Mikroampere
microamps Mikroampere * *(Plural)*
microcoded mikroprogrammiert
microcomputer Mikrocomputer
microcontroller Ein-Chip-Computer || Ein-Chip-Mikroprozessor
microelectronics Mikroelektronik
microfuse Feinsicherung * *super-rapid: superflinke*
micron Mikrometer || μ * *Dickenmaß 1/1000mm*
microorganisms Mikroorganismen
microprocessor Mikroprozessor
microprocessor controlled Mikroprozessor-gesteuert
microprocessor technology Mikroprozessortechnik
microprocessor-based control Mikroprozessorregelung
microprocessor-controlled mikroprozessorgesteuert
microprogram Mikroprogramm * *Vorschrift für die sequenzielle Abarbeitung eines Maschinenbefehls in* →*CISC-CPU*
microprogrammed mikroprogrammgesteuert
microscope Mikroskop
microscopic klein * *verschwindend ~* || kleinst || Kleinst- || verschwindend klein * *winzig*
mid Mitte * *mittler, Mittel-; mid of may: Mitte Mai; in the mid 19th century: in der ~ d.19 Jahrhunderts*
mid point Mitte * →*Mittelpunkt*
mid position Mittellage || Mittelstellung * *z.B. Tänzer*
mid-range Mittel * *in der Mitte rangierend, ~klasse-* || Mittelklasse- * *z.B. Produkt* || mittlerer Leistungsbereich
mid-sized mittelgroß

middle halb * *at the middle height: auf ~er Höhe* || Mitte || Mittel * *(Adj.)* || mittlerer * *in der Mitte liegend*
middle conductor Mittelleiter * *bei Gleichstrom*
middleman Vermittler * *z.B. Zwischenhändler*
middling mäßig * *(salopp) mittel~* || Mittel * *(Adj.) mittelmäßig* || mittlerer * *mittelmäßig*
midpoint Mittelpunkt
midpoint connection Mittelpunktschaltung * *Stromrichterschaltg.bei d.e. DC-Anschluss gleich dem Wechselstr.sternpkt. ist*
midrange Mittel * *(Adj.) im mittleren Bereich liegend* || Mittel- || mittlerer * *z.B. mittlere Entfernung*
might Gewalt * *zwingende Kraft* || Kraft * *Macht, Gewalt, Stärke, Kraft; with all ones's might: mit aller ~*
migrate eindringen * *hinein/abwandern* || migrieren * *(langsam hinein-) wandern* || wandern * *wegwandern, langsam hinein~ (auch techn.), ziehen, migrieren*
migration Migration
mildewy schimmelig * *['mildju:i] modrig, schimm(e)lig, brandig*
milestone Meilenstein
military militärisch
mill Fabrik * *in der Grundstoffindustrie (Papier, Stahl, Textil)* || fräsen || mahlen || Mühle
mill blower Mühlenluftgebläse
mill bus Fabrikbus * *in Papierfabrik, Walzwerk usw.*
mill entry section Bandeinlauf * *im Walzwerk*
mill motor Walzmotor || Walzwerkmotor || Walzwerksmotor
mill stand Walzgerüst * *(Walzwerk)*
mill-stand drive Walzantrieb
mill-stand motor Walzgerüst-Motor || Walzgerüstmotor || Walzmotor
millboard Pappe * *starke Pappe* || Wickelpappe * *geformte Pappe, Graupappe, Zellstoffpappe*
milliarde Milliarde * *(brit.)*
millimeter Millimeter * *(amerikan.)*
millimetre Millimeter * *(brit.)*
milling and boring machine Bohr- und -Fräswerk * *{Werkz.masch.}*
milling arbor Fräser-Dorn
milling center Fräszentrum * *{Werkz.masch.}*
milling cutter Fräser * *{Werkz.masch.}* || Fräskopf || Messerkopf * *zum Fräsen*
milling head Fräskopf
milling machine Fräsmaschine * *{Werkz.masch.} für Metall* || Fräswerk
milling plant Mahlanlage
milling stand Walzgerüst * *(Walzwerk)*
millstand drive Walzgerüst-Antrieb * *(Walzwerk)*
millstand motor Walzwerksmotor * *Walzgerüst-Motor*
millwide fabrikweit * *über die ganze Papierfabrik/ d. ganze Walzwerk (z.B. einheitl. Automatisierungssystem)*
MIME MIME * *'Multi-Purpose Internet Mail Extensions':Konvertiert 8Bit Code in 7Bit* →*ASCII Code (f.E-Mails)*
mimic nachbilden * *nachahmen*
mimic diagram Anlagenbild * *grafische Nachbildung der Anlage auf dem Bildschirm*
min. minimal
mind achten auf || ausmachen * *never mind: das macht nichts; would you mind if: macht es Ihnen etwas aus, wenn?* || beachten * *warnend; that: dass* || kümmern * *sich (be)~ um, (be)achten, Acht geben auf, sorgen für, sehen nach* || kümmern um || sich kümmern um || Sinn * *Verstand, Meinung*
mine abbauen * *Bodenschätze ~* || Bergwerk || Grube * *Bergwerk* || Mine * *Bergwerk; Sprengkörper*
mine cage Förderkorb

mine hoist Fördermaschine * Bergwerks-Förderaufzug || Schachtförderanlage
mine winder Förderanlage * Schachtförderanlage für Bergwerk || Fördermaschine * Haspelmaschine für Bergwerk || Schachtförderanlage * Förderaufzug || Schachtförderer
mine-type Bergwerks- || schlagwettergeschützt * für Untertage-Einsatz
mineral fiber Mineralfaser
mineral oil Erdöl
mingle mischen * sich ver~, verschmelzen, sich vereinigen/verbinden
mini Klein-
mini floppy disk Mini-Diskette * 5,25"-Diskette
mini PCB Kleinleiterplatte
mini PLC Kleinsteuerung * speicherprogrammierbare ~
mini- Miniatur-
miniature klein * ~st || Klein- || Kleinst-
miniature circuit breaker Sicherungsautomat
miniature motor Kleinmotor
miniature- Miniatur-
miniaturization Miniaturisierung || Verkleinerung
minidiskette Mini-Diskette * 5,25"-Diskette
minifuse Feinsicherung
minimization Minimierung
minimize minimieren || verkleinern * auf ein Minimum ~
minimized minimiert * for: bezüglich
minimized for harmonics oberschwingungsminimiert
minimum kleinst || kleinstmöglich * zahlen/wertmäßig || Mindest- || mindestens || Mindestmaß || minimal || Minimal- || Minimum * cut to a minimum: auf ein ~ reduzieren/beschränken
minimum configuration Minimalkonfiguration
minimum delay angle Gleichrichtertrittgrenze
minimum delay angle setting Vollaussteuerung * z.B. bei Phasenanschnittsteuerung
minimum dimension Mindestmaß * Längenmaß, Ausdehnung
minimum firing angle alpha g * Gleichrichtertrittgrenze || alpha-G || Gleichrichtertrittgrenze
minimum frequency Mindestfrequenz || Minimalfrequenz
minimum fusing current Sicherungsgrenzstrom
minimum gating angle Gleichrichtertrittgrenze
minimum maintenance requirements Wartungsfreundlichkeit
minimum quantity Mindestabnahmemenge
minimum requirements Mindestanforderung * regarding:an || Mindestanforderungen
minimum selector Minimalwertauswahl || Minimumauswahl || Minimumauswerter
minimum speed Mindestdrehzahl || Minimaldrehzahl
minimum value Minimalwert
minimum value selector Minimalwertauswahl || Minimumauswahl
mining Abbau * Bodenschätze || Bergbau || Bergwerks-
mining equipment Bergwerksausrüstung
mining hoist Schachtförderanlage
mining motor Bergbaumotor || Untertagemotor * Bergbaumotor
minor klein * geringfügig || kleiner * (Komparativ) geringer, untergeordnet, geringfügig, von geringerem Umfang || leicht * unbedeutend || niederprior * untergeordnet, rangniedriger || unter * klein(er), unbedeutend(er), geringfügig, ~geordnet (Rang)
minor control loop unterlagerter Regelkreis
minor matter Kleinigkeit
minority Minderheit || Minoritäts-
minority carrier Minoritätsladungsträger * bei Halbleitern

mint prägen * Münzen, Wörter, Begriffe usw. ~
minus abzüglich || minus * ['maines] || weniger * {Mathem.}
minus allowance Untermaß * zulässiges
minus symbol Minuszeichen
minute klein * winzig || Minute || umständlich * sehr genau || verschwindend klein * winzig
minuteness Kleinheit
minutes Protokoll * Verhandlungs-/Besprechungs~; enter in the minutes: ins ~ aufnehmen
minutes of meeting Bericht * Besprechungsbericht || Besprechungsprotokoll * Besprechungsbericht || Gesprächsprotokoll || Protokoll * Besprechungsbericht, Verhandlungs-/~; enter in the minutes: zu ~ nehmen
minutes of meetings Besprechungsbericht
mire Schlamm
mirror Spiegel || spiegeln * (auch figürlich) auch 'wider~'
mirror disk Spiegelplatte * redundante Festplatte (2 Platten mit gleichen Daten; siehe auch →RAID)
mirror image Gegenstück * identisches Teil, getreues Abbild || Spiegelbild
mirror in spiegeln an
mirror symmetric spiegelsymmetrisch
mirror symmetry Spiegelsymmetrie
mirror-image spiegelbildlich
mirror-inverted spiegelbildlich
mirrored spiegelbildlich
mis- Fehl- || schlecht * als Vorsilbe; z.B. mismanagement: ~es management || ver- * falsch, fälschlich(er weise)
mis-engineering Projektierungsfehler
misadjustment Verstellung * fehlerhafte/ungewollte ~
misalignment Dejustage || Fehlausrichtung || Fehljustierung || Fluchtfehler * Fehler in der Ausrichtung || Fluchtungsfehler || Verlagerung * Fluchtfehler, Fehler in der Ausrichtung || Versatz * Versatzfehler, Fehlausrichtung/Justierung
miscalculate verrechnen * sich
miscarry verloren gehen * z.B. Brief
miscellaneous diverse || Diverses || sonstig * vermischt || verschieden * vermischte || verschiedene * diverse || Verschiedenes
misconvergence Konvergenzfehler * ungleiche Positionierung der Farben bei Bildschirm
misfunction Fehlfunktion
mishandling Fehlbedienung
mishap Panne * (figürl.) unglücklich verlaufene Begebenheit, Unglück, Unfall, Panne
misinformation Fehlinformation
misinterpretation Fehlinterpretation * siehe auch →Missverständnis || Missverständnis * Fehlinterpretation
mismatch Unterschied * Nicht-Übereinstimmung, Nicht-Zusammenpassen
misorder durcheinander * (Subst.)
misplace verlegen * an die falsche Stelle, unauffindbar
misprint Druckfehler || Fehler * Druck~ (auch 'typographic error')
miss Fehlschlag || vergeben * Chance ~
miss the bus Anschluss verlieren * (figürl.)
missed vorbei * gefehlt, verfehlt
missing abwesend || daneben * verpassend; miss: vorbeigehen; miss the mark: danebentreffen/-schlagen || fehlend
mission Aufgabe * Reisezweck/auftrag, Sendung, taktischer Auftrag || Auftrag * Reisezweck, Sendung, taktischer Auftrag || Einsatz * Auftrag (für Dienstreise), taktischer Auftrag
misspelling Fehler * orthographischer ~

mist lubrication Nebelschmierung * *durch Ölnebel*
mistake Fehler * ~ *beim Rechnen, Schreiben usw.; Versehen, Missgriff, Irrtum* || versehen * *(Substantiv)* || Verstoß * Fehler || Verwechslung
mistake a person for another verwechseln * *jemanden mit jemandem* ~
mistaken irrtümlich
mistakenly irrtümlich * *(Adv.)* ~*erweise*
misty unklar * *neblig*
misunderstanding Missverständnis
misuse falsche Behandlung || Missbrauch * *falsche/ unkorrekte An/Verwendung*
miter Gehrung * *(amerikan.)* siehe 'mitre'; **miter joint**: ~*sverbindung*
miter gear Kegelrad * *(amerikan.)*
mitre Gehrung * *auch ~sfläche, ~sfugen gehren, auf ~ verbinden; mitre-welded: auf ~ geschweißt*
mitre cutting machine Gehrungsfräsmaschine * *{Holz}*
mitre gear Kegelrad || Winkelgetriebe * *z.B. Kegelradgetriebe*
mitre joint Gehrungsverbindung * *{Holz}*
mitre moulder Gehrungsfräsmaschine * *{Holz}*
mitre saw Gehrungssäge * *{Holz}*
mix durchsetzen * *durchmischen* || mischen || rühren * *mischen*
mix up verfälschen * *vermischen, durcheinandermischen, (völlig) durcheinander bringen* || vertauschen * *mischen* || verwechseln * *durcheinander bringen (with: mit)*
mix-up Verwechslung || Verwirrung * *(von Sachen)*
mixed gemischt || Misch-
mixed calculation Mischkalkulation
mixed friction Mischreibung
mixed up inkonsistent * *durcheinandergebracht*
mixer Mischer || Rührer || Rührwerk
mixer valve Mischventil
mixing Misch-
mixing chamber Mischkammer
mixing installation Mischanlage
mixing plant Mischanlage * *z.b. in der Gummi-/ Reifenindustrie*
mixing ratio Mischungsverhältnis
mixing up Vermaschung
mixing valve Mischventil
mixture Gemisch * *siehe auch →Mischung;* **mixture control**: ~*regelung* || Mischung
MLFB MLFB * *Abkürz. für 'Maschinenlesbare FabrikateBezeichnung' (SIEMENS-Bestellnummer)*
mnemonic mnemotechnisch * *auch menemotechnische Notation eines Maschinenbefehls in einer →Assemblersprache*
mnemotechnical mnemotechnisch
mobile beweglich * *z.B. auf Rädern* || fahrbar || mobil * *im Gegens. zu ortsfest* || tragbar * *nicht ortsfest* || verfahrbar
mobile applications mobiler Einsatz
mobile crane Autokran
mobile offshore drilling unit Bohrplattform
mobile phone Handy
mobility Beweglichkeit * *auch 'Mobilität'*
mobilize aktivieren * *Arbeitsweise, Methode* || Betrieb * *~sart (auch 'duty': Lastspiel bei einer elektr. Maschine usw.)* || Betriebsart || Form * *Art und Weise* || Mode * *Modus, Betriebsart* || Modus || Verfahren * *Art und Weise* || Weise
mock-up Attrappe || Modell * *~ in natürlicher Größe, Attrappe, Nachbildung* || Nachbildung * *Atrappe, Modell in natürlicher Größe*
modal number Ordnungszahl * *z.B. ~ von Oberwellen*
mode Art * *Art u.Weise, Modus, Betriebsart* || Art und Weise * *Wirkungsweise/Arbeitsweise, Methode* ||

mode change Betriebsartenänderung
mode of action Wirkungsweise
mode of control Regelart || Regelungsart
mode of dispatch Verpackungsart * *Angabe in Bestellformular*
mode of functioning Funktionsprinzip || Funktionsweise
mode of modulation Modulationsverfahren
mode of operation Arbeitsweise || Betriebsart || Funktionsprinzip * *Betriebsweise, Betriebsart* || Wirkungsweise
mode selection Betriebsartanwahl
mode selector Betriebsartenschalter
mode switch Betriebsartenschalter
model Ausführung * *Modell* || Form || ideal * *(auch Substantiv: Vorbild)* || Konzept * *Muster, Vorbild, Vorlage, Modell, Bauweise* || Modell * *auch mathematisches ~, Simulatons~ usw.* || Muster * *Vorbild, (verkleinerte) Nachbildung, Arbeitsmodell, Vorlage, Bauweise, Type* || mustergültig * *als Beispiel/ Muster dienend* || nachbilden * *durch ein Modell ~ (auch ein mathematisches)* || Schablone * *Modell, Muster* || Schema || Typ * *einer Produktreihe*
model description Modellbeschreibung
model line Modellreihe
model parameters Modellparameter * *für Rechenmodell*
modeling Modellbildung
modeller Konfigurator
modelling Nachbildung * *durch ein (mathemat.) Modell*
modem Modem * *Abk. f. 'Modulator/Demodulator': Gerät z.Übertragg. digitaler Nachrichten über Telefonleitg.*
moderate mäßig * *gemäßigt; in: in*
moderate price Preiswürdigkeit
moderation Maß * *Mäßigung*
modern fortschrittlich * *modern* || heutig * *modern* || modern || neu * *modern* || neuzeitlich || zeitgemäß * *modern*
modernisation Modernisierung
modernization Modernisierung * *(amerikan.)*
modernize modernisieren
modest anspruchslos * *auch 'bescheiden'*
modification Änderung || Modifikation || Überarbeitung * *Modifizierung (to/of:von)* || Umarbeitung * *Modifikation*
modification of the Um-
modification of the cabling Umverdrahtung
modifications Umbau * *(verbessernde) Änderungen z.B. an einem Gerät*
modified geändert * *modifiziert* || verändert * *modifiziert*
modified bearing life Modifizierte Lagerlebensdauer * *Wälzlagerlebensdauer bei bestimmten Werkstoff. u. Betriebsbedingungen*
modify abändern || ändern * *modifizieren* || korrigieren * *modifizieren* || überarbeiten * *modifizieren* || umarbeiten * *modifizieren* || umbauen * *modifizieren, ändern, verbessern* || umgestalten || verändern || verbessern * *umgestalten*
modify the cabling umverdrahten
modular Baustein- || modular
modular assembly Baustein * *Baugruppe in Modulbauweise*
modular assembly system Bausteinsystem * *modularer Aufbau bei Geräten, Maschinen usw.*
modular concept Modularität
modular construction modulare Bauweise
modular converter system modulares Umrichtersystem
modular design Baukastenprinzip * *modular entworfen/konstruiert* || Baukastensystem * *modular konstruiert/entworfen* || modulare Bauweise || modularer Aufbau || Modularität || Modularkonzept

modular layout modulare Bauweise
modular nature of the design Modularkonzept
modular packaging system Einbausystem
modular phase assembly Phasenbaustein
modular principle Baukastenprinzip || Modularität
modular system Baukastensystem || Bausteinsystem
modular terminal Anreihklemme || Reihenklemme
modular terminal block Anreihklemmenblock || Reihenklemme || Reihenklemmenblock
modular-assembly system Baukastensystem * einer Maschine, einer Konstruktion usw.
modularity Modularität
modulate modulieren
modulate upon aufmodulieren
modulation Aussteuerung * in der Mess-/Nachrichtentechnik usw. || Modulation
modulation frequency Modulationsfrequenz
modulation method Modulationsverfahren
modulation mode Modulationsart
modulator Steuersatz * (Pulsweiten-) Modulator bei Pulsumrichter
module Baugruppe || Baustein * Hardware/Software-Baugruppe || Einheit * Bau~, Baustein, Modul || Modul
module call Bausteinaufruf
module casing Kapsel * Baugruppengehäuse
module holder Kapsel * Baugruppenhalter
module library Modulbibliothek
module location Einbauplatz * für Baugruppe
module parameter diagram Modulparameterplan * (SIEMENS SIMADYN D (R))
module plug-in location Steckplatz
module width Teilung * Breiten-Raster von Einschubmodulen
modulus of elasticity Elastizitätskoeffizient || Elastizitätsmodul
moist feucht * with: von || nass * feucht
moist heat feuchte Wärme
moisten anfeuchten || befeuchten
moistening Befeuchtung
moistening unit Befeuchtungsgerät
moisture Feuchte || Feuchtigkeit
moisture condensation Betauung * Flüssigkeitskondensation
moisture condensation non-permissible Betauung nicht zulässig * siehe auch →keine Betauung || keine Betauung * Betauung nicht zulässig
moisture conditions Feuchtebeanspruchung * Betriebsbeanspruchung; high-: hohe
moisture content Feuchte * Feuchtegehalt (z.B. von Papier) || Feuchtegehalt
moisture control Feuchteregelung * z.B. bei Papiermaschine
moisture measuring instrument Feuchtigkeitsmessgerät
moisturization Befeuchtung
mold Form * (amerikan.) Guss-/Press~ || Gießen * (Verb; amerikan.) spritzgießen (Kunststoff, Glas usw.) || Schimmel
molding Formpressen * (amerikan.)
molding compound Pressmasse || Spritzgussmasse
moldy schimmelig
molest belästigen
molestation Belästigung
molten geschmolzen
molten mass Schmelze
molten material Schmelze
molybdenum Molybdän
molybdenum disulphide Molybdän-Disulfid * MoS2 (Schmiermittel, z.B. →Molykote)
Molycote Molykote
moment Augenblick * at the moment: im ~ || Moment * Augenblick || Zeitpunkt

moment of disengagement Ausrückmoment * einer (Sicherheits-) Kupplung
moment of inertia Massenträgheitsmoment * J[kg m²] Summe d.Produkte d.Einzelmassen u.d.Quadrat d.Entferng. z.Drehachse || Schwungmoment * Massenträgheitsmoment GD^2 in [kpm²]; SI-Einheit: J in [kgm²] || Trägheitsmoment
moment of inertia compensation Schwungmomentenkompensation || Schwungmomentkompensation
moment of momentum Drall
moment of sampling Abtastzeitpunkt
moment of truth Stunde der Wahrheit
momentarily momentan * (Adv.) für e. Augenblick, kurz, vorübergehend, flüchtig
momentary aktuell * momentan, augenblicklich || augenblicklich * vorübergehend || im Moment * momentan || kurzzeitig * vorübergehend || momentan * (Adj.) momentan, augenblicklich, vorübergehend, flüchtig, jed.Augenblick geschehend/mögl. || momentan * (Adj.) vorübergehend, flüchtig || nichtspeichernd * Signal ohne Selbsthaltung bei Relaissteuerung, Kontaktplan usw.
momentary impulse Wischimpuls
momentary power loss kurzzeitiger Netzausfall
momentary pulse Wischer || Wischimpuls
momentary switch Taster
momentary-contact rotary switch Schwenktaster
momentary-contact switch Taster
momentary-off type switch Taster * mit Öffnerkontakt
momentary-on type switch Taster * mit Schließerkontakt
momentous wichtig * sehr wichtig
momentum Moment * physikal. Impuls, "Moment einer Kraft", Triebkraft, Stosskraft, Wucht, Schwung, Fahrt || Schwungkraft * (auch figürl.) siehe auch →Schwung
monies Gelder * grant monies: ~ bewilligen
monitor -Wächter * Überwachungseinrichtung || beobachten * zur Überwachung ~ || Bildschirm || Meldegerät || Melder * Wächter || Monitor * Bildschirm, Überwachungseinrichtung || Überwachung * auf richtige Funktion ~ durch "Mithöhren" || Überwachung * als Vorrichtung || Überwachungseinrichtung || Wächter * Überwachungseinrichtung || Zustandsanzeige
monitor module Überwachung * als Baugruppe/Gerät
monitor program Monitorprogramm * Hilfsprogramm zum Testen von Anwenderprogrammen
monitor screen Bildschirm
monitoring beobachten * (Substantiv) || Beobachtung || Überwachung
monitoring circuit Überwachungsschaltung
monitoring device Kontrolleinrichtung
monitoring equipment Mithöranlage || Überwachungsgerät
monitoring function Beobachtungsfunktionen || Überwachungsfunktion
monitoring module Überwachungsbaugruppe
monitoring parameter Anzeigeparameter * Beobachtungsparameter || Beobachtungsparameter
monitoring program Monitorprogramm
monitoring relay Überwachungsrelais
monitoring signal Überwachungssignal
monitoring system Überwachungssystem
monitoring time Überwachungszeit * für Zeitüberwachung
mono- Ein- || Einzel-
monochrome monitor Schwarz-Weiß-Monitor || Schwarzweiß-Bildschirm
monoflop Impulsbildner * Zeitstufe || Zeitglied * Impulsbildner, Zeitstufe || Zeitstufe * monostable multivibrator: monostabiler Multivibrator

monopolize reißen * beherrschen, an sich ~, z.B. Gespräch, Markt
monopolize the market Markt an sich reißen || Markt beherrschen
monopolizing the market marktbeherrschend
monopoly Marktbeherrschung * *Monopol* || Monopol
monostable monostabil
monoticity Monotonität
monotone monoton * *stetig fallend oder steigend (ohne Wendepunkte)*
monotonic monoton
monotonous gleichförmig * *monoton (auch steigend/fallend), eintönig* || monoton
monotonousness Monotonität
monotony Monotonität * *stetiges Steigen oder Fallen ohne Wendepunkte*
monthly report Monatsbericht
monthly statement Monatsbericht
MOP Motorpotentiometer * *Abkürzg. für 'motor operated potentiometer'* || Motorpoti * *Abkürzung für 'MOtorized POtentiometer'* || Schwabbelscheibe
mordant Beize * *{Chemie}* || beizen * *{Textil}*
more mehr * *the more the better:je mehr desto besser; more than:* ~ *als; over/upwards of:* ~ *als (b.Zahlen)* || neu * weitere || weitere * *mehr*
more detailed information genauere Angaben
more detailed genauer * *(Komparativ) detaillierter, mehr ins Einzelne gehend* || näher * *detaillierter*
more detailed data genauere Angaben
more distant weiter * *entfernter*
more dynamically verstärkt * *(Adv.) energischer, tatkräftiger*
more energetically verstärkt * *(Adv.) energischer*
more extensive weitergehend * *umfassender*
more favourably priced preisgünstiger
more intensely verstärkt * *(Adv.)*
more intensive weitergehend * *stärker, heftiger, intensiver, verstärkt*
more precise näher * *genauer*
more quickly schneller * *(Adv.)*
more rapid schneller
more rarely seltener * *(Adv.)*
more recent neuer * *(Komparativ) z.B. Software-version*
more significant höherwertig * *Ziffer, Datenbit usw.*
more specific näher * *mehr ins Einzelne gehend, spezieller, genauer*
more strongly verstärkt * *(Adv.)*
more than über * *mehr als*
moreover außerdem * *außerdem, überdies, ferner* || daneben * *außerdem* || darüber hinaus || weiterhin * *ferner*
mortise fräsen * *{Holz} (brit.)* ~ *von Fugen/Nuten* || Fuge * *Falz, Fuge; Zapfenloch*
mortiser Stemmmaschine * *{Holz} (brit.)*
mortize fräsen * *{Holz} (amerikan.)* ~ *von Fugen/Nuten*
mortize-cutting machine Schlitzmaschine * *{Holz}*
mortizer Fräsmaschine * *{Holz} Fugen-/Nut-~* || Stemmmaschine * *{Holz} (amerikan.)*
mortizing machine Langlochbohrmaschine * *{Holz}* || Stemmmaschine * *{Holz} mit Meißeln arbeitend*
MOS MOS * *Abk. f. 'Metal-Oxyde Semiconductor'*
MOS transistor MOS-Transistor
MoS2 Molykote * →*Molybden-Disulfid (Schmiermittel)*
mosaic panel Mosaikwarte
mosaic-type control board Mosaikwarte * *Bedien-/Beobachtungssystem in der Verfahrenstechnik*

most meist * *the most: am ~en (Adv.)* || meistens * *am meisten*
most certainly ganz * *(Adv.)* ~ *gewiss*
most common am gebräuchlichsten
most commonly encountered gebräuchlichst * *am häufigsten anzutreffen*
most commonly used am häufigsten verwendet
most current neuest * *am aktuellsten, höchst aktuell*
most favourable optimal * *günstigst*
most important hauptsächlich || wichtigst
most important point wesentlicher Punkt * *wichtigster Punkt*
most important thing Hauptsache
most likely wahrscheinlichst
most lowest price preisgünstigst
most of meist * *most of the time: die ~e Zeit*
most of all meistens * *am meisten*
most of it meist * *das ~e*
most popular am häufigsten verwendet
most recent neuest * *jüngst, am kürzesten zurückliegend, frischest*
most recently zuletzt * *(Adv.) neuerdings, zeitlich am wenigsten zurückliegend, jüngst, frischest, neuest*
most significant bit höchstwertiges Bit
most significant digit höchstwertiges Zeichen * *einer Ziffernanzeige*
most widely used hauptsächlich verwendet
mostly bevorzugt * *(Adv.) hauptsächlich* || größtenteils || meistens * *meistenteils, meistens* || vorwiegend * *(Adv.) meist(ens)* || vorzugsweise || zum größten Teil
motherboard Grundleiterplatte || Hauptplatine * *trägt Zusatzbaugruppen ,z.B. bei PC* || Trägermodul * *Trägerleiterkarte, Hauptplatine*
motif Gegenstand * *(Leit)Motiv (auch künstlerisches), Leitgedanke,*
motion Bewegung || Fahrt * *Bewegung* || Gang * *Bewegung, set in motion: in* ~ *setzen/bringen* || Lauf * *Bewegung*
motion control Bewegungssteuerung
motion controller Bewegungssteuerung * *als Einrichtung/Gerät* || Lageregler * *Bahnkurvenregler* || Positionierregler
motion mechanism Antriebsmechanik * *Antriebsmechanismus*
motion process Bewegungsablauf
motional action Bewegungsablauf
motional process Bewegungsablauf
motivate motivieren
motivation Ansporn
motive Antrieb * *Beweggrund* || Gesichtspunkt * *Beweggrund* || Grund * *Beweg-* || Moment * *Beweggrund* || Ursache * *Beweggrund*
motive force Antriebskraft
motive power Antrieb * *Antriebskraft* || Antriebsenergie || Antriebskraft || Antriebsleistung
motor Motor * *(elektr.)*
motor actuator Stellantrieb * *motorisch betriebener Stellantrieb* || Stellmotor
motor attachments Motoranbauten * *z.B. Tacho, Bremse, Lüfter, Kupplung, Schutzdach*
motor back EMF EMK
motor base Motorsockel
motor blower Motorlüfter
motor braking heat Bremswärme im Motor
motor building Motorenbau
motor capacitor Motorkondensator * *z.B. für Einphasen-Wechselstrommotor*
motor capacity Motorleistung
motor carriage Triebwagen * *Straßenbahn~*
motor catalog Motorkatalog
motor circuit Motorkreis
motor circuit-breaker Motorschutzschalter * *siehe*

z.B. DIN VDE 0660 || Schutzschalter
* →*Motorschutzschalter*
motor commutated   maschinengeführt * *Stromrichter*
motor commutation   Maschinenführung * *Motortaktung eines Stromrichters (Fremdführung durch d. Motor)* || Maschinenkommutierung * *Lastkommutierg.e. (Thyristor)Stromrichters durch d.Läuferlage b.Synchronmotor*
motor connection   Motoranschluss
motor control   Motorregelung
motor control unit   Motorsteuergerät * *z.B. Sanftanlaufgerät*
motor controller   Antriebsstromrichter || Motorsteuergerät * *z.b. Sanftanlaufgerät*
motor current   Maschinenstrom * *Motorstrom*
motor current rating   Motornennstrom
motor data   Motordaten
motor direction   Drehrichtung * *eines Motors*
motor driven potentiometer   Motorpotentiometer
motor efficiency   Motor-Wirkungsgrad
motor fan   Motorlüfter
motor feeder   Motorabzweig || Motorzuleitung
motor feeder cable   Motoranschlusskabel || Motorkabel * *Anschlusskabel* || Motorleitung || Motorzuleitung
motor flux   Maschinenfluss * *bei Motor*
motor for operation with converter   Motor für Betrieb am Umrichter
motor frame   Achshöhe * *Motor-Baugröße* || Baugröße * *für Elektromotor (Achshöhe)* || BG * *Motor-Baugröße, Achshöhe* || Motorgehäuse
motor frame size   Motor-Baugröße || Motorgröße * *Baugröße, entspr. der Achshöhe*
motor housing   Motorgehäuse
motor identification   Motoridentifikation * *durch Selbstoptimierungsvorgang*
motor identification routine   Motoridentifikation * *automat. Ermittlung der Motor-Kenngrößen durch Antriebs-Stromrichter*
motor identification run   Motoridentifikation * *automat. Aufnahme der Motor-Kenngrößen durch Stromrichter*
motor in hazardous locations   EExe Motor
motor lamination   Motorbleche
motor laminations   Motorenbleche
motor lead   Motorleitung * *Zuleitung zum Motor* || Motorzuleitung
motor model   Maschinenmodell * *softwaremäßige Nachbildung eines Motors*
motor monitoring   Motorüberwachung
motor mount   Motoranbau * *Anbau/Montage (von Impulsgeber, Bremse usw.) an Motor*
motor noise   Motorgeräusch * *bei umrichtergespeistem AC-Käfigl.motor: siehe VDE 0530 Teil 1, Beiblatt 2*
motor nominal current   Motornennstrom
motor operated potentiometer   Motorpotentiometer
motor operating capacitor   Betriebskondensator
motor operation   Motorbetrieb || motorischer Betrieb
motor output   Motorleistung
motor output rating   Motornennleistung
motor power   Motorleistung
motor protecting relay   Motorschutzrelais
motor protecting switch   Motorschutzschalter * *(CEE 19)*
motor protection   Motorschutz * *Schutzeinrichtg. gegen thermische Überlastung e.Motors*
motor protection circuit breaker   Motorschutzschalter
motor protection exclusively by PTC thermistors   Alleinschutz * *{el. Masch.}*

motor protection relay   Motorschutzrelais
motor protective relay   Motorschutzrelais
motor protector   Motorschutzschalter
motor PTC resistor   Motorkaltleiter
motor PTC thermistor   Motorkaltleiter
motor rating   Motorbemessung * *{el.Masch.}* *elektromagnetische, thermische und mechanische Auslegung eines Motors* || Motorleistung || Motornennleistung
motor rpm   Motordrehzahl
motor shaft   Antriebswelle * *Motorwelle* || Motorwelle
motor side   motorseitig
motor size   Baugröße * *{el.Masch.} eines Motors (Achshöhe plus Buchstabe für Gehäuselänge; IEC72-1/2)* || Motor-Baugröße
motor specifications   Motordaten
motor speed   Motordrehzahl
motor speed meter   Motordrehzahl-Messinstrument
motor speedometer   Motordrehzahl-Messinstrument
motor starter   Anlasser || Motoranlasser || Motorstarter
motor starting-heat   Anlaufwärme im Motor * *{el.Masch.}*
motor supply cable   Motoranschluss * *Anschlusskabel* || Motoranschlusskabel || Motorkabel * *Anschlusskabel* || Motorleitung || Motorzuleitung
motor supply lead   Motoranschluss * *Leitung* || Motorzuleitung
motor thruster   Motordrücker * *Betätigg.selement f.Bremse/Kupplung/Schaltgetriebe m. groß. Stellkraft u.klein.Hub*
motor thrustor   Motordrücker
motor torque   Motormoment * *das vom Motor an der Welle abgegebene Drehmoment*
motor utilization   Motorausnutzung
motor vehicle   Kraftfahrzeug
motor winding   Motorwicklung
motor with brake   Bremsmotor
motor with thermally critical rotor   läuferkritischer Motor * *Läufer erreicht Grenztemperatur eher als Ständer*
motor with transversal flux   Transversalflussmotor
motor-back EMF   Gegen-EMK || Gegenspannung * *Gegen-EMK eines Motors*
motor-commutated   maschinengetaktet * *Motor gibt die Kommutierungszeitpunkte vor (→lastgeführter Stromrichter)*
motor-compressor   Motorverdichter * *Kompressor mit eingebautem Motor; hermetic: in e. gemeinsamen gasdichten Gehäuse*
motor-fixed cable   frei herausgeführtes Kabel * *{el. Masch.} aus d.Motor herausgeführte Wicklgs.-enden; kein Klemmenkasten*
motor-integrated inverter   Umrichter im Motor
motor-operated   elektromotorisch betrieben
motor-run capacitor   Betriebskondensator
motor-side   maschinenseitig * *motorseitig (z.B. ~er Stromrichter)* || motorseitig
motoring   motorisch
motoring operation   motorischer Betrieb
motorized   motorisch angetrieben
motorized impeller   Ventilator
motorized potentiometer   Motorpotentiometer || Motorpoti || Sollwerteinsteller * *motorischer* || Sollwertsteller * *motorischer*
motorized setpoint potentiometer   motorisches Sollwertpotentiometer
Motorola format   Motorola-Format * *Datenformat; höherwertig.Byte/Wort auf niedriger, niederwertig.auf höherer Adresse*
motorpot   Motorpotentiometer || Motorpoti

motorpotentiometer Motorpotentiometer
motorpower Motorleistung
motorpulley Trommelmotor * *Außenläufer-Motor z.B. für Bandförderer*
motto Motto * *Motto, Wahlspruch, Sinnspruch*
mould Form * *Guss~, Press~* ‖ fräsen * *[Holz]* ~ *von Formteilen/Kehlen/Leisten usw.* ‖ Gießen * *(Verb) spritzgießen (Kunststoff, Glas usw.)* ‖ Modell * Form ‖ profilieren * *[Holz]* ‖ Schimmel ‖ Werkzeug * *(Guss)Form*
mould press Formpresse * *[Holz] Formteilpresse*
mouldable verformbar * *(Kunststoff)*
moulded case Isolierstoffgehäuse ‖ Kompakt- * *im geschlossenen Gehäuse (Leistungsschalter usw.)*
moulded part Formteil * *(Kunststoff, Metall usw.)*
moulded-case gekapselt
moulded-case circuit braker Kompaktleistungsschalter * *in Isolierstoffgehäuse*
moulded-case circuit breaker Kompakt-Leistungsschalter * *mit Isolierstoffgehäuse*
moulder Fräsmaschine * *[Holz]* ‖ Kehlmaschine * *[Holz]*
moulding Formpressen * *(brit.)*
moulding compound Vergussmasse
moulding cutter Fräser * *[Holz]*
moulding machine Fräsmaschine * *[Holz] Fräsmaschine, Kehl(hobel-/fräs-)maschine* ‖ Kehlmaschine * *[Holz] zum Fräsen von Kehlen*
moulding resin Gießharz
moulding unit Fräseinheit * *[Holz] zum Profilieren*
mouldy schimmelig
mount anbringen * *montieren, befestigen (at: an/auf)* ‖ ansteigen * zunehmen ‖ aufbauen * errichten, aufstellen, montieren, anbringen, einbauen ‖ aufziehen ‖ befestigen ‖ bestücken ‖ einbauen * montieren ‖ Halterung ‖ montieren * *bestücken, anbringen* ‖ steigen ‖ Träger
mount in bearings lagern * *in Wälz/Gleitlager* ~
mount on befestigen
mount side by side aneinander reihen * *aneinander stoßend befestigen/montieren*
mountable montierbar
mountable side by side anreihbar
mounted angebracht * *montiert, befestigt (siehe auch →anbringen)* ‖ aufgebaut * *montiert* ‖ befestigt ‖ bestückt ‖ eingebaut ‖ montiert
mounted equipment Anbaugeräte
mounted on roller bearings rollengelagert
mounted underneath untenliegend * *unterhalb montiert*
mountes accessories Anbaugeräte * *z.B. Tacho, Bremse, Kupplung usw. an einem Motor*
mounting Anbau ‖ Aufbau * *Montage* ‖ Aufstellung * *Montage* ‖ Befestigung * *Montage* ‖ Befestigungs-* ‖ Bestückung ‖ Einbau * *Anbau, Befestigung, Bestückung* ‖ Lagerung * *Befestigung* ‖ Montage * *Anbringung, Bestückung, Einbau, Befestigung auf e.Unterlage, auch Leiterplattenbestückung*
mounting accessories Befestigungselemente * *als Zubehörsatz ("Satz loser Teile")* ‖ Befestigungsmaterial * *Montagezubehör*
mounting arrangements Montagemaße
mounting bolt Montagebolzen
mounting bore Anschraubbohrung
mounting bracket Halterung ‖ Montagebügel
mounting cutout Montageausschnitt
mounting depth Einbautiefe
mounting dimension Einbaumaß
mounting dimensions Befestigungsmaße * *für Trafos/Drosseln: siehe EN 60852-4* ‖ Montagemaße
mounting documentation Montageschriften
mounting element Befestigungselement

mounting elements Befestigungselemente
mounting feet Montagefüße
mounting flange Anbauflansch ‖ Befestigungsflansch
mounting frame Montagerahmen
mounting guidelines Aufbaurichtlinien
mounting hardware Befestigungsteile
mounting hole Befestigungsbohrung * *tapped mounting hole: ~ mit Innengewinde zur Schraubbefestigung* ‖ Montagebohrung ‖ Montagelochung
mounting instructions Einbauhinweise
mounting kit Anbausatz ‖ Einbausatz
mounting location Aufstellungsort ‖ Einbauort ‖ Einbauplatz
mounting lug Befestigungslasche
mounting material Befestigungsmaterial
mounting of components Bauelement-Bestückung ‖ Bauelemente-Bestückung
mounting opening Montageöffnung
mounting plate Einbauplatte * *Montageplatte* ‖ Montageplatte ‖ Tragblech
mounting position Bestückungsplatz ‖ Einbaulage ‖ Einbauplatz
mounting rack Einbaurahmen * *z.B. für Baugruppen*
mounting rail Befestigungsschiene * *Montageschiene* ‖ Halteschiene * *z.B. Installations-/Aufschnappschiene* ‖ Hutschiene * *Montageschiene zum Aufschnappen von Geräten* ‖ Installationsschiene ‖ Montageschiene * *z.B. Hutschiene* ‖ Profilschiene * *(Hutprofil-) Montageschiene, siehe auch →Installationsschiene* ‖ Tragschiene * *siehe auch →Installationsschiene*
mounting screw Befestigungsschraube
mounting set Einbausatz
mounting side by side Anreihmontage
mounting specification Einbauvorschrift
mounting strap Befestigungslasche
mounting style Befestigungsart
mounting support Halterung
mounting surface Einbaufläche ‖ Montagefläche
mounting technology Aufbautechnik
mounting tolerance Einbautoleranz
mounting version Einbauvariante
mounting width Einbaubreite
mountings Armatur
mouse Maus * *auch Computer~*
mouse click Knopfdruck * *at a mouse click: mit einem Mausklick*
mouse driver Maustreiber
mouse key Maustaste * *lefthand: linke; righthand: rechte*
mouse-click Mausklick
mouth Öffnung * *Mündung (einer Flasche, eines Tunnels usw.)*
MOV MOV * *Abkürzg. f. 'Metal-Oxide Varistor': Überspannungsschutzelement* ‖ Varistor * *Abkürzung für 'Metal-Oxide Varistor':Metalloxid-Varistor*
MOV surge suppressor Metalloxidableiter * *Abk. f. 'Metal Oxide Varistor'*
movable beweglich ‖ gängig * *gangbar, bewegbar verschiebbar, verstellbar* ‖ verschiebbar
movable property Betriebsvermögen * *bewegliches Vermögen (Maschinen, Geschäftsausstattung usw.)* ‖ bewegliches Vermögen * *Maschinen, Geschäftsausstattung usw.*
movables Betriebsvermögen * *bewegliches Vermögen (Maschinen, Geschäftsausstattung usw.)* ‖ bewegliches Vermögen * *Maschinen, Geschäftsausstattung usw.*
move bewegen * *auch sich; freely:frei; to and fro:- hin und her* ‖ Bewegung * *mit einer bestimmten Absicht* ‖ laufen * *sich bewegen* ‖ Schritt

* *(Schach-) Zug, Schritt, Maßnahme*; make the first move: den ersten ~ machen || Übergang * *Aufbruch, (Fort-) Bewegung, Schritt, Entwicklung, Maßnahme, Fortschreiten* || umziehen * *to:nach* || Umzug || verlegen * *bewegen, versetzen* || verschieben
move forward vorwärtsbewegen
move in a circle kreisen
move in reverse direction zurücklaufen
move into beziehen * *Gebäude, Wohnung, Stellung usw.* ~
move off abziehen
move time Verfahrzeit * *bei Positioniersteuerung*
move up aufrücken * →*aufschließen (to/with: zu)* || aufschließen * *zur Spitzengruppe* ~ *(to/with: zu)*
move upward steigen
moveable verschiebbar
moved away verzogen * *umgezogen*
movement Bewegung || Förderung * *Bewegung, z.B. von Luft durch e. Ventilator* || Gang * *Bewegung* || Handgriff * *Bedienungsgriff, Bewegung* || Lauf * *Bewegung* || Verlagerung * *Bewegung* || Werk * *z.B. ~ einer Uhr*
moving beweglich || verfahrbar
moving belt Förderband
moving column Fahrständer * *{Werkz.masch.}*
moving force treibende Kraft
moving in a straight line geradlinige Bewegung
moving parts bewegliche Teile
moving staircase Rolltreppe
moving target bewegtes Ziel
moving-coil instrument Drehspulinstrument
moving-iron type instrument Dreheiseninstrument
MOX arrester Metalloxid-Ableiter
MPP MPP * *Abkürzung für 'Maximum Power Point':Betriebspunkt maximaler Leistung bei Solaranlage* || Punkt der maximalen Leistung * *Maximum Power Point, z.B. bei Solar/Windkraftwerk*
MPP tracker MPP-Regler
MPPT MPP-Regler * *Abk. f.'Maximum Power Point Tracker':regelt auf d.Punkt max. Leistung b. Solaranlage*
MSB höchstwertiges Bit * *Most Significant Bit* || MSB * *Abkürzg.für 'Most Significant Bit':höchstwertiges Bit*
MSD höchstwertiges Zeichen * *Most Significant Digit; einer Ziffernanzeige* || MSD * *Abkürzg. für 'Most Significant Digit': Höchstwertiges Zeichen z.B. einer Ziffernanzeige*
MTBF MTBF * *MeanTime Between Failure:mittlere Ausfallzeit, Maß für die Zuverlässigkeit einer Einrichtung*
much höchst * *sehr* || viel * *much better: ~ besser; too much: zu ~*
mud Schlamm || Schmutz * *Schlamm*
mud pump Spülpumpe * *zum Ausspülen von (Öl-) Bohrlöchern*
muddiness Trübung
muddle durcheinander * *(Subst.) ~, Unordnung, Wirrwar, Unklarheit; make a muddle of sth.: ~bringen*
muddled unklar * *verworren*
muddy schmutzig * *schlammig* || trüb * *schlammig, trüb(e) (auch Licht usw.),* || unklar * *trüb*
muff Flansch * *Muffe, Stutzen, Flanschstück* || Muffe || Stutzen * *auch* →*Muffe,* →*Flanschstück*
muff coupling Schalenkupplung
muffle dämpfen * *(akustisch) Ton* ~
muffler Auspuff * →*Schalldämpfer,* →*Auspufftopf* || Auspufftopf * *Schalldämpfer* || Geräuschdämpfer || Schalldämpfer * *z.B. für Verbrennungsmotor, Kompressor usw.*
muffling Dämpfung * *Geräusche, Ton (siehe auch 'dämpfen')* || Schalldämpfung * *(amerikan.)*

mug Becher
multi Mehr-
multi drive Mehrmotorenantrieb
multi- Mehrfach- || Universal-
multi-axis Mehrachs-
multi-axis application Mehrachsanwendung
multi-axis control Mehrachssteuerung
multi-axis drive Mehrachs-Antrieb || Mehrachsantrieb
multi-axis speed controller Mehrachsendrehzahlregler
multi-blade frame saw Vollgattersäge * *{Holz}*
multi-buffer Wechselpuffer
multi-channel mehrkanalig
multi-channel control mehrkanalige Steuerung
multi-computing Parallelverarbeitung * *einer Aufgabe auf mehreren Rechnern gleichzeitig*
multi-conductor mehrdrähtig
multi-core mehradrig || mehrdrähtig || vieladrig
multi-cycle control Impulspaketsteuerung || Schwingungspaketsteuerung * *AC-Steller-Steuerverfahren m.Durchschaltg. ganz. Netzvollschwing.pakete* || Vollschwingungssteuerung * *bei Thyristorsteller: es wird e.ganze Anz. v.Netzschwingungen durchgelass.* || Vollwellensteuerung * *AC-Steller-Steuerverfahren mit Durchschaltung ganzer Netzvollschwingungspakete*
multi-disc rotor Mehrscheibenläufer
multi-drop busfähig * *Kommunikationsverbindung* || busförmig || busgekoppelt
multi-drop capability Vernetzbarkeit * *z.b. über Bussystem*
multi-drop communication busförmige Kommunikation
multi-drop connection Bus
multi-drop interface busfähige Schnittstelle
multi-faceted vielschichtig * *facettenreich*
multi-fingered mehrteilig * *z.B. Kohlebürsten*
multi-function Multifunktions-
multi-head Mehrspindel- * *z.B. Fräsmaschine*
multi-layered mehrlagig
multi-level mehrstufig * *z.B. Prozess, Vorgehen, Verfahren usw.*
multi-lingual mehrsprachig
multi-master capability Multimasterfähigkeit * *z.B. Bussystem*
multi-master capable multimasterfähig
multi-master operation Multimasterbetrieb * *bei Bussystem*
multi-microprocessor system Multi-Microcomputersystem
multi-motor drive Gruppenantrieb * *Antrieb mit mehreren Motoren z.B.an einem Stromrichter* || Mehrmotorenantrieb * *auch Betrieb von mehreren Motoren an einem Umrichter*
multi-motor drives Mehrmotorenantriebe
multi-motor operation Mehrmotorenbetrieb
multi-motor system Mehrmotorensystem
multi-motor-drive converter Gruppenumrichter * *Umrichter, der mehrere parallelgeschaltete Motoren speist*
multi-phase mehrphasig
multi-phase system Mehrphasensystem
multi-plate brake Lamellenbremse
multi-plate clutch Lamellenkupplung
multi-point connector Anschlussverteiler || Rangierverteiler * *Anschlussverteiler, Klemmenvervielfacher*
multi-pole mehrpolig
multi-pole switching Vielfachpolumschaltung
multi-purpose Mehrfunktions- * *Mehrzweck-* || Multifunktions- * *Mehrzweck-* || universal * *Vielzweck-* || Universal-
multi-quadrant Mehrquadrant- || Mehrquadrantenantrieb
multi-quadrant drive Mehrquadrantenantrieb

multi-quadrant operation Mehrquadrantbetrieb || Mehrquadrantenbetrieb
multi-range Mehrbereichs- * *Spannung, Messbereich usw., auch -Messinstrument* || Universal- * *Mehrbereichs-, z.B. Messgerät*
multi-range instrument Multimeter || Vielfach-Messinstrument || Vielfachinstrument || Vielfachmessinstrument
multi-segmented mehrteilig * *aus mehreren Segmenten bestehend, z.B. Schleifbürsten*
multi-shaft drive Mehrachs-Antrieb || Mehrachsantrieb
multi-shift operation Mehrschichtbetrieb
multi-speed mehrstufig * *(Schaltgetriebe) mit mehreren Übersetzungen*
multi-speed gearbox Schaltgetriebe || Wechselgetriebe
multi-spindle Mehrspindel- * *z.B. Werkzeugmaschine*
multi-spindle lathe Mehrspindeldrehmaschine
multi-stage Mehrstufen- || mehrstufig
multi-stage star-delta starting mehrstufiger Stern-Dreieck-Anlauf
multi-step inverter Mehrpunktwechselrichter
multi-tasking Parallelverarbeitung * *in einem Rechner*
multi-tier mehrzeilig * *z.B. Baugruppenträger*
multi-voltage Motor spannungsumschaltbarer Motor * *{el.Masch.}*
multi-way drilling machine Bohrzentrum
multi-wire mehradrig
multiblock wire-drawing machine Mehrblock-Drahtziehmaschine * *{Draht}*
multicast Multicast- * *an mehrere (Bus)Teilnehmer adressiert (z.b. Telegramm)*
multicomputing Multicomputing * *Mehrprozessortechnik*
multicycle control Schwingungspaketsteuerung || Vollschwingungssteuerung || Vollwellensteuerung
multidisc brake Lamellenbremse
multidisc clutch Lamellenkupplung
multidrop busförmig || busförmig verbinden
multidrop communication Buskopplung
multifaceted facettenreich
multifacetted facettenreich
multifarious vielfältig * *mannigfaltig*
multifunction Multifunktions-
multifunctional Mehrfunktions- || Multifunktions-
multigrade Mehrbereichs- * *Schmieröl, Schmierfett usw.*
multigrade grease Mehrbereichsfett
multigrade oil Mehrbereichsöl
multilayer Mehrlagen- || mehrlagig || mehrschichtig || Vielschicht-
multilayer capacitor Schichtkondensator * *(Keramik-) Vielschichtkondensator* || Vielschichtkondensator * *siehe auch →Schichtkondensator*
multilayer ceramic capacitor Keramik-Vielschichtkondensator
multilayer PCB Mehrlagen-Leiterplatte || Mehrlagenleiterplatte || Multilayer-Leiterplatte
multilayer printed board Mehrlagenleiterplatte
multilayered mehrschichtig
multimeter Multimeter || Universal-Messgerät || Vielfach-Messgerät || Vielfach-Messinstrument || Vielfachinstrument || Vielfachmessinstrument
multipartite mehrteilig * *vielteilig, mehrteilig*
multiple Mehr- || mehrfach || Mehrfach- || Mehrfaches || Sammel- || vielfach * *(auch Substantiv: das Vielfache)* || Vielfaches * *as a multiple of/as multiples of: als ~ von*
multiple assigned mehrfach belegt * *z.B. Klemme*
multiple axis operation Mehrachsbetrieb
multiple block drawing machine Mehrblockziehmaschine * *{Draht} Drahtziehmaschine*

multiple branch Sprungleiste
multiple compressor set Verbund-Kompressorsatz
multiple language mehrsprachig
multiple motor drive Mehrmotorenantrieb
multiple quadrant drive Mehrquadrantenantrieb
multiple quadrant operation Mehrquadrantenbetrieb
multiple- Verbund- * *Mehrfach-*
multiple-choice wählbar
multiple-disc brake Lamellenbremse * *Reibungsbremse mit mehreren Scheiben (abwechselnd eine ruhend, eine drehend)*
multiple-disc clutch Lamellenkupplung
multiple-disk brake Lamellenbremse || Mehrscheibenbremse * *siehe auch →Lamellenbremse*
multiple-disk clutch Lamellenkupplung
multiple-speed gearbox Schaltgetriebe
multiple-spindle Mehrspindel-
multiple-star-delta starting Mehrfach-Stern-Dreieck-Anlauf
multiplex mehrfach
multiplex operation Multiplexbetrieb
multiplexer Multiplexer * *['maltiplexe]* || Umschalter
multiplication Multiplikation || Vervielfältigung * *(allg.)*
multiplicity Vielfalt * *Vielfalt, Vielzahl, Menge, Mehrfachheit, Mehrwertigkeit* || Vielzahl
multiplied by mal * *multipliziert mit*
multiplier Multiplizierer
multiply multiplizieren * *by/with:mit; together: miteinander* || vermehren * *an Zahl ~/vergrößern* || vervielfachen * *auch 'sich ~'*
multiplying multiplizierend
multiprocessor capability Multiprozessorfähigkeit
multiprocessor mode Multiprozessorbetrieb * *Betriebsart*
multiprocessor operation Multiprozessorbetrieb
multiprocessor system Mehrprozessorsystem
multipurpose vielseitig * *~ verwendbar*
multisdisk- Mehrscheiben-
multisectional mehrteilig
multispeed gearbox Schaltgetriebe
multispeed motor polumschaltbarer Motor
multstage mehrstufig
multitude Menge * *Vielzahl* || Vielfalt * *Mannigfaltigkeit, Vielzahl; a multitude of: eine Vielzahl von* || Vielzahl * *a multitude of: eine ~ von/an*
multiturn potentiometer Spindelpotentiometer
municipal authorities Stadtwerke
muriatic acid Salzsäure
mush winding Träufelwicklung * *Wicklungsart bei Motor: Drähte werden von Hand in die Nut "eingeträufelt"*
mushroom button with pushlock and twist-release Pilzrasttaster mit Drehentriegelung
mushroom pushbutton Pilztaster * *z.B. Not-Aus Rasttaster*
must Notwendigkeit || sollen * *müssen*
musty schimmelig
mutual beidseitig * *→untereinander, gegen-/wechsel-/beiderseitig, gemeinsam* || gegeneinander * *gegenseitig, wechselseitig, beiderseitig, gemeinsam* || gegenseitig * *gegen-/wechselseitig, gemeinsam* || gemeinsam * *gegenseitig* || wechselseitig
mutual effect gegenseitige Beeinflussung
mutual inductance Gegeninduktivität || Hauptinduktivität * *Gegeninduktivität*
mutuality Gemeinsamkeit
mutually untereinander * *gegen-/wechselseitig* || zueinander * *gegenseitig*
MUX Multiplexer * *Abk. f. 'MUltipleXer'*
MV Mittelspannung * *Abkürzung f. 'Medium Voltage'*

MV breaker Mittelspannungsschalter * *Abkürzg.f. 'Medium Voltage': Mittelspannungs-*
mysterious geheim * ~*nisvoll, mysteriös*

N

N Mittelleiter * *im Drehstromsystem* || Mittelpunktleiter || Mittelpunktsleiter || Nullleiter
N conductor N-Leiter
n-conductive n-leitend * *Halbleiterschicht*
N-end BS * *Abkürzg. für 'Non-Drive end': Nicht-Antriebsseite*
N-schaft end N-Wellenende * *Non-drive side, Nicht-Antriebsseite, A-Seite eines Motors*
N.A. entfällt * *Abkürzg. f. 'Not Applicable': 'nicht anwendbar', in Formularen, Tabellen, Listen*
NAFTA NAFTA * *Abkürzg.f. 'North American Free Trade Agreement': Wirtsch.gemeinsch. zw. USA, Kanada u.Mexiko*
nail Nagel
nailer Nagelgerät
naked blank * *bloß* || bloß * *nackt*
name benennen * *einen Namen geben* || Benennung * *Name* || Bezeichnung * *Name* || Name || nennen * *mit einem Namen/Ausdruck* ~
name plate Firmenschild * *an Maschine, Gerät usw.* || Leistungsschild * *Typenschild* || Typenschild
name-plate Firmenschild * *an Maschine, Gerät usw.*
named crwähnt * *named above: oben genannt*
namely nämlich
nameplate Typenschild
names Namen
naming Bezeichnung * *Namensgebung*
NAMUR NAMUR * *Normen-Arbeitsgemeinschaft für Mess- u.Regeltechnik i.d. chem.Industrie (deut.Normengremium)*
NAMUR recommendation NAMUR-Empfehlung
nap aufrauen * *[Textil] Textilstof, Tuch usw.* ~ || Flor * *[Textil] Gewebeflor in der Weberei* || Noppe || rauen * *noppen, rauhen, z.b. von Textilstoff* ~
narrow beengt * *eng, räumlich beschränkt, knapp* || begrenzt * *eingeschränkt, beschränkt (Mittel, Verhältnisse, Platz, Raum usw.)* || beschränken * *eingeschränkt* || beschränkt * *eingeengt* || eng || schmal || verengen
narrow down einschnüren * *einengen*
narrow space enger Raum
narrow-band schmalbandig * *z.B. Frequenzband, Filter, Bandsperre*
narrow-profile schmal
narrower schmaler * *(Komparativ)*
nascent entstehen * *anfänglich/erstmalig entstehen*
Nassi-Schneidermann chart Nassi-Schneidermann-Diagramm * *[Computer]* || Struktogramm * *Nassi-Schneidermann-Diagramm*
national national
national economy Volkswirtschaft * *Wirtschaft eines Landes*
national language Landessprache
native natürlich * *angeboren*
natural Eigen- * *-Frequenz, -Kapazität, -Schwingung usw.* || natürlich
natural air circulation freie Konvektion * *z.B. bei Luftselbstkühlung von Leistungshalbleitern*
natural air circulation cooling Luftselbstkühlung * *von Leistungshalbleitern*
natural air cooling Selbstbelüftung * *Luftselbstkühlung*

natural commutation natürliche Kommutierung
natural convection freie Konvektion || Selbstkühlung * *durch natürliche Konvektion (ohne Lüfter)*
natural cooling freie Kühlung * →*Konvektionskühlung* || Selbstkühlung * *ohne Lüfter*
natural fiber Naturfaser
natural fibers Naturfasern * *(amerikan.)*
natural fibres Naturfasern * *(brit.)*
natural firing point natürlicher Zündzeitpunkt
natural frequency Eigenfrequenz
natural gas Erdgas
natural number natürliche Zahl
natural philosophy Physik
natural rubber Naturkautschuk
natural trigger instant natürlicher Zündzeitpunkt
natural-air cooled luftselbstgekühlt * *bei Leistungshalbleitern. Kühlung über freie Konvektion, kein Lüfter*
natural-air cooling Luftselbstkühlung * *von Leistungshalbleitern* || Selbstkühlung * →*Luftselbstkühlung von Leistungshalbleitern nur mit Kühlkörper, ohne Lüfter*
natural-convection selbstbelüftet
natural-convection cooled luftselbstgekühlt
natural-convection cooling Luftselbstkühlung * *von Leistungshalbleitern*
naturally selbstverständlich * *(Adv.) natürlich*
nature Art * *Beschaffenheit* || Beschaffenheit * *Art, Sorte, Beschaffenheit; Naturt; Veranlagung, Charakter* || Eigenschaft * *Beschaffenheit*
nautical See- * *nautisch, seemännisch*
naval Marine- || See- * *Marine-*
naval forces Marine * *Seestreitkräfte*
naval ship Marineschiff
nave Nabe || Radnabe
navy Marine * *Kriegsmarine* || See- * *Kriegsmarine-*
NC nicht belegt * *Abkürzung für 'Not Connected': nicht angeschlossen (z.B. an Schaltkreis, Stecker)* || numerische Steuerung * *Abk. für 'Numeric Control'*
NC contact Öffner * *Abkürzg.für 'Normally Closed contact'* || Öffnerkontakt * *Abkürzung für 'Normally Closed contact'* || Ruhekontakt * *Abkürzung für 'Normally Closed contact'*
NC control NC-Steuerung * *Abk.f. 'Numeric Control': numerische Positioniersteuerg. (für Werkzeugmaschinen usw.)*
NC programming language NC-Sprache * *Programmiersprache für Bewegungssteuerungen, z.B. nach DIN 66025*
NC programming system NC-Programmiersystem
NC-controlled NC-gesteuert || numerisch gesteuert
ND end B-Seite * *Abk. für 'Non-Drive end'*
ND shaft end BS-Wellenende * *Abkürzg.f. 'Non-Drive': nicht antreibend*
NDE B-Seite * *Abkürzg. f. 'Non-Drive End':nondrive end* || BS * *Abkürzg. für 'Non Drive End':- Nicht-Antriebsseite eines Motors (B-Seite)* || Nicht-Antriebsseite * *[el.Masch.] Abk. für 'Non-Drive End'*
near bei * *(räumlich)* || dabei * *nahe* || nah || nähern * *nähern, sich* ~ || um * *zeitlich ungefähr*
near at hand nahe liegend * *in der Nähe verfügbar*
near by dabei * *nahebei*
near it daneben * *in der Nähe liegend*
near to neben * *dicht* ~
nearby nahe liegend * *in der Nähe*
nearer näher
nearest nächst * *nächstgelegen* || nächster * ~ *bezüglich Entfernung, Beziehung; am nächsten gelegen* || nächstgelegen
nearly gegen * *ungefähr* || nahezu * *beinahe*
nearly backlash-free spielarm * *nahezu spielfrei*

nearly maintenance-free anspruchslos * *wartungsarm* || problemlos * *wartungsarm* || wartungsarm * *nahezu wartungsfrei* || wartungsfreundlich * *nahezu wartungsfrei*
nearness Nähe
nearness to the market Marktnähe
neat aufgeräumt * *ordentlich, sauber, übersichtlich* || ordentlich || sauber * *ordentlich, aufgeräumt, klar, übersichtlich, geschickt (Arbeit, Äußeres, Handschrift usw.)*
neatness Reinheit
NEC NEC * *Abkürzg. für 'National Electrical Code': amerikanische Errichtungsbestimmungen*
necessarily notwendigerweise
necessary erforderlich * *notwendig* || nötig || notwendig * *necessitate:* ~ *machen* || zwangsläufig
necessary condition notwendige Bedingung
necessitate erfordern || nötig machen || notwendig machen
necessity Notwendigkeit
need benötigen || brauchen || Notwendigkeit
needed erforderlich * *benötigt* || nötig * *benötigt* || notwendig * *benötigt*
needful notwendig
needle Zeiger * *Nadel, z.B. Messnadel eines Messinstruments*
needle bearing Nadellager
needle cage Nadelkäfig * *bei Nadellager*
needle roller bearing Nadellager
needle-nosed pliers Spitzzange
needle-shaped nadelförmig
needling machine Nadelfliesmaschine * *{Textil}* || Vernadelungsmaschine * *{Textil}*
needs Anforderung * *Erfordernisse, Bedürfnisse*
negated invers * *logisch invertiert* || invertiert * *Binärsignal*
negated signal invertiertes Signal * *Binärsignal*
negative negativ
negative air pressure Unterdruck * *Luftdruck*
negative edge hintere Flanke * *negative Flanke* || negative Flanke
negative effects nachteilige Auswirkungen
negative feedback Gegenkopplung
negative logic negative Logik * *niedriger Spannungspegel entpr. logisch "1"*
negative phase-sequence component Gegenkomponente * *{el.Masch.}*
negative phase-sequence system Gegensystem * *{el. Masch.}*
negative pole Minuspol
negative pressure Unterdruck
negative sign Minuszeichen
negative terminal Minuspol * *Anschluss an Batterie, Kondensator*
negative-going edge fallende Flanke * *negative Flanke eines Impulses* || negative Flanke
negative-sequence component Gegenkomponente * *{el. Masch.} Anteil mit entgegengesetzter Phasenfolge e. 3-Phasensystems*
negative-sequence network Gegensystem * *{el. Masch.}*
negatively charged negativ geladen
neglect übergehen * *jemanden ~;* →*nicht beachten, außer Acht lassen, missachten* || unter den Tisch fallen lassen * →*vernachlässigen* || vernachlässigen * *nicht beachten, z.B. einens Nebeneffekt ~* || verwerfen * *übergehen, außer acht lassen, übersehen, missachten*
neglecting unter Vernachlässigung von
neglegted ungeordnet * *ungepflegt, vernachlässigt*
negligence Fahrlässigkeit * *gross/culpable:grobe*
negligent fahrlässig * *rechtlich relevant* || unachtsam * *nachlässig*
negligible geringfügig * *unwesentlich, nebensächlich, vernachlässigbar, unbedeutend* || unwesentlich * *vernachlässigbar* || vernachlässigbar
negligibly small verschwindend klein
negotation Besprechung * *Verhandlung*
negotiate vereinbaren * *vertraglich* ~ || verhandeln
negotiation Unterhandlung * *siehe auch* →*Verhandlung* || Verhandlung
negotiations Verhandlungen
neighboring Nachbar-
neighbourhood Nähe * *Umgebung, Nachbarschaft*
neighbouring anschließend * *räumlich benachbart* || benachbart || Nachbar-
neither auch nicht * *neither can I: ich kann es auch nicht*
neither nor weder noch
NEMA NEMA * *Abk.für 'National Electrical Manufacturers Association' Normengremium in den USA ähnlich VDE*
neon tube Röhre * *Neonröhre*
nerveless ruhig * *nervenstark*
nest Satz * *ineinanderpassende Gegenstände* || schachteln * *Unterprogramme, Interrupts usw.*
nested geschachtelt * *z.B. Unterprogramme, Interrupts usw.*
nesting Schachtelung * *z.B. von Unterprogrammen, Interrupts*
nesting depth Klammertiefe || Schachteltiefe * *z.B. von Unterprogrammen, Interrupts* || Schachtelungstiefe
nesting level Klammerebene || Schachteltiefe * *Schachtelungsebene* || Schachtelungstiefe
nesting stack pointer Klammerstackpointer
net netto || Netto- || Netz || rein * *Gewinn*
net cost price Selbstkostenpreis
net data Nettodaten * *z.B. Nutzdaten eines Telegramms (ohne Telegrammrahmen)* || Nutzdaten * *Nettodaten*
net data length Nutzdatenlänge
net price Nettopreis
net weight Nettogewicht
Netherlands Niederlande
netting Gewebe * *Netz*
netting machine Flechtmaschine * *z.B. zur Herstellung von Drahtgeflecht*
netting wire Geflechtdraht
network Geflecht * *(auch figürl.)* || koppeln * *über ein Netzwerk* ~ || Netz * *Kommunikations~, Vertriebs~, Service~; Stromversorgungs~; (public: öffentliches)* || Netzwerk || Schaltnetz || Verbund * *Netz(werk)* || vernetzen * *together: miteinander*
network analysis Netzwerkanalyse
network card Netzwerkkarte * *z.B. Netzwerks-Anschaltungsbaugruppe in einem PC*
network interconnection Netzverbund * *Datennetz*
network node Netzknoten * *im Nachrichtennetz* || Netzwerkknoten
network protective relaying Netzschutz
networkable busfähig * *netzwerkfähig, vernetzungsfähig* || netzwerkfähig
networked gekoppelt * *über ein Datennetz* ~ || im Verbund || vernetzt
networked connection Vernetzung
networked operation Vernetzung
networking Kopplung * *Vernetzung* || Vernetzung * *Kommunikation*
neural net neuronales Netz
neural network neuronales Netz * *d.menschl. Gehirn nachempfundene Computerarchitektur m.verteilten Knoten (Neuronen)*
neutral Leerlauf * *bei Schaltgetriebe* || Mittelleiter * *im Drehstromsystem* || Mittelpunktleiter || Mittelpunktsleiter || MP * *neutrale Phase* || N-Leiter || neutral || Neutralleiter || Nullleiter || Sternpunkt
neutral box Sternpunktkasten * *{el.Masch.} z.B.*

zur Bildung des Sternpunktes in Hochspannungsmaschine
neutral busbar Sternpunktschiene
neutral conductor Mittelleiter * *im Drehstromsystem* || Mittelpunktleiter || Mittelpunktsleiter || N-Leiter || Neutralleiter || Nullleiter
neutral earthing Sternpunkterdung
neutral gear Leerlauf * *bei Schaltgetriebe*
neutral grouding Sternpunkterdung
neutral point Mittelpunkt * *(elektr.) eines Drehstromsystems* || Netzmittelpunkt || Nullpunkt * *Sternpunkt* || Sternpunkt
neutral position neutrale Lage || Nulllage || Nullstellung || Ruhelage * *z.B. einer Tänzerwalze* || Ruhestellung
neutral zone Totband
neutralization Nullung
neutralize abmagnetisieren || aufheben * *each other: sich gegenseitig ~; ausgleichen* || entkoppeln * *neutralisieren* || lahm legen || neutralisieren || symmetrieren
neutralizing winding Kompensationswicklung
never keinesfalls * *niemals* || nie
never before noch nie
never mind bitte * *es macht nichts*
never-lost unverlierbar
nevertheless dabei * *dennoch* || jedoch || nichtsdestoweniger || trotzdem * *(Adv)*
new neu
new condition Neuzustand
new development Neuentwicklung
new edition Neuauflage * *~ eines Druckerzeugnisses*
new orders Auftragseingang
new start Neustart
new start-up Neuinbetriebnahme
newcomer Einsteiger || Erstanwender * *→Neuling* || Lehrling * *Neuling* || Neuling
newest neuest
newly-developed neuentwickelt
newness Neuheit
news Kunde * *Nachricht* || Meldung * *Neuigkeit* || Nachricht * *Kunde, Neuigkeit(en), Zeitungsnachricht; good/bad: gute/schlechte*
newsprint Druckpapier * *für Zeitungen* || Zeitungsdruckpapier || Zeitungspapier
Newton-Meter Newton-Meter
next ander * *folgender* || anschließend * *(räuml. u. zeitl.) nächstfolgend, nächststehend, (gleich) neben, nächst-, gleich nach* || künftig * *Jahr, Woche* || nächst * *in der Reihenfolge* || nächster * *in zeitlicher Reihenfolge*
next after nächster * *(Präposition)*
next to bei * *(räumlich)* || nächster * *(Präposition)* || nahezu * *next to impossible: ~ unmöglich* || neben * *unmittelbar ~* || zunächst * *am nächsten gelegen*
next to it daneben * *nächstliegend, angrenzend*
next to last vorletzt || vorletzter
next to one another nebeneinander
next to the last zweitletzter
nexus Verknüpfung * *Zusammenhang, Verknüpfung*
NF NF * *Abkürzg. für 'Norme Francaise': französische Norm*
NFPA NFPA * *Abkürzg. für 'National Fire Protection Association': amerikan. Brandverhütungsgesellschaft*
nibble nibbeln * *Blechbearbeitung*
nibble-coded Nibble-codiert * *je 4 bits bilden eine Informationseinheit*
nibbling machine Aushaumaschine * *zur Blechbearbeitung* || Nibbelmaschine * *zur Blechschneidbearbeitung (z.B. die TRUMATIC (R) der Fa. Trumpf)*
nice sauber * *hübsch, nett*

Nice to see you Guten Tag
niche Nische * *[nitsch] (auch figürl.: Marktnische, zugewiesener Platz usw.); fill: füllen/besetzen*
Nichols diagram Nichols-Diagramm
nickel Nickel * *chrome-nickel steel: ~chromstahl*
nickel-plate vernickeln
nickel-plated vernickelt
night setback Nachtabsenkung * *bei Heizung usw.*
nimble beweglich * *(figürl.) flink, hurtig, gewandt, behend, nimble mind: rasche Auffassungsgabe, ~er Geist*
nip Anpress- || Anpress- * *Press-, Abkneif-* || klemmen * *kneifen, zwicken (off:ab-), klemmen (Maschine)* || Klemmpunkt * *Stelle, an der d.Geschwind. e.durchlaufend.Warenbahn zwangsmäßig konstantgehalten wird* || Klemmstelle * *Stelle, an d.die Geschwind. e.durchlauf. Warenbahn zwangsmäßig konstantgehalten wird* || Walzspalt
nip off abzwicken
nip pressure Liniendruck * *Anpressdruck an einer Klemmstelle (z.B. bei einer Papaierbahn)*
nip roll Anpresswalze * *Haltewalze* || Mangelwalze * *zum Durchfördern band-/plattenförmigen Materials*
nip roller Andruckwalze * *(An)Press-/Haltewalze*
Nipco roll Nipco-Walze * *[Papier]*
nipple Nippel || Stutzen * *kleiner ~*
nitrogen Stickstoff
Nm Nm * *Newtonmeter;1 Nm entspr. 0,101972 kpm entspr 0,7376 lbf ft(pound-force feet) enspr. 8,851 lbf in*
no kein * *(Adj.)* || Nummer * *Abkürzung*
no acknowledgement Quittungsverzug * *NAK bei serieller Kopplung*
NO contact Arbeitskontakt * *Abkürzg. für 'Normally Open contact'* || Schließer * *Abkürzung für 'Normally Open contact'* || Schließkontakt * *Abkürzung für 'Normally Open contact'*
no drift Driftfreiheit
no good wertlos
no less than mindestens * *vor Zahlen, in einer Vorschrift*
no longer nicht mehr
no longer used nicht mehr gebräuchlich
no moisture condensation keine Betauung
no more than nicht mehr als
no one kein * *(Subst.) keiner*
no operation Leerbefehl
no pains, no gains ohne Fleiß kein Preis
no signal flow nicht durchgeschaltet * *Zustand eines Strompfades in Kontaktplandarstellung eines SPS-Programms*
no, thank you danke * *bei Ablehnung*
no- -Freiheit || -los * *(vorangestellt)*
no-contact berührungslos * *kontaktlos* || kontaktlos
no-current interval stromlose Pause * *wird z. Sicherheit von Kommandostufe bei Momentenumkehr eingefügt*
no-cutting working spanlose Bearbeitung
no-load Leerlauf || Leerlauf- || unbelastet
no-load characteristic Leerlaufkennlinie * *bei [el. Masch.]:Verlauf v. Leerlaufstrom/-verlusten i. Abhängigkt. v.d.Spann*
no-load condition Leerlauf * *under: im/bei*
no-load current Leerlaufstrom
no-load DC voltage Leerlaufgleichspannung * *ideal no-load DC voltage: →ideelle Gleichspannung*
no-load interval Pausenzeit * *Zeitdauer ohne Belastung*
no-load losses Leerlaufverluste * *bei [el.Masch.]: Leistung, die eine Motor ohne Belastung an d.Welle aufnimmt* || Leerlaufverlustleistung
no-load measurement Leerlaufmessung
no-load operation Betrieb ohne Last || Leerlauf
no-load proof leerlauffest

no-load reversing frequency Leerumschalthäufigkeit * {el.Masch.} zuläss. Anzahl aufeinanderfolg. Reversiervorgänge ohne Last
no-load screw-down Leeranstellung * {Walzwerk} →Walzenanstellung während des "Leerweges"
no-load speed Leerlaufdrehzahl
no-load starting frequency Leeranlaufschalthäufigkeit * {el.Masch.} →Leerschalthäufigkeit || Leerschalthäufigkeit * {el.Masch.} zuläss. Anz. aufeinander folgender Anlaufvorgänge ohne Last
no-load starting time Leeranlaufzeit * benötigt e.Motor ohne Last z. Hochlaufen auf Betriebsdrehzahl
no-load status Leerlauf
no-load test Leerlaufversuch * z.b. zur Aufnahme der Kurve 'Leerlaufstrom als Funktion der Drehzahl' bei DC-Motor
no-load voltage Leerlaufspannung
no-loss verlustfrei
no-maintenance wartungsfrei
no-slip formschlüssig
no-voltage condition spannungsloser Zustand * in:in
noble nobel
noble metal Edelmetall
nobody kein * (Subst.) keiner
node Busteilnehmer || Knoten * auch eines Computernetzwerk; Teilnehmer an einem Bussystem || Teilnehmer * Knoten eines Datennetzes bzw. Kommunikationsbusses
node address Teilnehmeradresse * bei Bussystem (z.b. PROFIBUS)
nofab fabriklos * z.b. nofab company: Unternehmen ohne eigene Fertigungseinricht.
noise Einstreuungen || Elektromagnetische Störung * Störung || EMV * Störeinkopplung || Geräusch || Geräusche || Lärm || Rauschen * pink/random:rosa || Störsignal || Störung * elektromagnetische ~, Rauschen
noise addition of several sources Gesamtgeräusch mehrerer Quellen * {el.Masch.} Gesamt-Schalldruckpegel mehrerer Maschinen
noise current Störstrom
noise dampening Schalldämpfung
noise development Geräuschentwicklung
noise emission Geräuschemission * Grenzwerte für Motoren: DIN 57530/VDE 0530/IEC 34 jeweils Teil 9 || Geräuschentwicklung * Abgabe, Emission || Störabstrahlung || Störungsabstrahlung
noise fedback into the supply system Netzrückwirkungen
noise feedback Netzrückwirkungen
noise filter Entstörfilter
noise generator Störgenerator || Störsimulator * Störgenerator
noise immunity Störfestigkeit || Störsicherheit * gegen elektromagnetische Störungen || Störanungsfestikeit || Störunempfindlichkeit * gegen elektromagnetische Störungen
noise immunity class Störfestigkeitsklasse * siehe IEC 801-4/5
noise injected back into the line Netzrückwirkungen * Oberwell.-behaftete Belastg.d.Netzes durch taktenden Stromrichter
noise injection Störeinkopplung
noise insensitive störsicher * gegen elektromagnetische Störungen || störunempfindlich * gegen elektromagnetische Störungen
noise insensitivity Störsicherheit * gegen elektromagnetische Störungen || Störunempfindlichkeit * gegen elektromagnetische Störungen
noise insulation Schallisolierung
noise interference Störbeeinflussung * [inte'fi:rens] elektromagnetische
noise level Geräusch * ~pegel || Geräuschentwicklung * Geräuschspiegel || Geräuschniveau * low: niedriges; high: hohes || Geräuschpegel * Messmethode: siehe DIN 45633 u. 45635; Motorgeräusch: siehe DIN VDE 0530 || Geräuschstärke || Rauschpegel || Schallpegel * Geräuschpegel || Störpegel
noise level measurement Schallmessung * Schallpegelmessung
noise level test Geräuschmessung * {el.Masch.} siehe DIN 45635 Teil 10
noise limit Geräuschgrenzwert * {el.Masch.} max. erlaubter Schallleistungspegel (VDE 0530 Teil 5)
noise margin Störabstand
noise measurement Geräuschmessung * für elektr. Maschinen: siehe DIN 45635 || Schallmessung * siehe d. DIN IEC 651
noise muffling Lärmschutz * Lärmminderung/-Dämpfung
noise of the fan Lüftergeräusch
noise power level Schall-Leistungspegel || Schallleistungspegel
noise protection Lärmschutz || Schallschutz
noise pulse Störimpuls
noise radiation Störabstrahlung || Störaussendung * Stör(ab)strahlung || Störungsabstrahlung
noise reduction Geräuschminderung || Lärmminderung || Schallisolierung * Geräuschreduzierung
noise rejection Störungsunterdrückung || Störunterdrückung
noise resistant störunempfindlich * gegen elektromagnetische Störungen
noise sensitive störanfällig * anfällig gegen elektromagnetische Störungen || störempfindlich * gegen elektromagnetische Störungen
noise sensitiveness Störanfälligkeit * gegen elektromagnetische Störungen || Störempfindlichkeit * gegen elektromagnetische Störungen
noise sensitivity Störanfälligkeit * gegen elektromagnetische Störungen || Störempfindlichkeit * gegen elektromagnetische Störungen
noise signal Störsignal
noise simulator Störgenerator * Störsimulator || Störsimulator
noise spectrum Geräuschspektrum
noise suppression Geräuschdämpfung || Lärmdämpfung
noise suppression capacitor Funkentstörkondensator
noise suppression choke Entstördrossel
noise suppression equipment Entstörmittel || Entstörung * Entstörmittel, Entstörausrüstung
noise suppression filter Funkentstörfilter
noise suppression level Funkentstörgrad
noise suppression measures Entstörmaßnahmen
noise voltage Fremdspannung * Störspannung || Störspannung
noise voltage spike Störspitze
noise voltages Störspannungen
noise- Stör- * Geräusch, elektromagnetische Störung
noise-absorbing geräuschgedämpft
noise-corrupted verrauscht
noise-free störsicher * keine elektromagnetische Störungen aussendend/abstrahlend
noise-immune störsicher * gegen elektromagnetische Störungen
noise-insulating cover Schalldämmhaube
noise-level value Geräuschwert * [dB (A)]; Messung: siehe DIN 45635, Motorgeräusch: DIN VDE 0530
noise-reduced schallgedämpft
noise-reducing cover Geräuschdämmhaube || Schalldämmhaube
noise-reduction equipment Schallschutzeinrichtung

noiseless geräuschlos || leise * *geräuschlos/arm*
noisy laut * *lärmend, geräuschvoll* || marktschreierisch * *schreierisch, grell, aufdringlich, krakeelend* || verrauscht
nomenclature Begriffe || Fachausdrücke || Fachbegriffe || Fachsprache * *Fachbegriffe* || Terminologie
nominal Bemessungs- || Nenn- || nominell
nominal AC voltage Nennanschlussspannung * *Stromrichter*
nominal air gap Bemessungsluftspalt * *bei einer Bremse usw.*
nominal conductor cross-section anschließbarer Leiterquerschnitt
nominal current Nennstrom
nominal data Nenndaten
nominal diameter Nenndurchmesser * *z.B. bei Ventilator*
nominal input voltage Eingangs-Nennspannung || Nennanschlussspannung
nominal line speed Nenngeschwindigkeit * *Bahn-, Bandgeschwindigkeit*
nominal operating conditions Bemessungsbetriebsbedingungen
nominal range Nennbereich
nominal rating Nenndaten || Nennleistung * *z.B. ~ eines (Stromrichter)Geräts*
nominal rating conditions Nennbedingungen
nominal release travel Bemessungslüftweg * *bei einer Bremse*
nominal speed Nenndrehzahl || Nenngeschwindigkeit
nominal torque Nenndrehmoment || Nennmoment
nominal value Nennwert * *gerundeter Zahlenwert einer Größe, mit dem e. Einrichtung/Gerät gekennzeichnet wird*
nominal values Nenndaten
nominal velocity Nenngeschwindigkeit
nominal voltage Nennspannung
nominal voltages of supply systems Nennspannungen für elektrische Netze * *siehe IEC 38 (einphasig 120/240V, 3-phasig 400/480/690/1000V)*
nominal working point Nennpunkt
nominate aufstellen * *als Kandidaten* ~ || aufstellen * *benennen, z.B. Teammitglieder* || benennen * →*ernennen* || ernennen || nennen * *Kandidaten* ~
non aging alterungsbeständig
non compensated unkompensiert * *z.B. Motor, Tachometermaschine*
non obligatory unverbindlich * *ohne Verpflichtung, nicht (rechts)verbindlich*
non stop winder Rollenwechsler * *Wickler mit fliegendem Rollenwechsel*
non- -frei * *(vorangestellt)* || nicht- || un- * *nicht*
non-ageing alterungsbeständig
non-asbestos asbestfrei
non-audible unhörbar
non-backed-up ungesichert * *Daten usw.*
non-binding unverbindlich * *nicht bindend*
non-bridging contact unterbrechender Kontakt * *Wechselkontakt, der ohne Überbrückung arbeitet*
non-coincidence Antivalenz
non-committal unverbindlich * *neutral, zurückhaltend, nichts sagend (Stellungnahme usw.)*
non-condensing keine Betauung * *als Datenblattspezifikation* || nicht-kondensierend || ohne Betauung
non-conductive nicht leitend || nichtleitfähig
non-contact kontaktlos
non-contact proximity detector BERO * *Berührungsloser Abstandssensor, 'Analog-BERO'*
non-contacting berührungsfrei || berührungslos
non-contacting limit switch BERO * *berührungsloser Endschalter*

non-continuous area Lückbereich
non-continuous current period Lückdauer
non-controllable nicht steuerbar || ungesteuert * *nicht steuerbar, z.B.Brückenschaltung aus Dioden*
non-controlled ungesteuert * *z.B. Gleichrichter aus Dioden (nicht mit Thyristoren)*
non-corroding korrosionsbeständig || nicht rostend
non-cutting machining spanlose Fertigung
non-cutting working spanlose Fertigung
non-cyclic nichtzyklisch
non-delayed unverzögert
non-desirable unerwünscht * *nicht wünschenswert*
non-destroyable zerstörsicher
non-destructive unzerstörbar || zerstörungsfrei * *z.B. Materialprüfung*
non-destructive material testing nicht zerstörende Werkstoffprüfung
non-detachable fest montiert * *nicht demontierbar/lösbar*
non-disclosure agreement Geheimhaltungsvereinbarung
non-drive end B-Seite * *Nicht-Antriebsseite eines Motors* || BS * *B-Seite, Nicht-Antriebsseite eines Motors* || Nicht-Antriebsseite * *{el.Masch.}*
non-drive shaft end BS-Wellenende * *auf der Nicht-Antriebsseite eines Motors*
non-driven nicht angetrieben
non-driving shaft end zweites Wellenende * *auf Nicht-Antriebsseite (B-Seite) eines Motors*
non-electrical nichtelektrisch
non-electrical data nichtelektrische Größen
non-electrical quantities nichtelektrische Größen
non-encapsulated ungekapselt
non-equivalence Antivalenz * *z.b. logische Exklusiv-Oder Funktion*
non-existent nicht vorhanden
non-fatal error ungefährlicher Fehler
non-fatal fault ungefährlicher Fehler
non-ferrous Nichteisen-
non-ferrous metal Nichteisenmetall
non-floating potenzialbehaftet || potenzialgebunden
non-fused nicht abgesichert * *z.B. mit Schmelzsicherung*
non-inflammable unbrennbar
non-interacting rückwirkungsfrei
non-interchangeable unverwechselbar
non-inverting nichtinvertierend
non-isolated potenzialbehaftet || potenzialgebunden * *nicht potenzialfrei*
non-latching nichtspeichernd * *Signal ohne Selbsthaltung bei Relaissteuerung, Kontaktplan usw.*
non-linear nichtlinear * *highly: stark*
non-linearity Nichtlinearität
non-linearized value Rohwert * *nichtlinearer Messwert (z.B. von Temperatursensor)*
non-linting nicht fasernd * *nicht fusselnd, z.B. Lappen* || nichtfasernd * *z.B. Putzlappen*
non-locating bearing Loslager
non-lube ölfrei * *z.B. Pneumatikzylinder, Kompressor*
non-magnetic amagnetisch * *antimagnetisch*
non-normalized unnormiert
non-observance Nichtbeachtung * *Nichtbefolgung, z.B. einer Vorschrift*
non-obstructed unbehindert
non-oscillating schwingungsfrei
non-oscillatory schwingungsfrei
non-overlapping deckungsgleich * *nicht überlappend*
non-periodic nicht periodisch
non-periodic duty nichtperiodischer Betrieb
non-polar ungepolt * *z.B. Kondensator*
non-polarized ungepolt * *Kondensator, Relais usw.*

non-polarized capacitor ungepolter Kondensator
non-positive reibschlüssig
non-professional Anfänger * Laie || Einsteiger * Laie || Laie * 'Nichtprofi' || Neuling * Nicht-Profi
non-proprietary firmenneutral * nicht firmeneigen/spezifisch || firmenübergreifend * nicht firmeneigen/herstellerspezifisch || offen * nicht firmeneigen (z.B. Bussystem)
non-proprietary protocol offenes Protokoll * nicht-herstellerspezifisches Protokoll
non-pulsating current nichtlückender Strom
non-pulsating current area nichtlückender Bereich
non-pulsating-current operation nichtlückender Betrieb
non-recognizable command unbekannter Befehl
non-regenerative braking Verlustbremsung
non-regenerative load nicht-generatorische Last
non-reliably unzuverlässig * (Adv.)
non-return valve Rückschlagventil
non-reverse ratchet Rücklaufsperre * Maschinenlement, das m. Kraft- oder Formschluss e.Drehen i.e.Richtg. verhindert
non-rigid supply system weiches Netz * Netz mit hoher Kurzschlussleistung (z.B.über 1%)
non-rotating nicht umlaufend
non-salient pole rotor Vollpolläufer
non-scheduled außerplanmäßig
non-sinusoidal nichtsinusförmig
non-slip formschlüssig || schlupffrei
non-slip engagement Formschluss
non-slipping formschlüssig
non-solid flüssig * als Gegensatz zu "fest"
non-sparking design Non-sparking-Ausführung * Zündschutzart Typ 'N' nach BS 5000 Pt.16 (British Standard)
non-sparking fan nicht funkender Lüfter
non-stabilized unstabilisiert * z.B. Versorgungsspannung/Netzteil
non-standard anormal * nicht dem normalen Standard entsprechend || Sonder- || Spezial-
non-standard motor Transnormmotor * AC Motor, in d.Leistung über d.Normreihe (DIN 42669..81) liegend (über ca.200kW)
non-stationary nichtstationär
non-stop rewinder Wendewickler
non-taxable steuerfrei
non-tilting kippsicher * mechanisch
non-tracking quality Kriechstromfestigkeit
non-twist drehsteif
non-uniform ungleichförmig || ungleichmäßig
non-uniform converter connection teilgesteuerte Stromrichtschaltung * siehe IEC 146-1-1 u. EN 60146-1-1; z.B. für 1Q-DC-Stromrichter
non-ventilated selbstbelüftet * siehe →selbstgekühlt || selbstgekühlt * z.B. Motor ohne Lüfter, Wärmeabfuhr durch Abstrahlung an der Gehäuseoberfläche || unbelüftet * z.B. Motor ohne Lüfter (Wärmeabstrahlung über die Gehäuseoberfläche)
non-vibrating schwingungsfrei * (mechan.)
non-visible nicht sichtbar || unsichtbar
non-volatile nichtflüchtig || nullspannungsgesichert * nichtflüchtig bei Abschalten der Versorgungsspannung (Speicher)
non-volatile memory nichtflüchtiger Speicher || Permanentspeicher * nichtflüchtiger Speicher
non-volatile storage without backup battery batterielose Pufferung
non-wearing verschleißfrei
non-withdrawable fest eingebaut
non-woven material Vliesstoff
non-wovens Vliese * [Textil]
nonageing alterungsbeständig
nonaging alterungsbeständig

noncircular unrund * nicht kreisförmig
noncutting working spanlose Bearbeitung
none kein * (Subst.) none of us:keiner von uns
noninterchangeable unverwechselbar
nonlinear nichtlinear
nonlinearity Linearitätsfehler || Linearitätsfehler || Nichtlinearität
nonpositive kraftschlüssig
nonreactive resistance Wirkwiderstand
nonreturn valve Rückschlagventil
nonstop winder Wickler mit fliegendem Rollenwechsel
nonventilated selbstkühlend * unbelüftet, z.B. Motor
nonvolatile nichtflüchtig
nook Ecke * Nische
noose Schlaufe || Schlinge * zusammenziehbare
NOP Leerbefehl * Abkürzung für 'No OPeration'
NOR gate NOR-Stufe * invertierte ODER-Verknüpfung
norm Norm * Arbeitsnorm
normal gewöhnlich * normal || natürlich || normal
normal applications allgemeine Anforderungen * for normal applications: für allgemeine Anforderungen (z.B. Kondensatoren)
normal case Regelfall
normal DC excitation Gleichstrom-Normalerregung * für Magnetbremse
normal operating conditions normale Anforderungen * normale Betriebsbedingungen
normal position Ausgangsstellung * →Ruhestellung || Ruhestellung * auch eines Relais
normal requirements normale Anforderungen
normal room temperature Raumtemperatur
normal-blow mittelträge * Sicherung
normal-mode Gegentakt- * -Spannung, -Signal, - Unterdrückung
normal-mode rejection Gegentaktunterdrückung
normalization Normierung
normalization factor Normierfaktor || Normierungsfaktor
normalization parameter Normierungsparameter
normalize bewerten * normieren || normieren * normalisieren
normalized normiert
normally im Allgemeinen * normalerweise || normalerweise || sonst * normalerweise || standardmäßig * (Adv.) normalerweise
normally closed contact Öffner || Öffnerkontakt || Ruhekontakt
normally open contact Arbeitskontakt || Schließer || Schließerkontakt
normative Norm- * normativ, normende Kraft habend || normativ
North America USA und Canada
north pole Nordpol * auch bei einem Magnet
Norway Norwegen
Nos Nummern * Abkürzung für 'numbers'
nose Nase * Nase, Vorsprung, Schnabel, Mündung
not nicht * nicht
not accurately lückenhaft * (Adv.)
not affecting the costs kostenneutral
not airtight undicht * luftdurchlässig, nicht luftdicht
not allowed verboten * nicht erlaubt
not any kein * (Adj.)
not any longer nicht mehr
not applicable entfällt * in Formularen
not assigned nicht belegt * nicht (einer Funktion) zugeordnet
not at all nie * keineswegs
not bad ganz * (Adv.; salopp) ~ gut
not be provided fehlen * nicht vorgesehen/zur Verfügung gestellt sein

not before erst * *nicht früher als*
not binding freibleibend * *z.B. Preise, Kurse*
not clear unklar
not completed nicht durchgeschaltet * *nicht geschlossen (Stromkreis)*
not connected nicht belegt * *nicht angeschlossen* || nicht durchgeschaltet || unbeschaltet * *Anschluss, Klemme*
not correct unfachgemäß
not counting außer * *nicht mitgezählt*
not critical from a time perspective zeitunkritisch
not effective unwirksam * *nicht wirksam, nicht in Kraft*
not either auch nicht * *I cannot do it either: ich kann es auch nicht*
not exceed nicht größer sein als
not exceeding bis * *nicht größer als* || höchstens || unterhalb * *nicht überschreitend (Wert, Zahl usw.)*
not far from annähernd
not flat uneben
not in abwesend
not incur the expenses of einsparen * *die Kosten/ den Aufwand für etwas ~*
not involved unberührt * *(figürl.) nicht einbezogen*
not looked after unbeaufsichtigt
not maintained nichtspeichernd * *Signal ohne Selbsthaltung bei Relaissteuerung, Kontaktplan usw.*
not my cup of tea nicht nach meinem Geschmack * *(salopp)*
not on any account unter keinen Umständen
not only ... but also sowohl als auch
not operational außer Betrieb
not required entfällt * *nicht benötigt (for: bei/für)*
not requiring fast response zeitunkritisch * *keine schnelle Reaktion erfordernd*
not suitable ungeeignet
not switched through nicht durchgeschaltet
not tight undicht
not time-critical zeitunkritisch
not to be affected unberührt * *by: von Gesetz, Regel usw.*
not to fall within the scope of unberührt * *von Vorschrift. Gesetz usw. ~ bleiben*
not to mention that abgesehen davon dass
not to my taste nicht nach meinem Geschmack
not to scale nicht maßstäblich || nicht maßstabsgetreu
not too far down the road in nicht zu ferner Zukunft
not true to size nicht maßhaltig
not under mindestens * *vor Zahlen, in einer Vorschrift*
not uniform ungleichförmig
not until erst * *nicht bevor*
not waterproof undicht * *nicht wasserdicht*
not watertight undicht * *nicht wasserdicht*
not yet noch nicht
not-grounded erdfrei
notation Bezeichnung * *Aufzeichnung, Schreibweise, Notierung* || Darstellung * *Notation* || Form * *{Mathem.} Darstellungsart/Schreibweise* || Notation * *Schreibweise, Bezeichnung, (chem.) Formelzeichen; Notierung, Aufzeichnung* || Schreibweise
notch Ausschnitt * *Kerbe, Einschnitt, Aussparung, Nute* || Aussparung * *Kerbe, Einschnitt, Aussparung, Nute* || Einbruch * *z.B. Kommutierungseinbruch* || einkerben || Einschnitt * *Kerbe, Scharte, Loch* || Falz * *Auskehlung, Kerbe, Falz* || Kerbe || Nut || Nute || nuten || Raste * *Kerbe, Einschnitt, Raste (zum Einklinken)* || Rastung * *Rastnase*
notch effect Kerbwirkung
notched gezackt * *gekerbt* || gezahnt * *gekerbt*
notching Rastung

notching effect Kerbwirkung
note angeben * *notieren, aufschreiben* || Anmerkung * *Hinweis (schriftlicher)* || aufzeichnen * *notieren* || beachte || beachten * *kindly note: bitte beachten* || Bemerkung || Erläuterung * *Anmerkung* || Fußnote || Hinweis * *in Beschreibung; concerning: betreffend* || merken * *beachten, notieren, aufschreiben* || Nachricht * *Botschaft* || Vermerk
note down vermerken
noted bekannt * *berühmt; for: wegen*
notes for calculation Berechnungshinweise
notes on safety Sicherheitshinweise
noteworthy erwähnenswert * *especially noteworthy: besonders ~*
nothing kein * *(Subst.) kein*
nothing but nur * *nichts als*
notice beachten || Beachtung * *be noticed: ~ finden* || feststellen * *bemerken, beobachten, wahrnehmen, merken* || Meldung * *Mitteilung, Ankündigung, Nachricht, Notiz* || merken * *wahrnehmen* || Mitteilung || Nachricht * *Mitteilung* || Vermerk || wahrnehmen * *bemerken, beobachten, wahrnehmen, achten auf, mit Aufmerksamkeit behandeln*
notice of cancellation Kündigung * *~ eines Vertrages*
notice of defects Mängelrüge
notice of termination Kündigung * *~ eines Vertrages*
notice to leave Kündigung * *~ seitens des Arbeitnehmers*
notice to quit Kündigung * *~ seitens des Arbeitnehmers*
noticeable beachtlich || deutlich * *wahrnehmbar* || erheblich * *beachtlich* || erkennbar || fühlbar * *geistig* || merklich * *wahrnehmbar, beachtlich, bemerkenswert* || sichtbar * *wahrnehmbar* || wahrnehmbar
notifiable anzeigepflichtig
notification Bekanntgabe || Bekanntmachung || Verständigung * *Benachrichtigung*
notification of defect Mängelrüge
notification of delivery-discontinuation Abkündigung * *→Produkt~, Liefer~* || Produktabkündigung
notify bekanntmachen * *anzeigen, melden, amtlich mitteilen* || hinweisen * *of: auf* || informieren * *jemanden benachrichtigen, (amtlich) mitteilen, melden* || melden * *amtlich/offiziell/schriftlich ~* || mitteilen * *(amtlich/offiziell) mitteilen/bekannt geben*
notion Begriff * *Vorstellung* || Gedanke * *Gefühl, Ahnung* || Idee
notional gedanklich * *begrifflich, rein gedanklich, spekulativ, eingebildet, imaginär*
notwithstanding that trotzdem * *(Bindewort)*
novel neu * *~artig* || ungewöhnlich * *neuartig*
novelty Neuheit * *auch einer Erfindung*
novice Anfänger * *→Neuling* || Laie * *Neuling, Anfänger* || Lehrling * *Neuling* || Neuling * *→Anfänger, Neuling* || unerfahren * *Anfänger-, Neuling-*
novices Anfänger * *(Plural)*
NOVRAM nichtflüchtiger Speicher * *Abkürzung für 'nonvolatile random access memory'* || NOVRAM * *Abk.f. 'NOn Volatile Random Access Memory':nichtflüchtiger Schreib/Lesespeicher*
now and then gelegentlich * *(Adv.)* || manchmal * *hin und wieder* || sporadisch * *hin und wieder*
nowadays heutzutage || jetzig * *in der jetzigenZeit*
noxious schädlich * *gesundheitsschädlich*
nozzle Ausgießer * *Schnauze, Tülle, Rüssel, Mundstück (an Gefäßen usw.)* || Düse
NPP Kernkraftwerk * *Abk. f. 'Nuclear Power Plant'*
NRZ NRZ * *Abk. f. 'Non Return to Zero': Logikpegel mit Low-Signal ungleich Null*
NS chart Nassi-Schneidermann-Diagramm

* *(Computer)* || Struktogramm * *Nassi-Schneidermann Diagramm*
NTC Heißleiter
NTC resistor Heißleiter * *Abkürzg. für 'Negative Temperature Coefficient'*
NTC thermistor Heißleiter * *als Temperaturfühler*
nuclear Kern- * *Atom-*
nuclear power plant Kernkraftwerk
nuclear technology Kerntechnik
nucleus Kern * ~ *eines Teams, einer Partei; Atom~ usw.*
nudge stoßen * *(leise/heimlich) an~, stupsen, pochen*
nuisance Ärgernis * *Ärgernis, Plage, etwas Lästiges/Unangenehmes, Missstand, Unfug, Quälgeist, Landplage* || Belästigung || Mangel * →*Missstand* || Missstand
null Null
null and void Null * *Null und nichtig* || null und nichtig * *declare null and void: für ~ erklären* || ungültig * *null und nichtig; declare null and void: für ~ erklären*
null modem cable Nullmodem-Kabel * *RS232-Kabel mit gekreuzten Transmit- u. Receive-Adern*
nullify annullieren
number Anzahl * *Zahl* || betragen * *zahlenmäßig ~* || Exemplar * *einer Zeitschrift usw.* || Nummer || nummerieren * *consecutively: fortlaufend* || Zahl * *auch Anzahl (even:gerade, odd:ungerade)*
number consecutively durchnummerieren * *fortlaufend*
number of -Zahl * *(vorangestellt) Anzahl der*
number of conductors Aderzahl
number of different types Typenvielfalt
number of increments Impulszahl * *bei (Inkremental)Impulsgeber* || Pulszahl * *bei (Inkremental) Impulsgeber* || Schlitzzahl * ~ *eines (Inkremental-) Impulsgeber*
number of increments per revolution Pulszahl * *bei (Inkremental-) Impulsgeber* || Schlitzzahl * ~ *eines Impulsgebers* || Strichzahl * *eines Inkrementalgebers/Impulsgebers*
number of items Stückzahl
number of phases Strangzahl
number of pieces Stückzahl * *z.B. ~ bei einer Bestellung*
number of pole pairs Polpaarzahl
number of poles Polzahl
number of pulses per revolution Pulszahl * *Impulsgeber/Tacho* || Schlitzzahl * *Pulszahl pro Umdrehung bei Impulsgeber* || Strichzahl
number of slots per revolution Schlitzzahl * ~ *eines optischen Drehimpulsgebers* || Strichzahl * *eines optischen Drehimpulsgebers*
number of starter steps Stufenzahl * *{el.Masch.} Anzahl d.abschaltbaren Teilwiderstände eines Motoranlassers*
number of startings Schalthäufigkeit * *bei Motor; higher: erhöht*
number of teeth Zähnezahl * *bei Zahnrad und Zahnriemen-Scheibe*
number of turns Windungszahl
number of turns per unit length Windungszahl * *pro Längeneinheit*
number of types Typenvielfalt * *lower: geringere*
number representation Zahlendarstellung
number skills mathematische Fähigkeiten
number system Zahlensystem
numbering Nummerierung
numbers Nummern
numbers employed Beschäftigtenzahl
numeral Kennziffer * *Ziffer* || Zahl * *Ziffer, auch 'Betrag, Wert'* || Ziffer
numerator Dividend * *Zähler, das über dem Bruchstrich stehende* || Zähler * *(mathem.) das über dem Bruchstrich stehende*
numeric numerisch
numeric control NC-Steuerung
numeric controls numerische Steuerung
numeric display Ziffernanzeige
numeric function Rechenfunktion
numeric function block Rechenbaustein * *Funktionsbaustein*
numeric keyboard Zehntastatur || Zifferntastatur
numeric keypad Zehnertastatur || Zifferntastatur || Zifferntasten * *Zehner-/Zifferntastenfeld*
numeric keys Zifferntasten
numeric quantities numerische Größen
numeric switch numerischer Umschalter * *Funktionsbaustein* || Zahlenschalter
numeric task Rechenaufgabe * *für Rechner*
numeric value numerischer Wert || Zahlenwert
numerical numerisch || Zahlen-
numerical control numerische Steuerung
numerical control system numerische Steuerung
numerical data Wertangabe * *Zahlenangaben*
numerical date Wertangabe * *Zahlenangabe*
numerical input Zahleneingabe
numerical keypad Zehnertastatur
numerical notation Zahlendarstellung
numerical quantities numerische Größen
numerical value Zahlenwert
numerical value equation Zahlenwertgleichung
numerically numerisch * *(Adv.)*
numerically controlled NC-gesteuert || numerisch gesteuert
numerous zahlreich
NURB NURB * *{Werkz.masch.} Non-Uniform Rational B-Spline:Darstellg.e. Kurvenstücks durch Polynom 3. Ordng.*
Nuremberg Nürnberg
nut Mutter * *für Schraubverbindung*
nut running drive Schraubantrieb * *mit Kraftübertragung über Schraubspindel*
nut setting Spindelpositionierung * *bei Hub/Gewinde/Positionierspindel*
nut with two holes Zweilochmutter
NVRAM nichtflüchtiger Speicher * *Abkürzung für 'nonvolatile random access memory'*
Nyquist criterion Nyquist-Kriterium
Nyquist theorem Abtasttheorem

O

o Null * *(salopp, gesprochen)*
o with Umlaut ö
O-ring O-Ring * *Gummidichtring*
O-seal O-Ring * *Dichtring*
O.K. einwandfrei * *(auch Adv.) vermutl. Abkürzg. für 'all correct':alles in Ordnung* || funktionsfähig || funktionstüchtig || in Ordnung
O.K. key Quittiertaste
OA Operationsverstärker * *Abkürzg. f. 'Operational Amplifier'*
obey befolgen * *Vorschrift ~, Gesetz ~; gehorchen, Folge leisten*
object beanstanden * *to: etwas ~* || Einspruch einlegen || Gegenstand * *Objekt, Gegenstand, Absicht, Ziel, Zweck* || Objekt || Ziel * *Gegenstand, Absicht, Zweck* || Zweck * *Ziel*
object code Objektcode * *Ausgangsprodukt eines Hochsprache-Übersetzers/-Compilers*
object dictionary Objektverzeichnis * *{Network} beim* →*PROFIBUS*

object language Maschinensprache * *Objektsprache*
object library Objektverzeichnis * *{Network}* beim →*PROFIBUS*
object oriented objektorientiert * *z.B. Programmiersprache (Visual C usw.) für grafisch. Betriebssystem*
object program Objektprogramm * *{Computer} (übersetztes) Programm in Maschinensprache*
object to aussetzen * *etwas einzuwenden haben*
object to change without prior notice Änderungen vorbehalten * *ohne vorhergehende Benachrichtigung*
object-oriented programming objektorientierte Programmierung * *(z.B. für MS Windows; Gegensatz: prozedurale Programmierung)*
object-to-image ratio Abbildungsmaßstab
objection Ablehnung * *against/to: von* || Einspruch || Einwand || Reklamation * *Ein/Widerspruch*
objective Aufgabenstellung * *Zielsetzung* || Lernziel || sachlich * *objektiv, unparteiisch* || Ziel * *taktisches/strateg./unternehmerisches ~, ~setzg., Operations-/Angriffs-/Lern~* || Zielsetzung
objective of development Entwicklungsziel
obligate verpflichten * *vertraglich ~*
obligation Gebot * *Verpflichtung* || Pflicht * *Verpflichtung* || Verpflichtung * *undertake an obl.: eine ~ eingehen; be under obl.: unter ~ sein*
obligation to exercise due care Sorgfaltspflicht * *(juristisch)*
obligatory bindend * *auch verpflichtend, (rechts)verbindlich, obligatorisch* || obligatorisch * *on: für* || verbindlich * *bindend (rechts)verbindlich, obligatorisch*
oblige verpflichten * *durch Umstände, Stellung, Vertrag, Umstände, Gesetz ~*
obliged dankbar * *verpflichtet*
obliging entgegenkommend * *(figürl.) verbindlich, gefällig, zuvorkommend, entgegenkommend* || kulant * *{Handelswesen} gefällig, zuvor-/entgegenkommend; verbindlich*
oblique schief * *schief, schräg; oblique-angled:- schiefwinklig* || schräg * *schief, schräg*
oblique position Schieflage * *Schräglage*
oblique stroke Schrägstrich
oblique-angled schiefwinklig
obliquity Schräge
oblong hole Langloch
obscure unklar * *undeutlich* || unverständlich * *z.B.(Beweg-) Gründe* || verdecken * *verbergen* || verdeckt * *verborgen, unbekannt*
obscured verdeckt * *verborgen*
obscurity Unklarheit * *Undeutlichkeit, Unverständlichkeit*
observability Beobachtbarkeit * *{Regel.techn.}*
observance Beachtung * *Befolgung; z.B. einer Vorschrift* || Befolgung * *of:von* || Einhaltung * *of:- von*
observation Feststellung * *Beobachtung, Wahrnehmung* || Überwachung * *~ durch einen Menschen, Beobachtung* || Wahrnehmung * *Beobachtung(svermögen), Wahrnehmung, Überwachung*
observe beachten * *befolgen; z.B.Gesetze* || befolgen * *Vorschrift ~* || beobachten || betrachten * *beobachten* || einhalten || feststellen * *beobachten* || sich halten an || wahrnehmen * *bemerken, wahrnehmen (auch Termin), beobachten*
observe the timing Timing einhalten
observer Beobachter * *auch in der {Regel.techn.}*
observer matrix Beobachtermatrix * *{Regel.tech.}*
obsolete nicht mehr gebräuchlich || veraltet
obstacle Hindernis || Schwierigkeit * *Hindernis; put obstacles in a person's way: jdm. ~en machen* || Sperre * *Hindernis*

obstacle detection system Hinderniserkennungssystem * *für führerloses (Transport-) Fahrzeug*
obstruct beeinträchtigen * *behindern, verstellen, verstopfen* || behindern * *versperren, verstopfen, blockieren, behindern, nicht durchlassen, auch Sicht, Verkehr* || deaktivieren * *blockieren, hemmen, lahm legen, hindern, versperren* || erschweren * *hemmen, behindern, versperren, verstopfen, blockieren, verhindern* || verstellen * *versperren* || verstopfen * *hemmen* || zusetzen * *→verstopfen*
obstructed blockiert * *blockiert, gehemmt, lahmgelegt, versperrt*
obstruction Behinderung || Sperrung || Störung * *Behinderung* || Verstopfung
obstrusive marktschreierisch * *aufdringlich*
obtain bekommen * *Anstellung, Ware usw.* || beschaffen * *erlangen* || beziehen * *Ware ~* || erhalten * *erlangen, bekommen, erwerben, sich verschaffen* || erreichen * *erlangen, erhalten, sich verschaffen* || erzielen * *erlangen, erhalten* || gewinnen * *erhalten, (Bodenschätze, Zwischenprodukte usw.) gewinnen*
obtainable erreichbar * *ereichbar, erlangbar, erhältlich, zu erhalten*
obvious deutlich * *einleuchtend* || einleuchtend || nahe liegend * *(figürl.)* || offensichtlich || selbstverständlich * *offensichtlich*
obviously natürlich * *(Adv.) auf der Hand liegend, offensichtlich, einleuchtend, augenfällig, klar*
occasion Grund * *Anlass, Anstoß, Ursache, Gelegenheit* || Ursache * *Anlass* || Veranlassung * *Anlass, Veranlassung, Gelegenheit*
occasional gelegentlich * *gelegentlich, zufällig* || sporadisch * *gelegentlich* || zeitweise * *gelegentlich*
occasionally gelegentlich * *(Adv.)*
occult geheim * *Lehre, Geheimwissenschaft*
occupancy Belegung * *Inanspruchnahme (von Raum, Förderkapazität usw.), Innehaben, Besitz(ergreifung)*
occupation Arbeit * *Berufstätigkeit, Tätigkeit* || Einnahme * *Besetzung* || Tätigkeit * *Beschäftigung; Besitz(-nahme)*
occupy belegen * *besetzen, (für sich) beanspruchen, z.B. Speicherplatz* || besetzen * *Markt, (Sitz-) Platz, Land usw* || wegnehmen * *Raum, Zeit usw*
occupy oneself with befassen * *sich befassen mit* || sich beschäftigen mit * *vornehmen * *sich beschäftigen mit*
occur anfallen * *auftreten, sich ergeben, passieren* || auftreten * *vorkommen, eintreten, geschehen; z.B. Fehler* || eintreten * *auftreten, sich ereignen, geschehen* || entstehen * *eintreten, sich ereignen, vorkommen, auftreten* || erfolgen * *geschehen, auftreten* || geschehen || passieren * *sich ereignen* || vorkommen
occurrence auftreten * *(Substantiv) Vorkommen, Auftreten; Ereignis, Vorfall, Vorkommnis* || Ereignis * *Vorfall*
octagon bolt Achtkantschraube
octagonal achteckig
octal achtfach
octave band analysis Oktavanalyse * *Darstellung d.unbewerteten Schalldruckpegels in Frequenzbändern v. Oktavbreite*
octet Byte * *{Network} z.B. beim →PROFIBUS*
odd ausgefallen * *sonderbar, seltsam, merkwürdig, kurios* || übrig * *überschüssig, überzählig* || ungerade * *ungerazahlig* || ungewöhnlich * *sonderbar, seltsam, kurios, merkwürdig*
odd number ungerade Zahl
odd-numbered ungeradzahlig
odd-order harmonic ungeradzahlige Oberwelle
odd-order harmonics ungeradzahlige Oberschwingungen

oder example Bestellbeispiel
ODP ODP * *Abkürzung für 'Open Drip-Proof':offen, tropfwassergeschützt (Schutzart IP22 nach IEC und DIN)*
ODP enclosure IP22 * *Abk. f. 'Open Drip-Proof':offen, tropfwassergeschützt nach IEC und DIN*
ODS Hinderniserkennungssystem * *'Obstacle Detection System'; für führerloses (Transport-) Fahrzeug*
oe ö
OE spinning machine OE-Spinnmaschine * *{Textil}* || Open-End-Spinnmaschine * *{Textil}*
OEM OEM * *Original Equipment Manufacturer: stellt Maschinen o. Komponenten her u.liefert an* →*Endabnehmer* || Systemanbieter * *Abk. für 'Original Equipment Manufacturer': Systemhersteller, z.B. Maschinenbauer* || Systemhersteller * *Original System Manufacturer, stellt s. Systeme z.T.aus Zulieferteilen zusammen*
of aus || über * sprechen
of a different opinion über Kreuz * *(figürl.) unterschiedlicher Meinung*
of a sudden plötzlich * *(Adv.)* || ruckartig * *(Adv.)* plötzlich || schlagartig * *(Adv.)* plötzlich
of advantage vorteilhaft * *be of advantage: ~ sein*
of all kinds aller Art
of change Änderungs- * *(nachgestellt) z.B.: rate of change: ~geschwindigkeit*
of course selbstverständlich * *(Adv.)*
of different makes aller Art * *verschieder Marken/Hersteller*
of figures Zahlen-
of first priority vorrangig
of full value vollwertig
of good value preiswert
of high dielectric strength spannungsfest
of high quality hochwertig
of higher priority höherprior
of highest priority vorrangig
of inferior quality minderwertig * *Arbeit, Produkt usw.*
of interest interessant * *to:für* || interessierend
of itself selbst * *automatisch*
of late neuerdings
of low order niedriger Ordnung * *z.B. Oberschwingungen*
of lower priority niederprior
of market policy marktpolitisch
of modular design modular aufgebaut
of necessity notwendigerweise
of no account ohne Bedeutung
of no significance ohne Bedeutung
of no use unbrauchbar || wertlos * *nutzlos*
of numbers Zahlen-
of one's own eigen || eigener
of opposite polarity entgegengesetzt gepolt
of Portugal portugiesisch
of resistant material beständig * *be of resistant material: aus ~em Material bestehend*
of short duration kurzfristig * *kurz, nur kurz andauernd, von kurzer Dauer*
of smaller size kleiner * *(Komparativ) von kleiner Größe*
of space räumlich * *an Raum*
of such a kind solcher * *~ Art*
of such kind derartig
of that davon * *(Adv.)*
of that kind derartig
of the 1st order erster Ordnung
of the competition Mitbewerbs- || Wettbewerbs- * *die Konkurrenz/Mitbewerber betreffend*
of the day jeweilig * *die jeweilige Person*
of the essence von ausschlaggebender Bedeutung
of the same -gleich
of the same construction baugleich

of the same kind gleichartig
of the same rating baugleich * *mit gleichen Nenn-/Kennwerten/Typenschildangaben* || leistungsgleich
of the same type of construction baugleich
of the same value gleichwertig * *~ bezüglich des Geld-/Nutzwertes*
of this kind derartig
of this nature derartig * *a voltage notch of this nature: ein ~er Spannungseinbruch* || solcher * *a voltage notch of this nature: ein ~ Spannungseinbruch*
of this sort solcher * *~ Art*
of thought gedanklich * *depth of thought:gedankliche Tiefe*
of today heutig || jetzig * *heutig*
of top priority vorrangig * *to be processed with top priority: ~ zu bearbeiten*
of twist Verdreh- * *(nachgestellt)*
of use nützlich
of which dessen * *(nachgestellt) für Sachen*
off ab * *weg, entfernt, ab; washed off: abgewaschen; rubbed off: abgerieben* || abgeschaltet || aus || davon * *hinweg/-fort* || frei * *off-day:freier Tag* || Weg * *(Adv.) fort*
OFF condition Aus-Zustand
OFF delay Ausschaltverzögerung
off gauge nicht maßhaltig * *z.B. Draht, Blech*
off- außerhalb * *z.B. off-chip: außerhalb des integrierten Schaltkreises*
off-center außermittig * *(amerikan.)*
off-centre außermittig * *(brit.)*
off-chip extern * *außerhalb des integrierten Schaltkreises*
off-circuit spannungsfrei || spannungslos
off-cut Fehlschnitt * *z.B. bei Querschneider*
off-delay time Ausschaltverzögerung * *~szeit*
off-limit condition Grenzwertüberschreitung
off-load unbelastet
off-print Sonderdruck * *z.B. eines Zeitschriftenartikels*
off-size condition Maßabweichung * *als Zustand/Tatbestand*
off-state current Sperrstrom * *z.B. bei Diode, Thyristor, Transistor*
off-state voltage Sperrspannung * *z.B. Thyristor/Diode*
off-the-shelf ab Lager || ab Lager lieferbar || handelsüblich * *lagermäßig, von ~r Qualität; Teile* || standard * *handelsüblich* || über den Ladentisch * *aus dem Verkaufsregal*
OFF1 AUS1 * *Steuerbit im* →*PROFIBUS-Profil; Motor über Rampe stillsetzen, bei Drehz.Null Spannungsabschaltg*
OFF2 AUS2 * *Steuerbit im* →*PROFIBUS-Profil: Motor mögl.schnell spannungslos machen, Antrieb trudelt aus*
OFF3 AUS3 * *Steuerbit im* →*PROFIBUS-Profil: Motor m.max. mögl. Bremsmoment stillsetz.,dann Spanng. abschalt*
offcut Abschnitt * *das Abgeschnittene*
offence Ärgernis * *(brit.) Ärgernis, Anstoß, Vergehen, Verstoß, Delikt; give offence: ~ erregen* || Verstoß
offend verstoßen * *against: gegen*
offend against verletzen * *Vorschrift, Anstand ~* || verstoßen gegen
offense Ärgernis * *(amerikan.)* || Verstoß * *(amerikan.)*
offer anbieten || Angebot * *on offer:im Angebot* || Antrag * *Antrag; Vorbringen (e. Vorschlags, e. Meinung usw.)* || bieten * *anbieten, bereitstellen* || gewähren * *anbieten, sich bereit erklären zu (z.B. zu Sonderpreis, Vorteil)* || vermitteln * *Vorstellung, Eindruck, Bild ~*

office Büro || Geschäftsräume * *Büro* || Geschäftsstelle || Niederlassung
office apparatus Büromaschine
office automation Büroautomatisierung
office environment Büroumgebung
officer Beamter || Beauftragter * *mit einer offiziellen Aufgabe betraut*
official Beamter
official in charge Bearbeiter * *Sachbearbeiter* || Sachbearbeiter
official journal Amtsblatt
official journey Dienstreise
official trip Dienstreise
officially approved amtlich anerkannt
offline calender Einzelkalander * *einzeln stehendes Glättwerk, nicht in Papiermaschine integriert*
offline operation Offline-Betrieb
offload Ausladung * *Entladung von Frachtgütern* || entlasten
offloading equipment Ablegeeinrichtung
offroad Gelände- * *{Verkehr}*
offset Regelabweichung || Versatz * *z.B. Software: Basisadresse, Spuren eines Impulsgebers; time offset: zeitlicher ~* || verschieben * *bezüglich Phase/Spannung ~; auch 'ausgleichen'* || versetzt * *offset by 60°: um 60° ~; against/with respect to each other: gegeneinander*
offset address Adressversatz
offset adjustment Nullabgleich * *Nullpunktabgleich/-korrektur* || Offset-Abgleich
offset compensation Nullpunktabgleich || Nullpunktsabgleich || Offset-Abgleich * *Nullpunktkorrektur*
offset error Nullpunktfehler || Nullpunktsfehler || Nullpunktverschiebung * *Nullpunktfehler z.B. durch Drifteffekte* || Offset-Fehler
offset error suppression Nullpunktunterdrückung
offset press Offsetdruckmaschine
offset printing Buchdruck * *Offsetdruck, Druck auf Gummi, Platten-Druckverfahren* || Offsetdruck
offset printing machine Offsetdruckmaschine
offset printing paper Offsetpapier
offset printing press Offsetdruckmaschine
offshore offshore * *auf dem Meer (z.B. Ölplattform zum Abbau v. Bodenschätzen)*
offshore drilling platform Bohrplattform
Ohm's law Ohmsches Gesetz
ohmic Ohmsch * *purely ohmic: rein ~; ohmic voltage drop: ~er Spannungsabfall*
ohmic component Ohmscher Anteil
ohmic load Ohmsche Last
ohmic losses Ohmsche Verluste || Stromwärmeverluste
ohmic resistance Ohmscher Widerstand || Wirkwiderstand * *Ohmscher Widerstand*
oidal -förmig * *z.B. sinus~, trapez~, scheiben~*
oil einfetten * *mit Öl* || Öl || schmieren * *mit Öl*
oil bath Ölbad
oil bath lubrication Öltauchschmierung
oil carbon Ölkohle
oil charge Ölfüllung * *z.B. in Motor, Getriebe, Verdichter*
oil circulating pump Ölumlaufpumpe
oil circulation Schmierkreislauf * *Ölumlauf*
oil cooler Ölkühler
oil cup Schmierbüchse * *für Öl*
oil drag Lagerreibung
oil drain plug Ölablassschraube
oil duct Ölkanal
oil feed Ölversorgung * *Ölzufuhr*
oil filling Ölfüllung
oil filling plug Öleinfüllschraube
oil film Schmierfilm * *Ölfilm*
oil immersion Ölkapselung * *Schutzart 'o' nach EN50015 für Ex-geschützte el. Betriebsmittel*

oil inlet Öleinführung
oil level Ölstand
oil level gauge Füllstandsanzeiger * *für Öl* || Ölstandsanzeige * *~r*
oil level indication Ölstandsanzeige
oil level indicator Füllstandsanzeiger * *für Öl*
oil lubrication Ölschmierung
oil mist Ölnebel
oil passage Ölkanal
oil pressure Öldruck
oil pump Ölpumpe || Schmierpumpe * *Ölpumpe*
oil quantity Ölmenge
oil resistance Ölbeständigkeit
oil resistant ölbeständig
oil rig Bohrplattform * *zum Bohren nach Öl auf dem Lande oder im Meer*
oil seal friction Losbrechreibung * *augrund der anfänglichen "Klebewirkung" des Schmieröls*
oil seal ring Simmerring
oil sealing Ölkapselung * *Abdichtung gegen Öl*
oil shortage Ölmangel
oil spray Ölnebel * *Sprühöl*
oil supply Ölversorgung
oil supply pipe Ölleitung * *zur Schmierölversorgung*
oil tank Ölbehälter
oil vapor Öldampf
oil vapour Ölnebel
oil-cooled ölgekühlt * *z.B. Transformator, Gleitlager usw.*
oil-cooled starter Ölanlasser * *Motoranlasser mit ölgekühlten Anlasswiderständen*
oil-feed Ölleitung * *oil pressure feed: Öldruckleitung*
oil-feed pump Ölpumpe
oil-fired boiler Ölkessel
oil-immersed motor Unterölmotor * *in (Hydraulik-)Öl betriebener Motor, z.B. Hydraulikpumpenmotor*
oil-immersed transformer Öltransformator * *mit Öl als Kühlmittel*
oil-laden air ölgeschwängerte Luft
oil-lead Ölleitung
oil-lubricated ölgeschmiert
oil-proof bearing öldichte Lagerung * *Lagerg.mit einseitig. Dichtung geg. druckloses Öl (z.B. mit →Radialwellendichtring)*
oil-tight öldicht
oil-type capacitor ölgefüllter Kondensator
oiler Schmierbüchse * *für Öl*
oilfree vacuum pump ölfreie Vakuumpumpe
oiling Schmierung * *mit Öl*
oiltight Öldicht
oilway Ölkanal
oily schmierig * *ölig*
old alt * *(auch Substantiv: "das Alte")*
old hand Praktiker * *(salopp)*
old things alt * *(Substantiv) das Alte*
old ways alt * *(Substantiv) die alte (Art und) Weise*
old-fashioned veraltet * *altmodisch*
old-time alt * *zurückliegend*
omen Zeichen * *vorausahnenlassendes ~, An~*
omission Lücke * *Auslassung*
omit außer Acht lassen * *aus/unter/weglassen (from: aus/von), versäumen (object to do: etw.zu tun)* || auslassen * *weglassen* || übergehen * *auslassen* || weglassen
on an || auf || aus || bei * *(zeitlich)* || EIN || nach * *on arrival: ~ (der) Ankunft; on receipt: ~ Erhalt* || über * *~ etwas (Abhandlung, Vortrag usw.)*
on a large scale in großem Umfang
on a long-term basis auf lange Sicht
on a par with gleichwertig
on a regular basis regelmäßig
on account of auf Grund * *um ... willen, wegen* ||

über * *wegen* || wegen * *wegen*, um ... *willen; on his account: seinet~*
on behalf of im Auftrag von * *auch für jemanden*
on both sides beidseitig || zweiseitig * *auf beiden Seiten*
on call auf Abruf
on condition unter der Bedingung * *that: dass* || unter der Voraussetzung * *that:daß* || vorausgesetzt * *that: vorausgesetzt, dass*
on condition that unter der Bedingung dass || vorausgesetzt dass
on contract-specific basis auftragsgebunden
ON delay Einschaltverzögerung
on demand auf Anforderung
on easy terms günstig * *zu ~en Bedingungen*
on edge hochkant
on front frontseitig
on hand auf Lager * *bei der Hand, greifbar* || vorrätig
on his part seinerseits * *(Adv.)*
on load belastet * *unter Last* || unter Belastung || unter Last
on no account auf keinen Fall || keinesfalls || nie * *keinesfalls* || unter keinen Umständen || unter keiner Bedingung
on no condition keinesfalls
on occasion gelegentlich * *(Adv.) bei Gelegenheit*
on output side abtriebsseitig * *z.B. bei Getriebe, Durchtrieb usw.*
on principle grundsätzlich || im Prinzip * *aus Prinzip* || prinzipiell * *aus Prinzip*
on request auf Anforderung || auf Anfrage || auf Wunsch * *auf Anfrage*
ON resistance Durchlasswiderstand * *z.B. Feldeffekttransistor, FET-Schalter* || Einschaltwiderstand * *z.B. eines FET-Transistors im On-Zustand*
on sale auf dem Markt * *käuflich*
on site vor Ort * *an Ort und Stelle, auf der Baustelle/Anlage/beim Kunden*
on stock vorhanden * *auf Lager* || vorrätig
on that account deshalb * *als Folge davon, infolgedessen*
on that matter darüber * *bezüglich dieser Angelegenheit/Sache*
on that point darüber * *in dieser Hinsicht*
on the average durchschnittlich * *(Adv.) im Durchschnitt*
on the cathode side kathodenseitig
on the contrary dagegen * *indessen* || vielmehr * *im Gegenteil*
on the customer side kundenseitig
on the DC side auf der Gleichstromseite || gleichstromseitig
on the due date termingemäß || termingerecht
on the end of the line am Telefon * *am (anderen) Ende der Leitung*
on the fly Echtzeit * *während des Betriebes, ohne anzuhalten* || fliegend * *während des Betriebes, ohne anzuhalten* || während des Betriebes * *ohne anzuhalten*
on the inside innen * *(Adv.)*
on the left links || seitlich links
on the left hand side seitlich links
on the left side links
on the level of hinsichtlich
on the line side netzseitig
on the market auf dem Markt
on the motor side maschinenseitig * *motorseitig, z.B. Stromrichter* || motorseitig
on the next page umseitig * *auf der nächsten Seite*
on the occasion dabei * *anlässlich*
on the occasion of bei * *gelegentlich des ...*
on the one hand einerseits * *on the other hand: andererseits*
on the one side einerseits

on the other hand andererseits || dagegen * *andererseits*
on the outside außen * *(Adv.) außerhalb* || von außen * *außerhalb*
on the phone am Telefon
on the plant side anlagenseitig
on the reverse umseitig
on the reverse page umseitig * *auf der Rückseite*
on the right rechts || seitlich rechts
on the right side rechts
on the right-hand side seitlich rechts
on the shelf auf Lager
on the shop floor vor Ort * *direkt an/bei den Produktionseinrichtungen*
on the side -seitig * *on the motor side:motorseitig* || seitlich * *(Adv.)*
on the spot sofort
on the strength of laut * *kraft*
on the supply system am Netz
on the surface oben * *auf der Oberfläche*
on the system side anlagenseitig
on the telephone am Telefon
on the three-phase side drehstromseitig
on the understanding unter der Voraussetzung * *that: dass*
on the whole im Großen und Ganzen
on this day heutig
on this occasion hierbei * *bei dieser Gelegenheit*
on time pünktlich || rechtzeitig || sofort * *pünktlich* || sofortig * *pünktlich* || termingemäß * *pünktlich* || termingerecht * *pünktlich* || zeitgerecht * *(Adv.)* pünktlich, termingerecht, rechtzeitig
on top oben * *oben(auf); from top to bottom: von ~ nach unten; towards the top: nach ~*
on top of auf * *auf ... oben 'drauf'*
on top of each other aufeinander || übereinander
on top of it darüber
on top of one another aufeinander
on trial versuchsweise
on weekdays werktags
on workdays werktags
on working days werktags
on-board integriert * *auf der Leiterkarte*
on-board power supplies Bordnetzversorgung
on-board power supply Bordstromversorgung
on-call service Bereitschaftsdienst
on-chip integriert * *Zusatzfunktion auf einem integriertem Schaltkreis* || intern * *auf Halbleiterchip integriert*
on-condition eingeschalteter Zustand * *auch eines Thyristors*
ON-delay Ansprechverzögerung || Anzugsverzögerung * *Relais* || Einschaltverzögerung
ON-delay relay ansprechverzögertes Relais
on-direction Durchlassrichtung
on-line während des Betriebes * *ohne abzuschalten*
on-line operation Netzbetrieb * *Betrieb am Stromnetz oder Kommunikationsnetz*
on-line-help Online-Hilfe
on-load current Betriebsstrom
on-load operation Betrieb unter Last
on-off control Zweipunktregelung
on-off controller Zweipunktregler
on-off logic module Einschalt-Ausschalt-Logik
on-off logic sequencing Einschalt-Ausschalt-Logik
on-off switch Ein-Aus-Schalter
on-resistance Durchlasswiderstand
on-screen instruction Bildschirmanweisung
on-state Durchlasszustand * *z.B. Thyristor* || eingeschalteter Zustand * *z.B.Thyristor*
on-state characteristic Durchlasskennlinie * *(Thyristor)*
on-state current Durchlassstrom * *bei Diode, Thyristor, Transistor usw.*

on-state loading limit Durchlassbelastungsgrenze * *z.b. bei Diode, Thyristor*
on-state resistance Durchlasswiderstand * *eines elektronischen Schalters (Diode, Transistor, Thyristor usw.)*
on-state voltage Durchlassspannung * *(Thyristor)*
on-the-job training Ausbildung am Arbeitsplatz || Einweisung * *Vor-Ort-Schulung*
on-the-spot vor Ort
on/off switch Ein-Ausschalter
once einmal || einmalig * *einmal* || einzig * *(Adv.)* never once: nicht ein ~es Mal || mal * *einmal; once upon a time: es war einmal*
once again erneut * *(Adv.)* noch einmal
once more erneut * *(Adv.)* noch einmal, von neuem
oncoming entgegenkommend * *z.B. Fahrzeug, Verkehr*
oncoming valve Folgeventil * *das nächste zu zündende Ventil in e. Stromrichter*
one eins || einzig * *my one thought: mein ~er Gedanke* || erst * *(nachgestellt) step one: ~er Schritt* || man
one after another aufeinander * *nacheinander* || nacheinander
one after the other hintereinander
one and a half anderthalb || eineinhalb
one and the same derselbe * *ein und ~*
one another gegenseitig * *einander*
one beneath the other untereinander * *(räumlich)*
one by one aufeinander * *nacheinander* || einzeln * *(Adv.) einer nach dem andern, jeder für sich* || hintereinander
one on top of the other aufeinander
one stop shop Komplettanbieter * *(salopp)* || Vollanbieter * *bietet alles aus einer Hand*
one stop shopping alles aus einer Hand
one thing eins * *das eine*
one upon the other aufeinander * *aufeinander, gegenseitig; matching: ~ abgestimmt* || übereinander || zueinander * *gegenseitig, →aufeinander*
one with another untereinander * *miteinander*
one's own eigener || selbst * *eigen*
one- Einfach- || Einzel-
one-for-one eins-zu-eins
one-half Hälfte * *one-half the: die ~ des*
one-hand Einhand * *z.B. ~bedienung, ~hobel usw.*
one-man operation Einmannbedienung
one-off production Einzelfertigung
one-part einteilig
one-piece einteilig
one-quadrant Einquadrant-
one-quadrant drive Einquadrantenantrieb
one-quadrant operation Einquadrantbetrieb
one-shot Impulsbildner * *Zeitstufe* || Zeitstufe * *Impulsbildner, Monoflop, monostabiler Multivibrator*
one-sided einseitig * *(auch figürlich)*
one-sidedness Einseitigkeit * *(auch figürlich)*
one-way Einweg-
one-way clutch Freilauf
one-way converter Einfachstromrichter * *Einquadrant-Stromrichter*
one-way rectification Einweggleichrichtung
one-way rectifier Einweggleichrichter
onerous schwierig * *lästig, beschwerlich* || ungünstig * *lästig, beschwerlich*
ongoing heutig
online Echtzeit * *z.B. bei laufender Maschine* || in Betrieb * *während des Betriebes, in Echtzeit* || während des Betriebes
online operation Online-Betrieb
only allein * *nur* || bloß * *nur* || einzig * *the only one: der ~e* || erst * *bloß* || nur
only intermittently zeitweise * *(Adv.) mit Unterbrechungen, nur zeitweise*

onshore onshore * *auf dem Lande (z.B. Abbau v. Bodenschätzen)*
onward an * *(Adv.)*
onwards ab * *seit; from version 3.2 onwards:ab Version 3.2; from this time onwards: ab diesem Zeitpunkt*
OOP objektorientierte Programmierung * *(Computer) Abkürz. für 'Object-Oriented Programming'*
Oops Huch * *(salopp)*
Op Amp Operationsverstärker * *Abkürzg.f. 'Operational Amplifier'*
op code Befehlscode * *in Binärdarstellung*
opacity Opazität
OpAmp Operationsverstärker * *Abkürzg.f. 'Operational Amplifier'*
OpAmp board Universalverstärkerbaugruppe
opcode Befehl * *(Computer) Computerbefehl in Binärdarstellg.* || Befehlscode * *in Binärdarstellung*
opcode decoder Operationscode-Decoder * *Funktionsblock in der CPU eines Computers*
open auf * *offen* || aufschließen * *Tür usw. ~* || auftrennen || ausschalten * *(Kontakt) öffnen* || erschließen || firmenneutral * *offen, z.B. Bussystem* || firmenübergreifend * *offen(gelegt) z.B. Bussystem/protokoll* || frei * *offen, freimütig* || frei * *Markt, Feld, Himmel, Stelle usw.* || in Betrieb setzen * *eröffnen, Betrieb aufnehmen* || inbetriebnehmen * *eröffnen, Betrieb aufnehmen* || inbetriebsetzen * *eröffnen, Betrieb aufnehmen* || lösen * *öffnen* || offen || offenliegend * *offen (-gelegt) z.b. technische Lösung für andere Firmen* || öffnen || übersichtlich * *Gelände* || zugängig * *to:für; (auch figürl. z.B. für Gründe)* || zugänglich * *to: für; (auch figürl., z.b. für Gründe)*
open circuit Unterbrechung * *Stromkreis-~*
open collector offener Kollektor * *Transistorschaltung* || Open-Kollektor * *Transistorschaltung mit offenem Kollektor*
open communication offene Kommunikation * *allgemein zugängliche/firmenübergreifende Kommunikation (-shilfsmittel)*
open contact offener Kontakt
open design offene Ausführung
open drip-proof ODP * *Schutzart IP22 nach IEC und DIN ('offen, tropfwassergeschützt')*
open emitter Open-Emitter * *Transistorschaltung mit offenem Emitter*
open line Freileitung
open loop speed control Drehzahlsteuerung
open loop tension control indirekte Zugregelung * *ohne Tänzer oder Zugmessdose*
open loop tension regulator indirekte Zugregelung
open machine offene Maschine * *innengekühlte, durchzugsbelüftete Maschine (Schutzart IP23)*
open points Unklarheit * *offene Fragen*
open protocol offenes Protokoll * *z.B. Busprotokoll zur seriellen Kommunikation*
open the door to einlassen
open type offene Bauart
open up aufblenden * *aufblättern, öffnen (auch Anzeigefeld auf dem Bildschirm)* || aufschließen * *Markt/Bodenschätze usw. auf-/erschließen* || eröffnen * *z.B. Möglichkeiten* || erschließen * *z.B. neue Anwendungen, Märkte usw. ~*
open version offene Ausführung
open, self-ventilated innengekühlt * *innengekühlt mit Motorwelle-getriebenem Lüfter;Kühlart IC01/11/21/31 IEC34.6*
open, separately ventilated innengekühlt durch Fremdbelüftung * *z.B. Motorkühlarten IC05/06/17/37 nach DIN/IEC 34 Teil 6*
open, ventilated innenbelüftet || innengekühlt * *Motorkühlart*

open-cast mining Tagebau * *{Bergwerk}*
open-chassis unit Einbaugerät * *zum Schrankeinbau ohne besondere Schutzart, z.B. auf offenem Tragblech (IP20)*
open-circuit Leerlauf * *Schaltung ohne Last* || Leerlauf- || offen * *Kontakt* || unbeschaltet
open-circuit air cooling Innenkühlung * *Durchzugsbelüftung (Motorkühlart)*
open-circuit air-cooled durchzugsbelüftet * →*eigen-* oder →*fremdbelüfteter Motor mit Luftstrom durch die Maschine* || innengekühlt
open-circuit air-cooling Durchzugsbelüftung
open-circuit characteristic Leerlaufkennlinie
open-circuit cooled durchzugsbelüftet || innengekühlt
open-circuit cooling Durchzugsbelüftung * *z.B.* →*eigen-* oder →*fremdgekühlter Motor mit Luftstrom durch d.Maschine* || Innenkühlung * *z.B. Motor mit innen durchströmender Kühlluft, Kühlarten IC01,IC11, IC21,IC31*
open-circuit DC voltage Leerlaufgleichspannung
open-circuit monitoring Drahtbruchüberwachung
open-circuit operation Leerlauf * *Betrieb ohne Last*
open-circuit pressurized enclosure Überdruckkapselung mit ständiger Durchspülung * *Zünd/Explosionsschutzart*
open-circuit proof leerlauffest * *z.B. für Stromsignal 0-20mA*
open-circuit ventilated durchzugsbelüftet * *Motorkühlart mit Luftstrom durch das Gehäuse (Schutzart z.B. IP23)* || innengekühlt
open-circuit ventilation Durchzugsbelüftung || Innenkühlung * *Motorkühlart mit innen durchströmender Kühlluft (Durchzugsbelüftung)*
open-close condition Schaltstellung * *eines Schalters, Relais, Schützes (z.B. opt. angezeigte)*
open-end rotor spinning machine OE-Spinnmaschine * *{Textil}*
open-end spinning machine OE-Spinnmaschine * *{Textil} im Gegensatz zur* →*Ringspinnmaschine* || Open-End-Spinnmaschine * *{Textil}*
open-end wrench Gabelschlüssel
open-ended spanner Gabelschlüssel
open-loop and closed-loop control functions - Steuer- und Regelfunktionen
open-loop control steuern * *im Gegensatz zum 'regeln'* || Steuerung
open-loop control functions Steuerungsfunktionen
open-loop control section Steuerteil
open-loop controlled gesteuert * *nicht geregelt* || ungeregelt * *gesteuert*
open-loop controlled variable Steuergröße
open-loop frequency control Frequenzsteuerung * *im Gegensatz zur Regelung (closed-loop)*
open-minded vorurteilsfrei
open-pit mine Tagebau * *{Bergbau} ~-Grube, ~- Bergwerk*
open-pit mining Bergwerks- * *Tagebau-*
open-type machine offene Maschine * *{el.Masch.} z.B. Motor mit* →*Innenkühlung*
open-up eröffnen || erschließen * *auch Absatzgebiete, Märkte usw. ~*
opencast working Tagebau * *{Bergbau}*
opened geöffnet * *auch Schaltkontakt*
opening Abschaltung * *Kontakt, Stromkreis* || Aussparung * →*Öffnung* || Beginn * *~ einer Verhandlung, Schul~ usw.* || Durchbruch * *Öffnung* || Durchlass * *Öffnung* || Fenster * *Öffnung* || Inbetriebnahme * *Eröffnung* || Inbetriebsetzung * *Aufnahme des Betriebes (of:von einer (Fabrik)Anlage usw.)* || Marktlücke || Öffnung || Spannweite * *eines Schraubenschlüssels*
opening for the cooling air Kühlluftöffnung
opening hours Öffnungszeiten

opening in the front panel Frontplattenaussparung
opening roll Auflösewalze * *{Textil} bei Open-End-Spinnmaschine*
opening time Ansprechzeit * *eines Öffnerkontakts* || Ausschaltzeit * *eines Relais usw.*
openly laut * *(Adv.) offen*
operable betriebsbereit * *siehe auch 'betriebsfähig'* || funktionsfähig
operand Operand || Parameter * *einer (SPS-) Anweisung*
operand area Operandenbereich * *bei SPS*
operand identifier Operandenkennzeichen
operate Ansprechen * *(Verb) Relais, Sicherung, Überspannungsableiter* || anziehen * *Relais, Bremse usw.* || arbeiten * *z.B.Antrieb/Verstärker/Gerät/Einrichtung* || auslösen || bedienen * *auch 'handhaben, betätigen'* || betätigen * *z.B.Schalter/Bremse; auch bedienen,laufen lassen,handhaben,einwirken (on/upon:auf)* || betreiben * *Maschine, Anlage ~ (at: bei (e. Betriebspunkt, Frequenz usw.))* || einwirken * *upon:auf* || gehen * *bei Maschine usw.* || handhaben * *Maschine, Werkzeug usw.* || wirken * *arbeiten, in Betrieb sein, funktionieren, laufen (on:auf)*
operate on wirken auf
operate the clutch kuppeln * *die Kupplung betätigen*
operate time Anzugszeit * *Relais*
operate voltage Anzugsspannung * *Relais*
operated -betreiben || betätigt || betrieben
operated through spindle gearbox spindelgetrieben
operating Bedienungs- || Betriebs- * *z.B. Strom, Spannung, Bedingungen*
operating a motor from a converter Betrieb eines Motors am Umrichter
operating accuracy Schaltgenauigkeit * *z.B. einer Kupplung/Bremse*
operating airgap Arbeitsluftspalt * *z.B. bei Kupplung, Bremse*
operating altitude Aufstellhöhe
operating and monitoring Bedienen und Beobachten
operating area Arbeitsbereich
operating as a generator generatorisch * *~ arbeitend*
operating assets Betriebsvermögen * *Betriebskapital, betriebsnotwendiges Kapital*
operating at the nominal working point Nennbetrieb * *am Nenn-Arbeitspunkt*
operating availability Verfügbarkeit * *relative ~ eines Geräts/einer Anlage/einer Maschine*
operating characteristics Betriebseigenschaften || Betriebsverhalten || Betriebswerte
operating condition Betriebsbedingung || Betriebszustand * *in:in*
operating conditions Einsatzbedingungen * *most severe: härteste*
operating connection Betriebsschaltung * *z.B. Stern-, Dreieckschaltung*
operating control Betriebsüberwachung
operating control panel Bedienfeld
operating costs Betriebskosten
operating current Arbeitsstrom * *z.B. bei Bremse, Magnetventil*
operating cycle Arbeitsgang || Arbeitszyklus || Schaltspiel * *Lastspiel, Betätigungsspiel, siehe auch EN50032 und DIN IEC 147-1D*
operating data Betriebsdaten
operating data acquisition Betriebsdatenerfassung
operating device Bedienelement
operating display Betriebsanzeige
operating element Bedienteil
operating enable Betriebsfreigabe * *Steuerkommando*

operating energy Antriebsenergie * *für Betätigungsantrieb*
operating environment Betriebssystem * *Umgebung*
operating environments Betriebsverhältnisse
operating error Bedienungsfehler
operating expenses Betriebskosten
operating experience Betriebserfahrung || Betriebserfahrungen
operating frequency Betriebsfrequenz || Schaltfrequenz * *von Lastspielen/Schaltspielen, siehe EN50032 und DIN IEC147-1D* || Schalthäufigkeit * *bei einem Last/Schalt/Betätigungsspiel*
operating from a three-phase system Betrieb am Drehstromnetz
operating hours Betriebsstunden || Betriebstunden
operating instructions Anleitung * *Betriebsanleitung* || Bedienungsanleitung || Beschreibung * *Betriebsanleitung* || Betriebsanleitung || Handbuch * *Betriebsanleitung*
operating lever Betätigungshebel
operating life Lebensdauer
operating lifetime Lebensdauer
operating limit Ansprechgrenze
operating mechanism Antrieb * *~/Betätigungsteil eines Leistungsschalters usw.*
operating memory Arbeitsspeicher
operating mode Betriebsart
operating noise Betriebsgeräusch
operating panel Bedienstation
operating parameter Betriebsparameter
operating period Betriebszeit
operating personnel Bedienungspersonal || Maschinenpersonal
operating point Arbeitspunkt || Betriebspunkt || Betriebszustand * *Betriebs/Arbeitspunkt* || Nennpunkt * *Betriebspunkt* || Schaltpunkt
operating position Arbeitsstellung || Betriebsstellung
operating pressure Betriebsdruck * *z.B. in Hydrauliksystem*
operating principle Funktionsprinzip || Funktionsweise || Wirkungsweise * *Funktionsprinzip*
operating principles Anwendungsrichtlinien
operating quadrant Betriebsquadrant
operating range Aktionsradius * *Fahrzeug* || Ansprechbereich || Arbeitsbereich
operating regime Fahrweise * *Art und Weise der Maschinenführung durch das Personal*
operating sequence Lastspiel * *"Programm" einer Arbeitsmaschine* || Schaltfolge
operating span Reichweite
operating specification Betriebswert * *z.B. Nennwert bei Bemessungsbedingungen*
operating speed Arbeitsgeschwindigkeit || Betriebsdrehzahl
operating speed range Betriebsdrehzahlbereich
operating staff Bedienungspersonal
operating state Betriebszustand
operating station Bedienstation
operating status Betriebszustand * *(Plural: operating statuses)*
operating statuses Betriebszustände
operating system Ablaufsteuerung * *Betriebssystem* || Betriebssystem * *Software. die den Ablauf u. die Resourcen-Zuteilung für Anwenderprogramme verwaltet* || System-Software || Systemsoftware * *Betriebssystem*
operating temperature Betriebstemperatur || Schaltpunkt * *Schalt-Temperatur, z.B. eines Thermoschalters*
operating test Funktionsprüfung
operating time Ausschaltzeit * *einer Schmelzsicherung usw.* || Einschaltdauer || Laufzeit * *einer Maschine*

operating unit Bediengerät
operating value Bedienwert * *bei SPS* || Betriebswert
operating voltage Betriebsspannung
operation -Wirkung * *Wirken, Wirkung(-weise), Betrieb, Tätigkeit,Lauf, Arbeitweise/-gang, Bedieng., Betätigg.* || Arbeitsweise || Bedienung || Befehl * *[Computer | SPS] Operation* || Betätigung * *Bedienung* || Betrieb * *Betreiben; out of op.:außer ~; start op.: ~ aufnehmen; go into op.:in ~ gehen* || Einwirkung || Funktion * *Operation (eines Rechners)* || Gang * *Betrieb* || Geschäft * *~stätigkeit eines Unternehmens* || Lauf * *Betrieb* || Operation || Schaltspiel * *bei Relais usw.* || Vorgang * *Arbeits(vor)gang, Prozess, Betätigung, Wirken, Tätigkeit* || Wirkung * *Tätigkeit*
operation at high delay angles herabgesteuerter Betrieb * *niedriger Aussteuergrad bei Thyristorbrücke*
operation at light loads Teillastbetrieb
operation characteristic Arbeitskennlinie
operation check Betriebskontrolle
operation code Befehlscode * *in Binärdarstellung*
operation condition Betriebsbedingung
operation data Betriebsdaten
operation execution Operationsausführung
operation execution time Befehlsausführungszeit * *[SPS]* || Operationszeit * *Ausführungszeit einer (SPS-) Befehls*
operation from a 60 Hz supply Betrieb am 60 Hz-Netz
operation from ungrounded supply Betrieb am ungeerdeten Netz * *z.B. eines Stromrichters*
operation in perfect synchronism Winkelgleichlauf
operation log Betriebsprotokoll * *für eine Produktionsanlage/Maschine* || Protokoll * *Betriebs~ z.B. für eine Produktionsanlage, Maschine*
operation panel Bediengerät * *Bedienfeld*
operation set Operationsvorrat * *Befehlsvorrat einer SPS*
operation speed Betriebsdrehzahl
operation stoppage Betriebsunterbrechung
operation test Funktionsprüfung
operation time Lüftzeit * *einer Bremse mit Ruhestromprinzip*
operation under rated conditions Nennbetrieb * *unter Nennbedingungen*
operational betrieblich || Betriebs- || betriebsbedingt * *betriebsmäßig vorkommend* || betriebsfähig * *betriebsmäßig * den Betrieb betreffend, während des Betriebs auftretend* || betriebstechnisch || betriebstüchtig * *siehe auch →betriebsfähig* || einsatzbereit * *betriebsfähig, arbeitend* || in Betrieb * *~ befindlich, betriebsfähig* || operativ
operational amplifier Operationsverstärker
operational behaviour Betriebsverhalten
operational brake Betriebsbremse * *im Gegensatz zur →Haltebremse*
operational data acquisition Betriebsdatenerfassung
operational dependability Betriebssicherheit
operational display Betriebsanzeige
operational duration Betriebsdauer
operational economics Wirtschaftlichkeit * *z.B. einer Maschine/Anlage*
operational efficiency Wirtschaftlichkeit
operational enable Betriebsfreigabe * *Steuerkommando*
operational example Bedienungsbeispiel
operational panel Bedienfeld
operational reliability Betriebssicherheit * *Zuverlässigkeit*
operational safety Betriebssicherheit

operational sequence Bedienablauf || Betriebsablauf
operational status Betriebszustand * *einer Maschine/Anlage*
operational test Funktionskontrolle * →*Funktionsprüfung* || Funktionsprüfung * siehe auch →*Funktionskontrolle*
operations Bedienung || Geschäft * *~stätigkeiten eines Unternehmens*
operative control Bedienung
operator Arbeiter * ~ *an der Maschine* || Bedienelement * *(amerikan.)* || Bediener || Bedienfeld * *(amerikan.)* || Bediengerät * *(amerikan.)* || Bedienungs- || Bedienungselement * *(amerikan.)* || Bedienungsmann || Betreiber || Wärter
operator communication Bedienung * *bei SPS usw.*
operator communication and observation Bedienen und Beobachten
operator console Bedienpult
operator control Bedienung || Bedienungs-
operator control and monitoring Bedienen und Beobachten
operator control and visualization functions Bedien- und Beobachtungsfunktionen
operator control element Bedienelement || Bedienteil || Bedienungselement || Betätigungselement * *zur Bedienung*
operator control functions Bedienfunktionen
operator control interface Bedienoberfläche
operator control panel Bedienfeld || Bedientableau * *siehe auch* →*Bedientafel* || Bedientafel || Gerätebedienfeld
operator control station Bedienplatz || Bedienstation
operator control unit Bedieneinheit || Bediengerät
operator convenience Bedienungskomfort || leichte Handhabbarkeit * *durch Bedienungsmann*
operator desk Bedienpult
operator device Bedienelement || Bedienungselement
operator devices Bedienungselemente
operator friendliness Bedienkomfort
operator input Bedienung * *Eingabe*
operator inspection Selbstprüfung * *Werkstückprüfung durch den Maschinenbediener selbst (z.B. bei e. Werkz.masch.)*
operator interaction Benutzereingriff * *Bedienereingriff*
operator interface Bedienoberfläche || Bedienschnittstelle || Bedienungsoberfläche || Bedienungsschnittstelle
operator intervention Bedieneingriff || Bedienereingriff
operator mistake Bedienungsfehler
operator panel Bedieneinheit * *Bediengerät* || Bedienfeld || Bediengerät * *Bedienfeld, z.B. für SPS* || Bedientafel
operator prompting Bedienerführung * *z.B. am Bildschirm*
operator safety Personensicherheit
operator side Bedienseite * *einer Maschine*
operator station Bedieneinheit || Bediengerät * *z.B. Bedienkasten/Pult*
operator unit Bediengerät
operator's control element Bedienungselement
operator's control elements Bedienungselemente
operator's guide Bedienungsanleitung
operator's panel Bedienstation || Bedientafel
operator's position Arbeitsplatz * *z.B. eines Maschinenführers*
operator's station Bedieneinheit || Bedienstation || Bedienungs- und Beobachtungsstation
operators guide Bedienungsanleitung
operators staff Bedienpersonal || Bedienungspersonal

opinion Ansicht * *Meinung; in my opinion: meiner ~ nach; be of the opinion that: der Meinung sein, dass* || Beurteilung * *Meinung; of:über*
opportune günstig * *Moment; opportunity: ~e Gelegenheit* || passend * *Zeit, Gelegenheit usw.* || zeitgemäß
opportunity Gelegenheit || Möglichkeit * *gute Gelegenheit*
oppose entgegengerichtet sein * *zu etwas* || entgegensetzen * *auch entgegentreten,sich widersetzen* || widersetzen * *sich ~, angehen gegen, bekämpfen*
opposed entgegengesetzt * *to:zu* || gegensätzlich * *entgegengesetzt, gegensätzlich, gegenüberliegend*
opposed to gegen * *~sätzlich*
opposing entgegengesetzt * *unvereinbar (to:mit), gegenüberliegend* || gegensätzlich * *gegenüberliegend, entgegengesetzt, opponierend* || gegenüberliegend || unvereinbar * *to:mit*
opposing current Gegenstrom
opposite ander * *gegenüberliegend* || anders * *(Adj.) gegenüberliegend, z.B.Seite* || entgegen * *~gesetzt (gerichtet), gegenüberliegend* || entgegengesetzt || Gegen- * *entgegengesetzt, in entgegengesetzter Richtung* || Gegensatz * *of: zu* || gegensätzlich || Gegenteil * *of:von* || gegenteilig || gegenüber * *to:von* || gegenüberliegend || nebenstehend * *in Dokument nebenan abgebildet* || umgekehrt * *entgegengesetzt, gegenteilig, widersprechend; the opposite holds good for: d. ~e gilt für* || vor * *gegenüber*
opposite direction Gegenrichtung
opposite in direction entgegengesetzt gerichtet
opposite number Entsprechung * *Gegenstück* || Gegenstück || Gesprächspartner * *Partner am Telefon, Gegenspieler*
opposite sign entgegengesetztes Vorzeichen
opposition Einspruch * *z.B. gegen Patentanmeldung* || Einwand || Gegensatz * *Widerspruch; in opposition to/as opposed to: im Gegensatz zu* || Widerspruch * *Wider-/Einspruch z.B. gegen Patentanspruch* || Widerstand * *Opposition, Wider-/Einspruch*
oppress unterdrücken * *bedrücken*
opt wählen * *sich entscheiden, optieren (for: für, between: zwischen)*
opt for wählen * *sich entscheiden/optieren für*
optical optisch
optical coupler Optokoppler
optical fibre Lichtleiter || Lichtwellenleiter * *(brit.)*
optical fibre interface Lichtleiteranschluss
optical industry optische Industrie
optical inspection Sichtprüfung
optical isolation Potenzialtrennung * *über Optokoppler/Lichtleiter*
optical isolator Optokoppler
optical sensor Lichtvorhang * *lichtempfindlicher Sensor*
optical waveguide Lichtwellenleiter
optics Optik
optimal optimal
optimally optimal * *(Adv.)*
optimisation Optimierung * *(brit.)*
optimistic optimistisch
optimistically optimistisch * *(Adv.)*
optimization Optimierung * *(amerikan.) siehe DIN 19236*
optimization guidelines Optimierungsrichtlinien
optimization routine Optimierungslauf
optimization run Optimierungslauf
optimize optimieren
optimized -optimiert || optimiert * *for: bezüglich*
optimized pulse patterns optimierte Pulsmuster * *bei Pulsumrichter*
optimizing Optimierung

optimizing compiler optimierender Compiler
optimizing edger Besäumautomat * *{Holz}*
optimizing the controller Regleroptimierung
optimum best * *optimal, am ~en* || *optimal* || Optimum
optimum of magnitude Betragsoptimum * *Regleroptimierg. so das Betrag d.Amplitud.gangs gleich 1 über mögl.weit.Frequ.*
option Alternative * *Wahlmöglichkeit; have no option:keine Alternative haben* || Erweiterung * *wählbare, zusätzlich erhältliche* || Möglichkeit * *(zusätzliche) Wahl~, Option* || Option || Wahl * *freie Wahl* || Zubehör * *wahlweise erhältliches* || Zusatz * *wählbarer ~, (gegen Aufpreis) zusätzlich lieferbarer ~* || Zusatz-
option board Optionsbaugruppe * *Leiterkarte* || Zusatzbaugruppe * *Leiterplatte*
option module Optionsbaugruppe || Zusatzbaugruppe
option- optional
optional als Option || freigestellt || optional || optionell * *als Option* || wählbar || wahlweise
optional extras Zubehör * *wahlweise erhältliches*
optionally optionell * *(Adv.)* || wahlweise * *(Adv.)*
opto-cable Lichtwellenleiter
opto-coupled potenzialfrei * *über Optokoppler gekoppelt*
opto-coupler Optokoppler
opto-isolated potenzialfrei * *über Optokoppler gekoppelt* || potenzialgetrennt * *über Optokoppler*
opto-isolation Potenzialtrennung * *über Optokoppler*
opto-isolator Optokoppler
optocoupler Optokoppler
optoelectronic optoelektronisch
optoelectronics Optoelektronik
optoelectronics and electro-optics Optoelektronik
or beziehungsweise || bzw || ODER || verodern
OR gate ODER-Glied
OR operation ODER-Verknüpfung
OR'd ODER-verknüpft || verodert * *logische Funktion*
oral mündlich
orally mündlich * *(Adv.)*
orbital Bahn- * *umlaufend* || umlaufend
orbital sander Schwingschleifer * *{Holz}*
order anordnen * *befehlen* || Anordnung * *Befehl* || Anweisung * *Befehl, Auftrag, Instruktion* || Auftrag * *Bestellung, Befehl, Instruktion; by order of: im ~ von; be in the order backlog: in ~ sein* || beauftragen || Befehl * *Auftrag* || bestellen || Bestellung * *Auftrag, Bestellung; give/place orders with: Bestellung bei ... aufgeben* || Folge * *Reihen~* || Größenordnung * *Art, Rang, Klasse* || Ordnung * *ordentlicher Zustand, Reihenfolge; second/third order: zweiter/dritter ~* || Ordnungszahl * *~ einer Oberschwingung* || Rangfolge || Reihenfolge * *alphabetic(al): alphabetische; rising/ascending:aufsteigende* || Vorschrift * *Befehl*
order address BZ-Empfänger * *Bestelladresse*
order administration Auftragsabwicklung || Bestellabwicklung
order blank Bestellformular * *(amerikan.)* || Bestellschein * *(amerikan.)* || Bestellzettel * *(amerikan.) Bestellformular* || BZ * *(amerikan.) Bestellformular*
order code Bestellnummer
order data Bestellhinweise
order designation Bestellbezeichnung
order entry Auftragseingang * *Bestelleingang, Eintreffen von Bestellungen* || Bestelleingang * *Zugang/Eintreffen von Bestellungen*
order form Bestellformular || Bestellschein || Bestellzettel * *Bestellformular* || BZ * *Bestellformular*

order form receiver BZ-Empfänger * *Bestellzettel-Empfänger*
order handling Auftragsabwicklung
order inflow Auftragslage
order intake Auftragseingang
order No Bestellnummer || BZ-Nummer * *Bestellnummer*
order No suffix Bestellnummerergänzung
order No. Artikel-Nr * *Bestellnr.* || MLFB
Order Nos. Bestellnummern * *Abkürzg. f. 'Order Numbers'*
order number Ordnungszahl * *~ einer Oberschwingung*
order number supplementary code Bestellnummerergänzung
order numbers Bestellnummern
order of harmonics Ordnungszahl * *~ von Oberschwingungen*
order of magnitude Größenordnung
order picker Regalbediengerät * *für (Hochregal-) Lager* || Regalförderzeug * *in (Hochregal-) Lagersystem*
order picking system Kommissioniergerät * *z.B. in Hochregallager*
order procedure Bestellabwicklung
order processing Auftragsabwicklung || Auftragsbearbeitung || Bestellabwicklung
order reference Auftragskennzeichen
order situation Auftragslage
order suffix Bestellnummerergänzung
order transaction Auftragsabwicklung
order-driven manufacturing auftragsbezogene Fertigung
order-picking Kommissionierung * *Zusammenstellung von Lieferungen im Auslieferungslager*
order-related auftragsspezifisch
ordered bestellt
ordered from Bestellort * *zu bestellen bei* || Bestellort ist
orderer Auftraggeber * *Besteller* || Besteller
ordering Bestellung * *Auftragserteilung, das Bestellen*
ordering code Bestellschlüssel
ordering data Bestellangaben || Bestelldaten
ordering example Bestellbeispiel
ordering information Bestellangaben || Bestellhinweise
ordering instructions Bestellhinweise
ordering location Bestellort
ordering of spare parts Ersatzteilbestellung
ordering procedure Bestellabwicklung
orderly geordnet * *(Adj. und Adv.; auch figürl.)* || ordentlich || ordnungsgemäß
orders Ordnungszahlen * *z.B. v. Oberschwingungen*
orders in hand Auftragsbestand || Auftragspolster * *Auftragsbestand*
orders placed Bestelleingang
orders received Auftragseingang
orders secured Geschäftsabschluss
ordinal Ordnungszahl
ordinal index Ordnungszahl
ordinal number Ordnungszahl
ordinarily in der Regel
ordinary gebräuchlich * *gewöhnlich* || gewöhnlich * *gewöhnlich (siehe auch →in gewohnter Weise)* || normal * *gewöhnlich* || üblich * *gewöhnlich*
ordinate Ordinate || Y-Achse * *Ordinate*
ordinate number Ordinalzahl
ore Erz
ore crusher Erzmühle
ore dressing Erzaufbereitung
ore mill Erzmühle
ORed ODER-verknüpft || verodert * *logische Funktion*

organization Strukturierung * *Organisation*
ORGALIME ORGALIME * *Abk.f.'ORGanisme de Liason des Industries MEtalliques Européen': Maschinenbau/Metallverein*
organic organisch * *(Chemie)*
organically organisch * *(Adv.) (Chemie)*
organization Apparat * ~ *einer Verwaltung* || Einrichtung || *Organisation*
organization block Organisationsbaustein * *{SPS}*
organization chart Organisationsplan
organizational organisatorisch
organizational structure Organisation * *Organisationsstruktur (z.B. eines Unternehmens)*
organizational unit Organisationseinheit
organize aufstellen * *Einheit, Streikräfte usw. ~; (arrange: anordnen, arrangieren)* || einrichten * *gründen, ins Leben rufen, gestalten, organisieren* || gliedern * *organisieren* || organisieren || zusammenfassen * *in ein System bringen; organize into groups: in Gruppen ~*
organized gegliedert * *organisiert*
organizer Veranstalter * *z.B. einer Messe, Tagung usw.*
organizing organisatorisch
orient ausrichten * *(Lage, geistig) ausrichten, (sich) orientieren, sich ~* || informieren * *sich informieren, sich orientieren* || orientieren * *to: an*
orientable Schwenk- * *im Winkel einstellbar*
orientation Ausrichtung * *(auch figürl.) Orientierung (towards: an), Ausrichtung, auch eines Materialgefüges* || Orientierung * *auch von Fasern, Gewebe, Material* || Ortung || Positionierung || Richtung * *Einstellung*
orientative value Richtwert * *(nur) zur Orientierung gedachter Wert*
oriented orientiert * *ausgerichtet (auch figürl.), orientiert, informiert*
orifice Öffnung * Mündung
origin Entstehung * *Ursprung, Quelle, Herkunft* || Herkunft || Nullpunkt * *eines Koordinatensystems* || Ursprung * *originate in/from: seinen ~ haben in; auch ~ eines Koordinatensystems*
original alt * *ursprünglich* || ursprünglich
original document Ursprungsdokument
original manufacturer's equipment Ursprungserzeugnisse
original state jungfräulicher Zustand * *Urzustand, ursprünglicher Zustand*
originality Neuheit * *Ursprünglichkeit*
originally anfangs * *ursprünglich* || eigentlich * *(Adv.) ursprünglich* || von Haus aus
originate entstehen * *stammen, herrühren (from/in:aus)* || stammen * *seinen Ursprung haben (in: in); entstehen, entspringen, von jdm. ausgehen (from: von)* || zurückgehen * *seinen Ursprung haben (in/from:von einer Sache/Person)*
orthodox herkömmlich * *anerkannt, üblich,* →konventionell
OS Betriebssystem * *Abkürzung für 'Operating System'*
OSB board OSB-Platte * *{Holz} Abk.f. 'Oriented Structural Board'*
oscillate atmen * *(hin und her) schwingen* || flattern * *schwingen* || oszillieren || pendeln * *oszillieren, schwingen, (hin und her) schwanken* || rattern * *schwingen* || schwanken * *schwingen* || schwingen * *periodisch*
oscillating Dreh- * *(sich) hin- und herdrehend* || Hin- und Herbewegung * *schnelle ~* || pendeln * *(Substantiv) Schwingen* || Schwenk- * *hin- und herschwingend*
oscillating chisel Schwingmeißel * *{Holz}*
oscillating circuit Schwingkreis
oscillating motion pendeln * *(Substantiv) Hin- u. Herbewegung z.Erleichtern d.Gangwechsels bei Schaltgetriebe*
oscillating torque Pendelmoment * *{el.Masch.} führt zu Drehschwingungen/Drehzahlschwankungen (b.kleiner Drehzahl)*
oscillating velocity Schwinggeschwindigkeit * *bei Schwingung-/Rütteltest [mm/s]*
oscillation Pendelung * *z.B. bei Synchronmotor; kleine Drehzahlvariationen z.leichten Schalten v.Schaltgetriebe* || Schwankung * *Schwingung* || Schwingung * *(elektr.) periodische; forced: erzwungene; free: freie; damped: gedämpfte*
oscillation damping Pendeldämpfung
oscillation excitation Schwingungsanregung * *Anregung einer elektr. Schwingung*
oscillation of the controlled variable Regelschwingungen * *der Regelgröße*
oscillation suppression Pendeldämpfung
oscillations Schwingungen * *periodische*
oscillator Oszillator
oscillogram Oszillogramm
oscillogram trace Oszillogramm
oscillographic curve Oszillogramm
oscillographic record Oszillogramm
oscilloscope Oszillograf || Oszillograph || Oszilloskop
oscilloscope trace Oszillogramm
oscilloscopic record Messschrieb * *von einem Oszilloskop* || Schrieb * *von einem Oszilloskop*
oscilloscopic trace Messschrieb * *von einem Oszilloskop; make: aufnehmen/-zeichnen* || Oszillogramm || Schrieb * *von einem Oszilloskop*
osmosis Osmose
Ossanna's circle diagram Ossannakreis * *Stromortskurve*
ostensible scheinbar * *scheinbar, angeblich, vorgeblich* || virtuell * *scheinbar, vorgetäuscht*
other ander || anders * *(Adj.)* || sonstig || weitere * *zusätzliche, andere*
other company Fremdfirma
other countries Ausland
other manufacturer Fremdfirma || Fremdhersteller
other SIEMENS sales divisions Verbund * *SIEMENS-interne ~vertriebspartner in anderen Unternehm.bereichen*
other than andere als || außer * *anders als, verschieden von*
others übrig * *the others: die ~en, die anderen*
otherwise anders * *(Adv.)* || sonst * *ansonsten* || übrig * *~ens*
OTP OTP * *Abk.f. 'One-Time Programmable': nur einmal programmierbar, nicht löschbar (Festwertspeicher)*
ought to sollen * *im Sinne von Verpflichtung (sittliche)*
oust verdrängen * *aus Stellung, Amt ~*
out außen || hervor
out of aus || außer * *außer (Betrieb, Dienst usw.)* || hervor * *hervor aus*
out of balance Exzentrizität * *Rundlauffehler, z.B. bei Motorwelle (siehe DIN 42955)*
out of date veraltet
out of doors außen * *im Freien*
out of focus unscharf
out of function außer Betrieb * *defekt*
out of operation außer Betrieb || stillstehend * *außer Betrieb*
out of order ausgefallen * *defekt* || betriebsunfähig || defekt * *nicht in Ordnung, defekt* || gestört * *nicht in Ordnung, defekt* || kaputt * *nicht in Ordnung, defekt*
out of phase phasenverschoben * *by 90 degrees: um 90 Grad* || verschoben * *periodische Signale gegeneinander; by 90 ° (vorangestellt): um 90° gegeneinander*

out of place unpassend * *unangebracht, deplatziert*
out of round Schlag * *Unrundheit* || unrund
out of service außer Betrieb
out of stock nicht auf Lager || vergriffen
out of tolerance außerhalb der Toleranz
out of true unrund
out-dated nicht mehr gebräuchlich
out-moded veraltet * *altmodisch*
out-of-round Unrundheit
outage Ausfallzeit
outcome Resultat
outdated veraltet
outdistance überholen * *übertreffen*
outdoor Freie * *das Freie, im Freien* || im Freien * *designed for outdoor services: geeignet für den Betrieb ~*
outdoor installation Außenraumaufstellung * *Aufstellung/Installation im Freien*
outdoor plant Freiluftanlage
outdoor- Freiluft- * *outdoors: im Freien*
outdoors außen * *im Freien* || Freiluft- * *for outdoor use: für ~aufstellung/Verwendung im Freien* || im Freien
outdrive Abtrieb
outer Außen || äußerer || *äußeres * am äußeren Rand befindlich* || überlagert * *Regelkreis*
outer bearing cap Schlussdeckel * *äußerer Lagerdeckel*
outer circuit Außenkreis * *z.B. Kühlkreislauf*
outer conductor Außenleiter
outer control loop äußerer Regelkreis || überlagerte Regelung || überlagerter Regelkreis
outer controller überlagerter Regler
outer diameter Außendurchmesser * *z.B. eines Kabels*
outer dimensions äußere Abmessungen
outer loop control überlagerte Regelung
outer ring Außenring
outer sheath Außenmantel * *eines Kabels*
outer surface Außenfläche || äußere Oberfläche
outfit Apparatur * *Gerät(e), Ausstattung, Werkzeug(e), Ausrüstung, Ausstattung*
outflow Auslauf * *von Flüssigkeit*
outgoing abgehend || abzulösendes * *zu löschendes (Ventil in Stromrichterschaltung)* || ausgehend * *zu löschendes Ventil i.Stromrichterschaltung* || stromabgebendes * *zu löschendes (Ventil in Stromrichterschaltung)*
outgoing cable Ableitung * *elektr. Kabel/Leitung*
outgoing feeder Abgang * *Ausleitung/Abzweig z.B. zur Leistungsversorgung*
outgoing leads Ableitung * *z.B. in einer Schaltanlage*
outgoing message gehende Meldung
outgoing section Abgang * *Stromkreis*
outgoing shaft Abtriebswelle * *bei Getriebe* || Ausgangswelle * *eines Getriebes*
outgoings Ausgaben * *Geldausgaben*
outing Ausflug
outlay Kosten * *(Geld-) Auslage; initial outlay: Anschaffungs-/einmalige ~*
outlet Abgang * *Herausführung; Anschluss für Stromverbraucher* || Ablauf || Vorrichtung || Abzug * *Auslass* || Ausgang * *Auslass* || Auslass * *outlet valve: Auslassventil* || Auslauf * *Auslass* || Austrittsöffnung * *z.B. für Kühlluft* || Durchlass * *Auslass* || Öffnung * *Auslass* || Steckdose * *Netzsteckdose*
outlet box Anschlussdose
outlet duct Auslasskanal * *z.B. für Kühlluft*
outlier Ausreißer
outline andeuten * *umreißen, in groben Zügen darstellen, skizzieren* || darlegen * *in Umrissen, in groben Zügen* || darstellen * *in Umrissen, in groben Zügen ~* || entwerfen * *in groben Zügen ~* || erläu-

tern * *in Übersichtsform* || Kontur || Überblick geben || Übersicht * *auch Abriss, Umriss, Ausblick, Darstellung in groben Zügen* || Umriss || zeichnen * *(auch figürl.) skizzieren, umreißen* || zusammenfassen * *Übersicht/Überblick geben* || Zusammenfassung * *Abriss, Übersicht, Überblick*
outline dimensions Außenabmessungen
outline drawing Maßbild * *Umrisszeichnung* || Umrisszeichnung
outlines Kurzbeschreibung
outpace überholen * *übertreffen*
outperform leistungsfähiger sein als || überbieten * *bezüglich der Leistung(sfähigkeit) ~* || überholen * *an Leistung(sfähigkeit) übertreffen* || überlegen * *an Leistungsfähigkeit ~ sein im Vergleich zu* || übertreffen * *in der Leistungsfähigkeit ~*
output Abgabeleistung || abgeben * *Leistung usw. ~* || abgegeben || abgegebene Leistung * *z.b. eines Stromrichters/Motors* || Abtrieb * *eines Getriebes* || Abtriebs- * *z.b. eines Getriebes, einer Kupplung usw.* || Ausbeute * *(Ertrag)* || Ausgang * *z.B. Digital-,Analog-, eines Stromrichters/Motors/Getriebes* || ausgeben * *z.B. ein Signal auf eine Digital/Analogausgabe* || ausgegeben * *z.b. Signal,Leistung* || getrieben * *Maschinenteil* || Herstellung * *Ausstoß* || Leistung * *Ausgangs~ z.B.eines Stromrichters oder Motors (Ausgangs~, Wellen~)*
output board Ausgangsbaugruppe * *Leiterkarte*
output capacitor Ausgangskondensator
output capacity Ausgangsleistung
output choke Ausgangsdrossel
output circuit Ausgangsschaltung
output class Leistungsklasse * *Stromrichter, Motoren*
output command Ausgabebefehl
output current Ausgangsstrom
output derating Leistungsherabsetzung * *d. Nennleistg.z.B. b. Motor/Umrichter aufgr.erschwerter (Umgeb.-)Bedingungen*
output device Ausgabegerät
output end Abtriebswelle * *z.B. einer Kupplung/eines Getriebes*
output filter Ausgangsfilter * *z.B. für Stromrichter (siehe auch 'Sinusfilter' u. 'dV-dt-Filter')*
output filter reactor Ausgangssiebdrossel * *z.B. für Pulsumrichter*
output flag Ausgangsmerker
output flange Abtriebsflansch * *z.B. eines Getriebes, Durchtriebs usw.*
output frequency Ausgangsfrequenz
output ihibit Befehlsausgabe-Sperre * *"BASP" bei SPS (Sperren der Digital-/Analogausgaben)*
output impedance Ausgangsimpedanz || Innenwiderstand * *Ausgangsimpedanz*
output increase Leistungssteigerung * *z.B. eines Stromrichters*
output inhibit Ausgabesperre * *z.B. für Digital-/Analogausgaben* || BASP * *Befehlsausgabe-Sperre für Digital-/Analogausgaben z.B. bei SPS*
output level Ausgangspegel
output margin Leistungsreserve * *bei Motor/Umrichter*
output module Ausgabebaugruppe || Ausgangsbaugruppe
output per employee Produktionswert je Beschäftigten
output power Abgabeleistung || Ausgangsleistung || Leistung * *abgegebene ~ (eines Motors, Stromrichters usw.)*
output quantity Ausgangsgröße * *Betrag/Höhe des Ausgangssignals*
output range Leistungsbereich * *z.B. einer Motoren- oder Stromrichterbaureihe*
output rating Ausgangsnennleistung || Leistung

output reactor

* *z.B. Ausgangsnenn~ eines Stromrichters* || Nennleistung * *z.b. Ausgangs~ eines Stromrichters*
output reactor Ausgangsdrossel * *z.B. für Pulsumrichter*
output reserves Leistungsreserven * *z.b. Stromrichter, Motor*
output resistance Ausgangswiderstand || Innenwiderstand * *Ausgangswiderstand einer Signal- oder Spannungsquelle*
output rise Leistungssteigerung * *Steigerung der Ausgangsleistung, z.B. bei Motor, Umrichter usw.*
output shaft Abtrieb * *Abtriebswelle eines Getriebes, einer Kupplung usw.* || Abtriebswelle * *eines Getriebes/einer Kupplung* || Antriebswelle * *Ausgangswelle (eines Getriebes)*
output side Abtriebsseite
output signal Ausgangsgröße * *Ausgangssignal* || Ausgangssignal
output smoothing reactor Ausgangssiebdrossel
output speed Abtriebsdrehzahl * *eines Getriebes*
output stage Ausgangsstufe * *z.B. eines Leistungsverstärkers* || Endstufe * *z.B. für Stromrichter*
output terminal Ausgangsklemme
output torque Abtriebsdrehmoment * *eines Getriebes* || Abtriebsmoment * *eines Getriebes usw.* || Antriebsdrehmoment * *am Ausgang e. Getriebes/Motors/Kupplung*
output value Ausgangsgröße * *Ausgangswert*
output variable Ausgangsgröße
output voltage Ausgangsspannung
output voltage range Ausgangsspannungsbereich
output voltage ratio Aussteuerung * *Verhältnis max. zu aktueller Ausgangsspannung bei Stromrichter*
output word Ausgangswort * *bei SPS*
outrun überholen * übertreffen || übertreffen * *in der Leistung ~*
outset Beginn
outside außen * *außerhalb; from the outside: von ~; outside diameter: ~durchmesser; auch "nach (dr)außen"* || Außenseite || äußerer * *außen; outside diameter:Außendurchmesser* || außerhalb * *the value lies outside: der Wert liegt außerhalb; siehe auch →außen* || Fremd- || nach außen || Umgebung * *das Außenliegende, die Außenwelt*
outside cable diameter Kabel-Außendurchmesser
outside capital Fremdkapital
outside diameter Außendurchmesser
outside edge Außenkante
outside funds Fremdkapital
outside Germany Ausland * *außerhalb Deutschlands*
outside make Fremdfabrikat
outside manufacture Fremdfertigung
outside production Fremdfertigung
outside the außerhalb des
outside the permissible range außerhalb des zulässigen Bereiches
outside the permitted limits außerhalb der erlaubten Grenzen
outside the permitted range außerhalb des erlaubten Bereichs
outside wall Außenwand
outstanding außerordentlich * →*hervorragend* || ausgezeichnet || exzellent * *hervor-/herausragend* || herausragend * *(figürl.) hervor-/überragend, hervorstechend, prominent, offen stehend (Rechnung usw.)* || hervorragend * *herausragend* || sehr gut * *hervorragend* || überdurchschnittlich || überragend
outstanding debts Rückstände * *Sculden*
outstanding feature Haupteigenschaft * *herausragendes Merkmal*
outstandingly hervorragend * *(Adv.)*
outstrip überholen * *übertreffen*

outward äußerer
outwards außen * *nach außen hin*
oval oval
oval turning lathe Ovaldrehmaschine
oven Ofen * *Backofen*
over abgeschlossen * *vorbei (zeitlich)* || gegenüber * *(dar)über, mehr (verglichen mit); over against:-gegenüber, im Gegensatz zu* || in Abhängigkeit von * *auf der X-Achse aufgetragene Variable einer Kurvendarstellung* || über * *auch betrags/zahlen/zeitmäßig; over the complete range: ~ den gesamten Bereic* || Über- || übrig * *übrig, über; have something over: etwas ~ haben* || vorbei * *(zeitlich)*
over a longer time period über längere Zeit
over and above it darüber hinaus * *(figürl.)*
over it darüber || darüber hinaus * *darüber hinweg*
over short duration kurzzeitig
over that darüber
over the complete range über den gesamten Bereich
over the long term langfristig
over them darüber * *(Plural)*
over- Erhöhung * *(Vorsilbe) z.B.: overvoltage: Spannungs~ (über die Nenn-/erlaubte Spannng.hinaus)*
over-all gesamt * *über-alles*
over-all loop gain Schleifenverstärkung * *eines Regelkreises*
over-all project Gesamtprojekt
over-critical überkritisch
over-excitation Übererregung
over-excited übererregt * *z.B. Synchronmaschine*
over-hauling load hängende Last
over-proportional überproportional
over-synchronous braking generatorische Bremsung * *[el.Masch.] Bremseffekt bei Asynchronmotor wenn Drehzahl > Synchr.drehz.*
over-temperature Übertemperatur
over-travel Nachlauf * *meist unerwünschtes Weiterlaufen e. Antriebs nach Erreichen der Ziel-/Halteposition*
over-voltage arrester Überspannungsableiter
overall allgemein * *umfassend* || ganz * *über alles (z.B. Länge)* || Gesamt- * *über alles* || gesamter || pauschal * *über alles, Gesamt-* || Summen- * *Gesamt-, Über-Alles-* || über alles * *gesamt* || über-alles
overall aim Prämisse * *übergeordnetes Ziel; with the overall aim to reduce costs: unter d. ~ der Kostensparung*
overall control system Leitsystem
overall costs Gesamtkosten
overall efficiency Gesamtwirkungsgrad
overall load capacity Gesamtbelastbarkeit
overall output rating Gesamtbelastbarkeit
overall plant Gesamtanlage
overall quantity Gesamtmenge
overall reset Urlöschen * *z.B. bei SPS*
overburden überlasten * *(figürlich)*
overcharge überlasten || übersteuern
overcome beheben * *Fehler ~* || beseitigen * *Hindernisse, Problem usw. ~* || in den Griff bekommen * *Herr werden, lösen, meistern (z.B. e. Problem)* || meistern * *Herr werden, lösen (z.B. Problem), überwinden* || überbrücken * →*überwinden* || überstehen * *(Schwierigkeiten usw.) überwinden, bestehen* || überwinden * *(auch figürl.) Problem, Schwierigkeit, Hemmungen, Widerstand, Reibung, Lastmoment usw. ~*
overcurrent Überstrom
overcurrent capability Überstrombelastbarkeit
overcurrent capacity Überstrom * *Überlastbarkeit, Überlastfähigkeit* || Überstrombelastbarkeit
overcurrent cut-off Überstromabschaltung
overcurrent limiting Überstrombegrenzung

overcurrent protection Überstromschutz
overcurrent relay Überstromrelais
overcurrent release Überstromauslöser
overcurrent releasing device Überstromauslöser * zum thermischen Motorschutz
overcurrent trip Überstromabschaltung * Störabschaltung || Überstromauslöser || Überstromauslösung
overcurrent tripping Überstromauslösung
overcurrent tripping unit Überstromauslöser
overdimension überdimensionieren
overdimensioned überdimensioniert
overdimensioning Überdimensionierung
overdrive übersteuern * über ein anderes Signal hinweg ~
overdue überfällig * verspätet
overexcitation Übererregung * z.B. z. beschleunigten Momentenrichtungswechsel durch Feldumkehr
overexcited übererregt
overextend übernehmen * sich ~
overflow Überlauf
overhang Ausladung * z.b. einer Werkzeugmaschine || Vorsprung * Überhang || Wickelkopf * z.B. eines Motors
overhang bracket Wickelkopfabstützung
overhang winding Wickelkopf
overhaul instand setzen * überholen || instandsetzen * überholen || prüfen * überholen || Reparatur * Überholung || Revision * technische Überholung || überholen * nachsehen, ausbessern, instand setzen, technisch überprüfen, wieder i.Ordnung bringen || Überholung * Überholung, gründliche Überprüfung; complete overhaul: ~ von Grund auf || untersuchen * überholen || warten * überholen || Wartung * Überholung
overhaul work Inspektion * →Überholungsarbeiten || Überholungsarbeiten
overhauling load hängende Last * am Seil hängend
overhead Freiluft- * Leitung || oben * ~liegend, ~gesteuert, Hoch-, Frei- || obenliegend * Frei-, Ober-, Hoch-, obengesteuert, obenliegend (z.B. Nockenwelle) || Ober- * obenliegend || Overhead-Folie || Overheadfolie
overhead chart Overhead-Folie || Overheadfolie * mit grafischer/tabellarischer Darstellung
overhead conductor Fahrleitung * {Bahn}
overhead conveyor Hängebahn * Fördersystem || Hängeförderer
overhead costs Gemeinkosten
overhead crane Hallenkran * mit obenliegenden Laufschienen
overhead expenses Gemeinkosten || Unkosten * allgemeine Kosten, (All)Gemeinkosten
overhead line Freileitung
overhead power transmission line Freileitung
overhead router Oberfräse * {Holz}
overhead saw Portalsäge * {Holz}
overhead slide Overhead-Folie || Overheadfolie
overhead transparency Overhead-Folie || Overheadfolie
overhead travelling hoist Katzfahrwerk * {Hebezeuge} mit Hebewinde
overhead-cabin railway H-Bahn
overhead-travelling gear Katzfahrwerk * {Hebezeuge}
overheads Fixkosten * Gemeinkosten || Gemeinkosten || Unkosten * allgemeine Kosten, (All)Gemeinkosten
overheat überhitzen || Überhitzung || Übertemperatur * Überhitzung
overheating Übererhitzung || Übererwärmung || Überhitzung
overheating protector Temperaturwächter

overhung fliegend * fliegend angeordnet (Montage, Lagerung usw.), freitragend, überhängend
overhung mounting fliegender Anbau
overlap überlappen || Überlappung * auch zeitlich || überschneiden * überlappen
overlap angle Überlappungswinkel
overlap interval Überlappungszeit
overlap period Überlappungszeit
overlap time Überlappungszeit
overlapping Überdeckung * Überlappung || übergreifend || überlappend * auch zeitlich || Überlappung
overlapping edge band Umleimer * {Holz} um die Kante herumgewickelter ~
overlay Überdeckung || überlagern * überdecken
overlay inhibit Überdeckungssperre * am Bildschirm
overleaf umseitig * see overleaf : siehe ~
overleap überspringen
overload überbelasten || Überlast || überlasten * by: um || Überlastung * Überlast || übersteuern * z.B. Messgerät ~
overload capability Überlastbarkeit * dynamic: dynamische; für E-Motoren nach EN 60034 (2 min mit 1,5-fach. Strom) || Überlastfähigkeit || Überlastmöglichkeit
overload capacity Überlastbarkeit * wert/betragsmäßig; el.Maschinen nach VDE0530 für 2 min m. 1,5fach.Strom belastb. || Überlastfähigkeit
overload condition Überlast * Überlast-Zustand/-Betrieb
overload factor Betriebsfaktor || Servicefaktor || Überlastbarkeit * Überlastbarkeitsfaktor (max. Leistung/Strom zu Nennleistung/Strom) || Überlastfaktor * Verhältnis 'zulässige Motordauerleistung/Motornennleistung'
overload margin Überlastfähigkeit * Bereich der ~
overload protection Überlastschutz
overload rating Überlastbarkeit * spezifizierte Überlastfähigkeit
overload relay Motorschutzrelais * Überlast-Schutzrelais || Überlastrelais
overload slipping clutch Sicherheitskupplung * Überlast verhindernde Rutschkupplung
overload-capable überlastbar || überlastfähig
overloadable überlastbar || überlastfähig
overloaded überlastet
overloading Überbeanspruchung || Überlastung || Übersteuerung * z.B. ~ eines Messgeräts
overlook übergehen * übersehen, nicht beachten
overmagnetization Übermagnetisierung
overmodulate übersteuern
overmodulation Übersteuerung * z.B. ~ eines Regelkreises
overmodulation capacity Regelreserve
overpaint überlackieren
overpower überwinden
overpressure Überdruck * übermäßiger Druck; pressurized: unter ~ stehend
overpressure test überdrücken * (Substantiv) Überdruckprüfung || Überdruckprüfung
overrange Überschreitung * des Werte-/Zahlen-/Messbereichs
overrate überdimensionieren
overrated überdimensioniert
override höherprior sein || sich hinwegsetzen über || übersteuern * über ein anderes Signal hinweg ~ ||
override circuit Ablöseschaltung
override control ablösende Regelung || Ablöseregelung * siehe auch →ablösende Regelung
override controller ablösender Regler
overriding control ablösende Regelung
overrule die Oberhand gewinnen über || verwerfen * Antrag/Entscheidung ~

overrule circuit Ablöseschaltung
overrule control ablösende Regelung
overrun überholen * *überlaufen, z.B. Softwarezyklus*
overrun control ablösende Regelung
overrunning clutch Freilauf
oversea Übersee
overseas Übersee
overseas branch Auslandsniederlassung
overshoot Nachlauf || nachlaufen * *überschwingen, überschießen* || Überhöhung * *überschwingen* || überschwingen * *(auch Substantiv)* || Überschwinger || Überschwingweite * *z.B. einer Sprungantwort*
overshoot amplitude Überschwingweite
overshoot capability Regelreserve * *(Regel.-techn.) über d.Nennwert hinausgehender Bereich des Istwerts/Stellgröße*
oversight versehen * *(Substantiv)*
oversize großzügig dimensioniert || überdimensionieren || Übermaß * *bezüglich des Abmaßes*
oversized großzügig dimensioniert || überdimensioniert || übergroß
oversizing Überdimensionierung
overspeed durchgehen * *eines Motors (Überdrehzahl)* || überdrehen * *mit Überdrehzahl fahren* || Überdrehzahl || Übergeschwindigkeit
overspeed monitor Drehzahlwächter * *Überdrehzahlüberwachung* || Überdrehzahlauslöser
overspeed protection Überdrehzahlschutz
overspeed relay Überdrehzahlauslöser
overspeed release Überdrehzahlauslöser
overspeed switch Fliehkraftschalter * *Überdrehzahlauslöser/-schalter*
overspeed test Schleuderprüfung * *z.B. mit 120% der max Drehzahl eines Motors (nach VDE 0530)*
overspeed test speed Schleuderdrehzahl * *für Schleuderprüfung eines Motors, normalerweise mit 1,2-mal max. Drehzahl*
overspeed trip Überdrehzahlauslöser
overspeed tripping unit Drehzahlwächter * *Auslösegerät zum Überdrehzahlschutz*
overstate ansetzen * *zu hoch ansetzen*
overstep überschreiten * *Maß, Grenzlinie ~*
overstressing Überbeanspruchung
oversynchronous übersynchron
oversynchronous braking übersynchrone Bremsung
overtake überholen * *Fahrzeug im Verkehr ~, Mitbewerber ~*
overtemperature Übertemperatur
overtight überspannen * *z.B. Riemen, Kette*
overtime charges Überstundenzuschlag
overtime hour Überstunde
overtime work Überstunden * *work overtime: ~ machen*
overtone Oberwelle * *(salopp)*
overtorque Momentenüberlast || Überlast * *zu hohes Drehmoment*
overtravel Nachlauf * *~en eines Antriebs über das Ziel hinaus* || nachlaufen * *über die Zielposition hinaus ~*
overview Überblick * *in Beschreibung, Vortrag usw.* || Übersicht * *of: über; Überblick; provide an overview: eine ~ geben; fast overview: Schnell~* || Zusammenfassung * *Übersicht*
overview diagram Übersichtsdiagramm
overview of operations Operationsübersicht * *Übersicht über SPS-Befehlssatz*
overview representation Übersichtsdarstellung * *z.B. eines SPS-Programms*
overvoltage Überspannung
overvoltage category Überspannungskategorie * *kennzeichnet Isolations-/Überspanngs.festigkeit nach VDE0110*

overvoltage condition Überspannung * *Auftreten/Vorhandensein einer ~*
overvoltage handling capacity Überspannungsfestigkeit
overvoltage protection Überspannungsschutz * *allg. Überspannungsverhinderung, Schutz gegen Zerstörung*
overvoltage release Überspannungsauslöser
overvoltage spike Spannungsspitze * *Überspannungsspitze*
overvoltage strength Überspannungsfestigkeit * *siehe z.B. DIN VDE 0160*
overvoltage surge Überspannungsspitze
overvoltage transient Spannungsspitze * *kurzzeitige Überspannung*
overvoltage trip Überspannungsabschaltung * *Fehlerabschaltung auf Grund unzulässig hoher Spannung* || Überspannungsauslöser
overvoltage tripping unit Überspannungsauslöser
overwhelm überhäufen * *(with: mit)* || überschütten * *überhäufen, überwältigen, überschütten, erdrücken*
overwhelming überwiegend * *Mehrheit*
overwind wickeln * *~ von oben* || wickeln von oben
overwork Überarbeitung * *zu viel Arbeit*
overworking Überbeanspruchung * *(figürl.)*
overwrite überschreiben * *z.B. Speicherinhalt, Parametersatz*
ower absenken
owing to auf Grund * *infolge, wegen, zurückzuführen auf* || dadurch * *dadurch dass, wegen* || dank || wegen * *infolge, wegen, dank; be owing to: zurückzuführen sein auf/... zuzuschreiben sein*
owing to the fact that dadurch * *dadurch dass, wegen; auf Grund d.Tatsache ,dass*
owing to this infolgedessen
owing to which infolgedessen
own besitzen * *innehaben, beitzen; auch Eigenschaft, Talent usw. ~* || eigen || eigener
own consumption Eigenverbrauch
own manufacture Eigenfertigung
own production Eigenfertigung
owner Besitzer * *Eigentümer; change hands: den Besitzer wechseln* || Inhaber * *Eigentümer*
owner of a business Geschäftsinhaber
owner's equity Eigenkapital
oxide Oxid || Oxyd
oxide film Patina * *Oxidschicht*
oxidize oxidieren || rosten * *oxidieren*
oxygen Sauerstoff
oxygen cutter Schneidbrenner
oxygenize oxidieren * *mit Sauerstoff verbinden*
oz g * *Unzen; 1 g entspr. 0,0352740 oz*
ozon injector Ozonbegasungsanlage * *zur Wasserbehandlung*

P

P component P-Anteil
P controller Proportionalregler
P gain P-Verstärkung
p-conductive p-leitend * *Halbleiterschicht*
P-controller P-Regler
P-element Proportionalglied
P-gain Proportionalverstärkung
p.a. pro Jahr * *Abk. für 'Per Annum'*
p.c. board Flachbaugruppe
p.d. pro Tag * *Abk. für 'Per Diem'*
P.O. box Postfach
P.O.B Postfach * *post-office box*

p.u. Prozent * *Abkürzung für 'per unit' (1% entspricht 0,01 p.u.)*
p.u. quantity Bezugsgröße * *diejenige Größe, die das "Einheitsnormal" darstellt*
pace Geschwindigkeit * *Fahrt, Tempo, Schwung, Marsch~ z.b. eines Fahrzeugs* || Schritt * *keep pace: ~ halten; pace-maker: ~macher* || Tempo * *Gangart; set/increase the pace: das ~ angeben/ steigern*
pace-maker Schrittmacher
pace-setter Vorreiter
pacer Schrittmacher * *set the pace for: ~ sein für*
pack -Paket * *Pack, Bündel, Ballen, Schub, Packung* || abdichten || Ballen || binden * *Ballen usw. ~* || füllen * *mit Schmierfett* || packen * *ein/ver~* || verpacken
pack up packen * *ein~* || verpacken
pack-up with beipacken * *mit/dazu einpacken*
package -Paket * *Pack(ung), (Bau-) Einheit, Gebinde* || Baueinheit || Einheit * *Bau~* || Gehäuse * *z.b. für Halbleiterbauelement, integrierten Schaltkreis usw.* || Paket * *großes ~; siehe auch →-Paket; (auch figürl.: Daten-/Aufgabenpaket usw.)* || verpacken * *Einzelstück; besonders 'maschinell verpacken'*
package carton Verpackungskarton
packaging Verpackung * *die ~, das Verpacken; seaworthy packaging: seegemäße ~* || Verpakkungs-
packaging density Packungsdichte
packaging equipment Verpackungsmaschinen
packaging machine Verpackungsmaschine
packaging material Packmittel || Packstoff * *z.b. für Verpackungsmaschine*
packaging means Packmittel * *z.b. für Verpakkungsmaschine*
packaging recycling Verpackungs-Recycling
packaging strap Verpackungsband * *streifen-/reifenförmig*
packaging system Aufbausystem || Aufbautechnik || Einbausystem
packaging technology Verpackungstechnik
packaging thread Verpackungsband * *Schnur, Kordel, Bindfaden*
packed verpackt
packed for sea transport seemäßig verpackt
packed gland Stopfbuchsverschraubung
packed weight Verpackungsgewicht * *Gewicht inklusive Verpackeung*
packer Packprogramm
packer program Packprogramm
packet Paket * *kleines ~*
packing Füllung || Verpackung * *die Hülle, das Hüllmaterial*
packing base Verpackungsboden
packing box Verpackungskarton
packing machine Einpackmaschine
packing material Verpackung * *~smaterial* || Verpackungsmaterial
packing of grease Fettfüllung
packing paper Packpapier
packing ring Manschette
packing unit Verpackungseinheit
packing washer Dichtscheibe
packpress machine Packpresse * *Verpackungsmaschine*
pact Abkommen * *Pakt, strategisches Abkommen*
pad ausfüttern * *polstern* || Polster * *Polster, Kissen (als Schutz gegen Stöße), Wulst, Bausch; oil pad: Schmierkissen* || Pratze * *{el.Masch.} zur Motorbefestigung bei Bauformen IM B30, IM V30 u.IM V31 (3/4-Punktaufhängg.)*
pad covering Gleitbelag
pad saw Stichsäge
pad-mounting Pratzenanbau * *{el.Masch.} von*

Motoren ohne Anbaufüße/Flansch; Bauformen IM B 30, IM V30 und IM V31
padding Polster * *Polsterung, Wattierung*
padding elements Polstermittel * *beim Verpacken*
padding materials Polstermittel * *{Verpack.- techn.}*
paddle Schaufel * *z.B. bei Mixer*
page Blatt * *Seite in einem Dokument/Druckerzeugnis* || Seite * *~ in Schiftstück/auf dem Bildschirm*
page frame Kachel * *Speicherbereich im Rückwandbus einer SPS (bei SIEMENS SIMATIC (R) S5)*
page number Seitennummer
page printer Drucker * *Seitendrucker*
paging Umbruch * *{Druckerei} Seiten~*
paging machinery Verpackungsmaschinen
pain threshold Schmerzschwelle
pains Mühe * *Mühe. Bemühungen, Quälerei* || Sorgfalt * *Mühe, Bemühung; take pains:sich Mühe geben/anstrengen*
painstaking sorgfältig * *gewissenhaft, eifrig, rührig*
paint anstreichen || Farbanstrich || Farbe * *Lack, Mal~ usw.* || Lack || lackieren || streichen * *anstreichen*
paint finish Anstrich || Farbanstrich || Lackierung
paint mist Farbnebel
paint shop Lackiererei
paint spraying system Farbspritzanlage
paint system Anstrich
paint-application system Lackiergerät
paint-brush Pinsel
paintability Lackierbarkeit
painting Anstrich || Lackierung
pair Paar
pair of compasses Zirkel
pair of dividers Zirkel
pair of pliers Zange
pair of poles Polpaar
pair of scales Waage
pair of steps Leiter * *Stehleiter*
paired zusammen * *paired with: gepaart mit*
palette Palette * *Brett zum Mischen von Farben; Farbmodell/Farbauswahlfunktion in der Computergrafik*
pallet Palette * *Stapel~, Transport~* || Transportplatte * *Palette*
pallet conveyor Palettentransportanlage
pallet loading machine Bepalettiermaschine || Palettierer * *{Verpack.technik}* → *Bepalettiermaschine* || Palettiermaschine * *Bepalettiermaschine, Palettenbelademaschine*
pallet unloading machine Entpalettiermaschine
palleting machine Palettiermaschine
palletiser Palettierer * *(brit.) {Verpack.technik}*
palletising equipment Palettieranlage * *(brit.)*
palletizer Palettierer * *(amerikan.) {Verpack.technik}* || Palettiermaschine * *Verpackungsmaschine*
palletizing Palettierung * *Aufstapeln von Teilen auf Palette*
palletizing machine Bepalettiermaschine * *siehe auch* → *Palletiermaschine* || Palettierer * *{Verpack.technik}* → *Palettiermaschine* || Palettiermaschine * *Verpackungsmaschine*
palletizing robot Palettierroboter * *{Verpack.technik}*
palletizing system Palettieranlage * *(amerikan.)*
palm-oil Schmiergeld
palpable fühlbar * *körperlich fühl/greifbar (auch figürl.: augenfällig, deutlich)* || greifbar * *(figürl.): augenfällig, handgreiflich, fühl-/greifbar; impalpable: nicht ~*
PAM winding Rawcliff-Wicklung * *für polumschaltbaren Motor ähnlich Dahlanderschaltung (siehe 'PAM-Wicklung')*

parameter assignment device Parametriereinrichtung
parameter assignment module Parametrierbaugruppe * *bei SPS*
parameter assignment unit Parametriereinrichtung
parameter change report Spontanmeldung * (→*PROFIBUS-Profil*)
parameter change rights Bedienhoheit * (→*PROFIBUS-Profil*) || Parametrierhoheit * (→*PROFIBUS-Profil*) || PKW-Bedienhoheit * (→*PROFIBUS-Profil*)
parameter characteristics Kennzeichen * (→*PROFIBUS-Profil*)
parameter data PKW * (→*PROFIBUS-Profil; Parameter/Kennung/Wert*)
parameter description Parameterbeschreibung * (*auch* →*PROFIBUS-Profil*)
parameter identification Parameteridentifikation * *[Regel.techn.]*
parameter identifier Parameterkennung * (→*PROFIBUS-Profil*)
parameter index Subindex * (→*PROFIBUS-Profil*)
parameter interface Parameterschnittstelle
parameter memory Parameterpeicher
parameter number Parameternummer
parameter print-out Parameterausdruck * *auf einen Drucker*
parameter sensitivity Parameterempfindlichkeit
parameter set Parametersatz
parameter setting Parametereinstellung || Parametrierung
parameter storage Parameterpeicher
parameter type Parametertyp
parameter value Parameterwert
parameter variation Parameterschwankung
parameter-change-flags array Nummernliste geänderter Parameterwerte * (→*PROFIBUS-Profil*)
parameterization Parametrierung
parameterization device Parametriereinrichtung
parameterization interface Parameterschnittstelle * *Parametrierschnittstelle*
parameterization panel Parametriereinheit
parameterization software Parametrier-Software
parameterization unit Parameterbedieneinheit || Parametriereinheit
parameterization work Parametrierung * *Tätigkeit des Parametrierens z.B. durch e. Techniker*
parameterize parametrieren
parameterized parametriert
parameterizing Parametrierung
parameterizing device Parametriereinrichtung
parameterizing program Parametrierprogramm
parameterizing software Parametrier-Software
parameterizing tool Parametrierwerkzeug
parameterizing unit Parametriereinheit
parameters of the controlled system Streckenparameter
paramount überragend * *(figürl.)*
parantheses Klammern * *als Schriftzeichen*
parasitic Fremd- * *parasitär (Strom usw.)* || parasitär || parasitärer Effekt || schädlich * *parasitär, störend* || Stör- * *parasitär* || störend * *parasitär, schädlich* || unerwünscht * *parasitär, schädlich, störend*
parasitic current Störstrom
parasitic effect parasitärer Effekt
parasitic phenomenon parasitärer Effekt
parasitic voltage Fremdspannung || Störspannung
parasitic voltages Fremdspannungen
parasitics Einstreuungen
parcel Paket
parcel out einteilen * *verteilen in Paketen, paketieren*

pardon bitte * *wie bitte?*
parent company Muttergesellschaft
parenthesis Klammer * *[pe'renthisis] Schriftzeichen '(', ')'; (Plural: parentheses); paranthesize: i.~n setz.*
parenthesize einklammern * *[pe'renthisais]*
parity Parität * *Datensicherung*
parity bit Parität * *Paritätsbit* || Paritätsbit || Parity-Bit * *Datensicherungsbit, ergänzt ein Datenwort/-Byte auf e. gerad- oder ungeradzahlige Zahl*
parity check Paritätskontrolle || Paritätsprüfung
parity checking Paritätsprüfung
parity error Paritätsfehler
parity generation Paritätserzeugung
parking brake Haltebremse
parry abwehren * *Stoß usw. (auch figürl.)*
part -Anteil * *Teil* || Anteil * *Teil, Pflicht; do one's part: seinen ~ leisten* || Artikel || Ausschnitt * *Teil* || Bestandteil * *Einzelteil; essential part: wesentlicher ~* || dabei * *take part in: teilnehmen an, dabei sein* || Einzelteil * *Teil* || Komponente || Rolle * *(figürl.) (An-) Teil; play a part: eine ~ spielen* || Teil * *Bau~, Artikel* || teilen * *sich ~* || Werkstück
part delivered by a subcontractor Zulieferteil
part delivered by an external manufacturer Zulieferteil
part drawing Teilzeichnung
part load Teillast
part No Bestellnummer * *Artikelnummer, Ersatzteilnummer*
part No. Artikel-Nr
part number Teilenummer
part program Teileprogramm * *[NC-Steuerung]*
part shipment Teillieferung
part subject to wear Verschleißteil
part- Teil-
part-delivery Teillieferung
part-load condition Teillastbetrieb
part-load operation Teillastbetrieb
part-speed Teildrehzahlen * *under/during part-speed operation:bei Teildrehzahlen*
part-time Teilzeit-
part-time student Werkstudent
part-winding starting Teilwicklungsanlauf * *[el.-Masch.] Anlassen e. Käfigläufermot. mit nur einem Teil d. Ständerwicklg.*
partial einseitig * *parteiisch* || partiell || Teil- * *teilweise; An/Bruchteil* || teilweise * *partiell*
partial converter Teilstromrichter
partial delivery Teillieferung
partial discharge Glimmentladung * →*Teilentladung* || Sprühentladung || Teilentladung * *Mikrolichtbögen im Dielektrikum bei Überschreiten d. Durchschlagfestigkeit*
partial function Teilfunktion
partial load Schwachlast * *Teillast* || Teillast * *under partial load condition: bei ~*
partial load range Teillastbereich
partial solution Teillösung
partial vacuum Unterdruck
partial view Teilansicht
partial-load operation Teillastbetrieb
partially teilweise * *(Adv.)*
participant Teilnehmer * *an Kurs, Besprechung usw.*
participant in a course Kursteilnehmer || Lehrgangsteilnehmer
participate betätigen * *sich betätigen (in:an/bei)* || beteiligen * *sich beteiligen, teilhaben (in: an/bei)* || teilnehmen * *in: an*
participation Beteiligung * *in: auf/bei* || Teilnahme * *in: an*
participator Teilhaber * *Teilnehmer (in:an)* || Teilnehmer * *in:an*

particle Korn || Partikel || Teil * Partikel, Körper, ~chen
particle board Spanplatte * *(Holz)*
particle size Korngröße
particles Späne * *(Holz)* Ausgangsmaterial bei d. Spanplattenherstellung
particles of mass Masseteilchen
particular besonders || Einzelheit * Einzelheit, besonderer/näherer Umstand; with full particular:- mit allen ~ || einzeln * besonder, speziell || entsprechend * besonder, einzeln, speziell, jeweilig || jeweilig * einzeln, individuell, eigentümlich, speziell, besonder || Näheres || schwierig * eigen, wählerisch || Sonder- * besonders || speziell * besonders || Umstand || vor allem * besonders
particular point Einzelheit * besonderer Punkt
particularity Besonderheit || Eigenart * Besonderheit, Umstand || Eigenheit || Eigenschaft * Eigenheit, Umstand || Einzelheit * Besonderheit, besonderer Umstand, Eigenheit, eigentümliche Beschaffenheit || Umstand * besonderer
particularize beschreiben * genau ~
particularly besonders * *(Adv.)* in besonderem Maße || insbesondere * *(Adv.)* || sehr * besonders
particulars Einzelheiten || Information * nähere Auskünfte
parting line Fuge * →Trennfuge || Trennfuge
partition aufteilen || Fach * Abteil, auch in Schrank, Aktentasche usw. || Partition * logisches Laufwerk (z.B. C:, D: usw. auf Festplatte) || Teilung * (Auf-, Ver-)Teilung, Trennung, Absonderung
partition off teilen * absondern
partitioning Aufteilung * in Teile
partitions Schottung * durch Trennwände abbgeschottete Teile
partly teilweise * *(Adv.)* zum Teil || zum Teil
partner Gesellschafter || Partner || Teilhaber * an einer Firma || Teilnehmer * Partner, Teilhaber, 'Kompagnon', Gesellschafter
partnership Assoziation * Partnerschaft || Beteiligung * geschäftliche ~, z.B. an einem Unternehmen || Partnerschaft || Personengesellschaft
parts Material * Teile (z.B. Ersatzteile)
parts at hazardous voltage levels berührungsgefährdete Teile
parts department Ersatzteillager
parts inventory Lagerhaltung * von Artikeln, Geräten usw.
parts list Stückliste
parts manufacture Vorfertigung
parts production Vorfertigung
parts stocking Lagerhaltung * von Artikeln, Geräten usw.
parts under voltage spannungsführende Teile
party Partner * Vertragspartner, Vertragspartei || Seite * Partei || Teilnehmer * (Vertrags)Partei, Teilnehmer/haber, Beteiligter, Fraktion
party to the transaction Geschäftspartner * bei Geschäftsverhandlungen
party who issues the order Auftraggeber * *(Rechtswesen)*
pass annehmen * Gesetz || Arbeitsgang || beschließen * eine Vorlage ~, einen Vorschlag ~ || durchfahren || Durchgang * Arbeitsgang, Durchlauf, Passage || Durchlauf || durchreichen || Lauf * Durchlauf || laufen * sich hindurchbewegen, hindurch/hinüber-laufen/führen, hindurchlassen || leiten * (hindurch)führen/leiten, weiterreich./ leiten, (Nachricht be)fördern, durchlass., dirigieren || passieren || Stich * *(Walzwerk)* || überholen || verlaufen * Zeit || weitergeben
pass away verstreichen * Zeit
pass by überholen * vorbeigehen, vorübergehen *(auch zeitlich)*

pass into übergehen * ~ in
pass mark Passmarke
pass on abgeben * weitergeben (to: an) || übergehen * weitergehen (to:zu) || weitergehen * to: an || weiterreichen
pass over nicht beachten * übergehen (to: zu) || passieren || übergehen * to: zu; auch 'übersehen', →'nicht beachten' (in silence: stillschweigend) || überleiten || überschreiten * überqueren || übertragen * überleiten, überführen, übertragen,
pass through durchfließen || durchführen * (durch etwas) hindurchführen || durchlaufen || durchreichen || gehen * gehen (hin)durch || hindurchführen || passieren
pass through zero Nulldurchgang
pass to übergehen * von Besitz || zur Verfügung stellen * an jdn./etwas weitergeben
pass to a person's discount berechnen * jemandem in Rechnung stellen
pass to a persons credit abführen * jemandem gutschreiben
pass to account verrechnen * verbuchen
pass-bills Begleitpapiere
pass-on weitergeben
passage Durchgang * für Personen, Flüssigkeiten, Gase || Durchlass || Gang * Durch~, Verbindungsgang || Kapitel * Abschnitt, (Text-) Passage || Öffnung * Durchlass || passieren * (Substantiv) || Übergang || Weg * Durchgang
passage through zero Nulldurchgang
passenger service Personenbeförderung
passenger traffic Personenbeförderung || Personenverkehr
passenger transport Personenbeförderung
passenger transportation Personenverkehr * (amerikan.)
passing Annahme * eines Gesetzes || vorübergehend * auch flüchtig, vergehend
passing contact Wischer * Wischkontakt || Wischkontakt
passing in Einschleusung
passing over Überleitung
passing through Durchführung * einer Leitung
passing-on Weitergabe
passionate eifrig
passivate passivieren
passivated passiviert
passivation Passivierung * z.B. Schutzauftrag aus Glas bei Hybridbauteil
passive passiv
passive components passive Bauelemente
passive resistance Übergangswiderstand
passivity Passivität
password Codewort || Passwort || Schlüsselwort * dient als Zugriffssperre für Unbefugte
past alt * (Substantiv) das Alte, die Vergangenheit || bisherig * ehemalig, vergangen || vorbei * (räumlich und zeitlich) || vorherig * vergangen, ehemalig
past it darüber hinaus
paste anleimen * →kleben || Brei || Druckfarbe * *(Druckerei)* bei Siebdruck || einfügen * Inhalt von Zwischenablage, Textpuffer usw. "einkleben" *(Cut and Paste-Funktion)* || kleben * z.B. Papier || Klebstoff * Kleister || Paste * breiige Masse, Paste, Brei, Kleister, Klebstoff, Glasmasse, Druck~ (bei Siebdruck)
paste on ankleben
paste over verkleben
paste up verkleben
paste-board Karton * starke ~pappe
pasteboard Pappe
pasteboard box Karton * Schachtel
pasted geklebt
pasted lined board kaschierte Pappe
paster Rollenwechsler * Rollenwechsel- u. -Ankle-

beeinrichtung; flying paster: ~ *mit fliegendem Rollenwechsel*
pasting Klebung || Verleimung * *(Papier)*
patch Balkon * *behelfsmäßige Modifikation eines Programms durch Ansprung einer ausgelagerten Befehlssequenz* || rangieren || Rucksack * *behelfsmäßige Programmänderung durch Ansprung einer provisorischen Befehlssequenz*
patchy lückenhaft * *zusammengestoppelt, Flickwerk; (auch Kenntnisse, Wissen usw.)*
patent Patent * *Erfindungsurkunde (for:auf); apply for:anmelden; infringe on:verletzen; pending:angemeldet* || patentieren * *take out a patent for a thing: sich etw. patentieren lassen*
patent infringement Patentverletzung
patentable patentierbar
patented patentiert
path Bahn- * *[NC-Steuerg.]* || Lauf * Weg || Pfad || Programmzweig || Weg * *(auch figürl.) Pfad, Signal-, Leistung-; position: Lage* || Zweig
path control Bahnsteuerung * *[NC-Steuerung] Position entlang einer Kontur Steuern*
path interpolation Bahninterpolation * *bei* →NC-Steuerung
pathway Bahn
patina Patina
patron Förderer * *Gönner, Förderer, Schirmherr*
patronize fördern * *als Gönner* ~
pattern Anordnung * *System* || Lehre * *Muster, Vorlage, Modell* || Modell * *(Gieß-) Muster* || Muster * *Impuls~, Bit~, Vorlage, Modell, Design, Motiv, Vorbild* || Ordnung * *Anlage* || Probe * *Muster* || Schablone * *Modell, Muster* || Schema * *Muster, Anordnung* || Struktur * *Gestalt(ung), Anlage, Muster, Schema, Gesetzmäßigkeit(en), (wirtsch./ soziale) Struktur*
pattern maker Modellbauer
pattern making Modellbau
pause aussetzen * *unterbrechen, pausieren* || Pause
pause interval Pausenzeit
pave the way for in die Wege leiten * *den Weg bahnen für, in die Wege leiten*
pawl Klaue * *Sperrhaken, Sperrklinke, Klaue* || Klinke * *Sperr*~ || Sperrklinke * *Sperrhaken, Sperrklinke, Klaue*
pawl stop Rücklaufsperre * *durch Sperrklinke/ Spreizhaken realisiert*
pay attention to achten auf * *siehe auch* →beachten || beachten || kümmern * *beachten*
pay damages Schadenersatz leisten
pay duty on verzollen
pay for itself bezahlt machen * *sich* ~
pay no heed to außer Acht lassen
pay off amortisieren
pay out abrollen * *Kabel* || nachlassen * *z.B. Papierbahn entspannen*
pay over abführen * *to: Geldbetrag an ...* ~
pay-off Abwickler * *z.B. für Metallband, Kabel, Draht* || Abwicklung * *Abwickeleinrichtung, z.B. für Blechband, Kabel* || Spuler * *Ab~/Abwickler, z.B. für Blechband, Kabel*
pay-off reel Abhaspel * *in Walzstraße/Blechbehandlungsanlage/Draht-/Kabelmaschine* || Abwickler * *Abhaspel in Walzwerk/Blechbehandlungsanlage*
pay-off stand Abwickler * *Abwickelmaschine, z.B. für Kabel* || Abwicklung * *Abwickler, z.B. für Kabel; Rollenständer*
pay-off station Abwickler * *Abwickelstation, z.B. für Kabel* || Abwicklung * *Abwickelstation, z.B. für Kabel*
pay-out reel Abwickeltrommel
payable zahlbar * *zu zahlen*
payback Amortisation || Amortisierung

paying lohnend * *einträglich, rentabel* || wirtschaftlich * *ertragreich, sich auszahlend*
payload Nutzlast
payment Honorar || Lohn * *Bezahlung* || Zahlung * *offer as payment: in* ~ *geben*
payment in advance Vorauszahlung
payment of damages Schadenersatz * *Schadensatzzahlung; liable to payment of damages: schadenersatzpflichtig* || Schadenersatzzahlung || Schadensersatzzahlung
payoff Vorteil * *Investitionsrückfluss, Profit, das 'Sich-Auszahlen'*
payoff reel Abhaspel * *in Walz-/Blech-/Draht-/Kabelstraße* || Abwickelrolle
payroll Gehaltsliste
PC PC
PC board Flachbaugruppe || Karte * *Leiterkarte* || Leiterkarte || Platine * *gedruckte Leiterkarte* || Print
PC board format Leiterplattenformat
PC board terminal Printklemme
PC track Leiterbahn
pc-board Leiterplatte
PCB Baugruppe * *Leiterkarte* || Flachbaugruppe || Karte * *Leiterkarte; Abkürz. für 'printed circuit board'* || Leiterkarte * *Abkürzg. für 'Printed Circuit Board'* || Leiterplatte || Platine * *Abk. für 'Printed Circuit Board': gedruckte Leiterkarte* || Print
PCB assembly machine Bestückungsautomat * *für Leiterkarten*
PCB holder Einzelkartenhalterung
PCD Lochkreisdurchmesser * *Abk. für 'Pitch Circle Diameter'*
pcs Stück * *Abkürzg. für 'Pieces'.*
PDMC motor permanenterregter Gleichstrommotor * *Abk. für 'Permanent-Magnet DC motor'*
PE Schutzleiter * *Abkürzung für 'Protective Earth': Schutzerde*
PE conductor Schutzleiter * *Abkürzung für 'Protective Earth': Schutzerde*
PE rail PE-Schiene * *Abk. für 'Protective Earth'*
PE terminal Schutzleiteranschluss * *Abk. f. 'Protective Earth'*
peak -Spitze || maximal * *Spitzen-* || Scheitel- * *z.B. -Wert, -Spannung* || Spitze * *(auch elektr.)* || Spitzen- * *kurzzeitig vorkommender Maximalwert*
peak brake power Spitzenbremsleistung
peak current Spitzenstrom || Stoßstrom
peak demand Spitzenbedarf
peak factor Scheitelfaktor
peak hours Spitzenlastzeit
peak inrush current Rush-Strom * *Stromspitze b. Aufbau eines Feldes (z.B. Motor, Trafo usw.)*
peak inrush torque Rush-Moment * *Momentenspitze z.B. bei Motorstart u. Drehrichtungsumkehr*
peak inverse voltage Spitzensperrspannung * *eines Halbleiters*
peak load Lastspitze || Spitzenbelastung || Spitzenlast
peak output Spitzenleistung * *z.B. Stromrichter, Motor*
peak overload Lastspitze * *Über*~
peak performance Spitzenleistung * *höchste Leistungsfähigkeit*
peak power Spitzenleistung * *[kW]*
peak starting current Anlassspitzenstrom * *[el. Masch.]*
peak temperature Spitzentemperatur
peak torque Stoßdrehmoment
peak transient torque Stoßmoment
peak value Scheitelwert || Spitzenwert
peak voltage Scheitelspannung || Spitzenspannung
peak-to-peak Spitze-Spitze
peak-to-peak value Spitze-zu-Spitze-Wert
peak-to-peak voltage Spitze-Spitze-Spannung
peaked spitz

pect voraussetzen * *erwarten*
pecularities typische Merkmale * *Eigentümlichkeiten*
peculiar eigenartig || komisch * *eigenartig, absonderlich, komisch*
peculiarity Eigenart || Eigenheit || Eigenschaft * *Eigenart*
pedal Pedal * *work the pedals: in die ~e treten*
pedestal Bock * *Untergestell, Sockel, Bock,* →*Lagerbock* || Lagerbock * *(Lager)Bock, Untergestell, Sockel, Postament, Säulenfuß, Grundlage (figürl.)* || Sockel * *Untergestell, (Lager)Bock, Säulenfuß, ~ eines Bauwerks*
pedestal bearing Stehlager * *auf Lagerbock/Untergestell angeordnet*
pedestal-type Ständer- * *{Werkz.masch.}*
pedestral Sockel * *Untergestell, (Lager)Bock, Säulenfuß, ~ eines Bauwerks*
peel off ablösen * *(sich) blättrig ~ (z.B. Farbe); (Klebeetikett usw.) abziehen; abschälen* || abziehen * *abschälen, z.B. Schutzfolie*
peeling knife Rundschälmesser * *{Holz} z.B. zur Furniererzeugung* || Schälmesser * *{Holz}*
peeling lathe Schälmaschine * *{Holz} zum Erzeugen von Furnier*
peeling machine Schälmaschine * *z.B, für Baumstämme*
peeling off Ablösung * *(Sich-) Abschälen, Abblättern, Abbröckeln*
peer Koppelpartner * *bei Datenkopplung ohne ausgeprägten Master*
peer to peer network Peer-to-Peer-Netzwerk * *Netzwerk mit gleichberechtigten PC's(jeder ist gleichzeitig Server u.Client)*
peer-to-peer connection serielle Kopplung zwischen gleichberechtigten Partnern
Peer-to-Peer link Peer-to-Peer-Schnittstelle * *Verbindg.zw. gleichberecht. Partnern,z.B. Stromrichtern (o.Master)* || Peer-to-Peer-Verbindung * *seriell. Verbindg.zw. gleichberecht. Partnern, z.B. Stromrichtern (o.Master)*
peerless unvergleichlich
peforated roll Lochwalze
peg Dübel
peg teeth Wolfszahnung * *{Holz} bei Säge*
pelletizer Granulator * *zur Herstellung von Kunststoffgranulat* || Granuliermaschine
pelletizing machine Granuliermaschine * *zur Herstellung von Kunststoffgranulat*
pellets Granulat * *kleine Kügelchen; z.B. zur Kunststoff-Verarbeitung*
pencil Bleistift || Stift * *Zeichen~, Schreib~*
pencil drawing Bleistiftskizze * *Bleistiftzeichnung*
pencil scetch Bleistiftskizze
pend anstehen * *z.B. Interrupt, Warnung*
pending anstehend * *z.B. Meldung, Warnung, Interrupt*
pending message anstehende Meldung
pending signal anstehende Meldung
pendulum Pendel
pendulum control Pendeldämpfung * *z.B. bei Kran*
pendulum generator Pendelgenerator || Pendelmaschine * *zur Drehmomentmessung*
penetrate durchdringen * *auch Markt/geistig; auch eindringen, durchbohren, durchstoßen, geistig ergründen* || eindringen * *durchdringen, eindringen (in) (auch geistig), durchbohren, durchstoßen*
penetration Durchbruch || Durchführung * *(Kabel)-Eintritt, Durchbohrung, Durchtritt*
pep ankurbeln * *(amerikan.; salopp) Pepp geben*
per durch * *per, durch, laut, gemäß, pro, je, für* || entsprechend * *laut, gemäß* || je * *pro* || pro
per annum pro Jahr
per cent Prozent

per day täglich * *pro Tag*
per diem pro Tag
per mill Promille
per mille Promille
per thousand Promille
per unit Prozent * *der Nenngröße* || prozentual * *Anteil des Nennwerts* || relativ * *bezogen auf Bezugsgröße, welche "1" oder "100%" entspricht* || Teile des Nennwerts
per unit value Bezugswert * *auf den sich andere Angaben beziehen (Bezugswert entspr. 1 oder 100%)*
per-unit bezogen * *auf eine Nenngröße/Bemessungsgröße*
per-unit torque bezogenes Drehmoment
perceive merken * *wahrnehmen* || wahrnehmen * *empfinden, wahrnehmen, (be)merken, spüren, verstehen, erkennen, begreifen*
percent Prozent * *(amerikan.)*
percent impedance uk * *siehe auch* →*Kurzschlußspannung (bei Drossel, Trafo)*
percentage Anteil * *Prozent~* || bezogen * →*prozentual* || Prozent * *Prozentanteil, Prozentsatz* || Prozentanteil || Prozentsatz
percentage impedance Kurzschlusspannung * *z.B. bei Drossel/Trafo (siehe auch 'uk')*
percentage ripple Welligkeit * *Anteil der Welligkeit in Prozent*
percentage value Prozentwert
percentage voltage drop prozentualer Spannungsabfall
percental bezogen * →*prozentual* || prozentual * *percentage: prozentualer Anteil*
percental voltage drop prozentualer Spannungsabfall
perceptibility Erkennbarkeit
perceptible deutlich * *wahrnehmbar* || erkennbar * →*wahrnehmbar* || fühlbar * *geistig; wahrnehmbar* || sichtbar * *wahrnehmbar* || wahrnehmbar
perception Wahrnehmung * *(sinnliche, geistige) Wahrnehmung, Empfindung*
perception threshold Hörgrenze * *Wahrnehmungsschwelle* || Hörschwelle * *Wahrnehmungsschwelle* || Wahrnehmungsschwelle
percussion drill Schlagbohrer || Schlagbohrmaschine
percussion drilling Schlagbohren
perfect abrunden * *perfektionieren* || ausarbeiten * *vollständig ~* || ausgezeichnet || einwandfrei * *fehlerfrei, vollendet, vollkommen* || fehlerfrei || ideal || mustergültig || optimal || perfekt || perfektionieren * *of technical perfection/technically perfect:technisch perfektioniert* || sehr gut || verfeinern * *vervollkommnen, perfektionieren* || völlig * *vollkommen* || vollkommen * *perfekt,makellos*
perfected ausgereift * *technische Lösung, Konstruktion usw.*
perfectibilism Perfektionismus
perfecting engine Kegelmühle
perfectionism Perfektionismus
perfectionist Perfektionist
perforate lochen || peforieren || perforieren || stanzen * *lochen*
perforation Lochung
perforator Locher * *auch im Büro*
perform ausführen * *durchführen, z.B. eine Aufgabe* || durchführen || leisten * *verrichten, Leistung bringen, leistungsfähig sein; perform well: gute Leistung erbringen* || leistungsfähig sein || versehen * *Pflichten ~*
perform the pressure test abdrücken * *Druckprüfung durchführen (z.B. für Pressluftkessel, Rohrsystem usw.)*
performance Ausführung * *einer Verrichtung* || Betriebsverhalten || Durchführung || Leistung

* ~*sfähigkeit, ~svermögen; perform well: gute ~ erbringen* || Leistungsfähigkeit || Qualität * *Leistungsfähigkeit* || Verarbeitungsleistung * *Leistungsfähigkeit* || Verhalten * *Leistungsfähigkeit, Betriebsverhalten*
performance boost   Leistungsschub * *Schub bezüglich der Leistungsfähigkeit* || Leistungssteigerung * *(salopp) Steigerung der Leistungsfähigkeit*
performance characteristics   Leistungsmerkmale
performance data   Leistungsdaten
performance feature   Leistungsmerkmal
performance goodness   Regelgüte
performance level   Funktionsklasse || Leistungsklasse * *Leistungsfähigkeit*
performance range   Leistungsbereich * *Leist.fähigkeit (upper/high-end:oberer), (medium/midrange:mittlerer)*
performance rise   Leistungssteigerung * *Steigerung der Leistungsfähigkeit*
performance scope   Leistungsumfang
performance test   Funktionsprüfung
performance value   Betriebsgüte * *Produkt (Wirkungsgrad mal cos phi) eines Motors*
performing better   hesser * *in der Leistungsfähigkeit überlegen sein*
perhaps   etwa * *vielleicht* || eventuell * *(Adv.) vielleicht* || möglicherweise * *vielleicht* || unter Umständen * *vielleicht*
perilous   gefährlich || lebensgefährlich
perimeter   Peripherie * *Umkreis (z.B. einer Stadt) Peripherie* || Umfang * *(geometr.) Umkreis*
period   Abschnitt * *Zeit~* || Dauer * *Zeitraum* || Lauf * *over the period of:im Laufe einer Zeitspanne* || Periode || Periodendauer || Phase * *Zeitraum* || Zeit * *Zeitraum/Spanne/Dauer, Frist, Periode* || Zeitabschnitt || Zeitdauer || Zeitraum || Zeitspanne
period between lubrications   Fettgebrauchsdauer
period of development   Entwicklungszeit
period of no current   stromlose Pause * *z.B.durch Kommandostufe erzeugt*
period of no torque   momentenfreie Pause || momentfreie Pause
period of oscillation   Schwingungsdauer
period of storage   Lagerungszeit
period of time   Zeit * *~spanne* || Zeitspanne * *in: nach e. Zeitspanne*
periodic   periodisch || regelmäßig * *in ~en Zeitabständen*
periodic duty   Aussetzbetrieb || periodischer Betrieb || Schaltbetrieb * *sich wiederholendes Lastspiel, siehe auch 'Aussetzbetrieb'*
periodic intermittent duty with electrical braking   periodischer Aussetzbetrieb mit elektrischer Bremsung * *[el.Masch]* →*Betriebsart S5*
periodic nature   periodischer Verlauf * *Art, Beschaffenheit*
periodical   periodisch || regelmäßig * *(zeitlich)* || wiederkehrend || wiederkehrend * *periodisch* || Zeitschrift * *regelmäßig erscheinende ~*
periodical installation guides   Montageschriften * *regelmäßig erscheinende ~*
periodical technical information   Informationsdienst * *regelmäßig erscheinender techn. ~*
periodically   periodisch * *(Adv.)*
peripheral   Peripherie- || Peripheriebaustein * *z.B. eines Mikroprozessors* || Peripheriegerät * *eines Computers* || Umfangs-
peripheral board   Peripheriebaugruppe * *Leiterkarte*
peripheral byte   Peripheriebyte * *bei SPS*
peripheral device   Peripheriegerät * *eines Computers*
peripheral force   Umfangskraft
peripheral memory   Peripherspeicher

peripheral module   Peripheriebaugruppe || Peripheriemodul
peripheral signal   Peripheriesignal
peripheral speed   Umfangsgeschwindigkeit
peripheral unit   Peripheriegerät
peripheral velocity   Umfangsgeschwindigkeit
peripherals   Peripherie * *I/O-Baugruppen für SPS, (Prozess-)Rechner* || Peripherie * *I/O-Baugruppen/ Geräte für SPS/(Prozess-)Rechner* || Peripheriegeräte * *z.B. Drucker, Bildschirm bei SPS*
peripheric   Umfangs- * *peripheric holes: ~bohrungen*
peripheric holes   Umfangsbohrungen
periphery   Peripherie || Rand * *Umkreis, Peripherie*
permanent   beständig * *dauerhaft* || bleibend * *(fort-) dauernd, bleibend, permanent, ständig, dauerhaft, Dauer-* || Dauer- || dauerhaft * *bleibend, (fort-) dauernd, ständig, Dauer-, dauerhaft, massiv* || endgültig * *dauernd* || fest * *dauernd, unveränderlich; z.B. Wohnsitz, Struktur, Beschaffenheit* || gleich bleibend * *(fort)dauernd, bleibend, permanent, ständig, ständig* || permanent || ständig * *fortdauernd, bleibend, permanent, dauerhaft* || unlöslich * *unlösbar*
permanent control error   bleibende Regelabweichung
permanent damage   bleibender Schaden
permanent error   bleibender Fehler || Restfehler * *bleibender Fehler* || stationärer Fehler * *bleibender Fehler*
permanent forcing   zwangssteuern * *(Substantiv) Festvorgabe von SPS-Ein-/Ausgängen zum Test*
permanent lubrication   Dauerschmierung
permanent magnet   Dauermagnet
permanent magnet motor   Dauermagnetmotor || Permanentmotor
permanent memory   Permanentspeicher * *nichtflüchtiger Speicher (z.B. EEPROM, NOVRAM, Flash-EPROM usw.)*
permanent mounting   Festeinbau
permanent offset error   bleibende Regelabweichung
permanent operation   Dauerbetrieb
permanent output current   Dauer-Ausgangsstrom
permanent processing   Permanentverarbeitung
permanent service   Dauerbetrieb || Dauerlast * *Dauerbelastung*
permanent temperature   Beharrungstemperatur * *im eingeschwungenen Zustand*
permanent-field   dauermagneterregt || permanenterregt
permanent-field excitation   Permanenterregung
permanent-field machine   permanenterregte Maschine
permanent-field motor   permanenterregter Motor
permanent-field rotor   permanenterregter Läufer
permanent-field synchronous motor   permanenterregter Motor * *z.B. SIMOSYN (R) Motor von SIEMENS* || permanenterregter Synchronmotor * *z.B. Siemosyn-Motor von SIEMENS* || Siemosyn-Motor * *z.B. Siemosyn-Motor(R) von SIEMENS*
permanent-magnet   Dauermagnet || Permanentmagnet
permanent-magnet brake   Dauermagnetbremse || Permanentmagnetbremse
permanent-magnet DC motor   permanenterregter Gleichstrommotor
permanent-magnet excited   dauermagneterregt || permanenterregt
permanent-magnet synchronous motor   permanenterregter Synchronmotor
permanent-split capacitor   Betriebskondensator * *Phasenschieberkondensator z.Betrieb e. Drehstrommotors an 1-phasigem Netz*
permanently   dauerhaft * *(Adv.)* || ständig * *(Adv.)*

permanently assigned fest belegt
permanently defined fest definiert
permanently excited dauermagneterregt
permanently excited machine permanenterregte Maschine
permanently fixed fest montiert
permanently flexible dauerelastisch
permanently installed fest verlegt * Kabel, Leitungen usw.
permanently lubricated dauergeschmiert
permanently mounted fest montiert
permanently-excited permanenterregt
permeability Durchlässigkeit || magnetische Leitfähigkeit || Permeabilität
permeable durchlässig || permeabel
permeance Durchlässigkeit || Leitwert * magnetischer ~ || magnetischer Leitwert
permissible erlaubt * zulässig || zulässig
permissible noise level Störabstand * zulässiger Störpegel
permissible number of starts in succession Anlasszahl * {el.Masch} zulässige Anz. aufeinanderfolg. Startvorgänge
permissible overload Überlastbarkeit * zugelassene Überlast (short-time: kurzzeitige)
permissible range zulässiger Bereich
permissible temperature Grenztemperatur * zulässige Temperatur
permissible thermal load Wärmebelastbarkeit
permissible variation Toleranz * zulässige Abweichung || Toleranz * erlaubte Schwankungsbreite
permission Genehmigung * Erlaubnis; give a person the permission:jdm. die Erlaubnis erteilen
permit erlauben * gestatten || Erlaubnis || ermöglichen * gestatten, erlauben || Freigabe * Erlaubnis, auch Aus-/Einfuhrerlaubnis || Genehmigung * behördliche Zulassung || gestatten
permit access Zugriff freigeben
permits Begleitpapiere * Zollpapiere
permitted erlaubt || zugelassen * erlaubt, gestattet || zulässig * zugelassen, erlaubt
permittivity Dielektrizitätskonstante
perpendicular senkrecht * (geometr.) einen rechten Winkel bildend, lotrecht (to: zu)
perpendicular position Senkrechte
perpendicularity Rechtwinkligkeit
perpetual permanent
persecution Verfolgung * drangsalierende Verfolgung
persevere verharren * beharren (in), ausdauern, aushalten (bei), fortfahren (mit), festhalten (an)
persevering beständig * beharrlich, ausdauernd
persist bleiben * verharren (in: bei), beharren (in: auf; z.B. Meinung), weiterarbeiten (with: an) || verharren * in:bei/auf (z.B. einer Meinung)
persistance of vision Stroboskopeffekt
persistent andauernd * ständig, nachhaltig, anhaltend, beharrlich, ausdauernd, hartnäckig || beständig * ausdauernd, beharrlich || bleibend * nichtflüchtig (Speicherinhalt), ständig, anhaltend, beharrlich, ausdauernd, hartnäckig || dauerhaft * hartnäckig
person responsible Bearbeiter * for:für
person responsible for quality Qualitätsbeauftragter
person responsible for the product Produktverantwortlicher
person to talk to Gesprächspartner
personal assistant Assistent * persönlicher ~
personal autobiography Lebenslauf
personal computer PC
personal data Angaben zur Person
personal injury Körperverletzung * Personenschaden

personal injury or death Personenschaden * Verletzung oder Tod
personal protection Personenschutz
personal protection gear Personenschutzeinrichtung
personally selbst
personify verkörpern * personifizieren, versinnbildlichen
personnel Belegschaft * Personal || Personal
personnel head Personalchef
personnel management Personalwesen
personnel safety Personensicherheit
personnel schedule Dienstplan
perspective Blickwinkel * (auch figürlich) || Grund * Blickwinkel, Perspektive || Perspektive * (auch figürl.: 'Ausblick') || Sichtweise
persuade überzeugen * of:von, that: dass; auch 'überreden'
persuasive überzeugend * überredend, überzeugend
pertain beziehen * to: betreffen, gehören zu
pertain to betreffen
pertaining to betreffend || für * betreffend, gehörend zu
pertinence Relevanz * Erheblichkeit || Wichtigkeit * Erheblichkeit
pertinent angemessen * passend, schicklich, sachdienlich, einschlägig || dazugehörig * (zur Sache) gehörig, einschlägig, sachdienlich, passend, angemessen || diesbezüglich * to this:in dieser Sache || einschlägig || gehörig * zur Sache gehörend, sachdienlich, passend, angemessen || relevant * erheblich, zutreffend || sachgemäß || sachgerecht || zugehörig * zur Sache gehörig, sachdienlich, (zu)gehörig, (to: zu), angemessen, passend
pervasive beherrschend * durchdringend, überall vorhanden/anzutreffen, beherrschend
pervious durchlässig * durchlässig, durchdringbar, gangbar (to:für), zugänglich, offen, undicht
perviousness Durchlässigkeit
pessimistic pessimistisch
pessimistically pessimistisch * (Adv.)
pester belästigen * plagen
PET PET * Abk. für 'PolyEthyleneterephThalate': Polyethylentheraftalat (für Kunststoffflaschen usw.)
petition Antrag * Gesuch, Eingabe, Petition
petrochemical petrochemisch
petrol Benzin * (brit.)
petrol engine Benzinmotor
petrolatum Vaseline
petroleum Erdöl || Petroleum * (allg.)
petty geringfügig * Betrag, Vergehen || leicht * unbedeutend (Fehler, Vergehen usw.)
PG gland PG-Verschraubung * Abk.f.'Panzerrohr-Gewinde'-Verschraubg. (Kabeldurchführg. m.'Quetschgummi-Dichtg.')
Pg. Seite * Abkürzung für 'Page'
pH-value pH-Wert
pharmaceutical industry pharmazeutische Industrie
pharmacy Apotheke
phase Außenleiter || Phase || Stadium * siehe auch →Phase || Strang * Phase eines Wechselstromsystems || Stufe * Phase, auch Entwicklungsphase || Takt || Zeit * Phase, Stadium || Zustand * Phase
phase angle Phasenlage * Phasenwinkel || Phasenverschiebungswinkel || Phasenwinkel
phase angle control Anschnitt * Phasenanschnitt (steuerung) || Anschnittsteuerung * Phasenschnittsteuerung || Phasenanschnitt || Phasenschnittsteuerung * z.B.für netzgeführten Stromrichter,Thyristorsteller usw.
phase angle variation Phasenanschnitt
phase asymmetry Phasenunsymmetrie

phase characteristic Phasenkennlinie
phase coincidence Phasengleichheit
phase commutation Phasenfolgelöschung
phase comparison Phasenvergleich
phase conductor Außenleiter || Hauptleiter * *Außenleiter im Drehstromnetz*
phase control Anschnittsteuerung || Ansteuerung * *z.b. Phasenanschnittsteuerung* || Phasenanschnittsteuerung
phase converter Phasenwandler
phase current Außenleiterstrom || Phasenstrom || Strangstrom
phase delay Phasenverschiebung * *Phasen-Nacheilung* || Phasenwinkel
phase difference Phasenabweichung || Phasenverschiebungswinkel
phase dip Phaseneinbruch * *kurzzeitiger*
phase displaced phasenverschoben
phase displacement Phasenabweichung || Phasendrehung * *Phasenverschiebung* || Phasenverschiebung || Phasenverschiebungswinkel || Phasenwinkel
phase displacement angle Phasenverschiebungswinkel || Phasenwinkel
phase effects on the system Netzrückwirkungen
phase failure Phasenausfall || Phasenbruch
phase failure monitoring Phasenausfallüberwachung
phase imbalance Phasenunsymmetrie
phase intersection point Phasenschnittpunkt
phase lag Phasennacheilung || Phasenverschiebung * *Phasen-Nacheilung* || Phasenverzögerung
phase lead Phasenverschiebung * *Phasenvoreilung* || Phasenvoreilung || Voreilwinkel * *Phasen-Voreilwinkel*
phase loss Phasenausfall || Phasenbruch
phase loss detection Phasenausfallerkennung
phase margin Phasenreserve * *{Regel.techn.}* Phasenrand, Phasenreserve
phase of operation Arbeitsgang * *in a single operation: in einem* ~
phase position Phasenlage
phase reactor Strangdrossel * *in d. wechselstromseitigen Zuleitg. z. Stromrichter*
phase relation Phasenlage
phase relationship Phasenlage
phase response Phasengang || Phasenkennlinie * *→Phasengang*
phase reversal Drehfeldumschaltung
phase rotation insensitive Drehfeld-unabhängig * *Netzphasen können beliebig angeschlossen werden* || drehfeldunabhängig * *Netzphasen können beliebig angeschlossen werden*
phase sequence Drehfeld * *Phasenfolge: clockwise: Rechts-, counterclockwise:Links-* || Phasenfolge
phase sequence commutation Phasenfolgelöschung * *Kommutierungsart b.I-Umrichter: Speisung je 2er Wicklungsstränge i.Folge*
phase sequence identification Drehfelderkennung
phase sequence monitoring Drehfeldüberwachung || Phasenfolgeüberwachung || Phasenüberwachung
phase sequence reversal Drehfeldumkehr
phase sequence tolerant Drehfeld-unabhängig || drehfeldunabhängig
phase shift Phasenverschiebung
phase shifter Phasenschieber
phase shifting Phasenschieben || Phasenverschiebung
phase short-circuit Phasenkurzschluss || Phasenschluss
phase symmetry Phasensymmetrie
phase to phase short circuit Phasenschluss
phase to phase voltage Netzspannung * *Drehstrom*
phase unbalance Phasenunsymmetrie
phase unsymmetry Phasenunsymmetrie

phase velocity Phasengeschwindigkeit
phase voltage Leiterspannung * *z.B. eines Transformators* || Phasenspannung * *Strangspannung* || Strangspannung
phase winding Phasenwicklung * *z.B. eines Trafos, einer Drossel*
phase-angle controller Winkelregler * *z.B. bei USV*
phase-control ansteuern * *bei Phasenanschnittsteuerung*
phase-control reactive power Steuerblindleistung * *{Stromrichter} auf Grund der Phasenanschnittsteuerung*
phase-delayed phasenverschoben * *verzögert*
phase-displacement transformer Schwenktrafo * *z. Phasendrehung z.B. z. Erzeugung d. Synchron.spanng. e. Steuersatzes*
phase-failure protection Phasenausfallschutz
phase-locked phasenstarr
phase-segregated phasengetrennt
phase-separated phasengetrennt
phase-sequence commutation Phasenfolgelöschung * *Thyristorlöschung jew. durch Zündg.d.Folgethyristors ü. Löschkondensator*
phase-sequence indicator Drehfeldmesser
phase-shift network Phasenschieber
phase-shifted phasenverschoben || phasenverschoben * *with respect to: gegenüber*
phase-shifted by 90 degrees um 90 Grad versetzt * *z.B. Ausgangssignale eines inkrementellen Impulsgebers*
phase-shifted transformer Doppelstocktrafo * *{Stromrichter} für 12-pulsige Stromrichterschaltung usw.*
phase-to-ground voltage Phasenspannung || Strangspannung
phase-to-neutral voltage Phasenspannung * *zwischen L1, L2, L3 und N* || Sternspannung * *zw. Phase und Mittelpunktsleiter* || Strangspannung * *Phasenspannung zw. L1, L2, L3 und jeweils N*
phase-to-phase short-circuit Phasenkurzschluss
phase-to-phase voltage Außenleiterspannung * *verkettete Spannung* || Dreieckspannung || Leiterspannung || verkettete Spannung
phasor Vektor || Zeiger * *Vektor*
phenomenal phänomenal
phenomenon Effekt * *(Physikal.)* Phänomen, Erscheinung || Ereignis || Erscheinung * *Natur-* || Phänomen
Pho. Tel
phone anrufen * *phone somebody about: jdn. anrufen wegen; be on the phone: am Telefon sein* || Telefon * *answer the phone: den ~hörer abnehmen; bo on the phone: am ~ sein*
phonometer Schallpegelmesser
phosphate phosphatieren
photo Bild * *Foto* || Foto
photo copier Fotokopierer
photocell Fotozelle
photocoupler Optokoppler
photograph Bild * *Fotografie* || Foto
photoresistor Fotowiderstand
phototransistor Fototransistor
photovoltaic photovoltaisch * *z.B. Solar-*
phrase Redewendung * *set/stock phrase: (fest)stehende Redensart/-wendung* || Satz * *Redewendung, Ausdruck, Phrase, kurzer* ~
physical physikalisch
physical channel physikalischer Kanal
physical fundamental principles physikalische Grundlagen
physical science Physik
physically segregated räumlich getrennt
physician Arzt
physics Physik

PI-controller PI-Regler * *[pie 'ai ...]*
pick off abgreifen * *z.B. Spannung von einem Potentiometer, Messsignal usw.* || abnehmen * *z.B. Strom von einem Stromabnehmer/Schleifkontakt ~/ abgreifen*
pick up abholen || Ansprechen * *(Verb) Relais* || anziehen * *Relais* || auffangen * *Neuigkeiten usw.* || aufheben * *vom Boden* ~ || aufnehmen * *z.B. vom Boden, Signal durch e. Funkempfänger/e. Messaufnehmer usw.* ~ || holen * *ab*~ *(auch Informationen)* || lernen * *aufschnappen*
pick-and-place machine Bestückungsautomat
pick-off Abgriff * *z.B. an Potentiometer, Schiebetrafo* || Schleifer * *Abgriff, z.B. an einem Potentiometer*
pick-off gear wheel Wechselrad
pick-up annehmen * *aufnehmen, (Telefonanruf usw.)* annehmen || Geber || Greifer * *Aufnehmer*
pick-up current Ansprechstrom * *Relais*
pick-up delay Ansprechverzögerung * *Relais*
pick-up felt Abnahmefilz * *in Papiermaschine (übernimmt das vom Sieb kommende Papier)*
pick-up point Aufnahmeort * *z.B. für Kran*
pick-up time Ansprechzeit * *Relais*
pick-up voltage Anzugsspannung * *Relais*
picking of ordered items kommissionieren * *(Substantiv) im Auslieferungslager*
picking warehouse Auslieferungslager * *Kommissionierlager* || Kommissionierlager
pickle Beize || beizen || Metall || entzundern
pickling plant Beizanlage * *z.B. für Blech*
pickup Abnehmer * *Abtastgerät, Messwertaufnehmer, Aufnahmeapparatur,* || Abtaster || Geber * *(Messwert-) Aufnehmer*
pickup current Anzugsstrom * *Relais*
pickup delay Ansprechverzögerung * *z.B. Relais* || Anzugsverzögerung * *Relais* || Einschaltverzögerung * *Relais, Schütz*
pickup power Ansprechleistung * *Relais*
pickup time Anzugszeit * *Relais*
pickup value Ansprechwert * *z.B. Relais*
pickup voltage Anzugsspannung * *Relais*
pickup winding Anzugswicklung * *Relais*
pickup-delayed anzugsverzögert || anzugverzögert
pictograph Piktogramm
pictorial bildlich * *bildmäßig, Bild-, Bilder-*
pictorial marking Piktogramm
picture Abbildung * *Bild* || beschreiben * *anschaulich* || Bild * *allg.* || Foto * *take a picture of: fotografieren, ein Foto machen von* || grafisch darstellen
picture formatting Bildaufbau
picture recognition Bilderkennung
picture refresh memory Bildwiederholspeicher
PID controller PID Regler || PID-Regler * *Regler mit Proportional-, Integral- u.Differenzialanteil*
piece Exemplar * *(Einzel-)Stück* || Stück * *(Einzahl)* || Teil * *Stück Ware*
piece of information Information * *(Singular) Einzel-/Teil*~
piece of wood Holz * ~*stück*
piece on ansetzen * *anstücken (to:an)*
pieces Stück * *(Mehrzahl)*
pierce lochen * *pierce holes into: Löcher in etwas* ~*; durchbohren/-dringen/-stechen*
piezo Piezo-
piezo-electric piezoelektrisch
piggyback Huckepack- * *auch Leiterkarte, IC usw.*
piggyback board Aufsteckkarte * *Subleiterkarte* || Huckepackbaugruppe * *Leiterkarte*
pigment Farbe * *Farbkörper*
pile -Paket * *Stapel, Stoß, Haufen* || aufstapeln || Haufen * *Haufen, Stapel, Stoß, (auch Geld-) Menge/Masse* || Stapel || stapeln

pile-wire loom Rutenwebstuhl * *{Textil} z.B. zur Teppichherstellung*
piler Stapler * *Vorrichtung zum Aufeinanderstapeln von Teilen*
piling device Stapler
piling-up Anhäufung
pillar Säule * *(auch figürlich)* || Ständer * *Pfeiler, Ständer, z.B. bei Bohrmaschine*
pillar terminal Buchsenklemme
pillar-type drilling machine Säulenbohrmaschine
pillow block Lagerbock * *Stehlager, Lagerbock*
pilot Führungs- || leiten * *steuern, (durch)lotsen (auch figürl.), führen, lenken, leiten*
pilot device Bedienungselement * *z.B. Taster, Schalter, Meldeleuchte*
pilot drive Leitantrieb
pilot equipment Pilotanlage
pilot installation Pilotanlage
pilot lamp Anzeigelampe || Kontrollleuchte || Meldeleuchte || Signallampe
pilot pin Zentrierstift
pilot plant Pilotanlage
pilot project Pilotprojekt
pilot train Vorzug * *{Bahnen}*
pilot version Pilotversion
pimpling Blasenbildung
pin Anschluss * ~*stift* || Bolzen || Dorn * *Nadel, Stift, Bolzen* || Nagel || Pin * *z.B. IC-/Steckerbeinchen* || Stift || verstiften * *pinned fitting: Verstiftung* || Zapfen * *Bolzen, Stift, Dorn; pin with thread/ theaded pin: Gewinde*~
pin assignment Anschlussbelegung * *Stecker* || Pinbelegung * *z.B. eines Steckers/ICs*
pin compatible anschlusskompatibel * *IC, Stecker*
pin coupling Bolzenkupplung
pin designation Pinbezeichnung * *z.B. bei einem IC*
pin joint Zinkenverbindung * *{Holz}*
pin plug Stiftdübel * *zum Einschlagen von Stahlstiften in harte Wände*
pin-end connector Aderendhülse * *Stiftkabelschuh*
pin-type face spanner Zweilochmutterndreher
pinch klemmen * *(ein)klemmen, quetschen, einengen, einzwängen, drücken, kneifen, zwicken* || Quetsch- || quetschen * *kneifen* || Stanze
pinch off abzwicken * *auch figürlich*
pinch roll Andruckrolle || Andruckwalze || Transportrolle * *z.B. in Walzwerk/Stranggießanlage* || Treiber * *{Walzwerk} Band-Transportrolle* || Treibrolle * *{im Walzwerk}* || Treibwalze * *{Walzwerk}*
pinch roll gap Treiberspalt * *Walzwerk*
pinching Einschnürung * *Einzwängung, Einzwicken; auch von Ladungsträgern in Halbleiter*
pinhole Nadelloch * *kleines Loch, z.B. in Papier*
pinion Ritzel || Zahnrad * *Ritzel, Antriebs(kegel)rad* || Zahnritzel
pinion cage Planetenträger
pinion drive Ritzelantrieb
pink noise Zufallsrauschen * *rosa Rauschen*
pinning Anschlussanordnung * *IC, Stecker* || Anschlussbelegung * *IC, Stecker* || Benadelung || Pinbelegung * *eines IC*
pinning diagram Anschlussanordnung * *IC- oder Stecker-Anschlussbild* || Anschlussbelegung * *IC- oder Stecker-Anschlussbild* || Pinbelegung * *IC-Anschlussbild*
pinout Anschlussanordnung * *IC, Stecker* || Anschlussbelegung * *IC, Stecker* || Belegung * ~ *von IC-Anschlüssen, Steckerstiften usw.* || Pinbelegung * *eines IC*
pinpoint bestimmen * *genau festlegen* || festlegen * *auf den Punkt genau* ~ || feststellen * *auf den Punkt genau* ~ || genau festlegen
pintle Drehbolzen * *z.B. bei Kabel-Auf-/Abwickler*

(greifen beidseitig in die Achse d. Kabeltrommel ein)
pioneering   bahnbrechend * *z.B. Erfindung* || zukunftsweisend * →*bahnbrechend*
pioneering accomplishment   Pionierleistung
pioneering achievement   Pionierleistung
pioneering feat   Pionierleistung * *Großtat, Kunst/ Meisterstück*
pioneering performance   Pionierleistung
pioneering technical accomplishment   technische Pionierleistung
pioneering technical achievement   technische Pionierleistung
pioneering technical feat   technische Pionierleistung * *Großtat, Kunst/Meisterstück*
pioneering technical performance   technische Pionierleistung
pioneering work   Pionierleistung
pipe   leiten * *durch ein Rohr* ~ || Leitung * *Rohrleitung* || Leitungsrohr * *water pipe: ~ für Wasser; gas pipe: ~ für Gas* || Rohr * *Rohrleitung, Leitungs~* || Rohr- || Röhre * *Leitungsrohr, Rohrleitung*
pipe adapter   Rohranschluss * *Anschluss (-stutzen)/ Zwischenstück*
pipe bend   Rohrkrümmer
pipe bending system   Rohrbiegeanlage
pipe burst   Rohrbruch
pipe clamp   Rohrschelle
pipe clip   Rohrschelle * *meist aus Kunststoff*
pipe connection   Rohranschluss
pipe coupling   Muffe * *Rohrverbindung* || Rohranschluss
pipe cutter   Rohrabschneider
pipe diameter   Rohrdurchmesser
pipe heat exchanger   Röhrenwärmetauscher
pipe line   Rohrleitung * *Fernleitung*
pipe penetration   Rohrdurchführung
pipe size   Rohrdurchmesser * *Rohrquerschnitt* || Rohrquerschnitt
pipe socket   Rohrmuffe || Rohrstutzen
pipe system   Rohrnetz * *open:offenes*
pipe union   Muffe * *Rohrverbindung*
pipe ventilated   Rohranschluss * *pipe ventilated machine: Maschine mit Rohranschluss für Kühlmittel*
pipe wrench   Rohrzange
pipe-ventilated   fremdbelüftet über Rohranschluss
pipeline   Pipeline * *Rohrleitung für Öl, Gas usw.* || Rohrleitung * *Fernleitung*
pipeline network   Rohrleitungsnetz
pipework   Rohrleitungen
piping   Rohr * *~material* || Rohrleitung * *Rohrleitung(snetz), Rohrverlegung* || Rohrleitungen || Rohrleitungsnetz || Rohrsystem * *Verrohrung* || Verrohrung
piping system   Rohrsystem
pistol   Pistole * *z.B. Wire-Wrap-~, Schrumpf~, Heißluft~*
piston   Kolben
piston compressor   Hubkolbenverdichter || Kolbenkompressor || Kolbenverdichter
piston rod   Kolbenstange || Pleuelstange * *Kolbenstange*
pit   anfressen * *an/zerfressen, Löcher/Grübchen/ Narben/Vertiefgn. bilden z.B. i.Lager, Zylinderlaufbahn* || fressen * *Grübchen/Narben/Löcher bilden; siehe auch* →*anfressen und* →*zerfressen* || Grube * *Bergwerk, Schacht; Montagegrube* || Schacht * *im Bergwerk* || Vertiefung * *Grube, Grübchen, Narbe* || zerfressen * *an-/zerfressen, Grübchen/Löcher/Narben bilden*
pitch   Abstand * *in einem Raster; z.B. auf dem Bildschirm, Zähne auf Zahnstange usw.* || Einschnürung * *magnet.* || Gewindesteigung * *achsparalleler Abstand, Gewindesteigung* || Steigung * *Gewinde~ (achsparaller Abstand der Gewindegänge), Gewindeteilung* || Teilung * *(Zahnrad-, Kommutator-, Nuten-, Lochteilung usw.)*
pitch circle   Teilkreis * *z.B. bei Zahnrad*
pitch circle diameter   Lochkreisdurchmesser * *Kreis, auf dem Bohrungen für die Befestigung von Masachinenteilen liegen* || Teilkreisdurchmesser * *z.B. eines Zahnrades*
pitch diameter   Teilkreisdurchmesser * *z.b. bei Zahnrad*
pitch of screw   Spindelsteigung
pitch-circle diameter   Wirkdurchmesser * *Teilkreisdurchmesser, z.b. eines Zahnrades*
pitting   Grübchenbildung * *Verschleiß-/Narbenbildung: Verschleiß-/Korrosionseffekt, z.B. in einem Lager* || Riefenbildung * *Lochfraß, Grübchenerosion*
PIV speed variator   PIV-Getriebe * *Markenname eines mechanisch stufenlos verstellbaren Getriebes*
pivot   Angelpunkt * *Drehzapfen* || anlenken * *dreh/ schwenk/klappbar lagern* || Drehpunkt * *(auch figürlich) einer Zapfenverbindung, eines Scharniers, eines Hebels* || Drehzapfen || Gelenk * *Dreh-/ Angelpunkt, Drehzapfen* || lagern * *drehbar ~, z.B. mittels Scharnier ~* || schwenken * *(ein)schwenken, sich drehen (upon/on:um), drehbar lagern* || Zapfen * *Drehzapfen eines Scharnieres/Gelenks*
pivot bearing   Zapfenlager
pivot pin   Lagerzapfen
pivot point   Angelpunkt * *Drehpunkt*
pivoted   angelenkt || drehbar * *drehbar gelagert* || Klapp- * *mit Scharnier/Zapfen versehen* || klappbar * *mit Zapfen/Scharnier versehen* || schwenkbar || schwenkbar * *drehbar gelagert*
pivoted lever   Schwenkhebel
pivoting   Schwenk-
pivoting drive   Schwenkwerksantrieb
pixel   Bildpunkt * *[Computer] z.B. auf einem Monitor* || Pixel * *[Computer] Bildpunkt am Monitor*
pixel frequency   Pixelfrequenz * *[Computer] Pixelanzahl x Bildwiederholrate x 1.3 bei Monitor*
pixel rate   Pixelfrequenz
place   anordnen * *örtlich; platzieren* || aufgeben * *order:Bestellung* || Ort * *Platz, Ortschaft* || placieren * *z.B.Produkt am Markt;(auf)stellen/setz.; (sich) postieren, unterbring.,(Bestellg.) aufg.* || Platz * *be in/out of place: am/nicht am ~ sein* || platzieren * *z.B. Produkt am Markt; (auf)stell./setz.; (s.)postieren; unterbring.; (Bestellg.)aufgeb.* || setzen * *(auf-)stellen, setzen, legen, placieren, postieren, jdn. ernennen/in e.Amt ein~* || Sitz * *Ort, Platz; place of residence: Wohn~; place of business: Geschäfts~* || Stelle * →*Platz,* →*Ort; auch Dezimal~, ~ in einer Bestellnummer usw.* || stellen || unterbringen * *(auch figürlich; z.B. Person, Aufträge usw. ~)* || vergeben * *place with: Auftrag usw. ~ an*
place after the decimal point   Nachkommastelle
place an order   Auftrag erteilen * *with: an*
place an order for   bestellen
place considerable significance on   besonderen Wert legen auf
place demands   Anforderungen stellen * *place extreme demands on: hohe ~ an*
place into   versetzen * *in eine Lage/einen Zustand*
place of application   Angriffspunkt * *einer Kraft usw.*
place of business   Geschäftssitz || Niederlassung
place of delivery   Lieferort
place of deposit   Ablage * *Aufbewahrungsort, ~ort*
place of deposition   Ablage * *Aufbewahrungs-/Ablageort*
place of event   Veranstaltungsort

place of principal business Hauptgeschäftsstelle
place of work Arbeitsplatz * *Stelle* || Platz * →*Arbeitsplatz*
place orders disponieren * *Aufträge erteilen*
place preceding the decimal point Vorkommastelle
place requirements Anforderungen stellen * *on: an*
place together zusammenstellen * *an einen gemeinsamen Ort ~/vereinigen*
placed untergebracht
placed into versetzt * *in eine Lage/einen Zustand ~*
placement Bestückung * →*SMD-Bauteile auf Leiterkarte*
placing Bestückung * →*SMD-Bauteile auf Leiterkarte*
placing of order Auftragserteilung
placing of orders Disposition * *Auftragserteilung*
plain anspruchslos * *einfach, schlicht* || deutlich * *klar, leicht verständlich, einfach, schlicht* || einfach * *klar, leicht verständlich, schlicht* || glatt || klar * *leicht verständlich, einfach,schlicht* || Plan * *(Adj.) glatt* || rein * *Wahrheit*
plain bearing Gleitlager
plain language display Klartextanzeige
plain text Klartext
plain text description Klartextangabe * *Klartext-Beschreibung/-Bezeichnung*
plain text display Klartextanzeige
plain text specification Klartextangabe
plain-text message Klartextmeldung
plainly frei * *(Adv.) unumwunden*
plainness Einfachheit * *Einfachheit, Schlichtheit, Deutlichkeit, Klarheit, Offenheit, Ehrlichkeit*
plaintext display Klartextanzeige
plaintext message Klartextmeldung
plait falten * *(Textil) Matte, Stoff, Geflecht usw.) falten, (ver)flechten, gefaltet ablegen* || Geflecht * *aus Textil* || Zopf- * *Zopf, Flechte, (Haar-/Stroh-, Textil-) Geflecht, Falte: (ver)flechten (Verb)*
plaiting machine Flechtmaschine * *z.B. zur Herstellung von Moniereisen-Geflechtmatten*
plan auslegen * *planen, projektieren, entwerfen* || Disposition * *Entwurf, Anlage, Plan* || einteilen * *planen* || gestalten || Lageplan * *of:von* || Plan || planen || Planung || projektieren * *planen* || System * *Plan* || vorhaben * *planen* vorhaben * *(Substantiv) Plan* || vorsehen * *planen*
plan ahead disponieren * *vorausplanen*
plan in advance vorausplanen
plan view Grundriss
planar eben * *Oberfläche*
plane abrichten * *(Holz) mit Hobel(maschine)* || eben * *plan, flach* || Ebene || hobeln || Plan * *(Adj) auch 'flach, eben'* || planieren
plane No. Flug Nr
plane of separation Trennfläche || Trennfuge
plane surface Planfläche
plane-parallel planparallel
planer Hobelmaschine
planer knife Hobelmesser * *(Holz)*
planet carrier Planetenträger * *in einem* →*Planetengetriebe*
planet gear Planetengetriebe || Planetenrad
planet pinion Planetenrad * *bei Planetengetriebe*
planet wheel Planetenrad * *bei Planetengetriebe*
planetary gear Planetengetriebe
planetary gear unit Planetengetriebe
planetary gearing Planetengetriebe
planetary variable speed drive Planetenverstellantrieb
planetary-roller screw Satellitenrollspindel
planing bench Hobelbank * *(Holz)*
planing blade Hobelmesser * *(Holz)*
planing knife Hobelmesser * *(Holz)*

planing machine Hobelmaschine
planing mill Hobelwerk * *(Holz)*
planing system Hobelwerk * *(Holz)*
planing tool Hobelstahl * *Schneidwerkzeug für Hobelmaschine*
planish planieren * *Metall ~, glätten, (ab)schlichten, (Holz) abhobeln, (Metall) glatt hämmern/polieren*
plank Bohle
planned vorgesehen * *geplant*
planner Projekteur
planning Einteilung * *Planung* || Planung * *be in the planning/under planning: in ~ sein; planning phase: ~sphase* || Projektierung
planning and design Projektierung
planning engineer Projekteur
planning example Projektierungsbeispiel
planning guide Projektierungsanleitung || Projektierungshandbuch
planning phase Planungsphase || Projektierungsphase
planning sheet Projektierungsblatt
planning stage Planungsphase || Projektierungsphase
planning task Planungsaufgabe
plant Anlage * *z.B. Produktionsanlage* || Betrieb * *Maschinen(anlage)* || Einrichtung * *Anlage* || Werk * *Fabrikanlage*
plant availability Anlagenverfügbarkeit
plant code Anlagenkennzeichen
plant conditions Anlagenbedingungen || Anlagenverhältnisse || Betriebsverhältnisse
plant construction Anlagenbau || Anlagentechnik || Industrieanlagenbau
plant control level Leitebene
plant data Anlagendaten
plant design Anlagenprojektierung
plant designation Anlagenkennzeichen
plant designation system Anlagenkennzeichnungssystem
plant diagram Anlagenbild || Anlagenschema
plant display Anlagenbild * *auf Bildschirm*
plant downtime Anlagenstillstand
plant engineer Anlagenprojekteur
plant engineering Anlagenbau || Anlagenprojektierung || Anlagentechnik
plant fault Anlagenfehler
plant frequency Anlagenfrequenz
plant identification Anlagenkennzeichen * *siehe DIN 40719 Teil 2*
plant identifier Anlagenkennzeichen
plant maintenance Betriebserhaltung || Betriebsunterhaltung
plant management Betriebsleitung
plant manufacturer Anlagenbauer
plant outage Anlagenstillstand
plant parameter Anlagenparameter
plant personnel Anlagenpersonal
plant requirements Anlagenerfordernisse
plant schematic Anlagenschema
plant shutdown Anlagenstillstand
plant side Anlagenseite
plant signals Anlagensignale || Kundensignale
plant standstill Anlagenstillstand
plant status Anlagenzustand
plant stoppage Anlagenstillstand
plant supervision Betriebsüberwachung
plant technology Anlagentechnik
plant uptime Verfügbarkeit * *~ einer Anlage*
plant-projecting Anlagenprojektierung
plant-side anlagenseitig
plant-specific anlagenspezifisch
plasma-jet cutting machine Plasmaschneidmaschine * *Werkzeugmaschine*
plaster over verkleben

plasterboard Gipskartonplatte || Gipsplatte
plastic Kunststoff || plastisch
plastic bag Plastikbeutel
plastic belt Kunststoffriemen
plastic film Kunststoff-Folie || Kunststofffolie || Plastikfolie
plastic film industry Folienindustrie * *Kunststofffolie*
plastic film machine Folienmaschine * *für Kunststofffolie*
plastic film production Folienherstellung * *Kunststofffolie* || Folienproduktion
plastic foil Kunststoff-Folie || Kunststofffolie || Plastikfolie
plastic granulate Kunststoffgranulat
plastic granules Kunststoffgranulat
plastic material Kunststoff
plastic optocable Kunststoff-Lichtleiter || Lichtleiter * *Plastik-Lichtleiter*
plastic pellets Kunststoffgranulat
plastic preform Kunststoff-Halbzeug
plastic wood Holz * *flüssiges ~*
plasticiser Weichmacher
plasticizer Weichmacher
plastics Kunststoff
plastics industry Kunststoffindustrie
plastics machine Kunststoffmaschine
plastics processing Kunststoffverarbeitung
plate Grobblech || Groblech || Lamelle * *auch in einer Kupplung/Bremse* || Lamelle * *Scheibe einer Mehrscheiben-/Lamellenkupplung/-bremse* || Platte * *(Glas/Metall/Druck)Platte, Tafel, Scheibe, Grobblech, Blechtafel, (Batterie)Elektrode* || Scheibe || Schild * *z.B. Namen/Firmen/Tür/Typenschild* || Tafel * *Platte, z.B. Blech~* || Tafelblech * *Grobblech, dickes Blech* || Wand * *Vorder~, Rück~*
plate assembly Lamellenpaket * *einer Mehrscheiben- bzw. Lamellenkupplung/-bremse*
plate capacitor Plattenkondensator
plate edge Blechkante
plate glass Flachglas || Spiegelglas
plate heat exchanger Plattenwärmetauscher
plate rolling mill Grobblechwalzwerk
plate shears Blechschere * *für Grobblech* || Metallschere || Schere * *für Grobblech*
plate spring Blattfeder
plate stack Lamellenpaket * *einer Mehrscheiben- bzw. Lamellenkupplung/-bremse*
plate turner Blechwender * *im Walzwerk*
plate-cutting machine Blechschere
plate-shearing machine Blechschere
plated through-hole Durchkontaktierung * *in Leiterkarte*
platen Druckplatte * *{Druckerei} auch Drucktiegel, Druckzylinder in Druckmaschine* || Drucktiegel * *{Druckerei} →Druckplatte (z.B. einer Bogendruckmaschine)* || Druckzylinder * *{Druckerei} einer Rotationsdruckmaschine* || Tiegel * *{Druckerei} Drucktiegel/-Platte für Bogendruck*
platen-printing Tiefdruck
platform Bühne * *Arbeits~, Steuer~* || Plattform * *(auch figürlich: Hard-/Software/Entwicklungsplattform)* || Tisch * *Plattform*
platform conveyor Plattenband * *~förderer* || Plattenbandförderer
plating Überzug * *Plattierung, Metallüberzug*
platinum Platin
plausibility Plausibilität
plausibility check Plausibilitätsabfrage || Plausibilitätskontrolle
plausible einleuchtend || plausibel * *make plausible: ~ machen; siehe auch →einleuchtend* || sinnvoll * *plausibel*

play Spiel * *come into play: ins ~ kommen* || Spielraum * *Spiel, siehe auch 'Spiel'*
play a decisive role entscheidende Rolle spielen
play a key role Schlüsselposition einnehmen
play one's cards well geschickt * *geschickt handeln*
PLC AG * *programmable logic controller: Speicherprogrammierbare Steuerung* || Automatisierungsgerät * *Abk.f.'Programmable Logic Controller': speicherprogrammierbare Steuerung* || programmierbare Steuerung * *Abk.f. 'Programmable Logic Controller': Speicher~* || Speicherprogrammierbare Steuerung * *SPS * programmable logic controller: Speicherprogrammierbare Steuerung (EN 61131-3, IEC 1131-3)* || Steuerung * *Abkürzung für 'Programmable Logic Controller': speicherprogrammierbare ~ (PLC)*
PLC Ladder programming language Kontaktplan * *Programmiersprache für SPS*
PLC system Steuerungssystem * *SPS-System*
PLD PLD * *Abk. f. 'Programmable Logic Device': programmierbarer Logikbaustein* || programmierbarer Logikbaustein * *Abk. f. 'Programmable Logic Device'*
pleasant angenehm
please bitte * *anfragend*
please inquire auf Anfrage * *'bitte fragen Sie an'*
pleased zufrieden * *zufrieden (with: mit), erfreut (at: über), angenehm berührt*
pleasing the eye formschön
plentiful reichlich * *reich(lich), im Überfluss (vorhanden)*
plenty Überfluss
plenty of reichlich * *(vor Substantiv)* || viel * *reichlich* || viele
plethora Überfluss * *Überfülle, Übermaß, Zuviel (of: an)*
pliable biegsam || nachgiebig * *(auch figürl.) biegsam, geschmeidig, nachgiebig, fügsam, leicht zu beeinflussen*
pliers Zange
PLL PLL-Schaltung * *Phase-Locked Loop,phasenstarre Regelschleife:gleiche Phasenlage b.Aus-/Eing.pulsen*
plot aufzeichnen * *Messkurve, Messwerte usw. ~ mit Mess-/Registriereinrichtung* || Aufzeichnung * *Messkurve* || Bild * *aufgezeichnete Messkurve* || darstellen * *grafisch, zeichnerisch ~* || Diagramm * *Kurvendiagramm, Messschrieb* || Kurvendarstellung * *durch ein registrierendes Messgerät usw.* || Messschrieb || plotten || zeichnen * *einen Plan anfertigen von, etwas in e. Plan ein~, etwas planen/ entwerfen*
plotted geplottet
plotter Kurvenschreiber * *digital (punktweise) angesteuert* || Plotter
plug Anschlussstecker || Dübel || stecken * *z.B. einen Stecker/eine Steckbaugruppe ~* || Stecker || Stopfen * *Stöpsel, Dübel, Zapfen, Propfen, Verschlussschraube* || Verschluss * *Stöpsel* || Verschlussschraube * *z.B. für Kabeleinführung, Ölablass usw.* || Verschlussstopfen * *z.B. Blindstopfen für Kabeleinführung in Klemmenkasten* || verstopfen * *abdichten (Loch usw.)*
plug braking Gegenstrombremsung * *e. Asynchronmotors durch Drehfeldumpolg. (Vertausch. 2er Phasenanschlüsse)*
plug connection Steckanschluss || Steckeranschluss || Steckverbindung
plug connector Messerleiste || Stecker || Steckverbinder || Stiftleiste || Stiftstecker || Übergangsstecker
plug housing Steckergehäuse
plug in anschließen * *mit Stecker* || einstecken || Steck-

plug on aufstecken
plug socket Steckdose || Stecksockel
plug switch Klinkenstecker
plug up abdichten * Loch ~ || verstopfen * abdichten (Loch usw.)
plug with earthing contact Schukostecker
plug-and-screw terminal Schraub-Steck-Klemme || Schraubsteck-Klemme || Schraubsteckklemme
plug-connector housing Steckergehäuse
plug-in anschlussfertig * steckbar || einsteckbar || Steck- || steckbar
plug-in board Steckmodul * Leiterkarte, Flachbaugruppe
plug-in cable Steckkabel || Steckleitung
plug-in card holder Einzelkartenhalterung * für Steckkarte || Steckkartenhalter * →Einzelkartenhalterung
plug-in compatible anschlusskompatibel * Stecker-kompatibel
plug-in connection Anschluss * Steck~ || Steckanschluss || Steckeranschluss * Anschluss mittels Stecker || Steckverbindung
plug-in connector Steckkupplung || Steckverbinder
plug-in jumper Brückenstecker || Steckbrücke
plug-in location Steckplatz
plug-in module Baugruppe * Steck~ || Einschub || Steckbaugruppe || Steckmodul
plug-in PCB Steckmodul * Leiterkarte
plug-in power supply Steckernetzteil
plug-in socket Stecksockel
plug-in station Steckplatz
plug-in terminal Steckklemme * siehe auch 'Schraubsteckklemme'
plug-in type steckbar
plug-type brush Köcherbürste
plug-type terminal Steckklemme
plugboard Kreuzschienenverteiler
pluggable connection Steckanschluss
pluggable screw terminal Schraub-Steck-Klemme || Schraubsteck-Klemme || Schraubsteckklemme
plugged verschlossen * durch Stöpsel/Blindstopfen ~ (z.B. Kabeleinführung)
plumber Monteur * ~ für sanitäre Anlagen
plumber's solder Lötzinn
plunge Absturz * Sturz, Stürzen, Ein/Untertauchen || eintauchen * (intransitiv, aktiv) into: in || Sturz * ~ ins Wasser || tauchen * plötzlich/kräftig ~
plunge in eintauchen
plunger Kolben * Tauchkolben (auch in der Hohlglasherstellung), Tauchbolzen, Tauchspule || Stößel
plunger coil Tauchspule
plunger pump Plungerpumpe * Tauchpumpe
plus Pluspunkt || samt
plus or minus plus-minus
plus point Pluspunkt
plus symbol Pluszeichen
plush nobel * luxuriös, plüschig
ply Lage * Faserstoff-Lage, Sperrholz-Schicht, Falte, (Garn-) Strähne; Hang, Neigung
ply with zusetzen * jdn. mit Fragen usw. →belästigen
plywood Furnierholz || Sperrholz
PM motor permanenterregter Motor * Permanent Magnet-erregt.Motor; Geg.satz:wound field motor:Mot. m.Feldwicklg.
pneumatic Druckluft- * Pneumatik * [nju'mätik] || pneumatisch * [nju'mätik]
pneumatic brake Druckluftbremse
pneumatic clutch Druckluftkupplung
pneumatic cylinder Druckluftzylinder * [nju'mätik ...] || Pneumatikzylinder * [nju'mätik ...]
pneumatic piston Pneumatikkolben * [nju'mätik 'pisten]
pneumatic tool Druckluftwerkzeug
pneumatic valve Pneumatikventil
pneumatically pneumatisch * (Adv.) [nju:'mätikälli]
pneumatically actuated pneumatisch betätigt
pneumatically actuated brake Druckluftbremse
pneumatically operated Druckluft- * [nju'mätik ...] || pneumatisch * mit Druckluft betrieben || pneumatisch betätigt * [nju'mätikälli...]
pneumatically operated brake Druckluftbremse
pocket einstecken * in die Tasche stecken, einstecken, einheimsen, mitgehen lassen
pocket calculator Taschenrechner
pocket grinder Pressenschleifer * zum Holzschleifen
pocket guide Tabellenheft * Kurzübersicht/Kurzbetriebsanleitung im Format DIN A5 oder kleiner
pocket knife Taschenmesser
pocket lamp Taschenlampe
pocket reference Tabellenheft
poineer bahnbrechend
point anspitzen || Argument * (springender) Punkt || Ort * Stelle || Platz * (amerikan.) Stelle || Punkt || Spitze || stehen * auf Mssßinstrument ~/angezeigt sein (to: auf e. Wert) || Stelle * Punkt || Stützpunkt * eines Polygonzuges || zeigen * gerichtet sein; to the rear: nach hinten; upwards: nach oben
point at hinweisen * ~ auf || hinweisen auf || zeigen * deuten auf
point discharge Punktentladung
point fixing Pratzenanbau * [el.Masch.] Dreipunktbefestigung e. fuß-/flanschlosen Motors an 3 Pratzen am Gehäuse
point of action Angriffspunkt
point of application Angriffspunkt * z.B. einer Kraft
point of attack Angriffspunkt
point of contact Anstoß * mechanisch
point of diversion Weiche * Umleitungspunkt, Ableitstelle, Umlenkungsposition, Weiche
point of engagement Angriffspunkt * Eingriffspunkt
point of intersection Schnittpunkt
point of load application Lastangriff * ~spunkt
point of main effort Schwerpunkt
point of sampling Abtastzeitpunkt
point of time Augenblick || Zeitpunkt
point of view Aspekt * Gesichtspunkt, Blickwinkel; from the...point of view:aus dem Blickwinkel der... || Gesichtspunkt * from the ... point of view: aus/unter dem ~ des ... betrachtet || Standpunkt * from this point of view: von diesem ~ aus betrachtet
point out herausstellen * (figürl.), z.B. Gedanken, Vorzüge || hervorheben * herausstreichen || hinweisen auf * point out that/to: darauf ~, dass || nachweisen || zeigen * deuten auf, darlegen
point out that hinweisen * darauf ~, dass
point to hinweisen auf || stehen * zeigen auf (z.B. Zeigerinstrument auf einen Wert)
point up fugen * verfugen, verstreichen, Fugen glatt streichen
point-blank bündig * geradeheraus, unverblümt, klipp und klar
point-focal punktuell * in der Optik
point-like punktförmig * point of light: ~ Lichtquelle
point-to-point Punkt-zu-Punkt-
point-to-point connection Punkt-zu-Punkt Verbindung || Punkt-zu-Punkt-Kopplung || Punkt-zu-Punkt-Verbindung
point-to-point control Positioniersteuerung || Punktsteuerung * (Numer. Steuerg.) || Streckensteuerung * Positioniersteuerung
point-to-point positioning Punktsteuerung * (Numer. Steuerg.)
point-to-point wire connection Einzelanschluss

* **Anschlusstechnik** ohne mehrpoligen Stecker bei SPS
pointed scharf * *spitz* || spitz
pointed pliers Spitzzange
pointed teeth Spitzzahnung * *(Holz)* bei Säge
pointer Tipp * *Fingerzeig, Tipp* || Zeiger * ~ *eines Messgeräts, Adress~ auf eine Variablenadresse, indirekte Adresse*
pointer deflection Zeigerausschlag * ~ *eines Messgerät*
pointer instrument Zeigerinstrument
pointing the way to the future zukunftsweisend
points Eigenschaft * *good/bad: gute/schlechte* || Punkte || Weiche * *work the points: die ~ stellen*
Poland Polen
polar gepolt * *z.B. (Elektrolyt-) Kondensator*
polarity Polarität || Polung || Vorzeichen * *eines (Analog-) Signals*
polarity evaluation Vorzeichenauswertung
polarity inverting Vorzeichenumkehr
polarity reversal Umpolung || Umschwingen * *Wechsel der Polarität* || Verpolung * *Umpolung* || Vorzeichenumkehr || Vorzeichenumschaltung * *einer Spannung, eines Signals*
polarization Polung * *Polarisation*
polarize Polen * *(Verb)*
polarized gepolt * *z.B. (Elektrolyt-) Kondensator* || verpolungssicher * *unsymmetrisch (Stecker)*
polarized capacitor gepolter Kondenstor
pole Mast || Nullstelle || Pol * *auch eines Magneten* || Polen * *(Verb)* || Stange * *Pfosten, Pfahl, Mast, (Telegrafen-, Zelt-) stange, Stab, (Wagen-) Deichsel*
pole assignment Polvorgabe * *(Regel.techn.)*
pole changing polumschaltbar * *n-speed pole-changing: n-fach ~*
pole count Polzahl
pole head Polschuh
pole pair Polpaar
pole pair number Polpaarzahl
pole pitch Polteilung * *(el.Masch.) Abstand der Polmitten zweier benachbarter ungleichnamiger Pole*
pole shape Polform * *z.B. einer el. Maschine*
pole shoe Polschuh
pole terminal Polklemme
pole-amplitude modulation Pol-Amplituden-Modulation * *(el. Masch.) PAM-Schaltg. für Schleifringläufer m. Stromumkehr in 1 Wicklg*
pole-changing Polumschaltung * *Umschaltg. d. Ständerwicklg. e.AC-Motors z.Drehzahländerung (Dahlander/PAM-Schaltg.)*
pole-changing motor polumschaltbarer Motor
pole-changing multispeed motor polumschaltbarer Motor
pole-changing switch Polumschalter * *für polumschaltbaren Motor (z.B. in Dahlanderschaltung)*
pole-count ratio Polzahlverhältnis
policy Methode * *Verfahrensweise* || Politik || Richtung * *Politik* || Strategie * *Taktik, Verfahrensweise* || Taktik * *Verfahrensweise, Taktik* || Verfahrensweise * *Richtlinie, Schema*
poling Polung
polish Glätte * *Politur* || polieren
polished blank * *poliert, geputzt* || geschliffen * *(figürl.) Vortrag usw.*
polisher Poliermaschine
polishing drum Putztrommel
polishing machine Poliermaschine
polishing paste Polierpaste
political economy Volkswirtschaft || Volkswirtschaftslehre
poll abfragen * *(zyklisches) Abfragen "auf Änderung" durch einen Rechner* || Abruf || abrufen

poll list Umlaufliste * *z.b. zum Aufruf der Slave-Teilnehmer vom Master bei Bussystem*
poll list for slave stations Busumlaufliste
polling Abfrage * *Wählen, zyklische Abfr. (im Computer), einen Partner nach d.anderen* || Abruf
polling list Umlaufliste * *z.B. bei Bussystem*
pollutant Schadstoff * *in Luft, Flüssen usw.*
pollute verschmutzen * *Wasser, Umwelt usw.* || verunreinigen
polluted verschmutzt * *Wasser, Umwelt usw.*
pollution Umweltverschmutzung || Verschmutzung || Verunreinigung
pollution degree Verschmutzungsgrad
pollution protection Immissionsschutz
poly Mehr-
polyamide Polyamid
polycarbonate Polycarbonat
polycristalline polykristallin
polydisc filter Scheibenfilter * *Mehrscheibenfilter*
polydisc- Mehrscheiben-
polyester Polyester
polyethylene Polyäthylen
polygon Polygon * *aus Geradenstücken bestehende Kurve* || Polygonzug
polygon curve Kurvenzug * *→Polygonzug* || Polygonzug * *aus Geradenstücken bestehende Kurve*
polygon curve block Polygonbaustein * *Funktionsbaustein*
polygon curve characteristic Kennlinienbaustein * *Polygonzug-Funktionsbaustein*
polygon-based interpolation block Polygonzug * *Funktionsbaustein*
polygonal course Polygonzug
polyline Polygonzug
polyline function block Kennlinienbaustein * *Polygonzug-Baustein* || Polygonbaustein * *Funktionsbaustein*
polymer Polymer
polynom Polynom-
polynom notation Polynomform
polynominal Polynom || Polynom-
polyphase mehrphasig
polyphase commutator machine Drehstrom-Kommutatormaschine
polyphase commutator motor Drehstrom-Kommutatormotor || Drehstrom-Nebenschlussmotor || Drehstromkommutatormotor
polyphase current Mehrphasenstrom
polyphase machine Drehfeldmaschine * *Oberbegriff für alle Drehstrommaschinen*
polyphase motor Drehfeldmotor * *Oberbegriff für alle Drehstrommotoren* || Drehstrommotor * *drei- und mehrphasig*
polyplanetary gear Polyplanetengetriebe
polypropylene Polypropylen
polystyrene Polystyrol
polyurethane Polyurethan
pony Klein- * *z.B. pony motor: →Hilfsmotor, Anwurfmotor*
pony motor Anwurfmotor || Hilfsmotor * *Anwurf-/Hilfsmotor (pony enspr. "klein, Zwerg-, Mittelklasse, Rangier-, Vorstreck-")*
pool gemeinsam benutzen * *Resourcen in einer Interessengemeinschaft vereinen* || Verbund * *Interessen/Arbeitsgemeinschaft, Ring* || vereint * *(Kräfte, Resourcen, Kapital usw.) vereinen/gemeinsam nutzen* || vereinigen * *Kräfte, Resourcen usw. ~* || zusammenfassen * *Resourcen, Material bündeln/gemeinsam benutzen*
pooling gemeinsame Benutzung * *von Resourcen in einer Interessengemeinschaft*
poor dürftig * *→ungenügend, →mangelhaft, →schwach, →unbefriedigend* || gering * *unzureichend, mangelhaft* || knapp * *mangelhaft, unzureichend, ungenügend, schlecht* || lahm * *dürftig,*

poor lubrication 660

mangelhaft, jämmerlich, schwach || mangelhaft * *ungenügend, unzureichend, (auch Zeugnisnote)* || mäßig * →*dürftig* || minderwertig * *Qualität usw.* || nicht ausreichend * siehe auch →*ungenügend* || schlecht * *armselig,wertlos, unzureichend. (Geschäft, Qualität, Schwingdämpfg., Aussicht.,Entschuldigg.)* || schmal * *ungenügend* || schwach * *schlecht, unzureichend* || unbefriedigend * siehe auch *'ungenügend'* || ungenügend * *schwach, schlecht (siehe auch 'mangelhaft')* || unzureichend * *armselig, schlecht, mager, mangelhaft, jämmerlich*
poor lubrication Ölmangel * *schlechte Schmierung*
poor terminal connection Wackelkontakt * *an (Klemmen)Anschluss*
pop up aufblenden * *sich entfalten, aufspringen (z.B. Menü auf dem Bildschirm)* || auftauchen * *(salopp) plötzlich auftauchen*
pope reel Poperoller
pope reel winder Poperoller * *Oberflächenwickler (d.h. kein Achswickler; meist mit einer Andruck-/Reibwalze)*
pope roller Poperoller
popular allgemein * *allgemein verbreitet* || beliebt * *with: bei* || gebräuchlich * *populär, allgemein beliebt, weit verbreitet* || verbreitet * *beliebt; with: bei* || verständlich * *allgemein verständlich* || weit verbreitet * *beliebt*
popularity Beliebtheit
populated bestückt * *(salopp) ein vorbereiteter Einbauplatz, z.B. für Speicherbänke im Rechner*
porcelain insulator Lüsterklemme
pore filler Spachtel * *{Holz}* ~masse
pore over wälzen * *Bücher* ~
pores Poren
porosity Durchlässigkeit || Porosität
porous durchlässig * *porös* || undicht * *porös*
porous-bronze bearing Sinterbronzelager
port Durchlass || Hafen || portieren * *z.B. Software-programm von einer Zielhardware auf eine andere* || Portierung * *{Computer} Software auf eine andere Hardware- oder Betriebssystemplattform* || Schnittstelle * *(Signal-) Anschluss, z.B. serielle Schnittstelle*
port of destination Zielhafen
portability Portabilität * *Übertragbarkeit von Programmen von einem Rechnertyp auf den anderen* || Portierbarkeit * *z.B. von Software von e. Hardwareplattform/Betriebssystem auf e. andere(s)* || Übertragbarkeit * *z.B. von Software auf eine andere Hardware- oder Betriebssystemplattform*
portable beweglich * *tragbar* || Hand- * *tragbar* || portabel || portierbar * *übertragbar (z.B. Software von einer Hardwarebasis auf eine andere)* || tragbar * *in der Hand ~ vom Gewicht/Volumen her, z.B. Gerät, Rechner* || übertragbar * *Programm auf einen andern Rechner*
portable control operator Handeingabegerät
portable electric tool Elektrowerkzeug
portable tachometer Handtacho || Tachometer * *Hand~*
portable unit Handgerät
portal Tor * *Portal, Tor (auch figürl.), (Haupt-)Eingang*
portal robot Portalroboter
portal saw Portalsäge * *{Holz}*
portfolio Aktenkoffer * *Aktenmappe* || Aktentasche * *Mappe* || Mappe
portion Teil * *An~, Bruch~*
portion out teilen * *aus~ (in Portionen)*
portray beschreiben * *anschaulich ~*
Portugal Portugal
Portuguese portugiesisch

pose stellen * *(Frage)* ~, *(Behauptung) auf*~, *(Anspruch) erheben, hin-/ausgeben (as: als)*
posh nobel * *(salopp) piekfein, todschick, fesch*
position anordnen * *in die (richtige) Lage bringen, aufstellen, positionieren* || einstellen * *in die richtige Lage bringen* || Lage * *räumlich und figürlich; be in the position:in der ~ sein; auch bei ~regelung* || Platz * *Stellung, Lage* || platzieren * *anbringen, in die richtige Lage bringen, lokalisieren* || Position * *Lage, auch bei {NC-Steuerung}* || Stand * *(auch figürl.)* || Standort || Stelle || stellen * *auch Schalter* || Stellung * *örtliche/taktische Lage, berufliche Position (of:als), Rang* || Zustand * *Lage*
position actual value Wegistwert
position calculator Positionierrechner
position control Wegregelung
position control loop Lageregelkreis
position control module Positionierbaugruppe
position controller Lageregler || Positionier-Regler || Positionierrechner || Positionierregler || Wegregler
position counter Positionszähler
position decoder Wegerfassung * *Decodierer*
position detection Wegerfassung
position detector Lagegeber * *über Schaltkontakt*
position deviation Positionsfehler
position encoder Lagegeber * *absolute/incremental:absoluter/inkrementeller; rotary/linear:Dreh-/Linear-* || Weggeber || Wegmessgeber
position error Lageabweichung * *{Regel.techn.} I NC-Steuerg}.* || Positionsfehler
position feedback Wegistwert
position finding Ortung
position loop Lageregelkreis
position measurement Wegerfassung
position measuring Lagemessung
position measuring system Lagemesssystem
position of rest Ruhestellung * *z.B. bei einem Schütz*
position reference Lagesollwert || Wegsollwert
position reference value Lagesollwert || Positionssollwert || Wegsollwert
position sensing Lageerfassung || Lagemessung * *Lageerfassung* || Positionserfassung || Wegerfassung
position sensor Lagegeber
position setpoint Lagesollwert || Positionssollwert * *siehe auch* →*Lagesollwert* || Wegsollwert
position signal Lageistwert
position switch Positionsschalter
position transducer Lagegeber || Wegaufnehmer * *Lage/Positionsgeber* || Weggeber
position value Lageistwert
positional variance Lageabweichung * *Toleranz in einer mechan. Konstruktion*
positioner Positionierantrieb * *als Stellantrieb z.B. für Proportionalventil* || Stellantrieb * *Stellgerät m.Positionsregelg. für proportional wirkend. Stellglied (z.B. Stellventil)* || Steller * *Stellgerät* || Stellgerät * *mit Lage/Positionsverstellung*
positioning Anordnung * *das In-Die-Richtige-Lage-Bringen, Aufstellung* || Positionier- || Positionieren || Positionierung
positioning accuracy Positioniergenauigkeit
positioning axis Positionierachse
positioning block Positionierdatensatz
positioning board Positionierbaugruppe * *Leiterkarte*
positioning control Positioniersteuerung || Positionsregelung
positioning control loop Positionier-Regelung
positioning controller Positionierbaugruppe || Positioniersteuerung
positioning data set Verfahrdatensatz || Verfahrsatz
positioning drive Positionierantrieb || Stellantrieb

positioning element Stellglied * *zur Lageeinstellung (z.B. Stellventilantrieb)*
positioning error Positionierfehler || Positionsfehler * *Positionierfehler*
positioning module Positionierbaugruppe
positioning motor Stellmotor
positioning PCB Positionierbaugruppe * *Leiterkarte*
positioning range Positionierbereich
positioning record Positionierdatensatz
positioning roller Positionier-Rollgang || Positionierrollgang || Rollgang * *Positionierrollgang*
positioning roller table Positionier-Rollgang || Positionierrollgang
positioning set Positionierdatensatz
positioning speed Verstellgeschwindigkeit * *Positioniergeschwindigkeit*
positioning table Positioniertisch
positioning task Positionieraufgabe
positioning time Positionierzeit || Stellzeit
positioning tolerance Positioniertoleranz
positioning valve Regelventil * *Stellventil* || Stellventil
positive formschlüssig || positiv * *zustimmend, Plus- (positive pole: Pluspol), unumstößlich, eindeutig, zustimmend, bejahend* || zwangsläufig * *Zwangs-, zwangsläufig, formschlüssig; bestimmt, definitiv, ausdrückl., fest, tatsächl.*
positive component Mitkomponente
positive direction Wirkrichtung * *Wirkungsrichtung*
positive edge positive Flanke
positive feedback Mitkopplung
positive logic positive Logik * *hoher Spannungspegel entspricht logisch "1"*
positive pole Pluspol
positive pressure Überdruck
positive sign Pluszeichen
positive terminal Pluspol * *Anschluss an Batterie, Kondensator*
positive-action zwangsgeführt
positive-displacement blower Kapselgebläse
positive-displacement pump Kolbenpumpe
positive-going edge positive Flanke
positive-sequence component Mitkomponente * *symmetrischer Anteil am Dreiphasensystem*
positive-sequence system Mitsystem * *siehe →Mitkomponente*
positively charged positiv geladen
positively driven zwangsgeführt
positively driven opening contact sicher öffnender Kontakt * *zwangsgeführter (Relais-) Kontakt z.B. für Sicherheitsfunktion*
positively-opening zwangsgeführt * *Öffnerkontakt, z.B. an einem Überwachung-/Schutzrelais*
possess besitzen * *auch eine Eigenschaft ~* || haben * *besitzen* || verfügen über * *besitzen, im Besitz haben, beherrschen*
possessor Besitzer || Inhaber * *Besitzer,Inhaber*
possibility -Möglichkeit * *z.B.: connecting possibility: Anschluss~* || Möglichkeit * *possibility of doing: ~ etwas zu tun*
possibility of -Möglichkeit * *of doing: ~ etwas zu tun*
possible etwaig * *möglich* || eventuell || möglich * *as fast as possible: so schnell wie ~; make possible: ~ machen* || möglichst * *the smallest possible: kleinstmöglich* || potenziell * *möglich*
possible applications Einsatzmöglichkeiten
possible combinations Kombinationsmöglichkeiten
possible counter-measure Abhilfemöglichkeit
possible explosion hazards Zündgefahren
possibly etwa * *möglicherweise* || eventuell * *(Adv.)*. möglicherweise || möglicherweise || unter Umständen * *möglicherweise* || womöglich * *wenn möglich*
post Arbeitsstelle || Ständer * *Pfahl, Pfosten, Ständer, Stab*
post code Postleitzahl
post processor Postprozessor * *{Computer}*
post production Nachbaufertigung * *~ von ausgelaufenen Geräten für den (Ersatzbedarf usw.)*
post- Nach-
post-event history Nachgeschichte
post-machine nachbearbeiten * *{Werkz.masch.}*
post-mortem diagnostics Fehldiagnose * *nachträgliche Diagnose der Fehler-Vorgeschichte*
post-mortem history Nachgeschichte * *nach einer Fehlerabschaltung aufgezeichnete Messdaten*
post-office box Postfach
post-optimization Nachoptimierung
post-production phase Nachbaufertigung * *~ nach Fertigungseinstellung eines Produkts*
post-ramp hinter dem Hochlaufgeber
post-treatment Nachbehandlung
postal address Postanschrift
poster Plakat || Werbeplakat
postforming saw Postformingsäge * *{Holz}*
postion control Lageregelung
postmortem review Fehleranalyse * *rückblickende Analyse nach Auftreten eines (fatalen) Fehlers*
postpone aufschieben * *zeitlich* || aussetzen * *verschieben* || verlegen * *zeitlich verschieben* || verschieben * *zeitlich ~, Termin ~*
postprocessor Postprozessor * *{Computer}*
pot eingießen * *z.B. in Becher* || Potentiometer || Poti || vergießen * *z.B. Baugruppe mit Gießharz*
potential künftig * *eventuell später möglich, potenziell* || Leistungsfähigkeit * *auch Wirkungsvermögen,innere Kraft* || möglich * *potenziell* || möglicherweise * *potenziell* || Möglichkeit * *Potenzial, Entwicklungs~(en)/-Reserven, Kraftvorrat* || Potenzial || potenziell
potential barrier Potenzialschwelle * *bei Halbleitern*
potential bonding Potenzialausgleich * *über ~sleitung usw.*
potential compensation Potenzialausgleich
potential compensation cable Potenzialausgleichsleitung
potential customer potenzieller Kunde
potential difference Differenzspannung * *Potenzialdifferenz* || Potenzialdifferenz
potential drop Spannungsabfall
potential energy potenzielle Energie
potential equilization Potenzialausgleich * *z.B. zwischen weit entfernten Anlagenteilen*
potential equilization cable Potenzialausgleichsleitung
potential explosive explosionsgefährdet
potential gradient Potenzialdifferenz * *el. Spannungsanstieg/-unterschied zwischen zwei Punkten* || Spannungsgefälle
potential isolation Potenzialtrennung
potential threshold Potenzialschwelle
potential-free potenzialfrei
potentiality Möglichkeit * *Entwicklungs~*
potentially hazardous area explosionsgefährdeter Bereich * *siehe DIN 57165/VDE 0165*
potentiometer Drehwiderstand || Potentiometer * *[po'tenschemiete]* || Poti
potted gekapselt * *vergossen* || vergossen * *z.B. Baugruppe*
potting material Vergussmasse || Vergussmaterial * *siehe auch →Vergussmasse*
Pound Pfund * *brit. Währung* || Pfund * *brit. Gewichtseinheit (1 pound entspr. 453,6 g)*
pour Gießen * *(Verb) Flüssigkeit ~* || strömen * *Regen usw.*

pour in einfüllen * *eingießen*
pour out leeren * *aus~*
powder Pulver || Sinter- || Staub * *Puder*
powder filling Sandkapselung * *Schutzart 'q' nach EN50017 für Ex-geschützte el. Betriebsmittel*
powder form pulverförmig
powder metallurgy Pulvermetallurgie
powder-coated pulverbeschichtet
power ansteuern * *mit Leistung ansteuern, Leistung zuführen* || antreiben * *mit Antriebskraft versehen* || Befugnis || Einfluss * *Macht* || Gewalt * *of/over:- über; judicial power:richterliche Gewalt* || Kraft * *Macht; motive power: treibende ~* || Leistung || Leistungs- || Mandat || Potenz * *Kraft, Stärke; auch mathematisch: power of:zur Potenz von* || speisen * *mit Energie versorgen* || Stärke
power actuator Leistungssteller * *Stellglied* || Leistungsstellglied
power amplifier Leistungssteller * *Leistungsverstärker* || Leistungsverstärker
power back-up Energiepufferung || Leistungspufferung
power balancing Energieausgleich || Energieaustausch * *Energieausgleich*
power boost Leistungssteigerung
power break Abkantpresse * *motorisch angetriebene Abkantbank zum Biegen von Blechen*
power budget Leistungsbilanz * *elektr.*
power burner Brenner * *für Heizungszwecke*
power cable Leistungskabel || Starkstromkabel || Starkstromleitung
power cabling Leistungsverdrahtung
power capacitor Leistungskondensator
power circuit Energiekreis
power company Elektrizitäts-Versorgungsunternehmen || Elektrizitätsgesellschaft || Energieversorgungsunternehmen || EVU || Stromversorgungsunternehmen || Versorgungsunternehmen * *Stromversorgungsunternehmen*
power connection Leistungsanschluss || Leistungsverbindung || Netzanschluss
power connections Leistungsverbindungen
power consumption Leistungsaufnahme || Stromaufnahme || Stromverbrauch
power contactor Leistungsschütz
power control Leistungsregelung
power controller Leistungsstellglied * *Steller* || Steller * *Leistungs~*
power controller unit Leistungssteller
power controlling element Leistungsstellglied * *z.B. Stromrichter*
power converter Stromrichter
power converter unit Stromrichtergerät
power cord Anschlusskabel * *flexible Stromversorgungsleitung, z.B. für Kleingerät* Anschlussleitung * *flexible Netzanschlussleitung* || Netzkabel * *flexible Netzanschlussleitung, z.B. für Kleingerät* || Netzleitung * *z.B. für mobiles Gerät*
power correction capacitor Kompensationskondensator * *zur Blindleistungskompensation*
power costs Energiekosten
power current Starkstrom
power demand Leistungsbedarf * *{el.Masch.} tatsächl.erforderl. Antriebsleistung, muss < Motorbemess.leistg. sein* || Strombedarf * *bezüglich der Stromversorgung*
power density Leistungsdichte
power diode Leistungsdiode
power dissipation Verlustleistung
power distribution Energieverteilung || Verteilung * *zur elektr. Energieversorgung*
power distribution installation Energieversorgungsanlage
power down ausschalten * *Stromversorgung usw.* ~

power drain Leistungsaufnahme
power efficiency Wirkungsgrad * *%uales Verhältnis Ausgangs-/Eingangsleistung; high-efficiency: mit hohem ~*
power electronics Leistungselektronik
power engineering elektrische Energietechnik || Energietechnik || Starkstromtechnik
power equalizing bus gemeinsamer Zwischenkreis * *zum Energieausgleich*
power factor cos phi * *Leistungsfaktor (Verhältnis Wirk-/Scheinleistung)* || Leistungsfaktor * *cos phi, Verhältnis Wirkleistung/Scheinleistung; unity power factor: ~ von eins*
power factor compensation Blindleistungskompensation || Blindstromkompensation || Kompensation * *Blindleistungs-Kompensation*
power factor compensation equipment Blindstromkompensationseinrichtung || Kompensationsanlage || Kompensationseinrichtung
power factor of the fundamental wave Grundschwingungsleistungsfaktor
power fail Netzausfall * *Stromversorgungsausfall, z.B. eines Rechners*
power failure Netzausfall || Netzspannungsausfall || Spannungsausfall * *Ausfall der Versorgungsspannung* || Spannungseinbruch * *Stromausfall* || Stromausfall
power flow Leistungsfluss * *grafisches Schema zur Darstellung der Wirkleistung und der Verluste*
power frequency Netzfrequenz
power generation Energieerzeugung || Stromerzeugung
power house Zentrale * *Schalthaus*
power input Leistungsaufnahme || Stromaufnahme
power interruption Spannungseinbruch * *Unterbrechung der Stromversorgung*
power lead Anschlusskabel * *Stromversorgungskabel* || Anschlussleitung * *für Stromversorgung*
power level Leistungspegel
power limit Grenzleistung
power line Starkstromleitung
power line carrier control Rundsteuerung
power line frequency Netzfrequenz
power line sag Netzspannungseinbruch
power loss Netzausfall || Spannungsausfall * *Ausfall der Stromversorgung* || Spannungseinbruch * *Ausfall der Stromversorgung* || Stromausfall || Verlustleistung || Verlustwärme * *→Verlustleistung*
power loss dissipation Verlustleistung * *als Wärme abgegebene ~*
power loss ride-through Netzausfallüberbrückung * *Motor bleibt nicht stehen und läuft nach Netzwiederkehr weiter*
power losses Verlustleistung
power measurement Leistungsmessung
power meter Leistungsmessgerät
power module Endstufe * *Leistungsstufe*
power of hoch * *zur Potenz von*
power of press Presskraft
power off abschalten * *die Versorgungsspannung ~* || ausschalten * *die Stromversorgung ~* || Netz-Aus
power on bestromen || einschalten * *die Stromversorgung* || Netz-Ein
power outage Spannungsausfall
power outlet Leistungsabgang
power output Abgabeleistung || Leistungsabgang
power output element Leistungsstellglied
power pack Netzgerät || Netzteil
power plant Kraftwerk
power plug Stecker * *für Starkstrom*
power rail Stromschiene * *in der Kontaktplandarstellung eines SPS-Programms*
power range Leistungsbereich * *elektr.Leistung; high:oberer; lower:unterer*

power rating Leistung * *Nenn/Typenwert* || Nennleistung
power ratio Leistungsverhältnis
power recovery Energierückspeisung
power rectifier Netzgleichrichter
power requirement Leistungsbedarf
power reserve Leistungsreserve
power reserves Leistungsreserve || Leistungsreserven
power restoration Netzspannungswiederkehr || Netzwiederkehr
power section Leistungsteil
power set Leistungsteil
power shovel Bagger * *Schaufelbagger*
power shutdown Spannungsabschaltung * *der Stromversorgung*
power spectrum Leistungsspektrum
power stabilization Pendeldämpfung * *im Stromversorgungsnetz*
power stage Endstufe * *Leistungs-Endstufe*
power stages Leistungsabstufungen
power station Kraftwerk || Zentrale * *Kraftwerk*
power station process control Kraftwerkleittechnik
power supply Energieversorgung || Netzgerät || Netzteil || Speisequelle || Stromversorgung
power supply capacity Anschlussleistung
power supply company Elektrizitäts-Versorgungsunternehmen || Elektrizitätsgesellschaft || EVU || Stromversorgungsunternehmen
power supply module Stromversorgung * *als Funktionsmodul*
power supply unit Netzgerät || Stromversorgungseinheit
power supply utility EVU
power supply variations Netzspannungsschwankungen * *bezogen auf die Netz-Versorgungsspannungen*
power switch Ein-Aus-Schalter * *für Stromversorgung* || Ein-Ausschalter * *für die Stromversorgung* || Leistungsschalter || Netzschalter
power synchro-tie Arbeitswelle * *ein Funktionsprinzip d. elektr. Welle mit Asynchron-Schleifringläufermotoren*
power system Netz * *Stromversorgungssystem*
power system management Netzleittechnik
power system protection Netzschutz
power systems control Netzleittechnik
power take-off Abtrieb * *Kraftausleitung* || Zapfwellenantrieb * *z.B. an Traktor*
power tool Elektrowerkzeug
power train Antriebsstrang
power transformer Leistungstransformator * *siehe IEC 76* || Netztransformator
power transistor Leistungstransistor
power transmission Kraftübertragung
power transmission element Antriebelement * *Kraftübertragungselement* || Kraftübertragungselement
power transmission engineering Antriebstechnik * *Techn. d.mechan. Kraftübertragg. (Getriebe, Kupplungen,Bremsen usw.)*
power transmission line Starkstromleitung
power transmission technology Antriebstechnik * *Technik d.mechan. Antriebselemente, z.B. Bremsen, Kupplgn., Getriebe*
power transmitting element Kraftübertragungselement
power unit Leistungsteil
power up einschalten * *an Spannung legen, hochfahren*
power utility Elektrizitäts-Versorgungsunternehmen * *EVU* || Elektrizitätsgesellschaft * *EVU* || Energieversorgungs-Unternehmen * *siehe auch 'EVU'* || Energieversorgungsunternehmen * *EVU* ||

EVU * *Energieversorgungsunternehmen* || Stromversorgungsunternehmen || Versorgungsunternehmen * *Stromversorgungsunternehmen*
power yield Leistungsausbeute
power-down logic Ausschaltlogik
power-down sequence Ausschaltreihenfolge
power-down sequence control Ausschaltsteuerung || Ausschaltsteuerwerk
power-factor correction Blindleistungskompensation
power-flow indication ssignalflußanzeige * *Anzeige des durchgeschalteten Leistungsstrangs*
power-light system Netz * *Hausstromversorgung*
power-loss factor Verlustfaktor
power-measuring test-bench Leistungsprüfstand
power-on pulse Richtimpuls * *wird beim Einschalten d. Stromversorgung erzeugt*
power-on reset Einschalt-Reset * *wirkt beim Einschalten der Versorgungsspannung* || Richtimpuls
power-on routine Anlaufprogramm
power-supply transformer Netztransformator
power-take-off shaft Zapfwellenantrieb * *Zapfwelle, z.B. an Traktor*
power-to-weight ratio Leistungsgewicht
power-up current surge Einschaltstromstoß
power-up procedure Initialisierung * *e. Rechners beim Einschalt. d. Versorgungsspannng.*
power-up sequence Einschaltfolge || Einschaltreihenfolge
power-up sequencing control Einschaltlogik || Einschaltsteuerung
power-weight ratio Leistungsgewicht
powered angetrieben
powered screw driver Schrauber * *für Handbetrieb* || Schraubmaschine * *Hand~*
powered trolley Motorkatze * *{Hebezeuge}*
powerflow readout ssignalflußanzeige * *Anzeige des durchgeschalteten Leistungsstrangs*
powerful kräftig * *tatkräftig, leistungsfähig, kraftvoll, mächtig* || kraftvoll || leistungsfähig || leistungsstark || massiv * *(figürlich) kraftvoll, mit Macht, wirksam, wuchtig* || stark * *mächtig, leistungsfähig* || wirkungsvoll * *höchst wirksam*
powers Können
PPP PPP * *'Point-To-Point Protocol': Datenübertr.protokoll im INTERNET über →MODEM u. Wählleitung (→SLIP)*
PPR Impulse pro Umdrehung * *Abkürzg. für 'Pulses Per Revolution' (z.B. bei Impulsgeber)*
PPR count Pulszahl * *Abkürzung für 'Pulses Per Revolution':Impulszahl pro Umdrehg.* || Schlitzzahl * *Abk.f. 'number of Pulses Per Revolution': Anzahl Impulse pro Umdrehung* || Strichzahl * *Abkürzung für 'Pulses Per Revolution':Impulse pro Umdrehung*
PR Öffentlichkeitsarbeit * *Abk. für 'Public Relations'*
practicability Ausführbarkeit || Durchführbarkeit || Möglichkeit * *Ausführbarkeit*
practicable anwendbar * *ausführbar* || durchführbar * *durch-/ausführbar; möglich, tunlich, brauchbar, anwendbar* || möglich * *durchführbar* || praktikabel || praktisch * *durchführbar* || praxisgerecht || realisierbar * *→durchführbar*
practical konkret * *tatsächlich, praktisch, handgreiflich* || praktikabel * *praktisch* || praktisch * *praktisch, tatsächlich, durchführbar; experience: Erfahrung, example:Beispiel* || sinnvoll * *praktikabel, praxisbewährt* || zweckmäßig
practical course Praktikum
practical exercises praktische Übungen
practical experience Betriebserfahrungen || Einsatzerfahrungen
practical man Praktiker
practical training Praktikum || praktische Übungen

practical training course Praktikum
practical-minded praktisch * *praktisch veranlagt/ denkend* || sachlich * *praktisch denkend*
practically nahezu * *praktisch* || praktisch * *(Adv.) so gut wie; in der Praxis* || sinnvoll * *(Adv.)*
practically oriented praxisorientiert
practically proven value Daumenwert * *in der Praxis bewährter Wert* || Richtwert * *praxisgerechter Wert*
practically-proven value Erfahrungswert
practice Arbeitsweise * *einer Person/Abteilung usw.* || Ausbildung * *Übung* || Kurs * *Übung* || Lehrgang * *Übung* || praktizieren || Praxis * *in practice:in der ~; put into praxis: in die ~ umsetzen* || Schulung * *Übung* || Schulungskurs * *Übung* || Übung || Übung * *'practice makes perfect': '~ macht den Meister'*
practice-oriented praxisgerecht
practise betreiben * *ausüben, (Beruf, Geschäft usw.) betreiben, tätig sein als/in, praktizieren* || lernen * *üben* || üben * *ein~, sich ~ in, jdn. schulen*
practise tactical evolutions manövrieren * *taktisch ~*
practised praktisch * *geübt*
practitian Praktiker
practitioner Praktiker
pragmatical value Daumenwert || Erfahrungswert || Richtwert * *plausibler Richtwert*
pre- Vor- || voraus- || vorher-
pre-alarm Vorwarnung
pre-assembled fertig montiert * *vormontiert*
pre-assembled cable konfektionierte Leitung * *z.B. Steckleitung* || konfektioniertes Kabel * *z.B. Steckkabel*
pre-assign vorbesetzen * *vorbelegen, vorläufige Zuordnung herstellen*
pre-assignment Vorbelegung
pre-calculate vorausberechnen
pre-charge vorladen
pre-charging Vorladung
pre-charging circuit Vorladekreis * *zur Aufladg. d. Zwischenkreiskondensatoren bei Umrichter* || Vorladeschaltung
pre-charging contactor Vorladeschütz
pre-charging device Vorladeeinrichtung * *z.B. z. Aufladen des Zwischenkreises bei Umrichter*
pre-compressive stress Druckvorspannung
pre-configured vorbereitet * *vorkonfiguriert, vorprojektiert* || vorprojektiert
pre-control vorsteuern * *bekannte Führungs- oder Störgröße direkt auf d. Stellgröße e. Regelkreises aufschalten* || Vorsteuerung
pre-control component Vorsteueranteil
pre-determine vorherbestimmen
pre-drill vorbohren
pre-drilled vorgebohrt
pre-dry vortrocknen
pre-eminence Vorrang
pre-engineered vorbereitet * *konstruktiv vorbereitet, vorprojektiert* || vorprojektiert
pre-event history Vorgeschichte
pre-fabrication Vorfertigung
pre-header Vorkopf * *bei Funktions-/Datenbaustein einer SPS*
pre-process vorverarbeiten
pre-processing Vorverarbeitung * *auch von Daten*
pre-project study Vorstudie * *zu einem Projekt* || Voruntersuchung * *zu einem Projekt*
pre-release tests Freigabeuntersuchung * *vor der Freigabe durchgeführte Tests*
pre-selectable vorwählbar || wählbar * *vor~*
pre-selection Vorauswahl || Vorwahl
pre-set voreingestellt * *z.B. Einstellparameter im Herstellerwerk*
pre-set speed Drehzahl-Festsollwert || Festdrehzahl

* *fester Drehzahlsollwert* || Festsollwert * *für Drehzahl*
pre-set value Festsollwert
pre-setting Voreinstellung
pre-standard Vornorm
pre-start-up Vorinbetriebnahme
pre-tax vor Steuern * *ohne Abzug der Steuern*
pre-tensioning Vorspannung
pre-warning Vorwarnung
pre-wired anschlussfertig
preamplifier Vorverstärker
preamplifying Vorverstärkung
preassemble vormontieren
preassembled vorkonfektioniert * *z.B. Kabel* || vormontiert
preassembled cable set vorkonfektioniertes Kabel
preassembly Vormontage
preassignment Vorbelegung
prebore vorbohren
prebored vorgebohrt
precalculate vorausberechnen || vorherbestimmen * *vorausberechnen*
precalculation Vorausberechnung
precarious schwierig * *misslich*
precaution Vorkehrung || Vorschrift * *Vorsichtsmaßregel* || Vorsichtsmaßnahme || Vorsichtsmaßregel
precautional measures Sicherheitsgründe * *Vorsichtsmaßnahmen*
precautionary measure Vorsichtsmaßnahme
precautions Vorsorge * *take precautions: ~ treffen*
precedence Vorrang * *take precedence of: ~ haben vor*
preceding davorliegend || vorhergehend || vorherig
preceding valve abgelöstes Ventil * *ausgehendes Ventil beim Kommutierungsvorgang eines Stromrichters*
preceed vorausgehen * *vorangehen (zeitlich früher)*
precharge circuit Vorladeeinrichtung * *zum sanften Aufladen der Zwischenkreiskondensatoren eines Umrichters* || Vorladeschaltung * *z.B. zum langsamen Aufladen d. Z.kreiskondensatoren b. Spannungszwischenkreisumr.*
precharging Vorladung
precharging unit Vorladeeinrichtung * *z.B. Begrenzen d. Einschaltstroms von Umrichter-Zwischenkreiskondensatoren*
precheck Vorprüfung
precious kostbar || wertvoll
precious metal Edelmetall
precipice Absturz * *Abgrund (auch figürlich)*
precipitation Fällung * *{Physikl Chemie}*
precise bündig * *genau* || detailliert * *genau* || eigentlich * *genau* || exakt * *präzise, klar, genau, exakt, korrekt, peinlich genau* || feinfühlig * *exakt, →genau* || genau * *präzise, exakt, genau, klar, korrekt, klar umrissen, richtig (Betrag, Augenblick usw.)* || präzise * *scharf * *genau*
precisely genau * *(Adv.)* || gerade * *(Adv.)*
precisely placed gezielt * *genau plaziert*
preciseness Genauigkeit
precision Genauigkeit || Präzision || Präzisions-
precision bench lathe Mechanikerdrehmaschine
precision engineering Feinmechanik * *als Fachrichtung*
precision gear unit Präzisionsgetriebe
precision grinding machine Präzisionsschleifmaschine
precision measuring instrument Feinmessgerät || Feinmesszeug
precision mechanics Feinmechanik
precision of -Genauigkeit * *(vorangestellt)*
precision tool Präzisionswerkzeug

precision-gauge block Endmaß * *zur Messung von Durchmessern usw.*
precocious vorzeitig * *frühzeitig, vor der (normalen) Zeit, frühzeitig entwickelt, frühreif*
precondition Voraussetzung * *Vorbedingung* || Vorbedingung
preconditioning time Anwärmzeit * *Messgerät*
preconfigure vorprojektieren
precontrol Vorsteuerung
precontrol angle Vorsteuerwinkel
precontrol signal Vorsteuersignal
precontrol value Vorsteuerwert
precrusher Vorbrecher * *[Holz] z.B. in der Spanplattenherstellung*
precut vorkonfektioniert * *Kabel*
precut cable konfektioniertes Kabel || vorkonfektioniertes Kabel
predecessor Vorgänger
predesigned vorprojektiert
predestined prädestiniert
predetermine vorausberechnen * *vorherbestimmen* || vorherbestimmen
predetermined vordefiniert * *z.B. Software-Regelbaustein* || vorgegeben * *vorherbestimmt, festgesetzt, genau definiert* || vorherbestimmt
predetermined breaking point Sollbruchstelle
predial vorwählen * *Vorwahlnummer am Telefon*
predicament Lage * *(missliche) Lage; Kategorie*
predict vorausberechnen * *vorher/voraussagen* || vorhersagen || vorhersehen * *vorhersagen*
predictable vorhersehbar * *voraussagbar*
prediction Aussage * *Vorhersage, Vor~* || Vorausberechnung * *Vorhersage* || Vorhersage
predictive prädikativ || prädiktiv || vorausschauend * *prädiktiv, z.B. Regler*
predictive controller prädiktiver Regler * *"vorausschauender" Regler*
predisposition Anfälligkeit * *to:für/gegen*
predominance Dominanz * *over: über*
predominancy Dominanz
predominant beherrschend * *vorherrschend, überwiegend, vorwiegend; überlegen, das Übergewicht innehabend* || überwiegend || überwiegend * *über-/vorwiegend, vorherrschend* || vorherrschend || vorwiegend || weitgehend * *überwiegend, vorwiegend, voherrschend, überlegen*
predominantly vorwiegend * *(Adv.)*
predominate dominieren * *vorherrschen* || herrschen * *vor~* || überwiegen * *vorherrschen* || vorherrschen
predrill vorbohren
predrilled vorgebohrt
prefab house Fertighaus
prefab house construction Fertighausbau
prefabricated fertig * *vorgefertigt* || vorgefertigt || vorkonfektioniert * *z.B. Kabel* || vormontiert
prefabricated cable konfektioniertes Kabel || vorkonfektioniertes Kabel
prefabricated part Fertigteil || Halbzeug
preface Vorwort * *meist vom Autor selbst*
preferable bevorzugt * *vorzuziehen(d), to: gegenüber* || Vorzugs- * *vorzuziehend* || vorzuziehen
preferable operating position Vorzugslage * *räumliche Anordnung eines Bauteils, Einbaulage*
preferably bevorzugt * *(Adv.) vorzugsweise, lieber, am besten* || vorzugsweise
preference Präferenz || Vorzug * *Bevorzugung (above/over/to: vor), Vorliebe* || Vorzugs-
preference authorization Präferenzberechtigung * *(beim Zoll)*
preference eligible präferenzberechtigt * *[SPS]*
preferential bevorzugt * *bevorzugt, bevorrechtigt, Vorzugs-* || Vorzugs- * *Vorzugs-, bevorzugt, bevorrechtigt*

preferential price Sonderpreis * *Vorzugspreis* || Vorzugspreis
preferentially bevorzugt * *(Adv.) vorzugsweise* || vorzugsweise
preferred bevorzugt * *bevorzugt, Vorzugs-* || Vorzugs- * *bevorzugt*
preferred orientation Vorzugsrichtung
preferred state Vorzugslage * *eines Speichers (Flip Flop usw.)*
preferred type Vorzugstyp
prefix Präfix
preform Halbzeug * *vorgeformtes Teil (z.B. aus Kunststoff)*
preformed coil Formspule * *[el.Masch.]*
preheat anwärmen * *vorwärmen* || vorwärmen
preheat period Anwärmzeit
preheating Vorheizung
preheating furnace Vorwärmofen * *z.B. bei Blechstraße*
prejudice beeinträchtigen * *jemandes Rechte ~* || Nachteil * *Schaden, Nachteil* || Schaden * *Nachteil, Schaden, Beeinträchtigung* || schädigen * *benachteiligen*
prejudicial schädlich * *nachteilig*
preliminary Vor- * *vorläufig, vorbereitend* || Vorab- || vorläufig || vorübergehend * *vorläufig*
preliminary contact Vorendschalter * *bei Positioniersteuerung, Aufzug usw.* || Vorkontakt
preliminary examination Voruntersuchung
preliminary stage Vorstufe
preliminary test Vorprüfung
preliminary work Vorarbeiten
preload anstellen * *Lager z.B. mit Federring zur Vermeidung von Axialspiel* || verspannen * *→vorspannen* || vorspannen * *(mechan.) z.B. eine Feder, ein Wälz-/Gleitlager in Axialrichtung zur Spielvermeidung*
preloaded vorgespannt
preloaded bearing angestelltes Lager * *Lager m.axialer Vorspanng.,die.e.Vorzugslage bewirkt u.so Vibrationen vermeidet* || vorgespanntes Lager * *zur Vermeidung von Axialspiel*
preloading Vorbelastung || Vorspannung * *z.B. durch Feder zur Vermeidung von Axialspiel i. e. Lager*
prelubricated dauergeschmiert
prelubricated bearing Lager mit Dauerschmierung
premachine Vorarbeiten * *(Verb) mit (Werkzeug-) Maschine vorarbeiten*
premagnetize vormagnetisieren
premature vorzeitig * *frühzeitig, verfrüht, voreilig, vorschnell, übereilt*
premature material fatigue Materialfehler * *vorzeitige Materialmüdung*
premise Prämisse * *Prämisse, Voraussetzung; das Oben Erwähnte* || Voraussetzung
premises Gelände * *Grundstück, Häuser nebst Zubehör, Lokal(ität), Räumlichkeiten*
premium Beitrag * *(Versicherungs-) Prämie* || Bonus || Preis * *Bonus, Prämie, Belohnung; be at a premium: hoch im Kurs stehen/sehr gesucht sein*
preparation Aufbereitung * *Aufbereitung (von Sollwerten, Erz, Kraftstoff usw.), Vorbehandlung, Vorbereitung* || Ausarbeitung * *Abfassung (z.B. e.Dokuments), Aufbereitung, Vorbereitung, Ausfüllen (z.B. Formulare)* || Bearbeitung * *Akten usw. ~* || Vorbereitung * *in preparation: in ~; preparatory to: als ~ zu/für*
preparation of offer Angebotsausarbeitung
preparation roll Avivagerolle * *zum Befeuchten der Fasern bei Kunstfaser-Herstellung* || Präparationswalze * *[Textil] z.B. in der Chemiefaserherstellung*
preparation time Vorlauf * *Vorbereitungszeit*

preparatory vorbereitend * *preparatory to:als Vorbereitung zu/für*
preparatory period Vorlauf * *Vorbereitungszeit*
preparatory time Vorbereitungszeit
preparatory work Vorarbeiten * *vorbereitende Arbeiten*
prepare anfertigen || aufbereiten * vorbereiten, zurecht/fertigmachen, ausarbeiten, anfertigen, *(Schriftstück)* abfassen || aufstellen * *Liste usw.* ~ || ausarbeiten * entwerfen || bearbeiten * ausarbeiten || bereitstellen || in die Wege leiten * *vorbereiten* || vorbereiten * *for:auf/zu; being prepared:in Vorbereitung befindlich*
prepared vorbereitet
preparing Aufbereitung * Bereitmachen, Präparieren, Zu-/Herrichten, Ausarbeitung *(auch von Dokumenten)*
preparing machine Vorbereitungsmaschine * *z.B. in d. Textilherstellung/Spinnerei*
preponderance Überlegenheit * *Übergewicht*
preponderant dominant * →*überwiegend,* →*entscheidend* || überwiegend || vorwiegend
preponderate überwiegen
prepress level Druckvorstufe * *(Druckerei)*
preprocess vorverarbeiten
preprocessing Vorverarbeitung
preproduction Vorfertigung
prerequisite Voraussetzung * *Vorbedingung, (erste) Voraussetzung* || Vorbedingung
prescaler Vorteiler * *für Frequenzen*
prescribe vorschreiben
prescribed vorgeschrieben * *angeordnet*
prescribed by law gesetzlich vorgeschrieben
prescribed period Frist * *vorgeschriebener Zeitraum*
prescription Vorschrift
preselect anwählen || vorwählen
preselection Vorauswahl || Vorwahl
presence Anliegen * *Vorhandensein* || Anwesenheit * *auch 'Vorhandensein'* || Vorhandensein * *auch 'Anwesenheit'*
present aktuell * *gegenwärtig* || aufweisen || augenblicklich * *gegenwärtig* || ausstellen * *präsentieren, (erstmalig) zeigen* || bisherig || dabei * *anwesend* || darstellen || derzeitig || einbringen * *präsentieren* || heutig * *gegenwärtig* || jetzig * *bestehend* || momentan * *(Adj.) gegenwärtig* || übergeben * *überreichen, bringen, einreichen, schenken* || verkörpern * *darstellen* || vorführen * *präsentieren* || vorhanden || vorstellen * *präsentieren*
present at einführen * *eine Person in einen neuen Kreis* ~
present-time jetzig
presentation Einführung * *Vorstellung* || Präsentation || Übergabe * *von Dokumenten* || Vorführung * *Präsentation* || Vorstellung * *Präsentation*
presentation case Vorführkoffer
presentation guidelines Referentenleitfaden
presenter's guidelines Referentenleitfaden
presently derzeit * *gegenwärtig* || im Augenblick * *(Adv.)* || zurzeit * *momentan, gegenwärtig*
preservation Aufrechterhaltung * *Bewahrung, Schutz, Konservierung, Vorbeugung* || Erhaltung * *Bewahrung, Vorbeugung, Konservierung* || konservieren * *das Konservieren,die Konservierung* || Konservierung || Schonung * *Erhaltung*
preservation of value Werterhaltung
preserve aufheben * *aufbewahren* || erhalten * *bewahren, aufrecht~, beibehalten* || konservieren || schonen * *erhalten* || schützen * *erhalten*
preset eingestellt * *voreingestellt, vorgegeben* || Festsollwert || Vorbesetzung || voreingestellt || voreinstellen || Voreinstellung || vorgegeben * *vorbesetzt, fest*
preset breaking point Sollbruchstelle

preset counter Vorwahlzähler
preset frequency Festfrequenz * *Frequenz-Festsollwert*
preset option Voreinstellung * *bei SPS-Programmiergerät*
preset speed Drehzahl-Festsollwert || Drehzahlsollwert * *Festsollwert*
preset the factory values urladen * *Parameter, Datensätze usw.* ~
preset value Festsollwert || Festwert * *vorbesetzter Wert* || Setzwert * *fest vorbesetzter Wert, z.B. Festsollwert* || Sollwert * *Fest~* || Vorbesetzung * *vorbesetzter Wert*
presetting einrichten * *Maschine usw.* ~ || Vorbesetzung || Voreinstellung || Vorwahl * *Vorbesetzung* ~
preside leiten * *vorsitzen, over a meeting: eine Versammlung* ~
president Direktor || Hauptgeschäftsführer || Leiter * *Direktor* || Vorsitzender || Vorstandsvorsitzender
press antippen * *Taste* ~ || betätigen * *Taste* || drängen * *(auch figürl.)* || drücken * *auch eine Taste* || Presse- * *siehe auch* →*Anpress-,* →*Quetsch-* || Presse * *Pressmaschine; Druck~; die Druckmedien, Zeitung; auch Trocken~ bei Papiermaschine* || pressen || Verlag
press against anpressen
press agent Pressereferent
press brake Abkantpresse * *zum Biegen von Blech*
press break Abkantpresse
press conference Pressekonferenz
press down anpressen * *niederdrücken*
press drive Pressenantrieb
press force Presskraft
press hard zusetzen * *press a person hard: jdm. zusetzen*
press machine Presse * *Pressmaschine*
press pad Niederhalter
press release Pressemitteilung
press roll Presswalze * *z.B. in Papiermaschine*
press safety valve Pressensicherheitsventil
press section Pressenpartie * *in Papiermaschine*
press the key Taste betätigen
press together zusammendrücken || zusammenpressen
press tool Stanzwerkzeug
press-fit eingepresst || Einpress- || kraftschlüssig
press-in Einpress-
press-in stator core Einschubstator * *{el.Masch.} Ständerblechpaket, das außerh.d. Motors gewickelt wird*
press-pack thyristor cell Scheibenthyristor
press-rolled board Wickelpappe
pressboard Pressspanplatte * *{Holz}* || Spanplatte * *{Holz}*
presser Quetsch-
pressing dringend * *auch dringlich* || Pressung
pressing on roller Druckwalze * *Andruckwalze*
pressing tool Presszange
pressing-on roller Andruckwalze
pressure Anpress- || Druck * *low/light/high: niedriger/leichter/hoher* || Pressung || Überdruck
pressure above atmospheric Überdruck * *überatmosphärischer Druck*
pressure accumulator Druckakkumulator * *in Hydrauliksystem*
pressure closed-loop control Druckregelung
pressure compensating air reservoir Druckausgleichsbehälter * *in einem Druckluftsystem*
pressure connector Quetschverbinder
pressure contact Druckkontakt * *z.B. bei Scheibenthyristoren*
pressure control Druckregelung
pressure controller Druckregler
pressure difference Druckdifferenz || Druckerhö-

hung * *Differenzdruck, Druckdifferenz* || Überdruck
pressure drop Differenzdruck * *Druckabfall* || Druckabfall
pressure equalizing reservoir Druckausgleichsbehälter
pressure finger Druckhebel * *kurzer Hebel*
pressure fit Presspassung
pressure gauge Druckmesser || Druckmessgerät
pressure increase Druckerhöhung * *z.B. bei Lüfter, Ventilator*
pressure lever Druckhebel
pressure line Druckleitung
pressure load Druckbeanspruchung
pressure lubrication Ölumlaufschmierung
pressure meter Druckmesser || Druckmessgerät
pressure of rolling Walzdruck
pressure oil Drucköl * *in Hydrauliksystem*
pressure pad Niederhalter
pressure pick-up Druckgeber
pressure piston Druckkolben * *z.B. in einem pneumat./hydraul. Aktuator/Zylinder*
pressure plate Druckplatte * *z.B. bei einer Kupplung*
pressure reducer Druckminderer || Druckminderungsventil || Druckminderventil
pressure reducing valve Druckminderungsventil || Reduzierventil
pressure regulating valve Druckbegrenzungsventil || Druckminderungsventil
pressure regulator Druckregler
pressure release Entlüftung * *~ zum Druckausgleich*
pressure reservoir Druckspeicher * *z.B. in Hydrauliksystem*
pressure ring Druckring * *z.B. bei einer Reibkupplung*
pressure roll Anpresswalze
pressure sensor Druckaufnehmer || Druckgeber
pressure spring Druckfeder
pressure strength Druckfestigkeit
pressure surge Druckstoss
pressure switch Druckschalter || Druckwächter
pressure tank Druckbehälter
pressure to cut costs Kostendruck
pressure transducer Druckaufnehmer * *Drucksensor* || Drucksensor
pressure transmitter Druckgeber
pressure tube Druckrohr
pressure-cylinder Druckzylinder * *z.B. Pneumatikzylinder*
pressure-forge stauchen * *in der Schmiedetechnik*
pressure-proof druckfest
pressure-reducing valve Druckminderer || Druckminderventil
pressure-relief joint Reißnaht * *[el.Masch.] Sollbruchstelle in einem lichtbogensicheren Klemmenkasten*
pressure-relief valve Überdruckventil
pressure-resistant druckfest
pressure-sensitive druckempfindlich
pressurized enclosure druckfeste Kapselung * *Zünd-/Explosionsschutzart* || Überdruckkapselung * *Schutzart'p' n.EN50016 f.Ex-geschützte el.-Betriebsmittel; Zündschutzgasfüllg.*
pressurized enclosure with leakage compensation Überdruckkapselung mit ständiger Durchspülung * *Zünd/Exschutzart mit Ausgleich d.Schutzgasverluste*
pressurizing gas Zündschutzgas * *zum Explosionsschutz z.B. von Motoren*
presswork Druck * *[Druckerei] Druckvorgang*
prestigious feature Alleinstellungsmerkmal * *mit Prestige behaftetes Merkmal*
prestigious object Vorzeigeobjekt

prestress vorspannen * *(mechan.)*
prestressed vorgespannt
prestressing Vorbelastung * *vorangehende Beanspruchung* || Vorspannung * *z.B. bei Riementrieb*
prestressing factor Vorspannfaktor * *z.B. Verhältn. treibende Umfangskraft/Vorspannquerkraft bei Riementrieb*
presumable vermutlich || voraussichtlich * *(Adj.) vermutlich, mutmaßlich, wahrscheinlich*
presumably vermutlich * *(Adv.)*
presume voraussetzen * *annehmen, vermuten, schließen*
presuppose bedingen * *→voraussetzen* || voraussetzen * *voraussetzen, zur Voraussetzung/als Bedingung haben, erfordern* || Voraussetzung * *(Verb) zur ~ haben*
presupposition Voraussetzung
pretend angeben * *etw.falsches vorgeben; protzen*
pretended so genannt * *angeblich* || virtuell * *vorgetäuscht, scheinbar*
pretension vorspannen * *(mechan.)*
pretext Entschuldigung * *Ausrede, Vorwand*
pretreatment Vorbehandlung
pretty sauber * *hübsch* || sehr * *(salopp) ganz schön, beträchtlich, ziemlich, einigermaßen* || ziemlich
preultimate vorletzt || vorletzter
prevail Anliegen * *(vor)herrschen, maß-/ausschlaggebend sein. maßgebend/gültig sein* || dominieren * *das Übergewicht haben, die Oberhand gewinnen (over, against), maß-/ausschlaggebend sein* || durchsetzen * *erfolgreich sein; auch Erzeugnis am Markt ~* || herrschen * *vor~, herrschen, maß-/ausschlaggebend/verbreitet sein* || überwiegen * *over:(gegen)über* || vorherrschen
prevailing bisherig || jetzig * *vorherrschend* || überwiegend || vorherrschend
prevalent häufig * *vorherrschend, überwiegend, weit verbreitet* || überwiegend || verbreitet * *(vor)herrschend, überwiegend, weit ~* || vorherrschend || weit verbreitet * *auch 'vorherrschend, überwiegend'*
prevent verhindern || vermeiden * *verhindern*
preventable vermeidbar * *verhindern, "verhütbar"*
prevention of accidents Unfallschutz || Unfallverhütung
preventive maintenance vorbeugende Wartung || vorsorgliche Wartung
previous bisherig * *vorherig* || früher * *vorhergehend* || vorher * *(Adj.) vorherig* || vorhergehend || vorherig
previous commercial training Vorkenntnisse
previous knowledge Vorkenntnisse
previous model Vorgängermodell
previous to vor * *früher als, zeitlich ~*
previous version bisherige Ausführung
previously bereits * *zuvor* || bisher * *vorher, früher; than previously:als bisher* || vorher * *(Adv.)*
previously mentioned vorstehend genannt * *siehe auch →obengenannt, →obig*
prewarning signal Vorwarnsignal
prewind vorwickeln * *z.B. Spule vor dem Einziehen in einen Motor bei Motorenfertigung*
prewire vorverdrahten
prewired anschlussfertig
prewound vorgewickelt * *z.B. fertig gewickelte Spule zum Einziehen in einen Motor*
price bewerten * *preislich* || Kosten * *Preis* || Preis * *at a lower price: zu einem niedrigeren ~*
price adaption Preisanpassung
price adapter Aufpreis || Mehrpreis * *Preisaufschlag* || Preisaufschlag
price agreement Preisvereinbarung
price aim Preisziel
price barrier Preishürde

price cut

price cut Preissenkung
price cutting Preissenkung
price deduction Preisnachlass
price development Preisentwicklung
price difference Preisunterschied
price drop Preisrückgang || Preisverfall
price erosion Preisverfall
price fluctuation Preisschwankung
price framework Preisgefüge || Preisgestaltung * *Preisgefüge* || Preisstruktur
price freeze Preisstopp
price gap Preisschere
price generation Preisbildung
price hurdle Preishürde
price increase Preissteigerung
price level Preisniveau
price list Preisliste
price per item Stückpreis
price per piece Stückpreis
price pressure Preisdruck
price recession Preisrückgang
price reducing preissenkend
price reduction Bonus * *Preisreduzierung* || Preisnachlass || Preisreduzierung * *by: um* || Preissenkung || Rabatt * →*Preisnachlass*
price revision Preisanpassung
price scissors Preisschere
price segment Preissegment * *upper: oberes; lower: unteres*
price sensation Preisschlager
price spiral Preisschraube
price stop Preisstopp
price structure Preisgefüge || Preisgestaltung * *Preisgefüge* || Preisstruktur
price support Preisstützung
price supporting Preisstützung
price war Preiskampf * *throat-cutting: existenzgefährdender* || Preiskrieg
price-competitive preisgünstig * *~ verglichen mit Mitbewerbsprodukten*
price-performance ratio Preis-Leistungsverhältnis
price-raising preistreibend
pricelist Preisliste
pricing Preis * *~stellung, ~gestaltung, for: für* || Preisbildung || Preisgestaltung * *Preisstellung, Preiskalkulation* || Preisstellung * *for:für* || Preisvereinbarung
pricing agreement Preisvereinbarung
pricing discipline Preisdisziplin
pricing policy Preisgestaltung
pricing pressure Preisdruck * *high: großer*
pricing sheet Preisblatt
pricing structure Preisgestaltung || Preisstruktur
prick Stich * *mit Nadel usw.*
prickle Dorn * *Stachel*
pricy teuer * *(salopp)*
primaries Primäranschlüsse * *z.B. eines Trafos*
primarily erstens * *an erster Stelle* || hauptsächlich * *(Adv.)* || in erster Linie
primary Haupt- * *primär* || hauptsächlich * *primär* || primär * *auch Trafowicklung* || Primärwicklung
primary coat Grundanstrich * *z.B. mit Vorstreichfarbe*
primary cooling air circulation innerer Kühlluft-Kreislauf
primary energy Primärenergie
primary mover Kraftmaschine
primary voltage Primärspannung
primary winding Primärwicklung
primary-switched power supply primärgetaktetes Schaltnetzteil * *auf dem Netzseite (vor dem Übertrager) getaktetes Schaltnetzteil*
prime ansaugen * *durch Pumpe* || best * *in prime condition: im ~en Zustand* || grundieren * *mit Vorstreichfarbe streichen* || grundlegend * *primär, ~,*

wichtigst, erst(klassig); Primzahl; anlassen (Pumpe), grundieren || Haupt- * *erst, wichtigst, wesentlichst, of prime importance: von größter Wichtigkeit*
prime cost Anschaffungskosten || Selbstkostenpreis
prime importance Vordergrund * *höchste Bedeutung/Wichtigkeit (of:von)*
prime mover Antriebsmaschine || Arbeitsmaschine || Kraftmaschine * *wandelt Wasserkraft, Brennstoffe, Dampf, Windenergie usw. in mechan.Energie um* || Triebwagen * *Straßenbahn~*
primer Fibel * *z.B. Einfachbetriebsanleitung/Nachschlagewerk* || Grundierung * *Grundierfarbe/-lack* || Handbuch * *Fibel, Leitfaden, Anfangslehrbuch* || Leitfaden * *Fibel, (Anfänger-) Lehrbuch, Leitfaden*
principal Chef * *Direktor* || Direktor || Geschäftsinhaber * *Direktor, Chef, Vorsteher, Geschäftsherr* || grundlegend * *hauptsächlich* || Haupt- * *hauptsächlich, Erst-, Haupt-, Stamm-, Chef-* || hauptsächlich || Leiter * *Chef, Direktor, Vorsteher* || Vorstand * *Einzelperson* || wesentlich * *erst, hauptsächlich, Haupt-; Stamm-*
principal application Hauptanwendung
principal arm Hauptzweig * *{Stromrichter}*
principal characteristic Haupteigenschaft
principal office Hauptgeschäftsstelle
principle Gesetz * *Naturgesetz, naturwissenschaftliches ~* || Gesetzmäßigkeit * *Prinzip, Grundsatz, Regel, (Natur)Gesetz* || Grundsatz || Prinzip * *on: aus; in: im; work on the principle of: nach dem ~ des ... arbeiten* || Regel * *Prinzip* || Satz * *Prinzip*
principle of operation Arbeitsweise || Funktionsprinzip
print drucken || Platine * *Leiterkarte*
print buffer Druckspeicher * *für Computerdrucker*
print out ausdrucken
print shop Druckerei
print-out Ausdruck * *durch einen Drucker* || ausdrucken
printability Bedruckbarkeit
printable character abdruckbares Zeichen
printed board Leiterplatte
printed circuit gedruckte Schaltung
printed circuit board Baugruppe * *Leiterkarte* || Flachbaugruppe || Leiterkarte || Leiterplatte
printed circuit track Leiterbahn
printed conductor Leiterbahn
printed form Formularvordruck || Vorlage * *gedruckte*
printed product Druckerzeugnis * *{Druckerei}*
printed publication Druckschrift
printed sheet Druckbogen
printed-board holder Einzelkartenhalterung
printed-circuit board Print
printed-circuit module Flachbaugruppe * *Leiterkarte*
printer Drucker * *auch Beruf*
printer interface Druckerschnittstelle
printer port Druckerschnittstelle
printhead Druckkopf * *eines Computerdruckers*
printing Buchdruck || Druck * *{Druckerei} Druckvorgang*
printing blanket Drucktuch * *{Druckerei}*
printing color Druckfarbe * *{Textildruck}*
printing company Druckerei
printing element Druckwerk
printing engine Druckwerk * *bei Laserdrucker*
printing factory Druckerei
printing firm Druckerei
printing form Druckform * *{Druckerei}*
printing industry Druckindustrie
printing ink Druckerschwärze || Druckfarbe * *{Druckerei}*

668

printing lever Druckhebel * *[Druckerei]* bei Bogendruckmaschine
printing machine Druckmaschine
printing mechanism Druckwerk
printing office Buchdruckerei || Druckerei
printing out Registrierung * *auf Drucker*
printing paper Druckpapier
printing plant Buchdruckerei * *(amerikan.)*
printing plate Druckplatte * *[Druckerei]*
printing press Druckerpresse
printing quality Druckbild
printing roller Druckwalze * *[Druckerei]*
printing shop Druckerei
printing style Druckbild
printing trade Druckgewerbe || grafisches Gewerbe
printing unit Druckeinheit * *[Druckerei]* || Druckwerk
printout Ausdruck * *von Drucker*
prior vorherig * *früher, älter (to:als); prior patent:älteres Patent; prior use:Vorbenutzung*
prior to vor * *früher als*
prioritize priorisieren * *Prioritäten zuteilen, z.B. Interrupts*
prioritized priorisiert * *z B Interrupts*
priority dringend * *vordringlich* || Dringlichkeit * *Vor~ (top: höchste, lowest: niedrigste), Priorität* || Dringlichkeitsstufe * *top:höchste* || Priorität * *low:niedrige; high:hohe: top:höchste* || Schwerpunkt * *Vorrangigkeit* || Vorrang * *Vorrang, Vordringlichkeit (over: gegenüber, top: höchsten)*
priority assignment Priorisierung * *z.B. von Interrupts, Buszugriffen usw.*
priority class Dringlichkeitsstufe
priority level Prioritätsebene
priority scheduling Priorisierung * *z.B. von Interrupts, Buszugriffen usw.*
priority sequence Prioritätenfolge || Prioritätenreihenfolge
prism Prisma
prismatic prismatisch
private eigen * *persönlich, privat, allein, vertraulich, geheim* || geheim * *vertraulich* || geschlossen * *private party: ~e Gesellschaft* || intern * *vertraulich*
private area Freiraum * *(figürlich)*
private power system Inselnetz
privately geheim * *(Adv.) im Geheimen*
privately owned capital Eigenkapital
privilege Befugnis * *durch Vorrecht/Privileg* || Konzession * *Privileg*
privileged bevorrechtigt
probability Wahrscheinlichkeit
probability distribution Wahrscheinlichkeitsverteilung
probability of error Fehlerwahrscheinlichkeit
probable vermutlich * *wahrscheinlich* || voraussichtlich * *(Adj.) vermutlich* || wahrscheinlich * *(Adj.)*
probably voraussichtlich * *(Adv.) wahrscheinlich* || wahrscheinlich * *(Adv.)*
probation Probe * *Bewährungsprobe*
probationer Praktikant * *Angestellter auf Probe, Probekandidat*
probe Sonde || Tastkopf * *z.B. für Oszilloskop*
probing head Tastkopf
problem Aufgabe || Frage * *Problem* || Problem || Schwierigkeit * *Problem* || Störung * *Problem*
problem description Problembeschreibung
problem solver Problemlösung
problem-free problemlos
problematical schwierig * *problematisch*
procedural prozedural * *z.B. Programmiersprache (Gegensatz: →objektorientierte Programmiersprache)*

procedure Ablauf || Arbeitsvorgang || Art und Weise * *Verfahren* || Prozedur * *auch Rechner-(Unter)-Programm* || Unterprogramm || Verfahren * *Vorgehensweise, Ablauf, Prozedur, Verfahrensweise* || Verfahrensweise || Vorgang * *Vorgehensweise, Ablauf* || Vorgangsweise || vorgehen * *(Substantiv) Vorgehens-/Verfahrensweise* || Vorgehensweise || Weg * *Verfahrensweise*
procedure-oriented prozedural
proceed handeln * *verfahren* || verlaufen * *Vorgang* || vorgehen * *verfahren; by stage:s chrittweise; according to/as described in:gemäß (z.B.Betriebsanltg.)*
proceed on fortsetzen
proceeding Vorgehensweise
proceedings Protokoll * *Sitzungs-/Tätigkeitsbericht(e)*
proceeds Einnahme * *(Plural) Erlös* || Gegenwert * *Erlös* || Gewinn * *Erlös*
process abarbeiten || Ablauf || abwickeln * *(figürl.) einen Vorgang, einen Auftrag usw. ~; behandeln, einem Verfahren unterwerfen* || aufbereiten * *in Poduktions/Verarbeitungsschritt* || bearbeiten * *verarbeiten, bearbeiten* || entwickeln * *Film ~* || herstellen * *verarbeiten* || Methode || Prozess *" Vorgang; Produktions~,Herstellungs~* || verarbeiten || veredeln * *Rohstoffe* || Verfahren || Verfahrensweise || Vorgang * *Ablauf, z.B. Herstellungs-, Produktions-*
process as an order Auftrag bearbeiten * *siehe auch →Auftragsabwicklung*
process automation Prozessautomation || Prozessautomatisierung
process automation system Prozessautomatisierungssystem
process chain Prozesskette
process communication keyboard Prozessbedientastatur * *bei SPS*
process computer Prozessrechner
process control Prozessautomation || Prozessautomatisierung * *Prozessleitung/-leitsystem* || Prozessführung || Prozessleitsystem || Prozesssteuerung
process control and instrumentation technology Prozessleittechnik
process control computer Prozessrechner * *Rechner m. schnell. Echtzeitbetriebssystem u.Digital/ Analog E/A-Funktionen*
process control level Prozessleitebene
process control loop Prozessregelkreis
process control system Prozessleitsystem || Prozesssteuerungssystem
process control technology Prozessleittechnik
process controller Technologieregler || verfahrenstechnischer Regler * *für Temperatur, Durchfluss, Druck, Feuchte, Flächengewicht usw.*
process data Betriebsdaten * *Prozessdaten* || Prozessdaten || PZD * *(→PROFIBUS-Profil; Abkürzg. für 'ProZessDaten')*
process data control Prozessdatenführung
process display Anlagenbild * *auf Bildschirm*
process display sytem Visualisierungssystem
process downtime Prozessunterbrechung
process engineering Verfahrenstechnik
process flowchart Technologieschema
process heat Prozesswärme
process I/O image Prozessabbild * *Abbild des Zustandes der Ein/Ausgänge bei einer SPS*
process identification Prozessidentifizierung
process image Abbild * *Prozessabbild* || Prozessabbild * *z.B. bei SPS*
process image table Prozessabbild * *gespeichertes Abbild des Prozess-Ein/Ausgänge bei SPS*
process input image Prozessabbild der Eingänge * *bei SPS*
process instrumentation Prozessmesstechnik

process interface module ssignalformer * *Prozeß-signalformer (z.B. binäre/analoge Ein/Ausgabebaugruppen)*
process interrupt Prozessalarm || Prozessinterrupt
process machine Verarbeitungsmaschine
process management level Prozessleitebene
process model Prozessmodell
process monitoring Prozessbeobachtung || Prozessüberwachung
process network Anlagenverbund
process of calculation Rechenvorschrift * *Rechengang*
process of iteration iterativer Vorgang
process of printing Druckvorgang
process output image Prozessabbild der Ausgänge
process peripherals Prozessperipherie * *bei SPS*
process quantity Prozessgröße
process regulator Technologieregler
process signal Prozesssignal
process status report Prozesszustandskontrolle * *bei SPS*
process steam Prozessdampf
process systems Prozessleittechnik
process technology Verfahrenstechnik
process variable Prozessgröße
process visualisation Prozessvisualisierung * *(brit.)*
process visualization Prozessbeobachtung || Prozessvisualisierung * *(amerikan.)*
process-dependent anlagenabhängig || technologieabhängig
process-oriented technologisch || verfahrenstechnisch
process-oriented board Technologiebaugruppe * *Leiterkarte*
process-oriented controller Technologieregler
process-oriented function technologische Funktion
process-oriented functions technologische Funktionen
process-oriented PCB Technologiebaugruppe * *Leiterkarte*
process-related anlagenspezifisch || technologisch * *z.B. Kenngrößen* || verfahrenstechnisch
process-related module Technologiebaugruppe
process-specific anlagenspezifisch || technologisch
processed verarbeitet
processed quantity Verarbeitungsgröße * *in der Software*
processed variable Verarbeitungsgröße * *in der Software*
processing Abarbeitung || Abwicklung * *Erledigung* || Aufbereitung * *Verarbeitung* || Bearbeitung * auch Akten, Werkstück usw. ~ || Behandlung || Verarbeitung * *Programm, Daten, Signale, Nachrichten; Herstellungsprozess*
processing capacity Rechenleistung
processing depth Verarbeitungstiefe
processing line Verarbeitungsstraße
processing machine Bearbeitungsmaschine || Verarbeitungsmaschine
processing of order Auftragsabwicklung
processing of orders Auftragsbearbeitung || Bestellabwicklung
processing of quotations Angebotsbearbeitung
processing of telegrams Telegrammbearbeitung
processing power Rechenleistung || Verarbeitungsleistung
processing rate Abarbeitungsgeschwindigkeit || Bearbeitungsgeschwindigkeit
processing sequence Abarbeitungsreihenfolge || Bearbeitungsreihenfolge
processing speed Abarbeitungsgeschwindigkeit || Bearbeitungsgeschwindigkeit || Rechengeschwindigkeit * *Verarbeitungsgeschwindigkeit, z.B. eines*

Rechners || Verarbeitungsgeschwindigkeit * *z.B. Rechner*
processing time Bearbeitungszeit || Durchlaufzeit * *für Bestellungen usw.* || Verarbeitungszeit * *in einem Computer*
processing unit Verarbeitungsanlage * *z.B. in einer Fabrik*
processor Prozessor * *z.B. Mikroprozessor* || Steuerwerk
processor board Prozessormodul * *als Leiterkarte* || Prozessormodule * *Leiterkarte, Flachbaugruppe*
processor crash Prozessorabsturz * *(Computer)*
processor loading Prozessorauslastung || Rechnerauslastung * *Prozessorauslastung*
processor module Prozessormodul
processor number Prozessor-Nummer * *bei Mehrprozessorsystem*
processor PCB Prozessormodul * *als Leiterkarte*
processor performance Rechenleistung
processor power Prozessorleistung
processor program Prozessorprogramm * *z.B. beim Antriebsregelsystem SIMADYN D (R) der Fa. SIEMENS*
processor utilization Prozessorauslastung || Rechnerauslastung * *Prozessorauslastung*
proclamation Bekanntgabe * *Verkündigung* || Bekanntmachung * *Verkündigung*
procure anschaffen * *besorgen* || beschaffen * *auch 'sich beschaffen, jemandem etwas beschaffen'* || beziehen * *Ware* ~ || vermitteln * *be-/verschaffen*
procurement Beschaffung || Bezug * *von Ware*
produce abgeben * *erzeugen* || ausüben * *erzeugen* || erzeugen || fertigen || herstellen || tragen * *hervorbringen (Gewinn, Zinsen, Früchte)* || verursachen
producer -Erzeuger || Hersteller || Produzent * *z.B. in der Film/Werbeindustrie*
product Erzeugnis || Fabrikat * *Produkt* || Produkt * *Erzeugnis* || Ware * *Erzeugnis*
product announcement Produktankündigung
product brochure Produktschrift * *Werbebroschüre, Werbeblatt, Farbfaltblatt*
product data base Fabrikatedatenbank
product description Erzeugnisbeschreibung || Produktbeschreibung
product designation Erzeugnisbezeichnung || Fabrikatebezeichnung || Produktbezeichnung || Typbezeichnung || Typenbezeichnung
product discontinuation Produktabkündigung * *Einstellung der Fertigung/Lieferung eines Produkts*
product documentation Produktschrift
product family Produktfamilie || Produktreihe * *Produktfamilie*
product feature Produktmerkmal
product from another manufacturer Fremdfabrikat
product group Produktgruppe
product highlight Produktmerkmal * *herausragendes*
product idea Produktidee
product identification code Erzeugnis-Identnummer
product information Produktinformation || Prospekt * *Produktinformation*
product label Marke * *eines ~nartikels*
product liability Produkthaftung
product liability law Produkthaftungsgesetz * *in Deutschland: Paragraph 823 BGB*
product line Produktpalette || Produktreihe || Produktspektrum * *Produktlinie*
product management Produktbetreuung
product manager Produktbetreuer || Produktverantwortlicher
product maturity Produktreife

product overview Produktübersicht
product planning Produktplanung
product profile Produktsteckbrief
product programme Produktpalette * *Produktprogramm* || Produktprogramm || Produktspektrum * *Produktprogramm*
product range Baureihe || Fertigungsprogramm * *Produktpalette* || Lieferprogramm || Produktionsprogramm || Produktpalette || Produktprogramm || Produktreihe || Typenspektrum * *Produktpalette/-Reihe*
product returned Rückware
product safety Produktsicherheit
product series Produktreihe
product spectrum Produktspektrum
product summary Kurzbeschreibung * *eines Produkts*
product support Produktbetreuung || Service * *Kundenunterstützung für ein Produkt*
product testing Produkterprobung
production Bau || Erzeugung || Fertigung || Herstellung
production and sales profit Fertigungs- und Vertriebsergebnis
production break Produktionsunterbrechung
production capacity Produktionskapazität
production center Fertigungszentrum * *(amerikan.)*
production centre Fertigungszentrum * *(brit.)*
production control Fertigungssteuerung || Produktionsführung || Produktionsleittechnik
production control computer Fertigungsleitrechner
production control level Leitebene * *Produktions-Leitebene*
production control system Fertigungsleitsystem || Produktionsleitsystem
production control systems Produktionsleittechnik
production control technology Produktionsleittechnik
production costs Fertigungskosten || Herstellkosten || Herstellungskosten
production data acquisition Betriebsdatenerfassung
production efficiency Produktivität
production engineering Fertigungsplanung || Fertigungstechnik
production equipment Produktionsmittel
production facilities Fertigungsanlagen || Fertigungseinrichtungen || Produktionsstätte
production facility Fertigungsbetrieb || Produktionsstätte || Werk * *Produktionsanlage,-stätte*
production flow Fertigungsfluss
production host computer Fertigungsleitrechner
production increase Produktionssteigerung
production line Fertigungsstraße || Fließband || Produktionslinie
production machine Arbeitsmaschine
production machinery Fertigungsmaschinen
production management Produktionsleittechnik
production management and control Fertigungssteuerung
production mix Produktpalette
production on a just-in-time basis Kommissionsfertigung
production outage Produktionsausfall
production part Werkstück
production planning Arbeitsvorbereitung * *Fertigungsplanung* || Auftragsvorbereitung * *Fertigungsplanung* || Fertigungsplanung || Fertigungsvorbereitung
production plant Fabrik * *Produktionsanlage* || Fertigungsanlage || Fertigungsbetrieb || Produktionsanlage

production process Fertigungsprozess || Produktionsablauf || Produktionsprozess
production programme Produktpalette
production quality Fertigungsqualität
production range Produktpalette
production rate Durchsatz * *Fertigung* || Produktionsgeschwindigkeit
production scheduling Arbeitsvorbereitung * *Fertigungsvorbereitung* || Fertigungsplanung || Fertigungsvorbereitung
production sequence Produktionskette
production spectrum Fertigungspalette * →*Fertigungsprogramm* || Fertigungsprogramm || Fertigungsspektrum
production speed Arbeitsgeschwindigkeit * *einer Produktionsmaschine/-linie/-anlage* || Produktionsgeschwindigkeit
production stage Fertigungsreife || Produktionsstufe
production standstill Anlagenstillstand
production step Arbeitsgang
production stoppage Produktionsunterbrechung
production technique Fertigungsverfahren
production technology Fertigungstechnik || Fertigungstechnologie || Produktionstechnik
production time Durchlaufzeit * *in der Fertigung*
production unit Betrieb * *produzierender* ~
productive nutzbar * *Gewinn bringend*
productivity Produktivität
productivity improvement Produktivitätssteigerung
productivity increase Produktivitätserhöhung
products -Güter || -Zeug || Waren * *Erzeugnisse*
profession Fach * *Beruf, Fach*
professional fachgerecht || Fachmann * *'Profi'; qualified: qualifizierter* || fachmännisch || Spezialist * *Profi*
professional association Berufsgenossenschaft * *Fachvereinigung*
professional book Fachbuch
professional organisation berufsständische Organisation
professional school Fachhochschule
professional training Fachausbildung
professionally fachmännisch * *(Adv.)* || sachgemäß * *(Adv.)*
professor Professor
PROFIBUS PROFIBUS * *Abkürzg. für 'PROcess FIeld BUS':firmenübergreifender Feldbus nach EN 50170,*
PROFIBUS profile 'Variable-Speed Drives' PROFIBUS-Profil 'Drehzahlveränderbare Antriebe' * *DIN 19245(PROFIBUS) u.VDI/VDE-Richtl.3689 (Profil)*
proficiency Fachkönnen
proficient erfahren * *tüchtig* || tüchtig * *erfahren, geübt*
profile Form * *Profil* || Kontur * *Profil, Seitenansicht, Kontur* || kopierfräsen * *kopier-/fassonfräsen, fassonieren* || Profil * *(auch figürlich u.Anwenderprofil für Bussystem (z.B.PROFIBUS, DRIVECOM))* || Steckbrief * *z.B. eines Produkts* || Verlauf * *Profil,Kontur, Seitenansicht, Längs-/Querschnitt; i.Profil/Quer-/Längsschnitt darstellen*
profile cutting Profilschneiden
profile grinder Profilschleifmaschine * *{Werkz.-masch.}*
profile milling Profilfräsen
profile moulder Fassonfräsmaschine * *{Holz}*
profile rail Profilschiene
profile rolling Profilwalzen
profile sander Profilschleifmaschine * *{Holz}*
profile sanding Profilschliff * *{Holz}*
profile section Aufriss * *Seitenriss* || Seitenriss * *in techn. Zeichnung*

profile shaping   Profilfräsen * *(Holz)*
profile view   Seitenansicht * *siehe auch*
→*Seitenriss* || Seitenriss * *in techn. Zeichnung*
profile-wrapping plant   Profilummantelungsanlage * *(Holz) in der Möbelindustrie*
profiled   profiliert * *z.B. Draht, Stahl, Blech, Holz*
profiler   Kopierfräsmaschine
profiling lathe   Fassondrehmaschine * *(Holz)*
profiling machine   Profilmaschine
profit   Ausbeute || Ergebnis * *Profit, Gewinn, Geschäftsergebnis* || Ertrag * *Gewinn, Profit* || Gewinn * *Profit* || Profit || Überschuss * *Gewinn* || Vorteil * *Gewinn*
profit center   selbstbilanzierender Bereich
profit margin   Gewinnspanne || Verdienstspanne || Vertriebsspanne
profit-and-loss statement   Gewinn- und Verlustrechnung * *(amerikan.)*
profitability   Wirtschaftlichkeit * *Rentabilität*
profitable   förderlich * *vorteilhaft, nützlich (to: für)*; Gewinn bringend, einträglich, lohnend, rentabel || günstig * *vorteilhaft, einträglich, profitabel* || gut * *profitabel; business:Geschäft* || lohnend * *profitabel, Gewinn bringend* || nutzbar * *Gewinn bringend* || nützlich * *Gewinn bringend, profitabel, vorteilhaft* || vorteilhaft * *to:für* || wirtschaftlich * *Ertrag bringend*
profiteer   Schieber * *Profithai*
proforma   Proforma-
proforma-invoice   Proforma-Rechnung * *(für den Zoll)*
profound   tief * *Wissen, Kenntnisse*
profoundness   Tiefe * *(figürlich) Wissen, Kenntnisse, Erfahrung*
profundity   Tiefe * *(figürlich)* ~ *von Wissen, Kenntnissen, Erfahrung*
profusion   Überfluss
prognosis   Prognose || Vorhersage
program   einstellen * *Parameter usw.* ~ || parametrieren || Programm * *Software~, Produkt~ usw.* || programmieren
program abort   Programmabbruch
program block   Programmbaustein * *bei SPS*
program breakdown   Programmabsturz
program capacity   Programmspeicherplatz * *betrifft dessen Größe (large:groß)*
program check   Ablaufkontrolle * *(SPS)* || Bearbeitungskontrolle * *bei SPS*
program code   Programmcode
program counter   Befehlszähler
program crash   Programmabsturz
program debugging   Programmtest
program element   Programmelement * *bei SPS (bei Kontaktplandarstellung)*
program execution   Programmablauf * *Programmausführung, Programmabarbeitung* || Programmausführung || Programmbearbeitung * *Ausführung*
program flow   Ablauf * *Programmablauf* || Programmablauf
program flowchart   Programmablaufplan
program header   Programmkopf
program heading   Programmkopf
program input   Programmeingabe * *z.B. SPS, CNC*
program library   Programmarchiv || Programmbibliothek
program memory   Programmspeicher
program module   Programmbaustein * *bei SPS*
program organization   Programmorganisation
program overview   Programmübersicht * *über Liefer/Verkaufs/Herstell/Softwareprogramm*
program run   Programmablauf
program section   Programmabschnitt
program sequence   Programmablauf * *z.B. in einer sequenziellen Steuerung*
program storage   Programmspeicher

program testing   Programmtest
programm section   Programmteil
programmable   einstellbar * *digital* || konfigurierbar || parametrierbar || programmierbar || speicherprogrammierbar * *SPS* || wählbar * *programmierbar, parametrierbar*
programmable analog input   analoger Wahleingang
programmable analog output   analoger Wahlausgang
programmable controller   AG * *speicherprogrammierbare Steuerung* || Automatisierungsgerät * *AG, SPS, speicherprogrammierbare Steuerung* || programmierbare Steuerung || Speicherprogrammierbare Steuerung || SPS
programmable input   Wahleingang
programmable logic device   PLD || programmierbarer Logikbaustein
programmable output   Wahlausgang
programmable signal   Wahlsignal
programmable terminal   Wahlklemme
programme   Produktpalette * *Programm lieferbarer Produkte* || Programm * *Fertigungs~, Verkaufs~, Schulungs~, Kurs~ usw.* || Serie * *(Verkaufs/Produkt-) Programm*
programmed   parametriert
programmer   Parameterbedieneinheit * *tragbare* ~ || PG * *Programmiergerät für SPS* || Programmierer || Programmiergerät * *(SPS)*
programmer port   PG-Schnittstelle * *Anschluss (-stecker) z.B. an einer SPS zum Anschluss eines Programmiergeräts*
programming   Parametrierung || Programmierung
programming adapter   Programmieradapter
programming adaptor   Programmieradapter * *z.B. zum "Schießen" von EPROMs/EPROM-Moduln/PLDs usw.*
programming console   Programmierplatz * *für SPS*
programming device   Programmiergerät
programming engineer   Programmierer
programming interface   Programmierschnittstelle
programming knowledge   Programmierkenntnisse
programming language   Programmiersprache
programming port   Programmierschnittstelle * *(für SPS)*
programming skills   Programmierkenntnisse
programming software   Programmiersoftware
programming station   Programmierplatz
programming unit   PG * *Programmiergerät für SPS* || Programmiergerät * *(SPS)*
programming unit interface   PG-Anschaltung
progress   Anstieg * *Fortschritt, (Weiter-)Entwicklung, Fortschreiten, Umsichgreifen* || Fortschritt * *in:in; make progress:Fortschritte machen* || Verlauf * *Fortschreiten,* ~ *eines Vorgangs*
progression   Progressivität || Reihe * *mathematische* ~
progressive   ansteigend * *fortlaufend, zunehmend, progressiv* || fortschrittlich || modern * *fortschrittlich* || progressiv || zukunftsorientiert * →*fortschrittlich*
progressive cavity pump   Exzenterschneckenpumpe
progressive jog   schnell tippen * *Kommando für Antrieb* || Tippen schnell
prohibited   verboten
project   Aktion * *Projekt, Plan, (Bau)Vorhaben, Entwurf* || Anlage * *in Planung befindliches Projekt* || entwerfen * *planen, entwerfen* || herausragen * *vorspringen, vorstehen, herausragen, vorragen (over: über)* || hervorstehen || Plan * *Vorhaben* || planen * *entwerfen, projektieren* || Projekt || projektieren * *entwerfen, planen* || projizieren || ragen * *project beyond: hinaus~ über* || überstehen * *vorspringen, vorstehen, (her)vorragen (over: über),*

(her)vortreten (lassen) || vorhaben * *(Substantiv) Projekt*
project beyond hinausragen über
project director Projektleiter
project documents Planungsunterlagen
project engineer Anlagenprojekteur || Projekteur
project engineering Anlagenprojektierung
project leader Projektleiter
project management Projektierung * *Abwicklung eines Projekts* || Projektleitung || Projektmanagement
project manager Projektleiter
project manual Planungsunterlagen * *Handbuch* || Projektierungsunterlagen * *Projekthandbuch*
project over überragen * *hervorstehen über*
project planning Projektierung
project planning engineer Projekteur
project processing Projektbearbeitung
project stage Ausbaustufe
projecting herausragend * *vorspringend, vorstehend* || hervorragend * →*herausragend (z.B. Maschinenteil)* || vorstehend
projection Ausladung * *das Nach-Außen-Ragen eines Teils*, →*Vorsprung* || Nase * *vorspringender Teil, Vorsprung* || Vorsprung * *vorspringender Teil, Überhang*, →*Ausladung*
projection method Projektionsmethode * *bei technischen Zeichnungen*
prolong aktualisieren * *Vertrag* || erneuern * *verlängern, z.b. Vertrag* || verlängern * *zeitlich ~, z.B. Lebensdauer, Frist*
prolongation Frist * *Aufschub*
prolonged länger * *(zeitlich) verlängert, anhaltend; for a prolonged time. für ~e Zeit*
prominent vorstehend
prominent features typische Merkmale * *hervorstechende Merkmale*
promiscuously durcheinander * *(Adv.) wahllos*
promise Zusage || zusichern * *versprechen; promise a thing to a person:jdm. etw. versprechen*
promised date Terminzusage
promising günstig * *viel versprechend*
prommer EPROM-Programmiergerät * *(salopp)* || Prommer * *(salopp) Programmiergerät für EPROMs usw.*
promote fördern * *Vorhaben/Plan/berufliches Fortkommen usw. unterstützen* || unterstützen * *fördern, unterstützen, (im Rang) befördern, werben für, organisieren*
promoter Veranstalter
promotion Akquisition * *Verkaufsförderung* || Aufstieg * *Beförderung* || Beförderung * *beruflich, im Rang* || Förderung * *Verkaufsförderung, Werbung, Befürwortung, Unterstützung, berufliche (Be)Förderung* || Promotion * *(Verkaufs-) Förderung* || Unterstützung * *Förderung, Befürwortung, Beförderung*
promotion article Werbeartikel
promotion campaign Werbeaktion || Werbefeldzug
promotion expert Werbefachmann * *(amerikan.)*
promotion item Werbeartikel
promotion material Akquisitionsunterlagen
promotion matter Werbematerial
promotional akquisitorisch * *werblich, Werbe-* || werblich
promotional budget Werbeetat
promotional items Werbemittel
promotional literature Werbeschriften
promotional material Werbematerial * *zur Verkaufsförderung* || Werbeunterlagen * *verkaufsfördernde Unterlagen*
promotional-arguments booklet Argumentenheft * *mit Nennung der Produktvorzüge zur Vetriebsunterstützung*
promotionally effective werbewirksam

promotive förderlich * *of: für*
prompt Aufforderung * *Eingabe-Aufforderung durch einen Computer auf d.bildschirm* || führen * *[Computer] den Anwender ~ durch ein dialoggesteuertes Programm* || pünktlich * *unverzüglich, prompt, sofortig* || rasch * *sofortig* || rechtzeitig * *pünktlich* || schlagartig || schnell || sofortig || umgehend * *pünktlich*
prompt handling schnelle Erledigung
prompting message Aufforderungstext * *auf einem Computerbildschirm*
prompting text Aufforderungstext * *auf einem Computerbildschirm*
promptly sofort * *prompt, pünktlich*
promt Eingabeaufforderung * *eines Betriebssystems am Bildschirm (z.B. "C:\>" bei MS-DOS)*
prone anfällig * *anfällig sein (to:für)*
prone to breakdown störanfällig || störempfindlich * *zum Ausfall neigend*
proneness Anfälligkeit * *to:für, gegen*
prong Zinke
pronounced ausgeprägt * *deutlich, stark ~, sichtlich, ausgesprochen, bestimmt, entschieden*
proof Beleg * *Beweis* || beständig || Beweis * *of:für* || imprägnieren * *Textilien usw. (wasser-) dicht machen, imprägnieren* || korrekturlesen || Nachweis || Probe * *Beweis (mathematisches)* || sicher * *geschützt (from: gegen/vor; against: vor)*
proof-of-concept phase Erprobungsphase
proofing press Andruckmaschine * *[Druckerei]*
proofread korrekturlesen
proofreading korrekturlesen * *(Substantiv)*
prop Strebe * *Stütze, Stempel, Stützbalken, Stützpfahl, Stelze* || Stütze * *Pfahl, Stempel, Strebe, Stützbalken* || versteifen || Versteifung * *Stütze*
propaganda Werbung * *Reklame, Werbung*
propagate ausbreiten * *sich ~, z.B. Ladungsträger i.Halbleiter, Signale, Verlustwärme usw.* || fortleiten * *fortpflanzen (Ton, Licht usw.), aus-/verbreiten* || fortpflanzen * *auch 'sich fortpflanzen, verbreiten, ausbreiten'* || vermehren * *fortpflanzen*
propagated verbreitet * *Lehre, Meinung usw.* || weit verbreitet * *Lehre, Meinung usw.*
propagation Ausbreitung * *das Sich-Ausbreiten/ Fortpflanzen, z.B. von Ladungsträgern, Signalen, Verlustwärme*
propagation delay Schaltverzögerung * *Signallaufzeit*
propagation delay time Schaltverzögerung * *beim Durchschalten von Signalen z.B. durch Logikgatter* || Signallaufzeit
propel antreiben * *Fahrzeug*
propeller Schraube * *Schiffs~, Flugzeug~*
propeller shaft Gelenkwelle * *bei Fahrzeug, Schiff usw.* || Kardanwelle * *bei Fahrzeug, Schiff usw.*
propelling drive Fahrantrieb * *Fahrzeug, Bagger usw.* || Fahrzeugantrieb
propensity Neigung * *(figürl.) Hang, Vorliebe (to/ for:zu)*
proper angemessen || eigentlich * *genau, richtig* || exakt * *ordentlich, ordnungsgemäß* || geeignet || genau * *ordnungsgemäß, angemessen, korrekt, einwandfrei, genau* || gut * *angebracht* || ordnungsgemäß || passend * *richtig, passend, geeignet, angemessen, ordnungsgemäß, zweckmäßig* || richtig * *gehörig, passend* || sachgemäß || sachgerecht || sorgfältig * *ordnungsgemäß, genau, passend, geeignet* || wahr * *eigentlich, richtig* || zuständig * *maßgebend, passend, richtig* || zweckmäßig
properties Eigenschaften
property -Eigenschaft || Beschaffenheit * *Eigenschaft* || Besitz || Eigenschaft * *Eigenart, Merkmal, Werkstoffeigenschaft, Vermögen(z.B.Isolationsvermögen)* || Eigentum || Merkmal * *Eigenschaft*
property damage Sachschaden

propitious günstig * *geeignet, vorteilhaft (to: für); z.B. Moment*
proportion anpassen * *im Verhältnis ~* || Anteil * *(An)Teil, Verhältnis, Ausmaß, Umfang* || bemessen * *bemessen (to:nach), anpassen, i.d.richtige Verhältn. bringen, verhältnismäßig verteilen* || Maß * *(Größen)Verhältnis* || Verhältnis
proportion of ingredients Mischungsverhältnis
proportional proportional * *to/with: zu; inversely/ directly: umgekehrt/direkt; be ~/in proportion to:~ sein zu* || prozentual || verhältnismäßig
proportional band Proportionalbereich
proportional coefficient Proportionalbeiwert
proportional control Proportionalsteuerung
proportional element Proportionalglied
proportional factor P-Verstärkung || Proportionalverstärkung
proportional gain P-Verstärkung || Proportionalbeiwert * *Proportionalverstärkung* || Proportionalverstärkung || Verstärkungsfaktor * *P-Verstärkung*
proportional plus integral controller PI-Regler
proportional plus integral-action controller PI-Regler
proportional term P-Anteil || Proportionalanteil
proportional to -proportional
proportional to speed drehzahlproportional
proportional to the -proportional
proportional to the voltage spannungsproportional
proportional to time zeitproportional
proportional valve Proportionalventil * *mit stetig veränderbarer Öffnung (z.B. für Pneumatik, Hydraulik usw.)*
proportional-action coefficient Proportionalbeiwert
proportional-action controller P-Regler || Proportionalregler
proportionality Proportionalität
proportionally proportional * *(Adv.) change proportionally with: sich ~ ändern mit*
proportionate gleichmäßig * *ebenmäßig*
proportionate with entsprechend * *im Verhältnis*
proportioning device Dosiereinrichtung
proportioning plant Dosieranlage || Dosiermaschine * *Dosieranlage*
proportioning pump Dosierpumpe
proportioning screw Dosierschnecke
proportioning system Dosieranlage
proposal Antrag * *Vorschlag* || Vorschlag
propose einbringen * *vorschlagen* || planen * *vorhaben* || vorhaben * *beabsichtigen, sich vornehmen* || vorschlagen
proposition Antrag || Satz * *Lehr~*
propound aufstellen * *Lehre ~, Theorie ~*
proprietary firmeneigen || firmenintern * *z.B. technische Lösung* || firmenspezifisch * *siehe auch →firmeneigen* || hauseigen * *→firmenspezifisch* || herstellerspezifisch
proprietor Besitzer * *Eigentümer* || Geschäftsinhaber || Inhaber * *Eigentümer*
propshaft Antriebswelle * *eines Fahrzeugs/Schiffs (z.B. Kardanwelle)*
propulsion Antrieb * *Antriebs-/Fortbewegungskraft, besonders für Fahrzeug; auch figürlich* || Schiffsantrieb
propulsion drive Schraubenantrieb * *für Schiff* || Wellenantrieb * *für Schiff*
propulsion force Antriebskraft * *bei Schiff, Fahrzeug, Linearmotor*
propulsion motor Antriebsmotor * *bei Fahrzeug* || Propellermotor * *zum Antrieb eines Schiffes*
propulsion power Antriebsleistung * *bei Schiff, Fahrzeug, Linearmotor*
pros für * *(Substantiv) the pros and cons:das Für und Wider*

prospect Perspektive * *(figürl.) Aussicht, Ausblick, Vor(aus)schau, Zukunftsaussicht*
prospective künftig * *voraussichtlich* || voraussichtlich * *(Adj.) in Aussicht stehend, (zu)künftig* || zukünftig * *Person*
prospectus Produktinformation * *Werbeprospekt* || Produktschrift * *Werbeprospekt* || Prospekt * *Werbeprospekt*
protect schonen * *schützen, erhalten* || schützen * *against: gegen, from:v or, protect oneself from: sich ~ vor* || sichern || wahrnehmen * *Interesse ~*
protect against verhindern * *schützen gegen*
protect by fuse absichern * *mit Sicherung ~*
protect by patent patentieren
protected beständig * *geschützt* || geschützt || sicher * *geschützt*
protected against abgesichert * *geschützt gegen*
protected against accidental contact berührgeschützt || berührungsgeschützt
protected against corrosion korrosionsgeschützt
protected against damaging zerstörsicher
protected against earth faults erschlußfest
protected against ground faults erdschlussfest || erdschlusssicher
protected against polarity reversal verpolungssicher * *z.B. Digitaleingang, Stromversorgungsanschluss*
protected against twisting verdrehsicher
protected against wire breakage drahtbruchsicher || drahtbruchsicher
protecting schützend
protecting device Schutzgerät
protecting ring Schutzring
protecting tube Schutzrohr
protection -Schutz || Absicherung * *gegen Gefahr, Störung und Zerstörung* || Schonung * *Erhaltung, Schutz* || Schutz * *against: gegen, from: vor, legal: rechtlicher* || Schutzart || Sicherheit * *Schutz*
protection against accidental contact Berührungsschutz * *siehe z.B. DIN VDE 0106 und 0113)*
protection against contact with live or moving parts Berührungsschutz
protection against corrosion Korrosionsschutz
protection against dew Betauungsschutz
protection against dust Staubschutz
protection against dust-ignition Staubexplosionsschutz * *Schutz gegen Zündung explosionsfähiger Stäube (DIN VDE 0165)*
protection against explosion Explosionsschutz || Explosionsschutzart * *siehe IEC-Publikation No.79, EN50014 bis 50020* || Schlagwetterschutz
protection against false polarity Verpolschutz
protection against firedamp Explosionsschutz * *Schlagwetterschutz* || Schlagwetterschutz
protection against polarity reversal Verpolungsschutz
protection against splashing water Spritzwasserschutz
protection against water jets Strahlwasserschutz * *z.B. Schutzart IP55, IP56*
protection circuit Schutzbeschaltung
protection class Schutzart * *Schutzklasse (siehe DIN 40050 und DIN VDE 0106 Teil 1)* || Schutzklasse * *z.B. für elektr. Antriebe in Haushaltsgeräten nach VDE 0730 o. Explosionsschutzkl.*
protection component Schutzeinrichtung * *Bauteil, Gerät*
protection cover Schutzkappe
protection device Schutzeinrichtung
protection equipment Schutzeinrichtung
protection function Schutzfunktion
protection hood Schutzhaube
protection mark Schutzzeichen
protection mode Schutzart

protection pipe Schutzrohr
protection shield Schutzdach * *z.B über senkrecht eingebauten Motor, damit keine Fremdkörper i.d.Lüfter fallen*
protection tube Schutzrohr
protection type Schutzart
protective Schutz- || Sicherheits- * *Schutz-*
protective break down Durchschlagsicherung * *{el. Masch.} Überspannungsableiter z.Schutz v.Meskreisen i.Hochspannungsmasch.*
protective cap Schutzkappe
protective casing Schutzgehäuse
protective circuit Schutzbeschaltung
protective circuit breaker Schutzschalter
protective clothing Schutzkleidung
protective coating Schutzanstrich || Schutzbeschichtung
protective conductor Schutzleiter
protective conductor connection Schutzleiteranschluss
protective conductor terminal Schutzleiterklemme * *Erdungsklemme*
protective connection Schutzleiterverbindung
protective cover Schutzabdeckung || Schutzdach || Schutzhaube
protective device Schutzeinrichtung
protective earth Schutzerde || Schutzerdung || Schutzleiter * *'PE', Schutzerde*
protective earth conductor Schutzleiter * *siehe z.b. VDE 0100 Teil 200*
protective earth connection Schutzleiteranschluss
protective earth terminal Schutzleiteranschluss * *Klemme, Schraube*
protective element Schutzelement
protective equipment Schutzeinrichtung
protective features Schutzfunktionen
protective firing Schutzzünden || Schutzzündung * *von Thyristoren bei Überschreiten der max. zulässigen Sperrspannung*
protective function Schutzfunktion
protective gas Schutzgas
protective grid Schutzgitter
protective grille Schutzgitter
protective ground Schutzleiter
protective ground connector Schutzleiter
protective ground potential Schutzleiterpotential
protective hood Schutzhaube
protective housing Schutzgehäuse
protective interlock Schutzverriegelung
protective interlocking Schutzverriegelung
protective logic Schutzverriegelung
protective measure Schutzmaßnahme
protective motor starter Motorschutzschalter || Schutzschalter * *→Motorschutzschalter*
protective neutral conductor Mittelleiter * *mit Schutzleiterfunktion*
protective package Schutzverpackung
protective painting Schutzlack
protective relay Schutzrelais
protective relaying Schutz * *~ durch Überwachungs- u. ~einrichtungen, z.B. Netz~, Maschinen~*
protective release Schutzabschaltung
protective roof Schutzdach
protective separation sichere Trennung * *siehe DIN 0100, Tl. 410 u.DIN VDE 0106, Tl. 101*
protective sheath Schutzmantel * *für Kabel*
protective shield Schutzschild
protective top cover Schutzdach * *z.B. über Motor b. Senkrechtmontage, damit keine Fremdkörper i.d.Lüfter fallen*
protective trip Schutzabschaltung * *auf Grund eines Fehlers, einer Grenzwertüberschreitung usw.*
protector -Wächter * *Schutzeinrichtung/-Gerät || Schutz * als Bauteil/Vorrichtung; edge protector: Kanten~ || Schutzelement*

protest Reklamation * *Einspruch* || versichern * *beteuern*
protestation Versicherung * *Beteuerung*
protocol Protokoll * *z.B. ~ für se(rielle) Datenübertragung* || Prozedur * *Kommunikationsprotokoll*
protocol converter Protokollumsetzer
prototype Baumuster || Entwicklungsmuster * *Prototyp* || Modell * *Prototyp* || Muster * *Prototyp* || Prototyp
prototype plant Pilotanlage
prototype test Typprüfung * *~ der Erstausführung; für Motor: siehe VDE 0530*
prototype test certificate Baumusterprüfbescheinigung
prototype test certification Baumusterprüfbescheinigung
protracted langwierig * *in die Länge gezogen*
protrude hervorstehen || überstehen * *heraussehen, (her)vorstehen, →hervorragen, hervortreten (lassen), herausstrecken*
protruding herausragend * *hervorstehend, →hervorragend (z.B. Maschinenteil)* || hervorragend * *→herausragend (z.B. Maschinenteil)* || vorstehend
protrusion Ausladung * *Hervorstehen/-treten/-springen (Maschinenteil), vorstehender Teil, Vorwölbung/-sprung*
protude herausragen * *heraustehen, (her)vorstehen, herausragen, hervortreten*
protude over überragen * *herausragen/hervorstehen über*
provable nachweisbar
prove ausfallen * *Ergebnis usw.* || belegen * *beweisen* || bewähren * *they have proven themselves:sie haben sich bewährt* || beweisen || erproben || erweisen * *sich erweisen (to be: als)* || herausstellen * *sich ~/erweisen (to be:als)* || nachweisen * *beweisen* || zeigen * *sich ~, herausstellen*
proven bewährt || erprobt * *['pru:ven]*
provide anordnen * *sorgen für, vorsehen* || anschaffen * *besorgen* || ausrüsten || ausstatten || beistellen * *beliefern/versorgen (with:mit), z. Verfügung stellen, bereitstellen* || beliefern * *with: mit* || bereitstellen || beschaffen * *sich* || bieten * *sorgen für, bereitstellen, z. Verfügung stellen, (be)liefern (with: mit), befriedigen* || erstellen || leisten * *liefern* || liefern * *be-~/versehen (with:mit); z. Verfügg./bereitstellen,auch Strom/Spanng./Leistg./Energie* || sicherstellen * *Vorsorge treffen, sich sichern (against: vor/gegen); bereit-/zur Verfügung stellen* || sorgen für * *beschaffen* || versehen * *with: mit* || versorgen * *with: mit* || voraussetzen * *provided that:vorausgesetzt, dass; sicherstell., vorsehen/-schreiben, Vorsorge treffen* || Vorsorge treffen * *against: gegen* || zur Verfügung stellen * *bereitstellen; versehen, versorgen, ausstatten(with:mit); sorgen für,liefern*
provide a clear insight to transparent machen
provide advice and assistance mit Rat und Tat zur Seite stehen
provide an overview Überblick geben
provide for sorgen für || vorsehen * *~ für, sorgen für; the law provides that: das Gesetz sieht vor, dass*
provide from own sources beistellen * *durch den Kunden usw.*
provide information informieren * *to somebody: jdn.*
provide security for absichern * *z.B. Kredit ~*
provide with anlegen * *versorgen mit*
provided ausgestattet || unter der Bedingung * *that: dass (kann auch weggelassen werden)* || unter der Bedingung dass || vorausgesetzt * *that: dass (kann auch weggelassen werden)* || vorausgesetzt dass ||

vorgesehen * *for a thing: etwas; provided that:* ~, *dass* || vorhanden * *zur Verfügung gestellt* || zur Verfügung gestellt
provided that unter der Bedingung dass || vorausgesetzt dass
providence Vorsorge * *Vorsorge, (weise) Voraussicht*
provider Anbieter * *Lieferant, Anbieter (auch von Online-Diensten und Netzzugängen)* || Lieferant
providing Versorgung * *Ausstattung, Vorsorge, Bereitstellung*
province Fach * *(Wissens)Gebiet, Fach*
provision Bedingung * *in Vertrag; auch 'Vorbehalt'* || Festlegung * *Bestimmung, Vorschrift, Vorkehrung,* || Versorgung * *Beschaffung, Bereitstellung, Vorkehrung* || Voraussetzung * *Vorsorge, Vorkehrung, Maßnahme; provided that:vorausgesetzt, dass* || Vorkehrung || Vorrat * *of: an* || Vorschrift * *(gesetzl.) Bestimmg.,Vorschrift; come within th.prov.s of th.law: u.d.gesetzl. Best.fal.*
provisional behelfsmäßig || provisorisch * *vorläufig, einstweilen, behelfsmäßig* || Übergangs- * *provisorisch, vorläufig* || vorläufig || vorübergehend * *provisorisch* || Zwischen- * *vorläufig, provisorisch, behelfsmäßig*
provisional solution Notlösung * *behelfsmäßige Lösung* || Zwischenlösung * *behelfsmäßige Lösung*
provisions Vorsorge * *Vorkehrung/-sorge, Maßnahme; make provisions:~ treffen; provide against: ~ treffen gegen*
provisions of the agreement Vertragsbedingungen
provoke aufbringen * *erzürnen, provozieren*
prowess Können * *Tüchtigkeit*
proximity Nähe * *auch '(An-) Näherungs-'*
proximity effect Stromverdrängungseffekt
proximity limit switch BERO * *Endschalter*
proximity switch BERO * *Näherungsschalter* || berührungsloser Näherungsschalter || Endschalter * *Näherungsschalter* || Näherungsschalter
proximity to the market Marktnähe
PS PS * *Pferdestärken; 1 PS entspr. 0,73550 kW entspr. 0,9863 hp*
pt 1 * *Engl. Pint; 1l entspr. 1,7598 pt*
PT 100 gauge PT100-Messfühler
PT1 element PT1-Glied
PT1-block PT1-Glied * *Funktionsbaustein*
PT1-function PT1-Glied
PT100 temperatur sensor PT100-Messfühler * *Widerstandsthermometer aus Platindraht (PT), d.bei 0° e.Wid. v. 100 Ohm hat*
PTB PTB * *Physikalisch-Techn. Bundesanstalt, amtl .deut. Behörde für d.Mess- und Eichwesen i.Braunschweig*
PTC Kaltleiter * *Positive Temperature Coefficient, Widerst. dess. Ohmwert s.m.steig. Temperatur erhöht* || PTC * *Positive Temperature Coefficient, Widerstand dessen Ohmwert sich mit steigend. Temperatur erhöht* || Thermistor
PTC resistor Kaltleiter * *Widerstand mit 'Positive Temperature Coefficient': positivem Temperaturkoeffizient* || PTC-Widerstand
PTC sensor Kaltleiter * *als Temperaturfühler*
PTC thermistor Kaltleiter * *bei Verwendung als Temperaturfühler* || PTC-Widerstand * *als Temperaturfühler verwendet*
PTC-thermistor Kaltleitertemperaturfühler
pto wenden * *Abk. für 'please turn over': bitte (Buchseite) wenden*
pub Wirtschaft * *Kneipe*
public distribution system öffentliches Netz
public holidy Feiertag * *offizieller*
public house Wirtschaft * *Wirtshaus*
public network EVU-Netz || öffentliches Netz
public relations Öffentlichkeitsarbeit || Werbung * *Werbe-/Presse-/Öffentlichkeitsarbeit*
public secondary distribution system öffentliches Niederspannungsnetz
public servant Beamter * *Berufsgruppe*
public supply öffentliches Netz * *Energieversorgungsnetz*
public supply system Versorgungsnetz
public utility Betrieb * *Versorgungs~*
publication Bekanntgabe || Bekanntmachung || Druckschrift || erscheinen * *(Substantiv) eines Buches (forthcoming: im ~ begriffen; when published: bei ~)* || Publikation || Veröffentlichung * *auch Schrift, Buch, Aufsatz, Broschüre*
publicity Werbung * *Reklame, Werbung*
publicity appeal Werbewirksamkeit
publicity article Werbeartikel
publicity gift Werbegeschenk
publicity leaflet Werbeblatt * *Faltblatt,* →*Prospekt* || Werbeprospekt || Werbeschrift * *Faltblatt,* →*Prospekt*
publish auflegen * *Buch usw. ~; reprint/republish: wieder~* || bekanntmachen * *öffentlich* || drucken * *veröffentlichen* || herausgeben * *Druckerzeugnis ~* || veröffentlichen * *Buch, Aufsatz/Artikel*
published data Datenblattangabe * *vom Hersteller veröffentlichte Daten*
publisher Herausgeber * *z.B. eines Buches/einer Zeitschrift* || Verlag
publishers Verlag
publishing house Verlag
puffing marktschreierisch
puffy bauschig
pug mill Kollergang * *Walzen-Mahlwerk*
pull abziehen || anziehen * *(allg.) auch Bremse* || Griff * *Zieh-~* || reißen * *ziehen* || Werbewirksamkeit || ziehen || Zug * *Ziehen, Zerren, Zug, Schubkraft, Anzug, Ruck, Anziehungskraft*
pull down abbrechen * *Gebäude ~* || abreißen * *Gebäude* || reißen * *zu Boden ~*
pull flush with floor level bündigziehen * *{Aufzüge}*
pull in anziehen * *Relais, elektromagnet. Bremse usw.* || einfahren * *{Bahnen}* || einziehen * *z.B. (Motor)wicklung in eine Nut*
pull magnet Zugmagnet
pull off abreißen * *auseinander reißen (aktiv)* || abziehen * *z.B. Kupplung ~* || anziehen * *abfahren, sich in bewegung setzen*
pull out of step außer Tritt fallen * *Kippen (z.B. Synchronmotor)* || Kippen * *(Verb) ~ einer el. Maschine*
pull relief Zugentlastung
pull roll S-Walze * *Zugwalze* || Zugwalze
pull the plug Geldhahn zudrehen
pull the trigger abdrücken * *Gewehr usw.*
pull through durchziehen
pull-in einschalten * *anziehen, z.B. Schütz, Relais*
pull-in technique Einziehtechnik * *Einziehen vorgefertigter Spulen in Motor (Herstellverfahren)*
pull-in value Ansprechwert * *z.B. Relais*
pull-in winding Einziehwicklung * *{el.Masch.}* *vorgewickelte Spulen aus Runddrähten, die maschinell eingezogen werden*
pull-off Abzug * *z.B. Antrieb i.Kabelmaschine, in d. d.Kabel geklemmt u.mit konstanter Geschw. geförd.wird*
pull-off capstan Scheibenabzug * *konstant angetriebene Scheibe in Kabelmaschine, an die d.Kabel angepresst wird*
pull-out Kippen * *(auch Verb) z.B. ~ einer Asychronmaschine*
pull-out protection Kippschutz * *z.B. Asynchronmotor*
pull-out torque Kippmoment * *Synchronmotor und Asynchronmotor*
pull-through winding Durchziehwicklung * *Wick-*

lungsart bei Motor: Drähte werden einzeln durch die Nut hindurchgezogen
pull-up torque Sattelmoment * *[el.Masch.]* bei *Käfigläufermotor (siehe VDE 0530 Teil 12)*
pulled in angezogen * *Relais*
puller Abziehvorrichtung || Abziehwerkzeug
puller screw Abdrückschraube
puller tool Abziehvorrichtung * *Abziehwerkzeug; z.B, für Lager, Zahnrad* || Abziehwerkzeug * *z.b. für Lager, Zahnrad*
pulley Riemenscheibe || Rolle * *Zug~ für Seil, Riemen, Kran, Flaschenzug, Sägeband* || Scheibe * →*Riemen~* || Seilscheibe
pulley block Flaschenzug
pulley diameter Riemenscheibendurchmesser * *bei Riementrieb*
pulley end Antriebsseite * *Seite, an der die Riemenscheibe angebracht ist*
pulley wheel Riemenscheibe
pulling aid Ziehhilfe
pulling dog Zange * *z.B. als Ziehwerkzeug beim Drahtziehen*
pulling down Abbruch * *Gebäude*
pulling grip Ziehgriff * *an Steckbaugruppe*
pulling handle Ziehgriff * *an Steckbaugruppe*
pulling lever Ziehhilfe * *Hebel* *z.B. an Flachbandleitung-Stecker*
pulling out of step Kippen * ~ *einer Synchronmaschine*
pullout torque Kippmoment * *Synchron- und Asynchronmotor*
pulp Brei * *Zellstoff-/Papier~* || Faserstoff * *{Papier} (Faser)Zellstoff} {Halbstoff * für die Papierherstellung (z.B. Zellstoff)* || Holzstoff * *{Papier}* || Zellstoff * *Ausgangsmaterial zur Papiererzeugung; Zellstoff, (Zellstoff-) Brei*
pulp and paper industry Papierindustrie * *Papier- und Zellstoffindustrie*
pulp arrical Stoffauflauf * *{Papier} Das Auflaugen des Stoffes auf das Sieb bei e. Papiermaschine*
pulp bleaching Zellstoffbleiche
pulp board Zellstoffkarton
pulp digester Zellstoffkocher
pulp drying machine Zellstofftrockenmaschine
pulp engine Holländer * *{Papier}*
pulp factory Zellstofffabrik
pulp mill Zellstofffabrik
pulp refiner Holzstoffrefiner * *{Papier} zur Zellstofferzeugung*
pulp vat Bütte * *Zellstoffbütte in der Zellstoff-/Papierherstellung*
pulp wood grinding zerfasern * *{Papier} (Subst.)*
pulper Holländer * *{Papier} Breimühle in der Zellstoffindustrie* || Pulper * *{Papier} Breimühle ("Holländer") zur Zellstofferzeugung* || Stofflöser * *für Zellstoff*
pulpwood Faserholz
pulsate pulsen || schlagen * *rythmisch stoßend*
pulsated gepulst
pulsating getaktet * *pulsierend, gepulst* || pulsierend || stoßweise * *(auch figürlich)*
pulsating current lückender Strom || Lückstrom || Stromlücken
pulsating DC operation Lückbetrieb
pulsating-current operation Lückbetrieb
pulsation Pulsation || Pulsieren || Schwingung * *Pulsieren* || Vibration * *pochend, schlagend*
pulsation factor Schwingungsgehalt
pulsation torque Momentenschwankungen * *harte Schläge verursachend*
pulse -pulsig || Impuls || impulsförmig || pulsförmig || stoßartig
pulse amplifier Impulsverstärker
pulse amplitude Impulsamplitude

pulse blocking Impulslöschung * *Impulssperre* || Impulssperre
pulse capacitor Impulskondensator * *mit hoher Impulsbelastbarkeit* || Stoßkondensator
pulse conditioning Impulsaufbereitung
pulse conditioning circuit Impulsformer
pulse contracting block Impulsverkürzer * *Funktionsbaustein*
pulse control Pulssteuerung
pulse control factor Aussteuergrad || Aussteuerungsgrad
pulse converter Pulsumrichter
pulse count Pulszahl * *z.B. Impulsgeber* || Schlitzzahl * *Anzahl der Schlitze in der Schlitzscheibe eines (Inkremental-) Impulsgeber*
pulse counter Impulszähler
pulse delay block Einschaltverzögerung * *Funktionsbaustein*
pulse diagram Impulsdiagramm
pulse dialing Pulswählverfahren * *Telefonwählverfahr.:für jede Ziffer wird e.entsprechende Zahl v.Pulsen übermittelt*
pulse disable Impulssperre
pulse displacement Impulsverschiebung
pulse distribution Impulsverteilung
pulse distribution module Impulsverteiler
pulse distributor Ringverteiler * *verteilt d. Zündimpulse auf d. einz. Thyristoren e. Stromrichters*
pulse divider Impulsteiler || Impulsuntersetzer
pulse duration Impulsbreite || Impulsdauer || Impulslänge
pulse duration modulation Pulsbreitenmodulation * *Pulsdauermodulation*
pulse duty cycle Puls-Pausenverhältnis * *Tastverhältnis*
pulse duty factor Tastverhältnis * ~ *einer Impulskette* || Tastverhältnis
pulse edge Impulsflanke
pulse edge evaluation Erstabfrage * *Flankenerfassung*
pulse enable Impulsfreigabe
pulse enabling Impulsfreigabe
pulse encoder Drehimpulsgeber || Impulsgeber * *z.B. an Motor* || Impulstacho * *Pulsgeber* || Inkrementalgeber || Pulsgeber
pulse encoder evaluation Impulsgeberauswertung
pulse encoder simulation Impulsgebernachbildung * *Ausgabe v.Impulsketten durch e. Resolver-/Sin-Cos-Geber-Auswerteschaltung*
pulse evaluation Impulsauswertung || Pulsauswertung
pulse frequency Impulsfrequenz || Pulsfrequenz || Taktfrequenz * *Pulsfrequenz, z.B. einer Thyristor-Stromrichterschaltg.*
pulse generating Impulsbildung * *als Funktion*
pulse generator Impulsbildner * *Impulsgenerator* || Impulsbildung * *als Einrichtung/Schaltung* || Impulsgeber * *allg.* || Impulsgenerator
pulse group Impulsgruppe
pulse handling capacity Impulsbelastbarkeit * *von Kondensatoren*
pulse inhibit Impulssperre
pulse input Impulseingang
pulse inverter Pulswechselrichter
pulse length Impulslänge
pulse number Pulszahl * *Anz.d.Kommutierungen in e.Stromrichter je Periode*
pulse number per revolution Schlitzzahl * ~ *eines Impulsgebers* || Strichzahl * *eines Impuls-/Winkelschrittgebers*
pulse packet Impulspaket
pulse pattern Impulsmuster || Impulsraster || Pulsmuster * *optimized: optimiertes (bei Pulsumrichter)* || Pulsraster
pulse period Impulsperiode

pulse rate Impulsfrequenz || Pulsfolgefrequenz
pulse repetition period Impulsperiode
pulse resistor Pulswiderstand
pulse run Impulsgruppe
pulse scaler Impulsteiler
pulse sensing Impulserfassung
pulse shape Impulsart || Impulsform
pulse shaper Impulsformer
pulse shaping Impulsaufbereitung * *Impulsformung* || Impulsformung * *z.B. Aufsteilung* || Pulsformung
pulse shift Impulsverschiebung
pulse strength Impulsbelastbarkeit * *von Kondensatoren* || Impulsfestigkeit * *von Kondensatoren*
pulse stretching block Impulsverlänger * *Funktionsbaustein*
pulse suppression Impulslöschung * *Impulsunterdrückung* || Impulssperre * *Impulsunterdrückung*
pulse suppression module Zündsperrbaugruppe * *z.b. für Thyristor-Leistungsteil*
pulse tach Digitaltacho || Drehzahlgeber * *Impulsgeber* || Drehzahlistwertgeber * *Impulsgeber als Tachometer* || Impulsgeber * *['pals 'täk] Impulstacho* || Impulstacho || Pulsgeber * *Impulstacho* || Tacho * *Impulstacho* || Tachometer * *Impulsgeber*
pulse tacho Digitaltacho || Impulstacho || Pulstacho
pulse tachometer Digitaltacho || Drehzahlgeber * *Impulsgeber* || Drehzahlistwertgeber * *Impulsgeber als Tachometer* || Impulsgeber * *['pals 'täkomiete]* || Impulstacho || Pulsgeber * *Impulstachometer*
pulse technique Pulsverfahren * *für Pulsumrichter*
pulse track Impulsspur * *z.B. eines Impulsgebers*
pulse train Impulsfolge || Impulskette || Impulspaket * *Impulskette* || Kettenimpuls * *Impulskette, z.B. zur Thyristorzündung i.Ggs. z. Kettenimpuls*
pulse transformer Impulsübertrager || Übertrager * *Impuls~ (für Thyristoren usw.)* || Zündübertrager * *[Stromrichter]*
pulse waveform Impulsart || Impulsform
pulse waveshape Impulsform
pulse width Impulsbreite || Impulsdauer || Impulslänge * *Impulsbreite* || Pulsbreite
pulse width modulation Pulsbreitenmodulation
pulse width modulator Pulsbreitenmodulator
pulse-controlled converter Pulsumrichter * *mit konstanter Zwischenkreisspannung*
pulse-edge modulation Flankenmodulation * *z.B. b. Wechselrichter (Pulsweitenmodul. mit nichtäquidistanten 0>1-Flanken)*
pulse-forced response Impulsantwort
pulse-pattern Pulsverfahren * *Pulsmuster, z.B. für Pulsumrichter*
pulse-shaped impulsförmig || pulsförmig
pulse-width modulated pulsbreitenmoduliert
pulse-width modulated inverter Pulswechselrichter
pulse-width modulation Pulsweitenmodulation
pulsed braking resistor Bremswiderstand * →*Pulswiderstand* || Pulswiderstand
pulsed load Wechsellast
pulsed resistor Pulswiderstand * *z.B. Bremswiderstand für Frequenzumrichter*
pulses per revolution Impulse pro Umdrehung
pulsing Taktung * *bei Stromrichter, Leistungsimpulse erzeugend*
pulverize mahlen * *zu Pulver* || stoßen * *pulverisieren* || zerkleinern * *zu Pulver zermahlen*
pulverized coal blower Kohlenstaubgebläse
pulverizer Mühle * *Zerkleinerer, Pulverisiermühle, Mahlanlage*
pulverizer mill Mahlanlage
pulverizing equipment Mahlanlage
pulverizing plant Brechanlage
pumice Bimsstein

pumice block Bimsstein * *z.b. zum Reinigen des Kommutators bei Gleichstrommaschine*
pump Pumpe || pumpen
pump and fan drives Strömungsmaschinen-Antriebe
pump characteristic Pumpenkennlinie * *Drehmoment bzw. Leistg. über der Drehzahl (bei Kreiselpumpe: M ~ n^2, P ~ n^3)*
pump delivery Förderkapazität * *einer Pumpe*
pump drive Pumpenantrieb
pumped medium Fördermedium * *(Plural: "media") durch eine Pumpe gefördertes ~*
pumped-storage machine Pumpspeicherwerksmaschine
pumped-storage power station Pumpspeicherkraftwerk
pumping head Förderhöhe * *einer Pumpe*
pumping line Seil * *für Bohrpumpenantrieb*
pumping station Pumpwerk
punch ankörnen || Körnung * *An~ mit einem Körner(-schlag)* || lochen * *auch Papier(bögen)* || stanzen || Stanzstempel || Stanzwerkzeug * *Stanzstempel (Oberteil)* || Stempel * *zum Stanzen* || stempeln * *auch Angabe auf Typenschild ~* || stoßen * *(aus)-stanzen, stempeln, durchschlagen, lochen*
punch out ausstanzen
punched gelocht * *mit ausgestanzten Löchern* || gestanzt
punched hole Lochung * *gestanztes Loch*
punched part Stanzteil
puncher Locher * *auch im Büro*
punching Lochung || Stanzteil
punching die Matrize * *Stanzmatrize (Unterteil)* || Stanzmatrize || Stanzwerkzeug * *Stanzmatrize (Unterteil)*
punching machine Stanze || Stanzmaschine
punching press Stanze * *Stanzpresse* || Stanzmaschine || Stanzpresse
punching sheet Stanzblech
punching shop Stanzerei
punching tool Stanze
punchings Stanzabfälle
punctate punktförmig
punctiform punktförmig
punctual präzise * *pünktlich* || pünktlich || sofortig * *pünktlich* || umgehend * *pünktlich*
punctually rechtzeitig * *(Adv.) pünktlich* || sofort * *pünktlich*
punctuate punktieren
puncture Durchschlag * *Isolationsversagen, (Ein-) Stich, Loch, Reifenpanne; puncture-proof: durchschlagsicher*
purchase anschaffen || Beschaffung * *Anschaffung, Einkauf* || Bezug * *Erwerb von Ware* || erkaufen * *(auch figürl.) at the expense of: mit* || Erwerb * *Kauf* || erwerben * *käuflich ~* || Kauf * *complete a purchase: einen ~ abschließen* || kaufen * *erwerben*
purchase cost Anschaffungskosten
purchase costs Investitionskosten * *Anschaffungskosten*
purchase order Bestellschein * *Bestellung, ausgefülltes Bestellformular* || Bestellung || Bestellzettel * *ausgefüllter Bestellzettel, Bestellung* || BZ * *Bestellung, ausgefülltes Bestellformular*
purchaser Abnehmer * *Käufer* || Besteller * *Abnehmer, Käufer* || Käufer || Kunde * *Käufer*
purchasing Einkauf * *das Einkaufen* || Kauf * *das ~en*
purchasing department Einkauf * *Einkaufsabteilung* || Einkaufsabteilung
purchasing management Einkauf
purchasing manager Einkäufer * *in einem Unternehmen*
purchasing power Kaufkraft

purchasing schedule Disposition * ~ des Waren-Einkaufs
pure netto * rein || rein * pur, unvermischt, bloß; pure P controller:reiner P-regler || sauber * rein, nicht verunreinigt
pure india-rubber Naturkautschuk
pure water Reinwasser
purely lediglich || rein * (Adv.)
purely defined unscharf
purely ohmic rein Ohmsch
purely resistive rein Ohmsch
pureness Reinheit
purge beseitigen * säubern, reinigen, entleeren || Beseitigung || Reinigung
purification Reinigung * auch von Abwässern usw.
purification plant Kläranlage || Klärwerk
purified water Reinwasser
purify reinigen * Luft
purifying agent Reinigungsmittel
purifying machine Reinigungsmaschine * zum →Abscheiden von Fremdmaterial
purity Reinheit
purple violett
purpose Ziel * Zweck || Zweck
purposeful zielführend || zielgerichtet
purposes Zwecke
purposive gezielt * Maßnahmen usw. || zielgerichtet * zweckdienlich, zweckmäßig || zweckmäßig * zweckmäßig, zweckdienlich
pursuant to entsprechend * gemäß, zufolge, entsprechend, laut (einer Vorschrift, einem Gesetz usw.) || gemäß * gemäß, entsprechend, laut (z.B. e. Norm, Vorschrift, Gesetz) || in Übereinstimmung mit * Vorschrift
pursue betreiben * Beruf, Geschäft, Hobby, Studien, Angelegenheit usw. ~ || dranbleiben * eine Idee, Politik usw. verfolgen || fortsetzen || verfolgen * Ziel, Zweck, Plan, Weg, Kurs ~
pursuit Verfolgung * Verfolgung (-sjagd)
push anstoßen || drängen || schieben * stoßen, schieben (into:in), stecken, drängen || Schub * Stoß || stoßen * stoßen, schieben, treiben, drängen
push away verdrängen
push belt Schubgliederband * geschobene Kette zur Kraftübertragg., stufenlos verstellbares Übersetz.verhältnis
push open aufschieben * z.B. einen Schieber öffnen
push out ausstoßen
push rod Schubstange
push-button Drucktaste
push-lock rastend * "einrastend, wenn man es drückt", z.B. Rasttaste
push-on contact Steckhülse * "weiblicher" Steckkontakt zum Aufstecken
push-over Umstürzen
push-pull Gegentakt || Gegentakt- * -Verstärker, -Eingang, -Ausgang
pushbutton Druckknopf || Drucktaste * ['puschbatn] || Knopf * Drucktaster || Taster * Druckknopf
pusher Aufdrückvorrichtung || Schieber
pusher tool Aufdrückvorrichtung || Aufziehvorrichtung * Aufdrückvorrichtung, z.B. für Lager
pushing away Verdrängung * mit Gewalt
pushing forward Verfolgung * das Vorantreiben
pushing rod Stößel * siehe auch →Schubstange
pushlock-button Rasttaster
pushlock-twist-release button Rasttaster mit Drehentriegelung * z.B. Not-Aus Pilztaster
put setzen || stellen
put before vormachen * Brett usw.
put down festhalten * schriftlich ~ || herabsetzen * heruntersetzen, z.B. Preise, Ausgaben; hinstellen/setzen
put in einfügen
put in brackets einklammern

put in charge beauftragen
put in effect umsetzen * zur Wirksamkeit bringen
put in its proper place einordnen * an der richtigen Stelle ~
put in order aufräumen * ordnen || ordnen || regeln * ordnen
put in parentheses einklammern
put into versetzen * in eine Lage/einen Zustand || versetzt * in eine Lage/einen Zustand ~
put into bags füllen * in Beutel ~
put into force in Kraft setzen
put into operation in Betrieb setzen || in Gang setzen * Maschine ~, inbetriebsetzen || inbetriebnehmen * z.B. Elektromotor || inbetriebsetzen
put into service in Betrieb setzen || inbetriebnehmen * z.B. eines vom Lager entnommenen Motors
put into stock einlagern
put into the archives archivieren
put off aufhängen * Mäntel, Garderobe ablegen || aufschieben * verlegen * zeitlich verschieben (to: auf) || verschieben * zeitlich ~
put on ansetzen || anziehen * Kleider ~ || eingefallen * z.B. Bremse
put on its hinges einhängen * in Scharnier ~ (z.B. Tür)
put on market auf den Markt bringen
put on sale auf den Markt bringen
put out ausmachen * Feuer/Licht usw. ~ || mitteilen * herausgeben, z.B. offizielle Information einer Firma
put out of action abschalten || deaktivieren * außer Betrieb/Gefecht setzen
put out of operation außer Betrieb nehmen || außer Betrieb setzen
put over stülpen * auf/überstülpen
put through durchschalten * Signal usw. ~ || durchsetzen * durch/ausführen || durchstellen * (amerikan) jemanden am Telefon weitervermitteln
put to anlegen
put to account nutzen || nützen * nutzen
put together zusammensetzen
put up ansetzen * (mathem.) Gleichung || aufstellen
put up with hinnehmen * sich gefallen lassen, in Kauf nehmen || Kauf * etwas (mit) in ~ nehmen
putting in order Ordnung * das In-~-Bringen
putting into operation Inbetriebnahme * eine Maschine/Anlage || Inbetriebsetzung
putting into service Inbetriebsetzung
putty verkitten
puzzle out herausbekommen * Rätsel usw. ~
puzzling schwierig * question:Frage
PVC-insulated cable PVC-isolierte Leitung
PWM Pulsbreitenmodulation
PWM converter Pulsumrichter * mit Pulsbreitenmodulation
PWM frequency Pulsfrequenz * bei Pulsweitenmodulation
PWM Inverter Frequenzumformer * Um-/Wechselrichter mit Pulsweitenmodulation || Frequenzumrichter * Pulse Width Modulation, mit Pulsbreitenmodulation || Pulsumrichter * Wechselrichter mit Pulsbreitenmodulation || Pulswechselrichter * mit Pulsbreitenmodulation
PWM modulation Pulsweitenmodulation
pylon bracing Tragrohr
pyrolysis Pyrolyse

Q

Q factor Gütefaktor
q'ty Anzahl * *Kurzform für 'quantity'*
q'ty. Menge * *Abkürzg. für 'quantity': Anzahl*
qty. Stück * *(Abkürzung) ~zahl*
quad vierfach || Vierfach-
quadrangular quadratisch * *viereckig, vierseitig* || viereckig
quadrant -Quadrant || Quadrant * *['kwodrent]*
quadratic quadratisch * *mathematisch*
quadrature Phasenverschiebung * *~ um 90°* || um 90 Grad versetzt * *z.B. Rechteck-Impulssignale eines Impulsgebers*
quadrature pulses Rechteckimpulse * *2 um 90° versetzte Impulsketten m. Tastverhältnis 1:1 (z.B. e.Inkrementalgebers)*
quadripole Vierpol
quadruple vervierfachen * *z.B. Impuls-/Inkrementalgebersignale, Produktion usw. ~* || Vierfach-
qualification Abschluss * *~ einer (Berufs)Ausbildung* || Berechtigung * *Befähigung (for:zu)* || Eignung * *~ einer Person (for: zu)* || Qualifikation || Voraussetzung * *an Wissen/Können/Ausbildung*
qualification test Eignungstest * *für Bewerber, Personal* || Zulassungsprüfung * *z.B. bei einem Kunden zur Zulassung eines Lieferanten*
qualified fähig * *qualifiziert, geeignet, befähigt (for: für)* || kompetent * *qualifiziert, befähigt, fähig* || qualifiziert * *geeignet, befähigt (for:für/zu), berechtigt* || tüchtig * *qualifiziert, geeignet, befähigt (for: für)*
qualify absichern * *eine Aussage ~*
qualify as a university lecturer habilitieren
quality Ausführung * *Qualität* || Eigenschaft || Güte * *auch eines Kondensators/Filters* || Qualität * *poor/low:schlechte, high: gute (siehe ISO 9000 ... 9004 (EN 29000 ... 29004))* || Sorte * *Qualität*
quality assurance Gütesicherung || Qualitätssicherung * *siehe ISO 9000...9004 (EN 29000...29004)*
quality assurance system Qualitätssicherungssystem
quality audit Qualitätsaudit
quality certificate Qualitätszertifikat
quality class Prüfklasse
quality control Qualitätskontrolle * *"Qualitätslenkung/-steuerung"* || Qualitätssicherung * *(für die Antriebstechnik relevant: DIN ISO 90001)* || Qualitätsüberwachung
quality control regulations Qualitätssicherungsbestimmungen || Qualitätssicherungsrichtlinien
quality control system Qualitätsleitsystem
quality control technology Qualitätsleittechnik
quality defect Qualitätsmangel
quality factor Güte * *eines Kondensators/Filters* || Gütefaktor * *cos Phi x Eta (Leistungsfaktor x Wirkungsgrad) z.B. eines El.-Motors*
quality feature Qualitätsmerkmal
quality grade Qualitätsstufe
quality inspection Qualitätskontrolle * *Qualitätsprüfung*
quality loss Qualitätseinbuße * *Qualitätsverlust* || Qualitätsverlust
quality management Qualitätsmanagement * *siehe ISO 9000 ... 9004 (EN 2900 ... 29004)* || Qualitätssicherung * *siehe ISO 9000...9004 (EN 29000...29004)*
quality management system Qualitätssicherungssystem
quality of cut Schnittqualität
quality officer Qualitätsbeauftragter
quality regulations Qualitätssicherungsbestimmungen
quality specifications Abnahmevorschrift
quality standard Qualitätsstand
quality status Qualitätsstand
quality surveillance Qualitätsüberwachung
quality tests Qualitätskontrolle
quality-conscious qualitätsbewusst
quantified wertmäßig * *quantifiziert, quantitativ bestimmt*
quantify bestimmen * *quantitativ ~* || ermitteln * *quantitifizieren, quantitativ bestimmen*
quantitative wertmäßig * *['kwontitetiw] quatitativ, Mengen-*
quantities Größen || Stückzahlen
quantity Anzahl * *auch Stück Ware* || Betrag * *Menge* || Größe * *Menge, Höhe, Pegel, Wert e.Signals, auch Einflussgröße, physikal.Größe, Informationsmenge* || Maß * *Menge* || Menge * *Anzahl, Menge, Betrag (Plural: quantities)* || Stück * *~zahl* || Stückzahl * *for a quantity of: bei einer ~ von; (high-) volume quatities: hohe ~en*
quantity attribute Größenattribut * *(→PROFIBUS-Profil)*
quantity discount Mengenrabatt * *grant:gewähren*
quantity of grease Fettfüllmenge
quantization Quantisierung
quantization error Quantisierungsfehler * *z.B. auf Grund der beschränkten Auflösung eines Analog/Digital-Wandlers*
quarter viertel
quartz Quarz
quartz crystal Quarz || Schwingquarz
quartz generator Quarzgenerator
quarz crystal unit Schwingquarz * *als Komplettbauteil (im Gehäuse, mit Beschaltung usw.)*
quasi nahezu * *gleichsam, gewissermaßen, sozusagen*
quasi-continuous quasikontinuierlich * *z.B. digitaler Regler mit hoher Abtastrate* || quasistetig * *z.B. digitaler Regler mit hoher Abtastrate*
quasi-continuous-action controller quasikontinuierlicher Regler * *Digitalregler mit schneller Abtastung*
quatation phase Angebotsstadium
quench löschen * *Funken, Asche, Koks usw. ~*
quenching Löschung * *~ von Funken, Halbleiterventilen usw.*
quenching element Funkenlöschglied * *zur Beschaltung von Induktivitäten (z.B. bei Relais, elektromagnet. Bremse usw.)*
queries Rückfragen
query abfragen * *aus/befragen* || Frage * *(besonders 'anzweifelnde, unangenehme') Frage* || Nachfrage * *Rückfrage, anzweifelnde Frage* || nachfragen * *(aus/be)fragen, zweifelnd fragen, beanstanden, in Frage stellen, in Zweifel ziehen* || Rückfrage
question Anfrage || Frage * *ask/settle/pose a question: e.~ stellen/klären/aufwerfen; FAQ: →oft gestellte ~n* || Rückfrage * *Frage*
question mark Fragezeichen
questionnaire Fragebogen
questions Rückfragen * *Fragen*
queue Reihe * *Schlange* || verketten * *eine Schlange bilden* || Warteschlange * *[Computer] [kju:] z.B. für Rechnerbefehle/Tasks*
quick rasch || schnell
quick breaking Schnellabschaltung * *Stromkreis*
quick charge Schnellladung
quick coupling Schnellkupplung
quick lifting Schnellabhebung * *quick lifting of the roll: Walzen~*
quick lifting of the roll Walzenschnellabhebung
quick selection Schnellauswahl
quick stop aus * *Kommando z.schnellstmögl. Still-*

setzen e. Antriebs a.d. Momentengrenze *(AUS3 i.→PROFIBUS)* || AUS3 * Schnellhalt *(max.-Bremsmom.)*, Steuerbit im →*PROFIBUS-Profil 'Drehzahlveränderb. Antriebe'* || Schnellhalt
quick stopping Schnellabschaltung * *Motor, Maschine* || Schnellbremsung
quick-acting flink * *Sicherung* || schnell * *z.B. Sicherung* || Schnell- * *schnell arbeitend/auslösend, z.B. Schalter, Sicherung* || schnellansprechend * *z.B. Sicherung* || unverzögert * *schnell arbeitend (z.B. Schalter)*
quick-acting tripping device Schnellauslöser
quick-action Schnell- * *schnell arbeitend/auslösend, z.B. Schalter, Sicherung*
quick-action clamping device Schnellspanneinrichtung
quick-action fixture Schnellspanneinrichtung
quick-break switch Schnellschalter
quick-connect terminal Steckhülse * *Leitungsanschluss-Element*
quick-disconnect screw terminal Schraubsteckklemme
quick-disconnect screw terminal block Schraubsteckklemmenblock
quick-disconnect screw terminal strip Schraubsteckklemmenblock
quick-reference manual Schnellübersicht * *Kurzanleitung*
quick-release schnellansprechend * *Schutz/Überwachungseinrichtung*
quick-release catch Schnappverschluss || Schnellverschluss
quick-release clamping system Schnellspannsystem
quick-release coupling Schnellkupplung || Schnelltrennkupplung
quick-release lock Schnellverschluss
quick-response dynamisch * *hochdynamisch, schnell reagierend* || hochdynamisch
quick-response drive hochdynamischer Antrieb
quick-selection table Schnellauswahltabelle
quick-stopping time Schnellhaltzeit
quicker schneller
quickly growing schnell wachsend
quiescent current Ruhestrom * *[kwai'esnt] bei Halbleitern*
quiet geräuscharm || leise * *still* || ruhig * *still*
quiet running Laufruhe * *ruhiger Lauf* || ruhiger Lauf * *z.B. Motor*
quietness Laufruhe
quintessence Inbegriff * *Kern, Quintessenz, höchste Vollkommenheit*
quit aufhören * *(amerikan.)* || gehen * *aus dem Dienst ~* || verlassen * *gänzlich ~*
quite ganz * *(Adv.) völlig; quite another thing: etw.~ anderes; not quite th.same thing: nicht ~ dasselbe* || rein * *(Adv.) gänzlich* || völlig * *(Adv.)*
quite apart from the fact that abgesehen davon dass * *ganz abgesehen davon dass*
quite the contrary umgekehrt * *genau das Gegenteil*
quittance Quittung * *[Wirtschaft] ; auch Vergeltung, Bezahlung/Erlassen (einer Schuld)*
quiver zittern
quota Anteil * *Quote* || soll * *(Subst.) (Produktions-, Liefer-) Ziel; fixed: festgesetztes; production qu.: Produktionsziel*
quotation Angebot * *Preis/Verkaufs/Lieferangebot; generate a quotation: ~ ausarbeiten/erstellen* || Berechnung * *Preisstellung* || Preisangebot || Zitat
quotation stage Angebotsstadium
quotation-marks Gänsefüßchen
quote anbieten * *Preis ansetzen* || anführen * *zitieren* || angeben * *z.B. eine Größe in einer Maßeinheit, Schreibweise ~; in Newton: in Newton* || Angebot || ansetzen * *(Preis) angeben, berechnen* || nennen * *anführen*
quote from entnehmen * *zitieren*
quotient Quotient

R

RD FuE * *Research and Development:Forschung und Entwicklung*
R-C network RC-Glied
R.H. Luftfeuchtigkeit * *Abk. f. 'Relative Humidity': relative Luftfeuchtigkeit* || relative Luftfeuchtigkeit * *Abk. f.'Relative Humidity'*
r.m.s. Effektivwert
r.p.m. Drehzahl * *in [upm]* || upm
r.p.m. of the driven side Abtriebsdrehzahl
R/D converter Resolver-Digital-Umsetzer
rabbet falzen * *zusammenfugen* || Fuge * *Fuge, →Falz, Nut, Falzverbindung; rabbet plane: Falzhobel* || fugen * *einfügen, (zusammen)fugen [Holz], falzen*
rabble kratzen * *Metall*
race Laufbahn * *auch Kugel-/Rollenlaufbahn in Wälzlager* || Laufring * *in Wälzlager* || Rille * *Lauf-/Gleitbahn z.B. für Kugellager*
raceway Führungsschiene || Kabelpritsche || Laufring * *in Kugellager*
rack Baugruppenträger || Gerüst * *Gestell, Gerüst* || Gerüst * *Gestell, Gerüst, Schalt~; open frame: offenes Schalt~* || Gestell || Gestell * *auch 'Schaltgerüst'* || Halter * *Gerüst, Baugr.träger, Einb./Stützrahmen, Gestell/Ständer für Zeitschriften, Prospekte usw.* || Zahnstange
rack coupling Rahmenkopplung
rack link Rahmenkopplung * *Datenkopplung zwischen Baugruppenträgern; serial: serielle; parallel: parallele*
rack store Regallager
rack system Aufbausystem
rack-and-pinion drive Zahnstangenantrieb
rack-and-pinion gear Zahnstangengetriebe
rack-and-pinion railway Zahnradbahn
rack-interfacing module Rahmenkopplungsmodul * *zur Datenkopplung zwischen Baugruppenträgern*
racking vehicle Regalförderzeug
racy schnell * *(salopp) rasant*
radial radial
radial capacity diagram Querkraftdiagramm * *[el. Masch.] Verlauf d.zulässigen Querkraft über d. Länge des Wellenstummels*
radial clearance Lagerspiel * *~ in Radialrichtung* || Radialspiel
radial eccentricity Rundlauffehler
radial eccentricity tolerance Rundlauftoleranz * *erlaubter Konzentrizitätsfehler z.B. bei Motorwelle (siehe DIN 42955)*
radial fan Radialgebläse || Radiallüfter
radial force Querkraft * *Radialkraft (z.B. auf eine Motorwelle wirkend)* || Radialkraft
radial lead radial bedrahtet
radial line Stichleitung
radial load Radialkraft * *radiale Belastung* || Radiallast
radial misalignment Wellenversatz * *paralleler Fluchtfehler*
radial movement Radialbewegung
radial play Radialspiel
radial run-out Rundlaufabweichung

radial runout Rundlaufabweichung || Rundlauffehler
radial sealing ring Radialdichtring || Simmerring * *Radialdichtring*
radial shaft seal Radialdichtring || Simmerring || Wellendichtring
radial shaft seal ring Radialwellendichtring * *siehe DIN 3760* || Wellendichtring
radial shaft sealing ring Simmerring || Wellendichtring * *Radial-Wellendichtring*
radial slide bearing Radialgleitlager
radial slots Radialnuten * *z.B. an Kupplungs-/Bremslamelle*
radial-flow fan Radiallüfter
radial-lead radial bedrahtet * *z.B. Kondensator, dessen beiden Anschlussdrähte nach unten herausschauen*
radially radial * *(Adv.)*
radially leaded radial bedrahtet
radian Bogenmaß * *Maßeinheit für Winkel (1 rad entspr. 360 Grad/2 x Pi entspr. 57 Grad 17 min)* || Radian * *Winkelmaß/Bogenmaß; 1 rad entspr. 360 Grad/2 x Pi entspr. 57 Grad 17 min* || Radiant * *siehe* →*Radian*
radian frequency Kreisfrequenz
radiate abstrahlen
radiated emissions Störstrahlung
radiated heat Strahlungswärme
radiated noise Störstrahlung * *Störabstrahlung (siehe z.b. EN55011)*
radiation Abstrahlung * *Strahlung* || *Einstrahlung* || *Strahlung*
radiation influences Strahlungseinflüsse
radiation noise Störstrahlung
radiation protection Strahlenschutz
radiation treatment Bestrahlung * *Strahlenbehandlung*
radiator Heizung * *Heizkörper* || Kühler * *in Flüssigkeits-Kühlkreislauf*
radical radikal
radical change Umbruch * *radikale Änderung*
radically grundlegend * *(Adv.) grundlegend, von Grund auf, radikal* || völlig * *(Adv.)* →*grundlegend, von Grund auf, radikal*
radii Radien
radio and television interference suppression - Funkentstörung
radio frequency generated interference Funkstörung
radio frequency interference suppression Funk-Entstörung || Funkentstörung
radio inteference limit value Funkstör-Grenzwert
radio interference Funkstörung * *hochfrequente Störg.,die d.Funkempfang stören kann* || Funkstörung
radio interference filter Funk-Entstörfilter || Funkentstörfilter
radio interference level Funkstörgrad * *class B radio interference level: ~ B* || Funkschutzgrad
radio interference limit Funkstörgrenzwert
radio interference suppression Entstörung * *Funkentstörung* || Funk-Entstörung || Funkentstörung * *bei Elektromotor: siehe VDE 0875* || Störungsunterdrückung * *Funkentstörung* || Störunterdrückung * *Funkentstörung*
radio interference suppression capacitor Entstörkondensator * *zur Funkentstörung*
radio interference suppression devices Funkentstörmittel
radio interference suppression filter Funkentstörfilter
radio interference suppression level Funkstörgrad * *siehe VDE 0871, EN 55011*
radio interference suppression reactor Funkentstördrossel

radio interference voltage filter Funk-Entstörfilter || Funkentstörfilter
radio interference voltage test Störspannungsprüfung
radio receiver Funkempfänger
radio reception Funk-Empfang || Funkempfang
radio-controlled ferngesteuert * *funkferngesteuert*
radio-frequency Hochfrequenz-
radio-interference level Funkstörgrad * *siehe z.B. DIN VDE 0875 Teil 11 und EN 55011*
radio-interference suppressed funkentstört
radioactive radiation radioaktive Strahlung
radiograph röntgen || Röntgenbild
radius Radius || Reichweite * *Umkreis, Wirkgs.-/Einflussbereich; rad. of action: Aktionsradius, Fahrbereich (Verkehr)* || Strahl * *(Mathem.)* || Umfang * *Umkreis, Wirkungs/Einflussbereich*
radius of action Aktionsradius * *Reichweite* || Reichweite
radius of curvature Krümmungsradius
radius of gyration Trägheitsradius
radius of influence Reichweite
radix Basis * *Grundzahl einer potenzierten Größe*
rag Lappen * *z.B. Reinigungslappen*
rag bolt Maschinenfußschraube * *Steinanker/Bartbolzen z.B. zur Motorbefestigung im Maschinenfundament* || Steinschraube * *z.B. zur Motorbefestigung (auch "Steinanker, Bartbolzen")*
ragged gezackt
rags Hadern * *Lumpen (zur Papierherstellung)* || Lumpen
RAID RAID * *'Redundant Array of Inexpensive Disks": Speicher aus mehreren jeweils redundanten Festplatten* || Spiegelplattensystem * *'Redundant Array of Inexpensive Disks': Speicher mit redundanten Festplatten*
rail Leiste * *Schiene; guide rail:Führungsschiene* || Schiene * *guide rail: Führungs~*
rail car Fahrzeug * *(Bahn)* || Triebwagen * *(Bahn)*
rail guide Schienenführung
rail section Schienenprofil
rail traction motor Bahnmotor
rail vehicle Schienenfahrzeug
rail-car Triebwagen * *(Bahn)*
rail-car for local transportation Nahverkehrs-Triebwagen
rail-guided schienengebunden
rail-mount housing Aufschnappgehäuse * *zum Aufschnappen auf* →*Installationsschiene*
railed vehicle Schienenfahrzeug
railing Geländer
railroad Bahn * *(amerikan.) Eisen~* || Bahn- * *(Verkehr)*
railroad vehicle Schienenfahrzeug
rails Geländer
railway Bahn * *Eisen~* || Bahn- * *(Verkehr)*
railway signalling systems Eisenbahnsignaltechnik
rainfall Niederschlag * *in Form von Regen*
raise anheben * *emporheben, erhöhen* || aufbringen * *Geld* || aufheben * *hochheben; Maßnahme usw. ~* || bauen * *errichten* || beschaffen * *z.B. Kapital* || erhöhen * *steigern (auch Preise)* || heben * *auch durch Kran* || heraufsetzen || heben * *Kommando, Taste* || lüften * *(an)heben* || Steigerung * *Erhöhung, Steigung (auch Straße), Aufbesserung (z.B. Gehalt)*
raise button Höher-Taste || Höher-Taster * *z.B. für Motorpotentiometer*
raise key Höher-Taste
raise objection Einspruch erheben
RAISE pushbutton Höher-Taste
raised erhöht * *angehoben, gesteigert*
raised-head screw Linsenkopfschraube

raising Steigerung * Erhöhung, (Ver)Stärkung, Vergrößerung, Vermehrung
raising roller Kratzwalze * {Textil} bei →Kratzenraumaschine
rake Rechen * (Substantiv und Verb)
RAL RAL * Deut.Normeninstitut f. Gütesicherg. u. Kennzeichng., früher ReichsAusschus für Lieferbedingungen
RAL color RAL-Farbe * Farbton, der durch eine 4-stellige (internat. anerkannte) RAL-Nummer gekennzeichnet ist
RAM Arbeitsspeicher * RAM-Speicher || RAM * Abkürz. für 'Random Access Memory': Schreib/Lesespeicher || Stößel || stoßen * rammen
RAM memory Arbeitsspeicher * RAM-Speicher || Lese-Schreib-Speicher || Schreib-Lesespeicher * Abk. f. 'Random-Access Memory':Speicher mit wahlfreien Zugriff
rammer Stößel
ramming machine Rüttler * Stoßramme
ramp Anstiegsvorgang * entlang einer Rampe || Hochlaufgeber || Rampe
ramp back to zero stillsetzen * entlang einer Rampe auf Drehzahl Null fahren (lassen)
ramp begin rounding Anfangsverrundung
ramp delay Anstiegsverzögerungszeit
ramp down herunterfahren * entlang einer Rampe || herunterlaufen * entlang einer Rampe
ramp down to zero auf Drehzahl Null herunterfahren * entlang einer Rampe || stillsetzen * entlang einer Rampe ~
ramp enabling Hochlaufgeberfreigabe
ramp end rounding Endverrundung
ramp from run mode abfallende Hochlaufgeber-Rampe
ramp function Rampenfunktion
ramp function generator Anstiegsfunktion * Rampenbaustein, Hochlaufgeber-Funktionsbaustein
ramp function generator tracking Hochlaufgebernachführung
ramp generator Anstiegsbegrenzer * Hochlaufgeber || Hochlaufgeber
ramp response Anstiegsantwort * eines Regelkreises bei Rampe als Eingangsgröße
ramp response time Anstiegsverzögerung * bei Rampe als Eingangsgröße || Anstiegsverzögerungszeit * bei Rampe als Eingangsgröße
ramp start rounding Anfangsverrundung * bei Hochlaufgeber
ramp stop aus * Kommando zum Stillsetzen eines Antriebs über Hochlaufgeber (AUS1 im →PROFIBUS-Profil) || AUS1 * Steuerbit im →PROFIBUS-Profil; Motor über Rampe stillsetzen, dann elektr. AUS
ramp time Hochlaufzeit * Oberbegriff für Hoch/Rücklaufzeit
ramp to run mode ansteigende Hochlaufgeber-Rampe * für Drehzahlsollwert || ansteigende Rampe * e. Drehzahlhochlaufgebers
ramp up beschleunigen * entlang einer Rampe || hochlaufen * entlang einer Rampe ~
ramp-down verzögern * entlang einer Rampe ~ || zurücklaufen * entlang einer (Hochlaufgeber-) Rampe ~
ramp-down time Ablaufzeit * Rücklaufzeit eines Hochlaufgebers || Rücklaufzeit || Tieflaufzeit * eines Hochlaufgebers
ramp-function generator Hochlaufgeber
ramp-up hochfahren * entlang einer Rampe || Hochlauf * über eine (~geber-)Rampe
ramp-up integrator Hochfahrintegrator
ramp-up time Hochlaufzeit
ramping Hochlaufzeit
ramping up Hochfahrt * über eine (Hochlaufgeber-) Rampe

ramping-down Rücklauf * along a ramp: entlang einer Hochlaufgeber-Rampe
ramping-up Hochlauf * entlang Rampe
random beliebig * wahlfrei, wahllos, zufällig,Zufalls...,aufs Geratewohl || sporadisch * wahllos/von Fall zu Fall auftretend (z.B. Fehler), Zufalls- || wahlfrei * beliebig (z.B. Möglichkeit, auf alle Speicherplätze belieb. m.gleicher Geschw.zuzugreifen) || willkürlich * beliebig, wahllos, zufällig (statistisch gesehen) || zufällig * (statistisch) wahllos
random access wahlfreier Zugriff * mit gleicher Zugriffszeit auf alle Daten (to: auf)
random failure Zufallsausfall
random module insertion steckplatzunabhängig * Steckbaugruppe kann in e. beliebigen Einbauplatz eingeschoben werden
random noise Zufallsrauschen
random sample Stichprobe * Zufallsstichprobe
random test Stichprobenkontrolle || Stichprobenprüfung
random-number generator Zufallsgenerator
random-wound winding Träufelwicklung
range -Bereich || Baureihe || Bereich * in the range from x to y/in the x to y range: im ~ von x bis y; low/high: klein/großer || bewegen * in einer Spanne (z.B.Preise) || erstrecken || Hub * Bereich || Messbereich || Reichweite * bei Funkübertragg. usw..; long/medium range: große/mittl. ~; outrange: an ~ übertreffen || Reihe * Kollektion, Angebot, Auswahl, Bereich || Serie * Bau-/Produktreihe || Umfang * Bereich, Reichweite
range of adjustment Regelbereich
range of application Anwendungsbereich || Anwendungsgebiet * Anwendungsbereich || Einsatzbereich
range of applications Einsatzbereiche || Einsatzgebiete || Einsatzspektrum
range of models Modellreihe
range of products Lieferprogramm * lieferbare Produkte
range of rated voltages Bemessungsspannungsbereich * {el.Masch.}
range selection Bereichsumschaltung
range-up to erreichen * im (Leistungs- usw.) Bereich
rank einordnen * im Rang ~ (among:unter) || Position * Rang || rangieren * im Rang; before:vor; with mit/unter || stehen * im Rang ~ || Stufe * Rang
ranking Rangfolge
rap klopfen
rapid rapid || rapide || schnell || Schnell-
rapid advance Schnellgang
rapid braking Schnellbremsung
rapid excitation Schnellerregung * z.B. bei elektromagnet. Bremse/Kupplung mit erhöhter Spannung und Vorwiderstand
rapid feed Eilgang * {NC-Steuerung}
rapid movement Schuss * rasende Bewegung
rapid shutdown Schnellabschaltung
rapid taverse Eilgang * {NC-Steuerung}
rapid traverse Schnellgang
rapidly rapid * (Adv.) || rapide * (Adv.) || schnell * (Adv.)
rare selten
rare-earth magnet Seltene-Erden-Magnet * z.B. aus →Samarium-Kobalt (SmCo) für d. Läufer e. permanenterregten E-motors || Seltenerdmagnet
rash unvorsichtig * übereilt, unüberlegt
rasp Feile * grobe || kratzen * Geräusch || schaben
ratchet Knarren- * mit Sperrklinke versehen, z.B. Schraubenschlüssel/-dreher || Rechen
rate angeben * with: einen Wert für eine Kenngröße e.Produkts ~/zusichern || ansetzen * abschätzen || auslegen * z.B. Stromrichter/Motor ~ || bemessen

rate as

* *z.B. Stromrichter/Motor (bezüglich des Nennwertes) auslegen* || bemessen * *Leistung bemessen; abschätzen* || *beurteilen* * *Leistung (e. Mitarbeiters, von Produkten im Vergleich usw.), bewerten, (Arbeits)Qualität* || bewerten * *klassifizieren* || dimensionieren * *z.B. Stromrichter/Motor auslegen* || Frequenz * *Häufigkeit des Auftretens von Ereignissen* || Geschwindigkeit * *Ereignisse/Menge pro Zeiteinheit* || Grad * *Quote, Maßstab, Verhältnis, Ziffer* || Häufigkeit * *Ereignisse pro Zeiteinheit* || Leistung * *z.B. Motor~* || Maß * *Verhältnis* || Menge * *~ pro Zeiteinheit* || Rate * *Ereignisse/Wert pro Zeiteinheit* || Stand * *Markt-Kurs/-Preis* || Steilheit || Strom * *Menge pro Zeiteinheit* || Tarif * *freight rate: Frachttarif*
rate as gelten als
rate limitation Anstiegsbegrenzung
rate multiplier Frequenzteiler * *Frequ.multiplizierer (lässt nur e.bestimmten %-Anteil d.Eingangsimpulse passieren)* || Impulsteiler * *Impulsmultiplizierer (läßt einen einstellbaren Anteil d. Eingangsimpulse passieren)*
rate of assessment Steuersatz * *{Wirtschaft}*
rate of change Änderungsgeschwindigkeit || Steilheit * *z.B. einer Signalflanke*
rate of exchange Kurs * *Wechsel~*
rate of flow Durchflussmenge * *pro Zeiteinheit*
rate of interest Prozent * *Prozent Kapitalzins* || Zins * *Prozent Kapital~*
rate of rise Änderungsgeschwindigkeit * *Anstiegsgeschwindigkeit* || Anstiegsgeschwindigkeit || Anstiegssteilheit || Anstiegszeit || Steilheit * *z.B. einer Impulsflanke*
rate of voltage rise Flankensteilheit * *eines Spannungsverlaufes* || Spannungsanstieg * *Anstiegsgeschwindigkeit (du/dt)*
rate of wear Abnutzung * *bezogen auf die Betriebs/Nutzungszeit*
rate of yield Gewinn * *prozentualer ~ (Ergebnis bezogen auf den Umsatz)*
rate per hour Stundensatz
rate plate Typenschild * *Leistungsschild*
rate regulator Mengenregelung * *Mengenregeler*
rate time Vorhaltezeit
rate-of-change limiting Ruckbegrenzung
rate-of-rise Anstiegssteilheit
rated bemessen * *(Adj.)* || Bemessungs- * *Nenn-* || dimensioniert * *von der Baugröße her* || Nenn- * *z.B. rated current: Nennstrom*
rated AC current Nennwechselstrom
rated AC insulation voltage Nennisolationswechselspannung
rated air gap Bemessungsluftspalt * *bei einer Bremse usw.*
rated alternating current Nennwechselstrom
rated apparent power Bemessungsscheinleistung
rated armature current Ankernennstrom || Bemessungs-Ankerstrom
rated armature voltage Ankernennspannung || Bemessungs-Ankerspannung
rated conductor size Nenn-Anschlussquerschnitt
rated connecting capacity Nenn-Anschlussquerschnitt
rated connection cross-section Nenn-Anschlussquerschnitt
rated contact loading Kontaktbelastbarkeit
rated continuous current Nenndauerstrom
rated converter current Umrichternennstrom
rated converter supply voltage Gerätenennanschlussspannung * *eines Stromrichters*
rated current Bemessungsstrom || Nennstrom
rated current of converter Gerätenennstrom * *eines Stromrichters*
rated DC current Bemessungsgleichstrom || Nenngleichstrom

rated DC voltage Nenngleichspannung
rated direct current Nenngleichstrom
rated duty Bemessungsbetrieb * *{el.Masch.}* Betrieb mit allen vom Hersteller festgelegten Bemessungsgrößen
rated field current Bemessungsfeldstrom
rated frequency Nennfrequenz
rated fusing current Sicherungsnennstrom
rated input power Nennantriebsleistung * *eines Getriebes*
rated insulation voltage Isolationsbemessung * *~sspannung einer Wicklung* || Nennisolationsspannung
rated load Bemessungslast * *at rated load conditions: bei ~* || Nennlast * *under: bei* || Vollast * *Nennlast*
rated load torque Nenndrehmoment * *Nenn-Lastmoment*
rated motor amps Motornennstrom
rated motor current Motor-Nennstrom || Motornennstrom
rated motor frequency Motornennfrequenz
rated motor impedance Motor-Nennimpedanz
rated motor output Motornennleistung
rated motor power Motornennleistung
rated motor slip Motornennschlupf * *((Synchrondrehzahl-Motornenndrehzahl)/Synchrondrehzahl) x 100%* || Nennschlupf * *((Synchrondrehzahl-Motornenndrehz.)/Synchrondrehzahl) x 100%*
rated motor speed Motornenndrehzahl
rated motor torque Motornennmoment
rated motor voltage Motornennspannung
rated output Bemessungsleistung * *z.B. eines Motors, Stromrichters usw.* || Nennleistung * *z.B. e. Stromrichters; Nennwellenleistung e.Motors (siehe DIN 42673, IEC72 u.VDE 0530)*
rated output capacity Ausgangsnennleistung * *auch eines Stromrichters*
rated point Nennpunkt
rated power Bauleistung * *z.B. eines Transformators* || Bemessungsleistung || Nennleistung
rated power factor Nennleistungsfaktor
rated quantity Bemessungsgröße
rated slip Motornennschlupf * *ca. 10% bei kleinen bis ca 0,3% bei großen Asynchronmotoren* || Nennschlupf * *b. kleinem bzw. großem Asynchronmotor ca. 10% bzw. 0,3% (s. auch →Motornennschlupf)*
rated slip frequency Schlupfnennfrequenz
rated speed Bemessungsdrehzahl || Betriebsdrehzahl * *Nenndrehzahl z.B. eines Motors* || Nenndrehzahl || Nenngeschwindigkeit
rated supply voltage Nennanschlussspannung * *z.B. für Motor, Umrichter*
rated thermal capacity Wärmekennwert * *spezifizierte Wärmebelastbarkeit einer Kupplung, Bremse*
rated torque Bemessungsmoment || Nenndrehmoment || Nennmoment
rated transformer voltage Trafo-Nennspannung
rated value Bemessungsgröße || Bemessungswert || Nennwert
rated values Nenndaten
rated voltage Bemessungsspannung * *für Motoren: siehe VDE 0530 und EN 60034-1* || Nennspannung || Normspannung * *siehe z.B. DIN 40030 für DC-Motoren*
rated wire gauge Nenn-Anschlussquerschnitt
rated wire range Nenn-Anschlussquerschnitt
rather beziehungsweise * *eher* || bzw * *eher* || einigermaßen * *ziemlich* || etwas * *(Adv.)* || reichlich * *(Adv.) ziemlich* || vielmehr || ziemlich
rather long länger * *ziemlich lang*
rating Auslegung || Belastbarkeit * *Nennleistung/-Strom/-Drehmoment* || Bemessung * *festgelegter Nennwert* || Bemessungsdaten * *continuous rating:*

Bemessungsdaten für Dauerbetrieb || Beurteilung * *Leistung (auch eines Mitarbeiters), Wert* || Bewertung * *(Ab)Schätzung, Beurteilung, (Zeugnis)-Note, Leistungsbeurteilung e.Person* || Dimensionierung || Leistung * *z.b. Nenn~ eines Stromrichters, Stellung im Produkt/~sspektrum* || Nennleistung || Nennwert
rating class   Belastungsklasse * *Betriebsart eines Stromrichters (definierte Lastspiele), siehe DIN 41756* || Betriebsart * *Belastungsklasse/Betriebsart für Stromrichter, siehe DIN 41756*
rating code designation   Leistungskennzeichen * *für Thyristor-/Diodensatz nach DIN 41752*
rating criterium   Auslegungskriterium * *für die Leistung (z.b. des Motors/Stromrichters; Plural: criteria)*
rating data   Nenndaten
rating of enclosure   Schutzart
rating of machines for variable values   Bemessung für Maschinen mit veränderlichen Größen * *[el.Masch.]Bem.größen dürf. mehrere Werte annehm.*
rating of machines with several speeds   Bemessung fur Maschinen mit mehreren Drehzahlen * *[el.Masch.] es gibt mehrere Bemessungsdrehzahlen*
rating plate   Leistungsschild * *für Motor siehe IEC 34-1 und VDE 0530* || Typenschild
rating plate data   Typenschildangaben || Typenschilddaten
rating plate indications   Typenschildangaben || Typenschilddaten
rating range   Leistungsbereich * *Bereich der Nennleistung(en), z.b. einer Stromrichter-Baureihe*
ratings   Leistungsdaten || Nenndaten
ratio   -Verhältnis || Anteil * *Verhältnis* || Quotient * *Verhältnis* || Rate * *Verhältniszahl* || Relation * *Verhältnisfaktor* || Relationsfaktor || Übersetzung * *(~s-) Verhältnis, z.b. Getriebe/Trafo/Stromwandlerübersetzung* || Übersetzungsverhältnis * *Trafo; one-to-one ratio: Übersetzungsverhältnis 1:1* || Verhältnis * *of...to:von...zu*
ratio control   Verhältnisregelung * *auch bei Winkelgleichlaufregelung mit 'Impulsübersetzung'*
ratio factor   Relationsfaktor
ratio of the number of poles   Polzahlverhältnis * *z.B. bei Maschinensatz*
rational   rationell || vernünftig * *vernunftsmäßig, rational, vernunftbegabt*
rationalization   Rationalisierung * *rationalization measures: ~smaßnahmen*
rationalization measures   Ratiomaßnahmen
rationalizing process   Rationalisierung * *~sprozess*
rattle   flattern * *rattern, klappern (z.B.Relais)*, rasseln, rütteln || Klang * *eines Papierwickels (beim 'Draufklopfen' hörbar)* || rattern
raw   roh * *unbearbeitet, wie aus der Natur gewonnen* || Roh-
raw coal   Rohkohle
raw data   Rohdaten
raw material   Rohstoff
raw signal   Rohsignal || Rohwert * *Rohsignal*
raw water   Rohwasser
raw-signal encoder   Rohsignalgeber || Sinus-Cosinus-Geber * *Rohsignalgeber (gibt keine Rechteckimpulse, sondern sin-/cos-Signale aus)*
Rawcliff winding   PAM-Wicklung * *nur 1 Wicklg.i. polumschaltb.Motor; Stromumkehr i.einz.Spulengruppen ändert Drehz.* || Rawcliff-Wicklung * *für polumschaltbaren Motor ähnlich Dahlanderschaltung (siehe 'PAM-Wicklung)*
ray   Strahl
rayon staple   Zellwolle
rays   Strahlung * *Strahlen*
raytracing   Raytracing * *Verfahren z.b.Berechnung*

v. *3D-Bildern unter Berücksichtigg. des Lichteinfalls (Schatten)*
RC circuit   RC-Beschaltung || Schutzbeschaltung * *RC-Beschaltung*
RC damping element   RC-Löschkombination * *z.b. zur Beschaltung einer Schützspule z. Vermeidung von Spannungsspitzen*
RC element   RC-Beschaltung || RC-Glied
RC snubber   RC-Beschaltung * *als →TSE-Beschaltung einer Thyristorschaltung*
RC snubber circuit   TSE-Beschaltung || TSE-Schutzbeschaltung
RC surge suppression element   RC-Löschkombination
RCD   Fehlerstromschutzeinrichtung * *Abk. für Residual Current Device; siehe auch →FI-Schutzschalter*
RDC   RDC * *[Papier] Recommended Drive Capacity: empfohl. Antriebsleistg. (im stationär.Betrieb nach TAPPI)*
re-   erneut * *als Vorsilbe für Verb* || erneutes * *Vorsilbe für Substantiv/Verb* || Nach- || neu * *als Vorsilbe (er~t, wieder-); auch 'um-'* || Neu- || Rück- * *wieder (zurück)* || Um- * *anders herum, neu, anders, wieder, zurück* || wieder- * *Vorsilbe für Substantiv/Verb* || zurück-
re-adjust   nachregeln
re-adjustable   nachstellbar
re-arrange   umarrangieren
re-arrangement   Umbau * *Anders-Anordnen*
re-balance   nachwuchten
re-balancing   Nachwuchtung
re-boot   neustarten * *einen Computer ~*
re-calculate   nachrechnen
re-charging current   Umladestrom * *z.b. bei einer langen Leitung auftretend*
re-check   Kontrolle * *nochmalige ~* || überprüfen * *nochmalig ~*
re-closure   Wiedereinschalten * *z.B. Stromkreis nach Wartungsarbeiten*
re-cooling system   Rückkühlanlage * *z.b. bei wassergekühltem Stromrichter*
re-design   Redesign
re-designate   umbenennen
re-edit   bearbeiten * *Buch ~* || überarbeiten * *Schriftstück ~/neu bearbeiten/neu herausgeben* || umarbeiten * *Schriftsatz ~*
re-editing   Überarbeitung * *Neubearbeitung eines Druckerzeugnisses usw.*
re-employ   wiederverwenden
re-enter   neu eingeben || wiedereintreten
re-equip   umrüsten * *z.B. Werkzeugmaschine*
re-equipping time   Umrüstzeit * *z.B. für eine (Textil-)Maschine*
re-establish   wiederherstellen * *z.B. einen Zustand, eine Datenverbindung*
re-establishing   Wiederherstellung * *einer Verbindung usw.*
re-import   wiedereinführen * *(aus dem Ausland)*
re-initialization   Wiederanlauf * *Programm*
re-measure   nachmessen * *erneut messen*
re-occur   wiederkehren * *wieder auftreten, wieder vorkommen*
re-ordering   Nachbestellung * *das Nachbestellen*
re-organization   Umorganisation
re-organize   umorganisieren
re-reeler   Umroller || Vorroller
re-set   umbauen * *(Maschine, Anlage usw.) umrüsten, neu einrichten*
re-start   neustarten || Wiederanlauf || Wiedereinschalten * *Neustart*
re-start lockout   Wiedereinschaltsperre * *für Antrieb*
re-stretch   nachspannen * *Riemen, Kette usw.*
re-tighten   nachspannen * *Riemen, Kette usw.*

re-usable Mehrweg-
re-usable packaging Mehrwegverpackung * *(Verpack.technik)*
re-use wiederverwenden
re-wire umverdrahten
reach anfahren * erreichen || erreichen * *z.B. einen Wert, ein Ziel* || erstrecken || erzielen * *Verständigung usw.* ~ || reichen * *to: bis* || Reichweite * *within reach/near at hand: in ~; medium reach: mittlere ~; out of reach: außer ~*
reach of screw Einschraubtiefe
reach of supply Reichweite * *~ von Lagerbeständen (stocks are exhausted: die Vorräte sind erschöpft)*
reachable erreichbar * *z.B. am Telefon*
reached erreicht
reaching erreichen * *(Subst.) Erreichung*
react antworten * reagieren || reagieren * *on:auf*
reactance Blindwiderstand || Reaktanz * *Blindwiderstand, induktiver Widerstand*
reaction -Wirkung * *Gegenwirkung* || Antwort * *Reaktion* || Reaktion * *to:auf* || Rückwirkung || Wirkung * *(Gegen)Reaktion*
reaction force Rückstellkraft
reaction on system Netzrückwirkung
reaction time Reaktionszeit
reactivate reaktivieren
reactive Blind- * *(elektr.)*
reactive current Blindstrom
reactive current compensation plant Blindstromkompensationsanlage || Kompensationsanlage * *Blindstrom~* || Kompensationseinrichtung
reactive current compensation system Blindstromkompensationsanlage
reactive power Blindleistung
reactive power compensation Blindleistungskompensation
reactive power consumption Blindleistungsaufnahme
reactive power due to the phase-angle control Steuerblindleistung * *~ auf Grund der Phasenanschnittsteuerung*
reactive power for commutation Kommutierungsblindleistung
reactive-current compensation Blindstromkompensation * *durch Kondensatoren in der Nähe induktiver Verbraucher*
reactive-free rückwirkungsfrei
reactive-power compensation equipment Kompensationsanlage * *zur Blindleistungskompensation*
reactive-power compensation system Kompensationsanlage
reactive-power component Blindleistungsanteil
reactive-power demand Blindleistungsbedarf
reactor Drossel * *größere* || Drosselspule * *siehe VDE 0532; größere* || Spule * *(elektr.) Drossel (eher größere)*
read ablesen * *Skala* || anzeigen * *Ablesewert an einem Messgerät ~* || gelesen || lauten * *Inhalt, Worte, Text; vor Doppelpunkt* || lesen
read access Lesezugriff * *z.B. auf Speicher*
read cycle Lesezyklus
read in einlesen
read off ablesen * *Messgerät, Anzeigewert usw.* ~
read out ablesen * *Messgerät, Anzeigewert usw.* ~ || ausgegeben * *ausgelesen* || auslesen * *z.B. Daten* ~
read the proofs korrekturlesen * *den Vorabzug/die Druckfahnen* ~
read-off Ablesung * *eines Mess/Anzeigewerts*
read-only schreibgeschützt
read-only memory Festwertspeicher || Nur-Lese-Speicher

read-only parameter Anzeigeparameter * *Nur-Lese-Parameter* || Beobachtungsparameter * *Nur-Lese-Parameter*
read-out Anzeige * *Ablesewert* || auslesen
readability Ablesbarkeit * *z.B. einer Anzeige* || Sichtbarkeit * *(Ab)Lesbarkeit*
readable lesbar * *lesenswert, leserlich, lesbar*
readaption Umarbeitung
readiness Bereitschaft * *for:zu*
readiness for action Einsatzbereitschaft
readiness for delivery Liefereinsatz
readiness for production Fertigungsreife
readiness for service Einsatzbereitschaft
reading Anzeige * *Ablesung e.Instruments* || Anzeigewert || Ausschlag * *Anzeigewert eines Messgeräts* || Messung * *Ablesung, Ablesung* || Messwert * *Ablesewert*
readjust nachstellen * *Lager usw.*
readout Ablesung * *von e. Messgerät* || Anzeige * *Ablesewert, z.B. von einem Messinstrument*
readout value Anzeigewert
ready bereit * *be ready for:bereit sein zu/für* || betriebsbereit || fertig * *bereit*
ready for -bereitschaft
ready for action einsatzbereit
ready for connection anschlussfertig
ready for energizing Einschaltbereit
ready for operation betriebsbereit || einsatzbereit
ready for service einsatzbereit
ready for use funktionsfähig * *gebrauchstüchtig/-fertig, funktionsbereit*
ready on hand greifbar * *zur Hand*
ready to go anfahrbereit || betriebsbereit || startbereit
ready to hand griffbereit || zur Hand
ready to help hilfsbereit
ready to run ablauffähig * *z.B. Softwareprogramm* || anfahrbereit || lauffähig * *z.B. Softwareprogramm* || startbereit
ready to send Sendebereitschaft * *als Signalname (z.B. "RTS" bei RS232-Schnittstelle)*
ready to start anfahrbereit || Einschaltbereit * *Motor* || Einschaltbereit * *Motor, Maschine*
ready-built fertig * *vorgefertigt*
ready-made konfektioniert || vorfabriziert
ready-to-connect anschlussfertig
ready-to-go lauffähig * *z.B. Maschine*
ready-to-run betriebsbereit || einsatzbereit * *startbereit*
ready-to-run unit Komplettgerät * *direkt lauffähiges Gerät*
ready-to-start betriebsbereit * *Motor*
ready-to-switch-on Einschaltbereit
ready-wired anschlussfertig
real eigentlich * *wirklich* || existent || natürlich * *real size:natürliche Größe* || ordentlich * *wirklich* || rein * *eigentlich, tatsächlich, wirklich, wahr* || sachlich || tatsächlich * *wirklich* || wahr * *echt*
real component Realteil * *(Mathem.) bei komplexen Zahlen*
real estate Grund * *~ u. Boden*
real life Praxis * *in real life: im wirklichen/praktischen Leben* || Wirklichkeit
real resistance Wirkwiderstand
real service tatsächlicher Betrieb
real time echtzeitfähig * *z.B. Rechner, Betriebssystem*
real time clock Systemzeit * *Echtzeituhr*
real time processing Echtzeitverarbeitung
real-time Echtzeit * *Rechnerfunktionen mit schnellen u.vorherbestimmbaren (deterministischen) Reaktionszeiten*
real-time clock Echtzeit * *Echtzeituhr* || Echtzeituhr || Zeitgeber * *Echtzeituhr*
real-time control Echtzeitregelung

real-time operating system Echtzeit-Betriebssystem || Echtzeitbetriebssystem
real-time quantity Echtzeitgröße
realistic realistisch || sachlich * nüchtern || sinnvoll * realistisch, wirklichkeitsnah, sachlich
reality Realität || Wirklichkeit
realizable realisierbar
realization Ausführung * Verwirklichung eines Plans, Konzepts || Durchführung * Realisierung, Verwirklichung, Ausführung || Realisierung || Umsetzung * Realisierung || Zustandekommen * Realisierung, Verwirklichung, Aus/Durchführung
realize ausführen * Plan/Konzept ~/verwirklichen || bewerkstelligen || durchsetzen * realisieren || erfassen * erkennen || erkennen * sehen, begreifen, erfassen, wahrnehmen || erreichen * realisieren, wahr machen || erzielen * Gewinn usw. ~ || herstellen * realisieren || merken * erkennen || realisieren || umsetzen * realisieren
realized realisiert
really eigentlich * (Adv.) tatsächlich || tatsächlich * (Adv.) wirklich || wirklich * (Adv.)
realm Gebiet * Reich, Sphäre, Bereich, (Fach-)~
realtime-capable echtzeitfähig
reamer Dorn * Reib~, Reibahle, Räum~, Räumahle || Räumer * Reib-/Räumahle/Räumnadel/-plutte zur Metallbearbeitung || Reibahle
reaming bit Reibahle * zum Einsetzen in Werkzeughalter
reap profitieren * profitieren von, ernten
rear hinter * (Adj.) || Hinter- * hinten angeordnet, hintenliegend || Rück- * Hinter-, hinten angeordnet, hintenliegend || Rückseite * on/at the rear: auf der ~ || rückseitig * rückwärtig * auf der Rückseite (befindlich)
rear drum Tragwalze 1 * eines Doppeltragwalzenrollers
rear panel Rückseite * Platte/Wand
rear panel connector Rückwandstecker
rear panel wiring Rückwandverdrahtung * z.B. eines Baugruppenträgers
rear shaft zweites Wellenende * hinteres Wellenende
rear view Ansicht von hinten || Rückansicht
rear-panel Rückwand * z.B. eines Schaltschranks, Geräts usw.
rear-side rückseitig
rearmost hinten * am Ende
rearrange umgestalten * anders/neu anordnen || umrangieren * anders anordnen || umsetzen * neu anordnen || umstellen
rearwards nach hinten * nach hinten zu
reason Grund * Ursache, Vernunfts~; for this reason: aus diesem ~; not unreasonably: nicht ganz ohne ~ || Sinn * Grund || Ursache * Grund || Veranlassung * Grund || Zweck * Grund
reasonable angemessen * vernünftig || annehmbar * vernünftig || ausreichend * angemessen, zumutbar, annehmbar || hinreichend * angemessen, zumutbar, annehmbar || sinnvoll * vernünftig, angemessen || tragbar * zumutbar || vernünftig * vernunftsgemäß, angemessen || vertretbar * annehmbar, angemessen, tragbar, vernünftig || zumutbar
reasonably priced preisgünstig
reasons Erklärung * Gründe || Gründe * for/because of technical reasons: aus technischen ~n
rebalance nachwuchten
rebate Abzug * Rabatt || Falz * {Holz} || falzen * zusammenfügen || fugen * einfügen, (zusammen)-fugen {Holz}, falzen || Preisnachlass * Rabatt || Rabatt
rebating cutter Falzhobel * {Holz}
rebound prellen * zurückprallen/schnellen
rebound crusher Prallbrecher

rebuild umbauen * umbauen, wiederaufbauen, wiederherstellen (figürl.)
rebuilding Umbau
recalculation Umrechnung * z.B. einer Gleichung
recall Abberufung || abrufen || reaktivieren * Person ~ (to service: zum Dienst) || Rückruf * z.B. am Telefon
recapitulate zusammenfassen * noch einmal ~; rekapitulieren
recast umgestalten
receipt Beleg * [ri'ßiet] Quittung || bescheinigen * z.B. mit Hilfe einer Quittung ~ || Empfang * [re'-ßiet] Erhalt (von Sendungen, Briefen usw.) || Erhalt * [ri'ßiet] Empfang e. Sendung, Brief || Quittung * [ri'ßiet] Empfangsbescheinigg., Qittung(on/against:gegen); sales receipt:Kassenzettel
receipt of orders Auftragseingang * [ri'ßiet] Empfangen, Erhalt von Aufträgen || Auftragserhalt * [ri'ßiet ...]
receipt of payment Zahlungseingang
receipts Einnahme * (Plural) [ri'ßiets] Geschäftseinnahmen || Zugang * Einnahmen
receive annehmen || aufnehmen * empfangen || bekommen || empfangen || erhalten * empfangen, bekommen || übernehmen * empfangen
receive block Empfangsbaustein * Funktionsbaustein
receive channel Empfangskanal
receive character Empfangszeichen
receive data Empfangsdaten
receive direction Empfangsrichtung
receive note Eingangsvermerk * z.b. auf einem Brief
receive the order Auftrag erteilt bekommen
Received Eingangsvermerk * z.B. auf einem Brief
received characters Empfangszeichen
receiver Abnehmer * Empfänger || Eingangs- || Empfänger || Empfangsbaustein
receiver block Empfangsbaustein * Funktionsbaustein
receiver machine Empfängermaschine * bei elektr. Welle
receiving aerial Empfangsantenne
receiving channel Empfangskanal
recent kürzlich * (Adj.) || neu * neu, jung, frisch, modern; kürzlich/vor kurzem/unlängst (geschehen/entstanden)
recently kürzlich * (Adv) || neuerdings
receptacle Behälter || Buchse * (Netz-)Steckdose, (Kaltgeräte-/Bananenstecker-) Buchse || Buchsenstecker || Gefäß * Behälter, Gefäß; Steckdose || Halterung * Aufnahme (für Stecker, Behälter ..), Fach, Fassung, (Sammel-) Gefäß, Steckdose || Steckdose || Steckplatz * Aufnahmeöffnung/-schacht, z.b. für Speichersubmodul bei SPS
reception Anmeldung * Empfangsbüro || Annahme * Entgegennahme || Aufnahme * Empfang || Empfang * z.B. Radioempfang, Empfang einer Nachricht, eines Telegramms
recess absetzen * aussparen, einsenken, vertiefen, ausbuchten, zurücksetzen || Ausschnitt * Aussparung, Vertiefung, Einschnitt || Aussparung * Aussparung, Vertiefung, Einschnitt || Ecke * Aussparung, Vertiefung, Einschnitt, Nische, das Innere || einlassen * aussparen, einsenken, vertiefen (z.B. Fläche für Möbelbeschlag) || Nische * Vertiefung, Aussparung, Einschnitt, (versteckter) Winkel, Nische || Vertiefung * Aussparung, Einschnitt, Ausbuchtung, Einsenkung
recession Geschäftsrückgang * Rezession * Wirtschaftskrise || Rückgang * Rezession, (Konjunktur)Rückgang
recession in prices Preisrückgang
recharge erneuern * wieder auffüllen; with grease: Fett ~ || umladen * z.B. Kondensator

rechargeable aufladbar || wiederaufladbar
recipe Rezept || Rezeptur
recipe administration Rezeptverwaltung
recipe control Rezeptsteuerung || Rezeptursteuerung
recipe for success Erfolgsrezept
recipe handling Rezeptverwaltung
recipe management Rezeptverwaltung
recipient Empfänger
reciprocal gegenseitig * wechsel-/gegenseitig, umgekehrt, reziprok || reziprok || umgekehrt * reziprok, umgekehrt, wechsel-/gegenseitig || wechselseitig
reciprocal action Wechselwirkung * gegenseitige
reciprocal in proportion umgekehrt proportional
reciprocal value Kehrwert || Reziprokwert
reciprocally gegeneinander * gegenseitig, wechselseitig, umgekehrt, reziprok || reziprok * (Adv.)
reciprocate sich hin und herbewegen * z.B. Kolben
reciprocating Hin- und Herbewegung
reciprocating compressor Hubkolbenverdichter || Kolbenkompressor || Kolbenverdichter
reciprocating engine Kolbenmaschine
reciprocating pump Kolbenpumpe
reciprocating saw Gattersäge
reciproce pendeln * sich hin- und herbewegen (z.B. Kolben)
recirculate zurücklaufen * Flüssigkeit in Kreislauf
recirculated air Umluft
recirculation Rücklauf * Flüssigkeit, Gas
reckless fahrlässig
recklessness Fahrlässigkeit
reckon rechnen * (be-/er-) rechnen; reckon in: ein~; reckon over: nach~.; reckon with: ~ mit
reckon out ausrechnen * (auch figürl.) || errechnen
reckon with rechnen * mit etwas Zukünftigem ~
reclaim wiedergewinnen * aus Altmaterial ~, regenerieren; zurückbringen/-führen (from: von/aus; to: zu)
reclamation Reklamation
reclosing lockout Wiedereinschaltsperre * für einen freigeschalteten Stromkreis z.B. bei Wartungsarbeiten
recognize anerkennen * as:als; auch durch eine Normenbehörde || erkennen * auch 'wahrnehmen, sehen'
recognized approbiert * anerkannt z.B. bei Normenbehörde UL, CSA usw.
recoil zurücklaufen * zurückwickeln/spulen (Blech, Draht, Kabel, Seil)
recoiler Aufhaspel * Wiederaufhaspel, z.B. bei Metallband-Behandlungsanlage || Aufwickler * Wiederaufhaspel für Blech, Draht, Kabel
recombination Rekombination * Neutralisierung von Ladungsträgern in Halbleiter
recommend empfehlen || vorschlagen * empfehlen
recommendation Empfehlung || Hinweis * Empfehlung || Vorschlag * Empfehlung
recommended empfohlen || nötig * empfohlen || Vorzugs- * empfohlen
recommended value Daumenwert || Leitwert * empfohlener Wert, Richtwert || Richtwert * empfohlener Wert
recommissioning Wiederinbetriebnahme
recompilation Rückübersetzung * von Hochspracheprogrammen
recompile rückübersetzen * von Hochspracheprogrammen
recompiler Rückübersetzer * für Hochspracheprogramme
reconcile in Übereinstimmung bringen * with/to:in Einklang bringen/abstimmen mit || vereinigen * in Einklang bringen
recondition erneuern * (wieder) instand setzen *

überholen, erneuern || instand setzen * überholen || instandsetzen * überholen || überholen * techn. ~
reconditioning Erneuerung || Instandsetzung * (Wieder-) Instandsetzung, Überholung, Erneuerung || Reparatur * Überholung || Überholung || Umbau * z.b. einer Maschine
reconnect rangieren * umverbinden || umklemmen || umrangieren * Signale, Leitungen || umschalten * umverbinden/klemmen
reconnection Wiedereinschalten * (auch Stromkreis)
reconsider überlegen * noch einmal ~
reconstruct rekonstruieren || umbauen * neu konstruieren, umbauen, wiederaufbauen/-herstellen, umformen, rekonstruieren
reconstruction Rekonstruktion || Umbau
record anzeigen * registrieren, aufzeichnen || aufnehmen * aufzeichnen, z.B. Messwerte, Kennlinie || aufzeichnen * Messwerte ~, Mitschrift/Niederschrift erstellen || Aufzeichnung || Aufzeichnung * das Aufgezeichnete || Datensatz * [Computer | NC-Steuerg.] in einer Datenbank (-Tabelle) oder in einem NC-Programm || eintragen || erfassen * aufzeichnen, statistisch erfassen || festhalten * in Wort/Bild/Ton ~ || Messschrieb * Aufzeichnung z.B. von einem Messschreiber || Nachweis * Unterlage || Protokoll * Verhandlung~; record in protocol: zu ~ nehmen || protokollieren || Registrierung * Aufzeichnung von Messwerten, Betriebsdaten usw. || Satz * data record: Daten~ (auch Verfahr~ bei NC-Steuerung) || Schrieb * von einem Messschreiber || vermerken
record chart Messschrieb * aufgezeichnetes Messkurvenblatt || Schrieb * aufgenommenes Kurvenblatt
record in writing festhalten * schriftlich ~
recorder Schreiber * Aufzeichnungsgerät, registrierendes Messgerät
recording -Erfassung * Aufzeichnung || Aufnahme * Aufzeichnung, z.B. von Signalen, eines Protokolls (auch 'plot': Kennlinie aufnehmen) || Aufzeichnung || Erfassung * Aufzeichnung || Protokollierung * Aufzeichnung || Registrierung * Aufzeichnung von Messwerten, Betriebsdaten usw.
recording instrument Registriergerät * z.B. Linienschreiber
recording of the field characteristic Feldkennlinienaufnahme
recording unit Registriereinrichtung * zum Aufzeichnen von Messwerten/Betriebsdaten
recover eintreiben * Zahlung ~ || gewinnen * aus Altmaterial usw. ~ || retten * (Güter, Geld, Stoff usw.) wiedergewinnen/-erlangen; (Verluste) wiedergutmachen/ersetzen || rückgewinnen || rückspeisen * wiedergewinnen, z.B. Bremsenergie || wiedererlangen || wiedergewinnen || wiederkehren * z.B. Spannung nach Spannungseinbruch || zurückliefern * zurückgewinnen
recovery Erholung || Rückgewinnung || Wiedererlangung || Wiedergewinnung * Rückgewinnung || Wiederherstellung * Rückgewinnung, Erholung || Wiederkehr * z.B. Spannung nach Spannungseinbruch
recovery flow Rückfluss * von Energie usw.
recovery time Erholzeit || Freiwerdezeit * braucht ein Thyristor beim Umkommutieren,um richtig zu sperren || Schonzeit
recovery voltage wiederkehrende Spannung
recruit einstellen * (Personal) rekrutieren
rectangle Rechteck
rectangular impulsförmig * rechteckförmig || pulsförmig * rechteckförmig || Rechteck- || rechteckförmig || rechteckig || rechtwinklig || viereckig
rectangular bars Flacheisen
rectangular design Rechteckbauweise

rectangular pulse Rechteckimpuls
rectangular pulse shape Rechteckimpulsform
rectangular pulses Rechteckimpulse
rectangular signal Rechtecksignal
rectangular wire Profildraht * *mit rechteckförmigem Profil*
rectangular-shaped rechteckförmig
rectification Gleichrichtung * *half-wave/one-way: Einweg~; fullwave/two-way: Zweiweg-/Vollwellen~*
rectification limit Gleichrichtertrittgrenze
rectified gleichgerichtet
rectified average value Gleichrichtmittelwert
rectified mean value Gleichrichtwert
rectified value Gleichrichtwert
rectifier Gleichrichter
rectifier and regenerative feedback unit Ein-Rückspeiseeinheit
rectifier block Gleichrichterblock
rectifier bridge Brückengleichrichter || Gleichrichterbrücke
rectifier diode Gleichrichterdiode * *DIN 41781/82*
rectifier operation Gleichrichterbetrieb
rectifier range Gleichrichterbetrieb * *bei Umkehrstromrichter*
rectifier roll Lochwalze * *in Papiermaschine*
rectifier set Gleichrichtersatz
rectifier stability limit alpha g * *Gleichrichtertrittgrenze* || alpha-G || Gleichrichtertrittgrenze * *minimaler Steuerwinkel, oft 30 Grad bei 4Q-Stromrichter*
rectifier unit Einspeiseeinheit * *netzseitiger Stromrichter für mehrere Wechselrichter an gemeins. Zwischenkreis*
rectifier/regenerative feedback unit Ein-Rückspeiseeinheit
rectify beheben * *berichtigen, korrigieren, richtigstellen, verbessern* || beseitigen * *(Fehler, Störung usw.) berichtigen, korrigieren, verbessern, richtigstellen, beseitigen* || gleichrichten
rectifying Gleichrichtung * *das Gleichrichten*
rectifying front end Einspeiseeinheit
rectifying unit Einspeiseeinheit * *netzseitiger Gleichrichter z.B. zur Speisung mehrerer Wechselrichter*
rectilinear geradlinig
recur wiederkehren * *sich wiederholen*
recurrence Wiederkehr * *periodische*
recurrence formula Rekursionsformel
recurrence frequency Folgefrequenz
recurrent periodisch * *wiederkehrend, sich wiederholend, ~ auftretend* || wiederholt * *wiederkehrend, sich wiederholend, periodisch auftretend* || wiederkehrend
recurring wiederkehrend * *in der Reihenfolge ~*
recursive rekursiv
recursive algorithm rekursiver Algorithmus
recycle wiederaufbereiten || wiedergewinnen || wiederverwerten
recycled Recycling- * *aus Altmaterial gewonnen, z.B. ~Papier*
recycled paper Altpapier
recycling Recycling || Rückgewinnung * *of raw materials: von Rohstoffen* || Wiederaufbereitung || Wiedergewinnung * *dem Zyklus wieder zugeführt* || Wiederverwertung
recycling of raw materials Rohstoffrückgewinnung
red rot
red figures Verlust * *rote Zahlen; run into red: in die ~zone gelangen*
red heat Rotglut
red hot glühend heiß || rot glühend
red nose Lötspitze * *(salopp)*
red-tape bürokratisch

redeem amortisieren * *(Anleihe usw.)*
redefine neu definieren
redelivery Rücksendung || Zurücksendung
redelivery of goods Warenrücksendung * *das Zurücksenden*
redemption Auslösung * *Abzahlung/-lösung, Tilgung, Freikauf, Wiedergutmachung, Ausgleich, Wiederherstellung*
redesign Redesign || überarbeiten * *umkonstruieren, anders auslegen, "umentwickeln", umbauen* || Überarbeitung * *Umkonstruktion, Entwicklungsänderung, Umdimensionierung* || umbauen * *umentwickeln* || umgestalten
redress beseitigen * *Übel ~* || Beseitigung
reduce abbauen * →*reduzieren,* →*verringern* || abschwächen || absenken * *reduzieren* || aufheben * *in einer Bruchrechnung* || auflösen * *(mathemat.) Bruch, Gleichung ~* || erniedrigen * *reduzieren* || herabsetzen * *verringern; auch Geschwindigkeit, Preis ~* || kürzen * *{Mathem.} Bruch* || reduzieren * *by:um* || senken * *reduzieren (by:um)* || vereinfachen * *Formel* || verengen || verkleinern * *reduzieren; auch Maßstab/Zeichnung* || vermindern * *reduzieren* || verringern || zerkleinern * *{Holz}* || zurückführen * *to: auf eine Regel usw.~* || zurücknehmen * *reduzieren*
reduce in size verkleinern * *in den Abmaßen ~*
reduce the size verkleinern
reduce to fibers zerfasern
reduce to small pieces zerkleinern
reduced reduziert || Teil- * *reduziert* || vereinfacht * *Formel*
reduced number of types Typenreduzierung
reduced output voltage niedrige Aussteuerung
reduced output voltage ratio niedrige Aussteuerung
reduced prices Preissenkung * *gesenkte Preise*
reduced scale verkleinerter Maßstab
reduced voltage starter Sanftanlaufgerät * *für Drehstrommotoren*
reduced voltages Teilspannungen * *reduzierte Spannungen*
reduced-load interval Pausenzeit * *Zeitdauer mit abgesenkter Belastung*
reducer Getriebemotor * *mit Untersetzungsgetriebe* || Reduzierstück || Zerkleinerer * *{Holz}*
reducer motor Getriebemotor * *mit Untersetzungsgetriebe*
reducing Reduktion || Reduzier- || Senkung * *Reduzierung*
reducing adapter Reduzierstück
reducing to zero Nullung
reduction Abbau * *Reduzierung* || Abminderung || Abschwächung || Abschwächung * *Reduzierung* || Absenkung * *Reduzierung* || Herabsetzung * *auch ~ von Preis/Geschwindigkeit* || Minderung * *Reduzierung* || Reduktion || Reduzierung || Rückgang * *Reduzierung* || Senkung * *Reduzierung (Preise usw.)* || untersetzt * *Getriebe* || Untersetzung * *v. Getriebe* || Verkleinerung || Verminderung || Verringerung
reduction factor Reduktionsfaktor * *Verhältn.Motornennstrom/Stromtragfähigkeit d.i. Klemmenkasten anschließb.Leitgn.*
reduction gear Reduktionsgetriebe || Reduziergetriebe || Untersetzungsgetriebe * *zur Drehzahlreduzierung*
reduction gearbox Untersetzungsgetriebe
reduction gearing Reduziergetriebe
reduction in price Preisnachlass
reduction in prices Preissenkung
reduction in size Verkleinerung
reduction of prices Preisnachlass || Preissenkung
reduction of time Zeitverkürzung
reduction ratio Getriebeübersetzung * *bei Unter-

PAM-connection PAM-Schaltung * *für polumschaltbaren Motor (Pol-Amplituden-Modulation)*
PAM-winding PAM-Wicklung * *Abk.f.'Pole Amplitude Modulation':Polamplit.modulat. b. polumschaltb.Mot m. 1 Wicklg.*
pamphlet Broschüre || Druckschrift || Heft * *Druckschrift, Broschüre* || Produktinformation * *Druckschrift, Broschüre, Merkblatt* || Produktschrift * *Druckschrift, Broschüre, Merkblatt* || Prospekt * *Druckschrift, Broschüre, Merkblatt*
pan Pfanne
pan grinder Kollergang * *Mahlwerk mit schweren Walzen, die auf einer waagerechten Kreisbahn rollen*
pancake motor Flachmotor * *{el.Masch.}* || Scheibenläufermotor * *(salopp)*
pane Scheibe * *Fenster~*
panel Feld * *~ in einer Schaltanlage* || Grundplatte || Platte * *(Schalt)Tafel* || Tafel * *Holz~, Blechplatte, auch zur Wandverkleidung* || Wand
panel aperture Frontplattenausbruch || Frontplattenöffnung
panel beam saw Plattenaufteilsäge * *{Holz}*
panel cutout Frontplattenausbruch || Schalttafelausschnitt * *z.B. zum Einbau eines Schalttafelinstruments*
panel cutting saw Plattenaufteilsäge * *{Holz}*
panel dividing saw Plattenaufteilsäge * *{Holz}*
panel meter Einbau-Messgerät * *Schalttafel-Instrument* || Messgerät * *Einbau/Schalttafel-Instrument* || Messinstrument * *Einbau/Schalttafel-Instrument* || Schalttafel-Messinstrument || Schalttafelinstrument
panel mounting Schalttafeleinbau
panel operator Bedienfeld * *(amerikan.)*
panel saw Plattenaufteilsäge * *{Holz}*
panel sizing circular saw Formatkreissäge * *{Holz}*
panel sizing plant Plattenaufteilanlage * *{Holz}* Zuschneidemaschine für Holzplatten
panel sizing saw Plattenaufteilsäge * *{Holz}*
panel sizing system Plattenaufteilanlage * *{Holz}* Zuschneidemaschine
panel-mount Schalttafelmontage * *für Schalttafelmontage*
panel-mounting Schalttafelmontage * *→Durchsteck-Montage z.B. bei/von Messgeräten*
panel-mounting measuring instrument Schalttafel-Messinstrument
panel-sizing circular saw Plattenaufteilsäge * *Plattenaufteil-Kreissäge*
paper Arbeit * *schriftliche Ausarbeitung* || Aufsatz * *Abhandlung, Referat, Aufsatz* || Ausarbeitung * *schriftliche ~* || Beitrag * *zu Kongress, Fachzeitschrift usw.* || Papier * *auch Dokument,Urkunde, Wert~; identity papers: Ausweis~; filigreed paper: ~ mit Wasserzeich.* || Unterlage * *Papier, Dokument, schriftliche Ausarbeitung* || Zeitschriftenaufsatz * *Abhandlung, Fachaufsatz, Referat*
paper break Papierabriss
paper clip Klammer * *Büro~*
paper core Wickelhülse * *zum Aufwickeln von Papier*
paper feed Papiervorschub * *bei Linienschreiber, Drucker usw.*
paper finishing line Papierveredelungsanlage
paper for web printing Rollendruckpapier
paper grade Papiersorte
paper guide roll Papierleitwalze
paper inspection machine Inspektionsmaschine * *→Sortierroller für Papier* || Sortierroller * *-wickelt Papier in beide Richtungen hin u.her z.Aussortieren v.Schadstellen*
paper machine Papiermaschine

paper making machine Papiermaschine
paper mangle Kalander * *Glättwerk für Papier*
paper manufacture Papierherstellung
paper mill Papierfabrik
paper processing machine Papierverarbeitungsmaschine
paper reel Tambour * *{Papier}*
paper tension Papierzug
paper web Papierbahn
paper-making industry Papierindustrie
paperboard Karton * *~agenpappe*
paperclip Büroklammer
paperweight Papiergewicht
parabolical parabelförmig
paradigm Beispiel * *Paradigma, (Muster)Beispiel*
paraffin Petroleum * *für Heiz- u. Leuchtzwecke*
paragon Inbegriff * *Muster, Vorbild, Ausbund* || Muster * *Vorbild, Muster, Ausbund*
paragraph Absatz * *in einem Text* || Kapitel * *z.B eines Buches* || Kapitel * *Absatz, Abschnitt*
parallel parallel || parallel schalten
parallel actuator Parallel-Hubspindeltrieb
parallel bus Parallelbus
parallel bus interface Parallelbuskopplung
parallel capacitor Parallelkondensator
parallel circuit Parallelschaltung
parallel configuration Parallelschaltung * *parallele Anordnung, auch von Stromrichtern, Motoren usw.*
parallel connected parallelgeschaltet
parallel connection Nebenschluss * *Parallelschaltung* || Parallelschaltung || Parallelverbindung
parallel coupling Parallelschnittstelle
parallel drive Gruppenantrieb * *bei Parallelschaltung von Motoren an einem Stromrichter*
parallel interface Parallelschnittstelle
parallel jackscrew Parallel-Hubspindeltrieb
parallel link parallele Schnittstelle || Parallelschnittstelle
parallel operation Parallelbetrieb
parallel processing Parallelverarbeitung * *mehrerer Programme quasi-gleichzeitig in einem Rechner*
parallel resistor Nebenwiderstand * *parallelgeschalteter Widerstand* || Parallelwiderstand
parallel running motors Parallebetrieb von Motoren * *{el.Masch.}mehrere Mot. arbeit.auf denselb. Beweg.vorgang e.Arb.maschine*
parallel subrack interfacing parallele Rahmenkopplung
parallel to it daneben * *gleichzeitig*
parallel winding Nebenschlusswicklung
parallel working Parallelbetrieb
parallel-axis achsparallel
parallel-motors drive Mehrmotorenantrieb * *mit mehreren Motoren an einem Stromrichter*
parallel-shaft wellenparallel
parallel-tuned inverter Schwingkreis-Wechselrichter
paralleled parallelgeschaltet
parallelling Parallelschaltung || Parallelschaltung * *das Parallelschalten*
parallelism Parallelität
paralyse Stillstand * *Lahmlegung, Lähmung; be paralysed: lahmgelegt/unwirksam/zum ~ gekommen sein*
paralyze lahm legen * *lähmen, paralysieren* || lähmen
parameter Größe * *Einflussgröße, Parameter* || Kenngröße || Kennwert || Operand * *Parameter (bei SPS)* || Parameter * *[pe'rämite]*
parameter adjustment Parametereinstellung
parameter assignment Parametrierung || Versorgung * *SPS*

reduction stage 690

setzungsgetriebe || Übersetzung * ~*sverhältnis eines Untersetzungsgetriebes* || Übersetzungsverhältnis * ~ *eines Untersetzungsgetriebes* || Untersetzung * *Untersetzungsverhältnis* || Untersetzungsverhältnis * *z.B. eines Getriebes*
reduction stage   Getriebestufe
redundancy   Redundanz * *one-out-of-three redundancy: 1-aus-3-~*
redundant   hochverfügbar * *redundant aufgebaut* || redundant
reed   Kamm * *[Textil] in der Weberei*
reed contact   Schutzgaskontakt
reed relay   Schutzgaskontaktrelais || Schutzgasrelais
reel   Poperoller || Rolle * *Spule, Haspel, Rolle (mit Hülse)* || schwanken * wanken, taumeln || Spule * *Webspule, Rad; ~ bei Auf/Abwickler, Webstuhl, Film, Tonband* || Tambour * *Maschinenrolle für/mit Papier* || taumeln * wanken, taumeln || wackeln * taumeln || Walze * *Haspel, Winde, Rolle*
reel change   Rollenwechsel || Tambourwechsel * *Wechsel der Papierrolle in Papiermaschine an der Aufrollung*
reel cutter   Rollenschneider || Rollenschneidmaschine
reel cutting machine   Rollenschneider || Rollenschneidmaschine
reel drum   Poperoller * *Andruckwalze, Stützwalze* || Stützwalze * *für Oberflächenwickler, z.B. bei Rollenschneidmaschine* || Umfangswickler * *Stütz-/Tragwalze*
reel drum winder   Oberflächenwickler
reel off   abwickeln * *Material ~*
reel splicer   Rollenwechsler * *Auf-/Abwickler mit fliegendem Rollenwechsel*
reel up   aufrollen
reel-change control   Rollenwechselsteuerung * *z.B. zum Tambourwechsel bei Papiermaschine*
reel-drum winder   Poperoller
reel-up   Aufroller || Poperoller || Roller * *Wickler für Papier (meist Umfangswickler)*
reeler   Aufrollung * *Aufwickler* || Haspel || Oberflächenwickler || Poperoller || Roller * *Wickler für Papier (meist Umfangswickler)* || Umfangswickler
reeling drum   Poperoller || Tambour
reeling drum exchange   Tambourwechsel * *Wechsel volle gegen leere Papierrollle z.B. am Ende einer Papiermaschine*
reeling machine   Wickelmaschine
reeling up   Aufrollung
reentrant   reentrant * *wiedereintrittsfähig, Fähigkeit eines Programms, sich selbst unterbrechen zu können*
reestablishing force   Rückstellkraft
Ref   Betreff * *Abkürzung für 'reference': Abkürzung in Briefkopf* || soll * *Kurzzeichen für 'reference': Sollwert*
refeed   zurückspeisen
refer   verweisen * *refer to: (sich) beziehen auf/~ auf/~ an* || weiterleiten * *Antrag, Fall, Vorgang (to:an)*
refer above   siehe oben
refer to   beziehen * *(sich) ~ auf* || beziehen auf * sich ~ || Bezug nehmen auf * *verweise/hinweisen auf* || erstrecken * *(figürlich) sich ~ auf* || hinweisen * *jdn. auf etw. ~; verweisen auf* || hinweisen auf || nachschlagen * *z.B. in Buch/Betriebsanleitung/Nachschlagewerk ~* || sich beziehen auf || siehe * *z.B.Verweis auf anderes Kapitel* || vergleiche * *siehe auch* || verweisen auf || zurückgreifen auf * *sich wenden an, befragen (Uhr, (Wörter-)Buch usw.)*
reference   Beziehung || Bezug * *['referens] (figürlich) Bezugnahme* || Bezugs- * *reference line: Bezugslinie; reference frequency: Bezugsfrequenz* ||

Hinweis * *['referens] Bezug, (Ver-) Weis (to:auf); with reference to: mit ~ auf* || Kennung || Referenz || Sollwert || Vergleichs- || Verweis * *Hinweis, Referenz, Verweisung (to:auf)* || verweisen || Zeichen * *Akten~ z.B.in einem Brief, our reference: unser ~* || Zeugnis * *Führungs~ für Angestellte*
reference book   Benutzerhandbuch * *mit Beschreibung aller/Bezugnahme auf alle Funktionen* || Nachschlagewerk * *auch zu einem technische Hard- oder Softwareprodukt*
reference bus   M-Leiter * *Schiene* || M-Schiene
reference cascade   Sollwertkaskade
reference chain   Sollwertkaskade * *Sollwertkette*
reference conductor   M-Leiter
reference control   Sollwertführung
reference diode   Referenzdiode
reference frequency   Bezugsfrequenz || Referenzfrequenz
reference input   Sollwerteingang
reference instant   Bezugszeitpunkt
reference integrator   Hochlaufgeber * *Sollwertintegrator* || Sollwertintegrator * *Hochlaufgeber*
reference list   Belegungstabelle * *z.B. für Ein-/Ausgänge bei SPS*
reference manual   Beschreibung * *ausführliches/umfassendes Handbuch* || Handbuch * *technisches Referenz~* || technische Beschreibung * *technisches Referenzhandbuch*
reference mark   Markierung * *Einstell/Eichmarke*
reference marker pulse   Referenzimpuls * *Nullimpuls eines Dreh-Impulsgebers*
reference number   Aktenzeichen || Kennziffer
reference point   Bezugspunkt * *z.B. für Potenzial/Signal* || Referenzpunkt
reference position   Referenzpunkt * *bei Lageregelung*
reference position pulse   Referenzimpuls * *für Lageregeregelung*
reference potential   Bezugspotenzial * *z.B. für ein(e) Signal(klemme)* || Bezugsspannung || Referenzspannung * *Bezugsspannung für Signal (Masse)*
reference profile   Bezugsprofil * *einer Verzahnung*
reference quantity   Bezugsgröße
reference service-conditions   Betriebs-Bezugsbedingungen
reference setting   Sollwertvorgabe
reference signal   Bezugssignal
reference sound power   Bezugsschallleistung * *[el.Masch.] Bezugsgröße P0 für Schallleistungen (10^-12 W)*
reference sound pressure   Bezugsschalldruck * *[el. Masch.] Eff.wert p0 d. Schallwechseldrucks an d.Hörschwelle bei 1kHz*
reference supply   Referenzspannungsquelle * *z.B. zum Anschluss eines Sollwertpotentiometers*
reference terminal   Bezugsklemme * *0V*
reference to   Bezug nehmen auf * *in Dokument/Buch Verweise anbringen auf*
reference transfer function   Führungsübertragungsfunktion * *[Regel.techn.]*
reference value   Bezugsgröße || Bezugswert || Führungsgröße * *Sollwert* || Sollwert
reference value cascade   Sollwertkaskade
reference value memory   Sollwertspeicher
reference value source   Sollwertquelle
reference variable   Bezugsgröße || Führungsgröße
reference voltage   Bezugsspannung * *Bezugspotenzial* || Referenzspannung * *genaue Spannung, ein Spannungsnormal darstellend*
reference voltage generator   Konstantspannungsquelle * *Referenzspannungsgeber*
reference voltage level   Bezugsspannung * *Bezugspotenzial*

reference voltage source Referenzspannungsquelle
reference-point approach Referenzfahrt
referred bezogen * *to:auf*
referred to bezogen auf *||* genannt * *z.b. mit einem Fachausdruck; be referred to as:bezeichnet werden als*
referred to a common signal ground potenzialbehaftet * *auf d. gleiche Masse-Bezugspotenzial bezogen*
referred to ground massebezogen
referred voltage drop Kurzschlussspannung
referring to bezogen auf * *Bezug haben auf*
referring to this diesbezüglich
refill auffüllen * *nachfüllen ||* füllen * *nachfüllen ||* Mine * *zum Nachfüllen (z.B. für Kugelschreiber) ||* nachfüllen * *wieder füllen, nach-/auffüllen ||* Patrone * *zum Nachfüllen*
refine aufbereiten * *z.B. in e. Produktionsschritt, auch 'verfeinern, ausklügeln' ||* ausarbeiten * *verfeinern, 'ausklügeln', aufbereiten ||* mahlen * *Zellstoff ||* verbessern * *verfeinern "ausklügeln", veredeln, raffinieren ||* veredeln * *verfeinern ||* verfeinern * *(auch figürl.: Methode, Arbeit, Stil, Geschmack usw.)*
refined raffiniert * *verfeinert, z.B. Zucker, Erdöl*
refinement Veredelung * *Verfeinerung*
refiner Mahlanlage * *zur Zellstofferzeugung ||* Raffineur * *Fein-Mahlwerk für Holzstoff/Hackschnitzel zur Zellstoffherstellung (→Refiner) ||* Refiner * *zum Feinmahlen v.Holz-/Zellstoff; conical refiner: Kegel~; disc refiner: Scheiben~*
refinery Raffinerie
refinish nachbearbeiten
refitable nachrüstbar
reflect entsprechen * *widerspiegeln, sich widerspiegeln, seinen Niederschlag finden in, s.auswirken auf ||* reflektieren *||* spiegeln * *reflektieren (auch figürl.) ||* widerspiegeln * *(auch figürl.)*
reflect on betrachten * *sinnend ~ ||* spiegeln an * *z.B. den Anforderungen ||* überlegen * *bedenken*
reflection Gedanke * *Betrachtung ||* Reflektion *||* Reflexion *||* Überlegung * *Nachdenken*
reflective reflektierend
reflow soldering Aufschmelzlöten * *von →SMD-Bauelementen ||* Reflowlöten * *Aufschmelzlöten z.B. von →SMD-Bauelementen*
reflux Rückfluss * *von Kapital usw.*
reform formieren * *z.B. Elektrolytkondensator ||* umgestalten * *verbessern, reformieren*
reforming Nachformierung * *eines Kondensators nach langer Lagerung*
reforming process Nachformierung * *langames Aufladen e. Kondensators nach langer Lagerung*
refract brechen * *{Optik} Lichtstrahl ||* reflektieren * *{Optik} brechen, ablenken*
refraction Ablenkung *||* Brechung * *{Optik} Lichtstrahl; plane/angle of action: ~sebene/-Winkel; refractive index: ~szahl*
refractive index Brechungszahl
refrain from Abstand nehmen von
refresh aktualisieren *||* auffrischen * *auch Daten, Prozessabbild, Ein/Ausgänge ||* Auffrischung *||* erfrischen *||* erneuern * *auffrischen, z.B. Daten, Farben*
refresh buffer Bildwiederholspeicher
refresh memory Bildwiederholspeicher
refresh rate Bildwiederholfrequenz * *Bilder/sec bei Monitor (ab ca. 70 Hz kein Flimmern mehr sichtbar)*
refresh time Auffrischzeit
refresher course Auffrischkurs *||* Kurs * *Auffrischungs~ ||* Lehrgang * *Auffrischungskurs ||* Schulungskurs * *Auffrischungskurs*
refrigerant Kältemittel * *für Kälteanlage*

refrigerant compressor Kältemittelverdichter * *bei Kältemaschine*
refrigerating capacity Kälteleistung * *einer Kältemaschine (DIN 8928, ISO 9303)*
refrigerating compressor Kältekompressor
refrigerating system Kältemaschine
refrigeration and air conditioning Kälte- und Klimatechnik
refrigeration technology Kältetechnik
refrigerator Kältemaschine
refrigerator system Kälteanlage
refuge Schutz * *Zuflucht*
refund Erstattung *||* Rückerstattung * *Geld ||* vergüten * *Ausgaben/Auslagen ~ ||* zurückerstatten * *Ausgaben, Kosten, Geld*
refunding Rückerstattung * *Geld*
refurbish modernisieren *||* überholen * *renovieren, aufpolieren, überholen (z.B. Maschine)*
refurbishing Überholung * *z.B. ~ einer Maschine*
refusal Ablehnung *||* Absage * *Ablehnung ||* Zurückweisung
refuse Abfall * *Abfall, Ausschuss, Abraum {Bergwerk}, Müll- ||* ablehnen *||* abweisen * *ablehnen ||* Annahme verweigern *||* beanstanden * *Waren, Lieferung usw. ~; refuse acceptance: Annahme verweigern ||* nicht beachten * *zurückweisen ||* verwerfen * *ablehnen, zurückweisen, abweisen ||* zurückweisen * *ablehnen; refuse to accept: Annahme verweigern*
refuse to accept Annahme verweigern
regain rückgewinnen * *wiedergewinnen (z.B. Information) ||* wiedererlangen
regard achten auf * *beachten, berücksichtigen, respektieren; disregard: nicht achten auf, nicht beachten ||* Beachtung * *Berücksichtigung; ~ schenken (Verb); disregard: keine ~ schenken ||* berücksichtigen * *beachten, berücksichtigen, betrachten, respektieren; disregard: nicht ~ ||* Berücksichtigung * *with (due) regard to: unter (angemessener) ~ von; disregard: nicht berücksichtigen ||* betrachten * *as: als ||* Bezug * *with regard to:in Bezug auf ||* kümmern * *angehen, betreffen ||* Rücksicht * *with regard to: mit ~ auf*
regard as highly unlikely ausschließen * *für höchst unwahrscheinlich halten*
regarding betreffend *||* bezüglich * *bezogen auf, wenn man ... betrachtet ||* hinsichtlich *||* in Hinblick auf * *betreffend ||* wegen * *betreffend ||* zu * *betreffend*
regarding the -lich * *z.B. regarding the price: preis~*
regarding the price preislich
regardless unabhängig * *regardless of whether...or not:unabhängig davon, ob...oder nicht*
regardless of unabhängig von * *~ Tatsachen/Umstand; ohne Rücksicht auf ||* ungeachtet
regardless of the direction of rotation drehrichtungsunabhängig *||* unabhängig von der Drehrichtung
regardless of the fact that unabhängig davon ob
regardless of the speed direction drehrichtungsunabhängig
regardless of whether unabhängig davon ob * *or not:oder nicht*
regardless whether unabhängig davon ob * *or not:oder nicht*
Regards Hochachtungsvoll
Regards, mit freundlichen Grüßen *||* mit vorzüglicher Hochachtung
regenerate regenerieren *||* rückgewinnen *||* rückspeisen * *z.B. (Brems-) Energie ins Netz ||* wiedergewinnen *||* zurückliefern * *z.B. Bremsenergie ins Netz ~ ||* zurückspeisen * *z.B. Bremsenergie ins Netz ~*

regenerating　Energierückspeisung || generatorisch * zurückspeisend
regenerating front end　Ein-Rückspeiseeinheit
regenerating rectifying unit　Ein-Rückspeiseeinheit * Umkehrstromrichter zur Zw.kreispeisg. f.(mehrere) Wechselrichter
regenerating unit　Rückspeiseeinheit
regeneration　Energierückspeisung || Rückspeisung * z.b. von Bremsenergie ins Netz || Wiedergewinnung
regeneration back to the mains　Netzrückspeisung
regeneration into the system　Netzrückspeisung
regeneration of energy　Energierückspeisung * back to the mains: →Netzrückspeisung
regeneration of energy back into the mains　Netzrückspeisung
regenerative　generatorisch * zurückspeisend || regenerativ || rückspeisefähig * generatorischer Betrieb möglich || Vierquadrant- * rückspeisefähig || zurückspeisend * Bremsenergie ins Netz
regenerative braking　generatorische Bremsung * mit Rückspeisung der Bremsenergie ins Netz || generatorisches Bremsen || Nutzbremsung * mit Rückspeisg.d.Bremsenergie ins Netz bzw.Energierückgewinnung || Rückspeisung * von Bremsenergie ins Netz
regenerative converter　Umkehrstromrichter * rückspeisefähiger Stromrichter
regenerative energy　generatorische Energie * wird z.b. beim Abbremsen eines Antriebs frei
regenerative energy feedback　Energierückspeisung * z.b. vom Motor ins Netz beim Bremsen
regenerative feedback　Energierückspeisung || Netzrückspeisung * z.B. von Bremsenergie || Rückspeisung * z.B. von Bremsenergie ins Netz
regenerative feedback into the supply　Netzrückspeisung
regenerative feedback unit　Netzrückspeise-Einheit || Netzrückspeiseeinheit * zur Rückspeisung v. Bremsenergie ins Netz
regenerative front end　Ein-Rückspeiseeinheit
regenerative mode　Wechselrichterbetrieb * Rückspeisebetrieb (ins Netz zurück) eines Stromrichters
regenerative rectifier unit　Ein-Rückspeiseeinheit * zur Zw.kreispeisg. v.Antriebswechselrichtern m.Bremsenergierückspeisg.
region　Region
regional　regional
regional office　ZN * (SIEMENS-) →Zweigniederlassung || Zweigniederlassung
register　anmelden * Lehrgang, Tagung || belegen * sich eintragen lassen || buchen || eintragen * amtlich ~ (z.B. Warenzeichen), sich amtlich ~ lassen; with: bei || erfassen * statistisch ~, registrieren, anmelden, verzeichnen, (handelgerichtlich usw.) eintragen || Liste * amtliche || Register * Wortspeicher, Eintragungsbuch, (Inhalts-) Verzeichnis, Registriervorrichtg., Druckmarken-~ || registrieren * z.B. Betriebs- oder Fehlerdaten || Zwischenspeicher * ~ für Datenwort
register accuracy　Registerhaltigkeit * {Druckerei}
register area　Wertespeicher * {SPS}
register control　Registerregelung * {Druckerei} sorgt für passgenaues Übereinanderdrucken der Farbwerke/Passermarken
register error　Registerfehler * {Druckerei}
register mark　Druckmarke * {Druckerei} || Registermarke * {Druckerei}
registered　zugelassen * amtlich eingetragen/registriert
registered design or pattern　Gebrauchsmuster
registered master　Meister * im Handwerk
registered office　Hauptsitz * (eingetragener/gerichtlicher) Firmensitz

registered trade-mark　eingetragenes Warenzeichen
registration　Anmeldung * auch zu Seminar, Kongress || Erfassung * statistische ~ || Registrierung || Zulassung
registration form　Anmeldeformular * z.B. für Schulungskurs
registration of a utility or design　Gebrauchsmuster-Eintragung || Gebrauchsmustereintragung
regreasable　nachschmierbar
regrease　nachschmieren
regreasing　Fetterneuerung * Nachschmieren || nachschmieren * (Subst.) || Nachschmierung
regreasing device　Nachschmiereinrichtung * z.B. für Elektromotor
regression　Rückgang * Rückbildung, Rückentwicklung, Rückschritt || Rückschritt
regret　bedauern
regular　fest * z.b. Kunden || gewöhnlich * üblich || gleichmäßig * gleichmäßig || laufend * Geschäfte usw.; regelmäßig (z.B. Wartung); geregelt, geordnet, vorschriftsmäßig || normal * üblich || ordentlich * regelrecht || ordnungsgemäß || regelmäßig * auch zeitlich || routinemäßig || ständig * regtelmäßig (Einkommen usw.) || turnusmäßig
regular design　Normalausführung * übliche Bauart
regular service procedure　Routinewartung
regularity　Gesetzmäßigkeit * Regelmäßigkeit, Gleichmäßigkeit, Stetigkeit || Gleichmäßigkeit || Regelmäßigkeit
regularly　laufend * (Adv.) regelmäßig, vorschriftsmäßig, geordnet, pünktlich, normal || regelmäßig * (Adv.)
regularly recurring　turnusmäßig
regulate　einrichten * regulieren, anpassen, einstellen (Gerät) || regeln * regeln, regulieren || regulieren
regulated　geregelt || regelmäßig * geordnet
regulating element　Stellorgan
regulating section　Regelteil || Reglerteil * siehe auch →Regelteil
regulating unit　Stellgerät
regulating valve　Stellventil
regulating variable　Stellgröße
regulation　Anweisung * Vorschrift || Bestimmung * Vorschrift, Norm; legal/statutory regulation: gesetzliche ~ || Festlegung * Regelung, Verordnung, Verfügung, Vorschrift, Satzung, Statuten || Gesetz * gesetzliche Vorschrift || Regelung * →Bestimmung, →Vorschrift, Verfügung || Regulierung || Spannungsabfall * z.B. einer Spannungsquelle bei Laständerung || Vorschrift
regulation accuracy　Regelgenauigkeit * Abweichung d. Regelgröße v. eingestellten (Soll)Wert
regulator　Regler
regulator enable　Reglerfreigabe
regulator output　Reglerausgang
regulator slipring　Regelschleifring * {el.Masch.} mit dauernd aufliegenden Bürsten zur Drehz.regelg. durch Vorwiderst.
regulator-slipring motor　Regelschleifringläufer-Motor * {el.Masch.}
regulatory body　Behörde * Vorschriften erlassende ~ || Normenbehörde
regulatory control　Regelung
rehearing　Revision * {Rechtswesen} Wiederaufnahme
reimburse　vergüten * Ausgaben ~
reimbursement　Erstattung || Rückerstattung * Auslagen, Kosten
reinforce　ertüchtigen * verstärken, kräftigen || verstärken * mechanisch ~/fester machen || versteifen * verstärken
reinforced　verstärkt * mechanisch stärker/kräftiger

gemacht; auch 'strengthened'; siehe auch →*verstärken*
reinforced bearing design   verstärkte Lagerung
reinforced design   verstärkte Ausführung
reinforcement   Bewehrung || Verstärkung * *mechanisch* || Versteifung * *Verstärkung, Armierung*
reinforcing mesh   Armiereisen * *in Mattenform*
reinforcing steel   Armiereisen
reinstall   wiedereinbauen
reinstate   erneuern * *Patent* ~
reinstatement   Erneuerung
reiteration   Erneuerung * *Wiederholung*
reject   ablehnen * *auch* →*abweisen* || abweisen * *ablehnen, abweisen* || Annahme verweigern || beanstanden * *Waren, Lieferung usw.* ~ || nicht beachten * →*ablehnen,* →*verwerfen* || verwerfen * *ablehnen* || zurückweisen * *ablehnen, abweisen*
rejection   Ablehnung * *Ablehnung, Zurückweisung, Verwerfung, Einwand* || Absage || Unterdrückung || Zurückweisung
rejection band   Bandsperre
rejection system   Ausleitsystem * *zum Aussondern fehlerhafter Produkte aus einer Transportlinie*
rejector circuit   Bandsperre * *Sperrkreis* || Sperrkreis * *siehe auch 'Bandsperre'*
rejector filter   Sperrfilter
rejects   Ausschuss * *nicht verwendbare/zurückgewiesene Teile*
rejuvenate   runderneuern * *(figürl.)*
relate   beschreiben * *erzählend* ~
relate to   beziehen * *sich* ~ *auf, z.B. auf eine Nenngröße* || beziehen auf || zuordnen * *in Beziehung setzen/bringen zu; i.Zusammenhang bringen mit; verbinden mit*
related   -nah || bezogen * *to:auf* || dazugehörig * *in Beziehung/Zusammenhang stehend* || verwandt * *to: mit* || zugehörig || zusammengehörig * *verwandt, in Beziehung/Zusammenhang miteinander stehend* || zusammenhängend * *verwandt, in enger Beziehung stehend*
related to   bezogen auf
related to time   zeitabhängig * *zeitbezogen*
relating thereto   diesbezüglich
relating to   bezüglich
relating to market policy   marktpolitisch
relating to space   räumlich * *auf den Raum/Platz bezogen* || räumlich
relation   Beziehung * *between:zwischen; to: zu/mit* || Relation * *Verhältnis, Beziehung* || Verhältnis * *Beziehung* || Zusammenhang * *Beziehung, Verhältnis, kausaler* ~
relational   relational * *z.B. Datenbank mit Querbeziehungen zwischen den Datensätzen*
relational operation   Vergleichsfunktion * *Computerbefehl*
relational operator   Vergleichsoperator
relationship   Abhängigkeit * *z.B. mathematische Funktion; between:zwischen; to:zu/mit* || Beziehung * *between: zwischen; to: zu/mit* || Verwandtschaft * *between:zwischen; to:zu/mit* || Zusammenhang * *Beziehung, Verhältnis, Rechtsverhältn., Verwandtschaft; between...and: zwischen...und*
relative   -Verhältnis * *vorangestellt* || relativ * *(Adj.) to:zu* || verhältnismäßig
relative air humidity   relative Luftfeuchtigkeit
relative density   spezifisches Gewicht
relative humidity   relative Feuchtigkeit || relative Luftfeuchtigkeit
relative motion   Relativbewegung
relative short-circuit power   Kurzschlussleistungsverhältnis
relative speed   Relativdrehzahl
relative thermal life expectancy   bezogene thermische Lebensdauererwartung * *[el. Masch.]* Lebensdauerveränder. b.veränd. Betriebsbeding.
relative to   bezogen auf || bezüglich * *relativ zu* || gegenüber * *relativ zu, bezogen auf*
relatively   relativ * *(Adv.)* || verhältnismäßig * *(Adv.)*
relax   entspannen * *Muskeln, Nerven, Geist* ~; *sich* ~ || nachlassen * ~ *von Tätigkeit, Anspannung*
relay   Relais || weitergeben * *(allg.) weitergeben; mit Relais steuern*
relay chatter   Relaisflattern
relay coil   Relaisspule
relay contact   Relaiskontakt
relay contact rating   Relais-Kontaktbelastung
relay equivalent   Hilfsmerker * *bei SPS in Kontaktplandarstellung* || Merker * *bei Kontaktplandarstellung eines SPS-Programms*
relay logic output   Relaisausgang
relay oscillation   Relaisflattern
relay output   Relaisausgang
relay rating   Relais-Kontaktbelastung
relay rattling   Relaisflattern
relay symbology   Relaissymbolik * *bei Kontaktplandarstellung eines SPS-Programms*
relays   Relais * *Mehrzahl*
release   -Schutz * *durch abschaltende Überwachungseinrichtung* || -Wächter * *Auslösegerät* || abfallen * *Relais, Schütz usw.* || aufheben * *Verriegelung* ~, *Sperre* ~ || Ausgabestand || Auslösegerät * *Auslöser, Auslösevorrichtung* || auslösen * *freigeben, freiklinken, z.B. Überwachungseinrichtung/ Schnappverschluss, Kameraverschluss* || Auslöser * *Auslöse-/Schutzeinrichtung* || ausrücken * *Kupplung, Bremse* || Demontage * *"Entschnappen" eines Geräts von einer Montage-Hutschiene* || entriegeln || Entriegelung * *Freigabe, mechanische* ~ || Entriegelungs- || Freigabe * *z.B. eines Produktes zur Fertigung oder Lieferung; durch Schutzvorrichtung* || freigeben * *ein Produkt* ~; *для sale: zum Vertrieb* || lösen * *Bremse, Kupplung, Griff, Befestigungsschraube usw.* ~ || lüften * *Bremse öffnen* || Lüftung * *Bremse* || öffnen * *freigeben, (Bremse, Kupplung usw.) öffnen* || Softwareversion
release apron   Schlingenleitblech * *[Walzwerk]*
release button   Feigabetaste
release device   Auslösegerät
release for delivery   Lieferfreigabe
release for general availability   Lieferfreigabe
release for launching into the market   Vertriebsfreigabe
release for sale   Vertriebsfreigabe || zum Vertrieb freigeben
release force   Auslösekraft
release spring   Ausrückfeder * *z.B. in Kupplung/ Bremse* || Druckfeder * *einer Kupplung, Bremse usw.* || Kupplungsdruckfeder
release the clutch   auskuppeln * *eine Kupplung*
release time   Lüftzeit * *einer Bremse*
release travel   Lüftweg * *bei einer Bremse*
released   ausgelöst
releasing device   Auslösegerät
releasing temperature   Auslösetemperatur * *eines thermischen Auslösegerätes*
relentless   unaufhaltsam * *anhaltend, unbarmherzig, schonungslos, hart*
relevance   Relevanz * *Erheblichkeit* || Wichtigkeit * *Erheblichkeit*
relevancy   Relevanz
relevant   anwendbar * *einschlägig* || betreffend || betroffen * *relevant, einschlägig* || diesbezüglich * *to this:in dieser Sache* || einschlägig || infragekommend || interessierend * *sachdienlich, von Bedeutung/Belang* || maßgebend * *bestimmend; z.B. Bestimmung, Vorschrift* || maßgeblich * *bestimmend* || relevant * *erheblich, zutreffend (to:für)*

reliability Betriebssicherheit * →*Zuverlässigkeit* || Sicherheit * *Zuverlässigkeit* || Störsicherheit * *Betriebssicherheit, Zuverlässigkeit* || Verfügbarkeit * *Zuverlässigkeit (high level of: hohe)* || Zuverlässigkeit * siehe auch →*Betriebssicherheit*
reliability in operation Betriebssicherheit * *Zuverlässigkeit*
reliable betriebstüchtig * *zuverlässig* || funktionssicher * *zuverlässig* || gut * *zuverlässig* || problemlos * *zuverlässig* || sicher * *verlässlich, zuverlässig* || störsicher * *betriebssicher, zuverlässig* || zuverlässig
reliable contacting Kontaktsicherheit
reliable isolation sichere Trennung
reliably sicher * *(Adv.) zuverlässig* || zuverlässig * *(Adv.)*
relief Abhilfe * *afford relief:Abhilfe schaffen* || Behebung || Entlastung
relief grinding Hinterschleifen
relief groove Freistich * *{Werkz.masch.} "Hinterdrehung" an Drehteil*
relief pressure valve Überdruckventil
relief valve Überdruckventil
relief valve jet Druckbegrenzungsventil * *Überdruckventil*
relieve ablösen * *Schicht, Wache, Einheit usw.* ~ || befreien * *of:von; Sorge, Last usw.* || entlasten * *befreien (of/from:von)* || entspannen * *z.B. eine Feder* ~ || erleichtern * *erleichtern, entlasten, befreien (of:von)* || hinterdrehen * *z.B. Drehteil auf Drehmaschine*
relieve one another ablösen * *sich (gegenseitig)* ~ *(at:bei)*
relieving cut Freistich * *mit Drehmaschine*
relieving lathe Hinterdrehmaschine
relinquish verzichten
relocate umziehen || verlegen * *versetzen (an einen anderen Ort/Platz)*
relocation Umzug
relubricate nachschmieren
relubricating device Nachschmiereinrichtung * *z.B. für Wälz-/Gleitlager (bestehend aus Schmiernippel und Rohrleitung)* || Nachschmiereinrichtung
relubrication nachschmieren * *(Subst.)*
relubrication cycle Schmierfrist
relubrication device Nachschmiereinrichtung
relubrication interval Nachschmierfrist
reluctance magnetischer Widerstand
reluctance motor Reluktanzmotor * *Drehstrom-Synchronmotor mit Käfigläufer (ohne besondere Erregung)*
reluctant widerstrebend * *widerwillig, widerstrebend, zögernd; be reluctant to do: s. sträuben, etw.zu tun*
rely verlassen * *vertrauen (on:jdm./etwas)*
rely on verlassen * *sich* ~ *auf*
remachine nachbearbeiten * *{Werkz.masch.}*
remagnetize aufmagnetisieren
remain bleiben || verbleiben || verharren * *in: bei/in; z.B. in einer bestimmten Stellung*
remain effective gültig bleiben * *wirksam bleiben*
remain unchanged stehen bleiben * *unverändert bleiben*
remain untouched stehen bleiben * *unberührt* ~
remain valid gültig bleiben
remainder Rest * *auch mathematisch; Restbestand*
remainders of cable materials Leitungsreste
remaining restlich || übrig * ~*geblieben* || verbleibend
remaining distance Restweg * *{NC-Steuerung}*
remanence Remanenz || Rest- * *restlich, übrig(-bleibend); z.B. Spannung, Feld*
remanence relay Remanenzrelais
remanence voltage Remanenzspannung

remanent remanent * *magnet.*
remanent circuit Remanenzschaltung || Selbsthalteschaltung
remanent contact Selbsthaltekontakt
remanent field Restfeld
remanent relay Haftrelais
remanent voltage Restspannung * *Remanenz*
remark Anmerkung || Bemerkung * *on:über* || Hinweis * *Anmerkung; on: zu*
remarkable außerordentlich * *bemerkenswert* || beachtlich * *bemerkenswert* || bedeutend * *bemerkenswert* || höchst * *bemerkenswert*
remedy Abhilfe * *take remedial measures: Abhilfe schaffen, Abhilfemaßnahmen ergreifen* || Abhilfemaßnahme * *(Plural: remedies)* →*Abhilfe* || beseitigen * *Nachteil, Unrecht, Fehler, Schaden* ~ || Beseitigung * *Abhilfe, Behebung, (Problem-) Beseitigung; siehe auch* →*Fehlerbeseitigung* || Hilfsmittel * *(Gegen)Mittel, Abhilfe*
remember behalten * *im Gedächtnis* || merken * *remember something: sich etwas* ~
reminder Mahnung * *Mahnbrief, Erinnerung*
remnant Rest * *Rest, Überbleibsel, letzter/kläglicher Rest, Spur, (Stoff- usw.) Rest*
remnants -Reste * *Rest, Überbleibsel, letzter/kläglicher Rest, Spur, (Stoff- usw.) Rest*
remodel modernisieren * *vom äußeren Aussehen her* || umarbeiten * *gänzlich* ~ || umbauen * *umbilden, umbauen, umformen, umgestalten* || umgestalten
remodeling Umbau * *an e. Gerät*
remodelling Umarbeitung
remote dezentral * *entfernt angeordnet, ferngesteuert* || entfernt || Fern- * *z.B. remote control: Fernsteuerung* || Fernbedien- || Vor-Ort-
remote control externe Steuerung * *Fernsteuerung, z.B. über seriellen Bus* || Fernbedienung * *als Funktion* || Fernsteuerung
remote controllability Fernsteuerbarkeit
remote data transmission Datenfernübertragung
remote diagnostics Ferndiagnose
remote display Fernanzeige
remote I/O dezentrale Peripherie * *entfernt angeordnete Ein-/Ausgaben, elektronische Klemmenleiste* || intelligente Klemmenleiste
remote I/O module intelligente Klemmenleiste * *entfernt vom Rechner angeordnete Ein/Ausgaben*
remote I/O terminal block elektronische Klemmenleiste
remote I/O unit dezentrales Peripheriegerät
remote indication Fernanzeige
remote IO module intelligente Klemmenleiste
remote measurement Fernmessung
remote operating unit Einzelbediengerät * *Vor-Ort-*~
remote operator control Fernbedienung
remote operator control unit Fernbediengerät
remote programming Fernprogrammierung * *bei SPS*
remote-controlled ferngesteuert
removable abnehmbar || abziehbar || ausziehbar * *herausnehmbar* || lösbar * *abnehmbar*
removable screw terminal Schraubsteckklemme
removable screw-terminal Schraubsteckklemme
removable terminal abziehbare Klemme * *z.B. Schraubsteckklemme* || Schraubsteckklemme * *abziehbare Klemme*
removal Abfuhr || Abnahme * *Demontage* || Abtransport * *Fortschaffen, Wegräumen* || Ausräumung * *z.B. von Ladungsträgern beim Abschalten e. Thyristors* || Behebung * *auch* ~ *von Fehlern* || Entnahme || Verlagerung || Verlegung
removal by suction Absaugung

remove abbauen * abnehmen, abmontieren, ausbauen || abnehmen * lösen || abtragen || abziehen || abziehen * z.B.Lager/Kupplung/Zahnrad ~ || ausbauen * abbauen (z.B. Maschinenteil) || ausräumen * z.B. Ladungsträger b. Abschalten e. Thyristors || ausräumen * Möbel usw. (auch figürl.: Bedenken usw. ~) || ausschleusen || beheben * Fehler ~ || beseitigen || beseitigen * Gegner ~ || entfernen * beseitigen || entnehmen * herausnehmen, entfernen (from:von/aus) || fortführen * entfernen || herausnehmen * from:aus etw. heraus || umziehen * to:nach || verlagern * to:nach || wegnehmen
rename umbenennen
rend durchreißen * entzweireißen
render abwerfen * Gewinn/Profit usw. ~ || bieten * möglich machen, ermöglichen, wiedergeben || darstellen * wiedergeben, interpretieren || leisten * Dienst || übergeben * z.B. Gewinn, Festung || zurückgeben * zurückgeben/erstatten, z.B. Gewinn
render difficult erschweren * schwer/schwierig machen; render more difficult: noch schwerer machen
render first aid Erste Hilfe leisten
render ineffective lahm legen * ineffektiv machen
render prominent hervorheben * become prominent: sich hervorheben (from:aus)
rendering compact Straffung * (figürl.)
renew aktualisieren || erneuern * auch Vertrag ~; auch 'wiederholen' || verlängern * Vertrag ~
renewal Erneuerung * renewal rate: ~srate
renewed erneut * erneut, erneuert || neu * aufgefrischt, frisch, er~ert
rengineer anpassen * umprojektieren, umkonstruieren
renormalization Umnormierung
renounce verzichten * ~ auf
renovate erneuern * wiederherstellen, erneuern
renovation Erneuerung
renowned bekannt * nahmhaft, berühmt
rent leihen * (ver)mieten, (amerikan.) aus~, (amerikan.) sich etwas ~, vermietet werden (at/for: zu) || verleihen * z.B.Auto || vermieten * (besonders i. Amerikan.)
rental car Leihwagen || Mietwagen
rental costs Leihgebühren * Mietkosten
rental fee Leihgebühren
renunciation Verzicht * of:auf
reorganization Umbau * (figürlich) Um/Neuorganisation || Umorganisation
reorganize umbauen * (figürl.) anders organisieren/anordnen, umorganisieren || umgestalten * neu/umorganisieren || umstrukturieren * umorganisieren
repair beheben * Schaden ~ || instand setzen * reparieren || instandsetzen || Instandsetzung * Reparatur || Reparatur * under: in; have repaired: in ~ geben/reparieren lassen || reparieren * have repaired: reparieren lassen/in Reparatur geben || wiederherstellen * reparieren
repair department Reparaturstelle * Abteilung
repair facility Reparaturstelle
repair price Reparaturpreis
repair report Reparaturbericht
repair service Reparaturdienst
repair shop Reparaturwerkstatt
repair station Reparaturplatz
repair switch Reparaturschalter
repair work Instandsetzung * Reparaturarbeiten
repair works Instandsetzungsarbeiten
repaired funktionsfähig * wieder in Ordnung, wiederhergestellt, repariert
repairing Instandsetzung * Reparatur
repairing of faults Störungsbeseitigung
repairs Instandsetzungsarbeiten * to:an
reparable reparabel || reparaturfähig

repared spot Reparaturstelle
repay zurückerstatten * Ausgaben, Kosten
repayment Erstattung
repeat erneuern * wiederholen || reproduzieren * wiederholen || wiederholen || Wiederholung
repeat factor nutzen * (Subst.) Anz.d.Teilstücke je Fläche (bei Leiterplattenfertigung) od. je Längeneinheit || Wiederholfaktor
repeat number Wiederholungszahl
repeat size Rapport * Abstand, in dem sich ein Muster wiederholt (bei Tapeten, Bodenbelägen, Textildruck)
repeat-order Nachbestellung
repeatability Konstanz * Reproduzierbarkeit || Reproduzierbarkeit * Wiederholbarkeit || Wiederholbarkeit || Wiederholgenauigkeit
repeatable reproduzierbar * wiederholbar
repeated erneut || mehrfach * wiederholt || wiederholt
repeatedly mehrfach * (Adv.) wiederholt || wiederholt * (Adv.)
repeater Busverstärker || Leitungsverstärker * inmitten der Leitung zum Erreichen einer größeren Reichweite || Repeater * Leitungsverstärker zum Erreichen größerer Entfernungen bei Datenübertragung || Verstärker * Zwischenverstärker für Kommunikationsleitung
repercussion Rückwirkung * Auswirkung
repertoire Vorrat * Repertoire, Bestand
repertory Vorrat * Repertoire, Bestand
repetition Durchlauf * Wiederholzyklus || Wiederholung
repetition frequency Folgefrequenz
repetition rate Folgefrequenz
repetitive periodisch * ~ auftretend/wiederkehrend || Wiederhol- || wiederholt * →wiederkehrend || wiederkehrend
repetitive cycle Wiederholzyklus
replace ablösen * ersetzen || austauschen * ersetzen || auswechseln * ersetzen || erneuern * auswechseln || ersetzen * by: durch || substituieren * ersetzen; by:durch || tauschen * aus~, z.B. im Reparaturfall || vertreten * ersetzen
replaceable austauschbar || easily:leicht || auswechselbar
replacement Ablösung * Ersetzen || Austausch * Ersetzen || Austausch- || Erneuerung || Ersatz * for:für || Ersatz- * (Aus-) Tausch || Substitution * Ersetzen, Ersatz || Tausch * Ersetzen, Austausch (z.B. bei Reparatur/eines defekten Geräts) || Wechsel * Austausch z.B. e. defekten Teils
replacement board Ersatzbaugruppe * Leiterkarte
replacement brush Ersatzbürste
replacement carbon brushes Ersatzkohlebürsten
replacement module Ersatzbaugruppe
replacement of the printed-circuit board Baugruppentausch * Tausch der Leiterkarte
replacement part Ersatzteil * Austauschteil
replacement product Nachfolgeprodukt
replacement type Ersatztyp || Nachfolgetyp
replacement unit Austauschgerät || Tauschgerät
replacement value Ersatzgröße * für eine nicht direkt messbare Größe
replacing Tausch * das Austauschen, Ersetzen (z.B. von Verschleiß-/Ersatzteilen)
replacing of software Softwaretausch * Austausch
replenish auffüllen * Vorräte usw. (siehe auch →füllen) || füllen * wieder(auf)füllen (Vorräte usw.) || nachfüllen * (wieder) auffüllen, ergänzen (with: mit)
replenishment Ergänzung * Auffüllung, Ersatz, Ergänzung
replica Abbild * genaue Nachbildung || Duplikat * Nachbau || Modell * genaue Nachbildung || Nachbildung * genaue ~

reply Antwort * *to: auf* || antworten * *to: auf* || Rückmeldung * *Antwort*
reply channel Antwortkanal
reply specifier Antwortkennung * (→*PROFIBUS-Profil; für Parameter-Schreib-/Leseaufträge*)
reply to beantworten
report bekanntmachen * berichten || Bericht || melden * *berichten, Bericht erstatten, Nachricht geben; dienstlich/amtlich melden/anzeigen* || Meldung || Mitteilung * *Bericht* || Protokoll * *Bericht* || Vortrag * *Bericht, Referat* || Zeugnis * *Schul~*
report back rückmelden || zurückmelden
report of loss Verlustmeldung
report of repair Reparaturbericht
reportable anzeigepflichtig
reporter Referent
reporting Protokollierung
reporting back Rückmeldung * *~ bei Rückkehr*
reporting system Meldesystem
repose liegen * *ruhen, verweilen, schlafen, beruhen (on: auf)*
reposition umsetzen * *z.B. Bauteil an eine andere Position*
represent bedeuten * *kennzeichnen* || darstellen * *allg.* || repräsentieren || verkörpern * *darstellen, verkörpern* || vertreten * *jemanden/etwas ~ (Firma, Partei usw.)* || vorstellen * *darstellen, präsentieren*
representation Darstellung * *z.B. Zahlendarstellung* || Vertretung * *Repräsentanz*
representation abroad Auslandsvertretung
representation method Darstellungsform
representation mode Darstellungsart
representation type Darstellungsart
representative Niederlassung || repräsentativ || typisch * *kennzeichnend (of:für), repräsentativ (Auswahl, Querschnitt i.Sinne d.Statistik)* || Vertreter * *Repräsentant, auch Handels~; local: örtlicher* || Vertretung * *z.B. Industrie~, Handels~, Verkaufsniederlassung, Vertriebsbeauftragter*
representative office Firmenvertretung || Industrievertretung
representative value typischer Wert
repress verdrängen * *(psycholog.)*
repression Verdrängung * *(psychologisch)*
reprimand Verweis * *Tadel*
reprint Sonderdruck * *z.B. eines Zeitschriftenartikels*
reprocess wiederaufbereiten
reprocessing Aufbereitung * *Wiederaufbereitung*
reproduce nachbilden * *kopieren, wiedererzeugen, nachbilden* || reproduzieren || vervielfachen * *nachbilden*
reproducibility Reproduzierbarkeit || Wiederholgenauigkeit
reproducible reproduzierbar
reproduction Modell * *Nachbildung* || Nachbau || Nachbildung * *Reproduktion, Nachbildung, Wiedererzeugung, Fortpflanzung, Wiedergabe* || Vervielfältigung
reproof Verweis * *Vorwurf, Tadel*
repulse abwehren || abweisen * *Angriff* || zurückweisen * *Angriff*
reputation Name * *Ruf, Ansehen*
request anfordern * *bitten, ersuchen; auch Zugriffsrechte, Speicherplatz usw. ~* || Anforderung * *Wunsch, Ersuchen, Nachfrage, Bitte; (for: von/um)* || Anfrage * *Anforderung/Frage; (up)on request: auf ~* || Aufforderung || Auftrag * *Anforderung* || Wunsch * *Bitte; on request: auf ~; by popular request: auf allgemeinen* || wünschen * *bitten, ersuchen um; request sth. from someone: jdn. um etwas ersuchen*
request channel Auftragskanal
request specifier Auftragskennung * (→*PROFIBUS-Profil; für Parameter-Schreib/Leseaufträge*)

requested angefordert || wunschgemäß * *angefordert*
require beanspruchen * *z.B. Mühe, Sorgfalt, Zeit, Platz ~* || benötigen || brauchen * *→benötigen* || erfordern * *benötigen* || fordern || Kosten * (Verb) benötigen, in Anspruch nehmen || nötig haben || verlangen * *benötigen, erfordern* || voraussetzen * *erfordern, benötigen, verlangen,fordern, wünschen*
required erforderlich * *benötigt, gefordert (to:für)* || gefordert || gewünscht * *erforderlich, benötigt, gefordert* || nötig * *benötigt* || notwendig * *benötigt*
required memory capacity Speicherbedarf
requirement Anforderung * *regarding: bezüglich; depending on individual requirements: je nach ~; auch e.Norm* || Anspruch * *(An)Forderung, Bedingung; meet all requirements:allen Ansprüchen gerecht werden* || Bedarf * *auch an Stromversorgung; as required:nach Bedarf* || Bedingung * *Anforderung* || Forderung * *for:nach; auch 'Anforderung'* || Notwendigkeit * *~ einer Sache* || Voraussetzung * *Bedingung, (An)Forderg., Erfordernis; meet the requirements: die ~gen erfüllen*
requirement for energy Energiebedarf
requirements Anforderungen * *meet the requ.:die ~ erfüllen;be subject to stringent requ.:hohe ~ stellen (Aufgabe)*
requiring a considerable amount of capital kapitalintensiv
requiring fast response zeitkritisch * *{SPS|Computer} kurze Reaktionszeiten erfordernd*
requisite nötig * *eforderlich, notwendig* || notwendig * *erforderlich, ~*
rereeler Umroller || Vorroller
resale price Verkaufspreis * *Wieder~*
resaw Nachschnittsäge * *{Holz}* || Spaltsäge * *{Holz}* || Trennkreissäge * *{Holz}* || Trennsäge * *{Holz}*
rescaling Umnormierung
reschedule aufschieben * *Termin* || neu festsetzen || verlegen * *Termin* || verschieben * *(Termin) neu festsetzen* || vertagen
rescheduling Verschiebung * *Termin~, Neufestlegung eines Termins*
research Forschung
research facility Forschungsstätte
research institute Forschungsinstitut
research lab Forschungslabor
research laboratory Forschungslabor
researcher Wissenschaftler * *Forscher*
reseller Fachhändler || Händler * *Wiederverkäufer* || Wiederverkäufer
resemblance Ähnlichkeit * *to/between: Ähnlichkeit mit/zwischen*
resemble ähneln * *ähnlich sein/sehen, gleichen* || ähneln || gleichen * *ähnlich sein*
resembling ähnlich * *ähnlich sehend/seiend, ähnelnd*
reserve belegen * *einen Platz usw. ~* || bestellen * *(amerikan.) reservieren (auch Zimmer)* || Ersatz- || Reserve || reservieren * *Zimmer, Flug, Restaurant-Tisch; auch '~ lassen'* || Vorrat * *Reserve*
reserve capacity Reserve * *Leistungs/Kapazitätsreserve*
reserve factor Sicherheitsfaktor
reserved reserviert || verschlossen * *(figürlich) reserviert*
reserved space Freiraum * *reservierter/freigehaltener Platz; Platz, der nicht belegt werden darf*
reservoir Behälter * *Vorrats-/Zwischen-/Ausgleichs~, Sammel-/Staubecken, Bassin; reservoir of: Vorrat an*
reservoir capacitor Speicherkondensator
reset Entriegelung * *logische ~, Rückstellen* || löschen * *den Inhalt eines Speichers ~, ein Signal ~*

|| Neustart || rücksetzen || Wiederanlauf * *Rücksetzen* || zurücksetzen
reset button Löschtaste || Reset-Taster || Rückstellknopf
reset jack Reset-Buchse
reset output Ausgang rücksetzen
reset pulse Löschimpuls * *Richt/Rücksetzimpuls* || Richtimpuls * *Rücksetzimpuls* || Rücksetzimpuls
reset socket Reset-Buchse
reset time Nachstellzeit
resetting Umbau * *Umstellung, (Neu-)Einstellung, Umrüstung, Nachjustierung, Rückstellung*
resetting button Rückstellknopf
resetting force Rückstellkraft
resharpen schärfen * *nach~*
reside liegen * *liegen/ruhen bei, (inne)wohnen, untergebracht sein (in: in), ansässig sein*
resident program speicherresidentes Programm
residental district Wohngebiet
residential area Wohngebiet
residential settlement Wohngebiet * *Wohnsiedlung*
residual remanent * *Rest-; z.b. remanent field: Restfeld/remanentes Feld* || Rest * *z.B. Feld, Magnetisierung* || übrig * *Rest-*
residual current Reststrom * *bei Relais/Schütz: führt gerade noch nicht z. Anziehen*
residual current device Fehlerstromschutzeinrichtung
residual error bleibender Fehler * *Restfehler* || Restfehler
residual field Restfeld
residual inductance Restinduktivität
residual magnetism Remanenz * *Restmagnetismus* || Restmagnetismus
residual ripple Restwelligkeit
residual torque Leerlaufdrehmoment * *Rest-Drehmoment einer Kupplung/Bremse in geöffnetem Zustand* || Leerlaufmoment * *Rest-Drehmoment einer Kupplung/Bremse in geöffnetem Zustand*
residual unbalance Restunwucht
residual voltage Restspannung * *führt bei Relais/Schütz gerade noch nicht zum Anziehen*
residuary übrig * *Rest-*
residue Ablagerung * *Rückstand, Überbleibsel* || Rückstand * *Rest*
resign gehen * *aus dem Amt ~* || verzichten * *~ auf*
resignation Kündigung * *~ seitens des Arbeitsgebers; period of notice: ~sfrist* || Verzicht
resilience Elastizität || Federung || Nachgiebigkeit * *elastische ~, z.B. einer Feder*
resiliency Rückstellkraft * *durch Federwirkung*
resilient federnd
resilient preloading disc Federring * *Federscheibe für axiale Vorspannung eines Lagers* || Federscheibe * *zur axialen Vorspannung eines Lagers*
resin Harz
resin boiling Harzkochung * *i.B. in der Zellstoff-Herstellung*
resin-bounded paper Hartpapier
resin-encapsulated gießharzisoliert
resist Schutzlack || standhalten * *einer Person oder Sache ~* || widersetzen * *widerstehen, Widerstand leisten, sich streuben gegen* || widerstehen
resistance Beständigkeit || Festigkeit * *Widerstandskraft/-fähigkeit, Beständigkeit, (Biegungs-/Stoß-/Verschleiß-) ~* || Widerstand * *elektr. Größe; (auch figürl.); low/high: klein/groß* || Widerstandsfähigkeit * *to: gegen* || Widerstandswert
resistance against ground-faults Erdschlussfestigkeit
resistance gauge Messwiderstand * *Widerstandsthermometer*
resistance heating Widerstandsheizung
resistance in the cold state Kaltwiderstand
resistance in the hot state Warmwiderstand
resistance method Widerstandsverfahren * *zur Ermittlg. d. Wicklungstemp. e. el.Masch. durch Messg. d. Wickl.widerstands*
resistance per unit length Leitungswiderstand * *Widerstandsbelag, Widerstand pro Längeneinheit* || Widerstandsbelag * *Widerstand pro Längeneinheit bei Kabel, Leitung*
resistance speed-control Widerstandssteuerung * *[el. Masch.] Drehz.veränderung e. Schleifringläufermot. mit Vorwiderständen*
resistance thermometer Widerstandsthermometer * *el. Thermometer, nutzt d.Temperaturabhängigkeit d.Leiterwiderstandes*
resistance to -Festigkeit
resistance to ageing Alterungsbeständigkeit
resistance to aging Alterungsbeständigkeit
resistance to corrosion Korrosionsbeständigkeit || Korrosionsfestigkeit
resistance to creepage currents Kriechstromfestigkeit
resistance to earthquakes Erdbebenfestigkeit
resistance to flow Strömungswiderstand
resistance to heat Temperaturbeständigkeit
resistance to soldering heat Lötwärmebeständigkeit
resistance to wear Haltbarkeit * *Verschleißfestigkeit* || Verschleißfestigkeit
resistance welding Widerstandsschweißen
resistance when cold Kaltwiderstand
resistance when hot Warmwiderstand * *z.B. einer Lampe (ist höher als der Kaltwiderstand)*
resistance wire Widerstandsdraht
resistant beständig * *widerstandsfähig (against:gegen)*
resistant against -fest
resistant against cyrogen frigenfest
resistant against frigen frigenfest
resistant to -fest * *(vorangestellt) widerstandsfähig/beständig gegen*
resistant to ageing alterungsbeständig
resistant to aging alterungsbeständig
resistant to arcing faults störlichtbogenfest
resistant to fracture bruchfest
resistant to obsolence zukunftssicher * *nicht veraltend*
resistant to switching transients schaltfest
resistive Ohmsch
resistive component Ohmscher Anteil
resistive load Ohmsche Last || Widerstandslast
resistive voltage drop Ohmscher Spannungsabfall
resistor Widerstand * *Bauteil*
resistor paste Widerstandspaste * *in der Hybridtechnik*
resistor starter Widerstandsanlasser
resistor temperature detector Widerstandsthermometer
resistor value Widerstandswert
resolution Ansprechempfindlichkeit * *Auflösung* || Auflösung || Beschluss * *Entscheidung* || Stufensprung * *Auflösung*
resolvable lösbar * *Gleichung*
resolve lösen * *Konflikt ~*
resolver Drehmelder || Resolver * *Drehmelder als Lage/Drehzahlgeber mit sinusförmiger Erregg. u. sin/cos-Signalen am Ausgang*
resolver/digital converter Resolver-Digital-Umsetzer
resonance Resonanz
resonance circuit Resonanzkreis
resonance dampening Resonanzdämpfung
resonance damping Resonanzdämpfung
resonance factor Resonanzfaktor * *relative Amplitudenerhöhung bei Resonanz*

resonance frequency Resonanzfrequenz
resonance point Resonanzpunkt
resonance sharpness Resonanzfrequenzüberhöhung
resonance speed Resonanzdrehzahl
resonant circuit Schwingkreis
resonant circuit capacitor Schwingkreiskondensator
resonant circuit inverter Schwingkreiswechselrichter
resonant converter Resonanzumrichter
resonant frequency Resonanzfrequenz
resonant oscillation Resonanzschwingung
resonant speed Resonanzdrehzahl
resonant-pole inverter Resonanzkreisumrichter || Resonanzumrichter
resort to zurückgreifen auf * auch 'Zuflucht nehmen zu'
resource Hilfsmittel * (Hilfs)Quelle/Mittel/Geldmittel, Aktiva
resourceful findig
resources Hilfsmittel * (Plural) (Hilfs-) Quellen/ Mittel, Geldmittel || Kapital || Leistungsfähigkeit * zur Verfügung stehende Leistung || Mittel * Mittel, Geld~, Reichtümer, Resourcen, Hilfsquelle(n), Hilfs~
resp bzw * Abkürzung für 'respectively'.
resp. beziehungsweise
respect Beziehung * in this/some/every respect: in dieser/mancher/jeder ~ || Rücksicht * with respect to: hinsichtlich || schonen * Eigentum, Rechte usw.~
respectable beachtlich * ansehnlich, (recht) beachtlich, respektabel || ordentlich * achtbar, beachtlich
respected gut * gut angesehen
respective betroffen * jeweilig || einschlägig * entsprechend || entsprechend * jeweilig, betreffend || infragekommend || jeweilig * (Adj.)
respectively beziehungsweise * nachgestellt || bzw * nachgestellt || jeweilig * (Adv.)
respiration Atmung * Ausdehnung und Schrumpfung bei Temperaturänderungen
respond Ansprechen * (Verb) z.B. Regelung auf Sollwertänderung; Messgerät || antworten * von einer Mess- oder Regeleinrichtung; to: auf; with: mit || ausregeln * ansprechen, reagieren auf; respond to:e. Störgröße ausregeln || reagieren * to:-auf
respond to beantworten
responded angesprochen * z.B. Regelung, Messung, Überwachung
response Änderungsgeschwindigkeit * bezüglich der Reaktion eines Systems auf einen Stimulus || Ansprechen * z.B. Messinstrument, Überwachung, Begrenzung, Regeleinrichtung || Antwort * ~ einer Mess- o. Regeleinrichtung (z.B. Sprung-/Anstiegs~) || Dynamik || Reaktion * Ansprechen, Antwort, Erwiderung; to:auf || Regeldynamik || Regeleigenschaften || Regelverhalten || Stellungnahme || Übergangsfunktion || Übertragungsverhalten
response characteristic Dynamik * dynam. Eigenschaften || Regelverhalten || Übergangsfunktion * in der Regelungstechnik || Übertragungsfunktion * z.B. eines Regelkreises (siehe auch 'Sprungantwort') || Übertragungsverhalten
response characteristics Regeleigenschaften
response delay Ansprechverzögerung * z.B. einer Kupplung/Bremse || Ansprechverzug * →Ansprechverzögerung z.B. einer Bremse, Kupplung
response level Ansprechwert
response message Antworttelegramm * Inhalt || Folgetelegramm * Antworttelegramm
response rate Ansprechgeschwindigkeit

response requirements dynamische Anforderungen
response sensitivity Ansprechempfindlichkeit
response telegram Antworttelegramm
response temperature Ansprechtemperatur
response threshold Ansprechempfindlichkeit * Ansprechschwelle || Ansprechgrenze * Ansprechschwelle || Ansprechpegel * Ansprechschwelle || Ansprechschwelle * z.B. e. Mess/Regel/Überwachungs/Schutzeinrichtung || Ansprechwert * Ansprechschwelle
response time Anschwingzeit || Ansprechzeit || Antwortzeit || Einschwingzeit || Reaktionszeit
response to disturbances Störverhalten * Reaktion eines Regelkreises bei Einprägung von Störgrößen
response to process-side and mains-side changes Störverhalten * Reaktion e.Regelg. auf Störgrößen u. Netzschwankungen
response to setpoint changes Führungsverhalten * {Regel.techn.}
response to step changes Sprungantwort
response to temperature changes Temperaturgang
response value Ansprechwert
response voltage Ansprechspannung
responsibility Aufgabe * Verantwortlichkeit || Haftung * Verantwortlichkeit || Kompetenz * Zuständigkeit (for: für) || Pflicht * Verantwortung || Sorgfaltspflicht * Verantwortung || Verantwortlichkeit || Verantwortung * carry: tragen; under sole responsibility: unter alleiniger ~
responsible kompetent * zuständig, verantwortlich (for:für) || selbstständig * verantwortlich || verantwortlich * for:für || zuständig * verantwortlich; z.B. Abteilung, Bearbeiter
responsible person Bearbeiter * zuständiger Bearbeiter || Sachbearbeiter
responsive dynamisch * schnell ansprechend, mit hoher Dynamik (z.B. Regelung)
responsiveness Ansprechempfindlichkeit || Dynamik
rest Auflage * Auflage, Stütze, Lehne || Auflagefläche * Auflage, Stütze || aufliegen * ruhen, sich stützen, lehnen (upon: auf); sich stützen/beruhen auf || basieren * beruhen, sich stützen ((up)on:auf) || liegen * ruhen || Pause * Ruhe~, auch einer Maschine || Rest || Stillstand * Ruhe, Ruhepause, Pause || übrig * for the rest: im ~en; the rest: die ~en || übrig * der Rest, das ~e; the rest of: das restliche/ ~e; for the rest: im ~en
rest and de-energized Stillstand mit stromlosen Wicklungen * {el.Masch.}Stillstand ohne Zufuhr el.o.mechan. Energie,"Pause"
rest period Pausenzeit || Ruhepause * auch in einem Lastspiel || Stillstandszeit
restart erneut starten || Neustart * Rechner, Motor usw. || Warmstart || Wiederanlauf * Motor, Rechner usw.; flying restart: ~ einer laufenden Maschine nach Netzausfall || Wiederaufschalten || Wiedereinschalten * (Subst. und Verb)
restart address Anfangsadresse
restart disabling Wiedereinschaltsperre * für Antrieb
restart on the fly Fangen * Aufschaltung auf laufenden Motor || Fangschaltung
restart time delay Wiedereinschaltverzögerung * z.B. zur Vermeidung zu hoher Schalthäufigkeit
restart-on-the-fly capability Fangschaltung
restart-on-the-fly circuit Fangschaltung
restarting Wiederanlauf
restarting at residual voltage Wiedereinschalten bei Restspannung * {el.Masch.} Einschalten e. nach Abschaltg. noch laufenden Motors
resting snugly satt anliegend * against:an

restitution Rückerstattung || Wiederherstellung * ~ *eines Rechts, des alten Zustandes usw.*
restock auffüllen * *Lager* || füllen * *Lager wieder auf~*
restoration -Wiederkehr * *z.B. Netzspannung nach einem Einbruch* || Ergänzung * *Wiederherstellung, Instandsetzung* || Erneuerung || Instandsetzung || Wiederherstellung || Wiederkehr * *z.b. Spannung nach Spannungseinbruch*
restoration of healthy conditions Bereinigung * *(figürl.)*
restore instand setzen * *wiederherstellen* || instandsetzen || wiederherstellen || zurückerstatten || zurückgeben * *zurückerstatten/bringen/geben/stellen*
restoring force Rückstellkraft
restoring spring Rückstellfeder
restrain behindern * *zurückhalten, i.Schranken halten, Einhalt gebieten; restrain s.o. from:jmd. abhalten von* || beschränken * *einengen* || zurückhalten * ~, *Einhalt gebieten, abhalten (to do:von), zügeln, unterdrücken (Gefühl), beschränken*
restrained beengt || beschränkt * *eingeengt*
restraint Gewalt * *Zwang, Beschränkung, Hemmnis, Einschränkung*
restrict begrenzen || beschränken * *to:auf*
restricted begrenzt * *eingeschränkt* || beschränkt * *to: auf* || eingeschränkt * *(to:auf)* || geheim * *Vermerk auf Dokumenten*
restricted place Platzmangel
restricted space beengter Platz
restriction Bedingung * *Einschränkung* || Begrenzung * *Einschränkung* || Beschränkung || Einschränkung
restrictor Drossel * *(mechan., strömungstechn.) z.b. in Hydrauliksystem*
restructure umorganisieren * *umstrukturieren* || umstrukturieren
result -Wirkung * *Ergebnis* || anfallen * *sich ergeben* || auftreten * *sich ergeben, resultieren, anfallen* || Ausgang * *Ergebnis* || auswirken * *Ergebnis* || Auswirkung * *Ergebnis* || Bilanz * *(figürl.)* || entstehen * *resultieren (from:aus)* || Erfolg * *(End)Ergebnis* || ergeben * *in:sich ~; from:sich ~ aus; to:- Ergebnis einer Berechnung* || Ergebnis || Fazit * *draw one's conclusions from something:das Fazit aus etwas ziehen* || Folge * *Resultat* || Folgeerscheinung || führen * *resultieren (in: zu)* || herauskommen * *resultieren* || Resultat || resultieren * *in: in; from (the fact that): aus* || Wirkung * *Ergebnis*
result bit Ergebnisanzeige * *bei SPS*
result from hervorgehen aus * *sich als Folge ergeben aus* || sich ergeben * ~ *aus* || sich ergeben aus
result in bewirken * *als Resultat haben, resultieren in* || ergeben || führen zu || zur Folge haben
result of logic operation Verknüpfungsergebnis * *RLO:VKE bei SPS*
resultant resultierend * *sich ergebend*
resume fortfahren * *wiederanfangen, wieder (an der alten Stelle) aufsetzen* || fortführen * *wieder fortsetzen* || fortsetzen * *wieder fortsetzen nach Unterbrechung* || wieder anfangen || wieder aufnehmen || wiederanfangen
resumé Resümee || Rsumee || Zusammenfassung * *Resümee, Zusammenfassung*
resumption Fortsetzung * *Wieder~, Wiederaufnahme* || Wiederaufnahme * *z.B. einer Tätigkeit, von Zahlungen usw.*
resumption of power supply Netzwiederkehr
resumption of service Wiederaufnahme des Betriebes
retail price Einzelpreis * *Einzelhandelpreis, Ladenpreis, Kleinmengenpreis* || Verkaufspreis * *Endverbraucher-/Einzelhandelspreis, Kleinmengen-Preis für Wiederverkäufer*
retail store Einzelhandelsgeschäft

retailer Einzelhändler || Händler * *Einzelhändler* || Verkäufer * *Einzelhändler, Wiederverkäufer* || Wiederverkäufer
retain anhalten * *beibehalten, halten (z.B. I-Anteil, Durchmesserrechner), bewahren* || aufrechterhalten * *beibehalten* || behalten * *auch im Gedächtnis, Speicher usw.* || beibehalten || erhalten * *aufrecht~* || festhalten * *einbehalten* || halten * *zurück-/fest~, (zurückbe-) halten, sichern, stützen, befestigen, (Wasser usw.) stauen* || zurückhalten * *zurückbehalten, einbehalten, beibehalten, (Eigenschaft/Posten/i.Gedächtnis) behalten*
retained condition gedrückter Zustand * *z.B. bei einer Taste*
retaining bolt Befestigungsschraube
retaining bracket Halteblech * *z.B. für PC-Karte*
retaining element Befestigungselement * *Halteelement*
retaining frame Halterahmen * *z.B. für Leiterkarte*
retaining hole Befestigungsbohrung
retaining nut Befestigungsmutter
retaining ring Sicherungsring || Sprengring
retaining screw Befestigungsschraube * *zum Festhalten/Niederhalten; verhindert das Herunterfallen*
retaining slot Befestigungsschlitz * *z.B. in Montageblech oder Spulen-/Trafo-Befestigungswinkel z.Schraubbefestig.*
retaining unit Halterung
retaining washer Sicherungsscheibe
retard verzögern * ~, *verlangsamen, bremsen, hemmen, nachstellen (Zündg.), verspäten, auf-/zurückhalten* || zurückhalten * *verzögern*
retardation Verzögerung * *Verzögerung, Verlangsamung, Verspätung, Aufschub*
retardation torque Verzögerungsmoment
retarder Retarder * *(hydraul.) Strömungsbremse für LKWs, Schienenfahrzeuge usw.; Gleisbremse* || Strömungsbremse * *hydraul. ~/Retarder für LKWs, Schienenfahrzeuge usw.*
retention Erhaltung * *Bewahrung, Festhalten, Beibehaltung, Zurückhalten, Einbehaltung* || Retention * *Zurück(be)haltung* || zurückhalten * *(Substantiv) Zurückhaltung, Einbehaltung, Beibehaltung, Festhalten, Merken, Bewahrung*
retention force Abzugskraft * *eines Steckers*
retention time Verweilzeit
retentive nichtflüchtig || nullspannungsgesichert * *retentive flag: Haftmerker bei SPS* || remanent * *Merker/Kontakt bei Kontaktplan/Relaissteuerung*
retentivity Remanenz
reticulation Geflecht * *Netz(werk)*
retouch überarbeiten
retract einziehen * *(Fahrwerk, Klauen usw.) ~, (Behauptung usw.) zurückziehen/widerrufen, zurücktreten* || zurücknehmen * *Vorwürfe, Urteil, Kritik usw. ~*
retracting zurückziehend * *z.B. Kennlinie*
retraining Weiterbildung
retreat rudernerneuern * *(Reifen)*
retrieval Rückgewinnung || Wiederherstellung
retrieve abholen * *heraus/zurückholen/fischen, wiedererlangen* || auskoppeln * *(wieder) herausholen, "herausfischen"; (from: aus)* || entnehmen * *herausholen* || gewinnen * *(Wieder-) gewinnen/-erlangen, (Verlust usw.) wettmachen, der Vergessenheit entreißen* || rekonstruieren * *wiederherstellen* || retten * *wiedergewinnen/-erlangen* || rückgewinnen * *wiederfinden, wiederbekommen, (sich etwas) zurückholen, (Fehler usw.) wiedergutmachen*
retrigger nachtriggern
retro- Rück- * *zurück, rückwärtsgerichtet*
retro-fitting Modernisierung * *nachträgliches Einbauen neuer Komponenten*

retroact rückwirken || zurückwirken
retroaction Rückwirkung
retroaction-free rückwirkungsfrei
retroactive effect Rückwirkung * zurückwirkender *Effekt*
retrofit anbauen * *nachträglich anbauen, nachrüsten* || einbauen * *nachrüsten, nachträglich einbauen* || nachrüsten * *nachträglich einbauen* || nachträglich einbauen || umbauen * *neue Teile einbauen/nachrüsten*
retrofit assembly Nachrüstsatz
retrofit set Nachrüstsatz || Umbausatz
retrofitting Umbau * *Einbau/Nachrüstung neuer Teile*
retrogression Rückgang * *Rückwärtsgehen* || Rückschritt
return abgeben * *wieder ~, zurückgeben* || Einnahme * *Geschäftseinnahme* || Ertrag * *Ertrag, Einnahme, Verzinsung, Gewinn; yield/bring returns: Nutzen abwerfen, ertragreich s.* || Rückfluss * *auch von Kapital, Investitionen usw.* || Rückgang * *Rückweg* || Rücklauf * *Umkehr* || Rücksendung || Rücksprung * *in e. Programm* || Umkehr * *(auch figürl.) to: zu* || umkehren || Wiederkehr * *auch von Spannung/Leistung/Versorgung* || wiederkehren || zurückerstatten || zurückgeben || zurückgehen * *wiederkehren* || zurücklaufen * *umkehren, zurücklaufen* || zurückliefern * *z.B. Vertriebsergebnis/Profit/Bremsleistung ans Netz ~* || zurückschalten || zurücksenden || Zurücksendung
return conductor Rückleitung
return conveyor Werkstückrückführung
return flow Rücklauf * *Flüssigkeit*
return journey Rückfahrt
return letter Antwort * *~schreiben, ~brief* || Stellungnahme * *Antwortschreiben*
return line Rückleitung * *für ein Signal*
return motion Rücklauf
return of energy Energierückspeisung * *z.B.bei Nutzbremsung (to:in)*
return of energy to the supply system Energierückspeisung ins Netz
return of investment Ertrag * *Nutzbringung des eingesetzten Kaptitals*
return on equity Eigenkapitalrendite * *Verhältnis (Gewinn/eingesetztes Kapital) x 100%*
return on investment Amortisation || Amortisierung || Kapitalrückfluss * *einer Investition* || Rendite
return pressure Rückstellkraft * *Rückdruck*
return spring Rückholfeder || Rückstellfeder
return thanks danken
return travel Rückfahrt * *auch eines Maschinenteils/Krans usw.*
return trip Rückfahrt
returnable packaging Mehrwegverpackung * *{Verpack.technik}*
returned good Rückware * *z.B. zur Reparatur/Gutschrift rückgesandtes Teil* || Warenrücksendung * *zurückgesandte Ware*
returned product Rückware
returned products form Retourenbegleitschein * *zur Rücklieferung defekter Geräte*
returns Gewinn * *Ertrag, Verzinsung, Gewinn, Rückfluss* || nutzen * *(Substantiv) Ertrag*
reuse Wiederverwendung
rev rückwärts * *Abkürzung für 'reverse'*
rev./min upm
revamp Umbau * *(salopp) Renovierung, Herausputzen, "Aufmotzen"* || umbauen * *(salopp) neu herausputzen, aufpolieren, 'aufmotzen', herrichten, renovieren*
reveal erschließen * *offenbaren, enthüllen, erkennen lassen, aufdecken* || zeigen * *(figürl.) offenbaren, enthüllen, zeigen, erkennen lassen, aufdecken, verraten; Fensterrahmen*
revenge Quittung * *(figürl.) Revanche*
revenue Einnahme * *~ des Staates* || Ertrag * *Geldeinnahmen* || Gerichtsstand * *Gerichtsort*
reversal -Umkehr || Invertierung * *Umkehr, z.B. der Drehrichtung* || Reversieren || Umkehr * *z.B. Drehrichtungs~, Drehmomenten~, Drehrichtungs~* || Umschaltung * *Umkehr*
reversal braking Gegenstrombremsung * *eines Asynchronotors durch Umpolung des Drehfeldes*
reversal of magnetization Ummagnetisierung
reversal point Umkehrpunkt
reversal time Umsteuerzeit
reverse entgegengesetzt * *Bewegung* || entgegengesetzt gerichtet * *Bewegung* || Gegenteil * *of:von* || Reversieren * *(Verb)* || Rück- * *umgekehrt, in umgekehrter Richtung* || Rückseite * *andere,umgekehrte Seite* || rückwärts * *Drehbewegung* || umgekehrt * *z.B. Richtung* || umkehren || umsetzen * *{Bahnen} Zug/Waggon in die Gegenrichtung umrangieren* || vertauschen * *Phasen* || wechseln * *umkehren* || wenden || zurück * *umgekehrt*
reverse action Rücklauf
reverse blocking Rücklaufsperre
reverse conveyor Rücklaufeinrichtung * *{Fördertechnik}*
reverse current Gegenstrom || Rückstrom || Sperrstrom * *(Diode, Thyristor usw.)*
reverse diode Rückspeisediode
reverse direction Rückwärtsrichtung || umgekehrte Richtung
reverse direction of rotation Linksdrehrichtung || Rückwärtsdrehrichtung
reverse documentation Rückdokumentation
reverse movement Rücklauf * *auch "reversing, reverse operation"*
reverse polarity Verpolung
reverse recovery time Sperrverzugszeit * *z.B. einer Diode*
reverse rotation Rückwärtsdrehrichtung
reverse running Linkslauf || Rücklauf * *Lauf in Rückwärtsrichtung*
reverse running prevention Rücklaufsperre
reverse side Rückseite * *andere,umgekehrte Seite* || Rückseite * *of this sheet:dieses (Papier)Blattes*
reverse speed direction Rückwärtsdrehrichtung
reverse stroke Rücklauf * *(Kolben-) Hub in Rückrichtung*
reverse the polarity umpolen
reverse thread Linksgewinde * *reverse threaded: mit ~ versehen sein*
reverse torque Rückwärtsdrehmoment
reverse translation Rückübersetzung
reverse voltage Sperrspannung * *z.B. v. Diode, Transistor*
reversed verkehrt * *verkehrt herum*
reversible reversibel || umkehrbar * *z.B. Drehrichtung* || umschaltbar * *z.B. Motor mit umschaltbarer Drehrichtung*
reversible connection Umkehrschaltung * *4-Quadrant Stromrichter*
reversible converter Umkehrstromrichter
reversible drive Reversierantrieb || Umkehrantrieb
reversible motor Umkehrmotor * *in der Drehrichtung*
reversible tip cutter Wendeschneidplatte * *{Werkz.masch.}*
reversing Drehrichtungsumkehr || Reversier- * *z.B. Walzwerk* || Reversieren || Umsteuern
reversing amplifier Umkehrverstärker
reversing connection Zu- und Gegenschaltung
reversing contactor Wendeschütz * *in Ständer-, Anker- oder Feldkreis zur Drehrichtungsumkehr eines Motors*

reversing drive Reversierantrieb || Umkehrantrieb
reversing gear Wendegetriebe * z.b.
reversing gearbox Umkehrgetriebe || Wendegetriebe * z.b. im Schiffbau
reversing logic module Kommandostufe
reversing motor Reversiermotor
reversing pole Wendepol
reversing process Umschaltvorgang * *Drehrichtungs/Momentenrichtungsumkehr* || Umsteuervorgang * *Momentenumkehr bei 4Q-Stromrichter*
reversing rolling motor Umkehr-Walzmotor
reversing starter Wendestarter
reversing switch Wechselschalter || Wendeschalter * *zur Motor-Drehrichtungsumkehr durch Tausch von Ständer-/Anker- oder Feldleitungen*
reversing time Reversierzeit * *z.B. in Walzgerüst* || Umsteuerzeit * *zur Drehzahl- oder Momentenumkehr benötigte Zeit*
review Besichtigung * *prüfende* ~ || Besprechung * *Buchbesprechung, Zwischenbesprechung, Entwicklungsbesprechung* || Beurteilung * *kritische Besprechung, Rezension (z.b. eines Buches), Kritik* || Durchsprache || durchsprechen || Kritik * *Besprechung* || prüfen * *nach~, über~ (Entwicklungsvorhaben, Entscheidung, Schriftstück), inspizieren, revidieren* || Prüfung * *Revision, Über~ (Buch, Gerichtsurteil, Entscheidung, Entwicklungsvorhaben usw.)* || Revision * *Überprüfung (Buch, Gerichtsurteil, Entscheidung, Entwicklungsvorhaben usw.)* || überarbeiten * *Schriftstück, Druckerzeugnis usw.* ~ || Überblick geben || überprüfen * *Druckschrift überarbeiten, Ablauf eines Projektes* ~ || Überprüfung * *Revision, ~ (Entwickl.vorhaben, Entscheidg., Gerichtsurteil), (Buch-) Rezension* || Übersicht * *Besprechung, Rückblick, Kritik, kritische Besprechung, auch 'eine ~ geben über'*
revisal Revision * *~ eines Vertrages*
revise ändern * *revidieren, überarbeiten (Buch usw.), (wieder/nochmals) durchsehen, überprüfen* || bearbeiten * *überarbeiten* || korrekturlesen * *revidieren; nochmals durchlesen, die zweite Korrektur lesen; Umbruch neu erstellen* || korrigieren * *revidieren, korrekturlesen, Druckerzeugnis/Dokument überarbeiten* || überarbeiten * *Gerät, Dokument usw.* ~ || umarbeiten * *Buch/Schriftstück* ~ || verbessern * *revidieren, z.B. Buch, Beschreibung, Software*
revised überarbeitet * *z.B. Buch, Software, Druckschrift*
revised edition Bearbeitung * *überarbeitete Auflage eines Druckerzeugnisses*
revision Bearbeitung * *~ eines Druckerzeugnisses; Überarbeitung* || Revision * *Revision, Durchsicht, Überarbeitung, Korrektur, verbesserte Ausgabe/ Auflage, Ausgabestand* || Überarbeitung * *~, Durchsicht, Korrektur, verbesserte Ausgabe e. Dokuments, Buchs, Softwareprogramms* || Umarbeitung
revision date Ausgabedatum * *von Softwareprogrammen, Druckerzeugnissen usw.*
revival Erneuerung
revive erneuern * *(sich) neu beleben*
revoke aufheben * *widerrufen* || zurücknehmen * *widerrufen*
revolution Drehung * *[rewol'uschen] Umdrehung* || Kreislauf || Rotation || Umdrehung * *[rewol'u:schen]* || Umlauf * *(Um)Drehung*
revolution counter Drehzahlmesser
revolution indicator Drehzahlmesser
revolutionary bahnbrechend * *revolutionär*
revolutionize revolutionieren
revolutions Umdrehungen || Umdrehungen pro Minute
revolutions per minute Umdrehungen pro Minute || upm

revolutions/minute upm
revolve drehen * *sich* ~ || kreisen || rotieren || sich drehen || umlaufen
revolving Schwenk- || schwenkbar * *drehbar* || umlaufend
revolving-cylinder roaster Trommelofen * *Röst-/Abschwelofen*
reward Belohnung || Lohn * *Entgelt, (gerechter) Lohn, Belohnung, Vergütung*
rewarding lohnend * *Gewinn bringend, auch figürlich*
rewind Rücklauf * *Zurückspulen/Wickeln* || umrollen * *z.B. eine Papierbahn* || umspulen || umwickeln * *von einer Rolle auf die andere* || wickeln * *wiederauf~ (z.b. beim Rollenschneider)* || zurücklaufen * *zurückwickeln/spulen*
rewinder Aufwickler * *Wiederaufwickler, auch Umroller* || Rollenschneider * *Umroller, Wiederaufroller* || Umroller || Vorroller * *Umroller, Wiederaufwickler*
rewire umverdrahten
rewiring Umverdrahtung || Verdrahtungsänderung
rewrite überschreiben * *z.b. einer Speicherzelle, Texteingabe auf Bildschirm* || umarbeiten * *Schriftstück/Softwareprogramm* ~
rewriting Umarbeitung * *Schriftsatz, Softwareprogramm*
RFI filter Funk-Entstörfilter * *z. Unterdrückung von 'Radio Frequency Interference'* || Funkentstörfilter || Netzfilter * *Abk.f. 'Radio-Frequency Interference suppression filter': Funkentstörfilter*
RFI level Funkstörgrad
RFI reactor Entstördrossel || Funkentstördrossel
RFI suppressed funkentstört
RFI suppression Entstörung * *Unterdrückung von 'Radio Frequency Interference': Funkentstörung* || Funk-Entstörung * *Unterdrückung von 'Radio Frequency Interference'* || Funkentstörung * *Unterdrückung von 'Radio Frequency Interference':- Funkfrequenz-Störungen* || Störunterdrückung * *Funkentstörung*
RFI suppression capacitor Entstörkondensator * *zur Funk→entstörung* || Funkentstörkondensator
RFI suppression class Funkentstörgrad * *(Abk.f. 'Radio Frequency Interference') siehe VDE 0871, 0875*
RFI suppression filter Funk-Entstörfilter * *z. Unterdrückung von 'Radio Frequency Interference'* || Funkentstörfilter * *z. Unterdrückg.v. 'Radio Frequency Interference': Funkfrequenzstörungen*
RFI suppression reactor Entstördrossel * *Funkentstördrossel* || Funkentstördrossel
RFI-Emission Störaussendung * *Radio Frequency Interference, elektromagnetische Störabstrahlung*
rheostatic braking generatorisches Bremsen * *mit Bremswiderstand* || Widerstandsbremsung * *z.B. bei Drehstrom- und 1Q-Gleichstromantrieb*
rheostatic starter Widerstandsanlasser
rib Lamelle * *an Kühler* || Rippe * *auch Kühlrippe eines Motors/Kühlkörpers*
rib cooling Rippenkühlung * *Kühlung über Kühlrippen (z.B. Motor)*
rib frame motor rippengekühlter Motor
rib-cooled rippengekühlt * *z.B. Motor*
rib-cooled motor rippengekühlter Motor
ribbed gerippt * *ribbed frame; geripptes (Motor-) Gehäuse* || verrippt * *z.B. mit Kühlrippen versehen (Motor, Kühlkörper usw.)*
ribbed heat sink Rippenkühlkörper
ribbed housing Rippengehäuse * *z.B. Motorgehäuse mit Kühlrippen*
ribbing Riefenbildung || Verrippung
ribbon cable Bandkabel || Flachbandkabel || Flachbandleitung || Flachkabel || Flachleitung || Signalbandkabel

ribbon cable connection Bandkabelverbindung
ribbon cable connector Flachbandstecker || Flachleitungs-Steckverbinder || Flachsteckverbinder * *für Flachleitung*
ribbon saw Bandsäge
ribbon-cable flat connector Flachleitungs-Steckverbinder
riddle Sieb * ~ *für Sand usw.*
ride fahren * *mit einem (beliebigen) Beförderungsmittel; auch laufen/gleiten/schweben/aufliegen auf* || Fahrt * *mit Wagen* || mitlaufen auf * *z.B. Tastrolle auf Warenbahn, Druckwalze auf Wickel*
ride comfort Fahrkomfort
ride-through Überbrückung * *eines Netzspannungseinbruchs bei Antrieb, ohne Abzuschalten*
ride-through interval Überbrückungszeit * *Netzausfall/Einbruchzeit, die überbrückt werden muss/kann*
rider roll Druckwalze * *z.B. obere 3.Rolle bei Doppeltragwalzenroller,dient auch zur Durchmessererfassung*
riding characteristics Fahreigenschaften * *eines Fahrzeugs/Aufzugs usw.*
riding comfort Fahrkomfort * *eines Fahrzeugs/ Aufzugs usw.*
riding direction Fahrtrichtung * *bei Zug, Bus, Aufzug usw.*
rift Schlitz * Spalt || Spalt * *Spalt(e), Ritze, Sprung, Riss*
rig Verspannung * ~ *aus (Draht)Seilen, Takelage*
rigging Verspannung * *mit Tauen*
right Befugnis || Berechtigung * *(An)Recht (to: zu), (Rechts)Anspruch, Berechtigung* || gut * *richtig* || rechts * *right from: ~ von; turn right: nach ~ drehen/abbiegen; on the right: ~* || richtig || zutreffend
right angle rechter Winkel * *be at right angles to each other:um 90 Grad gegeneinander versetzt sein*
right angle gear Winkelgetriebe * *Ein- und Ausgangswelle bilden e.rechten Winkel*
right away sofort
right from the start von Anfang an
right hand rechte Seite || rechtes * *(Adj.)* || Rechts- * *rechts befindlich*
right hand side rechte Seite || rechts * *~seitig; to the right of: ~ von*
right in front vorn * *ganz ~*
right next to direkt neben
right of access Zugangsberechtigung || Zugriffsberechtigung
right stop Rechtsanschlag * *z.B. eines Potis*
right up from below unten * *von ~ an*
right- Rechts-
right-angeled senkrecht * →*rechtwinklig*
right-angled rechtwinklig
right-hand rechtsgängig * *z.B. Schraube, Gewinde*
right-hand thread Rechtsgewinde
right-handed thread Rechtsgewinde
right-justified rechtsbündig * *z.B. in der Daten-/Textverarbeitung*
righthand rechts * *zur Rechten (stehend usw.), ~ angeordnet, ~gängig, ~läufig, Rechts-* || rechtsseitig
righthand side rechte Seite
rightmost rechts * *das am weitesten ~ befindliche*
rigid fest * *starr, stabil* || formschlüssig * *starr* || formstabil || robust * *standfest, formfest, stabil* || starr * *(auch figürl.)* || steif || streng * *unnachgiebig, ohne Ausnahme* || unempfindlich * *robust*
rigid coupling Formschluss * *starre Kopplung* || starre Kupplung
rigid foam Hartschaum
rigid shaper Starrfräse * *{Holz}*
rigid-spindle moulder Starrfräse * *{Holz} im Gegensatz zur* →*Schwenkfräse*

rigid-spindle shaper Starrfräsmaschine * *{Holz} im Gegensatz zur* →*Schwenkfräsmaschine*
rigidity Qualität * *Stand/Formfestigkeit, Stabilität* || Steifigkeit
rigidly coupled starr gekoppelt || starr gekuppelt
rigidly cyclic streng zyklisch
rigidly engaged formschlüssig
rigorous konsequent * *(peinlich) genau, strikt* || streng || strikt * *peinlich genau, strikt*
rigorous analytical process gründliche Analyse
rim bördeln || Kante * *Rand, Randwulst, Felge* || Rand * *~ eines Rings, einer Scheibe, einer Vertiefung; Absatz*
RINA RINA * *Abkürzg.f. 'Registro Italiano Navale': italienische Schiffsbauvorschriften*
ring Kreis * *Ring* || Reifen || Ring
ring binder Aktenordner * *Ringbuch, dünner Aktenordner*
ring buffer Ringpuffer || Ringspeicher * *Ringpuffer* || Umlaufpuffer || Umlaufspeicher
ring bus Ringbus
ring clamp Ringschelle
ring clip Ringschelle
ring counter Ringzähler
ring fan Ringventilator
ring gear Zahnkranz
ring gear chuck Zahnkranzbohrfutter * *für Bohrmaschine*
ring lug Ringöse * *z.B. Hebeöse am Motor*
ring spindle Ringspindel * *{Textil} in Ringspinnmaschine zur Garnerzeugung*
ring spinning machine Ring-Spinnmaschine * *Textilmaschine* || Ringspinnmaschine * *{Textil} (i.Gegensatz zur* →*Open-End-Spinnmaschine)*
ring wound transformer Ringkerntrafo
ring-around Umschwingen * *eines Thyristors*
ring-around arm Umschwingkreis * *sorgt für d.Kommutierung i.e.selbstgeführten Stromrichter*
ring-around circuit Umschwingkreis
ring-around reactor Umschwingdrossel * *unterstützt d. Kommutierg. in e. selbstgeführten Stromrichter*
ring-around thyristor Umschwingthyristor
ring-back arm Rückschwingzweig
ring-back thyristor Rückschwingthyristor
ring-type current transformer Durchsteckwandler * *Stromwandler*
ring-type transducer Durchsteckwandler * *z.B. Strom-Messwandler*
ring-wound transformer Ringkern-Transformator
ringed Ring-
ringlike Ring- || ringförmig
rinse spülen
rinse machine Spülmaschine
rinse out ausspülen
rinsing machine Ausspritzapparat * *z.B. zur Reinigung von Flaschen*
rip aufteilen * *{Holz} auf-/zertrennen, (der Länge nach) auseinandersägen* || schneiden * *(zer-) schlitzen, längs~ (z.B. Holzfurnier), der Länge nach auf~; rip up: aufschlitzen*
rip saw Leistensäge * *{Holz} Längsteilsäge*
rip up aufreißen
rip-cut saw Tiefenschnittsäge * *{Holz}*
ripeness Reife * *(auch figürl.) siehe auch* →*Ausgereiftheit*
ripping chisel Stechbeitel
ripping-tooth chain Spitzzahnkette * *{Holz} für Kettensäge*
ripple -Welligkeit * *z.B. von elektr. Spannung, Drehmoment usw.* || Überlagerung * *Welligkeit (z.B. auf einer Gleichspannung)* || Welligkeit * *z.B. von Spannung/Strom/Drehmoment*
ripple content Oberschwingungsanteil || Wechselspannungsanteil * *Welligkeits-Anteil* || Welligkeit

* *Effektivwert d.Wechselspannungsanteils e.Gleichspannung; DIN 41750 Tl.4 u.41755 Tl.1*
ripple control Rundsteuerung
ripple control pulse train Rundsteuertelegramm
ripple control receiver Rundsteuerempfänger
ripple control system Rundsteueranlage || Rundsteuerung * *Rundsteueranlage*
ripple control transmitter Rundsteuersender
ripple current überlagerter Wechselstrom * *Strom(rest)welligkeit*
ripple current capability Wechselstrombelastbarkeit * *z.B. eines Kondensators*
ripple rate Oberwellenanteil
ripple telecontrol signal Rundsteuertelegramm
ripple voltage Brummspannung * *Spannungswelligkeit* || überlagerte Wechselspannung * *Spannungs(rest)welligkeit*
rippled geriffelt || gewellt
RISC RISC * *'Reduced Instruction Set Computer' mit einfachen aber schnellen u.kurzen Befehlen (→CISC)*
RISC processor RISC-Prozessor * *Reduced Instruction Set Computer: Zentraleinh. mit wenigen aber schnellen Befehlen*
rise ansteigen * *Gelände, Spannung, Strom usw.; auch 'zunehmen'* || Anstieg || anziehen * *von Preisen usw.* || Aufschlag || *Erhöhung, Hochsetzung (Preis, Kurs usw.)* || aufsteigen * *Fläche, Gelände usw.; zunehmen* || Aufstieg || Entstehung * *Ursprung, Entstehung, Anlass* || Erhöhung * *Anstieg* || Schwung * *Auf~* || steigen * *Zahlenwert, Wasserspiegel, Temperatur, Preise, Kurse (to:bis); in die Luft ~* || Steigerung * *(Auf/An)Steigen, Aufstieg (auch figürl.), Aufschwung, Zuwachs, Zunehmen* || Steigung * *Anstieg* || zunehmen
rise delay Anstiegsverzögerung
rise in prices Preissteigerung
rise of -Anstieg
rise of temperature Temperaturanstieg
rise time Anregelzeit * *benötigt d.Regelgröße nach e.Sollwertsprung z. erstmaligen Erreichen d. Sollwerts* || Anstiegszeit * *eines Impulses*
rise up auftauchen
rising ansteigend * *Gelände, Fläche; zunehmen* || aufsteigend * *Gelände, Fläche; zunehmen* || steigend * *z.B. Flanke, Kennlinie*
rising edge ansteigende Flanke * *eines Impulses* || positive Flanke * *aufsteigende Flanke*
rising prices Preissteigerung
risk Gefahr * *Wagnis* || gefährden * *aufs Spiel setzen, riskieren* || Risiko * *take: eingehen, incur: übernehmen, at one's own risk:auf eigenes ~* || Wagnis
risk capital Wagniskapital
risk of injury or death Personengefährdung
risk of personnel injury Personengefährdung
risk transfer Gefahrenübergang * *{Wirtschaft}* Übergang des Risikos vom Hersteller auf den Käufer
risk-free risikolos
risky gefährlich * *riskant*
RIV test Störspannungsprüfung * *Abk. für 'Radio Interference Voltage'*
rival konkurrieren || konkurrierend
rival firm Konkurrenzfirma || Wettbewerber * *Konkurrenzfirma*
rivals Konkurrenz * *Marktrivalen* || Mitbewerb * *Marktgegner* || Mitbewerber * *(Markt)Gegner*
rivet Niet || Niete || nieten || vernieten
RLO Verknüpfungsergebnis * *Abkürzg.f. 'Result of Logic Operation'*
rms effektiv * *Effektiv- (Wert, Spannung, Leistung usw.)* || Effektivwert * *Abkürzg. für 'Root Mean Square value':quadrat. Mittelwert, Effektivwert*
RMS control Effektivwertregelung * *'Abkürzg. für*

'Root Mean Square':Quadratischer Mittelwert, Effektivwert
RMS current Effektivstrom || Effektivwert des Stroms
RMS ripple factor Welligkeit * *Effektivwert des Wechselanteils*
rms value Effektivwert || Mittelwert * *Effektivwert (d.h. quadratischer ~)*
RMS voltage Effektivspannung || Effektivwert der Spannung
road Straße || Weg * *Straße; road to success: ~ zum Erfolg*
road transport Straßentransport
roadway construction Straßenbau
roar rattern
robot Automat * *Handhabungsautomat, Roboter* || Roboter
robot arm Roboterarm
robot control Robotersteuerung
robot controller Robotersteuerung
robot technology Robotertechnik
robotics Robotertechnik
robust kräftig || robust * *kräftig, stark, stabil (auch Parameterunempfindlichkeit e.Regelung), widerstandsfähig* || stark * *kräftig* || störunempfindlich * *robust* || widerstandsfähig
robust control robuste Regelung
robustness Robustheit * *Stabilität (auch e. Regelung), Stärke, Festigkeit* || Störunempfindlichkeit * *Robustheit* || Widerstandsfähigkeit
rock rütteln || schwanken * *erbeben, sich wiegen, schaukeln* || schwingen || wackeln * *schwanken*
rock crusher Steinbrecher || Steinmühle
rocker Kipphebel || Schaltknebel * *an Handschalter* || Schaltwippe * *an Handschalter* || Schwinghebel || Wippe
rocker arm Kipphebel
rocker rarm Pendelarm
rocking Schwankung * *Schaukeln, Wiegen* || Schwenk- * *hin- und herschwenkend/-schaukelnd*
rod Rundstab || Stab * *Stange* || Stange
rod bearing Pleuellager
rod drawing die Stangenziehstein
rod heater Heizstab
rod-stop Bund * *Anschlag*
ROE Eigenkapitalrendite * *Abkürzg. für 'Return On Equity': Verhältnis Gewinn/eingesetztes Kapital*
role Funktion * *(figürlich) Rolle, Part* || Rolle * *(figürl.) play a (significant/decisive) role: eine (wesentliche/entscheidende) ~ spielen*
roll Liste * *Gehalts~, Lohn~, Steuer~* || Rolle * *Rolle, Walze; live roll: angetriebene ~* || Rolle * *auch Bildschirmanzeige (up:hoch; down:hinunter/rückwärts)* || Walze || walzen || wälzen * *auch 'sich ~'*
roll adjustment Walzenanstellung * *{Walzwerk}*
roll change Rollenwechsel
roll changing Walzenwechsel
roll cutter Rollenschneidmaschine
roll diameter Rollendurchmesser
roll down rückwärts rollen * *auf dem Bildschirm*
roll exchange Rollenwechsel * *Austausch einer (abgenutzten) Rolle/Walze* || Walzenwechsel * *Austausch*
roll feed Walzenvorschub
roll force Walzkraft
roll grinding machine Walzenschleifmaschine
roll heating Walzenheizung * *z.B. bei Papiermaschine, Kalander usw.*
roll off abrollen * *mit der Hand*
roll on edge stauchen * *{Walzwerk}*
roll out Markteinführung
roll pressure Walzdruck

roll shell Walzenmantel * z.B. bei Siebsaugwalze einer Papiermaschine
roll shifting Walzenverschiebung * Walzwerk
roll slitter Rollenschneider || Rollenschneidmaschine || Umroller * →Rollenschneider
roll stand Rollenständer * Abwickler || Walzgerüst
roll tightness Wickelhärte
roll transfer Rollenwechsel * bei Wickler
roll turning lathe Walzendrehmaschine
roll up rollen * (sich) aufrollen, z.B. Papier || vorwärts rollen * auf dem Bildschirm
roll up the shirt sleeves Ärmel hochkrempeln * (auch figürl.)
roll wear Walzenverschleiß
roll's tension characteristic Wickelhärtenkennlinie * mit d. Durchm. abnehm.~ verhind.Teleskopieren u.Quetschg. b.Aufwickler
rollback lock Rücklaufsperre
rollback-stop Rücklaufsperre * mechanische ~
rollbar Rollbalken * am Bildschirm || Schiebebalken * am Bildschirm
rolled edge Walzkante
rolled iron Walzeisen
rolled plate Blech * (dickes) Walz~ || Walzblech
rolled steel gewalzter Stahl || Walzstahl
rolled wire Walzdraht
roller Druckwalze * im Walzwerk || Rolle * Walze || Walze * (Stoff-, Garn-)Rolle, Zylinder, Druckwalze (in Druckmaschine), Lauf/Gleit/Führungsrolle
roller adjustment drive Walzenverstellvorrichtung
roller bearing Rollenlager
roller cage Rollring * Lagerkäfig
roller chain Rollenkette
roller conveyor Rollenbahn * {Fördertechnik} Rollenförderer (live: angetriebene); buffer roller conveyor: Stau-~ || Rollenbahnförderer || Rollenförderer || Rollgang * Transportrollgang || Transportrollgang
roller covering Walzenbezug * {z.B. in Druckerei}
roller finishing machine Glattwalzmaschine * {Werkz.masch.}
roller forming machine Profil-Walzmaschine
roller freewheel Freilauf * Klemmrollen-Freilauf || Klemmrollenfreilauf
roller lever Rollenhebel
roller screw Rollspindel
roller table Rollentisch * {z.B. Holz} Zuführeinrichtung || Rollgang || Transportrollgang
roller thrust bearing Rollendrucklager
roller transporter Transportrollgang
roller trestle Rollenbock * z.B. für Langholz im Sägewerk
roller-coaster Achterbahn
roller-table drive motor Rollgangsmotor
roller-table motor Rollgangsmotor * {el.Masch.} Antriebsmotor für Rollgänge in Walzwerken
rolling barrel Putztrommel
rolling bearing Wälzlager
rolling direction Walzrichtung
rolling friction rollende Reibung || Rollreibung * z.B. eines Laufrades am Fahrzeug
rolling gap Walzspalt
rolling gap control Walzspaltregelung
rolling line Walzstraße
rolling load Walzkraft
rolling mill Walzwerk
rolling mill drive Walzwerksantrieb
rolling mill line Walzstraße
rolling mill motor Walzmotor || Walzwerkmotor || Walzwerksmotor
rolling motor Walzmotor
rolling ring Rollring
rolling speed Walzgeschwindigkeit
rolling stand Walzgerüst * (Walzwerk)

rolling stock Fahrzeuge * Fuhrpark || Fuhrpark || Schienenfahrzeug * (Plural) Schienenfahrzeugpark
rolling table Rollgang
rolling-contact bearing Wälzlager
rolling-elemement bearing Wälzlager
rolling-mill main drive Walzwerk-Hauptantrieb
rollstand Abrollung * Rollenständer/Gerüst; shaftless: achslos
ROM Festwertspeicher * Abkürzg. für 'Read Only Memory':Nur-Lesespeicher || Nur-Lese-Speicher
roof Dach * roof section: Dachteil eines Schaltschranks
roof chamber Dachhaube * z.B. eines Schaltschranks
roof cover assembly Dachhaube * Über einem Schaltschrank angebrachte ~ zum Schutz der Kühlluft-Austrittsöffnung
roof fan Dachventilator
roof section Dachaufsatz * z.B. eines Schaltschranks; siehe auch →Dachhaube || Dachblech * z.B. eines Schaltschranks || Dachhaube * Dachteil (z.B. eines Schaltschranks)
roof-mounted fan Dachventilator
roofed-in überdacht
roofed-over überdacht
room Platz * Raum; make room/way for: ~ machen für || Raum * Zimmer
room of move Spielraum
room reservation Zimmerbestellung
room to move Entscheidungsspielraum * →Spielraum || Spielraum
room utilization Platzausnutzung * Raumausnutzung || Platznutzung * Raumausnutzung || Raumausnutzung || Raumnutzung
room-saving Platz sparend * raumsparend || raumsparend
roominess Geräumigkeit
roomy geräumig
root Wurzel * auch mathematisch; square/cubic root: zweite/dritte Wurzel
root diameter Fußkreisdurchmesser * eines Zahnrades
root directory Stammverzeichnis * Hauptverzeichnis eines Comuterlaufwerks (z.B. einer Festplattenpartition)
root locus diagram Wurzelortskurve
root mean square Mittelwert * Effektivwert (d.h. quadratischer ~)
root-mean-square value Effektivwert
rope Seil * Tau, Strick, Strang; auch bei Aufzug, ~bahn, Kran; one-strand rope: einträmmiges ~ || Tau
rope drive Seilantrieb
rope drum Seiltrommel * auch für Drahtseil
rope pulley Seilscheibe
rope stretching Seildehnung * {Aufzüge}
rope suspension Seilaufhängung * {Aufzüge}
rope vibrations Seilschwingungen * unerwünschter Effekt bei Aufzug
rope-tightening weight Spanngewicht * (kleines) Seil-Spanngewicht bei Aufzug
ropeway Seilbahn
ropiness Zähigkeit * ~ von Flüssigkeiten
ropy zäh * Flüssigkeit
rotary Rotations- || rotatorisch || rotierend || Rund- * drehbar, sich drehend || umlaufend
rotary axis Rundachse
rotary cement kiln Zementdrehofen
rotary compressor Rotationsverdichter
rotary converter Umformer * rotierender ~ (Maschine(nsatz) z. Änderung der Frequenz, Spannung oder Phasenzahl)
rotary cutter Rotationsschneider

rotary cutting machine Rundschälmaschine * *[Holz]* || Schälmaschine * *Rundschälmaschine*
rotary encoder Drehgeber * *digitaler*
rotary filter Drehfilter
rotary furnace Trommelofen
rotary grinding machine Rundschleifmaschine * *[Werkz.masch.]*
rotary incremental encoder Drehimpulsgeber
rotary indexing table Drehtisch * *Positionier~*
rotary kiln Drehofen * *z.B. zur Zementherstellung*
rotary knob Drehknopf
rotary latch Drehverschluss * ~ *mit Verriegelung*
rotary matrix Drehmatrix
rotary motion Drehbewegung || Umdrehung * *Drehbewegung*
rotary movement Rotation
rotary photogravure Tiefdruck * *Rollentiefdruck*
rotary press printing Rotation * *Rotationsdruck* || Rotationsdruck * *[Druckerei]*
rotary printing machine Rotationsdruckmaschine * *[Druckerei]*
rotary printing press Rotationsdruckmaschine * *[Druckerei]*
rotary pulse encoder Drehimpulsgeber
rotary pump Kreiselpumpe || Schraubenpumpe * →*Kreiselpumpe*
rotary rheostat Drehwiderstand
rotary shaft seal Radialdichtring * *Wellendichtring* || Radialwellendichtring
rotary switch Drehschalter || Zahlenschalter * *Drehschalter*
rotary table Drehteller * *Drehtisch* || Drehtisch * *z.B. bei Werkzeug-/Verpackungsmaschine* || Rundtisch * *Drehtisch, z.B. bei Förder-/Verpackungseinrichtung, Werkzeugmaschine*
rotary table winder Rundtisch-Wickler
rotary transducer Drehgeber
rotary transfer station Drehtisch * *[Fördertechnik]*
rotary transformer Drehtrafo || Drehtransformator || Stelltrafo * *Drehtrafo* || Stelltransformator * *Drehtrafo*
rotary-table milling machine Rundtisch-Fräsmaschine
rotary-vane vacuum pump Drehschieber-Vakuumpumpe
rotatable drehbar || schwenkbar * *drehbar*
rotate drehen * *rotate freely/unobstructed: (sich) frei/ungehindert* ~ || kreisen || rotieren || schwenken * *um eine Achse drehen* || sich drehen || umlaufen
rotate freely back to zero austrudeln * *bis Drehzahl Null*
rotating Dreh- * *sich drehend* || drehend || rotierend || Schwenk- * *drehbar* || umlaufend
rotating belt Drehband
rotating converter rotierender Umformer
rotating disk Drehteller
rotating exciter Erregermaschine * *z.B. bei Leonard-Umformer*
rotating field Drehfeld * *umlaufendes Feld*
rotating field monitoring Drehfeldüberwachung
rotating frequency converter rotierender Frequenzumformer * *zum Erreichen extrem hoher Antriebsdrehzahlen bei Asynchronmotor*
rotating key Schwenktaster
rotating mass Schwungmasse
rotating masses rotierende Masse || rotierende Massen || Schwungmasse * *rotierende Masse*
rotating movement Drehbewegung
rotating parts drehende Teile || rotierende Teile
rotating rectifier excitation RG-Erregung * *'Rotierender (Dioden)Gleichrichter',i.Rotor integrierten (→Stromr.motor)*
rotating table Drehtisch
rotating-field direction Drehfeldrichtung

rotating-field machine Drehfeldmaschine
rotating-field reversal Drehfeldumkehr
rotating-rectifier excitation RG-Erregung * *bürstenlose Erregg. über i.rotor integrierten (Dioden)-gleichrichter*
rotation Drehung || Rotation || Umdrehung || Wechsel * *regelmäßiger Austausch*
rotation angle Drehwinkel
rotation arrow Drehrichtungspfeil * *zur Kennzeichnung d.Drehsinns v. Maschinen, die nur in eine Drehrichtg. laufen*
rotation lock Drehrichtungssperre
rotation reversal Drehrichtungsumkehr
rotation setting Drehrichtungsvorgabe
rotational Dreh-
rotational accuracy Rundlaufgenauigkeit
rotational axis Drehachse
rotational direction Drehrichtung
rotational irregularity Ungleichförmigkeit * *einer Drehbewegung*
rotational motion Drehbewegung
rotational speed Drehzahl * *in upm (i. Gegensatz zu speed i. Sinne v.(Bahn)Geschwindigkeit)*
rotational speed of the shaft Wellendrehzahl
rotational twist Verdrehung
rotational vector Drehvektor
rotational velocity Drehgeschwindigkeit
rotational-direction switch Drehrichtungsumschalter
rotogravure press Tiefdruck-Rollendruckmaschine * *[Druckerei]*
rotogravure printing Tiefdruck * *Rollentiefdruck*
rotor Anker * *Rotor* || Läufer * *[el.Masch.] drehender Teil einer elektr.Maschine* || Polrad || Rotor
rotor angle Polradlage || Polradwinkel || Rotorlage
rotor body Läuferkörper
rotor braking heat Bremswärme im Läufer
rotor core Läuferblechpaket || Läuferpaket
rotor current Läuferstrom
rotor displacement angle Polradwinkel
rotor flux Ankerdurchflutung * *Rotordurchflutung* || Läuferdurchflutung || Rotorfluss
rotor frequency Läuferfrequenz
rotor hub Läufernabe * *bei Elektromotor*
rotor lamination Läuferblechpaket || Läuferblechung
rotor laminations Läuferbleche
rotor locking device Läuferhaltevorrichtung * *Transportsicherung für Motor*
rotor losses Läuferverluste
rotor position Polradlage || Rotorlage
rotor position encoder Polradlagegeber
rotor position sensor Polradlagegeber || Rotorlagegeber
rotor position transmitter Rotorlagegeber
rotor power Drehfeldleistung * *bei Motor*
rotor resistance Läuferwiderstand
rotor shaft Rotorwelle
rotor shaft angle encoder Rotorlagegeber
rotor shipping brace Läuferhaltevorrichtung * *Transportsicherung für Motor*
rotor speed Läuferdrehzahl
rotor standstill voltage Läuferstillstandsspannung * *[el. Masch.] an d. Schleifringen b.off. Läuferkreis i. Stillstand gemessen* || Läuferstillstandsspannung * *z.B. b.Schleifringläufer, der im Stillstand k. höchste Läuferspanng. hat*
rotor starting heat Anlaufwärme im Läufer
rotor time constant Rotorzeitkonstante
rotor voltage Läuferspannung
rotor winding Ankerwicklung * *Läuferwicklung einer elektr. Maschine* || Läuferwicklung
rotor-cage Läuferkäfig * *eines Käfigläufermotors*
rotor-fed läufergespeist
rotten faul * *ver~t*

rough annähernd || grob || oberflächlich * *ungefähr, grob* || rau || roh * *rau, grob, roh (z.B. Entwurf), ungehobelt, unbearbeitet, im Rohzustand* || überschlägig * *z.B. Schätzung, Rechnung, Wert* || überschläglich * *grob; estimation:Berechnung/Abschätzung* || überschlagsmäßig * *grob; estimation:- Berechnung/Schätzung* || Vor- * *Grob-*
rough and ready formula Faustformel || Faustregel * *Faustformel*
rough draft Bild * *Grob-/Entwurfsskizze* || Vorentwurf
rough drawing Bild * *Roh-/Grobskizze* || Handskizze * *Grobzeichnung*
rough estimation Grobabschätzung
rough formula Faustformel || Faustregel * *Faustformel*
rough industrial use rauer Industrieeinsatz
rough material Vormaterial * *z.B. im Walzwerk*
rough plate Grobblech
rough pulse Grobimpuls
rough service rauer Betrieb || Schwerlastbetrieb
rough service conditions rauer Betrieb * *raue Betriebsbedingungen*
rough setting Grobjustierung
rough sketch Handskizze * *Grobskizze, flüchtige Skizze*
rough strip Vorband * *Grundmaterial im Walzwerk*
rough up rauen * *z.B. Textilstoff* ~
rough value Anhaltswert || Daumenwert || Näherungswert || Richtwert * *überschlägiger Wert*
rough-draw skizzieren
rough-drill vorbohren
rough-drilled vorgebohrt
roughen aufrauen
roughing stand Vorgerüst * *zum Grobwalzen im Walzwerk*
roughly etwa * *grob, annähernd, ungefähr* || ungefähr * →*etwa, annähernd*
roughly speaking etwa * *grob gesprochen, etwa, ungefähr*
roughness Rauheit || Rauhigkeit * *Bendtsen roughness:* ~ *nach Bendtsen*
round abgerundet * *Zahl* || abrunden || geschlossen * *(figürlich) Arbeit, Leistung usw.* || herum || kreisförmig || Kreislauf || rund * *abgerundet (auch Zahl), rund geformt* || Rund- * *z.B. round iron:* ~*eisen* || Umfang * *be 10 inches round: einen* ~ *von 10 inches haben* || verschleifen * *verrunden, z.B. Signale* || voll * *prall*
round bar Rundstab
round cable Rundkabel || Rundleitung
round core Rundleiter * *{Kabel}*
round corner cutting Rundschneiden
round down abrunden * *eine Zahl nach unten* || runden * *abrunden, to:auf*
round iron Rundeisen
round log merchandizing plant Rundholzplatz
round off abrunden * *eine Zahl* || aufrunden * *eine Zahl nach oben oder unten* || runden * *z.B. gebrochene Zahlen auf/abrunden; to:auf*
round rod Rundstab
round steel Rundeisen
round timber Rundholz
round up abrunden * *eine Zahl nach oben* || aufrunden * *eine Zahl nach oben (to:auf)* || runden * *aufrunden, to:auf*
round-bar rotor Rundstabläufer * *{el.Masch.}*
round-frame motor Rundmotor
round-head srew Rundkopfschraube
roundabout Karussel
rounded verrundet
rounded ramp verrundete Hochlaufgeberrampe
rounding Abrundung || Rundung * *auch mathemath.* || Verrundung

rounding down Rundung * *(Zahl) Abrunden (nach unten)*
rounding function Verrundung * *bei Hochlaufgeber*
rounding off Abrundung || Rundung * *(Zahl) Auf/Abrunden (nach oben oder unten)*
rounding off error Rundungsfehler
rounding time Verrundungszeit * *z.b. bei einem Hochlaufgeber mit Verrundung*
rounding up Rundung * *(Zahl) Aufrunden (nach oben)*
rounding-off Verrundung * *z.B. Hochlaufgeber*
rounding-off of the ramp generator Hochlaufgeber-Verrundung
roundness Rundung * *abgerundete Stelle*
rounds Rundstab
roundwood yard Rundholzplatz * *{Holz} in Sägewerk, Papier-/Zellstofffabrik usw.*
rout-out buiscuit machine Stranggebäckmaschine
route fördern * *be~, leiten, weiterleiten* || führen * *leiten, befördern, verlegen (Leitung, Leiterbahn)* || Kurs * *Route* || leiten * *Verkehr/Signalverkehr* ~ *(over:über)* || rangieren * *Signale* || Reiseweg || Richtung * *(Fahr)Weg* || Strecke * *Fahrstrecke,* →*Weg* || Transportweg || Verbindungsweg || verlegen * *leiten, auch verlegen von Kabel/Leitungen* || Weg * *Reise~, Fahrtroute/-Weg,* →*Strecke* || weiterleiten * *Aufträge auf d. Dienstweg, Material/ Transportgüter, Datenpakete in Rechnernetz* ~ || zuführen * *Leitung, Signal* ~
route separated getrennt verlegen
route through hindurchleiten
router Oberfräse * *{Holz}* || Oberfräsmaschine * *{Holz}*
router head Fräskopf * *{Holz}* einer Oberfräse
routine Gang * *gewohnheitsmäßiger* ~ || laufend * *routinemäßig, z.B. Geschäft* || Routine * *Software-Unterprogramm, gleich bleibendes Verfahren, gewohnheitsmäßiger Gang* || routinemäßig * *(Adj.) siehe auch* →*turnusmäßig* || standardmäßig * *routinemäßig* || turnusmäßig * *routinemäßig*
routine test Stückprüfung * *jedes gefertigten Exemplars (i.Gegensatz zur Typprüfung); für Motor: siehe VDE 0530*
routinely routinemäßig * *(Adv.)* || standardmäßig * *(Adv.) routinemäßig*
routing Entflechtung || Verlegung * *Art und Weise der Leitungs~*
routing center Fräszentrum * *{Holz}*
routing controller Positioniersteuerung
routing machine Oberfräse * *{Holz}*
routing of cables Kabelverlegung * *Kabelführung, Art und Weise der Kabelverlegung* || Leitungsverlegung * *legen, führen; Art der Leitungsverlegung*
routing unit Fräseinheit * *{Holz}*
row Reihe * *(Häuser/Sitz usw.) Reihe, Zeile* || Textzeile || Zeile * *Reihe (z.B. in Tabelle, Baugruppenträger)*
royalties Honorar * *eines Buchautors usw.*
royalty Lizenzgebühr
royalty fees Lizenzgebühr * *auch 'Patentgebühr'*
RPM Drehzahl * *Abkürzung für 'Revolutions Per Minute':Drehzahl in [upm]* || min-1 * *Abkürzg.f. 'Revolutions Per Minute': Umdrehungen pro Minute* || Umdrehungen * *Abkürzung für 'Revolutions Per Minute':* ~ *pro Minute* || Umdrehungen pro Minute * *Abkürzg. f. 'Revolutions Per Minute'* || upm * *revolutions per minute [rewol'u:schen ...] : Umdrehungen pro Minute*
rpm control Drehzahlregulierung
rpm preselection Drehzahlvorwahl
RS232 RS232 * *'Recommended Standard':empfohlene Norm f.serielle Computerschnittstelle m.* ±*12V-Signalen (V24)*
RS232 interface V24-Schnittstelle

RS232 link V24-Schnittstelle
RS232 port V24-Schnittstelle * *Anschluss*
RS485 RS485 * *"Recommended Standard": Norm f.serielle Daten-Schnittstelle m.5V-Differenzsignal, IEC 1158-2*
RTD Temperaturfühler * *Abk.f. 'Resistance Temperature Detector':Widerstands-~ (NTC od. PTC)* || Temperatursensor * *Abk.f. 'Resistance Temperature Detector': temp.abh. Widerstand (Heiß-/Kaltleiter)*
rub anschleifen * *(ab)schleifen; with emery:abschmirgeln* || aufscheuern * *abreiben, reiben an, (ab)scheuern, (ab)schaben, abschleifen* || reiben || schaben * *abrubbeln* || scheuern * *abreiben, reiben an, (ab)scheuern, (ab)schaben, abschleifen* || schleifen * *(ab)schleifen (with emery:abschmirgeln), (ab)feilen*
rub down anschleifen || schleifen * *anschleifen, z.B. vor einem Farbauftrag*
rubber Gummi || Kautschuk
rubber band Gummiband || Gummiring * *Gummiband*
rubber bush Gummiring * *z.B. Dichtring für Kabeldurchführungs- (PG-) Verschraubung*
rubber cylinder Druckzylinder * *[Druckerei] Gummi-Druckzylinder einer Druckmaschine*
rubber injection molding system Gummi-Spritzanlage
rubber insulated gummiisoliert * *cable: Leitung*
rubber insulated cable Gummischlauchleitung
rubber kneader Gummikneter
rubber pad Gummiklotz * *z.B. in elastischer Kupplung* || Gummiklötzchen * *z.B. in Kupplung*
rubber roller Gummiwalze
rubber rolling mill Gummiwalzwerk
rubber seal Gummidichtung
rubber sealing ring Gummidichtring
rubber sealing washer Dichtring * *Gummidichtring* || Gummidichtring
rubber sleeve Gummimanschette || Gummitülle
rubber-bushed pin coupling Gummibolzenkupplung * *zur Welle-Welle-Verbindg. (kann kleine Fluchtfehler ausgleichen)*
rubber-bushing Gummibuchse
rubber-metal vibration damper Schwingmetall
rubber-tired gummibereift * *(amerikan.)*
rubber-tired front-end loader Radlader
rubber-tyred gummibereift * *(brit.)*
rubber-tyred shovel Radlader
rubberized gummiert
rubbish Krampf * *(salopp)* Müll, Abfall, Blödsinn, Quatsch, Schund, Ausschuss || Krempel * *(salopp)* Plunder, Kram
rubric Rubrik
rubustness Stabilität * *auch Regelung*
rucksack Rucksack
rude rau * *grob*
rugged kräftig * *→robust, grob, stark, ungehobelt* || massiv * *robust, stark, stabil* || rau || robust * *stark, stabil* || stabil * *stark, stabil, →robust, grob, ungehobelt* || strapazierfähig || widerstandsfähig
rugged design Robustbauform || Robustheit * *robuste Konstruktion*
rugged industrial application rauer Industrieeinsatz
ruggedized design Robustbauweise || Robustheit * *robuste Konstruktion* || verstärkte Ausführung
ruggedized model Robustbauform
ruggedness Festigkeit * *Robustheit* || Langlebigkeit * *Verschleißfestigkeit, Zerstörsicherheit, Robustheit* || Qualität * *Robustheit, Zerstörsicherheit, Verschleißfestigkeit, Stabilität* || Robustheit * *Zerstörsicherheit, Verschleißfestigkeit* || Stabilität * *der Bauart usw.* || Unempfindlichkeit * *Robustheit*

ruin Sturz * *Untergang, Ruin* || verschandeln || zerstören || Zerstörung
rule Anordnung * *Vorschrift* || Beherrschung || Gebot * *Vorschrift, Regel* || Gesetz * *Vorschrift, Regel* || Grundsatz * *Regel; make it a rule: es (sich) zum ~ machen* || herrschen * *over: über* || Norm * *Regel* || Regel * *(auch mathem., physikal.) rule of thumb: Daumen~* || Regelung * *→Regel, Normalfall, Richtschnur, Grundsatz, Spielregel, Richtlinie, Verhaltensmaßregel* || Vorschrift * *Regel, Dienst~*
rule for success Erfolgsrezept
rule of thumb Daumenregel * *siehe auch →Faustformel* || Faustformel * *Faustregel* || Faustregel
ruler Lineal
ruling Entscheidung || Regelung * *~ durch Vereinbarung/Vorschrift usw.*
rumbling noises Dröhnen
rummage for suchen * *wühlend, kramend ~*
run abarbeiten * *ein Programm* || ablaufen * *z.B. Programm, Prozedur* || ablaufen lassen * *z.B.ein Programm/Projekt* || arbeiten * *laufen (Maschine)* || bedienen * *fahren, laufen lassen, führen* || betreiben * *Maschine, Anlage, Unternehmen, Verkehrslinie usw. ~* || Betrieb * *~urt einer SPS, einer Maschine usw. (z.B. als Lampenbeschriftung)* || durchführen * *z.B. einen Versuch* || Durchgang * *Durchlauf, Folge, Arbeitsgang* || Durchlauf * *siehe auch 'Durchgang'* || fahren * *laufen, (be/durch)fahren, laufen lassen (Maschine, Fahrzeug)* || gehen * *bei Maschine usw.* || Lauf * *auch Durchlauf, Durchgang* || laufen * *smooth/swift:ruhig/schnell (auch Zeit); auch Softwareprogramm (on: auf (einem Computer))* || lauten * *Inhalt, Worte* || leiten * *betreiben, (Betrieb) führen/leiten, (Maschine/Anlage) laufen lassen* || verarbeiten * *Daten in e. Computerprogramm ~, eine durchlaufende Warenbahn in e. Maschine ~* || verkehren * *Verkehrsmittel* || Verlauf * *Ablauf* || verlaufen * *Grenze, Weg, Leiterbahn, Farben usw.*
run 24 hours a day durchlaufen * *rund um die Uhr ~*
run after nacheilen || nachlaufen * *hinterherlaufen*
run aground auffahren * *(Schiff) auf Grund fahren*
run as follows Wortlaut * *folgenden ~ haben*
run away durchgehen * *eines Motors (Überdrehzahl)*
run back zurücklaufen
run continuously durchlaufen * *ohne Unterbrechung/kontinuierlich ~*
run down ablaufen * *Uhr* || Auslauf * *{el.Masch.}* || auslaufen * *z.B. Motor* || verbrauchen * *z.B. Batterie*
run hot heißlaufen
run idle im Leerlauf fahren || leerlaufen
run in einfahren * *neues Auto, Getriebe usw. ~* || Erwärmungsprüfung * *Erwärmungslauf zur Erkennung von Frühausfällen*
run in no-load condition im Leerlauf fahren
run in synchronism gleichlaufen
run into anfahren * *rammen* || auffahren * *auf etwas ~* || eintreiben
run mode Betriebsart
run non-stop durchlaufen * *ohne Unterbrechung ~*
run off abfließen || ablaufen * *Flüssigkeit*
run off the iron abstechen * *Hochofen ~*
run on auffahren * *Schiff*
run one's own business selbstständig sein
run out ablaufen * *eines Zeitgliedes* || auslaufen * *Produktion, Motor usw.*
run out of true unrund laufen
run past vorbeilaufen
run position Arbeitsstellung
run reversely zurücklaufen
run the show Laden schmeißen

run through ablaufen * *Gegend* || durchströmen
run time Laufzeit * *e. Transportbandes; Zeit, während der ein Rechnerprogramm abläuft; Länge eines Films usw.* || Rechenzeit * *während des Ablaufs eines Programms*
run unsmoothly unrund laufen
run untruly unrund laufen
run up anfahren * *hochfahren* || anlaufen * *z.B. eine Maschine* || betragen * *zahlen/wertmäßig* ~ || Hochfahrt * *z.B. auf End-/Betriebsgeschwindigkeit/-Drehzahl* || hochlaufen * *z.B. eines Motors*
run up to betragen
run-down test Auslaufversuch * *[el.Masch.]* Aufnahme des zeitl. Drehzahlverlaufs nach Abschalten
run-time Ablaufzeit * *eines Programms*
run-time optimized laufzeitoptimal * *(Adj.)*
run-up anfahren * *(auch Substantiv)* || hochfahren || Hochlauf
run-up characteristic Hochlaufkennlinie
run-up delay Hochlaufverzögerung
run-up integrator Hochlaufintegrator
runability Laufeigenschaften
runaway durchgehen * *(Substantiv) z.B. eines Motors* || Instabilität * *Durchgehen (eines Motors usw.)*
rung Segment * *Strompfad in der Kontaktplandarstellung eines SPS-Programms* || Strompfad * *in Relaissteuerungs-/Kontaktplan (KOP)*
runnable ablauffähig * *z.B. Programmcode auf einem Rechner* || lauffähig * *z.B. Programmcode auf einem Rechner*
runner Laufrad * *Turbine* || Schlaufe
running Betrieb || Gang * *Laufen (smooth: ruhig, swift: schnell)* || hintereinander * *3 days running:3 Tage hintereinander; 5 times running: 5-mal hintereinander* || in Betrieb || Lauf * *smooth/swift:ruhiger/schneller Lauf; auch Maschine/Motor* || laufend * *im Laufen befindlich*
running at constant speed Konstantfahrt
running at crawl speed Kriechbetrieb
running capacitor Betriebskondensator * *für Einphasenbetrieb eines Drehstrommotors (Steinmetzschaltung)* || Betriebskondensator
running connection Betriebsschaltung * *für Motor/Trafo, z.B. Stern- od. Dreieckschaltung*
running costs Betriebskosten
running direction Drehrichtung || Fahrtrichtung || Laufrichtung
running down auslaufen * *(Substantiv)*
running frequency Frequenzistwert
running friction Gleitreibung
running gear Fahrwerk
running hours Betriebsstunden
running idle leerlaufend || nicht angetrieben * *leerlaufend*
running in reverse direction Rücklauf * *Lauf in umgekehrte Richtung*
running in synchronism Gleichlauf
running load Betriebslast * *eines Motors*
running noise Betriebsgeräusch
running on Nachlauf * *"Nachdieseln" eines Verbrennungsmotors*
running smoothness Laufruhe * *geräusch- und schwingungsarme Bewegung mit gleichmäßiger Winkelgeschwindigkeit*
running test Laufprüfung * *z.B. für Motor nach DIN 51806*
running time Betriebsdauer || Einschaltdauer * *z.B. bei Motor* || Fahrzeit || Laufzeit * *einer Maschine, eines Maschinenteils*
running up Hochlauf
running wheel Laufrad * *z.B. bei Lüfter*
running-up time Anlaufzeit * *[el.Masch.] Dauer eines Anlaufs v. Einschalten bis zum stationären Betrieb*

running-up torque Hochlaufmoment * *Dehmoment, das zum Hochlaufen benötigt wird*
runtime Ablaufzeit * *eines Programms* || Laufzeit * *Zeit, während der ein Computer ein Programm abarbeitet (during: zur)*
runtime environment Ablaufumgebung * *Hard-/Softwareumgebung, auf der eine Software ablaufen soll (→Zielsystem)*
runup Hochlauf * *auch ~ eines Rechners*
rupture Ansprechen * *(Verb) Sicherung* || brechen * *reißen, ab~* || reißen
rupture joint Sollbruchstelle
rupture test Zerreißprüfung
rupturing capacity Ausschaltleistung * *e. Leistungsschalters, e. Sicherung* || Schaltleistung * *Aus~, Ab~(auch einer Sicherung)*
rush Nachfrage * *rush for: starke* ~ *nach* || Schuss * *rasende Bewegung*
rush current Rush-Strom * *Stoßstrom*
rush sequence Eilablauf * *bei SPS*
rush torque Rush-Moment * *Stoßmoment*
rust Rost * *Eisenoxid* || rosten
rust attack Rostfraß
rust formation Rostbildung
rust protected rostgeschützt
rust-preventing agent Rostschutzfarbe
rust-proofing Rostschutz
rust-resistant rostgeschützt
rusting Rostbildung
rustless nicht rostend || rostfrei
rustproof nicht rostend || rostfrei || rostgeschützt
rustproof coating Rostschutzfarbe
rusty rostig
rythm Takt * *Rythmus*

S

S ramp verrundete Hochlaufgeberrampe * *in Form eines 'S' abgerundete Rampe*
SH Abtast-Halteglied * *Abk.f. 'Sample and Hold element'*
S-curve S-Kurve * *abgerundete Hochlaufgeber-Rampe*
S-curve rounding Hochlaufgeber-Verrundung * *Rampe hat die Form des Buchstabens "S"* || Verrundung * *e.Hochlaufgeber-Rampe, sodass e. S-förmige Kurve entsteht*
S-motor S-Motor * *energiesparender SIEMENS-Sondermotor mit hohem Wirkungsgrad*
S-roll S-Rolle * *Walzen mit großer Umschlingung, erzwingen konstante Geschwind.* || S-Walze * *1 von 2 S-förmig angeordneten Walzen m.großer Umschlingung z.Fördern bandförmig. Materials*
S-shaped S-förmig * *z.B. Hochlaufgeber-Rampe (mit Verrundung)*
sack Beutel * *(amerikan.) Sack, Beutel, Tüte* || füllen * *in Säcke* ~ || Sack * *auch 'Beutel, Tüte'*
sacrifice opfern * *(auch figürl.) Opfer bringen, mit Verlust verkaufen; Verzicht,Opfer, Aufopferung, Verlust* || verletzen * *Gesetz, Vorschrift* ~
sacrifice of time Zeitaufwand * *Zeitverlust, Zeiteinbuße, "Zeitopfer"*
saddle Schelle * *mit zwei Befestigungslappen für Rohr*
safe sicher * *from: vor; be on the safe side: auf der ~en Seite sein/~gehen*
safe from finger-touch fingersicher * *siehe VDE 0106 Teil 100*
safe isolation sichere Trennung

safe load Tragfähigkeit * ~ von Seil, Hebezeug, Brücke
safe-to-touch berührungsgeschützt || berührungssicher * siehe auch →berührungsgeschützt
safeguard Schutz * ~vorrichtung, Vorsichtsmaßnahme (against: gegen), Sicherheitsklausel; sichern, schützen || sichern * sichern (auch Daten), schützen, (Interessen) wahrnehmen || wahrnehmen * sicherstellen von Interessen usw.
safeguarding Wahrnehmung * ~ der Interessen usw.
safety Sicherheit * Personen~ , ~ von Wertpapieren; for safety reasons: ~shalber || Sicherheits-
safety against damage Zerstörsicherheit
Safety Authority Berufsgenossenschaft
safety class Schutzklasse
safety clearance Sicherheitsabstand
safety coupling Sicherheitskupplung * beginnt bei Überschreiten eines bestimmten Drehmoments zu rutschen
safety device Schutzvorrichtung || Sicherheitseinrichtung
safety disconnection Freischaltung * VDE 0100 Teil 200 || Sicherheitsabschaltung * Stromkreis
safety distance Sicherheitsabstand
safety earth connection Schutzleiteranschluss
safety earth terminal Schutzleiteranschluss * Klemme, Schraube
safety equipment Schutzausrüstung * persönliche || Sicherheitseinrichtung || Sicherheitseinrichtungen
safety factor Sicherheitsfaktor
safety fuse Schmelzsicherung
safety glass Sicherheitsglas
safety glasses Schutzbrille
safety goggles Schutzbrille * allseits geschlossene
safety ground Schutzerde
safety hazard Unfallgefahr
safety in operation Betriebssicherheit
safety information Warnhinweise * in Betriebsanleitung
safety instructions Sicherheitshinweise
safety interlock Sicherheitsverriegelung
safety isolation Freischaltung * VDE 0100 Teil 200
safety limit-switch Sicherheitsendschalter
safety margin Sicherheitsabstand * wertmäßig
safety notes Gefahrenhinweise || Sicherheitshinweise
safety officer Sicherheitsbeauftragter
safety precaution Sicherheitsmaßregel || Vorsichtsmaßregel * Sicherheitshinweis/Vorschrift/Maßregel/Vorkehrung
safety precautions Sicherheitshinweise * Vorsichtsmaßregeln
safety reasons Sicherheitsgründe * for: aus
safety regulation Sicherheitsvorschrift || Unfallverhütungsvorschrift
safety regulations Sicherheitsbestimmungen || Sicherheitsvorschriften || Unfallverhütungsvorschriften * Sicherheitsvorschriften
safety requirements Sicherheitsanforderungen || Sicherheitsbestimmungen
safety rule Unfallverhütungsvorschrift
safety rules Unfallverhütungsvorschrift * (Plural)
safety running features Notlaufeigenschaften
safety separation sichere Trennung
safety shutdown Sicherheitsabschaltung * Antrieb
safety slipping clutch Rutschkupplung * Sicherheits-~ zur Drehmomentbegrenzung || Sicherheits-Rutschkupplung * zur Drehmomentenbegrenzung || Sicherheitskupplung * →Schlupfkupplung
safety standard Sicherheitsnorm * für elektr. Antriebssysteme: z.B. EN 50178, EN 60204-1
safety standards Unfallverhütungsvorschrift * (Plural)
safety supervisor Sicherheitsbeauftragter

safety temperature Grenztemperatur
safety valve Sicherheitsventil * siehe auch →Überdruckventil
safety-oriented sicherheitsgerichtet
safety-related sicherheitsrelevant
safety-related purposes Sicherheitsgründe * for: aus
safety-slip clutch Sicherheits-Rutschkupplung
safety-tested sicherheitsgeprüft
sag absacken * z.B.Last am Kran || absinken * zusammensacken, herunterhängen || Durchbiegung * Durchhang, Durch-/Absacken, Senkung || Durchhang || durchhängen || durchsacken || Einbruch * Absacken z.b. der Netzspannung || Senkung * Mauer, Decke usw
sag control Durchhangregelung * z.B. in für Metallband in Schlingengrube oder in Kabelummantelungsanlage
sagging Biegung * Durchhang, Durchbiegung || Durchbiegung * z.b. eines stark belasteten Regalbodens
sail abgehen * Schiff (for: nach) || auslaufen * [Schifffahrt] || fahren * mit Schiff
sailing Auslauf * [Schifffahrt] das ~en eines Schiffes
salary Gehalt * Einkommen || Lohn * Gehalt
sale Vertrieb
saleable marktfähig
sales Absatz * von Produkten/Waren auf dem Merkt || Umsatz || vertrieblich
sales agency Firmenvertretung * Handelsvertretung, →Industrievertretung || Industrievertretung || Verkaufsbüro || Vertretung * Handels~, Industrie~
sales agreement Verkaufsvereinbarung || Vertriebsvereinbarung
sales aid Akquisitionshilfe
sales and marketing department Vertriebsabteilung
sales and marketing manager Vertriebsleiter
sales brochure Werbeschrift
sales channel Vertriebskanal || Vertriebsweg
sales company Vertriebsgesellschaft
sales department Vertrieb * ~sabteilung || Vertriebsabteilung || Vetriebsabteilung
sales engineer Vertriebsingenieur || Vertriebsmann * Vertriebsingenieur
sales expense Vertriebskosten * at great sales expense: mit hohen ~
sales literature Prospekte
sales manager Vertriebsleiter || Vertriebsmann * leitender
sales margin Handelsspanne || Verdienstspanne || Vertriebsspanne
sales network Vertrieb * ~snetz || Vertriebsnetz || Vertriebsorganisation * Vertriebsnetz || Vertriebssystem * Vertriebsnetz
sales office Vertretung * Vertriebsbüro || Vertriebsbüro || Vertriebsniederlassung
sales organization Vertriebsorganisation
sales personnel Vertriebspersonal
sales price Verkaufspreis
sales profit Vertriebsergebnis
sales profit margin Vertriebsergebnis * prozentuales ~
sales program Fertigungsprogramm * lieferbare Producte || Produktpalette * verkaufbare Produkte
sales programme Fertigungsprogramm * Verkaufsprogramm
sales promotion Akquisition || Werbung * Verkaufsförderung
sales promotion articles Werbemittel
sales region Vertriebsregion
sales release Vertriebsfreigabe * z.B. eines neuentwickelten Produkts
sales representative Vertreter * Verkäufer || Ver-

triebsniederlassung * *Handelsvertretung, Handelsvertreter*
sales territory Verkaufsgebiet
sales trip Dienstreise * *zu Verkaufsgesprächen bei Kunden* || Kundenbesuch * *Geschäftsreise eines Vertriebsmannes zu Verkaufsgesprächen*
sales trips Kundenbesuche * *Geschäftsreisen zu Kunden zwecks Verkaufsgesprächen*
sales turnover Umsatz
sales volume Umsatz
sales/marketing akquisitorisch * *Marketing-, vertrieblich*
sales/marketing Akquisition || Vertrieb || vertrieblich
sales/marketing activities Akquisition
sales/marketing department Vetriebsabteilung
sales/marketing goals vertriebliche Zielsetzung
sales/marketing resources Vertriebshilfsmittel
salesclerk Verkäufer * *(amerikan.)*
salesman Verkäufer * *Vertriebsmann* || Vertriebsmann
salient ausgeprägt * *herausragend, (her)vorspringend, hervorstechend, ins Auge springend* || herausragend * *(her)vorspringend, hervorstechend, ins Auge springend* || scharf * *hervorspringend* || vorstehend * *(her)vorspringend/-stehend* || wesentlich * *hearusragend, hervorstechend, ins Auge springend; salient point: springender Punkt*
salient features typische Merkmale * *herausragende Merkmale*
salient pole Schenkelpol * *ausgeprägter, hervorstehender Pol z.B. am Läufer e.Synchronmaschine (Ggs.: →Vollpol-)*
salt air salzhaltige luft
salt-laden air salzhaltige luft
salvage Altmaterial * *verwertbares ~*
salvage reel cutter Fehlrollenschneider * *(Papier)*
salvage winder Fehlrollenschneider * *(Papier)*
Samarium-Cobalt Samarium-Kobalt || SmCo * *Seltene-Erden-Permanentmagnetmaterial; z.B. für Motorläufer*
Samarium-Cobalt magnet Samarium-Kobalt-Magnet
same dito || gleich * *ähnlich, identisch*
same as above siehe oben * *wie oben (angegeben)* || wie oben
same old alt * *unverändert, das gleiche alte*
sample Abfrage * *Abtastung eines Sinalzustands* || Abfrageergebnis * *Abtastwert einer Messung* || abfragen * *Messgröße/wert abtasten* || abtasten * *Regler* || Abtastung * *(Abfragen eines) Augenblickwert(s)/Messwert(s)* || bemustern || erfassen * *(Messwert) abtasten* || Exemplar * *Einzelstück* || Muster * *Probe, (Stück-/Typen-/Vorlage- usw.)* Muster || Probe * *Muster* || Probe- * *Muster* || Stichprobe
sample and hold element Abtast-Halteglied
sample interval Abtastintervall
sample-and-hold Abtast-und-Halte Verstärker || Abtast-und-Halte-Glied || Abtast-und-Halteverstärker * *Abtast-und-Halteglied* || Abtast-und-Halteglied
sample-and-hold amplifier Abtast-und-Halte Verstärker || Abtast-und-Halte-Verstärker
sample-and-hold circuit Abtast-und-Halte-Glied
sample-hold amplifier Abtast-und-Halte-Verstärker
sampled abgetastet * *(Signal)*
sampled signal Abtastsignal
sampled-data system Abtastsystem
sampling Abtastung * *z.B. eines Reglers* || Auswahl * *z.B. der wichtigsten Produkte in e. Übersichtskatalog* || Erfassung * *Abtastung* || Stichprobennahme
sampling control Abtastregelung

sampling controller Abtastregler
sampling cycle Abtastzyklus
sampling error Abtastfehler
sampling frequency Abtastfrequenz || Abtastgeschwindigkeit * *Abtastfrequenz*
sampling inspection Stichprobenkontrolle || Stichprobenprüfung
sampling instant Abtastzeitpunkt
sampling interval Abtastintervall || Abtastzeit || Abtastzyklus * *Abtastintervall, Tastperiode* || Tastperiode * *Abtastintervall*
sampling period Abtastperiode || Tastperiode
sampling point Abtastpunkt
sampling pulse Abtastimpuls
sampling rate Abtastfrequenz * *Abtastrate* || Abtastgeschwindigkeit * *Abtastrate* || Abtastrate
sampling system Abtastsystem
sampling test Stichprobenkontrolle || Stichprobenprüfung
sampling theorem Abtasttheorem
sampling time Abtastzeit || Zykluszeit * *Abtastzeit bei Regelsystem*
sampling time interval Abtastzeitintervall
sampling time monitoring Abtastzeitüberwachung * *erkennt Überlauf eines Softwarezyklus* || Zyklusüberwachung * *erkennt Überlauf eines Softwarezyklus*
sanction billigen * *sanktionieren, billigen (auch nachträglich), gutheißen, dulden, bindend machen*
sand schleifen * *schmirgeln (z.B. Holz ~ mit Sandpapier/Bandschleifer usw.)*
sand blasting Sandstrahlen
sand-blasting blower Sandstrahlgebläse
sand-blasting machine Sandstrahlgebläse
sandblast blower Sandstrahlgebläse
sandblast unit Sandstrahlgebläse
sandblaster Sandstrahlgebläse
sandblasting machine Sandstrahlgebläse || Sandstrahlmaschine
sander Schleifmaschine * *(Holz) mit Sandpapier arbeitend* || Winkelschleifer
sanding Schleif- * *(Holz) mit Schmirgel arbeitend*
sanding beld Schleifband * *(Holz) mit Sandpapier*
sanding disc Schleifscheibe * *(Holz) mit Schmirgelpapier*
sanding drum Schleifzylinder * *(Holz)*
sanding dust Schleifstaub * *(Holz)*
sanding paper Schleifpapier
sandpaper Sandpapier
sandwich Schicht- * *aus mehreren ~en bestehend* || Verbund- * *Mehrschicht-*
sandwich brush Schichtbürste * *Kohlebürste für Motor/Generator*
sap Splint * *(Holz) Splint(holz)*
satellite Planetenrad * *bei Planetengetriebe*
satellite gear Planetenrad || Satellitenrad * *bei Planetengetriebe* || Satellitenzahnrad * *bei Planetengetriebe*
satellite plant Zweigbetrieb
satisfaction Zufriedenheit * *to my greatest satifaction: zu meiner größten*
satisfactorily zufrieden stellend * *(Adv.)*
satisfactory dankbar * *befriedigend* || einwandfrei * *zufrieden stellend* || genügend * *befriedigend, zufrieden stellend* || günstig * *befriedigend* || zufrieden stellend
satisfied zufrieden * *~gestellt, befriedigt*
satisfy einhalten * *z.B. Bedingung/Gleichung* || erfüllen * *z.B. eine Bedingung/Formel/Gleichung/Vorschrift ~* || genügen * *einer Bedingung/Vorschrift ~*
sattelite carrier Planetenträger
sattelite- Insel-
saturable sättigbar * *induktives Bauteil*
saturable reactor Saugdrossel * *(Stromrichter)*

saturable-core reactor Sättigungsdrossel
saturate durchsetzen * *sättigen* || sättigen
saturated geflutet * *gesättigt, z.B. Diode* || gesättigt
|| Satt-
saturated steam boiler Sattdampfkessel
saturation Sättigung || Übersteuerung * *Sättigung*
saturation degree Aussteuerung * *Drossel*
sauce beizen * *Tabak*
save abspeichern * *retten* || retten || schonen * *Kräfte, Vorrat usw.* ~ || sparen * *Geld, Kräfte, Mühe, Zeit usw.* ~
save one's strength schonen * *seine Kräfte sparen*
save up einsparen
saving Einsparung || Ersparnis * *in/of: an* || sparsam
saving of energy Energieeinsparung
saving of time Zeitersparnis || Zeitgewinn
savings Einsparung * *(Plural) cost saving(s): Kosteneinsparung* || Einsparungen
saw nähen || Säge || sägen
saw bench Kreissäge
saw blade Sägeblatt
saw carriage Sägewagen * *{Holz}*
saw filer Sägenfeilmaschine * *{Holz} zum Schärfen von Sägen*
saw setting machine Schränkmaschine * *{Holz} zum Schärfen von Sägen*
saw web Sägeblatt
saw-blade Sägeblatt
saw-tooth Sägezahn
saw-tooth voltage Sägezahnspannung
sawdust Sägemehl * *{Holz}*
sawing machine Sägemaschine
sawmill Sägewerk
sawn timber Schnittholz
sawtooth generator Sägezahngenerator * *erzeugt sägezahnförmige Dreieckimpulse*
say lauten
say again wiederholen * *verbal*
scaffold Gerüst * *{Bautechnik} Bau/Montage*~
scalable skalierbar * *in Stufen ausbaubar/vergrößerbar*
scalar Skalar * *(auch Adj.)*
scalar field Skalarfeld
scalar product skalares Produkt
scalar quantity Größe * *skalare Größe* || skalare Größe
scalar variable skalare Größe
scale bewerten * *skalieren* || Einteilung * *~ in Grade* || gewichten * *skalieren, normieren eines Signals* || Maß * *~stab; on a large scale: in großem ~e* || Maßstab * *on a reduced/larger scale: in verklein./vergrößertem ~; on a large scale: in großem ~* || normieren * *auf eine Bezugs/Nenngröße* || Skala || Tarif * *für Löhne, Steuern usw.* || Umfang * *large-scale: in großem ~* || Waage
scale division Skalenteil
scale down verkleinern * *Zeichnung/Maßstab*
scale off blättern * *abblättern* || entzundern
scale removal entzundern * *(Substantiv)*
scaled abgestuft || gewichtet * *skaliert* || normiert
* *auf eine Bezugsgröße bezogen (%, 10V, 20 mA usw.)*
scaler Untersetzer * *z.B. Frequenzteiler für Impulsketten*
scales Waage
scaling Abgleich * *~ auf Nennwert (z.B. Tachometersignal), Normierung, Gewichtung* || Bewertung
* *eines Signals mittels Anpassungsfaktor* || Gewichtung * *Skalierung, Normierung* || Normierung
|| Skalierung
scaling factor Anpassfaktor * *Normier/Skalierfaktor* || Bewertungsfaktor * *Skalierfaktor* || Normierfaktor * *Gewichtungsfaktor* || Normierung
* *(→PROFIBUS-Profil)*

scaling information Skalierung * *Skalierdaten, Skalierwerte*
scalloped gezackt
scamp pfuschen
scamped work Pfusch
scan Abfrage * *auf einen Signalzustand* || abfragen
* *abtasten, z.B. einen Signalzustand ~* || abtasten ||
beobachten * *dauernd ~, systematisch ~* || erfassen
* *abtasten (z.B. Messwert)*
scan rate Abtastfrequenz * *Abtastrate* || Abtastgeschwindigkeit * *Abtastrate* || Abtastrate
scan result Abfrageergebnis
scan time Abtastzeit || Zykluszeit * *Abtastzeit z.B. bei SPS/Regelsystem*
scan time monitor Zykluszeitüberwachung * *als Einrichtung/Gerät*
scan time monitoring Zykluszeitüberwachung
scandal Ärgernis * *skandalöses Ereignis, Schande, (öffentliches) Ärgernis*
scanner Abtaster * *optischer/berührungsloser, z.B. für Barcode* || Messanlage * *berührungslos arbeitende ~*
scanner system Messanlage * *{Papier} z.B. zum berührungsfreien Messen von Feuchte, Flächengewicht, Dicke usw.*
scanning Abfrage * *eines Signalzustandes/Impulses usw.; Abtastung v. Messwerten in e. Digitalsystem* || Abtastung * *z.B. optisch*
scanning frequency Abtastfrequenz
scanning head Abtastkopf || Tastkopf * *Abtastkopf (z.B. für Duckmarken)*
scanning of actual value Istwertabfrage
scanning speed Traversiergeschwindigkeit * *z.B. bei berührungsloser Messanlage in der Papierindustrie*
scanty dürftig * *spärlich* || knapp * *dürftig, kärglich, spärlich, knapp, unzureichend, beengt (Raum usw.)*
scarce knapp * *spärlich, selten, rar; scarce commodities: Mangelwaren; make oneself scarce:sich rar machen*
scarcity Mangel * *Knappheit*
scarf Lasche * *Laschung; scarf joint: Blattfuge/ Falzverbindung/Verlaschung*
scatter streuen * *statistisch~, physikalisch~* || Streuung * *statistische ~* || verteilen * *verstreuen, bestreuen (with: mit), verbreiten, ~; well-scattered: gleichmäßig verteilt*
scattering Streuung * *statistische ~, physikal. ~*
scenario Konfiguration * *→Konstellation* || Konstellation || Szenarium
schedule ansetzen * *Termin* || einteilen * *zeitlich ~, durch Zeit/Terminplan ~* || Einteilung * *zeitliche ~* || festlegen * *Termin ~ (on: auf)* || Liste * *Verzeichnis, Aufstellung, Tabelle, (Arbeits/Stunden)Plan* ||
Plan * *Zeitplan* || planen * *zeitlich/terminlich ~ (auch Personaleinsatz)* || Terminplan || verwalten
* *(zeitlichen) Ablauf von Programmen ~* || vorausplanen * *planen, festlegen* || vorsehen * *(zeitlich) planen* || Zeitplan
schedule a meeting Termin anberaumen * *für Besprechung*
schedule confirmation Terminzusage
schedule for courses Kursplan * *Zeitplan, Kursprogramm*
schedule orders disponieren * *zeitliche Verteilung von Aufträgen planen*
schedule promise Terminzusage
schedule tentatively andenken * *provisorisch einplanen, ungefähr zeitlich planen (for: für)*
scheduled listenmäßig * *in Liste/Katalog aufgeführt/eingetragen; festgelegt, geplant* || voraussichtlich * *(zeitlich ein)geplant* || vorgesehen
* *(zeitlich) geplant*
scheduled condition Sollzustand

scheduled date   Termin * *geplanter Zeitpunkt*
schedules   Termine * *in Zeitplan* || Terminplan
scheduling   Disposition * *(zeitliche) Planung* || Einteilung * *zeitliche ~* || Planung * *zeitliche ~, Termin~* || Terminplanung || Zeitplanung
schematic   schematisch * *[ski:'mätik]* || schematische Darstellung * *[ski:'mätik]* || Stromlaufplan * *auch 'schematische Darstellung, Übersicht, Schema'*
schematic circuit diagram   Übersichtsschaltplan
schematic diagram   Blockschaltbild || Prinzipschaltbild || Prinzipschaltplan || Schaubild * *Übersichtsdarstellung* || schematische Darstellung * *siehe auch →Übersichtsschaltbild* || Stromlaufplan || Übersichtsplan || Übersichtsschaltbild || Übersichtsschaltplan
schematic diagrams   Schaltungsunterlagen
schematic representation   schematische Darstellung
schematically   schematisch * *(Adv.)*
schematize   schematisieren
scheme   Aktion * *Plan, Projekt, Programm, Komplott, Intrige* || Anordnung * *Schema* || Entwurf * *Schema, System, Anlage, Übersicht, schemat.-Darstellg., Aufstellung, Plan, Projekt, Programm* || Konzept * *Schema, System, Plan, Entwurf* || Plan * *Schema, Anlage, System, →Entwurf* || Schema
Schmitt trigger element   Schmitt-Trigger * *Schwellwertschalter mit Hysterese*
scholarship   Wissen * *Gelehrsamkeit*
schooling   Ausbildung * *Unterricht, Ausbildung* || Schulung * *Unterricht, Ausbildung*
Schottky barrier diode   Schottky-Diode
Schottky diode   Schottky-Diode
schredder   Zerkleinerer * *für Akten, Lumpen, Abfälle usw.*
science   Lehre * *Wissenschaft* || Wissenschaft
science and research   Wissenschaft * *~ und Forschung*
science of business   Betriebswirtschaft * *Betriebswirtschaftslehre*
science of transport   Verkehrstechnik * *als Wissenschafts-/Forschungsgebiet*
scientific   wissenschaftlich
scientist   Wissenschaftler
scissor lift   Hebebühne
scissors   Schere * *allg.*
scoop   Baggereimer * *Schöpfeimer/-Kelle, (Wasser-) Schöpfer, Schippe* || Schaufel * *zum Schöpfen, auch bei ~bagger* || Tropfenverteiler * *in der Hohlglasherstellung*
scope   Anwendungsbereich * *Geltungsbereich z.B. für Vorschrift* || Freiheit * *Spielraum* || Geltungsbereich || Oszillograf || Oszillograph * *(salopp)* || Oszilloskop || Rahmen * *(Geltungs-) Bereich* || Spielraum * *Spanne; leave/allow someone scope: jdm. ~ lassen* || Umfang * *Anwendbarkeits/Geltungsbereich, Inhalt, beinhaltete Funktionen*
scope of applications   Einsatzspektrum * *wide: breites*
scope of delivery   Lieferumfang
scope of functions   Funktionsumfang
scope of supply   Lieferumfang * *be in the scope of supply: im ~ enthalten sein*
scope of tasks   Aufgabenumfang
score   Ergebnis * *erzielte Punkte, Treffer(zahl), Ergebnis, (Be-)Wertung, Spielstand* || erzielen * *Erfolg/Treffer ~* || Nut * *Kerbe, Rille (siehe auch "Rille")* || Rille * *Kerbe, Rille*
scored   geriffelt * *(ein)gekerbt, gerillt*
scorer   Nutmesser * *Kerbmesser, z.B. in der Buchbinderei*
scoring   Riefenbildung
scour   aufscheuern * *scheuern, schrubben, polieren* || entzundern || scheuern * *scheuern, schrubben, polieren*

SCR   Thyristor * *silicon controlled rectifier*
SCR type high speed fuse   Halbleiter-Sicherung * *zum Thyristorschutz* || Halbleitersicherung * *zum Schutz von Thyristoren* || Thyristorsicherung * *silicon-controlled rectifier:zum Thyristorschutz*
SCR type high speeed fuse   superflinke Sicherung * *~ zum Thyristorschutz*
scrap   Abfall || Altmaterial || Ausschuss * *auch Schrott* || schlecht * *Schrott-* || Schrott * *auch Eisen/Stahlschrott* || verschrotten
scrap paper   Schmierpapier
scrape   kratzen * *schaben* || schaben
scrape off   abkratzen || abschaben
scraper   Kratzer * *Schaber* || Räumer * *zur Wasseraufbereitung* || Schaber || Spachtel * *Malerwerkzeug* || Streichmesser * *bei Streich-/Papierbeschichtungsmaschine usw.* || Ziehklinge * *{Holz}*
scraper bar   Rakel * *Schaber, Kratzmesser, Ziehklinge*
scraper plate   Kratzblech
scrapping   Verschrottung
scratch   kratzen * *scratchy noise: kratzendes Geräusch* || Kratzer * *Kratzer, Schramme* || Riss * *Schramme, Ritz* || schaben * *kratzen*
scratch flag   Schmiermerker * *bei SPS*
scratch off   abkratzen
scratch pad memory   Zwischenspeicher * *schneller "Notizblock"-~*
scratching-type washer   Kratzscheibe * *z.B. zur Herstellung einer sicheren Erdverbindung im Schaltschrank*
screech   pfeifen * *kreischen (z.B. Lager), gellend/durchdringend schreien* || quietschen * *kreischen (z.B. Bremse), gellend schreien*
screen   abschirmen * *Abschirmung* || Bild * *Bildschirm~* || Bildschirm || Bildschirmseite || Monitor * *Bildschirm* || Schirm * *Abschirmung, Bildschirm* || Schutz * *Abschirmung* || schützen * *abschirmen* || Sichtgerät * *Monitor, z.B.für Personal Computer* || Sieb * *Gitter~ (z.B. für Sand, Zellstoff usw.); auch beim →Siebdruck* || sieben * *durch ein mechanisches Sieb (durch)sieben*
screen flicker   Bildschirmflimmern
screen form   Bildschirmmaske || Maske * *~/Formular auf dem Bildschirm*
screen mask   Maske * *Bildschirm~*
screen page   Bildschirmseite
screen printing   Siebdruck * *Druckverfahren*
screen shot   Bildschirmfoto * *Hardcopy eines Monitorbildes*
screen structure   Bildaufbau * *Anordnung der Grafikelemente auf d. Bildschirm*
screened   abgeschirmt * *screened enclosure: ~es Gehäuse* || geschirmt * *siehe auch →abgeschirmt*
screened cable   abgeschirmte Leitung
screening   Schirmung * *z.B. einer Leitung*
screening connection   Schirmanschluss
screening cover   Abschirmhaube
screening machine   Siebmaschine * *{Holz} für zur Trennung der Holzschnitzel in der Spanplattenherstellung*
screening plate   Abschirmblech
screening system   Siebsystem * *zum Trennen von Partikeln (z.B. zerklnertes Holz)*
screw   Schnecke * *z.B. bei Extruder, Schneckengetriebe* || Schraube * *wood screw: Holz~; auch Schiffs-~; countersunk screw: eingelassene ~/ Senk~* || schrauben || Spindel * *Schraub~, Gewindestange, Hub~* || verschrauben * *siehe auch →festschrauben (to: mit)*
screw bolt   Schraubenbolzen
screw cap   Schraubdeckel || Schraubverschluss * *Schraubdeckel* || Überwurfmutter
screw clamp   Schraubzwinge
screw closure   Schraubverschluss

screw compressor Schraubenkompressor || Schraubenverdichter
screw connection Schraubverbindung || Verschraubung
screw conveyor Schnecke * *Schneckenförderer, Förderschnecke* || Schneckenförderer
screw die Gewindeschneideisen || Schneideisen
screw driver Schraubendreher || Schraubenzieher
screw fitting Schraubbefestigung || Verschraubung * *z.B. ~ für Rohrverbindung, Kabeldurchführung*
screw fixing Schraubbefestigung
screw gear Schneckengetriebe || Schneckenrad || Schneckenradgetriebe
screw gland Stopfbuchsverschraubung
screw head Schraubenkopf
screw hole Schraubenloch
screw home festschrauben || verschrauben * *festschrauben (am Einbauplatz)*
screw in einschrauben
screw in place festschrauben * *am vorgesehenen Einbauplatz ~* || verschrauben * *am vorgesehenen Einbauplatz* →*festschrauben*
screw interlocking Schraubverriegelung * *z.B. für (Sub-D) Stecker*
screw lead Spindelsteigung
screw locking varnish Schraubensicherungslack
screw locking washer Schraubensicherungsring
screw off abschrauben || aufschrauben * *abschrauben*
screw on anschrauben * *siehe auch* →*festschrauben* || aufschrauben * *anschrauben* || festschrauben || verschrauben
screw out abschrauben * *herausschrauben* || herausschrauben
screw plug Verschlussschraube
screw pump Kreiselpumpe * →*Schraubenpumpe* || Schneckenpumpe || Schraubenpumpe
screw retaining ring Schraubensicherungsring
screw spindle Gewindespindel
screw tap Gewindebohrer
screw terminal Klemme * *Schraubklemme* || Schraubanschluss * *z.B. zum Herstellen der Kontaktierung an einem Bauteil* || Schraubklemme
screw tight festschrauben
screw together verschrauben * *zusammenschrauben*
screw union Schraubverbindung
screw-down Walzenanstellung * *{Walzwerk}*
screw-down motor Anstellmotor * *{Walzwerk} zur Walzenanstellung*
screw-down operation Walzenanstellung * *{Walzwerk}*
screw-driving equipment Schrauber
screw-in thyristor Schraubthyristor
screw-lock varnish Schraubensicherungslack
screw-locking device Schraubensicherungselement
screw-plate Kluppe * *{Werkz.masch.} (amerikan.) an Drehbank*
screw-setting machine Schraubmaschine * *z.B. in der Holzindustrie*
screw-type metering system Schneckendosierer
screw-type proportioning system Schneckendosierer
screw-type sealing machine Schraub-Verschließmaschine * *in der Lebensmittel-/Verpackungsindustrie*
screw-type terminal Schraubklemme
screwdriver bit Schraubwerkzeug * *Schraubeinsatz*
screwdriver tool Schraubwerkzeug
screwdriving unit Schrauber
screwed verschraubt
screwed conduit entry PG-Verschraubung
screwed connection Schraubverbindung

screwed gland Verschraubung * *~ einer Kabeldurchführung*
screwed joint Schraubverbindung || Verschraubung
screwed pin Gewindestift
screwed pipe joint Schraubverbindung * *von Rohren*
screwed rod Bolzenschraube || Gewindestange
screwed spindle Gewindespindel
screwing and threading machines Gewindeherstellungsmaschinen
screwing chuck Gewindeschneidkopf
screwing down Anstellung * *{Walzwerk}* →*Walzenanstellung in Walzgerüst* || Walzenanstellung * *{Walzwerk}*
screwing machine Schrauber * *automatic screwing machine: Schraubautomat*
screwstock Kluppe * *zum Festschrauben eines Teils*
scribe anreißen * *mit Reißnadel* || anzeichnen * *anreißen (z.B. mit Reißnadel)*
scriber Reißnadel
scroll blättern * *z.B. über Bildschirmseiten* || rollen * *(Bildschirm)Anzeige, die Ausschnitt e.größeren Bildes ist, up:hoch; down;hinunter* || Schnecke * *Schnecke, Spirale, Schnörkel; z.B. ·· bei (Schrauben-) Kompressor*
scroll corner cutting Formschneiden
scroll down hinunterrollen * *Anzeige, Bildschirm*
scroll up hinaufrollen * *Anzeige, Bildschirm*
scrollbar Rollbalken * *auf dem Bildschirm* || Schiebebalken * *am Bildschirm*
scrub scheuern * *schrubben, scheuern, reinigen*
scrupulous gewissenhaft * *peinlich genau* || sorgfältig * *(über-) gewissenhaft, peinlich genau, ängstlich, vorsichtig*
scrupulousness Sorgfalt * *Gewissenhaftigkeit*
scrutinize prüfen * *genau prüfen/untersuchen/ansehen/studieren* || untersuchen * *genau ~*
scrutiny Prüfung * *genaue Untersuchung, prüfender Blick* || Überprüfung * *genaue Untersuchung, prüfender Blick* || Untersuchung * *genaue*
sculptured surface Freiformfläche * *z.B. zur Bearbeitung mit Werkzeugmaschine*
sea See-
sea freight Seefracht
sea level Meereshöhe * *above sea level:über dem Meeresspiegel*
sea transport Seetransport
sea water-proof seewasserfest
seal abdichten * *versiegeln, dichtmachen* || dichten * *abdichten* || Dichtung || schließen * *hermetisch ver~, versiegeln, durch Kleben/Schweißen/Vergießen ~* || vergießen || verschließen * *hermetisch ~, z.B. durch Kleben, Schweißen, Vergießen* || Verschluss * *Dichtung* || versiegeln
seal curling bördeln * *(Substantiv) Verschließen durch Bördelung, z.B. Kondensator, Dose usw.* || verkapseln * *(Substantiv) verschließen durch Bördelung (Dose, Becherkondensator usw.)*
seal face Dichtfläche
seal lip Dichtlippe * *z.B. bei einem Dichtring*
seal ring Dichtring
seal-in circuit Selbsthalteschaltung
seal-in contact Selbsthaltekontakt
sealant Dichtmasse || Dichtmittel * →*Dichtmasse* || Dichtungsmittel
sealed abgedichtet || gekapselt * *abgedichtet, hermetisch dicht*
sealed bearing gedeckeltes Lager * *Rillenkugellager mit eingebauten Lagerdichtungen*
sealed enclosure geschlossene Kapselung
sealed membrane keyboard Folientastatur * *mit hoher Schutzart*
sealing Abdichtung || Dichtelement || Schutzart * *Abdichtung* || Selbsthaltung || Verguss

sealing agent Dichtmasse || Dichtungsmasse
sealing bar Schweißbacke * *(Verpack.techn.)* z.B. zum Verschließen von Kunststoff-Schlauchbeuteln
sealing compound Dichtmasse || Dichtungsmasse || Vergussmasse
sealing disc Dichtscheibe
sealing disk Dichtscheibe
sealing element Dichtelement || Dichtungselement
sealing end Endverschluss * *an einem Leistungskabel* || Kabelendverschluss * *Schmutz-/Feuchteschutz am Kabelende an der Verzweigung d. Einzelleiter*
sealing face Dichtfläche
sealing home Selbsthaltung * *be sealed home: in ~ gehen*
sealing in Selbsthaltung
sealing kit Dichtungssatz
sealing lip Dichtlippe
sealing machine Verschließmaschine * *(Verpack.-techn.)* z.B. für Flaschen, Tuben, zur Verpackung von Lebensmitteln usw.
sealing materials Verschließmittel * *(Verpack.-techn.)*
sealing ring Dichtring || Simmerring
sealing rubber Dichtgummi
sealing strip Dichtpropil * *z.B. für Schranktür*
sealing surface Dichtfläche
sealing test Dichtigkeitsprüfung
sealing washer Dichtring || Dichtscheibe * *z.B. Gummidichtscheibe in PG-Verschraubung*
seam Fuge * →*Naht* || Naht
seam welding Bahnschweißen
seam-rolling machine Sickmaschine
seam-welding machine Nahtschweißmaschine
seaming machine Falzmaschine * *zur Blechbearbeitung*
seamless nahtlos * *(auch figürl.)*
seaport Hafen * *Seehafen*
search Suche * *in search/quest of: auf der ~ nach* || suchen * *for: nach* || Suchlauf
search for reference Referenzfahrt * *Anfahren d.Referenzpunktes bei Positioniersteuerung* || Referenzpunktfahren * *bei Positioniersteuerung*
search function Suchlauf
search key Suchbegriff
search operation Suchvorgang
search sequence Suchvorgang * *aufeinander folgende Suchvorgänge z.B.zum Fangen e. Motors nach Spannungseinbruch*
season altern * *z.B. durch Wettereinflüsse/künstlich ~* || Jahreszeit || Kampagne * *Zuckerrübenkampagne* || lagern * *altern, ab~ lassen (z.B. Holz)* || trocknen * *~ durch Lagerung* || voraltern
season crack Alterungsriss
seasonable zeitgemäß
seasoned erfahren * *langgedient*
seasoned wood Holz * *abgelagertes ~*
seasoning Alterung || Lagerung * *Alterung, das Ablagern (z.B. von Holz)*
seasoning kiln Trockenofen
seat Arbeitsplatz * *~ im Büro, z.B. an einem Computer* || Auflage * *Sitz, Auflager, Fundament* || Auflagefläche || aufliegen * *seinen Sitz haben, liegen, sitzen* || Bund * *Auflager* || lagern * *Maschine in Maschinenbettung ~* || Platz * *Sitz~* || Sitz * *~gelegenheit, Ventil~, mechan. ~fläche, Niederlassung; registered seat: Geschäfts~*
seating Lagerung * *Sitz*
seating face Auflagefläche
seawater-resistant seewasserbeständig
seaworthy package seemäßige Verpackung
seaworthy packaging seemäßige Verpackung
secluded abgeschlossen * *abgesondert*
second ander * *zweiter* || anders * *(Adj.) zweit* || Sekunde || unterstützen * *beistimmen, sekundieren* || zweit

second grade zweite Wahl
second order Nachbestellung
second shaft end zweites Wellenende
second shaft extension zweites Wellenende
second source Zweithersteller
second-hand alt * *gebraucht, aus zweiter Hand* || gebraucht
second-order zweiter Ordnung
second-to-none konkurrenzlos * *unvergleichlich*
secondaries Sekundärwicklungen
secondary Neben- * *zweitrangig, untergeordnet, sekundär; z.b. secondary aspect: ~aspekt* || Niederspannungsseite * *eines "Abwärtsstrafos"* || sekundär || Sekundärdeite * *z.B. eines Transformators* || Sekundärseite * *eines Trafos* || sekundärseitig || Sekundärwicklung || untergeordnet || unterlagert * *untergeordnet*
secondary aspect Nebenaspekt || Nebenpunkt * *Nebenaspekt*
secondary control loop unterlagerter Regelkreis
secondary costs Folgekosten || Nebenkosten
secondary course weiterführender Kurs * *Aus-/Weiterbildungskurs*
secondary current control unterlagerte Stromregelung
secondary damage Folgeschaden
secondary distribution system Niederspannungsnetz
secondary effect Nebenerscheinung
secondary fault Folgefehler
secondary gear Abtriebsstufe * *bei Getriebe*
secondary phenomenon Begleiterscheinung * *Sekundäreffekt, Nebeneffekt* || Folgeerscheinung * *Nebeneffekt* || Nebeneffekt * *Begleiterscheinung* || Nebenerscheinung
secondary processing Nachbearbeitung
secondary pulse Zweitimpuls * *z.B bei Steuersatz für Thyristorsatz*
secondary sequencer Unterkette * *bei SPS*
secondary switched power supply sekundärgetaktetes Schaltnetzteil * *hinter dem Übertrager getaktetes Schaltnetzteil*
secondary voltage Läuferstillstandsspannung * *bei Schleifringläufermotor* || Sekundärspannung
secondary winding Sekundärwicklung * *Transformator*
secondary windings Sekundärwicklungen
secondary-side sekundärseitig
secondly zweitens
seconds range Sekundenbereich * *in the :im ~*
secret geheim * *in secret: im Geheimen; top/most secret: streng ~; secret agreement: Geheimabkommen* || intern * *geheim* || verborgen * *geheim* || vertraulich * *geheim; top secret: streng geheim*
secretariat Geschäftsstelle
secretary Sekretärin
secretly geheim * *(Adv.) ins~*
section Abschnitt * *Kapitel, Abschnitt/Teil einer Maschine/Anlage, Bereich* || Abteilung * *Ausschnitt * Teil* || Bereich * →*Abschnitt einer Maschine/Anlage usw.* || Feld * *auch ~ in Schaltanlage, Aufbausystem usw.* || Kapitel * *Abschnitt* || Pfad * *Teil* || Profil * *z.B. von stangenförmigem Material (Stahl, Rohr usw.)* || Schnitt * *Schnitt(Bild), Profil* || Schnittbild || Strecke * *Abschnitt* || Teil * *auch ~abschnitt, ~element* || Teil- * *Teilabschnitt, (Teil-)Element* || Wandstärke || Zweig * *Teil*
section by section abschnittweise * *(Adv.)*
section leader Gruppenführer
sectional drive Antriebsstaffel * *auch 'Antrieb innerhalb einer Antriebsstaffel* || Gruppenantrieb * *ein Antrieb in einem Mehrmotorenverbund* || Mehrmotorenantrieb * *auch: ein Antrieb im Mehrmotorenverbund*
sectional drives Mehrmotorenantriebe

sectional view Ansicht im Schnitt || Schnitt
* *Schnittbild-Ansicht in Zeichnung* || Schnittbild
sectional warper Schärmaschine * *{Textil}*
sectionalize unterteilen * *in Abschnitte (räumlich)*
~
sectionally abschnittweise * *(Adv.)*
sector Ausschnitt * *Kreisausschnitt* || Bereich
* *Sektor, Teil* || Branche * *Bereich* || Geschäftszweig || Sektor * *auch auf der Festplatte*
sector motor Sektormotor * *mit im Kreissektor angeordnetem "Käfigläufer" z.B. bei Pressen u. Steinbrechern*
sector of industry Industriebereich || Industriezweig
sector-oriented branchenorientiert
secure befestigen * sichern || beschaffen * erlangen || fest * *sicher, geschützt (from:vor), ~ (Grundlage, Hoffnung, Schaubverbindung), gesichert (Existenz)* || feststellen * *sichern; in position: in (s)einer Lage* || fixieren * *mechanisch sichern* || schützen * *sichern* || sicher * *geschützt (from; vor), ge~t, fest, gewiss* || sichern * *~ (against: gegen; on/ by: durch), Sicherheit bieten, befestigen, (fest ver-)schließen* || sicherstellen * *~ (auch i.Handelswesen), sichern (on/by: durch; z.b.Hypothek), Sicherheit bieten*
secure bus sichere Schiene * *(USV)*
secure with pins verstiften * *mit Stiften sichern*
secured geschützt * *gesichert*
securing Sicherstellung || Sicherung * *das Sichern, das Sicher-Machen (auch Datenübertragung, Transportlast usw.)*
securing screw Sicherungsschraube
securities Papier * *(Plural) Wert~e (z.B. Aktien, Anleihen)*
security Sicherheit * *from/against: vor; auch {Wirtschaft}: security for payment*
security access code Zugriffsschlüssel * *z.B. Codewort zur Vermeidung unberechtigten Zugriffs*
security code Sicherheitscode
security printing machine Wertpapierdruckmaschine
sediment Ablagerung * *{Geologie}* || Niederschlag * *Ablagerung* || Rückstand * *Bodensatz* || Satz * *Boden~*
see empfangen * *Zutritt gewähren* || erkennen * *(er/ein/nach/ab)sehen, erkennen, entnehmen, herausfinden, erleben, verstehen* || siehe
see above siehe oben
see also siehe auch
see below siehe unten
see you later Auf Wiedersehen * *bis später*
seek Positionierung * *Schreib-/Lesekopf in Computer-Laufwerk* || suchen
seeming scheinbar * *anscheinend* || virtuell * *scheinbar*
seemingly scheinbar * *(Adv.) scheinbar; es scheint, dass*
seen from the load von der Last her gesehen
seep ausschwitzen
sef-adapting selbsteinstellend * *selbstanpassend*
sef-tuning selbsteinstellend * *z.B. Regler*
segment Abschnitt * *Abschnitt, Teil, Segment* || Ausschnitt * *Kreisausschnitt* || Lamelle * *eins Kommutators* || Segment * *~ eines Kreises, eines Kommutators, einer Sieben~-Anzeige, Speicher~ usw.* || Teilabschnitt * *auch ~ eines Computernetzwerks*
segment display Segmentanzeige
segmentation Segmentierung
segregate absondern || teilen * *trennen,(sich) absondern, ausschneiden, (sich) abspalten, s.chem. abscheiden,abgesondert* || trennen * *ab~, isolieren, absondern, sich abspalten/absondern*
segregated getrennt * *getrennt, abgesondert, abge-*

spaltet, isoliert || isoliert * *getrennt, abgesondert, abgespaltet, isoliert*
segregated-loss method Einzelverlustverfahren * *zur Bestimmung der Nenndaten einer elektr. Maschine (siehe VDE 0530,Teil 2)*
segregation Trennung * *Absonderung*
seismic test Erdbebenprüfung
seismic withstand capability Erdbebenfestigkeit
seismic withstandibility Erdbebenfestigkeit
seize erfassen * *greifen (auch figürl.)* || fassen * *(er-)greifen, packen, fassen, befallen* || festfressen * *sich festfressen* || fressen * *(sich) fest~* || greifen || nützen * *(Gelegenheit usw.) ergreifen* || pakken * *fassen*
seize up fressen * *(sich) fest~*
seize upon reißen * *an sich ~*
seldom selten
select anwählen || auswählen * *from:aus* || einstellen * *auswählen* || selektieren || Umschaltung || vorwählen || wählen
select input Wahleingang * *z.b. binäre/digitale/ analoge Eingangsklemme*
select menu Auswahlmenü
select output Wahlausgang * *z.B analoge/digitale/ binäre Ausgangsklemme*
select parameter Auswahlparameter
select possibilities Auswahlmöglichkeiten
select signal Wahlsignal
select switch Wahlschalter
select terminal Wahlklemme
select window Auswahlfenster * *am Bildschirm*
selectable anwählbar || umschaltbar * *wählbar* || wählbar
selectable analog input analoger Wahleingang
selectable analog output analoger Wahlausgang
selectable signal Wahlsignal
selected ausgesucht || gewählt || selektiert
selecting Anwahl || Wahl * *das Auswählen*
selection Ansteuerung * *Auswahl* || Anwahl || Auswahl || Entscheidung * *Auswahl* || Umschaltung * *Aus/Anwahl z.B. über Schalter* || Wahl * *Auswahl*
selection capabilities Auswahlmöglichkeiten
selection criterion Auswahlkriterium * *(Plural: criteria)*
selection data Auswahldaten
selection features Auswahlkriterien
selection key Anwahltaste
selection list Auswahlliste
selection of the rotational direction Drehrichtungsauswahl || Drehrichtungswahl
selection of the speed direction Drehrichtungsauswahl || Drehrichtungswahl
selection sheets Auswahlliste * *Auswahlblätter*
selection switch Auswahlschalter || Vorwahlschalter || Wahlschalter
selection table Auswahltabelle
selective gezielt * *selektiv, auswählend, Auswahl-, trennscharf* || selektiv || wahlweise
selective advertising gezielte Werbung
selectivity Selektivität * *bei Schutzeinrichtung/system, Radio usw.*
selector Umschalter || Wahlschalter
selector button Anwahltaste
selector switch Umschalter * *Mehrfach-Umschalter* || Wahlschalter
selenium Selen
self selbst
self adapting selbstanpassend
self calibration Selbstabgleich * *eines Messsystems*
self circulation Eigenkühlung
self commutation Selbstführung * *für Stromrichterschaltung*
self diagnosis Selbstdiagnose
self test Selbsttest

self tuning routine Optimierungslauf
self ventilated eigengekühlt * *with fan: mit Lüfter*
self ventilation Eigenkühlung * *Motorkühlart* →*"eigenbelüftet" (Lüfter durch Motorwelle ange trieben)*
self- Eigen- * *Selbst-* || Selbst-
self-acting automatisch * *selbsttätig* || selbsttätig
self-adaption selbstanpassend
self-adherent selbstklebend
self-adhering selbstklebend
self-adhesive selbstklebend
self-aspirating selbstansaugend * *Verbrennungsmotor usw.*
self-bypassing ramp function generator ablösender Hochlaufgeber * *überbrückt sich nach erfolgtem Hochlauf*
self-centering selbstzentrierend
self-cleaning selbstreinigend
self-clocked eigengetaktet
self-clocked converter selbstgetakteter Stromrichter
self-commutated selbstgeführt
self-commutated converter selbstgeführter Stromrichter * *benöt. kein. ext.Kommutier.spanng. (Ggs.:last-/netzgef.;DIN41750 Tl.1)*
self-commutating selbstgeführt
self-contained abgeschlossen * *(in sich) geschlossen, unabhängig, selbstständig, autark* || autark * *in s.geschlossen, selbstständig (arbeitend), unabhängig* || eigenständig * *z.B. Gerät* || geschlossen * *in sich* ~ || selbstständig * *autark, eigenständig, unabhängig, in sich geschlossen; z.B. Gerät, Maschine* || unabhängig * *(in sich) geschlossen, selbstständig (arbeitend), autark,*
self-contained unit Komplettgerät * *autarkes, eigenständiges, selbstständig arbeitendes, in sich geschlossenes* ~
self-controlled selbstgesteuert * *z.B. Stromrichter*
self-controlling eigenverantwortlich
self-cooled eigenbelüftet * *selbstgekühlt (unüblicher Ausdruck für Motorkühlart)* || eigengekühlt * *(für Motorkühlart unüblicher Ausdruck)*
self-cooling Eigenkühlung * *z.B. eines Motors mit am Läufer angebrachten oder von ihm angetriebenen Lüfter* || selbstgekühlt || Selbstkühlung
self-diagnostic Selbstdiagnose
self-diagnostics Selbstüberwachung * *Selbstdiagnose*
self-discharge Selbstentladung
self-documenting selbstdokumentierend
self-employed freiberuflich * *be self-employed: freiberufl. tätig sein*
self-employed person Freiberufler
self-evident selbsterklärend || selbstverständlich
self-evident fact Selbstverständlichkeit
self-excitation Selbsterregung
self-explanatory selbsterklärend
self-extinguishing selbstverlöschend * *z.B. Brennbarkeitsstufe*
self-healing selbstheilend * *Kondensator nach Spannungsüberbeanspruchung/Durchschlag* || Selbstheilung * *z.B. von Kondensatoren nach einem Durchschlag des Dielektrikums*
self-holding contact Selbsthaltekontakt
self-inductance Eigeninduktivität || Selbstinduktion || Selbstinduktivität
self-induction Selbstinduktion
self-learning selbstlernend
self-locking sebstsperrend || selbsthemmend * *z.B. Gewinde,Getriebe (kein Rücklauf mögl. bei Wegbleiben o.Umkehr d.Antriebsmoments)* || selbstsichernd * *z.B. Schraube, Mutter* || selbstsperrend
self-locking screw Sperrzahnschraube * *flat-headed:mit Flachkopf*
self-lubricating selbstschmierend

self-monitoring Selbstüberwachung
self-moving selbstfahrend
self-nesting selbstzentrierend * *z.B. bei Steckverbindung*
self-optimization Selbstoptimierung
self-optimizing selbstoptimierend
self-priming selbstansaugend * *Pumpe*
self-propelled selbstfahrend
self-recuperating selbstheilend * *z.B. Kondensator nach Spannungüberbeanspruchung/Durchschlag*
self-regulation Ausgleich * *selbsttätig eintretender* ~
self-reliant selbstständig * *selbstbewusst, selbstsicher*
self-restoring selbstheilend * *Isolierung usw.*
self-sealing selbstdichtend
self-setting Selbsteinstellung * *z.B. Regler*
self-starting Selbstanlauf || selbstanlaufend * *Motor usw.*
self-styled so genannt * *angeblich/vorgeblich*
self-supervision Selbstüberwachung
self-supplying autark * *selbstversorgend*
self-supporting freitragend || selbstständig || selbsttragend
self-supporting construction freitragende Konstruktion
self-synchronous system elektrische Welle
self-tapping screw Blechschraube || selbstschneidende Schraube
self-test Selbsttest
self-testing Selbsttest
self-tuning Selbstabgleich || selbstoptimierend
self-ventilated eigenbelüftet * *mit durch die Motorwelle angetriebenem Lüfter* || eigengekühlt * *z.B. Motor mit vom Läufer angetriebenem Lüfterrad* || selbstbelüftet
self-ventilation Eigenbelüftung * *{el.Masch.} (veraltet)* →*Eigenkühlung*
selfcooling selbstkühlend * *z.B. Motor*
selftest Selbsttest * *power-on selftest: b.Einschalten d.Versorgungsspanng. autom. durchgeführter* ~
selftuning Optimierungslauf || Selbstoptimierung
sell absetzen * *verkaufen* || gehen * *sich verkaufen (Ware)* || verkaufen || vertreiben * *verkaufen*
sell below cost schleudern * *unter Preis ver*~
sell for Kosten * *(Verb) für ... verkauft werden*
sell underhand verschieben * *unter der Hand verkaufen, z.B. auf d. Schwarzmarkt* ~
sell well gehen * *sich glänzend verkaufen*
seller Verkäufer
selling Absatz * *das Verkaufen*
selling price Verkaufspreis
selling tool Akquisitionshilfe
sellotape Tesafilm * *Klebeband aus durchsichtiger dünner Kunststoff- oder Zellstofffolie*
selsyn Drehgeber || elektrische Welle
selvage Kante * *~ eines Tuches, Webkante*
semantic Semantik * *auch einer Programmiersprache*
semester Semester
semi- Halb- * *z.B. semi-public company: ~öffentliche Gesellschaft; semi-annual: ~jährlich* || Zwischen- * *Halb-*
semi-automatic halbautomatisch
semi-axial-flow fan Halbaxiallüfter * *Mischung aus Axial- und Radiallüfter (drehricht.abh. Lüfter zur Motorkühlg.)*
semi-chemical pulp Zellstoff * *Halbzellstoff*
semi-controlled halbgesteuert * *z.B. Brückenschaltung, zur Hälfte aus Thyristoren, zur anderen Hälfte aus Dioden*
semi-fluid grease Fließfett
semi-portal crane Halbportalkran
semi-skilled worker Arbeiter * *angelernter* ~

semicolon Semikolon || Strichpunkt
semiconductor Halbleiter || Halbleiterbauelement
semiconductor component Halbleiterbauelement
semiconductor device Halbleiterbauelement
semiconductor fuse Halbleitersicherung
semiconductor protecting fuse Halbleiterschutz-Sicherung || Halbleitersicherung
semiconductor protective fuse Halbleiterschutz-Sicherung
semiconductor relay Halbleiter-Relais
semiconductors Halbleiter * *(Plural)*
semifinished product Halbzeug
semifluid dickflüssig
seminar Lehrgang * *(Universitäts-) Seminar* || Seminar * *Universität*
semiportal crane Halbportalkran
semiproduct Halbzeug
semiskilled worker angelernter Arbeiter
send abordnen || entsenden || schicken * *Brief, Post, Botschaft, Person (to:nach/zu; for:nach)* || senden * *Geld, Brief, Nachricht usw.* ~ || versenden || zuschicken * *to: jemandem*
send back zurücksenden
send for bestellen * *kommen lassen*
send off abschicken || entsenden
send samples to bemustern
send your order to Bestellung an
sender Absender * *z.B. eines Briefs* || Sender * *Ab~*
sending Sendung * *Absendung*
sending samples Bemusterung
senior Chef * *besonders Rang/Dienstälterer* || leitend * *Angestellter, Ingenieur usw.* || ober * *(Substantiv) Chef-* || Ober- * *Chef-, Vorgesetzter, (rang)höher*
senior clerk Büroleiter * *bei Behörde, Bank*
senior engineer Chef * *rang-/dienstälterer Ingenieur* || Gruppenführer * *Ober-/Chefingenieur*
senior executive leitender Angestellter
senior management oberer Führungskreis
sensational eklatant
sense abfragen * *Messwert/Signalzustand usw.* ~ || abtasten * *Messwert von Messaufnehmer, Sensor* || erfassen * *Messwert (auch von Messgeber)* || Gefühl || Sinn * *Sinn, Verstand, Vernunft,Bedeutg., Richtg., Gefühl;in a narrow/broad sense: i.eng./weiterem~e*
sense of length Längsrichtung
sense of rotation Drehrichtung * *Drehsinn* || Drehsinn
sense wire Fühlerleitung * *z.B. b.Netzteil nach d.4-Drahtprinzip z.Kompensation d. Leitungs-Spanngs.abfälle*
sensibility Empfindlichkeit * *to:für/gegen; Empfindlichkeit, Empfänglichkeit, Sensibilität, Empfindsamkeit*
sensible bewusst * *bei Bewusstsein; be sensible of: sich einer Sache ~ sein; fühl-/spürbar, vernünftig* || fühlbar || sinnvoll * *vernünftig; make sense: ~ sein/ Sinn machen* || spürbar * *fühlbar*
sensing -Erfassung * *~ von Messwerten/Signalen von Messgebern usw.* || Abtastung * *das Abtasten, Messen, Erfassen eines Messwerts* || Erfassung * *von Messwerten* || Erkennung * *eines Signals/Zustandes*
sensing device Sensor
sensing element Sensor || Signalgeber
sensing probe Tastkopf * *z.B. für Näherungsschalter*
sensing range Erfassungsbereich * *für Messgeber, Näherungsschalter usw.*
sensing system Erfassungssystem * *für Messwerte*
sensitive anfällig * *to:gegenüber* || empfindlich * *allg.; auch leicht zerstörbar* || feinfühlig * *sensibel* || heikel * *Thema, Punkt usw.* || sensibel
sensitive to -empfindlich * *(vorangestellt)*

sensitive to pressure druckempfindlich
sensitivity Abhängigkeit * *Empfänglichkeit, Empfindlichkeit* || Anfälligkeit * *Empfindlichkeit, Sensibilität, Sensitivität (to:für/gegen)* || Ansprechempfindlichkeit || Ansprechwert * *kleinster erfassbarer Messwert* || Empfindlichkeit * *auch 'Sensibilität, Sensitivität, Veranlagung' (to:für/gegen)*
sensitivity level Ansprechschwelle * *z.B. einer Schutzeinrichtung*
sensor Abtaster || Aufnehmer * *Sensor, Mess~* || Fühler * *Messgeber* || Geber * *Messgeber* || Messaufnehmer || Messfühler || Messgeber * *auch Messfühler* || Messwertaufnehmer || Sensor || Signalgeber
sensor feedback Geberrückführung
sensor in the ambient Außenfühler * *Außen-Temperaturfühler für Heizungszwecke*
sensorless geberlos * *ohne Messgeber (arbeitend)*
sensorless control geberlose Regelung * *z.B. Drehzahlregelung ohne Drehzahl-Messgeber/Tacho*
sensors Sensorik
sentence Satz * *~ aus Worten*
separate abtrennen || auftrennen * *(mechan.)* || einzeln * *getrennt* || extra * *getrennt* || Fremd- || gesondert * *getrennt; to be ordered separately: ~ zu bestellen* || getrennt * *voneinander* || räumlich getrennt || selbstständig * *getrennt, eigenständig* || separat || speziell * *gesondert* || teilen * *absondern* || trennen * *(ab)trennen (from:von), (ab)spalten, absondern, unterscheiden zwischen, zentrifugieren* || vereinzeln * *absondern, abtrennen, ausscheiden* || zerlegen * *in seine physikalische Bestandteile ~*
separate cooling Fremdkühlung * *z.B. Motor durch fremdangetriebenen Lüfter oder Flüssigkeitskühlg*
separate excitation Fremderregung
separate fan Fremdlüfter
separate fan unit Fremdlüfter * *inklusive Lüfterzusatzaggregate (Antriebsmotor, Filter usw.)*
separate feeding Vereinzelung * *{Verpack.techn.}*
separate field Fremderregung
separate out abscheiden * *abtrennen* || ausgliedern
separate print Sonderdruck * *z.B. eines Zeitschriftenartikels*
separate ventilation Fremdbelüftung * *with duct connection: mit Luftzu-/abfuhr über Rohrschluss* || Fremdkühlung * *Motorkühlart m.Luftkühlung (z.B. durch nicht v.d.Motorwelle angetrieb. Lüfter)*
separate ventilation with duct-connection Fremdbelüftung über Rohranschluss * *z.B. Motorkühlarten IC05/06/17/37 nach IEC 34 Teil 6*
separate ventilation with pipe-connection Fremdbelüftung über Rohranschluss
separate-field fremderregt
separate-field excitation Fremderregung
separate-source voltage-withstand test Wicklungsprüfung * *mit Fremdspannung*
separated Insel- || räumlich getrennt
separated windings getrennte Wicklungen * *{el. Masch.} galvan. voneinand. getrennte Wicklungen bei polumschaltbarem Motor*
separately allein * *jedes Teil für sich* || einzeln * *(Adv.) getrennt*
separately cooled fremdbelüftet * *fremdgekühlt* || fremdgekühlt * *Motorkühlg. durch getrennt angetriebenen Lüfter (Kühlwirkg.unabh.von Motordrehz.)*
separately driven getrennt angetrieben
separately excited fremderregt * *z.B. Gleichstrommaschine ohne Permanenterregung* || getrennt erregt * *Motor*
separately fan-ventilated fremdbelüftet
separately ventilated fremdgekühlt * *→fremdbelüftet*

separately-driven fan Fremdlüfter * *fremdangetriebener Lüfter (sitzt nicht auf Motorwelle)*
separately-driven fan motor Fremdlüftermotor * *Motor mit Fremdlüfter*
separately-excited DC machine fremderregte Gleichstrommaschine
separately-excited DC motor fremderregter Gleichstrommotor
separately-ventilated fremdbelüftet * *Motorkühlart; with duct connection: mit Rohranschluss*
separately-ventilated with pipe-connection - fremdbelüftet über Rohranschluss
separating character Trennzeichen
separating filter Weiche * *z.B. Impuls~*
separating layer Zwischenlage * *~ zum Trennen aufeinander liegender Lagen*
separating line Trennfuge || Trennlinie
separation Lösung * *~ einer Beziehung* || Teilung * *Trennung* || Trennung * *auch von Stoffen/Bestandteilen*
separation paper Beilaufpapier * *Papierzwischenlage* || Papierzwischenlage * *z.B. in der Blech-/Gummiverarbeitung, auch in Kondensator*
separator Abscheider * *z.B. oil separator: Ölabscheider; centrifugal separator: Zentrifugal~* || Separator * *Trennanlage (z.B. mit Zentrifuge)* || Sichter * *trennt Stoffe z.B. durch Sieben, Zentrifugieren usw.* || Trennzeichen || Zwischenlage * *z.B. Wicklungs~ (aus Hostaphan usw.)*
separator inserting machine Gefachesteckmaschine * *Verpackungsmaschine, packt Abtrennpappen in Karton*
separator piece Distanzstück * *zum Trennen/Auseinander halten von Teilen*
sequel Folge * *~zeit; in the sequel: in der ~; ~erscheinung, (Aus)Wirkung, (Aufeinander-) Folge* || Folgeerscheinung || Sequenz * *(Aufeinander-) Folge, Folgeerscheinung, (Aus)Wirkung, Konsequenz; in the sequel: in d.Folge* || Verlauf * *(Aufeinander-) Folge, weiterer ~*
sequence Abfolge || Ablauf * *['si:kwens] Abfolge* || Aufeinanderfolge * *['siequens]* || Folge * *['siequens] Aufeinander~, Ab~* || Rangfolge || Reihe * *(Aufeinander-) Folge, ~nfolge, Reihe, Serie; in sequence: der ~ nach* || Reihenfolge * *alphabetic(al): alphabetische; inverse: umgekehrte* || Sequenz * *Aufeinanderfolge, Reihenfolge, Reihe, Serie; Escape sequence: Steuersequenz (für Drucker)* || Staffel * *(Reihen)Folge; z.B. Antriebe* || Verlauf * *~ von Zeit* || Vorgang * *(Ab-) Folge*
sequence block Schrittbaustein * *bei SPS*
sequence cascade Ablaufkette
sequence cascade control Ablaufkettensteuerung
sequence cascade element Kettenelement * *in Ablaufkette*
sequence chart Ablaufdiagramm
sequence control Ablaufkette || Ablaufsteuerung * *['si:kwens]* || Ein-Ausschaltlogik || Schaltwerk * *Ablaufsteuerung* || Steuerwerk
sequence control chart Ablaufplan
sequence control module Zustandssteuerwerk
sequence control system Ablaufsteuerung * *bei SPS*
sequence controller Ablaufsteuerung
sequence logic Ein-Ausschaltlogik
sequence message system Meldefolgesystem
sequence of assembling Montagefolge
sequence of functional activities Funktionsablauf
sequence of installation Montagefolge
sequence of motion Bewegungsablauf
sequence of movement Bewegungsablauf
sequence processor Schaltwerk * *elektronisches Ablaufsteuerwerk*
sequence step Ablaufschritt || Schrittelement * *in einer Ablaufkette bei SPS*

sequenced aufeinander folgend * *in Reihe angeordnet*
sequencer Ablaufkette || Ablaufsteuerung || Zustandssteuerwerk
sequencer cascade Ablaufkette
sequencer step Ablaufschritt
sequencing Verkettung * *das Aufeinanderfolgen-Lassen*
sequencing control Ablaufsteuerung || Schaltwerk * *Ablaufsteuerung* || Steuerwerk
sequencing controller Ablaufsteuerung * *als Einrichtung/Gerät*
sequential sequenziell || squentiell
sequential control Ablaufkette * *Ablaufsteuerung* || Ablaufkettensteuerung || Ablaufsteuerung || Folgesteuerung || Schaltwerk * *Ablaufsteuerung* || Schrittsteuerung
sequential control element Ablaufglied
sequential errors Folgefehler * *(Plural)*
sequential gating Folgesteuerung * *{Stromrichter}*
sequential logic element Ablaufglied
sequential logic stage Kommandostufe
SERCOS SERCOS * *Abk.f. 'SErial Realtime COmmunication System':schnell. Kommunikat.system für Numerikantriebe*
serial Reihen- || seriell
serial communication serielle Kopplung || serielle Schnittstelle * *Kommunikation, Datenaustausch* || serielle Verbindung * *Kommunikation*
serial communications serielle Kommunikation
serial connection serielle Schnittstelle * *Verbindung*
serial data transmission serielle Datenübertragung
serial interface serielle Kopplung || serielle Schnittstelle
serial interface board Kommunikationsbaugruppe * *Leiterkarte für serielle Kommunikation* || Schnittstellenbaugruppe * *serielle Schnittstellenbaugruppe (Leiterkarte)*
serial link serielle Kopplung || serielle Schnittstelle * *serielle Verbindung* || serielle Verbindung
serial networking serielle Vernetzung
serial No. Fertigungsnummer * *Seriennummer* || Seriennummer
serial number Fabrikationsnummer * *Seriennummer* || Fabriknummer * *Seriennummer*
serial port serielle Schnittstelle * *Anschluss*
serial production Serienfertigung
serial start-up Folgestart * *→zeitlich versetztes Einschalten* || zeitlich versetztes Einschalten * *z.B. bei Verbund-→Verdichtersatz zur Begrenzung des Anlaufstroms*
serial test Serienprüfung
serial transmission serielle Übertragung
series Baureihe || Folge * *Reihe, Serie* || Reihe * *auch Bau~, a series of: eine ~ von; connected in series: in ~ geschaltet* || Serie || Vor- * *in Serie geschaltet* || Vorschalt-
series availability Serieneinsatz * *Verfügbarkeit in Serienstückzahlen*
series capacitor Reihenkondensator
series characteristic Reihenschluss-Kennlinie * *[el.Masch.] Momentenkennlinie*
series circuit Reihenschaltung
series connection Hintereinanderschaltung || Reihenschaltung || Serienschaltung
series inductance Reiheninduktivität || Serieninduktivität || Vorschaltdrossel * *Vorschaltinduktivität* || Vorschaltinduktivität * *siehe auch →Vorschaltdrossel*
series inductor Vorschaltdrossel * *in Reihe geschaltete ~*
series manufacturing Serienherstellung
series of measurements Messreihe
series of products Baureihe

series production Serienfertigung || Serienherstellung || Serienproduktion * *Serienherstellung*
series reactor Vorschaltdrossel * *in Reihe geschaltete* ~ || Vorschaltinduktivität * *Vorschalt-Leistungsdrossel*
series resistance Reihenwiderstand || Serienwiderstand * *als physikal. Größe, z.B. im Ersatzschaltbild*
series resistor Reihenwiderstand || Serienwiderstand * *als Bauteil* || Vorwiderstand
series resonant circuit Saugkreis * *Serienresonanzkreis*
series stabilizing winding Hilfsreihenschlusswicklung * *in DC-Motor*
series transformer Vorschalttrafo * *siehe auch 'Vorschalttransformator'* || Vorschalttransformator || Vortrafo * *siehe auch →Vortransformator* || Vortransformator * *in Reihe geschalteter Trafo*
series winding Reihenschlußwicklung
series-connected in Reihe geschaltet || nachgeschaltet * *in Reihe geschaltet* || vorgeschaltet * *in Reihe geschaltet*
series-mode Gegentakt- * *-Spannung, -Signal, -Unterdrückung*
series-mode rejection Gegentaktunterdrückung
series-produced serienmäßig
series-produced type Serienausführung || Serienprodukt
series-wound machine Reihenschlussmaschine * *Feldwicklung in Reihe mit Ankerwicklung*
series-wound motor Hauptschlussmotor || Reihenschlussmotor * *Feldwicklung in Reihe mit Ankerwicklung*
serious erheblich * *Schaden, Aufwand* || ernst zu nehmend || gefährlich * *ernst, schwerwiegend* || gravierend
seriously hazardous environment Ex-Bereich
seriousness Wichtigkeit * *Ernst*
serrated gezackt || gezahnt * *gezackt*
serrated lock washer Fächerscheibe * *Schraubensicherungsscheibe*
serrated washer Kratzscheibe * *Zackenscheibe, Fächerscheibe* || Zackenscheibe
servants Personal * *Beamte, Angestellte (öffentlicher Dienst); Dienstboten*
serve bedienen * *z.B. einen Markt/Kunden/Gäste, versorgen, dienen, nützen* || dienen * *as: als; for: (da)zu; it serves the purpose of: es dient dazu, ...* || erfüllen * *serve the needs/demands: Bedürfnisse/Anforderungen* ~
serve as dienen als
serve one's apprenticeship lernen * *seine Lehre/gewerbliche Ausbildung machen (with:bei)*
server Server * *Zentralcomputer i.e. 'Client-Server Netzwerk', verwaltet d. gemeinsame (Dokumenten)Datenbank*
service bedienen * *softwaremäßig* || Bedienungs- || Betrieb * *Dienst (auch e.Maschine/Fahrzeugs), Betrieb (in: in; out of: außer)* || Dienst * *Dienstleistung, Dienstleistungseinrichtung, ~ eines Feldbussystems usw.* || Dienstleistung || Einsatz * *Dienst (auch von Maschinen)* || Instandhaltung || Kundendienst || Leistung * *Dienst~* || pflegen || Prüfung * *Kundendienst, Wartung* || Service || Service durchführen || überholen * *technisch nachsehen, warten, überholen* || Überprüfung * *Kundendienst, Wartung* || Verkehr * *Verkehrsdienst; withdraw from service: aus dem Verkehr ziehen* || warten * *Wartungsdienste leisten, z.B. an Maschine* || Wartung
service aid Servicehilfe * *Hilfsmittel*
service assignment Serviceeinsatz * *von Servicepersonal*
service brake Betriebsbremse

service cable Anschlusskabel * *für Hausanschluss* || Anschlussleitung * *Hausanschluss*
service call Serviceeinsatz
service capability Wartungsfreundlichkeit * *Wartbarkeit*
Service Center Service-Abteilung * *(amerikan.)* || Serviceabteilung * *Servicezentrum (Abteilungsbezeichnung)*
Service Centre Service-Abteilung || Serviceabteilung * *Servicezentrum (Abteilungsbezeichnung)*
service characteristic Belastungskennlinie * *[el. Masch.]* →*Betriebskennlinie, Belast.abhängigkeit v. Drehzahl, Strom usw.* || Betriebskennlinie * *[el. Masch.] Belastungsabhängigkeit v.Drehzahl, Strom, Schlupf, Leistung usw.*
service company Dienstleistungsbetrieb
service condition Betriebsbedingung * *[el.Masch.] Betriebsverhältnisse, die der Festlegung d.Betriebsart zugrundeliegen*
service conditions Betriebsverhältnisse
service contract Wartungsvertrag
service cover Bedienungsklappe || Serviceöffnung * *Wartungsklappe* || Wartungsöffnung * *Wartungsklappe*
service department Kundendienst * *Abteilung* || Service-Abteilung || Serviceabteilung || Wartungsdienst
service device Servicegerät
service effort Serviceaufwand
service engineer Inbetriebnahme-Ingenieur * *Service/Wartungsingenieur* || Inbetriebnahmeingenieur || Service-Ingenieur || Service-Techniker
service factor Betriebsfaktor * *Betriebsbeiwert/Servicefaktor: Verhältn. zuläss. Motordauerleistg./Motornennleistg.* || Service Factor * *zuläss.Überlast/Bemess.leistg. e.durchzugsbelüft. Elektromotors (NEMA MG1-14.36)* || Servicefaktor * *Verhältn. zulässige Motordauerleistung/Motornennleistung* || Überlastfaktor * *Service/Betriebsfaktor*
service functions Servicefunktionen
service fuse Anschlusssicherung
service hours Betriebsstunden
service instructions Servicehilfe * *Service-/Wartungsanleitung* || Wartungsanleitung
service interface Service-Schnittstelle
service interval Inspektionsintervall || Serviceintervall || Wartungsintervall
service life Gebrauchsdauer || Haltbarkeit * *Lebensdauer (von Verschleißteilen)* || Lebensdauer * *Gebrauchsdauer, z.B. von Schmierfett, Kohlebürsten, Kontakten, Verschleißteilen* || Nutzungsdauer || Standzeit * *eines Verschleißteils, z.B. Kohlebürsten, Lager*
service life of grease Fettgebrauchsdauer
service location Servicestelle
service mission Serviceeinsatz * *von Servicepersonal*
service network Service-Netz || Servicenetz || Servicesystem * *Servicenetz*
service opening Montageöffnung * →*Wartungsöffnung* || Serviceöffnung || Wartungsöffnung
service personnel Servicepersonal
service provider Dienstleistungsbetrieb
service purposes Servicezwecke
service staff Servicepersonal
service station Reparaturwerkstatt
service support Servicehilfe
service technician Service-Techniker
service unit Servicegerät
service vehicle Servicefahrzeug
service voltage Betriebsspannung * *z.B. eines Netzes*
service-call report Einsatzbericht * *von Servicepersonal*

service-friendliness Servicefreundlichkeit
service-friendly servicefreundlich
service-proven ausgereift * *betriebs-/praxiserprobt*
serviceable betriebsbereit * →*betriebsfähig* || *nützlich* * *dienlich* || praktikabel * *zweckdienlich, brauchbar, verwendbar, nützlich* || praktisch * *zweckdienlich, brauchbar, verwendbar*
servicing Instandhaltung || Service * *das 'Service-Leisten', ~leistung, Erbringen einer ~leistung, Warten* || Unterhaltung * *Wartung* || Wartung * *z.B. an Maschine*
servicing cover Bedienungsklappe * *siehe auch* →*Serviceöffnung* || Inspektionsklappe * *Wartungsklappe; siehe auch* →*Wartungsöffnung* || Serviceöffnung * *Wartungsklappe/-Deckel*
servicing function Servicefunktion
servicing program Serviceprogramm
serving zuständig * *~ Vertriebsniederlassung usw.*
servo amplifier Servoregler * →*Servoverstärker* || Servoverstärker * *Leistungs- und Regelteil z.Ansteuerung eines (kleinen)* →*Servomotors*
servo controller Servoregler
servo drive Servoantrieb || Stellantrieb * *(schneller) Positionierantrieb*
servo driver Servoverstärker
servo motor Servomotor * *hochdynamischer Motor für Positionieraufgaben/Bewegungssteuerung* || Stellmotor
servo technique Servotechnik
servo valve Regelventil || Servoventil || Stellventil
servo-amplifier Stromrichtergerät * *zur Ansteuerung eines Servomotors bzw.eines Positionierantriebs*
servo-control Regelung * *~ mit Stellantrieb*
servo-controlled drive geregelter Antrieb
set -Paket * *Satz* || -Satz * *gleicher Teile z.B. Steuersatz, Thyristorsatz* || ansetzen * *a date: einen Termin* || ansteuern * *e. Ein/Ausgang mit High-Signal (z.B. bei SPS)* || aufstellen * *Beispiel ~* || eingestellt || einrichten * *Werkzeugmaschine* * *Einstell-* * *z.B. ~schraube* || einstellen * *z.B. Parameter, Sollwert ~* || festlegen * *Termin ~* || feststellen * *festsetzen, bemessen, festlegen* || Folge * *~ von zusammengehörigen Teilen* || regulieren * *einstellen, setzen* || Reihe * *Gruppe* || Satz * *~ von Teilen* || Schar * *von Kurven* || schränken * *[Holz] Säge scharf machen durch "Verbiegen" der Zähne* || Senkung * *Fundament usw.* || setzen * *z.B. Ausgang/Signal/Parameter/Flipflop/Merker ~* || soll * *Kurzzeichen für 'setpoint': Sollwert*
set a high standard hohe Ansprüche stellen * *of:an*
set about angreifen * *(figürlich) Aufgabe usw.*
set afoot einfädeln * *(figürl.)*
set aside annullieren * *Urteil*
set back belasten * *finanziell mit einem Betrag von ... ~*
set conditionally begrenzt setzen * *bei SPS*
set down ablegen * *absetzen* || absetzen * *auf eine Unterlage* || abstellen * *auf eine Unterlage absetzen*
set going betätigen * *in Gang setzen* || bewegen * *in Bewegung setzen* || in Bewegung setzen || in Gang setzen
set in concrete einbetonieren
set in motion betätigen * *in Gang setzen* || bewegen * *in Bewegung setzen* || in Bewegung setzen || in Gang setzen
set in operation in Betrieb setzen || inbetriebnehmen || inbetriebsetzen
set input conditionally Eingang begrenzt setzen * *bei SPS*
set key Übernahme-Taste
set of curves Kurvenschar
set of cutters Fräsersatz * *{Holz}*

set of overhead transparencies Foliensatz
set of rated values Nenndaten
set off against verrechnen * *gegeneinander aufrechnen*
set on ansetzen
set on edge verkanten
set on foot in die Wege leiten || in Gang setzen * *(figürlich) in Gang bringen, auf den Weg bringen*
set one's facts against widersetzen
set out aufführen * *in einer Liste* || feststellen * *(ausführlich) darlegen, angeben*
set output Ausgang setzen
set permanently to fest einstellen auf
set pulse Setzimpuls
set screw Einstellschraube
set speed Drehzahlsollwert
set standards Maßstäbe setzen
set the temperature temperieren * *die Temperatur (richtig) einstellen*
set to work inbetriebnehmen * *ingangsetzen*
set to zero nullsetzen
set up aufbauen * *aufstellen, montieren, gründen, etablieren, kräftigen, wiederherstellen* || aufstellen * *auch Maschine ~* || einführen * *Einrichtungen, Methode usw. ~* || einstellen * *einstellen, regulieren; z.B. Parameter, Potentiometer* || Einstellung * *v.Parametern; auch f.Messgerät (Ozilloskop, Logikanalysator) u.PC-Konfiguration (BIOS)* || gründen * *z.B. eine Firma* || montieren * *aufstellen* || zusammenstellen * *Mannschaft usw. aufstellen/etablieren*
set value eingestellter Wert || Sollwert
set vibrating versetzen * *in Schwingungen ~* versetzt * *in Schwingungen ~*
set-screw Feststellschraube
set-up aufrichten * *von der Horizontalen in die Vertikale* || einrichten * *Maschine usw. ~* || Einrichtung || einsetzen * *z.B. eine Kommission, ein Gremium* || Einstellung * *z.B. ~ von Einstellparametern* || Inbetriebnahme || Voreinstellung * *auch Vorgabe der Hardwareparameter für das BIOS-System eines PCs*
set-up mode einrichten * *(Substantiv) Betriebsart bei NC-Steuerung und SPS*
set-up time Rüstzeit || Vorlaufzeit
set-value input Sollwertvorgabe
setback Rückschritt
setpoint Führungsgröße * *Sollwert* || Sollwert
setpoint adjuster Sollwerteinsteller || Sollwertgeber * *z.B. zur Drehzahlsollwertvorgabe* || Sollwertsteller
setpoint adjustment Sollwerteinstellung
setpoint cascade Sollwertkaskade || Sollwertkette || Sollwertstaffel
setpoint chain Sollwertkette
setpoint channel Sollwertkanal
setpoint computer Sollwertrechner
setpoint conditioning Sollwertaufbereitung
setpoint control Sollwertsteuerung
setpoint distribution Sollwertverteilung
setpoint enabling Sollwertfreigabe
setpoint entering Sollwertvorgabe
setpoint filtering Sollwertglättung || Sollwertsiebung
setpoint generation Sollwertaufbereitung * *Sollwerterzeugung* || Sollwertbildung
setpoint generator Sollwertgeber
setpoint indication Sollwertanzeige
setpoint indicator Sollwertanzeige * *Sollwert-Anzeigegerät*
setpoint input Sollwerteingang || Sollwertvorgabe
setpoint integrator Sollwertintegrator * *Hochlaufgeber*
setpoint issuing Sollwertvorgabe

setpoint limit Sollwertgrenze
setpoint limiting Sollwertbegrenzung
setpoint link Sollwertstaffel
setpoint polarity Sollwertpolarität
setpoint potentiometer Sollwertpotentiometer
setpoint preparation Sollwertaufbereitung
setpoint processing Sollwertaufbereitung
setpoint reduction Sollwertabminderung
setpoint response Führungsverhalten * *[Regel.- techn.]*
setpoint setting Sollwerteinstellung || Sollwertvorgabe
setpoint setting device Sollwerteinsteller || Sollwertsteller
setpoint smoothimg Sollwertglättung
setpoint source Sollwertgeber * *Sollwertquelle* || Sollwertquelle
setpoint transmitter Sollwertgeber
setpoint value Sollwert
setpoint value of speed Drehzahlsollwert
setpoint value setting Sollwertvorgabe
setpoint-actual value deviation Soll-Istwert-Abweichung
setpoint-actual value difference Regelabweichung * *Soll-Ist-Differenz* || Soll-Ist-Abweichung || Soll-Ist-Differenz
setpoint-actual value monitoring Soll-Ist-Überwachung || Soll-Istwert-Überwachung
setscrew Abdrückschraube || Feststellschraube || Klemmschraube || Stiftschraube * →*Feststellschraube*, →*Madenschraube*
setter Einrichter * *Maschinen~* || Maschineneinrichter
setting Eingabe * *per Bedienung* || Einrichtung * *Einstellung, z.B. einer Maschine* || Einstell- * *z.B. ~schraube,~knopf, ~ring* || Einstellung * *~ von Parametern, Potentiometern* || Einstellwert || Stellung * *eines Schalters, Parameters, Sollwerts, Potentiometers usw.* || Verstellung * *durch Bedienpersonal, Stellantrieb usw.* || Vorgabe * *z.B. eines Sollwerts*
setting accuracy Einstellgenauigkeit
setting apart Abzweigung
setting error Einstellfehler
setting gauge Einstellehre
setting instrument Einstellgerät
setting machine Schränkmaschine * *[Holz] zum Schärfen von Sägen*
setting of rotational direction Drehrichtungsvorgabe
setting possibilities Einstellmöglichkeiten
setting procedure Einstellvorschrift * *Vorgehensweise zum Einstellen*
setting range Einstellbereich * *z.B. für Parameter* || Einstellungsbereich || Sollwertbereich
setting ring Einstellring
setting rule Einstellregel * *siehe auch* →*Einstellvorschrift* || Einstellvorschrift
setting speed Verstellgeschwindigkeit
setting the course Weichenstellung * *Vorgabe des Kurses (for:auf/in Richtung)*
setting time Rüstzeit || Umrüstzeit
setting up Einstellarbeiten || Montage * *Aufstellen*
setting value Einstellwert || Setzwert * *zu setzender Wert,z.B. für Integrator, Speicher)*
setting-basin Klärbecken
setting-up Einrichtung * *Gründung*
setting-up time Rüstzeit
settle abschließen * *endgültig ~* || ausräumen * *(figürl.) Meinungsverschiedenheit usw. ~* || beseitigen * *Auseinandersetzng/Streit aus dem Weg räumen* || einigen * *on:sich einigen auf* || entscheiden * *endgültig* || erledigen * *Geschäft; Problem lösen* || klären || kümmern * *klären, regeln, erledigen, aus der Welt schaffen (Problem)* || lösen * *Problem ~* || ord-

nen * *abschließend* || regeln * *eine Vereinbarung treffen, sich vergleichen, einrichten* || senken * *sic ~ (Fundament usw.)* || setzen * *(Erdreich, Fundament, Flüssigkeit, Bodensatz)*
settle down permanently Fuß fassen
settled abgeschlossen * *endgültig abgeschlossen* || geregelt * *Problem, Verfahren usw.*
settlement Abkommen || Abmachung || Abrechnung * *of accounts: von Konten* || Abwicklung * *Erledigung* || Behebung * *Problem ~, Fehler ~* || Bereinigung * *siehe auch 'Fehlerbereinigung'* || Erledigung || Übereinkunft * *Vergleich, Schlichtungsergebnis* || Vergleich * *Einigung; out-of-court settlement: außergerichtlicher ~* || Zahlung * *~ einer Schuld*
settling Klärung
settling process Einschwingvorgang
settling tank Absetzbecken * *bei der Wasseraufbereitung*
settling time Ausregelzeit * *verstreicht nach.- e.sollwertsprung bis d.regelgröße endgültig i.toleranzbereich bleibt* || Einschwingdauer || Einschwingzeit
SEV SEV * *Abk.f. 'Electrotechnical Institute of Switzerland*'· *Schweizer. Elektrot. Verein (Normengremium)*
seven-segment display Siebensegment-Anzeige
sever lösen * *abtrennen, Beziehungen ~*
several mehrere || verschieden * *mehrere*
several times mehrfach * *(Adv.) mehrere Male*
severally einzeln * *(Adv.) getrennt, einzeln*
severance pay Auslösung * *Trennungsgeld*
severe beträchtlich * *ernsthaft, schwer, schlimm (Verluste, Kritik usw.)* || erheblich * *schwer, schlimm (z.B. Verluste, Kritik)* || groß * *(Kälte, Schaden, Nachteil, Unwetter usw.)* || hart * *streng* || hoch * *Schaden, Strafe, Gebühr* || schwer * *ernst, hoch (Kosten, Steuer, Gebühr, Strafe)* || stark * *schwer, schlimm (z.B. Verlust, Verschleiß, Fehlfunktion, Krankheit, Wetter)* || streng
severe carelessness grobe Fahrlässigkeit
severe environments erschwerte Umgebungsbedingungen
severely streng * *(Adv.)*
severity Härte * *(figürl.) Strenge* || Schärfegrad * *z.B. beim Funkstörgrad, bei Schwingfestigkeitsprüfung*
severity class Schärfegrad
severity level Schärfegrad
seviceable betriebsfähig * *siehe auch* →*betriebsbereit*
sewage Abwasser
sewage sludge Klärschlamm
sewage treatment plant Kläranlage * *zur Abwasserbehandlung* || Klärwerk
sewage work Klärwerk
sewer Rinne * *Wasserablauf~*
sewing cotton Zwirn
sewing machine Nähmaschine
SF Service Factor * *Abk. für* →*Service Factor*
SF6-insulated SF6-isoliert
SG Dehnungsmessstreifen * *Abkürzung für 'strain gauge'* || DMS * *Abkürzung für 'Strain Gauge':- Dehnungsmessstreifen*
sh tn t * *short ton, US-ton: 1 sh tn entspr. 907,2 kg; 1000 kg entspr. 1,1023 short tons*
shabby abschaben * *abgeschabt, z.B. Textilstoff* || dürftig * *schäbig* || schmutzig * *schäbig*
shaded-pole motor Spaltpolmotor * *1-phas. AC- Klein-Induktionsmotor m.Käfigläufer, Kurzschlussring erzeugt Hilfsphase*
shadow memory Schattenspeicher
shadow RAM Schatten-RAM * *(schneller) Speicher, in dem eine Kopie der Daten gehalten wird* || Schattenspeicher * *z.B. in* →*NOVRAM*

shady   faul * *verdächtig*
shaft   Achse || Röhre * *Schacht* || Schacht || Schaft * *einer Säule usw.* || Welle * *(mechan.)* Maschinenteil
shaft assembly   Wellenstrang
shaft axis   Wellenachse
shaft bearing   Achslager
shaft block   Läuferhaltevorrichtung * *Transportsicherung für Motor*
shaft bore   Wellenbohrung * *z.B. in einem Lagerschild*
shaft current   Wellenstrom * *elektr. Strom durch (Motor-) Welle (unerwünschter Effekt)*
shaft deflection   Wellendurchbiegung
shaft diameter   Wellendurchmesser
shaft encoder   Impulsgeber * *Inkrementalgeber* || Impulstacho * *Winkelschrittgeber* || Inkrementalgeber
shaft end   Wellenende * *z.B. einer Motorwelle (free: freies)* || Wellenstummel || Wellenzapfen * *Wellenende*
shaft exit   Wellendurchtritt * *Motor*
shaft expansion   Wellendehnung * *Längsdehnung einer Welle auf Grund von Erwärmung*
shaft extension   Wellenende * *für elektr. Maschine: siehe DIN 748/IEC 74, Teil 3 u.IEC 72* || Wellenstummel || Wellenzapfen * *Wellenende*
shaft extension pointing upwards   Wellenende nach oben
shaft generator   Wellengenerator * *im Schiff*
shaft height   Achshöhe * *{el.Masch.} Abstand zw. d. Längsachse der Welle u. der Fußauflagefläche (DIN 42939)* || Ah * *Achshöhe AH* || Wellenhöhe
shaft journal   Wellenzapfen
shaft keyway   Wellennut
shaft load   Wellenbelastung || Wellenlast
shaft loading   Wellenbelastung
shaft misalignment   Wellenversatz * *Fehlausrichtung der Welle(n)*
shaft mounted gearbox   Aufsteckgetriebe
shaft nut   Wellenmutter
shaft offset   Wellenversatz
shaft output   Wellenleistung * *z.B. eines Motors*
shaft packing   Wellendichtung
shaft position   Wellenlage * *horizontal: horizontal; vertical: vertikal* || Winkellage * *einer Welle*
shaft power   Wellenleistung
shaft revolutions   Wellenumdrehungen
shaft run-out   Rundlauf * *{el.Masch} Komzentrizität d.Wellenendes bezogen auf d.Flansch (DIN 42955, IEC 72-1)*
shaft rupture   Wellenbruch
shaft sag   Wellendurchbiegung
shaft seal   Wellenabdichtung || Wellendichtring || Wellendichtung
shaft seal ring   Wellendichtring * *siehe auch →Simmerring*
shaft sealing   Wellendichtung
shaft shoulder   Wellenbund || Wellenschulter
shaft stub   Wellenstummel
shaft tachometer   Achstacho
shaft test   Wellentest * *{el.Masch.} Festigkeitsprüfung des Wellenmaterials*
shaft torque   Wellenmoment
shaft train   Wellenstrang
shaft voltage   Wellenspannung * *{el.Masch.} unerwünschte Spannung zw. Welle u. Motorgehäuse, erzeugt →Lagerstrom*
shaft-driven winder   Achswickler
shaft-end guard   Wellenschutz * *Kappe als Transportschutz des Wellenendes*
shaft-flange squareness   Planlauftoleranz * *zwischen Wellenende und Flanschfläche*
shaft-mounted fan   Eigenlüfter * *durch (Motor-)welle selbst angetriebener Lüfter* || Eigenlüfter

* *z.B. durch die Welle angetriebener ~ bei eigenbelüftetem Motor*
shaft-mounted speed reducer   Aufsteckgetriebe * *als Untersetzungstriebe*
shaft/hub connection   Welle-Naben-Verbindung
shafting   Wellenstrang
shaftless   achslos * *z.B. Papierroller* || längswellenlos || LWL * *längswellenlos*
shake   rütteln || schütteln || schwanken * *schütteln, erbeben* || wackeln || zittern
shaking   Erschütterung * *das Schütteln* || Vibration * *Schütteln*
shaky   unzuverlässig * *Methode usw.*
shall   sollen * *im Sine von Gebot, Pflicht*
shank   Schaft * *einer Schraube, eines Werkzeugs, eines Schlüssels usw.*
shank cutter   Schaftfräser * *{Werkz.masch.}*
shape   Ausprägung * *(auch figürl.) Form, Gestalt* || Form * *Gestalt, Form (auch e. Impulses); take shape:Form annehmen; be in bad shape:nicht i.form sein* || fräsen * *form~* || Gestalt * *take shape: ~ annehmen* || gestalten * *in Umrissen ~ (z.b. ein Programm, Ideen); formen* || hobeln || Profil * *Form, Gestalt, Umriss* || profilieren || Variante * *(auch figürlich) Form, Gestalt, Umriss* || verformen * *in e.Form bringen, bearbeiten* || Verlauf * *~ eines Impulses/Signals, einer Kurve*
shape of the curve   Kurvenform
shape of the lamination   Blechform * *Form/Querschnitt der Magnetbleche in einem Blechpaket* || Blechschnitt * *Form der Magnetbleche in einem Blechpaket (z.B. e. Motorläufers)*
shaped component   Formteil * *{Holz}*
shaped part   Formteil * *{Holz}*
shaped workpiece   Formteil * *{Holz}*
shaper   Fräsmaschine * *{Holz} Form~* || Stoßmaschine * *{Werkz.maschine}*
shaping   Realisierung * *Ausgestaltung, Ausformung, Ausprägung*
shaping machine   Hobelmaschine || Stoßmaschine * *{Werkz.masch.}*
shaping press   Formpresse * *{Holz}*
shaping unit   Egalisierapparat * *{Holz} zum Schärfen von Sägen*
share   Aktie * *shares are at a premium: die ~n stehen gut; prospects are fine: die ~n stehen gut (figürl.)* || Anteil * *(An)Teil, Geschäfts/Kapital/Aktien~ e. Unternehmens, Teilhaberschaft* || Aufteilung * *z.B. ~ e. Last, ~ v. Last/Drehmoment auf mehrere Motoren; (An)Teil, Betrag, Beteiligg.* || Beitrag * *Beitragsanteil* || Beteiligung * *Anteil* || gemeinsam benutzen * *teilen* || teilen * *(sich) teilen (with: mit); teilhaben (lassen) (an:an)* || verteilen * *unter sich teilen*
share forces   zusammenwirken * *Kräfte vereinigen*
share in profits   beteiligen * *am Gewinn beteiligt sein*
share opinion   übereinstimmen * *share his opinion: mit ihm ~*
shared   gemeinsam * *~ benutzt, untereinander aufgeteilt*
shared forces   Verbund * *gemeinsame Kräfte*
shareholder   Aktionär || Gesellschafter || Teilhaber * *z.B. Aktionär*
sharer   Teilnehmer * *Teilhaber*
sharing   Aufteilung * *untereinander* || gemeinsame Benutzung
sharing of experience   Erfahrungsaustausch
sharp   scharf * *Messer, Zähne, Kurve, (auch figürl.: rau, herb)*
sharp bend   Knick * *in Kurve*
sharp-edged   scharfkantig
sharp-pointed   scharf * *spitz*
sharpen   anspitzen || schärfen
sharpener   Schärfmaschine

sharpening machine Schärfmaschine
sharpness Schärfe * *auch einer Bildschirm-Wiedergabe*
shave schaben * *dünne Späne abschälen, z.B. mit Schabeisen/Werkzeugmaschine*
shavings Schneidabfälle || Span || Späne
shear Abschneidmaschine * →*Schere* || Blechschere || Metallschere || Schere * →*Metall-/*→*Blech~* || Schub * *Scherkraft* || Schubkraft * *Querschub, Scherkraft*
shear off abscheren || abschneiden
shear section Scherquerschnitt * *einer (Blech-) Schere [in mm²]*
shear strength Scherfestigkeit * *[in N/mm²]*
shear stress Scherbeanspruchung || Scherbelastung || Scherspannung
shearing force Querkraft || Scherkraft || Scherspannung || Schubkraft * *Querschub, Scherkraft*
shearing resilience Scherspannung
shearing strain Scherspannung
shearing strength Abscherfestigkeit
shearing stress Scherspannung
shears Blechschere || Metallschere || Schere * →*Metall-/*→*Blech~, auch als Maschine; flying shears: —fliegende ~*
sheath Armierung * *z.B. eines Kabels* || Mantel * *eines Kabels*
sheathe umhüllen * *ummanteln, überziehen, armieren, z.B. Kabel* || ummanteln * *einhüllen, armieren, z.B. Kabel*
sheathing Schalung * *[Bautechnik]* || Umhüllung || Ummantelung * *z.B. von Kabel*
sheave Rolle * *Scheibe, Rolle* || Scheibe * →*Seil~* || Seilscheibe * *Rillenscheibe, Leitrolle, Auflaufhaspel*
shed abwerfen * *abstoßen (Wasser, Hörner usw.), abwerfen (auch Last), verbreiten (Geruch, Laune), ablegen* || Werkshalle * *Schuppen, kleine Werkhalle*
shed the load Last abwerfen
sheer bloß * *bloß, →rein, nichts als, nackt; sheer nonsense: nichts als/bloß Unsinn* || rein * →*bloß, rein, nicht als, by sheer force: mit bloßer/nackter Gewalt*
sheer force Gewalt * *nackte Gewalt*
sheet Blatt * *Papier, Holzfurnier* || Blech || Bogen * *Papier~* || Feinblech || Fell * *Zwischenprodukt bei der Gummiherstellung* || Platte * *Metall, Glas usw.*
sheet delivery unit Bogenausleger * *für Bogendruckmaschine*
sheet feeder Bogenanleger * *[Druckerei] für Bogendruckmaschine*
sheet glass Tafelglas
sheet length Format * *bei Querschneider* || Formatlänge * *bei* →*Querschneider*
sheet metal Blech || Stahlblech * *(dünnes) Metallblech, Feinblech*
sheet metal bending machine Blech-Rundbiegemaschine
sheet metal section Blechprofil
sheet metal working Blechbearbeitung * *Fein~*
sheet metal working machine Blechbearbeitungsmaschine
sheet of instructions Merkblatt
sheet piling device Bogenableger * *für Papierbögen* || Bogenstapler * *für fertig geschnittene oder bedruckte Papierbögen*
sheet size Format * *eines Papierbogens usw.*
sheet steel Blech * *dünnes Stahl~* || Stahlblech * *dünnes Blech, Feinblech*
sheet thickness Blechstärke
sheet width Formatbreite * *bei* →*Längsschneider*
sheet working Blechbearbeitung

sheet-fed offset press Bogenoffsetmaschine * *[Druckerei]*
sheet-fed press Bogendruckmaschine * *[Druckerei] druckt vorgeschnittene Papierbögen (i. Ggs. z. Rollendruckmaschine)*
sheet-fed printing machine Bogendruckmaschine
sheet-fed printing press Bogendruckmaschine
sheet-iron shears Blechschere
sheet-metal cover Abdeckblech
sheet-section machine Blech-Formmaschine * *erzeugt Formprofile*
sheet-straightening machine Blech-Richtmaschine
sheet-working machine Blechbearbeitungsmaschine
sheeter Querschneider * *Bogenschneider für Papier*
sheeting machine Querschneider
shelf Fach * *Regalbrett* || Regal
shelf life Haltbarkeit * *~ von Lebensmitteln, Schmierstoffen usw.* || Lagerfähigkeit * *z.B. von Bauteilen*
shell Hülse * *Schale, Kapsel, Gehäuse, das (bloße) Äußere* || Schale
shell-type grab Schalengreifer * *[Hebezeuge] bei Greiferkran*
shell-type motor Einbaumotor * *[el.Masch.]*
shelter Schutz * *Zuflucht, seek/take: suchen* || Schutzdach * *auch 'Schuppen, Schutz, Unterstand, Anbau, Schuppen'* || schützen * *gegen Wetter usw. ~; from:vor*
sheltered geschützt * *überdacht; in sheltered areas: in ~en Räumen* || überdacht
shield abschirmen * *from: gegen* || Abschirmung || Schild * *Schutzschild, Lagerschild usw.* || Schirm * *Abschirmung, mechan. Schild* || Schutz * *Abschirmung (auch figürl.)* || schützen * *abschirmen (auch figürl.)*
shield connection Schirmanschluss
shield cover Abschirmhaube
shield plate Abschirmblech || Abschirmplatte
shield winding Schirmwicklung * *bei Transformator*
shielded abgeschirmt || geschirmt || geschützt * *abgeschirmt*
shielded cable abgeschirmte Leitung
shielding Schirmung
shielding cable abgeschirmte Leitung
shift Ausweg * *Notbehelf* || Hilfsmittel * *Ausweg, Hilfsmittel, Notbehelf, Kniff, List, Ausflucht* || schalten * *Getriebe* || Schicht * *Arbeitsschicht* || Schicht- * →*arbeit* || schieben * *verschieben* || umschalten * *hinüberschieben* || umschalten * *Getriebe* || verlagern * *auch 'sich verlagern'* || Verlagerung * *(figürl.)* || verlegen * *verschieben, versetzen; (auch figürlich: Termin usw.)* || verschieben * *auch 'sich ~'* || Verschiebung || verstellen || Verzerrung * *Verschiebung (z.B. Farbverschiebung)* || wechseln * *Platz/Lage/Szene ~*
shift claw Schaltgabel
shift fork Schaltgabel
shift key Vortaste || Zweitfunktionstaste
shift latch Schieberverriegelung * *z.B. an (Sub-D) Stecker zur Arretierung u. Zugentlastung*
shift left linksschieben
shift operation Schichtbetrieb || Schiebeoperation
shift register Schieberegister
shift report Schichtprotokoll * *z.B. auf Meldedrucker ausgegebenes Protokoll einer Arbeitsschicht*
shift right rechtsschieben
shift through durchschieben
shift turnover Schichtwechsel
shift work Schichtarbeit
shifted verschoben
shifting Verlagerung || Verlegung || Verschiebung

shifting the firing pulses to the inverter range Wechselrichterschieben * *[Stromrichter]*
shifting-mode averaging gleitende Mittelwertbildung
shilly-shally schwanken * *unentschlossen sein, zaudern*
shim Ausgleichscheibe * *z. Reduzierg./Einstellg. d. Axialspiels, Unterlegscheibe z. Motorausrichtg. usw.* || Ausgleichsscheibe * *z. Reduzierg.d. Axialspiels, Unterlegscheibe z. Ausrichtg. e. Motors usw.* || Keil * *Klemmstück, Ausgleichsscheibe*
shine aufleuchten || leuchten * *scheinen, leuchten, strahlen*
shining blank
ship ausliefern || liefern * *ausliefern; ship back: zurück~* || schicken * *Ware versenden/(aus)liefern* || Schiff || versenden * *Fracht*
ship deck Schiffsdeck
ship load Frachtgüter * *bei Schiffstransport*
ship motor Schiffsmotor
ship's propulsion Schiffsantrieb
ship's propulsion system Schiffsantrieb
ship-building standards Schiffbauvorschriften
ship-lifting device Schiffshebewerk
shipment Abtransport * *Versand, Verladung* || Auslieferung || Lieferung * *Versand; auch: gelieferte Waren* || Sendung * *Absendung, Auslieferung von Gütern* || Transport || Verladung || Versand
shipped geliefert * *ausgeliefert, versandt*
shipping Auslieferung || See- * *Seefahrts-* || Versand
shipping address Lieferadresse || Versandanschrift
shipping agent Spediteur
shipping brace Transportsicherung
shipping company Spedition
shipping costs Versandkosten || Versandspesen
shipping cycle Lieferzeit
shipping damage Versandschaden
shipping date Auslieferdatum || Ausliefertermin || Auslieferungsdatum || Auslieferungstermin || Liefertermin
shipping documents Lieferpapiere || Versandpapiere
shipping note Lieferschein
shipping package Versandpackung
shiver Splitter
shivers Splitter * *(Plural)*
shock -Sturz || Erschütterung || Ruck * *Stoß* || Schlag * *z.B. elektrischer; get a shock: einen ~ bekommen* || Stoß * *mechanischer*
shock absorber Schwingungsdämpfer * *Stoßdämpfer* || Stoßdämpfer
shock absorbing stoßdämpfend
shock factor Stoßfaktor * *kennzeichnet d.Zahl d.Lastzyklen je Zeiteinheit (maßgebl. für Getriebeauslegung usw.)*
shock frequency Stoßfrequenz
shock load Lastspitzen * *(mechan.)* || Stoßlast
shock overload torque Stoßdrehmoment
shock protection Berührungsschutz * *gegen Stromschlag*
shock pulse measurement Stoßimpulsmessung * *z.B. z.Verschleißmessg. v.Wälzlagern durch Ermittlg. d. Körperschallspektr.*
shock resistance Schockfestigkeit || Stoßfestigkeit * *(mechan.)*
shock stressing Schockbeanspruchung || Stoßbeanspruchung * *mechanische ~* || Stoßbelastung
shock test Stoßprüfung * *für explosionsgeschützte Geräte (DIN EN 50014)*
shock tolerance Stoßfestigkeit
shock-free erschütterungsfrei
shock-hazard protection Berührungsschutz * *gegen Personengefährdung/elektr. Schlag* || Berührungssicherheit * *gegen elektr. Schläge*
shock-proof stoßfest * *(mechan.)*
shock-type stoßartig
shocked betroffen * *bestürzt*
shockproofness Schockfestigkeit * *Festigkeit gegen wiederholten mechan. Stoß*
shoe brake Backenbremse
shoot Gleitbahn * *Rutsche, Rutschbahn* || Rutsche * *Rutsche, Rutschbahn, Kipprinne*
shoot down Absturz * *zum ~ bringen*
shoot-through Wechselrichterkippen
shooting Schuss * *Emporschießen*
shooting-through Durchlegieren * *z.B. bei Thyristor*
shop Abteilung * *Werkstatt, Fertigungsabteilung* || Geschäft * *Laden* || Geschäftsräume * *Laden* || Halle * *Werkshalle* || Werkshalle || Werkstatt
shop assistant Verkäufer * *Ladengehilfe*
shop floor Fertigungsbereich
shop foreman Meister || Werkstattleiter
short abkürzen || knapp * *of money: an Geld; of cash: bei Kasse; be in short supply:~ sein; run short: ~ werden* || kurz * *räumlich und zeitlich ~* || Kurz- || kurzschließen * *(salopp)* || Kurzschluss || überbrücken || verkürzen
short circuit to core Eisenschluss * *[el.Masch.] unerwünschter galvan. Kontakt zw. Blechen in e. Blechpaket*
short description Kurzbeschreibung
short designation Kurzbezeichnung * *Kurzbezeichnung* || Kurzzeichen * *Kurzbezeichnung*
short distance Spanne * *räumlich*
short duration Kürze * *(zeitlich)*
short form Kurz- || Kurzform * *(figürl.) in short form: in ~*
short form designation Kurzangabe * *Kurzbezeichnung*
short infos Kurzangaben
short interruption Kurzunterbrechung
short motor Kurzmotor
short out überbrücken * *elektr. kurzschließen*
short price Selbstkostenpreis
short pulse Kurzimpuls
short space of time Spanne * *zeitlich*
short term Kurzbezeichnung
short term operation Kurzzeitbetrieb
short time operation Kurzzeitbetrieb
short time working Kurzarbeit
short version Kurzversion
short-circuit kurzschließen || Kurzschluss || überbrücken * *kurzschließen*
short-circuit braking Kurzschlussbremsung * *Motorklemmen werden zum Bremsen kurzgeschlossen*
short-circuit breaking current Kurzschlussausschaltstrom * *eines Leistungsschalters*
short-circuit cage Kurzschlusskäfig
short-circuit capability Kurzschlussfestigkeit
short-circuit capacity Kurzschlussfestigkeit * *auch bei einer Sicherung*
short-circuit characteristic Kurzschlusskennlinie
short-circuit clearing Kurzschlussfortschaltung
short-circuit condition Kurzschlussfall
short-circuit current Kurzschlussstrom
short-circuit making current Kurzschlusseinschaltstrom * *eines Leistungsschalters*
short-circuit power Kurzschlussleistung
short-circuit proof kurzschlussfest || kurzschlusssicher
short-circuit protected kurzschlusssicher
short-circuit protection Kurzschlussschutz || Kurzschlusssicherheit * *Kurzschlussschutz*
short-circuit resistant kurzschlussfest
short-circuit ring Dämpferring || Kurzschlussring * *[el.Masch.] ringförmige Verbindung der Käfigstäbe bei Käfigläufer*
short-circuit strength Kurzschlussfestigkeit

short-circuit withstandability Kurzschlusssicherheit * *Kurzschlussfestigkeit*
short-circuit-proof termination kurzschlussfester Anschluss * *{el.Masch.}masch.seitiger Kurzschl.führt nicht z.KS i. Klemm.kasten*
short-circuited kurzgeschlossen
short-circuiting contactor Überbrückungsschütz * *z.B. f. Zwischenkreis-Ladewiderstand e.Umrichters, USV, Sanftanlaufgerät*
short-circuiting relay Überbrückungsrelais * *z.B. bei Sanftanlaufgerät*
short-cycle process Kurztaktverfahren
short-dated kurzfristig * *mit nahegelegenem Stichdatum*
short-distance communication Nahverkehr * *Nahverkehrsverbindung*
short-distance passenger traffic Personennahverkehr
short-distance traffic Nahverkehr
short-form catalog Kurzkatalog
short-haul passenger service Personennahverkehr
short-haul rail-car Nahverkehrs-Triebwagen
short-haul transport system Nahverkehrssystem
short-period kurzfristig
short-shift work Kurzarbeit
short-stroke Kurzhub-
short-term kurzfristig || Kurzzeit- || kurzzeitig
short-term overload kurzzeitige Überlastung
short-time Kurzzeit- || kurzzeitig
short-time duty Betriebsart S2 * *{el.Masch.}* Kurzzeitbetrieb || Kurzzeitbelastung * *Betriebsart* || Kurzzeitbetrieb * *{el.Masch.} Betriebsart S2*
short-time duty type Kurzzeitbetrieb * *Betriebsart S2 nach VDE 0530/DIN 41756 für Motor/Stromrichter*
short-time interruption Kurzzeitunterbrechung
short-time life Kurzzeitlebensdauer * *{el.Masch.} begrenzte Lebensdauer von Motoren mit extrem hohen Temperaturen*
short-time loading Kurzzeitbelastung
short-time loading capacity Kurzzeitbelastbarkeit
short-time operation Kurzbetrieb || Kurzzeitbetrieb
short-time rating Kurzzeitleistung * *spezifizierte/zulässige ~ (bei d. Spezifikation der Überlastfähigkeit)*
shortage Engpass * *Knappheit, Mangel (of:an)* || Knappheit || Mangel * *Knappheit (of: an)* || Verknappung
shortcoming Fehler * *Unzulänglichkeit, Pflichtversäumnis, Verknappung* || Mangel * →*Fehler* || Nachteil * *Unzulänglichkeit, Fehler, Mangel* || Unzulänglichkeit * *auch 'Fehler, Mangel, Pflichtversäumnis, Fehlbetrag'*
shortcut Abkürzung * *(mnemotechnisches) Kürzel*
shorted kurzgeschlossen
shorten abkürzen || kürzen || verkürzen * *auch 'sich ~'*
shortened Kurz- || kurzgeschlossen
shortening Verkürzung * *(zeitl. und räumlich)*
shorter näher * *(Weg usw.)*
shortest kürzest * *zeitlich und räumlich* || nächster * *kürzest*
shortly Kürze * *in Kürze*
shortness Kürze * *(räumlich und zeitlich)*
shot Pressung * *Schub einer Pressmaschine* || Schuss
should sollen * *Möglichkeit*
should the occasion arise eventuell * *(Adv.) gegebenenfalls* || gegebenenfalls
shoulder Ansatz * *Ansatzstufe* || Bund * *Vorsprung, Schulter* || Kröpfung || Vorsprung * *Schulter, Vorprung, (mechan.) Ansatz z.B. an Maschinenteil*
shoulder housing Bundgehäuse * *Gehäuseabform z.B. einer Kupplung/Bremse*

shove drängen || schieben * *(beiseite) schieben/stoßen* || Schub * *Stoß* || stoßen * *schieben (z.B. beiseite)*
shovel Schaufel * *allg.; shovel dredger: ~bagger*
shovel plate Flanschplatte * *in Form einer Schaufel, z.B. zur Verbindung v. Motor und Getriebe* || Flanschwanne * *schaufelförmige Flanschplatte*
show abbilden * *zeigen (in e. Abbildung)* || angeben * *ausweisen* || anzeigen * *Ablesewert auf Messgerät ~; show no reading: nichts ~* || auffahren * *in einer Liste, Grafik* || aufweisen || ausstellen * *auf Messe ~* || Ausstellung || beweisen || darstellen * *veranschaulichen, z.B.in Bild/Tabelle/Anhang* || nachweisen || sichtbar machen || sichtbar werden || zeigen * *auch darstellen (Bild/Tabelle/Anhang)*
show off hervorheben * *herausstreichen*
show-room Vorführraum
showcase ausstellen * *in einem Schaukasten, auf einem Messestand usw.* || Vitrine * *für Werbe-/Ausstellungs-/Verkaufszwecke* || zeigen * *zur Schau stellen, ausstellen (auf Messe, in Vitrine, in Ausstellungsraum usw.)*
showdown Stunde der Wahrheit * *Aufdecken der Karten, spannender Höhepunkt, entscheidende Kraftprobe*
showing round Führung * *~ in einer Ausstellung, bei einer Besichtigung*
shown aufgeführt * *in Liste, Grafik usw.* || dargestellt * *z.B. in einer Zeichnung, einer Tabelle*
shown by hervorgehen aus * *z.B. aus einer Abbildung/einem Kapitel ~*
shown on a larger scale vergrößert dargestellt
shown on a reduced scale verkleinert dargestellt
showroom Ausstellungsraum
shred schneiden * *schnitzeln* || Streifen * *Schnipsel*
shredder Häcksler * *z.B. für Holzabfälle*
shrink abnehmen * *schrumpfen* || aufziehen * *mit Hitze ~, aufschrumpfen* || einlaufen * *{Textil} schrumpfen; unshrinkable/Sanforized: nicht ~d* || schrumpfen || schwinden * *schrumpfen* || verkleinern * *schrumpfen; z.B. Halbleiterstrukturen durch verbesserten Herstellprozess*
shrink disc Schrumpfscheibe * *zur Welle-Naben-Verbindung*
shrink disk Schrumpfscheibe * *zur Wellenverbindung*
shrink film Schrumpffolie
shrink film wrapping Schrumpffolienverpackung
shrink on aufschrumpfen * *z.B. Aufziehen eines Lagers mit Wärme* || aufziehen * *mit Hitze ~, aufschrumpfen*
shrink sleeving Schrumpfschlauch
shrink tube Schrumpfschlauch
shrink tubing Schrumpfschlauch
shrink wrapping Folienschrumpfverpackung || Schrumpffolienverpackung
shrink-film wrapping Folienschrumpfverpackung
shrink-pressure disc Schrumpfscheibe
shrinkable film Schrumpffolie
shrinkable tube Schrumpfschlauch
shrinkage Schrumpfung || Schwund * *Schrumpfung*
shrinkhole Lunker
shrinking Schrumpf- || Schrumpfung || Verkleinerung * *Schrumpfen, z.B. von Halbleiterstrukturen durch verbesserten Herstellprozess*
shrinking film Schrumpffolie
shrunk-on aufgeschrumpft
shrunk-on ring Schrumpfring
shrunk-sleeve-insulated schrumpfschlauchisoliert
shuffle table Schlitten * *Schlepptisch*
shunt ableiten * *Strom* || Nebenwiderstand || parallel schalten || rangieren * *{Bahnen}* || Shunt || überbrücken || verschieben * *Eisenbahn*
shunt across parallel schalten * *zu etwas ~*
shunt arrangement Parallelschaltung

shunt capacitor Parallelkondensator
shunt characteristic Nebenschlusskennlinie * {el. Masch.}
shunt circuit Nebenschluss
shunt converter Shuntwandler
shunt motor Nebenschlussmotor
shunt out überbrücken
shunt resistor Bürde * Bürdenwiderstand zur Strommessung || Bürdenwiderstand * zur Strommessung || Nebenwiderstand || Parallelwiderstand
shunt transducer Shuntwandler * Stromwandler zur (potenzialfreien) Strommessung mit Shunt-Widerstand
shunt winding Nebenschlusswicklung
shunt wound DC motor Gleichstrom-Nebenschlussmotor
shunt-wound machine Nebenschlussmaschine
shunt-wound motor Nebenschlussmotor
shut geschlossen * verschlossen || schließen * (ver)schließen, zumachen; auch figürl.: shut one's eyes to: seine Augen ver~ vor || verschließen * auch figürl.: shut one's eyes to: die Augen ~ vor || verschlossen
shut down abschalten * z.B. Antrieb stillsetzen || abstellen || außer Betrieb nehmen || außer Betrieb setzen || ausschalten || deaktivieren * stillsetzen || herunterfahren * stillsetzen || stillsetzen
shut down after fault Störabschaltung
shut down on fault Störabschaltung
shut-down Abschaltung
shut-down brake Stillsetzbremse
shut-down control Ausschaltsteuerung || Ausschaltsteuerwerk
shut-down sequence Ausschaltreihenfolge
shut-down sequencing Ausschaltsteuerung || Ausschaltsteuerwerk
shut-off valve Absperrventil
shutdown abschalten * (Substantiv) Abschaltung || Abschaltung || Betriebsunterbrechung || stillsetzen || Stillsetzung || Stillstand * ~ einer Machine/Anlage
shutdown control Ausschaltsteuerung
shutdown period Betriebspause
shutdown sequence Ausschaltfolge
shutdown threshold Abschaltschwelle
shutter Jalousie || Schalung * {Bautechnik} || Verschluss * an (Foto)Kamera; shutter release: Verschlussauslösung
shuttering panel Schalungsplatte * {Holz} für Betonbau
shuttle Greifer * ~ zum Hin- und Herbewegen || pendeln * sich hin- und herbewegen; shuttle between: ~ zwischen (Fahrzeug)
shuttle car Verschiebewagen * z.B. zur Verteilg. d. Güter zu/von Regalbediengerät(en) i. Hochregallager
side Seite * on the side: auf der ~; aside: zur ~, bei~, abseits, seitwärts || Seiten- || seitlich || Wand * Seiten~ || Zusatz- * -Nutzen, -Abkommen, -Effekt usw.
side aspect Nebenaspekt || Nebenpunkt * Nebenaspekt
side benefit Folgeerscheinung * vorteilhafter Nebeneffekt || Zusatznutzen
side by side nebeneinander
side by side with neben
side cover Seitendeckel
side dresser Egalisierapparat
side effect Folgeerscheinung * Nebeneffekt || Nebenerscheinung || Schmutzeffekt * Nebeneffekt (siehe auch dort)
side elevation Seitenriss * in techn. Zeichnung
side frame Seitenrahmen
side guide Seitenführung * z.B. Bandführungs-Einrichtung im Walzwerk
side guiding Seitenführung * Funktion bei durchlaufendem Bandmaterial
side issue Nebenaspekt || Nebenpunkt
side ledge Seitenführung * in Form einer (Anschlag-) Leiste
side letter Zusatzvereinbarung * zu einem Vertrag
side panel Seitenteil * seitliche Wand/Blech/Platte/Holz
side plate Dichtscheibe * eines Walzlagers || Seitenwand
side rod Schubstange * bei Kurbelantrieb
side section Seitenprofil
side view Ansicht von der Seite || Seitenansicht
side wall Seitenteil * Seitenwand, auch eines Gerätes
side- Neben- * z.B. side letter: schriftliche ~vereinbarung
side-channel compressor Ringverdichter
side-channel vacuum pump Seitenkanal-Vakuumpumpe
side-effect Begleiterscheinung * Nebeneffekt || Nebeneffekt
side-panel Seitenwand * z.B. eines Schaltschrankes
side-shuttle Querausschleuser * bei Fördersystem
side-slip schleudern * Fahrzeug usw.
side-wall Seitenwand
sidelay register Nebenregister * {Druckerei}
sideslip schleudern * ins Schleudern kommen (Fahrzeug usw.)
sideward seitwärts
sidewards seitwärts
sideways seitwärts
Siemosyn-motor Siemosyn-Motor * (R) permanenterregter Synchronmotor von SIEMENS
sieve Sieb * Durchwurf, Rätter, Sieb
sifter Sichter * Siebeeinrichtung
sight Ansicht || ausmachen * beobachten, sichten, zu Gesicht bekommen, anvisieren
sight-glass Schauglas * z.B. zur Ölstandskontrolle || Sichtfenster * z.B. zur Ölstandskontrolle
sightseeing Besichtigung * sightseeing tour: Besichtigungsrundgang/-tour
sign Merkmal * Zeichen || Schild * (Schrift-) Zeichen, (Aushänge-) Schild || verpflichten * on: zu Arbeitsleistungen usw. ~ || Vorzeichen || Zeichen * Symbol, Wink, Kenn~, Hinweis~
sign inversion Vorzeichenumkehr
sign inverter Invertierer * Negierer, Vorzeichenumkehr || Negierer * Vorzeichenumkehrer
sign of burning Brandstelle * z.B. am Kommutator e. Gleichstrommaschine nach Bürstenfeuer
sign reversal Vorzeichenumkehr || Vorzeichenumschaltung
sign-board Firmenschild
sign-post Schild * Wegweiser
signal melden * signalisieren, z.B. Zustand, Fehler || Meldung * Signal || Signal * audible:akustisches, visible:sichtbares/optisches || signalisieren || Zeichen * Signal
signal acquisition Signalerfassung
signal adaption Signalanpassung
signal back rückmelden || zurückmelden
signal cable Signalleitung
signal change Signalwechsel
signal characterizer Kennliniengeber
signal common Bezugssignal * gemeinsames Bezugssignal, "Wurzel" || Masse * Gemeinsame Bezugsspannung, "Wurzel"
signal comparator Grenzwertmelder
signal conditioner Anpassverstärker || Signalumsetzer * zur Signalaufbereitung || ssignalformer
signal conditioning Signalanpassung * Signalaufbereitung || Signalaufbereitung || Signalumformung * Signalaufbereitung

signal conditioning circuit Signalanpassung * *Schaltung*
signal connection Signalanschluss
signal conversion Signalumformung
signal converter Messumformer || Signalumsetzer || Signalwandler
signal designator Signalbezeichner
signal dispatcher Signalverteiler
signal edge Signalflanke * *negative/positive going: fallende/steigende*
signal evaluation Signalauswertung
signal exchange Signalaustausch
signal flow durchgeschaltet * *Zustand eines Strompfades in der Kontaktplandarstellung eines SPS-Programms* || Signalfluss
signal flow chart Signalflussplan
signal generating signalerzeugend
signal generation Signalbildung
signal generator Messgenerator
signal ground Signalmasse
signal input Signaleingabe
signal interchange Signalaustausch
signal interpolation Signalauswertung * *Interpolation*
signal isolator Trennverstärker * *zur potenzialgetrennten Signalübertragung*
signal lamp Meldelampe || Meldeleuchte || Signallampe
signal lead Signalleitung
signal level Signalpegel
signal level adaption Pegelanpassung * *für ein Signal*
signal line Signalleitung
signal matching Signalanpassung
signal output Signalausgabe
signal path Signalpfad
signal pattern Signalmuster
signal power Übertragungsleistung
signal pre-processing Signalvorverarbeitung
signal preparing Signalaufbereitung
signal processing Signalverarbeitung
signal propagation delay Signallaufzeit
signal range Signalbereich
signal ribbon cable Signalbandkabel
signal sequence Signalverlauf * *(Ab)Folge*
signal status Signalzustand * *(Plural: signal statuses)*
signal status display Signalzustandsanzeige
signal statuses Signalzustände
signal tracing Signalverfolgung
signal tracking Signalverfolgung
signal transducer Signalumsetzer * *besonders 'Messumformer'* || Signalwandler * *z.B. Messwandler*
signal transfer Signalaustausch
signal transformation Signalumformung
signal variation Signalverlauf * *Änderung*
signal waveform Signalverlauf * *Wellen/Kurvenform (meist für periodisches Signal)*
signal-to-noise ratio Störabstand
signaled gemeldet
signaling Melde-
signaling block Meldebaustein * *Funktionsbaustein*
signaling contact Meldekontakt
signaling function block Meldefunktionsbaustein
signaling relay Melderelais
signaling system Meldesystem
signalize melden
signalled gemeldet
signaller Meldegerät
signalling Melde- || Meldung * *das Melden* || Signalisierung
signalling block Meldebaustein

signalling brush Meldebürste * *z.B. zur Bürstenverschleißmeldung in DC-Motor*
signalling contact Meldekontakt
signalling device Meldeeinrichtung
signalling relay Melderelais
signalling system Meldesteuerung || Meldesystem
signature Unterschrift
signboard Schild * *Firmen/Aushängeschild*
signed vorzeichenbehaftet
significance Bedeutung * *becoming increasingly significant: an ~ gewinnen* || Relevanz * *Wichtigkeit, Bedeutung* || Wertigkeit || Wichtigkeit * *Wichtigkeit; Bedeutung, (tieferer) Sinn*
significant deutlich * *bedeutsam, spürbar* || erheblich * *bedeutend, bedeutsam* || fühlbar * *deutlich, spürbar* || merklich * *bedeutsam, bedeutend* || nennenswert || relevant * *wichtig, bedeutsam* || spürbar || wesentlich * *deutlich*
significantly deutlich * *(Adv.)*
significantly expanded stark erweitert
signify andeuten * *an-/bedeuten, kundtun, ankündigen* || bedeuten
silencer Auspufftopf * *Schalldämpfer* || Geräuschdämpfer || Schalldämpfer
silencing Schalldämpfung || Schallisolierung * *als Funktion*
silent geräuscharm * *leise* || geräuschlos || leise * *ruhig, geräuschlos* || ruhig * *ruhig, geräuschlos*
silent running ruhiger Lauf
silicate glass Silikatglas
silicon Silizium
silicon carbide Siliziumkarbid
silicon controlled rectifier Thyristor
silicon crystal Siliziumkristall
silicon diode Siliziumdiode
silicon rectifier Siliziumgleichrichter
silicon semiconductor Siliziumhalbleiter
silicon slice Siliziumscheibe
silicon wafer Siliziumscheibe
silicone Silikon
silicone grease Silikonfett
silicone insulated silikonisoliert
silicone-free silikonfrei * *z.B. Öl, Kabelisolierung usw.*
silk screen Siebdruck * *auf Leiterkarte usw. aufgedruckter ~*
silk screening Siebdruck
silk-screen frame Siebdruckrahmen * *(Druckerei)*
sill Schwelle * *Türschwelle, Schwellbalken*
silo Bunker * *→Silo für Getreide usw.* || Silo * *ensilage: in ein ~ einlagern*
siluminium Silumin * *verschleißfeste, korrosionsbeständige Alulegierg.mit ca. 12% Siliziumanteil*
silver-coated versilbert
silver-plated versilbert
similar ähnlich * *to: zu/wie* || gleichartig * *ähnlich* || verwandt * *ähnlich (to: zu, mit)*
similar to analog zu * *ähnlich*
similarity Ähnlichkeit * *Ähnlichkeit, Gleichartigkeit*
SIMM SIMM * *(Computer) Single-Inline Memory Module': Speichermodul mit einreihigen Anschlusspins*
simple anspruchslos * *einfach* || einfach * *leicht* || leicht * *einfach* || natürlich * *einfach* || problemlos * *einfach, leicht* || unkompliziert * *einfach*
simple operations Bedienungsfreundlichkeit || einfache Bedienung
simple operator control panel Einfachbedienfeld
simplest einfachst
simplest case einfachster Fall * *in the simplest case: im einfachsten Fall*
simplicity Einfachheit * *to simplify matters/for reasons of simplicity/to save trouble:d. Einfachheit halber*

simplification Vereinfachung * *to simplify matters: zur ~*
simplified vereinfacht
simplify erleichtern * *vereinfachen* || vereinfachen
simply auf einfache Art und Weise || bloß * *nur, einfach* || nur * *einfach* || unbürokratisch * *(Adv.)*
simulate nachbilden * *simulieren, nachbilden* || simulieren
simulated operation Simulationsbetrieb
simulation Nachbildung * *Simulation, Nachahmung* || Simulation
simulation module Simulationsbaugruppe
simulation program Simulationsprogramm
simulator program Simulationsprogramm
simultaneitity Gleichzeitigkeit
simultaneity factor Gleichzeitigkeitsfaktor
simultaneous gleichzeitig * *simultaneously occurring:gleichzeitig auftretend* || zusammen * *gleichzeitig*
simultaneously zugleich * *(Adv.) gleichzeitig*
since seit * *von .. an; siehe auch →ab; since then/ from that time: seitdem*
sincere wahr * *aufrichtig*
Sincerely yours, mit freundlichen Grüßen || mit vorzüglicher Hochachtung
sine sinus || sinusförmig
sine cycle Sinusperiode
sine modulation Sinusmodulation
sine period Sinusperiode
sine wave Sinusperiode * *Sinusschwingung, Sinuswelle* || Sinusschwingung || Sinuswelle
sine-cosine encoder Rohsignalgeber * *Drehzahl-/ Lagegeber mit phasenverschobenen sin-Signalen* || Sinus-Cosinus-Geber * *Drehzahl-/Lagegeber mit phasenverschobenen Sinussignalen*
sine-modulated sinusmoduliert
sine-wave sinusförmig
sine-wave filter Sinusfilter
sinewave encoder Sinus-Cosinus-Geber * *Drehzahl-/Lagegeber mit sin- und cos-Signalen*
sinewave-filter Sinusfilter * *zur Erzeugung sinusförmiger Spannungen am Ausgang eines Umrichters*
single einfach * *einmal vorhanden* || Einzel- || einzeln || einzelner || einzig * *einzeln; not a single person: kein ~er*
single axis drive Einachs-Antrieb
single block drawing machine Einblockziehmaschine * *(Draht) Drahtziehmaschine*
single clocking Eintakt- * *z.B. Netzteil*
single digit einstellig
single drive Einzelantrieb
single drives Einzelantriebe
single ended auf Masse bezogen * *z.B. Analogsignal (im Gegensatz zum Differenzialsignal)*
single fault Einfachfehler * *bei SPS*
single feeder Stichleitung
single licence Einzellizenz
single license Einzellizenz * *(amerikan.)*
single motor drive Einzelantrieb * *mit nur einem Motor*
single packing Einzelverpackung
single phase einphasig
single quad 1Q
single quadrant 1Q
single scan Einzelzyklus * *bei SPS*
single sound Einzelton * *hörbare Schallschwingung mit einer einzigen Frequenz*
single step Einzelschritt * *(Computer\NC-Steuerung) Ablauf e. Programms mit Stopp nach jeder Anweisung zum Test*
single unit production Einzelfertigung * *Ein-Stück-Fertigung*
single- Ein- || Einfach-

single-axis positioning Einachs-Positionierung || Einachspositionierung
single-batch production Einzelfertigung
single-board computer Einplatinencomputer
single-chance fault Einfachfehler * *bei SPS*
single-channel control einkanalige Steuerung * *(SPS)*
single-chip computer Ein-Chip-Computer
single-chip processor Ein-Chip-Mikroprozessor
single-circuit controller Zweipunktregler * *z.B. zur Temperaturregelung*
single-conductor eindrähtig
single-core eindrähtig * *einadrig (Kabel)*
single-disc Einscheiben- * *Kupplung, Bremse usw.*
single-disc clutch Einscheibenkupplung
single-disk brake Einscheibenbremse
single-drum surface winder Poperoller
single-end einseitig
single-ended einseitig * *z.B. Antrieb* || massebezogen * *Signal (im Gegensatz zum Differenzialsignal)* || radial bedrahtet
single-ended analog input Masse-bezogener Analogeingang * *im Gegensatz zum Differenzeingang*
single-ended drive einseitiger Antrieb
single-ended termination einseitiger Anschluss * *z.B. bei einem Kondensator*
single-face Einflächen- * *z.B. Bremse, Kupplung*
single-head saw Einkopfsäge * *(Holz)*
single-inlet einseitig saugend * *Ventilator*
single-layer einlagig
single-layer winding Einschichtwicklung * *Motor-Wickl.art: jede Nut enth.nur Drähte einer Spule; Einziehtechnik-geeignet*
single-line einpolig * *Darstellung im Übersichtsschaltbild mit nur einer Phase des Drehstromsystems* || einzeilig * *auf Anzeige oder im (Aus)Druck*
single-line diagram Prinzipschaltplan * *in einpoliger Darstellung* || Übersichtsschaltbild * *in einpoliger Darstellung*
single-minded konsequent * *zielbewusst, zielstrebig* || zielgerichtet * *nur dies eine im Sinn habend*
single-mindedness Konsequenz * *Zielstrebigkeit; Aufrichtigkeit*
single-motor drive Einmotorenantrieb || Einzelantrieb * *z.B. mit nur einem Motor am Stromrichter*
single-part einteilig
single-PCB receptacle Einzelkartenhalterung
single-phase einphasig
single-phase AC controller Wechselstromsteller * *einphasig*
single-phase AC motor Wechselstrommotor * *einphasiger ~*
single-phase bridge connection B2-Schaltung * *(Stromrichter)* || Einphasenbrückenschaltung
single-phase capacitor motor Einphasenmotor * *mit Kondensator zur Erzeugung einer Hilfsphase (Steinmetzschaltung)* || Kondensatormotor * *m. Phasenschieberkondensator z.Erzeugg. e.Hilfsphase; Steinmetzschaltg.*
single-phase induction machine Wechselstromasynchronmaschine
single-phase motor Einphasenmotor * *(el.Masch.) Asynchronmotor mit Phasenschiebung der Spannung über Kondensator(en)*
single-phase operation Einphasenbetrieb * *z.B. eines Motors*
single-plate clutch Einscheibenkupplung
single-pole einpolig
single-pole representation einpolige Darstellung
single-quadrant Einquadrant-
single-quadrant drive Einquadrantenantrieb
single-quadrant operation Einquadrantbetrieb || Einquadrantenbetrieb
single-reduction einstufig * *bei Untersetzungsgetriebe*

single-row einzeilig * *Baugruppenträger usw.*
single-sided einseitig
single-speed eintourig * *z.B. Motor mit Festdrehzahl*
single-spindle Einspindel- * *z.B. Werkzeugmaschine*
single-spindle moulder Tischfräsmaschine * *{Holz}*
single-stage einstufig
single-throw contact Ein- oder Ausschaltkontakt * hat nur 2 Anschlüsse (kann entweder nur schließen oder nur öffnen)
single-tier einzeilig * *Baugruppenträger usw.*
single-tube heat exchanger Einfachrohrkühler
single-turn coil Spule mit einer Windung
single-user licence Einzellizenz
single-user license Einzellizenz * *(amerikan.)*
single-wave rectification Einweggleichrichtung
single-wave rectifier Einweggleichrichter
single-way Einweg- * *z.B. bei Stromrichterschaltung*
single-way connection Einwegschaltung * *{Stromrichter}*
singlehanded allein * *ohne Hilfe*
singly einzeln * *(Adv.)*
singular außerordentlich * *einzigartig, außergewöhnlich, einmalig*
singular solution Insellösung
singularity Singularität
sink absacken || absinken * *sinken* || Senke * *data sink: Daten~* || senken * *sich ~, sinken* || sinken || versenken
sink into einlassen
sinking Senkung * *(Ver)Sinken*
sinking depth Senktiefe
sinor Raumzeiger
sinor diagram Raumzeigerdiagramm
sinter Sinter- || sintern * *Erz, Stahl usw.*
sinter bronze Sinterbronze
sintered gesintert || Sinter-
sintered bearing Sinterlager * *Gleitlager*
sintered bronze bearing Bronze-Sinterlager
sintered lining Sinterbelag * *Reibbelag bei Kupplung, Bremse*
sintered material Sintermaterial
sintered metal Sintermetall
sintered sleeve bearing Sinterlager
sintering Sinter-
sinusodially evaluated sinusbewertet * *z.B. Modulationsverfahren für Umrichter*
sinusodially weighted sinusbewertet
sinusoidal sinusförmig
sinusoidal filter Sinusfilter * *aufwändiges Filt.am Pulsumrichterausg., mindert Motor-Spanngs.beanspr. u.Erwärmg.*
sinusoidal oscillation Sinusschwingung * *sinusförmige Schwingung*
sinusoidal output filter Sinusausgangsfilter * *siehe auch 'Sinusfilter'*
sinusoidal quantity sinusförmige Größe
sinusoidal wave Sinuswelle
sinusoidal waveform Sinuswellenform
sinusoidal-weighted sinusbewertet
sinusoidally modulated sinusmoduliert
siphon Faltenbalg
sirup Sirup * *z.B. Zuckerrübensirup*
sirupy dick * *zähflüssig*
sit setzen * *sich (auch zu Tisch) ~*
sit fast festsitzen
sit together zusammensetzen * *sich (z.B. zu einer Besprechung) ~*
site Anlage * *Baustelle, Industriegelände* || Aufstellungsort || Baustelle || Platz * *Lage (Bau)Platz* || Standort
site altitude Aufstellhöhe * *above sea level: über dem Meeresspiegel* || Aufstellungshöhe || Betriebshöhe
site condition Betriebsbedingung * *~ am Aufstellort, auf der Anlage*
site conditions Anlagenbedingungen
site configuration Anlagenkonfiguration
site elevation Aufstellungshöhe
site manager Bauleiter
site-specific anlagenspezifisch
SITOR set SITOR-Satz * *(R) SIEMENS-Thyristorsatz*
sitting Sitzung * *einzelne Sitzung, z.B.beim Arzt*
situated in front of davorliegend
situation Lage * *Situation* || Stand * *Lage* || Zustand * *Lage*
six-phase motor Sechsphasenmotor * *{el.Masch.}* AC-Motor mit zwei Drehstromwicklungen in insgesamt 6 Wicklungssträngen
six-pulse 6-pulsig || Sechspuls- || sechspulsig
six-pulse bridge connection sechspulsige Brückenschaltung
sizable umfangreich * *siehe 'sizeable'*
size Abmessung * *Größe* || aufteilen * *{Holz} zuschneiden; nach Größe sortieren* || auslegen * *bemessen, for: für* || Baugröße || bemessen * *Größe festlegen, nach Größe ordnen* || BG * *Sicherung, Trafo usw.* || dimensionieren * *Größe festlegen* || Feinheit * *{Textil} von Garn* || Format * *Größe, z.B. eines Papierbogens (untrimmed size: Rohformat)* || Format || Größe * *Umfang, Format; to DIN: nach DIN* || leimen * *Papier, Stoff usw. steifen* || Maß * *Größe* || Ort || planieren * *{Buchbinderei}* || Umfang * *Größe (auch figürlich)* || Volumen * *Größe*
size down verkleinern * *Fläche, (Bau-) Volumen verringern (z.B. durch Neukonstruktion)*
size of conductor Leiterquerschnitt
size of the market Marktvolumen
size press Leimpresse * *zur Papierbearbeitung*
size-cutting optimization Zuschnittoptimierung * *{Holz}*
sizeable umfangreich * *ziemlich groß, ansehnlich, beträchtlich*
sized -groß * *in ... Größe* || ausgelegt * *(in Wortzusammensetzg.) von...Größe, -dimensioniert, z.B.full/over/under/small-sized* || dimensioniert * *(i. Wortzusamm.setzg.) von...Größe, -dimensioniert; z.B. full-/under-/over-/small-*
sized paper geleimtes Papier
sizer Plattenaufteilanlage * *{Holz} Zuschneidemaschine*
sizing Dimensionierung || Leimung * *Steifung von Papier* || Sortierung
sizing machine Schneidemaschine * *{Holz} Zu~ für Holzplatten*
sizing press Stauchpresse * *{Walzwerk}*
skeleton agreement Rahmenabkommen
skeleton diagram schematische Darstellung
sketch Bild * *Skizze* || entwerfen * *flüchtig, grob ~* || Skizze || skizzieren * *(auch figürl.)* || zeichnen * *(auch figürl.) skizzieren*
sketch map Lageplan * *z.B. Landkartenskizze*
sketchy lückenhaft * *z.B. Kenntnisse*
skew schief * *schief, abgeschrägt, verdreht* || schräg * *schief, abge~t, verdreht*
skew-winding Schrägwicklung * *z.B eines Motor-Ankers*
skewed schief * *abgeschrägt,verdreht* || schräg * *abge~t,verdreht*
skewed armature Ankernutschrägung * *schräg genuteter Anker* || Schrägschlitzanker * *zur Verminderung der Nutwelligkeit (u.damit d. Momentenwelligkeit) eines Motors*
skewed commutator Gewindekommutator * *geschräger Kommutator bei DC-Maschine*

skewing Schräglage * *das 'Sich-Schräg-Legen'*
ski-lift Skilift
skid schleudern * *sich drehen (Auto usw.)*
skidding Holzrücken * *{Agrikultur}*
skilful geschickt * *at:zu, in:bei; geschickt, gewand, geübt, sachkundig; be skilful at:s.verstehen auf* || tüchtig * *geschickt*
skill Fachkönnen || Fähigkeit * *eines Menschen* || Fertigkeit || Können * *Fertigkeit*
skilled erfahren * *in:in; geübt, Fertigkeiten besitzend, geschickt (meist für Facharbeiter)*
skilled trade Handwerk * *auch 'handwerklicher Beruf'*
skilled worker Facharbeiter || gelernter Arbeiter
skim überdrehen * *z.B. Kommutator auf Drehmaschine* ~
skim down abdrehen * *an Drehmaschine leicht ~ zum Beseitigen der Unebenheiten*
skim over Streifen * *(Verb) über etwas hingleiten*
skimpy dürftig
skin abisolieren || Schicht * *Haut, äußerer Überzug*
skin effect Skin-Effekt * *Stromverdrängungseffekt* || Stromverdrängungseffekt * *Verdrängung des Stroms ins Leiteräußere auf Grund der Wirbelströme*
skin-effect losses Stromverdrängungsverluste
skin-pass edging stand Egalisiergerüst * *{Walzwerk} im Warmwalzwerk*
skin-pass stand Egalisiergerüst * *{Walzwerk} im Kaltwalzwerk*
skip auslassen * *überspringen, auslassen, übergehen* || aussetzen * *überspringen, überschlagen* || Sprung * *Überspringen (Seiten, Programmbefehle usw.)* || übergehen * *überspringen* || überspringen * *Schritt/Rechnerbefehl(e) usw. weglassen; skip over something: etwas ~/übergehen*
skip frequency Ausblendfrequenz * *ausgeblendeter Frequ.sollwert f.Frequenzumrichter zur Vermeidg.mechan. Resonanz* || Ausklammerfrequenz * *siehe auch →Ausblendfrequenz,* →Ausblendband
skip frequency zone Ausblendband * *zur Ausblendung von (mechan.) Resonanzfrequenzen im Sollwert* || Frequenzausblendband || Frequenzausklammerungsband || Frequenzbandausklammerung
sky-rocket ansteigen * *(salopp) jäh/raketenartig ~*
slab Bramme
slack Durchhang || Lose * *Durchhängen einer Warenbahn* || Schlaufe * *Lose, Durchhang* || Spiel * *Lose/Sack z.B. in einer Warenbahn*
slack control Aufholen-Nachlassen * *{Papier}* || Durchhangregelung
slack payout nachlassen * *z.B. bei Papierbahn*
slack take-up aufholen * *Durchhang, Lose aufholen, z.B. in Papierbahn*
slack-pay-out button Nachlassen-Taste * *z.B. bei Papiermaschine*
slack-rope control Schlaffseilregelung * *bei Greiferkran*
slack-take-up button Aufholtaste * *z.B. bei Papiermaschine*
slacken abfallen * *Geschwindigkeit* || entspannen * *Seil ~* || lockern || nachlassen * *lockern*
slackening control Schlaffseilregelung * *bei Greiferkran*
slackness Langsamkeit * *Unlust, Saumseligkeit, Trägheit, Nachlässigkeit* || Spiel * *toter Gang*
slackness control Schlaffseilregelung * *bei Greiferkran*
slant Schräge
slanting schief * *schräg* || schräg
slash herabsetzen * *stark kürzen, zusammenstreichen* || kürzen * *stark ~, zusammenstreichen, deutlich verringern (Gelder, Zeitaufwand, Gehalt*

usw.) || Schrägstrich * *Zeichen auf Bildschirm, Drucker*
slat Latte * *schmale ~* || Leiste * *~ aus Holz, Metall usw., Rippe, Lamelle, Jalousie-Element* || Plattenband * *Förder-/Vorschubeinrichtung*
slat conveyor Plattenbandförderer
slat feed Plattenbandvorschub * *Plattenbandförder/-Vorschubeinrichtung*
slate vorschlagen * *(vorläufig) aufstellen, (für eine Posten usw.) ~; be slated for: vorgesehen sein für*
slave Busteilnehmer * *passiver* || Folge- * *(Gegensatz: 'master': Leit-)* || Slave
slave address Teilnehmeradresse * *Slave-Teilnehmeradresse bei Master/Slave-Bussystem*
slave axis Folgeachse
slave drive Folgeantrieb || Nebenantrieb
slave interface Slave-Anschaltung * *für Bus*
slave station Busteilnehmer * *passive Station, Unterstation* || Teilnehmer * *Slave-Teilnehmer (z.B. an seriellem Bus)*
sleckness Glätte
sleeper Schwelle * *{Bahnen} Eisenbahn~*
sleeve Buchse * *Muffe, Hülse, Buchse, Manschette* || Hülse * *Hülse, Buchse, Büchse, Muffe, Manschette, Spann~* || Manschette || Muffe * *Muffe, Buchse, Manschette, Kabelverbindung* || Tülle * *Manschette, Muffe*
sleeve bearing Gleitlager
sleeve terminal Buchsenklemme
sleeved shaft Welle mit Abtriebsbuchse
slender dünn * *schlank* || schlank * *slenderize:- schlank machen* || schmal * *schlank*
slew schwenken * *herumdrehen, (herum)~*
slew rate Änderungsgeschwindigkeit * *Anstiegsgeschwindigkeit eines Signals* || Anstiegsgeschwindigkeit * *Halbleitertechn.: eines Impulses* || Anstiegssteilheit * *Spannungssteilheit * *Anstiegsgeschwindigkeit der Spannung, z.B bei Operationsverstärker* || Steilheit * *Anstiegsgeschwindigkeit, z.B. einer Impulsflanke (z.B. in [mV/µs])*
slew rate limiting Anstiegsbegrenzung * *Begrenzung der Anstiegsgeschwindigkeit (einer Spannung)*
slew round schwenken * *herumschwenken/drehen, im Kreis herumschwenken, baumeln lassen*
slewable schwenkbar
slewing Schwenk- || schwenkbar * *z.B. Kran*
slewing arm Schwenkarm
slewing crane Schwenkkran
slewing drive Schwenkwerk * *{Hebezeuge} bei Kran, Bagger* || Schwenkwerksantrieb * *Kran*
slewing gear Drehwerk * *{Hebezeuge}* →Schwenkwerk eines Krans || Schwenkwerk * *{Hebezeuge} bei Kran, Bagger*
slice Scheibchen * *Scheibe*
slick pfiffig * *flott, raffiniert, geschickt* || sauber * *(salopp) flott, raffiniert, geschickt, pfiffig*
slidable verschiebbar
slide Führung * *mechanische ~, Schlitten* || Gleitbahn || gleiten || Overhead-Folie * *Dia* || Rutsche || rutschen || gleiten, (aus)rutschen || schieben * *gleiten lassen* || Schieber || Schlitten * *gleitendes Teil, Führung, Schieber* || verschieben * *gleiten lassen*
slide bearing Gleitlager
slide caliper Kluppe * *Gabelmaß*
slide gauge Schieblehre
slide in einschieben || stecken * *einschieben*
slide rail Spannschiene * *zum Spannen eines Riementriebs z.B. durch Verschieben des Motors*
slide rule Rechenschieber
slide switch Schiebeschalter
slide transformer Stelltrafo || Stelltransformator
slide valve Schieber * *Steuerschieber, Schieberventil*
slide-in Einschub || einsteckbar * *einschiebbar*

slide-in fan unit Lüfter-Einschub || Lüftereinschub
slide-in module Einschub * *Einschubmodul*
slide-in unit Einschub
slide-way Führungsbahn * *z.B. für Verfahrschlitten bei Werkzeugmaschine*
slideable verschiebbar
slider Abgriff * *Schleifer an Potentiometer, Schiebetrafo usw.* || Schleifer * *Mittelabgriff am Potentiometer, Schiebetrafo usw*
slider pick-off Schleiferabgriff * *z.B. an Potentiometer, Stelltrafo*
slideway Führungsbahn || Gleitbahn
sliding ausziehbar || gleitend || verfahrbar * *verschiebbar, gleitend* || verschiebbar
sliding bearing Gleitlager
sliding block Gleitstein * *z.B. an einer mechanisch betätigten Kupplung*
sliding caliper Messschieber * →*Schieblehre* || Schieblehre
sliding carriage Schlitten
sliding closure Gleitverschluss * *in der Verpakkungstechnik*
sliding coat Gleitbelag
sliding contact Schleifer * *Schleifkontakt an Potentiometer usw.* || Schleifkontakt
sliding door Schiebetür
sliding fit Schiebesitz * *{Masch.bau} Passung, die ein Verschieben ermöglicht*
sliding friction Gleitreibung || rollende Reibung
sliding ring Gleitring
sliding sleeve Schiebemuffe
sliding speed Gleitgeschwindigkeit
sliding surface Gleitfläche || Lauffläche * *Gleitfläche*
sliding table Schiebetisch * *{Werkz.masch.}*
sliding-action contact Schleifkontakt
sliding-mode mean value generation gleitende Mittelwertbildung
sliding-rotor motor Verschiebeläufermotor * *(Hebezeug)Motor m. integr. Bremse, Lüftg.durch Verschiebg. d.konisch. Läufers*
sliding-table circular saw Formkreissäge * *{Holz}*
sliding-table saw Formatkreissäge * *{Holz}*
sliding-type mean value generation gleitende Mittelwertbildung
slight dünn * *schlank* || etwas * *gering(fügig), leicht, unbedeutend* || gering * *leicht, unbedeutend* || geringfügig * *leicht, gering, unbedeutend, schwach, ein wenig, oberflächlich* || leicht * *gering, z.B. Überdruck* || leise * *kaum merkbar*
slightly etwas * *(Adv.) gering(fügig), leicht, unbedeutend* || geringfügig * *(Adv.)*
slim dünn * *schlank* || flach * *schlank, flach* || schlank * *(auch Verb: schlank machen)* || schmal * *schlank, flach, dünn*
slim-line flach * *in schmaler/flacher Baugröße* || schmal * *in schlanker Bauform*
slime Schlamm * *schleimiger* ~ || Schleim
slimline dünn * *in schlanker Bauart* || schlank
sling Schlaufe || schleudern * *mit einer Schleuder(-vorrichtung)* ~, verspritzen, katapultieren
slinger Schleuderscheibe
slip ausrutschen || durchrutschen * *z.B. Kupplung, Bremse* || Fehler * *leichter* ~*, Flüchtigkeits~ beim Rechnen, Schreiben usw.* || gleiten * *(aus)gleiten* || rutschen * *rutschen (auch Kupplung, Bremse), (aus)gleiten* || schieben * *gleiten lassen* || Schlupf * *bei Asynchronmaschine: (Synchrondrehz. - Läuferdrehz.)/Synchrondrehz.; auch eines Riemens* || schlupfen || schlüpfen || SLIP * *'Serial-Line Internet Protocol': DÜ-Protokoll für Internet-Verbindgn. über* →*Modem (jetzt* →*PPP)* || verschieben * *gleiten lassen*

slip a person's memory entfallen * *dem Gedächtnis* ~
slip additive Gleitmittel * *als Zusatz(mittel)*
slip by verstreichen * *verrinnen (Zeit)*
slip clutch Schlupfkupplung
slip compensation Schlupfkompensation
slip control Schlupfregelung || Schlupfsteuerung * *{el.Masch.}*
slip core control Wickelhärtensteuerung * *Zugreduzierg. b.steig.Durchm.; verhind.Beschädigg. d.Innenlagen b.Aufwickler*
slip detecting Schlupferfassung
slip force Rutschkraft
slip frequency Schlupffrequenz * *Produkt Netzfrequenz x Schlupf*
slip friction Schlupfreibung
slip loss Schlupfverluste
slip losses Schlupfverluste
slip on aufschieben * *ein Teil auf ein anders "draufschieben"* || aufstecken * *ein Teil auf ein anderes*
slip power Schlupfleistung
slip regulation Schlupfregelung
slip regulator Schlupfregler
slip ring Schleifring * *zur Stromzuführung auf sich drehende Maschinenteile (z.B. Läufer e. Asynchronmaschine)*
slip ring motor Schleifringläufermotor * *z.B. Asynchronmot.m. Schleifkontakt z.Einschleif.e. ext. Anlasswiderstands* || Schleifringmotor
slip rotor Schlupfläufer || Widerstandsläufer
slip sensing Schlupferfassung
slip speed Schlupfdrehzahl
slip torque Schlupfmoment
slip-free schlupffrei
slip-frequency voltage Schlupfspannung
slip-on geared motor Aufsteck-Getriebemotor
slip-power recovery Schlupfleistungsrückgewinnung
slip-ring joint Schleifringübertrager * *m.Schleifkontakten z.Übertragg. v. Strom/Signalen auf drehend. Maschinenteil*
slip-synchronous schlupfsynchron
slippage Schlupf * *Durchrutschen*
slipper clutch Rutschkupplung
slipperiness Glätte * *Schlüpfrigkeit*
slippery glatt * *schlüpfrig, glatt, glitschig, aalglatt, gerissen (Person), heikel (Thema)* || rutschig
slipping gleitend
slipping clutch Rutschkupplung || Schlupfkupplung || Sicherheitskupplung * →*Schlupfkupplung*
slipping movement Schlupfbewegung
slipping period Rutschzeit * *einer Kupplung/Bremse*
slipping speed Rutschdrehzahl * *einer Kupplung*
slipping time Rutschzeit * *z.B. einer Kupplung, Bremse*
slipping torque Rutschdrehmoment * *einer Kupplung/Bremse* || Rutschmoment * *bei einer Kupplung/Bremse*
slipring clutch Schleifringkupplung * *elektromagnet. Kupplung mit magnet. durchfluteten Kupplungsscheiben/-lamellen*
slipring induction motor Asynchron-Schleifringläufermotor
slipring rotor Schleifringläufer
slipway Gleitbahn
slit Fuge * →*Schlitz* || längsgeteilt * *slit strip: längsgeteiltes Blech-/Walzband* || Schlitz * *z.B. zum Entweichen von Kühlluft* || schneiden * *der Länge nach* ~ || Spalt * *Schlitz* || Trennfuge * →*Fuge,* →*Schlitz*
slit width Schlitzbreite
slitter Kreismesser * *z.B. zum Längsschneiden von Papier* || Längsschneider * *siehe auch*

→*Rollenschneider* || Messer * Kreis~, z.B. zum Längsschneiden von Papier, Blech usw. || Rollenschneider * z.B. Längsschneider für Papier, Folie usw. || Schneidmaschine * →*Längsschneider*
slitter rewinder Rollenschneider || Rollenschneidmaschine || Umroller * →*Rollenschneidmaschine*
slitter unit Schneidpartie * Quer- oder Längsschneider für bandförmiges Material
slitter winder Rollenschneidmaschine
slitter-winder Rollenschneider || Rollenschneidmaschine
slitterwinder Rollenschneider || Rollenschneidmaschine
slitting Längsschneiden
slitting line Längsteilanlage * für Walzband, Blech usw.
sliver Band * (Textil) Flor~, Kammzug in der Spinnerei || Span * Splitter, Span, Stückchen
slogan Motto * Schlagwort, Werbespruch, Slogan || Werbespruch
slope abfallen * Gelände usw. || abschrägen || Abschrägung || ansteigen * Gelände || Flanke * schräg abfallende/steigende || Gefälle || Neigung * geneigte Fläche || Rampe * Neigung, Schräge, Gefälle, Steigung, Anstieg, Rampe || Schräge * schräge Fläche, Gefälle || Senkung * Neigung || Steigung * Neigung
slope down abflachen * sich ~
sloping geneigt * abschüssig || schief * abfallend, geneigt, ansteigend || schräg * abfallend
sloping loop channel Schlingenkanal
sloping position Schräglage
slot Einbauplatz * zum Einschieben einer Steckbaugruppe/Leiterkarte || Längsbohrung * (Holz) || Nut * z.B. im Motorläufer || Öffnung * Schlitz || Schlitz || Steckplatz * Einschub-Platz z.B. für Steckbaugruppe || stoßen * stanzen, (Nuten) stoßen
slot bracket Slotblech * für Personalcomputer-Einbaukarte
slot cover Steckplatzabdeckung * für Baugruppenträger zum Abdecken eines nicht bestückten Steckplatzes
slot form Nutform
slot insulation Nutisolierung * (el.Masch.) Flächenisolierung in den Nuten zw. Blechpaket und den Leitern
slot lining Nutauskleidung * z.B. Motornut mit Isolierstoff
slot mortiser Langlochfräsmaschine * (Holz) (brit.)
slot shape Nutform * z.B. des Läufers einer el. Maschine
slot side Nutwand
slot thermometer Nutthermometer * (el.Masch.) PT100-Widerstandsthermometer, das in die Nuten eingebaut ist
slot wedge Nutkeil * (el.Masch.) Nutverschlusskeil: Halbrund- o.Trapezprofilstab z.Verschluss d. Wicklgsnuten
slot width Schlitzbreite
slot-boring machine Langlochbohrmaschine * (Holz)
slot-mortizer Langlochfräsmaschine * (Holz) (amerikan.)
slot-termination wedge Nutverschlußkeil * z.B. für Wicklungsnut in →*Hochgeschwindigkeitsläufer eines Motors*
slothful faul * träge
slots nuten * (Subst.) z.B. eines Motors
slots of the core Nuten des Blechpakets * (el. Masch.) zur Aufnahme der Wicklung
slotted genutet || geschlitzt
slotted disc Schlitzscheibe
slotted hole Schlitz
slotted-head screw Schlitzschraube

slotter Stoßmaschine * (Werkz.masch.) slotting tool: Stoßmeißel
slotting Nut-
slotting machine Stoßmaschine * (Werkz.masch).
Slovacia Slowakei
slow langsam
slow down abbremsen * verzögern || abnehmen * an Geschwindigkeit ~ || absenken * Drehzahl, Leistungsfähigkeit usw. ~ || herunterbremsen || nachlassen * ~ des Tempos || verlangsamen || zurücklaufen * Geschwindigkeit/Drehzahl reduzieren
slow of comprehension langsam * geistig
slow speed shaft Abtriebswelle * bei Untersetzungsgetriebe
slow-acting langsam wirkend || träge * Sicherung
slow-down bremsen * langsamer machen/werden || herunterfahren * verlangsamen || herunterlaufen * verlangsamen || verzögern * langsamer machen, Geschwindigkeit/Drehzahl verringern
slow-down ramp Verzögerungsrampe
slow-motion Langsamkeit * Zeitlupentempo
slow-moving langsam * sich langsam bewegend || langsamlaufend
slow-moving machine Langsamläufer
slow-running langsamlaufend
slow-speed langsam * mit langsamer Geschwindigkeit
slowing down Abbremsung * Verzögerung || Dämpfung * Nachlassen || Rücklauf * Verringerung der Geschwindigkeit
slowly langsam * (Adv.)
slowness Langsamkeit || Trägheit * Langsamkeit (auch geistige)
sluable Schwenk- || schwenkbar * z.B. Kran
sludge Schlamm * Schlamm, Matsch, feuchter Bodensatz, Klärschlamm; Treibeis
slue schwenken * herumdrehen/-schwenken
sluggish langsam * träge || träge * träge, langsam, schwerfällig
sluggishness Stagnation * geschäftliche, wirtschaftliche ~ || Trägheit * Langsamkeit, Trägheit, Schwerfälligkeit
slump Fall * (Börsen-/Preis) Sturz, Geschäfts/Produktionsrückgang || fallen * plötzlich ~
slump in prices Preissturz
slurry Brei * Aufschlämmung || Dickstoff * Schlamm, Matsch
slush Dickstoff * Schlamm, Matsch
slush pump Schlammpumpe * z.B. im Bergbau
slush-money Schmiergeld
sly pfiffig * schlau, verschlagen, listig || raffiniert * schlau, verschlagen, listig
small gering || klein * small letter: ~buchstabe; small motor: ~motor; keep/grow small: ~ halten/ ~er werden || Klein- || schmal * gering
small and medium-sized companies mittelständische Firmen
small box Kästchen
small company mittelständische Firma * kleine Firma
small gift Aufmerksamkeit * kleines Geschenk
small load Schwachlast
small motor Kleinmotor
small parts Kleinteile
small parts store Kleinteilelager
small PC board Kleinleiterplatte
small plate Plättchen
small token Aufmerksamkeit * kleines Mitbringsel
small trade Handwerk
small-band analysis Schmalbandanalyse * bei der Geräuschmessung
small-footprint klein * mit ~er Grundfläche/Stellfläche

small-scale klein * *mit/in ~em Umfang/Maßstab, z.b. Unternehmen, Integrationsgrad*
smaller kleiner * *(Komparativ) bezüglich Umfang*
smallest kleinst
smallest possible kleinstmöglich * *Größe, Volumen*
smallness Kleinheit
smarmy schmierig * *(salopp) schmeichlerisch, ölig, kriecherisch*
smart elegant || geschäftstüchtig * *gewieft* || intelligent
smart solution elegante Lösung
smash brechen * *zertrümmern; to pieces: in Stücke* || Sturz * *lauter ~* || zusammenbrechen * *(salopp) zusammenkrachen, Bruch machen, zu Schrott fahren* || Zusammenbruch * *Vernichtung, Ruin (auch finanzieller)*
smashing prima * *(salopp) toll, sagenhaft, umwerfend* || toll * *(salopp) sagenhaft, umwerfend, prima*
SmCo Samarium-Kobalt || SmCo * *Samarium-Kobalt*
SMD oberflächenmontierbares Bauelement * *Abk. f. 'Surface-Mount Device'* || SMD * *Abkürzg. f. 'Surface-Mount Device': oberflächenmontierbares Bauteil*
SMD-type capacitor SMD-Kondensator
SMD-type resistor SMD-Widerstand
smelt Schmelze
smeltery Schmelze * *[Metallurgie] Schmelzanlage*
smelting house Hüttenwerk * *Schmelzhütte*
smelting line Schmelzstraße * *im Hüttenwerk*
smith's shop Schmiede
smokestack Schornstein * *Fabrik~*
smooth filtern * *glätten* || geglättet || geläufig * *fließend* || geräuscharm * *bei Motor; smooth running: ~er/geschmeidiger Lauf* || geschmeidig || glatt || glätten || gleichmäßig * *ruhig laufend* || leise * *weich, zügig, glatt, reibg.slos, geräuscharm, geschmeidig, z.B. Umrichter m.hoher Taktfrequ.* || reibungslos * *glatt, reibungslos (auch figürl.), zügig, flüssig, geschmeidig* || ruckfrei * *weich* || ruhig * *weich, zügig, reibungslos, geschmeidig, geräuscharm z.B.Umrichter m.hoher Taktfrequ.* || rund * *ruckfrei, geschmeidig, weich (z.B. Lauf einer Maschine/eines Motors usw.)* || sanft * *Bewegung, Lauf eines Maschine/eines Motors usw.* || sieben * *glätten* || stoßfrei * *weich, sanft* || unbürokratisch * *reibungslos* || verschleifen * *z.B. Signale in Form eines weichen Übergangs ~* || weich * *auch Regelcharakteristik, Kennlinie, Laufeigenschaften eines Antriebs* || zügig * *reibungslos*
smooth action Leichtgängigkeit
smooth operation Laufruhe || Rundlauf * *geschmeidiger Lauf, z.B. eines Motors*
smooth running Laufruhe * *Rundlauf, weicher/geschmeidiger Lauf* || ruhiger Lauf * *z.B. eines Motors* || Rundlauf * *ruhiger/guter/weicher Lauf (mit kleiner Drehmomentwelligkeit)*
smooth running characteristics Rundlauf * *~eigenschaften* || Rundlaufeigenschaft || Rundlaufeigenschaften || Rundlaufverhalten
smooth running properties Rundlaufeigenschaften
smooth start weiches Anfahren
smooth starting Sanftanlauf * *sanfter Anlauf* || sanfter Anlauf || weiches Anlaufen
smooth transition ruckfreier Übergang || weicher Übergang
smooth-running laufruhig
smooth-running properties Rundlaufgüte
smoothed geglättet || verrundet
smoothed ramp verrundete Hochlaufgeberrampe
smoothening Verschliff * *weicher Übergang*
smoother Spachtel * *Schmierkelle (make smooth: spachteln)*

smoothing Beruhigung || Filterung * *(elektr.) Glättung* || Glättung || Siebung || Vergleichmäßigung || Verrundung * *z.b. einer Rampe*
smoothing capacitor Glättungskondensator
smoothing choke Glättungsdrossel
smoothing element Glättungsglied * *Glättungsglied* || Siebglied * *Glättungsglied*
smoothing of reference value Sollwertglättung
smoothing press Offsetpresse * *zur Papierverarbeitung*
smoothing reactor Ankerdrossel * *Glättungsdrossel* || Glättungsdrossel * *z.B. i. Ankerkreis eines DC Motors* || Glättungsinduktivität * *[Stromrichter]*
smoothly moving leichtgängig
smoothly running leichtgängig
smoothness Glätte * *Glätte, Reibungslosigkeit, Geschmeidigkeit, glatter Fluss, Eleganz, Gewandheit*
smoothness of running Laufruhe
smoulder schwelen
smouldering temperature Glimmtemperatur * *Temp., bei der explosionsfähige Staub-Luftgemische explodieren können*
SMT SMD * *Abkürzg. f. 'Surface-Mount Technology':Oberflächenmontagetechnik* || SMD-Technik * *Surface Mount Technology: Oberflächenmontage*
smut Schmutz * *(meist figürl.)*
snap abreißen * *zerspringen, zerreißen, entweigehen* || brechen * *(zer)springen, (zer)reißen, entzweigehen* || knicken * *Zweig usw. ~* || reißen * *brechen* || schnappen * *z.b. auf eine Installationsschiene*
snap in einschnappen
snap into place aufschnappen * *auf den Einbauplatz ~* || einrasten
snap lid Schnappdeckel
snap lock Schnappverschluss || Schnellverschluss * *Schnappverschluss*
snap on aufschnappen * *snap onto: ~ auf* || aufstecken * *aufschnappen*
snap onto aufschnappen * *z.B. auf eine Installations/Hutschiene*
snap ring Sprengring
snap- einrastend
snap-fitting cover Schnappdeckel
snap-in fastener Schnappverschluss
snap-mount aufschnappen * *montieren per Aufschnappmontage*
snap-on Aufschnapp-
snap-on mounting Aufschnappmontage
snapping einrastend
snappy rasch * *(salopp) flott, forsch, fix, schwungvoll, schmissig*
snatch auffangen || Griff * *(auch figürl.) schneller ~ (at:nach)* || reißen * *wegschnappen*
snatch block Führungsrolle * *zum Regeln des Drahtdurchhangs* || Seilscheibe * *kleine ~, z.B. an Flaschenzug*
sneak current Kriechstrom
snippers Saitenschneider
snips Blechschere * *Handblechschere* || Metallschere * *Hand→blechschere*
snooping device Mithöranlage * *hört passiv am Speicher (z.B.auch Cache-Speicher) mit*
snub ablaufen * *snub a person off:jemanden ablaufen lassen*
snubber TSE-Beschaltung * *(salopp)*
snubber capacitor Überspannungsschutzkondensator
snubber circuit RC-Beschaltung * *zum Überspannungsschutz, z.B. TSE-Beschaltung bei Thyristorsatz* || Schutzbeschaltung * *gegen Überspannung(sspitzen), z.B. TSE-Beschaltung bei Thyristorsatz* || TSE-Beschaltung || TSE-Schutzbeschaltung || Überspannungs-Schutzbeschaltung

snubber circuitry TSE-Beschaltung
snubber network TSE-Beschaltung * *RC-Beschaltg. geg. d. 'TrägerSpeicherEffekt' b.Abkommutier. e. Thyristors* || TSE-Schutzbeschaltung * *RC-Beschaltg. gegen d. 'TrägerSpeicherEffekt' b. Abkommutieren e. Thyristors*
snubber roll Andruckwalze * *z.B. in Walzstraße* || Druckwalze * *z.B. in Walzstraße*
snug satt anliegend
snug fit Passsitz
so also * *(Adv.).* || deshalb || infolgedessen || so * *als, wie, folglich, also, auf diese Art und Weise; so that: sodass* || somit
so far bisher * *bis jetzt*
so on and so forth usw * *und so weiter und so fort*
so that sodass
so to speak gewissermaßen || gleichsam * *sozusagen*
so-called so genannt
SO2 Schwefeldyoxid
soak aufquellen * *einweichen* || imprägnieren * *durchtränken/-nässen/-feuchten, ~ (with:in); durchtränkt werden, sich vollsaugen* || tränken * *sich vollsaugen, durchtränkt werden, durchtränken, aufsaugen, imprägnieren*
soak in eindringen * *einer Flüssigkeit*
soaked nass * *tropf~*
soaking furnace Ausgleichsofen * *Walzwerk*
sociability Verträglichkeit * *im menschlichen Umgang*
sociable verträglich * *im menschlichen Umgang*
social climbing Aufstieg * *sozialer ~*
society Genossenschaft
socket Buchse * *(elektrisch)* || Fassung * *für Lampe, IC usw.* || Hülse * *Steck~, Muffe, Rohransatz, Fassung* || Sockel * *Lampen~, IC-~, Steck~, Rohransatz, Muffe, Fassung* || Steckdose || Tülle
socket connector Buchsenleiste || Buchsenstecker || Federleiste
socket outlet Steckdose
socket outlet with earthing contact Schukosteckdose
socket screw Zylinderschraube
socket screw wrench Inbusschlüssel || Zweilochmutterschlüssel
socket spanner Inbusschlüssel || Steckschlüssel
socket wrench Inbusschlüssel || Steckschlüssel
socket-head cap screw Inbusschraube
socle Sockel * *~ eines Bauwerks*
soda pulp Zellstoff * *Natron-/Sulfatzellstoff*
soda shop Sodahaus * *z.B. in Papierfabrik*
soft leise * *(seiden)weich, sanft, gedämpft, sacht* || ruhig * *(seiden)weich, sanft, gedämpft* || sanft || weich
soft braking Sanftauslauf * *z.B. mit Hilfe eines Dreh-/Wechselstrellers*
soft foam Weichschaum
soft magnetic weichmagnetisch
soft start weiches Anfahren
soft starter Sanftlasser || Sanftanlaufgerät * *{el. Masch.} Leistgs.steller, d.durch Spannungsverringerg. e.Sanftanlauf ermöglicht*
soft starting Sanftanlauf * *z.B.für Asynchronmotor*
soft starting unit Sanftanlaufgerät
soft stopping Sanftauslauf * *eines Motors z.B. mit Drehstromsteller*
soft-iron instrument Dreheiseninstrument * *Weicheiseninstrument (misst Effektivwert)* || Weicheiseninstrument * *misst Effektivwert*
soft-solder löten * *weich~*
soft-start sanftanlassen * *Motor*
soft-starting circuit Sanftanlauf-Schaltung
soft-stop Sanftauslauf
soft-torque starting Sanftanlauf * *{el.Masch.} Anlauf v.Käfigläufermotoren mit Anlasswiderstand o. Sanftanl.gerät*
soften dämpfen * *Licht ~* || nachlassen * *milder werden*
softening agent Weichmacher
softness Weichheit
softstart Sanftanlauf
software Programm * *Software~* || Software || softwaremäßig
software block Software-Modul * *Softwarebaustein* || Softwarebaustein * *Funktionsbaustein* || Softwaremodul * *Softwarebaustein*
software bug Programmfehler
software change Softwareänderung
software company Ingenieurbüro * *Softwarehaus* || Software-Haus || Softwarehaus
software designer Entwickler * *Softwareentwickler* || Programmierer || Software-Entwickler || Softwarebearbeiter * *Entwickler* || Softwareentwickler
software developer Programmierer * *Software-Entwickler* || Software-Entwickler || Softwareentwickler
software development Softwareentwicklung
software engineer Programmierer * *Ingenieur*
software function block Softwarebaustein * *Funktionsbaustein*
software manufacturer Software-Haus || Softwarehaus
software modification Softwareänderung
software module Software-Modul || Softwarebaustein || Softwaremodul
software package Programmpaket || Softwarepaket
software release Software-Version || Softwarestand
software solution Softwarelösung * *complete software solution: komplette ~*
software specialist Programmierer * *Spezialist*
software switch Software-Weiche || Softwareweiche || Weiche * *Software~*
software tool Softwarewerkzeug
software upgrade Softwaretausch * *auf eine neuere Version*
software version Software-Version || Softwarestand || Softwareversion
softwire verbinden * *per Software "verdrahten"* || verknüpfen * *per Software "verdrahten"*
softwood Holz * *Nadel~, Weich~* || Nadelholz * *Weichholz*
soil Boden * *Erde, Grund* || Grund * *Erdboden* || verschmutzen || verunreinigen * *beschmutzen, besudeln, verunreinigen, beflecken; schmutzig werden*
soil cultivation Bodenbearbeitung * *{Agrikultur}*
soiled schmutzig * *beschmutzt* || verschmutzt
solar radiation Sonneneinstrahlung
solder Lot || löten * *weich~* || Lötzinn || verlöten * *Weichlöten*
solder bath Lötbad
solder bridge Lötbrücke
solder connection Lötverbindung
solder in einlöten * *z.B. Lötbrücke, Bauteil auf Lötstützpunkte ~*
solder joint Lötstelle * *dry/faulty solder joint: kalte Lötstelle* || Lötverbindung
solder jumper Lötbrücke
solder lug Lötfahne || Lötöse || Lötstützpunkt * *→Lötfahne*
solder paste Lötpaste
solder pin Lötstift || Lötstützpunkt
solder post Lötstift
solder resist Lötstoplack || Lötstopplack
solder side Lötseite * *einer Leiterkarte*
solder tag Lötfahne || Lötstützpunkt
solder wire Lötdraht
solder-post Lötstützpunkt * *Pfosten*
soldered connection Lötverbindung

soldered in eingelötet
soldered joint Lötstelle || Lötverbindung
soldering Weichlöten
soldering bath Lötbad
soldering compound Lötmasse
soldering flux Flussmittel
soldering gun Lötkolben * *mit Pistolengriff* || Lötpistole
soldering iron Lötkolben
soldering jumper Lötbrücke
soldering lug Lötöse
soldering method Lötverfahren
soldering paste Lötfett || Lötmasse || Lötpaste
soldering pin Lötstift
soldering post Lötstützpunkt * *Pfosten*
soldering tag Lötfahne
soldering technique Lötverfahren
soldering terminal Lötstützpunkt
solderless lötfrei * *z.B. Verbindungstechnik*
sole einzig * *alleinig*
solely einzig * *(Adv.)* ~ *und allein* || lediglich || nur * *lediglich*
solenoid Antrieb * *Magnet~* || Drehmagnet || Magnet * *Betätigungsspule bei Ventil, Relais usw.*; Hubmagnet || Magnetventil || Relaisspule * *Betätigungsspule* || Spule * *Betätigungs~ (z.B. Magnetventil)*
solenoid actuator Magnetantrieb * *elektromagnet. Stellantrieb* || Stellantrieb * *Magnetantrieb z.B. für Ventil*
solenoid assembly Magnetkörper * *z.B. in einer Magnetbremse/Magnetventil*
solenoid coil Magnetspule * *einer Betätigungseinrichtung; bei Relais, Magnetventil, elektromagnet. Bremse usw.*
solenoid component Magnetteil * *z.B. bei einer elektromagnet. Bremse* || Megnetteil * *z.B. bei einer elektromagnet. Bremse*
solenoid operated drive Magnetantrieb * *z.B. ~ für Relais*
solenoid section Megnetteil * *bei einer elektromagnet. Bremse*
solenoid valve Elektromagnetventil || Magnetventil
solenoid-operator Magnetantrieb * *z.B.~ für Relais*
soleplate Sohlplatte * *z.B. für das Aufstellen einer elektr. Maschine*
solid Dickstoff * →*Feststoff* || eindrähtig * *steifer Leiter (keine Litze)* || fest * *von ~er Beschaffenheit, massiv* || Festkörper || Feststoff || kompakt * *fest, stabil, massiv, derb, geschlossen, einheitlich* || massiv * *fest, kräftig, stabil, gediegen, aus Vollmaterial* || räumlich * *(geometr.)* || stabil * *fest, robust* || ungeteilt || voll * *massiv, aus Vollmaterial* || Voll- * *z.B. aus Vollmaterial, ganz mit Material ausgefüllt*
solid cylinder Vollzylinder
solid foreign bodies Fremdkörper * *(Plural)* kornförmige/feste ~
solid friction Festkörperreibung
solid line durchgezogene Linie
solid matter Feststoffe
solid particles Festkörper
solid pulse Einzelimpuls * *z.B. zur Thyristor-Zündung i. Ggs. zum Kettenimpuls*
solid rotor Massivläufer * *ungeblechter ~ (→Turboläufer), hat eine höhere Grenzdrehzahl als e. geblechter Läufer*
solid rubber Vollgummi
solid shaft Vollwelle * *z.B. eines Getriebes*
solid state fester Zustand || kontaktlos * *in Halbleitertechnik ausgeführt (z.B. elektron. Relais)*
solid state electronics Halbleiterelektronik
solid state power component Leistungshalbleiter
solid substances Feststoffe

solid wire Volldraht * *im Gegensatz zum Litzenleiter*
solid wire connecting lead Anschlussdraht * *steifer Draht, kein Litzenleiter*
solid wood Massivholz
solid wood panel Massivholzplatte
solid wood working Massivholzbearbeitung
solid woodworking Massivholzverarbeitung
solid-born noise Körperschall
solid-pole rotor Vollpolläufer
solid-state relay Halbleiterrelais || kontaktloses Relais
solid-state switch kontaktloser Schalter * *Halbleiterschalter*
solidify härten * *fest werden (lassen), erstarren*
solidity Festigkeit * *kompakte/massive Struktur/Beschaffenheit; Dichtigkeit, Gediegenheit, Zuverlässigkeit*
solids Feststoffe * *granular solids: kornförmige ~*
solitary einzeln * *isoliert, einsam* || Insel-
solitary operation Inselbetrieb
soluble lösbar * *Problem, Aufgabe, Gleichung; Löslichkeit von Substanzen in Flüssigkeit*
solution Lösung * *Problem~, Realisierung, chemische ~* || Realisierung * *Problemlösung* || Variante * *Lösung* || Version * *Lösung*
solvable lösbar * *Gleichung*
solve auflösen * *Rätsel, Gleichung, Klammer usw. ~; for: nach* || lösen * *Problem, Aufgabe, (Differenzial-) Gleichung ; overcome a problem: ein Problem ~* || realisieren * *lösen (Problem, Aufgabe)*
solvent Lösungsmittel || zahlungsfähig
solvent-free lösungsmittelfrei
some etwas * *(Adj.)* || mehrere || Paar * *(Adj.) ein paar, einige*
something etwas
somewhat einigermaßen || etwas * *(Adv.) ein wenig, ein bisschen*
sonar Ultraschall * *nach dem Echoprinzip funktionierend*
sonorous laut * *stark klingend*
sooner früher * *zeitiger*
soot Ruß
sophisticate verfeinern * *(figürl.)*
sophisticated anspruchsvoll * *hochentwickelt, raffiniert* || ausgeklügelt || filigran * *(figürl.)* || komfortabel * *ausgeklügelt, ausgefeilt, raffiniert, hochentwickelt, anspruchsvoll, exquisit* || kompliziert * *hochentwickelt, hochgestochen, raffiniert, anspruchsvoll*
sophistication Komfort * *"Ausgefeiltheit", "Ausgeklügeltheit", Kultiviertheit*
sore aufscheuern * *Haut, Wunde usw.*
sorry Entschuldigung * *(gesproch.) um Verzeihung bittend; (I'm) sorry (about that): ~ (f.d.Ungemach)*
sort ordnen * *sortieren* || Sorte || sortieren
sorting Sortierung
sorting device Sortiereinrichtung
sorting equipment Sortieranlage || Sortiereinrichtung
sorting machine Sortiermaschine
sorting plant Sortieranlage
sorting system Sortieranlage
soul Sinn * *Seele*
sound anhören * *sich ~ (well: gut, badly: schlecht)* || Klang || lauten || Schall
sound absorber Geräuschdämpfer || Schalldämpfer
sound absorption Schalldämpfung
sound attenuation Schalldämpfung
sound conducted through solids Körperschall
sound dampening Schalldämpfung
sound insulation Schalldämpfung
sound intensity Lautstärke

sound level

sound level Geräuschpegel * *Schallpegel* || Schallpegel
sound level meter Schallpegelmesser
sound power level Schall-Leistungspegel * *z.B. für el. Maschinen: siehe DIN VDE 0530 und DIN 45635* || Schalldruckpegel * *Schallleistungspegel, siehe z.B. EN 50144* || Schalleistungspegel * *ergibt s. aus Schalldruckpegel u.Messfläche; für El.motoren: siehe DIN VDE 0530* || Schallleistungspegel * *siehe z.B. EN 50144*
sound pressure Schalldruck * *Effektivwert des Schallwechseldrucks in [N/m²]*
sound pressure level Schalldruckpegel * *[dB (A)] Messung: DIN 45635, EN 50144; Motorgeräusch: DIN VDE 0530, DIN 21680* || Schallleistungspegel * *siehe z.B. EN 50144*
sound propagation Schallausbreitung
sound protection Schallschutz
sound reflection Schall-Reflexion || Schallreflexion
sound source Schallquelle
sound transmission Schallübertragung
sound-absorbing schalldämpfend || schallgedämpft * *schalldämpfend*
sound-damping element Schalldämmelement
sound-deadening schalldämpfend
sound-insulated schallgedämpft
sound-insulating schalldämpfend
sound-proof schalldicht || schallisoliert
sound-proofed schallgedämpft
sound-proofing element Schalldämmelement
source Hand * *Quelle; from a single source: aus einer ~/von einem Anbieter* || Herkunft || Quelle * *auch ~ eines Signals; Quellcode eines Programms* || Quellsprache * *Quellsprachprogramm(listing)* || Quellsprache-Code * *(salopp)* || Ursprung * *Quelle, Ursprung, Herkunft (auch einer Nachricht)*
source code Quellcode || Quellsprache-Code || Quellsprachecode
source language Quellsprache
source level Quellspracheebene
source of supply Bezugsquelle || Lieferquelle
source reference Quellhinweis * *für Signal/Variable in Stromlaufplan/Blockschaltbild/Programmlisting*
source resistance Innenwiderstand * *Ausgangswiderstand*
source selection Quellenauswahl
source signal Eingangssignal * *Quellsignal* || Quellsignal
sourness Säure
south pole Südpol * *auch bei einem Magnet*
space Abstand * *(auch Verb: in einen definierten ~ bringen)* || Fläche * *Raum, Platz, Stelle, Zwischenraum* || Leerzeichen || Raum * *Fläche, Platz, Zwischen~, Stelle, Lücke, Zeit~, Abstand, Dimension im Gegensatz zu 'Zeit'* || Raum- || räumlich anordnen * *auch räumlich einteilen, mit/in Zwischenräumen anordnen* || Zeit * *~raum* || Zeitraum * *Zeitspanne* || Zwischenraum * *Leerzeichen zwischen Buchstaben*
space bar Leertaste || Zwischenraumtaste
space charge Raumladung
space heater Stillstandsheizung
space heating Stillstandsheizung
space key Leertaste
space liable to contain explosive atmospheres explosionsgefährdeter Raum
space of time Zeit * *~raum* || Zeitraum || Zeitspanne
space polygon Raumpolygon
space problem Platzproblem
space request Platzbedarf
space requirement Platzbedarf || Raumbedarf
space requirements Platzbedarf
space saving Platzeinsparung || Platzersparnis

space that is kept free Freiraum * *freigehaltener Raum*
space utilization Platzausnutzung || Platznutzung || Raumausnutzung * *Platzausnutzung* || Raumnutzung * *Platz-/Raumausnutzung*
space vector Raumvektor || Raumzeiger
space vector diagram Raumzeigerdiagramm
space vector modulation Raumzeigermodulation
space-saving Platz sparend || raumsparend
spaced versetzt || *auf Abstand gesetzt, spaced at 60°: um 60° ~*
spacer Abstandhalter || Abstandshalter * *auch in einer Batterie* || Abstandsstück * *siehe auch →Abstandshalter* || Distanzhalter || Distanzstück * *Abstandshalter* || Zwischenlage * *~ zum Abstand-Halten*
spacer bush Distanzhülse
spacer ring Ausgleichscheibe * *Distanzring, Abstandsscheibe* || Ausgleichsscheibe * *Distanzring, Abstandsscheibe* || Distanzring
spacer sleeve Distanzbuchse
spacer washer Abstandshalter * *Abstandsscheibe* || Ausgleichscheibe * *Abstandsscheibe* || Ausgleichsscheibe * *Abstandsscheibe* || Distanzscheibe
spacial Raum- || räumlich * *räumlich, Raum-*
spacing Abstand * *(einzuhaltender, vorgesehener, vorzusehender) Abstand* || Teilung * *Abstandsteilung* || Zwischenraum
spacing layer Zwischenlage * *~ zum Abstand-Halten*
spacious breit * *ausgedehnt* || geräumig || groß * *geräumig*
spaciousness Geräumigkeit
spade drill Spitzbohrer
spade terminal Kabelschuh * *offener ~ (schaufelförmig)*
Spain Spanien
span erstrecken * *sich ~ über* || Spanne || Spannweite * *eines Schraubenschlüssels* || Strecke * *Spanne* || Zeitspanne
Spanish spanisch
spanner Schlüssel * *Schraubenschlüssel* || Schraubenschlüssel
spanner gap Schlüsselweite * *eines Schraubenschlüssels*
spanner opening Schlüsselweite * *eines Schraubenschlüssels*
spanner surface Schlüsselfläche * *zum Ansetzen eines Schraubenschlüssels*
spar Holm * *z.B. bei Flugzeug, Schiff*
spare Ersatz- || knapp * *kärglich, dürftig, sparsam; a bare mile: eine ~ Meile; bare majority: ~ Mehrheit* || Reserve * *(Adj.) Ersatz-* || Reserve- * *z.B. Ersatzteil* || sparen * *Kosten, Mühe ~*
spare brushes Ersatzkohlebürsten
spare converter Reserveumrichter * *als Ersatzteil*
spare module Ersatzbaugruppe
spare part Ersatzteil || Teil * *Ersatz~*
spare parts Ersatzteile || Zubehörteile * *Ersatz-/Reserveteile*
spare parts catalog Ersatzteilkatalog
spare parts inventory Ersatzteilhaltung
spare parts list Ersatzteilliste
spare parts order Ersatzteilbestellung
spare parts service Ersatzteildienst
spare parts stockage Ersatzteilhaltung
spare parts stocking Ersatzteilhaltung
spare parts store Ersatzteillager
spare parts supply Ersatzteilversorgung
spare parts warehouse Ersatzteillager * *beim Hersteller/Lieferanten*
spare time Freizeit
spares Ersatzteile
spares back-up Ersatzteilhaltung
spares inventory Ersatzteilhaltung

spares supply Ersatzteilversorgung
spark Funken
spark discharge Funkenentladung
spark erosion Funkenerosion || Senkerodieren
spark gap Funkenstrecke
spark ignition Zünddurchschlag
spark ignition capacity Zünddurchschlagsvermögen * *bestimmt die Explosionsgruppe z.B. von Motoren*
spark over Ansprechen * *(Verb) Überspannungsableiter* || überspringen * *Funken*
spark quenching Funkenlöschung
spark suppression Funkenlöschung
spark suppression capacitor Funkenlöschkondensator
spark-erosion machine Funkenerosionsmaschine * *{Werkz.masch.}*
spark-over Funkenüberschlag || Überschlag
spark-quenching capacitor Funkenlöschkondensator
sparking Funkenbildung || Funkenüberschlag
sparking under the brushes Bürstenfeuer
sparkless funkenfrei * *z.B. Kommutierung einer Gleichstrommaschine*
sparkover level Ansprechpegel * *eines (Überspannungs-) Ableiters*
sparks Funken * *(Plural)*
spasmodic sprunghaft * *sprunghaft (auch Markt), krampfartig*
spatial Raum- || räumlich * *im Gegensatz zu 'zeitlich'*
spatial distribution räumliche Verteilung
spatial separation räumliche Trennung
spatially separated räumlich getrennt
spatula Spachtel * *~werkzeug*
SPDT contact Umschaltkontakt * *Abk. für 'Single-Pole Double-Throw contact: einpoliger ~*
speaker Referent * *Vortragender, Redner* || Sprecher || Vortragender * *Redner*
speaking am Telefon * *(salopp) 'am Apparat'*
spec Spezifikation * *Kurzform für 'specification'*
special anormal * *spezial, Sonder-* || besondere || besonders * *speziell* || einzeln * *besonder* || separat || Sonder- || Spezial- || speziell
special alloy Sonderlegierung
special application Sonderanwendung
special branch Spezialgebiet
special case Sonderfall
special coating Sonderanstrich
special construction Sonderanfertigung || Sonderausführung * *Sonderkonstruktion*
special demands besondere Anforderungen
special design Sonderanfertigung || Sonderausführung || Sonderbauart || Spezialauslegung
special driver Sondertreiber * *Softwaretreiber z.B. für Kommunikationsbaugruppe*
special excerpt Sonderdruck * *z.B. eines Zeitschriftenartikels*
special feature Besonderheit * *besonderes Merkmal* || Spezifikum
special features besondere Merkmale
special field Fachgebiet * *siehe auch →Fach*
special function Zusatzfunktion * *Spezial-/Sonderfunktion*
special grease Sonderfett
special knowledge Sachkenntnis || Spezialkenntnisse * *siehe auch →Fachwissen* || Spezialwissen
special machines Sondermaschinen
special measure Sondermaßnahme
special measures Sondermaßnahmen
special motor Sondermotor
special option Sonderoption
special periodical Fachzeitschrift
special price Sonderpreis || Vorzugspreis

special pricing agreement Sonderpreisvereinbarung
special production Sonderanfertigung
special property Spezifikum
special purpose Sonderzweck
special requirement Sonderwunsch
special requirements besondere Anforderungen
special rotor Sonderläufer * *{el.Masch.}*
special steel Sonderstahl
special subject Fachgebiet || Spezialgebiet
special tool Spezialwerkzeug
special training Fachausbildung
special version Sonderausführung
special wish Sonderwunsch
special-purpose Sonder- || Spezial- * *für einen speziellen Verwendungszweck*
special-purpose design Sonderbauart
special-purpose machines Sondermaschinen
specialist Experte || Facharbeiter || Fachingenieur || Fachmann || Inbetriebnahme-Ingenieur * *Fachmann* || Sachverständiger || Spezialist || Techniker * *Spezialist*
specialist area Spezialgebiet
specialist book Fachbuch
specialist dictionary Fachwörterbuch
specialist engineer Fachingenieur || Fachmann || Spezialist
specialist literature Fachliteratur
specialist publication Fachzeitschrift
specialist publications Fachliteratur
specialist service department Fachorientierte Leitstelle
specialist team Expertenteam
speciality Fach * *Spezial~, Spezialgebiet, Spezialität* || Fachgebiet * *Spezialfach, Spezialgebiet, Spezialität* || Spezial- || Spezialität || speziell * *Spezial-, besonders* || Spezifikum * *Spezialität*
specialization Spezialisierung
specialized book Fachbuch
specialized dealer Fachhändler
specialized dealers Fachhandel
specialized literature Fachliteratur
specialized manufacturing Sonderfertigung
specialized trade Fachhandel
specialty Fachgebiet * *(amerikan.)* || Spezifikum * *(amerikan.) Spezialität*
species Sorte
specific besonders || bestimmt * *spezifisch, speziell, bestimmt, eigen(tümlich). besonders, wesentlich, definitiv, präzise* || bestimmter * *spezifisch, speziell, eigen(tümlich), typisch, kennzeichnend, besonderer* || bezogen * *spezifisch* || gezielt || speziell * *spezifisch, speziell, bestimmt, typisch, kennzeichnend, eigen(tümlich), sonder, genau* || spezifisch
specific advertising gezielte Werbung
specific gravity spezifisches Gewicht
specific heat spezifische Wärme
specific loading Strombelag
specific volume spezifisches Volumen
specific weight spezifisches Gewicht
specifically speziell * *(Adv.) speziell, besonders, definitiv*
specification Angabe * *(technische) Spezifikation* || Anweisung * *Festlegung* || Bestimmung * *Spezifikation, Vorgabe* || Festlegung * *Spezifikation, Festlegung* || Lastenheft || Liste * *detaillierte Aufstellung* || Spezifikation || Vorgabe * *Aufgabenbeschreibung* || Vorschrift * *Spezifikation, Festlegung*
specification data technische Daten
specification draft Normentwurf
specifications Angaben * *z.B. in einem Datenblatt* || Lastenheft || technische Angaben || technische Daten
specifics Einzelheiten

specified angegeben * z.B. in Datenblatt ~, spezifiziert || aufgeführt * im Einzelnen || bestimmt * festgelegt || festgelegt * spezifiziert || spezifiziert || vorgegeben || vorgeschrieben * spezifiziert
specified application bestimmungsgemäße Verwendung
specified condition Soll-Zustand || Sollzustand
specified point Bezugspunkt
specifier Kennung
specify angeben * im Einzelnen ~, spezifizieren || aufführen * im Einzelnen ~ || aufstellen * Kosten ~ || bemessen * bestimmen, (i.einzelnen) festsetz., genaue Angaben machen, spezifizieren,(einzeln) angeben || beschreiben * spezifizieren, vorschreiben || festlegen * spezifizieren, genau angeben/vorgeben || spezifizieren || vorgeben || vorschreiben * spezifizieren, festlegen
specimen Exemplar * Einzelstück || Muster * Probestück || Probe * Prüfstück || Prüfling
speck Spritzer * Spenkel, Tupfer, Fleckchen, Pünktchen
spectacular eindrucksvoll * spektakulär || spektakulär
spectra Spektrum * (Plural)
spectrum Spektrum
spectrum analyzer Spektrumanalysator
speech Vortrag * Rede; make a speech:eine Rede halten
speed Bahngeschwindigkeit || Drehzahl || Fahrt * Tempo; at full speed: in voller ~; gather speed: ~ aufnehmen || Gang * Getriebe~ (first/second/top: erster/zweiter/höchster) || Geschwindigkeit * Drehzahl, Bahn~ || Getriebestufe * Gang || Tempo * Geschwindigkeit
speed actual value Drehzahlistwert
speed actual value encoder Drehzahlistwertgeber * digitaler
speed actual value evaluation Drehzahlistwerterfassung
speed actual value evaluator Drehzahlistwerterfassung * Baustein, Einrichtung
speed adaption Drehzahlanpassung
speed adjustment Drehzahlanpassung || Drehzahlregelung * Drehzahl(ein)stellung || Drehzahlverstellung
speed cascade Sollwertkaskade * für Geschwindigkeits-Sollwerte
speed changing Drehzahländerung
speed comparator Drehzahlgrenzwertmelder
speed constancy Drehzahlkonstanz
speed constant Drehzahlkonstante * eines Elektromotors [upm/V] (Kehrwert der →Spannungskonstanten)
speed control Drehzahlregelung || Drehzahlregulierung || Drehzahlsteuerung || Drehzahlverstellung || Geschwindigkeitsregelung
speed control error Drehzahlabweichung * Regelabweichung im Drehzahlregelkreis
speed control potentiometer Drehzahlsollwertpotentiometer
speed control range Drehzahlhub * Drehzahlstellbereich || Drehzahlstellbereich * Regelbereich (wide: großer) || Drehzahlverstellbereich * Drehzahlregelbereich || Stellbereich * der Drehzahl
speed controller Drehzahlregler || Geschwindigkeitsregler
speed controller input Drehzahlreglereingang
speed controller output Drehzahlreglerausgang
speed correction Drehzahlkorrektur
speed correction signal Geschwindigkeitskorrektursignal * z.B. von einem Tänzerlage- oder Zugregler
speed demand Drehzahlsollwertvorgabe * Drehzahlanforderung, geforderte Drehzahl
speed deviation Drehzahlabweichung

speed difference Drehzahldifferenz || Geschwindigkeitsdifferenz * relative: relative ~; z.B. zw. Antrieb 1 und 2: [(v2 - v1)/v1] x 100%
speed differential relative Geschwindigkeitsdifferenz * (v2-v1)/v1) x 100% zw. 2 Antrieben bei durchlaufender Warenbahn
speed direction Drehrichtung * siehe auch →Drehsinn
speed direction reversal Drehrichtungs-Umkehr
speed droop Drehzahlstatik * Abfall d. Drehzahl bei Belastungssteigerung, z.b. durch weich gemachten Drehz.regler
speed drop Drehzahlabfall || Drehzahleinbruch
speed encoder Drehzahlgeber * digitaler
speed error Drehzahlabweichung * Regelabweichung
speed feedback Drehzahlistwert
speed feedback adjustment Drehzahlistwertabgleich || Drehzahlistwertanpassung
speed feedback evaluation Drehzahlistwerterfassung
speed feedback loss Tachofehler * Wegbleiben des Drehzahlistwertsignals
speed fluctuation Drehzahländerung * Schwankung
speed fluctuations Drehzahlschwankungen
speed follower Folgeantrieb * erhält seinen Drehzahlsollwert vom Leitantrieb || Nebenantrieb * Folgeantrieb, fährt mit einer vom Leitantrieb vorgegebenen Geschwindigkeit
speed hunting Drehzahlschwankungen * Pendeln, Instabilität
speed limit Drehzahlgrenze
speed limit monitor Drehzahlgrenzwertmelder
speed loop control Drehzahlregelung
speed loop output Drehzahlreglerausgang
speed master Klemmstelle * Antrieb, der die Geschwindigkeit einer Produktions-/Verarbeitungsstraße vorgibt || Leitantrieb * gibt die Geschwindigkeit einer Verarbeitungs-/Produktionsstraße vor
speed match Drehzahlanpassung || Drehzahlgleichlauf || Geschwindigkeitsgleichlauf
speed measurement Drehzahlerfassung || Drehzahlmessung
speed meter Drehzahl-Messgerät
speed monitor Drehzahlrelais || Drehzahlwächter
speed of adjustment Verstellgeschwindigkeit
speed of current alternation Stromänderungsgeschwindigkeit
speed of lift Hubgeschwindigkeit
speed of response Dynamik || Reaktionsgeschwindigkeit
speed of the driven side Abtriebsdrehzahl
speed pre-selection Drehzahlvorwahl
speed ramp Hochlaufgeber * für Drehzahl, Geschwindigkeit
speed range Drehzahlhub || Drehzahlstellbereich || Drehzahlverstellbereich || Stellbereich * der Drehzahl
speed ratio Drehzahlrelation || Drehzahlverhältnis || Geschwindigkeitsrelation || Übersetzungsverhältnis * Getriebe
speed ratio control Drehzahlverhältnisregelung
speed reducer Getriebe * →Untersetzungsgetriebe || Reduktionsgetriebe * siehe auch 'Reduziergetriebe' || Reduziergetriebe || Untersetzungsgetriebe
speed reduction Getriebeuntersetzung || Untersetzung
speed reduction ratio Übersetzungsverhältnis * ~ eines Untersetzungsgetriebes
speed reference Drehzahlsollwert
speed reference setting Drehzahlsollwertvorgabe
speed reference value Drehzahlsollwert

speed regulation Drehzahländerung * *bei Laständerung* || Drehzahlregelung
speed regulation characteristic Drehzahlverhalten * *Drehzahl i.Abhängigkeit v. Lastmoment, gespiegelte Drehmom.kennl..*
speed regulator Drehzahlregler
speed reversal Drehrichtungsumkehr || Drehsinnumkehr * *siehe auch* →*Drehrichtungsumkehr*
speed rise time Drehzahlanregelzeit
speed search Drehzahlsuche * *zum Aufschalten eines Stromrichters auf einen laufenden Motor* || Fangen * *'Suchen',erlaubt Zuschalt.e. Frequ.umrichters auf lauf. Motor z.B.nach Spanngs.einbr.*
speed sensing Drehzahlerfassung || Drehzahlistwerterfassung
speed sensor Drehzahl-Messgeber || Drehzahlgeber || Drehzahlistwertgeber || Tacho || Tachogenerator * *Drehzahlgeber* || Tachometer
speed setpoint Drehzahlsollwert
speed setpoint setting Drehzahlsollwertvorgabe
speed setting Drehzahlsollwertvorgabe || Drehzahlverstellung
speed setting potentiometer Drehzahlsollwertpotentiometer
speed setting range Drehzahlstellbereich || Drehzahlverstellbereich
speed signal Drehzahlistwert
speed signal adaption Drehzahlistwertabgleich || Drehzahlistwertanpassung
speed slaving Drehzahlgleichlauf
speed stability Drehzahlkonstanz || Drehzahlstabilität
speed synchronism Drehzahlgleichlauf || Geschwindigkeitsgleichlauf
speed threshold Drehzahlschwelle || Drehzahlschwellwert
speed transducer Drehzahlgeber
speed transmitter Drehzahlgeber
speed trim Drehzahlistwertabgleich || Drehzahlistwertanpassung || Drehzahlkorrektur * *durch Korrektur-Zusatzsollwert (z.B. von Tänzerlage- oder Zugregler)*
speed trim signal Geschwindigkeitskorrektursignal * *z.B. von einem Tänzerlage- oder Zugregler*
speed undershoot Drehzahleinbruch
speed up aktivieren * *(Arbeit usw.) beschleunigen* || beschleunigen * *schneller machen/bearbeiten* || erhöhen * *die Geschwindigkeit ~* || fördern * *beschleunigen*
speed variation Drehzahländerung || Drehzahlregelung * *Veränderung der Drehzahl* || Drehzahlverstellung
speed variator Getriebe * *stufenlos verstellbar, z.B. PIV-Getriebe* || Regelgetriebe * *stufenlos regelbares ~* || stufenlos verstellbares Getriebe || Verstellgetriebe * *stufenlos verstellbares (z.B.* →*PIV-Getriebe)*
speed-controlled drehzahlgeregelt
speed-controlled coupling Anlaufkupplung * *[el. Masch.] Kupplung, die dem Motor einen Leerlauf ermöglicht*
speed-defining drehzahlbestimmend
speed-dependent drehzahlabhängig
speed-dependent current limiting drehzahlabhängige Strombegrenzung * *z.Einhaltg. d. Kommutierungs-Grenzkurve*
speed-dependent field weakening drehzahlabhängige Feldschwächung * *erlaubt bei Kranhubwerk schnelle Fahrt bei leichter Last*
speed-output diagram Drehzahl-Leistungsdiagramm * *z.B. bei DC-Motor*
speed-power diagram Drehzahl-Leistungsdiagramm * *z.B. bei DC-Motor*
speed-search suchen * *(Subst.) Suchen d.aktuell. Motordrehz. nach Spanngs.einbruch z. Zuschal-*
t.e.Umrichters || Suchen-Fangen * *Umrichter schaltet nach Netzspannungseinbruch wieder auf d.laufenden Motor auf*
speed-sychronous drehzahlsynchron
speed-synchronized drehzahlsynchron
speed-torque characteristic Drehmomentkennlinie * *Abhängigkeit des Drehmoments von d. Drehzahl* || Drehzahl-Drehmomentenkennlinie || Momentenkennlinie * *[el.Masch.] Verlauf des Drehmoments über der Drehzahl*
speed-torque curve Drehmomentkennlinie
speed-versus-time characteristic Fahrkurve * *Geschwindigkeitskurve über der Zeit*
speedometer Drehzahl-Messgerät || Motordrehzahl-Messinstrument
speedy beschleunigt * *zügig* || rasch * *zügig* || schnell * *zügig* || umgehend * *zügig* || zügig
spend aufwenden || verbrauchen * *Geld, Zeit usw.; (on: für; in doing: bei)* || verwenden * *aufwenden*
sphere Kreis * *Wirkungs~*
sphere of action Geschäftsbereich * *(allg./unternehmerischer) Tätigkeitsbereich* || Geschäftsgebiet * *(allg./unternehmerischer) Tätigkeitsbereich*
spheroidal graphite iron Sphäroguss
spider Drehkreuz * *z.B. bei Non-Stop-Wickler* || Läuferkörper
spider arm Tragbalken
spigot Zapfen * *Zapfen, (Leitungs-/Fass-) Hahn, Muffenverbindung*
spigot face Zentrierfläche
spike Dorn * *Spitze* || Nadelimpuls * *Spitze* * *Nadelimpuls*
spikes -Spitzen * *Nadelimpulse*
spillage Überlauf * *Flüssigkeit*
spin Drall || drehen * *(sich schnell) drehen* || mahlen * *z.B. Räder im Schlamm* || schleudern * *herumwirbeln, schnell drehen, schleudern. trudeln lassen, spinnen* || spinnen * *[Textil]Fäden, Garn usw; spin wool into yarn: Wolle zu Garn spinnen* || trudeln * *z.B. freilaufender Motor* || zentrifugieren * *zentrifugieren, schnell drehen, trudeln lassen, spinnen (Fäden)*
spin around kreisen
spin-dry schleudern * *Wäsche ~*
spindle Achse || Spindel * *z.B. ~ bei Werkzeugmaschine*
spindle bearing Zapfenlager
spindle collar Spindelhals * *z.B. einer Bohrmaschine*
spindle drive Spindelantrieb
spindle measuring system Spindelmesssystem * *[Werkz.masch.]*
spindle moulder Fräsmaschine * *[Holz]* || Tischfräse * *[Holz]* || Unterfräsmaschine * *[Holz]*
spindle orientation Spindelorientierung || Spindelpositionierung * *Ausrichtung*
spindle positioning Spindelpositionierung
spindle speed Spindeldrehzahl * *[Werkz.masch.] Textil]*
spindle-actuated potentiometer Spindelpotentiometer
spindle-nut gear Spindel-Mutter-Getriebe
spindly dünn * *mager*
spinneret Spinndüse
spinning duct Spinnschacht * *[Textil] bei der Chemiefaserherstellung*
spinning extruder Spinnextruder * *[Textil] zur Chemiefaserherstellung*
spinning fiber Spinnfaden * *Spinnfaser*
spinning frame motor Spinnmotor
spinning machine Drückmaschine * *"Drück-Drehbank" zur Blechformung (z.B. zur Erzeugung von Hohlgefäßen)* || Spinnereimaschine || Spinnmaschine * *Textilmaschine*
spinning nozzle Spinndüse

spinning pump Spinnpumpe
spinning section Spinnstelle * *{Textil}* *in Chemiefasermaschine*
spinning speed Spinngeschwindigkeit * *{Textil} bei Spinnmaschine*
spinning thread Spinnfaden
spinning unit Spinnstelle * *z.Chemiefaserherstellg., bestehend aus Spinnpumpe, Kühlschacht, Verstreckwerk, Wickler*
spinning/drawing/winding machine Spinn-Streck-Spulmaschine * *zur Erzeugung von Chemiefasern*
spiral Drall || schneckenförmig || schrauben * *up/down: hoch/herunter~ (Preise usw.), sich spiralförmig nach oben/unten bewegen* || schraubenförmig || Spirale || spiralförmig
spiral conveyor Förderschnecke || Schneckenförderer
spiral gear Schneckengetriebe || Schneckenradgetriebe
spiral grooved spiralgerillt
spiral grooves Spiralrillen * *z.B. an Kupplungs-/Bremslamelle*
spiral spring Schraubenfeder || Spiralfeder
spiral toothed schrägverzahnt
spirit Spiritus
spirit level Wasserwaage
spirit of enterprise Unternehmergeist
spirit of moving forward Aufbruchstimmung * *'zu neuen Ufern'*
spiritual geistig * *unkörperlich*
splash spritzen * *be~ (mit Wasser usw.)* || Spritzer
splash lubrication Tauchschmierung
splash-proof schwallwassergeschützt || Spritzwasser-geschützt || spritzwassergeschützt
splashing water Spritzwasser
splatter spritzen * *(be-/umher-) spritzen; beschmutzen, sprenkeln; planschen*
splendid ausgezeichnet * *großartig* || brillant
splice Klebestelle * *bei Abrollung/Abwickler* || Rollenwechsel * *Schneidevorgang bei (Nonstop-)Wickler* || Spleiß * *Rollenwechsel Klebe/Schneidvorgang bei fliegender Auf/Abwicklung* || Spleißstelle
splice box Muffe * *Kabelverbindungs~*
splice control Rollenwechselsteuerung * *z.B. in Papierbearbeitungmaschine*
splice tracking Klebestellenverfolgung * *z.B. in Papierbearbeitungs-/Druckmaschine*
splicer Rollenwechsler * *Auf-/Abroller mit fliegendem Rollenwechsel* || Wickler mit fliegendem Rollenwechsel
spline Feder * *Passfeder für Keilnut* || Federkeil * *Passfeder* || Keilnut * *~ einer Keilverzahnung* || Keilwelle * *Keilwelle, Nutwelle, Welle mit Keilverzahnung* || Kurve * *Kurvenlinie in d. Computergrafik* || Kurvenlineal || Kurvenlinie * *in der Computergrafik* || Nocken * *Keil, (Pass-) Feder; längliches, dünnes Stück* || Nutkeil * *Passfeder* || Passfeder * *Nutkeil* || Spline * *Kurvendarstellung im CAD- und Numerikbereich durch Polynom 3. Ordnung*
spline shaft Keilwelle || Vielkeilwelle
splined shaft Keilwelle * *mit →Federkeil versehene Welle*
splinter Span * *Splitter* || Splitter || splittern
split aufteilen * *(auf)spalten* || geteilt * *aufgespalten, (in zwei Hälften) geteilt (z.B. Gehäuse)* || reißen * *zer~, zer-/aufspalten, zerteilen, schlitzen; sich aufspalten/trennen/teilen* || spalten || splittern || teilen * *aufspalten* || unterteilen * *aufspalten, sich aufspalten* || zweiteilig * *in (zwei) Teile aufgespalten*
split brush Zwillingsbürste * *Kohlebürste für Motor/Generator*

split open aufreißen
split pin Splint * *geschlitzter ~* || Splint * *Spreizsplint*
split plug Bananenstecker * Büschelstecker
split taper sleeve Spannhülse * *konische Hülse mit Schlitzen*
split up aufgliedern || aufspalten || zerlegen * *(figürlich, mathemat.) in Bestandteile aufspalten*
split-cage motor Spaltrohrmotor * *Nassläufer-Pumpenmotor mit Schutzrohr im Luftspalt zwischen Läufer u. Stator*
split-pin hole Splintloch
split-pole motor Spaltpolmotor
split-second blitzschnell * *(Adj.) in Bruchteilen einer Sekunde*
split-up aufteilen * *aufspalten; be splitted up in(-to): sich aufteilen in*
splitting Plättchen
splitting saw Spaltsäge * *{Holz}*
splitting up Aufteilung * *into subassemblies:in Unterbaugruppen*
spoil verschandeln
spoilage Abfall
spoiled schlecht * *verdorben, ruiniert, kaputtgegangen, vernichtet*
spoke Speiche * *z.B. eines Rades*
spokesman Sprecher * *offizieller ~ einer Firma, Behörde usw.*
spontaneous plötzlich * *spontan*
spontaneous message Spontanmeldung * *(→PROFIBUS-Profil)*
spontaneously frei * *(Adv.) von selbst, unwillkürlich, selbsttätig*
spool Spule * *Garn-/Spinn-/Draht-/Faden~, Rolle, Haspel* || Tambour
spooler Spuler * *~ bei Drahtziehmaschine*
sporadic sporadisch || stoßartig * *in Stößen/Schüben/unregelmäßig auftretend*
sporadically sporadisch * *(Adv.)* || stoßartig * *(Adv.)* || stoßweise * *sporadisch auftretend*
spot Ort * *Fleck* || Platz || Punkt * *Fleck, Punkt, Platz* || Stelle * *Punkt, Fleck; hot spot: heiße ~; thick spot: Verdickung*
spot checking Stichprobenprüfung
spot welding Punktschweißen
spotted fleckig * *mit Schmutz, Rostflecken, Pusteln, Pickeln befleckt/getüpfelt/gesprenkelt*
spout Rohr * *Einfüll-, Schütt~* || Röhre * *Einfüll-/Schüttröhre, (Gieß-) Tülle, Abfluss-/Speirohr* || spritzen * *heraus~* || Tülle * *Gießröhre*
sprag Klemmkörper * *keilförmiger Klemmkörper, z.B. bei Rücklaufsperre, Freilauf* || Sperrklinke * *Spreiz-/Sperrklinke, Bremsklotz*
sprag clutch Rücklaufsperre * *Sperr-/Spreizzahnkupplung, verhindert Rücklaufen des Antriebs (ähnl. Freilauf)*
sprag freewheel Freilauf * *Klemmkeil-/Klemmkörper-Freilauf* || Klemmkörperfreilauf * *mit Klemmkeilen*
spray sprengen * *z.B. Wasser ~, mit dem Schlauch ~* || spritzen * *(ver-)sprühen, zerstäuben, besprühen, be~, spritzlackieren, sprengen* || sprühen
spray gun Spritzpistole
spray nozzle Sprühdüse
spray oil lubrication Sprühölschmierung
spray pipe Düsenrohr
spray tin Spraydose
spray water Sprühwasser
spraying booth Lackierkabine || Spritzstand * *Lackierkabine*
spraying robot Lackierroboter || Spritzroboter * *zum Lackieren*
spraying water Spritzwasser || Sprühwasser
spread auftragen * *verstreichen (z.B. Klebstoff)* || ausbreiten || spreizen || streuen * *statistisch~* ||

Streuung * *statistische ~* || verstreichen * *Farbe/ Salbe/Auftrag ~* || verteilen * *Farbe, Auftrag; (auch figürlich) over: über e. Zeitraum; evenly: gleichmäßig*
spread out   ausbreiten
spread-sheet   Tabellenblatt
spread-sheet program   Tabellenkalkulationsprogramm
spreaded   verbreitet
spreader   Spreize * *Spreize, Abstandshülse, "Auseinanderhalter", Zerstäuber, Spritzdüse*
spreader roll   Breitstreckwalze * *mit gebogener Achse zur Vermeidg.v. Falten bei Papiermaschine*
spreading   Ausbreitung || Häufung * *Verbreitung*
spreading knife   Streichmesser
spreading roller   Auftragswalze
spreadsheet   Tabellen-Arbeitsblatt * *[Computer] in einer Tabellenkalkulations-Software*
spring   Brunnen * *Quelle* || Feder * *z.B. Spiralfeder*
spring back   zurückfedern
spring band   Federband
spring bellows   Federbalg
spring bolt   Federbolzen
spring characteristic   Federkennlinie
spring constant   Federkonstante
spring curve   Federkennlinie
spring disc   Federscheibe
spring force   Federkraft
spring from   hervorgehen aus * *stammen von*
spring loaded brake with electromagnetic release   elektromagnetisch gelüftete Federkraftbremse
spring lock washer   Federring * *Feder-(Unterleg)-Scheibe* || Federscheibe * *Unterlegscheibe zur Schraubensicherung*
spring pressure   Federdruck
spring rate   Federkonstante
spring return pressure   Federrückdruck
spring shim   Federscheibe
spring suspension   Federung * *an einem Fahrzeug*
spring tension   Federkraft
spring up   entstehen
spring washer   Federring * *→Federscheibe, Unterlegscheibe mit axialer Federwirkung* || Federscheibe * *Unterlegscheibe mit axialer Federwirkung*
spring-actuated brake   Federdruckbremse
spring-applied   federbelastet * *z.B. Bremse/Kupplung*
spring-applied brake   Federdruckbremse
spring-coiling machine   Federwickelmaschine
spring-energized seal   Simmerring * *Gummidichtring mit innenliegender Spiralfeder, dichtet Wellenaustritt*
spring-loaded bearing   angestelltes Lager * *axial federnd vorgespanntes Kugellager*
spring-loaded brake   Federdruckbremse || Federkraftbremse * *zum Öffnen muss Strom fließen (Ruhestromprinzip)*
spring-loaded coupling   Federkupplung
spring-loaded sealing ring   Simmerring * *Gummiring m.innenliegender Spiralfeder, dichtet durch Anpressen an Welle*
spring-operated brake   Federdruckbremse
spring-pressure clutch   Federdruckkupplung
springiness   Elastizität
springing   Federung
springs   Federung * *(Plural)*
springy   federnd
sprinkle   sprengen * *z.B. Wasser ~* || spritzen * *besprengen*
sprinkling system   Beregnungsanlage
sprocket   Kettenrad
sprocket wheel   Kettenrad
SPS   Einbauplatz * *Abk. f. 'Standard Packaging Slot': →Standard~ (SEP, EP)* || EP * *Abk. für 'Standard Packaging Slot': Standard-Einbauplatz*

|| SEP * *Abk. f. 'Standard Packaging Slot': Standard-Einbauplatz* || Standardeinbauplatz * *Abk. für 'Standard Packaging Slot'*
spun   ersponnen * *{Textil} z.b. Chemiefasern*
spur   Stichleitung
spur gear   Stirnrad || Stirnradgetriebe
spur gearing   Stirnradgetriebe
spur line   Stichleitung * *z.b. im Netz*
spur rack   Zahnstange
spurious   falsch * *falsch, unecht, Fehl- (z.b. Messung)*
spurious measurement   Fehlmessung
spurious pulse   Wischer * *Wischimpuls* || Wischimpuls
spurious signal   Fehlsignal * *'unechtes' (Mess)Signal*
spurious tripping   Fehlauslösung * *z.B. eines Überwachungsgeräts*
spurt   spritzen * *(heraus)spritzen (Wasser, Flüssigkeit)*
sq   Quadrat * *Abkürzung für 'Square':Quadrat (sqmm: ~millimeter, mm^2 usw.)*
sqeeze   zusammendrücken * *(zusammen)drücken, auspressen/-quetschen, schröpfen*
squad leader   Gruppenführer * *Truppführer*
square   abgleichen * *Konten ~* || besäumen * *(Holz)-Platten u.* || Kästchen * *auf Rechenpapier usw.* || Potenz * *zweite Potenz* || Quadrat * *increase/vary with the square of: mit dem ~ der ... anwachsen/ändern* || quadratisch * *Fläche u. mathematisch* || quadrieren || Rechteck- * *quadratisch* || übereinstimmen * *übereinstimmen, passen* || viereckig || Vierkant
square bracket   Klammer * *eckige ~*
square brackets   eckige Klammern || rechteckige Klammern
square root   Quadratwurzel || zweite Wurzel
square root function   Radizierer
square speed/torque characteristic   quadratische Drehzahl-Drehmoment-Kennlinie
square speed/torque relationship   quadratische Drehzahl-Drehmoment-Kennlinie
square timber   Kantholz
square torque/speed characteristic   quadratische Drehmoment-Drehzahl-Kennlinie * *z.B. Lastkennlinie e.Ström.maschine (Pumpe, Lüfter usw.)*
square torque/speed relationship   quadratische Drehmoment-Drehzahl-Kennlinie
square type of construction   Viereckbauweise * *z.B. bei Elektromotor*
square-frame   Viereckbauweise * *z.B. bei Elektromotor*
square-frame design   Viereckbauweise * *z.B. eines Elekromotors*
square-head bolt   Vierkantschraube
square-law   quadratisch * *~e Gesetzmäßigkeit, Kennlinie (z.B. →Lastkennlinie bei Lüfter-/Pumpenantrieb)*
square-law characteristic   Lüfterkennlinie * *quadratische Kennlinie $M\sim n^2$* || quadratische Kennlinie
square-law counter-torque   quadratisches Gegenmoment * *z.B. beim Antrieb einer Strömungsmaschine (Pumpe, Lüfter)* || quadratisches Lastmoment * *Gegenmoment*
square-law load torque   quadratischer Momentenverlauf * *quadratische Lastkennlinie, z.B. bei Lüfterantrieb* || quadratisches Gegenmoment * *Lastmoment* || quadratisches Lastmoment * *z.B. bei Strömungsmaschine (Pumpen, Lüfter)*
square-law load torque characteristic   quadratische Drehmomentenkennlinie * *z.B. bei Strömungsmaschinen (Pumpen/Lüfter)* || quadratische Momentenkennlinie * *z.B. Lastkennlinie von Strömungsmasch. (Pumpen, Lüftern ...)*

square-root extractor Radizierer
square-wave blockförmig * *rechteckförmig, z.B. Impulse, Spannung, Strom, Modulation* || Rechteck- * *Impulse, Modulation usw.*
square-wave current Blockstrom * *rechteckförmiger Strom ;z.B. z. Ansteuerg. bürstenloser Servomotoren*
square-wave generator Rechteckgenerator
square-wave signal Rechtecksignal * *normalerweise mit Tastverhältnis 1:1*
squared hoch 2 * *(vorangestellt) sqared speed:- Drehzahl hoch 2* || quadriert
squared characteristic quadratische Kennlinie
squared timber Kantholz
squareness Rechtwinkligkeit
squarewave pulses Rechteckimpulse * *mit Tastverhältnis 1:1*
squaring quadrieren * *(Substantiv)*
squaring element Quadrierer
squaring-up Eckverbindung * *{Holz}*
squash quetschen * *(zu Brei) zerquetschen, zus.-drücken, flach-/breitschlagen, zerdrückt/-quetscht werden*
squeak quietschen
squeal pfeifen * *kreischen, grell/schrill schreien, quieken* || quietschen
squeegee Anpress- * *Gummiquetsch-* || Gummiquetschwalze || Gummiwalze * *['skwiedjie] →Gummiquetschwalze, Gummischrubber* || Quetschwalze || *['skwiedjie] (Gummi-) Quetschwalze* || Rechen
squeegee roll Gummiquetschwalze * *['skwiedjie 'roul]* || Quetschwalze
squeeze Engpass * *wirtschaftlicher, finanzieller ~, Klemme* || klemmen * *drücken, quetschen, zwängen* || pressen * *(zusammen-) drücken, (her-)aus~, (her-)ausquetschen (auch figürl.)* || quetschen * *(zus.-) drücken, (aus-) pressen, (aus-/heraus-) quetschen, zwängen (into: in), erpressen*
squeeze together zusammendrücken || zusammenpressen
squeezing Quetsch-
squezzing Anpress- * *Abquetsch-, Press-*
squirrel cage rotor Käfigläufer
squirrel-cage induction motor Asynchron-Käfigläufermotor || Käfigläufermotor
squirrel-cage motor Käfigläufermotor
squirrel-cage rotor Kurzschlussläufer
squirrel-cage winding Käfigwicklung * *{el. Masch.} Kurzschlusswicklung bestehend aus den Stäben des Läuferkäfigs*
squirt spritzen * *(hervor-/heraus-/be-) spritzen, hervorsprudeln; auch Substantiv: Strahl, Spritze* || Strahl * *(Wasser-) Strahl, Spritze*
SR motor SR-Motor * *Abkürz. für 'Switched Reluctance motor'*
SRAM SRAM * *Static Random Access Memory, statischer Schreib/Lesespeicher, braucht keine Auffrischzyklen*
sreening Sortierung * *~ unterschiedl. großer Partikel, z.B.d.Holzspäne durch Siebung i.d. Spanplattenindustrie*
srewed Gewinde- * *mit (Außen-) Gewinde versehen*
stability Beständigkeit || Festigkeit * *auch einer Konstruktion; Beständigkeit, Stand~* || Haltbarkeit || Kippsicherheit * *siehe auch →Kippsteifigkeit* || Konstanz * *Kurzzeitkonstanz (Einflussgrößen: Last, Frequenz, Netzschwankungen)* || Konstanz * *z.B. eines Kondensators* || Robustheit * *Stabilität* || Stabilität || Widerstandsfähigkeit
stability against alternating load Wechsellastfestigkeit * *z.B. einer Sicherung*
stability against pulsed load Wechsellastfestigkeit * *z.B. einer Sicherung*

stability criterion Stabilitätskriterium * *(Plural: 'criteria')*
stability limit Stabilitätsgrenze
stability limits Trittgrenzen * *Aussteuergrenzen für netzgeführten Stromrichter*
stability problems Stabilitätsprobleme
stability time Standzeit * *Lager, Fett usw.*
stabilization Beruhigung || Dämpfung * *Stabilisierung* || Stabilisierung
stabilize dämpfen * *Schwingungen ~* || stabilisieren
stabilized geregelt * *Stromversorgung, Versorgungsspannung* || stabilisiert * *z.B. Stromversorgung*
stabilized condition eingeschwungener Zustand * *thermal: thermisch ~*
stabilized-current source Konstantstromquelle
stabilizer Stabilisator || Stabilisierung * *stabilisierendes (Konstruktions-) Element*
stabilizing Stabilisierung
stabilizing post regulator Längsregler * *z.B. Spannungsregler mit Längstransistor b.stabilisierter Stromversorgg.*
stabilizing series winding Hilfs-Reihenschlusswicklung * *kompensiert Lastabhängigk. d.Drehz. e. DC-Motors bei konst. Ankerspanng.* || Kompoundwicklung * *kompensiert Lastabhängigk. d. Drehzahl e.DC-Motors bei konst. Ankerspanng.*
stable beständig * *dauerhaft, stabil (auch Börse, Nachfrage)* || dauerhaft * *fest* || fest * *dauerhaft, stabil, auch (Geschäfts)Beziehung* || kippsicher || konstant * *unveränderlich, stabil, dauerhaft, konstant* || stabil * *auch für Regelung* || unveränderlich * *stabil*
stable at no load leerlauffest * *Antrieb*
stable operation stabiler Betrieb
stable region stabiler Bereich
stable temperature Endtemperatur
stack -Paket * *aus gleichartigen Teilen, (auch Software-) Paket* || -Satz * *Stapel* || aufstapeln || Kellerspeicher * *in Rechner (Stapelspeicher, Stack)* || lagern * *auf Stapel ~* || Regal || Schichtsystem * *System der 7...8 ISO-Schichten bei Feldbus* || Schornstein || setzen * *Stapel* || Stack * *Stapelspeicher (zum Ablegen von Rücksprungadressen und Registerinhalten); LIFO-Speicher* || Stapel || stapeln || Stapelspeicher
stack conveyor Stapelband * *Transportsystem*
stack pointer Stapelzeiger * *Adresse des aktuellen Elements eines Stapelspeichers (stack)*
stacked übereinander * *~gestapelt*
stacked-film capacitor Schichtkondensator * *Kunststoff-~, kein Keramik-Viel~*
stacker Ableger * *Einrichtung zum Ablegen von Produkten (z.B. von bedruckten Papierbögen) auf einen Stapel* || Bogenableger * *Stapler; z.B. für Papierbögen* || Stapelanlage || Stapler * *z.B. Einrichtung zum Ablegen von Papierbögen auf Stapel*
stacker crane Regalbediengerät * *für Hochregallager* || Regalförderer || Regalförderzeug * *Bediengerät für Hochregallager* || RGB * *Abk. f. 'ReGalBediengerät'*
stacker truck Regalbediengerät * *Förderzeug für Hochregallager* || Regalförderzeug * *für Hochregallager*
stacking factor Eisenfüllfaktor * *{el.Masch.} →Stapelfaktor* || Stapelfaktor * *{el.Masch.} Verhältnis d.reinen Eisenlänge e. Blechpakets zu dessen Gesamtlänge*
stacking system Stapelanlage
staff ausstatten * *mit Personal* || Belegschaft || Mitarbeiter * *(Plural)* || Mitarbeiterschaft || Personal || Stab * *Mitarbeiterstab*
staff manager Personalchef
staff reduction Personalabbau
staffed ausgestattet * *mit Personal*

stage Abschnitt * *Stufe (eines Prozesses usw.)* || Ausbaustufe || Bühne * *im Theater* || Phase * *Stufe* || Stadium * *early: frühes (siehe auch →Stufe)* || Strecke * *Teil~* || Stufe * *(elektronische) Schalt~, Entwicklungsstadium, Ausbau~* || Zeit * *Stadium*
stage of development Entwicklungsstufe
stagger schwanken * *wanken, schwanken, taumeln, torkeln* || staffeln * *versetzt anordnen; z.B. Kohlebürsten, Arbeitszeit* || taumeln * *Schwanken, wanken, ~, torkeln, unsicher werden (figürl.), ins Wanken bringen, verblüffen* || versetzen * *versetzt anordnen, staffeln, auf Lücke setzen* || versetzt anordnen
staggered gegeneinander versetzt || gestaffelt * *gegeneinander versetzt angeordnet (auch zeitlich)* || versetzt * *~ angeordnet, staffeln, auf Lücke setzen*
stagnancy Stagnation
stagnant stagnierend * *z.B. Geschäft, Markt*
stagnate stagnieren * *z.B. Geschäft, Markt*
stagnated stagnierend * *z.B. Geschäft, Markt*
stagnation Stagnation || Stillstand * *~ von Geschäften, der Entwicklung usw.*
stain Beize * *{Holz} Holz~, Färbemittel* || beizen * *{Holz} Textil} Holz beizen, Glas bemalen, Stoff färben; Flecken verursachen, beflecken* || färben * *Glas, Papier usw.*
stained fleckig
staining plant Beizanlage * *{Holz}*
stainless nicht rostend * *Stahl* || rostfrei * *Stahl*
stainless steel Edelstahl || nichtrostender Stahl
stair-step treppenförmig
staircase function Treppenfunktion
staircase step Treppenstufe
stake nieten
stale air Abgas * *verbrauchte/abgasgeschwängerte Luft* || Abluft * *Abgase, verbrauchte Luft*
stale air extraction Entlüftung * *Abfuhr der Abgase, z.B. bei Tunnel*
stale air extraction for road tunnels Tunnellüftung * *Absaugung der Abluft bei Straßentunnels*
stall abschalten * *stehenbleiben* || abstellen * *Motor* || abwürgen * *Motor* || anhalten * *abwürgen, blockieren, auslaufen lassen* || aussetzen * *Motor* || blockieren * *aussetzen (Motor), sich festfahren, stecken bleiben* || Kippen * *z.B. ~ eines Asynchronmotors* || stehen bleiben * *(Motor) kippen (AC-Motor), absterben, stecken bleiben, abgewürgt werden* || Stillstand
stall limit Kippgrenze * *z.B. bei Asynchronmotor*
stall prevention Kippschutz * *"Kipp-Verhinderung" z.B. bei Asynchronmotor*
stall protection Blockierschutz || Kippschutz * *z.B. Asynchronmotor* || Kippsicherheit * *Kippschutz (bei Asynchronmotor usw.)*
stall stability Kippsicherheit * *Sicherheit gegen Kippen (bei Asynchronmotor usw.)*
stall tension Stillstandszug * *meist kleiner als Betriebszug, z.B. um b.DC-Motor den Kommutator zu schonen*
stall time Blockierzeit
stall torque Kippmoment
stall-protected kippsicher * *(elektr.) mit Kippschutz versehen, z.B. Motor mit Umrichter*
stall-resistant kippsicher
stalled blockiert * *abgewürgt, festgefahren, steckengeblieben*
stalled-motor protection Blockierschutz * *für Motor*
stalling torque Kippmoment * *Asynchronmotor (größtes stationäres Drehmoment unter Bemessungsbedingungen)*
stamp aufprägen || einprägen || prägen * *auf~, stanzen, pressen, stempeln* || stanzen || Stempel || stempeln * *z.B. auf Typenschild ~*

stamping Stanzteil
stamping machine Stanze || Stanzmaschine
stamping tool Stanzwerkzeug
stand bestehen * *sich behaupten* || Bock || Gerüst * *z.B. Walz~* || halten * *aus~* || Messestand || Stand * *Messe~, Gestell, ~ort, Platz; Ausrüstung* || standhalten * *aushalten; a test: einer Prüfung ~* || Stativ * *Ständer* || stehen * *(auch figürl.: stand for: ~ für)* || Stelle * *Standort* || verkraften * *ertragen, aushalten*
stand alone eigenständig
stand by in Bereitschaft stehen
stand duty Standdienst * *auf einer Messe/Ausstellung*
stand firm halten * *stand~*
stand in need of benötigen
stand of rolls Walzgerüst * *(Walzwerk)*
stand off from Abstand nehmen von * *Abstand halten/sich entfernt halten von*
stand out herausragen * *(figürl.) from: aus* || hervorstehen
stand out for sich auszeichnen durch
stand out from hinausragen über * *(figürl.)*
stand proud herausragen * *(figürl.) of: über*
stand still stehen * *still~* || stillstehen
stand up for eintreten * *für jdn. einstehen*
stand-alone autark * *allein arbeitend* || Einzel- * *→autark*
stand-by Not- * *Reserve* || Reserve * *auch als Vorsilbe; z.B. Umrichter, Maschine* || Reserve- * *z.B. Umrichter, Maschine*
stand-by generating set Notstromaggregat
stand-by mode Bereitschaft * *Betriebsart*
stand-by power supply Notstromversorgung
stand-off Abstandhalter || Abstandshalter || Distanzhalter
stand-up bestehen * *widerstehen, z.B.einer aggressiven Substanz*
standard einheitlich || gewöhnlich * *standardmäßig* || listenmäßig || Maß * *Eich-* || Muster * *Richtschnur* || Niveau * *(figürl.) not up to standard: unter dem ~* || Norm || Norm- || normal * *Abmessungen usw.* || Normvorschrift || Regel * *Norm* || serienmäßig * *(Adj.)* || standard || Standard- || standardmäßig * *(Adj.)* || Stufe * *Niveau* || üblich * *normal* || Vorschrift * *Norm* || Vorzugs- * *Standard*
standard asynchronous motor Norm-Asynchronmotor
standard atmospheric conditions Normalklima * *nach IEC*
standard block Standardbaustein * *Funktionsbaustein*
standard catalog listenmäßig
standard commercially available part Normteil * *im freien Handel erhältliches Teil*
standard component Messnormal || Standardkomponente
standard configured package Standardprojektierung * *(z.B. mit SIEMENS-Regelsystem SIMADYN D (R) erstellt)*
standard configured system Standardprojektierung * *z.B. bei freiprojektierbarem Regelsystem SIMADYN D (R) von SIEMENS*
standard configuring Standardprojektierung
standard design Normalausführung * *z.B. Motor* || Serienausführung || Standardausführung * *Standardbauart* || Standardmodell || Standardtyp * *Standardbauart*
standard deviation mittlere Abweichung || Standardabweichung * *(Wahrscheinlichkeitsrechnung)* || Streuung * *(statistische) Standardabweichung*
standard dimensions Standardmaße
standard excitation normalerregt * *Motor*
standard function package Standardprojektierung

standard induction motor   744

* *(z.B. erstellt mit SIEMENS-Regelsystem SIMA-DYN D (R))*
standard induction motor   Norm-Asynchronmotor
standard motor   Normmotor * *meist Käfigläufer-Asynchronmotor i.genormt. Baugröße nach IEC72, DIN42673 und 42677*
standard operator panel   Standardbedienfeld
standard option   Standarderweiterung
standard part   Normteil
standard peripherals   Standardperipherie * *bei SPS*
standard plug-in station   Standard-Einbauplatz * *(bei SPS) Rastermaß für Steckbaugruppen-Breite* || Standardeinbauplatz * *SEP bei SPS (Maß für die Breite einer Steckbaugruppe)*
standard product   Serienprodukt
standard range   Normreihe
standard series   Normreihe
standard sheet   Normblatt
standard size   Normgröße
standard solution   Standardlösung
standard specification   Normvorschrift
standard type   Grundtyp * *Normalausführung z.B. eines Geräts* || Kerntyp || Serienausführung || Standardausführung * *Standardmodell* || Standardmodell || Standardtyp
standard type of construction   Standardtyp * *standardmäßige Bauform*
standard types   Vorzugstypen
standard version   Standardausführung * *Normalversion*
standard voltage   Normspannung
standard voltages   Normspannungen * *Netzspannungen nach DIN IEC 38, z.B. 230V, 400V, 690V, 1000V usw.*
standard-design   serienmäßig * *in der serienmäßigen Ausführung*
standardised   standardisiert
standardization   Normung || Standardisierung || Vereinheitlichung
standardize   normen || normieren * *normen* || schematisieren || standardisieren || vereinheitlichen
standardized   einheitlich * *genormt* || genormt || standardisiert
standardized element   Normteil
standby   Bereitschaft || Ersatz- * *in Reserve/Bereitschaft stehend* || Hilfs- * *Reserve-* || Reserve-
standby converter   Reserveumrichter * *siehe auch →Synchronisierzusatz*
standby supply   Pufferung * *bei Netzspannungsausfall*
standing   Dauer- || Stellung * *Ansehen*
standing vibration   stehende Schwingung * *(mechan.)*
standing wave   stehende Welle
standoff   Abstandshalter || Stehbolzen * *als Abstandshalter*
standstill   Drehzahl Null * *Stillstand* || Stillstand * *at: im; come to a standstill: zum ~ kommen* || Stillstands-
standstill brake   Haltebremse * *→Stillstandsbremse* || Stillstandsbremse
standstill excitation   Stillstandserregung * *z.B. reduz. Feldstrom b. Gleichstrommotor, um Überhitzung zu vermeiden* || Stillstandsfeld
standstill field   Stillstandserregung || Stillstandsfeld * *z.B. bei DC-Motor: Absenkung d.Feldstroms i.Stillstand, um Überhitzg. zu vermeiden*
standstill heating   Stillstandsheizung * *z.B. bei Getriebe für möglichst geringe Reibung bei Neuanlauf*
standstill logic   Stillstandslogik
standstill monitoring   Stillstandsüberwachung
standstill voltage   Stillstandsspannung
staple   Faser * *{Textil}* Woll~, Rohwolle, Faser; Fadenlänge-/Qualität; *of short staple: kurz~ig/-staplig* || Heftklammer || Klammer * *Heft~* || Stapel * *(auch Fadenlänge/-qualität)*
staple fiber   Zellwolle
staple gun   Tacker
staple length   Faserlänge * *{Textil} z.B. in der Spinnrei; of short staple: kurzstaplig/mit kurzer Faden-/~*
stapler   Heftmaschine * *auch im Büro* || Klammergerät * *für Papier, Holz usw.; Tacker*
star   Stern * *Schaltgruppe z.B. für Trafo/Motor ; in star connection: in ~ geschaltet*
star connection   Sternschaltung * *z.B. Motor/Trafo; star-connected: in ~*
star point   Nullpunkt * *Sternpunkt* || Sternpunkt
star point resistor   Sternpunktwiderstand
star voltage   Phasenspannung * *Sternspannung* || Sternspannung * *zw. Phase u. Mittelpunktsleiter*
star-connected   in Stern geschaltet
star-delta connection   Stern-Dreieck-Schaltung || Stern-Dreieckschaltung
star-delta starting   Stern-Dreieck-Anlauf * *Sternschaltung reduziert Anlaufstrom (und -moment) auf 1/(Wurzel 3)*
star-delta switch   Stern-Dreieckschalter * *zum Starten eines Drehstrommotors mit verringertem Anlaufstrom*
star-double star starting   Stern-Doppelstern-Anlauf * *mit Dahlander bzw. PAM-Schaltung (siehe dort)*
starch   Stärke * *{Chemie}* Stärke(pulver), Stärkemehl/-Kleister
starchy   steif * *gezwungen*
start   abgehen * *Zug usw.* || abreisen * *for: nach* || anfahren * *losfahren, starten (Maschine, Fahrzeug)* || Anfang || anlassen || Anlauf || anlaufen || Anstoß * *(auch Verb)* || anwerfen || aufnehmen * *starten, the operation: den Betrieb ~* || Beginn || beginnen || Einsatz * *das Beginnen* || hochfahren * *anfahren* || in Betrieb setzen || in Bewegung setzen || in die Wege leiten || in Gang setzen || inbetriebnehmen || inbetriebsetzen
start address   Anfangsadresse
start bit   Startbit * *kennzeichnet den Beginn eines Zeichens bei serieller Übertragung (normalerweise 0-Pegel)*
start character   Startzeichen * *kennzeichnet z.B. das Ende eines Datentelegramms (häufig STX entspr. 02hex)*
start control   Anfahrsteuerung
start converter   Anfahrumrichter
start diameter   Anfangsdurchmesser * *bei Wickler*
start doing something   ansetzen * *mit etwas den Anfang machen*
start from ground zero   bei Null anfangen
start into a running motor   aufschalten auf einen laufenden Motor
start of course   Kursbeginn * *eines Aus/Weiterbildungskurses*
start of production   Produktionsbeginn
start of series production   Serieneinsatz || Serienstart * *siehe auch →Serieneinsatz*
start of shipping   Liefereinsatz
start of volume production   Serieneinsatz * *Beginn der Serienproduktion*
start of winding   Wicklungsanfang
start preconditioning circuit   Einschaltverriegelung * *~ für Motor, Maschine*
start program   Anlaufprogramm
start running   anlaufen
start up   anwerfen || anziehen * *Motor*
start value   Anfangswert
start winding   Hilfswicklung * *Anlaufwicklung in e. Motor*
start working   in Betrieb gehen

start-stop control   Ein-Ausschaltlogik || Ein-Ausschaltsteuerung || Einschalt-Ausschalt-Logik
start-stop logic   Ein-Ausschaltlogik || Einschalt-Ausschalt-Logik
start-stop sequencing   Ein-Ausschaltlogik || Ein-Ausschaltsteuerung * *Ablaufsteuerung zum Starten u. Anhalten e. Motors* || Einschalt-Ausschalt-Logik
start-up   anfahren * *Anlage, Maschine, Kraftwerk, Reaktor* || Anfangs- || Anlauf || anlaufen || Anlaufprogramm || IBN * *Inbetriebsetzung* || IBS * *Inbetriebsetzung* || in Betrieb setzen * *inbetriebnehmen* || Inbetriebnahme || inbetriebnehmen || inbetriebsetzen * *inbetriebnehmen* || Inbetriebsetzung
start-up block   Anlaufbaustein * *bei SPS*
start-up control   Einschaltlogik || Einschaltsteuerung * *z.b. Motor* || Einschaltsteuerwerk * *z.B. für Antrieb*
start-up costs   Inbetriebnahmeaufwand
start-up engineer   Inbetriebnahme-Ingenieur || Inbetriebnahmeingenieur || Inbetriebnehmer || Inbetriebsetzer ± *Inbetriebnahmeingenieur*
start-up frequency   Anfahrfrequenz
start-up guide   Inbetriebnahmeanleitung
start-up procedure   Inbetriebnahme
start-up program   Anlaufprogramm
start-up sequence   Einschaltfolge || Einschaltreihenfolge * *z.B. Motor*
start-up steps   Inbetriebnahmeschritte
start-up time   Inbetriebnahmezeit
start-up torque   Anfahrmoment
started   aufgenommen * *Betrieb*
starter   Anlasser * *während d.Hochlaufs zur Energie/Momenten-Entlastg.i.d.Motorkreis geschaltete Widerstände*
starting   ab * *starting in November:ab November; starting immediately:ab sofort* || anfahren * *(Substantiv)* || Anfangs- || Anlauf || Anlauf- || Aufnahme * *Beginn*
starting against high-inertia load   Schweranlauf || Schwerlastanlauf
starting aid   Anlaufhilfe || Starthilfe * *z.B. für Motor*
starting alarm   Anfahrwarnung
starting behavior   Anlaufverhalten
starting behaviour   Anlaufverhalten
starting by pole-changing   Polumschaltung als Anlassverfahren * *[el.Masch.] therm. schonendes Anfahren b. polumschaltbarem Motor*
starting cage   Anlaufkäfig * *in einem Läufer eines (Synchron-) Drehstrommotors (z.B. SIEMENS SI-MOSYN-Maschine (R))*
starting capacitor   Anlaufkondensator
starting characteristic   Anlaufverhalten
starting choke   Anlassdrossel
starting circuit-breaker   Motorschutzschalter
starting clutch   Anlaufkupplung * *reduziert das Anlaufdrehmoment beim Starten eines Motors*
starting condition   Anfahrbedingung
starting conditions   Anlaufbedingungen || Startbedingungen
starting control   Anlaufsteuerung
starting converter   Anfahrumrichter || Hochfahrumrichter * *übergibt den Motor nach erfolgtem Hochlauf ans Netz oder e.* →*Betriebsumrichter*
starting coupling   Anlaufkupplung * *für AC-Motor zum schnellen Hochlauf u. thermisch. Entlastg. (reduz. Anlaufmoment)*
starting current   Anfahrstrom || Anlaufstrom || Anzugsstrom * *eines Motors (gemessener Effektivwert II d.Stroms bei festgebremstem Läufer)* || Einschaltstrom * *~ eines Motors beim Starten*
starting cycle   Anlassvorgang || *Zyklus* || Anlauf * *Anlaufvorgang*
starting drawbar pull   Anzug * *Anzugskraft bei Fahrzeug*

starting duration   Anlaufdauer * *Zeit zum Hochlaufen z.B. eines Motors*
starting duty   Anlaufart * *[el.Masch.]*
starting equipment   Anlasshilfe * *z.b. für einen Motor (Softstarter, Anlaßwiderstand usw.)*
starting frequency   Anfahrfrequenz || Anlasshäufigkeit * *[el.Masch.] zuläss. aufeinander folgende Anz. d. Anlassvorgänge mittels Anlasser* || Schalthäufigkeit * *[el.Masch.] Anzahl period. wiederkehrender Schalthandlungen (Motorstarts je h)*
starting friction   Losbrechreibung
starting friction torque   Anlaufreibmoment
starting heat   Anlaufwärme * *Verlustwärme, die beim Anlaufen eines Motors entsteht*
starting inhibit circuit   Anlaufsperre
starting interlock   Einschaltverriegelung * *~ für Motor, Maschine*
starting load   Anlassschwere * *[el.Masch.]*
starting load factor   Anlassschwere * *[el.Masch.] mittlerer Anlassstrom/Läufernennstrom*
starting lockout   Anlaufsperre
starting logic   Einschaltlogik || Einschaltsteuerung * *Einschaltlogik*
starting motor   Anwurfmotor * *[el.Masch.] An wurfs-Hilfsmotor*
starting operation   Anlassvorgang
starting peformance   Anlaufverhalten * *für Käfigläufermotoren: siehe DIN 57530, VDE 0530/IEC 34 jeweils Teil 12*
starting performance   Anlaufeigenschaften * *eines Motors* || Anlaufgüte * *Anlaufeigenschaften e. Drehstrommotors (gewünscht:mögl. niedriger Anlaufstrom)*
starting period   Anlaufphase
starting point   Ausgangspunkt
starting point of non-continuous current   Lückgrenze
starting position   Ausgangspunkt || Ausgangsstellung
starting power   Anzug * *Anzugskraft, Anziehvermögen bei Fahrzeug/Motor*
starting process   Anlassvorgang
starting properties   Anlaufeigenschaften * *eines Motors*
starting quality   Anlaufgüte * *Anlaufeigenschaften e. Drehstrommotors*
starting reactor   Anlassdrossel * *[el.Masch.] 3phasige ~ mit Eisenkern, wird beim Anlassen vor d. Wicklg. geschaltet.* || Vorschaltdrossel * *als Anlaufhilfe für Motorstart*
starting resistor   Anfahrwiderstand || Anlasswiderstand
starting sequence   Anlassvorgang || Einschaltfolge || Einschaltreihenfolge * *z.B. Motor*
starting sequence control   Anfahrsteuerung || Einschaltsteuerwerk * *z.B. für Motor*
starting sequencing control   Einschaltlogik || Einschaltsteuerung
starting time   Anlasszeit * *[el.Masch.]* || Anlaufzeit * *benötigt ein Motor zum Anfahren d. Nenndrehz. aus d. Stillstand heraus*
starting torque   Anfahrdrehmoment || Anfahrmoment || Anlaufdrehmoment || Anlaufmoment * *z.B.Drehmoment e.Asynchronmotors bei Drehzahl Null* || Anzugsmoment * *Anlaufmoment eines Motors (kleinstes gemessen. Drehmomente e.festgebremsten Motors)* || Einschaltmoment * *Drehmoment*
starting tough loads   Schweranlauf || Schwerlastanlauf
starting tractive effort   Anzug * *Anzugskraft b. Fahrzeug*
starting tractive torque   Anzugsdrehmoment * *Bahn* || Anzugsmoment * *Bahn*

starting transformer Anlasstransformator * *für Sanftanlauf eines Motors*
starting unit Anlaufgerät
starting value Startwert
starting voltage Anlaufspannung * *wird für Motor häufig erhöht z. Überwindg. d.Losbrechmoments*
starting winding Anlaufwicklung || Hilfswicklung * *Anlaßwicklung in e. Motor*
starting-duty rating Anlassschwere * *{el.Masch.} Leistungsaufnahme beim Anlassen/Leist.aufn. im Bemessungspunkt*
starting-up Inbetriebsetzung
starting-up current Anfahrstrom
starting-up sequence Anlaufvorgang
starting-up temperature rise Übertemperatur bei Anlauf * *{el.Masch.}*
startled betroffen * *bestürzt, erschrocken*
startup Anlauf || Inbetriebnahme
state angeben * *angeben, darlegen, dar-/feststellen; z.B. Daten, (Sonder)Wünsche (i.Bestellformular)* || aufführen * *in einer Liste* || aufstellen * *Problem, Regel* ~ || behaupten * *erklären* || erklären * *(aus-)sagen, vorbringen/-tragen, (Einzelheit usw.)* angeben, behaupten, erwähnen, bemerken || feststellen * *konstatieren, erklären, behaupten, erwähnen, bemerken, anführen, darlegen, aussagen* || Stand * Zus~ || Zustand * *z.B. Betriebs~; steady: stationärer; solid: fester*
state company Staatsunternehmen
state control Zustandsregelung
state controller Zustandsregler
state estimation Istwertschätzung || Zustandsschätzung
state graph Zustandsfolgediagramm
state information Zustandsmeldung
state machine Ablaufsteuerung * *elektronisches oder softwaremäßig realisiertes Zustandssteuerwerk* || Steuerwerk * *elektronisches Zustandsfolge-Steuerwerk* || Zustandssteuerwerk * *{Computer}*
state observer Zustandsbeobachter
state of affairs Umstand * *Lage der Dinge*
state of development Entwicklungsstand
state of equilibrium Gleichgewichtszustand
state of flux Strukturwandel
state of the art Stand der Technik * *the present:der gegenwärtige*
state signal Zustandsmeldung
state space Zustandsraum
state transition diagram Zustandsfolgediagramm
state variable Zustandsgröße
state-of the art fortschrittlich * *modern, dem neuesten techn. Stand entsprechend*
state-of-the-art modern * *d.neuest. Stand d. Technik entsprechend, auf d.neuest. Entwicklungsstand befindl.* || neuzeitlich * *dem neuesten technischen Stand entsprechend* || zeitgemäß
state-sequence diagram Zustandsfolgediagramm
stated angegeben * *z.B. in einer Beschreibung* || aufgeführt * *in einer Liste*
statement Abrechnung * *Kosten-* || Angaben || Anweisung * *Computerbefehl, besonders für Hochsprache* || Aussage || Befehl * *{Computer} Anweisung in Hochsprache* || Erklärung * *Aussage, Feststellung (auch offizielle)* || Feststellung * *Aussage*
statement execution time Operationszeit * *Ausführungszeit eines (SPS-) Befehls*
statement list Anweisungsliste * *{SPS} Programmiersprache für Steuerungen mit mnemotechn. Befehlen nach DIN 19239* || AWL * *Anweisungsliste als Programmiersprache für SPS*
statement of contents Inhaltsübersicht
statement-list notation Listenform * *{SPS|Computer} Anweisungsliste als Quellsprachenotation*
static statisch
static accuracy statische Genauigkeit

static bypass switch Netzrückschalteinrichtung * *überbrückt e. unterbrech.freie Stromversorgg. (→USV) nach Netzwiederkehr*
static clutch torque übertragbares Drehmoment * *statisch* ~ *einer Kupplung (gilt nicht für Schaltvorgänge)*
static converter Stromrichter * *(veraltet) nicht rotierender Stromrichter (i. Gegens. z.B. zum Leonard-Umformer)* || Umformer * *nicht rotierender (i. Gegens. zum Leonhard-Umformer)*
static discharge Entladung statischer Elektrizität
static electricity statische Elektrizität
static error statischer Fehler * *Fehler im eingeschwungenen Zustand*
static feedback Statik * *nachgebende Reglerrückführung; siehe auch →Kennlinienneigung*
static friction Haftreibung || Losbrechreibung
static friction torque Losbrechmoment
static losses statische Verluste
static torque Stillstandsmoment || übertragbares Drehmoment * *z.B. über eine Kupplung (ist kleiner als das →schaltbare Drehmoment)*
static-torque motor Stillstandsmotor
statical statisch
statics Statik * *{Bautechnik}*
station -Platz * *z.B. Bedienplatz* || -Werk * *z.B. Pump~* || Anlage * *(Funk/Forschungs)Station, Kraftwerk* || Busteilnehmer || Haltestelle || Platz * ~ *zum Bedienen, Reparieren usw.* || Position * *geographische* ~ || Station || Teilnehmer * *z.B. Gerät an einem seriellen Bus*
station address Teilnehmeradresse * *z.B. bei Bussystem*
station number Teilnehmernummer * *bei Bussystem*
station time Taktzeit * *{Verkehr} bei Zug, Bus usw.*
stationary fest * *orts-~, stationär* || feststehend * *ortsfest* || nicht umlaufend || ortsfest || Papier * ~-*waren* || stationär * *auch ortsfest* || stillstehend
stationary application stationärer Einsatz
stationary core Statorpaket
stationary mounted fest eingebaut
stationary operation stationärer Betrieb
stationary precision statische Genauigkeit
stationary wave stehende Welle
stationary-mounted fest montiert * *fest eingebaut*
stationery printing machine Formulardruckmaschine * *(stationery: Schreib-/Papierwaren, Büro-/Schreibmaterial, Briefpapier)*
statistical statistisch
statistical analysis statistische Auswertung
statistical evaluation statistische Auswertung || statistische Berechnung
statistically statistisch * *(Adv.)*
stator Ständer * *Stator e. elektr. Maschine* || Stator
stator braking heat Bremswärme im Ständer
stator core Ständerblechpaket || Ständerpaket
stator current Ständerstrom
stator frame Ständergehäuse * *{el.Masch.}* || Statorgehäuse * *eines Motors*
stator frequency Ständerfrequenz
stator housing Ständergehäuse * *{el.Masch.}*
stator lamination Statorpaket
stator laminations Ständerbleche
stator resistance Ständerwiderstand
stator starting heat Anlaufwärme im Ständer * *{el.Masch.}*
stator voltage Ständerspannung
stator winding Ständerwicklung * *{el.Masch.} Wicklung im feststehenden Teil einer el.Masch.*
stator yoke Ständerjoch
stator-critical ständerkritisch * *{el.Masch.}*
stator-critical motor ständerkritischer Motor * *Ständer thermisch kritischer als der Läufer*

stator-fed ständergespeist
stator-resistance starting circuit Kusa-Schaltung * *Abk.f. 'KUrzschluss-SAnftanlaufschaltg.' mit Serienwiderst.*
stature Größe * *Gestalt*
status Eigenschaft || Position * *Stand* || Stand * *(Zu)Stand, (Gültigkeits- usw.)* Datum || Status || Zustand * ~ *eines Signals/Binärein/ausgangs (Plural: statuses)* || Zustand * *rechtlicher, politischer* ~
status as supplied Auslieferungszustand
status bit Statusbit || Zustandsbit || Zustandsmeldung * *Zustandsbit*
status check Statusabfrage
status description Zustandsbeschreibung
status display Statusanzeige
status indication Statusanzeige || Zustandsanzeige
status information Statusinformation
status message Zustandsmeldung
status output Binärausgang * *für (Betriebs-) Zustandssignal* || Digitalausgang * *für (Betriebs-) Zustandssignal*
status processing Statusbearbeitung * *bei SPS*
status register Zustandsregister
status report Statusmeldung
status scan Statusabfrage * *Test-/Diagnosefunktion bei SPS*
status signal Zustandsmeldung
status up until now bisheriger Stand
status word Statuswort || Zustandswort
statuses Zustände
statutory gesetzlich vorgeschrieben || vorgeschrieben * *(gesetzlich) vorgeschrieben, →gesetzlich, satzungsgemäß*
statutory regulation gesetzliche Bestimmung
Stauffer lubricator Schmierbüchse * *Staufferbüchse* || Staufferbüchse
stay bleiben || Strebe * *Stütze, Strebe, Verspannung, (mechan.) Anker, Korsett* || verankern * *absteifen, verankern, ab-/verspannen; hemmen, aufhalten* || verspannen * *ab-/verspannen, →verankern* || Versteifung * *Stütze, Strebe, Verspannung, Verankerung* || verweilen
stay-bolt Stehbolzen
stay-put rastend
STD code Vorwahlnummer * *(engl.) Abkürzg. für 'subscriber trunk dialing'*
steadfast beständig * *treu*
steady beständig * *ständig, stetig (auch Börse, Nachfrage usw.)* || fest * *von Markt, Kursen usw.* || gleich bleibend * *gleich bleibend/mäßig, ausgeglichen, fest, stabil* || gleichförmig * *→unveränderlich, gleich bleibend, gleichmäßig, stetig, regelmäßig* || gleichförmig * *unveränderlich, fest, stabil, stetig, regelmäßig* || gleichmäßig * *stetig, gleich bleibend, unveränderlich, ausgeglichen, fest, stabil* || laufend * *stetig, ständig, regelmäßig* || stabil * *gleichmäßig, stetig, ohne Abweichung* || ständig * *stetig, gleich bleibend/mäßig, unveränderlich* || stationär * *gleich bleibend* || stetig * *gleichmäßig, unerschütterlich*
steady light Dauerlicht * *z.B. einer Diagnose-Leuchtdiode*
steady state ausgeregelter Zustand || Beharrungszustand * *stationärer/eingeschwungener Zustand* || Gleichgewichtszustand * *stationärer/eingeschwungener Zustand* || stationärer Zustand
steady-state eingeschwungener Zustand || stationär * *z.B. im eingeschwungenen Zustand/Beharrungszustand* || statisch * *im stationären/eingeschwungenen Zustand*
steady-state accuracy Konstanz * *Genauigkeit im eingeschwungenen Zustand* || statische Genauigkeit
steady-state condition eingeschwungener Zustand || stationärer Betrieb * *als Betriebszustand* || stationärer Zustand

steady-state control deviation bleibende Regelabweichung
steady-state deviation bleibender Fehler * *Abweichung im stationären Zustand* || stationärer Fehler * *Abweichung im stationären Zustand*
steady-state error bleibender Fehler * *Fehler im stationären Zustand* || stationärer Fehler
steady-state operation stationärer Betrieb
steady-state temperature Beharrungstemperatur * *im eingeschwungenen Zustand/therm. Gleichgewicht* || Endtemperatur * *im thermisch eingeschwungenen Gleichgewichtszustand*
steady-state value Endwert * *im eingeschwungenen Zustand* || stationärer Wert
steady-state variation stationärer Verlauf
steadying Stabilisierung
steam Dampf
steam boiler Dampfkessel
steam generator Dampferzeuger
steam mains Sammelschiene * *zur Dampfversorgung*
steam moisture Dampffeuchte
steam nozzle Dampfdüse
steam pressure Dampfdruck
steam turbine Dampfturbine
steam-heated dampfbeheizt
steel Stahl * *(Bau)Stahl; siehe z.B. DIN 17006, DIN 17100 (z.B. ST37 mit Zugfestigkeit 370 N/mm^2)*
steel band Stahlband
steel base Stahlfundament * *aus Stahlteilen geschweißtes Maschinenfundament*
steel belt conveyor Stahlbandförderer * *{Fördertechnik}*
steel conduit thread PG * *verschraubte Kabeldurch/einführung (Abk.f.'Panzerrohr-Gewinde'-Verschraubung)* || PG-Verschraubung * *verschraubte Kabelein/-durchführung*
steel converter Konverter * *zur Stahlerzeugung*
steel cord Stahlcord * *{Reifen}*
steel frame Stahlgehäuse * *{el. Masch.}*
steel measuring tape Stahlbandmaß
steel mill Walzwerk * *Stahlwalzwerk*
steel pin Stahlnagel || Stahlstift
steel plate Blech * *dickes Stahl~* || Stahlblech * *dickes Blech, Grobblech*
steel sheet Blech * *{el.Masch.} Elektro~, →Magnetblech*
steel spring Stahlfeder
steel strip Stahlband
steel tape measure Stahlbandmaß
steel works Stahlwerk
steelmaking plant Stahlwerk * *Stahlwerk* || Stahlwerk
steelworks Hüttenwerk * *Stahlwerk* || Stahlwerk
steep aufquellen * *durchtränken (in/with: in), einweichen, eintauchen, imprägnieren* || imprägnieren * *eintauchen, durchtränken, imprägnieren (in/with: in)* || Lauge || steil
steep switching edges steile Schaltflanken
steep-fronted pulse Steilimpuls * *mit steiler Anstiegsflanke, z.B.zur Thyristoransteuerung*
steep-lead-angle thread Steilgewinde
steepen aufsteilen * *Impulsflanken usw.*
steepness Steilheit
steer leiten * *steuern, lenken, dirigieren*
steer clear of vermeiden * *umschiffen*
steering Lenkung * *eines Fahrzeugs*
steering committee Leitungskreis
steering gear Steuer * *{Schifffahrt} Ruder*
steering knuckle Achsschenkel * *{Auto}*
Steinmetz connection Steinmetz-Schaltung * *Einphasenbetrieb eines Drehstrommotors mit Betriebskondensator*
stem Dorn * *Stängel, Stiel, (Ventil-) Schaft, (Aufzieh-) Welle* || Schaft * *(Ventil-) Schaft, Stiel, Stän-*

stencil

gel, Stamm, Zapfen || stammen * *(amerikan) (from: von)*
stencil   Druckform * *{Druckerei}* || Schablone * *Zeichen/Malschablone*
stenter   Spannrahmen * *{Textil}* für Tuch
stentering machine   Spannrahmen * *{Textil}* Verstreckwerk mit Wärmebehandlung für Textilstoffe || Verstreckwerk * *{Textil}* Spannrahmen
step   Arbeitsgang || Ausbaustufe || Schritt * *step by step: ~ für ~; take small steps: kleine ~e machen; in steps of: in ~en von* || Schwelle * *Stufe* || Sprung * *~ eines Signals* || sprungartig || Stufe || stufen || Takt * *out of step: außer ~*
step back   Rückschritt
step by step   Schritt für Schritt || schrittweise * *Schritt für Schritt* || stufenweise * *(Adv.)*
step change   Sprung * *~artige Änderung* || sprungartige Änderung * *in:einer Größe* || Stoß * *stufenförmiger Sprung (z.B. als Stimulus z.Aufzeichnung d. Regler-Sprungantwort)* || Stufensprung * *Stimulationssignal für Regleroptimierung (Aufnahme der Sprungantwort)*
step change in load   Laststoß * *Lastsprung*
step change in speed   Drehzahlsprung
step criterion   Fortschaltbedingung * *(Plural: criteria) bei einer Ablaufsteuerung* || Fortschaltkriterium * *(Plural: criteria) in Ablaufsteuerung* || Weiterschaltbedingung * *(Plural: criteria) bei Ablaufsteuerung*
step down   heruntertakten || heruntertransformieren || schrittweise reduzieren || stufenweise reduzieren || transformieren * *heruntertransformieren* || untersetzen * *Trafo, Getriebe*
step enabling condition   Fortschaltbedingung * *bei Schrittsteuerung* || Weiterschaltbedingung
step flag   Schrittmerker
step function   Sprung * *~funktion* || Sprungfunktion
step function generator   Sprungfunktionsgenerator * *z.B. Funktionsbaustein zur Regleroptimierung*
step height   Stufenhöhe
step in   eingreifen * *(figürlich)* || eintreten * *hin~*
step logic element   Ablaufglied
step range   Schrittweite || Stufenhöhe * *Schrittweite* || Stufensprung * *Schrittweite* || Stufung * *Schrittweite*
step response   Sprungantwort
step rise time   Anregelzeit * *bei Sprungantwort*
step size   Schrittweite
step transition   Schrittfortschaltung * *bei Ablaufsteuerung/-kette*
step up   ankurbeln * *Produktion usw. ~* || hochtakten || hochtransformieren || schrittweise erhöhen || steigern * *Produktion usw. steigern/→hochtransformieren* || stufenweise erhöhen || transformieren * *hochtransformieren* || umsteigen * *(figürl.) to: auf eine bessere Lösung usw.*
step width   Schrittweite
step-by-step   stufenweise * *(Adj.)*
step-by-step controller   Proportionalregler * *z.B. zur Temperaturregelung*
step-change signal   Sprung * *Stimulus für Regleroptimierung (zur Aufnahme der ~antwort)*
step-down gear   Untersetzungsgetriebe
step-down gearing   Reduziergetriebe
step-down regulator   Abwärtsregler * *Schaltregler für Schaltnetzteil mit Ausgangsspanng. < Eing.-spanng.*
step-down transformer   Vorschalttrafo * *zur Spannungsreduzierung* || Vorschalttransformator * *zur Spannungsreduzierung* || Vortrafo * *zur Spannungsreduzierung* || Vortransformator * *zur Spannungsreduzierung*
step-forced response   Sprungantwort || Übergangsfunktion * *Sprungantwort*
step-up controller   Hochsetzsteller

step-up converter   Hochsetzsteller * *~-Stromrichter*
step-up gear   Übersetzungsgetriebe * *Hochsetzgetriebe*
step-up gearing   Getriebe * *Hochsetzgetriebe* || Übersetzungsgetriebe * *Hochsetzgetriebe*
step-up regulator   Aufwärtsregler * *(Schalt-) Regler (z.B. für Schaltnetzteil) mit Ausgangsspanng. > Eingangsspanng.*
stepless   kontinuierlich * *stufenlos* || stufenlos
steplessly variable   kontinuierlich * *stufenlos verstellbar*
stepped   stufenweise || treppenförmig
stepped characteristic   Treppenfunktion
stepper motor   Schrittmotor
stepping down   Untersetzung * *Getriebe, Trafo usw.*
stepping motor   Schrittmotor
stepping through   Schrittweises Durchschalten
stepping through the functions   Schrittweises Durchschalten * *bei SPS*
stepping transformer   Stufentransformator
steps   Leiter * *Stehleiter* || Weg * *by legal steps: auf gerichtlichem ~e*
steps ahead   Schritte voraus
stepwise   schrittweise || stufenweise
stereoscopic   räumlich * *dreidimensional (in der Optik)*
stereotype   Druckplatte * *{Druckerei}*
stick   ankleben * *angeklebt sein* || bleiben * *to something: bei etwas (z.B. Meinung)* || fressen * *Lager* || haften * *(an-) kleben; to: an* || kleben || Knüppel * *Holz~, Steuer~ usw.* || verharren * *to:bei*
stick fast   stecken * *festsitzen*
stick in someone's mind   merken * *sich einprägen*
stick magazine   Stangenmagazin * *z.B. für Bauteile bei der Leiterplattenbestückung*
stick out   aus dem Rahmen fallen
stick to   kleben * *kleben bleiben* || sich halten an
stick together   aneinander hängen * *aneinandergehängt/-geklebt sein* || verkleben * *kitten*
sticked together   verbacken * *miteinander verklebt, aneinander festsitzend*
sticker   Aufkleber || Haftbild || Klebebild
sticker diagram   Klebebild
sticky   heikel * *→schwierig, heikel, kritisch, zäh* || schmierig * *klebrig* || schwierig * *schwierig, heikel, kritisch; zäh, klebrig; schwül, stickig*
stiction   Haftreibung
stiff   stark * *(figürl.) z.B. Wind, Dosis, Gegner, Wettbewerb, Konkurrenz* || starr * *steif* || steif * *(auch figürlich) zäh, dick, stark, formell, steif, starr*
stiff competition   harter Wettbewerb
stiff system   starres Netz
stiffen   versteifen
stiffened   verstärkt * *versteift; high-capacity: mit ~er Tragfähigkeit (z.B. Lager)*
stiffening corrugation   Sicke * *zur Versteifung*
stiffening element   Versteifung
stiffness   Steifigkeit
still   jedoch || noch || noch immer
stimulate   ankurbeln * *(figürl.)* || anregen * *stimulieren* || antreiben * *stimulieren* || stimulieren
stimulation   Anregung
stimulus   Impuls * *(Plural:stimuli) Anstoß, (An-) Reiz, Stimulus, Antrieb, Ansporn* || Stimulus
stipulate   angeben * *festsetzen, vereinbaren* || ausmachen * *ausbedingen, (vertraglich) vereinbaren, festsetzen* || festlegen * *festsetzen, vereinbaren* || vereinbaren * *festsetzen, vereinbaren, ausbedingen, zur Bedingung machen (for:etwas)* || vorschreiben * *festsetzen, abmachen, vertraglich festlegen*
stipulated   obligatorisch * *zur Bedingung gemacht, (vertraglich) vereinbart*

748

stipulation Festlegung * (vertragliche) Abmachung, Übereinkunft, Festsetzung, Klausel,Bedingung || Übereinkunft * vertragliche ~
stir bewegen * unruhig; auch 'sich bewegen' || Bewegung * unruhige ~, Aufregung, Aufruhr; Betriebsamkeit, reges Treiben || rühren || umrühren
stir up aufwirbeln || umrühren
stirrer Rührwerk || Umrührer
stirring zündend * erregend, aufwühlend, mitreißend, bewegt, bewegend; z.B. Ereignis, Rede, Idee, Vision
stirring unit Umrührer
stitch Stich * mit Nähnadel usw.
STL AWL * Abkürzung für 'STatement List' Anweisungsliste als Programmiersprache für SPS
stock Aktie * (amerikan.) || auf Lager legen || einlagern * vorrätig haben, führen || Lager * Waren-, Ersatzteillager; supply from stock:ab Lager liefern; keep on stock:auf Lager haben || Lagerbestand || lagermäßig * z.B. stock motor: ~er Motor || Material * (Papier-/Faser-/Dick-) Stoff in der Prozeßtechnik; (Füll-)Gut, Material, Brühe || Stoff * Dick-/Dünn-/Grundstoff z.B. in d. Zellstoff-/Papierherstellung || Vorrat
stock clerk Lagerist
stock conveyor Flurförderzeug * Lagerförderzeug || Lagerförderzeug
stock in trade Betriebskapital
stock inventory Lagerbestand * ~saufnahme, festgestellter ~
stock list Lagerliste
stock of spare parts Ersatzteillager
stock on hand Lagerbestand
stock preparation Stoffaufbereitung * Herstellg.d. Papier-Grundmaterials z.weiteren Verarbeitg. i.d. Papiermaschine
stock preparation and equipment Stoffaufbereitung * (Papier) Vorstufe der Papiererzeugung
stock product lagerhaltiges Produkt * ab Lager lieferbares/lagermäßiges Produkt
stock pump Stoffpumpe * zum Fördern von Dick-/ Dünn-/Zellstoff in der Zellstoffaufbereitung
stock up auf Lager legen || einlagern * auf Lager legen, speichern
stock valve Stoffventil * in Zellstoff-/Papierherstellung
stock-pile Halde * Vorratshaufen
stockage Lagerhaltung
stocked lagermäßig || vorrätig * auf Lager liegend, vorrätig gehalten
stocking system Lagersystem
stockkeeper Lagerist
stockkeeping Lagerhaltung
stockkeeping of spares Ersatzteilhaltung
stockman Lagerist * (amerikan.)
stockyard Lagerplatz * im Freien
stone crusher Steinbrecher || Steinmühle
stop abbrechen * aufhören || abschalten || abstellen * Maschine || anhalten || Anschlag * mechan. End- || aufhören || aus * Halt-Kommando z.B. für Antrieb || aussetzen * (Tätigkeit, Maschinenlauf usw.) unterbrechen, anhalten; stoppage:das Aussetzen/Anhalten || einstellen * beenden || HALT || Haltestelle * für Straßenbahn, Bus usw. || Pause || Raste * Anschlag, Sperre, Hemmung, Klappe || Schluss * Beendigung; stop it: ~ damit! || stehen * anhalten || stehen bleiben || stillsetzen || stillstehen * anhalten || Stopp || stoppen
stop bit Stoppbit * kennzeichnet das Ende eines Zeichens bei serieller Übertragung (meist logisch "1")
stop control Ausschaltsteuerung * Antrieb || Ausschaltsteuerwerk * für Antrieb usw.
stop face Mittenanschlag * an einer (Schraub)-Klemme

stop interval Pausenzeit
stop plate Anschlagplatte * z.B. für Säge
stop position Halteposition
stop station Halteposition
stop time at start Anlaufverzögerung * Wartezeit bevor e.Antrieb nach d.Startbefehl hochläuft
stop valve Absperrventil
stopcock Absperrhahn
stopgap Hilfsmittel * Notbehelf, Lückenbüßer, Überbrückung, →Notlösung || Lückenbüßer || Notlösung * Notbehelf, Lückenbüßer, Überbrückung(slösung)
stopover Unterbrechung * ~ der Fahrt
stoppage Betriebsstörung || Einstellung * Anhalten, Stillstand, Betriebsstörung, Stockung, Verstopfung || Sperrung
stopper Stopfen * Stöpsel, Pfropfen (z.B. für Flasche), Absperrvorrichtung || Stopper || Verschluss * ~ einer Flasche usw.
stopping Abschaltung * Maschine || Stillsetzung || Verstopfung
stopping accuracy Haltegenauigkeit * (Aufzüge)
stopping distance Bremsweg
stopping place Haltestelle
stopping point Haltestelle * für Straßenbahn, Bus, Aufzug usw.
stopping position Haltestelle * (Aufzüge)
stopping sequence Ausschaltreihenfolge * Motor
stopping sequence control Ausschaltsteuerwerk * für Antrieb ~
stopping sequencing control Ausschaltsteuerung * für Antrieb
stopping time Auslaufzeit * z. Stillsetzen e. Antriebs || Bremszeit * zum Stillsetzen benötigte Zeit
storage Aufbewahrung || Einlagerung || Lager * Speicher || Lagerung * Ein-~/Lagerhaltung in einem Speicher/Lager || Speicherung || Warenspeicher
storage area Lagerplatz
storage battery Akkumulator * ladbare Batterie
storage bin Silo
storage capacitor Speicherkondensator
storage container Lagerbehälter
storage element Speicherglied
storage expenses Lagerkosten || Lagerungskosten
storage life Lagerfähigkeit * z.B. von Bauteilen
storage location Lagerplatz || Speicherzelle
storage medium Datenträger
storage place Lagerplatz
storage silo Lagersilo
storage system Lagersystem
storage tank Sammelbehälter * für Flüssigkeiten
storage temperature Lagertemperatur || Lagerungstemperatur
storage/retrieval sstem Lagersystem
store abspeichern || auf Lager halten * auf Lager nehmen || aufheben * lagern || einlagern || Geschäft * Laden || Geschäftsräume * Laden || hinterlegen * speichern, Daten in einer Datei hinterlegen || Lagerbestand || Lagerhalle || lagern * speichern, auch in Waren-/Ersatzteillager || Magazin * Speicher || speichern * abspeichern || unterbringen * lagern || Vorrat
store in a non-volatile mode nichtflüchtig abspeichern
store temporarily zwischenspeichern
store up einlagern
stored gespeichert
storehouse Lagerhalle
storekeeping Lagerhaltung
storey Etage || Stockwerk
storing Speicherung
storing in program library Programmarchivierung

story Etage * *(amerikan.)* ‖ Stockwerk * *(amerikan.)*
stout dick * *umfangreich*
stove Ofen * *Ofen, Herd, Brenn-/Trockenofen*
straddle spreizen * *Beine/sich ~, mit gespreizten Beinen stehen/gehen; straddle carrier: Containertransporter*
straight gerade * *straight line:gerade Linie* ‖ Geradeaus-
straight away sofort
straight in-line Geradeaus-
straight knife Langmesser
straight line gerade * *(Substantiv) [Mathematik]* ‖ gerade Linie ‖ Strahl * *[Mathem.]* ‖ Strecke * *(geometr.)*
straight-forward einfach * →*unkompliziert, direkt, geradlinig, gerade(her)aus* ‖ unkompliziert * *(Personen u. Sachen)* →*einfach, direkt, offen,gerade(her)aus, freimütig, aufrichtig*
straight-line Geradeaus- * *[Draht]*
straight-lined geradlinig
straight-lined motion geradlinige Bewegung
straight-sided press Zweiständerpresse
straight-through transformer Durchsteckwandler
straighten ausrichten * *gerade ~* ‖ richten * *geradebiegen durch Richtmaschine/Richtpresse usw.* ‖ strecken * *geradebiegen, (auf)richten*
straighten up aufräumen * *(amerikan.) z.B. Zimmer*
straightener Richtmaschine
straightening machine Richtmaschine * *zum Planbiegen von Blech, Geradebiegen von Draht, Rohren usw.*
straightening plate Richtplatte * *z.B. zum Glattrichten von Blech*
straightforward geradlinig * *(figürlich)* ‖ übersichtlich * *unkompliziert, einfach* ‖ unbürokratisch * *unkompliziert, einfach, geradlinig*
strain Anstrengung * *Strapaze* ‖ beanspruchen * *ermüden* ‖ Beanspruchung * *(starke) Inanspruchnahme, Belastung, Last* ‖ Belastung * *Anstrengung* ‖ durchlassen * *filtern* ‖ pressen * *(durch ein Sieb) seihen* ‖ schleudern * *Honig usw. ~; auch 'extract'* ‖ Spannung * *(mechan.) (verformende) Spannn., Verdehng.,Zerrg., Belastg.,Beanspruchg., Inanspruchnahme* ‖ Zerrung
strain gage Dehnungsmessstreifen * *(amerikan.)*
strain gauge Dehnungsmessstreifen ‖ DMS * *Dehnungsmessstreifen*
strain hardening Verfestigung * *Aushärtung*
strain relief Zugentlastung
strain relief bushing Zugentlastungsstutzen * *Durchführungshülse mit Kabel-Klemmvorrichtung zur Zugentlastung*
strain-relief bushing Durchführungshülse * *mit Zugentlastungsfunktion, z.B. für Kabel*
strainer Filter * *(mechan.) Sieb, Seiher, Filter* ‖ Sieb * *~ für Flüssiges*
strand Ader * *Litze(nleiter)* ‖ Draht * *Litze(nleiter)* ‖ drehen * *verseilen (Kabel, Seil)* ‖ Faser * *Strähne, Gewebe* ‖ Strang * *aus Metall, eines Seils/Taus* ‖ Trum * *einer Kette, Eines Riemens; siehe auch* →*Obertrum* ‖ Trumm * *einer Kette/eines Riemens; siehe auch* →*Trum* ‖ verseilen * *Seil, Kabel, Litzen eines Leiters usw.*
strand insulation Drahtisolierung * *für Motorwicklung und elektr. Wickelgüter*
stranded feindrähtig * *mit Litzenleiter* ‖ mehrdrähtig * *(Kabel, Leitung) mit Litzenleiter (kein starrer Draht)*
stranded conductor feindrähtiger Leiter ‖ Litzenleiter
stranded copper Kupferlitze
stranded wire Litzendraht ‖ Litzenleiter * *Litzendraht*

stranding cage Verseilkorb * *bei (Kabel-/Draht)Verseilmaschine*
stranding line Verseilanlage * *z.B. für Kabel*
strange eigenartig ‖ fremd ‖ komisch * *seltsam*
strangle abwürgen ‖ einschnüren * *strangulieren, (er-)würgen* ‖ würgen
stranglehold Griff * *würgender ~*
strap Bügel ‖ Lasche * *Riemen (-Stück), Gurt, Band, Strippe, Steg, Bandeisen* ‖ Leiste * *Lasche* ‖ Pratze * *Bügel*
strap brake Bandbremse
strapping machine Umreifungsmaschine * *Verpackungsmaschine z.Umreifen von Kartons/Paletten mit Kunststoff- o.Stahlband*
stratagem Manöver * *(figürlich)*
strategic strategisch
strategical strategisch
strategy Strategie
stratified vielschichtig * *geschichtet, schichtförmig*
stray Streu-
stray field Streufeld
stray flux Streufluss
stray loss torque Zusatzverlustmoment * *eines Motors auf Grund der Streuverluste (VDE 0530, Teil 2)*
stray losses Streuung * *magnet. Streuverluste* ‖ Zusatzverluste * *(magnet.) Streuverluste*
streak Streifen * *Streif(en) Strich, (Licht-) Streifen, Strahl*
stream strömen * *allg., auch Personen*
stream out ausströmen
streamer Magnetbandlaufwerk * *zur Datensicherung/Datenarchivierung*
streamline optimieren * *stromlinienförmig machen* ‖ verbessern * *modernisieren, wirkungsvoller/zügiger/zweckmäßiger/reibungsloser gestalten*
streamlined modern * *modernisiert, fortschrittlich, rationell* ‖ stromlinienförmig
streamlined wire Profildraht * *mit stromlinienförmigem Profil*
streamlining Feinanpassung ‖ Optimierung * *z.B. eines Prozesses, einer Organisation* ‖ Straffung * *einer Organisation, von Abläufen usw.*
streching unit Streckwerk
street car Straßenbahn * *(amerikan.) Straßenbahntriebwagen, evtl. mit Anhänger*
strength Festigkeit * *auch einer Konstruktion; Bruch/Zerreiß~, Stärke, Kraft* ‖ Kraft * *Stärke; gather strength: Kräfte sammeln* ‖ Robustheit * *Stärke, auch einer mechanischen Konstruktion* ‖ Stärke * *(auch figürlich: starke Seite(n) einer Person/eines Unternehmens)* ‖ Widerstand * *~skraft eines Materials* ‖ Widerstandsfähigkeit
strength class Festigkeitsklasse * *eines Werkstoffs, einer Schraube usw.*
strengthen ertüchtigen * *stärker machen* ‖ verstärken * *(mechan.u. figürl.) fester machen, (be-/ver-) stärken (auch zahlenmäß.),stärker werden*
strenuous mühsam * *anstrengend*
stress Anstrengung * *Strapaze* ‖ beanspruchen * *belasten, (Material, Maschinenteil, Bauelement) beanspruchen* ‖ Beanspruchung ‖ belasten * *beanspruchen* ‖ Belastung * *Beanspruchung* ‖ hervorheben * *betonen* ‖ hinweisen auf * *betonen/darauf hinweisen, dass* ‖ Spannung * *(mechan.) Beanspruchung, Spannung., Dehnung, Belastung* ‖ Zug * *Beanspruchung*
stress test Belastungstest * *Beanspruchungsprüfung (mit Temperatur-/Schwingungsbeanspruchung)*
stressability Beanspruchbarkeit ‖ Belastbarkeit * *Beanspruchbarkeit* ‖ Festigkeit * *Belastbarkeit, z.B. thermische, mechanische ~*
stressable belastbar * *beanspruchbar*
stressed belastet * *beansprucht*

stressing Beanspruchung * *von Material*
stretch anziehen * spannen || ausdehnen || ausweiten * *ausweiten* || dehnen || Dehnung * *elastische ~* || erstrecken * *auch 'sich ~' (from ... up to: von ... bis)* || recken || reichen * *stretch from ... to: ~/sich erstrecken von ... bis* || spannen * *strecken, spannen, straff ziehen, dehnen* || straff spannen || strecken || verlängern * *strecken* || verstrecken
stretch at break Bruchdehnung
stretch film Stretchfolie * *{Verpack.techn.}*
stretch roll Breitstreckwalze || Spannwalze
stretch-wrapping machine Folienwickler * *{Verpack.techn.} z.Verpacken v. Gegenständen, Paletten usw. i.Stretchfolie*
stretcher leveller Richtmaschine * *für Feinblech*
stretching Ausdehnung * Strecken || Dehnung * *elastische ~* || Spann- || Verstreckung
stretching device Spannvorrichtung * *zum Stramm-/Langziehen*
stretching factor Streckfaktor * *bei der Chemiefasern usw.*
stretching line Verstreckanlage
stretching machine Streckwerk
stretching ratio Streckfaktor * *bei Chemiefasern usw.* || Verstreckverhältnis
stretching roll Breitstreckwalze
stretching system Verstreckanlage * *z.B. für Textilfasern*
stretching unit Verstreckwerk
stretching-up Verstreckung * *z.B. von Textilfäden*
strewing Streuung * *Aus-/Bestreuen, z.B. Sand, Werbematerial*
stricken betroffen * *with:von; heimgesucht, schwer betroffen*
strict ausdrücklich * *Befehl usw.* || genau * streng || konsequent * *streng, durchgehend* || scharf * *streng* || streng * *scharf, bestimmt, genau; z.B. Reihenfolge, Anwendg. e.Prinzips* || strikt
stricter requirements erhöhte Anforderungen
strictly streng * *(Adv.) strictly confidential: vertraulich; adhere strictly to:sich ~ an etw. halten* || strikt * *(Adv.)*
strictly confidential vertraulich * *streng ~*
strictly speaking betrachten * *genau betrachtet* || eigentlich * *(Adv.) genaugenommen*
strike anschlagen || anstoßen || schlagen || stoßen * schlagen || Streik || treffen * *(auf/an/ein)schlagen, zufällig ~/entdecken*
striker clutch Schaltgabel
striker fork Schaltgabel * *z.B. für mechan. betätigte Kupplung*
striking Aufschlag * *Auftreffen* || eindrucksvoll || eklatant * *bemerkenswert, auffallend, eindrucksvoll, schlagend, überraschend, verblüffend, treffend*
string Schnur * *Bindfaden* || String * *{Computer} Zeichenkette, Folge von Textzeichen* || Zeichenfolge * *von Text-/ASCII-Zeichen usw.*
string together aneinander reihen || verknüpfen * *(figürl.) miteinander ~*
stringent knapp * *knapp (Geld), gedrückt (Geldmarkt), streng* || scharf * *streng, zwingend, bindend (Gesetz, (An)Forderung, Regel)* || streng * *zwingend, bindend (z.B. Maßahme, Regel, Vorschrift, (An)Forderung)*
stringent demands hohe Anforderungen
stringent operating conditions schwerste Bedingungen
stringent requirements hohe Ansprüche
strip abisolieren || abmanteln * *the insulation: Kabel ~* || abreißen * *Werksanlage, Maschine* || auseinandernehmen || Band * *Walzband, Metallband, Bandmaterial* || Leiste * *Streifen, Holzleiste* || Streifen * *kurzes/schmales Stück; Papierstreifen, Filmstreifen*

strip accumulator Bandspeicher * *für Metallband, wird zur Überbrückung der Coilwechselzeit gefüllt, dann geleert* || Warenbahnspeicher * *für Walzband*
strip annealing line Bandglühanlage
strip board Latte
strip breakage Bandriss * *{Walzwerk}*
strip chart Messschrieb * *z.B. Papierstreifen v. Linienschreiber*
strip entering Bandeinlauf * *Walzwerk*
strip finishing line Bandbearbeitungsmaschine * *Nachbearbeitungsstraße für Walz-/Blechband*
strip finishing plant Bandbehandlungsanlage * *für Metallband*
strip head Bandkopf * *{Walzwerk} Beginn des Walzbandes*
strip loop control Schlingenregelung * *{Walzwer-Sicherstellung e. definierten Durchhangs des Blechs i.d. Schlingengrube*
strip mill Bandwalzwerk || Flachwalzwerk * *zur Erzeugung von Metallband*
strip off Streifen * *(Verb) abstreifen*
strip processing line Bandbehandlungsanlage * *für Metallband*
strip speed Bandgeschwindigkeit * *z.B. von Walzband*
strip storage Warenspeicher * *für durchlaufendes Blechband (z.B. zur Überbrückung der Coil-Wechselzeit)*
strip tail Bandende * *{Walzwerk} Ende des Walzbandes*
strip tension Bandzug
strip thickness Banddicke * *z.B. von Walzband*
strip width Bandbreite * *{Walzwerk} ~ eines Walzbandes*
strip working machine Bandbearbeitungsmaschine * *für Blechband*
strip-levelling machine Bandrichtmaschine * *→Richtmaschine zum Planbiegen von Blechband*
strip-off absetzen * *Kabelisolierung*
strip-steel Stahlband
strip-tensioning control Bandzugregelung * *in Walzstraße* || Zugregelung * *Bandzugregelung in Walzstraße*
strip-type bandförmig
stripe Streifen
stripper Abisolierwerkzeug || Abstreifer * *z.B. in der Gießerei*
stripper roll Wendewalze * *{Textil}*
stripping Abbau
stripping length Absetzlänge * *Abisolierlänge, entlang derer das Kabel von der Isolierung befreit wird*
stripping tool Abisolierwerkzeug
strive streben * *after/towards: nach*
strive for anstreben * *streben nach, (erbittert) ringen um, sich mühen um*
strobe effect Stroboskopeffekt
stroboscope Stroboskop
stroboscopic effect Stroboskopeffekt
stroke Anschlag * *Schlag; auch bei Schreibmaschine* || anzeichnen * *mit Strich(en)* || Hub * *e. Werkzeugmaschine, ~spindel, Kompressors, Kolbens in Verbrennungsmotor* || Schlag * *Hieb, Hub* || Schrägstrich || Strich * *Pinsel-/Feder-/Kennzeichnungs-*
stroke circular saw Hubkreissäge * *{Holz}*
stroke length Hublänge * *z.B. bei Hubspindeltrieb*
stroke volume Hubvolumen * *eines (Hydraulik-) Kolbens*
strong heftig * *stark* || kräftig || stabil * *stark, robust* || stark
strong point Stärke * *starke Seite*
strongly dringend * *(Adv.) advise/warn strongly: ~*

abraten || sehr * *(Adv.) nachdrücklich, kräftig, stark*
structogram   Nassi-Schneidermann-Diagramm * *{Computer} Struktogramm* || Struktogramm
structural   Bau- || Struktur-
structural change   Strukturwandel
structural element   Konstruktionselement
structural lumber   Bauholz
structural material   Baustoff
structural part   Konstruktionsteil
structural problem   Strukturproblem
structural timber   Bauholz
structural-diagram oriented   strukturbildorientiert
structure   Anlage * *z.B. eines Romans* || Anordnung * *Aufbau* || Aufbau * *Struktur, Gefüge* || Aufgliederung * *Aufbau* || Beschaffenheit * *Struktur* || Gefüge || gliedern * *strukturieren* || Konstruktion * *Aufbau, Bauart* || Struktur || strukturieren || stukturieren || System * *Struktur, Gefüge* || Zusammensetzung * *Struktur, Gefüge*
structure diagram   Strukturbild
structure diagram oriented   strukturbildorientiert
structure-born noise   Körperschall
structured   gegliedert * *strukturiert* || geordnet || strukturiert
structured chart   Nassi-Schneidermann-Diagramm * *{Computer} Struktogramm* || Struktogramm
structured programming   strukturierte Programmierung * *z.B.nach Dijkstra, Nassi Schneidermann (z.B. direkte Sprünge verboten)*
structuring   Strukturierung
struggle   Einsatz * *Kampf, Mühen* || streben * *for: nach*
struggle against   widersetzen
strut   Strebe * *Strebe, Stütze, Spreize* || versteifen || Versteifung * *Strebe* || verstreben || Verstrebung
strutting   Verstrebung
stub axle   Achsschenkel * *{Auto}*
stub line   Stichleitung
stub-end feeder   Stichleitung * *im Versorgungsnetz*
stud   Bolzen * *Steh~, Stift(-Schraube), Zapfen* || Schraube * *~ ohne Kopf, Stift~, Stehbolzen* || Stehbolzen * *Stehbolzen, Stift, Zapfen; mittels Schraubenbolzen sichern* || Stiftschraube || Zapfen * *Stift, Zapfen, Stehbolzen, Pfosten*
stud board   Bolzenbrett * *zum Leitungsanschluss im Motorklemmenkasten*
stud bolt   Gewindestift * *Stehbolzen* || Stiftschraube
stud terminal   Bolzenklemme * *zum Anschluss von Leitungen*
stud torque   Anzugsdrehmoment * *für Schrauben* || Anzugsmoment * *beim Anziehen einer Schraube*
stud-casing thyristor   Schraubthyristor
stud-mounting thyristor   Schraubthyristor
stud-type thyristor   Schraubthyristor
student   Kursteilnehmer || Lehrgangsteilnehmer || Lernender * *Student, Kursteilnehmer* || Student || Teilnehmer * *an Lehrgang, Kurs, Seminar*
studies   Prüfung * *(wissenschaftliche) Untersuchung, Erforschung; genaue prüfende Betrachtung* || Überprüfung * *(wissenschaftliche) Untersuchung/Erforschung; genaue, prüfende Betrachtung*
studious   eifrig
study   befassen * *sich prüfend befassen mit* || lernen * *studieren* || prüfen * *(sorgfältig) untersuchen, erforschen, ~d ansehen, lesen* || Studie * *of: über*
study group   Arbeitsgemeinschaft || Arbeitskreis
stuff   -Zeug || abdichten || füllen * *vollstopfen* || Krempel * *(salopp) Zeug* || Material || vollstopfen * *(voll-) stopfen; stuff up: ver-/zustopfen/mit Fett imprägnieren/abschmieren* || Zeug
stuff and nonsense   Krampf * *(figürl.) (salopp) Quatsch, Unsinn, Blödsinn*
stuffer yarn   Füllgarn * *{Kabel}*

stuffing   Füllung || Stopfen * *Stopfbuchse, Füllung, Füllmaterial*
stuffing box   Stopfbüchse
stumbling-block   Hindernis * *(figürl.) Stolperstein, Stein des Anstoßes, Hindernis (to: für)* || Stolperstein * *(figürl.) Stolperstein, Stein des Anstoßes, →Hindernis (to: für)*
sturdiness   Robustheit * *Kräftigkeit, Robustheit, Standhaftigkeit*
sturdy   kräftig * *→stabil, →robust* || robust * *→stabil, kräftig, standhaft (auch figürl.), massiv, fest* || stabil * *kräftig, stabil, standhaft, fest, →massiv*
style   Art * *Stil, Typ, Art u.Weise, Manier* || Art und Weise * *Stil, Typ, Manier* || Ausführung * *Art, Typ, Art und Weise, Manier, (Bau-) Stil* || Form || Mode || nennen * *betiteln, benennen, bezeichnen, anreden mit/als* || Weise
styling   Design * *äußeres Design, stilistische Formgebung*
stylish   nobel * *fein "gestylt"*
Sub D connector   Sub D Stecker || Subminiatur D-Stecker
sub menu   Untermenü
sub task   Teilaufgabe
sub-   Teil- * *Unter-, untergeordnet* || Unter- * *untergeordnet*
sub-assembly   Baueinheit * *Baugruppe* || Baugruppe
sub-average   unterdurchschnittlich
sub-board   Sub-Leiterkarte
sub-critical   unterkritisch
Sub-D connector   Cannon-Stecker || D-Sub Stecker || SUB-D Stecker || Sub-D-Stecker * *auch "Cannon-Stecker", Grundfläche hat die Form eines "D"*
sub-division   Geschäftszweig * *in einer Unternehmensorganisation*
sub-group   Untergruppe
sub-miniature motor   Kleinstmotor
sub-module   Submodul
sub-office   Zweigbüro || Zweigniederlassung * *Zweigbüro*
sub-quantity   Untermenge
subcontracting industries   Zulieferindustrie
subcontracting industry   Zulieferindustrie
subcontracting part   Zulieferteil
subcontracting parts   Zulieferung * *Zulieferteile*
subcontractor   Lieferant * *Unterlieferant* || Unterlieferant || Zulieferant * *Unterlieferant* || Zulieferer * *Unterlieferant* || Zulieferfirma
subcontroller   unterlagerter Regler
subcritical   unterkritisch
subdivide   aufgliedern || aufschlüsseln || aufteilen * *unterteilen* || einteilen * *unterteilen, in/into:in* || gliedern * *unterteilen (into: in)* || unterteilen
subdivided   gegliedert * *unterteilt, unter~ (into: in)* || unterteilt * *into:in*
subdivision   Aufgliederung || Einteilung * *Unterteilung* || Untergliederung || Unterteilung
subdue   unterwerfen
subfunction   Teilfunktion * *Unterfunktion* || Unterfunktion * *siehe auch →Teilfunktion*
subharmonic   Schwingung * *subharmonische ~*
subharmonics   Schwingungen * *subharmonische ~*
subject   abhängig * *to:von Zustimmung usw.* || aussetzen * *unterwerfen, unterziehen, subject to: (etw. Hitze usw.) ~* || Betrifft * *'Thema' im Kopf eines Berichts, Briefes* || durchführen * *unterwerfen, unterziehen; someth.is subjected to: an etwas wird ... durchgeführt* || Fach * *Lehr/Schul/Studien~, Unterrichtsthema* || Gegenstand * *(Gesprächs- usw.) Gegenstand, Thema, Stoff, Objekt, Subjekt, Substanz* || Gesprächsthema || Sachgebiet || Thema * *Bericht, Brief, Besprechung, Tagung, Vortrag*

subject index Sachnummer ‖ Schlagwortverzeichnis * *Sachverzeichnis*
subject of conversation Besprechungsthema
subject to -behaftet * *(vorangestellt) z.B. subject to wear: verschleiß~* ‖ -pflichtig * *(vorangestellt) Gebühr, Abgabe, Provision usw.* ‖ abhängig von * *von Zustimmung*
subject to abrasion verschleißbehaftet * *sich abreibend*
subject to change freibleibend * *Änderungen unterworfen*
subject to change without notice Änderungen vorbehalten * *ohne Benachrichtigung d. Betroffenen*
subject to change without prior notice Änderungen vorbehalten * *kann geändert werden ohne vorherige Benachrichtigung*
subject to duty zollpflichtig
subject to taxation steuerpflichtig
subject to wear verschleißbehaftet
subject-matter Gegenstand * *Inhalt*
submerge eintauchen * *untertauchen, eintauchen, überschwemmen, unterdrücken, übertönen* ‖ untertauchen * *untersinken, untertauchen, überschwemmen, unter Wasser setzen, übertönen*
submergence Eintauchtiefe * *Untertauchtiefe, Eintauchung*
submerging Überflutung * *das Versenken*
submersible motor-pump Tauchmotorpumpe
submersible pump Tauchpumpe
submersible-pump motor Tauchpumpenmotor
submersion Überflutung * *Untertauchung* ‖ untertauchen * *(Subst.) Ein-/Untertauchen, Überschwemmung*
subminiature kleinst ‖ Kleinst-
Subminiature D connector Sub D Stecker ‖ Subminiatur D-Stecker
submission Antrag * *Unterbreitung, Vorlage*
submit to unterwerfen * *einer Prüfung, einem Schiedsgericht usw.; auch 'sich ~'*
submodule Modul * *Submodul, z.B. Speichermodul, "Huckepack-Kärtchen" usw.* ‖ Submodul
subordinate Mittel * *(Adj.) im Rang* ‖ untergeordnet ‖ unterlagert
subordinate control loop unterlagerter Regelkreis
subordinate controller unterlagerter Regler
subordinated unterlagert * *~ angeordnet, z.B. Regelkreis (to: einer Einrichtung)*
subordinated control circuit unterlagerter Regelkreis
subordinated controller unterlagerter Regler
subrack Baugruppenträger
subrack coupling Rahmenkopplung * *zwischen Elektronik-Baugruppenträgern*
subrack expansion Rahmenkopplung
subrack interface board Rahmenkopplungsmodul * *als Leiterkarte ausgeführte Anschaltung zur Datenkopplung zw. Baugr.trägern*
subrack interfacing Rahmenkopplung
subrack link Rahmenkopplung
subroutine Hilfsprogramm * *Unterprogramm* ‖ Unterprogramm
subscriber Abonnent * *auch eines Telekommunikations-Dienstes* ‖ Teilnehmer * *Telefonanschluss; ~ am Fernsehnetz*
subscriber's line Anschlussleitung * *Telefon-Endanschlussleitung*
subscriber's station Anschluss * *Telefonapparat*
subscript Index * *tiefgestellte(s) alphanumerisches Zeichen*
subscription Beitrag * *Mitgliederbeitrag*
subsequent anschließend * *zeitlich; to: an* ‖ Folge- * *nachfolgend* ‖ folgend * *nach~, später, nachträgl., Nach-, i. Anschluss an; subs'ly described:i. ~den beschrieben* ‖ nach * *in der zeitlichen Reihenfolge (~-)folgend; subs. to: später als/~, in An-*

schluss an ‖ nachfolgend * *(Adj.)* ‖ nachgeschaltet * *nachfolgend* ‖ nachträglich * *später* ‖ später * *nachfolgend*
subsequent processing Nachverarbeitung ‖ Weiterverarbeitung
subsequent telegram Folgetelegramm
subsequent to im Anschluss an
subsequently danach * *anschließend* ‖ im Folgenden * *subsequently described:im Folgenden beschrieben* ‖ nachfolgend * *(Adv.) z.B. in einer Beschreibung*
subsequently connected nachgeschaltet
subsequently listed folgend * *nach~ aufgeführt (in e. Aufzählung, Liste)* ‖ nachfolgend aufgeführt * *in Aufzählung, Liste*
subset Untermenge
subside abklingen * *(figürlich) einsinken, absacken, sich setzen* ‖ absinken * *z.B. Boden* ‖ aufhören * *allmählich*
subsidence Senkung * *(Erd-/Fundament-) Senkung, Absinken, Nachlassen, Abflauen (auch figürl.)*
subsidiary Tochtergesellschaft ‖ Zweigbüro * *Filiale, Tochtergesellschaft* ‖ Zweigniederlassung * *Filiale, Tochtergesellschaft*
subsidiary company Tochtergesellschaft
subsidies Subvention ‖ Subventionierung
subsidization Preisstützung * *Subventionierung* ‖ Subventionierung
subsidize stützen * *subventionieren* ‖ subventionieren
subsidy Unterstützung * *staatliche Gelder*
substance Inbegriff * *Wesen, Substanz, das Wesentliche, wesentlicher Inhalt/Bestandteil, Kern, Gehalt* ‖ Stoff * *Substanz* ‖ Substanz * *(auch figürlich: Grundlage)*
substantial beträchtlich * *wesentlich (Fortschritt, Unterschied usw.), wesentlich, substanziell, namhaft* ‖ erheblich * *substanziell, z.B.Sachschaden* ‖ grundlegend ‖ maßgebend * *beträchtlich* ‖ maßgeblich * *beträchtlich* ‖ substanziell * *wesentlich * in der Substanz, auch 'beträchtlich, substanziell, grundlegend'* ‖ wirklich * *wesentlich*
substation Unterstation ‖ Unterwerk
substation transformer Übergabetrafo * *in der Energieverteilung*
substitute Ersatz * *Ersatz-, ~mann, ~produkt* ‖ Ersatz- * *ersetzend, Austausch-, als Ersatz/Stellvertreter dienend* ‖ ersetzen * *ersetzen/austauschen/ an die Stelle setzen (for:durch/gegen/von)* ‖ setzen * *anstelle von ...: ~; er~* ‖ substituieren * *for:durch* ‖ Vertreter * *Stell~, Ersatzmann*
substitution Ablösung * *Substitution* ‖ Austausch * *Substitution, ersatzweise Verwendung, Ersatz-* ‖ Ersatz * *Substitution. ersatzweise Verwendung, Ersatz-, Ersetzung (durch neue Technik usw.)* ‖ Substitution
substitution operation Substitutionsbefehl * *bei SPS*
substitution variable Ersatzgröße
substract abziehen * *subtrahieren*
substraction Abzug * *Substraktion; when substracting: nach ~ von*
substrate Substrat ‖ Trägerschicht * *Trägermaterial, Unter-/Grundlage, Unterschicht, Träger, Medium, Grundschicht, Substrat*
substratum Trägerschicht * *Trägermaterial*
subsumption Unterbegriff * *(gemeinsame) Einordnung, Zusammenfasssung (under:unter)*
subsynchronous untersynchron * *converter cascade:Stromrichterkaskade*
subsynchronous converter cascade untersynchrone Stromrichterkaskade * *I-Umrichter f.Schleifr.läufer-Async.mot.(Ständ.am Netz; 1..20MW)*

subtilize verfeinern * *(figürl.)*
subtle raffiniert * *schlau*
subtract subtrahieren
subtraction Subtraktion
subtractor Subtrahierer
subtropical climate subtropisches Klima
suburban public transportation Personennahverkehr
suburban rail car Nahverkehrs-Triebwagen
suburban railway S-Bahn
suburban traction vehicle Nahverkehrs-Triebwagen
suburban transportation system Nahverkehrssystem * *siehe auch →Personennahverkehr*
suburbian traffic Nahverkehr
subvention Subvention
subway Tunnel * *Unterführung* || U-Bahn
succeed aufeinander folgen * *one another: einer auf den anderen* || durchsetzen * erfolgreich sein
succeed to übernehmen * *in ein Amt nachfolgen*
succeed with durchsetzen * erfolgreich
succesor Nachfolger * *im Amt*
success Erfolg * *great: großer/voller; successfully/with success. mit ~; be unsuccessful: keinen ~ haben*
success factor Erfolgsfaktor
successful erfolgreich * *in (doing): in/bei*
succession Abfolge || Aufeinanderfolge * *['saxeschen]* || Folge * *Aufeinander~; continuous: ununterbrochene/fortlaufende* || Reihenfolge || Wechsel * *Aufeinanderfolge*
succession to Übernahme * *Amt, Erbschaft usw.*
successive aufeinander folgend
successive approximation stufenweise Näherung * *z.B. schnelles Verfahren für A/D-Umsetzer*
successively hintereinander || nacheinander * *sukzessive*
successor product Nachfolgeprodukt
suceed in meistern * *(mit Gerundium) erfolgreich sein zu ...*
such derartig || solch || solcher * *such: solch einer*
such a so * *so ein* || solch
such as unter anderem
such as that solch * *(nachgestellt)*
such as this derartig * *(nachgestellt)* || solch * *(nachgestellt)*
such like this solch * *(nachgestellt)*
such that in einer Art und Weise dass
suchlike solcher * *~lei*
suck in ansaugen * *z.B. Kühlluft, Frischluft, Flüssigkeit, Wasser, Gas*
suck off absaugen
suction Absaugung || Ansaugung * *z.B. durch Pumpe, Lüfter* || Saug-
suction air Ansaugluft
suction box Absauggerät || Saugkasten * *in Siebpartie einer Papiermaschine*
suction capacity Ansaugleistung
suction couch roll Saugwalze * *Siebsaugwalze in der Siebpartie e. Papiermaschine* || Siebsaugwalze * *dient zur Entwässerung in Siebpartie einer Papiermaschine*
suction cup Saugnapf
suction filter Ansaugfilter * *für Pumpe, Hydrauliksystem usw.* || Saugfilter * *(mechan.)*
suction lift Ansaughöhe || Förderhöhe * *Ansaughöhe einer Pumpe* || Saughöhe * *einer Pumpe*
suction line Saugleitung * *z.B. eines Kompressors*
suction roll Saugwalze * *z.B. in der Siebpartie einer Papiermaschine*
suction side Ansaugseite * *Pumpe, Lüfter*
suction valve Ansaugventil
sudden plötzlich * *(Adj.)* || schlagartig || sprungartig
sudden change Sprung * *~artige/plötzliche Änderung* || sprungartige Änderung || Stoß * *schlagartige Änderung; sudden change in torque: Drehmomenten~*
sudden change in torque Drehmomentenstoß
sudden drop -Sturz * *z.B. 'sudden temerature drop': Temperatursturz*
sudden fall -Sturz || Absturz * *plötzlicher ~*
sudden fall in prices Preissturz
sudden load Stoßlast
sudden load change Laststoß
sudden load variation Laststoß
sudden loading Laststoß
sudden rise Schub * *plötzlicher Anstieg (Preise, Kosten usw.)*
sudden temperature drop Temperatursturz
sudden variation Sprung * *~artige Änderung* || sprungartige Änderung
suddenly auf einmal * *plötzlich* || mit einem Mal * *plötzlich* || plötzlich * *(Adv.)*
sue verklagen * *gerichtlich belangen, verklagen (for:auf); sue out:(Gerichtsbeschluss usw.) erwirken*
suffer erfahren * *erleiden* || leiden * *(from: an/unter) suffer severely from: schwer/stark ~ an/unter* || zulassen * *geschehenlassen, dulden*
suffer injury Schaden erleiden
suffice ausreichen * *genügen, hinreichen, reichen, (aus)reichen, jdm. genügen* || genügen * *ausreichen sein, ausreichen, hinreichen* || reichen * *genügen, aus~d/hin~d sein*
sufficient ausreichend || genügend || hinreichend
sufficient condition hinreichende Bedingung
sufficiently ausreichend * *(Adv.)* || genügend * *(Adv.) reichend, hinreichend*
suffix Anhängsel * *an ein Wort,eine Bestellbezeichnung,einen Variablennamen* || ergänzen * *durch eine angehängte Zusatzangabe, z.B. an Bestellnummer* || Ergänzung * *an Bestell-/Codenummer* || Zusatz * *z.B. an Bestellnummer*
sugar Zucker
sugar centrifuge Zuckerzentrifuge
sugar crystals Zuckerkristalle
sugar factory Zuckerfabrik
sugar industry Zuckerindustrie
sugar mill Zuckerfabrik || Zuckermühle
suggest anregen * *vorschlagen* || vorschlagen
suggestion Anregung * *Vorschlag* || Veranlassung * *Anregung, Vorschlag; at my suggestion: auf meine ~* || Vorschlag
suggestion of improvement Verbesserungsvorschlag
suggestive sinnvoll * *gehaltvoll, anregend*
suit anpassen * *to: an* || passen * *anpassen (to:an), passen zu, sich eignen zu/für, entsprechen, kleiden* || passen zu
suitability Eignung * *Eignung, Angemessenheit (to: für, zu)*
suitability test Eignungstest * *für anzuschaffendes Investitionsgut*
suitable anwendbar * *geeignet* || entsprechend * *passend* || geeignet * *passend* || günstig * *passend, zuträglich* || gut * *passend, geeignet* || passend * *suitable for use with: geignet für die Verwendung bei/mit* || richtig * *geeignet* || sachgemäß || zugehörig * *dazu passend* || zweckmäßig
suitable for entsprechend * *passend*
suitable for harzadous duty explosionsgeschützt
suitable to entsprechend * *passend*
suited geeignet * *passend; well/especially suited: gut/besonders geeignet (to/for:für)* || passend
sulfur Schwefel
sulphate Sulfat * *{Chemie} Salz der Schwefelsäure*
sulphide Sulfid * *{Chemie} Salz des Schwefelwasserstoffs*
sulphite Sulfit * *{Chemie} Salz der schwefligen Säure*

sulphite pulp Zellstoff * *Sulfitzellstoff*
sulphur dioxide Schwefeldyoxid * *['salfur 'daioxaid]* SO2
sum Summe || summieren
sum block Addierer * *Funktionsbaustein* || Summierer * *Funktionsbaustein*
sum up addieren * *auf-, zusammenaddieren* || rechnen * *zusammen~* || summieren * *auf~* || wiederholen * *zusammenfassen* || zusammenfassen * *noch einmal ~*
summarize Überblick geben || zusammenfassen * *gedrängt darstellen, z.B. einen Textes*
summarized zusammengefasst * *kurz dargestellt*
summarizing sheet Zusammenstellung * *Übersichts-Blatt*
summary Inhaltsübersicht || Resümee || Rsumee || Summen- * *(in) gedrängte(r) Übersicht, summarisch, Übersichts-, abgekürzt, Schnell-* || Überblick * *Zusammenfassung, Übersicht, Abriss, Inhaltsangabe* || Übersicht * *Zusammenfassung, Abriss, Inhaltsangabe, tabellarische ~ (of: über)* || Zusammenfassung
summary of operations Operationsübersicht * *Übersicht über SPS-Befehlssatz*
summary outline Übersicht * *Kurz-Zusammenfassung*
summation Summierung
summator Addierer || Summierer || Summierpunkt
summed up zusammengefasst * *kurz dargestellt*
summing amplifier Addierverstärker || Summierverstärker
summing element Addierer || Additionsglied || Summierer || Summierglied
summing junction Summierpunkt
summing point Addierpunkt || Addierstelle * *(Regel.techn.)* || Summationspunkt || Summierpunkt || Summierstelle
summing pulses Summenimpulse
summing up zusammenfassend
summon einberufen * *z.B. Versammlung ~*
sun gear Sonnenrad * *in einem →Planetengetriebe*
sun radiation Sonneneinstrahlung
sun wheel Sonnenrad * *bei →Planetengetriebe*
sundries Diverses * *Diverses, Verschiedenes, diverse Unkosten, allerlei Dinge*
sundry diverse * *(Adj.) verschiedene, diverse, allerlei, allerhand*
super Spitzen- * *überragend, höchst* || super * *(salopp) großartig* || Über-
super fast superflink * *Sicherung*
super-fast acting superflink * *Sicherung*
super-rapid superflink
superb hervorragend
supercalender Satinierkalander * *[Papier] Presse zur Herstellung hochkalandrierten Glanz-Papiers* || Superkalander * *Mehrstufig angeordnete Glättzylinder zum Erzeugen hochglatten (satinierten) Papiers* || superkalandrieren * *satinieren, hochfein glätten von Papier durch Glättzylinder*
supercharge vorverdichten
superconducting supraleitend
superconductive superleitend || supraleitend
superconductor Supraleiter
superelevation Überhöhung * *von Schienen usw.*
superficial oberflächlich * *überschläglich* || überschlägig * *oberflächlich* || überschläglich * *oberflächlich, flüchtig* || überschlagsmäßig * *oberflächlich, flüchtig*
superfluous überflüssig * *unnötig; render superfluous: ~ machen* || übrig * *überflüssig*
superheat überhitzen
superheating Übererhitzung || Übererwärmung || Überhitzung
superimpose * überlagern * *on:etwas/über etwas*
superimposed Leit- * *übergeordnet, überlagert* ||

übergreifend * *darüberliegend/gelegt, überlagert* || überlagert * *on: über etwas; auch Wechselspannungsanteil*
superimposed alternating current überlagerter Wechselstrom
superimposed alternating voltage überlagerte Wechselspannung
superimposed controller Leitregler * *überlagerter Regler* || überlagerter Regler
superimposed controls überlagerte Regelung
superimposition Überlagerung
superintend inspizieren * *beaufsichtigen* || überwachen * *beaufsichtigen*
superintendence Beaufsichtigung
superintendent Leiter * *auch Vorsteher,Direktor*
superior besser * *über dem Durchschnitt stehend,* überragend || exzellent * *hervorragend* || herausragend * *hervorragend* || hervorragend * *bezüglich der Leistung (-sfähigkeit), Qualität usw.* || höchst * *überlegen* || ober || sehr gut * *hervorragend* || überlegen * *(Adj.) funktionell überragend, i.Vorteil; (to:gegenüber/verglichen mit; in:an/bezügl.)* || überragend || vorzüglich || Vorzugs- * *hervorragend, überlegen (auch zahlenmäßig), bevorrechtigt, erlesen*
superiority Überlegenheit
superlative brillant || Spitzen- * *überragend, höchst*
superordinate übergeordnet
superordinated überlagert
superpose überlagern
superposed überlagert
superposition Überlagerung
superposition method Überlagerungsverfahren * *zur Ermittlung von Temperaturen*
supersede ablösen * *einem Vorgänger nachfolgen* || aufheben * *ersetzen* || ersetzen * *etwas/jdm. ersetzen/ablösen (by:durch), etw. abschaffen, beseitigen* || verdrängen * *als Nachfolger ~*
supersession Verdrängung * *(figürl.) Ablösung, Ersatz (durch Nachfolger)*
supersynchronous übersynchron
supervene dazwischenkommen * *auch unvermutet eintreten*
supervise kontrollieren * *siehe auch →inspizieren* || überwachen * *beaufsichtigen* || verwalten * *überwachen*
supervision Beaufsichtigung || Betreuung * *Beaufsichtigung* || Überwachung * *Aufsicht*
supervision panel Steuertafel
supervisor Kontrolleur || Werkmeister
supervisory Leit- * *aufsichtsführend, Leitsystem* || übergeordnet * *aufsichtsführend, Leit-(System)* || übergreifend * *aufsichtsführend, leitend*
supervisory board Aufsichtsrat
supervisory computer Leitrechner
supervisory controller überlagerter Regler
supervisory system Leitsystem || Überwachungssystem
supper Abendessen
supplement Anhang * *Ergänzung, Zusatz, Nachtrag, Anhang* || Beiblatt * *to:für* || ergänzen * *auch durch einen Anhang, z.B. an eine Bestellnummer* || Ergänzung * *Nachtrag, Anhang, Zusatz, das Ergänzte* || Nachtrag * *z.B. zu einem Katalog* || Zusatz * *Ergänzung*
supplement sheet Beiblatt * *z.B. zu einer Norm/einer Betriebsanleitung* || Zusatzblatt
supplementary ergänzend || Hilfs- || nachträglich * *ergänzend* || Neben- * *Zusatz-, Hilfs-, Ersatz-* || optional || Zusatz- || zusätzlich
supplementary board Zusatzbaugruppe * *Leiterplatte*
supplementary circuit Nebenstromkreis
supplementary device Hilfsmittel * *Hilfsvorrichtung, Gerät, Behelf*

supplementary equipment Hilfsmittel
supplementary features zusätzliche Merkmale
supplementary function Zusatzfunktion
supplementary information ergänzende Hinweise || Zusatzangaben
supplementary measures Zusatzmaßnahmen
supplementary module Zusatzbaugruppe
supplementary reference Hilfssollwert
supplementary reference value Hilfssollwert || Zusatzsollwert
supplementary setpoint Hilfssollwert || Zusatzsollwert * *Hilfssollwert*
supplementary sheet Beiblatt
supplementary unit Zusatzgerät
supplementary value Zusatzwert
supplementation Ergänzung * *Ergänzung, Nachtragen, Nachtrag, Zusatz*
suppliable lieferbar
supplied ausgestattet || geliefert
supplied from Lieferort * *lieferbar von*
supplier Anbieter * *Lieferant* || Auftragnehmer * *Lieferant* || Hersteller * *Lieferant* || Lieferant || Lieferer * *Lieferant* || Lieferfirma || Zulieferant || Zulieferer || Zulieferfirma
suppliers Lieferwerk
suppliers ref. Auftragskennzeichen
supplies Lagerbestand
supply abgeben * *power: Leistung* || abgeben * *z.B. Spannung/Strom/Leistung* ~ || anliefern || Anschluss * *Gas-/Wasser~* || ausliefern || ausrüsten || ausstatten * *with:mit* || beistellen || sorgen für || beschaffen * *liefern* || Bezug * *von Ware* || einspeisen * *Einspeisung* * *Versorgung* || erstellen || leisten * *liefern* || liefern * *versorgen mit; auch Ausgangssignal/Strom/Spannung/Leistung/Energie* ~ || Lieferung || Netz * *Versorgung, Einspeisung* || Netz- || speisen || Speisung
supply air Zuluft
supply cable Anschlusskabel * *Stromversorgungskabel* || Anschlussleitung * *für Stromversorgung, Motor, Stromrichter usw.* || Netzleitung || Netzzuleitung || Versorgungsleitung || Zuleitung * *z.B. (Netzanschluss)Leitung für Motor, Stromrichter*
supply connection Netzanschluss || Netzeinspeisung
supply dip Netzeinbruch * *kurzzeitiger ~, aus d. Sicht eines Verbrauchers/Motors/Stromrichters*
supply dip buffering Netzeinbruchüberbrückung
supply dip ride-through Netzeinbruchüberbrückung
supply failure Netzausfall * *aus d.Sicht eines Verbrauchers z.B. Motor/Stromrichter* || Netzfehler || Spannungsausfall
supply fault Netzstörung * *aus der Sicht eines Verbrauchers/Motors/Stromrichters*
supply feed Netzeinspeisung
supply feeder Netzeinspeisung || Netzzuleitung * *Einspeisung*
supply fluctuations Netzschwankungen || Netzspannungsschwankungen * *bezogen auf die NetzVersorgungsspannung*
supply frequency Netzfrequenz || Speisefrequenz
supply frequency adaption Netzfrequenzanpassung * *z.B. bei Steuersatz e. netzgeführten Stromrichters*
supply frequency tracking Netzfrequenznachführung * *z.B. bei netzgeführtem Stromrichter*
supply half wave Netzhalbwelle
supply interruption Netzausfall
supply lead Netzzuleitung * *Leitung, Anschluss (-leitung)*
supply line Netz * *Versorgungs~ vom Verbraucher aus gesehen* || Speiseleitung || Versorgungsleitung || Zuleitung
supply location Lieferort
supply logistics Lieferlogistik
supply monitoring Netzüberwachung
supply network Netz
supply pipe Anschlussleitung * *Anschlussrohr (z.B. in Wasserversorgungs-, Pneumatiksystem usw.)*
supply samples bemustern * *of:mit*
supply source Versorgungsquelle
supply symmetry Netzsymmetrie
supply system Netz * *Versorgungs-/Speisesystem; on the supply system: am* ~
supply system feeder Netzeinspeisung * *zu Verbrauchern*
supply terminal Anschlussklemme * *für Stromversorgung*
supply terminals Einspeiseklemmen
supply unit Einspeiseeinheit * *z.B. z. Erzeugen der Zwischenkreisspannung für Antriebswechselrichter*
supply voltage Anschlussspannung * *z.B. eines Stromrichters/Motors* || Leiterspannung || Netzanschlussspannung || Netzspannung * *Eingangsspannung für Stromrichter* || Speisespannung || Versorgungsspannung
supply voltage dip Netzeinbruch * *kurzzeitiger* ~
supply voltage dip buffering Netzeinbruchüberbrückung
supply voltage fluctuations Netzspannungsschwankungen
supply voltage range Netzspannungsbereich
supply voltage zero crossing Netznulldurchgang
supply with liefern * *supply someone with: jdm. etwas (zu)~*
supply zero crossover Netznulldurchgang
supply-side eingangsseitig * *Netz-/Versorgungs(-spannungs)-seitig*
supply-system Netz * *Versorgungs-/Speisesystem; on the supply-system: am* ~
supply-system neutral point Netzmittelpunkt
supply-system voltage dip Netzspannungseinbruch
supplying factory Lieferwerk
supplying samples Bemusterung
supplying works Lieferwerk
support abstützen || Abstützung || Auflage * *Stütze* || Aufrechterhaltung || Beratung * *Unterstützung* || Beteiligung * *Unterstützung (of: an/von)* || betreuen * *unterstützen* || Betreuung * *Unterstützung* || Bock || erhalten * *unterstützen* || fördern * *unterstützen* || Förderung * *Unterstützung* || Gestell || halten * *stützen* || Halter * *Stütze* || Halterung || lagern * *(unter)stützen, Maschine(nteil)* ~/betten || Lagerung * *Stützung, Aufhängung* || Strebe * *Stütze, Träger, Ständer, Strebe, Absteifung, Bettung, Stativ* || Stütze
support disk Tragscheibe
support frame Bock
support leg Stütze * *Stützbein*
support module Trägermodul
support plate Tragblech
support point Stützpunkt * *eines Polygonzuges*
support value Stützpunkt * *eines Polygonzuges/Kennlinienbausteins* || Stützwert * *Stützpunkt einer aus Geradenstücken entstehenden Kennlinie (Polygonzug)*
supporting Stütz- || Stützung * *auch von Preisen usw. (siehe auch →Subventionierung)* || Subventionierung * *Stützung (auch von Preisen, einer Währung usw.)*
supporting arm Tragarm
supporting capacitor Stützkondensator * *z.Stützen d. Stromversorgg. auf Elektron.leiterkarte, 'schluckt' Schaltspitzen* || Stützkondensator
supporting drum Stützwalze
supporting element tragendes Teil

supporting insulator Klemmenträger * *für eine (Schraub-) Klemme* || Stützisolator * *für eine Schraubklemme*
supporting module Trägermodul
supporting plate Montageplatte * *Trägerplatte* || Tragblech
supporting rail Tragschiene
supportive measures unterstützende Maßnahmen
suppose voraussetzen * *als Notwendigkeit/gegeben/mögl.voraussetz./annehmen;* || vorstellen, vermuten,glauben
supposed angenommen * *(Adj.) vermutet* || vermutlich
supposition Gedanke * *Mutmaßung* || Voraussetzung
suppress abbauen * *abschaffen* || ausblenden * *unterdrücken* || dämpfen * *unterdrücken* || deaktivieren * *unterdrücken* || totlegen * *unterdrücken* || unterdrücken || weglassen * *unterdrücken*
suppress interference entstören
suppressed unterdrückt
suppressed circulating current mode kreisstromfreier Betrieb
suppressed zero unterdrückter Nullpunkt
suppression -Sperre * *Unterdrückung* || Dämpfung * *Unterdrückung* || Sperre * *Unterdrückung* || Unterdrückung
suppression band Ausblendband
suppression choke Störschutzdrossel
suppression frequency Ausklammerfrequenz * *unterdrückter Frequenzsollwert z. Umrichters zur Vermeidung mechan.Resonanzen*
suppressor circuit Schutzbeschaltung * *z.B. Überspannungs-Schutzbeschaltung, TSE-Beschaltung bei Thyristorsatz* || TSE-Beschaltung || Überspannungs-Schutzbeschaltung
suppressor network Schutzbeschaltung
supreme höchst, oberst, größt, äußerst
supressor Funkenlöschglied
surcharge Aufpreis || Gebühr * *zusätzliche ~, Zuschlag, zusätzliche Belastung (Konto)* || Mehrpreis * *→Preisaufschlag* || Preisaufschlag || Zuschlag * *Aufschlag, (Preis-) Zuschlag*
sure sicher * *gewiss*
surety Sicherheit * *Gewissheit*
surface Fläche * *Ober~; machined surface: bearbeitete ~; mating surfaces: aufeinander arbeit. ~n* || Oberfläche
surface area Fläche * *(geometr.) Ober~* || Oberfläche
surface coating Oberflächenbeschichtung
surface cooling Oberflächenkühlung * *{el.Masch.} Ableitg. der Wärme über d. geschloss. äußere Oberfläche,VDE 0530 T.6*
surface engineering Oberflächentechnik
surface finish Oberflächenbearbeitung * *End/Nachbearbeitung, Veredelung* || Oberflächenveredelung
surface finishing Oberflächenbearbeitung * *Nach-/Feinbearbeiten der Oberfläche*
surface grinding machine Flachschleifmaschine * *{Werkz.masch.}*
surface integral Flächenintegral
surface leakage current Kriechstrom
surface machining Oberflächenbearbeitung * *mit Werkzeugmaschine*
surface metre per minute Umfangsgeschwindigkeit * *in m/min*
surface mining Bergwerks- * *Tagebau-* || Tagebau * *{Bergwerk}*
surface mounting Oberflächenmontage
surface pattern Oberfläche * *Gestaltung-/Aussehen der Oberfläche*
surface planer Abricht-Hobelmaschine * *{Holz}* || Abrichthobel * *{Holz}* || Abrichthobelmaschine * *{Holz}*

surface planing machine Abrichthobelmaschine
surface pressure Anpressdruck || Anpresskraft || Flächendruck
surface protection Oberflächenschutz
surface quality Oberflächenqualität * *z.B. v. Blech, Papier usw.*
surface sizing Oberflächenleimung * *von Papier*
surface speed Bahngeschwindigkeit * *Oberflächengeschwindigkeit* || Umfangsgeschwindigkeit * *Oberflächengeschwindigkeit*
surface tachometer Anlegetacho * *zur Messung der Bahngeschwindigkeit* || Bahntacho * *Anlegetacho* || Tachometer * *Bahn~, Anlege~, Messrad*
surface treatment Oberflächenbearbeitung || Oberflächenbehandlung || Oberflächentechnik
surface velocity Umfangsgeschwindigkeit * *Oberflächengeschwindigkeit*
surface ventilation Oberflächenkühlung * *z.B. über Kühlrippen bei Motor*
surface winder Kontaktwickler * *z.B. Druckwalzen/Doppeltragwalzenroller (m. Andruckrolle i Gegens. z. Achswickler)* || Oberflächenwickler * *Druckwalzenprinzip: Zug wird durch Anpresswalze erzeugt (i.Ggs.z. Achswickler)* || Umfangswickler || Wickler * *Oberflächenwickler, z.B. Stützwalzenroller, Doppeltragwalzenroller*
surface-cooled oberflächenbelüftet || oberflächengekühlt || selbstbelüftet || selbstgekühlt * *oberflächengekühlt, z.B. Motor ohne Lüfter (f.Motor wenig gebräuchlicher Ausdruck)* || unbelüftet * *oberflächengekühlt (wenig gebräuchlicher Ausdruck für Motorkühlart)*
surface-driven winder Umfangswickler
surface-mount Auf-Putz
surface-mount device oberflächenmontierbares Bauelement
surface-sized paper oberflächengeleimtes Papier
surface-treated oberflächenbehandelt
surfacer Spachtel * *{Holz} Spachtelmasse*
surge ansteigen * *vorübergehend/stoßartig ~, z.B. Spannung, (Luft-) Druck* || Anstieg * *vorübergehender, plötzlicher, hochbrandender ~ (z.B. Spannungs~, Druck~)* || Spannungsstoß || Stoß * *plötzlicher Anstieg, Spannungs~*
surge absorber diode Freilaufdiode * *z.B. parallel zu Relaisspule, vermeidet Überspanng.b.Schalten induktiver Last* || Überspannungsschutzdiode
surge arrester Überspannungsableiter
surge current Stoßstrom
surge diverter Überspannungsableiter
surge immunity Zerstörsicherheit * *gegen Überspannungen geschützt*
surge impedance Wellenwiderstand
surge protector Überspannungsableiter || Überspannungsschutz * *Überspannungsschutzelement*
surge strength Überspannungsfestigkeit * *Festigkeit gegen Spannungsspitzen*
surge strength class Überspannungskategorie * *kennzeichnet Überspannungsfestigkeit nach VDE 0110 Teil 2*
surge stressability Stoßfestigkeit
surge stressing Stoßbeanspruchung * *durch Spannungs/Stromstöße*
surge suppresion Überspannungsschutz * *gegen Spannungsspitzen*
surge suppression circuit TSE-Beschaltung * *RC-Beschaltg. geg. d. 'TrägerSpeicherEffekt' b.Abkommutieren e.Thyristors*
surge suppression network TSE-Beschaltung * *RC-Beschaltg. geg. d. 'TrägerSpeicherEffekt' b. Abkommutier. e.Thyristors* || TSE-Schutzbeschaltung
surge suppressor Ableiter * *z.B.Überspannungsableiter* || Schutzbeschaltung * *Überspannungs-Schutzbeschaltung, Überspannungsableiter* ||

surge voltage Überspannungsableiter || Überspannungsschutz * *Überspannungsableiter*
surge voltage Spitzenspannung * *Höhe von Spannungsspitzen, z.B. nach IEC*
surge withstand capability Stoßspannungsfestigkeit
surge-proof schaltfest * *z.B. Kondensator mit hoher Spitzenstrombelastbarkeit* || spannungsfest * *robust bezüglich Spannungsspitzen (z.B. bei Kondensatoren)* || stoßfest * *(elektr.) gegen Stoßspannungen*
surge-protection capacitor Überspannungsschutzkondensator
surge-suppressor components TSE-Beschaltungselemente * *verhindern den TSE- (Träger-Speicher-) -Effekt bei e. Thyristorschaltg.*
surging ansteigend * *plötzlich, z.B. Druck, Spannung; hochbrandend*
surmount überstehen * *überwinden* || übersteigen || überwinden * *Schwierigkeiten usw. ~*
surpass hinausgehen über || übersteigen * *Erwartungen usw. ~* || übertreffen * *überholen*
surplus Überfluss * *Überschuss* || überflüssig * *überschüssig* || Überschuss || überschüssig * *überschüssig, Überschuss-, Mehr-; surplus weight: Mehr-/Übergewicht*
surprise unerwartet * *überraschend, Überraschungs-*
surprising unerwartet * *überraschend*
surrender aushändigen || zurückgeben * *übergeben, ausliefern, aushändigen; to:an*
surreptitious geheim * *unerlaubt, erschlichen, betrügerisch, heimlich, verstohlen*
surround umgeben * *umgeben, umringen (auch figürl.)*
surrounding umgebend
surrounding air Außenluft * *z.B. bei Kühlsystem*
surroundings Nähe * *Umgebung* || Umgebung * *das Umgebende, der umgebende Raum z.B. einer Stadt, einer Maschine*
surveillance Beaufsichtigung || Überwachung * *Beaufsichtigung, genaue Betrachtung/Prüfung*
survey Anfrage * *Umfrage* || Aufstellung * *Übersicht* || beobachten * *genau betrachten, sorgfältig prüfen, mustern, überblicken, abschätzen, begutachten* || Studie * *→Untersuchung* || Überblick || Übersicht * *auch Überblick, Gutachten, Begutachtung, (Prüfungs-)Bericht* || überwachen * *genau betrachten, inspizieren, prüfen, begutachten* || Umfrage || Untersuchung * *Übersicht* || Untersuchung * *Prüfung, (Ab-) Schätzung, Begutachtung, Umfrage, Überblick, Aufnahme (v.Tatbeständen)* || Zusammenstellung * *Übersicht, Überblick (of: über)*
survey diagram Übersichtsplan || Übersichtsschaltbild
survey report Gutachten
surveying Messung * *Vermessung, Aufnahme, Prüfung, Begutachtung*
survive überstehen * *überleben*
susceptibility Anfälligkeit * *auch 'Empfindlichkeit, Empfänglichkeit' (to:für/gegen)* || Empfindlichkeit * *to:für/gegen; Anfälligkeit, Empfindlichkeit, Empfänglichkeit* || Suszeptibilität
susceptibility to breakdown Störanfälligkeit * *zum Ausfallen neigend* || Störempfindlichkeit * *zum Ausfallen neigend*
susceptibility to failure Störanfälligkeit * *Fehleranfälligkeit*
susceptibility to faults Störanfälligkeit * *Ausfall-Anfälligkeit*
susceptibility to interference Störanfälligkeit * *gegen elektromagnetische Störungen* || Störempfindlichkeit * *gegen elektromagnetische Störungen*
susceptible anfällig * *auch 'empfindlich, empfänglich' (to:für/gegen)* || empfindlich * *auch anfällig, empfänglich, zugänglich (to:gegen/für)*
susceptible to faults störempfindlich * *fehleranfällig* || störungsanfällig
susceptible to interference störanfällig * *anfällig gegen elektromagnetische Störungen* || störempfindlich * *gegen elektromagnetische Störungen*
susceptible to trouble störanfällig || störempfindlich * *fehleranfällig*
suscribe to beziehen * *z.B. Zeitschrift ~*
suspend aufhängen * *from:an* || aufheben * *zeitweilig~, vorläufig ~* || aussetzen * *Zahlung usw. ~* || einhängen * *into: in (Konstruktionselement, Bauteil usw.)* || unterbrechen * *vorübergehend anhalten*
suspended load hängende Last * *am Kran usw.*
suspended matters Schwebestoffe
suspended railway Hängebahn
suspension Aufhängung || Befestigung * *Aufhängung, Federung* || Einstellung * *vorübergehende Beendigung, Aufschieben* || Schwebe- || Suspension * *Flüssigkeit mit ungelösten, gleichmäßig verteilten (Schweb-) Stoffen/Partikeln* || Unterbrechung * *Aussetzung, Hemmung, vorübergehende Aufhebung, Aufschieben*
suspension cable Hängekabel
suspension of liability Haftungsausschluss
suspension pipe Tragrohr
suspension point Aufhängepunkt
suspension sifter Schwebesichter * *Sortieranlage (z.B. für Holzspäne)*
sustain aufrechterhalten * *Brauch, Lehre, Urteil ~* || erleiden * *Schaden usw. ~* || standhalten * *einem Angriff ~* || stützen * *stützen, tragen, (aufrecht)erhalten in Gang halten, (Interesse) wachhalten, unterhalten*
sustain injury Schaden erleiden
sustained andauernd * *anhaltend, Dauer-, ungedämpft, aufrechterhalten* || durchgängig * *aufrecht erhalten, durchgehalten, anhaltend; ungedämpft* || verbleibend
sustained forcing zwangssteuern * *(Substantiv) Festvorgabe von SPS-Ein-/Ausgangswerten zum Test*
swage Gesenk || stauchen * *{Holz} Schärfmethode für Sägen*
swager Stauchapparat * *{Holz} zum Schärfen von Sägen*
swaging machine Stauchmaschine * *{Holz} zum Schärfen von Sägen*
swaging unit Stauchapparat * *{Holz} zum Schärfen von Sägen*
swamp überhäufen * *(with: mit (Arbeit, Vorwürfen, Anrufen, Aufträgen usw.))* || überschwemmen
swank nobel * *(salopp) elegant, schick, protzig*
swap Austausch * *gegenseitiger ~; z.B. Nibbles in einem Byte* || austauschen * *(miteinander ver-)tauschen, z.B. Bytes in einem Datenwort, Daten in einem Speicher* || tauschen * *miteinander ver~, aus~, z.B. zwei Bytes in e. Datenwort/Briefmarken* || vertauschen * *vertauschen, z.B. Bytes in einem Wort*
swapfile Auslagerungsdatei * *zur vorübergehenden Auslagerung von RAM-Daten auf d.Festplatte*
swarm Menge * *Schwarm*
sway beeinflussen * *beherrschen* || Gewalt * *Herrschaft (over:über)* || schaukeln * *sich wiegen, schwanken* || schwenken * *schwenken, (sich) neigen, schaukeln, (sich) wiegen* || schwingen * *schwanken, schaukeln (sich) wiegen*
sweat out ausschwitzen
Sweden Schweden
Swedish schwedisch
sweep amplitude Wobbelamplitude
sweep away mitreißen
sweep bandwidth Wobbelbandbreite

sweep frequency Wobbelfrequenz * z.B. bei Funktionsgenerator
sweep frequency generator Wobbelgenerator
sweep generator Wobbelgenerator
sweep width Wobbelbandbreite
sweeping machine Kehrmaschine
swell aufquellen * anschwellen || prima * (salopp) totschick, piekfein, stinkvornehm, feudal
swelled bauschig
swelling Rundung * Wölbung, Ausbauchung, Beule
swift schnell * rasch, zügig, flüchtig (Bekanntschaft, Zeit), geschwind, eilig, flink, geschickt; Haspel || zügig
swill spülen
swing Ausladung * eines Schwenkkrans || klappen * schwenken || pendeln * (hin und her) schwingen/ schwenken/baumeln/pendeln lassen || schleudern * (hin- u.her-) schwingen, s. drehen, (herum-) schwenken, schlenkern, durchdrehen,anwerfen || schwanken * schwingen, sich wiegen || Schwenk- * swing-out: ausschwenkbar || schwenken * schwingen, schwenken, ein/ausschwenken lassen || schwingen * sich wiegen, pendeln || Schwung * Schwung, Schwingen, Schwingung, ~weite, Ausschlag (Pendel); in full swing: in vollem Gang
swing away ausschwenken
swing out ausschwenken
swing over ausschwenken
swing-frame Schwenkrahmen
swing-out Klapp- * ausschwenkbar || klappbar * ausschwenkbar || schwenkbar * ausschwenkbar
swing-out mechanism Schwenkvorrichtung * Ausschwenkmechanismus
swinging Schwenk-
swinging device Schwenkvorrichtung
swinging frame generator Pendelmaschine * zur Drehmomentmessung
swinging lever Schwenkarm
swinging stator dynamometer Pendelgenerator
swinging-frame dynamometer Pendelgenerator || Pendelmaschine * zur Drehmomentvorgabe/messung
swinging-frame generator Pendelgenerator * zur Drehmomentmessung
swinging-frame grinding machine Pendelschleifmaschine
swinging-frame machine Pendelmaschine * el.- Masch. mit ausschwingbar gelagertem Ständer zur Drehmomentmessung
swinging-stator dynamometer Pendelmaschine * zur Drehmomentmessung
switch rangieren * {Bahnen} (akerikan.) switcher: Rangierlokomotive || schalten || Schalter || umschalten || Weiche * work/shift the switches: die ~ stellen (auch figürl.)
switch back zurückschalten
switch board Montageplatte * z.B. für Mess- u.Bediengeräte, Schalttafel || Schalttafel
switch device Schaltgerät
switch disconnecter Lasttrenner || Lasttrennschalter
switch enclosure Schaltergehäuse
switch frequency Taktfrequenz * für Leistungsteil
switch group Schaltwerk * bei elektromechan. Schalter
switch in einschalten * hineinschalten, einlegen, aktivieren || zuschalten
switch into aufschalten
switch off abschalten || Abschaltung || abstellen || ausschalten || löschen * z.B. Thyristor ~
switch off delay Ausschaltverzögerung
switch on anschalten * einschalten || einschalten || leuchten * Licht einschalten
switch out abschalten || ausmachen * Licht usw. ~ || ausschalten

switch over übergehen * to: zu etwas || umschalten
switch over point Ablösepunkt * z.b. für Feldschwächung
switch over voltage Ablösespannung * für Feldschwächung
switch position Schalterstellung || Schaltstellung || Schaltzustand
switch room Schaltraum
switch state Schaltzustand
switch status Schaltzustand
switch status indicator Schaltzustandsanzeige * z.B. an Näherungsschalter, Schütz
switch through durchschalten
switch to aufschalten
switch to HIGH auf HIGH schalten
switch to LOW auf LOW schalten
switch-disconnector Lasttrenner || Lasttrennschalter
switch-fuse unit Schalter-Sicherungseinheit
switch-in aufschalten * (auch Substantiv) || Zuschaltung * (Substantiv) z.B. (Zusatz)Sollwert || zuschalten * (Substantiv) || Zuschaltung
switch-in delay Ansprechverzögerung
switch-in threshold Einschaltschwelle
switch-mode Schalt- * z.B. -Netzteil
switch-mode power supply getaktetes Netzteil || Schaltnetzteil
switch-off and on again aus- und wiedereinschalten
switch-off delay Abfallverzögerung
switch-off logic Ausschaltlogik
switch-off operation Abschaltvorgang
switch-on einschalten
switch-on and switch-off control Ein-Ausschaltlogik || Ein-Ausschaltsteuerung || Einschalt-Ausschalt-Logik
switch-on and switch-off logic Ein-Ausschaltlogik
switch-on command Einschaltbefehl
switch-on control circuit Einschaltsteuerwerk
switch-on delay Ansprechverzögerung * Einschaltverzögerung || Einschaltverzögerung
switch-on duration ED * →Einschaltdauer || Einschaltdauer
switch-on inhibit Einschaltsperre
switch-on ratio Einschaltverhältnis * Verhältnis Einschaltdauer zur Gesamt-Schaltperiode z.B. bei Bremswiderstand
switch-ON readiness Einschaltbereit * (Zustandsbit im →PROFIBUS-Profil) || Einschaltbereitschaft
switch-on sequencing Einschaltlogik * →Einschaltsteuerwerk
switch-on-sequence Einschaltvorgang
switch-over logic Kommandostufe
switch-over of operating mode Betriebsartenwechsel
switch-over time Umschaltzeit
switch-selectable umschaltbar * mit Schalter wählbar
switch-through führen * ein Signal durchschalten
switch-through to schalten auf * ein Signal führen auf
switchable umschaltbar
switchboard Schalttafel || Steuertafel
switchboard gallery Warte
switched abschaltbar || geschaltet
switched energy Schaltarbeit * bei Kupplung, Bremse
switched off abgeschaltet || ausgeschaltet * siehe →ausschalten
switched on eingeschaltet
switched reluctance motor geschalteter Reluktanzmotor || SR-Motor
switched through durchgeschaltet
switched-in eingefallen * z.B. Schützkontakt

switched-mode

switched-mode getaktet * *Netzteil, Netzgerät*
switched-mode power supply primärgetaktetes Schaltnetzteil * *Schaltnetzteil* || Schaltnetzteil
switchgear Schaltgerät || Schaltgeräte
switchgear and controlgear Schaltgeräte
switchgear cabinet Schaltschrank
switchgear cubicle Schaltschrank
switchgear frame Schaltgerüst
switchgear house Schalthaus * *Schaltanlage*
switchgear manufacture Schaltanlagenbau
switchgear room Schaltraum
switchgear system Schaltanlage
switchgear units Schaltgeräte
switching Schalt- || Schaltung * *Schaltvorgang, das Schalten*
switching action Schalthandlung || Schaltung * →*Schalthandlung*
switching capability Schaltvermögen
switching capacity Kontaktbelastbarkeit * *Schaltvermögen* || Kontaktbelastung * *Schaltvermögen, Kontaktbelastbarkeit* || Schaltleistung || Schaltvermögen
switching characteristic Schaltcharakteristik || Schaltverhalten
switching characteristics Schaltverhalten
switching command Schaltbefehl
switching condition Schaltungszustand
switching controller Schaltregler
switching current Schaltstrom
switching cycle Schaltspiel
switching delay time Schaltverzugszeit * *z.B. eines Leistungsschalters*
switching diode Schaltdiode
switching edge Schaltflanke * *(extremely steep: extrem steil)*
switching frequency Schaltfrequenz * *z.B. für PWM-Ansteuerung eines Leistungsteils* || Taktfrequenz * *für Leistungsteil*
switching hysteresis Schalthysterese
switching instant Schaltpunkt * *Zeitpunkt*
switching level Schaltpegel
switching lever Schalthebel * *bei elektr. Schalter*
switching losses Schaltverluste || Schaltverluste * *z.B. bei Thyristor, Transistor*
switching mode Schaltbetrieb * *bei Wechselstromsteller*
switching noise Schaltgeräusch * *einer Bremse/Kupplung*
switching of negligible currents annähernd stromloses Schalten
switching off Abschaltung
switching operation Schalthandlung || Schaltspiel || Schaltung * *Schaltvorgang (z.B. einer mechan. Kupplung; innerhalb eines Lastspiels usw.)* || Schaltvorgang
switching pattern Impulsmuster
switching performance Schaltcharakteristik * *Schaltverhalten*
switching point Schaltpunkt
switching process Schaltvorgang
switching rate Schalthäufigkeit * *~ von Kontakten*
switching regulator Schaltregler * *z.B. in Schaltnetzteil*
switching sequence Schaltfolge || Schaltvorgang * *Schaltfolge*
switching signal Schaltbefehl
switching spikes Schaltspitzen
switching state Schaltzustand
switching station Schaltanlage
switching time Schaltzeit
switching to ground gegen M schaltend * *z.B. Open Collector-Digitalausgabestufe*
switching to P potential gegen P schaltend * *z.B. Open Emitter-Digitalausgabestufe* || P-schaltend * *geg.pos. Versorg.panng. schaltend (z.B.Open Emitter-Digitalausg.stufe)*
switching transient Schaltspitze || Schaltvorgang * *(elektr.) auf Energieversorgungsleitung/-netz*
switching value Ansprechwert * *Relais, Schalter*
switching voltage Schaltspannung
switching-in Aufschaltung * *z.B. eines Zusatzsollwertes* || zuschalten * *(Substantiv)*
switchover Umschaltung
switchover sequence Umschaltvorgang
switchplant Schaltanlage
Switzerland Schweiz
swivel Dreh- * *Dreh-, Schwenk-, dreh/schwenkbar (z.B. über Gelenk, Drehzapfen)* || Schwenk- * *z.B. Achse, Hebel, Arm* || schwenken * *auf einem Zapfen schwenken, drehen*
swivel arm Schwenkarm
swivel frame Schwenkrahmen
swivel unit Schwenkwerk * *{Hebezeuge}*
swivel- Gelenk-
swivel-mounted schwenkbar * *schwenkbar gelagert*
swiveling Schwenk-
swiveling angle Schwenkwinkel
swiveling mechanism Schwenkvorrichtung
swivelling drehbar * *dreh/schwenkbar (z.B. über Gelenk, Drehzapfen)* || Schwenk- || schwenkbar
swivelling arm Schwenkarm
swivelling gear Schwenkwerk * *{Hebezeuge}*
swung-out type ausklappbar
symbiosis Symbiose
symbol Formelzeichen * *für {el.Masch}: siehe z.B. DIN 1304 Teil 7* || Kurzzeichen || Sinnbild || Symbol || Zeichen * *Symbol; auch Plus~, Minus~, Frage~ usw.*
symbolic symbolisch
symbolic name symbolischer Name
symbolical symbolisch
symbolical addressing symbolische Adressierung * *{Computer} durch e. mnemotechn. Namen, nicht durch eine physikal. Adresse*
symbolism Symbolik
symbology Symbolik * *auch die '(Bedeutung der) definierten Symbole'*
symmetric gleichmäßig * *ebenmäßig* || symmetrisch
symmetric components symmetrische Komponenten * *Zerlegung e.unsymmetr. Drehsromsystems in* →*Mit-, Gegen- und Nullkomponente*
symmetrical gleichmäßig * *ebenmäßig* || symmetrisch
symmetrical optimum symmetrisches Optimum * *Regleroptimierung,sodass Regelfläche gleich null ist (hohes Überschwingen)*
symmetrically split interpole winding symmetrisch geteilte Wendepolwicklung
symmetrize symmetrieren
symmetry Symmetrie * *(auch von Spannungen und Strömen)* || Symmetrierung * *Symmetrie*
sympathetic oscillation angeregte Schwingung
sympathetic vibration angeregte Schwingung * *mechanische*
sympathize with bedauern * *a person:jemanden*
sympathy Verständnis * *Mitgefühl*
symposium Symposium
sync mark Synchronisiermarke * *Signal, das den Referenzpunkt bei Winkel-/Weg-/Gleichlaufregelung vorgibt*
sync voltage Synchronisierspannung * *Kurzform für 'synchronizing voltage'*
synchro Drehmelder * *drehender Umformer, wandelt Wellenwinkelpositionen i.sinusförm. Signal um u.umgekehrt*
synchro control Gleichlaufregelung

synchro system   elektrische Welle * *Gleichlaufeinrichtung mit Asynchron-Schleifringläufermotoren*
synchro-generator   Drehmelder * *Drehgeber für elektrische Welle*
synchro-system   Ferndrehwelle * *Ausführungsform d. elektr. Welle mit Asynchron-Schleifringläufermotoren*
synchronisation   Synchronisierung
synchronisation control   Gleichlaufsteuerung
synchronisation controller   Gleichlaufregler
synchronising voltage   Synchronisierspannung
synchronism   Gleichlauf || Synchronismus || Synchronlauf
synchronism deviation   Gleichlauffehler || Gleichlaufschwankungen * *Gleichlauffehler*
synchronization   Synchronisation || Synchronisierung
synchronization block   Synchronisierstufe
synchronization circuit   Synchronisierstufe
synchronization control   Gleichlaufregelung
synchronization module   Synchronisierbaugruppe * *z.B. für Steuersatz e. netzgeführten Stromrichters*
synchronization torque   Synchronisierungsmoment * *bei Synchronmaschine*
synchronization voltage   Synchronisierspannung
synchronize   abgleichen * *in Übereinstimmung bringen (auch Datenbestände), abstimmen, synchronisieren* || in Übereinstimmung bringen * *with: mit* || synchronisieren
synchronized   synchronisiert
synchronized asynchronous motor   synchronisierter Asynchronmotor * *nach erfolgt.Hochl. Umschaltg.in Synchronbetr.(Läufer an Gleichsp.)*
synchronized drives control   Gleichlaufregelung
synchronized drives controller   Gleichlaufregler
synchronized induction motor   synchronisierter Asynchronmotor * *(el.Masch.)*
synchronizing   Synchronisier- || Zusammenschalten * *zum Synchronlauf*
synchronizing board   Synchronisierbaugruppe * *Leiterkarte zur Synchronisierung e.fremdgeführten (Thyristor-)Stromrichters*
synchronizing control   Gleichlaufregelung || Winkelgleichlaufregelung * *Gleichlaufregelung*
synchronizing controller   Gleichlaufregler
synchronizing mark   Synchronisiermarke
synchronizing option   Synchronisierzusatz * *z.B. z.Synchronisierg. Umrichter-Umr. (Reserveumr.)/ Umr.-Netz (Hochfahrumr.)*
synchronizing pulse   Synchronimpuls || Synchronisierimpuls
synchronizing transformer   Synchronisiertrafo * *erzeugt Synchronisiersignal, z.B. für netzgeführten Stromrichter* || Synchronisiertransformator * *erzeugt Synchronisiersignal z.B. für netzgeführten Stromrichter*
synchronizing voltage   Synchronisierspannung
synchronous   gleichzeitig || synchron
synchronous 'servo' motor   Servomotor
synchronous accuracy   Gleichlaufgenauigkeit || Gleichlaufgüte
synchronous control   Gleichlaufregelung
synchronous control loop   Gleichlaufregelung * *Gleichlaufregelkreis*
synchronous frequency converter   Synchron-Frequenzumformer * *rotierender*
synchronous machine   Synchronmaschine * *el. AC-Masch. mit Erregg. durch Dauermagnete oder DC-Feldwicklg. i.Läufer*
synchronous motor   Synchronmotor
synchronous operation   Gleichlauf || Synchronlauf
synchronous reactance   Ankerreaktanz
synchronous running   synchroner Lauf || Synchronismus * *synchroner Lauf* || Synchronlauf
synchronous speed   Drehfelddrehzahl || Synchron-

drehzahl * *(el.Masch.)Drehz.d. Drehfelds bei. AC-Motor [upm]: 120 x Netzfrequ./Polzahl* || synchrone Drehzahl * *e. AC-Motors in [upm]: (120-mal Netzfrequenz/Polzahl)*
synchronous speed control   Gleichlaufregelung
synchronous starting   Frequenzanlauf * *(el.-Masch.) Anlassen eines Asynchronmotors mit stetig wachsender Frequenz*
synergetic effect   Synergie * *Synergieeffekt* || Synergieeffekt
synergic effect   Synergie * *Synergieeffekt* || Synergieeffekt
synergism   Synergie
synergistic effect   Synergie * *Synergieeffekt* || Synergieeffekt
synergy   Synergie
synonym   Inbegriff * *Synonym, sinnverwandtes/bedeutungsgleiches Wort*
synonymous   gleich * *with: gleichbedeutend mit* || gleichbedeutend * *bedeutungsgleich, sinnverwand, gleichbedeutend (with: mit)*
synopsis   Inhaltsübersicht * *kurze* || Überblick * *Zusammenfassung, Übersicht, Abriss* || Übersicht * *Zusammenfassung, Abriss, vergleichende Zusammenschau (meist in einem Buch)* || Zusammenstellung * *Zusammenfassung, Übersicht, Abriss, (vergleichende) Zusammenschau*
syntax   Syntax * *auch einer Programmiersprache (Schreibweise von Anweisungen, Befehlen)*
syntax check   Syntaxprüfung * *z.B. ob ein Quellspracheprogramm den Syntaxregeln einer Programmiersprache entspricht*
syntax element   Sprachmittel * *(Computer) Syntaxelement*
syntax error   Syntaxfehler
synthesis   Synthese
synthetic   künstlich * *synthetisch, künstlich hergestellt* || synthetisch
synthetic fiber   Chemiefaser * *(amerikan.)* || Kunstfaser
synthetic fiber industry   Chemiefaserindustrie * *(amerikan.)*
synthetic fibre   Chemiefaser * *(engl.)*
synthetic fibre industry   Chemiefaserindustrie * *(brit.)*
synthetic fibres   Kunstfasern
synthetic material   Kunststoff
synthetic oil   Synthetiköl || synthetisches öl
synthetic resin   Kunstharz
syringe   spritzen * *(aus-/ab-/be-/ein-) spritzen*
system   Anlage * *(Energieversorgungs-) System, Anlage, Aggregat* || Apparatur || Aufbau * *Gefüge, System* || Gefüge || Methode || Netz * *Strom~, Energieversorgungssystem* || Ordnung * *System* || Schema * *Anordnung, System* || Strecke * *Regel~* || System * *systematize: in ein ~ bringen/~atisieren* || Systematik * *Aufbau* || Verfahren || Verfahrensweise * *Schema, System*
system architecture   Systemarchitektur
system bus   Systembus
system call   Systemaufruf * *(Computer)*
system cernel   Systemsoftware * *(Betriebs-) Systemkern*
system checkout   Systemprüfung || Systemtest
system clock   Systemtakt
system component   Systemkomponente
system configuration   Projektierung * *einer Anlage* || Systemkonfiguration
system connection   Netzanschluss
system control   Systemsteuerung
system control equipment   Regeleinrichtung
system crash   Systemabsturz * *auch ~ auf Grund eines Softwarefehlers*
system current   Netzstrom
system data   Systemdaten

system data area Bereich-System * *[SPS] Datenbereich für Systemdaten*
system data memory Betriebssystem-Speicherbereich * *bei SPS*
system design Anlagenplanung
system designer Anlagenprojekteur || Systementwickler
system developer Systementwickler
system deviation Regelabweichung || Regeldifferenz || Regeldifferenz * →*Regelabweichung* || Soll-Ist-Abweichung || Soll-Ist-Differenz * →*Regelabweichung* || Soll-Istwert-Abweichung
system display Anlagenbild * *auf einem Bildschirm*
system efficiency Anlagenwirkungsgrad
system engineering Anlagenprojektierung
system environment Systemumgebung
system error Systemfehler
system fault level Netzkurzschlussleistung
system fault power Netzkurzschlussleistung
system features Systemmerkmale
system frequency Netzfrequenz
system hum Netzbrumm
system impedance Netzimpedanz
system in reading Lesart
system infeed Netzeinspeisung * *ins (Stromversorgungs-) Netz*
system integration Systemintegration
system integrator Systemintegrator * *integriert für* →*OEM o.* →*Endabnehmer Systeme aus Komponenten verschied. Hersteller*
system library Systembibliothek
system management Systemmanagement
system matrix Systemmatrix
system noise Netzrückwirkungen
system of coordinates Koordinatensystem
system of designation Kennzeichnungssystem
system parameter Systemparameter
system parameters Streckenparameter
system performance Systemleistung
system perturbation Netzrückwirkung * *Beeinflussg.d.Netzspanng.durch Stromrichter/Motor*
system perturbations Netzrückwirkungen
system power factor Netzleistungsfaktor
system pressure Systemdruck * *in Hydrauliksystem*
system program Systemprogramm
system reaction Netzrückwirkung * *Beeinflussung der Netzspannung durch Stromrichter/Motor*
system recovery Netzwiederkehr
system software Dienstprogramme || System-Software
system solution Systemlösung
system specialist Systemspezialist
system start up engineer Inbetriebnahmeingenieur || Inbetriebnehmer
system start-up Inbetriebnahme * *Rechner/SPS softwaremäßig*
system supplier Systemanbieter || Systemintegrator * *Systemanbieter*
system supply impedance Netzimpedanz
system technology Anlagentechnik
system test Systemtest
system tick Systemtakt
system torque Anlagenmoment
system transfer Netzumschaltung * *{el.Masch.} Umschaltg. e. laufenden AC-Motors auf e.anders Netz (z.B. im Kraftwerk)*
system transfer data Systemtransferdaten
system vendor Systemhersteller * *Systemanbieter*
system voltage Anschlussspannung * *Netzspannung* || Netzspannung
system voltage dip Netzeinbruch * *kurzzeitig ~*
system voltage dip buffering Netzeinbruchüberbrückung

System-Based Drive Technology Antriebstechnik mit System * *SIEMENS-Slogan*
system-compatible systemgerecht
system-inherent systembedingt
system-overlapping systemübergreifend
system-side systemseitig
system-specific anlagenspezifisch
systematic geordnet || schematisch * *systematisch*
systematic person Systematiker
systematical systematisch || zielgerichtet * *systematisch*
systematically systematisch * *(Adv.)*
systematics Systematik
systematization Systematisierung
systematize systematisieren
systemize systematisieren
systems -Technik || Technik * *Systeme*
systems architecture Systemarchitektur
systems cable Netzleitung * *z.B. Motor-/Stromrichterzuleitung*
systems department Anlagenabteilung
systems engineering Anlagentechnik
systems from other manufacturers Fremdsysteme
systems integrator Systemintegrator
systems software Systemsoftware

T

t t * *metrische Tonne (entspr. 1000 kg)*
T section T-Profil
T-element T-Glied * *z.B. Tiefpass aus zwei Widerständen und einem Kondensator*
T-head bolt Hammerschraube * *für Schienenbefestigungssystem (siehe auch* →*T-Nut,* →*Nutstein)*
T-nut Nutenstein * *in Nut eingreifende Befestigungsmutter (z.B. in Montageschienensystem)* || Nutmutter * *für Schienenbefestigungssystem (siehe auch* →*Nutstein,* →*T-Nut,* →*Hammerschraube)* || Nutstein * *in* →*T-Nut eingreifende Befestigungsmutter*
T-slot Hammernut * *zur variablen Schraubbefestigung in einem Profilelement (mit* →*Nutsteinen)* || T-Nut * *z.B. ~ für variable Schraubbefestigung mit "Hammerschrauben"*
T-switch Weiche * *Daten~, z.B. für parallele/serielle Datenleitungen*
T.C. Zeitkonstante
T.C.T. tool HM-Werkzeug * *Abk. für 'Tungsten Carbide Tipped' mit Hartmetallplättchen bestückt*
t6-time t6-Zeit * *{el.Masch.} zuläss. Einschaltdauer e.AC-Motors mit blockiert.Läufer u. 6-fach. Bemess.strom*
tab Flachstecker * *Messerkontakt-Steckzunge* || Lappen * *Streifen, Lappen (z.B. aus Blech), Zipfel, Nase an Maschinenteil* || Nase * *Lappen, Zipfel, Schildchen, Nase, Öse, Fahne, Flachstift, Vorsprung* || Tabulator * *(salopp) Taste, die der Schreibmarke um mehrere Zeichenabstände weiterbewegt*
tab connector Faston-Stecker || Flachstecker || Flachsteckkontakt * *Flachstecker* || Messerkontakt * →*Flachstecker*
tab receptacle Flachstecker * *"weiblicher" Flachstecker, Flachsteckhülse* || Flachsteckhülse * *Gegenstück zur Flachstecker-Messerzunge* || Steckhülse * *Gegenstück für Flachstecker/Messerkontakt*
tab washer Schraubensicherungsblech * *mit hochgezogenen Laschen* || Sicherungsblech * *mit hochgebogenen Lappen zur Schraubensicherung*

table Aufstellung * *Tabelle* || Liste * *Tabelle, tabellarische Aufstellung* || Tabelle || Tisch || Übersicht * *(tabellarische) Aufstellung* || Verzeichnis * *Tabelle* || Zusammenstellung * *Tabelle*
table adjustment Tischverstellung * *{Werkz.-masch.}*
table measuring system Tischmesssystem * *(in der Werkzeugmaschinentechnik)*
table moulding machine Unterfräsmaschine * *{Holz}*
table of contents Inhalt * *z.B. Inhaltsverzeichnis am Anfang eines Buches* || Inhaltsverzeichnis
table roll Registerwalze * *in Papiermaschine/Papierverarbeitungsmaschine*
table value Tabellenwert
table-guide plate Tischleitblech * *z.B. bei Richtmaschine*
tabular tabellarisch
tabular form Tabellenform * *in a tabular form: in ~*
tabularly tabellarisch * *(Adv.)*
tabulate angeben * *in Tabelle ~* || aufführen * *in einer Tabelle* || auflisten * *Listen-/Tabellenförmig zusammenstellen* || zusammenstellen * *in einer Tabelle ~*
tabulated tabellarisch
tabulation Aufstellung * *tabellarische ~*
tabulator Tabulator * *Taste an Computer-Tastatur, die die Schreibmarke um mehrere Leerzeichen weiterbewegt*
tach Drehzahlgeber * *(salopp) Kurzform für 'tachometer'* || Tacho * *[täk]* || Tachometer * *[täk]*
tacho Drehzahlgeber || Tacho * *['täko]* || Tachogenerator || Tachometer * *['täko]*
tacho break-down Tachobruch
tacho failure monitoring Tachoüberwachung
tacho loss Tachoausfall * *Ausfall des Tachosignals z.B. durch Leitungsbruch* || Tachobruch * *Wegbleiben des Tachosignals z.B. durch Leitungsbruch* || Tachofehler * *z.B. durch Drahtbruch der Tacholeitung* || Tachostörung * *Ausfall des Tachosignals*
tacho loss monitoring Tachoüberwachung
tacho rotor Tacholäufer
tacho-generator Drehzahlgeber || Tachodynamo * *Gleichstrom- (manchmal auch Drehstrom-) Generator zur Drehzahlmessung* || Tachogeber || Tachogenerator || Tachomaschine
tacho-switch Drehzahlrelais || Drehzahlwächter
tachogenerator Drehzahlgeber || Tacho || Tachogenerator || Tachomaschine || Tachometerdynamo
tachometer Drehzahlgeber || Drehzahlistwertgeber || Drehzahlmesser || Tacho * *['täkomiete]* || Tachometer * *['täkomete]*
tachometer adaption Tachoanpassung
tachometer coupling Tachokupplung || Tachometerkupplung
tachometer failure Tachoausfall || Tachobruch * *Tacho-Fehler*
tachometer fault Tachostörung
tachometer generator Tachogenerator || Tachomaschine
tachometer interruption Tachobruch * *Unterbrechung des Tachosgnals/der Tacholeitung* || Tachostörung * *Tachobruch, Unterbrechung in der Tacholeitung*
tachometer interruption monitoring Tachoüberwachung
tachometer mounting Tachoanbau
tachometer rotor Tacholäufer
tachometer scaling Drehzahlanpassung * *siehe auch →Drehzahlistwertanpassung* || Drehzahlistwertabgleich || Drehzahlistwertanpassung
tachometer stator Tachoständer
tachometer voltage Tachospannung

tachometric relay Drehzahlrelais || Drehzahlwächter
tack Stift * *(Heft)Zwecke*
tackle angreifen * *(figürlich) e. Aufgabe usw. angreifen* || fertig werden mit || Flaschenzug || gewachsen sein * *fertig werden mit, lösen (z.B. Aufgabe, Problem)* || im Griff haben * *fertig werden mit (Aufgabe, Problem usw.)* || packen * *(an)packen, in Angriff nehmen, fertig werden mit* || überwinden * *(Aufgabe, Problem) lösen, fertig werden mit, etwas "packen"*
tackle block Flaschenzug
tackle with beherrschen * *fertig werden mit, lösen*
tactful feinfühlig * *taktvoll*
tactfulness Takt * *~gefühl*
tactical taktisch
tactics Taktik
tactile fühlbar * *mit Hand/Finger*
tactile touch Druckpunkt * *an Taste/Tastatur*
tag kennzeichnen * *mit Etikett/Kabelbezeichner versehen, (Ware usw.) auszeichnen*
taggered auf Lücke
tail Schwanz * *auch Stromschwanz, Rücken eines Impulses usw.*
tail rope Unterseil * *{Aufzüge | Seilbahn}*
tail wind Aufwind * *(auch figürl.) with tail wind: im ~ (z.B. Geschäfte)*
tailor anpassen * *to: an (Bedürfnisse, Kundenwunsch usw.)*
tailor-made maßgeschneidert
tailored zugeschnitten * *to: z.B. auf die Kundenbedürfnisse ~*
tailored to the market requirements marktgerecht
tailored to the needs of the market marktgerecht
tailstock Mitnehmer * *Reitstock bei Drehbank* || Reitstock * *bei Drehmaschine*
Taiwan Taiwan
take annehmen * *z.B. Form/Gestalt/Farbe; take various forms:verschiedene Formen annehmen* || aufwenden * *pains:Mühe* || beanspruchen * *z.B. Mühe, Sorgfalt, Zeit, Platz ~* || befolgen * *Ratschläge ~* || entnehmen * *e. Probe* || ergreifen * *auch Maßnahmen* || führen * *take a signal to the terminal: ein Signal auf die Klemme ~* || hinnehmen * *Kosten * (Verb) Mühe, Zeit usw. kosten* || tragen * *mitnehmen* || übernehmen * *Führung, Befehl, Risiko ~*
take a course verlaufen * *einen Verlauf nehmen*
take a message ausrichten * *Botschaft ~*
take a reading from ablesen * *Messwert von einem Messgerät*
take action eingreifen * *(figürlich)* || handeln * *in Aktion treten*
take advantage of ausnutzen * *vorteilhaft* || nutzbar machen || nutzen * *vorteilhaft ~* || nutzen * *vorteilhaft ~; take full advantage of: die Vorteile voll aus~*
take aim zielen * *zielen, als/aufs Ziel nehmen (at:-auf)*
take anti-noise measures entstören * *Entstörmaßnahmen ergreifen*
take as a basis zugrunde legen * *for:für* || zugrundelegen
take away wegnehmen
take back wiederholen * *zurückholen* || zurücknehmen
take benefit profitieren * *from: von/an*
take care beachten * *warnend* || vorsehen * *sich ~*
take care of achten auf || betreuen || kümmern * *achten/Acht geben auf, sorgen für, ~ um, erledigen* || kümmern um || schonen * *pfleglich behandeln* || sich kümmern um || sorgen für * *dafür sorgen, dass* || übernehmen * *sorgen für*
take care that achten auf * *darauf achten, dass*

take care! Vorsicht bitte * *vorsichtig behandeln*
take countermeasures Gegenmaßnahmen ergreifen
take down herabsetzen * *herunternehmen*
take down in writing festhalten * *schriftlich genau ~*
take effect anschlagen * *wirken, z.B. Arznei, Abhilfemaßnahme* || auswirken * *Auswirkung haben* || wirken * *wirksam werden, (Ein-) Wirkung/Einfluss haben; have an effect on: (ein)wirken auf*
take EMI-suppression measures entstören * *Entstörmaßnahmen ergreifen*
take exception to beanstanden
take for granted voraussetzen * *als bekannt/gesichert; stillschweigend voraussetzen*
take from entnehmen * *(auch figürlich) wegnehmen; auch Angabe aus einer Tabelle/e.Buch usw. ~*
take hold of fassen * *zu ~ bekommen*
take in ansaugen * *auch Luft* || hereinnehmen * *Geld, Ware, Auftrag ~* || zuführen * *hereinlassen (Flüssigkeit, Gas usw.)*
take in hand vornehmen * *beginnen, in die Hand nehmen*
take in stock hereinnehmen * *auf Lager nehmen*
take into account beachten * *berücksichtigen* || berücksichtigen || in Anschlag bringen * *leave out of account: nicht ~* || in Betracht ziehen
take into consideration berücksichtigen * *in Erwägung ziehen* || in Betracht ziehen
take into consultation konsultieren
take its rise entstehen
take legal action verklagen
take measures Maßnahmen ergreifen
take no notice of nicht beachten
take note of beachten
take notice of beachten || kümmern * *beachten*
take off ablenken * *Aufmerksamkeit, Gedanken usw.* || abnehmen * *lösen* || aufsteigen * *Flugzeug* || Aufstieg * *Flugzeug*
take on annehmen * *z.B. eine Farbe* || erledigen * *(Arbeit)* annehmen, übernehmen
take one's choice wählen * *seine Wahl treffen*
take one's doctor's degree promovieren
take one's place einordnen * *sich ~*
take out herausführen * *z.B. Signal aus einem Gerät* || herausnehmen * *from: aus etw. heraus*
take over übernehmen
take over from ablösen * *übernehmen von z.B. von Vorgänger*
take over the control Führung übernehmen * *(→PROFIBUS-Profil)*
take overlap überlappen
take part beteiligen * *sich beteiligen (in: an/bei)* || teilnehmen * *in: an*
take part in teilen * *teilhaben/nehmen an*
take place eintreten * *stattfinden* || erfolgen * *stattfinden* || passieren * *sich ereignen* || stattfinden
take possession of Übernahme * *Besitz ergreifen von, in Besitz nehmen* || übernehmen * *Besitz ~*
take precautions Vorsorge treffen
take special account of berücksichtigen * *(einem Umstand) Rechnung tragen* || Rechnung tragen * *siehe auch →berücksichtigen*
take the line Standpunkt vertreten
take the measurement messen * *ab~* || vermessen * *abmessen*
take the measurements of durchmessen * *Messwerte (z.B. eines Bauelements) aufnehmen*
take the place of ablösen * *einem Amtsvorgänger nachfolgen*
take the responsibility Verantwortung übernehmen
take the view Standpunkt vertreten * *that: dass*
take the view that Ansicht vertreten

take the wrong one verwechseln
take turns ablösen * *sich (gegenseitig) ~ (at:bei)*
take unpayed time off work unbezahlten Urlaub nehmen
take up aufheben || aufnehmen * *vom Markt, auch "take"* || wegnehmen * *Raum, Zeit usw.*
take-off starten * *Flugzeug*
take-off assistance Starthilfe * *z.B. Flugzeug*
take-off roll Auslaufwalze * *z.B. bei Textilmaschine*
take-off system Abnahmesystem * *Abtransport/Abfuhr von Teilen/Material aus einer Maschine*
take-up aufholen * *durchhängendes Material aufholen, z.B. in Papierbahn* || Aufwickler * *z.B. für Kabel* || Aufwicklung * *z.B. für Kabel* || beanspruchen * *(Platz, Zeit usw.) ~, in Anspruch nehmen, beanspruchen* || einnehmen * *(Platz, Zeit usw.) beanspruchen, in Anspruch nehmen, beanspruchen* || Spuler * *Auf-/Aufwickler, z.B. für Blechband, Kabel*
take-up machine Aufspulmaschine
take-up speed Abzugsgeschwindigkeit * *z.B. in Textilmaschine*
take-up stand Aufwickler * *Aufwickler, z.B. für Kabel* || Aufwicklung * *Aufwickelmaschine, z.B. für Kabel*
take-up station Aufwickler * *Aufwickelstation, z.B. für Kabel* || Aufwicklung * *Aufwickelstation, z.B. für Kabel*
take-up winder Aufwickler
taken aback betroffen * *verblüfft, überrascht*
taken for granted selbstverständlich * *take a thing for granted: etwas für ~ halten*
takeover Übernahme
takeover circuit Ablöseschaltung
taking charge of Übernahme * *Verwaltg., Aufsicht, Obhut, Verantwortg., Leitg., Besitz, Posten*
taking into account unter Berücksichtigung von
taking out Entnahme
taking over Übernahme
taking place Zustandekommen * *Stattfinden*
takings Einnahme * *(Plural) Geschäftseinnahmen*
takt to pieces for reutilization ausschlachten
talk Besprechung * *Gespräch* || Gespräch || Unterhaltung * *Gespräch* || verhandeln * *besprechen* || Vortrag * *Gespräch*
talker Sprecher
talks Verhandlung * *Gespräche*
tall groß * *~ von Wuchs (auch Baum, Haus usw.)*
tally passen * *zusammenpassen* || Rechnung * *{Wirtschaft} Rechnung, Ab-~, Gegen-~, Kontogenbuch (eines Kunden); Zählstrich, Coupon* || übereinstimmen * *übereinstimmen, entsprechen, aufgehen, stimmen (with:mit)*
tambour Rollstange * *für Papier* || Tambour * *Wickeldorn für/mit Papier* || Wickeldorn * *für/mit Papier*
tambour starter Tambourstarter * *{Papier}*
tan beige * *gelbbraun, hellbraun*
tandem Doppel- * *aus 2 gleichen Teilen bestehend (z.B. Kompressor)* || Kaskade || Tandem || Tandem- || Zweifach- * *in Reihen/Kaskadenanordnung, hintereinander angeordnet*
tandem arrangement Kaskadierung
tandem connection Kaskadenschaltung
tandem motor Tandem-Motor * *{el.Masch.} 2 Motoren i. 1 Gehäuse hintereinand. eingebaut, kleiner Durchm. u.Trägheit*
tangent Tangente
tangential tangential
tangential force Tangentialkraft || Umfangskraft * *Tangentialkraft*
tangential stress Scherspannung
tangible fühlbar * *körperlich* || fühlbar * *greif-/fühlbar, (figürl.) handfest* || greifbar * *(figürl.) greif-*

bar, fühlbar, real || konkret * *fühlbar, handgreiflich*
tangible property   Sachvermögen
tank   Behälter * *für Flüssigkeiten* || Bütte || Gefäß * *für Flüssigkeiten* || Wanne * *Flüssigkeitsbehälter*
tank filling and discharge control   Behältersteuerung * *für Flüssigkeiten*
tankage   Vorrat * ~ *an flüssigen Gütern*
tanked   voll * *(salopp) betrunken*
tanning vat   Gerbfass
tantalum   Tantal
tantalum capacitor   Tantalkondensator
tantamount   gleichbedeutend * *to: mit; be tantamount to: etwas gleichkommen*
tap   Abgriff * *Anzapfung, z.B. an Spule, Transformator* || Abzweig * *Stichleitung* || anzapfen * *Trafo/Drossel/Bierfass* ~ || Anzapfung * *tapped: mit ~ versehen* || auskoppeln * *z.B. Energie* || Gewinde bohren || Gewinde- * *Gewlndebohr-, Gewindebohrer-* || gewindebohren || Gewindebohrer || klopfen * *sanft klopfen* || Schlag * *leichter ~, Klaps* || schlagen * *leicht ~, klopfen/pochen (an, gegen), beklopfen, klopfen mit*
tap bolt   Stiftschraube
tap line   Stichleitung
tap off   abstechen * *Schlacke/Stahl ~* || abzweigen * →*anzapfen*
tap-off point   Abgriff * *Abgreifpunkt*
tape   Band * *schmales Band, Streifen, Textil/Klebeband* || Gurt * *z.B. für gegurtete Bauteile* || Streifen * *Band* || umwickeln * *mit Band ~/binden/heften*
tape drive   Bandlaufwerk * *für Magnetbandgerät*
tape measure   Bandmaß || Messband
tape packaging   Gurtung * *für elektr. Bauteile*
taper   abnehmen * *sich verjüngen; z.B. b. hoher Drehzahl abgesenkte Strombegrenzung, abnehmend Wickelhärte* || Abschrägung || Kegel * *verjüngtes Teil* || kegelförmig || kegelig || Konizität || spitz zulaufen || verjüngen * *sich ~, spitz zulaufen, konisch sein* || Verjüngung * *Konizität, z.B. abnehmende Wickelhärte b. hoher Aufwickler*
taper control   Wickelhärtensteuerung
taper current limit   drehzahlabhängige Strombegrenzung * *Stromabsenkung b. hoher Drehz. schützt Kommutator b.DC-Motor*
taper off   auslaufen * *allmählich ~ lassen* || spitz zulaufen * *auch 'sich verjüngen'* || verjüngen * *sich ~*
taper roller bearing   Kegelrollenlager
taper sleeve   Konus * *hohler Konus*
taper tension   abnehmende Wickelhärte * *Zugreduzierung m.steig. Durchm. bei Aufwickler z.Schonen d. Innenlagen* || Wickelhärte * *mit steigendem Durchmesser abnehmender Zug (zur Schonung d.Innenlagen bei Aufwickler)* || Zugreduzierung * *abnehmende Wickelhärte b. steigend. Durchmesser e.Aufwicklers zur Innenlagenschonung*
taper tension characteristic   Wickelhärtenkennlinie * *sorgt für abnehmende Wickelhärte beim Aufwickler*
taper tension control   Wickelhärtensteuerung * *z.B. abnehmende Wickelhärte beim Aufwickler; taper:Verjüngung*
taper tension curve   Wickelhärtenkennlinie * *sorgt für abnehmende Wickelhärte beim Aufwickler*
taper tension reduction   abnehmende Wickelhärte * *beim Aufwickler*
tapered   degressiv || kegelförmig || kegelig || konisch || spitz * ~ *zulaufend; taper (off): ~ zulaufen/sich verjüngen* || spitz zulaufend || verjüngt * *konisch, spitz zulaufend*
tapered current limit   drehzahlabhängige Strombegrenzung * *Stromabsenkung bei hoher Drehzahl z. Kommutatorschutz*
tapered socket   Konus
tapered-roller bearing   Kegelrollenlager

tapering   Verjüngung
taping   Bandage
taping machine   Banderoliermaschine * *{Verpack.-techn.} zum Verpacken mit (Klebe-) Streifen*
tapped   angezapft
tapped blind hole   Gewindesackloch || Sackbohrung * *Gewindesackloch* || Sackloch * *Gewinde-Sackloch*
tapped center hole   Zentrierbohrung * *(amerikan.) als Gewindebohrung ausgeführt*
tapped centre hole   Zentriergewinde * *z.B. am Wellenende eines Motors*
tapped hole   Gewinde * *Gewindeloch* || Gewindebohrung || Gewindeloch || Innengewinde
tapped variable inductor   Anzapfdrossel
tapping   abzweigend * *Leitung usw.* || Anzapfung
tapping bar   Räumeisen
tapping box   Abzweigdose
tapping die   Schneideisen * *Gewindebohrwerkzeug*
tapping point   Anzapfung
tapping screw   Blechschraube
tar   Teer
tardy   langsam * *säumig*
tare weight   Taragewicht
target   Aufgabenstellung * *Ziel(setzung)* || soll * *(Substantiv) (Produktions-, Liefer-) Ziel* || Ziel * *Leistungs-, Produktions~, Soll, ~scheibe* || Ziel- || zielen * *zielen auf* || Zielsetzung
target curve   Fahrkurve * *{Aufzüge}*
target date   Termin * *Zieldatum* || Zieltermin
target group   Zielgruppe * *für Werbung, Marketing, Schulung, Dokumentation usw.*
target hardware   Zielhardware * *auf der eine Software letztendlich ablaufen soll*
target market   Zielmarkt
target position   Lagesollwert * *Zielposition* || Positionssollwert || Zielposition
target specification   Pflichtenheft
target system   Zielsystem * *Hard-/Softwareystem, auf dem e. Software ablaufen soll (siehe auch* →*Ablaufumgebung)*
target tracking system   Zielsteueranlage * *{Fördertechnik} Hebezeuge}*
targeted   ausgerichtet * *auf ein Ziel ~*
tariff   Tarif * *(mit 'ff')*
tarnish   anlaufen * *Metall*
task   Arbeit * *Aufgabe* || Aufgabe || Aufgabenstellung || Auftrag * *Aufgabe* || Einsatz * *Aufgabe*
task definition   Aufgabenstellung
task distribution   Arbeitsteilung
task execution level   Ablaufebene * *in einem Echtzeit-Computersystem*
task manager   Aufgabenverwalter
task scheduler   Aufgabenverwalter * *in einem (Multitask)-Betriebssystem*
task solution   Aufgabenlösung
task-specific   anwendungsspezifisch
taste   Sinn * *Vorliebe, for: für*
tax   Steuer || steuern * *{Wirtschaft} (Substantiv) after tax: nach Steuern*
tax advisor   Steuerberater
tax authority   Steuerbehörde
tax burden   Steuerbelastung
tax consultant   Steuerberater
tax expert   Steuerberater
tax notice   Steuerbescheid
tax preparer   Steuerberater
tax rate   Steuersatz * *{Wirtschaft}*
tax severely   hohe Ansprüche stellen * *an etwas/jemanden*
taxi   Taxi
taxonomy   Systematik * *systematische Einordnung (z.B. Botanik, Zoologie)*
taylor   zuschneiden * *auch ein Produkt to:auf eine Anwendung/auf Kundenbedürfnis*

taylored maßgeschneidert
taylored to suit nach Maß
TC tipped hartmetallbestückt * *{Werkz.masch.} Abkürzg. für 'Tungsten-Carbide tipped'*
TCP-IP TCP-IP * *Transmission-Control Protocol/ Internet Protocol: Datenübertr.protokollfamilie f.Internet..*
TCT HM- * *{Werkz.masch.} Abkürzg. für 'Tungsten Carbide Tipped': hartmetallbestückt (Werkzeug)*
TDO Querrecke * *Abk.f. 'Transversal Directed Orientation': Breitstreckeinrichtung in Folienmaschine*
TDR Zeitrelais * *Abk. für 'Time-Delay Relay'*
teacher Ausbilder * *Lehrer* || Kursleiter * *Lehrer* || Schulungsleiter * *Lehrer*
teaching Unterricht
teaching aids Schulungsunterlagen * *Lehrmittel*
team Arbeitsgemeinschaft || Mannschaft || Stab * *Stab von Experten usw.*
team leader Gruppenführer
team spirit Teamgeist
team-work Zusammenspiel * *von Personen in einer Gruppe*
teamwork Kooperation || Teamarbeit * *by:in* || Zusammenarbeit * *Teamarbeit* || Zusammenspiel * *gemeinsame Zusammenarbeit im Team*
TEAO vollkommen geschlossen * *'Tot. Enclosed Air-Over': ~e Motorschutzart mit Gehäuse-Außenbelüftung*
tear durchreißen || reißen * *tear off: for~; tear out of: aus etwas heraus~; tear in two: entzwei~*
tear asunder durchreißen * *entzweireißen*
tear down abreißen * *niederreißen*
tear in two durchreißen * *entzweireißen*
tear off abreißen * *außeinanderreißen (aktiv)*
tear open aufreißen
tear the lead reißen * *die Führung an sich ~*
tear up aufreißen
tear-off closure Abreißverschluss * *{Verpack.-techn.}* || Aufreißverschluss * *{Verpack.techn.}*
tear-open strip Aufreißstreifen * *{Verpack.techn.}*
tearing resistance Durchreißwiderstand * *z.B. von Papier*
tearing strength Reißfestigkeit
teasel Krempel * *{Textil}*
technical sachlich * *(rein) technisch* || technisch * *for technical reasons: aus ~en Gründen*
technical academy Fachhochschule
technical advice Beratung * *technische ~*
technical adviser Fachberater
technical and trade literature Fachpublikationen
technical and trade publications Fachpublikationen
technical book Fachbuch
technical college Fachhochschule
technical data technische Angaben * *technische Daten* || technische Daten
technical demands technische Anforderungen
technical description technische Beschreibung
technical designer Konstrukteur
technical dictionary Fachwörterbuch
technical director Leiter * *technischer Leiter*
technical documentation technische Dokumentation * *siehe DIN-Entwürfe 8418, 66055 und VDI-Richtlinie 4500*
Technical Editorial Office technische Redaktion * *Abteilungsbezeichnung*
technical features technische Merkmale
technical information technische Angaben
Technical Inspectorate TÜV * *Techn. Überwachungsverein*
technical journal Fachzeitschrift * *technische ~*
technical knowledge Fachkenntnisse || Fachwissen || Kompetenz * *siehe 'Fachwissen'* || Vorkenntnisse * *Fachwissen* || Wissen * *praktische Erfahrung; wide: groß; extensive: umfangreich*
technical literature Fachliteratur
technical magazine Fachzeitschrift * *technische ~*
technical periodical Fachzeitschrift
technical publication Fachzeitschrift
technical requirements technische Anforderungen || technische Voraussetzungen
technical school Fachschule
technical specifications technische Angaben || technische Daten
technical support Beratung * *technische Unterstützung*
technical term Fachausdruck || Fachterminus || Vokabel * *Fachausdruck*
technical terms Begriffe * *techn. Fachbegriffe* || Fachbegriffe || Fachtermini
technical textiles technische Textilien
technical training Weiterbildung * *technische*
technical university Technische Hochschule
technically knowledgeable fachkundig
technician Monteur * *service technician: Kundendienst~* || Techniker
technique -Technik * *Arbeitsverfahren, Technik, Methode, Art der Ausführung, Geschicklichkeit* || Methode || Technik * *Verfahren* || Verfahren
technique of measurement Messmethode * *Messtechnik, Messverfahren* || Messtechnik * *i. einem konkreten Anwendungsfall, Messverfahren*
technological technologisch
technological advance technischer Fortschritt
technological analysis Aufgabenklärung * *technologische ~*
technological closed-loop controls technologische Regelungen
technological function technologische Funktion
technological supplementary unit Technologiezusatz
technologist Technologe
technology -Technik || Technik * *Wissenschaft, Technologie* || Technologie || technology
technology board Technologiebaugruppe * *Leiterkarte*
technology controller Technologieregler
technology functions Technologiefunktionen || technologische Funktionen
technology module Technologiebaugruppe
technology parameter Technologieparameter
technology-dependent technologieabhängig
tedious lästig * *langweilig, öde, ermüdend*
teething trouble Kinderkrankheiten * *auch von techn. Geräten, Verfahren und Einrichtungen*
TEFC eigengekühlt * *'Totally Enclosed Forced Ventilated': vollgekapselte Motorschutzart m. Fremdlüftg.* || IP44 * *Abkürzg. für 'Totally Enclosed Fan Cooled': vollk. geschlossen u.lüftergekühlt (Motorschutzart)* || oberflächenbelüftet * *'Totally Enclosed Fan-Cooled': oberflächengekühlt mit Lüfter (Motor-Kühlart)* || oberflächengekühlt * *'Totally-Enclosed Fan-Cooled': oberfläch.gekühlt m.Fremdlüfter (Motorkühlart)* || TEFC * *Abkürzg.f. 'Totally Enclosed Fan Cooled':vollk. geschlossen u.lüftergekühlt (Schutzart IP44)* || vollkommen geschlossen * *'Totally Enclosed Fan Cooled': ~e Motorschutzart IP44 mit Eigenbelüftung*
Teflon Teflon
Tel. Tel
tele- Fern- * *drahtlos*
tele-sale Telefonverkauf
telecommunication Nachrichtenübertragung || Telekommunikation
telecommunication power supply Fernmeldestromversorgung
telecommunications Fernmeldetechnik

telecommunications system Fernmeldeanlage ||
Fernmeldesystem
telecommunications technology Nachrichtentechnik
telecontrol Fernsteuerung * z.B. drahtlos || Fernwirktechnik
telecontrol station Fernwirkstation
telegram Telegramm * auch Datentelegramm
telegram data transfer Telegrammverkehr * auf Bus, serieller Verbindung usw.
telegram failure Telegrammausfall
telegram failure monitoring Telegrammausfallüberwachung
telegram failure monitoring time Telegrammüberwachungszeit
telegram failure time Telegramm-Ausfallzeit || Telegrammausfallzeit
telegram frame Telegrammrahmen * Aufbau eines Datentelegramms: Plätze für Nettozeichen plus Vorspann plus Nachspann
telegram header Telegrammkopf
telegram sequence Telegrammfolge
telegram timeout Telegramm-Ausfallzeit * Ansprechen der Telegramm-Zeitüberwachung
telegram timeout monitoring Telegrammausfallüberwachung * Zeitüberwachung der Telegrammpausen
telemetering Fernmessung * drahtlose ~ || Telemetrie * drahtlose Messwertübertragung
telemetry Fernmessung * drahtlose ~ || Telemetrie * drahtlose Messdatenübertragung
telephone Telefon
telephone answering machine Anrufbeantworter
telephone call Anruf
telephony Fernsprechtechnik * als Fachgebiet
teleprint Fernschreiben
telescope schieben * ineinander || Teleskop || verschieben * ineinander ~
telescopic ausziehbar || Teleskop-
telescoping Teleskopieren * ineinander verschieben; z.B. beim Aufwickler: innere Lagen werden herausgepresst
teleservice Fernservice * über Telefon/Modem
teleservice diagnostics Ferndiagnose * z.B. über Modem
teletype Fernschreiber
teletype message Fernschreiben
teletype writer Fernschreiber
tell about mitteilen
temper anlassen * Stahl || Härte * ~ von Stahl || härten * Stahl || temperieren || tempern * Wärmebehandlung/Glühen zum Ändern des Gefüges || vergüten * Wärmebehandlung zur Änderung d. Gefüges o. d. Oberflächeneigensch. von festen Werkstoffen
temper mill Dressiergerüst * zur Nachbehandlung in Walzstraße/Bandbehandlungsanlage || Dressiergerüst * in Blechstraße || Nachwalzgerüst * Dressiergerüst in Walzwerk zur Nachbehandlung
temperature Temperatur
temperature change Temperaturänderung
temperature class Temperaturklasse * Zündgruppe T1..6 nach EN50014 bis -20, kennzeichn. jew. Gase ähnl. Zündtemperatur || Wärmeklasse
temperature closed-loop control Temperaturregelung
temperature coefficient Temperaturbeiwert || Temperaturgang * Temperaturkoeffizient || Temperaturkoeffizient
temperature compensation Temperaturkompensation || Wärmeausgleich
temperature controller Temperaturregler
temperature decrease Temperaturabnahme
temperature dependence Temperaturabhängigkeit

temperature dependency Temperatureinfluss * Temperatureinfluss
temperature distribution Temperaturverteilung
temperature drop Temperaturabnahme * Temperaturabfall
temperature equilibrium Temperaturgleichgewicht
temperature evaluation Temperaturauswertung
temperature excursion Temperaturgang
temperature fluctuations Temperaturschwankungen
temperature gradient Temperaturgefälle
temperature increase Temperaturerhöhung * siehe auch →Übertemperatur
temperature limit Grenztemperatur * {el.Masch.} höchstzulässige Dauertemperatur (VDE 530 Teil 18.32, DIN 46416)
temperature loop controller Temperaturregler
temperature monitor Temperaturwächter
temperature monitoring Temperaturüberwachung
temperature range Temperaturbereich
temperature regulator Temperaturregler
temperature response Temperaturgang
temperature rise Erwärmung * Temperaturerhöhung/Anstieg (z.B. gegenüber der Umgebungs/Kühlmitteltemp.) || Temperaturanstieg || Übertemperatur * Temp.erhöhg. gegenüb. Vergleichspkt. (Umgebgs./Kühlmitteltemp.); f.Motor: VDE0530 || Wärmeklasse * Kurzform für 'temperature rise class'
temperature rise class Wärmeklasse * kennzeichnet max. zuläss. Übertemperaturen (el.Masch.: IEC 43, VDE 0530)
temperature rise test Erwärmungsprobe || Erwärmungsprüfung * z.B. für Elektromotor (siehe VDE 0530 Teil 1)
temperature rise-time Erwärmungszeit * {el. Masch.} benötigt e. blockierter ex.gesch. Motor z.Erreichen d.Grenztemp.
temperature sensing Temperaturerfassung
temperature sensitivity Temperaturgang
temperature sensor Temperaturfühler || Temperaturgeber || Temperatursensor || Thermofühler
temperature stability Temperaturgang * Unempfindlichkeit gegenüber Temperaturänderungen/-schwankungen
temperature transmitter Temperaturgeber
temperature-compensated temperaturkompensiert
temperature-rise limit Grenzerwärmung || Grenztemperatur * Grenzübertemperatur || Grenzübertemperatur
tempering furnace Härteofen
tempering steel Vergütungsstahl
template Lehre * Schablone || Schablone * Bohr/Schneid/Gußschablone
temporarily kurzfristig * (Adv.) vorübergehend || kurzzeitig * (Adv.) zwischenzeitlich, vorübergehend || momentan * (Adv.) vorläufig, temporär, behelfsmäßig, provisorisch || vorübergehend * (Adv.)
temporary gelegentlich * zeitweilig || kurzzeitig * zwischenzeitlich, vorübergehend || momentan * vorübergehend, zeitweise, provisorisch, zeit-/einstweilig || provisorisch * vorübergehend || temporär || vorerst * zeitweilig, vorübergehend || vorläufig vorübergehend * vorübergehend, zwischenzeitlich || Zwischen- * vorübergehend || zwischenzeitlich
temporary duty Kurzzeitbetrieb
temporary overvoltage Spannungserhöhung * vorübergehende
temporary storage Zwischenspeicher || Zwischenspeicherung
tenacious zäh * tenacity: Zähfestigkeit/Zähigkeit
tenacity Festigkeit * Zähigkeit || Zähigkeit * auch 'Zähfestigkeit', (auch figürl.: 'Verbissenheit'))

tenacy Härte * Zähigkeit, Zugfestigkeit, Zähfestigkeit
tend pflegen * Pflegehandlung ausüben || streben * zu (einer Richtung) hin ~, (to(wards): nach) || tendieren
tend to neigen zu
tendency Neigung * Tendenz (towards: zu) || Richtung * Tendenz || Tendenz * downward/upward tendency: ~ nach unten/oben || Trend * Tendenz (to: zu/für)
tender anbieten || schonend * pfleglich, zart, weich, empfindlich, heikel, bedacht (of: auf)
tender invitation Ausschreibung * für Angebotsabgaben
tenet Satz * Grund-~, Lehr~
tenon Zapfen * {Holz} (auch Verb: 'verzapfen')
tenon cutting machine Zapfenschneidmaschine * {Holz}
tenon-cutting machine Zapfenfräsmaschine * {Holz}
tenoner Zapfenfräsmaschine * {Holz} || Zapfenschneidmaschine * {Holz}
tenoning disc Zapfenschneidscheibe * {Holz}
tenoning machine Zapfenfräse * {Holz} im Zimmereihandwerk || Zapfenschneidmaschine * {Holz}
tenor Wortlaut * genauer
tens digit Zehnerstelle
tensible strength Streckfestigkeit
tensile energy absorption Zerreißarbeit
tensile force Zugkraft
tensile load Zugbeanspruchung || Zugspannung
tensile strength Bruchkraft * z.B. von Papier || Bruchwiderstand * z.B. von Papier || Reißfestigkeit || Zugfestigkeit
tensile stress Zugbeanspruchung || Zugspannung * →Zugbeanspruchung
tensiometer Zugmesseinrichtung
tension dehnen || spannen * dehnen, strecken, spannen (z.B. Riemen, Feder, Kette) || Spannung * (mechan.) Bahnspannung, Zug einer Warenbahn || Zug * Zugkraft || Zugkraft * Zug || Zugspannung * Zug
tension actual value Zugistwert
tension brake Abwickelbremse || Wickelbremse * z. Aufbringen d.Zugmoments bei Abwickler (magnet./mechan./hydraul./pneumat.Bremse)
tension control Zugkraftsteuerung || Zugregelung
tension controller Zugkraftregler || Zugregler
tension dynamometer Zugkraftmessgeber
tension fluctuations Zugschwankungen
tension force Zugkraft
tension leveller Richtmaschine * Streckricht-Maschine || Streckricht-Maschine || Streckrichtmaschine
tension load Spannkraft * elastische ~
tension measuring roll Zugmesswalze
tension meter Zugkraftmessgerät
tension pulley Spannrolle * für Riementrieb
tension regulator Zugregler
tension roll Spannwalze || Zugwalze
tension roller Spannrolle * für Riementrieb || Zugwalze
tension sensing device Zugmesseinrichtung
tension setting potentiometer Zugsollwertpotentiometer
tension spring Spannfeder || Zugfeder
tension taper abnehmende Wickelhärte * beim Aufwickler
tension test Zerreißprüfung
tension testing machine Zerreißprüfmaschine
tension transducer Zugkraftmessgeber || Zugmessdose || Zugmesseinrichtung
tension variations Zugschwankungen
tension-levelling machine Richtmaschine * Streck-

richt-Maschine || Streckricht-Maschine || Streckrichtmaschine
tension-relieve entspannen * z.B. Zugfeder, Zugspannung ~
tensional Zug- * Spann(ungs)-, Dehn(ungs)-
tensional force Spannkraft * elastische ~
tensional stress Zugbeanspruchung * tension-free: ohne ~
tensionally kraftschlüssig
tensioner Spanner * für Kette, Gurt, Sägeblatt usw. || Zugwerk * z.B. S-Walzengruppe in Folienmaschine
tensioning Spann- || spannen * (Substantiv) z.B. ~ eines Riemens
tensioning element Spannelement * für Kette, Gurt, Riemen, Seil, Sägeblatt usw.
tensioning stand Spannungsständer * {Textil} z.B. in Faserverstreckanlage
tensioning unit Spannvorrichtung * für Riemen, Kette, Seil usw.
tentative Probe- * versuchsweise || provisorisch * vorläufig, versuchsweise, probehalber || unverbindlich * Versuchs-, Probe-, vorläufig, provisorisch || vorläufig * versuchsweise || vorübergehend * versuchsweise
tentative standard Vornorm
tentatively unverbindlich * (Adv.) versuchsweise, probehalber || versuchsweise || zunächst * versuchsweise
tenth of one percent Promille
TENV TENV * Abkürz.f. 'Totally Enclosed Non Ventilated':vollkomm. geschloss. o.Lüfter (z.B.Schutzart IP54) || vollkommen geschlossen * 'Totally Enclosed Non Vented': ~e Motorschutzart IP54 ohne Kühlung
TENV enclosure IP54 * Abk.f. 'Totally Enclosed Non Ventilated':vollk.geschloss. ohne Lüfter;Motorschutzart
TEPV vollkommen geschlossen * 'Tot.Enclosed Pipe Vented': ~e Motorschutzart mit Kühlung über Rohranschluss
term -Anteil * in einem Algorithmus || Anteil * Teil einer Formel || Ausdruck * Fachausdruck,Teil einer Formel || Begriff * Ausdruck || benennen * mit einem Fachausdruck ~ || Benennung * Fachausdruck || bezeichnen * mit einem Wort/Ausdruck || Bezeichnung * (Fach-)Ausdruck || Fachausdruck || Name * Fachausdruck || nennen * mit einem (Fach-) Ausdruck be~ || Term * in einer Formel || Termin * Frist || Vokabel * (Fach-) Ausdruck || Wort * Ausdruck
term arranged Abmachung * Einzelpunkt einer Abmachung
term of delivery Lieferfrist || Liefertermin
term of shipment Lieferart * z.B. Unterscheidung Luft/Seefracht
termed genannt * mit e.Fachausdruck bezeichnet; termed as:bezeichnet als
terminal Anschluss * Klemme, Schraub~ || Anschlussklemme || Anschlusspunkt || Anschlussstelle || Endgerät * in Nachrichtenübertragungssystem || Klemme * (elektr.) || Pol * (Batterie-) Anschluss || Terminal
terminal arrangement Klemmenanordnung
terminal assignment Anschlussanordnung * Klemmenbelegung || Anschlussbelegung || Klemmenanordnung * Zuordnung der Klemmen(nummern) zu bestimmten Funktionen || Klemmenbelegung
terminal bar Anschlussschiene
terminal block Anschlussleiste || Klemmblock || Klemmenblock || Klemmenbrett || Klemmenleiste * Klemmenblock || Klemmenstein || Klemmleiste * Klemmenblock
terminal board Anschlussplatte || Klemmenbrett * z.B. im Motorklemmenkasten || Klemmenplatte

* *z.B. in einem (Motor-) Klemmenkasten* || Rangierverteiler * *Klemmenbrett*
terminal bolt Klemmenbolzen * *z.b. im Motorklemmenkasten*
terminal box Anschlusskasten || Klemmenkasten || Klemmkasten
terminal box base plate Klemmenkastensockel
terminal box housing Klemmenkastengehäuse || Klemmenkastenoberteil
terminal box position Klemmenkastenlage * *z.B. oben, links rechts bei einem Elektromotor*
terminal bracket Klemmbügel * *Teil in einer Schraubklemme*
terminal capacity Anschlussquerschnitt * ~ *der anschließbaren Leitungen bei Klemme*
terminal clamp Anschlussschelle
terminal clip Klemmbügel * *Teil einer Schraubklemme* || Schaltbügel * →*Klemmbügel*
terminal compartment Klemmenfach * *Klemmenanschlussraum in einem Gerät*
terminal connection Anschluss * ~ *an Klemme*
terminal connection diagram Anschlussplan * *für Klemmenanschlüsse* || Belegungsplan * *(Klemmen-) Anschlussplan* || Klemmenbelegungsplan
terminal designation Anschlussbezeichnung * *Klemmenbezeichnung; für E-Motoren:siehe DIN 57530/VDE 0530/IEC 34 jew.Teil 8* || Klemmenbezeichnung * *für E-Motoren: siehe DIN 57530/ VDE 0530/IEC 34 jew. Teil 8*
terminal diagram Belegungsplan * *(Klemmen-) Anschlussplan*
terminal end Anschlussseite
terminal expansion Klemmenerweiterung
terminal face Anschlussfläche
terminal housing Anschlusskasten || Klemmenkasten
terminal identifier Klemmenbezeichner
terminal impedance Anschlussimpedanz
terminal insulator Klemmenbrett || Klemmenstützer * *in Klemmenkasten usw.*
terminal jumper Klemmenbrücke || Schaltbügel * *Klemmenbrücke*
terminal layout Klemmenanordnung * *(siehe DIN46289, Leistungsklemmen: z.B. DIN 46206 T.2, DIN46223, DIN46260/64/65)*
terminal link Schaltbügel * *Leiterstück zur Überbrückg.v. KLemmen z.B. in Motorklemmenkasten* || Schaltbügel * *Klemmenbrücke z.B. in (Motor-) Klemmenkasten z.Herstellung d. gewünschten Schaltgruppe*
terminal lug Anschlussfahne || Anschlusslasche
terminal marking Anschlussbezeichnung || Klemmenbezeichnung
terminal module Klemmenmodul
terminal No Klemmennummer
terminal number Klemmennummer
terminal plate Anschlussplatte
terminal post insulator Klemmenstützer * *z.B. im Motor-Klemmenkasten*
terminal resistance Anschlusswiderstand
terminal screw Anschlussschraube || Klemmenschraube || Klemmschraube * *Klemmenschraube* || Kontaktschraube * ~ *für Kabelanschluss/an einer Anschlussklemme*
terminal strip Anschlussleiste || Klemmenblock * *Klemmenleiste* || Klemmenleiste || Klemmleiste || Reihenklemmenblock * *Klemmenleiste*
terminal stud Anschlussbolzen
terminal voltage Klemmenspannung
terminal with selectable function Wahlklemme
terminal-compatible anschlusskompatibel * *Klemmen-kompatibel*
terminals Anschlussklemmen
terminate abbrechen * *abschließen, beenden* || ablaufen * *z.B. Frist,Vertrag* || abschließen * *z.B. eine Leitung mit Wellenwiderstand ~* || aufheben * *Vertrag ~* || beenden * *abschließen* || beschließen * *beenden* || lösen * *Vertrag ~* || schließen * *beenden, auch Softwareprogramm* || verlassen * *beenden*
terminate and stay resident program speicherresidentes Programm
terminated abgeschlossen * *beendet, aufgehoben (z.B. Vertrag), abgeschloss. (Leitung m.Wellenwiderstand)*
terminating connector Abschlussstecker * *zum Leitungsabschluss*
terminating resistor Abschlusswiderstand
termination Ablauf * *eines Vertrages* || Abschluss * *auch Leitungs~ z.b. durch Stecker oder Bus~stecker* || Anschluss * *besonders für einen flächigen ~punkt (auch in Motorklemmenkasten)* || Beendigung
termination point Anschlusspunkt || Anschlussstelle
termination technology Anschlusstechnik
terminator Abschluss * *Leitungs~, Bus~widerstand, ~Stecker* || Abschlussstecker * *zum Leitungsabschluss*
terminology Begriffe * *Fachausdrücke* || Fachbegriffe || Fachsprache * *Namen* * *Terminologie, Fachausdrücke* || Terminologie
termite protection Termitenschutz * *relevant z.B. bei Langzeitlagerung von el.Motoren*
terms Begriffe * *Fachausdrücke* || Fachbegriffe || Größenordnung * *in terms of minutes:in einer ~ von Minuten* || Nomenklatur * *Fachausdrücke*
terms and conditions of warranty Garantiebedingungen || Gewährleistungsbedingungen
terms of contract Vertragsbedingungen
terms of delivery Lieferbedingungen
terms of guarantee Garantiebedingungen || Gewährleistungsbedingungen
terms of payment Zahlungsbedingungen
terms of sale Verkaufsbedingungen
terms of trade Geschäftsbedingungen
terraced stufenförmig * *Gelände usw.* || treppenförmig
terrific toll * *(salopp) fantastisch, gewaltig*
territory Gebiet * *räumliches ~*
tesioning cap Spannkappe
test Abfrage * *Prüfung* || abnehmen || erproben || Erprobung || Nachprüfung * *Prüfung* || Probe * *Erprobung* || Probe- * *Versuchs-* || prüfen || Prüfung || Test || testen || überwachen * *prüfen* || untersuchen * *for:auf* || Untersuchung || Versuch * *conduct a test: e. Versuch durchführen* || Versuchs-
test adapter Prüfadapter || Testmodul
test apparatus Prüfgerät
test arrangement Versuchsanordnung
test bay Prüffeld
test bed Prüfstand
test bench Prüfstand
test block Testbaustein * *Funktionsbaustein*
test certificate Kontrollbescheinigung * *Prüfbescheinigung* || Prüfbescheinigung || Prüfschein || Prüfungsschein * →*Prüfbescheinigung* || Prüfzertifikat || Prüfzeugnis
test certificate for explosion-protected apparatuses Prüfbescheinigung für explosionsgeschützte Betriebsmittel * *VDE 0171/02.61, DIN EN 50014,16,18-21*
test condition Prüfbedingung
test data Messdaten
test department Prüfabteilung || Prüffeld * *Abteilung*
test duration Prüfdauer * *z.B. bei Isolationsprüfung* || Prüfzeit * *z.B. bei Isolationsprüfung*
test engineering Prüftechnik
test equipment Prüfeinrichtung || Prüfgerät

test for correct functioning Funktionsprüfung
test installation Pilotanlage
test jack Hülse * *Prüfsteck~* || Prüfbuchse || Testbuchse
test jacket Testbuchse
test lab Prüffeld
test label Prüfschild
test laboratory Prüffeld
test mark Prüfzeichen
test method Prüfverfahren
test mode Testbetrieb
test model Modell * *Versuchs~* || Muster * *Versuchsmodell*
test module Testbaugruppe
test object Prüfling
test on selected samples Stichprobenkontrolle
test operation Testbetrieb
test overpressure Prüfüberdruck
test pattern Prüfmuster
test piece Prüfling
test planning Prüfplanung
test plug Prüfstecker
test point Messpunkt
test procedure Prüfverfahren
test pulse Prüfimpuls
test purposes Prüfzwecke * *for: für*
test receptacle Hülse * *Hülse/Verbindungsstück z.B. für Bananenstecker*
test report Prüfbericht || Testbericht
test result Prüfergebnis || Testergebnis
test rig Prüfstand
test routine Testroutine
test run Probelauf || Testlauf
test sample Prüfling
test schedule Prüfplan
test sequence Prüfreihenfolge
test severity Prüfschärfe || Schärfegrad * *Prüfschärfe*
test signal Testsignal
test socket Buchse * *Prüfbuchse* || Hülse * *Prüfsteck~* || Messbuchse * *Prüfbuchse* || Prüfbuchse || Testbuchse
test specification Prüfvorschrift
test specimen Prüfling
test stand Prüfstand
test torque Prüfdrehmoment
test voltage Messspannung * →*Prüfspannung* || Prüfspannung * *für Isolationsprüfungen; für Niederspannungsmotoren: siehe VDE 0530*
test-bed assembly Prüfaufbau
testabilty Prüfbarkeit
testbed Prüfstand
tested geprüft
testimonial Zeugnis * *Führungs~*
testing Erprobung
testing aid Testhilfe
testing appliance Prüfgerät
testing department Prüffeld
testing device Prüfgerät
testing equipment Prüfgeräte
testing machine Prüfmaschine
testing mark Prüfzeichen
testing method Prüfverfahren
testing of material Materialprüfung
testing officer Abnahmebeauftragter * *offizieller*
testing panel Testfeld * *z.B. an Mikrocomputersystem*
testing procedure Prüfverfahren
testing rate Prüfgeschwindigkeit
testing result Prüfergebnis || Testergebnis
testing stage Erprobungsphase
testing station Prüffeld
TEUC vollkommen geschlossen * *'Totally Enclosed Unit Cooled': ~e Motorschutzart mit Wärmetauscher*

text Text || Wortlaut
text display unit Textanzeigegerät
text editor Textverarbeitungsprogramm * *einfaches (z.B.z.Eingabe v.Quellprogrammen) ohne Formatierungsmöglichkt.*
text entry Texteingabe * *in Rechner*
text line Textzeile
text processing Textverarbeitung
text processor Textverarbeitung * *~sprogramm* || Textverarbeitungsprogramm || Textverarbeitungssystem
text string Zeichenfolge * *Text*
textile Gewebe || Stoff * *Textil* || Textil || textiler Stoff
textile application Textilanwendung
textile finishing Textilveredelung
textile industry Textilindustrie
textile loom Webstuhl
textile machine Textilmaschine
textile manufacture Textilherstellung
textile mill Textilfabrik
textile motor Textilmotor * *{el.Masch.} m. Schutz geg. Faserablagerg. (niedr. Kühlluftgeschw., kein Fettaustritt)*
textile stenter Spannrahmen * *{Textil} Verstreckwerk*
textiles Textilien
texts Texte
texture Gefüge * *(auch figürl.) Struktur, Beschaffenheit, Gefüge, Gewebetextur, Maserung* || Struktur * *Textur, Maserung, Struktur, Gefüge, Oberflächenbeschaffenheit*
texturing machine Texturiermaschine * *{Textil} raut glatte Chemiefasern auf und macht sie bauschig wie Naturfasern*
texturizing machine Texturiermaschine * *{Textil} raut glatte Chemiefasern auf u.macht sie bauschig wie Naturfasern*
than previously als bisher
than up to now als bisher
thank danken * *for:für*
thank you danke
thank you letter Dankschreiben
thank you very much danke * *dankeschön*
Thank's, fine Guten Tag * *Antwort hierauf*
thankful dankbar * *in Wort und Tat*
thanks dank * *(Substantiv)* || danke
thanks to dadurch * *dank* || dank
that so * *(Adv., salopp) solch, so (sehr), dermaßen; not that big: nicht so groß*
that can be shifted verschiebbar
that goes without saying selbstverständlich * *(Adv., nachgestellt) das ist ~*
that is nämlich
that is to say nämlich
that is why deshalb
that makes sense das ist sinnvoll
that way dadurch * *(örtlich) auf diesem Weg* || so * *(Adv.) auf diese Weise*
that which derjenige * *(auf eine Sache bezogen)*
that's why deshalb * *deswegen*
THD Klirrfaktor * *Abkürz. für 'total harmonic distortion'*
the am * *(in Zusammenhang mit Superlativ) z.B. the fastest: am schnellsten* || je * *the lower the price the greater the demand: je niedriger der Preis desto höher die Nachfrage*
the best am besten * *(Adv.)*
the bulk of it meist * *der Großteil, die große Masse*
the customary alt * *(Substantiv) das Alte/Gewohnte*
the day after tomorrow übermorgen
the fact is that tatsächlich * *(einleitend)*
the following is generally valid grundsätzlich gilt

the following is valid   folgendes gilt
the greater number   meist * die Mehrzahl
the greater part   meist * das ~e, der größte Teil
the greater the   umso größer je
the last   Letzteres
the last one   Letzteres
the majority of   meist * die Mehrheit/Mehrzahl von
the moment   sobald
the number of times   wie oft * so viel wie || wievielmal
the one who   derjenige
the other way round   umgekehrt
the question is if   handeln * es handelt sich darum, ob
the same   dasselbe || derselbe || dito || gleich * (Adv.) || unverändert * keep the same:unverändert beibehalten
the same as   genauso wie * (Adv)
the same as before   unverändert
the same phase relationship   Phasengleichheit
the same phase sequence   Phasengleichheit * z.B. zwischen Elektronik- und Leistungteilanschluss e. Stromrichters
the year round   ganzjährig
theme   Gegenstand * Thema || Thema * Stoff; ~ e. wissenschaftlichen oder Examensarbeit; Musik~
then   dabei * dann, anlässlich || dann
theorem   Satz * Lehr~, Theorem
theoretical   theoretisch
theoretical analysis   theoretische Untersuchung
theoritically   theoretisch * (Adv.)
theory   Lehre * Theorie || Theorie
therapy   Behandlung * ~ zur Besserung
there   dabei * anwesend
thereby   dabei * dadurch, auf diese Weise, daran, davon, nahe dabei || dadurch * auf diese/solche Weise || damit * dadurch || davon * thereby affected: davon betroffen
therefore   also || daher || deshalb * als Folge davon, infolgedessen || deswegen * →daher || so * folglich
thereof   davon * (Adv.) z.B. i. Zusammenhang mit Zahlenangabe
thereupon   anschließend * (Adv.) danach, (zeitlich) anschließend || danach * anschließend || daraufhin * zeitlich || hierbei * darauf, hierauf, danach, daraufhin, darum || somit * daraufhin, demzufolge, darauf, hierauf, danach, darum
therewith   hierbei * damit, darauf, hierauf, danach, daraufhin, demzufolge, darum || hiermit * damit, darauf, hierauf, daraufhin, demzufolge || somit * daraufhin, demzufolge, darum, damit
thermal   thermisch || Wärme-
thermal absorption capacity   Wärmeaufnahmefähigkeit
thermal bimetal contact   Thermokontakt
thermal capacity   Wärmebelastbarkeit || Wärmekapazität
thermal circuit breaker   Thermoschalter
thermal class   Wärmeklasse * kennzeichnet max. zuläss. Temp. i. el. Gerät nach IEC 85 (z.B. Y, A, E, B, F, H)
thermal classification   thermische Klassifizierung * Kennzeichn. d.→Wärmeklasse (früher: Isolierstoffklasse) e.Isoliersystems || Wärmeklasse
thermal conduction   Wärmeleitung
thermal conductivity   Wärmeleitfähigkeit
thermal contact   Thermokontakt
thermal cycling   Temperaturbeanspruchung * durch wechselnde Temperaturen/Temperaturzyklen
thermal economy   Wärmehaushalt
thermal EMF   Thermospannung * Abk.f. 'Electro-Motive Force'
thermal equilibrium   thermisches Gleichgewicht
thermal expansion   Wärmeausdehnung || Wärmedehnung * Längendehnung bei Erwärmung (Eisen 0.1%/100K; Kupfer 0.15 %/100K; Alu 0.2 %/100K)
thermal image   Abbild * thermisches Abbild
thermal impedance   Wärmewiderstand
thermal insulation   Wärmedämmung
thermal limiting value   Wärmegrenzwert
thermal load   Wärmebelastung * high: hohe
thermal loadability   thermische Belastbarkeit * z.B. eines Isoliersystems im Motor
thermal loading   Wärmebelastung
thermal losses   Verlustwärme
thermal model   thermisches Abbild
thermal motor protection   thermischer Motorschutz * {el.Masch.} Maschinenschutzeinrichtung z.Vermeidg. unzuläss. hoher Temp.
thermal motor time constant   thermische Motorzeitkonstante
thermal overload   thermische Überlastung
thermal overload capacity   thermische Überlastbarkeit
thermal overload protection   thermischer Überlastschutz
thermal power station   Wärmekraftwerk
thermal protection   Temperaturschutz || thermischer Schutz * siehe IEC 34 Teil 11 für in Motor eingebaute Schutzelemente
thermal protector   Temperaturwächter * Baueinheit von Temperaturfühler und Stromkreisunterbrecher
thermal raise class   Wärmeklasse
thermal rating   Wärmebelastbarkeit * spezifizierte Nenn-~, z.B. einer Kupplung/Bremse
thermal relay   Motorschutzrelais
thermal release   Temperaturwächter * Auslösegerät zum Übertemperaturschutz
thermal replica   thermisches Abbild
thermal reserve   thermische Reserve
thermal resistance   thermischer Widerstand || Wärmewiderstand
thermal rise time constant   thermische Zeitkonstante
thermal sensing device   Temperatursensor
thermal sensor   Temperaturfühler || Temperatursensor
thermal shock   Temperatursturz
thermal stability   Temperaturbeständigkeit || thermische Stabilität
thermal steady state   thermischer Beharrungszustand
thermal stress   Temperaturbeanspruchung || thermische Belastung || Wärmebelastung * Wärmebeanspruchung
thermal switch   Thermoschalter || Thermostat
thermal time constant   thermische Zeitkonstante * Zeit,nach d. e. Körper bei e.Sprung d. Umgeb.temp. 63% d.Endtemp. erreicht
thermal trip unit   Thermoauslöser * thermisch verzögerter (Motorschutz-) Schalter
thermally   thermisch * (Adv.)
thermally critical   läuferkritisch * {el.Masch.} Maschine, bei der die Temperaturgrenze zuerst i.Läufer erreicht wird
thermally delayed overload protection   thermisch verzögerter Überlastschutz
thermally insulated   wärmeisoliert
thermally utilized   thermisch ausgenutzt
thermally-critical-stator   ständerkritisch * {el.-Masch.} zuläss. Grenzübertemperatur wird zuerst im Ständer erreicht
thermistor   Heißleiter || Temperaturfühler * Heiß- oder Kaltleiter (NTC oder PTC) || Temperatursensor * →Heiß- oder →Kaltleiter (PTC/NTC Widerstand) || Thermistor || Widerstandsthermometer

* *Heiß/Kaltleiter-Temperaturfühler (mit NTC-, PTC-Widerstand)*
thermistor contact Thermokontakt
thermo sensor Thermofühler
thermo switch Thermoschalter
thermo-lubricant Wärmeleitpaste
thermo-mechanical thermomechanisch
thermocouple Temperatursensor
* →*Thermoelement* || Thermoelement * *Temperaturmessgeber bestehend aus 2 verschieden miteinand. leitend verbunden.Metallen* || Thermopaar
* →*Thermoelement zur Temperaturmessung*
thermodynamic thermodynamisch
thermoeletric voltage Thermospannung
thermoforming tiefziehen * *von Kunststoff (z.B. für Verpackungszwecke, Yogourtbecher, Armaturentafeln)*
thermolube Wärmeleitpaste
thermometer Temperaturmesser || Temperaturmessgerät
thermometer method Thermometerverfahren
* *zur Ermittlung d.Temperatur oder Übertemperatur an d. Oberfläche m.Thermometer*
thermoplastic Thermoplast || verformbar * *warm verformbar*
thermoplastic material Thermoplast
thermoplastic resin Thermoplast
thermostability Temperaturbeständigkeit
thermostat Temperaturregler || Thermoschalter || Thermostat
thermostatic cut-out Thermoschalter
thermoswitch Thermoschalter
these days heutzutage
thesis Dissertation || Doktorarbeit || Satz * *These, Behauptung, Streit~*
thick dick || steif * *dickflüssig* || trüb * *dick(flüssig), neblig, trübe (auch Wetter)*
thick mash Dickstoff * *Dickmaische*
thick mud Dickstoff * *Dickschlamm*
thick spot Verdickung * *dicke Stelle*
thick-film circuit Dickschichtschaltung
thick-film technology Dickschichttechnik * *z.B. zur Herstellung v. Hybridbauteilen auf Keramikträger*
thicken eindicken
thickener Dickungsmittel || Eindicker || Räumer * *Absetzbehälter zur Wasseraufbereitung*
thickening Eindickung || Verdickung
thickly liquid zäh * *zähflüssig*
thickness Dicke || Stärke * *Dicke* || Wandstärke
thickness gauge Dicken-Messanlage || Dickenmessanlage || Dickenmesser || Dickenmessgeber || Dickenmessgerät || Dickenmessung * *Dickenmessanlage/-Einrichtung*
thickness of pipe wall Wandstärke * *eines Rohres*
thickness of wall Wandstärke
thicknessing machine Dickenhobelmaschine * *{Holz}*
thicknessing planer Dickenhobel * *{Hobel}* || Dickenhobelmaschine * *{Holz}*
thin dünn || schmal * *dünn*
thin metal sheet Feinstblech
thin planing knife Streifenhobelmesser * *{Holz}*
thin sheet Feinstblech
thin sludge Dünnschlamm
thin-film technology Dünnfilmtechnik
thin-slab Dünnbramme
thin-walled dünnwandig
thing etwas * *(Substantiv) das Etwas* || Gegenstand * *Ding, (Haupt)Sache, Angelegenheit*
think annehmen * *vermuten* || finden * *denken* || glauben * *vermuten*
think out ausdenken
think over überlegen * *bedenken*
think up ausdenken * *(amerikan.)*

thinly-liquid dünnflüssig
thinner Verdünner
third dritt || Dritt- * *-klassig, -bester usw.* || Drittel * *one third: ein ~; two-thirds: zwei ~; a third of the amount: ein ~ des Betrages*
third from last drittletzt || drittletzter
third last drittletzt || drittletzter
third part Drittel * *der dritte Teil*
third party Dritt- * *Dritthersteller, dritte Kraft, dritte Partei/Person* || Drittfirma || Drittthersteller || Fremdfirma
third-band analysis Terzanalyse * *schmalbandige Messung des Geräuschspektrums*
third-party fremd * *von einem anderen Hersteller/Fremdhersteller*
third-party company Drittfirma
third-party computer Fremdrechner
third-party manufacturer Fremdhersteller
third-party product Fremdfabrikat
third-party vendor Drittanbieter
thirdly drittens
this das vorliegende || der vorliegende || die vorliegende
this day's heutig
this is why deshalb * *aus diesem Grunde* || so * *deshalb, aus diesem Grunde*
this side up oben * *auf Versandkiste*
this way so * *(Adv.) auf diese Weise*
Thomson measuring bridge Thomson-Messbrücke
thorn Dorn * *allg. (auch figürl.)*
thorough eingehend * *gründlich, sorgfältig, eingehend, genau; vollkommen, völlig, echt, durch und durch* || gründlich * *vollständig* || völlig
* →*gründlich* || vollständig * *vollkommen, völlig, durch und durch, gänzlich, gründlich*
thorough-going konsequent * *gründlich*
thoroughgoing eingehend * *kompromisslos, durch und durch* || kompromisslos * *durch und durch, extrem, gründlich*
those present Gesprächspartner * *(Plural) die Anwesenden* || Teilnehmer * *(Plural) die Anwesenden*
though jedoch * *(am Satzende) aber, dennoch, trotzdem, allerdings* || obwohl * *obwohl, obgleich, obschon, (je)doch; even though:wenn auch/selbst wenn; as though:als ob* || trotzdem * *(Adv., nachgestellt)*
thought Gedanke || Überlegung * *Gedanke; on second thoughts: bei näherer ~*
thought horizon geistiger Horizont
thousand tausend
thousand fold tausendfach
thousand millions Milliarde
thousand times tausendfach
thousands digit Tausenderstelle
thread aufführen * *der Papierbahn in einer Papiermaschine* || einfädeln || einführen * *z.B. eine Materialbahn ~* || einziehen || Faden || Faser * *Faden* || Gang * *Gewinde(gang)* || Garn * *Faden, Zwirn, Garn* || Gewinde * *right-/lefthand(ed): rechts/linksgängiges* || Gewinde- || Gewindegang || Gewindeschneiden * *(Verb) Gewinde schneiden in, mit Gewinde versehen* || kriechen * *mit Einzieh/Schleichdrehzahl fahren* || schleichen * *mit Einzug/Schleichdrehzahl fahren* || Schnur * *Faden, Zwirn, Garn* || Windung * *Gewindegang* || Zwirn
thread break Fadenbruch
thread connection Gewindeanschluss
thread cutting Gewindeschneiden
thread hobbing Gewindeschneiden * *mit Wälzfräser*
thread in einziehen * *z.B. Warenbahn in eine Maschine*
thread measured in inces Zollgewinde
thread pitch gauge Gewindelehre

thread reference Kriechsollwert * *Sollwert für Einziehgeschwindigkeit/-Drehzahl*
thread rolling Gewindewalzen
thread rolling machine Gewindewalzmaschine
thread size Gewindegröße
thread speed Einrichtdrehzahl * *Einziehdrehzahl/-geschwindigkeit* || Einziehgeschwindigkeit || Kriechgeschwindigkeit * *Kriechgeschwindigkeit, z.B. zum Materialeinziehen* || Schleichdrehzahl * *Einziehdrehzahl zum Material-Einziehen* || Schleichgang * *Einziehgeschwindigkeit*
threadbare abschaben * *abgeschabt, z.B. Textilstoff*
threaded Gewinde- * *mit Gewinde versehen*
threaded adapter nipple Gewindestutzen * *kleiner*
~
threaded bolt Gewindebolzen
threaded bore Gewindebohrung
threaded cable entry PG-Verschraubung * *für abgedichtete Kabeleinführung*
threaded cable entry hole PG-Verschraubung * *Kabeldurchführung mit Verschraubung*
threaded closure machine Schraub-Verschließmaschine * *in der Verpackungsindustrie*
threaded flange Gewindestutzen * *Gewindeflansch*
threaded hole Gewindebohrung
threaded joint Verschraubung * ~ *ohne Mutter*
threaded journal Gewindezapfen * *Dreh-/Lager-/ Wellenzapfen*
threaded pin Gewindestift || Gewindezapfen
threaded pipe Gewindestutzen
threaded pipe adapter Gewindestutzen * *für Rohranschluss*
threaded rod Gewindestange || Gewindestift * *Gewindestange*
threaded spindle Gewindespindel
threaded spud Gewindestutzen * *zum Anbau eines Rohrs*
threaded stem Gewindezapfen
threaded stud Gewindezapfen
threaded-stud terminal Bolzenklemme * *zum Anschluss von Leitungen*
threaded-type commutator Gewindekommutator * *bei DC-Maschine*
threading Einführung * ~ *einer Wahrenbahn in eine Maschine (z.B. Papier, Folie, Blechband, Textil usw.)* || Gewinde- * *Gewinde-herstellend* || Kriechbetrieb || Riefenbildung * *an Kommutator*
threading and tapping machines Gewindeherstellungsmaschinen
threading capstan Einziehscheibe * *erste →Ziehscheibe einer Drahtziehmaschine (gleich hinter der Abwicklung)*
threading die Schneideisen * *zum Gewindeschneiden*
threading lathe Gewindedrehbank
threading machine Gewindeschneidmaschine
threading operation Einziehbetrieb * *Einziehen der Materialbahn mit langsamer (Kriech-) Geschwindigkeit*
threading procedure Einziehvorgang * *für eine Warenbahn*
threading speed Einziehgeschwindigkeit || Kriechgeschwindigkeit * *Einziehgeschwindigkeit* || Schleichdrehzahl
three phase leads Drehstromzuleitung
three phase load Drehstromverbraucher
three shift working Dreischichtarbeit
three-dimensional dreidimenional || plastisch * *dreidimensional (bildliche Darstellung usw.)* || räumlich * *dreidimensional*
three-dimensional measuring machine Koordinaten-Messmaschine
three-phase Drehstrom- || dreiphasig

three-phase AC controller Drehstromsteller
three-phase AC drive Drehstromantrieb
three-phase AC power controller Drehstromsteller * *(Stromrichter) AC-AC-Stromrichter mit Phasenanschnittsteuerung*
three-phase bridge Drehstrombrücke * *z.B. Gleichrichterschaltung*
three-phase bridge circuit Drehstrombrückenschaltung
three-phase bridge connection B6-Schaltung * *(Stromrichter)* || Drehstrombrückenschaltung
three-phase busbar Drehstromsammelschiene
three-phase commutator motor Drehstrom-Nebenschlussmotor
three-phase commutator motor with shunt characteristic Drehstrom-Nebenschlussmotor
three-phase commutator shunt-type machine Drehstromnebenschlussmaschine
three-phase commutator shunt-type motor Drehstromnebenschlussmotor
three-phase current Drehstrom
three-phase distribution Drehstromverteilung
three-phase drive Drehstromantrieb
three-phase induction machine Drehstromasynchronmaschine
three-phase induction motor Drehstrom-Asynchronmotor || Drehstrommotor
three-phase motor Drehstrommotor * *dreiphasig*
three-phase power converter Drehstromsteller
three-phase power distribution Drehstromverteilung
three-phase shunt wound machine Drehstrom-Kommutatormaschine
three-phase shunt-wound motor Drehstrom-Kommutatormotor || Drehstrom-Nebenschlussmotor || Drehstromkommutatormotor
three-phase side Drehstromseite * *on the:auf der*
three-phase slipring motor Drehstrom-Schleifringläufermotor
three-phase supply cable Drehstromzuleitung
three-phase supply system Drehstromnetz
three-phase synchronous machine Drehstromsynchronmaschine
three-phase system Drehstomnetz || Drehstromnetz || Drehstromsystem
three-phase transformer Drehstromtransformator
three-phase voltage Drehspannung * *im dreiphasigen System*
three-phase winder Drehstromwickler * *z.B. mit Spezial-Wickelmotor der Fa. Lenze*
three-phase winding Drehstromwicklung
three-point controller Dreipunktregler
three-point fixing Dreipunktaufhängung * *(el. Masch.) eines Motors mit →Pratzenanbau* || Dreipunktbefestigung || Pratzenanbau * *(el.Masch.) →Dreipunktaufhängung*
three-position controller Dreipunktregler
three-position switch Dreistellungsschalter
three-pulse Dreipuls-
three-shift operation Dreischichtbetrieb
three-stage dreistufig
three-step action control Dreipunktregelung
three-step action controller Dreipunktregler
three-step control Dreipunktregelung
three-step controller Dreipunktregler
three-term controller Dreipunktregler
three-way Dreiwege-
three-way valve Dreiwegeventil
three-winding transformer Dreiwicklungstrafo * *siehe auch →Doppelstocktrafo* || Dreiwicklungstransformator
threshold Grenze * *Schwelle* || Schwelle || Schwellwert
threshold current Ansprechstrom
threshold of hearing Hörgrenze || Hörschwelle

threshold of indication Anzeigeschwelle
threshold of pain Schmerzgrenze || Schmerzschwelle
threshold value Schwellwert
threshold voltage Ansprechspannung || Schleusenspannung * *(Diode, Transistor)* || Schwellenspannung || Schwellspannung
throat Kehle * *{anatomisch}*
throat-cutting existenzgefährdend * *Wettbewerb usw.*
throb Pulsation * *Pochen, Klopfen, Hämmern, (Puls-) Schlag, Erregung*
throrougly eingehend * *(Adv.)* gründlich
throttle Drossel * *(mechan.)* ~ventil, ~klappe || drosseln || Gas * *Drosselklappe; open the throttle: Gas geben (auch figürl.); throttle down: Gas wegnehmen* || herunterbremsen * *(auch figürl.: Computer usw.)* || würgen
throttle back zurücknehmen * *drosseln (Gas beim Motor usw.)*
throttle valve Drossel * *(mechan.)* Drosselventil/klappe
throttling element Drosselorgan * *(mechan.) z.B. in Rohrleitung*
throttling losses Drosselverluste * *bei Durchflussregelung über Schieber/Drosselventile*
through aus || bis einschließlich || durch * *durch hindurch, z.B. Stromfluss; auch mittels* || mittels || über * *durch, auch 'mittels'* || um * *rotate through 90 Degrees: ~ 90 Grad drehen; shifted through 90 Degrees: ~ 90 Grad versetzt* || vermittels
through it dadurch * *auf solche Weise*
through oversight irrtümlich * *(Adv.)* versehentlich || versehentlich * *auf Grund eines Flüchtigkeitsfehlers*
through there dadurch * *(örtlich)* da hindurch
through which wobei * *wodurch*
through-feed Durchlauf- * *z.B. ~-Presse für Spannplatten usw.*
through-feed press Durchlaufpresse * *{Holz}*
through-hardened durchgehärtet
through-hole Durchgangsbohrung || Durchgangsloch || Durchsteck-
through-hole mounting Durchsteckmontage || Einsteckmontage || Wanddurchführungs-Montage
through-hole type bedrahtet * *Bauelement zur Bestückung auf Leiterkarte*
through-hole version bedrahtete Ausführung * *Bauelement zur Bestückung auf Leiterkarte*
through-shaft durchgehende Welle
through-wiring Durchgangsverdrahtung
throughout ganz * *(Adv.)* ~ durch
throughout the world weltweit
throughput Durchsatz || Menge * *Durchsatz (~ pro Zeiteinheit)*
throughput rate Durchsatz * *Rate*
throughput time Durchlaufzeit * *z.B. in einer Fertigungslinie*
throw Hub * *~ eines Exzenters, eines Kolbens, einer Presse usw.* || Kröpfung * *z.B. einer Kurbelwelle, eines Exzenters, einer Presse* || schleudern * *werfen, schleudern, zuwerfen, (Hebel, Schalter) umlegen, throw out: auskuppeln/-rücken* || Sprung * *~ in einer Wicklung*
throw in einrücken * *Gang, Kupplung usw.*
throw into gear einrücken * *Kupplung usw.* || in Gang setzen * *Maschine ~*
throw off abwerfen * *the load: Last ~*
throw off the load Last abwerfen
throw open to the public zugänglich machen * *der Öffentlichkeit ~*
throw out auskuppeln * *auskuppeln, ausrücken* || ausrücken * *Kupplung*
throw out of gear ausrücken * *außer Eingriff bringen*

throw over umschalten
throw-away Einweg-
throw-out Auswerfer
thru bis * *(salopp) bis einschließlich* || bis einschließlich * *(amerikan.)*
thru-the-door mounting Durchsteckmontage * *(amerikan.) in einen Türausbruch montiert/montierbar*
thrust Axialkraft || Axiallast || Axialschub || schieben * *vorwärtsstoßen* || Schub * *~ in Axialrichtung, Schiebekraft* || Schubkraft || Schwung * *Schub(-kraft)* || stoßen * *heftig ~, schubsen*
thrust aside verdrängen
thrust bearing Traglager
thrust force Schubkraft
thrust out ausstoßen
thrust rod Schubstange
thrust spindle Hubspindel * *Schubspindel*
thrust washer Druckscheibe * *zur →Vorspannung eines Lagers*
thrusting aside Verdrängung * *mit Gewalt*
thrustor Motordrücker
thsnd. tausend * *(Abkürzung)*
thud Aufschlag * *dumpfer Aufprall*
thumb wheel switch Zahlenschalter * *Daumenradschalter*
thumb-wheel switch Codierschalter * *Daumenradschalter, Zifferneinsteller*
thumbwheel switch Zahlenschalter * *Daumenradschalter*
thus also * *(Adv), auch auf diese Weise* || dadurch * *somit, folglich, demgemäß, so, folgendermaßen* || so * *dadurch, deshalb, daher* || somit
thus far bisher
thyristor Thyristor * *['theiriste] siehe DIN 41786/87*
thyristor AC controller Thyristorsteller * *Wechselstromsteller*
thyristor assemblies Thyristorsätze
thyristor assembly Thyristorsatz
thyristor block Thyristorbaustein
thyristor branch fuse Zweigsicherung * *in Thyristor-Brücke*
thyristor check Thyristorcheck
thyristor controller Thyristorsteller
thyristor converter Thyristor-Stromrichter
thyristor cubicle Thyristorschrank
thyristor fed DC drive thyristorgesteuerter Gleichstromanstrieb
thyristor module Thyristorbaustein || Thyristorblock || Thyristormodul
thyristor power controller Thyristorsteller
thyristor protection Thyristorschutz
thyristor reversing controller Thyristor-Umkehrsteller
thyristor set Thyristorsatz
thyristor sets Thyristorsätze
thyristor snubbers TSE-Beschaltung * *Schutzbeschaltung für Thyristoren*
thyristor stack Thyristorsatz
thyristor wafer Thyristortablette
tick Takt * *~ eines Rechners*
ticklish gefährlich * *heikel* || heikel
tidiness Reinheit
tidy ordentlich || sauber * *reinlich, ordentlich* || säubern * *Zimmer usw. in Ordnung bringen, säubern, richten; tidy up:aufräumen*
tidy up aufräumen * *z.B. Zimmer* || reinigen * *putzen*
tie binden * *(figürl.) hemmen,fesseln, lahm legen, (Produktion usw.) stillegen, (Resourcen) blockieren* || verknüpfen * *zusammenbinden*
tie in ankoppeln || einbinden * *(auch figürl.)*
tie to einbinden * *jdn. verpflichten (zu), jdn. binden an*

tie together   verknüpfen * *zusammenknüpfen*
tie up   binden * *Geldmittel ~* || lähmen * *Verkehr usw.*
tier   Etage * *Reihe, z.B. eines Baugruppenträgers* || Lage * *auch 'Reihe'; in tiers: ~nweise/in Reihen übereinander* || Reihe * *Reihe, Lage, Zeile (z.B. eines Baugruppenträgers); in tiers: in ~n/Lagen übereinander* || Zeile * *Reihe, Lage, Zeile in Baugruppenträger*
tight   dicht || eng * *Termin(plan), Etat, Geldmittel usw.* || fest * *fest(sitzend), stramm, straff, (an)gespannt, dicht (gedrängt)* || knapp * *eng (Termin), dicht (gedrängt), kritisch, mulmig, angespannt (Marktlage), geizig* || kompakt * *gedrängt (auch Programmcode), dicht (gedrängt), eng, prall(voll), straff* || straff * *Riemen, Terminplan usw.* || stramm * *straff; fit tightly: ~ sitzen* || undurchlässig
tight corners   schwerzugängliche Stellen
tight-fitting   satt anliegend
tighten   anpressen * *fest-/anziehen (Schraube, Zügel usw.), spannen (Feder, Gurt, Kette), straffen (Seil)* || anziehen * *Schraube; in diagonal-opposite sequence: über Kreuz* || fest anziehen || festdrehen * *Schraube* || festziehen || spannen * *straffen, z.B. Riemen* || straff spannen
tighten fully   fest anziehen * *z.B. Schraubverbindung*
tighten the belt   sparen * *(salopp) den Gürtel enger schnallen*
tighten the screw   schrauben * *fest~, fester an~*
tighten up   anziehen * *Schraube ~* || festschrauben * *(Schraube) festziehen* || festziehen
tightened   angezogen * *Schraube*
tightener   Spannrolle * *für Riementrieb*
tightening   Anpress- * *zum Straffen* || Spann- || Straffung * *(mechan.)*
tightening pulley   Spannrolle * *z.B. an einem Riementrieb*
tightening screw   Spannschraube
tightening torque   Anziehdrehmoment * *für Schrauben; siehe →Anzugsmoment* || Anzugsdrehmoment * *für Schrauben* || Anzugsmoment * *Schraube* || Anzugsmoment * *zum Anziehen von Schrauben*
tightness   Dichtheit * *z.B. gegen Luft, Wasser; test for leaks: auf ~ prüfen*
tile   Feld * *~ in der Wartentechnik*
till now   bisher
tilt   Kippen * *(Verb) mechanisch (um-/ab-)~* || Neigung * *Kipplage* || schrägstellen * *kippen* || schwenken || verkanten * *schrägstellen, neigen, kippen*
tilt-free   kippsicher * *mechanisch*
tilt/swivel stand   Schwenk-Neigefuß * *für Computer-Monitor*
tilted   schiefwinklig
tilting   Neigung * *Kipplage* || Schwenk- * *z.B. Fräser* || schwenkbar
tilting rigidity   Kippsteifigkeit * *mechan. Steifigkeit, die das (Um)Kippen verhindert*
tilting table   Kipptisch
tilting-spindle moulder   Schwenkfräse * *{Holz}*
tilting-spindle shaper   Schwenkfräsmaschine * *{Holz} zum Erzeugen von Profilen*
timber   Bauholz || Holz * *Nutzholz; (brit.) Schnitt~; round timber: Rund~* || Rundholz || Schnittholz * *(brit.)*
timber and woodworking industry   Holzindustrie
timber conveyor   Rundholzförderer
timber crafter   Abbundanlage * *{Holz}*
timber yard   Holzplatz * *im Sägewerk usw.* || Rundholzplatz * *{Holz} im Sägewerk, Papier-/Zellstofffabrik*
timbre   Klang * *Klangfarbe*
time   bemessen * *zeitlich bemessen* || einteilen * *zeitlich ~* || mal * *(Subst.) for the first/last time:* *zum ersten/letzten Mal; this time: dieses ~; two times:zwei~* || messen * *Zeit ~* || Zeit * *only one at a time: nur einer zur ~, jeweils nur einer; save time: ~ sparen* || Zeitpunkt || Zeitwert * *bei SPS*
time and data register area   Wertespeicher * *{SPS}*
time axis   Zeitachse
time base   Zeitablenkung * *(Oszilloskop) Zeitbasis* || Zeitachse || Zeitraster
time beam   Zeitstrahl * *Zeitachse für Zeit-/Projektplan*
time characteristic   zeitlicher Verlauf
time check   Zeitüberwachung
time constant   Zeitkonstante
time consuming   zeitraubend
time controlled   zeitgesteuert
time delay   Verzögerungszeit * *Zeitverzögerung* || Zeitverzögerung
time derivative   Änderungsgeschwindigkeit * *Differenzialquotient nach der Zeit*
time domain   Zeitbereich * *{Regel.techn.}*
time element   Zeitglied
time for one cycle   Spieldauer * *eines Lastspiels*
time gap   Zeitverschiebung * *Lücke*
time generator   Zeitgeber * *für Uhrzeit*
time grading   Zeitselektivität * *z.B. von Schutz-/Überwachungseinrichtungen*
time in between   Zwischenzeit
time information   Zeitangaben
time instant   Zeitpunkt
time interrupt   Weckalarm
time interval   Zeitabstand || Zeitspanne
time lag relay   Zeitrelais
time lapse   Zeitablauf
time limit   Zeitlimit
time monitoring   Zeitüberwachung
time of day   Uhrzeit
time of oscillation   Periodendauer
time of service   Gebrauchsdauer
time of year   Jahreszeit
time off   Freizeit
time out   Zeitüberschreitung * *Überschreitung einer Überwachungszeit*
time overlap   Zeitüberschneidung
time period   Zeitraum
time range   Zeitbereich
time reasons   Zeitgründe * *for: aus*
time register   Zeitwertspeicher * *bei SPS*
time resolution   Zeitauflösung
time response   zeitliches Verhalten || Zeitverhalten * *bezüglich der Reaktionszeit*
time saver   Zeitsparmaßnahme
time schedule   Terminplan || Zeitplan * *tight:eng/knapp*
time segment   Zeitabschnitt
time shift   Zeitverschiebung
time slice   Zeitscheibe
time slope   Zeitrampe
time slot   Zeitabschnitt * *Zeitscheibe für Softwareprogramme* || Zeitscheibe * *Rechnerzyklus*
time slot manager   Zeitscheibenverwalter
time slot scheduler   Zeitscheibenverwalter
time specification   Zeitangabe
time spent   Zeitaufwand * *aufgewendete Zeit*
time stamp   Zeitmarke * *→Zeitstempel, z.B. bei Diagnose-/Tracespeicher* || Zeitstempel * *z.B. Echtzeitgabe für Daten in einem Messwert-/Diagnose-/Tracespeicher*
time supervision   Zeitüberwachung
time sweep   Zeitablenkung * *(Oszilloskop)*
time switch   Zeitschalter
time tag   Zeitmarke
time to market   Entwicklungszeit * *bis ein Produkt marktfähig ist*
time value   Zeitwert

time-consuming langwierig * *zeitaufwändig* || zeitaufwändig
time-critical zeitkritisch
time-delay relay Verzögerungsrelais || Zeitrelais
time-delayed zeitverzögert
time-dependant zeitabhängig
time-dependent zeitabhängig
time-grade staffeln * *zeitlich ~; z.B. bei Schutzeinrichtungen*
time-lag träge * *Sicherung* || Verzögerung * *Hinterherhinken* || zeitverzögert * *relay:Relais*
time-lag relay Verzögerungsrelais
time-limit äußerster Termin * *Frist* || Frist * *vorgegebener Zeitpunkt* || Termin * *Frist*
time-limited zeitlich begrenzt
time-proportional zeitproportional
time-saving aufwandsarm * *zeitsparend* || zeitsparend
time-variant zeitabhängig
timely rechtzeitig || termingerecht || zeitgemäß
timeout Ablauf der Überwachungszeit || Zykluszeitüberschreitung
timeout monitoring Zeitüberwachung
timer Impulsbildner * *Zeitglied* || Wecker * *Zeitglied in der Digital-/Rechnertechnik* || Zähler * *Zeit~* || Zeitgeber || Zeitglied || Zeitstufe
timer interrupt Weckalarm
timer relay Zeitrelais
times mal * *ein vielfaches (Multiplikation); three times that of: dreimal so viel wie*
timing Timing || Zeiteinstellung || zeitlicher Ablauf || Zeitpunkt * *gewählter ~, Wahl des ~s*
timing chart Timing-Diagramm
timing diagram Impulsdiagramm || Timing-Diagramm
timing relay Zeitrelais
tin Dose * *Blechdose* || konservieren * *in Büchsen*
tin solder Lötzinn || Zinnlot
tin-coated verzinnt
tine Zinke
tinfoil Stanniol
tinge färben
tinman's solder Lötzinn
tinner's snips Blechschere * *Handblechschere*
tiny klein * *winzig*
tip Hinweis * *Tipp, Fingerzeig* || Spitze * *(mechan.)* || Tipp || Trinkgeld
tip cutter Schneidplatte * *{Werkz.masch.}* hartmetallbestücktes Schneidwerkzeug
tip of tooth Zahnspitze * *bei Zahnrad usw.*
tipping Klapp- || klappbar
tips Trinkgeld
tire ermüden * *ermatten* || Reifen * *(amerikan.)* Rad-/Auto-/Gummireifen
tire tread Laufstreifen * *eines Reifens*
tiring mühsam * *ermüdend*
tissue machine Tissuemaschine * *{Papier}* zur Herstellung von Hygienepapier
title Berechtigung * *Rechtsanspruch, Rechtstitel, Recht, Berechtigung* || Überschrift
title of course Kurstitel * *eines Schulungskurses*
title sheet Deckblatt * *z.B. eines Buches*
title to a debt Forderung * *Schuldforderung*
titre size Titer * *{Textil}* Fadendicke
TN system TN-Netz * *geerdetes Netz mit Schutzleiter-Verbindung zwischen den Sternpunkten (VDE 0100 Teil 410)*
to an || auf * *(mit Akkusativ)* || bis || für * *gegenüber* || gegen * *bei Spannungen, z.B. voltage to neutral: Sternspannung (~ den Sternpunkt)* || nach * *zu einem Punkt hin, in Richtung; to the right: ~ rechts*
to a certain extent einigermaßen
to a limited extent eingeschränkt * *(Adv.) in eingeschränktem Maße*
to all practical purposes praktisch * *(Adv.) so gut wie*
to and fro hin und her
to be discontinued abgekündigt
to be explained by zurückzuführen auf
to be on the safe side auf alle Fälle * *zur Sicherheit, sicherheitshalber, lieber* || sicherheitshalber * *um sicherzugehen* || zur Sicherheit * *um sicherzugehen*
to be ordered separately getrennt zu bestellen
to begin with erstens * *zunächst einmal* || zunächst * *erstens*
to correct dimensions maßhaltig * *(Adv.) z.B. cut to correct dimensions: ~ geschnitten*
to each other zueinander
to ground auf Masse bezogen || gegen Erde * *Spannung, Kapazität usw.*
to make sure sicherheitshalber * *um sicherzugehen*
to one another zueinander
to scale maßstäblich
to schedule fristgerecht * *termingerecht* || termingemäß || termingerecht
to some extent einigermaßen || gewissermaßen * *in gewissem Maße* || teilweise * *in gewissem Ausmaß* || zum Teil * *zu einem gewissen Ausmaß*
to tell the truth eigentlich * *(Adv.) offen gesagt*
to that end deshalb * *für diesen Zweck*
to the back hinten * *nach hinten*
to the best advantage vorteilhaft * *(Adv.) aufs vorteilhafteste*
to the credit of zugunsten
to the front nach vorne * *zur Vorderseite*
to the left links * *nach ~; to the left of: ~ von* || nach links
to the point bündig * *kurz und bündig*
to the rear hinten * *nach hinten; shifted towards the rear: nach ~ versetzt* || nach hinten * *zur Rückseite hin*
to the right nach rechts || rechts * *nach ~; to the right of: ~ von*
to the side seitlich
to what extent inwieweit * *bis zu welchem Ausmaße*
to-do list Erledigungsliste
to-pinpoint accuracy punktgenau * *(Adv.)*
today's heutig
together aneinander * *(Adv.)* || gemeinsam * *(Adv.)* || gleichzeitig || zueinander * *aneinander* || zugleich || zusammen
toggle hin- und herschalten || Kippen * *(Verb) – eines Bits, eines Flipflop usw.* || umklappen * *zwischen zwei Zuständen* || umschalten * *hin- und herschalten*
toggle press Kniehebelpresse
toggle switch Kippschalter
toggle-lever press Kniehebelpresse
toggled invertiert * *Binärsignal ('umgeklappt')*
toggling Umschaltung * *Hin- und Herschalten zwischen zwei Zuständen*
toilet Toilette * *(besonders i. Amerika)*
toilsome mühsam * *anstrengend*
token Token * *Sendeberechtigung bei einer seriellen Kommunikation* || Zeichen * *Anzeichen, Merkmal, Beweis, (umlaufende) Zugriffsberechtigg. für Bus-Kommunikation*
token passing Token Passing * *Weitergabe der Sendeberechtigung z.B. von einem Busteilnehmer zu e.anderen* || Token-Weitergabe * *Weitergabe der Sendeberechtigung (am seriellen) Bussystem* || Weitergabe der Sendeberechtigung
token-passing Tokenweitergabe * *Weitergabe der Zugriffsberechtigung von Teilnehmer zu Teilnehmer bei Bussystem*
tolarance zone Toleranzfeld

tolerable annehmbar * *leidlich* || tolerierbar || zulässig * *tolerierbar, zulässig, erträglich, leidlich*
tolerance Abweichung * *zulässige* ~ || Genauigkeit * *Toleranz* || Maßabweichung * *zulässige* ~ || Toleranz * *out of: außerhalb der; close: enge* || zulässige Abweichung
tolerance band Toleranzband || Toleranzbereich * *Toleranzband* || Toleranzfeld * *Toleranzband*
tolerance compensation Toleranzausgleich * *z.B. bei Werkzeugmaschine*
tolerance limit Toleranzgrenze
tolerance on fit Passung * *Passungstoleranz*
tolerance range Toleranzband * *Toleranzbereich* || Toleranzbereich
tolerance ring Toleranzring
tolerance window Toleranzband * *Toleranzfenster*
tolerances Passung
tolerant attitude Toleranz * *tolerante(r) Standpunkt, Geisteshaltung, Einstellung*
tolerate gestatten * *dulden, tolerieren, zulassen* || hinnehmen * *tolerieren* || tolerieren || zulassen * *geschehenlassen*
tolerated stress Grenzbeanspruchung
toleration Toleranz * *Duldung, Tolerierung*
toll Zoll * *Straßen-/Wege-/Brücken~, Standgeld, Wegegeld*
toll-free kostenlos * *nicht gebührenpflichtig (z.B. Telefongespräch)*
ton t * *in Großbritannien: 1 Brit. ton entspr. 1016 kg; in USA: 1 US-ton entspr. 907,2 kg* || Tonne * *Gewichtseinheit; siehe auch →t (1 Brit. ton entspr. 1016 kg; 1 US-ton entspr. 907,2 kg)*
tone Klang
tone dialing Frequenzwählverfahren || Tonwählverfahren * *Telefonwählverfahren: für jede Ziffer wird e.entsprechende Tonfrequenz übertragen*
tongs Zange * *große*
tongue Feder * *{Holz} bei einer Holzverbindung* || Lasche * *Zunge*
tongue and groove jointing Nut-und-Feder-Verbindung * *{Holz}*
tongue roll Oberwalze
tonnage Tonnage || Tragfähigkeit * ~ *eines Schiff*
tonne t * *metrische Tonne (entspr. 1000 kg)* || Tonne * *Gewichtseinheit "metrische Tonne" (entspricht 100 kg)*
too auch * *(nachgestellt)* || ebenfalls || zu * *too small: zu klein*
tool -Mittel * *Werkzeug* || -Zeug * *Werkzeug* || bearbeiten * *Metall ~, spanabhebend ~, z.B.auf Dehbank* || Hilfsmittel * *(auch figürlich) Werkzeug im weitesten Sinne* || Instrument || Mittel * *Hilfs~, Werkzeug (auch figürl.)* || Werkzeug * *auch Software~; tool kit: ~atz; power tool: Elektro~ (auch Heimwerkergerät usw.)*
tool bag Werkzeugtasche
tool bar Werkzeugleiste * *auf dem Bildschirm in einer grafischen Bedienoberfläche*
tool bit Drehling * *Werkzeugeinsatz*
tool breakage Werkzeugbruch
tool change Werkzeugwechsel
tool change system Werkzeugwechselsystem * *{Werkz.masch.}*
tool changeover Werkzeugwechsel
tool changer Werkzeugwechseleinrichtung || Werkzeugwechsler * *{Werkz.masch.}*
tool compensation Werkzeugkorrektur * *{NC-Steuerung} kompensiert Werkzeugverschleiß*
tool dressing Werkzeugzurichtung * *{Werkz.-masch.} schärfen, geradeschleifen (Schleifscheibe usw.)*
tool engineer Werkzeugbauer
tool grinder Werkzeugschleifmaschine
tool grinding machine Werkzeugschleifmaschine
tool holder Werkzeughalter

tool kit Werkzeugkasten * *Satz von Werkzeugen* || Werkzeugsatz || Werkzeugtasche * *Satz*
tool life Standzeit * *Werkzeug*
tool machine Werkzeugmaschine
tool manufacture Werkzeugbau
tool measuring instrument Werkzeugmessgerät
tool offset Werkzeugkorrektur * *{NC-Steuerung} kompensiert Werkzeuglänge (Nullpunktkorrektur)*
tool sharpening machine Werkzeugschleifmaschine * *zum Schärfen*
tool steel Werkzeugstahl
tool-changing system Werkzeugwechselsystem * *{Werkz.masch.}*
toolbox Werkzeugkasten
toolholder Halter * *Werkzeughalter*
tooling Bearbeitung * *spanabhebende* ~ || Bestückung * *einer Werkzeugmaschine mit Werkzeug(en)*
toolmaker Schlosser * *Werkzeugmacher* || Werkzeugbauer * *Werkzeugmacher*
toolmaking Werkzeugbau
tools Arbeitsmittel
toolsetter Maschineneinrichter
tooth Zahn * *auch an Zahnrad* || Zinke * *z.B eines Kamms*
tooth clutch Zahnkupplung
tooth construction Verzahnung
tooth engagement Zahneingriff * *bei Zahnrädern usw.*
tooth flank Zahnflanke * *z.B. an einem Zahnrad*
tooth meshing Zahneingriff * *bei Zahnrädern usw.*
tooth profile Zahnflanke || Zahnprofil * *z.B. eines Zahnrades*
tooth shape Zahnform * *z.B. an einem Zahnrad*
tooth system Verzahnung
tooth thickness Zahndicke * *z.B. bei Zahnrad*
tooth tip Zahnspitze * *bei Zahnrad usw.*
tooth tip clearance Zahnspitzenspiel * *bei Zahnrad*
tooth-clutch Zahnkupplung
tooth-flank backlash Zahnflankenspiel
toothed gezahnt || verzahnt * *externally: außen, internally: innen*
toothed belt Zahnriemen
toothed rack Zahnstange
toothed-belt pulley Zahnriemenscheibe
toothing Verzahnung
top Aufsatz * *Oberteil* || Dach || Deckel * *screw top: Schraub~* || höchst * *z.B. top priority: ~ Priorität, top speed: Höchstgeschwindigkeit/Drehzahl* || maximal * *Spitzen~* || oben * *ober(st)es Ende, Oberteil, Spitze* || Ober- * *obenliegend, oben angesiedelt (auch figürl.)* || Oberteil * *Spitze * (figürl.) einer Hierachie/Reihenfolge usw.* || Spitzen~ || Zopf-
top assembly Dachaufsatz * *eines Schaltschranks*
top bezel plate Kopfleiste * *an Schaltschrank*
top cable Oberseil * *{Aufzüge}*
top cover Dachblech || Deckel * *obenliegender* ~
top cover sheet Dachblech
top down von oben nach unten
top edge Oberkante
top face Oberseite * *obere Fläche (einer Platte, eines Brettes usw.)*
top housing cover Gehäusedeckel * *obere Abdeckung*
top management Führungsebene * *obere* ~
top part Oberteil
top performance Spitzenleistung * *beste Leistungsfähigkeit/Ausführung/Durchführung/Verrichtung/Hervorbringung*
top performer Leistungsträger * *z.B. Mitarbeiter, Arbeitskraft (~ bezüglich d. Leistungsfähigkeit/Arbeitsleistung)*
top plan view Ansicht von oben
top plate Dachblech * *aus dickerem Blech*
top priority höchste Priorität

top roll Oberwalze
top roller Oberwalze * *auch z.B. in der Flachglas-Industrie*
top rope Oberseil * *{Aufzüge}*
top sheet cover Dachblech
top side Oberseite
top speed Höchstgeschwindigkeit || Maximaldrehzahl || Maximalgeschwindigkeit || volle Geschwindigkeit
top surface Oberfläche
top up auffüllen * *(Öl usw.) nachfüllen* || nachfüllen * *Isolier-/Kühl-/Schmiermittel/Öl usw. auf den regulären Füllstand ~*
top view Draufsicht * *Sicht von oben*
top-hat rail Hutschiene * *zur Aufschnappmontage*
top-heaviness Kopflastigkeit
top-heavy kopflastig
top-mounted aufgebaut * *oben montiert, auf der Oberseite 'draufgebaut'* || *montiert* || oben montiert || obenliegend * *oben montiert*
top-mounted fan Aufbaulüfter
top-notch prima * *(salopp) prima, erstklassig* || Spitzen- * *(salopp) spitzenmäßig, →prima*
top-of-the-market führend * *am Markt*
top-priority hochprior || wichtigst * *von höchster Priorität, hochprior*
top-quality erstklassig
top-quality product Spitzenprodukt
top-roll motor Obermotor * *Walzwerk*
topic Gegenstand * *Thema, Gegenstand* || Gesprächsthema * *(of conversation)* || Kapitel * *Thema, Gegenstand* || Punkt * *Einzelthema* || Thema * *auch Gegenstand, Besprechungs~; current topics: aktuelle ~n*
topic of conversation Besprechungsthema
topical aktuell * *von aktueller Bedeutung/aktuellem Interesse, z.B. Frage, Problem, Film*
topmost höchst * *höchst, oberst*
topology Topologie * *Beziehg. einzeln. Komponenten bezügl.d. räuml./örtlich. Anordnung oder d. Datenflusses*
topple Kippen * *(Verb) (um)~ durch Kopflastigkeit*
torch Brenner * *Schweißbrenner* || Taschenlampe * *Stablampe*
torodial Ring- * *z.B. Trafo, Stellwiderstand usw.*
torodial core Ringbandkern * *z.B. bei Trafo*
toroidal ringförmig
toroidal coil winder Ringwickelmaschine
toroidal transformer Ringkern-Transformator || Ringkerntrafo
toroidal-core current transformer Ringkernstromwandler
torque Drehmoment * *['to:k] transmittable: übertragbares* || Moment * *[to:k] Drehmoment*
torque accuracy Momentengenauigkeit
torque actual value Momentenistwert
torque application Kraftangriff * *rotatorisch*
torque arm Drehmomentenstütze || Momentenstütze
torque block Momentenschale * *z.B. Stromregler, Kommandostufe und Steuersatz*
torque boost Drehmomentanhebung
torque bracket Drehmomentenstütze || Momentenstütze
torque build-up Drehmomentaufbau || Drehmomentenaufbau
torque calculator Drehmomentenrechner * *z.B. bei Motor-Prüfstand* || Drehmomentrechner * *z.B. bei Motor-Prüfstand*
torque causing slipping Schlupfmoment
torque change Momentenwechsel * *Momentenänderung*
torque changeover Momentenumschaltung
torque characteristic Drehmomentenverlauf ||

Drehmomentverlauf * *z.B. über der Drehzahl* || Momentenverlauf
torque class Läuferklasse * *kennzeichnet Momentenkennlinie von SIEMENS-Asynchronmotoren (besonders beim Anlauf)* || Momentenklasse * *bezeichnet mit 'KLxx' bei Motor*
torque constant Drehmomentkonstante * *eines Elektromotors [Nm/A]*
torque control Drehmomentenregelung || Drehmomentregelung
torque controller Momentenregler
torque converter Drehmomentenwandler || Drehmomentwandler
torque direction Drehmomentenrichtung || Drehmomentrichtung || Momentenrichtung
torque direction change Momentenwechsel * *Wechsel der Momentenrichtung*
torque direction reversal Momentenumkehr
torque fluctuations Momentenschwankungen
torque inrush Rush-Moment * *Momentenüberhöhung bei Start u.drehrichtungsumkehr eines Motors*
torque limit Momentengrenze * *along the torque limit: an der ~*
torque limiter Drehmomentbegrenzer * *siehe auch →Rutschkupplung, →Sicherheits-Rutschkupplung* || Drehmomentenbegrenzer * *z.B. →Rutschkupplung* || Sicherheitskupplung * *zur Drehmomentbegrenzung*
torque limiting Drehmomentbegrenzung || Momentenbegrenzung
torque loadability Momentenbelastbarkeit * *für el.Motor: siehe z.B. VDE 0530 Teil 1, Beiblatt 2 und Teil 2*
torque meter Drehmoment-Messeinrichtung || Drehmomentenwaage || Momentenwaage * *Drehmomenten-Messeinrichtung*
torque motor Drehfeldmagnet * *Drehmomentmotor mit hohem Stillstandsmoment* || Drehmagnet * *{el.Masch.}* →*Drehmomentmotor, Drehfeldmagnet* || Drehmomentmotor * *Motor mit hohem Stillstandsmoment ('Drehfeldmagnet')*
torque peak Drehmomentenspitze
torque range Drehmomentbereich
torque rating Nenndrehmoment
torque reaction Reaktionsmoment
torque reference Momentensollwert
torque reversal Drehmomentenumkehr || Drehmomentenwechsel || Momentenumkehr || Momentenwechsel
torque ripple Drehmomentenwelligkeit || Drehmomentwelligkeit || Momentenschwankungen * *Momentenwelligkeit* || Momentenwelligkeit
torque rise time Momentenanregelzeit
torque rod Schubstange * *bei Kurbelantrieb*
torque setpoint Momentensollwert
torque shell Momentenschale * *(z.B. beim SIEMENS-Regelsystem SIMADYN D (R))*
torque spanner Drehmomentenschlüssel || Drehmomentschlüssel
torque surge Momentenstoss
torque transmission Kraftübertragung * *rotatorisch*
torque when generating generatorisches Moment
torque when motoring motorisches Moment
torque wrench Drehmomentenschlüssel
torque-direction changeover Momentenumschaltung * *Umschaltung der Momentenrichtung*
torque-free interval momentenfreie Pause || momentfreie Pause
torque-generating momentenbildend
torque-limiting clutch Sicherheitskupplung * *zur Drehmomentbegrenzung*
torque-overload clutch Rutschkupplung * *rückt*

bei Überschreiten d. Nenndrehmoments aus (dient als Überlastungschutz)
torque-producing momentenbildend
torque-speed characteristic Drehmoment-Drehzahlkennlinie || Dremoment-Drehzahlkennlinie * siehe auch →Lastkennlinie
torque-transmission component Übertragungselement * zur Drehmomentübertragung, z.B. von el.- Motor auf die Arbeitsmasch.
torque/speed characteristic Drehmoment-Drehzahlverlauf || Drehzahl-Drehmoment-Kennlinie
torque/speed chart Drehmoment-Drehzahl-Ebene * Koordinatensystem
torsion Torsion || Verdreh- || Verdrehung || Verwindung
torsion angle Verdrehwinkel
torsion lock Verdrehschutz * z.b. für Kabelschuhe in Motorklemmenkasten
torsion resistance Torsionssteifigkeit
torsional Verdreh-
torsional angle Verdrehwinkel * bei Torsionsbeanspruchung
torsional backlash Verdrehspiel
torsional rigidity Drehsteifigkeit || Verdrehsteifigkeit * siehe auch →Drehsteifigkeit
torsional stiffness Drehsteifigkeit || Torsionssteifigkeit || Verdrehsteifigkeit
torsional vibration Drehschwingung || Torsionsschwingung
torsional vibrations Drehschwingungen || Torsionsschwingungen
torsionally flexible drehelastisch * z.B. Kupplung
torsionally rigid drehstarr * coupling: Kupplung || torsionssteif
torsionally stiff drehstarr * z.B. Kupplung || drehsteif * z.b. Kupplung
torsionally-rigid verwindungssteif
TORX screw Torx-Schraube
total betragen * mehrere Posten zusammen; sich belaufen (to:auf) || ganz * gesamt, ganz, völlig, gänzlich || gesamt * über-alles,vollkommen || Gesamt- || Gesamtanzahl || gesamter || insgesamt * gesamt, total || Summe * Gesamtsumme || Summen- * Gesamt-, Über-Alles- || über alles * gesamt || völlig * vollständig || vollständig * gesamt
total amount Gesamtbetrag || Gesamtmenge * Gesamtbetrag/summe
total failure Totalausfall
total harmonic distortion Klirrfaktor
total load Gesamtbelastung
total loss Totalausfall
total number Gesamtanzahl || Gesamtzahl * siehe auch →Gesamtanzahl
total operating I2t value Ausschalt-I2t-Wert * Für Sicherung: Summe aus Schmelz- und Lösch-I2t-Wert
total price Gesamtpreis
total to ausmachen * in der Summe ~, sich belaufen auf
total up summieren * (sich) auf~ (to: auf)
totality Inbegriff * Summe, Gesamtheit, Vollständigkeit || Summe * (figürl.) || Vollständigkeit
totalize summieren * auf~
totally voll * (Adv.) vollkommen || völlig * (Adv.)
totally enclosed voll geschlossen * Gehäuse || vollgeschlossen * hohe Schutzart || völlig geschlossen * Gehäuse, z.B. einer elektr. Maschine || vollkommen geschlossen * Schutzart
totally enclosed fan-cooled oberflächenbelüftet * Fremdlüfter bläst üb.Motoroberfläche; Motorkühlart IC0041 n.DIN IEC 34.6 || oberflächengekühlt * Fremdlüfter bläst üb.Motoroberfläche; Kühlart IC0041 nach DIN IEC 34.6
totally enclosed forced-ventilated oberflächenbelüftet * →fremdbelüftet, Luftstrom ü.d. Motor-

oberfläche || oberflächengekühlt * →fremdbelüftet, Luftstrom über die Maschinenoberfläche
totally enclosed machine geschlossene Maschine * {el.Masch.} Maschine mit geschlossenem Gehäuse
totally enclosed non-ventilated oberflächenbelüftet * lüfterlose Wärmeabfuhr über die Oberfläche des Motorgehäuses || oberflächengekühlt * Wärmeabfuhr ohne Lüfter über die Oberfläche des Motorgehäuses
totally enclosed self-ventilated oberflächenbelüftet * →eigenbelüftet, Luftstrom ü.d. Motoroberfläche (z.B. IP54)
totally enclosed type geschlossene Bauart * IP44
totally enclosed type of enclosure geschlossene Bauart * IP44 || IP54
totally enclosed type of protection geschlossene Bauart * IP 44
totally-enclosed self-ventilated oberflächengekühlt * →eigenbelüftet, Luftstrom ü.d.Oberfläche d.Maschine(z.B IP 54)
touch anfassen || angreifen * anfassen, (Kapital) angreifen || anreißen * Thema || antippen || befassen * (auch figürl.) || berühren || stoßen * berühren || Streifen * (Verb; auch figürl. (z.B. ein Thema)) berühren || tangieren
touch of a button Knopfdruck * at the: auf ~
touch pad keyboard Folientastatur
touch pad panel Folientastatur * Tastenfeld
touch up überarbeiten
touch-tone dialing Frequenzwählverfahren * Telefonwählverfahren: für jede Ziffer wird e.bestimmte Tonfrequenz übertragen || Tonwählverfahren || Tonwählverfahren * Telefonwählverfahren (→Frequenzwählverfahren, →Pulswählverfahren)
touching Berührung
tough aufwändig * schwierig, unangenehm, ein saures Stück Arbeit darstellend || hart * zäh, widerstandsfähig, robust, schwierig, unangenehm, 'bös', eklig, grob || mühsam * schwierig || robust * stark, robust, widerstandsfähig || schwierig * unangenehm, zäh, übel, 'bös', 'ein saures Stück Arbeit darstellend' || zäh * zäh, hart, widerstandsfähig, robust, hartnäckig; toughen: zäher machen
tough environmental conditions raue Umgebungsbedingungen
tough environments raue Umgebungsbedingungen
tough load schwere Belastung
toughness Härte * (figürl.) Zähigkeit, Brutalität, Aggressivität || Zähigkeit
tour Besichtigung * Rundgang, Rundreise (of: durch) || Fahrt * Reise || Rundgang
tour of the city Stadtrundfahrt
touting Kundenfang * aufdringlich, durch üble Tricks
tow schleppen * (ab-) schleppen, ins Schlepptau nehmen, am Haken haben
tow stacker Kabelzusammenführung * {Textil} z.B. in Faserverstreckanlage
toward in Richtung auf
towards auf * (mit Akkusativ) || entgegen * Richtung || gegen * örtlich, zeitlich, in Richtung auf; towards zero speed: ~ Drehzahl Null || in Richtung auf || nach * in Richtung; towards the top ~ oben; towards the bottom: ~ unten zu || um * zeitlich ungefähr || zu * Richtung
towards each other gegeneinander
towards one another gegeneinander
towards the bottom nach unten * nach unten zu
towards the front nach vorne * nach vorne zu || vorderer Bereich * im vorderen Bereich
towards the inside nach innen
towards the outside nach außen
towards the top nach oben
tower above hinausragen über * (figürl.)

towering achievement Meilenstein * *in der technischen Entwicklung*
town Ort * *Stadt*
toxity Giftigkeit || Toxität
toy Spielzeug
trace beschreiben * *eine Bahn usw.* ~ || Idee * Spur || Leiterbahn || Oszillogramm * *Aufzeichnung eines Oszillokop-Bildes* || suchen * *aufspüren (Fehler, Ursache, Spur usw.)* || Trace * *(meist triggerbarer) Speicher zum Aufzeichnen von Signalen über eine gewisse Zeit* || verfolgen * *eine Spur* ~
trace back zurückverfolgen
trace buffer Trace-Speicher || Tracepuffer || Tracespeicher
trace memory Diagnosespeicher * *speichert Signalverlauf* || Trace-Speicher || Tracespeicher
trace recording Traceaufzeichnung
traceability Verfolgbarkeit * *z.B. eines Produkts in der Vertriebskette zum Kunden*
traceable nachweisbar
traceable to zurückzuführen auf
tracing Aufzeichnung * *in einem Messwert/Diagnosespeicher*
track Bahn * *Pfad, Spur, auch Transport~ in Transferstraße* || Lauf * Spur || Leiterbahn || nachführen * *e. Größe e.anderen (to:auf); z.B. e. Prozessmodell, e. Hochlaufgeber auf d. Istwert usw.* || nachverfolgen || Pfad * *Spur, Weg, Route, Fährte* || Spur * *auch eines Impulsgebers* || Strang * *Schienenstrang* || verfolgen * *nach~* || Weg * *Spur; be on the right track: auf d. richtigen Spur/d. richtigen* ~ *sein*
track alignment Nachlauf * *Versatz zw. Radauflagepunkt und Lenkachse eines Fahrzeugrades*
track back zurückführen * *to: auf eine Ursache usw.* ~
track control Bahnsteuerung
track curve Fahrkurve
track displacement Spurversatz * *Impulsgeber*
track drive Raupenantrieb * *z.B. bei Bagger*
track offset Spurversatz * *Impulsgeber*
track-ID bit Spurkennbit * *bei SPS*
track-laying drive Raupenantrieb * *z.B. beim Bagger*
tracker Nachführung * *als Einrichtung/(Funktions-) Baustein*
tracking Nachführung * *eine Größe einer anderen* || Verfolgung * *Nachverfolgung eines Signals/Exemplars/(Verwaltungs-) Vorgangs/von Material*
tracking back Verfolgung * Zurückverfolgung
tracking control Bahnfolgeregelung
tracking error Schleppabstand || Schleppfehler * *z.B. bei einer Lageregelung*
tracking function Nachführung * *als Funktion*
tracking mode Nachführbetrieb
tracking system Verfolgungssystem * *z.B.für Klebe-/Schweiß-/Ausschussstellen b. Rotationsdruck, Blechverarbeit.straße*
traction Bahn- * *{Verkehr}* || Traktion || Zug * *~kraft, ~leistung {Bahn}, Ziehen, Bodenhaftung, Griffigkeit (Reifen), Fortbewegung*
traction converter Bahnumrichter
traction drive Bahnantrieb * *siehe VDE 0535 und IEC Publ.411* || Fahrantrieb || Bahnantrieb || Fahrzeugantrieb * *Traktionsantrieb, Bahnantrieb* || Traktionsantrieb
traction gear Fahrwerk * *bei Kran, Bagger usw.*
traction motor Antriebsmotor * *Fahrmotor bei Schienenfahrzeug* || Bahnmotor * *Traktionsmotor* || Fahrmotor * *{Bahn}*
traction power Zugkraft * *{Bahnen} eines (Schienen-) Fahrzeugs*
traction power supply Bahnstromversorgung
traction supply Bahnstromversorgung
traction supply system Fahrleitung * *{Bahn}*

traction vehicle Triebwagen
tractive power Zugkraft * *{Bahnen}*
trade Branche * *Gewerbe, Geschäftszweig* || Fach * *Geschäftszweig, Gewerbe, Branche, Handel* || Fachgebiet * *siehe auch →Fach* || Geschäft * *Handel* || Gewerbe || Handel * *Geschäftsverkehr* || handeln * *Handel treiben (with: mit einer Person, in: mit Waren)* || Handwerk * *(auch figürlich: z.B. e.Politikers) follow:betreiben* || Verkehr * *Handelsverkehr*
trade and industry Wirtschaft * *gewerbliche* ~
trade and technical literature Fachliteratur
trade association Berufsgenossenschaft * *Wirtschaftsverband* || berufsständische Organisation || Verband * *Fachvereinigung von Unternehmen*
trade barriers Handelsschranken
trade conditions Handelsbedingungen
trade fair Fachmesse || Messe * *Fachmesse, Industriemesse*
trade journal Fachzeitschrift * *~ für Industrie, Handel und Handwerk*
trade literature Fachbücher || Fachliteratur
trade magazine Fachzeitschrift * *~ für Industrie, Handel und Handwerk*
trade mark Marke * *Warenzeichen, Handels~; registered trade mark: eingetragenes Warenzeichen*
trade press Fachpresse * *siehe auch →Fachliteratur*
trade rivalry Konkurrenzkampf
trade secret Geschäftsgeheimnis
trade term Fachausdruck * *~ in der Geschäftswelt/ Branche/Gewerbe*
trade terms Fachausdrücke || Fachbegriffe
trade-mark Warenzeichen * *registered:eingetragenes*
trade-off Kompromiss * *make a trade-off: Kompromiss schließen; auch 'Absprache, Handel, Geschäft'*
trade-union Gewerkschaft
tradeoff Kompromiss
trader Händler
traders Händler * *(Mehrzahl)*
trading company Handelsgesellschaft || Vertriebsgesellschaft * *Handelsgesellschaft*
trading post Handelsniederlassung
traditional herkömmlich || klassisch * *wie gewohnt* || konventionell * *wie gewohnt*
traffic Verkehr
traffic engineering Verkehrstechnik
traffic system Verkehrssystem
traffic systems Verkehrstechnik
traffic-routing system Verkehrsleitsystem
trail nachlaufen * *hinter sich herziehen, (auf der Spur) verfolgen, (nach-) schleppen* || schleppen * *hinter sich herziehen*
trailer Anhänger * *Fahrzeug* || Nachspann * *z.B. eines Datentelegramms, Datensatzes*
trailing angehängt * *hinten/am Ende angehängt* || fallend * *hintere (Impulsflanke usw.)* || Hinter- || Rück-
trailing cable Schleppkabel || Schleppleitung
trailing capability Schleppfähigkeit * *eines Schleppkabels*
trailing edge abfallende Flanke * *Rückflanke eines Impulses* || fallende Flanke * *Rückflanke eines Impulses* || hintere Flanke || negative Flanke * *Rückflanke*
train ausbilden * *schulen* || Bahn * *Eisen~, Zug* || einarbeiten * *someone for a job: jemanden in eine Aufgabe* ~ || Kette * *Kolonne (von Fahrzeugen, Personen usw.), Impulskette, Reihe (von Ereignissen, Gedanken usw.)* || Strang * *Kette, Folge; drive train: Antriebs~* || üben * *schulen, ausbilden, jdm. etwas beibringen* || Zug * *Eisenbahnzug*
trained qualifiziert * *geschult, geübt, ausgebildet*

trained on the job angelernt * *Arbeiter*
trained worker Facharbeiter
trainee Informand * *Praktikant, Auszubildender* ||
Lernender * *bei beruflicher Aus/Weiterbildung; siehe auch 'Lehrling'* || Nachwuchskraft || Praktikant
trainer Ausbilder * *Schulungsleiter* || Schulungsleiter
training Ausbildung * *Schulung, innerbetriebliche/fachliche Aus-/Weiterbildung* || Einarbeitung || Einweisung * *Schulung, Ausbildung, Üben* || Schulung || Übung || Unterricht
training aids Lehrmittel
training costs Ausbildungskosten * *für berufliche Aus/Weiterbildung*
training course Kurs * *Lehrgang* || Lehrgang * *Schulungskurs* || Schulungskurs || Schulungsseminar || Seminar * *Lehrgang*
training documentation Kursunterlagen * *Schulungsunterlagen* || Schulungsunterlagen
training facility Ausbildungsstätte * *für berufliche/betriebliche Aus- und Weiterbildung*
training for the next generation Nachwuchsförderung
training material Schulungsunterlagen
training objectives Lernziel
training of new recruits Nachwuchsförderung
training period Einarbeitungszeit
training personnel Lehrkräfte * *für berufliche Aus/Weiterbildung*
training phase Lernphase
training program Schulungsprogramm
training requirements Ausbildungsaufwand * *z.B. zum Lernen der Maschinenbedienung*
training seminar Lehrgang || Schulungsseminar
training shop Lehrwerkstatt
training system Ausbildungssystem * *für berufliche Aus/Weiterbildung*
training workshop Lehrwerkstatt
trajectory Bahn * *{NC-Steuerung} Verfahr-/-kurve im 2-/3-dimensionalen Raum bei Positioniersteuerung usw.* || Bahn- * *{NC-Steuerg.}* || Bahnkurve
trajectory error Schleppfehler * *Abweichung von der vorgegebenen Bahnkurve bei Positionier-/Bahn-/NC-Steuerung*
tram Bahn * *Stadt~, Straßen~* || Straßenbahn
trans-standard motor Transnormmotor * *AC Motor, in der Leistung über der Normreihe liegend (über ca.200kW)*
transact abschließen * *einen Handel ~* || abwickeln * *Geschäfte ~*
transact business Geschäfte tätigen
transaction Abschluss * *~ eines Geschäfts* || Abwicklung * *~ einer Bestellung usw.* || Geschäft * *Unternehmung, Verhandlung,Abwicklung* || Handel * *Geschäft, Abwicklung, Verhandlung* || Transaktion || Verhandlung
transactions Verhandlungen
transceiver Leitungstreiber-Empfänger || Sender-Empfänger * *Abk.f. 'TRANSmitter/reCEIVER':Zusammenfassung von Sende- u. Empfangsbaustein(en)*
transducer Aufnehmer * *Mess~/Geber* || Geber * *Messumformer* || Messaufnehmer || Messfühler * *Messgeber* || Messgeber * *[trans'dju:se]* || Messumformer || Messwandler || Messwertaufnehmer || Umformer * *Messumformer* || Umsetzer * *(Mess) Umformer* || Wandler
transductor Transduktor * *Transduktor (-drossel), (vormagnetisierte) Regeldrossel*
transfer ablösen * *übergeben, übertreten (to:zu/an/nach)* || Anweisung * *Zahlungsüberweisung* || Transfer || transferieren || Transport * *von Daten* || Übergabe || Übergang * *Wechsel,* Übertragung, Transfer || übergeben * *übertragen, beschaffen, wechseln* || Überleitung * *Übertragung, Übergang, Transfer, Wechsel; to:zu* || übertragen * *auch elektr. Energie, Telegramm ~* || Übertragung || Überweisung || Verbindungs- || verlagern || Verlagerung || verlegen * *versetzen* || versetzen * *Mitarbeiter an einen anderen Ort/Platz* || weitergeben * *transferieren, übertragen, übergeben* || weiterleiten
transfer buffer Übergabepuffer
transfer characteristic Übertragungskennlinie
transfer circuit Ablöseschaltung
transfer control ablösende Regelung
transfer distance Übertragungsentfernung
transfer element Übertragungsglied
transfer error Übertragungsfehler
transfer function Übergangsfunktion || Übertragungsfunktion
transfer key Übernahme-Taste
transfer line Transferlinie * *Transferstraße* || Transferstraße
transfer operation Transferfunktion * *bei SPS*
transfer point Ablösepunkt
transfer position Übergabestelle
transfer power Übergabeleistung
transfer ratio Übertragungsfaktor
transfer resistance Übergangswiderstand
transfer signal Übertragungssignal
transferable übertragbar
transferred übertragen * *(passiv)* || versetzt * *(z.B. Mitarbeiter an einen anderen Ort)*
transferring Übertragung
transform transformieren * *(auch mathem. u. physikal.)* || Transformierte || umformen || umgestalten
transformation Transformation || Umrechnung * *einer Gleichung* || Umsetzung
transformation matrix Transformationsmatrix
transformation of coordinates Koordinatentransformation
transformation ratio Transformationsverhältnis || Übersetzung * *Übersetzungsverhältnis von Trafo, Strom-/Spannungswandler usw.* || Übersetzungsverhältnis * *Trafo, Strom/Spannungswandler usw.*
transformer Transformator || Übertrager * *Transformator f.Signale; pulse transformer: Impuls~ (z.B. z. Ansteuerung v.Thyristoren)* || Wandler
transformer capacity Trafoleistung || Transformatorleistung
transformer connection Schaltgruppe * *eines Transformators* || Transformatorschaltung
transformer having a loadable neutral Transformator mit belastbarem Sternpunkt
transformer incoming feeder Trafoeinspeisung
transformer-isolated potenzialfrei * *durch Übertrager potenzialgetrennt*
transgress überschreiten * *übertreten, z.B. Gesetz, Vorschrift*
transgression Überschreitung
transient -Spitze * *schlagartige Erhöhung* || Ausgleichsvorgang * *make/break transient: ~ bei Stromkreis-Schließung/-Unterbrechg.* || Ausgleichsvorgang || dynamisch * *vorübergehend, kurzzeitig; transient error: dynamische Regelabweichung* || Einschwingvorgang || flüchtig * *vorübergehend* || Kurz- * *vorübergehend, bedingt durch Einschwing-/Ausgleichsvorgang* || momentan * *(elektr., mechan.) z.B. Stoss, Einbruch, Erhöhung* || Spitzen- * *schlagartige, vorübergehende Erhöhung* || transient * *vorübergehend/in einem Übergang auftretend* || vorübergehend * *zeitlich; flüchtig; transient deviation: ~e Abweichung* || vorübergehende Abweichung
transient condition Einschwingvorgang
transient current Stoßstrom
transient deviation vorübergehende Abweichung

transient effect Ausgleichsvorgang
transient function Übergangsfunktion
transient interruption Kurzunterbrechung
transient phenomenon Ausgleichsvorgang
transient protection Überspannungsschutz * *gegen Spannungsspitzen*
transient pulse Wischer * *Kurzzeitimpuls*
transient reaction Ausgleichsvorgang
transient recorder Transientenrecorder
transient recovery Ausgleichsvorgang
transient response Sprungantwort || Übergangsfunktion * *Übergangsverhalten* || Übergangsverhalten
transient suppressor Überspannungsableiter
transient time Einschwingzeit
transient torque Stoßmoment
transients -Spitzen * *vorübergehende Erhöhungen/Einbrüche*
transistor Transistor * *(Abk.f. TRANSfer resistOR') transistorized: transistorisiert, mit ~en ausgestattet*
transistor chopper Transistorsteller * *{Stromrichter} z.B. für Servo-DC-Antriebe*
transistor output Transistorausgang
transistor PWM converter Transistorpulsumrichter * *mit pulsweitenmodulierter Ansteuerung*
transistor switch Transistorschalter
transition Schrittfortschaltung * *bei Ablaufsteuerung/-kette* || Übergang || Überleitung * *Übergang*
transition condition Weiterschaltbedingung
transition period Übergangszeit
transition phase Übergangsphase
transition point Eckpunkt * *Übergangspunkt* || Stoßstelle
transition resistance Übergangswiderstand
transition speed Übergangsdrehzahl
transition time Übergangszeit
transition-sensitive impulse Wischimpuls
transition-type circuit Ablöseschaltung
transition-type controller ablösender Regler
transitional Übergangs-
transitory vorübergehend
translate übersetzen * *into:in* || übertragen * *Grundsätze/Bewegung usw. ~ (into:in; to: auf); transl. into action: in die Tat umsetzen* || umsetzen * *→übertragen; translate ideas into action: Gedanken in die Tat ~*
translation Übersetzung * *z.B. in eine andere Landessprache; auch 'Übertragung, Auslegung'*
translation run Übersetzungslauf
translator Übersetzer
translatory motion geradlinige Bewegung
translucence Durchlässigkeit
translucent durchlässig * *~ für Licht*
transmission Abtrieb * *Kraft-/Drehmomentübertragung* || Getriebe * *Transmission, Übersetzung (-sgetriebe)* || Sendung * *Nachrichtenübermittlung/Übertragung* || Transmission || Übertragung * *z.B. von Daten, Zündimpulsen, Kraft, Drehmoment, Bewegung* || Verbindungs- * *zur Kraft-/Datenübertragung* || Weitergabe * *auch von Daten*
transmission belt Treibriemen
transmission channel Sendekanal
transmission data integrity Übertragungssicherheit
transmission direction Übertragungsrichtung
transmission distance Übertragungsentfernung || Übertragungsstrecke * *Entfernung/Länge der Übertragungsstrecke*
transmission element Antriebselement * *Kraftübertragungs-Element, z.B. Kette, Ritzel, Riemen, Getriebe* || Übertragungselement || Verbindungsglied * *mechanical: mechanisches* || Verbindungselement * *zur Kraft-/Datenübertragung*

transmission engineering Getriebetechnik * *als Fachrichtung*
transmission error Übertragungsfehler
transmission factor Durchlässigkeit * *otische*
transmission frequency Übertragungsfrequenz
transmission gear Vorgelege * *vor dem 'eigentlichen' Getriebe angeordnetes Zwischengetriebe*
transmission gearing Übersetzungstriebe
transmission length Übertragungsstrecke * *Länge der Übertragungsstrecke*
transmission line Übertragungsleitung || Übertragungsstrecke
transmission medium Übertragungsmedium * *z.B. Bus(system)*
transmission method Übertragungsverfahren
transmission mode Übertragungsverfahren
transmission of igniting flame Zünddurchschlag
transmission of internal ignition Zünddurchschlag
transmission path Übertragungsstrecke
transmission protocol Übertragungsprotokoll * *zur Datenübertragung*
transmission rate Übertragungsgeschwindigkeit * *z.B. bei serieller Kommunikation* || Übertragungsrate * *z.B. Baudrate bei serieller Datenübertragung*
transmission ratio Übersetzungsverhältnis * *Getriebe ("i" entspr. Getriebe-Eingangs- zu Ausgangsdrehzahl)*
transmission reliability Übertragungssicherheit
transmission security Übertragungssicherheit
transmission speed Übertragungsgeschwindigkeit
transmission time Übertragungszeit
transmit durchlassen * *Licht ~* || schicken * *übermitteln* || senden * *Nachricht/Signal über Funk, Draht, serielle Schnittstelle usw. ~* || übertragen * *Daten, Nachrichten, Signale, Kraft, Drehmoment, Kraft ~* || weitergeben || weiterleiten * *übertragen (auch Kraft), übersenden, übermitteln, (ver)senden*
transmit block Sendebaustein * *Funktionsbaustein*
transmit channel Sendekanal
transmit character Sendezeichen
transmit data Sendedaten
transmit direction Senderichtung
transmit request Sendeauftrag * *Sendeanforderung*
transmit task Sendeauftrag
transmittable übertragbar * *z.B. Drehmoment, Datenmenge*
transmittable torque übertragbares Drehmoment * *z.B. über eine Kupplung ~ (ist kleiner als das →schaltbare Drehmoment)* || übertragbares Moment
transmittal Sendung * *Nachrichtenübermittlung* || Übertragung * *auch Kraft, Drehmoment*
transmitter Absender * *einer Nachricht, von Daten* || Geber * *Messgeber* || Sendebaustein || Sender
transmitter block Sendebaustein * *Funktionsbaustein*
transmitter-machine Gebermaschine * *bei elektrischer Welle*
transmitting element Übertragungselement
transom Holm * *Querbalken*
transparency Durchsichtigkeit || Overhead-Folie || Overheadfolie || Transparentfolie * *für Overhead-Projektion* || Transparenz || Übersichtlichkeit
transparent transparent || übersichtlich * *klar (im Stil), durchsichtig, offen, ehrlich*
transparent package Klarsichtverpackung
transplant versetzen * *transplantieren, z.B. Baum*
transplanted versetzt * *(z.B. Baum)*
transport Abtransport * *Ab/Antransport, Beförderung, Versand, Verschiffung* || fördern * *transportieren* || Förderung * *Transport, Beförderung* || För-

derung * *Transport* || tragen * *befördern* || Transport || Verkehr * *von Gütern und Personen*
transport authority   Verkehrsunternehmen * *Staatliches/städtisches* ~
transport belt   Transportriemen
transport company   Spedition || Verkehrsbetrieb * *siehe auch* →*Verkehrsunternehmen* || Verkehrsunternehmen
transport container   Transportbehälter
transport damage   Transportschaden
transport documents   Transportpapiere * *siehe auch* →*Begleitpapiere*
transport equipment   Transporteinrichtung
transport into place   einfahren
transport park   Fuhrpark
transport path   Transportweg
transport plate   Transportplatte * *z.B. in Transferstraße*
transport roller table   Transportrollgang
transport system   Transportsystem * *auch bei serieller Kommunikation* || Verkehrssystem
transport systems   Verkehrstechnik
transport tarif   Transporttarif
transport track   Transportbahn * *in Transferstraße, z.B. Rollenbahn*
transport unit   Transporteinheit
transport vehicle   Transportfahrzeug
transportable unit   Transporteinheit
transportation   Transport * *(amerikan.)* || Verkehr * *(amerikan.) von Gütern und Personen*
transportation company   Verkehrsunternehmen
transportation roller   Tragwalze
transportation vehicle   Transportfahrzeug
transpose   versetzen * *mit etwas vertauschen, umstellen, umsetzen*
transposed   versetzt * *mit etwas vertauscht, umgesetzt, umgestellt*
transposition   Umsetzung
TRANSVEKTOR control   feldorientierte Regelung * *(R) Variante des Erfinders dieses Verfahrens, der Fa. SIEMENS* || Transvektor-Regelung * *Vektorregelung (Warenzeichen der Fa. SIEMENS)* || Transvektorregelung * *Vektorregelung (Warenzeichen d. Fa. SIEMENS)* || Vektor-Regelung * *(R) Variante des Erfinders dieses Verfahrens, der Firma SIEMENS* || Vektorregelung * *(R) Variante der Firma SIEMENS, des Erfinders dieses Verfahrens*
transversal   diagonal * *schräg, quer- (verlaufend), sich überschneidend; (to:zu)* || quer * *querverlaufend (to:zu), Quer-, schräg, diagonal* || Quer- || querliegend * *querverlaufend* || schräg * *quer hindurchgehend* || transversal
transversal-directed   Quer- * *in ~richtung*
transversal-directed orientation   Querrecke * *Breitstreck-Einrichtung in Folienmaschine*
transverse   diagonal * *schräg, quer- (verlaufend), sich überschneidend; (to:zu)* || quer * *quer(verlaufend), z.B. Kraft; schräg, diagonal; (to:zu)* || quergerichtet || querliegend
transverse belt   Querband * *in Fördersystem*
transverse drive   Querantrieb
transverse force   Querkraft || Schub * *Querkraft*
transverse load   Querbeanspruchung || Querkraft * *Balastung in Querrichtung*
transverse motion   Querbewegung
transverse orienter   Querrecke * *Breitstreckeinrichtung bei Kunststoff-Folienmaschine*
transverse profile   Querprofil * *Querverteilg.z.B. v. Flächengewicht, Feuchte,Dicke, Temperat. v. bandförm.Material*
transverse section   Querschnitt * *in technischer Zeichnung*
transverse stress   Querbeanspruchung
transverse travel   Querbewegung

transverse-flow fan   Querstromlüfter
tranversal direction   Querrichtung
trap circuit   Fangschaltung * *z.B. zum "Dingfest-Machen" eines Fehlers oder eines selten auftretenden Ereignisses*
trapezium function   Trapezfunktion
trapezoidal   trapezförmig
trapezoidal thread   Trapezgewinde
travel   Anfahrt || anreisen || bewegen * *sich (hin- und her-) bewegen, laufen; auch Maschinenteil, z.B. Kolben* || Bewegung * *eines Maschinenteils, auch 'Lauf,Hub'* || fahren * *reisen (by:mit), sich bewegen (auch Maschinenteil), laufen (auch Maschinenteil, Kolben)* || Fahrt * *Reise, Bewegung, Lauf, Hub (e. Kolbens usw.), Hin-und-her-Bewegung* || Hub * ~ *e. Maschinenteils, Stellventils, Positionierung* || Lauf * *Bewegung,Hub z.B.eines Kolbens* || laufen * *sich (hin- und her) bewegen, z.B.Kolben, Maschinenteil* || wandern || Weg * ~ *eines Machinenteils*
travel agency   Reisebüro
travel direction   Fahrtrichtung
travel distance   Fahrstrecke || Verfahrstrecke || Verfahrweg
travel drive   Fahrantrieb || Fahrwerk || Fahrwerksantrieb * *z.B. eines Krans*
travel forward   vorwärtsbewegen * *sich* ~
travel height   Hubhöhe * *[Aufzüge]*
travel length   Verfahrlänge * *z.B. eines Positionierantriebs*
travel operation   Fahrbetrieb
travel past   vorbeifahren
travel path   Verfahrstrecke
travel range   Verfahrbereich * *bei Positionierantrieb* || Verfahrweg * *Verfahrbereich (von ... bis)*
travel time   Fahrzeit
travel transducer   Wegaufnehmer
traveling gantry   Portalkran
traveling speed   Fahrgeschwindigkeit
travelled distance   Verfahrweg * *ver-/gefahrener/zurückgelegter Weg*
travelling   verfahrbar
travelling crab   Laufkatze
travelling data block   Verfahrdatensatz || Verfahrsatz
travelling direction   Fahrtrichtung * *Verfahrrichtung* || Verfahrrichtung
travelling distance   Verfahrstrecke || Verfahrweg * *z.B. bei einer Positioniersteuerung*
travelling drive   Fahrantrieb * *Kran* || Fahrwerksantrieb * *Kran*
travelling field   Wanderfeld
travelling field motor   Wanderfeldmotor * *z.B. Linear/Sektormotor mit linear/im Kreissektor angeordn. "Käfigläufer"*
travelling gear   Fahrwerk * *z.B.eines Krans*
travelling period   Verfahrzeit
travelling speed   Fahrgeschwindigkeit * *z.B. Kran, Förderzeug, Fahrzeug, Positionierantrieb*
traversable   schwenkbar * *(aus-) schwenkbar*
traverse   changieren * *Quer-/Hin- und Herverlegen beim Aufwickeln von Textilfäden, Kabel, Draht usw.* || Polygon || schwenken * *(seitwärts) schwenken, (sich) drehen, durchqueren* || Strebe * *Quer~* || traversieren * *sich hin- und herbewegen* || verlegen * *in Querrichtung hin- und herbewegen, z.B. Kabel bei der Aufwicklung*
traverse cam   Kehrgewindewelle * *z.B. zur* →*Changierung von Textilfäden bei der Aufwicklung*
traverse control   Changiersteuerung
traverse movement   Querbewegung
traverse unit   Changierung * *Einrichtg. z.seitlichen Hin- u.Herbewegen z.B. v. Textilfäden/Kabel b.d.Aufwicklg.*

traversing Changierung * seitliche Hin- und Herbewegung z.B. von Textilfäden/Kabel bei der Aufwicklung || Verlegung * das Hin- und Herbewegen in Querrichtung, z.B. von Kabel, Draht, Faden bei d. Aufwicklung
traversing control Changiersteuerung
traversing device Changiereinrichtung * {Textil}
traversing direction Verfahrrichtung * bei Positionierantrieb
traversing drive Changierantrieb * in Textilfaserverarbeitung || Fahrantrieb * Kran || Fahrwerksantrieb * Kran
traversing gear Fahrwerk * Kran
traversing profile Fahrkurve
traversing program Verfahrprogramm * {NC-Steuerung}
traversing record Verfahrdatensatz || Verfahrsatz
traversing speed Traversiergeschwindigkeit
traversing time Verfahrzeit
traversing unit Changiereinrichtung * {Textil} || Verlegung * Einrichtung zum Hin- u.Herbewegen von Kabel beim Aufwickeln
tray Mulde * Bottich || Schale * flache Schale, Tablett, Ablagekasten, Einsatz (i. Koffer usw.)
treacherous unzuverlässig * trügerisch
treacherousness Unzuverlässigkeit
tread Lauffläche * eines Rades, Reifens usw. || Lauffläche * eines Rades/(Gummi)Reifens; Spurweite || Laufstreifen * Lauffläche/Profil eines Reifens
treat Ansprechen * (Verb) Thema usw. behandeln || bearbeiten * behandeln || behandeln || handeln * (figürl.) ~ von/über || verarbeiten * behandeln || verhandeln * for: wegen
treat a person with indulgence schonen * jemanden ~d behandeln
treat separately entkoppeln * getrennt behandeln
treat with bearbeiten * behandeln
treatise Aufsatz * Abhandlung || Untersuchung * Abhandlung
treatment Aufbereitung * z.B. von Wasser || Bearbeitung * (allg.) auch Akten-~, Werkstück-~, chemische ~ usw. || Behandlung * ~ von Material || Verarbeitung
tree Baum
tree-type baumförmig
tree-type structure Baumstruktur
treenail Dübel
trellised gitterförmig
tremble zittern
tremendous enorm * gewaltig, ungeheuer, "toll", kolossal || gewaltig * enorm || prima * (salopp) kollossal, toll, gewaltig || riesig
trend Bewegung * Tendenz || Linie * Tendenz || Neigung * Tendenz (towards: zu) || Richtung * Entwicklung || Tendenz || Trend || Verlauf * Tendenz || zeitlicher Verlauf * Tendenz
trend analysis Trendanalyse
trend diagram Kurvendarstellung
trend display Kurvenanzeige
trend of the market Marktentwicklung
trend-setting zukunftsweisend
trendy modern * im Trend
TRESPAPHAN Polypropylen * Markenname (z.B. als Isolierfolie usw.)
trestle Bock * Gestell, Gerüst, Bock, Schragen || Bühne * Gestell, Gerüst, Bock
Triac Triac * Zweirichtungs-Thyristortriode, DIN 41787 Tl.2
trial Erprobung || Probe * Erprobung || Prozess * Strafverfahren || Prüfung * Erprobung || Überprüfung * Erprobung || Verhandlung * strafrechtliche ~ || Versuch * Versuch (of:mit), Erprobung, Probe, Prüfung
trial phase Erprobungsphase

trial run Probelauf
trial shipment Probelieferung
trial version Pilotversion
triangle Dreieck || Dreieck-
triangle signal Dreiecksignal
triangular Dreieck- * z.B. triangular wave: periodisches ~ssignal || dreieckförmig
triangular modulation Dreieckmodulation
triangular shaped dreieckförmig * z.B. Impulse, Kennlinie
triangular signal Dreiecksignal
triangular wrench Dreikantschlüssel
triblet Dorn * Ausweite~
trick Kniff * Kunstgriff, List || Kunstgriff || Manöver * trickreiches ~ || Trick
trickle charge Erhaltungsladung * einer Batterie || Pufferbetrieb * Erhaltungsladung für Batterie
trickle charging Erhaltungsladung * einer Batterie/eines Akkus
trickled gesintert
tricky kompliziert * (salopp) verzwickt || pfiffig * raffiniert, verschlagen, durchtrieben, trickreich || schwierig * (salopp) verzwickt || vielschichtig * (salopp) verzwickt, kompliziert, heikel
tried-and-tested bewährt || praxiserprobt
trifling gering * unbedeutend || geringfügig * unbedeutend, oberflächlich || klein * unbedeutend || leicht * unbedeutend || minimal * (figürl.) unbedeutend, geringfügig, trivial || unbedeutend
trigger Abzug * bei Gewehr usw. || ansteuern * Stromrichter || anstoßen * auslösen || Auslöse- || auslösen * (auch figürl.) ein Ereignis || Auslöser
trigger block Triggerbaustein * Funktionsbaustein
trigger circuit Ansteuerschaltung * für Thyristoren
trigger condition Triggerbedingung * bei Logikanalysator, Störspeicher/schreiber, Registriergerät
trigger delay angle Steuerwinkel
trigger equipment Steuereinrichtung * An~ für Stromrichter (erzeugt die Ansteuer-/Zünd-/Gateimpulse)
trigger event Trigger-Ereignis
trigger instant Triggerzeitpunkt
trigger logic Ansteuerlogik * z.B. für Thyristor
trigger module Ansteuerbaugruppe * Steuersatz für Thyristor-Stromrichter
trigger pulse Ansteuerimpuls * z.B. für Thyristor || Steuerimpuls || Zündimpuls
trigger pulse blocking Impulslöschung * für Thyristor/Transistor-Zündimpulse || Impulssperre * für Thyristoren
trigger pulse transformer Zündimpulsübertrager
trigger set Steuersatz * Zündimpulserzeugung
trigger time instant Triggerzeitpunkt
triggerable zündfähig * z.B. Thyristor
triggered ausgelöst
triggering Ansteuerung * Thyristor-Stromrichter || Anstoß ~d * Auslösung || Triggerung
triggering instant Zündzeitpunkt * für Thyristor
triggering power Ansteuerleistung * zur Ansteuerung von Thyristoren, Pneumatikventilen usw. || Steuerleistung * z.B. zur Ansteuerung eines Thyristors
triggering signal Ansteuersignal * für Transistor, Leistungstransistor usw.
trigonometrical function trigonometrische Funktion || Winkelfunktion
trim -Anpassung * Abgleich || Abgleich || abgleichen || abrichten * Schleifscheibe usw. || Anpassung * Abgleich || besäumen * {Holz} || beschneiden * z.B. Papierbogen, Papierbahn (seitlich) || justieren || kappen * an den Enden abschneiden || richtige Lage * z.B. einer Tänzerrolle || richtige Stellung || Stutzen * (Verb) || trimmen * i.d. richtige Lage

bringen, z.B. e.Tänzer durch Tänzerlageregler, der die Drehz. korrigiert
trim in einpassen
trim pot Trimmpotentiometer || Trimmpoti
trim potentiometer Trimmpotentiometer
trim signal Zusatzsollwert * *z.B. als Stellsignal von einem überlagerten Regler*
trim-pot Abgleichpotentiometer
trimmed abgeglichen
trimmer Abgleichkondensator || Abgleichpotentiometer || Abgleichwiderstand || Trimmer || Trimmkondensator || Trimmpotentiometer
trimmer capacitor Abgleichkondensator || Trimmkondensator
trimming capacitor Abgleichkondensator || Trimmkondensator
trimming machine Abbundanlage * *{Holz}*
trimming pot Abgleichpotentiometer
trimming potentiometer Abgleichpotentiometer
trimming press Abkantpresse
trimming resistor Abgleichwiderstand
trimming shear Besäumschere * *zum seitlichen Abschneiden der Blechränder in Blech-/Walzstraße* || Saumschere * *zum seitlichen Abschneiden des Walzbandes in Blechstraße*
trimmings Randstreifen * *(Plural) abgeschnittene Stücke; z.B. von einer Papierbahn, einem Papierbogen*
trip -Abschaltung * *auf Grund eines Fehlers* || - Wächter * *Auslösegerät, schaltet bei Grenzwertüberschreitung ab* || abschalten * *Störabschaltung* || Abschaltung * *z.B. auf Grund eines Fehlers oder des Ansprechens einer Überwachung* || Ansprechen * *(Verb und Substantiv) Überwachungseinrichtung* || Ausflug || Auslöse- * *Ansprechwert bei Überwachungseinrichtung* || auslösen * *z.b. Überwachungseinrichtung, Störabschaltung* || Auslösung * *einer Überwachungs/Schutzeinrichtung* || Fahrt * *Reise; official trip: Dienstreise* || fallen * *Sicherung, Schutzschalter usw.* || Störabschaltung || Zwangsabschaltung * *auf Grund eines Fehlers*
trip current Auslösestrom * *einer Überwachungs-/Schutzeinrichtung*
trip element Auslöser * *Schutzeinrichtung*
trip gear Sperrung * *Vorrichtung*
trip point Auslösepunkt * *z.B. zum Ansprechen einer Überwachung, eines Grenzwertmelders usw.*
trip report Reisebericht
trip unit Auslösegerät
trip-free ohne Fehlerabschaltung || störsicher * *ohne Störabschaltungen* || störunempfindlich * *ohne Störabschaltungen*
triple dreifach || verdreifachen
triple- drei- || dreifach-
triple-pulse Dreipuls- || dreipulsig
tripod Stativ
tripping Abschaltung * *durch Überwachungskontakt* || Auslösung * *einer Überwachungseinrichtung* || Störabschaltung
tripping characteristic Auslösekennlinie * *z.B. eines Motorschutzschalters* || Schaltcharakteristik * *z.B. eines Auslösegerätes/Überwachungsrelais*
tripping control Auslösegerät
tripping curren Auslösestrom * *einer Überwachungs-/Schutzeinrichtung*
tripping device Auslösegerät || Auslöser * *Schutzeinrichtung*
tripping relay Schutzschalter
tripping sequence Abschaltvorgang * *Ablauf einer Störabschaltung (Abschaltung auf Grund e. Fehlers)*
tripping speed Abschaltdrehzahl * *Auslösedrehzahl einer Drehzahlüberwachung*
tripping temperature Auslösetemperatur * *einer Temperaturüberwachung*

tripping threshold Abschaltschwelle * *für Fehlerabschaltung/Auslösung*
tripping time Ansprechzeit * *z.B. e. Auslösegeräts, Überwachungseinrichtung, Leistungsschalters* || Auslösezeit * *eines Auslösegeräts* || Ausschaltzeit * *einer Überwachungseinrichtung*
tripping torque Abschaltdrehmoment
tripping unit Auslösegerät * *Schutz/Überwachungseinrichtung/-Relais z.B. für Thermofühler im Motor* || Auswertegerät * *für Überwachungszwecke (Auslösegerät), z.b. Temperaturüberwachung*
trivial geringfügig * *unbedeutend, gering, trivial, banal, oberflächlich* || unbedeutend
trolley Katze * *{Hebezeuge}* || Laufkatze * *z.b. bei Kran (→Katzfahrwerk)* || Wagen * *(zweirädriger) Wagen, Laufkatze, Förder~, Draisine*
trolley block Flaschenzug * *mit eingebauter Laufkatze* || Laufkatze * *mit eingebautem Flaschenzug*
trolley carriage Katzfahrwerk * *{Hebezeuge}*
trolley cross travel gear Katzfahrwerk * *{Hebezeuge}*
trolley drive Katzfahrwerk * *{Hebezeuge}*
trolley hoist Katzfahrwerk * *{Hebezeuge} mit Hebewinde* || Laufkatze * *mit Hebezeug*
trolley travel gear Katzfahrwerk * *bei Portalkran*
tropic proofing Tropenschutz
tropical climate tropisches Klima
tropical climates tropische Klimata
tropicalized tropenfest
tropicalized insulation tropenfeste Isolierung * *z.B. einer Motorwicklung*
tropicalized package Tropenschutzverpackung
tropically humid tropisch feucht
tropics-proof tropenfest
tropics-proof insulaton tropenfeste Isolierung
trouble belästigen * *stören* || Fehler * *Störung, Defekt, Verdruss, Schwierigkeit(en)* || kümmern * *s.bemühen (to do:zu tun),s. Mühe/Umstände machen; th. doesn't trouble me:d. kümm. mich nicht* || Mühe * *Mühe, Plage, Last, Belästigung, Verdruss, Scherereien, Unannehmlichkeiten* || Panne || Problem * *Schwierigkeiten, Scherereien, Störung, Defekt, Unannehmlichkeiten* || Schwierigkeit * *Schwierigkeiten; give trouble: einer Person ~en machen* || stören * *belästigen, beunruhigen* || Störung * *auch 'Unannehmlichkeiten, Schwierigkeiten, Scherereien, Plage, Belästigung'* || Unannehmlichkeit * *~en*
trouble proneness Störanfälligkeit * *Fehleranfälligkeit* || Störempfindlichkeit * *Fehleranfälligkeit*
trouble-free mühelos || problemlos * *auch mühelos* || störunempfindlich || störungsfrei
trouble-prone störanfällig || störempfindlich * *fehleranfällig*
troublefree fehlerfrei
troubleshooting Diagnose * *→Fehlersuche, →Fehlerbeseitigung* || Fehlerbehebung * *Fehlerbeseitigung* || Fehlerbeseitigung || Fehlersuche || Störungsbeseitigung || Störungssuche
troublesome aufwändig * *mühsam* || hart * *mühevoll* || lästig * *beschwerlich* || mühsam * *mühevoll* || schwierig * *mühsahm, lästig, beschwerlich* || störend * *unangenehm* || umständlich * *unbequem* || unangenehm * *verdrießlich, lästig, beschwerlich* || unverständlich * *→unbequem*
trough Mulde * *Bottich, Trog, Mulde, Wanne, Rinne* || Rinne * *Trog, Mulde, Wanne, Rinne, Kanal, Wellental, Tiefdruck*
truck Lastauto || Lastwagen || LKW * *trucking: LKW-Transport; light/heavy-duty truck:leichter/schwerer LKW*
truck-trailer unit LKW * *Lastzug, LKW mit Anhänger*
trucking Straßentransport

true abrichten * z.B. Schleifscheibe ~, einpassen, einschleifen, abziehen, nachschleifen || ausrichten || eigentlich * *wirklich, echt* || einpassen || genau * *echt* || rund * *(ruckfreier) Lauf eines Motors ohne Rundlauffehler/Momentenwelligkeit* || wahr || wirklich * *echt, wahr, wahrhaftig* || zutreffend
true output Wirkleistung
true power Wirkleistung
true running Rundlauf || Rundlaufverhalten
true running characteristics Rundlaufgüte * *Gleichmäßigkeit der Drehbewegung, besonders bei kleinen Drehzahlen*
true to gauge maßhaltig
true to scale maßstäblich * *maßstabsgerecht* || maßstabsgerecht
true to size maßhaltig
truism Selbstverständlichkeit * *Binsenwahrheit*
truly wirklich * *(Adv.)*
truncate abbrechen * *Rechenvorgang/Programmablauf ~; abrunden* || beenden * *Programmablauf, Digitalisierg., Rechenalgorithmus abbrechen/-schneiden (Restfehler mögl.)*
truncated abgebrochen * *z.B. Programmablauf/Rechenalgorithmus usw.*
truncated cone Kegelstumpf
truncation Abbruch * *eines Rechenvorganges*
truncation error Restfehler * *Rundungsfehler, ~ bei Rechenalgorithmus (Reihenentwicklung usw.)* || Rundungsfehler * *bei Rechnerprogramm*
trunk Bus * *(amerikan.) Kommunikations(haupt)bus, Hauptnervenstrang* || Fern- * *z.B. -Kabel, -Leitung, -Straße, -Verbindung* || Hauptnervenstrang * *(amerikan.) eines Kommunikationsnetzes* || Verbindungs- * *in der Telekommunikation*
trunk line Hauptlinie * *{Bahnen} Hauptstrecke/-Linie z.B. bei der Eisenbahn*
trunnion Lagerzapfen * *Schildzapfen* || Ritzel
trunnion seat Zapfenlager
truss binden * *bündeln, schnüren (up: fest), zusammen~* || bündeln * *bündeln, schnüren (up: fest), zusammen→binden* || Gerüst * *Gitterwerk, Gerüst, Träger {Bautechnik}, Fachwerk*
trust bauen * *vertrauen (in:auf)* || Vertrauen * *(auch Verb) in:auf*
trustworthiness Glaubwürdigkeit || Sicherheit * *Vertrauen (-swürdigkeit)*
truth Wirklichkeit
truth table Wahrheitstabelle
try erproben || prüfen * *erproben* || verhandeln * *Strafrecht* || Versuch * *(salopp)*
try out erproben
try-out Erprobung || Versuch * *Bemühung*
trying schwierig * *kritisch, unangenehm, nervtötend*
TSR program speicherresidentes Programm * *Terminate a.Stay Resident program,bleibt nach Beendig.i.Hauptspeicher* || TSR-Programm * *Abk.f. "Terminate and Stay Resident": verbleibt auch nach Beendigung im Speicher*
TT system TT-Netz * *auf Einspeise- u.Verbraucherseite geerdeter Sternpunkt, jedoch ohne Schutzleit. dazwischen.*
TTL logic TTL-Logik
TTL signal TTL-Signal
TTY Fernschreiber * *Abk.f. 'TeleTYpe writer': Fernschreiber (mit serieller Stromschleifen-Schnittstelle)*
TTY interface 20mA-Schnittstelle
TTY link 20mA-Schnittstelle
TTY port 20mA-Schnittstelle * *Anschluss*
tub Bütte || Fass * *Bottich*
tube Hülse * *Röhre, Tubus, Tube, Kanal* || Leitung * *Rohr(leitung), Schlauch* || Rohr * *Röhre; seamless: nahtlos* || Rohr- || Röhre * *Schlauch * Rohr (-leitung), Schlauch, Luftschlauch, Röhre* || Tube || Tubus || Tunnel * *U-Bahn-Tunnel* || U-Bahn
tube cap Tubenhütchen
tube cooling Röhrenkühlung * *{el.Masch.} Kühlart mit geschloss. Primär-Röhrenkühlkreislauf (IC511/611/516)*
tube cutter Rohrabschneider
tube diameter Rohrdurchmesser
tube drawing die Rohrziehstein
tube mill Rohrwalzwerk
tube nozzle Anschlussstück * *für Rohr*
tube straightening machine Rohrrichtmaschine
tube-bending machine Rohrbiegemaschine
tube-forming machine Rohrwalzmaschine
tubiform röhrenförmig
tubing Rohr * *~material, Ver~ung, ~anlage* || Rohrleitung * *Röhrenmaterial, Rohr, Röhrenanlage, Rohrleitung(ssystem)*
tubular hohl * *röhrenförmig* || Rohr- * *röhrenförmig* || röhrenförmig || rohrförmig || tubusförmig
tubular bag Schlauchbeutel * *für Verpackungszwecke*
tubular heater Heizstab
tubular jacket Schutzrohr * *tubusförmige Ummantelung* || Tubus * *tubusförmige Umhüllung/Ummantelung*
tubular lamp Soffitte
tubular-bag filling machine Schlauchbeutel-Füllmaschine
tubulature Tubus
tubus Tubus
tug reißen * *zerren (an), ziehen an*
tumble Fall * *Sturz* || fallen * *hinfallen, purzeln, stolpern (over: über), taumeln, Saltos machen, plötzlich ~* || Sturz * *Fall, Sturz, Purzelbaum, Salto* || taumeln * *schleudern, Purzelbäume schlagen, durcheinanderwerfen, durchwühlen, umstürzen*
tumble box Fallschacht * *{Metallurgie}*
tumbler Becher || Putztrommel || Wäschetrockner
tumbler screening machine Taumelsiebmaschine * *{Holz} zur Trennung von Holzschnitzeln unterschiedlicher Größe*
tun Fass * *großes Fass Tonne* || Tonne
tune abgleichen * *einstellen, abstimmen, anpassen, optimieren, in Übereinstimmung/Resonanz bringen* || abstimmen * *z.B. Radio, Schwingkreis; to a frequency:auf e. Frequenz* || anpassen * *einem Zweck ~* || einstellen * *auch abstimmen (to:auf)* || optimieren * *fein anpassen, eine Regelung ~*
tune out entkoppeln * *z.B. bei Rundfunkempfänger*
tungsten Wolfram
tungsten carbide Hartmetall * *Wolframkarbid* || Wolframkarbid * *Hartmetall*
tungsten carbide drill bit Hartmetallbohrer * *Bohreinsatz*
tungsten filament lamp Glühlampe
tungsten-carbide tipped hartmetallbestückt * *{Werkz.masch.}*
tuning Abstimmung * *eines Schwing-/Regelkreises usw.* || Anpassung || Optimierung * *z.B. eines Reglers*
tuning capacitor Abgleichkondensator
tuning of the regulator Regleroptimierung
tunnel Gang * *~ unter Tage* || Rohr * *Tunnel* || Röhre * *Tunnel* || Tunnel
tunnel diode Tunneldiode
tunnel terminal Buchsenklemme
tunnel ventilation Tunnelbelüftung
turbid trüb * *dick(flüssig), trübe, schlammig, verschwommen (figürl.)*
turbidity Trübung
turbidness Trübung
turbine leerlaufen * *mitgezogen werden (Pumpenantrieb)* || Turbine
turbine control Turbinenregelung

turbine pump Kreiselpumpe
turbine-generator set Turbogeneratorsatz
turbo compressor Turbokompressor
turbo coupling Turbokupplung
turbo fan Turboventilator
turbo- Schnell-
turbo-compressor Turboverdichter
turbo-cooling Turbokühlung
turbo-rotor Turboläufer * *Hochgeschwindigkeitsläufer, häufig als ungeblechter →Massivläufer ausgeführt*
turbocompressor Turboverdichter
turboset Turbosatz * *Turbinen-/Generatorsatz zur Erzeugung elektr. Energie*
turbulence Turbulenz
turks head roll Türkenklopfwalze
turn Biegung || drehen * *auch mit Drehmaschine* ~ || Drehung || Kurve * *Biegung, Kurve, Drehung, Kehre, Krümmung, Wendung, Umkehr* || stülpen * *inside out:das Innere nach außen; upside down:- das Obere nach unten* || umbiegen * *abbiegen; up/down: nach oben/unten* || Umdrehung || Umlauf || verdrehen || Verlauf * *~einer Sache* || wenden * *about/round:um-* || Windung * *einer Spule*
turn aside abbiegen || ablenken || umlenken * *zur Seite lenken*
turn away ablenken
turn back umkehren || umlenken * *zurücklenken*
turn down abdrehen * *an der Drehmaschine* || umklappen * *nach unten* || umlegen * *Gegenstand von der Vertikalen in die Horizontale* || verwerfen * *Vorschlag usw.* ~
turn left abbiegen * *nach rechts* ~
turn off abbiegen * *in; auch Straße usw.; into: in.* || ablenken || abschalten || abstellen * *(Wasser, Gas)* abdrehen, *(Licht, Radio usw.)* ausschalten, *(Schlag usw.)* abwenden/-lenken || ausschalten || löschen * *z.B. Thyristor*
turn on anschalten * *einschalten* || einschalten || zünden * *z.B. Thyristor*
turn out ablaufen * *Ausgang nehmen* || ausfallen * *Ergebnis usw. (well: gut; badly: schlecht)* || ausmachen * *Licht usw.* ~ || erweisen * *turn out to be difficult: sich als schwierig* ~ *(siehe auch →herausstellen)* || herausstellen * *sich* ~
turn out to be sich herausstellen als
turn over wälzen * *auch Gedanken usw.* ~ || weitergeben * *die Sache an:the matter over to* || wenden * *Buchseite; please turn over/P.T.O./pto:bitte wenden*
turn over the leaves blättern * *in Buch*
turn right abbiegen * *nach rechts* ~
turn round umlenken * *herum-/zurücklenken*
turn to account nutzbar machen
turn up schrauben * *in die Höhe* ~ || zeigen * *plötzlich erscheinen*
turn-off löschen * *(Substantiv) das Löschen* || Löschung * *z.B.* ~ *eines Thyristors*
turn-off angle Löschwinkel
turn-off arm Löschzweig
turn-off current Abschaltstrom * *z.B. bei Thyristor*
turn-off thyristor Abschaltthyristor
turn-off time Abschaltzeit * *Thyristor* || Ausschaltzeit * *eines Transistors, Thyristors usw.* || Freiwerdezeit * *Abschaltzeit eines Thyristors* || Löschdauer * *eines Thyristors*
turn-on loss Einschaltverlustleistung
turn-on resistance Einschaltwiderstand * *z.B. eines FET-Transistors*
turn-on time Einschaltzeit
turn-table rollers Drehrollen * *"Um-Die-Ecke"- Rollgang (z.B. im Walzwerk)*
turnable drehbar

turnblade tool Wendeplattenwerkzeug * *{Werkz.- masch.| Holz}*
turned part Drehteil
turned-over edge Falz
turner Wender
turner bar Wendestange * *{Druckerei} im Falzapparat*
turning Biegung || Schwenk- * *sich drehend*
turning away Ablenkung
turning back Umkehr
turning cylinder Wendewalze
turning lathe Drehbank || Drehmaschine * *{Werkz.- masch.} Drehbank*
turning machine Drehmaschine * *{Werkz.masch.| Holz}*
turning off Ablenkung
turning point Wendepunkt
turning round Umdrehung
turning tool Drehstahl * *für →Drehbank*
turning unit Drehvorrichtung || Wendevorrichtung * *board turning unit: ~ für Holzplatten*
turnkey schlüsselfertig
turnkey erection schlüsselfertige Erstellung * *z.B. einer Fabrik*
turnover Umsatz || Volumen * *Umsatz* || Wechsel * *Hinüber~*
turnover station Wendestation * *{Werkz.maschine|Holz}*
turnover unit Drehvorrichtung * *Wendevorrichtung (zum Umdrehen von Teilen, Werkstücken, Holzplatten usw.)* || Wendevorrichtung * *für Holzplatten usw.*
turns Umdrehungen * *Drehungen*
turnstile Drehkreuz * *zur Zugangskontrolle an Durchgängen/Toren*
turntable Drehtisch
turpentine Terpentin * *['törpentain]*
turquoise türkis
turret Drehkreuz * *Drehkreuz bei Wickler mit fliegendem Rollenwechsel* || Karussel * *Karussel- (Drehbank), Wende-/Drehkreuz- (Wickler); vertical turr.boring machine: ~drehbank* || Revolver- * *z.B. turret lathe: ~drehbank*
turret butt Drehkreuz * *Drehkreuz-Auf/Abrollung bei Wickler mit fliegend. Rollenwechsel (butt: aneinander fügen)* || Revolverkopf * *→Drehkreuz bei Wickler mit fliegendem Rollenwechsel*
turret butt splice Drehkreuz-Rollenwechsler * *für fliegenden Rollenwechsel bei Auf/Abwickler* || Revolverkopf-Ankleber * *für fliegenden Rollenwechsel bei Auf/Abwickler*
turret head Revolverkopf * *z.B. an Drehmaschine*
turret lathe Revolverdrehmaschine
turret winder Drehkreuz * *Drehkreuzwickler* || Drehkreuzwickler * *Nonstop-Wickler mit fliegendem Rollenwechsel; automatic: automatischer*
turret-head drilling machine Revolverbohrmaschine
turtleback Wölbung * *(schildkrötenförmig) gewölbte Fläche*
tutor Kursleiter * *Studienleiter*
tutorial Seminar
TÜV TÜV * *authorized inspection agency:Technischer Überwachungsverein*
twelve-pulse 12-pulsig || zwölfpulsig
twelve-pulse converter configuration zwölfpulsige Stromrichterschaltung
twice doppelt * *(Adv.)* || zweimal * *twice as large: ~ so groß*
twice as doppelt * *twice as much: doppelt so viel*
twice the doppelt * *wert/zahlenmäßig; twice the amount:der doppelte Betrag/das Doppelte des Betrages*
twin doppel || doppelt * *Zwillings-* || zwei-

twin converter Doppelstromrichter || Zweifachstromrichter
twin drive Doppelantrieb * z.B. Walzmotor-Paar || Zwillingsantrieb * z.B. Walzmotor-Paar
twin drum winder Doppeltragwalzenroller
twin motor Doppelmotor * {el.Masch.} 2 konzentrische ineinandergebaute Käfigläufermotoren (→doppelte Drehz.)
twin- Doppel- * zweifach, Zwillings- || Zweifach-
twin-face Zweiflächen- * z.B. Bremse, Kupplung
twin-face brake Zweiflächenbremse
twine Faden * gezwirnter || Garn * gezwirntes ~ || Schnur * Bindfaden || Zwirn * Spinnereigarn
twinkle blinken * blitzen, glitzern, funkeln, blinken, zwinkern || glühen * glimmen || leuchten * blinken
twinkling blinkend * glimmend
twist Drall || drehen || Drehung * Ver~ || krümmen || schrauben * drehen || verbiegen * auch 'sich verbiegen' || verdrehen * against: gegenüber || Verdrehung || verdrillen || vertauschen * drehen || zwirnen * {Textil} Fäden usw
twist drill Spiralbohrer
twist switch Schwenktaster
twist-lock tie Kabelbinder * mit Knebel-Drehverschluss
twisted gedreht || gekrümmt || krumm * verdreht, verbogen || verdrillt * z.B. Kabel; in pairs: paarweise
twisted in pairs paarweise verdrillt * Kabel
twisted pair paarweise verdrillt * Kabel || paarweise verseilt
twisted pair cable Zweidrahtleitung * verdrillte Zweidrahtleitung
twisted pairs paarweise verdrillt * (Plural) Kabel
twisted thread Zwirn
twisted yarn Zwirn * Garn
twisted-pair cable verdrillte Zweidrahtleitung
twister Kabelverseilanlage || Kabelverseilvorrichtung || Verseilanlage * z.B. für Kabel
twisting Verbiegung * Verdrehung || Verdreh- || Verdrehung || Zopf-
twisting cage Verseilkorb * bei (Kabel-/Draht-) Verseilmaschine
twisting line Verseilanlage * z.B. für Kabel
twisting machine Verlitzmaschine * für Kabel
two zweit * (nachgestellt) z.B. step two: zweiter Schritt
two drum rewinder Doppeltragwalzenroller
two drum surface winder Doppeltragwalzenroller * Umfangswickler mit 2 Tragwalzen (Tragw. 1 drehz., Tragw.2 momentengeregelt)
two layer paper Duplexpapier
two wire system Zweidrahtsystem * für serielle Datenübertragung
two- Duo- || zwei- || Zweifach-
two-cannel zweiphasig * Impulsgeber, Inkrementalgeber
two-column press Zweisäulenpresse
two-conductor cable Zweidrahtleitung * z.B. für ein serielles Datenübertragungsverfahren nach RS485
two-core zweiadrig
two-cycle engine Zweitaktmotor
two-drum hoist Zweitrommelwinde
two-high Duo- * Walzstraße, Walzwerk
two-level controller Zweipunktregler * z.B. bei Füllstandsregelung
two-part zweiteilig
two-phase zweiphasig
two-phase operation Zweileiterbetrieb * am Drehstromnetz
two-piece zweiteilig
two-point controller Zweipunktregler
two-pole zweipolig
two-port Vierpol

two-position action controller Zweipunktregler
two-quadrant Zweiquadrant-
two-quadrant operation Zweiquadrantbetrieb
two-side zweiseitig
two-sided zweiseitig
two-sidedness Zweiseitigkeit
two-speed gearbox Zweiganggetriebe
two-speed pole-changing zweifach polumschaltbar
two-stage zweistufig * auch bei Getriebe
two-stage controller Dreipunktregler * z.B. Temperaturregler
two-state controller Zweipunktregler
two-step zweistufig
two-step action controller Zweipunktregler
two-step control Zweipunktregelung
two-stroke engine Zweitaktmotor
two-terminal circuit Zweipol
two-terminal network Zweipol
two-terminal-pair network Vierpol
two-tier zweizeilig
two-way zweiseitig * Zweiweg-
two-way contact Umschaltkontakt || Wechselkontakt || Wechsler || Wechslerkontakt
two-way converter Umkehrstromrichter
two-winding transformer Zweiwicklungstrafo
two-wire cable Zweidrahtleitung * z.B. für serielle Verbindung nach RS485
two-wire connection Zweidraht-Verbindung || Zweileiteranschluss
two-wire operation Zweidrahtbetrieb
twofold doppelt * (Adj.)
twosome Paar * aus zwei Teilen (bestehend); (auch figürl.)
type Art * Typ, Modell, Ausführung, Baumuster || Ausführung * Typ || Bauart || Bauform || Bauweise || eingeben * über Tastatur ~ || Form * Art, Typ || Modell * Typ, Bauart || Muster * Bautype || Qualität * Sorte, Bauart || Sorte || Typ || Variante * Typ, Bauform || Version * Typ, Art, Bauform
type change Sortenwechsel
type code Typenbezeichnung * Kurz~, z.B. in Bestellnummer || Typenschlüssel
type designation Typbezeichnung || Typenbezeichnung * bei el.Masch. z.B. entspr. d. Baugrößenbezeichng. nach IEC 72
type in eingeben * über Tastatur ~ || Taste betätigen * über Tastatur eingeben
type key Typenschlüssel * in Bestellnummer usw.
type of bearing Lagertyp * Gleit -/Wälzlager usw.
type of construction -Bauweise || Ausführung * Bauart || Bauart * Bauform || Bauform * für elektr. Maschinen: siehe DIN IEC 34, Teil 7 u. DIN 42950 || Baureihe * →Bauart || Bauweise * Bauform || Modell * Bauart || Sorte * Bauform || Typ * Bauform, Bauart || Variante * Bauweise || Version * Bauweise
type of construction and mounting arrangement IM * {el.Masch.} 'International Mounting' (VDE 0530 T.7)
type of cooling Kühlart * für elektr. Maschinen: siehe DIN IEC 34, Teil 6
type of data Parametertyp
type of enclosure Schutzart
type of grease Fettsorte
type of housing Gehäuseform
type of load Betriebsart * Belastungsklasse
type of modulation Modulationsart || Modulationsverfahren
type of mounting Befestigungsart
type of operation Nennbetrieb * Betriebsart z.B. für Motor nach VDE 0530 oder Stromrichter nach DIN 41756 || Nennbetriebsart * Betriebsart z.B. für Motor nach VDE 0530 oder Stromrichter nach DIN 41756
type of parameter Parameterart || Parametertyp

type of protection Explosionsschutzart || Schutzart * *(siehe auch 'IP') kennz.Schutz geg. Fremdkörper (IEC34-5/VDE0530/EN60034 Tl.5)* || Zündschutzart * *siehe EN 50014 und 50016 bis 50021*
type of protection E EX I Schlagwetterschutz * *Ex-Schutzart nach EN50014*
type plate Typenschild
type series Baureihe
type setting Satz * *{Druckerei}*
type spectrum Typenspektrum
type summary Typenübersicht
type test Typenprüfung || Typprüfung
type-of-construction code Bauform-Kennziffer
type-tested typgeprüft
types of converter connections Stromrichterschaltungen * *Arten von Stromrichterschaltungen*
typical charakteristisch * *of:für* || typisch
typical features typische Merkmale
typical properties typische Merkmale * *Eigenschaften*
typified gekennzeichnet * *charakterisiert*
typify kennzeichnen * *charakterisieren*
typographic drucktechnisch
typographic error Druckfehler
typographical drucktechnisch
typographical error Druckfehler
typography Buchdruck * *Buchdruckerkunst, (Buch-) Druck* || Buchdruckerei * *Buchdruckerkunst, (Buch-) Druck*
tyre Reifen * *(brit.) Rad-/Auto-/Gummireifen*

U

U packing ring Dichtmanschette
U section U-Profil
u with umlaut ü
U-bolt U-Bügel
U-seal Dichtmanschette || Manschette * *→Dichtmanschette*
U-shaped bracket U-Bügel
U-shaped washer U-Bügel * *Unterlegscheibe*
U.K. Großbritannien * *United Kingdom: England, Schottland, Wales u.Nordirland*
U.S. Vereinigte Staaten
U.S.A. USA
UART UART * *'Universal Asynchronous Receiver/Transmitter': parallel<>seriell-Wandler f. asynchr.Schnittst.*
ue ü
ugly unangenehm * *garstig, widerwärtig*
uk Kurzschlussspannung || uk * *Kurzschlusspann. z.B.Drossel (Eing.spann. bei d.bei kurzgeschloss. Ausgang der Nennstrom fließt)*
UK gal 1 * *Engl.Gallonen; 1 l entspr. 0,2200 Engl.Gallonen*
UK pt 1 * *Engl. Pint; 1 l entspr 1,7598 UK pt*
UL UL * *Abkürzg.für 'Underwriters Laboratory': USA-Normengremium der Feuerversicher*
UL approved UL approbiert
UL listed UL approbiert
UL recognized UL approbiert
ultimate End- * *äußerst, (aller)letzt, endlich, End- (Ergebnis, Verbraucher usw.)* || letzt * *äußerst, aller~, endgültig, schließlich, Höchst-, Grenz-* || letzter * *äußerst, allerletzt, schließlich, endgültig* || schließlich || Spitzen- * *ultimativ*
ultimate buyer Endabnehmer * *einer Ware* || Endkunde * *Endabnehmer einer Ware*
ultimately letztendlich || schließlich * *(Adv.) zuletzt* || zuletzt * *schließlich*

ultra high-speed response höchste Dynamik
ultra-clean hochrein * *sauber*
ultra-modern hochmodern
ultra-precise hochgenau
ultra-pure hochrein * *mit geringem Anteil an Fremdstoffen*
ultrasonic Ultraschall-
ultrasonic motor Ultraschallmotor
ultrasonic transducer Ultraschallgeber
ultrasonics Ultraschall
umlauted a ä
umlauted o ö
umlauted u ü
un Ent- * *Vorsilbe*
un- -los * *(vorangestellt)* || un- * *nicht*
unaffected natürlich * *ungekünstelt* || unbeeinflusst * *unbeeinflusst/beeindruckt; by:von* || unempfindlich * *unbeeinflusst, unbeeindruckt (by: durch; to: gegen)*
unaided selbst * *ohne fremde Hilfe* || selbstständig * *ohne Beistand/Hilfe*
unalloyed unlegiert
unalloyed steel unlegierter Stahl
unalterable unabdingbar * *unabänderlich, unveränderlich* || unveränderlich * *unabänderlich, unveränderlich*
unaltered unverändert
unambiguous eindeutig * *unzweideutig* || unzweideutig
unanimous übereinstimmend * *einstimmig*
unanimously geschlossen * *(Adv.) einstimmig*
unanticipated unerwartet
unapproachable unzugänglich
unarranged ungeordnet
unassailable einwandfrei * *unanfechtbar*
unassisted allein * *ohne Hilfe*
unassuming anspruchslos * *schlicht*
unattended automatisch * *unbeaufsichtigt, ohne Beteiligung von Bedienpersonal* || unbeaufsichtigt
unauthorized nichtautorisiert || unberechtigt
unauthorized person Unbefugter
unavoidable unvermeidbar || unvermeidlich
unbalance Unsymmetrie || Unwucht * *nicht ausgeglichene Fliehkräfte (z.B. U in [gmm] entspr. u (Unwuchtmasse [g]) x r [mm])*
unbalance measuring equipment Unwuchtmessgerät
unbalanced einseitig * *unausgeglichen* || ungleichmäßig
unbalanced load Schieflast || ungleiche Belastung
unbalanced state Unwucht
unbeatable konkurrenzlos * *unschlagbar* || unschlagbar
unbelievable unglaublich
unbiased sachlich * *unparteiisch*
unbiassed sachlich * *unparteiisch (mit einem oder zwei "s" richtig)*
unblocking Freigabe
unbroken lückenlos * *unbeeinträchtigt, unvermindert, ungebrochen, gesamt, heil*
unburden entlasten
unbureaucratic unbürokratisch
unbureaucratically unbürokratisch * *(Adv.)*
uncertain unbestimmt * *unsicher* || unzuverlässig * *unsicher*
uncertainty Unklarheit * *Unsicherheit* || Unschärfe * *(figürl.)* || Unsicherheit * *Ungewissheit, Unvorhersehbarkeit* || Unzuverlässigkeit
unchallenged unbestritten
unchangeable gleich bleibend * *unveränderlich* || unveränderlich
unchanged alt * *unverändert* || unverändert * *remain unchanged:unverändert bleiben*
unchanging beständig * *unveränderlich*
uncheckable unaufhaltsam

unchecked unkontrolliert * *ungeprüft*
unclarity Unklarheit
unclutter entwirren * *Ordnung in etwas bringen, Wirrwar "geradeziehen"*
uncluttered aufgeräumt * *z.B. Bildschirmdarstellung, Bedienungselement usw.*
uncoil abrollen * *Blech, Draht, Kabel, Seil* || abwickeln * *Blech, Kabel, Draht von einer Haspel* ~
uncoiler Abhaspel * *Abwickler für Blech, Draht, Kabel* || Abwickelhaspel || Abwickler * *Abhaspel für Blech, Draht, Kabel oder Seil* || Abwicklung * *Abhaspel für Blech, Draht, Kabel*
uncomfortable lästig * *unbequem, unkomfortabel*
uncommon außerordentlich * *außergewöhnlich* || ungewöhnlich * *it's uncommon for someone to do: es es für jemanden ungewohnt ... zu tun* || unüblich
uncompensated unkompensiert * *z.B. Motorwicklung*
uncompensated motor unkompensierter Motor * *ohne* →*Kompensationswicklung*
uncomplicated einfach * *unkompliziert* || unkompliziert
uncompromising kompromisslos
uncompromisingly kompromisslos * *(Adv.)*
unconditional unbedingt * *unbedingt, bedingungslos (auch Programmsprung), uneingeschränkt, vorbehaltlos*
unconditional jump absoluter Sprung || unbedingter Sprung * *im Programm*
unconditionally unbedingt * *(Adv.) bedingungslos*
uncontested unbestritten
uncontrollable unkontrollierbar * *nicht unter Kontrolle zu halten, nicht regelbar*
uncontrolled unbeaufsichtigt || uneingeschränkt || ungeregelt || ungesteuert * *z.B. Brückenschaltung, Ventil, Gleichrichter* || unkontrolliert
uncontrolled drive ungeregelter Antrieb
uncontrolled rectifier ungesteuerter Gleichrichter
uncontrolled valve ungesteuertes Ventil
uncouple abkoppeln || auskoppeln * *auskuppeln, entkoppeln* || auskuppeln * *entkuppeln/koppeln* || entkoppeln * *aus/los-koppeln/kuppeln*
uncovered bloß * *unbedeckt*
undamaged unbeschädigt
undamped unbeschaltet * *Schütz/Relaisspule mit RC-Glied oder Diode zum Überspannungsschutz*
undamped oscillation ungedämpfte Schwingung
undampened ungedämpft * *oscillation: Schwingung*
undecided schwankend * *unentschlossen*
undefined undefiniert
undelayed unverzögert
undependability Anfälligkeit * *Unzuverlässigkeit* || Unzuverlässigkeit
undependable anfällig * *unzuverlässig* || unzuverlässig
under im Rahmen von * *im Rahmen eines Vertrages, einer Vorschrift usw.* || laut * *kraft* || unter * *auch ~ einem Namen*
under certain circumstances unter Umständen
under conditions of bei * *z.B. bei bestimmten Betriebsbedingungen*
under consideration anstehend * *Entscheidung usw.*
under consideration of unter Berücksichtigung von
under hot running condition betriebswarm * *Verbrennungsmotor usw.*
under it darunter
under load unter Belastung || unter Last
under locked-rotor conditions festgebremst * *Motor*
under no circumstances nie * *unter keinen Umständen* || unter keinen Umständen
under no-load condition leerlaufend

under partial load condition bei Teillast
under preparation in Vorbereitung
under that darunter
under the condition unter der Bedingung
under the condition that unter der Bedingung dass
under them darunter * *(Plural)*
under these circumstances unter diesen Umständen
under voltage unter Spannung
under worst case conditions im schlechtesten Fall * *unter den ungünstigsten Umständen*
under- ungenügend * *(Vorsilbe) z.B. underpaid: unterbezahlt*
under-deck installation Unterdeck-Aufstellung * *z.B. v. El.motoren auf Schiffen unt.Deck (Temp.-/Feuchte-/Rüttelbelastg.!)*
under-powered untermotorisiert
under-voltage Unterspannung
undercoat Grundanstrich
undercut Freistich * *{Werkz.masch.} mit Drehmaschine, an Drehteil* || unterbieten * *Preis* ~ || Unterschnitt * *Unterschnitt, Unterhöhlung, {Werkz.-masch.}* →*Freistich*
undercut saw Kappsäge * *{Holz}*
underdimension unterdimensionieren
underestimate unterschätzen * *Fähigkeiten usw.*
underflow Unterlauf * *bei Rechenoperation*
undergo unterliegen * *ausgesetzt sein* || unterziehen * *sich unterziehen, unterzogen werden*
undergraduate Student
underground geheim * *unerlaubt* || unter Tage
underground mining Bergwerks- * *Untertage-* || Untertage || Untertagetechnik
underground railway U-Bahn
underground train U-Bahn
underline anstreichen * *Fehler, Textstelle* ~ || herausstellen * *unterstreichen, herausstreichen, betonen; z.B. in einem Bericht* || unterstreichen * *Text mit einer Linie* ~
underliner Schonschicht * *z.B. bei der Papier- oder Gummiverarbeitung*
underlying zugrundeliegend
undermentioned unten erwähnt || unten stehend
undermine schwächen * *unterminieren, z.B. Gesundheit*
underneath darunter || unter || unterhalb * *unter*
underrate unterschätzen
underscore unterstreichen * *(auch figürlich: betonen)* || Unterstrich * *Zeichen '_' auf Tastatur/Bildschirm/Drucker*
undersell schleudern * *zu billigen Preisen ver~* || unterbieten * *die Konkurrenz* ~
underselling Preisunterbietung
undershoot Einbruch * *Unterschwinger* || unterschwingen * *(auch Substantiv)* || Unterschwinger
underside unten * *from the underside: von* ~ || Unterseite
undersize unterdimensionieren || Untermaß
understand Verständnis * *Verständnis haben für* || verstehen
understandable anschaulich * →*verständlich* || verständlich
understanding Übereinkunft || Vereinbarung * *Vereinbarung, Übereinkunft* || Verständnis * *Verstehen, Einsicht; show understanding for a person: jdm. Verständnis entgegenbringen* || verstehen * *(Subst.) Verstehen, Verständnis, Verständigg., Vereinbarg., Übereinkunft, Abmachung* || Wissen * *Verstehen, Intelligenz, Verständnis, Verstand, Vereinbarung, Verständigung*
understate ansetzen * *zu niedrig ansetzen*
understood verstanden
undertake annehmen * *Auftrag* || durchführen * *(Sache) in die Hand nehmen, (Aufgabe, Verantwortung) übernehmen, (Reparatur)* ~ || überneh-

men * *Arbeit, Verantwortung* ~ || verpflichten * sich ~ *(to do:zu tun), sich verbürgen (that: dass), Risiko/Verantwortg.* übernehmen || vornehmen * *(Aufgabe) übernehmen/beginnen, (Sache) in die Hand nehmen, unternehmen*
undertaking Übernahme * *einer Aufgabe, Verpflichtung, Garantie* || Zusage * *Verpflichtung*
undervalue unterschätzen
undervoltage Unterspannung
undervoltage monitor Unterspannungsüberwachung * *als Schaltung/Einrichtung*
undervoltage monitor module Unterspannungsüberwachung * *als Baugruppe/Gerät*
undervoltage monitoring Unterspannungsüberwachung
undervoltage monitoring module Unterspannungsüberwachung * *als Baugruppe/Gerät*
undervoltage protection Unterspannungsschutz
undervoltage release Unterspannungsauslöser
undervoltage sensing Unterspannungserfassung
undervoltage trip Unterspannungsabschaltung * *Fehlerabschaltung auf Grund unzulässig kleiner Spannung*
underwater motor Unterwasser-Motor
underwind wickeln * ~ *von unten* || wickeln von unten
undesigned unabsichtlich * *unbeabsichtigt* || unbeabsichtigt || versehentlich * *unbeabsichtigt*
undesirable lastig * *unerwünscht* || unerwünscht
undesirably unerwünscht * *(Adv.)*
undesired überflüssig * *unerwünscht, nicht gewünscht* || unerwünscht || ungeplant * *unerwünscht*
undiminished unvermindert * *(Adj.) nicht reduziert*
undisputably unbestritten * *(Adv.)*
undisputed unbestritten
undivided einheitlich * *ungeteilt* || ganz * *ungeteilt* || ungeteilt * *(auch figürl.)*
undo löschen * *rückgängig machen* || rückgängig machen * *z.B. Computereingabe* || widerrufen * *Computereingaben usw.* →*rückgängig machen*
undoubted zweifellos
undoubtedly zweifellos * *(Adv.)*
undue überflüssig * *übermäßig, übertrieben, unangemessen, unzulässig* || übermäßig * *unnötig* || unzulässig
undulatory wellenförmig * *wellenförmig, Wellen-*
unearthed ungeerdet
unearthed system ungeerdetes Netz
uneasiness Unbehagen * *körperl./geistiges* ~, *(innere) Unruhe, Unbehaglichkeit*
uneconomic unwirtschaftlich
unedged unbesäumt * *[Holz]*
UNEL UNEL * *ital. Norm entspr. DIN*
unenclosed ungekapselt
unending langwierig * *endlos*
unequal ungleich || ungleichförmig
unequalled unübertroffen
uneven uneben || ungeradzahlig || ungleichmäßig || unterschiedlich * *schwankend*
uneven running unruhiger Lauf
unevenness Ungleichförmigkeit
unexcelled unübertroffen
unexpected unerwartet
unexpectedly unerwartet * *(Adv.)*
unexperienced unerfahren
unfailing unvermeidbar || unvermeidlich
unfair competition unlauterer Wettbewerb
unfamiliar fremd * *ungewohnt, nicht vertraut* || unbekannt * *nicht vertraut*
unfavorable ungünstig * *(amerikan.)*
unfavourable nachteilig * *ungünstig* || ungünstig * *auch nachteilig (z.B. Bedingungen), nicht dienlich*
unfilled orders Auftragsbestand

unfinished roh * *nicht veredelt, nicht fertigbearbeitet* || unbearbeitet * *ohne Endbehandlung*
unfit ungeeignet * *for:zu*
unfold erschließen * *(sich) entfalten, ausbreiten, öffnen, enthüllen, offenbaren*
unforeseen unerwartet * *unvorhergesehen, unerwartet* || unvorhergesehen || Unvorhergesehenes
unformatted unformatiert * *z.B. Text ohne Angaben von Schriftarten, Absatzgestaltung usw.*
unfortunately leider || unglücklicherweise
ungeared getriebelos
ungeared drive Direktantrieb * *ohne Getriebe*
ungrounded erdfrei || ungeerdet
ungrounded supply ungeerdetes Netz * *z.B. Netzanschluss für Stromrichter/Motor (from:an)*
ungrounded supply network ungeerdetes Netz * *aus der Sicht eines Verbrauchers (z.B. Motor, Stromrichter usw.)*
ungrounded system IT-Netz || ungeerdetes Netz
unguarded ungesichert
unhampered freizügig
unhang abhängen
unhealthy inkonsistent
unhook abhängen * *vom Haken*
unhurried langsam * *ohne Eile*
unhurt intakt
UNI UNI * *Abkürzg. für 'Ente Nazionale Italiano di Unificazione': italienisches Normenbüro*
unidirectional Einquadrant- * *mit einer Drehrichtung* || mit nur einer Drehrichtung || nur in einer Richtung || Zweiquadrant- * *nur in einer Drehrichtung arbeitend*
unidirectional fan drehrichtungsabhängiger Lüfter
unidirectional flux Gleichfluss * *magnetic: magnetischer* ~
unification Vereinheitlichung
unified durchgängig * *vereinheitlicht, z.B. e. techn. Lösung (bei e. Produktfamilie)* || durchgehend * *durchgängig* || einheitlich * *ver~t*
unified system concept Durchgängigkeit
uniform durchgängig * *einheitlich, gleichmäßig, übereinstimmend* || einheitlich || geschlossen * *einheitlich* || gleichartig * *einheitlich* || gleichförmig * *gleichmäßig; uniform in texture:* ~ *strukturiert (Oberfläche)*
uniform converter connection vollgesteuerte Stromrichterschaltung * *siehe IEC 146-1-1 und EN 60146-1-1*
uniform movement gleichförmige Bewegung
uniformity Durchgängigkeit * *z.B. einer techn. Lösung über eine Produktfamilie* || Gleichmäßigkeit
uniformity of components Gleichteiligkeit * *von Bauteilen*
unify vereinheitlichen
unilateral einseitig * *Vertrag usw.*
unilateral drive einseitiger Antrieb
unimpaired unvermindert * *(Adj.) ungeschwächt, ungeschmälert, unvermindert, unverhindert*
unimplemented nicht vorhanden * *nicht implementiert/realisiert*
unimportant gering * *unbedeutend, unwichtig* || geringfügig * *unwichtig, nicht von Gewicht* || unwesentlich * *unwichtig*
uninfluenced unbeeinflusst
unintelligible unverständlich
unintentional unabsichtlich * *unbeabsichtigt* || unbeabsichtigt || ungewollt || versehentlich * *unabsichtlich*
unintentional effect Folgeerscheinung * *ungewollter Effekt, unbeabsichtigte Folge* || Nebeneffekt * *ungewollter/unbeabsichtigter Effekt* || Schmutzeffekt * *ungewollter/unbeabsichtigter Effekt*
unintentionally ungewollt * *(Adv.)*
uninterruptable power supply unterbrechungsfreie Stromversorgung || USV

uninterrupted lückenlos || unterbrechungsfrei * *z.B. Stromversorgung, Produktion, Betrieb* || zügig * *ohne Unterbrechung*
uninterrupted duty unterbrechungsfreier Betrieb
uninterrupted operation Durchlaufbetrieb
uninterrupted power supply unterbrechungsfreie Stromversorgung || USV
uninterruptible unterbrechungsfrei * *nicht unterbrechbar, z.B. Stromversorgung*
uninterruptible power supply unterbrechungsfreie Stromversorgung || USV * *Unterbrechungsfreie Stromversorgung*
union Anschlussstück || Arbeitsgemeinschaft * *Vereinigung* || Gewerkschaft || Muffe * *Rohrverbindung* || Verband * *Vereinigung, Verbindung, Bund* || Verschraubung * *Rohrverbindung*
Union of German Technical Engineers VDE * *Verein Deutscher Elektrotechniker*
unique einmalig || einzig * *~artig; be unique: ~ darstehen* || einzigartig || unvergleichlich * *einzigartig*
unique feature Alleinstellungsmerkmal
unison Übereinstimmung * *Einklang, Gleichklang, Übereinstimmung; in unison with: in Einklang mit*
unit Aggregat || Einheit * *Gerät, Maß~* || Gerät || Maßeinheit
unit area Flächeneinheit * *energy per unit area: Energie pro ~*
unit interfacing Gerätekoppling || Gerätekopplung
unit of -Einheit * *Maßeinheit*
unit of measure Einheit * *Maß~* || Größenindex * *(→PROFIBUS-Profil)* || Maßeinheit || Umrechnungsindex * *(→PROFIBUS-Profil)*
unit of time Zeiteinheit
unit pack Verpackungseinheit
unit pressure Flächendruck * *Druck pro Flächeneinheit, z.B. in einem Gleitlager*
unit price Einzelpreis || Stückpreis
unit pulse Dirac-Funktion || Dirac-Impuls * *Sprungantwort eines Differenzierers auf "Einheitssprung"* || Einheitsstoß
unit ramp Anstiegsvorgang * *Anstiegsfunktion mit Anstiegszeit 1 sec ('Einheitsrampe')*
unit replacement Gerätetausch
unit step Einheitsprung || Einheitssprungfunktion
unit volume Bauvolumen * *eines Geräts* || Stückzahl * *Gesamt~/verkaufte ~ eines Produkts*
unitarily einheitlich * *(Adv.)*
unite vereinen * *verbinden, (sich) vereinigen, sich anschließen, in sich vereinigen (Eigenschaften usw.)* || vereinigen || zusammenfassen * *miteinander verbinden* || zusammenschließen * *(sich) vereinigen*
united geschlossen * *vereint*
United Kingdom Großbritannien * *England, Schottland, Wales u.Nordirland*
unitized construction Baukastenprinzip * *Baukastensystem, System aus abgeschlossenen Baueinheiten* || Baukastensystem
units Geräte
unity Einheit * *zusammengehörende ~* || eins * *unity power factor: Leistungsfaktor von eins*
unity delay-angle setting Vollaussteuerung * *(Stromrichter)*
universal allgemein * *universell, (all-) umfassend* || Allzweck- || durchgängig * *(all)umfassend, universal, allgemein(gültig), allgemein üblich* || universal * *Universal-* || universell
universal AC/DC motor Allstrommotor
universal amplifier Universalverstärker
universal current Allstrom
universal diode Universaldiode
universal joint Gelenk * *Kardangelenk* || Kardangelenk
universal motor Allstrommotor * *mit Kommutator* || Universalmotor * *mit Kommutator für Gleich- und Wechselstromanschluss in Hauptschlussschaltg.*
universal relay Allstrom Relais || Allstromrelais
universal shaft Gelenkwelle
universal type of contruction Universalbauform
universal validity Allgemeingültigkeit
universal-joint shaft Gelenkwelle
universality Allgemeingültigkeit || Durchgängigkeit * *Allgemeingültigkeit, Universalität, Vielseitigkeit*
universally valid allgemein gültig
university Hochschule || Universität
university degree akademischer Grad || Hochschulabschluss
university lecturer Pofessor
university of technology Technische Hochschule
unknown fremd * *unbekannt* || unbekannt
unlatch rücksetzen * *ein in Selbsthaltung befindliches Relais, z.B. in Kontaktplandarstellg. bei SPS* || speichernd rücksetzen * *z.B. Relais*
unlatch output Ausgang rücksetzen * *in Kontaktplandarstellung eines SPS-Programms*
unlawful rechtswidrig || unzulässig * *ungesetzmäßig*
unlegal nicht erlaubt * *nicht gesetzmäßig, illegal*
unless wenn nicht
unlike im Gegensatz zu * *a/the:zu* || ungleich * *unähnlich* || verschieden * *unähnlich*
unlikely unwahrscheinlich
unlimited unbegrenzt || uneingeschränkt * *unbegrenzt* || unendlich || ungehindert * *unbegrenzt*
unload entladen * *Fahrzeug, Waggon, Ladung, Gewehr usw. ~* || entlasten || löschen * *entladen*
unloaded unbelastet
unloaded position Nulllage * *ohne Belastung* || Nullstellung * *ohne Belastung* || Ruhestellung * *ohne Belastung*
unloading Ausladung * *das Entladen* || Entlastung
unloading equipment Entladeeinrichtung
unloading facility Entladeeinrichtung
unloading system Entladeeinrichtung
unlock aufschließen * *Tür usw. ~* || entriegeln * *freigeben* * *Sperre aufheben* || freischalten * *Software ~ (z.B. durch Eingabe eines Freischaltcodes)* || öffnen * *Schloß, Sperre ~*
unlocking Entriegelung * *mechanische oder steuerungstechnische ~*
unmachined roh * *nicht (fertig)bearbeitet (Werkstück auf Werkz.maschine)* || unbearbeitet * *(durch Werkzeugmaschine)*
unmanned unbemannt
unmatched unübertroffen
unmesh ausrücken * *Zahnräder usw. außer Eingriff bringen*
unmethodical unsystematisch
unnecessary überflüssig * *unnötig* || unnötig
unnoticed unbemerkt * *by:von*
unobjectionable einwandfrei
unobstructed frei * *unversperrt, ungehindert, ohne Hindernisse*
unoccupied leer * *unbelegt, unbesetzt* || nicht belegt * *z.B. Einbauplatz für Zusatzbaugruppe*
unpack entpacken * *eingepackte Produkte, Daten ~*
unpacked unverpackt
unpacking machine Auspackmaschine
unparalleled unvergleichlich
unparalleled feature Alleinstellungsmerkmal * *einzigartiges Merkmal*
unpermissible nicht erlaubt
unplanned ungeplant
unpleasant unangenehm * *unangenehm, unerfreulich*

unpleasantness Unannehmlichkeit
unplug abstecken * *Stecker* || abziehen * *Stecker/ Stöpsel* ~ || ziehen * *Stecker/Steckbaugruppe* ~
unpolluted sauber * *nicht verunreinigt*
unpotted unvergossen * *z.B. Baugruppe*
unpretending anspruchslos
unpretentious anspruchslos
unproductive loop Untätigkeitsschleife
unprofessional unsachgemäß
unprofitable unwirtschaftlich
unprotected offenliegend * *ungeschützt* || ungesichert * *ungeschützt*
unqualified unberechtigt * *unqualifiziert, unberechtigt, nicht approbiert; ungeeignet, untauglich, unbefähigt* || uneingeschränkt * *unbedingt, uneingeschränkt, ausgesprochen* || ungeeignet * *untauglich, ungeeignet, unbefähigt (auch Person)* || unqualifiziert
unquestioning unbedingt * *bedingungslos, blind; z.B. Anhänger, Glaube, Vertrauen*
unravel auflösen * *entwirren* || entwirren * *auch 'sich enträtseln/~'*
unrecognized unzulässig * *nicht anerkannt*
unreel abrollen
unregulated ungeregelt * *z.B. Stromversorgung*
unregulated power supply ungeregelte Stromversorgung || ungeregeltes Netzteil
unreliability Anfälligkeit * *Unzuverlässigkeit* || Störanfälligkeit * *Unzuverlässigkeit* || Störempfindlichkeit * *Unzuverlässigkeit* || Unzuverlässigkeit
unreliable anfällig * *unzuverlässig* || empfindlich * *unzuverlässig* || störfällig * *unzuverlässig* || störempfindlich * *unzuverlässig* || unzuverlässig
unreliable supply Schwachnetz * *unsicheres Netz (mit häufigen Spannungseinbrüchen, Unterbrechungen)*
unreliable supply system schwaches Netz * *unsicheres Netz (mit häufigen Spannungseinbrüchen/ Unterbrechungen)*
unreserved unbedingt * *uneingeschränkt (z.B. Billigung), vorbehaltlos, rückhaltlos, völlig, offen, freimütig*
unrestrained frei * *unbehindert*
unrestricted freizügig || uneingeschränkt * *ungehindert* || ungehindert
unrivalled konkurrenzlos
unsafe gefährlich * *unsicher* || unzuverlässig * *unsicher*
unsatisfactory mangelhaft * *unbefriedigend* || unbefriedigend || ungenügend * *unbefriedigend*
unsaturated ungesättigt
unscreened ungeschirmt * *z.B. Kabel/Leitung*
unscrew abschrauben || aufschrauben * *abschrauben* || lösen * *Schraube, Mutter ~ (abschrauben auch 'undo')*
unscrew machine Entschraubmaschine
unseasonable unpassend * *zur Unzeit*
unsecured ungesichert
unserviceable unbrauchbar * *nicht betriebsfähig, z.B. Maschine*
unsettled ungeordnet
unsharpness Unschärfe
unshielded ungeschirmt * *z.B. Kabel/Leitung*
unsigned ohne Vorzeichen
unskilled worker Arbeiter * *ungelernter* ~ || Hilfsarbeiter || ungelernter Arbeiter
unsmooth running unruhiger Lauf || unrunder Lauf
unsmoothed ungeglättet
unsnarl entwirren * *(auch figürl.)*
unsolder ablöten || auslöten || entlöten
unsplit ungeteilt
unstable instabil || schwankend * *veränderlich, instabil*
unstable condition Instabilität

unstable mains schwankendes Netz
unsteadiness Unsicherheit * *Schwanken* || Unstetigkeit
unsteady unstetig
unsteady running unruhiger Lauf
unstoppable unaufhaltsam
unsuitable unangemessen || ungeeignet || unpassend
unsuited unpassend
unsurmountable unüberwindbar * *Probleme, Schwierigkeit*
unsurpassed einmalig * *unübertroffen* || unübertroffen
unsymmetric unsymmetrisch
unsymmetrical ungleichmäßig
unsymmetry Unsymmetrie
unsystematic unsystematisch
unthrustworthiness Unzuverlässigkeit
untight undicht
until bis * *(zeitlich)*
until further notice bis auf weiteres
until now bisher
until you are informed otherwise bis auf weiteres * *(an jemanden gerichtet) bis Sie weitere Nachrichten erhalten*
untimely unpassend * *zur Unzeit*
untouched unberührt
untoward ungünstig * *widrig, ungünstig, unglücklich (Umstand usw.), widerspenstig, ungefügig* || widrig * *widrig, ungünstig, unglücklich (Umstände usw.), ungefügig, widerspenstig*
untreated water Rohwasser
untrue unrund
untrue running unruhiger Lauf * *unrunder Lauf* || unrunder Lauf || Unrundheit * *unrunder Lauf*
unused jungfräulich * *unbenutzt, ungebraucht, (noch) nicht verwendet/beansprucht*
unused opcode nicht interpretierbarer Befehl * *nicht ausführbarer Maschinenbefehl* || unbekannter Befehl * *[Computer] vom Befehlsdecoder als unbekannt zurückgewiesener (Maschinen-) Befehl*
unusual außergewöhnlich * *→ungewöhnlich, außergewöhnlich, ungewohnt, selten, äußerst* || außerordentlich * *ungewöhnlich, außergewöhnlich, ungewohnt, selten, äußerst* || ausgefallen * *unüblich* || ungewöhnlich * *unüblich*
unusually außergewöhnlich * *(Adv.)*
unvalid ungültig
unveil vorstellen * *erstmalig präsentieren, enthüllen, den Schleier fallen lassen*
unverifiable unkontrollierbar * *nicht nachprüfbar*
unvulcanized rubber Kautschuk
unwanted unerwünscht * *ungewünscht* || ungewollt * *→unerwünscht* || ungewünscht * *siehe auch 'unerwünscht'*
unwelcome unangenehm * *unwillkommen* || unerwünscht
unwind abrollen * *abwickeln* || Abrollung || abwickeln * *Material v. einem Abwickler* ~ || Abwickler || Abwickelung * *Material-Abrollung/Abwicklung/ Abwickler* || wickeln * *ab~*
unwind reel Abwickelrolle
unwind stand Abroller || Abrollung * *Abroller* || Abwickler * *→Rollenständer* || Rollenständer * *am Anfang einer Rotationsdruckmaschine* || Rollenträger * *Abwickler*
unwinder Abroller || Abwickler || Abwicklung * *Material-Abroller/Abwickler*
unworkable unbrauchbar * *Plan usw.*
unwrapping tool Abwickelwerkzeug * *für Wire-Wrap Verbindung*
unzip dekomprimieren * *(salopp) Daten* || entpacken * *(salopp) Daten* ~
up an * *(Adv.)* || auf * *(Adv.) hinauf; up and down: auf und ab* || oben * *in die Höhe, nach* ~, *hinauf, herauf, aufwärts, empor*

up pulses Vorwärtsimpulse
up to bis * *maximal, bis hinauf zu* || bis zu
up to and including bis * *bis einschließlich; z.B. Versionsnummer, Baugröße* || bis einschließlich
up to date laufend * *be up to date: auf dem ~en sein* || zeitgemäß
up to expectations wie erwartet
up to now bisher
up to standard vollwertig
up until bis * *(zeitlich) bis (zu)*
up until now bisher * *bis jetzt*
Up-/Down button Höher-Tiefer-Taster
up-date aktualisieren
up-down counter Vor-Rückwärtszähler
up-down interpreter Vor-Rückwärtsauswertung * *Funktionsbaustein* || Vorwärts-Rückwärts-Auswerter * *Funktionsbaustein* || Vorwärts-Rückwärtsauswertung * *Funktionsbaustein*
up-time Verfügbarkeit * *~sdauer, Dauer ungestörten Betriebs* || Verfügbarkeitsdauer
up-to-date aktuell * *modern, gängig, dem letzten Stand entsprechend* || auf dem Laufenden || fortschrittlich * *dem neuesten Stand entsprechend* || modern * *auf dem Stand der Technik*
up-to-the-minute aktuell * *in letzter Minute verfügber (Informationen, Nachrichten usw.)*
up-wind Aufwind
up/down counter Vor Rückwärtszähler
update aktualisieren || Aktualisierung || Änderung * *neue Version* || auf den neuesten Stand bringen * *Software* || auffrischen * *z.B. einen Puffer, Speicher*
updating Aktualisierung
upgradability Ausbaubarkeit * *bezüglich der Leistungsfähigkeit; Höchrüstbarkeit* || Ausbaufähigkeit * *bezüglich der Leistungsfähigkeit*
upgradable ausbaubar || ausbaufähig * *hoch/aufrüstbar* || erweiterungsfähig * *hochrüstbar* || nachrüstbar * *hochrüstbar*
upgrade auf den neuesten Stand bringen || aufbereiten * *Kohle* || aufrüsten * *hochrüsten (Hard-/Software)* || Aufrüstung || ausbauen * *auf/hochrüsten* || ersetzen * *durch eine neue/leistungsfähigere Version* ~ || ertüchtigen * *auf/hochrüsten* || erweitern * *in der Funktion hochrüsten* || Erweiterung * *Aufrüstung (Hard- und Software)* || Erweiterungs- | Funktionserweiterung * *Hochrüstung* || hochrüsten * *Hard- und Software ~; from: von; to: auf* || Hochrüstung || nachrüsten * *hochrüsten (Hardoder Software)*
upgrade kit Aufrüstsatz
upgrade set Nachrüstsatz * *Hochrüstsatz*
upgradeable ausbaubar
upgrading Aufbereitung * *von Kohle, Erz usw.* || Hochrüstung
upheaval Umbruch * *politischer/wirtschaftlicher/gesellschaftlicher ~*
uphill operation Bergfahrt
uphold aufrechterhalten * *Brauch, Lehre, Urteil ~* || tragen * *stützen*
upholster ausfüttern * *polstern*
upholstery Polster
upkeep Erhaltung * *~ von Bauten, Anlagen usw.* || Instandhaltung || Unterhaltung * *Instandhaltung* || Wartung * *Instandhaltung, Unterhalt*
upload auslesen * *herauflagen von Daten/Parametern aus einem Gerät heraus* || einlesen * *herauflesen von Daten/Parametern aus einem Gerät heraus*
upon an || auf || aus || bei * *(zeitlich)*
upon command auf Kommando
upon request auf Anfrage
upper oben * *ober, höher; at upper left: links ~* || ober || Ober- * *oberes* || oberes

upper digit höherwertige Stelle * *z.B.an Zahlenschalter oder Ziffernanzeige*
upper limit obere Grenze || Obergrenze
upper limit value oberer Grenzwert
upper motor Obermotor * *Walzwerk*
upper part oberer Teil || Oberteil
upper performance level oberer Leistungsbereich * *bezüglich der Leistungsfähigkeit*
upper performance range hoher Leistungsbereich * *bezügl. der Leistungsfähigkeit* || oberer Leistungsbereich * *bezügl. der Leistungsfähigkeit*
upper rope Oberseil * *{Aufzüge}*
upper-left links oben * *(Adjektiv)*
uppermost höchst * *oberst, höchst, ganz oben, obenan, zuoberst*
upread auslesen * *herauflesen von Daten/Parametern aus einem Gerät heraus* || einlesen * *herauflesen von Daten/Parametern aus einem Gerät heraus*
upright gerade * *Haltung usw.*
upright projection Aufriss * *Zeichnung*
upright shaft Stehwelle
UPS unterbrechungsfreie Stromversorgung * *Abkürzg. für 'Uninterruptible Power Supply'* || USV * *Abk.f. 'Uninterruptible Power Supply': Unterbrechungsfreie Stromversorgung*
UPS system USV-Anlage * *Abk. f. 'Uninterruptible Power Supply': unterbrechungsfreie Stromversorgung*
upset stauchen * *hot/cold upsetting: Heiß-/Kaltstauchen* || stören * *durcheinander bringen*
upshot Ausgang * *Ergebnis* || Endergebnis || Fazit * *(End)Ergebnis, Ende, Ausgang, Fazit; sum something up:das Fazit aus etw. ziehen*
upside down falsch herum * *die Oberseite nach unten*
upstairs nach oben * *in e. Haus* || oben * *~ im Haus*
upstream davorliegend * *im Signal-/Leistungs-/Energiefluss* || Vor- * *im Signal-/Energiefluss vorgeschaltet* || vorgeschaltet * *im Signal/Leistungsfluss ~* || Vorschalt- * *im Signal/Energiefluss vorgeschaltet*
upstream of vor * *im Signal-/Daten-/Energie-/Kühlmedien-/Materialfluss ~gelagert*
uptake Aufnahme * *(auch figürl.: Waren vom Markt usw.)*
uptimes Verfügbarkeitsdauer
upturn Aufschwung * *economic: wirtschaftlicher ~*
upward aufwärts || oben * *aufwärts (Tendenz)*
upward control Aufwärtssteuerung * *beim Aufzug*
upwards auf * *(Adv.) hinauf* || nach oben || oben * *aufwärts, nach ~; with the shaft end pointing upwards: mit dem Wellenende nach ~*
upwards compatible aufwärtskompatibel
urge antreiben * *(figürlich) vorwärtstreiben, anspornen, drängen, nötigen* || drängen * *(figürl.) auch zur Eile ~* || zusetzen * *drängend, mahnend*
urgency Dringlichkeit || Notwendigkeit * *Dringlichkeit*
urgent dringend * *be in urgent need of: ~ brauchen* || notwendig * *dringlich*
urgent case dringender Fall
urgently dringend * *(Adv.)* || unbedingt * *(Adv.) dringend*
URL URL * *{Network} Abk. für 'Universe Resource Locator': Adresse im World Wide Web des Internet*
US gal 1 * *amerikan.Gallonen; 1 l entspr. 0,2642 US Gallonen*
usability Anwenderfreundlichkeit || Benutzerfreundlichkeit || einfache Bedienbarkeit || Handhabbarkeit
usable brauchbar * *siehe auch →verwendbar* || einsatzbereit * *einsatzfähig* || einsetzbar * *verwendbar; universally: universell* || verwendbar

usage Behandlung * *Gebrauch, Verwendung, Behandlungsweise* || Form * *Brauch, Gepflogenheit, Usus* || Praxis * *Brauch*
USART USART * *Universal Synchronous/Asychronous Receiver/Transmitter: Schaltg. f.sychr./ asynchr.Schnittst.*
use anwenden * *for: für, zu; benutzen, verwenden* || Anwendung * *Gebrauch* || Anwendungsfall * *Verwendung(szweck), Verwendbarkeit, Gebrauch, Benutzung* || aufwenden * *anwenden* || benutzen * *share: gemeinsam* ~ || Benutzung || Einsatz * *Verwendung; in use/at work: im* ~; *be put to use: zum* ~ *kommen* || einsetzen * *verwenden, for:für* || Führung * ~ *eines Titels* || Gebrauch * *Anwendung, Gebrauch; in use: in* ~ || nutzen || nützen * *(be) nutzen* || verwenden || Verwendung * *Verwendung(szweck), Verwendbarkeit, Zweck, Nutzen, Nützlichkeit*
use a person's name as a reference beziehen * *sich auf jemanden* ~
use as basis zugrundelegen
use commonly gemeinsam benutzen
use force Gewalt anwenden
use in a wide range of applications breiter Einsatz
use up abnutzen
use with thyristor converters Stromrichterbetrieb * *Betrieb z.B. eines (Gleichstrom-) Motors am Thyristor-Stromrichter*
used alt * *gebraucht* || gebraucht || verwendet
used oil Altöl
useful förderlich * *nützlich* || gut * *nützlich* || nutzbar || nützlich || praktisch * *zweckmäßig, nützlich* || sinnvoll * *zweckdienlich*
useful data Nutzdaten * *z.B. bei einem seriellen Telegramm*
useful horsepower Nutzleistung
useful life Standzeit
useful output Nutzleistung * *nutzbare Ausgangsleistung; z.B. eines Motors, Stromrichters*
useful output power Nutzleistung
useful power Nutzleistung
useful signal Nutzsignal
useful value Rechenwert
useless überflüssig * *nutzlos, unnötig* || unbrauchbar || wertlos * *nutzlos*
user Anwender || Benutzer || Betreiber || Verbraucher * *Anwender, Benutzer*
user advantages Anwendernutzen
user advice Anwenderberatung
user benefits Anwendernutzen
user convenience Anwenderfreundlichkeit || Bedienkomfort * *Anwenderfreundlichkeit* || Benutzerfreundlichkeit
user convenient anwenderfreundlich * *bequem*
user data Nutzdaten
user friendly anwenderfreundlich
user interaction Benutzereingriff
user interface Anwender-Schnittstelle || Anwenderoberfläche || Bedienoberfläche || Bedienungsoberfläche || Benutzeroberfläche || Oberfläche * *Anwenderoberfläche einer Software/eines Rechnersystems*
user interrupt Anforderungsalarm * *bei SPS*
user layer Anwenderebene * *z.B. bei einem Bussystem*
user manual Anleitung * *Anwenderhandbuch* || Benutzerhandbuch
user memory Anwenderspeicher * *für Anwenderprogramme verwendbarer Speicher, z.B. in einer SPS*
user needs Anwenderbedürfnisse
user program Anwenderprogramm
user programmable wählbar * *vom Anwender programmierbar*
user prompt Eingabeaufforderung

user prompting Bedienerführung
user support Anwenderunterstützung
user terminal block Kundenklemmenleiste
user's guide Anwendungsrichtlinien || Benutzeranleitung
user's instructions Anwendungshinweise
user's manual Benutzerhandbuch || Handbuch * *Anwender~*
user's needs Anwenderbedürfnisse
user-configurable frei projektierbar * *vom Anwender projektierbar* || freiprojektierbar
user-definable frei belegbar * *z.B. Tasten*
user-defined Anwender-definiert || benutzerdefiniert
user-friendliness Anwenderfreundlichkeit || Bedienkomfort * *Anwenderfreundlichkeit* || Bedienungsfreundlichkeit * *Anwenderfreundlichkeit* || Benutzerfreundlichkeit
user-friendly anwenderfreundlich || komfortabel * →*anwenderfreundlich, leicht zu bedienen* || problemlos * *anwenderfreundlich*
user-oriented anwenderorientiert
user-specific anwendungsspezifisch
usual gängig * *gebräuchlich* || gebräuchlich * *with:bei* || gewöhnlich * *gebräuchlich* || herkömmlich * →*üblich* || normal * *gebräuchlich* || üblich
usual in the market marktüblich
usual in trade handelsüblich
usually gewöhnlich * *(Adv.)* || im Allgemeinen * *gewöhnlich* || meistens * *gewöhnlich, normaler/ gebräuchlicherweise* || normalerweise * *gebräuchlicherweise* || sonst * *für gewöhnlich*
usurp reißen * *an sich* ~, *usurpieren; z.B. Macht, Markt*
UTE UTE * *Abkürzg. für 'Union Technique de l'Electricité': französ. elektrotechn. Vereinigung*
utilities Dienstprogramme
utility Hilfsprogramm || nutzen * *(Substantiv) Nutzen (to:für), Nützlichkeit, Nützliches, nützliche Einrichtung* || Versorgungsunternehmen
utility company Versorgungsunternehmen
utility-model patent Gebrauchsmuster
utilizable nutzbar
utilization Anwendung * *Nutzbarmachung* || Auslastung * *Ausnutzung* || Ausnutzung || Einsatz * *Nutzbarmachung* || Nutzung || Verwendung * *Nutzbarmachung, (Aus)Nutzung, (Nutz)Anwendung*
utilize anwenden * *sich zunutze machen, ausnutzen* || ausnutzen * *nutzbar machen; under-utilize:- schlecht ausnutzen* || benutzen * *nutzbar machen* || in Anspruch nehmen * *(aus)nutzen* || nutzbar machen || nutzen * *(aus)nutzen* || nützen * *nutzen* || verwenden * *ausnutzen, nutzbar machen*
utilized ausgenutzt
utmost größt * *äußerst, höchst, größt, entlegenst, fernst* || größtmöglich || letzt * *do one's utmost: sein ~es hergeben/bestes tun; to the utmost: bis zum ~en*
utter höchst * *äußerst, höchst, völlig*
UV eraser Löscheinrichtung * *zum Löschen von EPROM-Modulen* || Löschlampe * *UV-Löschlampe für EPROM-Speicher* || UV-Löscheinrichtung || UV-Löschgerät * *zum Löschen von EPROMS*
UV-erasable UV-löschbar * *z.B. Speicherbausteine*

V

V U * *voltage:* Spannung
V- Dreieck-
V-belt Keilriemen
V-belt drive Keilriemenantrieb
V-belt pulley Keilriemenscheibe
V-connection V-Schaltung * *Messschaltung zur Stromistwerterfassung mit 2 Stromwandlern und Diodenbrücke*
V-ring V-Ring
v.v. umgekehrt * *Kurzform für 'vice versa'*
V/F characteristic Spannungs-Frequenz-Kennlinie || U-f-Kennlinie
V/F pattern U-f-Kennlinie
V/F-converter U-f-Wandler
VAC VAC * *Volt Wechselspannung*
vacancy Arbeitsplatz * *freier ~/Stelle*
vacant frei * *Stelle, Leitung* || leer * *unbesetzt* || offen * *Stelle*
vacuum absaugen * *mit Staubsauger usw.* || Unterdruck || Vakuum
vacuum bottle Vakuumröhre
vacuum circuit breaker Vakuum-Leistungsschalter || Vakuumschalter * *Hochspannungsschalter*
vacuum cleaner Staubsauger
vacuum contactor Vakuumschütz
vacuum cup Saugfuß
vacuum cup clamping device Werkstückansauggerät * *[Werkz.masch.| Holz]*
vacuum deposition Vakuumbeschichtung
vacuum lifting device Vakuumhebegerät
vacuum lifting equipment Vakuumhebegerät
vacuum off absaugen * *mit Unterdruck*
vacuum pump Vakuumpumpe
vacuum tube Vakuumröhre
vacuum-fluorescent display Vakuum-Fluoreszenz-Anzeige * *selbstleuchtende Ziffernanzeige*
vague unbestimmt * *vage* || ungenau * *(figürl.) vage* || unklar * *vage*
vagueness Unklarheit * *Verschwommenheit, Unbestimmtheit, Undeutlichkeit, Unverständlichkeit*
valence Wert * *~igkeit, Verhältniszahl, mitwirkende Kraft, Faktor* || Wertigkeit
valid gültig * *validate/legalize: ~ machen; remain valid: ~ bleiben* || zulässig * *→gültig*
validatable validierbar
validate abnehmen * *zertifizieren* || bestätigen * *für (rechts)gültig erklären, bestätigen (z.B. durch Prüfungsbehörde), rechtswirksam machen*
validity Berechtigung * *Rechtskraft, Rechtsgültigkeit, Gültigkeit, Gültigkeitsdauer* || Gültigkeit
validity period Gültigkeitsdauer
valuable kostbar || teuer * *wertvoll* || wertvoll * *(auch figürl.) be of value to s.o.: jdm. wertvoll sein*
valuation Abschätzung * *Bewertung, (Ab-) Schätzung, Wertbestimmung, Taxierung, Veranschlagung* || Berechnung * *Bewertung* || Bewertung * *(Ab)-Schätzung, Wertbestimmung* || Schätzung
value abschätzen * *bewerten* || ansetzen * *abschätzen, bewerten* || bewerten * *(allg. at:auf; by:nach)* || Gegenwert || nutzen * *(Substantiv) Gegenwert, Gegenleistung, Wert; Gewicht (eines Wortes/einer Meinung usw.)* || rechnen * *bewerten* || veranschlagen * *bewerten, ansetzen* || Wert || Wertigkeit * *Wert*
value added Mehrwert
value added tax Mehrwertsteuer
value buffer Wertepuffer
value characteristic Betragskennlinie
value indication Wertangabe * *z.B. bei Meldesystem*
value information Wertangabe
value of correcting variable Stellwert * *Wert der Stellgröße*
value of manipulated variable Stellwert * *Wert der Stellgröße*
value performer Leistungsträger * *Produkt mit gutem Preis-/Leistungsverhältnis*
value range Wertebereich
value retention Werterhaltung
value-added tax Mehrwertsteuer
value-adding Wertschöpfung
valued bewertet * *at: auf, by: nach* || gewichtet * *bewertet*
valueless wertlos
valve Röhre * *Elektronenröhre* || Schieber * *Ventil* || Ventil * *auch Halbleiter~, Thyristor~; safety-valve: Sicherheits~*
valve arm Stromrichterzweig || Ventilzweig
valve block Ventilinsel * *zusammenfasste Einheit aus mehreren (Pneumatik-) Ventilen*
valve control Ventilansteuerung
valve device Ventilelement
valve direction Ventilrichtung
valve seal Ventilverschluss * *z.B. für Verpackung*
valve-side ventilseitig
valves and fittings Armaturen
vamp reparieren * *flicken, reparieren, herrichten, zurechtschustern, "aufmotzen"*
vane Flügel * *z.B. eines Lüfters* || Lüfterflügel || Ventilatorflügel * *with adjustable pitch: einstellbarer ~* || Ventilatorschaufel
vane-axial fan Axialventilator
vanish schwinden * *ver~* || verschwinden
vapor Dampf * *(amerikan.)*
vapor-deposited aufgedampft
vaporization Eindampfung
vaporize verdampfen
vaporizer Verdampfer
vapour Dampf * *(brit.)* || Dunst
vapour cooling Siedekühlung
vapour extractor Dunstabsauganlage
vapours Dämpfe
VAr Blindleistung * *'reactive' Volt-Ampere*
VAr compensation Blindleistungskompensation
VAr compensator Blindleistungskompensationsanlage || Blindstromkompensationseinrichtung || Kompensationsanlage * *Blindleistungs-Kompensationseinrichtung* || Kompensationseinrichtung * *Blindleistungs-Kompensationseinrichtung*
VAr component Blindleistungsanteil
variable änderbar || Faktor * *veränderlicher ~* || Größe * *sich ändernde mathem./physikal. Größe, Variable* || regelbar * *infinitely:stufenlos* || stellbar || unterschiedlich * *schwankend* || variabel || Variable || veränderbar * *variabel* || veränderlich || verstellbar
variable capacitor Drehkondensator
variable frequency inverter Frequenzumrichter
variable quantity Größe * *variable Größe*
variable resistor Drehwiderstand || Potentiometer || veränderbarer Widerstand
variable speed drehzahlveränderbar
variable speed gearing Regelgetriebe
variable transformer Drehtransformator * *Stelltrafo* || Regeltrafo * *siehe auch →Stelltrafo* || Regeltransformator * *siehe auch →Drehtransformator* || Stelltrafo || Stelltransformator
variable type Datentyp * *Variablentyp*
variable-frequency drive drehzahlveränderbarer Antrieb * *Drehstomsntrieb* || frequenzgestellter Antrieb
variable-ratio transformer Stelltrafo || Stelltransformator

variable-speed drehzahlgeregelt || drehzahlgesteuert || drehzahlstellbar || drehzahlveränderbar
variable-speed belt drive Riemenverstellgetriebe
variable-speed drive drehzahlgeregelter Antrieb || drehzahlveränderbarer Antrieb || geregelter Antrieb || Regelantrieb * *geregelter/ drehzahlveränderbarer Antrieb* || regelbarer Antrieb
variable-speed drives drehzahlveränderbare Antriebe * *siehe VDI/VDE 2185*
variable-speed gearing Verstellgetriebe
variable-speed slipring motor Regelschleifringläufer-Motor * *{el.Masch.}*
variables Größen * *Variablen*
variably variabel * *(Adv.)*
variance Varianz
variant Variante || Version * *Variante*
variation Abweichung || Änderung * *Veränderung, Wechsel, Schwankung* || Schwankung || Ungleichförmigkeit * *Schwankung, Abweichung* || Unterschied * *Veränderung, Abweichung, Schwankung, Wechsel* || Veränderung * *Variierung, Variation; in:in/von; to:an* || Verlauf * *z.B. ~ einer Funktion* || Verstellung
variation from rated system voltage Netzspannungsabweichung * *vom Nennwert*
variation in load Belastungsänderung || Laständerung * *Lastschwankung* || Lastschwankungen
variation in time zeitliches Verhalten
variation of -Änderung
variation of temperature Temperaturänderung
variation with time zeitlicher Verlauf * *Änderung mit der Zeit* || Zeitverlauf * *zeitlicher Verlauf*
variations Schwankungen
variations in load Belastungsänderungen
variator Verstellgetriebe
varied unterschiedlich * *abwechslungsreich, mannigfaltig, verschieden(artig), varriiert* || verschieden * *wechselnd* || verschiedenst * *verschiedenartig, mannigfaltig* || vielfältig * *verschieden(artig), mannigfaltig, abwechslungsreich, bunt*
variety Sorte * *Abart* || Sortiment * *of:an* || Variante * *Abart, Spielart* || Version * *Abart* || Vielfalt * *Mannigfaltigkeit, Auswahl, Anzahl, Vielseitigkeit; a wide variety of: eine große ~ an* || Vielzahl * *Vielfalt, Mannigfaltigkeit, Anzahl, Auswahl; a wide variety of: eine große Vielfalt an* || Vorrat * *Anzahl, Reihe, Vielfalt, Auswahl*
variety of different types Typenvielfalt
various diverse || mehrere * *verschiedene* || verschieden * *diverse* || verschiedene * *diverse* || verschiedene * *→diverse* || vielfältig * *verschieden(artig), mehrere, verschiedene*
varistor Überspannungsableiter || Varistor * *spannungsabhängiger Widerstand, z.B. zum Überspannungsschutz*
varnish Lack * *auf Ölbasis* || lackieren * *auch mit Isolierlack*
varnish-insulated wire Lackdraht
varnished copper wire Kupferlackdraht
varnishing roller Lackierwalze * *{Druckerei}*
vary abweichen || ändern * *variieren, verändern, sich verändern (with:mit)* || atmen * *variieren, sich ändern; auch Preise, Kurse* || bewegen * *sich ändern* || schwanken * *variieren, sich ändern; auch Preise, Kurse* || steuern * *durch einen Bereich (hin-) durchsteuern, variieren* || variieren * *verändern, abwechseln, sich verändern* || verstellen * *variieren, auch verändern, wechseln* || wechseln * *variieren, (sich) verändern, verstellen*
vary with the square of quadratisch abhängig sein von
varying abweichend || schwankend * *Wert, Größe, Signal, Anzahl* || ungleich * *unterschiedlich * schwankend* || veränderlich || wechselnd * *(sich) verändernd*
varying duty Betrieb mit wechselnder Belastung
varying load Wechsellast || wechselnde Belastung
varying loads Lastschwankungen
varying-voltage winding Weitbereichswicklung * *für e. Bemessungsspannungsbereich ausgelegte Wicklg. e. Drehstrommotors*
vaseline Vaseline
vast breit * *ausgedehnt* || groß * *weit* || überwiegend * *Mehrheit* || weit * *weit, ausgedehnt, unermesslich, riesig*
vat Bütte || Fass * *Bottich* || Mehrwertsteuer * *Abk. für 'Value-Added Tax'*
vault Wölbung
vaulted gewölbt
VCB Vakuumschalter * *Abkürzg.f. 'Vacuum Circuit Breaker': Hochspannungs-~*
VCO U-f-Wandler * *Abkürzg. für 'Voltage Controlled Oscillator':spannungsgesteuerter Oszillator*
VCR Videorecorder * *Abkürzung für 'Video Cassette Recorder'*
VDC VDC * *Volt Gleichspannung*
VDE VDE * *Abkürzg. für 'Verein Deutscher Elektrotechniker': Union of German Electrical Engineers*
VDE regulation VDE-Bestimmung * *elektrotechn. Sicherheitsbestimmung des →VDE (für el.Masch.: VDE 0560, IEC 34)* || VDE-Vorschrift
VDE-regulation VDE-Bestimmung
VDMA VDMA * *'Verband Deutscher Maschinen- u. Anlagenbau': German Machinery Plant Manufacturer's Assoc.*
VDR Varistor * *Abkürzg. für 'Voltage Dependent Resistor': spannungsabhängiger Widerstand*
VDU Bildschirmgerät * *Abk. für 'Video Display Unit'* || Monitor * *Abk. f. 'Video Display Unit': Bildschirmanzeigegerät* || Sichtgerät * *video display unit*
VDW VDW * *Abk.f.'Verein Deutscher Werkzeugmaschinenfabriken e.V.': German Machine Tool Builders Associat.*
vector Vektor || vektoriell || Zeiger * *Vektor, Raum~*
vector addition geometrische Summe * *Zeigeraddition*
vector circulation Vektordrehung
vector control feldorientierte Regelung * *für Asynchronmotor* || Vektor-Regelung || Vektorregelung
vector controlled vektorgeregelt
vector diagram Vektordiagramm || Zeigerbild || Zeigerdiagramm
vector field Vektorfeld
vector group Schaltgruppe * *z.B. eines Transformators*
vector modulation Vektormodulation
vector product vektorielles Produkt
vector quantity Größe * *vektorielle Größe* || vektorielle Größe
vector rotation Vektordrehung
vector rotator Vektordreher
vector variable vektorielle Größe
vectorial vektoriell
vectorial multiplication Vektormultiplikation
veer schleudern * *sich drehen (auch Fahrzeug), umspringen, umschwenken (figürl.), wenden,drehen, schwenken*
vehemence Gewalt * *Wucht*
vehement heftig
vehicle Fahrzeug * *allg. für Landfahrzeug*
vehicle control Fahrzeugsteuerung
vehicle drive Fahrzeugantrieb
vehicle engine Kraftfahrzeugmotor
vehicle engineering Fahrzeugbau * *als (Ingenieur-) Fachrichtung*

vehicle industry Fahrzeugbau * *als Industriezweig*
vehicle pool Fuhrpark
vehicle washing system Fahrzeugwaschanlage
vehicles Fahrzeuge
velocity Geschwindigkeit * *Oberflächen~, Bahn~*
velocity control Geschwindigkeitsregelung
velocity correction signal Geschwindigkeitskorrektursignal
velocity limiter Anstiegsbegrenzer
vendor Anbieter * *Verkäufer, Lieferfirma* || Hersteller * *Verkäufer, Lieferer* || Lieferant * *Verkäufer* || Lieferfirma * *Verkäufer* || Verkäufer * *(Firma)*
veneer Furnier || furnieren
veneer clipper Furnierschere
veneer cutting machine Schälmaschine * *{Holz} Rundschälmaschine zur Erzeugung von Furnier*
veneer jointing guillotine Furnierschneidemaschine * *{Holz} zum Schneiden von lückenlos anfügbaren Furnierblättern*
veneer knife Furniermesser * *{Holz}*
veneer peeler Furnierschälmaschine * *{Holz}*
veneer peeling lathe Furnierschälmaschine * *{Holz}*
veneer saw Furniersäge * *{Holz}*
veneer slicer Furniermessermaschine * *{Holz}*
veneering Furnierung
veneering press Furnierpresse * *{Holz}* || Überfurnierpresse * *{Holz}*
veneers Furnierholz
vent belüften || entlüften * *z.B. ölgekühlten Trafo ~* || Entlüftung || Luftauslass || Öffnung * *Lüftung*
vent hole Entlüftungsbohrung
vent plug Entlüftungsschraube
ventilate belüften || entlüften || lüften
ventilated luftgekühlt
ventilated by duct system fremdgekühlt über Rohranschluss
ventilating duct Luftkanal
ventilating fan Ventilator
ventilation Belüftung || Bewetterung * *Belüftung im Bergbau; mine ventilation: Gruben~* || Entlüftung || Luftkühlung || Lüftung || Ventilation
ventilation circuit Luftführung * *Kreislauf*
ventilation duct Belüftungskanal || Lüftungskanal
ventilation ductwork Lüftungskanäle
ventilation fan Belüftungsventilator
ventilation opening Belüftungsöffnung || Kühlluftöffnung
ventilation slot Lüftungsschlitz
ventilation system Belüftung || Belüftungsanlage || Lüftungsanlage * *in der Klimatechnik*
ventilator Ventilator
venting Entlüftung
venting roof Lüftungsdach
venting screw Entlüftungsschraube
venting slot Lüftungsschlitz
venture Wagnis
venture capital Wagniskapital
verbose ausführlich * *ausführlich, vollständig, wortreich, geschwätzig, langatmig*
verbosely ausführlich * *(Adv.) in voller Ausführlichkeit*
verdigris Grünspan
verifiable nachweisbar
verification Nachprüfung || Prüfung * *Nach~ (auf Richtigkeit)* || Überprüfung * *Nachprüfung (auf Richtigkeit)*
verification of operation Funktionsprüfung * *Nachprüfung auf Einhaltung d. Spezifikationen* || Nachprüfung * *Funktionsprüfung*
verify belegen * *beweisen* || bescheinigen * *beglaubigen* || beweisen * *verifizieren, nachweisen* || kontrollieren * *die Richtigkeit ~, die Richtigkeit/Korrektheit nachweisen* || nachprüfen * *verifizieren, auf die Richtigkeit hin prüfen, die Richtigkeit feststellen* || prüfen * *auf Richtigkeit ~* || überprüfen * *verifizieren, auf Richtigkeit/Echtheit untersuchen, die (fehlerfrei) Funktion ~*
veritable wahr * *echt*
vernier Feineinsteller
vernier pulse Feinimpuls
versatile anpassungsfähig * *['wörsetail] vielseitig* || beweglich * *(figürl.) wendig, vielseitig* || komfortabel * *vielseitig (verwendbar), wandelbar* || vielseitig * *['wörsetail]*
versatile in application vielseitig einsetzbar
versatility Verwendung * *vielseitige Verwendung/Verwendbarkeit, flexible Einsetzbarkeit* || Vielseitigkeit
versed versiert * *well versed: gut bewandert*
version -Ausführung || Ausführung * *Version* || Ausgabestand || Ausprägung || Variante * *Version* || Version * *auch eines Softwareprogramms*
version adaption Versionsanpassung
version matching Versionsanpassung
version-number Versionsnummer * *z.B. einer Software*
versus gegen * *juristisch z.B. Prozessgegner* || gegenüber * *verglichen mit* || in Abhängigkeit von * *bei Kurve/Kennlinie/Funktion (z.B. Größe auf der X-Achse)* || über * *z.B. Kurve/Funktion über der X-Achse*
vertex Eckpunkt * *Scheitelpunkt* || Scheitelpunkt * *(auch mathemat.); (Plural: vertices)*
vertical längs * *vertikal, senkrecht, lotrecht* || senkrecht || vertikal
vertical accumulator Schlingenturm * *senkrecht. Warenspeicher (für Kabel, Walzband) z.Überbrück.d. Rollenwechsels*
vertical conveyor Senkrechtförderer
vertical frequency Bildwiederholfrequenz || Vertikalfrequenz * *{Computer} bei einem Monitor*
vertical plan Aufriss * *Zeichnung*
vertical plane Senkrechte
vertical position Senkrechte
vertical raceway Schacht * *senkrechter (Kabel)-Schacht*
vertical section Aufriss * *senkrechter Schnitt*
vertical shaft Königswelle * *senkrecht angeordnete Welle* || Stehwelle * *siehe auch →Königswelle*
vertical trunking Schacht * *senkrechter (Kabel)-Schacht*
vertical type senkrechte Bauform
vertical-shaft type senkrechte Bauform * *Motor, Getriebe usw.*
vertical-stack warehouse Hochregallager
vertically senkrecht * *(Adv.) vertically above each other; ~/rechtwinklig aufeinander* || vertikal * *(Adv.)*
verve Schwung * *Schmiss*
very ganz * *(Adv.) sehr* || höchst * *sehr* || sehr
very able man Könner
very easy kinderleicht
very functional funktionsgerecht * *siehe auch →zielführend*
very much sehr * *(Adv.)*
very quick acting superflink * *Sicherung*
very quick acting fuse superflinke Sicherung
very well hervorragend * *(Adv.)*
VESA VESA * *Abk.f. 'Video Electronics Standards Association': Normengremium für Bildschirmstandards*
vessel Gefäß || Schiff * *allg. Wasserfahrzeug*
vestige Rest- * *-Überrest, -Überbleibsel, -Spur*
veto Einspruch einlegen * *gegen etwas*
vex stören * *ärgern*
vexation Ärgernis * *Verdruss, Ärger, Plage, Qual, Belästigung, Schikane*
VFC Spannungs-Frequenz-Umsetzer * *Voltage-To-Frequency Converter* || Spannungs-Frequenz-

Wandler * *Abkürzg. f. 'Voltage-to-Frequency Converter'*
VFD drehzahlveränderbarer Antrieb * *Abk.f. 'Variable-Frequ. Drive': AC-Antrieb mit veränderbarer Frequenz*
VGA monitor VGA-Monitor
VGA screen VGA-Monitor
VHDL VHDL * *Abk.f. 'Very Highly integrated circuits Description Language': Beschreibungssprache für ASICs*
via über * *mit Hilfe von, mittels, auf dem Wege* ~
viable durchführbar * machbar, *lebensfähig* || lebensfähig * *(auch figürl.)* || realisierbar * machbar
vibrate rattern * *vibieren, pulsieren, zittern, beben* || rütteln || schütteln || schwingen * *mechanisch* || vibrieren || zittern * *vibrieren*
vibrating conveyor Schwingförderer
vibrating machine Rüttler * *vibrierend*
vibrating table Rütteltisch
vibration Erschütterung || Schwingung * *mechanische* ~ || Vibration
vibration absorber Schwingungsdämpfer * *z.B. zur Aufhängung von Maschinen*
vibration amplitude Schwingamplitude * *einer mechanischen Schwingung* || Schwingungsamplitude * *Schwingwegamplitude einer mechanischen Schwingung* || Schwingwegamplitude * *einer mechan. Schwingung*
vibration characteristics Schwingungseigenschaften * *von mechan. Schwingungen* || Schwingungsverhalten * *mechanisches*
vibration control Schwingungsdämpfung * *mechan.; über Regel-/Steuereinrichtung*
vibration damper Schwingungsdämpfer * *gegen mechan. Schwingungen, z.B. zur Motoraufstellung/-hängung*
vibration damping Schwingungsdämpfung * *mechan.*
vibration excitation Schwingungsanregung * *Anregung einer mechan. Schwingung*
vibration fatigue test Dauerschwingprüfung
vibration feeder Rütteltopf * *{Verpack.techn.}*
vibration frequency Schwingfrequenz * *~ einer mechan. Schwingung*
vibration level Schwingstärkestufe * *für Elektromotoren: siehe z.B. DIN ISO 2373*
vibration measurement Schwingungsmessung
vibration measuring Schwingungsmessung
vibration meter Schwingungsmesser || Schwingungsmeßgerät
vibration motor Rüttlermotor || Vibrationsmotor * *{el. Masch.} Motor erzeugt gewollte Schwingbewegung durch eingebaute Unwuchtscheibe*
vibration resistance Rüttelfestigkeit || Schwingfestigkeit || Vibrationsfestigkeit
vibration resistant rüttelsicher * *rüttelfest*
vibration roller Vibrationswalze
vibration screening machine Vibrationssiebmaschine * *{Holz} z.B. zum Trennen von Holzspänen in der Spanplattenherstellung*
vibration severity Schwingstärke * *mechan.; für el.Maschinen: siehe DIN ISO 2373, IEC34 Tl.14, DIN45660 u. VDI2065* || Schwingungsstärke * *mechan.*
vibration severity grade Schwingstärke * *~stufe, gemess. Eff.wert d.→Schwinggeschwindigkeit,DIN ISO3945* || Schwingstärkestufe * *Klassifiz.stufe für →Schwingstärke; el.Masch.: VDE0530 T.14, VDI 2056*
vibration speed Schwinggeschwindigkeit
vibration strength Schüttelfestigkeit
vibration stress Rüttelbeanspruchung
vibration stressing Rüttelbelastung || Schwingungsbeanspruchung * *mechanisch*

vibration table Rütteltisch * *auch zur Rüttelprüfung*
vibration velocity Schwinggeschwindigkeit * *Augenblickswert d.Geschw. e.mech.Schwingg., DIN ISO3945, DIN5483, VDE0530 T.1*
vibration-damped schwingungsisoliert
vibration-free erschütterungsfrei
vibration-free operation Laufruhe
vibration-related design schwingungstechnische Auslegung
vibration-related dimensioning schwingungstechnische Auslegung
vibration-resistant rüttelfest
vibration-safe rüttelfest * *rüttelsicher* || rüttelsicher
vibrational behavior Schwingungsverhalten * *mechanisches*
vibrational load Schwingungsbeanspruchung
vibrations Schwingungen * *Mechanische*
vibrations from external sources immitierte Schwingungen * *mechanische* ~ || immitierte Schwingungen * *mechanische* ~
vibrator Rüttler
vibrator conveyor Schwingbandförderer
vibratory sander Vibrationsschleifer * *{Holz}*
vibratory stressability Schwingfestigkeit
vibro-stable schwingsteif || schwingungssteif
vibrometer Schwingungsmesser || Schwingungsmessgerät
vibrostability Rüttelfestigkeit || Schwingfestigkeit
vice Schraubstock
vice versa umgekehrt * *dasselbe* ~
vice- Vize-
vicinity Nähe * *Umgebung*
vicinity of seconds Sekundenbereich * *in the: im* ~
video board Grafikkarte
video card Grafikkarte
video display unit Bildschirmgerät
video driver Grafiktreiber
video memory Bildschirmspeicher
video terminal Sichtgerät
videograbber Videograbber * *{Computer}* →*Framegrabber mit einer Digitalisierrate von 25 Bildern (Frames) pro sec*
view anschauen * *auch auf dem Bildschirm; (auch figürl.)* || ansehen * *siehe auch* →*anschauen* || Ansicht * *optische; Blick; side view: Ansicht von der Seite* || Berücksichtigung * *Hinblick; with a view of: in Hinblick auf; in view of: unter* ~ *von* || betrachten * *(auch figürl.)* || beurteilen * *betrachten* || Beurteilung * *Ansicht, Auffassung, Urteil, in my view:meines Erachtens* || Einstellung * *Ansicht (of:über)* || Perspektive * *(auch figürl.). Ansicht, Aussicht, Ausblick (of/over: auf), Überblick (of: über)* || Standpunkt * *take the view that:auf dem* ~ *stehen, dass* || Übersicht * *Überblick, Ausblick*
view from the left Ansicht von links
view from the right Ansicht von rechts
viewed gesehen * *right/left viewed: von rechts/ links aus* ~
viewing beobachten * *(Substantiv)*
viewpoint Aspekt * *siehe 'Gesichtspunkt'* || Gesichtspunkt || Standpunkt
vigilance Aufmerksamkeit * *Wachsamkeit*
vigor Härte * *(amerikan., figürl.) Nachdruck, Stärke, Tatkraft, Vitalität, Strenge, Härte, Rauheit* || Kraft * *(Körper-/Geistes-) Kraft, Vitalität, Energie, Nachdruck, Wirkung,* →*Härte*
vigour Härte * *(brit., figürl.) Nachdruck, Stärke, Tatkraft, Vitalität* || Stärke * *Tatkraft*
VIK VIK * *Verband d. Industriellen Energie- u. Kraftwirtsch., legt techn.Anforderungen für AC-Motoren fest*
violate brechen * *Vertrag usw.* ~ || verletzen * *Regel, erlaubte Grenze, Begrenzung, Vorschrift, Gesetz* ~ || verstoßen * ~ *gegen* || verstoßen gegen

violate a limit Grenzwert überschreiten
violation Abweichung * *Verletzung einer Vorschrift* || Verstoß * *Zuwiderhandlung (of:gegen)*
violation of the time limit Zeitüberschreitung
violence Gewalt * *Gewalttätigkeit*
violent heftig * *Sturm usw.* || stark * *heftig*
virgin jungfräulich * *(figürl.) jungfräulich, unberührt* || unberührt * *in der Natur*
virtual Kunst- * *scheinbar* || scheinbar * *virtuell* || virtuell * *(techn/physikal.) nur scheinb. vorhanden; (umgangssprachl.) faktisch, praktisch,eigentlich*
virtual junction temperature Ersatzsperrschichttemperatur * *bei Halbleiterbauelement*
virtually nahezu * *eigentlich, praktisch, im Grunde genommen* || praktisch * *(Adv.) so gut wie, nahezu, fast gänzlich* || virtuell * *(Adv.)*
virtually-continuous quasistetig
virtue Vorzug * *Tugend, gute Eigenschaft, Wirksamkeit*
virus Virus * *auch Computer~*
virus infection Virenbefall * *auch durch Computerviren*
virus-free virenfrei * *auch "frei von Computerviren"*
viruses Viren * *auch Computer~*
vis--vis gegenüber
viscid dick * *zähflüssig, dickflüssig, klebrig* || dickflüssig
viscometer Viskosimeter * *zum Messen der Zähigkeit von Flüssigkeiten*
viscosimeter Viskosimeter
viscosity Viskosität || Zähigkeit * *~ von Flüssigkeiten*
viscosity class Viskositätsklasse * *bei Schmieröl usw.*
viscosity grade Viskosität * *Viskositätsstufe (von Öl usw.)*
viscous dickflüssig || konsistent * *Schmiermittel* || zäh * *Flüssigkeit*
viscous friction Gleitreibung || Viskosereibung * *Gleitreibung*
vise Schraubstock * *(amerikan.)*
visibility Sichtbarkeit || Visualisierung * *die "Sichtbarmachung" als Funktion*
visible deutlich * *sichtbar* || merklich * *sichtbar, offensichtlich, deutlich* || offensichtlich || sichtbar || wahrnehmbar * *sichtbar* || wirklich * *sichtbar, festgestellt*
visible at a glance überschaubar
visibly offensichtlich * *(Adv.)*
vision Vision * *Seherblick, Weitblick, Phantasie, Vision, Traum-/Wunschbild* || Vorstellung * *Vision*
visionary visionär
visit Besichtigung || inspizieren * *besuchen, aufsuchen, inspizieren, in Augenschein nehmen*
visited betroffen * *by:von; heimgesucht, befallen*
visiting card Visitenkarte
visual optisch * *Signal usw.*
visual check optische Kontrolle || Sichtkontrolle || Sichtprüfung
visual contact Sichtkontakt
visual indicator Sichtmelder
visual inspection optische Kontrolle || Sichtprüfung
visual signal Lichtsignal
visualization beobachten * *(Substantiv) Visualisierung* || Beobachtung * *Visualisierung* || Visualisierung || Visualisierungssystem
visualization functions Beobachtungsfunktionen * *Visualisierung, z.B. über Bildschirm*
visualization system Visualisierungssystem
visualize grafisch darstellen * *visualisieren* || sichtbar machen || visualisieren || vorstellen * *(sich) ein Bild machen, (sich) bildlich vorstellen, visualisieren*

vital entscheidend * *grundlegend wichtig* || wesentlich * *unerlässlich* || wichtig * *unerlässlich*
vital point wesentlicher Punkt
vitality Schwung * *Vitalität*
vivid anschaulich * *deutlich, lebendig, intensiv, klar, lebhaft*
VLSI hochintegriert * *Abkürzg.für 'Very Large Scale Integration': (IC) mit sehr hohem Integrationsgrad* || VLSI * *Abkürzg. für 'Very Large Scale Integration': hochintegriert*
VOB VOB * *Abk. für 'VerfügungsanOrdnung für Bauleistungen': German Construction Contract Procedures*
vocabulary Wörterbuch * *Vokabelverzeichnis, Wörterbuch, Vokabular, Wortschatz*
voice Sprache * *Stimme,Ton*
void leer * *bedeutungslos* || leeren * *ent~* || Lücke * *(figürl.) freier Raum, Leere, Lücke, Gefühl der Leere; fill the void: die ~ schließen* || ungültig * *(rechts)unwirksam, ungültig, nichtig, frei, unbesetzt*
volatile flüchtig * *z.B.Speicher*
volontary selbst * *von ~, freiwillig*
voltage Spannung * *(elektr.)*
voltage asymmetry Spannungs-Unsymmetrie
voltage at which field weakening starts Ablösespannung * *Einsatzpunkt f. Feldschwächung*
voltage boost Spannungsanhebung * *z.B. zur Erhöhung d.Anfahrmoments e.Motors, um d.Losbrechreibg. zu überwinden*
voltage class Spannungsklasse
voltage code Spannungskennziffer * *Bestellnummer-Ergänzung bei SIEMENS-Motor*
voltage control Spannungssteuerung
voltage controlled oscillator U-f-Wandler * *spannungsgesteuerter Oszillator*
voltage converter Spannungswandler
voltage dependent resistor Varistor
voltage dip kurzzeitiger Spannungseinbruch || kurzzeitiger Stromausfall || Netzeinbruch * *kurzzeitiger ~* || Spannungseinbruch * *kurzzeitiger*
voltage disconnect Spannungsfreischaltung
voltage disconnection Spannungsfreischaltung
voltage divider Spannungsteiler
voltage drop Spannungsabfall || Spannungseinbruch
voltage endurance Spannungsfestigkeit
voltage failure Spannungsabfall || Spannungsabfall * *irregulärer Spannungsausfall/-einbruch* || Spannungsausfall
voltage feedback drive EMK-geregelter Antrieb * *tacholoser Betrieb eines DC Motor*
voltage fluctuation Spannungsschwankung
voltage fluctuations Spannungsschwankungen
voltage follower Impedanzwandler * *Spannungsfolger (Operationsverstärkerschaltung zur Signalentkopplung)* || Spannungsfolger * *Impedanzwandler mit Verstärkung 1 (mit Operationsverstärker(n))*
voltage gradient Spannungssteilheit
voltage increase Spannungsüberhöhung
voltage induced on circuit interruption Abschaltspannung * *beim Ausschalten e.induktiven Last*
voltage level Spannung * *(elektr.) Spannungshöhe* || Spannungspegel
voltage limitation Spannungsbegrenzung
voltage monitoring Spannungsüberwachung
voltage path Spannungspfad
voltage peak Spannungsspitze
voltage peaks Spannungsspitzen
voltage phase control Spannungsanschnittsteuerung
voltage phasor Spannungszeiger
voltage range Spannungsbereich
voltage rate-of-rise Spannungssteilheit

voltage ratio Übersetzungsverhältnis * *Trafo*
voltage recovery Spannungsrückkehr || Spannungswiederkehr * *nach Spannungseinbruch* || Wiederkehr der Spannung
voltage reduction Spannungsabsenkung
voltage reference Referenzspannung * *Referenzspannungsquelle, auch als Bauelement* || Referenzspannungsquelle * *auch als Bauelement*
voltage regulation Spannungsabfall * *z.B. einer Spannungsquelle bei Laständerung*
voltage regulator Spannungsregler
voltage reserve Spannungsreserve
voltage restoration Spannungswiederkehr
voltage ripple Spannungswelligkeit
voltage rise Spannungsanstieg || Spannungserhöhung || Spannungsüberhöhung
voltage sag Spannungsabfall * *'Absacken' der Spannung* || Spannungseinbruch * *'Einsacken' der Spannung über längere zeit*
voltage sensing Spannungserfassung
voltage simulation model Spannungsmodell * *Rechenmodell, z.B. für Drehstrommaschine*
voltage source Spannungsquelle
voltage source converter Spannungszwischenkreis-Umrichter
voltage source DC-link Spannungszwischenkreis
voltage source DC-link converter Spannungszwischenkreisumrichter
voltage spike Spannungsspitze * *Nadelimpuls, z.B. Störimpuls*
voltage spikes Spannungsspitzen * *Nadelimpulse*
voltage stabilization Spannungsstabilisierung
voltage stabilizer Spannungskonstanthalter
voltage stressing Spannungsbeanspruchung * *bei umrichtergespeisten Drehstrommotoren: siehe VDE 0530 Teil 1, Beiblatt 2*
voltage suppressor Spannungsunterdrücker
voltage surge Spannungsstoß || Stoßspannung || Überspannung * *kurzzeitige ~, ~sspitze*
voltage surge diverter Überspannungsableiter
voltage surge suppressor Überspannungsableiter
voltage switched Schaltspannung
voltage tap Spannungsanzapfung * *an einem Trafo usw.*
voltage to frequency converter U-f-Wandler
voltage to neutral Sternpunktspannung || Sternspannung
voltage tolerance Spannungstoleranz
voltage transducer Spannungswandler * *Messwandler*
voltage transformer Spannungswandler
voltage unbalance Spannungsunsymmetrie
voltage unsymmetry Spannungsunsymmetrie
voltage variation Spannungsänderung || Spannungsschwankung
voltage withstand capability Spannungsfestigkeit
voltage zero Nulldurchgang * *Nulldurchgang der Spannung*
voltage-clocked maschinengetaktet * *Umrichtertaktfrequ. stellt sich abhäng.v. d.Maschinenspanng. ein (Synchronmasch.)*
voltage-conversion motor spannungsumrüstbarer Motor * *{el.Masch.} z.B. Bergwerksmotor, d. sich von 500 auf 1000V umrüsten läßt*
voltage-current characteristic Strom-Spannungs-Kennlinie
voltage-dependent spannungsabhängig
voltage-dividing resistor Spannungsteiler-Widerstand
voltage-frequency characteristic Spannungs-Frequenz-Kennlinie || U-f-Kennlinie
voltage-frequency control characteristic U-f-Kennlinie
voltage-frequency conversion Spannungs-Frequenz-Wandlung

voltage-frequency converter Spannungs-Frequenz-Wandler
voltage-isolating potenzialtrennend
voltage-proof spannungsfest
voltage-source DC link converter Spannungszwischenkreis-Umrichter
voltage-time-area Spannungs-Zeit-Fläche
voltage-to-frequency converter f-u-Wandler || Spannungs-Frequenz-Umsetzer || Spannungs-Frequenz-Wandler
voltage/frequency characteristic Frequenz-Spannungs-Kennlinie
voltage/frequency curve Frequenz-Spannungs-Kennlinie
voltageless potenzialfrei * *ohne Spannung*
voltmeter Spannungsmesser || Voltmeter
volts meter Voltmeter
volume Bauvolumen || Größe * *Rauminhalt, Höhe des Umsatzes usw.* || Größenordnung * *Umfang, Inhalt, Masse* || Maß * *Raummenge, Volumen* || Raum * *~inhalt* || Stückzahl || Umfang * *Rauminhalt* || Umsatz || Volumen * *(auch figürlich), Gesamtbetrag, Stückzahl; low:kleines/niedriges*
volume control Volumenregelung * *z.B.mechan. "Hubraumänderg." bei Schraubenkompressor zur Variation d. Kälteleistung*
volume expansion Ausdehnung * *des Volumens z.B. bei Erwärmung*
volume flow Volumenstrom
volume flow meter Durchflussmesser * *Volumendurchfluss-Messgerät* || Volumendurchfluss-Messgerät
volume flow rate Volumenstrom * *durchströmendes Volumen pro Zeiteinheit, z.B. bei Lüfter [Kubikmeter/h]*
volume integral Volumenintegral
volume of data Datenmenge
volume of orders received Auftragseingang * *geldmäßiger Betrag, Volumen*
volume production Serienproduktion * *siehe auch →Großserienfertigung*
volume proportion Mengengerüst
volume scenario Mengengerüst
volumetric volumetrisch
volumetric capacity Förderkapazität * *z.B. eines Lüfters* || Fördermenge * *z.B. Pumpe*
volumetric mass Dichte
volumetric metering filling machine Dosier-Füllmaschine * *in der Verpackungstechnik*
voluminous dick * *voluminös, umfangreich* || umfangreich * *körperlich*
voluntarily frei * *(Adv.) freiwillig*
vote abstimmen * *bei einer Wahl*
vouch garantieren * *['wautsch] bürgen; for: (sich ver-) bürgen für; that: dafür bürgen, dass*
vouch for bescheinigen * *beglaubigen* || gewährleisten * *['wautsch] (sich ver-) bürgen für*
vouch that gewährleisten * *['wautsch] dafür bürgen, dass*
voucher Beleg * *['wautscher] (Rechnungs-) Beleg, Quittung, Gutschein, Eintrittskarte, Verrechnungsscheck* || Nachweis * *['wautscher] Beleg* || Quittung * *['wautscher] Beleg, Gutschein*
voucherless ordering beleglose Bestellung
voyage Fahrt * *Seereise*
Vp-p Volt Spitze-Spitze
Vpp Volt Spitze-Spitze
VRAM VRAM * *Video-RAM (Speicher auf Grafikkarte, wird v.Grafik-Chip beschrieben und von Video-DAC gelesen)*
vs gegenüber * *Kurzform für 'versus': verglichen mit*
VSD drehzahlveränderbarer Antrieb * *Abk.f. 'Variable-Speed Drive'*
vulcanize vulkanisieren

vulcanizing agent Vulkanisiermittel
vulnerable anfällig * *to:gegenüber*
VVIDD VVIDD * *'Video Interface for Digital Displays':* Norm der →VESA für dig.Bildschirmansteuerung *(LCD...)*
vying konkurrierend * *wetteifernd; vying for the contract:* um den Auftrag ~

W

W x H x D B x H x T || Breite x Höhe x Tiefe
w/o ohne * *Abkürz. für 'without'*
wafer Siliziumscheibe
wafer-thin hauchdünn
waffle pattern Waffelmuster * *z.B. an Kupplungs-/Bremslamelle*
wage agreement Tarifvertrag
wages Lohn || Löhne
waggon Wagen * *{Bahnen} Güter~*
wait warten * *for:* auf
wait for one's cycle to come around warten bis man dran ist
wait time Wartezeit
waiting period Lieferzeit || Wartezeit * *Wartefrist*
waive verzichten * ~ *auf, auskommen ohne, sich e. Rechtes/Vorteils begeben*
walk gehen * *zu Fuß* ~ || Weg * *Gang*
walkaround tour Begehung
walking beam Hubbalken * *z.B. im Walzwerk*
wall Wand
wall bracket Wandhalter
wall entrance Durchführung * *Wanddurch/einführung von Leitungen* || Kabeldurchführung * *Wandeinlass*
wall mounted wandbefestigt
wall mounting Schalttafelmontage || Wandbefestigung * *eines Geräts* || Wandmontage * *eines Geräts; wall-mount: für ~*
wall outlet Steckdose
wall plug Dübel * *Wanddübel* || Steckdose
wall socket Steckdose
wall thickness Wanddicke || Wandstärke
wall-mount housing Wandgehäuse
wall-mounted casing Wandgehäuse
wall-mounted distribution board Wandverteiler
wallet Hülle * *Futteral*
WAN WAN * *Abkürzg.f. 'Wide Area Network':- Weitbereichs-Computernetz*
want benötigen * *wünschen* || suchen * *wünschen, haben wollen* || wünschen * *wollen, wünschen*
want of clearness Unklarheit
wanting unvollkommen * *fehlend, mangelnd; be in wanting of:z.wünschen übr.lassen bezügl./nicht gerecht werden*
ward off abwehren * *Stoß usw.(auch figürl.)*
Ward-Leonard converter Leonardsatz || Leonardumformer
Ward-Leonard set Leonardsatz || Leonardumformer || Ward-Leonard-Umformer
Ward-Leonard-Ilgner converter Leonard-Schwungradumformer
Ward-Leonard-Ilgner set Ilgner-Umformer
ware Ware * *Ware(n), Artikel, Erzeugnis(se)*
warehouse aufheben * *['wärhaus]* einlagern, auf Lager halten || einlagern || Lager * *für Waren, Produkte* || Lagerhalle || lagern * *['wärhaus] aufbewahren, auf Lager halten*
warehouse charges Lagerkosten * *[wärhaus] an externe Unternehmen zu zahlende* || Lagerungskosten * *[wärhaus ...] an dritte zu zahlende*

warehouse costs Lagerkosten * *[wärhaus]* || Lagerungskosten * *[wärhaus ...]*
warehouse disposition Lagerdisposition * *[wärhaus]*
warehouse list Lagerliste
warehousing Einlagerung || Lagerhaltung * *[wärhausing]* || Lagerung * *[wärhausing]* Ein~, i.Lager Aufbewahren, Auf-Lager-Halten
warhouse magazinieren * *in einem Lager einlagern*
warm erwärmen * *auch 'sich erwärmen'* || warm
warm condition betriebswarmer Zustand
warm operating condition warmer Betriebszustand
warm press Warmpresse
warm restart Fortsetzungsstart * *Warmstart* || Warmstart * *eines Rechners oder Antriebs* || Wiederanlauf * *Warmstart*
warm up anwärmen
warm-up lamp current Anlaufstrom * *für Lampe* || Lampen-Kaltstrom * *erhöhter Lampenstrom während der Erwärmungsphase nach d.Einschalten*
warm-up period Anwärmzeit
warmed air erwärmte Luft || Kühlluft * ~, *nachdem sie ihre "Kühlarbeit" verrichtet hat (am ~-Austritt)*
warming Erwärmung * *Wärmung*
warming up period Anwärmzeit
warn warnen * *of/against:* vor; *warn someone against doing: jdn. davor ~, etw. zu tun*
warning Lehre * *Warnung; let it be a warning to you: lasse Dir dies zur Warnung dienen* || warnend || Warnung || Warnungsmeldung
warning channel Warnungskanal
warning information Warnhinweis
warning instruction Warnhinweis
warning label Warnschild * *Warn-Aufkleber*
warning light Warnleuchte
warning message Warnungsmeldung
warning plate Warnschild
warning sign Warnschild || Warnzeichen
warning signal Warnung * *Warnsignal* || Warnungsmeldung
warning symbol Warnzeichen
warp Kette * *{Textil} Kette(nfäden) in der Weberei; warp and woof:* ~ *und Schuss* || Krümmung * *Verziehen, Ver~, Verwerfung (von Holz usw.)* || sich verziehen * *{Holz}* || verbiegen * *{Holz}* || verwerfen * *{Holz} sich* ~ || verziehen * *sich* ~ *(Holz usw.)*
warp beam Kettenbaum * *{Textil}*
warp knitting Kettenwirken * *{Textil}*
warp knitting machine Kettenwirkmaschine * *{Textil}* || Wirkmaschine * *{Textil}*
warped verzogen * *{Holz}*
warping Verbiegung * *Verziehen, Verwerfen, Krümmen, z.B. Holz*
warping creel Spulengatter * *{Textil} in der Weberei*
warping frame Schärmaschine * *{Textil}*
warping machine Schärmaschine * *{Textil}*
warping speed Särgeschwindigkeit * *{Textil} in Schärmaschine*
warrant Befugnis || Berechtigung * *Vollmacht, Befugnis* || garantieren * *auch 'zusichern, haften für, bestätigen'* || gewährleisten * *garantieren, (sich ver)bürgen (für), haften für, bestätigen* || haften * *garantieren; warrant a thing: für etwas* ~ || rechtfertigen
warranted garantiert
warranty Berechtigung * *Berechtigung, Vollmacht, Rechtsgarantie* || Garantie * *des Verkäufers* || Gewährleistung * ~ *des Verkäufers; under: unter/in der*
warranty case Gewährleistungsfall

warranty claim Gewährleistungsanspruch * *warranty credits:Erstattung/Gutschrift von Gewährleistungsansprüchen*
warranty disclaimer Gewährleistungsausschluss * ~*erklärung*
warranty exclusion Gewährleistungsausschluss
warranty interval Garantiefrist || Gewährleistungsfrist
warranty period Garantiezeit || Gewährleistungsfrist || Gewährleistungszeit
warranty procedure Gewährleistungsverfahren
warranty repair Garantiereparatur || Gewährleistungsreparatur
wash aufbereiten * *Erz* || spülen || waschen
wash one's hand austreten * *Bedürfnis verrichten*
wash out ausspülen
washer Ausgleichscheibe * *Unterlegscheibe* || Ausgleichsscheibe * *Unterlegscheibe* || Reinigungsmaschine * *zum Waschen* || Ring * *Unterlegscheibe, Dichtring* || Scheibe * →*Unterleg~* || Unterlegscheibe || Waschmaschine * *für Nicht-Textilien, z.B. für Behälter in der Getränkeindustrie*
washing Aufbereitung * *Waschen, Wässern, Nasswäsche (auch von Erz, Schlamm)* || Wäsche
washing machine Waschmaschine
washing plant Wäscherei
washroom Toilette
wastage Verlust * *Material~, Verschleiß, Abfall, Abgang* || Verschleiß * *Verbrauch*
waste Abfall * *zum Wegwerfen* || Ausschuss * Schrott, verdorbenes Material || schlecht * *Abfall-, unbrauchbar, abgängig, verschlissen* || unbrauchbar * *Ausschuss/Abfall* || Verlust * *Abfall, Abgang*
waste disposal Abfallentsorgung || Entsorgung * *Abfallentsorgung*
waste edges Randstreifen * *z.B. bei Papiermaschine*
waste gas Abgas
waste heat Abwärme
waste of energy Mühe * *vergebliche ~*
waste of time Mühe * *vergebliche ~* || Zeitverschwendung
waste paper Altpapier
waste water Abwasser
waste water purification Abwasserreinigung
waste water treatment Abwasserbehandlung
waste wool Putzwolle
waste-basket Papierkorb
waste-heap Halde * *z.B. für Erz, Abraum usw.*
waste-paper basket Papierkorb
watch beobachten * *genau ~* || betrachten * *(aufmerksam) beobachten*
watch over schützen * *bewachen* || überwachen * *aufpassen auf*
watchdog Watchdog * *Zeitüberwachung eines zyklischen Vorgangs (Zeitglied, muss zyklisch getriggert werden)*
watchdog module Überwachungsbaugruppe * *überwacht den zyklischen Programmablauf e. Rechners/SPS*
watchdog monitor Funktionsüberwachung * *für Rechner/SPS usw.*
watchdog relay Überwachungsrelais * *mit eingebauter Überwachungseinrichtung*
watchdog timer Zeitüberwachung * *Zeitglied zur Überwachung d. zyklischen Programmablaufes eines Rechners* || Zykluszeitüberwachung * *Überwachungs-Zeitglied, das anspricht, wenn es nicht zyklisch getaktet wird*
watchfulness Aufmerksamkeit * *Wachsamkeit*
water spritzen * *wässern*
water cock Wasserhahn
water connection Wasseranschluss
water cooling Wasserkühlung

water drain Wasserablauf
water drain hole Wasserablaufbohrung
water filling Wasserfüllung
water hose Wasserschlauch
water inlet Wassereinlass * *water inlet temperature: ~temp.; b.wassergekühlt. Wärmetauscher f. El.motor 5...25 °C*
water jet Wasserstrahl
water jet cutting Wasserstrahlschneiden * *{Werkz.masch.}*
water jets Strahlwasser * *protection against water jets: Schutz gegen ~ (siehe →IP65)*
water level closed-loop control Wasserstandsregelung
water outlet Wasserablass || Wasserablauf * *Wasserauslass/-austritt*
water pipe Wasserleitung
water preparing plant Wasseraufbereitungsanlage
water protection Wasserschutz
water pumping station Wasserwerk
water service Wasserversorgung
water supply Wasseranschluss * *Wasserversorgung* || Wasserversorgung
water supply installation Wasserwerk
water supply plant Wasserwerk
water treatment Wasseraufbereitung
water utility Wasserwerk * *Wasserversorgungsunternehmen*
water works Wasserwerk
water cooled wassergekühlt
water-cooling system Wasserkühlung
water-gauge Wasserwaage
water-repellent wasserabweisend
water-to-water heat exchanger Wasser-Wasser-Rückkühlanlage * *bei einem geschlossenen Wasserkühlkreislauf*
waterchiller Kaltwassersatz * *Kältemaschine zur Erzeugung kalten Wassers*
watercooled wassergekühlt
waterleaf paper ungeleimtes Papier
waterproof abdichten * *gegen Wasser* || wasserdicht
watertight wasserdicht
waterwheel generator Wasserkraftgenerator
waterwork Wasserwerk
waterworks Wasserwerk
watery dünnflüssig
watt W * *Einzahl* || Watt
wattless current Blindstrom
wattmeter Leistungsmesser
watts W * *Mehrzahl*
wave schwanken * *schwanken, sich wiegen, sich hin und herbewegen; (to and fro:hin und her)* || Schwingung * *(elektr./magn.) Welle* || Welle * *elektrische ~/Schwingung; Wasser-, Angreifer usw.; standing: stehende*
wave crest Wellenberg
wave form Kurvenform * *einer periodischen Größe*
wave resistance Wellenwiderstand * *z.B. einer Leitung*
wave soldering Wellenlöten
wave-form Kurvenform * *einer periodischen Größe*
waveform Verlauf * *~ eines periodischen Signals* || Wellenform
waveform generator Frequenzgeber * *Funktionsgenerator* || Frequenzgenerator * *Funktionsgenerator* || Funktionsgenerator
wavelength Wellenlänge
waver schwanken * *zaudern, unschlüssig sein*
waviness -Welligkeit * *z.B. von Papier* || Welligkeit * *z.B. von Papier*
waving Schwankung * *Schwenken, Hin- und Herbewegung*

wavy wellenförmig
wavy setting machine Schränkmaschine * *(holz)* Wellen~ zum Scharfmachen von Sägen
wax Wachs
way Art und Weise * *(Lösungs/Realisierungs) Weg; in this way: auf diese ~* || Gang * *Weg* || Richtung * *Weg* || Variante * *Art und Weise, (Lösungs)-Weg* || Version * *Art und Weise* || Weg * *Fuß/Fahr~, Art und Weise, Richtung; stand in the way: im ~ stehen; midway: in der Mitte des ~es* || Weise
way back Rückfahrt * *on the way back: auf der ~/ dem Rückweg*
way bills Begleitpapiere
way of doing Methode
way of implementation Ausprägung * *Art u.Weise der Ausführung/Durchführung* || Variante * *Art und Weise der Aus/Durchführung* || Version * *Art und Weise der Aus/Durchführung*
way of realization Ausprägung * *Art u.Weise der Realisierung/Verwirklichung/Ausführung* || Variante * *Art u. Weise d. Realisierung/Verwirklichg./Aus-/Durchführung* || Version * *Art und Weise der Realisierung/Ausführung/Durchführung*
way of representation Darstellungsart
way out Ausweg * *auch 'Ausflucht'*
way-bills Begleitpapiere
waybill Frachtbrief
we reserve the right to make modifications Änderungen vorbehalten * *wir behalten uns das Recht auf Änderungen vor*
we would appreciate wir wären dankbar für
weak dünn * *Flüssigkeit* || schwach
weak network schwaches Netz
weak phase Schwächephase * *wirtschaftliche*
weak point Schwäche * *Schwachpunkt, schwache Seite* || Schwachpunkt || Schwachstelle
weak supply schwaches Netz || Schwachnetz
weak supply network Schwachnetz
weaken abschwächen * *z.B. Motorfeld ~* || angreifen * *schwächen* || dämpfen * *(ab)schwächen* || nachlassen * *schwächer werden* || schädigen * *schwächen* || schwächen * *(auch figürlich)*
weakened geschwächt
weakening Abschwächung * *z.B. Motorfeld* || Schädigung * *→Schwächung, →Abschwächung* || Schwächung * *(siehe auch 'Abschwächung')*
weakest point analysis Schwachstellenanalyse
weakness Fehler * *Schwäche, Mangel* || Schwäche
wealth Fülle * *of ideas/knowledge:an Ideen/Wissen* || Überfluss * *Reichtum, Fülle (of:an/von)*
wealth of experience Erfahrung * *~sschatz*
wealth of orders Auftragspolster
wealth of technical knowledge Wissen * *Erfahrungsschatz*
wear abnutzen || Abnutzung || Abrieb * *Verschleiß* || Alterung * *Verschleiß* || Beanspruchung * *Verschleiß* || Verschleiß * *Abnutzung (high: großer/ hoher)* || verschleißen
wear and tear Verschleiß * *Abnutzung*
wear characteristics Verschleißverhalten
wear compensation Verschleißausgleich * *z.B. an einer Kupplung/Bremse*
wear down ermüden * *abnutzen* || verschleißen * *sich 'abrubbeln'*
wear indication Verschleißanzeige * *z.B. für Brems-/Kupplungbelag*
wear indicator Verschleißanzeige * *Verschleißanzeiger, z.B. für Kupplungs-/Bremsbelag* || Verschleißanzeiger * *z.B. für Bremsbelag, Kohlebürste usw.*
wear of the friction face Reibflächenverschleiß * *bei Bremse/Kupplg.; specific: spezif. (Verschleißvolumen pro Reibarbeit)*
wear off abschaben * *abnutzen*
wear out ablaufen * *verschleißen* || abnutzen || altern * *verschleißen* || auslaufen * *(Masch.bau) Lager usw.* || verschleißen * *abnutzen, sich verschleißen*
wear resistance Abriebfestigkeit
wear up abnutzen || verbrauchen * *abnutzen*
wear-out Verschleiß
wear-resistant abriebbeständig || verschleißfest || verschleißfrei * *verschleißfest*
wearfree verschleißfrei
wearing verschleißbehaftet
wearing out verschleißbehaftet
wearing part Verschleißteil
wearisome langwierig * *ermüdend*
wearless verschleißfrei
wearout failure Verschleißausfall
weary schwach * *müde, matt, erschöpft (with:von/vor)*
weather factors Witterungseinflüsse * *weatherproofing measures: Maßnahmen gegen ~*
weather protected design wettergeschützte Ausführung * *z.B. Schutzart IPW24 für Motor*
weather protection Klimaschutz || Wetterschutz
weather resistant wetterfest
weathering Alterung
weatherproof wettergeschützt
weatherproofing Klimaschutz
weatherproofing measures Maßnahmen gegen Witterungseinflüsse
weave weben
weaver's loom Webstuhl
weaving loom Webstuhl
weaving machine Webmaschine || Webstuhl * *Webmaschine*
weaving mill Webwarenfabrik
web Bahn * *Material~ (aus Papier, Kunststoff, Textil)* || Gewebe || Warenbahn * *vor allem durchlaufende Papier/Textil/Kunststoffbahn*
web accumulator Warenbahnspeicher
web break Bahnriss
web breakage Bahnriss || Papierabriss
web force Zugkraft * *einer Warenbahn (Papier, Textil, Kunststoff)*
web guide Bahnführung * *Einrichtung zur seitl. Führung einer durchlaufenden Warenbahn*
web offset printing press Rollenoffsetmaschine * *(Druckerei)*
web speed Bahngeschwindigkeit * *einer Warenbahn (Papier, Textil usw)* || Warenbahngeschwindigkeit * *besonders Papier, Textil, Kunststofffolie* || Warengeschwindigkeit * *besonders einer Papier-/Textil-/Kunststofffolienbahn*
web storage Warenspeicher * *für durchlauf. Papier/Textil/Kunststoffbahn, z.B. z.Überbrückg. d. Rollenwechselzeit*
web tach Tachometer * *Bahn~*
web tension Bahnspannung * *Zug in einer Materialbahn* || Bahnzug * *Zug in einer Materialbahn (z.B. Papier, Textil)* || Papierzug || Warenbahnzug * *Zugkraft* || Warenzug * *Zugkraft* || Zugspannung * *Zugkraft einer Warenbahn*
web tension control Bahnspannungsregelsystem * *meist Abwickler mit Magnetbremse/Kupplung*
web tensioner Bahnspannungsregelsystem * *meist Abwickler mit Magnetbremse/Kupplung*
web tensioning unit Bahnspanngerät
web transfer Abnahme * *einer Papierbahn vom Sieb in d. Papiermaschine*
web velocity Bahngeschwindigkeit * *einer durchlaufenden Papier/Textil/Kunststoffbahn*
web viewer Bahnbeobachtungsanlage * *für durchlaufende Warenbahn (z.B. in Druckmaschine)*
web-fed offset press Offset-Rollendruckmaschine * *(Druckerei)*
web-fed printing machine Rollendruckmaschine || Rotationsdruckmaschine * *(Druckerei)*

web-fed printing press Rollendruckmaschine * *{Druckerei} im Gegensatz zur →Bogendruckmaschine; verarbeitet Papierrollen* || Rotationsdruckmaschine * *{Druckerei}*
web-glazing calender Satinierkalander * *{Papier} zum Zusammenpressen und glatt/glänzend machen von Papier*
web-speed tachometer Bahntacho || Messrad * →*Bahntacho*
web-spreading roller Breitstreckwalze
wedge Keil
weekdays werktags
weft Schuss * *{Textil} beim Wirkvorgang in der Weberei*
weigh aufliegen * *lasten ((up)on:auf)* || prüfen * *sorgsam ab-/erwägen* || wägen
weigh filler Abfüllwaage * *{Verpack.technik}*
weigher Waage * *{Verpack.techn} Abwiegeeinrichtung* || Wägeeinrichtung
weighing Verwiegung
weighing machine Wägeeinrichtung
weighing system Wägesystem
weight Belastung * *Gewicht* || Gewicht * *low: geringes* || gewichten || Wertigkeit
weight saving Gewichtsersparnis
weighted bewertet * *gewichtet* || gewichtet
weighting Bewertung * *Gewichtung* || Eingriff * *Gewichtung, auch eines Signals* || Gewichtung || Weitigkcit * *Gewichtung; (Umsatz-) ~ eines Kunden*
weights and measures Maß * *~e und Gewichte*
weighty wichtig * *gewichtig; z.B. Gründe, Argumente*
welcome angenehm * *willkommen* || bitte * *als Antwort auf "Danke!"* || empfangen * *freundlich ~*
welcome address Grußwort
weld Naht * *z.B. Schweißnaht* || schweißen || Schweißnaht || Schweißung
welding Schweißung || Verschweißen
welding cycle Schweißtakt
welding equipment Schweißgerät
welding gun Schweißzange
welding machine Schweißmaschine
welding operation Schweißung
welding plant Schweißerei
welding seam Schweißnaht
welding shop Schweißerei
welding technology Schweißtechnik
welding torch Schweißbrenner
welding transformer Schweißtrafo
well Brunnen || Schacht * *für Licht usw.*
well cared for gepflegt
well devised durchdacht * *gut durchdacht (z.B. Plan, Konstruktion, techn. Lösung)*
well disposed geneigt * *(figürl.) be well disposed towards somebody/something: jemandem/zu etwas ~ sein*
well disposed towards geneigt * *(figürl.) jdm./zu etwas ~*
well done sehr gut * *(anerkennend)*
well kept gepflegt || ordentlich
well organized übersichtlich
well proven erprobt * *wohl erprobt, gut bewährt*
well suited gut geeignet
well thought-out gut durchdacht
well up aufsteigen * *Gefühl*
well versed erfahren * *bewandert (in:in)*
well weighed durchdacht
well-aimed gezielt
well-balanced ausgewogen * *gut/wohl ~*
well-calculated gezielt
well-conducted gutgeführt
well-considered wohlüberlegt
well-contrived ausgeklügelt
well-devised gut durchdacht * *z.B. Plan*

well-directed zweckmäßig
well-established bewährt * *gut eingeführt, eine Referenz darstellend, unzweifelhaft*
well-founded fundiert * *z.B. Kenntnisse, Auskunft*
well-known bekannt * *gut bekannt*
well-patronized gutgeführt * *Firma*
well-placed gezielt * *gut plaziert*
well-priced kostengünstig
well-proven bewährt * *gut bewährt*
well-reasoned durchdacht || gut durchdacht
well-rounded abgerundet * *Leistung, Stil, Bildung*
well-stocked gut sortiert * *Händler, Lager usw.*
well-timed rechtzeitig
well-weighed gut durchdacht * *gut ausgewogen*
welt Falz * *Rollsaum in der Metallbearbeitung, Klempnerei* || falzen * *in der Klempnerei*
wet anfeuchten || befeuchten || feucht * *nass* || nass * *dripping wet: tropf~; become/get wet: ~ werden; wet: (Verb) ~machen*
wet end Nasspartie * *vorderer Teil einer Papiermaschine* || Siebpartie * *nasses Ende, Entwässerungspartie einer Papiermaschine*
wet press Nasspresse * *zur Papierentwässerung in Papiermaschine* || Presse * *{Papier} Nass~ in der Papierindustrie*
wet running Nasslauf * *Kupplung, Bremse*
wet section Nasspartie * *vorderer Teil einer Papiermaschine*
wet strength Nassfestigkeit * *von Papier*
wet-rotor Flüssigkeitsläufer * *Nassläufer (bei Pumpe)* || Nassläufer * *bei Pumpenantrieb: Läufer läuft in der zu fördernden Flüssigkeit*
wet-rotor motor Nassläufermotor * *bei Pumpenantrieb: Läufer dreht sich in der zu fördernden Flüssigkeit*
wet-treatment Nassbehandlung
wettability Benetzbarkeit
wetting Befeuchtung * *nass machen* || Benetzung
wetting agent Netzmittel * *Benetzungsmittel*
wetting-up mode einrichten * *(Substantiv) Betriebsart bei NC-Steuerung und SPS*
what so * *was für ein*
what is more außerdem || darüber hinaus
whatever etwaig
Wheatstone measuring bridge Wheatstone-Messbrücke
wheel Rad || rollen * *auf Rädern* || Scheibe * *Schleif~ usw.*
wheel drive Radantrieb * *bei Fahrzeug; wheeldriven: mit ~*
wheel frame Drehgestell * *{Bahn} Radgestell*
wheel nave Radnabe
wheel set Radsatz * *{Bahn}*
wheel stand Schleifbock
wheel-frame Radsatz * *{Bahn} Radgestell*
wheel-mounted front-end loader Radlader
wheelbase Achsabstand * *Radabstand*
wheeled loader Radlader
when wenn
when generating generatorisch * *(Adv.) im Energie-erzeugenden Zustand*
when looking at the drive end mit Blick auf die Antriebsseite
when motoring motorisch
when requested auf Wunsch
when required bei Bedarf
when service work becomes necessary Servicefall * *im ~*
when there is a chance gelegentlich * *(Adv.) bei (passender) Gelegenheit*
where wenn * *in Fällen,in denen* || wobei
where time is crucial zeitkritisch * *(nachgestellt) wo (Rechen-, Reaktions-) Zeit ausschlaggebend ist*
whereas dagegen * *während* || während * *Gegensatz*

whereby wobei * *wodurch, wobei (gleichzeitig)*
wherever possible womöglich * *wo immer es möglich ist*
whether ob * *wether...or not: ob..oder nicht*
which can be -bar
which can be connected anschließbar * *was angeschlossen werden kann*
which can be neglected vernachlässigbar
which can be pivoted schwenkbar
which can be rotated drehbar * *(nachgestellt)*
which can be switched off abschaltbar
which can be turned off abschaltbar * *z.B. Thyristor (→GTO)*
while dagegen * *während* || *obwohl* * *wenn auch, obwohl* || während * *Gegensatz*
whilst dagegen * *während* || *obwohl* * *wenn auch, obwohl* || während * ~ *demgegenüber*
whisper quiet flüsterleise || geräuscharm * *flüsterleise*
white liquor Weißlauge
white water Rückwasser * *Siebwasser in Papiermaschine*
white-collar worker Angestellter
whiteness Weiße || Weißgrad * *z.B. von Papier*
whole ganz * *vollkommen, vollständig, heil, unversehrt; whole number: ~e Zahl* || Gesamt- * *das ganze* || konsistent * *ganz, unversehrt* || sämtliche * *ganz* || voll * *ganz (z.B. Betrag)* || vollständig
whole number ganze Zahl
wholesale Großhandel
wholesale dealer Großhändler
wholesaler Großhändler
wholesalers Fachgroßhandel || Großhandel * *Großhändler*
wholscale in großem Umfang
whose dessen * *für Personen, umgangssprachlich auch für Sachen*
wicked schlecht * *boshaft, verworfen*
wide breit || weit * *weit, ausgedehnt, breit, ~gehend, umfassend, reich (Erfahrung usw.), ~ offen* || weitestgehend * *weitgehend, z.B. Vollmacht* || weitgehend * *Vollmacht, Befugnisse usw.*
wide planer Breithobelmaschine * *{Holz}*
wide range weiter Bereich
wide-band breitbandig
wide-frame drawing press Breitziehpresse
wide-range Weitbereichs-
wide-ranging umfangreich * *weit reichend, weitgehend* || weit reichend
wide-spread bekannt * *weit verbreitet* || beliebt * *weit verbreitet*
wide-spreaded verbreitet * *(weit) verbreitet* || weit verbreitet
wide-strip Breitband * *im Walzwerk*
wide-strip line Breitbandstraße * *Walzwerk*
widely stark * *(Adv.)* weit, stark *(verstreut, streuend, abweichend, bekannt, verschieden usw.)* || vielfach * *(Adv.)*
widely differing verschiedenst * *z.B. widely differing applications: ~e Anwendungen*
widely distributed weit verbreitet
widely held verbreitet * *(Ansicht usw.)* || weit verbreitet * *Ansicht*
widely used weit verbreitet * *häufig verwendet*
widen erweitern * *ausweiten (auch figürlich)* || vergrößern * *verbreitern*
wider höher * *Bereich usw.*
widespread verbreitet || weit verbreitet
width Breite * *['with]* || Größe * *Geräumigkeit, Weite*
width across flats Schlüsselweite * *eines Schraubenschlüssels*
width drawing machine Breitstreckmaschine * *z.B. zur Herstellung von Kunststofffolien*
width drawing roll Breitstreckwalze

width gauge Breitenmessung * *Breitenmessanlage/-Einrichtung*
width of machine Maschinenbreite
width over flats Schlüsselweite * *eines Schraubenschlüssels*
width x height x depth Breite x Höhe x Tiefe
width x heigth x depth B x H x T
wield handhaben * *(Macht, Einfluss) ausüben, (Waffe, Werkzeug) handhaben, führen, schwingen*
willow Reißwolf
willowing machine Reißwolf
win abbauen * *Bodenschätze* ~ || bekommen * *Auftrag, Preis* || erwerben * *Achtung, Vertrauen usw.* ~ || Gewinn * *Sieg* || gewinnen * *(Sieg, Preis, Unterstützg. usw.) ~; zu Auftrag usw. gelangen; s. durchsetzen, siegen*
win the order Auftrag erteilt bekommen * *gegen Konkurrenz*
win through durchsetzen * *erfolgreich sein*
winch Winde * *Winde, Haspel, Kurbel*
winch drive Hubwerk * *Seil-/Windenantrieb* || Windenantrieb
wind aufwickeln || Aufwicklung * *Aufwickelstelle* || aufziehen * *Feder* ~ || spannen * *(Spiral)Feder* ~ || wickeln * *auch auf-* || Wickelstation
wind around umschlingen
wind off abwickeln * *Material von einer Rolle/Spule/Haspel/Trommel* ~
wind on umwickeln
wind over umwickeln
wind stand Aufrollung * *Aufroller* || Wickelstation
wind tunnel Windkanal
wind up abwickeln * *Geschäfte usw.* ~ || aufwickeln
wind-driven generator Windgenerator
wind-screen wiper Scheibenwischer
wind-up Abwicklung * *(amerikan.) das Abwickeln von Geschäften usw.*
windage Luftwiderstand || Windeinfluß
windage loss Luftreibung * *Reibung auf Grund des Luftwiderstandes*
windage losses Luftwiderstandsverluste
windage noise Lüftergeräusch * *Luftrauschen*
winder Aufroller || Aufwickler || Aufwicklung * *Aufwickler* || Spuler * *Wickler* || Wickier
winder drive Wickelantrieb || Wicklerantrieb
winder station Wickelstation
winder with flying splice Wickler mit fliegendem Rollenwechsel
winding Abwicklung * *das Abwickeln von Material von e.Rolle/Spule/Trommel/Haspel* || Wickel || Wicklung * *eines Motors/Trafos usw.; tapped: unterteilte* ~ *mit Anzapfungen*
winding calculator Wickelrechner * *berechnet d. Durchmesser b. Wickler u. gibt das d.Sollzug entspr. Drehmoment vor*
winding characteristic Wickelhärte * *Wickelhärtenkennlinie (Zugkraftverlauf) abhängig vom Durchmesser* || Wickelhärtenkennlinie * *m. d.Durchmess. abnehm. ~ verhindert Beschädigg. d.Innenlagen bei Aufwickler*
winding circuit Wicklungsstrang
winding connection Schaltart * *Schaltgruppe z.B. Stern/Dreieck eines Motors/Trafos* || Schaltgruppe * *Verschaltung der Wicklungen, z.B. Stern/Dreieck bei Motor/Trafo*
winding density Wickelhärte
winding density control Wickelhärtensteuerung
winding diameter Wickeldurchmesser
winding drive Wickelantrieb
winding end Wicklungsende
winding factor Wicklungsfaktor * *Reduzierung d. in verteilten Spulen induz. Spanng gegenüber konzentrierter Anordng.*
winding gear Auf-Abwickler

winding head Spulkopf
winding insulation Wicklungsisolation
winding lead Wicklungszuleitung * z.b. einer Motorwicklung
winding losses Wicklungsverluste
winding machine Aufspulmaschine || Förderanlage * Haspel für Bergwerk || Fördermaschine * im Bergbau || Wickelmaschine
winding off Abwicklung * das Abwickeln von Material von e. Rolle/Spule/Haspel
winding overhang Wickelkopf * {el.Masch.} aus d. Blechpaket herausragender Teil e.Wicklung, z.b. eines Motors
winding phase Wicklungsstrang * eines Drehstromsystems
winding pressure Wickelhärte
winding renewal Wicklungserneuerung
winding resistance Wicklungswiderstand
winding section Wicklungsteil
winding shaft Wickelwelle
winding sleeve Wickelhülse * z.B. Papphülse zum Wickeln von Papier, Folie, Textilien usw.
winding slot Wicklungsnut
winding speed Wickelgeschwindigkeit
winding station Wickelstation
winding system Wicklungssystem * z.B. einer elektr. Maschine
winding technology Wickeltechnik
winding temperature Wicklungstemperatur
winding temperature rise Wicklungserwärmung
winding up Aufwicklung * das Aufwickeln
winding-head insulation Wickelkopfisolierung * Isolierg. nebeneinanderliegend. Spulen im →Wickelkopf durch →Phasentrenner
winding-type current transformer Durchsteckstromwandler * ohne Primärwicklung
winding-type transformer Durchsteckwandler * ohne Primärwicklung
winding-up Abwicklung * das Abwickeln von Geschäften usw.
windlass Winde * Winde, Förderhaspel, Ankerspill
windmill leerlaufen * mitgezogen werden (z.B. Lüfterantrieb) || trudeln * (salopp) mitgezogen werden, ohne Antriebsmoment freilaufen (z.B. Motor)
windmilling austrudeln * (Substantiv; salopp) || leerlaufend * (salopp) trudelnd, 'mitgezogen werdend' (z.B. von der Speisequelle getrennter Motor)
window Fenster * auch auf dem Bildschirm
window discriminator Fensterdiskriminator * Grenzwertmelder m. 2 Schwellen (erkennt "im oder außerhalb Fenster")
windows technology Fenstertechnik * Bedienoberfläche für Softwareprogramm
windshield wiper Scheibenwischer
wing Flügel * auch eines Gebäudes; lend wings to a person: jdm. ~ verleihen
wing nut Flügelmutter
winning lohnend * gewinnend (auch figürlich), siegreich, einnehmend
winning post Ziel * ~ im Sport
wipe contact Schleifkontakt
wiper Abstreifer || Rakel * Abstreifer || Schleifer * an Potentiometer, Schiebetrafo usw. || Schleifkontakt || Wischer * Scheibenwischer an Fahrzeug
wiper contact Schleifkontakt || Wischkontakt
wiper motor Wischermotor * Scheiben~
wire Ader || anschließen * verdrahten || Draht * solid: massiver, steifer || Leitung || Sieb * ~ bei Papiermaschine; wire section: ~partie; wire side: ~seite || verdrahten
wire annealing Drahtglühen
wire break Leitungsbruch
wire breakage Drahtbruch
wire breakage monitoring Drahtbruchüberwachung

wire breakage sensing Drahtbrucherkennung
wire brush Drahtbürste
wire bundle Kabelbaum
wire coiler Drahthaspel
wire cross section Anschlussquerschnitt
wire drawing Drahtziehen * Drahtherstellung mittels →Drahtziehmaschine
wire drawing mill Drahtzieherei
wire end Drahtende
wire gauge Anschlussquerschnitt || Drahtdicke || Leitungsquerschnitt * AWG: American Wire Gauge: amerikanische Maßeinheit für Leitungsquerschnitt || Querschnitt * für Kabel
wire grate Drahtgitter
wire grating Drahtgitter
wire guide Sieblaufregler * in Papiermaschine
wire industry Drahtindustrie
wire mesh Drahtgeflecht
wire mill Drahtwalzwerk
wire part Siebpartie * in Papiermaschine
wire range Anschlussquerschnitt
wire rod Walzdraht
wire rod finishing block Draht-Fertigblock * in Drahtwalzwerk
wire roll Siebleitwalze * in Papiermaschine
wire rope Draht * Drahtseil || Drahtseil || Seil * Draht~ || Stahlseil
wire section Siebgruppe * bei Papiermaschine || Siebpartie * Entwässerungspartie in Papiermaschine, hinter dem Stoffauflauf angeordnet
wire size Leitungsquerschnitt
wire straightener Drahtrichtmaschine
wire straightening machine Drahtrichtmaschine
wire strand Draht * (Draht)Litze, Litzendraht, Drahtseil || Drahtlitze
wire stripper Abisolierwerkzeug || Abisolierzange
wire suction roll Siebsaugwalze * {Papier} zum Entwässern des Papierbreis in d. Siebpartie e. Papiermaschine
wire up anschalten * mit Draht || verschalten * verdrahten
wire wheel Drahthaspel
wire winder Haspel * Draht-Auf-/Abwickler
wire-break monitoring Drahtbruchüberwachung
wire-cutter Schere * Draht~
wire-drawing machine Drahtziehmaschine
wire-end ferrule Aderendhülse
wire-guide roll Siebregulierwalze * in Papiermaschine
wire-lattice Gitter * Drahtgitter
wire-lattice pallet Drahtgitterpalette || Gitterboxpalette
wire-up anschließen * hinverdrahten || verdrahten
wire-wound potentiometer Drahtpotentiometer
wire-wound resistor Drahtwiderstand
wire-wrap gun Wickelpistole * zum Herstellen von Wire-Wrap Verbindungen
wire-wrap technique Drahtwickeltechnik * Verdrahtung über Wire-Wrap-Stifte mit Wickelpistole
wireless drahtlos
wireless reception Funk-Empfang || Funkempfang
wireless-controlled ferngesteuert * funkferngesteuert
wiretap Anzapfung * ~ einer Daten-/Telefonleitung
wirewrap post Wirewrap-Stift
wiring Anschluss * Verdrahtung || Beschaltung * z.B. mit Bauelementen || Verdrahtung || Verkabelung
wiring accessories Installationsmaterial
wiring area Anschlussraum
wiring backplane Busplatine
wiring block Rangierung * zur Leitungsrangierung z.B. Rangierklemmen bei d.Leitungseinführung i. Schaltschrank || Rangierverteiler * zur Leitungs-

verteilung z.B. bei der Leitungseinführung i. Schaltschrank
wiring costs Verdrahtungsaufwand || Verdrahtungskosten * siehe auch →*Verdrahtungsaufwand*
wiring diagram Anschlussbild || Schaltschema * *Verdrahtungsplan* || Verdrahtungsplan
wiring diagrams Schaltungsunterlagen * *Verdrahtungspläne*
wiring distance Leitungslänge
wiring duct Kabelkanal || Leitungskanal
wiring example Anschlussbeispiel
wiring facilities Anschlussmöglichkeiten
wiring hole Kabeldurchführung * *Loch*
wiring layout Bauschaltplan
wiring manual Schaltbuch || Schaltungsunterlagen * *Schaltbuch*
wiring method Anschlusstechnik
wiring modification Verdrahtungsänderung
wiring overhead Verdrahtungsaufwand
wiring space Anschlussraum
wiring technique Anschlusstechnik
wise sinnvoll * *klug*
wish Anliegen * *Wunsch* || Wunsch * *with the best wishes:* mit den besten Wünschen || wünschen
wishful thinking Wunschdenken
with -behaftet * *(vorangestellt)* || bei * z.B. Abteilung, Firma, Person || mit || mittels || über * *deal with:* handeln ~/von (Abhandlung, Vortrag, Buch, Film usw.)
with a bang schlagartig * *(Adv.)*
with a high level of hoch * *mit einem hohen Grad an/von*
with a light heart leicht * *~en Herzens*
with a low harmonic content oberschwingungsarm
with a phase-angle displacement phasenverschoben
with a step function sprunghaft
with a wide voltage range Weitspannungs-
with abundant resources aus dem Vollen
with added force verstärkt * *(Adv.) mit mehr Energie/Tatkraft*
with additives legiert * *Schmieröl*
with an effort mühsam * *(Adv.)*
with an interference fit kraftschlüssig
with associated -behaftet * *(vorangestellt) z.B. with associated losses: verlust~*
with associated losses verlustbehaftet
with battery backup batteriegepuffert
with being wobei * *with the thyristor being blocked:* ~ der Thyristor gesperrt ist
with blazing speed blitzschnell * *(Adv.; salopp) verteufelt schnell*
with brushless excitation permanenterregt * *mit bürstenloser Erregung*
with bus capability busfähig
with care sorgfältig * *(Adv.)*
with care! Vorsicht bitte * *Aufschrift auf Kiste usw.*
with contacts kontaktbehaftet
with costs kostenpflichtig
with difficulty kaum * *mit Mühe* || mühsam * *(Adv.)*
with drop-out delay abfallverzögert
with due regard to unter Berücksichtigung von * *unter gebührender Berücksichtigung von*
with explosion-proof enclosure explosionsgeschützt
with focussed energy verstärkt * *mit konzentrierten Kräften*
with free components frei beschaltbar * *z.B. mit Bauelementen auf Leiterkarte ("Spielwiese")*
with free convection konvektionsgekühlt * →*selbstgekühlt* || selbstgekühlt * *konvektionsgekühlt (Leistungshalbleiter/-teil)*

with full details detailliert || im Einzelnen * *mit allen Einzelheiten*
with galvanic isolation potenzialfrei || potenzialgetrennt
with great ease mit Leichtigkeit
with high accuracy hochgenau
with increased efforts verstärkt * *(Adv.) mit ~en/erhöhten Anstrengungen*
with it hiermit
with large surface area großflächig * *(Adv.) z.B. Auflagefläche*
with lightning speed blitzschnell * *(Adv.)*
with low current displacement stromverdrängungsarm
with low harmonic content oberschwingungsarm
with low line supply stressing netzschonend
with low losses verlustarm * *(Adv.)*
with low motor stressing motorschonend
with low stressing schonend
with menu prompting menügeführt
with might and main Gewalt * *mit aller Gewalt*
with millimeter accuracy millimetergenau
with minimum environmental impact umweltfreundlich * *(Adv.)*
with more energy verstärkt * *(Adv.) mit mehr Energie/Tatkraft*
with more power verstärkt * *(Adv.) mit mehr Energie/Tatkraft*
with natural air cooling eigenbelüftet * *luftselbstgekühlt (z.B. Leistungshalbleiter, für Motor unüblich)* || selbstbelüftet * *mit Konvektionskühlung, z.B. Leistungsteil mit Kühlkörper ohne Lüfter*
with necessity zwangsläufig * *(Adv.)*
with no -los * *(vorangestellt)*
with no housing gehäuselos
with OFF delay abfallverzögert
with ON delay anzugsverzögert || anzugverzögert
with open-circuit cooling durchzugsbelüftet * *Motorkühlart mit Luftstrom durch das Gehäuse (niedrige Schutzart)* || innenbelüftet || innengekühlt * →*durchzugsbelüftet (Motorkühlart)*
with operator prompting dialoggesteuert
with optimal ride times fahrzeitoptimiert * *{Aufzüge}*
with optimized execution time laufzeitoptimal * *(Adv.)*
with optimum run time laufzeitoptimal * *(Adv.)*
with overlap überlappend * *auch zeitlich*
with preview action vorausschauend * *z.B. Regler*
with pull-in delay anzugsverzögert
with radial leads radial bedrahtet
with ratio selection umschaltbar * *Trafo*
with reference to bezogen auf
with regard to bezüglich || unter Berücksichtigung von
with respect to bezogen auf * *bezüglich* || bezüglich || gegen * *bei Spannungen* || gegenüber * *hinsichtlich, bezüglich, in Anbetracht von, verglichen mit* || hinsichtlich || mit Rücksicht auf
with respect to each other gegeneinander * *aufeinander bezogen* || untereinander * *aufeinander bezogen, →gegeneinander* || zueinander * *aufeinander bezogen, in Bezug ~*
with rib cooling rippengekühlt
with shaft-mounted fan eigengekühlt * *Motorkühlart mit durch die Motorwelle angetriebenem Lüfterrad*
with sign vorzeichenbehaftet
with special emphasis verstärkt * *(Adv.) mit besonderem Nachdruck*
with sufficient accuracy hinreichend genau
with the aid of mit Hilfe von * *einer Sache*
with the correct sign vorzeichenrichtig
with the cube of mit der dritten Potenz von
with the exception of abgesehen von * *mit Ausnah-*

me von || außer * *mit Ausnahme von* || ausgenommen || bis auf
with the help of anhand * *mit Hilfe von* || mit Hilfe von * *einer Person* || mittels || vermitteln
with the power on unter Spannung
with the rotor locked festgebremst * *Motor*
with the same -gleich * *z.B. with the same output rating: leistungs~*
with the shortest possible length kürzestmöglich * *bezogen auf die Länge*
with the square of the mit dem Quadrat des * *vary with the square of the voltage:sich mit d.Quadrat d. Spannung ändern*
with thermally critical rotor läuferkritisch
with this hierbei || hiermit
with undiminished violence unvermindert * *(Adv.)* *mit unverminderter Kraft/Gewalt*
with unidirectional leads radial bedrahtet * *nach IEC 286*
with version x and more recent versions ab Version x * *ab inklusive Version x*
with zero backlash spielfrei * *z.B. Getriebe*
withdraw abführen * *wegnehmen, abziehen* || abziehen * *Personal, Wälzlager, Stecker, Steckbaugruppe usw.* ~ || stornieren || wegnehmen * *ein Kommando/Signal* ~ || widerrufen * *(Auftrag, Aussage usw)* ~, *zurückziehen, zurücktreten, (Geld, Personal) abziehen* || ziehen * *z.B. eine Steckbaugruppe* ~ || zurücknehmen * *Vorwürfe, Urteil, Kritik usw.* ~ || zurückziehen
withdraw from service aus dem Verkehr ziehen
withdrawable abziehbar
withdrawable fan Lüftereinschub
withdrawal Abzug || Entnahme * *von Geld usw.* || Widerruf
withdrawal device Ausziehvorrichtung
withdrawal force Abzugskraft
withdrawal from the contract Rücktritt vom Vertrag
withdrawer screw Abdrückschraube
withdrawing force Abziehkraft
withhold festhalten * *einbehalten, zurückhalten, abhalten (withhold someone from: jdn.* ~ *von), vorenthalten* || zurückhalten * *vorenthalten*
within innerhalb * ~ *eines Bereiches, within a period:* ~ *einer Periode*
within a short time kurzfristig * *(Adv.)*
within everybody's grasp verständlich * *allgemein verständlich; difficult to grasp:schwer verständlich*
within reach erreichbar * *(örtlich) within easy reach: leicht* ~ || greifbar * *in Reichweite* || zur Hand * *in Reichweite, greifbar*
within reason tragbar * *im Rahmen des Tragbaren* || zumutbar * *im Rahmen des Zumutbaren*
within the im Rahmen von
within the boundaries of im Rahmen von * *innerhalb der Grenzen von*
within the bounds of possibility so weit möglich
within the framework of im Rahmen von * *innerhalb des Gefüges/der Struktur*
within the limits of im Rahmen von * *innerhalb der Grenzen von*
within the scope of im Rahmen von * *innerhalb des (Geltungs)Bereichs von*
within wide bounds weitgehend * *(Adv.) in weiten Grenzen*
without -los * *(vorangestellt)* || außen * *außerhalb; from without: von außen* || ohne || von außen * *außerhalb*
without a gap lückenlos
without a hitch reibungslos * *(Adv.) reibungslos, glatt, ungestört, ungehindert, ruckfrei*
without any difficulties problemlos || unbürokratisch * *(Adv.) problemlos*

without any problems reibungslos * *problemlos*
without assistance selbst * *ohne fremde Hilfe* || selbstständig * *ohne Hilfe/Unterstützung*
without circulating current kreisstromfrei
without cogging ruckfrei
without competition konkurrenzlos
without compromise kompromisslos
without delay unverzögert * *(Adv.)*
without doubt unbestritten * *(Adv.)*
without enclosure gehäuselos
without extra kostenlos * *ohne zusätzliche Kosten*
without extra cost kostenlos * *ohne zusätzliche Kosten*
without liability ohne Gewähr
without overshoot überschwingungsfrei
without problems problemlos
without replacement ersatzlos
without sliprings schleifringlos
without steps stufenlos
without supervision unbeaufsichtigt
without time delay verzögerungsfrei
without trouble mühelos || problemlos * *auch mühelos*
withstand halten * *widerstehen* || standhalten * *aushalten* || widerstehen
withstand capability -Festigkeit
withstandibility -Festigkeit
wiz Könner * *(salopp) Kurzform für 'wizzard': Zauberer*
wizard Assistent * *kleiner Helfer (Hilfe-Funktion in Softwareprogramm), Zauberer, Hexenmeister* || prima * *erstklassig, prima*
wizzard Könner * *(salopp) Zauberer*
wobble flattern * *z.B. Räder* || Schlag * *Unrundheit* || schlagen * *unrund laufen, wackeln* || schwanken * *wackeln*
wobble amplitude Wobbelamplitude * *für Wobbelgenerator bei Textil-Changierantrieb*
wobble generator Wobbelgenerator * *sorgt f.guten Spulenaufbau bei Textil-Changierantrieb durch period.Zusatzsollwert*
wobbling frequency Wobbelfrequenz
wobbling unit Wobbeleinrichtung * *{Textil}* || Wobbelgenerator * *erzeugt period.Zusatzsollw.(z.B.Dreieck) f.Textil-Changierantrieb> gut. Spulenaufbau*
wobbulation Wobbelung
wobbulation equipment Wobbeleinrichtung * *{Textil} für Changierantrieb beim Aufwickeln von Fäden*
wobbulation frequency Wobbelfrequenz
wobbulation generator Wobbelgenerator * *für Changierantrieb i.Textilmaschine z.Fadenaufwikkeln mit gutem Spulenaufbau*
wonder about sich Gedanken machen über
wood Holz * *(made) of wood/wooden: aus* ~; *wood pile: ~stapel; wood yard: ~platz*
wood chips Holzspäne
wood containing holzhaltig * *(Papier)*
wood grinder Holzschleifer * *zur Zellstofferzeugung*
wood machinery Holzbearbeitungsmaschinen
wood panel Holzplatte
wood peeling machine Entrindungsmaschine * *Holzschälmaschine* || Holzschälmaschine || Rundschälmaschine * *für Baumstämme*
wood plate Holzplatte
wood pulp Zellstoff * *aus Holz gewonnener* ~
wood screw Holzschraube
wood working Holzbearbeitung
wood-based material Holzwerkstoff
wood-handling yard Holzplatz * *z.B. in Zellstoffabrik, Sägewerk*
wood-wool Holzwolle

wood-working machine Holzbearbeitungsmaschine
wooden Holz- || steif * (figürlich) hölzern
wooden crate Holzkiste
wooden stick Holzstäbchen
woodfree holzfrei * Papier; fine paper/woodfree paper: ~es Papier
woodfree paper Feinpapier * holzfreies Papier
woodpulp Holzstoff * (Papier) Zellstoffrohprodukt
woodwool Holzwolle
woodworking Holzbearbeitung
woodworking industry Holzbearbeitungsindustrie
woodworking machinery Holzbearbeitungsmaschinen
woodworking trade Holzbearbeitungsindustrie || Holzhandwerk
woodyard Holzplatz
woof Schuss * (Textil) in der Weberei; warp and woof: Kette und ~
wool Faser * metallische Faser, Schlackenfaser || Wolle * Schurwolle
woolen rag Wolllappen
woolly unklar * verschwommen, wolkig (z.B. Gedanken)
wooly unklar * siehe 'woolly'
word Ausdruck || Begriff * Wort || Vokabel || Wort * auch Datenwort || Zusage
word processing Textverarbeitung
word processor Textverarbeitungsprogramm || Textverarbeitungssystem || Wortprozessor * bei SPS
wording Wortlaut
work Arbeit * auch als physikalische Größe in [Ws] oder [J] || arbeiten || bearbeiten * Metall ~, spanlos ~ || bedienen * arbeiten an, betreiben (Fabrik, Maschine), arbeiten lassen || betätigen * sich betätigen (as:als) || betreiben * Maschine, Anlage ~ || einwirken * on: auf eine Person ~ || Fabrik * Werk, Fabrik, Betrieb || funktionieren * Art u. Weise d. Funktionierens; this is how it works: so funktioniert es || gehen * bei Maschine usw.; doesn't work: geht nicht || laufen * Maschine || verformen * bearbeiten || Werk * Arbeitsergebnis, Aufgabe, (Fabrikations)Betrieb || Werkstück
work at bearbeiten || sich beschäftigen mit
work clothes Arbeitskleidung
work for betreiben * hinarbeiten auf
work force Belegschaft || Mitarbeiterschaft || Personalausstattung
work hard hart arbeiten
work hour Arbeitsstunde
work in einarbeiten * by extra work: einen Rückstand durch Zusatzarbeit
work in shifts ablösen * bei der Arbeit ~
work into einarbeiten * sich in eine Sache ~; neue Erkenntnisse in eine Arbeit ~ || einbringen * hineinarbeiten
work load Arbeitsbelastung * bezügl. der Menge der zu leistenden Arbeit || Auslastung
work loose lockern * sich lockern
work madly hart arbeiten * (salopp) "wie verrückt" arbeiten
work of friction Reibarbeit || Reibungsarbeit
work of reference Nachschlagewerk * auch zu einem technischen Hard- oder Softwareprodukt
work off abarbeiten
work oneself to the bones überarbeiten * sich ~/ überanstrengen
work out ausarbeiten * auch herausarbeiten || auswirken || bearbeiten * ausarbeiten || erarbeiten * herausarbeiten, 'in die Mache nehmen' || erarbeiten * →ausarbeiten || herausarbeiten * auch ausarbeiten || herausbekommen * Rätsel, Antwort usw. ~ || rechnen * aus~
work over umarbeiten

work rate Arbeitsleistung * einer Maschine
work schedule Arbeitsplan
work scheduling Arbeitsvorbereitung
work station Arbeitsplatz * ~ für Bedienpersonal einer Maschine, Anlage usw.
work through durcharbeiten * z.B. Lehrbuch
work together zusammenarbeiten || zusammenwirken
work up aufbereiten * ausarbeiten, entwickeln, erweitern (into:in), (Thema) bearbeiten, sich einarbeiten || ausarbeiten * erweitern (into: zu), entwickeln; (Thema usw.) ~, bearbeiten, s. einarbeiten, aufbauen
work well klappen * gut funktionieren
work-force Mitarbeiter * (Plural) Belegschaft, Mitarbeiterschaft
workability Durchführbarkeit
workable durchführbar * durch-/ausführbar (Plan, Vorhaben usw.); bearbeitungsfähig || einsatzbereit * einsatzfähig || funktionierend * →funktionsfähig || realisierbar || verformbar * (Metall)
workaround Ausweg * Notbehelf || Fehlerbehebung * provisorische Maßnahme zur Umgehung/ Behebung e. Fehlers || Fehlerbeseitigung * provisorische Fehlerumgehung || Notlösung * ~, die das eigentliche Problem umgeht
workbench Werkbank
workdays Werktage
worker Arbeiter || Kraft * Person, Arbeits~
worker participation Mitbestimmung * der Mitarbeiterschaft
worker roll Arbeitswalze * (Textil) in →Kardenmaschine
worker student Praktikant * Werkstudent
worker's council Betriebsrat
workforce Mannschaft * Belegschaft, Mitarbeiterschaft || Mitarbeiterschaft || Personal * Mitarbeiterschaft, Belegschaft
workhorse Arbeitspferd * (auch figürl.) robustes, erprobtes Produkt
working Bearbeitung * spanlose ~ || Bedienung || Betrieb * Arbeit(sweise) || Betriebs- || funktionsfähig || funktionstüchtig || in Betrieb || Verformung * non-cutting working: spanlose ~ || Wirkungsweise
working angle Arbeitswinkel * z.B. einer Bremse
working area Arbeitsbereich * Bereich in dem jemand arbeitet (Sicherheitszone) || Arbeitsplatz * Arbeitsbereich
working atmosphere Arbeitsklima || Betriebsklima
working capital Betriebskapital
working chart Arbeitsplan * als Diagramm auf Papier
working committee Arbeitskreis || Gremium * siehe auch →Arbeitskreis
working conditions Betriebsbedingungen || Einsatzbedingungen || Umgebungsbedingungen * →Betriebsbedingungen
working current Arbeitsstrom * siehe z.B VDE 0160
working days Arbeitstage || Werktage
working disk Arbeitsdiskette
working document Arbeitsunterlage
working group Arbeitsgemeinschaft || Arbeitsgruppe * siehe auch →Arbeitskreis, Gremium || Arbeitskreis
working hours Arbeitszeit || Dienstzeit * Arbeitszeit || Mannstunden
working life Lebensdauer || Standzeit
working load Betriebslast
working material Betriebsstoff
working materials Arbeitsmittel
working memory Arbeitsspeicher
working method Arbeitsweise
working moral Einsatz * Arbeitsmoral

working order Schuss * *get into working order: in ~ bringen*
working out Ausarbeitung * *Ausarbeitung, Entwicklung*
working over Umarbeitung
working part Werkstück
working place Arbeitsplatz * *z.B. Schreibtisch, Werkbank*
working point Betriebspunkt
working pool Arbeitsgemeinschaft
working pressure Anpressdruck * *in Druckluftsystem, für Pneumatikkolben usw.* || Betriebsdruck
working process Arbeitsgang
working roller table Arbeitsrollgang * *{Walzwerk}* || Rollgang * *Arbeitsrollgang*
working space Arbeitsplatz * *Arbeitsbereich*
working speed Arbeitsgeschwindigkeit
working student Werkstudent
working temperature Betriebstemperatur
working title Arbeitstitel * *eines Buches, Projekts usw.*
working width Arbeitsbreite * *einer Maschine zur Verarbeitung von durchlaufenden Warenbahnen (z.B. Papiermaschine)*
workman Arbeiter * *Hand~* || Handwerker
workmanlike fachmännisch * *Arbeit*
workmanship Ausführung * *Verarbeitungsgüte, (handwerkliche) Qualität* || Fachkönnen * *handwerkliches ~* || Verarbeitung * *qualititative Ausführung* || Werk * *kunstvolles Stück Arbeit*
workpiece Werkstück * *z.B. bei der Metallverarbeitung*
workpiece accuracy Fertigungsgenauigkeit * *{Werkz.masch.}*
workpiece measuring Werkstückmessung
workplace Arbeitsplatz * *Arbeitsplatz, Arbeitsbereich*
works -Werk || Anlage * *Produktionsanlage* || Betrieb * *Werk(e)* || Fabrik * *Werk(sanlagen), Betriebe, Fabrikationsanlagen* || Fertigungsbetrieb || Werk * *Fabrik; ex works: ab ~*
works management Betriebsleitung * *Werksleitung*
works manager Leiter * *Werksleiter*
works order number Werknummer
works standard sheet Werknormblatt
workshop Betrieb * *Werkstatt, Fertigungs/Werkhalle, Werk, Betrieb* || Halle * *Werkshalle* || Kurs * *~ mit praktischen Übungen* || Lehrgang * *mit praktischen Übungen* || Schulungskurs * *~ mit praktischen Übungen* || Seminar * *Schulungskurs mit praktischen Übungen* || Werkhalle || Werkshalle || Werkstatt
workshop drawing Konstruktionszeichnung
workshop hall Halle * *Werkshalle* || Werkhalle || Werkshalle
workstation Arbeitsplatzrechner || Workstation * *leistungsfähiger Rechner mit hoher Grafikleistung, oft vernetzt, z.B. für →CAD und →DTP*
workstation computer Arbeitsplatzrechner
world Welt
world (without Europe) Übersee * *von Europa aus gesehen*
world around us Umwelt
world class level Weltmarktniveau
world market Weltmarkt
worldwide weltweit
worldwide company weltweit tätiges Unternehmen
worldwide market share Weltmarktanteil
worm Gang * *Gewinde(gang)* || Gewinde || Schnecke * *Schneckenrad* || Windung * *Gewinde*
worm conveyor Schneckenförderer
worm drive Schneckengetriebe * *Schnecken(an)trieb*

worm gear Schneckengetriebe || Schneckenradgetriebe
worm gearbox Schneckenradgetriebe
worm gearing Schneckengetriebe
worm out herausbekommen * *Geheimnis usw. ~*
worm reducer Schneckengetriebe * *Schneckenuntersetzungsgetriebe*
worm wheel Schneckenrad
worm's eye view Ansicht von unten
worm-gear spindle Gewindespindel
worm-spur gear unit Schnecken-Stirnradgetriebe
worn alt * *verschlissen*
worn out alt * *verschlissen*
worn-out part Verschleißteil * *verschlissenes Teil*
worry about kümmern * *sich ~ um* || kümmern um * *sich Gedanken machen über* || sich Gedanken machen über * *sorgend* || sich kümmern um * *sich Gedanken machen über*
worse schlechter * *(Komparativ); get/grow worse: schlechter werden*
worsted Garn * *wollenes ~, Kammgarn, Woll-, Kammgarnstoff; worsted wool: Kammgarnwolle*
worth Wert * *innerer/kultureller ~*
worth highlighting herausragend * *wert, hervorgehoben zu werden*
worth mentioning erwähnenswert || nennenswert
worth the money kostengünstig * *preiswert/würdig* || preisgünstig * *preiswert* || preiswert * *seinen Preis wert*
worthless faul * *(figürl.) Wechsel, Versprechungen usw.* || schlecht * *wertlos* || wertlos * *(auch figürlich) worthless stuff: ~es Zeug*
worthwhile dankbar * *lohnend*
worthwile lohnend * *der Mühe wert*
worthy of note erwähnenswert
worthy of special mention erwähnenswert
would sollen * *Vermutung, Wahrscheinlichkeit*
wound gewickelt
wound capacitor Wickelkondensator
wound core Bandkern
wound field motor Motor mit Feldwicklung * *Gegensatz: PM motor:permanenterregter Motor*
wound goods Wickelgüter
wound material Wickel * *aufgewickeltes Material* || Wickelgut * *von Auf-/Abwickler gewickeltes Material*
wound rotor Schleifringläufer * *eine Wicklung tragender Läufer (kein Käfigläufer)*
wound-core transformer Bandkerntrafo
wound-rotor motor Schleifringläufermotor * *Motor mit Läuferwicklung (i.Gegensatz zum Käfigläufer)*
wound-rotor open-circuit voltage Läuferstillstandsspannung * *bei Schleifringläufermotor*
woven fabric Gewebe
woven material Gewebe
woven-bronze brush Bronzegewebebürste * *z.Stromübertragg. auf bewegte Teile (el.magn. Bremse/Kupplg.) b. Naslauf (in Öl)*
wow and flutter Gleichlaufschwankungen * *bei Magnetbandberät usw.*
wrangler Schieber * *Betrüger*
wrap around umschließen * *umhüllen, umwickeln* || umschlingen || umwickeln
wrap up einwickeln || packen * *ein~, einwickeln* || umhüllen * *in:mit* || verpacken * *einwickeln*
wrap-around labelling machine Ummantelungs-Etikettiermaschine
wrap-spring brake Federbandbremse
wraparounder Wickelmaschine * *{Verpack.-techn.} Umwickelmaschine*
wrapper Hülle || Wickelmaschine * *{Verpack.-techn.}*
wrapper insulation material Flächenisolierstoff

* *z.B. zur Isolation von Wicklungen für Motor, Transformator usw.*
wrapping   Ummantelung || Verpackung * ~ *aus Papier, Karton oder Folie*
wrapping around   umschließend * *z.B. ein Kabel durch eine Schelle*
wrapping machine   Einschlagmaschine * *{Verpack.techn.}* || Einwickelmaschine * *{Verpack.techn.}* || Verpackungsmaschine * *Einwickelmaschine (z.B. für Folienverpackung)* || Wickelmaschine * *{Verpack.techn.} zum Umwickeln mit Verpackungsmaterial (z.B. Folie, Papier)*
wrapping paper   Packpapier
wrapping roll   Wickelrolle * *Haspelhilfsantrieb im Walzwerk*
wreck   zerstören
wreckage   Zerstörung
wrench   Schlüssel * *Schraubenschlüssel* || Schraubenschlüssel || verdrehen
wrench opening   Schlüsselweite * *eines Schraubenschlüssels*
wretched   schlecht * *erbärmlich, miserabel, schlecht, dürftig, unangenehm, ekelhaft, scheußlich*
wrinkle   Falte
wrist axis   Handachse * *eines Roboters*
wrist-pin   Kolbenbolzen
write   ausarbeiten * *schriftlich* || schreiben * *auch ein Programm, be~ einer Speicherzelle usw.*
write a software program   programmieren
write access   Schreibzugriff * *z.B. auf Speicher*
write cycle   Schreibzyklus
write off   amortisieren * *abschreiben*
write protection   Schreibschutz * *z.B. für Diskette, Parameter*
write up   erstellen * *etwas Schriftliches ~*
write-off   Abschreibung
write-protected   schreibgeschützt
write-read memory   Schreib-Lesespeicher
writeable   änderbar * *beschreibbar, z.B. Spicherzelle, Datei, Parameter*
writing paper   Schreibpapier
written   schriftlich
wrong   falsch || Fehl- || verkehrt
wrong delivery   Falschlieferung
wrong measurement   Fehlmessung
WWW   WWW * *'World-Wide Web': multimediafähiger Dienst innerhalb des INTERNET-Datennetzes (→HTML, →HTTP)*
WxHxD   BxHxT * *width x height x depth: Breite x Höhe x Tiefe*
wye   Stern * *elektr. Schaltgruppe z.B. für Trafo/Motor; wye-connected: im ~ geschaltet*
wye connection   Sternschaltung
wye-delta connection   Stern-Dreieckschaltung

X

x-axis   X-Achse
X-coordinate   Abzisse
x-deflection   x-Ablenkung
X-ograph   Röntgenbild
X-ray   röntgen
X-ray photograph   Röntgenbild
X-Y plotter   X-Y-Schreiber
X-Y recorder   X-Y-Schreiber
Xtal   Quarz || Schwingquarz * *(salopp)*

Y

y-axis   Y-Achse
Y-connected   in Stern geschaltet
Y-connection   Sternschaltung * *z.B. Motor/Trafo*
Y-coordinate   Ordinate
Y-delta connection   Stern-Dreieckschaltung
yank   reißen * *(amerikan.) ruckartig ~* || Ruck * *(salopp)*
yankee cylinder   Glättzylinder * *großer Trocken- und Glättzylinder in Papiermaschine* || Trockenzylinder * *großer Trocken-/Glättzylinder bei Papiermaschine (bis 6 m Durchmesser)* || Yankee Zylinder || Yankee-Zylinder
yankee dryer   Glättzylinder * *großer Trocken- und Gättzylinder in Papiermaschine* || Yankee Zylinder * *großer Trocken- und Glättzylinder in Papiermaschine*
yard   -Hof * *z.B. Bauhof, Holzplatz* || Hof * *Hof(raum), (Arbeits-/Bau-/Stapel-) Platz* || Lager * *~hof (im Freien; z.B. für Holz)* || Platz * *Hof(raum), (Bau-/Stapel-) Platz (im Freien)*
yard-stick   Zollstock
yarn   Filament || Garn
yarn beam   Kettenbaum * *{Textil}*
yarn breakage   Fadenbruch || Fadenriss
yarn warping tension   Fadenschärspannung * *{Textil} in →Trockenmaschine in →Schärmaschine*
yaw   abweichen * *gieren, ~ (from: von; z.B. ungewolltes ~/-kommen e.Schiffs v.Kurs auf Grund d.Seegangs)* || gieren * *(um d.Hochachse) gieren/scheren (Flugzeug, Kran); vom Kurs abkommen (Schiff); abweichen* || schwanken * *→gieren (Schiff, Flugzeug, Kran); abweichen (figürl.; from: von)*
yd   * *yard; 1 m entspr. 1,0936 yd*
year of construction   Baujahr
year of manufacture   Baujahr
yearly   Jahres-
yeast   Hefe
yellow   gelb
yes, thank you   bitte * *Bitte sehr!*
yet   dabei * *dennoch* || jedoch
yield   Ausbeute * *Ertrag, Ernte, Ausbeute, Gewinn, Ergiebigkeit, Metallgehalt (von Erz)* || ergeben * *(Resultat,Ertrag, Ausbeute) ~; ein-/er-/hervorbringen; (Gewinn, Zinsen usw.) hervorbringen* || gewähren * *zugestehen, einräumen, hergeben* || Gewinn * *Ertrag* || Rendite * *→Gewinn, →Ertrag* || tragen * *ein-/hervorbringen (Ernte, Gewinn, Früchte, Zinsen, Resultat)*
yield point   Streckgrenze * *Fließ-/Streck-/Nachgiebigkeitsgrenze*
yield rate   Ausbeute * *Gut/Schlecht-Verhältnis, z.B. bei einem Produktionsprozess*
yield strength   Elastizitätsgrenze || Streckfestigkeit
yielding   nachgiebig * *nachgebend, dehnbar, biegsam; (figürl.) gefügig, nachgiebig, zum Nachgeben bereit*
yieldingness   Nachgiebigkeit * *Dehn-/Biegsamkeit; (figürl.) Nachgiebigkeit, Gefügigkeit*
yoke   Joch * *Magnetjoch, Poljoch; z.B. eines Trafos (wicklungsloser Teil e. magnet. Kreises)* || Ständer * *(Magnet-) Joch*
you are welcome   bitte * *als Antwort auf "Danke!"* || danke * *nichts zu danken!*
Young's modulus   E-Modul * *Elastizitätsmodul* || Elastizitätsmodul
youngster   Stift * *(salopp) Lehrling, junger Bursche, Junge, Knirps*
Yours faithfully   Hochachtungsvoll
Yours respectfully   Hochachtungsvoll
Yours sincerely   Hochachtungsvoll
Yours truly   Hochachtungsvoll

Z

Z-bearing Z-Lager * mit Wellenabdichtungsfunktion
Z-transform Z-Transformation
Z-type bearing Z-Lager * mit Wellenabdichtungsfunktion
Zener barrier Zenerbarriere
Zener diode Zenerdiode
Zener voltage Zenerspannung
zeo adjust Nullabgleich
zero Null || Nullpunkt || nullsetzen
zero adjust Nullpunktabgleich
zero balance Nullabgleich
zero balancing Nullabgleich || Nullpunktabgleich || Nullpunktsabgleich || Offset-Abgleich
zero center meter Nullpunkt-Mitte Instrument
zero component Nullkomponente
zero crossing Nulldurchgang
zero crossing of current Stromnulldurchgang
zero crossover Nulldurchgang
zero current Stromnulldurchgang
zero current flow interval stromlose Pause * wird z. Sicherheit v.Kommandostufe eingefügt bei Momentenwechsel
zero current signal Stromnullmeldung
zero delay angle alpha g || alpha-G || Gleichrichtertrittgrenze
zero delay angle setting Vollaussteuerung * z.B. bei Phasenanschnittsteuerung; at: bei
zero delay firing angle setting Vollaussteuerung * bei Phasenanschnittsteuerung
zero displacement Nullpunktverschiebung
zero error Nullpunktfehler || Nullpunktsfehler
zero insertion force socket Nullkraftsockel * für ICs
zero maintenance Wartungsfreiheit
zero mark Nullmarke || Nullmarke * Nullmarke
zero offset Nullabgleich * Nullpunktkorrektur (bei NC) || Nullpunktverschiebung
zero passage Nulldurchgang * z.B. einer Funktion/ Kurve/Spannung
zero point Nullpunkt
zero position Nullstellung || Ruhestellung
zero pulse Nullimpuls
zero reference point Nullpunkt * Bezugspunkt, Koordinaten-Nullpunkt (z.B. bei NC)
zero reference voltage M M-Potenzial
zero shift Nullpunktverschiebung
zero speed Drehzahl Null * at: bei || Stillstand * Drehzahl Null
zero speed monitor Stillstandsüberwachung * als Gerät/Einrichtung
zero speed sensing Drehzahl-Null-Erfassung
zero speed signal Drehzahl-Null-Meldung
zero suppression Nullpunktunterdrückung
zero- -frei * (vorangestellt)
zero-backlash spielfrei
zero-current stromlos
zero-current condition stromloser Zustand
zero-current interval stromlose Pause * z.b. durch Kommandostufe erzeugt
zero-divergence field quellenfreies Feld
zero-marking pulse Nullimpuls
zero-sequence system Nullsystem * bei Drehstromsystem
zero-speed monitor Drehzahlwächter * Stillstandsüberwachung, spricht bei Drehzahl Null an || Stillstandswächter * löst bei Drehzahl Null aus
zero-speed monitoring Stillstandsüberwachung * als Funktion
zero-speed relais Bremswächter
zero-speed relay Drehzahlwächter * Stillstandsüberwachungsrelais, löst bei Drehzahl Null aus || Stillstandsüberwachung * Stillstandsrelais || Stillstandswächter * Stillstandsrelais
zero-speed switch Drehzahlwächter * Stillstandsüberwachung, spricht bei Drehzahl Null an || Stillstandsüberwachung * Stillstandswächter || Stillstandswächter * löst bei Drehzahl Null aus
zero-torque interval drehmomentfreie Pause || drehmomentlose Pause || momentenfreie Pause || momentenlose Pause || momentfreie Pause
zero-voltage interval spannungsfreie Pause || spannungslose Pause || Spannungslücke
zeroing Nullung
ZIF socket Nullkraftsockel * Abk. f. 'Zero Insertion Force': Sockel ohne Steckkraft für ICs
zinc verzinken || Zink
zinc diecast Zinkdruckguss
zinc plated verzinkt
zinc-coat verzinken
zinc-plate verzinken
zip Anzug * (salopp) Anzugskraft bei Fahrzeug usw.
zip code Postleitzahl
zone Bereich * [soun] Gebiet, Zone, Gegend, Raum, Bereich || Gebiet * Zone, (territorialer) Bereich || Zone
zone-classification Zoneneinteilung * of hazardous areas: in explosionsgefährdeten Bereichen (DIN VDE 0165)
zoom ansteigen * (salopp) jäh ~ || Detaildarstellung || stufenlos vergrößern || vergrößern * Maßstab stufenlos verändern || zoomen
zoom effect Lupeneffekt
zooming Lupenfunktion || stufenlose Maßstabsveränderung
ZVEI ZVEI * Abkürzg.f. 'Zentralverband der Elektrotechnischen Industrie'
μ μ
μop Mikrobefehl * (salopp) Teilbefehl, in den ein Maschinenbefehl in einer →CISC-CPU zerlegt wird

Groß, Hans; Hamann, Jens; Wiegärtner, Georg
Elektrische Vorschubantriebe in der Automatisierungstechnik
Grundlagen, Berechnung, Bemessung

2000, 334 Seiten, 89 Abbildungen, 21 Tabellen,
14,3 cm x 22,5 cm, Hardcover
ISBN 3-89578-058-8
DM 98,00 / € 50,11 / sFr 89,00

Das Buch bietet eine umfassende Einführung in die physikalischen und technischen Grundlagen der Regelungs- und Antriebstechnik mit Schwerpunkt auf Berechnung und Bemessung von elektrischen Vorschubantrieben in der Automatisierungstechnik.

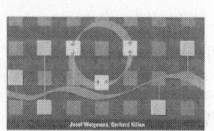

Weigmann, Josef; Kilian, Gerhard
Dezentralisieren mit PROFIBUS-DP
Aufbau, Projektierung und Einsatz des PROFIBUS-DP mit SIMATIC S7

2., überarbeitete und erweiterte Auflage, 2000, 223 Seiten,
90 Abbildungen, 70 Tabellen, 17,3 cm x 25 cm, Hardcover
ISBN 3-89578-123-1
DM 89,00 / € 45,50 / sFr 80,00

Neben Basiswissen zum Thema PROFIBUS konzentriert sich das Buch hauptsächlich auf die Projektierung des PROFIBUS-DP mit STEP 7 Version 5, erläutert unterschiedliche Möglichkeiten zum Datenaustausch mit Anwenderprogrammen und gibt wertvolle Tipps für die Inbetriebnahme und Störungssuche. Eine Reihe von praktischen Anwendungsbeispielen auf Basis der Speicherprogrammierbaren SIMATIC-Steuerungen helfen Anwendern, die Theorie in die Praxis umzusetzen.

Müller, Jürgen
Regeln mit SIMATIC
Praxisbuch für Regelungen mit SIMATIC S7 und SIMATIC PCS7

2000, 149 Seiten, 110 Abbildungen
17 Tabellen, 17,3 cm x 25 cm, Hardcover
ISBN 3-89578-147-9
DM 88,00 / € 44,99 / sFr 80,00

Durch die praxisnahe Beschreibung der Regelungstechnik anhand SIMATIC S7 und PCS7 erhalten die Leser in diesem Buch wertvolle Anregungen und Hilfestellungen für Projektierung und Inbetriebnahme regelungstechnischer Anwendungen.

Brich, Peter; Hinsken, Gerhard; Krause, Karl-Heinz
Echtzeitprogrammierung in Java
Automatisieren im Mikrosekundenbereich

2000, ca. 130 Seiten, ca. 30 Abbildungen,
17 cm x 25 cm, Hardcover
ISBN 3-89578-153-3
DM 88,00 / € 44,99 / sFr 80,00

Mit diesem Buch erhalten Leser eine Einführung in Echtzeit-Programmierung mit Java auf SICOMP-Rechnern. Es zeigt, wie mit Java in der Automatisierungstechnik wirtschaftlich Echtzeitfähigkeit und Anbindung an das Internet realisiert werden können.

Bezner, Heinrich (Bearbeiter)
Fachwörterbuch industrielle Elektrotechnik, Energie- und Automatisierungstechnik
Dictionary of Electrical Engineering, Power Engineering and Automation
Teil 1 Deutsch-Englisch   Part 1 German-English

4., überarbeitete und erweiterte Auflage, 1998,
608 Seiten, 14,3 cm x 22,5 cm, Hardcover
ISBN 3-89578-077-4
DM 128,00 / € 65,45 / sFr 113,00

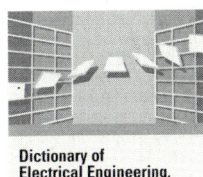

Fachwörterbuch industrielle Elektrotechnik, Energie- und Automatisierungstechnik
Dictionary of Electrical Engineering, Power Engineering and Automation
Teil 2 Englisch-Deutsch   Part 2 English-German

4., überarbeitete und erweiterte Auflage, 1998,
532 Seiten, 14,3 cm x 22,5 cm, Hardcover
ISBN 3-89578-079-0
DM 128,00 / € 65,45 / sFr 113,00

Dieses Standardwerk enthält in Teil 1 (Deutsch-Englisch) 67.000 und in Teil 2 (Englisch-Deutsch) 54.000 Fachbegriffe hauptsächlich aus den Bereichen Energieerzeugung, -übertragung und -verteilung, Antriebs-, Schaltgeräte- und Installationstechnik, Leistungselektronik, Meß-, Analysen- und Prüftechnik sowie Automatisierungstechnik.